Hemp Diseases and Pests: Management and Biological Control, 2nd edition

Hemp Diseases and Pests: Management and Biological Control, 2nd edition

John M. McPartland, University of Vermont

With contributions (chapter reviews) by
Ernest C. Bernard, University of Tennessee
Marguerite Bolt, Purdue University, Indiana
Kadie Britt, University of California, Riverside
Robert C. Clarke, BioAgronomics Group, California
Whitney Cranshaw, Colorado State University
Heather Grab, Penn State University
Gianpaolo Grassi, CREA Institute, Italy
Steven T. Koike, TriCal Diagnostics, California
Punya Nachappa, Colorado State University
Zamir K. Punja, Simon Fraser University, British Columbia
Ethan Russo, CReDO Science, Washington
Lindsey Thiessen, USDA APHIS, North Carolina
David P. Watson, HortaPharm B.V., Amsterdam

plus all growers and researchers who shared their experiences

CABI

CABI is a trading name of CAB International

CABI
Nosworthy Way
Wallingford
Oxfordshire OX10 8DE
UK

Tel: +44 (0)1491 832111
E-mail: info@cabi.org
Website: www.cabi.org

CABI
200 Portland Street
Boston
MA 02114
USA

Tel: +1 (617)682-9015
E-mail: cabi-nao@cabi.org

The views expressed in this publication are those of the author(s) and do not necessarily represent those of, and should not be attributed to, CAB International (CABI). Any images, figures and tables not otherwise attributed are the author(s)' own. References to internet websites (URLs) were accurate at the time of writing.

CAB International and, where different, the copyright owner shall not be liable for technical or other errors or omissions contained herein. The information is supplied without obligation and on the understanding that any person who acts upon it, or otherwise changes their position in reliance thereon, does so entirely at their own risk. Information supplied is neither intended nor implied to be a substitute for professional advice. The reader/user accepts all risks and responsibility for losses, damages, costs and other consequences resulting directly or indirectly from using this information.

CABI's Terms and Conditions, including its full disclaimer, may be found at https://www.cabidigitallibrary.org/terms-and-conditions.

A catalogue record for this book is available from the British Library, London, UK.

ISBN-13: 9781836990321 (hardback)
9781836990338 (paperback)
9781836990345 (ePDF)
9781836990352 (ePub)

DOI: 10.1079/9781836990352.0000

Commissioning Editor: Rebecca Stubbs
Editorial Assistant: Emma McCann
Production Editor: James Bishop

Typeset by Straive, Pondicherry, India
Printed in the USA

Contents

Part Five: Control of Diseases and Pests

Preface to the second edition

HD&P2 provides updated information and a new layout. Chapters are now organized by parts of the plant affected—roots, shoots, flowering tops, etc. This organization follows *Garden Insects of North America* (Cranshaw and Shetlar, 2018). Some insects and fungi attack several plant parts, such as *Botrytis cinerea*. They'll be featured in one chapter and excerpted elsewhere. *Botrytis*, for example, gets featured in Chapter 9 (Flowering top diseases) and cross-referenced in Chapter 10 (Leaf and stalk diseases) and Chapter 8 (Insects injuring seedlings or roots).

HD&P2 is written for the same audience as *HD&P1*—it's a hybrid text—a meticulous review for crop scientists, and a practical guide for growers. It's what made *HD&P1* successful. The technical jargon is minimalized, but some is unavoidable. We make room for "historical" information, because much research was published over a century ago, prior to prohibition. The internet has arisen since *HD&P1.0* was published, but *HD&P1* continues to sell, because articles on the internet often lack perspective, seldom look at the big picture, and indubitably provide sales-oriented information.

New research is reported here, such as phylogenetic studies of *Bipolaris* fungi (Fig. 1.12), the aphids *Phorodon* and *Myzus* (Fig. 4.11), and *Fusarium oxysporum* (Fig. 11.10). New species are named, such as *Neofusicoccum marconii* (Fig. 11.18) and *Dothiorella cannabis* (Fig. 11.20). We overhauled the synonymy of *Grapholita delineana* (section 6.2) and host associations in the *Ostrinia* spp. complex (section 6.3).

The goal of *HD&P2* remains the same as the first edition: growing a healthy crop using sustainable methods. Compared with 20 years ago, when *HD&P1* was written, maintaining sustainability has become more tenuous. Acreage has skyrocketed and attracted big money. Big money aims at short-term capital gains and stockholder dividends—often at odds with sustainable methods. At the same time, sustainable methods have become imperative: the past 20 years have seen global population growth, diminished arable land and water resources, and unpredictable climate change.

Some things haven't changed in the past 20 years. *Cannabis* is still prohibited in many countries around the world. The preface to *HD&P1* was written in Canadian English (with some Québécois French); Canada remains a shining light in the Western Hemisphere. We look around the USA, at the supersized bread and circus of a failing empire, and remain baffled by our federal government's continuing and costly efforts to restrict *Cannabis* cultivation.

One of our co-authors has died. I will be re-writing my Preface soon to reflect his passing.

Acknowledgements

HD&P2 builds on the original 2000 edition. At that time we acknowledged 30 colleagues who graciously reviewed parts of the manuscript. Their contributions have been carried forward in this edition. Financial sponsorship by GW Pharmaceuticals and Koppert Biocontrols allowed *HD&P1* to happen, and without *HD&P1* there'd be no *HD&P2*. CABI editor Rebecca Stubbs had the vision to bring *HD&P1* to print. Unique in the publishing world, we have the pleasure of her continued collaboration, 25 years later. *HD&P2* has more photographs than *HD&P1*; most were provided by contributors, but we thank many others who provided images.

Last but not least we thank our partners and families for their support and understanding while we worked on this project, instead of cleaning the kitchen. John McPartland dedicates this book to Patty Pruitt, who also provided copyediting skills, photographs, a sounding board, and love.

How to use this book

You need not begin this book at page one. If you tentatively identified a pest or pathogen infesting your plants, look up its name in the index. See if the problem in hand matches its description in the book. Proceed to read about the problem and its control measures. Control measures for common pests and pathogens are highlighted with charts, illustrations, and explicit instructions.

If you face an unknown assailant, turn to the identification keys in Chapter 1. If you dislike keys, most common pests can be diagnosed by the "picture-book method"—flip through the illustrations until you recognize your problem.

This book has no separate glossary; technical terms in this text are defined within it. If a technical term is unfamiliar to you, use the index to locate the term's definition. Technical terms appearing in **bold print** indicate they are being defined in that paragraph. We define technical terms in Chapters 1, 2, and 3, to build the language needed for pest identification using pest morphology (form and structure).

To control diseases and pests, we consider chemicals a last resort. We prefer cultural techniques, mechanical methods, and biological controls. Culture techniques alter the farmscape, making it less favorable for pests and disease organisms (see Chapter 19). Mechanical methods utilize traps, barriers, and other ingenious techniques (Chapter 19). Biological controls employ beneficial organisms to subdue pests and pathogens (Chapter 20). Although we discourage the use of chemicals, technical information regarding their use is presented in Chapter 21. This information is presented in the spirit of harm-reduction, not as a green light for chemical abuse. Use as little as possible.

1 Principles of plant protection

Abstract
Plant protection begins with the "crop damage triangle" concept. Diagnosing problems requires a knowledge of signs and symptoms, and the identification of causal agents. Tools for identification are based on morphological or molecular methods. Regarding the latter there are two approaches: DNA barcoding (the "best-hit" method), or the phylogenetic "backbone tree" method. A case study is provided that compares the two. Integrated pest management (IPM) applies control methods to all three sides of the crop damage triangle. We compare and contrast IPM with organic farming, biodynamic farming, and sustainable agriculture. The IPM pyramid is a useful guide for choosing IPM strategies—cultural and mechanical methods at the base of the pyramid, followed by biological control, and lastly chemicals. *Everyone* benefits by shifting from pesticides to biocontrol—consumers, cultivators, *Cannabis*, and the Earth. Different types of identification keys are explained and exampled. We finish with a brief history of *Cannabis* crop protection.

1.1 Introduction

Hemp (*Cannabis sativa* L.) is a complex crop to write about, partly because it's three-in-one. **Fiber-type crops** supply phloem and xylem cells, best known as bast fibers and woody hurds, respectively. **Seed-type crops** furnish fruits (i.e., seeds, achenes, grain) for food and feed. Crushed fruits yield seed oil and seed cake—the latter a protein-rich pulp after oil is expelled. **Drug-type crops** are grown for inflorescences ("flowers") that yield a vivid spectrum of phytochemicals.

Turner *et al.* (1980) tallied 420 phytochemicals in *Cannabis*—the 420 plant. Hanuš *et al.* (2016) cataloged 120 cannabinoids, the best known being tetrahydrocannabinolic acid (**THCA**) and cannabidiolic acid (**CBDA**). When THCA and CBDA are exposed to heat or light or time, they release CO_2 and tetrahydrocannabinol (**THC**) and cannabidiol (**CBD**). More heat or light or time causes THC to dehydrogenate into cannabinol (**CBN**) (Fig. 1.1). The K_i numbers (inhibition constants) in Fig. 1.1 are measures of binding affinity at human cannabinoid receptor 1 (hCB_1), discussed in section 18.4.

Cannabis produces about 120 **terpenoids**—primarily monoterpenoids ($C_{10}H_{16}$ template) and sesquiterpenoids ($C_{15}H_{24}$ template). Terpenoids include terpenes (pure hydrocarbons) and their oxygenated derivatives, which form alcohols, ethers, aldehydes, ketones, and esters. Terpenoids have insect-repellent properties and therapeutic applications. Additionally, *Cannabis* biosynthesizes flavonoids, phytosterols, indoles, spiroindans, and stilbenoids—a "phytochemical treasure trove" (Mechoulam 2005).

Cannabis is naturally **dioecious**, with separate female (pistillate) and male (staminate) individuals. **Monoecious** (intersexual) individuals have both kinds of flowers. Monoecious females (XX genotype) have been bred to provide fiber and seed crops. They have staminate flowers scattered amidst pistillate flowers. Abnormal **hermaphroditic** (bisexual) flowers have both stamens and pistils in the same flower structure.

Drug-type plants can be grown with or without seeds. Seeded plants include Moroccan and Afghani crops processed into *hashīsh*. Marijuana (considered a pejorative in some circles) was usually seeded in the 1960s and 1970s, such as Mexican brick weed. Plants without seeds—sinsemilla or *gañjā*—are females grown in the absence of male pollen. Traditionally, male plants were identified at puberty and removed from fields. Today, males are excluded by sowing "feminized" seeds, or by growing clones cut from female "mother plants."

In the USA, *Cannabis* acreage has approached what it was prior to prohibition: 146,065 acres were planted in 2019 (Mark *et al.*, 2020). In comparison, 204,124 acres were planted in 1943, during the "War Hemp" effort (US Treasury, 1944). Compared with 5 years ago, when the first edition of *Hemp Diseases and Pests* (*HD&P*) was written, crops are now being grown over a broader geographic range, and for more uses. The USA farm bills of 2014 and 2018 loosened regulations and acreage soared. This many-sided expansion has exposed *Cannabis* to new disease pathogens and insect pests. Many diverse scenarios—indoor versus outdoor; fiber, seed, or drug crops; high- or low-density—require unique approaches for controlling diseases and pests.

1.1.1 Placed in perspective

Diseases and pests reduce food security at household, national, and global levels. Oerke (2006) assessed global losses in six major crops (cotton, maize, wheat, soybean, rice, potatoes), and estimated 12.5% losses to disease pathogens, 10.8% losses to animal pests, and 8.8% to weeds—a total of 32.1%. In addition to these field losses, postharvest storage losses added another 10%.

Savary *et al.* (2019) revisited these estimates, using a different methodology. They estimated total losses—from diseases and pests—at 30.5% for rice, 22.6% for maize, 21.5% for wheat, 21.4% for soybean, and 17.2% for potato. These losses vary regionally, with greater losses in regions that face greater food insecurity, such as Africa and India. Crops become more vulnerable as we shift from small-scale, diverse farms to large-scale, genetically uniform monoculture production.

© John M. McPartland 2025. *Hemp Diseases and Pests* 2nd Edition (J.M. McPartland)
DOI: 10.1079/9781836990352.0001

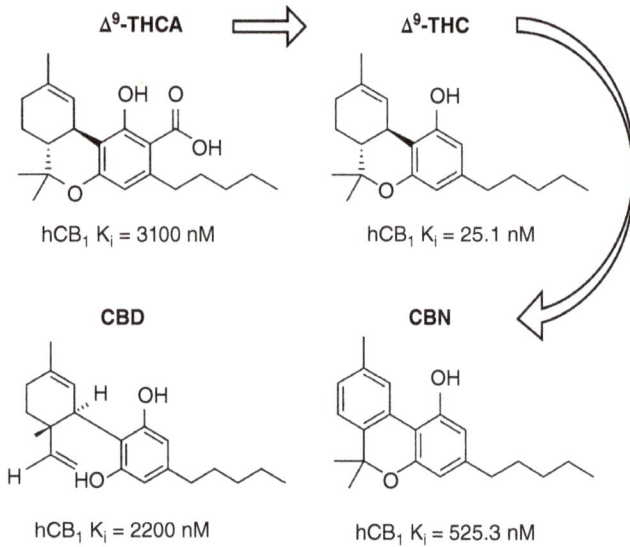

Fig. 1.1. Some cannabinoids

Consumers want healthy products—no discolored fiber, rancid seed oil, or aflatoxin-laden flowers. But we don't want pesticide-tainted products, either. *Everyone* benefits by shifting from chemical control to biological control—consumers, cultivators, *Cannabis,* and the Earth. Non-toxic pest control is the keystone for sustainable agriculture. Sustainability requires a shift in agronomic consciousness, and hemp can lead the way.

The hemp expert Dewey (1914) claimed "hemp has no enemies." Wright (1918) declared that "hemp is, as yet, entirely free from attacks by insects and diseases." Those rosy impressions have ebbed. Cranshaw *et al.* (2019) tallied 75 insects and mites attacking hemp crops in the USA. An inventory of disease pathogens listed 88 fungi, 11 nematodes, eight viruses, eight parasitic plants, and six bacteria (McPartland, 1992).

Insects, by definition, do not cause disease. Disease is caused by *continued or persistent irritation*, instigated by fungi, nematodes, bacteria, parasitic plants, and viruses. Insects cause *transient irritations*. They cause injury, not disease. **Entomologists** study insects and mites, and **plant pathologists** study disease pathogens. There are also **weed scientists**, who study undesirable plants "growing out of place."

Some diseases do not involve pests or pathogens. **Abiotic** (non-living) causes of diseases include nutrient deficiencies and imbalances, light deficiencies and imbalances, and environmental toxins. Scott and Punja (2022) emphasized that determining whether a disease is caused by an organism or by an abiotic factor is the first step in the diagnosis of a problem.

1.2   The crop damage triangle

Delving into plant protection begins with the "crop damage triangle" (Fig. 1.2). All three sides of the triangle are required for crop damage: *Side 1*, a susceptible host; *Side 2*, a pest or pathogen; and *Side 3*, an environment conducive to crop damage. Damage is limited by the triangle's shortest side—either an absence of pests and pathogens, or a health environment, or a

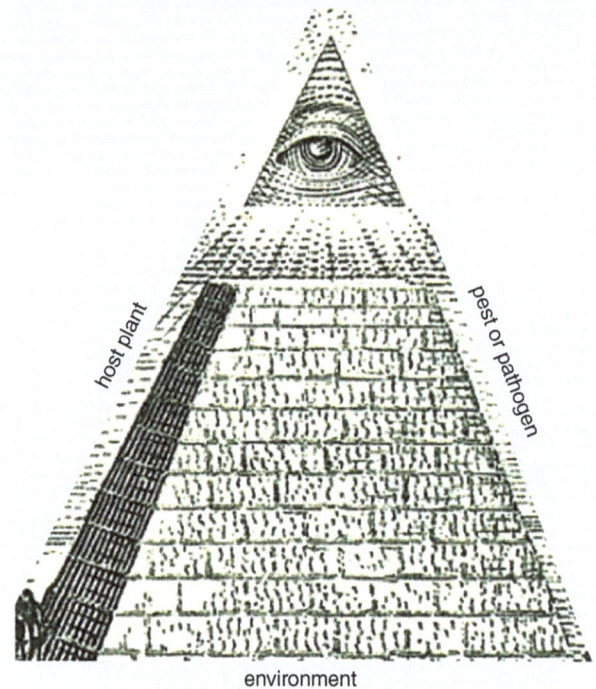

Fig. 1.2. Crop damage triangle

resistant host. Our control methods manipulate one or more sides of the crop damage triangle.

1.2.1   Environment

The base of the crop damage triangle is manipulated by cultural and mechanical methods. **Cultural methods** are basic agronomic practices, such as tilling the soil and keeping glasshouses clean. They *prevent* crop damage, rather than cure problems. **Mechanical methods** reduce crop damage by physical means. They can be as simple as hand-picking insects off plants. Mechanical methods can be preventive (e.g., bird netting) or curative (pruning away fungus-infected branches).

Indoor environments are easily manipulated. To control gray mold, lower the humidity and increase the temperature. To control spider mites, do the opposite. Indoor cultivation was a result of prohibition, so growers could evade detection. In today's legal milieu, some regulatory authorities still prefer indoor cultivation, because of enhanced security and surveillance. Indoor cultivation reduces crop losses from bad weather, and allows year-round cropping.

Indoor facilities may rely entirely on artificial light (**growrooms**), or sunlight supplemented with artificial light (**glasshouses,** i.e., greenhouses, hothouses). Glasshouses are open to the exterior environment (light, atmosphere, temperature, humidity), and semi-permeable to flying insects and fungal spores. Growrooms are mostly closed to the exterior environment; pests and pathogens are more easily excluded. Biocontrol organisms released in a growroom cannot fly away. However, when a pest or pathogen penetrates a growroom, an epidemic can wipe out the crop.

Outdoor environments offer higher yields, with lower costs, and less environmental impact. Outdoor drug-type crops in low-latitude countries may corner the market after global prohibition becomes history. Growroom energy costs will become prohibitive in the future, not to mention their expensive reliance on other external inputs, such as substrates, fertility, and water. If the market continues to trend away from artisanal flowers (best grown indoors) towards extractable concentrates (whose crops can be grown anywhere), the Global South will become the source of cannabinoids worldwide.

Outdoor environments can be manipulated to control pests and pathogens. We pull weeds, irrigate plants, rotate crops. *Site selection* influences pests and pathogens. Sites should be selected with an eye toward neighboring crops—if European corn borers are a problem, do not plant near maize. Also consider the previous season's crop—expect white root grubs if rotating after sod. Soil characteristics are critical factors in site selection. For instance, sites with heavy clay soil can result in high mortality from root diseases. Crop density (plants/ m²) alters the microenvironment around plants. Fiber-type plants sown at high density often suppress weeds but may become predisposed to gray mold.

1.2.2 Pests and pathogens

The second side of the triangle is controlled by cultural and mechanical methods, as well as biological and chemical controls. Biological control reduces pest and pathogen problems through the use of their natural enemies—predators, parasites, or pathogens (see section 20.6). Chemical control uses naturally occurring chemicals or biorational synthetic pesticides (see section 21.1). Agronomists often cite "the two Es" of pest and pathogen control: *exclusion* and *eradication*.

Exclusion led the US government to establish the Plant Quarantine Act in 1912—to exclude foreign pests and pathogens from the continent. Island ecosystems face a greater threat from foreign pests and pathogens. New Zealand, for example, has very strict biosecurity requirements regarding the importation of crop seeds. NZ's regulations for importing *Cannabis* germplasm relied upon the first edition of *HD&P* (Thomson, 2019). Farmers exclude local pests and pathogens in three ways: (1) cleaning all equipment before entering a field or indoor operation; (2) pasteurizing soil or growing media before bringing it into a glasshouse or growroom; and (3) using certified seed that is disease and pest-free.

Eradication requires the elimination of pests and pathogens once they have arrived. Some problems can be eradicated by cultural and mechanical means (e.g., starving pests via crop rotation, heat treatment of seeds, pruning infected branches). *Total eradication* usually requires the use of pesticides. But spraying one pest may increase the population of another pest, resulting in more crop damage. Eradication becomes an unrealistic and self-defeating goal. Eradication is not a concept embraced by practitioners of integrated pest management (section 1.3).

1.2.3 Host

The third side of the disease triangle is *Cannabis*, which can be selected and bred for resistance to pests and pathogens.

Breeding involves the mixing of genes and the selection of hybrids with resistant traits. Walking through an infested field, some plants may show resistance. This field variation has provided the foundation for many crop breeding efforts. Even a glasshouse full of clones may give rise to somatic mutations (Adamek *et al.*, 2022) or somaclonal variation (Chandra *et al.*, 2017). Variation itself varies, between crops grown from seeds versus clonal cuttings, and varies between crops grown from regular seeds versus feminized seeds (Lapierre *et al.*, 2023).

To breed resistance, we often turn to unique landraces or cultivars. A **landrace** or "traditional variety" refers to a population of plants locally adapted to a particular environment. A landrace has been mass-selected by local farmers for certain traits; in the case of *Cannabis*, for fiber, seed, or flower yields. A landrace expresses a phenotype that reproduces consistently, but it nevertheless harbors abundant genetic diversity. Plant breeders improve landraces by selectively breeding them into cultivars. A single landrace can be inbred and selected into a cultivar, or two landraces can be crossbred into a hybrid cultivar (Fig. 1.3).

A **cultivar** is defined as "an assemblage of plants that (a) has been selected for a particular character or combination of characters, (b) remains distinct, uniform, and stable in these characters when propagated by appropriate means" (Brickell, 2016). Cultivar names are placed in single quotation marks when they meet provisions established by the *International Code of Nomenclature for Cultivated Plants* (*ICNCP*) (Brickell, 2016). *ICNCP* provisions include valid publication of the name in an agricultural journal, typification ("Nomenclatural Standard"), and priority. 'FINOLA' is today's best-known *Cannabis* cultivar, judging from the number of internet search hits.

Breeders of drug-type *Cannabis* use "strain" names, most famously "Skunk #1". Small (2015) pointed out that strains are conceptually identical to cultivars, but almost no strains have met *ICNCP* requirements for cultivar recognition (thus double quotation marks). A database currently lists 31,481 strains (Seedfinder, 2024), a stupefying proliferation of what Doyle (2007) called *ganjanyms*.

Cannabis diversity can be compressed into a dichotomy: fiber-type plants or drug-type plants. This dichotomy *approximates* a pair of subspecies, *C. sativa* subsp. *sativa* and *C. sativa* subsp. *indica*, respectively (Small and Cronquist 1976). Hybrids between the subspecies have been created, for example most CBD-type plants (Fig. 1.3).

Drug-type plants, *C. sativa* subsp. *indica*, came to be divided into two categories, "Indica" and "Sativa". This vernacular taxonomy arose after Anderson (1980) published a line drawing of the plants (Fig. 1.4). "Indica" and "Sativa" are inconsistent with formal botanical nomenclature, as first noted in *HD&P* (McPartland *et al.*, 2000). Formally they are *C. sativa* subsp. *indica* var. *afghanica* and *C. sativa* subsp. *indica* var. *indica*, respectively (McPartland and Small, 2020).

"Indica" (var. *afghanica*) applies to plants with broad leaflets, short and compact habit, early maturation, a THC%/ CBD% ratio <7, and the presence of sesquiterpene alcohols (guaiol, β-eudesmol, γ-eudesmol). The landrace ancestors of "Indica" came from Central Asia, primarily Afghanistan and Turkestan.

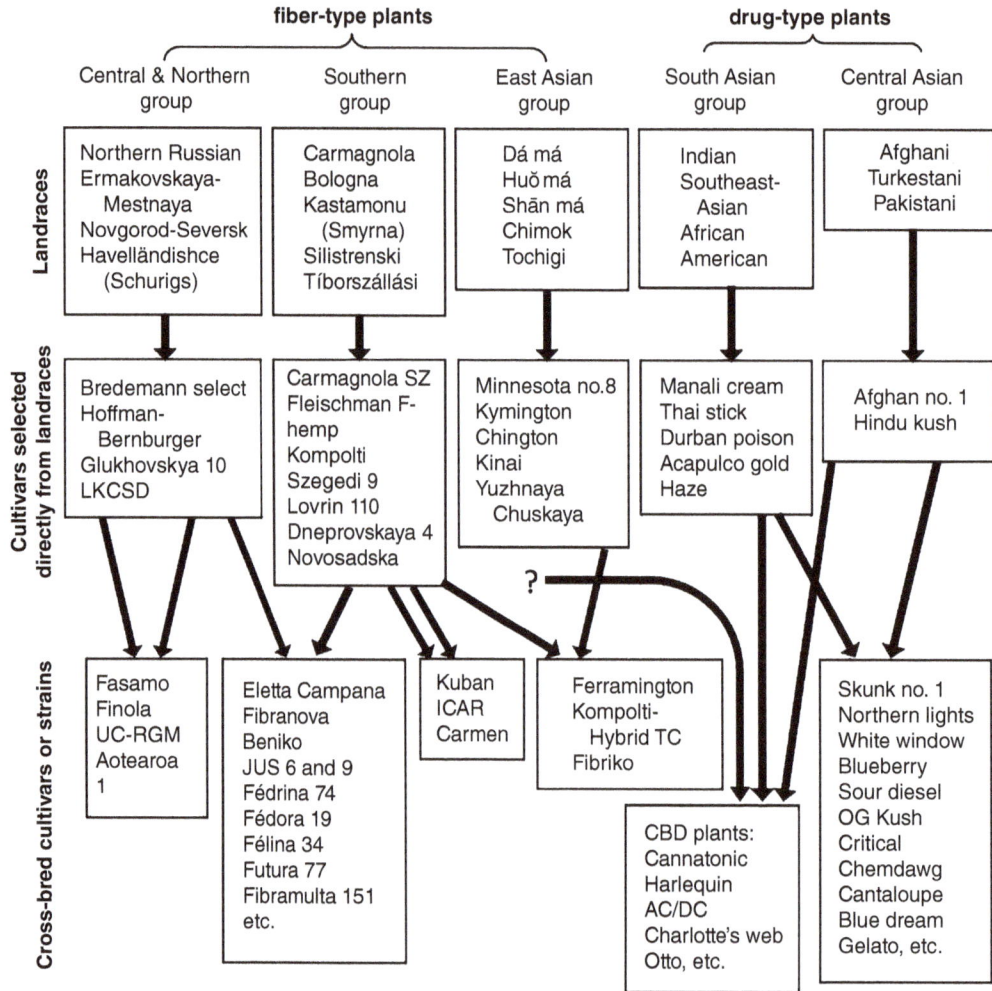

Fig. 1.3. Examples of landraces and cultivars, classified by population type (adapted from de Meijer, 1995)

Fig. 1.4. Vernacular taxonomy coined by Anderson (1980) (courtesy of Harvard University Herbaria and Botany Libraries)

"Sativa" (var. *indica*) applies to plants with narrow leaflets, tall and diffuse habit, late maturation, a THC%/CBD% ratio ≥7, and the absence of sesquiterpene alcohols. The landrace ancestors of "Sativa" originated in South Asia (India), with prehistoric distribution to Southeast Asia and Africa (de Meijer and van Soest, 1992).

The vast majority of drug-type plants available today are "Indica–Sativa" hybrid *ganjanyms* (Fig. 1.3). Identical strain names have been applied to different genetics due to counterfeiting, and different strain names have been applied to the same genetics due to breeder theft (Sawler *et al.*, 2015). Hazekamp *et al.* (2016) proposed jettisoning all strain names in favor of simply listing cannabinoid and terpenoid profiles in plants, "from cultivar to chemovar."

Widespread crossbreeding over the past 50 years has endangered traditional drug-type landraces, a phenomenon called *extinction through introgressive hybridization* (Wiegand, 1935). The original landraces are worth preserving as breeding stock, and not hybridized into homogeneity. They differ in their resistance to many pests and pathogens. We highlight this factor throughout the book.

The above information pertains to domesticated *Cannabis*. In addition, "wild-type" *Cannabis* grows around the world—a potentially rich source of resistance genes untapped by *Cannabis* breeders. Wild-type *Cannabis* was tagged with vernacular "Ruderalis" (Fig. 1.4). "Ruderalis" germplasm was first offered for sale by Schoenmakers (1986), who highlighted its autoflowering character (see section 2.1.4).

Lastly, bioengineering can transfer genes from other species into the *Cannabis* genome, creating a **genetically modified organism** (GMO). The transfer of genes across species boundaries creates **transgenic** organisms. This stimulates great debates, discussed in Chapter 22 (available on the web). Here we briefly touch on techniques.

To create a transgenic organism, useful DNA is identified and isolated, then incorporated into a **vector**, and the vector is inserted into the host. If the recombinant DNA integrates with the host genome, this results in stable expression. The genetic modification is permanent. Reproduction of the GMO yields genetically modified progeny. The most common vector is *Agrobacterium tumefaciens*. This bacterium can insert a segment of its DNA into plants, known as the Ti (tumor-inducing) plasmid. Genetic engineers have deleted the genes that cause tumors, and added novel genes to the plasmid. *A. tumefaciens* naturally infects *C. sativa* (Lopatin, 1936).

The world's first fully regenerated transgenic *Cannabis* was created for disease resistance. MacKinnon *et al.* (2000) utilized the gene for polygalacturase-inhibiting protein (PGIP) obtained from raspberry (*Rubus idaeus*). PGIP confers resistance to gray mold caused by *Botrytis cinerea*. MacKinnon and colleagues incorporated the raspberry PGIP gene into two fiber-type cultivars, 'Fedora 19' and 'Felina 34'. In Fig. 1.5, a transgenic seedling with resistance to *B. cinerea* appears on the right, and the four control plants on the left are susceptible. When the transgenic plants were inoculated with *B. cinerea*, they showed 39% fewer lesions than control plants transformed with an empty *A. tumefaciens* vector (MacKinnon, 2003). The GMO genie is out of the bottle.

Genome editing involves deleting or modifying a short stretch of DNA in an existing genome, rather than inserting a foreign gene. The tweak may resemble a naturally-occurring mutation. As such, plants with edited genomes create less societal concern than transgenic GMOs. Two genome editing

Fig. 1.5. GMO *Cannabis*. Photograph from MacKinnon *et al.* (2000) (courtesy of Steve Millam)

technologies have been used, **RNAi** (RNA interference) and **CRISPR-Cas** systems. One RNAi-based technology, called virus-induced gene silencing, has been applied to *Cannabis* (Schachtsiek *et al.*, 2019; Zhang *et al.*, 2021; Alter *et al.*, 2022).

1.3 Integrated pest management (IPM)

IPM integrates all sides of the crop damage triangle to control pests and diseases. IPM began as a solution to ecological and economic problems posed by pesticides (Stern *et al.*, 1959). IPM replaces the concept of *eradication* with the concept of *coexistence*. IPM practitioners (IPMers) recognize pests and pathogens as integral elements of human-made agroecosystems (Savary *et al.*, 2019). IPMers can coexist with pests, as long as pests remain below economically damaging levels. What constitutes "economically damaging" varies from plant to plant. A gardener entering a product in a flower contest may consider damage by a single budworm intolerable. Fiber crops, in contrast, forbear many budworms before economic thresholds are reached.

IPM integrates ideas from conventional agriculture and **organic farming**. Organic farming is defined by the *National Organic Standards Board* as "an ecological farm management system that promotes and enhances biodiversity, biological cycles, and soil biological activity. It is based on minimal use of off-farm inputs and on management practices that restore, maintain and enhance ecological harmony."

Organic concepts were put forth by **Johann Goethe** (1749–1832), who wrote about hemp in his research, diaries, and autobiography (Goethe, 1812, 1815, 1884, 2007). **Rudolf Steiner** (1861–1925) popularized Goethe's approach as "Anthroposophical thought" (Steiner, 1897). Steiner delivered a series of lectures that laid the foundation for an alternative agriculture, later known as **biodynamic farming**.

Steiner (1924a,b) championed the use of animal manure and vegetable compost. Being a mystic, he threw in some astrology, and postulated the primacy of certain herbs for composting, such as yarrow, valerian, stinging nettle, dandelion, chamomile, oak bark, and horsetail. The latter, *Equisetum arvense*, he recommended as a field spray to control fungal diseases. Horsetail is rich in silica, which does indeed control *Cannabis* powdery mildew (Dixon *et al.*, 2022b; Scott and Punja, 2022).

Composting as most people practice it (chopped waste material spread in layers and kept moist, with periodical turnings) was popularized by Sir **Albert Howard** (1873–1947). He served as the "Imperial Economic Botanist" in India and studied composting methods there (Howard and Wad, 1931; Howard, 1943). He applied his methods to many crops, but not to *Cannabis*. "The attention of the Imperial Economic Botanist has been drawn to the need of a biological study of the *ganja* plant, but at present he is fully employed on other and more important work" (Howard, 1906). After he retired, his successors conducted *ganja* studies (Shaw, 1926, 1931, 1932; Bose, 1930).

Steiner's protégé **Ehrenfried Pfeiffer** (1899–1961) coined the word biodynamic (Pfeiffer, 1938) and mainstreamed Steiner's concepts. "Pfeiffer's role could be characterised as having carried forward Anthroposophical thoughts of agriculture from mystery to muck" (Paull, 2011). Muck as in compost.

Pfeiffer advocated border plantings of hemp to repel mole crickets and cabbage moths, an idea he got from Steiner. Pfeiffer emigrated from Switzerland in 1940 and settled in Kimberton, Pennsylvania.

Forty-five miles away, **Jerome I. Rodale** (1898–1971) founded the Rodale Organic Gardening Experiment Farm in 1940. He was inspired by Albert Howard's work, and he travelled to Kimberton to share ideas with Pfeiffer. Rodale published *Organic Farming and Gardening* magazine. The first issue had articles by Howard (1942) and Pfeiffer (1942), and Rodale (1942) described how tobacco fertilized with manure yielded cigars with a better aroma than tobacco fertilized with chemical fertilizers. Rodale (1945) summarized organic farming concepts in *Pay Dirt*, with an introduction by Howard. His son **Robert Rodale** (1930–1990) began to edit our first book on *Cannabis* diseases and pests (McEno, 1991), but died in a car crash.

William Albrecht (1888–1974) at the University of Missouri linked natural soil fertility with healthy crops and healthy people. He threw down the gauntlet in 1940: "NPK formulas, as legislated and enforced by State Departments of Agriculture, mean malnutrition, attack by insects, bacteria and fungi, weed takeover, crop loss in dry weather, and general loss of mental acuity in the population, leading to degenerative metabolic disease and early death" (quoted in Storl, 2013). Here's another pithy quote: "The use of [pesticide] sprays is an act of desperation in a dying agriculture. It is not the over-powering invader we must fear, but the weakened condition of the victim" (Albrecht, 1975).

Organic farmers focus on soil fertility much more than IPMers. Natural soil fertility has many benefits (see section 17.2). Unfortunately, the USDA has hollowed out the term "organic farming," and turned it into a marketing tool. National USDA certification standards, in effect since 2002, are weaker than standards previously established by individual states. Organic farmers and IPMers differ regarding pesticides. Organic farmers eschew most synthetic pesticides, as well as many natural pesticides (see *The National List* in section 21.4). IPMers, in contrast, tend to use any legal pesticide that works. But IPMers reject the *conventional* approach of spraying pesticide, which is rigidly based on the calendar date. Instead, IPMers spray on a schedule determined by three aspects: (1) pest monitoring; (2) climate monitoring; and (3) the presence of beneficial organisms. With IPM, *careful observation* replaces the brute force of conventional chemical warfare. IPMers closely monitor crop conditions, biocontrol organisms, the weather, and *all* pests in the area, not just single target species. IPM is pest management for the information age.

To manage complexity, IPM methods are arranged in a hierarchy, depending on pest populations and crop density. The primary IPM strategy is *selectivity*. A control method should *selectively* kill pests and not beneficial organisms. *Selective* timing and *selective* treatment applied to *selective* plants (not the entire field) minimize collateral damage.

Sustainable agriculture was defined in the USA Farm Bill of 1977 as:

> "an integrated system of plant and animal production practices having a site-specific application that will, over the long-term—
> (A) satisfy human food **and fiber** needs [emphasis added];
> (B) enhance environmental quality and the natural resource

base upon which the agriculture economy depends; (C) make the most efficient use of nonrenewable resources and on-farm resources and integrate, where appropriate, natural biological cycles and controls; (D) sustain the economic viability of farm operations; and (E) enhance the quality of life for farmers and society as a whole" (7 US Code § 3103).

IPM and the use of biological controls are implicit in the definition. Sustainability is really a moral obligation, given the harmful effects on humans and the Earth if unsustainable methods continue to be used.

The year before the USA Farm Bill of 1977, **Wes Jackson** (1936–) founded the Land Institute in Salina, Kansas. "The plowshare may well have destroyed more options for future generations than the sword" (Jackson, 1980). His breeding research on perennial crop plants grown in polyculture systems (modeled after prairie ecosystems) has provided a *modus operandi* for sustainable and even regenerative agriculture.

Regenerative agriculture was coined by Robert Rodale. "Under the banner of sustainability we are, in effect, continuing to hamper ourselves by not accepting a challenging enough goal. I am not against the word sustainable, rather I favor regenerative agriculture" (Rodale, 1985). Thus, regenerative agriculture seeks to *reverse* the harmful effects of unsustainable methods, particularly through groundwater conservation and topsoil regeneration. Rodale Institute (2014) reframed their regenerative arguments around climate change, available online.

1.3.1   IPM steps

IPMers proceed in five steps: (1) monitoring plant health; (2) monitoring the environment; (3) deciding the proper IPM intervention; (4) implementing the intervention; and (5) post-intervention reassessment.

Monitoring plant health prior to the arrival of a problem is essential. Monitoring methods vary from casual hearsay between neighbors to daily quantitative trap sampling. Monitoring efforts must be increased after a problem appears. The effort should match the severity of the problem. The larger the crop, the greater is the monitoring effort. Be sure to inspect plants in hard-to-see spots, like centers of crop fields or glasshouse "hotspots" located near doors and window vents. Record the date, time, and location of any crop damage observed. Estimate the number of pests, either qualitatively ("many") or quantitatively ("average 10 aphids per leaf"). Record the temperature, humidity, and time of day. Mark infested plants with a bright-colored pole or flag so they can be easily found.

Identifying the causal agent at this stage can be challenging; we elaborate upon identification methods in the next subsection. The degree of damage can be estimated with **visual scales** (Fig. 1.6). The American plant pathologist Nathan Cobb first devised visual scales. Cobb worked with disease damage, but his scales have also been used for estimating insect damage (Tehon and Stout, 1930).

Environmental monitoring keeps an eye on the weather. Weather affects the severity of diseases and pests, and the efficacy of control methods. For example, if cutworms are a problem, cool, wet weather causes them to proliferate.

Wet weather hampers cutworm biocontrol by *Trichogramma* wasps, but enhances biocontrol by *Steinernema* nematodes. Environmental monitoring is easier in growrooms and glasshouses, because indoor growers have better control over their environments. Small, battery-powered sensors provide data on sunlight, air temperature, soil moisture, fertilizer levels, and more. New ones have bluetooth connectivity, and send reports to your smart phone.

Implementing IPM strategies comes next. The person in charge of monitoring pests and the environment should also be the decision-maker who implements control strategies. If not, then the monitor and the decision-maker must closely communicate. Similarly, if the decision-maker and implementer are separate, communication is key.

The **IPM pyramid** is a useful guide for choosing IPM strategies (Fig. 1.7). As the pyramid illustrates, preventive strategies are the foundation of IPM methods. Moving up the pyramid, to biological and finally to chemical controls, the degrees of intervention and toxicity increase.

Post-intervention monitoring follows implementation. In the "good old days," farmers sprayed DDT and were done,

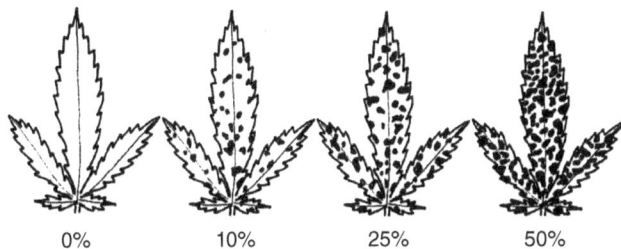

Fig. 1.6. Visual scale for damage assessment, with affected area covering 0% to 50% of the surface (McPartland *et al.*, 2000)

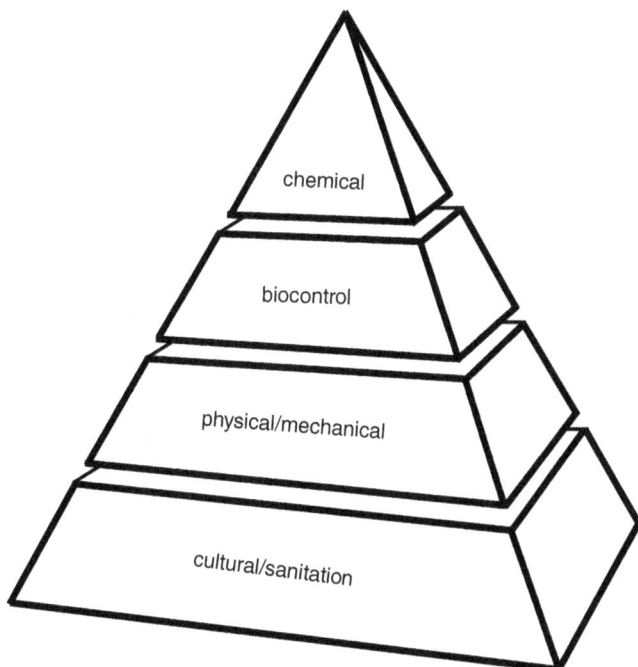

Fig. 1.7. IPM pyramid

knowing pests were dead. Not anymore. Today, monitoring pests and pathogens must continue *after* a control intervention. Post-intervention monitoring provides feedback for evaluating the effectiveness of the IPM program. Biocontrol does not kill all pests, so the *percentage* of control must be estimated.

1.4 Signs and symptoms

A **sign** is visual evidence of the causal organism itself, such as a throbbing mass of spider mites and their webbing. Excrement is a sign: **honeydew** (a sticky liquid excreted from sap-sucking insects), **fecal spotting** (dark stains from small insects), and **frass** (solid excreta of large caterpillars and grasshoppers). Slugs leave mucus trails. Signs of fungi are usually cryptic little black dots—their reproductive structures. When you look at powdery mildew, you're seeing a sign, not a symptom—the fungus itself.

A **symptom** is a detectable change in the plant. Symptoms include leaf yellowing, loss of turgor (wilting), and swelling in branches or the stalk. Symptoms of spider mites begin as small, light-colored leaf spots. Symptoms caused by fungi include leaf spots and stem cankers. You are not actually seeing the fungus, but rather the symptom it is causing.

Nutritional imbalances produce symptoms, not signs, and notoriously non-specific. Symptoms caused by drought (wilting) can be confused with symptoms caused by fungi, insects, and nematodes—anything that attacks roots. In general, symptoms are less useful for identifying the cause of a problem than are signs. We provide a list of common signs and symptoms in Table 1.1.

Ferentinos *et al.* (2019) used artificial intelligence (AI) to diagnose three fungal diseases, two insect pests, and three nutritional deficiencies, based on signs and symptoms. They trained an image-processing computer with *Cannabis* images downloaded from the internet. They reported 91% accuracy; the most difficult problems to diagnose were powdery mildew and potassium deficiency.

1.4.1 Arthropods

Stippling consists of small, pale specks and spots left by sap-sucking insects. They may coalesce into larger chlorotic spots (pale green or white). The stippling pattern may be random (e.g., from spider mites) or linear, in parallel with leaf veins (from thrips). **Shotholes** are small round holes in leaves, caused by leaf-chewing insects and some fungi. **Notching** is like a shothole, but along the margin of the leaf, caused by leaf-chewing insects. The cuts may be semicircular or angular. **Cupping and curling** are responses to damage during leaf expansion. They are characteristic of sap-sucking insects.

Folding and rolling is caused by caterpillars that roll over the margins of leaves, and feed within the fold. Some spin silk to web together margins. **Skeletonizing** is caused by leaf-chewing insects that selectively feed on the delicate leaf tissue between veins, leaving behind a skeleton of leaf veins. Chewing insects also bore into branches and stalks (**stem**

Table 1.1. Common signs and symptoms, with their most common causes

Sign or symptom	Causes
No seedlings	Old seed, cold soil, damping off fungi, seed-eating insects
Wilting	Too little or too much H_2O, nutritional imbalances, leaf-sucking insects, root insects, wilt fungi, wilt bacteria, nematodes
Spindly stalks	Not enough light, too much yellow light, deficient N, K, or calcium
Lumpy stalks	European corn borers, hemp borers, beetle grubs, canker fungi, stem nematodes
Disfigured roots	Soil fungi, nematodes, broomrape, grubs, maggots, rodents
Missing limbs or tops	Deer, cattle, rodents, humans
Missing seeds in flowering tops	Birds, budworms, hemp borers
Chlorotic (pale green or yellow) leaves	Deficient N, poor pH, nematodes, bacteria, leaf-sucking insects, herbicides
Leaves with large discolored spots	Leaf-sucking insects, leaf fungi, leafminers, excess fertilizer
Leaf margin discoloration	Excess fertilizer, deficient K, dry air, brown blight
Leaf stippling	Mites, thrips, leafhoppers, whiteflies, plant bugs
Leaf shotholes	Caterpillars, flea beetles, grasshoppers, fungi, bacteria
Leaf margin notching	Caterpillars, beetles, grasshoppers, flea beetles, weevils, sawflies
Leaf margin cupping and curling	Thrips, aphids, bugs, fungi, herbicides, cold injury
Leaf folding and rolling	Hemp borers and other caterpillars
Leaf skeletonizing	Hemp borers, budworms, other caterpillars, flea beetles, Japanese beetles, sawflies

borers), roots (**root borers**), or even into the narrow spaces inside leaves (**leafminers**). **Seed eaters** may puncture seeds and suck out the contents, or break open seeds and shell out the contents, or eat seeds whole.

1.4.2   Nematodes

Nematodes, with one exception, infest roots. Their above-ground symptoms mimic those of root injury—stunting, chlorosis, and **insipient wilting** (drooping of leaves during midday with recovery at night). Below-ground symptoms can be distinctive, including **root knots** or **root galls**, which are irregular lumps and bumps in feeder roots. Signs of nematodes—the culprits themselves—are generally under a millimeter in size. Our singular exception, the stem nematode, *Ditylenchus dipsaci*, invades stems and branches. It produces symptoms of swelling, twisting, and curling of infested plants.

1.4.3   Fungi

Fungi cause many types of symptoms. They may obstruct plant xylem, causing a **wilt** (drooping of leaves or shoots), or **leaf spots** (localized necrotic lesions of dead and collapsed cells), **blights** (non-localized, general necrosis of leaves, flowers, and shoots), and **blotches** (irregular necrotic leaf lesions with indistinct chlorotic margins, often accompanied by wilting). If wilting becomes permanent, whole branches suffer **dieback**—an extensive blight and necrosis, often beginning at branch tips. **Cankers** are localized, well-defined necrotic lesions on stalks or branches; cankers may also cause wilting and dieback. **Rot** is a brown liquifying necrosis, indicating complete tissue destruction. A **root rot** indicates necrosis and collapse of part or all of the root system. **Crown rot** involves the crown

(the transition zone between the root and stalk). **Damping off** is a rapid collapse of young seedlings.

Some signs are easily recognizable. **Rust fungi** produce pustules of spores that look like rust spots on leaves, especially under magnification (Fig. 1.8B). **Powdery mildew** arises on leaves as a thin covering of powdery, white spores. **Black mildew** and **sooty mold** appear as black growths on leaves; the latter is associated with aphid honeydew. These characteristic signs can narrow the differential diagnosis down to a handful of species.

1.4.4   Bacteria

Symptoms caused by bacteria are similar to those caused by fungi. They include chlorosis, necrosis, rosettes, wilting, leaf spots, blotches, blights, cankers, diebacks, root rots, crown rots, and wilts. A soil-borne species, *Agrobacterium tumefaciens*, causes hyperplasia and hypertrophy of plant cells, called a **crown gall**—a mutant, cancer-like growth. Phytoplasmas are a special group of bacteria that produce two unique symptoms. **Witches' broom** is a shortening of internodes, with a proliferation of axillary buds into branches, causing a bushy appearance. **Phyllody** arises in flowering tops, where floral organs (bracts, stamens) transform into small leafy structures, adding to the bushy appearance.

Bacteria are microscopic, but their signs can be seen. Place a leaf with a leaf spot in a moisture chamber. The chamber can be a petri dish with moist filter paper (off to one side, not touching the leaf). Humid conditions cause bacteria to ooze from infected surfaces, into tiny puddles of viscid fluid. Bacteria that colonize xylem (wilt organisms) can be visualized by the "strand test": slice a stalk, approximate the ends, then gently pull apart. Bacteria oozing from cut xylem vessels will produce mucous strands. Alternatively, suspend a sliced stalk in a beaker of water, and look for heavier-than-water bacterial slime flowing down in a cloudy stream from the cut end.

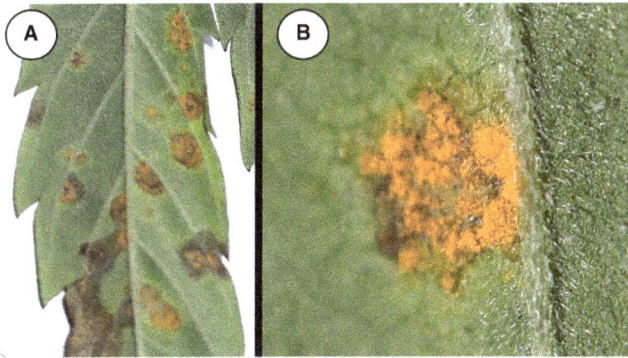

Fig. 1.8. Sign of a rust fungus: **A.** macro; **B.** micro (courtesy John Bruce)

1.4.5 Viruses and viroids

Common symptoms caused by viruses and viroids include **stunting** of entire plants, and **chlorosis**, which is a yellowing of leaves. Chlorosis over the whole leaf is called **virus yellows**. Chlorosis may form in circular patterns as **ring spot**, in striped patterns as **hemp streak**, or appear randomly over the leaf as **virus mosaic**. Viruses may also cause **hypertrophy** and **hyperplasia** of leaves and flowers, which is a distorted enlargement and thickening of tissues. The lamina and leaf margins of infected plants can also become distorted and appear as **wrinkle leaf**.

1.5 Identifying causal agents

Identifying the cause of crop damage can be difficult. A seasoned entomologist or plant pathologist may identify the species at first glance. For the rest of us, several items of diagnostic equipment come into play. These items can be purchased from sources listed by BIRC (Bio-Integral Resource Center) (https://www.birc.org). For difficult-to-find tools, such as collecting tubes and aspirators, go to Gempler's (https://gemplers.com).

The ultimate tool for identifying pests and pathogens is an expert. The USA government has employed consultants since 1854, when Asa Fitch was hired as an entomologist. The State of New York hired the first plant pathologist, Joseph C. Arthur, in 1882. Today the US Department of Agriculture (USDA) maintains a network of county extension agents across the country; find your local office on the web.

Legalization has emancipated extension agents, heretofore unwilling or unable to work with *Cannabis* growers. *Cannabis*-specific extension publications are appearing, and even a book-length expert's guide, *Diagnosing hemp and cannabis crop diseases* (Wang, 2021). With the crop out of the closet, *Cannabis* has attracted plant pathologists and entomologists at universities. Research money has flowed into the field, enabling the adoption of new technology, previously unaffordable. Money has also attracted "disease consultants" wielding new technology, but often lacking the prerequisites needed for proper interpretation of results. We elaborate on this below, in the section on molecular techniques.

To identify causal agents, diagnosticians traditionally used morphological characters. Microscopes came into play. Biochemical tests were employed, which necessitated *in vitro* culturing of bacteria and fungi. The first edition of *HD&P* relied on these methods (see section 1.5.1). We called ourselves "unapologetic morphologists," but no longer. We've gone molecular (see section 1.5.2).

1.5.1 Tools for morphological identification

Light microscopes (LMs) range from single-lens "magnifying glasses" to compound LMs using two or more lenses. **Digital microscopes** use a camera with magnified optics that displays an image on a screen. Several people can observe the image at the same time, which can be captured and saved to a digital file. Digital microscopes are widely available and affordable.

Specialists enhance LM optics by using polarized light or specific wavelengths of light (fluorescence LMs). The latter can be enhanced with fluorescent stains, such as DAPI, a nucleic acid stain that fluoresces in ultraviolet light. High-tech fluorescence LMs, fitted with pinhole apertures, are known as **confocal microscopes**. They capture multiple images at different depths in a sample, and reconstruct the images with computer software, which yields good representations of 3-D shapes. The latest iteration of this technology is light-sheet microscopy (Stelzer *et al.*, 2021). Livingston *et al.* (2019) published fabulous photos of *Cannabis* trichomes using confocal laser-scanning LM.

LMs are limited to magnifications <2000×, with a resolution of about 200 nanometers (nm), due to light diffraction. Electron microscopes (EMs) can magnify up to 10,000,000× with 50 picometer resolution. EMs use electrons instead of light, and electromagnets instead of glass lenses. **Transmission electron microscopes** (TEMs) transmit electrons though extremely thin sections of specimens. **Scanning electron microscopes** (SEMs) bounce electrons off the surface of a specimen and collect the backscatter. SEMs magnify less than TEMs, but they can image specimens with a greater depth of field. SEMs produce images that are good representations of 3-D shapes (Fig. 1.9).

Morphological identification: fungi

To identify fungi by morphology requires an examination of spore-producing structures. Often these structures can be dissected directly from *Cannabis* and viewed under a microscope (Fig. 1.10). Morphology easily identifies fungi to genus, but often falls short at the species level. Spore size and shape in a species may vary, making identification challenging. "Immature" fungi, which lack spore-producing structures, defy morphological identification.

Fungi may need to be isolated and cultured on **artificial media** in petri plates (Fig. 1.10). Artificial media can be natural or synthetic. Natural media include potato dextrose agar, corn meal agar, and V-8 juice agar. Synthetic media include

Czapek-Dox agar and Sabourad agar. Synthetic media contain defined amounts of carbohydrates, nitrogen, and vitamin sources, and can be duplicated with precision every time they are made. Adding antibiotics to media (e.g., ampicillin, streptomycin, chloramphenicol) will prevent bacterial contaminants from overrunning the fungal culture.

Morphological identification: bacteria

Morphological characters for identifying bacteria are limited. Under LM, bacteria are seen as spherical, rod-shaped, or rarely spiral-shaped. SEM adds little information beyond LM microscopy. TEM can detect flagella placement, and the absence of a cell wall (in phytoplasmas). **Gram staining** is a simple LM technique that divides bacteria into two major groups: "gram-negative" bacteria stain red or pink, and "gram-positive" bacteria stain purple.

Fig. 1.9. *Aspergillus*, conidiophore with conidia (SEM, 800×)

Invariably, bacteria must be cultured in artificial media. Nutrient agar is a general-purpose medium for growing bacteria. A variant, known as plate count agar (PCA) or standard methods agar (SMA), is used to count "total" or viable bacteria in a *Cannabis* sample. Bacterial counts are done for consumer protection, rather than diagnosing plant disease, because human pathogens may contaminate *Cannabis* and cause disease outbreaks (see section 18.3.7).

Morphological identification: insects

Most insects can be collected with tweezers or a hand trowel, plus a flashlight (many are nocturnal). An aspirator or venturi suction trap is useful for collecting small, mobile insects. A penknife may be needed to extract recalcitrant individuals from stalks. Smashed insects are difficult to identify, so a collection jar keeps them incarcerated for closer scrutiny. Insects can be knocked out of foliage with muslin sweep nets or beating sheets. Baited insect traps allow you to monitor pests 24/7 (see section 19.6.1).

Measuring morphological traits to speciate insects and mites can be difficult and time consuming. Phenotypic plasticity often complicates morphological identification. A magnifying lens (10× to 16×) is usually needed. Some species require microscopic examination of their genitalia, after chemical clearance with potassium hydroxide.

Immature insects that lack genitalia—larvae, caterpillars, grubs, maggots, etc.—are harder to identify than adults, and the immatures are often causing the damage on plants. Perversely, many identification guides illustrate only adults. With care, captured larvae can be nurtured into adulthood for proper identification.

Morphological identification: nematodes

There are several ways to extract nematodes from soil or plant tissues, and they vary in efficacy (EPPO, 2013). The Baermann funnel method is simple and produces a high yield of nematodes, but takes 2–5 days. Soil or plant samples are chopped up

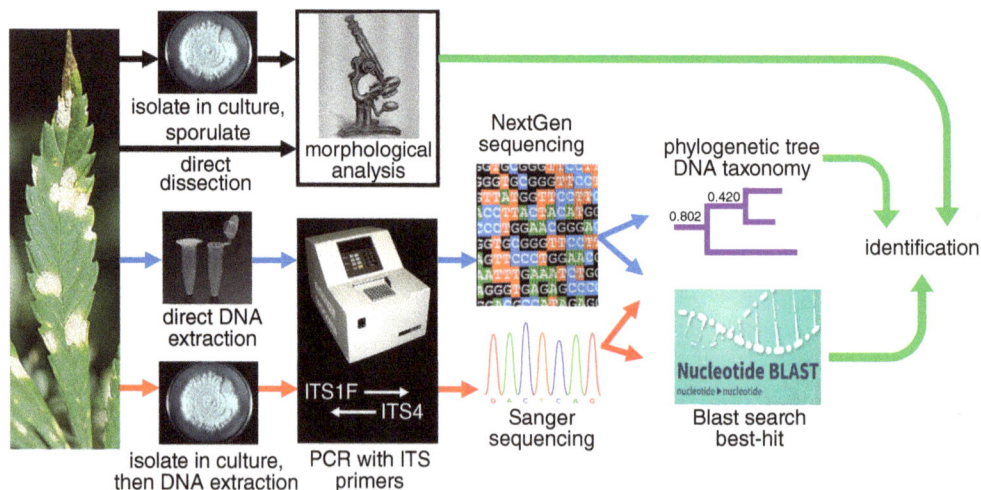

Fig. 1.10. Flowchart for identifying unknown fungi

and placed in a funnel lined with a filter (cheesecloth or a paper towel). The funnel is filled with water. Nematodes leave the sample, pass through the filter, and sink to the bottom of the funnel. This only works with motile nematodes.

Motile nematodes can also be extracted with flotation and sieving methods. Add 200 mL soil to 5 liters of water in a bucket, and stir vigorously. Allow the soil to settle, and pour the supernatant through a series of three sieves with 50 μm apertures. Then wash material collected on the sieves into a Baermann funnel.

Centrifugal flotation mixes a soil sample in a suspension of water and a solute (sugar or $MgSO_4$) with a specific gravity greater than nematodes. The nematodes will float, and the heavier soil particles will sink. Kaolin is added, which forms a visible white layer between the suspension (with the nematodes) and the heavier particles. Centrifugal flotation is more expensive, but is fast, and delivers a clean nematode suspension. It is the only way to extract both motile and immotile nematodes. Several less common extraction methods are also detailed by EPPO (2013), available online.

Simply inspecting roots can identify some nematodes to genus, such as root knot nematodes (genus *Meloidogyne*). Using morphology to identify nematodes to species is time-consuming, labor-intensive, and requires specialized expertise. It entails a microscopic study of the head, anal shield, or reproductive organs (Mai *et al.*, 1996). Nematodes are the most difficult animals to identify to species level.

Morphological identification: viruses

And then there are viruses. Simply determining their presence can be difficult. Seeing them requires a TEM microscope. Samples for TEM are "negative stained" using electron-opaque heavy metal salts (e.g., uranyl acetate) or gold particles conjugated with antibodies for viral glycoproteins. This gives the TEM image a dark background, which contrasts with the surface features of virus particles (see Fig. 14.3).

Morphology observed with a TEM—particle shape and size—can classify viruses to family level (Hull, 2014). Most *Cannabis* viruses are helical: either rigid rod-shapes (e.g., tobacco mosaic virus), or flexuous shapes (e.g., lettuce chlorosis virus). Icosahedral viruses are roughly spherical (e.g., cannabis cryptic virus), and some consist of multiple icosahedral particles joined together (e.g., cucumber mosaic virus). Bacilliform viruses have rounded ends separated by a short tubular section (e.g., alfalfa mosaic virus).

1.5.2 Tools for molecular identification

DNA biotechnology is replacing microscopes and petri plates. In the first edition of *HD&P*, monoclonal antibody-based **ELISA tests** (enzyme-linked immunosorbent assays) were the rage. ELISA tests can instantly *confirm* the identify of a fungus causing disease (e.g., *Botrytis cinerea*, *Rhizoctonia solani*, and *Sclerotinia sclerotiorum*). But ELISA tests are specific to individual species, and therefore require prior knowledge of a suspected pathogen. **Lateral flow immunoassay** (LFIA) kits have replaced ELISA tests. They are simpler, cheaper, and quicker. LFIA kits are used in roadside tests for THC (Plouffe and Murthy, 2017) and in home pregnancy tests. LFIA kits are antibody-based, like ELISA, so they only confirm the presence of a suspected organism.

Methods that utilize the **polymerase chain reaction** (PCR) are taking over. PCR technology is relatively new. According to Chen *et al.* (2020), the first publication that used PCR-based methods to identify a *Cannabis* pathogen (McPartland and Cubeta, 1997) was followed by a 12-year gap before the next study. In the past decade, however, dozens have been published.

PCR makes millions of copies of a specific DNA sequence, thereby amplifying a DNA fragment from the target organism into a larger sample amenable to analysis. Small oligonucleotide fragments called **primers** are chosen to target specific DNA sequences for amplification. The choice of primers makes PCR-based identification quite flexible. At one end of the spectrum, primers can be designed to detect a single species. At the other end of the spectrum, **universal primers** can amplify a targeted sequence of all organisms in a given phylum (fungi, bacteria, nematodes, insects).

DNA amplified with PCR can then be sequenced. There are two types of sequencing methods (Fig. 1.10). Older **Sanger sequencing** reads one fragment or sequence of DNA at a time using DNA polymerase. High-throughput **next-generation sequencing** (NextGen or NGS) can read thousands or millions of sequences at the same time, using parallel arrays of DNA polymerase. The NextGen fragments are then reassembled, based on overlapping segments, using various software programs (e.g., Sequencher, Bioedit, Geneious).

As illustrated in Fig. 1.10, PCR can amplify fungal DNA sampled directly from a diseased leaf, or amplify DNA after the fungus has been isolated in pure culture. A diseased leaf typically harbors multiple fungi: the disease pathogen, airborne molds, and other non-pathogenic fungi. This poses a problem for Sanger sequencing, which sequences only one DNA fragment at a time. In the presence of multiple fungi, Sanger cannot distinguish which fragment came from which organism. The sequence reads will overlap, yielding undecipherable data. Therefore, isolating the pathogen in pure culture is required for Sanger sequencing (Fig. 1.10).

NextGen sequencing can independently sequence DNA fragments from multiple fungi at once, so no culturing is necessary. This results in a GIGO error: NextGen will amplify *every* fungus in a *Cannabis* specimen—which sequence belongs to the actual pathogen? The first *Cannabis* study that used NextGen technology yielded an astonishing 75 species from flowering tops (McKernan *et al.*, 2016a). More than 80% of the species had never been reported as *Cannabis* pathogens. Many of them were no doubt endophytes rather than pathogens (see section 13.4). Two reported species were clearly not plant pathogens—*Trichophyton rubrum* (ringworm fungus) and *Candida albicans* (candidiasis yeast). They came off the hands of people trimming buds.

The dichotomy between morphological and molecular techniques is particularly acute for diagnosing fungal diseases. When encountering a diseased *Cannabis* specimen, Rivedal *et al.* (2022) decided whether to use morphological or molecular techniques based on plant symptoms, as follows.

1. Symptoms of wilting, leaf chlorosis and necrosis, flower and seed browning, and stem or root discoloration, or with visible signs of fungal or oomycete mycelia: *isolate in culture and identify morphologically.*
2. Symptoms of leaf mosaic, leaf cupping, leaf curling, shortened internodes, and plant stunting: *extract DNA and amplify.*

Real-time PCR (qPCR) monitors the amplification of a targeted DNA molecule while it happens, by coupling PCR with fluorescent probes or DNA-binding dyes. Commercial kits use qPCR for the detection of a predefined target. For example, if *Botrytis cinerea* is suspected, test the diseased plant tissue with a kit that restricts PCR amplification to a DNA sequence specific to *B. cinerea.*

Real-time PCR has a clear utility in testing cannabis products destined for human consumption. Fungal contamination is a public health issue, because fungi can cause allergies, asthma, or opportunistic infections. Several companies sell multiplex qPRC tests, which detect a predefined panel of the most frequent offenders (e.g., *Aspergillus flavus*, *A. fumigatus*, *A. niger*) (see Fig. 18.6). As a field-deployable technology, qPCR has difficulties (e.g., alternating thermal cycles). This led to a new generation of field tests based on loop-mediated isothermal amplification (LAMP). LAMP primers amplify DNA at a constant temperature, and the amplicon can be detected via fluorescence or simple color changes. Fernandez i Marti *et al.* (2023) developed primers for a LAMP assay to detect hop latent viroid (HLVd), a serious threat to the *Cannabis* industry.

DNA barcodes

Running through a series of qPCR assays for predefined targets is too expensive, and the options are limited. Instead, a segment of the unknown organism's DNA can be sequenced, and compared with a reference library of known DNA sequences. These sequences are known as **DNA barcodes**. The premise of DNA barcoding is the same as supermarket scanners—they read the black stripes of the UPC barcode to identify an item against the scanner's reference UPC database.

The field procedure is easy: extract and purify DNA from diseased plant tissue with a commercial kit, such as the DNeasy Plant Kit (by Qiagen). To extract insect DNA, commercial kits include the Qiagen DNeasy Blood and Tissue DNA purification kit. Purified DNA is brought into the lab and amplified with PCR, using universal primers that target the organism's phylum (fungi, bacteria, nematodes, insects). Between the primers, nucleotide variation is often sufficient to discern individual species.

Insect barcode

The DNA barcode for animals (including insects and us) derives from a mitochondrial gene, cytochrome *c* oxidase I (*COI* or COX1). The entire gene has ~1500 base pairs (bp), but the *COI* barcode uses a 648 bp segment. This is an exceedingly small sample of an animal's total DNA. The human genome contains 3 billion bp, so the barcode-to-genome ratio is 5×10^6. The *Oxford English Dictionary* contains 350 million letters, and its ISBN barcode has ten characters, a barcode-to-content ratio of 4×10^7 (Schindel, 2005).

Paul Hebert, who spearheaded DNA barcoding, proposed a 2.7% difference between two *COI* sequences as the threshold for flagging distinct species (Hebert *et al.*, 2004). Any two humans differ by only one or two nucleotides in their *COI* barcodes. My *COI* barcode (obtained from 23andMe®) differs from the Neanderthal barcode at six nucleotides (Fig. 1.11). Six nucleotides represent a difference of only 0.99%—less than the 2.7% difference proposed by Herbert. Indeed, Neanderthals are the same species as us, *Homo sapiens neanderthalensis.* In contrast, the difference between my barcode and the chimp barcode is 8%—different species, as well as different genera (*Pan troglodytes*).

With the *COI* barcode of an unknown insect in hand, its sequence can be compared with those in a reference database. The largest databases are **GenBank** and **BOLD** (Barcode of Life Database). Both databases contain sequences for about 318,000 arthropod species (with more every year). Searching the databases is simple. The National Center for Biotechnology Information (NCBI) provides free BLAST services (https:// blast.ncbi.nlm.nih.gov). **BLAST** (Basic Local Alignment Search Tool) is an algorithm that compares the unknown *COI* barcode with sequences in NCBI's database (GenBank). Submitting the unknown "query" sequence to NCBI's BLAST server will determine its closest match (best-hit) in GenBank.

The metric quantifying a best-hit is "sequence identity"— the extent to which the unknown query sequence has identical nucleotides at the same positions as the best-hit sequence. If the unknown query shares ≥97.3% sequence identity (i.e., <2.7% difference) with its best-hit, you have a match.

Tahir *et al.* (2018) collected insect pests in rice fields and compared the accuracy of morphological identification to that of barcoding. Morphological methods failed to identify immature caterpillars in the Noctuidae, Crambidae, and Hesperiidae families. *COI* sequencing was performed on 59 insects, of which only 48 specimens were successfully sequenced. *COI* sequencing resolved the identities of 46 specimens that shared ≥97.3% sequence identity with a species in the GenBank database. Two insects had *COI* sequences with no close resemblance to anything in the database.

Meiklejohn *et al.* (2019) compared the accuracy of the GenBank and BOLD databases. They sequenced *COI* barcodes from 17 insects—curated reference specimens in a museum. GenBank outperformed BOLD, with 53% and 35% correct identifications, respectively. Paul Hebert's team at BOLD took offense, and examined the possible causes of this low success rate. Pentinsaari *et al.* (2020) determined that Meiklejohn's misidentification arose in sequencing, and not in the BOLD sequence library.

Fungal barcode

The barcode for fungi is the internal transcribed spacer (ITS) region (Schoch *et al.*, 2012). The ITS region includes genes for

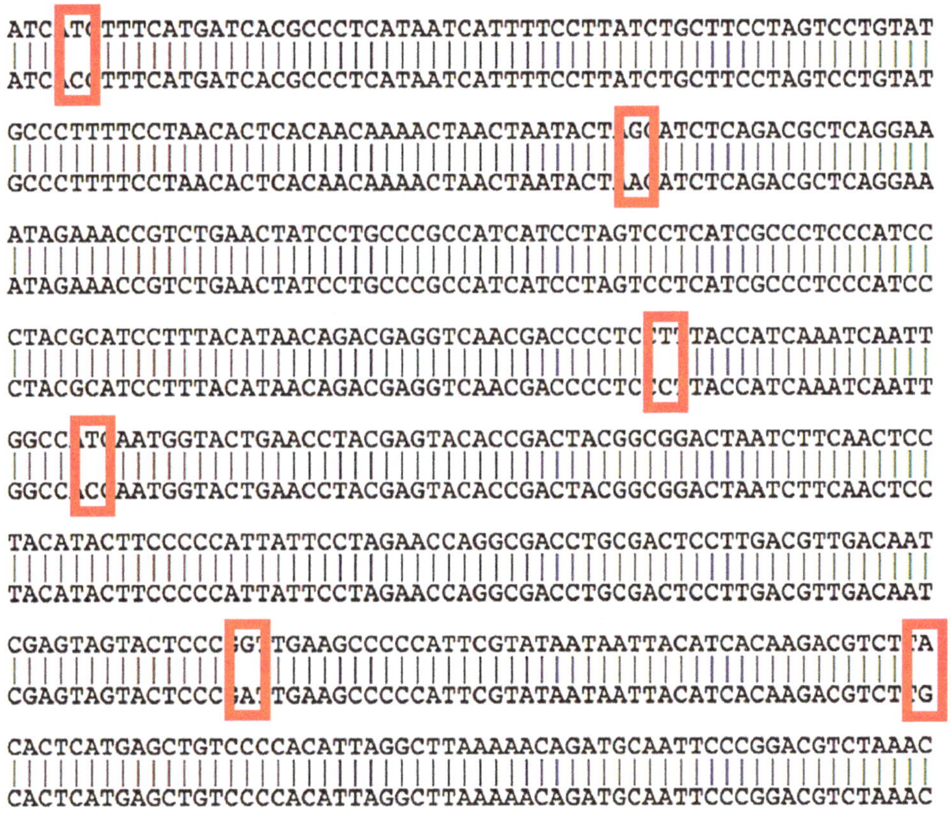

Fig. 1.11. Pairwise comparison of human and Neanderthal *COI* barcodes, with six non-identical nucleotide sites highlighted

all or parts of three ribosomal subunits, 18S, 5.8S, and 25–28S, and most importantly, the highly variable spacers between them: *ITS1* (between 18S and 5.8S) and *ITS2* (between 5.8S and 25–28S). With the ITS sequence of an unknown fungus in hand, there are two approaches to fungal identification: the BLAST best-hit method, or the phylogenetic "family tree" method.

BLAST best-hit method

This approach compares the unknown ITS sequence with its closest match in GenBank. A decade ago, GenBank contained 172,000 fungal ITS sequences, and 56% were linked with a Latin binomial, representing 15,500 species (Schoch *et al.*, 2012). This is an exceedingly small sample of the Kingdom Fungi, with 120,000 described species (Hawksworth and Lücking, 2017). Thus, the metrics quantifying a best-hit have been relaxed for fungi: a sequence identity ≥97%, within a "query coverage" (percentage of the unknown query sequence that overlaps with the best-hit) of ≥80% (Raja *et al.*, 2017).

Meiklejohn *et al.* (2019) compared the accuracy of the GenBank and BOLD databases. They sequenced barcodes from 16 curated reference specimens of mushroom fungi (*Amanita, Psilocybe, Inocybe, Conocybe, Clitocybe, Gyromitra*

spp.). The two databases performed equally poorly, with only 57% correct identifications. Why? The sequences deposited at GenBank are a crowd-sourced endeavor, and may be misidentified (Raja *et al.*, 2017). If an unknown's BLAST best-hit lands on a misidentified GenBank sequence, the unknown will be assigned an incorrect identification.

We tested the accuracy of the BLAST best-hit method using *Bipolaris, Drechslera,* and *Curvularia* spp. *B/D/C* fungi are difficult to identify by morphological methods (see Fig. 10.34). We searched GenBank for sequences annotated as *Cannabis* pathogens and identified as *B/D/C* spp. There were six. Assuming that sequences of other *B/D/C* spp. may have been misidentified in GenBank, we conducted a BLAST search of GenBank for fungi with similar sequences (> 90% identity), that were annotated as *Cannabis* pathogens. This search retrieved three additional accessions, which were identified in GenBank as *Cochliobolus* sp. (a synonym of *Bipolaris* sp.) or *Alternaria* spp.

We compared the putative identities of nine retrieved sequences with their best-hits in the Genbank database (Table 1.2). In three of the nine cases, the identity of the best-hit in GenBank did not match the putative identity assigned to the *Cannabis* pathogen by the original author (mismatches highlighted in red font in Table 1.2). One was way off: the best-hit of a *Cochliobolus* sp. was a *Stemphylium* sp.

Phylogenetic method

The other identification method places an unknown ITS sequence in a phylogenetic "backbone tree" of ITS sequences from related organisms, to find its closest relative. Hyde *et al.* (2014) provided "backbone trees" for 25 groups of plant pathogenic fungi. Their reference sequences consisted of authenticated accessions of reliably identified species. Many other "backbone trees" of plant pathogenic fungi have been made available since then.

The phylogenetic method is more robust than the BLAST best-hit method, but it is more complicated. It requires multiple sequence alignment (MSA), phylogeny estimation software, and tree-building tools. In the resulting phylogenetic tree, branching patterns reflect how species evolved from a series of common ancestors.

Phylogenetics, like the BLAST best-hit method, uses sequence variation among different species to identify taxa. Phylogenetics, however, uses an *explicitly statistical framework*. The unknown ITS sequence and its closest relative will share a common ancestor in a "sister" relationship. Robustness of the relationship is quantified by "bootstrap support," a statistical measure ranging from 0.0 (completely random) to 1.0 (a strongly supported relationship).

We tested the phylogenetic method using the same nine sequences in Table 1.2. The backbone tree was constructed from 15 authenticated accessions (Zhang and Berbee, 2001; Manamgoda *et al.*, 2012, 2014). The analysis was performed online, using the software platform at MABL (www.phylogeny. fr). MUSCLE 3.8.31 performed the multiple sequence alignment (MSA). Gblocks 0.91b trimmed ambiguously aligned

positions from the MSA. Sequences with excessive lengths in the MSA were trimmed by hand. PhyML 3.1 analyzed the MSA in a phylogenetic framework using maximum likelihood. TreeDyn provided tree visualization.

Results with the phylogenetic method were largely consistent with the BLAST best-hit method. For example, "*Bipolaris* sp." whose best-hit was *Curvularia spicifera* (Table 1.2) fell into the *Curvularia* clade (Fig. 1.12). "*Cochliobolus* sp." KX641964, whose best-hit was *Stemphylium* sp. (Table 1.2), did not fall into an identifiable clade (Fig. 1.12), because we did not include *Stemphylium* spp. in the backbone tree. "*Cochliobolus* sp." KX641967, whose best-hit was *Bipolaris sorokiniana* (Table 1.2), fell into the *Bipolaris* clade (Fig. 1.12). Bootstrap support values >60% are shown in branches of the tree.

In summary, there are pitfalls when using ITS sequences as a stand-alone method for identifying *Cannabis* pathogens. Shortfalls lie in incorrect identification, and in the proper interpretation of results. Incorrect identification can be minimized by sequencing other genes in addition to ITS, such as elongation factor 1α (*EF-1α*), β-tubulin (*TUB*), calmodulin (*CAL*), or actin (*ACT*). The proper interpretation of results requires training in plant pathology. "Disease consultants" wielding this new technology, but lacking prerequisite training, report nonsense like ringworm fungi as *Cannabis* pathogens.

Some of the species in Table 1.2, such as *Bipolaris* sp. KY781820 and *Drechslera* sp. KX641972, were probably endophytes rather than pathogens. Punja *et al.* (2019) used ITS to sequence several fungi from *Cannabis*, and then they conducted pathogenicity tests to sort endophytes from pathogens.

Table 1.2. Testing the putative identity of *Cannabis* fungi in a BLAST best-hit analysis (mismatches in red)

Putative identity, geographical location of specimen, GenBank number	Best hit in GenBank Species identification GenBank number	Identity %, query coverage %
Bipolaris sp. USA MK477544	Bipolaris sp. KX512317	100, 99
Bipolaris sp. Pakistan KY781820	Curvularia spicifera MH010905	100, 100
Bipolaris sp. China MH881227	Bipolaris victoriae MN856279	100, 98
Curvularia pseudobrachyspora USA MT072019	Curvularia aeria MT470608	99.77, 100
Drechslera sp. Canada KX641972	Pyrenophora tritici-repentis (=Drechslera tritici-repentis) KM011994	97.71, 95
Cochliobolus sp. Canada KX641964	Stemphylium sp. MN153952	100, 100
Cochliobolus sp. Canada KX641967	Cochliobolus sativus (=Bipolaris sorokiniana) KC311473	99.91, 100
Alternaria sp. Canada KX641959	Alternaria alternata MT453271	100, 100
Alternaria sp. Canada KX641966	Alternaria alternata MT453271	99.81, 100

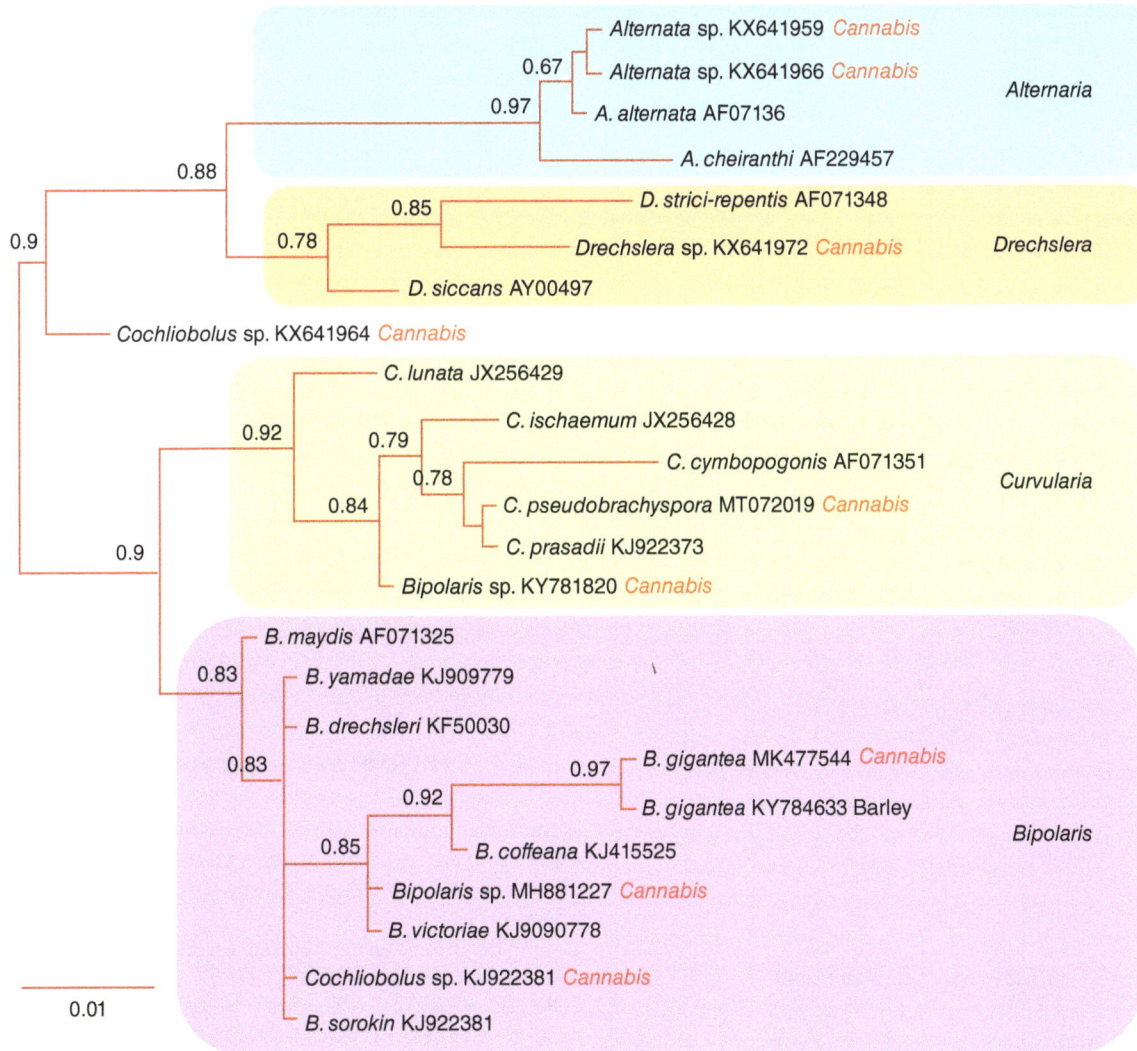

Fig. 1.12. Phylogram generated from a maximum likelihood analysis of ITS sequences of nine *Cannabis* fungi plus reference sequences

Bacterial barcode

The bacterial barcode is based on the 16S rRNA gene sequence. Some *Cannabis* pathogens, such as *Pseudomonas syringae* and *Xanthomonas campestris*, have wide host ranges, so these species are further divided into categories called **pathovars** with narrower host ranges. Identifying a pathovar formerly involved laborious host susceptibility testing with living plants. With the advent of 16S rRNA gene sequencing, the classification of pathovars infecting *Cannabis* has undergone complete overhauls (see Chapter 15).

Testing consumer products for bacterial contamination usually involves artificial culture in petri plates, which can be considered slow and sometimes inaccurate. Because of these shortcomings, and to accelerate turn-around time, some laboratories use qPCR assays with primers for 16S rRNA. McKernan *et al.* (2015) compared qPCR and three petri plate- or film-based detection systems. They tested 17 *Cannabis* samples obtained from dispensaries. Six samples tested positive with

qPCR, five samples tested positive with Simplate Biocontrol Systems™, four samples tested positive with 3M Petrifilm™, and one sample tested positive with BioLumix™. Drawbacks to qPCR include cost, and the method's indifference to living or dead DNA.

Nematode barcode

The *COI* barcode in nematodes performs poorly compared with *COI* in insects, so other sequences are jockeying for primacy, including *COII*—the second subunit of cytochrome *c* oxidase. Another candidate, the LSU (large subunit) rRNA sequence, includes the three-gene array of 5S, 5.8S, and 28S (including their internal spacers). The SSU (small subunit) rRNA sequence consists of the 18S gene.

Kiewnick *et al.* (2014) required a multi-locus approach to identify *Meloidogyne* at species level. They combined SSU rDNA with either *COI*, *COII*, or LSU rDNA. Powers *et al.* (2018) designed a *Meloidogyne*-specific primer set for *COI*, and tested it on 322 specimens. Most of them could be identified to species, but several specimens had *COI* sequences with no resemblance to anything in GenBank. Nematodes are the most difficult animals to identify to species level.

Viral barcode

A universal barcode for viruses is impossible, because their genetic material is all over the place. Two-thirds of plant virus genomes are (+)-strand ssRNA, but the remaining third consists of (–)-strand ssRNA, ssDNA, or dsDNA replicating by reverse transcription (Hull, 2014). PCR primers for the detection of *specific* viruses are available, for example cannabis cryptic virus (Ziegler *et al.*, 2012; Righetti *et al.*, 2018).

Instead of barcodes, whole viral genomes can be sequenced using NextGen sequencing. Plant virus genomes are small. Satellite viruses, which require a helper virus for replication, have genomes as small as 1 kb (the *COI* barcode is 0.648 kb). The largest viral genomes are under 20 kb. In comparison, bacterial genomes range from 130 to 14,000 kb. Barba *et al.* (2014) reviewed NextGen approaches for detecting and identifying viruses in plants.

Serological procedures and fluorescent antibody tests provide rapid and simple methods for identifying viruses. However, they are specific to individual viruses, so they only confirm the presence of a suspected pathogen. ELISA kits are available for widespread viruses (e.g., alfalfa mosaic virus, Arabis mosaic virus, cucumber mosaic virus, lettuce chlorosis virus), but these viruses rarely infest *Cannabis*. Righetti *et al.* (2018) developed a qPCR assay to detect cannabis cryptic virus.

1.5.3   Identification keys

Identifying pests and pathogens is not as hard as we might make it sound. With practice, it becomes easy. We frequently encounter the same pests and pathogens around the world.

Identification keys are great tools for diagnosing common crop problems. The best keys are host-plant specific, so we provide a few for *Cannabis*. There are basically two types of identification keys: synoptic and dichotomous. Synoptic keys rely on pattern recognition, while dichotomous keys are structured decision trees.

In addition to *Cannabis*-specificity, keys can also be devised for specific stages of growth (seedlings versus flowering plants), different plant parts (flowers, leaves, stalks, roots), and indoor versus outdoor crops. A series of synoptic keys are presented in Table 1.3, to help identify frequently encountered problems.

1.5.4   Proving pathogenicity

Koch's postulates are four criteria designed for establishing a causative relationship between a microbe and a disease. Examining diseased *Cannabis* may reveal the presence of *many*

microbial organisms. This is especially true when using molecular DNA methods. Which one of the microbes is the pathogen? Koch's postulates consist of the following: (1) isolate a microbe from a diseased organism; (2) grow the microbe in pure culture (in a petri dish); (3) the cultured microbe causes disease when introduced into a healthy organism; (4) reisolate the microbe from the diseased organism (Koch, 1891).

The second postulate is problematic for viruses, which cannot be grown in pure culture; they require host cells to grow and reproduce. Because of this, Koch's postulates could not be used to document the SARS-CoV-2 virus as the cause of Covid-19 disease, which led to crackpot denialism. Another test of causality was applied: a Covid-19 challenge study, in which healthy young volunteers were exposed to the SARS-CoV-2 virus and developed the disease (Killingley *et al.*, 2022). Alternative tests of causality have been applied to *Cannabis* viruses (see section 14.2).

The second postulate cannot be fulfilled with biotrophic fungi, which cannot grow in pure culture, such as powdery mildew fungi. For powdery mildew fungi on *Cannabis*, causality has been demonstrated by tapping conidia (spores) off infected leaves onto healthy plants, and watching for disease development (e.g., Pépin *et al.*, 2018; Farinas and Peduto Hand, 2020).

1.6   History of *cannabis* crop protection

Our work builds on earlier efforts by hundreds of men and women. Here we highlight a few previous researchers and their work. For a more complete history, see the two-part series by McEno (1987, 1988).

The first insects reported on *Cannabis* were leaf-eating omnivores: locusts (Justell, 1686) and *Autographa gamma* (Réaumur, 1736). The first *Cannabis*-specific insect was *Phorodon cannabis* (Passerini, 1860). Giovanni Passerini (1816–1893), a botanist and entomologist at Parma, is the only person who has identified a new insect species on *Cannabis* (*P. cannabis*) as well as a new fungus (*Ophiobolus cannabinus*, Passerini 1888). The first mite was *Tetranychus urticae* (Cuboni, 1889). However, the first *Cannabis* pest of any sort was a bird. In AD 544, Jiǎ Sīxié, an agronomist in Shāndōng, coined the name 麻雀 (*má què*, hemp bird) for the seed-eating Eurasian tree sparrow, *Passer montanus* (Jiǎ, 1962).

The first pathogen described on *Cannabis* happens to be the physically largest one: a root-parasitic plant, *Phelipanche ramosa* (Mattioli, 1554). The first fungus described on *Cannabis* was made by an American, **Lewis David von Schweinitz** (1780–1834) (Fig. 1.13B). Schweinitz (1832) found the fungus near Winston-Salem, and named it *Sphaeria cannabis*. He was an interesting character—a cigar-smoking minister, and the first American to earn a PhD. He named other hemp diseases, and made one of the first sightings of feral hemp growing in North America (Schweinitz, 1836).

In Europe, the earliest reported fungi were leaf pathogens: *Ascochyta cannabis* (Lasch, 1846), *Erysiphe communis* (Westendorp, 1854), and *Depazea cannabis* (Kirchner, 1856). *Septoria cannabis* was described in Belgium (Westendorp, 1857), and three weeks later in Germany (Rabenhorst, 1857). Our modern-

Table 1.3. Six "top 10" lists of common disease and pest problems, indexed with section numbers in the text

Category	Section
Seed and seedling problems	
Cutworms	8.3
Crickets	8.5
Damping-off oomycetes and fungi	12.2–12.3
Rodents	16.6
Slugs and snails	16.4
Flea beetle	5.4
Birds	16.5
Insufficient light or water	17.5–17.7
Overwatering	17.7.3
Old or genetically sterile seed	17.8
Flower and leaf problems, outdoors	
Gray mold	9.2
Aphids	4.6
Yellow leaf spot	10.2
Nutritional diseases	17.3
Budworms and Armyworms	5.2–5.3
Flea beetles	5.4
Downy mildew	10.5
Plant bugs	4.14
Hemp leaf spot	10.10
Deer, rabbits and rodents	16.6
Stem and branch problems, outdoors	
Eurasian hemp borers	6.2
European corn borers	6.3
Beetle & weevil grubs	6.8–6.10
Gray mold	9.2
Sclerotinia canker and crown rot	10.3
Fusarium canker	10.4
Rhizoctonia sore shin	11.6
Anthracnose	10.8
Striatura ulcerosa (bacteria)	15.2
Stem nematodes	15.9
Root problems, outdoors	
Root knot nematodes	15.7
Pythium root rot	11.2
Fusarium crown and root rot	11.3
Southern blight	11.5
Rhizoctonia sore shin and root rot	11.6
Broomrape	16.1
Rodents	16.6
White root grubs	8.3
Wire worms	8.4
Root maggots	8.8
Whole-plant problems, outdoors	
Fusarium wilt	11.4
Twig blight	11.7
Charcoal rot	11.8
Nutritional diseases	17.3
Overwatering	17.7.3
Virus, viroid and phytoplasma diseases	14.2–14.7
Dodder	16.2
Russet mites	4.2
Aphids	4.6
Armyworms	5.3

Table 1.3. Continued.

Category	Section
Common indoor problems	
Spider mites	4.2
Russet mites	4.2
Aphids	4.6
Thrips	4.7
Whiteflies	4.8
Root aphids	8.2
Powdery mildew	9.1
Gray mold	9.2
Virus, viroid & phytoplasma diseases	14.2–14.7
Nutritional diseases	17.3

Fig. 1.13. Outstanding *Cannabis* researchers: **A.** Vavilov; **B.** Schweinitz; **C.** Charles; **D.** Haberlandt; **E.** Kirchner; **F.** Dewey holding a male plant growing next to a female plant. Photos courtesy USDA archives, except (B) (*Popular Science Monthly*, April 1894, frontispiece), (D) (*Wiener Landwirtschaftliche Zeitung*, 2 November 1878, p. 1), and (F) (*Journal of Heredity* 1916, p. 326)

day plague, gray mold (caused by *Botrytis cinerea*), was first described on *Cannabis* by Hazslinszky (1877).

Carlo Berti Pichat (1799–1878) first illustrated a fungus infesting hemp, *Septoria cannabis*. He was educated to become a politician, and became one (going into exile after the failed Italian insurrection against Austria in 1848). But his passion was agronomy, and living in Bologna meant hemp. Berti Pichat

(1866) wrote a 103-page review of hemp cultivation and diseases and pests, which we cite throughout this book.

Jean-Henri Fabre (1823–1919) is France's best-known entomologist. Fabre's family grew hemp and his grandmother spun hemp thread (Fabre, 1897). As a boy, he dove into hemp fields to catch goldfinches (Fabre, 1918). In his series of popular books on science, he frequently used hemp as a metaphor. Fabre (1913) compared web-spinning by spiders to rope-makers paying out hemp while walking backwards. In a model of the solar system, Mercury is the size of a hemp seed (Fabre, 1872). In a trippy portrayal of time-compression, needed to appreciate natural phenomena, Fabre (1914) wrote about "this sublime phantasmagoria of the grain of hemp which in a few hours has been transmuted into the finest cloth."

Fabre (1876a) wrote a book about the plant world and dedicated a chapter to hemp, with a nice illustration of the plant. He proposed that *Cannabis* with its *odeur vireuse* was native to the East Indies. Fabre (1876b) highlighted countries that grew hemp; in India they grew hemp to smoke *hatchisch*, and he passed no judgement. After all, "match sticks are made of hemp."

Friedrich J. Haberlandt (1826–1878) (Fig. 1.13D) became an agronomy professor in Vienna, after he quit working as a lawyer. Unsung in history books, Haberlandt is quoted throughout our book. He was interested in hemp seeds—their chemistry (Haberlandt, 1860a), and their ability to germinate at low temperatures (Haberlandt, 1874, 1875). Haberlandt (1879) estimated the total leaf area of a hemp plant, its transpiration rate, and the *Wärmesummen* (heat-sum) needed to grow fiber-type hemp.

Haberlandt (1860b) was a pioneer in biogeography, but his book's few copies were destroyed in WWII. All we have is a fragment—he described wild hemp in Hungary (Haberlandt, 1861). Haberlandt (1877) initiated research on gender inheritance in hemp. Lastly, Haberlandt (1878) published a 20-page analysis on the load-bearing capacity of hemp fiber. He died that year, too young, from a botched surgery.

Oskar Kirchner (1851–1925, Fig. 1.13E) wrote his dissertation on Theophrastus, and noted that the ancient Greek botanist did not mention *Cannabis* (Kirchner, 1874). Many authors still erroneously say that Theophrastus did. Kirchner (1883) elaborated on Haberlandt's study of seedlings growing at low temperatures; hemp could grow as low as low as 0.75°–1.0°C. Kirchner and Eichler (1900) studied wild-type *Cannabis* growing in southwestern Germany. Then Kirchner turned to plant pathology and entomology, and wrote about all kinds of *Cannabis* problems—fungi and insects, as well as nematodes, bacteria, and parasitic plants (Kirchner, 1890, 1906). His artwork was admired by many; his classic lithograph of *Septoria cannabis* (Kirchner and Boltshauser, 1898) was plagiarized many times.

Lyster Hoxie Dewey (1865–1944) (Fig. 1.13F) had a long career in the USDA, beginning with taxonomic work on the Gramineae. He wrote about medicinal herbs, and deplored the over-harvesting of wild goldenseal (Dewey, 1905). His ecological perspective was precocious—Dewey explained how destroying our native prairie enabled tumbleweed (Russian thistle, *Salsola australis*) to spread across midwestern rangelands. Dewey's contemporaries, in contrast, believed tumbleweed was a Russian plot to destroy American agriculture (Young, 1991). Dewey led fiber-plant investigations at the USDA from 1899 to 1935. He became a champion of hemp, dedicating his energies and talents to the advancement of *Cannabis*.

Dewey imported seeds from all over the world, from fiber *and* drug plants, and evaluated them on American soil. For decades he grew hemp at two sites (Fig. 1.14): Arlington Farms (ironically now where the Pentagon sits), and across the Potomac River in Potomac Flats (now the Lincoln Memorial). An ecologist at heart, Dewey made paper out of hemp hurds instead of tree pulp. "There seems to be little doubt that the present wood supply cannot withstand indefinitely the demands placed upon it ... Our forests are being cut three times as fast as they grow" (Dewey and Merrill, 1916). Unfortunately, Dewey lived to see his hemp efforts undone by Harry Anslinger's anti-marihuana propaganda.

Dewey was not expert in pests and pathogens. For these problems he collaborated with **Vera K. Charles** (1877–1954) (Fig. 1.13C). Dr Charles was one of the first women appointed to a professional position in the USDA. She worked at Arlington Farms and Potomac Flats (Fig. 1.14), and collected *Cannabis* specimens from across the country. Charles collaborated with **Flora W. Patterson** (1847–1928), the first female mycologist hired by the USDA. Patterson spearheaded passage of the Plant Quarantine Act of 1912. Together they coauthored state-of-the-art research on a new fungus attacking hemp (Charles and Jenkins, 1914).

Dewey's counterpart in Russia was **Nikolai I. Vavilov** (1887–1943) (Fig. 1.13A). Among the great scientists of the 20th century, Vavilov has been lionized as an international statesman of agriculture and plant genetics (Medvedev, 1969). Vavilov collected *Cannabis* from around the globe, including three trips to Central Asia, which he considered the center of *Cannabis* diversity.

Vavilov's research was terminated by political action, like Dewey's. But Vavilov himself was terminated. After he published *The Origin of the Cultivation of our Primary Crops, in Particular of Cultivated Hemp*, Vavilov locked horns with T.D. Lysenko. Lysenko fabricated nonsense genetic theories based on Marxist doctrine, and became a powerful toady of Stalin. Lysenko had Vavilov arrested. Shortly before Vavilov died in one of Stalin's gulags, he wrote for the ages, "We shall go to the

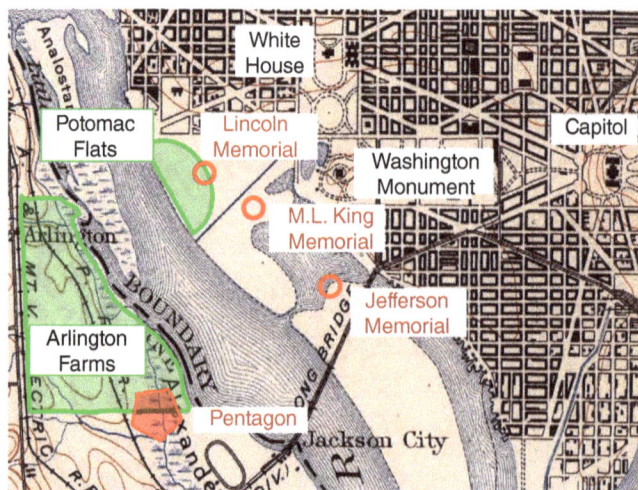

Fig. 1.14. Location of Arlington Farms and Potomac Flats. US Geological Survey map, Washington quadrangle, 1900 edition. Future buildings labeled in red

pyre, we shall burn, but we shall not renounce our convictions" (Medvedev, 1969).

Stalin also executed the phytopathologist **Veniamin S. Bakhtin** (1888–1937), on trumped-up charges of counter-revolutionary activity. Bakhtin coined a new species, *Macrosporium cannabinum,* and we reproduce his artwork (see Fig. 10.38). Many other *Cannabis* researchers died during WWII, including Kirchner and Klebahn in Germany, Curzi in Italy, Guilliermond in France, Lange in Denmark, and Komarov and Tranzschel in the USSR.

One survivor was **Kurt Röder** (1909–1990). He researched a range of *Cannabis* problems—fungi, viruses, insect vectors—before his Berlin laboratory was bombed in 1945. After the war Röder developed fungicides for Schering (including maneb and zineb), and patiently answered McPartland's questions in the 1980s.

Italy became a center for hemp research in the 1950s, thanks to the tireless efforts of Floriano Ferri and Gabriele Goidànich in Bologna, and Carmine Noviello in Napoli. Eastern Europe got busy, with publications from Poland, Romania, Bulgaria, the former Yugoslavia and USSR, and especially Hungary. Hungarian research has been led by Barnabas Nagy (an entomologist) and Ivan Bòsca (an agronomist).

In China, entomology and plant pathology have combined the old and the new: time-honored customs and modern methods. Two plant pathologists pulled China into modernity: Dài Fānglán (戴芳澜, 1893–1973) and Dèng Shūqún (邓叔群, 1902–1970). Both of them earned doctorates at Cornell University, and studied hemp diseases. Both of them were purged during China's Cultural Revolution, and died in rural poverty.

The study of pests and pathogens became schizophrenic during the 1960s and 1970s, thanks to the rise of anti-*Cannabis* biocontrol research funded by the USA. While Europeans tried to kill pests and pathogens of hemp, USA researchers used the same pests and pathogens to kill marijuana. Arthur McCain, a biocontrol researcher at UC-Berkeley, said this schizophrenic era ended when his research grants were cancelled by the Carter administration (Zubrin, 1981).

Meanwhile, illicit *Cannabis* growers began hiring disease and pest consultants—Frank, Rosenthal, Clarke, McEno, and the Bush Doctor. They published in gray journals such as *High Times* and *Sinsemilla Tips*. With miniscule budgets and simple tools, they accumulated data. We still cite their work.

The 1980s and 1990s saw a resurrection of industrial hemp in Western Europe. This sparked new monographs on hemp pests (Spaar *et al.*, 1990; Gutherlet and Karus, 1995). New disease research came from Holland (de Meijer, 1993, 1995; Kok *et al.*, 1994; Van der Werf, 1994). In the USA, McPartland (1983, etc.) initiated phytopathology research at the University of Illinois. The university has long history of hemp research (Tehon and Boewe, 1939; Hackleman and Domingo, 1943; Tehon, 1951; Gerdemann, 1955; Boewe, 1963; Haney and Bazzaz, 1970; Haney and Kutscheid, 1973, 1975), and groundbreaking cannabinoid research (summarized by Adams, 1942).

The USDA's anti-*Cannabis* research in the 1970s had a silver lining: after the program was declassified in 1986, McPartland gained access to its archives—about 15 linear feet of world literature on diseases and pests. The USDA's nest egg hatched the first edition of *HD&P* (McPartland *et al.*, 2000). His co-authors, Robert Clarke and David Watson, were Amsterdam-based agronomists. They sent problems to McPartland; he'd make diagnoses and recommend control measures; they'd trial the controls and report what worked. Clarke and Watson are cited throughout this text. It's a collaboration 40 years along (imagine that in academia).

Around the same time, a new generation of scientists saw dollar signs in anti-*Cannabis* biocontrol (e.g., Sands, 1991; Tiourebaev *et al.*, 1998, US Patents 6673746, 6403530). They shuttled government-funded research to Sands's private corporation, Ag/Bio Con. Public outrage (Fields, 1998) and scientific scrutiny (McPartland and West, 1999) helped curtail the research. Sands collected *Cannabis* pathogens in Kazakhstan and Tajikistan, sometimes without obtaining clearance from the State Department (FOIA documents released to Jeremy Bigwood). Rules were bent in those days. *Cannabis* specimens collected in Nepal and India were smuggled home, resulting in new discoveries (McPartland and Hughes, 1994; McPartland and Cubeta, 1997).

As stated by Kovalchuk *et al.* (2020), legalization was required before "true scientists" could work with *Cannabis*. Canada legalized fiber-type *Cannabis* cultivation in 1994. Canadians soon published work on pests and pathogens (e.g., Scheifele *et al.*, 1997; Bains *et al.*, 2000; Small *et al.*, 2007; Feeney and Punja, 2017; Pépin *et al.*, 2018; Punja, 2018).

Everything changed after the USA Farm Bills of 2014 and 2018 legalized fiber-type and CBD crops, respectively. Back in the 1990s, McPartland proposed editing a *Hemp Compendium* for the American Phytopathological Society, but academicians spurned collaboration. While writing the first edition of *HD&P*, university researchers ridiculed us as stoners, shunned and underfunded. What a difference legalization has made—researchers are chasing dollars, which have flowed into the field.

Entomologists started publishing research (Britt *et al.*, 2017; Cranshaw *et al.*, 2018; Bolt and Skidmore, 2019; Cranshaw *et al.*, 2019) as did plant pathologists (Beckerman *et al.*, 2017; Thiessen and Schappe, 2019; Koike *et al.*, 2019; Gauthier, 2020). This book will be the last of its kind: a comprehensive review. With *Cannabis* legalization, the scientific literature is expanding in too many directions to be captured in one tome. The future belongs to subspecialty texts (e.g., Wang 2021) and summary compendia (e.g., the APS *Hemp Compendium*, forthcoming).

(Prepared by J. McPartland; revised by H. Grab)

References

Adamek K, Jones AMP, Torkamaneh D. 2022. Accumulation of somatic mutations leads to genetic mosaicism in cannabis. *Plant Genome* 15: e20169.

Adams R. 1942. Marihuana. *Bulletin New York Academy of Medicine* 18: 705-730.

Albrecht WA (Walters C, *ed*). 1975. *Albrecht's foundational concepts. The Albrecht papers, vol. 1*. Acres USA, Austin, Texas.

Alter H, Peer R, Dombrovsky A, Flaishman M, Spitzer-Rimon B. 2022. Tobacco rattle virus as a tool for rapid reverse-genetics screens and analysis of gene function in *Cannabis sativa* L. *Plants* 11(3): e327.

Anderson LC. 1980. Leaf variation among *Cannabis* species from a controlled garden. *Harvard University Botanical Museum Leaflets* 28(1): 61–69.

Bains PS, Bennypaul HS, Blade SF, Weeks C. 2000. First report of hemp canker caused by *Sclerotinia sclerotiorum* in Alberta, Canada. *Plant Disease* 84: 373.

Barba M, Czosnek H, Hadidi A. 2014. Historical perspective, development and application of next-generation sequencing in plant virology. *Viruses* 6: 106–136.

Beckerman J, Nisonson H, Albright N, Creswell T. 2017. First report of *Pythium aphanidermatum* crown and root rot of industrial hemp in the United States. *Plant Disease* 101: 1038–1039.

Berti Pichat CB. 1866. "La pianta della canapa," pp. 396-499 in *Istituzioni scientifiche e tecniche, ossia Corso teorico e practico de Agricoltura, Libri XIII, Volume Quinto*. Presso l'Union Tipografico-Editrice, Torino.

Boewe GH. 1963. Host plants of charcoal rot disease in Illinois. *Plant Disease Reporter* 47: 753–755.

Bolt M, Skidmore A. 2019. IPM in hemp: managing pests in a "new" crop. *Entomology Today*, Available at https://entomologytoday.org/2019/09/05/hemp-ipm-managing-pests-new-crop/

Bose RD. 1930. A study of sex in the Indian hemp. *Agricultural Journal of India* 25: 495–507.

Brickell CD, *ed*. 2016. *International Code of Nomenclature for Cultivated Plants, Ninth edition*. International Society for Horticultural Sciences, Leuven, Belgium.

Britt KE, Byrd J, Kuhar T, Fisk J. 2017. *Facts about industrial hemp*. Virginia Cooperative Extension, no CSES-196NP. Available at: https://vtechworks.lib.vt.edu/bitstream/handle/10919/83956/CSES-196.pdf?sequence=1

Charles VK, Jenkins AE. 1914. A fungous disease of hemp. *Journal of Agricultural Research* 3: 81–84.

Chen Y, Tang XY, Gao CS, Li ZM, *et al*. (4 other authors). 2020. Molecular diagnostics and pathogenesis of fungal pathogens on bast fiber crops. *Pathogens* 9: e223.

Cranshaw WS, Halbert SE, Favret C, Britt KE, Miller GL. 2018. *Phorodon cannabis* Passerini (Hemiptera: Aphididae), a newly recognized pest in North America found on industrial hemp. *Insecta Mundi* 662: 1–12.

Cranshaw W, Schreiner M, Britt K, Kuhar TP, McPartland JM, Grant J. 2019. Developing insect pest management systems for hemp in the United States: a work in progress. *Journal of Integrated Pest Management* 10(1): e26.

Cuboni G. 1889. Nota dei casi di malatti de vegetali presentati alla R. Stazione di patologia vegetale di Roma durante i mesi di agosta e settember 1889. *Bollettino di Notizie Agrarie* 11: 1942–1949.

de Meijer EPM. 1993. Evaluation and verification of resistance to *Meloidogyne hapla* Chitwood in a *Cannabis* germplasm collection. *Euphytica* 71: 49–56.

de Meijer EPM. 1995. Fiber hemp cultivars: a survey of origin, ancestry, availability and brief agronomic characteristics. *Journal International Hemp Association* 2(2): 66-73.

de Meijer EPM, van Soest LJM. 1992. The CPRO *Cannabis* germplasm collection. *Euphytica* 62: 201–211.

Dewey LH. 1905. Golden seal. *Hunter-Trader-Trapper* 10(6): 21.

Dewey LH. 1914. "Hemp," pp. 283-347 in: *USDA Yearbook 1913*. United States Department of Agriculture, Washington, DC.

Dewey LH, Merrill JL. 1916. *Hemp hurds as paper-making material*. USDA Bulletin No. 404. Washington, DC.

Dixon E, Leonberger K, Amsden B, *et al*. (6 other authors). 2022. Suppression of hemp powdery mildew using root-applied silicon. *Plant Health Progress* 23: 260–264.

Doyle R. 2007. "The transgenic involution," pp. 70–82 in Kac E, *ed. Signs of life*. Massachussetts Insititute of Technology, Cambridge, Massachusetts.

EPPO (European and Mediterranean Plant Protection Organization). 2013. PM 7/119 (1) Nematode extraction. *EPPO Bulletin* 43(3): 471–495.

Fabre JH. 1872. *Astronomie élémentaire*. C. Delagrave, Paris.

Fabre JH. 1876a. *La plante: leçons à mon fils sur la botanique*. C. Delagrave, Paris.

Fabre JH. 1876b. *Géographie*. C. Delagrave, Paris.

Fabre JH. 1897. *Souvenirs entomologiques, sixième série*. C. Delagrave, Paris

Fabre JH (Teixeira de Mattos A, *trans*). 1913. *The life of the spider*. Dodd, Mead & Co., New York.

Fabre JH (Mial B, *trans*). 1914. *Social life in the insect world*. The Century Co., New York.

Fabre JH (Teixeira de Mattos A, *trans*). 1918. *The wonders of instinct*. The Century Co., New York.

Farinas C, Peduto Hand F. 2020. First report of *Golovinomyces spadiceus* causing powdery mildew on industrial hemp (*Cannabis sativa* L.) in Ohio. *Plant Disease* 104: 2727.

Feeney M, Punja ZK. 2017. "The role of *Agrobacterium*-mediated and other gene-transfer technologies in *Cannabis* research and product development," pp. 343–363 in Chandra S, *et al*., eds. *Cannabis sativa: Botany and Biotechnology*. Springer International Publishing, Cham, Switzerland.

Ferentinos KP, Barda M, Damer D. 2019. "An image-based deep learning model for *Cannabis* diseases, nutrient deficiencies and pests identification," pp. 135–145 in Oliveira PA, *et al*., eds. *Progress in artificial intelligence*. Springer International, Cham, Switzerland.

Fernandez i Marti A, Parungao M, Hollin J, *et al*. (5 other authors). 2023. A novel, precise and high-throughput technology for viroid detection in *Cannabis* (MFDetect™). *BioRxiv Preprint*, https://doi.org/10.1101/2023.06.05.543818

Fields G. 1998. U.S. might enlist fungi in drug war. *USA Today* 17(28) (22 Oct 1998): 1.

Gauthier N. 2020. "Considering crop rotations and the potential for carry-over," p. 25 in Gauthier N, Leonberger K, Bowers C, *eds. Science of hemp: production and pest management*. University of Kentucky Agricultural Experiment Station, no. SR-112, Lexington, KY. Available at: https://plantpathology.ca.uky.edu/files/sr112.pdf

Gerdemann JW. 1955. Relation of a large soil-borne spore to phycomycetous mycorrhizal infections. *Mycologia* 47: 619–632.

Goethe JW von. 1812. *Zür Farbenlehre*. Geistinger, Wien.

Goethe JW von. 1815. *Aus meinem Leben: Dichtung und Wahrheit, Dritter Theil*. Geistinger, Wien.

Goethe JW von (Schmitz LD, *ed*). 1884. *Miscellaneous travels of J. W. Goethe*. George Bell & Sons, London.

Goethe JW von (Albrecht W, *ed*). 2007. *Tagebücher: Band V,1 Text (1813–1816)*. Springer-Verlag, Stuttgart.

Gutherlet V, Karus M. 1995. *Parasitäre Krankheiten und Schädlinge an Hanf (Cannabis sativa L.)*. Nova Institute, Köln.

Haberlandt FJ. 1860a. Einige Bemerkungen über Secretions-Erscheiimngen an Pflanzen im Allgemeinen, und über Ausscheidungen an gequellten Sämereien im Besonderen. *Oesterreichische Botanische Zeitschrift* 10: 118–121.

Haberlandt FJ. 1860b. *Die wichtigsten Culturpflanzen und Unkräuter nach ihren Standorten zusammengestellt.* Alexander Czéh, Ungarisch-Altenburg, Hungary.

Haberlandt FJ. 1861. Von Keszthely nach Tiliany. *Oesterreichische Botanische Zeitschrift* 11: 10–19.

Haberlandt F. 1874. Die oberen und unteren Temperaturgrenzen für die Keimung der wichtigeren landwirtschaftlichen Sämereien. *Landwirtschaftlichen Versuchs-Stationen* 17: 104–116.

Haberlandt F. 1875. Ueber die untere Grenze der Keimungstemperatur der Samen unserer Getreidepflanzen. *Wissenschaftlich-Praktische Untersuchungen auf dem Gebiete des Pflanzenbaues* 1: 109–117.

Haberlandt FJ. 1877. Welche Einflüsse bedingen das Geschlecht der Hanfpflanzen? *Fühlings Landwirtschaftliche Zeitung* 26: 881.

Haberlandt FJ. 1878. Versuche über die Tragfähigkeit und Elasticität der Bastbänder gerösteter Hanfpflanzen. *Forschungen auf dem Gebiete der Agriculturphysik* 1: 415–434.

Haberlandt F. 1879. *Der allgemeine landwirtschaftliche Pflanzenbau.* Faesy & Fick, Vienna.

Hackleman JC, Domingo WE. 1943. Hemp: an Illinois war crop. *University of Illinois Agricultural Experiment Station Circular No. 547.* Urbana, Illinois.

Haney A, Bazzaz FA. 1970. "Some ecological implications of the distribution of hemp (*Cannabis sativa* L.) in the United States of America," pp. 39–48 *in* Joyce CRB, Curry SH, eds. *The botany and chemistry of Cannabis.* J & A Churchill, London.

Haney A, Kutscheid BB. 1973. Quantitative variation in the chemical constituents of marihuana from stands of naturalized *Cannabis sativa* L. in East-Central Illinois. *Economic Botany* 27: 193–203.

Haney A, Kutscheid BB. 1975. An ecological study of naturalized hemp (*Cannabis sativa* L.) in east-central Illinois. *American Midland Naturalist* 93: 1–24.

Hanuš LO, Meyer SM, Muñoz E, Taglialatela-Scafati O, Appendino G. 2016. Phytocannabinoids: a unified critical inventory. *Natural Product Reports* 33: 1357.

Hawksworth DL, Lücking R. 2017. Fungal diversity revisited: 2.2 to 3.8 million species. *Microbiology Spectrum* 5: 79–95.

Hazekamp A, Tejkalová K, Papadimitriou S. 2016. *Cannabis*: from cultivar to chemovar II—a metabolomics approach to *Cannabis* classification. *Cannabis and Cannabinoid Research* 1(1):202–215.

Hazslinszky F. 1877–78. *Polyactis infestans*, (nov. spec.). *Grevillea* 6: 77.

Hebert PD, Stoeckle MY, Zemlak TS, Francis CM. 2004. Identification of birds through DNA barcodes. *PLoS Biology* 2: e312.

Howard A. 1906. "Summary of the report of the imperial economic botanist," pp. 53-54 in *Annual Report of the Imperial Department of Agriculture for the year 1904–05.* Government Central Press, Calcutta.

Howard A. 1942. The Indore method of composting. *Organic Farming and Gardening* 1(1): 6–8.

Howard A. 1943. *An agricultural testament.* Oxford University Press, Oxford, UK.

Howard A, Wad YD. 1931. *The waste products of agriculture.* Oxford University Press, Oxford UK.

Hull R. 2014. "Replication of plant viruses," pp. 341–421 in Hull R, ed. *Plant virology 5th edn.* Academic Press, Amsterdam.

Hyde KD, Nilsson RH, Alias SA, *et al.* (37 other authors). 2014. One stop shop: backbone trees for important phytopathogenic genera: I. *Fungal Diversity* 67: 21–125.

Jackson W. 1980. *New roots for agriculture.* Friends of the Earth, San Francisco.

Jiă SX. 1962. *A preliminary survey of the book Ch'i min yao shu : an agricultural encyclopaedia of the 6th century.* Science Press, Beijing.

Justell RRS 1686. An extract of a letter written from Aramont in Languedoc near Avignon, giving an account of an extraordinary swarm of grasshoppers in those parts. *Philosophical Transactions* 16: 147–149.

Kiewnick S, Holterman M, van den Elsen S, *et al.* (3 other authors). 2014. Comparison of two short DNA barcoding loci (COI and COII) and two longer ribosomal DNA genes (SSU & LSU rRNA) for specimen identification among quarantine root-knot nematodes (*Meloidogyne* spp.) and their close relatives. *European Journal of Plant Pathology* 140: 97–110.

Killingley B, Mann A, Kalinova M, *et al.* (28 other authors). 2022. Safety, tolerability and viral kinetics during SARS-CoV-2 human challenge. *Nature Medicine.* https://doi.org/10.1038/s41591-022-01780-9

Kirchner L. 1856. *Depazea cannabis. Lotos* 6: 183.

Kirchner O. 1874. Die botanischen Schriften des Theophrast von Eresos. *Jahrbücher für classische Philologie, Suppl.* 7: 451–539

Kirchner O. 1883. Ueber das Längenwachstum von Pflanzenorganen bei niederen Temperaturen. *Beiträge zur Biologie der Pflanzen* 3: 335–364.

Kirchner O. 1890. *Die Krankheiten und Beschädigungen unserer landwirtschaftlichen Kulturpflanzen.* Eugen Ulmer, Stuttgart.

Kirchner O. 1906. "Hanf, *Cannabis sativa* L.," pp. 319–323 in *Die Krankheiten und Befehädigungen uhferer landwirtschaftlichen Kulturpflanzen.* E. Ulmer, Stuttgart.

Kirchner O, Boltshauser H. 1898. Blattflecken und blattminen an hanf. *Atlas der Krankheiten und Beschädigungen. Serie 3, Tafel 20.* Verlag von Ulmer, Stuttgart.

Kirchner O, Eichler J. 1900. *Exkursionsflora für Württemberg und Hohenzollern.* Eugen Ulmer, Stuttgart.

Koch R. 1891. "Über bakteriologische Forschung,' pp. 35–47 in *Verhandlung des X Internationalen Medicinischen Congresses, Band 1.* A. Hirschwald, Berlin.

Koike ST, Stangbellini H, Mauzey SJ, Burkhardt A. 2019. First report of sclerotinia crown rot caused by *Sclerotinia minor* on hemp. *Plant Disease* 103: 1771.

Kok CJ, Coenen GCM, de Heij A. 1994. The effect of fibre hemp (*Cannabis sativa* L.) on selected soil-borne pathogens. *Journal of the International Hemp Association* 1(1): 6–9.

Kovalchuk I, Pellino M, Rigault P, *et al.* (13 other authors). 2020. The genomics of *Cannabis* and its close relatives. *Annual Review of Plant Biology* 71: 713–739.

Lapierre É, de Ronne M, Boulanger R, Torkamaneh D. 2023. Phenotypic characterization of a diverse population of *Cannabis sativa* for agronomic, morphological, and biochemical traits. *Preprints.*

Lasch WG. 1846. *Ascochyta cannabis. Klotzchii Herbarium vivum mycologicum, Editio I*, exsiccatus no. 1509.

Livingston SJ, Quilichini TD, Booth JK, *et al.* (7 other authors). 2019. *Cannabis* glandular trichomes alter morphology and metabolite content during flower maturation. *Plant Journal* 101: 37–56.

Lopatin MI. 1936. Поражаемость растений возбудителем корневого рака растений *Bacterium tumefaciens* Sm. A. Town. Микробиология 5: 716–724.

MacKinnon L. 2003. *Genetic transformation of Cannabis sativa Linn: a multi purpose fibre crop.* Doctoral thesis, University of Dundee, Scotland.

MacKinnon L, McDougall G, Aziz N, Millam S. 2000. Progress towards transformation of fibre hemp. *Scottish Crop Research Institute Annual Report 2000/2001.* Scottish Crop Research Institute, Dundee, pp. 84–86.

Mai WF, Mullin PG, Lyon HH, Loeffler K. 1996. *Plant-parasitic nematodes: a pictorial key to genera.* Cornell University Press, Ithaca, New York.

Manamgoda DS, Cai L, McKenzie EHC, Crous PW, *et al.* (5 other authors). 2012. A phylogenetic and taxonomic re-evaluation of the *Bipolaris - Cochliobolus - Curvularia* complex. *Fungal Diversity* 56: 131–144.

Manamgoda DS, Rossman AY, Castlebury LA, Crous PW, *et al.* (3 other authors). 2014. The genus *Bipolaris*. *Studies in Mycology* 79: 221–288.

Mark TB, Shepherd J, Olson DW, *et al.* (3 other authors). 2020. *Economic viability of industrial hemp in the United States: a review of state pilot programs.* USDA Economic Research Service, Economic Information Bulletin no. 217.

Mattioli PA, *ed.* (Dioscorides P). 1554. *Commentarii in sex libros Pedacii Dioscoridis de medica material.* Vincent Valgrisum, Venice.

McEno J. 1987. History of *Cannabis* plant pathology, part I. *Sinsemilla Tips* 7(3): 37–39.

McEno J. 1988. History of *Cannabis* plant pathology, part II. *Sinsemilla Tips* 7(4): 42–44.

McEno J, *ed.* 1991. *Cannabis ecology: a compendium of diseases and pests.* Amrita Press, Middlebury, Vermont.

McKernan K, Spangler J, Zhang L, *et al.* (4 other authors). 2015. Cannabis microbiome sequencing reveals several mycotoxic fungi native to dispensary grade cannabis flowers. *F1000Research* 4: 1422.

McKernan K, Spangler J, Zhang L, *et al.* (6 other authors). 2016. *Cannabis* microbiome sequencing reveals *Penicillium paxilli* and the potential for paxilline drug interactions with cannabidiol. *Proceedings of the 26th Annual Symposium on the Cannabinoids.* International Cannabinoid Research Society, Research Triangle Park, NC, pg. Pl-28.

McPartland JM. 1983a. Fungal pathogens of *Cannabis sativa* in Illinois. *Phytopathology* 72: 797.

McPartland JM. 1983b. *Phomopsis ganjae* sp. nov. on *Cannabis sativa. Mycotaxon* 18: 527-530.

McPartland JM. 1992. The *Cannabis* pathogen project: report of the second five-year plan. *Mycological Society of America Newsletter* 43(1): 43.

McPartland JM, Cubeta MA. 1997. New species, combinations, host associations and location records of fungi associated with hemp (*Cannabis sativa*). *Mycological Research* 101:853-857.

McPartland JM, Hughes S. 1994. *Cannabis* pathogens VII: a new species, *Schiffnerula cannabis. Mycologia* 86: 867-869.

McPartland JM, Small E. 2020. A classification of endangered high-THC cannabis (*Cannabis sativa* subsp. *indica*) domesticates and their wild relatives. *PhytoKeys* 144: 81–112.

McPartland JM, West D. 1999. Killing *Cannabis* with mycoherbicides. *Journal of the International Hemp Association* 6(1): 1, 4–8.

McPartland JM, Clarke RC, Watson DP. 2000. *Hemp diseases and pests – management and biological control.* CABI Publishing, Wallingford, UK.

Mechoulam R. 2005. Plant cannabinoids: a neglected pharmacological treasure trove. *British Journal of Pharmacology* 146: 913–915.

Medvedev Z. 1969. *The rise and fall of T.D. Lysenko.* Columbia University Press, New York.

Meiklejohn KA, Damaso N, Robertson JM. 2019. Assessment of BOLD and GenBank—their accuracy and reliability for the identification of biological materials. *PLoS ONE* 14(6): e0217084.

Oerke EC. 2006. Crop losses to pests. *Journal of Agricultural Science* 144: 31–43.

Passerini G. 1860. *Gli afidi con un prospetto dei generi ed alcune specie nuove italiane.* Tipografia Carmignani, Parma.

Passerini G. 1888. Diagnosi di funghi nuovi. Nota III. *Atti della Reale Accademia dei Lincei. Rendiconti di classe di Scienze Fisiche, Matematiche e Naturale* 4(2): 55–66.

Paull J. 2011. Biodynamic agriculture: the journey from Koberwitz to the new world, 1924–1938. *Journal of Organic Systems* 6(1): 27–41.

Pentinsaari M, Ratnasingham S, Miller SE, Herbert PDN. 2020. BOLD and GenBank revisited – Do identification errors arise in the lab or in the sequence libraries? *PLoS ONE* 15: e0231814.

Pépin N, Punja ZK, Joly DL. 2018. Occurrence of powdery mildew caused by *Golovinomyces cichoracearum sensu lato* on *Cannabis sativa* in Canada. *Plant Disease* 102: 2644.

Pfeiffer EE. (Heckel F, *trans*). 1938. *Bio-dynamic farming and gardening: soil fertility renewal.* Anthroposophic Press, New York.

Pfeiffer EE. 1942. Introduction to the bio-dynamic techniques. *Organic Farming and Gardening* 1(1): 8–9.

Plouffe BD, Murthy SK. 2017. Fluorescence-based lateral flow assays for rapid oral fluid roadside detection of cannabis use. *Electrophoresis* 38: 501–506.

Powers T, Harris T, Higgins R, Mullin P, Powers K. 2018. Discovery and identification of *Meloidogyne* species using COI DNA barcoding. *Journal of Nematology* 50: 399–411.

Punja ZK. 2018. Flower and foliage-infecting pathogens of marijuana (*Cannabis sativa* L.) plants. *Canadian Journal of Plant Pathology* 40: 514–527.

Punja ZK, Collyer D, Scott C, Lung S, Holms J. Sutton D. 2019. Pathogens and molds affecting production and quality of *Cannabis sativa* L. *Frontiers in Plant Science* 10: e1120.

Rabenhorst GL. 1857. *Spilosphaeria cannabis. Klotzchii Herbarium vivum mycologicum, Editio II,* exsiccatus no. 559.

Raja HA, Miller AN, Pearce CJ, Oberlies NH. 2017. Fungal identification using molecular tools: a primer for the natural products research community. *Journal of Natural Products* 80: 756–770.

Réaumur RA de. 1736. *Mémoires pour servir a l'histoire des insectes, tome second.* Imprimerie Royale, Paris.

Righetti L, Paris R, Ratti C, *et al.* (8 other authors). 2018. Not the one, but the only one: about *Cannabis cryptic virus* in plants showing 'hemp streak' disease symptoms. *European Journal of Plant Pathology* 150: 575–588.

Rivedal HM, Funke CN, Frost KE. 2022. An overview of pathogens associated with biotic stresses in hemp crops in Oregon, 2019 to 2020. *Plant Disease* 106: 1334–1340.

Rodale JI. 1945. *Pay dirt.* Devin-Adair Co., New York.

Rodale R. 1985. "The past and future of Regenerative Agriculture," pp. 312–317 in Edens TC, Fridgen C, Battenfield SL, *eds. Sustainable agriculture & integrated farming systems: 1984 Conference proceedings.* Michigan State University Press, East Lansing.

Rodale Institute. 2014. *Regenerative organic agriculture and climate change.* Rodale Institute, Kutztown, Pennsylvania.

Sands DC. 1991. *Interim report: Cannabis sativus.* January/February Report, Cooperative Agreement 58-3K47-9-036. Boseman, Montana.

Savary S, Willocquet L, Pethybridge SJ, *et al.* (3 other authors). 2019. The global burden of pathogens and pests on major food crops. *Nature Ecology & Evolution* 3(3):1.

Sawler J, Stout JM, Gardner KM, *et al.* (5 other authors). 2015. The genetic structure of marijuana and hemp. *PLoS ONE* 10(8): e0133292.

Schachtsiek J, Hussain T, Azzouhri K, Kayser O, Stehle F. 2019. Virus-induced gene silencing (VIGS) in *Cannabis sativa* L. *Plant Methods* 15: e157.

Scheifele G, Dragla P, Pinsonneault C, Laprise JM. 1997. *Hemp (Cannabis sativa) research report, Kent County, Ontario, Canada.* Ridgetown College Agricultural Technology, Ridgetown, Canada.

Schindel D. 2005. *DNA barcoding the Consortium for the Barcode of Life.* Available at: http://dimacs.rutgers.edu/archive/Workshops/Barcode/slides/slides.html

Schoch CL, Siefert KA, Huhndorf S, *et al.* (153 other authors). 2012. Nuclear ribosomal internal transcribed spacer (ITS) region as a universal DNA barcode marker for fungi. *Proceedings of the National Academy of Sciences USA* 109: 6241–6246.

Schoenmakers N. 1986. *The Seed Bank 1986/1987 catalogue.* Drukkerij Dukenburg Printers, Nijmegen, The Netherlands.

Schweinitz LD de. 1832. Synopsis fungorum in American Boreali media degentium. *Transactions of the American Philosophical Society, New Series* 4: 141–316.

Schweinitz LD de. 1836. Remarks on the plants of Europe which have become naturalized in a more or less degree in the United States. *Annals of the Lyceum of Natural History of New York* 3: 148–155.

Scott C, Punja Z. 2022. "Management of diseases on Cannabis in controlled environment production," pp. 231–251 in Zheng YB, *ed. Handbook of Cannabis production in controlled environments.* CRC Press, Boca Raton, Florida.

Seedfinder. 2024. *Seedfinder online database.* Available at: https://en.seedfinder.eu/

Shaw FJF. 1926. "Report of the Imperial Economic Botanist," pp. 13–25 in *Scientific Reports of the Imperial Institute of Agricultural Research, Pusa, 1925–26.* Government of India Central Publication Branch, Calcutta.

Shaw FJF. 1931. "Report of the Imperial Economic Botanist," pp. 10–33 in *Scientific Reports of the Imperial Institute of Agricultural Research, Pusa, 1929–1930.* Government of India Central Publication Branch, Calcutta.

Shaw FJF. 1932. "Report of the Imperial Economic Botanist," pp. 58–78 in *Scientific Reports of the Imperial Institute of Agricultural Research, Pusa, 1930–1931.* Government of India Central Publication Branch, Calcutta.

Small E. 2015. Evolution and classification of *Cannabis sativa* (marijuana, hemp) in relation to human utilization. *Botanical Review* 81: 189–294.

Small E, Cronquist A. 1976. A practical and natural taxonomy for *Cannabis. Taxon* 25(4): 405-435.

Small E, Marcus D, Butler G, McElroy AR. 2007. Apparent increase in biomass and seed productivity in hemp (*Cannabis sativa*) resulting from branch proliferation caused by the European corn borer (*Ostrinia nubilalis*). *Journal of Industrial Hemp* 12(1): 15–26.

Spaar D, Kleinhempel H, Fritzsche R. 1990. *Öl- und Faserpflanzen.* Springer-Verlag, Berlin.

Steiner R. 1897. *Goethes Weltanschuung.* E. Felber, Weimar, Germany.

Steiner R. 1924a. *Geisteswissenschaftliche Grundlagen zum Gedeihen der Landwirtschaft.* R. Steiner Verlag, Dornach.

Steiner R. 1924b. "Report to members of the Anthroposophical Society after the Agriculture Course, Dornach, Switzerland, June 20, 1924," pp. 1–12 in Creeger CE, Gardner M, *trans, ed. 1984. Spiritual foundations for the renewal of agriculture.* Biodynamic Farming and Gardening Association, Junction City, Oregon.

Stelzer EHK, Strobl F, Chang BJ, *et al.* (4 other authors). 2021. Light sheet fluorescence microscopy. *Nature Reviews Methods Primers* 1: e73.

Stern VM, Smith RF, van den Bosch R, Hagen KS. 1959. The integration of chemical and biological control of the spotted alfalfa aphid: the integrated control concept. *Hilgardia* 29: 81–101.

Storl WD. 2013. *Culture and horticulture: the classic guide to biodynamic and organic gardening.* North Atlantic Books, Berkeley, California.

Tahir HM, Noor A, Mehmood S, Sherawat SM, Qazi MA. 2018. Evaluating the accuracy of morphological identification of insect pests of rice crops using DNA barcoding. *Mitochondrial DNA Part B* 3(2): 1220–1224.

Tehon LR. 1951. *The drug plants of Illinois.* Illinois Natural History Survey Circular 44, Urbana, Illinois.

Tehon LR, Boewe GH. 1939. Charcoal rot in Illinois. *Plant Disease Reporter* 23: 312–321.

Tehon LR, Stout GL. 1930. Epidemic diseases of fruit trees in Illinois 1922–1928. *Bulletin of the Illinois Natural History Survey* 18(3): 415–503.

Thiessen LD, Schappe T. 2019. First report of *Exserohilum rostratum* causing foliar blight of industrial hemp (*Cannabis sativa*). *Plant Disease* 103: 1414.

Thomson P. 2019. *Import health standard: seeds for sowing.* Ministry for Primary Industries, Wellington, New Zealand.

Tiourebaev KS, Pilgeram AL, Anderson TW, Baizhanov MK, Sands DC. 1998. *Fusariun oxysporum* f. sp. *cannabina* as a promising candidate for biocontrol of *Cannabis* sp. *Phytopathology* 88: S89.

Turner CE, ElSohly MA, Boeren EG. 1980. Constituents of *Cannabis sativa* L. XVII. A review of the natural constituents. *Journal of Natural Products* 43: 169–234.

Van der Werf HMG. 1994. *Crop physiology of fibre hemp (Cannabis sativa L.).* Doctoral thesis, Wageningen Agricultural University, Wageningen, the Netherlands.

Wang SH. 2021. *Diagnosing hemp and cannabis crop diseases.* CABI, Wallingford UK.

Westendorp GD. 1854. *Les Cryptogames.* I.S. Van Doosselaere. Gand, Belgium.

Westendorp GD. 1857. *Septoria cannabina. Bulletin l'Académie Royal de Belgique Ser.* II, 2(7): 576.

Wiegand KM. 1935. A taxonomist's experience with hybrids in the wild. *Science* 81: 161–166.

Wright 1922 p2, **NIR** (refs and RSC 1918 and 1947 only)

Wright AH. 1918. *Wisconsin's hemp industry.* Bulletin 293, Agricultural Experiment Station of the University of Wisconsin, Madison.

Young JA. 1991. Tumbleweed. *Scientific American* 264(3): 82–87.

Zhang GJ, Berbee ML. 2001. *Pyrenophora* phylogenetics inferred from ITS and glyceradehyde-3-phosphate dehydrogenase gene sequences. *Mycologia* 93: 1048–1063.

Zhang XY, Xu GC, Cheng CH, *et al.* (10 other authors). 2021. Establishment of an *Agrobacterium*-mediated genetic transformation and CRISPR/Cas9-mediated targeted mutagenesis in hemp (*Cannabis sativa* L.). *Plant Biotechnology Journal* 19: 1979–1987.

Ziegler A, Matoušek J, Steger G, Schubert J. 2012. Complete sequence of a cryptic virus from hemp (*Cannabis sativa*). *Archives of Virology* 157: 383–385.

Zubrin R. 1981. The fungus that destroys pot. *War on Drugs* June 1981: 61–62.

Abstract

Five requirements: light, moisture, air, warmth, soil nutrients. Light has three parameters: intensity (measured as quantity or energy), spectrum (wavelength), and photoperiod (daylength, latitude). Effects of these parameters upon crop yields—particularly cannabinoids—have unsolved facets (e.g., effects of supplemental ultraviolet light). Plant hydration can be measured in various ways (water potential, total leaf saturation, relative water content) as can soil moisture (field capacity, volumetric water content). Plants vary in water use efficiency and tolerance of vapor pressure deficits. Warmth is best quantified by growing degree days (°Cd). Plants vary in their needs for O_2 and CO_2 depending on temperature and moisture. Soil texture depends upon its *organic* phase (particulate organic material) and its *inorganic* phase (sand, silt, or clay). Hydroponic growth media also vary in texture. This chapter introduces macronutrients (N-P-K, calcium, sulfur and magnesium) and micronutrients (trace elements), and their effects on growth and cannabinoid content.

Introduction

Hemp is touted as an Earth-friendly crop, but its needs impact the environment. Montford and Small (1999) compared hemp's "environmental friendliness" with that of other crops. They assessed 25 criteria, including cultivation parameters (energy consumption, soil impact, annual versus perennial, needs for irrigation, fertilizer, pesticides), impacts on biodiversity (monocrop tendency, crop rotation suitability, soil microbiome, wildlife impact), and the costs of processing crops into products (energy, air and water pollution). Fiber-type hemp scored above average in friendliness (Fig. 2.1).

Cherrett *et al.* (2005) calculated the "ecological footprint" of producing a ton of textile fiber. They compared hemp, cotton, and polyester, in regards to energy expenditures and CO_2 generation. Factoring these and water needs, for both cultivation and processing, yielded a bottom-line "ecological footprint" (smaller footprint the better): organic hemp 1.5; conventional hemp 1.6; organic cotton 2.2; polyester 2.2; conventional cotton 3.0.

Drug-type *Cannabis* impacts the environment more than fiber crops (Fig. 2.1). Mills (2011, 2012) scrutinized the energy costs of growrooms, and rendered a dismal assessment. He accounted for lighting, air exchange, dehumidification, space heating, CO_2 supplementation, drying the crop, and other parameters. Total electrical usage of growrooms came to 200 watts/ft². This is on a par with high-end computer data centers. Mills estimated that indoor operations burned about $5 billion worth of electricity annually—about 1% of Canadian power consumption. Given the carbon footprint of electrical grid power sources, a single cannabis cigarette generates about two pounds of CO_2.

Plants photosynthesize energy (as sugar), hence into other raw materials that they need, and products that we consume. Plants have five requirements to perform this evolutionary miracle: light, moisture, air, warmth, and soil nutrients.

2.1   Light

Darwin (1880) described *Cannabis* as a sun-loving heliotrope, whose leaves move in response to solar tracking. Exposure to open sky is the most important ecological parameter influencing seed production in wild-type *Cannabis* (Haney and Kutscheid, 1975). When plants are under no limiting factors (water or nutrients), light intensity directly correlates with crop yields: Potter (2009) calculated a light intensity-to-yield correlation of $r^2 = 0.92$. He quantified yield as the weight of dried flowering tops per m² of grow space.

Half the sun's radiation reaching the Earth's surface is visible light, with a **wavelength spectrum** from 420 to 740 nanometers (nm). We revisit wavelength many times in this chapter. Visible light is flanked by ultraviolet radiation (UV, 100–420 nm), and infrared radiation (IR, 700–1000 nm). Artificial light can *supplement* sunlight (in glasshouses) or *replace* sunlight (in growrooms), but the intensity and spectrum of sunlight is hard to replicate (see Fig. 17.7). All artificial light sources have shortcomings—incandescent, fluorescent, mercury vapor (MV), metal halide (MH), high-pressure sodium (HPS), and light-emitting diode (LED).

2.1.1   Measuring light

Light intensity can be measured by quantity or by energy. **Light quantity** (brightness) is a function of photon flux density. Light quantity is what the exposure meter in a camera measures. The brightness cast by a candle upon a square foot (1 ft²) of surface, 1 ft away, equals 1 foot-candle or 1 **lumen** (lm).

John H. Schaffner (1866–1939) employed 2400 lm in the world's first "indoor grow" (Schaffner, 1926). He was a botanist at Ohio State University who published eight *Cannabis* articles, including one in *Science*. Schaffner used two 100-watt (100 W) Mazda™ incandescent light bulbs, each rated at 1200 lm (Fig. 2.2). Apparently 2400 lm was insufficient, because

DOI: 10.1079/9781836990352.0002

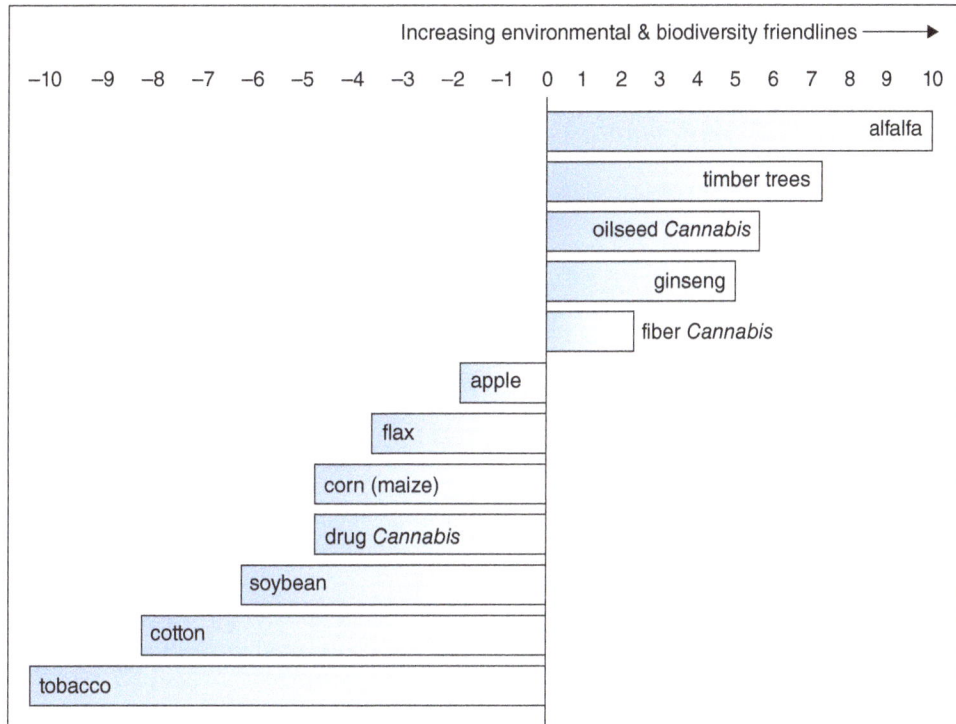

Fig. 2.1. Environmental friendliness scale (adapted from Montford and Small, 1999)

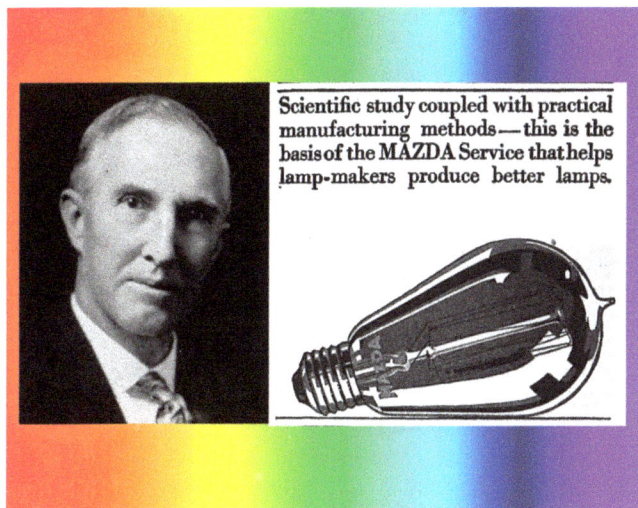

Fig. 2.2. World's first indoor grower and his light bulb. Photo credits: Schaffner portrait in Waller (1941), open access journal; Mazda advertisement in *Popular Science Monthly* 90(1917): 108.

Schaffner (1928) said the stressed plants were attacked by "scale insects, mealy bugs, and aphids."

In the metric world, 1 lm/m² equals 1 **lux** (lx). Sunlight atop Mauna Loa in Hawai'i at midday hits 100,000 lx. A cloudy day at midday in Vermont drops to 4000 lx. Indoor light in a home averages 150 lx. A full moon emits 0.25 lx. Schaffner

used 2400 lm to illuminate an area 3 ft in diameter, which equals 280 lx. Sáringer and Nagy (1971) maintained *Cannabis* under 600 lx, but did not flower the plants.

Heslop-Harrison (1956) flowered plants with high-pressure mercury vapor lamps emitting 3500–4000 lx. Paris *et al.* (1975) used fluorescent and incandescent lamps emitting 14,000–18,000 lx. Bazzaz *et al.* (1975) measured the **light saturation point** of *Cannabis*—the light intensity at which photosynthesis reaches its maximum, where more light has no effect on photosynthesis. They determined that *Cannabis* reaches a light saturation point at 120,000 lx.

Photon flux density (PFD) quantifies the number of moles of photons impacting a surface area. A photon is a mass-less particle, a quantum of the electromagnetic field. Photosynthesis is a quantum process: 8 to 10 photons (quanta) of light will fix one carbon molecule, when the wavelength of those photons falls into the photosynthetically active spectrum (Emerson, 1958).

McCree (1972b) defined the portion of PFD used by plants, in a wavelength spectrum of 400–700 nm, as **photosynthetic photon flux energy**. PPFD is measured in units of $\mu mol/m^2/$second (s). Chandra *et al.* (2015) determined a PPFD of 2000 $\mu mol/m^2/s$ as optimal for *Cannabis* photosynthesis. It bears mentioning that 2000 $\mu mol/m^2/s$ is full sunlight atop Mauna Loa, and not economically feasible in terms of energy costs.

Eaves *et al.* (2020) estimated that most growers used 600 $\mu mol/m^2/s$, based on the "industry standard" of one 1060 W HPS bulb placed 76 cm above 1.48 m² of plant area. Eaves reported a positive correlation ($r^2 = 0.87$, 0.90, and 0.94 in three trials) between flower yield and PPFD under a range of intensities between 200 and 1498 μmol/m²/s. Substituting HPS lights with LEDs did not alter yields, but plants under LEDs produced shorter internodes and denser flowers. Plants under LEDs reached peak maturity about 5 days earlier than plants under HPS lights.

Light energy is the other way to measure light—its ability to do work (e.g., drive photosynthesis). Energy is measured in kilocalories per hour imparted per square meter, or watts per square meter (**W/m²**). Sunlight entering the Earth's atmosphere imparts 1350 W/m². By the time sunlight reaches the Earth's surface, its energy has dropped to 1000 W/m²—at noon, on a clear summer day, near the equator. On a cloudy day sunlight dissipates to 100 W/m². Moonlight exerts 0.01 W/m².

Cannabis seedlings can be grown under 96 W/m² of mixed incandescent and fluorescent lighting (McPartland, 1984). At the dawn of the indoor era, Frank and Rosenthal (1978) suggested flowering *Cannabis* under a minimum of 215 W/m². These are measures of *total* light energy, however, which is not the best metric for measuring effects on plants. McCree (1972a) defined **photosynthetically active radiation** (PAR) as the portion of light energy used by plants (wavelength spectrum 400–700 nm), quantified as W/m² PAR. Potter (2009) flowered *Cannabis* under 70 W/m² PAR. Potter and Duncombe (2012) reported photosynthetic activity leveling off above 100 W/m² PAR.

Light energy depends on color. Color is a function of wavelength. Energy increases in proportion to wavelength. Short-wave light has less energy than long-wave light. For example, purple light (short wavelength 420 nm) requires 130 milliwatts to generate 1 lumen of brightness, whereas yellow light (long wavelength 570 nm) needs only 1.4 milliwatts to generate 1 lumen. Converting from brightness (lumens or lux) to energy (W/m²) is not simple, since most light represents a mix of wavelengths.

2.1.2   Wavelength

Plants, like people, prefer certain colors. Specific wavelengths are critical for plant growth. Plants look green because chlorophyll absorbs blue and red wavelengths, and reflects green wavelengths. The absorbance spectra of chlorophyll A and chlorophyll B are illustrated in Fig. 2.3. Chlorophyll A absorbance peaks at 430 and 663 nm, and chlorophyll B peaks at 453 and 642 nm.

Plants have receptors that sense light at different wavelengths. **Phytochromes** are sensitive to red light (650–700 nm) and far-red light (700–740 nm). They regulate seed germination and chlorophyll synthesis, promote stalk elongation, and trigger flowering. **Cryptochromes** are sensitive to blue light (420–520 nm). They regulate phototropism (directional growth toward a light source), decrease stalk elongation, and regulate circadian rhythms. **Phototropins** are sensitive to blue light (420–520 nm); they regulate phototropism and stomatal activity.

Cryptochromes and phototropins are also sensitive to light in the ultraviolet spectrum. UV wavelengths include UV_A (420–315 nm), UV_B (315–280 nm), and UV_C (280–100 nm). Although UV light damages nucleic acids and proteins in plants (and people), small quantities have benefits, such as improving photosynthetic

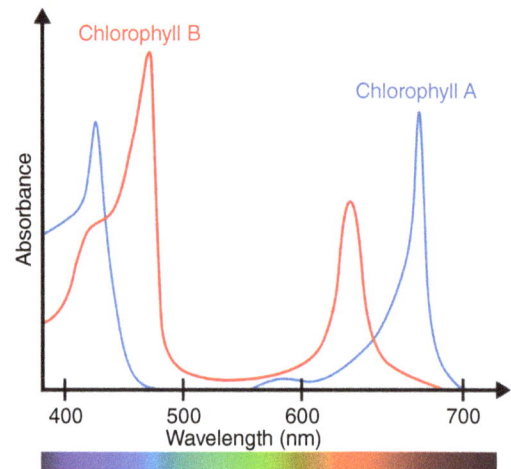

Fig. 2.3. Chlorophyll absorption spectra (courtesy of M0tty, Wikipedia Commons)

efficiency, and promoting resistance to pests and pathogens. Cryptochromes and phototropins respond to UV_A. The UVR8 (UV resistance locus 8) photoreceptor responds to UV_B. UVR8 regulates the production of compounds that protect against UV_B light, such as flavonoids (and possibly cannabinoids, see below). UVR8 also regulates pathogen resistance and stomatal activity.

2.1.3   Cannabinoids and light

Cannabinoids can be measured by total yield (mg/plant or mg/grow area) or by potency (THC% or CBD% in dried plant material). Balduzzi and Gigliano (1985) first analyzed the relationship between cannabinoids and light. They measured light intensity at four levels of illumination (25, 50, 75, and 100% natural solar light) and correlated it with THC% and CBD% in female flowers.

Three key studies (Table 2.1) measured *Cannabis* yield and cannabinoid potency at two levels of light intensity, low and high. (1) Potter (2009) tested high-THC and high-CBD cultivars and measured dried flowering tops per m² of grow space. He measured cannabinoid yield as potency (THC%, CBD%), or as g/grow area, and measured light intensity as W/m² PAR. (2) Potter and Duncombe (2012) grew seven high-THC hybrid strains, and measured light intensity as W/m² (rather than W/m² PAR). (3) Vanhove *et al.* (2011) grew four high-THC hybrid strains, measured light intensity as W/m², and measured dried flowering tops per plant or per m² of grow space.

In only one study (Potter, 2009) did *potency* significantly increase at higher light intensity. Potter measured W/m² PAR, the other two studies measured W/m². He combined HPS and MH lamps, but so did Vanhove *et al.* (2011), whereas Potter and Duncombe (2012) used only HPS lamps. In all three studies, *yield* of flowering tops increased (g/m²), as did *yield* of THC or CBD (g/m²). Potter and Duncombe explained that THC yield/m² increased despite THC% not changing because the flower-to-leaf ratio increased with higher light intensity.

Table 2.1. Light intensity, flower yield, and cannabinoid yield

Potter, 2009	Low intensity, 25 W/m² PAR	High-intensity, 70 W/m² PAR
High-THC plants	Dried flowers 188 g/m²	Dried flowers 397 g/m²
	11% THC, or 23 g/m²	16% THC, or 62 g/m
High-CBD plants	Dried flowers 157 g/m²	Dried flowers 251 g/m²
	6% CBD, or 9 g/m²	11% CBD, or 28 g/m²
Potter and Duncombe, 2012	Low intensity, 270 W/m²	High intensity, 600 W/m²
High-THC plants	Dried flowers 422 g/m²	Dried flowers 544 g/m²
	14.46% THC, or 61.2 g/m²	14.49% THC, or 78.4 g/m²
Vanhove et al., 2011	Low intensity, 400 W/m²	High intensity, 600 W/m²
High-THC plants	Dried flowers 11.65 g/plant	Dried flowers 20.08 g/plant
	or 210.5 g/m², 12.6% THC	or 361.7 g/m², 12.8% THC

A series of studies by Suman Chandra and colleagues documented that **photosynthetic rate (P_N)** increased with light intensity. P_N correlates with biomass yield. Chandra et al. (2008) measured P_N in a Mexican landrace under four levels of brightness (500, 1000, 1500, 2000 PPFD) and five temperatures (20, 25, 30, 35, 40°C). Photosynthesis peaked (P_N = 24.6 µmol/m²/s) at 30°C and 1500 PPFD. Greater light intensity (2000 PPFD) at 30°C caused photosynthesis to decline. At 25°C, photosynthesis reached its highest level at 2000 PPFD, but less than the 30°C peak (P_N = 22.5).

Chandra et al. (2015) measured P_N in four high-THC/low-CBD strains (Mexican, K2, MX, W1) under five levels of brightness (400, 800, 1200, 1600, 2000 PPFD) at 25°C. Like their 2008 study, P_N increased linearly with increased PPFD. All four strains topped out at 2000 PPFD, but the magnitude of P_N increase varied: W1, 25.54; MX, 23.70; K2, 21.7; Mexican, 19.82.

Magagnini et al. (2018) measured the effects of light *spectrum* on cannabinoid potency. They flowered a high-THC strain under HPS, or two types of LED lights—AP673L or NS1—all at identical intensity (450 µmol/m²/s at canopy height). The three light sources differed in their percentage of PAR and their ratio of red to far-red light (R/FR, i.e., 650–670/720–740 nm). HPS bulbs emitted 96% PAR, primarily in the 500–600 nm spectrum, with R/FR ratio of 2.8. AP673L emitted 93% PAR, primarily in the 600–700 nm spectrum, with R/FR ratio of 6.07. NS1 emitted 94% PAR, and spanned the 400–700 nm spectrum, with R/FR ratio of 10.05.

They ran two trials, with slightly different results (Table 2.2). HPS yielded less THC% than either of the LEDs (LEDs did not statistically differ from each other). Similar trends were seen for CBD, THCV, and CBG. Regarding flower weight, HPS plants had greater yields than the LEDs, statistically significant in trial 1, but not significant in trial 2. Plants grown under HPS were taller than LED plants (tall height is not a desired characteristic indoors). Magagnini attributed this to HPS lamps having a low R/FR ratio, therefore triggering stem elongation through phytochrome sensors. Differences in THC% were attributed to cryptochrome and phototropin receptors, which signal in response to blue light in the LEDs.

Hawley et al. (2018) measured cannabinoid and terpenoid content in plants exposed to three spectra: control lighting (source not stated, but their Conviron AT60 growth chamber used fluorescent and incandescent bulbs), or control lighting supplemented with sub-canopy LED lighting, either red-blue

Table 2.2. Plant responses to different wavelengths, from Magagnini et al. (2018)

Light	THC in flowers	Flower weight	Plant height
HPS trial 1	9.5%	32.7 g	67.4 cm
HPS trial 2	11.7%	26.6 g	79.2 cm
AP6 trial 1	12.8%	24.3 g	58.3 cm
AP6 trial 2	15.4%	23.1 g	54.5 cm
NS1 trial 1	15.4%	26.2 g	58.7 cm
NS1 trial 2	16.5%	22.8 g	54.8 cm

(RB) or red-green-blue (RGB). In the lower canopy, RGB significantly increased concentrations of alpha-pinene and borneol. Either RG or RGB significantly increased concentrations of THCA compared with control. In the upper canopy, RGB significantly increased concentrations of alpha-pinene, limonene, myrcene, and linalool, but not THCA.

Rodriguez-Morrison et al. (2021a) criticized as too simplistic the studies that measured PAR (Potter, 2009; Vanhove et al., 2011; Potter and Duncombe, 2012) or PPFD (Chandra et al., 2008, 2011, 2015; Eaves et al., 2020). However, their model, using "light integrals" (total light integral, daily light integral) is overly complex for an average grower or even an expert (S. Chandra, pers. comm., 2022).

Rodriguez-Morrison et al. (2021a) made some concrete observations: flower yield increased linearly from 116 to 519 g/m² (i.e., 4.5 times higher) as PPFD increased from 120 to 1800 µmol/m²/s. CBD yield (mg/m²) also increased at the same rate—due to increased inflorescence yield, and not because of increased CBD potency. Conversely, terpenoid potency *did* increase, by 25%, as PPFD increased from 120 to 1800 µmol/m²/s. Lastly, they reported that augmented light intensity increased the *density* of dried apical buds: 0.0893 g/cm³ at PPFP 120, to 0.115 g/cm³ at PPFP 1800 (i.e., 1.3 times greater).

Danziger and Bernstein (2021) showed that low cannabinoid concentrations in inflorescences at the bottom of plants correlated with low light penetration, and that increasing light penetration by selective defoliation increased cannabinoid concentrations. They worked with a high CBD cultivar, grown in a naturally lit glasshouse. Interestingly, they found that cannabinoid content in the apical primary inflorescence (the "cola") was actually lower than concentrations in the tops of upper branches and mid-level branches.

Ultraviolet light

UV light may affect cannabinoid production. Based on a literature review, Pate (1983) proposed that plants upregulated THC production as an adaptive response to stress in locations with high UV_B (high altitude, low latitude locations). This adaptive response (i.e., over evolutionary time) has been tested in acclimation studies (i.e., responses to UV light in a single growing season): Fairbairn and Liebmann (1974) grew Nepalese plants in a greenhouse, and the upper leaves contained 2.4% THC; supplementing with two 400 W UV_{A-B} bulbs for 4 h/day increased potency to 2.96% THC. They did not run stats to see if this difference was statistically significant.

Lydon *et al.* (1987) grew Jamaican plants in a greenhouse, supplemented with UV_B for 6 h/day for 40 days. Floral tissues grown in ambient greenhouse light contained 2.5% THC, plants supplemented with 6.7 kJ/m^2 UV_B contained 2.8% THC, and plants supplemented with 13.4 kJ/m^2 UV_B contained 3.2% THC. They did not test the statistical significance of these differences. They also grew fiber-type plants of Czechoslovakian origin, and reported no change in CBD levels under the different light regimes, although they did not present data.

Potter and Duncombe (2012) and Cervantes (2015) concluded that such a small increase in THC did not warrant human exposure to UV_B, whereas Short (2013) recommended using UV_B. Zhang and Björn (2009) speculated that chalcone synthase, a UV_B-regulated enzyme, may be responsible for UV_B-stimulated cannabinoid biosynthesis.

Marti *et al.* (2014) detached leaves from *Cannabis* plants, placed them on damp blotting paper in a petri plate, and exposed them to a UV_C lamp (253 nm), delivering 0.18 kJ/min, for 10 min, 13 cm above the leaves. After 48 h of exposure, they found no change in cannabinoid content compared with control plants. There was a significant increase in cannabisin D (a lignanamide unique to *Cannabis*), and non-*Cannabis*-specific stilbenes, spirans, cinnamic acids, and flavonoids.

In the study by Magagnini described above (Table 2.2), one type of LED bulb, NS1, emitted more UV_A radiation than the other LED bulb (AP673L), but THC content did not significantly differ. Giupponi *et al.* (2020) cultivated CBDA-dominant 'Kompolti' plants in Italy, at two sites, 1200 m and 130 m altitude (same latitude). High-altitude plants yielded CBDA 9.96%, and low-altitude plants yielded 6.82%, a significant difference they attributed to increased UV light exposure. However, other environmental parameters (temperature, rainfall) were not reported.

Rodriguez-Morrison *et al.* (2021b) exposed flowering plants to UV_{C-B} (range 275–305 nm) for 60 days. Plants received a range of UV photon flux densities, between 0.01 to 0.80 μmol/m^2/s at canopy height. Increased UV exposure did not change THC concentration in one strain ("Low Tide"), whereas in another strain ("Breaking Wave"), THC concentration increased with UV exposure ($r^2 = 0.33$), but the increase was marginal. Their study documented damage caused by UV radiation—foliar chlorosis and necrotic leaf spots, curled leaflet margins, stigma browning, and premature senescence. Increased UV exposure correlated negatively with leaf size, plant height, growth index, and bud yield. They concluded that UV supplementation provided no commercial benefits. Any increase in THC potency was offset by reduced biomass. Given the the costs of supplemental UV light (equipment and energy), and health risks to employees, they did not recommend it.

2.1.4   Daylength

Most experts describe *Cannabis* as a **short-day plant**. It flowers in autumn, when the photoperiod drops below ~12–14 h/day. The connection between photoperiod and flowering in the plant world was first demonstrated by Tournois (1912), and he used *Cannabis*. Tournois actually described *Cannabis* as a **long-night plant**, and he was correct—flowering is induced when night time crosses a threshold of ~10–12 h in autumn. The darkness must be uninterrupted: a short burst of light during the night will prevent flowering. Conversely, interruption of the light period by even a long dark period will not prevent flowering.

Hoffmann (1961) studied the influence of photoperiod on three fiber-type landraces, (from Finland, Italy, and Japan) plus a "wild-type" from Nūristān in Afghanistan (Table 2.3). He counted the number of days between seedling emergence and flowering. Under natural daylength in Berlin, northern Finnish plants began flowering in 3–4 weeks, independent of photoperiod. The others began flowering after daylength <12 h/day. When daylength was artificially shortened to 10 h/day, it hastened flowering time in Italian and Japanese landraces. Hoffmann surmised that when it comes to daylength response, latitude is destiny.

Hoffmann's "latitude is destiny" was demonstrated in a study by Zhang *et al.* (2018b). They genotyped 52 *Cannabis* accessions—mostly from China—and their phylogenetic tree segregated the accessions into three haplogroups. To a large degree, the three haplogroups segregated by latitude: low, medial, and high (<30 °N, 30–40 °N, >40 °N). Thus latitude (hence photoperiod sensitivity) generated a phylogenetic signal … and Zhang utilized "neutral markers" in the chloroplast genome. Neutral markers should not confer fitness advantages involved in latitude acclimation.

Table 2.3. Number of days between seedling emergence and flowering, in four types of *Cannabis* under different day lengths (from Hoffmann, 1961)

Provenance, Latitude	Normal daylength[a]	Short daylength[a]
Finland 60–63 °N	24–30	22–27
Italy 41–45 °N	86–108	30–37
Japan 35–44 °N	47–74	27–34
Nūristān 35 °N	158	158

[a]Days to flowering

Autoflowering or day-neutral plants are indifferent to daylength, and begin to flower due to intrinsic mechanisms. Hoffmann's Finnish and Afghani plants were autoflowering plants—they did not respond to daylength manipulation (Table 2.3). Small (2015) stated that autoflowering is more common in plants at the latitudinal extremes of the *Cannabis* geographic range. At high latitudes, the growing season is so short that plants would not have time to mature seeds if they waited until days began to shorten (see section 2.6). Close to the equator, daylength is 12–13 h all year round, without a change in daylength.

Autoflowering plants with short vegetation periods and short stature have become immensely popular among growers of THC- and CBD-yielding crops. Clarke and Merlin (2016) proposed that "all modern autoflowering cultivars likely have a common ancestor, possibly 'FINOLA'." Callaway (2004a) originally described 'FINOLA' as "early-flowering." But research suggests 'FINOLA' is autoflowering (Petit *et al.*, 2020a; Schilling *et al.*, 2022).

Green (2005) contended that "Ruderalis" was the source of autoflowering genetics that went into Sasha the Joint Doctor's "Lowryder," the first widely marketed autoflowering strain, bred in 2002. "Ruderalis" was first sold by Schoenmakers (1986). He also marketed the first autoflowering hybrid strain, "Shady Lady", a cross between "Ruderalis" and "Afghani #1".

Much *Cannabis* research has focused on two factors that govern flowering time—the switch-point photoperiod and photoperiod sensitivity. **Switch-point photoperiod (SPS)** is the number of hours of daylight in autumn that trigger the initiation of flowering. The SPS is sometimes called the critical photoperiod, although that term also refers to the number of hours of daylight above which flowering never occurs.

Borthwick and Scully (1954) first measured SPS in a glasshouse supplemented with artificial lights. They determined that a photoperiod of ≤14 h initiated flowering in 'Kentucky' hemp. The SPS of many fiber- and drug-type varieties has been measured (e.g., Heslop-Harrison and Heslop-Harrison, 1967; Van der Werf *et al.*, 1994; Struik *et al.*, 2000; Amaducci *et al.*, 2008a; Sengloung *et al.*, 2009; Hall *et al.*, 2014).

Petit *et al.* (2020a) conducted the most detailed study. They grew 123 accessions at three different latitudes (Rovigo in Italy, 45 °N; Sarthe in France, 48 °N; Westerlee in Holland, 53 °N), and measured three time intervals: (1) length of vegetation period (VEG, measured in days); (2) beginning of flowering time (FL-B, measured in growing degree days, °Cd, see section 2.3.1 for this metric); and (3) time to full flowering (FL-F, °Cd). Regardless of latitude, some accessions exhibited early flowering, notably 'FINOLA': mean VEG: 32.63 days, mean FL-B: 469.03 °Cd, mean FL-F 639.83 °Cd. Other accessions flowered late, regardless of latitude, notably Chinese accessions such as 'Yúnmá 5'—VEG: 117.78 days, FL-B: 2097.42 °Cd, FL-F 3440.87 °Cd.

Photoperiod sensitivity (PS) reflects variation in flowering response to suboptimal photoperiods, which contributes to variation in flowering time. Three research groups have modeled PS (Lisson et al., 2000; Amaducci et al., 2008a; Zhang et al., 2021). They highlighted the interacting roles of photoperiod and temperature, with temperature measured in "thermal time" or growing degree days (°Cd).

Because flowering time is so important for fiber yield and cultivar adaptation to new locations, there is considerable interest in identifying the genes responsible. Two key genes, first identified in *Arabidopsis*, are **Flowering Locus T** (*FT*) and **CONSTANS** (*CO*). These are genes for transcription factors, which are proteins that bind to regulatory sequences and thereby promote or silence other genes.

Petit *et al.* (2020b) detected QTL (quantitative trait loci) markers governing flowering time by testing 123 accessions cultivated at three different latitudes. After mapping the QTLs to the "Purple Kush" genome, they identified "candidate genes" for the QTLs. These included *FT* and *CO*, as well as other genes that code for transcription factors (*Flowering locus D*, *Agamous*, *Bed*, *bZIP*, *Floricaula*, *Leafy*, *Vrn1*, and *Squamosa*), plus genes involved in light perception (*Cry1*, *PhyA*, *phyE*, *Spa1*, *Uvr8*), and a circadian timekeeper gene (*Xap5*).

Whole-genome studies have revealed that *Cannabis* has multiple copies of both *FT* and *CO* (Panerio, 2020). Pan *et al.* (2021) identified no fewer than 13 *CONSTANS*-like genes. Chen *et al.* (2022) proposed that alterations in these genes were responsible for differences in flowering time between wild-type plants and fiber-type cultivars.

Autoflowering genes differ from early flowering genes. Toth *et al.* (2022) conducted elegant breeding studies linked with whole-genome analyses. They identified a gene locus on *Cannabis* chromosome 1 associated with early flowering, which they named *Early1*, with the candidate gene *Casein kinase 1-like* (*CKL*). Elsewhere on chromosome 1 they identified a gene locus associated with autoflowering, which they named *Autoflower1*. Candidate genes for *Autoflower1* included *APETALA2* and *UPF2*, as well as Nuclear transcription Factor Y subunit B-1 (*NFYB1*).

Toth *et al.* (2022) genotyped several cultivars that expressed the wild-type allele for *Autoflower1* (i.e., photoperiod-sensitive plants), and subjected them to continuous light. As expected, most did not flower ('Carmagnola', 'Puma', 'RN16'), but some did ('CFX-2', 'Picolo', 'Henola'). This suggests that additional genetic mechanisms impart photoperiod insensitivity. Phylos Biosciences (2021) also searched for genes responsible for autoflowering, and also implicated *APETALA2* and *UPF2*.

Daylength and cannabinoids

Daylength affects cannabinoid synthesis. Potter (2009) induced flowering by exposing plants to short days, either 13, 12, or 11 h of light. In terms of yield and cannabinoid content, there was no significant difference between 13 and 12 h, but yields decreased in plants grown in the 11 h regimen. Potter compared 6, 8, and 10 weeks of short days, and THC yield peaked at 10 weeks, although there were differences between clone lines.

De Backer *et al.* (2012) showed that THC content peaked 5–6 weeks after flowering began. Aizpurua-Olaizola *et al.* (2016) reported that THC content peaked after 9 weeks of short days in high-THC clones, and THC+CBD content peaked at 11 weeks in mixed THC/CBD clones. Richins *et al.* (2018) measured THC content in three "medical" strains after the initiation of a 12 h flowering regime. THC content in "Sour Willie" peaked at 60 days, "Bohdi Tree" peaked at 62 days, and "Blue Dream" peaked at 74 days.

Yang *et al.* (2020) measured the THC:CBD ratio in flowers as they matured. The THC:CBD ratio increased in three photoperiod-sensitive CBD strains ("Cherry Blossom", "Cherry×T1", "Cherry Wine"), but in two autoflowering CBD strains ("Pipeline", "Maverick"), the THC:CBD ratio decreased as flowers matured. Photoperiod-sensitive strains became "hot" (THC ≥0.3%) 4 weeks after flowering began, whereas in autoflowering strains this did not happen until 6–7 weeks after flowering began. Toth *et al.* (2022) attributed the reduced THC:CBD ratio to the conversion of THC to CBN in autoflowering plants, which entered senescence earlier.

Daylength and sexual expression

The female-to-male ratio, normally 1-to-1, shifts under artificially shortened daylength. McPhee (1924) fitted a glasshouse with black-out shutters; shortening daylength to 5 h shifted the ratio to 1.3-to-1. Furthermore, monoecious (intersex plants) cropped up. McPhee (1925) subjected Dewey's dioecious 'Chington' to short days, which induced intersex plants. When pollen from female plants with intersex male flowers was crossed with normal female flowers, the resulting seeds produced all-female plants. Today these are known as **feminized seeds**, often induced with plant hormones (e.g., gibberellin, silver thiosulfate), and now a major industry.

Frank and Rosenthal (1978) self-pollinated intersex plants to generate feminized seeds—the earliest report we could find since McPhee. Lapierre *et al.* (2023) cast a cold eye on feminized seeds: plants grown from feminized seeds yielded 4.74% less THCA than plants grown from regular seeds, possibly because they matured earlier (104 days versus 109 days).

Hoffmann (1972) also shifted the normal ratio towards female plants by shortening daylength. The feminizing effects of short days could be reversed by treating plants with gibberellin. Treating plants with an auxin (1-naphthylacetic acid) did the opposite, further increasing the percentage of female plants. Mohan Ram and Jaiswal (1972) first noted that gibberellin induced the formation of male flowers on female plants.

Van der Werf *et al.* (1994) reported that 'Kompolti Hybrid TC' grown under continuous light shifted its sexual expression from 1:1 to a preponderance of females. Conversely, Schilling *et al.* (2022) reported that 'FINOLA' grown under continuous light shifted its sexual expression from 1:1 to a preponderance of males.

2.2   Moisture

Cannabis needs water for photosynthesis and transpiration. **Transpiration** is the movement of water from roots to leaves, and the evaporation of water from leaf surfaces through pores called stomata. Transpiration rate is a function of plant size (i.e., leaf area) and the rate of transpiration through stomata. Leaf area is large in *Cannabis*, compared with other annual plants. Haberlandt (1879) estimated a leaf area of 9607.6 cm² in a mature plant. He also estimated the transpiration rate: 7.22 g

$H_2O/cm^2/h$—the highest of seven species studied (green beans were lowest, 3.85).

Quantitative estimates of *Cannabis* water consumption vary in the literature. This is understandable, given the wide variety of *Cannabis* phenotypes, and the crop's wide geographic range. Duke (1982) collated rainfall data in regions where fiber-type hemp was cultivated, and he found a huge range: from 310 mm to 4030 mm rainfall per year, with a mean of 970 mm. For better estimates, see section 17.7.

Wild-type *Cannabis ruderalis* is more drought resistant than domesticated plants (Janischevsky, 1924). Kubešova *et al.* (2010) considered *Cannabis ruderalis* one of the most drought-resistant weeds in the Czech Republic. Haney and Bazzaz (1970) stated that wild-type plants in Illinois—descendants of Chinese fiber-type cultivars—had regained drought tolerance after several decades of growing wild. Drake (1970) estimated that wild-type plants consumed four times less water than domesticated drug-type plants.

Drug-type plants vary in their drought resistance, and this may be due to where they evolved. The geographic locations of wild-type drug plants were mapped by McPartland and Small (2020), based on examinations of herbarium specimens (Fig. 2.4). They differentiated between wild-type "Sativa"- and "Indica"-type plants. These geographic locations were then overlaid with a map of floristic regions, based on Djamali *et al.* (2012): the Indian floristic region (green-colored area), the Irano-Turanian floristic region (red area), and the Saharo-Sindian region (lilac area). "Sativa"-type wild types—green triangles—mostly occurred in the wet and monsoonal Indian floristic region. "Indica"-type wild types—red circles—mostly occurred in the arid Irano-Turanian floristic region. This correlates with observations by Cervantes (2015), who stated that "Sativa" plants consume more water than "Indica" plants.

Plant hydration is expressed as **water potential** (ψ), in units of megaPascals (MPa) or bars. Measures of ψ in plants are always negative numbers, and become more negative with dehydration. Caplan *et al.* (2017b) measured ψ in *Cannabis*: adequately watered plants averaged –0.9 to –1.0 MPa (= –9 to –10 bars). At –1.5 MPa, leaves began to wilt. During dry summer months, ψ routinely dropped to –1.2 MPa. There's no visible stress at –1.2 MPa (no wilting), but plants become predisposed to fungal pathogens at –1.2 MPa (McPartland and Schoeneweiss, 1984).

Percentage of **total leaf saturation** (TLS) is another measure of plant hydration. *Cannabis* growth peaks at 85–93% TLS (Slonov and Petinov, 1980). Bahador *et al.* (2017) measured **relative water content** (RWC) in leaves, using the formula (FW–DW)/(TW–DW) ×100, where FW is fresh weight; DW is dry weight; TW is turgid weight (cut leaves saturated in water for 3 h). In fully irrigated *Cannabis* plants, RWC = 60%. In plants given only 40% of their water requirements, RWC dropped to 34%.

Soil water is measured different ways (Kirkham, 2014). The **wilting point** (WP) is defined as the moisture content of soil (expressed as a percentage of the dry weight) when a plant reaches the permanent wilting point (it cannot recover). **Field capacity** (FC) is the amount of water content held in field soil after excess water has drained away. At 100% FC, the water and air contents of the soil are considered optimal for crop growth.

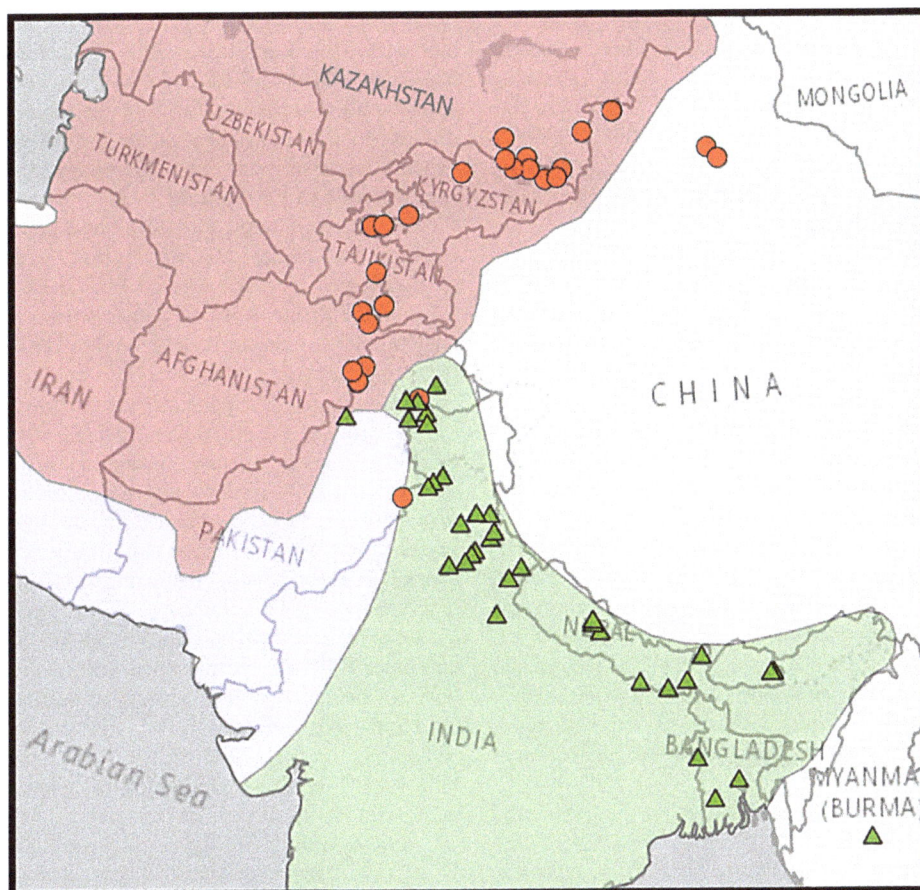

Fig. 2.4. Distribution of wild-type herbarium specimens in floristic zones (McPartland and Small, 2020)

Above 100%, the soil is saturated or even waterlogged. Desert soil might drop to 0% FC. *Cannabis* tolerates fairly low FC percentages, thanks to its extensive root system. But *Cannabis* grows poorly in saturated soil. *Cannabis* growth peaks at 80% FC (Slonov and Petinov 1980).

Volumetric water content (θ) is the ratio of water volume to soil volume, expressed as a percentage. The θ of soil at FC and WP is site specific, and varies due to soil texture, organic matter, and temperature. Several *Cannabis* studies reported θ at FC and WP, respectively: 45% and 5% (Faux *et al.*, 2013); 27.3% and 9.45% (Angelini *et al.*, 2014); 35% and 12% (Babaei and Ajdanian, 2020). Guo *et al.* (2011a,b) considered a θ of 15% as the minimum for growing 'Yúnmá 1', a drought-sensitive cultivar. In a follow-up study of 'Yúnmá 1', they identified 1292 genes in drought-stressed plants that were down- or up-regulated, many in the abscisic acid (ABA) signaling pathway (Gao *et al.*, 2018).

Caplan *et al.* (2019) quantified *Cannabis* wilt by measuring "relative leaf angle." They selected for measurement fully expanded leaves on a side-branch from the first internode. The acute angle between the center of the middle leaflet and the stem from which it originates was measured with a protractor at midday. The angle in wilted, drought-stressed plants was 50% greater than well-hydrated turgid plants. WP was −1.5 MPa (compared with −1.0 MPa in controls), and the potting

substrate's volumetric moisture content was 5.3% (33.3% in controls).

Water use efficiency (WUE) is defined as the amount of carbon assimilated as biomass per unit of water used by the crop. WUE can be approximated as the ratio of plant dry matter (in grams) to the amount of water consumed (in liters). Lisson and Mendham (1998) measured WUE in 'Kompolti' cultivated in various degrees of water deficit. Predictably, WUE increased as water was withheld, ranging from 2.4 to 3.1 g/L. WUE increased with increasing N input (plants irrigated with 30, 80, 160, 240, or 320 mg/L), but plateaued at 240 mg/L (Saloner and Bernstein, 2021).

Chandra *et al.* (2008) calculated WUE as a ratio of photosynthetic rate (P_N) to transpiration rate (E). They measured WUE in a Mexican landrace exposed to various CO_2 levels (350, 450, 550, 650, 750 ppm) and light intensities (500, 1000, 1500, 2000 µmol/m²/s). WUE increased linearly with CO_2 levels, from 4.64 (at 350 ppm) to 9.81 (at 750 ppm). At 30°C and ambient CO_2 (350 ppm), WUE increased with light intensity, peaking at an intensity of 1500 (WUE 4.8), and dropping at 2000 (WUE 3.8).

Chandra *et al.* (2011) calculated WUE at two temperatures, 20°C and 40°C. Predictably, WUE decreased as temperature increased, because of increased transpiration. They also calculated WUE in three drug-types (Mexican, MX, W1) and four fiber-types ('Felina 34', 'Kompolti', 'Zolo 11', 'Zolo 15'). Optimal WUE was greater in drug-types than fiber-types, but not statistically significant. Chandra *et al.* (2015) calculated WUE in four drug-types (Mexican, K2, MX, W1) under five levels of brightness (400, 800, 1200, 1600, 2000 PPFD) at 25°C. WUE in W1, MX and HPM increased with PPFDs up to 2000, whereas in K2 the highest WUE was observed at 1600 PPFD, and decreased at 2000 PPFD.

Cosentino *et al.* (2013) calculated the WUE of 'Futura 74' in four irrigation regimens, ranging from fully watered to water-stressed conditions. WUE increased as water was withheld, ranging from 2.73 to 3.45 g/L. Tang *et al.* (2018) added precision by measuring the ratio of canopy CO_2 consumption (rather than yield dry matter) to canopy transpiration, a metric called canopy **photosynthetic water-use efficiency** ($PWUE_C$). They analyzed 'Futura 75' in fully watered and water-stressed conditions, and obtained $PWUE_C$ values between 4.0 and 7.4 mmol CO_2 /mol H_2O.

Babaei and Ajdanian (2020) calculated the WUE of 47 ecotypes of Iranian *Cannabis* in fully watered and water-stressed conditions. The WUE in fully watered plants had a wide range, from 4.9 to 0.4. Water-stressed plants ranged from 4.8 to 0.2. Ecotypes varied in their response to water stress, so Babaei and Ajdanian calculated a **Tolerance Index** (TOL), measured as yield when fully watered minus yield when water-stressed. They used TOL to identify ecotypes suitable for cultivation in arid areas.

Low-level deficits may *increase* cannabinoid content and yield, if carefully controlled. To optimize potency, Frank (1988) said crops should be watered sufficiently to maintain growth, but no more; soil should dry out between waterings. Pate (1993) reviewed older literature hinting at this phenomenon.

Caplan *et al.* (2017b) demonstrated this experimentally. Beginning in week 7 of the flowering period, they withheld daily watering until wilting appeared. Flower yield, measured as grams dry weight per growing area, averaged 232 g/m²; well-watered control plants averaged only 178 g/m². Yields increased despite the fact that leaf photosynthetic rate (P_N) in water-deficit plants was 42% lower than control plants. THCA% (dried flowers) averaged 5.3% in water-deficit plants, and 4.7% in control plants. THCA yield per growing area averaged 11.0 g/m² in water-deficit plants, and 7.7 g/m² in control plants. Measures of CBD concentrations and yield showed the same trends.

Caplan *et al.* (2019) grew plants in Pro-Mix HP Mycorrhizae substrate, fertigated as previously reported. Thirty-nine days into the flowering period, fertigation was withheld, and wilting was observed after 11 days. WP in wilted plants was –1.5 MPa, compared with –1.0 MPa in controls. Net photosynthetic rate (Pn) in leaves differed: 13.2 μmol/m²/s in controls, 7.7 in drought-stressed plants. All plants resumed normal fertigation until harvest 54 days into the flowering period. THCA% significantly differed (controls 4.7%, drought-stressed 5.3%) as did CBDA% (controls 9.1%, drought-stressed 10.3%). Drought had substantial effects on cannabinoid yield per unit growing area, for THCA (controls 7.7 g/m², drought-stressed 11.0 g/m²) and CBDA (controls 15 g/m², drought-stressed 22 g/m²).

Chronic high-level deficits, however, may *decrease* cannabinoid content. Campbell *et al.* (2019) cultivated 13 European fiber-type cultivars in Colorado, under dry conditions (irrigation applied only to seedlings, if wilting was observed), versus wet conditions (full irrigation all season long, at 100% of the ET_0 rate). In dry and wet conditions, THC% averaged 0.136% and 0.158%, respectively. CBD% averaged 1.431% and 2.244%, respectively.

Besides soil water and plant hydration, cultivators must account for *atmospheric* water; that is to say, **humidity**. *Cannabis* grows best at a relative humidity (RH) of 40–80% (Frank, 1988), but RH >60% promotes gray mold and powdery mildew. Therefore, RH of 40–60% is optimal during flowering, to avoid disease.

Pate (1993) reviewed literature that suggests climates with low RH produce plants with greater cannabinoid content. Bouquet (1950) observed that plants growing in coastal Lebanon, exposed to humid air from the Mediterranean, produced less resin than plants growing in Lebanon's arid interior. Paris *et al.* (1975) grew plants in a growroom under either 50% or 80% RH, and the former produced a higher percentage of THCA and THC.

Vapor pressure deficit (VPD) is more accurate than RH at describing conditions leading to leaf transpiration. VPD combines RH with temperature, and quantifies the difference between the actual and maximum amounts of water the air can hold for a given temperature. VPD and RH have an inverse relationship: a high VPD means the air has a high capacity to hold water (i.e., low humidity), while a low VPD means the air is near saturation (i.e., high humidity).

The optimal VPD for *C. sativa* growth is relatively high. Frank (1988) observed that *Cannabis* grows best at 75°F at RH 40–80%, which we convert to a VPD of 1.8–1.2 kPA. Purdy (2018) directly measured VPD, and reported optima of 0.8–1.1 kPa in the vegetative stage, 1.0–1.4 during early flowering, and 1.3–1.5 during late flowering. If VPD is too high (RH too low), it may induce wilt and necrosis of leaf tips. If VPD is too low (RH too high), it encourages fungal pathogen growth.

2.3 Temperature

Temperature measurements were possible after Fahrenheit invented a mercury thermometer in 1714. He proposed the Fahrenheit scale. In 1742 Celsius proposed a different 100-to-zero scale. He died two years later, and his student Linnaeus inverted the scale to what we use today. "We" does not include the USA, Liberia, Belize, and the Cayman Islands, where Fahrenheit is still used.

Cannabis tolerates a wide range of temperatures. This enables the species to inhabit a wide range of latitudes, from the equator to the Arctic Circle, and a wide range of altitudes, from sea level to 3977 m above sea level. Duke (1982) summarized worldwide data from 50 reports, and found that fiber-type hemp does best in regions with a mean temperature of 14.3°C (range 5.6–27.5°C). Wild-type plants are more cold-tolerant (Janischevsky, 1924; Scholz, 1957). Shaikhislamova *et al.* (2006) considered *Cannabis ruderalis* one of the most cold-tolerant weeds in Russia.

Dewey (1914) said that hemp requires 4 months free from killing frosts, for the production of fiber, and 5.5 months for full seed maturity. He suggested that hemp grows best where temperatures are in the range of 60–80°F (15.6–26.7°C) during the months of May through August. He added that young seedlings and mature plants can endure light frosts, of short duration, with little injury—better than oats or maize.

Temperature optima (T_{opt}) can be measured in controlled indoor environments. Cool temperatures cause metabolism to slow, and high temperatures cause plants to expend energy in transpiration, or shut down photosynthesis altogether. Malloch (1922) estimated a T_{opt} of 29.4°C for 'Ferramington', a Dewey cultivar of mixed Chinese and Italian heritage.

Bazzaz et al. (1975) estimated T_{opt} by measuring plant CO_2 production, in plants of tropical provenance (from Jamaica and Panama) and temperate provenance (Nepal and Illinois). Plants from Jamaica, Panama, and Nepal were drug-type plants, and the Illinois plants were feral, likely of Chinese origin. In a "warm chamber" (32°C daytime, 23°C at night), the T_{opt} of plants of tropical provenance (Jamaica 30°C, Panama 27.5°C) was higher than plants of temperate provenance (Nepal 25°C, Illinois 23°C). CO_2 production (hence photosynthesis) in all four populations was about equal at their respective peaks. In a "cool chamber" (26°C daytime, 23°C at night), Jamaica and Panama again peaked at higher temperatures than Nepal and Illinois, but CO_2 production in Jamaica and Panama was lower than Nepal and Illinois in the cool chamber.

Chandra et al. (2011) determined T_{opt} for photosynthetic rate (P_N), by measuring gas exchange in leaf samples (CO_2, H_2O). They reported that T_{opt} was higher in drug-type plants than in fiber-type plants (Table 2.4), but mean T_{opt} for these groups, 30°C and 27.5°C respectively, was not significantly different. Peak P_N did not correlate with drug- or fiber-type characterizations.

Supplementing plants with CO_2 increases their T_{opt}. Frank (1988) recommended growing CO_2-enriched plants at 10°F above normal. Cervantes (2015) suggested growing CO_2-enriched plants at 5°F above normal. Chandra et al. (2008) showed that photosynthetic rate increases with CO_2 levels, and Chandra et al. (2011) showed that photosynthetic rate goes up with temperature, so it is clear that increases in both CO_2 and temperature markedly increase photosynthetic rate.

2.3.1 Growing degree days

Friedrich Haberlandt (see Fig. 1.13) introduced a new thermal metric to the agronomy world, Wärmesummen (heat-sum), which measures heat accumulation over a growing season.

Measuring the daily temperature in a locality, added day-by-day, is powerfully predictive: it can assess the suitability of a location for a particular crop, predict when a crop might mature in a location, determine the best timing for fertilizer application, and compare a particular season with "normal" or average for a location, to predict that season's crop yield.

Haberlandt recognized that plant growth required a minimal temperature, and this base temperature varied between plant species. He determined T_{BASE} experimentally, and determined a T_{BASE} for Cannabis of 0°C or 1°C. For other crops, T_{BASE} ranged from 0°C to 15°C. Measuring Wärmesummen begins the day a crop is planted. Wärmesummen is now called **growing degree days** (GDD), measured in units of °Cd (or in the USA, °Fd). Mean daily temperature is calculated as the daily maximum + minimum divided by two. For example, a day with a high of 20°C and a low of 10°C (and a T_{BASE} of 0°C) contributes $(20° + 10° \div 2) - 0° = 15$ °Cd towards the GDD.

Haberlandt (1879) determined that hemp required 2600–2900 °Cd to reach maturity. He didn't give the hemp's provenance, but he worked in Vienna, so the germplasm was probably Hungarian, and Hungarians were growing Italian landraces by then. Haberlandt calculated GDD in 36 crop plants, and only rice and tobacco needed more GDD to reach maturity than hemp (sunflower equaled hemp).

Marquart (1919) calculated 1800–2000 °Cd for fiber crops, and 2200–2800 °Cd for seed crops. Again, no provenance, but Marquart worked in Landsberg an der Warthe (now Gorzów Wielkopolski, Poland), so he worked with a Middle-European landrace. Using a T_{BASE} of 1°C, Bòcsa and Karus (1997) determined that Middle and Southern European dioecious cultivars needed 1900–2000 °Cd for male flower maturity, and 2700–3000 °Cd for seed maturity. Monoecious 'Felina 34' needed only 1600–1700 °Cd to reach maturity.

Struik et al. (2000) measured GDD to maturity at three different locations: In Italy, dioecious 'Carmagnola' required 2660 °Cd, and monoecious 'Félina 34' and 'Futura 77' needed more, 2923 °Cd (they were growing south of their latitudinal optimum). In the Netherlands, 'Félina 34' and 'Futura 77' averaged 2126 °Cd. In the United Kingdom, 'Félina 34' and 'Futura 77' averaged 1584–1653 °Cd. Pagnani et al. (2018) grew 'FINOLA' in an Italian greenhouse, and it required only 848.5 °Cd to flower.

Several groups measured GDD required by seeds to emerge from soil. See section 17.7.4 for more. GDD can also predict the emergence and progression of insect pests, and determine the best timing for biocontrol release or pesticide application. The T_{BASE} for the European corn borer is 10°C.

2.4 Atmosphere

Plants, like all living things, require oxygen (O_2). Unlike animals, they also need carbon dioxide (CO_2) to drive photosynthesis. The photosynthetic equation is:

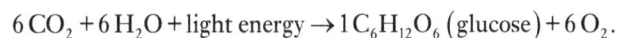

$$6\,CO_2 + 6\,H_2O + \text{light energy} \rightarrow 1\,C_6H_{12}O_6\,(\text{glucose}) + 6\,O_2.$$

Plants provide their own O_2 through the process of photosynthesis.

Table 2.4. T_{opt} and P_N from Chandra et al. (2011)

Plant, type[a]	T_{opt} (°C) at peak P_N	Peak P_N
Mexican, d	35	26.42
W1, d	30	22.32
MX, d	25	21.62
'Zolo 11', f	30	24.83
'Zolo 15', f	30	23.93
'Kompolti', f	25	20.23
'Felina 34', f	25	22.76

[a]d, drug-type; f, fiber-type

Several researchers discovered pieces of the photosynthesis puzzle, and some used hemp in their experiments. Senebier (1782) observed that hemp stalks after harvest turned from green to white when exposed to sunlight. Yet this did not happen to living plants, so they must utilize sunlight to stay green. Senebier (1788) showed that both sunlight and air were required to turn harvested hemp stalks from green to white.

Saussure (1804) added H_2O to the equation. He determined that the weight of O_2 produced by plants, plus the organic matter formed in plants, was considerably larger than the weight of CO_2 consumed. He surmised that the difference must have been supplied by water. He also demonstrated that plants obtained their carbon from CO_2 in the atmosphere, not through uptake from the soil. Lastly, he noted that plants converted some O_2 to CO_2 in metabolic processes, which he demonstrated with germinating hemp seeds.

Plants have evolved different photosynthetic pathways to assimilate CO_2, known as C_3 and C_4 carbon fixation (Hatch and Slack, 1966). *Cannabis* is a **C_3 plant**, like rice, wheat, barley, and 80% of angiosperm species. Examples of C_4 plants include maize, sugarcane, and sorghum—tropical grasses. C_3 plants thrive in ecosystems with moderate temperatures and sunlight intensity, with plentiful groundwater. They grow with difficulty in very hot or arid conditions, because of their high transpiration rates: these conditions cause their stomates to close (to reduce water loss), but this reduces CO_2 levels available for photosynthesis.

C_3 plants require higher CO_2 levels than C_4 plants. Atmospheric CO_2 can be the limiting factor for *Cannabis* photosynthesis. Adding supplemental CO_2 to the growroom produces larger *Cannabis* plants with greater yields. Frank (1988) reported peak growth at CO_2 levels between 1500 and 2000 ppm (=1.5–2.0%). This CO_2 level is 3.6–4.8 times greater than the current atmospheric concentration, which is around 413 ppm (*ca.* 2024).

Cervantes (2015) suggested a CO_2 optimum of 1200 ppm, and given other cost limitations (electrical costs for increased light intensity and temperature), he suggested "the biggest bang for the buck" at 700–900 ppm. Chandra *et al.* (2008) measured photosynthetic rate (P_N, as μmol CO_2/m^2/s) in a Mexican variety, at five CO_2 levels (250, 350, 450, 550, 650, 750 ppm). P_N increased linearly with increased CO_2 levels, from 12.48 to 36.80 μmol CO_2/m^2/s.

Elevated CO_2 modifies resistance to disease pathogens. Zhou *et al.* (2019) grew *Arabidopsis thaliana* (a C_3 plant) under ambient (450 ppm) and elevated (800 ppm) CO_2 levels. Elevated CO_2 led to *increased* resistance to necrotrophic *Botrytis cinerea*, and *decreased* resistance to hemi-biotrophic *Pseudomonas syringae*. Disease severity caused by soil-borne *Fusarium oxysporum* and *Rhizoctonia solani* did not change. Disease caused by biotrophic powdery mildew fungi worsened under elevated CO_2 in all crop species studied (Bencze *et al.*, 2013; Itagaki *et al.*, 2015; Khan and Rizvi, 2020).

2.5 Soil chemistry

Soil chemistry is the most complex feature a grower must manipulate. By "soil" we include **hydroponics**, which is soil chemistry

without the soil. Prior to chemistry taking its present scientific form in the 18th century, specific types of soil were recognized as favorable for hemp. The Roman agronomist Columella (AD 4–70) stated, "Hemp demands a rich, manured, well-watered soil, or one that is level, moist, and deeply worked" (Columella, 1941). The anonymous text *Geoponika* (Γεωπονικα, "*Farm work*") was compiled for the Byzantine emperor Constantine VII (AD 904–959). It stated that hemp grew best in valley land and soil that is consistently humid (Beckh, 1895).

Olivier de Serres (1539–1619), France's "founder of agronomy," recommended *grasse* soil (loam) for a *chenevière* or hemp seed garden (de Serres, 1600). English author Gervase Markham (1568–1637) stated that the best soil for *hempe* was "a rich mingle earth of clay and sand." Pure clay was too heavy, and pure sand "too barren, too hot, and too light" (Markham, 1615). Henry (1771) wrote, "There is no soil in which Hemp will thrive so well as in fine, fat, rich, black-moulded sand … the mould should be deep and easily penetrated, for the fibres of the roots are soon checked, and if impeded are no longer capable of performing their proper offices."

Antoine Lavoisier (1743–1794) co-founded modern chemistry. Lavoisier (1770) debunked the "phlogiston theory" of Van Helmont (1648), who stated that rainwater "transmuted" into soil as part of Aristotle's "elemental cycle" (earth, air, fire, water). Furthermore, water was not an "element," but a compound of two elements, which he called "oxygen" and "hydrogen" (Lavoisier, 1777). Lavoisier established a "model farm" to experiment with crop manuring, and devised chemical tests to detect adulterated tobacco (Lavoisier, 1862). In 1787 he bemoaned France's shortage of hemp when the country's soil was so fit for its cultivation, and he called for spinning mills in the countryside so farmers would gain value-added product (Lavoisier, 1893). The following year he promoted hemp as a premier source of paper, and protested the restriction of river-retting (Lavoisier, 1868). He could have accomplished more but he was guillotined during the French Revolution.

Albrecht Thaer (1752–1828), a German physician/botanist, promoted the "humus theory" of plant nutrition. Thaer believed that plants depended on humus-derived carbon (C), hydrogen (H), oxygen (O), and nitrogen (N). Thaer harbored phlogistonist sentiments, and described a vital force (*Lebenskraft*) that enabled plants to transmute C, H, O, and N into potassium (K) and other elements. In his famous text, Thaer (1812) stated that *Cannabis sativa* required "a strong, humus-rich soil, which has a moist location, and is loose."

Thaer influenced Michał Oczapowski (1788–1854) (Fig. 2.5A), who wrote Poland's first agronomy textbook. Oczapowski (1848) penned a remarkable two-page chapter, *Land Appropriate for Hemp*. He recommended "moderately moist, quite deep, medium-plump soil." Lowlands between hills were good for hemp, because rain washed humus particles down to the lowlands. The humus should not be acidic, otherwise decay arises in roots. In clay soil, hemp only grew well in dry years, and it produced woody stalks.

Thaer's student Carl Sprengel (1787–1859) turned soil fertility into agricultural chemistry. Sprengel (1828) identified six essential soil nutrients, now called plant macronutrients: nitrogen (N), phosphorus (P), potassium (K), calcium (Ca), sulfur (S), and magnesium (Mg). Sprengel recognized that if one of the six is deficient, then plant growth will be poor, even if all the other macronutrients are abundant.

Sprengel's concept became the *Law of the Minimum*, which was popularized by Justus von Liebig (1803–1873). Liebig (1840) noted that some crops can be grown for years without depleting the soil, such as peas and clover. Other plants—Liebig listed hemp first—needed fertilizer supplementation. He became the "founder of the fertilizer industry." Liebig's "patent manure" contained all of Sprengel's nutrients except one: Liebig (1840) argued that plants got all the N they needed from atmospheric ammonia, constantly washed into the soil from rain.

Jean-Baptiste Boussingault (1802–1887) (Fig. 2.5B) experimentally proved some of Liebig's hypotheses, and disproved

Fig. 2.5. *Cannabis* nutritionists: **A**. Oczapowski; **B**. Boussingault; **C**. Tibeau; **D**. Drake. Photos courtesy: (A) Wikimedia Commons, (B) Boussingault (1900), (C) Stephen Lowers at https://www.findagrave.com, (D) portrait by Terry Rutledge, reproduced with permission from Drake (1970)

others. Peas and clover did not deplete the soil because they fixed atmospheric N, and restored N to the soil (Boussingault, 1841). The primacy of N came to Boussingault's mind back in 1832, when he discovered that Peruvians fertilized soil with guano.

Boussingault and Payen (1842) measured N% in a wide variety of fertilizers, including *tourteaux de chènevis* (hempseed cake—after oil has been pressed from seed). They converted raw data to comparative values, setting the N% of dried farm manure at 100%. Guano equaled 31.4% and *tourteaux de chènevis* equaled 40.8%. Boussingault (1851) turned his attention to phosphate. He measured phosphate content in a number of substances, including hemp seed: 3.20%.

Boussingault (1857) showed how N interacted with phosphate by growing hemp in nutrient-deficient soil (calcined sand) and adding selective nutrients (Table 2.5). In one pot he sowed 7 seeds (weighing 2.89 grains (gr)) and added calcium pyrophosphate ($Ca_2P_2O_7$). In a second pot he sowed 7 seeds (weighing 2.89 gr) and added ammonium carbonate (($NH_4)_2CO_3$). In a third pot he sowed 5 seeds (weighing 2.05 gr) and added both $Ca_2P_2O_7$ and potassium nitrate (KNO_3).

By the 50th day, plants fertilized with $Ca_2P_2O_7$ weighed 4.722 gr, only 1.65-fold greater than their initial seed weight (Table 2.5). Plants fertilized with ($NH_4)_2CO_3$ weighed 4.14-fold greater than their initial seed weight. Plants fertilized with both N (KNO_3) and P ($Ca_2P_2O_7$) weighed 14.17-fold greater than their initial seed weight.

Plants fertilized with both N and P flowered a week earlier than plants in other fertilizer treatments, and both males and females were taller than the other plants (Table 2.5). Hemp plants fertilized with only $Ca_2P_2O_7$ were extremely stunted (Fig. 2.6).

Boussingault (1857) disproved Liebig's belief that N wasn't needed for fertilizer: "These last results prove that, in order to contribute actively to plant productions, the basic calcium phosphates, the alkaline salts, ought to be associated with a substance that can furnish nitrogen." Boussingault's name has all but disappeared from modern literature, compared with Liebig's fame (Google hits 49,500 and 903,000, respectively), but Boussingault was the better scientist. And he was quite interested in hemp: Boussingault (1851) gave advice for optimal soil, proper plowing, and suggested a seeding rate of 450 liters/ha.

Soubeiran and Girardin (1851) analyzed N content in seed cake. High N content made seed cake an attractive food supplement for cattle. Seed cake could also be used as fertilizer, so they also measured phosphate content. Soubeiran and Girardin analyzed seed cake from 16 plants, and *tourteaux de chènevis* had the second highest N (6.2%, second to carnation: 7.0%), and the highest phosphate content (7.1%).

Table 2.5. *Cannabis* growth in three fertilizer regimes (Boussingault, 1857)

Fertilizer added	Dried weight at harvest, increase over seed weight	Flowering dates, comments
$Ca_2P_2O_7$	4.73 gr × 1.65-fold	50th day; ♀ 1.5" tall, ♂ 5.5" tall
($NH_4)_2CO_3$	11.86 gr × 4.14-fold	49th day; ♀ 4.0" tall, ♂ 5.5" tall
KNO_3 and $Ca_2P_2O_7$	28.98 gr × 14.17-fold	43rd day; ♀ 7.5" tall, ♂ 12.0" tall

Fig. 2.6. Stunted *Cannabis* plants irrigated with only $Ca_2P_2O_7$ (Boussingault, 1857)

Liebig mentored many agricultural chemists, such as Wilhelm Knop (1817–1891). In a letter to Knop, Leibig disparaged "nitrogenists" such as Boussingault (Finlay, 1992). However, Knop became a nitrogenist. Knop (1865) developed a nutrient solution based on soil analysis, and grew plants in "water culture" without soil. Knop's solution birthed hydroponics. Knop (1868) analyzed the elements present in plant seeds, including hemp seeds, and added silicon (Si), sodium (Na), and chlorine (Cl) to the list of essential elements.

Sister Mary Tibeau (1886–1970) (Fig. 2.5C) conducted the first macronutrient study on *Cannabis*. Tibeau was a Catholic nun from Cedar Rapids. Her Master's Thesis at the University of Iowa sorted the *Law of the Minimum* for hemp. She published her research a year before the USA prohibited *Cannabis* cultivation. Tibeau (1936) chose *Cannabis* because "it shows marked sexual dimorphism and grows well under laboratory conditions." She grew control plants in pots with quartz sand, supplemented with Knop's solution. She tested plants for nutrient deficiencies or toxicities by growing them in Knop's solution modified by a single nutrient—either absent, or 8-fold greater concentration than Knop's solution. Tibeau recorded the responses of plants to nutrient manipulation over time. She described and illustrated symptoms of nutrient deficiencies, including photomicrographs of stalks and leaves in cross-section.

Bill Drake (1941–) (Fig. 2.5D) rediscovered Tibeau's research. Drake (1970) summarized, "Sister Mary is every grower's spiritual Mother Superior." Drake expanded on Tibeau's work, and inaugurated the homegrown era with his groundbreaking *Cultivator's Handbook of Marijuana*. We cite his work frequently, because he also made some of the first observations and recommendations regarding hemp diseases and pests. Cockson *et al.* (2019) replicated Tibeau's study, without citing her. They grew plants in pots with silica sand, and used Hoagland's solution instead of Knop's. They withheld single

nutrients, or added tenfold greater concentrations than Hoagland's solution. We detail their work below.

2.5.1   Soil texture

The texture of soil depends upon both its organic phase and its inorganic phase. These phases bind soil into aggregates. Aggregate structure determines the amount of air space, water dynamics, and nutrient retention in the soil.

The **organic phase** of soil is known as humus, or technically as **particulate organic material** (POM). POM consists of partially decomposed plant matter, living and dead insects, and microorganisms. POM provides a source of nutrients, and keeps nutrients from leeching away. As POM decomposes, the nutrients become available to plants. Dewey (1914) commented that humus "appears to be more necessary for hemp than for most other farm crops."

The **inorganic phase** (minerals) depends on particle size, which is classified as sand, silt, or clay. **Sand** consists of relatively large particles, from 2.0 to 0.05 mm in diameter. Sandy soil feels gritty. **Silt** consists of particles from 0.05 to 0.002 mm in diameter, with a floury feel. **Clay** particles are smaller than 0.002 mm, invisible under light microscopes. Wet clay soil "slicks out," developing a continuous ribbon when pressed between the thumb and finger.

Coarse-textured soil is mostly sand, which does not compact. It drains well, often too well. To improve the water-holding capacity of sand, add organic material (e.g., coir) or synthetic soil conditioners (e.g., perlite). Fine-textured soil is mostly clay. Clay compacts easily, drains poorly, becomes puddled, and dries into large hard clods. To improve clay soil, add organic material or sand. Adding vermiculite is a poor choice here, because it decomposes into a clay sludge. A loose loam soil is half solid (45% mineral, 5% organic) and half space.

Loam—a mixture of sand, silt, and clay—is the best soil for *Cannabis*, first stated by de Serres (1600). The USDA classifies different types of loam in a "soil texture triangle" (Fig. 2.7). Dewey (1914) thought a clay loam was best for hemp; other experts have suggested silt loam, sandy loam, or sandy clay loam. Perhaps *Cannabis* is the best expert regarding its optimal soil texture: Haney and Kutscheid (1975) surveyed the habitats of wild-type *Cannabis* in Illinois, analyzed their soil, and mapped them onto the USDA's soil texture triangle (Fig. 2.7). Most of the wild-type plants grew in loam or sandy loam soils.

Hydroponic growth media also vary in texture. Rockwool and coir consist of filaments, perlite and vermiculite consist of globules. Peat moss (decayed *Sphagnum*) has lost popularity to coir; peat accumulations have not been extracted sustainably; peatlands worldwide are now endangered wetlands. Storm (1987) de-emphasized peat moss in the first book about *Cannabis* hydroponics.

Rockwool is made from super-heated rock spun into a dense cotton-candy texture. The narrow air spaces in rockwool exert strong capillary action. Rockwool easily becomes water-saturated, which reduces root zone oxygen, and predisposes roots to diseases. Its high pH requires conditioning.

Coir is made of coconut fiber. Caplan *et al.* (2017b) tested two coir products: U2 had a higher water-holding capacity than U2-HP; consequently U2 held more water and less oxygen in the root zone than U2-HP. Plants grown in U2-HP compared with U2 yielded 11% higher dry-weight flowers, 20% higher THC concentration, and 57% higher THC yield per plant. A well-oxygenated root zone is essential.

Perlite is volcanic glass, typically made from super-heated obsidian, which expands into spheres. It is very porous (so light that it floats), with good water drainage. **Vermiculite** is a silicate mineral that expands when heated into granules with an accordion shape. It is similar to perlite, but with a higher cation-exchange capacity, meaning it retains nutrients longer.

2.5.2 Soil nutrients

Plant **macronutrients** were discovered by Sprengel (1828): nitrogen, phosphorus, potassium (N-P-K), as well as calcium

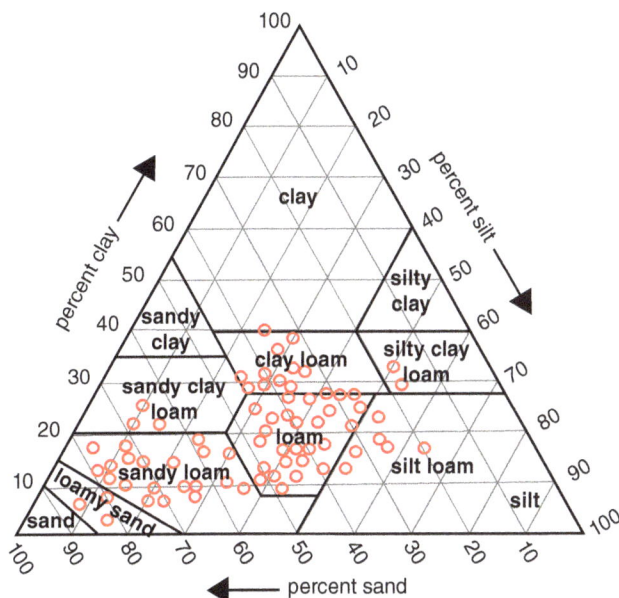

Fig. 2.7. Locations of feral hemp sites, mapped onto the USDA's soil texture triangle (adapted from Haney and Kutscheid, 1975)

(Ca), sulfur (S), and magnesium (Mg). The inorganic phase provides K, Mg, and Ca, and the organic phase provides N, P, and S. The forms and functions of macronutrients are summarized in Table 2.6.

Physiological studies regarding macronutrients are detailed below. For symptoms caused by imbalanced macronutrients and micronutrients, see section 17.3.3. Universal fertilizer rates adopted by conventional agronomists (e.g., "100 kg N/ha") are rarely applicable to *Cannabis*, because different rates pertain to different end products. Crops grown for cannabinoids are sown at a low density (often 10 plants/m^2), lower than crops sown for seeds (30 plants/m^2), or for fiber (50–70 plants/m^2).

Early experimental agronomists simultaneously added N, P, and K, so the effects of N alone cannot be dissected out. Jaffa (1891) attempted to dissect these data, by measuring the elemental content in plants. He extrapolated these data to the "amount of soil ingredients withdrawn from one acre" (Table 2.7). Hemp extracted more nutrients from the soil than flax, about the same as cotton, and less than ramie.

Csókás (1914) dissected N-P-K data using a different approach. To field soil he added 35 kg N/ha, 59 kg P$_2$O$_2$/ha, and 188 kg K$_2$O/ha. Then he measured nutrient uptake in fiber-type plants: N uptake 139 kg/ha, P$_2$O$_2$ uptake 36 kg/ha, and K$_2$O uptake 253 kg/ha. His results were likely skewed by heavy K$_2$O amendment.

Dewey (1914) argued that fiber crops removed few nutrients from the field if stalks were dew-retted *in situ* and fiber extracted in the field by portable machines. Fiber removed from the field consists of cellulose, $(C_6H_{10}O_5)_n$, with little N-P-K. His argument was supported by data from Jaffa (Table 2.7, right-hand column). In 800 lb of fiber, Dewey reported a mineral ash content of only 13.1 lb, compared with 30.7 lb in an equivalent 70 bushels of maize.

Cockson *et al.* (2019) estimated macronutrient requirements in a drug-type strain by growing plants with a complete fertilizer (Hoagland's solution) versus a solution deficient in one macronutrient. At the onset of visual symptoms of deficiency, they harvested above-ground biomass and weighed it after drying (Table 2.8). The time course for the onset of visual symptoms varied, which explains the variable weights of their

Table 2.6. Macronutrients required for plant growth, and their forms and functions

Nutrient	Forms taken up by plant	Nutrient function
Nitrogen N	NH_4^+ NO_3^-	A component of amino acids (proteins), nucleic acids (DNA and RNA), enzymes, coenzymes, cell membranes, and chlorophyll
Phosphorus P	$H_2PO_4^-$ HPO_4^{---}	A component of sugar phosphates (ATP), nucleic acids, lipids, and coenzymes, promotes root formation and flowering
Potassium K	K^+	The primary intracellular cation and a major enzyme catalyst, fuels the "hydrogen pump" and drives stomatal movement
Calcium Ca	Ca^{++}	Cements the middle lamella in cell walls and regulates N metabolism, promotes healthy root and stem development
Sulfur S	SO_4^{---}	Required constituent of amino acids, enzymes and coenzymes, involved in the formation of vitamins
Magnesium Mg	Mg^{++}	The core of chlorophyll, regulates P metabolism, and may activate enzymes that synthesize THC

Table 2.7. Amount of soil ingredients withdrawn from one acre of hemp, from Jaffa (1891); numeric values in pounds

	Whole plant	Leaves	Stems	Clean fiber
Weight per acre	5975	1975	3000	1000
N	62.74	ND	ND	ND
K_2O	101.30	56.46	44.44	0.40
P_2O_3	33.23	18.70	12.91	1.61
$CaCO_3$	130.70	98.62	24.86	7.22
H_2SO_4	6.14	4.46	1.63	0.05
Na_2O	1.95	1.40	0.47	0.08
MgO	17.71	11.55	5.50	0.65
SiO_2	12.66	11.36	0.75	0.49
Cl	0.86	0.31	0.50	0.01

ND = no data

Table 2.8. Effects of macronutrients on mean plant weight (g), from Cockson et al. (2019)

	N	P	K
Control	13.46	20.00	44.20
Deficient	6.77[a]	15.44[b]	32.27[a]
	Ca	S	Mg
Control	20.00	20.00	20.00
Deficient	16.63[b]	18.01[b]	18.01[b]

[a]Statistically significant difference between control and deficient plants
[b]No significant difference between control and deficient plants

control plants. Cockson and colleagues provided excellent descriptions of deficiency symptomology, with color photographs (see section 17.3).

Nitrogen (N)

Cannabis is a nitrophile. It likely evolved that way from coevolving with animals. Vavilov (1926) noted that wild-type *Cannabis* thrived in soil manured by N-rich "excrements of wild animals." Animals excrete N-P-K into feces and urine, and those elements are required by plants. Humans excrete 11 g N/day (Rose *et al.*, 2015), and cattle excrete 104 g N/day (Dong *et al.*, 2014).

Early experimental agronomists did not dissect out N/P/K data. Jaffa tried (Table 2.7), but regarding N, he did not report data for leaves, stems, and clean fiber; the amount of N from whole plants (i.e., whole *above-ground* plant) seems far too low.

In the past 30 years, many studies on fiber-type crops have shown that amending soil with N increases biomass, fiber yield, and seed yield (e.g., Van der Werf and van den Berg, 1995; Van der Werf *et al.*, 1995a; Iványi *et al.*, 1997; Struik *et al.*, 2000; Amaducci *et al.*, 2002; Iványi and Izsáki, 2009; Vera *et al.*, 2010; Prade *et al.*, 2011; Tang *et al.*, 2018).

Fiber-type hemp smothers weeds with its dense canopy. But for this to happen, the crop needs "start-up nitrogen" to establish its canopy. Callaway (2004a) recommended that N be added to soil *before* the crop is sown. He argued that dwarf seed

cultivars such as 'FINOLA' required less N than conventional cultivars because of reduced biomass.

Iványi *et al.* (1997) monitored N uptake by 'Kompolti' over the course of a season, and 79% of N uptake occurred in the first 4–8 weeks (as did 77% of K_2O). During the period of intensive nutrient consumption, N uptake was 3.4 kg/ha per day, and K_2O uptake was 3–6 kg/ha per day. Uptake rate of P_2O_5 was even throughout the season, 0.25–0.64 kg/ha per day.

However, supplementing soil with excess N—out of balance with P and K—produces poor fiber that lacks strength (Berger, 1969). Van der Werf *et al.* (1995a) measured bark fiber percentage in plants supplemented with either 80 or 200 kg N/ha, and bark fiber decreased from 35.5% to 34.0%, respectively.

Excess N may skew sex expression: Heyer (1884) studied gender ratios in dioecious hemp, and reported that "rich" soils shifted the gender ratio towards females. Subsequent studies reported conflicting results, reviewed by Heslop-Harrison (1957). In monoecious hemp, Arnoux and Mathieu (1966) reported that N supplementation "masculinized" plants, by producing a greater percentage of male flowers.

Van der Werf and Van den Berg (1995) assessed gender in crops grown at either 80 or 200 kg N/ha, over two seasons. In the first season, the female-to-male ratio was 1.07 at 80 kg, and 1.72 at 200 kg. In the second season this difference was not as evident, with ratios of 1.42 and 1.45, respectively. The authors concluded that the gender ratio shift was due to N causing females to grow more robustly, and slender male plants were eliminated through self-thinning.

A case of diminishing returns was demonstrated by Tang *et al.* (2018). They measured the ratio of canopy N consumption to canopy transpiration, a metric called **canopy nitrogen use efficiency** (NUE_C). They grew 'Futura 75' in a sealed growroom, and applied urea at three rates: no fertilizer, 1.0 g N per container, or 2.0 g N per container (equivalent to a field application of 0, 60, or 120 kg N/ha). With increased N supplementation, N content in the canopy increased, but NUE_C decreased, from 0.46 to 0.44.

Bevan *et al.* (2021) grew high-THC "Gelato" to investigate the interactive effect of N, P, and K on plant yield (dry weight flowers). During the flowering stage they used a range of concentrations (N 70–290 mg/L, P 20–100 mg/L, K 60–340 mg/L) in 15 different hydroponic solutions. The various combinations of nutrients were statistically analyzed to create contour plots using a central composite design. Yield did not significantly respond to K in the tested range. When K = 200 mg/L, highest flower biomass was obtained with N = 194 mg/L and P = 59 mg/L (Fig. 2.8).

Regarding seed crops, yield should increase with N supplementation, because hemp seeds are N-rich (22.5% protein (Wirtshafter, 1995). Aubin *et al.* (2015) tested the effects of N amendment on 'Anka' and 'CRS-1'. On average, seed yields increased from 519 kg/ha (no added N) to 1340 kg/ha (200 kg N/ha). Vera *et al.* (2010) grew 'FINOLA' and 'Crag' in fields supplemented with five rates of urea, up to 200 kg N/ha. Seed yield for 'Crag' ranged from 584 kg/ha (no added N) to 875 kg/ha (at 175 kg N/ha). 'FINOLA' was more responsive, from 548 kg/ha (no added N) to 1100 kg/ha (at 200 kg N/ha).

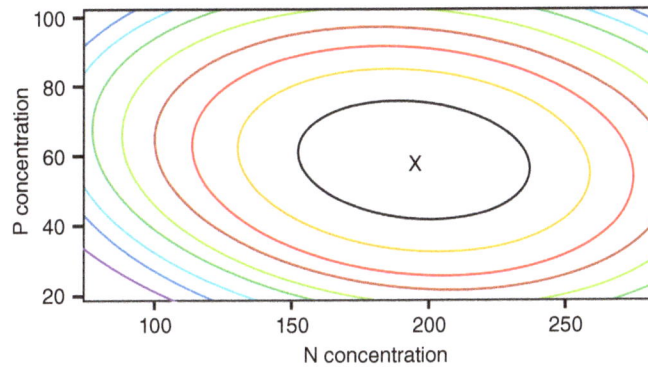

Fig. 2.8. Contour plot of N × P (adapted from Bevan *et al.*, 2021)

Phosphorus (P)

Some studies regarding P are hard to decipher; it's not clear if they are reporting P/ha or P_2O_5/ha, which are different measures (25 kg P/ha equals 57 kg P_2O_5/ha). Jaffa (1891) found that hemp extracted less P than N or K (Table 2.7). Csókás (1914) found that fiber-type hemp required less P_2O_2 than N or K. Bòcsa and Karus (1997) stated that fiber-type hemp needed P, in part to effectively utilize soil N, but also to improve the tensile strength and elasticity of bast fiber cells. Gauca *et al.* (1977) grew fiber-type hemp in field soil supplemented with P_2O_5 at different rates. Fiber yields increased with the application of P_2O_5, but they reported diminishing returns at rates above 40 kg/ha.

Iványi *et al.* (1997) grew 'Kompolti' in field soil supplemented with P_2O_5 between zero (control) and 1000 kg P_2O_5/ha. They reported no significant increase in dry biomass yield. Iványi (2005) grew fiber-type hemp in field soil supplemented with P_2O_5 at different rates. Stalk yields increased from 8.99 t/ha (P_2O_5 at 158 kg P/ha), to 11.95 t/ha (P_2O_5 at 267 kg P/ha). Iványi and Izsáki (2009) grew fiber-type hemp in a field supplemented with various levels of superphosphate fertilizer, resulting in soil P_2O_5 ranging from 158 kg/ha (ambient soil levels) to 267 kg/ha. Stem yield increased from 8.99 t/ha to 11.95 t/ha.

Seed crops require ample P (Callaway, 2004a). Berzak (1975) supplemented a Caucasus landrace with a combination of P+K (60 kg/ha of both P_2O_5 and K_2O), either with or without N (60 kg/ha). The seed yield with P+K+N was 640 kg/ha, the yield with N alone was 530 kg/ha, and the yield with P+K alone was 370 kg/ha. This suggests N is more important than K+P for seed yields.

Vera *et al.* (2010) grew 'FINOLA' and 'Crag' in fields supplemented with four rates of superphosphate, up to 80 kg P/ha. 'FINOLA' was more responsive to P (seed yield gain of 2.14 kg/ha for every additional kg of fertilizer P/ha) than 'Crag' (seed yield gain 0.94 kg/ha for every additional kg of fertilizer P/ha). Yield gains, however, were less than yield gains seen with N supplementation. Aubin *et al.* (2015) tested the effects of added P on 'Anka' and 'CRS-1', by supplementing soil with up to 100 kg P/ha. They found little change in seed yield.

Potassium (K)

Some studies aren't clear if they are reporting K/ha or K_2O/ha, which are different measures. Jaffa (1891) and Csókás (1914)

showed that fiber-type hemp required ample K_2O. Tibeau (1936) found the response of *Cannabis* to K was second only to N. Bòcsa and Karus (1997) stated that K promoted thickening of cell walls, and increased fiber yield and tensile strength.

Iványi (2005) grew fiber-type hemp in field soil supplemented with K_2O at different rates. Stalk yields slightly increased from 5.38 t/ha (K_2O at 215 kg/ha), to 5.95 t/ha (K_2O at 465 kg/ha). Iványi and Izsáki (2009) grew fiber-type hemp in a field supplemented with various levels of potassium chloride (KCl), resulting in soil K_2O ranging from 290 kg/ha (ambient soil levels) to 490 kg/ha. Stem yield increased from 11.46 t/ha to 12.51 t/ha—which was marginal, compared with their results with N or P supplementation. Finnan and Burke (2013) grew three fiber-type cultivars in field soil supplemented with K at five rates from 0 to 150 kg K/ha. Their results suggest that the crop needed about 70 kg K/ha.

Deng *et al.* (2019) conducted a central composite design with four factors: planting density, N application, P application, and K application rate. They grew 'Yunma 1' in field conditions with five levels set for each factor, a total of 36 combinations. Factors for optimal fiber yield (2000 kg fiber/ha) were a density of 329,950–371,500 plants/ha, N application of 251–273 kg/ha, P application of 85–95 kg/ha, and K application of 212–238 kg/ha.

Seed crops may not need K amendment if sufficient levels are in the soil (Callaway, 2004a). Seed yields should increase with K, because hemp seeds have a high K content, 463–2841 mg/100 g seed (Mihoc *et al.*, 2012). However, Aubin *et al.* (2015) tested the effects of added K on 'Anka' and 'CRS-1', by supplementing soil with up to 200 kg P/ha. They found little change in seed yield.

Calcium (Ca)

Jaffa (1891) found that hemp extracted a fair amount of Ca from the soil (Table 2.7). Plants grown in conditions of Ca deficiency grow slowly, but, when Ca is added, recover fairly rapidly (Tibeau, 1936). Plants fed a nutrient solution deficient in Ca showed symptoms late. Their weight at the onset of visual symptoms was not significantly different than control plants (Table 2.8) (Cockson *et al.*, 2019).

Under conditions of maximal stalk yield, Iványi (2011) measured mineral content in leaves—2.4–3.0% for Ca; only two nutrients were higher: N (5–6%) and K (2.7–3.3%). Landi (1997) analyzed four dioecious fiber-types for Ca content, which was greater in leaves (5.12%) than in flowers (2.87%). Hemp seeds have a relatively high Ca content, 144–955 mg/100 g seed (Mihoc *et al.*, 2012), so it seems likely that Ca supplementation would increase yield, but no studies have been done.

Magnesium (Mg)

Jaffa (1891) found that hemp extracted a fair amount of Mg from the soil (Table 2.7). Plants grown in conditions of Mg deficiency were stunted, and when Mg was added, recovery was slow (Tibeau, 1936). Plants fed a nutrient solution deficient in Mg showed symptoms late. Their weight at the onset of visual symptoms was not significantly different than control plants (Table 2.8) (Cockson *et al.*, 2019).

Under conditions of maximal stalk yield, Iványi (2011) measured mineral content in leaves—only 0.6–0.8% for Mg, compared with 5–6% for N. Landi (1997) analyzed fiber-type varieties for Mg content, which was greater in leaves (0.59%) than in flowers (0.50%). Hemp seeds have a relatively high Mg content, 237–694 mg/100 g seed (Mihoc *et al.*, 2012), so it seems likely that Mg supplementation would increase seed harvest, but no studies have been done.

Sulfur (S)

Jaffa (1891) found that hemp extracted a low amount of H_2SO_4 from the soil (Table 2.7). Plants fed a nutrient solution deficient in S showed relatively minor symptoms. Their weight at the onset of visual symptoms was not significantly different than control plants (Table 2.8) (Cockson *et al.*, 2019).

Vera *et al.* (2010) tested the effects of sulfur amendment on 'FINOLA' and 'Crag'. They conducted a field trial on S-deficient soil (ambient soil S ranged from 19 to 30 kg/ha), and added potassium sulfate at rates of 10, 20, or 30 kg S/ha. They reported no increase in either biomass or seed yield, but did not present data. "These results indicate that hemp may not have as high an S requirement as canola and timothy, or that hemp may have a more efficient root system, and hence may be more tolerant to soils relatively deficient in S."

Micronutrients

Micronutrients or "trace elements" are also essential for plant growth: iron (Fe), zinc (Zn), boron (B), copper (Cu), manganese (Mn), chlorine (Cl), and molybdenum (Mo). "Trace" is an accurate word—a plant needs a million N atoms for every Mo atom (Jones, 1998). The forms and functions of micronutrients are summarized in Table 2.9. Some researchers argue that other elements are plant micronutrients, such as nickel, cobalt, sodium, vanadium, titanium, and silicon (Jones, 1998).

Symptoms of micronutrient deficiency are described in section 17.3.3. Composted manure contains adequate micronutrients. Maintaining organic matter and proper soil pH is usually sufficient for most micronutrients.

Micronutrient *toxicity* is commonly seen in gardens that are over-zealously fertilized. Soldatova and Khryanin (2010) added copper sulfate ($CuSO_4$) at a trace concentration of 10^{-10} M, or zinc sulfate ($ZnSO_4$) at a trace concentration of 10^{-9} M. Treated plants showed a shift from the normal 1:1 ratio of females-to-males towards females. They also tested a non-micronutrient, lead nitrate, $Pb(NO_3)_2$, at 10^{-9} M, which shifted the ratio towards males.

Table 2.9. Micronutrients required for plant growth, and their forms and functions

Nutrient	Forms taken up by plant	Nutrient function
Iron Fe	Fe^{++} Fe^{+++}	Occurs in respiratory enzymes and a catalyst of chlorophyll formation and perhaps THC synthesis
Boron B	$B(OH)_4^-$ BO_3^{---}	Appears in enzymes and regulates K and Ca metabolism, promotes root development and prevents tissue necrosis from excess oxygen
Manganese Mn	Mn^{++} Mn^{+++}	A component of photosynthetic enzymes, perhaps THC-synthesis enzymes, involved with N and Fe metabolism
Copper Cu	Cu^{++}	Catalyzes respiratory and photosynthetic enzymes, involved in cell wall formation and lignification
Molybdenum Mo	MoO_4^{--}	Serves as a metal component of enzymes required for utilization of N
Chlorine Cl	Cl^-	Major intracellular anion, activates photosynthesis, involved with K in the regulation of osmotic pressure
Zinc Zn	Zn^{++} $Zn(OH)_2$	Required for DNA and protein synthesis and formation of auxin and other growth hormones
Cobalt Co	Co^{++}	Enhances the growth of organisms involved in symbiotic N fixation, constituent of vitamin B_{12}
Vanadium V	V^+	Promotes chlorophyll synthesis, functions in oxidation–reduction reactions
Silicon Si	$Si(OH)_4$	Forms enzyme complexes that act as photosynthesis regulators, plays a role in the structural rigidity of cell walls
Sodium Na	Na^+	Involved in regulation of osmotic pressure

2.5.3 Soil nutrients and potency

Here we cover N-P-K basics. See section 17.4 for nuanced topics of nutrient supplementation, such as NH_4/NO_3 ratios, and organic versus chemical fertilizers.

Optimizing cannabinoid *potency* (THC%, CBD%) is the goal of many growers. Or the goal is cannabinoid *yield*, a product of potency × flower biomass. Nutritional optimization for biomass may come at a cost of lower potency, and vice versa. The decoupling of potency and biomass is due to two factors: (1) nutrient stress may trigger plants to produce more cannabinoids as a stress response; and (2) optimized nutrition for biomass may decrease cannabinoid potency through a dilution effect.

The latter factor was demonstrated by Caplan *et al.* (2017b). They grew plants during the flowering stage in two coir-based substrates (U2 with high water-holding capacity, U2-HP less so), and used liquid organic fertilizer (2.00N–0.87P–3.32K) at five rates (delivering 117, 234, 351, 468, and 585 mg N/L). Biomass yield (floral dry weight, g/plant) increased with increased fertilizer rate. At the same time, THCA% decreased (Fig. 2.9). They attributed this to a dilution effect. No symptoms of nutritional deficiency were seen at the lowest rate.

Suboptimal nutrition (the first factor listed above) was tested by Tanney *et al.* (2023). They grew high-CBD plants in a peat–coir–perlite substrate for 4 weeks under 18/8 h L/D and then flowered them under 12/12 h L/D for 8 weeks. One set of plants received suboptimal nutrition for the first 6 weeks, then optimal nutrition. Another set of plants received optimal nutrition throughout. Plants that received suboptimal nutrition were more potent than optimized plants, CBDA 8.2% and 7.2%, respectively, and THCA 5.6% and 4.6%, respectively.

Anderson *et al.* (2021) grew high-CBD strains in a peat-based substrate, fertigated with Peters Professional 20-20-20 at six N rates (0, 50, 150, 300, 450, and 600 ppm).

Nutrition status was gauged with a SPAD meter, which indirectly measures chlorophyll content. Plants at 0 ppm N were

nutrient deficient, plants at 50 ppm N were well-fertilized, and higher fertilization rates were superfluous. Lack of fertilizer (0 ppm N) yielded plants with the least biomass (Fig. 2.10). Greatest biomass was achieved at 50 ppm N, and progressively declined with higher fertilization rates. Decline was attributed to the accumulation of salts within the root zone; *Cannabis* does not tolerate salinity. Biomass generally correlated with CBD%, except for disparities at 0 ppm (↑ CBD% due to nutrient stress) and at 600 ppm (↑ CBD% due to salinity stress).

Massuela *et al.* (2023) explored nutrient use efficiency (NUE), which refers to the ratio of nutrient inputs (fertilizers) to nutrient outputs in harvested materials. A higher NUE can be achieved either by increasing nutrient uptake or by increasing nutrient re-mobilization into harvested materials. The former is measured by nutrient uptake efficiency (NUpE): the assimilation of soil nutrients into plants. The latter is measured by nutrient utilization efficiency (NUtE): the translocation of nutrients from sources (roots, leaves) to sinks (flowers, seeds).

NUpE did not significantly change when plants were fertilized at low rates (160 mg/L) versus high rates (240 mg/L), and this was true for N, P, and K. NUtE significantly changed when plants were fertilized at low rates versus high rates, for N ($p = 0.0011$) and K ($p = 0.0106$), but not for P ($p = 0.546$).

NUtE increases when nutrient uptake is limited during flowering, and the nutrient translocation is evidenced by fan leaf turning yellow during flowering. Nutrient translocation from sources to sinks can be quantified by measuring N content (Fig. 2.11). In leaves, N content rises (NUpE) and falls (NUtE) as the season progresses, whereas in flowers, N content steadily increases (NUpE + NUtE). Translocation even occurred at the highest fertilization rate (N at 240 mg/L), indicating a preference of the plant to shift N to flowers. However, under nutrient stress (N 80 mg/L), translocation began earlier (65 days after planting) compared the highest fertilization rate (83 DAP). Massuela and colleagues also explored differences between mineral and organic fertilizers (section 17.4).

Many experts advise decreasing N and increasing P and K during the flowering stage to increase THC% (Frank and Rosenthal, 1978; Frank, 1988; Cervantes, 2015). Benefits of decreasing N during flowering are demonstrated in Fig. 2.9. Evidence supports the practice of increasing P during

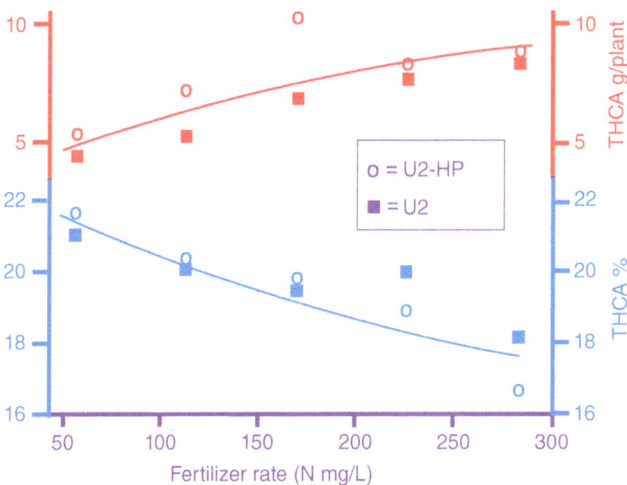

Fig. 2.9. THCA biomass and percent (adapted from Caplan *et al.*, 2017b)

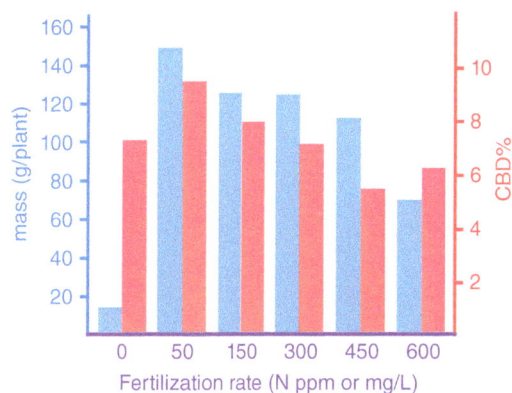

Fig. 2.10. Biomass and CBD% (adapted from Anderson *et al.*, 2021)

flowering, but nowhere near the rates espoused by companies that sell "bloom formulas" and "bud food." Evidence for increasing K during flowering is less robust. Excessive P and K is a waste of money and their runoff leads to the eutrophication of water bodies. To prevent runoff, the disposal or treatment of nutrient wastewater from indoor *Cannabis* cultivation is regulated in some jurisdictions (Bevan *et al.*, 2021).

Coffman and Gentner (1977) conducted the first experimental study regarding soil N-P-K and THC content—both THC potency and THC yield (Table 2.10). They grew Afghani plants and provided N as ammonium nitrate, NH_4NO_3; P as superphosphate, $Ca(H_2PO_4)_2$; and K as potassium chloride, KCl. They applied N at three rates: "low" (no fertilizer added, 0.23 ppm in soil), "medium" (25 ppm added), and "high"

(50 ppm added). Three P rates were "low" (no fertilizer added, 15 ppm P_2O_5 in soil), "medium" (50 ppm added), and "high" (150 ppm added). Three K rates were "low" (no fertilizer added, 35 ppm K_2O in soil), "medium" (50 ppm added), and "high" (150 ppm added). Their study has shortcomings: They harvested plants young, only 50 days old, when males started to flower, and combined male and female *leaves* to measure THC content. Their base soil was deficient in magnesium and plants showed symptoms of Mg deficiency. Their base soil contained toxic levels of manganese, which had deleterious effects on plants.

Nitrogen (N) and THC

Two types of studies have examined the relationship between soil N and cannabinoid potency. One type is observational: researchers study populations of plants grown in different soils, measure soil nutrients and plant cannabinoids, and search for correlations. The other type is experimental: researchers supplement plants with exact amounts of nutrients, and measure cannabinoids.

In an observational study of wild-type *Cannabis* in Illinois, Haney and Kutscheid (1973) found a positive correlation between soil N and THC content. Latta and Eaton (1975) conducted a similar study in Kansas, and did not report any correlation, positive or negative, between leaf tissue N and THC content. Coffman and Gentner (1975) grew an Afghani landrace in 11 different soils; they measured soil nutrients, measured the same elements in leaves, and looked for correlations. They found a positive correlation between leaf N and THC content.

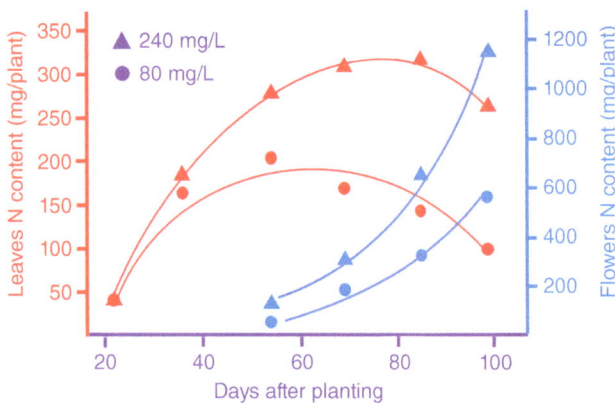

Fig. 2.11. Nitrogen content (adapted from Massuela *et al.* 2023)

Table 2.10. Effects of low, medium, or high N-P-K, from Coffman and Genter (1977)[a]

N	P	K	Weight (g, dry)	THC% (ppm)	THC yield (mg/plant)
L	L	L	0.59	4676	2.7
L	M	L	2.91	4707	13.7
L	H	L	4.55	8948	40.4
L	L	M	1.15	4972	5.6
L	L	H	0.91	4516	4.0
M	L	L	0.71	4052	2.8
M	M	L	3.96	4788	19.0
M	H	L	3.41	**9251**	31.4
M	L	M	1.06	3362	3.6
M	L	H	0.75	5768	0.4
H	L	L	0.75	4537	0.3
H	M	L	5.33	2434	12.8
H	H	L	6.58	**9310**	**61.2**
H	L	M	1.42	4681	6.7
H	L	H	0.67	5339	3.6
L	M	M	3.41	8768	30.0
L	M	H	5.32	7885	**42.0**
L	H	M	**5.54**	**9472**	**52.6**
L	H	H	**5.67**	4968	28.3
M	M	M	5.34	7146	37.9
M	M	M	**5.44**	4891	26.6

[a]Top three yields in each column are highlighted

In an experimental study, Coffman and Gentner (1977) measured the highest THC%, and second-highest total THC yield, in plants amended with N-P-K at low-high-medium (Table 2.10). Bòcsa et al. (1997) reported a negative correlation between soil N and THC content. When 'Kompolti Hybrid TC' was supplemented with 150, 450, and 600 mg N/kg soil, leaf THC content dropped from 275 to 260 to 225 (GC-FID peak areas). Adding N significantly increased plant height and weight, so decreased THC% was presumably a dilution effect.

Caplan et al. (2017a) tested the effects of fertigation (fertilizer added to water irrigation) on a high-THC strain during the vegetative phase. They grew plants in two coir (coconut fiber)-based substrates: U2 had a higher water-holding capacity than U2-HP. Fertigation was applied at five N rates (117, 234, 351, 468, and 585 mg/L), using a liquid organic fertilizer (4.0 N, 1.3 P, 1.7 K). No differences in biomass yield were found between the two substrates. Pooled data from both substrates showed that the highest yield was achieved at a rate that supplied 389 mg N/L (interpolated from yield-fertilizer responses).

Turning to the flowering phase, Caplan et al. (2017b) grew a high-THC strain in U2 and U2-HP, and tested fertigation at five N rates (117, 234, 351, 468, and 585 mg/L) using a different liquid organic fertilizer (2.00 N, 0.87 P, 3.32 K). In both substrates, THCA% decreased linearly with increasing fertilizer rates: in U2-HP dropping from 21.6% to 16.7%, in U2 dropping from 21.0% to 18.1% (Fig. 2.9). This was a dilutional effect, because biomass increased with increased fertilization rate, with greater biomass in U2-HP. To maximize both THCA% and biomass yield in UP-HP, they recommended a rate supplying N at 212–261 mg/L. In UP-2, a rate supplying N at 283 mg/L maximized yield, although a lower rate was desirable to increase THCA%.

Pagnani et al. (2018) grew 'FINOLA' in pots with peat-based compost potting soil, vermiculite, and perlite, which served as controls. Another set of pots were fertilized at a rate equivalent to 80 kg N/ha. Female plants were harvested when flowering began. Dried weight (leaves and flowers) of controls averaged 2.04 g/plant; fertilized plants averaged 3.03 g/plant. THC+CBN% of control plants averaged 0.176%, fertilized plants averaged 0.182% (significant, p <0.01). CBD% of control plants averaged 1.899%, fertilized plants averaged 2.065% (significant, p <0.01). Pagnani and colleagues also tested the effects of growth-promoting rhizobacteria, see section 13.3.

Saloner and Bernstein (2021) applied nitrogen (80% nitrate NO_3^-, 20% ammonium NH_4^+) to 'Annapurna' at the onset of flowering under a 12:12 light regimen. Irrigation solutions contained 30, 80, 160, 240, or 320 mg/L (ppm). Plant growth was restricted at the lower rates (30 and 80 mg/L), demonstrating N deficiency with symptoms of reduced flower weight and leaf chlorosis. However, THCA and CBDA concentration was greatest at 30 mg/L (THCA 8%, CBDA 8%), with a linear dose-response down to 320 mg/L (THCA 5%, CBDA 5%). Total CBDA yield is a product of flower weight × CBDA%, so they suggested a rate of 160 mg/L to hit that sweet spot. They also found optimal growth during the vegetative period at 160 mg/L (Saloner and Bernstein, 2020).

James et al. (2023) tested the effects of N on female clones of high-CBD "BaOx" under field conditions. Urea–ammonium nitrate was split-applied, 10 and 28 days after transplanting clones, at five rates (0, 56, 122, 168, 224 kg N/ha). Baseline N in soil was not measured because N leaches from the rooting zone between growing seasons in North Carolina. Plants were harvested at 50% amber trichome stage and dried in forced-air tobacco curing barns at 66°C for 2–3 days. Inflorescence cuttings, 15–20 cm long (flowers and leaves) were dried a second time at 54°C for at least 24 h until they reached a constant weight.

Inflorescence yield showed an increasing response at lower N rates that plateaued at higher rates. It reached a plateau of 3384.5 kg/ha, with an inflection point for optimal yield at 86.8 kg N/ha. CBD% showed a shallow bell-shaped response curve rather than a plateau: 11.33% at 0 kg N/ha, peaking 12.11% at 115 kg N/ha, dropping to 11.42% at 224 kg N/ha. Total CBD yield (inflorescence yield × CBD%) plateaued at 389.34 kg/ha, with an inflection point for optimal yield at 84.2 kg N/ha. The authors noted that this optimal N rate to maximize CBD yield was lower than rates needed to maximize seed yields in the study by Vera et al. (2010), with inflection points of 180 kg N/ha for 'Crag' and 189 kg N/ha for 'FINOLA'.

Massuela et al. (2023) grew high-CBD plants in a wood fiber-and-perlite substrate and fertigated at three N rates (80, 160, 240 mg/L) through the vegetative and flowering stages. Flower dry matter at the three rates increased linearly from 16.1, 22.6, to 26.3 g/plant. CBD% at the three rates showed a downward trend, 5.29%, 5.36%, and 4.98%. CBD yield (potency × flower biomass) at the three rates followed the dry matter trend: 989, 1389, and 1450 g/plant. Thus, fertilization at 160 mg/L produced 95% of the CBD yield of 240 mg/L, while receiving one-third less nutrients. Massuela used a SPAD meter (which indirectly measures chlorophyll content) at season's end, which showed that plants fed 80 mg/L were deeply deficient, plants fed 160 mg/L were slightly deficient, and plants fed 160 mg/L were well-fertilized. They proposed that the controlled induction of nutrient stress during flowering can increase cannabinoid yield.

Song et al. (2023) showed that a decrease in the plant's ratio of C (carbon) to N by N-limitation induced a shift in the plant metabolism towards lower production of N-containing metabolites and higher production of metabolites without N. They fertilized plants with 30, 80, 160, 240, or 320 mg N/L and measured metabolites. Elevation of N resulted in an increase in N-containing compounds (chlorophylls and most amino acids, and total biomass) and decreased the content of cannabinoids, phenols, and flavonoids.

Phosphorus (P) and THC

In an observational study of wild-type *Cannabis* in Illinois, Haney and Kutscheid (1973) reported a negative correlation between soil P and THC content, but that appears to be a typo—their Fig. 1 showed a positive correlation. In another observational study, Coffman and Gentner (1975) grew a drug-type Afghani landrace in 11 different soils; they reported a strong positive correlation (r = 0.88) between leaf P content and THC content. They hypothesized that phosphate plays a role in cannabinoid synthesis, because geranyl-pyro*phosphate* is a cannabinoid precursor.

In an experimental study (Coffman and Gentner, 1977) highest THC yield (mg/plant) was seen in plants supplemented with high P, high N, and low K (Table 2.10). Plants supplemented with high P, low N, and medium K yielded the highest potency (THC%). Berstein *et al.* (2019b) supplemented a high-THC strain with P, and then analyzed mineral content in plants to see where the P was translocated. Most of it translocated to flowers, with lower concentrations in leaves and stalks. However, Berstein *et al.* (2019a) found that increased P supplementation (40 ppm P_2O_5 versus 108 ppm K_2O) did not increase THC content.

Shiponi and Bernstein (2021a) measured plants during the vegetative stage, and they produced as much biomass when fertilized with 30 mg/L as those supplied with 90 mg/L. Shiponi and Bernstein (2021b) measured a high-THC strain and a mixed THC/CBD strain during the flowering stage, at five levels of P fertigation (5, 15, 30, 60, 90 mg/L). Inflorescence dry weight increased with P in both strains up to 30 mg/L and then plateaued. Inflorescence THCA% and CBDA% in both strains were greatest at 5 mg/L and then decreased due to dilutional effects of greater biomass. Cannabinoid yield (cannabinoid% x biomass) was greatest at 30 mg/L.

Cockson *et al.* (2020) supplemented high-CBD "BaOx" in a peat moss–perlite substrate with Hoagland's solution at six P rates (3.75, 7.50, 11.25, 15.0, 22.50, 30.0 mg/L). They reported a significant increase in apical bud weight up to 22.50 mg/L. CBDA% peaked at 11.25 mg/L, and then slightly declined. The same group used "BaOx" at higher rates (15, 60, 120, 180 mg/L) and limited their analysis to biomass production (Veazie *et al.*, 2021). Rates above 15 mg/L did not significantly increase total above-ground biomass. Foliar P concentrations increased linearly up to 180 mg/L—no doubt contributing to the harsh smoke and black, greasy ash characteristic of over-fertilized herb (section 17.4).

Westmoreland and Bugbee (2022) grew a high-CBD strain in a peat–vermiculite substrate with nutrient solutions at three P rates (25, 50, 75 mg/L). Dried inflorescence yield (flowers plus sugar leaves) slightly decreased (680 g/m² at 25 mg/L versus 610 g/m² at 75 mg/L) but was not statistically significant. CBD% was essentially unchanged across all treatments. Nutrient partitioning (fraction of P in inflorescences versus leaves) decreased with increased P supplementation from 65% to 25%. Phosphorus use efficiency (PUE) predictably decreased with increased P supplementation.

Leachate P (into drainage water) increased 12-fold in response to the threefold increase in P input. The authors summarized the literature regarding P fertilization as follows. During the vegetative phase, 15 mg/L was sufficient in greenhouse conditions with ambient CO_2. During the flowering phase, studies showed no benefit of P above 25 mg/L (their study), above 30 mg/L (Shiponi and Bernstein, 2021b) or even above 11.25 mg/L (Cockson *et al.*, 2020). For growers that supplement with of CO_2, which increases photosynthetic rate, they proposed a theoretical P demand of 24 mg/L. Anything more becomes leachate.

Potassium (K) and THC

An observational study of wild-type *Cannabis* in Illinois by Haney and Kutscheid (1973) reported a negative correlation between soil K and THC content. Similar observational studies did not report any correlation, positive or negative, between K and THC content, in Kansas soils (Latta and Eaton, 1975) and Maryland soils (Coffman and Genter, 1975).

In an experimental study, Coffman and Gentner (1977) grew Afghani plants in soil supplemented with various amounts of N, P, or K (Table 2.10). Plants supplemented with high levels of K did not appreciably increase THC yields unless accompanied by moderate P, and low to moderate N. Berstein *et al.* (2019b) supplemented a high-THC variety with K, and K translocated to flowers more than stalks (the difference was not as great as seen with P).

Yep and Zheng (2021) grew a high-CBD strain in a peat moss substrate fertigated with aquaponic solutions at three K levels (75, 113, 150 mg/L) during the flowering stage. Increasing K levels increased the harvest index (marketable inflorescence to shoot weight) by up to 22% compared with the control. Total inflorescence weight increased from 49.7 to 68.0 g/plant, but not statistically significant in a pooled *t* test.

Saloner and Bernstein (2022b) tested the effects of five levels of K fertigation (15, 60, 100, 175, and 240 mg/L) during the flowering stage. In a high-THC strain, THCA content in apical inflorescences decreased from 17.5% (K = 15 mg/L) to 15.0% (K = 240 mg/L). In a mixed THC/CBD strain, THCA content decreased from 8% (K = 15 mg/kg) to 5.5% (K = 240 mg/L), and CBDA content dropped from 7.2% (K = 15 mg/kg) to 5.8% (K = 240 mg/L). This was probably a dilutional effect, because inflorescence dry weight increased with K supplementation: In the high-THC strain, from 24.0 g (K = 15 mg/L) to 38.7 g (K = 240 mg/L). Likewise the mixed THC/CBD strain increased from 36.7 g (K = 15 mg/L) to 57.3 g (K = 240 mg/L). Interestingly, at K = 15 mg/L, where cannabinoid potency was greatest, plants showed symptoms of K deficiency: leaf chlorosis. The authors suggested applying K = 60 mg/L to optimize cannabinoid yield (potency × biomass).

James *et al.* (2023) tested the effects of K supplementation on high-CBD "BaOx" plants under field conditions. K_2SO_4 was applied 10 days after transplanting, at five rates (0, 46, 93, 139, 185 kg K/ha). Untreated soil contained a baseline K of 99.2 mg/kg. Plants were harvested at 50% amber trichome stage. K supplementation did not significantly increase biomass (dried inflorescences) or CBD% over baseline K in soil.

Other nutrients and THC

Calcium (Ca) content in soil correlated with THC content in an observational study of wild-type *Cannabis* by Haney and Kutscheid (1973). A similar study in Kansas did not report any correlation, positive or negative (Latta and Eaton 1975). Coffman and Gentner (1975) grew plants in 11 different soils. Ca by itself did not correlate with THC%, but the combination of Ca and magnesium correlated with THC% ($r = 0.74$). Radosavljevic-Stevanovica *et al.* (2014) analyzed soil associated with seized *Cannabis* in Serbia. They reported a positive correlation between Ca and THC content.

Coffman and Gentner (1977) grew Afghani plants in soil supplemented with various amounts of N, P, or K. Their P supplement contained Ca–superphosphate, $Ca(H_2PO_4)_2$. They reported a positive correlation between soil Ca and plant size, but a negative correlation between leaf Ca and THC content. Berstein *et al.* (2019b) supplemented a medicinal *Cannabis* variety with Ca, and more of it translocated to flowers than to stalks.

Magnesium (Mg) content in soil did not correlate with THC content in an observational study of wild-type *Cannabis* by Haney and Kutscheid (1973). A similar study by Latta and Eaton (1975) reported a significant positive correlation between leaf Mg and THC content, $r = 0.66$. Radosavljevic-Stevanovica *et al.* (2014) analyzed soil associated with seized *Cannabis* in Serbia. They reported a positive correlation between Mg and THC content.

Coffmann and Genter (1975) grew plants in 11 different soils, and reported a significant negative correlation between soil Mg and THC content, $r = -0.64$. They subsequently grew plants supplemented with various fertilizers, and reported a positive correlation between Mg and plant size, but no correlation, positive or negative, between soil Mg and THC content.

Morad and Bernstein (2023) grew a mixed THC/CBD strain in perlite fertilized with balanced nutrition at five Mg rates (2, 20, 35, 70, 140 mg/L) and studied the effects on plants during the vegetative stage. Plant development and function were optimal at the 35 and 70 rates. Plants at the 2 and 20 rates were stunted and showed Mg deficiency symptoms. Plants at the 140 rate showed Mg toxicity symptoms.

Micronutrients have not received much attention. Yep and Zheng (2021) grew a high-CBD strain in a peat moss substrate fertigated with an aquaponic solution (AP) or AP augmented with micronutrients (AP+M):

AP (mg/L): Fe^{3+} 0.56, Cu^{2+} 0.01, B^{3+} 0.02, Zn^{2+} 0.48, Mo^{3+} 0.0, Mn^{2+} 0.02.

AP+M (mg/L): Fe^{3+} 1.96, Cu^{2+} 0.03, B^{3+} 0.28, Zn^{2+} 0.56, Mo^{3+} 0.01, Mn^{2+} 0.42.

Additional micronutrients did not have any effect on vegetative growth; total inflorescence yields increased: AP = 43.2 g/plant, AP+M = 49.7 g/plant, but not statistically significant in a pooled *t* test. Additional micronutrients (AP+M) increased leaf levels of micronutrients compared with AP.

2.6 Altitude and latitude

The *Cannabis* **altitude record** was set by Waddell (1905). He came upon "rank plants of Indian hemp, 6 feet high" growing near Gyantse in Tibet, 250 km southwest of Lhasa. He recorded the altitude as 12,000 ft above sea level, but Gyantse is actually 3977 m (13,050 ft). Either way, this holds the record. In second place, Sharma (1983) reported wild-type *Cannabis* growing at 3700 m in the Indian Himalaya.

The European record is less, because the landscape is lower. Jaccard (1895) recorded *Cannabis* at 1593 m (5226 ft) near Bellwald, Switzerland, not far from the Matterhorn. The European record, however, is not in the Alps, but in the Caucasus mountains in southern Russia. That range is taller than the Alps, topping out

with Mount Elbrus (5642 m, 18,510 ft)—compared with Mont Blanc in the Alps (4808 m, 15,774 ft).

Portiere (2012) stated that *Cannabis ruderalis* grew in the North Caucasus "Upland xerophyte belt," between 1200 and 1800 m. Chadaeva (2020) gave a specific location and altitude: she collected *Cannabis ruderalis* in a soil-disturbed pine forest near the rural settlement of Elbrus, on the flanks of Mount Elbrus, at 1775 m (5824 ft). South of the Caucasus, in the Transcaucasus highlands of Armenia, Gabrielian and Zohary (2008) said that *Cannabis sativa* grew on dry stony hills between 700 m and 2000 m. Their distribution map shows a collection near the headwaters of the Pambak River (2200 m, 7217 ft). But that is in Asia, so it doesn't count as a European record.

In North America, *Cannabis* has been cultivated as high as 2500 m (8200 ft) near Gypsum, Colorado (Brenner, 2019). Glasshouse crops have been grown higher, at 3094 m (10,150 ft), in Leadville, 60 miles from Gypsum (Walsh, 2019).

The primary environmental constraint for high-altitude growth is temperature. But another environmental constraint for Europe versus Asia is latitude: Gyantse in Tibet is relatively low, at 28°55' N, whereas Elbrus village is 43°15' N, and Bellwald is 46°26' N. Gypsum is at 39°39' N.

Latitudinal limits to growth, like altitude limits, are partially a function of temperature. High-latitude locations also tend to have nutrient-deficient soils with permafrost, and less rainfall (yet flooding during thaws). Latitudinal limits are inexorably linked to daylength, and daylength is linked to flowering time, which we introduced earlier (section 2.1.3). As latitude increases, so do annual fluctuations in daylength (Fig. 2.12). At the North Pole, the "midnight sun" lasts from late March to late September. But it's a cold light, because the sun isn't high in the sky: a solar elevation of only 23.5° on the Summer Solstice (versus 90° overhead at noon the same day along the Tropic of Cancer at 23.44 °N).

The autumnal equinox (September 23) marks the moment when the Sun crosses the Earth's equator, and the length of day is around 12 h. Daylength less than 12 h triggers flowering in *Cannabis*. At higher latitudes, daylength shortens rapidly after

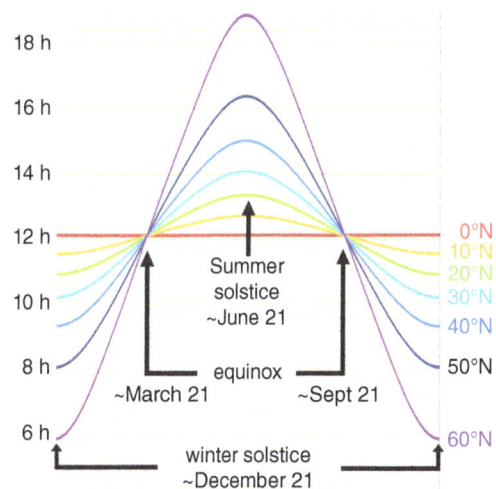

Fig. 2.12. Length of day as a function of latitude in the northern hemisphere

the equinox (Fig. 2.12), solar elevation drops precipitously, and frost hits flowering plants before they can mature seeds. Thus Billings (1987) stated that temperature is the primary latitudinal constraint—cold air and soil, strong and chilling winds, and a thin life zone between permafrost and the wind. Billings demonstrated the effects of latitude by estimating the relative richness of vascular floras: the Southeastern United States at 25–37 °N has 5557 plant species, compared with the Canadian Arctic Archipelago at 61–83 °N with 340 plant species.

The *Cannabis* latitude record was set by Schübeler (1875), who grew plants at 68°7' N in Lofoten, Norway. This is north of the Arctic Circle (66°33'47.5" N). Lofoten is warmed by the far reaches of the Gulf Stream, imparting perhaps the world's greatest temperature anomaly relative to latitude. Schübeler also mentioned high-latitude hemp crops at Sodankylä in Finland (67°30'), Haparanda in Sweden (65°50'), and Reykjavík in Iceland (64°08'). At the other end of the latitude spectrum, drug-type plants grow at the equator, 0°, across South America (Colombia to Brazil), across Africa (Congo to Kenya), and in Indonesia (Sumatra, Borneo).

Cannabis exhibits a latitudinal "regression to the mean." Plants from northern latitudes when transplanted south will flower earlier, with shortened stature (Barbieri, 1952; Yao *et al.*, 2007; Cosentino *et al.*, 2012; Petit *et al.*, 2020a). Plants from southern latitudes when transplanted north will flower later, with increased growth (de Meijer and Keizer, 1996; Amaducci *et al.*, 2008a; Hu *et al.*, 2012; Salentijn *et al.*, 2015; Petit *et al.*, 2020a).

Many fiber-type cultivars were developed at 35–55 °N latitude; they perform poorly in equatorial locations, due to early flowering. Many drug-type landraces were developed at equatorial locations, and perform poorly in high latitudes, due to late flowering and early frost. Southern drug-type strains failed to reach maturity at 55°57' N (Christison, 1850), 53°34' (Bredemann, 1952), 51°30' (Fairbairn and Liebmann, 1974), 48°17' (Fournier, 1981), 45°12' (Small and Beckstead, 1973), and even 35°21' (Turner *et al.*, 1979).

Latitude became an issue when Sir Joseph Banks, England's great hemp advocate, imported germplasm from southern China. Banks was prodded by Keane Fitzgerald, who obtained 40 seeds from a "General Elliot" in Canton (Guǎngzhōu, 23°08' N). The plants grew to "an amazing size" in London (51°30.5'), 14 ft tall, but frost killed them before they set seed (Fitzgerald, 1782). Banks obtained a pound of seed smuggled out of Canton in 1784 (Faujas de Saint Fond, 1799). Banks's crop suffered the same fate—killed by frost before setting seed. Banks gave germplasm to French naturalists, hoping that plants might fully ripen in the south (Faujas de Saint Fond, 1799). Faujas distributed germplasm to 11 recipients across France; all reported that Chinese hemp grew luxuriously. But frost killed them before they set seed, except in southern France—at Montélimar (44°33') and Drôme (44°45').

Barbieri (1943) reported latitude making a difference within Italy. When northern varieties ('Carmagnola', 'Bolognese', 'Ferrarese') were grown south around Napoli, they showed *nanismo*, "dwarfing," and *prefioritura*, "bolting" or "pre-flowering." This also happened when French and Hungarian varieties were cultivated in southern Italy (Cosentino

et al., 2012), and when northern Chinese varieties were cultivated in southern Nánjīng (Yao *et al.*, 2007).

2.7   Elicitors and effectors

Elicitors are signaling molecules that induce physiological changes in plants. They attach to receptor proteins located on cell membranes, and trigger responses in metabolic pathways (i.e., cannabinoid biosynthesis), pest and pathogen resistance, and even the female-to-male ratio. Elicitors may be *intrinsic*, such as plant hormones, or *extrinsic* (foreign signaling molecules), such as chitosan.

Salicylic acid (SA) is a plant hormone with many functions in plant growth and development, such as seed germination, vegetative growth, and flowering. Spraying SA on plants at the initiation of flowering increased *THCAS* gene expression and THC content (Jalali *et al.*, 2019). SA is an ingredient in some commercial "bloom formulas."

SA induces **systemic acquired resistance** in plants. SAR is a systemic "whole plant" response that arises following a localized exposure to a pathogen or pest. Ross (1961) coined SAR after he inoculated the lower leaves of tobacco plants with tobacco mosaic virus (TMV). Seven days later, leaves distal to the inoculation site reacted when injected with TMV. They walled-off the infection by producing small necrotic lesions, now called a **hypersensitivity response** (HR).

White (1979) discovered that SA induces SAR, allegedly after his mother suggested he test the effects of acetylsalicylic acid (aspirin) on plants, because she took aspirin for her own health problems. SA-inducible defenses are generally effective against biotrophic pathogens, which derive nutrients from living host tissues (e.g., the powdery mildew fungus, *Golovinomyces cichoracearum*). SA-inducible defenses are less effective against necrotrophic pathogens, which first destroy host cells, then feed on the contents (e.g., the gray mold fungus, *Botrytis cinerea*). A product with SA, indole-3-butyric acid, and chitosan (Consensus®) designed to induce SAR was approved for use on *Cannabis* (EPA, 2023).

SA also induces resistance against abiotic stresses, such as cold, drought, heavy metals, and UV radiation. Shi *et al.* (2009) induced *Cannabis* resistance to heavy metal (cadmium) toxicity by soaking seeds in SA prior to planting. Gonsior *et al.* (2004) improved resistance against the parasitic plant *Phelipanche ramosa* by watering plants with BTH, a derivative of SA. SA-mediated defenses are induced by insects with piercing-sucking mouthparts, such as aphids and whiteflies. Conversely, insects with chewing mouthparts trigger jasmonic acid-induced plant defenses (Thaler *et al.*, 2012).

Jasmonic acid (JA) is a hormone that mediates **induced systemic resistance** (ISR) independent of the SA pathway. ISR is "a state of enhanced defensive capacity developed by a plant when appropriately stimulated" (van Loon *et al.*, 1998). JA-inducible defenses are generally effective against insects with chewing mouthparts (Thaler *et al.*, 2012) and necrotrophic pathogens (e.g., *Botrytis cinerea*). ISR can be induced by spraying plants with JA derivatives (jasmonates).

ISR can be induced via **plant growth-promoting rhizobacteria** (PGPRs). PGPRs are non-pathogenic bacteria living in the soil around roots, including *Bacillus* and *Pseudomonas* spp. (see section 13.3). Gonsior *et al.* (2004) induced ISR in *Cannabis*, which improved resistance against the parasitic plant *Phelipanche ramosa*. They induced ISR by treating plants with nonpathogenic *Pseudomonas* sp., or with extracts from a seaweed, *Ascophyllum nodosum*. Romeo® is a product designed to induce ISR, and approved for use on *Cannabis* by Colorado DOA (2023). Romeo® contains "cerevisane" (cells walls of *Saccharomyces cerevisiae* strain LAS117), labeled for use against fungal pathogens.

Flores-Sanchez *et al.* (2009) exposed cell cultures (leaf explants of "Skunk #1") to three hormone elicitors: jasmonic acid (JA), methyl jasmonate (MeJA), and salicylic acid (SA); as well as five extrinsic elicitors: *Citrus* fruit pectin, silver nitrate ($AgNO_3$), cobalt chloride ($CoCl_2$), nickel sulfate ($NiSO_4$), and sodium alginate from brown algae. The cell cultures did not produce cannabinoids, even after elicitation treatments. However, a number of other metabolites were modulated, including some amino acids, organic acids, and sugars. The same research group revisited cell culture responses to JA and *Citrus* fruit pectin: Peč *et al.* (2010) found that JA increased the production of tyrosol, an antioxidant metabolite, while *Citrus* fruit pectin increased saturated fatty acid content.

Gibberellins (GAs) regulate key processes such as stem elongation, fiber content, flower development, leaf senescence, and seed dormancy. Today we know dozens of GAs, which are modified terpenoids. The third to be discovered, gibberellic acid (GA_3), has received the most attention.

Atal (1959) applied 100 ppm gibberellin (no GA specified) as a foliar spray to seedlings of Punjabi provenance. Female plants produced male and hermaphroditic flowers. Atal and Sethi (1961) applied the same treatment, and examined fiber content in stalks. The bark fiber-to-wood ratio increased in GA-treated plants (37%) compared with controls (33%). Treated plants compared with controls produced fiber cells that were longer (3.8 cm versus 0.43 cm) and wider (24 μm versus 18 μm), and the cell walls were thicker (7.6 μm versus 4.7 μm).

Herich (1960) soaked 'Rastislavická' seeds in gibberellin (no GA specified), and shifted the female-to-male ratio toward females. The ratio in control plants was 0.93, shifting to 1.22 with GA at 10 ppm. However, at 100 ppm, the ratio reverted to 0.96. Heslop-Harrison and Heslop-Harrison (1961) found no gender shifts in plants treated with GA_3. Hoffmann (1972) used GA_3 to shift the ratio towards males.

Holenarasipur Y. Mohan Ram (1930–2018) published eight *Cannabis* papers in his trailblazing research at Delhi University. "One of the principal reasons why we chose *Cannabis* is that it grows abundantly and naturally in India. We have few competitors because western scientists might be behind bars if they cultivated it!" (Mohan Ram, 2002). Mohan Ram and Jaiswal (1972) induced male flowers on female plants with GA_3. This was confirmed by Galoch (1978).

Frank (1988) popularized the use of GA_3 to induce male flowers (and pollen) on female plants, and then cross that pollen with normal female flowers to create feminized seeds. Cervantes (2015) recommended spraying plants with 25–100 ppm GA_3 for the creation of feminized seeds—now a major industry.

Auxins coordinate phototropism, geotropism, apical dominance, and other growth responses by stimulating cell elongation. Auxins promote root initiation in callus cell cultures, and induce wound responses, fruit development, and seed dormancy. Five endogenous auxins have been found, of which **indole-3-acetic acid** (IAA) is the most important. The synthetic auxin **1-naphthylacetic acid** (NAA) is widely used. **Indole-3-butric acid** (IBA) is an ingredient in many horticultural plant rooting products, and approved for use on *Cannabis* (EPA, 2023).

Heslop-Harrison (1956) induced female flowers on male plants with NAA; at high doses NAA reduced total flowering response. Hoffmann (1972) shifted the gender ratio towards females with NAA. Galoch (1978) dabbed IAA (25, 50, and 100 ppm) on shoot apices of fiber-type 'LKCSD.' Increasing concentrations step-wise decreased overall flowering. Male plants increased their percentage of female flowers, female plants showed no change. Galoch provided photomicrographs of male flowers transforming into hermaphrodites, and then females, in four stages. Chailakhyan and Khryanin (1978a) applied IAA to seedlings through the root system. They tested 'USO-6,' an incompletely-monoecious cultivar—control plants were 34% monoecious, 37% female, and 29% male. IAA-treated seedlings matured into 60% monoecious, 40% females, and no males.

Abscisic acid (ABA) was originally believed to play a major role in leaf abscission, but this is only true in some species. ABA regulates responses to environmental stress (drought, heat, cold tolerance) and immune responses to fungal pathogens.

Mohan Ram and Jaiswal (1972) sprayed female plants with ABA, which did not alter the gender, but when ABA was added to GA_3, it reduced the induction of male flowers by GA_3. Galoch (1978) also reported no effect with ABA, but it antagonized the effects of GA_3 and IAA. Galoch (1980) measured endogenous hormone levels in plants, presumably fiber-type 'LKCSD.' Male plants produce higher levels of GAs than female plants, whereas female plants produce higher levels of ABA and auxins than males.

Ethylene plays roles in plant growth, leaf senescence, fruit ripening, and responses to environmental stress and fungal pathogens. Mohan Ram and Jaiswal (1970) used ethephon (metabolized into ethylene by plants) to induce female flowers on male *Cannabis* plants. High doses caused **epinasty**—twisted, malformed growth of branches and leaves. Galoch (1978) also reported a feminizing effect on male plants with ethylene.

Ethylene synthesis in plants can be inhibited by a number of substances, thus inducing male flowers on female plants. Mohan Ram and Sett (1979) did this with cobalt chloride, applying 50 to 100 μm/plant. Aminoethoxyvinylglycine (AVG) also induced male flowers on female plants, and when AVG was applied to male plants sprayed with ethephon, the feminization effect of ethephon was markedly curtailed (Mohan Ram and Sett, 1982a).

Silver nitrate ($AgNO_3$) is a potent inhibitor of ethylene. Sarath and Mohan Ram (1979) used 100 μm/plant, applied to the shoot tip of female plants in 10 μm drops for 10 days. All 10 plants produced male flowers. In comparison, only half the plants produced male flowers when treated with GA_3 at the same dosage. Furthermore, GA_3 caused marked shoot elongation, not seen with $AgNO_3$, and induced male flowers were smaller than those induced by $AgNO_3$.

Mohan Ram and Sett (1982b) compared $AgNO_3$ to silver thiosulfate (STS, $Ag(S_2O_3)_2$). In their hands STS was more effective than $AgNO_3$ at inducing male flowers in female plants; "this work is useful for producing seeds that give rise to only female plants." STS has become the industry standard for producing feminized seeds. Two comparative studies showed that GA_3 induced male flowers in only 2.2–2.5% of treated inflorescences, whereas a variety of STS products resulted in 21–54% male flower induction (Flajšman et al., 2021; Owen et al., 2023). The latter authors emphasized that STS is light-sensitive, and they recommended a shelf-stable commercial product.

The effects of elicitors upon cannabinoid content have yielded contradictory results. Spraying ABA on plants had different effects depending on the plant stage when it was applied. In flowering plants of Iranian provenance, THC content increased to 7% in ABA-treated plants, compared with 5% in controls (Mansouri et al., 2009a). In contrast, THC and CBD content decreased in vegetative-stage plants (Mansouri and Asrar, 2012). ABA-treated vegetative plants showed shifts in terpenoid content: more monoterpenoids and less sesquiterpenoids, compared with control plants (Mansouri and Asrar, 2012).

Spraying GA_3 on flowering plants decreased THC in a dose-related manner in bract tissue: controls, 3.4% THC; 50 μM GA_3, 3.0%; 100 μM GA_3, 2.0%. This dose response was not seen in lower leaves: controls, 2.8%; 50 μM GA_3, 1.25%; 100 μM GA_3, 2.0% (Mansouri et al., 2009b). In vegetative plants, GA_3 did the opposite: controls, 2.4% THC; 100 μM, 7.0% (Mansouri et al., 2011).

Spraying SA on plants caused an upregulation of the *THCAS* gene and an increase in THC content, but *CBDAS* expression decreased (Jalali et al., 2019). In another study, also using plants of Iranian provenance, SA treatment increased both THC and CBD content (Mirzamohammad et al., 2021). This result was replicated in Spain (Garrido et al., 2022). Spraying the JA analog methyl jasmonate (MeJA) on female flowers increased THC content in weeks 1–3 but not week 4 (Apicella et al., 2022). A similar study that measured both THCA and CBDA observed increases in both (Garrido et al., 2022).

MacWilliams et al. (2023) measured endogenous levels of SA, JA, and ASA in plants in response to 20 days of feeding by cannabis aphids (*Phorodon cannabis*). This caused a robust increase in levels of JA and ASA, and especially SA. Yet no significant changes were seen in cannabinoid content compared with controls.

Effectors are molecules secreted by pathogens. They can either trigger or compromise immune responses in plants. Flores-Sanchez et al. (2009) tested the reaction of *Cannabis* cell cultures to fungal effectors extracted from yeast (*Saccharomyces cerevisiae*), *Botrytis cinerea*, and *Pythium aphanidermatum*. These effectors modulated some *Cannabis* metabolites, but not cannabinoids.

Mansouri and Salari (2014) treated plants with mevinolin, an effector produced by several fungi. Mevinolin decreased THC content, yet increased CBD content. Mansouri and Rohani (2014) treated plants with 2-chloro-ethyl-trimethyl-ammonium chloride (Cytocel), a plant growth retardant that inhibits gibberellin biosynthesis. They found gender differences: females showed no reduction in growth, and flowers showed an increase in THC content at a dosage of 500 mg/L. Male plants treated with 500 mg/L were shorter than controls, with a decrease in THC content.

Ascorbic acid (AsA, vitamin C) upregulates induced systemic resistance (ISR) in plants. Soltan and Dadkhah (2022) sprayed AsA on female plants, which upregulated gene expression of *THCAS*, *CBDAS*, *PT*, and *OLS* compared with controls. THC and CBD content increased in just 3 days (greater at a dose of 0.5 mM than 0.3 mM or 1 mM).

Wood et al. (2022) searched for *Cannabis* genes involved with effector-triggered immunity to pathogens and pests. They identified *SRFR1*, whose homolog in *Arabidopsis thaliana* triggers resistance to *Pseudomonas syringae*, *Fusarium oxysporum*, *Botrytis cinerea*, and the budworm *Spodoptera exigua* (Son et al., 2021). Interestingly, Wood and colleagues determined that *SRFR1* had undergone positive selection in wild-type *Cannabis* plants but not in domesticated cultivars, "an adaptation to natural environments during feralization."

2.8 Putting it all together: crop yields over the years

We extracted some historical reports of crop yields from the literature, for comparative purposes. This summary is by no means comprehensive. We wanted to show how yields have changed over the years, in regards to fiber, seeds, and flowering tops.

2.8.1 Fiber yields

Measuring crop yield (harvest weight per unit of land) is relatively new, arising during the agricultural revolution of the past three centuries. Early-modern books on hemp did not provide crop yield statistics—not in France (e.g., Duhamel, 1747; Marcandier, 1758), Italy (Baruffaldi, 1741; Targioni Tozzetti, 1802), Germany (Baur, 1794; Vogelmann, 1840; Dosch, 1850), or Great Britain (Chomel and Bradley, 1725; Miller, 1768).

Suddenly two reports were published in 1788 (Table 2.11). William Hinton (1760–1805), an Oxford-trained doctor of divinity, grew Chinese hemp on a small plot (322 yd²), and extrapolated to per-acre yields. The other report, by an anonymous author in Suffolk, we attribute to Arthur Young (1741–1820), who edited the journal. Young was a gentleman farmer, traveler, and editor of texts on agriculture and economics (see Fig. 17.6). His estate, Bradfield Hall in Suffolk, was surrounded by hemp. Other anonymous articles mentioning hemp are now attributed to him (e.g., Young, 1768, 1772). Later he signed his name (e.g., Young 1797, 1799, Table 2.11).

Table 2.11. Fiber yields, early reports. Authors' original estimates in **bold** print, with our conversions added (1 lb/acre = 1.12 kg/ha)[a]

Source	lb/acre	kg/ha
Young (1788), Suffolk, England, average **36–38 stone/acre**, but up to **60 stone/acre**	504–532 up to 840	564–596 up to 941
Hinton (1788), Norfolk, England, Chinese hemp, **95.5 stone/acre**	1337	1497
Young (1797), Suffolk, England, **45 stone/acre**	630	706
Young (1799), Lincolnshire, England **35 stone/acre**	490	549
Traill (1828), Kumaon Himalaya, India, **4 *maunds* per *pucca bigha***	527	590
Clay (1830), Lexington, Kentucky	**600–1000**	672–1120
Rham (1842), Flanders, Netherlands	**350**	392
Rowlandson (1849), Lincolnshire, **60–70 stone/acre**; after a crop of beans, **90–100 stone/acre**	840–980 1260–1400	941–1098 1411–1568
Hamm (1854), F= Maine-et-Loire, France; B= giant hemp of Bologna; E= "extraordinary European yield"	F= **390** B= **600** E= **700–800**	F= 437 B= 672 E= 784–896
Gasparin (1848), France, Si= simultaneous harvest of males+females for fiber; Se= separate harvests of males for fiber and females for seed	Si= 696 Se= 615	Si= **780** Se= **689**
Bouché and Grothe (1884), Austro-Hungarian Empire: 39,184 ha yielded 20,500,000 kg	460	**515**
Bouché and Grothe (1884), Chinese hemp grown in Algeria	1339–1429	1500–1600
U.S. Census (1889), 25,054 acres nationwide yielded 11,511 tons	919	1029

[a]Notes: 1 *pucca bigha* = 0.253 ha, 1 *maund* = 37.32 kg, 1 *seer* = 1/40 *maund*

Dewey (1914) said fiber yields in the USA averaged 1000 lb/acre (1120 kg/ha). He listed average European yields per acre: France 662 lb, Italy 622 lb, Hungary 504 lb, Russia 358 lb. Lojacono (1910) said Piedmonte hemp yielded a total of 11,105 quinti (1 quintale = 100 kg) across 936.4 ha, or 1186 kg/ha. Cavazza (1930) said Piedmonte hemp yielded 7000 quinti across 900 ha, or 778 kg/ha. Bruna (1955a) gave a Piedmonte yield of 800 kg/ha.

Experimental studies reported high fiber yields. Haring (1922) conducted field trials of Dewey's new cultivars at the California Agricultural Experiment Station in Davis: 'Kymington' 1745 kg/ha, 'Chington' 1685 kg/ha, 'Ferramington' 1965 kg/ha, 'Tochimington' 2373 kg/ha. In an early study of monoecious hybrids, Sengbusch (1956b) reported an early-maturing line yielded 2300 kg/ha, a mid-maturing line yielded 2400 kg/ha, and a late-maturing line yielded 2200 kg/ha. His flagship dioecious cultivar, 'Fibrida', yielded 2500 kg/ha.

Modern reports offer other metrics: yields of freshly cut stalks, dried stalks, extracted bast fiber, or primary bast fiber. Dewey (1914) converted these metrics. Per acre, green freshly cut stalks yielded 15,000 lb, dried stalks yielded 10,000 lb, dried stalks after retting yielded 6000 lb, and clean fiber extracted from retted stalks yielded 1000 lb, which consisted of 75% primary fiber and 25% tow.

Cromack (1998) compared five cultivars under two different sowing densities. Stem dry matter was highest in dioecious 'Uniko B' (12,900 kg/ha) and 'Kompolti' (12,700 kg/ha), compared with monoecious 'Futura 77' (11,300 kg/ha), 'Fédora 19' (10,600 kg/ha), and 'Félina 34' (10,400 kg/ha). Struik *et al.* (2000) grew several cultivars under various conditions (nitrogen, plant density, harvest date) in three countries (UK, Holland, Italy). Stem dry matter was highest in 'Carmagnola' (16,300 kg/ha), 'Futura' (16,100 kg/ha), and 'Kompolti' (15,100 kg/ha).

2.8.2 Seed yields

Seed yields are often measured by volume—the bushel. Dewey (1914) said a bushel of hemp seeds weighed 44 lb (19.96 kg), and he reported an average yield in the USA of 16–18 bushels/acre (788–887 kg/ha). In Lincolnshire, England, Young (1799) reported a seed yield of 16 strikes/acre. The strike is an old unit of volume with various meanings, often equivalent to 1.5 bushels, which converts Young's yield to 1183 kg/ha.

Also in Lincolnshire, Rowlandson (1849) reported yields of 20–22 bushels/acre (986–1084 kg/ha), and following a crop of nitrogen-fixing beans, 28–30 bushels (1380–1478 kg/ha). In the Flanders region, Rham (1842) reported yields of 30–35 bushels/acre (1478–1725 kg/ha). Allegret (2013) said French hemp back in 1818 yielded 6 hectoliters per ha, about 340 kg/ha. Also in France, Gasparin (1848) reported 293 kg/ha.

Traill (1828) reported seed yields in Himalayan hemp, from plants that produce small, semi-domesticated seeds: 48 *seers/pucca bigha*, about 177.0 kg/ha. Southern Chinese landraces produce large seeds; in England, Hinton (1788) grew Chinese hemp on a small plot (322 yd^2), and extrapolated a per-acre yield of 11 bushels, two pecks, and half a pint, or about 586 kg/ha. Mangin (1859) cultivated *chanvre géant de la Chine* in Algeria, which yielded 1470 kg/ha. Chinese agronomists in Yúnnán bred 'Yún Má No.1' from local landraces, which yielded 1500 kg/ha^2 (Guo *et al.*, 2011b).

In Italy's Piedmont region, Lojacono (1910) reported a harvest of 1305 quinti seed. If that harvest came from the same 936.4 ha that he reported for fiber, it's a yield of only 139.4 kg/ha. Cavazza (1930) gave a Piedmonte yield of 1500 quinti on 900 ha, or 166.7 kg/ha. Bruna (1955a) gave a Piedmonte yield of 800 kg/ha seed, equal to its fiber yield.

Bòcsa (1999) said that modern monoecious cultivars potentially yield 1200–1500 kg seed/ha. However, in agronomic trials, monoecious 'Fédora', 'Fédrina', 'Félina', 'Futura', and

'Fasamo' yielded 450–675 kg/ha (Deleuran and Flengmark, 2004). Callaway (2004) claimed that 'FINOLA' yielded nearly 1700 kg/ha. However, a 3-year trial in Canada reported an average 'FINOLA' yield of only 619 kg/ha (Vera *et al.*, 2004). Vogl *et al.* (2004) compared yields in 20 cultivars (one dioecious, the rest monoecious) in small plots (24 m²) cultivated under organic conditions. Yields ranged from a low of 53.8 g/m² (= 538 kg/ha) for dioecious 'Kompolti' to a high of 143.0 g/m² (= 1430 kg/ha) for monoecious 'Fedora 19'.

2.8.3 Flower yields

Flower yields are measured per plant or per growing area. Various types of flower products have been measured, and comparing them is a little apples-to-oranges: Should the weight of seedless *gañjā* be equated to that of seeded marijuana? Other flower products include "manicured material" and "botanical raw material," with variable amounts of leaf material.

Yield per plant

By this metric, outdoor plants yield more than indoor plants. Outdoor plants have more room to spread out, and often have a longer growing season than indoor plants. In British India's famed *Gañjā Mahal*, Prain (1893) reported yields of 2.5 ounces (71 g) per plant of dried, manicured *gañjā*. In comparison, DEA (1992) claimed that an outdoor plant yielded one pound (454 g) of dry-weight marijuana. The DEA's high estimate enabled them to inflate statistics in their eradication reports. Sentencing guidelines in USA court cases were more extravagant—they estimated a yield of 2.2 lb (1000 g) per plant (US Congress, 1996). In the same report, DEA-funded researchers at the University of Mississippi reported yields of only 295–412 g per plant (US Congress, 1996).

Indoor yields according to Toonen *et al.* (2006) average 33.7 g of manicured bud per plant. They surveyed a variety of conditions in 77 growrooms, most of which used Dutch "sea of green" methods. On average, the growrooms employed high plant density (15 plants/m²) and a short growth cycle (plants matured in 3 months), so the average plant height was only 2 ft (0.61 m).

Vanhove *et al.* (2011) compared indoor yields at two plant densities. At a density of 16 plants/m², yields averaged 17.6 g/plant. At a higher density of 20 plants/m², yields dropped to 14.5 g/plant. Averages were calculated from four different cultivars; the winner was "Big Bud," yielding 28.0 g/plant when planted at 16 plants/m². Vanhove *et al.* (2011) subsequently decreased the density to 12 plants/m², and yields increased to 52.0 g/plant. The winner was "Silver Haze #9," with an average of 62.0 g/plant.

Yield per unit area

By this metric, indoor plants yield more than outdoor plants. Yields for outdoor crops appear in Table 2.12. They vary a great deal. *Gañjā* yields in British India were measured in non–standard units, such as the *bigha*, which equaled 0.33 acre in Bengal, 0.25 acre in Punjab, 0.20 acre in Himachal Pradesh. Some of these British estimates for outdoor *gañjā* surpassed indoor yields by some modern growers.

The estimate by Peacock (1871) in Table 2.12 was ballpark. He presented data from Rājshāhi in Bengal, soon to be known as *Gañjā Mahal*. He averaged the annual quantity of *gañjā* produced from 1854 to 1870, divided by the area under cultivation. The most trustworthy estimate in British India was by Kerr (1877), who made a detailed study of *gañjā* in *Gañjā Mahal*.

Stockberger (1915), a USDA crop specialist, cultivated *Cannabis indica* near Timmonsville, South Carolina. He grew well-fertilized plants, removed all males, and waited until females grew "heavy and sticky," yet his yield was relatively low. The estimate of *kif* yields by Mikuriya (1967) must have included stems and seeds.

Indoor plants are fed more light, more nutrients, and perhaps supplemental CO_2. UN-ODC (2008) estimated that indoor hydroponic crops yielded up to 15 times more than outdoor crops per unit area. But they vastly overestimate indoor yields (Table 2.13). Perhaps their estimate was inflated by the concept that two to four crops can be grown per year indoors, compared with one crop per year outdoors. This may also explain the inflated estimate by Backer *et al.* (2019); one of their facility sites, 0.56 ha, reported yielding 20,000 kg, or 3588 g/m², which is impossible, unless they summed multiple harvests per year.

2.8.4 *Hashīsh* yields

Hashīsh is a mechanical (not chemical) extract of resin from flowering tops. Resin is encapsulated within gland heads atop capitate stalked glandular trichomes. **Sieved** *hashīsh* consists of gland heads mostly intact, whereas the gland heads are ruptured in **hand-rubbed** *charas*. We would expect lower yields in hand-rubbed *charas* (see section 18.5). Comparing yield data is bedeviled by the issue of *hashīsh* quality (Clarke, pers. comm., 2023), quality and quantity being inversely related (see below).

Traill's estimate in Table 2.14 was probably accurate—parallel to his accurate estimates of fiber and seed yields. Huddleston reported yield per *beesee* (*bísí*), a nebulous unit of land area. A *bísí* equaled an area sown with 20 *nális* of seed, or an area plowed by two yoke of bullocks in one day (Traill, 1828). The best estimate of a *bísí* is about 4800 square yards, or 0.99 acres. Clarke's estimate—the highest for hand-rubbed *charas*—was obtained from modern cultivated plants. Clarke (1998) reported that skilled Nepalis could rub 5–10 g per day of high-quality *charas*, or 50 g per day of commercial-grade *charas*. Turning to sieved *hashīsh*, Vavilov's estimate is way too high. His data came through interpreters.

Gland heads atop capitate stalked glandular trichomes average 70–100 μm in diameter in a dried state. To selectively sieve gland heads, plant material should pass through a screen with a pore size of 135 μm, and be retained on a screen with a pore size of 50 μm (Clarke, 1998). Sieving through larger pore sizes results in a mixture of gland heads, leaf fragments, and other detritus.

Hashīsh "quality" is the percentage of *hashīsh* composed purely of gland heads, and this affects yields. Clarke (1998) used two metrics of quality. **Extraction percentage** is the weight ratio of extracted gland heads to original plant material. For example, the extraction percentage of sieved Moroccan *hashīsh* is around 1%. The second metric, **extraction ratio**, is the inverse: the weight of plant material to the weight of resin powder. An extraction ratio of 100 equals an extraction percentage of 1%.

Table 2.12. Outdoor crop yield estimates. Authors' original estimates in **bold** print, with our conversions added (1 lb/acre = 1.12 kg/ha)[a]

Source	lb/acre	kg/ha
Peacock (1871), *gañjā* in Bengal, **8.0 *maunds* per *bigha***	1997	2237
Anonymous (1874), *gañjā* in Bengal, **5.5–9.5 *maunds* per *bigha***	1369–2365	1533–2649
Kerr (1877), in Bengal, round ga*ñjā* **3.8 *maunds* per *bigha*** (RG), flat g*añjā*	RG: 946	RG: 1060
5.4 *maunds* per *bigha* (FG)	FG: 1344	FG: 1505
Indian Hemp Drugs Commission (1894) **12–16 maunds** *gañjā* per acre, India	988–1317	1107–1475
Mair (1898), round *gañjā* and flat *gañjā* in Bengal (BG), flat *gañjā* in Bombay (BB).	BG: **1333**	BG: 1493
	BB: **400**	BB: 448
Stockberger (1915), dried tops, South Carolina	**400–500**	448–560
Cherian (1932), pressed flowers, India	**246–820**	276–902
Small *et al.* (1975), manicured marijuana, Canada	**1800**	2160
Potter (2009), "botanical raw material" (BRM) floral and foliar mixture, **500 g/m²**	4464	5000
Mikuriya (1967) **2 kg/m²** of marketable *kif*, Morocco	17,857	20,000
UN-ODC (2005b), *brut* (raw) *kif* in Morocco, dryland (D), irrigated (I)	D= 669.6	D= **750**
	I= 1133.9	I= **1270**
UN-ODC (2008), worldwide average	688	**770**

[a]Notes: 1 *pucca bigha* = 0.253 ha, 1 *maund* = 37.32 kg, 1 *seer* = 1/40 *maund;* conversion factors: 1 *maund/bigha* = 249.7 lbs/acre; 1 lb/acre = 1.12 kg/ha

Table 2.13. Indoor crop yield estimates. Authors' original estimates in **bold** print, with our conversions added (1 lb/acre = 1.12 kg/ha)

Source	lb/acre	kg/ha
Frank and Rosenthal (1978), "smoking material," **one ounce per ft²**	2723	3049
Alexander (1989), **1.5 ounce per ft²**	4574	4084
D. Watson (pers. comm. 1998), flowering tops in a greenhouse with supplemental light, **250 g per m²**	2232	2500
Toonen *et al.* (2006), female flowering bud, **505 g per m²**	4509	5050
Knight *et al.* (2010), female flowering bud, **275 g per m²**	2455	2750
Potter and Duncombe (2012), floral material, **544 g per m²**	4857	5440
Vanhove *et al.* (2012), female flowers, **627 g per m²**	5598	6270
UN-ODC (2008), hydroponic gardens	12,500– 26,786	**15,000– 30,000**
Backer *et al.* (2019) mean of 61 Canadian production sites, calculated from reported yield of dried female flowers divided by facility size, **1219.7 g per m²**	10,882	**12,197**

Table 2.14. Yield estimates of hand-rubbed *charas* and sieved *hashīsh*, with our conversions added (1 lb/acre = 1.12 kg/ha)[a]

Source	lb/acre	kg/ha
Traill (1828), rubbed *charas* from cultivated or feral Himalayan hemp, Kumaon District, **4 seers per *pucca bigha***	13.17	14.75
Huddleston (1841), rubbed *charas* from cultivated or feral Himalayan hemp, Garhwal District, **3 seers per *beesee* (bísí)**	6.22	6.97
Clarke (2007) rubbed *charas* from cultivated plants in Nepal, **up to 200 g/100m²**	17.86	20.0
Vavilov and Bukinich (1929), sieved *hashīsh* in Afghanistan, **30 kg per 1/16 ha**	429	480
UN-ODC (2005b) sieved *hashīsh* in Morocco, dryland (D), irrigated (I)	D= 18.88	D= **21.15**
	I= 31.97	I= **35.81**
UN-ODC (2010b), sieved *hashīsh* in Afghanistan	129.5	**145**
UN-ODC (2010b), sieved *hashīsh* in Morocco	35.71	**40**
Chouvy and Macfarlane (2018), sieved *hashīsh* in Morocco	62.5	**70**

[a]Notes: 1 pucca bigha = 0.253 ha, 1 maund = 37.32 kg, 1 seer = 1/40 maund; conversion factors: 1 maund/bigha = 249.7 lb/acre; 1 lb/acre = 1.12 kg/ha

The lower the *hashīsh* quality, the higher is the *hashīsh* yield. Clarke (1998) reported an extraction percentage of 1% in Moroccan *hashīsh*, whereas UN-ODC (2005b) reported a Moroccan extraction percentage of 2.82%. However, their estimate was ballpark: they divided a *hashīsh* yield of 21.15 kg (dryland, Table 2.14) by a *kif* yield of 750 kg (dryland, Table 2.12) × 100.

2.8.5 THC yield

THC yield can be reported in terms of weight (in g or kg) per grow area (in m² or ha). The more traditional metric is THC potency (percentage of Δ^9-THC in dried female flowering tops). Drug warriors claim that THC potency has risen dangerously—causing an escalation in abuse, dependence, and schizophrenia. Alarm over increased potency is not new, but has been voiced repeatedly (e.g., Walton, 1938; Maugh, 1975). Nearly 150 years ago, Fronmüller (1882) stated that *Cannabis indica* potency "has increased significantly over the years," a fourfold increase in the previous 30 years. Fronmüller believed that "a more careful selection of the plants, a more careful preparation, for example production in India itself may be the cause."

Addiction specialists proclaimed, "Marijuana has increased 1,400 percent in potency since 1970," which Mikuriya and Aldrich (1988) refuted as "patently absurd." The Director of the US Office of National Drug Control Policy said that "superweed" was 10 or 20 times stronger than a generation ago (Walters, 2002). In fact, "superweed" was available 50 years ago. In the UK, "Thai sticks" averaged 17% THC in 1975 (Baker *et al.*, 1980b). Bazzaz *et al.* (1975) reported THC levels of 13.2% and 12.9% in plants from Jamaica and Nepal. In San Francisco, Perry (1977) analyzed "Thai sticks" with 9–10% THC, and sinsemilla with 14% THC.

Sober analytical chemists in the USA reported a mean fivefold increase in THC between 1979 and 2008, from 1.58% to 8.8% (ElSohly *et al.*, 2000; Mehmedic *et al.*, 2010). In the UK, Potter *et al.* (2008) also estimated a fivefold increase in Δ^9-THC in recent decades. This might be compared to the fivefold difference in caffeine between green tea and filtered coffee—not a dangerous thing. Between 2008 and 2017, mean THC content in seedless flowering tops nearly doubled, from 8.8% to 17.1% (Chandra *et al.*, 2019).

De Meijer (2004) may have been first to report THC yields from an agronomic perspective: kg/ha. He predicted that a *Cannabis* crop producing 500 g of flowering tops/m² with 20% THC would yield nearly 100 g of THC/m², or equivalent to 1 kg/ha, "impressive for a secondary metabolite." Actual yields were lower than that, up to 78 g THC/m² (see Table 2.1). This yield pales in comparison with that of nicotine. A crop of flue-cured *Nicotiana tabacum* produces 2500 kg of leaves/ha with a nicotine content of 4%, or a nicotine yield of 100 kg/ha (Mitchell, 1999).

Most jurisdictions require labeling of THC and CBD potency for consumers. However, there is controversy as to whether all reported results are legitimate. To investigate this concern, Jikomes and Zoorob (2018) obtained data from the Washington State Liquor and Cannabis Board, for all *Cannabis* tested for THC and CBD content between 2014 and 2017. They found interlab differences in cannabinoid content, even after controlling for plausible confounds. One laboratory in particular reported consistently higher values than the other five labs, across all products tested. This was likely due to the economic benefits of "cannabinoid inflation"—labs that consistently reported higher THC content attracted producers who require testing. This illegitimate activity was also seen among labs in Michigan, where blinded "round-robin" audits of identical products revealed consistent discrepancies (J. Crookston, pers. comm., 2022).

2.8.6 Essential oil yield

Essential oil (EO) is an aromatic liquid extracted from a plant by steam distillation, hydrodistillation, solvent extraction, or vaporization. The EO of *Cannabis* consists primarily of a mixture of **terpenoids**, which are volatile, readily outgas, and release the characteristic odor of *Cannabis*.

Ross and ElSohly (1996) demonstrated the ephemeral nature of EO yields. They steam-distilled a "high potency hybrid" cultivated indoors under lights, and quantified EO content as a percentage of "marijuana bud" weight. Freshly collected flowering tops yielded 0.29% v/w essential oil. Buds air-dried at room temperature for a week yielded 0.20%, a loss of 31%. Month-old buds air-dried and stored in a paper bag yielded 0.16%, a loss of 45%. After 3 months in a paper bag, buds yielded 0.13%, a loss of 55%.

Comparisons between EO studies can be apples-to-oranges, due to different extraction methods employed by different studies. But it appears that fiber-type and drug-type plants yield similar EO concentrations, whether dioecious or monoecious. Fiber-type cultivars were hydrodistilled by Bertoli *et al.* (2010), who quantified EO content as a percentage of fresh inflorescences. They reported a mean EO content of 0.20% in dioecious plants (e.g., 'Carmagnola' and 'C.S.') and a mean of 0.13% in monoecious 'Felina 34' and 'Codimono'. Nissen *et al.* (2010) steam-distilled inflorescences, and reported the opposite: more EO in monoecious 'Futura' (0.31%) than dioecious 'Carmagnola' and 'Fibranova' (0.23–0.26%).

(Prepared by J. McPartland; revised by E. Russo)

References

Aizpurua-Olaizolo O, Sodaner U, Öztürk E, *et al.* (5 other authors). 2016. Evolution of the cannabinoid and terpene content during the growth of *Cannabis sativa* plants from different chemotypes. *Journal of Natural Products* 79: 324–331.

Alexander T. 1989. A report on five strains from SSSC. *Sinsemilla Tips* 8(4): 21–23.

Allegret S. 2013. "The history of hemp," pp. 4–26 in Bouloc P, Allegret S, Arnaud L, *eds. Hemp: industrial production and uses.* CABI International, Wallingford, UK.

Amaducci S, Colauzzi M, Bellocchi G, Venturi G. 2008a. Modelling post-emergent hemp phenology (*Cannabis sativa* L.): theory and evaluation. *European Journal of Agronomy* 28: 90–102.

Anderson SL, Pearson B, Kjelgren R, Brym Z. 2021. Response of essential oil hemp (*Cannabis sativa* L.) growth, biomass, and cannabinoid profiles to varying fertigation rates. *PLoS ONE* 16: e0252985.

Angelini LG, Tavarini S, Cestone B, Beni C. 2014. Variation in mineral composition in three different plant organs of five fibre hemp (*Cannabis sativa* L.) cultivars. *Agrochimica* 58(1): 1–18.

Anonymous. 1874. Culture of gunjah in Bengal. *Journal of Applied Science* 5 (Feb 1): 19.

Apicella PV, Sands LB, Ma Y, Berkowitz GA. 2022. Delineating genetic regulation of cannabinoid biosynthesis during female flower development in *Cannabis sativa*. *Plant Direct* 6: e142.

Arnoux M, Mathieu G. 1966. Sur quelques problèmes posés par la sélection et l'amélioration de la monoecie chez le chanvre (*Cannabis saliva* L.). *Fibra* 11(2): 51–63.

Atal CK. 1959. Sex reversal in hemp by application of gibberellin. *Current Science* 28: 408–409.

Atal CK, Sethi JK. 1961. Increased fibre content in hemp (*Cannabis sativa*) and sunn (*Crotalaria juncea*) by application of gibberellins. *Current Science* 30: 177–179.

Aubin MP, Seguin P, Vanasse A, Tremplay GF, Mustafa AF, Charron MB. 2015. Industrial hemp response to nitrogen, phosphorus, and potassium fertilization. *Crop, Forage & Turfgrass Management* 1: 0159.

Babaei M, Ajdanian L. 2020. Screening of different Iranian ecotypes of *Cannabis* under water deficit stress. *Scientia Horticulturae* 260: e108904.

Backer R, Schwinghamer T, Rosenbaum P, *et al.* (8 other authors). 2019. Closing the yield gap for *Cannabis*: a meta-analysis of factors determining *Cannabis* yield. *Frontiers in Plant Science* 10: 495.

Bahador M, Tadayon MR, Rafie-Alhoseini M, Salehi MH. 2017. Changes of canopy temperature and some physiological traits of hemp (*Cannabis sativa*) under deficit water stress and zeolite rates. *Environmental Stresses in Crop Sciences* 10: 269–279.

Baker PB, Bagon KR, Gough TA. 1980b. Variation in the THC content in illicitly imported cannabis products. *Bulletin on Narcotics* 32(4): 47–84.

Balduzzi A, Gigliano GS. 1985. Influenza dell'intensità luminosa sull'accumulo di cannabinoli in *Cannabis sativa* L. *Atti dell' Istituto Botanico e del Laboratorio Crittogamico Dell-Università di Pavia, Series 7*, 4: 89–92.

Barbieri R. 1943. Aspetti e soluzione del problema del semi di canapa in Campania. *Annali della Facoltà di agraria di Portici della R. Università di Napoli* 15: 85–127.

Barbieri R. 1952. La "prefioritura" della canapa in Campania nell'annata 1952. *Agricoltura Napoletana* 19(7): 5–16 and 19(9): 5–29.

Baruffaldi G. 1741. *Il canapajo di Girolamo Baruffaldi libri VIII. Con le annotazioni.* Lelio dalla Volpe, Bologna.

Baur G. 1794. *Stallfütterung, Klee-, Hanf-, Flachs- und Grundbirnbau : sammt verschiedenen nützlichen Landwirthschafts-Gegenständen.* Rieger, Augsburg.

Bazzaz FA, Dusek D, Seigler DS, Haney AW. 1975. Photosynthesis and cannabinoid content of temperate and tropical populations of *Cannabis sativa*. *Biochemical Systematics and Ecology* 3: 15–18.

Beckh H, *ed.* 1895. *Geoponica sive Cassiani Bassi scholastici de re rustica ecologae.* Teubner, Lipsiae.

Bencze S, Vida G, Balla K, Varga-László E, Veisz O. 2013. Response of wheat fungal diseases to elevated atmospheric CO_2 level. *Cereal Research Communications* 41: 409–419.

Berger J. 1969. *The world's major fibre crops: their cultivation and manuring.* Centre d'Etude de l'Azote, Zurich.

Berstein N, Gorelick J, Zerahia R, Koch S. 2019a. Impact of N, P, K, and humic acid supplementation on the chemical profile of medical cannabis (*Cannabis sativa* L). *Frontiers in Plant Science* 10: 736.

Berstein N, Gorelick J, Koch S. 2019b. Interplay between chemistry and morphology in medical cannabis (*Cannabis sativa* L.) *Industrial Crops & Products* 129: 185–194.

Bertoli A, Tozzi S, Pistelli L, Angelini LG. 2010. Fibre hemp inflorescences: from crop-residues to essential oil production. *Industrial Crops and Products* 32: 329–337.

Berzak IA. 1975. Effect of nitrogen fertilizers on yield of hemp on leached chernozem soils in the central cis-Causcasus region. *Khimiya v Sel'skom Khozyaistve* 13: 30–31.

Bevan L, Jones M, Zheng YB. 2021. Optimisation of nitrogen, phosphorus, and potassium for soilless production of *Cannabis sativa* in the flowering stage using response surface analysis. *Frontiers in Plant Science* 12: e764103.

Billings WD. 1987. Constraints to plant growth, reproduction, and establishment in Arctic environments. Alpine Research 19: 357–365.

Bòcsa I. 1999. "Genetic improvement: conventional approaches," pp. 153–184 in: Ranalli P, *ed., Advances in Hemp Research.* Haworth Press, Binghamton, New York.

Bòcsa I, Karus M. 1997. *Der Hanfanbau: Botanik, Sorten, Anbau und Ernte.* C.F. Müller, Heildelberg. English translation: *The cultivation of hemp: botany, varieties, cultivation, and harvesting.* Hemptech, Sebastopol, California.

Bòcsa I, Máthé P, Hangyel L. 1997. Effect of nitrogen on tetrahydrocannabinol (THC) content in hemp (Cannabis sativa L.) leaves at different positions. *Journal of the International Hemp Association* 4: 80–81.

Borthwick HA, Scully NJ. 1954. Photoperiodic responses of hemp. *Botanical Gazette* 116: 14–29.

Bouché CB, Grothe H. 1884. *Ramie, Rheea, Chinagras und Nesselfaser.* J. Springer, Berlin.

Bouquet J., 1950. Cannabis. *Bulletin on Narcotics* 2: 14–30.

Boussingault JB. 1841. De la discussion de la valeur relative des assolements, par les résultats de l'analyse élémentaire. *Annales de Chimie et de Physique, Troisième Série* 1: 208–246.

Boussingault JB. 1851. *Economie rurale considérée dans ses rapports avec la chimie, la physique et la meteorology, deuxième edition, Tome Premier.* Béchet Jeune, Paris.

Boussingault JB. 1857. Alimentation des plantes. Influence du phosphate de chaux du engrais sur la production vegetale. *Journal d'Agriculture Pratique, quatrième série,* 8: 441–449.

Boussingault JB, Payen A. 1842. Mémoire sur les engrais. *Annales de Chimie et de Physique, Troisième Série* 6: 449–464.

Bredemann G. 1952. Weitere Beobachtungen bei Züchtung des Hanfes auf Fasergehalt. *Der Züchter* 22: 257–269.

Brenner D. 2019. A high alpine weed grower's unusual harvest. *High Country News.* Available at: https://www.hcn.org/articles/agriculture-blind-pot-farmer-cultivates-marijuana-at-high-altitude-with-zero-carbon-footprint

Bruna T. 1955a. "Canapa Piemontese e canapa Carmagnola," pp. 286–369 in Mattioli E, *ed. Aspetti e problemi della canapicoltura Italiana*. Ramo Editoriale Delgi Agricoltori, Rome.

Callaway JC. 2004a. Hemp seed production in Finland. *Journal of Industrial Hemp* 9(1): 97–103.

Callaway JC. 2004b. Hempseed as a nutritional resource: an overview. *Euphytica* 140: 65–72.

Campbell BJ, Berrada AF, Hudalla C, Amaducci S, McKay JK. 2019. Genotype x environment interactions of industrial hemp cultivars highlight diverse responses to environmental factors. *Agrosystems, Geosciences & Environment* 2: 180057.

Caplan D, Dixon M, Zheng YB. 2017a. Optimal rate of organic fertilizer during the vegetation-stage for *Cannabis* grown in two coir-based substrates. *HortScience* 52: 1307–1312.

Caplan D, Dixon M, Zheng YB. 2017b. Optimal rate of organic fertilizer during the flowering stage for *Cannabis* grown in two coir-based substrates. *HortScience* 52: 1796–1803.

Caplan D, Dixon M, Zheng YB. 2019. Increasing inflorescence dry weight and cannabinoid content in medical cannabis using controlled drought stress. *HortScience* 54: 964–949.

Cavazza L. 1930. *La canapa gigante di Carmagnola*. Tipografia Sacerdote, Alba.

Cervantes J. 2015. *The Cannabis encyclopedia*. Van Patten Publishing, Vancouver, Washington.

Chadaeva VA. 2020. "Чужеродные виды растений национального парка «Приэльбрусье» (Кабардино-Балкарская Республика, Центральный Кавказ)," pp. 304–309 in Ishmuratova MM, *ed. Всероссийской научно-практической конференции «Актуальные вопросы охраны биоразнообразия на заповедных территориях»*, Bashkir State University, Ufa, Russia.

Chailakhyan MK, Khryanin VN. 1978a. The influence of growth regulators absorbed by the root on sex expression in hemp plants. *Planta* 138: 181–184.

Chandra S, Lata H, Khan IA, ElSohly MA. 2008. Photosynthetic response of *Cannabis sativa* L. to variations in photosynthetic photon flux densities, temperature and CO_2 conditions. *Physiology and Molecular Biology of Plants* 14: 299–306.

Chandra S, Lata H, Khan IA, ElSohly MA. 2011. Temperature response of photosynthesis in different drug and fiber varieties of *Cannabis sativa* L. *Physiology and Molecular Biology of Plants* 17(3): 297-303.

Chandra S, Lata H, Mehmedic Z, Khan IA, ElSohly MA. 2015. Light dependence of photosynthesis and water vapor exchange characteristics in different high Δ^9-THC yielding varieties of *Cannabis sativa* L. *Journal of AppliedResearch on Medicinal and Aromatic Plants* 2: 39–47.

Chandra S, Lata H, Khan IA, ElSohly MA. 2017. "*Cannabis sativa* L.: botany and horticulture," pp. 79–100 in Chandra *et al.*, eds. *Cannabis sativa L.—botany and biotechnology*. Springer International, Cham, Switzerland.

Chandra S, Radwan MM, Majumdar CG, *et al.* (3 other authors). 2019. New trends in cannabis potency in USA and Europe during the last decade (2008–2017). *European Archives of Psychiatry and Clinical Neuroscience* 269: 5–15.

Chen X, Guo HY, Zhang QL, *et al.* (10 other authors). 2022. Whole-genome resequencing of wild and cultivated cannabis reveals the genetic structure and adaptive selection of important traits. *BMC Plant Biology* 22: e371.

Cherian MC. 1932. Pests of ganja. *Madras Agricultural Journal* 20: 259–265.

Cherrett N, Barrett J, Clemett A, Chadwick M, Chadwick MJ. 2005. *Ecological footprint and water analysis of cotton, hemp, and polyester*. Stockholm Environmental Institute, Stockholm.

Chomel N (Bradley R, *trans, ed*). 1725. *Dictionaire oeconomique; or, the family dictionary*. D. Midwinter, London.

Chouvy PA, Macfarlane J. 2018. Agricultural innovations in Morocco's cannabis industry. *International Journal of Drug Policy* 58: 85–91.

Christison A. 1850. On *Cannabis indica*, Indian hemp. *Transactions and Proceedings of the Botanical Society of Edinburgh* 4: 59–69.

Clarke RC. 1998. *Hashish!* Red Eye Press, Los Angeles, California.

Clarke RC. 2007. Traditional *Cannabis* cultivation in Darchula District, Nepal—seed, resin and textiles. *Journal of Industrial Hemp* 12(2): 19–42.

Clarke RC, Merlin MD. 2016. *Cannabis* domestication, breeding history, present-day genetic diversity, and future prospects. *Critical Reviews in Plant Sciences* 35: 293–327.

Clay H [as H.C.]. 1830. "Hemp," pp. 226–236 in *Western Agriculturist, and Practical Farmer's Guide*. Robinson and Fairbank, Cincinnati.

Cockson P, Landis H, Smith T, Hicks K, Whipker BE. 2019. Characterization of nutrient disorders of *Cannabis sativa*. *Applied Sciences* 9: 4432.

Cockson P, Schroeder-Moreno M, Veazie P, *et al.* (5 other authors). 2020. Impact of phosphorus on *Cannabis sativa* reproduction, cannabinoids, and terpenes. *Applied Sciences* 10: e7875.

Coffman CB, Genter WA. 1975. Cannabinoid profile and elemental uptake of *Cannabis sativa* L. as influenced by soil characteristics. *Agronomy Journal* 67: 491–497.

Coffman CB, Genter WA. 1977. Responses of greenhouse-grown *Cannabis sativa* L. to nitrogen, phosphorus, and potassium. *Agronomy Journal* 69: 832–836.

Colorado DOA (Department of Agriculture). 2023. *Pesticide use in Cannabis production information*. Available at: https://ag.colorado.gov/plants/pesticides/pesticide-use-in-cannabis-production-information.

Columella LIM (Ash HB, *trans.*). 1941. *On agriculture: with a recension of the text and an English translation, Vol* 1. Harvard University Press, Cambridge, Massachusetts.

Cosentino SL, Testa G, Scordia D, Copani V. 2012. Sowing time and prediction of flowering of different hemp (*Cannabis sativa* L.) genotypes in southern Europe. *Industrial Crops and Products* 37: 20–33.

Cosentino SL, Riggi E, Testa G, Scordia D, Copani V. 2013. Evaluation of European developed fibre hemp genotypes (*Cannabis sativa* L.) in semi-arid Mediterranean environment. *Industrial Crops and Products* 50: 312–324.

Cromack HTH. 1998. The effect of cultivar and seed density on the production and fibre content of *Cannabis sativa* in southern England. *Industrial Crops and Products* 7: 205–210.

Csókás G. 1914. A kender tápanyagfelvétele és hatása a rost mennyiségére és minőségére. *Kisérletügyi Közlemények* 17: 64–120.

Danziger N, Bernstein N. 2021. Shape matters: plant architecture affects chemical uniformity in large-size medical *Cannabis* plants. *Plants* 10: e1834.

Darwin CR (assisted by Darwin F). 1880. *The power of movement in plants*. John Murray, London.

De Backer B, Maebe K, Werstraete AG, Charlier C. 2012. Evolution of the content of THC and other major cannabinoids in drug-type *Cannabis* cuttings and seedling during growth of plants. *Journal of Forensic Sciences* 57: 918–922.

de Meijer EPM. 2004. "The breeding of *Cannabis* cultivars for pharmaceutical end uses," pp. 55-69 in Guy G, Robson R, Strong K, Whittle B, eds. *The Medicinal Use of Cannabis*. Royal Society of Pharmacists, London.

de Meijer EPM, Keizer LCP. 1996. Patterns of diversity in *Cannabis*. *Genetic Resources and Crop Evolution* 43: 41–52.

de Serres O. 1600. *Théâtre d'agriculture et mesnage des champs*. Jamet-Métayer, Paris.

DEA. 1992. *Cannabis yields*. US Department of Justice, Drug Enforcement Administration. Washington, DC.

Deleuran L, Flengmark PK. 2004. Yield potential of hemp (*Cannabis sativa* L.) cultivars in Denmark. *Journal of Industrial Hemp* 10: 19–31.

Deng G, Du GH, Yang Y, Bao YN, Liu FH. 2019. Planting density and fertilization evidently influence the fiber yield of hemp (*Cannabis sativa* L.). *Agronomy* 9: 368.

Dewey LH. 1914. "Hemp," pp. 283–347 in: *USDA Yearbook 1913*. United States Department of Agriculture, Washington, DC.

Djamali M, Brewer S, Breckle SW, Jackson ST. 2012. Climatic determinism in phytogeographic regionalization—a test from the Irano-Turanian region SW and Central Asia. *Flora* 207: 237–249.

Dong RL, Zhao GY, Chai LL, Beauchemin KA. 2014. Prediction of urinary and fecal nitrogen excretion by beef cattle. *Journal of Animal Science* 92: 4669–4681.

Dosch J. 1850. *Deutschlands flachs und hanf-bau*. A. Emmerling, Freiburg.

Drake B. 1970. *The cultivator's handbook of marijuana, Revised ed.* Agrarian Reform Co., Eugene, Oregon.

Duhamel du Monceau HL. 1747. *Traité de la fabrique des manoeuvres pour les vaisseaux, ou, l'art de la corderie perfectionné*. L'Imprimerie Royale, Paris.

Duke JA. 1982. "Ecosystematic data on medicinal plants," pp. 13–23 *in* Atal CK, Kapur BM, eds. *Utilization of medicinal plants*. United Printing Press, New Delhi.

Eaves J, Eaves S, Morphy C, Murray C. 2020. The relationship between light intensity, cannabis yields, and profitability. *Agronomy Journal* 112: 1466–1470.

ElSohly MA, Ross SA, Mehmedic Z, *et al.* (3 other authors). 2000. Potency trends of delta-9-THC and other cannabinoids in confiscated marijuana from 1980-1997. *Journal of Forensic Science* 45: 24–30.

Emerson R. 1958. The quantum yield of photosynthesis. *Annual Review of Plant Physiology* 9: 1–24.

EPA (Environmental Protection Agency). 2023. *Pesticide products registered for use on hemp*. Available at: https://www.epa.gov/pesticide-registration/pesticide-products-registered-use-hemp.

Fairbairn JW, Liebmann JA. 1974. The cannabinoid content of *Cannabis sativa* L grown in England. *Journal of Pharmacy and Pharmacology* 26: 413–419.

Faujas de Saint Fond B. 1799. *Travels in England, Scotland, and the Hebrides, Vol.* 1. James Ridgway, London.

Faux AM, Dray X, Lambert R, *et al.* (3 other authors). 2013. The relationship of stem and seed yields to flowering phenology and sex expression in monoecious hemp (*Cannabis sativa* L.). *European Journal of Agronomy* 47: 11–22.

Finlay MR. 1992. *Science, practice and politics: German agricultural experiment stations in the nineteenth century*. Doctoral thesis, Iowa State University, Ames, Iowa.

Finnan J, Burke B. 2013. Potassium fertilization of hemp (*Cannabis sativa*). *Industrial Crops and Products* 41: 419–422.

Fitzgerald K. 1782. Experiments with Chinese hemp seed. *Philosophical Transactions of the Royal Society of London* 72: 46–49.

Flajšman M, Slapnik M, Murovec J. 2021. Production of feminized seeds of high CBD *Cannabis sativa* L. by manipulation of sex expression and its application to breeding. *Frontiers in Plant Science* 12: e718092.

Flores-Sanchez IJ, Pec J, Fei J, Choi YH, Dusek J, Verpoorte R. 2009. Elicitation studies in cell suspension cultures of *Cannabis sativa* L. *Journal of Biotechnology* 143: 157–168.

Fournier G. 1981. Les chimiotypes du chanvre (*Cannabis sativa* L.) Intérêt pour un programme de selection. *Agronomie* 1: 679–688.

Frank M. 1988. *Marijuana grower's insider's guide*. Red Eye Press, Los Angeles.

Frank M, Rosenthal E. 1978. *Marijuana grower's guide*. And/Or Press, Berkeley, California.

Fronmüller F. 1882. Gerbsaures cannabin. *Allgemeine medizinische Central-Zeitung* 51: 833–834; *Zeitschrift des Allgemeinen österreichischen Potheker-Vereines* 20: 457–459.

Gabrielian E, Zohary D. 2008. Wild relatives of food crops native to Armenia and Nakhichevan. *Flora Mediterranea* 14: 5–80.

Galoch E. 1978. The hormonal control of sex differentiation in dioecious plants of hemp (*Cannabis sativa*). *Acta Societatis Botanicorum Poloniae* 47: 153–162.

Galoch E. 1980. The hormonal control of sex differentiation in dioecious plants of hemp (*Cannabis sativa*). *Acta Physiologiae Plantarum* 2: 31–39.

Gao CS, Cheng CH, Zhao LN, *et al.* (8 other authors). 2018. Genome-wide expression profiles of hemp (*Cannabis sativa* L.) in response to drought stress. *International Journal of Genomics* 2018: 3057272.

Garrido J, Rico S, Corral C, *et al.* (4 other authors). 2022. Exogenous application of stress-related signaling molecules affect growth and cannabinoid accumulation in medical cannabis (*Cannabis sativa* L.). *Frontiers in Plant Science* 13: e1082554.

Gasparin AEP. 1848. *Cours d'agriculture , tome quatrième*. Librairie Agricole de la Maison Rustique, Paris.

Gauca C, Les M, Goncearu V, Ionescu A, Craciun E. 1977. Influenta masurilor agrofitotehnice asupra productiei la cinepa pentru fibre. *Cercetari Agronomice in Moldava* 1977(2): 105–108.

Giupponi L, Leoni V, Pavlovic R, Giorgi A. 2020. Influence of altitude on phytochemical composition of hemp inflorescence: a metabolomic approach. *Molecules* 25: e1381.

Gonsior G, Buschmann H, Szinicz G, Spring O, Sauerborn J. 2004. Induced resistance: an innovative approach to managed branched broomrape (*Orobanche ramosa*) in hemp and tobacco. *Weed Science* 52: 1050–1053.

Guo HY, Guo MB, Hu XI, *et al.* (5 other authors). 2011a. Industrial hemp variety 'Yúnmá No.1' seed and stalk high yield cultivation model. *Southwest China Journal of Agricultural Sciences* 24(3): 888–895.

Guo Y, Wang YF, Qiu CS, *et al.* (3 other authors). 2011b. Preliminary study on effects of drought stress on physiological characters and growth of different hemp cultivars (*Cannabis sativa* L.). *Plant Fiber Sciences in China* 2011(5): 235-239.

Haberlandt F. 1879. *Der allgemeine landwirtschaftliche Pflanzenbau*. Faesy & Fick, Vienna.

Hall J, Bhattaral SP, Midmore DJ. 2014. The effects of photoperiod on phenological development and yields of industrial hemp. *Journal of Natural Fibers* 11: 87–106.

Hamm W. 1854. Die culture des hanfs. *Agronomische Zeitung* 9: 97–103 (also *Die Grundzüge der Landwirthschaft, Zweiter Band*. Vieweg & Sohn, Braunschweig).

Haney A, Bazzaz FA. 1970. "Some ecological implications of the distribution of hemp (*Cannabis sativa* L.) in the United States of America," pp. 39–48 *in* Joyce CRB, Curry SH, eds. *The botany and chemistry of Cannabis*. J & A Churchill, London.

Haney A, Kutscheid BB. 1973. Quantitative variation in the chemical constituents of marihuana from stands of naturalized *Cannabis sativa* L. in East-Central Illinois. *Economic Botany* 27: 193–203.

Haney A, Kutscheid BB. 1975. An ecological study of naturalized hemp (*Cannabis sativa* L.) in east-central Illinois. *American Midland Naturalist* 93: 1–24.

Haring CM. 1922. "Report of the Director of the Agricultural Experiment Station," pp. 15–187 in *Report of the College of Agriculture and the Agricultural Experiment Station of the University of California*. University of California Press, Berkeley, California.

Hatch MD, Slack CR. 1966. Photosynthesis by sugar cane leaves. A new carboxylation reaction and the pathway of sugar formation. *Biochemical Journal* 101: 103–111.

Hawley D, Graham T, Stasiak M, Dixon M. 2018. Improving *Cannabis* bud quality and yield with subcanopy lighting. *HortScience* 53: 1593–1599.

Henry D. 1771. *The complete English farmer*. F. Newbery, London.

Herich G. 1960. Gibberellin and sex differentiation of flowering plants. *Nature* 188: 599–600.

Heslop-Harrison J. 1956. Auxin and sexuality in *Cannabis sativa. Physiologia Plantarum* 9: 588–597.

Heslop-Harrison J. 1957. The experimental modification of sex expression in flowering plants. *Biological Review* 32: 1–52.

Heslop-Harrison J, Heslop-Harrison Y. 1961. Studies on flowering plant growth and organogenesis. IV. Effects of gibberellic acid on flowering and the secondary sexual difference in stature in *Cannabis sativa. Proceedings of the Royal Irish Academy, Section B* 61: 219–231.

Heslop-Harrison J, Heslop-Harrison Y. 1967. "*Cannabis sativa* L.," pp. 205–226 in Evans LT, *ed. The induction of flowering*. Cornell University Press, Ithaca, New York.

Heyer F. 1884. Untersuchungen über das Verhältnis des Geschlechts bei einhäusigen und zweihäusigen Pflanzen. *Berichte Physiologischen Laboratorium und der Versuchsanstalt des Landwirtschaftlichen Instituts der Universität Halle* 1(5): 5–81

Hinton W. 1788. Papers in agriculture. *Transactions of the Society, instituted at London, for the Encouragement of Arts, Manufactures, and Commerce* 6: 105–110.

Hoffmann W. 1961. "Hanf, *Cannabis sativa* L.," pp. 204–263 in Kappert H, Rudorf W, Roemer T, *eds. Handbuch der Pflanzenzüchtung, 2nd Edition, Vol.* 5. Paul Parey, Berlin.

Hoffmann W. 1972. A környezeti tényezők hatása a kender ivaranak alakulására. *Rostnövények* 1972: 61–80.

Hu XL, Guo HY, Liu XY, *et al.* (4 other authors). 2012. 云南工大麻品种在黑江大麻安岭地区的适性研究 (Study on adaptability of Yunnan industrial hemp varieties in Daxinganling region of Heilongjiang province). *Southwest China Journal of Agricultural Sciences* 25: 838–841.

Huddleston H. 1841. Report on hemp cultivation, etc. in British Gurhwal. *Transactions of the Agricultural & Horticultural Society of India* 8: 260–278.

Indian Hemp Drugs Commission. 1894. *Report of the Indian Hemp Drugs Commission, 1893–94*. Government Central Printing Office, Simla.

Itagaki K, Shibuya T, Tojo M, Endo R, Kitaya Y. 2015. Development of powdery mildew fungus on cucumber leaves acclimatized to different CO_2 concentrations. *HortScience* 50: 1662–1665.

Iványi I. 2005. Relationship between leaf nutrient concentrations and yield of fibre hemp (*Cannabis sativa* L.). *Cereal Research Communications* 33: 97–100.

Iványi I. 2011. Relationship between leaf nutrient concentrations and the yield of fibre hemp (*Cannabis sativa* L.). *Research Journal of Agricultural Science* 43: 70–76.

Iványi I, Izsáki Z, van der Werf HGM. 1997. Influence of nitrogen supply and P and K levels of the soil on dry matter and nutrient accumulation of fiber hemp (*Cannabis sativa* L.). *Journal of the International Hemp Association* 4: 84–89.

Jaccard H. 1895. *Catalogue de la flore Valaisanne*. H. Georg, Genève et Lyon.

Jaffa ME. 1891. *Composition of the ramie plant*. Bulletin of the University of California Agricultural Experiment Station, Bulletin No. 94. Berkeley, California.

Jalali S, Salami SA, Sharifi M, Sohrabi S. 2019. Signaling compounds elicit expression of key genes in cannabinoid pathway and related metabolites in cannabis. *Industrial Crops & Products* 133: 105–110.

James MS, Vann MC, Suchoff DH, *et al.* (4 other authors). 2023. Hemp yield and cannabinoid concentrations under variable nitrogen and potassium fertilizer rates. *Crop Science* 63: 1555–1565.

Janischevsky DE. 1924. Форма конопли на сорных местах в Юго-Восточной России (A form of cannabis in wild areas of south-eastern Russia). Ученые записки Саратовского государственного университета имени Н. Г. Чернышевского 2(2): 3–17.

Jikomes N, Zoorob M. 2018. The cannabinoid content of legal cannabis in Washington State varies systematically across testing facilities and popular consumer products. *Scientific Reports* 8: e4519.

Jones JB. 1998. *Plant nutrition manual*. CRC Press, Boca Raton, Florida.

Kerr HC. 1877. Report of the cultivation of, and trade in, ganja in Bengal. *British Parliamentary Papers* 66: 94–154.

Khan MR, Rizvi TF. 2020. Effect of elevated levels of CO_2 on powdery mildew development in five cucurbit species. *Scientific Reports* 10: e4986.

Kirkham MB. 2014. *Principles of soil and plant water relations*. Academic Press, Amsterdam.

Knight G, Hansen S, Conner M, *et al.* (3 other authors). 2010. The result of an experimental indoor hyroponic *Cannabis* growing study using the 'screen of green' (ScOG) method: yield, tetrahyrocannabinol (THC) and DNA analysis. *Forensic Science International* 202: 36–44.

Knop W. 1865. Quantitative Untersuchungen über die Ernährungsprozeß der Pflanzen. *Landwirtschaftlichen Versuchs-Stationen* 7: 93–107.

Knop W. 1868. *Lehrbuch der Agricultur-Chemie, Ester Band*. H. Haessel, Leipzig.

Kubešova M, Moravcová L, Suda J, Jarošík V, Pyšek P. 2010. Naturalized plants have smaller genomes than their non-invading relatives: a flow cytometric analysis of the Czech alien flora. *Preslia* 82: 81–96.

Landi S. 1997. Mineral nutrition of *Cannabis sativa* L. *Journal of Plant Nutrition* 20: 311–326.

Lapierre É, de Ronne M, Boulanger R, Torkamaneh D. 2023. Phenotypic characterization of a diverse population of *Cannabis sativa* for agronomic, morphological, and biochemical traits. *Preprints*.

Latta RP, Eaton BJ. 1975. Seasonal fluctuations in cannabinoid content of Kansas marijuana. *Economic Botany* 29: 153–163.

Lavoisier A. 1770. Sur la nature de l'eau. *Mémoires de l'Académie Royale des Sciences* 1770: 73–82, 90–107.

Lavoisier A. 1777. Mémoire sur la combustion en général. *Mémoires de l'Académie Royale des Sciences* 1777: 535–547.

Lavoisier A. 1862. "Résultats de quelques expériences d'agriculture, et réflexions sur leurs relations avec l'économie politique," pp. 812-823 in *Oeuvres de Lavoisier, Tome II*. Imprimerie Nationale, Paris.

Lavoisier A. 1868. "Relative à l'essai des chiffons," pp. 699–700 in *Oeuvres de Lavoisier, Tome Iv*. Imprimerie Impériale, Paris.

Lavoisier A. 1893. "Instruction sur l'agriculture pour les assemblées Provinciales," pp. 203-215; "Mémoire sur les encouragements qu'il est necessaire d'accorder à l'agriculture (1787)," pp. 216-226; in *Oeuvres de Lavoisier, Tome VI*. Imprimerie Nationale, Paris.

Liebig J. 1840. *Die organische Chemie in ihrer Anwendung auf Agricultur und Physiologie*. Friedrich Vieweg und Sohn, Braunschweig, Germany.

Lisson S, Mendham N. 1998. Response of fiber hemp (*Cannabis sativa* L.) to varying irrigation regimes. *Journal of the International Hemp Association* 5(1): 9–15.

Lisson SN, Mendham NJ, Carberry PS. 2000. Development of a hemp (*Cannabis sativa* L.) simulation model 1. General introduction and the effect of temperature on the preemergent development of hemp. *Australian Journal of Experimental Agriculture* 40: 405—411.

Lojacono A. 1910. *La canapa di Carmagnola*. Vicenzo Bona, Torino.

Lydon J, Teramura AH, Coffman CB. 1987. UV-B radiation effects on photosynthesis, growth and cannabinoid production of two *Cannabis sativa* chemotypes. *Photochemistry and Photobiology* 46: 201–206.

MacWilliams J, Peirce E, Pitt WJ, Schreiner M, *et al.* (4 other authors). 2023. Assessing the adaptive role of cannabidiol (CBD) in *Cannabis sativa* defense against cannabis aphids. Frontiers in Plant Science 14: e1223894

Magagnini G, Grassi G, Kotiranta S. 2018. The effect of light spectrum on the morphology and cannabinoid content of *Cannabis sativa* L. *Medical Cannabis and Cannabinoids* 2018(1): 19–27.

Mair W. 1898. Indian hemp. *The Pharmaceutical Era* 20: 281–282.

Malloch WS. 1922. Value of the hemp plant for investigating sex inheritance. *Journal of Heredity* 13: 277–283.

Mangin A. 1859. "Chanvre," pp. 594–599 in *Dictionnaire universel théorique et practique du commerce et de la navigation, Tome Premier.* Guillaumin et Cie, Paris.

Mansouri H, Asrar Z. 2012. Effects of abscisic acid on content and biosynthesis of terpenoids in *Cannabis sativa* at vegetative stage. *Biologia Plantarum* 56: 153–156.

Mansouri H, Rohani M. 2014. Response of *Cannabis sativa* L. to foliar application of 2-chloro-ethyl-trimethyl-ammonium chloride. *Iranian Journal of Plant Physiology* 5: 1225–1233.

Mansouri H, Salari F. 2014. Influence of mevinolin on chloroplast terpenoids in *Cannabis sativa. Physiology and Molecular Biology of Plants* 20: 273–277.

Mansouri H, Asrar Z, Szopa A. 2009a. Effects of ABA on primary terpenoids and Δ^9-tetrahydrocannabinol in *Cannabis sativa* L. at flowering stage. *Plant Growth Regulation* 58: 269–277.

Mansouri H, Asrar Z, Mehrabani M. 2009b. Effects of gibberellic acid on primary terpenoids and Δ^9-tetrahydrocannabinol in *Cannabis sativa* at flowering stage. *Journal of Integrative Plant Biology* 51(6): 553–561.

Mansouri H, Asrar Z, Amarowicz R. 2011. The response of terpenoids to exogenous gibberellic acid in *Cannabis sativa* at flowering stage. *Acta Physiologiae Plantarum* 33: 1085–1091.

Marcandier P. 1758. *Traite du Chanvre.* Jean-Luc Nyon, Paris.

Markham G. 1615. *Countrey contentments in two bookes, the second intituled the English Huswife.* Roger Jackson, London.

Marquart B. 1919. *Der Hanfbau.* Paul Parey, Berlin.

Marti G, Schnee S, Andrey Y, *et al.* (4 other authors). 2014. Study of leaf metabolome modifications induced by UV-C radiations in representative *Vitis, Cissus,* and *Cannabis* species by LC-MS based metabolomics and antioxidant assays. *Molecules* 19: 14004–14021.

Massuela DC, Munz S, Hartung J, Nkebiwe PM, Graeff-Hönninger S. 2023. Cannabis Hunger Games: nutrient stress induction in flowering stage—impact of organic and mineral fertilizer levels on biomass, cannabidiol (CBD) yield and nutrient use efficiency. *Frontiers in Plant Science* 14: e1233232.

Maugh TH. 1975. An escalation of potency. *Science* 190: 867.

McCree K. 1972a. The action spectrum, absorptance and quantum yield of photosynthesis in crop plants. *Agricultural Meteorology* 9: 191–216.

McCree K. 1972b. Test of current definitions of photosynthetically active radiation against leaf photosynthesis data. *Agricultural Meteorology* 10: 443–453.

McPartland JM. 1984. Pathogenicity of *Phomopsis ganjae* on *Cannabis sativa* and the fungistatic effect of cannabinoids produced by the host. *Mycopathologia* 87: 149–153.

McPartland JM, Schoeneweiss DF. 1984. Hyphal morphology of *Botryosphaeria dothidea* in vessels of unstressed and drought-stressed *Betula alba. Phytopathology* 74: 358–362.

McPartland JM, Small E. 2020. A classification of endangered high-THC cannabis (*Cannabis sativa* subsp. *indica*) domesticates and their wild relatives. *PhytoKeys* 144: 81-112.

McPhee HC. 1924. The influence of environment on sex in hemp, *Cannabis sativa* L. *Journal of Agricultural Research* 28: 1067–1081+plate.

McPhee H. 1925. The influence of environment on sex in hemp. *Journal of Agricultural Research* 31: 935–943.

Mihoc M, Pop G, Alexa E, Radulov I. 2012. Nutritive quality of Romanian hemp varieties (*Cannabis sativa* L.) with special focus on oil and metal contents of seeds. *Chemistry Central Journal* 6: e122.

Mikuriya TH. 1967. Kif cultivation in the Rif mountains. *Economic Botany* 21: 231–234.

Mikuriya TH, Aldrich MR. 1988. Cannabis: Old drug, new dangers. The potency question. *Journal of Psychoactive Drugs* 20(1): 47–55.

Miller P. 1768. *The gardeners dictionary, 8th edition.* J & J Rivington, London.

Mills E. 2011. *Energy up in smoke. The carbon footprint of indoor cannabis production.* Available at: http://evan-mills.com/energy-associates/Indoor.html.

Mirzamohammad E, Alirezalu A, Alirezalu K, Norozi A, Ansari A. 2021. Improvement of the antioxidant activity, phytochemicals, and cannabinoid compounds of *Cannabis sativa* by salicylic acid elicitor. *Food Science and Nutrition* 9: 6873–6881.

Mitchell TG. 1999. *Prospects for augmenting nicotine content of tobacco products.* Document code 3107, British American Tobacco Co. Available at: http://tobaccodocuments.org/youth/CgNcBAT00000000.Rg1.html

Mohan Ram HY. 2002. A passion for plant life. *Journal of Biosciences* 27: 651–664.

Mohan Ram HY, Jaiswal VS. 1970. Induction of female flowers on male plants of *Cannabis Sativa* L. by 2-chloroethanephos-phonic acid. *Experientia* 15: 214–216.

Mohan Ram HY, Jaiswal VS. 1972. Induction of male flowers on female plants of *Cannabis sativa* by gibberellins and its inhibition by abscisic acid. *Planta* 105: 263–266.

Mohan Ram HY, Sett R. 1979. Sex reversal in the female plants of *Cannabis sativa* by cobalt ion. *Proceedings of the Indian Academy of Sciences* 88B: 303–308.

Mohan Ram HY, Sett R. 1982a. Modification of growth and sex expression in *Cannabis sativa* by aminoethoxyvinylglycine and ethephon. *Zeitschrift für Pflanzenphysiologie* 105: 165–172.

Mohan Ram HY, Sett R. 1982b. Induction of fertile male flowers in genetically female *Cannabis sativa* plants by silver nitrate and silver thiosulphate anionic complex *Theoretical and Applied Genetics* 62: 369–375.

Montford S, Small E. 1999. A comparison of the biodiversity friendliness of crops with special reference to hemp (*Cannabis sativa* L.). *Journal of the International Hemp Association* 6(2): 53–63.

Morad D, Bernstein N. 2023. Response of medical cannabis to magnesium (Mg) supply at the vegetative growth phase. *Plants* 12: e2676.

Nissen L, Zatta A, Stefanini I, *et al.* (4 other authors). 2010. Characterization and antimicrobial activity f essential oils of industrial hemp varieties (*Cannabis sativa* L.). *Fitoterapia* 81: 413–419.

Oczapowski M. 1848. *Gospodarstwo wiejskie, Tom VI.* S.H. Merzbacha, Warsaw.

Owen LC, Suchoff DH, Chen H. 2023. A novel method for stimulating *Cannabis sativa* L. male flowers from female plants. Plants 12: e3371.

Pagnani G, Pellegrini M, Galieni A, *et al.* (8 other authors). 2018. Plant growth-promoting rhizobacteria (PGPR) in *Cannabis sativa* 'Finola' cultivation: an alternative fertilization strategy to improve plant growth and quality characteristics. *Industrial Crops and Products* 123: 75–83.

Pan G, Li Z, Yin M, *et al.* (9 other authors). 2021. Genome-wide identification, expression, and sequence analysis of CONSTANS-like gene family in *Cannabis* reveals a potential role in plant flowering time regulation. *BMC Plant Biology* 21: 142.

Panerio RJG. 2020. *Hedging your bets: the importance of flowering time and agronomic practise in the search for the ideal Australian hemp seed cultivar.* Master's thesis, Southern Cross University, Lesmore, Australia.

Paris M, Boucher F, Cosson L. 1975. The constituents of *Cannabis sativa* pollen. *Economic Botany* 29: 245–253.

Pate DW. 1983. Possible role of ultraviolet radiation in evolution of *Cannabis* chemotypes. *Economic Botany* 37: 396–405.

Pate DW. 1993. Possible role of ultraviolet radiation in evolution of *Cannabis* chemotypes. *Economic Botany* 37: 396–405.

Peacock FB. 1871. "Letter from F.B. Peacock, Esq., Officiating Secretary to the Board of Revenue, Lower Provinces, to the Officiating Secretary to the Government of Bengal, Revenue Department," pp. 67–72 in Godley A, *ed. East India (consumption of ganja).* Her Majesty's Stationery Office, London.

Peč J, Flores-Sanchez IJ, Choi YH, Verpoorte R. 2010. Metabolic analysis of elicited cell suspension cultures of *Cannabis sativa* L. by ^{1}H-NMR spectroscopy. *Biotechnology Letters* 32: 935–941.

Perry DC. 1977. Street drug analysis and drug use trends 1969-1975. Part II. *PharmChem Newsletter* 6(4): 1–3.

Petit J, Salentijn EMJ, Paulo MJ, *et al.* (10 other authors). 2020a. Genetic variability of morphological, flowering, and biomass quality traits in hemp (*Cannabis sativa* L.). Frontiers in Plant Science 11: e102.

Petit J, Salentijn EMJ, Paulo MJ, Denneboom C, Trindade LM. 2020b. Genetic architecture of flowering time and sex determination in hemp (*Cannabis sativa* L.): a genome-wide association study. *Frontiers in Plant Science* 11: e569958.

Phylos Biosciences. 2021. *Autoflowering markers. International Patent WO 2021/097496 A2.* World Intellectual Property Organization, Geneva, Switzerland.

Portiere H. 2012. *Флора и ботаническая география Северного Кавказа (Flora and botanical geography of the North Caucasus).* Association of Scientific Publications KMK, Moscow.

Potter DJ. 2009. *The propagation, characterisation and optimisation of Cannabis sativa L. as a phytopharmaceutical.* Doctoral thesis, King's College, London.

Potter DJ, Duncombe P. 2012. The effect of electrical lighting power and irradiance on indoor-grown cannabis potency and yield. *Journal of Forensic Science* 57: 618-622.

Potter DJ, Clark P, Brown MB. 2008. Potency of delta 9-THC and other cannabinoids in cannabis in England in 2005: implications for psychoactivity and pharmacology. *J.ournal of Forensic Science* 53: 90–94.

Prade T, Svensson SE, Andersson A, Mattson JE. 2011. Biomass and energy yield of industrial hemp grown for biogas and solid fuel. *Biomass & Bioenergy* 35: 3040–3049.

Prain D. 1893. *Report on the cultivation and use of gánjá.* Bengal Secretariat Press, Calcutta.

Purdy T. 2018. *Vapor pressure deficit and HVAC system design.* Application Note 28, Desert Aire, Milwaukee, Wisconsin.

Radosavljevic-Stevanovica N, Markovic J, Agatonovic-Kustrin S, Razic S. 2014. Metals and organic compounds in the biosynthesis of cannabinoids: a chemometric approach to the analysis of *Cannabis sativa* samples. *Natural Product Research* 28: 511–516.

Rham WL. 1842. Agriculture of the Netherlands, part II. *Journal of the Royal Agricultural Society of England* 3: 240–263.

Richins RD, Rodriguez-Uribe L, Lowe K, Ferral R, O'Connell MA. 2018. Accumulation of bioactive metabolites in cultivated medical *Cannabis*. *PLoS One* 13: e0201119.

Rodriguez-Morrison V, Llewellyn D, Zheng Y. 2021a. Cannabis yield, potency, and leaf photosynthesis respond differently to increasing light levels in an indoor environment. *Frontiers in Plant Science* 12: e456.

Rodriguez-Morrison V, Llewellyn D, Zheng Y. 2021b. Cannabis inflorescence yield and cannabinoid concentration are not increased with exposure to short-wavelength ultraviolet-B radiation. *Frontiers in Plant Science* 12: e725078.

Rose C, Parker A, Jefferson B, Cartmell E. 2015. The characterization of feces and urine: a review of the literature to inform advanced treatment technology. *Critical Reviews in Environmental Science and Technology* 45: 1827–1879.

Ross AF. 1961. Systemic acquired resistance induced by localized virus infections in plants. *Virology* 14: 340–358.

Ross SA, ElSohly MA. 1996. The volatile oil composition of fresh and air-dried buds of *Cannabis sativa*. *Journal of Natural Products* 59: 49–51.

Rowlandson T. 1849. On hemp. *Journal of the Royal Agricultural Society of England* 10: 172–182.

Salentijn EMJ, Zhang QY, Amaducci S, Yang M, Trindade LM. 2015. New developments in fiber hemp (*Cannabis sativa* L.) breeding. *Industrial Crops and Products* 68: 32–41.

Saloner A, Bernstein N. 2020. Response of medical cannabis (*Cannabis sativa* L.) to nitrogen supply under long photoperiod. *Frontiers in Plant Science* 11 e1517.

Saloner A, Bernstein N. 2021. Nitrogen supply affects cannabinoid and terpenoid profile in medical cannabis (*Cannabis sativa* L.). *Industrial Crops and Products* 167: e113516.

Sarath G, Mohan Ram HY. 1979. Comparative effect of silver ion and gibberellic acid on the induction of male flowers on female *Cannabis* plants. *Experientia* 35: 333–334.

Sáringer G, Nagy B. 1971. The effect of photoperiod and temperature on the diapause of the hemp moth (*Grapholita sinana* Feld.) and its relevance to the integrated control. *Proceedings, 13th International Congress of Entomology, Moscow* 1: 435–436.

Saussure NT de. 1804. *Recherches chimiques sur la vegetation.* Nyon, Paris.

Schaffner JH. 1926. The change of opposite to alternate phyllotaxy and repeated rejuvenations in hemp by means of changed photoperiodicity. *Ecology* 7: 315–325.

Schaffner JH. 1928. Further experiments in repeated rejuvenation in hemp and their bearing on the general problem of sex. *American Journal of Botany* 15: 77–85.

Schilling S, Melzer R, Dowling CA, Shi JQ, Muldoon S, McCabe PF. 2022. A protocol for rapid generation cycling (speed breeding) of hemp (*Cannabis sativa*) for research and agriculture. *Plant Journal* 113: 437–445.

Schoenmakers N. 1986. *The Seed Bank 1986/1987 catalogue.* Drukkerij Dukenburg Printers, Nijmegen, The Netherlands.

Scholz H. 1957. Der wilde Hanf als Ruderalpflanze Mitteleuropas. *Verhandlungen des Botanischen Vereins für die Provinz Brandenburg* 83(97): 61–64.

Schübeler FC. 1875. *Die Pflanzenwelt Norwegens. Ein Beitrag zur Natur-und Culturgeschichte Nord-Europas.* A.W. Brøgger, Christiania.

Senebier J. 1782. *Mémoires physico-chymiques, sur l'influence de la lumière solaire, tome troisime.* Chirol, Geneva.

Senebier J. 1788. *Mémoires physico-chymiques, sur l'influence de la lumière solaire, nouvelle edition, tome second.* Chirol, Geneva

Sengbusch RV. 1956b. *Die Züchtung von monözischen und diözischen faserertragreichen Hanfsorten.* Lecture notes, Kortrijk, Belgium. Available at: http://pubman. mpdl.mpg.de/pubman/item/escidoc:37261:8/component/escidoc:49364/sengbusch_248_PDFA.pdf

Sengloung T, Kaveeta L, Nanakorn W. 2009. Effect of sowing date on growth and development of Thai hemp (*Cannabis sativa* L.). *Kasetsart Journal (Natural Science)* 43: 423–431.

Shaikhislamova EF, Suyundukov YT, Mirkin BM. 2006. Statistical analysis of factors determining the composition of segetal communities in the Bashkir Transural Region. *Russian Journal of Ecology* 37: 284–287.

Sharma GK. 1983. Altitudinal differentiation in *Cannabis sativa* L. *Science and Culture* 49(2): 53–55.

Shi GR, Cai QS, Liu QQ, Wu L. 2009. Salicylic acid-mediated alleviation of cadmium toxicity in hemp plants in relation to cadmium uptake, photosynthesis, and antioxidant enzymes. *Acta Physiologiae Plantarum* 31: 969–977.

Shiponi S, Bernstein N. 2021a. Response of medical cannabis (*Cannabis sativa* L.) genotypes to P supply under long photoperiod: functional phenotyping and the ionome. *Industrial Crops and Products* 161: e113154.

Shiponi S, Bernstein N. 2021b. The highs and lows of P supply in medical cannabis: Effects on cannabinoids, the ionome, and morpho-physiology. *Frontiers in Plant Science* 12: 657323

Short DJ. 2013. Proposing a new classification: 'Cannabis indoor.' *O'Shaughnessy's* Winter/Spring 2013, p. 60.

Slonov LKh, Petinov NS. 1980. The content of nucleotides and the ATPase activity in hemp leaves in relation to the water supply. *Fiziologiya Rastenii* 27(5): 1095–1100.

Small E. 2015. Evolution and classification of *Cannabis sativa* (marijuana, hemp) in relation to human utilization. *Botanical Review* 81: 189–294.

Small E, Beckstead HD. 1973. Common cannabinoid phenotypes in 350 stocks of *Cannabis*. *Lloydia* 36: 144–146.

Small E, Beckstead HD, Chan A. 1975. The evolution of cannabinoid phenotypes in *Cannabis*. *Economic Botany* 29: 219–232.

Soldatova NA, Khryanin VN. 2010. The effects of heavy metal salts on the phytohormonal status and sex expression in marijuana. *Russian Journal of Plant Physiology* 57: 96–100.

Soltan K, Dadkhah B. 2022. Studies of the major gene expression and related metabolites in cannabinoids biosynthesis pathway influenced by ascorbic acid. *Planta Medica* 9: e116.

Son GH, Moon JY, Selake RH *et al.* (5 other authors). 2021. Conserved opposite functions in plant resistance to biotrophic and necrotrophic pathogens of the immune regulator *SRFR1*. *International Journal of Molecular Science* 22: e6427.

Song C, Saloner A, Fait A, Berstein N. 2023. Nitrogen deficiency stimulates cannabinoid biosynthesis in medical cannabis plants by inducing a metabolic shift towards production of low-N metabolites. *Industrial Crops and Products* 202.

Soubeiran E, Girardin J. 1851. Examen comparatif des tourteaux des grains oléagineuses. *Journal d'Agriculture Pratique, troisième série* 2: 89–95.

Sprengel C. 1828. Von den Substanzen der Ackerkrume und des Untergrundes. *Journal für Technische und Ökonomische Chemie* 2: 423–474, 3: 42–99, 313–352, and 397–421.

Stockberger WW. 1915. Drug plants under cultivation. *USDA Farmer's Bulletin No. 663*. Washington, DC.

Storm D. 1987. *Marijuana hydroponics*. And/Or Books, Berkeley, California.

Struik PC, Amaducci S, Bullard MJ, *et al.* (3 other authors). 2000. Agronomy of fibre hemp (*Cannabis sativa* L.) in Europe. *Industrial Crops and Products* 11: 107–118.

Tang K, Fracasso A, Struik PC, Yin X, Amaducci S. 2018. Water- and nitrogen-use efficiencies of hemp (*Cannabis sativa* L.) based on whole-canopy measurements and modeling. *Frontiers in Plant Science* 9: 951.

Tanney CAS, Lyu DM, Schwinghamer T, *et al.* (3 other authors). 2023. Sub-optimal nutrient regime coupled with *Bacillus* and *Pseudomonas* sp. inoculation influences trichome density and cannabinoid profiles in drug-type *Cannabis sativa*. *Frontiers in Plant Science* 14: e1131346.

Targioni Tozzetti O. 1802. *Lezioni di agricoltura specialmente Toscana del dottore Targioni Tozzetti. Tomo II*. Guglielmo Piatti, Firenze.

Thaer A. 1812. *Grundsätze der rationellen Landwirthschaft, Vierter Band*. G. Reimer, Berlin.

Thaler JS, Humphrey PT, Whiteman NK. 2012. Evolution of jasmonate and salicylate signal crosstalk. *Trends in Plant Science* 17: 260–270.

Tibeau ME. 1936. Time factor in utilization of mineral nutrients by hemp. *Plant Physiology* 11: 731–747.

Toonen M, Ribot S, Thissen J. 2006. Yield of illicit indoor cannabis cultivation in the Netherlands. *Journal of Forensic Science* 51: 1050-1054.

Toth JA, Stack GM, Carlson CH, Smart LB. 2022. Identification and mapping of major-effect flowering time loci *Autoflower1* and *Early1* in *Cannabis sativa* L. *Frontiers in Plant Science* 13: e991680.

Tournois J. 1912. Influence de la lumière sur la floraison du houblon japonais et du chanvre déterminées par des semis haitifs. *Comptes rendus hebdomadaires des séances de l'Académie des sciences* 155: 297–300.

Traill GW. 1828. Statistical sketch of Kamaon. *Asiatick Researches* 16: 137–236.

Turner CE, Cheng PC, Lewis GS, Russell MH, Sharma GK. 1979. Constituents of *Cannabis sativa*. XV: botanical and chemical profile of Indian variants. *Planta Medica* 37: 217–225.

UN-ODC. 2005b. *Maroc. Enquête sur le cannabis 2004*. UN-ODC, Vienna.

UN-ODC. 2008. *World Drug Report*. United Nations Office on Drugs and Crime. Vienna International Center, Vienna.

UN-ODC. 2010b. *Afghanistan Cannabis survey 2009*. United Nations Office on Drugs and Crime. Vienna, Austria.

US Congress. 1996. "Marijuana use in America," *Hearings before the Subcommittee on Crime of the Committee on the Judiciary, House of Representatives, 104th Congress, 2nd session*. Government Printing Office, Washington, DC.

Van der Werf HMG, van den Berg W. 1995. Nitrogen fertilization and sex expression affect size variability of fiber hemp (*Cannabis sativa* L.). *Oecologia* 103: 464–470.

Van der Werf HMG, Haasken HJ, Wijlhuizen M. 1994. The effect of daylength on yield and quality of fibre hemp (*Cannabis sativa* L.). *European Journal of Agronomy* 3: 117–123.

Van der Werf HMG, van Geel WCA, van Gils LJC, Haverkort AL. 1995a. Nitrogen fertilization and row width affect thinning and productivity of fibre hemp (*Cannabis sativa* L.). *Field Crops Research* 42: 27–37.

Van Helmont JB. 1648. *Ortus Medicinae*. Elzevirium, Amsterdam.

van Loon LC, Bakker PAHM, Pieterse CMJ. 1998. Systemic resistance induced by rhizosphere bacteria. *Annual Review of Phytopathology* 36: 453–483.

Vanhove W, Van Damme P, Meert N. 2011. Factors determining yield and quality of illicit indoor cannabis (*Cannabis* spp.) production. *Forensic Science International* 212: 158–163.

Vanhove W, Surmont T, van Damme P, de Ruyver B. 2012. Yield and turnover of illicit indoor cannabis (*Cannabis* spp.) plantations in Belgium. *Forensic Science International* 220: 265–270.

Vavilov NI. 1926. The origin of the cultivation of "primary" crops, in particular cultivated hemp. *Bulletin of Applied Botany and Plant-Breeding* 16(2): 221–233.

Vavilov NI, Bukinich DD. 1929. Konopli. Труды по прикладной ботанике, генетике и селекции 33 (Suppl.): 380–382. Supplement published separately as: Земледельчиский Афганистан (*Agricultural Afghanistan*). Publications of the All-Union Institute of Applied Botany and New Cultures at the SNK., Leningrad.

Veazie P, Cockson P, Kidd D, Whip B. 2021. Elevated phosphorus fertility impact on cannabis sativa 'BaOx' growth and nutrient accumulation. *International Journal of Innovative Science, Engineering & Technology* 8: 345–351.

Vera CL, Malhi SS, Raney JP, Wang ZH. 2004. The effect of N and P fertilization on growth, seed yield and quality of industrial hemp in the Parkland region of Saskatchewan. *Canadian Journal of Plant Science* 84: 939–947.

Vera CL, Malhi SS, Phelps SM, May WE, Johnson EA. 2010. N, P and S fertilization on industrial hemp in Saskatchewan. *Canadian Journal of Plant Science* 90: 179–184.

Vogelmann V. 1840. *Der hanfbau im Großherzogthum Baden*. G. Brann, Karlsruhe.

Vogl CR, Lissek-Wolf G, Surböck A. 2004. Comparing hemp seed yields (*Cannabis sativa* L.) of an on-farm scientific field experiment to an on-farm agronomic evalutation under organic growing conditions in lower Austria. *Journal of Industrial Hemp* 9: 37–49.

Waddell L. 1905. *Lhasa and its mysteries: with a record of the expedition of 1903–1904*. John Murray, London.

Waller AE. 1941. Professor John Henry Schaffner. *Ohio Journal of Science* 41: 251–286.

Walsh C. 2019. A matter of altitude: growing cannabis up in the mountains. *MMJ Daily*. Available at: https://www.mmjdaily.com/article/9107267/us-co-a-matter-of-altitude-growing-cannabis-up-in-the-mountains/

Walters JP. 2002. The myth of harmless marijuana. *Washington Post*, May 1, 2002, p. A25.

Walton RP. 1938. *Marihuana, America's new drug problem*. J.B. Lippincott Co., Philadelphia.

Westmoreland FM, Bugbee B. 2022. Sustainable *Cannabis* nutrition: elevated root-zone phosphorus significantly increases leachate P and does not improve yield or quality. *Frontiers in Plant Science* 13: e1015652.

White RF. 1979. Acetylsalicylic acid (aspirin) induces resistance to tobacco mosaic virus in tobacco. *Virology* 99: 410–412.

Wirtshafter D. 1995. Nutrition of hemp seeds and hemp seed oil," pp. 546–555 in: *Bioresource Hemp, 2nd edition*. Nova-Institute, Köln, Germany.

Wood P, Price N, Matthews P, McKay JK. 2022. Genome-wide polymorphism and genic selection in feral and domesticated lineages of *Cannabis sativa. GGG* 2022: jkac209.

Yang R, Berthold E, McCurdy CR *et al.* (3 other authors) 2020. Development of cannabinoids in flowers of industrial hemp (*Cannabis sativa* L.) – a pilot study. *Journal of Agricultural and Food Chemistry* 68: 6058–6064.

Yao QJ, Xiong YN, Peng F, He SL, Xia B. 2007. 不同生态类型大麻品种在南京引种的生育表 (Growth and developmental character of different ecotypic hemp varieties naturalized in Nanjing). *Plant Fiber Sciences in China* 29: 270–275.

Yep B, Zheng YB. 2021. Potassium and micronutrient fertilizer addition in a mock aquaponic system for drug-type *Cannabis sativa* L. cultivation. *Canadian Journal of Plant Science* 101: 341–352.

Young A. 1768. *The farmer's letters to the people of England*. W. Nicoll, London.

Young A. 1772. *Political essays concerning the present state of the British Empire*. Strahan and Cadell, London.

Young A, ed. 1788. On the culture and manufacture of hemp, in Suffolk, by a manufacturer. *Annals of Agriculture and Other Useful Arts* 10: 377–389.

Young A. 1797. *General view of the agriculture of the County of Suffolk*. Macmillan, London.

Young A. 1799. *General view of the agriculture of the County of Lincoln*. W. Bulmer & Co., London.

Zhang WJ, Björn LO. 2009. The effect of ultraviolet radiation on the accumulation of medicinal compounds in plants. *Fitoterapia* 80: 207–218.

Zhang QP, Chen X, Guo HY, *et al.* (6 other authors). 2018. Latitudinal adaptation and genetic insights into the origins of *Cannabis sativa* L. *Frontiers in Plant Science* 9: e1876.

Zhou YL, van Leeuwen SK, Pieterse CMJ, Bakker PAHM, van Wess SCM. 2019. Effect of atmospheric CO_2 on plant defense against leaf and root pathogens of Arabidopsis. *European Journal of Plant Pathology* 154: 31–42.

3 Taxonomy and ecology of pests and pathogens

Abstract

This chapter was formerly called "the Seven Wonders" (McEno, 1991), enumerating viruses, bacteria, protozoans, fungi, oömcetes, plants, and animals—the seven kingdoms whose members interact with *Cannabis*. Today a different taxonomy is recognized, but the fundamentals—classification and nomenclature—remain the same. In light of evolution, taxonomy becomes *systematics*, and molecular systematics (based on DNA sequencing) have given rise to new classification schemes. The members of each kingdom share synapomorphies (e.g., unique morphologies) derived through evolution and shaped by ecological forces. At the interface of evolution and ecology lies an unanswered question: for what purpose does *Cannabis* produce cannabinoids? We summarize worldwide research regarding the effects of THC and CBD upon other organisms. Several coevolutionary hypotheses are proposed.

3.1 Introduction

Taxonomy includes **classification** (the identification and categorization of organisms) and **nomenclature** (the naming and describing of organisms). Adding a fourth dimension—time—turns taxonomy into **systematics**: the evolutionary relationships among living things. Classification, in the light of evolution, becomes **phylogenetics**: the genealogical study of relationships among individuals and groups in a nested hierarchy.

Evolution is inherently linked with **ecology**; ecological relationships provide the context for evolutionary change. Ecology was coined by Haeckel (1866), "the relationship of the organism to the environment." By "environment" Haeckel included the physical habitat (sunlight, moisture, soil) as well as other organisms.

From our perspective, crop agronomy is applied ecology; IPM is the management of ecosystems. For instance, Darling (1910) meticulously investigated the ecology of malaria-transmitting insects in the Panama Canal Zone. "If you wish to control mosquitoes, you must learn to think like a mosquito," he reportedly said (Paul, 1955).

At the interface of ecology and evolution lies an unanswered question: for what purpose does *Cannabis* produce cannabinoids? THC's evolutionary *raison d'être* has many hypotheses, two leading ones being: (1) protection against other organisms (the effects of THC and CBD upon other organisms is a subtheme of this chapter); and (2) protection against harsh environmental conditions (UV radiation, drought, high temperatures, nutrient deficiency). The upregulation of cannabinoid production (and glandular trichome growth) has been linked with genes involved with generalized stress tolerance, such as proteins in the MYB and HD-ZIP families (Liu *et al.*, 2021; Singh *et al.*, 2021; Ma *et al.*, 2022; Haiden *et al.*, 2022).

Cannabinoids and terpenoids are produced in **glandular trichomes**. *Stalked* glandular trichomes (Fig. 3.1) consist of a gland head atop a stalk. They predominate in flowering tops. *Sessile* glandular trichomes, which lack stalks, predominate in lower leaves. The contents of gland heads (resin heads) provide a *chemical* defense against herbivores. Many researchers have documented the lethal effects of cannabinoids and terpenoids against insects, mites, and nematodes. It's been a topic of recent reviews (McPartland and Sheikh, 2018; Bernard *et al.*, 2022; Ona *et al.*, 2022). We present highlights later in this chapter.

Potter (2009) measured a monoterpenoid-to-sesquiterpenoid ratio of 4:1 in stalked glandular trichomes. Sessile glandular trichomes had nearly the reverse ratio. Potter proposed that *monoterpenoids*, with lower viscosity, worked best against insects in flowering tops. Russo (2011) extended Potter's idea, and proposed that *sesquiterpenoids*—bitter and gut-altering—repel vertebrate herbivores, who feed primarily on lower leaves. Jin *et al.* (2020) measured a sesquiterpenoid-to-monoterpenoid ratio of 6.5:1 in lower fan leaves. Flowering tops expressed a reverse ratio, albeit less lop-sided: a monoterpenoid-to-sesquiterpenoid ratio of 1.9:1.

Cystolith trichomes also cover the plant (Fig. 3.1). They resemble pointed needles of glass and are in fact heavily silicified. They contain calcium carbonate crystals at their base. Cystoliths provide a *mechanical* defense against herbivores. They impede the mobility of smaller herbivores, and may damage the mouthparts of larger ones. Soft-bodied insects impale themselves on cystolith spikes—they may remain impaled and die, or suffer a morbid series of impale-and-escape episodes (Levin, 1973).

Glandular trichomes also provide a mechanical defense against herbivores, independent of any chemical repellence provided by their contents. Gland heads in living plants are friable; they burst easily and release their fluids. The fluids oxidize and polymerize into a gummy substance, which mechanically disables small insects. Potter (2009) photographed aphids (*Aphis gossypii*) glued to trichome "flypaper." The aphids struggled for a while and then died (see Fig. 4.15).

Nelson *et al.* (2019) proposed that *Cannabis* trichomes entrap pest species. The entrapped pests, in turn, attract predators, who scavenged the bodies, thereby increasing predator populations, and further reducing plant damage. On the negative

DOI: 10.1079/9781836990352.0003

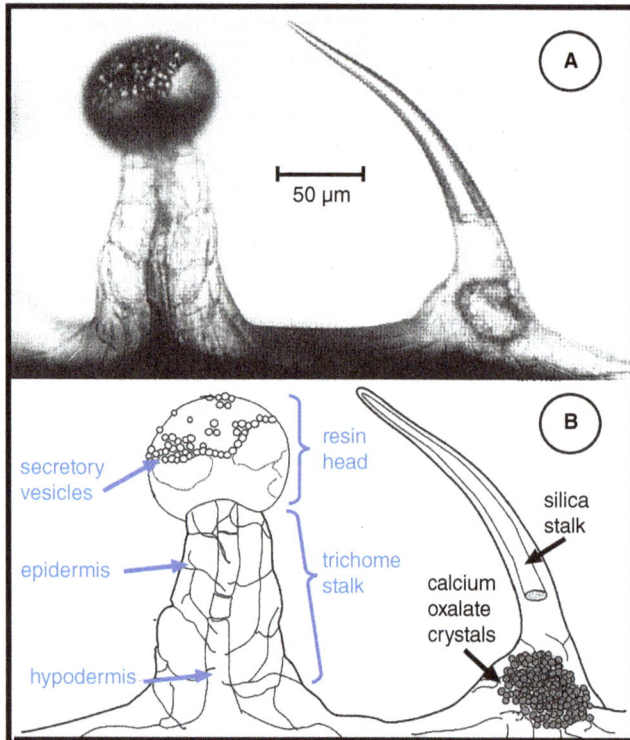

Fig. 3.1. Capitate stalked glandular trichome (left), and cystolith trichome (right): **A**. photomicrograph; **B**. diagram. Reproduced with kind permission from Potter (2004).

side, trichomes may impair the movement of beneficial predators and parasitoids (Kashyap *et al.*, 1991; Lemay *et al.*, 2022).

Ecologists and their agronomic siblings—IPMers—must understand a complex web of ecological relationships, summarized as parasitism, mutualism, and competition. Animals have a **parasitic** relationship with *Cannabis*. There are exceptions, such as *Homo sapiens*, that have a **mutualistic** relationship with *Cannabis*. Mutualism occurs when *both* species benefit from their interaction. *Homo sapiens* nurtures the plant's growth (cultivation) and disperses its seeds (zoochory); in turn, the plant provides *Homo sapiens* with fiber, oil, and medicines. *Cannabis* has a **competitive** relationship with other plants. They compete for the raw materials of photosynthesis—soil, sunlight, and water.

Any two species' interactions may depend on the influence of a third species. For instance, studies show that a fungal infection may protect a plant from subsequent insect infestations. When two insects meet on a plant, they compete—especially in the hollow stalk of a *Cannabis* plant.

But insects don't have to meet to compete: Plants damaged by leaf-chewing insects are subsequently avoided by leafminer insects (Faeth, 1986). Damaged plants contain more defense chemicals in their remaining leaves, which makes them less desirable to subsequent pests. Leafminers that feed on previously damaged leaves suffer greater parasitism than leafminers who feed on unchewed leaves. This is because **parasitoids** (parasites of pests) use chewed leaves as clues to locate their hosts. Amazingly, leaves damaged by mechanical causes do not attract parasitoids, unless oral secretions from leaf-eating insects are added (Turlings *et al.*, 1990).

The final outcome of any interaction is often arbitrated by the environment, or the microclimate. A pest may flourish on a plant's lower leaves, shaded and protected, but not survive in the harsh environment near a plant's apex. This is especially true in *Cannabis*, where flowering tops accumulate THC and CBD—chemicals with pesticidal activity.

3.2 Taxonomy

Taxonomy—classification and nomenclature—ultimately has two functions: (1) to identify and classify organisms based on consensus pattern recognition; and (2) to construct a stable nomenclature for effective storage, retrieval, and communication of information (Athreya and Hopkins, 2021).

Taxonomy is vital for communication. Lewis Carrol described a place without taxonomy in *Through the Looking Glass*, when Alice entered a dark wood "where things have no names" (Carroll, 1871). She found it impossible to have a conversation. In the *Analects*, when Confucius was asked what would be his first action when administrating a government, he replied, "I would begin by defining the names of things. If names of things are not properly defined, words will not correspond to facts" (Lau, 2000).

This seems straightforward, but massive conceptual issues underlie classification and nomenclature. A 300-year debate argues over whether taxonomies are real things in nature (**essentialism**) or arbitrary human constructs (**nominalism**). Linnaeus was an essentialist—he classified the unchanging order of life created by God. Darwin was a nominalist. He could not reconcile the continuous process of evolution with the discrete concept of species. "I look at the term species as one arbitrarily given for the sake of convenience to a set of individuals closely resembling each other ... I was much struck how entirely vague and arbitrary is the distinction between species and varieties" (Darwin, 1859).

After Alice left the dark wood, she came across Tweedledum and Tweedledee. They drew Alice's attention to the Red King sleeping under a nearby tree. Tweedledee proposed that Alice was a figment of the Red King's dream, and upon awaking she would cease to exist. Upset, she replied, "I *am* real!" As it turned out, Alice herself had been dreaming. "I do hope it's *my* dream, and not the Red King's! I don't like belonging to another person's dream" (Carroll, 1871).

Modern taxonomists have dreamed up systems that Athreya and Hopkins (2021) characterized as "intellectual"—organisms are classified regardless of the practical use or relevance of the system. In contrast, "folk taxonomies" of non-literate indigenous societies are strictly utilitarian. "Folk species" are classified in around 500 "folk genera" and higher ranks, related to how human memory stores information (Raven *et al.*, 1971). Folk taxonomies can be quite fine-grained: Schultes (1986) discovered that Amazonian shamans distinguish between seven kinds of *Banisteriopsis caapi* (an ingredient in ayahuasca) based on their psychoactivity; to Schultes they were the same morphologically.

Intellectual systems are designed for storage and retrieval of information, whereas folk taxonomies are designed for communication (Raven *et al.*, 1971). In this book we seek a sweet spot between the two. For example, we deconstruct overly intellectual *Ostrinia* host plant associations (see sections 6.3 to 6.5). Strictly DNA-based classification has completely destabilized the names of species, especially fungal species, and we critique these name changes where appropriate.

The naming of species is governed by strict rules: the *International Code of Zoological Nomenclature* (Ride, 1999) and the *International Code of Nomenclature for Algae, Fungi, and Plants* (Turland, 2018). They are legalistic documents, updated periodically, and have become rather complicated. Both codes have pre-Darwinian roots (Strickland, 1843; de Candolle, 1867) with a holdover Linnaean (essentialist) method: a "type specimen" must be designated when naming a new species.

Typological practice has a utilitarian aspect: new samples can be compared with the type specimen to judge morphological variability. But this phenetic criterion is anthemic to DNA-based taxonomists. Phillips *et al.* (2013) suggested "disregarding" fungal species that lacked DNA data. They designated "epitypes" (specimens in culture) rather than type specimens to fit their new taxonomic concepts and species names.

Two fundamental precepts guiding rules of nomenclature are the **valid publication** of taxa, and the principle of **priority** (sanctioning the oldest valid taxon available). Rules were made to be broken: the taxon *Cannabis sativa* L. 1753 was sanctioned by the *International Code of Nomenclature for algae, fungi, and plants* (*ICN*), despite the fact that Barbaro (1516) coined the binomial name earlier, and should have priority. As our mentor von Arx (1987) said, "nonspecialists have difficulties in understanding the code and adhering to its provisions."

From a practical standpoint, consistent and stable Latin names are essential for effective communication. Common names often vary; Latin names can be exact. Exact identification is especially crucial when you use biocontrol methods, which work on specific pests. Applying biocontrol against a misidentified pest is a waste of time and money. Sloppy identification is permissible only when using sledgehammer pesticides.

3.3 Kingdoms

Many of us learned a two-kingdom taxonomy in school. Everything was crammed into the plant or animal kingdoms. Here we describe six kingdoms: **Vira**, little bits of bad news wrapped in a protein coat; **Monera**, the prokaryotes, including bacteria, phytoplasmas, and actinomycetes; **Protoctista**, unicellular eukaryotes, including protozoans, algae, slime molds, and oomycetes; **Fungi**, molds, mildews, smuts, etc.; **Plantae**, e.g., *Cannabis*; and **Animalia**, such as nematodes, mollusks, insects, and vertebrates. See Fig. 3.2 for an illustration of *Cannabis* parasites representing these kingdoms. Kingdoms are subdivided into Phyla, then Classes, Orders, Families, Genera, and Species ("King Phillip Came Over From Germany Slowly").

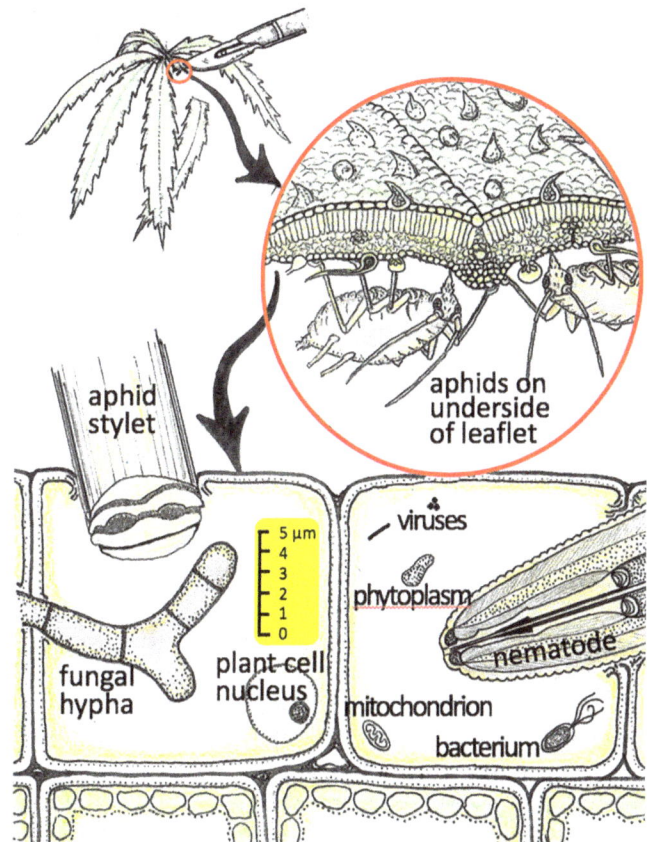

Fig. 3.2. Shapes and sizes of some organisms associated with *Cannabis* (adapted from Agrios 2005)

3.3.1 Kingdom Vira

Conceptually, the Virus kingdom resides within the other kingdoms. Viruses only replicate in connection with a living host. Viruses cannot "grow," they do not eat, there is no sex. They are complex molecules, entities between chemicals and life. Viruses contain DNA or RNA, which encode information for the reproduction of identical entities. Viruses cause disease by re-programing their host's metabolic machinery, causing host cells to produce foreign (viral) proteins.

Plant viruses are transmitted (vectored) by aphids, as the aphids move from diseased plants to healthy plants. Viruses are also vectored by leafhoppers, whiteflies, mites, mealybugs, thrips, and other insects with sucking mouthparts. Some plant viruses can replicate in their insect vectors. Insect-specific viruses may have switched hosts millions of years ago, when insects fed on plants. One such insect killer, the nuclear polyhedrosis virus (NPV), is sold as a pesticide.

Viruses and their vectors have a mutualistic relationship. Viruses induce a change in plant metabolism—a kind of premature senescence—which makes plant sap more nutritious for sap-sucking insects (Kennedy *et al.*, 1959). The insects, in turn, transmit the viruses to new hosts. Nematodes, fungi, and parasitic plants occasionally vector viruses. Viruses commonly spread through cloning (vegetative propagation) of infected "mother plants." Viruses can be transmitted through seeds of infected plants. Viruses also spread plant-to-plant via root grafts, or between leaves rubbing in the wind, or by workers moving among diseased and healthy plants. Tobacco mosaic virus (TMV) can even be transmitted by smoking infected cigarettes near uninfected plants.

Until recently, no plant viruses were known to cause disease in people; likewise, plants cannot catch the flu. However, a metagenomic analysis of human feces discovered the pepper mild mottle virus (PMMV) in 12 of 18 fecal samples (66.7%) collected from healthy individuals (Zhang *et al.*, 2006). PMMV is a pathogen of *Capsicum* (chili peppers). Zhang infected peppers with fecal-borne PMMV. Colson *et al.* (2010) recovered PMMV from commercial sources of hot sauce (e.g., Tabasco). Some people who tested positive for anti-PMMV antibodies suffered from fever, abdominal pain, and pruritus.

Cannabinoids can inhibit the replication and pathogenicity of viruses, at least in experimental *in vitro* studies. This has been shown with herpesvirus (Blevins and Dumic, 1980; Lancz *et al.*, 1990,1991), poliovirus 1 (Braut-Boucher *et al.*, 1985), influenza A (Buchweitz *et al.*, 2008), simian immunodeficiency virus (Molina *et al.*, 2011), HIV-1 (Williams *et al.*, 2014; DeMarino *et al.*, 2022), hepatitis C (Lowe *et al.*, 2017), and Covid-19 (Van Breemen *et al.*, 2022; Nguyen *et al.*, 2022).

Most of these anti-viral studies used reagent-grade THC. However, plant extracts were also utilized, including aqueous extracts. Aqueous extracts *do not* test the effects of cannabinoids, because *cannabinoids and most terpenoids are not very water soluble* (Table 3.1). Only 2.8 mg of THC dissolves into a liter of water at 23°C. That's why water pipes work. THCA's carboxylic acid likely makes it more water soluble, but its solubility has not been measured. Monoterpenes that are pure hydrocarbons (abbreviated M-h in Table 3.1) are not very water soluble. Hydrocarbon sesquiterpenes (S-h) are downright hygrophobic. Conversely, oxygenated monoterpenoids (M-oh) show 60-fold greater solubility than hydrocarbon monoterpenes.

The active ingredients in aqueous *Cannabis* extracts are water-soluble proteins, carbohydrates, flavonoids, anthocyanins, chlorophylls, and minerals. Contrarily, cannabinoids and terpenoids are highly soluble in polar solvents (ethanol, methanol, butanol) and non-polar solvents (supercritical CO_2, petroleum ether, hexane).

Terpenoids and cannabinoids can be extracted by steam distillation or hydrodistillation. Steam distillation passes steam through a bed of plant material in a closed system. Volatile compounds are carried away in the steam, condensed and separated. Hydrodistillation is a variation of steam distillation, where plant material is soaked in water, then boiled, and volatile compounds are carried away in the vapor, condensed and separated.

Malingré *et al.* (1975) estimated that steam distillation extracted 3.3% of cannabinoids in plant material, and 75% of terpenoids in plant material. Hydrodistillation extracts a much higher percentage of cannabinoids: a group in Italy extracted 0.1% CBD from steam-distilled 'Felina 32' (Benelli *et al.*, 2018b). Their hydrodistilled extract contained 1.69–1.89% CBD (Bertoli *et al.*, 2010). That's essentially all of the CBD in 'Felina 32': dried flowering tops average 1.4–1.5% CBD (Callaway, 2008; Ingallina *et al.*, 2020).

3.3.2   Kingdom Monera

The Monera comprises organisms with a prokaryotic cell organization—having no nuclear membrane. Antoni van Leeuwenhoek discovered bacteria in 1683, shortly after he invented the microscope. Leeuwenhoek temporarily blinded himself by observing the ignition of gunpowder through his invention. Thereafter he turned to more pedestrian materials, such as the scum on his teeth, where he discovered bacteria.

Unique prokaryotes that outgas methane have been recognized since Smit (1930) named *Methanosarcina* (*Zymosarcina*) *methanica*. They were considered bacteria until Woese and Fox (1977) analyzed 16S ribosomal RNA sequences of *Methanosarcina*, and they were more closely related to animals and plants than to bacteria. Woese named a new domain, **Archaea**, separate from bacteria, for *Methanosarcina* and related organisms. Archaeans are extremophiles—they produce methane, hold records for surviving high temperatures (130°C for a short time), extreme acid (pH of –0.2), extreme salinity (five times greater NaCl than seawater), and nuclear war—they can survive 30 kGy gamma radiation (Pikuta *et al.*, 2007). Archaean are rarely pathogens, and often exist as mutualists within animals, such as our own methane-producing guts.

The **Bacteria** domain contains a varied lot of organisms: Phylum Pseudomonadota (i.e., Proteobacteria—many human and *Cannabis* pathogens), Phylum Cyanobacteria ("blue-green algae" with chlorophyll), Phylum Spirochaetota (causes of Lyme disease, syphilis, and leptospirosis), Phylum Actinomycetota (*Mycobacterium*, which causes tuberculosis, and *Streptomyces*, which kills *Mycobacterium* with streptomycin), and Phylum Mycoplasmatota (mycoplasmas that infect humans, phytoplasmas that infect *Cannabis*).

Small bacteria are the size of large viruses. But bacteria are far more complex than viruses. Bacteria, like us, are enclosed by a cell membrane composed of proteins, carbohydrates, and lipids. Unlike us (but like plant cells), bacteria also have a cell wall (except for mycoplasmas and phytoplasmas). Unlike organisms in other kingdoms, bacteria have only one chromosome, and it is circular,

Table 3.1. Water solubility

Compound	Solubility (mg/L)
THC[a]	2.8
CBD[b]	12.6
α-Pinene (M-h)[c]	5.0
Limonene (M-h)[c]	20.4
Myrcene (M-h)[c]	29.9
(E)–β-caryophyllene (S-h)[d]	0.05
trans-α-Bergamotene (S-h)[d]	0.03
Linalool (M-oh)[d]	1559
α-Terpineol (M-oh)[d]	1889
Carveol (M-oh)[d]	2931

[a]Garrett and Hunt (1974), [b]Stella *et al.*, 2021, [c]Fichan *et al.* (1999), [d]Rao and McLements (2012)

without a nuclear membrane. Bacteria communicate with each other by passing genes between themselves.

Bacteria are extremely prolific. Some double their numbers every 20 minutes. If unchecked, one such bacterium could multiply into a colony covering the Pentagon in 16.50 hours. Bacteria are very hardy. *Bacillus cereus* can revert to a spore stage which survives boiling water, frozen water, or no water. Bacteria inhabit practically every habitat imaginable, including us. The bacterial cells within a human body outnumber our own cells, although estimates of 10-to-1 are overblown (Sender *et al.*, 2016). Bacteria in us weigh about 3 lb (1.36 kg)—the same as our brain.

Thomas Burrill first proposed that bacteria cause disease in 1878. He studied "fire blight" of pear trees at the University of Illinois. Burrill predated Pasteur, but Pasteur studied bacteria in *people*, so he got all the glory. *Cannabis*-pathogenic bacteria share some generalities: they are usually rod-shaped bacilli, gram-negative, aerobic, motile (moving with propeller-like flagellae), enter plants through small wounds or other openings, generally have a narrow host range, and rarely harm humans (see section 15.1).

Interest has exploded regarding bacteria in the phyllosphere and rhizosphere of *healthy* plants. The **phyllosphere** refers to above-ground portions of plants, including stems, leaves, flowers, and seeds. The **rhizosphere** refers to below-ground parts of the plants. Some rhizosphere bacteria can fix atmospheric nitrogen for plant use. We dedicate Chapter 13 to phyllosphere and rhizosphere bacteria.

Bacteria also provide benefits by killing phytophagous insects. *Bacillus thuringiensis* (BT) is a well-known insect killer. Spray BT on plants, and insects die after they eat BT-coated leaves. Another insect killer, *Xenorhabdus nematophilus*, lives within nematodes, which serve as its vector. You can purchase nematodes with *X. nematophilus*, and mix them into soil. The nematodes find insects, penetrate them, and then release *X. nematophilus*. The bacteria kill the insects, and the nematodes feed off the cadavers. Quite a delivery system. Such teamwork.

Cannabis extracts are antibacterial, at least in experimental *in vitro* studies (Table 3.2). These studies assayed antibacterial activity with agar diffusion assays, measured as zone diameter

Table 3.2. Antibacterial activity of *Cannabis* extracts

Citation	Extract	Results, gram-positive (+) or gram-negative (−) bacteria
Ferenczy 1956	Whole seeds pressed into agar	Active against (+) *Bacillus cereus*
Chailakhyan *et al.* 1957	Water extracts, leaves, stems, roots	Active against (−) *Rhizobium leguminosarum*; root > shoot > leaf extracts. Hemp ranked 4th out of 15 non-leguminous crop plants
Ferenczy *et al.* 1958	"Organic solvent" extract, flowering tops	Active: (+) *B. cereus, B. subtilis*, and *Staphylococcus aureus*; inactive: (−) *Escherichia coli, Pseudomonas aeruginosa, Salmonella paratyphi, Shigella* sp.
Kabelík *et al.* 1960	Ethanol/ethyl ether extract, leaves and shoots	Active: (+) *B. subtilis, B. anthracis, S. aureus, Streptococcus haemolyticus, S. pneumoniae, Corynebacterium diphtheriae, Diplococcus pneumonia*, (+/−) *Mycobacterium tuberculosis*; inactive: (−) *E. coli, P. aeruginosa, Salmonella typhi, Shigella dysenteriae, Proteus vulgaris*
Krejčí 1961	Ethanol extract, flowering tops	Active: (+) *B. cereus*
Radošević *et al.* 1962	Ethanol extract, resin	(+) *B. cereus*, extracts from fiber-type plants (Germany, Spain, Cyprus) more potent than drug-type plants (Brazil, Greece, Yugoslavia)
Veliky and Genest 1972	Ethanol extract of *Cannabis* cell cultures	Active: (+) *S. aureus, Bacillus megaterium*, gram (−) *E. coli*; inactive: (−) *P. aeruginosa*
Veliky and Latta 1974	Ethanol extract, four *Cannabis* cell lines	Active: (+) *B. megaterium*; less active: (+) *S. aureus*, (−) *E. coli*, (−) *P. aeruginosa*; inactive: (+/−) *Mycobacterium smegmatis*
Braut-Boucher *et al.* 1985	Hexane and chloroform extracts	"Very strong" against bacteria (unnamed species)
Vijai *et al.* 1993	Aqueous extract, leaves	Active: (−) *Erwinia carotovora*
Isahq *et al.* 2015	Various extracts, *Cannabis indica* leaves, stems, or seeds	Active: *S. aureus, B. cereus, Klebsiella pneumoniae, Proteus mirabilis*; extract potency: leaf > stem > seed, and (generally) methanol > *n*-butanol = chloroform = ethyl acetate > aqueous
Anjum *et al.* 2018	Various extracts of leaves	Active: greatest inhibition was seen in ethanol extracts against (−) *E. coli*, but aqueous extracts worked best against (+) *S. aureus* and (−) *P. aeruginosa*
Muscarà *et al.* 2021	Hexane extraction after hydrodistillation	Active: (+) *S. aureus* MIC = 4.88 µg/mL, several methicillin resistant *S. aureus* strains MIC = 2.44–4.88 µg/mL

of inhibition (mm); or in serial dilutions, which enabled the calculation of **minimum inhibitory concentration** (MIC).

The studies in Table 3.2 are apples-to-oranges: aqueous extracts contain a paucity of terpenoids (EOs) and cannabinoids, whereas solvent extracts contain these substances (see Table 3.1 and text below).

Few studies have field-tested antibacterial activity. In Ukraine, an aqueous extract of wild hemp, called "cansatine," was sprayed on other crop plants to protect them from bacteria (Zelepukha, 1960; Zelepukha et al., 1963). The extract worked best against gram-positive (gram+) *Corynebacterium* (*Clavibacter*) species, worked okay against gram-negative (gram–) *Xanthomonas* species (including *X. campestris*), and was least effective against gram+ *Bacillus* and gram– *Pseudomonas*, *Erwinia*, *Bacterium*, and *Agrobacterium* species (Bel'tyukova, 1962).

Cannabis essential oils (EOs, a.k.a. terpenoids) were tested for antibacterial activity by four groups. Fournier et al. (1978) extracted EOs from French monoecious hemp (FH), Moroccan *hashīsh* (MH), and Mexican drug plants (MD). EOs worked best against gram+ *Staphylococcus aureus* and *S. faecalis*, FH > MH > MD. They likely extracted cannabinoids in the EOs, because relative efficacy of the three extracts corresponded to CBD content. The EOs worked equally well against *Mycobacterium smegmatis* (gram+/–). Gram– *E. coli* and *P. fluorescens* were resistant.

Novak et al. (2001) prepared five EOs from five European fiber cultivars, and tested activity against 21 bacteria. Only five bacteria were sensitive to at least four of five EOs: gram+ *Bacillus subtilis*, *Brevibacterium linens*, *Brochothrix thermosphacta*, and gram– *Acinetobacter calcoaceticus* and *Beneckea* (*Vibrio*) *natriegens*. They did not run statistics, but our one-way ANOVA calculated $F = 1.153$, no significant difference between the five EOs at $p < 0.05$. Six other bacteria were sensitive to at least one EO: gram+ *S. aureus* and *Micrococcus luteus*, and gram– *E. coli*, *Aeromonas hydrophila*, *Flavobacterium suaveolens*, and *Yersinia enterocolitica*.

Nissen et al. (2010) compared EOs from 'Carmagnola', 'Fibranova', and 'Futura'. All three EOs inhibited gram+ *Enterococcus hirae*, *Enterococcus faecium*, and *Streptococcus salivarius*, well below the threshold limit (2.00% v/v). In contrast, only 'Futura' gave satisfactory results against *Clostridium* spp. 'Futura' gave best results against six species of gram– *Pseudomonas*, including the plant pathogens *P. syringae*, *P. fluorescens*, *P. savastonoi*, and *P.* (*Erwinia*) *carotovorum*. 'Carmagnola' and 'Fibranova' were above the threshold limit except for *P. savastonoi* and *P. carotovorum*.

Satyal and Setzer (2014) used hydrodistillation to extract cannabinoids and terpenoids from wild plants in Nepal. They summarized antibacterial activity as "essential no effect": (+) *B. cereus* MIC = 625 µg/mL, *S. aureus* = 2500 µg/mL, *P. aeruginosa* = 2500 µg/mL. Zengin et al. (2018) tested EOs against several multidrug-resistant microbial strains. Against 15 strains of *Helicobacter pylori*, MIC values ranged from 8 to 64 mg/mL. Against *Staphylococcus aureus* (four strains), the mean MIC value was 8 mg/mL.

Pure cannabinoids are antibacterial. Krejčí and Šantavý (1955) isolated a compound they named *kannabidiolovou* (cannabidiolic acid, CBDA). *Kannabidiolovou* formed a zone of inhibition around *Staphylococcus aureus*. Schultz and Haffner (1959) inhibited gram+ *S. aureus* and *B. subtilis* with a dilute solution of CBD (1:100,000). Gram– *E. coli* required a more concentrated solution of CBD (1:1000).

After a series of tests with extracts, Ferenczy et al. (1958) concluded that "the antibacterial activity is ever proportionate to the intensity of the hashish reaction." In contrast, Radoševič et al. (1962) stated that "antibiotic activity decreases with ... the increase of hashish activity." They made ethanol extracts from eleven different varieties, including drug-type plants (from Brazil) and fiber-type plants (from Europe and Canada), and assayed their effects on *B. subtilis*. They concluded that CBDA was the primary antibiotic agent.

Gal et al. (1969) tested CBDA as a fruit juice preservative, and reported its effectiveness against gram+ *B. cereus*, *Lactobacillus plantarum*, and *Leuconostoc mesenteroides*. Klingeren and Ham (1976) added THC and CBD to culture media at various doses and measured MICs. Gram+ *S. aureus*, *Streptococcus pyonenes*, and *S. faecalis* were sensitive with MICs of 1–5 µg/ml (CBD was slightly more potent than THC). Gram– *E. coli*, *Salmonella typhi*, and *Proteus vulgaris* were resistant at 100 µg/ml.

Turner and ElSohly (1981) tested cannabichromene (CBC). An agar diffusion screen demonstrated that gram+ *B. subtilis* and *S. aureus* were sensitive, whereas gram+/– *Mycobacterium smegmatis* and gram– *E. coli* and *P. aeruginosa* were resistant. Subsequent MIC tests gave values of 0.39 µg/ml for *B. subtilis*, and 1.56 µg/ml for *S. aureus*. ElSohly et al. (1982) tested cannabigerol (CBG) alongside CBD and THC in the agar diffusion screen test, quantified as width of the inhibition zone (in mm). Results were identical for *S. aureus* and *B. subtilis*: CBG = 25, CBD = 10, THC = 5. For *M. smegmatis*, CBG = 25, CBD = 10, THC = 2. *E. coli* and *P. aeruginosa* were resistant.

Harpaz et al. (2021) created a panel of genetically modified strains of *E. coli* to determine the mode of toxicity of cannabinoids: THC and THCA produced genotoxic effects (DNA damage), while CBD, CBDV, CBC, and CBG induced cytotoxic or oxidative damages.

In summary, terpenoids and cannabinoids impair the growth and survival of many bacteria. Perhaps that's why only ten bacterial species cause disease in *Cannabis*. In contrast, Woolhouse and Gaunt (2007) counted 541 human pathogenic species.

3.3.3 Kingdom Protoctista

This group comprises eukaryotic organisms that are not fungi, plants, or animals. The Protoctista is a paraphyletic assemblage of similar-appearing but diverse creatures, mostly unicellular. Some protists are animal-like, such as *Amoeba* and bacteria-eating *Paramecium*. Some are plant-like and have chlorophyll, such as *Karenia brevis*, the cause of Red Tide in Florida. Diatoms are photosynthetic microalgae—a major source of the planet's oxygen. The shells of dead diatoms constitute diatomaceous earth, an organic control against aphids.

Protists cause human disease (e.g., *Trypanosoma brucei*, sleeping sickness; *Leishmania* spp., leishmaniasis; *Plasmodium* spp., malaria) as well as plant diseases (e.g., *Phytomonas leptovasorum*, phloem necrosis of *Coffea*). Against *Trypanosoma brucei*, Nok *et al.* (1994) reported *in vitro* lethal effects by a cannabis petroleum ether extract. It also worked *in vivo*; mice infected by *T. brucei* were injected with extract (50 mg/kg/day), and cured in 5 days. Nok assumed that the extract contained cannabinoids, but he extracted seeds, which do not contain cannabinoids.

Anti-*Leishmania* activity has been documented in several studies. A team in Mississippi tested individual compounds in *Cannabis* and determined IC_{50} values (µg/mL) against *Leishmania donovani*: cannflavin B, 5.0 µg/mL; cannabigerolic acid, 12.0 µg/mL (Radwan *et al.*, 2008a); cannflavin A, 4.5 µg/mL; cannflavin C, 17.0 µg/mL (Radwan *et al.*, 2008b); 5-acetyl-4-hydroxycannabigerol, 10.7 µg/mL (Radwan *et al.*, 2009). They also tested ten THC derivatives—relatively potent, in a range of 0.5 to 4.5 µg/mL, but never tested THC itself (Osman *et al.*, 2018).

A hexane-extracted EO showed little activity against *L. donovani* (>40 µg/mL) but a subfraction (β-caryophyllene, α-humulene, caryophyllene oxide) yielded 7.47 µg/mL (Wanas *et al.*, 2016). Meghini *et al.* (2021) treated *Leishmania tropica*-infected mice by injecting them with steam-distilled EOs (2.5 mL/kg) from three fiber-type cultivars (Eletta campana, Futura 75, Carmagnola selezionata). "Percentage cure rate" was 96%, 84%, and 75%, respectively, compared with 95% in amphotericin B-treated mice (control mice 0%).

Anti-*Plasmodium falciparam* activity was also explored by the Mississippi group, both a chloroquine-sensitive strain and a chloroquine-resistant strain, with the following IC_{50} values by *Cannabis* compounds: 6-prenylapigenin, 2.8 and 2.0 µg/mL, respectively (Radwan *et al.*, 2008b), two cannabichromene derivatives, 7.2 and 4.0–6.7 µg/mL, respectively (Radwan *et al.*, 2009), and eight THC derivatives, range 0.16–4.76 and 0.20–4.50 µg/mL, respectively (Osman *et al.*, 2018).

De Sousa *et al.* (2021) found that THC inhibited the growth of chloroquine-sensitive and chloroquine-resistant strains of *P. falciparam* (IC_{50} = 0.79 and 0.72 µg/mL, respectively), more potently than CBD (4.1 µg/mL chloroquine-sensitive, other strain not tested). THC inhibited β-haematin formation (IC_{50} = 11.3 µg/mL), thereby increasing the concentration of free haem in the parasite, killing it through oxidative stress.

Akinola *et al.* (2018) reported a significant difference in the survival rate of *Plasmodium berghei*-infected mice fed a diet augmented with *C. sativa* (leaves, twigs, and seeds, ratio 6:3:1). Survival was 15.2 days with 40% *C. sativa* and 9.4 days with 1% *C. sativa*, compared with 7.6 days in controls, and 17.2 days for chloroquine-treated mice.

Slime molds are fungus-like protists. Some "individuals" can exist as either one multicellular aggregation or a collection of single-celled organisms. If there is enough food around, single cells go about their business, growing and dividing like amoebae. But if starved, they aggregate into clumps and crawl off like a slug. Finding better conditions, the slug erects a tall stalk topped by spores. The spores blow off, revert to amoebae, and go their separate ways. Hrebenyuk (1984) reported a slime mold species, *Didymium clavus*, climbing hemp stalks.

Oomycetes go by many names: phycomycetes, algal fungi, water molds, and pseudofungi. They can be difficult to differentiate from true fungi. Oomycetes have cellulose in their cell walls (like plants), instead of the chitin found in fungi. Two Oomycete genera, *Pseudoperonospora* and *Pythium*, comprise several species that infect *Cannabis*. *Pseudoperonospora* species cause downy mildew, a blue-white felt that forms on the undersides of leaves. *Pythium* species cause damping off, a rapid collapse of small seedlings. *Cannabis* EO was tested against the Oomycete *Phytophthora infestans*; but the results "were not promising" (Mediavilla and Steinemann, 1997).

Harvey (1925) invented a technique to isolate Oomycetes. He steam-sterilized *Cannabis* seeds, and floated the seeds in pond water as Oomycete "bait." Thanks to Harvey's technique, there are scores of reports of Oomycetes infesting *Cannabis* seeds. None of these organisms cause problems unless you store seeds in pond water. Oomycetes act as biocontrol agents against mosquito larvae (e.g., *Lagenidium*, *Leptolegnia*, *Aphanomyces* spp.). Researchers have grown them in liquid culture with hemp seed extract (Domnas *et al.*, 1982; Nakumusana, 1986; Patwardhan *et al.*, 2005).

3.3.4 Kingdom Fungi

Fungi are not plants—they lack chlorophyll and do not engage in photosynthesis. They lack roots. Instead, they produce threadlike tubular filaments termed **hyphae**. Hyphae usually contain **septa**, which are incomplete cross-walls. The body of a fungus, its collection of hyphae, is called a **mycelium**. Fungi have chitin in their cell walls, whereas plants have cellulose. Chitin is something fungi share with animals—chitin constitutes the exoskeleton of lobsters, crabs, and insects. Fungi store reserve energy as glycogen, as do animals (plants store energy as starch).

Based on a combination of ultrastructural and biochemical characters, Cavalier-Smith (1987) proposed that fungi are more closely related to animals than plants. This has been confirmed with protein and DNA sequences (Baldauf and Palmer, 1993). Nevertheless, the *International Code of Nomenclature* (*ICN*) for naming fungi is shared with plants (Turland, 2018), and not animals (Ride, 1999).

Fungi can be unicellular or multicellular. They contain one, two, or many nuclei per cell. Some are mobile. Most are sexual. They are everywhere we look. In this aspect they are more successful than insects—fungi have conquered the seas, but we find no insects in marine environments. Hawksworth and Lücking (2017) estimated there are 2.2 to 3.8 million species of fungi. With 120,000 currently accepted species, only about 3% to 8% have been identified to date.

Fungi have no digestive system, so they absorb (not ingest) nutrients. Fungi grow into food, secrete enzymes that cause digestion to occur around them, then absorb the nutrients. Fungi reproduce a variety of ways. Yeasts reproduce by **budding** (like bacteria), but most fungi reproduce via **spores**. There are several types of spores. Fungal spores produced by **mitosis** are genetically haploid, i.e. $1n$ (like our sperm and ova), and the spores are called **conidia**. Unlike our sperm and ova, conidia can germinate directly into whole $1n$ organisms.

The hyphae of two 1n organisms can intertwine and fuse, forming **diploid** (2n) organisms. This is fungal sex. Sometimes the hyphae fuse, but their nuclei remain separate—forming **dikaryotic** (1+1n) organisms. The hyphae of all these organisms—haploid, diploid, and dikaryotic—are identical in external appearance. Only their nuclei know for sure. Diploids (2n) can produce spores by **meiosis**. These spores may form at the site of haploid fusion (at the zygote, hence zygospores). Or the spores arise distally, in sacs (ascospores) or on club-like basidia (basidiospores).

Fungi infesting *Cannabis* produce spores in assorted sizes and shapes. Figure 3.3 illustrates *Alternaria alternata* (A.a.), *Aspergillus flavus* (A.f.), *Athelia rolfsii* (A.t.), *Botrytis cinerea* (B.c.), *Curvularia* sp. (C.s.), *Diaporthe ganjae* (D.g.), *Epicoccum nigrum* (E.n.), *Fusarium oxysporum* (F.s.), *Lasiodiplodia theobromae* (L.t.), *Ophiobolus anguillides* (O.a.), *Penicillium chrysogenum* (P.c.), *Pestalotiopsis* sp. (P.s.), *Pseudoperonospora cannabina* (P.c.), along with a human red blood cell (RBC). **Pier Andrea Saccardo** (1845–1920) developed a system for classifying fungi by spore shape and color. His massive 25-volume *Sylloge Fungorum* (1882–1931) is full of research on *Cannabis* pathogens. *Sylloge* remained the primary system of classification prior to molecular DNA analysis. Saccardo (1879) also conducted research on *Cannabis* sexuality, and the plant's response to different soil types.

Fungi cause more plant disease than the rest of Earth's organisms combined. The scientific literature lists over 420 Latin names of fungi associated with *Cannabis*. Many names on that list are saprophytes, not pathogens, such as "hemp retting" fungi. Other names on that list are taxonomic synonyms. The fungus causing gray mold, for instance, masquerades under seven different Latin names (McPartland, 1995d). Other species cited in the literature are misidentifications

(McPartland, 1995a). After a name-by-name review, McPartland (1992) determined the 420 Latin names that represent 88 species of *Cannabis* pathogens. They are presented in Chapters 9 to 13.

Saprophytes live off already-dead material. They benefit us by decomposing organic matter and releasing nutrients back to the soil. Saprophytic fungi ruin our food and overrun leather, cotton, and paper in damp places. They ret and rot hemp fiber, and colonize poorly stored seeds and flowers.

Some fungi "go both ways," living as saprophytes of dead plants *and* parasites of living plants. Termed **facultative parasites** (FPs), they normally act saprophytically, but can attack living hosts. Many root-rot fungi and damping-off fungi fall into this category. Some FPs exist as epiphytes or endophytes in the phyllosphere or rhizophere (for these terms, see the above section on bacteria). They may be commensal or mutualistic, or flip into a parasitic lifestyle. Some phyllosphere fungi are beneficial to *Cannabis* (see section 13.4).

Mycorrhizae are rhizosphere endophytes. They extend hyphae into the deep soil, drawing water and minerals back to their host's roots. In return, the host supplies the mycorrhizae with photosynthetically-derived carbohydrates. Mycorrhizae also protect roots from nematodes, insects, and other fungi. The basic science is covered in section 13.2, and the commercial application of mycorrhizal fungi is detailed in section 20.6.

Some *Cannabis* endophytes raise concern because of their potential health hazards to human producers and consumers. These fungal contaminants produce liver-harming mycotoxins, precipitate asthma attacks, and cause opportunistic lung infections. We illustrate a rogues' gallery in Fig. 18.6.

Over 1000 species of fungi are known to infect and kill insects (Shang *et al.*, 2015). *Cordyceps sinensis* is a famous one; it is also esteemed as a valuable remedy in Chinese medicine. Many fungi are marketed for insect biocontrol. Other "friendly" fungi feed on other fungi or trap soil nematodes (see section 20.2.3).

Fungal taxonomy and nomenclature

Taxonomy gets complicated here. Even Linnaeus, a genius taxonomist, found the fungi frustrating. He placed some in *Chaos*, a genus he created for animals. Other fungi he placed in *Tremella*, a genus of algae. Out of 5247 plant species in *Species Plantarum* (Linnaeus, 1753), only 89 are fungi (Farlow, 1910).

Fruiting bodies are multicellular reproductive structures that give rise to fungal spores (Fig. 3.4). Some fruiting bodies give rise to *asexual* spores: A **sporangium** (sp) is a spherical structure atop a long stalk. A **pycnidium** (py) is spherical or pear-shaped, and embedded within a leaf or branch. Its internal cavity is lined with **conidiophores**, which are special conidium-bearing hyphae. Mature conidia discharge from the pycnidium through an opening at the top. An **acervulus** (ac) is similar, but more lens-shaped, and its whole roof comes off, releasing conidia.

Some fruiting bodies give rise to *sexual* spores: A **perithecium** (pe) is shaped like a pycnidium, but gives rise to asci (singular ascus)—small sacs where meiosis takes place, giving rise to 2–8 ascospores. An **apothecium** (ap) is a cup-shaped structure atop a small stalk, capped by a layer of asci. Powdery

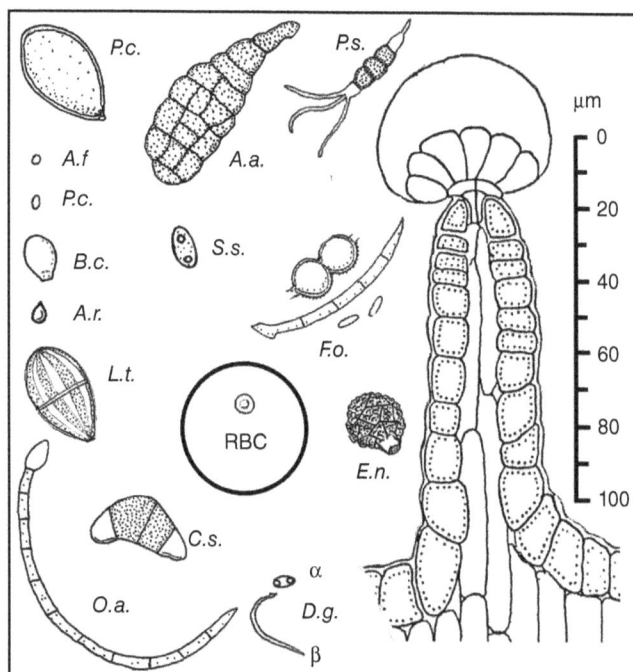

Fig. 3.3. Spores drawn to scale with a glandular trichome

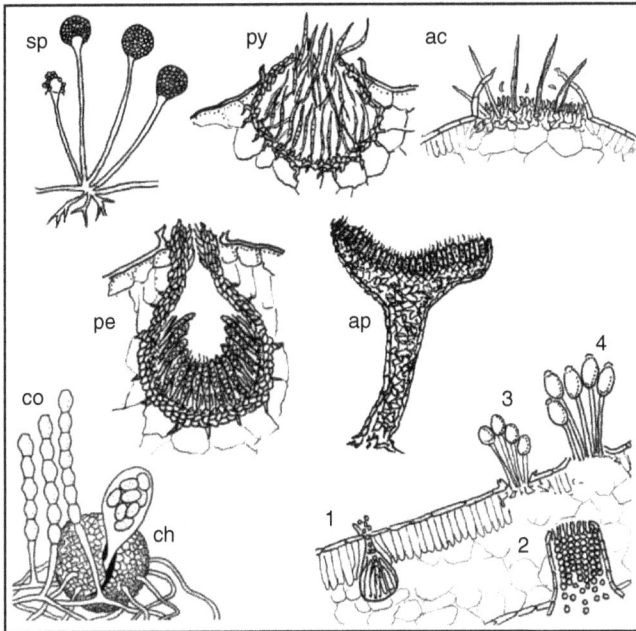

Fig. 3.4. Some fungal fruiting bodies

mildew fungi give rise to asexual **conidiophores** (co) with chains of conidia, as well as a sexual-stage **chasmothecium** (ch), a spherical structure filled with asci and ascospores.

Rust fungi (Order Pucciniales) produce several fruiting bodies in the course of a year (Fig. 3.4). (1) A flask-shaped **spermagonium** produces spermatia and "receptive hyphae" (both haploid). When spermatia from elsewhere fertilize a receptive hypha, they fuse into a dikaryon (binucleate, with two haploid nuclei per cell). (2) A cup-like **aecium** arises from dikaryotic mycelia, bearing chains of binucleate aeciospores. Aeciospores are dispersed by wind and infect another host. (3) **Uredinia** arise in pustules on the new host. They burst through the epidermis and form binucleate, rust-colored urediniospores. (4) **Telia** take the place of uredinia near the end of the growing season. Binucleate teliospores overwinter, undergo meiosis, and in the spring give rise to **basidia**. Basidia give rise to haploid basidiospores, which start the life cycle anew.

Many fungi also produce **chlamydospores** or **sclerotia**, which are haploid hyphae with hard, thickened walls. They survive conditions causing the rest of the fungus to die. After bad conditions pass, these survival units regenerate hyphae.

The sexual reproductive structure is called the **teleomorph**, and the asexual reproductive structure is called the **anamorph**. Many fungi are known only by their anamorph stage. The teleomorph may be rare, or the species has lost its ability to sexually reproduce. Two Frenchmen called these asexual organisms "imperfect fungi," and the term stuck.

In many species of fungi, the teleomorph and the anamorph were discovered by different scientists. Because these reproductive structures looked different, and did not seem related to each other, the scientists gave them different names. This was permissible until 2012. For example, the anamorph of the gray mold fungus is very common. *Botrytis cinerea* was named in 1801, at the dawn of mycological taxonomy.

Its teleomorph is relatively rare. It was originally named *Peziza fuckeliana* in 1866, and was not persuasively connected with *B. cinerea* until nearly a century later, after painstaking research (Groves and Loveland, 1953).

Thanks to DNA sequencing and molecular systematics, an anamorph with no known sexual stage can be linked to a genus whose teleomorph is known. For example, McPartland (1983b) coined *Phomopsis ganjae* for an anamorph with no known teleomorph. Gomes *et al.* (2013) sequenced genes from *P. ganjae*, and it formed a clade with fungi whose sexual states were *Diaporthe* spp. On this basis, they renamed the fungus *Diaporthe ganjae*. Given breakthroughs in molecular systematics, the International Botanical Congress adopted the principle of "one fungus, one name" (McNeill, 2012). The International Botanical Congress ruled that the "one name" would be decided by priority of publication—whichever genus name came first, regarding the anamorph versus teleomorph (Turland, 2018).

Wholesale renaming of fungi by molecular taxonomists has thrown nomenclature into turmoil. "One fungus, one name" turned into "one fungus, which name?" Hawksworth (2012) estimated that 10,000–12,000 fungi would change names. He cautioned against uncritical name changes, because "proof of holomorphy" (uniting anamorph and teleomorph names) is problematic with polyphyletic and paraphyletic genera. **Polyphyletic** organisms derive from more than one common evolutionary ancestor, and are not suitable for grouping together. A **paraphyletic** group descended from a common ancestor, but that group excludes some of the descendants.

In animal taxonomy, the author of a name change goes uncited. For instance, Linnaeus assigned the name *Sphinx atropos* L. to the death's head moth (see Fig. 7.2). Fifty years later, Jakob Laspeyres changed the name, and it is cited as *Acherontia atropos* (L.). In plant and fungal taxonomy, the author of a name change gets cited. For example, *Phomopsis ganjae* McPartland is now *Diaporthe ganjae* (McPart.) Gomes, Glienke & Crous.

The mycologist Curtis Gates Lloyd wrote: "In my opinion, the prevailing custom of attaching the names of writers to the names of plants serves no purpose unless it be to gratify the vanity of authors" (Lloyd, 1905). He critiqued names changes in articles entitled "Advertising pages" and "Juggled names." Molecular taxonomists have gone whole-hog into "advertising." Pedro Crous has advertised his name far more than any other mycologist. He conducted molecular research at breakneck speed, some of it sloppy, and he is no longer Director of the Westerdijk Fungal Biodiversity Institute.

DNA research has also thrown taxonomy into flux at higher taxonomic levels (Phylum, Class, Order). Indeed, one entire Phylum, the Deuteromycota (anamorphic fungi), is no longer formally recognized. However, it serves as a useful place to categorized anamorphic fungi that haven't yet hit the Advertising pages. The higher taxonomic levels recognized in this book are presented in Table 3.3, which we limited to *Cannabis* pathogens and biocontrol fungi. Thus Table 3.3 omits many otherwise-important fungi, such as the Order Agaricales (gilled mushrooms), as well as an entire phylum, the Chytridiomycota (chytrids).

Table 3.3. Taxonomy of some fungi associated with *Cannabis*

Phylum or Subphylum	Description
Mucoromycotina	Hyphae rarely have septa; teleomorph produces zygospores.
	Order Mucorales: produce a profuse mycelium, much of which is immersed in the host; they reproduce asexually by sporangia. They are saprophytic (as storage molds) or weakly pathogenic on *Cannabis*: *Mucor* and *Rhizopus*
Entomophthoromycota	Parasitic on insects, biocontrol agents against pests: *Erynia, Conidiobolus, Entomophthora*
Glomeromycota	Soil fungi forming mycorrhizal associations with plants. *Cannabis* symbionts: *Glomus*
Basidiomycota	Hyphae septate with clamp connections; teleomorph produces basidiospores.
	Class Pucciniomycetes: simple septa present; basidiospores borne on promycelia and teliospores
	Order Pucciniales: obligate parasites with complicated life cycles; the Rust fungi. *Cannabis* pathogens: *Aecidium, Uromyces, Uredo*
	Class Agaricomycetes: dolipore septa present; basidia and basidiospores borne on a hymenium (fertile layer lining a fruiting body)
	Order Atheliales (Amylocorticiales). *Cannabis* pathogens: *Athelia, Thanatephorus*
	Order Agaricales: the gilled mushrooms
Ascomycota	Hyphae septate; teleomorph produces ascospores.
	Subphylum Saccharomycotina (reproduce by budding and rarely produce ascocarps)
	Saccharomycetales: brewer's yeast *Saccharomyces cerevisiae*; human pathogen *Candida albicans*
	Subphylum Pezizomycotina (reproduce by fission)
	Order Dothideales: pseudothecia are uni- or multiloculate, lacking a hamathecium. *Cannabis* pathogens: *Leptosphaerulina, Schiffnerula, Botryosphaeria* (anamorphs *Diplodia, Dothiorella, Fusicoccum, Lasiodiplodia, Leptodothiorella, Macrophomina, Phyllosticta, Septoria*)
	Order Pleosporales, many endophytes. *Cannabis* pathogens: *Pleospora, Leptosphaeria, Ophiobolus, Didymella* (anamorphs *Phoma, Macrophoma, Ascocyta, Alternaria*)
	Order Capnodiales: mostly sooty molds, plus *Cannabis* pathogen *Mysospharella* (anamorphs *Cercospora, Pseudocercospora*)
	Order Eurotiales: storage fungi; *Talaromycetes, Eupenicillium, Emericella* (anamorphs *Aspergillus, Penicillium*)
	Class Leiotiomycetes:
	Order Erysiphales, powdery mildews. *Cannabis* pathogens: *Leveillula, Golovinomyces, Sphaerotheca*
	Order Helotiales, inoperculate discomycetes. *Cannabis* pathogen: Sclerotinia sclerotiorum
	Class Sordariomycetes
	Order Sordariales, mostly saprophytic; *Sordaria, Neurospora, Chaetomium, Thielavia*
	Order Glomerellales, darkly pigmented perithecia. *Cannabis* pathogen: *Glomerella* (anamorph *Colletotrichum*)
	Order Hypocreales, lightly pigmented perithecia. *Cannabis* pathogens: *Gibberella, Nectria* (anamorphs *Fusarium*); *Calonectria* (anamorph *Cylindrocladium*); *Hypocrea* (anamorphs *Verticillium, Acremonium*); biocontrol fungi (anamorphs *Trichoderma, Gliocladium*); *Myrothecium, Stachybotrys*
	Order Diaporthales, perithecia in stromatic tissue. *Cannabis* pathogens: *Diaporthe* (anamorph *Phomopsis*)
Deuteromycota	Teleomorph lacking; group no longer recognized, but useful for field identification.
	Hyphomycetes: mycelium bears conidia directly on special hyphae (conidiophores), conidiophores free or bound in tufts (coremia) or cushion-like masses (sporodochia). *Cannabis* pathogens: *Alternaria, Aspergillus, Botrytis, Cercospora, Cephalosporium, Cladosporium, Curvularia, Cylindrosporium, Epicoccum, Fusarium, Myrothecium, Penicillium, Periconia, Phymatotrichopsis, Pithomyces, Pseudocercospora, Ramularia, Sarcinella, Stemphylium, Thyrospora, Torula, Trichothecium, Ulocladium, Verticillium*
	Coelomycetes: conidia borne on conidiophores enclosed in pycnidia or acervuli. *Cannabis* pathogens: *Ascochyta, Botryodiplodia, Colletotrichum, Coniothyrium, Diplodina, Macrophomina, Microdiplodia, Phoma, Phomopsis, Phyllosticta, Rhabdospora, Septoria, Sphaeropsis*
	Agonomycetes: "mycelia sterilia," mycelium with no reproductive structures. *Cannabis* pathogens: *Rhizoctonia* species

Cannabis extracts are antifungal, at least in experimental *in vitro* studies. These studies have utilized aqueous or solvent extracts, which contain different compounds (see text below Table 3.1). Solvent extracts contain EOs and cannabinoids, whereas aqueous extracts do not.

- "Organic solvent" extract, flowering tops of 'Bologniensis', no effect on yeasts (*Saccharomyces, Rodotorula, Hansenula* spp.) or filamentous fungi (*Aspergillus, Penicillium* spp.), Ferenczy *et al.*, 1958
- Dried flowers and leaves mixed with *Pinus* seeds before planting (1:10 v/v) reduced *Fusarium* damping off in seedlings, 1.5% losses compared to 87% in control, Vysots'kyi, 1962
- Aqueous extract, "cansatine 4," inhibited *Candida*, Zelepukha *et al.*,1963
- Ethanol extract, *Cannabis* cell culture, did not inhibit *Candida*, Veliky and Genest 1972
- Aqueous or ethanol extracts, leaves, no effect on *Trichophyton rubrum* or *Aspergillus niger* (Gupta and Banerjee 1972).
- Ethanol extract, leaves, killed *Ustilago tritici* and *Ustilago hordei*, Misra and Dixit 1979
- Aqueous extract, leaves, inhibited *Aspergillus, Penicillium, Cladosporium, Drechslera, Fusarium, Cephalosporium, Rhizopus, Mucor* and *Curvularia* spp., Pandey 1982
- Ethanol extract, leaves, inhibited spore germination of *Neovassia indica*, Gupta and Singh 1983
- Ethanol extracts, leaves, inhibited spore germination of *Ustilago maydis* and *U. nuda*, Singh and Pathak 1984
- Aqueous extract, leaves, inhibited *Colletotrichum truncatum*, but not *Septoria glycines* or *Phoma exigua*, Kaushal and Paul 1989
- Dried leaves mixed into wheat straw compost suppressed *Fusarium solani*, Grewal 1989
- Aqueous extract, leaves, inhibited *Curvularia lunata* (greater inhibition with an ethanol extract), Upadhyaya and Gupta 1990
- Aqueous, ethanol, acetone, or chloroform extracts of leaf material, all inhibited *Aspergillus niger* and *Fusarium* sp., but no difference between the four extracts, Anjum *et al.* 2018
- Acetone extract of inflorescences, inhibited *Aspergillus flavus* in a dose-dependent manner, with an IC_{50} of 0.225 mg dry matter/mL. Extracts of leaf or stem material did not significantly reduce *A. flavus* growth, Al Khoury *et al.* 2021

Cannabis EOs (terpenoids) have been tested by six groups. Satyal and Setzer (2014) used hydrodistillation to extract EOs from wild plants in Nepal. They summarized antifungal activity as "essentially no effect": *Aspergillus niger* MIC = 625 μg/mL, *Candida albicans* = 1250 μg/mL. Kędzia *et al.* (2016) used steam distillation to extract EOs from fiber-type 'Beniko', and obtained similar results (reported in mg/mL rather than μg/mL): "relatively sensitive" yeasts (*Candida utilis, C. parapsilosis, C. lusitaniae, Geotrichum candidum, Rhodotorula mucilaginosa*) had MIC ≤1.5–12.5 mg/mL. "Less sensitive" yeasts (*Candida albicans, C. krusei*) had MIC ≥50.0 mg/mL. Zengin *et al.* (2018) tested the human pathogens *C. albicans, C. glabrata, C. krusei, Malassizia furfur*, and *M. sympodialis*, and found little inhibition (MIC ≥12.5 mg/mL).

Siwulski *et al.* (2012) tested three EOs on growth of the edible mushroom *Agaricus bisporus*. Hemp EO did not eliminate growth until a concentration of 0.55% was reached (less potent than sage EO or tea tree EO). Nissen *et al.* (2010) compared EOs from 'Carmagnola', 'Fibranova', and 'Futura' on the growth of six yeasts. None of them inhibited *Saccharomyces cerevisiae*, and only 'Futura' inhibited *Torulospora delbrueckii* and *Zygosaccharomyces bailii* at MIC values below 2.00% v/v. 'Carmagnola' inhibited *Kluyveromyces marxianus* and *Pichia membranaefaciens* more than 'Futura', while the opposite was true in *Schizosaccharomyces japonicus*.

Wanas *et al.* (2016) obtained a hexane-extracted EO from a "high-potency" strain that showed modest activity (IC_{50} 33.15 μg/mL) against *Cryptococcus neoformans*. However, a subfraction (β-caryophyllene, α–humulene, and caryophyllene oxide) was relatively potent: 3.85 μg/mL.

Pure cannabinoids are antifungal. Dahiya and Jain (1977) assayed THC and CBD against *in vitro* growth of 18 fungi. Generally, THC inhibited human pathogens more than CBD (e.g., *Microsporium* and *Trichophyton* spp.). Conversely, CBD inhibited plant pathogens more than THC (e.g., *Alternaria alternata, Curvularia lunata, Fusarium solani, Trichothecium roseum*). Two fungi were completely resistant to THC and CBD, *Aspergillus niger* and *Penicillium chrysogenum*. Interestingly, these two species are frequently isolated from moldy marijuana (McPartland and Pruitt, 1997).

McPartland (1984) extracted flowering tops with petroleum ether, then used thin-layer chromatography (TLC) to separate the extract's components. Eluted TLC plates were sprayed with *Phomopsis ganjae* spores suspended in a nutrient solution. Spore germination was inhibited at three positions on the plate (Rf 0, Rf 0.20, Rf 0.68). The latter two Rf values corresponded to CBD and THC, respectively.

Turner and ElSohly (1981) tested cannabichromene (CBC), and reported mild to moderate antifungal activity. MIC values were 25 μg/mL for both *Saccharomyces cerevisiae* and *Trichophyton mentagrophytes*; *Candida albicans* and *Aspergillus niger* were resistant. ElSohly *et al.* (1982) tested cannabigerol (CBG) alongside CBD and THC in the agar diffusion screen, quantified in width of the inhibition zone (in mm). For *T. metagrophytes*, CBG = 5, CBD = 4, THC = 3. For *S. cerevisiae*, CBG = 4, CBD = 2, THC = 1. For *C. albicans*, CBG = 3, CBD and THC = 2. *A. niger* was resistant.

3.3.5 Kingdom Plantae

Vascular plants—with xylem and phloem—comprise five classes: psilopsids, club mosses, horsetails, ferns, and seed plants. Seed plants divide into Gymnospermae (cyads, ginkgos, conifers) and Angiospermae (monocots and dicots). Among the Angiosperms we find *Cannabis* and some of its parasites and weed competitors.

Parasitic plants have little or no chlorophyll, and leach off other plants for food (and sometimes water). Parasitic plants send haustoria (modified roots) into the stems or roots of their host plants. More than 2500 species of parasitic plants are known around the world. Less than a dozen species leach off of *Cannabis*. Collectively they are known as dodder and broomrape (see section 16.1). To add injury to insult, parasitic plants may spread viruses to their hosts.

Weeds compete with *Cannabis* for sunlight, water, and nutrients. "Weed" is a human construct, "a plant out of place." *Cannabis* becomes a weed in steppe landscapes where humans are trying to grow other plants. Weeds have encroached upon crop fields since the dawn of agriculture. In 2012, weed-killing herbicides accounted for 58% of the pesticide market in the USA, compared with 25% for insecticides and 16% for fungicides. Worldwide, those percentages were 44%, 29%, and 26%, respectively (Atwood and Paisley-Jones, 2017).

Hemp, when densely sown, is considered a "smother crop." It competes successfully against many weeds for growth space. This was noted by early writers (e.g., Teodosi, 1553; Tusser, 1557; Markham, 1615; Rozier, 1787; see more in section 16.7). These authors credited the shade imparted by hemp's dense canopy to "choke out" weed growth.

Lotz *et al.* (1991) investigated this assertion. They monitored the nasty weed yellow nutsedge (*Cyperus esculentus*) growing under crops of hemp, maize (*Zea mays*), and barley (*Hordeum vulgare*). Nutsedge produced a mean of 0.2 tubers/plant under hemp, 25.1 under maize, 46.2 under barley, and 171.2 in fallow conditions. Yet hemp and maize provided the same amount of shade: only 5% of photosynthetically active light (PAR) above the canopies reached the nutsedge. Barley let in 30% of light.

Cannabis may impede other plants through **allelopathy**: the chemical inhibition of one plant by another, through the release of substances that inhibit seed germination or growth. The allelopathic firepower of *Cannabis* is a hotly debated topic (see section 16.7), and appraised in nearly 20 studies (Table 3.4).

Many of these studies used an overly simplistic bioassay: place weed seeds on filter paper, soak the filter paper with an

Table 3.4. Allelopathic effects of *Cannabis* against other plants

Source of inhibition	Results with test plants	Reference
Weeds irrigated with water leached from pots in which *Cannabis* grew	Inhibited growth of *Stellaria media,* whereas *Centaurea cyanus* and *Sinapsis arvensis* not affected	Stupnicka-Rodzynkiewicz 1970
Seeds exposed to aqueous extracts obtain from aerial parts, germination tested	Decreased seed germination of *Matricaria recutita* and *Lepidium sativum*	Stupnicka-Rodzynkiewicz 1970
Tubers soaked in aqueous extracts	Inhibited tuber germination of *Cyperus rotundus*	Srivastava and Das 1974
Six assays, including aqueous extracts, residual toxicity in soil, rainwater collected off plants	Equivocal results with seeds or seedlings of five crop species	Inam *et al.* 1989
Dry leaf leachates or fresh leaf leachates of female plants	Inhibited germination of *Parthenium hysterophorus,* dry > fresh	Singh and Thapar 2003
Seeds placed on filter paper soaked in aqueous extracts of shoots, stems, or roots	Inhibited germination of *Pisum sativum* and *Triticum aestivum;* shoots > roots > stems	Ulmer *et al.* 2010
Hydroalcoholic extract of flowering tops	Inhibited seed germination of *Avena fatua, Chenopodium album,* and *Amaranthus retroflexus*	Makkizadeh *et al.* 2011
Seeds placed on filter paper soaked in aqueous extracts from fresh flowers of fiber hemp (Białobrzeskie)	Inhibited monocots (*Triticum aestivum, Secale cereale*) more than dicots (*Brassica napus, Lupinus luteum*)	Pudełko *et al.* 2014
Seeds placed in agar containing dried, powdered leaves	Inhibited germination and growth of *Lactuca sativa*	Akhtar *et al.* 2013
Seeds placed on filter paper soaked in aqueous extracts of shoots or roots	Inhibited germination of *Pisum sativum* and *Triticum aestivum;* shoots = roots	Sharma and Devkota 2014
Seeds placed on filter paper soaked in aqueous extracts of shoots or roots	Shoot extracts inhibited germination of *Lactuca sativa;* root extracts had a stimulatory effect	Mahmoodzadeh *et al.* 2015
Plants grown in soil where hemp or sugar beets grew the previous year	No difference in germination of *Lupinus luteum, Pisum sativum,* or *Vicia faba;* but reduction in stem length of *P. sativum* and *V. faba* when previous crop was hemp	Olsson 2015
Seeds placed on filter paper soaked in essential oil	Inhibited germination of *Amaranthus retroflexus* and *Bromus secalinus* > *Zea mays* and *Echinochloa crus-galli* > *Avena sativa* and *Avena fatua*	Agnieszka *et al.* 2016
Seeds placed on filter paper soaked in essential oil	Inhibited growth of hypocotyl and epicotyl of *Allium cepa, Solanum lycopersicum* and *Daucus carota*	Konstantinović *et al.* 2018
Seeds placed on filter paper soaked in aqueous extracts of shoots or roots; seedlings sprayed with aqueous extracts	Inhibited germination of *Cicer arietinum,* shoots > roots; sprayed seedlings had reduced chlorophyll, NR activity and total protein content	Aarti 2018
Weed seedlings sprayed with aqueous leaf extracts (3% w/v)	Reduced the growth of eight weed species, but not as effectively as leaf extracts from *Moringa oleifera* or *Parthenium hysterophorus*	Gurmani *et al.* 2021

aqueous *Cannabis* extract, and see if the seeds germinate. A few studies used methods approaching naturalistic conditions: they placed weed seeds in soil irrigated with water coming off *Cannabis* plants, or placed weed seeds in soil collected from a hemp field. Naturalistic studies produced equivocal results, compared with simplistic bioassays that produced positive results.

A few studies merit elaboration: Inam *et al.* (1989) tested *Cannabis* allelopathy in simplistic and naturalistic assays. Their simplistic assay reduced seed germination in all test plants: *Trifolium resupinatum* (TR), *Trigonella foenum-graceceum* (TFG), *Brassica campestris* (BC), *Vigna mungo* (VM), and *Sorghum bicolor* (SB). They tested four naturalistic assays. (1) Watering seedlings in soil with *Cannabis* leaf extracts reduced the growth of TFG, BC, and SB, but not TR and VM. (2) Planting seeds in soil collected from a hemp field reduced radicle growth in only BC and VM. (3) Watering seeds in soil with rainwater collected off *Cannabis* leaves actually stimulated seed germination and/or radicle growth in all species except SB. (4) Watering seeds in soil with rainwater collected off *Cannabis*—and fourfold concentrated–reduced seed germination and/or radicle growth in all species except VM.

Pudełko *et al.* (2014) measured transcription levels of genes associated with isoflavonoid synthesis in yellow lupine (*Lupinus luteus*). These genes become upregulated in plants exposed to environmental stress. Exposing lupine seedlings to *Cannabis* aqueous extracts upregulated two of three genes they analyzed: chalcone synthase and isoflavone synthase.

Haney and Bazzaz (1970) speculated that terpenoids excreted by *Cannabis* may suppress the growth of surrounding vegetation. Three studies in Table 3.4 utilized extracts that contained terpenoids (hydroalcoholic extracts or essential oils), and they inhibited the germination and growth of test seeds.

Perhaps THC and CBD evolved as allelochemicals. Two correlative studies showed that wild-type *Cannabis* surrounded by competing weeds produced more THC than plants lacking competitive pressure (Haney and Kutscheid, 1973; Latta and Eaton, 1975). Correlation is not the same as causality. Causality could be inferred by the studies in Table 3.4, but most of them utilized aqueous extracts, which did not contain much THC or CBD (Table 3.1).

Two *in vitro* assays showed that THC and CBD (or THCA and CBDA) induced apoptosis and cellular necrosis in every plant tested—rice, soybean, *Arabidopsis thaliana*, and *Scutellaria baicalensis* (Morimoto *et al.*, 2007; Shoyama *et al.*, 2008). THCA and CBCA even cause necrosis in *Cannabis* cells, which might explain why the plant sequesters cannabinoids in non-cellular gland heads. Morimoto *et al.* (2007) referred to gland heads as "death capsules."

Small (2015b) argued against cannabinoids evolving as allelochemicals. Prior to flowering, *Cannabis* produces few glandular trichomes. Therefore THC would not provide allelopathic effects until late in the season, when competing plants have fully grown. Allelochemicals usually target the *seeds* or *seedlings* of competing plants.

Inam *et al.* (1989) implicated water-soluble compounds in *Cannabis* allelopathy: caffeic acid, ferulic acid, benzoic acid, and p-OH-benzoic acid. These compounds have been shown to produce allelopathic effects in other studies, so Inam and colleagues looked for them in *Cannabis*, and found them. Gurmani *et al.*

(2021) did not cite Inam *et al.* (1989), but extracted some of the same water-soluble compounds in their *Cannabis* allelopathy study: 4-OH-benzoic acid, p-OH-benzoic acid, 4-OH-phenyl acetic acid, p-coumaric acid, and β-resorcylic acid.

3.3.6 Kingdom Animalia

Dividing the animal kingdom into vertebrates and invertebrates is history. Based on embryological characters and DNA phylogenetics, animals are now split into Protostomia and Deuterostomia. **Protostomia** divide into **Ecdysozoa** (nematodes and arthropods) and **Lophotrochozoa** (snails, slugs, mussels, squids). Ecdysozoans are invertebrates that grow through ecdysis (molting their exoskeleton). Lophotrochozoans have lophophores (ciliated tentacles surrounding the mouth) and do not molt. The two groups are monophyletic, based on 18S ribosomal DNA evidence (Halanych *et al.*, 1995; Aguinaldo *et al.*, 1997).

Deuterostomia are characterized embryologically by the anus forming before the mouth, whereas the opposite is seen in Protostomia. The group may be paraphyletic (Kapli *et al.*, 2021). Deuterostomia divide into **Ambulacraria** (starfish, acorn worms) and **Chordata** (basal chordates and vertebrates). Ambulacraria and basal chordates (sea squirts and lancelets) do not have bones or vertebrae, but they profoundly differ from other invertebrates.

Herbivores (animals that feed on plants) are nitrogen-challenged. Animals are mostly protein, whereas plants consist mainly of carbohydrates. The difference is nitrogen. Animals consist of 7–14% nitrogen by dry weight. Plants rarely contain more than 7% nitrogen, except for *seeds*. Thus seeds become very attractive to nitrogen-starved herbivores. *Cannabis* seeds develop in bracts covered with trichomes (Fig. 3.1), and the trichomes contain THC and CBD, so it seems likely that THC and CBD evolved to deter herbivores.

Deter herbivores by what molecular mechanism? THC activates **cannabinoid receptors** (CBRs) in animals. CBRs evolved as receptors for *endogenous* cannabinoids produced by animals, known as **endocannabinoids** (eCBs), such as anandamide (AEA) and 2-arachidonylglycerol (2-AG). By activating CBRs, THC is a mimic of AEA and 2-AG, and elicits many kinds of behavioral responses—often repellency. Other plants with repellent properties work similarly: nicotine deters herbivores by activating their nicotinic acetylcholine receptors. Capsaicin works through capsaicin receptors.

Only some animals have CBRs, however. The dividing line between the haves and the have-nots occurred when Deuterostomia divided into Chordata (haves) and Ambulacraria (have-nots). McPartland and Glass (2003) proposed this hypothesis after they evaluated an array of animals for the presence of CBRs. The evolution of Chordata, hence the evolution of CBRs, occurred during the Cambrian explosion, 538.8 million years ago. So CBRs existed long before *Cannabis* evolved, about 27.8 mya (McPartland and Guy, 2010).

A hotly contested debate between haves and have-nots revolves around insects and nematodes. The presence or absence of CBRs in an animal can be appraised two ways. (1) The genome of an animal can be searched for orthologs (homologous genes in different species) that share identities with human CB_1 or CB_2. (2) The neuronal tissue of an animal can be tested for its ability to bind with [³H]CP55,940, a super-potent analog of THC.

Abbott (1990) reported "preliminary evidence" of a gene for CB_1 in the fruit fly *Drosophila melanogaster*. Howlett *et al.* (1992) tested [³H]CP55,940 in *D. melanogaster*, and "a low concentration of binding sites were discernible." Howlett *et al.* (2000) again detected [³H]CP55,940 binding sites in *D. melanogaster*, although the binding was not displaced by CB_1-specific SR141716A or CB_2-specific SR144528.

McPartland *et al.* (2001) found no high-affinity binding of [³H]CP55,940 in a panel of insects: *D. melanogaster*, *Apis mellifera*, *Gerris marginatus*, *Spodoptera frugiperda*, and *Zophobas atratus*. These negative binding results were confirmed by Elphick and Egertová (2005). The genome of *D. melanogaster* lacks orthologs of human CB_1 or CB_2 (McPartland *et al.*, 2006a).

The nematode *Caenorhabditis elegans* had its entire genome sequenced, and McPartland and Glass (2001) searched for orthologs of human CB_1 or CB_2. The gene with closest similarity, C02H7.2 (which encodes a receptor called **NPR-19**), shared only 23% sequence identity with human CB_1. Furthermore, its sequence revealed crippling substitutions at critical amino acid residues involved in the binding of THC to receptors. These results have been replicated (McPartland and Glass, 2003; Estrada-Valencia *et al.*, 2023).

NPR-19 has been revisited by others, who proposed that the receptor can bind AEA or 2-AG, and these endocannabinoids modulate nematode behavior (Oakes, 2018, Pastuhov *et al.*, 2016; Levichev *et al.*, 2023). This does not mean that NPR-19 has binding affinity for THC. The receptor binding pocket for THC differs subtly from that for AEA and 2-AG. Receptor binding studies using [³H]CP55,940 have not found high-affinity binding sites in nematodes (Elphick and Egertová, 2001), or binding sites with a relatively low percentage of specific binding (58%, McPartland *et al.*, 2006a; 30–35%, Guha *et al.*, 2020).

3.4   Phyla

3.4.1   Phylum Nematoda

Nematodes are extremely abundant—a fistful of soil may contain thousands of them. They occupy every earthly niche from mountaintop to sea bottom. Crop losses from nematodes tend to be underestimated because of their small size, and because their damage goes mostly unseen—in roots, underground. A few nematodes move above ground, including the first-discovered nematode. In 1743 Turberville Needham extracted a white fibrous material from diseased wheat seeds. To his amazement, the fibrous material began squirming when soaked in water. The "fibers" were matted larvae of *Anguina tritici*, the wheat gall nematode.

Most nematodes are dioecious, with males and females. Males are usually smaller than females. Males of one species, *Trichosomoides crassicauda*, are so small that they live in the female's uterus. Some nematodes are **hermaphroditic** (females with additional male gonads). Others eliminate males altogether, and reproduce **parthogenetically**.

The physical characteristics of typical plant-pathogenic nematodes are illustrated in Fig. 3.5. Nematodes are not related to earthworms. Built on a far simpler scale, nematodes have no respiratory or circulatory systems. Their simple nervous system can be described at the level of individual cells: *Caenorhabditis elegans*, for instance, has exactly 302 neurons. A complete wiring diagram of its nervous system has been compiled. Nematodes have few muscle cells, and thrash aimlessly. Unassisted, nematodes move barely 1 ft from where they hatched.

At least 20 species of nematodes attack *Cannabis*. Most are **polyphagous**—they feed on many plant species—but some are **oligophagous**, feeding on a narrower spectrum of plant species. None are monophagous on hemp. Nematodes feed with a hypodermic-like **stylet**, which resembles a hollow spear. Nematodes thrust their stylets into plant cells and suck out cell cytoplasm.

Nematodes that feed from outside roots as **ectoparasites** (e.g., *Paralongidorus maximus*) or they enter plants to feed as **endoparasites**. Endoparasites can be **sedentary**, remaining imbedded in roots (e.g., *Meloidogyne*, *Heterodera* spp.) or **migratory**, moving to above-ground plant parts (e.g., *Ditylenchus dipsaci*), or migrate from plant to plant (e.g., *Pratylenchus penetrans*).

Nematode populations are naturally culled by viruses, bacteria, fungi, protozoans, and other nematodes. Some plants secrete metabolites that suppress or kill nematodes. *Cannabis* appears to be one such plant. A few studies have shown that nematode populations are suppressed by hemp crops.

Kir'yanova and Krall (1971) rotated potatoes with a hemp crop to suppress the potato cyst nematode, *Globodera rostochiensis*. Kok *et al.* (1994) rotated hemp to suppress populations of the root knot nematode, *Meloidogyne chitwoodi*. Scheifele *et al.* (1997) assessed nematode populations before and after a hemp crop. Hemp suppressed soybean cyst nematodes (*Heterodera glycines*), but hemp increased populations of spiral nematodes (*Heliocotylenchus* or *Scutellonema* spp.) and root knot nematodes (*Meloidogyne* spp.).

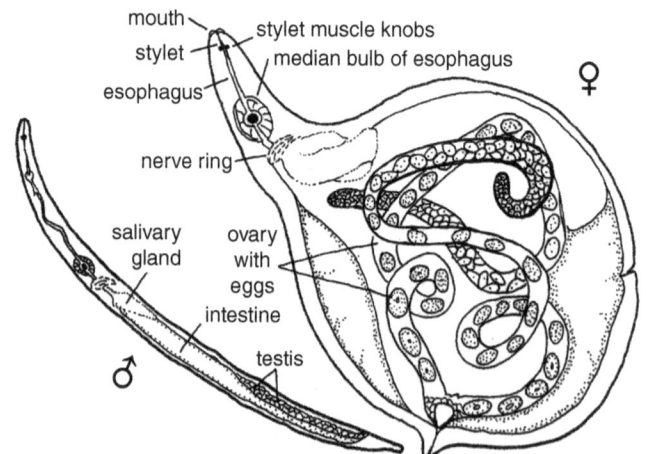

Fig. 3.5. Male and female nematodes showing important parts

Zhang *et al.* (2013) found that soil populations of *Heterodera glycines* declined after a hemp crop. Mateeva (1995) counted the number of "root knots" caused by *Meloidogyne* spp. and the number of *Meloidogyne* larvae (juveniles) in soil surrounding roots. His positive control was marigold (*Tagetes* sp.), long known to suppress *Meloidogyne* spp. (Steiner, 1941):

- cucumber: 56 root knots, 396 larvae
- tomato: 42 root knots, 318 larvae
- hemp: 5 root knots, 21 larvae
- marigold: 1 root knot, no larvae

Dried leaves and flowers of Indian hemp, mixed into potting soil, suppressed *M. incognita* populations (Khanna and Sharma, 1998; Haidar and Askary, 2011; Kayani *et al.*, 2012; Thakur, 2014; Ganaie and Khan, 2017). These results are paradoxical, because *M. incognita* can infest *Cannabis* (see section 15.7). Other nematode species have also been suppressed (Somvanshi and Gupta, 2003; Thakur, 2014). This has also been seen when *whole plants* were chopped into soil (Goswami and Vijayalakshmi, 1986).

Aqueous extracts of *Cannabis* leaves and flowers can kill some nematodes, although aqueous extracts contain little or no THC and CBD (Table 3.1). Aqueous extracts caused high mortality rates in *M. incognita* (Vijayalakshmi *et al.*, 1979; Nandal and Bhatti, 1983; Sharma, 1996; Singh and Singh, 2002; Saxena and Gupta, 2004; Adegbite, 2011; Mukhtar *et al.*, 2013; Shah *et al.*, 2018). The same was seen with *Meloidogyne javanica* (Bajpai and Sharma, 1992), and both *M. incognita* and *M. javanica* (Nazar and Nath, 1989).

Other nematode species have also been suppressed with aqueous extracts (Haseeb *et al.*, 1978; Khanna *et al.*, 1988; Grewal, 1989; Mojumder *et al.*, 1989; Pandey and Dwivedi, 2000). In one case, extracts made from *roots* proved effective (Ganaie and Khan, 2017). Aqueous extracts failed to show nematocidal activity against *Strongyloides papillosus* and *Haemonchus contortus* (Boyko and Brygadyrenko, 2019). Against *M. incognita*, a research group tested dried leaves in one study (Kayani *et al.*, 2012) and aqueous extracts in another study (Mukhtar *et al.*, 2013). Unfortunately, they did not comment on the relative efficacies of the two approaches.

Solvent extracts, with cannabinoids, have rarely been tested. A methanol extract of leaves imparted 89% mortality of *M. javanica* (Amini *et al.*, 2011). Laznik *et al.* (2020) tested ethanol extracts (EEs), but they assayed nematode chemotaxis, rather than mortality, in three entomophagic nematodes, *Steinernema carpocapsae*, *S. feltiae*, and *Heterorhabditis bacteriophora*. Overall, EEs were chemoattractants, especially EEs made from flowers, both fiber-type and CBD-type varieties. EEs were weakly repellent against one species, *S. carpocapsae*.

Satyal and Setzer (2014) used hydrodistillation to extract terpenoids and cannabinoids. The extract's LD_{50} against *Caenorhabditis elegans* was 232 µg/mL—relatively weak. In comparison, nicotine exerted an LD_{50} of 46.1 µg/mL (Yu and Potter, 2008).

Van Es-Remers *et al.* (2022) exposed *C. elegans* to THC extracts or CBD extracts. Both extracts elicited a biphasic reaction, depending on dosage: Extracts with high concentrations of THC or CBD (> 1 µg/mL) reduced the lifespan of *C. elegans*. Extracts with low concentrations (< 0.005 µg/mL), however, resulted in significant effects with respect to appetite (pharyngeal pumping activity), body oscillation, motility, and nervous system-related functions.

Van Es-Remers and colleagues correctly questioned whether CBRs were involved. THC and CBD were nearly equal in efficacy, and CBD does not activate CBRs, suggesting that other receptors were involved. As we explicated in text following Fig. 3.5, nematode receptors orthologous to CBRs are activated by eCBs like AEA and 2-AG, but have little or no binding affinity for THC.

Other receptors are possible. CBD is a potent agonist of the vanilloid receptor type one (TRPV1, EC_{50} = 3.35 µm), and CBD also increases AEA levels by inhibiting its breakdown (Bisogno *et al.*, 2001). AEA also activates TRPV1 (Ross, 2023), and *C. elegans* expresses TRPV1 orthologs (Ben Salem *et al.*, 2022)—CBD at TRPV1 might be one mechanism by which *Cannabis* repels nematodes. Shrader *et al.* (2020) found another mechanism: in their hands, CBD (30 µM) caused paralysis in 40% of *C. elegans*. They implicated dopamine D_2-like receptors, because mutant nematodes lacking these receptors were not paralyzed. CBD does indeed bind to vertebrate dopamine D_2 receptors (Seeman, 2016).

Nematodes are underground, so Bernard *et al.* (2022) highlighted *Cannabis* constituents produced in roots that might repel nematodes. They focused on phytosterols (β-sitosterol, camesterol, stigmasterol) and triterpenoids (friedelin and epifriedelanol), and reviewed studies on the nematicidal effects of these compounds. Latter *et al.* (1975) isolated from roots two spermidine alkaloids unique to *Cannabis* (cannabisativine and anhydrocannabisativine). Their nematicidal effects have not been analyzed.

3.4.2 Phylum Arthropoda

Phylum Arthropoda is huge—estimates of the number of species run from 1.2 million to 5–10 million; arthropods account for over 80% of all known living animal species (Ødegaard, 2000). At least six classes of arthropods are hemp herbivores. **Class Insecta** is the largest. The body of an insect is segmented into the head, thorax, and abdomen (Fig. 3.6).

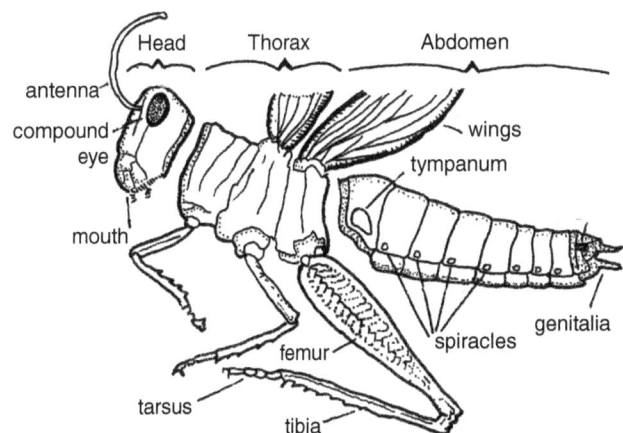

Fig. 3.6. Typical insect showing important parts

Externally, the head may contain one pair of compound eyes, one or more pairs of simple eyes (ocelli), one pair of segmented antennae, breathing tubes (tracheae), and mouthparts. The thorax may sport one or two pairs of wings, and three pairs of legs. Legs are segmented and jointed, with a hip (trochanter), upper leg (femur), lower leg (tibia), and a foot (tarsus). The abdomen may exhibit vestigial legs (prolegs, the fleshy unjointed stubs on caterpillars), tympana (thinned sections of abdomen which serve as "ears"), genitalia (modified as *stingers* in bees), and cerci (caudal appendages serving olfactory or tactile functions).

Internally, the head houses the brain, a blood vessel, and the esophagus. The thorax contains nerve ganglia ("subbrains"), the lower esophagus (including the crop), two pairs of spiracles (breathing apparati), the aorta, and muscles for locomotion. The abdomen contains the heart, digestive organs, more nerve ganglia, eight more pairs of spiracles, excretory organs, and the reproductive system.

In order to grow, an insect larva must periodically shed its external skeleton, while simultaneously building a new, larger exoskeleton. This process is called **molting**. All insects must molt repeatedly during their lifetime. The term **instar** refers to the development stage between each molt. Larvae typically pass through three to seven instars as they develop. The 1st instar emerges from the egg, molts into the 2nd instar, etc. The final stage is the sexually mature adult.

Metamorphosis refers to a relatively abrupt change in body structure. Among insects, there are two general patterns: simple metamorphosis and complete metamorphosis. **Simple metamorphosis** (i.e., incomplete) arises as a gradual difference between instars. These instars are called **nymphs**. Between molts, nymphs gradually alter their body proportions, change colors, add body segments, increase their head width, lengthen antennae, and perhaps add wing pads. Insects undergoing simple metamorphosis have three life stages: egg, nymph, and adult. Adults differ from nymphs by being sexually mature and, if they are winged, having functional wings.

Complete metamorphosis consists of four stages: egg, larva, pupa, and adult. In insects undergoing complete metamorphosis, the larval and adult stages look radically different. For example, caterpillars (larvae of moths and butterflies) may have up to 22 legs. After metamorphosis, the adults have only six legs, plus four wings. In contrast, maggots (larvae of flies) have no legs. After metamorphosis, the adults have six legs, plus two wings.

Complete metamorphosis takes place in a **pupa** (i.e., cocoon, chrysalis). Insects metamorphize from very active, ravenously eating larvae, to complete immobile, non-feeding pupae. Pupation may last weeks, months, or even years, depending on the insect species and temperature. Temperature modulates all stages of insect growth and reproduction, because they are cold-blooded. Regardless of the type of metamorphosis, further development of external structures ceases once insects molt to their ultimate adult form.

Cold temperatures may induce **winter dormancy** or **hibernation** in insects. Generally, only one stage will hibernate (egg, larva, pupa, or adult, depending on the species). Some insects go dormant *before* temperatures become unfavorable. This is called **diapause**, and is frequently triggered by short photoperiods in autumn. Insects may be **univoltine**, producing one generation per year, or **multivoltine**, producing several generations per year.

Insects have a range of mouthparts. **Piercing-sucking insects** have mouthparts that enable sucking of internal fluids. Some are human pests (mosquitoes, ticks, lice, fleas), and others feed on fellow insects (predatory mites, assassin bugs, pirate bugs). Herbivorous insects with piercing-sucking mouthparts include aphids, whiteflies, leafhoppers, and true bugs. They mostly feed on phloem sap, although some feed on xylem sap. Sap-suckers must absorb huge amounts of sap to obtain their required nutrients. They may suck fluids weighing 100–300 times their own body weight per day. Excess water and carbohydrates are excreted as honeydew.

Chewing insects have jaws (mandibles, and chelicera in arachnids) that chew, bite, pinch, burrow, or crush. Carnivorous chewers include human pests (bedbugs, chiggers, scabies mites, bot flies) and insect-preying biocontrol species (lady beetles, parasitoid wasps, lacewing larvae, aphid midges). Some herbivorous chewers, such as caterpillars and maggots, have jaws as larvae, which are lost in adults. Other, such as beetles, have jaws as larvae and adults.

Thrips are **rasping insects**. They have only half a jaw. The single mandible makes a wound in plant epidermal cells. Their maxillae, shaped as stylets, repeatedly thrust and withdraw, and they suck up cytoplasm leaking from the wound. Insects with **sponging mouthparts** (e.g., flies) and siphoning mouthparts (adult moths and butterflies) are not *Cannabis* pests.

Mostafa and Messenger (1972) listed 272 taxa of insects and mites associated with *Cannabis*. In this book we describe fewer organisms, about 200. Many insects and mites listed by Mostafa and Messenger were no doubt "incidental migrants," caught while wandering across hemp plants. However, some incidental migrants may occasionally feed on *Cannabis*, while they breed on neighboring plant species (Cranshaw *et al.*, 2019).

Currently, entomologists recognize about 30 orders of insects. Many are never encountered, because of their small size, scarcity, or restricted environments. A dozen orders known as *Cannabis* pests or biocontrol organisms are listed in Table 3.5.

DNA-based phylogenetic research has upended traditional insect taxonomy, but not to the same degree as fungal taxonomy. Traditionally, Order Hemiptera divided into suborder Heteroptera (true bugs) and suborder Homoptera (aphids, whiteflies, scale insects). A phylogenetic study based on 18S rRNA judged Homoptera paraphyletic (Von Dohlen and Moran, 1995), and split it into Sternorrhyncha (aphids, whiteflies, scale insects), Auchenorrhyncha (leafhoppers, treehoppers, cicadas), and Coleorrhyncha (moss bugs). More recent phylogenetics based on mDNA suggests that Homoptera maybe monophyletic after all (Song *et al.*, 2012).

Cannabis as insect killer

The number of studies investigating *Cannabis* as an insect killer far outnumber those concerning antibacterial, antifungal, allelopathic, or nematocidal effects. McPartland and Sheikh (2018) reviewed 88 publications. That review is available online;

Table 3.5. A synopsis of the insect Orders associated with *Cannabis*

Order	Examples	Characteristics
Orthoptera	Crickets, grasshoppers	Incomplete metamorphosis, 4 wings, chewing mouthparts
Dermaptera	Earwigs	Incomplete metamorphosis, 4 wings, chewing mouthparts
Isoptera	Termites	Incomplete metamorphosis, chewing mouthparts
Thysanoptera	Thrips	Incomplete metamorphosis, 4 wings, rasping-sucking mouthparts
Hemiptera (suborder Heteroptera)	True bugs	Incomplete metamorphosis, "half-wings" (part chitinous, part membranous), piercing-sucking mouthparts arising from front part of the head
Homoptera	Aphids, scales, whiteflies, leafhoppers	Incomplete metamorphosis, wings either chitinous or membranous (but uniform), piercing-sucking mouthparts arising from posterior part of the head
Neuroptera	Lacewings	Complete metamorphosis, 4 wings, chewing mouth parts in larvae and adults (larvae carnivorous, eating other insects)
Lepidoptera	Butterflies and moths	Complete metamorphosis, 4 wings, chewing mouth parts in larvae, siphoning mouthparts in adults; larvae lack compound eyes
Coleoptera	Beetles and weevils	Complete metamorphosis, 4 wings (the front pair hardened into a sheath), chewing mouth parts in larvae and adults; larvae lack compound eyes
Hymenoptera	Bees, wasps, ants, sawflies	Complete metamorphosis, 4 wings, chewing or reduced mouthparts in larvae and chewing-lapping in adults; larvae lack compound eyes
Diptera	Flies	Complete metamorphosis, 2 wings, chewing or reduced mouth parts in larvae and sucking-sponging in adults; larvae lack compound eyes.

here we cite some key findings. The 88 studies were grouped into five types of applications: companion planting ($n = 17$ publications), applying harvested plant material without any extraction ($n = 25$), aqueous extracts ($n = 20$), essential oil extracts (EOs, $n = 9$), and solvent extracts ($n = 17$). The first type of application, companion planting with living *Cannabis*, is a kind of biological control. We discuss that research in section 20.2.4.

Freshly harvested or dried plants have been used to repel pests since ancient times. Pliny the Elder (AD 23–79) said that "juice" squeezed from cannabis seed "drives out of the ears the worms and any other creature that has entered them" (Pliny, 1870). Pliny was cribbed by Culpeper (1652), who stated that "juice" squeezed from fresh hemp leaves dropped into ears "draweth forth earwigs."

Around AD 900, the Iraqi agronomist Ibn Wahshīyah fumigated crops with sulfur mixed with *šahdānaj* (hemp) to drive away locusts (Rogers, 1980). Around AD 950, the anonymous Byzantine author of *Geoponika* wrote, "If you lay a flowering branch of κανναβεως (hemp) near you when you go to sleep, κώνωπες (mosquitos) will not touch you" (Beckh, 1895). A passage in *Wù Lèi Xiāng Gǎn Zhì*, written by Zànníng *ca.* AD 980, recommended burning *má yè* (hemp leaves) to repel mosquitoes (Needham *et al.*, 1986).

This repellent quality was employed as a taxonomic character by the Italian naturalist Ulisse Aldrovandi. He assigned the adult and larval stages of mosquitos to different species, arguing that adult *Culex* was driven away by *Cannabis*, whereas larval *Intestinum aquaticum* thrived in *Cannabis* retting ponds (Aldrovandi, 1602). In Renaissance and early-modern Europe, between 1496 and 1803, a dozen authors recommended using fresh shoots of *Cannabis* to drive away human pests— mosquitoes, bedbugs, fleas, and clothes moths (McPartland and Sheikh, 2018).

Khare *et al.* (1974) used an "olfactameter" to show that powdered leaves repelled *Sitophilus oryzae* from stored grain. Kashyap *et al.* (1991) showed that powdered leaves repelled the potato tuber moth *Phthorimaea operculella*. Anecdotal reports of *Cannabis* leaf material repelling grain weevils date back to 1789, reviewed by McPartland and Sheikh.

Aqueous extracts repel insects, although aqueous extracts lack cannabinoids (Table 3.1). Piemontese (1555) boiled *semenza del canape* (hemp seeds) in seawater, and poured the decoction around the house to get rid of *pulici* (fleas). Buc'hoz (1775) killed underground nests of *courtillières* (mole crickets, *Gryllotalpa* sp.) by flooding them with water suffused with hemp seed oil. Experiments with aqueous extracts began with Mackiewicz (1962), who sprayed potato plants in the laboratory with a hemp extract. It had no repellent effect on ovipositing potato beetles, *Leptinotamus decemlineata*, and did not affect larval development.

Fifteen other studies tested the effects of aqueous extracts, and reported various degrees of repellency, antifeedant effects, oviposition deterrence, and egg and larva mortality (cited in McPartland and Sheikh). Few of these studies described their extraction techniques, whether they used infusions (either cold or heated), or boiled decoctions. No studies chemically analyzed the constituents in their aqueous extracts. Few studies had control arms, but several studies compared *Cannabis* to other plant extracts. Paradoxically, some studies reported efficacy against known *Cannabis* pests (e.g., budworm *Helicoverpa armigera*, cutworm *Spodoptera litura*).

Studies on EO (terpenoid) extracts were judged the most rigorous by McPartland and Sheikh, and the most recent—all were published in the 21st century. Most authors described their extraction technique (i.e., steam distillation or hydrodistillation). Few reported the content of their EOs, but among those who did, minor amounts of CBD or THC were present along with terpenoids. No studies had control arms, although several compared *Cannabis* with EOs from other plants, and in one case, a synthetic pesticide. Five studies showed efficacy against mosquito larvae. Three studies showed efficacy against *Cannabis* pests (*Spodoptera litura, Spodoptera littoralis,* and *Myzus persicae*). Benelli *et al.* (2018a) made a comparison between leaf EOs and flower EOs; the latter was more potent against three pest species.

Solvent extract studies were numerous (*n* = 17) but variable: extracts were prepared from ethanol, methanol, petroleum ether, or carbon tetrachloride. Although few studies chemically analyzed the contents of their extracts, it is safe to assume they contained cannabinoids, because of the solvents they used.

The first solvent extract study (Metzger and Grant, 1932) utilized a pharmaceutical USP ethanol extract, diluted to 1/64. It repelled adults of Japanese beetle, *Popillia japonica*. They tested extracts from 390 plant species, and only 56 showed any repellency. Five solvent extract studies showed efficacy against human pests (mosquitoes, ticks), and three showed efficacy against *Cannabis* pests (*Tetranychus urticae, Spodoptera litura, Spodoptera frugiperda*); see summaries in McPartland and Sheikh.

Two studies compared solvent extracts to aqueous extracts, and showed surprisingly little difference in efficacy. Rothschild and Fairbairn (1980) tested oviposition deterrence in the cabbage moth, *Pieris brassicae*, by spraying cabbage leaves with four types of extracts. Aqueous extracts containing terpenoids (cold extracts) were more effective than aqueous extracts lacking terpenoids (heated extracts, terpenoids boiled off). Ethanol extracts of purified THC or CBD were even less effective. THC exerted nominal deterrence, but CBD was actually an oviposition *attractant*.

Frischknecht and Waser (1980) added THC to sugar water (1 mg/mL), which did not repel ants (*Formica pratensis*). It induced subtle changes their behavior: ants made more foraging excursions away from the nest. *Cannabis* pollen is contaminated with THC and CBD (Paris *et al.*, 1975), but this does not hinder honey bees from collecting pollen (see section 20.4). Sirikantaramas *et al.* (2005) tested the effects of THCA on an insect cell culture, the Sf9 cell line derived from *Spodoptera frugiperda*. THCA at only 50 μM induced Sf9 cell death via apoptosis.

Mantzoukas *et al.* (2020) tested 4th-instar larvae of *Tribolium confusum, Oryzaephilus surinamensis,* and *Plodia interpunctella* on wheat, rice and maize seeds sprayed with ethanol-extracted "CBD oil" (3% CBD) at three different doses (15, 45, and 90 mg/mL). Mortality significantly increased as the dose increased. *T. confusum* larvae suffered 17–100% mortality on wheat, 17–93% on maize, and 26–83% on rice. *O. surinamensis* mortality ranged from 17% to 100% on wheat, 36–96% on maize, and 67–100% on rice. *P. interpunctella* mortality ranged from 16% to 76% on wheat, 13–60% on maize, and 33–63% on rice. Mortality was greater than assays conducted with neem oil (3% azadirachtin).

He *et al.* (2021) evaluated food preference and food intake in fruit flies, *Drosophila melanogaster*, exposed to phytocannabinoids (CBD, CBDV, CBC, CBG), endocannabinoids (AEA, 2-AG), and a synthetic analog of THC (CP 55,940). They showed that fruit flies possessed an ability to detect cannabinoids in food, and developed a slight preference for them over time. Paradoxically, all cannabinoids inhibited food intake. Direct comparisons between the cannabinoids were not presented.

Rather than cannabinoid extracts, Njoroge and Berenbaum (2019) tested seed oils for controlling mosquito larvae (*Aedes aegypti*) in rain barrels for storing drinking water. They compared seed oils from 13 crop plants, and hemp seed oil was the most lethal (LC_{50} = 348 ppm), followed by sesame seed oil (LC_{50} = 670 ppm). The authors hypothesized that linoleic acid content was responsible, forming a stable film that suffocated larvae.

Feeding choice studies

Rothschild *et al.* (1977) raised caterpillars of *Arctia caja* (a *Cannabis* pest) on fresh leaves of either high-CBD (Turkish) or high-THC (Mexican) plants. Larvae reared on high-THC leaves did not survive beyond the 3rd instar. Those fed high-CBD leaves pupated successfully. But in a feeding choice experiment, caterpillars showed a preference for high-THC leaves.

Park *et al.* (2019) conducted a feeding choice test with tobacco hornworm, *Maduca sexta*, a model organism used in herbivore studies (but not a *Cannabis* pest). Caterpillars preferred low-CBD plants to high-CBD plants; caterpillars reared on a CBD-infused artificial diet suffered reduced growth and increased mortality.

Fall armyworm (*Spodoptera frugiperda*) is another model organism used in herbivore studies, and a *Cannabis* pest. Bolt *et al.* (2021) subjected 3rd-instar larvae to a forced-feeding study with three fiber-type cultivars ('Futura 75', 'CFX-1', 'X-59'). Plants were cultivated with supplemental nitrogen—either 168 or 224 kg/ha. The ratio of total cannabinoids (THC, CBD) to nitrogen in foliage negatively correlated with larval performance (growth rate, leaf consumption, frass production).

Abendroth *et al.* (2023) reared *S. frugiperda* larvae on an artificial diet spiked with different concentrations of CBD. As CBD content increased (0, 5, 10, and 15%), food consumption, frass production, and larval growth decreased. On average, 20–30% of the CBD ingested was recovered in frass, suggesting that CBD was either metabolized or complexed with other compounds. Measurements of metabolic enzymes showed that β-glucosidase enzyme activity increased, and protease and cytochrome P450 enzyme activity decreased.

MacWilliams *et al.* (2023) tested the performance of cannabis aphids (*Phorodon cannabis*) on leaves of either low-CBD 'Tiborszallasi' or high-CBD 'Unicorn'. At maturity these cultivars produce 2–3% and 8–9% CBD, respectively, although the study utilized vegetative-stage 5-week-old plants. Aphid preference was determined by placing 50 winged aphids in a cage with three plants of each cultivar and counting aphids between 1 and 72 h after release. No preference was seen until 72 h, when an average of 19 aphids were counted on low-CBD plants and 14 on high-CBD plants.

Adult longevity was greater on low-CBD plants (10.19 days) than high-CBD plants (5.31 days). Net reproductive rate (number of offspring produced per female) was greater on low-CBD plants (39.7) than high-CBD plants (15.5). Next, they tested the effects of CBD on aphids by rearing them on an artificial diet or an artificial diet supplemented with 1 mM

CBD. The addition of CBD *increased* aphid fecundity and longevity (although the latter fell short of statistical significance). These results with dietary CBD suggested that other cannabinoids and/or metabolites (such as terpenoids) were responsible for the observations made on 5-week-old plants.

Stack *et al.* (2023) tested the performance of cabbage looper, *Trichoplusia ni*, a model herbivore (not reported on *Cannabis*). They crossed 'Carmagnola' × 'USO-31', and F$_2$ plants segregated into CBD-dominant, CBG-dominant, or cannabinoid-free populations. In a field experiment, cannabinoid-free genotypes showed greater herbivore damage than CBD- or CBG-dominant genotypes. Foliar cannabinoid concentrations inversely correlated with herbivore damage. A detached leaf bioassay showed that larvae consumed more leaf area of cannabinoid-free genotypes than of CBD-dominant genotypes, and the larvae gained more weight and had a higher survival rate. To isolate cannabinoids from confounding factors, an artificial diet was amended with CBDA and CBGA in a range of physiologically relevant concentrations. Increased concentrations of CBDA and CBGA decreased larval growth and survival.

Instead of testing the effects of cannabinoids on insects, several researchers have flipped the experiment and tested the effects of insect infestation upon cannabinoid production. As a defense against herbivory, cannabinoids and terpenoids can be either: (1) **constitutive chemicals**, which are routinely produced by plants; or (2) **inducible chemicals**, which are upregulated rapidly in response to herbivore feeding. Producing defense chemicals is expensive, so inducible chemicals are a cost-saving strategy—in the absence of herbivores, plants can focus their resources on growth and reproduction. Conversely, plant species or specific plant tissues that are regularly attacked will produce constitutive chemicals (Zangerl and Rutledge, 1996).

Tobacco hornworm (*Manduca sexta*) is a model organism for these kinds of studies. For example, tobacco plants attacked by *M. sexta* rapidly upregulate the production of terpenoids (Halitschke *et al.*, 2000), whereas nicotine production appears to be constitutive and does not upregulate (McCloud and Baldwin, 1997). Park *et al.* (2022) placed *M. sexta* in immature tops of a CBD strain. Twenty 3rd-instar larvae were placed in six hemp plants. After seven days of feeding, CBD content *decreased* from 755 μg/g (controls plants) to 194 μg/g. CBGA also decreased, while CBDA, THCA, and THC levels remained unchanged.

Different results were obtained with an actual *Cannabis* pest: Jackson *et al.* (2021) placed corn earworms (*Helicoverpa zea*) in mature tops of two CBD strains for 24 h in a growroom. CBD content in "Cherry Blossom" was higher in test plants (17%) than controls (8%). "Wife" yielded the same results, 12% vs 6%, respectively. In field trials the differences were less: "Cherry Blossom" 14% vs 8.4%; "Wife" 9% vs 5% (the latter not statistically different). Similar results were seen with THC in in a growroom experiment.

Pulkoski (2022) tested the effects of budworm feeding (*Chloridea virescens*, *Heliothis zea*) on cannabinoid content in high-CBD strains, and the results were inconsistent. In one experiment, concentrations of THCA (0.550%) and CBDA (12.424%) were higher after 14 days of *C. virescens* feeding,

compared with controls (THCA 0.483%, CBDA 10.743%). In the second experiment, there was no significant difference in cannabinoid content between controls and plants exposed to *C. virescens* or *H. zea*.

MacWilliams *et al.* (2023) measured nine cannabinoids in response to 20 days of feeding by cannabis aphids (*Phorodon cannabis*). They evaluated vegetative 8-week-old plants of either low-CBD 'Tiborszallasi' or high-CBD 'Unicorn'. Compared with control plants there was no change, except for an increase in THC (from 0.00014% to 0.00024%) in 'Tiborszallasi' (but no significant change in THCA). In contrast, 20 days of feeding caused a robust response in plant defense hormones, with increases seen in JA (jasmonic acid), ASA (abscisic acid), and especially SA (salicylic acid).

Three studies measured cannabinoids in response to defoliation as a proxy to insect-feeding damage. Britt (2021) artificially defoliated 'Felina 32' at 0%, 25%, 50%, or 75%, to represent increasing levels of chewing injury. Defoliation was carried out early-, mid-, or late-season (20, 40, or 60 days after planting). Cannabinoid concentrations (CBD%, THC%) remained stable regardless of the timing or degree of defoliation. Similarly, Park *et al.* (2022) found no significant differences in CBD, CBG, or THC levels after they punched holes in leaves (20% tissue loss) and measured cannabinoid levels 5 days later. Toth *et al.* (2021) used a weed whacker to damage 40–50% of foliage on lower and middle parts of three CBD strains. This was done late in the season, and cannabinoid content was measured for 3 weeks, which revealed no significant changes compared to controls.

Cannabinoid mechanism of action

As we explicated earlier, insects do not have CBRs. Nevertheless, THC exerts effects on insects. THC is a "promiscuous ligand," with weak activity at other receptors and enzymes (McPartland *et al.*, 2015; Russo and Marcu, 2017). CBD altered the electrophysiological responses of nerve ganglia in 3rd-instar *Manduca sexta* larvae (Park *et al.*, 2019). THC and CBD likely affect insects though other receptors and enzymes, which are also targeted by insecticides:

Acetylcholinesterase (AChE)

AChE is targeted by organophosphate pesticides. THC may inhibit the AChE enzyme (Eubanks *et al.*, 2006), or it may not (Moss *et al.*, 1975). CBD does not (Benelli *et al.*, 2018a). AChE is inhibited by *Cannabis* EOs: Smeriglio *et al.* (2020) hydrodistilled EOs from two fiber-type varieties, from China and Italy; they inhibited AChE, IC$_{50}$ = 74.64 and 57.31 μg/mL, respectively. Karimi *et al.* (2021) extracted EOs from flowering tops of Iraqi plants or from *charas* (*hashīsh*) purchased in Iran. They inhibited AChE 52.3% and 80.0%, respectively. In contrast, EO hydrodistilled from fiber-type 'Futura 75' only moderately inhibited AChE, IC$_{50}$ = 4.0 mg/mL (Benelli *et al.*, 2018a).

Gamma-aminobutyric acid (GABA) receptors

GABA receptors are targeted by cyclodienes (aldrin, dieldrin, chlordane) and fipronil. THC impacts vertebrate GABAergic neurons indirectly, through CB$_1$ receptors located in GABAergic neurons. However, an elegant study of knockout mice lacking CB$_1$ in GABAergic neurons showed that THC decreased locomotion, nociception, and body temperature (Monory *et al.*, 2007). These results might be relevant to insects, which also lack CB$_1$ in GABAergic neurons. THC and CBD inhibit GABA uptake (Banerjee *et al.*, 1975). GABAergic effects by *Cannabis* EOs have not been measured, but individual components, such as linalool, modulate GABA activity (Russo, 2011).

Other receptors

Voltage-gated sodium channels are targeted by pyrethroids and DDT. They are inhibited by THC (Turkanis *et al.*, 1991) and CBD (Okada *et al.*, 2005; Rimmerman *et al.*, 2013). Nicotinic acetylcholine receptors are targeted by neonicotinoids. They are blocked by terpenoids in *Cannabis* EOs, such as β-caryophyllene and β-eudesmol (Russo and Marcu, 2017).

Glutamate receptors are targeted by avermectins. Synthetic cannabinoids (CP55,940, WIN55,212-2) inhibit glutamatergic synaptic transmission (Shen *et al.*, 1996). Linalool in *Cannabis* EO exerts antiglutamatergic effects (Russo and Marcu, 2017).

Octopamine receptors in insects are equivalent to norepinephrine receptors in vertebrates. They are targeted by chlordimeform and amitraz, as well as EOs extracted from *Citronella*, *Pinus*, *Cedrus*, and *Eucalyptus* (Rattan, 2010). Octopaminergic activity by *Cannabis* EO or cannabinoids has not been assayed in insects. In nematodes, 2-AG activates the octopamine receptor OCTR-1 (Oakes, 2018).

Koch (2021) proposed a completely different mechanism. CBD-fed *Manduca sexta* larvae failed to molt, becoming constricted in their own exuviae leading to their death. GC-MS of exoskeletons in CBD-fed larvae revealed greater levels of metabolites involved in the production of N-β-alanyldopamine, required for exoskeleton sclerotization (cuticle hardening). Transcriptomic analysis revealed that expressed genes involved in exoskeleton development were highly unregulated in response to CBD administration.

Collectively, insect studies show that many constituents in *Cannabis* exert insecticidal activity. So how do pests that eat our *Cannabis* crops survive? Perhaps they intersperse marijuana meals with less toxic lunches on other plants. *Spilosoma obliqua*, for instance, feeds on *Cannabis* leaves and female flowers (Baloch and Ghani, 1972; Nair and Ponnappa, 1974). But when Deshmukh *et al.* (1979) force-fed *S. obliqua* caterpillars a pure *Cannabis* diet, they died after 20 days.

Kinder, gentler insects

Not all insects are crop pests. Some of them prey upon crop pests; we dedicate section 20.6 to these beneficial insects. Some insects are even domesticated, such as the silk moth (*Bombyx mori*), which now exists only in captivity. Centuries of selective breeding have turned *B. mori* into fat, sedentary, silk-spewing machines, unable to fly.

Insects and plants may interact in mutualistic relationships—the cooperative interaction between species. Darwin (1859) wrote, "Thus I can understand how a flower and a bee might slowly become, either simultaneously or one after the other, modified and adapted to each other." He viewed mutualism in the context of coevolution—when two or more species reciprocally affect each other's evolutionary modification of traits.

Cannabis is wind pollinated, so it has not evolved mutualistic relationships with pollinators. There is no **entomophily syndrome**, where flowers grow large and colorful, or offer nectar as insect rewards. However, *Cannabis* produces copious pollen, and its pollen is sought by bees (see section 20.4).

Insect-mediated **zoochory** (seed dispersal by animals) is another well-known example of mutualism. Zoochory is primarily mediated by birds and mammals, but insects play their part. Some plants have evolved a lipid-rich appendage on seeds, which attracts insects. Insects feed on the appendage and discard the seed in an intact and viable state, providing a seed dispersal service to the plant. Sernander (1906) coined the term **elaiosome** to describe lipid-rich appendages of seeds dispersed by ants.

Janischevsky (1924, 1925) described zoochory of *Cannabis ruderalis* by **Pyrrhocoris apterus** L., known as the red bug (Fig. 3.7A). Janischevsky said the bug was attracted to an elaiosome, which he located in the elongated base of wild-type seeds. He illustrated *P. apterus* carrying a seed by its piercing-sucking proboscis (Fig. 3.7B). The insect carried the seed "far distances," and then sucked oil out of the elaiosome. The rest of the seed was left intact and capable of germination. *P. apterus* is a well-known species with a Palearctic distribution, yet no one else has observed *P. apterus* carrying around hemp seeds. Small (1975d) dissected a number of wild-type accessions,

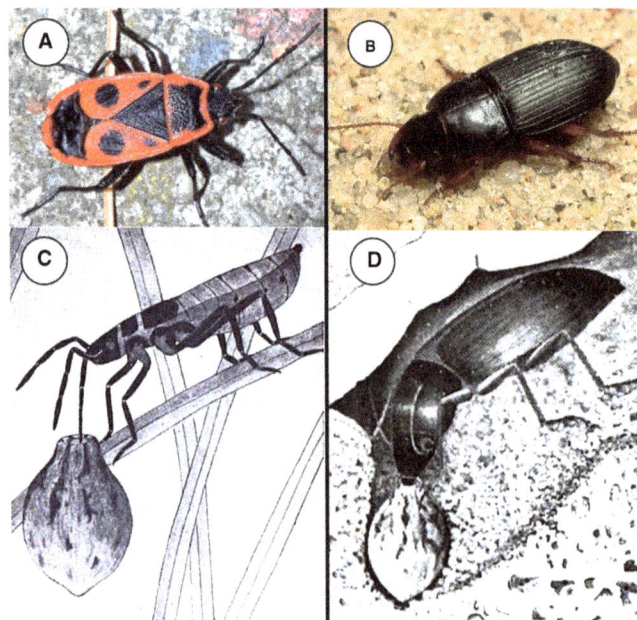

Fig. 3.7. Zoochory: **A**, **B**: *Pyrrhocoris apterus*; **C**, **D**: *Harpalus smaragdinus*. Sources: (A) André Karwath at Wikipedia Commons, (B) Janischevsky, 1924, (C) Alek Brzozowski at https://insektarium.net, (D) Janischevsky, 1925

including plants of Russian origin: "I have not been able to perceive much basal oil cell proliferation in most achenes of wild plants."

Janischevsky (1925) reported more cases of zoochory, by ground beetles such as **Harpalus smaragdinus** (Fig. 3.7C), as well as Harpalus (Ophonus) calceatus, H. hirtipes, H. servus, H. serripes, H. latus, H. flavicornis, H. smaragdinus, and Amara fulva. The beetles carried C. ruderalis seeds into their burrows (Fig. 3.7D). They chewed off the elaiosome but left the rest of the seed intact, which germinated and grew out of the shallow burrow. Brygadyrenko and Reshetniak (2014) presented a related species, H. rufipes, with a "cafeteria" of 15 different crop seeds, and hemp seed ranked 7th in preference.

Other Arthropods

Beyond insects (which have three pairs of legs), at least five other classes of arthropods are hemp herbivores: **Arachnida** (spiders, ticks, mites) with four pairs of legs; **Crustacea** (including pillbugs) with five to seven pairs of legs; **Symphyla** (garden centipedes) with 12 pairs of legs; **Chilopoda** (true centipedes) with one pair of legs per segment; and **Diplopoda** (millipedes) with two pairs of legs per segment, and many segments.

ARACHNIDA. Mites and ticks hatch from eggs with three pairs of legs. After their first molt, they gain their full complement of four pairs. Mites and ticks feed with piercing-sucking mouthparts. Perhaps the worst pest faced by indoor growers is the two-spotted spider mite, *Tetranychus urticae*. The hemp russet mite, *Aculops cannabicola,* a burgeoning quandary, was practically unknown 30 years ago.

Arachnids, like insects, do not have CBRs. Nevertheless, some studies show that *Cannabis* or cannabis extracts can repel or kill arachnids. Reznik and Imbs (1965) used powdered leaves to kill larvae of four ticks: *Ixodes redikorzevi, Haemaphysalis punctate, Rhipicephalis rossicus,* and *Dermacentor marginatus.* Surina and Stolbov (1981) rubbed the inner walls of honeybee hives with powdered leaves to control the honeybee mite, *Varroa jacobsoni.* Extracts reportedly repel or kill *Tetranychus urticae,* a serious *Cannabis* pest. This holds for aqueous extracts (Fenili and Pegazzano, 1974), essential oil extracts (Fiedler *et al.,* 2013; Górski *et al.,* 2016), or solvent extracts (Taisiya *et al.,* 2010).

Hayes *et al.* (2023) tested the performance of hemp russet mite (*Aculops cannabicola*) on detached leaves of low-CBD 'Elite' (2–3%) versus high-CBD 'Unicorn' (8–9%). Adult mites were transferred to leaves and maintained for 2 weeks. Mites performed better on low-CBD leaves compared with high-CBD leaves, in terms of development time (9.00 days vs 5.76 days) and number of offspring (19.24 vs 6.11). Next, they supplemented low-CBD 'Elite' with CBD by submerging the petiole of detached leaves in a CBD solution. Mites on CBD-supplemented leaves compared with normal leaves showed no differences in survival or fecundity—which suggests that constituents other than CBD were involved in reducing mite populations with the high-CBD cultivar in the first experiment.

Hayes *et al.* (2023) also flipped the experiment and tested the effects of *A. cannabicola* on cannabinoid production. In a 2-year field experiment, they measured cannabinoids in mite-infested plants compared with miticide-treated plants. Mite-infested plants produced significantly lower levels in 12 of 20 cannabinoids measured in one year, and 13 of 14 cannabinoids measured the second year.

Kostanda and Khatib (2022) tested the effects of *T. urticae* infestation on cannabinoid production. Clones of a mixed THC/CBD strain were grown for 2 months in vegetative phase (16 h light/day) and 2 months in flowering phase (12 h/day). They added mites midway through the vegetative stage. Thereafter plants were sampled five times: four during the vegetation phase (leaf tissue) and once during flowering ("flower tissue"). Compared with controls, cannabinoid response to herbivory was muted. CBDA and THCA levels never significantly differed during the five sample times. Other cannabinoids (CBD, THC, CBCA, CBC, CBG) increased once or twice during the five sample times, but no trends were seen. Terpenoid response, in contrast, was robust: in flowers, nearly all terpenoids showed significant differences between tests and controls: β-pinene, myrcene, limonene, ocimene, fenchone, fenchol, terpinolene, borneol, α- and β-caryophyllene, caryophyllene oxide, valencene, guaiol, and β-eudesmol.

Pulkoski and Burrack (2023) tested the effects of *T. urticae* infestation on cannabinoid production. Three weeks after flower initiation, "BaOx" plants (high-CBD/low-THC) were infested at low rates (50 mites/plant) or high rates (200 mites/plant). Mite populations exploded and plants rapidly declined, so cannabinoid levels were measured after just 7 days. A second experiment utilized lower infestation rates (low: 25 mites/plant; high: 125 mites/plant) and cannabinoid levels were measured at 0, 4, 7, 11, and 14 days after infestations. In both experiments, THC and CBD levels were significantly *lower* in infested plants than control plants (Fig. 3.8A). Higher densities of mites negatively correlated with THC and CBD levels (Fig. 3.8B). The authors concluded that cannabinoids were not inducible chemicals, but may serve as constitutive defensive chemicals.

CRUSTACEA. Crustaceans can be distinguished from insects and mites by their biramous limbs, which divide to form two branches. Having said that, biramous limbs are missing in some crustacean groups (e.g., Order Isopoda, the pillbugs). Pillbugs and sowbugs sometimes kill *Cannabis* seedlings (see section 16.4.3).

Among all the invertebrates, a crustacean was subject to the first experimental study on cannabinoids. Harry E. Warmke (1907–1985) led an effort to breed fiber-type *Cannabis* for "rope without dope" (Armagnac, 1943). The standard biological assay for measuring drug potency—the dog ataxia test—was not practical for analyzing hundreds of plants required by a breeding program. So Warmke turned to the water flea, *Daphnia* sp. Warmke and Davidson (1943) added serial dilutions of leaf extracts to a jar of water with *Daphnia,* to determine LD_{50} levels. Warmke and Davidson (1944) calibrated the *Daphnia* assay with purified cannabinoids available at that time: THC-acetate and CBN-acetate. Because these compounds were inactive in dogs, but killed *Daphnia,* they concluded that "the *Daphnia* assay is not specific for the marihuana drug."

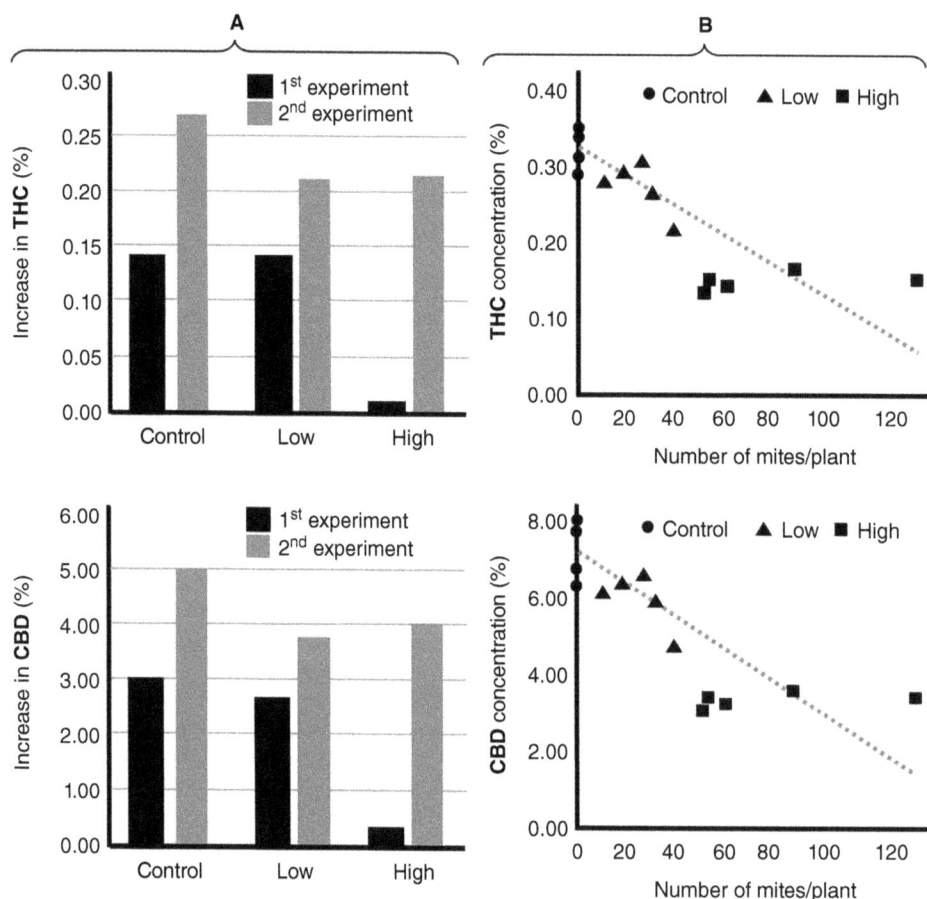

Fig. 3.8. Results by Pulkoski and Burrack (redrawn): **A.** Changes in cannabinoid content (final concentration – initial concentration) under low or high infestations compared with controls; **B.** Correlations between cannabinoid content and mite populations.

Warmke didn't designate a species, but he probably used *Daphnia magna*, a test animal for monitoring water quality (e.g., Ellis, 1940). The *D. magna* genome has been sequenced (Lee *et al.*, 2019), and a BLAST search found no genes resembling those of human CB_1 or CB_2 (http://wfleabase.org). McPartland *et al.* (2006a) searched for CBRs in a crustacean, the rock lobster (*Jasus edwardi*), using [3H]CP55,940], and found no specific binding. A lobster chef in Maine claimed that blowing cannabis smoke through a straw into a lobster tank sedated the animals. McPartland debunked the idea, saying that lobsters lacked CBRs, and THC was not very water soluble, "that's why water pipes work" (Grunewald, 2019).

Satyal and Setzer (2014) determined LD_{50} levels in another crustacean, *Artemia salina* (brine shrimp). An EO extract (terpenoids and cannabinoids) hydrodistilled from wild plants in Nepal was quite lethal: LD_{50} = 13.6 µg/mL. The LD_{50} in insects was much weaker: *Chaoborus plumicornis* (fly midge) 227 µg/mL, *Reticulitermes virginicus* (termite) 354 µg/mL, *Drosophila melanogaster* (fruit fly) 500 µg/mL.

LOPHOTROCHOZOA. Superphylum Lophotrochozoa includes Phylum **Annelida** (segmented worms, including leeches), Phylum **Platyhelminthes** (flatworms), and Phylum

Mollusca (molluscs). The annelid *Lumbricus terrestris* deserves praise as the "intestines of the earth," to quote Aristotle. Earthworms tunnels aerate soil and their manure constitutes one of the finest fertilizers available. Earthworms can be prolific—in prairie soil they weigh up to 6000 lb/acre (6725 kg/ha). But in adjacent crop fields, their numbers may plummet to 78 lb/acre (87.4 kg/ha). The difference is due to their dislike of pesticides and intolerance of plowing (Zaborski, 1998).

McPartland *et al.* (2006a) found no specific binding of [3H]CP55,940 in cerebral ganglia of *Lumbricus terrestris*. Nevertheless, earthworms are affected by *Cannabis* extracts. Mattioli (1554) learned from Italian fishermen, "A fresh decoction of hemp leaves, poured into earthworm holes, at once calls them up," which the fishermen used as bait. Kabelík *et al.* (1960) also described fishermen soaking the ground with an aqueous extract of macerated hemp leaves to force up earthworms. Duquénois (1950) exposed earthworms to a *Cannabis* aqueous extract, and they showed signs of excitation (twisting and wriggling), which subsequently diminished, followed by paralysis and death.

Platyhelminthes are stripped-down Lophotrochozoans, lacking specialized circulatory and respiratory organs. One group, the planaria, exhibit an extraordinary ability to regenerate lost body parts. High doses of THC or CBD are lethal to the planarian *Girardia tigrine* (Lenicque *et al.*, 1972). *Girardia* spp. exposed to a synthetic cannabinoid (WIN 55,212-2) showed stimulated motor activity (Buttarelli *et al.*, 2002), and abrupt cessation of WIN 55,212-2 caused a withdrawal reaction (Rawls *et al.*, 2006).

Phylum Mollusca is the second-largest invertebrate phylum after the Arthropoda. Terrestrial mollusks—snails and slugs—are familiar creatures, seemingly benign. But they can be nasty pests (see section 16.4.1). Bedini *et al.* (2016) found that *Cannabis* EO was toxic to the freshwater snail *Physella acuta*. A relatively low dose, 100 µL/L, killed 100% of the snails. Among sea creatures, Mollusca is the largest marine phylum in terms of species numbers, comprising about 23% of all named marine animals: squids, octopuses, clams, and mussels. Octopi have the largest nervous system among invertebrates. The genome of *Octopus bimaculoides* has been sequenced (Albertin *et al.*, 2015), and lacks any genes with homology to human CBRs (McPartland, BLAST search 2016).

3.4.3 Phylum Chordata

Subphylum **Vertebrata** numbers about 66,000 species, half of which are fish. Fish comprise three of the seven classes in Vertebrata. Of the remaining four classes of vertebrates, two do not normally feed on *Cannabis* (Amphibia, Reptilia), and the other two do (Aves, Mammalia). Fish do not normally feed on hemp seeds—not having access—*but they would if they could*. Platière (1785) caught fish in a river where hemp was retted, and they had swallowed hemp seeds, as well as some leaflets. Platière ate the fish, despite warnings against it, and said his meal was excellent.

Fishermen use hemp seed as hook bait or groundbait. Groundbait is thrown into water to attract fish to an area. Hemp seed groundbait was first used by the French angler Kresz (1818), and soon adopted by the English (Surtees, 1829). Parker (1930) dedicated several pages to hemp seed as groundbait. Parker also recommended hemp seed as hook bait, although presenting hard-shelled hemp seed to the hook can be difficult. Some anglers germinate hemp seeds, and then hook them when white root tips emerge from seed coats; they resemble water snails (Fairbairn, 1976). *Cannabis* seeds are also used in dough balls for carp fishing in Europe (Clarke, pers. comm. 1997).

Fish have CBRs (Yamaguchi *et al.*, 1996) and *Cannabis* elicits psychoactivity in them. Schultes (1973) proposed that *Cannabis* was domesticated by East Asians who used it as a "fish-stupefying plant." This was done in Iraq, documented long ago by Ibn Wahshīyah (AD 860–931). "All the fish in the water to swoon and rise to the surface" after *Cannabis* plants were thrown in the water (Levey, 1966). O'Shaughnessy (1839) mentioned that fish were susceptible to *Cannabis indica*. They "invariably and speedily exhibited the intoxicating influence of the drug."

With the rise of international prohibition regimes, forensic specialists searched for ways to gauge drug potency in police seizures. Some developed fish bioassays and added *Cannabis* extracts to fishbowls. Exactly what they were testing is hard to say, because THC is not very water soluble (Table 3.1).

Pierre Duquénois (1904–1986), a pharmacy professor at Strasbourg, used minnows, *Phoxinus laevis*. Duquénois (1939, 1950) added an aqueous maceration of *hashīsh* to a fish bowl (80 cc) with minnows, and observed a biphasic reaction: "At the beginning there is difficulty breathing, anxiety and an excitation phase: sudden movements, rapid swimming, merry-go-round movements and convulsive attacks. A depressive period follows: motor slowing, slowing of breathing, which becomes panting, loss of balance. The minnow falls to the bottom of the aquarium, remains on its side and dies after a time which is related to the concentration of the poison."

Duquénois also tested goldfish (*Carassius auratus*) and sticklebacks (*Gasterosterius leiurus*). He preferred fish assays over mammal assays because less test material was need to detect symptoms—as little as 0.5 mg in sticklebacks, compared with 25 mg in mice. In Italy, Carbonaro and Imbesi (1941) used goldfish to assay the activity of *C. indica* from various sources. They added progressively larger doses of ethanol-based extracts to a 360 mL bowl. Low doses caused hyperexcitability and rapid mouth and gill movements. Higher doses caused uncoordinated movements, balance disorders, and depression. At highest doses, fish moved to the bottom of the bowl in a peculiar vertical position, and remained motionless until they died.

In the search for "rope without dope", hemp breeders used fish for measuring drug potency. First was Brittain B. Robinson (1899–1969), who took over USDA research after Dewey retired. Robinson (1941) placed 2 mL of an acetone extract of *Cannabis* leaves in a liter of water containing goldfish, *Carassius auratus*, and measured survival time. Extracts from drug-type plants killed fish faster (Tunisia, 68 min) than extracts from fiber varieties ('Ferramington', 267 min; 'Kentucky', still alive at 24 h).

Harry Warmke of *Daphnia* fame (see above) modified Robinson's "goldfish test" by using Atlantic minnows, *Fundulus heteroclitus*. To four fishbowls he added increasing amounts of an acetone extract. A pair of minnows was placed in each bowl, and after 24 h he counted the number of dead fish (Warmke, 1942, 1944). He found that the deadly *Cannabis* "alkaloid" was more concentrated in upper leaves than lower leaves.

Studies in the 21st century have utilized zebrafish (*Danio rerio*), whose genome expresses CB_1 and a pair of CB_2 orthologs (McPartland *et al.*, 2007c). Adding THC to water (after dissolving it in 0.05% ethanol) caused normally top-swimming zebrafish to slow their movements and swim near the bottom, an anxiety-like behavior (Stewart and Kalueff, 2014). THC alters spatial learning and memory, slows locomotor activity and escape responses, and modifies social behavior (Ruhl *et al.*, 2014, 2015; Neeley *et al.*, 2018). These changes are detrimental in prey species, and these behavioral studies suggest fish dislike getting "high." When food pellets are spiked with the endocannabinoid 2-AG, minnows (*Phoxinus phoxinus*) spit them back out (Schaaf and Dettner, 2000).

Birds (Class Aves)

Birds are perhaps the most voracious seed predators of *Cannabis* (see section 16.5). Some birds pluck seeds from flowering tops (Order Passeriformes, perching birds), other birds glean seeds off the ground (Columbiformes, Galliformes).

A number of birds evolved around the same time and place that *Cannabis* evolved. Two genera of Columbiformes, *Columba* and *Streptopelia* (well-known hemp pests), evolved and diverged about 30 mya in Central Asia (Pereira *et al.*, 2007). Passerines came later; their fossil record in Central Asia dates to the Miocene, *ca.* 17 mya (Zelenkov and Kurochkin, 2012).

Plants have evolved a variety of seed defenses in response to seed predation. These include cryptic coloration for visual concealment, hardened seed coats for mechanical defense, and toxins for chemical defense. The cryptic-colored perianth of wild-type *Cannabis* provides camouflage against visually cued seed predators (Small, 1975d). The camouflage is ill-adapted against Passeriformes in flowering tops (Fig. 3.9), rather it would protect the seeds from ground-gleaning feeders (Columbiformes, Galliformes).

THC biosynthesis may have evolved as a deterrent to seed predation. Boaler (1919) suggested that birds crack open hemp seeds, eat the kernel and spit out the husk because "this outer shell contains some quantity—small, of course—of an irritant poison." Altered sensory perception renders animals susceptible to environmental hazards and predators. Acute toxicity is not the issue: the lethal dose of THC in pigeons is 180 mg/kg (McMillan and Dewey. 1971), the equivalent of a 70 kg person smoking 1260 joints with 10 mg THC in each. But 1% of that dose, 1.8 mg/kg, stopped pigeons from pecking for food. Pigeons dosed with a cannabis extract make errors in color perception (Siegel, 1969), or misjudge timing when food is available for a brief period (Siegel, 1989). At high doses they withdraw from social interactions and spend less time with their chicks (Siegel, 1989). Cannabinoids may hamper bird reproduction—they cause zebra finches to forget vocal learning, and disrupt song production (Soderstrom and Tian, 2006).

On the other hand, successful seed dispersal in many plants requires the participation of animals. It's a tradeoff—most seeds are destroyed by predation, but a few benefit from **zoochory**: the transport of seeds by animals. Plants invest energy in various adaptions for animal dispersal, such as adhesive mucus and a variety of hooks, spines, and barbs. *Cannabis* has

none of those, but its protein- and lipid-rich seeds are nutritious and very attractive to birds (see section 16.5).

Cannabis has adaptions for **endozoochory**: the transport of seeds via animal digestive systems. Seeds that escape beak and gizzard may undergo long-distance dispersal while traversing the bird gut, and then get deposited in a pearl of fertilizer (bird excrement). Plants adapted for "classical" endozoochory have seeds embedded in the fleshy pulp of an edible fruit or berry. Alternative traits are small, round achenes or nuts with smooth surfaces and hard pericarps (Ridley, 1930; Van der Pijl; 1982). These traits have adapted *Cannabis* for endozoochory (Spengler and Mueller, 2019).

Janischevsky (1924) made note of the "great strength of the pericarp" in wild-type *Cannabis ruderalis* seeds. "It is difficult to separate the [*C. ruderalis*] pericarp into two identical shells, such as one obtains from the fruits of the *cultivated variety* when they are cracked by birds." Small (2017) cited Janischevsky, and said the comparatively thick wall of wild-type seeds made it more difficult for herbivores to break open the seeds. Three years later he offered conflicting data. Naraine *et al.* (2020) measured the force needed to fracture seeds: 20.2 N in wild-types, and 27.4 N in fiber-type cultivars. However, Van der Meij and Bout (2006) needed only 12.16 N to crack the seed of a cultivar.

In *Origin of Species*, Darwin (1859) documented endozoochory of *Cannabis* seeds by carrier pigeons (*Columba livia domestica*) flying from France to England. Darwin did not quantify the survival rate of eaten seeds—he simply observed that some seeds germinated. An unpublished study by Eaton (cited in Small *et al.*, 2003) quantified hemp seed survival through the avian gut. Eaton found that bobwhite quail (*Colinus virginianus*) passed one viable seed for every 700 eaten. Mourning doves (*Zenaida macroura*) passed one viable seed per 12,400 seeds consumed.

Darwin actually documented *diplo*endozoochory, "two-phase dispersal," because hawks and owls preyed upon the pigeons in England. They disgorged pellets of indigestible material, from which some seeds of hemp germinated. Sanchis Serra *et al.* (2013) also described diploendozoochory. They found *Cannabis* seeds in the nest of an Egyptian vulture (*Neophron percnopterus*). The seeds came from carcasses the vulture carried to the nest.

Mammals (Class Mammalia)

Many mammalian herbivores also evolved in Central Asia, particularly in the Order Rodentia (rats, mice, hamsters, voles, squirrels), Order Lagomorpha (rabbits, hares, and pikas), Family Cervidae (deer) and Family Equidae (horses). They were early seed predators of *Cannabis*. The endozoochory tradeoff operates here. Vavilov (1926) hinted at endozoochory when he stated that wild-type *Cannabis* thrived in soil manured by grazing cattle, and in gorges and ravines where the dung of wild animals accumulated. Van der Pijl (1982) noted that agricultural fields manured with ruminant dung sometimes acquired mass infestations of weed species.

Lush foliage may attract mammals who accidently eat seeds "sufficiently small, tough, hard and inconspicuous to escape the molar mill" (Janzen, 1984). Spengler and Mueller

Fig. 3.9. Camouflaged wild-type *Cannabis* seed still embedded in the flower (courtesy Naraine, 2018)

(2019) proposed that several plants domesticated in Central or East Asia, with small and hard seeds—millet, buckwheat, hemp—were originally dispersed via endozoochory by "megafaunal ruminants" (animals ≥40 kg, including horses and deer). The color scheme of wild-type seeds, however, is poorly adapted against megafauna grazing seeds in foliage, but provides camouflage against ground-gleaning rodents and rabbits (Fig. 3.9).

McPartland and Naraine (2018) tested the hypothesis that mastication was the seed-killing step in mammalian endozoochory. They ran hemp seeds through an *in vitro* gastrointestinal system that lacked mastication but included all human digestive enzymes. Then they conducted a pair of *in vivo* tests, with variable amounts of mastication: A human ate 1000 hemp seeds mixed into granola breakfast cereal, chewed and swallowed normally. A canine wolfed down 1000 seeds mixed into dog food with minimal chewing. Feces were collected for daily installation into a germination bed. Seedling emergence was compared with a control plot, which yielded a 76% germination rate. In comparison: 59.5% germinated after passing through the *in vitro* digestion model, 13.6% after passing through canine gut (minimal mastication), and 1.3% after passing through human gut (maximal mastication). These survival rates are much higher than seeds passing through the avian gut.

THC as a deterrent to seed predation also operates here. Consuming a lethal dose would be difficult to achieve—the oral LD_{50} of THC in mice is huge: >21,600 mg/kg (Loewe, 1946). In contrast, the oral LD_{50} of aspirin in mice is 200 mg/kg. According to Ley (1843), O'Shaughnessy tried to kill to a dog with ½ oz of a potent extract. The extract was active in humans at ⅙ grain, therefore the dog received 1313 times an active dose. The dog became cataleptic, slept for 2 days, and then recovered and ate voraciously. Reports of bovines and horses consuming lethal amounts of *Cannabis indica* (Cardassis, 1951; Driemeier, 1997) seem implausible.

Campbell *et al.* (2019) suggested that cannabinoids repelled animals through "a combination of toxicity, repellent taste, and objectionable stickiness." Rats will readily self-administer cocaine, heroin, fentanyl, valium, etc., but they refuse to self-administer THC, unless they are food-deprived (reviewed in Justinova *et al.*, 2005). Mice will not consume THC unless it is laced into sweetened dough, and even then not at high doses (Smoker *et al.*, 2019).

Altered sensory perception is a hazard. Rats injected with a cannabis extract turn the wrong way in mazes (Carlini and Kramer, 1965). Siegel (1989) observed rats stripping seeds from *Cannabis* plants in Hawai'i, and then, "perhaps slowed by intoxication," mongooses caught and killed them. Dogs dosed with a cannabis extract develop muscular incoordination, lassitude, and dulled attention. This was calibrated into the "dog ataxia test," utilized by pharmaceutical companies who tested the potency of their *Cannabis indica* products from the 1890s until the 1940s (Loewe, 1944).

Ruminants (cow, deer, reindeer) and hindgut fermenters (horse, rabbit, pig, rodents) depend on gut microflora for digestion. The antibacterial activity of cannabinoids and terpenoids may harm them. THC modified the hindgut microflora in mice, at least in obese mice (Cluny *et al.*, 2015). A combination of THC+CBD altered the hindgut microflora in mice by inhibiting *Akkermansia muciniphila* (Al-Ghezi *et al.*, 2019). This probiotic bacterium occurs in many ruminants and hindgut fermenters, as well as us (Geerlings *et al.*, 2021). Terpenoids found in sagebrush inhibited the bacteria in deer rumen, resulting in decreased rumen motility and reduced food consumption (Nagy *et al.*, 1964).

(Prepared by J. McPartland; revised by S. Koike)

References

Aarti YD. 2018. The phytotoxic effect of aqueous extract of *Cannabis sativa* on the germination and growth of *Cicer arietinum*. *Research Journal of Pharmacy and Technology* 11: 5096–5100.

Abbott A. 1990. The switch that turns the brain on to cannabis. *New Scientist* 127: 31.

Adegbite AA. 2011. Effects of some indigenous plant extracts as inhibitors of egg hatch in root-knot nematode (*Meloidogyne incognita* race 2). *American Journal of Experimental Agriculture* 1(3): 96–100.

Agnieszka S, Magdalena R, Jan B, *et al.* (4 other authors). 2016. Phytotoxic effect of fiber hemp essential oil on germination of some weeds and crops. *Journal of Essential Oil Bearing Plants* 19: 262–276.

Aguinaldo AM, Turbeville JM, Linford LS, *et al.* (4 other authors). 1997. Evidence for a clade of nematodes, arthropods and other moulting animals. *Nature* 387: 489–493.

Akhtar MT, Shaari K, Verpoorte R. 2013. Biotransformation of tetrahydrocannabinol. *Phytochemistry Reviews* 15: 921–934.

Al Khoury A, Sleiman R, Atoui A, *et al.* (4 other authors). 2021. Antifungal and anti-aflatoxigenic properties of organs of *Cannabis sativa* L.: relation to phenolic content and antioxidant capacities. *Archives of Microbiology* 203: 4485–4492.

Anjum M, Zel-E-Arooj, Azem S, Rehman P, Khadim J. 2018. Evaluation of antimicrobial activity and ethnobotanical study of *Cannabis sativa* L. *Pure and Applied Biology* 7: 706–713.

Athreya S, Hopkins A. 2021. Conceptual issues in hominin taxonomy: *Homo heidelbergensis* and an ethnobiological reframing of species. *Yearbook of Physical Anthropology* 175(Suppl. 72): 4–26.

Atwood D, Paisley-Jones C. 2017. *Pesticides industry sales and usage. 2008-2012 market estimates*. Office of Pesticide Programs, US Environmental Protection Agency, Washington, DC.

Bajpai NK, Sehgal VK. 1993. Oviposition preferences, larval development, and survival of *Helicoverpa armigera* (Hübner) on chickpea and weed hosts at Pantnagar, India. *International Chickpea Newsletter* 29: 15–17.

Bajpai NK, Sharma VK. 1992. possible use of hemp (*cannabis sativa* l.) weeds in integrated control. *Indian Farmers' Digest* 25(12): 32, 38.

Baldauf SL, Palmer JD. 1993. Animals and fungi are each other's closest relatives: congruent evidence from multiple proteins. *Proceedings of the National Academy of Sciences USA* 90: 11558–11562.

Barbaro E, *ed*. (Dioscorides P). 1516. *In hoc volumin haec continentur, Ioannis Baptistae Egnatii in Dioscoridem ab Hermolao Barbaro tralatum annotamenta*. Aloisius & Franciscus Barbari, Venice.

Bel'tyukova KI. 1962. Чутливість фітопатогенних бактерій до кансатину 4 [Sensitivity of phytopathogenic bacteria to cansatine 4]. *Mikrobiolohichnyi Zhurnal (Kiev)* 24(5): 62–65.

Benelli G, Pavela R, Lupidi G, *et al.* (8 other authors). 2018a. The crop-residue of fiber hemp cv. Futura 75: from a waste product to a source of botanical insecticides. *Environmental Science & Pollution Research International* 25: 10515–10525.

Benelli G, Pavela R, Petrelli R, *et al.* (7 other authors). 2018b. The essential oil from industrial hemp (*Cannabis sativa* L.) by-products as an effective tool for insect pest management in organic crops. *Industrial Crops and Products* 122: 308–315.

Bernard EC, Chaffin AG, Gwinn KD. 2022. Review of nematode interactions with hemp (*Cannabis sativa*). *Journal of Nematology* 54: e2022-2. Available at: https://www.ncbi.nlm.nih.gov/pmc/articles/PMC8975275/

Bertoli A, Tozzi S, Pistelli L, Angelini LG. 2010. Fibre hemp inflorescences: from crop-residues to essential oil production. *Industrial Crops and Products* 32: 329–337.

Blevins RD, Dumic MP. 1980. The effect of delta-9-tetrahydrocannabinol on herpes simplex virus replication. *Journal of General Virology* 49: 427–431.

Braut-Boucher F, Cotte J, Fleury C, *et al.* (5 other authors). 1985. Exemple de variabilité spontanée mis en évidence par l'activité biologique des extraits tissulaires de *Cannabis sativa* L. *Bulletin de la Société Botanique de France* 132 (3/4): 149.

Buchweitz JP, Karmaus PW, Williams KJ, Harkema JR, Kaminski NE. 2008. Targeted deletion of cannabinoid receptors CB1 and CB2 produced enhanced inflammatory responses to influenza A/PR/8/34 in the absence and presence of delta9-tetrahydrocannabinol. *Journal of Leukocyte Biology* 83: 786–796.

Callaway JC. 2008. A more reliable evaluation of hemp THC levels is necessary and possible. *Journal of Industrial Hemp* 13: 117–144.

Carroll L. 1871. *Through the Looking Glass*. Macmillian, London.

Cavalier-Smith T. 1987. "The origin of fungi and pseudofungi," pp. 339-353 in Rayner *et al.*, eds. *Evolutionary biology of the fungi*. Cambridge University Press, Cambridge, UK.

Chailakhyan MK, Mehrabyan AA, Karapetyan NA. 1957. О бактерицидности небобовых растений по отношению к клубеньковым бактериям. *Proceedings of the National Academy of Sciences of the Armenian SSR* 25(2): 81–85.

Colson P, Richet H, Desnues C, *et al.* (8 other authors). 2010. Pepper mild mottle virus, a plant virus associated with specific immune responses, fever, abdominal pains, and pruritus in humans. *PLoS ONE* 5(4): e10041.

Dahiya MS, Jain GC. 1977. Inhibitory effects of cannabidiol and tetrahydrocannabinol against some soil inhabiting fungi. *Indian Drugs* 14(4): 76–79.

Darling ST. 1910. *Studies in relation to malaria*. Government Printing Office, Washington, DC.

Darwin CR. 1859. *On the origin of species*. John Murray, London.

de Candolle ALP. 1867. *Lois de la nomenclature botanique*. Masson et fils, Paris.

De Sousa ACC, Combrinck JM, Maepa K, Egan TJ. 2021. THC shows activity against cultured *Plasmodium falciparum*. *Bioorganic & Medicinal Chemistry Letters* 54: e128442.

DeMarino C, Cowen M, Khatkar P, *et al.* (12 other authors). 2022. Cannabinoids reduce extracellular vesicle release from HIV-1 infected myeloid cells and inhibit viral transcription. *Cells* 11: e723.

Domnas AJ, Fagan SM, Jaronski S. 1982. Factors influencing zoospore production in liquid cultures of *Lagenidium giganteum* (Oomycetes, Lagenidiales). *Mycologia* 74: 820–825.

Elphick MR, Egertová M. 2001. The neurobiology and evolution of cannabinoid signaling. *Philosophical transactions of the Royal Society of London. Series B, Biological sciences* 356: 381–408.

Elphick MR, Egertová M. 2005. The phylogenetic distribution and evolutionary origins of endocannabinoid signalling. Handbook of Experimental Pharmacology 168: 283–297.

ElSohly HN, Turner CE, Clark AM, ElSohly MA. 1982. Synthesis and antimicrobial activities of certain cannabichromene and cannabigerol related compounds. *Journal of Pharmaceutical Sciences* 71: 1319–1323.

Estrada-Valencia R, Eduardo de Lima M, Colonnello A, *et al.* (5 other authors). 2023. The endocannabinoid system in *Caenorhabditis elegans*. *Review of Physiology, Biochemistry and Pharmacology* 184: 1–31.

Faeth SH. 1986. Indirect interactions between temporally separated herbivors mediated by the host plant. *Ecology* 67: 479–494.

Farlow WG. 1910. A consideration of the "*Species plantarum*" of Linnaeus as a basis for the starting-point of the nomenclature of cryptogams. *American Naturalist* 44: 385–394.

Ferenczy L. 1956. Antibacterial substances in seeds. *Nature* 178: 639–640.

Ferenczy L, Gracza L, Jakobey I. 1958. An antibacterial preparatum from hemp (*Cannabis sativa*). *Naturwissenschaften* 45: 188.

Fichan I, Larroche C, Gros JB. 1999. Water solubility, vapor pressure, and activity coefficients of terpenes and terpenoids. *Journal of Chemical & Engineering Data* 44: 56–62.

Fournier G, Paris MR, Fourniat MC, Quero AM. 1978. Activité bactériostatique d'huiles essentielles de *Cannabis sativa* L. *Annales pharmaceutiques Françaises* 36: 603–605.

Gal IE, Vajda O, Bekes I. 1969. A kannabidiolsav néhány tulajdonságának vizsgálata élelmiszertartósítási szempontból. *Elelmiszervizsgalati Közlemenyek* 4:208–216.

Ganaie MA, Khan TA. 2017. Evaluation of some locally available plant leaves for the control of *Meloidogyne incognita* and *Rhizoctonia solani* disease complex. *Trends in Biosciences* 10: 8534–8539.

Garrett ER, Hunt CA. 1974. Physiochemical properties, solubility, and protein binding of delta-9-tetrahydrocannabinol. *Journal of Pharmaceutical Sciences* 63: 1056–1064.

Gomes RR, Glienke C, Videira SIR, *et al.* (3 other authors). 2013. *Diaporthe*: a genus of endophytic, saprobic, and plant pathogenic fungi. *Persoonia* 31: 1–41.

Goswami BK, Vijaylakshmi K. 1986. Effect of some indigenous plant materials and oilcake amended soil on the growth of tomato and root-knot nematode population. *Annals of Agricultural Research* 7: 363–366.

Grewal PS. 1989. Effects of leaf-matter incorporation on *Aphelenchoides composticola* (Nematoda), mycofloral composition, mushroom compost quality and yield of *Agaricus bisporus*. *Annals Applied Biology* 115: 299–312.

Groves JW, Loveland CA. 1953. The connection between *Botryotinia fuckeliana* and *Botrytis cinerea*. *Mycologia* 45: 415–425.

Guha S, Calarco S, Salomé Gachet M, Gertsch J. 2020. Juniperonic acid biosynthesis is essential in *Caenorhabditis elegans* lacking Δ6 desaturase (fat-3) and generates new ω-3 endocannabinoids. Cells 9: e2127.

Gupta SK, Banerjee AB. 1972. Screening of selected West Bengal plants for antifungal activity. *Economic Botany* 26: 255–259.

Gupta RP, Singh A. 1983. Effect of certain plant extracts and chemicals on teliospore germination of *Neovossia indica*. *Indian Journal of Mycology & Plant Pathology* 13(1): 116–117.

Gurmani AR, Khan SU, Mehmood T, Ahmed W, Rafique M. 2021. Exploring the allelopathic potential of plant extracts for weed suppression and productivity in wheat (*Triticum aestivum* L.) *Gesunde Pflanzen* 73: 29–37.

Haeckel E. 1866. *Generelle Morphologie der Organismen*, 2 vols. Georg Reimer, Berlin.

Haidar MG, Askary TH. 2011. Management of plant parasitic nematodes through botanicals and growth of sugarcane (*Saccharum officinarum* L.). *Annals of Plant Protection Sciences* 19: 433–436.

Haiden SR, Apicella PV, Ma Y, Berkowitz GA. 2022. Overexpression of *CsMIXTA*, a transcription factor from *Cannabis sativa*, increases glandular trichome density in tobacco leaves. *Plants* 11(11): e1519.

Halanych KM, Bacheller JD, Aguinaldo AMA, *et al.* (3 other authors). 1995. Evidence from 18S ribosomal DNA that the lophophorates are protostome animals. *Science* 267: 1641–1643.

Haney A, Bazzaz FA. 1970. "Some ecological implications of the distribution of hemp (*Cannabis sativa* L.) in the United States of America," pp. 39–48 *in* Joyce CRB, Curry SH, eds. *The botany and chemistry of Cannabis*. J & A Churchill, London.

Haney A, Kutscheid BB. 1973. Quantitative variation in the chemical constituents of marihuana from stands of naturalized *Cannabis sativa* L. in East-Central Illinois. *Economic Botany* 27: 193–203.

Harpaz D, Veltman B, Sadeh Y, *et al.* (3 other authors). 2021. The effect of cannabis toxicity on a model microbiome bacterium epitomized by a panel of bioluminescent *E. coli*. *Chemosphere* 263: e12841.

Harvey JV. 1925. A study of the water molds and pythiums occuring in the soils of Chapel Hill. *Journal of the Elisha Mitchell Scientific Society* 41: 151–164.

Hawksworth DL. 2012. Managing and coping with names of pleomorphic fungi in a period of transition. *IMA Fungus* 3(1): 15–24.

Hawksworth DL, Lücking R. 2017. Fungal diversity revisited: 2.2 to 3.8 million species. *Microbiology Spectrum* 5: 79–95.

Howlett AC, Evans DM, Houston DB. 1992. "The cannabinoid receptor," pp. 73-92 in Murphy L, Bartke A, eds. *Marihuana/cannabinoids neurobiology and neurophysiology*. CRC Press, Boca Raton, Florida.

Hrebenyuk NV. 1984. Распространенне грибов на тресте конопли [The occurrence of fungi on hemp stems]. *Микология и фитопатология* 18(4): 322–326.

Inam B, Hussain F, Bano F. 1989. *Canabis sativa* L. is allelopathic. *Pakistan Journal of Scientific and Industrial Research* 32: 617–620.

Ingallina C, Sobolev AP, Circi S, *et al.* (19 other authors). 2020. *Cannabis sativa* L. inflorescences from monoecious cultivars grown in Central Italy: an untargeted chemical characterization from early flowering to ripening. *Molecules* 25(8): e1908.

Isahq MS, Afridi MS, Ali J, *et al.* (3 other authors). 2015. Proximate composition, phytochemical screening, GC-MS studies of biologically active cannabinoids and antimicrobial activities of *Cannabis indica*. *Asian Pacific Journal of Tropical Disease* 5(11): 897–902.

Jin D, Xie Z, Chen, J. 2020. Secondary metabolites profiled in *Cannabis* inflorescences, leaves, stem barks, and roots for medicinal purposes. *Scientific Reports* 10: e3309.

Kabelík J, Krejcí Z, Santavy F. 1960. *Cannabis* as a medicament. *Bulletin on Narcotics* 12(3): 5–23.

Kapli P, Natsidis P, Leite DJ, *et al.* (6 other authors). 2021. Lack of support for Deuterostomia prompts reinterpretation of the first Bilateria. *Science Advances* 7: e2741.

Kashyap RK, Kennedy GG, Farrar RR. 1991. Behavioral response of *Trichogramma pretiosum* and *Telenomus sphingis* to trichome/methyl ketone mediated resistance in tomato. *Journal of Chemical Ecology* 17: 543–556.

Kaushal RP, Paul YS. 1989. Inhibitory effects of some plant extracts on some legume pathogens. *Legume Research* 12: 131–132.

Kayani MZ, Mukhtar T, Hussain MA. 2012. Evaluation of nematicidal effects of *Cannabis sativa* L. and *Zanthoxylum alatum* Roxb. against root-knot nematodes, *Meloidogyne incognita*. *Crop Protection* 39: 52–56.

Kędzia A, Kaniewski, R, Hołderna-Kędzia E, Opala B. 2016. Ocena działania eterycznego olejku konopnego (*Cannabis sativa* L.) wobec grzybów drożdżopodobnych. *Postępy Fitoterapii* 2016(4): 262–267.

Kennedy JS, Booth CO, Kershaw WJS. 1959. Host finding by aphids in the field. *Annals Applied Biology* 47: 424–444.

Khanna AS, Sharma NK. 1998. "Phytotherapeutic effect of some indigenous plants/ nematicides on *Meloidogyne incognita* infesting tomato," pp. 4–6 in Dhawan SC, Kaushal KK, eds. *Proceedings of National Symposium on rational approaches in nematode nanagement for sustainable agriculture*. Nematological Society of India Press, Delhi.

Kir'yanova ES, Krall EL. 1971. *Plant-parasitic nematodes and their control, vol. II*. Academy of Sciences of the USSR, Nauka Publishers, Leningrad.

Klingeren B van, Ham MT. 1976. Antibacterial activity of delta-9-tetrahydrocannabinol and cannabidiol. *Antonie van Leeuwenhoek* 42: 9–12.

Kok CJ, Coenen GCM, de Heij A. 1994. The effect of fibre hemp (*Cannabis sativa* L.) on selected soil-borne pathogens. *Journal of the International Hemp Association* 1(1): 6–9.

Konstantinović B, Vidović S, Stojanović A, *et al.* (5 other authors). 2018. "Allelopathic effect of essential oil of *Cannabis sativa* L. on selected vegetable species," pp. 1212–1215 in *Proceedings, IX International Scientific Agriculturla Symposium "Agrosym 2018."* Jahorina, Bosnia and Herzogovina.

Krejčí Z. 1961. K otázce látek s antibakteriálním a hašišovým účinkem v konopí (*Cannabis sativa* L.). *Časopis Lékařů Českých* 43: 1351–1354.

Krejčí Z, Šantavý F. 1955. Isolace dalších látek z listí indického konopí *Cannabis sativa* L. *Acta Universitatis Palackianae Olomucensis Facultatis Medicae* 6: 59–66.

Lancz G, Specter S, Brown HK. 1990. Suppressive effect of delta-9-tetrahydrocannabinol on herpes simplex virus infectivity *in vitro*. *Proceedings Society Experimental Medicine & Biology* 196: 401–404.

Latta RP, Eaton BJ. 1975. Seasonal fluctuations in cannabinoid content of Kansas marijuana. *Economic Botany* 29: 153–163.

Latter HL, Abraham DJ, Turner CE, Knapp JE, Schiff PL, Slatkin DJ (1975) Cannabisativine, a new alkaloid from *Cannabis sativa* L. root. *Tetrahedron Letters* 16, 2815–2818.

Lau DC, *trans*. 2000. *Confucius: the Analects*. Chinese University Press, Hong Kong.

Lemay J, Zheng YB, Scott-Dupree C. 2022. Factors influencing the efficacy of biological control agents used to manage insect pests in indoor cannabis (*Cannabis sativa*) cultivation. *Frontiers in Agronomy* 4: e795989.

Levichev A, Faumont S, Berner RZ, *et al.* (4 other authors). 2023. The conserved endocannabinoid anandamide modulates olfactory sensitivity to induce hedonic feeding in *C. elegans*. *Current Biology* 33: 1625–1639.

Levin DA. 1973. The role of trichomes in plant defense. *Quarterly Review of Biology* 48: 3–15.

Linnaeus C. 1753. *Species Plantarum, Tomus II*. Laurentii Salvii, Holmiae [Stockholm].

Liu YY, Zhu PP, Cai S, Haughn G, Page JE. 2021. Three novel transcription factors involved in cannabinoid biosynthesis in *Cannabis sativa* L. *Plant Molecular Biology* 106: 49-65.

Lloyd CG. 1905. "Advertising pages," pp. 19–20 in *Index of the Mycological Writings of C. G. Lloyd. Vol. I, 1898-1905*. Cincinnati, Ohio.

Lotz LAP, Groeneveld RMW, Habekotté B, Van Oene H. 1991. Reduction of growth and
reproduction of *Cyperus esculentus* by specific crops. *Weed Research* 31: 153–160.

Lowe HIC, Toyang NJ, McLaughlin W. 2017. Potential of cannabidiol for the treatment of viral hepatitis. *Pharmacognosy Research* 9: 116–118.

Ma G, Zelman AK, Apicella PV, Berkowitz G. 2022. Genome-wide identification and expression analysis of homeodomain leucine zipper subfamily IV (HD-ZIP IV) gene family in *Cannabis sativa* L. *Plants* 11(10): e1307.

Mahmoodzadeh H, Ghasemi M, Zanganeh H. 2015. Allelopathic effect of medicinal plant *Cannabis sativa* L. on *Lactuca sativa* L. seed germination. *Acta Agriculturae Slovenica* 105: 233–239

Makkizadeh TM, Farhoudi R, Rabii M, Rastifar M. 2011. Evaluation allelopathic effect of hemp (*Cannabis sativa* L.) on germination a growth of three kinds of weeds. *Crop Physiology (Iran)* 3(11): 77–88.

Malingré TM, Hendriks H, Batterman S, Bos R, Visser J. 1975. The essential oil of *Cannabis sativa*. *Planta Medica* 28: 56–61.

Markham G. 1615. *Countrey contentments in two bookes, the second intituled the English Huswife*. Roger Jackson, London.

Mateeva A. 1995. Use of unfriendly plants against root knot nematodes. *Acta Horticulturae* 382 (Feb): 178–182.

McEno J. 1991. "*The seven wonders*," pp. 13–29 in: McEno, *ed. Cannabis ecology: a compendium of diseases and pests*. Amrita Press, Middlebury, Vermont.

McNeill J, *chair*. 2012. *International Code of Nomenclature for Algae, Fungi, and Plants (Melbourne Code)*. Koeltz Scientific Books, Königstein, Germany.

McPartland JM. 1983b. *Phomopsis ganjae* sp. nov. on *Cannabis sativa*. *Mycotaxon* 18: 527–530.

McPartland JM. 1984. Pathogenicity of *Phomopsis ganjae* on *Cannabis sativa* and the fungistatic effect of cannabinoids produced by the host. *Mycopathologia* 87: 149–153.

McPartland JM. 1992. The *Cannabis* pathogen project: report of the second five-year plan. *Mycological Society of America Newsletter* 43(1): 43.

McPartland JM. 1995a. *Cannabis* pathogens VIII: misidentifications appearing in the literature. *Mycotaxon* 53: 407–416.

McPartland JM. 1995d. *Cannabis* pathogens XII: lumper's row. *Mycotaxon* 54: 273–279.

McPartland JM, Glass M. 2001. The nematocidal effects of *Cannabis* may not be mediated by cannabinoid receptors. *New Zealand Journal Crop & Horticultural Science* 29: 301–307.

McPartland JM, Glass M. 2003. Functional mapping of cannabinoid receptor homologs in mammals, other vertebrates, and invertebrates. *Gene* 312: 297–303.

McPartland JM, Guy GW. 2010. *THC synthase in Cannabis has undergone accelerated evolution and positive selection pressure. Proceedings of the 20th Annual Symposium on the Cannabinoids*. International Cannabinoid Research Society, Research Triangle Park, North Carolina, p. 43.

McPartland JM, Pruitt PP. 1997. Medical marijuana and its use by the immunocompromised. *Alternative Therapies in Health and Medicine* 3(3): 39–45.

McPartland M, Sheikh Ź. 2018. A review of *Cannabis sativa*-based insecticides, miticides, and repellents. *Journal of Entomology and Zoology Studies* 6: 1288–1299.

McPartland JM, Di Marzo V, De Petrocellis L, Mercer A, Glass M. 2001. Cannabinoid receptors are absent in insects. *Journal of Comparative Neurology* 436: 423–429.

McPartland JM, Agraval J, Gleeson D, Heasman K, Glass M. 2006a. Cannabinoid receptors in invertebrates. *Journal of Evolutionary Biology* 19: 366–373.

Mediavilla V, Steinemann S. 1997. Essential oil of *Cannabis sativa* L. strains. *Journal of the International Hemp Association* 4(2): 82-84.

Meghini L, Ferrante C, Carradori S, *et al.* (10 other authors). 2021. Chemical and bioinformatics analyses of the anti-leishmanial and anti-oxidant activities of hemp essential oil. *Biomolecules* 11: e272.

Misra SB, Dixit SN. 1979. Antifungal activity of leaf extracts of some higher plants. *Acta Botanica Indica* 7: 147–150.

Molina PA, Winsauer P, Zhang P, *et al.* (9 other authors). 2011. Cannabinoid administration attenuates the progression of simian immunodeficiency virus. *AIDS Research and human Retroviruses* 27: 585–592.

Morimoto S, Tanaka Y, Sasaki K, *et al.* (5 other authors). 2007. Identification and characterization of cannabinoids that induce cell death through mitochondrial permeability transition in *Cannabis* leaf cells. *Journal of Biological Chemistry* 282: 20739–20751

Mukhtar T, Kayani MZ, Hussain MA. 2013. Nematicidal activities of *Cannabis sativa* L. and *Zanthoxylum alatum* Roxb. against *Meloidogyne incognita*. *Industrial Crops and Products* 42: 447–453.

Muscarà C, Smeriglio A, Trombetta D, *et al.* (4 other authors). 2021. Phytochemical characterization and biological properties of two standardized extracts from a non-psychotropic *Cannabis sativa* L. cannabidiol (CBD)-chemotype. *Phytotherapy Research* 35: 5269–5281.

Nakumusana ES. 1986. Histopatological studies on the progress of infection of *Leptolegnia* sp. (SC-1) in *Anopheles gambiae* larvae exposed to zoospores in the laboratory. *Current Science* 55: 633–636.

Nandal SN, Bhatti DS. 1983. Preliminary screening of some weeds shrubs for their nematicidal activity against *Meloidogyne javanica*. *Indian Journal of Nematology* 13: 123–127.

Nazar MZH, Nath RP. 1989. Nematicidal effect of certain weed on root know nematode *Meloidogyne incognita* and *M. javanica* mixture. *Research Rajendra Agricultural University* 7(12): 95–96.

Nelson PN, Burrack HJ, Sorenson CE. 2019. Arthropod entrapment increases specialist predators on a sticky crop and reduces damage. *Biological Control* 137: e104021.

Nguyen LC, Yang D, Nicolaescu V, *et al.* (31 other authors). 2022. Cannabidiol inhibits SARS-CoV-2 replication through induction of the host ER stress and innate immune responses. *Science Advances* 8: eabi6110.

Nissen L, Zatta A, Stefanini I, *et al.* (4 other authors). 2010. Characterization and antimicrobial activity f essential oils of industrial hemp varieties (*Cannabis sativa* L.). *Fitoterapia* 81: 413–419.

Nok AJ, Ibrahim S, Arowosafe S, *et al.* 1994. The trypanocidal effect of *Cannabis sativa* constituents in experimental animal trypanosomiasis. *Veterinary and Human Toxicology* 36: 522–524.

Novak J, Zitterl-Eglseer K, Deans SG, Franz CM. 2001. Essential oils of different cultivars of *Cannabis sativa* L. and their antimicrobial activity. *Flavour and Fragrance Journal* 16: 259–262.

Oakes MD. 2018. *Uncovering cannabinoid signaling in C. elegans: a new platform to study the effects of medicinal Cannabis.* Doctoral thesis, University of Toledo, Ohio.

Olsson M. 2015. *Allelopatisk effekt av industrihampa som förfrukt till åkerböna, ärt och lupin.* Doctoral thesis. Sveriges Lantbruksuniversitet, Uppsala, Sweden.

Ona G, Balant M, Bouso JC, *et al.* (4 other authors). 2022. The use of *Cannabis sativa* L. for pest control: from the ethnobotanical knowledge to a systematic review of experimental studies. *Cannabis and Cannabinoid Research* 7: 365–387.

Osman AG, Elokely KM, Yadav VK, *et al.* (11 other authors). 2018. Bioactive products from singlet oxygen photooxygenation of cannabinoids. *European Journal of Medicinal Chemistry* 143: 983–996.

Pandey KN. 1982. Antifungal activity of some medicinal plants on stored seeds of *Eleusine coracana. Journal of Indian Phytopathology* 35: 499–501.

Park SH, Staples SK, Gostin EL, *et al.* (6 other authors). 2019. Contrasting roles of cannabidiol as an insecticide and rescuing agent for ethanol–induced death in the tobacco hornworm *Manduca sexta. Scientific Reports* 9: e10481.

Park SH, Pauli CS, Gostin EL, *et al.* (4 other authors). 2022. Effects of short-term environmental stresses on the onset of cannabinoid production in young immature flowers of industrial hemp (*Cannabis sativa* L.). *Journal of Cannabis Research* 4: e1.

Pastuhov SI, Matsumoto K, Hisamoto N. 2016. Endocannabinoid signaling regulates regenerative axon navigation in *Caenorhabditis elegans* via the GPCRs NPR-19 and NPR-32. *Genes to Cells* 21: 696–705.

Patwardhan A, Ghandhe R, Ghole V, Mourya D. 2005. Larvicidal activity of the fungus *Aphanomyces* (Oomycetes: Saprolegniales) against *Culex quinquefasciatus. Journal of Communicable Diseases* 37: 269–274.

Paul BD. 1955. *Health, culture, and community.* Russell Sage Foundation, New York.

Phillips AJL, Alves A, Abdollahzadeh J, *et al.* (4 other authors). 2013. The *Botryosphaeriaceae*: genera and species known from culture. *Studies in Mycology* 76: 51–167.

Pikuta EV, Hoover RB, Tang J. 2007. Microbial extremophiles at the limits of life. *Critical Reviews in Microbiology* 33: 183–209.

Potter DJ. 2004. "Growth and morphology of medical cannabis," pp. 17-54 in Guy G, Robson R, Strong K, Whittle B, *eds. The medicinal use of Cannabis.* Royal Society of Pharmacists, London.

Potter DJ. 2009. *The propagation, characterisation and optimisation of Cannabis sativa L. as a phytopharmaceutical.* Doctoral thesis, King's College, London.

Pudełko K, Majchrzak L, Narozna D. 2014. Allelopathic effect of fibre hemp (Cannabis sativa L.) on monocot and dicot plant species. *Industrial Crops and Products* 56: 191–199.

Radoševič A, Kupinič M, Grlič L. 1962. Antibiotic activity of various types of *Cannabis* resin. *Nature* 195: 1007–1009.

Radwan MM, Ross SA, Ahmed SA, *et al.* (3 other authors). 2008a. Isolation and characterization of new cannabis constituents from a high potency variety. *Planta Medica* 74: 267–272.

Radwan MM, ElSohly MA, Slade D, *et al.* (5 other authors). 2008b. Non-cannabinoid constituents from a high potency *Cannabis sativa* variety. *Phytochemistry* 69: 2627–2633.

Radwan MM, ElSohly MA, Slade D, *et al.* (3 other authors). 2009. Biologically active cannabinoids from high-potency *Cannabis sativa. Journal of Natural Products* 72: 906–911.

Rao J, McLements DJ. 2012. Impact of lemon oil composition on formation and stability of model food and beverage emulsions. *Food Chemistry* 134: 749–757.

Raven PH, Berlin B, Breedlove DE. 1971. The origins of taxonomy. *Science* 174: 1210–1213.

Ride WDL, *chair.* 1999. *International code of zoological nomenclature.* International Trust for Zoological Nomenclature, London.

Rodgers RH. 1980. Hail, frost, and pests in the vineyard: Anatolius of Berytus as a source for the Nabataean Agriculture. *Journal of the American Oriental Society* 100: 1-11.

Rozier F. 1787. "Mémoire sur la culture et rouissage du chanvre," pp. 1-136 in *Recueil de mémoires sur la culture et le rouissage du chanvre.* Chez les frères Périsse, Lyon.

Russo EB. 2011. Taming THC: potential cannabis synergy and phytocannabinoid-terpenoid entourage effects. *British Journal of Pharmacology* 163: 1344–1364.

Saccardo PA. 1879. Sulle cause determinant la sessualità nella canapa. *Bullettino della Società Veneto-Trentina di Scienze Naturali* 1879: 1–8.

Satyal P, Setzer WN. 2014. Chemotyping and determination of antimicrobial, insecticidal, and cytotoxic properties of wild-grown *Cannabis sativa* from Nepal. *Journal of Medicinally Active Plants* 3: 9–16.

Saxena R, Gupta S. 2004. Efficacy of aqueous extract of some medicinal plants against second stage juveniles of *Meloidogyne incognita. Current Nematology* 15: 51–53.

Scheifele G, Dragla P, Pinsonneault C, Laprise JM. 1997. *Hemp (Cannabis sativa) research report, Kent County, Ontario, Canada.* Ridgetown College Agricultural Technology, Ridgetown, Canada.

Schultes RE. 1986. Recognition of variability in wild plants by Indians of the northwest Amazon: an enigma. *Journal of Ethnobiology* 6: 229–238.

Schultz OE, Haffner G. 1959. Zur Kenntnis eines sedativen und antibakteriellen Wirkstoffes aus dem deutschen Faserhanf (*Cannabis sativa*). *Zeitschrift für Naturforschung* 14B: 98–100.

Sender R, Fuchs S, Milo R. 2016. Are we really vastly outnumbered? Revisiting the ratio of bacterial to host cells in humans. *Cell* 164: 337–340.

Shah NH, Dar AR, Qureshi IA, Afroza A, Wani MR, Lubna A. 2018. Control of root-knot disease of brinjal (*Solanum melongena* L.) by the application of leaf extracts of certain medicinal plants. *Indian Journal of Agricultural Research* 52: 444–447.

Shang YF, Feng P, Wang CS. 2015. Fungi that infect insects: altering host behavior and beyond. *PLoS Pathogens* 11(8): e1005037.

Sharma GC. 1996. Leaf extract and their action against *Meloidogyne incognita* (in vitro). *Journal of Phytological Research* 9: 155–157.

Sharma S, Devkota A. 2014. Allelopathic potential and phytochemical screening of four medicinal plants of Nepal. *Scientific World* 12(12): 56–61.

Shoyama Y, Sugawa C, Tanaka H, Morimoto S. 2008. Cannabinoids act as necrosis-inducing factors in *Cannabis sativa. Plant Signaling and Behavior* 3: 1111–1112.

Singh KV, Pathak RK. 1984. Effects of leaf extracts of some higher plants on spore germination of *Ustilago maydis* and *U. nuda*. *Fitoterapia* 55: 318–320.

Singh SP, Singh V. 2002. Evaluation of some plant extracts against root-knot nematode, *Meloidogyne incognita*. *Indian Journal of Nematology* 32: 226–227.

Singh NB, Thapar R. 2003. Allelopathic influence of *Cannabis sativa* on growth and metabolism of *Parthenium hysterophorus*. *Allelopathy Journal* 12: 61–70.

Singh KS, Cordeiro EMG, Troczka BJ, et al. (30 other authors). 2021. Global patterns in genomic diversity underpinning the evolution of insecticide resistance in the aphid crop pest *Myzus persicae*. *Communications Biology* 4: e847.

Siwulski M, Sobieralski K, Górski R, Lisiecka J, Sas-Golak I. 2012. Wpływ wybranych olejków eterycznych na wzrost grzybni i wiązanie owocników pieczarki *Agaricus bisporus* (Lange) Imbach. *Progress in Plant Protection* 52: 1217-1221.

Small E. 2015b. Response to the erroneous critique of my *Cannabis* monograph by R. C. Clarke and M.D. Merlin. *Botanical Review* 81: 306–316.

Smit J. 1930. *Die Gärungssarcinen: eine Monographie*. Gustav Fischer, Jena.

Somvanshi VS, Gupta MC. 2003. Effect of soil treatment with aromatic plants on population dynamics of *Tylenchorhynchus* and *Helicotylenchus* spp. and on maize plant growth. *International Journal of Nematology* 13: 195–200.

Srivastava PP, Das LL. 1974. Effect of certain aqueous plant extracts on the germination of *Cyperus rotundus* L. *Science & Culture* 40: 318–319.

Steiner G. 1941. Nematodes parasitic on and associated with roots of marigolds (*Tagetes* hybrids). *Proceedings of the Biological Society of Washington* 54: 31–34.

Stella B, Baratta F, Pepa CD, et al. (3 other authors). 2021. Cannabinoid formulations and delivery systems: current and future options to treat pain. *Drugs* 81: 1513–1557.

Strickland HE. 1843. Series of propositions for rendering the nomenclature of zoology uniform and permanent, being the report of a Commitee for the consideration of the subject, appointed by the British Association for the Advancement of Science. *London, Edinburgh and Dublin Philosophical Magazine and Journal of Science* 23(150): 108–124.

Stupnicka-Rodzynkiewicz E. 1970. Zjawiska allelopatii między niektórymi roślinami uprawnymi i chwastami. [Phenomena of allelopathy between some crop plants and weeds.] *Acta Agraria et Silvestria (Series Agraria)* 10(2): 75–105.

Teodosio G. 1553. *Medicinales epistolae LXVII*. Nicolaum Episcopium Juniorem, Basileae.

Thakur SK. 2014. Effect of green manuring and plant dry powder on soil properties and nematode infecting maize. *Agricultural Science Digest* 34: 56–59.

Turland NJ, ed. 2018. *International code of nomenclature for algae, fungi, and plants (Shenzhen code)*. Koeltz Scientific Books: Königstein, Germany.

Turlings TC, Tumlinson JH, Lewis WJ. 1990. Exploitation of herbivore-induced plant odors by host-seeking parasitic wasps. *Science* 250: 1251–1253.

Turner CE, ElSohly MA. 1981. Biological acitivity of cannabichromene, its homologs and isomers. *Journal of Clinical Pharmacology* 21: 283S– 291S.

Tusser T. 1557. *A hundredth good pointes of husbandrie*. Richard Tottel, London.

Ulmer A, Yousaf Z, Khan F, et al. (4 other authors). 2010. Evaluation of allelopathic potential of some selected medicinal species. *African Journal of Biotechnology* 9: 6194–62006.

Upadhyaya ML, Gupta RC. 1990. Effect of extracts of some medicinal plants on the growth of *Curvularia lunata*. Indian Journal of Mycology and Pathology 20: 144–145.

Van Breemen RB, Muchiri RN, Bates TA, et al. (4 other authors). 2022. Cannabinoids block cellular entry of SARS-CoV-2 and the emerging variants. *Journal of Natural Products* 85: 176–184.

Veliky IA, Genest K. 1972. Growth and metabolites of *Cannabis sativa* cell suspension cultures. *Lloydia* 35: 450–456.

Vijai P, Jalali I, Parashar RD. 1993. Suppression of bacterial soft rot of potato by common weed extracts. *Journal of the Indian Potato Association* 20: 206–209.

Vijayalakshmi K, Mishra SD, Prasad SK. 1979. Nematicidal properties of some indigenous plant materials against second stage juveniles of *Meloidogyne incognita*. *Indian Journal of Entomology* 41: 326–331.

Von Arx JA. 1987. Plant pathogenic fungi. *Nova Hedwigia Beihefte* 87: 1–288.

Vysots'kyi GA. 1962. Застосування фітонцидів з коноплі для контролю фузаріозу сосна. *Journal of Microbiology (Kiev)* 24(2): 65–66.

Wanas AS, Radwan MM, Mehmedic Z, et al. (3 other authors). 2016. Antifungal activity of the volatiles of high potency *Cannabis sativa* L. against *Cryptococcus neoformans*. *Record of Natural Products* 10: 214–220.

Williams JC, Appelberg S, Goldberger BA, et al. (3 other authors). 2014. Δ(9)-Tetrahydrocannabinol treatment during human monocyte differentiation reduces macrophage susceptibility to HIV-1 infection. *Journal of Neuroimmune Pharmacology* 9: 369–379.

Woese CR, Fox GE. 1977. Phylogenetic structure of the prokaryotic domain: the primary kingdoms. *Proceedings of the National Academy of Sciences USA* 74 (11): 5088–5090.

Woolhouse M, Gaunt E. 2007. Ecological origins of novel human pathogens. *Critical Reviews in Microbiology* 33: 231-242.

Zelepukha SI. 1960. Третя нарада з проблеми фітонцидів [Third conference on the problem of phytoncides]. *Mikrobiolohichnyi Khurnal (Kiev)* 22(1): 68–71.

Zelepukha SI, Rabinovich AS, Pochinok PY, Negrash AK, Kudryavtsev VA. 1963. Вивчення антимікробних властивостей препарату з конопль – кансантина 4 [Study of antimicrobial properties of preparations from hemp—cansantin 4]. *Mikrobiolohichnyi Khurnal (Kiev)* 25(2): 42–46.

Zengin G, Menghini L, Di Sotto A, et al. (13 other authors). 2018. Chromatographic analyses, *in vitro* biological activities, and cytotoxicity of *Cannabis sativa* L. essential oil: a multidisciplinary study. *Molecules* 23: 3266.

Zhang T, Breitbart M, Lee WH, et al. (7 other authors). 2006. RNA viral community in human feces: prevalence of plant pathogenic viruses. *PLoS Biology* 4(1): e3.

Zhang QY, Li ZL, Han BJ, et al. (3 other authors). 2013. Immediate responses of cyst nematode, soil-borne pathogens and soybean yield to one-season crop disturbance after continuous soybean in Northeast China. *International Journal of Plant Production* 7: 341–353.

4 Insects and mites that suck fluids from leaves and flowers

Abstract

This chapter lumps together insects with piercing-sucking mouthparts. The list is long: aphids, whiteflies, scales, leafhoppers, thrips, and true bugs (stink bugs, lygus bugs, capsid bugs). Here we include mites, which also suck. It's the largest chapter in the book, with the nastiest indoor pest (two-spotted spider mite, *Tetranychus urticae*), and other indoor nuisances: broad mites (*Polyphagotarsonemus latus*), whiteflies (*Trialeurodes vaporariorum, Bemisia tabaci*), thrips (no fewer than six genera), mealybugs (*Pseudococcus* spp.), and a sliding scale of scales (soft, giant, armored). Cannabis aphid (*Phorodon cannabis*) is possibly the most common outdoor problem—along with sisters *Phorodon humuli* and *Myzus persicae*, a phylogenetic topic we explore. The potato leafhopper (*Empoasca fabae*) is one of the most destructive *native* insects to North American agriculture. Leafhopper relatives include treehoppers, froghoppers, and planthoppers. Lastly there's a new pest with no clear solution, the hemp russet mite (*Aculops cannabicola*).

4.1 Introduction

The sap-sucking guild: insects with piercing-sucking mouthparts comprise the Paraneoptera, a monophyletic superorder (Johnson *et al.*, 2018). Subdividing Paraneoptera elicits debate, particularly within order Hemiptera—traditionally divided into Homoptera and Heteroptera. This tradition was upended by a phylogenetic study based on 18S rRNA (Von Dohlen and Moran, 1995), but a study based on mDNA has restored the tradition (Song *et al.*, 2012). Subgroups within Hemiptera employ different feeding strategies and give rise to various symptoms, as follows.

Homopterans (aphids, whiteflies, scales, leafhoppers) have long stylets and drill down to phloem vessels. This feeding strategy is nearly invisible until pest populations explode. **Heteropterans** (stink bugs and lygus bugs) are "lacerate-and-flush" feeders. This feeding strategy causes more destruction than other Paraneoptera, giving rise sometimes to browning and necrosis. **Thysanopterans** (thrips) have rasping mouthparts, and suck sap from parenchyma (epidermal) cells. They cause leaf stippling as well as cupping and curling of young leaves.

Mites are not insects, but they have piercing-sucking mouthparts. Their stylets are relatively short (100–150 µm) compared with phloem-sucking insects (up to 800 µm). Tetranychidae (spider mites) have a single and retractable stylet, whereas Eriophyidae (russet mites) have several stylets that are not retractable.

We present members of the sap-sucking guild in their order of economic impact. **Common names** are standards accepted by the Entomological Society of America (ESA, 2019). These standards may differ from common names used in Europe (Wood, 1989) or Australia (CSIRO, 2019). **Scientific names** (Latin binomials) can be unstable; we have included taxonomic synonyms that appeared in *Cannabis* literature ... as links to earlier publications.

4.2 Spider mites

Spider mites (Acari: Tetranychidae) are the most important family of plant-feeding mites. Over 1300 species occur worldwide, and most crops support one or more species. Primary sources on spider mites include Migeon and Dorkeld (2020) and Jeppson *et al.* (1975). Spider mites are generally small (<1.0 mm), have an oval shape, and are pale colored. Yellow or green are common colorations, but some are reddish, and dark spotting may be present. Colors vary within a species, and they change during **diapause**, a period when spider mites go into a dormant condition, and they become orange or reddish.

Spider mites feed on either side of a leaf, but more commonly on the underside, often concentrated near veins. Spider mite eggs are laid on the surface of the plant and are generally spherical. The first immature form that emerges from the egg is known as a **larva** and has only six legs. The larva molts into the next stage, the **protonymph**, which is eight-legged, the normal condition for most mites (except eriophyid mites). The protonymph transforms to a third immature stage, the **deutonymph**, that is similar in appearance but larger. The deutonymph molts into the adult form. Both male and females are produced, with males being slightly smaller, often with a more pointed abdomen. Spider mites can reproduce with a mixture of sexual and asexual reproduction, termed **arrhenotoky**, where females arise from fertilized eggs and males from unfertilized eggs. Populations on plants are typically skewed to higher proportions of females.

The common name "spider mite" refers to silk production by these mites, although silk is also produced by other kinds of mites and insects. Silk helps protect eggs and the colony from predators and effects of weather. Silk is used for locomotion, as silk trails can allow spider mites to move more easily within and between plants. Silk is also used to aid spider mites when they disperse on wind currents, known as ballooning.

DOI: 10.1079/9781836990352.0004

4.2.1 Signs and symptoms

Spider mites feed on plant cells using a **stylet**, a thin tube formed by two interlocking segments of their mouthparts known as chelicerae (Bensoussan *et al.*, 2016). The stylet penetrates cells of the mesophyll, from which mites extract fluids through a narrow canal in the center of the stylet. In reaching to the mesophyll, the stylet does not penetrate cells of the epidermis, but instead enters through stomates or between epidermal pavement cells. Enzymes introduced in saliva help exodigest organelles within the mesophyll cell, and cell contents are then extracted through the canal within the stylets. Because of the very small diameter of the food canal, only liquids or extremely tiny particles can be ingested.

In each feeding event the contents of a single cell are consumed. There is no immediately visible symptom, but feeding injuries cause cells to collapse, producing chlorotic spotting. These symptoms are often referred to as **stippling injuries** (Fig. 4.1A). Insects that feed on mesophyll cells also cause stippling (e.g., thrips, some leafhoppers). When high populations of spider mites are present, and damage becomes extensive, foliage may take on an overall change of color, often looking more grayish or bronze. Leaves and entire plants can be killed by spider mites during outbreaks. Plants appear bleached or brown, as if they have dried out.

Egg shells after hatching often remain attached to the leaf, as well as old "skins" (exuviae) discarded after molting. These can be helpful in diagnosing the presence of spider mites. Silk

produced by spider mites may also be useful in diagnosis, but silk often does not become visible until populations of mites are very high. When spider mites enter diapause, they migrate and cluster together at tips of leaves and flowering tops. Clusters of diapausing mites may grow as big as golf balls and contain thousands of individuals (Fig. 4.1B).

4.2a. Two-spotted spider mite (TSM)

Tetranychus urticae Koch 1886=*T. telarius* (L.) 1758, =*T. bimaculatus* Harvey 1893, =T. *cinnabarinus* (Boisduval) 1867

TSM is polyphagous and cosmopolitan, and the most economically important spider mite affecting crops worldwide. TSM feeds on plants in over 120 families (Migeon and Dorkeld, 2020). Its host range in glasshouses includes vegetables (e.g., tomatoes, cucumbers, eggplant, peppers), as well as flowers (e.g., rose, chrysanthemums, orchids). Outdoors in warm climates the pest attacks many field crops—maize, soybeans, legumes, and cotton. As far north as Canada the species causes problems in fruit trees, grapevines, hops yards, and berry crops.

TSM can be a key pest of *Cannabis* where ever it is grown, and may become very destructive when the crop is grown indoors. At the dawn of the indoor era, Stevens (1975) encountered TSM and proposed control measures. Since then, every indoor grower's guide has bemoaned the pest. "If spider mites become established in a garden room, they are nearly impossible to exterminate" (Cervantes, 2015).

Outdoor crops can also be damaged; the earliest reports of hemp damage were outdoors in Italy (Cuboni, 1889) and Germany (Kirchner, 1906; Blunck, 1920). Early North American records came from Illinois (Forbes and Hart, 1900) and California (Essig, 1915). In northeastern India, Fletcher (1917) wrote, "In Bengal the red spider is very bad on Indian hemp. This is the worst pest we have in Bengal on this plant." Cherian (1932) reported 30–50% losses in southern India. Even with adequate control measures, outdoor *gañjā* crops suffered considerable losses (Ayyar, 1963; Mukundan, 1964). TSM also infests wild-type plants in India (Choudhury and Mukherjee, 1972).

Early scientific names given to TSM depended on where it was found and its associated hosts. These names have since been synonymized under *T. urticae*. We list three synonyms that appear in older *Cannabis* literature. TSM has also been given many common names: "glasshouse spider mite," "simple spider mite," "common spinning mite," "red spider mite," and "carmine spider mite." The red and carmine labels are linked with the scientific name *T. cinnabarinus*. Studies by Auger *et al.* (2013) established that these are reddish form of *T. urticae*, and *T. cinnabarinus* is no longer considered to be a valid separate species.

Description: *Eggs* are spherical and 0.14 mm in diameter (Fig. 4.2A). Initially translucent to white, eggs turn a straw color just before hatching. Hatching *larvae* have six legs and two tiny red eye spots. *Protonymphs* become eight-legged and molt into *deutonymphs* (Fig. 4.2A), which later molt to the adult form. *Adult* females average 0.4–0.5 mm in length. Males are slightly smaller, with a less-rounded posterior. Typical coloration is straw-colored or pale green. Plum-red to brick-red

Fig. 4.1. Two-spotted spider mite: **A**. leaf stippling; **B**. webbing and clusters of diapausing mites (R. Clarke)

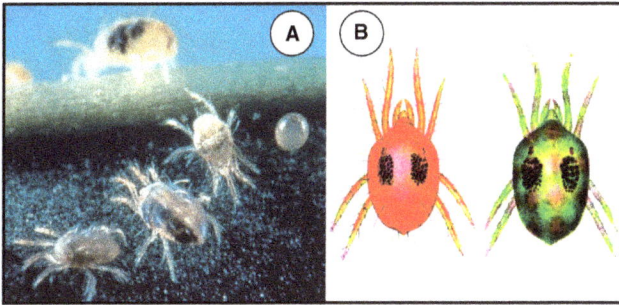

Fig. 4.2. Two-spotted spider mite: **A**. all life stages (courtesy Koppert Biocontrols); **B**. extremes in color variation (Ewing, 1914)

forms, the "carmine spider mite," are more common in warmer climates. Ewing (1914) illustrated extremes in color variation found among the descendants of a single female (Fig. 4.2B). Regardless of strain, when TSM enters diapause, females turn yellowish-orange to red. A pair of dark patches ("two-spotted") appear on the back of the mite. They are accumulations of waste, visible through the translucent body. The size of spots varies, and may be absent immediately following a molt (Fasulo and Denmark, 2009).

4.2.2 Life history and habits

TSM can reproduce sexually or asexually. Sex ratios may be skewed; Laing (1969) reported a 3:1 female-to-male ratio in a laboratory life history study. Females usually mate immediately after molting to the adult form. After a couple of days—the pre-oviposition period—females begin to lay eggs on host plant. Reports vary on the number of eggs produced. Laing (1969) found females typically laid 2–3 eggs each day over a period of 2–3 weeks. Shih *et al.* (1976) reported a higher number, averaging 143.9 eggs over 19 days.

Environmental factors affect the development, survival, and reproduction of TSM. Temperature is perhaps central; rate of development is closely associated with air temperature. For example, the life cycle (period when egg is laid until egg-laying adults are present) requires about 14 days at 21°C, but is halved to seven days at 29°C (Cloyd, 2011a). TSM prefers dry conditions; Hazan *et al.* (1973) reported an optimal relative humidity of 38% for the carmine mite strain in Israel.

TSM can reproduce continuously and year-round when conditions are favorable—as in indoor production. Outdoors in temperate climates, conditions preceding winter cause TSM to go into diapause. These conditions include decreased day length, exposure to cold temperatures, and decline of food plant quality or availability. These triggers may act alone, or interact in complex ways

Only adult females enter diapause, and adult females are the stage that survives between crop seasons. When in diapause, metabolism slows, and development and reproduction cease. Tolerance to stress increases, which may include the production of cryoprotectants for freeze tolerance. Diapausing TSM seeks sheltered locations such as crevices on bark, rocks,

or buildings (Bryon *et al.*, 2013). Diapause ceases, is "broken", when favorable conditions return, often a combination of rising temperatures and increasing day length. Some TSM populations are more likely to enter diapause than others, due to genetic differences (Takafuji *et al.*, 1991).

TSM develops mostly on leaves. Older leaves often show strongest symptoms of mite feeding injuries, but females tend to move to newer growth and lay eggs on emerging leaves (Cloyd, 2011a). Feeding damage results in loss of photosynthesis, and may accelerate water loss (De Angelis *et al.*, 1982). Seriously damaged leaves may be killed or abscise. As vegetative growth slows with flower production, feeding damage becomes concentrated on the developing flower buds. Cherian (1932) noted that spider mites were attracted to female flowers, and plants yielding the largest flowers were most heavily infested.

McPartland *et al.* (2000) published an infestation severity index for TSM (Table 4.1). Approaches to pest control are based on the infestation severity. Heavy infestations may cause allergies in workers (contact dermatitis, allergic rhinitis, asthma), as attested in many publications regarding other crops. Biotic stress induced by TSM causes leaves and flowers to upregulate production of cannabinoids (Kostanda and Khatib, 2022).

Other members of the spider mite family (Acari: Tetranychidae) have been identified as *Cannabis* pests, as follows.

4.2b. Citrus brown mite

Eutetranychus orientalis (Klein) 1936
Description: Eggs are subspherical, pale brown, 0.14 mm diameter. Larvae are light brown and six-legged. Nymphs emerge from molts with 8 legs, like adults. Adult females are round-bodied, greenish-brown, up to 0.5 mm long, and covered with short spines. Males are more slender, smaller, and redder than females.
Comments: Citrus brown mite normally attacks Citrus spp., but infests a wide variety of hosts, including Cannabis in Pakistan (Chaudhri et al., 1974), India (Dhooria, 1983; Gupta, 1985), and Greece (Stavraki, 2005). Upper leaf surfaces of infested plants turn yellow, then red-brown, then die. On Cannabis in Pakistan E. orientalis is preyed upon by the phytoseiid mite Neoseiulus longispinosus (Raza, 2008). A related species, Eutetranychus africanus Tucker 1926, was reported as "an important pest of hemp plants" in Thailand (Wasuwan et al., 2021). No one else has cited Cannabis as a host of E. africanus (Jeppson et al., 1975; Smith Meyer, 1987; Bolland et al., 1998).

Table 4.1. Infestation Severity Index for TSM

Light	Any mites seen
	Often no symptoms present
Moderate	<5 mites/leaf (not leaflet)
	Feeding patches present
Heavy	>5 mites/leaf
	Feeding patches coalescing
Critical	>25 mites/leaf
	Shriveled leaves and webbing

4.2c. Glover's spider mite

Tetranychus gloveri Banks 1900
Description: *Adult* females are red, males yellow-orange. *Eggs* are colorless when laid. *T. gloveri* is also known as the cotton red mite. Identification characteristics separating it from *T. urticae* are provided by Beard (2018).
Comments: Glover's spider mite has been recorded from over 110 plant species, including cotton, okra, beans, peas, celery, eggplant, and sweet potato (Jeppson *et al.*, 1975; Beard, 2018). Leaf injuries are described as rusty blotches and speckling. Glover's spider mite is primarily a tropical species, with most records from South America, Florida, and the Caribbean. Damage to hemp was observed in southern Florida (Lance Osborne, University of Florida, pers. comm. to WC, April 10, 2020) where populations reached levels that began to produce leaf loss. A mass release of the predatory mite *Phytoseiulus macropilis* was able to provide control.

4.2.3   Differential diagnosis

Leaf injury by TSM is perhaps most similar to injury produced by thrips. Thrips often produce somewhat larger and more silvery flecking wounds, and small dark fecal spots often mark the damaged areas. Unlike TSM, thrips do not leave behind egg shells or discarded "skins" on the leaf surface, which is a useful diagnostic sign.

Other mites (russet mites and broad mites) are much smaller than TSM. Their mouthparts are also much smaller, which limits feeding to epidermal cells, so they do not produce the chlorotic spotting that occurs when mesophyll cells collapse. Generalized off-colors (bronzing, graying), a brittleness of foliage, and distortions of new growth are common symptoms of russet mites and broad mites. Russet mites and broad mites do not make visible webbing.

Chlorotic spotting can be produced by some leafhoppers and lace bugs, which suck fluids from mesophyll cells. Leafhoppers produce larger leaf spots than those of spider mites. Populations of leafhoppers never approach those of spider mites, so the spotting is much less common, and more widely dispersed on leaf surfaces. Lace bugs will produce even larger areas of chlorotic spotting, but can be easily diagnosed by viewing the undersides of leaves. Dark fecal "tar spots" mark the area where lace bugs feed. Old egg shells and molted "skins" are also present on leaves, but these are much larger and more visible than those produced by TSM.

4.2.4   Cultural and mechanical control

Cultural and mechanical methods are elaborated in Chapter 19. For indoor production, the primary focus should be on prevention (exclusion). Holding new plants in a quarantine area, until determined to be pest-free, is a very important practice to prevent spider mites—and many other pests of the crop. If the growing facility is not well sealed and entry points exist that can allow mites to crawl indoors, these should be sealed. TSM may develop on many kinds of weeds that grow outside around the building, so a plant-free barrier should be maintained around the building.

If TSM is present in a building, then efforts are needed to prevent its spread from infested plants to plants that are clean of mites. These areas should be physically separated. Movements of people and plants should be curtailed between infested rooms and clean rooms. Attention should be given to wind currents generated by fans and air-conditioning. TSM is readily blown on air currents. Wind should be directed so that airborne mites do not flow to mite-free areas of the building. Implacable infestations prod small-scale growers to start over: remove everything from the growroom, and thoroughly disinfest the space. Cervantes (2015) wiped down walls and equipment with 5% bleach, and rented a wall-paper steamer to sterilize cracks in walls.

A program of continuous monitoring is central to any TSM management program. This is best done by regular visual inspection to detect leaf injury symptoms produced by TSM. Suspect leaves can be checked under magnification to detect live mites, old egg shells, or discarded "skins" to confirm the diagnosis.

Some conditions affecting plant growth can also affect TSM. High levels of nitrogen fertilization, and the succulent plant growth it promotes, improves food quality for TSM, increasing its reproduction. Key TSM nutrients (amino acids, salts) can become more concentrated in drought-stressed plants, which also improves food quality, and leaf temperatures often rise as stomates close to conserve water. TSM prefers high temperatures and low relative humidity. Hazan *et al.* (1973) advocated a high-humidity regime for controlling TSM (which unfortunately predisposes plants to gray mold).

Use of cultivars with resistance to TSM will be a useful management method in the future, but has not been developed. Anecdotal evidence does suggest that genetic resistance may exist, as growers sometimes have observed cultivars that are less or more prone to TSM problems. Cherian (1932) compared six varieties of Indian *gañjā* for resistance to TSM (carmine mite strain). The most resistant varieties unfortunately produced the smallest female flowers.

4.2.5   Biological control

Biological control organisms cited here are discussed at length in section 20.6. A mixture of *Phytoseiulus persimilis*, *Neoseiulus californicus*, and *Mesoseiulus longipes* provides excellent biocontrol for most situations. Wainwright-Evans (2017) reported best success with *Phytoseiulus persimilis*, *Amblyseius californicus*, and *Neoseiulus fallacis*. For outdoor crops, *Galendromus occidentalis* and *Galendromus pyri* have been used. Other predatory mites that eat spider mites in the absence of their primary hosts include *Neoseiulus cucumeris* and *Iphiseius degenerans*.

As soon as clones are rooted, immediately start them on a predatory mite program (Wainwright-Evans, 2017). Wainwright-Evans also reported success with the green lacewing *Chrysoperla rufilabris* and the predatory midge *Feltiella acarisuga*. The predatory beetle *Stethorus punctillum*, which specializes in *Tetranychus* spp., proved less helpful. Mahieu *et al.* (2019) preferred the predatory mites *P. persimilis* and *N. californicus*. Grammenos *et al.* (2021) reported success using a combination of *N. californicus* and *F. acarisuga*, released four times in April and May in a glasshouse in Greece.

For outdoor crops, monitor populations of naturally occurring predators and parasites. For indoor crops, *biocontrol must be established before spider mite populations explode*. If mite populations get a head-start, biocontrols never catch up. Predatory mites work best against low or moderate TSM infestations (Table 4.1). This is because predatory mites, such as *P. persimilis*, reproduce every seven days at 21°C—twice as fast as TSM. Populations of predators soon overpower a few stray pests (Fig. 4.3).

California DPR (2017) approved the use of two microbial biocontrols for treating TSM, the bacterium *Burkholderia* spp. strain A396, and the fungus *Cordyceps* (*Isaria*) *fumosorosea*. In fact they have limited usefulness against TSM. Optimal conditions for TSM (warm, dry) actually inhibit *I. fumosorosea*; Zemek *et al.* (2016) found a lack of efficacy in glasshouse conditions. *Burkholderia* spp. strain A396 has some efficacy against TSM (Cordova-Kreylos *et al.*, 2013). The fungus *Hirsutella thompsonii* works against TSM. A spider mite-specific fungus, *Neozygites floridana*, is under development.

To control citrus brown mite (*Eutetranychus orientalis*), Dhooria (1983) cited two biocontrols—a predator mite (*Amblyseius alstoniae*) and a thrips (*Scolothrips indicus*). *Neoseiulus californicus* feeds on *E. orientalis* but prefers TSM. Wasuwan *et al.* (2021) successfully controlled *Eutetranychus africanus* on hemp with the biocontrol fungus *Metarhizium* sp. strain BCC 4849. This strain from Thailand was much more effective than commercially available *Metarhizium anisopliae*.

4.2.6 Chemical control

TSM has long been one of the more difficult-to-manage of any agricultural pests, worldwide. Its capacity to develop resistance against a broad range of pesticides has produced special challenges. It is far easier to find and use pesticides that are completely ineffective for control of TSM under real-life growing conditions—or worse, to find and use pesticides that will create

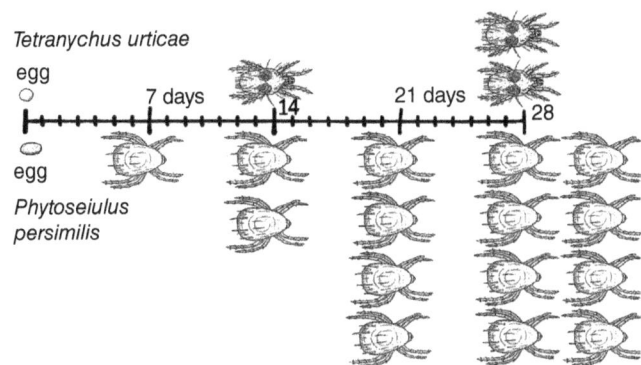

Fig. 4.3. A month of mites (redrawn from Olkowski *et al.*, 1991)

greater problems with spider mites—than to identify a pesticide that will actually provide grower benefit.

One insecticide that has been widely used on *Cannabis* that can aggravate TSM problems is the natural product pyrethrins, as well as its synthetic derivatives, known as pyrethroids. These are poor products for control of almost any spider mite, and one can expect TSM to be resistant to them. Pyrethrins and pyrethroids often will cause TSM to increase activity and disperse, accelerating its spread through a crop (Penman and Chapman, 1988).

Imidacloprid is another insecticide that has been used (illegally) in *Cannabis* production that has been shown to increase problems with TSM. In this case, imidacloprid and the related neonicotinoid chlothianidan have been shown to suppress plant defenses that help resist spider mites (Szczepaniec *et al.*, 2013; Ruckert *et al.*, 2018).

The status of whether a pesticide can be legally used on a *Cannabis* crop—or any crop—is something that is determined by regulatory agencies of governments that establish which pesticides can be safely used on crops and their manner of use. Each country has its own agencies that determine which pesticides are allowable for use within their borders. In the USA, that regulatory unit is the Environmental Protection Agency. However, because of the peculiar history of *Cannabis sativa* crops in the USA, long illegal under federal law, a unique situation has evolved where presently (2024) regulation of which pesticides are allowed to be used on *Cannabis* has devolved to individual states. This situation is further discussed in section 21.4.

Problems arising from illegal pesticide usage flared up when *Cannabis* crops were first legalized by some states within the USA. At that time many growers used pesticides that they found recommended through internet sources or from word-of-mouth. This resulted in extensive use of clearly illegal and possibly unsafe pesticides on *Cannabis*, such as abamectin, acequinocyl, bifenazate, and extazole for TSM control (McPartland and McKernan, 2017). When these uses became exposed in numerous media reports across the USA, public concerns about the safety of *Cannabis* products became widespread. The history of irresponsible practices used to manage TSM on *Cannabis* in the USA continues to cloud the reputation of the industry to this day.

Some pesticidal active ingredients are more easily registered for crop use. For example, the need to establish a tolerance level for a pesticide may be exempt "because the chemical is considered to be safe enough for the use described in the tolerance exemption that a maximum permissible level does not need to be established" (EPA, 2020). This includes active ingredients with some miticidal activity such as horticultural oils (mineral oils, plant seed oils), insecticidal soaps, sulfur, and some microbial-derived pesticides. Other pesticides are classified as "minimum risk" or FIFRA (Federal Insecticide, Fungicide, and Rodenticide Act) "Section 25(b)" pesticides that do not have to meet federal registration requirements (EPA, 2020). Examples of these are corn oil, sodium lauryl sulfate, potassium sorbate, and many essential oils (thyme, rosemary, garlic, peppermint).

Of these, sulfur has the longest history of use. Sulfur can be used to help control spider mites on some crops, although few sulfur products are legally sanctioned for indoor or glasshouse use in the USA. Growers have circumvented this by using sulfur burners, which vaporize sulfur pellets, and claim its use as a "fertilizer" rather than a pesticide, and therefore legal. Disregarding the obvious falsity of the value of sulfur fertilizer applied through this method and its dubious legality, this practice also poses serious potential hazards to humans, equipment, and natural enemies of spider mites. In high humidity, vaporized sulfur produces sulfuric acid, which causes chlorosis in *Cannabis* leaves and flowers (Goidànich, 1959). TSM resistance to sulfur is another consideration, and was first reported over a century ago (Parker, 1913b).

There have been numerous studies documenting adverse effects of sulfur on biocontrol organisms, including predatory mites that are used to suppress TSM (Anonymous, 2020). Much of this has involved observations of TSM population resurgence following sulfur use on fruit crops. James and Prischman (2010) reported the deleterious effects of sulfur on predatory mites in fields of hops, a *Cannabis* relative, which led to a rebound of TSM. In an early study of sulfur products, Cherian (1932) tested a lime–sulfur spray and a fish oil soap. *Gañjā* sprayed with lime–sulfur "was tested by veteran smokers who gave their verdict against it." *Gañjā* sprayed with fish oil soap passed the smoker's test.

Horticultural oils, including mineral oils and various oils extracted from seeds (e.g., neem, cottonseed, soybean, canola) act primarily by smothering. Decades of controlled field studies, on many crops, have demonstrated the effectiveness of "hort oils" against TSM. Ability to control the egg stage is a feature of hort oils that few other pesticides have been shown to provide.

Insecticidal soaps are less well studied, and are less commonly included in TSM management recommendations. Both hort oils and insecticidal soaps are purely contact pesticides with no residual effects. Thorough application that coats the spider mite is required for hort oils and soaps to control spider mites, and repeated applications are required. Some reports demonstrated that hort oils, and especially insecticidal soaps, can have adverse effects on predatory mites. However, oils and soaps lack residual effects, so susceptible natural enemies can be released on plants very shortly after application.

Azadirachtin is extracted from neem seed oil. Azadirachtin is useful as an insecticide, acting as an insect growth regulator, but no studies support its use as a miticide. Most azadiractin labels do not include spider mites as a controllable pest (although products marketed to *Cannabis* growers often do). Neem seed oil does have miticidal activity, similar to other hort oils.

Essential oils (EOs, i.e., terpenoids extracted from aromatic plants) have long attracted attention as natural products for insect and mite control. In recent years a great many pesticides using EOs have been developed, and they are exempt from pesticide registration (Section 25(b) pesticides in the USA). Many EO-based products are marketed to *Cannabis* growers, and some jurisdictions have allowed the use of EOs on *Cannabis* crops (e.g., California DPR, 2017). For some EO-based products (rosemary oil, thyme oil), some published studies show their ability to kill TSM under laboratory conditions (e.g., Cloyd *et al.*, 2009).

Full-scale field studies that have demonstrated efficacy against TSM are largely lacking for Section 25(b) products marketed for TSM control on *Cannabis* crops. It is of interest that some laboratory studies have shown that *Cannabis* extracts can kill TSM under laboratory conditions. These include EO extracts (Górski *et al.*, 2016), ethanol extracts (Taisiya *et al.*, 2010), and aqueous extracts (Fenili and Pegazzano, 1974).

Pesticides can be useful in managing TSM during early stages of clonal production, when used as prophylactic agents. Wainwright-Evans (2017) recommended dipping clones in horticultural oil or insecticidal soap. These prophylactics can be applied thrice: to mother plants before clones are cut, on clones several days after transplanting, and again on clones the day before flowering is induced.

The use of pesticides must also consider how well they integrate with biological controls. Summaries of pesticide compatibility with biocontrols have been developed by some major suppliers of biological control species used on indoor/glasshouse crops (e.g., Koppert, Biobest) and are publicly accessible. Some examples appear in Table 20.1.

4.3   Eriophyid mites

The **hemp russet mite** (HRM) belongs to a very large group of mites known as eriophyid mites or russet mites (Acari: Eriophyoidae), with over 5000 species (Zhang, 2017). Present taxonomy divides superfamily Eriophyoidae into three families: Diptilomiopidae, Phytoptidae, and Eriophyidae. The latter is the largest and most important family. Within the Order Acari (mites and ticks), only the spider mite family (Tetranychidae) is more damaging than the Eriophyidae family.

HRM and other eriophyid mites are minute, around 0.2 mm when full grown, with a unique body form that is elongated and tapered at their posterior end. Also unusual is that they possess only two pairs of legs, half the number typical of mites and other arachnids. Most eriophyids, including HRM, live on the surface of plants as leaf vagrants. They often produce plant injury in the form of bronzing, hence the names "rust mites" or "russet mites." A minority of eriophyids, known as "gall mites," live under the leaf surface, which induces various plant distortions and growths within which the mites reside.

There can be variations in the life history of eriophyid mites, but basically there are four life stages (Manson and Oldfield, 1996). Eggs are laid on plants, usually on the leaf surface or around buds. After egg hatch there are two immature stages, a larva followed by a nymph. They resemble the adult stage in general form, but are smaller. Both the larva and nymph feed on plants, with a short resting period at the end of each of these stages before they molt.

Adult males and females are very similar in appearance, unlike many other mites. Also dissimilar is that they do not directly mate. Instead the male will produce packets of sperm (spermatophores) that it places on the plant. The female moves to a spermatophore and takes it into the genital opening, fertilizing her eggs. Some eriophyid mites will produce a different female form (deuterogyne) that can survive through periods of adverse conditions when mites go dormant or into diapause.

This form may have slightly different appearance, with modifications of the exterior cuticle that can help it conserve water.

4.3.1 Signs and symptoms

HRM and other eriophyid mites feed on epidermal cells of the leaf surface. Injured cells may collapse and die, and when large numbers of mites are present this may result in various leaf discolorations: bronzing, russeting, or graying/silvering. The cardinal symptom by spider mites—chlorotic spotting (stippling)—is not caused by eriophyid mites, since their mouthparts cannot extend beneath epidermal cells. Eriophyid mite injury can induce changes in cells adjacent those where the mites feed, producing thickened cell walls. This results in increased thickness and/or brittleness of foliage (Petanović and Kielkiewicz, 2010a,b).

Symptoms caused by HRM were first illustrated by McPartland and Hillig (2003). Most feeding occurs on the underside of leaves, but HRM can be present on the upper leaf surface. Mites may crowd plants by the thousands, their bodies giving leaves a powdery beige appearance (Fig. 4.4A). High populations cause extensive damage to epidermal cells, producing the aforementioned leaf discolorations (Fig. 4.4B). As the quality of leaves degrades following extensive damage, HRM will migrate to new sites and settle on newly expanding leaves.

When HRM feeds on expanding leaves, symptoms of leaf curling may appear. A slight up-rolling along the margins of leaflets, particularly the base of the leaflet, is most often reported. Leaves may also cup downward. These differences in leaf curling patterns are due to HRMs feeding primarily on the upper leaf surface (producing an upward curl) or lower leaf surface (producing downward cupping), in a manner similar to leaf curling patterns produced by broad mite (Gerson, 1992). When HRMs congregate on petioles, the petioles become brittle, and leaves may break off.

On developing female flowers, HRMs selectively feed on pistils, which necrose and die prematurely. High populations reduce the size of buds and diminish resin production. Cannabinoid concentrations are reduced (Hayes et al., 2023). A few eriophyid mites are vectors of plant viruses, a subject reviewed by De Lillo et al. (2018) and Oldfield and Proeseler (1996). None of the viruses presently known from Cannabis are vectored by eriophyid mites.

4.3a. Hemp russet mite

Aculops cannabicola (Farkas) 1965
= *Vasates cannabicola* Farkas
HRM was discovered on fiber-type hemp in Hungary (Farkas, 1965). A couple of years later it was found in North America (Kansas) on wild-type hemp (Hartowicz et al., 1971). The damage it caused led to proposals for its deployment against illicit *Cannabis* as a biocontrol weapon (Eaton et al., 1972; Weber, 1978). HRM has also been reported from Iran (Panah and Namaghi, 2014). It infests both indoor and outdoor crops (McPartland and Hillig, 2003; Cranshaw et al., 2019). In a survey of California *Cannabis* growers (Wilson et al., 2019), 36% reported problems with HRM, somewhat higher than those reporting problems with spider mite (33%).

Description: HRM has a body form typical of most eriophyid mites; elongate-fusiform and wormlike, with only two pairs of legs (Fig. 4.5). The body is pale beige or very pale pink. *Adults* are 0.17–0.21 mm long—a quarter the size of *Tetranychus urticae*. Immature stages are similar in body form but smaller than the adults. Petanović et al. (2007) supplemented the original description by Farkas (1965) with line drawings of males and juvenile nymphs. *Eggs* are usually spherical and clear or translucent. They are small, ranging from 0.02 to 0.06 mm, but relatively large compared with the mother, perhaps 15–25% her body length.

Fig. 4.4. HRM injury: **A.** early, outdoors (W. Cranshaw); **B.** late, indoors (R.T. Villanueva)

4.3.2   Life history and habits

The biology of HRM is very little studied, but is likely similar to related eriophyid species that feed on other herbaceous hosts (Cranshaw *et al.*, 2019). The best studied of these is tomato russet mite (*Aculops lycopersici*), reviewed by Perring and Farrar (1986) and Perring (1996). The entire life cycle of that species (initially laid egg through first egg laying by the adult) is completed within 2 weeks at temperatures of 25°C. Egg production by tomato russet mite typically averages 1–2 dozen per female. Adults of tomato russet mite normally live about 3 weeks.

In some important aspects, the biology of tomato russet mite and HRM may differ. To date *Cannabis* is the only reported host for HRM, and attempts to establish it on related plants in the Cannabaceae family (hops, hackberry) have so far been unsuccessful (Cranshaw, unpublished data). In contrast, tomato russet mite has been found on tomato as well as pepper (*Capsicum*), jimsonweed (*Datura stramonium*), field bindweed (*Convolvulus arvensis*), and some small fruits (wild gooseberry, blackberry) (Perring, 1996). These non-crop hosts allow tomato russet mite to survive when tomato crops are absent.

Tomato russet mite is not reported to produce a special stage that allows extended survival during periods when host plants are not present (e.g., a deutogyne form). In contrast, HRM likely does produce a deutogyne form, because it is adapted to survival between seasons in climates with freezing winter weather, when host plants are not available. Studies are needed to determine how HRM survives during periods when living *Cannabis* plants are not present, including investigations into whether diapause forms exist.

HRM is capable of surviving and reproducing year-round on *Cannabis* crops grown indoors in continuous culture. Under such conditions the pest causes serious damage. Outdoors, HRM populations on hemp grown from indoor-grown transplants were observed to steadily increase from an average of 50/leaf in early June to 450/leaf through early September (Cranshaw *et al.*, 2019).

On their own, HRMs can crawl only very short distances, and immature stages are particularly immobile. However, adults are capable of some crawling, and may move to the edge of leaves where they can be picked up and carried on air currents. In enclosed areas, fans can quickly spread mites. Small breezes readily distribute HRMs through fields. The mites are easily collected on greased slides placed above the crop canopy of infested plants.

4.3.3   Differential diagnosis

Injuries produced by HRM are similar to those of broad mite (see next section). Both species feed on epidermal cells, and high populations cause leaves to become bronzed to gray and brittle. Both may also cause leaf distortion injuries, notably leaf curling. Leaf curling produced by broad mite may be more pronounced than that of HRM. Leaf thickening/brittleness and, particularly, brittleness of petioles is a symptom that is more likely associated with HRM.

Leaf curling can be an indication of infestation by HRM (or broad mite) but can never be used as a positive diagnosis. Not all plants produce leaf cupping symptoms in response to HRM feeding, and varietal differences are possible. Some hemp varieties produce similar leaf curling ("taco leaf") in the absence of mites. HRM will not produce the stippling (chlorotic spotting) produced by two-spotted spider mite or thrips. HRM does not produce silk. Diagnosis of HRM requires microscopic examination of leaves to detect the presence of the mites. Their small size, unique body shape and color, and slow movement on the plant all can be used to readily distinguish HRM from other mites on *Cannabis* plants. Koike (pers. comm.., 2020) encountered both two-spotted spider mites and HRM on the same plant—perfect for a size comparison (Fig. 4.6).

Fig. 4.5. *Aculops cannabicola* mounted on a microscope slide

Fig. 4.6. Two-spotted spider mite surrounded by HRMs (S. Koike)

Falcon-Brindis *et al.* (2023) estimated the size of HRM populations using photographs taken with a handheld digital microscope (×55), by imaging the underside of proximal and middle sections of the central leaflet. Mite counts were conducted either directly from the photograph or via the Adobe Photoshop counting tool. An infestation severity index, similar to that of spider mites (Table 4.1) has not yet been proposed.

4.3.4 Cultural and mechanical control

In glasshouses and growrooms, strict sanitation is key: start with clean plants. Make sure mother plants are mite-free. Propagating clones is a particularly vulnerable period. Closely check plants before introducing them into new rooms. The small size of HRM and an absence of visible symptoms until populations are large can allow crops to harbor non-detectable populations for months following an initial infestation. In Colorado, outdoor crops plagued by HRM likely resulted from plants initially infested in hoop house nurseries (Cranshaw *et al.*, 2019).

Quarantine questionable clones, and observe for HRM before moving clones into growrooms. It is absolutely essential to regularly scout for HRM in all grow situations. Once part of a growing operation is found to be infested, make efforts to prevent infestations of new (presumably clean) plantings. Limit visitors and staff. People and equipment can move from clean areas to infested areas, but the reverse movement must be restricted (Wainwright-Evans, 2017).

Mechanical controls have limited efficiency. Outdoor growers with heavy infestations have burned their fields (Cranshaw *et al.*, 2019). In a glasshouse with prized plants, Hillig killed HRM by enclosing potted plants in large plastic bags, and inflated the bags with CO_2 for 2 hours (McPartland and Hillig, 2003). Cranshaw (unpublished data, 2019) tried this method and found it unreliable, with high potential to cause plant injury. To kill 100% of HRM required >50% CO_2 for 24 h.

Hayes *et al.* (2023) tested the effects of hot-water treatments or surfactant dips on HRM populations. Clone cuttings rooted in rockwool were infested with mites, and then treated six ways. Treatment at 43°C for 15 min was the most effective (more so than 41°C, or for 10 min), with a 75% reduction in mite populations compared with untreated control. Next in effectiveness (68% reduction) was a surfactant treatment—dipping clones in Dr Bronner's Pure-Castile Soap (1%) for 1–2 seconds, just long enough to fully submerge the clone. None of the treatments affected rooting success, indicating that all treatments were well tolerated by the cuttings.

4.3.5 Biological control

In Colorado, Cranshaw *et al.* (2019) found that outdoor crops infested with HRM have very low populations of naturally occurring biocontrol organisms, such as predatory mites, minute pirate bugs (*Orius* spp.), or spider mite midges (*Feltiella acarisuga*). Mahieu *et al.* (2019) recommended the predatory mites *Amblyseius*

swirskii and *Amblyseius andersoni*. In fact, no commercially available predatory mites have been clearly demonstrated to be effective against HRM (Cranshaw *et al.*, 2019). Specific time-points in production, particularly during clonal propagation, might be favorable for their use. The biocontrol fungus *Beauveria bassiana* strain GHA (BoteGHA® ES) controlled HRM in a short-term 2-week study, whereas *Cordyceps* (*Isaria*) *fumosorosea* apopka strain (PFR97™) did not (Villanueva *et al.*, 2021). Other eriophyid mites infesting other host plants have been controlled with *Hirsutella thompsonii* (McCoy, 1996).

4.3.6 Chemical control

Controlling russet mites in other crops has been difficult. Even high-powered synthetics, off-limits in *Cannabis*, work poorly (Childers *et al.*, 1996; Van Leeuwen *et al.*, 2010). California DPR (2017) posted a list of approved pesticides, most of which don't work. Limited trials indicate that horticultural oils (e.g., SuffOil-X®) appear promising for suppressing HRM (Cranshaw *et al.*, 2019; Villanueva *et al.*, 2022c).

Britt and Kuhar (2020) trialed 11 products in a growroom infested with HRM. Vegetative-stage plants were sprayed until runoff using a manual-pump spray bottle, and surviving mites were counted 22 days later (Fig. 4.7). Control was provided by SuffOil-X®, Mammoth® (thyme oil), and PLP® (citronella, lemongrass, peppermint, cinnamon, garlic oils). They performed as well as synthetic pesticides off-limits in *Cannabis* (Agrimec®, Movento®). Next came Sulfur #1, Requiem™ (plant extract of *Chenopodium ambrosioides*), and Venerate® (heat-killed *Burkholderia* sp.). Lastly Grandevo® and M-Pede® were not significantly different than untreated control plants.

In a short-term 2-week study, the use of DeBug® Optimo (neem oil + azadirachtin) and Sil-Matrix® LC (potassium silicate) was not significantly different than plants sprayed with water (Villanueva *et al.*, 2021). Sil-Matrix® LC showed efficacy 3 days after spraying, but not at 2 weeks.

In a 2-year field trial, Hayes *et al.* (2023) tested the efficacy of sulfur (Microthiol Disperss®). Plants were sprayed early in the growing season (July 5 or July 15), or late in the season at the beginning of flowering (August 16 or 17), or a combination

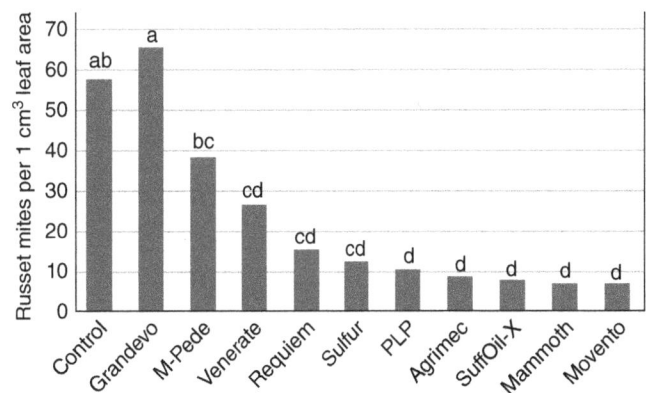

Fig. 4.7. Pesticides trial (adapted from Britt and Kuhar 2020). Columns with the same letter not significantly different

of both. Following dual application, HRM populations were reduced by 98% the first year (nearly eradicated), but only 43% the second year compared with untreated controls. With single applications, July treatments worked better than August treatments. Dual application was the only treatment that significantly increased yields (flowers and leaves) compared with control. Treatment also resulted in a significant increase in cannabinoid concentrations.

The effectiveness of sulfur (Microthiol Disperss®) was also demonstrated by Szczepaniec *et al.* (2023) in a field trial. Its control of HRM was on par with fenpyroximate, a synthetic acaricide, as well as rosemary oil. The latter was the most effective, a nice surprise. Mineral oil (SuffOil-X®), abamectin, and synthetic etoxazole were not effective. Szczepaniec and colleagues also conducted a greenhouse trial, where all the tested pesticides knocked down HRM populations.

New clones can be dipped in pesticides as a preventive measure. Wainwright-Evans (2017) recommended SuffOil-X®, but emphasized that clones should still be isolated for observation. Cranshaw (unpublished data, 2019) found that dipping clones in Sulfur #1 provided good control; SuffOil-X® and neem oil were less effective. Insecticidal soap worked poorly; pyrethrins and azadirachtin were totally worthless.

4.4   Tarsonemid mites

Tarsonemid mites (Acari: Tarsonemidae) are very small, often less than 0.3 mm long. They are nearly colorless or may have a pale brown tint. Their life cycle differs from that of spider mites; they have only four life stages: egg, larva (six-legged, feeding), nymph (eight-legged, non-feeding), and adult. Development can be very rapid; generation times can be completed in about 1 week.

Two species of tarsonemid mites are widely distributed and can be damaging to a wide range of crops plants: the cyclamen mite (*Phytonemus pallidus*) and the broad mite (*Polyphagotarsonemus latus*). Only the latter is known to affect *Cannabis* crops. Broad mite was first reported on *Cannabis* by Goff (1987) on outdoor crops in Hawai'i. Wainwright-Evans (2017) reported problems with indoor-grown crops. In a survey of California growers, 8% reported problems with broad mite (Wilson *et al.*, 2019). Broad mite causes problems in Dutch greenhouses (Mahieu *et al.*, 2019).

Broad mite has a worldwide distribution and wide host range. Crops listed by Fasulo (2016) include apple, avocado, cantaloupe, castor, chili, citrus, coffee, cotton, eggplant, grapes, guava, jute, mango, papaya, passion fruit, pear, potato, sesame, string or pole beans, tea, and tomato. The pest is common on outdoor crops in warmer, tropical climates. In temperate areas it primarily damages indoor-grown crops, including many ornamental and flowering plants (Cloyd, 2011b).

4.4.1   Signs and symptoms

Broad mites often concentrate on actively growing, meristematic tissues. Their minute size limits their ability to feed within the leaf, and feeding is restricted to cells of the epidermis. Growth distortions often result when mites feed on new growth, producing leaf curling or twisting (Fig. 4.8). This may be exacerbated by mites introducing toxins in saliva as they feed. High populations may occur on flowering *Cannabis*: leaves turn a yellow-copper color, with a thickened and distorted appearance; flower pistils may be damaged and they necrose and die prematurely (Wainwright-Evans, 2017).

4.4a.   Broad mite

Polyphagotarsonemus latus (Banks) 1904
≡ *Tarsenomus latus* Banks
Description: Adult females are elongate-oval, broad or swollen in profile, 0.2 mm long, and light yellow to amber or green in color with an indistinct, light, median stripe that forks near the back end of the body. Males are similar in color but lack the stripe, and 0.11 mm long. Eggs are about 0.08 mm wide, pale-colored, elongate-oval in shape, and pressed on the surface of the plant, so the bottom is flattened. Eggs are covered with scattered tufts of white hairs. The first-stage larva ranges from 0.1 mm to 0.2 mm, and is whitish. The second-stage nymph is clear and pointed at both ends (Fasulo, 2016).

4.4.2   Life history and habits

The life cycle of broad mite can be completed very quickly (Fasulo, 2016). Eggs hatch about 2–3 days after being laid, producing

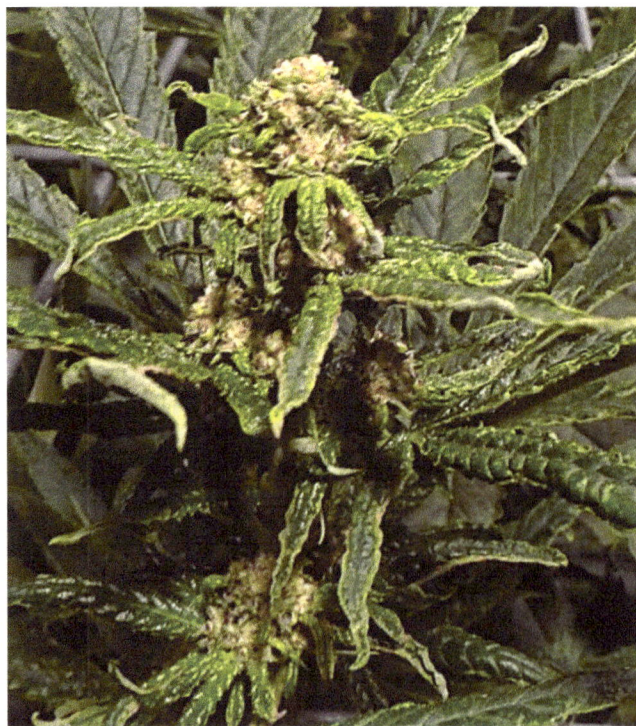

Fig. 4.8. Damage from broad mite, *Polyphagotarsonemus latus* (Baile Oakes)

a six-legged larva. The larva feeds for about 2–3 days, moving little. It then molts to a non-feeding quiescent stage (nymph) that occurs within the old skin of the larva. A day or two later the adult forms emerge. Adult females live 8–13 days, laying an average of about five eggs per day. Males are shorter lived, 5–9 days. Populations are skewed towards females, often in a 4:1 female-to-male ratio (Cloyd, 2011b).

Adult females move little, but males are quite active. They will often carry the female nymphs, and then mate with them when the adult female emerges. Much of the dispersal of broad mite occurs this way. Broad mites may also disperse in wind currents, on hands or clothing of humans working a crop, and reportedly attach to and are carried by adult whiteflies (Gerson, 1992).

Broad mites thrive in humid conditions of mild temperature, with population growth peaking around 25°C. Broad mites favor areas of lower light, and may be concentrated in shaded areas of plants. In field populations monitored year-round in southern Florida and Guadeloupe, populations were usually highest in April through August, with lowest numbers found from December through March. Broad mite does not produce a diapause condition, and thus reproduces continuously year-round.

4.4.3 Differential diagnosis

Injuries produced by broad mite are similar to those caused by hemp russet mite. Both feed on epidermal cells, and high populations cause leaves to become bronzed or gray and brittle. Growth distortion in the form of leaf curling can be produced by both species, but is often more pronounced by broad mite.

Diagnosis requires microscopic examination of the plant. Broad mites may be found on leaves, usually the underside, but many concentrate in new growth and can be difficult to see in the folds of emerging leaves. If adult stages can be found, broad mite and hemp russet mite can be readily distinguished, the latter having a much more elongate body form. The eggs of broad mite, flattened on the leaf with oval form and white tufts, may be the easiest way to identify this species (Wainwright-Evans, 2017).

4.4.4 Control measures

As with essentially all pests of indoor production, make strong efforts to prevent the introduction of broad mite into the facility. Broad mite is most likely to be introduced on plant material containing live stages. Broad mite can be introduced on hands or clothing after a person has recently worked with infested plants. The mobility of broad mite is more limited than that of spider mites, and the pest is unlikely to migrate indoors from outside plants. Once within a facility, dispersal will be largely through mites being carried on wind currents, or by people moving through and working with infested plants, then moving to non-infested plants.

Establishing a host-free period, when no live plants are present to support broad mite, can effectively eliminate this pest from an infested facility. Elimination is more easily achieved than with spider mites, as broad mites never go into a dormant diapause condition. Broad mite cannot survive long on tough, older leaves, and dies if the plant dries up or rots (Gerson, 1992).

Immersion of plants in warm water baths has potential to eliminate broad mite from cuttings and small plants. Cloyd (2011b) recommended treatment in water heated to 40–43°C, for 15–45 min. Extended exposures are needed to ensure that the water treatment fully reaches broad mite stages present in the crevices of buds and meristematic tissues. For biological control, Wainwright-Evans (2017) reported best success with *Neoseiulus cucumeris* and *Neoseiulus californicus*. Broad mites are also eaten by *Neoseiulus barkeri* (Cloyd, 2011b). Insecticidal soaps and horticultural oil may help, but good spray coverage is difficult for this tiny pest.

4.5 False Spider Mites

False spider mites or flat mites (Acari: Tenuipalpidae) have a flattened and somewhat elongate body. Most false spider mites are orange to red, and have short, stubby legs. They tend to be highly polyphagous; Childers *et al.* (2003b) provided host records for three species—collectively 928 plant species in 139 families, rivaling *Tetranychus urticae* for host range. Most reports of crop injury occur on citrus and other fruiting trees, shrubs, and orchids. A majority of false spider mites that cause significant crop damage are in the genus *Brevipalpus*, and five *Brevipalpus* spp. have been reported from *Cannabis sativa*.

4.5a. *Brevipalpus* spp.

1. Privet mite
a. *Brevipalpus obovatus* Donnadieu 1875
2. Other false spider mites
b. *Brevipalpus californicus* (Banks) 1904
c. *Brevipalpus phoenicis* (Geijskes) 1939
d. *Brevipalpus rugulosus* Chaudri *et al.* 1974
e. *Brevipalpus khalidae* Mansoor-ul-Hasan *et al.* 2005
Description: *Eggs* are bright red, 0.1 mm long. *Larvae* and *nymphs* are bright red. *Adults* are oval in outline, rather flattened, average 0.3 mm long. Most are reddish, but colors may vary from reddish green to dark green. Distinguishing these species requires a close look at their tarsi, and/or male genitalia. They are slow moving and can be more difficult to observe on plants than many other mites. False spider mites produce little or no visible silk. Discarded exuviae following molting remain present on plants that can be used in diagnosis.
Comments: Childers *et al.* (2003b) reported *B. obovatus* on *Cannabis* in the Middle East and South America. Gupta (1985) reported *B. obovatus* in India. *B. obovatus* is a polyphagous pest found in temperate and subtropical regions around the world.

Childers *et al.* (2003b) reported *B. californicus* and *B. phoenicis* on *Cannabis* in Asia. *B. californicus* is known as the omnivorous mite in the USA, and found worldwide. *B. phoenicis* also occurs globally, and is polyphagous. Gupta (1985) reported *B. rugulosus* on *Cannabis* in India. The species is limited to South Asia. Mansoor-ul-Hasan *et al.* (2005) described *Brevipalpus khalidae* as a new species, infesting *Cannabis* in Pakistan.

Brevipalpus spp. feed on fruit, leaf, stem and bud tissues. A toxic saliva is introduced while feeding. Descriptions of injury to *Cannabis* are lacking. In other crop species, symptoms are described as coppery or bronze foliage, with leaves prematurely dropping. Damage is greatest in drought-stressed plants (Hill, 1983).

4.5.1   Life history and control measures

Childers *et al.* (2003a) provide a general review of the biology and feeding injuries produced by *Brevipalpus* spp. Parthenogenesis is predominant and there are four active life stages: larva, protonymph, deutonymph, and adult. Development is slower than that of spider mites, with reported generation times typically ranging from 4 to 6 weeks. Adults live about a month; females lay 30–55 eggs.

Prevent false spider mite from entering indoor facilities: stringently inspect plant material entering a grow space, and quarantine clones. Predatory mites recommended for two-spotted spider mite do not provide economic control for false spider mite. In other crops, sulfur was used to control false spider mites (Hill, 1983).

4.6   Aphids

Aphids or "plant lice" are small, soft-bodied, sap-sucking insects. They tend to have pear-shaped bodies with relatively long legs and antennae. Their abdomens project a pair of tube-like **siphunculi** (i.e., cornicles or siphons) and a single, rear-pointed **caudum**. Most adults do not have wings (and are called **apterae**), but some do (**alatae**).

Aphid damage *increases* in moderately warm, moist growing seasons with gentle rain and little wind. Damage *decreases* in hot, dry weather (Parker, 1913a). In addition to direct feeding damage, aphids vector viruses. Aphids and viruses have a symbiotic relationship. Viruses induce changes in plant metabolism that make plant sap more nutritious for aphids (Kennedy, 1951). The aphids, in turn, transmit the viruses to new hosts (Kennedy *et al.*, 1959).

Family Aphididae comprises around 510 genera and 5000 species. Roughly half the species are in Subfamily Aphidinae—including all aphids on *Cannabis*. Aphidinae subdivides into two tribes, the Macrosiphini and Aphidini. Early mentions of aphids attacking hemp did not identify the species, in Italy (Scopoli, 1763) and in France (Marc, 1827; Vallot, 1836). The first named species was *Phorodon cannabis* (Passerini, 1860).

Eight aphid species reportedly attack *Cannabis*, although some of these host associations are questionable. Six are in Tribe Macrosiphini (*P. cannabis, P. humuli, Myzus persicae, Lipaphis erysimi, Aulacorthum solani, Uroleucon jaceae*) and two are in Tribe Aphidini (*Aphis fabae, A. gossypii*). Related to aphids are psyllids, or jumping plant lice (Homoptera:

Psylloidea). One psyllid has been reported on *Cannabis*: Labina (2008) found *Craspedolepta lineolata* Loginova 1962 infesting wild-type plants in the Altai steppes of Siberia. *C. lineolata* primarily infested *Artemisia* spp. and *Urtica cannabina*.

4.6.1   Signs and symptoms

Aphids congregate on the undersides of leaves, along petioles, and on small branches. Some aphid species prefer older, lower leaves (e.g., *Myzus persicae*), and some prefer younger, upper leaves (e.g., *Aphis fabae*). Some species infest flowering tops (e.g., *Phorodon cannabis*). Leaves may turn yellow, their blades curl, wilt, and look scorched (Fig. 4.9). Chronic, heavy infestations cause a decrease in plant height, and a significant loss of yield (Fedorenko *et al.*, 2016). One aphid is known to infest roots instead of above-ground parts: the rice root aphid, *Rhopalosiphum rufiabdominale* (see section 8.1a).

Honeydew exudes from the anus of feeding aphids. On warm afternoons, honeydew may be seen falling as a mist from severely infested plants (Parker, 1913a). Honeydew causes secondary problems: it attracts ants, and supports the growth of black sooty mold. Ants attack the predators of aphids, and must be eliminated for biocontrol to be effective. Sooty mold reduces plant photosynthesis and leaf transpiration. Honeydew residues might end up in CBD and THC products extracted with solvents (ethanol, butane).

4.6.2   Causal organisms

The *Cannabis* literature regarding aphids is filled with Latin synonyms. Some aphids are nearly impossible to tell apart,

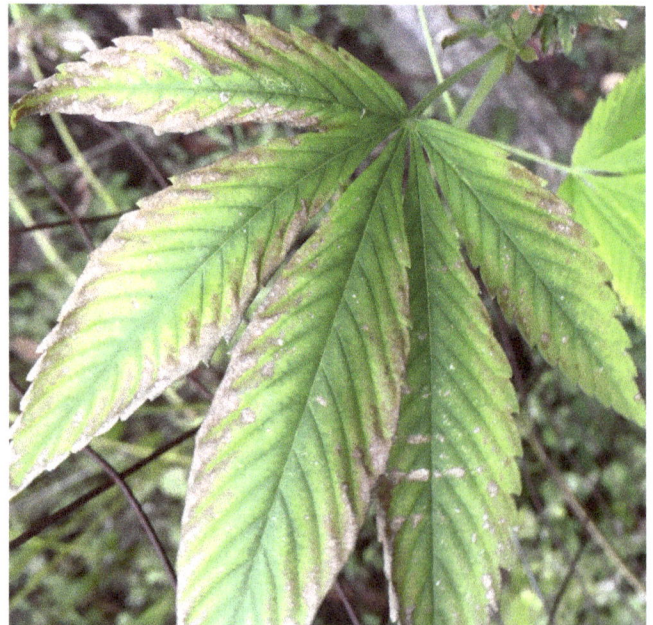

Fig. 4.9. Damage caused by *Phorodon cannabis*

such as *Phorodon cannabis* and *Phorodon humuli*. Some reports of *P. humuli* and *Myzus persicae* may represent misidentifications of *P. cannabis*. See the differential diagnosis section.

4.6a. Cannabis aphid

Phorodon cannabis Passerini 1860
≡*Aphis cannabis* (Passerini), ≡*Myzus cannabis* (Passerini), ≡*Paraphorodon cannabis* (Passerini), ≡*Diphorodon cannabis* (Passerini), =*Phorodon asacola* Matsumura 1917, =*Capitophorus cannabifoliae* Shinji 1924, =*Semiaphoides cannabiarum* Rusanova 1943

Cranshaw (2018) established the common name "cannabis aphid" for P. cannabis. It is called the hemp louse in Europe, and the bhāng aphid in India (Das, 1918). Matsumura (1917) said Japanese farmers called it the asa (hemp) iboabura (aphid), so he coined the Latin name Phorodon asacola.

Cranshaw *et al.* (2018) untangled the genus *Phorodon*. The genus now includes seven species: *P. cannabis*, *P. humuli* (on hops), *P. humulifoliae* (on hops), *P. japonensis* (on *Prunus*), *P. persifoliae* (on *Prunus*) *P. lanzhouense* (host unknown), and *P. viburni* (on *Viburnum*). Passerini (1860) named an eighth, *P. inulae* (on *Inula*), which Van der Goot (1915) moved to *Capitophorus* as *C. inulae*. *Capitophorus* was utilized by Shinji (1924) who coined *C. cannabifoliae*, which Miyazaki (1972) synonymized under *P. cannabis*, along with Matsumura's *P. asacola*.

Börner (1952) considered the bristles of *P. cannabis* with knobbed apices a unique character, and erected *Diphorodon* as a subgenus of *Phorodon*. Heinze (1960) raised the subgenus to generic status, coining *Diphorodon cannabis*. The species has also been placed in the genus *Paraphorodon* by Eastop and Hille Ris Lambers (1976). Then Eastop and Blackman (2005) returned *Paraphorodon cannabis* to *Phorodon*. Frank (1896) placed *P. humuli* and *P. cannabis* in the genus *Aphis*, which is where Schrank (1801) originally classified *P. humuli*. Frank was a good mycologist (he coined "mycorrhiza"), but not an entomologist.

Description: (see "Life History" for these terms) Immature *apterous* (wingless) *viviparae* are pale yellowish green, gradually enlarging as they go through four molts (Fig. 4.10A). Adult *apterae* have flattened, elongate-oval bodies, tapering towards ends, more than twice as long as broad, averaging 1.7–2.7 mm in length. They are pale yellow without stripes (Fig. 4.10A), to yellowish green with one to three longitudinal darker green stripes (Fig. 4.10B). Some take a pink hue (not shown). Dusky lateral stripes mark the abdominal segments (Das, 1918). Antennae are 1.1–2.2 mm long. Siphunculi are white, up to 0.8 mm long (nearly a third of the body length), cylindrical, and taper towards their tips. The caudum tapers evenly to its tip, a third the length of the siphunculi.

Heads of apterae are covered with microscopic bristles that have knobbed apices (flabellate setae). On the front of the head, between the antennae, are two finger-like projections called antennal tubercles; the tubercles are long (at least half as long as the first antennal segment) and covered with knobbed bristles. Between the tubercles a smaller midline knob arises,

Fig. 4.10. Cannabis aphids. See text for letters **A–D** (W. Cranshaw)

also bristled. Close-up color photos of the tubercles and bristles are available on the web (Oregon Department of Agriculture, 2017).

Alatae (winged) *viviparae* are slightly smaller than apterae, 1.6–1.8 mm long, and the flabellate character of the setae is less pronounced. They develop black-brown patches on heads and abdomens. The wings are large and broad, slightly smoky in appearance with dark veins, and held vertically over the abdomen when at rest (Fig. 4.10C).

Oviparous apterae differ little from viviparous apterae, except they lay eggs (Fig. 4.10C). *Eggs* are ovate, 0.7 mm long, yellow-green when first laid, but turn shiny black as they harden (Fig. 4.10D). The white egg in Fig. 4.10D was laid by a syrphid fly, an aphid predator. Eggs hatch into *fundatrices*, which are more oval than other apterae, but slightly shorter (1.8–2.4 mm), with long antennae.

Comments: *P. cannabis* is the most common aphid on *Cannabis*. It infests *C. sativa, C. indica,* and *C. ruderalis* (Müller and Karl, 1976). The species is native to temperate Eurasia, infesting fiber-type crops from Britain to Japan. *P. cannabis* is common in India, Pakistan, and Bangladesh (Das, 1918; Ghosh, 1974; Müller and Karl, 1976; Ghosh, 1994; Naumann-Etienne and Remaudière, 1995). *P. cannabis* has been reported on drug-type plants in Nepal (Das and Raychaudhuri, 1983), North Africa (Tuatay and Remaudiere, 1964), and Lebanon

(Remaudière and Talhouk, 1999). *P. cannabis* infests wild-type plants in Uzbekistan (Mukhamediev, 1979). Under the synonym *Semiaphoides cannabiarum* it occurs in Azerbaijan (Holman, 2009).

The species is new to North America, with "first record" claims in Colorado (Halbert, 2016) and Oregon (Oregon Department of Agriculture, 2017). In fact, Batra (1976) provided the first record in North America. The species now appears almost everywhere that *Cannabis* is cultivated in the USA and Canada (Cranshaw *et al.*, 2018, 2019). The Suction Trap Network (STN) was established in the Midwest to monitor populations of aphid pests. STN has trapped *P. cannabis* in Illinois, Indiana, Iowa, Minnesota, and Wisconsin (Lagos-Kutz *et al.*, 2018). Now that industrial hemp is federally legal, *P. cannabis* has become the first *Cannabis* pest that is reportable to USDA-APHIS (CFDA, 2019).

Although Pichat (1866) reported aphids infesting seedlings, mass infestations of *P. cannabis* do not appear until August (Müller and Karl, 1976). Earlier in the season, *P. cannabis* populations are limited to the underside of leaves. Exploding populations invade terminal shoots (Balachowsky, 1936). Durak *et al.* (2021) grew *Cannabis* in temperature-controlled growth chambers to measure *P. cannabis* demographic parameters at 20, 25, 28, and 30°C. The optimal temperature was 25°C, when insects took 6.4 days to reach reproductive age, with a reproductive period of 15.2 days, and an average fecundity of 112 nymphs per apterous female.

Cannabis aphid is particularly damaging to female flowers, noted by Passerini (1860) in his original description. Kirchner (1906) wrote that "it sits between female flowers and seeds, sucking plant sap." Crescini (1940) said *P. cannabis* reduced seed formation. He despised the aphid, because its honeydew damaged the cellophane bags that he used for trapping pollen for breeding research. The aphids proliferated in the protected environment of the pollen bags. Infested flowers may become distorted and hypertrophied (Villanueva and Kennedy, 2019). Cannabis aphid vectors the hemp streak virus (Goidànich, 1955), hemp mosaic virus, hemp leaf chlorosis virus (Ceapoiu, 1958), pea mosaic virus (Karl, 1971), cucumber mosaic virus, alfalfa mosaic virus (Schmidt and Karl, 1970), and potato virus Y (Pitt *et al.*, 2022).

4.6b. Hops aphid

Phorodon humuli (Schrank) 1801
≡*Aphis humuli* Schrank 1801, ≡*Myzus humuli* (Schrank)
Description: All life stages of hops aphid resemble *P. cannabis*, first noted by Passerini (1860). Müller and Karl (1976) illustrated differences in *apterae* (wingless forms): The head of *P. humuli* sports few bristles than *P. cannabis*. The few bristles on antennal tubercles of *P. humuli* have blunt or pointed apices, never knobbed. The midline knob on the top of the head of *P. humuli* is scarcely apparent. Cranshaw *et al.* (2018) illustrated these differences in a publication available on the internet. They also detailed other diagnostic differences (the shape of antennal projections and siphunculi), as well as more technical characters, such as the length ratio between antennal segment

III and the siphunculi, and the ratio between segment IV and caudal length.
Comments: Hops aphid overwinters on *Prunus* spp. (*P. avium*, *P. cerasus*, *P. persica*). In the spring several generations feed on *Prunus* spp. Winged alatae migrate to hops (*Humulus lupulus*) in the spring and summer, and produce wingless apterae whose populations can explode parthenogenetically. Hops aphid occurs in Europe, Central Asia, East Asia, North Africa, and North America. In California *P. humuli* is aided by a black ant, *Formica subsericea* (Parker, 1913a).

Cannabis has been reported as a host for *P. humuli* in Germany (Blunck, 1920; Flachs, 1936; Eppler, 1986), Japan (Takahashi, 1965; Higuchi and Masahisa, 1969; Moritsu, 1983), Korea (Okamoto and Takahashi, 1927), and Iran (Hojat, 1988).

Cranshaw *et al.* (2018) contended that these are misidentifications of *P. cannabis*. In host-transfer experiments, *P. humuli* populations nosedived when transferred to two drug-type strains of *Cannabis* (Carmichael, 2020). Vice versa, *P. cannabis* reportedly infests hops (Higuchi and Masahisa, 1969; Batra, 1976; Blackman and Eastop, 2006). However, transfer trials failed to establish *P. cannabis* on hops (Cranshaw, 2018). In a careful survey of Fraser Valley hops fields, no specimens of *P. cannabis* were found (Carmichael, 2020).

Anonymous (1940) proposed that *P. cannabis* was synonymous with *P. humuli*. Their morphology is similar, but their life cycles differ—*P. cannabis* has reduced to an autoecious condition, whereas *P. humuli* has a heteroecious life cycle. Genetic "barcode" comparisons suggest they are different species: Lagos-Kutz *et al.* (2018) sequenced the *COI* gene from two specimens of *P. cannabis*, and compared them with three previously sequenced *COI* records for *P. humuli*. The pairwise distance between them ranged from 4.24% to 4.56%. Pairwise distances > 2.6% represent different species, according to a large comparison of *COI* sequences in 274 aphid species (Coeur d'acier *et al.*, 2014).

4.6c. Green peach aphid

Myzus persicae (Sulzer) 1776
≡*Aphis persicae* Sulzer, ≡*Phorodon persicae* Sulzer
Description: *Apterae* (wingless forms) are green (sometimes yellow-green or pink), oval in outline, averaging 2.0–3.4 mm in length. They resemble *P. cannabis* but are about 25% larger. Tubercles between the antennae are short (less than half as long as the first antennal segment) and point inward. Siphunculi projecting from the abdomen are cylindrical with a slight swelling distal to the midpoint, and are more than twice as long as the caudum. The caudum is lightly bristled and constricted slightly at the midpoint. *Alatae nymphs* are often pink or red, and develop wing-pads. *Alatae adults* have black-brown heads, a black spot in the middle of their abdomens, and hold their wings vertically over their abdomens when at rest.
Comments: Genus *Myzus* currently comprises 55 species with a cosmopolitan distribution. In a molecular clock analysis based on three genes (*COI*, *COII*, *EF-1α*), Rebijith *et al.* (2017) included *M. persicae* and *M. varians*, which diverged

20.14 million years ago. Passerini (1860) coined the genus *Myzus,* for *M. persicae,* as well as genus *Phorodon,* for *P. cannabis* and *P. humuli.* For morphological differences between these genera, see the differential diagnosis section below. These species have switched places: Blunck (1920) transferred *P. humuli* to *Myzus* as *M. humuli.* Bodenheimer (1930) transferred *M. persicae* to *Phorodon* as *P. humuli.* The taxon *Myzus cannabis* (Passerini) turns up on websites, but we couldn't trace its source.

In a phylogenetic study of aphids whose entire genomes have been sequenced, Mathers *et al.* (2022) included several *Myzus* spp. and *P. humuli.* The analysis was based on the amino acid sequences of 928 conserved genes, using maximum likelihood implemented in FastTree. No bootstrap values were given, but "all branches received maximum support." A portion of their phylogenetic appears in Fig. 4.11A. They found that *P. humuli* nested within a clade of *Myzus* spp. This would mean that *Myzus* and *Phorodon* are monophyletic. Taxonomically, *Phorodon* has priority, having been described by Passerini a page before his *Myzus* description.

We wanted to replicate the study, but most of the genomes were unavailable, except for *Myzus persicae* strain 006—the original whole-genome sequence of that species (Singh *et al.,* 2021). Searching GenBank we found two genes, *COI* and *EF-1α,* conserved in several of Mathers's analyzed species. Our analysis used the software pipeline described for the phylogram in Chapter 1 (see Fig. 1.12). The resulting phylogenetic tree (Fig. 4.11B) showed that *P. cannabis* and *P. humuli* form a clade, sister to *M. persicae,* albeit with weak bootstrap support (0.634). Further research is required.

Green peach aphid attacks dozens of plant species, and now occurs worldwide (Spaar *et al.,* 1990). Like *P. humuli,* green peach aphid can overwinter on *Prunus* spp. The species is exceptionally restless; alatae repeatedly land on plants, probe briefly, then take off for other plants. They are "the most notorious vectors of plant viruses" (Kennedy *et al.,* 1959). Blattný *et al.*

(1950) described hemp plants with a viral infection concurrently infested by *M. persicae.* Schmidt and Karl (1969, 1970) also found hemp plants infested with viruses and *M. persicae.* They used *M. persicae* to experimentally transmit alfalfa mosaic virus and cucumber mosaic virus to hemp plants.

M. persicae has also been reported on *Cannabis* in India (Sekhon *et al.,* 1979; Kumar *et al.,* 2005), in Dutch glasshouses (Clarke, pers. comm., 1995), and on feral hemp in Illinois (Bush Doctor, 1985). Blackman and Eastop (2006) cited *M. persicae* on *Cannabis* but without a location record. Holman (2009) cited *M. persicae* on *Cannabis* in the Czech Republic, India, and Iran. Grammenos *et al.* (2021) reported *M. persicae* in a glasshouse in Greece. Some of these reports may be misidentifications; Villanueva and Kennedy (2019) found that *P. cannabis* was misidentified as *M. persicae* in Oregon. Pulkoski and Burrack (2023) transferred tobacco-reared *M. persicae* to *Cannabis;* they survived and reproduced on excised leaf discs maintained on agar, but did not survive on whole plants.

4.6d. Cotton aphid or melon aphid

Aphis gossypii Glover 1877
=*Aphis sativae* Williams 1891
Description: *Apterae* are rotund, nearly spherical in outline, dark green to black, 1–2 mm long, with prominent black siphunculi (Fig. 4.12). They turn a paler color when present in high densities, but their siphunculi remain black. The caudum is inconspicuous, with only two or three pairs of setae. *Alatae* (winged) parthenogenetic females are black on their head and thorax, with a green abdomen, 1.1–1.7 mm long. Egg-laying females are dark purplish green, with bodies more elongate than apterae. Their wings are tinted brown with dark-brown veins, and twice the length of their bodies.

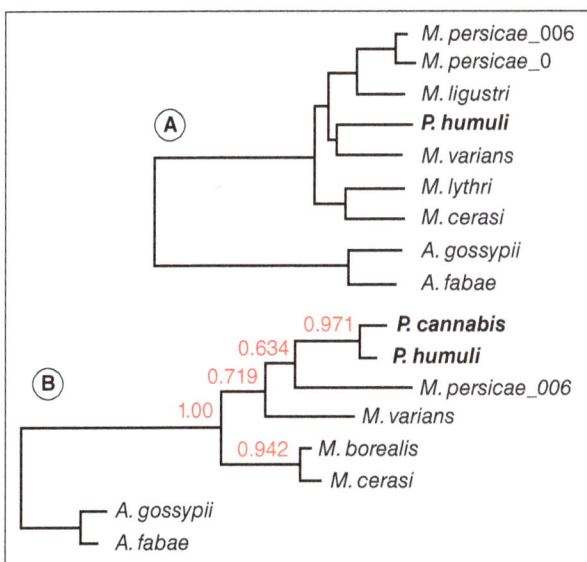

Fig. 4.11. Phylograms: **A**. portion of Mathers *et al.* (2022); **B**. *COI* and *EF-1α*

Fig. 4.12. Apterae of *Aphis gossypii* in male flowers (R. Clarke)

Comments: Cotton aphid lives around the world. It seriously damages cotton and cucurbit crops, but feeds on almost anything. The species prefers high temperatures; at 27°C the apterae mature in seven days. *A. gossypii* vectors over 50 plant viruses and is often ant-attended. The species is common on *Cannabis* in India (Cherian, 1932; Raychaudhuri, 1984; Ghosh, 1990; Singh *et al.*, 2014). The species has also been reported on *Cannabis* in South Africa (Dippenaar *et al.*, 1996), Hungary (Clarke, pers. comm., 1998) (Fig. 4.12), New Zealand (McPartland and Rhode, 2005), Serbia, Uzbekistan, and Morocco (Holman, 2009), and Greece (Grammenos *et al.*, 2021).

Williams (1891) coined *Aphis sativae* for aphids infesting hemp flowers in Nebraska. Williams (1910) gave a description: *Apterae* short and broad, 1.55 mm long, "very dark greenish brown to almost black." They had inconspicuous antennae tubercles, siphunculi black, robust, 0.22 mm long, with a short cadua. Winged forms slightly smaller, blackish-brown, with wings tinted brown with medium-sized brown veins, and cauda with bristles. "This cannot be *Phorodon cannabis*, as that species has porrect frontal tubercles and is of a much paler color." Davis (1911) examined Williams's *A. sativae* type specimens, preserved in a slide mount. He said they were badly mutilated, "an unrecognizable mass." Blackman and Eastop (2006) synonymized *A. sativae* under *A. gossypii*. Judging from Williams's description, we agree.

4.6e. Other aphids

The black bean aphid, *Aphis fabae* (Scopoli) 1763, infested fiber-type hemp in Poland (Heller *et al.*, 2006) and Ukraine (Pivtoraiko *et al.*, 2020). Holman (2009) cited location records in Sweden, the Czech Republic, Spain, and Kazakhstan. Blackman and Eastop (2006) cited the species on *Cannabis* without a location record. In England, *A. fabae* infestation was greater in male plants (11.9%) than female plants (1.3%), and the pest was absent in female-only sinsemilla crops (Potter, 2009). *A. fabae* resembles *Aphis gossypii* (Fig. 4.12), and shares that species' cosmopolitan and polyphagous habits. *A. fabae* is ant-attended, often the black ant *Lasius niger* L. 1758.

The glasshouse potato aphid, *Aulacorthum solani* (Kaltenbach) 1863, is cited by Blackman and Eastop (2006) without a location record. Holman (2009) cited the species on *Cannabis* in Czech Republic. This species has one of the broadest host ranges of any aphid in the world. *A. solani* superficially resembles *P. cannabis* and *P. humuli*.

The large knapweed aphid, *Uroleucon jaceae* (Linnaeus) 1758, was found on *Cannabis* in India (Raychaudhuri, 1984). *U. jaceae* is a large (2.5–3.5 mm long) reddish brown to brown-black aphid that infests Asteraceae in Europe and Asia. The mustard aphid, *Lipaphis erysimi* (Kaltenbach) 1843, reportedly infested Indian hemp in Delhi (Prasad, 1997). *L. erysimi* normally feeds on turnips and other cruciferous plants. It resembles *P. cannabis*. Gaujac (1810) claimed that hemp plants repelled an unidentified turnip aphid, probably *L. erysimi*. Cranshaw (pers. comm., 2019) contended that host associations of *Aulacorthum solani*, *Uroleucon jaceae*, and *Lipaphis erysimi* with *Cannabis* are likely erroneous.

4.6.3   Differential diagnosis

Symptoms from aphids (wilting, yellowing) can be confused with damage by spider mites and whiteflies. Turn over a leaf to see what lurks beneath. Young aphids can be confused with young whiteflies, young scales, even young tarnished plant bugs. Use a hand lens to distinguish aphids from these similar pests.

Differentiating between aphid species can be difficult. We previously described diagnostic differences between *P. cannabis* and *P. humuli*. A third look-alike is *M. persicae*. *M. persicae* is about 25% larger than *Phorodon* spp., its antennal tubercles are short and rounded, it lacks a midline knob between the tubercles, its siphunculi develop a slight swelling, and its caudum is slightly constricted. If apterae are dark green-black and lack antenna tubercles, suspect *A. gossypii*. But don't forget *A. fabae*, which can also be quite dark and lacks tubercles. The jet-black siphunculi of *A. gossypii* set it apart from *A. fabae*.

4.6.4   Life history and habits

The complicated life cycle of aphids requires special terminology. **Autoecious** aphids spin through their life cycle on a single host species, such as *P. cannabis* (Müller and Karl, 1976; Cranshaw *et al.*, 2018). Other aphids, such as *P. humuli*, *M. persicae*, and *A. fabae*, migrate between primary hosts and secondary hosts during the grow season. This life cycle is called **heteroecious**. The life cycle of *P. cannabis* can also be called **holocyclic**, or cyclical parthenogenesis—it alternates between asexual (parthenogenetic) reproduction and sexual reproduction (Fig. 4.13). Sexual morphs are rarely seen in some populations; they may be **anholocyclic** (producing only asexual females).

P. cannabis overwinters outdoors as shiny black eggs laid on hemp stalks and flowering tops (Müller and Karl, 1976). Cranshaw *et al.* (2018) proposed three overwintering reservoirs for eggs—debris from unharvested plants, volunteer seedlings, and nearby feral hemp. Field infestations may also arise from indoor-grown plants harboring aphids that are transplanted outdoors.

Each egg hatches into a **fundatrix** or "founding mother" (Fig. 4.13). Hatch time is temperature dependent. *P. cannabis* eggs hatch when the temperature reaches 22–26°C. Fundatrices multiply parthenogenically, which begins within a few days. They give birth **viviparously**—eggs hatch within their reproductive tract—so larvae, not eggs, are "born alive." Fundatrices bear 60–100 **fundatrigeniae**.

Fundatrices and fundatrigeniae are *apterous* (wingless) females. They begin giving birth to more live females (*apterous viviparae* or *apterae*). In late spring the first *alatae* (winged aphids) develop. At the end of the summer, special alatae called **sexuparae** give birth to about ten **sexuales**. Sexuales are either females or males, alatous or apterous. They mate and the females become **oviparous** and lay 5–10 eggs, which overwinter. Outoors in France, oviparous alatae emerged in October, and began laying eggs on stalks (Balachowsky, 1936). Indoors in Ontario, *P. cannabis* began laying eggs about 4 weeks after

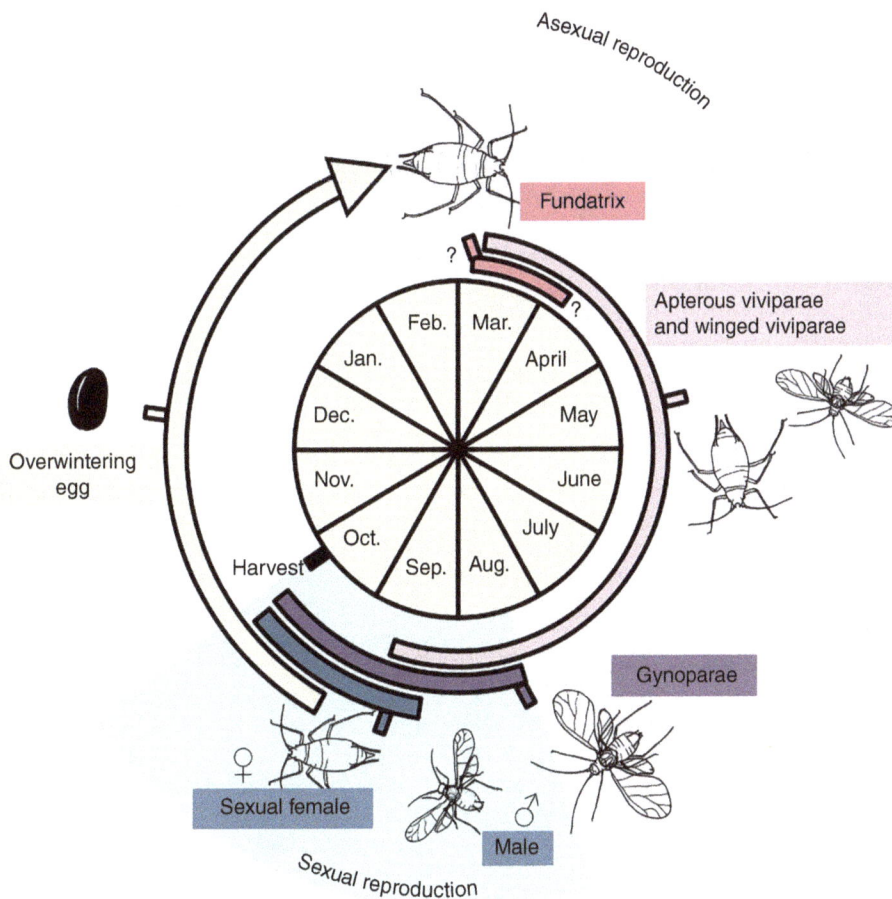

Fig. 4.13. Life cycle of *Phorodon cannabis* (courtesy M. Schreiner and E. Perice)

the glasshouse was switched from a 16/8 to a 12/12 light/dark day cycle. The change in photoperiod likely triggered the production of sexual forms (Cranshaw *et al.*, 2018).

A few researchers believe *P. cannabis* has a heteroecious life cycle. In India, Das (1918) could not find *P. cannabis* eggs on *Cannabis*, and proposed there may be an alternative host. Das found *P. cannabis* on two incidental hosts, *Cirsium* (*Cnicus*) *arvensis* and *Artemisia* spp. In Japan, Higuchi and Masahisa (1969) claimed that *P. cannabis* migrates to *Prunus communis* and *P. mume* (a host they said was shared by the hops aphid *P. humuli*, but perhaps they confused the two).

In aphids with heteroecious life cycles, the "primary" host is the one in which eggs overwinter. *M. persicae* and *P. humuli* overwinter as eggs on *Prunus* spp. *A. fabae* overwinters on *Euonymus* or *Virburnum* spp. Eggs hatch into fundatrices, which give birth viviparously to fundatrigeniae, which give birth to more fundatrigeniae. Thus fundatrices may live to see their great-great-great-granddaughters. Hill (1994) claimed that a single springtime fundatrix could give rise to up to 12 generations of aphids in one year—and theoretically 600,000 million offspring.

In late spring the first alatae (winged aphids) develop. They are called spring migrants and fly off to secondary hosts,

such as *Cannabis*. Aphids are weak fliers, with a flight speed of only 1.6–3.2 km/h in still air. To migrate, alatae of *P. humuli* fly straight up into a moving air mass. The wind governs their direction of flight. In this way they usually migrate 15–20 km (Dixon, 1985). They terminate migration by actively flying downward and settling on plants.

Once settled on *Cannabis*, alatae give birth to apterae; apterae undergo four molts in about 10 days to reach sexual maturity. Each aptera gives birth to 30–70 young. Crowded conditions or a lack of food induce more alatae, which fly off seeking unexploited *Cannabis*. They are called summer migrants. At the end of the summer, sexuparae (autumn migrants or return migrants) fly back to the primary host and give birth to sexuales. For *P. humuli* this migration peaks in late September (Dixon, 1985). Sexuales mate and females lay eggs, which overwinter.

4.6.5   Cultural and mechanical control

Prevention is the best control: Limit the movement of people and plants in a grow space. Hold new plants in quarantine until determined to be aphid-free. Screen all entrances to glasshouses to thwart winged females. Sanitation is essential, elaborated in section 19.2. Scout regularly, at least weekly, by closely inspecting undersides of young leaves. Incipient outbreaks can be "nipped in the bud" by roguing infested plants. Bag the rogued plants before carrying them through a grow space to dispose of them. Indoor problems get transplanted outdoors via infested mother plants and their clones (Cranshaw *et al.*, 2018).

Yellow sticky traps are invaluable for monitoring populations of winged aphids. See details regarding sticky traps in section 19.6.1. Traps will not "eliminate" aphids as advertised, simply because most aphids are not winged. Yellow sticky traps also kill beneficial insects, such as parasitic wasps, lady beetles, and hoverflies (Broughton and Harrison, 2012). Reflective material laid between plants (aluminum foil) confuses winged aphids, and can be useful for short plants. However, aluminum foil or any mulch that protects the soil may favor rice root aphids (Cranshaw and Wainwright-Evans, 2020).

In field situations, monitor meteorological conditions to forecast aphid outbreaks. Meteorological models based on temperature (degree days) can predict the spring migration of *P. humuli* to hops (Worner *et al.*, 1995). Predictive decisions regarding when to apply pesticides on hops have been based on temperature (Lorenzana *et al.*, 2013).

Balachowsky (1936) noted that oviparous alatae lay eggs on feral *Cannabis* plants, and urged their destruction. In Ukraine, field crops badly infested with *P. cannabis* are burned at the end of the season, and feral hemp is destroyed (Fedorenko *et al.*, 2016). Field residues of infested plants should be plowed under, to kill overwintering eggs (Cranshaw *et al.*, 2018). If aphids are ant-attended (e.g., *Aphis fabae*, *A. gossypii*), locate and destroy ant nests.

Environmental stress predisposes plants to aphids, so provide adequate moisture, light, and balanced nutrition. In the first-ever "indoor grow" under lights, Schaffner (1928) noted that stressed, under-illuminated plants were attacked by "scale insects, mealy bugs, and aphids." Aphids thrive on juicy plants excessively fertilized with nitrogen. Reduce nitrogen fertilization to the minimum amount needed.

McPartland *et al.* (2000) posted an infestation severity index (Table 4.2). The index is used for decisions regarding biological and chemical control, and has been adopted by others (Oregon Department of Agriculture, 2017).

4.6.6   Biological control

A bewildering array of biocontrols are available. Each has its own niche, requiring nuanced choices for each grow situation,

Table 4.2. Infestation Severity Index for aphids

Light	Any aphids seen
	Often no symptoms present
Moderate	<10 aphids/leaf (not leaflet)
	Feeding patches present
Heavy	11–50 aphids/leaf
	Feeding patches coalescing
Critical	>50 aphids/leaf
	Shriveled, discolored leaves and honeydew

sometimes by season. Mimicking nature is a good start: Cranshaw *et al.* (2018) identified naturally occurring predators of *P. cannabis* in Colorado: convergent lady beetle (*Hippodamia convergens*), multicolored Asian lady beetle (*Harmonia axyridis*), green lacewings (*Chrysopa oculata* and *C. plorabunda*), aphid midge (*Aphidoletes aphidimyza*), and several species of hover flies (Diptera: Syrphidae). Other generalist predators included *Orius insidiosus*, *Geocoris punctipes*, *Nabis alternatus*, and *Chlamydatus associatus*.

These naturally occurring biocontrols can be enhanced with conservation efforts (see section 20.3.3). In a study of plants harboring natural populations of lacewings in Xīnjiāng, Liu *et al.* (2021) ironically called for the conservation of *Cannabis sativa* (a common weed there) for "biological control services."

Stalked glandular trichomes in flowers interfere with the mobility of some predators. López Carretero *et al.* (2022) evaluated the mobility of *Aphidoletes aphidimyza* and *Chrysoperla carnea*. On leaves, covered by *non*-glandular trichomes, the larvae of both predators were able to move and prey upon aphids. But in flowers, *A. aphidimyza* larvae were impeded, and *C. carnea* larvae were slowed. The authors suggested employing *A. aphidimyza* during the vegetative stage and *C. carnea* during the flowering stage.

Cranshaw *et al.* (2018) infrequently observed natural parasitism by braconids (Hymenoptera: Braconidae). In Pakistan and India, Das (1918) reported parasitism of *P. cannabis* by two braconids, *Aphidius* sp. and *Lysiphebus* sp. Dey and Akhtar (2007) identified *Aphidius matricariae* parasitizing *P. cannabis* in northern India. McPartland and Rhode (2005) found *Aphidius* sp. parasitizing *A. gossypii* on *Cannabis* in New Zealand. Cranshaw (Fig. 4.14) found *P. cannabis* mummies in Colorado with exit holes suggestive of *Aphidius* sp.

Commercially available biocontrols are described in section 20.6. Predators work better than parasitoids at controlling established aphid infestations. Popular predators include green

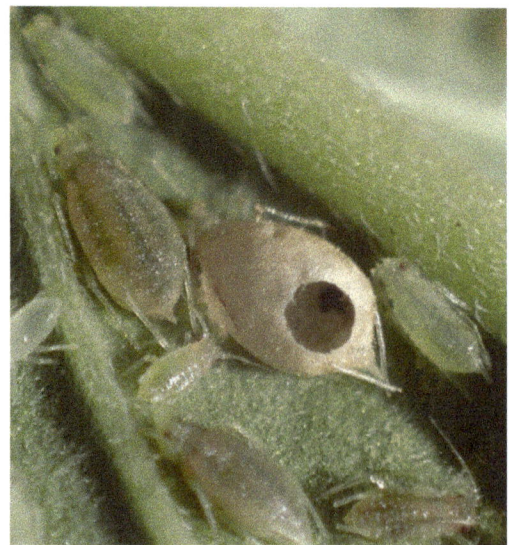

Fig. 4.14. *Phorodon cannabis* mummy, with exit hole made by *Aphidius* sp. (W. Cranshaw)

lacewings (*Chrysoperla carnea*), aphid midges (*Aphidoletes aphidimyza*), and pirate bugs (*Orius* spp.). Wild-harvested lady beetles such as *Hippodamia convergens* are no longer recommended; their populations are in decline (Harmon *et al.*, 2007). Other lady beetles are reared in commercial insectaries: *Cryptolaemus montrouzieri* will feed on aphids, but its primary hosts are mealybugs. *Stethorus* spp. may feed on aphids, but their primary hosts are spider mites. Predator release rates depend upon the infestation severity index (Table 4.2).

Parasitoid wasps control aphids when released prophylactically or for treating very light infestations. They can't control epidemics. The big three are *Aphidius matricariae, A. colemani*, and *Aphelinus abdominalis*. Mahieu *et al.* (2019) preferred *A. colemani* and *A. abdominalis* along with the predators *A. aphidimyza, Chrysopa carnea*, and the lady beetle *Adalia bipunctata*. Grammenos *et al.* (2021) reported success against *M. persicae* and *A. gossypii* using a combination of *Aphidius ervi, A. matricariae*, and *Chrysoperla carnea*, released weekly in April and May in a glasshouse in Greece.

Bacillus thuringiensis (Bt) doesn't work with members of the sap-sucking guild. Bt is sprayed on plant surfaces, and aphid stylets essentially bypass plant surfaces to penetrate phloem, so aphids won't ingest Bt. Amazingly this holds true when the Bt toxin is bioengineered into cotton. Aphids and other sap feeders are not affected by transgenic Bt cotton (Lu *et al.*, 2010).

Fungi work on contact, and directly penetrate aphid skin. They need not be ingested. CFDA (2019) recommended *Beauveria bassiana* for *P. cannabis*. California DPR (2017) approved the use of *Cordyceps* (*Isaria*) *fumosorosea* for treating aphids. Other fungi include *Lecanicillium* (*Verticillium*) *lecanii, Metarhizium anisopliae*, and *Pandora neoaphidis*. California DPR (2017) approved the use of *Burkholderia* spp. strain A396 for treating aphids on *Cannabis*. The bacterium is heat-killed (60°C for 2 h), and its thermostable toxins degrade the exoskeleton and interfere with molting and pupation.

Breeding for genetic resistance glitters with future potential—but thus far, only aphid-*susceptible* plants have been identified. Alexander (1990) reported that aphids "just love" *afghanica* landraces and *afghanica*-dominant hybrids. Hillig (2004) showed that *afghanica* landraces ("WLD biotype" or "Indica") produced significantly less **(E)-β-farnesene** (EBF) than South Asian landraces ("NLD biotype" or "Sativa").

EBF is an aphid alarm pheromone, secreted in droplets from cornicles in response to attack by predators or parasites. Volatized EBF causes aphids to stop feeding and disperse across plants—exposing them to biocontrols—or drop to the soil. *Cannabis* produces EBF in glandular trichomes. Kunert *et al.* (2010) showed that *constitutive* volatilization of EBF from plants does not repel aphids. However, the authors allowed that *pulsed emission* may repel aphids, for example when *M. persicae* causes glandular trichomes to burst and release EBF (Gibson and Pickett, 1983).

EBF production by *Cannabis* may explain its efficacy at protecting other aphid-susceptible crops growing nearby. See the end of section 20.3 for reports of hemp plants repelling four species of aphids as well as related phylloxera (*Daktulosphaira vitifoliae*).

Harand (2000) investigated hemp's protective effect against potato aphids. He used gas chromatography to analyze hemp plants. EBF was the primary terpenoid, followed by β-caryophyllene (BCP) and α-humulene (AHL). Dawson *et al.* (1984) made an interesting discovery with hops that is applicable to *Cannabis*: A dosage of 60 ng synthetic ECB caused 89% of *M. persicae* to disperse, yet the EO (essential oil) of hops, at a level equivalent to 900 ng of EBF, caused only 5% of *M. persicae* to disperse. They determined that other terpenoids in hops EO *inhibited* the effects of EBF—primarily BCP, but also AHL to a lesser degree. High levels of EBF relative to BCP in the glandular trichomes of alfalfa (*Medicago saliva*) also enhanced aphid deterrence (Mostafavi *et al.*, 1996).

Aphid species vary in the dosage of EBF required to induce dispersal. Montgomery and Nault (1977) found a 375-fold difference in sensitivity to EBF between *Myzuz persicae* (0.2 ng) and *Aphis fabae* (75 ng). They correlated EBF sensitivity with two traits: aphid aggregation density and whether the aphid was ant-attended. *P. cannabis* shares traits with EBF-sensitive *M. persicae*: it aggregates in medium-dense colonies, and is not attended by ants. Mondor *et al.* (2002) proposed that aphids with long cornicles use them to scent-mark aphid predators with EBF. *P. cannabis* has moderately long cornicles; it is probably sensitive to EBF.

Rather than chemical deterrence, Potter (2009) proposed that glandular trichomes imparted a mechanical deterrence. Gland heads on living plants are "touch-sensitive," burst easily, and release their fluids, which oxidize and polymerize into a sticky resin. Potter photographed nymphs of *Aphis gossypii* with their legs adhered to burst resin heads of glandular trichomes (Fig. 4.15). They struggled for a while and then died.

4.6.7 Chemical control

See introductory comments regarding pesticide regulations in the spider mite section, as well as in section 21.4. Direct all sprays at undersides of leaves. Hosing plants with plain cold water may knock back aphid populations (Villanueva and

Fig. 4.15. Ensnared nymphs of *Aphis gossypii* (David Potter)

Kennedy, 2019). For treating aphids, California DPR (2017) approved the use of aromatic plant oils (cinnamon, garlic, peppermint, rosemary, thyme), other plant oils (corn, soybean, cottonseed, sesame), horticultural oil (petroleum oil), potassium silicate and potassium sorbate, insecticidal soaps (potassium salts of fatty acids), citric acid, and azadirachtin (neem oil). For treating *P. cannabis*, CFDA (2019) listed azadirachtin, horticultural oil, insecticidal soap, and rosemary and peppermint oils.

Cloyd *et al.* (2009) compared several US Environmental Protection Agency (EPA) FIFRA Section 25(b) exempt pesticide products against *Myzus persicae* on garden chrysanthemums (*Dendranthema vestitum*)—the same ones they tested against spider mites (see above). None worked well. Historically, *P. cannabis* was controlled with nicotine (Balachowsky, 1936; Crescini, 1940; Ceapoiu, 1958). Spray dormant oil on neighboring *Prunus* species to kill overwintering eggs of *M. persicae* and *P. humuli*.

William Wilde (Oscar's father) rid glasshouses of aphids in London by fumigating *Cannabis indica* foliage (Wilde, 1840). Five experimental studies that tested cannabis extracts against aphids were reviewed by McPartland and Sheikh (2018), available online. The most rigorous study, by Benelli *et al.* (2018), killed *M. persicae* with an EO steam-distilled from flowering tops of the fiber-type cultivar 'Felina 32'. They calculated an LC_{50} dose of 3.5 mL/L. The EO contained 3% EBF.

In hops crops, Neve (1991) proposed that EBF-induced dispersal would increase aphid contact with insecticides or biocontrol organisms when administered together. This paradigm has been tested on other crops, with mixed results, depending on the aphid, host plant, insecticide, or biocontrol organism. High-powered synthetic pesticides are a losing proposition against aphids. Because of their prolific nature, aphids famously develop resistance to these pesticides. *M. persicae* resists a broader range of pesticides than nearly any other insect.

4.7   Thrips

Thrips (Thysanoptera: Thripidae) are small, slender insects. Adults have wings but are poor fliers. They prefer to jump, and spring quickly to safety when confronted. Among surveyed California growers, 25% reported problems with thrips—in third place behind hemp russet mites and two-spotted spider mites (Wilson *et al.*, 2019).

4.7.1   Signs and symptoms

Thrips rasp plant surfaces and suck up the exuded sap. This gives rise to pale speckles or streaks, colored gray, silver, or yellow (Fig. 4.16A). Damage begins on undersides of leaves, but soon moves topside. Heavily infested plants become covered with black specks of thrips excrement. Ultimately, leaves curl up and whole plants may wither, brown and die. McPartland *et al.* (2000) proposed an infestation severity index for thrips (Table 4.3). McCune *et al.* (2021) proposed lower limits for onion thrips (*Thrips tabaci*) on indoor *Cannabis*.

Fig. 4.16. Onion thrips, *Thrips tabaci*: **A.** typical leaf damage (W. Cranshaw); **B.** adult cleared and mounted on a microscope slide (S. Koike)

Table 4.3. Infestation Severity Index for thrips

Light	Any thrips seen
Moderate	Any thrips *damage* seen OR 2–5 thrips per leaf
Heavy	Thrips damage on many plants but confined to lower leaves OR 6–10 thrips per leaf
Critical	Thrips damage on growing shoots OR >10 thrips per leaf

4.7.2   Causal organisms

At least six genera of thrips can infest *Cannabis*. Additionally, a species formerly in the *Thrips* genus, **Chirothrips manicatus** (Haliday) 1836, infested *Cannabis* in western Siberia (John, 1924). Another renamed species, *Neohydatothrips signifer* (Priesner) 1932, has been reported on *Cannabis* in Columbia (Ravelo, 2020). Mahieu *et al.* (2019) reported *Echinothrips americanus* Morgan 1913 in Dutch greenhouses. This species was first described in Florida, and has spread to Europe and China.

Thrips are best identified to species under a compound microscope. They need to be captured, killed in ethyl alcohol, "cleared" overnight in 1–2% NaOH, then mounted on a microscope slide, ventral side up (Mound and Kibby, 1998).

4.7a.   Onion thrips

Thrips tabaci Lindeman 1888
Description: *Adults* vary in color from pale yellow to dark brown, are pointed at both ends, and grow to 1.2 mm in length. Key morphological features include antennae with seven segments, and the body segment directly behind the head (pronotum) bears two pairs of long hairs (setae) on the posterior edge.

Female wings are slender and fringed with delicate hairs (Fig. 4.16B). They are very active insects, especially when disturbed. *Larvae* are small, pale, wingless versions of the adults, barely visible without a hand lens. Pupae are rarely seen because they stay in the soil, but they look like larvae with wing buds. *Eggs* are white and kidney-shaped, oviposited in stems or leaves, and all but invisible to the naked eye.

Comments: Onion thrips attacks a wide variety of crops, and it vectors viral diseases (Koike *et al.*, 2006). The species reportedly transmits hemp streak virus (Ceapoiu, 1958) and the *Cannabis* pathogen Argentine sunflower virus (Traversi, 1949). Onion thrips have damaged hemp in Romania (Ceapoiu, 1958), Italy (Martorina, 2011), and Poland (Bakro *et al.*, 2018). Drug-type plants were infested in Iran (Bhatti *et al.*, 2009).

The first report in the USA was on feral hemp in Illinois (Stannard *et al.*, 1970). Onion thrips was the dominant thrips species recovered in a survey of survey of outdoor *Cannabis* pests in eastern Colorado (Schreiner, 2019), and central coastal California (Koike, pers. comm., 2020). In Colorado, onion thrips caused serious foliar damage during the indoor phase, but when plants were transplanted outdoor, the pest failed to reach damaging population levels (Cranshaw *et al.*, 2019). Onion thrips reduced yields in indoor-grown *Cannabis* in Québec, an average loss of 31% compared with control plants (McCune *et al.*, 2021). Furthermore, THC content decreased, from 19.3% in control plants to 17.6% in test plants.

4.7b. Marijuana thrips

Oxythrips cannabensis Knechtel 1923
Description: *Adults* are yellow, turning brown near their tails. *Females* reach 1.6 mm in length, and *males* are 1.3 mm. The body segment directly behind the head (pronotum) bears only one pair of long hairs (setae) on the posterior edge. *Larvae* are smaller, lighter in color, and wingless (Fig. 4.17).
Comments: This species appears to be host-specific on *Cannabis*; Stannard *et al.* (1970) coined the common name. It has been isolated from hemp in Romania, Hungary, Poland, Czech Republic, Russia, Siberia, Israel, and the USA

(Vierbergen *et al.*, 2010). Marijuana thrips has infested leaves and female flowers of feral hemp in Illinois (Stannard *et al.*, 1970) and Kansas (Hartowicz *et al.*, 1971). Figure 4.17 illustrates a sample from the Stannard collection. Hartowicz and colleagues proposed using *O. cannabensis* as a biocontrol agent against marijuana, and suggested the species could vector viruses. Fyodorov *et al.* (2021) recommended deploying *O. cannabensis* against wild hemp in eastern Siberia.

4.7c. Western flower thrips

Frankliniella occidentalis (Pergande) 1895
Description: *Adults* are different colors—some populations are pale yellow, others are darker brown. They have light-colored heads and dark eyes. The forewings are light-colored, with two longitudinal veins bearing two rows of delicate hairs (Fig. 4.18). Key morphological features include antennae with eight segments, and the body segment directly behind the head (pronotum) bears two pairs of long hairs (setae) on the anterior edge and three pairs of setae on the posterior edge. Females reach 1.3–1.7 mm in length, males are shorter (0.9–1.1 mm). *Larvae* are small, pale yellow, red-eyed, wingless versions of adults, feeding along large veins of leaflets. They develop into prepupal and pupal stages in soil (usually 1.5–2.0 cm deep). *Eggs* are kidney-shaped, 0.25 × 0.5 mm, and oviposited in leaves near main veins.
Comments: Western flower thrips (WFT) is California's gift to glasshouses around the world. WFT invaded Europe around 1983 (in Holland) and has spread throughout the continent. And it's still in California, infesting *Cannabis* (California DPR, 2017). The species has been described on CBD crops in Colorado, mostly associated with flowers and pollen (Cranshaw *et al.*, 2019), and indoors in Holland (Mahieu *et al.*, 2019) and Greece (Grammenos *et al.*, 2021).

Fig. 4.17. Adult female marijuana thrips

Fig. 4.18. *Frankliniella occidentalis* cleared and mounted on a microscope slide (S. Koike)

WFT infests at least 250 species of plants, and is attracted to flowers and buds. Adults can fly for only 24 h after emerging from pupae. At 25°C the life cycle is completed in 12–13 days. Females lay an average of 150 eggs (up to 300); mated females are more fecund than unmated ones. Females feeding on leaves and pollen are more fecund than females feeding solely on leaves.

Five other *Frankliniella* species have also been reported: *Frankliniella intonsa* (Tryborn) on *Cannabis* in western Siberia (John, 1924), *Frankliniella tritici* Fitch on feral hemp in Illinois (Stannard *et al.*, 1970), *Frankliniella fusca* (Hinds) on Mississippi marijuana (Lago and Stanford, 1989), *Frankliniella schultzei* (Tryborn) on Polish hemp (Bakro *et al.*, 2018), and *Frankliniella bicolor* Moulton on plants in Columbia (Ravelo, 2020).

4.7d. Indian bean thrips

Caliothrips indicus (Bagnall) 1913
≡ Heliothrips indicus Bagnall
Description: *Adults* are blackish brown, with brown-and-white banded forewings. The body segment directly behind the head (pronotum) has no long hairs (setae). Females reach 1.2 mm in length; males are shorter (0.9 mm) and lighter brown. Females occur in a 6:1 ratio to males. *Larvae* are small, pale, wingless versions of adults; they feed along large veins. *Eggs* are oval, white, and oviposited in the upper surfaces of leaves near main veins.
Comments: Indian bean thrips normally attacks flax and various legumes. The species is limited to the Indian subcontinent. *C. indicus* closely resembles the North American bean thrips, *Caliothrips fasciatus* (Pergande). Cherian (1932) considered Indian bean thrips a minor pest of young *C. indica* in India. Ananthakrishnan and Sen (1980) reported the species on *C. indica* all over India, as well as *Caliothrips graminicola* (Bagnall & Cameron) in Bengal and Kashmir. Mated females produce an average of 65 eggs, unmated females average 51 eggs. In tropical zones the species completes its life cycle in 11–14 days. They are very active insects. Even the pupae of Indian bean thrips can jump when disturbed.

4.7e. Greenhouse thrips

Heliothrips haemorrhoidalis (Bouché) 1833
Description: *Adults* have dark brown to black bodies covered by a reticular pattern, a red abdomen tip, light-colored legs, and glassy-appearing wings which look white against the black body, 1.3–1.8 mm long. The body segment directly behind the head (pronotum) has no long hairs (setae). *Larvae* are wingless, slender, and light brown to almost white in color.
Comments: Greenhouse thrips are tropical, and now found in warm glasshouses everywhere. The species has been identified on *Cannabis,* albeit not by thrips experts (McPartland, 1996b; Deardorff and Wadsworth, 2017; California DPR, 2017). Feeding by greenhouse thrips causes small dark specks or streaks on the undersides of leaves. Each female lays about

45 eggs. The *H. haemorrhoidalis* life cycle is completed in as little as 30–33 days at 26–28°C. This species pupates on foliage of host plants (Hill, 1994). Young larvae secrete a rectal liquid which gums up parasitoids and repels predators.

4.7f. Banded thrips

Aeolothrips fasciatus (Linneaus) 1758
Description: *Adults* have dark brown bodies and legs, with elongate abdomens that taper to a tip with a long ovipositor, 1.4–1.8 mm long (males smaller than females). The body segment directly behind the head (pronotum) has no long hairs (setae). Eyes are large and black. Forewings are fringed, dark with three distinct white-colored cross bands (Fig. 4.19). Antennae have nine segments and are also dark with a band. *Larvae* are elongate and yellow to orange.
Comments Banded thrips is polyphagous—it feeds on many plant hosts as well as other thrips, aphids, and whitefly larvae. The species occurs across temperate Eurasia as well as southern Africa, Australia, and New Zealand. Karadjova and Krumov (2015) reported banded thrips on *Cannabis* in Bulgaria. Damage was due to feeding as well as their egg-laying punctures. At 25°C the life cycle is completed in 15 days.

4.7.3 Differential diagnosis

Thrips damage resembles that of other sap-suckers, particularly spider mites. Thrips leave irregular-shaped lesions, which may eventually fill the spaces between leaf veins. Adult thrips can be distinguishable from other insects, thanks to their cigar-shaped bodies. Thrips nymphs can be confused with leaf-hopper nymphs, although thrips nymphs move faster. Differentiating adult thrips to species is difficult, and telling apart the nymphs is virtually impossible.

Fig. 4.19. *Aeolothrips fasciatus* cleared and mounted on a microscope slide (S. Koike)

4.7.4 Life history and habits

Outdoors, thrips overwinter in soil, on weeds, and on plant debris. Glasshouse thrips do not hibernate. Most thrips stir into activity when the temperature reaches 16°C. Thereafter, the warmer the temperature, the worse their damage. Females with ovipositors insert their eggs into leaf and stem tissues. Those lacking ovipositors usually lay eggs in cracks and crevices. Eggs hatch in 3–10 days under favorable conditions.

First and second instars feed voraciously, and become full grown in less than a month. Third and fourth instars (called *pre-pupae* and *pupae*) are generally inactive, do not feed, and mostly go underground. After pupation they develop wings (in some species males do not have wings). Males are rare and reproduction is usually parthenogenic. Both mated and unmated females produce eggs; virgin females generally produce only females. In the field, four generations arise per year, while up to eight generations arise indoors.

4.7.5 Cultural and mechanical control

Indoors, sanitation and exclusion are key. Cover air vents and windows with fine mesh, and make sure doorframes shut tight. Keep glasshouse perimeters as weed-free as possible. Inspect plant material prior to bringing it into a grow space. Adult thrips jump and scurry when disturbed; to detect thrips, shake foliage or flowers over a piece of white paper. Anything that jumps onto the paper can be picked up with a small paintbrush and placed in a vial with alcohol for identification.

Thrips can be caught with colored sticky traps. Blue and yellow traps are commercially available; blue is more attractive to thrips (Broughton and Harrison, 2012).

However, blue also attracts beneficial insects; enhance selectivity by adding semiochemicals, such as Thripline® (for *F. occidentalis*), or Lurem-TR® (for *F. occidentalis* and *T. tabaci*). See section 19.6.1 for details on sticky traps.

In thrips-infested glasshouses, remove all plant residues after harvest, and heat the house for several days, to starve any remaining thrips. Water-stressed plants are particularly susceptible to thrips damage, so keep plants well-hydrated. Reflective material laid between plants (aluminum foil) confuses flying thrips, and might be useful when plants are short.

4.7.6 Biological control

Concrete floors in modern glasshouses are one reason why thrips are a problem. In soil-floored glasshouses, watering plants on slatted benches kept floors damp, which encouraged a naturally occurring soil fungus, *Entomophthora thripidum*. The fungus infects thrips when they drop to the ground to pupate. With no damp soil, there is no fungus, and no natural biocontrol.

Many predators are available. Outdoors, *Aeolothrips intermedius* is a naturally occurring predatory thrips that preys on *T. tabaci* in *Cannabis* fields (Seczkowska, 1969). *Ceranisus menes* is a Eulophid wasp with a world-wide distribution that parasitizes thrips larvae (*F. occidentalis* and *T. tabaci*). The species is famous for its introduction into Hawai'i to eliminate *T. tabaci* (Sakimura, 1937). Indoors, choices of predatory mites include *Neoseiulus cucumeris*, *Iphiseius degenerans*, *Neoseiulus barkeri*, *Amblyseius swirskii*, and *Stratiolaelaps scimitus*. Mahieu *et al.* (2019) preferred *Ambyseius cucumeris*, *A. swirskii*, and *Hypoasis miles*.

Predatory bugs include *Deraeocoris brevis* and *Orius* spp. *D. brevis* was taken off the market in 1998 due to decreased demand, but it is native to the USA, so don't kill the little brown mirids. Several *Orius* spp. are commercially available. Less effective thrips predators include lacewings and lady beetles. Most thrips predators cannot be mixed together, except for a combination of *Orius* spp., *Neoseiulus* spp., and *Stratiolaelaps scimitus*. Grammenos *et al.* (2021) reported success against *F. occidentalis* using a combination of *Amblyseius swirskii* and *Orius laevigatus*, released weekly in April and May in a glasshouse in Greece.

Nematodes (*Steinernema* and *Heterorhabditis* spp.) may parasitize thrips pupating in soil, but will not reproduce in thrips cadavers. Parasitoid wasps include *Thripoctenus javae* and *Ceranisus menes*. They are members of the Eulophidae, a wasp family that has the unique ability to parasitize the Thysanoptera (thrips family). The commercial availability of Eulophid wasps is spotty.

Turning to fungi, Hussey and Scopes (1985) controlled onion thrips and other species with the Mycotal™ strain of *Lecanicillium* (*Verticillium*) *muscarium*. They also controlled onion thrips with *Beauveria bassiana*, spraying twice at 10-day intervals with a 0.25% water suspension of Boverin™. California DPR (2017) approved *Beauveria bassiana* and *Cordyceps* (*Isaria*) *fumosorosea*. Thrips are killed by *Metarhizium anisopliae*, which has a wide host range, and *Metarhizium album*, which specifically targets thrips. California DPR also approved *Burkholderia* spp. strain A396, which is a heat-killed bacterium. In Dutch glasshouses, Mahieu *et al.* (2019) recommended *Beauveria bassiana* (ATCC74040 and GHA strains), *Lecanicillium muscarium* strain Ve6, and *Paecilomyces fumosoroses* (strain Fe9901).

Potter (2009) observed onion thrips totally immobilized by sticky resins released from burst glandular trichomes. Perhaps this provides some thrips deterrence, and perhaps it's why Foy (1851) observed Egyptian farmers planting *Cannabis* in onion fields. They may have been companion planting; onions don't have many pest problems, but onion thrips is a major pest in that crop.

4.7.7 Chemical control

Give clones and seedlings a complete dunk in horticultural oil or insecticidal soap before moving them into a grow space. California DPR (2017) offered a list of pesticides for controlling thrips, identical to the approved list for aphids (see above). Direct all sprays at undersides of leaves, and apply at midday, when thrips are most active.

Cloyd *et al.* (2009) compared several biorational pesticides against western flower thrips on *Gerbera* plants. A product that contained 0.05% spinosad provided the best control (100% mortality). Spinosad is a natural substance produced by the soil bacterium *Saccharopolyspora spinosa*. EPA Section 25(b) exempt pesticide products—the same ones Cloyd tested against spider mites (see above)—did not provide sufficient control (<30% mortality). Cherian (1932) repelled thrips from *gañjā* with a mix of fish emulsion and soap sprays. Historically, nicotine controlled thrips in *Cannabis* (Ceapoiu, 1958). Taisiya *et al.* (2010) tried controlling western flower thrips with a *Cannabis* ethanol extract. After 5 days, <20% of 2nd-instar larvae died. Their *Cannabis* extract worked much better against spider mites.

4.8   Whiteflies

Whiteflies (Homoptera: Aleyrodidae) primarily cause problems in glasshouses and growrooms. They resemble tiny moths, but are neither moths or flies. Whiteflies are related to aphids, scales, and leafhoppers. They damage plants by sucking sap, and they vector plant viruses, including the hemp streak virus (Ceapoiu, 1958) and the *Cannabis* pathogen lettuce chlorosis virus (Hadad *et al.*, 2019).

4.8.1   Signs and symptoms

Whiteflies produce few initial symptoms. This allows their populations to build until the garden is suddenly engulfed by a massive infestation. Symptoms from whiteflies resemble aphid damage—leaves with yellow, white, or silver spots and stipples. In heavy infestations plants lose vigor, leaves droop, plants wilt, and sometimes die. Leaves become glazed with honeydew. Larvae and adults congregate on undersides of leaves, out of sight to the casual observer. When infested plants are shaken, a billowing cloud of whiteflies fills the air for several seconds before they resettle. McPartland *et al.* (2000) proposed an infestation severity index for whiteflies (Table 4.4).

4.8.2   Causal organisms

Three species of whiteflies infest *Cannabis*. The greenhouse whitefly, like its name implies, is primarily an indoor problem. The sweetpotato whitefly causes problems indoors and outdoors. The silverleaf whitefly has plagued growers in the southern USA, indoors and outdoors. Whiteflies were the most commonly collected insects in a survey conducted in southern California (Britt *et al.*, 2022).

4.8a.   Greenhouse whitefly

Trialeurodes vaporariorum (Westwood) 1856
Description: *Eggs* are 0.2 mm long, pale yellow, oval or football-shaped, with short stalks anchoring them to the leaf. Eggs turn purple-gray to brown-black before hatching. First-instar *larvae* are tiny 0.3 mm long "crawlers," almost transparent, oval in outline, nearly flat, and radiate a halo of short waxy threads from their body (Fig. 4.20A). Subsequent instars lose their legs and resemble immature scale insects, reaching 0.7 mm in length. Late in the fourth stage, larvae change from transparent to an off-white color, by secreting an extra layer of wax to pupate within. *Pupae* project long body hairs as well as a halo of short wax threads from their palisade-like perimeter (Fig. 4.21). *Adults* rarely measure over 1 mm in length. Their wings are off-white, have rounded contours, and are held flat over their abdomens almost parallel to the leaf surface. The wings and body may become covered with a white dust or waxy powder (Figs 4.20A, 4.21).
Comments: The greenhouse whitefly has been reported in the USA (Frank and Rosenthal, 1978; Frank, 1988; McPartland, 1996b; Cervantes, 2015; California DPR, 2017; Wang, 2021). The pest is also a problem in Italy (Martorina, 2011), Holland (Mahieu *et al.*, 2019), and Greece (Grammenos *et al.*, 2021). In temperate countries it attacks many glasshouse crops, and lives outdoor in warmer climates around the world. Zuefle (2021) found *T. vaporariorum* outdoors, in New York State, infesting CBD crops for 3 weeks in July. *T. vaporariorum* has approximately 860 host plant species.

4.8b.   Sweetpotato (or tobacco) whitefly

Bemisia tabaci (Gennadius) 1889
= *B. gossypiperda* Misra & Lamba 1929

Table 4.4. Infestation Severity Index for whiteflies

Light	Any thrips seen when plant is shaken
Moderate	5–10 adults/plant, seen on more than one plant
Heavy	11–20 adults/plant, seen on many plants
Critical	>20 adults/plant OR sooty mold and leaf discoloration present OR a few adults on all plants

Fig. 4.20. Whiteflies: **A.** *Trialeurodes vaporariorum* (W. Cranshaw); **B.** *Bemisia tabaci* (USDA)

T. vaporariorum

B. tabaci

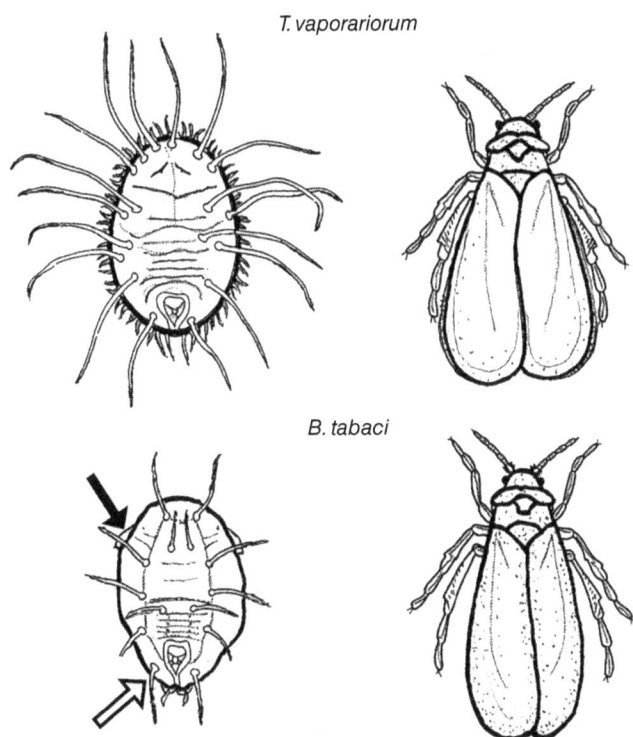

Fig. 4.21. Comparing pupae and adults of two whitefly species

Description: *Eggs* are 0.2 mm long, white, oval, with short stalks anchoring them to the leaf. As eggs mature they turn yellow or amber. *Larvae* are oval, flat, with segmentations, and light green to nearly colorless. *Pupae* have reddish colored eyespots, their bodies are pale, marked by a caudal groove, and slightly pointed in the rear. Margins of pupae taper to the leaf surface, they lack long hairs, and project only a few short wax threads. They grow an anterior wedge of wax at tracheal folds (Fig. 4.21, solid arrow), and a posterior wax fringe which extends lateral to the caudal setae (Fig. 4.21, hollow arrow). *Adults* are light beige to yellow, with longitudinal striations and angled (not rounded) wing tips, held in a tent-like position over their abdomens (Fig. 4.20B).

Comments: Sweetpotato whitefly is a pest of field crops in warm climates, and indoor crops in temperate climates. Outdoors, the species infests *Cannabis* crops in southern Europe (Sorauer, 1958), Egypt (Azab *et al.*, 1970), Brazil (Flores and Silberschmidt, 1958), and Pakistan (Attique *et al.*, 2003). Indoors, sweetpotato whitefly causes problems in the USA (Cervantes, 2015; California DPR, 2017; Cranshaw *et al.*, 2019), Holland (Mahieu *et al.*, 2019), and Greece (Grammenos *et al.*, 2021).

4.8c. Silverleaf whitefly

Bemisia argentifolii (Bellows & Perring) 1994
= *B. tabaci* poinsettia strain, or biotype B, or biotype MEAM1
Description: *B. tabaci* and *B. argentifolii* are nearly identical. *B. argentifolii* pupae have narrower wedges of wax at tracheal

folds, the posterior wax fringe does not extend lateral to caudal setae, and they have one fewer pairs of hairs. The adults are slightly larger than those of *B. tabaci* (Bellows *et al.*, 1994).

Comments *B. argentifolii* appeared in the USA around 1986. It was originally called the *B. tabaci* poinsettia strain in Florida. It is also called *B. tabaci* biotype MEAM1 (Middle East Asia Minor 1). Summer rains in Florida dampen the pest's damage there, but within 5 years it spread to southern California. It can feed on anything the Imperial Valley has to offer, with resistance to almost all pesticides. As a consolation, *B. argentifolii* transmits fewer viruses than *B. tabaci*. Wu *et al.* (2002) reported *B. argentifolii* on Chinese hemp, and California DPR (2017) also reported the species. In Israel, Hadad *et al.* (2019) proved that silverleaf whitefly could transmit the *Cannabis* pathogen LCV (lettuce chlorosis virus).

4.8.3 Differential diagnosis

Injuries caused by whiteflies can be confused with those of aphids, scales, and hoppers. Look for the insects. Immobile whitefly larvae may be confused with immature scale insects. Whitefly species can be difficult to tell apart. Eggs of *T. vaporariorum* are white and turn purple or brown before hatching, while *Bemisia* eggs turn yellow. Their pupae differ in color, hairiness, and the angle of their edges: *Bemisia* pupae taper smoothly to the leaf surface, whereas edges of *T. vaporiorum* pupae drop at a 90° angle to the leaf surface and look like they were made with a cookie cutter.

T. vaporariorum adults are slightly larger and lighter colored than *Bemisia* adults, and hold their wings differently. *B. tabaci* and *B. argentifolii* are genetically different—their DNA sequences are as different from each other as they are from *T. vaporariorum*. They are biologically distinct (*B. argentifolii* is more fecund and produces more honeydew), and morphologically different (Bellows *et al.*, 1994). *B. argentifolii* also differs from *B. tabaci* by the ability of its larvae to induce "silverleaf" symptoms (Costa *et al.*, 1993). Wu *et al.* (2002) used genetic markers (RAPD-PCR) to differentiate *B. tabaci* and *B. argentifolii* (which they called biotype B). *B. argentifolii* was collected from hemp in Xīnjiāng Region; *B. tabaci* was collected from other plants.

4.8.4 Life history and habits

Indoor populations reproduce year-round. Outdoor populations overwinter as eggs. Females of all three species lay ≥100 eggs. Eggs are laid on undersides of leaves near the top of plants, often clustered in a circular pattern. Eggs hatch after 7–10 days. First-instar larvae crawl around plants searching for suitable feeding sites. Once feeding begins, larvae settle in one spot to suck sap, and pupate there. Larvae take 2–4 weeks to reach adulthood, depending on temperature. Adults live 4–6 weeks. Generations often overlap, so all stages may be found together. Reproduction rate is dependent on temperature and the host plant.

4.8.5 Cultural and mechanical control

Sanitation is supreme. Follow instructions given for aphids: clean clones, screen openings, limit visitors, monitor crops, and rogue infested plants. Mahieu *et al.* (2019) emphasized that mother plants must be scrupulously clean of the greenhouse whitefly.

Lay aluminum foil on the ground between plants to confuse flying adults. This is less effective on taller plants. Yellow sticky traps for catching adult whiteflies are widely available (see section 19.6.1). Outdoors, do not plant *Cannabis* near whitefly magnets: eggplant, sweet potato, tobacco, or cotton.

4.8.6 Biological control

Early whitefly infestations are hard to detect, so growers should release biocontrols before pests are detected. The best prophylactics are wasps—*Encarsia formosa* and *Eretmocerus eremicus* (Mahieu *et al.*, 2019). Predators include *Delphastus pusillus*, *Geocoris punctipes*, *Dicyphus hesperus*, and *Macrolophus pygmaeus*. Release rates depend on the Severity Index (Table 4.4). Second-line agents are pirate bugs (*Orius* spp.), green lacewings, and lady beetles. The predatory mite *Amblyseius swirskii* feeds on eggs and young nymphs. Grammenos *et al.* (2021) reported success against *T. vaporariorum* and *B. tabaci* using a combination of *Encarsia formosa* and *Eretmocerus eremicus*, released weekly in April and May in a glasshouse in Greece.

California DPR (2017) approved two fungi for controlling whiteflies on *Cannabis*: *Beauveria bassiana* and *Cordyceps* (*Isaria*) *fumosorosea*. Mahieu *et al.* (2019) recommended *Beauveria bassiana* (ATCC74040 and GHA strains). *Metarhizium anisopliae* kills whiteflies, but has a wide host range. Fungi with narrower host ranges might be more sensible. Mycotal®, a product made from *Lecanicillium muscarium*, specifically targets whiteflies and thrips. *Aschersonia aleyrodis* specifically parasitizes whitefly larvae (*T. vaporariorum*, *B. tabaci*). The fungus was commercialized by Florida's agricultural department for sale to citrus growers, but no longer. California DPR also approved *Burkholderia* spp. strain A396, which is a heat-killed bacterium.

4.8.7 Chemical control

See introductory comments regarding pesticide regulations in the spider mite section, as well as in section 21.4. Direct all sprays at undersides of leaves. Small plants in pots can be dipped. Three or four applications (at weekly intervals) are required to kill whiteflies emerging from eggs that escaped initial applications. For treating whiteflies, California DPR (2017) offered a list of pesticides, identical to the approved list for aphids. They won't eliminate the pests, but will slow them down. Insecticidal soap and horticulture oil are compatible with some biocontrol organisms. So is neem oil, which is effective against whitefly nymphs.

Liu and Stansly (2000) tested several surfactants against nymphs of silverleaf whitefly on tomato. Sodium lauryl sulfate (SLS), an EPA Section 25(b) exempt pesticide, killed 50% at a concentration of 10 mL/L. Synthetic surfactants did better: Silwet L-77 (silicone-polyester copolymer, 1.0 g/L), killed 99.7%, and APSA80 (80% alkyl aryl alkoxylate, free fatty acids, 1.0 g/L) killed 87.7%. They also tested insecticidal soap (M-Pede®, 20 mL/L), which killed 72.1%, and horticulture oil (Sunspray oil, 5 mL/L), which killed 74.7%.

Cloyd *et al.* (2009) compared several EPA Section 25(b) exempt pesticide products against silverleaf whitefly on poinsettia plants—the same ones they tested against spider mites (see above). None provided sufficient control. The Rodale crowd recommended ryania and spinosad (Yepsen, 1976).

4.9 Leafhoppers

Leafhoppers constitute an enormous family of plant-feeding insects (Hemiptera: Cicadellidae), with over 20,000 species described worldwide (Dietrich, 2013). Leafhoppers are often the most species-rich insect family found on *Cannabis* plants outdoors. Lago and Stanford (1989) cataloged 19 leafhopper species in Mississippi, and Schreiner and Cranshaw (2020) reported 14 species in Colorado. Few leafhoppers, however, reproduce on the crop (Cranshaw *et al.*, 2019). Many found on foliage appear to be transient adults, associated with neighboring weeds or crops. Leafhopper sightings on *Cannabis* often do not describe any damage—the insects possibly did not even feed on the crop.

Leafhoppers have piercing-sucking mouthparts, but can feed in several different ways (Backus *et al.*, 2005). Most feed on fluids from phloem, in a manner similar to aphids, causing very little cell injury and producing little, if any, visible symptoms. These leafhoppers excrete honeydew, but since leafhoppers never approach the high populations often seen in aphids, and because they because move around plants much more actively, honeydew is less commonly observed. Some phloem-feeding leafhoppers can transmit viruses or phytoplasmas to plants. Beet curly top virus causes a serious hemp disease and is transmitted by *Neoalitursus tenellus*, described below.

Leafhoppers in the Cicadellinae subfamily ("sharpshooters") are xylem feeders (Wilson *et al.*, 2009). They do not produce any external symptoms, and none of them likely produce any significant plant injury. Some vector xylem-limited bacteria, but none of these bacteria are reported on hemp. Some members of subfamily Typhocybinae feed on the contents of mesophyll cells. The wounds produced by this type of feeding result in small light flecking of foliage, known as stippling. Mesophyll feeders do not produce honeydew but will often excrete small dark droplets known as tar spots.

About two dozen species of leafhoppers and planthoppers worldwide use a lacerate-and-flush feeding strategy, including a few species on *Cannabis*. This causes a serious injury known as **hopperburn**. The dynamic interaction between wounding, leafhopper saliva, and plant responses is termed the "hopperburn cascade" (Backus *et al.*, 2005). Hopperburn often results in leaf yellowing and curling, even necrosis (Fig. 4.22). This is typically first evident along leaf margins. Hopperburn injuries can result in dramatic suppression of photosynthesis and decrease yields, sometimes in the absence of foliar symptoms.

Leafhoppers have a simple metamorphosis pattern of development. Eggs are inserted into leaves, leaf veins, or stems. Upon egg hatch the immature nymphs feed on their host plants. There are four nymphal stages that do most of the feeding. They then molt a final time, producing a winged adult form. Leafhoppers have well-developed hindlegs that allow them to jump. When disturbed, adults will often use their legs to propel them into flight, and the adults are often observed after brushing against plants. Nymphs may jump, but many remain on leaves and may be seen crawling rapidly, sometimes crab-like, across leaves.

The taxonomy of leafhoppers is constantly in flux and the great majority of species are thought to be undescribed. Three relevant subfamilies of Family Cicadellidae described here include Typhlocybinae (the majority of those below, with potato leafhopper as a major problem), Cicadellinae, and Deltocephalinae. The taxonomy used below is based on the web-accessible *World Auchenorrhyncha Database* (Dmitriev, 2022).

4.9.1 Causal organisms: Typhlocybinae

4.9a. Potato leafhopper

Empoasca fabae (Harris) 1841
= E. mali LeBaron 1853, = E. flavescens Gillette 1898, not E. flavescens (Fabricius) 1794
Description: *Adults* are green with faint white spots on their head, large eyes, a row of six to eight white spots on the thorax behind the head, prominent spines on the hind legs, and 3.5–4 mm in length. Seen from above they appear wedge-shaped, with heads 1 mm wide, tapering to acute-angled wingtips (Fig. 4.23A). *Nymphs* are similar in shape but generally smaller, pale green, and lack wings. *Eggs* are white to pale white, slender, elongate, 0.9 mm long (Fig. 4.23B).
Comments: The first clear report of potato leafhopper damage to hemp was made by Dudley (1920). He reported hopperburn on hemp in New Hampshire by *Empoasca mali*, a name widely used a century ago that has since been synonymized as *E. fabae*. Dudley (1926) reiterated that hemp with hopperburn grew near potato fields. Poos and Wheeler (1943) reared *E. fabae* on hemp, as well as *Empoasca recurvata* DeLong, but without location records. Potato leafhopper produced classic hopperburn symptoms in Indiana (Bolt, 2019; Cranshaw, pers. observ., 2019). It has also been observed in upstate New York (Zuefle, 2020, 2021).

As *E. fabae* is thought originally to have been restricted to areas of North America east of the Rocky Mountains (Chasen *et al.*, 2014), an early report of *E. fabae* damaging hemp in California (Essig, 1915) was possibly a misidentification, perhaps of *Empoasca mexara* Ross and Moore, which is the predominant native species in that area associated with alfalfa. Within the past decade *E. fabae* has been confirmed as present in California.

An unidentified *Empoasca* sp. was found commonly on hemp in eastern Colorado (Schreiner and Cranshaw, 2021). It bred on the crop, and feeding resulted in small light flecking wounds on foliage, not hopperburn. *Empoasca* sp. was relatively common in an insect survey conducted in southern California (Britt *et al.*, 2022).

Life history and habits

Potato leafhopper is one of the most destructive *native* insects to North American agriculture. It is a major pest of potato and alfalfa but can damage more than 100 plant species, including legumes, hops, and maple. It is perhaps the most important of all hopperburn-producing leafhoppers worldwide (Hartzell, 1923; Backus *et al.*, 2005). Recent reviews of the species include Chasen *et al.* (2014) and Capinera (2020). Adults lay an average of 35 eggs in petioles and leaf veins. Up to four generations arise each year. In potato crops the presence of 10 nymphs per 100 mid-plant leaves is the threshold necessitating control measures (Howard *et al.*, 1994).

Potato leafhopper is described as an "annual circular migrant across multiple generations" in eastern North America (Taylor and Shields, 2018). During the winter, adult stages retreat to regions along the Gulf of Mexico and

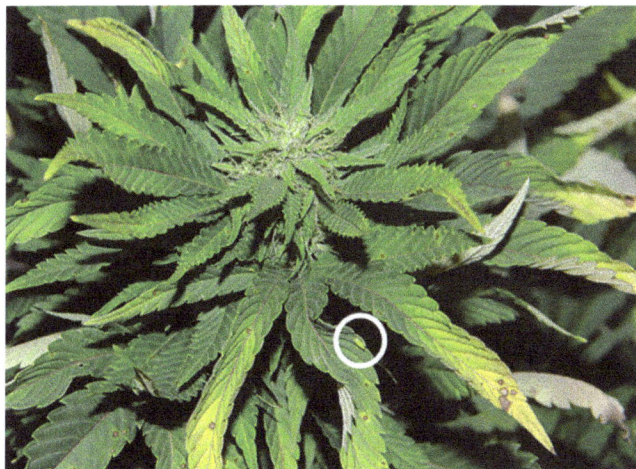

Fig. 4.22. Hopperburn caused by the potato leafhopper *Empoasca fabae*, with a nymph circled (M. Bolt)

Fig. 4.23. Potato leafhopper: **A**. adult (Robert Webster, Wikimedia Commons); **B**. nymph on *Cannabis* (W. Cranshaw)

sustain themselves primarily on pines while in a reproductive diapause condition. In spring they resume reproduction and development on a wider range of plants,
producing two or three generations in early spring. As host
plants decline in quality in overwintering areas, potato leafhopper adults disperse northward assisted by wind (Pienkowski
and Medlar, 1964). Multiple generations may be produced in
these northern locations until late in summer, when an increasing proportion goes into the diapause condition, suspends
reproduction, and then makes a reverse migration to overwintering areas in the southern USA.

Potato leafhopper belongs to subfamily Typhlocibinae, and
several other species from this subfamily have been reported
on *Cannabis*, such as *Edwardsiana flavescens*, *Kyboasca bipunctata*, and *Emelyanoviana mollicula*. These include leafhoppers
that produce leaf flecking/stippling symptoms or produce
hopperburn injuries. Many of these were originally identified
as *Empoasca* spp. but are now placed in segregate taxa.

4.9b. *Edwardsiana flavescens* (Fabricius) 1794

= *Empoasca flavescens* (Fabricius), ≠ *E. flavescens* Gillette
1898 (=*E. fabae* Harris)
Comments *Adults* are 6 mm in length, slender, yellowish-green. The head is pointed and yellow. They resemble a
pale version of *E. fabae*. This species causes hopperburn in
many crops, including cotton, hops, and tea plantations in
India. Spaar *et al.* (1990) reported *E. flavescens* infesting
European hemp, although this species name has been applied
to what are now recognized as several species. It is sometimes
known as the hemp leafhopper but its association with hemp
needs confirmation.

4.9c. *Kyboasca bipunctata* (Oshanin) 1871

= *Empoasca bipunctata* (Oshanin), = E. punctum Haupt
1912
Comments *Adults* are 3.3–3.6 mm in length, slender, yellowish-green. The head is broad with faint white spots. On the
thorax behind the head are two longitudinal orange bands separated by a median white band. Wings are yellow-green anteriorly, tapering to smoky-colored ends. This Eurasian native has
been introduced into the USA and is now widespread. It primarily attacks alfalfa, cotton, hollyhock, and elms. *K. bipunctata* has been reported on hemp in Italy (Poos and Wheeler,
1943), Denmark (Ossiannilsson, 1981), France (Giustina,
1989), Turkey (Lodos and Kalkandelin, 1983), and Uzbekistan
(Zakhvatkin, 1953).

4.9d. *Emelyanoviana mollicula* (Boheman) 1895

Comments *Adults* are 3.2–3.6 mm in length, slender, pale
yellowish-green, no spots on head. Wings are lightly banded,
yellow-green anteriorly, tapering to smoky-colored ends. This

species is a Eurasian native, not yet in the USA, and associated
with a range of herbaceous plants. *E. mollicula* has been reported on hemp in alpine Italy (Vidano, 1965), Denmark
(Ossiannilsson, 1981), France (Giustina, 1989), and Uzbekistan
(Zakhvatkin, 1953).

4.9e. Other typhlocibine leafhoppers

Empoasca decipiens Paoli 1930 has been cited on *Cannabis* in
Europe (Ossiannilsson, 1981; Dmitriev, 2022). This species has
a wide host range and it causes hopperburn (Backus *et al.*,
2005). Gottwald (2002) reported *Cannabis* as a new host of
Eupteryx atropunctata Goeze 1778 in Germany. Humphries
and Florentine (2019) described *Austroasca alfalfae* (Evans)
1940 causing hopperburn in Australian hemp. Asian records
include *Asymmetrasca uniprossicae* (Sohi), originally named
Empoasca uniprossicae on *Cannabis* in India (Sohi, 1977).
Amrasca biguttula (Ishida) 1913 (= *Empoasca bipunctata*,
E. devastans) has been described on hemp in India (Sohi, 1983).
Elbelus tripunctatus Mahmood 1967 has been described on
Chinese hemp (Cao *et al.*, 2019).

4.9.2 Causal organisms: Cicadellinae and Deltocephalinae

Beyond the Typhlocibinae, other leafhoppers have caused
problems. Among them are xylem-feeding Cicadellinae in the
genus *Graphocephala*, and phloem-feeding Deltocephalinae in
the genus *Neoaliturus*. The phloem feeders are a worry; the
xylem feeders raise concerns but rarely cause damage.

4.9f. Candystriped leafhopper

Graphocephala coccinea (Foerster) 1771
Description: *Adults* 8–10 mm in length, slender, with yellow,
pointed heads; wings reflect alternate bands of magenta and
green with yellow margins. *Nymphs* are yellow to green.
Comments: This pretty insect has been collected on wild
hemp in Kansas and Illinois (Hartowicz *et al.*, 1971; McPartland,
1996b) as well as drug-type plants (Frank and Rosenthal, 1978;
Frank, 1988; Cervantes, 2015). Many websites have photographs of *G. coccinea* on *Cannabis*; none show damage. A related species, *Graphocephala versuta* (Say) 1830, the "versute
sharpshooter," was the most frequently encountered leafhopper out of 19 species in Mississippi marijuana (Lago and
Stanford, 1989). The same study reported other Cicadellinae
sharpshooters: *Neokolla hieroglyphica* (Say) 1830, *Cuerna
costalis* (Fabricius) 1803, *Draeculacephala balli* (Van Duzee)
1915, and *Draeculacephala portola* (Ball) 1927.

4.9g. Beet leafhopper

Neoaliturus tenellus (Baker) 1896
=*Circulifer tenellus* (Baker)

Description: *Adults* 3.0–4.0 mm in length. Their coloration and patterning are highly variable, with some indistinct striping of the wings (Fig. 4.24). Forms produced in the first spring generation tend to be greenish. Under high summer temperatures the coloration may be yellow. With cooler temperatures the overall coloration darkens and dark patches develop on the wings. Overwintering forms are dark. Only adult forms are found on hemp. *Nymphs* are wingless and generally pale green-brown but do not develop on *Cannabis*.

Comments: Beet leafhopper is native to the Middle East but has become much more widespread and is presently an important agricultural pest in western North America. The species feeds on a wide variety of hosts, but breeding occurs on relatively few plants, such as Russian thistle (*Salsola tragus*) and kochia (*Bassia scoparia*). Beet leafhopper is a migratory species, and can travel hundreds of kilometers from overwintering sites. The migrations and host plants used in various regions were summarized by Cook (1967) and in the Pacific Northwest by Rondon and Oppedisano (2020). Typically, 3–4 overlapping generations are produced on various host plants used through late spring and summer.

Beet leafhopper adults migrate into hemp and feed, but do not lay eggs and develop on the crop. No significant plant injury appears to be produced. However, the species is capable of vectoring viruses and phytoplasmas, notably beet curly top virus (see section 14.5) and *Spiroplasma citri* (section 14.8). Beet leafhopper is the only known vector of beet curly top virus (Bennett, 1967), and it causes serious hemp disease in western North America (Giladi *et al.*, 2020; Hu *et al.*, 2021; Chiginsky *et al.*, 2021; Rivedal *et al.*, 2022). A short period of time, 4 hours, must pass after the leafhopper has ingested the virus before it can transmit the virus to healthy plants in subsequent feedings.

Beet leafhopper adults may be present for only a brief time within hemp. Their detection in a hemp field is difficult, requiring collection, usually by sweep net. More importantly they subsequently require correct identification. As beet leafhopper superficially resembles other non-damaging leafhoppers that occur in the crop, it often must be examined under magnification. Better assessment of local populations of beet leafhopper can be done indirectly, by sampling favored host plants on which the insects breed, such as Russian thistle and kochia.

The incidence of beet leafhopper, and the associated infections of beet curly top, vary greatly from season to season due

largely to climatic conditions. Perhaps most important are the effects of weather on populations of weedy plants that support overwintering populations of the leafhopper and sustain reservoirs of the virus. Temperature, rainfall, and wind patterns directly affect survival, reproduction, and migration.

4.9.3 Others leafhoppers

The most common leafhopper species reported by Schreiner and Cranshaw (2021) in Colorado hemp was *Aceratagallia uhleri* (Van Duzee) 1894, and it was one of only a few leafhoppers confirmed to reproduce within the crop. No symptoms were observed to be produced by this insect and it is unlikely to have any effect on hemp growth and yield. Two other species in the same family, Agalliinae, were reported by Lago and Stanford (1989) in Mississippi: *Ceratagallia sanguinolenta* (Provancher) 1872, and *Agallia constricta* (Van Duzee) 1894.

Graminella nigrifrons (Forbes) 1885, the blackfaced leafhopper, was reported by Lago and Stanford (1989) as "common" from May through September. This leafhopper is well-known in the southeastern USA, and develops on a variety of plants, particularly grains and grasses. *Scaphytopius frontalis* (Van Duzee) 1890, the yellowfaced leafhopper, was also reported by Lago and Stanford (1989) as a common species, most abundant in June. A less common species, *Scaphytopius acutus* (Say) 1830, is known to be capable of transmitting several phytoplasmas that cause disease in other crops (Weintraub and Beanland 2006).

Outside North America, Batra (1976) reported *Olidiana indica* (Walker) 1851 (= *Jassus indicus*) as the most common cicadellid on Indian *Cannabis*. In China's Hénán Province, *Macropsis cannabis* was a new *Macropsis* spp. described by Wei and Cai (1998).

4.9.4 Differential diagnosis

Young leafhopper nymphs may resemble aphids. They move faster than aphids, however. The stippling injury produced by some leafhoppers is somewhat similar to feeding injury caused by true bugs. Thrips and spider mites also produce light flecking wounds, but the size of the spotting is much smaller and they leave characteristic signs on the leaf such as fecal spots (thrips), egg shells and cast skins (spider mites). Leafhoppers may also leave old skins on the leaf after molting, like spider mites and aphids, but leafhopper skins are larger and different in shape. Hopperburn symptoms can be confused with sun scald or nutrient imbalances.

4.9.5 Cultural and mechanical control

With a few exceptions, leafhoppers cause little damage and do not require control measures. Catch adult leafhoppers with yellow sticky traps for monitoring purposes. Sweep net sampling is effective for collecting adult leafhoppers, particularly

Fig. 4.24. Beet leafhopper (W. Cranshaw)

migratory species. Although adult leafhoppers are highly mobile and may disperse over long distances, there may be some site considerations that can decrease risk of leafhopper colonization. One example would be the proximity of hemp to alfalfa, within which large populations of potato leafhoppers often develop. Potato leafhoppers disperse from alfalfa into nearby hemp fields, particularly following alfalfa cutting. Eliminate weeds that serve as breeding sites for beet leafhoppers (Russian thistle and kochia).

Reflective foil was shown to reduce the colonization of common bean, *Phaseolus vulgaris*, by potato leafhopper (Wells *et al.*, 1984). Hemp grows considerably taller than bean, however, and fiber crops form a closed canopy. Breeding for resistance is a possibility. Chasen *et al.* (2014) reviewed the status of developing alfalfa resistant to potato leafhopper—specifically, cultivars with increased leaf "hairiness." Increased trichome density and the presence of certain kinds of glandular trichomes have two effects on potato leafhopper—increasing mortality of the first stage nymphs, and reducing the settling on plants of adults that visit. Preliminary observations of hopperburn injury to hemp in Indiana suggested that symptoms varied among hemp populations (Cranshaw, pers. obs., 2019). Cultivar differences in susceptibility to potato leafhopper injury in other crops is well documented.

4.9.6   Biological and chemical control

A review of natural enemies of potato leafhopper by Chasen *et al.* (2014) included many common generalist predators (e.g., larvae of green lacewings, damsel bugs, minute pirate bugs), egg parasitoids, and in rare situations fungal parasites. However, Chasen and colleagues concluded that none of the natural enemies played a "crucial role in suppressing potato leafhopper populations on alfalfa."

Anagrus atomus is a mymarid wasp that parasitizes the eggs of leafhoppers. It is commercially available in Europe, supplied as parasitized eggs in bottles. Given that many leafhoppers in hemp fields do not even breed on the crop and are transitory adults that originate outside of the plantings, such as beet leafhopper, biological controls are likely to have even fewer effects on their populations in hemp. Damage is rarely enough to justify pesticides. Indoor leafhoppers panicked Frank (1988) and Cervantes (2015) into spraying with pyrethrum.

4.10   Treehoppers and Froghoppers

Treehoppers (Hemiptera: Membracidae) are closely related to leafhoppers but can be distinguished by the area behind the head (pronotum) becoming very enlarged and extending over the back of the insect. In some treehopper species the pronotum forms a broad "helmet", giving the insects a humpbacked form. Most treehoppers lay eggs on woody plants. Females use a strong, serrated ovipositor to insert eggs into stems and branches. The physical damage produced during egg laying (oviposition injury) is often more serious than injuries produced by feeding.

"Membracidae treehoppers" have been described on feral hemp in Kansas (Hartowicz *et al.*, 1971; Eaton *et al.*, 1972). They could have been either *Spissistilus festinus* or *Stictocephala bisonia*, both common in Kansas, and described below.

4.10a. Threecornered alfalfa hopper

Spissistilus festinus (Say) 1830
Description: *Adults* are robust, wedge-shaped, light green and 6–7mm long. They get their name from the hardened triangular (three-cornered) area over the thoracic area as seen from above. *Nymphs* in early stages are pale green or straw colored, and darken as they grow to yellow-brown and in late stages progressively greener. They have spine-like extensions that project from the back of the insect, a feature shared with some other treehopper nymphs, notably "buffalo treehoppers" discussed below. Nymphs are very active and will quickly crawl to hide behind the leaf or stem when disturbed.
Comments: Lago and Stanford (1989) reported *S. festinus* as among the most common insects present in Mississippi marijuana. They found nymphs developing on *Cannabis* but did not make note of any damage. Threecornered alfalfa hopper can develop on over 20 plant species, but reproduces best on legumes—alfalfa, soybean, and peanut (Beyer *et al.*, 2017). Winter is normally spent in the form of an adult female in reproductive diapause. In the spring, clovers and vetch are common host plants for the first generation. Eggs are inserted into stems, often near the soil line. In some plants threecornered alfalfa hopper will make a series of feeding probes, ringing the stem and disrupting the phloem, producing a girdling.

4.10b. Buffalo treehopper

Stictocephala bisonia Kopp & Yonke 1977
Description: *Adults* are green, 6–8 mm long, and the body is broadly triangular when viewed from above (Fig. 4.25). The pronotum is broad, tinged with brown, and has two "horns" along the side that are widely spread. *Nymphs* are tiny, pale

Fig. 4.25. Buffalo treehopper, *Stictocephala bisonia* (W. Cranshaw)

green, humpbacked and characteristically have long forward pointing spines in the front of the body.

Comments: "Buffalo treehopper" is the common name for several *Stictocephala* spp. in North America. Rosenthal (2012) collected *S. bisonia* on *Cannabis* in California. *S. bisonia* is easily confused with *Stictocephala* (*Ceresa*) *bubalus* (Fabricius) 1794, a taxon that Jerald DeWitt assigned to buffalo treehoppers on feral hemp in Illinois (McPartland, pers. obs., 1982). Specimens collected from drug-type plants near Buffalo, NY (Hillig, pers. comm., 1993) and in southern Indiana (Clarke, pers. comm., 1985) were assigned to *S. bubalus* (McPartland *et al.*, 2000). Dennis (1952) identified buffalo treehoppers collected from feral hemp in Wisconsin as *Stictocephala diceros* (Say) 1824.

Nymphs hatch late in the spring from overwintering eggs laid in tree branches. They drop to the ground and suck sap from many species of herbaceous plants. They become adults by August. The real damage is done by egg-laying adult females—by slicing stems with their knife-sharp ovipositers to lay eggs. One generation arises per year. Adults are shy and fly away with a loud buzzing noise. Buffalo treehoppers live throughout North America and cause economic damage in orchards and tree nurseries.

Micrutalis calva (Say) 1831, the honeylocust treehopper, was collected from *Cannabis* in Mississippi (Lago and Stanford, 1989) and Colorado (Schreiner and Cranshaw, 2021), in both cases being "uncommon." Honey locust trees (*Gleditsia* spp.) are a common host, but *M. calva* is polyphagous and nymphs develop on herbaceous plants (Nixon and Thompson, 1987). Other collections of treehoppers from *Cannabis* include *Echenopa* (= *Campylenchia*) *latipes* (Say) 1824, reported as "uncommon" by Lago and Stanford (1989), and a *Stictolobus* sp. was listed as "rare" by Schreiner and Cranshaw (2021).

4.10c. Meadow spittlebug

Philaenus spumarius (L.) 1758
Description: *Adults* are 5–7 mm long and are highly variable in patterning, ranging from very pale to mottled brown to entirely black. Weaver and King (1954) illustrated eight different color patterns most commonly found in Ohio. *Nymphs* resemble tiny pale green frogs, 6 mm in length, with small black antennae. They usually occupy the crotches of small branches, and hide in a froth of excreted spittle (Fig. 4.26).

Comments: The spittle is due to *P. spumarius* feeding on xylem fluid, which is an extremely dilute food source. Nymphs ingest sap at a rate ten times their body weight per hour. They digest what little is available, then excrete >99% of the fluid as water mixed with a mucilaginous material that produces small bubbles. Weaver and King (1954) listed over 100 host plants. Goidànich (1928) reported the species on hemp in Italy, and R. Clarke found it in Hungary (Fig. 4.26). Spittlebugs have been found on feral and cultivated plants in Oregon, Illinois, New Jersey, and Vermont (McPartland, pers. obs.).

Nymphs feed in the spring, and go through five stages during which they may move and resettle on different plants or areas of the plant. They molt into adults from late May to late June and escape from the spittle mass. Females lay eggs in

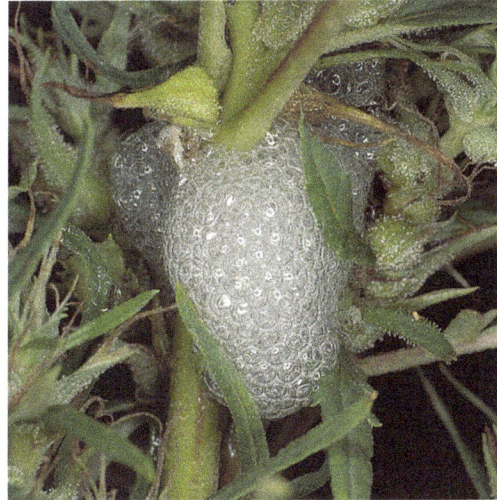
Fig. 4.26. Spittle froth made by *Philaenus spumarius* (R. Clarke)

August and September. One generation arises per year, the eggs overwinter. Feeding by meadow spittlebug causes little, if any, injury to plants. High populations of nymphs are capable of producing stunting of first cuttings of alfalfa, and modestly reduce fruit size in strawberries. In Europe, spittlebugs have recently gained importance, as they are able to transmit the xylem-limited bacterium *Xylella fastidiosa* that produces olive quick decline syndrome, a serious disease (Cornara *et al.*, 2018).

Lago and Stanford (1989) listed another spittlebug, *Clastoptera xanthocephala* Germer (Hemiptera: Clastopteridae), among the more common insects found in outdoor-grown marijuana in Mississippi. This spittlebug is sometimes known as the "sunflower spittlebug" and nymphs have been reported to develop on sunflower (*Helianthus*), ragweed (*Ambrosia*), goldenrod (*Solidago*), and *Chrysanthemum morifolium*.

4.10.1 Control measures

Damage from treehoppers and froghoppers is rarely sufficient to justify control measures. Elude spittlebugs with crop rotation—stay away from their favorite hosts, alfalfa and clover. To avoid treehoppers, do not plant near elm trees or orchards.

4.11 Planthoppers

Planthoppers (superfamily Fulgoroidea) are a large group of insects that include at least 20 families. Although some may superficially resemble leafhoppers or treehoppers, all planthoppers share a unique set of certain physical features, notably details of the structure and location of the antennae and unique wing venation. Planthoppers are found nearly worldwide, but the majority of the 12,000 species are found in Africa and the tropical Americas. Few cause serious crop injuries, but some are important in the transmission of plant viruses. Most records of planthoppers on *Cannabis* are from Asia.

4.11a. Spotted lanternfly

Lycorma delicatula (White) 1845
Description: *Adults* have light brown forewings with black spots. When resting, the crimson hindwings are partially visible through the semi-translucent forewings, giving the lanternfly a pink cast. Females are 22–27 mm long, males 21–22 mm long. *Nymphs* are mostly black with white spots, but the fourth and final instar has red markings. *Egg casings* are covered by a brownish gray waxy secretion, and usually contain fewer than 50 eggs each.
Comment: Spotted lanternfly is indigenous to China. Its preferred host is tree of heaven (*Ailanthus altissima*), but it has >170 known host plants, and damages fruit crops, grape vines, as well as herbaceous crops. EPPO (2016) and Liu (2019) cited two Chinese publications (unavailable to us) that recorded *L. delicatula* on Chinese hemp (Zhou, 1946; Zhou *et al.*, 1985). The species was first detected in the eastern USA in 2014. It has been found on hemp, first in Delaware (Chris Tipping, pers. obs., 2019). Scouting in Pennsylvania hemp fields revealed ≤2 adults per plant, with peaks in early October (Leach and Briggs, 2021). Leach and Briggs placed fourth-instar nymphs on caged, potted hemp plants; forced feeding on this single host resulted in low-to-medium survivorship, 32–43%.

4.11b. Other planthoppers

Geisha distinctissima Walker 1858 (Hemiptera: Flàtidae) has been reported on hemp in China, Korea and Japan (Takahashi, 1919; Clausen, 1931; Shiraki, 1952; Sorauer, 1958). *Ricania japonica* Melichar 1898 (Hemiptera: Ricaniìdae) has also been reported on hemp in East Asia (Takahashi, 1919; Clausen, 1931; Shiraki, 1952; Sorauer, 1958). Both species are polyphagous, and the latter is now known from Georgia, Crimea, Turkey, and Bulgaria.

Individual records of planthoppers on *Cannabis* include *Eurybrachys tomentosa* Fabricius 1775 (Hemiptera: Fulgóridae) as a minor pest in India (Cherian, 1932), and the new species *Stenocranus qiandainus* Kuoh 1980 (Hemiptera: Delphácidae) on hemp in China (Kuoh, 1980). *Scorlupella montana* Becker 1865 (Hemiptera: Issidae) was found on *Cannabis ruderalis* in the Russian steppe near Rostov (Gnezdilov, 2010). The passionvine hopper, *Scolypopa australis* (Walker) 1851 (Hemiptera: Ricaniidae) occurs on *Cannabis* in New Zealand (McPartland and Rhode, 2005). Adults of a *Melanoliarus* sp. (Hemiptera: Cixiidae) have been regularly found in surveys of hemp fields in eastern Colorado (Schreiner and Cranshaw, 2021). *Flatormenis proxima* (Walker) 1851 (as *Anormenis sepentrionalis*, Hemiptera: Flatidae) was reported as "common" in Mississippi (Lago and Stanford, 1989).

4.11.1   Control methods

Areas infested with invasive species such as spotted lanternfly must enact quarantine measures. Scrape egg masses into rubbing alcohol. Brown sticky traps capture nymphs, but adults can muscle out of the adhesive. The primary host of spotted lanternfly, *Ailanthus altissima*, is also an invasive and should be extirpated.

4.12   Mealybugs

Mealybugs (Hemiptera: Pseudococcidae) are closely related to soft scales (Coccidae), but can be distinguished by having the body covered with powdery wax. Many species possess waxy threads that protrude from the body, often as "tails." Mealybugs extract fluids from phloem, like aphids, and may excrete sticky honeydew. Mealybugs retain their legs throughout life and may be found on either above- or below-ground parts of plants.

Mealybugs live outdoors in warm habitats; in temperate climates they are an indoor problem. Reports on *Cannabis* began after the advent of indoor cultivation. A century ago, in the first-ever "indoor grow" that utilized electric lights, mealybugs infested plants stressed by insufficient light (Schaffner, 1928). Indoor grow consultants described control measures for generic "mealybugs" (Frank and Rosenthal, 1978; Puritch, 1982; Frank, 1988). McPartland *et al.*, (2000) began identifying them to species. All species reported from *Cannabis* are highly polyphagous pests.

4.12a. Longtailed mealybug

Pseudococcus longispinus Targioni-Tozzetti 1867
=Pseudococcus adonidum (L.) Zimmerman 1948, nec Coccus adonidum L. 1767
Description: *Adult females* are broadly oval, about 3 mm long, yellowish or slightly grayish. Two long wax filaments grow from the tip of the abdomen, which may exceed the length of the body, although these "tails" may break off. A row of conspicuous thin wax spikes arises along the sides of the body (Fig. 4.27a). *Adult males* are tiny, somewhat gnat-like, lack mouthparts, and are rarely noticed. *Nymphs* hatch immediately upon oviposition—there is no visible egg stage. The first stage nymph, known as a "crawler," is minute, lacks the waxy material that develops in later stages, and is highly mobile. The

Fig. 4.27. Mealybugs: **A.** classic longtailed mealybug (Comstock 1904); **B.** probable citrus mealybug (Vijay Cavale, Wikipedia Commons)

next two nymphal stages increase greatly in size, and increasingly resemble the adult form.

Comments: Longtailed mealybug can develop on 170 genera of plants, including avocado, citrus, grapes, pear, persimmon, and pineapple (García Morales *et al.*, 2016). The species also infests many commonly grown houseplants, including indoor *Cannabis* (McPartland *et al.*, 2000; Rosenthal, 2012; Cervantes, 2015). The life cycle and biology of this pest were reviewed by Byron and Gillette-Kaufman (2016). The length of time to complete a life cycle is highly dependent on temperature, but may be completed in about a month under favorable conditions. Since no dormant forms are produced, all life stages may be present year-round with multiple generations being produced in a year. Crawlers can be wind-blown or transported by animals (humans, birds, etc.); it is likely that infestations of *Cannabis* crops are due to dispersal of crawlers onto new plants.

4.12b. Citrus mealybug

Planococcus citri (Risso) 1813

Description: *Adults* are sexually dimorphic, like those of P. longispinus. Adult females resemble P. longispinus, except they lack the long wax filaments "tails."

Comments Citrus mealybug can develop on 240 genera of plants, including many economically important crops such as citrus (outdoors) and tomatoes and houseplants (indoors) (García Morales *et al.*, 2016). Rosenthal (2012) named this species on *Cannabis*, and Cervantes (2015) provided photographic evidence. A mealybug photographed on *Cannabis* outdoors in India appears to be P. citri (Fig. 4.27b). Citrus mealybug was identified on *Cannabis* by entomologists in Uruguay (Da Silva *et al.*, 2020). The life cycle and biology of this pest is similar to that of *P. longispinus*, except the eggs have a longer incubation period, 2–10 days. Depending on temperature, up to ten overlapping generations arise annually.

4.12c. Other mealybugs

Cervantes (2015) identified the pink hibiscus mealybug, *Maconellicoccus hirsutus* Green 1908, and provided credible photographic evidence. This species occurs in most tropical areas of Eurasia, and has hitchhiked to the Caribbean, Mexico, and California. It infests 230 genera of plants, including herbaceous crops (hibiscus, cotton, okra) and tree species (citrus, coffee, mango) (García Morales *et al.*, 2016).

Varshney (1992) cited the striped mealybug, *Ferrisia virgata* (Cockerell) 1893, on *Cannabis* in India. Striped mealybug is highly polyphagous, reported from 211 genera of plants (García Morales *et al.*, 2016), and has spread around the globe. The paired longitudinal stripes on its body become obscured with wax over time, and its paired wax filaments "tails" resemble those of the longtailed mealybug, albeit not as long and thin. Cotton mealybug, *Phenacoccus solenopsis* (Tinsley) 1898, has been reported from *Cannabis* in India (Mruthyunjana, 2012) and Pakistan (Abbas, 2010). It is polyphagous, infesting 210 genera of plants (García Morales *et al.*, 2016). Cotton

mealybug is native to MesoAmerica and the southwestern USA, but has become invasive in many areas of the world (Tong *et al.*, 2019).

4.12.1 Control measures

Exclusion from glasshouses and growrooms is key. Adult mealybugs can be removed by hand in small-scale situations. The predatory lady beetle *Cryptolaemus montrouzieri* is known as the "mealybug destroyer," but also eats young scales and aphids. Two mealybug-selective parasitoids are commercially available, *Anagyrus pseudococci* and *Leptomastix dactylopii*. Both are compatible with *Cryptolaemus montrouzieri* (see section 20.6). The fungus *Lecanicillium muscarium* kills mealybugs, but not as effectively as aphids and whiteflies. Damage is rarely heavy enough to justify chemicals. Frank (1988) daubed mealybugs with cotton swabs dipped in rubbing alcohol. Puritch (1982) killed 97% of mealybugs on *Cannabis* with insecticidal soap.

4.13 Scales

Scale insects (Infraorder Coccomorpha) are mostly closely related to aphids (Aphididae). Coccomorpha comprises about 20 families—many in taxonomic flux—and about 8000 species. Scales infesting *Cannabis* belong to three families: **Coccidae** (soft scales) is a large family with >1200 species. Soft scales secrete wax that forms a covering that is attached to and covers much of the body. They feed on fluids in phloem, and their honeydew may be incorporated into the scale covering. All soft scales recorded from *Cannabis* are highly polyphagous. Many have nearly worldwide distributions, and include the most common scale insects found in greenhouses and indoor plantings.

Monophlebidae (giant scales) is a smaller family, with 265 species. Giant scales are so-named because they are larger than other scale insects, up to 10 mm in length. They are primarily found in tropical areas. Adult females retain legs through life and have conspicuous antennae. Some species, but not all, make conspicuous egg sacs.

Diaspididae (armored scales) is the largest family of scale insects, with nearly 2700 species. Armored scales produce a hard covering, known as a test. The test is made of wax, proteins, and old cast skins (exuviae) that encloses the body and is often either circular or elongate in form. Only first-stage nymphs (crawlers) are mobile. After they have settled to feed and then molt, armored scales remain in place for the remainder of their life, with eggs laid within the test. Armored scales feed on the contents of individual cells and do not access the fluids of the phloem like most scale insects.

4.13a. Brown soft scale (Coccidae)

Coccus hesperidum L. 1758

Description: *Adult* females are oval, and flat or slightly convex., 2.5–4 mm long (Fig. 4.28A). Younger adults are yellowish-green to

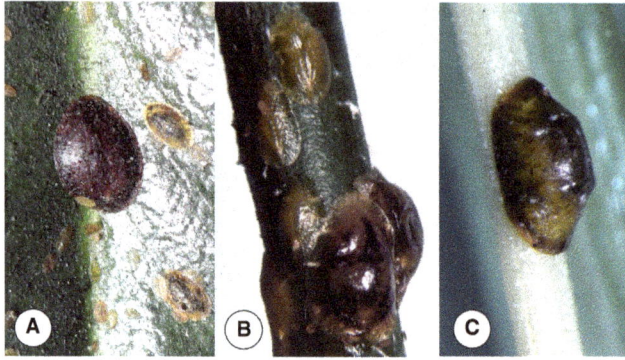

Fig. 4.28. Soft scales: **A.** *Coccus hesperidum* (W. Cranshaw); **B.** *Parthenolecanium corni* (W. Cranshaw); **C.** *Saissetia coffeae* (courtesy USDA)

yellowish-brown, flecked with brown spots. Older adults often darken in color. First-instar *nymphs* (crawlers) are oval, flat, yellow-brown, 0.4–0.5 mm long, with two eye spots, with legs and antennae projecting beyond the edge of the body. Later stages become immobile, feeding in place.

Comments: Brown soft scale has been reported on *Cannabis* in Croatia (Schmidt, 1956), Florida (Hamon and Williams, 1984), and Nicaragua (Maes, 2002). In a Kentucky glasshouse, *C. hesperidum* infested stems and leaves of CBD-dominant plants, and killed a test plant in two weeks (Villanueva *et al.,* 2020). The species is polyphagous, recorded on 385 genera of plants, and best known as a pest of citrus (García Morales *et al.,* 2016). Outdoor populations are restricted to subtropical and tropical regions. Brown soft scale can thrive under protected conditions in temperate areas, and is often the most common scale insect that occurs on indoor plants in the USA.

Brown soft scale produces asexually, and hatches eggs internally, giving live birth. Crawlers are sluggish and usually settle near their mother, but these are the most mobile stage. Villanueva *et al.* (2020) attributed a glasshouse *Cannabis* infestation to wind-borne spread of crawlers—the windows were open. Alternatively the scales survived a 4-month fallow period following a previous crop of ornamental plants. Multiple generations can be produced annually, and often all stages are present on plants. The insect is often detected by the large amount of sticky honeydew it excretes. The honeydew is forcibly expelled and may become deposited several centimeters from the insect. Cranshaw (2020) discussed its management on indoor plants in Colorado.

4.13b. European fruit lecanium (Coccidae)

Parthenolecanium corni (Bouché) 1844
=*Lecanium corni* Bouché, =*Lecanium robinarium* Douglas
Description: *Adult* females are oval to round and may be moderately convex. Size is variable, from 3 mm to 5 mm, and they are dark brown in color, with rough dermal surface, sometimes with black stripes and spots. A woolly white ring appears under the edge of the body as eggs are produced and hatch. Young adult females are yellow to brown in color, and have a longitudinal band surrounded by black mottling that runs down the center of the back (Fig. 4.28B).

Comments: European fruit lecanium has been recorded on *Cannabis* in Germany (Kovačević, 1929), Russia (Borchsenius, 1957), Turkey (Toros, 2002), Italy (Marotta, 1987), and Sweden and Central Asia (Mostafa and Messenger, 1972). The species is distributed around the Northern Hemisphere, cited on 120 species of plants, and best known as a pest of fruit and nut trees (García Morales *et al.,* 2016). European fruit lecanium produces a single generation per year. In temperate climates, winter is spent in the stage of a dormant adult female, and egg production occurs in late spring. As eggs hatch the first-stage crawlers disperse to settle on leaves, where they feed and develop through summer. At the end of the growing season they crawl to twigs, where the adult female stage occurs and subsequently remains through winter.

4.13c. Hemispherical scale (Coccidae)

Saissetia coffeae (Walker) 1852
=*Saissetia hemisphaerica* (Targioni) 1867.
Description: *Adult* females have an oval body, 2–3 mm diameter, which becomes strongly convex, appearing helmet-like (Fig. 4.28C). Young adult females are yellowish, with a faint H-shaped ridge that disappears at maturity. Older adult females become brown, more convex and hardened. As eggs begin to be produced, white silky wax protrudes from beneath the abdomen. *Eggs* are light purple in color, stored under body of the mother.

Comments: Hemispherical scale may be the unnamed brown scale that Frank (1988) described on stems of indoor *Cannabis* in California. McPartland *et al.* (2000) identified *S. coffeae* on glasshouse plants in Michigan. Cervantes (2015) photographed *S. coffeae* on a *Cannabis* branch. Hemispherical scale is highly polyphagous, infesting 315 genera of plants. It develops outdoors in the tropics and subtropics. In temperate areas it is one of the most common scale insects present in glasshouses and other protected indoor growing sites. Females reproduce parthenogenetically and may produce six or more generations in a year.

4.13d. Other soft scales (Coccidae)

Varshney (1992) reported two species in India. *Ceroplastes actiniformis* Green 1896 is a polyphagous species from semi-tropical India, Sri Lanka, and Indonesia. *Parasaissetia nigra* (Nietner) 1861, known as the nigra scale, is highly polyphagous and has become distributed worldwide. Joshi *et al.* (2021) reported another soft scale in India, *Ceroplastes cirripediformis* Comstock 1881. This polyphagous pest has been recorded from 118 genera of plants. It originated in North America but now occurs in southern Europe and most recently in India.

4.13e. Cottony cushion scale (Monophlebidae)

Icerya purchasi Maskell 1878
Description: *Adult* females are bright orange-brown, yellow or brown, oval in shape, about 7 mm long. The most distinctive feature is the very large waxy ovisac (egg sack) that

the female produces as eggs are laid. It is white, becomes longitudinally fluted, and ultimately exceeds the body in length. When the ovisac reaches full size, the body of the scale tilts so it appears to be standing on its head. The ovisac contains 300–1000 oblong red eggs. *Nymphs* are red-bodied with black legs, prominent antennae, and congregate under leaves along the midribs.

Comments: Cottony cushion scale infests *Cannabis* in Europe (Sorauer, 1958) and Iran (Afchar, 1937; Bodenheimer, 1944; Beccari, 1959). The latter author described the infestation "of considerable economic importance." Cranshaw (unpublished data, 2018) identified *I. purchasi* from photographs of *Cannabis* in Hawaii. Cottony cushion scale has a wide host range, 193 genera of plants (García Morales *et al.*, 2016). Cottony cushion scale is hermaphroditic and able to self-fertilize. Eggs hatch within the ovisac and first-stage nymphs (crawlers) emerge. Dispersal to new plants almost always occurs by movement of crawlers, which can be blown by wind and carried by birds and other animals. All stages of larvae remain mobile, unlike most scale species. Typically four generations arise annually where temperatures remain warm, fewer in cooler areas. During feeding they excrete large amounts of honeydew which, in turn, may be colonized by sooty molds.

4.13f. Mango scale (Monophlebidae)

Drosicha mangiferae (Stebbing) 1903
Description: Also known as mango mealybug or giant mealybug, adult females are segmented, elliptical in shape, 12–18 mm long, and tan colored, but a fine covering of minute hairs makes them gray-white in color—and look like mealybugs.

Comments: Prathapan (1992) dedicated an entire article to *D. mangiferae* infesting Cannabis in India. Nymphs and females sucked sap from inflorescences, young leaves, and fruit peduncles. Infested inflorescences dried up. Mango mealybug is native to South Asia; it infests mango and 16 other genera of plants (García Morales *et al.*, 2016).

4.13g. Armored scales (Diaspididae)

Armored scales live on perennial plants, usually woody plants, and their appearance on *Cannabis* would result from dispersal of the crawlers from primary hosts onto nearby plants. Five armored scales appear in the *Cannabis* literature.

Dekle (1976) cited two in Florida: *Chrysomphalus dictyospermi* (Morgan) 1889 is polyphagous and a serious pest of *Citrus* worldwide. The lesser snow scale, *Pinnaspis strachani* (Cooley) 1899, has even a wider host range, and also a worldwide distribution. Maes (2002) cited both species on *Cannabis* in Nicaragua.

Red scale, *Aonidiella aurantii* (Maskell) 1879, was reported on *Cannabis* in India (Varshney, 1992). It a notorious citrus pest with a wide host range. White peach scale, *Pseudaulacaspis pentagona* (Targioni-Tozzetti) 1886, was reported on Japanese hemp (Shiraki, 1952). The species originated in East Asia but now infests all of subtropical Asia, southern Europe and southeastern USA. San Jose scale, *Comstockaspis perniciosus* (Comstock) 1881, has been collected from *Cannabis*

in India (Rahman and Ansari, 1941) and Russia (Borchsenius, 1966). The species originated in the Far East, but has spread to every continent except Antarctica. It is a major pest of fruit trees. *C. perniciosus* has notoriety as the first documented case of insecticide resistance (Melander, 1914).

4.13.1 Differential diagnosis

Soft scales (Coccidae) are hard to speciate, they look similar, with overlapping size ranges (Fig. 4.28). Giant scales (Monophlebidae) have a more unique appearance, particularly the two species on *Cannabis*. Adults of armored scales (Diaspididae) are characterized by their waxy, somewhat concentrically ringed, "stuck-on" appearance. Although most scales produce honeydew like aphids and whiteflies, the insects themselves look completely different.

4.13.2 Control measures

Exclusion is key for indoor situations. Practice good sanitation—"crawlers" are carried into glasshouses via people and tools, and can blow in through open windows. Monitor carefully, for both insects and their honeydew—early detection improves the ease of managing scales.

Control of cottony cushion scale by the vedalia lady beetle, *Novius* (*Rodolia*) *cardinalis*, was the first biocontrol success story. The pest was unwittingly introduced into California from Australia. It posed a serious threat to the citrus industry until the USDA imported a mere 129 vedalia lady beetles in 1888. *Lindorus lophanthae*, "the scale destroyer," preys on soft scales, but will also eat the nymphs of hard scales in a pinch. The related species *Rhyzobius forestieri* preys on soft scales, and will also eat mealybug nymphs and hard scale nymphs. Predation by *Chrysoperla carnea* seems less effective.

Turning to parasitoids, soft scales are attacked by two encyrtid wasps, *Metaphycus helvolus* and *Microterys nietneri*. The latter species is particularly effective against *Coccus hesperidum*, but commercial availability is spotty. Larvae parasitize the scales, but adults of both *M. helvolus* and *M. nietneri* can host feed—they puncture young scales and suck them dry. *Encarsia formosa* may parasitize soft scales, but much prefers whiteflies. Hard scales are parasitized by *Aphytis melinus* and *Comperiella bifasciata*. The fungus *Lecanicillium muscarium* (formerly called *Verticillium lecanii*) kills scales and mealybugs, but not as effectively as aphids and whiteflies.

Insecticides control first-instar crawlers, but are less effective on later stages. Cranshaw (2020) recommended insecticidal soaps and horticultural oils. They need to be reapplied regularly to kill later-hatched crawlers.

4.14 True Bugs

True bugs belong to Hemiptera suborder Heteroptera. They are the largest group of insects with incomplete (simple) metamorphosis, comprising 90 families, 6000 genera, and 42,000 species. Heteropterans have "hetero" wings, which are half hard (chitinous), and half diaphanous (membranous).

Within the sap-sucking guild, Heteropterans utilize a unique "lacerate-and-flush" feeding strategy. Laceration (vigorous stylet movements) and flush (copious, toxic saliva) kills cells around the feeding site. They penetrate parenchyma, as well as deeper mesophyll cells, and even xylem and phloem vessels. Their feeding strategy causes more feeding site damage than other sucking insects. In addition to the usual spots and stippling, leaves can become distorted and necrotic, and flowers may abort. Heteropterans can lacerate-and-flush their way into immature seeds, turning them brown and lumpy. Adding insult to injury, some bugs inject a toxic saliva as they feed.

Heteropterans tend to be polyphagous. Species with multiple generations per year often switch host plants over the course of the growing season. *Cannabis* may only be infested during part of the year (Fig. 4.29).

The best approach for describing Heteropterans is by families, of which there are four that infest *Cannabis*: **Pentatomidae**, shield bugs or stink bugs; **Miridae**, mirids or capsid bugs; **Lygaeidae**, "seed bugs;" and **Rhopalidae**, scentless plant bugs.

4.14.1   Pentatomidae

Insects in this family are known as shield or stink bugs. "Shield" refers to the body shape of adults, which resembles a heraldic shield when viewed from above. "Stink" refers to their ability to release a pungent odor when disturbed or crushed. The family contains over 4700 species, including many important crop pests.

4.14a. Southern green stink bug

Nezara viridula (Linnaeus) 1758

Description: *Adults* are easily recognized by their shield-like shapes, 13 mm long. Green stink bugs are not always green; they may turn a russet color before diapause. McPartland also found white colormorphs with spots in Illinois (Fig. 4.30A). *Eggs* are barrel-shaped, pale yellow, laid in dense batches of 50–60 on undersides of leaves, which resemble tiny honeycombs. *Nymphs* are oval, bluish-green with red markings.

Comments: *N. viridula* feeds on leaves and flowering tops of hemp in Europe (Ferri, 1959a; Sorauer, 1958; Dippenaar *et al.*, 1996), Australia (Jobling and Warner, 2001; Humphries and Florentine, 2019), feral hemp in the USA (Hartowicz *et al.*, 1971; Paulsen, 1971; Batra *et al.*, 1981; McPartland, 1996b), and drug-type plants in India (Rao, 1928; Cherian, 1932; Nair and Ponnappa, 1974) and Pakistan (Ullah, 2018).

Besides direct feeding damage, the bug also vectors diseases. Paulsen (1971) cultured the brown blight fungus, *Alternaria alternata*, from *N. viridula* feeding on feral hemp. *N. viridula* probably originated in Ethiopia, but is now found worldwide. It is polyphagous, with a preference for brassicas

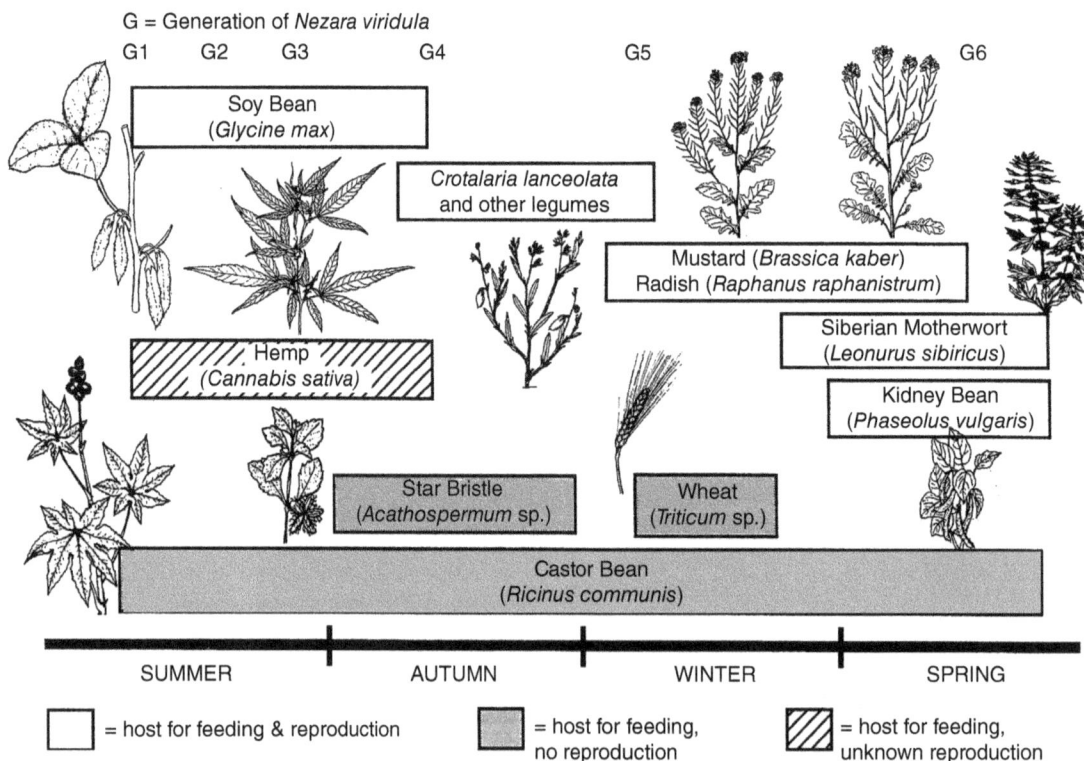

Fig. 4.29. Sequence of plant hosts used by successive generations of *Nezara viridula,* in a semi-tropical climate where six generations arise per year (adapted from Panizzi, 1997)

Fig. 4.30. Stink bugs. **A**. *Nezara viridula* colormorphs (McPartland); **B**. *Thyanta custator* (W. Cranshaw); **C**. *Chlorochroa ligata* (W. Cranshaw); **D**. *Halyomorpha halys* (K. Britt).

and legumes (especially soybeans). Adults overwinter in above-ground debris. Females lay up to 250 eggs. In warm weather total development time averages 22–32 days and adults live for 20–60 days (Panizzi, 1997). Three generations may arise per year in temperate zones, but in tropical regions they breed continuously, and switch hosts over the course of a year (Fig. 4.29).

4.14b. Uhler's stink bug

Chlorochroa uhleri (Stål) 1872
Description: *Adults* are bright green to olive, with abdominal edging in yellow, and three diagnostic light spots at the base of the scutellum, antennae almost entire black; big and broad, 12–16 mm long.
Comments: Uhler's stink bug is the most common stink bug on hemp in the western USA (Cranshaw *et al.*, 2019). The species is related to the conchuela bug, *Chlorochroa ligata* (Say) 1832, which also occurs on hemp in the western USA (Cranshaw *et al.*, 2019). Conchuela bug is the biggest stink bug out there, up to 18 mm long. Adults are green with a yellow band around their bodies; the marginal band turns red in autumnal brown colormorphs (Fig. 4.30B). Uhler's stink bug is polyphagous, feeding on Russian thistle, wheat, barley, alfalfa, and vegetable crops. Conchuela bug feeds on weeds in the spring, then moves to early-fruiting sorghum, and then to cotton.

4.14c. Redshouldered stink bug

Thyanta custator (Fabricius) 1803
Description: *Adults* are light green with varying amounts of reddish coloration across the "shoulders" (scutellum behind the head), with black spots on the ventral abdomen (Fig. 4.30C). The autumnal colormorph is brown, 10–11 mm long. *Nymphs*

are brownish in color. *Eggs* are gray, round, flat-topped, and laid in clusters.
Comments: *T. custator* is common on hemp in the southwestern USA (Cranshaw *et al.*, 2019). It feeds on beans, wheat, maize, and fruit trees, but economic damage is minor (Cranshaw and Shetlar, 2018).

4.14d. Marmorated stink bug

Halyomorpha halys (Stål) 1855
Description: *Adults* are larger than most stink bugs, 12–17 mm in length. Their marmorated (variegated) colors include brown, black, beige, and dark red. Key features include dark and light bands on margins of the abdomen, and light bands on antennae (Fig. 4.30D). *Nymphs* have an orange-red abdomen with black spots. *Eggs* are light green or light blue, 1 mm diameter, and laid in clutches of 25–28 eggs.
Comments: The marmorated stink bug, native to East Asia, arrived in Allentown (Pennsylvania) in 1999. The species has emerged as a major pest of vegetables and fruit trees. It's a nuisance to homeowners, because it moves into houses to overwinter. Britt *et al.* (2019) reported it feeding on seeded tops of *Cannabis* in Virginia. Eggs were found on plants, and caged nymphs successfully matured to adulthood on a diet of hemp alone. Zuefle (2020) observed it in upstate New York, and now it occurs on *Cannabis* throughout the USA (Munir *et al.*, 2023).

Marmorated stink bug has invaded Europe. In the Thessaloniki region of Greece, *Cannabis* was the third most-common host, in terms of egg-laying, after *Prunus armeniaca* and *Phaseolus vulgaris* (Andreadis *et al.*, 2021). Registered insecticides have not been effective against marmorated stink bug. A natural enemy was identified in China, *Trissolcus japonicus* (Ashmead), and USDA researchers initiated the long testing process required to import a foreign biocontrol. Then they discovered *T. japonicus* was already here (Buffington *et al.*, 2018).

4.14e. Pigeonpea stink bug

Dolycoris indicus Stål 1876
Description: *Adult* bodies and wings yellow-brown and covered with dark brown or black speckles, abdominal margins with brown and yellow stripes, covered with fine hairs, 10–11 mm long.
Comments: Pigeonpea stink bug is common in South Asia, where it destroys *gañjā* (Cherian, 1932; Nair, 1975). It has spread to Europe, and infests hemp (Rataj, 1957; Sorauer, 1958; Batra *et al.*,1981). A related species, *Dolycoris baccarum* L. 1758, infests hemp in Finland (Vappula, 1965), Germany (Patschke *et al.*, 1997; Gottwald, 2002), and Ukraine (Fedorenko *et al.*, 2016).

4.14f. Other stink bugs

The green stink bug, *Chinavia hilaris* (Say) 1832, fed upon Mississippi marijuana, and laid eggs in the crop (Lago and

Stanford, 1989). *C. hilaris* looks like *N. viridula* (and was formerly classified in *Nezara*) but is larger (13–19 mm long), with abdomen margins marked by black notches, and black bands on antennal segments. *C. hilaris* is broadly distributed across the USA. A second report of *C. hilaris* comes from Missouri, where Akotsen-Mensah (2021) trapped stink bugs in hemp fields using Pherocon® stink bug lures. He also caught brown stink bugs (*Euschistus* spp.). In Colorado, Cranshaw *et al.* (2019) caught the brown stink bug *Euschistus servus* (Say) 1832; the rice stink bug *Oebalus pugnax* Fabricius 1775; and the twice-stabbed stink bug *Cosmopepla lintneriana* Kirkaldy 1909.

The painted bug, *Bagrada hilaris* (Burmeister) 1835, feeds on *Cannabis* in Pakistan (Palumbo *et al.*, 2016). *B. hilaris* arrived in southern California in 2008. Since then it has spread to Nevada, Arizona, Texas, New Mexico, and Mexico. *B. hilaris* has caused widespread damage to organically grown crops in California, mowing down crucifer crops, as well as maize, wheat, and pearl millet.

4.14.2   Miridae

Miridae are called mirids or capsid bugs. The family contains over 10,000 species—largest in the Heteroptera. They often have a hunched look, because of the shape of the prothorax, which carries the head bent down. Some mirids are notorious crop pests. Other species are predatory and beneficial insects.

4.14g. Chinese cotton mirids

a. *Apolygus lucorum* Meyer-Dür 1843
b. *Adelphocoris suturalis* (Jakovlev) 1882
Descriptions: *A. lucorum* adults a uniform greenish-yellow in color, with brown wing tips, oval in shape, and 5–6 mm long, with small heads and strong femora (Fig. 4.31A). *A. suturalis* adults have elongate-oval bodies, a mix of yellow and grayish brown with brownish-black wing tips, 5.5–7.5 mm long, with small heads and strong femora.
Comments: Mirids have emerged as major pests in China, an unforeseen side effect of transgenic Bt cotton. The Bt toxin kills cotton bollworm (*Helicoverpa armigera*), but is ineffective against members of the sap-sucking guild (Lu *et al.*, 2010). Hagenbucher *et al.* (2013) demonstrated an interesting mechanism: damage by *H. armigera* induces plant synthesis of terpenoids. Suppression of *H. armigera* by Bt cotton means that fewer terpenoids are produced, thus enhancing the crop's susceptibility to other pests, such as mirids and aphids.

A. lucorum in Bt cotton has spread to other hosts. Pan *et al.* (2013, 2015) surveyed its host preferences. Out of 242 plant species, *Cannabis* ranked seventh. *A. lucorum* adults preferred feeding on hosts in flower, and switched hosts according to the seasonal succession of flowering plants. Out of six survey years, *Cannabis* led the survey one year, with a peak of 14 adults per m² during flowering in August. Host preference by *A. lucorum* is guided by volatile compounds. Pan *et al.* (2018) identified seven compounds in common among the top 18 host plants. *Cannabis* produced all seven. A subset of four elicited positive

behavioral responses in Y-tube olfactometer bioassays: m-xylene, butyl acrylate, butyl propionate, and butyl butyrate.

A. suturalis has spread to other hosts. Lu *et al.* (2011) searched for overwintering hosts. They found eggs on 126 plant species, and then monitored plants for successful nymph emergence. *Cannabis* averaged 83% emergence. They also searched for volatile compounds shared by 11 key hosts. Adults showed behavioral responses to compounds, n-butyl ether, butyl acrylate, butyl propionate, and butyl butyrate. *Cannabis* produced three out of four (Xiu *et al.*, 2019).

4.14h. Potato capsid

Closterotomus norvegicus (Gmelin) 1790
=*Calocoris bipunctatus* Fabricius 1779
Description: *Adults* are elongate, dull greenish-yellow in color, with two small black dots marking their pronotums, thinly covered with black and fine yellow hairs, and reach 6–7 mm in length (Fig. 4.31B).
Comments: Potato bug may be the most common mirid on European hemp. It infests leaves and flowers in Italy (Goidànich, 1928; Ragazzi, 1954; Ferri, 1959a), Germany (Flachs, 1936; Gutberlet and Karus, 1995; Patschke *et al.*, 1997), and the Middle East (Arzone *et al.*, 1990). The species originated in Europe, but now occurs across North America.

4.14i.   Other mirids

Common nettle bug, *Liocoris tripustulatus* (Fabricius) 1781, normally infests nettles in Europe. It was observed feeding on pollen in Dutch greenhouses (Clarke, pers. comm., 1998). Potter (2009) encountered *L. tripustulatus* adults feeding in flowering tops of outdoor crops in England. *L. tripustulatus* adults are dark brown, with a yellow-orange V-shaped scutellum. Wingtips are also yellow-orange (Fig. 4.31C).

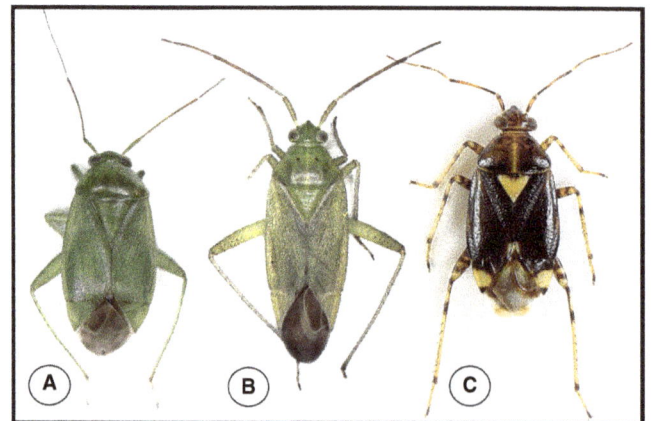

Fig. 4.31. Mirids (all courtesy Wikimedia Commons): **A.** *Apolygus lucorum* (B. J. Schoenmakers); **B.** *Closterotomus norvegicus* (B. J. Schoenmakers); **C.** *Liocoris tripustulatus* (Janet Graham)

Fourlined plant bug, *Poecilocapsus lineatus* (Fabricius) 1798, is native to North America east of the Rocky Mountains, and feeds on many herbaceous plants. It has been reported on *Cannabis* in Pennsylvania (Wheeler and Miller, 1981), upstate New York (K. Hillig, pers. comm., 2008; Zuefle, 2020, 2021), and Vermont (Hazelrigg and Maia, 2019). The adults have distinctive light green or yellow wings with four black linear stripes, and orange heads.

Tomato bug, *Engytatus modestus* (Distant) 1893, was commonly collected in an insect survey of *Cannabis* crops in southern California (Britt *et al.*, 2022). Alfalfa plant bug, *Adelphocoris lineolatus* (Goeze) 1778, was collected on Ukrainian hemp (Fedorenko *et al.*, 2016). They collected only nine adults, compared with 457 collections of *Lygus pratensis*. They also collected *Polymerus cognatus* (Fieber) 1858 and *Trigonotylus ruficornis* Geoffroy 1785. Henry (2015) collected *Ceratocapsidea complicata* (Knight) 1927 from *Cannabis* near Washington, DC. Kaur *et al.* (2012) encountered *Liocoris myittae* Distant 1904 on *Cannabis* in the Punjab.

Batra (1976) collected two mirids from *C. indica* in northern India: *Paracalocoris* sp. and *Rhopalus* sp. Lago and Stanford (1989) reported several mirids in Mississippi marijuana, led by the aforementioned *Lygus lineolaris*. Also abundant was the garden fleahopper, *Microtechnites bractatus* (Say) 1832. They also reported the cotton fleahopper *Pseudatomoscelis seriatus* (Reuter) 1876, *Neurocolpus nubilus* (Say) 1832, and *Keltonia sulphurea* (Reuter) 1907.

4.14.3 Lygaeidae

Lygaeidae or "seed bugs" differ from mirids by the presence of ocelli, or simple eyes. The family is smaller than the two aforementioned bug families, comprising around 1000 species, but its members are well-represented on *Cannabis*.

4.14j. Tarnished plant bug

Lygus lineolaris (Palisot de Beauvois) 1818
Description: *Adults* are flattened and oval in outline, 4–7 mm in length, mostly greenish-brown but irregularly mottled by reddish-brown coloring, with a distinct yellowish triangle or "V" located on their backs. Their triangular wingtips are characteristically yellow, tipped by a black dot (Fig. 4.32A). *Nymphs* when young look like yellow-green aphids, but are more active. Final-instar nymphs have green bodies marked by four black dots on the thorax, one dot on the abdomen, long antennae, and red-brown legs with indistinct stripes. *Eggs* are elongate and curved.
Comments: Tarnished plant bug has been collected on wild hemp in Illinois (McPartland, unpublished data 1981), cultivated hemp in Manitoba (Moes, pers. comm., 1995), Mississippi marijuana (Lago and Stanford, 1989), and hemp crops in eastern Colorado, Tennessee, and Virginia (Cranshaw *et al.*, 2019). The species was the most abundant insect in Tennessee hemp fields (Seals, 2019). It was relatively common in an insect survey conducted in southern California (Britt *et al.*, 2022).

Tarnished plant bug has been reported on *Cannabis* in Europe (Goidànich, 1928; Rataj, 1957; Sorauer, 1958). Ferri (1959a) illustrated a mass of nymphs and adults practically covering a female flowering top. The pest primarily infests alfalfa, canola, cotton, and vegetable crops.

Adults overwinter in soil, weeds, and crop stubble. Females emerge in spring, and insert eggs into stems and petioles of crop plants and weeds. Nymphs molt five times, gradually taking on an adult appearance. The life cycle takes only 3–4 weeks in warm climates, permitting 3–5 generations each year. In northern extremes only one generation arises per year. Tarnished plant bug infestations increase in hot dry weather. They are like human thieves: nowhere to be seen, until just before harvest time, and then they're all over the plants.

14.4k. Other *Lygus* spp.

The western tarnished plant bug, *Lygus hesperus* Knight 1917, is found west of the Rockies (Cranshaw *et al.*, 2019), where it infests flowering tops (Fig. 4.32B). *L. hesperus* is polyphagous like eastern *L. lineolaris*, but on fewer host species, 142 and 385 respectively. *L. hesperus* and *L. lineolaris* have emerged as primary pests of cotton, because Bt cotton, engineered for lepidopteran pests, does not affect them. *Lygus elisus* Van Duzee 1914 is often the dominant *Lygus* sp. in Colorado *Cannabis* (Cranshaw *et al.*, 2019). Adults of this species have a uniformly

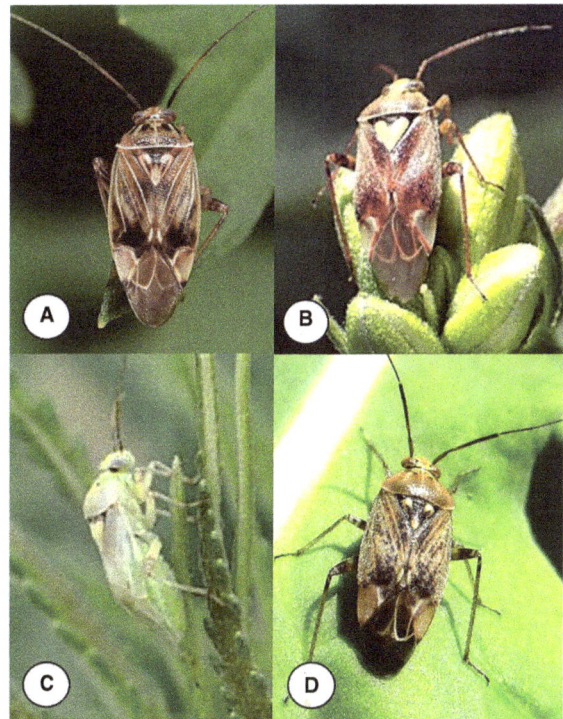

Fig. 4.32. *Lygus* spp.: **A**. *L. lineolaris* (C. Butler, Wikimedia Commons); **B**. *L. hesperus* (W. Cranshaw); *L. elisus* (W. Cranshaw); *L. rugulipennis* (B. J. Schoenmakers, Wikimedia Commons)

green abdomen, compared with *L. hesperus* and *L. lineolaris* (Fig. 4.32C). *L. elisus* is common west of the Rockies, and a major pest of alfalfa.

The European tarnished plant bug, ***Lygus rugulipennis*** Poppius 1911, is a pest in Germany (Patschke *et al.*, 1997; Gottwald, 2002), Romania (Trotuş and Naie, 2008), and Ukraine (Pivtoraiko *et al.*, 2020). It resembles its North American counterparts (Fig. 4.32D). The meadow bug, ***Lygus pratensis*** L. 1758, infests *Cannabis*, alfalfa, and cotton in Uzbekistan (Daminova, 2017). In a survey of fiber-type crops in Ukraine, Fedorenko *et al.* (2016) collected 457 specimens of *L. pratensis* (the second most common pest), and 84 specimens of *L. rugulipennis*. Conversely, *L. pratensis* was a minor problem in Germany (Gottwald, 2002).

Taylor (1947) reported two *Lygus* spp. breeding on *Cannabis* in Uganda, ***Lygus arboreus*** Taylor 1947, and ***Lygus nairobiensis*** Poppius 1912. In Turkey, ***Lygus saxatilis*** (Scopoli) 1763 has infested hemp (Lodos *et al.*, 1999). ***Lygus equestris*** L. 1758 has been found in Iran (Linnavuori, 2011). Shilenkov and Tolstonogova (2006) reported unidentified *Lygus* spp. constituting 50% of insects captured in a survey of Siberian *Cannabis ruderalis*.

4.14l. False chinch bugs

a. *Nysius ericae* Schilling 1895
b. *Nysius raphanus* Howard 1872
Description: *Adults* are small, 3–5 mm long, with dark gray-brown bodies and lighter colored translucent wings with dark markings (Fig. 4.33A). *Nymphs* are reddish-brown versions of adults without wings, darkening as they grow older. *Nysius* spp. are difficult to speciate without a reference specimen.
Comments *N. ericae* has been reported in Europe (Gilyarov, 1945; Sorauer, 1958), and *N. raphanus* in Colorado (Cranshaw *et al.*, 2019). Adults fly south to winter quarters for hibernation. They migrate north in the spring. Females lay several hundred eggs on sheaths of grasses. Nymphs limit their

damage to grasses and weedy hosts. Adults feed on a wide variety of crops, sometimes in alarming swarms. They inject a toxin while feeding which may cause chlorotic spots.

4.14.4 Rhopalidae

Rhopalidae or scentless plant bugs lack the well-developed scent glands present in most Heteroptera. The family is small, with about 240 species. Rhopalids live principally on trees (e.g., the ubiquitous boxelder bug, *Boisea trivittata* Say), but some cause problems in herbaceous plants.

4.14m. Hyaline grass bug

Liorhyssus hyalinus (Fabricius) 1794
Description: *Adults* have yellow-brown to reddish bodies, with a dark abdomen, 6.5-7.5 mm long (Fig. 4.33B). They can be distinguished by the length of their hyaline wings, which extend well beyond their abdomen.
Comments Hyaline grass bug has been reported on *Cannabis* in Europe (Hradil *et al.*, 2007). Cranshaw *et al.* (2019) encountered high populations in Colorado, and all life stages were found in Virginia. The species was collected in an insect survey conducted in southern California (Britt *et al.*, 2022). Additionally, the Rhopalid *Arhyssus lateralis* (Say) 1825 has been found on Mississippi marijuana (Lago and Stanford, 1989). Unidentified *Arhyssus* spp. occur in Colorado hemp, possibly spreading from neighboring weeds (Cranshaw *et al.*, 2019).

4.14.5 Other true bugs <or 4.14n>

Pyrrhocoris apterus L. 1758 (Hemiptera: Pyrrhocoridae) was collected from *Cannabis ruderalis* in Russia by Janischevsky (1924, 1925). He described *P. apterus* practicing zoochory (see Fig. 3.7). Batra (1976) collected another member of the Pyrrhocoridae from *C. indica* in northern India: the red cotton stainer, ***Dysdercus cingulatus*** (Fabricius) 1775. Red cotton stainer is a serious cotton pest, but is polyphagous, and occurs in South Asia, Southeast Asia, and northern Australia.

Four species of assassin bugs (Hemiptera: Reduviidae) have been identified: Ullah (2018) collected ***Scipinia horrida*** Stål 1843 in Pakistan. In Colorado, Schreiner (2019) collected the spined assassin bug, ***Sinea diadema*** (Fabricius) 1776, which was fairly common, and eggs found on plants indicated the species breeds in hemp. ***Zelus tetrancanthus*** Stål 1862 and ***Phymata americana*** Merlin 1930 were less common. Assassin bugs prey on other insects, and are generally beneficial.

4.14.6 Differential diagnosis

The primary Heteroptera families infesting *Cannabis*—Pentatomidae and Miridae—are easy to differentiate, at least as adults. Young Heteropteran nymphs might be confused with

Fig. 4.33. Seed and scentless bugs. **A**. *Nysius raphanus;*
B. *Liorhyssus hyalinus* (W. Cranshaw)

young leafhoppers or planthoppers. Very young bugs look like aphids, but lack cornicles and move much faster. Short, black bugs can be confused with beetles.

4.14.7 Cultural and mechanical control

To control true bugs, practice good sanitation, eliminate crop residues after harvest, and remove weeds within and around fields. Small-acreage farmers historically removed bugs by hand (Palumbo *et al.*, 2016). Adult *Lygus* bugs can be shaken off plants, best done early in the morning. *Lygus lineolaris* is attracted to white sticky traps (see section 19.6.1 for details).

Bugs are highly mobile, and migrate from host to host. Learn what sequence of hosts they infest, carefully monitor those hosts, and watch for dispersal from alternative hosts to *Cannabis* (e.g., Fig. 4.29). Watch for sudden infestations of *Lygus* species after neighboring alfalfa is cut for hay, forcing the bugs to migrate. Selecting plants for bug resistance is possible. So far there's only one report: Jobling and Warner (2001) found that a cultivar of Chinese provenance ('INSX') suffered less damage from *N. viridula* than two European cultivars ('Futura 77' and 'Kompolti').

4.14.8 Biological control

Bug nymphs are eaten by their predatory brethren: the big-eyed bug *Geocoris punctipes*. Assassin bugs also feed on bug nymphs, and some species naturally breed in hemp: *Sycanus collaris* (Batra, 1976) and *Sinea diadema* (Schreiner, 2019). Parasitoids decrease bug populations: *Trichopoda pennipes* (a tachinid parasitoid of squash bugs and southern green stink bugs), and *Ooencyrtus submetallicus* (an egg parasitoid of stink bugs).

Anaphes iole is a mymarid wasp that parasitizes the eggs of tarnished plant bugs (*Lygus hesperus* and *L. lineolaris*). It is native to North America, works best at 13–35°C, and is commercially available. *Peristenus digoneutis* is a braconid wasp that parasitizes *Lygus rugulipennis* in Europe. The USDA established a population in New Jersey that controls *L. lineolaris*. The scelionid wasp *Trissolcus basalis* parasitizes eggs of stink bugs, particularly *Nezara viridula*. Strains from southern Europe were successfully introduced into California.

4.14.9 Chemical control

Prior to the days of synthetic pesticides, ryania or sabadilla controlled green stink bugs, tarnished plant bug, chinch bugs, and other heteropterans (Metcalf *et al.*,1962). Young nymphs may be susceptible to soap. Interestingly, growers of cocoa (*Theobroma cacao*) in Cameroon repelled the mirids *Sahlbergella singularis* and *Distantiella theobroma* with aqueous extracts of banga, *Cannabis sativa* (Coulibaly *et al.*, 2002).

Britt and Kuhar (2019) tested three insecticides against naturally occurring populations of tarnished plant bug in a Virginia hemp field. The baseline population was low, however, so insecticides did not yield statistically significant results. Trends could be seen: Closer® (a xylem-mobile systemic nAChR agonist) and Besige® (a ryanoid + pyrethroid combo) worked better than Thuricide® (Bt-kurstaki strain).

(Prepared by J. McPartland; revised by W. Cranshaw & S. Koike)

References

Abbas G. 2010. *Taxonomy ecobiology and management of mealybug on cotton in Pakistan.* Doctoral thesis, University of Agriculture, Faisalabad, Pakistan.

Abbasi I, Lopo de Queiroz AT, Kirstein OD, *et al.* (13 other authors). 2018. Plant-feeding phlebotomine sand flies, vectors of leishmaniasis, prefer *Cannabis sativa*. *Proceedings of the National Academy of Sciences USA* 115: 11790–11795.

Afchar DJ. 1937. *Les insects nuisibles aux arbres fruitiers en Iran.* Ministry of Agriculture, Teheran. 112 pp

Akotsen-Mensah C. 2021. "Industrial hemp IPM updates in Missouri," p. 16 in *Second annual Science of Hemp Conference*. Auburn University, Auburn, Alabama.

Alexander T. 1990. The micro-security garden. *Sinsemilla Tips* 9(1): 39–41.

Ananthakrishnan TN, Sen S. 1980. *Taxonomy of Indian Thysanoptera.* Zoological Survey of India, Calcutta.

Andreadis SS, Gogolashvili NE, Fifis GT, Navrozidis EI, Thomidis T. 2021. First report of native parasitoids of *Halyomorpha halys* (Hemiptera: Pentatomidae) in Greece. Insects 12: e984.

Anonymous. 1940. *Lezioni al corso di perfezionamento per la stima del tiglio di canapa.* Faenza-Fratelli Lega Editori, Rome.

Anonymous. 2020. *Sulfur.* University of California/Integrated Pest Management Program. Pesticide Information Fact Sheet Series. Available at: http://ipm.ucanr.edu/TOOLS/PNAI/pnaishow.php?id=67

Arzone A, Vidano C, Alma A. 1990. Vineyard agro-ecosystem Heteroptera in the Mediterranean region. *Scopolia* Suppl. 1990(1): 101–107.

Attique MR, Rafiq M, Gaffar A, Ahmad Z, Mohyuddin AI. 2003. Hosts of *Bemisia tabaci* (Genn.) (Homoptera: Aleyrodidae) in cotton areas of Punjab, Pakistan. *Crop Protection* 22: 715–720.

Auger P, Migeon A, Ueckermann EA, Tied L, Navajas M. 2013. Evidence for synonymy between *Tetranychus urticae* and *Tetranychus cinnabarinus* (Acari, Prostigmata, Tetranychidae). *Acarology* 53: 383–415.

Ayyar TVR. 1963. *Handbook of economic entomology for South India.* Controller of Stationery and Print, Madras.

Azab AK, Megahed MM, El-Mirsawe HD. 1970. On the range of host-plants of *Bemisia tabaci* (Genn.) (Homoptera: Aleyrodidae). *Bulletin de la Société Entomologique d'Égypte* 54: 319–326.

Backus EA, Serrano MS, Ranger CM. 2005. Mechanisms of hopperburn: an overview of insect taxonomy, behavior, and physiology. *Annual Review of Entomology* 50: 125–151.

Bakro F, Wielgisz K, Bunalski M, Jedrycka M. 2018. An overview of pathogen and insect threats to fibre and oilseed hemp (*Cannabis sativa* L.) and methods for their biocontrol. *IOBC-WPRS Bulletin* 136: 9–20.

Balachowsky AS. 1936. "*Phorodon cannabis* Pass. (Hem. Aphididae)," pp. 1429–1430 in *Les insects nuisibles aux plantes cultivées, vol. 2*. Busson, Paris.

Batra SW. 1976. Some insects associated with hemp or marijuana (*Cannabis sativa* L.) in northern India. *Journal of the Kansas Entomological Society* 49: 385–388.

Batra SW, Coulson JR, Dunn PH, Boldt PE. 1981. *Insects and fungi associated with Carduus thistles (Compositae)*. USDA Technical Bulletin No. 1616, Washington, DC.

Beard JJ. 2018. "Fact Sheet – *Tetranychus gloveri* Banks 1900," in *Spider mite species of Australia (including key exotic southeast Asian pest species). V 1.0*. Accessed April 12, 2020 at https://keys.lucidcentral.org/keys/v3/spider_mites_australia/index.html

Beccari F. 1959. Entomofauna Persiana: primo elenco di insetti nocivi. *Rivista di Agricoltura Subtropicale e Tropicale* 53: 67–93.

Bellows TS, Perring TM, Gill RG, Headrick DH. 1994. Description of a species of *Bemisia* (Homoptera: Aleyrodidae). *Annals Entomological Society of America* 87: 195–206.

Benelli G, Pavela R, Petrelli R, *et al.* (7 other authors). 2018. The essential oil from industrial hemp (*Cannabis sativa* L.) by-products as an effective tool for insect pest management in organic crops. *Industrial Crops and Products* 122: 308–315.

Bennett CW. 1967. Epidemiology of leafhopper-transmitted viruses. *Annual Review of Phytopathology* 5: 87–108.

Bensoussan N, Santamaria ME, Zhurov V, Diaz I, Grbić M, Grbić V. 2016. Plant-herbivore interaction: dissection of the cellular pattern of *Tetranychus urticae* feeding on the host plant. *Frontiers in Plant Science* 7: e1105.

Beyer BA, Srinivasan R, Roberts PM, Abney MR. 2017. Biology and management of the threecornered alfalfa hopper (Hemiptera: Membracidae) in alfalfa, soybean, and peanut. *Journal of Integrated Pest Management* 8(1): e10.

Bhatti JS, Alavi J, Strassen R, Telmadarraiy Z. 2009. Thysanoptera in Iran 1938-2007: an overview. *Thrips* 7-8: 7–172.

Blackman RL, Eastop VF. 2006. *Aphids on the world's herbaceous plants and shrubs, vol. 1*. John Wiley and Sons, Chichester UK. Available at: http://www.aphidsonworldsplants.info/index.htm

Blattný C, Osvald CV, Novak J. 1950. Virosy a z viros podezřelé zjevy u konopí. *Ochrana Rostlin* 23: 5–9.

Blunck H. 1920. Die niederen tierischen feinde unserer gespinstpflanzen. *Illustrierte landwirtschaftliche Zeitung* 40: 259–260.

Bodenheimer FS. 1930. Die Schadlingsfauna Palastinas. *Monographien zur Angewandten Entomologie* 10: 1–438.

Bodenheimer FS. 1944. Note on the Coccidea of Iran, with descriptions of new species (Hemiptera-Homoptera). *Bulletin de la Société Fouad 1er d'Entomologie* 28: 85–100.

Bolland HR, Gutierrez J, Flechtmann CHW. 1998. *World catalogue of the spider mite family*. Brill, Leiden.

Bolt M. 2019. Insect pests continue to attack hemp. *Purdue Cooperative Extension Service, Pest & Crop Newsletter* 27: 1–2.

Borchsenius NS. 1957. Fauna of USSR, Homoptera, Coccidae. *Trudy Zoologicheskogo Instituta, Akademiia nauk SSSR* 66(9): 365.

Borchsenius NS. 1966. Каталог щитовок (Диаспидоидеа) мировой фауны. Nauka, Moscow.

Börner CB. 1952. Europae centralis Aphides. *Mitteilungen der Thüringischen Botanischen Gesellschaft* 3: 1–484.

Britt KE, Kuhar TP. 2019. Evaluation of insecticides to control insect pests on hemp in Virginia, 2018. *Arthropod Management Tests* 44(1): 1–2.

Britt KE, Kuhar TP. 2020. Evaluation of miticides to control hemp russet mite on indoor hemp in Virginia, 2019. *Arthropod Management Tests* 45(1): 1–2.

Britt KE, Pagani MK, Kuhar TP. 2019. First report of brown marmorated stink bug (Hemiptera: Pentatomidae) associated with *Cannabis sativa* (Rosales: Cannabaceae) in the United States. *Journal of Integrated Pest Management* 10(1): 1–3.

Britt KE, Meierotto S, Morelos V, Wilson H. 2022. First year survey of arthropods in California hemp. *Frontiers in Agronomy* 4: e901416.

Broughton S, Harrison J. 2012. Evaluation of monitoring methods for thrips and the effect of trap colour and semiochemicals on sticky trap capture of thrips (Thysanoptera) and beneficial insects (Syrphidae, Hemerobiidae) in deciduous fruit trees in Western Australia. *Crop Protection* 42: 156–163.

Bryon A, Wybouw N, Dermauw W, *et al.* 2013. Genome wide gene-expression analysis of facultative reproductive diapause in the two-spotted spider mite *Tetranychus urticae*. *BMC Genomics* 14: e815.

Buffington ML, Talamas EJ, Hoelmer KA. 2018. Team *Trissolcus*: integrating taxonomy and biological control to combat the brown marmorated stink bug. *American Entomologist* 64: 224–232.

Bush Doctor. 1985. Aphids or plant lice. *Sinsemilla Tips* 5(2), 22–23.

Byron MA, Gillett-Kaufman JL. 2016. *Longtailed mealybug*. Featured Creatures, University of Florida. Available at: http://entnemdept.ufl.edu/creatures/fruit/MEALYBUGS/longtailed_mealybug.htm

California DPR (Department of Pesticide Regulation). 2017. Legal pest management practices for Cannabis growers in California. Available at: https://www.cdpr.ca.gov/docs/county/cacltrs/penfltrs/penf2015/2015atch/attach1502.pdf

Cao YH, Dmitriev DA, Dietrich CH, Zhang YL. 2019. New taxa and new records of Erythroneurini from China (Hemiptera: Cicadellidae: Typhlocybinae). *Acta Entomologica* 59: 189–210.

Capinera JL. 2020. *Handbook of vegetable pests, 2nd edition*. Academic Press, New York.

Carmichael E. 2020. *A survey of aphid species and their associated natural enemies in Fraser Valley hop fields and an exploration of potential alternative summer hosts of the damson-hop aphid, Phorodon humuli (Hemiptera: Aphididae)*. Master's thesis, Virginia Polytechnic Institute, Blacksburg, Virginia.

Ceapoiu N. 1958. *Cinepa, studiu monografic*. Editura Academiei Republicii Populare Romine. Bucharest.

Cervantes J. 2015. *The Cannabis encyclopedia*. Van Patten Publishing, Vancouver, Washington.

CFDA (California Department of Agriculture). 2019. *California pest rating profile for Phorodon cannabis Passerini: cannabis aphid*. Available at: https://blogs.cdfa.ca.gov/Section3162/wp-content/uploads/2019/07/PRP2019-Phorodon-cannabis_Profile_ADA.pdf?fbclid=IwAR0UG1W5iHUcVcSRsBu5jv5nsFG6MK8RWvVYzNpNuTWmwf05KkTpv3l622I

Chasen EM, Dietrich C, Backus EA, Cullen EM. 2014. Potato leafhopper (Hemiptera: Cicadellidae): ecology and integrated pest management focused on alfalfa. *Journal of Integrated Pest Management* 5(1): A1–A8.

Chaudhri WM, Akbar S, Rasool A. 1974. *Taxonomic studies of the mites belonging to the families Tenuipalpidae, Tetranychidae, Tuckerellidae, Caligonellidae, Stigmaeidae and Phytoseiidae.* University of Agriculture Lyallpur, Pakistan.

Cherian MC. 1932. Pests of ganja. *Madras Agricultural Journal* 20: 259–265.

Chiginsky J, Langemeier K, MacWilliams J, et al. (6 other authors). 2021. First insights into the virus and viroid communities in hemp (*Cannabis sativa*). *Frontiers in Agronomy* 3: e778433.

Childers CC, Easterbrook MA, Solomon MG. 1996. "Chemical control of eriophyioid mites," pp 695–728 in Lindquist et al., eds. *Eriophyoid mites*. Elsevier Science SV, Amsterdam.

Childers CC, French JV, Rodrigues JC. 2003a. *Brevipalpus californicus, B. obovatus, B. phoenicis,* and *B. lewisi* (Acari: Tenuipalpidae): a review of their biology, feeding injury and economic importance. *Experimental and Applied Acarology* 30: 5–28.

Childers CC, Rodrigues JC, Welbourn C. 2003b. Host plants of *Brevipalpus californicus, B. obovatus,* and *B. phoenicis* (Acari: Tenuipalpidae) and their potential involvement in the spread of viral diseases vectored by these mites. *Experimental and Applied Acarology* 30: 29–105.

Choudhury AKS, Mukherjee AB. 1972. Wild plants as alternate hosts of red spider mite, *Tetranychus telarius* (Linnaeus) (Tetranychidae: Acarina). *Indian Journal of Entomology* 33(1): 108–110.

Clausen CP. 1931. *Insects injurious to agriculture in Japan.* USDA Circular no. 168, Washington, DC.

Cloyd RA. 2011a. *Twospotted spider mite. Management in greenhouses and nurseries.* Kansas State University Extension Publication MF-2997.

Cloyd RA. 2011b. *Broad mite and cyclamen mite. Management in greenhouses and nurseries.* Kansas State University Extension Publication MF-2938.

Cloyd RA, Galle CL, Keith SR, Kalschuer NA, Kemp KE. 2009. Effect of commercially available plant-derived essential oil products on arthropod pests. *Journal of Economic Entomology* 104: 1567–1579.

Coeur d'Acier AA, Cruaud E, Artige G, et al. (7 other authors). 2014. DNA barcoding and the associated PhylAphidB@se website for the identification of European aphids (Insecta: Hemiptera: Aphididae). *PLoS ONE* 9: e97620.

Comstock JH. 1904. *Manual for the study of insects.* Comstock Publishing Co., Ithaca, New York.

Cook WC. 1967. *Life history, host plants, and migration of the beet leafhopper in the western United States.* USDA-ARS Technical Bulletin 1365, Washington, DC.

Cordova-Kreylos AL, Fernandez LE, Koivunen M, et al. (3 other authors). 2013. Isolation and characterization of *Burkholderia rinojensis* sp. nov., a non-*Burkholderia cepacia* complex soil bacterium with insecticidal and miticidal activities. *Applied and Environmental Microbiology* 79: 7669–7678.

Cornara D, Bosco D, Fereres A. 2018. *Philaenus spumarius*: when an old acquaintance becomes a new threat to European agriculture. *Journal of Pest Science* 91: 957–972.

Costa HS; Ullman DE; Johnson MW; Tabashnik BE, 1993. Squash silverleaf symptoms induced by immature, but not adult, *Bemisia tabaci. Phytopathology* 83: 763–766.

Coulibaly O, Mbila D, Sonwa DJ, Adesina A, Bakala J. 2002. Responding to economic crisis in Sub-Saharan Africa: new farmer-developed pest management strategies in cocoa-based plantations in southern Cameroon. *Integrated Pest Management Reviews* 7: 165–172.

Cranshaw W. 2018. *Proposal form for new common name or change of ESA-approved common name: cannabis aphid, Phorodon cannabis Passerini.* Available at: https://www.entsoc.org/sites/default/files/files/Cannabis%20Aphid%20Common%20Name%20Proposal.pdf

Cranshaw W. 2020. *Brown soft scale—a common insect pest of indoor plants.* Colorado State University Extension Fact Sheet 5.599. Available at: https://extension.colostate.edu/docs/pubs/insect/05599.pdf

Cranshaw WS, Shetlar D. 2018. *Garden insects of North America, 2nd edn.* Princeton University Press, Princeton, New Jersey.

Cranshaw WS, Wainwright-Evans S. 2020. *Cannabis sativa* as a host of rice root aphid (Hemiptera: Aphididae) in North America. *Journal of Integrated Pest Management* 11(1): 15; 1–3

Cranshaw WS, Halbert SE, Favret C, Britt KE, Miller GL. 2018. *Phorodon cannabis* Passerini (Hemiptera: Aphididae), a newly recognized pest in North America found on industrial hemp. *Insecta Mundi* 662: 1–12.

Cranshaw WS, Schreiner M, Britt K, Kuhar TP, McPartland JM, Grant J. 2019. Developing insect pest management systems for hemp in the United States: a work in progress. *Journal of Integrated Pest Management* 10(1): 26; 1–10.

Crescini F. 1940. Formen von Hanf (*Cannabis sativa* L.). *Der Züchter* 12(5): 105–115.

CSIRO (Commonwealth Scientific and Industrial Research Organisation). 2019. *CSIRO Handbook of Australian insect names.* Available at: http://www.ces.csiro.au/aicn/intro.htm

Cuboni G. 1889. Nota dei casi di malatti de vegetali presentati alla R. Stazione di patologia vegetale di Roma durante i mesi di agosta e setterr 1889. *Bollettino di Notizie Agrarie* 11: 1942–1949.

Da Silva VCP, Kaydan MB, Basso C. 2020. Pseudococcidae (Hemiptera: Coccomorpha) in Uruguay: morphological identification and molecular characterization, with descriptions of two new species. *Zootaxa* 4894 (4): 501–520.

Daminova DB. 2017. Ecological features of Miridae bugs of cotton–alfalfa biocoenosis of Uzbekistan. *Journal of Novel Applied Sciences* 2017 (6-2): 52-55.

Das B (Van der Goot P, ed). 1918. The Aphididae of Lahore. *Memoirs of the Indian Museum* 6(4): 135–274.

Das BC, Raychaudhuri DN. 1983. Aphids (Homoptera: Aphididae) of Nepal. *Records of the Zoological Survey of India, Occasional Paper* no. 51. Zoological Survey of India, Calcutta.

Davis JJ. 1911. *Williams' "The Aphididae of Nebraska"; a critical review.* Contributions from the Department of Entomology, University of Nebraska, No. 8. Lincoln, Nebraska.

Dawson GW, Griffiths DC, Pickett JA, Smith MC, Woodcock CM. 1984. Natural inhibition of the aphid alarm pheromone. *Entomologia Experimentalis et Applicata* 36: 197–199.

De Angelis JD, Larson KC, Berry RE, Krantz GW. 1982. Effects of spider mite injury on transpiration and leaf water status in peppermint. *Environmental Entomology* 11(4): 975–978.

de Lillo E. Pozzebon A, Valenzano D, Duso C. 2018. An intimate relationship between eriophyoid mites and their host plants. *Frontiers in Plant Science* 9: e1786.

Deardorff D, Wadsworth K. 2017. *What's wrong with my marijuana plant?* Ten Speed Press, Berkeley, California.

Dekle GW. 1976. *Florida armored scale insects, revised second printing.* Florida Department of Agriculture and Consumer Services, Gainesville, Florida.

Dennis CL. 1952. The Membracidae of Wisconsin. *Transactions of the Wisconsin Academy of Sciences, Arts, and Letters* 42: 129–152.

Dey D, Akhtar MS. 2007. Diversity of natural enemies of aphids belonging to Aphidiinae (Hymenoptera: Braconidae) in India. *Journal of Asia-Pacific Entomology* 10: 281–296.

Dhooria MS. 1983. An outbreak of the citrus mite, *Eutetranychus orientalis* (Klein) in Delhi. *Pesticides* 17(11): 36.

Dietrich CH. 2013. Overview of phylogeny, taxonomy and diversity of the leafhopper (Hemiptera: Auchenorrhyncha: Cicadomorpha: Membracoidea: Cicadellidae) vectors of plant pathogens. *Proceedings of the International Symposium on Insect Vectors and Insect-Borne Diseases, Taiwan*, pp. 47–70.

Dippenaar MC, du Toit CLN, Botha-Greeff MS. 1996. Response of hemp (*Cannabis sativa* L.) varieties to conditions in Northwest Province, South Africa. *Journal International Hemp Association* 3(2): 63–66.

Dixon AFG. 1985. *Aphid Ecology*. Blackie/Chapman & Hall, New York.

Dmitriev DA. 2022. *World Auchenorrhyncha Database*. Available at: http://dmitriev.speciesfile.org/

Dudley JE. 1920. Control of the potato leafhopper (*Empoasca mali* Le B.) and prevention of "hopperburn." *Journal of Economic Entomology* 13: 408–425.

Dudley JE. 1926. *The potato leafhopper and how to control it*. USDA Farmers' Bulletin no. 1462. USDA, Washington, DC.

Durak R, Jedryczka M, Czjka B, *et al*. (3 other authors). 2021. Mild abiotic stress affects development and stimulates hormesis of hemp aphid *Phorodon cannabis*. *Insects* 12: e420.

Eastop VF, Blackman RL. 2005. Some new synonyms in Aphididae (Hemiptera: Sternorrhyncha). *Zootaxa* 1089: 1–36.

Eastop VF, Hille Ris Lambers D. 1976. *Survey of the world's aphids*. W. Junk, The Hague.

Eaton BJ, Hartowicz LE, Latta RP, *et al*. (3 other authors). 1972. Controlling wild hemp. *Kansas Agricultural Research Station, Report of Progress* 188: 1–10.

EPA (Envrironmental Protection Agency). 2020. *Need for tolerances and tolerance exemptions for minimum risk pesticides*. Available at: https://www.epa.gov/minimum-risk-pesticides/need-tolerances-and-tolerance-exemptions-minimum-risk-pesticides

Eppler A. 1986. Untersuchungen zur Wirtswahl von *Phorodon humuli* Schrk. I. Besiedelte Pflanzenarten. *Anzeiger für Schädlingskunde, Pflanzenschutz, Umweltschutz* 59: 1–8.

EPPO (European Plant Protection Organization). 2016. *Pest risk analysis for Lycorma delicatula*. EPPO, Paris.

ESA (Entomological Society of America). 2019. *Common names of insects database*. Available at: https://www.entsoc.org/common-names

Essig EO. 1915. *Injurious and beneficial insects of California, 2nd edition*. California State Commission of Horticulture, Sacramento, California.

Ewing HE. 1914. *The common red spider or spider mite*. Oregon Agricultural College Experimental Station Bulletin 121: 1–95.

Falcon-Brindis A, Villanueva RT, Viloria Z, Bradley CL. 2023. A novel approach to tally *Aculops cannabicola* (Acari: Eriophyidae) for field and laboratory studies. *Journal of Economic Entomology*.

Farkas H. 1965. "Family Eriophyidae, Gallmilben," p. 84 *in Die Tierwelt Mitteleuropas, Band III, Lief 3*. Quelle & Meyer, Leipzig.

Fasulo TR. 2016. *Broad mite. Featured creatures series*. University of Florida, Department of Entomology. Gainsville, Florida.

Fasulo TR, Denmark HA. 2009. *Twospotted spider mite. Featured creatures series*. University of Florida, Department of Entomology. Gainsville, Florida.

Fedorenko VP, Kabanetz VV, Kabanets VM. 2016. Шкідники конопель посівних. National Academy of Agrarian Science, Kiev.

Fenili GA, Pegazzano F. 1974. Metodi avanzati di lotta contro gli acari fitofagi. *Noti ed Appunti Sperimentali di Entomologia Agraria* 15: 33–41.

Ferri F. 1959. *Atlante delle avversità della canapa*. Edizioni Agricole, Bologna, Italia.

Fiedler Ż, Sosnowska D, Kaniewski R, Władyka-Przybylak M. 2013. Wykorzystanie kompozycji z olejku konopnego do ograniczania liczebności przędziorków (Tetranychidae). *Progress in Plant Protection* 53: 679–682.

Flachs K. 1936. Krankheiten und schädlinge unserer gespinstpflanzen. *Nachrichten über Schädlingsbekämpfung* 11: 6–28.

Fletcher TB. 1917. *Report of the Proceedings of the Second Entomological Meeting, Pusa*. Superintendent Government Printing, Calcutta.

Flores E, Silberschmidt K. 1958. Relations between insect and host plant in transmission experiments with infectious chlorosis of Malvaceae. *Anais da Academia Brasileira de Ciências* 30: 535–560.

Forbes SA, Hart CA. 1900. "The red spiders. The common red spider, *Tetranychus bimaculatus* Harv.," pp. 406-407 in *University of Illinois Agricultural Experiment Station, Bulletin No. 60*, Urbana, Illinois.

Foy F. 1851. "Haschisch,' pp. 325–331 in Tardieu A, *ed. Suplément au dictionnaire des dictionnaires de médecine français et étrangers*. G. Baillière, París.

Frank AB. 1896. *Die tierparasitären Krankheiten der Pflanzen*. Eduard Trewendt, Breslau.

Frank M. 1988. *Marijuana grower's insider's guide*. Red Eye Press, Los Angeles, California.

Frank M, Rosenthal E. 1978. *Marijuana grower's guide*. And/Or Press, Berkeley, California.

Fyodorov IA, Cherosov MM, Troeva EI. 2021. Конопля в Якутии: биология, распространение, меры борьбы. Nauka, Novosibirsk, Russia.

García Morales M, Denno BD, Miller DR, *et al*. (3 other authors). 2016. *ScaleNet: a literature-based model of scale insect biology and systematics*. Available at: http://scalenet.info.

Gaujac C. 1810. Suite de l'extrait d'un Mémoire de M. Gaujac, Propriétaire-Cultivateut à Dagny, près Coulommiers (Seine-et-Marne), sur la culture compare des plantes oléagineuses. *Bulletin Société d'Encouragement pour l'Industrie Nationale* 9: 247–252.

Gent DH, James DG, Wright LC, *et al*. (5 other authors). 2009. Effects of powdery mildew fungicide programs on twospotted spider mite (Acari: Tetranychiidae), hops aphid (Hemiptera: Aphididae), and their natural enemies in hops yards. *Journal of Economic Entomology* 102: 274–286.

Gerson U. 1992. Biology and control of the broad mite, *Polyphagotarsonemus latus* (Banks) (Acari: Tarsonemidae). *Environmental and Applied Acarology* 12: 163–178.

Ghosh AK. 1974. A list of aphids (Homoptera: Aphididae) from India and adjacent countries. *Journal of the Bombay Natural History Society* 71: 201–220.

Ghosh AK. 1994. Fauna of West Bengal, part-5. *Zoological Survey of India, State Fauna Series 3*. Zoological Survey of India, Calcutta.

Ghosh LK. 1990. A taxonomic review of the genus *Aphis* Linnaeus (Homoptera: Aphididae) in India. *Memoirs of the Zoological Survey of India* 17: 1–159.

Gibson RW, Pickett JA. 1983. Wild potato repels aphids by release of aphid alarm pheromone. *Nature* 302: 608–609.

Giladi Y, Hadad L, Luria N, Cranshaw W, Lachman O, Dombrovsky A, 2020. First report of beet curly top virus infecting *Cannabis sativa* L., in Western Colorado. *Plant Disease* 104: 1656-1656.

Gilyarov MS. 1945. [A new and dangerous insect pest of the seeds of kok-saghyz and krym-saghyz]. *Dopov. Akad. Nauk URSR* 1945 (1-2): 47–55.

Giustina W della. 1989. *Homoptères Cicadellidae, vol. 3. Compléments aux ouvrages d'Henri Ribaut*. Faune de France no. 73. Lechevalier, Paris.

Gnezdilov VM. 2010. New synonyms, combinations, and faunistic records of Western Palaeartic planthoppers of the Family Issidae (Homoptera, Fulgoroidea). *Entomological Review* 90: 1024–1030.

Goff ML. 1987. *A catalog of Acari of the Hawaiian islands*. Hawaii Institute of Tropical Agriculture and Human Resources, Honolulu.

Goidànich A. 1928. Contributi alla conoscenza dell'entomofauna della canapa. I. Prospetto generale. *Bollettino del Laboratorio di Entomolgia del R. Istituto Superiore Agrario di Bologna* 1: 37–64.

Goidànich G. 1955. *Malattie crittogamiche della canapa*. Associazione Produttore Canapa, Bologna-Naples.

Goidànich G. 1959. *Manual di patologia vegetale*. Edizioni Agricole, Bologna.

Górski R, Sobieralski K, Siwulski M. 2016. The effect of hemp essential oil on mortality *Aulacorthum solani* Kalt. and *Tetraychnus urticae* Koch. *Ecological Chemistry and Engineering* 23: 505–511.

Gottwald R. 2002. Entomologische Untersuchungen an Hanf (*Cannabis sativa* L.). *Gesunde Pfanzen* 54(5): 146–152.

Grammenos G, Kouneli V, Mavroeidis A, *et al*. (4 other authors). 2021. Beneficial insects for biological pest control in greenhouse *Cannabis* production. *Bulletin of University of Agricultural Sciences and Veterinary Medicine Cluj-Napoca, Horticulture* 78(2): 85–93.

Gupta SK. 1985. *Plant mites of India*. Government of India Press, Calcutta.

Gutberlet V, Karus M. 1995. *Parasitäre Krankheiten und Schädlinge an Hanf (Cannabis sativa)*. Nova Institut, Köln, Germany.

Hadad L, Luria N, Smith E, Sela N, Lachman O, Dombrovski A. 2019. Lettuce chlorosis virus disease: a new threat to *Cannabis* production. *Viruses* 11: e802.

Hagenbucher S, Wäckers FL, Wettstein FE, *et al*. (3 other authors). 2013. Pest trade-offs in technology: reduced damage by caterpillars in Bt cotton benefits aphids. *Proceedings of the Royal Society B* 280: 20130042.

Halbert S. 2016. *Phorodon cannabis*, hemp aphid, a new Western Hemisphere record. *TRI-OLOGY* 55(4): 6.

Hamon AB, Williams ML. 1984. *The soft scales of Florida (Homoptera: Coccoidea:Coccidae)*. Florida Department of Agriculture and Consumer Services, Gainesville, Florida.

Harand W. 2000. Hanf (*Cannabis sativa*) gegen Blattläuse. *Pflanzenschutz* 16(4): 8–9.

Harmon JP, Stephens E, Losey J. 2007. The decline of native coccinellids (Coleoptera: Coccinellidae) in the United States and Canada. *Journal of Insect Conservation* 11: 85–94.

Hartowicz LE, Knutson H, Paulsen A, Eaton BJ, Eshbaugh E. 1971. Possible biocontrol of wild hemp. *North Central Weed Control Conference, Proceedings* 26: 69.

Hartzell A. 1923. The Genus *Empoasca* in North America. *Proceedings of the Iowa Academy of Science* 30: 87–133.

Hayes CB, Carter O, MacWilliams, *et al*. (4 other authors). 2023. Biology and management of hemp russet mite (Acari: Eriophyidae). *Journal of Economic Entomology* 116: 1706–1714.

Hazan A, Gerson U, Tahori AS. 1973. Life history and life tables of the carmine spider mite. *Acarologia* 15: 414–440.

Hazelrigg A, Maia G. 2019. *Plant Diagnostic Clinic 2019*. University of Vermont Extension. Available at: https://www.uvm.edu/sites/default/files/Agriculture/2019_Report_Plant_Diagnostic_Clinic.pdf

Heinze K. 1960. Systematic der mitteleuropäischen Myzinae mit besonderer Berücksichtigung der im Deutschen Entolmologischen Institut befindlichen Sammlung Carl Börner (Homoptera: Aphidoidea Aphididae). *Beiträge zur Entomologie* 10(7–8): 744–842.

Heller K, Andruszewska A, Grabowska L, Wielgusz K. 2006. Ochrona lnu i konopi w Polsce i na świecie. *Progress in Plant Protection* 46: 88–89.

Henry TJ. 2015. Revision of the Ceratocapsine Renodaeus group: Marinonicoris, Pilophoropsis, Renodaeus, and Zanchisme, with descriptions of four new genera (Heteroptera, Miridae, Orthotylinae). *ZooKeys* 490: 1–156.

Herbert PD, Stoeckle MY, Zemlak TS, Francis CM. 2004. Identification of birds through DNA barcodes. *PLoS Biology* 2(10): e312.

Higuchi H, Masahisa M. 1969. A tentative catalogue of host plants of Aphidoidea in Japan. *Insecta Matsumurana, Supplement* 5: 1–66.

Hill DS. 1983. *Agricultural insect pests of the tropics and their control, 2nd ed.*, Cambridge University Press, Cambridge UK.

Hill DS. 1994. *Agricultural Entomology*. Timber Press, Portland, Oregon.

Hillig KW. 2004. A chemotaxonomic analysis of terpenoid variation in *Cannabis*. *Biochemical Systematics and Ecology* 32: 875–891.

Hojat H (Lampel G, *trans*). 1988. *A list of aphids and their host plants in Iran*. Shahid-Chamran University, Ahwaz, Iran.

Holman J. 2009. *Host plant catalog of aphids*. Springer, Dordrecht.

Howard RJ, Garland JA, Seaman WL, eds. 1994. *Diseases and pests of vegetable crops in Canada*. Entomological Society of Canada, Ottawa, Ontario.

Hradil K, Kment P, Rohacova M. 2007. New records of *Liorhyssus hyalinus* (Heteroptera: Rhopalidae) in the Czech Republic, with a review of its worldwide distribution and biology. *Acta Musei Moraviae, Scientiae biologicae* 92: 53–107.

Humphries T, Florentine S. 2019. Cultivation of low tetrahydrocannabinol (THC) *Cannabis sativa* L. cultivation in Victoria, Australia: do we know enough? *Australian Journal of Crop Science* 13: 911–919.

Hussey NW, Scopes N. 1985. *Biological pest control: the glasshouse experience*. Blandford Press, Poole, UK.

Jablonowski J. 1916. A szőlő és egyéb gazdasági növények paizstetvei. *Kísérletügyi Közlemények* 19: 161–288.

James DG, Prischmann D. 2010. "The impact of sulfur on biological control of spider mites in Washington State vineyards and hop yards," in Sabelis M. Bruin J, eds. *Trends in Acarology*. Springer, Dordrecht.

Janischevsky DE. 1924. Форма конопли на сорных местах в Юго-Восточной России. Ученые записки Саратовского государственного университета имени Н. Г. Чернышевского 2(2): 3–17.

Janischevsky DE. 1925. Энтомохория у *Cannabis ruderalis* Janisch. *Proceedings of the Saratov Society of Naturalists* 1(2-3): 39–49.

Jeppson LR, Baker EW, Keifer HH. 1975. *Mites injurious to economic plants*. University of California Press, Berkeley, California.

Jobling T, Warner P. 2001. New tropical industrial hemp. *Proceedings of the 10th Australian Agronomy Conference*, Hobart, Australia.

John O. 1924. Thysanopteren aus West-Sibirien. *Entomologische Mitteilungen* 13(1): 7–10.

Johnson KP, Dietrich CH, Friedrich F, *et al*. (24 other authors). 2018. Phylogenomics and the evolution of hemipteroid insects. *Proceedings of the National Academies of Science USA* 115: 12775–12780.

Joshi S, Bhaskar H, Aashiq Poon VS, *et al*. (10 other authors). 2021. Occurrence and spread of *Ceroplastes cirripediformis* Comstock (Hemiptera: Coccomorpha: Coccidae) in India. *Zootaxa* 5039(4): 561–570.

Kabanets VV, Fedorenko. 2014. Энтомокомплекс травостоя конопляного поля. *Защита и карантин растений* 2014: 30–34.

Karadjova O, Krumov V. 2015. Thysanoptera of Bulgaria. *ZooKeys* 504: 93–131.

Karl E. 1971. Neue Vektoren für einige nichtpersistente Viren. *Archiv für Pflanzenschutz* 7: 337–342.

Kaur H, Singh D, Suman V. 2012. Faunal diversity of terrestrial Heteroptera (Insecta: Hemiptera) in Punjab, India. Journal of Entomological Research 36: 177–181.

Kennedy JS. 1951. A biological approach to plant viruses. *Nature* 168: 890–894.

Kennedy JS, Booth CO, Kershaw WJS. 1959. Host finding by aphids in the field. *Annals Applied Biology* 47: 424–444.

Kirchner O. 1906. "Hanf, *Cannabis sativa* L.," pp. 319–323 in *Die Krankheiten und Befehädigungen uhferer landwirtschaftlichen Kulturpflanzen*. E. Ulmer, Stuttgart.

Koike ST, Kammeijer K, Bull CT, O'Brien D. 2006. First report of bacterial blight of romanesco cauliflower (*Brassica oleracea* var. *botrytis*) caused by *Pseudomonas syringae* pv. *alisalensis* in California. *Plant Disease* 90: 1511.

Kopp DD, Yonke TR. 1977. Taxonomic status of the buffalo treehopper and the name *Ceresa bubalus*. *Annals of the Entomological Society of America* 70: 901–905.

Kostanda E, Khatib S. 2022. Biotic stress caused by *Tetranychus urticae* mites elevates the quantity of secondary metabolites, cannabinoids and terpenes, in *Cannabis sativa* L. *Industrial Crops & Products* 176: e114331.

Kovačević Z. 1929. Über die wichtigsten schädlinge der kulturpflanzen in Slawonien und Bačka. *Verhandlungen der Deutsches Gesellschaft für Angewandte Entomology, 7. Mitgliederversammlung München*, pp. 33–41.

Kumar A, Parihar SBS, Ramkishore. 2005. Record of weed host plants of *Myzus persicae* (Sulzer). *Insect Environment* 11(2): 58–59.

Kunert G, Reinhold C, Gershenzon J. 2010. Constitutive emission of the aphid alarm pheromone, (E)-β-farnesene, from plants does not serve as a direct defense against aphids. *BMC Ecology* 10: e23.

Kuoh CL, Huang CL, Tian LX, Ding JH. 1980. New species and new genus of Delphacidae from China. *Acta Entomologica Sinica* 23: 413–426.

Labina ES. 2008. The jumping-lice (Homoptera, Psyllinea) fauna of the Republic of Altai. *Entomological Review* 88: 277–285.

Lago PK, Stanford DF. 1989. Phytophagous insects associated with cultivated marijuana, *Cannabis sativa*, in northern Mississippi. *Journal of Entomological Science* 24: 437–445.

Lagos-Kutz D, Potter B, DiFonzo C, Russell H, Hartman GL. 2018. Two aphid species, *Phorodon cannabis* and *Rhopalosiphum rufiabdominale*, identified as potential pests on industrial hemp, *Cannabis sativa* L., in the US Midwest. *Crop, Forage & Turfgrass Management* 4(1): 1–3.

Laing JE. 1969. Life history and life table of *Tetranychus urticae*. *Acarologia* 11: 32–42.

Leach HL, Briggs L. 2021. *Spotted lanternfly survivorship and damage to specialty agricultural crops*. Penn State Extension. Available at: https://extension.psu.edu/spotted-lanternfly-survivorship-and-damage-to-specialty-agricultural-crops-2021

Linnavuori RE. 2011. Studies on the Cimicomorpha and Pentatomomorpha (Hemiptera: Heteroptera) of Khuzestan and the adjacent provinces of Iran. *Acta Entomologica Musei Nationalis Prague* 51: 21–48.

Liu HP. 2019. Oviposition substrate selection, egg mass characteristics, host preference, and life history of the spotted lanternfly (Hemiptera: Fulgoridae) in North America. *Environmental Entomology* 48: 1452–1468.

Liu TX, Stansly PA. 2000. Insecticidal activity of surfactants and oils against silverleaf whitefly (*Bemisia argentifolii*) nymphs (Homoptera: Aleyrodidae) on collards and tomato. *Pest Management Science* 56: 861–866.

Liu YT, Liu B, Li H, Liu JM, Wang PL, Lu YH. 2021. Lacewing density and dynamics on different weeds in cotton-growing region of northern Xinjiang. *Chinese Journal of Biological Control* 37: 671–678.

Lodos N, Kalkandelen A. 1983. Preliminary list of Auchenorrhyncha with notes on distribution and importance of species in Turkey. XII. Famil Cicadellidae Typhlocybinae: Empoasini. *Türkiye Bitki koruma Dergisi* 7: 153–165.

Lodos N, Önder F, Pehivan E, *et al.* (5 other authors). 1999. *Faunistic studies on Lygaeidae (Heteroptera) of western Black Sea, central Anatolia, and Mediterranean regions of Turkey*. University of Ege, Izmir.

López Carretero P, Pekas A, Stubsgaard L, *et al.* (3 other authors). 2022. Glandular trichomes affect mobility and predatory behavior of two aphid predators on medicinal cannabis. *Biological Control* 170: e104932.

Lorenzana A, Hermoso-do-Mendoza A, Seco MV, Casquero PA. 2013. Population dynamics and integrated control of the dahson-hop aphid *Phorodon humuli* (Schrank) on hops in Spain. *Spanish Journal of Agricultural Research* 11: 505–517.

Lu YH, Wu KM, Jiang YY, Xia B, Li P, *et al.* 2010. Mirid bug outbreaks in multiple crops correlated with wide-scale adoption of Bt cotton in China. *Science* 328: 1151–1154.

Lu YH, Jiao ZB, Li GP, Wyckhuys KAG, Wu KM. 2011. Comparative overwintering host range of three *Adelphocoris* species (Hemiptera: Miridae) in northern China. *Crop Protection* 30: 1455–1460.

Maes JM. 2002. Insectos asociados al esparrago en Nicaragua. Available at: http://bio-nica.info/Biblioteca/MaesInsectosEsparrago.pdf

Mahieu CM, Biesebeek JD, Graven C. 2019. *Inventarisatie van gewasbescherming toepasbaar in de teelt van Cannabis binnen het "Experiment met een gesloten coffeeshopketen."* Rijksinstituut voor Volksgezondheid en Milieu, Rapport 2019-0232. Bilthoven, Netherlands.

Manson DCM, Oldfield GN. 1996. Life forms, diapause and seasonal development. *World Crop Pests* 6: 173–183.

Mansoor-ul-Hasan, Waqas W, Bashir F, Kwon YJ. 2005. Descriptions of two new species of the genus *Brevipalpus*(Acari: Tenuipalpidae) Donnadieu from Punjab, Pakistan. *Journal of the Acarological Society of Japan* 14(1): 13–15.

Marc JA 1827. "Notice sur la culture alternative du chanvre et du blé, dans le Département de la Haute-Saòne," pp. 86–88 in de Neufchâteau NF, *ed. Mémoire sur la manière d'étudier et d'enseigner l'agriculture*. Aucher-Éloy, Blois.

Marotta S. 1987. I Coccidi (Homoptera: Coccoidea: Coccidae) segnalati in Italia, con riferimenti bibliografici sulla tassonomia, geonemia, biologia e piante ospiti. *Bollettino del Laboratorio di Entomologia Agr Filippo Silvestri* 44: 97–119.

Martorina S. 2011. *Ottimizzazione dei principali fattori produttivi della Cannabis sativa L. destinata all'impiego*. Master's thesis, University of Ferrara, Italy.

Mathers TC, Mungford ST, Wouters RHM, *et al.* (5 other authors). 2022. Aphidinae comparative genomics resource. *Zenodo Preprint* before print.

Matsumura S. 1917. A list of the Aphididae of Japan, with description of new species and genera. *Journal of the College of Agriculture, Tohoku Imperial University* 7: 351–414.

McCoy CW. 1996. "Pathogens of eriophyid mites," pp. 481–492 in Lindquist EE, *et al.*, eds. *Eriophyoid mites*. Elsevier, Amsterdam.

McCune F, Morphy C, Eaves J, Fournier V. 2021. Onion thrips, *Thrips tabaci* (Thysanoptera: Thripidae), reduces yields in indoor-grown cannabis. *Phytoprotection* 101: 1–37.

McPartland JM. 1996a. A review of *Cannabis* diseases. *Journal International Hemp Association* 3(1): 19–23.

McPartland JM. 1996b. *Cannabis* pests. *Journal International Hemp Association* 3(2): 49, 52–55.

McPartland JM. 1999. "A survey of hemp diseases and pests," pp. 109–131 in Ranalli P, *ed. Advances in Hemp Research*, Haworth Press, New York.

McPartland JM, Hillig KW. 2003. The hemp russet mite. *Journal of Industrial Hemp* 8: 107–112.

McPartland JM, McKernan KJ. 2017. "Contaminants of concern in *Cannabis*: microbes, heavy metals and pesticides," pp. 457–474 in Chandra S, Lata H, ElSohly MA, eds. *Cannabis sativa: Botany and Biotechnology*. Springer International Publishing, Cham, Switzerland.

McPartland JM, Rhode B. 2005. New hemp diseases and pests in New Zealand. *Journal of Industrial Hemp* 10: 99–108.

McPartland JM, Sheikh Ź. 2018. A review of *Cannabis sativa*-based insecticides, miticides, and repellents. *Journal of Entomology and Zoology Studies* 6: 1288–1299.

McPartland JM, Clarke RC, Watson DP. 2000. *Hemp diseases and pests – management and biological control.* CABI Publishing, Wallingford, UK.

Melander AL. 1914. Can insects become resistant to sprays? *Journal of Economic Entomology* 7: 167–173.

Metcalf C, Flint W, Metcalf R. 1962. *Destructive and useful insects, 4th edn.* McGraw-Hill Book Co., New York.

Migeon A, Dorkeld F. 2020. *Spider Mites Web.* Available at: http://www.montpellier.inra.fr/CBGP/spmweb

Miyazaki M. 1972. A revision of the tribe Macrosiphini of Japan. *Insecta Matsumurana* 34(1): 1–247.

Mondor EB, Roitberg BD, Stadler B. 2002. Cornicle length in Macrosiphini aphids: a comparison of ecological traits. *Ecological Entomology* 27: 758–762.

Montgomery ME, Nault LR. 1977. Comparative response of aphids to the alarm pheromone, (E)-β-farnesene. *Entomologia Experimentalis et Applicata* 22: 236–242.

Moritsu M. 1983. 日本原色アブラムシ図鑑. Zenkoku Nōson Kyōiku Kyōkai, Tōkyō.

Mostafa AR, Messenger PS. 1972. *Insects and mites associated with plants of the genera Argemone, Cannabis, Glaucium, Erythroxylum, Eschscholtzia, Humulus, and Papaver.* Unpublished manuscript, University of California, Berkeley.

Mostafavi R, Henning JA, Gardea-Torresday J, Ray IM. 1996. Variation in aphid alarm pheromone content among glandular and eglandular-haired *Medicago* accessions. *Journal of Chemical Ecology* 22: 1629–1638.

Mound LA, Kibby G. 1998. *Thysanoptera: an identification guide, 2nd edn.* CSIRO Entomology, Canberra.

Mruthyunjana HM. 2012. *Bio-ecology and crop loss estimation of Phenacoccus solenopsis (Homoptera: Pseudococcidae) Tinsley on sunflower (Helianthus annuus L.).* Doctoral thesis, Raichur University of Agricultural Sciences, Raichur, India.

Mukhamediev AA. 1979. Тли Ферганской долины (*Aphids of the Fergana valley*). Izd-vo "FAN" Uzbekskoĭ SSR, Tashkent.

Mukundan TK. 1964. *Plant protection: principles and practice.* Asia Publishing House, Bombay.

Müller FP, Karl E. 1976. Beitrag zur kenntnis der bionomie und morphologie der hanfblattlaus, *Phordon cannabis* Passerini, 1860. *Beiträge zur Entomologie, Berlin* 26: 455–463.

Munir M, Leonberger K, Kesheimer K, *et al.* (14 other authors). 2023. Occurrence and distribution of common diseases and pests of US *Cannabis*: a survey. *Plant Health Progress.*

Nair MRGK. 1975. *Insects and mites of crops in India.* Indian Council of Agricultural Research, New Delhi.

Nair KR, Ponnappa KM. 1974. *Survey for natural enemies of Cannabis sativa and Papaver somniferum.* Commonwealth Institute of Biological Control, India Station Report, pp 39–40.

Naumann-Etienne K. & Remaudière G. 1995. A commented preliminary checklist of the aphids (Homoptera: Aphididae) of Pakistan and their host plants. *Parasitica* 51: 1–61.

Neve RA. 1991. *Hops.* Chapman & Hall, London.

Nixon PL, Thompson HE. 1987. Biology of *Micrutalis calva* (Homoptera: Membracidae) on honey locust. *Journal of the Kansas Entomological Society* 60: 272–279.

Okamoto H, Takahashi R. 1927. Some Aphididae from Corea. *Insecta Matsumurana* 1(3): 130–148.

Oldfield GN, Proesler G. 1996. "Eriophyoid mites as vectors of plant pathogens," pp. 259–276 in Lindquist EE, *et al.*, eds. *Eriophyoid mites.* Elsevier Science SV, Amsterdam.

Olkowski W, Daar S, Olkowski H. 1991. *Common-Sense Pest Control: Least-Toxic Solutions for Your Home, Garden, Pests and Community*, 1st edn., The Taunton Press, Newtown, Connecticut.

Oregon Department of Agriculture. 2017. *Pest alert: Cannabis or bhang aphid.* Available at: https://www.oregon.gov/ODA/shared/Documents/Publications/IPPM/CannabisAphidAlert.pdf

Orenzana A, Hermoso-de-Mendoza A, Seco MV, Casquero PA. 2013. Population dynamics and integrated control of the damson-hop aphid *Phorodon humuli* (Schrank) on hops in Spain. *Spanish Journal of Agricultural Research* 11: 505–517

Ossiannilsson F. 1981. *The Auchenorrhyncha (Homoptera) of Fennoscandia and Denmark, vol. 2.* Scandinavian Science Press, Klampenborg, Denmark.

Palumbo JC, Perring TM, Millar JG, Reed DC. 2016. Biology, ecology and management of an invasive stink bug, *Bagrada hilaris*, in North America. *Annual Review of Entomology* 61: 453–473.

Pan H, Lu Y, Wyckhuys KAG, Wu K. 2013. Preference of a polyphagous mirid bug, *Apolygus lucorum* (Meyer-Dür) for flowering host plants. *PLoS ONE* 8(7): e68980.

Pan H, Liu B, Lu Y, Wyckhuys KAG. 2015. Seasonal alterations in host range and fidelity in the polyphagous mirid bug, *Apolygus lucorum* (Heteroptera: Miridae). *PLoS ONE* 10(2): e0117153.

Pan HS, Lu YH, Xiu CL, *et al.* (7 other others). 2018. Volatile fragrances associated with flowers mediate host plant alternation of a polyphagous mirid bug. *Scientific Reports* 5: 14805.

Panah SR, Namaghi HS. 2014. New contribution to the eriophyoid (Acari: Eriophyidae) fauna of Iran. *Zoology and Ecology* 24: 352–354.

Panizzi AR. 1997. Wild hosts of pentatomids: ecological significance and role in their pest status on crops. *Annual Review of Entomology* 42: 99–122.

Parker WB. 1910. The life history and control of the hop flea beetle. *USDA Entomology Bulletin* 82(4): 32–58.

Parker WB. 1913a. The hop aphis in the Pacific region. *USDA Entomology Bulletin* 111: 9–39.

Parker WB. 1913b. The red spider on hops in the Sacramento valley of California. *USDA Entomology Bulletin* 117: 1–41.

Passerini G. 1860. *Gli afidi con un prospetto dei generi ed alcune specie nuove italiane.* Tipografia Carmignani, Parma.

Patschke K, Gottwald R, Müller J. 1997. Erste Ergebnisse phytopathologischer Beobachtungen im Hanfanbau im Land Brandenburg. *Nachrichtenblatt des Deutschen Pflanxenschutzdienstes* 49: 286–290.

Paulsen AQ. 1971. *Plant diseases affecting marijuana (Cannabis sativa).* Unpublished manuscript, Kansas State University.

Penman DR, Chapman RB. 1988. Pesticide-induced mite outbreaks: pyrethroids and spider mites. *Experimental and Applied Acarology* 4: 265–276.

Perring TM. 1996. "Vegetables," pp 611–618 in Lindquist EE, *et al.*, eds. *Eriophyoid mites.* Elsevier Science SV, Amsterdam.

Perring TM, Farrar CA. 1986. Historical perspective and current world status of the tomato russet mite (Aari: Eriophyidae). *Miscellaneous Publications of the Entomological Society of America* 63: 1–19.

Petanović R, Kielkiewicz M. 2010a. Plant-eriophyoid mite interactions: cellular biochemistry and metabolic responses induced in mite-injured plants. *Experimental and Applied Acarology* 51: 61–80.

Petanović R, Kielkiewicz M. 2010b. Plant-eriophyoid mite interactions: specific and unspecific morphological alterations. Part II. *Experimental and Applied Acarology* 51: 81–91.

Petanović R, Magud B, Smiljanić D. 2007. The hemp russet mite *Aculops cannabicola* (Farkas, 1960) (Acari: Eriophyoidea) found on *Cannabis sativa* L. in Serbia: supplement to the description. *Archives of Biological Sciences Belgrade* 59: 81–85.

Pichat CB. 1866. "La pianta della canapa," pp. 396–499 in *Istituzioni scientifiche e tecniche, ossia Corso teorico e practico de Agricoltura, Libri XIII, Volume Quinto*. Presso l'Union Tipografico-Editrice, Torino.

Pienkowski RL, Medler JT. 1964. Synoptic weather conditions associated with long-range movement of the potato leafhopper, *Empoasca fabae*, into Wisconsin. *Annals of the Entomological Society of America* 57: 588–591.

Pitt WJ, Kairy L, Villa E, Nalam VJ, Nachappa P. 2022. Virus infection and host plant suitability affect feeding behaviors of cannabis aphid (Hemiptera: Aphididae), a newly described vector of Potato Virus Y. *Environmental Entomology* 20: 1010.

Pivtoraiko VV, Kabanets VV, Vlasenko VA 2020. Шкідлива ентомофауна конопель посівних, *Cannabis sativa* L. *Plant Quarantine and Protection* 2020 (7-9): 20–25.

Poos FW, Wheeler NH. 1943. *Studies on host plants of the leafhoppers of the genus Empoasca*. USDA Technical Bulletin No. 850, Washington, DC.

Potter D. 2009. *The propagation, characterisation and optimisation of Cannabis sativa L. as a phytopharmaceutical*. Doctoral thesis, King's College, London.

Prain D. 1904. On the morphology, teratology and diclinism of the flowers of *Cannabis*. *Scientific Memoirs by Officers of the Medical and Sanitary Departments of the Government of India* 12: 51–82.

Prasad SK. 1997. Hemp *Cannabis sativa*, a new place to harbour mustard aphid, *Lipaphis erysimi*. *Indian Journal of Entomology* 59: 232.

Prathapan KD. 1992. Incidence of *Drosicha mangiferae* (Green) (Homoptera: Margarodidae) on *Cannabis sativa* L. (Cannabinaceae). *Entomon* 17: 139.

Pulkoski M, Burrack H. 2023. Assessing the impact of piercing-sucking pests on greenhouse-grown industrial hemp (*Cannabis sativa* L.). *Environmental Entomology*.

Puritch G. 1982. An inside look at insecticidal soap. *Sinsemilla Tips* 3(3): 34.

Ragazzi G. 1954. Nemici vegetali ed animali della canapa. *Humus* 10(5): 27–29.

Rahman KA, Ansari AR. 1941. Scale insects of the Punjab and North-West Frontier Province usually mistaken for San Jose scale. *Indian Journal of Agricultural Sciences* 11(5): 816–830.

Rao YR. 1928. *Administration report of the government entomologist for 1927–28*. Coimbatore Agricultural Research Institute, India.

Rataj K. 1957. Skodlivi cinitele pradnych rostlin. *Prameny literatury* 2: 1–123.

Ravelo EEE. 2020. *Riqueza y distribución geográfica de Thysanoptera: Thripinae con énfasis en Frankliniella Karny, en especies de plantas cultivadas y no cultivadas en las regiones Andina, Orinoquía y Caribe de Colombia*. Doctoral thesis, Universidad Nacional de Colombia, Bogotá.

Raychaudhuri DN. 1984. *Food plant catalogue of Indian Aphididae*. Graphic Printall Press, Calcutta.

Raza ABM. 2008. *Biodiversity of Amblyseius Berlese and Neoseiulus Hughs (Acarina: Phytoseiidae) of Punjab-Pakistan*. Doctoral Thesis, University of Agriculture, Faisalabad, Pakistan.

Rebijith KB, Asokan R, Hande HR, *et al.* (4 other authors). 2017. Reconstructing the macroevolutionary patterns of aphids (Hemiptera: Aphididae) using nuclear and mitochondrial DNA sequences. *Biological Journal of the Linnean Society* 121: 796–814.

Remaudière G, Talhouk AS. 1999. Les aphides du Liban et de la Syrie avec la description d'une nouvelle espece du genre *Brachyunguis* Das (Hemiptera, Aphididae). *Parasitica* 55: 149–183.

Rivedal HM, Funke CN, Frost KE. 2022. An overview of pathogens associated with biotic stresses in hemp crops in Oregon, 2019 to 2020. *Plant Disease* 106: 1334–1340.

Rondon SI, Oppedisano T. 2020. *Biology and management of beet leafhopper and purple top phytoplasma affecting potatoes in the Pacific Northwest*. Oregon State University Extension Service Publication EM9282. Corvallis, Oregon.

Rosenthal E. 2012. *Marijuana pest & disease control*. Quick American Publishing, Oakland, California.

Ruckert A, Allen LN, Ramirez RA. 2018. Combinations of plant water-stress and neonicotinoids can lead to secondary outbreaks of banks grass mite (*Oligonychus pratensis* Banks). *PLoS ONE* 13(2): e0191536.

Sakimura K. 1937. Introduction of *Thripoctenus brui* Vuillet, a parasite of *Thrips tabaci* Lind. from Japan to Hawai. *Journal of Economic Entomology* 30: 799–802.

Schaffner JH. 1928. Further experiments in repeated rejuvenation in hemp and their bearing on the general problem of sex. *American Journal of Botany* 15: 77–85.

Schlein Y, Muller G. 1995. Assessment of plant tissue feeding by sand flies (Diptera: Psychodidae) and mosquitoes (Diptera: Culicidae). *Journal of Medical Entomology* 32: 882–886.

Schmidt HE, Karl E. 1970. Ein beitrag zur analyse der virosen des hanfes under berücksichtigung der hanfplattlaus als virusvektor. *Zentralblatt für Baktteriologie, Parasitenkunde, Infektionskrankheiten und Hygiene, Abteilung* 2, 125: 16–22.

Schmidt L. 1956. Štitaste uši Hrvatske. *Zaštita Bilja* 36: 5–11.

Schrank F. von P. 1801. Blattlaus. Fauna Boica. *Durchgedachte Geschichten der in Baiern einheimischen und zahmen Thiere* 2: 102–140.

Schreiner M. 2019. *A survey of the arthropod fauna associated with hemp (Cannabis sativa L.) grown in eastern Colorado*. Master's thesis, Colorado State University, Fort Collins.

Schreiner M, Cranshaw W. 2021. Survey of the arthropod fauna associated with hemp (*Cannabis sativa* L.) grown in eastern Colorado. *Journal of the Kansas Entomological Society* 93: 113–131.

Scopoli GA. 1763. *Entomologia carniolica*. Trattner, Vindobonae.

Seals WC. 2019. *Insects associated with industrial hemp, Cannabis sp. L., in Tennessee and development of Helicoverpa zea (Boddie) on a hemp-based diet*. Master's thesis, University of Tennessee, Knoxville.

Seczkowska K. 1969. *Thrips tabaci* Lind. (Thysanoptera) jako wektor *Lycopersicum* virus 3 w Lubelskim Okregu Upraw Tytoniu Przemyslowego. *Annales Universitatis Mariae Curie-Sklodowska, sectio C Biologia* 24: 341–354.

Sekhon SS, Sajjan SS, Kanta U. 1979. A note on new host plants of greenpeach aphid, *Myzus persicae* from Punjab and Himachal Pradesh. *Indian Journal of Plant Protection* 7: 106.

Shih CT, Poe SL, Cromroy HL. 1976. Biology, life table, and intrinsic rate of increase of *Tetranychus urticae*. *Annals of the Entomological Society of America* 69: 362–364.

Shilenkov VG, Tolstonogova EV. 2006. Вредители конопли в Прибайкалье (*Cannabis* pests in the Baikal region). Бюллетень ВСНЦ СО РАМН (*Acta Biomedica Scientifica*) 2006(2): 159–160.

Shinji O. 1924. New aphids from Morioka. 動物学雑誌 36(431): 343–372.

Shiraki T. 1952. *Catalogue of injurious insects in Japan*. General Headquarters, Supreme Command Allied Powers, Tokyo.

Singh G, Singh NP, Singh R. 2014. Food plants of a major agricultural pest *Aphis gossypii* Glover (Homoptera: Aphididae) from India: an updated checklist. *International Journal of Lice Sciences Biotechnology and Pharma Research* 3(2): 1–26.

Singh KS, Cordeiro EMG, Troczka BJ, *et al.* (30 other authors). 2021. Global patterns in genomic diversity underpinning the evolution of insecticide resistance in the aphid crop pest *Myzus persicae*. *Communications Biology* 4: e847.

Small E. 1975. Morphological variation of achenes of *Cannabis*. *Canadian Journal of Botany* 53: 978–987.

Smith Meyer MKP. 1987. African Tetranychidae (Acari: Prostigmata)—with reference to the world genera. *Entomology Memoir, Department of Agriculture and Water Supply, Republic of South Africa* 69: 1–175.

Sohi AS. 1977. New genera and species of Typhlocybinae (Homoptera: Cicadellidae) from north-western India. *Oriental Insects* 11: 347–362.

Sohi AS, 1983. "The oriental Typhlocybinae with special reference to the pests of cotton and rice: a review," pp. 49–74 in *Proceedings of the 1st International Workshop on Leafhoppers and Planthoppers of Economic Importance*. Commonwealth Institute of Entomology, London.

Song N, Liang AP, Bu CP. 2012. A molecular phylogeny of Hemiptera inferred from mitochondrial genome sequences. *PLoS ONE* 7(11): e48778.

Sorauer P. 1958. *Handbuch der Pflanzenkrankheiten, Band 5*. Paul Parey, Berlin.

Spaar D, Kleinhempel H, Fritzsche R. 1990. *Öl- und Faserpflanzen*. Springer-Verlag, Berlin.

Stannard LJ, DeWitt JR, Vance TC. 1970. The marijuana thrips, *Oxythrips cannabensis*, a new record for Illinois and North America. *Transactions Illinois Academy of Science* 63: 152–156.

Stavraki E. 2005. Προκαταρτικές παρατηρήσεις επί της πληθυσμιακής διακύμανσης του *Eutetranychus orientalis* (Klein) σε εσπεριδοειδή του νομού Αττικής. Doctoral thesis, Kalamata Institute of Agricultural Technology, Kalamata, Greece.

Stevens M. 1975. *How to grow marijuana indoors under lights, 3rd edn*. Sun Magic, Seattle.

Szczepaniec A, Raupp MJ, Parker RD, Kerns D, Eubanks MD. 2013. Neonicotinoid insecticides alter induced defenses and increase susceptibility to spider mites in distantly related crop plants. *PLoS ONE* 8(5): e62620.

Szczepaniec A, Lathrop-Melting A, Janecek T, *et al.* (4 other authors). 2023. Suppression of hemp russet mite, *Aculops cannabicola* (Acari: Eriophyidae), in industrial hemp in greenhouse and field. *Journal of Economic Entomology* Epub ahead of print.

Taisiya D, Chermenskaya TD, Stepanycheva EA, Shchenikova AV, Chakaeva AS. 2010. Insectoacaricidal and deterrent activities of extracts of Kyrgyzstan plants against three agricultural pests. *Industrial Crops and Products* 32: 157–163.

Takafuji A, So P, Tsuno N. 1991. Inter- and intra-population variations in diapause attribute of the two-spotted spider mite, *Tetranychus urticae* Koch, in Japan. *Researches on Population Ecology* 33: 331–344.

Takahashi R. 1965. Some new and little known Aphididae from Japan (Homoptera). *Insecta Matsumurana* 28: 19–61.

Takahashi S. 1919. Notes on insects injurious to hemp, and on *Thyestes gebleri* Fald. *Konchu Sekae (Insect World) Gifu* 23(1): 20–24.

Taylor RAJ, Shields. 2018. Revisiting Potato Leafhopper, *Empoasca fabae* (Harris), migration: implications in a world where invasive insects are all too common. *American Entomologist* 64: 44–51.

Taylor THC. 1947. Some East African species of *Lygus*, with notes on the host plants. *Bulletin of Entomological Research* 38: 233–258.

Tong HJ, Ao Y, Li ZH, Wang Y, Jiang MX. 2019. Invasion biology of the cotton mealybug, *Phenacoccus solenopsis* Tinsley: current knowledge and future directions. *Journal of Integrative Agriculture* 18(4): 758–770.

Toros S. 2002. *Ankara ili ve çevresinde bulunan Coccoidea (kabuklubit ve koşnil) Türleri ve doğal düşmanlarının tespiti*. Ankara Üniversitesi, Turkey.

Traversi BA. 1949. Estudio inicial sobre una enfermeded del girasol (*Helianthus annuus* L.) en Argentina. *Lilloa* 21: 271–278.

Trotuş E, Naie M, 2008. Cercetări privind reducerea atacului unor agenţi patogeni şi dăunători specifici culturilor de cânepă prin tratamentul chimic al seminţei. *Lucrari Stiintifice, Universitatea de Stiinte Agricole Si Medicina Veterinara "Ion Ionescu de la Brad" Iasi, Seria Agronomie* 51(2): 219–223.

Tuatay N, Remaudiere G. 1964. Première contribution au catalogue des Aphididea (Hom.) de lat Turqui. *Revue de Pathologie Végétale et d'Entomologie agricole de France* 43: 243–278.

Ullah S. 2018. *Biosystematics of the true bugs (Heteroptera) of District Swat Pakistan*. Doctoral thesis, Hazara University Mansehra, Pakistan.

Vallot JN. 1836. Sur une Phrygane. *Mémoires de l'Académie des Sciences, Arts et Belles-lettres de Dijon* 1836(2): 237–246.

Van Der Goot P. 1915. *Beiträge zur Kenntnis den Holländischen Blattläuse*. Willink & Zoon, Haarlem.

Van Leeuwen T, Witters J, Nauen R, Duso C, Tirry L. 2010. The control of eriophyoid mites: state of the art and future challenges. *Experimental and Applied Acarology* 51: 205–224.

Vappula NA. 1965. Pests of cultivated plants in Finland. *Annales Agriculturae Fenniana* 1 (Suppl.): 7–239.

Varshney RK. 1992. *A check-list of the scale insects and mealy bugs of South Asia, part 1*. Zoological Survey of India, Calcutta.

Vavilov NI. 1926. происхождение культурной конопли и вооникновение культуры группы «первичных» растений. *Труды по прикладной ботанике, генетике и селекции* 16(2): 107–121.

Vidano C. 1965. A contribution to the chorological and oecological knowledge of the European Dikraneurini (Homoptera, Auchenorhyncha). *Zoologische Beitrage (Neue Folge)* 2: 343–367.

Vierbergen GB, Kucharczyk H, Kirk WDJ. 2010. A key to the second instar larvae of the Thripidae of the Western Palaearctic region (Thysanoptera). *Tijdschrift voor Entomologie* 153: 99–160.

Villanueva RT, Kennedy B. 2019. *Cannabis aphid found in hemp grown in greenhouses in western Kentucky*. Available at: https://kentuckypestnews.wordpress.com/2019/02/19/cannabis-aphid-found-in-hemp-grown-in-greenhouses-in-western-kentucky/

Villanueva RT, Gauthier NL, Ahmed MZ. 2020. First record of *Coccus hesperidum* L. (Hemiptera: Coccidae) in industrial hemp in Kentucky. *Florida Entomologist* 103(4): 514–515.

Villanueva RT, Viloria Z, Falcon-Brindis A, Bradley C. 2021. "Approaches to conduct balanced studies on hemp russet mites," p. 59 in *Second annual Science of Hemp Conference*. Auburn University, Auburn, Alabama.

Villanueva RT, Viloria Z, Ochoa R, Ulsamer A. 2022. *ENTFACT-162: Hemp russet mite, a key pest of hemp in Kentucky*. Available at: https://entomology.ca.uky.edu/ef162

Von Dohlen CD, Moran NA. 1995. Molecular phylogeny of the Homoptera: a paraphyletic taxon. *Journal of Molecular Evolution* 41: 211–223.

Wainwright-Evans S. 2017. Take control of mites on cannabis crops. *Greenhouse Grower*, Available at: https:// www.greenhousegrower.com/production/ insect-control/ take-control-of-mites-in-cannabis-crops/.

Wang SH. 2021. *Diagnosing hemp and cannabis crop diseases*. CAB International, Wallingford, UK.

Wasuwan R, Phosrithong N, Pomdonkoy B, *et al.* (12 other authors). 2021. The fungus *Metarhizium* sp. BCC 4849 as an effective and safe mycoinsecticide for the management of spider mites and other insect pests. *Insects* 13: e42.

Weaver CR, King DR. 1954. Meadow spittlebug. *Ohio Agricultural Experiment Station Research Bulletin* 741: 1–100.

Weber H. 1930. *Biologie der Hepipteren*. Julius Springer, Berlin.

Weber JB. 1978. Marihuana—weed that is a weed: Part I. *Weeds Today* 9(4): 5–6.

Wei CS, Cai P. 1998. Two new species of the genus *Macropsis* Lewis (Homoptera: Cicadellidae: Macropsinae) from China. *Entomotaxonomia* 20(2): 119–122.

Weintraub PG, Beanland L, 2006. Insect vectors of phytoplasmas. *Annual Review of Entomology* 51: 91–111.

Wells PW, Dively GP, Schalk JM. 1984. Resistance and reflective foil mulch as control measures for the potato leafhopper (Homoptera: Cicadellidae) on *Phaseolus* species. *Journal of Economic Entomology* 77: 1046–1051.

Wheeler AG, Miller GL. 1981. Fourlined plant bug (Hemiptera: Miridae), a reappraisal: life history, host plants, and plant response to feeding. *Great Lakes Entomologist* 14: 23–35.

Wilde WR. 1840. *Narrative of a voyage to Madeira, Teneriffe and along the shores of the Mediterranean, vol.* I. William Curry, Jun. & Co., Dublin.

Williams TA. 1891. *Host-plant list of North American Aphididae*. University of Nebraska Department of Entomology, Special Bulletin no. 1. Lincoln, Nebraska.

Williams TA. 1910. *The Aphididae of Nebraska*. University Studies of the University of Nebraska 10: 85–175.

Wilson H, Bodwitch H, Carah J, *et al.* (4 other authors). 2019. First known survey of cannabis production practices in California. *California Agriculture* 73: 119–127.

Wilson MR, Turner JA, McKamey SH. 2009. *Sharpshooter leafhoppers of the world (Hemiptera: Cicadellidae subfamily Cicadellinae)*. Available at: http://natural-history.museumwales.ac.uk/Sharpshooters

Wood A. 1989. *Insects of economic importance: a checklist of preferred names*. CAB International, Wallingford, UK.

Worner SP, Tatchell GM, Woiwod IP. 1995. Predicting spring migration of the damson-hop aphid *Phorodon humuli* (Homoptera: Aphididae) from historical records of host-plant flowering phenology and weather. *Journal of Applied Ecology* 32: 17–28.

Wu XX, Hu DX, Li ZX, Shen ZR. 2002. Using RAPD-PCR to distinguish biotypes of *Bemisia tabaci* (Homoptera: Aleyrodidae) in China. *Entomologia Sinica* 9(3): 1–8.

Xiu CL, Pan HS, Liu B, Luo ZX, Williams L, Yang YZ, Lu YH. 2019. Perception and behavioral responses to host plant volatiles for three *Adelphocoris* species. *Journal of Chemical Ecology* 45: 779–788.

Yepsen RB. 1976. *Organic plant protection*. Rodale Press, Emmaus, Pennsylvania.

Zakhvatkin AA. 1953. Сборник научных работ (*Collection of Scientific Works*). Publishing House of Moscow University, Moscow.

Zemek R, Kopačka M, Šimáčková K. 2016. Evaluation of *Isaria fumosorosea* efficacy for the control of spider mites. *IOBC-WPRS Bulletin* 120: 93–97.

Zhang ZG. 2017. Eriophyoidea and allies: where do they belong? *Systematic and Applied Acarology* 22: 1091–1095.

Zhou Y. 1946. 斑衣蜡蝉之研究 (Research on the variegated wax cicada). 昆虫与艺术 1(2): 31–54.

Zhou Y, Lu JS, Huang J, Wang SZ. 1985. 中国经济昆虫志第36册同翅目蜡蝉总科 (*China economic insects, vol. 36, Homoptera Cicada superfamily*). Science Press, Beijing.

Zuefle M. 2020. *Hemp pest survey 2020*. New York State Integrated Pest Management Program, Geneva, New York. Available at: https://ecommons.cornell.edu/ handle/1813/103765

Zuefle M. 2021. *Hemp pest survey 2021*. New York State Integrated Pest Management Program, Geneva, New York. Available at: https://hdl.handle. net/1813/111249

Abstract

The previous chapter treated insects that suck. This chapter features insects with chewing mouthparts; specifically, insects that chew on flowering tops. Budworms lead the list: *Helicoverpa armigera* in the Old World, and *Helicoverpa zea* in the New World. Also troublesome are marbled clover (*Heliothis viriplaca*) and tobacco budworm (*Chloridea virescens*). Budworms zero in on seeds, but they'll eat sinsemilla—anything with relatively high nitrogen content. Armyworms, like budworms, infest flowering tops, eat seeds, and damage leaves. Armyworms are so named because they sometimes march *en masse* across a crop field (*Spodoptera exigua, S. frugiperda, S. ornithogalli, Melanchra picta, Mamestra brassicae*). Armyworms and budworms are members of the Nocturid family, so they share some control measures. Last but not least comes the hemp flea beetle (*Psylliodes attenuatus*), causing double-trouble: late season adults feed on leaves and flowering tops, overwinter in soil, then emerge in the spring to feed on young seedlings.

5.1   Introduction

This chapter focuses on insects with *chewing* mouthparts that damage flowering tops. Lepidopterans lead the list in this chapter: **budworms** (*Helicoverpa zea, Helicoverpa armigera, Heliothis viriplaca, Chloridea virescens*) and **armyworms** (*Spodoptera exigua, Spodoptera frugiperda, Spodoptera ornithogalli, Melanchra picta, Mamestra brassicae*). Plus there's a Coleopteran, the hemp flea beetle (*Psylliodes attenuatus*).

Other Lepidopterans feed on flowers but cause greater damage elsewhere so we describe them in other chapters. *Grapholita delineana* is best known a stalk borer (see section 6.2) but can damage flowers and seeds (see Fig. 6.2). Cutworms notoriously cut down seedings (see section 8.2) but they sometimes infest flowers at season's end (see Fig. 8.4). Leaf defoliators may feed on flowers, such as the death's head moth (*Acherontia atropos*, see section 7.2b) and Japanese beetle (*Popillia japonica*, see section 7.4a). Sap suckers also harm flowering tops—see Chapter 4 for mites, aphids, thrips, etc.

5.2   Budworms

Budworms are caterpillars that preferentially feed on high-nitrogen plant parts, namely buds (flowers, fruits, seeds). Budworm damage on plants cultivated for flowering tops is greater than damage on seed crops, and much greater than plants grown for fiber (Britt *et al.*, 2021b). A single grower estimated a loss of over US$0.5 million in a CBD crop in 2016 (Cranshaw *et al.*, 2019).

Budworms may spin loose webs around flowering tops, wherein they feed and frass. They may drill deep inside floral clusters, where damage is not initially visible, until flowers are ruined. Budworms deep in the inflorescence may sever its axil, leading to necrosis of the entire flowering top. Dead tissue provides a nidus for gray mold disease, caused by *Botrytis cinerea* (Fig. 5.1).

Budworms are members of the superfamily Noctuoidea, a hugely successful radiation of five or six families and >70,000 species. Sorting species into correct families has taken centuries, and now with DNA data everything's in flux. Five budworms described here belong to Family Noctuidae. They share life histories. Noctuids usually overwinter as pupae in the soil, although in warm climates they may overwinter as caterpillars in cocoons. The moths are nocturnal—hence their family name. They fly at night, mate, and lay eggs on or near host plants. They may lay 1000 eggs or more, in masses or one at a time. Eggs hatch in a few days and larvae commence feeding. The larval period lasts 15–60 days, depending on latitude and temperature. The number of generations per year, from one to ten, also depends on latitude and temperature.

Moths from late-season generations lay their eggs in flowering tops. Some moths are good fliers and perform seasonal migrations between summer and winter ranges, exploiting favorably directed wind streams. The caterpillars are cannibalistic and feed on each other in crowded conditions, or if suitable vegetation is unavailable.

Budworm species are polyphagous; each species has many Latin names, assigned by entomologists who encountered them on different hosts in different parts of the world. We are indebted to taxonomists who collated these names, examined museum specimens, and sorted the synonyms. For example, Hardwick (1965) took nearly 250 pages to sort the "corn earworm complex."

5.2a.   Corn earworm or bollworm

Helicoverpa zea (Boddie) 1850
≡*Heliothis zea* (Boddie), =*Bombyx obsoleta* Fabricius 1793, ≠ *B. obsoleta* Fabricius 1775
Description: *Larvae* when newly hatched are pale yellow with dark longitudinal stripes, 1.5 mm long. They grow into stout caterpillars up to 40 mm in length. Mature caterpillars vary in color from green to light brown, or sometimes yellow, pinkish,

Fig. 5.1. *Helicoverpa zea* damage in tops: **A.** unseeded (W. Cranshaw); **B.** seeded (M. Bolt)

Fig. 5.2. *Helicoverpa zea*: **A–C**, larva color morphs; **D.** pupa; **E.** moth (all W. Cranshaw except E, courtesy BoldSystems)

or mostly black (Fig. 5.2A–C). Color depends upon food source; green is most common in *Cannabis* feeders. The head is light brown to orange; legs and thoracic plates are black. Narrow dark stripes often occur down the center of the back. Each side has a dark stripe above the spiracles, and a yellow or white stripe below the spiracles. They have numerous tiny spines on their skin (visible through a hand lens), in addition to the dozen or so longer bristles found on each segment, giving the skin a rough feel.

Pupae are buried in the ground, shiny mahogany-brown in color, 17–22 mm in length (Fig. 5.2D). *Adults* are variable in color. Forewings usually yellowish brown, with light and dark bands, often bearing a small dark spot centrally. Hindwings are creamy white basally, darker laterally, bearing a small dark spot centrally (Fig. 5.2E), wingspan 32–45 mm. *Eggs* are deposited singly, subspherical, shiny, radially ribbed, white when newly laid but darkening to tan or gray with time, 0.5–0.6 mm diameter.

Comments: Early reports of *H. zea* infesting US hemp were mistakenly identified as *Helicoverpa armigera* (Riley, 1871, 1885; Comstock, 1879; Mah, 1923), the latter being a Eurasian species. Bruner (1892) said the species infested "heads" of hemp in Nebraska, and Forbes (1904) said the species "feeds freely" on hemp in Illinois. Tietz (1972) correctly cited *H. zea* on North American *Cannabis*. *H. zea* has been reported in Colorado, Illinois, Indiana, Kentucky, Nevada, North Carolina, Tennessee, Vermont, Virginia, Wisconsin, and Alabama (Britt *et al.*, 2019; Cranshaw *et al.*, 2019; Seals, 2019; Ajayi and Samuel-Foo, 2021).

Corn earworm moths migrate to hemp after plants begin to flower. This happens in late summer, after corn (maize) has passed the green silk period and is no longer attractive for egg laying. Corn earworm is rarely seen in hemp prior to flowering, but once flowering begins, hemp becomes very attractive. The caterpillars chew into flower buds and seed heads.

The species is not a permanent resident above 40 °N (it cannot overwinter), but every summer the adults migrate as far north as Ontario (Howard *et al.*, 1994). In the north only one

generation occurs per year; two occur in northeastern states; three in the southeast, central Great Plains, and northern California, four to five in Louisiana and southern California; and six in southern Florida and southern Texas. The life cycle can turn in as little as 30 days. Females lay between 500 and 3000 eggs. More than 100 plant species are recorded as hosts. Ajayi and Samuel-Foo (2021) predicted a range expansion on *Cannabis* due to climate change.

Today *H. zea* is considered the greatest threat to flowering tops in the USA, due to a lack of effective monitoring and management tools (Britt *et al.*, 2021b). Despite its common name, Britt *et al.* (2019) found that corn earworm development on a pure hemp diet was equal to that on maize—from early instar larva to pupation. Caterpillars force-fed hemp leaves and seeds weighed less than caterpillars fed a balanced diet (Seals, 2019), and caterpillars fed nothing but *Cannabis* buds and leaves suffered a 19.1% mortality rate (Falcon-Brindis *et al.*, 2022).

Comstock (1879) reported that a Texan tried repelling egg-laying *H. zea* moths from a cotton field by planting a ring of hemp around the crop. He never published the results, but the effort likely failed, since hemp is a host. Pheromone traps designed to catch *H. zea* also attract **Heliothis phloxiphaga** Grote and Robinson 1867 (Hoffmann *et al.*, 1991). This pest species infests trees (*Ulmus, Malus, Prunus* spp.) but Munir *et al.* (2023) reported it feeding on *Cannabis*.

5.2b. Cotton bollworm or Old World bollworm

Helicoverpa armigera Hübner 1809
≡*Heliothis armigera* (Hübner), =*Heliothis obsoleta* (Fabricius), =*Chloridea obsoleta* (Fabricius), =*Bombyx obsoleta* Fabricius 1775, ≠*B. obsoleta* Fabricius 1793

Description: Larvae, pupae, and adult moths of *H. armigera* closely resemble those of *H. zea*, (compare Fig. 5.2E and Fig. 5.3A). See the "Differential diagnosis" section.

Comments: Passerini (1832) described cotton bollworm on flowering tops in Italy—one of the first pests reported on *Cannabis*. Accounts of *H. armigera* ruining hemp in Europe range from France to southern Russia (Goureau, 1866; Noel, 1917; Vinokurov, 1927; Rataj, 1957; Ceapoiu, 1958; Khamukov and Kolotilina, 1987; Matov and Kononenko, 2012). The species was "very rare" on hemp in Warsaw, "it is a southern species, appearing only briefly in northern Europe" (Romaniszyn, 1929).

H. armigera has damaged hemp in Japan (Matsumura, 1899; Shiraki, 1952), China (Li and Rahmann, 1997; Zhang *et al.*, 2001), Iran (Bagheri and Isfahani, 1989), South Africa (Dippenaar *et al.*, 1996), New Zealand (McPartland and Rhode, 2005), and Australia (Humphries and Florentine, 2019). Feral *Cannabis* harbors *H. armigera*, from which the pest migrates to hemp crops (Vörös, 1996).

Cotton bollworm is notably noxious in drug-type crops. **Harold Maxwell-Lefroy** (1877–1925) reported the species attacking drug-type plants in India (Theobald, 1904). Maxwell-Lefroy (1906) published a detailed report of *H. armigera* in India, including its impact on flowering tops of *gañjā*. He was a marvelous entomologist who invented new methods of mechanical, chemical, and biological control—including the discovery of *Encarsia formosa* (Maxwell-Lefroy, 1915). He died young while experimenting with an arsenic fumigant.

Fletcher (1917) considered *H. obsoleta* (= *H. armigera*) "a bad pest, eating the leaves and cutting the topshoots" in Pakistan. Cherian (1932) claimed that 100 caterpillars could eat a pound of *gañjā* per day. He considered cotton bollworm second only to spider mites in destructive capacity. Many others have cited this species on Indian hemp (Cotes, 1893; Alcock, 1903; Ballard, 1920; Rao, 1928; Nair and Ponnappa, 1974; Chandra, 2004; Singh Kriti *et al.*, 2014).

Bajpai and Sehgal (1993) conducted an oviposition preference test with seven host plants, and *Cannabis* ranked fourth. Parihar and Singh (1992) force-fed *H. armigera* a pure *gañjā*

diet and growth was slow, compared with growth on pigeon peas, chickpeas, or brassica. Kumar *et al.* (2006) reported a "growth index" of 1.25 when *H. armigera* fed on *bhāng*, lower than 4.43 when budworms fed on cauliflower.

Cotton bollworm produces 1–6 generations per year, depending on latitude and temperature. Moths emerge from pupae as late as June in northern Russia (Vinokurov, 1927). The species does its worst damage in the tropics, where its life cycle can be as short as 28 days (Hill, 1983). It attacks at least 60 cultivated and 67 wild hosts plants, spanning 34 plant families (Pogue, 2004). Cotton bollworm causes greatest damage on cotton, maize, tobacco, and chickpea. Larvae sometimes turn up in canned tomatoes. The common name "Old World bollworm" is now a misnomer, because *H. armigera* has spread to South America, and the species has been intercepted at the USA border (Pogue, 2004).

5.2c. Marbled clover

Heliothis viriplaca (Hüfanagel) 1766
≡*Heliothis dipsacea* (Linnaeus) 1767

Description: *Larvae* vary in color, from grayish green to dark brown, sometimes pinkish, with yellow-green heads, up to 25 mm long. Heads light green, pale brown, or pink. Dorsal stripes are bundles of multicolored bands—green, yellow, or rust-colored, often dark-edged; lateral stripes are lighter in color, green, yellow, or white. *Pupae* in a flimsy cocoon, on or below ground. *Adults* variable in color, 25–30 mm wingspan. Forewings yellowish brown, with light and dark bands and botches. Hindwings mostly dark brown to nearly black, with irregular central beige blotches and small beige blotches near outer edge (Fig. 5.3B).

Comments: Marbled clover infests hemp crops across Europe (Kirchner, 1906; Blunck, 1920; Vinokurov, 1927; Ceapoiu, 1958; Arcozzi *et al.*, 1980; Matov and Kononenko, 2012), and East Asia (Clausen, 1931; Shiraki, 1952). The pest also infests *Cannabis ruderalis* in Siberia (Shilenkov and Tolstonogova, 2006). It has reached Australia and infests hemp (Humphries and Florentine, 2019). Marbled clover is polyphagous, with >70 host plants, including >20 cultivated crops, particularly flax, cotton, alfalfa, and soybean. It occurs in cooler climates, producing 1–4 generations per year. Moths undertake long-distance migration on prevailing winds. Unlike nocturnal *Helicoverpa* moths, marbled clover flies by day as well as night.

5.2d. Tobacco budworm

Chloridea virescens (Fabricius) 1777
≡*Heliothis virescens* (Fabricius) 1777

Description: *Larvae* vary in color from light brown to green, and are nearly identical to those of *H. zea*. Mature larvae can be differentiated by tubercles (dark bumps) on their 1st, 2nd, and 8th abdominal segments. *H. zea* has microspines on the tubercles, and *C. virescens* does not (Neunzig, 1964). *Pupae* are indistinguishable from *H. zea* pupae. *Adult* forewings usually beige to creamy white, with three darker bands; hindwings are white with the distal margin bearing a darker band.

Fig. 5.3. Noctuids: **A.** *Helicoverpa armigera* (Amy Carmichael); **B.** *Heliothis viriplaca* (Dumi); **C.** *Spodoptera ornithogalli* (Jacy Lucier); **D.** *Melanchra picta* (Robert Bauenfeind) (all Wikimedia Commons except D (Bugwood.org))

Comments: *Heliothis virescens* was transferred to *Chloridea* based on microscopic morphological characters (protibial setae, male abdominal tergites) and multilocus genotype data (Pogue, 2013). *C. virescens* prefers tobacco and *H. zea* prefers maize, but their host ranges overlap—both thrive on maize, tobacco, tomato, pepper, soybean, alfalfa, cabbage, and many flower crops and weeds. To date *C. virescens* has only been reported on *Cannabis* in the southeast USA: North Carolina (Pulkoski and Burrack, 2020) and Virginia (Britt *et al.*, 2021b). The ratio of species in a hemp field consisted of 79% *H. zea* and 21% *C. virescens* (Pulkoski and Burrack, 2020).

C. virescens overwinters in the south, and gives rise to four or five generations. Adults are strong fliers and disperse north of the Mason–Dixon line but do not overwinter. Female moth longevity ranges from 25 days at 20°C to 15 days at 30°C, and they lay 300–500 eggs. Like *H. zea*, the moths are attracted to flowering plants and do not infest hemp prior to flowering. Hatching caterpillars chew into flower buds and seed heads. Pulkoski (2022) investigated the effects of *C. virescens* feeding upon THC and CBD levels, and the results were inconsistent.

5.2.1   Differential diagnosis

Differentiating *H. zea* from *H. armigera* is not easy, and requires a close look at male and female genitalia (Hardwick, 1965; Pogue, 2004). Geography separates them: *H. zea* is native to the Americas, while *H. armigera* is a Eurasian species—although *H. armigera* has established a beachhead in Brazil. Barbosa *et al.* (2016) compared the two species in Brazil; *H. armigera* larvae developed faster and were smaller than *H. zea* larvae.

Differentiating *H. zea* from *C. virescens* is also difficult. The wings of *C. virescens* are lighter than those of *H. zea*. The wings of *C. virescens* are held in an upright tent; *H. zea* holds its wings in a flat triangular shape. *Helicoverpa* and *Heliothis* spp. must be differentiated from leaf-eating Noctuids that stray into flowering tops. Some are armyworms, discussed below. The light-colored heads and presence of spines distinguish *Helicoverpa* larvae from some armyworms (e.g., *Spodoptera exigua*) and stalk borers that stray into flowering tops (*Ostrinia nubilalis*, *Grapholita delineana*).

5.3   Armyworms

"Armyworm" refers to caterpillars who sometimes march *en masse* across a crop field. The five armyworm species discussed below are Noctuids, like budworms. They infest flowering tops, eat seeds, and damage leaves. We chose to place them here rather than with leaf chewers in Chapter 7. Armyworms are polyphagous, and will likely gain prominence as the acreage of *Cannabis* increases in the USA. For example, soybean looper, *Chrysodeixis includens* (Walker), readily switches from soybean to hemp. Arey *et al.* (2022) reared *C. includens* larvae on soybean leaves compared with two CBD varieties, and they grew fatter on hemp, with higher reproductive rates.

5.3a.   Beet armyworm

Spodoptera exigua (Hübner) 1827
≡ *Laphygma exigua* Hübner

Description: *Larvae* when young are green, becoming variable gray-green or brown, with wavy lines down the dorsum, and pale lateral stripes along each side, up to 30 mm long (Fig. 5.4). Mature larvae have dark heads, and a dark spot on each side of the body above the second true leg. *Pupae* are dark brown, and found in the upper 10 mm of soil in a cell formed of soil and plant debris. *Adult* bodies are brown. Forewings are gray-brown with a yellow kidney-shaped spot in the mid-front margin; hindwings are white with a dark anterior fringe, wingspan 25–30 mm. *Eggs* are green (changing to gray), and laid in clusters of 50–300, in several layers, covered by moth hairs and scales.

Comments: Moths emerge in spring and deposit 500–900 eggs over 4–10 days. Eggs hatch in 2–5 days. Larvae feed for about 3 weeks, and pupate in half that time. In warm areas 4–6 generations arise per year. The species is native to India and China, where it feeds on *Cannabis* (Cherian, 1932; Chandra, 2004). Beet armyworm has spread around the world, and attacks hemp in Europe (Ceapoiu, 1958) and Africa (Jansen, 2008). Adult moths of *S. exigua* visit *Cannabis* flowers; Jia *et al.* (2023) found *Cannabis* pollen on the proboscis of migratory moths in northeast China, along with pollen from 33 other plant species.

The species reached California by 1876, and by 1924 it had spread to Florida. Cranshaw *et al.* (2019) found it in Colorado. Arey *et al.* (2022) reared *S. exigua* larvae through six instars to pupation; on two CBD varieties they developed faster than larvae reared on two other hosts (soybean, *Glycine max;* cowpea, *Vigna unguiculata*). Host plant choice did not have a significant impact on pupal weight or survivorship.

5.3b.   Fall armyworm

Spodoptera frugiperda (J.E. Smith) 1797
≡ *Laphygma frugiperda* J.E. Smith

Fig. 5.4. *Spodoptera exigua* larva (W. Cranshaw)

Description: *Larvae* when young are greenish with a black head. Older larvae greenish laterally, but the dorsum turns brown, with white subdorsal and lateral lines, elevated spots on the dorsum are dark in color and bear spines, and the head becomes reddish brown. The face is marked by a light-colored inverted "Y" mark. *Pupae* are reddish-brown, 14–18 mm long. *Adults* have a dark brown thorax and lighter brown abdomen. Forewings variegated brown-gray with triangular white spots at the tip and near the center of the wing; hindwings are iridescent silver-white-beige with a narrow dark border; wingspan 32–40 mm. *Eggs* are pale yellow to light green, laid in clusters of 100–200, usually in a single layer, covered by moth hairs and scales.

Comments: Fall armyworm is native to warmer regions of the Western hemisphere. The species primarily damages Poaceae (maize, wheat, sorghum, sugar cane) but can feed on nearly anything, from cotton to clover to cabbage. In the USA fall armyworm overwinters along the Gulf of Mexico. The moths are strong fliers and migrate north, to nearly all states east of the Rocky Mountains. Moths deposit 1500 eggs over 4–10 days. The life cycle and number of generations depend on latitude; in the tropics the life cycle may only take 30 days, with up to six generations per year; in northern areas, up to 90 days, and only a single generation.

Fall armyworm has spread to other warm regions around the globe. In India, an outbreak of *S. frugiperda* devastated maize crops (KRC Times, 2019), and one photo showed the pest feeding on seeded flowering tops of *Cannabis*. The pest invaded the Canary Islands, where it attacks fiber-type and drug-type *Cannabis* (Salami, 2020).

S. frugiperda has served as an experimental "lab rat." Vaughn *et al.* (1977) created an *in vitro* cell line called Sf9 (*Spodoptera frugiperda* 9). Genetic engineers have created Sf9 cells that express proteins from other species; Pettit *et al.* (1994) transfected the gene for cannabinoid receptor 1 (CB_1) into Sf9 cells. Sirikantaramas *et al.* (2004) transfected the *Cannabis* gene for THCA synthase into Sf9 cells.

Bolt *et al.* (2021) subjected third-instar larvae to a forced-feeding study with three fiber-type cultivars ('Futura 75', 'CFX-1', 'X-59'). Larvae showed greater consumption and frass production when fed 'X-59' compared with the other cultivars. In plants grown with supplemental nitrogen (either 168 or 224 kg/ha), larvae showed even greater consumption, frass production, and weight gain. Arey *et al.* (2022) reared larvae on leaves of two CBD varieties or other hosts (soybean, *Glycine max*; cowpea, *Vigna unguiculata*). Host plant did not have a significant impact on development time or survivorship, but pupal weight was highest in larvae reared on hemp.

5.3c. Yellowstriped armyworm

Spodoptera ornithogalli (Guenée) 1852
Description: *Larvae* are robust, gray-green to nearly black, with lateral stripes bright yellow or white, and a black head, up to 35 mm long (Fig. 5.5). *Pupae* in soil are reddish-brown, 18 mm in length. *Adults* have a robust beige body. Forewing color is a complex and highly contrasting pattern of brown, yellow, beige, and white. Hindwing color is white, with a light brown band near the fringes, 34–41 mm wingspan (Fig. 5.3C). *Eggs* laid in clusters of 200–300, subspherical, radially ribbed, greenish, 0.46–0.52 mm diameter.

Comments: Yellowstriped armyworm is polyphagous and feeds on many crop plants, including tomato and cotton. Generally 3–4 generations arise annually. The species has infested *Cannabis* in Mississippi (Lago and Stanford, 1989), Colorado (Cranshaw *et al.*, 2019), Virginia (K. Britt, pers. obs., 2020), Alabama (Conner *et al.*, 2020), Missouri (Akotsen-Mensah, 2021), and Kentucky (Falcon-Brindis *et al.*, 2022). Larvae primarily attack flowers and developing seeds, although they feed on foliage.

5.3d. Zebra caterpillar

Melanchra picta (Harris) 1841
Description: *Larvae* have a distinctive black and white "zebra" pattern, dorsal black stripe, upper lateral stripes black and bright yellow, with a reddish-brown head and orange feet, 30–35 mm long (Fig. 5.6). *Pupae* in soil are reddish-brown, 18 mm in length (Fig. 5.6). *Adults* have a robust, reddish-brown body (Fig. 5.3D). Forewings are reddish-brown with an indistinct gray spot, reniform-shaped. Hindwings white with a narrow reddish-brown at the fringes, 35–40 mm wingspan.

Comments: Zebra caterpillar primarily attacks flowers and developing seeds, although it feeds on foliage (Cranshaw *et al.*, 2019). The species occurs throughout the USA and southern Canada, is more common in the western states, but has attacked hemp in Vermont (Fig. 5.6A). It is a broad generalist feeder, causing problems in alfalfa, vegetable crops, and fruit trees. One generation arises per year in the west, two or more generations in the east.

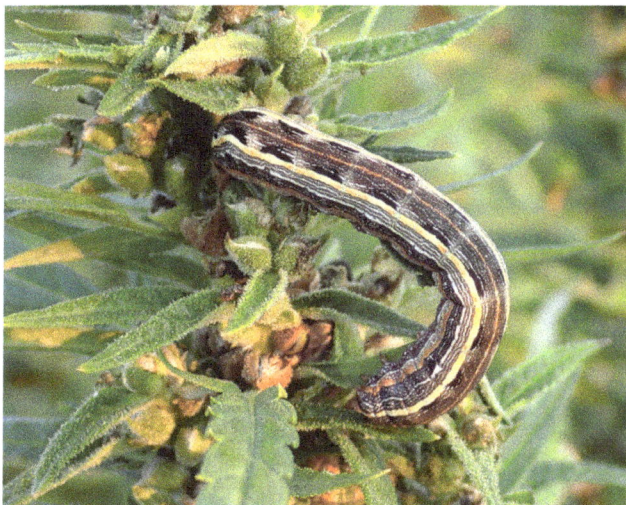

Fig. 5.5. *Spodoptera ornithogalli* larva (W. Cranshaw)

Table 5.1. Infestation severity index for budworms

Light	Any budworms or webs or bud damage seen
Moderate	Signs or symptoms in more than one plant, moths present
Heavy	Many worms and webbings seen, most buds in crop infested
Critical	Every plant with bud damage, moths common

Fig. 5.6. *Melanchra picta*: **A**. larva (John Bruce); **B**. adult emerging from pupa (W. Cranshaw)

5.3e. Cabbage moth

Mamestra brassicae (L.) 1758
≡*Barathra brassicae* L.

Description: *Larvae* are yellow-green when young, their dorsum darkening to a green-brown color by the fourth instar. They have yellow heads. Side stripes consist of a chain of dark semilunar spots. Rusty-brown setae cover the whole body, reaching 38–45 mm in length. *Pupae* are reddish-brown and glossy, up to 22 cm long. *Adults* are brown Noctuid moths, nearly indistinguishable from other species except for a curved dorsal spur on the tibia of the foreleg. *Eggs* are transparent (turning light yellow to brown), ribbed, and laid in batches of 50 or more, in single-layered geometric rows.

Comments: Cabbage moth is Palearctic in distribution and polyphagous. It overwinters as pupae. Adults emerge and lay eggs in May–June. Young larvae act as cutworms; later they eat leaves. Two generations arise per year. It attacks hemp in central Europe (Rataj, 1957; Ferri, 1959a), Siberia (Matov and Kononenko, 2012), and Japan (Matsumura, 1899; Takahashi, 1919; Harada, 1930; Clausen, 1931; Shiraki, 1952). In Japan the cabbage moth feeds like a budworm. A North American relative, the Bertha armyworm, *Mamestra configurata* Walker 1856, infested hemp in Manitoba (Moes *et al.*, 1999). It likely migrated from neighboring canola, a common host.

5.3.1   Cultural and mechanical control

McPartland *et al.* (2000) published an infestation severity index for budworms in general (Table 5.1). Britt *et al.* (2021b) specifically addressed corn earworm, based on data from sites

in Virginia, North Carolina, Alabama, Maryland, and Delaware. They scouted crops weekly, by sampling 30 random buds, beginning at the onset of bud formation. They recorded the number of larvae and the extent of damage on a 0–3 scale, where 0 was no damage; 1 was visual damage but bud still marketable; 2 was damage to 50% or less of bud material, rendering it unmarketable; 3 was damage to 50% or more of bud material. Peak of infestation occurred mid-September, when an average of 7.2 ±2.0 larvae was found per 30 buds. Regression analysis indicated that the number of larvae during the last week of August was a significant predictor of the damage rating, and densities greater than 2.5 larvae per 30 buds resulted in unacceptable crop damage.

Budworm pheromone traps, designed to catch male moths, are monitoring tools used in many crops. Armyworm pheromone traps are also available as monitoring tools (e.g., *Spodoptera frugiperda*, *Mamestra brassicae*).

Britt *et al.* (2021b) tested the efficacy of budworm pheromone traps to monitor corn earworm populations, at the locations where they assessed larval densities and bud injury. They used Mesh Scentry *Heliothis* pheromone traps, with a Hercon Luretape corn earworm lure attached to a string hanging below the bottom opening of the trap (see Fig. 21.5). Lures were replaced every 14 days, and trap catch recorded weekly from the onset of bud formation until harvest. Regression analysis found no significant relationship between moth trap catch and peak larval abundance. This suggests that weekly scouting for larvae, a time-consuming endeavor, is a more reliable predictor of potential crop damage than pheromone traps.

To discourage budworms and armyworms, keep weeds down within and around crops. That includes feral *Cannabis* (Vörös, 1996). Most budworms and armyworms overwinter in soil, and post-harvest plowing exposes them to the elements. Night-flying moths are attracted to light traps, useful for monitoring purposes (see Fig. 19.6). Small-acre farmers historically removed caterpillars by hand. Cherian (1932) shook infested plants every 10 days, and collected all caterpillars that fell to the ground. Tomato breeders have conferred some resistance against *H. zea* by selecting for plants that produce methyl ketones (Kashyap *et al.*, 1991). Several methyl ketones are produced by *Cannabis* (Turner *et al.*, 1980).

5.3.2   Biological control

Generalist predators of *Helicoverpa* eggs and young larvae include pirate bugs (*Orius* spp.), big-eyed bugs (*Geocoris* spp.), lady beetles (*Harmonia*, *Hippodamia* spp.), and lacewings (*Chrysoperla* spp.). All are commercially available (see section

20.6). Villanueva (2022b) observed naturally occurring *Orius* spp. cruising for *H. zea* eggs in Kentucky hemp fields.

Trichogramma **species** are minute wasps that parasitize the eggs of Lepidopterans. There are many species (see section 20.6). Khamukov and Kolotilina (1987) controlled cotton bollworm in hemp with *Trichogramma pretiosum* and *T. minutum*. They did not cite release rates. Smith (1996) said corn growers release a million *T. pretiosum* wasps per hectare, every 2–3 days while moths are laying eggs, for a total of 11–18 releases per season. *T. brassicae*, *T. dendrolimi*, and *T. confusum* have also controlled *Heliothis* spp. at release rates of 200,000–600,000 wasps/hectare (Li, 1994).

In the braconid wasp family, a pair of naturally occurring parasitoids have gained attention: *Microplitis croceipes* deposits eggs in larvae of *Helicoverpa zea*, *H. armigera*, and *H. virescens*. The wasp is attracted to volatile compounds released when its host feeds on plants, including several *Cannabis* compounds: β-caryophyllene, α-humulene, α-farnesene, and β-pinene (Morawo and Fadamiro, 2016). *Chelonus insularis* parasitizes eggs and young larvae of *Spodoptera frugiperda* as well as *Heliothis* spp. and other Noctuids. Naturally occurring *C. insularis* was the second most-common parasitoid of *S. frugiperda* in a Florida survey, and it was #1 in Central and South America (Meagher *et al.*, 2016). It has not been commercially developed, but its populations can be conserved by avoiding pesticides.

Tachinid flies parasitize *H. zea* in Kentucky hemp fields. Falcon-Brindis *et al.* (2022) found that 45% of *H. zea* larvae had tachinid eggs externally attached to their bodies (typically 1–5, but up to 22). Parasitized or unparasitized larvae (no visible eggs) were placed individually in plastic cups (25°C, 20% RH, photoperiod 12 L:12 D), and fed *ad libitum* with leaves and buds. Parasitized *H. zea* larvae died before pupating into moths at a higher rate (45.9%) than unparasitized larva (19.1%).

Adult tachinids emerging from *H. zea* were identified as *Winthemia rufopicta* (50% of larvae) and *Lespesia aletiae* (40% of larvae), with both tachinids emerging from a single host in 10% of cases. Adults emerged from larvae (90%) or pupae (10%). Larvae recognized egg-laying adult flies as enemies and assumed defensive postures (Fig. 5.7). Falcon-Brindis *et al.* (2022) also posted videos of the tachinids attacking *H. zea* larvae, available as supplemental files attached to their online publication.

Bacillus thuringiensis (Bt) *kurstaki* strain was traditionally used to control budworms, armyworms, and other lepidopteran pests. Bt *kurstaki* produces endotoxins in the Cry1A toxin family. Transgenic Bt maize (Cry1Ab, Yieldguard™) and cotton (Cry1Ac, Bollgard®) were both released in 1996. Predictably, widespread cultivation of these crops has resulted in Bt *kurstaki*-resistant pest populations.

Britt and colleagues tested several Bt *kurstaki* products (Thuricide®, Gemstar®, DiPel®, Javelin®, Deliver®, Crymax®), and they exerted only moderate efficacy against corn earworm (Britt and Kuhar, 2019, 2020; Britt *et al.*, 2021a). They achieved better control with products containing Bt *aizawai* (XenTari®, Agree®) which expresses additional endotoxins (Cry1C, Cry1D). Similarly, infestations of fall armyworm (*S. frugiperda*) were better controlled with Bt *aizawai* than Bt *kurstaki* (Salami, 2020).

Britt and colleagues also tested the *Helicoverpa zea* nuclear polyhedrosis virus (HzNPV, Gemstar®) against corn earworm. By itself, Gemstar® exerted only moderate efficacy (Britt and Kuhar, 2020). Better control was achieved with another HzNPV product, Heligen®. This control was statistically equivalent to a combination of Gemstar® and the fungus *Beauveria bassiana* strain GHA (Britt *et al.*, 2021a).

Pulkoski and Burrack (2023) tested several biocontrols against naturally occurring mixed infestations of *H. zea* (79%) and *C. virescens* (21%). Damage rating in untreated controls (1.6 on a 0–3 scale) was not statistically different than any treatments, although two came close: Gemstar® + *Beauveria bassiana* GHA (rating 0.8), and Heligen® + XenTari® (rating 0.8). Gemstar® has demonstrated efficacy against other budworms (*H. armigera*, *H. viriplaca*) in other crops. NPV viruses for armyworms are available (e.g., *Spodoptera exigua* NPV, *Autographa californica* NPV, *Mamestra brassicae* NPV); see section 20.6.

5.3.3 Chemical control

Britt and Kuhar (2019) tested Closer®, a xylem-mobile systemic nAChR agonist, and Besige®, which combines a ryanoid and a pyrethroid. They got marginally better results than with a Bt *kurstaki* product. Britt and Kuhar (2020) tested PyGanic® (botanical pyrethrins) and Entrust® (spinosad)—both were highly effective, statistically better than Bt *kurstaki* and Bt *aizawai* products. Britt *et al.* (2021a) tested Entrust® (spinosad) and Coragen® (chlorantraniliprole, targeting ryanodine receptors like caffeine). Both were highly effective, statistically better than any of the bacterial, viral, or fungal products listed above. Many budworm and armyworm species are resistant to pyrethrum and pyrethroids. Historically, Cherian (1932) sprayed plants with lead arsenate to control *H. armigera*; he estimated that 100 g of sprayed *gañjā* contained 2.0 mg of arsenic residue, "allowed as a medical dose."

Fig. 5.7. *Winthemia rufopicta* attacking *H. zea* larva (courtesy Armando Falcon-Brindis)

5.4   Hemp flea beetles

Beetles are the most common animals on Earth, and the most common beetles on *Cannabis* are flea beetles (Family Chrysomelidae). Hemp flea beetle, *Psylliodes attenuatus*, is especially damaging to flowering tops, but it also chews leaves and attacks roots and seedlings. It caused one of the first epidemics ever recorded on hemp: in 1843, flea beetles in Kursk Oblast, Russia, inflicted 15,000 Rubles in damage. The same season, flea beetles badly damaged crops in neighboring Oryol Oblast (Köppen, 1880).

Trotuş and Naie (2008) surveyed entomofauna in Romanian hemp for three seasons, and they considered *P. attenuatus* the worst pest—an average density of 113 adults/m². Fedorenko *et al.* (2016) considered *P. attenuatus* the most dangerous pest in Ukraine. In a sweep-net study conducted over several years, they collected 117 species of insects, and 80.8% of individuals were *P. attenuatus*. They showed that *P. attenuatus* is getting worse, due to global warming. They ran statistics on crop reports from 2000 to 2014, which correlated ($r = 0.72$) with hydrothermal coefficients in Ukraine.

Due to the damage that *P. attenuatus* can inflict, with feeding largely restricted to *Cannabis*, Mohyuddin and Scheibelreiter (1973) recommended the species for eradicating illegal marijuana. However, *P. attenuatus* has long been known to attack hops (Targioni Tozzetti. 1888), and has emerged as a serious pest of that crop in Europe (Mumm *et al.*, 2017). *P. attenuatus* has not been reported in North America. Pierce (1917) included *P. attenuatus* in a list of dangerous insects that would cause serious damage if accidently imported into the USA.

5.4.1   Signs and symptoms

Larvae and adults feed on plants, but adults cause more damage. Adults emerge from pupation in August, and feed on leaves and flowering tops. They gnaw holes in leaves (Fig. 5.8), and sometimes completely skeletonize leaves. Late-season beetles selectively infest female tops; they feed on flowers (Leprieur, 1864) and seeds (Silantyev, 1896). They nestle between developing seeds and their enclosing bracts; the seeds become wrinkled and do not germinate (Girard, 1881). Sometimes nearly the entire seed is eaten, with only the shell

Fig. 5.8. *Psylliodes attenuatus*: **A.** damage on foliage (Clarke); **B.** adult (courtesy Vera Vokotrub)

remaining (Fedorenko *et al.*, 2016). Beetles may bite through the flower base, terminating seed development (Dmitriev, 1935). Beetles deep in the inflorescence may sever the axil, leading to necrosis of the entire flowering apex.

Adults feed on tops until autumn, then burrow into soil for the winter. They stir back into activity in the spring, and feed on young seedlings. Damage to emerging hemp seedlings is a critical period of potential crop loss (Fedorenko *et al.*, 2016). Gnawing holes in cotyledonous leaves in dry conditions may kill the seedlings. Damage to seedlings reduces crop yields at the end of the season—stalk losses of 17.5–26.1%, and seed losses of 23.5–41.1% (Dmitriev, 1935). Guo *et al.* (2020) reported flea beetles feeding on seeds sown in soil before they germinated.

Adults mate in April, and females lay eggs in soil. Eggs hatch into grubs that bore into roots and the crown (root–stalk junction). Older grubs move back into the soil and feed externally on root hairs. Tölg (1913) said grubs are hard to see because they are very light-shy, and crawl away immediately when uncovered. Root damage is minimal compared with the devastation caused by adults, but may decrease plant growth.

5.4a.   Psylliodes attenuatus (Hoffmann) 1803

≡*Haltica attenuata* Hoffmann 1803, ≡*Psylliodes attenuata* Koch 1803 (according to various authors), ≡*Altica attenuata* Illiger 1807 (according to various authors), =*Macrocnema picicornis* Stephens 1831, =*Psylliodes vicina* Redtenbacher 1874, =*Psylliodes japonica* Jacoby 1885, =*Psylliodes attenuata* var. a. *elytris unicoloribus* Foudras 1860, =*Psylliodes attenuata* var. b. *thorace posticè irregulariter bidepresso* Foudras 1860, =*Psylliodes attenuata* var. A Weise 1888, =*Psylliodes attenuata* forma *brunneotestacea* Heikertinger 1926, *Psylliodes attenuatus* var. *picicornis* Tölg 1938, =*Psylliodes attenuata* forma *coerulea* Král 1947
Taxonomic notes: Entomologists across Eurasia repeatedly "discovered" the hemp flea beetle, and assigned different Latin names. *Psylliodes attenuatus* is based on the basionym *Haltica attenuata*, published in *Entomologische Hefte*. The title page of that text listed four authors: Hoffmann, Koch, Müller, and Linz. Some entomologists, not knowing which author described *Haltica attenuata*, cited the text, *Entomologische Hefte*, and not a specific author (e.g., Duftschmid, 1825; Dejean, 1837; Sturm, 1843; Redtenbacher, 1848; Maehler, 1850; Perris, 1852; Calwer, 1858; Foudras, 1860; Sella, 1864; Łomnicki, 1868).

Illiger (1807) suggested the author was Koch. Due to Illiger's oversized influence, many researchers accepted *Psylliodes attenuata* Koch. However, Wilhelm Koch (1771–1849) was a botanist and not an entomologist. Other researchers recognized Johann Jacob Hoffmann (1767–1814) as the author of *Haltica attenuata* (e.g., Du Val and Migneaux, 1854; Leprieur, 1864; de Marseul, 1866; Lethierry, 1871). Hoffmann was an entomologist. Most recently Bousquet (2016) recognized Hoffmann as the author, and we do too.

Two spellings are widespread, *P. attenuata* and *P. attenuatus*. Latin is the language of zoological nomenclature, so the species epithet must grammatically agree in gender with the genus name. Thus, when *Haltica attenuata* is transferred to the genus *Psylliodes*, the species epithet becomes *attenuatus*. Targioni Tozzetti (1888) first used that spelling. Most recent monographs use *attenuatus* (Aslan *et al.*, 1999; Warchałowski, 2000; Nadein, 2010), but Takizawa (2005) stuck with *P. attenuata*.

A species in Japan, *Psylliodes japonica* Jacoby, is recognized as a synonym of *P. attenuatus*. *Psylliodes* (*Macrocnema*) *apicalis* Stephens was synonymized under *P. attenuatus* by Spaar *et al.* (1990) and Nadein (2007), but not by Warchałowski (2000). Two species epithets, *vicina* and *picicornis*, were attributed a bronze tone imparted by tiny hairs on elytrae according to Heikertinger (1913), which he synonymized. Later he erected a superfluous name for an "aberrant specimen" with a brown-ish-yellow-red body, *P. attenuata* forma *brunneotestacea* (Heikertinger, 1926). Other subspecies or varieties of *P. attenuatus* have been erected, which we placed in synonymy under the species name.

Description: *Adult* body shape is oval in outline, short ("*attenuatus*") compared with most flea beetles, moderately convex, 1.6–2.6 mm long, males smaller than females. Color reflecting off elytrae looks black at a distance, but metallic green up close. Elytrae have tiny hairs that are colored bronze-gray, which impart a bronze tone (Fig. 5.8). The adult's head exhibits sharp outgrowths, crossed in an "x" between the eyes (Fig. 5.9). Antennae are long, half the body length. A line where the hardened forewings meet (elytral suture or stria) is deep and distinct, and the forewings are covered by longitudinal lines of fine punctuations. *Larvae* (grubs) are elongate–cylindrical, white with tiny spots, 3.5–4.5 mm long (Fig. 5.9). The head is brownish, round. The thorax has three pairs of legs. The abdomen has conspicuous bristles. *Pupae* are the length of adults; pupae start pearly white but slowly darken, beginning with the eyes, followed by other parts of the head, legs, and finally the elytra. *Eggs* are suboval, pale yellow, deposited singly around plant roots near the soil line.

5.4.2 Differential diagnosis

Eurasian *P. attenuatus* has a nearly identical sister species in North America, *Psylliodes punctulata*, that feeds on hemp (see section 7.4c). Like the similarity between *Helicoverpa armigera* and *H. zea*, it's largely geography that tells them apart. We found *COI* barcodes for *P. attenuatus* (GenBank KM440790.1) and *P. punctulata* (GenBank MW254880.1), and they differed by 15%—clearly different species.

Other foliage-infesting flea beetles must be differentiated, such as *Phyllotreta nemorum* (see section 7.4d). The aforementioned epidemics in Kurst and Oryol in 1843 were attributed to *Altica oleracea* L. and *Psylliodes hyoscyami* L., but Köppen (1880) noted these species are not hemp pests. The "*Altica oleracea*" epidemic spread from nettles to hemp in the springtime, which is consistent with the *P. attenuatus* life cycle. Species in other beetle families may infest flowering tops, such as the ubiquitous Japanese beetle, *Popillia japonica* (see section §7.4a). They're six times the size of *P. attenuatus*.

5.4.3 Life history and host range

Adults overwinter in the soil, and can survive when air temperatures dive to –25°C (Fedorenko *et al.*, 2016). Adults emerge at the end of March to April, when topsoil temperature reaches 12–16°C. Often this is before the emergence of hemp seedings. In the absence of hemp seedlings, *P. attenuatus* feeds on hops and nettle plants. After hemp seedlings emerge, flea beetles attack them (Bertolini, 1861). "As soon as seedlings are born they fear flea beetles" (Pichat, 1866). Goureau (1863) wrote, "by the time the hemp starts to grow and has reached 5 to 10 centimeters in height, we often notice that the young leaves are riddled with small round holes."

Adults migrate between 2 and 4 km (Fedorenko *et al.*, 2016). They mate in April, and egg-laying commences 10 days later. Eggs are laid in soil near host plants, between the soil line and a depth of 8 cm. Oviposition extends from late April to the end of July. Females lay an average of 2.6 eggs/day for a total of 55 eggs (Angelova, 1968). Dmitriev (1935) claimed up to 325 eggs are possible, depending on food availability and environmental factors. Eggs develop in 6–20 days, which depends on temperature and soil moisture, with 40% soil moisture optimal for development (David'yan, 2009).

Larvae have three instars and develop in 21–42 days, depending on temperature and soil moisture. The depth at which larvae feed varies with the nature of the soil and moisture conditions. They burrow deeper in loose, sandy soil than in loamy soil (Tölg, 1913). Pupation occurs in the soil and lasts 6–25 days, depending on conditions. Adults emerge in August. Diapause begins in September–October.

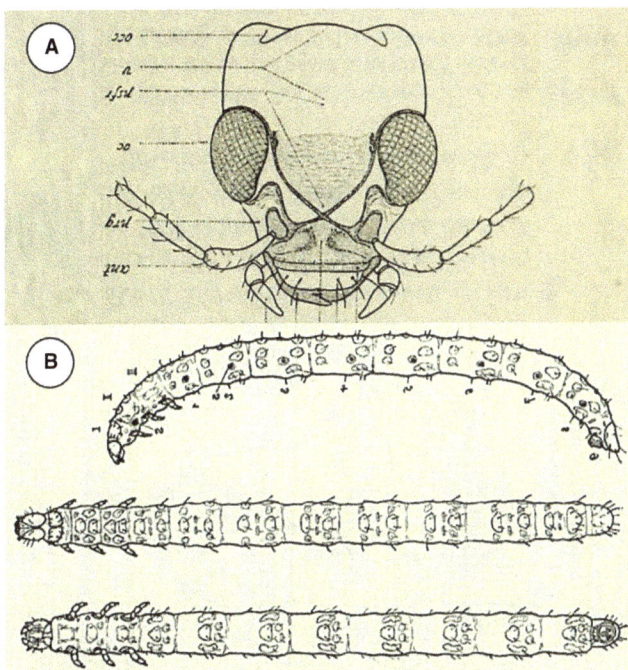

Fig. 5.9. *Psylliodes attenuatus*: **A**. head morphology (Heikertinger, 1913); **B**. three views of larva (Tölg, 1913)

Field infestations are patchy, with clusters of insects in some areas, and other areas free of insects. Margins of fields are particularly damaged (Shilenkov and Tolstonogova, 2006). Populations vary seasonally, with peak damage in late spring/ early summer (in seedlings) and the end of summer (during flowering). Fedorenko *et al.* (2016) proposed a pest infestation index (Table 5.2).

The geographic range of *P. attenuatus* extends from Ireland, Great Britain, France, and Spain in the west, south to Sardinia and the Greek islands, through Middle and Eastern Europe, the Caucasus, Turkey, Kazakhstan, Uzbekistan, Kyrgyzstan, across Siberia to Vladivostok, and all of Central Asia (Heikertinger, 1926; David'yan, 2009; Nadein, 2010). In Western Siberia and the Russian Far East, *P. attenuatus* is a "massive pest" of *Cannabis ruderalis* (Shilenkov and Tolstonogova, 2006). In northern China, Korea, and Japan, *P. attenuatus* was called *P. japonica*.

Guo *et al.* (2020) studied the genetic diversity of *P. attenuatus* in China. They used the *COI* sequence to genotype 281 specimens, collected in six provinces. Many haplotypes were restricted to specific provinces, whereas other haplotypes were distributed widely—indicative of long-distance dispersal, likely from humans shipping contaminated seeds. Their analysis suggested that *P. attenuatus* is an ancient species in China, with Shāndōng, Húnán, and especially Jílín province as centers of diversity.

P. attenuatus can only complete its development on hemp and hops, and perhaps nettles (Fedorenko *et al.*, 2016). Adults in spring commonly feed on nettles before *Cannabis* seedlings emerge (Silantyev, 1896). According to Dmitriev (1935) and Lukash (1938), adults lacking nettle and hemp seedlings can survive on weeds (*Chenopodium* spp., *Arctium* spp.) or seedlings of other crops (*Solanum tuberosum*, *S. lycopersicum*, *Beta vulgaris*, and *Phaseolus* spp.).

5.4.4 Cultural and mechanical control

Eliminate feral *Cannabis*, hops, and nettles (*Urtica dioica*) from the area. Gianpaolo Grassi provided a photo of badly infested hemp seedlings next to a fencerow full of nettles (Fig. 5.10). Eliminate alternative hosts in autumn and again in the spring when seedlings emerge. In badly infested crops, all crop residues and alternative hosts should be collected after harvest, followed by burning after the first frost (Fedorenko *et al.*, 2016). Alternatively, nettles can serve as a trap crop: eliminate all but a few plants. Flea beetles crowd on the remaining nettles, which can be burned or sprayed with pesticides.

Adults overwinter in the soil, so post-harvest plowing exposes them to the elements. Careful spring tillage prior to sowing seeds

accomplishes three goals: eliminating weeds, killing adults, and prepping the seed bed for robust seedling growth. Fertilize soil before sowing seeds, to accelerate seedling development so they can escape serious harm (David'yan, 2009).

European farmers sow seeds early, resulting in seedlings that are older, hardier, and able to withstand flea beetles emerging in the spring (Camprag *et al.*, 1996). Higher seedling rates dilute the damage caused by emerging adults. Sow all fields at the same time, rather than staggered over a period of time (Fedorenko *et al.*, 2016). At the other end of the season, early harvest of crops may escape late-season flea beetles, cut short their feeding time (thus reduce body fat stores), and destroy adults before they move to the soil. This is done in China (Clarke, 1995) and Ukraine (Fedorenko *et al.*, 2016).

Continuous cropping of hemp builds up populations of *P. attenuatus* (Fedorenko *et al.*, 2016). For crop rotations, Camprag *et al.* (1996) suggested planting a new hemp crop at least 0.5–1 km away from the previous year's hemp crop. Farmers in China rotate with other crops for one or two years (Guo *et al.*, 2020).

Use resistant varieties. Dmitriev (1935) reported that Chinese, Japanese, and Hungarian varieties were more resistant than Central Russian varieties. Serebryakova and Sizov (1940) also reported a latitudinal gradient: Northern Russian landraces were the most susceptible, followed by Middle Russian forms, and most resistant were Japanese, Turkish, and Far Eastern hemp, as well as Central Asian drug-type plants.

Research in the 1930s demonstrated resistance to *P. attenuatus* in two American cultivars, 'Kentucky' and 'Chinamington' (Grumm-Grzhimailo, 1962). These were Lyster Dewey's hybrids of Chinese ancestry. Bócsa (1954) created a *P. attenuatus*-resistant hybrid by crossing susceptible 'Kompolti' with a resistant Chinese landrace. Bócsa and Beke (1956) reiterated that two Chinese landraces cultivated in Hungary were resistant to *P. attenuatus*. Two Russian cultivars, 'Southern

Table 5.2. Infestation severity index for hemp flea beetles

Level 1	Damage to 5% of leaf surface
Level 2	Damage to 6–25% of leaf surface
Level 3	Damage to 26–50% of leaf surface
Level 4	Damage to 51–75% of leaf surface
Level 5	Damage to >76% of leaf surface

Fig. 5.10. Hemp and nettles (G. Grassi)

Krasnodar' and 'JUS-6' were selected for resistance to *P. attenuatus* in the 1950s (de Meijer 1999). Fedorenko *et al.* (2016) evaluated resistance in five Ukrainian cultivars, using a pest infestation index (Table 5.2). Least damaged cultivars were 'Glana' (mean point score 0.77) and 'Zolotoniski 15' (0.82). They were less affected than 'Glesia' (0.89), 'Glukhivskaya 51' (0.93), and 'Victoria' (1.01).

Flea beetles readily jump when disturbed; Heikertinger (1913) dissected the muscles and illustrated the mechanism. This behavior can be used against them. Jablonowski (1905) constructed a hollow cone, and coated the inside with sticky tar. Holding the cone with one hand, his other hand shook the plant, and the beetles jumped into the cone to their death. Heikertinger (1913) made the first European mention of "Tree Tanglefoot," an American product. He wrapped tanglefoot around a large, flat board, and killed flea beetles the same way as Jablonowski. Heikertinger perhaps got the tanglefoot idea from Parker (1910), who coated cymbals with tanglefoot to trap *Psylliodes punctulata*.

Flea beetles are attracted to white sticky traps, useful for monitoring pest populations. *P. attenuatus* is attracted to the terpenoids linalool and (E)-α-bergamotene, which can be added to traps (Mumm *et al.*, 2017). Monitor meteorological data. Pest density depends on weather conditions during hibernation, and on soil humidity during egg and larva stages (David'yan, 2009). *P. attenuatus* replication and crop damage is enhanced by warm, dry weather (Flachs, 1936; Fedorenko *et al.*, 2016). Adults are most active on warm sunny days; in overcast or wet weather they retreat to soil cracks, beneath clods of earth, or on the lower surface of leaves.

Cultural methods alone reduced damage from *P. attenuatus* but did not eliminate the pests (Pavlov, 1950). He estimated that complete elimination of crop debris killed only 10% of overwintering adults, thus he was a proponent of biological control.

5.4.5 Biological control

Soil-borne controls of flea beetle grubs include nematodes, *Heterorhabditis* and *Steinernema* spp. Grubs are killed by *Bacillus thuringiensis* var. *tenebrionis*. Leslie (1994) claimed that Bt-t also kills adult flea beetles. The fungus *Metarhizium anisopliae* has good efficacy against many Coleopteran species. *Beauveria bassiana* is a fungus with a broad host range, including Coleopterans. *M. anisopliae* and *B. bassiana* work best as soil drenches, but they can be sprayed on foliage in humid conditions.

Several braconid wasps parasitize *P. attenuatus*. Dmitriev (1935) identified *Perilitus lobili* Retzius and *Perilitus (Townesilitus) bicolor* Wesmael in Russia. Together they parasitized up to 60% of

P. attenuatus adults. Fedorenko *et al.* (2016) encountered these species in Ukraine. They noted that biological features limited wasp viability under natural conditions, and they raised doubts about the feasibility of commercial development. However, *T. bicolor* has been introduced into Canada for biological control (Wylie, 1988). Other *Townesilitus* and *Microcronus* spp. have parasitized *P. attenuatus*, see Table 20.6.

5.4.6 Chemical control

Heaviest infestations of *P. attenuatus* occur around the margins of hemp fields (Shilenkov and Tolstonogova, 2006). Aim chemical controls around the edges. Spinosad and cryolite kill a variety of flea beetles. Pyrethrins, ryania, and sabadilla are less effective. Historically, Heikertinger (1913) recommended Bordeaux mixture as a feeding deterrent and repellant. Dehtyaryov (1931) recommended tobacco dust and calcium arsenate. Combating hemp flea beetles with aerial dusting of calcium arsenate, using a classic Polikarpov Po-2A crop duster plane, appeared in the 1954 film Трубчевские коноплеводы (*Trubchevskie Hemp Growers*), available on YouTube.

Fedorenko *et al.* (2016) said that farmers in the 1980s and 1990s used diazinon, phosalone, and methyl parathion (all now banned), as well as pyrethroids. They got good results by coating hemp seeds with Cruiser® (thiamethoxam). They considered seed treatment more environmentally friendly than crop spraying, but thiamethoxam is a systemic neonicotinoid. Trotuş and Naie (2008) also tested seed treatment with a neonicotinoid, Gaucho® (imidacloprid), and the carbamate Oncol® (benfurcarb)—all banned on hemp in the USA, for good reasons (see section 21.8).

Farmers in China spray unnamed pesticides on emerging seedlings to prevent *P. attenuatus* from killing them (Guo *et al.*, 2020). Yepsen (1976) claimed that wood ashes repel flea beetles, but Heikertinger (1913) said wood ashes were useless. Heikertinger emphasized the fact that any pesticide must be reapplied frequently, practically daily, to kill beetles arriving from surrounding areas and infesting new foliage and flowers.

(Prepared by J. McPartland)

References

Ajayi OS, Samuel-Foo M. 2021. Hemp pest spectrum and potential relationship between *Helicoverpa zea* infestation and hemp production in the United States in the face of climate change. *Insects* 12: e940.

Akotsen-Mensah C. 2021. "Industrial hemp IPM updates in Missouri," p. 16 in *Second Annual Science of Hemp Conference*. Auburn University, Auburn, Alabama.

Alcock A, *ed.* 1903. Notes on insect pests from the entomological section, Indian Museum. *Indian Museum Notes* 5: 103–202.

Angelova R. 1968. Characteristics of the bionomics of the hemp flea beetle, *Psylliodes attenuatus* Koch. Растениевъдни науки 5(8): 105–114.

Arey NC, Lord NP, Davis JA. 2022. Evaluation of hemp (*Cannabis sativa*) (Rosales: Cannabaceae) as an alternative host plant for polyphagous Noctuid pests. *Journal of Economic Entomology* 115: 1947–1955.

Arcozzi L, Gelosi A, Pollini A. 1980. *Chloridea viriplaca* Hfn. (= *Heliothis dipsacea* L.): una nottua dannosa ai medicai. *Informatore Fitopatologico* 30: 21–22.

Aslan I, Gruev B, Özbek H. 1999. A preliminary review of the subfamily Alticinae (Coleoptera, Chrysomelidae) in Turkey. *Turkish Journal of Zoology* 23: 373–414.

Bagheri MR, Isfahani MN. 1989. خشخ ي شي از نون و ت افآ نون فوپدنايان گياهاي ديفم ناياياپدنان گياهاي ويورايي و مرتعي در اصفهانان. *Journal of Entomological Research, Islamic Azad University* 3: 119–131.

Bajpai NK, Sehgal VK. 1993. Oviposition preferences, larval development, and survival of *Helicoverpa armigera* (Hübner) on chickpea and weed hosts at Pantnagar, India. *International Chickpea Newsletter* 29: 15–17.

Ballard E. 1920. Note on *Heliothis* (*Chloridea*) *obsoleta*, Fb., as a pest of cotton. *Agricultural Journal of India* 15: 462–464.

Barbosa TAN, Mendes SM, Rodrigues GT, *et al.* (4 other authors). 2016. Comparison of biology between *Helicoverpa zea* and *Helicoverpa armigera* (Lepidoptera: Noctuidae) reared on artificial diets. *Florida Entomologist* 99: 72–76.

Berdrow H, *ed.* 1907. Lebensrätsel im Pflanzenreich. *Illustriertes Jahrbuch der Naturkunde* 5: 131–158.

Bertolini G. 1861. Descrizione di una nuova malattia della canepa nel Bolognese. *Memorie della Accademia delle Science dell'Istituto di Bologna* 12: 289–303.

Blunck H. 1920. Die niederen tierischen feinde unserer gespinstpflanzen. *Illustrierte landwirtschaftliche Zeitung* 40: 259–260.

Bócsa I. 1954. Kender heterózis-nemesítésieredmények. *Növénytermelés* 3: 301–316.

Bócsa L, Beke F. 1956. Néhány külföldi kenderfajta hazai termesztési és nemesítési értéke. *Növénytermelés* 5: 39–50.

Bolt M, Beckerman JL, Couture JJ. 2021. Agronomic management of industrial hemp alters foliar traits and herbivore performance. *Arthropod-Plant Interactions* 15: 137–151.

Bousquet Y. 2016. Litteratura Coleopterologica (1758-1900): a guide to selected books related to the taxonomy of Coleoptera with publication dates and notes. *ZooKeys* 583: 1–776.

Britt KE, Kuhar TP. 2019. Evaluation of insecticides to control insect pests on hemp in Virginia, 2018. *Arthropod Management Tests* 44(1): 1–2.

Britt KE, Kuhar TP. 2020. Laboratory bioassays of biological/organic insecticides to control corn earworm on hemp in Virginia, 2019. *Arthropod Management Tests* 45(1): 1–2.

Britt KE, Kuhar TP, Cranshaw W, *et al.* (20 other authors). 2021b. Pest management needs and limitations for corn earworm (Lepidoptera: Noctuidae), an emergent key pest of hemp in the United States. *Journal of Integrated Pest Management* 12(1): 34; 1–11.

Britt KE, Reed TD, Kuhar TP. 2021a. Evaluation of biological insecticides to managed corn earworm in CBD hemp, 2020. *Arthropod Management Tests* 46(1): 1–2.

Britt KE, Taylor S, Kuhar T. 2019. *Corn earworm, Helicoverpa zea, a pest of hemp, Cannabis sativa, in Virginia.* Virginia Cooperative Extension ENTO-328NP. Virginia Tech, Blacksburg, Virginia.

Bruner L. 1892. "Report of the entomologist," pp. 240–309 in *Annual Report, Nebraska State Board of Agriculture for the year 1891.* Lincoln, Nebraska.

Calwer CG. 1858. *Käferbuch.* Krais & Hoffmann, Stuttgart.

Cameron AM. 1902. Our insect pests. *Calcutta Review* 114: 291–300.

Camprag D, Jovanic M, Sekulic R. 1996. Štetočine konoplje i integralne mere suzbijanja. *Zbornik Radova* 26/27: 55–68.

Ceapoiu N. 1958. *Cinepa, studiu monografic.* Editura Academiei Republicii Populare Romine. Bucharest.

Chandra R. 2004. Status of medicinal plants with respect to infestation of insect pests in and around Chitrakoot, District-Satna (M.P.). *Flora and Fauna (Jhansi)* 10(2): 88–92.

Cherian MC. 1932. Pests of ganja. *Madras Agricultural Journal* 20: 259–265.

Clarke RC. 1995. Hemp (*Cannabis sativa* L.) cultivation in the Tai'an district of Shandong province, Peoples Republic of China. *Journal of the International Hemp Association* 2(2): 57, 60–65.

Clausen CP. 1931. *Insects injurious to agriculture in Japan.* USDA Circular no. 168, Washington DC.

Comstock JH. 1879. *Report upon cotton insects.* Government Printing Office, Washington DC.

Conner K, Kesheimer K, Sikora E, Kemble J. 2020. "Diseases and insects of hemp in Alabama," p. 21 in Gauthier N, ed. *Science of hemp: production and pest management.* University of Kentucky Agricultural Experiment Station no. SR-112. Lexington, Kentucky.

Cotes EC. 1893. A conspectus of the insects which affect crops in India. *Indian Museum Notes* 2(6): 145–176.

Cranshaw WS, Schreiner M, Britt K, Kuhar TP, McPartland JM, Grant J. 2019. Developing insect pest management systems for hemp in the United States: a work in progress. *Journal of Integrated Pest Management* 10(1): 26; 1–10.

Datta M, Chakraborti M. 1983. On a collection of flower flies (Diptera: Syrphidae) with new records from Jammu and Kashmir. *Records of the Zoological Survey of India* 81: 237–253.

David'yan GE. 2009. *Psylliodes attenuatus* (Koch)—hop flea beetle. *Interactive Agricultural Ecological Atlas of Russia and Neighboring Countries.* Available at: http://www.agroatlas.ru/en/content/pests/Psylliodes_attenuatus/index.html

De Geer C. 1776. *Memoires pour servir a l'histoire des insects, tome sixieme.* Hesselberg, Stockholm.

de Marseul SA. 1866. *L'abeille mémoires d'entomology, tome III.* Deyrolle, Paris.

de Meijer EPM. 1999. "*Cannabis* germplasm resources," pp. 133–151 in: Ranalli P, ed. *Advances in Hemp Research.* Haworth Press, New York.

Dehtyaryov M. 1931. Шкідники конопель та боротьба з ними. State Agricultural Inspectorate, Moscow.

Dejean PFMA. 1837. *Catalogue des Coléoptères, troisième édition.* Méquignon-Marvis père et fils, Paris.

Dippenaar MC, du Toit CLN, Botha-Greeff MS. 1997. Response of hemp (*Cannabis sativa* L.) varieties to conditions in Northwest Province, South Africa. *Journal of the International Hemp Association* 3(2): 63–66.

Dmitriev GV. 1935. Конопляный блошак (*Psylliodes attenuata* Koch.) в условиях правобережья Куйбышевского края. *Защита растений* 1935(5): 91–107.

Dong K, Dong Y, Luo YZ. 2004. 黑带食蚜蝇成虫室内的饲养方法. *Entomological Knowledge* 41(2): 157–160.

Du Val J, Migneaux J. 1854. *Genera des Coléopteres d'Europe.* Chez les auteurs, Paris.

Duftschmid CE. 1825. *Fauna Austriae, Dritter Theil.* Akademie Buchhandlung, Linz und Leipzig.

Dziock F. 2002. *Überlebensstrategien und Nahrungsspezialisierung bei räuberischen Schwebfliegen (Diptera, Syrphidae).* UFZ-Bericht, Leipzig.

Falcon-Brindis A, Stiremann JO, Viloria ZJ Villanueva RT. 2022. Parasitism of corn earworm, *Helicoverpa zea* (Boddie) (Lepidoptera: Noctuidae), by tachinid flies in cultivated hemp. *Insects* 13(6): e519.

Fedorenko VP, Kabanetz VV, Kabanets VM. 2016. *Шкідники конопель посівних.* National Academy of Agrarian Science, Kiev.

Ferri F. 1959a. *Atlante delle avversità della canapa.* Edizioni Agricole, Bologna, Italia.

Flachs K. 1936. Krankheiten und schädlinge unserer gespinstpflanzen. *Nachrichten über Schädlingsbekämpfung* 11: 6–28.

Flanders SE. 1930. Mass production of egg parasites of the Genus *Trichogramma.* Hilgardia 4: 465–501.

Fletcher TB. 1917. *Report of the Proceedings of the Second Entomological Meeting, Pusa.* Superintendent Government Printing, Calcutta.

Forbes SA. 1904. *The more important insect injuries to Indian corn.* University of Illinois Agricultural Experiment Station Bulletin No. 95. Urbana, Illinois.

Girard M. 1881. M. Maurice Girard lit les communications suivantes. *Bulletin des séances de la Société Entomologique de France* 1881 (Supplément): 131–132.

Gistel J. 1856. *Die mysterien der europäischen insectenwelt.* Dannheimer, Kempten.

Goureau C. 1863. Insectes nuisibles aux céréales et aux plantes fourragères. *Bulletin de la Société Centrale de l'Yonne* 1863: 71–88.

Goureau C. 1866. *Les insectes nuisibles a l'homme, aux animaux et a l'economie domestiques, 2nd supplement.* V. Masson, Paris.

Grumm-Grzhimailo AG. 1962. *В поисках растительных ресурсов мира : некоторые научные итоги путешествий Н.И. Вавилова.* Publishing House of the Academy of Sciences of the USSR, Leningrad.

Guo LT, Gao F, Cheng Y, *et al.* (5 other authors). 2020. Mitochondrial COI sequence variations within and among geographic samples of the hemp pest *Psylliodes attenuata* from China. *Insects* 11: e370.

Harada T. 1930. 大麻の害患難を大麻象鼻貴職に就きて. *Konchū Sekai (Insect World)* 34: 118–123.

Hardwick DF. 1965. The corn earworm complex. *Memoirs of the Entomological Society of Canada* 97 (Suppl. S40): 5–247.

Heikertinger F. 1913. *Psylliodes attenuata* Koch, der Hopfen- oder Hanf-Erdfloh. II. Teil. Morphologie und Bionomie der Imago. *Verhandlungen der Zoologisch-Botanischen Gesellschaft in Österreich* 63: 98–138.

Heikertinger F. 1926. Bestimmungstabelle der Halticinengattung *Psylliodes* aus dem palaarctischen Gebiete. II. Die helltarbigen Arten. *Koleopterologische Rundschau* 12: 101–138.

Hill DS. 1983. *Agricultural insect pests if the tropics and their control, 2nd edn.* Cambridge University Press, Cambridge, UK.

Hoffmann JJ, Koch JDW, Linz JM, Müller PWJ. 1803. *Entomologische Hefte enthaltend Beiträge zur weitern Kenntniss und Aufklärung der Insektengeschichete, Zweites heft.* Friedrich Esslinger, Frankfurt am Main.

Hoffmann MP, Wilson LT, Zalom FG. 1991. Area-wide pheromone trapping of *Helicoverpa zea* and *Heliothis phloxiphaga* (Lepidoptera: Noctuidae) in the Sacramento and San Joaquin valleys of California. *Journal of Economic Entomology* 84: 902–911.

Howard RJ, Garland JA, Seaman WL, eds. 1994. *Diseases and pests of vegetable crops in Canada.* Entomological Society of Canada, Ottawa, Ontario.

Humphries T, Florentine S. 2019. Cultivation of low tetrahydrocannabinol (THC) *Cannabis sativa* L. cultivation in Victoria, Australia: do we know enough? *Australian Journal of Crop Science* 13: 911–919.

Illiger K. 1807. *Magazin für Insektenkund, Sechster Band.* Reichard, Braunschweig.

Jablonowski J. 1905. *Útmutatás a komlót pusztító állatok irtására.* Részvénytársaság Nyomdája, Budapest.

Janischevsky DE. 1924. Форма конопли на сорных местах в Юго-Восточной России (A form of cannabis in wild areas of south-eastern Russia). *Ученые записки Саратовского государственного университета имени Н. Г. Чернышевского (Scientific Notes of the Saratov State University named after N. G. Chernyshevsky)* 2(2): 3–17.

Janischevsky DE. 1925. Энтомохория у *Cannabis ruderalis* Janisch. *Proceedings of the Saratov Society of Naturalists* 1(2-3): 39–49.

Jansen PCM. 2008. "*Cannabis sativa* L.," pp. 135–142 in Schmetzer GH, Gurib-Fakim A, eds. *Plant resources of tropical Africa.* Prota Foundation, Wageningen, Netherlands.

Jia HR, Wang TL, Li XK, *et al.* (5 other authors). 2023. Pollen molecular identification from a long-distance migratory insect, *Spodoptera exigua,* as evidenced for its regional pollination in eastern Asia. *International Journal of Molecular Sciences* 24: e758.

Kashyap RK, Kennedy GG, Farrar RR. 1991. Behavioral response of *Trichogramma pretiosum* and *Telenomus sphingis* to trichome/methyl ketone mediated resistance in tomato. Journal of Chemical Ecology 17: 543–556.

Khamukov VB, Kolotilina ZM. 1987. Расширяем применение биометода (We are extending utilization of the biological method). *Защита растений* 1987(4): 30–31.

Kirchner O. 1906. "Hanf, *Cannabis sativa* L.," pp. 319–323 in *Die Krankheiten und Befehädigungen uhferer landwirtschaftlichen Kulturpflanzen.* E. Ulmer, Stuttgart.

Köppen FP. 1880. *Die schädlichen Insekten Rußlands.* Kaiserlichen Akademie der wissenschaften, St Petersburg.

Kozhanchikov IV. 1956. *Fauna SSSR: Nasekomyl Cheshuekrylye* 3(2): 444. Zoologicheskii Institut Akademii Nauk, SSSR.

KRC Times 2019. *Insect outbreak caused major crop destruction in Mizoram.* Available at: https://www.krctimes.com/news/insect-outbreak-caused-major-crop-destruction-in-mizoram/

Kumar A, Kurl SP, Ramkishore. 2006. Alternate food plants of tomato fruit borer *Helicoverpa armigera* (Hubner) in Meerut. *Plant Archives* 6: 769–770.

Lago PK, Stanford DF. 1989. Phytophagous insects associated with cultivated marijuana, *Cannabis sativa,* in northern Mississippi. *Journal of Entomological Science* 24: 437–445.

Leprieur M. 1864. Notes sur quelques Coléopteres des environs de Colmar. *Bulletin de la Société d'Histoire Naturelle de Colmar* 5: 35–87.

Leslie AR. 1994. *Handbook of integrated pest management for turfgrass and ornamentals.* CRC Press, Boca Raton.

Lethierry L. 1871. Supplément au catalogue des Coléoptéres de la faune Belge. *Annales de la Société Entomologie de Belgique* 15: LIII–LX.

Li LY. 1994. "Worldwide use of *Trichogramma* for biological control in different crops," pp. 37-53 in Wajnberg E, Hassan SA, eds. *Biological control with egg parasitoids.* CAB International, Wallingford, UK.

Li SQ, Rahmann H. 1997. Baumwoll-Schädlingsmanagement in China. I: Schädlingsspektrum. *Zeitschrift für Pflanzenkrankheiten und Pflanzenschutz* 104: 611–621.

Łomnicki AM. 1868. "Wycieczka na Czarnogérę," pp. 132-152 in *Sprawozdanie komisyi fizyograficznej c.k. towarzystwa naukowego Krakowskiego.* Drukarni Uniwersytetu Jagiellońskiego, Kraków.

Lukash II. 1938. Продолжительность жизни конопляной блохи в весенний период в условиях голода. *Лен и конопля* 1938(2): 6–9.

Maehler FJ. 1850. *Enumeratio Coleopterorum circa Heidelbergam indigenarum.* E. Mohr, Heidelberg.

Mah LM. 1923. *Hemp-Cannabis sativa: an agronomic study.* Master's thesis, University of California.

Malinowska D. 1973. Larwy bzygowatych (Diptera) w koloniach mszyc na niektórych roslinach uprawnych. *Polskie Pismo Entomologiczne* 43: 607–613.

Malinowska D. 1979. Communities of aphidophagous syrphids (Diptera, Syrphidae) in the Lublin region. *Memorabilia Zoologica* 30: 37–62.

Marshall GAK. 1917. *Psyche cannabinella. Review of Applied Entomology, Series A (Agriculture)* 5: 159.

Matov AY, Kononenko VS. 2012. *Трофические связи гусениц совкообразных чешуекрылых фауны России (Lepidoptera, Noctuoidea: Nolidae, Erebidae, Euteliidae, Noctuidae).* Dalnauka, Vladivostok.

Matsumura S. 1899. 日本害蟲篇. Shōkabō, Tōkyō.

Maxwell-Lefroy HM. 1906. *Indian insect pests.* Government Printing Office, Calcutta.

Maxwell-Lefroy HM. 1915. The control of white fly and soft scale. *The Gardeners' Chronicle* 58: 154.

McPartland JM, Rhode B. 2005. New hemp diseases and pests in New Zealand. *J Industrial Hemp* 10(1): 99–108.

McPartland JM, Clarke RC, Watson DP. 2000. *Hemp diseases and pests – management and biological control.* CABI Publishing, Wallingford, UK.

Meagher RL, Neussly GS, Nagoshi RN, Hay-Roe MM. 2016. Parasitoids attacking fall armyworm (Lepidoptera: Noctuidae) in sweet corn habitats. *Biological Control* 95: 66–72.

Meigen JW. 1822. *Systematische Beschreibung der bekannten Europäischen zweiflügeligen Insekten, dritter theil.* Schultz-Wundermann'schen Buchhandlung, Hamm.

Mitra B, Banerjee D. 2007. Fly pollinators: assessing their value in biodiversity conservation and food security in India. *Records of the Zoological Survey of India* 107: 33–48.

Moes J, Sturko A, Przybylski R. 1999. "Agronomic research on hemp in Manitoba," pp. 300–305 in Janick J, ed. *Perspectives on new crops and new uses.* ASHS Press, Alexandria, Virginia.

Mohyuddin AI, Scheibelreiter GK. 1973. *Investigations on the fauna of Papaver spp. and Cannabis sativa.* Annual Report, Commonwealth Institute of Biological Control, Switzerland Station, pp. 32–33.

Morawo T, Fadamiro H. 2016. Identification of key plant-associated volatiles emitted by *Heliothis virescens* larvae that attract the parasitoid, *Microplitis croceipes*: implications for parasitoid perception of odor blends. *Journal of Chemical Ecology* 42: 1112–1121.

Mumm R, van Tol RWHM, Weihrauch F. 2017. Elucidation of the role of volatile compounds in the chemical communication of the hop flea beetle *Psylliodes attenuata. Proceedings of the Scientific-Technical Commission, International Hop Growers' Convention* 2017, pp. 60–64.

Munir M, Leonberger K, Kesheimer K, *et al.* (14 other authors). 2023. Occurrence and distribution of common diseases and pests of US *Cannabis*: a survey. *Plant Health Progress.*

Nadein KS. 2007. A review of the genus *Psylliodes* Latreille (Coleoptera, Chrysomelidae) of the fauna of Russia and neighboring countries: I. a key to subgenera, species-groups, and species. *Entomological Review* 87(3): 330–360.

Nadein KS. 2010. A review of the genus *Psylliodes* Latreille (Coleoptera, Chrysomelidae) of the fauna of Russia and neighboring countries: II. An annotated list of species. *Entomological Review* 90(8): 1035–1074.

Nair KR, Ponnappa KM. 1974. *Survey for natural enemies of Cannabis sativa and Papaver somniferum.* Commonwealth Institute of Biological Control, India Station Report, pp 39–40.

Neunzig HH. 1964. The eggs and early-instar larvae of *Heliothis zea* and *Heliothis virescens* (Lepidoptera: Noctuidae). *Annals of the Entomological Society of America* 57: 98–102.

Noel P. 1917. Les ennemis du chanvre (*Cannabis* Tourn.). *Bulletin du Laboratoire Régional d'Entomologie Agricole, Rouen* 1917(2): 11–12.

Paclt J. 1976. Der Hanf (*Cannabis sativa*)–eine unbekannte futterpflanze von *Aglope infausta* (L.) (Lep., Zygaenidae)? *Anzeiger für Schädlingskunde, Pflanzenschutz, Umweltschutz* 49(8): 116.

Parihar SBS, Singh OP. 1992. Role of host plants in development and survival of *Heliothis armigera* (Hubner). *Bulletin of Entomology (New Delhi)* 33: 74–78.

Parker WB. 1910. The life history and control of the hop flea beetle. *USDA Entomology Bulletin* 82(4): 32–58.

Passerini C. 1832. *Osservazioni e notizie relative alle larve pregiudizievoli alla pianta del Gran turco (Zea mays).* [No printer listed], Florence.

Pavlov IF. 1950. О создании условий, повышающих эффективность деятельности паразитов конопляной блохи. *Зоологический журнал* 29(3): 15–18.

Perris E. 1852. Second excursion dans les grandes-landes. *Annales de la Société Linnéenne de Lyon* 1852: 145–216.

Pettit DA, Showalter VM, Abood ME, Cabral GA. 1994. Expression of a cannabinoid receptor in baculovirus-infected insect cells. *Biochemical Pharmacology* 48: 1231–1243.

Pichat CB. 1866. "La pianta della canapa," pp. 396–499 in *Istituzioni scientifiche e tecniche, ossia Corso teorico e practico de Agricoltura, Libri XIII, Volume Quinto.* Presso l'Union Tipografico-Editrice, Torino.

Pierce WD. 1917. *A manual of dangerous insects likely to be introduced in the United States through importations.* Government Printing Office, Washington DC.

Pogue MG. 2004. A new synonym of *Helicoverpa zea* (Boddie) and differentiation of adult males of *H. zea* and *H. armigera* (Hübner) (Lepidoptera: Noctuidae: Heliothinae). *Annals of the Entomological Society of America* 97: 1222–1226.

Pogue MG. 2013. Revised status of *Chloridea* Duncan and (Westwood), 1841, for the *Heliothis virescens* species group (Lepidoptera: Noctuidae: Heliothinae) based on morphology and three genes. *Systematic Entomology* 38 (3): 523–542.

Pulkoski M. 2022. *Assessing the impact of arthropod pests on industrial hemp (Cannabis sativa L) to inform management.* Master's thesis, North Carolina State University, Raleigh.

Pulkoski M, Burrack H. 2020. *Corn earworm and tobacco budworm in industrial hemp.* NC State Extension Factsheet. Available at: https://content.ces.ncsu.edu/corn-earworm-and-tobacco-budworm-in-industrial-hemp

Pulkoski M, Burrack H. 2023. Evaluating efficacy of biological and reduced risk pesticides against heliothine caterpillars in hemp, 2021. *Arthropod Management Tests* 48(1): 1–2.

Rao YR. 1928. *Administration report of the government entomologist for 1927–28.* Coimbatore Agricultural Research Institute, India.

Rataj K. 1957. Skodlivi cinitele pradnych rostlin. *Prameny literatury* 2:1–123.

Redtenbacher L. 1848. *Fauna Austriaca. Die Käfer.* Carl Gerold, Wien.

Riley CV. 1871. Third annual report on the noxious, beneficial and other insects of the State of Missouri. *Annual Report of the Missouri Board of Agriculture* 3: 157–158.

Riley CV. 1885. *On the cotton worm together with a chapter on the boll worm.* Fourth Report of the US Entomological Commission, USDA. Government Printing Office, Washington, DC.

Romaniszyn J. 1929. *Fauna motyli Polski (Fauna Lepidopterorum Poloniae), Tom. VI.* Polska Akademija Umiejętności, Kraków, Poland.

Salami M. 2020. *Una nueva plaga del cannabis en Canarias: Spodoptera frugiperda oruga cogollera*. Available at: https://drgrow.home.blog/2020/08/23/una-nueva-plaga-del-cannabis-en-canarias-spodoptera-frugiperda-oruga-cogollera/

Schiner JR. 1862. *Fauna Austriaca. Die Fliegen (Diptera), Teil 1*. Carl Gerold's Sohn, Wien.

Schmid U. 1996. Rettet *Episyrphus balteatus* (De Geer, 1776). *Volucella* 2: 101–103.

Schmidt M. 1929. Blattlausfliegen (Syrphidae) als vorratsschädlinge. *Mitteilungen Gesellschaft für Vorratsschutz* 5(6): 80–81.

Scopoli GA. 1763. *Entomologia carniolica*. Trattner, Vindobonae.

Seals WC. 2019. *Insects associated with industrial hemp, Cannabis sp. L., in Tennessee and development of Helicoverpa zea (Boddie) on a hemp-based diet*. Master's thesis, University of Tennessee, Knoxville.

Sella E. 1864. Sopra alcuni Coleotteri che s'incontrano nel Biellese. *Atti della Società Italiana di Scienze Naturali* 7: 105–148.

Serebryakova TY, Sizov IA. 1940. "Cannabinaceae," pp. 1–53 in: Vavilov NI, Vul'f EV, eds., Культурная Флора СССР, Том V (*Cultural Flora of the USSR, Vol. 5*). State Publishing House of Collective-farm and State-farm Literture, Moscow-Leningrad.

Shilenkov VG, Tolstonogova EV. 2006. Вредители конопли в Прибайкалье. *Acta Biomedica Scientifica* 2006(2): 159–160.

Shiraki T. 1952. *Catalogue of injurious insects in Japan*. General Headquarters, Supreme Command Allied Powers, Tokyo.

Silantyev AA. 1896. Результаты исследования конопляной и свекловичной блохи, совершенного по поручению Министерства в 1895–96 годах. Bulletin of the Ministry of Agriculture and Government Estates, no. 49, St. Petersburg.

Singh Kriti J, Dar MA, Khan ZH. 2014. Biological and taxonomic study of agriculturally important Noctuid pests of Kashmir. *World Journal of Agricultural Research* 2: 82–87.

Sirikantaramas S, Morimoto S, Shoyama Y, *et al.* (4 other authors). 2004. The gene controlling marijuana psychoactivity. *Journal of Biological Chemistry* 279: 39767–39774.

Smith SM. 1996. Biological control with *Trichogramma*. *Annual Review Entomology* 41: 375–406.

Spaar D, Kleinhempel H, Fritzsche R. 1990. *Öl- und Faserpflanzen*. Springer Verlg, Berlin.

Sturm J. 1843. *Catalog der Kaefer-sammlung*. Kosten des Verfassers, Nürnberg.

Targioni Tozzetti AD. 1888. *Annali di Agricoltura 1888*. Tipografia dell'Arte della Stampa, Firenze.

Takahashi S. 1919. Notes on insects injurious to hemp, and on *Thyestes gebleri* Fald. *Konchu Sekae (Insect World) Gifu* 23(1): 20–24.

Takizawa H. 2005. A revision of the genus *Psylliodes* Latreille in Japan (Chrysomelidae: Alticinae). *Insecta Matsumurana* 62: 175–185.

Tenhumberg B. 1995. Estimating predatory efficiency of *Episyrphus balteatus* (Diptera: Syrphidae) in cereal fields. *Environmental Entomology* 24: 687–691.

Theobald FV. 1904. *Second report on Economic Zoology*. British Museum of Natural History, London.

Tietz HM. 1972. *An index to the described life histories, early stages and hosts of the Macrolepidoptera of the continental United States and Canada*. AC Allyn, Sarasota, Florida.

Tölg F. 1913. *Psylliodes attenuata* Koch, der Hopfen- oder Hanf-Erdfloh. I. Teil. Morphologie und Bionomie der Präimaginalstadien. *Verhandlungen der Zoologisch-Botanischen Gesellschaft in Österreich* 63: 1–25.

Trotuş E, Naie M, 2008. Cercetări privind reducerea atacului unor agenţi patogeni şi dăunători specifici culturilor de cânepă prin tratamentul chimic al seminţei. *Lucrari Stiintifice, Universitatea de Stiinte Agricole Si Medicina Veterinara "Ion Ionescu de la Brad" Iasi, Seria Agronomie* 51(2): 219–223.

Tubbs PK, ed. 2000. Precedence of names in wide use over disused synonyms or homonyms in accordance with Article 23.9 of the Code. *Bulletin of Zoological Nomenclature* 57: 6–10.

Turner CE, ElSohly MA, Boeren EG. 1980. Constituents of *Cannabis sativa* L. XVII. A review of the natural constituents. *Journal of Natural Products* 43: 169–234.

Vallot JN. 1836. Sur une Phrygane. *Mémoires de l'Académie des Sciences, Arts et Belles-lettres de Dijon* 1836(2): 237–246.

Vaughn JL, Goodwin RH, Tomkins GJ, McCawley P. 1977. The establishment of two cell lines from the insect *Spodoptera frugiperda* (Lepidoptera; Noctuidae). *In Vitro* 13: 213–217.

Villanueva RT. 2022b. *Minute pirate bugs were active in hemp fields*. Available at: https://kentuckypestnews.wordpress.com/2022/10/11/minute-pirate-bugs-were-active-in-hemp-fields/

Vinokurov GM. 1927. Larvae of the lucerne or flax noctuid (*Chloridea dipsacea* L.) as a pest of grain in the ear. *Bulletin Irkutsk Plant Protection Station* 1: 67–79.

Vörös G. 1996. A gyapottok-bagolylepke (*Helicoverpa armigera* Hübner) kártétele szőlőben. *Növényvédelem* 32: 229–234.

Warchałowski A. 2000. *Chrysomelidae Stonkowate (Insecta: Coleoptera), Część VII, Fauna Polski, Tom. 22*. Muzeum i Instytut Zoologii PAN, Warszawa.

Wylie HG. 1988. Release in Manitoba, Canada of *Townesilitus bicolor* (Hym.: Braconidae), an European parasite of *Phyllotreta* spp. (Col.: Chrysomelidae). *Entomophaga* 33: 25–32.

Yepsen RB. 1976. *Organic plant protection*. Rodale Press, Emmaus, Pennsylvania.

Zhang DF, Liu HL, Wang AL, Han DQ. 2001. Survey on occurrence and damage of cotton bollworm in Qianghai Province. *Plant Protection* 27: 22–25.

6 Insects chewing stalks and branches

Abstract

This chapter features the Eurasian hemp borer (EHB, *Grapholita delineana*) and European corn borer (ECB, *Ostrinia nubilalis*). Both species have nomenclature issues, so this chapter has a sizeable taxonomic component. EHB larvae that hatch late in the growing season change their tactics and feed on flowers and seeds, for which they are equally notorious. Other boring caterpillars include the common stalk borer (*Papaipema nebris*), larvae of ghost moths (*Endoclita excrescens*, *Hepialus humuli*), and carpenter moths (*Cossus cossus*, *Zeuzera multistrigata*). A beetle parade follows: longhorns (*Thyestilla gebleri*, *Agapanthia cynarae*), tumbling flower beetles (six *Mordellistena* spp.), and weevils or curculios (*Rhinoncus pericarpius*, *Ceutorhynchus rapae*, *Cardipennis rubripes*). The cultural and mechanical control of boring beetles and caterpillars shares many commonalities, but pest identification is required for species-specific biological control. Pesticides have little use—once larvae bore into stalks, surface sprays will not affect them.

6.1 Introduction

Two lepidopterans lead the list of stalk-boring insects: *Grapholita delineana* and *Ostrinia nubilalis*. Both species have nomenclature issues, so this chapter has a sizeable taxonomic component. *G. delineana* is mired in a swamp of synonyms, and its author citation is in question. *O. nubilalis*, according to a small cloister of taxonomists, only attacks corn (maize). They identify the hemp pest as another species: *O. scapulalis* in Europe and *O. furnacalis* in Asia. Hogwash, we say. Those species infest hemp, but so does *O. nubilalis*.

Populations of *O. nubilalis* have decreased worldwide due to the widespread adoption of transgenic maize that expresses the BT toxin. *G. delineana* appears to be a hemp specialist and its populations have increased due to the widespread adoption of legal *Cannabis*.

Other lepidopterans and coleopterans chew on stalks, but primarily cause problems elsewhere: See Chapter 5 for budworms (*Helicoverpa zea*, *Helicoverpa armigera*, *Heliothis viriplaca*) and the hemp flea beetle (*Psylliodes attenuatus*). Insects that dabble in stalks but primarily damage leaves are in Chapter 7. Members of the sap-sucking guild may congregate on stalks and branches; see Chapter 4.

6.1.1 Signs and symptoms

Newly hatched stalk borers feed on leaves. Then they switch to pith, and bore into branches. From there, they tunnel into the main stalk. Their damage is often hidden within. A tiny entrance hole may visible, sometimes extruding a slimy mix of sawdust and frass. Their tunnels may cut xylem vessels, which causes leaf flagging, wilting, and necrosis.

Tunneling may give rise to visible deformities in branches or stalks (Fig. 6.1). These swellings or galls are fusiform to globose in shape. Galls are structurally weak, and depending on the pest species, may cause stalks to snap. Photographs of badly infested fields (Ceapoiu, 1958; Nagy, 1959) show half the plants lodged (toppled) at odd angles. Not pretty pictures. Lodged plants become entangled in harvest machinery, with fiber yield losses up to 50% (Grigoryev, 1998).

6.2 Eurasian hemp borer (EHB, *Grapholita delineana*)

Young Eurasian hemp borer (EHB) larvae skeletonize leaves in early spring, move to leaf petioles, and then bore into branches. They bore a series of holes, eventually reaching the stalk, often entering at a branch node (Ovsyannikova and Grichanov, 2009). A fusiform-shaped gall may arise, with frass ejected from the bore hole (Fig. 6.1). Leaf flagging may occur adjacent to the stalk gall. Stalks rarely break; EHB tunnel length averages only 1 cm (Miller, 1982) or 1–2 cm long (Nagy, 1967). Dissection of galls often reveals feeding galleries but no larvae. Some galls have moth exit holes, and these may contain pupal casings (McPartland, 2002).

EHB larvae that hatch in late summer or autumn change tactics. Instead of boring into stems, they feed within folded leaves, bound together by strands of silk. Due to this shelter-building habit, EHBs are also known as "hemp leafrollers." They crawl into buds and feed on flowers and seeds (Fig. 6.2). Cranshaw *et al.* (2019) considered bud damage more significant than stalk damage.

Senchenko and Timonina (1978) illustrated damage to seeds, which were notched or hollowed out, leaving only a shell. Kryachko *et al.* (1965) described EHBs causing 80% losses of flowering tops in Russia. Bes (1978) reported 41% seed losses in unprotected Yugoslavian hemp. Smith and Haney (1973) estimated that each larva consumed an average of 16 seeds. Anti-marijuana researchers considered the EHB "an excellent biocontrol weapon" (Mushtaque *et al.*, 1973; Baloch *et al.*, 1974; Scheibelreiter, 1976). Forty larvae could kill a seedling 15–25 cm tall in 10 days (Baloch *et al.*, 1974). Ten larvae per plant crippled growth and seed production.

DOI: 10.1079/9781836990352.0006

6.2a. *Grapholita delineana* Walker 1863, Lepidoptera: Tortricidae

≡*Cydia delineana* (Walker), ≡*Laspeyresia delineana* (Walker), ≡*Enarmonia delineana* (Walker)

Fig. 6.1. Stalk damage by *Grapholita delineana*

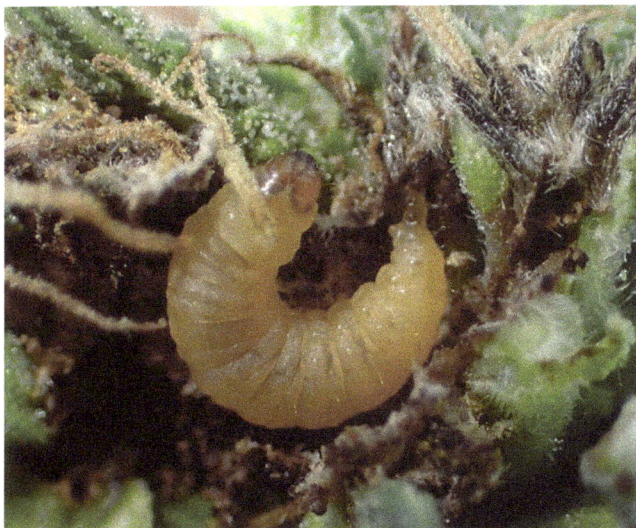

Fig. 6.2. Flower damage by EHB (W. Cranshaw)

=*Grapholita apicalana* Walker 1863, =*Grapholita sinana* Felder and Rogenhofer 1874, ≡*Cydia sinana* (F. & R.), =*Grapholitha tetragrammana* Staudinger 1880, =*Grapholita mundana* Christoph 1882, =*Grapholita terstrigana* Ragonot 1894, =*Laspeyresia quadristriana* Walsingham 1900, =?*Laspeyresia isacma* Meyrick 1907, ≡*Grapholita isacma* (Meyrick)

Description: *Larvae* cream-colored, to pinkish-orange, to pale brown, darkening with age, 8–12 mm long (Fig. 6.2). Several pale bristles per segment are barely visible. Heads yellow-red, with black ocelli, averaging 0.91 mm wide. Abdominal legs with 16–20 hooks form a uniserial crown. *Pupae* light brown, 5–7 mm in length, with two rows of spinules directed backward on dorsal side of 2nd–8th segments (Fig. 6.3A), sometimes covered in silken cocoons within stalks or in soil.

Adults are small with grayish to rusty-brown bodies (Figs 6.3B and 6.4). Forewings variegated in color, from light brownish-ochrous to dark brownish-red; with light/dark stripes along the anterior edge, and four slanting chevron-like stripes near the center. Hindwings grayish-brown, darkening from peripheral to lateral. Fringe on forewings and hindwings is prominent, gray-brown with a metallic shine (see Figs 6.3B and 6.4). Body length and wingspan average 6–7 and 10–15 mm, respectively, in females, and 5 mm and 9–13 mm, respectively, in males. *Eggs* are white to pale yellow, oval, 0.4 mm wide. In the spring, eggs are laid one-by-one or 2–3 at a time, either on the lower side of hemp leaves (60% of eggs), or the upper surface of hemp leaves (30%), or on stalks and petioles (10%). Moths of subsequent generations lay eggs in inflorescences.

6.2.1 Distribution

EHB has been given many Latin names by entomologists across Eurasia, in four genera: *Grapholita*, *Cydia*, *Laspeyresia*, and *Enarmonia*. The orthographic variant *Grapholitha* (with second *h*) often appears. Sorting EHB's geographic distribution is complicated by misidentifications in the literature. Early taxonomic names were assigned to free-flying adult moths, whose hosts were unknown. Systematic entomologists identified adults reared from larvae (e.g., Nagy, 1967; Miller, 1982), or compared

Fig. 6.3. EHB: **A.** pupa and frass in stalk; **B.** adult on leaf (W. Cranshaw)

field specimens to museum specimens (e.g., Clarke, 1958; Bradley, 1961; Danilevsky and Kusnetsov, 1968; Razowski, 2003).

Peering through our thicket of synonyms, East Asia appears to be the origin of EHB. It was originally described in China (Walker, 1863), and is still common there (Tsao, 1963; Wang and Rong, 1992; Liu and Li, 2002). It also occurs in Japan (Walsingham, 1900; Matsumura, 1931; Esaki *et al.*, 1932; Issiki, 1957; Asano, 2012; Jinbo *et al.*, 2014), Korea (Bae *et al.*, 2012; Byun *et al.*, 2012), and the Russian Far East (Christoph, 1882; Kuznetsov, 1988).

In South Asia, EHB occurs in Pakistan (Mushtaque *et al.*, 1973; Baloch *et al.*, 1974, 1975), India (Sankaran and Ramachandran Nair, 1973), and Nepal (Fig. 6.1, taken in Nepal). In Western and Central Asia, EHB occurs in Turkey, the Caucasus, Iran, Kyrgyzstan, and Kazakhstan (Ovsyannikova and Grichanov, 2009). In Europe, an EHB epidemic began in the 1960s–1970s, but evidence indicates the species reached Europe earlier. EHB reached North America by the 1940s.

No one has sorted the synonyms, so we will here. Walker (1863) erected two species names for moths collected in China: *Grapholita delineana* for a female specimen, and *Grapholita apicalana* for a male specimen. Fletcher (1932) recognized they were two sexes of one species, and synonymized *G. apicalana* under *G. delineana*. Bradley (1961) examined type specimens, and reiterated the synonymy. Felder and Rogenhofer (1874) coined *Grapholitha sinana* for a moth collected by Georg Frauenfeld in Shanghai. Rather than a description, they provided a color drawing (Fig. 6.4). *G. sinana* and *G. delineana* were considered distinct by Obraztsov (1959), whereas Danilevsky and Kusnetsov (1968) synonymized *G. sinana* under *G. delineana*, and this has been universally accepted.

Staudinger (1880) erected *Grapholitha tetragrammana* for a female moth collected by Maximilian Wocke at Bursa in Turkey, and a female moth collected by Julius Lederer at Gorgān in Iran. Rebel (1889) caught a moth in Vienna and sent it to Wocke, who identified it as *G. tetragrammana*. Hinneberg (1895) linked *G. tetragrammana* to a host plant: wild hops growing in Germany. Schütze (1931) detailed the life cycle of *G. tetragrammana* on hops. Meyrick (1937) compared Staudinger's specimens with the type of *G. sinana* by Felder and Rogenhofer, and considered them identical. Obraztsov (1959) synonymized *G. tetragrammana* under *G. sinana*, hence under *G. delineana* (Komai, 1999; Razowski, 2003).

Fig. 6.4. *Grapholita sinana* (Felder and Rogenhofer 1874)

Christoph (1882) coined *Grapholitha mundana* for moths collected in hemp fields in the Russian Far East. It has been synonymized under *G. delineana* (Danilevsky and Kusnetsov, 1968; Razowski, 2003). Ragonot (1894) erected *Grapholitha terstrigana* for a moth collected by Pierre Millère on "the Mediterranean coast." The species was synonymized under *G. delineana* (Fletcher, 1932), then under *G. sinana* (Obraztsov, 1959), then under *G. delineana* again (Danilevsky and Kuznetsov, 1968; Kuznetsov, 1989). Walsingham (1900) erected *Laspeyresia quadristriana* for moths collected by Pryer in Japan, and Leech in Fújiàn Province, China. Matsumura (1931) and Esaki *et al.* (1932) also used this taxon. It has been synonymized under *G. delineana* (Obraztsov, 1959; Danilevsky and Kusnetsov, 1968).

Meyrick (1907) erected *Laspeyresia isacma* for moths collected in the Khasi Hills (Assam, India). Then Meyrick (1908) synonymized *L. isacma* under *G. delineana*, noting that *G. delineana* occurred in China and India. Others supported the synonymy (Fletcher, 1932; Clarke, 1958; Obraztsov, 1959). Conversely, Bradley (1961) reinstated Meyrick's species, as *Grapholita isacma*. He compared Meyrick's specimen to the type of *G. delineana*, and found differences in genitalia: "in *isacma* the apical part of the aedeagus is slender and tapers to an acute point and is curved almost at right angles to the main part." Some agreed with Bradley (Obraztsov, 1967; Diakonoff, 1982; Komai, 1999), others re-synonymized *G. isacma* under *G. delineana* (Bae and Park, 1997; Bae *et al.*, 2012). The host plant of *G. isacma* has never been identified.

Grapholita tristrigana (Clemens) 1865 reportedly infested *Cannabis* in North America (MacKay, 1959; Smith and Haney, 1973; Haney and Kutscheid, 1975). Miller (1982) examined museum specimens of "*G. tristrigana*" on hemp, and they proved to be EHB. He collected larvae in stalks of feral hemp in Minnesota, reared them into adults, and identified them as EHB. The USA distribution of EHB, based on museum specimens, was Minnesota, Iowa, Missouri, Wisconsin, Illinois, Kentucky, and New York. The earliest specimens dated to 1943 (from Lexington, Kentucky, and Madison, Wisconsin). EHB likely arrived via contaminated European seed, imported during World War II's "war hemp" effort.

EHB emerged as a serious problem in southeastern Europe during the 1960s. It was first detected in 1960, in Ukraine (Kryachko *et al.*, 1965) and Russia (Danilevski and Kuznetsov, 1968). The species subsequently appeared in Romania by 1963 (Manolache *et al.*, 1966) and Hungary by 1964 (Nagy, 1967). EHB was reported next in Bosnia-Herzegovina (Bes, 1967), Armenia and Moldavia (Shutova and Strygina, 1969), Serbia and Montenegro (Lekic and Mihajlovic, 1971), Bulgaria (Gerginov, 1974), Greece (Vassilaina-Alexopoulou and Mourikis, 1976), Slovakia (Bako and Nitri, 1977), and Slovenia (Camprag *et al.*, 1996).

Barnabás Nagy (1921–2020), a legendary Hungarian entomologist (Fig. 6.5), added EHB to his portfolio of hemp pests. Nagy's career began in the mid-1950s. Forty years later he helped edit the first version of this book (McPartland, 1991). We cite his research on *G. delineana* (Nagy, 1967, 1979, 1980; Sáringer and Nagy, 1971), *Ostrinia nubilalis* (Nagy and Csehi, 1955; Nagy, 1958, 1959, 1976, 1986), and *Ceutorrhynchus rapae* (Nagy *et al.*, 1982). In a review of hemp pests, Nagy (1962) considered the hemp flea beetle (*Psylliodes attenuata*)

Fig. 6.5. Barnabás Nagy (courtesy Szabolcs Simó)

the most common problem, adding, "A whole army of insects with chewing and sucking mouth parts can live in the above-ground parts of hemp." Later he changed his mind and named EHB as the worst hemp pest in Hungary (Nagy, 1986).

Well before the 1960s epidemic, EHB infested European hops—masquerading under the names *G. tetragrammana* or *G. terstrigana* (Rebel, 1889; Ragonot, 1894; Hinneberg, 1895; Schütze, 1931). McPartland (2002) proposed that the European hemp epidemic began when a new Asian strain was introduced via contaminated Chinese hemp seed. He interviewed Ivan Bòcsa, who stated that Hungarian breeders imported several thousand metric tons of hemp seed from China in the 1950s. Sergey Grigoryev stated that the Vavilov Institute in Russia imported hemp seed from China in 1953. Janos Berenji said that Serbian bird seed manufacturers imported tons of Chinese hemp seed in the 1950s and 1960s. The new hemp strain in Europe rarely attacked hops (Nagy, 1967; Bes, 1967). Similarly, the Asian race did not feed on hops (Mushtaque *et al.*, 1973). Song *et al.* (2021) sequenced the mitochondrial genome of six *G. delineana* specimens collected in Hénán and Shāndōng provinces. Their phylogenetic analysis revealed that *G. delineana* is not monophyletic. New species names should be forthcoming.

6.2.2 Host range and differential diagnosis

EHB has a limited host range: *Cannabis sativa*, *Humulus lupulus*, and *Humulus japonicus* (= *H. scandens*). Knotweed (*Polygonum* spp.) is an alternative host, and there are probably others (Cranshaw *et al.*, 2019). Mushtaque *et al.* (1973) ran host range tests, and larvae failed to feed on Urticaceae and Asteraceae hosts.

EHB larvae are smaller than European corn borers (ECBs, *Ostrinia nubilalis*). EHB damage often occurs in the top one-third of plants (Nagy, 1967), whereas ECBs usually form galls in the lower three-quarters of plants (Nagy, 1959). In Romanian

hemp, EHBs occurred at a higher density than ECBs, averaging $10.2/m^2$ and $0.5/m^2$, respectively (Trotuş and Naie, 2008). ECBs and other boring caterpillars drill longer tunnels than EHB larvae. Weevils and curculios also bore into stems and form galls. Late-season hemp borers that infest buds may be confused with budworms.

6.2.3 Life cycle

EHBs overwinter in the larval stage, in crop stubble, neighboring weeds, and in stored seed (Shutova and Strygina, 1969). Depending on latitude, EHB is univoltine (one generation per year) or multivoltine (two or more generations per year). The univoltine life cycle occurs in northern Germany (Schütze, 1931), and the northern part of the Russian Far East (Kuznetsov, 1988). Two generations per year occur in Hungary (Nagy, 1979), Ukraine (Kryachko *et al.*, 1965), and the southern part of the Russian Far East (Kuznetsov, 1988). A partial third generation arises in Armenia (Shutova and Strygina, 1969), Moldova, southern Ukraine, and southernmost Russia (Ovsyannikova and Grichanov, 2009). Up to four generations overlap in Pakistan (Mushtaque *et al.*, 1973). Three or four generations arise in the USA (Cranshaw *et al.*, 2019).

Schütze (1931) described in detail the univoltine life cycle of EHB (as *G. tetragrammana*), to which we add data from Ovsyannikova and Grichanov (2009): overwintering larvae emerge when daily average temperatures reach 7°C, usually in early to mid-April. They skeletonize hemp leaves, and then bore into branches. Feeding duration is 21–33 days. Pupation begins in late April to June. Larvae pupate in stalks, unless temperatures are warm, and then pupate in soil. Moths emerge between May and July. They are day fliers; flight time and egg-laying continue for 8–14 days. Emerging larvae feed on upper leaves and flowering tops. Larvae diapause between August and October, induced by day length under 14 h. Temperature influences diapause—warm weather slows photoperiodic effects (Sáringer and Nagy, 1971). Larvae may overwinter in rolled-up leaves or in flowering tops.

Later-generation larvae gnaw holes in buds, flowers, and especially seeds, but may bore in stalks. They preferentially infest female flowers over males (Kryachko *et al.* 1965, Cotuna *et al.* 2020). Later generations often pupate within rolled leaves, bound together by strands of silk. They may also pupate within stalks (Nagy, 1967), or in soil under plant debris (Mushtaque *et al.*, 1973). Pupal development lasts 16–22 days. In bivoltine populations, moths of the first generation fly from May to mid-June, and the second generation flies from mid-July to mid-August (Kuznetsov, 1988). Eggs hatch in 5–6 days at 20–25°C (Kryachko *et al.*, 1965).

EHB moths are not strong fliers; Nagy (1979) conducted flight chamber experiments, and calculated flight speeds of 3.2–4.7 km/h (2.0–2.9 mph) at 21–24°C. Upon finding a hemp field, females land quickly, usually within 3 m of the field's edge. After mating, females lay 350–500 eggs. Adults live less than 2 weeks. Eggs hatch in 5–6 days at 22–25°C, or 3–4 days at 26–28°C. Out of 350–500 eggs, Smith and Haney (1973) estimated only 17 larvae survived to first instar.

6.2.4 Cultural and mechanical control

Scout for symptoms (stem galls, rolled leaves) and search for moths with sweep nets. Synthetic pheromone traps await commercial development. Nagy (1980) demonstrated proof-of-concept by trapping male moths with female sex hormones. Commercially available pheromones for other *Grapholita* spp. failed to lure EHBs (Cranshaw, pers. obs., 2019). He tested pheromones from the oriental fruitworm, *G. molesta*; cherry fruitworm, *G. packardi*; and lesser appleworm, *G. prunivora*.

Light traps might work. Komlódi (1970) could not catch EHB moths with "the standard light trap," but reported success with a mercury vapor lamp. Nagy (1980) reported a peak flight time for EHB males between 5 and 7 pm.

Eradicate "sanctuary stands" of feral hemp. Larvae may overwinter in neighboring weeds (Shutova and Strygina, 1969), so eliminate other weeds as well. This should be done in autumn, and again in the spring when weed seedlings emerge. All plant residues and weeds should be collected after harvest; Fedorenko *et al.* (2016) burned residues after the first frost. Nagy (1967) noted that larvae overwintered in crop stubble, which should be removed from the field. Shutova and Strygina (1969) recommended deep plowing to kill overwintering larvae and pupae in stubble and the soil.

Larvae may overwinter in harvested stalks (Nagy, 1979), so stalks should not be transported into uninfested areas. Larvae also overwinter in seed, so clean seed before storage. McPartland (2002) carefully threshed seed from feral hemp in Vermont, and at least one larva still snuck into the harvest. Seed quarantine measures to exclude *G. delineana* are in effect in New Zealand (McPartland *et al.*, 2004), Australia (Biosecurity Australia, 2010), and Thailand (Sutabtra, 2007).

Nagy (1967) recommended early harvesting, because problems get worse with delays. Early-maturing *Cannabis* varieties would be helpful, especially in areas with multivoltine populations. East Asia appears to be EHB's origin, so East Asian *Cannabis* may harbor resistance. Ree (1966) reported that Korean and Japanese landraces were more resistant to "stem borers" (likely EHB) than an Italian cultivar. Cotuna *et al.* (2020) evaluated resistance in four Romanian varieties, counting the average number of larvae per plant: 'Lovrin 110' was the most resistant, averaging 36.83 larvae/plant, followed by 'Armaca' with 39.0, then 'Silvana' with 43.5, and lastly 'Teodora' with 61.16.

6.2.5 Biological control

Ovsyannikova and Grichanov (2009) counted 22 hymenopteran species that naturally parasitized EHBs in Russia, killing up to 38.5% of larvae and 14.5% of pupae. In Hungary, Scheibelreiter (1976) counted 22 species of parasitoids, led by a *Scambus* sp. with a parasitism rate of 30%. In Pakistan, larvae were parasitized by a *Goniozus* sp. (Mushtaque *et al.*, 1973).

Peteanu (1972) controlled EHBs by releasing *Trichogramma evanescens* (egg parasitoids, see section 20.6) in Romania. *T. evanescens* also parasitized EHBs in India (Rajmohana and Peter, 2014). Peteanu (1980) tested three release rates for *T. evanescens*, 80,000, 100,000 or 120,000 wasps/ha. At the two lower rates, *T. evanescens* was equal in effectiveness to *Bacillus*

thuringiensis (Dipel® or Thuringin®). At the highest rate it was equal to spraying with fenitrothion, a synthetic pyrethroid.

Camprag *et al.* (1996) used *Trichogramma* wasps to control the 1st-generation infestation with 51–68% efficiency. They released 75,000–100,000 wasps/ha, and repeated the release one week later. They did not report which *Trichogramma* sp. they released. Commercially available *T. pretiosum* and *T. minutum* may work. *T. dendrolimi* is an egg parasitoid with a wide host range, including ECBs, budworms (*Helicoverpa* and *Heliothis* spp.), and *Grapholita molesta*. Smith (1996) controlled *G. molesta* by releasing *T. dendrolimi* at a rate of 600,000 wasps/ha, repeated every 5 days while moths were laying eggs.

In Illinois, Smith and Haney (1973) found up to 75% of EHB larvae infested by the tachinid fly *Lixophaga variablis* (Coquillett) and the braconid wasp *Macrocentrus delicatus* Cresson. Spraying with *Bacillus thuringiensis* (Bt) *kurstaki* (e.g., Dipel®) may be effective. But like spraying with pesticides, Bt must be applied while the larvae are still on the surface of leaves. Once the larvae bore into stems, surface sprays will not affect them.

6.2.6 Chemical control

Nagy (1979) described an "edge effect," where weakly flying EHB moths landed within the first 3 m around a hemp field. He recommended spraying this edge zone with pesticides as moths arrive. In severe infestations, the edge zone should be cut down and buried or burned. Kryachko *et al.* (1965) sprayed plants with high-powered pesticides (e.g., DDT, parathion), and results were never more than 50% effective. Trotuş and Naie (2008) knocked back EHB by spraying with three pesticides banned in the USA. Repeated applications were required for multivoltine populations. To prevent EHBs from spreading via contaminated seed, Kryachko *et al.* (1965) fumigated seed with methyl bromide to kill larvae. This tactic might work with less-toxic fumigants (e.g., CO_2) or seed disinfectants (bleach diluted to 10%).

6.3 European corn borer (ECB, *Ostrinia nubilalis*)

The European corn borer (ECB) is first of five *Ostrinia* spp. that attack hemp. *Ostrinia* taxonomy was overhauled by **Akira Mutuura** (1923–2004) and **Eugene Munroe** (1919–2008). Mutuura studied *Ostrinia* in Kyūshū before joining the Systematic Entomology Unit in Ottawa, Canada. Munroe led the Unit and specialized in *Ostrinia* spp. and other members of family Pyralidae (now the Crambidae).

Mutuura (1954) classified Japanese *Ostrinia* spp. by male genitalia and the size of their mid-tibiae. Mutuura and Munroe (M&M) (1970) expanded this to a world-wide classification system. M&M classified 20 *Ostrinia* spp. into three "species groups" based on the number of lobes on the uncus. The **uncus** is an extension of the tegumen (distal abdominal segment) that the male uses to hold the female during mating (Fig. 6.6). All *Ostrinia* spp. that infest *Cannabis* belong to the "trilobed uncus" group: three major pests (*O. nubilalis*, *O. scapulalis*, *O. furnacalis*) and two minors (*O. narynensis*, *O. kurentzovi*).

Within the "trilobed uncus" group, M&M differentiated species by the size of the mid-tibia on males: small (*O. nubilalis*, *O. furnacalis*), medium (*O. narynensis*, *O. kurentzovi*), or large (*O. scapulalis*). M&M also differentiated species by two other parts of male genitalia. The **juxta** provides a ventral support at the base of the aedeagus (penis). The **sacculus** provides anterio-lateral support to the aedeagus. Yang and Zhang (2011) analyzed these structures in the three majors (*O. nubilalis*, *O. scapulalis*, *O. furnacalis*). These species have a V-shaped juxta, with both anterior and posterior arms. They have an inflated sacculus with dorsal spines. Yang *et al.* (2021) provided schematic drawings (Fig. 6.6).

Female *Ostrinia* moths emit different **pheromones** to attract males. Some *Ostrinia* spp. emit a blend of Z- and E- isomers of 11-tetradecenyl acetate, others emit a blend of Z- and E- isomers of 12-tetradecenyl acetate (Fig. 6.6).

Prior to M&M splitting ECB into segregate species, others recognized that ECB could be split into groups, but within one species. Rataj (1957) proposed that "specialized races" of ECB were involved in hemp damage, a question "that has not been solved yet." Nagy (1958) described a new bivoltine "ecotype" of ECB responsible for hemp losses.

Damage by ECB is similar to damage caused by EHB. In early spring, young ECB larvae chew on hemp leaves, on the underside, near the midline vein (Nagy, 1959). Time spent on the leaves depends on relative humidity—young ECB larvae are very hygrophilic. Soon they gnaw into leaf petioles, and then bore into branches. After a week or two in branches, ECBs tunnel into stalks. Their bore holes extrude sawdust and frass, and predispose plants to fungal infection by *Macrophomina phaseolina* (McPartland, 1996b) and *Fusarium* spp. (Grigoryev, 1998).

ECBs may tunnel through xylem vessels and cause upstream wilting—sometimes just a solitary leaf, sometimes the entire plant above the bore hole. Stalks at tunnel sites may swell into galls, fusiform- or globose-shaped (Fig. 6.7A). The misshapen galls disrupt bast cell growth and thereby weaken fiber. Plants may snap at gall sites. ECB larvae "gnaw inside the stem until it is knocked down by the wind" (Heuzé, 1864).

Infested fields may turn into a maze of toppled stalks; see the mess photographed by Ceapoiu (1958) and Nagy (1959). Stalks may become twisted and curved, photographed by Zwölfer (1928). Twisted, curved, and broken stalks are difficult to harvest; Nagy (1959) illustrated a tangled harvesting machine. Between 5 to 12 ECBs can destroy a hemp plant (Emchuck, 1937). Camprag *et al.* (1996) reported ECB losses as high as 80–100%.

If tunneling kills the plant's apical meristem, lateral branching takes over, and the plant develops a shorter, bushier appearance (Ceapoiu, 1958). Small *et al.* (2007) proposed that bushier plants (Fig. 6.7B) may *increase* yields of flowering tops. However, ECB larvae can infest flowering tops, wherein they spin webs and scatter feces. They selectively feed on female flowers and immature seeds. Seed yield losses can range from 25% (Ceapoiu, 1958) to 40% (Camprag *et al.*, 1996). Vasilyev (1956) gave losses as 15% in seed crops and 25% in fiber crops.

6.3a. *Ostrinia nubilalis* (Hübner) 1796, Lepidoptera: Pyralidae

≡*Pyralis nubilalis* Hübner, ≡*Botys nubilalis* (Hübner), ≡*Pyrausta nubilalis* (Hübner), =*Pyralis nubilalis* Hübner, ≡*Botys silacealis*

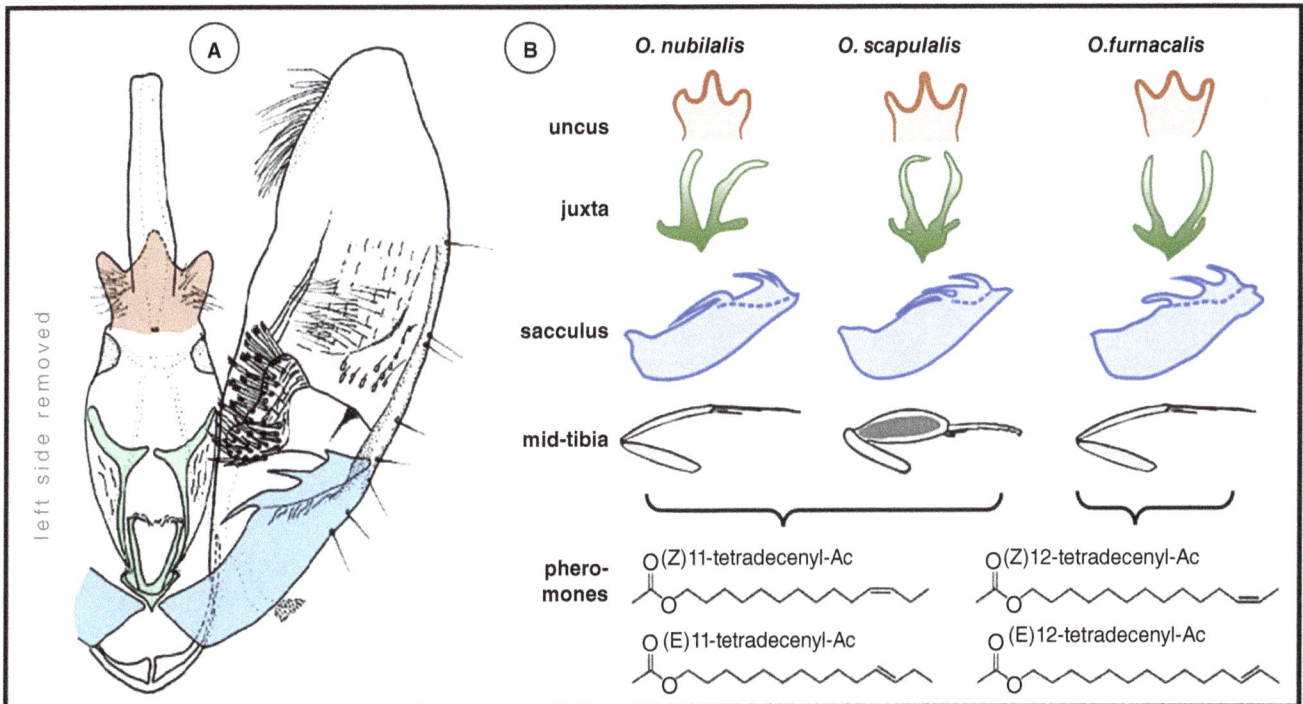

Fig. 6.6. Taxonomic characters: **A.** male genitalia (right half only, redrawn from Yang and Zhang, 2011); **B.** three *Ostrinia* spp. (redrawn from Yang *et al.*, 2021)

(Hübner), ≡*Pyrausta silacealis* (Hübner), =*Botys lupulinalis* Guenée 1854, =*Botys lupulina* Heinemann 1863

Taxonomic notes: Hübner (1796) assigned *Pyralis nubilalis* to a male moth and *Pyralis silacealis* to a female moth. Treitschke (1829) recognized that they represented two sexes of a single species. Marion (1957) transferred the species to *Ostrinia,* as *O. nubilalis.* The synonyms listed above are limited to *Cannabis* literature; Mutuura and Munroe (1970) listed a dozen others.

Description: *Larvae* are light brown with dark brown heads (Fig. 6.8A). The body has nine rows of dark brown spot-like plates, each with setae. Mature larvae 15–25 mm long. *Pupae* are reddish-brown, torpedo-shaped, 10–20 mm long, sometimes in flimsy cocoons (Fig. 6.8B). *Adults* are sexually dimorphic. Female moths are beige to dusky yellow, with irregular olive-brown bands running in wavy lines across their 25 mm wingspan.

Fig. 6.7. ECB damage: **A**. lodging (G. Grassi); **B**. branching (E. Small)

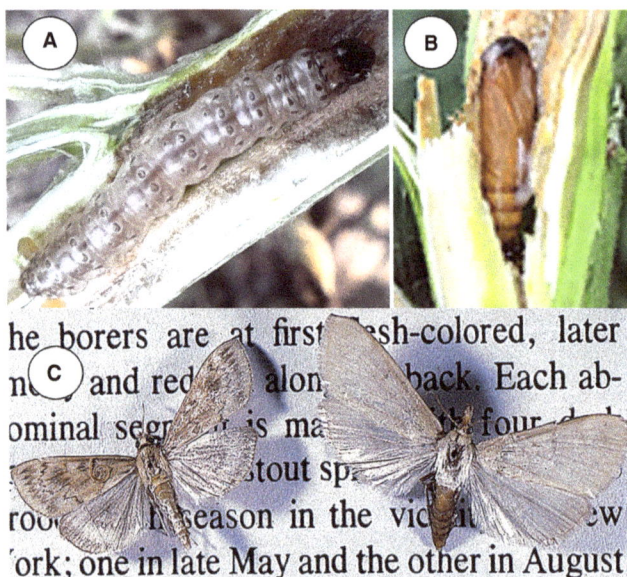

Fig. 6.8. *Ostrinia nubilalis*: **A**. larva (M. Zuefle); **B**. pupa (M. Bolt); **C**. moths: female (L), and male (R)

Males are slightly smaller and darker, also with olive-brown markings on their wings (Fig. 6.8C). Sacculus spines more uneven in size compared with Asian corn borer (ACB) (section 6.4) or adzuki bean borer (ABB) (section 6.5) with a long posterior spine (Fig. 6.6). Uncus height is taller than that of ACB or ABB with sharper lobe tips. *Eggs* are creamy-white, <1 mm in diameter, laid in a cluster (average 23 eggs), and overlap like tiles on a roof. Just before hatching, the dark brown head of the larva becomes visible within the creamy-white egg, the "blackhead" stage.

6.3.1 Life history

Mature larvae overwinter in crop stubble near the soil line. Springtime feeding begins when temperatures exceed 10°C. Larvae pupate for 2 weeks and emerge as moths in late May (or June, or even August in Canada and northern Europe). Females are night fliers, seeking out host plants to lay eggs. Long-range migrations are limited to dozens of kilometers, driven by wind (Dorhout *et al.*, 2008). Moths lay up to 500 eggs in 25 days (Taylor and Deay, 1950). Eggs are deposited on lower leaves of the most mature (i.e., earliest planted) hosts. Eggs of second-generation ECBs hatch in a week or less.

Larvae feed for about 3 weeks, then spin cocoons and pupate in hemp stalks (Fig. 6.8b). Moths emerge, mate, and repeat the life cycle. A hard freeze late in the year kills all but the most mature (5th-instar) larvae. One to four generations of *O. nubilalis* arise each year, depending on latitude and local weather. In western Europe, only one generation arises north of about 45° latitude. In the USA, only one generation arises north of a line from Maine to mid-Wyoming. Two generations arise between that latitude and a line that runs from Virginia to northern Colorado. South of there, three generations arise. Four generations arise in a belt from Georgia to southern Arkansas and Louisiana. Summers with high humidity and little wind favor egg laying, egg survival, and larval survival.

6.3.2 Distribution and host range

ECB is native to Europe. It has infested hemp as far north as Lancashire, UK (53.8 °N), "probably a casual immigrant only" (Meyrick, 1895). Recently it has expanded north to Minsk in Belarus (53.9 °N), which Bykovskaya and Samonov (2018) attributed to global warming. Migrants infest hemp up to 55 °N in Russia (Grigoryev, 1998). ECB ranges south to Mediterranean Africa (Morocco, Algeria, Tunisia, Libya, Egypt), and east to West Asia (Caucasus, Turkey, Syria, Lebanon, Israel, Iran), and to Central Asia (Uzbekistan, Xīnjiāng Region).

ECB has been recorded from 230 host species, basically any herbaceous plant with a stalk large enough to bore into. Its pestilence entered a new orbit when New World maize (corn) arrived in Europe. In the mid-1500s an acclimated maize variety was introduced to Italy and adopted throughout Europe (Brandolini and Brandolini, 2001; Rebourg *et al.*, 2003). An Italian entomologist made the first report of ECB infesting maize (Passerini, 1832).

Prior to maize arriving in Europe, ECB's crop hosts included hemp, hops, and millet—*Panicum miliaceum* (Nagy, 1986). ECB was the third insect pest ever described on *Cannabis*; Filippo Re (1806) described it in the Po River valley of Italy. He tentatively identified it as a *Pyralis* sp. Filippo Re (1813) found up to six larvae in a single hemp stalk; mature plants with large stalks were affected the worst. Bertolini (1842, 1844) named *Botrys silacealis* as a hemp pest, also in the Po River valley. Pichat (1866) stated that Bertolini and Filippo Re described the same species.

Claude Roberjot (1752–1799) reportedly described ECB earlier. Bosc (1809) wrote, "a caterpillar, which Roberjot first made known, lives in the interior of its [hemp's] stem and often kills it." Bosc added a footnote to Filippo Re (1813), stating that Roberjot had observed the species in 1796 near Mâcon in France. Heuzé (1860) said Roberjot described *pyrale du chanvre* in 1796. Roberjot published in *Mémoires de la Société Royale d'Agriculture de Paris*, and France's Bibliothèque Nationale—the ultimate repository of that journal—is missing the 1796 volume. Roberjot certainly knew about hemp, and experimented with the crop (Roberjot, 1787).

Guenée (1854) coined *Botys lupulinalis* for a pest of hops and millet in France. Köppen (1880) reported hemp damage by *Botys silacealis* in Taganrog Oblast, Russia. Robin and Laboulbène (1884) reported both these species on hemp, and synonymized them under *Botys nubilalis*. Heinemann (1863) coined *Botys lupulina* for ECB infesting hops and hemp in Germany. Taschenberg (1879) observed *Botys lupulina* infesting hemp in Germany. Robin and Laboulbène synonymized that taxon under *Botys nubilalis*.

In Germany, Zwölfer (1928) and Schlumberger (1941) noted that some ECB populations infested hemp and hops but not maize, whereas the reverse was true in other populations. Romaniszyn and Schille (1930) reported ECB as a pest of hemp and hops in Poland, with no mention of maize. In Ukraine, Karpova (1959) found that larvae collected from maize would not infest hemp, and larvae collected from hemp in neighboring Russia (Bryansk) would not infest maize.

ECB arrived in Boston around 1910. Vinal (1917) presumed ECB came from Europe, hidden in hemp stalks imported by a cordage manufacturer, probably the Charlestown Navy Yard, which operated the USA's largest rope walk. Vinal said that no hemp and hops grew near Boston, so ECB became established on maize. The pest has spread throughout the USA and Canada east of the Rocky Mountains. Two host-range studies were conducted in Boston after ECB arrived there. Caffrey and

Worthley (1927) grew test plants adjacent to a heavily infested maize field, and quantified infestations (Table 6.1). Their methodology was marred by variability in the number of test plants examined, but hemp was clearly the "winner." ECB caused severe losses in flint corn, whereas dent corn was less susceptible. Flint corn was the variety introduced into Italy in the mid-1500s, when ECB got its start on the new host (Brandolini and Brandolini, 2001).

Hodgson (1928) also surveyed host plants in the Boston area. Feral hemp was frequently infested, "the greater part of the interior of the plant is sometimes destroyed." ECB was bivoltine in Boston, and second-generation larvae were more commonly found on hemp than first-generation larvae, and ECB overwintered in hemp.

ECB damage didn't get much attention until the "war hemp" effort during WWII (Wilsie *et al.*, 1942; Robinson, 1943b). ECB nearly ruined Illinois's hemp crop (Hackleman and Domingo, 1943). Bush Doctor (1987b) reported ECB attacking feral hemp and drug-type *Cannabis* across the midwestern "corn belt." Since then, the adoption of Bt transgenic maize has suppressed ECB populations, but the pest still infests hemp in the USA (Cranshaw *et al.*, 2019) and Canada (Small *et al.*, 2007). ECB populations in the corn belt are rebounding on hosts other than transgenic maize (Lizotte, 2019).

Research on ECB in Europe up-ticked in the 1930s. A bibliography of East European literature (Brindley *et al.*, 1975) yielded many hits on hemp: Durnovo in Russia: up to 6% of crop stubble harbored overwintering larvae; Dmitriev in Russia: a survey of control measures; Laduizhenskaya in Russia: microclimate and ECB survival in hemp stalks; Silant'ev in Russia: the deleterious effects of pond retting on larval survival; Popov in Bulgaria: ECB hammering hemp; Emchuck in Ukraine: widespread damage to hemp crops; Laduizhenskaya in Russia: impact of ecological conditions on egg survival; Rumyantzev in Russia: the duration of external (leaf) feeding.

Another round began in the late 1950s. ECB caused "grave injury" in Czechoslovakia (Rataj, 1957), and attacked both fiber and seed crops in Romania (Ceapoiu, 1958). Hemp damage suddenly increased in southern Hungary near Szeged (Nagy and Csehi, 1955). The damage was due to a new bivoltine ecotype; Nagy (1958) mapped the distributions of bivoltine and univoltine ecotypes, and peak flight times for the moths. The number of generations depended on temperature and

Table 6.1. Host-plant choice experiment, Caffrey and Worthley (1927)

Host plant	Plants examined (number)	Plants w/ larvae (number)	Percent plants w/ larvae	Total number of larvae found	Average larvae per 100 plants
Hemp	10	10	100	150	1500
Hops	12	3	25	5	41.6
Hegari sorghum	84	70	83.3	218	259.5
Milo sorghum	420	254	60.4	444	105.7
Broomcorn sorghum	450	191	42.4	314	69.8
Japanese millet	200	27	13.5	38	19.0
Sudan grass	50	5	10.0	7	14.0
Cotton	656	22	3.4	67	3.5

photoperiod, as well as host plants—bivoltine populations in the south preferred hemp over maize, whereas univoltine populations fed on maize and not hemp.

Nagy (1959) described and photographed various types of injuries, ranging from leaf skeletonization, petiole injury, branch and stalk boring, and flower damage. Nagy (1976) conducted oviposition preference tests. A univoltine strain obtained from maize showed no significant preference between maize, hemp, *Artemisia vulgaris*, or *Amaranthus retroflexus*. A multivoltine strain from hemp preferred hemp over the other plants. In a feeding-choice test with first-instar larvae, the maize strain preferred *Artemisia* (40% of larvae) over maize (23%), hemp (17%), hops (18%), and other plants. Larvae of the hemp strain preferred hemp (72%), *Artemisia* (11%), maize (11%), and other plants.

Sáringer (1976) noted that ECB was univoltine on maize and multivoltine on hemp. Diapause was triggered by photoperiod, but modified by temperature. Sáringer correlated these environmental triggers to the relative abundance of host plants—either maize or hemp, which drove food choice. Sandru (1986) reported that overwintering larvae collected from hemp were shorter and weighed less than those from maize. Sandru attributed this disparity to differences in the nutritional value between the two host plants.

6.3.3   Cultural and mechanical control

Eradicate feral hemp, wild hops, and *Artemisia* growing near hemp fields, because these weeds serve as alternative hosts. Grigoryev (1998) emphasized that ECB moves from *Artemisia vulgaris* to hemp. ECB can switch from maize to hemp (Table 6.1), so avoid rotating hemp after maize.

Durnovo (1933) reported that 6% of hemp crop stubble harbored overwintering larvae. Vasilyev (1956) recommended destroying all crop residues, and plowing fields in the fall and the spring. Planting late in the spring may avoid first-generation larvae. Maize crops cultivated in organically managed soils suffered less ECB damage than crops cultivated with conventional fertilizers (Phelan *et al.*, 1996). Grigoryev (1998) reported less infestation in densely sown hemp fields (4–5 million plants/ha) than in sparse plantings (0.5 million plants/ha). Goidànich (1928) noted that larvae hiding in harvested stalks are destroyed if the stalks are pond-retted.

Synthetic pheromone traps are available for monitoring moth flights. Gschloessl *et al.* (2013) stated that ECB strains in hemp were attracted to (E)11-tetradecenyl acetate (Fig. 6.6). But Bengtsson *et al.* (2006) reported that a hemp-infesting strain was attracted to (Z)11-tetradecenyl acetate. The (E) and (Z) traps should be separated by at least 30 m, so that the pheromones do not interfere with each other.

Night-flying moths are attracted to light. Taylor and Deay (1950) determined that "blacklight" in the near-ultraviolet region (320–380 nm) worked best. Blacklights can be coupled with funnel traps, although the traps often get clogged by other insects attracted to the light. Blacklights can be surrounded by an electric grid to electrocute ECB (Hienton, 1974).

Breeding hemp for resistance began with Tkalich (1965) in Ukraine. Tkalich *et al.* (1983) reported some resistance in

'ЮCO-14' (YUSO-14, JUSO-14). Virovets and Lepskaya (1983) identified other resistant Ukrainian cultivars (YUSO-22, YUSO-25) and landraces from Yugoslavia, Turkey, France, and Italy. Grigoryev (1998) highlighted YUSO-25, YUSO-27, YUSO-25, Carmagnola (from Italy), Chenevis (from France), and Domaca (from Serbia).

6.3.4   Biological control

In the first description of ECB on hemp, Filippo Re (1813) noted that larvae were parasitized by a wasp that he tentatively identified as *Microgaster deprimator* Fabricius. The wasp crawled into hemp stalks at ECB entry holes, and deposited one egg in the body of each larva. "Let the farmer not be angry with this little animal; on the contrary, may he regard him as a friend and conservationist of hemp!"

Other researchers have studied ECB predators and parasitoids in hemp crops: Lesne (1920) reported that an outbreak in France in the 1880s "was brought to an end by natural enemies." Goidànich (1928) and Zwölfer (1928) studied ECB parasitoids in hemp growing in Italy and Germany. They reported several species in common—ichneumonid wasps, braconid wasps, and tachinid flies. One common species, the braconid *Macrocentrus cingulum*, has been imported and released in the USA (Fig. 6.9). According to Harvard *et al.* (2014), *M. cingulum* parasitizes ECB in North America but less successfully in Europe. It vectors the ECB pathogen *Nosema pyrausta* (see below), but is also infected by it.

A tachinid fly, *Lydella thompsonii* Herting, was introduced by the USDA in the 1920s. It became established, then crashed in the 1960s, possibly due to DDT. The USDA introduced more in the 1980s (Hoffmann and Frodsham, 1993). Adults resemble large, bristly houseflies. Females run up and down hemp stalks and lay eggs next to ECB entrance holes. Hatching larvae enter holes and bore into ECB larvae. Larvae pupate

Fig. 6.9. *Macrocentrus cingulum* laying an egg in *O. nubilalis*. Public domain, courtesy InsectBase, http://www.insect-genome.com/data/detail.php?id=157

within stalks, then emerge as adults. The life cycle can be as short as 20 days. *L. thompsonii* overwinters as maggots in ECB larvae.

Several *Trichogramma* spp. parasitize ECB eggs (see section 20.6). Marin (1979) scattered *T. minutum* pupae from an airplane across 5000 acres of hemp in Russia. Release rates were not given. Tkalich (1967) released 120,000–150,000 wasps/ha, timed when ECB oviposition began in hemp, and repeated when oviposition peaked. Manojlović (1984b) reported that *T. evanescens* parasitized 3.2–7.4% of ECBs infesting hemp, about the same as ECBs infesting maize or hops. Li (1994) said that out of 20 tested species, *T. evanescens* was the second-most effective species against ECBs. *T. ostriniae* is native to China, and was introduced into the USA to control ECB in maize. One USA producer is now marketing it for use in hemp: 20 cards (each with 8000 *T. ostriniae* pupae) per acre, released as soon as moths are first detected. Repeat weekly for 5 weeks.

ECB eggs are preyed upon by green lacewings (*Chrysoperla* spp.) and pirate bugs (*Orius* spp.). The fungus *Beauveria bassiana* kills ECB larvae (Cartwright, 1933). A protozoan species, *Nosema pyrausta* (Paillot), reduces ECB longevity and fecundity (Lewis *et al.*, 2009). Don Jackson of Nature's Control (pers. comm., 1997) suggested killing ECB with *Steinernema* and *Heterorhabditis* nematodes by injecting them into stalks.

Products made from *Bacillus thuringiensis* (Bt) (e.g., Thuricide®) have lost effectiveness due to resistance arising from widespread cultivation of transgenic Bt maize and cotton. Tembo and Pavuk (2011) sprayed either Bt or *Beauveria bassiana*, compared with *Trichogramma pretiosum* against ECBs on maize. *T. pretiosum* provided the least control (the wasps flew away from the treatment plots). *B. bassiana* performed better than Bt, and better than a combination of *B. bassiana* plus Bt. Surface sprays of Bt or chemicals will not affect ECB once it bores into stalks.

6.3.5 Chemical control

For treating choice plants, Frank (1988) used a large-bore syringe to inject stalk galls with horticultural oil. Clarke (pers. comm., 1995) retrofitted an aerosol can of pyrethrin with a spray nozzle and tube, and sprayed the pyrethrin into borer holes. After treatment, wipe away all frass from stems. New excrement indicates a need for repeated treatment.

Two botanicals on the USDA's *National List* of organic substances, neem and ryania, are effective against ECB larvae *… while they are still on leaf surfaces*. Tembo and Pavuk (2011) tested the efficacy of spraying spinosad on ECB, and it worked much better than spraying with Bt or *Beauveria bassiana*, or releasing *Trichogramma pretiosum*. Lizotte (2019) recommended spinosad for treating ECB in hops.

6.4 Asian corn borer

The corn borer in East Asia was traditionally identified as *O. nubilalis*. Then Mutuura and Munroe (1970) dropped the bomb: the primary maize pest in East Asia was *Ostrinia furnacalis* (ACB). *O. nubilalis* (ECB) did not exist in East Asia. This would negate all reports of ECB infesting hemp in East Asia.

Some of these reports of ECB on hemp in East Asia were under synonyms: Onuki (1903) named *Botys lupulinalis*. Others cited *Pyrausta nubilalis* (Takahashi, 1919; Harada, 1930; Clausen, 1931; Cartwright, 1933; Koo, 1940; Senda, 1951; Li, 1951; Sugawara *et al.*, 1953). Kuwayama (1930a) published a 142-page monograph on *P. nubilalis*. He synonymized several taxa under *P. nubilalis*, including *Botys lupulinalis* and *Spilodes kodzukalis* Matsumura 1897, the latter of which Mutuura and Munroe (1970) synonymized under *Ostrinia furnacalis*.

However, some careful studies have revealed ECB in East Asia. For example, Li *et al.* (2003) identified a specimen of *O. nubilalis* in Héběi, near Běijīng, based on mating studies and an examination of genitalia. It was feeding on *Cannabis*. Along the same lines, old reports of *O. nubilalis* in South Asia (e.g., India and Pakistan) (Hampson, 1896) need confirmation. Ditto old reports of *O. nubilalis* infesting hemp in the Russian Far East (Staudinger and Rebel, 1901).

6.4a. *Ostrinia furnacalis* (Guenée) 1854

≡*Botys furnacalis* Guenée 1854, =*Botys damoalis* Walker 1859, =*Botys salentialis* Snellen 1880, =*Spilodes kodzukalis* Matsumura 1897

Description: Larvae, pupae, and adult moths of *O. furnacalis* closely resemble those of *O. nubilalis*. Dissection of male genitalia is required for species identification (Fig. 6.6). But even this is difficult; both species have a "trilobate" uncus, and the lobes are relatively sharp-ended. The ACB sacculus has longer spines than ECB and the spines are less compressed than ABB (Fig. 6.6). Female sex pheromones differ between ACB and ECB which (Fig. 6.6).

Taxonomic notes: Guenée (1854) erected *Botys furnacalis* for a female moth collected in Australia. He said it resembled *B. lupulinalis* (= *O. nubilalis*), but the forewings and hindwings were a slightly different shape and color. Pryer (1885) identified *B. furnacalis* in Japan. Meyrick (1886) found *B. furnacalis* in Fiji. Semper (1896) found it in the Philippines and transferred it to *Pyrausta* as *P. furnacalis*.

Buligan (1929) quoted correspondence from Carl Heinrich, who examined *B. furnacalis* specimens in the Smithsonian Institution. Heinrich considered *B. furnacalis* a synonym of *P. nubilalis*. All of these collections were free-flying moths, and the host plant of *B. furnacalis* was unknown. Mutuura and Munroe (1970) transferred the species to *Ostrinia* as *O. furnacalis*, under which they synonymized *B. damoalis* and *B. salentialis*— pests of maize. Unlike Carl Heinrich, they differentiated *O. furnacalis* from *O. nubilalis* based on their genitalia.

6.4.1 Distribution and host range

ACB is distributed widely. In East Asia: Japan, Korea, the Russian Far East, and eastern China. In Southeast Asia: Thailand, Vietnam, Laos, Cambodia, Malaysia, Indonesia, New Guinea, Philippines, and Solomon Islands. ACB is also found in eastern Australia. Latitudinally, ACB ranges from Hēilóngjiāng Province (48° N) through the equator to Melbourne in Australia

(37.8 °S). In comparison, ECB's latitudinal range is from 53.9 °N (Belarus) to central Egypt (26.2 °N).

ACB now overlaps with ECB in the Yīníng area of northwestern Xīnjiāng Region (43.0 °N), after ACB invaded in the 1990s. Wang et al. (2017) detected gene flow from ACB into ECB in Yīníng, and documented the presence of hybrids. Li and Yang (2022) confirmed that Yīníng is a hybrid zone between ACB and ECB.

Frolov et al. (2007) claimed that ACB was a "maize specialist," akin to their equally-erroneous concept of ECB as a maize specialist in Europe (discussed below). ACB is not a maize specialist. Carefully identified ACB specimens have been reported on *Cannabis* in Korea (Park et al., 2017), and China (Lu et al., 1992; Cao and Deng, 2004; Li et al., 2009). Wang and Wang (2008) dedicated an entire article to ACB infesting hemp in Hēilóngjiāng Province—its life cycle, the damage it causes, and control measures.

Hattori and Mutuura (1987) examined the hosts of ACB in Japan, based on species identification in Mutuura and Munroe (1970). ACB had a wide host range: maize and seven other monocots, three Solanaceae spp., plus Asteraceae, Polygonaceae, Malvaceae, and Fabaceae spp.

Hattori and Mutuura were ambivalent about *Cannabis*, citing a study by Obi (1940) that probably evaluated ACB, but lacked voucher specimens. In Obi's study, *Cannabis* ranked third in host preference, behind maize and millet. ACB infested hemp stalks and flowering tops, and Obi provided three photos of damage. Ishikawa et al. (1999) described ACB as polyphagous.

Of course, ACB did not infest maize until the crop was introduced to East Asia by the Portuguese in 1579. What were the original host plants of ACB? Yang et al. (2021) included Cannabaceae among the ancestral hosts of ACB. Yang et al. (2011) conducted a genetic study of *Ostrinia* spp. in China (plus one sample from France), using the *COI* gene. One ACB specimen was collected from *Cannabis*. It fell into a clade of ACB specimens feeding on maize (Fig. 6.10). ACB is not a maize specialist.

6.4.2   Life history

The ACB life cycle echoes that of ECB. Mature larvae diapause over-winter in crop stubble near the soil line. Springtime

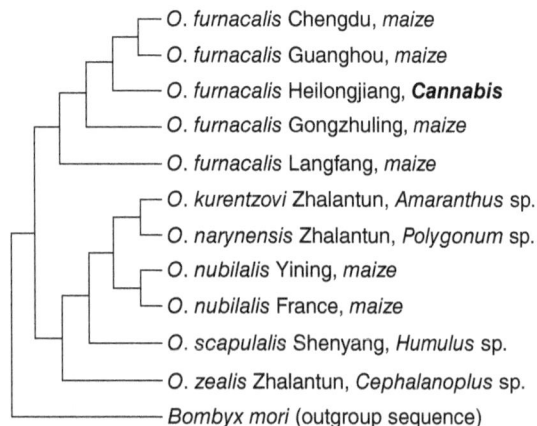

Fig. 6.10. Phylogenetic tree of *Ostrinia* spp. (adapted from Yang et al., 2011)

feeding begins when temperatures exceed 10°C. Larvae pupate for 2 weeks within stalks, and emerge as moths in late May. Moths are night fliers, and long-range migrations are limited to dozens of kilometers. They lay up to 1000 eggs on host plants in masses of 25–50 eggs. Eggs hatch in 3–10 days.

Cartwright (1933) conducted field work on *P. nubilalis* (probably ACB) in East Asia for 3 years. He mapped the number of generations arising in Japan, Korea, and China: 1 generation in Hokkaidō and Jílín Province, 2 generations in central Japan and Korea, 3 generations in Kyūshū, 4 generations in Okinawa, and 5 generations in Taiwan. Seven generations arise in southern China, and in the tropics up to 10.

Kuwayama (1930a) listed hemp, hops, maize, and millet as primary hosts of *P. nubilalis* (probably ACB). He cited six reports of *P. nubilalis* on *Cannabis* in Japan and Korea, including agricultural research stations in Tochigi, Niigata, and Chosen, plus his own observations. He photographed a plant with its stalk snapped at a stalk swelling, its flowering top flopped over and wilted. A ten-page summary was published in English (Kuwayama, 1930b).

Hemp infestations averaged 20–50% in Tochigi (Cartwright, 1933). Further south, in Kyūshū, the pest caused greater economic damage. In Sinŭiju (North Korea), there was "practically a complete loss of the crop." Cartwright reported that some populations attacked *Cannabis* and not maize. Koo (1940) also reported a Japanese strain that preferred *Cannabis* to maize, in both egg-laying and feeding damage.

6.4.3   Control of ACB

For cultural and mechanical control, see the section on ECB above. Wang and Wang (2008) added the novel method of flooding fields in early spring to drown overwintering ACB larvae.

Cartwright (1933) identified parasitoids attacking larvae of *P. nubilalis* (probably ACB) in Japan and Korea. Dominant species were the ichneumonid *Cremastus flavoorbitalis* (Cameron) in June, the braconid *Macrocentrus gifuensis* Ashmead in July, and the braconid *Microgaster tibialis* Nees in August. Highest rates of parasitization were by *M. gifuensis*—now a synonym of *Macrocentrus cingulum* (Fig. 6.9). All three species have been utilized for control of ACB as well as ECB. Cartwright identified a nematode, *Hexamermis meridionalis* Steiner (now *Hexamermis albicans* Siebold), parasitizing borers that infested hemp in Hiroshima.

Cartwright overlooked *Trichogramma* wasps, but they are tiny. Pang and Chen (1974) identified *Trichogramma ostriniae* parasitizing ACB eggs near Běijīng, and it has become an international star for controlling ACB (in China) and ECB (in Europe and the USA). *T. nubilale*, a North American parasitoid of ECB, has controlled ACB in China (see section 20.6). Products made from *Bacillus thuringiensis* (Bt), (e.g., Thuricide®) have lost effectiveness due to resistance arising from widespread cultivation of transgenic Bt maize and cotton. ACB is susceptible to the fungus *Beauveria bassiana* and the microsporidian *Nosema furnacalis*. Pesticide control of ACB on hemp has been effective (Senda, 1951; Sugawara et al., 1953), but those studies used chemicals now banned, such as BHC (lindane).

6.5   Adzuki bean borer

The common name ABB refers to adzuki bean, *Vigna angula-ris*, cultivated in East Asia, where ABB was first described. Walker (1859) coined *Botys scapulalis* for free-flying moths collected in China by G.T. Laye. Early Japanese research on this species fell under the epithet *varialis*, which unfortunately was used for two species: *Botys varialis* Bremer 1864 was described in the Russian Far East. Shibuya (1929) synonymized it under *Botys zealis* Guenée 1894, which Mutuura and Munroe (1970) treated as a subspecies, *Ostrinia zealis varialis*. Mutuura (1954) recombined Bremer's taxon as *Micractis varialis*, but he actually worked with a different species. Mutuura and Munroe (1970) synonymized *Micractis varialis* Mutuura under Walker's *B. scapulalis*, and transferred it to *Ostrinia* as *O. scapulalis*.

6.5a.   *Ostrinia scapulalis* (Walker) 1859

≡*Botys scapulalis* Walker, ≡*Pyrausta scapulalis* (Walker), =*Micractis varialis* Mutuura 1954

Description: Larvae, pupae, and adult moths of ABB closely resemble those of ECB and ACB. Male genitalia differ, but subtly. For example, ABB, ECB, and ACB have a "trilobate" uncus, and the lobes are relatively sharp-ended. The ABB sacculus compared with those of ECB and ACB has shorter spines that are more dorsally compressed (Fig. 6.6).

Mutuura (1954) noted a large mid-tibia in ABB (as *Micractis varialis*), which differentiated it from ECB and ACB (Fig. 6.6). This was repeated by Mutuura and Munroe (1970). Frolov (1981) argued that mid-tibia size was inherited in a simple Mendelian fashion, and ABB was polymorphic regarding mid-tibia size—he found ABB populations with small, medium or large mid-tibiae. Nevertheless, other taxonomists still use the large mid-tibia as a criterion (e.g., Yang *et al.*, 2021).

Researchers in France began studying two *O. nubilalis* populations that they considered races: a race that infested maize, and another that fed on *Artemisia* and hops. Later they referred to these as "*O. nubilalis* sensu Frolov" and "*O. scapu-lalis* sensu Frolov," respectively. These races showed genetic differences (Bourguet *et al.*, 2000) and the races were partially reproductively isolated from each other (Bethenod *et al.*, 2005). Thomas *et al.* (2003) found three other differences. (1) Moths of the *Artemisia*/hops race emerged from pupae 10 days earlier than moths of the maize race. (2) Female moths pro-duced sex pheromones with different E/Z isomeric ratios of 11-tetradecenyl acetate (Fig. 6.6). The *Artemisia*/hops race produced mostly (E)-11-14Ac, and the maize race produced mostly (Z)-11-14Ac. (3) Over-wintering mortality was higher in the *Artemisia*/hops race than the maize race, and this was attributed to parasitism by the braconid wasp *Macrocentrus cin-gulum* (Fig. 6.9).

In genetic studies of the genus *Ostrinia*, ABB is often sister to ECB (e.g., Fig. 6.10). Using eight microsatellite markers, Frolov *et al.* (2012) determined that ABB and ECB populations differed genetically, with a mean $F_{ST} = 0.051$ (fixation index, a measure of population differentiation). Note that Hey and Pinho (2012) proposed $F_{ST} = 0.35$ as a threshold for species

differentiation; pairs with lesser values (e.g., $F_{ST} = 0.051$) are identified as subspecies. European ABB and ECB populations interbreed in natural settings (Malausa *et al.*, 2007a). In a gen-etic study using *COI* and *COII* genes, Piwczyński *et al.* (2016) did not find distinct mtDNA haplotypes among larvae feeding on maize ("*O. nubilalis*") vs hops and *Artemisia* ("*O. scapularis*").

6.5.1   Distribution and host range

ABB is widely distributed. Mutuura and Munroe (1970) named six ABB subspecies based on their geographic distribu-tions (as well as subtle morphological differences): (1) *O. scap-ulalis* subsp. *scapulalis* in SE China (Shanghai, Nánjīng); (2) subsp. *pacifica* in Japan (Hokkaidō); (3) subsp. *subpacifica* in Japan (Yokohama); (4) subsp. *perpacifica* in Korea, the Russian Far East (Ussuri) and NE China (Hēilóngjiāng Province); (5) subsp. *assamensis* in NE India (Assam); (6) subsp. *rossica* in Crimea. Ishikawa *et al.* (1999) mapped a discontinuous Palearctic distribution for ABB: in Ukraine and European Russia, and then in China, Korea, and Japan. In contrast, Frolov *et al.* (2007) mapped a distribution between France and European Russia—and not in East Asia.

As mentioned above, Frolov (1981) considered mid-tibia size a poor criterion for species discrimination, and a group in France began differentiating two races of *O. nubilalis* by their host plants (Bourguet *et al.*, 2000; Bethenod *et al.*, 2005). Frolov and the French group joined forces: Frolov *et al.* (2007) proposed that host plant was the best criterion for distin-guishing species. ABB infested dicotyledonous plants (hemp, hops, *Artemisia* spp., *Xanthium* spp.), and ECB specialized in maize. They argued that *O. nubilalis* evolved from an ancestral species close to today's *O. scapulalis* by acquiring the ability to specialize on maize, an introduced host. However, a genetic study suggested that *O. nubilalis* and *O. scapulalis* began diverg-ing 75,000 to 150,000 years ago, long before the introduction of maize into Europe (Malausa *et al.*, 2007b).

Frolov's defining characteristic of ABB infesting hemp and not maize does not hold. In China's Xīnjiāng Region, ABB fed on both hemp and maize (Xu and Zhang, 1979). Larvae infested up to 78–90% of hemp plants, and 37% of stalks broke. Hattori and Mutuura (1987) examined five *Ostrinia* spp. in Japan, and ABB had the widest host range: maize and three other monocots, eight Asteraceae spp., six Polygonaceae spp., plus Solanaceae, Malvaceae, Geraniaceae spp., and *Cannabis* and *Humulus*. Ishikawa *et al.* (1999) described ABB as polyphagous. Yang *et al.* (2021) listed ABB host plants by family: Cannabaceae, Fabaceae, Gramineae, Polygonaceae, Chenopodiaceae, Euphor-biaceae, and Moraceae.

Frolov's notion that ABB alone and not ECB infests hemp is untenable. Manojlović (1984a) collected ECB larvae from maize and fed them other host plants. Larvae gained the most weight when feeding on maize, but did well on hemp, hops, and *Artemisia*. Fecundity (number of eggs per female) was greatest on maize (<500 eggs), but averaged 300 eggs on hemp and hops. Recall that when ECB was accidently introduced into the USA in hemp stalks, it immediately switched to maize (Vinal,

1917). When Caffrey and Worthley (1927) conducted a host-plant choice experiment, that same population switched back to hemp (Table 6.1).

6.5.2   Life history

ABB life cycle echoes that of ECB: larvae spend the winter in diapause. Calcagno *et al.* (2010) showed that ABB larvae overwinter in host stalks at locations higher than those of ECB larvae. Human selective pressure may play a role: ECB larvae overwinter low in crop stubble—a location near the soil line avoids harvesting equipment. ABB larvae often overwinter in *Artemisia* and other unharvested weeds. Early-instar larvae graze on leaves before boring into stalks, which is where they pupate. The moths are nocturnal and are attracted to light.

6.5.3   Control of ABB

For cultural and mechanical control see the section on ECB. Parasitoids of ABB include the braconid *Macrocentrus cingulum*, tachinid *Lydella thompsoni*, and the egg parasitoid *Trichogramma brassicae*. Thomas *et al.* (2003) reported that overwintering mortality was greater in borers overwintering in *Artemisia*/hops compared with maize, and they attributed greater mortality to parasitism by *M. cingulum* (Fig. 6.9). Pélissié *et al.* (2010) confirmed that parasitism rates and mortality by *M. cingulum* was greater in ABB than ECB. *Nosema pyrausta* is equally effective against ABB, ACB, and ECB (Grushevaya *et al.*, 2020). See comments on *Bacillus thuringiensis* (Bt) in the previous section.

A naturally occurring bacterium, *Wolbachia* sp., kills males or turns them into infertile pseudofemales. Tokarev *et al.* (2017) assessed the *Wolbachia* infection rate in *Ostrinia* larvae collected in Krasnodar: 2.9% in samples collected on maize, 14.3% in samples collected on *Artemisia*, and 65.8% in samples collected on hemp.

6.6   Other *ostrinia* species

Mutuura and Munroe (1970) named two new *Ostrinia* spp. in the trilobate uncus group that infest *Cannabis*. Literature on these species is limited.

6.6a.   Wild hemp stalk borer

Ostrinia narynensis Mutuura and Munroe 1970
Description: M&M segregated *O. narynensis* from *O. nubilalis* because of its larger mid-tibia, as well as a larger middle lobe of its trilobed uncus, and a "hair pencil" on the enlarged mid-tibia. They examined *O. narynensis* specimens collected in two places: near Fort Naryn in Kyrgyzstan, and the Malaya Almatinka River valley in Kazakhstan. They did not offer a host plant.
Comments: Frolov (1979) reported *O. narynensis* feeding on *Cannabis* near Krasnodar in the northern Caucasus. Lu *et al.*

(1992) identified *O. narynensis* feeding on hemp in China's Yúnnán Province. In a Kazakhstan population of *O. narynensis*, Frolov *et al.* (2012) found mid-tibia size a polymorphic character. Based on host plant preference (the larvae fed on hemp), Frolov *et al.* (2012) synonymized *O. narynensis* under *O. scapulalis*. However, a genetic study of *Ostrinia* spp. by Yang *et al.* (2011) showed that *O. narynensis* separated from a clade that included *O. nubilalis* and *O. scapulalis* (Fig. 6.10).

Similarly, in a study that analyzed whole mitochondrial genomes using Bayesian inference, *O. narynensis* separated from a clade that included *O. nubilalis* and *O. scapulalis* (Zhou *et al.*, 2020). Agropages (2018a) assigned the common name 麻秆野螟 (wild hemp stalk borer) to *O. narynensis*, and described its geographic distribution as Xīnjiāng Region and the former Soviet Union. Host plants were hemp, hops, *Xanthium* spp., and *Polygonum* spp.

6.6b.   Asada wild hop stalk borer

Ostrinia kurentzovi Mutuura and Munroe 1970
Description: M&M split *O. kurentzovi* away from *O. nubilalis* because of its larger mid-tibia. They noted that *O. kurentzovi* resembled *O. narynensis*, but *O. kurentzovi* occupied the Russian Far East and Manchuria, whereas *O. narynensis* occupied Central Asia.
Comments: In a phylogenetic analysis based on morphological characters, Yang *et al.* (2008) grouped *O. kurentzovi*, *O. narynensis*, and *O. nubilalis* in a clade, separate from *O. furnacalis*. In a genetic study using the *COI* gene, Yang *et al.* (2011) found that *O. kurentzovi* and *O. narynensis* were sister species (Fig. 6.10). Agropages (2018b) assigned the common name 麻田酒花秆野螟 (Asada wild hop stalk borer) to *O. kurentzovi*, describing its geographic distribution as Hēilóngjiāng, Inner Mongolia, and the Russian Far East. Agropages recorded its host plants as hemp, hops, *Xanthium* spp., and maize.

6.7   Other boring caterpillars

Larvae of six other Lepidopterans damage *Cannabis* stalks. These borers generally do not form galls. Infested plants appear stunted and unhealthy. Inspection of stalks reveals entrance holes, often exuding sawdust-like frass. Hollowed stalks may fall over.

The first two species (*Papaipema nebris*, *P. cataphracta*) are members of the Noctuidae, akin to budworms. The other four species are members of uncommon families.

6.7a.   Common stalk borer

Papaipema nebris (Guenée) 1852
=*Papaipema nitela* (Guenée) 1852
Description: *Young larvae* are reddish-brown with a thin, cream-colored dorsal stripe. A pair of side stripes run from the anal plate to about mid-body. *Mature larvae* slowly lose their stripes and the darker colors fade, leaving a uniform, dirty-gray color.

Heads and anal shields are brown; mandibles and true legs are black. They reach 45 mm in length and bore large exit holes (7 mm dia.). *Adults* have robust brown bodies and grayish brown wings with small white dots, wingspan 30 mm. *Eggs* are white, globular, ridged.

Life history and host range: Common stalk borer is an indiscriminate feeder. Decker (1931) listed 176 species of hosts in 44 plant families, including *Cannabis* in Iowa. Tietz (1972) also reported *Cannabis* infestation. Zuefle (2020) found it in upstate New York (Fig. 6.11). *P. nebris* occurs east of the Rocky Mountains, from southern Canada to the Gulf of Mexico. Larvae usually enter stems near the ground, bore for 25–50 cm, then exit in search of a new host. This restless habit multiplies crop losses. In August larvae enter soil to pupate. Moths emerge in September. Females prefer to deposit eggs on perennial grasses. *P. nebris* causes more crop damage in no-till or minimum-till fields, where eggs are laid in grassy weeds. Eggs hatch in spring. One generation arises per year.

6.7b. Burdock borer

Papaipema cataphracta (Grote) 1864
The burdock borer was reported on hemp by Tietz (1972). Burdock borers resemble common stalk borers (*P. nebris*), except that side stripes are not interrupted, and they are smaller. Larvae pupate in stems, not soil. One generation arises per year. The species is sympatric with common stalk borer.

6.7.1 Ghost moths

Ghost moths or swift moths, family Hepialidae, are considered ancient Lepidopterans that evolved in Gondwana. The males of some species fly swiftly to females. In other species, males hover over females, slowly rising and falling, primarily at dusk ("ghost" moths). Two species reportedly bore into hemp stalks: Japanese swift moth, and European ghost moth.

Fig. 6.11. *Papaipema nebris* in hemp (courtesy M. Zuefle)

6.7c. Japanese swift moth

Endoclita excrescens (Butler) 1877
≡*Hepialus excrescens* Butler 1877, =*Phassus satsamanis* Yazakai 1926
Description: *Larvae* are short and stout with dark brown dorsum and paler brown ventral surface and large black heads. *Adults* have greenish-gray forewings which become brownish after collection, with a lighter bands and black spots. Hindwings are dark brown, wingspan large (81–90 mm). *Eggs* are black-brown, almost globular, 0.8 mm diameter; up to 1000 eggs are laid per female.
Comments: *E. exrescens* is considered a pest of tobacco. It also attacks *Raphanus* (radish) and tree species (*Castanea, Quercus, Salix, Populus, Paulownia* spp.). The species feeds on hemp in Japan (Takahashi, 1919; Clausen, 1931; Shiraki, 1952). Its range also includes North and South Korea, northeast China, and the Russian Far East. Eggs hatch in spring, and larvae feed on stalks of herbaceous plants and then move into tree species. Moths emerge from late August to mid-October in Japan and are crepuscular, flying and mating at dusk.

6.7d. European ghost moth

Hepialus humuli (L.) 1758
Description: *Larvae* are whitish ochrous, maggot-like, with a reddish-brown head and black spiracles. *Female adults* have yellowish-buff forewings with darker linear markings, with brown hindwings, and a wingspan of 25–29 mm. *Male adults* are smaller, and typically have white or silver wings.
Comments: *H. humuli* is common throughout Europe, except for the far southeast. Larvae tunnel into roots and perhaps stalks of plants. They also cut down plants at ground level like cutworms. *H. humuli* attacks *Humulus* as well as *Cannabis*; grasses are its wild hosts (Carter, 1984).

6.7.2 Carpenter moths

Carpenter moths, family Cossidae, are so-called because of their habit of boring into trees and feeding on xylem and phloem. The larvae can migrate to new hosts as they develop. Two species reportedly bore into hemp stalks: goat moth and casuarina borer.

6.7e. Goat moth

Cossus cossus (L.) 1758
Description: *Larvae* are red-violet and grow 90 mm long. *Pupae* are reddish-brown, 38–50 mm long, covered with pieces of wood and debris. *Adults* are robust, brown-bodied, with olive-gray variegated wings reaching a span of 90 mm. Larvae are said to smell like goats (as will plant stalks, even after larvae exit to pupate in soil).
Comments: Goat moth produces smell secretions, hence the name. Ferri (1959a) described larvae attacking Italian hemp.

Moths emerge from pupae in June–July. Goat moth takes 3–4 years to complete its life cycle. *C. coccus* normally infests tree trunks, in Europe, central Asia, and north Africa.

6.7f. Casuarina borer

Zeuzera multistrigata Moore 1881
=*Zeuzera indica* Walker 1856, nec Z. indica Herrich-Schäffer 1854
Description: *Larvae* are predominantly white, with black heads and dark spots. *Adults* have predominantly white bodies, with three pairs of steel-blue spots on the thorax, and seven black bands across the abdomen. Wings are white with steel-blue spots and streaks, veins have an ochreous tinge, wings span 85 mm in females, 65 mm in males. Heads and antennae are black, lower legs blue, femora white. They resemble the Leopard moth (*Z. pyrina* L.).
Comments: Casuarina borer occurs across the Himalayan foothills from Dharmsala to Darjeeling. Baloch and Ghani (1972) described this species causing serious damage in Pakistan. Casuarina borer takes 2–3 years to complete its life cycle. *Z. multistrigata* normally feeds within the trunks of *Casuarina* and *Coffea* spp. (Sorauer 1958).

6.7.3   Control of other boring caterpillars

Follow general advice for ECBs, such as sanitation. Clearing weeds is particularly important for the pair of *Papaipema* species, *P. nebris* and *P. cataphracta*. Mow down grass around fields, and be sure to mow again in August, before moths emerge to lay over-wintering eggs. Infestations of *P. nebris* are common in crops grown in recently broken sod. Most Lepidopterans are killed by Bt. The two Noctuids (*P. nebris* and *P. ataphracta*) might be sensitive to nuclear polyhedrosis virus (NPV).

6.8   Longhorn beetles

Longhorn beetles comprise the Coleopteran family Cerambycidae. They begin our section on grubs—the larvae of beetles and weevils. "Longhorns" are named for their long antennae. The grubs feed within branches and stalks, ejecting excrement at intervals through frass holes. Stalks may snap at tunnel sites. Three species have been cited on *Cannabis*. Controls for longhorns are provided after the weevil section.

6.8a. Hemp longhorn beetle

Thyestilla gebleri (Faldermann) 1835
≡*Saperda gebleri* Faldermann, ≡*Thyestes gebleri* (Faldermann) Bates 1873, =*Thyestes funebris* Gahan 1888, =*Thyestilla lepesmei* Gilmour 1950, plus varietal names: *T. gebleri* var. *pubescens* Thompson 1864, *T. gebleri* var. *funebris* Gahan 1888, *T. gebleri* var. *nigrinus* Plavilstshikov 1915, *T. gebleri* var. *subuniformis* Breuning 1952, *T. gebleri* var. *transitiva* Breuning 1952

Description: *Larvae* are robust, cylindrical, creamy white, with prominent heads capped by a black dot, up to 20 mm long ("roundheaded borers"). *Pupae* are cylindrical, light brown, intermediate in size between grubs and adults. *Adults* are black with white stripes down the prothorax, elongate, cylindrical, up to 15 mm long, with striped antennae nearly as long as their bodies (Fig. 6.12).
Life history and host range: Faldermann (1835) described and illustrated *T. gebleri* collected by Alexander Bunge during his trans-Siberian expedition to northern China. No host plant was recorded; the given location was "northern China." Matsumura (1899) described *T. gebleri* infesting Japanese hemp. Others have characterized *T. gebleri* as the most destructive pest of hemp in Japan (Takahashi, 1919; Clausen, 1931). Shiraki (1952) gave it the common name hemp longhorn. Compilation Group (1972) reported *T. gebleri* in China, and illustrated its stalk damage.

Hemp longhorn beetle occurs in eastern Siberia, Mongolia, northern China, Korea, and Japan. Japanese populations plummeted after World War II, when hemp cultivation was restricted by the USA's Occupational authorities. For a while the species was listed in the *Red Data Book of Endangered Insects in Japan* (McPartland and Hillig, 2007). In the absence of *Cannabis* the species can live on *Artemisia* spp. (sagebrush), *Cirsium* spp. (thistle), and *Boehmeria nivea* (white ramie).

T. gebleri overwinters as grubs in plant debris. Grubs pupate in roots at the ends of tunnels, which they plug with fibrous fragments. Adults emerge in May or June. Females deposit eggs in hemp stalks, one or two at a time, "usually five inches below the first joint" (Takahashi, 1919). One generation arises per year.

6.8b. Thistle longhorn beetle

Agapanthia cynarae (Gyllenhal) 1817
≡*Saperda cynarae* Gyllenhal
=*Agapanthia cynarae* Germar 1825 (not 1817)
Description: *Larvae* are cylindrical, creamy white, with prominent heads, and reach 25 mm long. *Adults* are elongate, cylindrical, greenish-olive with a greenish-gray pubescence, a

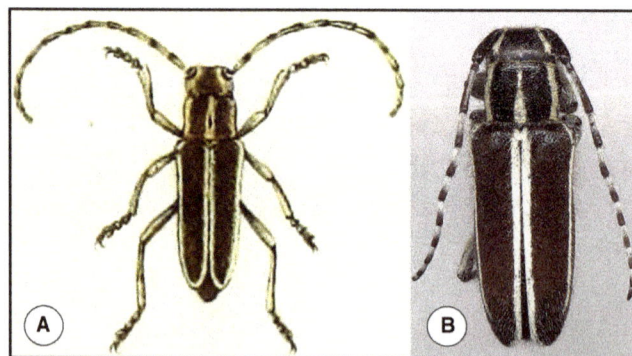

Fig. 6.12. Hemp longhorn: **A.** original drawing by Faldermann (1835); **B.** photograph courtesy Oleg Berlov, http://www.zin.ru/Animalia/Coleoptera/rus/thygebob.htm

black head with three yellow-orange bands, long striped antennae, 14–23 mm long.

Life history and host range: Thistle longhorn beetle bores into stalks of many herbaceous plants, especially thistles (*Carduus* and *Cirsium* spp.) and sunflowers (*Helianthus annuus*). Larvae overwinter in roots, pupate in spring, and adults are active in May and June. They feed on branches and stalks, and lay eggs there. *A. cynarae* is distributed around the Mediterranean—in southern Europe, Turkey, the Levant, and North Africa. Rataj (1957) said it infested hemp in Czech Republic. Fontaine (2019) cited it on *Cannabis* in Crete.

A related species, *Agapanthiola leucaspis* (Steven) 1817, was collected on hemp in central Turkey (Şabanoğlu 2013), and near Orenburg in southwest Russia (Shapovalov 2012). The beetles are metallic blue-green, with black heads, legs, and antennae, and smaller than *A. cynarae* (length 7–10.5 mm). *A. leucaspis* also bores into thistles and many other herbaceous plants. It ranges from Turkey to Transbaikalia in Siberia. An unspeciated *Agapanthia* sp. was collected in Turkey (Mohyuddin and Scheibelreiter, 1973).

6.9 Tumbling flower beetles

Tumbling flower beetles comprise the family Mordellidae. "Tumbling" is performed by adults, to escape predators while feeding in flowers. Adults feed on pollen, a mostly benign activity, although some chew on flowers. Grubs do the damage, by boring into branches and stalks. The adults are also called humpbacked beetles for their convex thorax, or pintail beetles due to their prominent abdominal tip. Mordellidae is a sizeable family with 115 genera and 2310 species. The six species reported on *Cannabis* belong to one genus, *Mordellistena*. They have attracted attention as potential biocontrol weapons against illicit *Cannabis* crops (Baloch and Ghani, 1972; Mushtaque *et al.*, 1973). Controls for Mordellidae are provided after the weevil section.

6.9.1 Symptoms

Depending on where eggs are laid, the grubs bore into narrow branches, petioles, and even central leaf veins. They soon bore into the central stem, and replace pith with fine borings and frass (Fig. 6.13). Stalks at feeding sites often swell. The sites are structurally weak and may snap, causing a wilt of distal plant parts. Damage peaks in late July and early August.

6.9a. Hemp humpback beetle

Mordellistena micans (Germar) 1817
≡*Mordella micans* (Germar)
Description: *Larvae* are cylindrical, legless, yellow, covered with fine bristles, 3–4 mm in length, and sport a dark red barb on their posterior end (Fig. 6.14B). *Pupae* have a red-brown dorsum and a yellow abdomen, 3 mm long (Fig. 6.14C). *Adults* are wedge-shaped, humpbacked, black on the upper surface

and covered with a dense pubescence of gray-brown hair, brown on the underside, 2.5–3.5 mm long (Fig. 6.14D). They have long strong hindlegs like flea beetles. The abdomen tapers to a barb which projects beyond the wings.

Life history and host range: Lindemann (1885) called *M. micans* the "hemp beetle." It infests hemp in Europe (Lindemann, 1885; Kirchner, 1906; Flachs, 1936; Sorauer, 1958; Camprag *et al.*, 1996; Pivtoraiko *et al.*, 2020), and North Africa and Syria (Sorauer, 1958). Larvae overwinter in crop stubble—lower stalks and roots. They pupate there in spring, and adults become active in May. Adults feed on flowers, and lay eggs on leaves. Larvae bore tunnels up to 6 cm long (Kirchner, 1906). One generation arises per year. In addition to hemp, *M. micans* adults tumble around in umbelliferous hosts, such as parsley and carrot.

6.9b. Sunflower humpback beetle

Mordellistena parvula (Gyllenhal) 1827
≡*Mordella parvula* Gyllenhal, =*Mordellistena pusilla* Redtenbacher 1844
Description: *Larvae* resemble those of *M. micans*, except *M. parvula* larvae lack the dark red barb. *Adults* also resemble

Fig. 6.13. Stalk damage by *Mordellistena micans* (R. Clarke)

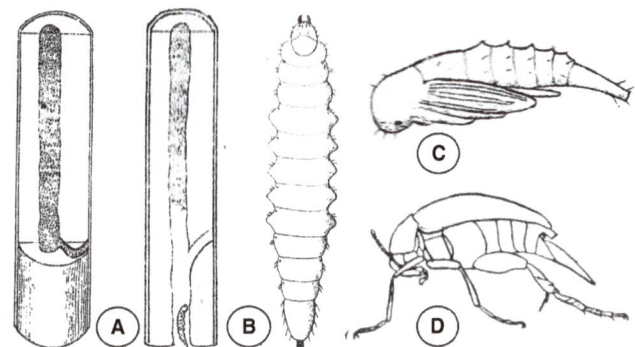

Fig. 6.14. Hemp humpback beetle: **A**. damage (stalk on right shows larva at base); **B**. larva; **C**. pupa; **D**. adult (Lindemann, 1885)

M. micans, except they are more slender and shorter (2.5–3.1 mm long). Emery (1876) made a careful comparison between the adults of *M. parvula* and *M. micans*.

Life history and host range: Krustev (1957) counted 10 grubs per plant on severely damaged crops in Bulgaria. Trotuş and Naie (2008) counted an average density of 1.1 adults/m² in Romanian hemp fields (the worst pest, *Psylloides attenuatus*, averaged 113/m²). Durnovo (1933) reported *M. parvula* as "abundant" in north Caucasus and Kyrgyzstan. In Kyrgyzstan its damage peaked in late July and early August; it infested 95% of early-sown hemp and 66–78% of hemp sown later in April and May. Damage to hemp has also been reported in Russia (Tkalich *et al.*, 1983) and Holland (Batten, 1986). The species ranges from Central Europe, through Iran and Kazakhstan, to possibly China. Besides hemp, the adults feed on pollen of Umbelliferae hosts. Larvae bore into *Artemisia*, *Achillea*, *Cirsium*, *Valerian*, and sunflower, *Helianthus annuus* (Batten, 1986).

Pivtoraiko (2022) studied the life history of *M. parvula* in northeastern Ukraine over the course of three seasons: Larvae overwintered within the lower stalks and the crown (between stalk and root). They stirred out of hibernation when the average daily temperature reached 10°C. Pupation began mid-May to early June, depending on weather conditions. Pivtoraiko calculated growing degree days (°Cd; see section 2.3.1); pupation began after 160 °Cd (T_{BASE} 10°C). Pupation lasted 13–18 days (average 170.6 °Cd) and adults emerged between the third week of May and the second week of June.

Mass flight of adults began mid-June to early July, and the flight period continued until early August, a total of 1.5–2 months—in synchrony with the availability of *Cannabis* pollen. Females laid eggs on upper parts of plants, beginning when the average daily temperature was 24.1–26.2°C. Eggs hatched after 15–20 days. Larvae bored into stalks and worked their way towards the base of the plant. The time period between hibernation ending in spring to diapause in autumn lasted from 64 to 103 days, and required an average of 1571 °Cd (T_{BASE} 10°C).

6.9c. Spike-bearing humpback

Mordellistena connata Ermisch 1969
Description: *Larvae* resemble those of *M. parvula*. *Adults* also resemble those of *M. parvula*, except for anatomical differences in the male's copulatory apparatus, illustrated by Ermisch (1969). The beetles are slightly larger than *M. parvula*, 3.3–3.5 mm long (not counting the pygidium "spike," which is last abdominal somite that extends beyond elytra). Pivtoraiko (2022) coined the common name. The Latin name may be a later homonym: Shelford (1913) cited "*Mordellistena connata* Lec." on plants in the Illinois prairie. "Lec." abbreviates LeConte, but we could not find this taxon in LeConte's *oeuvre*.

Life history and host range: Ermisch (1969) described *M. connata* as a new species in Poland, without naming a host, and identified specimens from Hungary, Albania, and France. Fedorenko *et al.* (2016) found both *M. connata* and *M. parvula* in hemp stalks in Ukraine. Pivtoraiko (2022) studied the life cycles of both species, which were nearly synchronous in their development (see above). *M. connata* was far less common. Sweep-net sampling, at peak flight time, yielded up to 285 *M. parvula* beetles per 100 net sweeps, compared with only 22 *M. connata* beetles.

6.9d. East Asian hemp humpback

Mordellistena comes Marseul 1876
=*Mordellistena cannabisi* Matsumura 1915
Description: *M. comes* lacks a clear diagnostic description; Marseul (1876) gave a rather generic description of the adults. Shiyake (1994) wrote, "This nominal species varies much in body size, ridges on hind leg, and male genitalia, so that it may in reality be a mélange of several species." Larvae as illustrated by Matsumura (1899) are shorter than most *Mordellistena* spp. (Fig. 6.15C).

Life history and host range: Marseul (1876) described *M. comes* in Japan, and several Japanese entomologists have reported it infesting hemp (Matsumura, 1899; Shinji, 1944; Kyokai, 1965). The taxon *Mordellistena cannabisi* is considered a synonym of *M. comes* (Kōno, 1928; Shiyake, 1994), although Sorauer (1958) synonymized *M. cannabisi* under *M. micans*. Takahashi (1919) considered *M. cannabisi* a serious pest in Japan, but Clausen (1931) deemed it "of less consequence." Matsumura (1899) illustrated considerable damage (Fig. 6.15). Compilation Group (1972) reported *M. cannabisi* in China, and illustrated adults and grubs, as well as damage to stalks and roots.

6.9e. South Asian hemp humpback

Mordellistena ghanii Franciscolo 1974
Description: *Larvae* are cylindrical, legs strongly reduced to tufts of spiny hairs (essentially legless), pale yellow, soft and fleshy, covered with fine bristles, 7.4 mm in length; heads sclerotized, pale brown, with darker brown mouthparts, and a

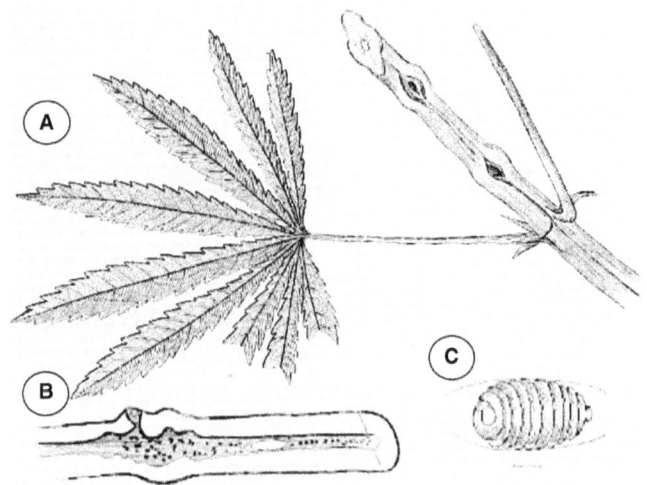

Fig. 6.15. *Mordellistena comes*: **A.** external damage; **B.** internal damage; **C.** larva (Matsumura, 1899)

darker and pointed tail end. *Adults* are moderately slender, parallel-sided, convex (humpbacked), uniformly black with a blond pubescence; males 3.74 mm and females 3.91 mm in length. Antennae with dull brick-brown proximal segments, legs yellowish-brown, heads black.

Life history and host range: Baloch and Ghani (1972) identified a *Mordellistena* sp. infesting *Cannabis* and *Artemisia* in Rawalpindi, Pakistan. They reviewed the literature, and found three *Mordellistena* spp. on *Cannabis*: *M. parvula* (see above), *M. cannabisi* (see above), and *M. gardneri* (see below). Ghani sent specimens in alcohol to Mario Franciscolo for identification.

Franciscolo (1974) named Ghani's specimens a new species, *M. ghanii*. Franciscolo provided an exemplary description (700 words long) of key diagnostic structures in *Mordellistena* adults, with many illustrations. He gave an equally long description of the larvae, a feature often missing in *Mordellistena* descriptions. In Franciscolo's opinion, if more good descriptions were published, larvae of *Mordellistena* might be identifiable to species. The key diagnostic features are: (a) the structure of the last abdominal segment, especially in the arrangement, number and form of tergal spines; (b) the form of the maxilla and maxillary palpus; and (c) the form of the palatum and lower labium.

Franciscolo provided biological notes supplied by Ghani: *M. ghanii* began boring in side branches in May, and ultimately entered the main stalk. Up to seven larvae were found in a single plant. Pupation took place within the stalk, and the adult exited by cutting a small hole. Two generations arose per year. One photograph showed an exit hole, and another photo, of a bisected stalk, showed larvae feeding in the narrow pith characteristic of *afghanica* plants in Pakistan.

6.9f. *Mordellistena gardneri* Blair 1930

Description: *Larvae* not described. *Adults* dark brown with yellow-brown mouth parts, yellow and brown antennae and legs, with silvery pubescence, narrow and elongate, 3.34 mm long.
Comments: Blair (1930) erected *M. gardneri* for beetles he found in Darjeeling, associated with alder trees, *Alnus nepalensis*. Beeson (1941) reported *M. gardneri* from *A. nepalensis*, *Artemisia vulgaris*, and *Cannabis sativa*. Beetles laid eggs at the angle of the leaf petiole and branch or stalk. Larvae bored into the pith of stalks, and the stalks sometimes snapped at gall sites. Franciscolo (1974) examined Blair's type specimens of *M. gardneri*, and the beetles differed from *M. ghanii*. He provided a new description of the adults.

6.9g. Rare Mordellidae

Additional humpbacks have solo citations in hemp literature: Goidànich (1928) reported *Mordellistena reichei* on Italian hemp. The species was originally described in France by Emery (1876), who said that *M. reichei* resembled *M. parvula*, differentiated by "the proportions of the pygidium and the posterior spurs." In Ukrainian hemp, Fedorenko *et al.* (2016) identified

Mordella holomelaena Apfelbeck 1914, in addition to aforementioned *M. parvula* and *M. connata*. The adults of *M. holomelaena* feed on pollen of many plants, the larvae bore in rotten hardwood logs. Schreiner (2019) captured unspeciated *Mordellistena* sp. and *Mordella* sp. in a survey of the arthropod fauna in eastern Colorado.

6.10 Weevils and curculios

Weevils and curculios belong to Family Curculionidae. Weevils are beetles with long curved snouts. They are also known as snout beetles. Curculios are weevils with even longer snouts. Their snouts sprout antennae and chewing mouthparts. About ten species of Curculionids attack *Cannabis*. The grubs feed on pith and cause a swelling at feeding sites. Some grubs feed on roots. The adults chew small holes in leaves, or they notch leaf margins. The holes and notches often become surrounded by chlorotic halos. When disturbed, adult weevils and curculios draw in their legs and antennae, drop to the ground, and play dead.

6.10a. Eurasian hemp weevil

***Rhinoncus pericarpius* (L.) 1758**
≡*Curculio pericarpius* L., =*Curculio castor* Fabricius 1792
Description: *Larvae* are white with dark brown-black heads; they soon grow plump, reaching 4–6 mm in length. *Adults* are broadly oval, dark reddish-brown to black, thinly covered with grayish-yellow hairs, length 3.5–4 mm. The scutellum sports a conspicuous light-colored spot; the beak is slightly longer than the head and slightly ridged (Fig. 6.16). Eggs are oval, white, less than 1 mm long.

Fig. 6.16. Hemp weevil, *Rhinocus pericarpius* (courtesy Udo Schmidt, Wikimedia Commons)

Life history and host range: Matsumura (1899) first cited *R. pericarpius* on Japanese hemp. Harada (1930) considered *R. pericarpius* the most injurious pest of hemp in Japan. Wang *et al.* (1995) labeled it a major pest in Ānhuī Province, China, whereas Clausen (1931) considered it less consequential. The species has been called the "hemp weevil" (Shiraki, 1952; Wang *et al.*, 1995), although it attacks other herbaceous plants: *Saponaria* spp., *Euphorbia* spp., *Rumex* spp., and *Polygonum* spp.

R. percarpius lives throughout temperate Europe and Asia, and now in North America. Grubs feed within stalks and cause galls. The galls are structurally weak and may snap. Grubs pupate in stalks. Adult beetles feed on leaves and overwinter. One generation arises annually. Lu *et al.* (2019) reported another *Rhinoncus* sp. on *Cannabis* in China, **Rhinoncus leucostigma** (Marsham) 1802. The species normally infests *Polygonum*, *Rumex*, and *Rheum* spp.

6.10b. Cabbage curculio

Ceutorhynchus rapae Gyllenhal 1837
Description: *Young larvae* have cylindrical white bodies and brown heads. *Mature larvae* slightly darken in color, become somewhat plump, and 4–6 mm in length (Fig. 6.17). They assume a C-shape when exposed. *Adults* are oblong to oval, gray to black, and covered with yellow to gray hair-like scales. The curved snout is slightly longer than the head and thorax; the snout is slender, cylindrical, finely punctuated with striae on the basal half, and antennae arise near its middle. Wing covers are marked by fine longitudinal ridges and rows of shallow punctuations (Fig. 6.17). Length 3.5–5.0 mm.
Life history and host range: *C. rapae* is called the "cabbage curculio" in the USA, but in Hungary it is called the "hemp curculio" (Nagy *et al.*, 1982). The species is native to west-central Europe, and was introduced into North America about

1855. *C. rapae* infests hemp in the Czech Republic (Rataj, 1957), Hungary (Nagy *et al.*, 1982), Serbia (Camprag *et al.*, 1996), Italy (Goidànich, 1928; Ferri, 1959a, 1961c; Tremblay, 1968), and Iran (Ghahari and Colonnelli, 2012). Nagy *et al.* (1982) reported up to 40% crop losses in Hungary.

Adults overwinter and emerge in April–May to mate. Females insert eggs into hemp stems when seedlings are 3–8 cm tall. As a result, only the lower 50 cm of mature plants contain pith-feeding grubs (Nagy *et al.*, 1982). Grubs leave exit holes in June to pupate in small cocoons just beneath the soil surface. Adults emerge in late June–July and feed on leaves. One generation arises per year. *C. rapae* primarily infests cruciferous crops.

6.10c. East Asian hemp weevil

Cardipennis rubripes (Hustache) 1916
≡*Ceutorhynchus rubripes* Hustache 1916
Description: Korotyaev (1980) erected the genus *Cardipennis* for three East Asian *Ceutorhynchus* spp., including *C. rubripes*. They exhibited unique characters: tibia lack mucrones (tooth-like processes); the funicle (mid-part of antenna) has seven segments; the slender rostrum (curved snout) is longer than the pronotum (thorax); the pronotum has an apical restriction; the tarsal claws have a large tooth (all illustrated by Huang *et al.*, 2018). Otherwise, *C. rubripes* adults look like those of *Ceutorhynchus* spp.: body plump, oval in outline; dorsum matt-black and covered with fine, brown or whitish hair-like scales, with a light-colored scutellar spot. Wing covers are marked by fine longitudinal ridges, finely punctate. Legs reddish-brown and densely covered with white and brown scales. Length 2.3–2.4 mm (Fig. 6.18).
Life history and host range: Hustache (1916) collected *C. rubripes* in Kyōto; he did not list a plant host. Korotyaev (1980) connected it with *Cannabis*, and transferred it to *Cardipennis*. The other two species that he transferred to *Cardipennis* infest *Humulus*: *C. shaowuensis* (Hustache) 1916 and *C. sulcithorax* (Voss) 1958.

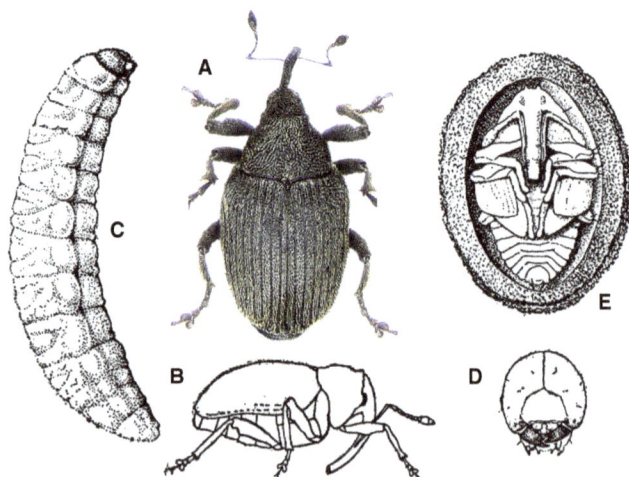

Fig. 6.17. Cabbage curculio: **A**. adult from above; **B**. adult from side; **C**. grub; **D**. head of grub from front; **E**. pupa. (Line drawings from Blatchley (1916); color photo courtesy Udo Schmidt, Wikimedia Commons)

Fig. 6.18. East Asian hemp weevil. (courtesy Runzhi Zhang (Huang *et al.*, 2018))

Shilenkov and Tolstonogova (2006) found *C. rubripes* on *Cannabis ruderalis* in the Baikal region of Siberia. It infested 3.5% of plants, surpassed only by the hemp flea beetle (*Psylliodes attenuatus*) with 10.6%. The authors described larvae boring into stalks, and adults chewing holes in flowers and developing seeds. Feeding was limited to *Cannabis*, and Shilenkov and Tolstonogova considered *C. rubripes* an excellent biocontrol against *Cannabis*. They collaborated with USDA scientists in California, who proposed weaponizing *C. rubripes* against illegal marijuana (Carruthers, 2004).

Huang *et al.* (2018) examined over 1500 *Cardipennis* specimens in museums. *C. rubripes* specimens came from Japan, Korea, Mongolia, Siberia, and many places in China (Inner Mongolia, Běijīng, Héběi, Hénán, Húnán, Ānhuī, Shānxī, Chóngqìng, Yúnnán, Gānsù). Filimonov (2012) collected *C. rubripes* from Chelyabinsk Oblast, east of the Ural Mountains, so the species may be moving towards Europe.

6.10d. Other weevils and curculios

Tremblay and Bianco (1978) reported the cauliflower weevil, *Ceutorhynchus pallidactylus* (Marsham) 1802, feeding on hemp. The life cycle of *C. pallidactylus* mimics that of *C. rapae* except *C. pallidactylus* eggs are laid on leaves. *C. pallidactylus* weevils are slightly smaller than *C. rapae* but otherwise identical; the grubs cannot be differentiated. *C. pallidactylus* also infests hemp in Iran (Ghahari and Colonnelli, 2012).

Tremblay (1968) reported three species on Italian hemp: *Ceutorhynchus pleurostigma* (Marsham) 1802, *Ceutorhynchus quadridens* (Panzer) 1795, and *Ceutorhynchus roberti* Gyllenhal 1837. He illustrated the morphological differences between them. Colonnelli (2004) cited three species on hemp: the turnip weevil *Ceutorhynchus assimilis* (Paykull) 1800, the cabbage stem weevil, *Ceutorhynchus pallidactylus* (Marsham) 1802, and *Ceutorhynchus anatolicus* Schultze 1900.

Mohyuddin and Scheibelreiter (1973) identified *Ceutorhynchus macula-alba* Herbst 1795 on Rumanian hemp. They tested the species as a biocontrol agent against marijuana. Goidànich (1928) cited seven other Curculiónidae on Italian hemp—*Ceutorhynchus sulcicollis* (Paykull) 1800, *Gymnetron labile* (Herbst) 1795, *Gymnetron pascuorum* (Gyllenhal) 1813, *Polydrusus sericeus* (Schaller) 1783, *Sitona humeralis* (Stephens) 1831, *Sitona lineatus* (L.) 1758, and *Sitona sulcifrons* (Thumberg) 1798.

Li (1985) tagged a species "the hemp weevil," but he only identified it to genus, *Ceuthorrhynchus* sp. It attacked crops in Gānsù Province. Li illustrated eggs, larvae, pupae, adults, and two great images of plant symptoms (stalk galls, plant stunting). He detailed its life history, meteorological conditions leading outbreaks, and control measures.

Other reports: the *Cannabis*-specific weevil *Lepropus sublateralis* Poorani in India (Poorani and Ramamurthy, 1997), the Asiatic oak weevil, *Cyretepistonmus castaneus* (Rolofs) on Mississippi marijuana (Lago and Stanford, 1989), the Ber weevil, *Xanthoprochilis faunus* Olivier 1807 in India (Cherian, 1932), and *Corigetus mandarinus* Fairmaire 1888 in Vietnam (Hanson, 1963).

Seed weevils, *Apion* spp., have been reported in India (Batra, 1976) and Italy (Goidànich, 1928). *Apion* spp. are called weevils but belong to the family Apionidae, not the Curculionidae. Conversely, *Thamurgus* spp. are called bark beetles (Scolytinae), but are now folded into the weevil family (Curculionidae). *Thamurgus caucasicus* Reitter 1887 has infested *Cannabis* stems in Astrakhan, Russia (Mandelshtam *et al.*, 2012).

The clover stem borer, *Languria mozardi* Latreille 1807, is a member of the Cucujoidae, not the Curculionidae, but we'll mention it here. Garman (1906) observed the grubs "eating out the stems of hemp" in Kentucky. Adults are eye-catching, with an orange-red head and thorax, and metallic black elytra, 4–9 mm long

6.10.1 Differential diagnosis

Adult longhorn beetles are big, and their "longhorns" give them away. Tumbling flower beetles (*Mordellistena* spp.) are the smallest beetles in hemp, and their humpbacks and pintails give them away. Weevils and curculios have unmistakable snouts. Grubs follow the same size trends: longhorn grubs are bigger than the others. Grubs of *Mordellistena* spp. are the smallest, and lack visible legs. Grubs of weevils and curculios are difficult to differentiate from flea beetle grubs.

Galleries tunneled by longhorns are of wider caliber than those of other boring beetles and weevils. Tunneling by *Mordellistena* and *Ceutorhynchus* spp. tends to cause a slight swelling in stalks. These swellings look similar to symptoms caused by stem-boring caterpillars.

6.10.2 Cultural and mechanical control

Apply methods described in the ECB section, above. Feral hemp and wild hops serve as alternative hosts for many boring beetles and weevils, so eradicate those hosts. The same goes for *Artemisia* (sagebrush) and *Cirsium* (thistle). Many boring beetles overwinter in plant debris and crop stubble. Destroy plant residues, and plow crop stubble in the fall and the spring. Many *Ceutorhynchus* spp. on hemp, especially *C. rapae*, primarily infest *Brassica* spp., so do not rotate after a cruciferous crop, and eliminate cruciferous weeds.

Alternatively, use *Brassica* spp. as a trap crop. Destroy trap crops after eggs are deposited on them. Collect adults by laying a tarp under infested plants and shaking the plants. Weevils and curculios fall to the ground and remain motionless long enough to be gathered for soup. No synthetic pheromone traps are commercially available for beetles and weevils. *C. rapae* and other *Brassica* pests can be caught in yellow water traps baited with *Brassica* volatiles (isothiocyanates). No breeding for resistance appears in the literature.

6.10.3 Biological and chemical control

Parasitoid wasps have been identified. *Trichomalus perfectus* (Walker) parasitizes several *Ceutorhynchus* spp. (Haye *et al.*, 2015). Biocontrol researchers are testing *Scambus pterophori*, a parasitoid of stem-boring beetle grubs (especially weevils).

Grubs of *Mordellistena* spp. in *Cannabis* stems were parasitized by three parasitoid wasps in Pakistan: *Tetrastichus* sp., *Rhaconotus* sp., and *Buresium* sp. (Baloch and Ghani, 1972). *Buresium naso* was reared from grubs of *Mordellistena* spp. in *Cannabis* stems in Pakistan (Bouček, 1983).

Entomopathogenic nematodes (*Steinernema scarabaei*, *Heterorhabditis bacteriophora*) may control beetles and weevils that overwinter in roots and plant debris. The biocontrol bacterium *Bacillus popilliae* is another option. In small operations, inject stalks with nematodes or *B. popilliae*. For chemical control, see recommendations in the *O. nubilalis* section. No surface sprays kill borers inside stalks.

(Prepared by J. McPartland)

References

Agropages. 2018a. 麻秆野螟. Available at: http://cn.agropages.com/Bcc/Bdetail-744.htm

Agropages. 2018b. 麻田酒花秆野螟. Available at: http://cn.agropages.com/Bcc/Bdetail-748.htm

Asano K. 2012. 名古屋大学東山構内の蛾類リスト. *Bulletin of the Nagoya University Museum* 28: 99–118.

Bae YS, Park KT. 1997. Systematic study of the genus *Grapholita* Treitschke (Lepidoptera, Tortricidae) from Korea. *Korean Journal of Biological Sciences* 1: 539–547.

Bae YS, Chae DY, Ahn NH. 2012. *Insect fauna of Korea, vol. 16, no. 3: Arthropoda: Insecta: Lepidoptera: Tortricidae: Olethreutinae Leafroller II*. National Institute of Biological Resources, Incheon, S. Korea.

Bako L, Nitri I. 1977. Pokusy s ochranou proti obalovaci konopnému (*Grapholitha sinana* Feld). *Len a Konopí* 15: 13–31.

Baloch GM, Ghani MA. 1972. *Natural enemies of Papaver somniferum and Cannabis sativa*. Annual report, Commonwealth Institute of Biological Control, Pakistan station, pp. 55–56.

Baloch GM, Mushtaque M, Ghani MA. 1974. *Natural enemies of Papaver spp. and Cannabis sativa*. Annual report, Commonwealth Institute of Biological Control, Pakistan station, pp. 56–57.

Batra SW. 1976. Some insects associated with hemp or marijuana (*Cannabis sativa* L.) in northern India. *Journal of the Kansas Entomological Society* 49: 385–388.

Batten R. 1986. Die Nederlandse soorten van de Keverfamilie Mordellidae. *Zoologische Bijdragen* 19: 1–37.

Beeson FC. 1941. *The ecology and control of the forest insects of India and the neighboring countries*. Government of India, New Delhi.

Bengtsson M. Karpati Z. Szöcs G, *et al*. (3 other authors). 2006. Flight tunnel responses of Z strain European corn borer females to corn and hemp plants. *Environmental Entomology* 35: 1238–1243.

Bertolini G. 1842. *De Botyde silacealis deque damno quo afficit Cannabin sativam L.* Emygdii ab Ulmo, Bonaniae, Italy.

Bertolini G. 1844. De *Botyde silaceali*, deque damno, quo afficit *Cannabim Sativam*. *Novi Commentarii Academiae Scientiarum Instituti Bononiensis* 6: 91–100.

Bes A. 1967. Konopljin savijac (*Grapholitha delineana* Walk.) nova stetocina konoplje. *Zastita Bilja* 18: 399–400.

Bes A. 1978. Prilog poznavanju izgledna osteçenja i stetnosti konopljinog savijaca — *Grapholitha delineana* Walk. *Radovi Poljoprivrednog Fakulteta Univerzita u Sarajevu* 26(29): 169–189.

Bethenod MT, Thomas Y, Rousset F, *et al.* (4 other authors). 2005. Genetic isolation between two sympatric host plant races of the European corn borer, *Ostrinia nubilalis* Hübner II—assortative mating and hostplant preferences for oviposition. *Heredity* 94: 264–270.

Biosecurity Australia. 2010. *Draft review of policy: importation of hops (Humulus species) into Australia*. Biosecurity Australia, Canberra.

Blair KG. 1930. Some new species of Indian Heteromera. *Entomologist's Monthly Magazine* 66: 177–181.

Blatchley WS. 1916. *Rhynchophora or weevils of north eastern America*. Nature Publishing Co., Indianapolis.

Bolt M. 2019. A rough season can lead to invaluable information on hemp production. *Purdue University Extension Entomology Pest & Crop Newsletter*. Available at: https://extension.entm.purdue.edu/newsletters/pestandcrop/article/a-rough-season-can-lead-to-invaluable-information-on-hemp-production/

Bosc LAG. 1809. "Chanvre," pp. 307–317 in *Nouveau cours complet d'agriculture théorique et pratique, tome 3*. Deterville, Paris.

Bouček Z. 1983. On *Buresium*, *Masneroma* (n.gen.) and some other Eurytomidae (Hymenoptera). *Entomologica Scandinavica* 14: 186–194.

Bourguet D, Bethenod M-T, Trouvé C, Viard F. 2000. Host-plant diversity of the European corn borer *Ostrinia nubilalis*: what value for sustainable transgenic insecticidal Bt maize? *Proceedings of the Royal Society of London Series B* 267: 1177–1184.

Bradley JD. 1961. Microlepidoptera from the Solomon Islands. *Bulletin of the British Museum (Natural History) Entomology* 10: 113–168.

Brandolini A, Brandolini A. 2001. Classification of Italian maize (Zea mays L.) germplasm. *Plant Genetics Resources Newsletter* 126: 1–11.

Brindley TA, Milton KM, Guthrie WD. 1975. European corn borer: a bibliography 1925-1973. *Iowa State Journal of Research* 49: 217–253.

Buligan CT. 1929. The corn borer, *Pyrausta nubilalis* Hübner (Pyralidae, Pyraustinae, Lepidoptera). *Philippine Agriculturist* 17: 397–450.

Bush Doctor. 1987b. European corn borers. *Sinsemilla Tips* 7(2): 45–47.

Bykovskaya AV, Samonov AS. 2018. Влияние гидротермических условий на ареал стеблевого кукурузного мотылька (*Ostrinia nubilalis* Hbn.) в Беларуси. *Защита Растений* 42: 201–208.

Byun BK, Lee BW, Bae KH, Lee KJ. 2012. A review of the genus *Grapholita* (Lepidoptera, Tortricidae) in North Korea. *Animal Systematics, Evolution and Diversity* 28: 291–296

Caffrey DJ, Worthley LH. 1927. *A progress report on the investigation of the European corn borer*. USDA Bulletin no. 1476. Government Printing Office, Washington, DC.

Calcagno V, Bonhomme V, Thomas Y, Singer MC, Bourguet D. 2010. Divergence in behaviour between the European corn borer, *Ostrinia nubilalis*, and its sibling species *Ostrinia scapulalis*: adaptation to human harvesting? *Proceedings of the Royal Society, Biological Sciences* 277: 2703–2709.

Camprag D, Jovanic M, Sekulic R. 1996. Štetočine konoplje i integralne mere suzbijanja. *Zbornik Radova* 26/27: 55–68.

Cao GC, Deng BZ. 2004. 特用玉米栽培实用技术. China Agricultural Press, Beijing.

Carruthers RI. 2004. *Classical biological control of narcotic plants*. Western Regional Research Center, Albany, California. Available at: https://reeis.usda.gov/web/crisprojectpages/0402468-classical-biological-control-of-narcotic-plants.html

Carter DJ. 1984. *Pest Lepidoptera of Europe, with special reference to the British Isles*. W. Junk, Dordrecht.

Cartwright WB. 1933. *Observations on the European corn borer and its major parasites in the orient*. USDA Circular no. 289. Washington, DC.

Ceapoiu N. 1958. *Cinepa, studiu monografic*. Editura Academiei Republicii Populare Romine. Bucharest.

Cherian MC. 1932. Pests of ganja. *Madras Agricultural Journal* 20: 259–265.

Christoph HT. 1882. Neue Lepidopteren des Amurgebietes. *Bulletin de la Société Impériale des Naturalistes de Moscou* 56(4): 405–436.

Clarke JFG. 1958. *Catalogue of the type specimens of Microlepidoptera in the British Museum (Natural History) described by Edward Meyrick. Vol. III. Tortricidae, Olethreutidae, Noctuidae*. British Museum, London.

Clausen CP. 1931. *Insects injurious to agriculture in Japan*. USDA Circular no. 168. Washington, DC.

Colonnelli E. 2004. *Catalogue of Ceutorhynchinae of the world, with a key to genera (Insecta Coleoptera Curculionidae)*. Argania Editio, Barcelona.

Compilation Group. 1972. 中国农作物病虫图谱. 第四分册, 棉麻病虫. Agricultural Publishing House, Beijing.

Cotuna O, Panda A, Sărăţeanu V, Durău C, Paraschivu M. 2020. Research regarding the attack of the Eurasian hemp moth *Grapholita delineana* Walker in several hemp varieties cultivated at SCDA Lovernin (Western Romania). *Life Science and Sustainable Development* 1: 13–19.

Cranshaw WS, Schreiner M, Britt K, Kuhar TP, McPartland JM, Grant J. 2019. Developing insect pest management systems for hemp in the United States: a work in progress. *Journal of Integrated Pest Management* 10(1): 26; 1–10.

Danilevsky AS, Kuznetsov VI. 1968. Листовёртки Tortricidae. Триба плодожорки Laspeyresiini. *Fauna of the USSR, Lepidopterous Insects* 5 (Part 1): 261–263.

Decker GC. 1931. *The biology of the stalk borer Papaipema nebris (Gn)*. Iowa State College Agricultural Experiment Station Research Bulletin no. 143, Ames, Iowa.

Diakonoff A. 1982. On a collection of some families of micro-Lepidoptera from Sri Lanka (Ceylon). *Zoologische Verhandelingen* 193: 3–124.

Dorhout DL, Sappington TW, Rice ML. 2008. Evidence for obligate migratory flight behavior in young European corn borer (Lepidoptera: Crambidae) females. *Environmental Entomology* 37: 1280–1290.

Durnovo ZP. 1933. "Итоги работ по кукурузному мотыльку и другим вредителям однолетних новых лубяных культур," pp. 85–106 in Болезни и вредители новых лубяных культур. Institute New Bast Raw Materials, Moscow.

Emchuck EM. 1937. Some data on the injurious entomofauna of the truck farms and orchards of the Desna river region. *Travaux de l'Institut de Zoologie et Biologie. Académie des Sciences d'Ukraine* 14: 279–282.

Emery MC. 1876. Essai monographique sur les Mordellides de l'Europe et des contrées limitrophes. *L'Abeille, Journal d'Entomologie*, 3e série 2: 1–128.

Ermisch K. 1969. Neue Mordelliden aus Europa, Nordafrika und dem Nahen Osten (Coleoptera, Mordellidae). 59. Beitrag zur Kenntnis der Mordelliden. *Entomologische Blätter für Biologie und Systematik der Käfer* 65: 104–115.

Esaki T, Uchida S, Ishii T. 1932. 日本昆蟲圖鑑 = *Iconographia Insectorum Japonicorum*. Hokuryūkan, Tokyo.

Faldermann F. 1835. Coleopterorum ab illustrissimo Bungio in China boreali, Mongolia et montibus Altaicis collectorum. *Mémoires de l'Académie Impériale des Sciences de St. Pétersbourg, 6e série* 2: 337–464.

Fedorenko VP, Kabanetz VV, Kabanets VM. 2016. Шкідники конопель посівних. National Academy of Agrarian Science, Kiev.

Felder R, Rogenhofer AF. 1874. Reise der Österreichischen Fregatte Novara um die Erde in den Jahren 1857, 1858, 1859. *Zoologischer Theil, Zweiter Band*, pl. CXXXVII, fig. 42.

Ferri F. 1959a. *Atlante delle avversità della canapa*. Edizioni Agricole, Bologna, Italia.

Ferri F. 1961c. Le avversità delle piante viste alla lente: canapa. *Progresso Agricolo (Bologna)* 7: 764–765.

Filimonov RV. 2012. "К фауне долгоносикообразных жуков (Coleoptera, Curculionoidea) памятника природы "Черный Бор" (Челябинская область). *Proceedings of the Orenburg branch of the Russian Entomological Society* 2012(2): 77–94.

Filippo Re. 1806. *Sopra le danneggiarono ne' due anni scorsi i Canapaj in alcunni Dipartimenti del Regno d'Italia*. G. Silvestri, Milano.

Filippo Re. 1813. Lettre sur les insectes nuisibles. *Annales de l'Agriculture Française* 53: 215–247.

Flachs K. 1936. Krankheiten und schädlinge unserer gespinstpflanzen. *Nachrichten über Schädlingsbekämpfung* 11: 6–28.

Fletcher TB. 1932. *Life-histories of Indian Microlepidoptera, second series: Alucitidae (Pterophoridae) Tortricina and Gelechiadae*. Imperial Council of Agricultural Research, Calcutta.

Fontaine P. 2019. Faune—853—*Agapanthia cynarae* (Gyllenhal, 1817). Available at: http://phrygana.eu/Fauna/Coleoptera/Cerambycidae/Agapanthia-cynarae/Agapanthia-cynarae.html

Franciscolo ME. 1974. New and little-known *Mordellistena* Costa from Pakistan and India (Coleoptera: Mordellidae). *Oriental Insects* 8: 71–84.

Frank M. 1988. *Marijuana grower's insider's guide*. Red Eye Press, Los Angeles.

Frolov AN. 1979. "The main trends of differentiation of corn borers *Ostrinia* Hübner," pp. 25–33 in *Proceedings of the All-Union Research Institute for Plant Protection. Questions of ecological physiology of insects and problems of plant protection*. Leningrad, USSR.

Frolov AN. 1981. Genetic analysis of 'large' tibia - the taxonomic character of brush-leg borer *Ostrinia scapulalis* Wlk. (Lepidoptera, Pyraustidae). *Genetika* 17: 2160–2166.

Frolov AN, Bourguet D, Ponsard S. 2007. Reconsidering the taxonomy of several *Ostrinia* species in the light of reproductive isolation: a tale for E. Mayr. *Biological Journal of the Linnean Society* 91: 49–72.

Frolov AN, Audiot P, Bourguet D, et al. (5 other authors). 2012. From Russia with lobe: genetic differentiation in trilobed uncus *Ostrinia* spp. follows food plant, not hairy legs. *Heredity* 108: 147–156.

Garman H. 1906. *Observations and experiments on clover, alfalfa and soy beans*. Kentucky Agricultural Experiment Station Bulletin no. 125. Lexington, Ky.

Gerginov L. 1974. *Grapholitha delineana* Walk., a new hemp pest in Bulgaria. *Rastenievud Nauk* 11(1): 147–154.

Ghahari H, Colonnelli E. 2012. Curculionoidae from Golestan Province, northern Iran. *Fragmenta Entomologica* 44: 101–161.

Goidànich A. 1928. Contributi alla conoscenza dell'entomofauna della canapa. I. Prospetto generale. *Bollettino del Laboratorio di Entomolgia del R. Istituto Superiore Agrario di Bologna* 1: 37–64.

Grigoryev SV. 1998. Survey of the VIR *Cannabis* collection I: the resistance of accessions to corn stem borer (*Ostrinia nubilalis* Hb.). *Journal of the International Hemp Association* 5(2): 72–74.

Grushevaya I, Ignatieva A, Tokarev Y. 2020. Susceptibility of three species of the genus *Ostrinia* (Lepidoptera: Crambidae) to *Nosema pyrausta* (Microsporidia: Nosematida). BIO Web of Conferences 21: e00040.

Gschloessl B, Beyne E, Audiot P, Bourguet D, Streiff R. 2013. De novo transcriptomic resources for two sibling species of moths: *Ostrinia nubilalis* and *O. scapulalis*. *BMC Research Notes* 6(1): e73.

Guenée AM. 1854. *Histoire naturelle des Insectes. Species général des Lépidoptères, Tome Huitième, Deltoïdes et Pyralites*. De Roret, Paris.

Hackleman JC, Domingo WE. 1943. *Hemp: an Illinois war crop*. University of Illinois Agricultural Experiment Station Circular No. 547. Urbana, Illinois.

Hampson GF. 1896. *The fauna of British India, including Ceylon and Burma. Moths–vol. 4*. Taylor and Francis, London.

Haney A, Kutscheid BB. 1975. An ecological study of naturalized hemp (*Cannabis sativa* L.) in east-central Illinois. *American Midland Naturalist* 93: 1–24.

Hanson HC. 1963. *Diseases and pests of economic plants of Vietnam, Laos and Cambodia*. American Institute of Crop Ecology, Washington DC.

Harada T. 1930. 大麻の害患難を大麻象鼻貴職に就きて. *Konchū Sekai* (Insect World) 34: 118–123.

Harvard S, Pélissier C, Ponsard S, Campan EMD. 2014. Suitability of three *Ostrinia* species as hosts for *Macrocentrus cingulum*: A comparison of their encapsulation abilities. *Insect Science* 21: 93–102.

Hattori I, Mutuura A. 1987. 日本産アワノメイガ属(Ostrinia)の種の同定と寄生植物. *Plant Protection* 41: 62–69.

Haye T, Mason PG, Gillespie DR, *et al*. (5 other authors). 2015. Determining the host specificity of the biological control agent *Trichomalus perfectus* (Hymenoptera: Pteromalidae): the importance of ecological host range. *Biocontrol Science and Technology* 25: 21–47.

Heinemann H von. 1863. *Die Schmetterlinge Deutschlands und er Schweiz, zweite abtheilung, band I, helft I*. Schwetschke und Sohn, Brunswick.

Heuzé G. 1860. *Les plantes industrielles, seconde partie*. Hachette et Cie, Paris.

Heuzé G. 1864. *L'agriculture de l'Italie septentrionale*. Hachette et Cie, Paris.

Hey J, Pinho C (2012) Population genetics and objectivity in species diagnosis. Evolution 66: 1413–1429.

Hienton TE. 1974. *Summary of investigations of electric insect traps*. USDA Technical Bulletin No. 1498, Washington, DC.

Hill DS. 1983. *Agricultural insect pests of the tropics and their control, 2nd edn*, Cambridge University Press, Cambridge, UK.

Hinneberg C. 1895. *Grapholitha tetragrammana* Stgr. und *Plutella incarnatella* Steudel, zwei bisher wenig beobachtete Kleinschmetterlingsarten. *Stettiner Entomologische Zeitung* 55: 345–357.

Hodgson BE. 1928. *The host plants of the European corn borer in New England*. USDA Bulletin no. 77. Washington, DC.

Hoffmann MP, Frodsham AC. 1993. *Natural enemies of vegetable insect pests*. Cooperative Extension Bulletin, Cornell University, Ithaca, NY.

Huang JH, Yoshitake H, Qin M, Zhang RZ, Ito H. 2018. Taxonomic study of the East Palaearctic genus *Cardipennis* Korotyaev (Coleoptera: Curculionidae: Ceutorhynchinae). *Transactions of the American Entomological Society* 144: 217–238

Hübner J. 1796. *Sammlung europäischer Schmetterlinge*. No publisher identified, Augsburg.

Hustache A. 1916. Synopsis des Ceuthorrhynchini du Japon. *Annales de la Société Entomologique de France* 85: 107–144.

Ishikawa Y, Takanashi T, Kim CG, *et al*. (3 other authors). 1999. *Ostrinia* spp. in Japan: their host plants and sex pheromones. *Entomologia Experimentalis et Applicata* 91: 237–244.

Issiki S. 1957. "Eucosmidae, Tortricidae," pg 54 in: Esaki T *et al*., eds. *Icones Heterocerorum Japonicorum in Coloribus Naturalibus*. Hoikusya Press, Osaka.

Jablonowski J. 1897. A kukoriczamoly életmódjáról és kártételeiről. *Rovartani Lapok* 4(8): 164–166.

Jablonowski J. 1930. The black locust-tree-scale, Lecanium robiniarum Dougl., and the European corn borer, *Pyrausta nubilalis* Hübn., a biological parallel," pp. 455–462 *in Proceedings, 4ᵗʰ International Congress of Entomology, Ithaca, August 1928*. G. Pätz, Wenzelsring.

Jinbo U, Arita Y, Nakajima H, *et al*. (3 other authors). 2014. 皇居の蛾類調査(2009–2013). *Memoirs of the National Science Museum Tokyo* 50: 129–237.

Karpova AI. 1959. Развитие и кормовые связи стеблевого мотылька Pyrausta nubilalis (Lepidoptera, Pyralidae) в новых районах возделывания кукурузы. *Entomological Review* 38: 724–733

Kirchner O. 1906. "Hanf, *Cannabis sativa* L." pp. 319–323 in *Die Krankheiten und Befehädigungen uhferer landwirtschaftlichen Kulturpflanzen*. E. Ulmer, Stuttgart.

Komai F. 1999. A taxonomic revision of the genus *Grapholita* and allied genera (Lepidoptera: Tortricidae) in the Palaearctic region. *Entomologica Scandinavica* (supplement) 55: 1–226.

Komlódi J. 1970. A kis kendermoly bioldgiájának vizsgálata és a védekezési kisérletek eredményei. *Növényvédelem* 6(8): 343–348.

Kōno H. 1928. Die Mordelliden Japans (Col.). *Transactions of the Sapporo Natural History Society* 10(2): 26–46.

Koo M. 1940. *Studies on Pyrausta nubilalis Hüber attacking the cotton plant*. Yamanashi Agricultural Experiment Station, Kofu, Japan.

Köppen FP. 1880. *Die schädlichen Insekten Rußlands*. Kaiserlichen Akademie der wissenschaften, St Petersburg.

Korotyaev BA. 1980. Материалы к познанию Ceutorhynchinae (Coleoptera, Curculionidae) Монголии и СССР. *Insects of Mongolia* 7: 107–282.

Krustev VP. 1957. A new pest of hemp, *Mordellistena parvula* Gyll., in Bulgaria. *Sofia Bulletin of Plant Protection* 5(1): 87-88.

Kryachko Z, Ignatenko M, Markin A, Zaets V. 1965. Внимание, Конопляная листоверт. *Zashchita Rastenii̇ ot Vreditelei̇ i Boleznei̇* 5: 51-54.

Kuwayama S. 1930a. アハノメイガに関する研究. *Hokkaido Experimental Stations Reports* 25: 1–142.

Kuwayama S. 1930b. "Some observations on the so-called European corn borer in Japan," pp. 100-109 *in Proceedings, 4ᵗʰ International Congress of Entomology, Ithaca, August 1928*. G. Pätz, Wenzelsring.

Kuznetsov VI. 1989. "Leaf-rollers (Lepidoptera, Tortricidae) of the southern part of the Soviet Far East and their seasonal cycles," pp. 57–249 in Kryzhanovskii OL, ed., *Lepidopterous fauna of the USSR and adjacent countries*. National Science Foundation, Washington, DC.

Kyokai NSB. 1965. 農林病害虫名鑑. Japanese Plant Protection Association, Tōkyō.

Lago PK, Stanford DF. 1989. Phytophagous insects associated with cultivated marijuana, *Cannabis sativa*, in northern Mississippi. *Journal of Entomological Science* 24: 437–445.

Lekic M, Mihajlovic L. 1971. *Grapholitha sinana* Felder (Tortricidae, Lepidoptera) opasna stetocina konoplje na podrucju Vojvodine. *Savremena Poljoprivreda* 19(3): 63–68.

Lesne P. 1920. Une ancienne invasion du "Botys du millet" (*Pyrausta nubilalis* Hb.) en France. *Bulletin de la Société de pathologie Végétale de France* 7(1): 15–16.

Lewis LC, Bruck DJ, Prasifka JR, Raun ES. 2009. *Nosema pyrausta*: its biology, history, and potential role in a landscape of transgenic insecticidal crops. *Biological Control* 48: 223–231.

Li B, Yang ZF. 2022. Multilocus evidence provides insight into the demographic history and asymmetrical gene flow between *Ostrinia furnacalis* and *Ostrinia nubilalis* (Lepidoptera: Crambidae) in the Yili area, Xinjiang, China. *Ecology and Evolution* 12: e9504.

Li CL. 1985. 大麻龟象发生规律及防治研究. *Zoological Research* 6(1): 87–93.

Li FS. 1951. 中国经济昆虫学,增订版. Self-published, Beijing.

Li GL, Li MZ, Liu ZH, Tang JH. 2009. 转Bt基因抗虫玉米的研究. *Biotechnology Bulletin* 2009(5): 9–13.

Li LY. 1994. "Worldwide use of *Trichogramma* for biological control in different crops," pp. 37–53 *in* Wajnberg E, Hassan SA, *eds. Biological control with egg parasitoids.* CAB International, Wallingford, UK.

Li WD, Chen SX, Qin JG. 2003. 亚洲玉米螟与欧洲玉米螟混生区的研究 (Identification of sympatric Asian and European corn borers). *Entomological Knowledge* 40(1): 31–35.

Lindemann KE. 1869. Нѣсколько словъ по поводу конопли, поврежденной насѣкомыми. *Русское Сельское Хозяйство* 3: 72–75.

Lindemann KE. 1885. *Конопляный жук (Mordellistena micans).* Ministry of Agriculture and Government Estates, St Petersburg.

Liu YQ, Li GW. 2002.中国动物志 昆虫纲 第二十七卷 鳞翅目:卷蛾科. Science Press, Beijing.

Lizotte E. 2019. *Pest alert: European corn borer in Michigan hop.* Michigan State University Extension. Available at: https://www.canr.msu.edu/news/pest-alert-european-corn-borer-in-michigan-hop

Lu JH, Huang JH, Colonnelli E, Yoshitake H, Bo C, Zhang RZ. 2019. A checklist of Chinese Ceutorhynchinae (Coleoptera; Curculionidae). *Annales Zoologici* 69(2): 201–240.

Lu MR, Yang BL, Lou YZ, Yang JQ, Sun YX. 1992. 亚洲玉米螟近缘种云南新记录. *Journal of Yunnan Agricultural University* 7(1): 43–45

MacKay, M. R. 1959. Larvae of the North American Olethreutidae (Lepidoptera). *Canadian Entomologist* 91(Suppl. 10): 72.

Malausa T, Dalecky A, Ponsard S, *et al.* (4 other authors). 2007a. Genetic structure and gene flow in French populations of two *Ostrinia* taxa: host races or sibling species? *Molecular Ecology* 16: 4210– 4222.

Malausa T, Leniaud L, Martin JF, *et al.* (6 other authors). 2007b. Molecular differentiation at nuclear loci in French host races of the European corn borer (*Ostrinia nubilalis*). *Genetics* 176: 2343–2355.

Mandelshtam MY, Petrov AV, Korotyaev BA. 2012. To the knowledge of the herbivorous scolytid genus *Thamnurgus* Eichhoff (Coleoptera, Scolytidae). *Entomological Review* 92: 329–349.

Manojlović B. 1984a. Uticaj biljke hraniteljke na težinu gusenica, plodnost i nosivost leptira kukuruznog plamenca (*Ostrinia nubilalis* Hbn., Lepidoptera, Pyralidae). *Zaštita Bilja* 35: 153–163.

Manojlović B. 1984b. Efektivnost *Trichograma evanescens* West. (Hymenoptera, Trichogrammatidae) u parazitiranju jaja kukuruznog plamenca na raznim biljkama hraniteljkama. *Zaštita Bilja* 35: 347–356.

Manolache C, Sandru I, Romascu E. 1966. Un nou daunator al culturilor de cinepa—molia cinepii (*Grapholitha delineana* Walk.—Lepidoptera—Tortricidae). *Probleme Agricole* 1966 (6): 68–72.

Marin AN. 1979. Биометод—на поля! *Защита растений* 1979(11): 24.

Marion H. 1957. Complément à la classification et nomenclature des Pyraustidae d'Europe. *Entomologiste* 13: 129–130.

Marseul SA de. 1876. *Mordellistena comes. Annales de la Société Entomologique de France* 6: 473–474 (also numbered 103–104).

Matsumura S. 1899. 日本害蟲篇. Shōkabō, Tōkyō.

Matsumura S. 1931. *Nihon konchū dai zukan; or, 6000 illustrated insects of the Japanese Empire.* [Publisher not identified], Japan.

McPartland JM. 1991. "Chapter 6: Insects and mites," pp. 116–168 in McEno J, *ed. Cannabis ecology: a compendium of diseases and pests.* Amrita Press, Middlebury, Vermont.

McPartland JM. 1996a. A review of *Cannabis* diseases. *Journal of the International Hemp Association* 3(1): 19–23.

McPartland JM. 1996b. *Cannabis* pests. *Journal of the International Hemp Association* 3(2): 49, 52–55.

McPartland JM. 2002. Epidemiology of the hemp borer, *Grapholita delineana* Walker, a pest of *Cannabis sativa* L. *Journal of Industrial Hemp* 7(1): 25–42.

McPartland J, Hillig K. 2007. Longhorn beetles and *Botryosphaeria*. *Journal of Industrial Hemp* 12(2): 123–133.

McPartland JM, Clarke RC, Watson DP. 2000. *Hemp diseases and pests – management and biological control.* CABI Publishing, Wallingford, UK.

McPartland JM, Cutler S, McIntosh DJ. 2004. Hemp production in Aotearoa. *Journal of Industrial Hemp* 9(1): 105–115.

Meyrick E. 1886. Descriptions of Lepidoptera from the South Pacific. *Transactions of the Entomological Society of London* 1886: 189–296.

Meyrick E. 1895. *A handbook of British Lepidoptera.* Macmillan, London.

Meyrick E. 1907. Descriptions of Indian Micro-Lepidioptera. *Journal of the Bombay Natural History Society* 18: 137–160.

Meyrick E. 1908. Descriptions of African micro-Lepidoptera. *Proceedings of the Zoological Society of London* 1908: 716–756.

Meyrick E. 1937. *Laspeyresia tetragrammana. Exotic Microlepidoptera* 5(5): 100.

Miller WE. 1982. *Grapholita delineana* (Walker), a Eurasian hemp moth, discovered in North America. *Annals of the Entomological Society America* 75(2): 184–186.

Mohyuddin AI, Scheibelreiter GK. 1973. *Investigations on the fauna of Papaver spp. and Cannabis sativa,* pp. 32–33 in Annual Report, Commonwealth Institute of Biological Control, Switzerland Station.

Mushtaque M, Baloch GM, Ghani MA. 1973. *Natural enemies of Papaver spp. and Cannabis sativa,* pp. 54–55 in Annual report, Commonwealth Institute of Biological Control, Pakistan station.

Mutuura A. 1954. Classification of Japanese *Pyrausta* group based on the structure of the male and female genitalia. *Bulletin of the Naniwa University, Series B, Agriculture and Biology* 4: 7–33.

Mutuura A, Munroe E. 1970. Taxonomy and distribution of the European corn borer and allied species: genus *Ostrinia* (Lepidoptera: Pyralidae). *Memoirs of the Entomological Society of Canada* S71: 1–112.

Nagy B.1958. "Vizsgalatok martonvásári és szegedi kukoricamoly populációkkal kapcsolatban," pp. 339–347 in Bajai J, *ed. Kukoricatermesztési Kísérletek.* Akadémiai

Nagy B. 1959. Kukoricamoly okozta elváltozások és károsítási formák kenderen. *Különlenyomat a Kísérletügyi Közlemények (Növénytermesztés)* 52(4): 49–66.

Nagy B. 1962. "Állati kártevők," pp. 70–73 in Mándy G, Bócsa I, *eds.* 1962. A kender. *Magyarország Kultúrflórája* 7(14): 1–101.

Nagy B. 1967. The hemp moth (*Grapholith sinana* Feld., Lepid.:Tortricidae), a new pest of hemp in Hungary. *Acta Phytopathologica Academiae Scientiarum Hungaricae* 2: 291–294.

Nagy B. 1976. Host selection of the European corn borer (*Ostrinia nubilalis* Hbn.) populations in Hungary. *Symposium Biologica Hungarica* 16: 191–195.

Nagy B. 1979. Different aspects of flight activity of the hemp moth, *Grapholitha delineana* Walk., related to intergrated control. *Acta Phytopathologica Academiae Scientiarum Hungaricae* 14: 481–488.

Nagy B. 1980. *Interspecific sex-phermone and sexual behavior of the hemp moth, Grapholitha delineana Walk.* Abstracts of Papers presented at the Conference on New Endeavours in Plant Protection, Budapest, September 2–5, 1980. Ohne Angabe, Budapest.

Nagy B. 1986. "European corn borer: historical background to the chages of the host plant pattern in the Carpathian basin," pg. 174–181 in *Proceedings of the 14th Symposium of the International Working Group on Ostrinia.* IOBC, Bejing.

Nagy B, Csehi É. 1955. Néhány megfigyelés a kenderen károsító kukoricamolyról. *Magyar Tudományos Akadémiai Agrártudományok Osztályának Közleményei* 8: 106–108.

Nagy B, Gulyás S, Pétchy I, Tărkányi-Szücs S. 1982. A kenderormányos (*Ceutorrhynchus rapae* Gyll.) kártételének magyarországi jelentkezése. *Növényvédelem* 18(7): 289–298.

Obi M. 1940. 棉に加害する粟野娯蛾に関する研究. *Yamanashi Prefectural Agricultural Experiment Station* 3: 1–81.

Obraztsov NS. 1959. Die Gattungen der Palaearktischen Tortricidae. II. Die Unterfamilie Olethreutinae 2. Teil. *Tijdschrift voor Entomologie* 102: 175–216.

Obraztsov NS. 1967. Die Gattungen der Palaearktischen Tortricidae. III. Addenda und Corrigenda. 2 Teil. *Tijdschrift voor Entomologie* 110: 13–64.

Onuki S. 1903. 実用昆虫学 *(Jitsuyō konchū-gaku).* Seibidō, Tōkyō.

Ovsyannikova EI, Grichanov IY. 2009. *AgroAtlas pests: Grapholita delineana Walker. — Eurasian hemp moth.* Available at: http://www.agroatlas.ru/en/content/pests/Grapholita_delineana/

Pang XF, Chen TL. 1974. *Trichogramma* of China (Hymenoptera: Trichogrammatidae). *Acta Entomologica Sinica* 17: 441–454.

Park CG, Seo BY, Jung JK, Kim HY, Lee SW, Seong KY. 2017. Forecasting spring emergence of the Asian corn borer, *Ostrinia furnacalis* (Lepidoptera: Crambidae), based on postdiapause development rate. *Journal of Economic Entomology* 110: 2443–2451.

Park KT, Lee BW, Bae YS, Lan HL, Byun BK. 2014. Tortricinae (Lepidoptera, Tortricidae) from Province Jilin, China. *Journal of Asia-Pacific Biodiversity* 7: 355–363.

Passerini C. 1832. *Osservazioni e notizie relative alle larve pregiudizievoli alla pianta del Gran turco (Zea mays).* [No printer listed], Florence.

Pélissié B, Ponsard S, Tokarev YS, *et al.* (10 other authors). 2010. Did the introduction of maize into Europe provide enemy-free space to *Ostrinia nubilalis*? Parasitism differences between two sibling species of the genus. *Ostrinia. Journal of Evolutionary Biology* 23: 350–361.

Peteanu Ş. 1972. Cercetări privind utilizarea viespii oofage *Trichogramma evanescens* Westw. in combaterea moliei cînepii (*Grapholitha delineana* Walker). *Analele Institutului de Cercetari pentru Cerealele si Plante Tehnice, Fundulea* 38: 317–322.

Peteanu Ş. 1980. Contributii la studiul combaterii biologice si integrate a moliei cinepii (*Grapholitha delineana* Walker). *Cereale si Plante Tehnice, Productia Vegetala* 32: 39–43.

Phelan PL, Norris KH, Mason JF. 1996. Soil-management history and host preference by *Ostrinia nubilalis:* evidence for plant mineral balance mediating insect-plant interactions. *Environmental Entomology* 25: 1329–1336.

Pichat CB. 1866. "La pianta della canapa," pp. 396–499 in *Istituzioni scientifiche e tecniche, ossia Corso teorico e practico de Agricoltura, Libri XIII, Volume Quinto.* Presso l'Union Tipografico-Editrice, Torino.

Pivtoraiko VV. 2022. Особливості розвитку шипоносок (Coleoptera: Mordellidae) в агроценозі конопляного поля у північно-східному лісостепу україни. *Bulletin of Sumy National Agrarian University, Agronomy and Biology Series* 47(1): 108–118.

Pivtoraiko VV, Kabanets VV, Vlasenko VA 2020. Шкідлива ентомофауна конопель посівних, *Cannabis sativa* L. *Plant Quarantine and Protection* 2020 (7-9): 20–25.

Piwczyński M, Pabijan M, Grzywacz A, *et al.* (3 other authors). 2016. High regional genetic diversity and lack of host-specificity in *Ostrinia nubilalis* (Lepidoptera: Crambidae) as revealed by mtDNA variation. *Bulletin of Entomological Research* 106: 512–521.

Poorani J, Ramamurthy VV. 1997. Weevils of the genus *Lepropus* Schoenherr from the Oriental region. *Oriental Insects* 31: 1–82.

Pryer H. 1885. Additions and corrections to a catalogue of the Lepidoptera of Japan. *Transactions of the Asiatic Society of Japan* 8: 22–68.

Ragonot EL. 1894. Notes synonymiques sur les Microlépidoptères et descriptions d'espèces peu connues ou inédites. *Annales de la Société Entomologique de France* 63: 161–226.

Rajmohana K, Peter A. 2014. "Review of the Hymenopteran parasitoids of Lepidoptera in India," pp. 27–47 in Rahman M, Anto M, *eds. Forest entomology.* Narendra, Delhi.

Rataj K. 1957. Skodlivi cinitele pradnych rostlin. *Prameny literatury* 2: 1–123.

Razowski J. 2003. *Tortricidae (Lepidoptera) of Europe. Vol. 2, Olethreutinae.* Frantiœsek Slamka, Bratislava.

Rebel H. 1889. Beiträge zur Microlepidopteren-Fauna Oesterreich-Ungarns. *Verhandlungen der Zoologisch-Botanischen Gesellschaft in Wien* 39: 293–326

Rebourg C, Chastanet M, Gouesnard B, Welcker C, Dubreuil P, Charcosset A. 2003. Maize introduction into Europe: the history reviewed in the light of molecular data. *Theoretical and Applied Genetics* 106: 895–903.

Ree JH. 1966. Hemp growing in the Republic of Korea. *Economic Botany* 20: 176–186.

Roberjot C. 1787. Mémoire sur un moyen propre à détruire les chenilles qui ravagent la vigne. *Mémoires de la Société Royale d'Agriculture de Paris, trimestre de printemps,* pp. 193–206.

Robin C, Laboulbène A. 1884. Sur les dégâts causés au maïs et au chanvre par les chenilles du *Botys nubilalis* Hübner. *Annales de la Sociéte Entomologique de France* 4: 5-16.

Robinson BB. 1943a. Seed treatment of hemp seeds. *Journal of the American Society of Agronomy* 35: 910–914.

Robinson BB. 1943b. *Hemp.* USDA Farmer's Bulletin No. 1935. Washington, DC.

Robinson BB. 1946. Dew retting of hemp uncertain west of longitude 95°. *Journal of the American Society of Agronomy* 38: 1106–1109.

Robinson BB. 1952. *Hemp (Revised Edition).* USDA Farmer's Bulletin No. 1935. Washington, DC.

Romaniszyn J, Schille F. 1930. *Fauna motyli Polski (Fauna Lepidopterorum Poloniae), tom. VII.* Polska Akademia Umiejętności, Kraków, Poland.

Şabanoğlu B. 2013. *İç anadolu bölgesi Cerambycıdae (Coleoptera) Familyası Üzerinde Sistematik Çalışmalar.* Doctoral thesis, Hacettepe Üniversitesi, Ankara, Turkey.

Sandru I. 1986. Sfredelitorul tulpinilor-*Ostrinia nubilalis* Hb. pe cînepă. Biologie, ecologie, etologie. *Probleme de Protecţia Plantelor* 14: 223–245.

Sankaran T, Ramachandran Nair K. 1973. *Survey for natural enemies of Cannabis sativa and Papaver somniferum.* Annual report, Commonwealth Institute of Biological Control, India station, pp. 48–49.

Sáringer G. 1976. Diapause-Versuche mit der ungarischen Population von *Ostrinia nubilalis* Hbn. (Lepid.: Pyraustidae). *Zeitschrift fur Angewandte Entomologie* 80: 426–434.

Sáringer G, Nagy B. 1971. The effect of photoperiod and temperature on the diapause of the hemp moth (*Grapholita sinana* Feld.) and its relevance to the integrated control. *Proceedings, 13th International Congress of Entomology* 1968: 435–436.

Scheibelreiter GK. 1976. *Investigations on the fauna of Papaver spp. and Cannabis sativa.* Annual report, Commonwealth Institute of Biological Control, European station, pp. 34–35.

Schlumberger O. 1941. Der Maiszünsler (*Pyrausta nubilalis*) als Hopfen- und Hanfschädling. *Nachrichtenblatt für den Deutschen Pflanzenschutzdienst* 21(3): 18–20.

Schreiner M. 2019. *A survey of the arthropod fauna associated with hemp (Cannabis sativa L.) grown in eastern Colorado*. Master's thesis, Colorado State University, Fort Collins.

Schütze KT. 1931. *Die Biologie der Kleinschmetterlinge unter besonderer Berücksichtigung ihrer Nährpflanzen und Erscheinungszeiten*. Verlag des Internationalen Entomologischen Vereins e.V., Frankfurt am Main.

Semper G. 1896. *Die Schmetterlinge der Philippinischen Inseln, Zweiter Band*. Kreidel, Wiesbaden.

Senchenko GI, Timonina MA. 1978. Конопля. Kolos Press, Moscow.

Senda C. 1951. 大麻のアワノメイガの防ぎ方-特に栽培條件による被害回避について. *Iwate Agricultural Extention* 3(4): 12–14.

Shapovalov AM. 2012. Жуки-усачи (Coleoptera, Cerambycidae) Оренбургской области: фауна, распространение, биономия. *Proceedings of the Orenburg branch of the Russian Entomological Society* 2012(3): 1–220.

Shelford VE. 1913. *Animal communities in temperate America*. University of Chicago Press, Chicago.

Shibuya J. 1929. On the known and unrecorded species of the Japanese Pyraustinae. *Journal of the Faculty of Agriculture, Hokkaido Imperial University* 25: 151–242.

Shilenkov VG, Tolstonogova EV. 2006. Вредители конопли в Прибайкалье. *Acta Biomedica Scientifica* 2006(2): 159–160.

Shinji O. 1944. "Mordellideae," pp. 431–434 in 虫癭と虫癭昆虫. Shun'yōdō, Tōkyō.

Shiraki T. 1952. *Catalogue of injurious insects in Japan*. General Headquarters, Supreme Command Allied Powers, Tokyo.

Shiyake S. 1994. On the hind tibial spurs in the genus *Mordellistena* (Coleoptera: Mordellidae). *Bulletin of the Osaka Museum of Natural History* 48: 9-22.

Shutova NN, Strygina SP. 1969. Конопляная моль [The hemp moth]. *Zashchita Rasteniĭ* 14(11): 49–50.

Small E, Marcus D, McElroy A, Butler G. 2007. Apparent increase in biomass and seed productivity in hemp (*Cannabis sativa*) resulting from branch proliferation caused by the European corn borer (*Ostrinia nubilalis*). *Journal of Industrial Hemp* 12(1): 15–26.

Smith GE, Haney A. 1973. *Grapholitha tristrigana* (Lepidoptera:Torttricidae) on naturalized hemp (*Cannabis sativa* L.) in east-central Illinois. *Transactions of the Illinois State Academy of Science* 66: 38–41.

Smith SM. 1996. Biological control with *Trichogramma*. *Annual Review of Entomology* 41: 375–406.

Song L, Shi YX, Zhang HF, et al. (3 other authors). 2021. Complete mitochondrial genome of the hemp borer, *Grapholita delineana* (Lepidoptera: Tortricidae): gene variability and phylogeny among *Grapholita*. *Journal of Asia-Pacific Entomology* 24(2): 250–258.

Sorauer P. 1958. *Handbuch der Pflanzenkrankheiten, Band 5*. Paul Parey, Berlin.

Spencer KA. 1966. Notes on the Oriental Agromyzidae – 4. *Stuttgarter Beiträge zur Naturkunde* 147: 1–15.

Staudinger O. 1880. Lepidopteren-fauna Kleinasiens's. *Horae Societatis Entomologicae Rossicae* 15: 159–435.

Staudinger O, Rebel H. 1901. *Catalog der Lepidopteren des Palaearctishen Faunengebietes, dritte auflaud*. Friedländer & Sohn, Berlin.

Sugawara H, Omori H, Oya T. 1953. 大麻髄虫アワノメイガの藥剤防除について. *Annual Report of the North Japan Institute of Disease and Pest Research* 4: 186–187.

Sutabtra T. 2007. *Notification of Ministry of Agriculture and Cooperatives Re: Specification of plant pests as prohibited articles under the Plant Quarantine Act*. Available at: https://www.ippc.int/static/media/files/reportingobligation/2014/11/quarantine_pest_2.pdf

Takahashi S. 1919. 大麻の害虫と大麻天牛に就きて. *Konchū Sekai (Insect World)* 23(1): 20–24.

Taschenberg EL. 1879. *Praktische Insekten-Kunde: oder, Naturgeschichte, Erster Theil*. Heinsius, Bremen.

Taylor JG, Deay HO.1950. Electric lamps and traps in corn borer control. *Agricultural Engineering* 31: 503–505.

Tembo RD, Pavuk DM. 2011. The impact of *Beauveria bassiana*, *Bacillus thuringiensis* (Bt) spray, *Trichogramma pretiosum* and spinosad on the Lepidoptera: Crambidae, European Corn Borer (*Ostrinia nubilalis*). *Journal of Agricultural Science and Technology A* 1: 678–692

Thomas Y, Bethenod MT, Pélozuelo L, Frérot B, Bourguet D. 2003 Genetic isolation between two sympatric host-plant races of the European corn borer, *Ostrinia nubilalis* Hübner. I. Sex pheromone, moth emergence timing. *Evolution* 57: 261–273.

Tietz HM. 1972. *An index to the described life histories, early stages and hosts of the Macrolepidoptera of the continental United States and Canada*. AC Allyn, Sarasota, Florida.

Tkalich PP. 1965. Устойчивость конопли к стеблевому мотыльку. *Защита Растений от Вредителей и Болезней* 11: 52.

Tkalich PP. 1967. "Выявление и применение энтомофагов против стеблевого мотылька на конопле," pp. 143–146 in *Возделывание и первичн обработка конопли*. Harvest press, Kiev.

Tkalich PP, Lepskaya PA, Goloborodko PA. 1983. Система защиты конопли. *Plant Protection* 1983(1): 46–49.

Tokarev YS, Yudina MA, Malysh JM, et al. (4 other authors). 2017. Встречаемость эндосимбиотической бактерии рода *Wolbachia* в природных популяциях *Ostrinia nubilalis* и *Ostrinia scapulalis* (Lepidoptera: Pyraloidea). *Ecological Genetics* 15(1): 44–49.

Treitschke F. 1829. *Die schmetterlinge von Europa, siebenter band*. Fleischer, Leipzig.

Tremblay E. 1968. Osservazioni sui Punteruoli della Canapa (Coleoptera Curculionidae). Note morfologiche, biologiche e lotta chimica. *Bollettino del Laboratorio di Entomologia Agraria "Filippo Silvestri" Portici* 26: 139-190.

Tremblay E, Bianco M. 1978. *I punteruoli del cavolfiore in Campania*. Note Divulgative No. 12. Istituto di Entomologia Agraria della Università di Napoli, Portici.

Trotuş E, Naie M, 2008. Cercetări privind reducerea atacului unor agenţi patogeni şi dăunători specifici culturilor de cânepă prin tratamentul chimic al seminţei. *Lucrari Stiintifice, Universitatea de Stiinte Agricole Si Medicina Veterinara "Ion Ionescu de la Brad" Iasi, Seria Agronomie* 51(2): 219–223.

Tsao C. 1963. Quarantine aspects of the hemp moth. *Acta Phytophylacica Sinica* 2: 371–377.

Vasilyev VP 1956. *Шкідники і хвороби сільськогосподарських рослин*. State Publishing House of Agricultural Literature of the Ukrainian SSR, Kiev.

Vassilaina-Alexopoulou P, Mourikis PA. 1976. A new insect pest of the hemp in Greece, *Cydia delineana* Walker. *Annales de l'Institut Phytopathologique Benaki* 11: 254–255.

Vinal SC. 1917. *The European corn borer, Pyrausta nubilalis Hübner: a recently established pest in Massachusetts*. Massachusetts Agricultural Experiment Station, Bulletin no. 178. Amherst, Massachusetts.

Virovets VG, Lepskaya LA. 1983. "Сортовая устойчивость конопли к стеблевому мотыльку," pp. 53-58 in Senchenko GI, ed. *Биологические особенности, технология возделывания и первичная обработка лубяных культур*. Institute of Bast Crops, Glukhov, Ukraine.

Walker F. 1859. *Botys scapulalis. List of the Specimens of Lepidopterous Insects in the Collection of the British Museum*, part XVIII: 657.

Walker F. 1863. *Grapholita delineana. List of the Specimens of Lepidopterous Insects in the Collection of the British Museum*, part XXVIII: 389.

Walsingham T. 1900. Asiatic Torticidae. *Annals and Magazine of Natural History, series seven* 6: 429–447.

Wang LN, Wang DK. 2008. 大麻田中玉米螟的危害及防治技术. *Heilongjiang Agricultural Sciences* 2008(6): 70–71.

Wang TK, Rong WG. 1992. A study on the moth (*Grapholitha delineana* Wal.) of hemp (*Cannabis sativa* L.). *Chung-kuo ma tso* [*China's Fiber Crops*] 1992(2): 26–29, 35.

Wang T, Shen G, Feng X, Guo D, Lui L. 1995. A study on the bionomics of the hemp weevil *Rhinoncus pericarpius*. *China's Fiber Crops* 17(1): 37–39.

Wang Y, Kim KS, Guo W, *et al.* (4 other authors). 2017. Introgression between divergent corn borer species in a region of sympatry: Implications on the evolution and adaptation of pest arthropods. *Molecular Ecology* 26: 6892–6907.

Wilsie CP, Dyas ES, Norman AG. 1942. *Hemp—a war crop for Iowa.* Agricultural Experiment Station, Iowa State College, Bulletin P49, Ames, Iowa.

Xu HR, Zhang WG. 1979. 新疆沙湾县紫玉米螟的观察初报. *Chinese Bulletin of Entomology* 1979(3): 120–122.

Yang RS, Wang ZY, He KL, Bai SX. 2008. 中国秆野螟属(Ostrinia)雄性外生殖器系统比较与分析. *Journal of Shenyang Agricultural University* 39(1): 33–37.

Yang RS, Wang ZY, He KL, Bai SX, Jiang YR. 2011. 中国秆野螟属昆虫线粒体COI玉基因遗传多样性及其分子系统学. *Journal of Nanjing Agricultural University* 34(5): 73–80.

Yang ZF, Zhang YL. 2011. Comparison of ultrastructure among sibling species of *Ostrinia* (Lepidoptera: Crambidae) from China. *Canadian Entomologist* 143: 126–135.

Yang ZF, Plotkin D, Landry JF, Storer C, Kawahara AY. 2021. Revisiting the evolution of *Ostrinia* moths with phylogenomics (Pyraloidea: Crambidae: Pyraustinae). *Systematic Entomology* 46: 827–838.

Yepsen RB. 1976. *Organic plant protection.* Rodale Press, Emmaus, Pennsylvania.

Zhou N, Dong YL, Qiao PP, Yang ZF. 2020. Complete mitogenomic structure and phylogenetic implications of the genus *Ostrinia* (Lepidoptera: Crambidae). *Insects* 11: e232.

Zuefle M. 2020. *Hemp pest survey 2020.* New York State Integrated Pest Management Program. Available at: https://ecommons.cornell.edu/handle/1813/103765

Zwölfer W. 1928. Bericht über die Untersuchungen zur Biologie und Bekämpfung des Maiszünslers (*Pyrausta nubilalis*) in Süddeutschland 1926. *Arbeiten aus der Biologischen Reichsanstalt für Land- und Forstwirtschaft* 15: 335–400.

Abstract

Caterpillars lead the list of insects that chew on leaves. Perhaps the worst are Noctuidae, related to budworms and armyworms (*Autographa gamma*, *Melanchra persicariae*). Members of other families pile on: Sphingidae (*Acherontia atropos*), Crambidae (*Loxostege sticticalis*), Torticidae (*Grapholita delineana*), and Erebidae (*Spilosoma obliqua*, *Spilosoma virginica*, *Arctia caja*). Another group of opportunistic generalists that feed on hemp are locusts and grasshoppers (*Locusta migratoria*, *Calliptamus italicus*, *Melanoplus bivittatus*, *Melanoplus differentialis*, and several minor species). Defoliating beetles include *Popillia japonica* and a panoply of leaf beetles (family Chrysomelidae)—foremost *Psylliodes attenuatus*, *Psylliodes punctulata*, *Phyllotreta nemorum*, *Chaetocnema concinna*, and *Systena frontalis*, but dozens of others. Leafminers are larvae that live and feed within leaves. Their taxonomy is in flux, but four species appear to be problems: *Liriomyza cannabis*, *Liriomyza strigata*, *Agromyza reptans*, and *Phyllotreta nemorum*. Gall midges (*Melanagromyza urticivora*) and sand flies (*Phlebotomus* spp.) round out the chapter, plus a grab-bag of sawflies, springtails, and earwigs.

7.1   Introduction

Holes chewed in leaves may look alarming, but a lot of defoliation is required before losses become significant. Britt (2021) investigated the impact of leaf loss on seed yield. She artificially defoliated 'Felina 32' at 0%, 25%, 50%, or 75%, to represent increasing levels of chewing injury. Defoliation was carried out early-, mid-, or late-season (20, 40, or 60 days post-planting). Plants were harvested *ca.* 90 days post-planting, and yields assessed as seed weight per plant. Yields at all levels of defoliation, at all dates, did not differ from control plants (at statistically significant levels). She repeated the study with three CBD strains, defoliated either 0% or 50%, at early-, mid-, or late-season. Removal of 50% leaf tissue did not result in yield loss (dry weight flowering tops) at a statistically significant level.

Insect herbivory differs from mechanical defoliation. Oral secretions from insects may stimulate plants to produce defensive chemicals, not seen in response to sterile mechanical wounds. Feeding damage by tobacco hornworms (*Maduca sexta*) induced the release of terpenoids from tobacco plants, as did mechanical wounds dabbed with oral secretions, but not mechanical wounds by themselves (Halitschke *et al.*, 2000). Similar studies with *Cannabis* have yielded inconsistent results (see section 3.3.6).

Caterpillars (Lepidopterans) lead the list of insects that chew leaves. Foremost are members of the Noctuidae. Some leaf-eating Noctuids cause more damage to flowering tops and seeds (see section 5.2). "Cutworms" may feed on leaves, but they're better known for cutting down seedings (see section 8.2). Next come Coleopterans: Japanese beetle and several species of flea beetles. The most notorious flea beetle, *Psylliodes attenuatus*, we cover in section 5.4 because of its damage to flowering tops and seeds. Weevils and curculios may feed on leaves, but they primarily damage stalks (see section 6.10). Orthopterans—grasshoppers and locusts—can defoliate plants down to their bones. Cranshaw *et al.* (2019) considered them the most important defoliators in arid Colorado. Lastly, we come to Dipterans (flies): leafminers and gall midges.

Leaf symptoms include **notching** at leaf margins; notches may have smooth or angular edges. **Shotholes** are small round holes cut into leaves. **Skeletonizing** describes the selective feeding on tissues between leaf veins. This type of feeding may leave only a skeleton (the midrib and largest veins). **Leaf rolling** is caused by caterpillars who fold over leaf margins and feed within their little shelter, webbed together with silk. **Leafmining** pertains to feeding *within* leaves, producing a network of tunnels.

7.2   Caterpillars

Caterpillars that defoliated plants, due to their high-profile damage, were among the earliest pests described on hemp. Dozens of species reportedly defoliate *Cannabis*. Many reports are erroneous. Caterpillars wander, and they may cross *Cannabis* while en route to a true host plant. For instance, web blogs implicate as a *Cannabis* pest the eastern tent caterpillar, *Malacosoma americanum* (Fabricius). It's an unlikely pest of hemp; the moths oviposit almost exclusively on fruit trees (apple, cherry, plum).

7.2a.   Silver Y moth

Autographa gamma (L.) 1758, Lepidoptera: Noctuidae
≡*Plusia gamma* (L.) ≡*Phytometra gamma* (L.) ≡*Noctua gamma* L.
Description: *Larvae* have bodies narrower towards the head and wider towards the posterior, greenish-yellow to olive green, with whitish dorsal lines and lateral pale stripes, reaching 24–40 mm (Fig. 7.1). Head with fine dark-brown spots, and black coloring on each lateral side. Larvae "loop" as they walk, having only three pairs of abdominal legs (absent on 3rd and 4th abdominal segments). *Pupae* are black and shiny, encased in a loose, web-like cocoon, 20 mm long, on plants or near the surface of soil. *Adults* are brown-bodied moths, forewings variegated brown-gray, each with a silver-colored "Y" marking

(or γ, Greek gamma), hindwings brownish-gray with darker veins and a broad blackish border, wingspan 30–45 mm (Fig. 7.1). *Eggs* white, round, ribbed, 0.5–0.6 mm in diameter.

Comments: Silver Y moth was the first hemp pest to be illustrated (Fig. 7.1). Réaumur (1736) called it *des arpenteuses à douze jambes* (the loopers with twelve legs), and added, "*Elles n'aiment que trop le chanvre*" (They love hemp only too much). An outbreak of the pest caused *La panique de 1735*. Farmers abandoned hemp crops in Auvergne and Burgundy. Later authors identified his species as *A. gamma* (Filippo Re, 1806; Curtis, 1860), as did the editor of a new edition of Réaumur's text (Réaumur, 2001).

Angelini (1827) described an outbreak of *A. gamma* in the Po Valley that affected many crops, including hemp. He compared it with devastation caused by locusts 2 years prior (Bendiscioli, 1825). The following year, an outbreak of *A. gamma* hit East Prussia, impacting hemp (Nördlinger, 1869). Another outbreak on hemp occured in 1833, in the Smolensk region of Russia (Köppen, 1880).

Larvae attack *Cannabis* seedlings (Kaltenbach, 1874; Kirchner, 1906), and leaves or flowers of mature plants (Freyer, 1830; Thon, 1838; Macquart, 1855; Gistel, 1856; Melicher, 1869; Taschenberg, 1879; DuBois, 1884; Del Guercio, 1903; Blunck, 1920; Goidànich, 1928; Ceapoiu, 1958; Nagy, 1962; Spaar *et al.*, 1990; Patschke *et al.*, 1997; Matov and Kononenko, 2012). Gottwald (2002) reported "remarkable population densities" in Germany, but 80% of larvae were parasitized by a wasp, *Copidosoma truncatellum* (Dalman) 1820.

A. gamma overwinters in Mediterranean Europe, the Middle East, and North Africa, where 4–5 generations arise annually. Migratory moths are strong fliers and ride airstreams north, all the way to the UK and Scandinavia. The species has been recorded on *Cannabis* as far north as Finland (Vappula, 1965), and east to the Lake Baikal region (Shilenkov and Tolstonogova, 2006). In northern France the caterpillars feed on leaves in June (Noel, 1917).

Migration paths vary annually, sometimes in huge swarms consisting of millions of individuals. Two generations or one generation will arise along the migration path. Females lay 500–1000 eggs on undersides of leaves, singly or in small groups. Young larvae skeletonize leaves; older larvae eat the entire leaf lamina, leaving only the midrib. In outbreak years they march like armyworms. *A. gamma* is polyphagous; its main hosts include

sugarbeet, vegetable crops, *Prunus*, *Rubus*, and *Sambucus*. In the past 30 years its numbers have declined, possibly due to climate change (Ellis, 2016).

7.2b. Death's head moth

Acherontia atropos (L.) 1758, Lepidoptera: Sphingidae
≡ *Sphinx atropos* L.

Description: *Larvae* are uniformly pale green after hatching. Older larvae have green and yellow chevrons with lateral black dots, although polymorphic brown forms are common; they are large, up to 120–130 mm. *Pupae* are smooth and glossy, red to reddish-brown. *Adults* show sexual dimorphism; female moths sport a brown-and-yellow skull pattern on their thorax (Fig. 7.2). Forewings are brown and speckled with yellow, amber, gray, and cream; hindwings are yellow with two brown stripes. They are large ("weight of a mouse," Young, 1997), wingspan 80–120 mm.

Comments: Death's head moth overwinters in Africa, Mediterranean Europe, and the Middle East. Summer migrants reach all of Europe. *A. atropos* has unique behaviors: adult moths shriek when disturbed, and dive into bee hives to rob honey. *A. atropos* is normally a pest of Solanaceae (potato, tomato, tobacco, eggplant, datura).

Gladis and Alemayehu (1995) found larvae devouring hemp leaves—midveins and all. Their feeding experiments showed that larvae preferred hemp to nightshade plants, and larvae completed development on hemp alone. The authors considered hemp a new host of *A. atropos*. In fact, *A. atropos* was one of the first and most frequently reported insects on hemp (Brez, 1791; Saint-Amans, 1798; Cuvier, 1798; Filippo Re, 1806; Godart, 1822; Poiret, 1827; Meigen, 1827; Macquart, 1855; Curtis, 1860; Kranz, 1860; Melicher, 1869; Kaltenbach, 1874; Kirchner, 1906; Goidànich, 1928; Dempsey, 1975).

7.2c. Beet webworm

Loxostege sticticalis (L.) 1761, Lepidoptera: Crambidae
≡ *Phlyctaenodes sticticalis* L.

Description: *Larvae* are yellowish-green, darkening with age from olive-green to nearly black, with a patchwork of lighter stripes and rows of darker marks, reaching 30 mm in length (Fig. 7.3). *Moths* are mottled brown, with straw-colored markings and a dark margin on the underside of their hindwings.

Fig. 7.1. *Autographa gamma* (drawings by Réaumur (1736); larva photo courtesy Charles Olsen, Bugwood.org; adult photo courtesy of Dumi (Wikipedia Commons))

Fig. 7.2. *Acherontia atropos* female (Didier Descouens, Wikimedia Commons)

Fig. 7.3. *Loxostege sticticalis* larva (W. Cranshaw)

Comments: Beet webworm ranges across temperate Eurasia and has been introduced into the USA and Canada. Larvae overwinter in silk-lined cells in the soil, pupating there in the spring. Moths emerge from March to June. Moths are strong night fliers, migrating hundreds of miles along river valleys with prevailing winds. Females lay 150 eggs in groups of 2–20 on undersides of leaves. The entire life cycle can be as short as 35 days, with up to three generations arising each year.

This polyphagous pest is also known as the meadow moth or steppe caterpillar, where it feeds on *Artemisia* spp. Young larvae skeletonize hemp leaves, often webbing several leaves together; older larvae eat entire leaves. *L. sticticalis* has been reported in Europe (Köppen, 1880; Kulagin, 1915; Ceapoiu, 1958; Nadejde *et al.*, 1975; Camprag *et al.*, 1996) and in Colorado (Cranshaw *et al.*, 2019).

7.2d. Dot moth

Melanchra persicariae (L.) 1761, Lepidoptera: Noctuidae
= *Polia persicariae* (L.), = *Mamestra persicariae* (L.)
Description: *Larvae* vary in color from pale gray-green to dark green or purple-brown. They have a light green head, pale dorsal line down their backs, a series of V-shaped marks, and a dorsal hump on segment eight. Setae (1–1.3 mm) cover the body, giving larvae a velvety texture, up to 40 mm in length. *Adults* are gray to bluish-black, with darker forewings, marked by two conspicuously white, kidney-shaped dots (hence the name); wingspan 38–50 mm. *Eggs* hemispherical, ribbed, whitish–green becoming pinkish-brown, laid singly or in small masses on undersides of leaves.
Comments: Dot moth spans Eurasia from Spain to Japan, north to Scotland and the Russian Far East, south to Italy, the southern Caucasus, Iran, and central China.
Moths emerge in June or July and promptly lay eggs. Eggs hatch in a week and larvae damage plants for up to 90 days before moving underground in autumn. In northern France the caterpillars fed on leaves in September and October (Noel, 1917). Only one generation arises per year.

Dot moth is polyphagous, but prefers *Urtica*, *Plantago*, *Salix*, and *Sambucus* spp. Vallot (1836) reported *M. persicariae* on hemp bearing seed; it was parasitized by a chalcid he identified to the genus *Eulophus* (likely *E. pennicornis* Nees 1834). Macquart (1855) reported the species "devastating" hemp, whereas Blunk (1920) rarely encountered *M. persicariae* in hemp. Gistel (1856) found *M. persicariae* feeding on seeds as well as leaves. Others have cited the species on hemp in Europe (Melicher, 1869; Kaltenbach, 1874; Leunis, 1877; Taschenberg, 1879; Kirchner, 1906; Rataj, 1957; Matov and Kononenko, 2012).

7.2e. Bihar hairy caterpillar

Spilosoma obliqua (Walker) 1855, Lepidoptera; Erebidae; Arctiinae
=*Diacrisia obliqua* (Walker)
Description: *Larvae* have a gray body with a dorsal white stripe and black side stripes. They are "woolly-bears" with tufts of long black hairs and yellow-orange hairs, alternating in bands down the body. *Adults* have orange abdomens and orange-brown wings, with a scattering of black dots, wingspan 25 mm.
Comments: Bihar hairy caterpillar is native to South Asia, from Pakistan to Burma. It overwinters as pupae beneath the soil surface. Females lay eggs in large conspicuous masses, up to 1000 at a time. In warm climates the life cycle takes only 30 days; 3–8 generations arise per year. Young larvae skeletonize leaves; older larvae completely consume leaves. *S. obliqua* sometimes infests flowering tops.

Fletcher (1917) first reported *S. obliqua* eating *Cannabis* leaves, in Bengal. The species attracted researchers seeking a biocontrol of *Cannabis* (Nair and Ponnappa, 1974; Baloch and Ghani, 1972). *Cannabis* was a preferred host in India (Dhaliwal, 1993). Deshmukh *et al.* (1979) offered larvae 154 different plant species, and found *Cannabis* among the six most preferred. Yet, when fed a diet of pure *Cannabis*, 50% of larvae died after 24 days. Sharma *et al.* (2011) repeated the experiment, and *S. obliqua* larvae fed a pure *bhang* diet died after 15 days.

7.2f. Garden tiger moth

Arctia caja (L.) 1758, Lepidoptera; Erebidae; Arctiinae
Description: *Larvae* have a black dorsum, rust-colored undersides, flecked with white spots on their flanks; these "woolly-bears" are quite hirsute, up to 60 mm long. *Adults* have a brown thorax and orange abdomen. Forewings are beige with brown spots, hindwings are rust to orange with iridescent blue spots, wingspan 45–65 mm.
Comments: Garden tiger moth is distributed in temperate climates worldwide in the Northern Hemisphere. Larvae hatch at the end of summer (from August to September), overwinter, re-emerge in spring, and finish growth by June. Moths are active from June to September. The larvae are polyphagous, and feed on hemp in Europe (Sorauer, 1958) and Japan (Matsumura, 1899; Harada, 1930; Shiraki, 1952). Rothschild *et al.* (1977) fed larvae a pure *Cannabis* diet, and they became stunted and died before maturity. The larvae preferred eating high-THC plants to high-CBD plants.

7.2g. Yellow woolly bear

Spilosoma virginica (Fabricius) 1798, Lepidoptera; Erebidae; Arctiinae
Description: *Larvae* are yellow but color-variable—from beige to red-brown, covered in long soft hairs of variable length, body up to 50 mm long (Fig. 7.4). *Pupae* glossy reddish-brown, in a cocoon, 16–20 mm long. *Adults* have white bodies with yellow markings on the abdomen, and white wings with black dots, wingspan 32–52 mm. *Eggs* yellow, round, laid on leaves in loose clusters, 0.6 mm in diameter.

Comments: Yellow woolly bear is widespread throughout North America, but more common in the east. Larvae feed on a wide range of garden, field, and ornamental crops. *S. virginica* has been collected on *Cannabis* in Mississippi (Lago and Stanford, 1989), Vermont (McPartland *et al.*, 2000), and Colorado (Cranshaw *et al.*, 2019). In upper New York state, *S. virginica* caused serious damage in a hoop house, where it was unable to disperse (Zuefle, 2020).

Adult moths can be confused with those of two other Arctiinae, but the larvae look different. Larvae of fall webworm, *Hyphantria cunea* (Drury) 1773, range from yellow to dark gray, with cream-colored side stripes, yellow spots, and they are not very hirsute. Furthermore, larvae congregate in webbed "tents." *H. cunea* has been reported infesting hemp in Serbia (McPartland *et al.*, 2000), as well as the Midwestern USA (M. Bolt, pers. obs., 2019). Larvae of saltmarsh caterpillar, *Estigmene acrea* (Drury) 1773, vary in color from dark brownish-black to rusty-brown, with a few rows of orange or black warts, and tufts of gray to rust-colored setae on each segment, up to 5 cm long. *E. acrea* is consistently encountered in Colorado hemp (Cranshaw *et al.*, 2019) and in the Midwestern USA (M. Bolt, pers. obs., 2020).

7.2h. Painted lady

Vanessa cardui (L.) 1758, Lepidoptera: Nymphalidae
≡ *Papilio cardui* L., ≡ *Cynthia cardui* (L.)
Description: *Larvae* are grayish-green to purple-gray, with yellow side stripes and tufts of spines, up to 32 mm long (Fig. 7.5). *Pupa* (chrysalis) is brownish-gray with cream-colored spots, hanging from host plant. *Adults* have a monarch-like coloration: body olivescent ochreous-brown, abdomen with ochreous bands. Forewings with orange, brown, and white patterns; hindwings orange with black dots, wingspan 50–70 mm. *Eggs* are light green, with 14–19 vertical ribs, and laid singly.
Comments: Painted lady is found on every continent except South America and Antarctica. The species overwinters in warmer areas, with mass migrations in the spring. In Europe it migrates from North Africa and the Mediterranean, all the way to the UK and Sweden. Larvae act as leaf rollers, webbing together the edges of leaves and feeding within. The species is polyphagous, preferring herbaceous plants in the families Asteraceae and Malvaceae. Larvae have fed on hemp in Germany (Kranz, 1860), Romania (Trotuş and Naie, 2008), Russia (Lvovsky and Morgun, 2007), Ukraine (Pivtoraiko *et al.*, 2020), Japan (Shiraki, 1952), Manitoba (Brooke, 2008), and Colorado (Cranshaw *et al.*, 2019).

Related Nymphalidae also infest hemp: the small tortoiseshell, *Aglais* (*Vanessa*) *urticae* (L.) 1758, has been reported in Japan (Shiraki, 1952) and Russia (Lvovsky and Morgun, 2007). It normally infests nettles. The comma, *Polygonia* (*Vanessa*) *c-album* (L.) 1758, was reported on *Cannabis* in Japan and Taiwan (Clausen, 1931) and the Russian Far East (Tuzov *et al.*, 2000), where it also infests *Humulus* spp. Shiraki (1952) reported three additional Nymphalidae on hemp: *Vanessa indica* (Herbst) 1794, *Araschnia burejana* (Bremer) 1861, and *Araschnia levana* (L.) 1758.

7.2i. Brown sweep

Sterrhopterix fusca (Haworth) 1809, Lepidoptera: Psychidae.
=*Psyche cannabinella* Doumerc 1860, ≡*Perianthosuta cannabinella* Doumerc 1860,
≡*Perianthophila cannabinella* Doumerc 1860
Description: *Larvae* slender and pale, pinkish white, with brown heads and legs, bristling with short brown hairs. Shortly after hatching, larvae weave a silken case, spindle-shaped or conical, 20–23 mm long and 6–7 mm wide, and cover it with fragments of plant material. Therein they feed and eventually pupate. *Adults* have a strong sexual dimorphism. Males have a gray-brown thorax and abdomen. Forewings gray with brown spots, bordered posteriorly with gray fringe; hindwings are grayish white with silver-white fringe, wingspan 18–25 mm. Females are small, wingless, and wormlike, with only rudimentary legs, and live in their leaf-covered "bags."
Comments: Doumerc (1860) described *Psyche cannabinella* infesting feral hemp in Luxembourg. Larvae preferred infesting female flowers to male flowers. The identity of Doumerc's species has been debated. Even Doumerc suggested *Psyche* was not correct, and proposed two new genera, *Perianthosuta* or *Perianthophila*. Dalla Torre and Strand (1929) moved the species to *Sterrhopteryx*, as *S. cannabinella*. Kozhanchikov (1956) synonymized the species under *Sterrhopterix fusca*. Paclt (1976) considered the species identical to *Aglaope infausta* (Linnaeus) 1758. Marshall (1917) suggested the species might be a

Fig. 7.4. *Spilosoma virginica* larva (W. Cranshaw)

Fig. 7.5. *Vanessa cardui* larva (W. Cranshaw)

Hymenopteron! Sobczyk (2011) and Arnscheid and Weidlich (2017) synonymized *P. cannabinella* under *Sterrhopteryx fusca*, a polyphagous species ranging from England to European Russia, south to Italy and Romania.

7.2j. Leafroller moths

Leafrollers (Lepidoptera: Torticidae) are named for shelter-building by the larvae, who feed within folded leaves, bound together by strands of silk. The Eurasian hemp borer, *Grapholita delineana*, is a Torticid best known as a stalk borer (see section 6.2), but it has been called "the hemp leafroller" (Fig. 7.6A). The flax tortrix, *Cnephasia asseclana* (Denis & Schiffermüller) 1775, infests flax and other herbaceous plants in Europe. It has been described on hemp under two of its synonyms: under *Cnephasia interjectana* by Fritzsche (1959), and under *Cnephasia wahlbomiana* by Goidànich (1928) and Spaar *et al.* (1990). Adults have light brown forewings with darker variegated patterns, white hindwings, thorax gray-brown, abdomen gray, wingspan 15–20 mm (Fig. 7.6B).

The pallid tortrix, *Clepsis pallidana* (Fabricius) 1776, infested hemp crops in China (Park *et al.*, 2014). *C. pallidana* is a polyphagous pest whose distribution spans Eurasia, from Ireland to Japan, south to Iran. Adults look like a pallid version of *G. delineana*, with yellowish forewings, white hindwings, thorax gray-brown, abdomen gray, wingspan 15–20 mm (Fig. 7.6C).

The obliquebanded leafroller, *Choristoneura rosaceana* (Harris) 1841, was collected from CBD hemp in Indiana, and raised in the lab for identification (Bolt, 2019). The species is polyphagous and native to North America. Since then it has been reported on *Cannabis* throughout the USA (Munir *et al.*, 2023). Adults have light brown forewings with darker brown oblique bands, hindwings are cream-colored, wingspan 11.5–14 mm (Fig. 7.6D).

7.2k. Assorted caterpillars

Tussock moths (family Erebidae, tribes Nygmiini and Lymantriini) have tussock-like tufts of urticating hairs that may cause painful reactions when handled. Cherian (1932) observed two tussocks (both Nygmiini) damaging Indian hemp, *Euproctis fraterna* Moore 1883 and *Somena scintillans* (Walker) 1856. The world's most notorious tussock, the spongy moth (formerly gypsy moth, *Lymantria dispar* (L.) 1758, Tribe Lymantriini), has been reported on hemp in Japan (Kojima, 1917; Schaefer *et al.*, 1987), as well as Italy (Goidànich, 1928).

Four members of family Noctuidae have been sighted rarely on hemp: Matsumura (1899) cited *Plataplecta consanguis* (Butler) 1879 on Japanese hemp. Shiraki (1952) referred to *P. consanguis* as "the hemp dagger moth." *P. consanguis* is now a synonym of *Acronicta pruinosa* (Grenée) 1852, with a geographic range spanning Japan, south to Indonesia and Sri Lanka. In Colorado, Cranshaw *et al.* (2019) encountered the variegated cutworm, *Peridroma saucia* (Hübner) 1808, cotton square borer, *Strymon melinus* (Hübner) 1818, and yellowstriped armyworm, *Spodoptera ornithogalli* (Guenée) 1852. The latter species primarily infests seeded tops (see section 5.3c).

7.2.1 Control measures

The importance of eliminating weeds from within and around crops cannot be overstated. Most leaf-eating caterpillars overwinter in soil or crop stubble, so post-harvest plowing exposes them to the elements. Night-flying moths are attracted to light, so light traps can be used for monitoring purposes (see Fig. 19.6).

Synthetic pheromone traps are available for some species as monitoring tools (e.g., *Autographa gamma*, *Spodoptera exigua*, *Loxostege sticticalis*, *Melanchra persicariae*, *Choristoneura rosaceana*). Small-acreage farmers historically removed pests by hand.

Fig. 7.6. Tortrix moths: **A**. *Grapholita delineana*; **B**. *Cnephasia asseclana* (S. Haapal, Wikimedia Commons); **C**. *Clepsis pallidana* (M. Virtal, Wikimedia Commons); **D**. *Choristoneura rosaceana* (J. Powell, CalPhotos)

Wasps in the genus *Trichogramma* (*T. evanescens*, *T. brassi-cae*, *T. dendrolimi*) parasitize the eggs of Noctuidae (see section 20.6). Leaf-eating caterpillars are susceptible to *Bacillus thuringiensis* (Bt, e.g., Thuricide®), although widespread resistance has arisen. Nuclear polyhedrosis viruses that kill specific armyworms (*Spodoptera exigua* NPV, *Autographa gamma* NPV) are approved for use on *Cannabis* (EPA, 2023). Pesticides are rarely needed. Two botanical pesticides on the USDA's *National List* of organic substances, neem (azadirachtin) and ryania, repel or kill caterpillars. Spinosad, not on the list, also works.

7.3 Grasshoppers and allies

The order Orthoptera includes locusts, grasshoppers, and crickets. Most of them are moderately large insects, with prominent eyes, jaws, and large hindlegs (see Fig. 3.6). They may be winged or wingless. In hemp, Orthopteran pests are opportunistic generalists.

Grasshoppers tend to produce angular notching in leaves. Heavy infestations may defoliate leaves down to mid-ribs. Cranshaw *et al.* (2019) considered grasshoppers the most important defoliators of hemp in the arid Mountain Time Zone (Colorado, New Mexico, Arizona, Wyoming, Idaho, Montana, Nebraska, Kansas, South Dakota, and Alberta). Unlike other defoliators, grasshoppers may chew through branches and stalks (Fig. 7.7). Swarming locusts can strip foliage down to bare branches; Clarke (pers. comm. 1994) saw this happen in a matter of minutes in western Africa.

Grasshoppers rarely attack seedlings, but there's always an exception: Clarke (pers. comm., 1994) saw grasshoppers in Mexico biting through stems of seedings to topple them. Grasshoppers rarely cause problems indoors. An exception: Villanueva (2022a) reported devastation in an indoor hoop house caused by the Carolina grasshopper, *Dissosteira carolina* (L.). It was a two-fold exception: they consumed seedlings.

Differentiating locusts from grasshoppers is like separating mushrooms from toadstools. Locusts are the swarming phase of short-horned grasshoppers in the family Acrididae. The distinction between locusts and grasshoppers is not phylogenetic,

Fig. 7.7. Grasshopper damage in Colorado (W. Cranshaw)

it is based on whether a species swarms under suitable environmental conditions.

Locusts are usually solitary, like grasshoppers. Crowding upon diminished food sources causes a surge in serotonin, which initiates a "gregarious state," the swarming stage. Swarming locusts undergo morphological changes—bigger wings, smaller bodies—better for a nomadic lifestyle. Uvarov (1921) first recognized that gregarious *Locusta migratoria* (L.) 1758 was identical to solitary *Locusta danica* (L.) 1767. Everyone prior to Uvarov, including Linnaeus, thought they were different species.

Locusts assailed all ancient agricultural civilizations; they appeared in the oldest Mesopotamian cuneiform texts and Egyptian hieroglyphics (Kaynar, 2019). Plagues of locusts loomed large in the *Hebrew Bible* and the *Qurān*. Theophrastus wrote about them *ca.* 300 BC in Greece. Chinese texts documented hundreds of locust plagues between 243 BC and AD 1911 (McNeill, 1976).

The first recorded insect plague on hemp involved locusts. Justell (1686) described a "grasshopper," gray colored, an inch long, swarming over Languedoc in southern France. "Their swarms were so thick that they covered the sun like a cloud, and were whole hours in passing." Justell's meager description of the locust does not permit us to identify it to species. Two candidates are *Locusta migratoria* and *Calliptamus italicus*, described below.

Despite the ubiquity of locusts—and their wide host ranges—they are rarely cited on *Cannabis*. Justell (1686) stated, "After having eaten up the corn, they attacked the vines, the pulse, the willows, and *lastly the hemp*, notwithstanding its great *bitterness*" (emphasis added). Justell's characterization of "bitter" hemp became a meme in locust literature, as we will see.

Three ancient Chinese texts reputedly claimed that locusts did not attack hemp: *Lǔ Shì Chūnqiū* (written *ca.* 239 BC), *Jìnshū* (*ca.* AD 648), and *Jīn Shǐ* (*ca.* 1344). These claims were cited by Western authors (e.g., Kolb, 2007; Clarke and Merlin, 2016). However, machine-readable versions of these Chinese texts are available, via Chinese Text Project (https://ctext.org), and we could not find these claims.

Nóng Shū, a famous treatise on agriculture written by Wáng Zhēn in AD 1313, also reputedly made this claim. The famous treatise was cited by Xú Guāngqǐ (1562–1633), who wrote *Complete Book of Agricultural Administration* (Xú, 2020), and Chén Fāngshēng (born 1642), who wrote *A Study of Locust Catching* (Chén, 2020). We accessed *Nóng Shū* via Chinese Text Project, and could not find that claim.

More recently, the entomologist Qiū Shìbāng (1911–2010) noted an absence of locusts in hemp fields (Qiū, 1996). He suggested that *Cannabis* closed its canopy early, so weeds eaten by locusts cannot grow. In northern Afghanistan, Shamonin (1964) stated that the locust *Dociostaurus maroccanus* (Thunberg) 1815 preferred cereals and pulses, and avoided feeding on *Cannabis sativa*, *Peganum harmala*, and *Artemisia* spp.

7.3a. Migratory locust

Locusta migratoria (L.) 1758, Orthoptera: Acrididae
≡*Gryllus migratorius* L., ≡*Pachytylus migratorius* (L.)
=*Gryllus danicus* L., ≡*Pachytylus danicus* (L.), ≡*Locusta danica* (L.)

Fig. 7.8. Locusts: **A.** *Locusta migratoria* (Jonathan Hornung, Wikimedia Commons); **B.** *Calliptamus italicus* (Kurt Kulac, Wikimedia Commons)

Description: *Adults* are elongated insects with big heads and blue mandibles (Fig. 7.8A). Elytra lengthy and shiny; wings colorless, without bands. Hind femora bluish-black. Adults vary by phase. The gregarious form is brownish-yellow, body length 30–45 mm, with larger wings and shorter hind femora than the solitary form. Solitary form is brown with various greenish tints depending on food source, body length 35–50 mm (male), 45–55 mm (female). *Larvae*: gregarious form develops bold black and yellow or orange stripes; solitary form is green or brown. *Egg pod* is a column of hardened pinkish-white froth containing 95–160 eggs (solitary form) or 40–80 eggs (gregarious form).

Comments: *L. migratoria* once ranged across nearly all temperate and tropical regions of the Eastern Hemisphere, short of boreal forests. The species overwinters in sandy grasslands (e.g., *Phragmites* reedbeds). Its populations have declined since the late 19th century. Like all locusts that breed in sandy alluvia, their ecosystem has been disrupted by drainage, irrigation, and farming. Eggs hatch in May; 5th-instar larvae become gregarious when >7000 individuals/m². Adults begin flying in July, and reproduce while migrating, depositing eggs in crowds—conditions leading to another swarm. Swarms travel 5–130 km/day.

Solitary *L. migratoria* feeds on wild grasses, but swarms will eat anything green, including *Cannabis*. Thon (1838) described an outbreak of *L. migratoria* in Bonțida, Siebenbürgen (present-day Cluj County, Romania) in 1780. Swarms "ate up the grain first, then attacked the vineyards, the vegetables, the trees, and finally even the bitter hemp." Rion (1843) reported an outbreak of *L. migratoria* in Valais, Switzerland. "Cereals, hay, flax, hemp, vegetable plants, everything was eaten to the roots."

In Russia, Keppen (1870) reported that *L. migratoria* destroyed hemp, "probably, only when there is a lack of other, more tasty food." Ingenitsky (1897) reported that *L. migratoria* avoided eating wild hemp and cultivated hemp in Semirechye (near Lake Balkhash in Kazakhstan). Conversely, Andreevsky (1878) reported 977 ha of hemp destroyed in Kursk Oblast. The Kingdom of Hungary passed a law in 1848 that mandated the extirpation of *L. migratoria,* which destroyed "grain, corn, all kinds of crops, garden crops, meadows, reeds, even bitter fibrous hemp" (Linzbauer, 1861).

Switzerland suffered an outbreak in 1875; Fruhstorfer (1921) mentioned hemp at two sites. At Erlach, along the sandy banks of Lake Biel, *L. migratoria* gutted grain fields, peas, potatoes, herbs, and hemp. At Fläsh, 215 km east of Erlach, swarms emerged from a sandy alluvial plain of the Rhine River. Oat and wheat fields were chewed to stubble, but hemp, maize, and potatoes were spared, "dozens of the brown fellows climbed on some hemp stems to sunbathe on the green canopy."

7.3b. Italian locust

Calliptamus italicus (L.) 1758, Orthoptera: Acrididae
≡*Gryllus italicus* L., =*Gryllus germanicus* Fabricius 1775, =*Acrydium italicum* Latreille 1804

Description: *Adults* are medium-sized and stocky, with well-developed elytra and wings (Fig. 7.8B). Hind femora dull red or mauve. Coloration varies: gray-brown to brownish-red, with spotted elytra, body length 14–32 mm (females), 10–22 mm (males). Individuals of gregarious phase are somewhat larger and their elytra and wings are longer than the ones of individuals of solitary phase. *Egg pod* is cylindrical, 22–42 mm long, with 20–50 eggs.

Comments: *C. italicus* ranges from Mediterranean Europe (France) to Central Asia (Kazakhstan). Egg-laying females prefer the sandy soils of *Artemisia* steppes. They lay multiple egg pods. The species is univoltine, with egg diapause in autumn and winter. Solitary *C. italicus* feeds on drought-resistant dicots like *Artemisia* spp. Swarming *C. italicus* devours nearly anything. Bendiscioli (1825) first reported *C. italicus* on *Cannabis* in Italy. It stripped leaves, flowers, and seeds. Villa (1839) still marveled over the devastation. In Russia's Voronezh Oblast, Lindemann (1892) observed young hemp plants mowed to the ground, but in older plants only leaves were eaten. In Lower Volga provinces, Semenov (1902) described an outbreak of *C. italicus* beginning in "weeds." Swarms soon switched to crops of peas, buckwheat, sunflower, flax, and hemp, the latter of which "was hard to completely destroy" by the locusts.

7.3c. Rocky Mountain locust

Melanoplus spretus (Walsh) 1866, Orthoptera: Acrididae
Description: *Adults* elongate, with elytra and wings longer than their bodies, body length 28 mm (females), 25 mm (males). Coloring dull brown with darker spots and stripes; parts of femora, head, and abdomen yellow-brown. *Egg pod* is curved-cylindrical, 25 mm long, with 20–35 eggs. *M. spretus*

is only known from its gregarious phase. It may be related to solitary *Melanoplus bruneri* Scudder 1897 or *Melanoplus sanguinipes* Fabricius 1798, based on genetic analysis of *M. spretus* specimens (Chapco and Litzenberger, 2004).

Comments: Migratory swarms of immense sizes were documented in the mid-1800s. The size of an 1875 swarm, estimated with the aid of telescopes and telegraphs, was 1800 miles (2900 km) long and 110 miles (180 km) wide—an estimated 3.5 trillion locusts (Lockwood, 2004). Then the species went extinct. The last living specimens were collected in 1902. Walsh (1868) recounted an 1820 outbreak of *M. spretus* in Missouri, which destroyed wheat, cotton, flax, and hemp crops. Riley (1877a) reported *M. spretus* in Nebraska, "devouring everything in some localities … even weeds, such as Jamestown weed and wild hemp."

7.3d. Other locust species

Schistocerca gregaria (Forsskål) 1775 is now the most feared locust in the world. The species devastates crops in Africa and southwest Asia. *S. gregaria* may have been the unidentified locust that Alcock (1903) described in Bengal, India—a two-mile-long swarm that destroyed most crop plants, but "did not attack the hemp crop." Rothschild *et al.* (1977) fed *S. gregaria* a pure diet of high-THC Mexican *Cannabis*, which proved lethal. In a feeding-choice test with various plants from India, *S. gregaria* refused to consume *Cannabis* (Manchanda *et al.*, 1982).

A related North American species, *Schistocerca lineata* (Scudder) 1899, occurs in Colorado hemp fields (Cranshaw, pers. obs., 2018). Thomas (1880) referred to the species by its synonym *Acridium emarginatum*, "I remember noticing a large number feeding on some hemp growing on the bank of the Missouri River."

The citrus locust, *Chondracris rosea* (DeGeer) 1773, attacked mature hemp crops in Taiwan (Sonan, 1940). *Dichroplus maculipennis* Blanchard 1851 is the most dangerous locust in temperate South America, where it is called *langosta brava*, the wild locust. Swarms have attacked fiber hemp in Chile (González *et al.*, 1973). Liebermann (1944) described "intense" damage on Chilean hemp, despite predation of *D. maculipennis* by birds and flesh flies (Diptera; Sarcophagidae).

7.3e. Two-striped grasshopper

Melanoplus bivittatus (Say) 1825, Orthoptera: Acrididae
Description: *Adults* large (30–55 mm long), light green to brown, with a pair of pale-yellow stripes running along the top of the body, from above the eyes to wing tips (Fig. 7.9A).
Comments: *M. bivittatus* is found coast to coast, particularly in the prairies of the USA and southern Canada. It is very common and sometimes destructive, feeding on a wide variety of herbaceous dicots. In hemp it has a habit of roosting on the crop late in the day, where it strips bark off branches and stalks and causes girdling injuries (Schreiner and Cranshaw, 2020) (Fig. 7.9A).

Fig. 7.9. Grasshoppers: **A**. *Melanoplus bivittatus*; **B**. Melanoplus differentialis (W. Cranshaw)

7.3f. Differential grasshopper

Melanoplus differentialis (Thomas) 1865, Orthoptera: Acrididae
Description: This polymorphic species ranges from bright green to brown, with dark chevrons along the hind femura, and yellow tarsi and antennae; females 34–50 mm, males 28–37 mm long (Fig. 7.9b).
Comments: *M. differentialis* is also found coast to coast, as common as *M. bivittatus* in the northern part of its range, but greatly outnumbers that species in the southern USA and Mexico. *M. differentialis* is polyphagous, preferring herbaceous dicots. The species is not migratory, but large populations can build into swarms and cause serious damage. It has infested hemp in Manitoba (McPartland *et al.*, 2000) and Colorado (Cranshaw *et al.*, 2019).

7.3g. Other grasshopper species

In addition to *M. bivittatus* and *M. differentialis*, Cranshaw *et al.* (2019) also reported the red-legged grasshopper, *Melanoplus femurrubrum* (De Geer) 1773, and the lakin grasshopper, *Melanoplus lakinus* (Scudder) 1878. *M. femurrubrum* "badly riddled" hemp in Kentucky (Garman, 1894a). McPartland *et al.* (2000) reported two other North American species: the sprinkled locust, *Chloealtis conspersa* Harris 1841 (in New York and Illinois), and the clearwinged grasshopper, *Camnula pellucida* (Scudder) 1862 (in Manitoba).

Grasshoppers that move through hemp, but do not damage it, may cause unwarranted alarm in growers. Cranshaw *et al.* (2019) exampled the chenopod specialist *Aeoloplides turnbulli* (Thomas) 1872, common in hemp along field edges where Russian thistle (*Salsola* spp.) is abundant. Siegel (1989) observed grasshoppers in a maize field that landed on hidden *Cannabis* plants; they jumped around erratically, attempting to clean off resin that adhered to antennae and tympanic membranes.

The elegant grasshopper, **Zonocerus elegans** (Thunberg) 1773, is less elegantly known as the stink grasshopper. It is named for its beautiful green, red, blue, and orange markings, and for its smelly defensive glands. *Z. elegans* is polyphagous with a predilection for poisonous food plants. Rothschild *et al.* (1977) fed *Z. elegans* nymphs a diet of pure *Cannabis* for 6 weeks, and most survived. The nymphs sequestered cannabinoids in their cuticle (as a predator deterrent?) and excreted the rest in frass.

In northern India, Batra (1976) reported the tobacco grasshopper, **Atractomorpha crenulata** Fabricius 1793 (Pyrgomorphidae), as the most common grasshopper. Other Asian pests include **Atractomorpha bedeli** Bolivar 1884 (Pyrgomorphidae) in Japan (Shiraki, 1952), **Hieroglyphus nigrorepletus** Bolivar (Acrídidae) in northern India (Roonwal, 1945), and **Chrotogonus saussurei** Bolivar 1884 (Acrídidae) in southern India (Cherian, 1932).

7.3.1 Differential diagnosis

Defoliation by grasshoppers can be confused with damage caused by caterpillars or leaf beetles. When grasshoppers chew through branches and stalks (Fig. 7.9), the injury resembles stalk borer damage.

7.3.2 Cultural and mechanical control

Discourage grasshoppers by eliminating grassy weeds. Be watchful when rotating after sod or pasture. In small gardens, keep hoppers away from tender seedlings with tent screens or row covers. When a locust swarm hit hemp in Bonțida, Romania, the town mobilized 1500 people, each of whom collected a sack of locusts every day, which were crushed, burned, or buried (Thon, 1838). Hemp has been turned against locusts: Valdrighi (1876) made torches of *canepa* stalks and "massacred" locusts. Salido y Estrada (1874) used "locust whips" topped with hemp cords or leather straps, to "make a great slaughter of them." In China, "every kind of oil is equally destructive to locusts, hemp oil being the best" (Riley, 1883). Cereal crops in Iraq were protected from locusts by fumigating hemp (*šāhdānaj*) and sulfur, according to a text written in the early 10th century (Rodgers, 1980).

7.3.3 Biocontrol and chemical control

Orthopterans are susceptible to *Metarhizium anisopliae*. A species segregated from *M. anisopliae*, named *Metarhizium acridum*, specifically infects locusts and grasshoppers. *Beauveria bassiana* has a wide host range, including Orthopterans. *Antonospora locustae* (formerly *Nosema locustae*) is back on the market, formulated as a bran bait for grasshoppers. *Entomophaga grylli*

is a wicked fungus (see Fig. 20.7). *Bacillus thuringiensis* (Bt) does not work against Orthopterans. Neem (azadirachtin) is a pesticide on the USDA's *National List* of organic substances. Pyrethrins may control grasshoppers. Some products combine pyrethrins with *Beauveria bassiana*. Yepsen (1976) used sabadilla against grasshoppers.

7.4 Leaf-eating beetles

Among Coleopterans, the most damaging defoliator is the hemp flea beetle (*Psylliodes attenuatus*). We cover this species in section 5.4 because of its damage to flowering tops. Adult weevils and curculios often notch leaves. Their larvae cause greater damage as stalk borers (see section 6.10). Many flea beetles feed on hemp. But first we turn to larger beasts—scarab beetles (family Scarabaeidae) and darkling beetles (family Tenebrionidae).

7.4a. Japanese beetle

Popillia japonica Newman 1838, Coleoptera: Scarabaeidae
Description: *Adults* are attractive, with robust metallic green bodies, and iridescent copper-colored wing covers, up to 15 mm long (Fig. 7.10). *Larvae* (grubs) live in soil; they are plump, white with a brown head, six dark legs, 15–25 mm long. They lie curled in a C-shaped position (see Fig. 8.7).
Comments: Japanese beetle is native to Japan, where natural enemies hold it in check. In Japan it primarily damages clover, with no mention of *Cannabis* (Clausen, 1931). A nursery in New Jersey accidently imported these pests in 1916. Since then, they have spread across North America, except for seven states west of the Rockies. *P. japonica* feeds on over 300 plant species, and hemp is not spared. Adult beetles notch and skeletonize foliage (Zuefle, 2021). They also feed on flowers (Cranshaw *et al.*, 2019). Japanese beetles are likely attracted to hemp pollen, as they are frequently seen in grain or fiber varieties during periods when an abundance of pollen is present on plants. The grubs live in soil and feed on roots (see Fig. 8.7).

Fig. 7.10. Japanese beetle (K. Britt)

The Chinese scarab beetle *Adoretus sinicus* (Burmeister) 1855 was accidently imported into Hawai'i, where it attacks over 250 species of plants. When David West established the first research plot of industrial hemp in Hawai'i, *A. sinicus* promptly became his #1 problem (D. West, pers. comm. 2002).

the soil line. Larvae feed on roots. A new generation of adults appears in the summer and feeds on leaves. Leaves become riddled by fine shot holes, then become skeletonized, and sometimes are completely defoliated down to midribs and major veins.

7.4b. Darkling beetle

Opatrum sabulosum (L.) 1761, Coleoptera: Tenebrionidae
Description: *Adults* are dark gray-brown to black, oval in shape, with a deep groove separating the thorax from abdomen. Wing covers (elytra) with longitudinal rows of tubercles, 7–10 mm long. *Larvae* are dark gray, flattened-cylindrical, 18 mm long.
Comments: Darkling beetle territory spans Europe, through Central Asia, to northwestern China. One generation arises per year, but they live 2–3 years. Adults emerge from soil in early spring and damage young seedlings; later they feed on leaves, and gnaw on seeds. Durnovo (1933) said damage peaks in May, especially in hemp fields adjacent to weedy steppe. Trotuş and Naie (2008) counted an average of 4 adults/m² in Romanian hemp (compared with 113/m² for *P. attenuatus*, the worst pest).

7.4.1   Leaf beetles

Species in the family Chrysomelidae are known as leaf beetles. The family includes more than 2500 genera and 50,000 species. Darwin reportedly said that the Creator must have an "inordinately fondness for beetles," because there are so many of them. We'll treat the Chrysomelidae by tribes, beginning with Tribe Alticini.

Tribe Alticini, the flea beetles, are so-called because they can jump like fleas using their enlarged hindleg femora. Many Alticini have the same life cycle: adults overwinter and emerge to feed on young plants (Fig. 7.11). Pichat (1866) wrote of hemp, "As soon as the seedlings are born they fear the *Altiche* or fleas of the earth." Adults mate and females lay eggs close to

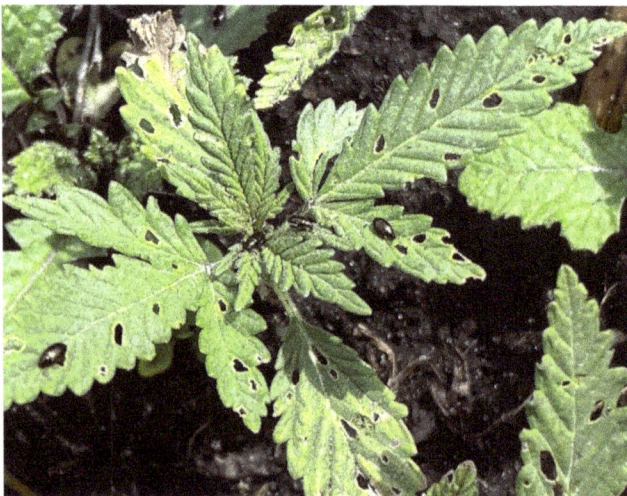

Fig. 7.11. Flea beetles chewing holes in seedlings (M. Bolt)

7.4c. Hop flea beetle

Psylliodes punctulata Melsheimer 1847
Description: Hop flea beetle closely resembles the hemp flea beetle, *P. attenuatus*. In fact, descriptions of *P. punctulata* (Chittenden, 1910) seem identical to *P. attenuatus* (Heikertinger, 1913). See the *P. attenuatus* description near Fig. 5.8. Hop and hemp flea beetles are differentiated by geography: North America and Eurasia, respectively. No one has made a side-by-side comparison. The genetic barcode (COI gene) of *P. attenuatus* has been sequenced, but not *P. punctulata*.
Comments: *P. punctulata* occurs coast to coast in the northern half of the USA and southern Canada. It is polyphagous and a serious pest in hops and cruciferous crops (Chittenden, 1910). It has been reported on *Cannabis* in Canada (Glendenning, 1927) and the USA (Clark *et al.*, 2004). The life cycle of *P. punctulata* mirrors *P. attenuatus*—one generation per year.

7.4d. Yellow-striped flea beetle

Phyllotreta nemorum (L.) 1758
≡*Haltica nemorum* L.
Description: *Adult* head and thorax metallic black, elytra black with tiny hairs that impart a bronze tone, with a broad yellowish stripe down each wing cover, 3.5 mm long. *Larvae* are yellow-bodied and brown-headed, 3–4 mm long.
Comments: Yellow-striped flea beetle has a Palearctic distribution. Like most *Phyllotreta* spp., it hammers cruciferous crops. But it frequently infests hemp (Nördlinger, 1869; Kirchner, 1906; Borodin, 1915; Goidànich, 1928; Kovačević, 1929). The lifecycle of *P. nemorum* departs from other members of Tribe Alticini: females lay eggs on the leaf surface, instead of the soil line, and larvae act as leafminers.

7.4e. Brassy flea beetle

Chaetocnema concinna (Marsham) 1802
Description: *Adult* body shape is oval in outline, 2–2.5 mm long, with greenish-black head, thorax, and elytra. The elytra are covered by longitudinal lines of fine punctuations. Tiny hairs are colored bronze-gray, which impart a bronze or brassy tone, especially after several days in sunshine. Sorauer (1958) pointed out that *C. concinna* is easily confused with the hemp flea beetle, *P. attenuatus*.
Comments: Brassy flea beetle spans northern Europe and Siberia; it reached New England by the 1970s and Canada's Maritime Provinces by the 1980s (Lesage and Majka, 2010). Its life cycle is consistent with Tribe Alticini in general.

C. concinna has been reported on hemp in Europe (Borodin, 1915; Kovačević, 1929) and North America (Clark *et al.*, 2004). In a pest survey of Romanian hemp, Trotuş and Naie (2008) counted an average of 41 adults/m². In their survey, *C. concinna* was second only to *P. attenuatus*. The species also infests hops; Ormerod (1890) called it "the hop flea." Davidyan (2006) said that ovipositing females targeted *Polygonum lapathifolium*, and larvae ate roots of *P. lapathifolium*, *Fagopyrum esculentum*, and *Cannabis sativa*.

7.4f. Red-headed flea beetle

Systena frontalis (Fabricius) 1801

Description: *Adult* body shape elongated, taping toward the head, with a shiny dark brown-black thorax and elytra, a dark-red head, and long banded antennae, 2.5–6.35 mm long. *Larvae* are off-white with a brown head, a cylindrical body ending with a fleshy projection, 5–10 mm long.

Comments: Red headed flea beetle is distributed in temperate North America and Europe, and feeds on a wide variety of plants. It commonly moves into hemp in Indiana (M. Bolt, pers. obs., 2020). In a survey conducted around Rochester, New York, Zuefle (2020) ranked *S. frontalis* the #1 cause of leaf-feeding damage. Symptoms included windowpane spots and skeletonizing injuries (Fig. 7.12). *S. frontalis* was first detected mid-June and was present through early September.

7.4g. Other Alticini

Flea beetles often move into hemp fields from other neighboring crops. We have compiled some other species in Tribe Alticini that appear in the literature.

Fig. 7.12. *Systena frontalis* and damage (courtesy M. Zuefle)

Genus *Altica*: In Europe, Melicher (1869) and Köppen (1880) identified *Altica oleracea* (L.) 1758 feeding on hemp leaves. Batra (1976) cited *Altica* sp. on *Cannabis* in India. Schreiner (2019) cited *Altica* sp. on *Cannabis* in Colorado. "*Haltica flavicornis* Baly" has been cited on Japanese hemp (Matsumura, 1899; Takahashi, 1919), but the genus *Haltica* is no longer recognized. Clausen (1931) wrote that "no information is available regarding *Haltica flavicornis*."

Genus *Podagrica*: *Podagrica aerata* Marsh 1802 infests hemp in Europe (Kaltenbach, 1874; Dempsey, 1975). The species has been renamed *Batophila aerata* (Marsham), the raspberry flea beetle. Rataj (1957) cited *Podagrica malvae* Illiger 1807. Hill (1983) reported a *Podagrica* sp. in Thailand.

Genus *Chaetocnema*: Kulagin (1915) described *Chaetocnema hortensis* (Geoffroy) 1785 attacking plants near Moscow; the species normally infests grasses. In Mississippi, Lago and Stanford (1989) identified *Chaetocnema denticulata* (Illiger) 1807, which normally infests grasses, and *Chaetocnema pulicaria* Melsheimer 1800, a pest of maize. Hartowicz *et al.* (1971) found a *Chaetocnema* sp. infesting wild hemp in Kansas.

Genus *Phyllotreta*: Borodin (1915) cited *Phyllotreta atra* (Fabricius) 1775 in Russia. In North America, Clark *et al.* (2004) cited *Phyllotreta lewisii* (Crotch) 1873. Schreiner (2019) identified *Phyllotreta pusilla* Horn 1889 in Colorado.

Genus *Disonycha*: In Mississippi, Lago and Stanford (1989) identified *Disonycha glabrata* (Fabricius) 1775, which was abundant one year, and absent the next. Schreiner (2019) identified *Disonycha triangularis* (Say) 1824 in Colorado.

Genus *Systena*: Garman (1906) cited *Sytena blanda* Melscheimer 1847 in Kentucky. In Mississippi, Lago and Stanford (1989) also identified *Sytena blanda*, which was common from May until August, and *Systena elongata* (Fabricius) 1798, which was abundant from May through September. *S. blanda* was identified in Colorado (Schreiner, 2019) and southern California (Britt *et al.*, 2022).

Genus *Neocrepidodera*: Goidànich (1928) identified *Neocrepidodera ferruginea* (Scopoli) 1763 infesting hemp in Italy.

7.4h. Tribe Luperini

Adults in Tribe Luperini skeletonize leaves, and larvae are root feeders. Many Luperini have hosts in the Cucurbitaceae, with evidence of a coevolutionary relationship (Jolivet *et al.*, 1994). *Cannabis* is an incidental host. The xylem pathogen *Erwinia tracheiphila*, an obligately insect-transmitted bacterium, has inserted itself into this relationship—but *Cannabis* is not susceptible to this pathogen.

Spotted cucumber beetle (a.k.a. southern corn rootworm), ***Diabrotica undecimpunctata howardi*** Barber 1947, is a serious pest of maize as well as cucurbits. It feeds on many alternative plants, including hemp and hops (Clark *et al.*, 2004). In Mississippi, Lago and Stanford (1989) described it as common, particularly from July through September. Spotted cucumber beetle was readily observed in Colorado (Schreiner and Cranshaw, 2020) and Virginia (Britt, pers. obs. 2020), and no significant damage was associated with its feeding. The species caused damage in New York, but no economic loss (Zuefle, 2021).

Striped cucumber beetle, *Acalymma vittata* (Fabricius) 1775, specializes in cucurbits, but adults can subsist on alternative hosts. Frank and Rosenthal (1978) and Alexander (1980) noted that stray individuals were commonly seen on *Cannabis* but rarely caused damage. Its western counterpart, the western corn rootworm, *Diabrotica virgifera* LeConte 1868, has also been reported on *Cannabis* (Clark *et al.*, 2004; Schreiner 2019). Banded cucumber beetle, *Diabrotica balteata* LeConte 1865, was collected in an insect survey in southern California (Britt *et al.*, 2022).

Golden beetle, *Monolepta dichroa* Harold 1877, has been identified on Japanese hemp (Yuasa, 1927; Shiraki, 1952). In addition to cucurbits and hemp, it feeds on hops and sweetpotato (Clausen, 1931). Redshouldered leaf beetle, *Monolepta australis* Jacoby 1882, has defoliated industrial hemp in Australia (Weeden, 2006; Hall *et al.*, 2014). Goidànich (1928) described *Luperus flavipes* (Lucas) 1849 on Italian hemp.

7.4i. Other leaf beetles

Rounding out the Chrysomelidae, several leaf beetles from other tribes have been identified on hemp. From Tribe Chrysomelini, Goidànich (1928) described *Chrysomela rossia* (Illiger) 1802 and *Gastrophysa polygoni* (L.) 1758 in Italy. From Tribe Lemini, Kulagin (1915) observed *Oulema melanopa* (Linnaeus) 1758 feeding on hemp near Moscow. From Tribe Eumolpini, Lago and Stanford (1989) identified two species by their old *Nodonota* names: *Brachypnoea tristis* (Olivier) 1808, and *Brachypnoea clypealis* (Horn) 1892; both were uncommon in Mississippi marijuana.

From Tribe Doryphorina, *Chrysolina fastuosa* (Scopoli) 1763 fed on *Cannabis* in Greece (Rozner and Rozner, 2014). Gorb and Gorb (2002) conducted an experiment on the ability of *C. fastuosa* to attach to leaf surfaces in 83 plant species, and they rated *Cannabis* among 29 plants with a "good" attachment surface. Gorb and Gorb attributed this to cystolith trichomes—which are supposed to serve as an herbivore repellent (see Fig. 3.1).

7.4j. Leaf-eating Curculionidae

Some weevils and curculios damage foliage. Most are known as stalk borers (see section 6.10). In India, Cherian (1932) reported the Ber weevil, *Xanthochelus faunus* Olivier 1807, defoliating *Cannabis*. Also in India, *Lepropus sublateralis* Poorani 1997 is reportedly a *Cannabis*-specific weevil (Poorani and Ramamurthy, 1997). Batra (1976) reported a straight-snouted weevil, *Apion* sp., in Indian hemp.

The gold dust weevil, *Hypomeces squamosus* (Fabricius) 1794, is a pest in Thailand (Hill, 1983). Adults notch leaf margins, and grubs live in the soil and feed on roots. Adults are gray, 10–15 mm long, and covered with a fine golden-green "dust" of hairs. Their wings fuse together in midline, rendering the adults flightless. In Vietnam, Hanson (1963) identified *Corigetus mandarinus* Fairmaire 1888.

Lastly, Coccinellidae (lady beetles) have been reported. In India, Cherian (1932) identified *Henosepilachna pusillanima* (Mulsant) 1850, using the old name *Epilanchna dodecastigma* (Wiedemann). Batra (1976) collected an *Epilanchna* spp. on *Cannabis* in northern India. The majority of coccinellids are considered beneficial insects, because they eat aphids.

7.4.2   Cultural and mechanical control

Many beetles overwinter in stubble or soil, so plowing fields in the fall will crimp their lifestyle. Given that so many Chrysomelidae feed on Cruciferae, eliminate cruciferous weeds. Alternatively, use crucifers as trap crops, like Chittenden (1910) suggested for the hop flea beetle, *Psylliodes punctulata*. Cruciferous plants lure the beetles away from hops (or hemp), where they can be destroyed. Follow instructions for the hemp flea beetle, *P. attenuatus* (see section 5.4). Wu *et al.* (2010) evaluated 21 Chinese landraces for resistance to flea beetles in the seedling stage, but never identified the species of flea beetles. Landraces from Shānxī and Inner Mongolia were the most resistant. A landrace from Guìzhōu was the most susceptible.

7.4.3   Biocontrol and chemical control

Leaf beetles are susceptible to the same biocontrols as the hemp flea beetle, *P. attenuatus* (see section 5.4). Grubs and overwintering adults in the soil can be killed by nematodes, *Heterorhabditis* and *Steinernema* spp. Chrysomelid beetle grubs are killed by *Bacillus thuringiensis* var. *tenebrionis*. *Bacillus popilliae* is a soil bacterium that kills *Popillia japonica*, the Japanese beetle. The fungi *Metarhizium anisopliae*, *Beauveria bassiana*, and *Cordyceps* (*Isaria*) *fumosorosea* have good efficacy against many Coleopteran species. A genus of parasitic wasps, *Townesilitus* spp., specializes in Chrysomelidae, particularly the genera *Phyllotreta* and *Psylloides*. See Table 20.5.

Chittenden (1910) sprayed hops with Bordeaux mixture to repel *Psylliodes punctulata*. Pyrethrins, ryania, sabadilla, and spinosad have been used. Foliar applications must be repeated to kill new beetles arriving from surrounding areas. Kill underground grubs with soil drenches. Metzger and Grant (1932) repelled Japanese beetles from test plants with pharmaceutical grade (USP, United States Pharmacopeia) *Cannabis indica* extract. They made extracts from 390 plant species; only 56 plants showed any repellency, including *C. indica*.

European researchers tested hemp as a repellent against a notorious Chrysomelid, the Colorado potato beetle, *Leptinotarsa decemlineata*. Stratii (1976) sowed hemp plants around a potato plot, and potato plants nearest the hemp were free of *L. decemlineata*. However, Mackiewicz (1962) grew hemp around and within a potato field, and found no effect on *L. decemlineata*. Both researchers also tested aqueous extracts; Mackiewicz (1962) found no effect, whereas Stratii (1976) claimed all adults and larvae died within 45 minutes. Stratii even claimed that waving flowering hemp plants over infested potato plants caused beetles to fall to the ground paralyzed.

7.5 Leafminers

Leafminer larvae live and feed within leaves. They are tiny and somewhat flattened, to fit their nearly two-dimensional niche. To gain fresh food, larvae tunnel through tissues like miniature coal miners. The tunnels look pale green to white, due to light reflecting off air in the mines (Fig. 7.13). As leafminers grow, their mines increase in width, and they leave behind a trail of frass. Their embedded lifestyle shields them from many natural enemies, and they avoid plant surface defense structures, such as glandular trichomes.

Leafminers are flies, members of their own family, the Agromyzidae, with around 2500 species. Kirchner (1906) identified one leafminer that was *not* a fly—larva of a beetle, *Phyllotreta nemorum* (adults chew on leaves, section 7.4d). *P. nemorum* larvae were easily differentiated from other leafminers—they had a distinct brown head and three pairs of brown legs. Agromyzid leafminers have no legs, and no distinct head—they are fly maggots.

Mine topography can help diagnose the leafminer species responsible. Diagnostic aspects of topography include the following: (1) the layer of leaf tissue being mined: the upper (palisade) parenchyma, the lower (spongy) parenchyma, or both layers; (2) the path taken: spiral (serpentine), linear (curvilinear), or tunneling in several directions (a blotch mine); (3) the trail of expelled frass: in continuous strips, or widely-spaced pellets, or one big dump; and (4) where larvae pupate: in the mine, or externally (in soil).

The first leafminer named on hemp, *Liriomyza strigata*, was reported by Brischke (1881) near Danzig (now Gdańsk, Poland). *L. strigata* is a well-known, polyphagous species, and Brischke identified *L. strigata* by mine topography alone—the insect was no longer present, having left the mine to pupate in soil.

The best-known *Cannabis* leafminer, *Liriomyza cannabis*, has a convoluted history, beginning with Martin Hering (1893–1967), a Berlin entomologist whose leafminer research spanned nearly 50 years. Hering (1921) encountered spiral-shaped mines on *Cannabis*. Based on mine topography, he identified *Liriomyza eupatorii*, which normally infested *Eupatorium* spp. Hering (1924) elaborated on the species' mine topography, and illustrated it twice, with different drawings (Hering, 1926, 1927).

Fig. 7.13. Leafminer damage (courtesy Fera, UK Crown Copyright)

De Meijere (1925) considered dubious any identifications based on mine topography. He identified leafminers by larva morphology (an equally dubious method), and reduced Hering's species on *Cannabis* to a new subspecies, *Liriomyza fasciola* subsp. *eupatorii*. Friedrich Hendel (1874–1936) identified species by adult morphology—including two hemp pests, *Phytomyza atricornis* (Hendel, 1926) and *Agromyza reptans* (Hendel, 1936). Turning to Hering's species, Hendel (1936) argued that mine topography differed: in *Cannabis* the initial spiral made 3–4 rotations before straightening; in *Eupatorium* the spiral was larger, and made 6–7 rotations. Next he compared adults emerging from *Cannabis* and from *Eupatorium*. They differed, so he erected a new species for the hemp pest, *Liriomyza cannabis*.

Hering (1937) accepted Hendel's argument, and dropped *L. eupatorii* from the list of leafminers on *Cannabis*. He provided a diagnostic key of three species based upon *Gangminen* (mine paths). Hering (1957) provided a revised key of four *Cannabis* leafminers, which in turn was revised by Ellis (2017) in Holland and Pitkin *et al.* (2019) in the UK. We provided a key of five species in Table 7.1.

Mine topography may change as a larva matures. A hatchling often mines in tight spirals or intestinal-shaped coils, until it gains strength and can cross leaf veins. Thereafter it plows ahead, laying linear or curvilinear tracks. The path may vary depending on where the egg is laid. If the egg of *L. cannabis* is laid on a leaf edge, the track is linear, following serrations along the leaf margin. If the egg is laid away from the edge, the initial track is laid in a spiral.

Hering (1957) provided line drawings of mine paths. Ellis (2017) and Pitkin *et al.* (2019) presented color photographs of mine paths. Spaar *et al.* (1990) drew a composite of several species's mine paths on a single *Cannabis* leaf, for comparative purposes, a method utilized here: Fig. 7.14A illustrates mines by *Phytomyza horticola*, based on Ferri (1959a). Figure 7.14B illustrates *Liriomyza cannabis*, based on Hering (1927, 1937). Figure 7.14C illustrates *Agromyza reptans*, based Buhr (1937). Figure 7.14D illustrates *Phyllotreta nemorum*, based on Kirchner (1906).

Leafminer damage is commonly encountered in the USA, but the causal species has not been identified with certainty. In a Colorado survey, 2.5–6.5% of insects caught in sweep nets were Agromyzid spp., identified to genus as *Agromyza* and *Liriomyza* spp. (Schreiner, 2019). Schreiner described leafminer tunnels having a "meandering serpentine form." No larvae were successfully reared to adults. An insect survey in California also recovered a fair number of unidentified Agromyzid spp. (Britt *et al.*, 2022).

7.5.1 Life history and habits

Agromyzid leafminers have similar life cycles. They overwinter as pupae. Adults emerge in spring to mate. In warm glasshouses leafminers do not hibernate; they breed continuously. Females drill eggs into leaves, one at a time, often clustered together. Adult flies feed on sap oozing from their ovipositor punctures. Some species lay up to 350 eggs (Hill, 1983). Eggs

Table 7.1. Topography of mines by four *Cannabis* leafminers

Mine path	Leaf layer mined	Frass trail	Pupal location
Liriomyza strigata produces a primary mine alongside the leaf midrib, with side branches radiating in a pinnate shape along leaf veins	Upper surface	In discontinuous pellets	Outside the mine
Liriomyza cannabis begins with a spiral (unless starting in leaf tip) and then becoming linear but tortuous	Uppersurface	Nearly continuous strips, "like strings of pearls" (Hering 1937)	Outside the mine
Phytomyza horticola elongated and meandering but relatively straight, initial part never spiraled	Upper or lower	Small, widely spaced pellets	Leaf, lower surface
Agromyza reptans initially follows the leaf margin, and then expands into an irregular brown blotch	Full thickness	In lumps or short pellets, never in long strings	Outside the mine
Phyllotreta nemorum short, wide tunnels with uneven edges, expanding to a blotch; larva exits mine to start elsewhere	Full thickness	In lumps or short pellets	Outside the mine

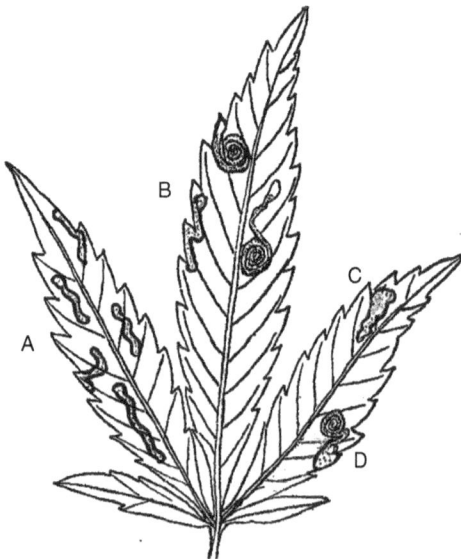

Fig. 7.14. Leafmines made by four species

hatch in 2–6 days. Maggots of most species undergo four molts, then pupate. Some species pupate in mines, and project two spiracular "horns" through leaf epidermis. Other species cut a slit in leaves and drop to the ground to pupate. Pupation takes 1 week to several months, depending on the species and season. Outdoors, 2–6 generations arise per year; in glasshouses the generations overlap, so all stages can be found at any time.

7.5a. Liriomyza cannabis Hendel 1931

Description: *Larvae* are legless cylindrical maggots without a head capsule, translucent white becoming yellowish, up to 2 mm long. The "cephalo-pharyngeal skeleton" takes the place of mandibles: a black, sclerotized pharynx with paired mouth hooks, whose morphology is used to speciate *Liriomyza* spp. (illustrated by de Meijere, 1937; Sasakawa, 1961). De Meijere

and Sasakawa also illustrated *L. cannabis* anterior and posterior spiracles (tracheal openings). *Pupae* form within the hardened last larval skin, with prominent transverse ridges, yellow turning brown, and no longer translucent. *Adults* resemble tiny houseflies, head yellow with large red eyes, thorax shiny gray-black with yellow markings, abdomen dark dorsally but sides yellow; body length and wingspan 1.5 mm long (Fig. 7.15).

Comments: Hering (1921, 1924) originally identified this species as *L. eupatorii*. Hendel (1931) differentiated *L. cannabis* from *L. eupatorii* based on external morphology, confirmed by Séguy (1934). Buhr (1937) conducted the first-ever transplantation of Agromyzid larvae. He transplanted *L. cannabis* larvae from *Cannabis* to *Eupatorium*, and transplanted *L. eupatorii* from *Eupatorium* to *Cannabis*. In both cases, the larvae mined, matured, and pupated. Nowakowski (1962) examined male genitalia in males reared from *Cannabis* and *Eupatorium*, and found no distinct differences. Nowakowski considered *L. cannabis* and *L. eupatorii* "sibling species," if not host races of a single species. However, male genitalia of *L. cannabis* and *L. eupatorii* are considered distinct by others (Sasakawa, 1961; Spencer, 1990). We await genetic studies; the COI sequence in *L. eupatorii* is available in public databases, but not the *L. cannabis* sequence.

L. cannabis has been reported in Germany (Hering, 1921, 1937; Spaar *et al.*, 1990), Finland (Hendel, 1931), France (Séguy, 1934), Italy (Ferri, 1959a), Japan (Sasakawa, 1961; Kyokai, 1965), Afghanistan (Mushtaque *et al.*, 1973), Turkey (Giray, 1980; Civelik, 2003), Great Britain (Collins *et al.*, 2016), and Holland (Mahieu *et al.*, 2019). In a Romanian field survey, *L. cannabis* density averaged 0.4 adults/m² (Trotuş and Naie, 2008). Two generations arise per year, and the species pupates in soil (Hendel, 1931).

7.5b. Liriomyza strigata (Meigen) 1830

≡Agromyza strigata Meigen 1830
Description: *Larvae* resemble those of *L. cannabis*: elongated cylindrical and soft, white becoming yellowish, but smaller (1.6 mm long). De Meijere (1937) noted differences between

Fig. 7.15. Adult *Liriomyza cannabis* (courtesy BioBest Group NV)

Fig. 7.16. A. *"Liriomyza strigata"* in *Cannabis* (Kirchner and Boltshauser, 1898); B. *Liriomyza strigata* in *Galeopsis* (courtesy Willem Ellis)

L. strigata and *L. cannabis* regarding the shape of the cephalo-pharyngeal skeleton. *Adult* external morphology resembles that of *L. cannabis*. In *L. strigata* the sides of the abdomen have narrow bands of yellow (in *L. cannabis* the sides are completely yellow), and *L. strigata* body length is longer, 2.0 mm (Séguy, 1934; Hendel, 1936). Spencer (1990) described differences in male genitalia between the two species.

Comments: *L. strigata* is polyphagous. Brischke (1881) named hemp as a host of *L. strigata* by mine topography: "bright, almost white, meandering tunnels with a green center, in which fine black feces form an irregular fecal line ... Maggots 2 mm long, greenish." A description of the mine and maggots by Kirchner (1890) clearly derived from Brischke. Kirchner and Boltshauser (1898) illustrated mines that were based on Brischke's written description (Fig. 7.16A).

Subsequently, Hering (1937) described very different mine topography by *L. strigata* on hemp—along the midrib, with side-tunnels along leaf veins. He reiterated this later (Hering, 1951, 1957). This distinctive topography is made by *L. strigata* in other hosts (Fig. 7.16B), and it's clearly not what Brischke (1881) described, or Kirchner and Boltshauser illustrated.

Later illustrations of mine topography by *L. strigata* on hemp (Ciferri and Brizi, 1955; Spaar *et al.*, 1990) were copied from Kirchner and Boltshauser. Other reports of *L. strigata* on *Cannabis* also appear derivative (Flachs, 1936; Anonymous, 1940; Rataj, 1957; Ferri, 1959a). Berlese (1924) said *L. strigata* occurred on Italian hemp. His Italian rival, Goidànich (1928), considered it a misidentification.

7.5c. *Phytomyza horticola* (Goureau) 1851

=*Phytomyza atricornis* Meigen 1838
Description: *Larvae* are elongate and cylindrical maggots, greenish-white, and somewhat larger than the aforementioned species, 2.5 mm long (Dempenwolf, 2001). *Pupae* in leaves, brown in color, with two cornices protruding from the lower epidermis. *Adults* have a black head, but the "face" (clypeus) is yellow. Thorax mostly black, abdomen black with yellow lateral lines, legs mostly black, with yellow "knees" (joint between tibia

and femur), 1.5–2.5 mm long (Ferri, 1959a). The wings of *Phytomyza* spp., unlike most other Agromyzids, lack crossveins in the posterior portions.

Comments: Highly polyphagous *P. horticola* has a common name: the pea leafminer. It has been recorded from 268 host genera, but most frequently in Fabaceae, Brassicaceae, and Asteraceae (Spencer, 1990). It is widely distributed across Eurasia and much of Africa. Hendel (1926) first reported *P. horticola* on hemp, under its synonym, *Phytomyza atricornis*. "Topside mine tunnel with pupal cradle at the end. Typical."

Ferri (1959a) described *P. atricornis* at length, with a drawing of the adult fly and mine topography in *Cannabis*, "irregular elongated tunnels." Adults appeared in late April, and mines appeared in the first week of May. Life cycle was 18–20 days, with 2–7 generations per year. Other European authors cited *P. horticola* on hemp under different synonyms (Hendel, 1932; Hering, 1937; Anonymous, 1940; Rataj, 1957; Spaar *et al.*, 1990). Griffiths (1967) stated that *P. horticola* also occurred on *Cannabis* in Japan and India. Parihar and Ramkishore (2002) reported *Cannabis* as an "alternative host" for *P. horticola* infesting potato crops in India.

7.5d. Agromyza reptans Fallén 1823

=*A. haplacme* Steyskal 1972
Description: *Larvae* are elongated cylindrical and soft, white becoming yellowish, and they are the largest leafminers on *Cannabis*, 4.6 mm long (Dempewolf, 2001). *Pupae* are reddish-brown. *Adults* have a dark-brown head, reddish-brown eyes, brownish-black thorax, brown abdomen, black femora, and tibiae and tarsi with hints of yellow, wing length 3.0–3.9 mm.

Comments: *A. reptans* is distributed in Europe and North America. The species is narrowly oligophagous, feeding on hosts in the Urticaceae: mostly *Urtica* spp., as well as *Parietaria* and *Laportea* spp. Buhr (1937) conducted transplantation experiments with *A. reptans*. Larvae in *Urtica* spp. were transplanted to *Cannabis* and *Humulus* spp., and they fed and developed normally. *A. reptans* infests hemp in northern Europe (Hendel, 1936; Hering, 1937, 1951; Patschke *et al.*, 1997; Černý, 2009).

7.5.2   Control of leafminers

Regarding hemp leafminers, Ferri (1959a) wrote, "The damage these insects cause is never very significant. In general, it is rare that they cause defoliation that compromises the vitality of the plant, only when the attacks are very early—young seedlings deprived of part of the assimilating apparatus can deteriorate, and therefore be easy prey for other parasites, especially fungi."

Leafminers sometimes tunnel into floral leaves, making the flowers unattractive to consumers when tunnels are filled with frass. Leafminers may require control if they get into glasshouses. Biological control is a good option; Ferri (1959a) noted that leafminer larvae in *Cannabis* were parasitized by the braconid wasp *Dacnusa areolaris* and the chalcid wasp *Closterocerus formosus*. Commercially available species include the braconid wasp *Dacnusa sibirica* and the chalcid wasp *Diglyhus isaea* (see section 20.6). Leafminer eggs are eaten by *Chrysoperla carnea*. Leafminers that pupate in soil can be culled by nematodes (*Heterorhabditis* and *Steinernema* spp.), as well as *Bacillus thuringiensis* var. *israelensis*. To control *L. cannabis*, Mahieu *et al.* (2019) preferred using *Diglyphus isaea* and the fungus *Beauveria bassiana* (ATCC74040 and GHA strains).

7.6   Assorted others

Other flies besides leafminers infest *Cannabis*, such as gall midges and sand flies. Plus there's a grab-bag of other insects that occasionally chew hemp leaves—sawflies, springtails, and earwigs.

7.6.1   Gall midges

At least one errant member of the Agromyzidae feeds on pith instead of mining leaves. The larvae cause **galls** in small branches or in flower peduncles and pedicels. In addition to Agromyzidae, flies in family Cecidomyiidae may act as gall midges. Mushtaque *et al.* (1973) reported an *Asphondylia* **sp.** forming galls in male flowers in Pakistan. "It does not appear to be a promising biocontrol agent, because pollination of hemp is almost complete by the time the infestation starts" (Baloch *et al.*, 1974). Batra (1976) collected a *Lestodiplosis* **sp.** on *Cannabis* in Pantnagar, northern India. He said it was phytophagous, but many *Lestodiplosis* spp. are carnivorous and prey on other insects.

7.6a.   Nettle midge

Melanagromyza urticivora Spencer 1966

=*Melanagromyza cannabini* Hering 1957, nomen nudum
Description: *Larvae* are small, white maggots without legs or distinct head capsules. *Pupae* pale yellowish-brown, 3 mm long (Fig. 7.17A). In order to breathe, pupae drill a pair of spiracles to the surface. Each spiracle is crowned with an ellipse of 14 bulbs, somewhat diagnostic of *M. urticivora* (Fig. 7.17B). *Adults* are black to faintly greenish, 1.5 mm long, wingspan 2.4 mm in males, to 3 mm in females (Fig. 7.17A).
Life history and host range: Spencer (1966) found larvae galling nettles (*Urtica* sp.) in Pakistan and coined *M. urticivora*. Mushtaque *et al.* (1973) found larvae galling *Cannabis* in Pakistan. Baloch *et al.* (1974) reported that *M. urticivora* only attacked *Cannabis* growing near *Urtica* spp. In a letter to Kenneth Spencer in 1957, Martin Hering described *Melanagromyza cannabini* larvae feeding within *Cannabis* stems. Spencer, who edited their correspondence, noted that *M. cannabini* was not formally described, a *nomen nudum* (Hering, 1968), but they are likely one and the same species.

7.6.2   Sand flies

The adults of some sand flies (Diptera: Psychodidae) have piercing-sucking mouthparts and feed on blood, similar to mosquitoes (Diptera: Culicidae) and black flies (Diptera: Simuliidae). Schlein and Muller (1995) showed that some blood-feeding sand flies also probe plants for sugar-based meals. They treated plants with calcfluor, a florescent dye that stains internal tissues of plants but not the outer epidermis. Then they housed **Phlebotomus papatasi** (Scopoli) 1786 overnight with labeled plants, and dissected their guts, which contained florescence. The guts also contained fructose, which suggested phloem penetration.

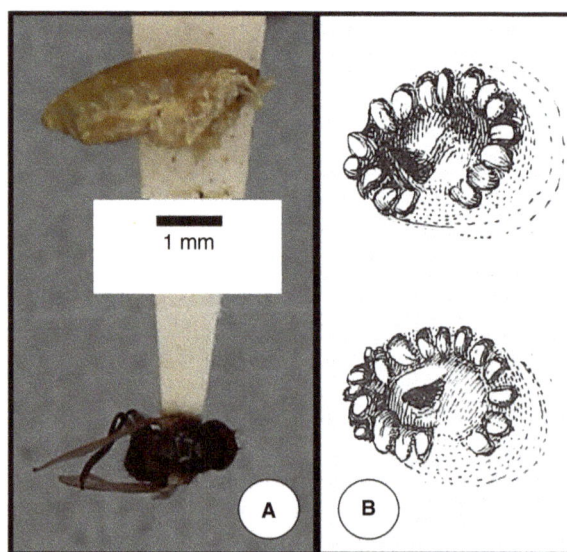

Fig. 7.17. Nettle midge: **A.** Spencer's type specimen; **B.** close-up of spiracles

To identify plants consumed by wild-caught sand flies, Abbasi *et al.* (2018) searched for plant DNA in them, using primers for chloroplast *rbcL* genes. They found *Cannabis* DNA in several *Phlebotomus* spp. around the world: *P. sergenti* Parrot 1917, collected in Palestine; *P. orientalis* (Parrot) 1936, collected in Ethiopia; *P. mongolensis* Sinton 1928, collected in Kazakhstan; and *Lutzomyia longipalpis* Lutz & Neiva 1912, collected in Brazil. The authors concluded that *Phlebotomus* spp. preferentially feed on *Cannabis*. This is bad news, because *Phlebotomus* spp. vector leishmaniasis, a terrible human disease. Many sand fly specimens with *Cannabis* DNA also tested positive for *Leishmania* spp.

7.6.3   Sawflies

7.6b.   Hemp sawfly

Trichiocampus cannabis Xiao & Huang 1987, Hymenoptera: Tenthredinidae
Description: *Larvae* resemble pale caterpillars, with light green heads that darken to black as they age. They have 6–8 pairs of hookless prolegs (caterpillars have five or fewer pairs of prolegs, bearing tiny hooks or crochets). They grow to 10 mm in length. *Adults* have a black head, black or rust-brown thorax, and yellowish-orange abdomen. Wingspan averages 16 mm, body length 6.0–6.8 mm (females), 4.7–5.3 mm (males). Vikberg (2013) provided a series of excellent photographs. *Eggs* are white, oblong, up to 1 mm long.
Comments: Sawflies are wasps. Xiao *et al.* (1986) offered the name *Cladius* (*Trichiocampus*) *cannabis*, and some texts still use that name (Belokobylskij and Lelej, 2017). Subsequently, Wang *et al.* (1987) shortened the trinomial name to *T. cannabis*. Earlier, Takeuchi (1949) mentioned "*Trichiocampus cannabis* Takeuchi, in litt.," but we could not find his "in litt." publication. Hemp sawfly is monophagous on *Cannabis*. Larvae skeletonize leaves, or gnaw holes and notch edges, and sometimes eat everything but midribs and main veins. The species overwinters in soil as mature larvae (Wang *et al.*, 1987) or as pupae (Shilenkov and Tolstonogova, 2006). Two generations arise per year.

Hemp sawfly occurs in Japan (Takeuchi, 1949), Ānhuī Province in China (Wang *et al.*, 1987), and the Russian Far East (Zhelochovtsev and Zinovjev, 1992). Shilenkov and Tolstonogova (2006) found *T. cannabis* defoliating *Cannabis ruderalis* in the Baikal region of Siberia. In their survey, counts of *T. cannabis* were second only to the hemp flea beetle. Vikberg (2013) authenticated museum specimens of *T. cannabis* collected in the Volga region of European Russia, and Gānsù Province in central China.

In China, 15–30% of *T. cannabis* were eaten by lady beetles, lacewings, and other predators, or parasitized by Ichneumonid wasps. In the Baikal region, larvae were parasitized by the tachinid fly *Bessa selecta* (Beigen) 1824. To control *T. cannabis* in Chinese hemp, Wang *et al.* (1987) tested carbamates and organophosphates.

Purslane sawfly, *Schizocerella pilicornis* (Holmgren) 1868, was reported on Mississippi marijuana. Lago and Stanford (1989) encountered many adults, but no larvae. Schreiner

(2019) recovered *S. pilicornis* in Colorado hemp, "only because purslane (*Portuacha oleracea*) is abundant in cultivated fields in Colorado." In Indiana, if there is purslane in a hemp field, *S. pilicornis* is there too (M. Bolt, pers. obs., 2019).

7.6.4   Springtails

Springtails (subclass Collembola) are small, simple insects, and quite abundant. They are so-named because a specialized appendage under their abdomen enables them to spring 10 cm (100 times their body length). They are seldom seen because of their tiny size and concealed soil habitats. Most springtails feed on decaying plant material, fungi, and bacteria. Some are garden pests. Two springtail species have been described from *Cannabis*. Many websites report springtails as pests, but not specific species. Springtails can be observed in the soil surrounding *Cannabis* plants, both indoors and outdoors.

7.6c.   Garden springtail

Bourletiella hortensis Fitch 1863, Collembola: Bourletiellidae)
Description: *Adults* have a round head tenuously attached to their globose body; they have black eyes and long antenna, no wings, up to 1 mm long. Their color varies from yellow to reddish-brown to dark green or black. They undergo a simple metamorphosis, so juveniles look much like sexually mature adults, except smaller.
Comments: *B. hortensis* lives near the soil surface. Populations can be estimated by placing a white card on the ground and gently disturbing soil. *B. hortensis* damages seedlings by nibbling at stems near the soil level and eating small holes in leaves. These injuries provide entry portals for pathogenic fungi. Lago and Stanford (1989) found them on seedlings up to 15 cm tall, with three or four individuals feeding around the same hole. They disappeared from older plants, and the authors attributed this to cannabinoids adversely affecting palatability.

7.6d.   Clover springtail

Sminthurus viridis L. 1758, Collembola: Sminthuridae
Description: *Adults* have round heads and globular bodies, long antenna, no wings, up to 3 mm long. Color varies from yellow to bright green.
Comments: *S. viridis* is also called the lucerne flea—it infests lucerne (alfalfa) and jumps like a flea. This Eurasian species has invaded Australia, where it infests hemp (Ditchfield, pers. comm., 1997). Feeding injury causes stippling in leaves.

7.6.5   Earwigs

Earwigs (Dermaptera: Forficulidae) are known for prominent forceps-like cerci at their tail end. They are abundant across Eurasia and the Americas. The genus *Forficula* comprises nearly 70 species, one of which has been cited on *Cannabis*.

7.6e. European earwig

Forficula auricularia L. 1758, Dermaptera: Forficulidae
Description: *Adults* are dark reddish-brown with flattened and elongated bodies, 10–15 mm long, with a pair of prominent cerci. Their hindwings when unfolded resemble a human ear, hence the name. *Nymphs* are pale brown, and their wings and cerci are much reduced or absent. *Eggs* are pearly-white and laid in masses underground.
Comments: *F. auricularia* is a beneficial species (it eats aphids and other pests), but the species is omnivorous, and can become a garden pest. Large males can inflict a painful pinch with their cerci. They are nocturnal. Earwigs overwinter as adults in small underground nests. Eggs are laid in late winter and females guard their brood until they hatch around May. Earwigs can fly, but not far.

Kovacs (1809) stated that *F. auricularia* "loves the sweet fruit hemp variety" in Hungary. Earwigs have killed seedlings in California (Frank, 1988), infested foliage in India (Batra, 1976), occupied stem galleries of European corn borers (Goidànich, 1928), and occurred in Colorado hemp fields (Schreiner, 2019). Culpeper (1652) used "juice" squeezed from *Cannabis* leaves to repel earwigs … from his patients' ears.

(Prepared by J. McPartland; revised by M. Bolt and K. Britt)

References

Abbasi I, Lopo de Queiroz AT, Kirstein OD *et al.* (13 other authors). 2018. Plant-feeding phlebotomine sand flies, vectors of leishmaniasis, prefer *Cannabis sativa*. *Proceedings of the National Academy of Sciences USA* 115: 11790–11795.

Alcock A, *ed.* 1903. Notes on insect pests from the entomological section, Indian Museum. *Indian Museum Notes* 5: 103–202.

Alexander T. 1980. Spring pests. *Sinsemilla Tips* 1(1): 7.

Andreevsky I.E. 1878. *Земский ежегодник за 1876 годъ*. Free Economic Society, St Petersburg.

Angelini B. 1827. Dei danni principalmente causati nel 1826 dalla *Noctua gamma* nella provincia veronese. *Biblioteca Italiana* 45: 57–61.

Anonymous. 1940. *Lezioni al Corso di Perfezionamento per la Stima del Tiglio di Canapa*. Faenza-Fratelli Lega Editori, Roma.

Arnscheid WR, Weidlich M. 2017. *Psychidae*. Brill, Leiden, The Netherlands

Baloch GM, Ghani MA. 1972. *Natural enemies of Papaver somniferum and Cannabis sativa*. Annual report, Commonwealth Institute of Biological Control, Pakistan station, pp. 55–56.

Baloch GM, Mushtaque M, Ghani MA. 1974. *Natural enemies of Papaver spp. and Cannabis sativa*. Annual report, Commonwealth Institute of Biological Control, Pakistan station, pp. 56–57.

Batra SW. 1976. Some insects associated with hemp or marijuana (*Cannabis sativa* L.) in northern India. *Journal of the Kansas Entomological Society* 49: 385–388.

Belokobylskij SA, Lelej AS. 2017. *Annodated catalogue of the Hymenoptera of Russia, volume* I. *Symphyta and Apocrita: Aculeata*. Zoological Institute of the Russian Academy of Sciences, St Petersburg.

Bendiscioli G. 1825. Alcuni cenni sull'Acridium italicum, sulla sua straordinaria moltiplicazione, e sui danni che areca alle campagne del commune di Poggio nella provincia di Mantova. *Biblioteca italiana ossia Giornale di Letteratura, Scienze ed Arti* 39: 241–248.

Berlese A. 1924. *Entomologia Agraria. Manuale sugli insetti nocivi alle piante coltivate*. Reale Stazione di Entomologia Agraria, Firenze.

Blunck H. 1920. Die niederen Tierirschen feinde unserer Gespinstpflanzen. *Illustrierte Landwirtschaftliche Zeitung* 40: 259–260.

Bolt M. 2019. *Influence of agronomic management strategies on hemp-insect interactions*. Master's thesis, Purdue University, West Lafayette, Indiana.

Borodin DI. 1915. Первый отчёт о деятельности энтомологического бюро и обзор вредителей Полтавской губернии. D. N. Podzemsy, Poltava, Ukraine.

Brez J. 1791. *Flore des insectophiles*. Wild et Altheer, Utrecht.

Brischke CGA. 1881. Die Blattminierer in Danzigs Umgebung. *Schriften der Naturforschenden Gesellschaft in Danzig, Neue Fogle* 5: 233–290.

Britt KA. 2021. *Insect pest management in hemp in Virginia*. Doctoral thesis, Virginia Polytechnic Institute and State University, Blacksburg.

Britt KE, Meierotto S, Morelos V, Wilson H. 2022. First year survey of arthropods in California hemp. *Frontiers in Agronomy* 4: e901416.

Brooke G. 2008. *National industrial hemp strategy*. Manitoba Agriculture, Food and Rural Initiative, Agriculture and Agri-Food Canada.

Buhr H. 1937. Parasitenbefall und pflanzenverwandtschaft. *Botanische Jahrbucher* 68: 142–198.

Camprag D, Jovanic M, Sekulic R. 1996. Štetočine konoplje i integralne mere suzbijanja. *Zbornik Radova* 26/27: 55–68.

Ceapoiu N. 1958. *Cinepa, studiu monografic*. Editura Academiei Republicii Populare Romine. Bucharest.

Černý M. 2009. Jizerských hor a Frýdlantska Agromyzidae. *Sborník Severočeského Muzea, Přírodní Vědy, Liberec* 27: 115–140.

Chandra R. 2004. Status of medicinal plants with respect to infestation of insect pests in and around Chitrakoot, District-Satna (M.P.). *Flora and Fauna (Jhansi)* 10(2): 88–92.

Chapco W, Litzenberger G. 2004. A DNA investigation into the mysterious disappearance of the Rocky Mountain grasshopper, mega-pest of the 1800s. *Molecular Phylogenetics and Evolution* 30: 810–814.

Chén FS. 2020. 捕蝗考. Chinese Text Project, Available at: https://ctext.org/wiki.pl?if=en&chapter=317644

Cherian MC. 1932. Pests of ganja. *Madras Agricultural Journal* 20: 259–265.

Chittenden FH. 1910. "The hop flea-beetle (*Psylliodes punctulata* Melsh.)," pp. 71–92 in *Some insects injurious to truck crops*. USDA Bureau of Entomology Bulletin No. 66, Washington, DC.

Ciferri R, Brizi A. 1955. *Manuale di patologia vegetale, Vol. III, Table 29*. Societa Editrice Dante Alighieri, Roma.

Civelek HS. 2003. Checklist of Agromyzidae (Diptera) of Turkey, with a new record. *Phytoparasitica* 31: 132–138.

Clark SM, LeDoux DG, Seeno TN, *et al.* (3 other authors). 2004. *Host plants of leaf beetle species occurring in the United States and Canada Coleoptera, Megalopodidae, Orsodacnidae and Chrysomelidae, excluding Bruchinae.* Special Publication of the Coleopterists Society, no. 2. Sacramento, California.

Clarke RC, Merlin MD. 2016. *Cannabis* domestication, breeding history, present-day genetic diversity, and future prospects. *Critical Reviews in Plant Sciences* 35: 293–327.

Clausen CP. 1931. *Insects injurious to agriculture in Japan.* USDA Circular no. 168, Washington, DC.

Collins DW, Gaunt A, Tschirnhaus M, Pye D, 2016. The first occurrence of *Liriomyza cannabis* Hendel (Diptera, Agromyzidae) in Great Britain. *Dipterists Digest* 23: 163–167.

Comes O. 1885. Sulla grillotalpa (*Gryllotalpa vulgaris*) e sul mezzo di combatterla. *Annuario della Regia Scuola superiore di agricoltura in Portici* 5: 55–63.

Cranshaw WS, Schreiner M, Britt K, Kuhar TP, McPartland JM, Grant J. 2019. Developing insect pest management systems for hemp in the United States: a work in progress. *Journal of Integrated Pest Management* 10(1): 26; 1–10.

Culpeper N. 1652. *The English physitian.* Peter Cole, London.

Curtis J. 1860. *Farm insects.* Blackie & Son, London.

Cuvier G. 1798. *Tableau élémentaire de l'histoire naturelle des animaux.* Baudouin, Paris.

Dalla Torre KW von, Strand E. 1929. *Psychidae. Lepidoterorum Catalogus pars 34.* W. Junk, Berlin.

Davidyan GE. 2006. *Chaetocnema concinna Marsh.* – mangold flea beetle. Interactive Atlas of Economic Plants and Pests in Russia and Neighboring Countries. Available at: http://www.agroatlas.ru/en/content/pests/Chaetocnema_concinna/

de Meijere JCH. 1925. Die Larven der Agromyzinen. *Tijdschrift voor Entomologie* 68: 195–293.

de Meijere JCH. 1937. Die Larven der Agromyzinen. Dritter Nachtrag. *Tijdschrift voor entomologie* 80: 167–243

Del Guercio G. 1903. Notizie e suggerimenti pratici per conoscere e combattere gli animali nocivi alle piante coltivate ed ai loro frutti. *Nuove Relazioni intorno ai Lavori della R.Stazione di Entomologia Agraria di Firenze, Serie Prima* 5: 1–208.

Dempewolf M. 2001. *Larvalmorphologie und Phylogenie der Agromyzidae (Diptera).* Doctoral thesis, Universität Bielefeld, Germany.

Dempsey JM. 1975. "Hemp," pp. 46–89 in *Fiber crops.* University of Florida Press, Gainesville.

Deshmukh PD, Rathore YS, Bhattacharya AK. 1979. Larval survival of *Diacrisia obliqua* Walker on several plant species. *Indian Journal of Entomology* 41(1): 5–12.

Dhaliwal JS. 1993. Role of some weeds in the carry-over of *Spilosoma obliqua* (Walker) to Egyptian clover (*Trifolium alexandrinum* L.). *Journal of Research, Punjab Agricultural University* 30(3-4): 168–170.

Doumerc A. 1860. Notice sur la teigne du chanvre (*Psyche cannabinella*). *Annales de la Société Entomologique de France Serie III* 8: 322–325.

DuBois A. 1884. *Les lépidoptères de la Belgique, tome troisième.* Muquardt, Merzbach et Falk, Bruxelles.

Durnovo ZP. 1933. "Итоги работ по кукурузному мотыльку и другим вредителям однолетних новых лубяных культур," pp. 85-106 in Болезни и вредители новых лубяных культур. Institute New Bast Raw Materials, Moscow.

Ellis WN. 2016. What is going on with the silver Y, *Autographa gamma*? *Entomologische Berichten* 76: 21–27.

Ellis WN. 2017. *Canabis hemp. Dichotomous table for leafminers.* Available at: https://bladmineerders.nl/host-plants/plantae/spermatopsida/angiosperma/eudicots/superrosids/rosids/fabids/rosales/cannabaceae/cannabis/

EPA. 2023. *Pesticide products registered for use on hemp.* Available at: https://www.epa.gov/pesticide-registration/pesticide-products-registered-use-hemp.

Ferri F. 1959a. *Atlante delle avversità della canapa.* Edizioni Agricole, Bologna, Italia.

Filippo Re. 1806. *Sopra le danneggiarono ne' due anni scorsi i Canapaj in alcuni Dipartimenti del Regno d'Italia.* G. Silvestri, Milano.

Flachs K. 1936. Krankheiten und schädlinge unserer gespinstpflanzen. *Nachrichten über Schädlingsbekämpfung* 11: 6–28.

Fletcher TB. 1917. *Report of the Proceedings of the Second Entomological Meeting, Pusa.* Superintendent Government Printing, Calcutta.

Frank M. 1988. *Marijuana grower's insider's guide.* Red Eye Press, Los Angeles.

Frank M, Rosenthal E. 1978. *Marijuana grower's guide.* And/Or Press, Berkeley.

Freyer CF. 1830. *Beiträge zur geschichte Europäischer Schmetterlinge mit Abbildungen nach der Natur, dritter band.* Berfasser, Augsburg.

Fritzsche R. 1959. Der schattenwickler (*Cnephasia wahlbomiana* L.) als schädling an lein und hanf. *Wissenschaftliche Zeitschrift (Mathematisch-Naturwissenschaftliche Reihe)* 8(6): 1117–1119.

Fruhstorfer H. 1921. Die Orthopteren der Schweiz. *Archiv für Naturgeschichte* 87(5-6): 1–262.

Fulton BB. 1924. *The European earwig.* Oregon Agricultural Experiment Station Bulletin no. 207. Corvallis, Oregon.

Garman H. 1894a. *Destructive locusts in Kentucky.* Kentucky Agricultural Experiment Station Bulletin no. 49. Lexington, KY.

Garman H. 1906. *Observations and experiments on clover, alfalfa and soy beans.* Kentucky Agricultural Experiment Station Bulletin no. 125. Lexington, Kentucky.

Giray H. 1980. *Türkiye' de Bitki Yaprakla-rında Galeri Açan Böcekler Faunasına Ait İlk Liste ile Bunların Konukçu ve Önem-lilerinin Galeri Şekilleri Hakkında Notlar.* Ege Üniversitesi Ziraat Fakültesi Yayınları, no. 374. İzmir, Turkey.

Gistel J. 1856. *Die mysterien der europäischen insectenwelt.* Dannheimer, Kempten.

Gladis T, Alemayehu N. 1995. Larven von *Acherontia atropos* L. (Lep., Sphingidae) neuerdings auch an Hanf (*Cannabis sativa* L.) – oder bislang übersehen? *Entomologische Nachrichten und Berichte* 39: 209–212.

Glendenning R. 1927. *The cabbage flea beetle and its control in British Columbia.* Pamphlet No. 80, New Series, Canada Department of Agriculture.

Godart JB. 1822. *Histoire naturelle des lépidoptères ou papillons de France, Tome troisième.* Crevot, Paris.

Goidànich A. 1928. Contributi alla conoscenza dell'entomofauna della canapa. I. Prospetto generale. *Bollettino del Laboratorio di Entomolgia del R. Istituto Superiore Agrario di Bologna* 1: 37–64.

González RH, Arretz P, Campos LE. 1973. *Catálogo de las plagas agrícolas de Chile.* Facultad de Agronomia, Universidad de Chile, Santiago.

Gorb EV, Gorb SN. 2002. Attachment ability of the beetle *Chrysolina fastuosa* on various plant surfaces. *Entomologia Experimentalis et Applicata* 105: 13–28.

Gottwald R. 2002. Entomologische Untersuchungen an Hanf (*Cannabis sativa* L.). *Gesunde Pflanzen* 54(5): 146–152.

Griffiths GCD. 1967. Revision of the *Phytomyza syngenesiae* group (Diptera, Agromyzidae), including species hitherto known as "*Phytomyza atricornis* Meigen." *Stuttgarter Beiträge zur Naturkunde* 177(3): 1–28.

Halitschke R, Keßler A, Kahl J, Lorenz A, Baldwin IT. 2000. Eco-physiological comparison of direct and indirect defenses in *Nicotiana attenuata*. *Oecologia* 124: 408–417.

Hall J, Bhattarai SP, Midmore DJ. 2014. Effect of industrial hemp (*Cannabis sativa* L.) planting density on weed suppression, crop growth, physiological responses, and fibre yield in the subtropics. *Renewable Bioresources* doi: 10.7243/2052-6237-2-1

Hanson HC. 1963. *Diseases and pests of economic plants of Vietnam, Laos and Cambodia*. American Institute of Crop Ecology, Washington, DC.

Harada T. 1930. 大麻の害患難を大麻象鼻貴職に就きて. *Konchū Sekai (Insect World)* 34: 118–123.

Hartowicz LE, Knutson H, Paulsen A, *et al*. 1971. Possible biocontrol of wild hemp. *North Central Weed Control Conference, Proceedings* 26: 69.

Heikertinger F. 1913. *Psylliodes attenuata* Koch, der Hopfen- oder Hanf-Erdfloh. II. Teil. Morphologie und Bionomie der Imago. *Verhandlungen der Zoologisch-Botanischen Gesellschaft in Österreich* 63: 98–138.

Hendel F. 1926. *Blattminenkunde Europas: I. Die Dipterenminen, Liefg. 1*. F. Wagner, Wien.

Hendel F. 1931. "Agromyzidae," pp. 1–320 *in* Lindner E, *ed. Die Fliegen der Paläearktishen Region, Band 6(2)*. Stuttgart.

Hendel F. 1936. "Agromyzidae," pp. 1–320 *in* Lindner E, *ed. Die Fliegen der Paläearktishen Region, Band 6(2)*. E. Nägele Stuttgart.

Hering M. 1921. Minenstudien II. *Deutsche Entomologische Zeitschrift* 1921: 123–147.

Hering M. 1924. Minenstudien IV. *Zeitschrift für Morphologie und Ökologie der Tiere* 2: 217–250.

Hering M. 1926. *Die Ökologie der blattminierenden Insektenlarven*. Borntraeger, Berlin.

Hering M. 1927. *Zweiflügler oder Diptera: Agromyzidae*. G. Fischer, Jena.

Hering EM. 1937. *Die Blattminen Mittel- und Nord-Europas*. Verlag Gustav Feller, Neubranddenburg.

Hering EM. 1951. *Biology of the leaf miners*. Junk Publ., Gravenhage, Netherlands.

Hering EM. 1957. *Bestimmungstabellen der Blattminen von Europa*. Junk Publ., Gravenhage, Netherlands.

Hering M (Spencer KA, *ed.*). 1968. *Briefe über Blattminierer*. Junk, The Hague.

Hill DS. 1983. *Agricultural insect pests of the tropics and their control, 2nd edn*. Cambridge University Press, Cambridge, UK.

Ingenitsky I. 1897. Вредныя насѣкомыя Семирѣчья. *Agriculture and Forestry* 1897: 177–197.

Jansen PCM. 2008. "*Cannabis sativa* L.," pp. 135–142 in Schmetzer GH, Gurib-Fakim A, *eds. Plant resources of tropical Africa*. Prota Foundation, Wageningen, Netherlands.

Jolivet PH, Cox ML, Petitpierre E, *eds*. 1994. *Novel aspects of the biology of Chrysomelidae*. Kluwer, Dordrecht.

Justell RRS. 1686. An extract of a letter written from Aramont in Languedoc near Avignon, giving an account of an extraordinary swarm of grasshoppers in those parts. *Philosophical Transactions* 16: 147–149.

Kaltenbach JH. 1874. *Die Pflanzenfeinde aus der Klass der Insekten*. Julius Hoffmann, Stuttgart.

Kaynar F. 2019. Locusts in the Hittite culture. *Tüba-Ar (Turkish Academy of Sciences Journal of Archaeology)* 25: 77–84.

Keppen TP. 1870. О саранчѣ и другихъ вредныхъ прямокрылыхъ изъ семейства Acridiodea, преимущественно по отношенію къ Россіи. *Proceedings of the Russian Entomological Society of St. Petersburg* 5: 1–351.

Kirchner O. 1890. *Die Krankheiten und Beschädigungen unserer landwirtschaftlichen Kulturpflanzen*. Carl Hammer, Stuttgart.

Kirchner O. 1906. "Hanf, *Cannabis sativa* L.," pp. 319–323 in *Die Krankheiten und Befehädigungen uhferer landwirtschaftlichen Kulturpflanzen*. E. Ulmer, Stuttgart.

Kirchner O, Boltshauser H. 1898. Blattflecken und blattminen an hanf. *Atlas der Krankheiten und Beschädigungen, Serie 3, Tafel 20*. Verlag von Ulmer, Stuttgart.

Kojima G. 1917. On pest control in Shimane Prefecture. *Forest Bulletin* 10: 997–998.

Kolb RT. 2007. "Kurze Einführung in die Bekämpfung agrarischer Schadinsekten im spätkaiserzeitlichen China (1368–1911)," pp. 191–230 in Engelken K, *et al.*, *eds. Beten, Impfen, Sammeln*. Universitätsverlag Göttingen.

Köppen FP. 1880. *Die schädlichen Insekten Rußlands*. Kaiserlichen Akademie der wissenschaften, St. Petersburg.

Kovacs AI. 1809. *Utasitás a' fåk betegségeiröl, gyogyitásairol és a' természet mivoltárol*. Maros Vásarheljen, Második

Kovačević Z. 1929. Ueber die wichtigsten schädlinge der kulturpflanzen in Slawonien und Backa. *Deutschen Gesellschaft für Angewandte Entomologie, 7th Mitgliederversammlung München, 31. Mai–2. Juni 1928*, pp. 33–41.

Kozhanchikov IV. 1956. Fauna SSSR: *Nasekomyl Cheshuekrylye* 3(2): 444. Zoologicheskii Institut Akademii Nauk SSSR.

Kranz JB. 1860. *Schmetterlinge um München: enthaltend: Tagfalter, Schwärmer, Spinner und Eulen*. Georg Franz, München.

Kulagin NM. 1915. Insects injurious to cultivated field plants in European Russia in 1914. *Bulletin of the Moscow Entomological Society* 1: 136–161.

Kyokai NSB. 1965. 農林病害虫名鑑. Japan Plant Protection Association, Tokyo.

Lago PK, Stanford DF. 1989. Phytophagous insects associated with cultivated marijuana, *Cannabis sativa*, in northern Mississippi. *Journal of Entomological Science* 24: 437–445

Lesage L, Majka CG. 2010. Introduced leaf beetles of the Maritime Provinces, 9: *Chaetocnema concinna* (Marsham, 1802) (Coleoptera: Chrysomelidae). *Zootaxa* 2610: 27–49.

Leunis J. 1877. *Synopsis der Pflanzenkunde, Zweite Abtheilung*. Hahn, Hannover.

Liebermann J. 1944. Sobre la importancia económica de las especies chilenas del género *Dichroplus* Stål (Orth. Acrid. Cyrtacanth.), con algunas consideraciones acerca de su biogeografia. *Revista Chilena de Historia Natural* 47: 241–247.

Lindemann KE. 1892. Итальянская саранча в Воронежской губерніи. *Сельскохозяйственные обзоры по Воронежской губернии* 1892: 1–38.

Linzbauer FX, *ed*. 1861. "Methodus exstirpandi Locustas. 8 Febr. 1848," pp. 876–894 in *Codex sanitario-medicinalis Hungariae, Tomus III, Sectio V*. Universitatis Hungaricae, Budae.

Lockwood JA. 2004. *Locust: the devastating rise and mysterious disappearance of the insect that shaped the American frontier*. Basic Books, New York.

Lvovsky A, Morgun D. 2007. Булавоусые чешуекрылые Восточной Европы. KMK Scientific Partnership, Moscow.

Mackiewicz S. 1962. Wpływ wsiewek konopi w ziemniakach i burakach na gęstość porażenia przez stonkę ziemniaczaną i mszyce. *Biuletyn Instytutu Ochrony Roślin* 16: 101–131.

Macquart J. 1855. Insectes nuisibles a l'agriculture. *Annuaire des cinq Départements de l'ancienne Normandie* 21: 460–486.

Mahieu CM, Biesebeek JD, Graven C. 2019. *Inventarisatie van gewasbescherming toepasbaar in de teelt van Cannabis binnen het "Experiment met een gesloten coffeeshopketen."* Rijksinstituut voor Volksgezondheid en Milieu, Rapport 2019-0232. Bilthoven, Netherlands.

Manchanda SK, Sachan GC, Rathore YS. 1982. Growth and development of *Schistocerca gregaria* Forsskål on various host plants. *Indian Journal of Entomology* 44: 273–279.

Marshall GAK. 1917. *Psyche cannabinella*. *Review of Applied Entomology, Series A (Agriculture)* 5: 159.

Matov AY, Kononenko VS. 2012. Трофические связи гусениц совкообразных чешуекрылых фауны России (Lepidoptera, Noctuoidea: Nolidae, Erebidae, Euteliidae, Noctuidae). Dalnauka, Vladivostok.

Matsumura S. 1899. 日本害蟲篇. Shōkabō, Tōkyō.

McNeill WH. 1976. *Plagues and people*. Anchor Books, Garden City, New York.

McPartland JM, Clarke RC, Watson DP. 2000. *Hemp diseases and pests – management and biological control.* CABI Publishing, Wallingford, UK.

Meigen JW. 1827. *Handbuch für Schmetterlingsliebhaber: besonders für Anfänger im Sammeln.* La Ruelle & Destez, Aachen.

Melicher LJ. 1869. *Skizze der nützlichen und schädlichen niederen Thiere.* Braumüller, Wien.

Metzger FW, Grant DH. 1932. *Repellency to the Japanese beetle of extracts made from plants immune to attack.* USDA Technical Bulletin no. 299, Washington, DC.

Moes J, Sturko A, Przybylski R. 1999. "Agronomic research on hemp in Manitoba," pp. 300-305 in Janick J, ed. *Perspectives on new crops and new uses.* ASHS Press, Alexandria, Virginia.

Munir M, Leonberger K, Kesheimer K, *et al.* (14 other authors). 2023. Occurrence and distribution of common diseases and pests of US *Cannabis:* a survey. *Plant Health Progress.*

Mushtaque M, Baloch GM, Ghani MA. 1973. *Natural enemies of Papaver spp. and Cannabis sativa.* Annual report, Commonwealth Institute of Biological Control, Pakistan station, pp. 54–55.

Nadejde M, Pasol P, Schmidt E. 1975. Cercetari asupra combaterii omizii de stepa (*Loxostege sticticalis* L.) prin tratamente cu volum ultraredus in conditiile anului 1975. *Lucrari Stiintifice, Institutul Agronomic "Nicolai Balcescu"* 1975: 59–66.

Nagy B. 1962. "Állati kártevők," pp. 70–73 in Mándy G, Bócsa I, eds. 1962. *A kender. Magyarország Kultúrflórája* 7(14): 1–101.

Nair KR, Ponnappa KM. 1974. *Survey for natural enemies of Cannabis sativa and Papaver somniferum.* Commonwealth Institute of Biological Control, India Station Report, pp 39–40.

Noel P. 1917. Les ennemis du chanvre (*Cannabis* Tourn.). *Bulletin du Laboratoire Régional d'Entomologie Agricole, Rouen* 1917(2): 11–12.

Nördlinger H. 1869. *Die kleinen Feinde der Landwirthschaft, zweite auflage.* J.G. Cotta, Stuttgart.

Nowakowski JT. 1962. Introduction to a systematic revision of the family Agromyzidae (Diptera) with some remarks on host-plant selection by these flies. *Annales Zoologici Warszawa* 20(8): 67–183.

Ormerod EA. 1890. *A manual of injurious insects, 2nd edn.* Simpkin, Marshall, Hamilton, Kent & Co., London.

Paclt J. 1976. Der Hanf (*Cannabis sativa*) – eine unbekannte futterpflanze von *Aglope infausta* (L.) (Lep., Zygaenidae)? *Anzeiger für Schädlingskunde, Pflanzenschutz, Umweltschutz* 49(8): 116.

Parihar SBS, Ramkishore. 2002. Carry over of leaf miner, *Phytomyza atricornis* Meigen through alternate host plant in potato. *Insect Environment* 8(1): 32–33.

Park KT, Lee BW, Bae YS, Lan HL, Byun BK. 2014. Tortricinae (Lepidoptera, Tortricidae) from Province Jilin, China. *Journal of Asia-Pacific Biodiversity* 7: 355–363.

Patschke K, Gottwald R, Müller R. 1997. Erste Ergebnisse phytopathologischer Beobachtungen im Hanfanbau im Land Brandenburg. *Nachrichtenblatt des Deutschen Pflanxenschutzdienstes* 49: 286–290.

Pettit DA, Showalter VM, Abood ME, Cabral GA. 1994. Expression of a cannabinoid receptor in baculovirus-infected insect cells. *Biochemical Pharmacology* 48: 1231–1243.

Pichat CB. 1866. "La pianta della canapa," pp. 396-499 in *Istituzioni scientifiche e tecniche, ossia Corso teorico e practico de Agricoltura, Libri XIII, Volume Quinto.* Presso l'Union Tipografico-Editrice, Torino.

Pitkin B, Ellis W, Plant C, Edmunds R. 2019. Key for the identification of the known mines of British insects (Diptera and non-Diptera) recorded on *Cannabis.* Available at: http://www.ukflymines.co.uk/Keys/CANNABIS.php

Pivtoraiko VV, Kabanets VV, Vlasenko VA 2020. Шкідлива ентомофауна конопель посівних, *Cannabis sativa* L. *Plant Quarantine and Protection* 2020 (7-9): 20–25.

Poiret JLM. 1827. *Histoire philosophique, littéraire, économique des plantes de l'Europe, tome quatrième.* Didot, Paris.

Poorani J, Ramamurthy, VV. 1997. Weevils of the genus *Lepropus* Schoenherr from the Oriental region. *Oriental Insects* 31: 1–82.

Qiū SB. 1996. 邱式邦文选 (*Selected works of Qiū Shìbāng*). China Agricultural Publishing House, Beijing.

Ragazzi G. 1954. Nemici vegetali ed animali della canapa. *Humus* 10(5):27-29.

Rataj K. 1957. Skodlivi cinitele pradnych rostlin. *Prameny Literatury* 2: 1–123.

Réaumur RA de. 1736. *Mémoires pour servir a l'histoire des insectes, tome second.* Imprimerie Royale, Paris.

Réaumur RA de (Albouy V, ed). 2001. *Réaumur Histoire des insects, Morceux choisis par Vincent Albouy.* Millon, Grenoble.

Riley CV. 1877a. *Ninth annual report on the noxious, beneficial, and other insects of the State of Missouri.* Jefferson City, Missouri.

Riley CV. 1877b. *The locust plague in the United States: being more particularly a treatise on the Rocky Mountain locust.* Rand, McNally & Co., Chicago.

Riley CV, ed. 1883. *Third report of the United States entomological commission.* Government Printing Office, Washington, DC.

Rion JA. 1843. Relations des ravages causés en Valaise parles sauterelles en 1837, 1838 et 1839. *Actes de la Société Helvétique des sciences naturelles réunie à Lausanne* 28: 118–131.

Rodgers RH. 1980. Hail, frost, and pests in the vineyard: Anatolius of Berytus as a source for the Nabataean Agriculture. *Journal of the American Oriental Society* 100: 1–11.

Roonwal ML. 1945. Notes on the bionomics of *Hieroglyphus nigrorepletus* Bolivar at Benares, United Provinces, India. *Bulletin of Entomological Research* 36: 339–341.

Rothschild M, Rowan MR, Fairbairn JW. 1977. Storage of cannabinoids by *Arctia caja* and *Zonocerus elegans* fed on chemically distinct strains of *Cannabis sativa. Nature* 266: 650–651.

Rozner I, Rosner G. 2014. Data to the leaf-beetle fauna of Greece (Coleoptera: Chrysomelidae). *Natura Somogyiensis* 24: 81–98.

Saint-Amans JF. 1798. *Philosophie entomologique.* Dugour, Paris.

Salido y Estrada A. 1874. *La langosta. Todo cuanto más notable se ha escrito, sobre la plaga.* Aguado, Madrid.

Sasakawa M. 1961. A study of the Japanese Agromyzidae (Diptera) part 2. *Pacific Insects* 3: 307–472.

Schaefer PW, Ikebe K, Higashiura Y. 1987. *Gypsy moth, Lymantria dispar (L.) and its natural enemies in the far east (especially Japan).* Delaware Agricultural Experimental Station Bulletin No. 476, Newark, Delaware.

Schlein Y, Muller G. 1995. Assessment of plant tissue feeding by sand flies (Diptera: Psychodidae) and mosquitoes (Diptera: Culicidae). *Journal of Medical Entomology* 32: 882–886.

Schreiner M. 2019. *A survey of the arthropod fauna associated with hemp (Cannabis sativa L.) grown in eastern Colorado.* Master's thesis, Colorado State University, Fort Collins.

Schreiner M, Cranshaw W. 2020. A survey of the arthropod fauna associated with hemp (*Cannabis sativa* L.) grown in eastern Colorado. *Journal of the Kansas Entomological Society* 93: 113–131.

Séguy E. 1934. *Diptères (Brachycéres). Faune de France tome 28*. LeChevalier et fils, Paris.

Selgnij S. 1982. Sun, seeds, soil and soul. *Sinsemilla Tips* 2(4): 18–21, 30.

Semenov DP, *ed.* 1902. Год в сельскохозяйственном отношеніи. Kirshbaum, St. Petersburgh.

Shamonin MG. 1964. *Марокканская и пустынная саранча в Афганистане*. Doctoral thesis, All-Union Scientific Research Institute for Plant Protection, Leningrad.

Sharma N, Pandey S, Khan MA. 2011. Larval duration and survival of *Spilarctia obliqua* (Walker) on different natural host plants. *Pantnagar Journal of Research* 9: 145–146.

Shilenkov VG, Tolstonogova EV. 2006. Вредители конопли в Прибайкалье. *Acta Biomedica Scientifica* 2006(2): 159–160.

Shiraki T. 1952. *Catalogue of injurious insects in Japan*. General Headquarters, Supreme Command Allied Powers, Tokyo.

Siegel RK. 1989. *Intoxication: life in pursuit of artificial paradise*. E.P. Dutton, NY.

Sirikantaramas S, Morimoto S, Shoyama Y, *et al.* (4 other authors). 2004. The gene controlling marijuana psychoactivity. *Journal of Biological Chemistry* 279: 39767–39774.

Sirikantaramas S, Taura F, Tanaka Y, *et al.* (3 other authors). 2005. Tetrahydrocannabinolic acid synthase, the enzyme controlling marijuana psychoactivity, is secreted into the storage cavity of the glandular trichomes. *Plant and Cell Physiology* 46: 1578–1582.

Smith RL, Olson CA. 1982. Confused flour beetle and other coleoptera in stored marijuana. *Pan-Pacific Entomologist* 58: 79–80.

Sobczyk T. 2011. *World catalogue of Insects, vol. 10. Psychidae (Lepidoptera)*. Apollo Books, Stenstrup, Denmark.

Sonan J. 1940. On the life history of the Citrus locust (*Chondracris rosea* DeGeer) in Formosa. *Formosan Agricultural Review* 36: 839–842.

Sorauer P. 1958. *Handbuch der Pflanzenkrankheiten, 26 volumes*. Paul Parey, Berlin.

Spaar D, Kleinhempel H, Fritzsche R. 1990. *Öl- und Faserpflanzen*. Springer Verlag, Berlin.

Spencer KA. 1966. Notes on the Oriental Agromyzidae – 4. *Stuttgarter Beiträge zur Naturkunde* 147: 1–15.

Spencer KA. 1990. *Host specialization in the world Agromyzidae (Diptera)*. Kluwer Academic Publishers, Dordrecht.

Stratii YI. 1976. Конопля и колорадский жук. *Защита растений* (*Plant Protection*) 1976(5): 61.

Takahashi S. 1919. Notes on insects injurious to hemp, and on *Thyestes gebleri* Fald. *Konchu Sekae (Insect World) Gifu* 23(1): 20–24.

Takeuchi K. 1949. A list of the food-plants of Japanese sawflies. *Transactions of the Kansai Entomological Society* 14(2): 47–50.

Taschenberg EL. 1879. *Praktische Insekten-Kunde: oder, Naturgeschichte, Erster Theil*. Heinsius, Bremen.

Thomas C. 1880. "The Acrididae of Illinois," pp. 73–140 in *Ninth report of the State Entomologist of the State of Illinois*, Springfield, Illinois.

Thon T. 1838. *Die Insekten, Krebs- und Spinnenthiere*. Eduard Eikenach, Leipzig.

Trotuş E, Naie M, 2008. Cercetări privind reducerea atacului unor agenţi patogeni şi dăunători specifici culturilor de cânepă prin tratamentul chimic al seminţei. *Lucrari Stiintifice, Universitatea de Stiinte Agricole Si Medicina Veterinara "Ion Ionescu de la Brad" Iasi, Seria Agronomie* 51(2): 219–223.

Tuzov VK, Bogdanov PV, Churkin SV, *et al.* (4 other authors). 2000. *Guide to the butterflies of Russia and adjacent territories, vol. 2: Libytheidae, Danaidae, Nymphalidae, Riodinidae, Lycaenidae*. Pensoft, Sofia.

Uvarov BP. 1921. A revision of the genus *Locusta* L. (= *Pachytylus* Fieb.), with a new theory as to the periodicity and migrations of locusts. *Bulletin of Entomological Research* 12: 135–163.

Valdrighi LF. 1876. *Le locuste e una loro invasione recente rilevata da note dell'Avv Luigi Savani*. Toschi, Modena.

Vallot JN. 1836. Sur une Phrygane. *Mémoires de l'Académie des Sciences, Arts et Belles-lettres de Dijon* 1836(2): 237–246.

Vappula NA. 1965. Pests of cultivated plants in Finland. *Annales Agriculturae Fenniana* 1 (Suppl.): 7–239.

Vaughn JL, Goodwin RH, Tomkins GJ, McCawley P. 1977. The establishment of two cell lines from the insect *Spodoptera frugiperda* (Lepidoptera; Noctuidae). *In Vitro* 13: 213–217.

Vikberg V. 2013. Hemp sawfly found in European Russia (Hymenoptera: Tenthredinidae, Nematinae, Cladiini). *Entomologica Fennica* 24: 172–178.

Villa A. 1839. Notizie intorno alle locuste. *Pictorial Cosmorama* 4: 247.

Villanueva RT. 2022a. *Carolina grasshopper consumed hemp seedlings in high tunnels*. Available at: https://kentuckypestnews.wordpress.com/2022/07/12/carolina-grasshopper-consumed-hemp-seedlings-in-high-tunnels/

Walsh BD. 1868. *First annual report on the noxious insects of the State of Illinois*. Prairie Farmer Co., Chicago.

Wang T, Cui L, Wan Z. 1987. 大麻叶蜂研究. *Acta Entomologica Sinica* 30: 407–413.

Weeden BR. 2006. *Evaluation of hemp and kenaf varieties in tropical and sub-tropical environments*. RIRDC Publication no. 06/091. Rural Industries Research and Development Corporation, Mareeba, Australia.

Wu FM, Li M, Wang B, Feng SH, Wang DQ. 2010. Experimental study on the resistance of Frutus cannabis germplasm. *Medicinal Plant* 1(5): 13–15.

Xiao GR, Zhou SZ, Huang XY. 1986. Two new species of sawflies from China (Hymenoptera: Pamphiliidae, Cephalcinae; Tenthredinidae: Nematinae). *Scientia Silvae Sinicae* 22(4): 356–359.

Xú GQ. 2020. 農政全書. Chinese Text Project, Available at: https://ctext.org/wiki.pl?if=en&chapter=733415

Yepsen RB. 1976. *Organic plant protection*. Rodale Press, Emmaus, Pennsylvania.

Young M. 1997. *The natural history of moths*. T & AD Poyser, London.

Yuasa H. 1927. Notes on the Japanese Chrysomelidae. 1. On the food plants of several species. *Kontyu* 2(2): 130–133.

Zhelochovtsev AN, Zinovjev AG. 1992. "Подотряд Symphyta – Сидячебрюхие," pp. 199–221 in Насекомые Хинганского заповедника, Часть 2. Dalnauka, Vladivostok.

Zuefle M. 2020. *Hemp pest survey 2020*. New York State Integrated Pest Management Program. Available at: https://ecommons.cornell.edu/handle/1813/103765

Zuefle M. 2021. *Hemp pest survey 2021*. New York State Integrated Pest Management Program. Available at: https://hdl.handle.net/1813/111249

Abstract

Roots have their specialist pests, and some subspecialize in indoor plants: fungus gnats (*Bradysia* spp.) and the rising menace of root aphids (*Rhopalosiphum rufiabdominale*). Outdoors expect white root grubs (*Popillia japonica, Melolontha melolontha, Melolontha hippocastani*), other scarab beetles (*Maladera holosericea, Heteronychus arator*), wireworms (*Agriotes* and *Selatosomus* spp.), mole crickets (*Gryllotalpa gryllotalpa, Gryllotalpa orientalis*), root maggots (*Delia* spp.), even ants and termites. Seedings have their specialist pests—primarily cutworms. They are plump greasy noctuid caterpillars. Cutworms are polyphagous, but four species cause the most damage: *Spodoptera litura, Agrotis ipsilon, Agrotis segetum,* and *Agrotis gladiaria.* Seedlings are also attacked by hemp flea beetles (*Psylliodes attenuatus*), but they are treated in section 5.4. Other flea beetles and crickets attack seedlings, but they are primarily leaf defoliators (Chapter 7). A careful combination of cultural and biological control successfully mitigates damage; pesticides are rarely needed (e.g., insecticidal baits).

Introduction

Healthy roots are essential for crop yields. "Deep in their roots, all flowers keep the light" (Roethke, 1968). Withal, root damage is underappreciated because it is underground and unseen. The root system of a large plant is an abundant resource for certain insects (Fig. 8.1). Roots of small seedlings offer little sustenance to insects, yet they are more sensitive to grievous harm. A hemp seedling's root system is not as well developed as its above-ground parts (Neuer, 1935). Root tips of a hemp seedling are under more stress than root tips of older plants—they are closer to the soil surface and exposed to greater temperature fluctuations (Hackenberg, 1908).

8.1   Root aphids

Root aphids (Hemiptera: Aphididae) are an emerging problem faced by indoor growers, regardless of growth substrate—rockwool, coconut coir, aeroponics, or soil (Cranshaw and Wainwright-Evans, 2020). In a survey of California growers, 5% had problems with root aphids (Wilson *et al.*, 2019). Symptoms above ground are not specific—a generalized decline and reduced growth rate. Winged aphids (alatae) trapped on glandular trichomes are often the first recognizable sign of a root aphid infestation. Aphid root wounds serve as portals of infection for soil fungi.

8.1a. Rice root aphid

Rhopalosiphum rufiabdominale (Sasaki) 1899
Description: *Apterae* (wingless) are elongate-oval in shape, olive to dark green, with rufous (reddish-brown) or yellowish

tints in the abdomen, darker cauda and cornicles, 2.0–2.6 mm long (Fig. 8.2A). *Alatae* (winged) are darker and smaller, with long wings (Fig. 8.2B).

Comments: A pest of rice in Asia, rice root aphid attacks other monocots (wheat, barley), as well as dicots (cotton, vegetable crops). The species reached the USA in the early 1900s, and by 1937 was infesting cotton crops (Blackman and Eastop, 2006). By the 1990s *R. rufiabdominale* surfaced as a pest of hydroponic vegetables in the USA and Europe.

Rice root aphid has a holocyclic life cycle in Asia, with various *Prunus* spp. serving as primary hosts. On *Cannabis* the pest has an anholocyclic life cycle. Alatae females emerge from soil and fly to new plants. Females give birth to live young on stems and foliage, and they migrate down to roots. Rice root aphid has been encountered on indoor plants in Michigan (Lagos-Kutz *et al.*, 2018), Colorado (Cranshaw *et al.*, 2019), California, Nevada, Oregon, Washington, Florida, and Maine (Cranshaw and Wainwright-Evans, 2020). The pest has become a plague in Dutch glasshouses (Mahieu *et al.*, 2019).

8.1.1   Control of root aphids

See section 4.6 for aphids in general. Sanitation is key: seedlings and clones brought into a facility must be aphid-free. Screen all glasshouse entrances to thwart winged females. Do not reuse pots or soil. Do not transplant infested plants outdoors, because *R. rufiabdominale* can live outdoors, at least in southern California. Eliminate grassy weeds from the vicinity of crops, and eliminate ant nests—*R. rufiabdominale* is protected by ants, who feed on its honeydew.

Aphids thrive on plants excessively fertilized with nitrogen. Aluminum foil placed atop soil is recommended for confusing many winged pests, but this may favor rice root aphid, by protecting the soil (Cranshaw and Wainwright-Evans, 2020). Yellow sticky

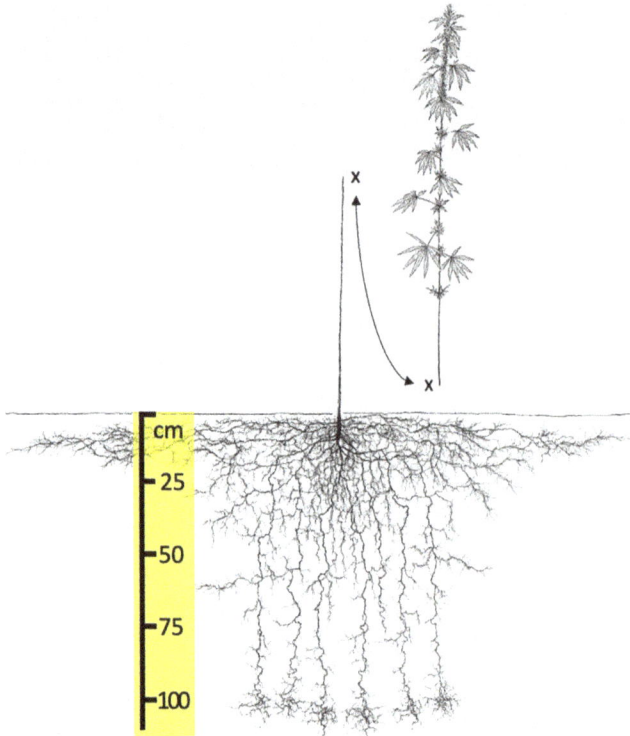

Fig. 8.1. Root system of a fiber-type plant (from Lichtenegger *et al.*, 2018, with kind permission; size scale highlighted from original)

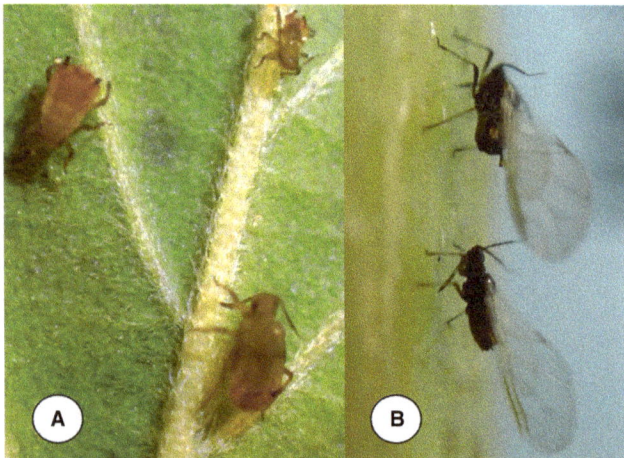

Fig. 8.2. Rice root aphid: **A**. nymphs; **B**. winged forms (W. Cranshaw)

traps attract aphids, including *R. rufiabdominale*. Yellow sticky traps are used for monitoring populations of winged adults, but on a large scale they can be used for mass-trapping (see Fig. 19.6).

Sticky traps devised for *R. rufiabdominale* can be disc-shaped for placement around plant stems over rockwool blocks or soil pots—complete with holes for drip irrigation tubes (Fig. 8.3). Sticky tape can also be wrapped around the stalk to capture alatae that emerge from soil and climb up the stalk before dispersal, or capture apterae as they climb down the stalk to infest roots.

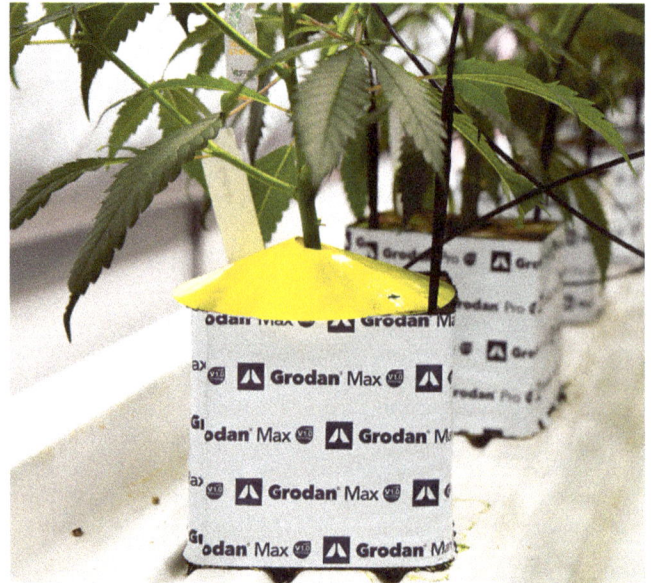

Fig. 8.3. Yellow sticky trap for root aphids (Horiver Disc™, courtesy Koppert Biocontrols)

Biocontrols for treating aphids are described in section 4.6. Above-ground predators and parasitic wasps work poorly against rice root aphid. Mahieu *et al.* (2019) preferred *Stratiolaelaps scimitus* (formerly *Hypoaspis miles*) and nematodes (*Heterorhabditis bacteriophora*, *Steinernema carpocapsae*, and *Steinernema feltitae*). They also recommended *Bacillus thuringiensis* subsp. *israelensis* strain AM65-52. Dutch growers use the fungus *Cordyceps fumosorosea* (formerly *Isaria fumosoroseus*) as a soil drench (W. Ravensberg, pers. comm. 2023).

Pace (2019) tested nematodes (*H. bacteriophora* and *S. feltitae*) and the fungus *Beauveria bassiana*. Rice root aphid could kick or rub nematodes off their bodies, preventing penetration. The fungus killed about 50% of aphids, and the survivors appeared weakened. Pace suggested using both: apply *B. bassiana*, wait 2–5 days, and then apply the nematodes. A combination of *B. bassiana* plus azadirachtin reduced rice root aphid in organic celery (Dara, 2015). The combination reduced populations by 62% compared with controls. Less effective were soil drenches of Grandevo® (*Chromobacterium subtsugae*, 29% reduction) or Venerate™ (*Burkholderia* spp., 24% reduction).

8.2   Cutworms

Cutworms (Lepidoptera: Noctuidae) mow down seedlings by chewing through delicate hypocotyls (stems). They eat enough to fell a seedling, and then move on. This wasteful mode of feeding results in disproportionate damage to crops. Older seedlings may not be completely severed; they tilt, wilt, and die. Cutworms are nocturnal and burrow underground shortly before dawn, usually within 25 cm of damaged plants. This distinguishes them from many leaf-feeding caterpillars. The culprits are rarely seen in action. Raking soil around plants will uncover them. When exposed to light, cutworms roll into tight spirals.

Larvae that hatch later in the season, lacking seedlings to cut, will climb plants to feed on leaves and even flowers. Some species accumulate in large numbers and crawl *en masse* across fields, earning a new name: armyworms. Armyworms are also called "climbing cutworms." See section 5.3 for armyworms that may also cut down seedlings (e.g., *Spodoptera exigua*, *Mamestra brassicae*).

Four cutworms commonly attack *Cannabis*, but cutworms are polyphagous, and as hemp is cultivated in more places, more cutworm species will appear on the crop. For example, the western bean cutworm, *Striacosta albicosta* (Smith), is a key pest of corn and beans, but it readily feeds on hemp. Difonzo (2022) reared *S. albicosta* on a diet of pure hemp inflorescences ('Grandi' fiber/grain cultivar); 86% survived to full grown larvae (6th instar) and 15% pupated and emerged as moths. In the field, larvae placed in cages with flowering plants lived in soil during the daytime and crawled into flowers at night. Late in the season *S. albicosta* larvae behaved like "climbing cutworms" and shelled seeds (Fig. 8.4).

8.2a. Black cutworm

Agrotis ipsilon (Hufnagel) 1766
=*Noctua ypsilon* Rottenburg 1776
=*Agrotis suffusa* var. *pepoli* Bertoloni 1874

Description: *Larvae* are plump, greasy, gray-brown, with dark lateral stripes and an indistinct pale gray band down their back, up to 45 mm long (Fig. 8.5). They have a dark head with two white spots. Close examination reveals many convex granules on their skin, resembling rough sandpaper. *Pupae* are brown, spindle-shaped, with flexible abdominal segments. *Adults* are typical Noctuids with a robust body, brown thorax, and beige to light brown abdomen. Forewings variegated brown, gray, beige, and some red tinges; each with an indistinct marking shaped like the letter Y (or Greek lowercase gamma). Hindwings whitish-gray with light brown margins, wingspan 40–50 mm. *Eggs* are round to conical, ribbed, yellow, laid either singularly or in clusters of 2–30 on plants close to the ground.

Comments: Black cutworm is distributed through the northern hemisphere in temperate regions. The number of generations per year depends on latitude, from one to four. Black cutworm overwinters in soil as pupae or mature larvae. Moths emerge in April or early May; they are strong fliers, and migrate north in springtime. Each moth lays up to 1800 eggs. Eggs hatch in 5–10 days. Larvae feed for 10–30 days before pupating, repeating their life cycle until a heavy frost. This pest is despised for its pernicious habit of cutting down many seedlings to satisfy its appetite. It feeds on almost any herbaceous plant.

In *A Tale of the Kentucky Hemp Fields*, Allen (1900) called cutworms "hemp's only enemy." Five years prior, Kentucky got hit by a plague of cutworms, and "hemp especially suffered" (Garman, 1895). Garman named several cutworm species, *A. ipsilon* among them. Black cutworm was probably the species that devastated *chanvre* (hemp) and *haricots* (green bean) in a French epidemic (Guérin-Méneville, 1846). They came out at night to gnaw at stems; 50 hectares were "completely devastated in the space of four days." Guérin-Méneville shipped larvae and pupae to another entomologist for identification, but they died *en route*.

Bertoloni (1874) described a new variety on hemp, *Agrotis suffusa* var. *pepoli*. Targioni Tozzetti (1876) synonymized the variety under *A. ipsilon*. Del Guercio (1903) called *A. ipsilon* "the Noctuid of grains, legumes, and hemp." Other Italians have reported *A. ipsilon* cutting down hemp (Berlese, 1924;

Fig. 8.4. Western bean cutworm, *Striacosta albicosta* (courtesy C. DiFonzo)

Fig. 8.5. Black cutworm (courtesy W. M. Hantsbarger, Bugwood.org)

Ragazzi, 1954), as have other Europeans (Nagy, 1962; Ebert, 2005; Matov and Kononenko, 2012; Rak Cizej and Policnik, 2018). The species has attacked *gañjā* in India (Hector, 1923; Sinha *et al.*, 1979).

8.2b. Common cutworm

Agrotis segetum (Denis & Schiffermüller) 1775
Description: *Larvae* are plump, greasy, cinnamon-gray to gray-brown with dark heads and faint dark dorsal stripes, up to 45 mm long. *Pupae* are smooth, shiny brown, 12–22 mm long and found in the soil. *Adults* are typical Noctuids with a robust body, brown thorax, and lighter brown abdomen. Forewing color variable, ranging from dark brown to pale buff, sometimes reddish tinged, with two darker spots on each wing: a circular one, and a larger, distal kidney-shaped one. Hindwings whitish-gray with light brown margins, wingspan 32–42 mm.
Comments: Common cutworm is distributed throughout Eurasia, as well as regions of Africa that are temperate to subtropical. In recent decades its numbers have decreased in forested areas, becoming restricted to the steppe zone (Matov and Kononenko, 2012). The number of generations per year depends on latitude, from one to five. In temperate regions the species overwinters in soil as larvae. Moths emerge in April; they are strong fliers, and migrate north in springtime. Females lay light yellow eggs haphazardly, on plants or soil, singly or groups of 2–20. Common cutworm is sometimes called the turnip moth; it is polyphagous. *A. segetum* has been reported on hemp in Russia (Durnovo, 1933; Matov and Kononenko, 2012) and elsewhere in Europe (Nagy, 1962; Bakro *et al.*, 2018; Rak Cizej and Policnik, 2018).

8.2c. Tobacco cutworm

Spodoptera litura (Fabricius) 1775
Description: *Larvae* are stout, smooth-skinned, with black heads and dull gray to gray-green bodies, partially covered with short setae, 40–50 mm long. They often have yellow dorsal and lateral stripes, between which runs a series of semi-lunar black marks (Fig. 8.6A). *Pupae* in soil, reddish-brown, 18–22 mm. *Adults* have robust bodies, light brown to orange. Forewings dark brown with lighter colored lines and stripes. Hindwings whitish with brown margins and venation, wingspan 28–38 (Fig. 8.6B). *Eggs* spherical, with 40 longitudinal ribs, green turning black, laid in batches covered with orange-brown hair-like scales from the female's body.
Comments: Tobacco cutworm is distributed in South Asia, East Asia, and Australasia. The species is nearly identical to *Spodoptera littoralis* (Boisduval) 1833. Only a close inspection of genitalia tells them apart; some authors consider them one species. *S. litura* lives in warm climates and rarely overwinters. Its life cycle can turn in 30 days; as many as eight generations arise per year (Hill, 1983). Females lay up to 300 eggs, in clusters under leaves of host plants.

    S. litura is polyphagous and damages tobacco, tomato, cotton, rice, and maize. Irrigated fields are attractive to ovipositing

Fig. 8.6. Tobacco cutworm: **A**. larva (Kenpei, Wikimedia Commons); **B**. adult (Birgit Rhode, Wikimedia Commons)

females. Cherian (1932) considered *S. litura* a "minor pest" of *gañjā*. In contrast, Nair and Ponnappa (1974) reported it defoliating plants; they considered the species a possible biocontrol weapon against illegal *Cannabis*. Chandra (2004) also reported *S. litura* infesting *gañjā*.

8.2d. Claybacked cutworm

Agrotis gladiaria (Morrison) 1874
Description: *Larvae* are gray with a distinctly paler dorsum, set with numerous small, flat, shining granules arranged down the back like pavement stones. The body tapers both anteriorly and posteriorly, up to 37 mm long. *Adults* resemble other Noctuids: robust brown bodies, forewings darker than hindwings, with variegated dark brown markings.
Comments: Claybacked cutworm lives east of the Rocky Mountains, becoming scarce south of Tennessee and Virginia. Overwintering larvae cause their greatest destruction from May to late June. After that they pupate underground. Moths emerge between September and October, mate, lay eggs, and repeat their life cycle (one generation per year). Claybacked cutworm is polyphagous and injures clover, beans, maize, and tobacco. Garman (1895) identified *A. gladiaria* among cutworms infesting hemp in Kentucky. Tietz (1972) cited it on hemp.

8.2.1 Differential diagnosis

Cutworm damage can be confused with injuries caused by crickets, slugs, and snails. Birds sometimes damage seedlings while searching for seeds. Post-emergent damping off, caused by fungi and oomycetes, causes seedlings to keel over. Cutworms must be differentiated from armyworms and other leaf-eating caterpillars.

8.2.2 Cultural and mechanical control

Keep weeds down within and around crops. *A. ipsilon* particularly targets weedy, poorly cultivated crops. Without specifying a species, Robinson (1943b) considered cutworms a problem in hemp rotated after pasture or sod. Dewey (1914) said cutworms caused losses in late-sown hemp, with less damage in plants sown at the proper time. Cutworms overwinter in the soil, so post-harvest plowing exposes them to the elements. Tilling of soil exposes cutworms to toads, birds, and small mammals that relish eating them. Night-flying moths are attracted to light traps, useful for monitoring purposes (see section 19.6.1). Pheromone traps for *A. ipsilon* are available. Mechanical barriers (plastic collars, pressed into soil around seedlings) are not particularly effective. Hector (1923) resorted to hand-picking *A. ipsilon* off *Cannabis* when pesticides didn't work.

8.2.3 Biological and chemical control

Hector (1923) used biocontrol, of sorts, against *S. litura*: he flood-irrigated fields, which forced cutworms to the soil surface, where birds preyed on them. The beneficial nematode *Steinernema carpocapsae* works well against mobile hosts like cutworms; *Steinernema feltiae* also kills cutworms. The bacterium *Bacillus thuringiensis* (Bt-kurstaki) can be effective. The fungus *Metarhizium rileyi* specifically kills Noctuid larvae. *Trichogramma* wasps are less effective, because cutworms come out at night.

Insecticidal baits are used against cutworms, but Hector (1923) tried this in a *gañjā* crop, and "poison bait proved useless." Terpenoids extracted from thyme (*Thymus vulgaris*) were highly toxic to *S. litura* and deterred feeding (Hummelbrunner and Isman, 2001). Terpenoids extracted from *Cannabis* showed moderate toxicity against *Spodoptera littoralis* (Benelli *et al.*, 2018). In contrast, *Cannabis* extracts showed relatively little activity against *S. litura* (meta-analysis of five studies) (McPartland and Sheikh, 2018).

8.3 White root grubs

White root grubs are larvae of scarab beetle (Coleoptera: Scarabaeidae). Antennae of adult scarabs end in flattened, palmate segments which unfurl like a hand-fan; this feature distinguishes them from other beetles. Larvae of four scarab beetles feed on *Cannabis* roots. All four pests are polyphagous, and commonly attack grasses.

Symptoms often become apparent when seedlings are 30–60 cm tall. Then seedlings wilt, yellow and die. Damage arises in patches across infested fields. Roots are slightly gnawed or entirely eaten away by grubs. The tasty grubs attract secondary problems: moles, skunks, and raccoons. Their digging indicates grubs are in the soil. Forbes (1894) tried mitigating damage by white root grubs attacking maize in Illinois by rotating with hemp. It didn't work. The four species (*Phyllophaga fusca* Frölich 1792, *P. inversa* Horn 1887, *P. futilis* LeConte 1850, *Cyclocephala immaculata* Oliver 1789) likely fed on hemp roots.

8.3a. Japanese beetle

Popillia japonica Newman 1838
Description: *Adults* feed on *Cannabis* leaves (see Fig. 7.10). *Grubs* are plump and white, with a brown head, six dark legs,

15–25 mm long. They lie curled in a C-shape when exposed to light (Fig. 8.7).
Comments: Adults lay eggs individually or in small clusters near the soil surface in late June and July. Hatchlings make a cell in the soil and feed on fine rootlets projecting into the cell. As they mature, grubs enlarge their cells, and consume progressively larger roots, usually within 5–10 cm of the soil surface. As winter approaches, they hibernate deeper in the soil, below frost line. In springtime when soil temperatures warm to ≥10°C, they break hibernation and move closer to the surface. Grubs resume feeding for a month, then pupate. *P. japonica* grubs are polyphagous. They cause a lot of damage to turfgrass.

8.3.1 European cockchafers

"Cockchafer" derives from German *cock* (in reference to size) + *chafer*, "beetle." More prosaic names refer to the adults' flying times, maybugs and June bugs. These names refer to *Melolontha* spp., with two reported on *Cannabis*, *M. melolontha* and *M. hippocastani*. Both are European species. *M. melolontha* is more common than *M. hippocastani*, and their ranges differ. *M. melolontha* has a more southern distribution, and predominates in open fields. *M. hippocastani* has a more northern distribution, and predominates at forest margins of deciduous and mixed upland forests.

Early *Cannabis* researchers used a third taxon, *Melolontha vulgaris*, which unfortunately is a synonym common to either species: *M. vulgaris* Fabricius 1775 is a synonym of *M. melolontha*, and *M. vulgaris* Olivier 1789 is a synonym of *M. hippocastani*. The cockchafer life cycle parallels that of the Japanese beetle (see section 7.4a), except the larval stage may last 3 years (*M. melolontha*) or 4 years (*M. hippocastani*). Larvae of both species damage the roots of fruit trees, pasture grasses, clover and alfalfa fields, and cereal and vegetable crops. Grubs of both species are big and nasty, and inflict a painful bite.

Fig. 8.7. White root grubs: **A.** Japanese beetle; **B.** European cockchafer (courtesy D. Cappaert)

8.3b. Common cockchafer

Melolontha melolontha (L.) 1758
=*Melolontha vulgaris* Fabricius 1775
Description: *Larvae* are large, C-shaped, whitish-yellow with a translucent lower abdomen, light brown heads, strong mandibles, and three pairs of light brown legs, 45–50 mm long (Fig. 8.7). *Adults* are hefty beetles, 24–32 mm long. Their small head is retracted into a black or red prothorax, covered with a dense pubescence of light hairs. The abdomen has a series of white triangles along its sides. Wing covers (elytrae) vary in color from red to brown, with longitudinal striations. The pygidium (last abdominal somite) is elongated, extends beyond elytrae, and tapers to a pointed tip.
Comments: Common cockchafer is distributed across Europe, from Portugal in the west, to Turkey and Ukraine in the east. It was formerly abundant, and a very destructive pest. Synchronized "mass flights" occur every 4 or 5 years, and a longer 30-year cycle peaks with massive infestations. *M. melolontha* populations drastically declined in the 20th century, due to widespread pesticide usage. With recent pesticide restrictions, their numbers have rebounded.

Ratzeburg (1839) reported a "mass flight" of *M. vulgaris* Fabricius (=*M. melolontha*) that devastated hemp in Germany. Exactly 30 years later, Nördlinger (1869) and Melicher (1869) reported *M. vulgaris* Fabricius in German hemp fields. In France, Anjubault (1862) reported *M. vulgaris* Fabricius damaging hemp plants, "their underground parts cut or gnawed, losing the hair and bark of their roots." Leunis (1877) reported problems with "*Melolontha vulgaris* L.", a non-existent taxon. He probably meant *M. melolontha* (= *M. vulgaris* Fabricius). Trotuş and Naie (2008) reported *M. melolontha* in Romanian hemp fields, with an average density of 2.3 specimens/m². Fedorenko *et al.* (2016) said damage to Ukrainian hemp by *M. melolontha* peaked in late May and June.

8.3c. Forest cockchafer

Melolontha hippocastani Fabricius 1801
=*Melolontha vulgaris* Olivier 1789
Description: *Larvae* are nearly indistinguishable from *M. melolontha*. *Adults* resemble *M. melolontha*, including their variability in color. *M. hippocastani* is slightly shorter and broader than *M. melolontha* (length 22–27 mm). Its pygidium tapers abruptly, ending in a small knob, whereas the *M. melolontha* pygidium narrows evenly and gradually to a pointed tip.
Comments: Forest cockchafer spans northern Eurasia, from the UK to the Russian Far East. Jäger (1874) and Kirchner (1906) reported two species damaging hemp, *M. hippocastani* and *M. vulgaris* (the latter probably *M. melolontha*). Goidànich (1928) reported *M. hippocastani* and *M. melolontha* attacking Japanese hemp.

8.3.2 Other Scarab beetles

8.3d. Maladera holosericea Scopoli 1772

Description and comments: *Larvae* are soft-bodied and fleshy, creamy-white with a brown head and hind segments, up to 4.0 mm long. *Adults* are dark brown, 7–9 mm in length, with elytra marked by rows of shallow punctations. Both adults and larvae overwinter in soil. Adults emerge in spring and are active at night. Larvae and adults have been observed in hemp fields in Yugoslavia (Camprag, 1961) and Korea (Anonymous, 1919). Shiraki (1952) cited a related species, *Maladera orientalis* (Motschulsky) 1857, on hemp in Japan.

8.3e. Heteronychus arator (Fabricius) 1794

Description and comments: *Larvae* are soft-bodied and fleshy, curved in a C-shape, with well-developed thoracic legs, reaching 35 mm long at maturity. *Adults* are black, rounded scarabs 15–20 mm long. *H. arator* occurs throughout tropical Africa and southern Asia. It is also found in Australia, where it is called the African Black Beetle, and infests hemp (Ditchfield, pers. comm. 1997).

8.3.3   Differential diagnosis

Above-ground symptoms caused by white root grubs (i.e., wilting) can be confused with symptoms caused by wireworms, root maggots, nematodes, root-rot fungi, or drought. Dig up roots for inspection. The difficult part is differentiating white root grubs from each other—they look quite similar. Luckily they share control measures.

8.3.4   Control of white root grubs

Dewey (1914) noted that white grubs cause the greatest damage in hemp rotated after sod. Keep weeds down within and around crops. Grubs overwinter in soil, so post-harvest plowing exposes them to the elements. Frequent tilling during the season exposes the grubs to birds and small mammals that eat them. To monitor adult populations, pheromone/food attractant traps are available for *Popillia japonica*, and male cockchafers are strongly attracted to light traps.

Bacillus popilliae kills grubs of *Popillia japonica*, and *Bacillus thuringiensis* var. *tenebrionis* kills cockchafer grubs. All white root grubs are killed by nematodes, *Heterorhabditis* and *Steinernema* spp. The fungi *Metarhizium anisopliae*, *Beauveria bassiana*, and *Cordyceps* (*Isaria*) *fumosorosea* have good efficacy against most beetles.

8.4   Wireworms

Wireworms are the grubs of click beetles (Coleoptera: Elateridae) that live in the soil. Wireworms usually scavenge dead organisms, but can become agricultural pests. They chew on freshly sown seeds, before or after germination, and gnaw on small roots throughout the season. Wireworms attack many crop plants, especially grasses (maize, wheat, etc.).

Agronomists in the 19th century believed that growing hemp for several consecutive years would *eliminate* wireworms

(Barnum, 1829; Wright, 1847; Sempers, 1894). This was even recommended in Kentucky (Garman, 1892). Saalas (1923) stated that hemp was the only crop in Finland that did not suffer damage from wireworms, it "destroyed the worms." This belief was abolished by Heinrich Neuer. He was a hemp breeder and paid attention to every seed sown in soil; he invented a single-seed planting drill for hemp breeding studies (Neurer et al., 1946). Neuer (1935) wrote, "the wireworm is known to be devastating, especially on *umgebrochenen Wiesen* [converted grassland], and experience has shown that it can cause extraordinary damage to hemp."

Flader and Neuer (1939) identified *Agriotes lineatus* and *Agriotes obscurus* damaging hemp seedlings grown on former grassland, especially during dry periods, "since the juicy hemp roots then provide good food." The damage occurred in patches in the field. Since then, more wireworm species have been cited on *Cannabis*. Some publications simply identify them by common name, or to genus, as *Agriotes* spp. (Nagy, 1962; El Bassam, 2010; Oprescu et al., 2019; Britt et al., 2020). Specific species cited on *Cannabis* include the following:

- *Agriotes lineatus* L. 1758 (in Germany: Flader and Neuer, 1939; Gutberlet and Karus, 1995; in Ukraine: Fedorenko et al., 2016; in Poland: Bakro et al., 2018)
- *Agriotes obscurus* L. 1758 (in Germany: Flader and Neuer, 1939; in Siberia: Cherepanov, 1957; in Romania: Trotuş and Naie, 2008)
- *Agriotes ustulatus* (Schaller) 1783 (in Romania: Trotuş and Naie, 2008)
- *Agriotes sputator* L. 1758 (in Ukraine: Fedorenko et al., 2016)
- *Selatosomus aeneus* (L.) 1758 (in Ukraine: Fedorenko et al., 2016)
- *Selatosomus latus* (Fabricius) 1792 (in Ukraine: Fedorenko et al., 2016)

Agriotes and *Selatosomus* spp. look similar. *Larvae* are slender, hard-bodied, and have a metallic sheen ("wireworms"), with many smooth-jointed segments (Fig. 8.8). They can be confused with centipedes or millipedes, but have only three pairs of legs, behind the head. The head is black, body is yellow brown, up to 25 mm long. *Adults* are light brown to black, with lined electra, cylindrical, 7.5–11 mm long. A spine on the prosternum can be snapped into a corresponding notch on the mesosternum, producing a click that launches the beetle into the air.

The life cycles of *Agriotes* and *Selatosomus* spp. are similar. They overwinter as larvae or adults in the soil. In early spring adults become active. Females lay eggs around the roots of plants. Adults live 10–12 months, spending most of the time in the soil. Larvae take 2–6 years to complete their development,

feeding on seeds or roots. They migrate vertically in the soil, depending on soil temperature and moisture. They dive deep during the winter and dry summer months, and are closest to the surface in the spring.

Fedorenko et al. (2016) considered wireworms the most harmful soil pests of hemp fields in Ukraine. Their quantitative survey unearthed *A. sputator* (16.1 larvae/m²), *A. lineatus* (3.9 larvae/m²), *S. aeneus* (1.4 larvae/m²), and *S. latus* (1.4 larvae/m²). Fedorenko proposed that global warming has increased wireworm damage in recent years. In a survey of Romanian hemp fields, Trotuş and Naie (2008) found average densities of 3.5 specimens/m² for *A. ustulatus*, and 1.3 specimens/m² for *A. obscurus*.

8.4.1 Control of wireworms

Greatest wireworm damage occurs in crops rotated after grassland or following sod. Several authors have warned against rotating hemp after maize or cereal crops (Flader and Neuer, 1939; Fedorenko et al., 2016; Oprescu et al., 2019). Prior to planting in a suspect field, survey the wireworm populations with a bait trap. Three weeks prior to sowing seed, dig a hole 10 cm deep and 20 cm wide, and place 120 cc (0.5 cup) of mixed maize/wheat seeds in the bottom. Backfill with loose soil, and cover area with plastic. Then dig up the bait 2 weeks later and search through seeds for wireworms. The presence of even a couple wireworms may spell trouble. This is particularly true in poorly drained soils.

For control, follow recommendations for white root grubs. Biocontrols recommended for white root grubs apply to wireworms. Gutberlet and Karus (1995) suggested *Metarhizium anisopliae*. Pesticides are problematic; wireworms can recover from sublethal poisoning. Van Herk et al. (2008) exposed *A. obscurus* to ten different insecticides. One insecticide was highly toxic to wireworms, the rest were not; seemingly moribund wireworms made a full recovery. Trotuş and Naie (2008) tested three insecticides (not currently permitted on *Cannabis*) against *A. obscurus* in hemp fields. Compared with controls (13.6% affected plants), the insecticides decreased infestation to 1.8–2.2%.

8.5 Crickets

Crickets (Orthoptera: Gryllidae) are related to grasshoppers and locusts but have shorter legs and antennae. Most species are black and produce familiar nocturnal chirping sounds. Cricket damage is uncommon and no methods for controlling crickets in *Cannabis* have been reported. When crickets sever seedlings at the soil line, their damage may be confused with cutworms.

8.5a. House cricket

Acheta domesticus (Linnaeus) 1758
Description: *Adult* body is not black but yellow-brown in color, with dark crossbars and light-colored heads, 16-21 mm long. *Nymphs* look like small adults.
Comments: *A. domesticus* is most likely native to Europe or southwest Asia, but now occurs wherever it is sold (for pet food

Fig. 8.8. Wireworm larvae of *Agriotes* sp. (courtesy M. Gulesci, Bugwood.org)

and fish bait). The species is omnivorous, and takes 2–3 months to complete its life cycle at 26–32°F (–3.3–0°C). Rataj (1957) reported *A. domesticus* infesting Czech hemp. It may have been the unidentified species that Selgnij (1982) reported attacking young *Cannabis* seedlings in California.

8.5b. Field crickets

Field crickets are the same size as house crickets, but darker in color—usually black. Their wings are less developed than house crickets; they can jump but not fly.

Del Guercio (1900b) reported the European field cricket, *Gryllus campestris* L. 1758, feeding on Italian hemp. Two former *Gryllus* spp. were reported in Italian hemp (Goidànich, 1928; Anonymous, 1940; Ragazzi, 1954). *Melanogryllus desertus* (Pallas) 1771, the steppe cricket, inhabits meadows in southern and eastern Europe, east to Kazakhstan. *Eumodicogryllus bordigalensis* (Latreille) 1804 (=*Gryllus chinensis* Weber 1801) is another meadows species with a similar geographic range. Both species killed sprouting seeds and young seedlings; in older plants they damaged leaves and delicate stems.

8.5c. Bush and tree crickets

Bush crickets or katydids (family Tettigoniidae) appear in the *Cannabis* literature. *Tettigonia cantans* (Fuessly) 1775 infested hemp in Italy (Goidànich, 1928). This species occurs throughout the Mediterranean region, and extends east to China. Goidànich also cited two less common species, *Bothrogonia ferruginea* (Fabricius) 1787 and *Tettigonia orientalis* Uvarov 1924. Cranshaw *et al.* (2019) encountered bush katydids, *Scudderia* spp., in Colorado. The striped meadow cricket *Oecanthus fasciatus* Fitch 1856 was a pest in Kentucky (Garman 1894b, 1904b). Garman (1894b) also cited the slender meadow katydid *Conocephalus fasciatus* De Geer 1773.

Tree crickets (Family Gryllidae, Subfamily Oecanthinae) frequent trees as well as meadows. Fruhstorfer (1921) named *Cannabis* as a host of the Italian tree cricket, *Oecanthus pellucens* (Scopoli) 1763. Garman (1904b) named *Cannabis* as a host of *Oecanthus fasciatus* Fitch 1856 in Kentucky. Cranshaw *et al.* (2019) encountered *Oecanthus* spp. in Colorado.

8.6   Mole crickets

Mole crickets (Orthoptera: Gryllotalpidae) live underground. They are the insect version of moles, adapted to subterranean lifestyles. The "mole" body plan includes a cylindrical body, velvety "fur," inconspicuous eyes, and broad, shovel-like forelimbs. Mole crickets dig extensive tunnel systems, the width of a finger. They often burrow in zig-zag paths, within a few centimeters of the surface. Females dig underground chambers wherein they lay 200 eggs and guard their young. Mole crickets are omnivorous and feed on roots, tubers, and small soil invertebrates. They live up to 3 years.

8.6a. European mole cricket

Gryllotalpa gryllotalpa (L.) 1758
=*Gryllotalpa vulgaris* Latreille 1804
Description: *Adult* body stout, elongated, dark brown with fine velvety hairs that impart a silky shimmer, undersides yellowish, and *big* for a cricket, 40–50 mm long (Fig. 8.9). Mesothorax broad (housing foreleg muscles); forelegs powerful and shovel-like, modified for digging. Elytra are half the length of the abdomen; wings folded into pleats. *Larvae* are smaller, pale versions of adults, without wings.
Comments: European mole cricket occurs in Europe, northern Africa, and western Asia. The species was accidently imported into New Jersey, around the same time as the Japanese beetle (Weiss, 1915). Unlike the latter, mole crickets have not spread very far—it's taken a century for them to spread to neighboring New York (Westchester County). European populations are declining, due to widespread pesticide usage and the drainage of damp meadows.

European mole cricket is polyphagous and its damage to roots can kill young seedlings. Damage to hemp has been recorded in Italy (Pichat, 1866; Comes, 1885; Del Guercio, 1900b; Berlese, 1902), Russia (Bogdanov-Kat'kov, 1930; Osmolovsky, 1976), Hungary (Nagy, 1962; Barna *et al.*, 1982), and Germany (Mühle, 1966; Gutberlet and Karus, 1995).

Hemp damage has not been reported in France. *Au contraire*, Olivier de Serres claimed that *sterpis* (mole crickets) are repelled by hemp plants. Everyone believed him, because he founded French agronomy. De Serres (1600) claimed that growing hemp in every fourth part of a garden would banish *sterpis* for years, because "its natural smell is strong." French *bio-dynamique* gardeners still espouse this technique (Pfeiffer, 1979), as has a Russian author (Olshevsky, 1970).

8.6b. Other mole crickets

Oriental mole cricket, *Gryllotalpa orientalis* Burmeister 1838, has been reported on *Cannabis* in Russian Turkestan (Arnoldi,

Fig. 8.9. Mole cricket, *Gryllotalpa gryllotalpa* (courtesy F.A. Rubén, iNaturalist.org)

1949) and China (Xu *et al.*, 2013). Clarke (pers. comm. 1994) reported mole crickets damaging *Cannabis* in Africa. African mole crickets could have been either *G. gryllotalpa* or *Gryllotalpa africana* Beauvois 1805. Clarke also reported mole crickets in California—possibly *Neocurtilla* (*Gryllotalpa*) *hexadactyla* Perty 1832. Control is rarely necessary. Flood irrigation and tillage get rid of them (Weiss, 1915). Buc'hoz (1775) killed underground nests of *courtillières* (mole crickets) by flooding them with water suffused with hemp seed oil. Mole crickets are susceptible to *Metarhizium anisopliae*. A related fungus, *Metarhizium acridum,* specifically infects Orthopterans.

8.7   Ants and termites

Ants and white ants (termites) sometimes curse *Cannabis* in warm climates. They tunnel into taproots and stalks. Ants and termites look somewhat similar, and they cause identical damage. But they are not related—ants join honeybees in the Hymenoptera, and termites join cockroaches in the Blattodea.

8.7.1   Ants

Ants (Hymenoptera: Formicidae) tunnel into taproots and also eat feeder roots (Fig. 8.10). Infested plants wilt and sometimes collapse. They are easily pulled from the ground, exposing part of the colony. All ages of plants are affected. Clarke (pers. comm. 1996) encountered leafcutter ants in Mexico. Leafcutter ants cut pieces of leaves and carry them off to underground nests. In severe cases leafcutter ants devoured entire plants up to 1 m tall, leaving only the roots. Leafcutter ants are nocturnal and can destroy several plants a night.

8.7a.   Tropical fire ant

Solenopsis geminata (Fabricius) 1804
Description: Worker ants have a large head and abdomen, which are separated by a narrow thorax and a petiole with two

Fig. 8.10. Root damage caused by ants in Italy (G. Grassi).

nodes. Fire ants are called "red ants" but they are copper-brown, with a lighter head and darker abdomen (almost black), 2.5–6 mm long. Mandibles are large, have teeth, and they bite. The abdomen is tipped with a venomous stinger.
Comments: Cherian (1932) cited *S. geminata* chewing *gañjā* seedlings and tunneling into roots of mature plants. *S. geminata* is likely native to lands bordering the Caribbean, but was found as early as 1851 in India, and occurs everywhere tropical, including the southeastern USA (Creighton, 1930).

8.7b.   Black and Red imported fire ants

BIFAs and RIFAs, native to South America, are among the worst invasive species in the world. They are attracted to disturbed soil created by humans. BIFA (*Solenopsis richteri*) and RIFA (*Solenopsis invicta*) have invaded the southeastern USA. They are difficult to differentiate from each other and from aforementioned *S. geminata*. Hung *et al.* (1977) provided an identification key for the three *Solenopsis* spp., available on the web.

Solenopsis richteri Forel 1909 was introduced into the USA, at the port of Mobile, Alabama, around 1918 (Creighton, 1930). *S. richteri* has largely been displaced by *S. invicta* in the USA, and is now restricted to southern Tennessee, northeast Mississippi, northwest Alabama, and the tidewater area of Virginia. *S. richteri* infested *Cannabis* roots in Virginia (Britt *et al.*, 2020).

Solenopsis invicta Buren 1972 also came through Mobile, in the late 1930s. It was discovered by an aspiring teenage entomologist—Edward O. Wilson (1929–2021), who became the world's foremost ant expert. Wilson thought it was a reddish subspecies of BIFA and called it *Solenopsis saevissima richeti* (Wilson, 1951). Skeptical of Wilson's taxonomy, Buren (1972) named the red ant a new species, *S. invicta*. Conner *et al.* (2020) reported *S. invicta* damaging hemp in Alabama. The ants build mounds near the base of the plant, strip the bark, and tunnel into stalks. Damaged plants wilt, turn yellow, and sometimes fall over. Young plants are especially at risk.

8.7c. Marauder ants

Carebara diversa (Jerdon) 1851 was collected from marijuana confiscated in New Zealand by Crosby *et al.* (1986), probably imported from Southeast Asia, where *C. diversa* is native. This species is in the same subfamily as fire ants (Myrmicinae), but occupies a different niche: marauder ants collect nectivorous plant materials and prey on other insects.

8.7.2   Termites

Termites (Blattodea: Termitidae) rival honeybees in social behavior. Termites have queens and kings, a division of labor, cooperative caring for offspring, and communicate through touch and smell (they are blind). Queens can live for 30–50 years, and their workers build colonies far larger than those of honey bees, with several million individuals. Termite mounds can be 10 m tall—the largest structures built by any non-human animal. In proportion to their tiny size, that's twice the size of our tallest human structure, Burj Khalifa in Dubai. Termites tunnel into taproots and upwards into stalks (Fig. 8.11). Infested plants wilt and collapse. Plants are easily pulled from the ground, exposing part of the colony.

8.7d. *Odontotermes obesus* (Rambur) 1842

Description: Worker termites have glistening bodies with no eyes. Heads are relatively large, taper toward the front, orange-brown in color. Heads of soldier termites have a pair of long, slender, saber-shaped mandibles. Thorax and abdomen are cream colored and nearly translucent, total length 4.4-5.4 mm. **Comments:** *O. obesus* is common in South Asia, where it chews away the roots of *Cannabis* (Cherian, 1932; Chandra, 2004). *O. obesus* feeds predominantly on woody soil debris, and builds mounds up to 3 m tall.

Alexander (1994) documented two cases of termite damage in Florida. In one case, the grower invited disaster by transplanting seedlings into planting holes mulched with wood chips. No species was identified, but it may have been the Formosan termite, *Coptotermes formosanus* Shiraki 1909, an invasive species taking over southeastern USA. Termites have chewed *Cannabis* roots as far north as Virginia (Britt *et al.*, 2020). Termite damage was also encountered in equatorial Africa (Fig. 8.11).

8.7.3   Cockroaches

Cockroaches are related to termites, so we mention them here. Del Guercio (1900b) cited the European cockroach, *Blatta orientalis* L. 1758, harming Italian hemp. Harming what, he didn't say, but seeds in soil seem likely. Dass Baba (1984) reported cockroaches eating seeds in peat pots.

8.7.4   Control of ants and termites

Flooding soil around *Cannabis* plants repelled ants and termites (Frank and Rosenthal, 1978). Frank (1988), a proponent of hydroponics, noted that ants and termites only cause problems in soil-rooted crops. Practice good sanitation: clean out stumps,

roots, and other termite-attractive debris. Rouge infested plants from fields, and destroy the local colony. Optimize nutrition and hydration, because ants and termites attack stressed plants.

Pseudacteon tricuspis is a wicked little phorid fly that controls fire ants. The species is native to South America and was imported into the USA in 1997 (Porter and Gilbert, 2005). Adult flies are brown with black heads, only 0.9–1.5 mm long. Females track ant semiochemicals from up to 50 m away, and make a dive-bombing oviposition strike (Fig. 8.12A). After the larva hatches, it moves into the ant's head, growing there until the head decapitates (Fig. 8.12B). Total development time is 4–12 weeks, depending on temperature.

Entomopathogenic fungi include *Beauveria bassiana* and *Metarhizium anisopliae*. The first species is added to ant baits, and the latter species is added to a product that specifically targets termites (BioBlast™). *Cannabis* extracts have been used against termites. Sattar *et al.* (2014) found that aqueous extracts of seeds were more effective than leaf extracts at killing 4th–5th-instar termites (*Microtermes obesi* and *Odontotermes lokanandi*). Clarke (pers. comm. 1996) stabbed an opening into subterranean colonies, and, depending on his mood, either drowned them with water, or poured kerosene into the colonies and burned them.

8.8   Root-damaging flies

Here we lump together root-damaging flies (Order Diptera): fungus gnats, crane flies, and root maggots.

8.8a. Fungus gnats

Bradysia spp., Diptera: Sciaridae
Description: *Larvae* (maggots) are slender, legless, with a black head and a smooth, white body, with semi-transparent skin revealing digestive tract contents, up to 5 mm long (Fig. 8.13). *Adults* are small and delicate, but heftier than mosquitos, with dark brown bodies 3 mm long, tiny heads with prominent eyes, long antennae, and relatively long legs and wings. Wings are dusky—Sciaridae are known as dark-winged fungus gnats.
Comments: In *Cannabis* literature, *Bradysia* spp. have not been identified to species. One website guessed the dark-winged fungus gnat, *Bradysia coprophila* (Linter) 1895.

Fig. 8.11. Termite damage in Africa (R. Clarke).

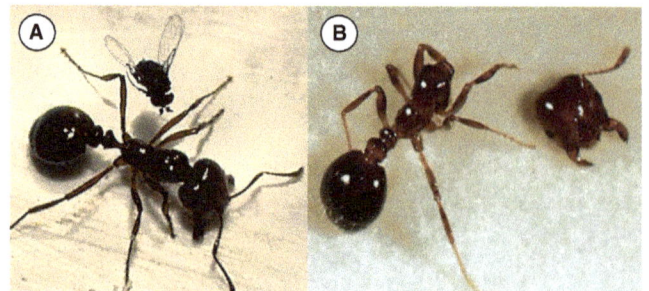

Fig. 8.12. *Pseudacteon tricuspis* (courtesy Porter and Gilbert, 2005)

Fungus gnats are a relatively new phenomenon. Plants growing in hydroponic systems and sterile potting media are more susceptible than plants growing in soil (Howard *et al.*, 1994). The earliest reports of *Bradysia* spp. in *Cannabis* are by Arnaud (1974) and Batra (1976). In Holland, Mahieu *et al.* (2019) referred to *Bradysia* spp. in *Cannabis* by their synonym *Sciara* spp. Glasshouse plants infested with *Bradysia* spp. continued to suffer outdoors (Zuefle, 2020). Larvae eat fine root hairs and gnaw into larger roots in heavy infestations. These injuries predispose plants to soil fungi. The adults become trapped in sticky flowers and are aesthetically displeasing. Females lay up to 200 eggs in soil or rockwool. New generations arise on a monthly basis in warm climates and glasshouses.

8.8b. European crane fly

Tipula paludosa Meigen 1830, Diptera; Tipulidae
Description: *Larvae* are known as leatherjackets, pink to grayish-black, with black heads, 25–38 mm long. *Adults* resemble gigantic mosquitoes, with gray-brown bodies up to 25 mm long, with long, delicate wings and legs, "daddy-longlegs with wings."
Comments: This Eurasian species was introduced into British Columbia and Washington state around 1955. Larvae live in soil and cause damage to pasture grasses. They feed on seedlings and roots. Bovien (1945) reported *T. paludosa* attacking Danish hemp crops. In Hungary, Nagy (1962) said *Tipula* damage was greatest in fertile, peaty soil. *T. paludosa* has also been reported in Germany (Mielke and Schöber-Butin, 2002), and Batra (1976) caught a *Tipula* sp. in India.

8.8.1 Root maggots

Root maggots (Diptera: Anthomyiidae) burrow into freshly sown seeds, which fail to germinate, leaving gaps in the seedling stand. The maggots also attack above-ground parts, chewing cotyledons and true leaves. They burrow into roots of older plants, which become honeycombed with slimy brown tunnels. Damaged roots often rot from invading soil fungi. Plants, if they survive, are yellow and stunted. They wilt during the day but recover at night.

8.8c. Seedcorn maggot

Delia platura Meigen 1826
=*Chortophila cilicruta* Rondani 1859, =*Pegomya fusciceps* Zetterstedt 1845
Description: *Larvae* are 5–7 mm long, white, with pointy heads and blunt posteriors, and the posteriors sport a pair of unforked tubercles. *Pupae* are spindle-shaped and brown. *Adults* look like half-sized house flies, 5 mm long, with gray pointed abdomens. They lay fusiform-shaped eggs covered by net-like surface ornaments.
Comments: Seedcorn maggot occurs on every continent except Antarctica. It causes greatest problems in temperate regions. The species feeds on decaying vegetable matter, but attacks a wide range of crops. Germinating seeds of maize and bean are particularly vulnerable to attack, especially in cold, wet spring conditions when seed germination is slow (Howard *et al.*, 1994). *D. platura* overwinters as pupae; adults emerge in spring.

Adults emerged from hemp fields on May 20 in Kentucky (Garman 1904a). Females lay 100 eggs, a few at a time, in disturbed soil. During the season *D. platura* pupates 2–4 cm under the soil near damaged plants; 2–5 generations arise per year. Garman (1904a) described the pest as "very destructive to hemp, destroying the young plants when 1.5–2 inches tall, an injury for which the cutworms get the blame commonly, since this insect is so minute that it escapes the observation of farmers."

D. platura has infested hemp in Japan (Harukawa and Kondo, 1930; Shiraki, 1952), the USA (Mah 1923; Reid 1940), and Europe (Rataj, 1957). Batra (1976) reported a *Delia* spp. in

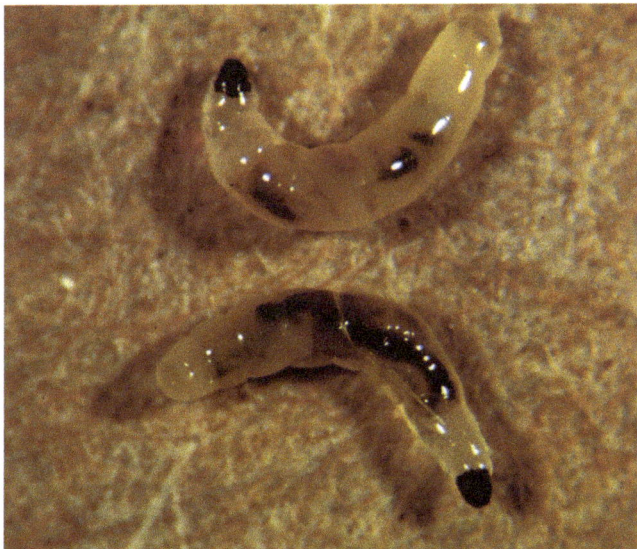

Fig. 8.13. Larvae of fungus gnats, *Bradysia* sp. (W. Cranshaw)

Fig. 8.14. Damage caused by seedcorn maggot, *Delia platura* (W. Cranshaw).

India. In a survey of Romanian pests, Trotuş and Naie (2008) recovered *D. platura* at an average density of 1.4 specimens/m².

maggots, don't plant near maize or *Brassica* crops. Eliminate grassy and cruciferous weeds from around crops.

8.8d. Cabbage maggot

Delia radicum (L.) 1758
=*Chortophila brassicae* Wiedemann 1817, =*Chortophila brassicae* Bouché 1833
Description: *Larvae* are spike-shaped, with pointed heads and wide flat posteriors, and the posteriors have paired tubercles that are forked at their apex. *Adults* have dark gray bristly bodies, 7 mm long, with black stripes on their thoraxes. They lay fusiform-shaped eggs covered by longitudinal striations.
Comments: Cabbage maggot is an Old World pest that was introduced into North America about 125 years ago. The species normally infests *Brassica* crops; larvae feed on roots and can completely destroy the root system. There's one citation on *Cannabis*: Goriainov (1914) claimed that *D. radicum* destroyed 40% of a hemp crop in Russia. A related species, the wheat root maggot, *Delia coarctata* (Fallén) 1825, was reportedly repelled from wheat by intercropped with *Cannabis* (Pakhomov and Potushanskii, 1977).

8.8.2   Cultural and mechanical control

Fungus gnats are an indoor problem. Practice good hygiene: don't allow organic detritus to stay wet, because the maggots can live on fungi. They also live on green algae growing on hydroponic media. Cover potting soil and hydroponic media with perlite to reduce algae. Avoid overwatering and excess humidity. Monitor adult populations with yellow sticky traps set at the base of plants (see Fig. 8.3).

Crane flies and root maggots are outdoor problems. For crane flies, avoid rotating hemp after grasslands. For root

8.8.3   Biological and chemical control

Aleochara bilineata is a staphylinid rove beetle native to North America that provides natural biocontrol against root maggots. Its larvae chew into maggot pupae and complete their life cycle therein, emerging as adults, that eat up to five maggots/day. The adults look like earwigs without rear pinchers.

The predator mite *Stratiolaelaps scimitus* (formerly *Hypoaspis miles*) preys on fungus gnat pupae. *Bacillus thuringiensis* var. *israelensis* kills the larvae of fungus gnats and crane flies. It is compatible with *Stratiolaelaps* mites and beneficial nematodes. The nematodes *Steinernema feltiae* and *Heterorhabditis* spp. show good efficacy against fungus gnat larvae (Lacey and Georgis, 2012), as well as outdoor root maggots and crane flies.

Mahieu *et al.* (2019) preferred using *S. scimitus*, *S. feltitae*, and *B.t. israelensis* strain AM65-52 to control fungus gnats. California DPR (2017) approved the use of BT and nematodes against fungus gnats. They also approved using soil drenches with azadirachtin (neem). A drench with insecticidal soap may knock back fungus gnats.

(Prepared by J. McPartland)

References

Alexander T. 1994. Florida reader relates. *Sinsemilla Tips* 5(1): 27.
Allen JL. 1900. *The reign of law, a tale of the Kentucky hemp fields.* Macmillan Co., New York.
Anjubault. F. 1862. Les Hannetons. *Bulletin de la Société d'Agriculture, Sciences et Arts de la Sarthe* 8: 529–561.
Anonymous. 1919. 害虫に関する調査 Governor-General's Office of Business Modeling, Korea.
Anonymous. 1940. *Lezioni al Corso di Perfezionamento per la Stima del Tiglio di Canapa.* Faenza-Fratelli Lega Editori, Roma.
Arnaud PH. 1974. Insects and mites associated with stored *Cannabis sativa* Linnaeus. *Pan-Pacific Entomologist* 50: 91–92.
Arnoldi LV, *ed.* 1949. Вредные животные Средней Азии : справочник. Publishing House of USSR Academy of Sciences, Moscow.
Bakro F, Wielgisz K, Bunalski M, Jedrycka M. 2018. An overview of pathogen and insect threats to fibre and oilseed hemp (*Cannabis sativa* L.) and methods for their biocontrol. *IOBC-WPRS Bulletin* 136: 9–20.
Barna J, Lencsepeti JS, Rközy P, Zsombokos G. 1982. *Mezögazdasági Lexikon, Vol 1 (A–K).* Mezögazdasági Kiadó, Budapest.
Barnum AW. 1829. "Letter from Gen. Barnum to the editor of the Vermont *Aurora*, Vergennes," p. 15–21 in Fairbanks E., *ed. Compilation of articles relating to the culture and manufacture of hemp in the United States.* Jewett and Porter, St. Johnsbury Vermont.
Batra SW. 1976. Some insects associated with hemp or marijuana (*Cannabis sativa* L.) in northern India. *Journal of the Kansas Entomological Society* 49: 385–388.
Benelli G, Pavela R, Petrelli R, *et al.* (7 other authors). 2018. The essential oil from industrial hemp (*Cannabis sativa* L.) by-products as an effective tool for insect pest management in organic crops. *Industrial Crops and Products* 122: 308–315.
Berlese A. 1902. La grillotalpa ed il modo seguito per combatteria a nola. *Bollettino di Entomologia Agraria* 9: 104–116, 128–140, 150–159, 185–189.
Berlese A. 1924. *Entomologia Agraria. Manuale sugli insetti nocivi alle piante coltivate.* Reale Stazione di Entomologia Agraria, Firenze.
Bertoloni A di. 1874. Al danno arrecato alla canapa, al formentone ec. *dall'Agrotis suffusa* Ochs. var. *pepoli* Bertol. nep. *Bollettino della Società Entomologica Italiana* 6: 139–146.
Blackman RL, Eastop VF. 2006. *Aphids on the world's herbaceous plants and shrubs, vol. 1.* John Wiley and Sons, Chichester, UK. Available at: http://www.aphidsonworldsplants.info/index.htm

Bogdanov-Kat'kov NN. 1930. *Краткий учебник теоретической и прикладной энтомологии, Издание второе.* State Publishing House, Moscow.

Bovien P. 1945. Plantesygdomme i Danmark 1944. *Tidsskrift for Planteavl* 50: 1–76.

Britt K, Fike J, Flessner M, Johnson C, Kuhar T, McCoy T, Reed D. 2020. *Integrated pest management of hemp in Virginia.* Virginia Cooperative Extension, Virginia Tech Bulletin ENTO-349NP, Blacksburg, Virginia.

Buc'hoz PJ. 1775. *Dictionnaire vétérinaire et des animaux domestiques.* Brunet, Paris.

Buren WF. 1972. Revisionary studies on the taxonomy of the imported fire ants. *Journal of the Georgia Entomological Society* 7: 1–26

California DPR (Department of Pesticide Regulation). 2017. *Legal pest management practices for Cannabis growers in California.* Available at: https://www.cdpr.ca.gov/docs/county/cacltrs/penfltrs/penf2015/2015atch/attach1502.pdf

Camprag D. 1961. Observations on the occurrence and injuriousness of *Maladera holosericea* Scop. with special reference to sugar-beet. Зборник Матице српске за природне науке 21: 122–131.

Chandra R. 2004. Status of medicinal plants with respect to infestation of insect pests in and around Chitrakoot, District-Satna (M.P.). *Flora and Fauna (Jhansi)* 10(2): 88–92.

Cherepanov AI. 1957. *ЖукиУщелкуны Западной Сибири.* Publishing House of the Academy of Sciences of the USSR, Novosibirsk.

Cherian MC. 1932. Pests of ganja. *Madras Agricultural Journal* 20: 259–265.

Comes O. 1885. Sulla grillotalpa (*Gryllotalpa vulgaris*) e sul mezzo di combatterla. *Annuario della Regia Scuola superiore di agricoltura in Portici* 5: 55–63.

Conner K, Kesheimer K, Sikora E, Kemble J. 2020. "Diseases and insects of hemp in Alabama," p. 21 in Gauthier N, *ed. Science of hemp: production and pest management.* University of Kentucky Agricultural Experiment Station no. SR-112. Lexington, Kentucky.

Cranshaw WS, Wainwright-Evans S. 2020. *Cannabis sativa* as a host of rice root aphid (Hemiptera: Aphididae) in North America. *Journal of Integrated Pest Management* 11(1): 15; 1–3

Cranshaw WS, Schreiner M, Britt K, Kuhar TP, McPartland JM, Grant J. 2019. Developing insect pest management systems for hemp in the United States: a work in progress. *Journal of Integrated Pest Management* 10(1): 26; 1–10.

Creighton WS. 1930. The New World species of the genus *Solenopsis. Proceedings of the American Academy of Arts and Sciences* 66: 39–151.

Crosby TK, Watt JC, Kistemaker AC, Nelson PE. 1986. Entomological identification of the origin of imported *Cannabis. Journal of the Forensic Science Society* 26(1): 35–44.

Dara SK. 2015. *Reporting the occurrence of rice root aphid and honeysuckle aphid and their management in organic celery.* E-Journal of Entomology and Biologicals. Available at: https://ucanr.edu/blogs/strawberries-vegetables/index.cfm?start=48

Dass Baba S. 1984. Truth, fact, and hearsay from the Big Island. *Sinsemilla Tips* 4(3): 12–13.

de Serres O. 1600. *Le theâtre d'agriculture.* Cher Jamet Metayer, Paris.

Del Guercio G. 1900b. La cronaca della R. Stazion Entomologica de Firenze dal 1886 al 1900. *Nuove relazioni intorno ai Lavori della R. Stazione di Entomologia Agraria di Firenze, Serie Prima* 3: 160–369.

Del Guercio G. 1903. Notizie e suggerimenti pratici per conoscere e combattere gli animali nocivi alle piante coltivate ed ai loro frutti. *Nuove Relazioni intorno ai Lavori della R.Stazione di Entomologia Agraria di Firenze, Serie Prima* 5: 1–208.

Dewey LH. 1914. "Hemp," pp. 283–347 in: *USDA Yearbook 1913.* United States Department of Agriculture, Washington, DC.

Difonzo CD. 2022. Western bean cutworm (Lepidoptera: Noctuidae) feeding and development on industrial hemp in the laboratory and field. *Great Lakes Entomologist* 55(1/2): 1–9.

Durnovo ZP. 1933. "Итоги работ по кукурузному мотыльку и другим вредителям однолетних новых лубяных культур," pp. 85–106 in Болезни и вредители новых лубяных культур. Institute New Bast Raw Materials, Moscow.

Ebert G. 2005. *Die Schmetterlinge Baden-Württembergs: Ergänzungsband.* Ulmer, Stuttgart.

El Bassam N. 2010. *Handbook of bioenergy crops.* Routledge, Abingdon, UK.

Fedorenko VP, Kabanetz VV, Kabanets VM. 2016. Шкідники конопель посівних. National Academy of Agrarian Science, Kiev.

Flader C, Neuer H. 1939. *Der deutsche Hanfbau. Eine Anleitung für die Praxis.* Paul Parey, Berlin.

Forbes SA. 1894. "Injuries by white grubs," pp. 109–146 in *Eighteenth Report of the State Entomologist on the noxious and beneficial insects of the State of Illinois.* Springfield, Illinois.

Frank M. 1988. *Marijuana grower's insider's guide.* Red Eye Press, Los Angeles, California.

Frank M, Rosenthal E. 1978. *Marijuana grower's guide.* And/Or Press, Berkeley, California.

Fruhstorfer H. 1921. Die Orthopteren der Schweiz. *Archiv für Naturgeschichte* 87(5-6): 1–262.

Garman H. 1892. *Some common pests of the farm and garden.* Kentucky Agricultural Experiment Station Bulletin no. 40. Lexington, Kentucky.

Garman H. 1894b. "The Orthoptera of Kentucky," pp. 23–30 in *Sixth Annual Report of the Kentucky Agricultural Experiment Station.* Frankfort, Kentucky.

Garman H. 1895. *Cutworms in Kentucky.* Kentucky Agricultural Experiment Station Bulletin no. 58. Lexington, Kentucky.

Garman H. 1904a. "The fringed anthomyian *Pegomya fusciceps*," pp. 45–46 in *Insects injurious to cabbage.* Kentucky Agricultural Experiment Station Bulletin no. 114, Lexington, Kentucky.

Garman H. 1904b. *On an injury to fruits by insects and birds.* Kentucky Agricultural Experiment Station Bulletin no. 116, Lexington, Kentucky.

Goidànich A. 1928. Contributi alla conoscenza dell'entomofauna della canapa. I. Prospetto generale. *Bollettino del Laboratorio di Entomolgia del R. Istituto Superiore Agrario di Bologna* 1: 37–64.

Goriainov AA. 1914. Вредители сельско-хозяйственных растений Рязанской губ. Ryazan Government Printing Office, Ryazan, USSR.

Guérin-Méneville FE. 1846. Sur une chenille de Noctua. *Revue Zoologique* 9: 267–268.

Gutberlet V, Karus M. 1995. *Parasitäre Krankheiten und Schädlinge an Hanf (Cannabis sativa).* Nova Institut, Köln, Germany.

Hackenberg H. 1908. *Über die Substanzquotienten von Cannabis sativa und Cannabis gigantea.* Doctoral thesis, Friedrich-Wilhelms-Universität, Bonn.

Harukawa C, Kondo S. 1930. 種蝿(*Hylemyia cilicrura* Rondani) に就いて (第一報). *Agricultural Research (Nōgaku kenkyū)* 14: 449–469.

Hector GP. 1923. "Report of the economic botanist to the government of Bengal," pp. 37–41 in *Annual Report of the Department of Agriculture, Bengal, for the year 1921–22.* Bengal Secretariat, Calcutta.

Hill DS. 1983. *Agricultural insect pests of the tropics and their control, 2nd edn.,* Cambridge University Press, Cambridge, UK.

Howard RJ, Garland JA, Seaman WL, eds. 1994. *Diseases and pests of vegetable crops in Canada.* Entomological Society of Canada, Ottawa, Ontario.

Hummelbrunner LA, Isman MB. 2001. Acute, sublethal, antifeedant, and synergistic effects of monoterpenoid essential oil compounds on the tobacco cutworm, *Spodoptera litura* (Lep., Noctuidae). *Journal of Agricultural and Food Chemistry* 49: 715–720.

Hung ACF, Barlin MR, Vinson SB. 1977. *Identification, distribution, and biology of fire ants in Texas.* Texas Agricultural Experiment Station Publication B-1185, College Station, Texas.

Jäger G. 1874. *Deutschlands Thierwelt nach ihren Standorten eingetheilt, Zweiter Band.* Kröner, Stuttgart.

Kirchner O. 1906. "Hanf, *Cannabis sativa* L.," pp. 319–323 in *Die Krankheiten und Befehädigungen uhferer landwirtschaftlichen Kulturpflanzen.* E. Ulmer, Stuttgart.

Lacey LA, Georgis R. 2012. Entomopathogenic nematodes for control of insect pests above and below ground with comments on commercial production. *Journal of Nematology* 44: 218–225.

Lagos-Kutz D, Potter B, DiFonzo C, Russell H, Hartman GL. 2018. Two aphid species, *Phorodon cannabis* and *Rhopalosiphum rufiabdominale*, identified as potential pests on industrial hemp, *Cannabis sativa* L., in the US Midwest. *Crop, Forage & Turfgrass Management* 4(1): 1–3.

Leunis J. 1877. *Synopsis der Pflanzenkunde, Zweite Abtheilung.* Hahn, Hannover.

Lichtenegger E, Kutschera L, Sobotik M. 2018. *Wurzelatlas der Kulturpflanzen gemäßigter Gebiete mit Arten des Feldgemüsebaues.* DLG-Verlag, Frankfurt am Main.

Mah LM. 1923. *Hemp-Cannabis sativa: an agronomic study.* Master's thesis, University of California.

Mahieu CM, Biesebeek JD, Graven C. 2019. *Inventarisatie van gewasbescherming toepasbaar in de teelt van Cannabis binnen het "Experiment met een gesloten coffeeshopketen."* Rijksinstituut voor Volksgezondheid en Milieu, Rapport 2019-0232. Bilthoven, Netherlands.

Matov AY, Kononenko VS. 2012. Трофические связи гусениц совкообразных чешуекрылых фауны России (Lepidoptera, Noctuoidea: Nolidae, Erebidae, Euteliidae, Noctuidae). Dalnauka, Vladivostok.

McPartland M, Sheikh Ź. 2018. A review of *Cannabis sativa*-based insecticides, miticides, and repellents. *Journal of Entomology and Zoology Studies* 6: 1288–1299.

Melicher LJ. 1869. *Skizze der nützlichen und schädlichen niederen Thiere.* Braumüller, Wien.

Mielke H, Schöber-Butin B. 2002. *Pflanzenschutz bei Nachwachsenden Rohstoffen, Zuckerrübe, Öl- und Faserpflanzen.* Parey, Berlin.

Mühle E. 1966. Insektenlarven als allgemeine wurzelschädlinge. *Phytopathologie und Pflanzenschutz* 2: 44–61.

Nagy B. 1962. "Állati kártevők," pp. 70–73 in Mándy G, Bócsa I, *eds.* 1962. A kender. *Magyarország Kultúrflórája* 7(14): 1–101.

Nair KR, Ponnappa KM. 1974. *Survey for natural enemies of Cannabis sativa and Papaver somniferum.* Commonwealth Institute of Biological Control, India Station Report, pp 39–40.

Neuer H. 1935. Der Anbau des Hanfes. *Berichte über Landwirtschaft: Sonderheft* 20: 16–30.

Neuer HV, Prieger E, Sengbusch RV. 1946. Hanfzüchtung. I. Die Steigerung des Faserertrages von Hanf. *Der Züchter (Zeitschrift für theoretische und angewandte Genetik)* 17/18(2): 33–39.

Nördlinger H. 1869. *Die kleinen Feinde der Landwirthschaft, zweite auflage.* J. G. Cotta, Stuttgart.

Olshevsky VD. 1970. Конопля против крота медведка обыкновенная. *Plant Protection* 15: 46.

Oprescu RM, Biris SS, Voicea I, Vladut V. 2019. Considerations on hemp cultivation technology. *Acta Technica Corviniensis* 12(3): 85–88.

Osmolovsky GE. 1976. *Определитель сельско-хозяйственных вредителей по повреждениям культурных растений.* Kolos, Leningrad.

Pace R. 2019. *Studying biocontrols for root aphids.* Cannabis Horticultural Association of Humbolt County. Available at: https://www.chascience.com/horticultural-blog/tag/Beauveria+bassiana

Pakhomov VI, Potushanskii VA. 1977. Озимая муха в Ульяновской области. *Plant Protection* 9: 18–19.

Pfeiffer E. 1979. *Fécondité de la terre: méthode pour conserver ou rétablir la fertilité du sol, le principe bio-dynamique dans la nature.* Triades, Paris.

Pichat CB. 1866. "La pianta della canapa," pp. 396–499 in *Istituzioni scientifiche e tecniche, ossia Corso teorico e practico de Agricoltura, Libri XIII, Volume Quinto.* Presso l'Union Tipografico-Editrice, Torino.

Porter SD, Gilbert LE. 2005. "Parasitoid case history: an evaluation of methods used to assess host ranges of fire and decapitating flies," pp. 634–650 in *International Symposium on biological control of arthropods.* USDA Forest Service, Morgantown, West Virginia.

Ragazzi G. 1954. Nemici vegetali ed animali della canapa. *Humus* 10(5): 27–29.

Rak Cizej M, Policnik F. 2018. Škodljivci *industrijske konoplje (Cannabis sativa L.) v Sloveniji.* Hmeljarski Bilten 25: 36–43.

Rataj K. 1957. Skodlivi cinitele pradnych rostlin. *Prameny Literatury* 2: 1–123.

Ratzeburg JTC. 1839. *Die Forst-Insecten oder Abbildung und Beschreibung der in den Wäldern Preussens und der Nachbarstaaten als schädlich oder nützlich bekannt gewordenen Insecten, ester theil.* Nicolai'schen Buchhandlung, Berlin.

Reid WJ. 1940. *Biology of the seed-corn maggot in the coastal plain of the South Atlantic states.* Technical Bulletin no. 723, USDA, Washington, DC.

Robinson BB. 1943b. *Hemp.* USDA Farmer's Bulletin No. 1935. Washington, *DC.*

Roethke T. 1968. The stony garden. *Poetry* 113(2): 104–108.

Saalas U. 1923. Studien über die Elateriden Finlands. *Annales Societatis Zoologicae Botanicæ Fennicæ Vanamo* 2: 121–168.

Sattar A, Naeem M, ul-Haq E. 2014. Efficacy of plant extracts against subterranean termites *i.e., Microtermes obesi* and *Odontotermes lokanandi* (Blattodea: Termitidae). *Journal of Biodiversity, Bioprospecting and Development* 1(2): 122.

Selgnij S. 1982. Sun, seeds, soil and soul. *Sinsemilla Tips* 2(4): 18–21, 30.

Sempers FB. 1894. *Injurious insects and the use of insecticides.* Atlee Burpee & Co., Philadelphia.

Shiraki T. 1952. *Catalogue of injurious insects in Japan.* General Headquarters, Supreme Command Allied Powers, Tokyo.

Sinha RP, Singh BN, Zafar IM, Choudhary RS. 1979. Ganja, *Cannabis sativa* L., a new host for cutworms in Bihar. *Indian Journal of Entomology* 41: 296.

Targioni Tozzetti A. 1876. *Annali del Ministero di Agricoltura, Industria e Commercio 1876, Agricoltura.* Eredi Botta, Roma.

Tietz HM. 1972. *An index to the described life histories, early stages and hosts of the Macrolepidoptera of the continental United States and Canada.* AC Allyn Publ., Sarasota, Florida.

Trotuş E, Naie M, 2008. Cercetări privind reducerea atacului unor agenţi patogeni şi dăunători specifici culturilor de cânepă prin tratamentul chimic al seminţei. *Lucrari Stiintifice, Universitatea de Stiinte Agricole Si Medicina Veterinara "Ion Ionescu de la Brad" Iasi, Seria Agronomie* 51(2): 219–223.

Van Herk WG, Vernon RS, Tolman JH, Ortiz Saavedra H. 2008. Mortality of a wireworm, *Agriotes obscurus* (Coleoptera: Elateridae), after topical application of various insecticides. *Journal of Economic Entomology* 101: 375–383.

Weiss HB. 1915. *Gryllotalpa gryllotalpa* Linn the European mole cricket in New Jersey. *Journal of Economic Entomology* 8: 500–501.

Wilson EO. 1951. Variation and adaption in the imported fire ant. *Evolution* 5: 68-79.

Wilson H, Bodwitch H, Carah J, *et al.* (4 other authors). 2019. First known survey of cannabis production practices in California. *California Agriculture* 73: 119–127.

Wright J. 1847. *On our present domestic condition and the cultivation of waste lands.* Dolman, London.

Xu J, Liu N, Zhang R. 2013. "Other pests—China," pp. 193–226 in Giordanengo P *et al., eds. Insect pests of potato.* Academic Press, Oxford, UK.

Yepsen RB. 1976. *Organic plant protection.* Rodale Press, Emmaus, Pennsylvania.

Zuefle M. 2020. *Hemp pest survey 2020.* New York State Integrated Pest Management Program. Available at: https://ecommons.cornell.edu/handle/1813/103765

Abstract

Chapter 9 begins a series of four chapters on fungal diseases. Two potentially catastrophic fungal diseases are addressed here. Gray mold caused by *Botrytis cinerea* can destroy flowering tops, so we feature it in this chapter, but it also damages stalks (Chapter 10) and seedlings (Chapter 12). Powdery mildew, primarily caused by *Golovinomyces* spp., badly infests flowering tops and leaves. Other fungi occasionally infect flowering tops but they cause greater problems in leaves (Chapter 10) or cause blights (Chapter 11). *Botrytis* is a necrotrophic pathogen—it kills plant cells and feeds on the dead. *Golovinomyces* is an obligate biotrophic pathogen—it must draw nutrients from living plant cells. *Botrytis* is an opportunist (attacking stressed plants) and relatively easy to control; the same cannot be said for *Golovinomyces*. Common names for *Cannabis* diseases were standardized by the American Phytopathological Society (McPartland, 1989, 1991). For an update see https://www.apsnet.org/edcenter/resources/commonnames/Pages/Hemp.aspx).

9.1 Powdery mildew

Powdery mildew fungi, Family Erysiphaceae, are the world's most familiar yet poorly known plant pathogens (Glawe, 2008). About 650 species infect nearly 10,000 species of angiosperms. Historically, powdery mildew was rarely reported in outdoor fiber-type varieties (Després, 1854; Polonio, 1863; Transhel *et al.*, 1933; Gitman and Boytchenko; 1934), and outdoor drug-type varieties (Doidge and Bottomley, 1931; Doidge *et al.*, 1953).

Today's epidemic is due to the intersectionality of indoor cultivation and new, susceptible drug-type hybrids. Inklings of the epidemic surfaced at the dawn of the indoor era (Stevens, 1975; McPartland, 1983). When indoor plants with powdery mildew are transplanted outdoors, the disease may dissipate (Szarka *et al.*, 2019; Tymon *et al.*, 2022). In a recent survey, Wilson *et al.* (2019) reported that 43% of California *Cannabis* growers were plagued by powdery mildew—more than any other disease or pests.

Powdery mildew fungi are obligate biotrophic pathogens. They draw nutrients from *living* cells and therefore must maintain host viability. They use **haustoria**—specialized feeding structures—to penetrate the host cell wall without disrupting the plasma membrane. Biotrophs slowly drain nutrients from the host, and gradually decrease plant fitness and yield (Williamson *et al.*, 2007). The host mounts a **hypersensitive response**, expressed as chlorosis and necrosis. The hypersensitive response combats fungal growth, but it compounds the damage.

9.1.1 Symptoms and signs

Powdery mildew appears on leaves that have stopped expanding. Early symptoms are nearly invisible—raised humps or blisters on leaf surfaces. From these areas, a white, dusty or powdery mycelium emerges. The mycelium may remain isolated in irregular colonies, or coalesce over the entire leaf, accompanied by chlorotic and necrotic areas. Infected plants remain alive for indefinite periods, or prematurely wilt and die. Some powdery mildew fungi are largely restricted to upper (adaxial) leaf surfaces, and some can grow on both leaf surfaces. Some powdery mildew fungi can extend onto petioles and branches, and into inflorescences (Fig. 9.1).

Erysiphaceae produce asexual spores—conidia—often in the absence of sexual spore structures (chasmothecia). The genus name *Oidium* was traditionally applied to Erysiphaceae known only by their conidia. *Oidium* posed a problem to taxonomists, who classified Erysiphaceae by their chasmothecia. Classification based on conidial characteristics has a long history (Sawada, 1927; Zaracovitz, 1965; Boesewinkel, 1980; Cook *et al.*, 1997). Informative characters include colony characteristics (endo- or exophytic mycelium), appressorium shape, conidiophore morphology (foot cell size and shape), conidial ontogeny (conidia produced singly or in chains), conidial size, shape, and wall surface, and the presence or absence of fibrosin granules within conidia (visible in 3% KOH as highly refractive, angular crystals a few micrometers long).

Conidial states have been reported from fiber-type plants in France (Després, 1854), Italy (Polonio, 1863), and Russia (Hirata, 1966). Reports of conidial states in South Africa (Doidge and Bottomley, 1931; Doidge, 1950) likely involved drug-type plants. Doidge collected her specimens in Potchefstroom, and Potchefstroom has long been a center of *Cannabis indica* cultivation (Davy, 1904).

Thanks to molecular methods, it is far easier to identify conidia-bearing fungi in the absence of their sexual state. At first, DNA methods identified powdery mildew fungi to genus but not species: *Golovinomyces* sp. in Kentucky (Gauthier, 2015), Massachusetts (McKernan *et al.*, 2016), and Washington (Couch *et al.*, 2017), and *Erysiphe* sp. in Nevada (Schoener and Wang, 2016).

Four species of powdery mildew have been identified on hemp: two *Golovinomyces* spp., *Leveillula taurica*, and *Podosphaera*

Fig. 9.1. Powdery mildew: **A.** leaf; **B.** inflorescence (from Punja *et al.*, 2019, with kind permission)

macularis. A fifth taxon was named as a new variety by Westendorp (1854a) in Belgium: *Erysiphe communis* var. *urticearum* Westendorp. McPartland (1995a) examined Westendorp's specimen, loaned from the National Herbarium of Belgium. No powdery mildew! The specimen was a *Phoma* sp. mixed with spider mite webbings and dead spider mites.

9.1a. *Golovinomyces ambrosiae* (Schwein.) Braun & Cook 2009

≡*Erysiphe ambrosiae* Schweinitz 1834
=*Golovinomyces spadiceus* (Berk. & Curtis) Braun 2012,
≡*Erysiphe spadicea* Berkeley & Curtis 1876
Description: Conidiophores upright, unbranched, cylindrical, hyaline or light brown, 120–140 μm in height. Conidia produced singly or in short chains of 2–4, hyaline, lacking fibrosin granules, ellipsoid-ovoid to cylindrical to barrel shaped, single-celled, 25–40 × 14–20 μm. Chasmothecia are subglobose or flattened on the side next to the leaf surface, black, smooth, 80–140 μm in diameter. Appendages flexuous, hyaline, usually unbranched, up to 200 μm in length. Asci numerous, obovoid-saccate with a short stalk, 40–70 × 25–35 μm. Ascospores broadly ellipsoid-ovoid, hyaline, unicellular, 15–25 × 10–15 μm.
Comments: *G. ambrosiae* was first cited on *Cannabis* in Nevada (Schoener and Wang, 2018; Wang, 2018). *G. ambrosiae* is distributed worldwide with a wide host range, including Cucurbitaceae (especially cucumbers and squash), Solanaceae (tobacco and potato), and Compositae (sunflower, gerbera, echinacea). The organism was identified in Kentucky as *G. spadiceus*, between 2014 and 2018, on fiber-, seed-, and drug-type plants, using ITS sequences (Szarka *et al.*, 2019; Dixon *et al.*, 2022b). Weldon *et al.* (2020) used ITS and 28S sequences to identify *G. spadiceus* infesting glasshouse-grown CBD-type strains in New York. Farinas and Peduto Hand (2020) used ITS sequences to identify *G. spadiceus* infesting glasshouse-grown drug-type plants in Ohio.

 G. ambrosiae and *G. spadiceus* cannot be resolved by their 28S and ITS regions (Takamatsu *et al.*, 2013). *G. spadiceus* was synonymized under *G. ambrosiae* by Qiu *et al.* (2020), based on a multilocus analysis (ITS, 28S, IGS, *TUB2, CHS1*). The correct nomenclature, *G. ambrosiae*, has been applied in Oregon (Wiseman *et al.*, 2021), Washington (Mihalyov and Garfinkel, 2021), New Jersey (Rajmohan *et al.*, 2022), Quebec (Brochu *et al.*, 2022), and British Columbia (Scott and Punja, 2020).

9.1b. Golovinomyces cichoracearum (DC) Heluta 1988

≡*Erysiphe cichoracearum* DC 1805
≡*Oidium asteris-punicei* Peck 1911, conidial state
Description: Compared with *G. ambrosiae*, conidiophores are taller (140–260 μm) and conidia are larger (29–37 × 12–17 μm). Chasmothecia compared with *G. ambrosiae* are broader (100–200 μm) with larger asci (48–90 × 28–37 μm) and larger ascospores (17–28 × 12–22 μm).
Comments: *G. cichoracearum* is distributed worldwide in temperate climates. Braun (1987) confined *G. cichoracearum* to powdery mildews on hosts in the Asteraceae (sunflower, gerbera, echinacea). But molecular identification has widened the host range to Cucurbitaceae (especially cucumbers and squash), Solanaceae (tobacco and potato), okra, and phlox. Using ITS sequences, Rodriguez *et al.* (2015) identified *G. cichoracearum* in an indoor hydroponic facility in British Columbia. *G. cichoracearum sensu lato* was reported in British Columbia and New Brunswick (Pépin *et al.*, 2018; Punja *et al.*, 2019). They labeled it *sensu lato* because its ITS sequence could not clearly distinguish it from *Golovinomyces ambrosiae*. Punja *et al.* (2019) reported the fungus infesting leaves, stalks, and inflorescences. Maymon *et al.* (2021) identified *G. cichoracearum sensu lato* in Israel. Its ITS sequence shared 100% identity with a Canadian isolate (Pépin *et al.*, 2018). In their strain, no chasmothecia were observed.

9.1c. *Leveillula taurica* (Lév.) Arnaud 1921

≡*Leveillula taurica* f. *cannabis* Jaczewski 1927
≡*Oidiopsis taurica* (Lév.) Salmon 1906, conidial state
Description: Conidiophores are often two-celled, upright, simple or occasionally branched, hyaline, emerging from host stomata, up to 250 μm in height. Conidia are borne singularly atop conidiophores, either cylindrical or navicular in shape, often with apical points, single-celled, hyaline, 50–79 × 14–20 μm (Fig. 9.2). Chasmothecia are globose, black, smooth, 135–250 μm in diameter, with numerous hyphal appendages (Fig. 9.2). Chasmothecial appendages hyaline to light brown, indistinctly branched, less than 100 μm in length. Chasmothecia contain around 20 asci, which are ovate, distinctly stalked, 2-spored, 70–110 × 25–40 μm. Ascospores cylindrical to pyriform, sometimes slightly curved, hyaline, unicellular, 24–40 × 12–22 μm.
Comments: *L. taurica* is distributed worldwide, mostly in warm and dry climates. In the USA it is restricted to western states. *L. taurica* has a wide host range, including Solanaceae, Cucurbitaceae, *Allium*, and *Gossypium* spp. Different races of *L. taurica* show host-specificity. Jaczewski (1927) assigned the

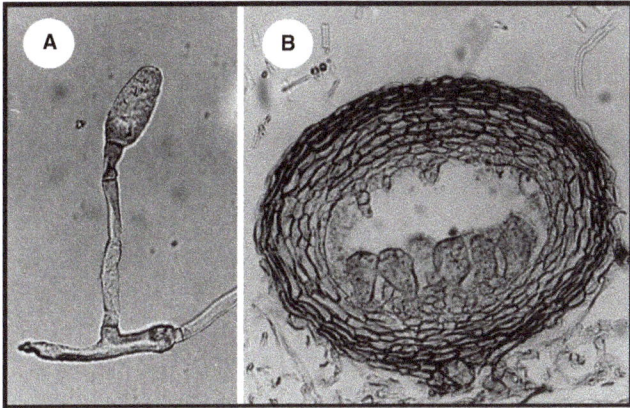

Fig. 9.2. *Leveillula taurica*: **A**. conidium and conidiophore; **B**. chasmothecium with asci

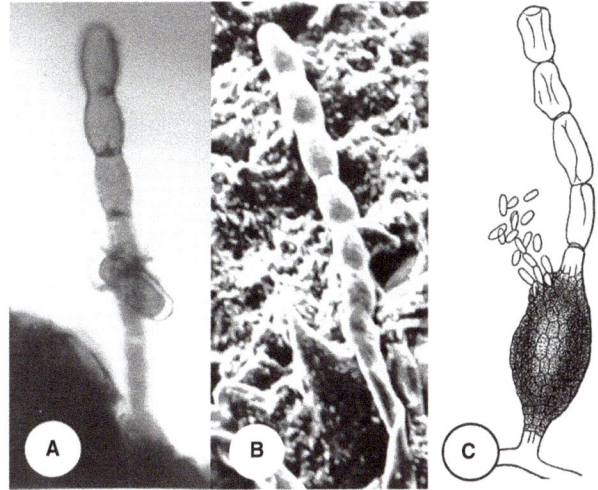

Fig. 9.3. Conidiophores and conidia of *P. macularis*: **A**. LM ×390; **B**. SEM ×390; **C**. conidiophore parasitized by *Ampelomyces quisqualis* (DeBary, 1887)

taxon *L. taurica* f. *cannabis* to a race infesting hemp in Turkestan. Within Turkestan, Zaprometov (1933) specified a place: the Ferghana Valley in Tashkent District. The fungus has also been reported in neighboring Uzbekistan (Hirata, 1966) and Kazakhstan (Rakhimova *et al.*, 2013). It also occurs on *Cannabis* in Europe: in Russia (Transhel *et al.*, 1933), France (Hirata, 1966), and Germany (Brandenburger, 1985), as well as Turkey (Kabaktepe *et al.*, 2015).

9.1d. *Podosphaera macularis* (Wallr.) Braun & Takamatsu 2000

≡*Sphaerotheca macularis* (Wallr.: Fr.) Lind 1913
≡*Oidium* sp. (subgenus *Fibroidium*), conidial state
Description: Conidiophores are upright, simple, hyaline, 50–100 μm in height. Conidia are produced in chains, hyaline (turning brown with age), ovate to barrel-shaped, single-celled, with a smooth outer wall, containing fibrosin granules (which disappear with age), averaging 30.2 × 14.0 μm (Fig. 9.3). Chasmothecia are globose, black, smooth, 60–125 μm in diameter, with few to many hyphal appendages (Fig. 9.4). Chasmothecial appendages are hyaline to brown, unbranched, flexous, tapering at ends, 300–500 μm in length. Chasmothecia contain one ascus, which is subglobose to broadly elliptical, indistinctly stalked, 8-spored, 50–90 × 45–75 μm. Ascospores are ellipsoidal to oval, hyaline, unicellular, 18–25 × 12–18 μm.
Comments: *P. macularis* occurs worldwide, mostly in temperate climates. It commonly parasitizes hops and strawberries, and was identified on drug-type plants in Illinois (Fig. 9.3A,B) (McPartland and Cubeta, 1997). This identification was based on conidial morphology—a key character being fibrosin granules, present in *Podosphaera* and *Sphaerotheca,* and absent in *Golovinomyces* and *Erysiphe* (Boesewinkel, 1980). Gauthier (2020a) dismissed the identification of *P. macularis* "as reported." Soon *P. macularis* was confirmed in Oregon (W. Mahafee, unpublished data, 2020), New York (Weldon *et al.*, 2020), Oregon again (Bates *et al.*, 2021), and Washington (Punja, 2022). In all these cases, *P. macularis* conidial morphology was confirmed by ITS and 28S sequencing.

Fig. 9.4. Chasmothecia of *P. macularis*: **A**. low magnification; **B**. high magnification (from Weldon *et al.*, 2020, with kind permission)

Weldon *et al.* (2020) cross-inoculated *P. macularis* from hops to hemp. Then they co-inoculated two sexually compatible isolates on *Cannabis*. Chasmothecia began forming within 18 days, visible under low magnification as brown–black dots (Fig. 9.4A). Under higher magnification they visualized one ascus per chasmothecium (Fig. 9.4B). Two varieties were susceptible: 'Anka' is an early-maturing, monoecious, fiber- and seed-type cultivar bred from Russian accessions. "Wild Horse" is an "Indica-dominant" hybrid of murky pedigree. A resistant variety, "TJ's CBD", is reportedly a "50% Indica, 50% Sativa" hybrid.

9.1.2 Differential diagnosis

Powdery mildew can be confused with symptoms of downy mildew, which produces a gray mycelium, usually on undersides of leaves. Powdery mildew has also been confused with pink rot, which looks white and powdery in its initial states (McPartland and Hillig, 2008).

Differentiating amongst Erysiphaceae is best accomplished by a morphological inspection of the sexual state, or by molecular methods. The fungi cause slightly different signs and symptoms: *Golovinomyces* spp. are the most aggressive; they colonize leaves, petioles, branches, stalks, and inflorescences. In contrast, *P. macularis* growth tends to be limited to upper (adaxial) leaf surfaces, and *L. taurica* usually grows on the undersides of leaves. Conidiophores of *Golovinomyces* spp. and *P. macularis* arise on the leaf surface, whereas *L. taurica* conidiophores emerge through stomata.

9.1.3 Disease cycle and epidemiology

Powdery mildew fungi overwinter as chasmothecia or as dormant mycelia in crop residues. Chasmothecia discharge ascospores in the spring, and dormant mycelia release conidia. Ascospores and conidia disperse by wind. Young plants become infected in May, but may take weeks to show symptoms. The fungi produce copious conidia, which spread by the slightest breeze to sites of secondary infections. Conidia of *P. macularis* do not spread very far, usually ≤2 m from host plants, whereas conidia of *G. cichoracearum* may disperse up to 200 km (Glawe, 2008). Glawe cited a study that estimated *G. cichoracearum* could disseminate 80 million conidia/ha per day in moderately infected crops.

L. taurica is adapted to xerophytic conditions; conidia can germinate in 0% relative humidity (RH). In contrast, *P. macularis* conidia germinate best at 100% RH, but tolerate RH down to 10–30%. Conidia of *Golovinomyces* spp. germinate best at 80–90% RH, but tolerate RH down to 20%. Optimal temperatures are 25°C for *L. taurica*, 18–24°C for *Golovinomyces* spp., and 15–20°C for *P. macularis*.

Disease development is favored by warm days without rain coupled with cool nights with saturated humidity (giving rise to dew). Under optimal conditions for fungal growth, conidia arise 1 week after infection and quickly create epidemics of secondary infections. Losses multiply as plants approach maturity. Low light intensity (indoors) or shaded areas (outdoors) increase disease severity, as does poor air circulation. Succulent plants treated with excess nitrogen suffer the greatest damage.

9.1.4 Cultural and mechanical control

Prevention is the best cure. Obtain clones from trusted sources, and place them in temporary quarantine for observation. Clones can be disinfected using plant tissue culture/explant methods (e.g., the Sigma-Aldrich protocol). Prune infected plants or discard them, and quarantine suspect plants.

Maintain high light intensity, and augment air flow. Slow oscillating fans (≤20 kph) promote air exchange adjacent to the phylloplane. The use of air filtration and purification systems can remove airborne conidia from indoor spaces. Managing relative humidity (RH) is essential. Ventilation expels moist air to the outside. Venting is not an option for growers who add CO_2 to growrooms; they must depend on dehumidifiers. Coordinate venting and dehumidification with diurnal transpiration cycles.

Rather than RH, **vapor pressure deficit** (VPD) has the greatest effect on host–parasite interactions (Jarvis *et al.*, 2002). VPD is a metric that measures the interaction of RH and temperature. VPD and RH have an inverse relationship: a high VPD means the air has a high capacity to hold water (i.e., low humidity), while a low VPD means the air is near saturation (i.e., high humidity). The optimal VPD for *C. sativa* growth is relatively high: 0.8–1.1 kPa in the vegetative stage, to 1.0–1.4 during early flowering, to 1.3–1.5 during late flowering (Purdy, 2018). Optimal VPD enhances plant vitality and discourages fungal growth. Cucumber plants suffer less *Podosphaera xanthii* damage when grown under high VPD rather than a low VPD regime (Itagaki *et al.*, 2014).

In outdoor settings, plant in wide rows that are oriented for optimal air circulation and sunlight penetration. Crops sown in dense stands must be monitored closely. Water plants in dry conditions, because drought stress increases disease. Irrigate at the soil level or use drip irrigation, and not overhead irrigation.

Indoor facilities with a history of powdery mildew should be completely emptied and sanitized, otherwise epidemics will arise in sequential crops. Remove all clippings, fallen leaves, and other debris, and decontaminate all equipment with hydrogen peroxide. Some facilities have a decontamination area near the entrance, where personnel change into work attire and sanitize their footwear.

Brais (2017) recommended a combination of air filtration and irradiation with ultraviolet light (UV-C, 100–280 nm) to reduce the inoculum load of powdery mildew conidia in *Cannabis* glasshouses. Scott and Punja (2020) exposed plants to UV-C by moving a hand-held light (CleanLight™ Pro) over and around plants, once daily for 28 days. This treatment reduced disease by 45.2% compared with controls.

Succulent plant growth induced by excess nitrogen predisposes plants to powdery mildew; use balanced nutrition (McPartland and Hillig, 2008). Dixon *et al.* (2022b) supplemented peat-based soilless growth media with silicon (wollastonite: calcium inosilicate, 24% Si) to control powdery mildew. Disease severity in the upper canopy was significantly reduced at 300 kg Si/ha, while 600 kg Si/ha was needed for the mid-canopy. Foliar sprays of potassium silicate (Silamol®) controlled powdery mildew at low disease pressures (Scott and Punja, 2022), but application of silicon to roots seems to be more effective.

No breeding for resistance has been done, but screening for disease has identified resistant cultivars (Cala *et al.*, 2019; Weldon *et al.*, 2020; Scott and Punja, 2020). Hops breeders selected for resistance to *P. macularis* within a decade of the pathogen arriving in the Pacific Northwest. Conversely, tobacco breeders have tried to select for resistance to *G. cichoracearum* since the 1920s (Lucas, 1975).

Powdery mildew has not been reported in India, so landraces in the land of monsoons appear to be resistant. In contrast, Stevens (1975) provided the first description of Afghani landraces in the USA, and he first bemoaned powdery mildew. Afghani ("Indica") landraces appear to be more susceptible than "Sativa" landraces of South Asian provenance—however, most drug-type strains have hybridized these landraces, and the hybrids are susceptible (McPartland *et al.*, 2000; McPartland and Hillig, 2008).

McKernan *et al.* (2020) searched for resistance genes in a resistant strain, "Jamaican Lion", reportedly "100% Sativa." They focused on the thaumatin-like protein (TLP) family. "Jamaican Lion" expressed increased copy numbers of one particular gene, *CsTLP1*, that was deleted in susceptible strains. McKernan cloned *CsTLP1* into *Escherichia coli*, which upregulated *β*-1,3-glucanase activity.

Mihalyov and Garfinkel (2021) discovered a resistance locus they named *PM1* by screening a Pacific Northwest germplasm collection for resistance to *G. ambrosiae*. The accession with greatest resistance, "PNW39," was not characterized, but their photograph suggests a broad-leafleted Afghani hybrid, which is surprising. Crossing "PNW39" with susceptible plants revealed that *PM1* was inherited in a dominant Mendelian fashion. ~~Soren~~ Seifi *et al.* (2025) discovered a second resistance locus, which they named *PM2*, by screening 510 *Cannabis* genotypes from across the globe. Conidiophore production was inhibited by 90% in a resistant line (WO3) compared with a susceptible line (AC). *PM2*, like *PM1*, was also inherited in a dominant Mendelian fashion.

Instead of resistance, Pépin *et al.* (2021) explored susceptibility, imparted by *Mildew Locus O* (*MLO*) genes. They searched for sequences resembling *Arabidopsis thaliana MLO1* to *MLO15* in the genomes of three drug-type plants and 'FINOLA'. The copy number varied: "CBDRx" (14 *CsMLO* genes), "Jamaican Lion" (*n* =17), "Purple Kush" (*n* =18), and 'FINOLA' (*n* =18). Inoculation of plants with *G. ambrosiae* resulted in transcriptional upregulation of two genes in particular, *CsMLO1* and *CsMLO4*. They searched for loss-of-function mutations of *CsMLO1* and *CsMLO4* in 32 *Cannabis* cultivars, and found none.

Stack *et al.* (2023) sequenced *CsMLO1* in a *G. ambrosiae* resistant strain ("FL 58") and a susceptible strain ("TJ's CBD"). *CsMLO1* in the resistant strain contained a 6.8 kb insertion in the *CsMLO1* sequence that caused improper mRNA splicing resulting in a premature stop codon. Despite the prediction that knockout of both *CsMLO1* and *CsMLO4* would be required for abolishing susceptibility, in "TJ's CBD" only *CsMLO1* knockout was required. The authors developed a PCR allele competitive extension (PACE) assay using *CsMLO1* sequences to distinguish resistant and susceptible genotypes.

Adhikary *et al.* (2022) analyzed transcript level responses of fiber-type 'X59' to infection by *G. ambrosiae*. As disease progressed, 910 genes upregulated, and 988 genes downregulated. Upregulated genes were related to *Mildew Locus O*, thaumatin-like protein, salicylic acid (particularly *PAD4* and *EDS1*), and transcription factors (WRKY and MYB).

9.1.5 Biological and chemical control

The superficial nature of powdery mildew fungi renders them susceptible to biological control agents, individually described in section 20.6. Anonymous (2008) controlled *Cannabis* powdery mildew with Serenade® (*Bacillus subtilis*), which is also recommended by state agencies (California DPR, 2017a; Colorado DOA, 2018). Pscheidt and Ocamb (2018) recommended Actinovate® AG (*Streptomyces lydicus*). Health Canada (2018) approved Prestop® (*Clonostachys rosea*). EcoSwing® (*Swinglea glutinosa* extract) is approved by New York.

Scott and Punja (2020) compared biological and chemical controls, applied weekly, to prevent and control powdery mildew. Best control was achieved with potassium bicarbonate (MilStop®) or Regalia® (an extract of *Reynoutria sachalinensis*). Next came *Bacillus subtilis* strain QST 713 (Rhapsody®) and *Bacillus amyloliquefaciens* strain F727 (Stargus®). Lower efficacy was observed with Actinovate® (*Streptomyces lydicus*), boric acid, orthosilicic acid, hydrogen peroxide, and neem oil.

Cala *et al.* (2021) compared biological and chemical controls, applied weekly, to control powdery mildew. Best control was achieved with *Bacillus mycoides* isolate J (LifeGard®), followed by a three-way tie between potassium silicate (Sil-MATRIX®), LifeGard® alternating with *Bacillus amyloliquefaciens* strain D747 (Double Nickel®), and Double Nickel® alone.

Akinrinlola and Hansen (2002) tested eight organic fungicides and biocontrols, compared with an off-limits systemic pesticide (Luna® Experience). Plants were inoculated with *G. ambrosiae* conidia 1 day before or 7 days after the first fungicide spray; additional applications were made at weekly intervals. Compared with control plants, three fungicides reduced disease by 96–100%: Luna® Experience, MilStop®, and Bonide® sulfur. Four reduced disease by 86–95%: Sil-MATRIX®, Regalia®, Exile™ (potassium salt of fatty acids), and Cinnerate™ (cinnamon oil and potassium oleate). Two reduced disease by 76–85%: Stargus® (*Bacillus amyloliquefaciens* strain F727) and Defguard™ (*B. amyloliquefaciens* strain D727).

In other crops, powdery mildew has been controlled with *Ampelomyces quisqualis* (see Fig. 9.3), *Sporothrix flocculosa*, *Trichoderma harzianum*, and *Verticillium lecanii*. All kinds of plant extracts have been tested against powdery mildew fungi. Masheva *et al.* (2014) tested five plant extracts to control *Podosphaera xanthii*, and their extract made from *C. sativa* showed the least efficacy.

Sulfur burners were recommended by Anonymous (2008), but they are illegal in USA glasshouses, can harm plants, and are hard on equipment. Vaporized sulfur kills predatory mites, leading to a surge in spider mites (James and Prischman, 2010). Health Canada does permit the use of vaporized sulfur to control powdery mildew, specifically Ascend™ Vaporized Sulphur by Agrotek. Mahieu *et al.* (2019) recommended a similar product in Holland.

Wettable sulfur is recommended by many (California DPR, 2017a; Colorado DOA, 2018; Health Canada, 2018; Pscheidt and Ocamb, 2018). Mahieu *et al.* (2019) recommended a range of wettable sulfurs for powdery mildew. The same authorities recommended potassium bicarbonate, sodium lauryl sulfate, horticultural oils (either canola or mineral oil), and neem oil (or its extracted limonoid, azadirachtin). Potassium bicarbonate is particularly useful. It is mildly fungitoxic and prevents infection by altering the pH of leaf surfaces.

Forensic analyses have found myclobutanil residues in samples from licensed dispensatories in California, Oregon, Colorado, and Canada (McPartland and McKernan, 2017). Heating myclobutanil produces hydrogen cyanide; it is not approved in tobacco. Myclobutanil is explicitly banned for use on *Cannabis* (California DPR, 2017b; Colorado DOA, 2018; Health Canada, 2018). Myclobutanil residues are the most common reason that Colorado has recalled cannabis products.

9.2   Gray mold

The gray mold fungus, *Botrytis cinerea*, attacks everything above the soil line. Its primary notoriety lies in bud rot. *B. cinerea* can also attack leaves (Thiessen *et al.*, 2020), or cause stem cankers, which Rataj (1957) considered "the most dangerous disease of this plant." *B. cinerea* can destroy seedings ("damping off"), which Flachs (1936) considered the gravest threat. *B. cinerea* can be a problem in clonal propagation chambers (McPartland and Thiessen, 2022).

Muth (1906) reported that seed-bearing plants "were attacked by a botrytis, presumably *Botrytis cinerea*." Seeds harvested from infected plants did not germinate. *B. cinerea* continues to attack buds after harvest if stored improperly. The fungus even ramifies through poorly processed *hashīsh* (section 18.3). Adding insult to injury, *B. cinerea* can metabolize THC (Binder and Meisenberg, 1978).

Gray mold afflicts fiber-, seed-, and drug-type plants, outdoors and indoors. The fungus attacks many crops around the world. It is ubiquitous in vineyards; some types of wine are made from "botrytized" grapes: *Sauterne, Trockenbeerenauslese,* and *Tokay* (*Aszú*). The conidia of *B. cinerea* are everywhere, including in air samples in bud trimming rooms (Couch *et al.*, 2017; Victory *et al.*, 2018). Workers exposed to spore-laden air may develop allergies, hypersensitivity pneumonitis, and even pulmonary *Botrytis* infections.

B. cinerea thrives in high humidity (65–70% RH) and cool to moderate temperatures (17–24°C or 62–75°F) (Mahmoud *et al.*, 2023). These conditions are pervasive in maritime climates (e.g., the Netherlands, the Pacific Northwest, New England). In these climates *B. cinerea* can reach epidemic proportions and destroy a *Cannabis* crop within a week (Barloy and Pelhate, 1962; Frank, 1988).

9.2.1   Symptoms and signs

Gray mold often looks brown instead of gray. The disease predominates in two places—flowering tops and stalks. Flower infestations tend to arise in grain- and drug-type plants, especially in large, moisture-retaining female buds. Infestations often begin *within* buds, so initial symptoms are not visible, unless the tissues are pried apart. Fan leaves wilt and turn yellow, then pistils begin to brown, followed by floral leaves, and the whole top turns brown (Fig. 9.5).

Fig. 9.5. Bud rot caused by *Botrytis cinerea*: **A**. field condition; **B**. individual plant (courtesy C. Vogl)

In high humidity, flowering tops become enveloped by a gray fuzz: a nimbus of conidia, liberated in a gray cloud by the slightest breeze.

Stalk rot is common in fiber-type plants (Patschke *et al.*, 1997), and *B. cinerea* was first reported on stalks (Hazslinszky, 1877–78). Diseased flowering tops were not reported until 60 years later by Trunov (1936), and he said the disease spread to flowering tops from stalks. Stalks lose color and become gray, and then develop brown cankers. In high humidity, cankers become covered by a gray-brown mantle of conidia (Fig. 9.6). The fungus releases pectinolytic and cellulolytic enzymes, and stalks may snap at canker sites. If a canker encircles the stem, everything above it wilts and dies. After the plants dies, sclerotia may form within its stalk.

Flachs (1936) first described *B. cinerea* as a cause of damping off. He photographed hemp seedlings covered by fuzzy *B. cinerea* mycelium. Botrytis blight affects vegetative plants before they've flowered. It begins as a yellowing of leaves in the upper canopy, followed by shriveling and necrotic lesions (Punja *et al.*, 2019). In high humidity a gray fuzzy growth covers leaves and branches (Thiessen, 2019; Gauthier, 2020a).

9.2.2   Causal organisms

In the heyday of morphologists splitting species, the genus *Botrytis* included around 100 species. Lumpers in the 20th century consolidated the list to 25 *Botrytis* spp. Since then, thanks to molecular hair-splitting, a dozen new *Botrytis* spp. have been named in the 21st century. In parallel fashion, four *Botrytis* species were reported on *Cannabis* a century ago, subsequently consolidated to one species (McPartland *et al.*, 2000), and now there are two or three.

Fig. 9.6. Stalk rot caused by *Botrytis cinerea*: **A**. macro view (G. Grassi); **B**. micro view (S. Koike)

9.2a. *Botrytis cinerea* Persoon 1794, Leotiomycetes, Sclerotiniaceae

=*Botrytis infestans* (Hazslinszky) Saccardo 1887, =*Botrytis felisiana* Massalongo 1899
Sexual state: *Botryotinia fuckeliana* (deBary) Whetzel 1945, ≡*Sclerotinia fuckeliana* (deBary) Fuckel 1870
Description: Conidiophores upright, gray-brown, branching near the apex, 5–22 μm in diameter. Conidia borne on conidiophore apex in botryose clusters, hyaline to yellow-gray (gray en masse), aseptate, round to ovoid, 8–14 × 6–10 μm. A botryose cluster is illustrated in Fig. 9.7A as a stereopair: view with a stereoscope or hold about 20 cm away, cross your eyes and align double vision into a single 3-D image. Microcondidia are uncommon, *Myrioconium*-like, arising from flask-shaped phialides along the mycelium, hyaline, oval, aseptate, 2.0–2.5 μm in diameter.

Apothecia arise from sclerotia on a 3–10 mm stalk, topped by a yellow-brown disc; disc flat to slightly convex, 1.5–5 mm in diameter, capped by a single layer of asci and paraphyses. Asci cylindrical, with long tapered stalks, 8-spored, 120–140 × 8 μm. Paraphyses hyaline, septate, filiform. Ascospores aseptate, usually uniguttulate, uniseriate, hyaline, ovoid to ellipsoid, 8.5–12 × 3.5–6 μm. Sclerotia hard, black, rough, plano-convex, irregularly round to elongate, 1–15 mm long (averaging 5 mm on *Cannabis* according to Flachs, 1936), sometimes arising in chains; cross-section of sclerotium reveals a thin black rind covering a hyaline interior.

Comments: *B. cinerea* attacks 586 genera of plants (Garfinkel *et al.*, 2019). Hazslinszky (1877–78) named *Polyactis infestans* on Hungarian hemp stems. Saccardo (1886) renamed Hazslinszky's species *Botrytis infestans*. Sorauer (1888) encountered *B. cinerea* on hemp in Germany, which he thought was identical

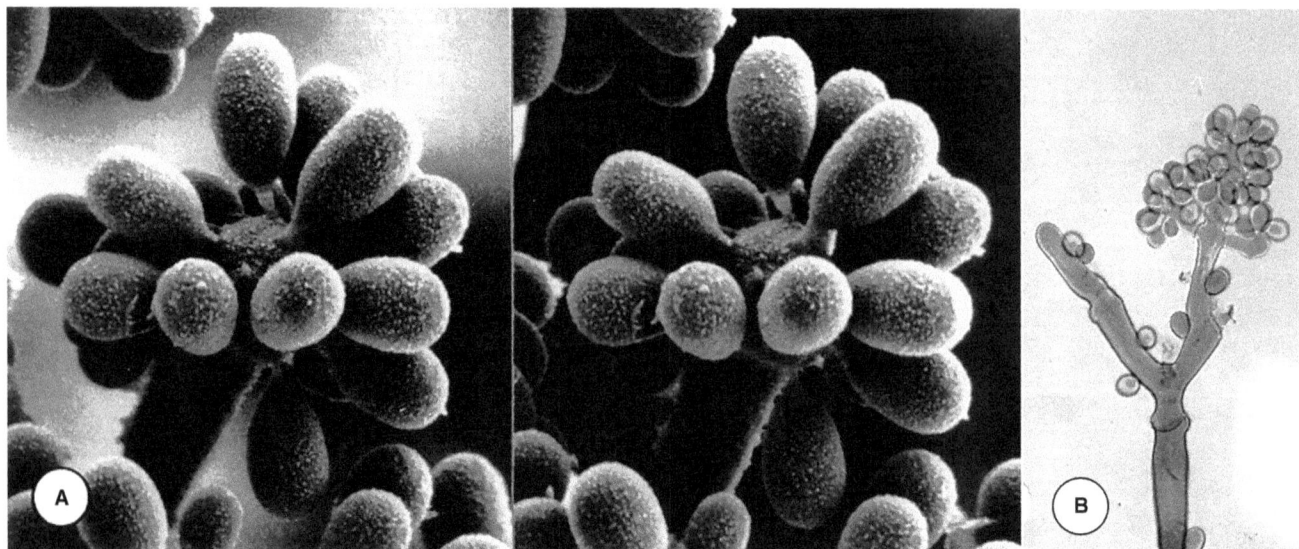

Fig. 9.7. Conidiophores and conidia of *Botrytis cinerea*: **A.** stereopair, SEM ×1250 (courtesy Merton Brown and Harold Brotzman); **B.** LM ×260 (courtesy Bud Uecker)

to *B. infestans*. Behrens (1891) considered *B. cinerea* a serious hemp disease, and connected it with the sexual state *Botryotinia fuckeliana*. Massalongo (1899) coined *Botrytis felisiana* for a fungus on Italian hemp stems. Behrens (1902) isolated *Botrytis vulgaris* from German hemp stems. Saccardo and Sydow (1902) noted a resemblance between *B. vulgaris* and *B. felisiana*. Smith (1900) synonymized *B. vulgaris* under *B. cinerea*. Lindau (1907) synonymized *B. felisiana* under *B. infestans*.

Rapaić von Ruhmwerth (1914) isolated *B. cinerea* from hemp stems, under which he synonymized *B. felisiana* and *B. infestans*. Old nomenclature lingered. In Russia, Gitman and Boytchenko (1934) said *B. infestans* caused stem disease, and *B. cinerea* infested leaves and tops. In Italy, Ferraris (1935) employed the taxa *B. vulgaris* and *B. cinerea* for flower and stem pathogens, respectively. In French hemp, Barloy and Pelhate (1962) considered *B. infestans* the cause of bud rot and stem canker; they differentiated it from *B. cinerea* by conidial size and color.

9.2b. *Botrytis pseudocinerea* Walker *et al.* 2011

Sexual state: *Botryotinia pseudofuckeliana* Walker *et al.* 2011
Description: *B. pseudocinerea* cannot be discriminated from *B. cinerea* by conidiophore and conidium morphology, or mycelial growth *in vitro*, sporulation rate, or germination rate. Ecological parameters (infection rates on tomato and bean, temperature optima) do not significantly differ. The two species were distinguished by gene sequences (*G3PDH*, *HSP60*, *MS547*) and fenhexamid fungicide sensitivity (*B. pseudocinerea* is resistant). Sexual crosses between *B. pseudocinerea* and *B. cinerea* individuals were not successful, demonstrating a lack of gene flow between the species.
Comments: *B. pseudocinerea* was discovered in French vineyards, living in sympatry with *B. cinerea* (Walker *et al.*, 2011). *B. pseudocinerea* was more abundant in springtime, on floral debris of grapes and blackberries. A molecular clock analysis of

gene sequences estimated that *B. pseudocinerea* and *B. cinerea* diverged 7–18 million years ago.

Garfinkel (2020) isolated both *B. pseudocinerea* and *B. cinerea* from CBD-type plants in Oregon suffering from bud rot and dieback. Out of 23 pure cultures, 20 were *B. cinerea*, two were *B. pseudocinerea*, and one was a third *Botrytis* sp. unrelated to *B. pseudocinerea* or *B. cinerea*. Completion of Koch's postulates demonstrated pathogenicity by all three isolates, with symptoms arising after 7 days. Punja and Ni (2021) recovered *B. pseudocinerea* from indoor crops in British Columbia.

9.2.3 Differential diagnosis

In flowering tops, gray mold may be confused with brown blight caused by *Alternaria alternata* (see section 11.10). *A. alternata* and *B. cinerea* look very different under the microscope. A microscope may also be needed to differentiate gray mold from Fusarium flower blight, caused by *Fusarium graminearum* (Thiessen, 2019). Powdery mildew looks like a white mold, not a gray mold.

On stems, many fungi cause cankers (e.g., anthracnose fungi and *Fusarium* spp.), but none grow as fast or thick as gray mold. Sclerotia formed by *B. cinerea* can be confused with sclerotia formed by *Sclerotinia sclerotiorum*, the cause of hemp canker. Generally, *B. cinerea* sclerotia arise within stalks, whereas *S. sclerotiorum* sclerotia arise within stalks or on the exterior surface. *B. cinerea* sclerotia are smaller than *S. sclerotiorum* sclerotia. If sclerotia produce apothecia, the two are easily distinguished under a microscope.

9.2.4 Disease cycle and epidemiology

Unlike fungi that cause powdery mildew, which are biotrophic and slowly drain nutrients from the host, *B. cinerea* is necrotrophic.

Necrotrophic pathogens first destroy host cells, then feed on the contents. As an assault strategy, *B. cinerea* triggers a host defense mechanism—programmed cell death. Then it exudes enzymes that destroy host cells and feeds on the contents (Williamson *et al.*, 2007).

Van Kan *et al.* (2014) described *B. cinerea* as "an endophyte gone rogue." It is capable of colonizing plants, without causing any disease or stress symptoms. When conditions are right, often during flowering, the fungus mounts an assault. *B. cinerea* turns the host's defense mechanism against it, and also squelches plant immune responses: *B. cinerea* transfers small RNA (sRNA) into host cells, which binds to genes involved with immune responses, and silences them (Weiberg *et al.*, 2013).

B. cinerea overwinters within dead stalks as sclerotia (Fig. 9.8A) or dormant mycelium (Fig. 9.8B). Pietkiewicz (1958) and Noble and Richardson (1968) documented seed-borne infection (Fig. 9.8C), which gives rise to infected seedlings the following spring (Fig. 9.8D). Sclerotia may persist for years in soil. Spring rains induce sclerotial germination, usually producing conidia (Fig. 9.8E), or apothecia with ascospores (Fig. 9.8F). Sclerotia or dormant mycelium may give rise to hyphae, which penetrate seedlings directly (Fig. 9.8G).

Conidia (rarely ascospores) are blown or splashed onto plants. High humidity or dew is needed for conidial germination. Optimal conditions for conidial *production* are 15–20°C and RH >65%; optimal temperatures for conidial *germination* are 20–24°C and RH >65% (McPartland and Thiessen, 2022). Conidia germinate on seedlings, and germ tubes directly penetrate the thin epidermis. On older plants the fungus infects epidermis damaged by insects (especially budworms), rough handling, frost, or even improper fertilization. During the summer, the fungus forms new conidia, which give rise to secondary infections. Cycles of secondary infections build to epidemics in mature plants (Fig. 9.8H).

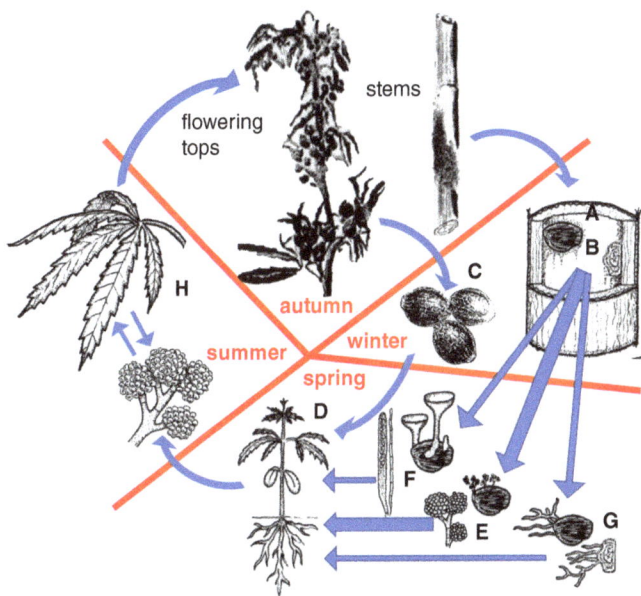

Fig. 9.8. Disease cycle of *Botrytis cinerea*: see text for description. Redrawn from Barloy and Pelhate (1962) and McPartland *et al.* (2000).

Fiber-type plants become more susceptible after canopy closure, about 60 days after sowing. Gray mold can be severe in high-density stands of fiber-type hemp. De Meijer *et al.* (1995) sowed four fields at rates of 20, 40, 80, and 140 kg seed/ha. These rates initially produced 104, 186, 381 and 823 seedlings/m². Gray mold decimated the high-density stands. By harvest, all fields yielded about 100 plants/m². Crop yield (stalk biomass) was actually greatest in crops sown at the lowest seedling rate.

Drug-type plants are most susceptible during flowering. *B. cinerea* often colonizes senescent flowers, and from this foothold it invades the rest of the plant. Watch for gray mold during periods of high humidity (>65% RH) and cooling temperatures, especially when dew point is reached. Indoors, planting density affects plants; 2 plants/m² suffered significantly more bud rot than 1 plant/m² (Mahmoud *et al.*, 2023).

9.2.5 Cultural and mechanical control

For prevention indoors, keep relative humidity <60%, keep light intensity high, and the temperature warm (>25°C) to inhibit conidial germination. For prevention outdoors, utilize row spacing as wide as commercially feasible. Neenan (1969) tested seeding rates in Ireland (whose maritime climate is perfect for *B. cinerea*) and disease incidence dropped off at seeding rates of ≤50 lb/acre (56 kg/ha). Air circulation can be improved outdoors by clearing away weeds, and improved indoors with ventilation fans. Avoid excess nitrogen and phosphorus, add calcium, and neutralize acid soils to enhance calcium absorption.

Sanitation: remove senescing plant debris (especially dead fan leaves) throughout the season. Trimming fan leaves also improves air circulation. Trimming should be done on a hot, sunny day, practicing good scissors sanitation—use sharp scissors, and dip in isopropyl alcohol between cuts. Dab prune sites with alcohol, or with biocontrol organisms. Poor scissors sanitation creates wounds that provide portals of infections.

Trunov (1936) noted that *B. cinerea* attacks males after they've pollinated (and lose vigor), and then spreads to female tops. Harvest males if possible. Female flowers should be harvested while resin glands are white or amber, not brown. Brown and wilted pistils are prime *B. cinerea* fodder. If scouting reveals *B. cinerea*, or the weather forecast looks bad, it's better to harvest early than never.

Breeding resistant plants is the ultimate solution. Fröier (1956) bred a Swedish monoecious fiber-type cultivar, 'Mona', with greater resistance to *B. cinerea* than existing German varieties. Van der Werf *et al.* (1995) noted that Hungarian 'Kompolti Hybrid TC' and French monoecious 'Fédrina 74' were more susceptible to gray mold than other fiber varieties. Dempsey (1975) said the Russian cultivars 'JUS-1' and 'JUS-7' were resistant, but these cultivars are no longer available (de Meijer, 1995).

Among drug-type plants, Clarke (1987) noted that the dense, tightly packed buds of Afghani landraces tend to hold moisture and easily rot. Afghani landraces evolved in arid conditions and have no resistance to gray mold. This unfavorable trait is often expressed in hybrids that are only a small percentage of *afghanica* heritage (McPartland *et al.*, 2000). In comparison,

"Sativa" landraces (Indian, Mexican, Colombian, Thai plants) are relatively resistant to bud rot.

Jin *et al.* (2021) analyzed 21 "Sativa–Indica" hybrid strains in a search for phenotypic characters that might correlate with a plant's THC/CBD ratio. Several phenotypic characters correlated: "wide leaflets, large and compact inflorescences, dense and resinous trichomes, and *B. cinerea* susceptibility." Mahmoud *et al.* (2023) determined that *B. cinerea* disease incidence correlated with apical inflorescences that were larger, heavier, and denser. Shape mattered: apical inflorescences that congealed into a single solid cola (typical of *afghanica*) were more susceptible than thready apical inflorescences (typical of Thai landraces). Lastly, disease incidence was higher in plants whose apical inflorescences had a greater number of "inflorescence leaves" (sometimes called subtending floral bracts, typical of *afghanica*).

The fight against gray mold continues after harvest. Flowering tops must be dried in dark rooms. Ultraviolet (UV-B) light is required for *B. cinerea* sporulation (Sasaki and Honda, 1985). Provide good air circulation, low humidity, and temperatures >25°C or even >30°C (Mahmoud *et al.*, 2023). Propane heaters and dehumidifiers are helpful. Rain-soaked crops pulled into drying sheds need higher temperatures for a few days and more ventilation fans. Websites offer "desperation methods" for quick-drying suspicious materials, such as drying buds in sunlight, or in ovens.

Mechanical wet trimming can be problematic; Punja (2021a) showed that fungal counts tripled in wet trimmed buds. In a survey of air samples, the relative abundance of *B. cinerea* conidia was highest in bud-stripping and hand-trimming rooms (59%), compared with 19% in drying rooms, and 18% in glasshouse air (Green *et al.*, 2018). Methods for monitoring the levels of airborne conidia were described by Mahmoud *et al.* (2023). Jerushalmi *et al.* (2020b) used ionizing radiation, e-beam irradiation, and cold plasma treatment to kill *B. cinerea* inoculum in harvested material (see section 18.3.8). Ruchlemer *et al.* (2015) assayed the effects of these treatments on THC content (section 18.3.8).

9.2.6   Biological and chemical control

Preventive sprays of biocontrols are useful, and individually described in section 20.6. *Bacillus subtilis* (Serenade®) is recommended by state agencies (California DPR, 2017a; Colorado DOA, 2018). *Clonostachys rosea* (Prestop®) was approved by Health Canada (2018), and it is used in Holland (Mahieu *et al.*, 2019). *Bacillus amyloliquefaciens* strain F727 (Stargus®) effectively controls disease (Scott and Punja, 2020), and the product includes *B. cinerea* on its label.

Balthazar *et al.* (2020, 2022b) tested the ability of PGPRs (plant growth-promoting rhizobacteria) to reduce *B. cinerea* in *Cannabis*. They tested *Bacillus velezensis*, *B. subtilis*, *Pseudomonas synxantha*, or *P. simiae*. First, Balthazar *et al.* (2020) inoculated soil with these PGPRs, which did not significantly control gray mold. Subsequently, Balthazar *et al.* (2022b) sprayed aboveground parts of plants with these PGPRs, and some performed well (see Fig. 13.6).

Punja and Ni (2021) tested *Trichoderma asperellum* and *Bacillus amyloliquefaciens* against isolates of *B. cinerea* that infested *Cannabis*, both *in vitro* and *in vivo*. When applied to detached inflorescences 48 h prior to pathogen inoculation, they visibly reduced pathogen development after incubation under high humidity for 7 days. Cala *et al.* (2021) inoculated plants with *B. cinerea* and treated them weekly with *Bacillus mycoides* isolate J (LifeGard®), potassium silicate (Sil-MATRIX®), and *Bacillus amyloliquefaciens* strain D747 (Double Nickel®). Disease severity was no different than in untreated controls.

Dixon *et al.* (2022a) screened eight biocontrols and three biorational pesticides for activity against *B. cinerea* and other fungal pathogens. They published a wealth of data, based on three *in vitro* screening methods, as follows.

1. *Amended media assays* added fungicidal products to potato dextrose agar, and measured the inhibition of fungal growth (Cueva®, copper soap; Procidic® 2, citric acid; SilMatrix®, potassium silicate; Trilogy®, extract of *Azadirachta indica*; LifeGard® WP, *Bacillus mycoides* isolate J).
2. *Exposure timing assays* were used with products with metabolite-producing active ingredients, and measured the inhibition of fungal spore germination, after exposure for 0 h, 2 h, or 4 h (Bio-Tam®, *Trichoderma asperellum* + *T. gamsii*; Cease®, *Bacillus subtilis* strain QST713; Double Nickel®, *Bacillus amyloliquefaciens* strain D747; Regalia®, extract of *Reynoutria sachalinensis*; Stargus®, *Bacillus amyloliquefaciens* strain F747 plus fermentation media).
3. *Inhibition zone assays* evaluated products using a dual-plating method to measure zone of fungal growth inhibition (Botrystop®, *Ulocladium oudemansii* strain U3; Cease®, *Bacillus subtilis* strain QST713; Double Nickel®, *Bacillus amyloliquefaciens* strain D747; Stargus®, *Bacillus amyloliquefaciens* strain F747 plus fermentation media).

Dixon and colleagues evaluated products in these three assays, applied at different rates, variable exposure times, and comparisons with Custodia® (azoxystrobin + tebuconazole fungicide). We simplify their results in Table 9.1. Cueva® (copper soap) and LifeGard® (*Bacillus mycoides*) produced the best results against *B. cinerea*.

Spray outdoor plants with potassium bicarbonate after rainfall. It is mildly fungitoxic and prevents infection by altering the pH of leaf surfaces. Bordeaux mixture knocks back *B. cinerea* in early stages, and prevents infections until the next rainfall washes the mixture off plants. *B. cinerea* has developed resistance to most synthetic fungicides.

9.3   Fusarium flower blight

We introduce the genus *Fusarium* here; the genus also features in Chapter 10 (stalk cankers), Chapter 11 (root rots, wilts, blights), and Chapter 12 (damping off). Dean *et al.* (2012) composed a "Top 10" list of the world's worst fungal pathogens, and *Fusarium graminearum* and *Fusarium oxysporum* were ranked fourth and fifth, respectively. Both species infest *Cannabis*.

Fusarium taxonomy and nomenclature have been difficult for centuries. Link (1809) described the first species, *Fusarium*

Table 9.1. Effects of biocontrols and biorational pesticides, abstracted from Dixon *et al.*, 2022a[a]

	Botrytis cinerea	*Bipolaris gigantea*	*Cercospora flagellaris*	*Septoria cannabis*
Bio-Tam®	I: 0, E: ++	I: 0	I: 0, E: 0	I: 0, E: 0
Botrystop®	I: 0	I: +	I: +	I: +
Cease®	I: +, E: ++	I: ++	I: +, E: +	I: +, E: +
Cueva®	A: +++	A: +++	A: +++	A: +++
Double Nickel®	I: +, E: ++	I: ++	I: +, E: 0	E: 0
LifeGard®	A: +++, E: ++	A: +++	A: ++, E: 0	A: +++, E: 0
Procidic® 2	I: 0, A: ++, E: 0	I: 0, A: ++	I: 0, A: ++, E: 0	I: 0, A: ++, E: +++
Regalia®	E: +		E: 0	E: +
SilMatrix®	A: +, E: 0	A: ++	A: +, E: 0	A: +, E: 0
Stargus®	I: +, E: +	I: ++	I: +, E: +	I: +, E: 0
Trilogy®	A: +,	A: +	A: +	A: +

[a]See text for full names of biocontrols and biorational pesticides
Assay abbreviations: A, amended media assay; E, exposure time assay; I, inhibition zone assay
0, no different than control; +, slight inhibition; ++, more inhibition; +++, most inhibition

roseum. After Link died, it was discovered that his herbarium contained three different species under that name (Gams *et al.*, 1997). *F. roseum* appears in *Cannabis* literature; it's always difficult to tell what the authors actually had on hand.

Classifying groups of *Fusarium* spp. is even more torturous. The challenge attracted Hans Wilhelm Wollenweber (1879–1949), who wrestled with *Fusarium* from 1908 to 1943. He was a Berlin mycologist who worked at the USDA in Washington DC from 1911 to 1914 (where he first wrote about *Fusarium* spp. on *Cannabis*). Returning to Germany, he was drafted and served for 4 years, becoming an anti-war pacifist. He preferred hiking (Reinking, 1950).

Initially Wollenweber recognized six "Types" of *Fusarium*, such as "*Fusarium solani*-Typ" (Appel and Wollenweber, 1910). Subsequently he divided the genus into six "Sections" based on the morphology of conidia and chlamydospores: *Elegans, Martiella, Discolor, Gibbosum, Roseum*, and *Ventricosum* (Wollenweber, 1913).

This system—expanded to 16 sections—culminated in a book, *Die Fusarien* (Wollenweber and Reinking, 1935). They were lumpers, for example reducing section *Martiella* to just *F. solani, F. javanicum*, and *F. coeruleum*, into which they lumped dozens of other *Fusarium* taxa as varieties of these three species. They ended up with 65 species, 55 varieties, and 22 forms. Prior to them, over 1000 species, varieties, and forms of *Fusarium* had been named. The record goes to *F. lateritium*, into which Wollenweber and Reinking subsumed 133 taxonomic synonyms (for the anamorph and teleomorph).

The baton passed to William C. Snyder (1904–1980) and Hans Nicholas Hansen (1891–1960), whose adjoining laboratories at UC Berkeley had a door connecting them. Snyder went to Berlin and studied under Wollenweber for a year. Snyder and Hansen (1940) considered Wollenweber's system too complex, "not sufficiently usable for the average worker who has need for it," and they offered a stripped-down version, beginning with section *Elegans* (see section 11.4). They continued the lumping process, collapsing all of section *Martiella* into one species, *F. solani*, with five forms (Snyder and Hansen, 1941). A series of articles (Snyder and Hansen, 1940, 1941, 1945) reduced

Wollenweber's 65 species to just nine. Noviello and Snyder (1962) published a seminal paper on *Fusarium* wilt of *Cannabis*.

Paul E. Nelson (1927–1996) established the Fusarium Research Center at Penn State. He published 140 papers on *Fusarium* as well as several books. Nelson *et al.* (1983) considered Snyder's system an oversimplification; they recognized 30 *Fusarium* spp. Although Nelson never published on *Cannabis*, he helped us identify *F. graminearum* shortly before he died (McPartland and Cubeta, 1997). This brief history of *Fusarium* taxonomy is elaborated by Nelson (1991).

The advent of DNA-based phylogenetics and "one fungus, one name" have thrown nomenclature into a tizzy (Geiser *et al.*, 2013). The clades in phylogenetic trees are now named **species complexes**, of which there are either 18 (Crous *et al.*, 2021) or 23 (Geiser *et al.*, 2021), depending on gene sequences and phylogenetic software.

Kerry L. O'Donnell (1946–) has utilized the USDA's NRRL culture collection in Peoria (*ca.* 93,000 strains of fungi and bacteria) to conduct many phylogenetic studies. O'Donnell (1996, 2000) applied a multilocus approach (LSU rRNA, ITS, *tub2*, *EF-1α*) to section *Martiella*. He recognized about 50 phylogenetic segregates, collectively called the ***Fusarium solani* species complex** (FsSC). In total he recognized about 300 phylogenetically distinct *Fusarium* spp. We've had the opportunity to work with him on *Cannabis* diseases (O'Donnell *et al.*, 2022).

In a review of *Fusarium* spp. that infest *Cannabis*, Gwinn *et al.* (2021) recognized 16 *Fusarium* spp. in six of the 23 species complexes proposed by Geiser *et al.* (2021). Here we do the same, expanding the list to 20 *Fusarium* spp. in eight species complexes (Table 9.2).

The incidence of Fusarium flower blight is increasing. It was only mentioned in passing as "black rot" by McPartland *et al.* (2000). In North Carolina, the disease was first called flower rot (Thiessen, 2019), then flower blight (Cochran *et al.*, 2020), where it represented 20% of all disease problems (Thiessen *et al.*, 2020). The disease was called bud rot in British Columbia (Punja *et al.*, 2019) and Tennessee (Hansen *et al.*, 2020). In Kentucky, the disease was called head blight (Gauthier, 2020a), and flower blight (Yulfo-Soto *et al.*, 2022).

Table 9.2. *Fusarium* spp. causing *Cannabis* diseases, organized by species complexes

Species complex	Chapter 9 Flower blight	Chapter 10 Stalk canker	Chapter 11 Root rots, wilts, blights	Chapter 12 Damping off
Fusarium sambucinum	F. graminearum F. sporotrichiodes	F. graminearum F. sporotrichiodes F. sambucinum F. culmorum	F. brachygibbosum	
Fusarium incarnatum-equiseti	F. equiseti		F. equiseti	
Fusarium solani	F. solani		F. solani, F. haematococca, F. javanicum F. lichenicola, F. petroliphilium, F. keratoplasticum, F. falciforme	F. solani
Fusarium fujikuroi	F. fujikuroi	F. fujikuroi F. proliferatum		
Fusarium tricinctum		F. avenaceum	F. tricinctum	
Fusarium oxysporum			F. oxysporum	F. oxysporum
Fusarium redolens			F. redolens	
Fusarium lateritium			F. lateritium	

9.3.1 Symptoms and signs

Symptoms can begin just 7 days after artificial inoculation (Yulfo-Soto *et al.*, 2022). Fan leaves develop dark, necrotic spots, surrounded by chlorotic halos. Entire leaves turn yellow and wilt. Flowers turn brown-black, with wilting of floral leaves and bracts. In humid conditions, abundant white to pink mycelium enshrouds the flowers (Fig. 9.9).

9.3.2 Causal organisms

9.3a. *Fusarium graminearum* Schwabe 1838, Sordariomycetes; Nectriaceae

Comments: *F. graminearum* is better known as a cause of stem cankers. See section 10.4 for a full description of this fungus. High humidity is required for the fungus to move into leaves and flowering tops. It sporulates rapidly, leading to epidemics of secondary infections. *F. graminearum* was identified as the cause of Fusarium flower blight in Illinois (Fig. 9.9B, McPartland *et al.*, 2000), North Carolina (Fig. 9.9A) (Thiessen, 2019; Cochran *et al.*, 2020; Thiessen *et al.*, 2020), Kentucky (Gauthier, 2020a; Yulfo-Soto *et al.*, 2022; Smith *et al.*, 2022), and Tennessee (Hansen *et al.*, 2020).

Voronin (1890, 1891) identified "*F. roseum*" infecting oats, other grasses, and hemp in the Russian Far East. The hemp pathogen was subsequently identified as *F. graminearum* (Vakhrusheva and Davidian, 1979). Ferri (1961b) demonstrated that "*F. roseum*" colonized hemp seeds. Yulfo-Soto *et al.* (2022) identified the seed pathogen as *F. graminearum*. Consumption of "*F. roseum*" infected seeds gave rise to "drunken bread" syndrome (Voronin, 1891). The syndrome is caused by trichothecene mycotoxins, such as trichothecene T-2, zearalenone, and deoxynivalenol (Gagkaeva *et al.*, 2021). An unspeciated *Fusarium* spp. caused bud blight in New York (Bergstrom *et al.*, 2019, 2020), and seeds harvested from the plants were contaminated

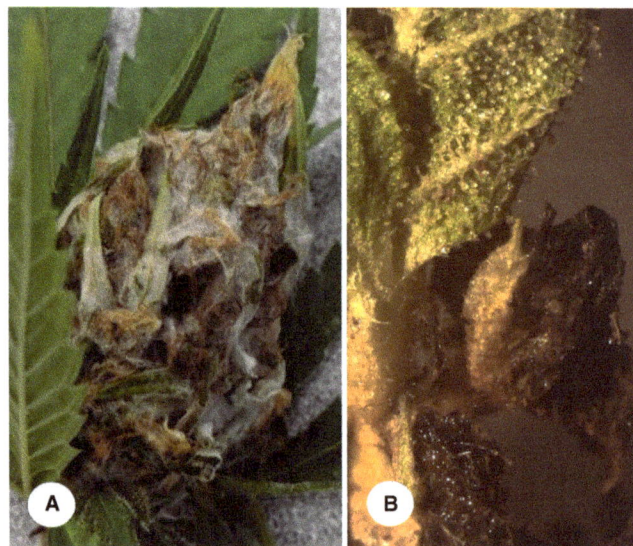

Fig. 9.9. Fusarium flower rot, caused by *Fusarium graminearum*: **A.** macro view (L. Thiessen); **B.** micro view

with deoxynivalenol in excess of 7 ppm—exceeding the FDA advisory of 1 ppm for food. Narváez *et al.* (2020) analyzed ten CBD gelatin capsules, and detected T-2 or zearalenone in 60% of them. Consuming small amounts of zearalenone or deoxynivalenol ("vomitoxin") can cause nausea, vomiting, diarrhea, headache, chills, and convulsions (Rippon, 1988).

9.3b. *Fusarium solani* (Martius) Saccardo 1881

Comments: *F. solani* is better known as a cause of root rot; see section 11.3 for a full description of this fungus. *F. solani* was isolated from leaves and flowers by Punja *et al.* (2019), Gauthier (2020a), and Jerushalmi *et al.* (2020a).

9.3c. *Fusarium equiseti* (Corda) Saccardo 1886

Comments: See section 11.3 for a full description of this fungus. *F. equiseti* is a soil-borne fungus, and causes head blight disease in wheat, barley, and maize. *F. equiseti* was isolated from leaves and flowers in British Columbia (Punja *et al.*, 2019), North Carolina (Cochran *et al.*, 2020; Thiessen *et al.*, 2020), Tennessee (Hansen *et al.*, 2020), Kentucky (Smith *et al.*, 2022), and Israel (Jerushalmi *et al.*, 2020a).

9.3d. *Fusarium sporotrichiodes* Sherbakoff 1915

Comments: *F. sporotrichiodes* was recovered from flowering tops in a drying room in British Columbia (Punja, 2021a), identified morphologically and by ITS sequencing. Pathogenicity was confirmed on living inflorescences. The species was recovered from flowering tops in Kentucky (Smith *et al.*, 2022). *F. sporotrichiodes* is a worldwide cause of maize kernel rot. Punja (2021a) also isolated *Fusarium proliferatum*, which is primarily a stalk pathogen (see section 10.4).

9.3e. *Fusarium fujikuroi* Nirenberg 1976

Comments: *F. fujikuroi* was observed producing chlorosis and necrosis of flowering tops in Virginia (Chawla *et al.*, 2021). Pinkish-white fuzzy growth was also observed on bracts of hemp seeds; seeds harvested from infected plants showed significantly reduced germination. Identification was confirmed by sequencing ITS, elongation factor 1-α, and β-tubulin. Pathogenicity was confirmed on 10-week-old plants in a humid chamber.

9.3.3 Differential diagnosis

Fusarium flower blight might be confused with gray mold. In high humidity, *Fusarium*-infested flowers may become shrouded by a pinkish mycelium, whereas flowers infested by *Botrytis* spp. turn gray. Microscopically, the conidia of *Fusarium* spp. and *Botrytis* spp. look very different. Differentiating between *Fusarium* spp. is best accomplished by molecular means.

9.3.4 Disease cycle and epidemiology

Fusarium flower blight is primarily a problem in drug-type crops grown outdoors, but it has also been reported indoors. Punja *et al.* (2019) demonstrated a temporal sequence where stalk infection gave rise to airborne conidia, which secondarily spread to foliar and flower infections on the same or adjacent plants. Thiessen (pers. comm., 2020) implicated ascospores as a source of primary inoculum.

9.3.5 Disease management

Follow guidelines for gray mold. *Bacillus subtilis* (Serenade®, Kodiak®) is labeled for biocontrol of *Fusarium* spp. as a soil application. So is *Burkholderia vietnamiensis* (Botrycid®), *Clonostachys rosea* (=*Gliocladium catenulatum*, Prestop®), *Pseudomonas chlororaphis* (Howler®), and *Fusarium oxysporum* Fo47 (Biofox C®, Fusaclean®).

9.4 Penicillium leaf and flower necrosis

In most cases, *Penicillium* spp. act as saprophytes that invade plants after harvest (Chapter 18). New research shows that *Penicillium* spp. can live as endophytes, and inhabit living plants (Chapter 13). Given the opportunity, they act as necrotrophic pathogens, and infest floral parts in drug-type crops.

9.4.1 Symptoms and signs

Symptoms begin as a "burning" of leaves (Babu *et al.*, 1977), or a browning and necrosis and wilting of female pistils (styles and stigmas). This is followed by browning and necrosis of floral leaves (Punja *et al.*, 2019). The styles and stigmas of drug-type sinsemilla plants, protected against pollination, grow quite large (Small and Naraine, 2016), which predisposes them to necrotrophic pathogens.

9.4.2 Causal organisms

Although a dozen *Penicillium* spp. are implicated in post-harvest decay (see section 18.3), only a couple are cited as pathogens of living plants. They are members of Class Eurotiomycetes, Family Trichocomaceae.

9.4a. Penicillium chrysogenum Thom 1910

=*Penicillium notatum* Westling 1911
Description: Conidial heads paint-brush-like, bearing conidia in well-defined columns up to 200 μm long. Conidiophores smooth, hyaline, up to 350 μm long × 3.0–3.5 μm wide. Branches (penicilli) usually biverticillate, measuring 15–25 μm × 3.0–3.5 μm. Metulae borne atop penicilli in clusters of 2–5, measuring 10–12 μm long × 2–3 μm wide. Phialides borne atop metulae in compact clusters of 4–6, flask-shaped, mostly 8–10 × 2.0–2.5 μm. Conidia elliptical to almost globose, smooth, pale yellow to pale green, mostly 3.0–4.0 × 2.8–3.5 μm (see Fig. 18.6).
Comments: *P. chrysogenum* occurs worldwide on a variety of substrates, including rotten fruits and vegetables, and salted food products. Dr Alexander Fleming discovered "penicillin" when this species floated into his laboratory. *P. chrysogenum* conidia are common air contaminants in damp or water-damaged buildings, and they cause allergies or asthma in sensitized individuals.

Babu *et al.* (1977) described *P. chrysogenum* causing "burning" symptoms in India. Suyal *et al.* (2011) analyzed airborne mycoflora over *Cannabis* fields in the Himalaya, and *P. chrysogenum* was the third most common species. Punja *et al.*

(2019) isolated *P. chrysogenum* from coconut (coco coir) fiber used as a hydroponic medium. In plants grown in that medium, Punja isolated *P. chrysogenum* as a leaf endophyte, and from the central pith of stems.

9.4b. *Penicillium olsonii* **Bainier & Sartory 1912**

Description: Conidia form in congested heads, bearing conidia in columns up to 300 µm long (Fig. 9.10). Conidiophores smooth, hyaline, up to 1000 µm long × 4–6 µm wide. Branches (penicilli) usually two-stage branched (terverticillate). Metulae borne atop penicilli in clusters of 2–5, 9–12 × 2–4 µm. Phialides borne atop metulae in compact clusters of 3–7, flask-shaped, 9–12 × 2–3 µm. Conidia subglobose to ellipsoidal, 3–4 × 2.5–3.0 µm, finely roughened.

Comments: *P. olsonii* occurs worldwide, but is less common than *P. chrysogenum*; it causes spoilage of tomatoes, beans, and even salami. Punja *et al.* (2019) found *P. olsonii* sporulating in the central hollow of stems. Inoculation of inflorescences with a conidial suspension resulted in abundant mycelial growth and sporulation on pistil surfaces. Jerushalmi *et al.* (2020a) isolated *P. olsonii* from floral tissues, but unsuccessfully complete Koch's postulates on detached inflorescences.

9.4c. **Other** *Penicillium* **spp.**

Punja *et al.* (2019) isolated **Penicillium copticola** Houbraken *et al.* 2011 from plants in British Columbia. Inoculation of inflorescences with a conidial suspension resulted in mycelial growth. Jerushalmi *et al.* (2020a) isolated **Penicillium citrinum** Thom 1910 and **Penicillium steckii** Zaleski 1927 from plants in Israel. They completed Koch's postulates for both species, using detached flowering tops.

Fig. 9.10. Sporulation of *Penicillium olsonii* on papillae of a stigma (courtesy Z. Punja)

Punja (2021a) isolated a slew of *Penicillium* spp. from dried inflorescences, including most of the aforementioned (*P. chrysogenum*, *P. olsonii*, *P. copticola*, *P. citrinum*). He also reported species new to *Cannabis*: *P. brevicompactum*, *P. corylophilum*, *P. daleae*, *P. expansum*, *P. glabrum*, *P. griseofulvum*, *P. manginii*, *P. oxalicum*, *P. pancosmium*, *P. sclerotiorum*, *P. simplicissimum*, *P. spathulatum*, *P. terrigenum*).

9.4.3   Differential diagnosis and management

Flower symptoms illustrated by Punja *et al.* (2019) are subtle, and could be confused with the natural dieback of pistils in mature flowers. Differentiating *Penicillium* spp. prior to the molecular era required *in vivo* culture on different media, checking for color changes in the media, and measuring metabolite production. *P. chrysogenum* conidia are smooth-walled, *P. olsonii* conidia are finely roughened.

Penicillium spp. flourish in damp organic matter, so all plant debris and paper products must be removed. *Penicillium* spp. present in unpasteurized coco fiber growing medium are potential sources of contamination (Punja *et al.*, 2019). Pasteurized products accrue microbial populations as plants mature. Rockwool might provide a safer medium. *Penicillium* spp. are weak pathogens and primarily attack stressed plants.

Farahmand *et al.* (2023) sprayed copper sulphate (100 mg/L, four weekly foliar applications) on plants inoculated with *P. olsonii*. The treatment significantly reduced conidia formation compared with controls, and qPCR analysis showed that *P. olsonii* levels were significantly lower.

9.5   Pink rot

The fungus causing pink rot, *Trichothecium roseum*, can infect everything above the soil line. In the past 25 years, it has been isolated from indoor drug-type crops. The first report of pink rot attacking flowering tops was in Amsterdam, on "Skunk no. 1" clones, where it caused considerable losses (McPartland *et al.*, 2000). Prior to that, *T. roseum* was described as a foliar pathogen in Pakistan (Nair and Ponnappa, 1974; Ponnappa, 1977).

Jerushalmi *et al.* (2020a) isolated *T. roseum* from plants, and fulfilled Koch's postulates on detached flowering tops. The fungus initially developed a white-colored mold, followed by the appearance of pink conidia. *T. roseum* has been isolated from hemp stems in Italy (Ghillini, 1951) and Iowa (Fuller and Norman, 1945). It also causes seed-borne infections in Italy (Ciferri, 1941; Ferri, 1961b) and Russia (Pospelov *et al.*, 1957).

9.5.1   Symptoms and signs

Pink rot is a bit of a misnomer. The fungus produces a *whitish* fuzz on leaves or flowering buds. Hyphae may form a mat on the surface of the leaf, or extend into a web that completely encases small flowers (Fig. 9.11). A pink tint arises with the production of conidia. In Amsterdam the fungus grew over branches and girdled them. Girdled limbs wilted. Ghillini (1951) and Fuller and Norman (1945) noted that the fungus ruins hemp fiber.

Fig. 9.11. Pink rot caused by *Trichothecium roseum* (R. Clarke)

9.5.2 Causal organism

9.5a. *Trichothecium roseum* (Persoon) Link 1809, Sordariomycetes; Hypocreales

=*Cephalothecium roseum* Corda 1838
Description: Colonies quickly turn dusty pink with conidia. Conidiophores upright, unbranched, often with 3 septa near the base, up to 2 mm long, 4–5 μm wide. Conidiogenesis basipetal, conidia often remain in contact with each other in zig-zag chains. Conidia ellipsoidal to pyriform with truncate basal scars, 2-celled with upper cell larger and rounder than lower cell, hyaline (pink *en masse*), with a thick smooth wall, 12–23 × 8–10 μm.

The initial white fuzz of pink rot has been confused with both gray mold and powdery mildew, as documented by McPartland and Hillig (2008). The fungi causing these diseases are easy to differentiate under a microscope or even a strong hand lens.

9.5.3 Disease cycle and epidemiology

T. roseum overwinters on crop debris or in the soil. It occurs worldwide as a saprophyte of stored foodstuffs, and turns up in termite nests, paper mill slime, sewage sludge, and hundreds of other substrates. In New Zealand *T. roseum* sporulated on pest excreta on hemp leaves (snail slime and aphid honeydew). It gained energy from these substrates and then invaded healthy plant tissue (McPartland and Rhode, 2005).

In optimal conditions for fungal growth, conidia arise about 1 week after infection, and create epidemics of secondary infections. Epidemics of *T. roseum* have occurred in cloning chambers, because conditions (relative humidity near 100%, little airflow) provide a perfect germination tank for *T. roseum* conidia.

9.5.4 Disease management

T. roseum flourishes in damp organic matter, so all plant debris and paper products must be removed. Cloning chambers must be kept scrupulously clean. Eliminate pollen and pest excreta lying upon the surface of plants. Get rid of aphids. Avoid overcrowding plants, overhead irrigation, and excess humidity. Evidence suggests that *T. roseum* may cause seed-borne infections, so seeds from infected plants should not be sown. Infected plants should not be consumed: *T. roseum* produces trichothecene metabolites, specifically T-2 toxin, which is neurotoxic and carcinogenic.

Afghani landraces and their hybrids (e.g., "Skunk no. 1") are particularly susceptible. McPartland (unpublished data, 1994) found *T. roseum* on herbarium specimens collected in Afghanistan by Schultes in 1971. "Immunizing" plants with beneficial endophytic fungi may provide biological control in the future. Kusari *et al.* (2013) isolated several naturally occurring endophytic fungi from *Cannabis*. The most dominant species, *Penicillium copticola*, strongly antagonized *T. roseum* in a dual culture assay. However, see comments regarding *P. copticola* in the previous section. No fungicides are effective, although Dahiya and Jain (1977) reported that pure THC and CBD inhibited the growth of *T. roseum*.

9.6 Sooty mold

Sooty mold grows on plant surfaces contaminated by "honeydew" produced by aphids and whiteflies. Symptoms of sooty mold consist of black or dark-brown fungal growth, primarily on the upper surfaces of leaves. The growth can vary from a fine sooty specking to a crust of black mold. Sooty mold indirectly harms plants by blocking sunlight and reducing photosynthesis. Plants with bad infestations are unsightly and their products lose market value.

Any darkly pigmented fungus that superficially colonizes plant surfaces has been called a sooty mold. Some mycologists limit the definition of sooty molds to members of the family Capnodiaceae (such as *Capnodium* spp.), of which none yet have been reported on *Cannabis*. *Aureobasidium pullulans* (de Bary) Arnard 1918 (Dothideomycetes; Dothioraceae) has been cited on *Cannabis* (Lentz, 1977; Hrebenyuk, 1984; Ondřej, 1991; Scott *et al.*, 2018). This ubiquitous black, yeast-like fungus grows anywhere that adequate humidity can support it.

Some darkly pigmented fungi growing on leaves are actually pathogens, and not sooty molds, such as *Alternaria alternata* (see section 11.10) and *Cladosporium herbarium* (see section 10.11). They establish a foothold on honeydew and then invade tissues. To control sooty mold means controlling insects that produce honeydew. And their ant masters. Once that's accomplished, remove the most infested leaves with judicious pruning. Hosing will not wash off the mold.

(Prepared by J. McPartland)

References

Adhikary D, El-Mezawy A, Khatri-Chhetri U, *et al.* (6 other authors). 2022. Transcriptome response of cannabis (*Cannabis sativa* L.) to the pathogenic fungus *Golovinomyces ambrosiae*. *BioRxiv.* doi 10.1101/2022.08.01.501243

Akinrinlola R, Hansen ZR. 2022. Efficacy of organic fungicides against hemp powdery mildew caused by *Golovinomyces ambrosiae* in a greenhouse in Tennessee. *Plant Disease.*

Anonymous. 2008. Miserable mold. *High Times* May 2008, p. 104.

Appel O, Wollenweber HW. 1910. Grundlagen einer Monographie der Gattung Fusarium (Link). *Arbeiten aus der Kaiserlichen Biologischen Anstalt für Land- und Forstwirtschaft* 8: 1–207.

Babu R, Roy AN, Gupta YK, Gupta MN. 1977. Fungi associated with deteriorating seeds of *Cannabis sativa* L. *Current Science* 46: 719–720.

Balthazar C, Cantin G, Novinscak A, Joly DL, Filion M. 2020. Expression of putative defense responses in Cannabis primed by *Pseudomonas* and/or *Bacillus* strains and infected by *Botrytis cinerea*. *Frontiers in Plant Science* 11: e572112.

Balthazar C, Joly DL, Filion M. 2022a. Exploiting beneficial *Pseudomonas* spp. for *Cannabis* production. *Frontiers in Microbiology* 12: e833172.

Balthazar C, Novinscak A, Cantin G, Joly DL, Filion M. 2022b. Biocontrol activity of *Bacillus* spp. and *Pseudomonas* spp. against *Botrytis cinerea* and other *Cannabis* fungal pathogens. *Phytopathology* 112: 549–560.

Barloy J, Pelhate J. 1962. Premières observations phytopathologiques relatives aux cultures de chanvre en Anjou. *Annales des Épiphyties* 13: 117–149.

Bates TA, Block MH, Wisemann MS, *et al.* (3 other authors). 2021. First report of powdery mildew caused by *Podosphaera macularis* on hemp in Oregon. *Plant Health Progress* 22: 567–569.

Behrens J. 1891. Ueber das auftreten des Hanfkrebses in Elsass. *Zeitschrift für Pfanzenkrankheiten* 1: 208–215.

Behrens J. 1902. Untersuchungen ueber die gewinnung der hanffaser durch natuerliche roestmethoden. *Centralblatt für Bakteriologie und Parasitenkunde* 11: 264–268, 295–299.

Bergstrom G, Starr J, Myers K, Cummings, J. 2019. *Diseases affecting hemp in New York*. Available at: https://hemp.cals.cornell.edu/resource/diseases-affecting-hemp-new-york-2019/

Bergstrom G, Starr J, Myers K, Cummings J. 2020. "An early view of diseases affecting hemp in New York," p. 10 in Gauthier N, *ed. Science of hemp: production and pest management*. University of Kentucky Agricultural Experiment Station no. SR-112. Lexington, Kentucky.

Binder M, Meisenberg G. 1978. Microbial transformation of cannabinoids: Part 2; A screening of different microorganisms. *European Journal of Applied Microbiology and Biotechnology* 5: 37–50.

Boesewinkel H. 1980. The morphology of the imperfect states of powdery mildews (Erysiphaceae). *Botanical Review* 46: 167–224.

Brais N. 2017. *Marijuana powdery mildew reduction using ultraviolet germicidal system*. Available at: https://sanuvox.com/wp-content/uploads/2017/06/FOOD-Marijuana-Powdery-Mildew-Reduction-using-Germicidal-Ultraviolet.pdf

Brandenburger W. 1985. *Parasitische Pilz an Gefäßpflanzen in Europa*. Gustav Fisher Verlag, Stuttgart.

Braun U. 1987. A monograph of the Erysiphales (powdery mildews). *Beihefte zur Nova Hedwigia* 89: 1–700.

Brochu AS, Labbe C, Bélanger RR, Edel Pérez-López E. 2022. First report of powdery mildew caused by *Golovinomyces ambrosiae* on *Cannabis sativa* L. (marijuana) in Quebec, Canada. *Plant Disease* 106: 2747.

Brown MF, Brotzman HG. 1979. *Phytopathogenic fungi: a scanning electron stereoscopic survey*. University of Missouri, Columbia, Missouri.

Cala A, Portilla N, Weldon B, *et al.* (9 other authors). 2019. Evaluating disease resistance in CBD hemp cultivars. Cornell Hemp. Available at: http://hemp.cals.cornell.edu

Cala A, Day C, Giles G, Snyder SI, Rose JKC, Smart C. 2021. *Biological fungicide efficacy against hemp powdery mildew and Botrytis gray mold*. Cornell Hemp. Available at: http://hemp.cals.cornell.edu

California DPR (Department of Pesticide Regulation). 2017a. *Legal pest management practices for marijuana growers in California*. California Environmental Protection Agency, Sacramento, California.

Chawla S, Geyer E, Dhakal R, Ren SX, Mersha Z. 2021. "Fusarium head and bud blight detected on hemp varieties in Virginia," p. 22 in *Second annual Science of Hemp Conference*. Auburn University, Auburn, Alabama.

Ciferri R. 1941. *Manuale di patologia vegetale*. Societa Editrice Dante Alighieri, Rome.

Clarke RC. 1987. *Cannabis evolution*. Master's thesis, Indiana University, Bloomington, Indiana.

Cochran S, Schappe T, Thiessen L. 2020. "*Fusarium* spp. on industrial hemp," p. 20 in Gauthier N, *ed. Science of hemp: production and pest management*. University of Kentucky Agricultural Experiment Station, no. SR-112, Lexington, Kentucky.

Colorado DOA (Department of Agriculture). 2018. *Pesticide use in Cannabis production information*. Available at: www.colorado.gov/pacific/agplants/pesticide-use-cannabis-production-information

Cook RTA, Inman AJ, Billings C. 1997. Identification and classification of powdery mildew anamorphs using light and scanning electron microscopy and host range data. *Mycological Research* 101: 975–1002.

Couch J, Victory K, Lowe B, *et al.* (5 other authors). 2017. *Evaluation of potential hazards during harvesting and processing Cannabis at an outdoor organic farm*. US Department of Health and Human Services, report no. 2015-0111-3271.

Crous PW, Lombard L, Sandoval-Denis M, *et al.* (124 other authors). 2021. Fusarium: more than a node or a foot-shaped basal cell. *Studies in Mycology* 98: e10116.

Dahiya MS, Jain GC. 1977. Inhibitory effects of cannabidiol and tetrahydrocannabinol against some soil inhabiting fungi. *Indian Drugs* 14(4): 76–79.

Davy JB. 1904. Alien plants spontaneous in the Transvaal. *South African Journal of Science* 2: 252–229.

De Bary A. 1887. *Comparative morphology and biology of the fungi, mycetozoa and bacteria*. Clarendon Press, Oxford.

de Meijer EPM. 1995. Fiber hemp cultivars: a survey of origin, ancestry, availability and brief agronomic characteristics. *Journal of the International Hemp Association* 2(2): 66–73.

de Meijer WJM, van der Werf HMG, Mathijssen EWJM, van den Brink PWM. 1995. Constraints to dry matter production in fibre hemp (*Cannabis sativa* L.). *European Journal of Agronomy* 4: 109–117.

Dean R, Van Kan JAL, Pretorius ZA, *et al.* (8 other authors). 2012. The Top 10 fungal pathogens in molecular plant pathology. *Molecular Plant Pathology* 13: 414–430.

Dempsey JM. 1975. "Hemp" pp. 46-89 in *Fiber crops*. University of Florida Press, Gainesville.

Després M. 1854. Le chanvre (*Cannabis sativa* L.) preserve la Vigne de l'*Oïdium*. *L'illustration Horticole* 1: 6–7.

Dixon E, Leonberger K, Szarka D, *et al*. (4 other authors). 2022a. Prescreening of biological and biorational fungicides against common hemp pathogens using *in vitro* analyses. *Plant Health Progress* 23: 250–255.

Dixon E, Leonberger K, Amsden B, *et al*. (6 other authors). 2022b. Suppression of hemp powdery mildew using root-applied silicon. *Plant Health Progress* 23: 260–264.

Doidge EM. 1950. The South African fungi and lichens to the end of 1945. *Bothalia* 5: 1–1094.

Doidge EM, Bottomley AM. 1931. *A revised list of plant diseases occurring in South Africa*. *Memoirs of the Botanical Survey of South Africa* 11: 1–78.

Doidge EM, Bottomley AM, van der Plank JE, Pauer GD. 1953. A revised list of plant diseases in South Africa. *South African Department of Agriculture Science Bulletin* No. 345. Pretoria. 122 pp.

Farahmand H, Robinson GI, Gerasymchuk M, Kovalchuk I. 2023. Copper sulphate inhibits *Penicillium olsonii* growth and conidiogenesis on *Cannabis sativa*. *Journal of Plant Pathology*.

Farinas C, Peduto Hand F. 2020. First report of *Golovinomyces spadiceus* causing powdery mildew on industrial hemp (*Cannabis sativa* L.) in Ohio. *Plant Disease* 104: 2727.

Ferraris T. 1935. *Parassiti vegetali della canapa*. Rivista Agricola, Rome.

Ferri F. 1961b. Microflora dei semi di canapa. *Progresso Agricolo (Bologna)* 7(3): 349–356.

Flachs K. 1936. Krankheiten und schädlinge unserer gespinstpflanzen. *Nachrichten über Schädlingsbekämpfung* 11: 6–28.

Frank M. 1988. *Marijuana grower's insider's guide*. Red Eye Press, Los Angeles.

Fröier K. 1956. Svalöfs samkönade Mona-hampa. *Allmänna Svenska Utsädesaktiebolaget Svalöf* 1956: 28–29.

Fuller WH, Norman AG. 1945. Biochemical changes involved in the decomposition of hemp bark by pure cultures of fungi. *Journal of Bacteriology* 50: 667–671.

Gagkaeva T, Orina A, Gavrilova O. 2021. Fusarium head blight in the Russian Far East: 140 years after description of the "drunken bread" problem. *PeerJ* 9: e12346.

Gams W, Nirenberg HI, Seifert KA, *et al*. 1997. Proposal to conserve the name *Fusarium sambucinum* (Hyphomycetes). *Taxon* 46: 111–113.

Garfinkel AR. 2020. Three *Botrytis* species found causing gray mold on industrial hemp (*Cannabis sativa*) in Oregon. *Plant Disease* 104: 2727.

Garfinkel AR, Coats KP, Sherry DL, Chastagner GA. 2019. Genetic analysis reveals unprecedented diversity of a globally-important plant pathogenic genus. *Scientific Reports* 9: e6671.

Gauthier N. 2015. *Management of powdery mildew begins with understanding the causal fungus*. Available at: http://www.kyhempdisease.com/powdery-mildew-of-hemp.html

Gauthier NW. 2020a. *Hemp disease in Kentucky: grasping the reality of yield loss*. Available at: https://www.uvm.edu/sites/default/files/Northwest-Crops-and-Soils-Program/2020%20HempConf%20Presentations/NicoleGauthier_Hemp_Disease_and_Yield_Loss_VT.pdf

Geiser DM, Aoki T, Bacon CW, *et al*. (66 other authors). 2013. One fungus, one name: defining the genus *Fusarium* in a scientifically robust way that preserves longstanding use. *Phytopathology* 103: 400–408.

Geiser DM, Al-Hatmi AMS, Aoki T, *et al*. (165 other authors). 2021. Phylogenomic analysis of a 55.1 kb 19-gene dataset resolves a monophyletic *Fusarium* that includes the *Fusarium solani* species complex. *Phytopathology* 111: 1064–1079.

Ghillini CA. 1951. I parassiti nemici vegetali della canapa. *Notiziario sulle Malattie delle Piante* 15: 29–36.

Gitman L, Boytchenko E. 1934. "*Cannabis*" pp. 45-53 *in* Справочник по болезням бовых лубианых культур. Institute New Bast Raw Materials, Moscow.

Glawe DA. 2008. The powdery mildews: a review of the world's most familiar (yet poorly known) plant pathogens. *Annual Review of Phytopathology* 46: 27–81.

Green BJ, Couch JR, Lemons AR, *et al*. (4 other authors). 2018. Microbial hazards during harvesting and processing at an outdoor United States cannabis farm. *Journal of Occupational and Environmental Hygiene* 15: 430–440.

Gwinn KD, Hansen Z, Kelly H, Ownley BH. 2021. Diseases of *Cannabis sativa* caused by diverse *Fusarium* species. *Frontiers in Agronomy* 3: e796062.

Hansen Z, Siegenthaler T, Akin E, Cartwright M, Kelly H. 2020. "Hemp extension in Tennessee; most common diseases of 2019," p. 27 in Gauthier N, *ed. Science of hemp: production and pest management*. University of Kentucky Agricultural Experiment Station no. SR-112. Lexington, Kentucky.

Hazslinszky F. 1877–78. *Polyactis infestans*, (nov. spec.). *Grevillea* 6: 77.

Health Canada. 2018. *Health Canada testing of Cannabis for medical purposes for unauthorized pest control products*. Available at: www.canada.ca/en/health-canada/services/drugs-medication/cannabis/licensed-producers/policies-directives-guidance-information-bulletins/testing-cannabis-medical-purposes-unauthorized-pest-control-products.html

Hirata K. 1966. *Host range and geographical distribution of the powdery mildews*. Niigata University Press, Japan. 472 pp.

Hrebenyuk NV. 1984. Распространенне грибов на тресте конопли (The occurrence of fungi on hemp stems). *Микология и фитопатология* 18(4): 322–326.

Itagaki K, Shibuya T, Tojo M, Endo R, Kitaya Y. 2014. Atmospheric moisture influences on conidia development in *Podosphaera xanthii* through host-plant morphological responses. *European Journal of Plant Pathology* 138: 113–121.

Jaczewski AA. 1927. *Карманный определитель грибов: Выпуск второй. Мучнисто-росяные грибы (Pocket determinant of fungi, vol. 2: powdery mildew fungi)*. Mycological Laboratory A. A. Jaczewski State Research Institute of Experimental Agronomy, Leningrad, p. 412.

James DG, Prischmann D. 2010. "The impact of sulfur on biological control of spider mites in Washington State vineyards and hop yards," pp. 477–482 *in:* Sabelis M, Bruin J, eds. *Trends in Acarology*. Springer, Dordrecht.

Jarvis W, Gubler WG, Grove GG. 2002. "Epidemiology of powdery mildews in agricultural ecosystems," pp. 169-199 in Belanger R, *et al*., eds., *The powdery mildews: a comprehensive treatise*. American Phytopathological Society, St Paul, Minnesota.

Jerushalmi S, Maymon M, Dombrovsky A, Freeman S. 2020a. Fungal pathogens affecting the production and quality of medical cannabis in Israel. *Plants* 9(7): e882.

Jerushalmi S, Maymon M, Dombrovsky A, Freeman S. 2020b. Effects of cold plasma, gamma and e-beam irradiations on reduction of fungal colony forming unit levels in medical cannabis inflorescences. *Journal of Cannabis Research* 2: e12.

Jin D, Henry P, Shan J, Chen, J. 2021. Identification of phenotypic characteristics in three chemotype categories in the genus *Cannabis*. *HortScience* 56: 481–490.

Kabaktepe S, Heluta VP, Akata I. 2015. Checklist of powdery mildews (Erysiphales) in Turkey. *Biological Diversity and Conservation* 8/3: 128–146.

Kusari P, Kusari S, Spiteller M, Kayser O. 2013. Endophytic fungi harbored in *Cannabis sativa* L.: diversity and potential as biocontrol agents against host plant-specific phytopathogens. *Fungal Diversity* 60: 137–151.

Lentz P. 1977. *Fungi and diseases of Cannabis. Unpublished manuscript, notes and collected literature.* National Fungus Collections, USDA, Beltsville, Maryland.

Lindau G. 1907. *Die Pilze Deutschlands, Oesterreichs und der Schweiz, VIII. Abteilung: Fungi imperfecti: Hyphomycetes.* Eduard Kummer, Leipzig.

Link HF. 1809. Observationes in ordines plantarum naturales. Dissertatio I. *Magazin der Gesellschaft Naturforschenden Freunde Berlin* 3(1): 3–42.

Lucas GB. 1975. *Diseases of tobacco, 3rd ed.* Biological Consulting Assoc., Raleigh, North Carolina.

Mahieu CM, Biesebeek JD, Graven C. 2019. *Inventarisatie van gewasbescherming toepasbaar in de teelt van Cannabis binnen het "Experiment met een gesloten coffeeshopketen."* Rijksinstituut voor Volksgezondheid en Milieu, Rapport 2019-0232. Bilthoven, Netherlands.

Mahmoud M, BenRejeb I, Punja ZK, Buirs L, Jabaji S. 2023. Understanding bud rot development, caused by *Botrytis cinerea*, on cannabis (*Cannabis sativa* L.) plants grown under greenhouse conditions. *Botany* 101: 200–231.

Masheva S, Lazarova T, Velkov N, Velichkov G. 2014. Botanical products against powdery mildew on cucumber in greenhouses. *Turkish Journal of Agricultural and Natural Sciences* 2014(2): 1707–1712.

Massalongo C. 1899. Funghi della Provincia di Ferrara, I Serie. *Atti della Accademia delle Scienze Mediche e Naturali in Ferrara* 72: 61–94.

Maymon M, Jerushalmi S, Freeman S. 2021. First report of *Golovinomyces cichoracearum sensu lato* on *Cannabis sativa* in Israel. *New Disease Reports* 42: 11.

McKernan K, Spangler J, Helbert Y, *et al.* (9 other authors). 2016. Cannabis microbiome sequencing reveals *Penicillium paxillin* and the potential for paxilline drug interactions with cannabidiol. *Proceedings of the 26th Annual Symposium on the Cannabinoids.* International Cannabinoid Research Society, Research Triangle Park, North Carolina, p. P1–28.

McKernan KJ, Helbert Y, Kane LT, *et al.* (12 other authors). 2020. Sequence and annotation of 42 *Cannabis* genomes reveals extensive copy number variation in cannabinoid synthesis and pathogen resistance genes. *BioRxiv Preprint*, Available at: https://doi.org/10.1101/2020.01.03.894428

McPartland JM. 1983. Fungal pathogens of *Cannabis sativa* in Illinois. *Phytopathology* 72: 797.

McPartland JM. 1989. Proposed list of common names for diseases of *Cannabis sativa* L. *Phytopathology News* 23(4): 46–47.

McPartland JM. 1991. Common names for diseases of *Cannabis sativa* L. *Plant Disease* 75: 226–227.

McPartland JM. 1995a. *Cannabis* pathogens VIII: misidentifications appearing in the literature. *Mycotaxon* 53: 407–416.

McPartland JM, Cubeta MA. 1997. New species, combinations, host associations and location records of fungi associated with hemp (*Cannabis sativa*). *Mycological Research* 101: 853–857.

McPartland JM, Hillig KW. 2008. Differentiating powdery mildew from false powdery mildew. *Journal of Industrial Hemp* 13: 78–87.

McPartland JM, McKernan KJ. 2017. "Contaminants of concern in cannabis: microbes, heavy metals and pesticides," in Chandra S, Lata H, ElSohly MA, *eds. Cannabis sativa: botany and biotechnology.* Springer International Publishing, Cham, Switzerland.

McPartland JM, Rhode B. 2005. New hemp diseases and pests in New Zealand. *Journal of Industrial Hemp* 10: 99–108.

McPartland JM, Thiessen LD. 2022. "Gray mold," in Gauthier NW, Thiessen LD, *eds. Compendium of Cannabis diseases.* APS Press, St Paul, Minnesota.

McPartland JM, Clarke RC, Watson DP. 2000. *Hemp diseases and pests – management and biological control.* CABI Publishing, Wallingford, UK.

Mihalyov PD, Garfinkel AR. 2021. Discovery and genetic mapping of PM1, a powdery mildew resistance gene in *Cannabis sativa* L. *Frontiers in Agronomy* 3: e720215.

Muth F. 1906. Untersuchungen über die Früchte des Hanfes (*Cannabis sativa* L.). *Jahresbericht der Vereinigung für Angewandte Botanik* 3: 76–121.

Nair KR, Ponnappa KM. 1974. *Survey for natural enemies of Cannabis sativa and Papaver somniferum.* Commonwealth Institute of Biological Control, India Station Report, pp 39–40.

Narváez A, Rodríguez-Carrasco Y, Castaldo L, Izzo L, Ritieni A. 2020. Ultra-high-performance liquid chromatography coupled with quadrupole orbitrap high-resolution mass spectrometry for multi-residue analysis of mycotoxins and pesticides in botanical nutraceuticals. *Toxins* 12: e144.

Neenan M. 1969. The cultivation of hemp in Ireland. *Fibra* 14: 23–33.

Nelson PE. 1991. History of *Fusarium* systematics. *Phytopathology* 81: 1045–1048.

Nelson PE, Toussoun TA, Marass WFO. 1983. *Fusarium species: an illustrated manual for identification.* Pennsylvania State University Press, University Park, Pennsylvania.

Noble M, Richardson MJ. 1968. *An annotated list of seed-borne diseases.* Commonwealth Mycological Institute, Kew, UK.

Noviello C, Snyder WC. 1962. Fusarium wilt of hemp. *Phytopathology* 52: 1315–1317.

O'Donnell K. 1996. Progress towards a phylogenetic classification of *Fusarium*. *Sydowia* 48: 57–70.

O'Donnell K. 2000. Molecular phylogeny of the *Nectria haematococca–Fusarium solani* species complex. *Mycologia* 92: 919–938.

O'Donnell K, McPartland JM, Thiessen LD. 2022. "Fusarium wilt," in Gauthier NW, Thiessen LD, *eds. Compendium of Cannabis diseases.* APS Press, St Paul, Minnesota.

Ondřej M. 1991. Výskyt hub na stoncích konopí (*Cannabis sativa* L.). *Len a Konopí (Sumperk, Czech Rep)* 21: 51–57.

Patschke K, Gottwald R, Müller R. 1997. Erste Ergebnisse phytopathologischer Beobachtungen im Hanfanbau im Land Brandenburg. *Nachrichtenblatt des Deutschen Pflanzenschutzdienstes* 49: 286–290.

Pépin N, Herbert FO, Joly DL. 2021. Genome-wide characterization of the MLO gene family in *Cannabis sativa* reveals two genes as strong candidates for powdery mildew susceptibility. *Frontiers in Plant Science* 12: e729261.

Pépin N, Punja ZK, Joly DL. 2018. Occurrence of powdery mildew caused by *Golovinomyces cichoracearum sensu lato* on *Cannabis sativa* in Canada. *Plant Disease* 102: 2644.

Pietkiewicz TA. 1958. Mikroflora nasion konopi. Przeglad literatury. *Roczniki Nauk Rolniczych Seria A* 77(4): 577–590.

Polonio F. 1863. "Sopra l'oidio della canape," pp. 43–44 in Bertoloni G, ed. *Rendiconto delle sessioni dell'Accademia delle scienze dell'Istituto di Bologna.* Gamerini e Parmeggiani, Bologna.

Ponnappa K M. 1977. New records of fungi associated with *Cannabis sativa. Indian Journal of Mycology and Plant Pathology* 7: 139–142.

Pospelov AG, Zapromatov NG, Domasheva AA. 1957. *Грибная флора Киргизской ССР, Вып. 1.* Academy of Sciences of the Kyrgyz Republic, Frunze.

Potter D. 2009. *The propagation, characterisation and optimisation of Cannabis sativa L. as a phytopharmaceutical.* Doctoral thesis, King's College, London.

Pscheidt JW, Ocamb CM, *eds.* 2018. *Pacific Northwest plant disease management.* Oregon State University, Corvallis, Oregon.

Punja ZK. 2021a. The diverse mycoflora present on dried cannabis (*Cannabis sativa* L., marijuana) inflorescences in commercial production. *Canadian Journal of Plant Pathology* 43: 88–100.

Punja ZK. 2021b. Epidemiology of *Fusarium oxysporum* causing root and crown rot of cannabis (*Cannabis sativa* L., marijuana) plants in commercial greenhouse production. *Canadian Journal of Plant Pathology* 43: 216–235.

Punja ZK. 2021c. First report of *Fusarium proliferatum* causing crown and stem rot, and pith necrosis, in cannabis (*Cannabis sativa* L., marijuana) plants. *Canadian Journal of Plant Pathology* 43: 236–245.

Punja ZK. 2021d. Emerging diseases of *Cannabis sativa* and sustainable management. *Pest Management Science* 77: 3857–3870.

Punja ZK. 2022. First report of the powdery mildew pathogen of hops, *Podosphaera macularis*, infecting marijuana (*Cannabis sativa* L.) plants under field conditions. *Canadian Journal of Plant Pathology* 44: 235–249.

Punja ZK, Ni L. 2021. The bud rot pathogens infecting cannabis (*Cannabis sativa* L., marijuana) inflorescences: symptomology, species identification, pathogenicity and biological control. *Canadian Journal of Plant Pathology* 43: 827–854.

Punja ZK, Collyer D, Scott C, Lung S, Holms J. Sutton D. 2019. Pathogens and molds affecting production and quality of *Cannabis sativa* L. *Frontiers in Plant Science* 10: e1120.

Punja ZK, Scott C, Lung S. 2022. Several *Pythium* species cause crown and root rot on cannabis (*Cannabis sativa* L., marijuana) plants grown under commercial greenhouse conditions. *Canadian Journal of Plant Pathology* 44: 66–81.

Purdy T. 2018. *Vapor pressure deficit and HVAC system design*. Application Note 28, Desert Aire, Milwaukee, WI. Available at: www.desert-aire.com/sites/default/files/28-AN-Vapor-Pressure-Deficit-and-HVAC-System-Design.pdf

Qiu PL, Liu SY, Bradshaw M, *et al*. (10 other authors). 2020. Multi-locus phylogeny and taxonomy of an unresolved, heterogeneous species complex within the genus *Golovinomyces* (Ascomycota, Erysiphales), including *G. ambrosiae*, *G. circumfusus* and *G. spadiceus*. *BMC Microbiology* 20: e51.

Rajmohan N, Price DC, Buckley RJ, *et al*. (7 other authors). 2022. First report of powdery mildew caused by *Golovinomyces ambrosiae* on industrial hemp in New Jersey. *Plant Disease* 105: 2733.

Rakhimova EV, Nam GA, Yermekova BL. 2013. Ключ для определения видов мучнисторосяных грибов Казахстана по семействам и родам питающих растений. *Turczaninowia* 16: 176–196.

Rapaić von Ruhmwerth R. 1914. Phytopathologische Beobachtungen in Debrecen (Ungarn). *Zeitschrift für Pflanzenkrankheiten* 24: 211-218.

Rataj K. 1957. Skodlivi cinitele pradnych rostlin. *Prameny literatury* 2: 1–123.

Reinking OA. 1950. Hans Wilhelm Wollenweber, 1879–1949. *Phytopathology* 40: 119–127.

Rippon JW. 1988. *Medical mycology, 3rd edn*. WB Saunders Co., Philadelphia, Pennsylvania.

Rodriguez G, Kibler A, Campbell P, Punja ZK. 2015. *Fungal diseases of Cannabis sativa in British Columbia, Canada*. American Phytopathological Society Annual Meeting, Poster 529-P.

Ruchlemer R, Amit-Kohn M, Raveh D, Hanuš L. 2015. Inhaled medicinal cannabis and the immunocompromised patient. *Supportive Care in Cancer* 23: 819–822.

Saccardo FA. 1886. *Syllogue Fungorum, vol. IV*. Typis Seminarii, Patavii.

Saccardo FA, Sydow P. 1902. *Syllogue Fungorum, vol. XVI*. Typis Seminarii, Patavii.

Sasaki T, Honda Y. 1985. Control of certain diseases of greenhouse vegetables with ultraviolet-absorbing vinyl film. *Plant Disease* 69: 530–533.

Sawada K. 1927. 分生子時代より觀たる台湾産粉病菌属. *Agricultural Department Report, Central Research Institute of the Taiwan Governor's Office* 24: 1–55

Schoener J, Wang SH. 2016. *Nevada Department of Agriculture, Plant Industry Division, December 12, 2016 Board Updates Plant Pathology*. Available at: http://agri.nv.gov/uploadedFiles/agri.nv.gov/Content/Administration/Board_of_Agriculture/2016(1)/BOA_2016_03_29/7a_plant_update_dec17_final.pdf

Schoener J, Wang SH. 2018. First detection of *Golovinomyces ambrosiae* causing powdery mildew on medical marijuana plants in Nevada. *Phytopathology* 108: S1.186.

Scott C, Punja ZK. 2020. Evaluation of disease management approaches for powdery mildew on *Cannabis sativa* L. (marijuana) plants. *Canadian Journal of Plant Pathology*. doi: 10.1080/07060661.2020.1836026

Scott C, Punja Z. 2022. "Management of diseases on Cannabis in controlled environment production," pp. 231–251 in Zheng YB, ed. *Handbook of Cannabis production in controlled environments*. CRC Press, Boca Raton, Florida.

Scott M, Rani M, Samsatly J, Charron JB, Jabaji S. 2018. Endophytes of industrial hemp (*Cannabis sativa* L.) cultivars: identification of culturable bacteria and fungi in leaves, petioles, and seeds. *Canadian Journal of Microbiology* 64: 664–680.

Seifi S, Leckie KM, Giles I, *et al*. (6 other authors). 2025. Mapping and characterization of a novel powdery mildew resistance locus (PM2) in *Cannabis sativa* L. *Frontiers in Plant Science* 16: 1543229.

Small E, Naraine SGU. 2016b. Size matters: evolution of large drug-secreting resin glands in elite pharmaceutical strains of *Cannabis sativa* (marijuana). *Genetic Resources and Crop Evolution* 63: 349–359.

Smith HS, Szarka D, Dixon E, *et al*. (5 other authors). 2022. Emerging *Fusarium* spp. causing head blight on hemp in Kentucky. *Plant Health Progress*.

Smith RE. 1900. *Botrytis* and *Sclerotinia*: their relation to certain plant diseases and to each other. *Botanical Gazette* 29: 369–407.

Snyder WC, Hansen HN. 1940. The species concept in *Fusarium*. *American Journal of Botany* 27: 64–67.

Snyder WC, Hansen HN. 1941. The species concept in *Fusarium* with reference to section *Martiella*. *American Journal of Botany* 28: 738–742.

Snyder WC, Hansen HN. 1945. The species concept in *Fusarium* with reference to section *Discolor* and other sections. *American Journal of Botany* 32: 657–666.

Sorauer P. 1888. *Die Schäden der einheimischen Kulturpflanzen*. Paul Parey, Berlin.

Stack GM, Cala AR, Quade MA, *et al*. (9 other authors). 2023. Genetic mapping, identification, and characterization of a candidate susceptibility gene for powdery mildew in *Cannabis sativa* L. *Molecular Plant-Microbe Interactions*.

Stevens M. 1975. *How to grow marijuana indoors under lights, 3rd edn*. Sun Magic, Seattle.

Suyal N, Upadhyaya ML, Gupta RC. 2011. Studies on air mycoflora over the field of *Cannabis sativa* L. in Almora hills. *Vegetos* 24: 183–190.

Szarka D, Tymon L, Amsden B, Dixon E, Judy J, Gauthier N. 2019. First report of powdery mildew caused by *Golovinomyces spadiceus* on industrial hemp (*Cannabis sativa* L.) in Kentucky. *Plant Disease* 103: 1773.

Takamatsu S, Matsuda S, Grigaliūnaitė B. 2013. Comprehensive phylogenetic analysis of the genus *Golovinomyces* (Ascomycota: Erysiphales) reveals close evolutionary relationships with its host plants. *Mycologia* 105: 1135–1152.

Thiessen L. 2019. *Hemp diseases in North Carolina*. Department of Entomology and Plant Pathology, North Carolina State University. Available at: https://growingsmallfarms.ces.ncsu.edu/wp-content/uploads/2019/03/Hemp-Diseases-Lindsey-Thiessen.pdf?fwd=no

Thiessen LD, Schappe T, Cochran S, Hicks K, Post AR. 2020. Surveying for potential diseases and abiotic disorders of industrial hemp (*Cannabis sativa*) production. *Plant Health Progress* 21(4): 321–332.

Thompson GR, Tuscano JM, Dennis M, *et al.* (8 other authors). 2017. A microbiome assessment of medical marijuana. *Clinical Microbiology and Infection* 23: 269–270.

Transhel V, Gutner L, Khokhryakov M. 1933. Новые виды грибных паразитов на новых лубяных растениях. (A list of fungi found on new bast crops). *Болезни и вредители новых лубяных культур* 4: 127–140.

Trunov GA. 1936. "Материалы по фитопатологическому изучению конопель," pp. 69–113 *in Collection of scientific papers on plant protection.* Ukrainian State Publishing Office, Kiev.

Tymon L, McPartland JM, Thiessen LD, Gauthier NW. 2022. "Powdery mildew," in Gauthier NW, Thiessen LD, *eds. Compendium of Cannabis diseases.* APS Press, St Paul, Minnesota.

Vakhrusheva TE, Davidian GG. 1979. Методические указания по инвентаризации болезней и микрофлоры льна и конопли. All-Union Vavilov Scientific Research Institute of Crop Cultivation, Leningrad.

Van der Werf HMG, Van Geel WCA, Wijhuizen M. 1995. Agronomic research on hemp (*Cannabis sativa* L.) in the Netherlands, 1987-1993. *Journal of the International Hemp Association* 2: 14–17.

van Kan JAL, Shaw MW, Grant-Downton RT. 2014. *Botrytis* species: relentless necrotrophic thugs or endophytes gone rogue? *Molecular Plant Pathology* 15: 957–961.

Victory KC, Couch J, Lowe B, Green BJ. 2018. Occupational hazards associated with harvesting and processing *Cannabis*—Washington, 2015–2016. *Morbidity and Mortality Weekly Reports* 67: 259–260.

Voronin MC. 1890. О пьяном хлебе в Южно-Уссурийском крае. *Ботанические записки* 3(1): 13–21.

Voronin MC. 1891. Über das "Taumelgetreide" in Süd-Ussurien. *Zeitschrift für Pflanzenkrankheiten* 1: 234–236.

Walker AS, Gautier A, Confais J, *et al.* (5 other authors). 2011. *Botrytis pseudocinerea*, a new cryptic species causing gray mold in French vineyards in sympatry with *Botrytis cinerea*. *Phytopathology* 101: 1433–1445.

Wang SH. 2018. *Industrial hemp crop diseases. What we've seen and what we know.* Nevada Department of Agriculture. Available at: http://agri.nv.gov/uploadedFiles/agrinvgov/Content/Plant/Seed_Certification/Industrial_Hemp/NV%20Industrial%20Hemp%20Pathology%20-%203-7-2018.pdf

Weiberg A, Wang M, Lin FM, *et al.* (5 other authors). 2013. Fungal small RNAs suppress plant immunity by hijacking host RNA interference pathways. *Science* 342: 118–123.

Weldon WA, Ullrich MR, Smart LB, Smart CD, Gadoury DM. 2020. Cross-infectivity of powdery mildew isolates originating from hemp (*Cannabis sativa*) and Japanese hop (*Humulus japonicus*) in New York. *Plant Health Progress* 21: 47–53.

Westendorp GD. 1854a. Quatrième notice sur quelques Cryptogames récemment découvertes en Belgique. *Bulletins de l'Académie Royale Belgique Series II*, 21: 229–236.

Williamson B, Tudzynski B, Tudzynski P, Van Kan JAL. 2007. *Botrytis cinerea:* the cause of grey mould disease. *Molecular Plant Pathology* 8: 561–580.

Wilson H, Bodwitch H, Carah J, *et al.* (4 other authors). 2019. First known survey of cannabis production practices in California. *California Agriculture* 73: 119–127.

Wiseman MS, Bates T, Garfinkel A, Ocamb CM, Gent DH. 2021. First report of powdery mildew caused by *Golovinomyces ambrosiae* on *Cannabis sativa* in Oregon. *Plant Disease* 105: 2733.

Wollenweber HW. 1913. Studies on the *Fusarium* problem. *Phytopathology* 3: 24–50.

Wollenweber HW, Reinking OA. 1935. *Die Fusarien.* Paul Parey, Berlin.

Yulfo-Soto GE, Smith H, Szarka D, *et al.* (3 other authors). 2022. First report of *Fusarium graminearum* causing flower blight on hemp (*Cannabis sativa*) in Kentucky. *Plant Disease* 106: 334.

Zaprometov N. 1933. Список болезней новых лубяных культур Средней Азии [List of diseases of new bast crops in Central Asia]. *Болезни и вредители новых лубяных культур* 3: 20–21.

Zaracovitiz C. 1965. Attempts to identify powdery mildew fungi by conidial characters. *Transactions of the British Mycological Society* 48: 553–558.

10 Leaf and stalk diseases

Abstract

The long list of fungi that cause these diseases seems overwhelming, and a proper diagnosis can be critical for control. Nearly 70 species are described here. The chapter is organized to constrain the maelstrom. Some fungi prefer leaves and cause conspicuous symptoms: rust fungi (primarily *Uredo kriegeriana*), downy mildews (four *Pseudoperonospora* spp.), and fungi that cause yellow spot (*Septoria* spp.), brown spot (*Phoma* and *Ascochyta* spp.), white spot (*Diaporthe* and *Phomopsis* spp.), and *Cercospora* spot. Others prefer stalks and present unique diseases: *Sclerotinia* and *Fusarium* cankers. Anthracnose by *Colletotrichum* spp. goes both ways, as do combo-stem–leaf spots by *Stemphylium*, *Leptosphaeria*, and *Cladosporium* spp. An emerging disease, hemp leaf spot, is caused by *Bipolaris gigantea* and allied species in the *Bipolaris–Curvularia–Exserohilum* complex. Keeping in mind 20 rarely encountered fungi listed at the end of the chapter, the long list becomes manageable.

10.1 Introduction

Some fungi cause leaf *and* stalk diseases, such as *Colletotrichum* and *Phomopsis* spp. But most have preferences: *Septoria* and *Cercospora* spp. prefer leaves, *Sclerotinia* and *Fusarium* spp. prefer stalks. Some fungi that cause leaf diseases are worse in flowering tops, see Chapter 9 for powdery mildew and gray mold. Leaves are wilted by root rotters and vascular pathogens; they're treated in Chapter 11. Leaves are also infected by viruses (Chapter 14) and bacteria (Chapter 15). Many fungi in this chapter attracted the attention of government-supported researchers who aimed to eradicate illegal *Cannabis* (Fig. 10.1).

Nicole Gauthier's group compared the "big three" causes of leaf diseases in Kentucky: *Septoria cannabis*, *Bipolaris gigantea*, and *Cercospora flagellaris*. In field plots, disease by *Septoria* appeared first, mid-July, and was most severe in lower to mid-canopies. Disease by *Bipolaris* began in late July, and was most severe in upper canopies. Disease by *Cercospora* developed in September and became severe in both upper and lower canopies (Gauthier, 2021).

Next came a deeper dive into epidemiology: Munir *et al.* (2024) inoculated field-transplanted female clones of six CBD-type cultivars, at two sites in two consecutive years. Disease severity varied across cultivar, location, and year; among those factors, cultivar choice played the largest role. Trump group cultivars ('Trump 1', 'Wife') were the most susceptible, while Otto II group cultivars ('Otto II', 'Endurance') were the least susceptible. 'BaOx' and 'Cherry Citrus' were intermediate in susceptibility.

Trump group cultivars produced denser canopies, providing a more conducive microclimate for disease progression. Surprisingly, leaf spot diseases minimally impacted floral biomass and had no effect on CBD yield, leading the authors to suggest that, "regardless of disease severity, leaf spot diseases may seldom warrant management."

10.2 Yellow leaf spot

More papers have been published about yellow leaf spot than any other disease in fiber-type hemp, plus five monographs (Watanabe and Takesawa, 1936; Ferri, 1959b; Punithalingam, 1980; McPartland, 1995c; Rahnama *et al.*, 2021). The disease is ubiquitous. The primary causal agent, *Septoria cannabis*, was placed on a list of pathogens to be excluded through quarantine in New Zealand (MAF Biosecurity, 2009) and Australia (Paini, 2011), but alas, it was already in Australia (Weeden, 2006).

Treating the number of peer-reviewed studies as a proxy for prevalence, yellow leaf spot causes problems in Russia ($n = 13$), Italy and USA (9 each), Germany, Hungary, and China (5 each), Romania (4), France, Japan, and Kazakhstan (3 each), Ukraine and Canada (2 each), and once each in Armenia, Australia, Belgium, Bulgaria, Chile, Czech Republic, Estonia, Iran, Latvia, Kyrgyzstan, Netherlands, Poland, Spain, Turkey, and Uzbekistan.

Drug-type plants have been infested in India (Patel, 1933; Pandotra and Husain, 1966; Bilgrami *et al.*, 1981), Pakistan (Mushtaque *et al.*, 1973; Ghani *et al.*, 1978; Fig. 10.1), and Nepal (McPartland *et al.*, 2000). When germplasm from those places was grown at the NIDA farm in Mississippi, the plants developed yellow leaf spot (Yu, 1973). Wild-type *Cannabis* has been affected in Hungary (Rapaić von Ruhmwerth, 1914), Russia (Shembel, 1927; Gamalitskaia, 1964), Kazakhstan (Kuzhantayeva, 1991), Kyrgyzstan (Bavlankulova *et al.*, 2013), and the USA (McPartland, 1983a).

Yellow leaf spot is sometimes relegated to the status of an unsightly nuisance. Rataj (1957) considered the disease "widespread but not likely to cause extensive damage." However, leaf photosynthesis is the engine driving crop yields, so yellow leaf spot can result in yield losses of fiber, flowers, and grain. In extreme cases, plants can lose 50–90% of leaves (Gauthier, 2020c). Surprisingly, yellow leaf spot did not affect CBD yields (Munir *et al.*, 2024). Effects on seed and fiber yields await epidemiology studies of that caliber.

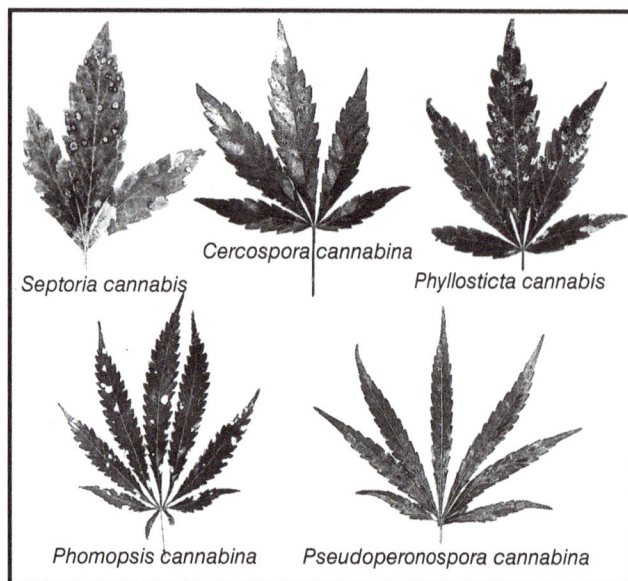

Fig. 10.1. Fungi causing leaf diseases in Pakistan (adapted from Ghani et al., 1978)

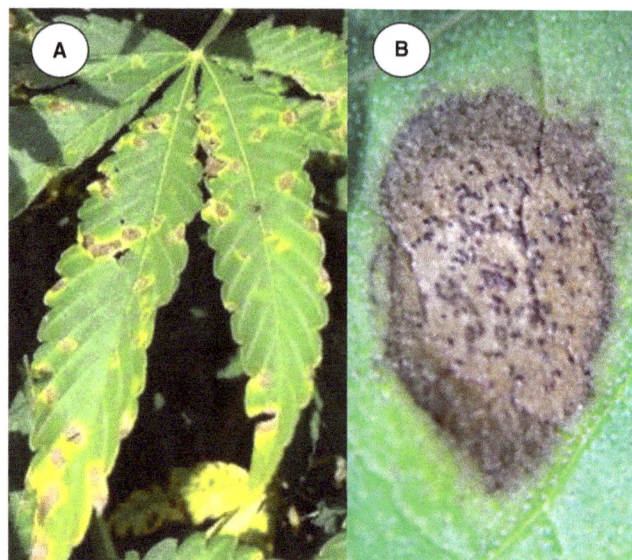

Fig. 10.2. Yellow leaf spot: **A.** macro; **B.** micro (courtesy N. Gauthier)

10.2.1 Symptoms and signs

Yellow spots appear in June, on lower leaves within the canopy. They consist of center lesions (ocher or gray-brown in color) surrounded by a yellow halo. Spots may remain small and round, but usually enlarge to irregularly polygonal shapes, their edges partially delineated by leaf veins (Fig. 10.2). Tiny black pycnidia arise within the center lesions (Fig. 10.2). Under high humidity, conidia ooze out of each pycnidium in a white cirrus—a columnar or tendril-like mass. In severe infections the leaves curl, wither, and fall off prematurely, defoliating the lower part of the plant. Coalescing damage led Ghani et al. (1978) to call the disease "Septoria leaf blight."

Septoria cannabis was the first fungus on *Cannabis* meriting a microscopic illustration, although Berti Pichat (1866) simply labeled it *le fungosità parassitiche*. Kirchner and Boltshauser (1898) illustrated yellow leaf spot in a classic lithograph, copied by others (Flachs, 1936; Ceapoiu, 1958; Barloy and Pelhate, 1962). Spots primarily appear on lower fan leaves, but they may arise on floral leaves of flowering tops (McPartland, 1995c), on stalks (Gitman and Boytchencko, 1934; Ferri 1959a,b), and cotyledons of seedlings (Ferri, 1959a,b).

10.2.2 Causal organisms

Genus *Septoria* (Dothideomycetes, Mycosphaerellaceae) contains about 2000 taxa, with species traditionally delimited by host specificity and morphology. Most *Septoria* spp. produce only asexual conidia. Sexual states linked to *Septoria* include *Sphaerulina* and *Mycosphaerella*, although these sexual states are also linked with other anamorphs. *Septoria* is a polyphyletic assembly of organisms (Verkley *et al.*, 2013; Quaedvlieg *et al.*, 2013). These DNA studies showed that some *Septoria* spp. on different hosts were genetically identical, suggesting that *Septoria* spp. may have wider host ranges than traditionally recognized.

Two *Septoria* spp. that cause *Cannabis* leaf spot, *S. cannabis* and *S. neocannabina*, are compared in Fig. 10.3. Images of pycnidia are the same scale (LM × 500), as are images of conidia (LM × 980). Recently a third species has been named, *S. cannabicola* (Ujat *et al.*, 2024).

10.2a. *Septoria cannabis* (Lasch) Saccardo 1884

≡*Ascochyta cannabis* Lasch 1846, =*Spilosphaeria cannabis* Rabenhorst 1857, =*Septoria cannabina* Westendorp 1857
Description: *Conidiomata* (pycnidia) epiphyllous, gregarious, immersed but eventually erumpent, globose to flask-shaped, averaging 90 μm in diameter (range 39–145 μm) (Rahnama *et al.*, 2021), peridium thick-walled *textura angularis-globulosa*, dark brown, ostiole round, 15 μm in diameter (Fig. 10.3A). *Conidiogenous cells* subglobose to ampulliform, simple, hyaline, holoblastic. *Conidia* hyaline, straight or curved, filiform, tapering towards apex, base truncate, 3–4 septate, (30–)42.2(–55) × 2.0–2.5 μm (Fig. 10.3C). *Microconidia* hyaline, aseptate, ovoid to ellipsoid, 2.5–3.5 × 1.5 μm.
Comments: *S. cannabis* nomenclature is complex and convoluted, explicated and emended by McPartland (1995c). Lasch's type specimen complicated the situation, because the specimen was collected from a plant with two diseases (discussed in the section on brown leaf spot). The taxon *S. cannabis* (Lasch) Saccardo requires nomenclatural conservation, because Saccardo's assistant, Aloysius Vido, made a perfectly valid recombination five years earlier (Vido, 1879).

McPartland (1995c) described conidiogenous cells as holoblastic and determinate (growth ceases after formation of a conidium), whereas Kuzhantayeva (1991) described them as holoblastic and indeterminate (growth continuing after conidial formation), with sympodial-type proliferation (conidiogenous cells grow beneath or beside a previously formed conidium). *S. cannabis* may produce microconidia. Kuzhantayeva described microconidia produced by holoblastic conidiogenous cells, a sort of *Asteromella* morph, whereas McPartland considered the microconidia a type of secondary conidia, budding off primary conidia.

Rahnama *et al.* (2021) provided excellent photos of symptoms caused by *S. cannabis,* along with photomicrographs. They compare morphology (pycnidial diameter, conidial length) in seven isolates from Kentucky. Three isolates were subjected to a robust seven-locus phylogenetic analysis (*TEF1, TUB, RPB2, LSU,* ITS, *ACT, CAL*). The isolates were genetically identical and formed a clade separate from 42 other *Septoria* spp. on other hosts.

10.2b. *Septoria neocannabina* McPartland 1995

≡*Septoria cannabina* Peck 1884 (not *Septoria cannabina* Westendorp 1857), =*Septoria cannabis* var. *microspora* Briosi & Cavara 1888

Description: *Conidiomata* (pycnidia) epiphyllous, gregarious, immersed but erumpent and nearly cupulate, globose, 66 µm in diameter (Fig. 10.3B), peridium thin-walled *textura angularis-globulosa*, honey-brown near the ostiole to almost colorless at the base, ostiole irregular, 20 µm in diameter. *Conidiogenous cells* short ampulliform to lageniform, simple, hyaline, holoblastic, up to 8 µm long. *Conidia* hyaline, usually curved, filiform, pointed at the apex with a truncate base, 1–3 septate, (20–)28.6(–30) × 1.0–2.0 µm (Fig. 10.3D).

Comments: Peck (1884) coined *Septoria cannabina* in New York. The label on Peck's type specimen indicates he originally called the fungus "*Septoria cannabina* Westendorp." Realizing

his specimen was different, Peck scratched out Westendorp's name and wrote "*Septoria cannabina* Peck." Since Peck's later homonym is illegitimate, McPartland (1995c) provided an avowed substitute.

S. cannabis and *S. neocannabina* can be distinguished by several characteristics. Leaf spots caused by *S. neocannabina* often have a dark reddish-brown border (Peck, 1884; Kirchner, 1906; Kirchner, 1966). *S. cannabis* pycnidia are large and thickwalled compared with *S. neocannabina* pycnidia (Fig. 10.3A,B, same scale). Ostioles in *S. cannabis* pycnidia are relatively small, whereas *S. neocannabina* ostioles may open to nearly half the diameter of pycnidia. Conidia of *S. cannabis* are longer and wider, and contain more septa than those of *S. neocannabina* (Fig. 10.3C,D, same scale).

Five authors proposed synonymizing the species (Martin, 1887; Gilman and Archer, 1929; Seymour, 1929; Transchel *et al.*, 1933; Punithalingam, 1980). Others considered the species distinct (Saccardo, 1884; Kirchner, 1906; Oudemans, 1920; Voglino, 1924; Gitman and Boytchenko, 1934; Savulescu, and Sandu-Ville, 1936; Ceapoiu, 1958; Barloy and Pelhate, 1962; Mándy and Bócsa, 1962; Kirchner, 1966; Teterevnikova-Babajan ,1987). None of them examined type specimens. McPartland (1995c) examined the types of Lasch, Rabenhorst, Westendorp, and Peck, as well as 60 other collections of *Septoria* spp. at the US National Fungus Collections (BPI).

Briosi and Cavara (1888) collected a fungus in Italy, noted its similarity to *S. cannabina* Peck, but named it a new variety, *S. cannabis* var. *microspora*. They illustrated cupulate conidiomata (Fig. 10.4A), also seen in Peck's fungus. Voglino (1924) reduced *S. cannabis* var. *microspora* to a synonym under Peck's taxon. In 1917 J.T. Rogers collected a "*Cylindrosporium* sp." in Lyster Dewey's research plots, which also proved to be *S. neocannabina* (McPartland, 1995c). Ostioles in mature pycnidia average 26 µm in diameter, oozing conidia (Fig. 10.4B.).

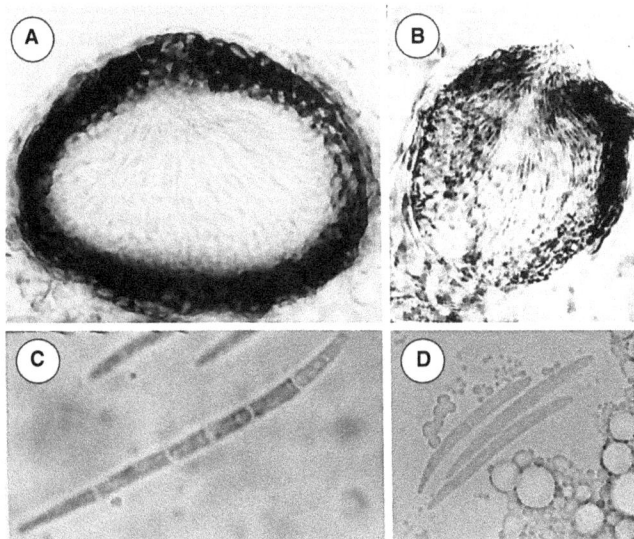

Fig. 10.3. Two *Septoria* species. *S. cannabis*: **A.** pycnidium, **C.** conidia; *S. neocannabina*: **B.** pycnidium, **D.** conidia

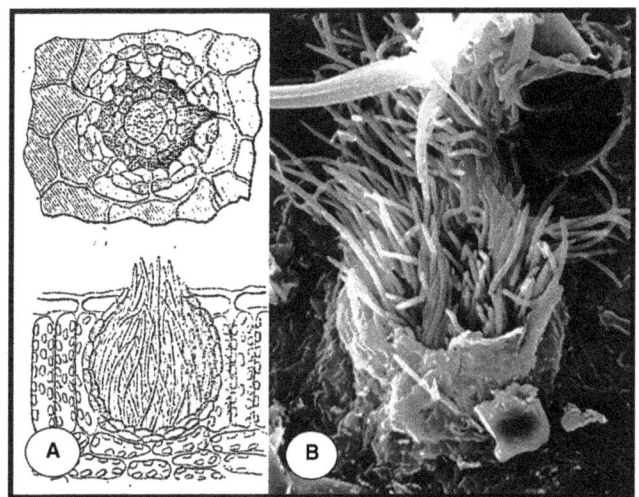

Fig. 10.4. *Septoria neocannabina*: **A.** Briosi and Cavara (1888); **B.** "*Cylindrosporium* sp." (SEM ×500)

10.2c. *Septoria cannabicola* Ujat & C. Nakashima 2024

Description: *Conidiomata* (pycnidia) amphigenous, mainly hypogenous, immersed or erumpent, globose to subglobose, 88–125 μm in diameter, peridium thin-walled *textura angularis*, pale brown to brown, ostiole 20–25 μm in diameter. *Conidiogenous cells* short ampulliform, hyaline, lining inner cavity of basal half of conidiomata, holoblastic, percurrently proliferating, 2.5–5 × 2–3 μm. *Conidia* hyaline, straight to slightly curved, cylindrical to obclavate, pointed at the apex with a truncate base, 0–4 septate, 30–40 × 1.0–2.5 μm.

Comments: Ujat *et al.* (2024) erected this species based on a seven-loci phylogenetic analysis, which differentiated it from three *S. cannabis* specimens genotyped in Kentucky by Rahnama *et al.* (2021). Ujat and colleagues isolated *S. cannabicola* in Japan. The fungus caused leafspots on lower leaves and early defoliation. Leaf spots were described as amphigenous, angular, yellow with indistinct border at early stages, later becoming brown, circular to irregular, surrounded by yellow halo, 2–8 mm.

The authors noted that earlier Japanese phytopathologists attributed the disease to *S. cannabis* (e.g., Watanabe and Takesawa, 1936). Inspection of Japanese herbarium specimens collected from 1896 to 1947 were consistent with *S. cannabicola*. The wide, nearly cupulate ostioles of *S. cannabicola* resemble those of *S. neocannabina*, but the pycnidia are much larger. Conidia of *S. cannabicola* are intermediate in size between *S. cannabis* and *S. neocannabina*.

10.2.3 Differential diagnosis

Yellow leaf spot can be confused with brown leaf spot (*Phoma* and *Ascochyta* spp.), whose lesions are smaller and darker. The lesions of white leaf spot (*Phomopsis* spp.) are paler than yellow leaf spot, and the dried lesions rarely flake away. Symptoms of Cercospora leaf spot (*Cercospora* and *Pseudocercospora* spp.) and downy mildew (*Pseudoperonospora* spp.) are similar, but their causal agents look quite different under the microscope.

S. cannabis sometimes sporulates on stalks (Gitman and Boytchencko, 1934; Ferri, 1959a,b; McPartland, 1995c), and may be confused with other fungi. Fautrey (1899) coined *Rhabdosora cannabina* for a fungus on hemp stalks; his description of *R. cannabina* resembles *S. cannabis* (see later in this chapter). *Jahniella bohemica* sporulates on hemp stalks, and its conidia look like those of *S. cannabis* (see later in this chapter).

10.2.4 Disease cycle and epidemiology

Septoria spp. overwinter as pycnidia in crop residue near the soil surface. Seedlings and young plants become infected in early spring. Conidia of *S. cannabis* germinate rapidly and infect plants by penetrating stomates (Ferri, 1959b). Optimal growth of *S. cannabis* occurs at 25°C, with an incubation period of 6–7 days between inoculation and early symptoms (Watanabe and Takesawa, 1936). Pycnidia arise on infected plants, and spew copious amounts of conidia. Conidia spread by splashed rain and give rise to epidemics in the summer.

Gauthier (2020c) proposed that neighboring weeds may serve as alternative hosts, providing a "green bridge" for survival from one season to the next. Barloy and Pelhate (1962) suggested seed-borne infection, but did not confirm this experimentally. Other studies of *Cannabis* seed-borne fungi have not implicated *Septoria* spp. (Pietkiewicz, 1958; Ferri, 1961b; Babu *et al.*, 1977). Mándy (1972) stated that *Septoria* disease was worse on female plants, because their longer lifespan led to more secondary infections.

10.2.5 Disease control

Sanitation is the cornerstone of control. Rogue infected plants and remove all infected crop residues. Crop residues have been burned in the field after harvest (Serzane, 1962; Punithalingam, 1980). Alternatively, crop residues can be tilled into the soil—burial suppresses *Septoria* inoculum. Crop rotation is an option, monocots are a good bet, although host range studies of *S. cannabis* and *S. neocannabina* have not been done. Eliminate weeds—they create a humid microclimate that enhances conidial germination, and some might serve as alternative hosts.

Decrease nitrogen fertilizer and increase P and K (Serzane, 1962). Dixon *et al.* (2022b) suggested adding silicon to peat-based growth media; they cited literature showing that silicon reduced *Septoria* disease in other crops, and they showed that silicon supplementation controlled powdery mildew. Avoid overhead irrigation. Monitor for disease during rainy seasons. Stay out of *Cannabis* fields when plants are wet, since conidia are easily spread by contact. No breeding for resistance has been done. Bòcsa (1958) said that inbred monoecious cultivars were susceptible to *Septoria* infection. Barloy and Pelhate (1962) identified other susceptible varieties: late-maturing fiber-types, and a landrace from Libya.

Dixon *et al.* (2022a) screened several biocontrols and biorational pesticides for *in vitro* efficacy against *S. cannabis* (see Table 9.1). Best results were obtained with Cueva® (copper soap), Procidic® (citric acid), and LifeGard® (*Bacillus mycoides*). Gyenis *et al.* (2003) needed to repeat applications every 7–14 days to control *Septoria* disease in poplar trees. Yu (1973) sprayed plants with Bordeaux mixture to control *S. cannabis* in NIDA's *Cannabis* garden in Mississippi. Schwartz *et al.* (2024) tested nine fungicides against *S. cannabis* at Cornell. A combination of Double Nickel® (*Bacillus amyloliquefaciens* strain D747) and CureZin® (copper + zinc) reduced disease severity, followed by CureZin® alone, but statistically they did not differ from controls. Other fungicides were nearly worthless (Double Nickel® alone, Regalia®, Cease®, Cueva®, MilStop® SP, PERpose Plus™, RootShield® Plus).

10.3 Sclerotinia canker and crown rot

Rataj (1957) considered this disease the #1 scourge of fiber-type hemp, "Hemp canker caused by Sclerotinia sclerotiorum … is the only disease capable of destroying the plants and has a devastating effect on the fibre." Termorshuizen (1991) ranked it #2 behind gray mold. The dank canopy enclosing densely sown fiber-type crops provides a perfect microclimate (de Meijer *et al.*, 1995). The disease has caused 40% crop losses in Nova Scotia (Hockey, 1927), and 60–70% losses in India (Sharma *et al.*, 2016).

This disease also damages crops grown for seed (Callaway and Laakkonen, 1996). Given proper meteorological conditions, *Sclerotinia* spp. have afflicted widely spaced plants grown for *kif* in Morocco (Anonymous, 1937), *gañjā* in India (Bilgrami *et al.*, 1981; Sharma *et al.*, 2016; Pathak *et al.*, 2017), and hoophouse sinsemilla in Canada (Punja and Ni, 2021). The sequence of countries reauthorizing hemp cultivation parallels the re-emergence of this disease: the USA during WWII (Buchholtz, 1943), and more recently in Holland (de Meijer *et al.*, 1995), Germany (Patschke *et al.*, 1997), Australia (Lisson and Mendham, 1995), Canada (Moes *et al.*, 1999; Bains *et al.*, 2000; Punja and Ni, 2021), New Zealand (McPartland and Rhode, 2005), the USA (Darby *et al.*, 2017; Koike *et al.*, 2019; Garfinkel, 2021), and Colombia (Rodríguez-Yzquierdo *et al.*, 2021).

10.3.1   Symptoms and signs

Symptoms usually begin in late summer, in late vegetative stages or flowering stages. The disease may arise near the soil line (Fig. 10.5A), or in the upper 2/3rds of the plant (Fig. 10.5B). Symptoms begin with a water-soaked lesion on the stalk. Cortical tissues beneath the lesion collapse, creating a pale brown canker. In high humidity, the canker becomes shrouded in a white, cottony mycelium. Scraping away the mycelium reveals a shredded appearance.

If the canker encircles the stalk, all foliage above the lesion wilts. Flowering, if it has begun, promptly ceases (Flachs 1936). Wilting is followed by necrosis, with leaves remaining attached to plants (Darby *et al.*, 2017). The fungus destroys bast fibers at the canker site. Stalks may snap and fall over. Harvested plants yield a higher percentage of tow than normal (Buchholtz,

1943). By September large black sclerotia arise on stalk surfaces, or within the hollow of dead stalks.

10.3.2   Causal organisms

Three *Sclerotinia* spp. are *Cannabis* pathogens. *S. sclerotiorum* has been cited, under six synonyms, in Australia, Canada, Colombia, China, Czech Republic, Finland, France, Germany, Holland, Italy, Morocco, New Zealand, Poland, Russia, Ukraine, and the USA. *S. minor* has been reported in California (Koike *et al.*, 2019). The third species, *S. fuckeliana*, is the sexual state of *Botrytis cinerea*—the cause of gray mold—and discussed in section 9.2.

10.3a.   *Sclerotinia sclerotiorum* (Lib.) deBary 1884, Leotiomycetes; Sclerotiniaceae

≡*Peziza sclerotiorum* Libert 1837, ≡*Whetzelinia sclerotiorum* (Lib.) Korf & Dumont 1972, =*Sclerotinia libertiana* Fuckel 1870, =*Sclerotinia kauffmanniana* (Tikhom.) Saccardo 1889, ≡*Peziza kauffmanniana* Tikhomirov 1868, =*Acoromorpha cannabis* Bertolini 1861

Description: *Sclerotium* hard, yellow becoming black, smooth, oblong, 5–13 × 3–7 mm. When growing on the stalk surface the sclerotium is globose, and becomes oblong within the stalk's hollow (Fig. 10.6A). Cross-section of the sclerotium reveals a dark rind surrounding a soft, white center. Each sclerotium produces one to seven apothecia Fig. 10.6B). *Apothecium* consists of a slender brown stipe up to 20 mm long, topped by a yellow-brown cup, usually 2–4 mm wide (Fig. 10.6C).

Fig. 10.5. *Sclerotinia sclerotiorum*: **A.** near base of plant; **B.** in upper part of plant (McPartland *et al.*, 2000)

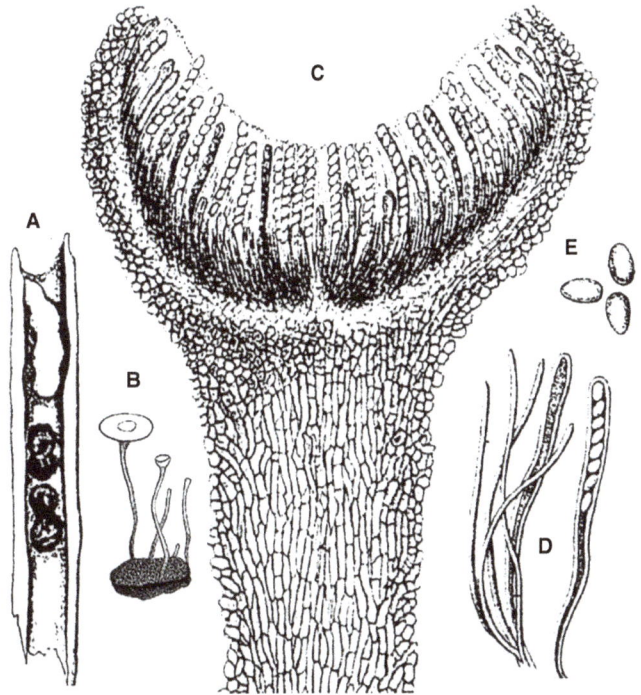

Fig. 10.6. *Sclerotinia sclerotiorum*. All from Tikhomirov (1868), except B by de Bary (1887) (see text for letters)

The apothecial hymenium is filled with a single layer of closely packed asci and paraphyses. *Asci* are narrow-cylindrical, 8-spored, bluing with Melzer's reagent, 110–140 × 6–10 μm (Fig. 10.6D). *Paraphyses* filiform, hyaline, same length as asci. *Ascospores* aseptate, ovoid to elongate, hyaline, usually biguttalate, 9–16 × 4–7 μm (Fig. 10.6E).

Comments: *S. sclerotiorum* attacks over 500 species of crops and weeds, mostly herbaceous dicots. A race attacking Russian hemp also attacked Jerusalem artichoke, potato, safflower, flax, and colza (Antonokolskaya, 1932). A German strain on hemp also infested colza, i.e., canola (Raabe, 1939). DNA data from a worldwide collection of *S. sclerotiorum* found that a haplotype isolated from *Cannabis* was shared by collections from canola, cabbage, tobacco, and kiwi fruit (Carbone and Kohn, 2001).

Early hemp researchers gave other names to *S. sclerotiorum*. Bertolini (1861a) assigned the name *Psoromorpha cannabis* to a fungus he subsequently named *Acoromorpha cannabis* (Bertolini, 1861b). He added that his colleague Luigi Botter had provisionally named the fungus *Spumaria cannabis*. Comes (1891) synonymized *A. cannabis* under *Peziza kauffmanniana*.

Peziza kauffmanniana was described on hemp by Tikhomirov (1868) in Smolensk, Russia. De Bary (1886) obtained a specimen from Tikhomirov, and said it was morphologically identical to *S. sclerotiorum*. Saccardo (1889), unaware of de Bary's work, transferred *P. kauffmanniana* to *Sclerotinia*, as *S. kauffmanniana*. Kohn (1979) reduced *S. kauffmanniana* to a synonym of *S. sclerotiorum*, unaware that de Bary had done so. Behrens (1891) identified *Sclerotinia libertiana* attacking hemp in France. Prillieux (1897) synonymized *S. kauffmanniana* under *S. libertiana*, and *S. libertiana* is now a synonym of *S. sclerotiorum*.

S. sclerotiorum was reported in India for the first time by Shaw and Ajerakar (1915) on several hosts, including *Cannabis*. They provided photographs of plant symptoms and microscopic images. The causal agent was presented to them as *Rhizoctonia napi*, but they doubted that was correct, and suggested a *Botrytis* sp. Joshi (1924) identified their fungus as *S. sclerotiorum*.

10.3b. *Sclerotinia minor* Jagger 1920, Leotiomycetes; Sclerotiniaceae

Description: *Sclerotium* hard, black, irregularly shaped, and small, 0.5–3 mm in length, always formed on stalk surfaces, embedded in mycelium. The sclerotium rarely sprouts apothecia, usually single. *Apothecium* consists of a slender yellow-brown stipe, topped by a beige or yellowish funnel-shaped cup, 3–5 mm wide. The apothecial hymenium is filled with a single layer of closely packed asci and paraphyses. *Asci* are narrow-cylindrical, 8-spored. *Ascospores* aseptate, elliptical-obovoid, hyaline, usually biguttalate, 8–17 × 5–7 μm.

Comments: *S. minor* attacks a wide range of vegetable crops, including lettuce, endive, escarole, and radicchio (Koike *et al.*, 2007). Koike *et al.* (2019) isolated it from field-grown hemp in California. A white to gray mycelium formed at the crown (base of stalk) in contact with soil, giving rise to sclerotia (Fig. 10.7). Tissue beneath the epidermis was necrotic. Roots were not affected. Disease incidence was low, approximately 1% of plants in the field. DNA extracted from the pathogen matched

Fig. 10.7. Crown rot caused by *Sclerotinia minor*, with penny for size comparison (S. Koike, TriCal Diagnostics)

sequences for *S. minor*. Koike and colleagues successfully performed Koch's postulates, proving pathogenicity.

10.3.3 Differential diagnosis

The gray mold fungus, *Botrytis cinerea*, causes stem cankers (see Fig. 9.6). In humid conditions, *B. cinerea* covers stems with a blanket of gray-brown mycelium and conidia, whereas *S. sclerotiorum* and *S. minor* produce a white mycelium. *S. sclerotiorum* sclerotia are larger than those of *B. cinerea* or *S. minor*. The rind covering *S. sclerotiorum* sclerotia consists of 3–4 layers of cells with thick black walls, whereas the rind of *B. cinerea* sclerotia is thinner, usually 1–2 cells thick. The internal medulla of both species is white, but *B. cinerea* is denser and more gelatinous. *S. minor* sclerotia are small and quite irregular in shape.

Wilt caused by *Sclerotinia* spp. can be mistaken for diseases caused by wilt pathogens (e.g., *Fusarium* spp., *Verticillium* spp.), or leaf blight fungi (*Alternaria* spp., *Botryosphaeria* spp., *Colletotricum* spp., *Pseudocercospora cannabina*). Stalk lesions superficially resemble those of Southern blight (*Sclerotium rolfsii*) or Rhizoctonia sore shin (*Rhizoctonia solani*). *S. rolfsii* and *R. solani* form sclerotia, but much smaller than *S. sclerotiorum* sclerotia. In dry conditions, symptoms of Sclerotinia canker and crown rot—lacking flocculose mycelia—may resemble those of Cladosporium stem canker. Molecular-based assays can detect *S. sclerotiorum* in plants and even in soil. The assays are based on antibody-based ELISA tests or real-time PCR.

10.3.4   Disease cycle and epidemiology

S. sclerotiorum and *S. minor* overwinter as sclerotia in plant debris or in soil. They survive periods of freezing and thawing, and survive dry heat up to 70°C. Springtime moisture and moderate temperatures (15–20°C) initiate sclerotial germination: *S. sclerotiorum* forms apothecia or infective hyphae, *S. minor* usually just the latter. Apothecia forcibly eject asci into the air, which are carried by wind to nearby plant surfaces. Ascospores germinate in the presence of moisture, and directly penetrate host epidermis or penetrate via wounds. Both species grow inter- and intracellularly through host parenchyma and cortex. *S. sclerotiorum* and *S. minor* thrive in cool, moist conditions, just like gray mold. *S. sclerotiorum* disease accelerates after canopy closure, especially in high-density stands (de Meijer *et al.*, 1995).

10.3.5   Cultural and mechanical control

Prevention is key. Once established, *S. sclerotiorum* and *S. minor* are hard to kill. Whenever possible, plant in wide rows that are oriented for optimal air circulation and sunlight penetration. Crops sown in dense stands must be monitored closely. Rogue infected plants, but wear gloves, because handling *S. sclerotiorum* may cause a contact photodermatitis. Irrigate at the soil level or use drip irrigation, and not overhead irrigation. To reduce sclerotial germination, allow the soil surface to completely dry between irrigation events. High nitrogen enhances this disease. Amend soil with calcium, and neutralize acidic soils to enhance calcium absorption (McPartland *et al.*, 2000). Ferraris (1935) suggested adding phosphate.

Best tilling practices are debatable. The number of viable sclerotia in soil declines in no-till conditions, but crop debris may harbor sclerotia. Alternatively, deep plowing of crop debris can be effective—sclerotia do not germinate if buried deeper than 6 cm underground (Lucas, 1975). Romashchenkov (1936) reported that *S. sclerotiorum* disease incidence was higher in densely sown crops (40%) than less dense stands (10–15%); mowing grass between rows reduced humidity and reduced disease incidence. Weeding is important, given the wide host range of *S. sclerotiorum* and *S. minor,* and their preference for humidity in weedy fields. Wild mustard, *Sinapsis arvensis*, provided the source of inoculum for hemp canker in Manitoba (Moes *et al.*, 1999). Rotate with grains and grasses for at least 3–4 years, kept weed-free, to starve the long-lived sclerotia.

S. sclerotiorum invades the seeds of many plants; seed-borne infection probably happens with *Cannabis*, although it has not been documented. Small sclerotia are the same size and density of hemp seeds, can pass through the threshing and screening process, and contaminate seed lots (Moes *et al.*, 1999). Serzane (1962) emphatically stated that seeds from infected crops should not be used. No breeding for resistant cultivars

has been done. Darby *et al.* (2017) compared infection rates (percentage of total plants) in eight cultivars: 'Katani' 10.2%, 'Grandi' 8.2%, 'CRS-1' 0.72%, 'Canda' 0.655%, 'CFX-2' 0.45%, 'Full sun' 0.35%, 'Anka' 0.27%, 'Delores' 0.13%. They reported a trend: shorter cultivars had higher infection rates than taller cultivars.

10.3.6   Biological and chemical control

Glasshouse soil can be dried for several days, and then soaked for 2–3 weeks. This encourages the growth of soil organisms that parasitize and kill sclerotia. Commercially available biocontrol fungi include *Trichoderma harzianum*, *Trichoderma virens*, and *Gliocladium catenulatum* (see section 20.6). Health Canada (2018) approved several products containing these fungi for use on medical cannabis. *Paraphaeosphaeria minitans* also shows efficacy. *Bacillus subtilis* (Serenade®, etc.) is a soil bacterium that colonizes plant roots and inhibits fungi. Strgulec *et al.* (2016) treated fiber crops with Serenade®, applied as a soil drench at a rate of 8 L/ha, up to six applications per season.

Fungicides for *S. sclerotiorum* are registered in other crops, but consistent control is rarely achieved. De Meijer *et al.* (1995) alternated vinclozolin (0.5 kg/ha) and iprodione (0.5 kg/ha), from June through August at 14-day intervals, and this did not provide consistent control.

10.4   *Fusarium canker*

We introduced the genus *Fusarium* (Sordariomycetes; Nectriaceae) in section 9.3. Stalk cankers are caused by seven *Fusarium* spp., classified in three *Fusarium* "species complexes" (see Table 9.2). Cankers usually arise in midseason-to-mature plants.

10.4.1   Symptoms and signs

Symptoms begin as water-soaked lesions on stalk epidermis, followed by epidermal chlorosis and necrosis. Stalks often swell at the lesion site, creating fusiform-shaped cankers that may split open. Dewey (1915) said cankers often appear at branch nodes, and foliage emerging from those nodes become wilted and necrosed, clinging to the plants. Bast fibers beneath the canker become shredded, and the stalk may snap at the lesion site. The canker rarely girdles the stalk, but if it does, all foliage above the canker wilts. Slicing open the canker reveals a reddish-brown discoloration. The discoloration may also be seen if the canker splits open (Fig. 10.8).

10.4.2   Causal organisms

Seven *Fusarium* spp. reportedly cause this disease. Only three species are common, and described below. The other four species are cited but spared further details. "*Fusarium roseum*" in *Cannabis* literature is a *nomen confusum*; the name has been applied to two species.

10.4a. *Fusarium graminearum* Schwabe 1838

=*Fusarium roseum* Link emended Snyder, Hansen & Oswald 1957, *pro parte*
Sexual state: *Gibberella zeae* (Schweinitz) Petch 1936, ≡*Sphaeria zeae* Schweinitz 1822 (not *Sphaeria zeae* Schweinitz 1832)
=*Gibberella saubinetii* (Durieu & Montagne) Saccardo 1879, ≡*Sphaeria saubinetti* Durieu & Montagne in Durieu 1849, =*Botryosphaeria dispersa* De Notaris 1863
Description: *Conidiogenous cells* hyaline, short, doliiform, phialidic, simple or branched dichotomously, proliferating through previous phialides, 10–14 × 3.5–4.5 µm. *Macroconidia* sickle-shaped, hyaline, with a well-marked foot cell, 3–7 septate, 25–30 × 3–4 µm. *Microconidia* absent. *Chlamydospores* rare, single or in chains, in hyphae or rarely in macroconidia, hyaline to brown, thick-walled, smooth or slightly rough surface, 10–12 µm in diameter. *Perithecia* superficial, clustered, blue-black, ovoid, with a rough tuberculate surfaced, 140–250 µm in diameter. *Asci* clavate, usually 8-spored (rarely 4–6), 60–65 × 8–11 µm, but rapidly deliquesce after ascospore production. *Ascospores* fusoid, hyaline to light brown, curved, 3-septate (20% 2-septate or less), 19–30 × 3–5 µm.
Comments: Saccardo (1873) cited *"F. roseum"* on Italian hemp, and gave its sexual state as *Botryosphaeria dispersa*. Subsequently, Saccardo (1883) cited *"F. roseum"* with the sexual state as *Gibberella saubinetii*. Many authors cite *G. saubinetii* on hemp (Cuboni and Mancini, 1886; Voglino, 1912; Oudemans, 1920; Wollenweber and Reinking, 1935; Miller *et al.*, 1960; Vakhrusheva and Davidian, 1979). *G. saubinetii* was synonymized under *G. zeae* by Shear and Stevens (1935), whereas

Petch (1936) synonymized it under *G. cyanogena* (see next species). The debate continues, and some mycologists consider *G. saubinetii* a legitimate species, but most agree with Shear and Stevens.

The sexual state *Gibberella zeae* was first cited on hemp in Romania (Ceapoiu, 1958). McPartland and Cubeta (1997) isolated *F. graminearum* from stalks of wild-type hemp in Illinois (Fig. 10.8), and identified the sexual state as *G. zeae*. It was homothallic and readily produced perithecia. They deposited a culture at the Fusarium Research Center, Pennsylvania State University (accession R-8965), where Paul Nelson confirmed the identification. Canker caused by *F. graminearum* has been reported in North Carolina (Thiessen, 2019), Virginia (Britt *et al.*, 2020), and Kentucky (Ricciardil and Smith, cited in Gwinn *et al.*, 2021). *F. graminearum* can also infect flowering tops (see section 9.3).

10.4b. *Fusarium sambucinum* Fuckel 1863

=*Fusarium sulphureum* Schlechtendahl 1824, =*Fusarium roseum* Link 1809
Sexual state: *Gibberella pulicaris* (Fries) Saccardo 1877, =*Gibberella cyanogena*(Desmazières) Saccardo 1883, =*Gibberella quinqueseptata* Sherbakoff 1928.
Description: *Conidiogenous cells* hyaline, cylindrical, forming palmately from metulae, phialidic, 12–20 × 3–5 µm. *Macroconidia* fusiform, hyaline, with a curved, pointed apical cell and a marked foot cell, usually 3–4 septate (range 0–6), 30–50 × 3.5-5.0 µm (Fig. 10.9). *Microconidia* absent. *Chlamydospores* rare, single or short chains within hyphae or macroconidia, smooth-walled, 8–10 µm in diameter. *Perithecia* superficial, scattered, blue-black, globose, peridium textura angularis-globulosa, with a distinct ostiole, 150–300 µm in diameter. *Asci* clavate, 8-spored, 65–90 × 12–20 µm. Pseudoparaphyses present but evanescent. *Ascospores* ellipsoid, hyaline, straight, slightly constricted at 3 septa, 20–25 × 5–7 µm (Fig. 10.9).
Comments: The taxon *F. sambucinum* has been conserved, under which fall the synonyms *F. roseum*, *F. sulphureum*, and

Fig. 10.8. Fusarium canker, with dot-like perithecia of the sexual state, *Giberella zeae* indicated by arrow

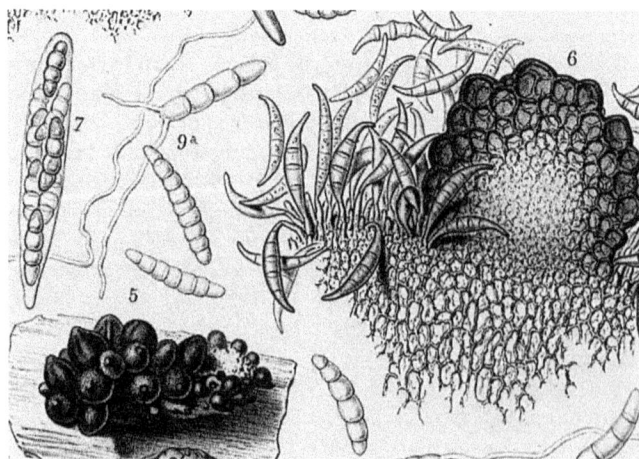

Fig. 10.9. *Fusarium sambucinum* and its sexual state *Gibberella pulicaris* (Tulasne and Tulasne, 1865)

their sexual states (Gams *et al.*, 1997). The sexual state *Gibberella pulicaris* was cited on hemp by Oudemans (1920). Wollenweber (1914) cited *Giberella cyanogena* on hemp, but erroneously, treating it as a synonym of *Gibberella saubinetii* (see previous species). In Czech hemp, Bartoš (1968) connected *F. sambucinum* with *G. pulicaris*. Lyster Dewey photographed diseased hemp stalks at Pierceton, Indiana (McPartland, pers. obs., USDA archives). He sent specimens to Flora Patterson, who diagnosed "*Fusarium* sp., and apparently none has been reported on this host" (Dewey, 1915). Sherbakoff (1928) re-examined the specimen, found the sexual state, and named it *Gibberella quinqueseptata*. McPartland (1995d) synonymized *G. quinqueseptata* under *G. cyanogena*, and Gams *et al.* (1997) synonymized *G. cyanogena* under *G. pulicaris*.

10.4c. *Fusarium proliferatum* (Matsushima) Nirenberg 1976

Sexual state: *Gibberella intermedia* (Kuhlman) Samuels *et al.* 2001
Description: *Conidiogenous cells* hyaline, short or long, and mono- or polyphialidic. *Macroconidia* rare to abundant, slender, rather straight, slightly sickle-shaped due to curved apical and foot cells, 3–5 septate, 19.5–50.5 × 2.5–4.5 μm. *Microconidia* abundant, ellipsoid to club-shaped with a flattened base, aseptate, 3.5–15.5 × 1.5–6.2 μm, commonly formed in chains (up to 10 microconidia) or small balls ("false heads"). *Chlamydospores* absent.
Comments: *F. proliferatum* attacks many crops, including maize, wheat, rice, onion, asparagus, tomato, soybean, banana, fig, mango, and pineapple. It produces a slew of mycotoxins, including fumonisin, furarin, fusaric acid, and beauvericin. *F. proliferatum* caused stem rot and pith necrosis in glasshouse-grown plants (Punja, 2021c). It also caused damping-off on rooted cuttings in propagation rooms. Punja isolated *F. proliferatum* from facilities in British Columbia, Ontario, New Brunswick, and California. In some cases, *F. oxysporum* was recovered with *F. proliferatum*. Punja confirmed pathogenicity with Koch's postulates, which showed that *F. proliferatum* caused more extensive stem colonization than

Fig. 10.10. Pith necrosis caused by *F. proliferatum*: **A.** cross-sectioned stems; **B.** longitudinally sectioned stems (healthy plant on right side) (courtesy Z. Punja)

F. oxysporum. F. proliferatum uniquely proliferated in pith and surrounding parenchyma cells, causing brown-black discoloration, with shredding of surrounding xylem tissues in advanced disease (Fig. 10.10). Breton *et al.* (2022) isolated *F. proliferatum* from plants in Québec.

10.4.3   Other reported fusaria

10.4d. *Fusarium culmorum* (W.G. Smith) Saccardo 1895

F. culmorum normally attacks cereal crops. It was reported from Argentinean *Cannabis* (Abiusso, 1954). No sexual state known. *F. culmorum* was recovered from plants in New Brunswick, identified using ITS sequences (Cormier, cited in Gwinn *et al.*, 2021).

10.4e. *Fusarium lateritium* Nees 1817

Sexual state: *Gibberella baccata* (Wallroth) Saccardo 1883
Pietkiewicz (1958) isolated this species from hemp seeds in Poland. It normally infests woody hosts.

10.4f. *Fusarium avenaceum* (Fries) Saccardo 1886

Sexual state: *Gibberella avenacea* R.J. Cooke 1967
F. avenaceum attacks a wide range of hosts. Zelenay (1960) reported *F. avenaceum* infecting *Cannabis* seedlings in a host-range experiment. Punja (2021d) recovered the fungus from indoor-grown drug-type plants and outdoor fiber-type plants with crown and root rots.

10.4g. *Fusarium fujikuroi* Nirenberg 1976

Sexual state: *Gibberella fujikuroi* (Sawada) Wollenweber 1931
F. fujikuroi was isolated from field-grown hemp in Virginia (Chawla, cited in Gwinn *et al.*, 2021). *F. fujikuroi* causes "foolish rice disease," where seedlings grow tall and spindly. These symptoms are due to fungal metabolites, named gibberellins after the *Gibberella* sexual state.

10.4.4   Differential diagnosis

The two most common causes, *F. sambucinum* and *F. graminearum*, produce macroconidia ≥30 μm and ≤30 μm long, respectively. Conidiogenous cells of *F. sambucinum* branch in a palm shape, whereas *F. graminearum* conidiogenous cells branch dichotomously or not at all. The presence of microconidia in *F. proliferatum* differentiate it from *F. graminearum* and *F. sambucinum*. Abundant microconidia, however, are also produced by *F. solani*, the cause of crown and root rot (see section 11.3). Molecular methods—sequencing ITS and *EF-1α* —are often needed for secure identification.

Fusarium spp. cause internal stem discoloration (reddish in *F. graminearum* and *F. sambucinum*, brown-black in *F. proliferatum*). This discoloration differentiates Fusarium canker from gray mold, hemp canker, and southern blight. Stem swelling caused by *Fusarium* spp. can be confused with stem swellings caused by European corn borer, hemp borer, and assorted beetles.

10.4.5   Disease cycle and epidemiology

Sexual states (perithecia) arise after plant death, and they overwinter in crop stubble or soil. In the spring, ascospores or conidia infect seedlings. *F. graminearum* can colonize hemp seeds (Ferri, 1961b; Yulfo-Soto *et al.*, 2022), and thereby cause seed-borne infections. Later in the season, *Fusarium* spp. invade roots via wounds created by nematodes and *Orobanche* parasites.

Fusarium conidia move in water droplets, so they spread best in damp conditions and heavy soils. Watch for *Fusarium* epidemics during seasons with above-average rainfall. *Fusarium* conidia easily spread to sites of secondary infection, arising at stalk nodes or stalk wounds (e.g., wounds caused by stalk-boring insects, wind, improper pruning, and from damage caused by dry, caking soil or salt damage). Rockwool blocks should not be allowed to dry out, because evaporated fertilizer salts may accumulate around the base of the stalk and favor infections.

All seven *Fusarium* spp. are distributed worldwide. *F. graminearum* and *F. avenaceum* predominate in cooler climates. *F. graminearum* is the most pathogenic of the bunch (Booth, 1971), and seems to be the most common causal agent in *Cannabis*. Most *Fusarium* spp. are facultative parasites, and exist as soil saprophytes until a parasitic opportunity presents itself.

10.4.6   Cultural and mechanical control

Booth (1971) characterized Fusarium canker as a "disease of accumulation." Sanitation is key: rogue and destroy infected plants, and eliminate weeds. Bury overwintering perithecia by plowing under all crop residues. Do not plant in heavy, wet soils, or low-lying areas, or fields with a history of *Fusarium* disease. Seed-borne infection is possible—do not save seeds from infected plants.

Use drip irrigation and avoid overhead irrigation. Avoid excess nitrogen; increase potassium and phosphorus. Choose KCl as a potassium supplement instead of K_2SO_4 or KNO_3. The latter may actually increase *Fusarium* disease (Elmer, pers. comm., 1990). Since *Fusarium* spp. have a wide host range and live saprophytically in soil, crop rotation is not effective. Where *F. graminearum* is an issue, do not rotate after wheat, maize, or sorghum. Dempsey (1975) reported that the cultivar 'Fibramulta 151' is resistant to *Fusarium* spp.

10.4.7   Biological and chemical control

Glasshouse soil can be dried for several days, and then soaked for 2–3 weeks. This encourages the growth of soil organisms that parasitize and kill *Fusarium* spp. Commercially available biocontrols include *Bacillus subtilis*, *Clonostachys rosea*, *Pseudomonas chlororaphis*, *Pseudomonas fluorescens*, *Pythium oligandrum*, *Streptomyces griseoviridis*, *Streptomyces lydicus*, and the *Trichoderma* sisters: *T. harzianum*, *T. virens*, and *T. asperellum*.

Encourage the growth of mycorrhizal fungi to control *Fusarium* spp. Products with *Glomus intraradices* or other mycorrhizal fungi are often combined with the aforementioned biocontrols. Vysots'kyi (1962) protected pine seedlings from *Fusarium* by mixing hemp leaves into soil. Spraying fungicides on plants has not proved practical. Neither has soil fumigation, except to control synergistic organisms such as nematodes.

10.5   Downy mildew

Downy mildew was considered worse than powdery mildew, until indoor cultivation brought powdery mildew to the fore. Zabrin (1981) proposed using downy mildew to destroy illegal *Cannabis* plantations: "A single infected plant introduced into Colombia or Jamaica during a wet season could cause complete devastation."

Downy mildew is caused by *Pseudoperonospora* spp. They are obligate biotrophic pathogens, and draw nutrients from living cells—a lifestyle shared by powdery mildew fungi. *Pseudoperonospora* spp., however, induce a strong hypersensitivity immune response in the host, resulting in considerable necrosis of host tissues. Taxonomically, *Pseudoperonospora* spp. are not related powdery mildew fungi—they are not even fungi. They belong to Class Oomycota, Family Peronosporaceae.

Taking the number of published reports as a proxy for prevalence, downy mildew causes problems in Italy ($n = 8$), USA ($n = 6$), Russia, Germany (4 each), France, Hungary, Pakistan (3 each), Canada, China, India, Japan, Poland, Portugal (2 each), and Bulgaria, Estonia, Kyrgyzstan, Latvia, Lithuania, Romania, Ukraine, and Uzbekistan (1 each). Downy mildew of hemp has been the subject of several monographs (Hoerner, 1940; Glażewska, 1971; Waterhouse and Brothers 1981; McPartland, 2000b; Choi *et al.*, 2005).

Stevenson (1926) included *Pseudoperonospora cannabina* in a list of "foreign plant diseases" posing a threat to US agriculture if accidently imported. McPartland and Cubeta (1997) found it on feral hemp in Illinois. *P. cannabina* infested CBD crops near Geneva in New York (Zuefle, 2020), where it ranked second in severity among foliar pathogens, behind only powdery mildew (Stack *et al.*, 2020). Another group in New York isolated the fungus (Giles *et al.*, 2023). *P. cannabina* has also been identified in Guelph, Ontario, 200 miles west of Geneva (Melzer and Shan, 2022).

10.5.1   Symptoms and signs

Symptoms arise as yellow leaf spots, of irregular size and angular shape, often limited by leaf veins. With age they turn brown (Fig. 10.11A). Opposite the lesions, on undersides of leaves, discoloration is darker brown (Fig. 10.11B). The downy mycelial growth on undersides of leaves appears a lustrous violet-gray color when wet with dew. Lesions enlarge and leaves become twisted and contorted. Leaves may wither, brown, and fall off. Whole plants and entire fields may follow this course.

10.5.2 Causal organisms

Exactly how many *Pseudoperonospora* spp. infect *Cannabis* remains to be determined. Four "suspects" have been named, representing half the entire *Pseudoperonospora* genus.

Pseudoperonospora cannabina (Otth) Curzi 1926 ≡*Peronospora cannabina* Otth 1869, ≡*Peronoplamopara cannabina* (Otth) Peglion 1917, ≡*Pseudoperonospora cannabina* (Otth) Hoerner 1940

Pseudoperonospora humuli (Miy. & Tak.) Wilson 1914 ≡*Peronoplasmopara humuli* Miyabe & Takahashi 1906, ≡*Plasmopara humuli* (Miy. & Tak.) Saccardo 1912, ≡*Peronospora humuli* (Miy. & Tak.) Skalický 1966, =*Pseudoperonospora celtidis* (Waite) Wilson var. *humuli* Davis 1910

Pseudoperonospora celtidis (Waite) Wilson 1907 ≡*Peronospora celtidis* Waite 1892, ≡*Plasmopara celtidis* (Waite) Berlese 1898, ≡*Peronoplasmopara celtidis* (Waite) Clinton 1904

Pseudoperonospora cubensis (Berk. & Curtis) Rostovzev 1903 ≡*Peronospora cubensis* Berkeley & Curtis 1869, ≡*Peronoplasmopara cubensis* (Berk. & Curtis) Berlese 1902

Hoerner (1940) thought the four taxa represented physiological races of a single species, *P. humuli*. Waterhouse and Brothers (1981) allied *P. cannabina* with *P. celtidis*. Shin and Choi (2003) and Choi *et al.* (2005) allied *P. humuli* with *P. cubensis*, whereas Mitchell *et al.* (2011) and Runge *et al.* (2011) said *P. humuli* and *P. cubensis* are different species. These opinions were based on morphological, pathological, and molecular taxonomic characters.

Jaczewski (1928) stated that *P. cannabina* exhibited a unique sporangiophore branching pattern (Fig. 10.12), and its sporangium shape was distinct from *P. humuli*. Waterhouse and Brothers (1981) thought that *P. cannabina* was morphologically similar to *P. celtidis*. Choi *et al.* (2005) considered *P. cannabina* and *P. celtidis* sufficiently distinct: the sporangia of *P. cannabina* did not stain with picronigrosin (the three other species did). *P. celtidis* uniquely contained callose plugs in its sporangiophores (the three other species did not), and *P. celtidis* sporangia had the greatest length/width ratio. Furthermore, *P. cannabina* rarely produced oospores, which was common in *P. celtidis*.

Choi *et al.* (2005) considered *P. humuli* and *P. cubensis* morphologically indistinguishable. Runge and Thines (2012) statistically analyzed 11 morphological characters in *P. humuli* and *P. cubensis*. The two species could be distinguished on their original hosts. Then Runge and Thines cross-inoculated *P. humuli* onto the hosts of *P. cubensis* (cucurbit species), and morphological differences diminished. See Table 10.1 for a summary of the four species.

Biotrophic organisms such as *Pseudoperonospora* often show host specificity—at the level of species, genus, or family.

Fig. 10.11. Symptoms caused by *P. cannabina*: **A.** upper leaf surface; **B.** underside of leaf (courtesy Herrmann and Smart, 2024)

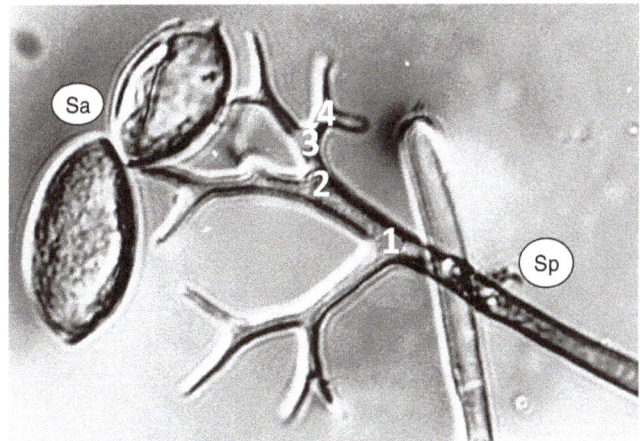

Fig. 10.12. *Pseudoperonospora cannabina*: Sa, two sporangia; Sp, sporangiophore; numbers indicate branching order

Table 10.1. Morphological comparisons of four *Pseudoperonospora* species

Character	P. cannabina	P. humuli	P. celtidis	P. cubensis
Sporangiophore length	160–350 × 5–8 μm	200–460 × 6–8 μm	200–320 × 6–8 μm	120–480 × 5–10 μm
Sporangiophore branching habit[a]	Dichotomous, 3–4 orders of branching	Dichotomous, 5–6 orders of branching	Dichotomous, 4–5 orders of branching	Dichotomous, 5–6 orders of branching
Sporangia size	23–36 × 12–24 μm	22–26 × 15–18 μm	20–38 × 14–16 μm	22–38 × 15–22 μm
Sporangia shape	Ovo-ellipsoid	Ovo-ellipsoid to obovate	Ellipsoid	Ovo-ellipsoid

[a]Example of branching order illustrated in Fig. 10.12

Cannabis and *Humulus* are sister species in the family Cannabaceæ. Hoerner (1940) showed that inoculum from *Humulus* could infect *Cannabis* cultivars ('Bologna,' 'Tochigi,' 'Kymington', etc.), as well as *Celtis* spp. (*C. occidentalis, C. sinensis, C. tournefortii, C. mississippiensis*). *Celtis* spp. are now placed in Cannabaceæ.

Glażewska (1971) reported that *P. humuli* sporangia could infect *Cannabis* (cultivar 'Krasnodarskaja' with 60% success, 'Jugoslowińskie' with 11.4%), and could also infect *Celtis occidentalis* (100%). Choi *et al.* (2005) and many others cite Salmon and Ware (1928) cross-inoculating *P. humuli* onto *Cannabis*, but Salmon and Ware did not. *P. cubensis* causes downy mildew in cucurbits (squash, pumpkin, cucumber, cantaloupe, watermelon). Mitchell *et al.* (2011) reported a low rate of infectivity of *P. cubensis* on hops, and a very low rate (one case) of *P. humuli* infecting cucurbits. Runge and Thines (2012) reported limited infectivity of *P. cubensis* on hops, and *P. humuli* could infect cucurbits.

Reports of *P. humuli* infecting *Cannabis* are problematic for three reasons. (1) *P. humuli* can be devastating—a combination of *P. humuli* and the powdery mildew fungus *Podosphaera macularis* caused New York State's hop industry to collapse a century ago (Summers *et al.*, 2015b). (2) Hop cultivation is currently expanding into many regions where hemp is grown. (3) Hop is a perennial plant, and therefore provides an overwintering site for *P. humuli*. Gent *et al.* (2019) genotyped 216 isolates of *P. humuli*, and demonstrated that subpopulations exist within the species. Cross-infectivity with *Cannabis* is hopefully limited to a subset of subpopulations. Beckerman (pers. comm., 2019) observed no downy mildew in a hemp field growing next to a hops field badly infested by *P. humuli*.

Four DNA studies have addressed this issue (Fig. 10.13). Choi *et al.* (2005) used the ITS region (ITS1, 5.8S rDNA, ITS2), and *P. cannabis* was basal. Repeating their study, but adding *P. urticae*, resulted in a similar tree (Choi *et al.*, 2017). Choi revisited the issue using a multilocus approach (*cox2*,

ypt1, ITS region) with more specimens (Runge *et al.*, 2011). Their tree was more complex, revealing several cryptic species. *P. cannabis* was basal to *P. celtidis*, but *P. celtidis* split into two phylogenetic lineages, within which fell one *P. humuli* specimen (not shown in Fig. 10.13).

In a focus on *P. humuli* and *P. cubensis*, Mitchell *et al.* (2011) published several gene trees, but one gene tree (ITS region) included our other two species of interest. *P. cannabis* and *P. celtidis* were sisters; sequences of *P. humuli* and *P. cubensis* formed a polytomy. When they used other loci (ITS+ β-tublin, *cox1-2* region), *P. humuli* and *P. cubensis* segregated (Fig. 10.13).

Summers *et al.* (2015a) conducted a whole-genome study, which clearly segregated *P. humuli* from *P. cubensis* (they did not include *P. cannabis* or *P. celtidis*). Summers *et al.* (2015b) developed an assay that differentiated *P. humuli* from *P. cubensis* based on an SNP in the *cox2* gene. Taken together, these studies indicate that *P. humuli* and *P. cubensis* are similar, but they exhibit genetic differences that support their retention as separate species. *P. cannabis* and *P. celtidis* clearly differ, and *P. cannabis* is the basal species among the four.

10.5.3 Differential diagnosis

Downy mildew can be confused with Cercospora leaf spot or brown leaf spot. Ferraris (1935) thought the leaf contortions caused by downy mildew resembled symptoms caused by *Ditylenchys dipsaci*, the stem nematode. Careful observation with a microscope or strong magnifying lens makes these problems easy to differentiate.

10.5.4 Disease cycle and epidemiology

Sporangia are spread by wind, and secondarily spread by splashing water. Sporangia germination requires a wet period (a heavy dew will do). They germinate into hyphae, or cleave into zoospores. The hyphae penetrate plant epidermis directly, whereas zoospores can only invade via stomata. Hyphae grow within the host intercellularly, sending haustoria into cells. Sporangiphores emerge from stomata and form sporangia, initiating secondary infections. Epidemics arise when warm, humid, cloudy days are followed by cool wet nights.

Gent and Ocamb (2009) created an algorithm for predicting the risk of *P. humuli* infection in hops. Variables included hours of relative humidity >80%, degree-hours of leaf wetness, and mean night temperature. Their algorithm was useful in regions with endemic disease, and provided disease management "action thresholds" early in the season, when disease prevalence was still relatively low.

P. humuli produces oospores in host tissues, which are overwintering structures. *P. cannabina* rarely forms oospores, with only one report in the literature (Waterhouse and Brothers, 1981). Nevertheless, Barloy and Pelhate (1962) blamed oospores as a source of infection. They noted that *P. cannabina* can persist endemically in a field, becoming progressively worse over the years. *P. cannabina* mycelium can penetrate developing

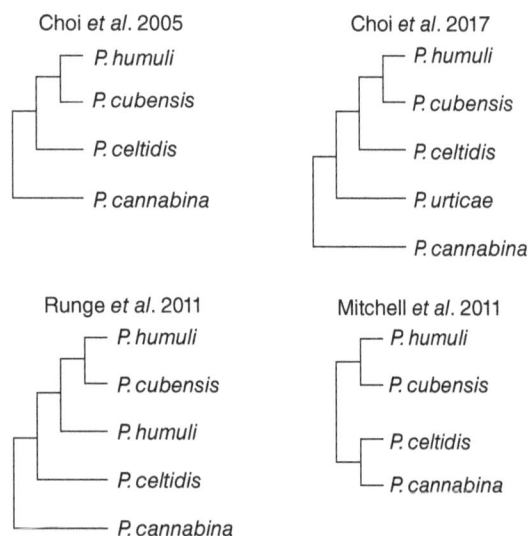

Fig. 10.13. Phylogenetic comparisons of four *Pseudoperonospora* spp.

seeds and lie dormant, initiating seed-borne infections the following year (Hall, 1991).

10.5.5   Control methods

Plant in wide rows that are oriented for optimal air circulation and sunlight penetration. Sanitation is important: rogue infected plants (including roots), and remove all diseased material. Assuming oospores exist, plow under all crop residues. Optimize plant nutrition, and do not over-use nitrogen. Crop rotation is a good option, rotate other crops for a minimum of 3 years (and not with hops).

Although *P. cannabina* is rarely reported in glasshouses or growrooms, follow the advice for glasshouse powdery mildew. Adjust heat and humidity to avoid dew formation in glasshouses. Indoor infection of clones will provide an overwintering bridge if they are transplanted outdoors the following spring.

Breeding for resistance is possible: McCain and Noviello (1985) cited two Italian hemp cultivars, 'Superfibra' and 'Carmagnola Selezionata', that were resistant to a strain of *P. cannabina* that destroyed all drug cultivars they tested. Stack *et al.* (2020) observed variable susceptibility in CBD cultivars, varying in disease incidence from 5% to zero. They have expanded resistance screening to 105 accessions, using detached leaf assays, whole plant ratings, and the Blackbird phenotyping system (Herrmann and Smart, 2024).

Biocontrol is limited. Lebeda and Cohen (2011) managed *P. cubensis* by spraying leaves with the mycoparasite *Pythium oligandrum*. These authors also suggested spraying leaves with an amino acid, β-aminobutyric acid, to induce host resistance. Ferraris (1935) controlled *P. cannabina* epidemics with copper sulfate. Be sure to spray the undersides of leaves.

10.6   Brown leaf spot and stem canker

BLS&SC is caused by at least eight *Phoma* and *Ascochyta* spp. (Dothideomycetes, Didymellaceae). BLS&SC rivals yellow leaf spot as the most common foliage problem in *Cannabis*. Sometimes the two diseases occur on the same plant (Fig. 10.14). BLS&SC is a "disease of attrition"—as leaf losses increase, crop yields decrease.

10.6.1   Symptoms and signs

Leaves low in the canopy begin developing beige or brown spots in late May or June. Spots remain circular, or form a straight edge along leaf veins. They average 5 mm in diameter, rarely up to 15 mm. Sohi and Nayar (1971) described spots coalescing into large irregular lesions. Shukla and Pathak (1967) described spots as brick red in color, also unusual. Pinpoint black pycnidia arise within the lesions. Depending on the species involved, lesions may remain intact (Fig. 10.14) or break apart, leaving small shot-holes in leaves (Fig. 10.15).

On stems, the spots begin as chlorotic areas. They quickly turn gray, beige, or brown, and elongate along the stem axis (Fig. 10.15). Pycnidia form within the lesions. Diseased plants are stunted, 30 cm shorter than healthy plants by mid-June (Röder, 1939).

10.6.2   Causal organisms

Identifying *Phoma* and *Ascochyta* to species is difficult. Some researchers sidestep the issue and simply list the cause as *Phoma* sp., in Lithuania (Brundza, 1933), Iowa (Fuller and Norman, 1944, 1945), Kansas (Paulsen, 1971), and India (Srivastara and Naithani, 1979). Researchers attempting to identify their pathogen to species have named at least 23 taxa on *Cannabis*. McPartland (1994a) sorted them out, and reduced the 23 taxa to eight species. Only two arise on a regular

Fig. 10.14. Brown leaf spot (*Neodidymelliopsis cannabis*) and yellow leaf spot (*Septoria cannabis*) on the same plant in Nepal

Fig. 10.15. Brown leaf spot (*Didymella glomerata*) and brown stem spot (*Boeremia exigua*)

basis: *Phoma cannabis* (now *Neodidymelliopsis cannabis*) and *Ascochyta arcuata*. Their morphological characters are described below. The other six species are described in Table 10.2 but spared further details.

10.6a. *Neodidymelliopsis cannabis* (Winter) Chen & Cai 2015

≡*Didymella cannabis* (Winter) von Arx 1962, ≡*Mycosphaerella cannabis* (Winter) Röder 1937, ≡*Mycosphaerella cannabis* (Winter) Magnus 1905, ≡*Sphaerella cannabis* Winter 1872
Anamorph: *Phoma cannabis* (Kirchner) McPartland 1994, ≡*Depazea cannabis* L. Kirchner 1856, =*Diplodina cannabicola* Petrak 1921, ≡*Diplodina parietaria* Brun. f. *cannabina* von Höhnel 1910, =*Phyllosticta cannabis* (Kirch.?) Speg. *apud* Saccardo 1884, =*Ascochyta cannabis* (Speg.) Voglino 1913
Description: *Conidiomata* (pycnidia) immersed then erumpent, globose to subglobose, peridium 3–4 cells thick (5–10 μm), *textura angularis-globulosa*, (65–)130(–180) μm in diameter on host substrate (Fig. 10.16). *Ostioles* round, sometimes slightly papillate. *Conidiogenous cells* enteroblastic, phialidic, ampulliform to short cylindrical, 2–6 μm tall. *Conidia* hyaline to light brown *en masse*, oval to ellipsoidal, sometimes biguttulate, 0- to 1-septate on host (nonseptate in culture), 3–8 × 2–3 μm on host substrate (3–10 × 2–4 μm in culture), adhering to form densely aggregated masses. *Chlamydospores* intercalary or terminal, usually globose to oval (in culture sometimes a dictyochlamydospore), single or in chains, brown, with smooth thick walls, 8–17 μm in diameter (Fig. 10.16).

Pseudothecia dark brown, immersed then erumpent, flattened on host substrate (stems), globose in culture, (90–)135(–180) μm in diameter on host substrate (to 190 μm in culture), averaging 140 asci per pseudothecium (Röder, 1937). *Ostioles* papilliform. *Asci* emerging parallel from hymenium, bituni-

cate, clavate to cylindrical, with a slight pedicle, 8-spored, 50–85 × 9–10 μm (Fig. 10.16). *Ascospores* hyaline, subovoid to oblong, submedially 1-septate with two cells unequal sized, constricted at septum, (11–)13.6(–17) × (4.5–)5.4(–8) μm. *Pseudoparaphyses* filamentous, 0- to 3-septate, 25–50 × 1 μm.

Comments: Taxonomy of the anamorph, *Phoma cannabis*, became entangled with *Septoria cannabis*. Lasch's type specimen of *Ascochyta cannabis* contains two different fungi. Spegazzini (1881) saw unicellular conidia (*Ascochyta* spp. are bicellular), and changed Lasch's taxon to *Phyllosticta cannabis* (Lasch) Spegazzini. That taxon was cited by Saccardo (1881). Subsequently, Saccardo (1884) examined Lasch's type, saw *Septoria*-type conidia, and changed Lasch's taxon to *Septoria cannabis* (Lasch) Saccardo.

Saccardo hypothesized that Spegazzini examined Kirchner's type specimen of *Depazea cannabis*, so Saccardo renamed Spegazzini's recombination, with a question mark, as "*Phyllosticta cannabis* (Kirchn?) Speg.," and listed *Depazea cannabis* as basionym. He was wrong—Spegazzini (1881) listed Lasch's taxon as basionym. Nevertheless, many authors cite "*Phyllostica cannabis* (Kirchn.) Speg.", including Farr (1973) and the Mycobank website. Voglino (1913) cited "*Phyllostica cannabis* (Kirchn.) Speg." but he saw bicellular conidia, so he changed the name to *Ascochyta cannabis* (Speg.) Voglino.

The sexual state, *Sphaerella cannabis* Winter, was originally described without mentioning its anamorph. Röder (1937) used single ascospore cultures to connect the sexual state with the anamorph, which he assigned to "*Phyllostica cannabis* (Kirchn.) Speg." Chen *et al.* (2015) conducted a multi-locus phylogenetic analysis (based on ITS, LSU, *rpb2* and *tub2*) of several dozen *Phoma* spp., most of which had *Didymella* teleomorphs. They included Röder's culture from 1937. It separated from *Didymella* in a clade that included two other *Phoma* spp. They renamed all three species with the new genus name *Neodidymelliopsis*.

Table 10.2. Characteristics of eight *Phoma* or *Ascochyta* species reported from *Cannabis*.

Species	Pycnidia diameter (μm)	Conidia characteristics (μm)	Chlamydospores (μm)	Sexual state
P. cannabis	65–(130)–180	1–2-celled, oval-ellipsoid, guttulate, 3–8 × 2–3	Globose to dictyo-form, 8–17	Neodidymelliopsis
A. arcuata	90–(135)–150	2-celled, short-cylindrical, guttulate, 9–28 × 3.5–8.0	Globose to oval, 8–12	Didymella
P. exigua	110–220	1–2-celled, ellipsoid-cylindrical, guttulate, 5–10 × 2–3.5	Absent	None known
P. glomerata	50–300	Usually 1-celled, ellipsoid-cylindrical, guttulate, 4.5–10 × 1.5–4	Dictyoform, 30–65 × 15–20	None known
P. herbarum	80–260	Usually 1-celled, ellipsoid, non-guttulate, 4.5–5.5 × 1.5–2.5	Absent	None known
P. nebulosa	125–185	1-celled, ellipsoid-cylindrical, polar guttules, 5–7 × 1.5–2.5	Absent	None known
A. prasadii	55–93	Usually 2-celled, oblong, non-guttulate, 7.5–10 × 1.5–2.5	Absent	None known
A. cannabina	120–130	1–2-celled, oval-oblong, non-guttulate, 7–8.8 × 3.2–4	Absent	None known

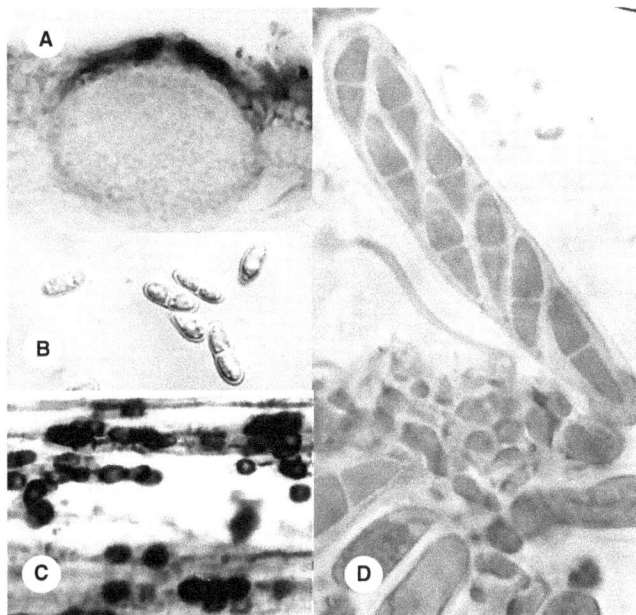

Fig. 10.16. *Neodidymelliopsis cannabis*: **A**. pycnidium (×200); **B**. conidia (×120); **C**. chlamydospores (×240); **D**. asci and ascospores (×830)

Older publications cited the cause of this disease as *Phyllosticta cannabis* (see Fig. 10.1). Fungi with *Phoma*-like unicellular conidia were traditionally placed in *Phyllosticta* if they occurred on leaves. Van der Aa (1973) restricted *Phyllosticta* to species with appendage-bearing conidia. McPartland (1994a) examined 25 herbarium specimens labeled "*Phyllosticta cannabis*." None were true *Phyllosticta* spp.; most were either *Neodidymelliopsis cannabis* or *Ascochyta arcuata*. Herbarium specimens of "*Phyllostica cannabis*" confidently reassigned to *Neodidymelliopsis cannabis* came from Europe (Belgium, Germany, Italy, Poland, Romania, Serbia, Russia, Ukraine), Asia (India, Pakistan, Nepal, Japan), and USA (Illinois, Kansas, Maryland, Wisconsin).

10.6b. *Didymella arcuata* Röder 1939

Anamorph: *Ascochyta arcuata* McPartland 1994
Not: *Ascochyta cannabis* Lasch 1846 (≡*Septoria cannabis* (Lasch) Saccardo)
Description: *Pseudothecia* brown-black, immersed or erumpent, globose on leaves and flattened on stems, 130–150 µm in diameter on host substrate (to 250 µm in culture). *Ostioles* round. *Asci* emerging parallel from hymenium, bitunicate, clavate to pyriform, with a slight pedicle, 8-spored, 50–80 × 9–14 µm. *Ascospores* hyaline, subovoid to oblong, submedially 1-septate with two cells unequal in size, constricted at septum, (16–)18.6(–22) × (4–)5(–6) µm. *Pseudoparaphyses* filamentous, 0- to 3-septate, 23–60 × 1 µm.
Conidiomata (pycnidia) immersed then erumpent, subglobose, unilocular, ostiolated, peridium *textura angularis-globulosa*, (90–)135(–150) µm in diameter on host substrate (up to 450 µm

in culture). *Conidiogenous cells* enteroblastic, phialidic, discrete, determinate, ampulliform to short cylindrical, (5–)8(–11) µm tall. *Conidia* hyaline to light brown *en masse*, ellipsoidal to short cylindrical, straight or curved, usually 1-septate (rarely 2-septate), not constricted at the septum, finely guttulate, 9–28 × 3.5–8.0 µm on host substrate; larger in culture (to 35 × 9 µm) (Röder, 1939). *Chlamydospores* intercalary, globose to oval, single or in short chains, brown, 8–12 µm in diameter.
Comments: Röder (1939) isolated a *Cannabis* fungus producing two-celled conidia (an *Ascochyta* sp.), and a sexual state belonging to *Didymella* sp. He named it *Didymella arcuata*, believing it was the sexual state of Lasch's fungus. Röder noted that Lasch's type specimen of *Ascochyta cannabis* was collected from a plant with two diseases. He accepted *Ascochyta cannabis* Lasch, and rejected *Septoria cannabis* (Lasch) Saccardo.

McPartland (1994a) disagreed: first, Saccardo's recombination has achieved common usage, whereas Lasch's basionym has not; second, Lasch's description (part of the taxon's prologue) described *Septoria* conidia and not *Ascochyta*; third, only one tiny leaf spot on the type specimen contains non-*Septoria* conidia, and these are not an *Ascochyta*, but a *Phoma* species. Thus, McPartland proposed an avowed substitute for the name of Röder's anamorph: *Ascochyta arcuata*.

McPartland (1994a) examined herbarium specimens from around the world; specimens confidently assigned to *D. arcuata* were limited to Europe: Germany, France, and Italy. A synopsis of differences between *P. cannabis* and *D. arcuata* appears in Table 10.2.

Röder submitted a culture of *D. arcuata* to CBS (no. 118.40). Crous and colleagues examined CBS 118.40 (Chen *et al.*, 2015). Phylogenetic analysis found no genetic difference in four genes (ITS, LSU, *rpb2* and *tub2*) between CBS 118.40 and a collection of *Didymella exitialis* (Morini) Müller *from wheat* (CBS 389.86). They synonymized CBS 118.40 and CBS 389.86, and coined the name *Neoascochyta exitialis* (Morini) Chen and Cai. We're dubious. Chen *et al.* (2015) stated that the host of CBS 118.40 was unknown. They added that *Neoascochyta* spp. do not produce chlamydospores; Röder (1939) described and illustrated *D. arcuata* chlamydospores.

10.6c. *Boeremia exigua* var. *exigua* (Desm.) Aveskamp *et al.* 2010

≡*Phoma exigua* Desmazieres 1849
=*Ascochyta phaseolorum* Saccardo 1878
=*Plenodomus cannabis* (Allescher) Moesz & Smarods in Moesz 1941, ≡*Phoma herbarum* f. *cannabis* Allescher 1901.
Comments: Allescher (1901) erected *Phoma herbarum* f. *cannabis* for a fungus he collected from dry hemp stalks in Bavaria. Moesz and Smarods (in Moesz, 1941) synonymized it under *Plenodomus cannabis,* collected from hemp stems in Latvia. Boerema (1970) synonymized *Phoma herbarum* f. *cannabis* under *Phoma exigua*. McPartland (1994a) and Boerema *et al.* (1996) added *Plenodomus cannabis* to the synonymy after examining the type specimen. Another synonym, *Ascochyta phaseolorum*, was cited on *Cannabis* by Sutton and Waterston (1966). McPartland isolated *B. exigua* from living plants in Nepal (see Fig. 10.15).

Domsch *et al.* (1980) considered *P. exigua* the most common pycnidial fungus in the world; it attacks over 200 plant genera, giving rise to over 100 synonyms. Aveskamp *et al.* (2010) conducted a phylogenetic analysis of 159 *Phoma* spp., and *P. exigua* fell into a clade with ten other *Phoma* spp. They named the clade *Boeremia* and renamed *P. exigua* as *Boeremia exigua*. Kaushal and Paul (1989) tested extracts of *Cannabis* on the growth of *P. exigua*, and it was not inhibited by the extracts.

10.6d. *Phoma herbarum* Westendorp 1852

Comments: *P. herbarum* was collected from hemp stalks in Italy (Saccardo, 1898), France (Brunard, 1899), Denmark (Lind, 1913), the Netherlands (Oudemans, 1920), and Japan (Kyokai, 1965; Kishi, 1988). *P. herbarum* is a well-known saprophyte; Lind (1913) stated he collected *P. herbarum* from dead hemp stalks. The other authors didn't specify whether their collections came from living or dead plants.

10.6e. *Didymella glomerata* (Corda) Chen and Cai 2015

≡*Phoma glomerata* (Corda) Wollenweber & Hochapfel 1936
≡*Coniothyrium glomeratum* Corda 1840
Comments: *P. glomerata* parasitizes at least 94 plant genera worldwide (Sutton, 1980). Although no *Didymella* sexual state has ever been found, Chen *et al.* (2015) renamed *P. glomerata* as *Didymella glomerata*, based a phylogenetic analysis, because *P. glomerata* fell into a clade with *Didymella* spp. McPartland (1994a) collected *P. glomerata* from leaves of a drug-type hybrid in Illinois, causing shotholes in leaves (see Fig. 10.15). The chlamydospores of *P. glomerata* are unique among *Phoma* spp., and can be mistaken with conidia produced by *Alternaria alternata* (Fig. 10.17).

10.6f. *Phomatodes nebulosa* (Persoon) Chen & Cai 2015

≡*Phoma nebulosa* (Persoon) Berkeley 1860
Comments: *P. nebulosa* is primarily saprophytic, isolated from herbaceous and woody plants. De Gruyter *et al.* (1993) isolated it from stems of "withering *Cannabis sativa*" in Holland. Crous and colleagues renamed *P. nebulosa* as *Phomatodes glomerata*, based a phylogenetic analysis, because *P. nebulosa* formed its own clade (albeit with weak bootstrap support, 53), so they coined a new genus for it (Chen *et al.*, 2015).

10.6g. *Ascochyta prasadii* Shukla & Pathak 1967

Shukla and Pathak (1967) coined this taxon for a *Cannabis* pathogen causing serious leaf disease in Udaipur, India. They claimed the morphology of *A. prasadii* differed from other

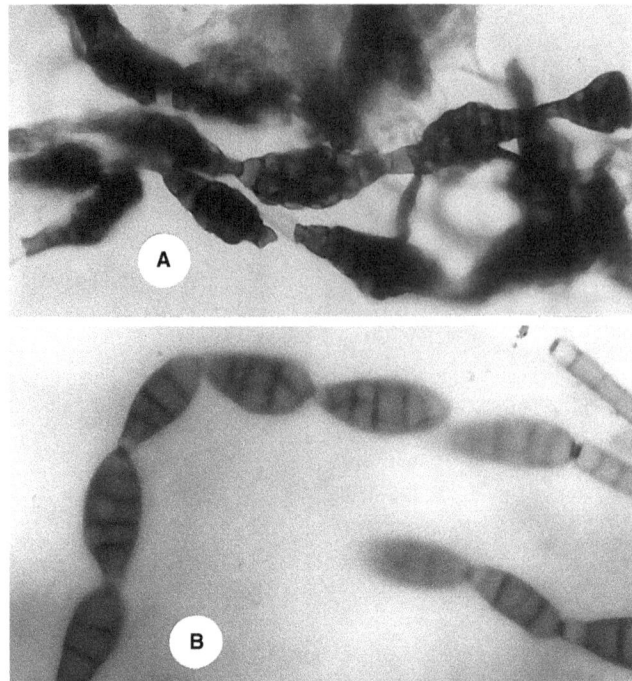

Fig. 10.17. *Didymella glomerata* chlamydospores (**A**, ×660), compared with *Alternaria alternata* conidia (**B**, ×490)

Ascochyta spp. described on *Cannabis*. The fungus caused unique brick red-colored leaf spots.

10.6h. *Ascochyta cannabina* E.I. Reichardt 1925

Reichardt (1925) coined *A. cannabina* for a fungus causing leaf spots on hemp growing in the botanical garden of the former Czar's summer palace near St Petersburg. He made no mention of previously described *Ascochyta* spp. on *Cannabis*. The size and shape of pycnidia and conidia fall into the range of *Neodidymelliopsis cannabis*. His taxon has slipped into obscurity; Mel'nik (1977) made no mention of *A. cannabina* in his 246-page tome of *Ascochyta* in the USSR.

10.6.3   Differential diagnosis

Yellow leaf spot and white leaf spot produce larger lesions than brown leaf spot, and they are lighter in color. When brown leaf spot lesions break up and fall out, the holes may resemble insect damage. Dobrozrakova *et al.* (1956) noted that lesions on stems resembled those caused by *Dendrophoma marconii* (see section 11.7) and *Microdiplodia abramovii* (later this chapter). Stem cankers may be confused with symptoms caused by *Fusarium*, *Coniothyrium*, *Leptodothiorella*, *Fusicoccum*, or *Phomopsis* spp. These fungi can be differentiated with a microscope.

More than a microscope may be needed to differentiate the eight *Phoma* and *Ascochyta* spp. (Table 10.2). Röder (1937,

1939), who cultured the two most common species (*N. cannabis* and *A. arcuata*), distinguished them by conidial size, shape, and septation. The conidia of *A. arcuta* produce septa, and the conidia of *P. cannabis* do not. *A. arcuata* conidia, *in vitro* and on the host, are longer than *P. cannabis* conidia. Chen *et al.* (2015) sequenced Röder's two cultures; *N. cannabis* and *A. arcuata* clearly segregated in their phylogenetic tree.

10.6.4 Disease cycle and epidemiology

Phoma and *Ascochyta* spp. overwinter as pycnidia and perithecia in crop residue near the soil surface. *P. cannabis*, *A. arcuata*, and *P. glomerata* also survive as chlamydospores in soil. The disease cycle mirrors that of *Septoria* spp. (see above). Infection occurs in early spring, but symptoms may take a month to manifest. Conidia are spread via splashed rain and wind-driven water, and build to epidemics by late summer. The optimal temperature for mycelial growth and spore germination is 19–22°C (Röder, 1939). Paulsen (1971) reported *Phoma* sp. growing on leaves, stalks, and "green seeds." Seed-borne infection has been documented in *P. exigua* (Sutton and Waterston, 1966).

10.6.5 Control methods

Follow cultural control methods described under *Septoria* spp. Brown spot is worse on hemp grown in peat soils (Rataj, 1957); avoid overuse of nitrogen fertilizer. Given some evidence of seed-borne infection (Sutton and Waterston, 1966; Paulsen, 1971), do not harvest seed from infected females. Many *Phoma* diseases explode in crops weakened by environmental stress. Be sure plants are well-maintained.

Biocontrol against brown leaf spot and stem canker is limited. Balthazar *et al.* (2022b) screened beneficial *Bacillus* and *Pseudomonas* spp. for *in vitro* activity against *Phoma* sp. isolated from *Cannabis* leaves. Two performed well: *B. velezensis* and *B. subtilis* (see Fig. 13.6). Kirchner (1966) used copper-based sprays such as Bordeaux mixture.

10.7 Cercospora leaf spot

This disease is caused by two *Cercospora* spp. and two species in the segregate genera *Pseudocercospora* and *Phaeomycocentrospora*. All four fungi belong to Family Mycosphaerellaceae, Class Dothideomycetes. All four fungi cause symptoms that begin on older leaves in the lower canopy and then spread upward.

10.7.1 Symptoms and signs

Spots begin as yellow specks on upper surfaces of leaves, and then turn light brown, sometimes dark brown, sometimes with distinct margins. Growth on the underside of the leaf may impart an olive color, but may appear yellow-brown rather than

olive. As symptoms advance, whole leaves may wilt, curl, and drop off.

Pseudocercospora cannabina leaf spots are irregularly shaped, slightly raised and stromatic, at first vein-delimited (Fig. 10.18), but eventually coalesce with other spots (Lentz *et al.*, 1974). *Cercospora cannabis* lesions usually remain relatively small and circular, with a dark margin and pale brown center (Fig. 10.19). Spots may enlarge to leaf veins, causing straight edges and somewhat rectangular shapes. Growth on the underside of the leaf may look yellow-brown rather than olive.

Cercospora flagellaris spots begin as small yellow dots, then enlarge to circular lesions 2–6 cm in diameter. They have a dark brown to purplish margin surrounded by a yellow halo; the center becomes pale brown to white. *Phaeomycocentrospora cantuariensis* causes large leaf spots, roundish to irregularly shaped, with a distinctive dark-brown edge and a light-brown to grayish center (Kauschitz and Plenk, 2022). Leaf spots illustrated by Bergstrom *et al.* (2019) lacked distinctive brown edges, and were capable of expanding across the leaf midrib.

10.7.2 Causal organisms

These species are not easily differentiated. Some authors simply list the cause as *Cercospora* sp., in India (McRae, 1934)

Fig. 10.18. Cercospora leaf spot caused by *Pseudocercospora cannabina*

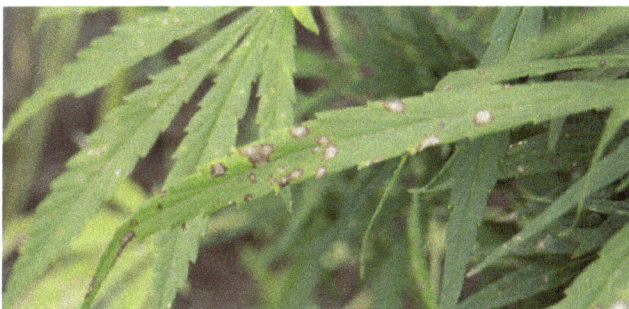

Fig. 10.19. Cercospora leaf spot caused by *Cercospora cannabis* (courtesy J. Beckerman)

and Chile (Viégas, 1961). Side-by-side, Vasudeva (1961) contrasted *P. cannabina* (from Godagari, Bangladesh) from *C. cannabis* (from Mysore, India). Lentz *et al.* (1974) examined herbarium specimens from around the world and showed that mistakes have been made in identification.

10.7a. *Pseudocercospora cannabina* (Wakefield) Deighton 1976

≡*Cercospora cannabina* Wakefield 1917, =*Helicomina cannabis* Ponnappa 1977
Description: *Conidiophores* emerge from stomata or erumpent through epidermis in fascicles, simple or sometimes branched, straight, septate, olivaceous brown, narrowing to a slender apex, up to 150 x 3.5-4.5 µm. *Conidia* with unthickened basal scars, cylindrical, slightly or strongly curved, colored dilutely olivaceous, usually 5–7-septate, 30–90 × 3.5–5.5 µm (Fig. 10.20).
Comments: Wakefield discovered *P. cannabina* in Uganda. Sydow and McRae (1929) collected it in Peshawar (Pakistan) and Godagari (Bangladesh). They described conidial lengths up to 120 µm. Petrak (1947) reported *P. cannabina* in China. Stevenson (1926) placed *P. cannabina* on a list of "foreign plant diseases" that posed a threat to US agriculture if accidently imported.

Lentz *et al.* (1974) documented the arrival of *P. cannabina* in the USA, at the NIDA farm in Mississippi where germplasm from around the world was cultivated. Lentz confirmed the identity of *P. cannabina* in herbarium specimens collected in Bangladesh, Cambodia, India, Pakistan, and China. Nair and Ponnappa (1974) identified *P. cannabina* in India, then Ponnappa (1977) described their isolate as a new species, *Helicomina cannabis*, unaware that Deighton (1976)

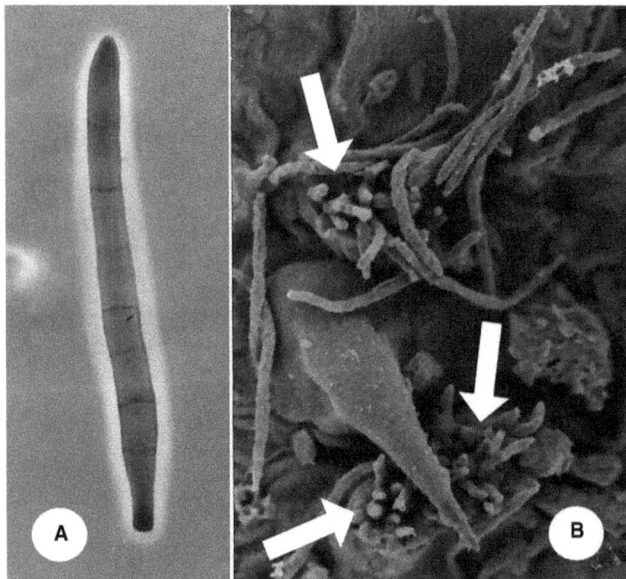

Fig. 10.20. *Pseudocercospora cannabina*: **A**. conidium (LM, ×1200); **B**. leaf surface showing three erumpent fascicles of conidiophores (arrows) with conidia (SEM, ×375)

had synonymized *Helicomina* under *Pseudocercospora*. McPartland (1995d) examined Ponnappa's type specimen, and *Helicomina cannabis* proved to be *P. cannabina*. Ghani and Basit (1976) added collection records in Punjab, Kashmir, and Khyber Pakhtunkhwa. Bush Doctor (1987) added Nepal. Crous and Braun (2003) added Korea and Azerbaijan.

10.7b. *Cercospora cannabis* Hara & Fukui *in* Hara 1925

≡*Cercosporina cannabis* Hara 1928, ≡*Cercospora cannabis* Hara & Fukui *in* Shirai & Hara 1927, ≡*Cercospora cannabis* (Hara) Chupp *apud* H.C. Green 1944
Description: *Conidiophores* emerge from stomata, either solitary or in fascicles of 2–12, simple, straight or occasionally 1–2 geniculate, septate, pale brown, not narrowing at the apex but broad and flat, 10–100 × 3.5–5.5 µm (up to 125 µm according to Phengsintham *et al.*, 2013). *Conidia* bear thickened detachment scars at the base, not cylindrical but tapering toward their tips, straight or slightly curved, hyaline, indistinctly multiseptate, 20–90 × 3–5 µm (83–125 µm long according to Phengsintham *et al.*, 2013).
Comments: *C. cannabis* was originally identified in Japan. It was described in the Soviet Union (Gitman and Boytchenko, 1934) and China (Teng, 1932; Tai, 1936), although Lentz *et al.* (1974) noted that both Chinese descriptions sounded more like *P. cannabina*. Next *C. cannabis* cropped up in Wisconsin (Greene, 1946), Missouri (Maneval, 1948), and southern India (Thirumalachar and Chupp, 1948). Herbarium voucher specimens from these sites were examined and confirmed by Lentz *et al.* (1974), who also recorded a find at the NIDA farm in Mississippi. Tai (1979) cited nine reports of *C. cannabis* in China. The species also occurs in Nepal (Crous and Braun, 2003) and Laos (Phengsintham *et al.*, 2013).

10.7c. *Cercospora flagellaris* Ellis & G. Martin 1882

=*Cercospora* cf. *flagellaris* apud Groenewald *et al.* (2013)
Description: Sporulation amphigenous. *Conidiophores* emerge from stomata or erumpent through epidermis, in 2–23 loose to dense fascicles, simple, straight or geniculate, pale brown to brown, not narrowing at the apex, size ranges widely, 250–735 × 2.5–6.5 µm. *Conidia* with thickened detachment scars at the base, cylindrical to acicular, sometimes obclavate, straight or slightly curved, hyaline, indistinctly multiseptated, (68–)100–200(–300) × 3–5.5 µm.
Comments: *C. flagellaris* is a worldwide pathogen, reported from phylogenetically-unrelated plant hosts spanning 24 families. Groenewald *et al.* (2013) introduced the concept *Cercospora* cf. *flagellaris*, where cf. (Latin *confer*, "compare with") equates with a question mark (?). They conducted a molecular study of 360 *Cercospora* cultures worldwide, of which 15 were morphologically similar to existing species, but taxonomic names could not be applied without doubt, receiving a "cf." designation. None of their cultures came from *Cannabis*.

Doyle *et al.* (2019) sequenced a *Cercospora* sp. isolated in Kentucky, plugged it into the phylogenetic tree by Groenewald *et al.* (2013), and identified *Cercospora* cf. *flagellaris* on hemp. Their isolate differed from descriptions of *Cercospora cannabis*, producing longer conidiophores (324.4–618.1 µm) and conidia (102.2–199.6 µm). They fulfilled Koch's postulates; inoculated plants showed symptoms within 21 days. Marin *et al.* (2020a) sequenced a *Cercospora* sp. in Florida, accessing the same databank, which also yielded *Cercospora* cf. *flagellaris.* Coolong (2020) cited *Cercospora* cf. *flagellaris* in Georgia.

10.7d. *Phaeomycocentrospora cantuariensis* (Salmon & Wormald) Crous et al. 2012

≡*Pseudocercospora cantuariensis* (Salmon & Wormald) Braun 1993,
≡*Cercospora cantuariensis* Salmon & Wormald 1923
Description: Sporulation amphigenous but primarily on underside of leaf. *Conidiophores* emerge from stomata or erumpent through epidermis, in loose fascicles, olivaceous brown, straight to slightly curved, unbranched, not geniculate, narrowing at the apex, proliferating sympodially, 0–3-septate, 30–140 × 7–20 µm. *Conidia* with a broad base (7–10 µm) without a thickened detachment scar, filiform to cylindrical, straight or slightly curved, subhyaline to pale olivaceous, 3–15 septate, (100–)140–300(–500) × 7–15 µm.
Comments: In a phylogenetic study this species formed its own clade, separate from other *Pseudocercospora* and *Cercospora* spp., so Crous *et al.* (2012) erected the genus *Phaeomycocentrospora* for it. Morphologically, they distinguished *P. cantuariensis* from *Pseudocercospora* spp. by its broad-based conidia, as well as hyaline hyphae and hyphopodia-like structures *in vitro. P. cantuariensis* is a worldwide pathogen of *Humulus* spp. (*H. lupulus* in Europe, *H. japonicus* in East Asia). Bergstrom *et al.* (2019) illustrated a *Cannabis* leaf spot caused by *Phaeomycocentrospora* sp. in New York. Kauschitz and Plenk (2022) isolated *P. cantuariensis* from hemp in Austria.

10.7.3   Differential diagnosis

Early symptoms of Cercospora leaf spot may be confused with yellow leaf spot, brown leaf spot, or even downy mildew. The differential diagnosis can be accomplished with a microscope. Side-by-side comparisons of *P. cannabina* and *C. cannabis* have been made (Chupp, 1953; Vasudeva, 1961; Lentz *et al.*, 1974, McPartland, 1995d). *P. cannabina* spreads diffusely over the leaf surface, has branched conidiophores, and colored conidia. *C. cannabis* growth is less diffuse, and the fungus produces conidiophores in fascicles, its conidia are hyaline and bear thickened detachment scars at the base.
Cercospora cf. *flagellaris* sporulation is amphigenous, and it produces big conidiophores and conidia. *P. cantuariensis* sporulation is also amphigenous; its conidia are even larger. Conidia of *P. cantuariensis* and *P. cannabina* are pale olivaceous, whereas conidia of *C. cannabis* and *Cercospora* cf. *flagellaris* are hyaline.

10.7.4   Disease cycle and epidemiology

Sexual states for most *Cercospora* and *Pseudocercospora* spp. are unknown. Crous and Braun (2003) described *Mycosphaerella*-like sexual states, but *Mycosphaerella* is now restricted to fungi that form *Ramularia* anamorphs. *Cercospora* and *Pseudocercospora* spp. overwinter in plant debris or in soil. Conidia form in the spring and spread by wind, splashing water, or farm hands. Unlike *Septoria* conidia, which are "sticky" and disseminate only over several meters, conidia of *Cercospora* and *Pseudocercospora* can be blown over greater distances (Howard *et al.*, 1994). High humidity or free water is required for conidial germination. Penetration occurs via stomates or wounds. Damage escalates rapidly in August, particularly during rainy summers after canopy closure; it has also been reported on glasshouse-grown cannabis where plants near doors are exposed to outside conditions (Gauthier *et al.*, 2022).

10.7.5   Control and management

Prevention is key. Follow directions for yellow leaf spot (*Septoria* spp.). *Cercospora* seed-borne infection has not been detected in *Cannabis* (Pietkiewicz, 1958; Ferri, 1961b; Babu *et al.*, 1977), although it is very common in other plants—therefore do not harvest seeds from infected plants. Crop rotation to control *C. flagellaris* is difficult; the fungus is a generalist pathogen. *P. cantuariensis* is a hops pathogen. Host-range studies have not been done for *P. cannabina* and *C. cannabis.*

Dixon *et al.* (2022a) screened several biocontrols and biorational pesticides for *in vitro* efficacy against *C. flagellaris* (see Table 9.1). Compared with three other pathogens, *C. flagellaris* was the hardest to control. Best results were obtained with Cueva® (copper soap). Even their positive control, Custodia® (azoxystrobin + tebuconazole fungicide), did not completely inhibit *C. flagellaris* conidial germination. Balthazar *et al.* (2022b) screened beneficial *Bacillus* and *Pseudomonas* spp. for *in vitro* activity against *Cercospora* sp. Two performed well: *B. velezensis* and *B. subtilis* (see Fig. 13.6). Yu (1973) managed *Cercospora* in NIDA's Mississippi *Cannabis* garden with sulfur dust and Bordeaux mixture.

10.8   Anthracnose

French authors combined Greek *anthrakas* "coal" and *nósos* "disease" to describe dark spots (anthracnose) on grape leaves (Fabre and Dunal, 1853). The modern association of anthracnose with *Colletotrichum* spp. was largely due to two American women. Effie A. Southworth (1860–1847) became the first female plant pathologist at the USDA. Southworth (1890) described *Colletotrichum* spp. causing anthracnose of cotton and hollyhock. Bertha Stoneman (1866–1943) earned her PhD at Cornell University based on research of anthracnose diseases. Stoneman (1898) described six diseases caused by *Colletotrichum* spp.

10.8.1 Symptoms and signs

Leaf symptoms begin as light green, water-soaked spots. Spots enlarge to circular or irregular shapes, with gray-white to tan-colored centers, and brownish margins surrounded by a yellow halo. Spots caused by *Colletotrichum coccodes* may become zonate. Centers may drop out, leaving shotholes in leaves. Spots sometimes coalesce into large necrotic regions, and leaves become puckered, twisted, and wilt (Fig. 10.21).

Stalk lesions initially turn beige or white. Then black, dot-like acervuli arise in the lesions, lending a salt-and-pepper appearance (Fig. 10.22). Affected stalks may swell slightly and develop cankers. The periderm peels off easy. Tissues beneath cankers become brittle, and stalks sometimes snap. Distal plant parts may wilt. Young plants die. Post-emergent damping-off of seedlings caused by *C. coccodes* was illustrated by Hoffmann (1959). Sunken cankers extended up the epicotyl and down to the radicle. Hoffmann also reported pre-emergent damping-off.

10.8.2 Causal organisms

Cannabis anthracnose is caused by three *Colletotrichum* spp. (Glomerellaceae; Sordariomycetes). *Colletotrichum* spp. also exist as non-disease-causing endophytes. Gautam *et al.* (2013) isolated *Colletotrichum* spp. from 2.5% of leaves and 1.4% of stalks of healthy plants in Himachal Pradesh. Jin *et al.* (2013) identified *Colletotrichum* as the dominant genus in a survey of endophytic fungi inhabiting hemp in Yúnnán Province.

10.8a. *Colletotrichum dematium* (Persoon) Grove 1918

≡*Vermicularia dematium* (Pers.) Fries 1849, =*Vermicularia dematium* f. *cannabis* Saccardo 1880
Description: *Conidiomata* (acervuli) round to elongated, up to 400 μm diameter, strongly erumpent through epidermis and exuding smoky-gray conidial masses with divergent setae. *Setae* smooth, stiff, rarely curved, usually 3–4-septate, 4–7.5 μm wide at the base tapering to sharpened apices, 60–200 μm long (Fig. 10.23). *Conidiophores* hyaline, cylindrical. Conidiogenous cells phialidic. *Conidia* hyaline, aseptate (becoming 2-celled during germination), smooth, thin-walled, guttulate, falcate and curved, 18–26 μm long and 3–3.5 μm wide in the middle and tapering to pointed apices (Fig. 10.24). *Appressoria* club-shaped to circular, medium brown, edge usually entire, 8–11.5 × 6.5–8 μm, often becoming complex.
Comments: *C. dematium* is a weak parasite on many hosts; von Arx (1957) listed 88 synonyms. *C. dematium* has been collected from hemp stems in Italy (Saccardo, 1880; Cavara, 1889), Hungary (McPartland *et al.*, 2000) (Fig. 10.23), and possibly Kansas (Hollandbeck, 2019). Saccardo (1880) coined a new subspecies, *Vermicularia dematium* f. *cannabis,* which he distributed as an exsiccatus. McPartland (1995d) examined an isotype at BPI, which proved morphologically indistinguishable from *C. dematium.*

Crous and colleagues sequenced 97 cultures of *Colletotrichum* spp., which formed 20 clades, of which 12 included cultures identified as *C. dematium* (Damm *et al.*, 2009). The clade with the most specimens (*n* = 8) they identified as *C. dematium sensu stricto*, and designated one culture (CBS

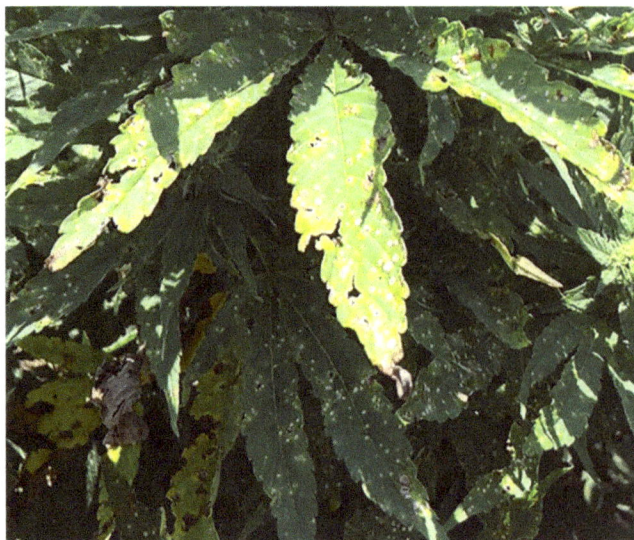

Fig. 10.21. Anthracnose caused by *Colletotrichum fioriniae* (courtesy of N. Gauthier)

Fig. 10.22. Anthracnose caused by *Colletotrichum dematium* (R. Clarke)

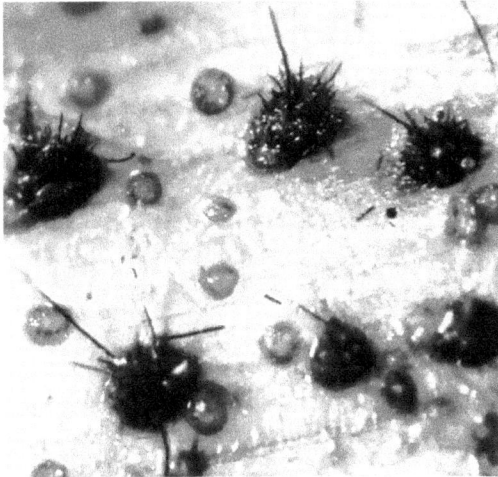

Fig. 10.23. *Colletotrichum dematium* on *Cannabis* (LM ×40, courtesy Bud Uecker)

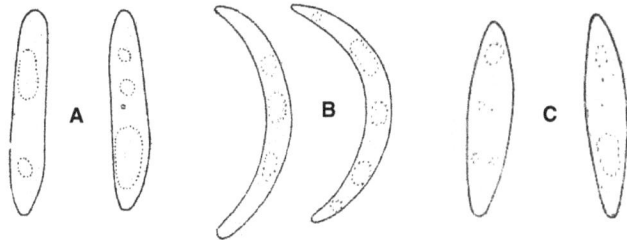

Fig. 10.24. *Colletotrichum* conidia, ×1500: **A**. *C. coccodes*; **B**. *C. dematium*; **C**. *C. fioriniae*

125.25) as the epitype of *C. dematium*. Their chosen epitype came from Killian (1926), who identified his fungus as *Vermicularia eryngii*. They noted that Persoon's holotype specimen (at the Leiden herbarium) lacked conidia, "but all structures observed, e.g. setae and conidiogenous cells, resemble those of the epitype," and they illustrated setae in the epitype. *C. dematium* produces ample setae (Persoon, 1801; Grove, 1918; von Arx, 1957; Sutton, 1980), as does *Vermicularia eryngii* (Killian, 1926). Thiessen and Gauthier (2022) misread Damm *et al.* (2009) and said, "*C. dematium* does not produce setae … Therefore, classification of the *Cannabis* pathogen previously reported as *C. dematium* deserves reexamination."

10.8b. *Colletotrichum coccodes* (Wallroth) Hughes 1958

=*Colletotrichum atramentarium* (Berk. & Broome) Taubenhaus 1916, ≡*Vermicularia atramentaria* Berkeley & Broome 1850
Description: *Conidiomata* (acervuli) on stems round or elongated, up to 300 µm in diameter, at first covered with epidermis; then dehiscing irregularly and exuding slimy conidia and bristling setae. *Setae* smooth, stiff, septate, slightly swollen and dark brown at the base and tapering to sharpened and paler apices, up to 100 µm long. *Conidiophores* hyaline, cylindrical,

occasionally septate and branched at their base. Conidiogenous cells phialidic, smooth-walled, hyaline, with a minute channel and periclinal thickening of the collarette. *Conidia* hyaline (honey-colored to salmon-orange *en masse*), aseptate, smooth, thin-walled, guttulate, straight, cylindrical to fusiform, with rounded apices, often with a slight median constriction, averaging 14.7 × 3.5 µm on *Cannabis* stems but ranging 16–22 × 3–4 µm in culture (Fig. 10.24). *Appressoria* club-shaped, medium brown, edge irregular to almost crenate, 11–16.5 × 6–9.5 µm, rarely becoming complex. Sclerotia are black, irregularly spherical, 150–250 µm in diameter.
Comments: *C. coccodes* has a wide host range, with over a dozen additional synonyms listed by von Arx (1957). Worldwide it causes serious damage in solanaceous crops (tomatoes, potatoes, eggplants). In fiber-type hemp, Hoffmann (1958, 1959) reported heavy losses in Germany. Conversely, Gitman (1968b) considered the pathogen of little importance in the USSR. Hoffmann (1958) completed Koch's postulates, and also infected hemp with *C. coccodes* isolates from potato. Ghani and Basit (1975) likely collected *C. coccodes* from Pakistani drug plants, described as "*Colletotrichum* sp. … cylindrical conidia with both ends rounded." A hemp specimen collected by V.K. Charles in Virginia (deposited at herb. BPI, the US National Fungus Collection Herbarium) also proved to be *C. coccodes* (McPartland, unpublished data, 1995).

10.8c. *Colletotrichum fioriniae* (Marcelino & Gouli) Pennycook 2017

≡Colletotrichum acutatum var. fioriniae Marcelino & Gouli 2008
Sexual state: Glomerella acutata Gueber & Correll 2001
Description: *Conidiomata* reduced to a cushion of pale brown, thick-walled, angular cells. *Conidiophores* formed directly on vegetative hyphae, up to 35 µm long. *Conidiogenous cells* phialidic, smooth-walled, hyaline to pale brown, cylindrical to ampulliform, with a minute channel and distinct periclinal thickening of the collarette, 4–12 × 3.5 µm. *Conidia* hyaline (orange *en masse*), aseptate, smooth-walled, straight, cylindrical to fusiform with acute ends, averaging 9–15 × 3.0–4.5 µm (Fig. 10.24). *Setae* absent. *Appressoria* solitary or in loose groups, pale to medium brown, smooth-walled, ellipsoidal, clavate to irregular outline, entire edge or undulate, 12–16 × 4.0–5.4 µm.
Comments: *C. fioriniae* is part of the *C. acutatum* species complex. Damm *et al.* (2012) separated *C. fioriniae* from *C. acutatum* based on culture characteristics and ITS sequences. They argued that the sexual state, *G. acutata*, is an interspecific hybrid between *C. fioriniae* and *C. acutatum*—evidence that these two taxa constitute a single biological species. *C. fioriniae* has a wide host range, including blueberry, strawberry, grape, apple, clover, and it has been isolated as an endophyte from 25 other species. It also parasitizes an insect, *Fiorinia externa* Ferris. Szarka *et al.* (2020a) found *C. fioriniae* causing leaf spots in Kentucky, confirmed its identity with *ACT*, *CH-1*, *GAPDH*, and *TUB2* sequences, and completed Koch's postulates.

10.8.3   Differential diagnosis

Anthracnose can be confused with blights, leaf spots, or stem cankers. The three *Colletotrichum* spp. on *Cannabis* produce conidia with different shapes (Fig. 10.24). *C. dematium* conidia are curved and falcate, widest in the middle, with pointed ends. *C. coccodes* conidia are straight and rather cylindrical, often with a slight median constriction, and have rounded ends. *C. fioriniae* conidia are straight and fusiform, with acute to pointed ends.

10.8.4   Disease cycle and epidemiology

Hoffmann (1958) found no evidence of seed-borne transmission. *Colletotrichum* spp. overwinter in plant debris or soil. In the spring they sporulate, and conidia splash onto seedlings. Conidia form appressoria which directly penetrate epidermal tissue or enter via stomates and wounds. *C. dematium* prefers a warmer optimum temperature (26–32°C) than *C. coccodes* or *C. fioriniae*. Expect anthracnose epidemics during cool damp weather, especially in heavy, poorly drained soils. Hoffmann (1958, 1959) noted more losses in nutrient-poor "bog" soils. Disease escalates in plants under stress from drought or frost damage. Anthracnose can rage in monocropped glasshouses, especially with hydroponic systems (Smith *et al.*, 1988). Conidia give rise to secondary infection, and spread via splashing water and wind-driven rain. Hoffmann (1958, 1959) and Cook (1981) described heaviest infections after plants have flowered, with males succumbing before females. Concurrent infection by the nematode *Heterodera schactii* or the fungus *Rhizoctonia solani* increases plant susceptibility to anthracnose (Smith *et al.*, 1988).

10.8.5   Control and management

Follow directions for yellow leaf spot (*Septoria* spp.). Practice good sanitation (gather and dispose of crop residues), and eliminate weeds (which may serve as alternative hosts). Hoffmann (1958) found *C. coccodes* on crop stubble, which he plowed under. Provide balanced nutrient and adequate hydration (no overhead irrigation). Avoid heavy soils. If *C. coccodes* is the causal agent, do not rotate after solanaceous crops. Stay out of *Cannabis* when plants are wet, since conidia spread by contact. Clarke (pers. comm., 1995) noted that Hungarian 'Uniko-B' growing in China was susceptible to *C. dematium*, while local Chinese landraces were resistant to the fungus. Biocontrol options include *Bacillus subtilis* (Serenade®), *Trichoderma harzianum* (or more likely *Trichoderma asperellum*, several products), *Burkholderia vietnamiensis* (Botrycid®), *Clonostachys rosea* (PreStop®), and *Pseudomonas chlororaphis* (Howler®). There are no effective chemical controls.

10.9   White leaf spot and atem canker

This disease was formerly called Phomopsis leaf spot and stem canker, but most *Phomopsis* spp. have been transferred to *Diaporthe* (Diaporthaceae; Sordariomycetes). Five species are involved: *D. ganjae* causes leaf spots that accrue over the growing season; the others cause stem cankers late in the growing season or post-harvest (*P. cannabina*, *D. eres*, *D. arctii*, *D. phaseolorum*).

10.9.1   Symptoms and signs

Leaf symptoms begin as pinpoint spots. Lesions enlarge, irregularly circular, and become slightly raised or thickened. They remain white or darken to a beige color (Fig. 10.25). Black pycnidia arise in concentric rings. In severe cases the spots coalesce, leaf tissue between spots becomes chlorotic, and the leaves drop off. The leaf symptoms attracted biocontrol researchers (see Fig. 10.1). Spots rarely arise in flowering tops; THC and CBD inhibit the growth of *D. ganja* (McPartland, 1984).

Stem symptoms begin in late summer, appearing as light-colored lesions with raised margins, about 3.0–3.5 cm long. Lesions form in "sub-apical" positions of the main stem, usually near branching nodes (De Corato, 1997). Plant parts distal to lesions wilt and dry early, within 10–15 days of first symptoms, followed by black pycnidia scattered through the lesion (Fig. 10.26). After the plant dies, pycnidia continue to form, and the stem surface blackens (Gitman and Malikova, 1933). In some species, a sexual state arises by November (Saccardo, 1897).

10.9.2   Causal organisms

Five *Diaporthe–Phomopsis* spp. have been reported from *Cannabis* (Table 10.3). The *Phomopsis* state is pycnidial, globose to subglobose, varying in size, solitary or clustered in groups,

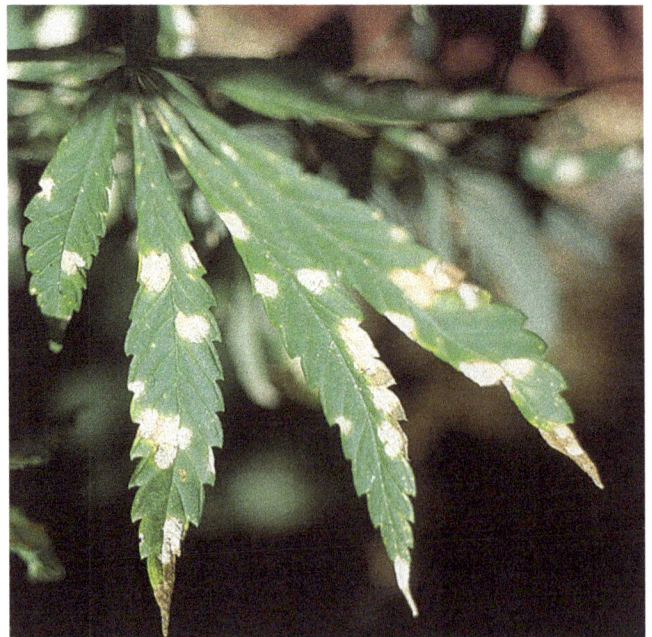

Fig. 10.25. White leaf spot caused by *Diaporthe ganjae*

immersed in a *pseudostroma* (a mass of fungal hyphae and host tissue) which in some cases is quite thin; upper peridium is black and stromatic; lower peridium is parenchymous and pale sooty-brown. *Conidiophores* cylindrical and slightly tapering towards the apex, straight to sinuous, simple or branched, septate. *Conidiogenous cells* enteroblastic, phialidic, cylindrical to obclavate, hyaline. Two types of conidia are commonly produced (Table 10.3): *Alpha-(α-)conidia* unicellular or 1-septate, hyaline, fusiform to ellipsoidal, with acute to rounded ends, often biguttulate. *Beta-(β-)conidia* unicellular, hyaline, filiform,

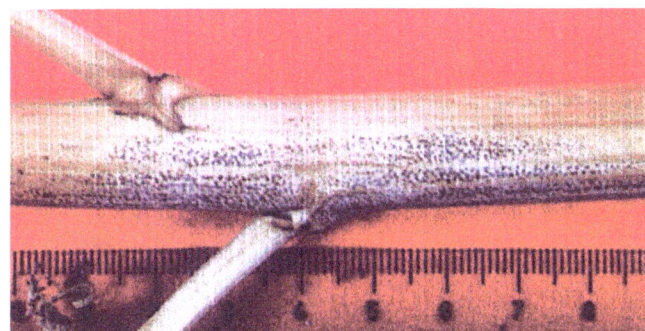

Fig. 10.26. Stem canker caused by *Phomopsis* sp. in Italy (courtesy Ugo De Corato).

usually curved or hooked, eguttulate, tapered or truncated towards ends. They do not germinate, except to produce a short lateral protuberance (Fig. 10.27D).

A *Diaporthe* sexual state may be present: *ascomata* are perithecial, subglobose or irregular, solitary or clustered in groups, brown to black, and immersed in a *pseudostroma*. which is widely effuse over the stem surface; its ventral margin may be delimited by a narrow dark-celled prosenchymatous "zone line," seen by sectioning the stem (Fig. 10.28A). *Perithecium* often has a tubular ostiolated beak, which emerges from the pseudostroma (Fig. 10.28B). *Paraphyses* occasionally produced, intermingled among asci, elongate, multiseptated, hyaline, smooth. *Asci* unitunicate, 8-spored, sessile, often with an apical refractive ring, elongate to cylindrical. *Ascospores* often biseriate within the ascus, ellipsoid to cylindrical, septate, hyaline, often guttulate, varying in size (Table 10.3).

10.9a. *Phomopsis cannabina* Curzi 1927

Comments: The description of *P. cannabina* in Table 10.3 comes from Curzi (1927). Curzi's herbarium collection is at

Table 10.3. Characteristics of five *Diaporthe–Phomopsis* spp. reported from *Cannabis*

Species	Perithecium	Asci and ascospores	Pycnidia	Conidia
P. cannabina	None known	None known	Scattered, eustromatic, oblong to lens-shaped, 300–450 × 100–150 μm	α-Conidia unicellular, fusiform, straight or curved, biguttulate, 7–8 × 2.5–3.0 μm. β-conidia not described
D. ganjae	None known	None known	Globose to elliptical, ostiolate, 120–220 × 120–300 μm	α-Conidia unicellular, fusiform to elliptical, biguttulate, usually 7–8 × 2.5 μm. β-conidia filiform, 16–22 × 1.0 μm
D. arctii	Globose to slightly flattened, 280–480 × 160–320 μm diam., with a long, tubular ostiolated beak	Asci clavate, 47–60 × 7–10 μm. Ascospores 1-septate when mature, fusoid-ellipsoidal, straight or curved, 12–15 × 2.5–4.0 μm	Subglobose to lens-shaped (long axis parallel with the stem), ostiolate, 250–450 × 100–190 μm	α-Conidia usually 1-septate, fusiform, straight or slightly curved, attenuated ends, biguttulate, 7–10 × 2–3 μm. β-conidia filiform-hamate, 18–25 × 1 μm
D. eres	Subglobose, clustered in groups, 200–300 μm diam., ostiolated (300–700 μm long)	Asci elongate to clavate, 48.5–58.5 × 7–9 μm. Ascospores 1-septate, often 4-guttulate, elliptical, 12.5–14.5 × 3-4 μm	Eustromatic, globose, 200–250 μm diam., ostiolate (200–300 μm long)	α-Conidia unicellular, ovate to ellipsoidal, base subtruncate, 6.5–8.5 × 3.4 μm. β-conidia filiform to hooked, 25 × 1.3 μm
D. phaseo-lorum	Globose, clustered in groups, 230–270 μm diam., ostiolated (210–430 μm long)	Asci ellipsoid, 34.5–36.0 × 7.5–8.5 μm. Ascospores ellipsoid, 1-septate, each cell biguttulate, 9.3–9.7 × 2.4–2.6 μm	Globose to irregular, 210 × 180 μm	α-Conidia unicellular, ellipsoidal to fusiform, biguttulate, 5.3–7.7 × 1.5–4.6 μm. β-conidia filiform, 10–18 μm × 0.5–1.8 μm

Pavia, and *P. cannabina* is missing (Dr V. Terzo, Curator, pers. comm.). No specimens exist at any other major Italian herbaria, including MIPV, PAD, POR, SIENA, TO (McPartland, 1983b). In the USSR, Gitman and Malikova (1933) described *P. cannabina* causing "stem blackening" after harvest in unfavorable storage conditions. Gitman and Boytchenko (1934) redescribed *P. cannabina* on stems, pycnidia 70–100 μm in diameter, α-conidia biguttulate, septate, 9.8–11.5 × 4.2 μm (which departs from Curzi's description). De Corato (1997) described a *Phomopsis* sp. on hemp stems in southern Italy with pycnidia globose, 500–800 μm in diameter, α-conidia 6–9 × 2–4 μm, *β*-conidia 17–23 × 1.2 μm. De Corato described this as a disease of living plants, and provided a color photograph of the symptoms (Fig. 10.26).

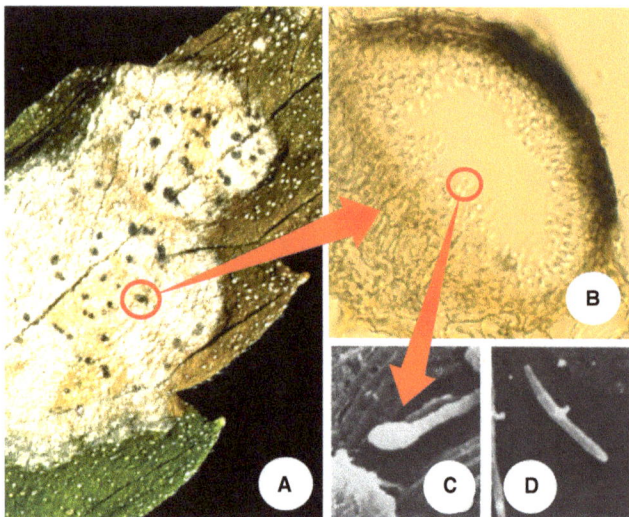

Fig. 10.27. *Diaporthe ganjae*: **A.** LM ×8; **B.** LM ×250; **C.** SEM ×1500; **D.** SEM ×1100 (McPartland, 1983b)

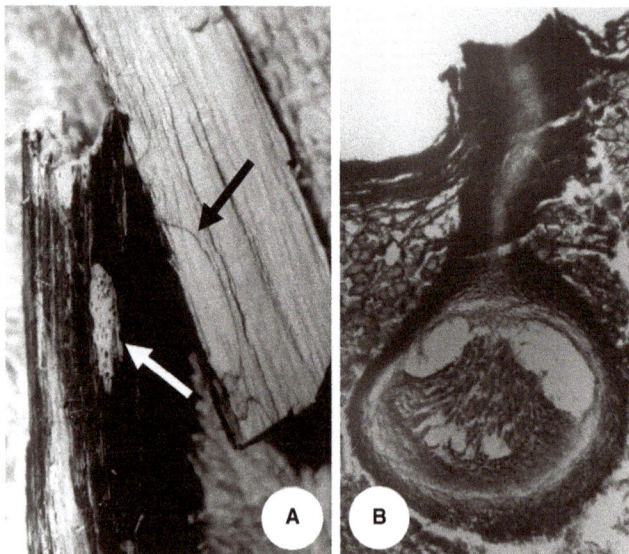

Fig. 10.28. *Diaporthe arctii* on *Cannabis*

10.9b. *Diaporthe ganjae* (McPartland) Gomes, Glienke & Crous 2013

Anamorph: *Phomopsis ganjae* McPartland 1983

Comments: A fungus causing white leaf spot in Illinois (Fig. 10.25) was provisionally identified as *P. cannabina* (McPartland, 1983a), but a closer look prompted McPartland (1983b) to erect *P. ganjae*. Cultures were submitted to ATCC and CBS. Examination of herbarium specimens revealed misidentifications of *P. ganjae*: in Pakistan as *Phoma* sp. (Ghani and Basit, 1976) or *Phomopsis cannabina* (Ghani *et al.*, 1978) (see Fig. 10.1), in India as *P. cannabina* (Sohi and Nayar, 1971; Gupta, 1985), and in Kansas as *Phyllosticta cannabis* (unpublished, Worthie H. Horr). *P. ganjae* has been reisolated in Kansas, with pycnidia arising in classic concentric rings (Hollandbeck, 2019).

P. ganjae has not produced a *Diaporthe* sexual state. Uecker (1989) featured *P. ganjae* in a *Phomopsis* workshop, and could not induce a *Diaporthe* state, despite his many tricks of the trade. Crous and colleagues transferred *P. ganjae* to *Diaporthe* based on a phylogenetic analysis of the CBS culture—*D. ganjae* was sister to *Diaporthe manihotia* (Gomes *et al.*, 2013). They erroneously described the species on "wilted, dead leaves" (compare Fig. 10.25) and "limited to the USA."

In other phylogenetic studies, *D. ganjae* was sister to a different species, *Diaporthe compacta* (Gao *et al.*, 2017; Dissanayake *et al.*, 2017; Yang *et al.*, 2018; Sukarno *et al.*, 2021; Cao *et al.*, 2022). Dissanayake *et al.* and Yang *et al.* repeated erroneous statements by Gomes *et al.* ("limited to the USA"). Dissanayake stated that that Gomes *et al.* described and illustrated *D. ganjae*, which they had not.

Crous and colleagues found a fungus that infected *Pyrus pyrifolia* and formed a clade with *D. ganjae,* so they identified the fungus as *D. ganjae* (Guo *et al.*, 2020). If their study utilized *D. manihotia* or *D. compacta* instead of *D. ganjae*, no doubt they would have identified the *Pyrus* fungus as one of those species. They said that Gomes *et al.* (2013) "discovered" *D. ganjae* on "dead leaves of *Cannabis sativa* in Illinois, USA." Adding incredulously: "This is the first description of its asexual morph." The asexual morph on *Pyrus* differed significantly from that on *Cannabis*: pycnidia were larger (229–634 μm in diameter), α-conidia were shorter (5.31–7.25 μm), and no *β*-conidia were observed, unlike *D. ganjae* on *Cannabis* (Fig. 10.27D).

Karani *et al.* (2022) found a fungus that infected *Berchemia discolor* in Kenya. In a phylogenetic study the Kenyan fungus was sister to *D. ganjae,* so that's what they called it. Inoculation studies showed that Kenyan *D. ganjae* caused tree cankers on croton (*Croton megalocarpus*), tamarind (*Tamarindus indica*), and olive (*Olea europaea*).

Serial magnifications of *Diaporthe ganjae* appear in Fig. 10.27: (A) multiple pycnidia in a leaf spot; (B) cross-section of pycnidium, with multiple α-conidia; (C) α-conidium, with a robust germination tube; (D) *β*-conidium, germinating a limited lateral protuberance.

10.9c. *Diaporthe arctii* (Lasch) Nitschke 1870

≡*Sphaeria arctii* Lasch 1846
=*Diaporthe tulasnei* Nitschke f. *cannabis* Saccardo 1897

Anamorph: *Phomopsis arctii* (Lasch) Traverso 1906
Comments: Curzi (1924) identified this fungus as *Phomopsis tulasnei* Saccardo 1906, collected in Porto d'Ascoli. Saccardo (1897) coined a new subspecies, *Diaporthe tulasnei* f. *cannabis*, which he distributed as an exsiccatus. Saccardo collected his specimens from Italian plants in November. Wehmeyer (1933) synonymized Saccardo's taxon under *D. arctii*. Unaware of this, McPartland (1995d) did the same, after examining a Saccardo isotype (Fig. 10.28).

D. arctii has a wide host range; Wehmeyer (1933) described it from 49 plant genera—one tree (*Phoenix* sp.), one grass species (*Phleum* sp.), and the rest herbaceous dicots. In a molecular study by Gomes *et al.* (2013), three specimens labeled *D. arctii* scattered across their phylogenetic tree, indicative of several cryptic species. The surface of infected stems turns black; shaving the surface exposes pockets of perithecia (white arrow in Fig. 10.28A). Splitting the stem reveals zone lines (black arrow in Fig. 10.28A). Sectioning a perithecium shows a long, tubular ostiole (Fig. 10.28B).

10.9d. *Diaporthe eres* Nitschke 1870

≡*D. cotoneastri* (Punithalingam) Udayanga *et al.* 2012
Anamorph: *Phomopsis oblonga* Desmazières (Traverso) 1906
Comments: *D. eres* causes stem cankers and leaf spots in diverse families of both woody and herbaceous plants. A phylogenetic study by Udayanga *et al.* (2014) demonstrated that *D. eres* consists of nine distinct species. *D. eres* in the strict sense was isolated from many woody plants, as well as herbaceous *Abutilon*, *Alliaria*, *Allium*, *Articum*, *Cucumis*, *Hordeum*, *Phaseolus*, and *Rumex* spp. Punja (2021d) added *Cannabis* to the list of *D. eres* hosts.

10.9e. *Diaporthe phaseolorum* (Cooke & Ellis) Saccardo 1882

≡*Sphaeria phaseolorum* Cooke & Ellis 1878
Anamorph: *Phomopsis phaseoli* (Desmazieres) Saccardo 1915
Comments: *D. phaseolorum* causes stem cankers on soybean and common bean, as well as sweet potato, peanut, cowpea, and other hosts. DNA analysis has revealed that *D. phaseolorum* is polyphyletic (Santos *et al.*, 2011; Gomes *et al.*, 2013). *D. phaseolorum* caused stem cankers in 3-month-old, field-cultivated Italian cultivars ('Eletta Campana', 'Carmagnola Selezionata', 'Tygra') growing in Florida (Marin *et al.*, 2021). Symptoms began on the main stalks with light-to-dark brown lesions of different sizes and shapes. The lesions coalesced into large necrotic areas and bore pycnidia. Koch's postulates proved pathogenicity. Phylogenetic analysis confirmed its identity.

10.9.3 Differential diagnosis

White leaf spot can be confused with yellow leaf spot (*Septoria* spp.) and brown leaf spot (*Phoma* and *Ascochyta* spp.). White leaf spot uniquely produces pycnidia in thickened, stromatic tissue. On leaves, pycnidia often form in concentric rings, compared with the random distribution seen in *Septoria*, *Phoma*, and *Ascochyta* spp. The stem canker can also be confused with disease caused by *Septoria*, *Phoma*, and *Ascochyta* spp.

Phylogenetic studies place *D. ganjae*, *D. eres*, *D. arctii*, and *D. phaseolorum* in different clades (Gomes *et al.*, 2013; Gao *et al.*, 2017; Dissanayake *et al.*, 2017); α-conidia of *D. ganjae* are longer, with a length/width ratio (7.7) greater than those of *P. cannabina* (Curzi: 2.7, De Corato: 2.5), *D. eres* (Udayanga: 3.0), or Saccardo's specimen of *D. arctii* (4.0).

10.9.4 Disease cycle and epidemiology

Diaporthe/Phomopsis spp. overwinter in crop debris. They sporulate in spring and infect seedlings. Seedlings infected with high inoculum loads are killed (Fig. 10.29). Leaf symptoms caused by *D. ganja* arise low in the canopy in early summer. Conidia give rise to secondary infections, and epidemics arise late in the season. Stem infections follow a later time course, getting under way in late summer. Infections arise on stems near branch nodes, probably entering through breaks in the epidermis. Stem cankers continue to develop after plant death, and some species produce a sexual state in autumn.

10.9.5 Control and management

Follow directions for yellow leaf spot (*Septoria* spp.). Practice good sanitation—gather and dispose of crop residues, and eliminate crop stubble. Scout for disease during wet seasons. Gitman and Malikova (1933) listed several resistant fiber varieties, which are no longer available. De Corato (1997) tested four fiber varieties: 'Yellow stem' (Hungarian 'Kompolti Sárgaszárú') was the most susceptible, followed by two Chinese varieties, 'Shan-ma' and 'Chain-chgo'. Most resistant was 'Foglia pinnatofida'. Biocontrol of *Phomopsis* spp. in other

Fig. 10.29. Healthy seedling flanked by plants inoculated with fungal pathogens

crops has utilized *Trichoderma atroviride* (Binab®), *Trichoderma virens* (GlioGard®), and *Bacillus subtilis* (Serenade®). Bordeaux mixture inhibits *Diaporthe/Phomopsis* species on other crops.

10.10 Hemp leaf spot

This is a new disease, coined by Szarka *et al.* (2020b), who named *Bipolaris gigantea* as the causal organism. We expanded the franchise, adding six related species in *Bipolaris*, *Curvularia*, and *Exserohilum*. Many species in these genera were formerly classified as *Drechslera* spp., or "graminicolous *Helminthosporium* species" (Sivanesan, 1987). They belong to Class Dothideomycetes, Order Pleosporales, Family Massarinaceae. Difficult to sort morphologically, molecular methods have revolutionized their identification. A phylogenetic analysis of the *Bipolaris–Curvularia–Exserohilum* complex appears in Fig. 1.12.

10.10.1 Symptoms and signs

Tiny light-colored spots appear in July, on lower leaves within the canopy. The spots enlarge to round lesions, colors ranging from dark brown to light tan with a darker margin (Fig. 10.30).

Fig. 10.30. Hemp leaf spot caused by *Bipolaris gigantea* (courtesy Z. Hansen)

The multicolored lesions are called "zonate leaf spots" in other hosts. By August spots enlarge to irregularly polygonal shapes, their edges partially delineated by leaf veins. Spots may fragment, leaving behind shot-holes. In warm and rainy summers, when conditions are right for the fungus, lesions coalesce and cause a necrotic blight, affecting fan leaves and sugar leaves, even bracts (Gauthier and Szarka, 2022). Entire plants may become affected, their growth stunted, with serious yield losses.

10.10.2 Causal organisms

Species in the *Bipolaris–Curvularia–Exserohilum* complex have multicellular conidia with distosepta (i.e., pseudosepta). Distosepta separate individual cells from each other, but do not extend to the outer cell wall of the conidium, like true septa.

10.10a. *Bipolaris gigantea* (Herald & Wolf) Lane *et al.* 2020

≡*Drechslera gigantea* (Herald & Wolf) Ito 1930
≡*Helminthosporium giganteum* Heald & Wolf 1911, *nomen illegitimum*
≠*Helminthosporium giganteum* Renault & Roche 1898

Description: Sporulation amphigenous. *Conidiophores* emerge from leaf tissue singularly or in pairs, simple, pale to dark brown, with 3–6 indistinct septa, base slightly bulbous, 200–375 × 7–11 μm. *Conidia* are large, straight and cylindrical, hyaline to pale yellow-brown, 3–6 distosepta, with a hilum (base) that is slightly truncate and protuberant, 200–350 × 19–27 μm (Fig. 10.31A). Conidia exhibit bipolar germination, producing multiple germ tubes from basal and apical cells, usually growing at angles not on axis with the conidium. Conidia may give rise to secondary *Dendryphion*-like microconidia in penicilliate chains, hyaline to pale yellow-brown, 1–3 pseudosepta, 12–28 × 4–6 μm.

Nomenclature notes: The basionym *H. giganteum* Heald & Wolf is a *nomen illegitimum*, because that taxon was previously applied to a different fungus, *H. giganteum* Renault & Roche. *International Code of Nomenclature* rejects later homonyms, such as Heald & Wolf's taxon, under Article 53.1 (Turland, 2018). Websites (*Mycobank*, *Index Fungorum*) as well as Lane *et al.* (2020) claim that Ito (1930) erected the species *Drechslera gigantea* Ito, but this is incorrect. He transferred *H. giganteum* Heald & Wolf to *Drechslera*, coining *Drechslera gigantea* (H. et W.) Ito. He changed the species epithet to agree grammatically with the generic name *Drechslera*. Recombinations based on a rejected basionym, such as *D. gigantea* (H. et W.) Ito and *B. gigantea* (Herald & Wolf) Lane *et al.*, are illegitimate (Turland, 2018). *Mycobank* and *Index Fungorum* changed the nomenclature to *B. gigantea* (Ito) Lane *et al.*, which is equally untenable. Someone needs to coin an avowed substitute, such as *Bipolaris neogigantea*, *nomen novum*.

Comments: *B. gigantea* is a worldwide pathogen with over 50 known hosts, mostly wild and weedy monocots; rare dicot hosts include *Teramnus* and *Zinnia* spp. (Szarka *et al.*, 2023).

B. gigantea is sufficiently pathogenic that several research groups evaluated it as a bioherbicide (e.g. Shabana *et al.*, 2010). That work has been shelved now that hemp has been identified as a crop plant susceptible to *B. gigantea*.

B. gigantea first gained attention as a *Cannabis* pathogen in Kentucky (Gauthier, 2020a; Szarka *et al.*, 2020b,c). It has been reported in Tennessee (Cartwright and Hansen, 2020), Virginia (Mersha *et al.*, 2020), Alabama (Conner *et al.*, 2020), and New York (Bergstrom *et al.*, 2020). Szarka (2021) reported isolations in 15 states but didn't list them. Devastating outbreaks have spread as far west as Kansas and Oklahoma (Gauthier and Szarka, 2022).

Szarka *et al.* (2020b) searched for alternative hosts around hemp fields. They found six, including *Acalypha ostryifolia*, a dicot weed common in Kentucky. No sexual state has been reported for this species. Populations have been identified with more than one mating type, and proto-perithecial-like structures have been observed (Szarka *et al.*, 2020c; Szarka, 2021; Szarka *et al.*, 2023). *B. gigantea* populations on hemp exist as haploids or "heteroploids" (either diploid or heterokaryotic). Heteroploids were distinguished from haploids by genome sizes, the number of *RPB2* and *TEF1* alleles, and the presence of one or both mating-type idiomorphs *MAT1-1* and *MAT1-2* (Szarka *et al.*, 2023).

B. gigantea conidia differ from other *Bipolaris* and *Drechslera* spp. in three characters: their "gigantic" size, strikingly cylindrical shape, and secondary *Dendryphion*-like microconidia (Drechsler, 1928). Because of this, Luttrell (1964) proposed removing *Drechslera gigantea* from *Drechslera*. Kenneth (1983) suggested erecting *Colossosporium* as a new genus for the species, but he never formally published the name. *B. gigantea* conidia have a hilum (base) that is slightly truncate and protuberant, consistent with *Bipolaris* and not *Drechslera* spp. (Alcorn, 1988). Phylogenetically, *B. gigantea* falls into a clade with other *Bipolaris* spp. (see Fig. 1.12), although the long branch length leading to *B. gigantea* indicates evolutionary divergence.

10.10b. *Bipolaris victoriae* (Meehan & Murphy) Shoemaker 1959

≡*Drechslera victoriae* (Meehan & Murphy) Subramanian & Jain 1966 ≡*Helminthosporium victoriae* Meehan & Murphy 1946
Description: *Conidiophores* emerge singularly or in small groups, simple, straight or flexuous, sometimes geniculate at upper part, pale to mid-brown, septate, 100–250 × 6–10 μm. *Conidia* are straight or slightly curved, broadly fusiform or obclavate fusiform, tapering towards rounded ends, pale to mid-brown, usually 7 distosepta (range 4–11), hilum slightly protuberant, 55–90 × 12–16 μm (Fig. 10.31B).
Comments: *B. victoriae* is a worldwide pathogen of oats and similar grain crops. Pei *et al.* (2018) reported "*Bipolaris* sp." causing hemp leaf spots in Yúnnán Province. They deposited a genetic sequence of its ITS region in GenBank (no. MH881227). We subjected the sequence to a BLAST "best-hit" analysis of GenBank, and it shared 100% identity with *B. victoriae* (see Table 1.2). Phylogenetically, their sequence fell into a clade with other *Bipolaris* species (see Fig. 1.12).

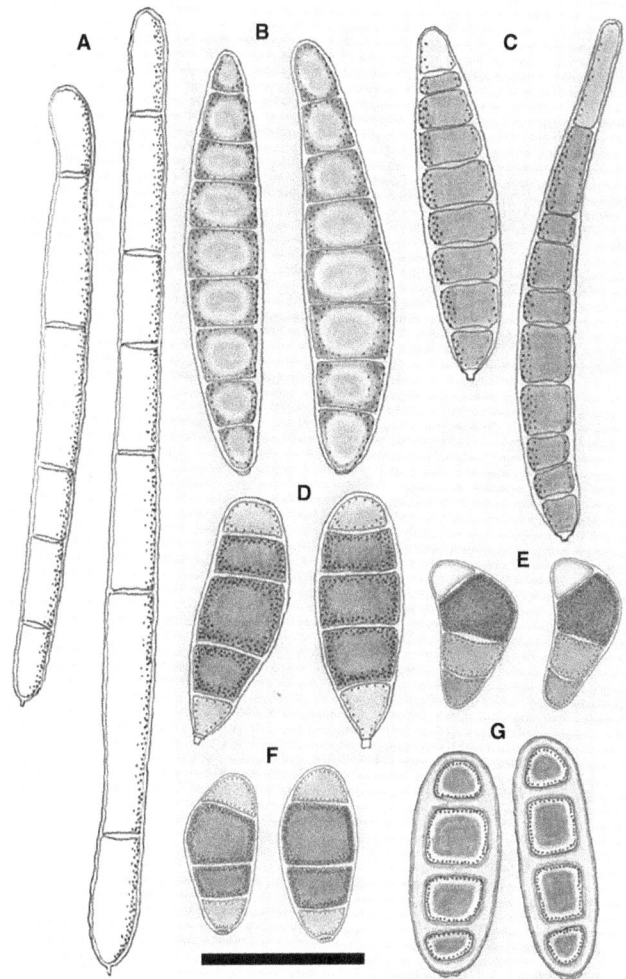

Fig. 10.31. Comparison of conidia: **A**. *Bipolaris gigantea*; **B**. Bipolaris victoriae; **C**. Exserohilum rostratum; **D**. Curvularia cymbopogonis; **E**. Curvularia lunata; **F**. Curvularia pseudobrachyspora; **G**. *Curvularia australiensis*. Scale bar = 25 μm, except *B. gigantea*

10.10c. *Curvularia lunata* (Wakker) Boedijn 1933

Sexual state: *Cochliobolus lunatus* [as *lunata*] Nelson & Haasis 1964.

Description: *Conidiophores* up to 650 μm long. *Conidia* with 3 distosepta, third cell from the base is curved and larger and darker brown than other cells, fourth cell (end cell) nearly hyaline, surface smooth to verruculose, 13–32 × 6–15 μm (see Figs 3.3 and 10.31E). *Pseudothecia* solitary or gregarious, black, ellipsoidal to globose with a tall beak, ostiolate, up to 700 μm in height including a beak 210–560 μm long, up to 530 μm in diameter. *Asci* bitunicate, cylindrical to clavate, with a short stipe, 1–8-spored, 160–300 × 10–20 μm. *Ascospores* filiform, 6–15 septate, hyaline, arranged either straight, or coiling in a helix within asci, 130–270 × 3.8–6.5 μm.

Comments: *C. lunata* is distributed worldwide in warmer climates. It attacks a variety of crops, including maize, millet, rice, and amaranth. Babu *et al.* (1977) recovered *C. lunata* from

Indian *Cannabis* seeds, and Shahzad and Ghaffar (1995) report the pathogen in Pakistan. Suyal *et al.* (2011) analyzed air mycoflora over *Cannabis* fields in the Himalaya, and frequently found *C. lunata*. Hansen *et al.* (2020) isolated *C. lunata* from hemp leaf spots in Tennessee. They also identified two other species based on ITS sequence data: *Curvularia americana* Da Cunha *et al.* 2014, and *Curvularia trifolii* (Kauffman) Boedijn 1933. They hesitated calling these *Cannabis* pathogens, pending completion of Koch's postulates.

10.10d. *Curvularia cymbopogonis* (Dodge) Groves & Skolko 1945

≡*Helminthosporium cymbogonis* C.W.Dodge 1942
Sexual state: *Cochliobolus cymbopogonis* Hall & Sivanesan 1972
Description: *Conidiophores* simple, septate, brown, up to 300 μm long. *Conidia* acropleurogenous, smooth, straight or curved, clavate to ellipsoidal, 4 (sometimes 3) distosepta, averaging 40–50 × 12–15 μm, middle cells dark brown, base cell and end cell paler (Fig. 10.31D), base cell obconical with a protuberant hilum (distinctive among *Curvularia* spp.). *Pseudothecia* scattered or aggregated in concentric zones on agar, black, globose with a long cylindrical beak, up to 575 μm diameter. *Asci* bitunicate, cylindrical with a short stipe, 8-spored, 210–275 × 15–23 μm. *Ascospores* filiform, 8–14-septate, hyaline, 195–420 × 3.5–4.5 μm.
Comments: McPartland and Cubeta (1997) isolated *C. cymbopogonis* from seeds in Nepal. *C. cymbopogonis* occurs worldwide, attacks barley, rice, sorghum, citronella grass, and other crops; it is frequently reported as a seed-borne pathogen.

10.10e. *Curvularia pseudobrachyspora* Marín, Cheewankoon & Crous 2017

Description: *Conidiophores* arise singly or in groups, septate, straight or flexuous, sometimes geniculate at upper part, size of cells rarely decreasing towards apex, sometimes branched, 110–420 × 3.5–6.0 μm. *Conidiogenous cells* smooth-walled, terminal or intercalary, proliferating sympodially, pale brown to brown, subcylindrical to slightly swollen, 7–24 × 4–6.5 μm. *Conidia* verruculose, mostly curved, ellipsoidal to obovoid, pale brown to brown, usually 3 distosepta, apical and basal cells paler than the middle cells, 21.5–27.0 × 8–14 μm (Fig. 10.31F).
Comments: Marín-Felix *et al.* (2017) discovered this species in Thailand, isolated from *Eleusine indica*. It resembled *Curvularia brachyspora* Boedijn 1933, but its ITS and *EF1-α* sequences differed. Marin *et al.* (2020b) found a *Curvularia* sp. causing leaf spots on hemp plants in Florida, with conidia measuring 20.3–31.7 × 9.5–15.9 μm. A BLAST search of its ITS and *EF1-α* sequences hit on *C. pseudobrachyspora*. In a phylogenetic analysis, their ITS sequence fell into a clade with other *Curvularia* species (see Fig. 1.12). Marin and colleagues noted that *Eleusine indica* is a common weed in Florida, and might serve as an alternative host.

10.10f. *Curvularia australiensis* (Ellis) Manamgoda, Cai & Hyde 2012

≡*Bipolaris australiensis* (Ellis) Tsuda & Ueyama 1981, ≡*Drechslera australiensis* M.B. Ellis 1971
Description: *Conidiophores* single, flexuous, geniculate, septate, smooth, cylindrical, reddish-brown, up to 235 × 6–15 μm. *Conidiogenous cells* with verruculose nodes. *Conidia* straight, ellipsoidal or oblong, rounded at the ends, reddish-brown to black-brown, usually 3 distosepta (rarely 4 or 5), 27–43 × 7.5–10 μm (Fig. 10.31G).
Comments: Shahzad and Ghaffar (1995) isolated *Bipolaris australiensis* from *Cannabis* in Pakistan. Based on a phylogenetic analysis of LSU, ITS, and *EF1-α*, Manamgoda *et al.* (2012) transferred this species from *Bipolaris* to *Curvularia*. The species attacks Poaceae (*Setaria, Chloris, Pennisetum*) but also infects humans.

10.10g. *Exserohilum rostratum* (Drechsler) Leonard & Suggs 1974

≡*Drechslera rostrata* (Drechsler) Richardson & Fraser 1968, ≡*Helminthosporium rostratum* Drechsler 1923
Sexual state: *Setosphaeria rostrata* Leonard 1976
Description: *Conidiophores* arising singly or in groups of 2–5 from stomata or between epidermal cells, flexuous, geniculate, olivaceous-brown, usually 5–7 distosepta, 40–180 × 6–8 μm. *Conidia* straight or slightly curved, ellipsoidal to obclavate-rostrate, olivaceous-brown; base with a protuberant, cylindrical hilum; apex with a rostral beak; 7–12 distosepta, mean dimensions 75.64 ± 8.31 × 15.61 ± 1.41 μm (Fig. 10.31C).
Comments: Conidial length is quite variable, Leonard (1976) surveyed the literature and reported a range of 15–190 μm. Drechsler gave a range of 32–184 μm in his original description. Variability was due to the length of the rostral apex.

E. rostratum has a wide host range—primarily Poaceae, but also dicots, as well as cats, dogs, horses, cattle, and humans (Sharma *et al.*, 2014). Thiessen and Schappe (2019) isolated *E. rostratum* from hemp in North Carolina. It caused foliar, stem, and floral blight symptoms. Lesions on leaves were round, brown to black, with dark margins (Fig. 10.32). Identification was confirmed by sequencing *rpb2* and the ITS gene region, and conducting a BLAST search and phylogenetic analysis. Pathogenicity was confirmed by fulfilling Koch's postulates. *E. rostratum* was isolated from 16% of all diseased samples submitted to the North Carolina Department of Agriculture (Thiessen *et al.*, 2020). Plenk *et al.* (2021) isolated an *Exserohilum* sp. from diseased hemp in Austria.

10.10.3 Endophytic species

Bipolaris–Curvularia spp. can behave as **endophytes**—commensal organisms that live within plants. Endophytes absorb nutrients from extracellular sources without causing disease. They may enhance plant growth, by improving the plant's ability to tolerate drought, or boost resistance to pathogens and

Fig. 10.32. Hemp leaf spot caused by *Exserohilum rostratum* (L. Thiessen)

pests. Under certain circumstances, however, endophytes may flip into pathogens (see Chapter 13).

Lubna *et al.* (2019) isolated a *Bipolaris* sp. colonizing healthy *Cannabis* plants in Pakistan. They considered the fungus a beneficial endophyte—it produced indole-3-acetic acid (IAA) and gibberellin (GA$_3$). Lubna deposited its ITS sequence in GenBank. In Table 1.2 we subjected the ITS sequence to a BLAST "best-hit" analysis, and Lubna's *Bipolaris* sp. shared 100% identity with **Curvularia spicifera** (Bainier) Boedijn 1909. When placed in a phylogenetic tree, the ITS sequence fell into a clade with other *Curvularia* spp. (see Fig. 1.12).

Scott *et al.* (2018) isolated endophytes from *Cannabis* in Quebec. They identified *Cochiobolus* sp. and *Drechslera* sp., and deposited ITS sequences in GenBank. A closer look at these sequences, using BLAST and a phylogenetic analysis, identified them as **Bipolaris sorokiniana** (Sacc.) Shoemaker 1959 and **Drechslera tritici-repentis** (Diedicke) Shoemaker 1962, respectively (see Table 1.2, Fig. 1.12).

10.10.4 Differential diagnosis

Leaf spots caused by *Bipolaris–Curvularia–Exserohilum* spp. superficially resemble lesions by other pathogens, such as *Phoma* and *Septoria* spp. But examination with a hand lens will reveal dark conidiophores sporulating on the surface of lesions, and no immersed conidiomata such as *Phoma* and *Septoria* pycnidia. The gigantic conidiophores and conidia of *B. gigantea* are visible without magnification. Lesions by *B. gigantea* and *E. rostratum* can expand into a blight, thereby resembling symptoms caused by *Alternaria*, *Stemphylium*, *Fusarium*, and *Colletotricum* spp.

10.10.5 Disease cycle and epidemiology

Fungi in the *Bipolaris–Curvularia–Exserohilum* complex overwinter as conidia or dormant mycelium in the dead leaves of grassy weeds, which sporulate in springtime with fresh conidia. Some species have sexual states that overwinter, releasing ascospores in springtime. Szarka *et al.* (2020c) investigated seed-borne infection by *B. gigantea*, which they considered insignificant. *Curvularia* spp. can be transmitted by seed-borne infection. Leaf spots arise low in the canopy in early summer. Airborne conidia give rise to secondary infections, and epidemics arise late in the season. Some species can cause an almost complete destruction of foliage (e.g., *B. gigantea*).

10.10.6 Control and management

Follow directions for yellow leaf spot (*Septoria* spp.), beginning with sanitation. Remove all infected crop residues. After harvest, till the soil to bury any remaining crop residues. Eliminate weeds; Szarka *et al.* (2020b) isolated *B. gigantea* from grassy weeds (*Eleusine* and *Microstegium* spp.) as well as a dicot host (*Acalypha ostryifolia*) growing near and within hemp crops; genetic sequences from these hosts were identical to *B. gigantea* infecting hemp.

Szarka *et al.* (2020c) reported greater susceptibility in CBD cultivars than in fiber-type or grain cultivars. High-THC cultivars are also highly susceptible to *B. gigantea* (Gauthier and Szarka, 2022). Dixon *et al.* (2022a) screened several biocontrols and biorational pesticides for *in vitro* efficacy against *B. gigantea* (see Table 9.1). Best results were obtained with Cueva® (copper soap) and LifeGard® (*Bacillus mycoides*). Biocontrol of other *Bipolaris* spp. has been accomplished with *Trichoderma viride* and *T. harzianum*.

10.11 Cladosporium stem and leaf spot

Several *Cladosporium* spp. cause this disease, members of Family Davidiellaceae, Class Dothideomycetes. Stem spots arise as dark areas, and elongate on axis with the stem. The patches turn velvety green-gray under humid conditions. *Cladosporium* spp. produce cellulolytic enzymes, which rot bast fiber. They can continue to rot fiber after the stems have been harvested (see Fig. 18.2). Leaf spots have been described as round to irregular in shape, covered by a green-gray mycelial mass. Diseased tissues turn necrotic and break up into ragged shot-holes.

10.11.1 Causal organisms

10.11a. *Cladosporium herbarum* (Persoon) Link 1816

≡*Dematium herbarum* Persoon 1794, =*Hormodendrum herbarum* auct.
Sexual state: *Mycosphaerella tassiana* (deNotaris) Johanson 1884
Description: *Conidiophores* cylindrical and straight (in culture) or nodose and geniculate (*in vivo*), unbranched until near the apex, smooth, pale to olivaceous brown, 36 µm in diameter and up to 250 µm long. *Conidia* blastic, produced sympodially

in simple or branched chains, ellipsoidal or oblong, rounded at the ends, thick-walled, distinctly verruculose, golden to olivaceous brown, with scars at one or both ends, 0–1 septate, 8–15 × 4–6 µm. *Pseudothecia* globose, black, scattered to aggregated, up to 160 µm in diameter. *Asci* bitunicate, subclavate, short stipitate, 8-spored, 35–90 × 15–30 µm. *Ascospores* hyaline to rarely pale brown, ellipsoid, usually 1-septate, slightly constricted at septum, 15–30 × 4.5–9.5 µm. *Paraphyses* not present.

Comments: *C. herbarum* is a common fungus found worldwide. Farr *et al.* (1989) found hosts in over 90 plant genera in the USA. It grows saprophytically on an assortment of nonplant substrates. *C. herbarum* attacks stalks of fiber-type plants, especially after plants enter senescence (Curzi and Barbaini, 1927; Ceapoiu, 1958; Hrebenyuk, 1984). Lentz (1977) reported the sexual state on *Cannabis* stems.

   C. herbarum may participate in the retting process (see Fig. 18.2). The fungus is not welcome because of its cellulolytic enzymes, which ruin bast fiber. Long ago, Behrens (1902, 1903) implicated *C. herbarum* in rotten, over-retted hemp, especially in *Tauröste* (dew retting) and *Winterlandröste* (winter retting or snow retting). Affected fibers stain a dark brown-black color and lose their tensile strength (Gitman and Malikova, 1933; Vakhrusheva and Davidian, 1979). Fibers stained by the fungus lose commercial value; Trunov (1936) reported a 24% loss in Ukrainian crops one year.

   C. herbarum also acts as a secondary invader of plants parasitized by *Podosphaera*, *Septoria*, or *Phomopsis* spp. (McPartland *et al.*, 2000). There are reports of *C. herbarum* attacking foliage (Curzi and Barbaini, 1927; Nair and Ponnappa, 1974; Ponnappa, 1977). Ferri (1961b) cultured *Cladosporium* sp. from Italian hemp seeds, and Babu *et al.* (1977) isolated *C. herbarum* from seeds in India that were surface-sterilized with 0.1% mercuric chloride.

10.11b. *Cladosporium westerdijkiae* Bensch & Samson 2018

Cladosporium cladosporioides (Fresenius) DeVries 1952
≡*Hormodendrum cladosporioides* (Fresenius) Saccardo 1880.
Description: *Conidiophores* straight or flexous, simple or once-branched, smooth or verruculose, pale to olivaceous brown, 20–135(–180) × 2.5–3.5 µm. *Ramo-conidia* an extension of conidiophores, cylindrical to oblong, smooth or rarely minutely verruculose, 12–45 × 2–4 µm. *Conidia* formed in simple or branched chains, pale olivaceous brown, mostly smooth-walled, rarely minutely verruculose, 1-celled; small terminal conidia globose to ovoid, average 3.3 × 1.0 µm; larger intercalary conidia ovoid to ellipsoid, average 7.1 × 2.1 µm.
Comments: *C. cladosporioides* is a common mold found worldwide, more dominant than *C. herbarum* indoors, but less common as a plant pathogen. Hrebenyuk (1984) isolated *C. cladosporioides* from hemp stalks in Ukraine. Suyal *et al.* (2011) analyzed air mycoflora over *Cannabis* fields in the Himalaya, and *C. cladosporioides* was the dominant species. Jerushalmi *et al.* (2020) isolated *C. cladosporioides* from foliar tissues in Israel, and fulfilled Koch's postulates.

Bensch *et al.* (2018) sequenced hundreds of *C. cladosporioides* cultures at CBS; their phylogenetic analysis prompted them to cleave four new species out of *C. cladosporioides*. One of the new species, *Cladosporium westerdijkiae*, was isolated from *Cannabis* by Punja *et al.* (2019). They exposed petri plates to greenhouse air, and recovered high counts of *C. westerdijkiae*. The fungus was also recovered from cut surfaces of stems that had been pruned. Flowering tops also yielded *C. westerdijkiae*.

   Two other *Cladosporium* spp. have been reported on *Cannabis*, once each. See Bensch *et al.* (2018) for morphological descriptions. Hrebenyuk (1984) isolated *Cladosporium resinae* (Lindau) DeVries 1955 from hemp stems in Ukraine. *C. resinae* has been isolated from petrochemical substrates such as jet fuel and facial creams; it hasn't been reported from herbaceous plants. Nair and Ponnappa (1974) identified *Cladosporium tenuissimum* Cooke 1878 from diseased leaves in India. The species normally acts as a saprophyte, isolated from dead leaves, stems, and other organic matter. Bensch *et al.* (2018) stated that *C. tenuissimum* is "confusable with *C. cladosporioides*."

10.11.2   Differential diagnosis

Stem spots caused by *Cladosporium* spp. can be confused with hemp canker, brown blight, and Stemphylium stem disease. Leaf symptoms have not been seen by us for sake of comparisons. *C. westerdijkiae* differs from *C. cladosporioides* in producing marginally shorter intercalary conidia (average 6.5 µm long), and forming shorter conidial chains (≤4, compared with ≤10 in *C. cladosporioides*) (Bensch *et al.*, 2018).

10.11.3   Disease cycle and epidemiology

Cladosporium spp. overwinter in crop debris. They are competent saprophytes and can grow anywhere, even at refrigerator temperatures. Warm humid conditions favor conidial production. A slight breeze detaches conidia and can carry them for miles. Airborne *C. herbarum* and *C. cladosporioides* conidia are major causes of "mold allergy" and asthma. Martyny *et al.* (2013) tested air samples in 30 indoor grows, and *Cladosporium* sp. ranked first. Spore counts at one site peaked at 500,000 spores/m^3 while plants were being harvested and removed. Opportunistic *Cladosporium* infections in humans range from eye ulcers to pulmonary fungus balls (Rippon, 1988).

   Cladosporium spp. exist as endophytes in *Cannabis*: Gautam *et al.* (2013) isolated *Cladosporium* sp. from 2.8% of leaves and 3.9% of stems of healthy plants in Himachal Pradesh, India. From an endophytic foothold, *Cladosporium* spp. can invade damaged tissues, such as flowering tops after they have been trimmed.

   Koivula *et al.* (2004) monitored mold counts in stems of a fiber-type crop ('USO-31'), twice a month for two successive seasons. Identified genera included *Cladosporium*, *Fusarium*, *Alternaria*, *Penicillium*, and *Mucor* spp. Colony counts (cfu/g) of sampled material increased steadily over the course of the season. Colony counts were greater in stems sampled from upper parts of plants than stems sampled from the base of plants. Fields treated three times with a fungicide (propiconazole) showed no appreciable differences in colony counts.

10.11.4 Control and management

Cladosporium spp. cannot be eliminated—they exist as endophytes (Gautam *et al.*, 2013), and their conidia are ubiquitous in air around *Cannabis* plants (Suyal *et al.*, 2011; Martyny *et al.*, 2013; Punja *et al.*, 2019). Remove all damp organic matter from grow spaces, especially paper products. Pruning wounds should be dabbed with *Bacillus subtilis* (Serenade®) or similar biocontrol organisms.

Punja *et al.* (2019) reported that *C. westerdijkiae* colony counts increased fourfold in trimmed buds subjected to mechanical trimming immediately after harvest ("wet trim"), due to the associated wounding of plant tissues. Dry trimming is a safer alternative. Babu *et al.* (1977) documented seedborne infection, so don't use seeds from infected plants. Fungicides are pointless; the fungus is difficult to eradicate. Indeed, Durrell and Shields (1960) found *C. herbarum* growing at Ground Zero in Nevada shortly after nuclear weapons testing.

10.12 Stemphylium leaf and stem spot

Rataj (1957) considered *Stemphylium botryosum* a dangerous pathogen in Czech hemp. *Stemphylium* sp. caused "considerate amounts" of leaf disease in Canadian hemp (McCurry and Hicks, 1925). Conversely, Barloy and Pelhate (1962) considered *S. botryosum* a secondary problem, only invading plants weakened by *Alternaria* spp. Leaf spotting by *S. botryosum* has also been reported in Denmark (Bovien *et al.*, 1943) and Germany (Spaar *et al.*, 1990). Steve Koike (pers. comm., 2021) identified a *Stemphylium* sp. causing leaf spots in California (Fig. 10.33).

Stem spots caused by *S. botryosum* have been reported in Ukraine (Hrebenyuk, 1984) and Czechoslovakia (Ondřej, 1991). A synonym, *Stemphylium cannabinum*, has been reported in Russia (Bakhtin and Gutner, 1933; Dobrozrakova *et al.*, 1956), Latvia (Serzane, 1962), Czechoslovakia (Ondřej, 1974), and Germany (Brandenburger, 1985).

10.12.1 Symptoms and signs

Spaar *et al.* (1990) referred to leaf symptoms as "brown fleck disease"—the spots seldom coalesce. Lesions arise as grayish-green, tan, or brown spots, often limited by leaf veins. Lesion centers may break apart, leaving shot-holes in leaves (Fig. 10.33). Sometimes the spots have darker margins (Fig. 10.34A). Disease may spread to stems (Hrebenyuk, 1984; Ondřej, 1991). Conidia (rarely perithecia) of the causal fungi form in necrotic plant tissue.

Fig. 10.33. Stemphylium leaf spot (S. Koike)

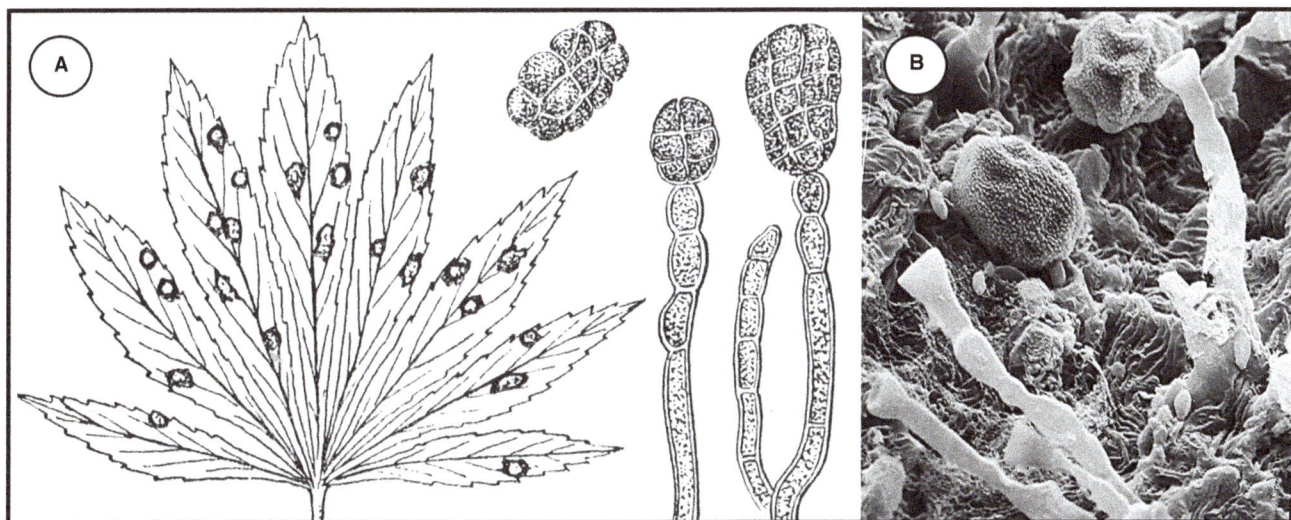

Fig. 10.34. Stemphylium leaf spot and its causal organism: **A.** redrawn from Bakhtin and Gutner (1933); **B.** Bakhtin and Gutner's type specimen (SEM ×750)

10.12.2   Causal organisms

Two *Stemphylium* spp. cause this disease. Both appear in *Cannabis* literature under piles of synonymous names. Genus *Stemphylium* currently consists of 28 "species-clades," each with many synonyms (Woudenberg *et al.*, 2017). They belong to Family Pleosporaceae, Class Dothideomycetes.

10.12a.   *Stemphylium botryosum* Wallroth 1833

=*Macrosporium cannabinum* Bakhtin & Gutner 1933, ≡*Stemphylium cannabinum* (Bakhtin & Gutner) Dobrozrakova *et al.* 1956, =*Tyrospora cannabis* Ishiyama 1936
Sexual state: *Pleospora tarda* Simmons 1986
Description: *Conidiophores* arise in clumps, straight or flexous, simple or occasionally branched at their base, 1–7-septate, cylindrical except for swelling at the site of spore formation, light olive-brown darkening at swollen apex, smooth becoming echinulate at swollen apex, 4–6 µm wide, swelling to 7–10 µm at apex, usually 25–85 µm long, but continued percurrent proliferation through 1–5 previously formed apices may extend conidiophore length to 120 µm (Fig. 10.34). *Conidia* initially oblong to subspherical, hyaline and verruculose; at maturity oblong to subdoliiform, densely echinulate, with 3 (sometimes 4) transverse septa and 2–3 (sometimes 4) complete or nearly complete longitudinal septa, with a single constriction at the median transverse septum, dilute to deep olive-brown, 30–35 × 20–25 µm (Fig. 10.34).

Pseudothecia large (700 µm in diameter), hard, black sclerotic bodies, with formation of asci occurring slowly, often after 8 months in culture. *Asci* subcylindrical, 8-spored, averaging 200 × 40 µm. *Ascospores* hyaline, oblong and constricted at 3 transverse septa when immature; when mature they exhibit a broadly rounded apex and almost flat base, turn yellowish-brown, develop 4 secondary transverse septa in addition to the 3 primary septa, form 1 or 2 longitudinal septa, and average 40 × 17 µm.

Comments: Bakhtin and Gutner (1933) described *Macrosporium cannabinum* near Leningrad (Fig. 10.34A). Bakhtin's research went no further; Stalin arrested him that year, and he was executed. Dobrozrakova *et al.* (1956) transferred the taxon to *Stemphylium*, coining *S. cannabinum*. Examination of Bakhtin and Gutner's type specimen of *M. cannabinum* revealed it to be *S. botryosum* (Fig. 10.34B). McPartland (1995d) sent a slide to Emory G. Simmons, who concurred with the identification.

Ishiyama (1936) cited *Macrosporium* sp. causing *Cannabis* leaf spots on Sakhalin Island. Ishiyama (1936b) then coined a new taxon for the fungus, *Tyrospora* [*Thyrospora*] *cannabis*. Ishiyama's type specimen has gone missing (pers. comm., curators at TI, TNS and KYO). Ishiyama's fungal description was identical to *S. botryosum*, so McPartland (1995d) added it to the synonymy. The genus *Thyrospora* is nomenclaturally illegitimate, and most *Thyrospora* spp. have been transferred to *Stemphylium*.

Garcia Rada identified "*Coniothecium* sp." causing leaf spots in Lima, Peru. He deposited a herbarium specimen at

BPI. It turned out to be *S. botryosum* (McPartland, 1995a). John T. Presley described "*Tetracoccosporium* sp." on leaves and stems of hemp in Beltsville, Maryland (unpublished 1955 data, USDA archives). No specimen survives, but Presley's description suggests *S. botryosum*. Termorshuizen (1991) mentioned *S. botryosum* colonizing hemp in Holland.

Traditionally, the sexual state of *S. botryosum* was misinterpreted as *Pleospora herbarum* (see next species). Careful research by Emory Simmons showed that *S. botryosum* and *P. herbarum* are two different species. Therefore, Simmons (1986) coined *P. tarda* for the sexual state of *S. botryosum*. For the asexual state of *Pleospora herbarum*, Simmons coined *Stemphylium herbarum*.

A phylogenetic study using ITS and *gapdh* showed that *S. herbarum* was sister to *Stemphylium vesicarium* (Câmara *et al.*, 2002). Both fungi were collected from the same host, alfalfa. Inderbitzin *et al.* (2009) used more sequences (ITS, *gapdh*, *tef1*, *vmaA-vpsA*), which showed the same clustering. They stated that *S. herbarum* and *S. vesicarium* might be synonymized by researchers who are wedded to the phylogenetic species concept ("However, other characters tell a different story") and they highlighted morphological differences between *S. vesicarium* and *S. botryosum*.

Graf *et al.* (2016) sequenced *S. vesicarium* and *S. botryosum* (ITS, *gapdh*, *cytochrome B* region). A large intron in *cytochrome B* (3 kb long) present in *S. botryosum* differentiated it from *S. vesicarium*. Crous and colleagues revisited *Stemphylium* using a four-locus approach (ITS, *gapdh*, *tef1*, *rpb2*), which clustered *S. vesicarium*, *S. botryosum*, and many other *Stemphylium* spp. Wedded to the phylogenetic species concept, Crous's team synonymized them (Woudenberg *et al.*, 2017). We do not.

10.12b.   *Stemphylium vesicarium* (Wallroth) E.G. Simmons 1969

≡*Macrosporium vesicarium* (Wallroth) Saccardo 1886, ≡ *Helmisporium vesicarium* Wallroth 1833, =*Stemphylium herbarum* E. G. Symmons 1986
Sexual state: *Pleospora herbarum* (Persoon) Rabenhorst 1857, ≡*Sphaeria herbarum* Persoon 1801, =*Pleospora allii* (Rabenhorst) Cesati & de Notaris 1863, ≡*Sphaeria allii* Rabenhorst 1846
Description: *Conidiophores* and conidia resemble those of *S. botryosum*; Simmons (1969) said *S. vesicarium* conidia (33.4 × 17.7 × 22 µm) have a length/width ratio (≥1.9) greater than *S. botryosum* conidia (≤1.0). Mature *S. vesicarium* conidia retain constrictions at three transverse septa, *S. botryosum* at only the median transverse septum.

Comments: see taxonomic comments above regarding *S. botryosum* versus *S. vesicarium*. For many other synonyms, see Woudenberg *et al.* (2017). The sexual state, *P. herbarum*, was isolated from hemp stems in Ukraine (Hrebenyuk, 1984) and in Czechoslovakia (Ondřej, 1991). However, both authors also cited *S. botryosum*—at that time considered the asexual state of *P. herbarum*, now considered a different species.

Jerushalmi *et al.* (2020) isolated *S. vesicarium* from foliar tissues in Israel, identified by ITS and *EF1-α* sequences. They did not perform Koch's postulates. Gwinn *et al.* (2023) isolated

S. vesicarium from products sold by licensed cannabis facilities in British Columbia.

Roberts and Punja (2022) isolated *S. vesicarium* from hemp seeds in British Columbia. The conidial length/width ratio (1.49) fell between Simmons's values for *S. vesicarium* and *S. botryosum*. Sequencing (ITS, *TUB*) confirmed its identity as *S. vesicarium*. They conducted pathogenicity tests on detached leaves, detached stems, scalpel-wounded seedings (14-day-old rooted clones), and scalpel-wounded flowering plants placed inside perforated polyethylene bags. In all cases, *S. vesicarium* caused leaf spots and stem cankers, but less than other pathogens in the study (*Neofusicoccum parvum*, *Lasiodiplodia theobromae*, *Alternaria alternata*).

10.12.3   Differential diagnosis

Stemphylium leaf spot can be confused with brown leaf spot and brown blight. The latter is caused by *Alternaria*, a "dematiaceous hyphomycete" like *Stemphylium*, with similar brown, multiseptated conidia (see Figs 10.17, 11.25). Conidiophores of the two genera differ: *Stemphylium* elongates via percurrent proliferation through previously-formed apices, *Alternaria* via geniculate, sympodial proliferation.

10.12.4   Disease cycle and epidemiology

S. botryosum and *S. vesicarium* occur in temperate and subtropical regions worldwide. *In vitro* growth of *S. vesicarium* peaked at 25°C, dropped off at 30°C, and plummeted at 35°C (Roberts and Punja, 2022). Both species attack many hosts. They overwinter in crop debris. Springtime infections primarily arise via air-dispersed conidia. During the growing season copious amounts of conidia form on diseased leaves. They cause a rapid build-up of inoculum and secondary infections.

10.12.5   Control and management

Follow directions for yellow leaf spot (*Septoria* spp.). Practice good sanitation—gather and dispose of crop residues, and eliminate crop stubble. Scout for disease during wet seasons. Bakhtin and Gutner (1933) tested susceptibility in several fiber-type landraces. Plants from Ukraine (Glukhov, i.e., Hlukhiv) were most resistant, plants from Russia (Oryl), Belarus, and France were intermediate, and Japanese landraces were most susceptible. They also showed that heavy fertilization with NPK predisposed plants to disease.

10.13   Chaetomium leaf and stem spot

Chaetomium spp. are common soil organisms. They aggressively decompose cellulose, including hemp and cotton fibers. Most *Chaetomium* spp. act as saprophytes, but reports of *Chaetomium* spp. on living *Cannabis* plants appear in the literature. Chandra

(1974) described leaf symptoms as light brown spots, which enlarged but remained limited by leaf veins. Spots gradually darkened, dried, and dropped out, leaving irregular holes in leaves. Chaffin *et al.* (2020) described symptoms as chlorotic lesions arising near leaf margins, which progressed to necrotic lesions with chlorotic halos.

10.13.1   Causal organism

10.13a.   *Chaetomium globosum* Kunze 1817

Description: *Ascomata* (perithecia) superficial, ostiolate, olivaceous in reflected light and brown-black in transmitted light; size reported from *Cannabis*: 240–313 µm tall, 174–233 µm wide (Wang *et al.*, 2016, reported 160–300 µm tall, 135–250 µm wide). Ascomatal hairs abundant, flexuous, undulate to loosely coiled. *Asci* fusiform or clavate, spore-bearing part 30–40 × 12–17 µm atop a stalk 15–25 µm long, 8-spored. *Ascospores* olivaceous brown when mature, limoniform, averaging 9.3 × 7.5 µm on *Cannabis* (Wang *et al.*, 2016, reported 8.5–10.5 × 5.5–6.5 µm). Anamorph absent.

Comments: *C. globosum* is ubiquitous, growing saprophytically on a wide variety of substrates. Its airborne spores cause allergies and asthma, and opportunistic infections in humans. Hrebenyuk (1984) isolated *C. globosum* from stems. Punja *et al.* (2019) isolated *C. globosum* as an endophytic colonizer of stem tissues. They completed Koch's postulates by inoculating the cut surface of a hemp stem with *C. globosum*, and recovered it up to 1 cm further down the stem. Chaffin *et al.* (2020) found *C. globosum* causing leaf spots in Tennessee, and completed Koch's postulates.

A phylogenetic analysis of 80 cultures by Crous and colleagues sliced *C. globosum* into 36 species (Wang *et al.*, 2016), so new names will be appearing in the *Cannabis* literature. Chandra (1974) described *Chaetomium succineum* causing leaf spots in India, but phylogenetic analysis has moved that species to the genus *Arxotrichum* (Crous *et al.*, 2018).

Chaetomium spp. overwinter as perithecia in soil or crop stubble. They are saprophytes and thrive on all sorts of organic material. Ascospores are water-dispersed. Disease by *Chaetomium* spp. can be managed by maintaining vigorous plants. Dead plants are another matter; any parts in contact with soil are subject to rapid attack. *Chaetomium* spp. are well-known hemp retting organisms (Chapter 11).

10.14   Rusts

Rust fungi (Basidiomycota; Pucciniales) are obligate pathogens that cause some of the world's worst crop diseases. Symptoms caused by rust fungi echo their common name: pockets of rust-colored dust arising on leaves (Fig. 10.35). Many rust fungi have complicated life cycles that require two unrelated hosts, called *heteroecious* rusts; others are *autoecious*, and complete their entire life cycle on a single host.

The rust life cycle involves several types of spore-producing structures. *Macrocyclic* rusts have five: spermagonia, aecia,

Fig. 10.35. Rust symptoms caused by *Uredo kriegeriana*: **A**. upper leaf surface; **B**. lower leaf surface (courtesy M. Cartwright)

uredia, telia, and basidia (see Fig. 3.4). *Demicyclic* rusts lack the uredinal state. *Microcyclic* rusts lack uredinal and aecial states, and sometimes spermatial states, thus possessing only telial and basidial states. Consequently, microcyclic rusts are always autoecious (basidiospores reinfect the same host species). In heteroecious rusts, the "primary" host supports uredinal and telial states. Primary hosts may be annual or perennial plants. The "alternate" host is always a perennial, and carries spermatial and aecial states.

10.14.1   Causal organisms

Four species of rust fungi appear in the *Cannabis* literature, but one is a typographical error: ***Melampsora cannabis***, a misspelling by Rataj (1957) of *Melanospora cannabis*, the cause of red boot disease. Rataj's typo has been misconstrued as hemp rust. McPartland (1995a) listed half a dozen authors who made that mistake.

The other three species—*Aecidium cannabis*, *Uredo kriegeriana*, and *Uromyces inconspicuus*—are each known from only one spore state. Each represents one stage of the macrocyclic life cycle (aecia, uredia, and telia, respectively). Conceivably they could be different stages of a single autoecious species. If that's true, the genus name *Uromyces* would have priority, particularly over *Uredo* (Aime *et al.*, 2018). More likely they're different species. *Aecidium cannabis*, on an annual like *Cannabis*, would need to be autoecious. *Uredo kriegeriana* and *Uromyces inconspicuous* could be autoecious or heteroecious; if the latter, what perennial plant would be the alternate host?

10.14a. *Aecidium cannabis* Shembel 1927

Description: *Aecia* arise in pale leaf spots on undersides of lower leaves, "globose" (cupulate?), orange-colored, gregarious,

360–400 μm deep and 300–340 μm in diameter. Aecial peridial cells rectangular, often rhomboidal, 32–40 × 18–24 μm, exterior walls 6–8 μm thick. *Aecidiospores* round to ellipsoidal, subhyaline, verrucose, single-celled, 24–28 × 20–24 μm, with an epispore 1.5–2.0 μm thick.

Comments: The author's name Шембель is often transliterated as Szembel. Shembel (1927) collected *A. cannabis* from *Cannabis ruderalis* growing near the Caspian Sea. Plants were concurrently infected by *Septoria cannabis*. Bavlankulova *et al.* (2013) reported the species in Kyrgyzstan.

10.14b. *Uredo kriegeriana* H. & P. Sydow 1902

Description: *Uredia* are borne on undersides of leaves in irregular yellow to ocher spots, either sparsely or gregariously distributed, initially subepidermal ("hidden in enclosing peridium") then erumpent and pin-cushion shaped, covered by yellow spores (Fig. 10.35). *Uredospores* subglobose to ellipsoidal, finely echinulate, orange on the inside, single-celled, 21–27 × 15–22 μm, bearing many germination pores.

Comments: The description comes from Sydow and Sydow (1902). Sydow and Sydow (1924) amended the description, "*Uredinia* appear in loose aggregation within leaf spots, the covering perideum consists of roundish to obtuse-angled thin-walled cells, hyaline and membranous, 1 μm thick and 12–17 μm long by 10–13 wide. *Uredinospores* contain a hyaline epispore 1.5 μm thick, and the germination pores are difficult to see." Father and son Sydow collected *U. kriegeriana* near Schandau, Germany. Stevenson (1926) placed *U. kriegeriana* on a list of "foreign plant diseases" that posed a threat to US agriculture if accidently imported. It's here. *U. kriegeriana* turned up simultaneously in Indiana and Virginia (M. Bolt, pers. comm. 2020), Tennessee (Hansen *et al.*, 2020), New York (Bergstrom *et al.*, 2020), and Vermont (J. Bruce, pers. comm. 2020) (see Fig. 1.8).

Traditionally, the generic name *Uredo* was applied indiscriminately to any uredinial morph. DNA phylogenetic studies have revealed that *Uredo* spp. span several families of rusts (e.g., Pucciniaceae, Pucciniastraceae, Cronartiaceae). With the advent of "1 fungus = 1 name," the need for the name *Uredo* no longer exists (Aime *et al.*, 2018). The genetic markers useful for identifying rust fungi—the LSU and SSU rRNA regions—were sequenced from specimens of *U. kriegeriana* from New York. Perhaps we'll have an answer soon. Or not: fewer than 3% of all rust species have sequences deposited at GenBank, so reliance on LSU and SSU for species-level identification is not recommended (Hyde *et al.*, 2014).

10.14c. *Uromyces inconspicuus* Otth 1869

Description: Leaf spots emerge on undersides of leaves. *Telia* inconspicuous, growing in punctiform tufts, containing few teliospores. *Teliospores* borne on short hyaline pedicels; spores dark brown-black, ellipsoid, verruciform, single-celled, with a small hyaline apex, 32 × 18 μm.

Comments: Otth (1869) described (but did not illustrate) *U. inconspicuous* near Bern, Switzerland, infesting hemp as well as potatoes, strawberries, and seven genera of weedy plant species. Such a wide host range is unlikely for a rust fungus. Saccardo and Sydow (1899) expressed their doubts: "*Quid sit haec species pantogena non liquet, certe dubia res*." Kuntze (1898) transferred this species to the genus *Caeomurus* as *C. inconspicuus* in his wholesale transfer of a hundred *Uromyces* spp. to *Caeomurus*. He likely never examined Otth's specimen. Genus *Caeomurus* is no longer recognized. In a monograph of Swiss rusts, Fisher (1904) thought Otth misidentified a hyphomycete fungus.

10.14.2 Differential diagnosis, disease control

True rusts should not be confused with "white rusts," which are related to downy mildews. Manandhar (pers. comm., 1986) reported seeing a white rust in Nepal, caused by an *Albugo* sp. Rusts are rare on *Cannabis*, and cause little harm. Misra and Dixit (1979) used ethanol extracts of *Cannabis* to kill *Ustilago hordei* and *Tilletia* (*Ustilago*) *tritici*, two smut fungi that are somewhat related to rusts.

10.15   Leptosphaeria leaf and stem spot

Three *Leptosphaeria* spp. (Leptosphaeriaceae; Dothideomycetes) have been cited on *Cannabis*. *L. cannabina* causes irregularly-shaped leaf spots, ocher to dull white with an ocher margin, 3–5 mm in diameter (Ferraris and Massa, 1912). *L. woroninii* was described on seeds (Negru *et al.*, 1972). *L. acuta* was described on stalks (Saccardo and Roumeguère, 1883).

10.15.1   Causal organisms

10.15a. *Leptosphaeria cannabina* Ferraris & Massa 1912

Description: *Pseudothecia* epiphyllous, globose, black, erumpent at maturity, ostiolate, 130–140 µm diameter, peridium "membranous"

[thin-walled?]. *Asci* clavate, straight or curved, apex rounded, base constricted to a short pedicle, 45–50 × 7–10 µm. *Paraphyses* not described. *Ascospores* honey-colored, fusiform, 3-septate, slightly constricted at septa, 19–20 × 5 µm.

Comments: Ferraris and Massa (1912) found *L. cannabina* causing leaf spots in wilted plants near Alba, Italy. The fungus was also reported in Russia (Gitman and Boytchenko, 1934; Gitman, 1935; Dobrozrakova *et al.*, 1956). Ferraris and Massa considered *L. cannabina* a "probable" teleomorph of *Septoria cannabis*. They did not confirm this in culture. Perhaps they found pycnidia of *S. cannabis* near *L. cannabina*, because *S. cannabis* is a ubiquitous hemp pathogen; proximity does not constitute a relationship.

10.15b. *Leptosphaeria woroninii* Docea & Negru *in* Negru, Docea & Szasz 1972

Description: *Pseudothecia* flask-shaped, brown-black, carbonaceous, immersed then erumpent, 150–180 µm diameter. *Asci* cylindrical to clavate, 8-spored, 58–100 × 18–20 µm. *Paraphyses* filiform, hyaline, with branching apices. *Ascospores* fusiform, slightly curved, 4–6-septate, slightly constricted at septa, guttulate, granular cytoplasm, yellow, 23–29 × 3.5–5.0 µm.

Comments: Negru *et al.* (1972) found *L. woroninii* on *Cannabis* growing near Cluj, Romania. No one has reported the fungus since then.

10.15c. *Leptosphaeria acuta* (Mougeot & Nestler) Karsten 1873

=*Leptosphaeria acuta* f. *cannabis* Roumeguère 1887

Description: *Pseudothecia* conical, surface smooth, glistening black, immersed then erumpent, ostiolate, peridium a thick-walled textura globulosa near ostiole thereafter radiating textura prismatica, 300–450 µm wide, beak 100–200 µm tall, topped by a 30–40 µm diameter ostiole (Fig. 10.36). *Asci* cylindrical, 8-spored, base constricted to a short pedicle, 90–140 × 7–11 µm (Fig. 10.36). *Paraphyses* filiform, hyaline, septate.

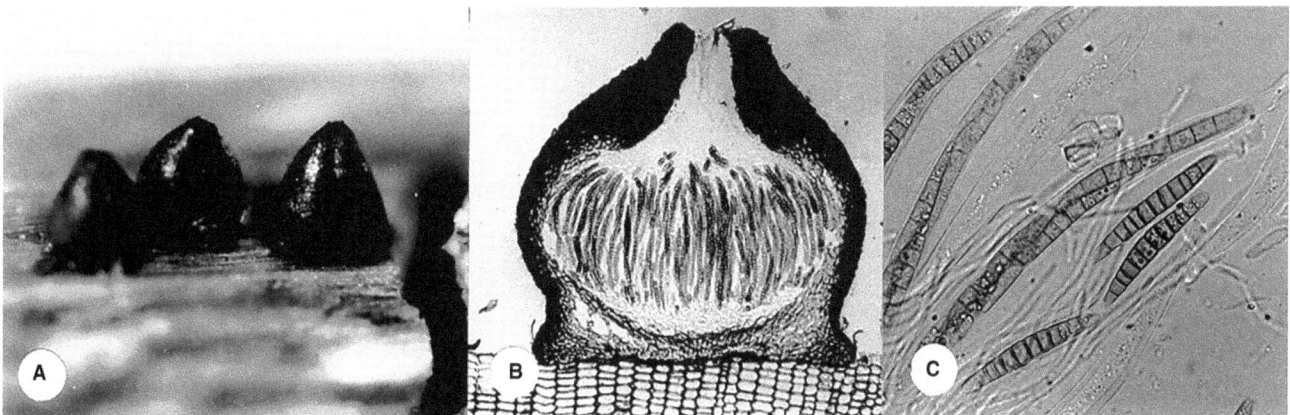

Fig. 10.36. *Leptosphaeria acuta*: **A**. pseudothecia on stem, LM ×40; **B**. sectioned pseudothecium showing asci and minute ascospores, LM ×105; **C**. ascospores, LM ×410 (courtesy Bud Uecker)

Ascospores pale yellow, guttulate, fusiform, usually 7-septate and constricted at central septum, 35–45 × 4–7 µm.

Comments: Saccardo and Roumeguère (1883) reported *L. acuta* on hemp stems. Subsequently, Roumeguère (1887) erected a new subspecies, *L. acuta* f. *cannabis*. Both of these publications were distributed as exsiccati, which consist of sets of actual fungi, rather than written descriptions and illustrations of fungi. Exsiccati are distributed to herbaria around the world, and they are considered legitimate "publications."

Specimens distributed in exsiccati are assumed to be uniform in content, but that's not always the case. When original specimens ran out, other specimens were substituted under the same name. Sometimes the substitutions differed from the originals (Stevenson, 1967). The exsiccatus of *L. acuta* f. *cannabis* in herbarium BPI is a specimen of *L. acuta* (Fig. 10.36), whereas the exsiccatus in herbarium BR is a specimen of *Jahniella bohemica* (see Fig. 10.38). On the exsiccatus label of *L. acuta* f. *cannabis*, Roumeguère designated Leige, Belgium, as the type location. *L. acuta* is usually restricted to *Urtica* species.

10.16   Assorted leaf and stem pathogens

The 16 fungi in this section have only been mentioned once or twice in the literature. Some of them lack herbaria-preserved voucher specimens, so their identification cannot be verified. They are briefly described and discussed, in alphabetical order:

10.16a. *Arthrinium phaeospermum* (Corda) M.B. Ellis 1965

Description: *Hyphae* hyaline to pale brown, smooth or verruculose, septate, 1–6 µm thick. *Conidiophore mother cells* short, lageniform, smooth or verruculose, 5–10 × 3–5 µm. *Conidiophores* mostly long, cylindrical, flexous, simple, septate, hyaline, 5–65 × 1–1.5 µm. *Conidia* borne on short sterigmata along the lengths of conidiophores, lens-shaped, golden-brown with a hyaline band around the perimeter, 8–12 µm diameter.

Comments: *A. phaeospermum* grows worldwide on sedges and reeds (*Carex*, *Glyceria*, *Phragmites* species). Chandra (1974) described *A. phaeospermum* causing leaf spots on *Cannabis* in India. Leaf spots initially are light-brown and circular. Spots gradually necrose and collapse, leaving irregular holes in the leaf.

10.16b. *Coniothyrium cannabinum* Curzi 1927

Description: *Pycnidia* scattered along stems, immersed then erumpent, spherical or flattened from above, ostiole somewhat sunken, sooty brown-black, peridium consisting of small parenchymous cells finely woven together, 90–120 × 60–90 µm. Conidiophores scarcely apparent. *Conidia* thick-walled, olive-brown, almost spherical to oval, with 1 large central guttule, 4–5 × 2.5–3.5 µm.

Comments: The description comes from Curzi (1927). Bestagno-Biga *et al.* (1958) described slightly larger conidia, 5–5.5 × 2.5 µm. Unfortunately, Curzi's type specimen is missing

(Curator Dr V. Terzo, pers. comm. 1987). *C. cannabinum* has been reported in Italy (Curzi, 1927; Bestagno-Biga *et al.*, 1958) and Russia (Gitman and Boytchenko, 1934). Hrebenyuk (1984) cited two other species, *Coniothyrium tenue* Diedicke and *Coniothyrium olivaceum* Bonorden, which may be misdeterminations (McPartland, 1995a). Fuller and Norman (1944) recovered an unspeciated *Coniothyrium* sp. from field-retted hemp in Iowa. No disease symptoms are described in any of these reports.

10.16c. *Grovesinia moricola* (I. Hino) Redhead 2014; Leotiomycetes; Sclerotiniaceae

≡*Cristulariella moricola* (I. Hino) Redhead 1980
≡*Sclerotinia* (*Botrytis*) *moricola* I. Hino 1929

Description: *Sporophore* is a pyramidal or cone-shaped structure, erect, white, generally hypophyllous, 330–650 µm tall and 100–180 µm wide at the base (Fig. 10.37, inset). The upper portion of the sporophore is considered an individual multi-celled conidium. Branches within the pyramidal head are short and compact, septate, di- or trichotomously branched, producing a series of hyaline globose cells ("microconidia") 2.5–4.0 µm in diameter. *Sclerotium* hard, black, round to irregularly shaped, 2–5 mm in diameter, giving rise to one or more apothecia. *Apothecia* comprise a slender stipe topped by an ochraceous-brown cup, 3–5 mm in diameter. Apothecium hymenium is filled with a single layer of closely packed asci and paraphyses. *Asci* stalked, 8-spored, 133–150 × 6–8 µm. *Ascospores* ellipsoid, hyaline, thick-walled, 10–12 × 4–5 µm.

Comments: *G. mor–cola* causes concentrically zonated leaf spots, alternating brown and beige (Fig. 10.37). Under a hand lens, the unique sporophores give it away. As lesions enlarge they coalesce, leading to premature defoliation. Sporophores detach easily from the leaf and disperse by splashing water.

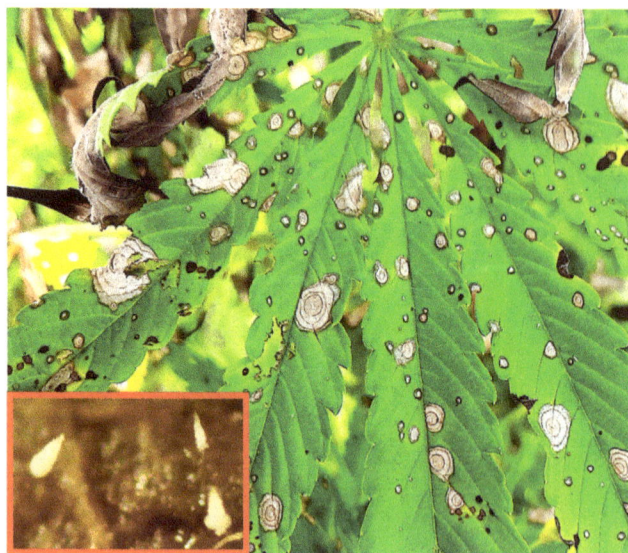

Fig. 10.37. *Grovesinia moricola*. Inset: microscopic sporophores (courtesy Marion Zuefle)

Sclerotia overwinter in plant debris or the soil. The species is distributed in temperate North American and East Asia, with a wide host range, >50 plants species in 36 plant families. Boxelder (*Acer negundo*) is a frequent host; Zuefle (2021) found *G. moricola* infesting *Cannabis* in New York State, and boxelder trees with the pathogen grew along a field edge. Disease severity averaged 15% in the region, although some growers had to harvest their hemp early and estimated that they lost nearly 50% of potential yield.

10.16d. *Cylindrosporium cannabina* Ibrahimov 1955

Description: *Acervuli* epiphyllous, gregarious, barely conspicuous, 55–96 µm wide. *Conidiophores* small, hyaline, cylindrical-acuminate, crowded together. *Conidia* unicellular, filiform, hyaline, curved, rarely straight, 18–54 × 1.2–2.0 µm.
Comments: Ibrahimov discovered this species causing leaf spots in Azerbaijan. Leaf spots are yellow-brown, 0.1–6 mm in diameter. Ibrahimov's illustration of cupulate acervuli in *C. cannabina* closely resembles the cupulate pycnidia of *Septoria cannabis* var. *microspora* illustrated by Briosi and Cavara (1888). Perhaps the two species are conspecific. McPartland (1995a) examined a "*Cylindrosporium* species" collected in 1917 by J.T. Rogers near Yarrow, Maryland (see Fig. 10.4). The specimen was a mix of *Septoria neocannabina* and *Pseudocercospora cannabina*. Rogers's erroneous identification is the sole basis of "*Cylindrosporium* sp." listed as a *Cannabis* pathogen in the USA (Miller *et al.*, 1960; Farr *et al.*, 1989). Ghani and Basit (1975) described *Cylindrosporium* sp. forming a blackish fluffy mycelium on undersides of leaves in Pakistan. Conidia were described as filiform, 2–4-septate, and bent at the apex.

10.16e. *Corynespora* sp.

Unspeciated *Corynespora* sp. caused "head blight" in Anui Province, China (Zeng *et al.*, 2014), and caused leaf spots in Kentucky (Gauthier, 2020a). Members of the genus *Corynespora* are difficult to speciate, and a phylogenetic study revealed that the genus is polyphyletic (Voglmayr and Jaklitsch, 2017). Zeng *et al.* (2014) tested nine fungicides to control *Corynespora* sp. on *Cannabis*, but none are approved for use on *Cannabis*.

10.16f. *Epicoccum nigrum* Link 1816

=*Epicoccum pururascens* Kunze ex Schlechtendahl 1824
Description: *Conidiomata* (sporodochia) 100–200 µm diameter. *Conidiophores* densely compacted, straight or flexuous, occasionally branched, colorless and smooth but turning pale brown and verrucose at the tip. *Conidia* formed singularly, monoblastically, globose to pyriform in shape, golden brown to dark brown, with a warted surface obscuring muriform septa which divide conidia into many cells (up to 15), conidia 15–25 µm diameter (see Fig. 3.3).

Comments: *E. nigrum* arises on leaves and stems, forming small dark pustules of fungal growth. Pustules reach a finite size (<2 mm) and become covered with black conidia. Pustules may form in small chlorotic spots, or on seemingly healthy green tissue. Leaf disease has been reported in India (Nair and Ponnappa, 1974; Ponnappa, 1977) and Maryland (McPartland *et al.*, 2000). Stem disease occurs in Europe (Hrebenyuk, 1984; Ondřej, 1991). *E. nigrum* also colonizes poorly stored marijuana (Punja *et al.*, 2023; Gwinn *et al.*, 2023). In Illinois *E. nigrum* acted as a saprophyte, colonizing male flowers after pollen release (McPartland *et al.*, 2000). The species grows worldwide on a wide range of plants, animals, and processed foodstuffs.

Jones *et al.* (2021) co-isolated *E. nigrum* with known fungal pathogens (*Bipolaris, Fusarium, Alternaria* spp.). Disease symptoms did not develop when *Cannabis* plants were inoculated with *E. nigrum* alone. Plants co-inoculated with *Bipolaris* and *E. nigrum* developed less disease severity than plants inoculated with *Bipolaris* alone. Dual cultures on potato dextrose agar (PDA) showed that *E. nigrum* inhibited *Bipolaris* growth, and the authors hypothesized that *E. nigrum* has potential to reduce diseases in hemp.

10.16g. *Jahniella bohemica* Petrak 1921

Description: *Pycnidia* unilocular, brown to black, immersed then erumpent, flattened subglobose, ostiolated, (310–)600(–850) µm in diameter, up to 400 µm tall, with a thick scleroplectenchymatic wall (50–60 µm thick, Fig. 10.38), peridium textura angularis. *Ostiolum* central, papillate, up to 110 µm tall. *Conidiogenous cells* holoblastic, discrete, determinate, ampulliform to short cylindrical, 4–14 × 2–5 µm. *Conidia* hyaline, filiform, straight or curved, base truncate, smooth-walled, finely guttulate, indistinctly (2–)3–4(–5) septate, (18.5–)45(–55) × 1.0–2.5 µm.
Comments: Saccardo and Roumeguère (1883) distributed exsiccati of *L. acuta* (Fig. 10.36), but some of their specimens substituted *J. bohemica* (McPartland and Hillig, 2006). They were collected from hemp stalks near Liege, Belgium. *J. bohemica* regularly infests figwort (*Scrophularia nodosa*) and garden loosestrife (*Lysimachia vulgaris*).

Jahniella spp. can be confused with *Septoria* spp., because their conidia look similar. *Jahniella* accommodates species with scleroplectenchymatous (thickened) pycnidial walls, whereas

Fig. 10.38. *Jahniella bohemica* with pycnidial wall stained red with iodine solution

Septoria spp. have pseudoparenchymatous (thin) pycnidial walls. Sutton (1980) noted the *Jahniella–Septoria* relationship was similar to that between *Plenodomus* (scleroplectenchymatous) and *Phoma* (pseudoparenchymatous).

The scleroplectenchymatous walls of *Plenodomus* spp. stain a red color when treated with iodine solution, whereas the pseudoparenchymatous walls of *Phoma* spp. do not (Boerema *et al.*, 1981). When *J. bohemica* was stained with iodine solution, its scleroplectenchymatous wall stained a deep red color, whereas the pseudoparenchymatous wall of *Septoria cannabis* did not (McPartland and Hillig, 2007).

10.16h. *Leptosphaerulina trifolii* (Rostrup) Petrak 1959

=*Pleosphaerulina cannabina* Gutner 1933
Description: *Mycelium* in culture becomes gray and only slightly floccose, pseudothecia arise in black crusts, produced in concentric rings of satellite colonies growing from the central inoculation point. *Pseudothecia* globose, brown, erumpent at maturity, ostiolate, non-paraphysate, thin-walled, 120–250 μm diameter. *Asci* bitunicate, ovate, 8-spored, 50–90 × 40–60 μm. *Ascospores* hyaline (becoming light brown at maturity), oval to ellipsoid, smooth, with 3–4 transverse septa and 0–2 longitudinal septa, 25–50 × 10–20 μm.
Comments: *L. trifolii* displays a wide variation of spore and pseudothecium sizes, resulting in a large synonymy. To that synonymy McPartland (1995d) added *Pleosphaerulina cannabina,* based on a study of Gutner's type specimen. *L. trifolii* causes symptoms consisting of round yellow-brown leaf spots 2–4 mm in diameter; pseudothecia appear as tiny black dots within the spots (Gutner, 1933). According to Gitman and Boytchenko (1934), pseudothecia form on both sides of leaves. *L. trifolii* also occurred on leaves in Nepal (Bush Doctor, 1987).

10.16i. *Leptosphaerulina chartarum* Roux 1986

Anamorph: *Pithomyces chartarum* (Berkeley & Curtis *apud* Berkeley) M. B. Ellis 1960.
Description: *Mycelium* dark olive-brown, septate, smooth to verruculose. *Pseudothecia* globose, dark olive-brown, immersed or superficial, with a large ostiole, 5–7 asci per pseudothecium, non-paraphysate. *Asci* bitunicate, short ovate but elongating with maturation, 8-spored, 100–150 × 60–100 μm. *Ascospores* hyaline to light brown, broadly ellipsoidal, smooth, usually 3 transverse septa and 1 longitudinal septum, slightly constricted at septa, 23–27 × 7–12 μm. *Conidiophores* straight or slightly curved, cylindrical, hyaline to subhyaline, holoblastic, 2.5–10 × 3–3.5 μm. *Conidia* borne singularly atop conidiophores, light to dark brown, broadly ellipsoidal, verruculose to echinulate, usually 3 transverse septa and 1 longitudinal septum, 18–29 × 10–17 μm.
Comments: The anamorph was isolated from leaves in India (Nair and Ponnappa, 1974; Ponnappa, 1977). No symptoms were described. The fungus attacks many plants. It produces an oligopeptide, sporidesmin, that causes eczema in mammals (Rippon, 1988).

10.16j. *Melanospora cannabis* Behrens 1891

Description: *Perithecia* globose, orange-red, averaging 210–230 μm diameter, ostiole atop a 60–90 μm perithecial neck. *Asci* described as swollen, no dimensions listed. Paraphyses not found. *Ascospores* black, elliptical, 22–26 × 15–17 μm. *Conidiophores* more or less erect, verticillately branched, multicellular, flask-shaped, narrowing at tips. *Conidia* arise successively at the apices of phialides, unicellular, forming in clusters or chains, red in color, 4.4 × 3.0 μm.
Comments: *M. cannabis* rapidly encases stems in thick red mats of mold, which Behrens (1891) called "red boot." Fiber extracted from these stems was soft, frail, stained a red color, and valueless. It's a secondary disease—red boot only arises on stems previously damaged by the canker-causing fungus *Sclerotinia sclerotiorum.* Behrens (1891) found *M. cannabis* unable to infect *Cannabis* in the absence of *S. sclerotiorum.* The red boot fungus parasitized *S. sclerotiorum,* destroying mycelia and sclerotia. Red boot has been reported in Germany (Behrens, 1891; Kirchner, 1906; Flachs, 1936) and Ukraine (Hrebenyuk, 1984). In a monograph on *Melanospora,* Douguet (1955) said *M. cannabis* was similar to *Melanospora zobelii* (Corda) Fuckel and *Melanospora fimicola* Hansen. Douguet did not examine the type specimen, however, and expressed regret over Behrens's failure to measure asci.

10.16k. *Microdiplodia abramovii* Nelen 1977

≡*Microdiplodia cannabina* Abramov 1939
Description: *Conidiomata* (pycnidia) scattered, thin-walled, papillate, brown, cells around the ostiole become black, 130–180 μm diameter. *Conidia* olive to pale brown, bicellular, with rounded ends, 8–11 × 4.0–4.5 μm.
Comments: Abramov (1939) found *M. cannabina* on stems of male plants near Vladivostok. Dobrozrakova *et al.* (1956) described it again in Russia. Abramov named the species but didn't describe it in Latin, rendering it a *nomen nudum.* Nelen (1977) provided a Latin description, and assigned a *nomen novum* to the *nomen nudum,* which he named after Abramov.

10.16l. *Paramyrothecium roridum* (Tode) Lombard & Crous 2016

≡*Myrothecium roridum* Tode 1790
Description: *Conidiomata* (sporodochia) sessile or slightly stalked, cushion shaped, light-colored to dark, diameter variable, 16–750 μm. *Conidiophores* hyaline, cylindrical, septate, with multiple branches bearing phialides in whorls. *Conidia* ellipsoid to elongate, subhyaline to olivaceous, aseptate, guttulate, 5–10 × 1.5–3.0 μm.
Comments: *M. roridum* causes "corky leaf spot." Symptoms begin as small brown leaf spots in July. Lesions enlarge and sometimes coalesce. Masses of slimy conidia turn spots black. Upon drying, they become corky and tough. Heavy infection leads to defoliation and death. *M. roridum* overwinters in the soil. The fungus is a weak pathogen and usually invades plants

via wounds. Although *P. roridum* attacks many crops around the world, it has only been reported on *Cannabis* in India (Nair and Ponnappa, 1974; Ponnappa, 1977) and Pakistan (Ghani and Basit, 1976). Hrebenyuk (1984) cited a closely related and easily mistaken fungus on Ukrainian hemp, *Albifimbria* (*Myrothecium*) *verrucaria* (Albertini & Schweinitz) Lombard & Crous.

10.16m. *Pestalotiopsis* sp.

Description: *Acervuli* epidermal, often epiphyllous, circular to oval, up to 200 µm in diameter. *Conidiophores* cylindrical, septate, occasionally branched, hyaline, up to 10 µm in length. *Conidiogenous cells* holoblastic, annellidic, hyaline, cylindrical. *Conidia* fusiform, 5-celled, smooth-walled, averaging 24–26 × 5.6 µm; basal cell hyaline, with a simple hyaline appendage (pedicel) averaging 6.6 µm in length; 3 median cells umber to olive-brown, thick-walled; apical cell hyaline, conical, with 3 or less commonly 2 simple hyaline appendages (setulae) averaging 17.1 µm in length (see Fig. 3.3).
Comments: McPartland and Cubeta (1997) found these cute conidia associated with leaf and stem smudge of *Cannabis* near Pokhara, Nepal. Paulsen (1971) isolated a *Pestalotia* sp. from leaves of feral hemp in Iowa. Vieto *et al.* (2022) isolated a *Pestalotiopsis* sp. from a canvas made of hemp.

10.16n. *Phyllachora cannabis* Hennings 1908

Description: *Perithecia* lie tightly bunched together, developing in a blackened plano-pulvinate pseudostroma under stem epidermis, perithecial wall membranous, brown, 16 µm thick, ostioles erumpent, 170 µm wide × 120 µm deep. *Asci* clavate with rotund apices, 8-spored, 45–55 × 10–12 µm. *Ascospores* in 1–2 rows, arranged at oblique angles to ascus axis, hyaline to grayish blue, oval to almost fusiform, non-septate, 12–17 × 5–6 µm.
Comments: Hennings (1908) discovered *P. cannabidis* on stems near Sao Paulo, Brazil. The perithecia of *Phyllachora* species merge together atop host epidermis to form a clypeus. Clypei are shield-like, shiny, and black, and cause "tar spots." Hennings described *Cannabis* tar spots having circular and angular shapes. Theissen and Sydow (1915) re-examined *P. cannabidis,* and said it belonged to the Clypeosphaeriaceae, not the Phyllachoraceae, because its erumpent ostioles freely penetrated the "pseudo-clypeus." Parberry (1967) included species with erumpent ostioles in *Phyllachora*. Parberry proposed that many of the 1100 *Phyllachora* taxa described on different hosts are synonymous.

10.16o. *Ramularia collo-cygni* Sutton & Waller 1988

Description: *Conidiophores* emerge from plant stomates, uniquely curl in the shape of a swan's neck or shepherd's crook, and produce up to five conidia. *Conidia* obovoid to ellipsoid,

with a verrucose surface, and an eccentrically placed basal hilum, 7–11 µm long × 3–6 µm wide.
Comments: *R. collo-cygni* is known as a late-season pathogen of barley in northern Europe and New Zealand; it also infects maize and other grasses. Scheuer and Bechter (2012) found it causing necrotic leaf spots of *Cannabis* in Austria. The morphology of *R. collo-cygni* is not typical of *Ramularia* spp., but in a phylogenetic analysis, Crous *et al.* (2000) showed that it clustered with other *Ramularia* spp.

10.16p. *Rhabdospora cannabina* Fautrey 1899

Description: *Pycnidia* appear on blackened stems, intensely aggregated, numerous, immersed then erumpent, surface woven and thinly reticulate, gray to black, papillate, with a round ostiole. *Conidia* variously curved, guttulate, 40–48 × 1.75–2.0 µm.
Comments: Fautrey (1899) discovered *R. cannabina* on hemp stalks near Semur, France. His description of *R. cannabina* resembles *Septoria cannabis*. The genus *Rhabdospora* was established to accommodate *Septoria*-like fungi growing on stalks instead of leaves. It's an artificial distinction; *S. cannabis* can grow on stalks as well as leaves (Gitman and Boytchencko, 1934; Ferri, 1959b). In a review of 11 *Rhabdospora* spp. by Sutton (1980), most were transferred to *Septoria*. The review didn't include *R. cannabina*. Fautrey's type specimen is missing (McPartland 1995c).

Hrebenyuk (1984) reported two other *Rhabdospora* spp. on hemp stems: *R. origani* (Brunand) Saccardo 1884, and *R. hypochoeridis* Allescher 1897. Both species produce conidia smaller than Fautrey described for *R. cannabina*.

10.16q. *Schiffnerula cannabis* McPartland & Hughes 1994

Anamorph: *Sarcinella* sp.
Description: *Mycelium* composed of brown hyphae, straight or sinuate, septate (cells mostly about 16 µm long), 4–5 µm thick, branching irregular. *Hyphopodia* numerous, alternate, hemispherical or subglobose, brown, 9.5 × 6.5 µm. *Ascomata* circular in outline, 37–43 µm diameter, containing 1–3 asci, wall at maturity mucilaginous and breaking up in water. *Asci* ellipsoid to subglobose, non-paraphysate, 8-spored, 23 × 21 µm. *Ascospores* massed into a ball, at first hyaline, later turning light brown, oblong, smooth, 1-septate, scarcely constricted, the cells unequal, 17.4 × 7.3 µm. *Conidiophores* indistinguishable from hyphae. *Conidiogenous cells* intercalary, short, cylindrical, integrated, monoblastic. *Conidia* opaquely black-brown, ellipsoid to subglobose, dictyoseptate, composed of 8–15 cells, bullate but smooth-walled, 31.9 × 27.5 µm (Fig. 10.39).
Comments: *Schiffnerula cannabis* produces copious amounts of *Sarcinella* conidia, but none has been observed to germinate (McPartland and Hughes, 1994). Their role in the life cycle is accordingly uncertain. *S. cannabis* covers the upper surface of leaves with a flat, gray-black, film-like

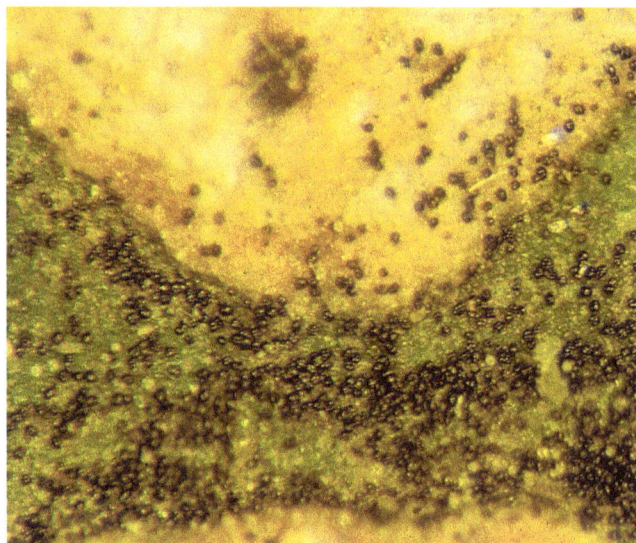

Fig. 10.39. Tiny black conidia of *Schiffnerula cannabis* next to an unrelated white leaf spot

growth. The common name of the disease is "black mildew" (McPartland *et al.*, 2000). Black mildew fungi, like powdery mildew fungi, are obligate pathogens. They tap host resources by immersing haustoria into host leaves. They should not be confused with sooty molds, which are entirely superficial and saprophytic. The tiny black conidia of *S. cannabis* in Fig. 10.39 are next to a white leaf spot with a single pycnidium of *Diaporthe ganjae*. Compare with Fig. 10.27 for a sense of scale.

(Prepared by J. McPartland; revised by L. Thiessen)

References

Abiusso NG. 1954. Fungitoxicidad de productos quimicos en ensayos de laboratorio. *Revista Eva Peron Universidad Nacional Facultad de Agronomia* 30: 149–161.

Abramov IN. 1939. *Болезни сельскохозяйственных растений Дальнего Востока*. Dalgiz, Khabarovsk.

Aime MC, Castlebury LA, Abbasi M, *et al.* (9 other authors). 2018. Competing sexual and asexual generic names in *Pucciniomycotina* and *Ustilaginomycotina* (*Basidiomycota*) and recommendations for use. *IMA Fungus* 9: 75–89.

Alcorn JL. 1988. The taxonomy of "*Helminthosporium*" species. *Annual Review of Phytopathology* 26: 37–57.

Allescher A. 1901. "*Phoma herbarum* forma *cannabis*," p. 330 in *Dr. L. Rabenhorst's Kryptogamen-Flora von Deutschland, Oesterreich und der Schweiz. Die Pilze, VI Abtheilung: Fungi imperfecti*. E. Kummer, Leipzig.

Anonymous. 1937. *Memento du service de la défense des végétaux*. Bureau de la Défense des Végétaux, no. 44, Rabat, Maroc.

Antonokolskaya MP. 1932. The races of *Sclerotinia libertiana* Fckl. on the sunflower and other plants. *Bulletin of Plant Protection, Leningrad* 5(1): 39–62.

Appel O, Wollenweber HW. 1910. Grundlagen einer Monographie der Gattung Fusarium (Link). *Arbeiten aus der Kaiserlichen Biologischen Anstalt für Land- und Forstwirtschaft* 8: 1–207.

Aveskamp MM, de Gruyter J, Woudenberg JHC, Verkley GJM, Crous PW. 2010. Highlights of the Didymellaceae: a polyphasic approach to characterise *Phoma* and related pleosporalean genera. *Studies in Mycology* 65: 1–60.

Babu R, Roy AN, Gupta YK, Gupta MN. 1977. Fungi associated with deteriorating seeds of *Cannabis sativa* L. *Current Science* 46(20): 719–720.

Bains PS, Bennypaul HS, Blade SF, Weeks C. 2000. First report of hemp canker caused by *Sclerotinia sclerotiorum* in Alberta, Canada. *Plant Disease* 84: 373.

Bakhtin VS, Gutner LS. 1933. Новая болезнь конопли. *Болезни и вредители новых лубяных культур* 3: 68–70.

Balthazar C, Novinscak A, Cantin G, Joly DL, Filion M. 2022b. Biocontrol activity of *Bacillus* spp. and *Pseudomonas* spp. against *Botrytis cinerea* and other *Cannabis* fungal pathogens. *Phytopathology* 112: 549.

Barloy J, Pelhate J. 1962. Premières observations phytopathologiques relatives aux cultures de chanvre en Anjou. *Annales des Épiphyties* 13: 117–149.

Bartoš J. 1968. *Ochrana rostlin*. Statni zemedelske nakl., Praha.

Bavlankulova KD. Mosolova SV, Kasymbekova ES. 2013. Микофлора лекарственных растений Кыргызстана. *Наука новые технологии* 2013(1): 114–124.

Behrens J. 1891. Ueber das auftreten des Hanfkrebses in Elsass. *Zeitschrift für Pfanzenkrankheiten* 1: 208–215.

Behrens J. 1902. Untersuchungen über die Gewinnung der Hanffaser durch natürliche Röstmethoden. *Centralblatt für Bakteriologie, Parasitenkunde, und Infektionskrankheiten, Zweite Abteilung* 8: 264–268, 295–299.

Behrens J. 1903. Ueber die taurotte von flachs und hanf. *Centralblatt für Bakteriologie, Parasitenkunde, und Infektionskrankheiten, Zweite Abteilung* 10: 524–530.

Bensch K, Groenewald JZ, Meijer M, *et al.* (6 other authors). 2018. *Cladosporium* species in indoor environments. *Studies in Mycology* 89: 177–301.

Bergstrom G, Starr J, Myers K, Cummings, J. 2019. *Diseases affecting hemp in New York*. Available at: https://hemp.cals.cornell.edu/resource/diseases-affecting-hemp-new-york-2019/

Bergstrom G, Starr J, Myers K, Cummings J. 2020. "An early view of diseases affecting hemp in New York," p. 10 in Gauthier N, *ed. Science of hemp: production and pest management*. University of Kentucky Agricultural Experiment Station no. SR-112. Lexington, Kentucky.

Berti Pichat C. 1866. "Capitolo XII. Della Canapa," pp. 395–499 in *Istituzioni scientifiche e tecniche, ossia Corso teorico e pratico di Agricoltura, Volume Quinto*. L'Union Tipografico-Editrice, Torino.

Bertolini G. 1861a. Descrizione di una nuova malattia della canepa nel Bolognese. *Rendiconto della sessioni dell'Accademia delle Scienze dell'Istituto di Bologna* 1861–1862: 31–35.

Bertolini G. 1861b. Descrizione di una nuova malattia della canepa nel Bolognese. *Memorie della Accademia delle Science dell'Istituto di Bologna* 12: 289–303.

Bestagno-Biga ML, Ciferri K, Bestagno G. 1958. Ordina mento artificiale dolle species del genere *Coniothyrium* Corda. *Sydowia* 12: 258–320.

Bilgrami K, Jamaluddin S, Rizwi MA. 1981. *Fungi of India: part II, host index and addenda.* Today & Tomorrow's Printers and Publishers, New Delhi.

Bócsa I. 1958. A kender beltenyésztésémek újabb jelenségei. *Növénytermelés* 7: 1–10.

Boerema GH. 1970. Additional notes on *Phoma herbarum. Persoonia* 6(1): 15–48.

Boerema GH, van Kesteren HA, Loerakker WM. 1981. Notes on *Phoma. Transactions of the British Mycological Society* 11: 61–74.

Boerema GH, Loerakker WM, Hamers MEC. 1996. Contributions towards a monograph of *Phoma* (Coelomycetes) – III. 2. Misapplications of the type species name and the generic synonyms of section *Plenodomus* (excluded species). *Persoonia* 16: 141–190.

Booth C. 1971. *The genus Fusarium.* Commonwealth Mycological Institute, Kew, UK.

Bovien P, Gram E, Hansen HR, Peterson JH, Weber A. 1943. Plantesygdomme i Danmark 1942. *Tidsskrift för Planteavl* 48: 1–90.

Brandenburger W. 1985. *Parasitische Pilz an Gefäßpflanzen in Europa.* Gustav Fisher Verlag, Stuttgart.

Breton AM, Dionne A, Hamel D, *et al.* (3 other authors). 2022. Maladies et problèmes abiotiques diagnostiqués sur les échantillons de plantes reçus en 2021 au laboratoire d'expertise et de diagnostic en phytoprotection du Ministère del l'Agriculture, des Pêcheries et de l'Alimentation du Québec. *Canadian Journal of Plant Pathology* 44(supp): S38–S51.

Briosi G, Cavara F. 1888. *Septoria cannabis var. microspora. I Funghi Parassiti Delle Piante Coltivate Od Utili*, exsiccatus no. 94.

Britt K, Fike J, Flessner M, Johnson C, Kuhar T, McCoy T, Reed D. 2020. *Integrated pest management of hemp in Virginia.* Virginia Cooperative Extension, Virginia Tech Bulletin ENTO-349NP, Blacksburg, Virginia.

Brunard P. 1899. Sphaeropsidees recoltees jusqu'a ce jor dans la Charente-Interieure. *Academie de la Rochelle, Sociètè des Sciences Naturelles de la Charente-Interieure* 26: 51–140.

Brundza K. 1933. Kai kurie parazitiniai grybeliai, surinkfi Liefuvoje 1927–1932. *Zemes Ukio Akademijos Leidinys* 1932: 199–208.

Buchholtz KF. 1943. *Local observations card file.* USDA archives, Beltsville, Maryland.

Bush Doctor. 1987. Roadtrip research. *Sinsemilla Tips* 6(4): 46–48.

California Department of Pesticide Regulation. 2017b. *Cannabis pesticides that cannot be used.* California Environmental Protection Agency, Sacramento, California.

Callaway JC, Laakkonen TT. 1996. Cultivation of *Cannabis* oil seed varieties in Finland. *Journal of the International Hemp Association* 3: 32–34.

Câmara MPS, O'Neill NR, van Berkum P. 2002. Phylogeny of *Stemphylium* spp. based on ITS and glyceraldehyde-3-phosphate dehydrogenase gene sequences. *Mycologia* 94: 660–672.

Cao LX, Luo D, Lin Wu, Yang Q, Deng XJ. 2022. Four new species of *Diaporthe* (Diaporthaceae, Diaporthales) from forest plants in China. *MycoKeys* 91: 25–47.

Carbone I, Kohn LM. 2001. A microbial population-species interface: nested cladistic and coalescent inference with multilocus data. *Molecular Ecology* 10: 947–964.

Cartwright ML, Hansen Z. 2020. *High demands for joint effort: disease in Tennessee hemp.* Available at: https://news.utcrops.com/wp-content/uploads/2020/01/2020-Madison-Co.-Hemp-Meeting.pdf

Cavara F. 1889. Materiaux de mycologie Lombarde. *Revue Mycologique* 11: 173–193.

Ceapoiu N. 1958. *Cînepa, Studiu monografic.* Editura Academiei Republicii Populare Romine. Bucharest.

Chaffin AG, Dee Me, Boggess SL, *et al.* (3 other authors). 2020. First report of *Chaetomium globosum* causing a leaf spot of hemp (*Cannabis sativa*) in Tennessee. *Plant Disease* 104: 1540.

Chandra S. 1974. Some new leaf-spot diseases from Allahabad. *Nova Hedwigia Beihefte* 47: 35–101.

Chen Q, Jiang JR, Zhang GZ, Cai L, Crous PW. 2015. Resolving the *Phoma* enigma. *Studies in Mycology* 82: 137–217.

Choi YJ, Hong SB, Shin HD. 2005. A re-consideration of *Pseudoperonospora cubensis* and *P. humuli* based on molecular and morphological data. *Mycological Research* 109: 841–848.

Choi YJ, Lee HB, Shin HD. 2017. *Pseudoperonospora urticae* occurring on *Urtica angustifuolia* in Korea. *Korean Journal of Mycology* 45: 160–166.

Chupp C. 1953. *A monograph of the fungus genus Cercospora.* Published by author, Ithaca, New York.

Colorado DOA (Department of Agriculture). 2018. *Pesticide use in Cannabis production information.* Available at: www.colorado.gov/pacific/agplants/pesticide-use-cannabis-production-information

Comes O. 1891. *Crittogamia agraria, volume unico.* Marghieri di Gius, Napoli.

Conner K, Kesheimer K, Sikora E, Kemble J. 2020. "Diseases and insects of hemp in Alabama," p. 21 in Gauthier N, *ed. Science of hemp: production and pest management.* University of Kentucky Agricultural Experiment Station no. SR-112. Lexington, Kentucky.

Cook A. 1981. *Diseases of tropical and subtropical field, fiber, and oil crops.* Macmillan Publ., New York.

Coolong T. 2020. *A preview of industrial hemp flower production in Georgia.* University of Georgia Cooperative Extension Bulletin no. 1530. Available at: https://secure.caes.uga.edu/extension/publications/files/pdf/B%201530_1.PDF

Crous PW, Braun U. 2003. *Mycosphaerella and its anamorphs: 1. Names published in Cercospora and Passalora.* Ponsen & Looyen, Wageningen, Netherlands.

Crous PW, Aptroot A, Kang J-C, Braun U, Wingfield MJ. 2000. The genus *Mycosphaerella* and its anamorphs. *Studies in Mycology* 45: 107–121.

Crous PW, Braun U, Hunter GC, *et al.* (5 other authors). 2012. Phylogenetic lineages in *Pseudocercospora. Studies in Mycology* 75: 37–114.

Crous PW, Wingfield MJ, Burgess TI, *et al.* (115 other authors). 2018. Fungal planet description sheets: 716-784. *Persoonia* 40: 239–392.

Crous PW, Lombard L, Sandoval-Denis M, *et al.* (124 other authors). 2021. Fusarium: more than a node or a foot-shaped basal cell. *Studies in Mycology* 98: e10116.

Cuboni J, Mancini V. 1886. *Synopsis mycologiae Venetae, secundum matrices.* Typis Seminarii, Patavii.

Curzi M. 1924. Sulla flora micologica delle March. *Atti dell'Istituto Botanico della Università e Laboratorio Crittogamico di Pavia (Serie III)* 1: 49–115.

Curzi M. 1927. De novis Eumycetibus. *Atti dell'Istituto Botanico della Università e Laboratorio Crittogamico di Pavia (Serie III)* 3: 203–208.

Curzi M, Barbaini M. 1927. Fungi aternenses. *Atti dell'Istituto Botanico della Università e Laboratorio Crittogamico di Pavia (Serie III)* 3: 147–202.

Damm U, Woudenberg JHC, Cannon PF, Crous PW. 2009. *Colletotrichum* species with curved conidia from herbaceous hosts. *Fungal Diversity* 39: 45–87.

Damm U, Cannon PF, Woudenberg JHC, Crous PW. 2012. The *Colletotrichum acutatum* species complex. *Studies in Mycology* 73: 37–113.

Darby H, Gupta A, Cubins J, Ruhl L, Ziegler S. 2017. *2016 industrial hemp planting date and variety trial.* University of Vermont Extension, online publication. Available at: http://www.uvm.edu/extension/cropsoil/wp-content/uploads/2016-Hemp-PDxVar-Trial.pdf

De Bary A 1886. Ueber einige Sclerotinien und Sclerotinienkrankheiten (fortsetzung). *Botanische Zeitung* 44: 449–461.

De Bary A. 1887. *Comparative morphology and biology of the fungi, mycetozoa and bacteria*. Clarendon Press, Oxford.

De Corato U. 1997. Le malattie della canapa in Basilicata. *Informatorie Fitopatologico* 47(5): 57–59.

De Gruyter J, Noordeloos ME, Boerema GH. 1993. Contributions towards a monograph of *Phoma* (Coelomycetes) – I. 2. Section *Phoma*. *Persoonia* 15: 369–400.

de Meijer WJM, van der Werf HMG, Mathijssen EWJM, van den Brink PWM. 1995. Constraints to dry matter production in fibre hemp (*Cannabis sativa* L.). *European Journal of Agronomy* 4: 109–117.

Deighton FC. 1976. Studies on *Cercospora* and allied genera, VI. *Mycological Papers* 140: 1–168.

Dempsey JM. 1975. "Hemp", pp. 46–89 in *Fiber crops*. University of Florida Press, Gainesville, Florida.

Dewey LH. 1915. *Local observations card file*. USDA archives, Beltsville, Maryland.

Dissanayake AJ, Phillips AJL, Hyde KD, Yan JY, Li XH. 2017. The current state of species in *Diaporthe*. *Mycosphere* 8: 1106–1156.

Dixon E, Leonberger K, Szarka D, *et al.* (4 other authors). 2022a. Prescreening of biological and biorational fungicides against common hemp pathogens using *in vitro* analyses. *Plant Health Progress* 23: 250–255.

Dixon E, Leonberger K, Amsden B, *et al.* (6 other authors). 2022b. Suppression of hemp powdery mildew using root-applied silicon. *Plant Health Progress* 23: 260–264.

Dobrozrakova TL, Letova MF, Stepanov KM, Khokhryakov MK. 1956. "*Cannabis sativa* L.," pp. 242–248 *in Определитель болезней растений*. State Publishing House for Plural-Economic Literature, Leningrad.

Domsch KH, Gams W, Anderson TH. 1980. *Compendium of soil fungi*. Academic Press, New York.

Douguet G. 1955. *Le genre Melanospora*. *Le Botaniste* 39: 1–313.

Doyle VP, Tonry HT, Amsden B, *et al.* (5 other authors). 2019. First report of *Cercospora* cf. *flagellaris* on industrial hemp (*Cannabis sativa*) in Kentucky. *Plant Disease* 103: 1784.

Drechsler C. 1928. Zonate eyespot of grasses caused by *Helminthosporium giganteum*. *Journal of Agricultural Research* 37: 473–492.

Durrell LW, Shields L. 1960. Fungi isolated in culture from soils of the Nevada test site. *Mycologia* 52: 636–641.

Fabre E, Dunal MF. 1853. *Observations sur les maladies régnantes de la vigne*. *Bulletin de la Société Centrale d'Agriculture du Département de l'Hérault* 40: 11–75.

Farr DF, Bills GF, Chamuris GP, Rossman AY. 1989. *Fungi on plants and plant products in the United States*. APS Press, St Paul, Minnesota.

Farr ML. 1973. *An annotated list of Spegazzini's fungus taxa, vol. I*. J. Cramer, Lehre, Germany.

Fautrey F. 1899. Espèces nouvelles de la Côte-d'Or. *Bulletin de la Société Mycologique de France* 15: 153–156.

Ferraris T. 1935. *Parassiti vegetali della canapa*. Rivista Agricola, Rome.

Ferraris T, Massa C. 1912. Micromiceti nuovi o rari per la flore micologica Italiano. *Annales Mycologici* 10: 285–302.

Ferri F. 1959a. *Atlante delle avversità della canapa*. Edizioni Agricole, Bologna, Italia.

Ferri F. 1959b. La Septoriosi della canapa. *Annali della sperimentazione agraria N.S.* 13: 6, Supplement p. CLXXXIX-CXCVII.

Ferri F. 1961b. Microflora dei semi di canapa. *Progresso Agricolo (Bologna)* 7(3): 349–356.

Fisher E. 1904. *Die Uredineen der Schweiz*. K.J. Wyss, Bern, Switzerland.

Flachs K. 1936. Krankheiten und schädlinge unserer gespinstpflanzen. *Nachrichten über Schädlingsbekämpfung* 11: 6–28.

Fuller WH, Norman AG. 1944. The nature of the flora on field-retting of hemp. *Proceedings of the Soil Science Society of America* 9: 101–105.

Fuller WH, Norman AG. 1945. Biochemical changes involved in the decomposition of hemp bark by pure cultures of fungi. *Journal of Bacteriology* 50: 667–671.

Gamalitskaia NA. 1964. *Микромицеты юго-западной части Центрального Тянь-Шаня*. Academy of Sciences of the Kyrgyz Republic, Frunze.

Gams W, Nirenberg HI, Seifert KA, *et al.* 1997. Proposal to conserve the name *Fusarium sambucinum* (Hyphomycetes). *Taxon* 46: 111–113.

Gao YH, Liu F, Duan WJ, Crous PW, Cai L. 2017. *Diaporthe* is paraphyletic. *IMA Fungus* 8(1): 153–187.

Garfinkel AR. 2021. First report of *Sclerotinia sclerotiorum* causing stem canker on *Cannabis sativa* L. in Oregon. *Plant Disease* 105: 2245.

Gautam AK, Kant M, Thakur Y. 2013. Isolation of endophytic fungi from *Cannabis sativa* and study their antifungal potential. *Archives of Phytopathology and Plant Protection* 46: 627–635.

Gauthier NW. 2020a. *Hemp disease in Kentucky: grasping the reality of yield loss*. Available at https://www.uvm.edu/sites/default/files/Northwest-Crops-and-Soils-Program/2020%20HempConf%20Presentations/NicoleGauthier_Hemp_Disease_and_Yield_Loss_VT.pdf

Gauthier N. 2020b. "Considering crop rotations and the potential for carry-over," p. 25 in Gauthier N, Leonberger K, Bowers C, *eds. Science of hemp: production and pest management*. University of Kentucky Agricultural Experiment Station, no. SR-112, Lexington, Kentucky.

Gauthier NW. 2020c. *Disease management guide to Septoria leaf spot in the field*. Available at: http://www.kyhempdisease.com/septoria-leaf-spot.html

Gauthier N. 2021. "Time of infection and host susceptibility of common leaf spots in Kentucky hemp," p. 31 in *Second annual Science of Hemp Conference*. Auburn University, Auburn, Alabama.

Gauthier NW, Szarka D. 2022. "Hemp leaf spot," in Gauthier NW, Thiessen LD, *eds. Compendium of Cannabis diseases*. APS Press, St Paul, Minnesota.

Gauthier NW, Price P, Doyle VP. 2022. "Cercospora leaf spot," in Gauthier NW, Thiessen LD, *eds. Compendium of Cannabis diseases*. APS Press, St Paul, Minnesota.

Geiser DM, Aoki T, Bacon CW, *et al.* (66 other authors). 2013. One fungus, one name: defining the genus *Fusarium* in a scientifically robust way that preserves longstanding use. *Phytopathology* 103: 400–408.

Geiser DM, Al-Hatmi AMS, Aoki T, *et al.* (165 other authors). 2021. Phylogenomic analysis of a 55.1 kb 19-gene dataset resolves a monophyletic *Fusarium* that includes the *Fusarium solani* species complex. *Phytopathology* 111: 1064–1079.

Gent DH, Ocamb CM. 2009. Predicting infection risk of hop by *Pseudoperonspora humuli*. *Phytopathology* 99: 1190–1198.

Gent DH, Adair N, Knaus BJ, Grünwald NJ. 2019. Genotyping-by-sequencing reveals fine-scale differentiation in populations of *Pseudoperonospora humuli*. *Plant Pathology* 109: 1801–1810.

Ghani M, Basit A. 1975. *Investigations on the natural enemies of marijuana, Cannabis sativa L. and opium poppy, Papaver somniferum L.* Annual Report, Commonwealth Institute of Biological Control, Pakistan station, Rawalpindi.

Ghani M, Basit A. 1976. *Investigations on the natural enemies of marijuana, Cannabis sativa L. and opium poppy, Papaver somniferum L.* Annual Report, Commonwealth Institute of Biological Control, Pakistan station, Rawalpindi.

Ghani M, Basit A, Anwar M. 1978. *Final report. Investigations on the natural enemies of marijuana, Cannabis sativa L. and opium poppy, Papaver somniferum L.* Commonwealth Institute of Biological Control, Pakistan station, Rawalpindi.

Giles G, Indermaur EJ, Gonzales-Giron JL, *et al.* (10 other authors). 2023. First report of downy mildew caused by *Pseudoperonospora cannabina* in New York. *Plant Disease* 107: 1638.

Gilman JC, Archer WA. 1929. The fungi of Iowa parasitic on plants. *Iowa State College Journal of Science* 3: 299–507.

Gitman LS. 1935. Список грибов и бактерий, зарегистрированных в СССР на новых лубяных культурах (A list of fungi and bacteria registered on new bast crops in the USSR). *За новое волокно (For New Fiber)* 6: 36–39.

Gitman LS. 1968b. Малоизвестные болезни конопли (Little-known hemp diseases). *Защита растений* 13(3): 44–45.

Gitman L, Boytchenko E. 1934. *"Cannabis"* pp. 45–53 *in Справочник по болезням бовых лубианых культур.* Institute New Bast Raw Materials, Moscow.

Gitman LS, Malikova TP. 1933. "Пораженность стеблей конопли по сортам," pp. 51–57 in *Болезни и вредители новых лубяных культур, вып. 3.* Institute of New Bast Raw Materials, Moscow.

Glażewska Z. 1971. Rośliny żywicielskie grzyba *Peronospora humuli* (Miy. et Tak.) Skal. *Pamiętnik Puławski, Prace Instytutu Uprawy, Nawożenia i Gleboznawstwa* 49: 191–204.

Gomes RR, Glienke C, Videira SIR, *et al.* (3 other authors). 2013. *Diaporthe:* a genus of endophytic, saprobic, and plant pathogenic fungi. *Persoonia* 31: 1–41.

Graf S, Bohlen-Janssen H, Miessner S, Wichura A, Stammler G. 2016. Differentiation of *Stemphylium vesicarium* from *Stemphylium botryosum* as causal agent of the purple spot disease on asparagus in Germany. *European Journal of Plant Pathology* 144: 411–418.

Greene HC. 1946. Notes on Wisconsin parasitic fungi, V & VI. *Transactions of the Wisconsin Academy of Science, Arts, and Letters* 36: 225–268.

Groenewald JZ, Nakashima C, Nishikawa J, *et al.* (6 other authors). 2013. Species concepts in *Cercospora*: spotting the weeds among the roses. *Studies in Mycology* 75: 115–170.

Grove WB. 1918. New or noteworthy fungi—part VI. *Journal of Botany, British and Foreign* 56: 340–346.

Guo YS, Crous PW, Bai Q, *et al.* (7 other authors). 2020. High diversity of *Diaporthe* species associated with pear shoot canker in China. *Persoonia* 45: 132–162.

Gupta RC. 1985. A new leaf spot disease of *Cannabis sativa* caused by *Phomopsis cannabina* Curzi. *Current Science* 54(24): 1287.

Gutner LS. 1933. Новые болезни конопли, бамии и испанского дрока. *Болезни и вредители новых лубяных культур* 3: 71–72.

Gwinn KD, Hansen Z, Kelly H, Ownley BH. 2021. Diseases of *Cannabis sativa* caused by diverse *Fusarium* species. *Frontiers in Agronomy* 3: e796062.

Gwinn KD, Leung MCK, Stephens AB, Punja ZK. 2023. Fungal and mycotoxin contaminants in cannabis and hemp flowers: implications for consumer health and directions for further research. *Frontiers in Microbiology* 14: e1278189.

Gyenis L, Anderson NA, Ostry ME. 2003. Biological control of *Septoria* leaf spot disease of hybrid poplar in the field. *Plant Disease* 87: 809–813.

Hall G. 1991. *Pseudoperonospora cannabina. Mycopathologia* 115: 223–234.

Hansen Z, Siegenthaler T, Akin E, Cartwright M, Kelly H. 2020. "Hemp extension in Tennessee; most common diseases of 2019," p. 27 in Gauthier N, *ed. Science of hemp: production and pest management.* University of Kentucky Agricultural Experiment Station no. SR-112. Lexington, Kentucky.

Heald FD, Wolf FA. 1911. New species of Texas fungi. *Mycologia* 3: 5–22.

Health Canada. 2018. *Health Canada testing of Cannabis for medical purposes for unauthorized pest control products.* Available at: www.canada.ca/en/health-canada/services/drugs-medication/cannabis/licensed-producers/policies-directives-guidance-information-bulletins/testing-cannabis-medical-purposes-unauthorized-pest-control-products.html

Hennings P. 1908. Fungi S. Paulenses IV. *Hedwigia* 48: 1–20.

Herrmann T, Smart C. 2024. *Hemp downy mildew.* Cornell Hemp. Available at: https://hemp.cals.cornell.edu.

Hockey JF. 1927. *Report of the Dominion field laboratory of plant pathology, Kentville, Nova Scotia.* Canada Department of Agriculture, pp. 28–36.

Hoerner GR. 1940. The infection capabilities of hop downy mildew. *Journal of Agricultural Research* 61: 331–334.

Hoffmann GM. 1958. Das auftreten einer anthraknose des hanfes in Mecklenburg und Brandenburg. *Nachrichtenblatt des Deutschen Pflanzenschutzdienstes NF* 12: 96–99.

Hoffmann GM. 1959. Untersuchungen über die anthraknose des hanfes (*Cannabis sativa* L.) *Phytopathologische Zeitschrift* 35: 31–57.

Hollandbeck G. 2019. Plant diseases found in industrial hemp. *Plant Disease in Kansas* 45(2): 1–3.

Howard RJ, Garland JA, Seaman WL, *eds.* 1994. *Diseases and pests of vegetable crops in Canada.* Entomological Society of Canada, Ottawa.

Hrebenyuk NV. 1984. Распространенне грибов на тресте конопли (The occurrence of fungi on hemp stems). *Микология и фитопатология* 18(4): 322–326.

Hyde KD, Nilsson RH, Alias SA, *et al.* (37 other authors). 2014. One stop shop: backbone trees for important phytopathogenic genera: I (2014). *Fungal Diversity* 67: 21–125.

Inderbitzin P, Mehta YR, Berbee MA. 2009. *Pleospora* species with *Stemphylium anamorphs:* a four locus phylogeny resolves new lineages yet does not distinguish among species in the *Pleospora herbarum* clade. *Mycologia* 101: 329–339.

Ishiyama T. 1936a. List of the fungi parasitic on agricultural plants in southern Sakhalin. *Journal of the Sapporo Society of Agriculture and Forestry* 28: 1–20.

Ishiyama T. 1936. ~~1936b~~ 南樺太産作物の寄生菌. (Parasitic fungi of South Sakhalin crops.) *Transactions of the Sapporo Natural History Society* 14(4): 297–308.

Ito S. 1930. On some new ascigerous stages of the species of *Helminthosporium* parasitic on cereals. *Proceedings of the Imperial Academy Japan* 6(8): 352–355.

Jaczewski AA. 1928. По вопросу о распространении *Pseudoperonospora humuli. Plant Protection* 5: 595–599.

Jerushalmi S, Maymon M, Dombrovsky A, Freeman S. 2020. Fungal pathogens affecting the production and quality of medical cannabis in Israel. *Plants* 9(7): e882.

Jin R, Liu FH, Yang MZ. 2013. 不同品种工业大麻的内生真菌多样性及其分布特征. *Journal of Yunnan University, Natural Sciences Edition* 35: 697–702.

Jones K, Lee M, Hatagan R, *et al.* (3 other authors). 2021. "Relationships between *Cannabis sativa* and *Epicoccum nigrum*," p. 39 in *Second annual Science of Hemp Conference.* Auburn University, Auburn, Alabama.

Joshi SD. 1924. The wilt disease of safflower. *Memoirs of the Department of Agriculture in India. Botanical* Series 13: 39–46.

Karani S, Njuguna J, Runo S, *et al.* (3 other authors). 2022. Molecular and morphological identification of fungi causing canker and dieback diseases on *Vangueria infausta* (Burch) subsp. *rotundata* (Robyns) and *Berchemia discolor* (Klotzsch) Hemsl in lower Eastern Kenya. *African Journal of Biotechnology* 21: 6–15.

Kauschitz J, Plenk A. 2022. Erstnachweis von *Phaeomycocentrospora cantuariensis* an *Cannabis sativa* in Österreich und Europa. *Stapfia* 113: 107–110.

Kaushal RP, Paul YS. 1989. Inhibitory effects of some plant extracts on some legume pathogens. *Legume Research* 12: 131–132.

Kenneth RG. 1983. Towards a more natural classification of so-called *Helminthosporium* on Gramineae," paper no. 546 in *Abstracts of papers: Fourth International Congress of Plant Pathology.* Australian Academy of Science, Canberra.

Killian C. 1926. Caractères morphologiques et culturaux du *Vermicularia eryngii* Corda (Fuck). *Bulletin trimestriel de la Société mycologique de France* 42: 51–61.

Kirchner HA. 1966. *Phytopathologie und Pflanzenschutz.* Akademie Verlag, Berlin.

Kirchner O. 1906. "Hanf, *Cannabis sativa* L.," pp. 319–323 in *Die Krankheiten und Befehädigungen uhferer landwirtschaftlichen Kulturpflanzen.* E. Ulmer, Stuttgart.

Kirchner O, Boltshauser H. 1898. Blattflecken und blattminen an hanf. *Atlas der Krankheiten und Beschädigungen,* Serie 3, Tafel 20. Verlag von Ulmer, Stuttgart.

Kishi K. 1988. *Plant diseases in Japan.* Zenkoku Nōson Kyōkai, Tōkyō.

Kohn L. 1979. A monographic revision of the genus *Sclerotinia. Mycotaxon* 9(2): 365–444.

Koike ST, Gladders P, Paulus AO. 2007. *Vegetable diseases, a color handbook.* Manson Publishing, London.

Koike ST, Stangbellini H, Mauzey SJ, Burkhardt A. 2019. First report of sclerotinia crown rot caused by *Sclerotinia minor* on hemp. *Plant Disease* 103: 1771.

Koivula MJ, Kymäläinen HR, Vanne L, *et al.* (3 other authors). 2004. Microbial quality of linseed and fibre hemp plants during growing and harvest seasons. *Agricultural and Food Science* 13: 327–337.

Kuntze O. 1898. *Revisio generum plantarum, pars III.* Arthur Felix, Leipzig.

Kuzhantayeva ZZ. 1991. Ультраструктура цитоплазмы и конидиогенез некоторых видов рода *Septoria* Sacc. *Bulletin of the Academy of Sciences of the Kazakh SSR* 1991(9): 67–70.

Kyokai NSB. 1965. 農林病害虫名鑑. Japan Plant Protection Association, Tōkyō.

Lane B, Stricker KB, Adhikari A, *et al.* (6 other authors). 2020. Large-spored *Drechslera gigantea* is a *Bipolaris* species causing disease on the invasive grass *Microstegium vimineum. Mycologia* 112: 921–931.

Lebeda A, Cohen Y. 2011. Cucurbit downy mildew (*Pseudoperonospora cubensis*)—biology, ecology, epidemiology, host-pathogen interaction and control. *European Journal of Plant Pathology* 129: 157–192.

Lentz P. 1977. Fungi and diseases of *Cannabis.* Unpublished manuscript, notes and collected literature. National Fungus Collections, USDA, Beltsville, Maryland.

Lentz P, Turner CE, Robertson LW, Genter WA. 1974. First North American record for *Cercospora cannabina,* with notes on the identification of *C. cannabina* and *C. cannabis. Plant Disease Reporter* 58: 165–168.

Leonard KJ. 1976. Synonymy of *Exserohilum halodes* with *E. rostratum,* and induction of the ascigerous state, *Setosphaeria rostrata. Mycologia* 68: 402.

Lind J. 1913. *Danish fungi.* Nordisk Forlag, Copenhagen.

Link HF. 1809. Observationes in ordines plantarum naturales. Dissertatio I. *Magazin der Gesellschaft Naturforschenden Freunde Berlin* 3(1): 3–42.

Lisson SN, Mendham NJ. 1995. Tasmanian hemp research. *Journal International Hemp* Association 2(2): 82–85.

Lubna SA, Khan AL, Waqas M, *et al.* (4 other authors). 2019. Growth-promoting bioactivities of *Bipolaris* sp. CSL-1 isolated from *Cannabis sativa* suggest a distinctive role in modifying host plant phenotypic plasticity and functions. *Acta Physiologiae Plantarum* 41: e65.

Lucas GB. 1975. *Diseases of tobacco, 3rd ed.* Biological Consulting Assoc., Raleigh, North Carolina.

Luttrell ES. 1964. Systematics of *Helminthosporium* and related genera. *Mycologia* 56: 119–132.

MAF Biosecurity 2009. *Importation of grains/seeds for consumption, feed or processing plant health requirements.* Ministry of Agriculture and Forestry (MAF), Wellington, New Zealand.

Manamgoda DS, Cai L, Hyde KD. 2012. A taxonomic evaluation of the holomorphic species complex: *Cochliobolus, Bipolaris* and *Curvularia* through multilocus phylogeny. *Fungal Diversity* 56: 131–144.

Mándy G. 1972. A kenderfajták ökólogiai vizsgálata. *Rostnövények* 1972: 113–126.

Mándy G, Bócsa I, *eds.* 1962. A kender. *Magyarország Kultúrflórája* 7(14): 1–101.

Maneval WE. 1948. A supplementary list of Missouri fungi. *University of Missouri Studies* 22(3): 1–65.

Marin MV, Coburn J, Desaeger J, Peres NA. 2020a. First report of *Cercospora* leaf spot caused by *Cercospora* cf. *flagellaris* on industrial hemp in Florida. *Plant Disease* 104: 1536.

Marin MV, Wang NY, Coburn J, Desaeger J, Peres N. 2020b. First report of *Curvularia pseudobrachyspora* causing leaf spots on hemp in Florida. *Plant Disease* 104: 3626.

Marin MV, Wang NY, Coburn JD, Desaeger J, Peres NA. 2021. First report of *Diaporthe phaseolorum* causing stem canker of hemp (*Cannabis sativa*). *Plant Disease* 105: 2018.

Marín-Felix Y, Senwanna C, Cheewangkoon R, Crous PW. 2017. New species and records of *Bipolaris* and *Curvularia* from Thailand. *Mycosphere* 8(9): 1556–1574.

Martin E. 1887. Enumeration and description of the Septorias of North America. *Journal of Mycology* 3: 37–41

Martyny JW, Serrano KA, Schaeffer JW, Dyke MV van. 2013. Potential exposures associated with indoor marijuana growing operations. *Journal of Occupational and Environmental Hygiene* 10: 622–639

Masheva S, Lazarova T, Velkov N, Velichkov G. 2014. Botanical products against powdery mildew on cucumber in greenhouses. *Turkish Journal of Agricultural and Natural Sciences* 2014(2): 1707–1712.

McCain AH, Noviello C. 1985. Biological control of *Cannabis sativa. Proceedings, 6th International Symposium on Biological Control of Weeds,* pp. 635–642.

McCurry JB, Hicks AJ. 1925. "Canadian plant disease survey," p. 22 in *Canada Department of Agriculture, Experimental Farms Annual Report No. 5,* Ottawa.

McPartland JM. 1983a. Fungal pathogens of *Cannabis sativa* in Illinois. *Phytopathology* 72: 797.

McPartland JM. 1983b. *Phomopsis ganjae* sp. nov. on *Cannabis sativa. Mycotaxon* 18: 527–530.

McPartland JM. 1984. Pathogenicity of *Phomopsis ganjae* on *Cannabis sativa* and the fungistatic effect of cannabinoids produced by the host. *Mycopathologia* 87: 149–153.

McPartland JM. 1989. Proposed list of common names for diseases of *Cannabis sativa* L. *Phytopathology News* 23(4): 46–47.

McPartland JM. 1991. Common names for diseases of *Cannabis sativa* L. *Plant Disease* 75: 226–227.

McPartland JM. 1994a. *Cannabis* pathogens X: *Phoma, Ascochyta* and *Didymella* species. *Mycologia* 86: 870–878.

McPartland JM. 1995a. *Cannabis* pathogens VIII: misidentifications appearing in the literature. *Mycotaxon* 53: 407–416.

McPartland JM. 1995c. *Cannabis* pathogens XI: *Septoria* spp. on *Cannabis sativa, sensu strico. Sydowia* 47: 44–53.

McPartland JM. 1995d. *Cannabis* pathogens XII: lumper's row. *Mycotaxon* 54: 273–279.

McPartland JM. 2000a. Data sheet: *Septoria cannabis. Crop Protection Compendium,* CAB International, Wallingford, UK.

McPartland JM. 2000b. Data sheet: *Pseudoperonospora cannabina. Crop Protection Compendium,* CAB International, Wallingford, UK.

McPartland JM, Cubeta MA. 1997. New species, combinations, host associations and location records of fungi associated with hemp (*Cannabis sativa*). *Mycological Research* 101: 853–857.

McPartland JM, Hillig KW. 2006. *Jahniella bohemica:* a rediscovered species that infests hemp stalks. *Journal of Industrial Hemp* 11(2): 97–108.

McPartland JM, Hillig KW. 2007. *Jahniella bohemica* Part II: mistaken identity with *Septoria cannabis* infesting hemp stalks. *Journal of Industrial Hemp* 12(1): 63–74.

McPartland JM, Hughes S. 1994. *Cannabis* pathogens VII: a new species, *Schiffnerula cannabis*. *Mycologia* 86: 867–869.

McPartland JM, Rhode B. 2005. New hemp diseases and pests in New Zealand. *Journal of Industrial Hemp* 10: 99–108.

McPartland JM, Small E. 2020. A classification of endangered high-THC cannabis (*Cannabis sativa* subsp. *indica*) domesticates and their wild relatives. *PhytoKeys* 144: 81–112.

McPartland JM, Clarke RC, Watson DP. 2000. *Hemp diseases and pests – management and biological control*. CABI, Wallingford, UK.

McRae W. 1934. Report of the imperial mycologist. *Scientific Reports of the Imperial Institute of Agricultural Research, Pusa* 1932–1933: 134–160.

Mel'nik VA. 1977. *Определитель грибов рода Ascochyta Lib.* Nauka, Leningrad.

Melzer M, Shan X. 2022. Diseases diagnosed on plant samples submitted to the plant disease clinic, University of Guelph in 2021. *Canadian Plant Disease Survey* 102: 30–37.

Mersha Z, Kering M, Dhakal R, Rahemi A, Ren SX. 2020. "Southern blight and foliar diseases of hemp in central Virginia," p. 35 in Gauthier N, ed. *Science of hemp: production and pest management*. University of Kentucky Agricultural Experiment Station no. SR-112. Lexington, Kentucky.

Miller PR, Weiss F, O'Brien MJ. 1960. *Index of plant diseases in the United States*. USDA Agriculture Handbook No. 165, Washington, DC.

Misra SB, Dixit SN. 1979. Antifungal activity of leaf extracts of some higher plants. *Acta Botanica Indica* 7: 147–150.

Mitchell MM, Ocamb CM, Grünwald NJ, Mancino LE, Gent DH. 2011. Genetic and pathogenic relatedness of *Pseudoperonospora cubensis* and *P. humuli*. *Phytopathology* 101: 805–818.

Moes J, Sturko A, Przybylski R. 1999. "Agronomic research on hemp in Manitoba," pp. 300–305 in Janick J, ed., *Perspectives on new crops and new uses*. ASHS Press, Alexandria, Virginia.

Moesz G. 1941. Neue Pilze aus Lettland. *Botanikai Közlemények* 38(1-2): 68–73.

Munir M, Smith H, Valentine T, *et al.* (9 other authors). 2024. Leaf spot disease development and its effect on yield of essential-oil producing hemp cultivars in Kentucky. *Plant Disease* 108: 1621–1631.

Mushtaque M, Baloch GM, Ghani MA. 1973. *Natural enemies of Papaver spp. and Cannabis sativa*. Annual report, Commonwealth Institute of Biological Control, Pakistan station, pp. 54–55.

Nair KR, Ponnappa KM. 1974. *Survey for natural enemies of Cannabis sativa and Papaver somniferum*. Commonwealth Institute of Biological Control, India Station Report, pp 39–40.

Narváez A, Rodríguez-Carrasco Y, Castaldo L, Izzo L, Ritieni A. 2020. Ultra-high-performance liquid chromatography coupled with quadrupole orbitrap high-resolution mass spectrometry for multi-residue analysis of mycotoxins and pesticides in botanical nutraceuticals. *Toxins* 12: e12020144.

Negru A, Docea E, Szasz E. 1972. De speciebus fungorum, semina germinantia inficientium novis et in Romania primum inventis. *Novosti Sistematiki Nizshikh Rastenii* 9: 167–170.

Nelen ES. 1977. Species fungorum pycnidialium novae e parte australi Orientis extremi. *Новости Систематики Низших Растений* 14: 103–106.

Ondřej M. 1974. Výskyt zajímavých imperfektních hub na lnu a konopí. *Len a Konopí* (Sumperk, Czech Rep) 12: 69–76.

Ondřej M. 1991. Výskyt hub na stoncích konopí (*Cannabis sativa* L.). *Len a Konopí* (Sumperk, Czech Rep) 21: 51–57.

Otth GH. 1869. Sechster Nachtrag zu dem in Nr. 15-23 der Mittheilungen enthaltenen Verzeichnisse schweizerischer Pilze. *Mitteilungen der Naturforschenden Gesellschaft in Bern* 654: 37–70.

Oudemans CAJA. 1920. *Enumeratio Systematica Fungorum, Vol. II*. Comitum, The Hague.

Paini D. 2011. *Final report CRC10184. Using likelihood of arrival and establishment to assess the world threat of invasive species*. Cooperative Research Centre for National Plant Biosecurity, Canberra, Australia

Pandotra VR, Husain A. 1966. Fungi on medicinal and aromatic plants in the north-west himalayas. *IV. Mycopathologia et Mycologia Applicata* 29: 155–160.

Parberry DG. 1967. Studies on graminicolous species of *Phyllachora*. *Australian Journal of Botany* 15: 271–375.

Patel MK. 1933. India: diseases in the Bombay Presidency. *International Bulletin Plant Protection* 7(11): 246.

Pathak D, Khan RU, Singh VP. 2017. Cultural and morphological variations among the isolates of *Sclerotinia sclerotiorum* (Lib.) de Bary causing Sclerotinia stem rot. *International Journal of Plant Protection* 10: 295–298.

Patschke K, Gottwald R, Müller R. 1997. Erste Ergebnisse phytopathologischer Beobachtungen im Hanfanbau im Land Brandenburg. *Nachrichtenblatt des Deutschen Pflanzenschutzdienstes* 49: 286–290.

Paulsen AQ. 1971. *Plant diseases affecting marijuana (Cannabis sativa)*. Unpublished manuscript, Kansas State University.

Peck CH. 1884. "Report of the botanist," pp. 125–164 *in Thirty-fifth annual report on the New York State Museum of Natural History*. Weed, Parsons & Co., Albany.

Pei W, Xu Y, Yang M, Yang M, Yu L. 2018. *First report of leaf spot of Cannabis sativa caused by Bipolaris sp. in Yunnan Province, China*. Available at: https://www.ncbi.nlm.nih.gov/nuccore/MH881227

Persoon CH. 1801. *Synopsis Methodica Fungorum, pars prima*. Henricus Dieterich, Göttingen.

Petch T. 1936. *Gibberella saubinetii* (Mont.) Sacc. *Annales Mycologici* 34: 257–260.

Petrak F. 1947. Plantae sinenses a Dre. H. Smith annis 1921-1922, 1924, et 1943 lectae XLII. Micromycetes. *Meddelelser från Göteborgs Botaniska Trädgård* 17: 113–164.

Phengsintham P, Chukeatirote E, McKenzie EHC, Hyde KD, Braun U. 2013. Monograph of Cercosporoid fungi from Laos. *Current Research in Environmental & Applied Mycology* 3: 34–158.

Pietkiewicz TA. 1958. Mikroflora nasion konopi. Przeglad literatury. *Roczniki Nauk Rolniczych Seria A* 77(4): 577–590.

Plenk A, Votzi J, Moyses A, Follak S. 2021. Erste Untersuchungen zum Auftreten von Pilzkrankheiten an Nutzhanf in Österreich. *Julius-Kühn-Archiv* 467: 213–214.

Ponnappa KM. 1977. New records of fungi associated with *Cannabis sativa*. *Indian Journal of Mycology and Plant Pathology* 7: 139–142.

Prillieux EE. *Maladies des plantes agricoles et des arbres fruitiers et forestiers, tome second*. Didot, Paris.

Punithalingam E. 1980. *Septoria cannabis*. CMI descriptions of pathogenic fungi and bacteria, no. 668. Commonwealth Mycological Institute, Kew, UK.

Punja ZK. 2021c. First report of *Fusarium proliferatum* causing crown and stem rot, and pith necrosis, in cannabis (*Cannabis sativa* L., marijuana) plants. *Canadian Journal of Plant Pathology* 43: 236–245.

Punja ZK. 2021d. Emerging diseases of *Cannabis sativa* and sustainable management. *Pest Management Science* 77: 3857–3870.

Punja ZK, Ni L. 2021. The bud rot pathogens infecting cannabis (*Cannabis sativa* L., marijuana) inflorescences: symptomology, species identification, pathogenicity and biological control. *Canadian Journal of Plant Pathology* 43: 827–854.

Punja ZK, Collyer D, Scott C, Lung S, Holms J. Sutton D. 2019. Pathogens and molds affecting production and quality of *Cannabis sativa* L. *Frontiers in Plant Science* 10: e1120.

Punja ZK, Li N, Lung S, Buirs L. 2023. Total yeast and mold levels in high THC-containing cannabis (*Cannabis sativa* L.) inflorescences are influenced by genotype, environment, and pre-and post-harvest handling practices. *Frontiers in Microbiology* 14: e1192035.

Quaedvlieg W, Verkley GJM, Shin HD, *et al.* (5 other authors). 2013. Sizing up *Septoria*. *Studies in Mycology* 75: 307–390.

Raabe A. 1939. Untersuchungen über pilzparasitäre Krankheiten von Raps und Rübsen. *Zentralblatt fur Bakteriologie, Parasitenkunde und Infektionskrankheiten* 2: 35–52.

Rahnama M, Szarka D, Hua Li H, *et al.* (4 other authors). 2021. Reemergence of Septoria leaf spot caused by *Septoria cannabis* on hemp in Kentucky. *Plant Disease* 105: 2286–2289.

Rapaić von Ruhmwerth R. 1914. Phytopathologische Beobachtungen in Debrecen (Ungarn). *Zeitschrift für Pflanzenkrankheiten* 24: 211–218.

Rataj K. 1957. Skodlivi cinitele pradnych rostlin. *Prameny Literatury* 2: 1–123.

Reichardt EI. 1925. Contribution a la flore mycologique du gouvernment Petrograd. *Bulletin de la Station Régionale Protectrice des Plants à Leningrad* 5: 46.

Rippon JW. 1988. *Medical mycology, 3rd ed.* W.B. Saunders Co., Philadelphia, Pennsylvania.

Roberts AJ, Punja ZK. 2022. Pathogenicity of seedborne *Alternaria* and *Stemphylium* species and stem-infecting *Neofusicoccum* and *Lasiodiplodia* species to cannabis (*Cannabis sativa* L., marijuana) plants. *Canadian Journal of Plant Pathology* 44: 250–269.

Röder K. 1937. *Phyllostica cannabis* (Kirchner?) Speg. eine nebeufruchtform von *Mycosphaerella cannabis* (Winter) n.c. *Zeitschrift für Pfanzenkrankheiten* 47: 526–531.

Röder K. 1939. Über einen neuen hanfschädiger, *Didymella arcuata* n. sp. und seine nebenfruchtformen. *Phytopathologische Zeitschrift* 12: 321–333.

Rodríguez-Yzquierdo GA, Patiño-Moscoso MA, Betancourt-Vásquez M. 2021. Caracterización fisiológica en plantas de *Cannabis* medicinal durante distintas etapas fenológicas bajo estrés biótico. *Agronomía Mesoamericana* 32: 823–840.

Romashchenkov DD. 1936. Конопля на Болотных почвах севера карельской АССР. *Bulletin of Applied Botany, Genetics, and Plant Breeding, series XV*, 5: 79–91.

Roumeguère C. 1887. "*Leptosphaeria acuta* f. *cannabis*," no. 4172 in *Fungi selecti exsiccati praecipue Galliae et Algeriae* Centurie XLII. *Revue Mycologique* 9: 146–155.

Runge F, Thines M. 2012. Reevaluation of host specificity of the closely related species *Pseudoperonospora humuli* and *P. cubensis*. *Plant Disease* 96: 55–61.

Runge F, Choi YJ, Thines M. 2011. Phylogenetic investigations in the genus *Pseudoperonospora* reveal overlooked species and cryptic diversity in the *P. cubensis* species cluster. *European Journal of Plant Pathology* 129: 135–146.

Saccardo FA. 1883. *Syllogue fungorum vol. II.* Typis Seminarii, Patavii.

Saccardo FA. 1884. *Syllogue fungorum, vol. III.* Typis Seminarii, Patavii.

Saccardo FA. 1897. *Diaporthe tulasnei f. cannabis.* Mycotheca Italica, century I, no. 91. Padua.

Saccardo PA. 1873. Mycologiae Venetae Specimen. *Atti della Società Veneto-Trentino di Scienze Naturali* 2: 53–264.

Saccardo PA. 1880. *Vermicularia dematium f. cannabis.* Mycotheca Veneta, century XVI, no. 1558. Padua.

Saccardo PA. 1881. Fungi Veneti novi vel critici v. mycologiae Veneti addendi. Series XII. *Michelia* 2: 241–301.

Saccardp PA. 1889. *Syllogue fungorum, vol. VIII.* Borntraeger, Berolini.

Saccardo PA. 1898. *Syllogue fungorum, vol. XIII.* Borntraeger, Berolini.

Saccardo PA, Roumeguère C. 1883. "*Leptosphaeria acuta*," no. 56 in Reliquiae Libertianae (Series III). *Revue Mycologique* 5: 233–239.

Saccardo PA, Sydow P. 1899. *Syllogue fungorum, vol. XIV.* Borntraeger, Berolini.

Salmon ES, Ware WM. 1928. Inoculation experiments with the downy mildews of the hop and nettle (*Pseudoperonospora humuli* (Miy. et Takah.) Wils. and *P. urticae* (Lib.) Salmon et Ware). *Annals of Applied Biology* 15: 352–370.

Santos JM, Vrandečić K, Ćosić J, Duvnjak T, Phillips AJL. 2011. Resolving the *Diaporthe* species occurring on soybean in Croatia. *Persoonia* 27: 9–19.

Savulescu T, Sandu-Ville C. 1936. Beitrag zur Kenntnis der Micromyceten Rumäniens. *Hedwigia* 75: 159–233.

Scheuer C, Bechter S. 2012. Pilzfunde aus dem Botanischen Garten Graz. *Mitteilungen des naturwissenschaftlichen Vereines für Steiermark* 142: 59–98.

Schwartz J, Gordon T, Stansell Z, Smart C. 2024. *Septoria leaf spot: resistance and management.* Cornell Hemp, available at https://hemp.cals.cornell.edu

Scott M, Rani M, Samsatly J, Charron JB, Jabaji S. 2018. Endophytes of industrial hemp (*Cannabis sativa* L.) cultivars: identification of culturable bacteria and fungi in leaves, petioles, and seeds. *Canadian Journal of Microbiology* 64: 664–680.

Serzane M. 1962. "Kanepju - *Cannabis sativa* L." pp. 366–369 *in Augu Slimibas, Praktiskie Darbi.* Latvijas Valsts Izdevnieciba, Riga.

Seymour AB. 1929. *Host index of the fungi of North America.* Harvard University Press, Cambridge, Massachusetts.

Shabana YM, Stiles CM, Charudattan R, Tabl AHA. 2010. Evaluation of bioherbicidal control of tropical signalgrass, crabgrass, smutgrass, and torpedograss. *Weed Technology* 24: 165–172.

Shahzad S, Ghaffar A. 1995. New records of soilborne root infecting fungi in Pakistan. *Pakistan Journal of Botany* 27: 209–216.

Sharma K, Goss EM, Dickstein ER, *et al.* (4 other authors). 2014. *Exserohilum rostratum*: characterization of a cross-Kingdom pathogen of plants and humans. *PLoS ONE* 9(10): e108691.

Sharma P, Meena PD, Razdan VK, Gupta V, Singh D. 2016. Worldwide new host record of *Sclerotinia sclerotiorum* on *Cannabis indica* in Jammu, India. *Journal of Oilseed Brassica* 7(2): 192–193.

Shaw FJF, Ajerakar SL. 1915. The genus *Rhizoctonia* in India. *Memoirs of the Department of Agriculture in India, Botanical Series* 7: 177–195.

Shear CL, Stevens NE. 1935. *Sphaeria zeae* (*Diplodia zea*) and confused species. *Mycologia* 28: 467–477.

Shembel SY. 1927. Микологические заметки. Новый вид ржавчины на конопле. *Commentarii Instituti Astrachanensis ad Defensionem Plantarum* 1(5-6): 59–60.

Sherbakoff CD. 1928. An examination of fusaria in the herbarium of the pathological collections, Bureau of Plant Industry, USDA *Phytopathology* 18: 148.

Shin HD, Choi YJ. 2003. A first check-list of Peronosporaceae from Korea. *Mycotaxon* 86: 249–267.

Shukla DD, Pathak VN. 1967. A new species of *Ascochyta* on *Cannabis sativa* L. *Sydowia* 21: 277–278.

Simmons EG. 1969. Perfect states of *Stemphylium*. *Mycologia* 61: 1–26.

Simmons EG. 1986. Perfect states of *Stemphylium*. II. *Sydowia* 38(1985): 284–293.

Sivanesan A. 1987. *Graminicolous species of Bipolaris, Drechslera, Curvularia, Exserohilum and their teleomorphs*. CAB International Mycological Institute, Kew, UK.

Smith IM, Dunez J, Phillips DH, Lelliott RA, Archer SA. 1988. *European handbook of plant diseases*. Blackwell Scientific Publications, Oxford, UK.

Snyder WC, Hansen HN. 1940. The species concept in *Fusarium*. *American Journal of Botany* 27: 64–67.

Snyder WC, Hansen HN. 1941a. The species concept in *Fusarium* with reference to section *Martiella*. *American Journal of Botany* 28: 738–742.

Snyder WC, Hansen HN. 1941b. The species concept in *Fusarium* with reference to section *Discolor* and other sections. *American Journal of Botany* 32: 657–666.

Sohi HS, Nayar SK. 1971. New records of fungi from Himachal Pradesh-III. *Research Bulletin of Panjab University* 22: 243–245.

Southworth EA. 1890. "Special subjects," pp. 406–408 in *Report of the Secretary of Agriculture, 1890*. Government Printing Office, Washington, DC.

Spaar D, Kleinhempel H, Fritzsche R. 1990. *Öl- und Faserpflanzen*. Springer-Verlg, Berlin.

Spegazzini C. 1881. Nova addenda ad mycologiam venetam. *Atti della Società Crittogamologica Italiana* 3(1): 42–71.

Srivastara SL, Naithani SC. 1979. *Cannabis sativa* Linn., a new host for *Phoma* sp. *Current Science* 48(22): 1004–1005.

Stack GM, Toth JA, Carlson CH, et al. (15 other authors). 2020. *Evaluation of high-CBD cultivars in New York State—results of 2020 Cornell hemp field trials*. Cornell University, Geneva and Ithaca, New York.

Stevenson JA. 1926. *Foreign plant diseases*. USDA Bureau of Plant Industry, Washington, DC.

Stevenson JA. 1967. Rabenhorst and fungi exsiccati. *Taxon* 16: 112–119.

Stoneman B. 1898. A comparative study of the development of some anthracnoses. *Botanical Gazette* 26: 69-120.

Strgulec M, Jesenko T, Škerbot I, et al. (4 other authors). 2016. *Tehnologija pridelave industrijske konoplje*. Kmetijsko gozdarska zbornica Slovenije, Ljubljana.

Sukarno N, Badia Ginting RC, Widyastuti U, et al. (5 other authors). 2021. Endophytic fungi from four Indonesian medicinal plants and their inhibitory effect on plant pathogenic *Fusarium oxysporum*. *Hayati Journal of Biosciences* 28: 152–171.

Summers CF, Gulliford CM, Carlson CH, et al. (5 other authors). 2015a. Identification of genetic variation between obligate plant pathogens *Pseudoperonospora cubensis* and *P. humuli* using RNA sequencing and genotyping-by-sequencing. *PLoS ONE* 10(11): e0143665.

Summers CF, Adair NL, Gent DH, McGrath MT, Smart CD. 2015b. *Pseudoperonospora cubensis* and *P. humuli* detection using species-specific probes and high definition melt curve analysis. *Canadian Journal of Plant Pathology* 37: 315–330.

Sutton BC. 1980. *The Coelomycetes. Fungi imperfecti with pycnidia, acervuli, and stromata*. Commonwealth Agricultural Bureau, London.

Sutton BC, Waterston JM. 1966. *Ascochyta phaseolorum*. C.M.I. Descriptions of Pathogenic Fungi and Bacteria No. 81. CABI Publications, London.

Suyal N, Upadhyaya ML, Gupta RC. 2011. Studies on air mycoflora over the field of *Cannabis sativa* L. in Almora hills. *Vegetos* 24: 183–190.

Sydow H, McRae W. 1929. Hyphomycetes Indiae orientalis Pars. I. *Annales de Cryptogamie Exotique* 2(3-4): 262–271.

Sydow H, Sydow P. 1902. Einige neue Uredineen. I. *Österreichische Botanische Zeitschrift* 52(5): 182–185.

Sydow P, Sydow H. 1924. *Monographia Uredinearum, Vol. 4*. Lipsiae Fratres, Borntraeger.

Szarka D. 2021. *Seeing double with Cannabis: heteroploidy populations of Bipolaris gigantea, causal agent of Bipolaris leaf spot*. Master's thesis, University of Kentucky, Lexington.

Szarka D, McCulloch M, Beale J, et al. (3 other authors). 2020a. First report of anthracnose leaf spot caused by *Colletotrichum fioriniae* on hemp (*Cannabis sativa*). *Plant Disease* 104(5): 1560.

Szarka D, Amsden B, Beale J, et al. (3 other authors). 2020b. First report of hemp leaf spot caused by a *Bipolaris* species on hemp (*Cannabis sativa*) in Kentucky. *Plant Health Progress* 21(1): 82–84.

Szarka D, Amsden B, Schardl CL, Gauthier N. 2020c. "Hemp leaf spot, new disease of hemp caused by *Bipolaris gigantea*," p. 43 in Gauthier N, ed. *Science of hemp: production and pest management*. University of Kentucky Agricultural Experiment Station no. SR-112. Lexington, Kentucky.

Szarka D, Gauthier NA, Rahnama M, Schardl CL. 2023. Seeing double on *Cannabis*: haploids and heteroploids of *Bipolaris gigantea* on hemp and other dicots. *Mycologia*.

Tai [Dai] FL. 1936. Notes on Chinese fungi. VII. 中國植物學會彙報 2(2): 45–66.

Tai [Dai] FL. 1979. *Sylloge fungorum sinicorum*. Science Press, Beijing.

Teng [Deng] SC. 1932. 南京真菌记载. II. *Contributions from the Biological Laboratory of the Science Society of China, Botanical Series* 8(1): 5–48.

Termorshuizen AJ. 1991. *Literatuuronderzoek over ziekten bij nieuwe potentiële gewassen*. IPO-DLO Rapport No. 91-08, Instituut voor Plantezektenkundig Onderzoek, Wageningen, The Netherlands.

Teterevnikova-Babajan DN. 1987. *Грибы рода Septoria в СССР*. Akademia Nauk Armjanskoi SSR, Erevan.

Theissen F, Sydow H. 1915. Die Dothideales. *Annales Mycologici* 13: 147–746.

Thiessen L. 2019. *Hemp diseases in North Carolina*. Department of Entomology and Plant Pathology, North Carolina State University. Available at: https://growingsmallfarms.ces.ncsu.edu/wp-content/uploads/2019/03/Hemp-Diseases-Lindsey-Thiessen.pdf?fwd=no

Thiessen LD Gauthier NW. 2022. "Anthracnose," in Gauthier NW, Thiessen LD, eds. *Compendium of Cannabis diseases*. APS Press, St Paul, Minnesota.

Thiessen LD, Schappe T. 2019. First report of *Exserohilum rostratum* causing foliar blight of industrial hemp (*Cannabis sativa*). *Plant Disease* 103(6): 1414.

Thiessen LD, Schappe T, Cochran S, Hicks K, Post AR. 2020. Surveying for potential diseases and abiotic disorders of industrial hemp (*Cannabis sativa*) production. *Plant Health Progress* 21(4): 321.

Thirumalachar MJ, Chupp C. 1948. Notes on some Cercosporae of India. *Mycologia* 40: 362–352.

Tikhomirov W. 1868. *Peziza Kauffmanniana*, eine neue, aus sclerotium stammende und auf hanf schmarotzende becherpilz-species. *Bullétin Société Imperiale Naturalistes Moscou* 41(2): 295–342.

Transchel V, Gutner L, Khokhryakov M, 1933. A list of fungi found on new cultivated textile plants. *Труды Института нового лубяного сырья* 4: 127–140.

Trunov GA. 1936. "Материалы по фитопатологическому изучению конопель," pp. 69–113 in *Collection of scientific papers on plant protection*. Ukrainian State Publishing Office, Kiev.

Turland NJ, ed. 2018. *International code of nomenclature for algae, fungi, and plants (Shenzhen code)*. Koeltz Scientific Books: Königstein, Germany.

Tulasne LR, Tulasne C. 1865. *Selecta fungorum carpologia, tomus tertius*. Imperiali Typographeo Excudebatur, Parisiis.

Udayanga D, Castlebury LA, Rossman AY, Chukeatirote E, Hyde KD. 2014. Insights into the genus *Diaporthe*: phylogenetic species delimitation in the *D. eres* species complex. *Fungal Diversity* 67: 203–229.

Uecker FA. 1989 "*Phomopsis*," Workshop on the taxonomy and identification of Coelomycetes, Mycological Society of America, 5 August 1989, Beltsville, Maryland.

Ujat AH, Konishi S, Kato Y, Tonami H, Nakashima C. 2024. *Septoria cannabicola*, a new species on *Cannabis sativa* in Japan. *Mycoscience* 65: myc623.

Vakhrusheva TE, Davidian GG. 1979. Методические указания по инвентаризации болезней и микрофлоры льна и конопли. All-Union Vavilov Scientific Research Insititute of Crop Cultivation, Leningrad.

van der Aa HA. 1973. Studies in *Phyllosticta* I. *Studies in Mycology* 5: 1–110.

van Kan JAL, Shaw MW, Grant-Downton RT. 2014. *Botrytis* species: relentless necrotrophic thugs or endophytes gone rogue? *Molecular Plant Pathology* 15: 957–961 (2014).

Vasudeva RS. 1961. *Indian Cercosporae*. Indian Council of Agricultural Research, New Delhi.

Verkley GJM, Quaedvlieg W, Shin HD, Crous PW. 2013. A new approach to species delimitation in *Septoria*. *Studies in Mycology* 75: 213–305.

Vido A. 1879. Repertorium mycologiae Venetae seu index alphabeticus fungorum in ditione Veneta hucusque cognitorum. *Michelia* 1(5): 553–619.

Viégas AP. 1961. *Indices de fungos da America do Sul*. Seçao de Fitopathologia. Campinas, Brazil.

Vieto S, Escudero-Leyva E, Avendaño R, *et al.* (6 other authors). 2022. Biodeterioration and cellulolytic activity by fungi isolated from a nineteenth-century painting at the National Theatre of Costa Rica. *Fungal Biology* 126: 101–122.

Voglino P. 1912. Pflanzenkrankheiten im Piemont. *Zeitschrift fur Pflanzenkrankheiten* 22: 153–155.

Voglino P. 1913. I funghi più dannosi alle piante osservati nella provincia di Torino e regioni vicine nel 1911. *Annali della Reale Accademia d'agricoltura di Torino* 55: 199–227.

Voglino P. 1924. *Patologia vegetale*. Unione Tipographico Editrice, Torino.

Voglmayr and Jaklitsch 2017. *Corynespora, Exosporium* and *Helminthosporium* revisited–new species and generic reclassification. *Studies in Mycology* 87: 43–76.

Von Arx JA. 1957. Die artender gattung *Colletotrichum* Cda. *Phytopathologische Zeitschrift* 29: 28–468.

Vysots'kyi GA. 1962. Застосування фітонцидів з коноплі для контролю фузаріозу сосна. Journal of Microbiology (Kiev) 24(2): 65–66.

Wang XW, Lombard L, Groenewald JZ, *et al.* (4 other authors). 2016. Phylogenetic reassessment of the *Chaetomium globosum* species complex. *Persoonia* 36: 83–133.

Watanabe T, Takesawa M. 1936. 大麻白星病の研究 (Studies on the leaf-spot disease of hemp). *Japanese Journal of Phytopathology* 6: 30–47.

Waterhouse GM, Brothers MP. 1981. The taxonomy of *Pseudoperonospora*. *Mycological Papers* 148: 1–28.

Weeden BR. 2006. *Evaluation of hemp and kenaf varieties in tropical and sub-tropical environments*. Rural Industries Research and Development Corporation, Publication no. 06/091, Kingston, Australia.

Wehmeyer LE. 1933. The genus *Diaporthe* Nitschke and its segregates. *University of Michigan Studies Scientific Series* 9: 1–349.

Wollenweber HW. 1914. Identification of species of *Fusarium* occurring on the sweet potato, *Ipomoea batatas*. *Journal of Agricultural Research* 2(4): 251–285.

Wollenweber HW, Reinking OA. 1935. *Die Fusarien*. Paul Parey, Berlin.

Woudenberg JHC, Hanse B, van Leeuwen GCM, Groenewald JZ, Crous PW. 2017. *Stemphylium* revisited. *Studies in Mycology* 87: 77–103.

Yang Q, Du Z, Tian CM. 2018. Phylogeny and morphology reveal two new species of *Diaporthe* from Traditional Chinese Medicine in Northeast China. *Phytotaxa* 336: 159–170.

Yu SL. 1973. *Fungal pathogens of Cannabis sativa grown in the USDA garden*. Unpublished manuscript, University of Mississippi, Oxford, Mississippi.

Yulfo-Soto GE, Smith H, Szarka D, *et al.* (3 other authors). 2022. First report of *Fusarium graminearum* causing flower blight on hemp (*Cannabis sativa*) in Kentucky. *Plant Disease* 106: 334.

Zabrin R. 1981. The fungus that destroys pot. *War on Drugs Action Reporter*, June 1981: 61–62.

Zelenay A. 1960. Fungi of the genus *Fusarium* occurring on seeds and seedlings of hemp and their pathogenicity. *Prace Naukowe, Instytutu Ochrony Roślin, Poznan* 2(2): 248–249.

Zeng XP, Wang HF, Chen Y, Yun X, Chen MC. 2014. "Preliminary report on indoor toxicity test of 9 fungicides on *Corynespora* sp. causing head blight of *Cannabis sativa*," pp. 538–540 in *Proceedings of the Chinese Society of Plant Protection in 2014, Fujian, China, 5–7 December, 2014*. Chinese Academy of Agricultural Sciences, Beijing.

Zhang GJ, Berbee ML. 2001. *Pyrenophora* phylogenetics inferred from ITS and glyceradehyde-3-phosphate dehydrogenase gene sequences. *Mycologia* 93: 1048–1063.

Zuefle M. 2020. *Hemp pest survey 2020*. New York State Integrated Pest Management Program. Available at: https://ecommons.cornell.edu/handle/1813/103765

Zuefle M. 2021. *Hemp pest survey 2021*. New York State Integrated Pest Management Program, Geneva, New York.

11 Root rots, wilts, and blights

Abstract

This chapter comprises a pot-pourri of diseases. Root rot is caused by oomycetes (*Pythium* and *Globisporangium* spp.) and fungi (primarily *Fusarium solani*). *Fusarium oxysporum* causes a wilt disease rather than root rot. One subspecies, *F. oxysporum* f. sp. *cannabis*, has been fingered as the *Cannabis* pathogen, but we provide phylogenetic evidence that it is actually *F. oxysporum* f. sp. *vasinfectum*. Blight disease is a general necrosis of leaves, flowers, and shoots—often accompanied by wilting. Some blight fungi received a taxonomic overhaul (*Botryosphaeria obtusa* is resurrected), and we describe two new taxa: *Neofusicoccum marconii* and *Dothiorella cannabis*. Don't overlook twig blight caused by *Lasiodiplodia theobromae* and *Neofusicoccum parvum*, as well as southern blight (*Athelia rolfsii*), charcoal rot (*Macrophomina phaseolina*), sore shin (*Rhizoctonia* spp.), verticillium wilt (*Verticillium* spp.), brown blight (*Alternaria* spp.), Texas root rot (*Phymatotrichopsis omnivora*), and black root rot (*Berkeleyomyces* spp.).

11.1 Introduction

Root rot looks like it sounds: roots turn mushy and brown. The normal white fuzz of root hairs melts away (Fig. 11.1A). Lesions may coalesce, easily seen in hydroponic culture (Fig. 11.1B). **Wilt** is also iconic: leaves lose turgor, go limp, and bend towards the ground. **Blight** is a sudden and extensive chlorosis and necrosis of leaves and flowers. **Dieback** refers to the progressive death of a branch that starts at the tip and works backwards. Sorting dieback from blight is a judgement call.

Blight may resemble a leaf spot disease whose spots have suddenly coalesced. Fungi that cause coalescing leaf spots in Chapter 10 (e.g., *Septoria*, *Bipolaris*, *Colletotrichum* spp.) could have been placed in this chapter. Some stalk diseases in Chapter 10 involve the vascular system and cause wilt symptoms (e.g., Sclerotinia canker and crown rot). Wilt can also be caused by nematodes (Chapter 15) or insects that injure roots (Chapter 8). Bacterial blights are detailed in Chapter 15. Gray mold of flowering tops is sometimes called Botrytis blight or bud rot (Chapter 9).

11.2 Pythium root rot

Pythium spp. have been called "water molds" for a long time (*wasser-schimmel*, Nees von Esenbeck, 1823). They produce **zoospores** that can swim through water and actively find plant hosts (see Fig. 12.4). Thanks to the rise of hydroponics, water molds have become leading causes of root rot. Many hydroponically grown vegetable crops have problems with *Pythium* spp. (Koike *et al.*, 2007). *Pythium aphanidermatum* and *Pythium ultimum* were identified early in the hydroponic era causing *Cannabis* root rot (McEno, 1990). Some names have changed—*Pythium ultimum* is now *Globisporangium ultimum*—but we'll keep the name Pythium root rot. Prior to hydroponics, these species were primarily maligned for causing "damping off" disease in seedlings (see section 12.2).

11.2.1 Symptoms and signs

When plants infected by *Pythium* or *Globisporangium* are pulled from soil, there's often a paucity of soil particles clinging to the roots, because of the loss of root hairs (Fig. 11.1A). Aboveground symptoms include stunted growth, due to root pruning (Fig. 11.2A). In older plants, water-soaked areas may arise at the base of the stem ("crown rot"), followed by a generalized leaf chlorosis and necrosis (Fig. 11.2B). Wilting may be present (Beckerman *et al.*, 2017) or not (Punja and Rodriguez, 2018). Plants may collapse and die. *Pythium* and *Globisporangium* spp. rarely produce hyphae on root surfaces. Razored root slices, mounted in lactophenol-cotton blue solution, may reveal oospores under ×125 or ×400 magnification (see Fig. 12.5D).

11.2.2 Causal organisms

Six *Pythium* spp. are associated with *Cannabis* root rot. A phylogenetic study reassigned two of them to the new genus *Globisporangium* (Uzuhashi *et al.*, 2010). All are members of Phylum Oomycota, Class Oomycetes, Family Pythiaceae. Isolating *Pythium* and *Globisporangium* spp. in pure culture is not easy. Non-specific nutrient media (potato dextrose agar, corn meal agar) become overrun by fungi or bacteria. "Water molds" can grow on "water agar" alone: 10 g of agar mixed into 500 ml water. This gives *Pythium* and *Globisporangium* spp. a selective advantage over fungi. Rose Bengal water agar is even more selective. Beckerman *et al.* (2017) isolated *Pythium aphanidermatum* from *Cannabis* with PARP agar, a selective medium with four compounds that suppress fungi and bacteria (pimaricin, ampicillin, rifampicin, pentachloronitrobenzene).

11.2a. *Pythium aphanidermatum* (Edson) Fitzpatrick 1923

Description: see section 12.2 and Fig. 12.5.

Fig. 11.1. *Cannabis* root rot by *Pythium aphanidermatum*: **A.** microscopic view (S. Koike); **B.** macroscopic view (courtesy of Punja and Rodriguez, 2018)

Fig. 11.2. Pythium root rot: **A.** root pruning by *P. dissotocum*: inoculated plant on left, control plant on right; **B.** leaf necrosis by *P. aphanidermatum* due to root death (courtesy of Punja and Rodriguez, 2018)

Comments: *P. aphanidermatum* has been recognized as a cause of damping off in seedlings for nearly a century (Galloway, 1937). Its portfolio has expanded to full-grown field plants (Beckerman *et al.*, 2018; Punja *et al.*, 2018), indoor plants grown hydroponically (Punja and Rodriguez, 2018; Punja *et al.*, 2019), and drip-irrigated plants (Hu and Masson, 2021). In all these studies, morphological identification was confirmed with ITS and/or *EF-1α* analysis. Disease causality was confirmed with pathogenicity tests.

11.2b. *Globisporangium ultimum* (Trow) Uzuhashi *et al.* 2010

≡*Pythium ultimum* Trow 1901
Description: see section 12.2 and Figs 12.4, 12.5.
Comments: *G. ultimum* is a well-known cause of damping off (section 12.2). It was first described as a cause of root rot in Argentina (Frezzi, 1956). *G. ultimum* causes root rot in field plants (Beckerman *et al.*, 2018) and plants grown hydroponically (Punja *et al.*, 2019; Pitman *et al.*, 2020) or in glasshouses (Thiessen *et al.*, 2020). In all these studies, morphological diagnosis was confirmed with ITS analysis, and disease causality was confirmed with pathogenicity tests.

11.2c. *Pythium myriotylum* Drechsler 1930

Description: see section 12.2.
Comments: Punja and Rodriguez (2018) recovered *P. myriotylum* from hydroponically grown plants in British Columbia. Morphological identification was confirmed with ITS and *EF-1α*; disease causality was confirmed with pathogenicity tests. Since then, *P. myriotylum* has been isolated in Connecticut (McGehee *et al.*, 2019), California (Pitman *et al.*, 2021), and Arizona (Hu, 2021). In all these reports, plants hit by *P. myriotylum* were growing in soilless peat-based potting mixes.

In North Carolina, Thiessen *et al.* (2020) recovered *P. myriotylum* from crown rot, and *G. ultimum* from root rot. Morphological identification was confirmed with ITS and *COI*; disease causality was confirmed with pathogenicity tests. Punja *et al.* (2022) obtained 43 isolates of *P. myriotylum* in a study of seven glasshouses, identified with ITS and *EF-1α*, and confirmed disease causality.

McGehee and Raudales (2021) obtained 21 isolates of *P. myriotylum* from plants cultivated in coconut coir and rockwool substrates, identified with ITS and *EF-1α*. Pathogenicity tests in seedlings demonstrated a loss in shoot and root biomass, with symptoms of chlorosis and wilting. The authors calculated the Area Under Disease Progress Curve, a quantitative summary of disease intensity over time—a first for *Cannabis*. All isolates were sensitive to mefenoxam, so the authors presumed that the isolates did not spread from other agricultural sites where mefenoxam resistance is common.

11.2d. *Globisporangium irregulare* (Buisman) Uzuhashi *et al.* 2010

≡*Pythium irregulare* Buisman 1927
Description: see section 12.2.
Comments: *G. irregulare* is a well-known cause of damping off (Chapter 12). It was first described as a cause of root rot in Connecticut (McGehee and Raudales, 2021), in plants cultivated in coconut coir and rockwool substrates. Disease causality was confirmed with pathogenicity tests—*G. irregulare* was less pathogenic than *P. myriotylum*, and was isolated less frequently.

11.2e. *Pythium dissotocum* Drechsler 1930

Description: *Sporangia* consist of elongated irregular swellings, simple or sparingly branched, 5–8 μm in diameter, scarcely distinguished from the mycelium. Sporangia produce long evacuation tubes (up to 1 mm in length) ending with swollen vesicles housing 10–75 zoospores, which measure 8–9 μm in diameter when encysted. *Oogonia* terminal or intercalary, spherical, smooth-walled, 12–29 μm in diameter. Each oogonium is fertilized by 1–2 antheridia (rarely 3–4). *Antheridia* swollen, sausage-shaped, curved, usually originating immediately below oogonia, but sometimes diclinous (originating at a distance from another hypha). *Oospores* aplerotic (space between the oospore wall and oogonium wall) though almost filling the oogonial cavity, smooth, 10–25 μm in diameter, with a moderately thick wall, 1.5–2 μm thick.
Comments: Punja and Rodriguez (2018) recovered *P. dissotocum* from hydroponically grown plants with root rot (Fig. 11.2). Morphological identification was confirmed with ITS and *EF-1α* analysis; disease causality was confirmed with pathogenicity tests. They recovered *P. dissotocum* more frequently than *P. myriotylum* or *P. aphanidermatum*. In a wider study, including greenhouses in British Columbia, Ontario, and California, *P. dissotocum* was the second most frequently recovered *Pythium*, behind *P. myriotylum* (Punja *et al.*, 2022).

11.2f. *Pythium catenulatum* Matthews 1931

Description: *Sporangia* consist of irregular swellings, catenulate (in a series of chains), often branched, scarcely distinguished from the mycelium. They give rise to a short evacuation tube ending with swollen vesicles housing 10–30 zoospores, which measure 6 × 8–11 μm when encysted. *Oogonia* terminal or intercalary, usually spherical, smooth-walled, 18–38 μm in diameter. Antheridia 1–12 to an oogonium, often 5–6, usually diclinous (originating at a distance from another hypha). *Oospore* nearly filling the oogonium, smooth, 16–26 μm in diameter, wall 1.5 μm thick.
Comments: Punja *et al.* (2019) recovered *P. catenulatum* from hydroponically grown plants with root rot in British Columbia, identified by ITS. Punja *et al.* (2022) demonstrated pathogenicity by *P. catenulatum; it caused less severe symptoms than other species in the study (P. myriotylum > G. ultimum > P. dissotocum > P. catenulatum).*

The organism was originally discovered by *Matthews (1931), who used hemp seed as "bait" to isolate oomycetes from pond water in North Carolina. Hemp seed bait has also isolated* P. *catenulatum from* river water in British Columbia (Thompson, 1979), irrigation water in Spain (*Sánchez and Gallego, 2000*), and reservoir water in Morocco (El Androusse et al., 2007).

11.2.3 Differential diagnosis

Symptoms of Pythium root rot could be confused with symptoms caused by root-rotting fungi, such as *Fusarium solani*. The causal organism needs to be isolated and identified. However, recovering *Pythium* or *Globisporangium* spp. from diseased plants may or may not indicate their role in crop damage, if other pathogens occur simultaneously in the same plants, such as *Fusarium* spp. or *Rhizoctonia solani* (Rothrock *et al.*, 2015). Causality can be confirmed by fulfilling Koch's postulates (e.g., Beckerman *et al.*, 2017, 2018; Punja *et al.*, 2018, 2019).

Pythium disease incidence and severity were placed in perspective by Punja *et al.* (2022), who isolated *Pythium* spp. from seven greenhouses located in British Columbia, Ontario, and California. The most common cause was *P. myriotylum* (43 isolates), followed by *P. dissotocum* (35 isolates), *P. catenulatum* (12 isolates), *P. aphanidermatum* (3 isolates), and *G. ultimum* (2 isolates). Inoculation studies showed *P. myriotylum* was the most pathogenic *(P. myriotylum > G. ultimum > P. dissotocum > P. catenulatum).*

11.2.4 Disease cycle and epidemiology

Recirculating hydroponic solutions can act as both a primary inoculum source and a dispersal mechanism, promoting the spread of *Pythium* and *Globisporangium* zoospores. For additional aspects of *Pythium* and *Globisporangium* epidemiology in soilless systems, see section 12.2.

Above-ground symptoms occur sooner in field crops than in hydroponically grown plants; the latter are constantly exposed to nutrient solutions, which partially compensates for root loss (Punja and Rodriguez, 2018). Field crops grown in low-lying or flood-prone soil are more susceptible. The incidence and severity of disease worsens when field crops are planted in early spring, in cool conditions, or in poorly draining soil (McPartland, 1996). Crusting of the soil surface after high-intensity rainfall predisposes to *Pythium* root rot (Beckerman *et al.*, 2017, 2018). Hu and Masson (2021) implicated drip irrigation in combination with plastic mulch.

11.2.5   Cultural and mechanical controls

Exclusion: Keep *Pythium* out of hydroponic systems. Before entering a facility, all clones and cuttings must be quarantined and inspected for infection. Other sources of inoculum include growing media, pipes and tubing, reservoir tanks, tools and equipment, as well as residues on workers' shoes. Good hygiene practices include restricted entry, footbaths at entry, and sanitation of all infrastructure and equipment. Don't reuse growing media.

Methods of disinfecting hydroponic solutions are reviewed in detail by Stewart-Wade (2011). Summaries of several methods are presented in section 19.2.3, including heat treatment, ozone, ultraviolet light, filtration systems, ionization, and chemicals.

Hydroponic systems with catastrophic problems should be dismantled and sanitized before another crop is introduced. Equipment amenable to heat treatment can be sterilized. Remove reservoir tanks and pipes from the room, and do a deep clean. Bleach down walls and floors. Reassemble, and "shock treat" the nutrient solution with disinfectants prior to adding plants.

Conditions that favor *Pythium* spp. include excessive nitrogen in the nutrient solution, low dissolved oxygen, and extreme temperatures that stress plants. Injury to roots or crowns provides an entry point for *Pythium*.

Pythium cannot be excluded from field soil—it's already there. Field crops should not be irrigated with water from ponds or streams. Pitman *et al.* (2020) reported a *Pythium* epidemic in a greenhouse whose irrigation system was supplied by a local stream. Other cultural and mechanical controls are presented in section 12.2.

11.2.6   Biological and chemical controls

Patel *et al.* (2014) tested several biocontrols against *P. aphanidermatum* in an *in vitro* comparison. *Bacillus subtilis* (e.g., Serenade®) did the best, followed by *Trichoderma harzianum* (e.g., Rootshield®), *Pseudomonas fluroscences* (e.g., Pseudon-F®, Ecomonas®), and *Trichoderma atroviride* (e.g., Binab®). Other effective biocontrols include *Bacillus cereus* (BioStart™ Defensor), *Streptomyces lydicus* (Actinovate®), *Streptomyces griseoviridis* (Mycostop®), and *Gliocladium catenulatum* (Prestop®).

One study tested the efficacy of four products against root rot caused by *P. myriotylum* in *Cannabis* (Scott, 2021; Scott and Punja, 2023). Young plants potted in a 3:1 mix of coco and perlite were drenched with biocontrols, followed a week later by *P. myriotylum* inoculation. Plants were assessed for disease severity 18 days later. Plants treated with *Gliocladium catenulatum* strain J1446 (Prestop®, Lalstop®) had the lowest disease

rating, followed by *Trichoderma harzianum* strain KRL-AG2 + *Trichoderma virens* strain G-41 (Rootshield® Plus WP), *Trichoderma asperellum* (Asperello®), and *Bacillus amyloliquefaciens* strain F727 (Stargus®).

Success with *Bacillus amyloliquefaciens* (Double Nickel 55®) against *Pythium*, *Rhizoctonia*, and *Fusarium* spp. was presented at a *Cannabis* symposium (Highland, 2020). Formulations of *B. amyloliquefaciens* for use in hydroponic systems are available (e.g., Hydroguard®). *Pythium oligandrum* (Polyversum®) is a hyperparasite that colonizes pathogenic *Pythium* spp. in the soil, as well as *Fusarium* spp., *Rhizoctonia solani*, *Botrytis cinerea*, and *Sclerotina sclerotiorum*. In fact, all the aforementioned biocontrols show some efficacy against these other pathogens. *P. oligandrum* also triggers the induction of defense-related reactions in plants, as do some of the other aforementioned biocontrols.

Fungicides effective against *Pythium* are not labeled for use on *Cannabis*. Metalaxyl appears on a list of pesticides by California DPR (2018) that explicitly cannot be used on *Cannabis* under any circumstances. Most *Pythium* populations are resistant to metalaxyl.

11.3   Fusarium crown and root rot

Fusarium spp. are soil-borne pathogens. They thrive in moisture levels bordering on the aquatic. The primary cause is *Fusarium solani*, a primitive beast—perhaps the oldest *Cannabis* pathogen. A molecular clock study estimated that *F. solani* diverged from its closest relative during the Paleocene, ~57 million years ago (mya). In comparison, other *Fusarium* spp. that infest *Cannabis* (*F. oxysporum*, *F. graminearum*, *F. sambucinum*) radiated during the early Miocene, 23–10 mya (O'Donnell *et al.*, 2013). *Cannabis* evolved around 28 mya (McPartland, 2018), back when *F. solani* was the only diverged species.

Lyster Dewey (see Fig. 1.13F) first described Fusarium root rot in his breeding plots at Arlington, Virginia. Dewey (1902) observed wilting in plants with rotted, red-stained roots. He gave a sample to Erwin F. Smith, who identified the cause as a *Fusarium* sp. Altogether, he reported this disease four times in Arlington (Dewey, 1902, 1917, 1919, 1924), a site with water-retaining clay soil that predisposes plants to root rot.

11.3.1   Symptoms and signs

Root rot affects the taproot and lateral roots, which show cortical decay and vascular discoloration. The discoloration often takes a reddish hue when caused by *Fusarium* spp. (Fig. 11.3). The root may become quite necrotic and have "a very disagreeable odor" when pulled from the ground (Dewey, 1919).

Crown rot—where the stem joins the root—is an upward extension of root rot (Fig. 11.4A). Discolored tissues include the epidermis and cortex, which may extend into vascular tissues. Reddish-brown internal decay is easily seen when the root is cross-sectioned (Fig. 11.4B). In humid conditions, a white or pink fluff of mycelium may cover the crown surface. Aboveground symptoms begin as a yellowing and drooping of leaves.

Next comes partial wilting, or complete wilting if crown rot girdles the stem. Wilting is worse in plants growing in alluvial soil (Barloy and Pelhate, 1962). Wilted leaves turn brown but stay attached to the plant (Dewey, 1924). Plants in all stages of development may be affected, including seedlings (Zelenay, 1960; Barloy and Pelhate, 1962; Punja and Rodriguez, 2018).

Fig. 11.3. Fusarium root rot caused by *Fusarium solani*

Fig. 11.4. Fusarium crown rot: **A.** subepidermal staining; **B.** internal decay (courtesy of Punja *et al.*, 2018)

11.3.2 Causal organisms

Fusarium taxonomy is difficult (see section 9.3). The genus belongs in Class Sordariomycetes, Family Nectriaceae, and consensus ends there. DNA-based phylogenetics now groups *Fusarium* spp. into **species complexes** (see Table 9.2). The eight *Fusarium* spp. causing this disease belong to four species complexes. We have organized them accordingly.

11.3.3 *Fusarium solani* species complex

11.3a. *Fusarium solani* (Martius) Saccardo 1881

≡*Neocosmospora solani* (Mart.) Lombard & Crous 2015
? Sexual stage: *Nectria cancri* Rutgers 1913, ≡*Hypomyces cancri* (Rutgers) Wollenweber 1914, ≡*Hypomyces haematococcus* var. *cancri* (Rutgers) Wollenweber 1930,
Description: *Conidiogenous cells bearing macroconidia* are hyaline, short, doliiform, frequently branched, and phialidic. *Macroconidia* fusoid, slightly curved, stout, broad, hyaline; with a foot-shaped basal cell and a pointed, somewhat beaked apical cell, 1–6 septate (usually 3), 40–100 × 5–8 μm (Fig. 11.5). *Conidiogenous cells bearing microconidia* hyaline, long, cylindrical, phialidic, up to 400 μm in length. *Microconidia* hyaline, ovoid to allantoid, sometimes becoming 1-septate, 8–16 × 2–4 μm. *Chlamydospores* usually in pairs, but sometimes single or in chains, globose to oval, smooth to rough walled, 9–12 × 8–10 μm.
Comments: Barloy and Pelhate (1962) first named *F. solani* as a *Cannabis* pathogen. McPartland *et al.* (2000) isolated the

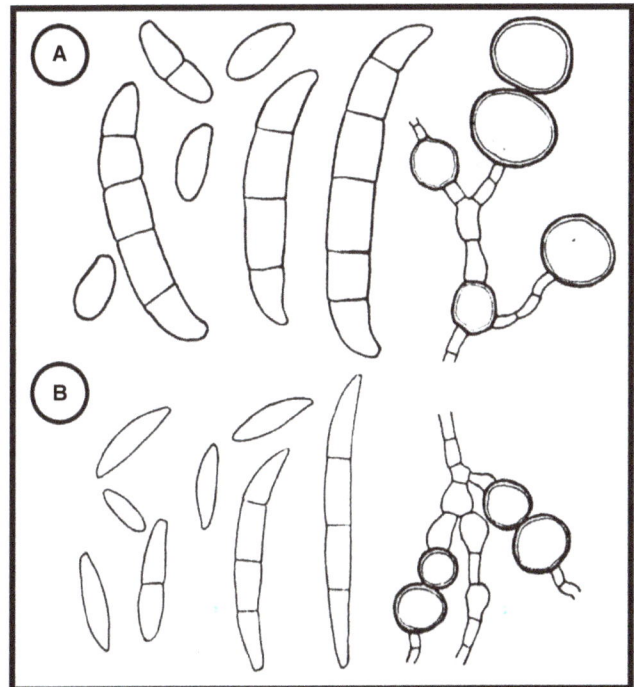

Fig. 11.5. Two fusaria isolated from *Cannabis*: **A.** *F. solani*; **B.** *F. oxysporum.* (redrawn from Barloy and Pelhate 1962)

fungus from wild hemp in Illinois; its identification was based on morphological characters in culture. Punja and Rodriguez (2018) found *F. solani* infecting roots of hydroponically grown *Cannabis* in British Columbia. Their morphological identification was confirmed with ITS and *EF-1α*. Sorrentino *et al.* (2019) isolated *F. solani* causing crown rot in Italy. They sequenced ITS and *EF-1α*, conducted a phylogenetic analysis, and their isolate belonged to Clade 3 of the FsSC. Jerushalmi *et al.* (2020a) isolated *F. solani* in Israel, and Breton *et al.* (2022) isolated it in Québec.

F. solani is ensnared in a web of synonyms, depending upon whether or not the species produces a sexual stage, and the identity of that sexual stage.

Wollenweber (1914) identified the sexual stage *Hypomyces cancri* from dead taproots of *Cannabis* in Dewey's Potomac Flats plots. Across the river in Arlington Farms, Flora Patterson isolated "a luxuriantly fruiting *Fusarium* sp. It may be connected with *Hypomyces cancri*" (Dewey, 1917). This report is the source of *H. cancri* cited on *Cannabis* in the USA (Seymour, 1929; Miller *et al.*, 1960; Westcott, 1990; Wang, 2021). Russian phytopathologists have also attributed hemp disease to *H. cancri* (Gitman and Boytchenko, 1934; Vakhrusheva and Davidian, 1979).

Hypomyces cancri was originally named *Nectria cancri* by Rutgers (1913), who isolated it from cacao plants in Java. Wollenweber (1914) transferred it to *Hypomyces* as *H. cancri*. Wollenweber did not name the anamorph, stating that it resembled *Fusarium radicicola* but its conidia had a pedicellate base (i.e., foot-shaped basal cell).

To the anamorph of *Nectria cancri* Rutgers (1913) assigned *Fusarium theobromae* Appel and Strunk 1904. Unlike Wollenweber, Rutgers described and illustrated microconidia along with macroconidia (Fig. 11.6), which is also seen in *F. solani*.

Subsequently, Wollenweber (1917) identified the anamorph of *H. cancri* as *Fusarium javanicum*, which he then reduced to a variety as *F. javanicum* var. *radicicola* (Wollenweber, 1931). Bilaĭ (1955) identified *F. javanicum* var. *radicicola* on dead stems and branches of *Cannabis* in Russia. Zelenay (1960) cited *F. javanicum* var. *radicicola* on *Cannabis* in Poland.

Meanwhile, Wollenweber (1930) reduced *Hypomyces cancri* to a variety: *Hypomyces haematococcus* var. *cancri*. Snyder and Hansen (1941) connected *Hypomyces haematococcus* with the anamorph *Fusarium solani*. Instead, Nalim *et al.* (2011) connected the sexual stage to a new anamorph taxon, *Fusarium haematococcum*.

Crous's team synonymized Rutgers's *Nectria cancri* with *Fusarium solani*, without mentioning Wollenweber's *Hypomyces cancri* (Lombard *et al.*, 2015). However, *Fusarium theobromae*—the anamorph that Rutgers connected with *Nectria cancri*—they considered a separate species, with no mention of the sexual stage (Sandoval-Denis *et al.*, 2019).

When Nalim *et al.* (2011) coined *Fusarium haematococcum* as the conidial stage of *Neocosmospora* (*Nectria*) *haematococca*, this left *F. solani* dangling, which prompted Geiser *et al.* (2013) to publish a petition to conserve the name *F. solani*, signed by 69 researchers. Schroers *et al.* (2016) created a lectotype and epitype of *F. solani*, thereby conserving the taxon. They stated that *F. solani* did not produce a sexual stage.

Sandoval-Denis *et al.* (2019) disagreed with the petition (thereby disagreeing with themselves, since co-author Crous signed the petition). They transferred *F. solani* to *Neocosmospora* as *N. solani*, removing the taxon from genus *Fusarium*. In response, O'Donnell *et al.* (2020) published "No to *Neocosmospora*." Geiser *et al.* (2021) published a robust 19-locus phylogeny (55.1 kg total sequence length), which removed all doubt about the inclusion of *F. solani* in the genus *Fusarium*. Crous *et al.* (2021) continued to place *F. solani* in *Neocosmospora*. They whined about Geiser publishing "a political generic concept, meant to assuage the concerns of plant pathologists and other applied scientists."

11.3b. Other species in the FsSC

Five other *Fusarium* spp. in the *F. solani* species complex have been named as causes of *Cannabis* root rot (see Table 9.2). They are noted briefly, without morphological descriptions.

Fusarium haematococcum Nalim, Samuels & Geiser 2011 is not a *Cannabis* pathogen. Its sexual stage has been cited in the *Cannabis* literature (Booth, 1971, as *Nectria haematococca*), back when that sexual stage was connected with *F. solani*. This concept traces back to Wollenweber (1930), who reduced *Nectria cancri* to a variety of *N. haematococca* (see above).

Fusarium lichenicola Massalongo 1903 was isolated from glasshouse plants in British Columbia (Punja *et al.*, 2021). Symptoms included stunted growth, crown discoloration, yellowing of lower leaves, and browning and decay of roots. Morphological identification was confirmed with ITS, *EF-1α*, and a phylogenetic analysis. Because *F. lichenicola* occurs in tropical climates, the authors proposed that it came from coconut coir imported into Canada. *F. lichenicola* was less

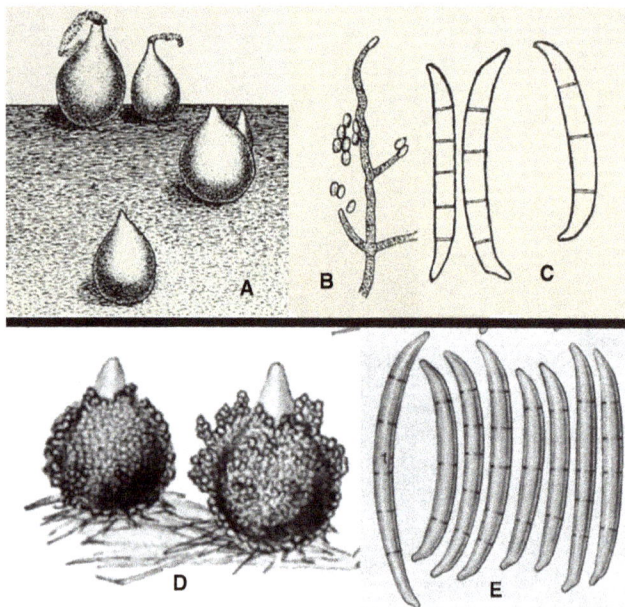

Fig. 11.6. *Hypomyces* (*Nectria*) *cancri*: Rutgers (1913): **A.** perithecia; **B.** microcondidia; **C.** macroconidia; Wollenweber (1914): **D.** perithecia; **E.** macrocondia

virulent than *F. solani* in pathogenicity tests. Breton *et al.* (2022) isolated *F. lichenicola* from rotten roots in Québec.

Fusarium falciforme (Carrión) Summerbell & Schroers 2002 was isolated from hydroponic crops in British Columbia (Punja and Rodriguez, 2018) and field-grown plants in California (Paugh *et al.*, 2022). In both studies, identification was confirmed with ITS and *EF-1α*, and Koch's postulates confirmed pathogenicity.

Fusarium petroliphilium (Chen & Fu) Geiser 2013 and *Fusarium keratoplasticum* Geiser 2013 were isolated from plants in Israel by Jerushalmi *et al.* (2020a). ITS and *EF-1α* were used for identification. Koch's postulates were not performed. *F. solani* was much more common.

11.3.4 *Fusarium incarnatum–equiseti* species complex

11.3c. *Fusarium equiseti* (Corda) Saccardo 1886

≡*Selenosporium equiseti* Corda 1838
Description: *Conidiogenous cells* monophialidic, occasionally with percurrent proliferation; phialides 9.6–21.6 × 3.0–4.0 µm. *Conidia* falcate, 3–7-septate, smooth, hyaline, with a unique apical cell that is tapered, curved, and described as "whip-like," 26.0–80.0 × 3.2–5.0 µm. *Chlamydospores* solitary or in chains, terminal or intercalary, smooth, hyaline to pale yellowish-brown, up to 12.8 µm wide.
Comments: *F. equiseti* is a cosmopolitan species with a wide host range, causing blight in cereal crops, cucurbits, and solanaceous species. Wang *et al.* (2019) analyzed 175 isolates in the *F. equiseti* species complex, using a multilocus approach. Within *F. equiseti* they detected 14 subclades, which they recognized as separate species—so new names are on their way.

Ondřej (1991) isolated *F. equiseti* from hemp stems in Czechoslovakia. Schoener *et al.* (2017a) isolated *F. equiseti* from fiber-type plants in Nevada. Morphological identification was confirmed with ITS sequencing. The same group isolated *F. equiseti* from drug-type plants in Nevada; Wang (2018) suggested *F. equiseti* may act as a secondary invader following *F. oxysporum* or *F. solani*. Punja *et al.* (2018) isolated *F. equiseti*, along with *F. oxysporum* and *F. solani*, from drug-type plants in California with symptoms of wilting and root rot. *F. equiseti* has also been isolated from leaves and flowering tops (see section 9.3). Breton *et al.* (2022) isolated *F. equiseti* from roots in Quebec.

11.3.5 Other *Fusarium* spp.

Three other fusaria have been isolated from *Cannabis* with root rot. They belong to other *Fusarium* species complexes (see Table 9.2), and are briefly noted here.

Fusarium tricinctum (Corda) Saccardo 1886, and *Fusarium redolens* Wollenweber 1913 were isolated in Nevada (Schoener *et al.*, 2017a; Wang, 2018). Morphological identification was confirmed with ITS sequencing.

Fusarium brachygibbosum Padwick 1945 was isolated from field-grown plants in California (Punja *et al.*, 2018).

They also isolated *F. solani*, *F. oxysporum*, and *F. equiseti*. Symptoms included chlorosis, wilting, and rot of vascular and cortical tissues. Morphological identification was confirmed with ITS, *EF-1α*, and a phylogenetic analysis. Plants in pots were inoculated with pure cultures poured into soil, and a wound was created at soil line with a scalpel. Test plants developed yellow leaves and stunting. After 5 weeks plants were depotted and root length measured: control 18 cm; *F. oxysporum* 12.5 cm; *F. brachygibbosum* 12.5 cm; *F. solani* 7 cm.

11.3.6 Differential diagnosis

On *Cannabis*, the macroconidia of *F. solani* and *F. oxysporum* (see Fusarium wilt, next) were differentiated by Barloy and Peltate (1962) (Fig. 11.5). *F. solani* has stouter, thicker macroconidia, and longer microconidiophores. The microconidia produced by *F. solani* and *F. oxysporum* are absent in species that cause stem canker (*F. graminearum*, *F. sambucinum*). Amplifying and sequencing ITS and *EF-1α* for a phylogenetic analysis is the best way of differentiating *Fusarium* spp.

Reddish root staining caused by *Fusarium* spp. is not caused by fungi responsible for gray mold, hemp canker, and southern blight. Those fungi produce a thicker mycelial mat on the plant surface, and sclerotia. Initial symptoms of Fusarium root rot (wilting) resemble symptoms caused by nematodes and some soil insects (e.g., root maggots, white root grubs), or wilting from insufficient hydration.

11.3.7 Disease cycle and epidemiology

F. solani overwinters in crop debris or in soil, as conidia, chlamydospores, or ascospores. Sexual states (perithecia) arise after plant death, and overwinter in crop stubble or soil. Perithecia are abundant in wet tropics and less common in temperate zones (Booth, 1971). The mode of infection is similar to *Fusarium* spp. described in section 10.4. *F. solani* acts synergistically with nematodes and parasitic plants that create root wounds. Barloy and Pelhate (1962) considered a combination of *F. solani* and broomrape (*Orobanche ramosa*) the greatest threat to *Cannabis* cultivation in southern France. Since *Fusarium* conidia move in water films and droplets, they do best in damp conditions and heavy clay soil. Watch for *Fusarium* epidemics during seasons with above-average rainfall. Like other fusaria, *F. solani* may also poses a threat to humans. The fungus can cause eye infections, with a majority of cases occurring in agricultural workers with antecedent corneal abrasions (Rippon, 1988).

11.3.8 Cultural and mechanical controls

Do not plant in heavy, wet soils, or fields with a history of *Fusarium* disease. Fusarium root rot usually appears after abundant rainfall in fields with heavy soils. Rogue affected plants as soon as possible. Control synergistic organisms such as nematodes and broomrape. Seed-borne infection is possible

(Zelenay, 1960); do not harvest seeds from infected plants. Since weeds serve as hosts for *F. solani*, removing weeds may reduce inoculum pressure (McPartland *et al.*, 2022).

For hydroponic systems, exclusion is key. Punja and Rodriguez (2018) recovered *F. solani* from recirculating hydroponic solution. See exclusion measures outlined in the previous *Pythium* section. Everything must be clean: growing media, pipes and tubing, reservoir tanks, tools and equipment. Hydroponic nutrient solution will disperse *Fusarium* conidia. Methods of nutrient solution disinfection also appear in the *Pythium* section—ozone injection, ionization, UV light, filtration systems, and chemicals.

No breeding for resistance has been done. Schoener *et al.* (2017a) reported that drug-type strains ("Fire OG", "Remedy CBD") were more susceptible than fiber-type cultivars ('Futura-75', 'Canda', 'Joey').

11.3.9   Biological and chemical controls

Biocontrol organisms are preventive—they suppress soil populations of pathogens—and not curative. Many biocontrols mentioned in the previous section on *Pythium* will also suppress *Fusarium* populations. Wang (2018) isolated *Trichoderma virens* from *Cannabis* roots and tested it against five *Fusarium* spp. *T. virens* was particularly efficacious against *F. solani*. A related species, *Trichoderma harzianum*, has been used against *F. solani* in other crop plants.

A bacterial endophyte isolated from *Cannabis* identified as *Pseudomonas orientalis* showed significant *in vitro* activity against *F. solani* (Scott *et al.*, 2018). A related species, *Pseudomonas stutzeri*, controlled *F. solani* in kidney beans (*Phaseolus vulgaris*). These species are related to *Pseudomonas fluorescens*, another useful biocontrol organism.

Spraying fungicides on *F. solani*-infected plants is not useful. Seed treatment can disinfect seeds suspected of harboring seed-borne infections. Mishra (1987) applied several fungicides to *Cannabis* seeds and controlled *F. solani*, but none are registered for the crop. Mixing dried *Cannabis* leaves into soil may suppress *Fusarium* spp. Grewal (1989) suppressed *F. solani* growth by mixing 3 kg of dried leaves into 137 kg of wheat straw compost. Dahiya and Jain (1977) reported that pure THC and CBD inhibited the growth of *F. solani*. Hydrodistilled essential oil of *C. sativa* showed little *in vitro* efficacy against *F. solani* (9.8% inhibition), compared with essential oil from other plants, such as *Putranjiva roxburghii*, 100%; *Ageratum houstonianum*, 80.5%; and *Ocimum canum*, 75.0% (Kumar, 2014). Similarly, an aqueous extract, lacking THC and CBD, was less effective than other plant extracts (Mahmood *et al.*, 2014).

11.4   Fusarium wilt

The pathogen causing this disease, *Fusarium oxysporum*, is a xylem inhabitant, rather than a cortical pathogen like *F. solani*. It plugs the plant's water-conducting tissues and causes a wilt (Fig. 11.7). *F. oxysporum* and *F. solani* have occurred simultaneously in the same field (Schoener *et al.*, 2017b; Jerushalmi

Fig. 11.7. Symptoms caused by *Fusarium oxysporum*

et al., 2020a). Barloy and Pelhate (1962) found them in the same plant (Fig. 11.5), as did Punja and Rodriguez (2018).

Hildebrand and McCain (1978) developed a method to mass-produce *F. oxysporum* inoculum. McCain wrote to President Nixon and explained how *F. oxysporum* could destroy marijuana crops worldwide (Shay, 1975). McCain claimed, "Just introduce a couple of pounds [of the fungus] into an area, and while it wouldn't have much of an effect the first year, in several years it would spread throughout the country with devastating results" (Zubrin, 1981). McCain said the project was suppressed by a conspiracy of Carter administration officials (Zubrin, 1981).

David C. Sands restarted mycoherbicide research aimed at "*Cannabis sativus*" (Sands, 1991). He sent *F. oxysporum* to the USDA in Beltsville, Maryland. They released the fungus in a field trial (!). It killed *Cannabis*, overwintered in the soil, and then killed plants the following year. Collaborators in Kazakhstan tested *F. oxysporum* in greenhouse experiments. The fungus proved difficult to contain, and spread to control plants (Anonymous, 1994). Nevertheless, they released it in field trials (Tiourebaev *et al.*, 2001). Sands's team created hyper-virulent mutants of *F. oxysporum* (Tiourebaev *et al.*, 2000; US Patent 6673746), and a special formulation designed for aerial deployment—hemp seeds coated with *F. oxysporum* conidia (US Patent 6403530). An attempt to release *F. oxysporum* in Florida met with public outrage (Fields, 1998) and scientific scrutiny (McPartland and West, 1999; McPartland and Nicholson, 2003). They argued that transposable elements ("jumping genes") in *F. oxysporum* f. sp. *cannabis* might transform non-pathogenic *F. oxysporum* into pathogens of other hosts. This has since been proven possible (Ma *et al.*, 2010).

11.4.1 Symptoms and signs

Initially, small, dark, irregular spots appear on lower leaves. Affected leaves suddenly become chlorotic. Wilt symptoms begin with an upward curling of leaf tips. Wilted leaves dry to a yellow-tan color, and hang on plants without falling off (Fig. 11.7). Stems may turn yellow-tan (Noviello and Snyder, 1962; Pellegrini *et al.*, 2021). Surviving plants are stunted (Tiourebaev *et al.*, 2001). Cutting into diseased stems reveals a reddish-brown discoloration of xylem tissue (Fig. 11.8). Pulled-up roots show no external symptoms. Barloy and Pelhate (1962) described the fungus wilting whole plants. Noviello and Snyder (1962) illustrated one-sided wilting.

11.4.2 Causal organisms

Members of the *Fusarium oxysporum* species complex (FoSC) are thought to comprise many clonal lineages, given that no sexual cycle has been discovered (O'Donnell *et al.*, 2022). *F. oxysporum* isolates may be morphologically identical, yet exhibit host-selective pathogenicity. Based on host ranges, Snyder and Hansen (1940) recognized 25 "biologic forms" within *F. oxysporum*. Any *Fusarium* causing a vascular wilt was lumped into *F. oxysporum*; for example, *F. vasinfectum* became *F. oxysporum* f. *vasinfectum*

"Biologic forms" became known as host-selective *formae speciales* (ff. spp.). Edel-Hermann and Lecomte (2019) listed 106 well-characterized *formae speciales*, together with 37 insufficiently documented ones, as well as 58 plant species susceptible to *F. oxysporum* where a *forma specialis* (f. sp.) has not been named.

Two *formae speciales* have been reported from *Cannabis*: *F. oxysporum* f. sp. *vasinfectum* (hereafter *F. o. vasinfectum*) and *F. oxysporum* f. sp. *cannabis* (*F. o. cannabis*). They are morphologically identical.

Mycelium in culture is usually floccose or felty, white to pink to purple, and grows abundantly. *Conidiophores* are branched or unbranched monophialides, usually short and barrel-shaped, in tufts of 1-4 atop metulae, 10–12 µm long. *Macroconidia* hyaline, 3–5-septate, straight to sickle-shaped, ends curved inward with a hooked apex and pedicellate base, as large as 45–55 × 3.5–4.5 µm (see Figs 3.3, 11.9). *Microconidia* hyaline, aseptate (rarely 1-septate), oval to cylindrical or kidney shaped, 5–16 × 2.2–3.4 µm (Fig. 11.9). *Chlamydospores* hyaline, thick-walled, with a rough or smooth surface, spherical, borne singularly or in pairs, formed atop conidiophores or intercalary within hyphae or macroconidia, 7–13 µm in diameter (Fig. 11.9).

11.4a. *Fusarium oxysporum* f. sp. *vasinfectum* (Atkinson) Snyder & Hansen 1940

≡ *Fusarium vasinfectum* Atkinson 1892

Comments: Selective pathogenicity has necessitated the subdivision of some *formae speciales* into **races**. *F. o. vasinfectum* consists of eight races (Edel-Hermann and Lecomte, 2019), with selective pathogenicity on hosts in the Malvaceae (*Gossypium*,

Fig. 11.8. Internal discoloration of pith tissue following inoculation by *Fusarium oxysporum* (courtesy Z. Punja)

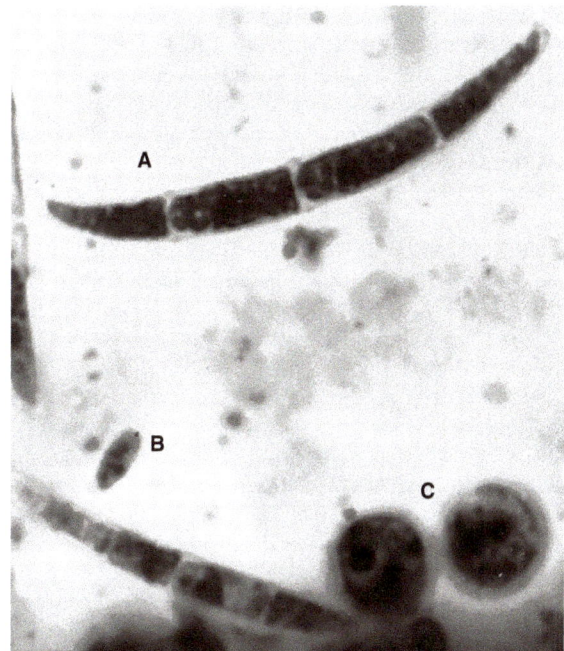

Fig. 11.9. *Fusarium oxysporum*: **A.** macroconidium; **B.** microconidium; **C.** chlamydospores (LM ×950; McPartland, 1983a)

Abelmoschus, Hibiscus spp.), Solanaceae (*Nicotiana, Physalis* spp.), Fabaceae (*Senna, Medicago, Glycine* spp.), and Asteraceae (*Tithonia* spp.). The race infesting *Cannabis* has not been determined; see discussion below.

Fusarium wilt of *Cannabis* was first described in Russia by Alexandra Ivanovna Raillo (1894–?). Raillo (1950) identified the pathogen as *F. vasinfectum*. She was a *Fusarium* expert, and pioneered the use of single spore culture methods. During her career, Raillo named seven new varieties of *F. oxysporum*. One variety, *Fusarium oxysporum* var. *pisi* (C. Hall) Raillo, she transferred from *F. vasinfectum*. Apparently she could perceive differences between *F. oxysporum* and *F. vasinfectum*.

Dobrozrakova *et al.* (1956) described hemp wilt symptoms caused by *F. vasinfectum*, as well as damping off in seedlings. The disease soon spread beyond Russia. The epidemic likely followed the seed supply. McPartland and Hillig (2004b) followed the paper trail to reconstruct the spread of *F. vasinfectum*: to Czechoslovakia (Rataj, 1957), Romania (Ceapoiu, 1958), Poland (Czyżewska and Zarzycka, 1961), and Latvia (Serzane, 1962).

Then *F. oxysporum* was identified as the causal organism, first in Poland (Zelenay, 1960) then France (Barloy and Pelhate, 1962). Mándy and Bòcsa (1962) implicated both *F. oxysporum* and *F. vasinfectum* in wilt disease and damping off in Hungary. The taxon *F. oxysporum* f. *cannabis* Bilaĭ appears on websites, but we could not trace its source publication. Vera I. Bilaĭ (1908–1994) was a Ukrainian mycologist who specialized in toxin-producing fungi. There's no mention of that taxon in her book *Fusaria: Biology and Systematics* (Bilaĭ, 1955).

11.4b. *Fusarium oxysporum* f. sp. *cannabis* Noviello & Snyder 1962

Comments: Noviello and Snyder (1962) isolated *F. oxysporum* f. sp. *cannabis* in Italy. It is still found there (Pellegrini *et al.*, 2021). Noviello and Snyder justified the new *forma specialis* saying that "the wilt disease and its pathogen have not been described." They apparently overlooked all of the aforementioned reports.

Some Russian authors adopted *F. o. cannabis* (Gitman, 1968b; Vakhrusheva and Davidian, 1979), although other Russians still cite *F. o. vasinfectum* (Serkov *et al.*, 2019). In North America, *F. o. cannabis* has been identified in Illinois (McPartland, 1983a) (Figs 11.7 and 11.9), Nevada (Schoener *et al.*, 2017b), California (Punja *et al.*, 2018), North Carolina (Thiessen, 2019; Cochran *et al.*, 2020), British Columbia (Punja and Rodriguez, 2018; Punja *et al.*, 2019), and Ontario (Punja, 2021b).

Fusarium wilt may have spread to Russia from Kazakhstan, which harbors vast stands of wild-type *Cannabis*. At least 125,000 ha of wild hemp grow in the Chuy River Valley alone (Tiourebaev *et al.*, 2001). Tiourebaev collected 125 isolates of *F. o. cannabis* throughout Kazakhstan, including the Chuy River Valley.

According to Gitman (1968b), Nikolay G. Zaprometov (1893–1983) first described *Fusarium* wilt of *Cannabis*, in a book we have not seen (Zaprometov, 1928). He worked at Lunacharsky near Tashkent, an experiment station established by *Novlubinstitut* (New Bast Institute). *Novlubinstitut* also established the Chuy Bast Crops Experiment Station in Kazakhstan, and began growing hemp in 1931 (L'dov, 1933). Zaurov (1948) bred a new variety of "southern hemp" at Chuy. His new variety was soon cultivated in southern Russia—a potential vector of Fusarium wilt into Europe.

Edel-Hermann and Lecomte (2019) recognized only one race of *F. o. cannabis*, although pathogenicity tests—the means of recognizing races—have been limited. Noviello and Snyder (1962) tested only *Cannabis*. Two research groups who aimed to weaponize *F. o. cannabis* conducted wider host-range studies (McCain and Noviello, 1985; Tiourebaev *et al.*, 2001). Both studies reported "no pathogenicity" in any plants tested other than *Cannabis* (Table 11.1). National Research Council (2011) criticized both studies, for a lack of clearly described testing procedures, and a lack of data other than "no pathogenicity."

McCain and Noviello (1985) tested plants at three sites. At one site, no pathogenicity was reported in any plants—not even *Cannabis*—which makes that trial a source of false-negatives. McCain and Noviello reported no infection in *Humulus lupulus*, stating that hops "are not known to be susceptible to a fusarium wilt disease anywhere in the world." They overlooked reports of *F. oxysporum* on *H. lupulus* in Germany (Rudolph, 1968; Leibelt and Senser, 1972), Poland (Solarska, 1981), and Australia (Sampson and Walker, 1982). Other reports have been published since then (Sabo *et al.*, 2002; Gryndler *et al.*, 2008; Ahren *et al.*, 2009), including the coining of *F. oxysporum* f. *humuli* Komarova (Danilova *et al.*, 2010).

McCain and Noviello demonstrated variable pathogenicity among *Cannabis* populations: A heavy inoculum load caused 20% mortality in 'Super Elite' (Italian fiber-type cultivar), 70% in feral hemp from Iowa (a Chinese fiber-type landrace), and 100% in "Iran" (a drug-type landrace). McCain and Noviello stated that three other Italian fiber-type cultivars were resistant, as was a fiber-type cultivar from Portugal. Drug-type plants from Mexico, Pakistan, Turkey, Thailand, India, Nepal, and South Africa were all susceptible.

11.4.3 Phylogenetics vs *formae speciales*

Cracks have formed in the *formae speciales* system of classification. Recent studies have shown that *formae speciales* on a single host species may not be genetically related (they have polyphyletic origins), and vice versa, *formae speciales* can jump host species.

We propose that some—if not all—isolates of *F. o. cannabis* belong to races of *F. o. vasinfectum*. Host range testing to identify the eight races of *F. o. vasinfectum* is well-established (Armstrong and Armstrong, 1960, 1968; Davis *et al.*, 2006). Key hosts in these determinations, such as *Nicotiana tabacum, Gossypium herbaceum, G. arboretum, G. barbadense,* and *Physalis alkekengi,* do not appear in Table 11.1.

Three non-susceptible plants in Table 11.1, *G. hirsutum, Medicago sativa,* and *Abelmoschus esculentus,* are not susceptible to *F. o. vasinfectum* races 3 or 4 (Armstrong and Armstrong, 1960). Thus, the results in Table 11.1 could be obtained with *F. o. vasinfectum* races 3 or 4.

Instead of *formae specialis*, Lombard *et al.* (2019) called for a classification system based on phylogenetics, to "clear the

Table 11.1.   Plants not susceptible to *F. oxysporum* **f. sp.** *cannabis* **in two host-range studies. Plant families in bold font include plant species used in host-range studies for susceptibility to the eight races of** *F. oxysporum* **f. sp.** *vasinfectum*

Plant family	McCain and Noviello (1985)	Tiourebaev *et al.* (2001)
Malvaceae	*Gossypium hirsutum, Abelmoschus esculentus*	
Solanaceae	*Lycopersicon esculentum, Solanum tuberosum, S. melongena, Capsicum frutescens*	*L. esculentum, S. tuberosum*
Fabaceae	*Medicago sativa, Glycine max, Phaseolus vulgaris, Pisum sativum, Arachis hypogaea, Cicer arietinum, Trifolium repens, Vicia sativa, V. faba, Vigna unguiculata*	*P. vulgaris, Alhagi pseudalhagi, Glycyrrhiza glabra*
Asteraceae	*Helianthus annus, Zinnia elegans, Chrysanthemum morifolium, Callistephus chinensis, Carthamus tinctorius, Lactuca sativa*	*Artemisia vulgaris, Tragopogon major*
Cannabaceae	*Humulus lupulus*	
Poaceae	*Triticum aestivum, Zea mays, Avena sativa, Hordeum vulgare, Oryza sativa, Sorghum bicolor*	*T. aestivum, Z. mays, Agropyron pectiniforme, Bromus inermis, Dactylis glomerata, Stipa dasyphylla*
Amaranthaceae	*Spinacia oleracea*	*Chenopodium* sp., *Atriplex alba, Ceratocarpus arenarius, Kochia prostrata*
Cucurbitaceae	*Cucumis melo, C. sativus, C, pepo*	
Brassicaceae	*Raphanus sativus, Brassica vulgaris, B. junicea, B. oleracea, B. rapa*	
Amaryllidaceae	*Allium cepa, A. sativum*	
Apiaceae	*Daucus carota, Anethum graveolens, Apium graveolens*	
Other families	*Linum usitatissimum, Morus alba, Papaver somniferum, Ipomoea batatas, Asparagus officinalis*	

taxonomic chaos." A phylogenetic study by Skovgaard *et al.* (2001) analyzed all eight races of *F. o. vasinfectum*—28 strains, all isolated from *Gossypium*. They used a multi-locus approach (*EF-1α*, NIR, PHO, mtSSU rDNA), but their phylogenetic signal came primarily from *EF-1α* and mtSSU rDNA. Their phylogram resolved four clades, which generally correlated with *F. o. vasinfectum* race and geographic origin.

In 2020 we compared Skovgaard's *EF-1α* sequences with *EF-1α* sequences for *F. o. cannabis* deposited in GenBank, as well as *EF-1α* sequences from other *formae speciales* available in GenBank. We utilized the platform at MABL, with the same software pipeline described for the phylogram in Fig. 1.12. The resulting phylogenetic tree (Fig. 11.10) showed that *F. o. cannabis* is polyphyletic. The original isolate from Italy (canna_Italy) is not genetically related to Canadian isolates (canna_ON, canna_ BC). Importantly, all *F. o. cannabis* sequences nested within the *F. o. vasinfectum* clade. The canna_Italy sequence grouped with races 4 and 7 (vasR4, vasR7), and the Canadian sequences were sister to a clade of races 3 and 5. This is consistent with our hypothesis that some isolates of *F. o. cannabis* might represent races 3 or 4 of *F. o. vasinfectum*.

O'Donnell *et al.* (2009) sequenced *EF-1α* and IGS from 850 isolates of *F. oxysporum*, including *F. o. cannabis* from Italy. The *EF-1α* and IGS sequences produced two different phylogenetic trees, and O'Donnell identified IGS as the source of topological incongruence. In the *EF-1α* tree, *F. o. cannabis* formed a polytomy with a *F. o. vasinfectum* sequence. Punja (2021b) sequenced *EF-1α* from 33 isolates of *F. o. cannabis* collected in

British Columbia and Ontario. His phylogenetic analysis showed that isolates from BC and Ontario belonged to separate clades (as seen in Fig. 11.10).

Jerushalmi *et al.* (2022) sequenced *EF-1α* from 11 isolates in Israel. Phylogenetic analysis revealed that the Israeli isolates were polyphyletic with each other, and with isolates from California, British Columbia, and Italy. Some of their isolates formed clades with *F. o. vasinfectum*, others did not. The "backbone tree" constructed by Jerushalmi differed from ours: fewer *F. o. vasinfectum* isolates (eight instead of our 13, all from *Gossypium*, races not identified), and more *formae speciales* from other hosts (33 instead of our 13).

Pellegrini *et al.* (2021) constructed a phylogenetic tree using ITS sequences. Their *F. oxysporum* isolate from *Cannabis* was sister to an isolate of *F. o. vasinfectum*. They conducted pathogenicity studies with *Medicago sativa* and *Capsicum annuum*, and no disease was seen, so they classified their *Cannabis* isolate as *F. o. cannabis*. However, *M. sativa* is not susceptible to *F. o. vasinfectum* races 3 and 4 (Armstrong and Armstrong, 1960), and the *Capsicum* pathogen is considered *F. o. capsici* or *F. o. radicis-capsici*, and not *F. o. vasinfectum* (Lomas-Cano *et al.*, 2016; Edel-Hermann and Lecompte, 2019).

Tang *et al.* (2022) reported the first case of Fusarium wilt caused by *F. oxysporum* in China (Hēilóngjiāng Province). They identified it by morphology, confirmed by sequencing ITS, *EF-1α*, and *RPG2*. No *formae speciales* were identified, but their *EF-1α* sequence was 100% identical to *EF-1α* sequences of *F. oxysporum* f. sp. *sesami* from China.

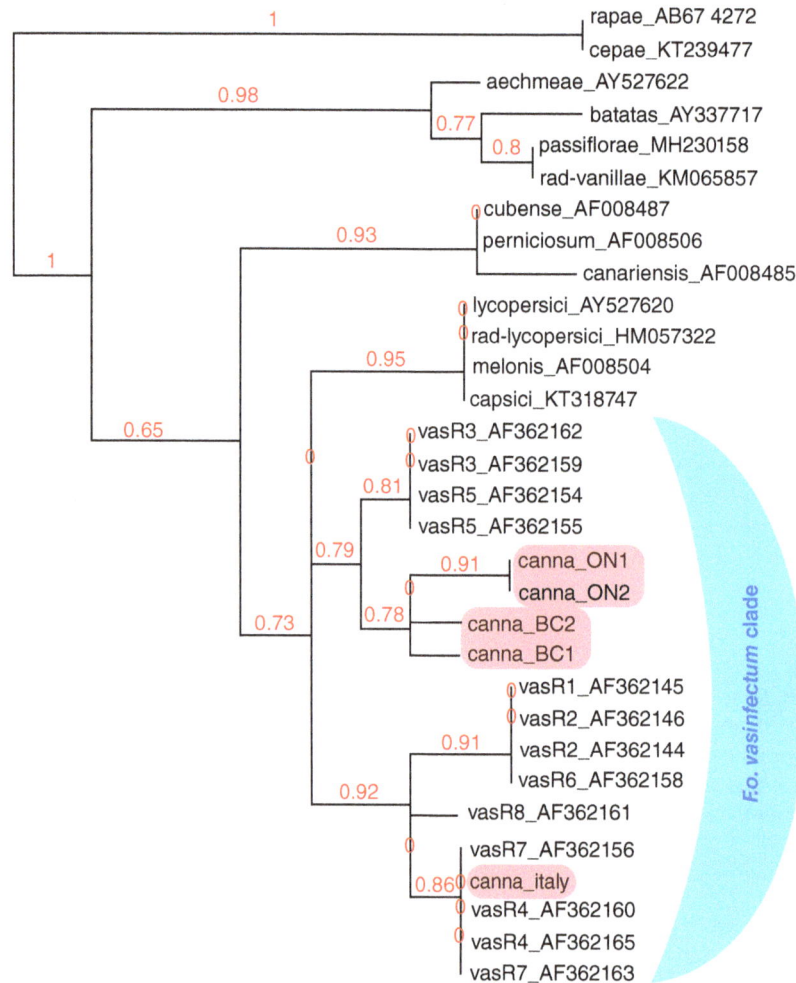

Fig. 11.10. Phylogenetic tree of five *F. o. cannabis* isolates with other *formae speciales*

11.4.4   Disease cycle

F. oxysporum can infect hemp seeds (Pietkiewicz, 1958; Zelenay, 1960; Noble and Richardson, 1968; Ghani *et al.*, 1977; Plăcintă and Murariu, 2016). Seed-borne infections lie dormant until seedlings sprout the following spring. This may give rise to damping off in seedlings (Barloy and Pelhate, 1962), or wilt in older plants.

F. oxysporum overwinters in soil or crop debris as chlamydospores. Chlamydospores produce hyphae in early spring, which infect seedlings by directly penetrating roots behind the root tip. In older "root-hardened" plants, hyphae must enter via wounds. Thus disease is more severe in fields harboring root-wounding nematodes or broomrape. In Hungary, Clarke (pers. obs., 1993) noted that a wet–dry–wet summer predisposed plants to a *Fusarium* epidemic. Lack of rain caused heavy clay soil to cake and crack, which wounded plant roots, allowing *Fusarium* to invade.

After hyphae penetrate roots, the fungus invades water-conducting xylem tissue. Microconidia arise in these vessels and flow upstream to establish a systemic infection. Hyphae and/or the host response occlude xylem vessels, interrupting water flow, and plants wilt. As plants die, chlamydospores are formed. Conidia arising on dead plants may be rain-splashed onto neighbors. The optimum temperature for fungal growth is 26°C (Noviello and Snyder, 1962). Disease symptoms do not become evident until July. In autumn, as temperatures cool, disease tapers off (Tiourebaev *et al.*, 2001).

At high inoculum levels (7000 propagules/g soil), 100% of plants died within 9 days of exposure (McCain and Noviello, 1985). At the lowest inoculum level (7 propagules/g soil), 50% of plants died after 47 days of exposure. Tiourebaev *et al.* (2001) reported that wilting began 2 weeks after exposure to a highly pathogenic strain.

In an indoor facility, Punja (2021b) demonstrated the spread of *F. o. cannabis* from stock (mother) plants to their clones. Mother plants with wilt symptoms yielded *F. o. cannabis* at a frequency of 70–100%. Cuttings obtained from non-symptomatic mother plants still showed pathogen recovery rates of 1–2% on potato dextrose agar.

Punja (2021b) proposed the aerial spread of inoculum, as well as through recycled hydroponic water. He conducted air sampling in an indoor facility, and recovered *F. o. cannabis* in areas used for rooting, vegetative propagation, and where mother plants were grown. Inoculum was present in coco substrate used to grow plants, and in hydroponic drainage water.

McGehee and Raudales (2021) isolated *F. oxysporum* from diseased plants growing in coconut coir and rockwool. They completed Koch's postulates, and calculated the 'area under disease' progress curve. In their hands, *F. oxysporum* was less pathogenic than *Pythium myriotylum*. Perhaps the *F. oxysporum* they isolated was not a primary pathogen—in their phylogenetic analysis, it was distant from *F. oxysporum* f. sp. *cannabis* isolated from Italy. Punja *et al.* (2019) isolated *F. oxysporum* from *Cannabis* that was growing as an endophyte in plants without symptoms.

11.4.5 Cultural and mechanical controls

See the preceding section on Fusarium crown and root rot. For hydroponic systems, exclusion is key. Sanitation is essential—Punja (2021b) isolated *F. oxysporum* in recirculating hydroponic solution, as well as in spaces dedicated to plant rooting, propagation, and production. Seed-borne infection is possible, so do not save seeds from infected plants.

Noble and Coventry (2005) suppressed *F. oxysporum* by adding compost to soil, 20% v/v. Adding compost and green manure to soil encourages the growth of natural *Fusarium* antagonists (Windels, 1997). Sterilization of compost resulted in a loss in disease suppression, indicating that the mechanism was predominantly biological.

In fields, avoid heavy, wet soils with a history of *Fusarium* disease. Flooding is injurious to hemp roots, and lack of moisture causes clay soil to crack, which damages plant roots (Noviello and Snyder, 1962). Zhalnina (1969) reported that "acid fertilizers" such as ammonium sulphate increased Fusarium wilt in hemp. She corrected this by adding lime. Control synergistic organisms such as nematodes and broomrape.

Crop rotation is possible. Czyżewska and Zarzycka (1961) reported a disease incidence of 15–20% in hemp that followed hemp, and only 6–10% in hemp that followed wheat. Two or more years may be required; McCain and Noviello (1985) could still detect *F. o. cannabis* in soil 16 months after inoculation. Punja (2021b) compared 14 drug-type *Cannabis* strains in pathogenicity studies, which did not reveal differences in susceptibility.

11.4.6 Biological and chemical controls

Trichoderma spp. have been used to prevent Fusarium wilt. Czyżewska and Zarzycka (1961) protected hemp plants by inoculating soil with *Trichoderma lignorum*. Wang (2018) demonstrated some efficacy with a strain of *Trichoderma virens* isolated from *Cannabis* roots.

A study tested the efficacy of five products against damping off by *F. oxysporum* in *Cannabis* (Scott, 2021; Scott and Punja, 2023). They tested cuttings rooted in rockwool blocks, in a damping-off paradigm (see section 12.3). Punja (2021b) recommended *T. harzianum* and *Gliocladium catenulatum*, applied early in the propagation stage and at transplantation of rooted cuttings. Serkov and Pluzhnikov (2012) used a seed treatment, *Bacillus megaterium* formulated with polyhydroxybutyrate (Albit®), which increased field germination of seeds by 18%. In other crops, *F. oxysporum* has been managed with most of the biocontrols listed in the previous *Pythium* section.

Pellegrini *et al.* (2021) controlled *F. oxysporum* f. sp. *cannabis* with PGPRs (plant growth-promoting rhizobacteria). Three species showed *in vitro* antagonistic activity (*Gluconacetobacter diazotrophicus*, *Herbaspirillum seropedicae*, *Burkholderia ambifaria*). The authors provided SEM photomicrographs of bacterial damage to *Fusarium* hyphae. They combined the three species for glasshouse trials. In a pre-emergence experiment, hemp seeds were soaked in a bacterial solution; in a post-emergence experiment, the bacterial solution was poured into soil at the base of seedlings. In both experiments, PGPRs protected plants inoculated with the pathogen. Even in the absence of the pathogen, PGPR-treated plants grew better than control plants.

Fusarium oxysporum Fo47 is a non-pathogenic strain useful for preventing disease by pathogenic *F. oxysporum*. Its discovery traces back to studies on "suppressive soils"—blessed farmland whose microflora naturally controls *Fusarium* pathogens (Smith and Snyder, 1971). Rouxel *et al.* (1979) isolated Fo47 from suppressive soil in Châteaurenard. When they heat-treated soil (killed Fo47), melons planted in the soil became diseased. When they reintroduced Fo47, melons stayed healthy. Fo47 colonizes plant roots but lives as a saprophyte. It suppresses pathogens by competing for nutrients in the soil, competing for infection sites on the root, reducing pathogen spore germination, parasitizing pathogen hyphae, and by inducing systemic resistance in the plant (Benhamou *et al.*, 2002). Kaur *et al.* (2010) collated a list of 26 studies where Fo47 controlled disease in over a score of crop plants. It is supplied as a wettable powder or granule (Biofox C®) or liquid (Fusaclean®). The dry formulations can be stored for months in cool conditions (8–11°C). It is applied as a seed treatment, mixed into potting soil, incorporated into drip irrigation, or as a furrow drench.

Regalia®, a fungicidal extract of giant knotweed (*Reynoutria sachalinensis*), was tested by Lung *et al.* (2019). Clone cuttings were inoculated with *F. oxysporum* and then rooted in rockwool cubes soaked with Regalia® (1% w/v). Treated cuttings showed 30% less wilting than positive controls. Treatment with RootShield® also significantly reduced wilting, but not Stargus®, Rhapsody®, or Prestop®.

Spraying wilted plants with fungicides is not useful. Seed treatment serves as a prophylactic, or can disinfect seeds suspected of harboring seed-borne infections. Two studies demonstrated efficacy with a total of ten fungicides on *Cannabis* seeds, but none are currently registered for the crop (Trotuş and Naie, 2008; Serkov and Pluzhnikov, 2012).

11.5   Southern blight

The cause of southern blight, *Athelia rolfsii* (formerly *Sclerotium rolfsii*, and perhaps *Agroathelia rolfsii*) attacks many plants, from corn to catnip. This "southern" disease arises in warm temperate regions and moist tropics. Disease in *Cannabis* was first reported in India (Hector, 1931; Uppal, 1933; Rao, 1977), where it still causes problems (Krishna, 1995).

Fiber-type plants have succumbed to *A. rolfsii* in South Carolina (Anderson, 1926; Weber, 1931), Tennessee (Hyre, 1944), Texas (Miller *et al.*, 1960), Italy (Ferri, 1961a), Japan (Kyokai, 1965; Kishi, 1988), and Ukraine (Hrebenyuk, 1984). More recently the disease has been reported in southern Italy (Pane *et al.*, 2007), Queensland, Australia (Hall, 2009), North Carolina (Thiessen *et al.*, 2020), Tennessee (Hansen *et al.*, 2020), Virginia (Mersha *et al.*, 2020; Amaradasa *et al.*, 2020), Georgia (Coolong, 2020), Alabama (Conner *et al.*, 2020), and Louisiana (Singh *et al.*, 2022).

11.5.1   Symptoms and signs

Symptoms begin as a dramatic wilting, accompanied by leaf yellowing. A light brown or tan lesion arises at the base of the stalk near the soil line. In warm weather, during the summer, symptoms progress to necrosis and sudden death (Fig. 11.11). Lesions sprout sclerotia, usually after death of the plant. The sclerotia are small and light brown; they resemble seeds of white mustard, *Synapsis alba* (Fig. 11.12). In humid conditions, the stalk lesion becomes enveloped in a cottony white mycelium. Rarely, a pale-brown hyphal mat radiates from the base of the stalk along the soil surface, producing the sexual stage.

11.5.2   Causal organism

11.5a. *Athelia rolfsii* (Sacc.) Tu & Kimbrough 1978

≡*Agroathelia rolfsii* (Sacc.) Redhead & Mullineux 2023, ≡*Corticium rolfsii* (Sacc.) Curzi 1932, anamorph: *Sclerotium rolfsii* Saccardo 1911

Fig. 11.11. Symptoms of southern blight (courtesy of Coolong, 2020)

Description: *Sclerotia* smooth or pitted, near spherical (slightly flattened below), at first amber (Fig. 11.12) then turning brown to black, usually 0.7–1 mm in diameter; cross-section of sclerotia reveals three cell layers—a darkly pigmented outer rind (3–4 cells thick), a middle cortex composed of dense, hyaline cells (4–10 cells thick), and an inner medulla, comprising loosely interwoven cells. *Hyphae* produce clamp connection at some septa, branching at acute angles. *Basidiocarps* resupinate, effused, loosely adherent to the stem surface. *Basidia* short, clavate, usually 4-spored. *Basidiospores* hyaline, smooth, tear-drop-shaped, 5.5–6.5 × 3.5–4.5 µm (see Fig. 3.3).

Comments: *A. rolfsii* (Agaricomycetes, Atheliaceae) is one of the few Basiodiomycetes that cause *Cannabis* disease. *A. rolfsii* is a facultative parasite, capable of living as a saprophyte in soil. Real-time PCR kits utilizing TaqMan and SYBR Green assays are available to detect *A. rolfsii* in soil. Recent studies have used molecular methods of identification (Amaradasa *et al.*, 2020; Thiessen *et al.*, 2020).

11.5.3   Differential diagnosis

Symptoms of southern blight may be confused with symptoms of hemp canker (caused by *Sclerotinia sclerotiorum*), charcoal rot (*Macrophomina phaseolina*), or Rhizoctonia sore shin (*Rhizoctonia solani*). Sclerotia of *S. sclerotiorum* and *R. solani* are larger and less symmetrical than those of *A. rolfsii*, and the sclerotia of *M. phaseolina* are smaller and irregular in shape. No other sclerotia consist of three distinct cell layers. Stem cankers and crown rots caused by *Fusarium* spp. may also be confused with southern blight, but no sclerotia are formed.

11.5.4   Disease cycle and epidemiology

A. rolfsii overwinters as sclerotia in soil and plant debris—either crop residues or weed hosts. Sclerotia can persist many

Fig. 11.12. Sclerotia of *Athelia rolfsii* on hemp (courtesy M. Cartwright)

years in soil. They spread by the movement of soil, infected plant material, and contaminated farm tools. Contamination of seeds is possible if stalks get pulled into threshing machines. Germination of sclerotia and mycelial growth is favored by warm temperatures (27–30°C), acidic soil (pH <6), and a well-aerated and moist environment (Mersha and Hansen, 2022). Sclerotia produce hyphal strands, which directly penetrate roots or enter via wounds.

The fungus grows inter- and intracellularly, primarily within the stem cortex. Optimum growth occurs around 30°C (Mersha *et al.*, 2020); at that temperature plants may die within 7–14 days, and *Cannabis* losses may reach 60% (Krishna, 1995). Basidiospores germinate best at 28°C, but are short-lived, and not very important in disease transmission. Hall (2009) reported greater disease in plants growing in lighter sandy soils. Recent reports come from field crops in plasticulture with drip irrigation (Amaradasa *et al.*, 2020; Coolong, 2020; Mersha *et al.*, 2020), which provides a warm, moist environment for *A. rolfsii* (Fig. 11.11).

11.5.5   Cultural and mechanical controls

Scout for disease in summer. Rogue infected plants and remove all crop residues. Eliminate weed hosts. After harvest, till the soil to bury sclerotia, which kills them—if buried deeper than 8 cm. Cyclical drying and wetting of soil induces sclerotia to germinate. In the absence of a host (i.e., fallow soil), germinated sclerotia die within 2 weeks (Lucas, 1975). Reduced moisture at ground level inhibits disease development, so remove weeds and limit irrigation. Given the daunting list of 189 host plants by Weber (1931), crop rotation has limited efficacy. Anecdotal observations indicate that disease is worse in hemp crops following unmanaged pastures (Mersha and Hansen, 2022).

Mersha *et al.* (2020) noted that temperatures ≥40°C were detrimental to *A. rolfsii*. Ristaino *et al.* (1991) killed sclerotia with soil solarization in North Carolina: they irrigated field soil, then covered it with clear polyethylene (0.025 mm thick, or 1 mil, i.e. 0.001 inches thick) for 6 weeks during the warmest months of summer. Soil temperatures reached 41–49°C at 10 cm soil depth. Krishna (1996) tested three *gañjā* cultivars from Uttar Pradesh (India) for resistance to *A. rolfsii*. Cultivar "Dwarf" was more resistant than cultivar "Tall", and "Medium" was intermediate. These lines might represent Afghani, Himalayan, and peninsular Indian landraces, respectively.

11.5.6   Biological and chemical control

Ristaino *et al.* (1991) combined soil solarization with thermotolerant *Gliocladium virens* to kill *A. rolfsii* sclerotia in soil. Combined treatment was no better than either treatment alone. Charirak *et al.* (2016) combined soil solarization with thermotolerant *Trichoderma harzianum*, and reported better control than either treatment alone.

Shrestha *et al.* (2018) mixed organic amendments with a high carbon-to-nitrogen ratio (10:1 to 40:1, e.g., wheat bran), at a rate of 8 g/kg soil. They called it anaerobic soil disinfestation (ASD), because it encouraged the growth of *Trichoderma* spp.,

Aspergillus spp., and other low-oxygen-tolerant fungi that parasitize *A. rolfsii* sclerotia.

Other biocontrols with proven efficacy against *A. rolfsii* sclerotia include *Albifimbria verrucaria* and *Talaromyces flavus* (see section 20.6). For chemical control, Krishna (1996) mixed urea into soil, at a rate of 100 kg/ha.

11.6   Rhizoctonia sore shin and root rot

Rhizoctonia disease in *Cannabis* was first described on drug plants in India (Shaw and Ajrekar, 1915), where *gañjā* crop losses may exceed 80% (Pandotra and Sastry, 1967). The disease also occurs in Pakistan (Shahzad and Ghaffar, 1995). Fiber-type plants have also been hammered, in Europe (Rayllo, 1927; Brandenburger, 1985; Ondřej, 1991), East Asia (Matuo, 1949; Tai, 1979), and the USA (Thiessen *et al.*, 2020; McPartland and Cubeta, 2022). In contrast, Trunov (1936) considered the disease of less consequence in Ukraine. Two causal agents have been identified—rare Basiodiomycetes that cause *Cannabis* disease, like *Athelia rolfsii*, but in a different family (Agaricomycetes, Atheliaceae).

11.6.1   Symptoms and signs

Leaf chlorosis is followed by wilting. Then a dark brown discoloration moves up the stalk from the roots below (Fig. 11.13). Several centimeters of stalk rot away, leaving a shredded "sore shin" appearance. Cortical tissues under the lesion may completely disintegrate (Pandotra and Sastry, 1967). Within 6–8 weeks young plants (<3 months old) topple over and die. Older plants often survive, but exhibit wilting and reduced growth (McPartland and Cubeta, 2022). Small black microsclerotia sometimes become visible in the shredded area. Occasionally a pale mat of basidiospores forms around the base of the stalk (Trunov, 1936).

Fig. 11.13. Symptoms of Rhizoctonia sore shin (courtesy Anna-Liisa Fabritius)

11.6.2   Causal organisms

11.6a.   *Rhizoctonia solani* Kühn 1858

=*Rhizoctonia napaeae* Westendorp & Wallays 1846.
Sexual stage: *Thanatephorus cucumeris* (Frank) Donk 1956, ≡*Hypochnus cucumeris* Frank 1883, =*Hypochnus solani* Prillieux & Delacroix 1891, =*Pellicularia filamentosa* (Patouillard) D.P. Rogers 1943, =*Corticium vagum* Berkeley & Curtis 1873
Description: *Hyphae* 5–12 µm wide, multinucleate, with dolipore septa, no clamp connections, at first colorless but rapidly becoming brown; branches form geometrically at 45° or 90° angles from parent hyphae. Hyphae become slightly constricted at branching points; a septum usually forms near the base of the branch (Fig. 11.14A). *Microsclerotia* are deep brown to black, smooth, somewhat flattened and irregular in shape, no differentiation between rind and medulla, 1–6 mm in diameter. *Basidiocarps* arise on stem surface, thin, effuse, discontinuous. *Basidia* are barrel-shaped, 10–25 × 6–19 µm, borne on imperfectly symmetrical racemes, usually with 4 sterigmata per basidium. Sterigmata are 6–36 µm long and bear basidiospores. *Basidiospores* hyaline, ellipsoid to oblong, flattened on one side, 5–14 × 4–8 µm.
Comments: *R. solani* has a wide host range, ~250 plant spp., and is global in distribution. Matuo (1949) demonstrated that *R. solani* isolated from diseased cotton seedlings also caused disease in *Cannabis*. Technically, *R. solani* is not a single species but a genetically diverse species complex. Ogoshi (1996) divided *R. solani* into 13 "anastomosis groups" (AGs), based on vegetative hyphal fusion and somatic recognition interactions. Each AG has a unique host range and disease cycle. The AG of *R. solani* responsible for *Cannabis* disease has not been determined.

Shaw and Ajrekar (1915) cited *Rhizoctonia napi* on *Cannabis* in India, an orthographic variant of *Rhizoctonia napaeae*, a synonym of *Rhizoctonia solani*. Matuo (1949) cited "*Corticium vagnum*" in Japan, i.e., *Corticium vagum*, another synonym. Stevenson (1926) placed *Rhizoctonia napi* on a list of *Cannabis* "foreign plant diseases" posing a threat to US agriculture if accidently imported.

Thiessen (2019) identified *R. solani* causing disease in roots and crowns of plants in North Carolina. Morphological identity was confirmed with ITS and 18S rRNA sequences. She was first to complete Koch's postulates of *R. solani* on *Cannabis*. *Cannabis* websites have sighted *R. solani* in California, Nevada, and Oregon. *R. solani* also causes damping off in *Cannabis* seedlings (Dippenaar *et al.*, 1996; Thiessen, 2019; Gauthier, 2020b). Dippenaar stated that *R. solani* caused both pre- and post-emergent damping off, and reduced the hemp stand to only a few plants.

11.6b.   Binucleate *Rhizoctonia* sp.

= *Rhizoctonia fragariae* Husain & McKeen 1963
= *Ceratobasidium* sp.
Description: This species is morphologically similar to *R. solani,* but has thinner hyphae (4–7 µm) and possesses only two nuclei ("binucleate") per hyphal cell (Fig. 11.14B).
Binucleate *Rhizoctonia* spp., like *R. solani*, have been divided into genetically separated "anastomosis groups" (AGs); but unlike *R. solani*, binucleate *Rhizoctonia* spp. often produce *Ceratobasidium* sexual stages (Ogoshi *et al.*, 1983).
Comments: McPartland and Cubeta (1997) isolated a binucleate *Rhizoctonia* sp. from the roots of "Skunk no. 1" in Amsterdam (Fig. 11.15). The binucleate condition was determined by staining with safranin-o and DAPI. The isolate belonged to Ogoshi's AG-G group, determined by PCR amplification using primers for the 25S rDNA region, and subjecting the PCR amplicon to digestion with *Hha* I, *Hpa* II, *Sau*3A I, and *Taq* I restriction endonucleases. According to Chen *et al.* (2020), this was the first study to use a PCR-based method to identify a *Cannabis* pathogen. Members of AG-G have a wide host range, including sugar beet, bean, pea, tomato, potato, melon, sunflower, peanut, strawberry, apple, and

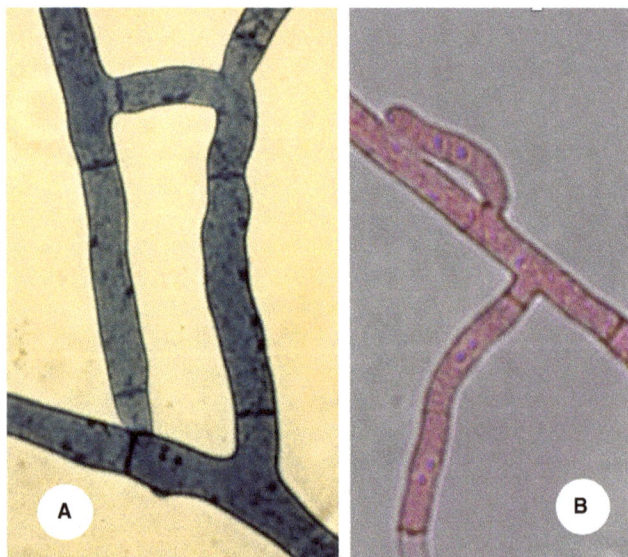

Fig. 11.14. Characteristic hyphae: **A**. *Rhizoctonia solani* (cotton blue stain); **B**. Binucleate *Rhizoctonia* sp. (safranin-o stain) (courtesy M. Cubeta)

Fig. 11.15. Root rot caused by a binucleate *Rhizoctonia* sp.

Rhododendron sp. Although several AGs have been assigned to specific *Ceratobasidium* sexual stages, AG-G has not (Gonzales *et al.*, 2016).

11.6.3 Differential diagnosis

Symptoms caused by *Rhizoctonia* spp. may resemble symptoms of hemp canker (caused by *Sclerotinia sclerotiorum*) and southern blight (*Athelia rolfsii*). These fungi, however, produce more prominent external hyphae and larger sclerotia than *R. solani* microsclerotia. The unique hyphae of *R. solani*, often branching at right angles, are easily differentiated from other fungi under a microscope.

11.6.4 Disease cycle and epidemiology

R. solani typically overwinters as microsclerotia, but mycelia can survive in plant debris. Microsclerotia germinate in early spring and produce hyphae. Hyphae enter roots near the soil line, either by direct penetration or through wounds. After penetration, *R. solani* produces cellulose-degrading enzymes (which disrupt xylem and cause wilting), and pectolytic enzymes (which cause cortical rot).

Disease symptoms increase under cool, damp conditions, or paradoxically when soil temperatures are elevated. Lucas (1975) simply explained that sore shin worsens at temperatures not optimal for the host. A strain of *R. solani* in India produced neither microsclerotia nor basidia—just hyphae (Pandotra and Sastry, 1967). A Ukrainian strain formed basidiospores (Trunov, 1936). Crosier (1968) stated that seed-borne transmission of *R. solani* occurred in *Cannabis,* and he cited Neergaard (1958), but Neergaard didn't include *Cannabis* in his study.

11.6.5 Cultural and mechanical controls

Do not plant in fields previously infested by *R. solani*; microsclerotia remain viable in soil for up to 6 years. Relatively high soil moisture and warm conditions that promote fungal growth rather than plant growth will favor disease development (McPartland and Cubeta, 2022). Many weeds serve as alternative hosts, so eliminate them. *R. solani* viability decreases in soils amended with high-carbon mulches, such as straw, corn stover, and even pine shavings (Lucas, 1975). Root-knot nematodes (*Meloidogyne* spp.) act synergistically with *R. solani* and must be controlled. Greater disease incidence occurred at a sowing rate of 250 seeds/m² compared with 150 seeds/m² (Dippenaar *et al.*, 1996). Breeding for resistance is a goal. Dippenaar *et al.* (1996) tested five hemp cultivars ('Fedora 19', 'Futura 77', 'Felina 34', 'Kompolti', and 'Secuini'), and all were equally susceptible to *R. solani*.

11.6.6 Biological and chemical controls

Several commercial biocontrol agents have shown efficacy against *R. solani* in other crop systems. *Bacillus subtilis, Gliocladium*

virens, and *Trichoderma harzianum* are permitted for use by *Cannabis* growers (California DPR, 2017). Endophytic bacteria cultured from above-ground parts of *C. sativa* inhibited *in vitro* growth of *R. solani* (Scott *et al.*, 2018). They identified the top performing species as *Pseudomonas fulva* and *P. orientalis.* Binucleate *Rhizoctonia* sp. was also significantly inhibited by *P. fulva* and *P. orientalis.*

Soil-borne pathogens such as *R. solani* are difficult to control with pesticides. Dippenaar *et al.* (1996) reported heavy losses despite spraying seedlings with Rhizolex™, a nasty mixture of thiram and tolclofos. Interestingly, a dozen studies have demonstrated that aqueous or ethanol extracts of *Cannabis* inhibit *in vitro* growth of *R. solani*, more so than most other plant extracts. This was confirmed in a field trial, where soil drenches of aqueous extracts (10% w/v) resulted in 82.1% reduction of *R. solani* in black gram, *Vigna mungo* (Kumar *et al.*, 2013).

11.7 Twig blight

Twig blight fungi invade at stalk nodes, whereas wilt fungi such as *Fusarium oxysporum* go through roots. Twig blight fungi penetrate vascular vessels (phloem, xylem), as well as tissues around vessels. They produce pectin- and hemicellulose-degrading enzymes—so they begin retting hemp stalks while plants are still alive.

Twig blight fungi are members of the Botryosphaeriaceae family, which consists of around 25 genera. Collectively they can be distinguished by their bitunicate or double-layered asci, with a visible apical chamber between the two layers (Fig. 11.16).

The ascocarp produced by Botryosphaeriaceae spp. is a **pseudothecium**—produced in a cavity within a **stroma** of fungal mycelium—but the stroma may be thin to nearly invisible, so the ascocarp resembles a **perithecium**, lacking a stroma. The conidia of Botryosphaeriaceae spp. vary in shape, septation, pigmentation, and development. A single species

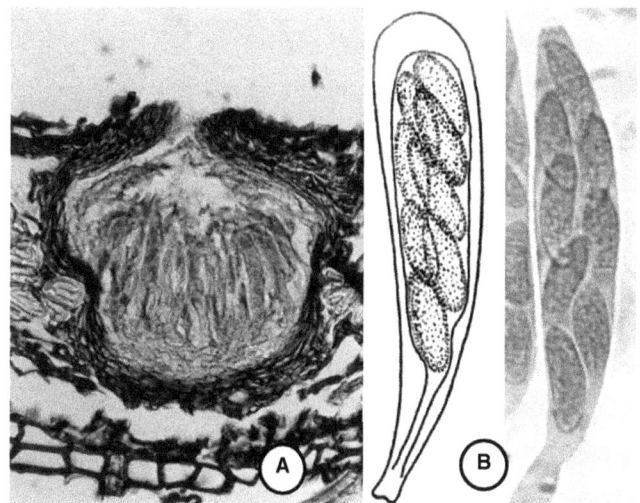

Fig. 11.16. *Neofusicoccum marconii*: **A**. vertical section through unilocular ascocarp; **B**. asci with ascospores (line drawing by Charles and Jenkins, 1914; photos by McPartland)

may produce two or more anamorph states, known as **synan-amorphs** (Hughes, 1979), often differentiated as "macroconidia" and "microconidia."

Taxonomy of Botryosphaeriaceae and the flagship genus *Botryosphaeria* was upended by a phylogenetic study based upon a single locus (LSU). Crous *et al.* (2006) sequenced 14 *Botryosphaeria* spp. among 115 Botryosphaeriaceae spp., and the 14 species scattered across the phylogenetic tree. They restricted *Botryosphaeria* to one clade with two species, *B. dothidea* and *B. corticis*. The other *Botryosphaeria* spp. were transferred to other genera, allied by their anamorphs (e.g., *Fusicoccum, Dothiorella*) or moved to a new genus, *Neofusicoccum*. The strict application of phylogenetics to delineate *Neofusicoccum* has been critiqued (Sakalidis *et al.*, 2011; Pem *et al.*, 2021). Crous continued shuffling names (Phillips *et al.*, 2013), based on phylogenetics and an inconsistent treatment of anamorph morphology (e.g., *Diplodia* and *Dothiorella*, discussed below).

Around 195 *Botryosphaeria* spp. have been named over the years. Crous and colleagues suggested "disregarding" all *Botryosphaeria* spp. that lacked DNA data (Phillips *et al.*, 2013). They altered genus concepts based on epitypes (specimens in culture) rather than holotypes (herbarium specimens). "Disregarding" species that lack DNA data applies to two species here.

11.7.1 Symptoms and signs

Twig blight begins as a wilting and drooping of leaves, usually later in the season. Wilted foliage browns and dies, but remains attached to plants (see the classic photograph in Charles and Jenkins, 1914, available online). Tips of young branches show symptoms first ("twig blight"). Within 2 weeks the entire plant may wilt and die. Stalks develop gray to beige lesions, elongated on axis with the stalk. Lesions darken to gray-brown, and bear black-dot pycnidia and/or pseudothecia. Lesions may continue to enlarge after plants are harvested (Gitman and Malikova, 1933).

11.7.2 Causal organisms

11.7a. *Dendrophoma marconii* Cavara 1888

Description: *Pycnidia* arise in gray stem spots, concealed by epidermis, flattened globose, ostiole slightly raised, 130–150 μm diameter (Fig. 11.17A). *Conidiophores* unbranched or widely dichotomous branching, septate, hyaline (Fig. 11.17B). *Conidia* pleomorphic, ovate-elliptical, teretiuscullis [tapering?], 1-celled, hyaline, 4.5–6.5 × 2–2.5 μm.

Comments: The description is by Cavara (1888). McPartland (1995b) examined Cavara's holotype (herb. Pavia), as well as exsiccati specimens (Briosi and Cavara's *I Funghi Parassiti delle Piante Coltivate od Utili*, no. 20; Roumeguère's *Fungi Selecti Exsiccati* no. 5064). Specimens of *D. marconii* were also collected from fiber-type plants in Austria and Michigan, and drug-type plants in Nepal. Conidiomata are pycnidial, stromatic, unilocular, scattered. Conidiophores are frequently

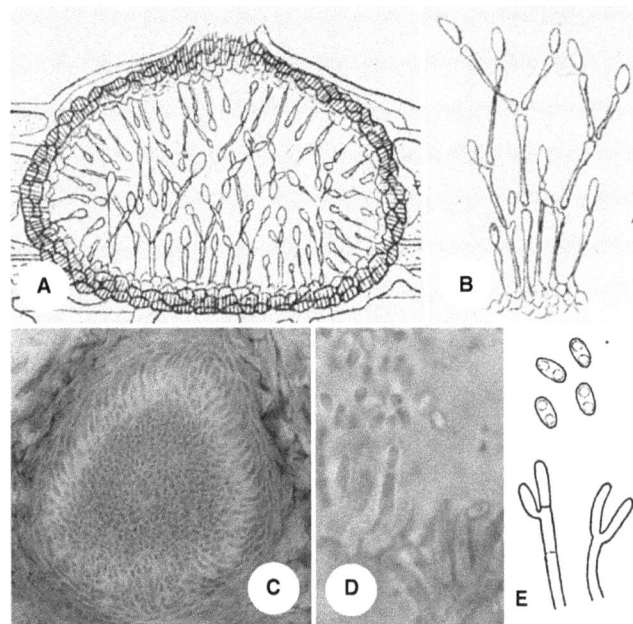

Fig. 11.17. *Dendrophoma marconii* (**A**, **B**, Cavara 1888) compared with *Leptodothiorella* morph of *Neofusicoccum marconii* (**C**, **D**, McPartland; **E**, Charles and Jenkins 1914)

multiseptate, widely branched, exhibit irregular monilioid swellings, up to 25 μm long. Conidiogenesis is probably holoblastic, but not visualized with certainty. Conidia arise at restrictions along the conidiophores. Conidia are pleomorphic and irregularly shaped, often resembling short sections of conidiophores.

11.7b. *Neofusicoccum marconii* (McPartl.) McPartl., comb. nov.

Basionym: *Fusicoccum marconii* McPartl. 1995, *Mycotaxon* 53: 421 ≡ *Leptodothiorella marconii* McPart. 1995 ≡ *Botryosphaeria marconii* (Cavara) Charles & Jenkins 1914, misapplied name

Holotype: BPI 600713

Description: *Ascomata* pseudothecioid or perithecioid in appearance, usually unilocular, globose, immersed then erumpent, 130–160 μm in diameter. Outer wall pale brown but darkening near the ostiole; inner wall hyaline and very thin. Interascal tissues (pseudoparaphyses) thin-walled, filiform, deliquescing. *Asci* bitunicate, 8-spored, clavate, with a short stalk, 80–90 × 13–15 μm. *Ascospores* fusoid to ellipsoid, aseptate, hyaline to pale green, 16–18 × 7–8 μm (Fig. 11.16). *Conidiomata of macroconidia* (*Fusicoccum*) pycnidial, same size and shape as ascomata. *Conidiophores* arise from inner wall, hyaline, smooth, simple or branched and septate near the base, 10–16 μm long, up to 4.0 μm wide at the base tapering to 2.5–3.0 μm at the tip. *Conidiogenous cells* at first holoblastic, later enteroblastic and phialidic. *Macroconidia* fusiform, single-celled, hyaline to glaucous, smooth-walled, base often truncate, 16–22 × 5–8 μm (Fig. 11.18). *Conidiomata of microconidia* (*Leptodothiorella*) pycnidial and smaller than conidiomata of macroconidia. *Conidiogenous cells* simple or rarely branched, lageniform to

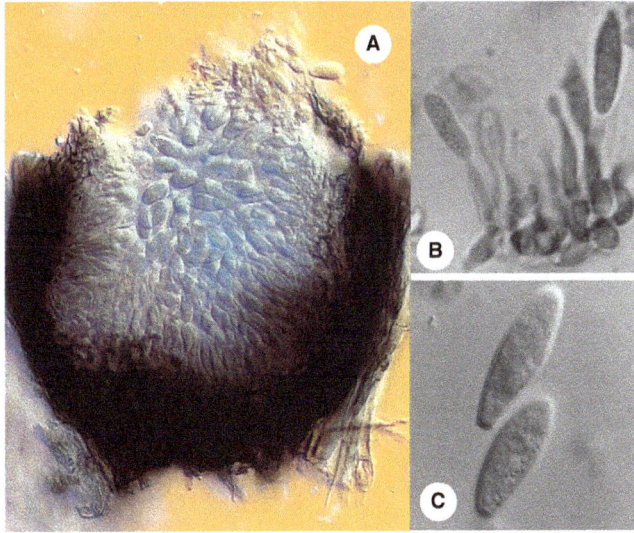

Fig. 11.18. *Fusicoccum* morph of *Neofusicoccum marconii*:
A. pycnidium stained with cotton blue; **B**. conidia and
conidiophores; **C**. conidia

cylindrical, 3–12 μm long, up to 3 μm wide at base tapering to 1 μm at apex, phialidic, integrated or discrete, with minute channel. *Microconidia* ellipsoid, single-celled, hyaline, often biguttulate and swollen towards each end, 3.0–4.0 × 0.5–2.5 μm (Fig. 11.17D,E).

Comments: In a notebook entry dated October 11, 1913, Lyster Dewey wrote about his research plot in Arlington Farms, Virginia. "A disease hitherto unknown in this country has nearly ruined our seed hemp" (www.lysterdewey.com). Dewey (1914) described symptoms; Charles and Jenkins (1914) reported *in vivo* and *in vitro* fungal morphology. Two anamorphs arose on living plants—first a microconidial one, "of the genus *Dendrophoma*," followed by a macroconidial one, "of the genus *Macrophoma*." The sexual morph formed on dead plants. The morphs arose in the same order *in vitro*: microconidia, macroconidia, then the sexual stage.

They named the fungus *Botryosphaeria marconii* (Cavara) Charles & Jenkins, believing it to be the sexual morph of Cavara's *Dendrophoma marconii*. Cavara's type specimen, however, differs from Charles and Jenkins's type specimen. *D. marconii* conidiophores are long, widely branched, and ramify throughout the pycnidial cavity (Fig. 11.17A,B), whereas the microconidial morph of Charles and Jenkins's fungus produces short conidiophores, usually unbranched, that line the pycnidial margin, filling the cavity with conidia (Fig. 11.17C,D,E). Condidiogenesis differs, as does conidial morphology.

Prior to the principle of "one fungus, one name" (McNeill, 2012), synanamorphs were often assigned separate taxonomic names. McPartland (1995b) assigned *Fusicoccum marconii* and *Leptodothiorella marconii* to macroconidial and microconidial morphs, respectively, in Charles and Jenkins's specimen. With the change to "one fungus, one name," the oldest epithet must be placed in the accepted genus (Hawksworth, 2012; May *et al.*, 2019). Since the epithet *Botryosphaeria marconii* was a misapplied name that Charles and Jenkins based on a different species

(Cavara's *D. marconii*), the oldest epithet becomes *Fusicoccum marconii*. Now the accepted genus is *Neofusicoccum*, and the necessary recombination is made here.

Crous *et al.* (2006) defined *Neofusicoccum* as a *Botryosphaeria*-like genus that forms both *Fusicoccum* and *Dichomera* synanamorphs. That definition does not hold: Phillips *et al.* (2013) recognized 22 *Neofusicoccum* spp., and 18 did not produce a *Dichomera* morph. Their taxonomic key of *Neofusicoccum* spp. relied on *Fusicoccum* morphology and host substrate. Their descriptions of *Neofusicoccum protearum*—its sexual morph, macroconidia, and microconidia—resemble large versions of the three morphs of *N. marconii* (see illustrations in Denman *et al.*, 2003).

Given the potential of mistaking *D. marconii* and *N. marconii*, all citations in the literature are suspect. For example, Anonymous (1943) cited *D. marconii* in Chile without a description. Barloy and Pelhate (1962) cited *D. marconii*, accompanied by an illustration, which clearly depicted long, dichotomous branching conidiophores—consistent with *D. marconii*. An illustration by Spaar *et al.* (1990) depicted short, unbranched conidiogenous cells—consistent with *N. marconii*. Several authors described both microconidia and macroconidia, consistent with *N. marconii* (Gitman and Boytchenko, 1934; Flachs, 1936; Petri, 1942; Mándy and Bòcsa, 1962).

Assuming we can trust reports of *D. marconii* in Italy, where it was first described, *D. marconii* seldom caused appreciative losses (Cavara, 1888; Ferraris, 1935). Barloy and Pelhate (1962) said the same in France. Assuming we can trust citations of *N. marconii* in Virginia, the pathogen caused serious losses, affecting between 66% and 95% of plants (Charles and Jenkins, 1914). In warm, moist weather, *N. marconii* killed a 10-ft-tall plant in 5 days (Dewey, 1914). Herbarium specimens at BPI show that *N. marconii* was collected seven times at Arlington Farms (or across the river in Potomac Flats) between 1902 and 1929.

11.7c. *Neofusicoccum parvum* (Pennycook & Samuels) Crous *et al.* 2006

≡*Botryosphaeria parva* Pennycook & Samuels 1985
≡*Fusicoccum parvum* Pennycook & Samuels 1985

Description: *Ascomata* pseudothecioid (rarely perithecioid in appearance), usually multilocular in botryose clusters, globose, immersed then erumpent, 150–250 μm in diameter. *Asci* bitunicate, 8-spored, clavate, with a short stalk, 75–150 × 17–21 μm. *Ascospores* broadly ellipsoidal to fusoid, hyaline, aseptate, occasionally becoming 1-septate, 18–23 × 8–10 μm. *Conidiomata* pycnidial, globose, unilocular, immersed then erumpent, 200–300 μm in diameter. *Conidiogenous cells* holoblastic, hyaline, cylindrical, proliferating percurrently to form 1–2 annellations, or proliferating at the same level to form phialidic periclinal thickenings. *Conidia of Fusicoccum morph* ellipsoidal, with apex round and base flat, unicellular, hyaline; old conidia becoming 1–2-septate and light brown with the middle cell darker than the terminal cells, 16.9–17.3 × 5.4–5.6 μm. *Conidia of Dichomera morph* (when present) subglobose to obpyriform, brown, apex obtuse, base truncate, 8.0–10.5 × 7–8(–9) μm, with muriform septation: 1–3 transverse septa, 1–2 longitudinal septa, and 1–2 oblique septa, almost *Alternaria*-like.

Comments: *N. parvum* has a wide host range with a global distribution. It causes stem canker and dieback of *Cannabis* in Italy (Alberti *et al.*, 2018) (Fig. 11.19) and Arkansas (Feng *et al.*, 2020). Both research groups extracted DNA, sequenced ITS and β-tubulin, and the sequences matched *N. parvum* sequences in GenBank. Both groups completed Koch's postulates. Roberts and Punja (2022) isolated *N. parvum* in British Columbia. Stems showed surface blackening and internal pith discoloration. They demonstrated pathogenicity with Koch's postulates. *N. parvum* was more pathogenic than *Lasiodiplodia theobromae*, *Stemphylium vesicarium*, and *Alternaria alternata* in inoculated seedlings. But in inoculated stem and flowers, *L. theobromae* edged out *N. parvum*. Six drug-type strains showed differences in susceptibility, ranging from "White Rhino" (84% diseased) to "Pink Kush" (32% diseased).

N. parvum differs from *N. marconii*; it is densely multilocular with a thick stroma. *N. marconii* is usually unilocular, with no stroma formed on the host, although a stroma is produced in culture (Charles and Jenkins, 1914). *N. parvum* ascomata, asci, and ascospores are larger than those of *N. marconii*. The *Fusicoccum* conidia of *N. parvum* and *N. marconii* are similar in shape, although those of *N. parvum* turn brown and develop septa—not seen in *N. marconii*. Their other synanamorphs radically depart from each other: *Leptodothiorella*-like versus *Dichomera*-like (if present).

11.7d. *Dothiorella cannabis* (Schwein.) McPartl., comb. nov.

Basionym: *Sphaeria cannabis* Schweinitz 1832, *Transactions of the American Philosophical Society* 4 (2): 222
Holotype: herb. PH, no. 00076340 (image at https://plants.jstor.org, underSphaeria cannabis)
Description: *Conidiomata* pycnidial, immersed then erumpent, stromatic, unilocular, globose, averaging 235 µm diameter. *Conidiogenous cells* hyaline, simple, cylindrical,

holoblastic, discrete, determinate, 8–14 µm in length, 3–4 µm in width. *Conidia* borne in mucilage, elliptical or broadly clavate, base often truncate and bordered by a scar, thick-walled, verruculose to almost smooth-walled, at first hyaline and unicellular with a large central guttule; becoming brown while still attached to conidiogenous cell; at length becoming 1-septate and biguttulate, 19–21 × 10–11 µm (Fig. 11.20).

Comments: *Sphaeria cannabis* was the first fungal pathogen reported on *Cannabis*. Schweinitz's publication has been dated 1831–1834 by various authors; Rogers (1944) argued for the middle of 1832. Schweinitz (1832) collected the specimen in Salem, North Carolina. Donald P. Rogers (pers. comm. 1984) collected a specimen in Champaign, Illinois, while consulting for Haney and Kutscheid (1975).

McPartland (1995b) examined Schweinitz's type specimen and compared it with the type of *Sphaeria obtusa* Schweinitz 1832, now *Botryosphaeria obtusa* (Schwein.) Shoemaker 1962. Anamorphs of both *S. cannabis* and *B. obtusa* could be accommodated in either *Dothiorella* or *Diplodina*. Phillips *et al.* (2005) established criteria differentiating these genera: *Diplodia* conidia are hyaline and unicellular, not darkening until after

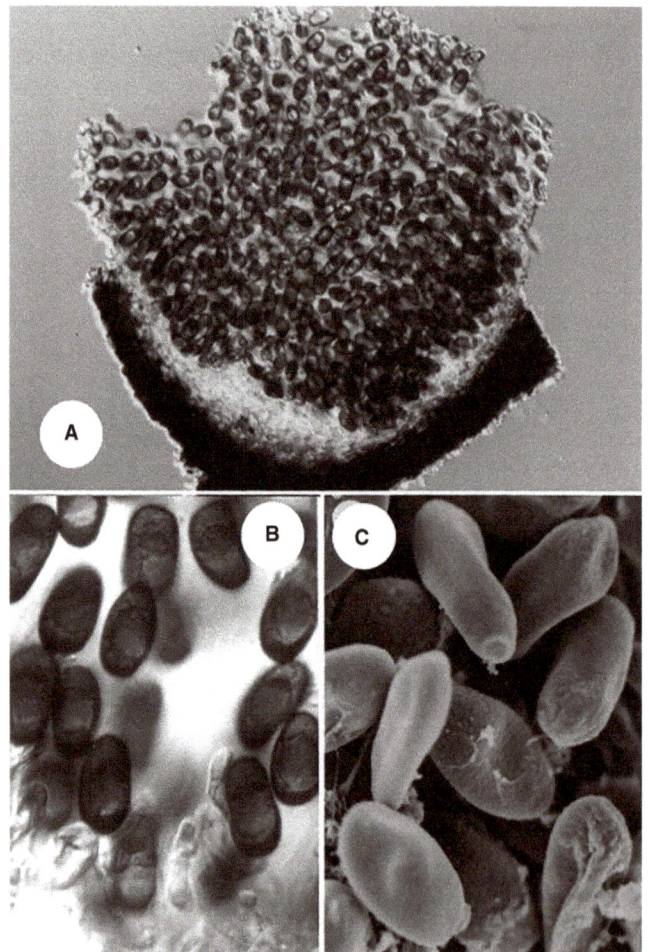

Fig. 11.20. *Dothiorella cannabis*: **A.** sectioned pycnidium with conidia in a mucilaginous mass; **B.** conidia and conidiophores; **C.** conidia visualized with scanning electron microscopy

Fig. 11.19. *Neofusicoccum parvum* causing dieback in Italy (courtesy G. Grassi)

discharge from the conidiomata, occasionally becoming 1-septate, and frequently formed through percurrent proliferation of conidiogenous cells. *Dothiorella* conidia become brown while still attached to conidiogenous cells, often becoming 1-septate, and percurrent proliferation is uncommon. *Sphaeria cannabis* agrees with *Dothiorella* (Fig. 11.20), and the recombination is made here.

Crous and colleagues erroneously synonymized *Botryosphaeria obtusa* under *Diplodia seriata* (Phillips *et al.*, 2007). They based this on a phylogenetic analysis and a study of "appropriate" herbarium specimens (which did not include Schweinitz's type specimen). Schweinitz's type of *Sphaeria obtusa* is not a *Diplodia*, but a *Dothiorella* sp., according to their own criteria—conidia become brown while still attached to conidiogenous cells (Fig. 11.21A). A microconidial morph is present in Schweinitz's type (Fig. 11.21B), and not present in *D. seriata* (Phillips *et al.*, 2007). Microconidia have been observed in other *Dothiorella* spp., such as *D. sarmentorum* (Phillips *et al.*, 2005).

Others have observed *B. obtusa* microconidia (Hesler, 1912; Shear and Stevens, 1924; Dodge, 1930; Morgan-Jones and White, 1987). Some of these authors cited *B. obtusa* under the synonym *Sphaeropsis malorum* Peck 1881. In Peck's specimen, macroconidia become brown while still attached to conidiogenous cells (Fig. 11.21C). Stevens (1933) said Peck's *S. malorum* specimens were indistinguishable from Schweinitz's *S. obtusa* specimen. Stevens examined Schweinitz's herbarium and synonymized 12 species under *B. obtusa*—but not *Sphaeria cannabis*.

11.7e. *Lasiodiplodia theobromae* (Patouillard) Griffon & Maublanc 1909

≡*Botryodiplodia theobromae* Patouillard 1892
=*Botryosphaeria rhodina* (Berkeley & Curtis) Arx 1970
Description: *Pycnidia* stromatic, simple or compound (uni- or multilocular), often aggregated, carbonaceous black, ostiolate, between 300 and 5000 µm in diameter. *Conidiogenous cells* hyaline, simple, cylindrical, sometimes septate, rarely branched, holoblastic, 5–15 × 3 µm. *Paraphyses* hyaline, cylindrical, septate, up to 50 µm long. *Conidia* at first ellipsoid, hyaline and thin-walled, then thick-walled, later developing a median septum, dark brown pigmentation, longitudinal striations and a truncate base, 20–30 × 10–15 µm (see Fig. 3.3). *Ascomata* perithecioid, stromatic, uni- or multilocular, gregarious. *Asci* bitunicate, clavate, 8-spored, 90–120 × 15–28 µm. *Ascospores* 1-celled, hyaline, ellipsoidal, 24–42 × 7–18 µm.

Comments: *L. theobromae* has a wide host range. McPartland (1995b) isolated *L. theobromae* from stem cankers on feral hemp in Illinois. Thiessen (2020) found *L. theobromae* in North Carolina, causing surface lesions as well as vascular discoloration in tissues beneath lesions. Stalks weakened and split (Fig. 11.22). Jerushalmi *et al.* (2020a) isolated *L. theobromae* from 11% of stems in Israel, using ITS sequences to confirm identification.

Roberts and Punja (2022) isolated *Lasiodiplodia theobromae* in British Columbia, using ITS sequences to confirm identification. Stems and flowers inoculated with *L. theobromae* showed greater symptoms than those by *Neofusicoccum parvum*, *Stemphylium vesicarium*, and *Alternaria alternata*. But in inoculated seedlings, *N. parvum* edged out *L. theobromae*. Alves *et al.* (2008) conducted a phylogenetic analysis of 25 *L. theobromae* isolates using ITS and EF1-α sequence data. They fell into three clades, so the authors erected new species names for two of the clades, *L. pseudotheobromae* and *L. parva*. Singh (2021) identified *L. pseudotheobromae* causing cankers on hemp in Louisiana.

11.7.3 Disease cycle and epidemiology

D. marconii, *N. marconii*, *D. cannabis*, *N. parvum*, and *L. theobromae* overwinter in crop debris. In the spring either conidia or ascospores are spread by splashing rain. Disease usually does not arise until mid-season or later. The sole report of disease in young plants, in Hungary by Mándy and Bòcsa (1962), was caused by *N. marconii*. Plants that survived to maturity produced fewer

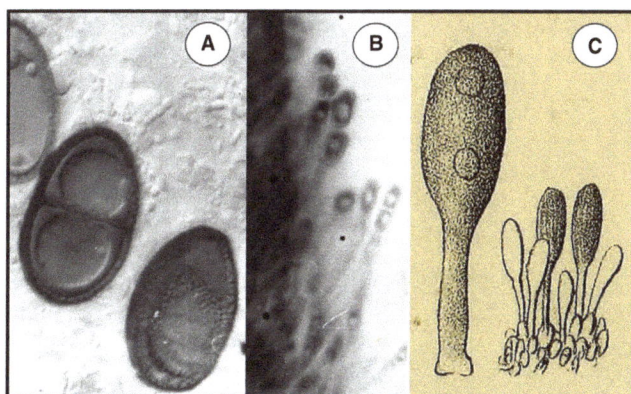

Fig. 11.21. *Botryosphaeria obtusa*: **A**, **B**, Schweinitz's type specimen; **C**, *Sphaeropsis malorum* (Peck 1881)

Fig. 11.22. Stalk splitting caused by *Lasiodiplodia theobromae* (L. Thiessen)

fibers, of poor quality. As a percentage of dry stalk weight, fiber yield fell to 1.8% (instead of normal 10–12%).

Botryosphaeriaceae spp. are classic facultative pathogens. They remain latent within their hosts, and become virulent when plants get stressed. McPartland and Schoeneweiss (1984) showed how *Botryosphaeria dothidea* shifted from a harmless commensal to a raging vascular pathogen in host plants exposed to drought stress. Host plants respond to stress, in part, by producing defensive compounds such as lignin. *L. theobromae* hijacks this response by utilizing the phenylpropanoid precursors of lignin for its own metabolism (Paolinelli-Alfonzo *et al.*, 2016).

Dewey (1914) stated that *N. marconii* appeared first on male plants and then spread to female plants. Trunov (1936) said the same about *D. marconii*. Male plants enter senescence prior to females. Wounding of stems provides a portal for infection. Dewey (1914) stated that an *N. marconii* epidemic followed a severe hailstorm. Stem abrasions in plants staked upright with twine led to *L. theobromae* infections in Virginia (C. Johnson, pers. comm., 2020). After pycnidia have formed on stems, epidemics of secondary infections arise in warm, damp conditions.

11.7.4   Disease management

Rogue infected plants and remove all plant residues. Eliminate weeds around and within crops. Weeds create a humid micro-environment, and may serve as alternative hosts. Avoid overhead irrigation. Monitor for disease during rainy seasons. Harvest fiber crops early, soon after male plants shed pollen and enter senescence (Charles and Jenkins, 1914; Ferraris, 1935; Trunov, 1936; Serzane, 1962). After stems are mowed down, damage may continue, so they must be dried quickly (Serzane, 1962; Mándy and Bòcsa, 1962). Dew retting may reactivate fungal growth.

Mándy and Bòcsa proposed a harsh crop rotation—do not plant hemp in infected fields for 8–10 years. Dempsey (1975) listed five Russian and Ukrainian fiber-type cultivars with resistance to *F. marconii*: 'Monoecious Central Russia', 'Odnodomnaja 2', 'USO-1' (i.e., YUSO-1, JUSO-1), and USO-1's parents—'USO-6' and 'Odnodomnaya Bernburga.' These cultivars are no longer available, although some are the parents of current cultivars, such as 'USO-14' (a selection from 'USO-1', de Meijer, 1995).

Biocontrol of other Botryosphaeriaceae (e.g., *Botryosphaeria dothidea*) has been accomplished with *Bacillus amyloliquefaciens*. Spraying these bacteria on pruning sites and other wounds should prevent colonization by twig blight pathogens. Bordeaux mixture has controlled *Botryosphaeria* spp. on other hosts, mostly by preventing secondary infections.

11.8   Charcoal rot

Charcoal rot is cause by *Macrophomina phaseolina*, another member of the Botryosphaeriaceae family. Instead of invading stems *à la* twig blight, *M. phaseolina* enters plants through roots. In fiber varieties, *M. phaseolina* has been reported in

Illinois (Tehon and Boewe, 1939), Missouri (Goodnight *et al.*, 2021), Italy (Goidànich, 1955; De Corato, 1997), Cyprus (Georghiou and Papadopoulos, 1957), Serbia (Acinovic, 1964), Spain (Casano *et al.*, 2018), and Australia (Poudel *et al.*, 2021).

From drug-type or CBD cultivars, *M. phaseolina* has been reported in Illinois (McPartland, 1983a), North Carolina (Thiessen, 2019), Iran (Mahdizadeh *et al.*, 2011), and Spain (Casano *et al.*, 2018). Wild-type plants have also been infected in Illinois (Boewe, 1963) and Hungary (Simay, 1990).

11.8.1   Symptoms and signs

Plants rapidly develop a systemic chlorosis and wilting, then necrose and die. Pith inside the stalk becomes punky and peppered with black microsclerotia. Roots become necrotic with a brown-violet discoloration. *M. phaseolina* kills plants approaching maturity; its disease is sometimes called "premature wilt." *M. phaseolina* also causes post-emergent damping-off in young seedlings (see Fig. 10.29).

11.8.2   Causal organism

11.8a.   *Macrophomina phaseolina* (Tassi) Goidànich 1947

≡*Macrophoma phaseolina* Tassi 1901, ≡*Tiarosporella phaseolina* (Tassi) van der Aa 1981, =*Macrophomina phaseoli* (Maublanc) Ashby 1927, ≡*Macrophoma phaseoli* Maublanc 1905; =*Rhizoctonia bataticola* (Taubenhaus) Briton-Jones 1925, ≡*Sclerotium bataticola* Taubenhaus 1913.

Description: *Pycnidia* solitary or gregarious, brown to black, subglobose, immersed in host tissue but becoming erumpent, ostiolate, averaging 180 μm in diameter. *Conidiophores* short, hyaline, simple (sometimes branched, according to Punithalingam, 1982). *Conidiogenous cells* at first holoblastic, becoming phialidic, with a minute collarette, lageniform to doliiform. *Conidia* hyaline, aseptate, obovoid to fusiform, thin-walled, smooth, often guttulate, 5–10 × 14–30 μm. Punithalingam described a conidial appendage visible after treatment with Leifson's flagella stain. The appendage is apical, cap-like or cone-shaped, formed from the eversion of an outer conidial sheath. *Microsclerotia* smooth, black, hard, oval to irregular in shape, averaging 44 × 75 μm on hosts, and 60–200 μm *in vitro*.

11.8.3   Differential diagnosis

The complete and rapid necrosis caused by charcoal rot has been mistaken for herbicide destruction. *Fusarium* and *Verticillium* cause wilt symptoms, but do not pepper the pith with sclerotia. *M. phaseolina* sclerotia are much smaller than those produced by *S. sclerotiorum* or *A. rolfsii*.

11.8.4 Disease cycle and epidemiology

M. phaseolina is distributed worldwide, and causes disease in more than 500 plant species. Maize and soybeans are particularly susceptible hosts. *M. phaseolina* is now considered a species complex, with several cryptic species recognized by molecular methods. An isolate on *Cannabis* in Australia, subjected to a five-locus phylogenetic analysis, fell into the clade *M. phaseolina sensu stricto* (Poudel *et al.*, 2021). An isolate on *Cannabis* in Spain, subjected to a four-locus analysis, also belonged to *M. phaseolina sensu stricto* (Viejobueno *et al.*, 2022).

 M. phaseolina overwinters as microsclerotia in the soil, and in infected plant debris. Microsclerotia persist in soil up to 3 years. They germinate when temperatures are warm, 28–35°C. Their germ tubes form appressoria that penetrate root epidermal cell walls, either directly or through wounds. Disease by *M. phaseolina*, typical of Botryosphaeriaceae spp., worsens under unfavorable conditions that stress the plant, particularly during hot, dry weather. The pathogen plugs xylem vessels with microsclerotia, and by toxin production and enzymatic action.

11.8.5 Disease management

Diagnostic real-time PCR kits utilizing TaqMan and SYBR Green assays are available to detect *M. phaseolina* in soil. Early planting results in earlier canopy closure, which reduces soil temperatures and therefore reduces the pathogenicity of *M. phaseolina*. Maintain good soil moisture with irrigation. Adequate levels of phosphorus and potassium reduce nutrient stress and encourage healthy plant growth.

 Crop rotation is difficult, given the wide host range of *M. phaseolina*; do not rotate after corn or beans. Casano *et al.* (2018) reported variable susceptibility in drug-type strains: "Sara" and "Aida" (disease incidence: 25.5% and 37.1%, respectively) were more susceptible than "Theresa" (3.8%), "Pilar" (3.2%), and "Juani" (2.7%).

 Plant growth-promoting rhizobacteria such as *Bacillus subtilis*, *Burkholderia cepacia*, *Paenibacillus lentimorbus*, and *Pseudomonas fluorescens* have controlled *M. phaseolina* in other crops. Biocontrol fungi such as *Trichoderma harzianum* and *T. viride* also show efficacy. No fungicides stop charcoal rot once it has begun.

11.9 Verticillium wilt

One or maybe two *Verticillium* spp. (Sordariomycetes; Plectosphaerellaceae) cause this disease in *Cannabis*. Both *Verticillium* spp. attack many plants. Reports of *Cannabis* verticillium wilt come from Europe, but both pathogens live worldwide.

11.9.1 Symptoms and signs

Leaves turn yellow along margins and between veins, then turn gray-brown. Lower leaves show symptoms first. Wilting may not arise until periods of drought stress. Slightly wilted plants often recover at night or after irrigation. These partial recoveries are transient as wilt becomes permanent. Dissection of diseased stems reveals a brownish discoloration of xylem tissue. If the fungus invades only a few xylem bundles, the plant may express only one-sided wilting. Verticillium wilt resembles Fusarium wilt. Wilting mimics symptoms caused by nematodes and root-boring insects (root maggots, white grubs), or drought.

11.9.2 Causal organisms

11.9a. *Verticillium dahliae* Klebahn 1913

Description: *Conidiophores* abundant, completely hyaline, with 3 or 4 whorled phialides arising at regular nodes along an upright branch which can reach 150 μm in height (Fig. 11.23A). *Phialides* hyaline, 19–35 μm long, 1.5 μm wide. *Conidia* arise singly but congregate in small droplets at tips of phialides (Fig. 11.23B), oval to ellipsoid in shape, hyaline, aseptate, $2.5–8 \times 1.5–3.2$ μm. *Microsclerotia* arise by lateral budding of a single hypha into long chains of cells, becoming dark brown to black, irregularly spherical to elongated, 15–100 μm in length (Fig. 11.23C).

Comments: Reports of *V. dahliae* on hemp come from Russian Turkestan (Vasiliev, 1933), Russia (Gitman, 1968b), Italy (Noviello, 1957), Czechoslovakia (Ondřej, 1991), Holland (Kok *et al.*, 1994), and Germany (Patschke *et al.*, 1997). Barnett *et al.* (2020) analyzed the soil microbiome associated with *Cannabis* crops in upstate New York. *V. dahliae* was present in soil immediately surrounding roots (the rhizosphere). *V. dahliae* attacks many cultivated, weedy and wild plants, in temperate zones and the tropics.

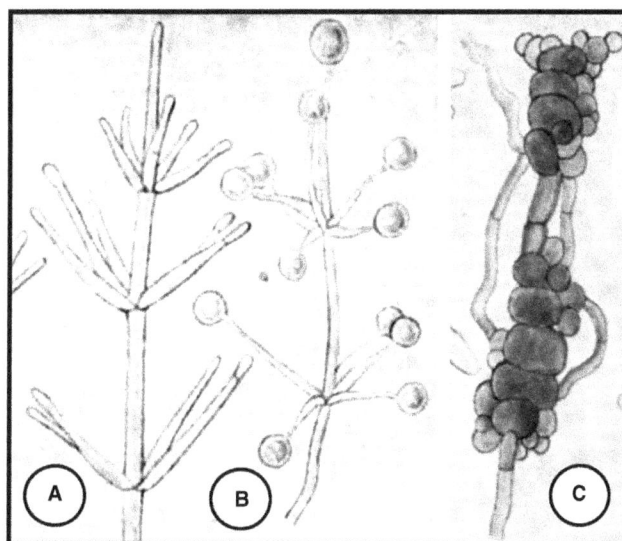

Fig. 11.23. *Verticillium dahliae*: **A.** conidiophore with conidia; **B.** same, with conidia in water droplets; **C.** microsclerotia (Carpenter, 1918)

11.9b. *Verticillium albo-atrum* Reinke & Berthier 1879

Description: *Conidiophores* abundant, mostly hyaline but with a darkened base (especially *in vivo*), with 2–4 phialides arising at regular intervals along an upright branch up to 150 µm tall. *Phialides* hyaline, variable, 20–30 (up to 50) µm in length and 1.5–3.0 µm wide. *Conidia* arise singly, ellipsoid to short-cylindrical, hyaline, usually single-celled but occasionally 1-septate, 3.5–10.5 × 2.5–3.5 µm. No microsclerotia formed.

Comments: *V. albo-atrum* has only been reported on hemp by Hrebenyuk (1984). Green (1951) conducted host range pathogenicity tests with *V. albo-atrum*. He inoculated nine crop plants and 21 "weed plants" in Indiana, *Cannabis* among the latter. *V. albo-atrum* could not infect healthy *Cannabis* seedlings, but grew saprophytically on sterilized stems.

11.9.3   Disease cycle and epidemiology

Verticillium spp. overwinter in soil as microsclerotia, and invade roots of seedlings in the spring. Once in roots, *Verticillium* spp. spread via xylem tissue; they block the movement of water and cause wilts. Verticillium wilt increases in soils rich in clay and soil moisture. Nematodes predispose plants to *Verticillium* infection. *V. albo-atrum* grows best at 23.5°C, and the optimum for *V. dahliae* is 21°C.

11.9.4   Disease management

Follow instructions for *Fusarium* wilt. Burying crop debris at the end of the season is not helpful, because *Verticillium* microsclerotia can live 75 cm deep in soil. Soil solarization works well against *V. dahliae*, but *V. albo-atrum* is heat-resistant. Organic material and green manures added to soil encourage the growth of natural *Verticillium* antagonists. Grewal (1989) reduced *Verticillium* growth by mixing compost with dried *Cannabis* leaves.

Monocotyledonous crops are not susceptible to *Verticillium* spp., so crop rotation is an option. But rotation must be long (4 years for *V. albo-atrum*, twice that for *V. dahliae*). Monocot crops must be weed-free, since almost all dicot weeds serve as alternative hosts. Root-knot nematodes (*Meloidogyne* spp.) should be eliminated. Breeding for resistance is possible—Kok *et al.* (1994) found partial resistance to *V. dahliae* in fiber cultivar 'Kompolti Hybrid TC'. *Pythium oligandrum* (Polyversum®), *Trichoderma atroviride* (Binab®), and *Streptomyces lydicus* (Actinovate®) are labeled for *Verticillium* control. Marois *et al.* (1982) screened 34 fungi as potential agents against *V. dahliae*, and *Talaromyces flavus* provided the best control. There is no chemical control.

11.10   Brown blight

Several *Alternaria* spp. (Dothideomycetes; Pleosporaceae) cause this disease in *Cannabis*. The most ubiquitous cause, *Alternaria alternata*, attacks plants as an opportunistic saprophyte in necrotic tissues, as well as a primary pathogen in leaves and floral tissues. *A. alternata* is also a common field-retting organism (Fuller and Norman, 1944), and even destroys poorly stored hemp textiles (Agostini, 1927).

11.10.1   Symptoms

Brown blight usually arises on mature plants late in the growing season. Symptoms begin as pale green or gray leaf lesions, on upper surfaces and lower surfaces. These irregularly circular spots may or may not develop chlorotic halos. Spots often coalesce into blight-like symptoms, and whole leaves turn brown. Frail necrotic tissue breaks up, resulting in irregular leaf perforations. Lesions may extend to leaf petioles and appear on stems. The fungus also infests female flowers (and seeds), turning them gray-brown.

Cheng *et al.* (2022) isolated three *Alternaria* spp. causing leaf spots in Gānsù Province, China. They inoculated the fungi on wounded or unwounded *Cannabis* leaves, and derived pathogenicity scores based on sizes of lesions (Fig. 11.24). Necrotic spots from *A. alternata* averaged 10–14.4 mm in diameter, *A. cannabina* averaged 5.5–8.5 mm, and *A. helianthi-inficiens* averaged 4.5–5.5 mm.

11.10.2   Causal organisms

11.10a. *Alternaria alternata* (Fries) Keissler 1912

=*Alternaria tenuis* Nees 1817

Description: *Conidiophores* simple or branched, straight or curved, septate, yellow to golden brown, up to 50 µm long, 3–6 µm thick. *Conidia* borne in chains, straight or slightly curved, obyriform to obclavate (rarely ellipsoid), yellow to golden brown, with 3–8 transverse and 1–2 longitudinal septa, tapering

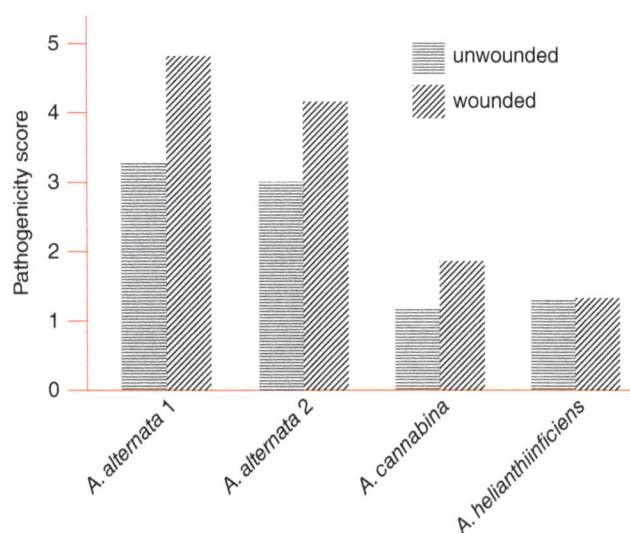

Fig. 11.24. Pathogenicity score of *Alternaria* spp. (adapted from Cheng *et al.*, 2022)

to a short beak (sometimes absent), overall length 20–63 µm, 7–18 µm at greatest width (see Figs 3.3, 10.17B).

Comments: *A. alternata* can live as a commensal endophyte: Gautam *et al.* (2013) isolated *A. alternata* from 4.6% of leaves and 1.4% of stems of healthy plants in Himachal Pradesh, India. Scott *et al.* (2018) isolated endophytes from healthy plants in Quebec. They identified a pair as "*Alternaria* sp." and deposited ITS sequences in GenBank. We analyzed the sequences and they proved to be *A. alternata* (see Table 1.2, Fig. 1.12).

A. alternata readily invades senescent tissues, such as necrotic fan leaves or wilted stamens and pistils (McPartland, 1983a). *A. alternata* caused leaf spots on drug-type plants in India (Nair and Ponnappa, 1974; Bhargava *et al.*, 1981; Singh *et al.*, 1998). Fiber-type plants also suffer leaf spots, in Poland (Zarzycka and Jaranowska, 1977), Russia (Vakhrusheva and Davidian, 1979), and Tasmania (Lisson and Mendham, 1995).

On feral hemp, Hartowicz *et al.* (1971) described *A. alternata* as a weak pathogen, attacking senescent or damaged tissue in Kansas. Conversely, on feral hemp in Illinois, Haney and Kutscheid (1975) considered it a serious pathogen, causing a blight and sometimes killing plants. They estimated that *A. alternata* destroyed 20.5% of developing seeds. The new generation has discovered *A. alternata* in Canada (Punja *et al.*, 2019), Tennessee (Hansen *et al.*, 2020), Virginia (Mersha *et al.*, 2020), and China (Tang *et al.*, 2021; Cheng *et al.*, 2022). In Israel, Jerushalmi *et al.* (2020a) isolated *A. alternata* from foliage and flowers in 18% of plants—more than any other fungus, including *Botrytis cinerea*.

A. alternata has caused seed-borne infections (Pietkiewitcz, 1958; Ferri, 1961b; Jian *et al.*, 2019). Plăcintă and Murariu (2016) isolated *A. alternata* from 2.5% of sampled seeds—the highest percentage of any fungi. Paulsen (1971) cultured the fungus from green stink bugs (*Nezara viridula*) feeding on *Cannabis* seeds. Suyal *et al.* (2011) analyzed air mycoflora over *Cannabis* fields in the Himalaya, and *A. alternata* was the second most common species. Roberts and Punja (2022) isolated *A. alternata* from seeds, stems, and flowers in British Columbia, and proved pathogenicity with Koch's postulates. Cigarettes made from *Alternaria*-infected tobacco produce a harsh and irritating smoke (Lucas, 1975). More significantly, *A. alternata* produces alternariol, a carcinogenic metabolite, which McPartland (1995f) implicated as a cause of esophageal cancer, the most common malignancy associated with cannabis smoking.

11.10b. *Alternaria solani* Sorauer 1896

=*Alternaria solani* (Ellis & Martin) Jones 1896, ≡*Macrosporium solani* Ellis & Martin 1882, =*Alternaria porri* (Ellis) Ciferri f. sp. *solani* Neergaard 1945

Description: *Conidiophores* simple, straight or curved, septate, pale to dark olive-brown, up to 100 µm long, 6–10 µm thick. *Conidia* borne singularly, straight or slightly curved, obyriform to obclavate, light olive-brown, with 8–11 transverse and 0–2 longitudinal septa, tapering to a long beak which is usually equal in length to the body, overall length 140–280 µm, 15–19 µm at greatest width.

Comments: Barloy and Pelhate (1962) cited *A. solani* on *Cannabis*, but the fungus they illustrated looks more like *A. alternata* than *A. solani*, and the leaf symptoms resemble those of *A. alternata* illustrated by Zarzycka and Jaranowska (1977). The synonym *A. porri* f. sp. *solani* was cited on *Cannabis* by Neergaard (1945). *A. solani* usually attacks solanaceous crops (potato, tomato), and *Brassica* (cabbage, cauliflower), but has a wide host range.

11.10c. *Alternaria cannabina*
Cheng & Deng 2020

? =*Alternaria cannabis* Yu 1973

Description: *Conidiophores* simple or branched, straight or curved, 2–5 septa, with 1–4 geniculate conidiogenous loci, 18–83 × 4–6 µm; secondary conidiophores occasionally arising from the apex of conidia, with 1–4 branches. *Conidia* borne solitary or in chains, straight or slightly curved, variously shaped (subglobose, ovoid, obclavate, or ellipsoid), tapering to a short beak (sometimes absent), light brown, 0–7 transverse septa and 0–2 longitudinal or oblique septa, 8–70 × 7–21 µm (Fig. 11.25B).

Comments: Yu (1973) isolated and described a new species, *Alternaria cannabis*, from drug-type plants growing at the University of Mississippi. Her description appeared in an unpublished manuscript, rendering it a *nomen nudum* according to the *ICN*. Cheng *et al.* (2022) erected *A. cannabina* for a fungus in Gānsù Province. Irregularly shaped dark brown spots were surrounded by chlorotic halos (Fig. 11.25A). *A. cannabina was less virulent than A. alternata in pathogenicity tests* (Fig. 11.24). Cheng and colleagues conducted a large phylogenetic analysis (ITS, TEF1, rpg2, ATPase) of 99 worldwide *Alternaria* spp., and *A. cannabina* was sister with species in a clade called *Alternaria* section *Infectoriae*, distant from *A. alternata* (in *Alternaria* section *Alternaria*), and *A. solani* (in *Alternaria* section *Porri*).

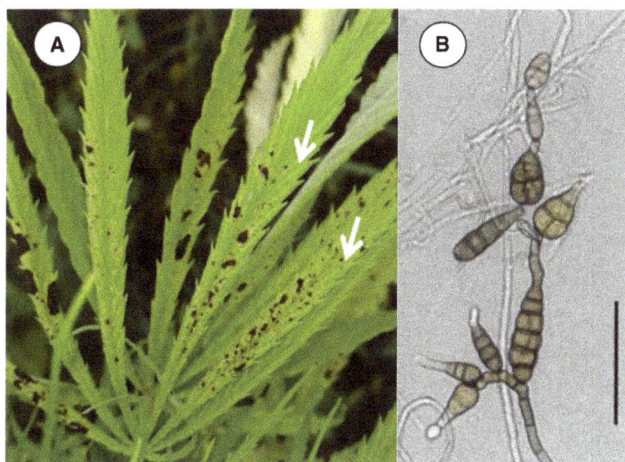

Fig. 11.25. Brown leaf blight by *A. cannabina*: **A**. symptoms; **B**. conidia, scale bar 50 µm (courtesy Deng Jianxin)

11.10d. Other *Alternaria* spp.

Cheng *et al.* (2022) also isolated *Alternaria helianthiinficiens* Simmons, Walcz & Roberts 1986 from leaf spots in Gānsù Province. They differentiated *A. helianthiinficiens* morphology from that of *A. alternata* by its larger conidia (body 65–110 × 19–24 μm and beak 14–138 μm) and fewer conidia in chains (only 1 or 2). Phylogenetic analysis placed their *Cannabis* isolate sister to *A. helianthiinficiens* isolates from sunflower (*Helianthus annuus*). *A. helianthiinficiens* was less virulent than other *Alternaria* spp. (Fig. 11.24).

Schoener *et al.* (2018a) observed fungal growth on *Cannabis* seedlings growing in a germination tray. They analyzed the seed stock from which the seedlings were sourced, by surface-sterilizing seeds with sodium hypochlorite and plating them on PDA agar. Fifty-nine percent of seeds sprouted fungi, from which 29 *Alternaria* isolates were obtained. Half the isolates produced white-gray colonies, and half produced gray-black colonies. Using ITS primers and a BLAST search of GenBank identified *Alternaria infectoria* Simmons 1986 (the white-gray colonies) and *Alternaria tenuissima* (Kunze) Wiltshire 1933 (the gray-black colonies).

Hrebenyuk (1984) isolated *Alternaria cheiranthi* (Libert) Bolle 1924 from retted hemp stems. He identified a total of 79 species, and *A. cheiranthi* was second most common (behind *Cladosporium cladosporioides*), yet ubiquitous *A. alternata* was not on his list. He may have misidentified *A. alternata* as *A. cheiranthi*.

11.10.3 Differential diagnosis

Microscopic characteristics separate *A. solani* from *A. alternata*: the former produces larger, darker conidia with longer beaks. *A. solani* produces solitary conidia whereas *A. alternata* often sporulates in chains. Conidia of both species are considerably larger than those of *A. alternata*. Conidia of *Alternaria* spp. could be confused with chlamydospores of *Didymella glomerata* (see Fig. 10.17).

11.10.4 Epidemiology and management

Alternaria spp. overwinter as conidia or mycelia in host-plant debris. *A. alternata* colonizes plants during warm, wet weather, but symptoms do not arise until late in the season. *Alternaria* conidia disperse by wind or splashed rain. A water film is required for conidial germination. Conidia germ tubes penetrate host epidermis directly, or enter via stomates and wounds. Lucas (1975) reported a tenfold increase in infection when *Alternaria* conidia were mixed with plant pollen. The pollen served as an energy source for conidia.

Disease management hinges on avoiding seed-borne infection. Do not harvest seeds from infected plants. In stored seed, *A. alternata* infection was reduced by decreasing storage temperature and relative humidity (Jian *et al.*, 2019). Surprisingly, seed infection was not significantly influenced by "dockage" in stored seed (i.e., chaff—leaf, bud or stem fragments, weed seeds, and other foreign materials).

Avoid overhead irrigation, and do not walk through fields of wet plants. Fertilize with low-nitrogen amendments; Lucas (1975) used a 3:18:15 NPK ratio. Several biocontrol products are marketed for *Alternaria* control: *Clonostachys rosea* (PreStop®), *Pythium oligandrum* (Polyversum®), *Streptomyces griseoviridis* (Mycostop®), and *Trichoderma* sp. (PlantShield®). Balthazar *et al.* (2022b) screened beneficial *Bacillus* and *Pseudomonas* spp. for *in vitro* activity against *A. alternata*. Two performed well: *B. velezensis* and *B. subtilis* (see Fig. 13.6). Yu (1973) controlled *Cannabis* brown blight with monthly sprays of Bordeaux mixture, which is also used by tobacco growers (Lucas, 1975).

11.11   Texas root rot

Texas root rot occurs in the southwestern USA and northern Mexico. The causal fungus belongs to Class Pezizomycetes, Family Rhizinaceae. It is limited to soils high in calcium carbonate, high pH, high temperatures, and low organic matter. The disease is prevalent in the blackland prairies of central Texas. In this region the fungus attacks over 2000 species of wild and cultivated dicots.

11.11.1   Causal organism

11.11a. *Phymatotrichopsis omnivora* (Duggar) Hennebert 1973

≡*Phymatotrichum omnivorum* Duggar 1916, =*Ozonium omnivorum* Shear 1907

Description: *Mycelium* forms thick-walled cruciform aerial setae and aggregating into subterranean cord-like funicles. *Conidiophores* borne directly on hyphae, hyaline, simple or branched, clavate to globose, hyaline, 20–28 × 15–20 μm. *Conidia* holoblastic, ovate to globose, smooth, thin-walled, hyaline, with a broad base exhibiting a detachment scar, 6–8 × 5–6 μm. *Sclerotia* borne on hyphae, ovate to globose, at first yellow but turning reddish-brown, 1–2 mm diameter, often aggregated in clusters reaching 10 mm diameter.

Comments: *P. omnivora* has attacked hemp in Texas, Oklahoma, and Arizona (Chester. 1941), causing 30–60% mortality rates (Taubenhaus and Killough, 1923), or up to 95% (Killough, 1920). The cruciform setae and funicles produced by *P. omnivora* are unique. *P. omnivora* grows in rope-like funicles a foot or two underground. The fungus invades roots in the spring. In damp conditions *P. omnivora* emerges from the ground and spreads via conidia. In dry conditions the fungus produces sclerotia in soil. These durable structures survive for years.

Deep tillage (0.3–0.6 m) exposes funicles to desiccation. Crop rotation with monocots must be of sufficient duration to outlast long-lived sclerotia. Mixing organic matter into soil encourages natural biocontrol organisms. Use composted animal manure or a green manure of legumes or grasses. No fungicides are effective. Moving from Texas is the best solution.

11.12 Black root rot

Two *Berkeleyomyces* spp. (Ceratocystidaceae; Sordariomycetes) cause this disease. They have wide host ranges and live around the world in temperate climates. Above-ground symptoms include wilting, stunting, and leaf chlorosis. Root symptoms begin with a dark brown-black discoloration within the root, followed by black lesions on the root surface. Internal discoloration may extend up from the root, through the crown, into the stem (Fig. 11.26A).

11.12.1 Causal organisms

Berkeley and Broome originally named the primary pathogen *Torula basicola*, which shuttled through several genus names, landing on *Thielaviopsis basicola*. Using DNA-based techniques, Nel *et al.* (2018) showed that *Thielaviopsis basicola* segregated from other *Thielaviopsis* spp., so they erected the genus *Berkeleyomyces*. Six isolates of the fungus split into two clades: *B. basicola* and the new species *B. rouxiae*. Both have been reported on *Cannabis*.

11.12a. *Berkeleyomyces basicola* (Berk. & Broome) Nel *et al.* 2017

≡*Thielaviopsis basicola* (Berk. & Broome) Ferraris 1912, ≡*Thielavia basicola* (Berk. & Broome) Zopf 1876, ≡*Torula basicola* Berkeley & Broome 1850, =*Chalara elegans* Nag Raj & Kendrick 1975

Description and comments: *Conidiophores* borne terminally or laterally on vegetative hyphae. *Conidiogenous cells*

phialidic, cylindrical, hyaline to subhyaline, 33–70 μm long, 4–6 μm wide at base and tapering toward apex. *Conidia* unicellular, cylindrical, hyaline, produced singly or in chains, 10–18 × 3.5–4.5 μm. *Chlamydospores* dark brown, club-shaped chains of spores held together by an outer membrane, 25–65 × 10–12 μm (Fig. 11.26B).

Yarwood (1981) isolated this species, as *Chalara elegans*, from *Cannabis* roots in California. Koike (pers. obs., 2021) isolated *B. basicola* from field plants in California. Thiessen *et al.* (2020) cited *Thielaviopsis* sp. in North Carolina, and Joshi *et al.* (2022) identified *B. basicola* in British Columbia. This species acts as a hemibiotroph: early in the disease cycle it drains nutrients from living host cells (the biotrophic stage), then switches to a necrotrophic stage and destroys host cells. The fungus is soil-borne, but it has little saprophytic ability in soil without a host—Hood and Shew (1997) actually classified it as an obligate parasite. *B. basicola* has a wide host range, and lives around the world in temperate climates.

11.12b. *Berkeleyomyces rouxiae* Nel *et al.* 2017

Description and comments: *B. rouxiae* cannot be distinguished morphologically from *B. basicola*. DNA sequence data are needed to separate them. Both *B. rouxiae* and *B. basicola* have wide host ranges; differentiating the two species may be important in terms of pathogenicity and host susceptibility. Rahnama *et al.* (2022) isolated *B. rouxiae* from hemp in Kentucky. It caused 30% dieback in clones potted in artificial media. Culturing the fungus yielded both endoconidia and chlamydospores. Koch's postulates confirmed pathogenicity. Dumigan *et al.* (2023) isolated *B. rouxiae* from the surface of diseased roots in an aeroponic facility in British Columbia. They sequenced the entire genome of *B. rouxiae*, a first. Koch's postulates demonstrated that roots became stunted and discolored just 14 days after inoculation.

11.12.2 Differential diagnosis, disease cycle

Above-ground symptoms caused by *Berkeleyomyces* spp., such as wilting, stunting, and leaf chlorosis, can be confused with all kinds of causes. The chlamydospores found in roots and stems are pathognomonic (Fig. 11.26B).

Chlamydospores or conidia germinate in soil and penetrate root tissues. Spores move easily in soil water, and can spread via transplants. The disease is more severe in cool soils (17–23°C) that are below the optimum range for the plant. Wet soil plays a role, and soil pH greater than 5.5 favors the fungus. Fungus gnats vector *B. basicola*.

The chlamydospores are built for the long haul and hard to eradicate. Pots and trays with diseased plants should not be reused; keep equipment and work areas clean of old potting media. Maintain soil pH <5.5, control fungus gnats, and supplement growing medium with mycorrhizae.

Fig. 11.26. Black root rot: **A**. black discoloration in center of stem; **B**. chlamydospores in central pith (S. Koike)

11.13 Basidio root rot

Here we group three basidiomycete members of Class Agaricomycetes. The three species have only been reported once each on *Cannabis*, so their descriptions are truncated here. The first two belong to Family Atheliaceae; their basidiocarps are effuse, thin, white to buff, and inconspicuous: 2–6 cm long and 1–3 cm wide, 75–150 µm thick in cross-section. Both species produce globose sclerotia, composed of dark brown parenchymatous cells, 0.2–1.0 mm in diameter (larger in culture, 1–5 mm in diameter). The two taxa may represent the same species (Adams and Kropp, 1996). The third species belongs to Family Physalacriaceae; its basidiocarps are true mushrooms, with caps and stipes.

11.13a. *Athelia epiphylla* Persoon 1822

=*Hypochnus centrifugus* (Léveillé) Tulasne 1861, ≡*Corticium centrifugum* (Léveillé) Bresadola 1903. Anamorph: *Fibularhizoctonia centrifuga* (Léveillé) Adams & Kropp 1996, ≡*Rhizoctonia centrifuga* Léveillé 1843
Endo (1931) recovered *A. epiphylla* from *Cannabis* in Japan. The fungus decays forest leaves, rots wood, and causes fisheye decay of stored apples. It is common in Eurasia, less so in North America.

11.13b. *Athelia arachnoidea* (Berkeley) Jülich 1972

≡*Hypochnus arachnoideus* (Berkeley) Bresadola 1903, ≡*Corticium arachnoideum* Berkeley 1844. Anamorph: *Fibularhizoctonia carotae* (Rader) Adams & Kropp 1996, ≡*Rhizoctonia carotae* Rader 1948
Lentz (1977) cited *A. arachnoidea* on *Cannabis*. This fungus is common in North America and Scandinavia. It grows on forest leaf humus, parasitizes lichens, and causes a cold-storage disease of carrots (*Daucus carota* L.).

11.13c. *Armillaria gallica* Marxmüller & Romagnesi 1987

Description: Basidiocarp cap convex to nearly flat, yellow with brownish scales, 3–10 cm in diameter, gills running down the stem; stipe beige becoming brown, 4–7 cm long, 1–3 cm thick, often attached to black rhizomorphs. Basiodiospores smooth, hyaline to yellowish, ellipsoid with an apiculus.
Comments: *A. gallica* was isolated from *Cannabis* with root rot, growing in a forest surrounded by *Quercus* spp. in California (Feng *et al.*, 2022). Above-ground symptoms included leaf necrosis and wilting, followed by death within 24 h. White mycelial fans formed at the base of the stem. Identification was confirmed with *EF-1α* and ITS region sequences. Pathogenicity tests showed wilting 8 weeks after inoculation. *A. gallica* gained international attention when Anderson *et al.* (2018) used genomic DNA to determine that a single individual living in a Michigan forest covered 75 ha, weighed about 440 tons (the weight of three blue whales), and was 2500 years old.

11.14 Dematophora root rot

The causal agent of this disease is the only member of Class Sordariomycetes, Family Xylariaceae that infects *Cannabis*. Above-ground symptoms include stunting, wilting, and early plant death. Below-ground signs include white spheres of mycelium forming on root surfaces, and a feather-like mycelium growing within roots. Death generally occurs 2–3 weeks after the appearance of first symptoms (Sorrentino *et al.*, 2021).

11.14a. *Dematophora necatrix* Hartig 1883

=*Rosellinia necatrix* Prillieux 1904
Description: *Hyphae* brown, septate, forming pear-shaped swellings at ends of cells next to septa. *Conidiophores* bound together in an upright synnemata, which is 40–400 µm thick and 1–5 mm tall, often dichotomously branched towards the apex. *Conigiogenous cells* polyblastic, integrated and terminal or discrete and sympodial. *Conidia* light brown, ellipsoid, 1-celled, smooth, 3–4.5 × 2–2.5 µm. *Rosellinia* morph *perithecia* densely aggregated, globose, brown-black, ostiolate, 1000–2000 µm diameter. *Asci* unitunicate, cylindrical, long-stalked, 8-spored, 250–380 × 8–12 µm. *Ascospores* fusiform, straight or curved, single-celled, brown, 30–50 × 5–8 µm, with a longitudinal slit. *Paraphyses* filiform.
Comments: *D. necatrix* causes "white root rot" of hemp in Japan (Kyokai, 1965; Tsuge and Sato, 2020) and Italy (Sorrentino *et al.*, 2021). The fungus has a wide host range and lives in temperate and semi-tropical areas worldwide. *D. necatrix* is a hardy soil fungus, difficult to eradicate once established. *Trichoderma harzianum* infests and kills *D. necatrix*.

(Prepared by J. McPartland; revised by Z. Punja; D.L. Hawksworth made helpful suggestions regarding *Neofusicoccum marconii*)

References

Acinovic M. 1964. The occurrence of *Macrophomina phaseolina* on some agricultural crops and morphological and ecological properties of the parasite. *Savremena Poljoprivredna (Novi Sad)* 12: 55–66.

Adams GC, Kropp BR. 1996. *Athelia arachnoidea*, the sexual state of *Rhizoctonia carotae*, a pathogen of carrot in cold storage. *Mycologia* 88: 459–472.

Agostini A. 1927. Observazioni informa a due ifomiceti saprofiti dannosi di tessuti di canapa. *Atti della Reale Accademia dei Fisiocritici* 1(3): 25–33.

Ahren W, Berthold C, Böck G, *et al.* (11 other authors). 2009. Ursachen des Hopfensterbens. *Journal für Kulturpflanzen* 61: 375–379.

Alberti I, Prodi A, Nipoti P, Grassi G. 2018. First report of *Neofusicoccum parvum* causing stem and branch canker on *Cannabis sativa* in Italy. *Journal of Plant Diseases and Protection* 125: 511–513.

Alves A, Crous PW, Correia A, Phillips AJL. 2008. Morphological and molecular data reveal cryptic speciation in *Lasiodiplodia theobromae*. *Fungal Diversity* 28: 1–13.

Amaradasa BS, Turner A, Lowman S, Mei C. 2020. First report of southern blight caused by *Sclerotium rolfsii* in industrial hemp in southern Virginia. *Plant Disease* 104: 1563.

Anderson JB, Bruhn JN, Kasimer D, *et al.* (3 other authors). 2018. Clonal evolution and genome stability in a 2500-year-old fungal individual. *Proceedings of the Royal Society B* 285: e20182233.

Anderson PJ. 1926. *Check list of diseases of economic plants in the United States.* USDA Department Bulletin no. 1366. Washington, DC.

Anonymous. 1943. Principales enfermedades de origen parsitario que fueron objeto de consulta en el semester Julio-Decembre de 1942. *Boletín de Sanidad Vegetal Santiago* 2(2): 152–155.

Anonymous. 1994. *Alma Ata fusaria pathogenicity testing in Cannabis sativa. USDA report, May, 1994.* Document released under FOIA to Jeremy Bigwood.

Armstrong GM, Armstrong JM. 1960. *American, Egyptian, and Indian cotton-wilt fusaria. Their pathogenicity and relationship to other wild fusaria.* USDA Technical Bulletin no. 1219, Washington, DC.

Armstrong GM, Armstrong JK. 1968. Formae speciales and races of *Fusarium oxysporum* causing a tracheomycosis in the syndrome of disease. *Phytopathology* 58: 1242–1246.

Balthazar C, Novinscak A, Cantin G, Joly DL, Filion M. 2022b. Biocontrol activity of *Bacillus* spp. and *Pseudomonas* spp. against *Botrytis cinerea* and other *Cannabis* fungal pathogens. *Phytopathology* 112: 549–560.

Barloy J, Pelhate J. 1962. Premières observations phytopathologiques relatives aux cultures de chanvre en Anjou. *Annales des Épiphyties* 13: 117–149.

Barnett SE, Cala AR, Lansen JL, *et al.* (5 other authors). 2020. Evaluating the microbiome of hemp. *Phytobiomes Journal* 4: 351–363.

Beckerman JL, Nisonson H, Albright N, Creswell T. 2017. First report of *Pythium aphanidermatum* crown and root rot of industrial hemp in the United States. *Plant Disease* 101: 1038.

Beckerman JL, Stone J, Ruhl GE, Creswell T. 2018. First report of *Pythium ultimum* crown and root rot of industrial hemp in the United States. *Plant Disease* 102: 2045.

Benhamou N, Garand C, Goulet A. 2002. Ability of nonpathogenic *Fusarium oxysporum* strain Fo47 To induce resistance against *Pythium ultimum* infection in cucumber. *Applied and Environmental Microbiology* 68: 4044–4060.

Bhargava SN, Pandey RS, Shukla DN, Dwivedi DK. 1981. Two new leaf spot diseases of medicinal plants. *National Academy Science Letters* 4(7): 276–277.

Bilaĭ VI. 1955. *Фузарии. Биология и систематика.* Ukrainian Academy of Sciences, Kiev.

Boewe GH. 1963. Host plants of charcoal rot disease in Illinois. *Plant Disease Reporter* 47: 753–755.

Booth C. 1971. *The genus Fusarium.* Commonwealth Mycological Institute. Kew, UK.

Brandenburger W. 1985. *Parasitische Pilz an Gefäßpflanzen in Europa.* Gustav Fisher Verlag, Stuttgart.

Breton AM, Dionne A, Hamel D, *et al.* (3 other authors). 2022. Maladies et problèmes abiotiques diagnostiqués sur les échantillons de plantes reçus en 2021 au laboratoire d'expertise et de diagnostic en phytoprotection du Ministère del l'Agriculture, des Pêcheries et de l'Alimentation du Québec. *Canadian Journal of Plant Pathology* 44(supp): S38–S51.

California DPR. 2018. *Cannabis pesticides that cannot be used.* California Department of Pesticide Regulation. Available at: https://www.cdpr.ca.gov/docs/cannabis/cannot_use_pesticide.pdf

Carpenter CW. 1918. Wilt diseases of okra. *Journal of Agricultural Research* 12: 529–546.

Casano S, Hernández-Cotan A, Marín-Delgado M, *et al.* (4 other authors). 2018. First report of charcoal rot caused by Macrophomina phaseolina on hemp (*Cannabis sativa*) varieties cultivated in Southern Spain. *Plant Disease* 102: 1665–1666.

Cavara F. 1888. Appunti di Patologia Vegetale. Alcuni funghi parassiti di piante coltivate. *Atti dell'Istituto Botanico della Università e Laboratorio Crittogamico di Pavia* (Serie II) 1: 425–436.

Ceapoiu N. 1958. *Cînepa, Studiu monografic.* Editura Academiei Republicii Populare Romine. Bucharest.

Charirak P, Saksirirat W, Joglogy S, Saepaisen S. 2016. Integration of soil solarization with chemical and biological control of stem rot disease of Jerusalem Artichoke. *Journal of Pure and Applied Microbiology* 10: 2531–2539.

Charles VK, Jenkins AE. 1914. A fungous disease of hemp. *Journal of Agricultural Research* 3: 81–84.

Chen Y, Tang XY, Gao CS, *et al.* (5 other authors). 2020. Molecular diagnostics and pathogenesis of fungal pathogens on bast fiber crops. *Pathogens* 9: e223.

Cheng H, Zhao L, Wie X, *et al.* (4 other authors). 2022. *Alternaria* species causing leaf spot on hemp (*Cannabis sativa*) in Northern China. *European Journal of Plant Pathology* 162: 957–970.

Chester K. 1941. *Cotton root rot or Texas root rot.* Oklahoma Agricultural Experiment Station Circular No. 53, Stillwater, Oklahoma.

Cochran S, Schappe T, Thiessen L. 2020. "Fusarium spp. on industrial hemp," p. 20 in Gauthier N, *ed. Science of hemp: production and pest management.* University of Kentucky Agricultural Experiment Station, no. SR-112, Lexington, Kentucky.

Conner K, Kesheimer K, Sikora E, Kemble J. 2020. "Diseases and insects of hemp in Alabama," p. 21 in Gauthier N, *ed. Science of hemp: production and pest management.* University of Kentucky Agricultural Experiment Station no. SR-112. Lexington, Kentucky.

Coolong T. 2020. *A preview of industrial hemp flower production in Georgia.* University of Georgia Cooperative Extension Bulletin no. 1530. Available at: https://secure.caes.uga.edu/extension/publications/files/pdf/B%201530_1.PDF

Corda ACI. 1838. *Icones fungorum hucusque cognitorum, Tomus II.* J.G. Calve, Prague.

Crosier WF. 1968. *Rhizoctonia solani* in seeds of small grains and other plants. *Proceedings of the Association of Official Seed Analysts* 58: 111–117.

Crous PW, Slippers B, Wingfield MJ, *et al.* (7 other authors). 2006. Phylogenetic lineages in the Botryosphaeriaceae. *Studies in Mycology* 55: 235–253.

Crous PW, Lombard L, Sandoval-Denis M, *et al.* (124 other authors). 2021. Fusarium: more than a node or a foot-shaped basal cell. *Studies in Mycology* 98: e10116.

Czyżewska S, Zarzycka H. 1961. Ergebnisse der bodeninfektionsversuche an *Linum usitatissimum*, *Crambe alyssinica*, *Cannabis sativa* und *Cucurbita pepo* var. *oleifera* mit einigen *Fusarium*-Arten. *Instytut Ochrony Roslin, Reguly, Polen* 41: 15–36.

Dahiya MS, Jain GC. 1977. Inhibitory effects of cannabidiol and tetrahydrocannabinol against some soil inhabiting fungi. *Indian Drugs* 14(4): 76–79.

Danilova YS, Treyvas LY, Trofimova NA. 2010. Практические оценки грибной и бактериальной инфекции в ризосфере хмеля обыкновенного (*Humulus lupulus* L.). *Agricultural Science of Euro-North-East* 2(17): 19–22.

Davis RM, Colyer PM, Rothrock CS, Kochman JK. 2006. Fusarium wilt of cotton: population diversity and implications for management. *Plant Disease* 90: 692–703.

De Corato U. 1997. Le malattie della canapa in Basilicata. *Informatorie Fitopatologico* 47(5): 57–59.

de Meijer EPM. 1995. Fiber hemp cultivars: a survey of origin, ancestry, availability and brief agronomic characteristics. *Journal International Hemp Association* 2(2): 66–73.

Dempsey JM. 1975. "Hemp," pp. 46–89 in *Fiber crops*. University of Florida Press, Gainesville.

Denman S, Crous PW, Groenewald JZE, *et al.* (3 other authors). 2003. Circumscription of *Botryosphaeria* species associated with Proteaceae based on morphology and DNA sequence data. *Mycologia* 95: 294–307.

Dewey LH. 1902. *Local observations card file*. USDA archives, Beltsville, Maryland.

Dewey LH. 1914. "Hemp," pp. 283–347 in: *USDA Yearbook 1913*. United States Department of Agriculture, Washington, DC.

Dewey LH. 1917. *Local observations card file*. USDA archives, Beltsville, Maryland.

Dewey LH. 1919. *Local observations card file*. USDA archives, Beltsville, Maryland.

Dewey LH. 1924. *Local observations card file*. USDA archives, Beltsville, Maryland.

Dippenaar MC, du Toit CLN, Botha-Greeff MS. 1996. Response of hemp (*Cannabis sativa* L.) varieties to conditions in Northwest Province, South Africa. *Journal of the International Hemp Association* 3(2): 63–66.

Dobrozrakova TL, Letova MF, Stepanov KM, Khokhryakov MK. 1956. "*Cannabis sativa* L.," pp. 242-248 *in* Определитель болезней растений. State Publishing House for Plural-Economic Literature, Leningrad.

Dodge BO. 1930. Breeding albinistic strains of *Monilia* bread mold. *Mycologia* 22: 9–38.

Dreschsler 1952. Production of zoospores from germinating oospores of *Pythium ultimum* and *Pythium debaryanum*. *Bulletin of the Torrey Botanical Club* 79: 431–450.

Dumigan C, Maddock S, Bray-Stone D, Deyholos MK. 2023. Hybrid genome assembly of *Berkeleyomyces rouxiae*, an emerging *Cannabis* fungal pathogen causing black root-rot in an aeroponic facility. *Plant Disease*.

Edel-Hermann V, Lecomte C. 2019. Current Status of *Fusarium oxysporum formae speciales* and races. *Phytopathology* 109: 512–530.

El Androusse A, El Aissami A, Rahouti M, *et al.* (3 other authors). 2007. First record of three species of *Pythium* from Moroccan waters. *Acta Botanica Malacitana* 32: 35–40.

Endo S. 1931. The host plants of *Hypochnus centrifugus* (Lév.)Tul. ever recorded in Japan. *Transactions of the Tottori Society of Agricultural Science* 3: 254–270.

Feng C, Villarroel-Zeballos MI, Ficheau PF, *et al.* (3 other authors). 2020. First report of *Neofusicoccum parvum* causing dieback and canker disease on hemp in the United States. *Plant Disease* 104: 3075.

Feng X, Baumgartner K, Dubrovsky S, Fabritius A-L. 2022. First report of root and crown rot caused by *Armillaria gallica* in California. *Plant Disease* 106: 3215.

Ferraris T. 1935. *Parassiti vegetali della canapa*. Rivista Agricola, Rome.

Ferri F. 1959b. La Septoriosi della canapa. *Annali della sperimentazione agraria N.S.* 13:6, Supplement pp. CLXXXIX–CXCVII.

Ferri F. 1961a. Sensibilitá di*Sclerotium rolfsii* avari fungicidi. *Phytopathologia Medeterranea* 3: 139–140.

Ferri F. 1961b. Microflora dei semi di canapa. *Progresso Agricolo (Bologna)* 7(3): 349–356.

Ferri F. 1961c. Le avversità delle plante viste alla lente: Canapa. *Progresso Agricolo (Bologna)* 7: 764–765.

Fields G. 1998. U.S. might enlist fungi in drug war. *USA Today* 17(28) (22 Oct 1998): 1.

Flachs K. 1936. Krankheiten und schädlinge unserer gespinstpflanzen. *Nachrichten über Schädlingsbekämpfung* 11: 6–28.

Frezzi MJ. 1956. Especies de *Pythium* fitopatogenas identificadas en las República Argentina. *Revista de Investigaciones Agrícolas Buenos Aires* 10 (2): 113–241.

Fuller WH, Norman AG. 1944. The nature of the flora on field-retting of hemp. *Proceedings of the Soil Science Society of America* 9: 101–105.

Galloway LD. 1937. Report of the imperial mycologist. *Annual Science Report of the Agricultural Research Institute, New Delhi* 1935-1936: 105–111.

Gautam AK, Kant M, Thakur Y. 2013. Isolation of endophytic fungi from *Cannabis sativa* and study their antifungal potential. *Archives of Phytopathology and Plant Protection* 46: 627–635.

Gauthier N. 2020b. "Considering crop rotations and the potential for carry-over," pg. 25 in Gauthier N, Leonberger K, Bowers C, *eds. Science of hemp: production and pest management*. University of Kentucky Agricultural Experiment Station, no. SR-112, Lexington, Kentucky.

Geiser DM, Aoki T, Bacon CW, *et al.* (66 other authors). 2013. One fungus, one name: defining the genus *Fusarium* in a scientifically robust way that preserves longstanding use. *Phytopathology* 103: 400–408.

Geiser DM, Al-Hatmi AMS, Aoki T, *et al.* (165 other authors). 2021. Phylogenomic analysis of a 55.1 kb 19-gene dataset resolves a monophyletic *Fusarium* that includes the *Fusarium solani* species complex. *Phytopathology* 111: 1064–1079.

Georghiou GP, Papadopoulos C. 1957. *A second list of Cyprus fungi*. Cyprus Department of Agriculture Technical Bulletin TB-5. Nicosia, Cyprus.

Ghani M, Basit A, Anwar M. 1977. *Investigations on the natural enemies of marijuana, Cannabis sativa L. and opium poppy, Papaver somniferum L.* Commonwealth Institute of Biological Control, Pakistan station, Rawalpindi.

Gitman LS. 1968b. Малоизвестные болезни конопли. *Plant Protection* 13(3): 44–45.

Gitman L, Boytchenko E. 1934. "*Cannabis*," p. 45–53 *in* Справочник по болезням бовых лубяных культур. Institute New Bast Raw Materials, Moscow.

Gitman LS, Malikova TP. 1933. "Пораженность стеблей конопли по сортам," pp. 51–57 in *Болезни и вредители новых лубяных культур, вып. 3*. Institute of New Bast Raw Materials, Moscow.

Goidànich G. 1955. *Malattie crittogamiche della canapa*. Associazione Produttore Canapa, Bologna-Naples.

Gonzalez D, Rodriguez-Carres M, Boekhout T, *et al.* (5 other authors). 2016. Phylogenetic relationships of *Rhizoctonia* fungi within the Cantharellales. *Fungal Biology* 120: 603–619.

Goodnight KM, Tian P, Bissonnette KM. 2021. "Disease firsts in the first year of hemp production in Missouri," p. 33 in *Second annual Science of Hemp Conference*. Auburn University, Auburn, Alabama.

Green RJ. 1951. Studies on the host range of the *Verticillium* that causes wilt of *Mentha piperita* L. *Science* 113: 207–208.

Grewal PS. 1989. Effects of leaf-matter incorporation on *Aphelenchoides composticola* (Nematoda), mycofloral composition, mushroom compost quality and yield of *Agaricus bisporus*. *Annals Applied Biology* 115: 299–312.

Gryndler M, Krofta K, Gryndlerová H, *et al.* (3 other authors). 2008. Potentially dangerous fusarioid microorganisms associated with rot of hops (*Humulus lupulus* L.) plants in field culture. *Plant, Soil & Environment* 54: 149–154.

Hall J. 2009. Industrial hemp and sugarcane rotation—not just a pipe dream. *Australian Sugarcane* 2009, June-July, pp. 21–22.

Haney A, Kutscheid BB. 1975. An ecological study of naturalized hemp (*Cannabis sativa* L.) in east-central Illinois. *American Midland Naturalist* 93: 1–24.

Hansen Z, Siegenthaler T, Akin E, Cartwright M, Kelly H. 2020. "Hemp extension in Tennessee; most common diseases of 2019," p. 27 in Gauthier N, *ed. Science of hemp: production and pest management*. University of Kentucky Agricultural Experiment Station no. SR-112. Lexington, Kentucky.

Hartig R. 1883. *Rhizomorpha (Dematophora) necatrix* n. sp. *Untersuchungen aus dem Forstbotanischen Institut zu München* 3: 95–135.

Hartowicz LE, Knutson H, Paulsen A, Eaton BJ, Eshbaugh E. 1971. Possible biocontrol of wild hemp. *Proceedings of the North Central Weed Control Conference* 26: 69.

Hawksworth DL. 2012. Managing and coping with names of pleomorphic fungi in a period of transition. *IMA Fungus* 3(1): 15–24.

Hector GP. 1931. "Report of the economic botanist to the government of Bengal," pp. 35–44 in *Annual Report of the Department of Agriculture, Bengal, for the year 1930–31*. Bengal Secretariat, Calcutta.

Hesler LR. 1912. The New York apple tree canker. *Proceedings of the Indiana Academy of Science* 1911: 325–339.

Highland B. 2020. "The use of Double Nickel 55 WDG and Double Nickel LC fungicide/bactericides in organic plant disease programs for fruit and vegetable production," p. 29 in Gauthier N, *ed. Science of hemp: production and pest management*. University of Kentucky Agricultural Experiment Station, no. SR-112, Lexington, Kentucky.

Hildebrand DC, McCain AM. 1978. The use of various substrates for large scale production of *Fusarium oxysporum* f. sp. *cannabis* inoculum. *Phytopathology* 68: 1099–1101.

Hood ME, Shew HD. 1997. Reassessment of the role of saprophytic activity in the ecology of *Thielaviopsis basicola*. *Phytopathology* 87: 1214–1219.

Hrebenyuk NV. 1984. Распространенне грибов на тресте конопли (The occurrence of fungi on hemp stems). *Микология и фитопатология* 18(4): 322–326.

Hu JH. 2021. First report of crown and root rot caused by *Pythium myriotylum* on hemp (*Cannabis sativa*) in Arizona. *Plant Disease* 105: 2736.

Hu JH, Masson R. 2021. First report of crown and root rot caused by *Pythium aphanidermatum* on industrial hemp (*Cannabis sativa*) in Arizona. *Plant Disease* 105: 2257.

Hughes SJ. 1979. Relocation of some species of *Endophragmia* auct. with notes on relevant generic names. *New Zealand Journal of Botany* 17: 139–188.

Hyre RA. 1944. Tennessee plant disease survey, 1943. *Plant Disease Reporter suppl.* 145: 249–254.

Jerushalmi S, Maymon M, Dombrovsky A, Freeman S. 2020a. Fungal pathogens affecting the production and quality of medical cannabis in Israel. *Plants* 9: e882.

Jerushalmi S, Maymon M, O'Donnell K, Freeman S. 2022. Members of the Fusarium oxysporum complex causing wilt symptoms in medical cannabis in Israel, Italy and North America comprise a polyphyletic assemblage. *Plant Disease*.

Jian FJ, Al-Mamun MA, White NDG, *et al.* (3 other authors). 2019. Safe storage time of FINOLA® hemp (*Cannabis sativa*) seeds with dockage. *Journal of Stored Products Research* 83: 34–43.

Joshi V, Burlakoti P, Babu B. 2022. Diseases/symptoms diagnosed on commercial crop samples submitted to the British Columbia Ministry of Agriculture, Food and Fisheries (BCMAFF) Plant health Laboratory in 2021. *Canadian Plant Disease Survey* 102: 9–16.

Kaur R, Kaur J, Singh RS. 2010. Nonpathogenic *Fusarium* as a biological control agent. *Plant Pathology Journal* 9: 79–91.

Killough DT. 1920. *Unpublished letter of November 16, from USDA Substation No. 5, Temple, TX to USDS headquarters*. USDA archives, Beltsville, Maryland.

Kishi K. 1988. *Plant diseases in Japan*. Zenkoku Nōson Kyōkai, Tōkyō.

Koike ST, Gladders P, Paulus AO. 2007. *Vegetable diseases, a color handbook*. Manson Publishing, London.

Kok CJ, Coenen GCM, de Heij A. 1994. The effect of fibre hemp (*Cannabis sativa* L.) on selected soil-borne pathogens. *Journal of the International Hemp Association* 1(1): 6–9.

Krishna A. 1995. Chemical control of root rot of ganja. *Current Research—University of Agricultural Sciences (Bangalore)* 24(6): 99–100.

Krishna A. 1996. Resistance in ganja cultivars to *Sclerotium rolfsii*. *Indian Journal of Mycology & Plant Pathology* 26: 307.

Kumar N. 2014. Search of a natural remedy for control of fusarial wilting of shisham (*Dalbergia sissoo* Roxb). *International Journal of Engineering Research and Science and Technology* 3(1): 34–44.

Kumar S, Garkoti A, Tripathi HS, *et al.* (3 other authors). 2013. Management of *Rhizoctonia* web blight of urdbean through botanicals and animal products. *Annals of Plant Protection Sciences* 21: 342–344.

Kyokai NSB. 1965. 農林病害虫名鑑. Japan Plant Protection Association, Tokyo.

L'dov N. 1933. Культура лубяных растений в Советской Азии—За индустр (Culture of bast plants in Soviet Asia—for industry). *За индустриализацию Советского Востока* 1933(2): 103–128.

Leibelt W, Senser F. 1972. Die Welke des Hopfens und ihre Bekampfung. *Hopfen-Rundschau* 23: 19–25.

Lentz P. 1977. *Fungi and diseases of Cannabis*. Unpublished manuscript, notes and collected literature. National Fungus Collections, USDA, Beltsville, Maryland.

Lisson SN, Mendham NJ. 1995. Tasmanian hemp research. *Journal International Hemp Association* 2(2): 82–85.

Lomas-Cano T, Boix-Ruiz A, de Cara-García M, *et al.* (4 other authors). 2016. Etiological and epidemiological concerns about pepper root and lower stem rot caused by *Fusarium oxysporum* f. sp. *radicis-capsici* f. sp. nova. *Phytoparasitica* 44: 283–293.

Lombard L, van der Merwe NA, Groenewald JZ, Crous PW. 2015. Generic concepts in Nectriaceae. *Studies in Mycology* 80: 189–245.

Lombard L, Sandoval-Denis M, Lamprecht SC, Crous PW. 2019. Epitypification of *Fusarium oxysporum* – clearing the taxonomic chaos. *Persoonia* 43: 1–47.

Lucas GB. 1975. *Diseases of tobacco, 3rd edn.* Biological Consulting Assoc., Raleigh, North Carolina.

Lung SM, Betz EC, Roberts AJ, Punja ZK. 2019. *Infection of Cannabis sativa cuttings by Fusarium oxysporum and F. proliferatum and investigation into potential biofungicide control*. Poster presentation, Canadian Phytopathological Society.

Ma LJ, van der Does HC, Borkovich KA, *et al.* (64 other authors). 2010. Comparative genomics reveals mobile pathogenicity chromosomes in *Fusarium*. *Nature* 464: 367–373.

Mahdizadeh V, Safaie N, Aghajani MA. 2011. New hosts of *Macrophomina phaseolina* in Iran. *Journal of Plant Pathology* 94 (4, Supplement): S4.70.

Mahmood R, Naseer M, Shahid AA, Usmani A, Ali A. 2014. Screening of indigenous weed extracts against *Fusarium solani* with an emphasis to soil fertility-related microbial activities. *Journal of Food, Agriculture and Environment* 12: 958–962.

Mándy G, Bòcsa I. 1962. A kender (*Cannabis sativa* L.). *Magyarország kultúrflórája 14.* Akadémiai Kiadó, Budapest.

Marois JJ, Johnston SA, Dunn MT, Papavizas GC. 1982. Biological control of *Verticillium* wilt of eggplant in the field. *Plant Disease* 66: 1166–1168.

Matthews VD. 1931. *Studies on the genus Pythium.* University of North Carolina Press, Chapel Hill.

Matuo T. 1949. On the "sore shin" disease of grown-up cotton and several other fibre plants. *Annals of the Phytopathological Society of Japan* 13: 26–28.

May TW, Redhead SA, Bensch K, *et al.* (4 other authors). 2019. Chapter F of the International Code of Nomenclature for algae, fungi, and plants as approved by the 11th International Mycological Congress, San Juan, Puerto Rico, July 2018. *IMA Fungus* 10: e21.

McCain AH, Noviello C. 1985. Biological control of *Cannabis sativa. Proceedings, 6th International Symposium on Biological Control of Weeds* 1985: 635–642.

McEno J. 1990. Hydroponic IPM. *The Growing Edge* 1(3): 35–39.

McGehee CS, Raudales RE. 2021. Pathogenicity and mefenoxam sensitivity of *Pythium, Globisporangium,* and *Fusarium* isolates from coconut coir and rockwool in marijuana (*Cannabis sativa* L.) production. *Frontiers in Agronomy* 3: e706138.

McGehee CS, Apicella P, Raudales R, *et al.* (3 other authors). 2019. First report of root rot and wilt caused by *Pythium myriotylum* on hemp (*Cannabis sativa* L.) in the United States. *Plant Disease* 103: 3288.

McNeill J, *chair.* 2012. *International Code of Nomenclature for Algae, Fungi, and Plants (Melbourne Code).* Koeltz Scientific Books, Königstein, Germany.

McPartland JM. 1983a. Fungal pathogens of *Cannabis sativa* in Illinois. *Phytopathology* 72: 797.

McPartland JM. 1995b. *Cannabis* pathogens IX: anamorphs of *Botryosphaeria* species. *Mycotaxon* 53: 417–424.

McPartland JM. 1995f. Mycotoxin alternariol: letter to the editor. *Townsend Letter for Doctors* No. 139: 88.

McPartland JM. 1996. A review of *Cannabis* diseases. *Journal of the International Hemp Association* 3: 19–23.

McPartland JM. 2000c. Data sheet: *Fusarium oxysporum* f.sp. *cannabis. Crop Protection Compendium,* CAB International, Wallingford, UK.

McPartland JM. 2018. *Cannabis* systematics at the levels of family, genus, and species. *Cannabis and Cannabinoid Research* 3: 203–212.

McPartland JM, Cubeta MA. 1997. New species, combinations, host associations and location records of fungi associated with hemp (*Cannabis sativa*). *Mycological Research* 101: 853–857.

McPartland JM, Cubeta MA. 2022. "Rhizoctonia sore shin and root rot," in Gauthier NW, Thiessen LD, *eds. Compendium of Cannabis diseases.* APS Press, St Paul, Minnesota.

McPartland JM, Hillig KW. 2004b. Fusarium wilt. *Journal of Industrial Hemp* 9(2): 67–77.

McPartland JM, Nicholson J. 2003. Using parasite databases to identify potential nontarget hosts of biological control organisms. *New Zealand Journal of Botany* 41(4): 699–706.

McPartland JM, Schoeneweiss DF. 1984. Hyphal morphology of *Botryosphaeria dothidea* in vessels of unstressed and drought-stressed *Betula alba. Phytopathology* 74: 358–362.

McPartland JM, West D. 1999. Killing *Cannabis* with mycoherbicides. *Journal of the International Hemp Association* 6(1):1, 4–8.

McPartland JM, Clarke RC, Watson DP. 2000. *Hemp diseases and pests – management and biological control.* CABI Publishing, Wallingford, UK.

McPartland JM, Punja ZK, Thiessen LD. 2022. "Fusarium crown and root rot," in Gauthier NW, Thiessen LD, *eds. Compendium of Cannabis diseases.* APS Press, St Paul, Minnesota.

Mersha Z, Hansen Z. 2022. "Southern blight," in Gauthier NW, Thiessen LD, *eds. Compendium of Cannabis diseases.* APS Press, St Paul, Minnesota.

Mersha Z, Kering M, Dhakal R, Rahemi A, Ren SX. 2020. "Southern blight and foliar diseases of hemp in central Virginia," p. 35 in Gauthier N, *ed. Science of hemp: production and pest management.* University of Kentucky Agricultural Experiment Station no. SR-112. Lexington, Kentucky.

Miller PR, Weiss F, O'Brien MJ. 1960. *Index of plant diseases in the United States.* USDA Agriculture Handbook No. 165, Washington, DC.

Mishra D. 1987. Damping off of *Cannabis sativa* caused by *Fusarium solani* and its control by seed treatment. *Indian Journal of Mycology and Plant Pathology* 17(1): 100–102.

Morgan-Jones G, White JF. 1987. Notes on Ceolomycetes. II. Concerning the *Fusicoccum* anamorph of *Botryosphaeria ribis. Mycotaxon* 30: 117–125.

Nair KR, Ponnappa KM. 1974. Survey for natural enemies of *Cannabis sativa* and *Papaver somniferum.* Commonwealth Institute of Biological Control, India Station Report, pp 39–40.

Nalim FA, Samuels GJ, Wijesundera RL, Geiser DM. 2011. New species from the *Fusarium solani* species complex derived from perithecia and soil in the Old World tropics. *Mycologia* 103: 1302–1330.

National Research Council. 2011. *Feasibility of using mycoherbicides for controlling illicit drug crops.* National Academies Press, Washington, DC.

Neergaard P. 1945. *Danish species of Alternaria and Stemphylium.* Munksgaard, Copenhagen.

Neergaard P. 1958. Infection of Danish seeds by *Rhizoctonia solani* Kuehn. *Plant Disease Reporter* 42: 1276–1278.

Nees von Esenbeck CG (Carus CG, *ed*). 1823. Pythium. *Nova Acta Physico-medica Academiae Caesareae Leopoldino-Carolinae Naturae Curiosum* 11(2): 515–516.

Nel, WJ, Duong TA, Wingfield BD, Wingfield MJ, de Beer ZW. 2018. A new genus and species for the globally important, multi-host root pathogen *Thielaviopsis basicola. Plant Pathology* 67: 871–882.

Nobel M, Richardson MJ. 1968. *An annotated list of seed-borne diseases.* Commonwealth Mycological Institute, Kew, UK.

Noble R, Coventry E. 2005. Suppression of soil-borne plant diseases with composts: a review. *Biocontrol Science and Technology* 15: 3–20.

Noviello C. 1957. Segnalazione di *Verticillium* sp. su *Cannabis sativa. Ricerche, Osservazioni e Divulgazioni Fitopatologiche per la Campania ed il Mezzogiorno* 13-14: 161–163.

Noviello C, Snyder WC. 1962. Fusarium wilt of hemp. *Phytopathology* 52: 1315–1317.

O'Donnell K, Gueidan C, Sink S, *et al.* (23 other authors). 2009. A two-locus DNA sequence database for typing plant and human pathogens within the *Fusarium oxysporum* species complex. *Fungal Genetics and Biology* 46: 936–948.

O'Donnell K, Rooney AP, Proctor RH, *et al.* (12 other authors). 2013. Phylogenetic analyses of RPB1 and RPB2 support a middle Cretaceous origin for a clade comprising all agriculturally and medically important fusaria. *Fungal Genetics and Biology* 52: 20–31.

O'Donnell K, Al-Hatmi AMS, Aoki T, *et al.* (34 other authors). 2020. No to *Neocosmospora*: Phylogenomic and practical reasons for continued inclusion of the Fusarium solani species complex in the genus *Fusarium. mSphere* 5: e00810–20.

O'Donnell K, McPartland JM, Thiessen LD. 2022. "Fusarium wilt," in Gauthier NW, Thiessen LD, *eds. Compendium of Cannabis diseases.* APS Press, St Paul, Minnesota.

Ogoshi A. 1996. "Introduction—the genus *Rhizoctonia*," pp. 1–9 in: Sneh B. *et al.*, eds. *Rhizoctonia species: Taxonomy, molecular biology, ecology, pathology and disease control.* Springer, Dordrecht.

Ogoshi A, Oniki M, Araki T, Ui T. 1983. Studies on the anastomosis groups of binucleate Rhizoctonia and their perfect states. *Journal of the Faculty of Agriculture, Hokkaido University* 61: 244–260.

Ondřej M. 1991. Výskyt hub na stoncích konopí (*Cannabis sativa* L.). *Len a Konopí* (Sumperk, Czech Rep) 21: 51–57.

Pandotra VR, Sastry KSM. 1967. Wilt: a new disease of hemp in India. *Indian Journal of Agricultural Science* 37: 520.

Pane A, Cosentino SL, Copani V, Cacciola SO. 2007. First report of southern blight caused by *Sclerotium rolfsii* on hemp (*Cannabis sativa*) in Sicily and southern Italy. *Plant Disease* 91: 636.

Paolinelli-Alfonzo M, Villalobos-Escobendo JM, Rolshausen P, *et al.* (4 other authors). 2016. Global transcriptional analysis suggests *Lasiodiplodia theobromae* pathogenicity factors involved in modulation of grapevine defensive response. *BMC Genomics* 17: e616.

Patel JK, Joshi KR, Prajapati HN, Jage NP. 2014. Bait techniques for isolation of *Pythium aphanidermatum* causing damping-off of chilli from soil and efficacy of bio-agents in *vitro. Trends in Biosciences* 76: 474–476.

Patschke K, Gottwald R, Müller R. 1997. Erste Ergebnisse phytopathologischer Beobachtungen im Hanfanbau im Land Brandenburg. *Nachrichtenblatt des Deutschen Pflanxenschutzdienstes* 49: 286–290.

Paugh KR, Del Castillo Múnera J, Swett CL. 2022. First report of *Fusarium falciforme* (FsSC 3+4) causing rot of industrial hemp (*Cannabis sativa*) in California. *Plant Disease* 106: 1723.

Paulsen AQ. 1971. *Plant diseases affecting marijuana (Cannabis sativa).* Unpublished manuscript, Kansas State University.

Peck CH. 1881. "Report of the botanist," pp. 24–57 *in* in *Thirty-fourth annual report on the New York State Museum of Natural History.* Weed, Parsons & Co., Albany.

Pellegrini M, Ercole C, Gianchino C, *et al.* (3 other authors). 2021. *Fusarium oxysporum* f. sp. *cannabis* isolated from *Cannabis sativa* L.: *in vitro* and *in planta* biocontrol by a plant growth promoting-bacteria consortium. *Plants* 10: e2436.

Pem D, Jeewon R, Hyde KD, *et al.* (4 other authors). 2021. Species concepts of Dothideomycetes: classification, phylogenetic inconsistencies and taxonomic standardization. *Fungal Diversity* 109: 283–319.

Petri L. 1942. Rassegna dei casi fitopatologici osservati nel 1941. *Bollettino della R. Stazione di Patologia vegetale di Roma, N.S.* 22(1): 1–62.

Phillips AJL, Alves A, Correia A, Luque J. 2005. Two new species of *Botryosphaeria* with brown, 1-septate ascospores and *Dothiorella* anamorphs. *Mycologia* 97: 513–529.

Phillips AJL, Crous PW, Alves A. 2007. *Diplodia seriata*, the anamorph of "*Botryosphaeria*" *obtusa. Fungal Diversity* 25: 141–155.

Phillips AJL, Alves A, Abdollahzadeh J, *et al.* (4 other authors). 2013. The *Botryosphaeriaceae*: genera and species known from culture. *Studies in Mycology* 76: 51–167.

Pietkiewicz TA. 1958. Mikroflora nasion konopi. Przeglad literatury. *Roczniki Nauk Rolniczych Seria A* 77(4): 577–590.

Pitman TL, Philbrook RN, Vetterli MR, Warren JG. 2020. First report of *Pythium ultimum* causing crown rot in greenhouse grown *Cannabis sativa* (L.) in California. *Plant Disease* 105: 1230.

Pitman TL, Philbrook RN, Warren JG. 2021. First report of *Pythium myriotylum* causing root rot in *Cannabis sativa* in California. *Plant Disease* 105: 3766.

Plăcintă DD, Murariu D. 2016. Fungus evaluation from seeds germplasm before medium and long term storage. *Cercetări Agronomice în Moldova* 49: 71–82.

Poudel B, Shivas RG, Adorada DL, *et al.* (9 other authors). 2021. Hidden diversity of *Macrophomina* associated with broadacre and horticultural crops in Australia. *European Journal of Plant Pathology* 16: 1–23.

Punithalingam E. 1982. Conidiation and appendage formation in *Macrophomina phaseolina* (Tassi) Goid. *Nova Hedwigia* 36: 249–290.

Punja ZK. 2021a. The diverse mycoflora present on dried cannabis (*Cannabis sativa* L., marijuana) inflorescences in commercial production. *Canadian Journal of Plant Pathology* 43: 88–100.

Punja ZK. 2021b. Epidemiology of *Fusarium oxysporum* causing root and crown rot of cannabis (*Cannabis sativa* L., marijuana) plants in commercial greenhouse production. *Canadian Journal of Plant Pathology* 43: 216–235.

Punja ZK. 2021c. First report of *Fusarium proliferatum* causing crown and stem rot, and pith necrosis, in cannabis (*Cannabis sativa* L., marijuana) plants. *Canadian Journal of Plant Pathology* 43: 236–245.

Punja ZK. 2021d. Emerging diseases of *Cannabis sativa* and sustainable management. *Pest Management Science* 77: 3857–3870.

Punja ZK, Ni L. 2021. The bud rot pathogens infecting cannabis (*Cannabis sativa* L., marijuana) inflorescences: symptomology, species identification, pathogenicity and biological control. *Canadian Journal of Plant Pathology* 43: 827–854.

Punja ZK, Rodriguez G. 2018. *Fusarium* and *Pythium* species infecting roots of hydroponically grown marijuana (*Cannabis sativa* L.) plants. *Canadian Journal of Plant Pathology* 40: 498–513.

Punja ZK, Scott C, Chen S. 2018. Root and crown rot pathogens causing wilt symptoms on field-grown marijuana (*Cannabis sativa* L.) plants. *Canadian Journal of Plant Pathology* 40: 528–551.

Punja ZK, Collyer D, Scott C, Lung S, Holms J. Sutton D. 2019. Pathogens and molds affecting production and quality of *Cannabis sativa* L. *Frontiers in Plant Science* 10: e1120.

Punja ZK, Ni L, Roberts A. 2021. The *Fusarium solani* species complex infecting cannabis (*Cannabis sativa* L., marijuana) plants and a first report of *Fusarium* (*Cylindrocarpon*) *lichenicola* causing root and crown rot. *Canadian Journal of Plant Pathology* 43: 567–581.

Punja ZK, Scott C, Lung S. 2022. Several *Pythium* species cause crown and root rot on cannabis (*Cannabis sativa* L., marijuana) plants grown under commercial greenhouse conditions. *Canadian Journal of Plant Pathology* 44: 66–81.

Rahnama M, Szarka D, Boyadjieva L, Ward Gauthier NA. 2022. First report of black root rot (*Berkeleyomyces rouxiae*) on greenhouse hemp (*Cannabis sativa*) in Kentucky. *Plant Disease* 106: 2534.

Raillo AI. 1950. *Грибы рода фузариум.* Publishing House of Agricultural Literature, Moscow.

Rao VG. 1977. Diseases of fibre crops in India. *Sydowia* 30: 164–185.

Rataj K. 1957. Skodlivi cinitele pradnych rostlin. *Prameny Literatury* 2: 1–123.

Rayllo AN. 1927. Опыты искусственного заражения грибком *Hypochnus solani* Prill. et Delacroix. *Materials on Mycology and Phytopathology, Leningrad* 6(1): 166–179.

Rippon JW. 1988. *Medical mycology, 3rd edn.* W.B. Saunders Co., Philadelphia, Pennsylvania.

Ristaino JB, Perry KB, Lumsden RD. 1991. Effect of solarization and *Gliocladium virens* on sclerotia of *Sclerotium rolfsii*, soil microbiota, and the incidence of southern blight of tomato. *Phytopathology* 81: 1117–1124.

Roberts AJ, Punja ZK. 2022. Pathogenicity of seedborne *Alternaria* and *Stemphylium* species and stem-infecting *Neofusicoccum* and *Lasiodiplodia* species to cannabis (*Cannabis sativa* L., marijuana) plants. *Canadian Journal of Plant Pathology* 44: 250–269.

Rogers DP. 1944. On the dates of publication of Schweinitz's synopsis. *Mycologia* 36: 526–531.

Rothrock CS, Avanzato MV, Rupe JC. 2015. "Pythium seed rot, damping-off, and root rot," pp. 76–79 in Hartman GL, *et al.*, eds. *Compendium of soybean diseases and pests, 5ᵗʰ edn.* APS Press, St Paul, Minnesota.

Rouxel F, Alabouvette C, Louvet J. 1979. Recherches sur la résistance des sols aux maladies. IV. Mise en évidence du rôle des *Fusarium* autochtones dans la résistance d'un sol à la fusariose vasculaire du melon. *Ann. Phytopathol.* 11: 199–207.

Rudolph E. 1968. Untersuchungen zur Welkekrankheit des Hopfens (*Humulus lupulus* L.). *Zeitschrift für Pflanzenkrankheiten und Pflanzenschutz* 75: 401–412.

Rutgers AAL. 1913. The fusariums from cankered cacao-bark and *Nectria cancri nova* species. *Annales du Jardin Botanique de Buitenzorg* 27: 59–63.

Sabo J, Durić T, Jasnić S. 2002. *Fusarium* fungi as a pathogen causing hop wild. *Plant Protection Science* 38: 308–310.

Sakalidis ML, Hardy GESJ, Burgess TI. 2011. Use of the Genealogical Sorting Index (GSI) to delineate species boundaries in the *Neofusicoccum parvum-Neofusicoccum ribis* species complex. *Molecular Phylogenetics and Evolution* 60: 333–344.

Sampson PJ, Walker J. 1982. *An annotated list of plant diseases in Tasmania.* Department of Agriculture, Tasmania, Australia.

Sánchez J, Gallego E. 2000. *Pythium* spp. present in irrigation water in the Poniente region of Almería (south-eastern Spain). *Mycopathologia* 150: 29–38.

Sandoval-Denis M, Lombard L, Crous PW. 2019. Back to the roots: a reappraisal of *Neocosmospora*. *Persoonia* 43: 90–185.

Sands DC. 1991. *Interim report: Cannabis sativus.* January/February Report, Cooperative Agreement 58-3K47-9-036. Boseman, Montana

Schneider K. 1987. Gene scientist freed germs in 1984 tests. *New York Times* Sept. 2, A1, D19.

Schoener J, Wilhelm R, Rawson R, Schmitz P, Wang S. 2017a. Five *Fusarium* species associated with root rot and sudden death of industrial hemp in Nevada. (Abstr.) *Phytopathology* 107: S5.98.

Schoener J, Wilhelm R, Rawson R, Schmitz P, Wang SH. 2017b. First detection of *Fusarium oxysporum* and *F. solani* causing wilt of medical marijuana plants in Nevada. *Phytopathology* 107: S5.98.

Schoener J, Wilhelm R, Wang S. 2018a. Molecular identification of *Alternaria* species associated with imported industrial hemp seed. *Phytopathology* 108: S1.185

Schroers HJ, Samuels GJ, Zhang N, *et al.* (3 other authors). 2016. Epitypification of *Fusisporium* (*Fusarium*) *solani* and its assignment to a common phylogenetic species in the *Fusarium solani* species complex. *Mycologia* 108: 806–819.

Schweinitz LD de. 1832. Synopsis fungorum in America Boreali media degentium. *Transactions of the American Philosophical Society*, New Series, 4: 141–316.

Scott CA. 2021. *Disease management approaches for Cannabis sativa L.* Master's thesis, Simon Fraser University, British Columbia, Canada.

Scott CA, Punja ZK. 2023. Biological control of *Fusarium oxysporum* causing damping-off and *Pythium myriotylum* causing root and crown rot on cannabis (*Cannabis sativa* L.) plants. *Canadian Journal of Plant Pathology* 45: 238–252.

Scott M, Rani M, Samstatly J, Charron JB, Jabaji S. 2018. Endophytes of industrial hemp (*Cannabis sativa* L.) cultivars: identification of culturable bacteria and fungi in leaves, petioles, and seeds. *Canadian Journal of Microbiology* 64: 664–680.

Serkov VA, Pluzhnikov II. 2012. Эффективность предпосевной обработки семян однодомной конопли посевной (Efficiency of presowing seed treatment in monoecious hemp). Достижения науки и техники АПК 2012(2): 46–46.

Serkov VAS, Belousov RO, Alexandrova MR, Davydova OK. 2019. Новый сорт конопли посевной Милена. *International Agriculture Journal* 5: 16–19.

Serzane M. 1962. "Kaņepju – *Cannabis sativa* L." pp. 366–369 *in Augu Slimibas, Praktiskie Darbi.* Latvijas Valsts Izdevnieciba, Riga.

Seymour AB. 1929. *Host index of the fungi of North America.* Harvard University Press, Cambridge, Massachusetts.

Shahzad S, Ghaffar A. 1995. New records of soilborne root infecting fungi in Pakistan. *Pakistan Journal of Botany* 27: 209–216.

Shay R. 1975. Easy-gro fungus kills pot among us. *The Daily Californian*, March 14, p. 3.

Shaw FJF, Ajrekar SL. 1915. The genus *Rhizoctonia* in India. *Memoirs of the Department of Agriculture in India, Botanical Series* 7: 177–195.

Shear CJ, Stevens NE. 1924. *Botryosphaeria* and *Physalospora* on currant and apple. *Journal of Agricultural Research* 28: 589–598.

Shrestha U, Dee ME, Ownley BH, Butler DM. 2018. Soil disinfestation reduces germination and affects colonization of *Sclerotium rolfsii* sclerotia. *Phytopathology* 108: 342–351.

Simay EI. 1990. Adatok a *Macrophomina phaseolina* (Tassi) Goid. gazdanövényköréhez Magyarországon. *Növénytermelés* 39(1): 23–27.

Singh R. 2021. "Two new diseases of industrial hemp in Louisiana," p. 51 in *Second annual Science of Hemp Conference*. Auburn University, Auburn, Alabama.

Singh R, de Souza M, Burks T, Price T. 2022. First report of southern blight of industrial hemp caused by *Athelia rolfsii* in Louisiana. *Plant Health Progress* 23.

Singh RV, Singh AK, Ahmad R, Singh SP. 1998. "Influence of agronomic practice on foliar blight, and identification of alternate hosts in the rice-wheat cropping system," pp. 346–348 in Duveiller E. *et al.*, eds. *Helminthosporium blights of wheat: spot blotch and tan spot.* CIMMYT, Mexico City.

Skovgaard K, Nirenberg HI, O'Donnell K, Rosendahl S. 2001. Evolution of *Fusarium oxysporum* f. sp. *vasinfectum* races inferred from multigene genealogies. *Phytopathology* 91: 1231–1237.

Smith SN, Snyder WC. 1971. Relationship of inoculums density and soil types to severity of *Fusarium* wilt of sweet potato. *Phytopathology* 61: 1049–1051.

Snyder WC, Hansen HN. 1940. The species concept in *Fusarium. American Journal of Botany* 27: 64–67.

Snyder WC, Hansen HN. 1941. The species concept in *Fusarium* with reference to section Martiella. *American Journal of Botany* 28: 738–742.

Solarska E. 1981. Więdnięcie infekcyjne chmielu (*Humulus lupulus* L.) uprawianego na. Lubelszczyźnie. *IUNG Puławy, seria R* 156: 23.

Sorrentino R, Rergamo R, Battaglia V, *et al.* (3 other authors). 2019. Characterization and pathogenicity of *Fusarium solani* causing foot rot on hemp (*Cannabis sativa* L.) in southern Italy. *Journal of Plant Diseases and Protection* 126: 585–591.

Sorrentino R, Baldi GM, Battaglia V, *et al.* (3 other authors). 2021. First report of white root rot of hemp (*Cannabis sativa* L.) caused by *Dematophora necatrix* in Campania region (southern Italy). *Plant Disease* 105: 3299.

Spaar D, Kleinhempel H, Fritzsche R. 1990. *Öl- und Faserpflanzen.* Springer-Verlag, Berlin

Stevens NE. 1933. Two apple black rot fungi in the United States. *Mycologia* 25: 536–548.

Stevenson JA. 1926. *Foreign plant diseases.* USDA Bureau of Plant Industry, Washington, DC.

Stewart-Wade SM. 2011. Plant pathogens in recycled irrigation water in commercial plant nurseries and greenhouses: their detection and management. *Irrigation Science* 29: 267–297.

Suyal N, Upadhyaya ML, Gupta RC. 2011. Studies on air mycoflora over the field of *Cannabis sativa* L. in Almora hills. *Vegetos* 24: 183–190.

Tai [Dai] FL. 1979. *Sylloge fungorum sinicorum.* Science Press, Beijing.

Tang LL, Song XX, Zhang LG, Wang J, Zhang SQ. 2021. First report of leaf spot on industrial hemp (*Cannabis sativa*) caused by *Alternaria alternata* in China. *Plant Disease* 105: 3294.

Tang LL, Fan C, Yuan HM, Wu GW, Zhang SQ. 2022. First report of Fusarium wilt of industrial hemp (*Cannabis sativa*) caused by *Fusarium oxysporum* in northeast China. *Plant Disease* 106(12): 3205.

Taubenhaus J, Killough DT. 1923. *Texas root rot of cotton and methods of its control*. Texas Agricultural Experiment Station Bulletin No. 307, College Station, Texas.

Tehon LR, Boewe GH. 1939. Charcoal rot in Illinois. *Plant Disease Reporter* 23: 312–321.

Thiessen L. 2019. *Hemp diseases in North Carolina*. Department of Entomology and Plant Pathology, North Carolina State University. Available at: https://growingsmallfarms.ces.ncsu.edu/wp-content/uploads/2019/03/Hemp-Diseases-Lindsey-Thiessen.pdf?fwd=no

Thiessen L. 2020. "Industrial hemp disease pressures in North Carolina," p. 45 in Gauthier N, *ed*. *Science of hemp: production and pest management*. University of Kentucky Agricultural Experiment Station no. SR-112. Lexington, Kentucky.

Thiessen LD, Schappe T, Cochran S, Hicks K, Post AR. 2020. Surveying for potential diseases and abiotic disorders of industrial hemp (*Cannabis sativa*) production. *Plant Health Progress* 21(4): 321.

Thompson TA. 1979. *Some aspects on the taxonomy, ecology, and histology of Pythium pringsheim species associated with Ficus distichus in estruaries and marine habits of British Columbia*. Master's thesis, University of Arizona, Tucson.

Tiourebaev KS, Nelson S, Sands DC, *et al.* (4 other authors). 2000. "Amino acid excretion enhances virulence of bioherbicides," pp. *295–299 in Proceedings of the X International Symposium on biological control of weeds*. Montana State University, Bozeman, Montana.

Tiourebaev KS, Semenchenko GV, Sands DC, *et al.* (5 other authors). 2001. Biological control of infestations of ditchweed (*Cannabis sativa*) with *Fusarium oxysporum* f. sp. *cannabis* in Kazakhstan. *Biocontrol Science and Technology* 11: 535–540.

Trotuş E, Naie M, 2008. Cercetări privind reducerea atacului unor agenţi patogeni şi dăunători specifici culturilor de cânepă prin tratamentul chimic al seminţei. *Lucrari Stiintifice, Universitatea de Stiinte Agricole Si Medicina Veterinara "Ion Ionescu de la Brad" Iasi, Seria Agronomie* 51(2): 219–223.

Trunov GA. 1936. "Материалы по фитопатологическому изучению конопель," pp. 69-113 *in Collection of scientific papers on plant protection*. Ukrainian State Publishing Office, Kiev.

Tsuge T, Sato T, *eds*. 2020. 日本植物病名目録. Japanese Society for Plant Pathology. Available at: https://www.ppsj.org/pdf/mokuroku/mokuroku202001.pdf

Turland NJ, *chair*. 2018. *International Code of Nomenclature for Algae, Fungi, and Plants (Shenzhen Code)*. Koeltz Scientific Books, Königstein, Germany.

Uppal BN. 1933. India: diseases in the Bombay presidency. *International Bulletin of Plant Protection* 6(5): 103–104.

Uzuhashi S, Tojo M, Kakishima M. 2010. Phylogeny of the genus *Pythium* and description of new genera. *Mycoscience* 51: 337–365.

Vakhrusheva TE, Davidian GG. 1979. *Методические указания по инвентаризации болезней и микрофлоры льна и конопли*. All-Union Vavilov Scientific Research Insititute of Crop Cultivation, Leningrad.

Vasiliev AA. 1933. Увядание *лубяных культур* в условиях Средней Азии. *Болезни и вредители новых лубяных культур* 3: 22–24.

Viejobueno J, de los Santos B, Camacho-Sanchez M, *et al.* (3 other authors). 2022. Phenotypic variability and genetic diversity of the pathogenic fungus *Macrophomina phaseolina* from several hosts and host specialization in strawberry. *Current Microbiology* 79: e189.

Wang MM, Chen Q, Diao YZ, Duan WJ, Cai L. 2019. *Fusarium incarnatum-equiseti* complex from China. *Persoonia* 43: 70–89.

Wang SH. 2018. *Industrial hemp crop diseases. What we've seen and what we know*. Nevada Department of Agriculture. Available at: http://agri.nv.gov/uploadedFiles/agrinvgov/Content/Plant/Seed_Certification/Industrial_Hemp/NV%20Industrial%20Hemp%20Pathology%20-%203-7-2018.pdf

Wang SH. 2021. *Diagnosing hemp and cannabis crop diseases*. CABI, Wallingford, UK.

Weber GF. 1931. Blight of carrot caused by *Sclerotium rolfsii* with geographic distribution and host range of the fungus. *Phytopathology* 21: 1129–1140.

Westcott C (Horst RK, revised). 1990. *Wescott's plant disease handbook, 5th edn*. Springer Science, New York.

Windels CE. 1997. "Altering community balance: organic amendments, selection pressures, and biocontrols," pp. 282-300 *in* Andow DA, *et al.*, *eds. Ecological interactions and biological control*. Westview Press, Boulder, Colorado.

Wollenweber HW. 1914. Identification of species of *Fusarium* occurring on the sweet potato, *Ipomoea batatas*. *Journal of Agricultural Research* 2: 251–285.

Wollenweber HW. 1917. Fusaria autographice delineata. *Annales Mycologici* 15: 1–56.

Wollenweber HW. 1930. *Hypomyces haematococcus var. cancri. Fusaria Autographice Delineata* 3: 829.

Wollenweber HW. 1931. Fusarium-Monographie. Fungi parasitici et saprophytici. *Zeitschrift für Parasitenkunde* 3: 269–516.

Wollenweber HW, Reinking OA. 1935. *Die Fusarien*. Paul Parey, Berlin.

Wollenweber HW, Sherbakoff CD, Reinking OA, Johann H, Baley AA. 1925. Fundamentals for taxonomic studies of *Fusarium*. *Journal of Agricultural Research* 30: 833–844.

Yarwood CE. 1981. The occurrence of *Chalara elegans*. *Mycologia* 73: 524–530.

Yu SL. 1973. *Fungal pathogens of Cannabis sativa grown in the USDA garden*. Unpublished manuscript, University of Mississippi, Oxford, Mississippi.

Zaprometov NG. 1928. *Материалы по микофлоре Средней Азии, вып. II*. Novolubinstitut, Tashkent.

Zarzycka H, Jaranowska B. 1977. Alternarioza konopi. *Acta Agrobotanica* 30: 419–421.

Zaurov EI. 1948. Полиморфизм признаков пола у южной конопли и его использование в селекции. *Reports of the Academy of Sciences of the Uzbek SSR* 1948(6): 19–27.

Zelenay A. 1960. Grzby z rodzaju Fusarium wystepujace nanasionach i siewkach konopi oraz ich patogenicznosé. *Prace Naukowe, Instytutu Ochrony Roślin, Poznan* 2(2): 248–249.

Zhalnina LS. 1969. Porazhennost' konopli fuzariozom pri dlitel' nom primenenii udobrenii. *Khimiya v Sel'skom Khozyaistve* 7(11): 33–34.

Zubrin R. 1981. The fungus that destroys pot. *War on Drugs* 1981(June): 61–62.

12 Damping-off diseases of seedlings

Abstract

Damping off is a seedling disease caused by fungi and fungus-like organisms. The latter comprise oomycetes, a distinct phylogenetic lineage. Their primary phenotypic departure from fungi consists of cellulose, rather than chitin, in their cell walls. Five *Pythium* and *Globisporangium* spp. cause most the damage. Fungal offenders include *Botrytis cinerea*, *Rhizoctonia solani*, *Fusarium solani*, *Fusarium oxysporum*, and *Alternaria alternata*. Damping off comes in two flavors: pre-emergent (seeds or seedlings die *before* they emerge from the soil) or post-emergent (seedlings die *after* soil emergence). Understanding the disease cycle and epidemiology is critical. Cultural control is key: avoid heavy, slow-draining soil, and sow seed under conditions favorable for germination. Prevent disease in hydroponic systems with proper sanitation. Many biological controls are commercially available. Fungicides can be applied as seed dressings—historically the first use of fungicides on *Cannabis*.

12.1   Introduction

Damping off arises when seeds struggle to germinate in cold, wet soil. It is a disease of seeds *or* seedlings, before *or* after germination. Damping off primarily appears in field crops, subject to meteorological whims, although damping off may arise in fecklessly managed glasshouses. The disease befalls crops seeded directly into fields (fiber-type and grain-type cultivars). It seldom strikes crops grown from transplanted seedlings (drug-type cultivars).

12.1.1   Symptoms and signs

Pre-emergent damping off kills seeds or seedlings *before* they emerge from the soil. Seeds may die before they germinate. Or they germinate, and the radicle (rootlet) develops a water-soaked lesion—a wet, sunken, darkened, translucent area. A water-soaked lesion can be seen at the tip of the radicle in Fig. 12.1. The lesion spreads, growth ceases, and the radicle and seed discolor and rot. Affected plants fail to break through the surface of the soil.

Post-emergent damping off hits *after* seedlings have emerged from the soil. Young seedlings develop a tan to brown lesion at the soil line, and then wilt and topple over (Fig. 12.2). The rootlet and hypocotyl (part of the stem between rootlet and cotyledons) may discolor and rot. In older seedlings (with up to eight pairs of true leaves), growth ceases, the leaves turn pale yellow and wilt, and eventually the plants fall over. When seedlings are removed from the ground and examined, much of their affected root tissue sloughs off. Infected roots lose their root hairs and darken to a tan, beige, or brown color (see Fig. 11.1A).

Seedlings may develop fungal leaf spots (Kirchner, 1906). Leaf spots by themselves are *not* damping off. *Trichothecium roseum* and *Septoria cannabis* can attack cotyledons. Since stems remain untouched and seedlings stay upright, this is not damping off.

12.1.2   Causal organisms

People formerly believed damping off was caused by excess soil moisture, hence the name. Damping off is caused by fungi as well as oomycetes ("water molds")—two groups that look similar, but differ: oomycetes have cellulose in their cell walls (like plants), instead of the chitin found in fungi. Oomycete hyphae are usually non-septate, whereas fungal hyphae are usually septate. Oomycetes are diploid, whereas fungi are haploid or dikaryotic during the major part of their life cycle.

The sublime weirdness of oomycetes was appreciated by John R. Raper (1911–1974), who studied their sexual lifecycle for 40 years. "Sex, quite aside from its recreational possibilities, has profound and far-reaching biological significance" (Raper, 1959). He named new species, such as *Achlya bisexualis* and *A. ambisexualis*, and grew them on halved hemp seeds floating in water (Fig. 12.3). "*A. bisexualis* and *A. ambisexualis* are capable of a sexual ambivalence in which maleness and femaleness are determined in each mating by common consent of the mated" (Raper, 1959).

12.2   Oomycetes

Two genera of oomycetes, *Pythium* and *Phytophthora*, "specialize" in damping off. Curiously, no *Phytophthora* spp. have been identified in *Cannabis* peer-reviewed literature. Aqueous extracts of *Cannabis sativa* have been used to control *Phytophthora* disease in cocoa, *Theobroma cacao* (Coulibaly *et al.*, 2002). *Phytophthora nicotianae* kills tobacco, and rotating hemp with tobacco breaks the disease cycle (Adedokun *et al.*, 2021).

The genus *Pythium* is a highly divergent group. Uzuhashi *et al.* (2010) divided the genus into five monophyletic clades. Accordingly, Uzuhashi dispersed members of the genus into *Pythium* (*sensu stricto*) and four new genera: *Globisporangium*, *Ovatisporangium*, *Elongisporangium*, and *Pilasporangium*.

© John M. McPartland 2025. *Hemp Diseases and Pests* 2nd Edition (J.M. McPartland)
DOI: 10.1079/9781836990352.0012

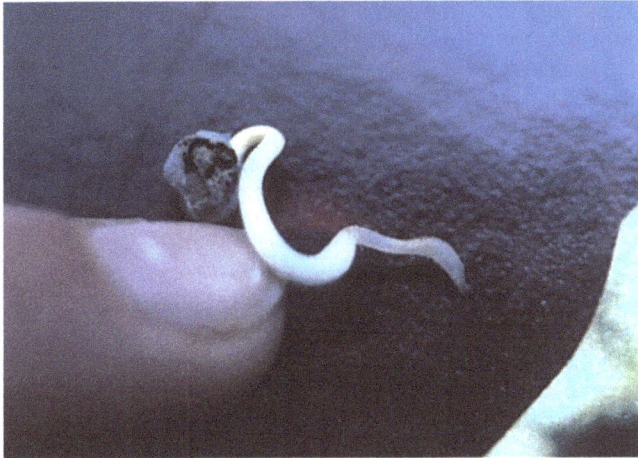

Fig. 12.1. Pre-emergence damping off

Fig. 12.2. Post-emergence damping off

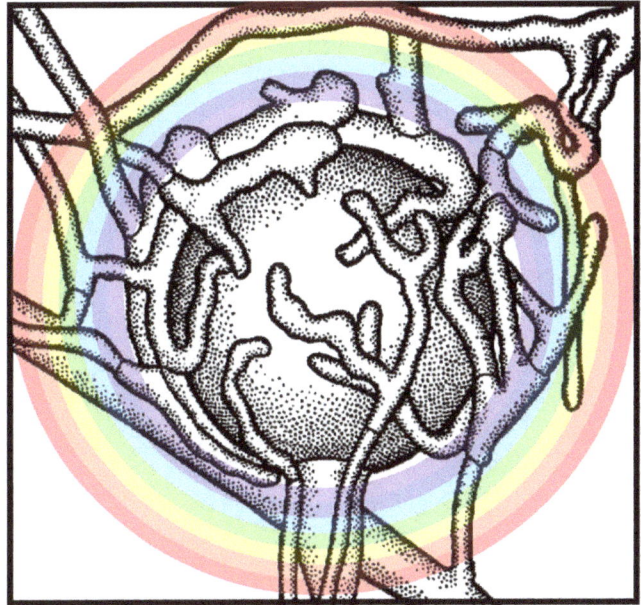

Fig. 12.3. *Achlya ambisexualis* having sex (redrawn from Raper, 1939)

zoospores with flagella (Fig. 12.4). Flagella enable them to swim, and actively find plant hosts.

12.2a. *Globisporangium ultimum* (Trow) Uzuhashi *et al.* 2010

≡*Pythium ultimum* Trow 1901

Description: *Sporangia* are spherical to doliiform, 20–29 × 14–28 μm, and germinate only by germ tubes. *Oogonia* smooth, spherical, formed terminally, 19–23 μm in diameter (Fig. 12.4). Each oogonium is fertilized by one antheridium (rarely 2–3 antheridia). *Antheridia* swollen, sausage-shaped, curved, originating immediately below oogonia. Each fertilized oogonium forms a single oospore. *Oospores* aplerotic, globose, 17–20 μm in diameter, wall often 2 μm thick, or more. Oospores germinate as germ tubes or (rarely) they produce an elongate emission tube, 5–135 μm long, with a vesicle at its end. Vesicles cleave into 8–15 zoospores (Fig. 12.4).

Comments: *G. ultimum* attacks hundreds of plants and occurs worldwide. Damping off has been reported in central Europe (Marquart, 1919; Schultz, 1939) and Argentina (Palmucci, 2015). The pathogen also causes root rot in older plants, in Argentina (Frezzi, 1956) and the USA (Beckerman *et al.*, 2018; Punja *et al.*, 2018; Gauthier, 2020b). *G. ultimum* also appears in studies using hemp seed bait (e.g., Ghaderian *et al.*, 2000; Pettitt *et al.*, 2002).

12.2b. *Pythium aphanidermatum* (Edson) Fitzpatrick 1923, Oomycota; Pythiaceae

Description: *Sporangia* mature into lobate, elongated, irregular swellings, branched or unbranched, 50–1000 × 2–20 μm,

In addition to the five species described below, dozens of other oomycetes have been isolated from hemp seeds, but not to worry: Harvey (1925) discovered that hemp seeds served as excellent "bait" to attract oomycetes. He steam-sterilized seeds and suspended them in pond water. In this fashion, many species new to science have been described from hemp seeds—dozens of *Achlya*, *Allomyces*, *Olpidium*, *Phytophthora*, and *Pythium* spp. (reviewed in McPartland *et al.*, 2000). These organisms shouldn't cause problems unless you store hemp seeds in pond water.

Oomycetes also cause root rot in older plants growing in hydroponic systems (McEno, 1990; Punja *et al.*, 2019), rooted in rockwool media, and in field crops (McPartland, 1996; Beckerman *et al.*, 2017, 2018; Punja *et al.*, 2019). See section 11.2. In these production systems, oomycetes can spread rapidly. They produce sometimes scarcely distinguished from the mycelium (Fig. 12.5A). *Sporangia* germinate into emission tubes of variable lengths, 2–5 µm diameter, and form vesicles at their ends (Fig. 12.5B). *Vesicles* cleave into 100 or more zoospores. *Zoospores* reniform, laterally biflagellate, 7.5 × 12 µm (Fig. 12.5C). *Oogonia* spherical, formed terminally (rarely intercalary), 22–27 µm diameter. Each oogonium is fertilized by 1–2 antheridia. *Antheridia* are formed on hyphae adjacent to oogonia, terminally or intercalary, doliiform to broadly clavate, 9–11 × 10–15 µm. Each fertilized oogonium forms a single oospore. *Oospores* aplerotic, globose, 20–22 µm in diameter, wall 1–2 µm thick (Fig. 12.5D).

Comments: *P. aphanidermatum* has a wide host range and occurs worldwide.

Damping off has been reported in Romania (Ceapoiu, 1958) and India (Galloway, 1937). *P. aphanidermatum* causes root rot, wilting, and mortality in older plants (Beckerman *et al.*, 2017; Schoener *et al.*, 2018b; Punja *et al.*, 2018, 2019). *P. aphanidermatum* appears in studies that use hemp seed bait (e.g., Rodríguez *et al.*, 1999).

12.2c. *Globisporangium debaryanum* (Hesse) Uzuhashi *et al.* 2010

≡*Pythium debaryanum* Hesse 1874, Oomycota; Pythiaceae
Description: *Globisporangium debaranum* is a legitimate name, but its identification is problematic. Hesse's description was based on a mixture of organisms, thus its morphological descriptions in the literature are suspect (Dreschler, 1953). Plaats-Niterink (1981) stated that most "*Pythium debaryanum*" citations are misidentifications of either *G. ultimum* (see above) or *G. irregulare* (see below). We defer a morphological description of *G. debaryanum* to the description of *G. ultimum*.

Comments: Two phylogenetic studies using ITS sequences showed *G. debaryanum* clustering with *G. ultimum* (Paulitz and Adams, 2003; Lévesque and de Cock, 2004). Of course, molecular studies are only as good as the identification of their accessions (identified by morphology). Uzuhashi *et al.* (2010)

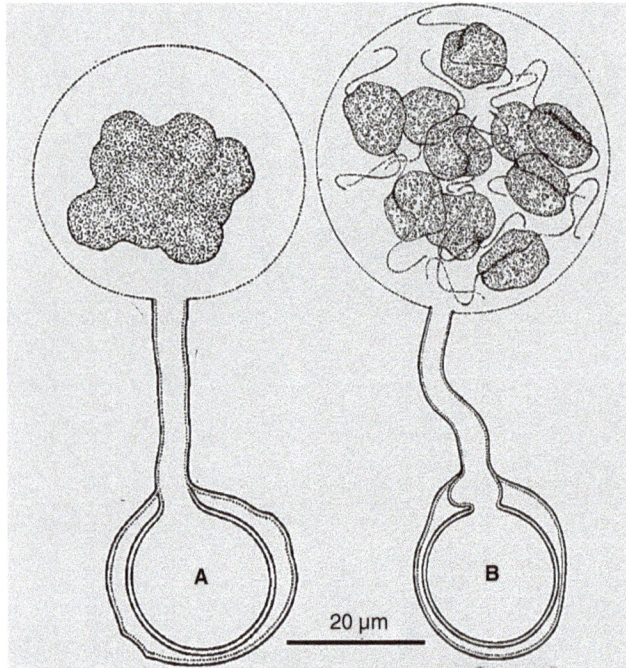

Fig. 12.4. Production of zoospores by *Globisporangium ultimum*: **A.** oospore producing a germ tube and vesicle with immature zoospores; **B.** vesicle with ten mature, flagellated zoospores (redrawn from Drechsler, 1952)

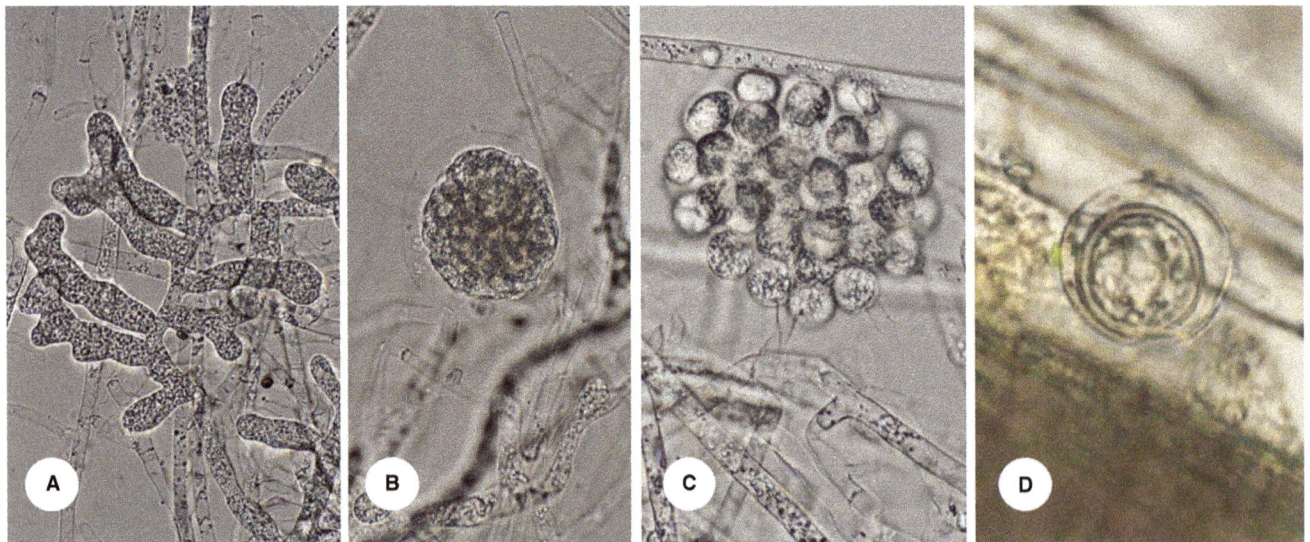

Fig. 12.5. *Pythium aphanidermatum*: **A.** lobate sporangia; **B.** vesicle cytoplasm differentiating into zoospores; **C.** zoospore cluster within vesicle; **D.** oogonium and thick-walled oospore in root tissue. Various magnifications (S. Koike)

transferred *P. debaryanum* to *Globisporangium*, although they did not include this species in their phylogenetic analysis.

G. debaryanum is the most commonly cited cause of damping off in Europe (Kirchner, 1906; Seda, 1935; Flachs, 1936; Schultz, 1939; Rataj, 1957; Ceapoiu, 1958; Serzane, 1962; Kirchner, 1966; Vakhrusheva and Davidian, 1979; Barna *et al.*, 1982; Bócsa and Karus, 1997; Trotuş and Naie, 2008). Many of these citations are probably misidentifications of *G. ultimum*, less so *G. irregulare*.

12.2d. *Globisporangium irregulare* (Buisman) Uzuhashi *et al.* 2010

≡*Pythium irregulare* Buisman 1927
Description: *Sporangia* spherical, terminal or intercalary, 10–20 µm in diameter, rarely produced. *Oogonia* spherical, smooth or sometimes with irregular, finger-like projections, 16–21 µm in diameter. *Antheridia* stalked, sometimes branched, monoclinous or hypogynous, 1–2 per oogonium. *Zoospores* reniform, laterally biflagellate, 7 × 10 µm. *Oospores* mostly aplerotic, occasionally plerotic, globose, 15–18 µm in diameter, wall mostly 1–1.5 µm thick.
Comments: Gauthier *et al.* (2020b) cited the species causing damping off in Kentucky, and commented that *G. irregulare* was not aggressive. McGehee and Raudales (2021) isolated *G. irregulare* from plants with root rot, and inoculated seeds in petri plates. Their germination rate was ~30%, indicating aggressive virulence. However, the researchers removed the seed coat with a scalpel and forceps, "to ensure uniform germination." *G. irregulare* has been isolated from baited hemp seed (Sinobas *et al.*, 1999; Sánchez *et al.*, 2000).

12.2e. *Pythium myriotylum* Drechsler 1930

Description: *Sporangia* terminal or intercalary, filamentous, consisting of undifferentiated and inflated lobulate or digitate elements of variable length and mostly 7–17 µm wide. *Oogonia* globose, smooth-walled, terminal or intercalary 26–32(–35) µm in diameter. *Antheridia* stalked and branched, diclinous or occasionally monoclinous, 3–6 per oogonium. *Zoospores* reniform, laterally biflagellate, 10 × 12 µm. *Oospores* aplerotic, colorless or yellowish, 20–27 µm in diameter, wall up to 2 µm thick.
Comments: *P. myriotylum* is a newcomer to the *Cannabis* world. Punja *et al.* (2018) isolated *G. myriotylum* from older plants with root rot (see section 11.2). Gauthier *et al.* (2020b) cited the species causing pre-emergent damping off in Kentucky. It has also been isolated in North Carolina (Thiessen, pers. comm., 2020).

12.3 Fungi

Several fungi (not oomycetes) cause damping off. They are listed below, with full descriptions provided elsewhere, under

the diseases they cause in mature plants (e.g., gray mold, sore shin, Fusarium wilt).

12.3a. *Fusarium solani* (Martius) Saccardo 1881

12.3b. *Fusarium oxysporum* Schlechtendal 1824

Comments: For morphological descriptions of *F. solani* and *F. oxysporum*, see sections 11.3 and 11.4. Both species have been implicated in damping off (Zelenay, 1960; Barloy and Pelhate, 1962; Mishra, 1987; Patschke *et al.*, 1997). Mishra (1987) reported *F. solani* causing 78% mortality in *Cannabis* seedlings (44% as post-emergence, 34% as pre-emergence). *F. solani* has caused damping off in Kentucky (Gauthier, 2020b) and North Dakota (Khan *et al.*, 2023). Pellegrini *et al.* (2021) reported *F. oxysporum* reducing seed germination by 45% compared to control plants. Seedlings that survived were damaged and stunted.

Barloy and Pelhate (1962) illustrated different symptoms caused by these two fungi (Fig. 12.6). Seedlings with *F. solani* developed red discolored roots, whereas seedlings with *F. oxysporum* became enveloped in a pink mass of hyphae. *Fusarium* spp. also cause cotyledon drop (Rataj, 1957). Studies of seed-borne infection often identify *F. oxysporum* (Pietkiewicz, 1958; Zelenay, 1960; Noble and Richardson; 1968, Ghani *et al.*, 1977; Plăcintă and Murariu, 2016).

Zelenay (1960) named two other taxa as pathogens of hemp seeds and seedlings, which he called *Fusarium avenaceum* var. *herbarum* (Corda) Saccardo 1886 and *Fusarium javanicum* var. *radicicola* Wollenweber 1931. Both these taxa are problematic.

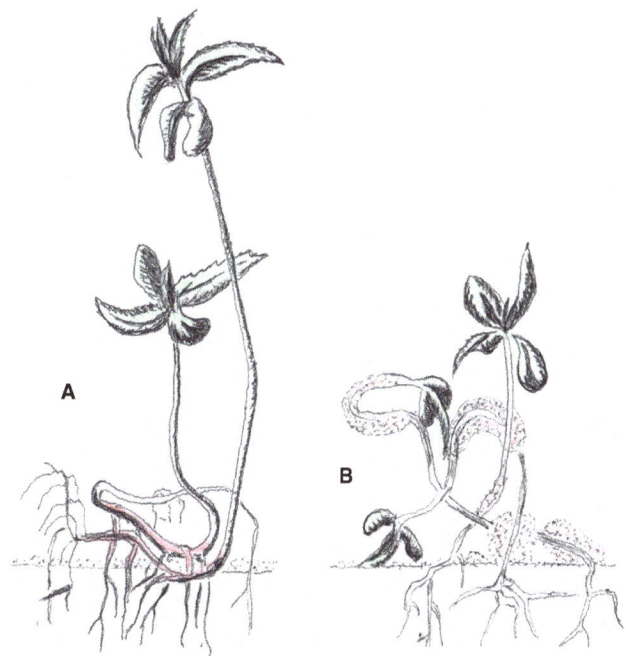

Fig. 12.6. *Fusarium* damping off: **A.** *Fusarium solani*; **B.** *Fusarium oxysporum* (redrawn from Barloy and Pelhate, 1962)

Saccardo (1886) coined *F. avenaceum* (Fries) Saccardo with no subspecies name. Sandoval-Denis *et al.* (2019) lumped *F. javanicum* var. *radicicola* into *F. solani*.

12.3c. *Rhizoctonia solani* Kühn 1858

Comments: For a morphological description see section 11.6. Dippenaar *et al.* (1996) reported heavy losses by *R. solani* despite spraying seedlings with a fungicide (Rhizolex™). Gauthier (2020b) also reported damping off by *R. solani*, in Kentucky. *R. solani* does not require excess moisture, unlike most damping-off fungi.

12.3d. *Botrytis cinerea* Persoon 1794

Comments: For a morphological description see section 9.2. Flachs (1936) first described *B. cinerea* causing damping off in hemp. He photographed post-emergent seedlings covered by fuzzy *B. cinerea* mycelium. The pathogen spreads by seed-borne infection in *Cannabis* (Pietkiewicz, 1958; Noble and Richardson, 1968; Patschke *et al.*, 1997). Expect epidemics in seeds harvested from females infested by gray mold.

Confirmation that *B. cinerea* is the primary pathogen causing damping off is difficult. This fungus is an aggressive secondary invader that colonizes tissues damaged by other factors. *Cannabis* seedlings that are declining due to post-emergent damping off caused by a *Pythium* sp. can be readily colonized by *B. cinerea* and show extensive gray sporulation. In this case, concluding that the disease is Botrytis damping off would be a misdiagnosis.

12.3e. *Alternaria* spp.

Comments: Several *Alternaria* spp. cause brown blight in mature plants late in the growing season (see section 11.10). *A. alternata* has caused seed-borne infections (Pietkiewitcz, 1958; Ferri, 1961b; Plăcintă and Murariu, 2016; Jian *et al.*, 2019). Haney and Kutsheid (1975) estimated that *A. alternata* destroyed 20.5% of seeds developing on plants. Schoener *et al.* (2018a) isolated two other species, *A. infectoria* and *A. tenuissima*, on *Cannabis* seedlings growing in a germination tray. Sorrentino *et al.* (2023) used ITS sequencing to identify *Alternaria rosae* Simmons & Hill causing pre-emergent damping off in Italy. They fulfilled Koch's postulates.

12.4 Olpidiomycota

In a DNA study of root endophytes in China, Wei *et al.* (2021) isolated an ITS sequence corresponding to *Olpidium brassicae* (Woronin) Dang 1886. This species is an obligate parasite that infects the epidermal cells of roots. *O. brassicae* parasitism is often symptomless, but in some hosts it causes damping off in young seedlings and leaf chlorosis in older seedlings. This occurs when soil humidity is high, in poorly drained clay soils in cool conditions. The disease status of *O. brassicae* was not explicated by Wei *et al.* (2021).

O. brassicae infects many cultivated and non-cultivated plants, and it causes problems in cabbage and tobacco seedlings. The species has a worldwide distribution. Primary infections occur via flagellated zoospores, and the species overwinters in soil as thick-walled "resting spores." *O. brassicae* is a minor pathogen, mainly of seedbeds. Its significance lies in its ability to transmit plant viruses. It is a primitive beast, formerly classified in the *Chytridiomycota*. A phylogenetic analysis by Chang *et al.* (2021) indicated that the ancestor of *Olpidium* spp. diverged from Chytrids around 680 million years ago—the closest living organism to non-flagellated, terrestrial fungi.

12.4.1 Differential diagnosis of damping off

Absentee seedlings in a field plot may be due to many causes. The seeds could have been dead before they went into the ground. Birds and rodents dig them up. Some underground insects (wireworms and root maggots) kill seeds before they emerge from soil. Post-emergence damping off can be incorrectly attributed to frost, heavy rains, or other environmental causes. Cutworms cause seedlings to topple over.

Pythium and *Globisporangium* spp. seldom produce visible hyphae on diseased roots. Signs of these oomycetes may be observed as thick-walled oospores under 125× or 400× magnification. Make free-hand razor sections through a diseased root, and mount in lactophenol-cotton blue solution (Fig. 12.5D).

Fungi may produce visible structures if the humidity is high. *Botrytis cinerea* forms a fuzz of gray-white growth of hyphae and spores, and *Fusarium oxysporum* produces pinkish-white hyphae. Seedlings killed by *Fusarium solani* develop reddish root discoloration (Fig. 12.6). In seedlings killed by *Rhizoctonia solani*, tiny bits of soil dangle from roots of a seedling pulled from the ground—this diagnostic sign is attributed to the pathogen's coarse, clinging, and relatively thick hyphae.

A presumptive diagnosis can be made by carefully removing a diseased seedling from the soil or rooting medium. Lightly grasp the root between your thumb and forefinger. Then pull the root away from the stem. If the outer layer of the root (epidermis and cortex) slips away leaving only a thin inner cylinder (endodermis and stele), you likely have an oomycete problem.

Differentiating amongst *Pythium* and *Globisporangium* spp. is difficult. Isolating them in pure culture is not easy—petri plates become overrun by fungi or bacteria. See section 11.2 for a description of selective media for growing *Pythium* and *Globisporangium* spp. Damping-off diagnosis can be further complicated due the presence of *Pythium* spp. that are non-pathogenic to *Cannabis*, but nevertheless present in soil and rooting media. They can grow in roots damaged by true *Cannabis* pathogens, and are accidently isolated in culture along with the pathogenic species.

Recent studies have used molecular methods: isolate the organism in culture, extract DNA, perform PCR with ITS primers, and then use BLAST to compare the amplified sequence with sequences of known *Pythium* spp. deposited at GenBank (Beckerman *et al.*, 2017, 2018; Punja *et al.*, 2019).

12.4.2 Disease cycle and epidemiology

In the field, the incidence and severity of *Pythium* and *Globisporangium* disease worsens in early spring plantings under cool conditions, in low-lying or flood prone regions (McPartland, 1996). Disease is typically worse in field locations where the soil has a high clay content, has a hard pan or other obstruction in the soil profile, and where irrigation water or rain water does not drain well. Crusting of the soil surface after high-intensity rainfall creates a *Pythium* playground (Beckerman *et al.*, 2017, 2018). Rataj (1957) said *Pythium* often arises in "faulty or mechanically damaged seeds, or from seeds that are suffering from some physiological injury."

P. aphanidermatum is the "warm weather *Pythium*," with optimal growth at 32°C. Three-week-old seedlings inoculated with *P. aphanidermatum* at 25°C began showing symptoms 18 days after inoculation (Beckerman *et al.*, 2017). After 31 days, all inoculated plants displayed symptoms of chlorosis, crown rot, wilt, dieback, and eventually death. *G. ultimum* lives in temperate regions, and attacks plants in cooler temperatures (12–20°C). Two-week-old seedlings inoculated with *G. ultimum* at 25°C began showing symptoms within the first week post-inoculation. After 42 days, 80–100% of plants were infected (Beckerman *et al.*, 2018).

Oospores, which are the resilient survival structures of *Pythium* and *Globisporangium* spp., develop in diseased roots and allow the pathogens to persist over time (from season to season), sometimes for years. Oospores in crop residues, after the residues decompose, can spread the pathogen within or between fields when they are transported by equipment or by flood events.

Damp soil and proper temperatures induce oospores to germinate and form sporangia and release zoospores. They can swim short distances (≤1 cm) with flagella but also spread from plant to plant via surface, sub-surface, or flood waters. Zoospores encyst on root surfaces, germinate, and penetrate the roots. Hyphae grow intercellularly in root tissues, and form haustoria to penetrate host cells.

Disease epidemiology for indoor production and soilless systems is quite different (Rossouw, 2015). For hydroponic *Cannabis* the continuous presence of water in the root system greatly favors *Pythium* and *Globisporangium* spp. If these species are introduced into a hydroponic system, zoospores and other propagules can be rapidly disseminated, resulting in widespread root rot.

Other indoor rooting media (peat moss-based mixes, rockwool, foam) tend to remain wet for long periods, especially if irrigation schedules are not carefully monitored. Overly wet rooting media will favor *Pythium* and *Globisporangium*. Controlled environments maintained in indoor-grown *Cannabis*, lacking high or low temperatures, tend to favor the development of water molds and other damping-off pathogens.

12.4.3 Cultural and mechanical control

Avoid planting in wet soil in low places. In slow-draining soil, plant in raised beds (10–15 cm high). Consider lightening the soil with perlite or other amendments. Vermiculite is not recommended; it retains water and eventually decomposes into a clay sludge, which exacerbates soil heaviness.

Delay sowing when soil is wet and cool. Basically, plant under conditions favorable for hemp seed germination and emergence. Sow high-quality seed, and avoid planting deeper than 3 cm. Do not over-irrigate crops, and do not water the soil immediately after sowing—it will pack soil around seeds. Use well water or municipal water, not water pumped out of a pond or stream.

Crop rotation is thwarted by the wide host range of *Pythium* spp. and the long survival of oospores in soil. No cultivars with resistance have been reported. Correct soils deficient in phosphorus, potassium, and especially calcium (Seda, 1935; Ceapoiu, 1958). Etiolation predisposes plants to damping off, and is prevented by high light intensity. Do not plant seeds harvested from female plants infected with *B. cinerea* and *Fusarium* spp., since these pathogens can be spread by infested seeds.

For indoor production, sanitation measures are critical. Bench tops should be cleaned and sanitized between crops. Trays, flats, pots, and other containers should not be re-used, or if re-used should be thoroughly washed and sanitized. Re-using rockwool or foam materials is risky, because such rooting media cannot be completely sanitized and may harbor inoculum.

Prevent disease in hydroponic systems by keeping nutrient tanks clean, removing senescent material, maintaining oxygenation, and using an enzyme product (e.g., Hygrozyme®) to facilitate the breakdown of dead plant matter in circulating systems. Fungus gnats (*Bradysia* spp.) transport zoospores, and must be eliminated. Between crops, clean all hydroponic system components and disinfect with hydrogen peroxide or UV-C irradiation. Ensure that water tanks remain free of water mold inoculum. Hydroponic nutrient solutions can be disinfected; see section 19.2.3.

12.4.4 Biological control

In soil-based glasshouses and growrooms, Noble and Coventry (2005) controlled damping off by adding compost to soil, at a moderate rate (20% v/v). The treatment suppressed *Pythium ultimum*, *Rhizoctonia solani*, and *Fusarium oxysporum*. The mechanism was predominantly biological, because sterilization of compost resulted in a loss in disease suppression.

Many commercially available biological controls have been approved for use on *Cannabis* by regulatory agencies (California DPR, 2017; Colorado DOA, 2023; Health Canada, 2023). It's a bewildering array of products—many consist of different strains of the same biocontrol species. For example, Colorado approved four strains of *Bacillus amyloliquefaciens*: D747 (Double Nickel 55® and many others), F727 (Stargus®), ENV503 (BellaTrove Companion® Maxx WP), and MBI600 (Serifel® NG). *B. amyloliquefaciens* is a PGPR (plant growth-promoting rhizobacterium) that colonizes roots and suppresses *Pythium*, *Fusarium*, and *Rhizoctonia* spp. Other PGPRs effective against damping-off organisms—and approved for use on *Cannabis*—include *Bacillus subtilis*, *Bacillus megaterium*, and *Pseudomonas chlororaphis* (all described in section 20.6).

Damping off can also be suppressed by fungi, rather than PGPRs. These include *Gliocladium catenulatum* and *G. roseum* (now considered subspecies of *Clonostachys rosea*), *Trichoderma harzianum*, *T. virens*, *T. asperellum*, *Streptomyces lydicus*, and *S. griseoviridis*. Again, an array of strains for most of these species is available, and combinations of the species have been commercialized (section 20.6).

Which to choose? Specific biocontrols show greater efficacy against specific damping-off pathogens. For example, Patel *et al.* (2014) compared the efficacy of several biocontrols against *P. aphanidermatum*. But that was an *in vitro* study—the best evaluation would be an *in vivo* study with *Cannabis* seedlings.

One study tested the efficacy of five products against damping off by *F. oxysporum* in *Cannabis* seedlings (Scott, 2021; Scott and Punja, 2023). Cuttings rooted in rockwool blocks were drenched with biocontrols, followed 48 h later by *F. oxysporum* inoculation. Two weeks later, seedlings treated with *Gliocladium catenulatum* strain J1446 (Prestop®, Lalstop®) had the lowest disease rating. The other products resulted in intermediate but significant disease reduction, nearly equal in efficacy: *Bacillus subtilis* strain QST 713 (Rhapsody® ASO), *Bacillus amyloliquefaciens* strain F727 (Stargus®), *Trichoderma asperellum* (Asperello®), and *Trichoderma harzianum* strain KRL-AG2 + *Trichoderma virens* strain G-41 (Rootshield® Plus WP).

The same study (Scott, 2021; Scott and Punja, 2023) tested the efficacy of these products against *Pythium myriotylum*, which causes damping off, but they tested older plants in a root-rot paradigm (see section 11.2). *Pythium oligandrum* (Polyversum®) is a hyperparasite that colonizes pathogenic fungi in the soil, including other *Pythium* spp., *Fusarium* spp., *Rhizoctonia solani*, *Botrytis cinerea*, and *Sclerotina sclerotiorum*. It also triggers the induction of defense-related reactions in plants.

Sands *et al.* (2022) investigated Harpin$_{EA}$, a protein from the bacterium *Erwinia amylovora* known to activate immune responses in plants. Seeds were soaked for 40 h in a Harpin$_{EA}$ solution (0.5 g/L) and placed on water agar in a petri plate. They were exposed to *P. aphanidermatum* for 7 days, and then seedling length was measured. Seedlings treated with Harpin$_{EA}$ solution averaged 5.3 cm long; untreated seedlings averaged 2.0 cm long. Treated seedlings showed upregulation of the resistance gene *CsERF1*.

12.4.5 Chemical control

Fungicide treatments of seed-borne infections have a long history. Lorenz Hiltner discovered that *Fusarium* diseases are seed-borne, existing as mycelia within the seed coat. He was a pioneer in rhizosphere ecology (see section 13.1). Hiltner (1910) invented Fusariol®, the first seed dressing for treating fungal diseases. Fusariol® did not penetrate the seed coat (*Fusarium*'s lair), but coating a seed with Fusariol® prevented infection of the root and shoot as they emerged from the seed. Fusariol® consisted of 50% mercuric chloride with binding agents (Hiltner and Gentner, 1915).

Hiltner and Gentner (1916) published the first controlled trial of a fungicide on *Cannabis*, complete with photographs. They sowed 100 hemp seeds into soil infested by *Fusarium*, *Penicillium*, *Mucor*, and other fungi. In the control plot, only 22 plants emerged and grew (Fig. 12.7A). In a test plot sown with 100 Fusariol®-treated seeds, 66 plants emerged and grew (Fig. 12.7B).

Bayer Corporation released **Uspulun™** (hydroxyphenylmercury chloride) as a seed dressing in 1914. Two bigwigs tested Uspulun™ on hemp: Nathanael Pringsheim, the famous botanist, was not impressed. He soaked hemp seeds in Uspulun™ for 15 h, which did not rid the seeds of fungi (Pringsheim, 1928). Gustav Bredemann, the famous hemp breeder, found that Uspulun™ significantly improved the germination rate of seeds (Bredemann, 1922). Others used Uspulun™ to improve hemp's "germination power" (Burk, 1923; Lucke, 1925; Heuser, 1927; Stephan, 1929). The famous German botanist Anneliese Niethammer (1901–1983) conducted several studies on the effects of Uspulun™ on hemp seeds (Niethammer, 1930a,b,c).

Uspulun™ was a liquid, and "wet dressings" are difficult to apply—they must dry quickly, otherwise seeds will germinate. Bayer launched a dry dressing in 1929, **Ceresan™** (ethylmercury chloride). It was marketed as a healthy, wholesome alternative to wet dressing. Ceresan™ birthed a new industry: contract seed treatment, where farmers paid seed companies to supply them with dressed seeds.

Fungal diseases of *Cannabis* were controlled with Ceresan™ and other organomercurials in Europe (Flachs,

Fig. 12.7. The first controlled trial of a fungicide on hemp (Hiltner and Gentner, 1916)

1936; Röder and Krüger, 1942; Andrén, 1946, 1947; Ferri, 1961b, Ferri and Noviello, 1963; Kotev and Georgieva, 1969).

Hemp became an essential "war crop" in the USA during WWII, giving rise to a seed shortage. The government funded studies on Ceresan™ and other mercurials (Robinson, 1943a; Haskell *et al.*, 1943; Wilsie *et al.*, 1944; Koehler, 1946; Wilsie and Reddy, 1946). After the war, US corporations appropriated German patents, and DuPont grabbed Bayer's patent for Ceresan™. DuPont released New Improved Ceresan™ and Ceresan M™, both labeled for use on hemp (Leukel, 1948). The hazards of environmental contamination with mercury led to the phasing out of organomercurials during the 1980s. Mercurial pesticides were finally banned in 1992, due to adverse toxicology (Maude, 1996).

The first non-mercurial fungicide applied to hemp seeds was **Thiram™**, a sulfur carbamate (Robinson, 1943a). Andrén (1946, 1947) tested thiram and other fungicides to control damping off by *B. cinerea* in *Cannabis*. Trotuş and Naie (2008) treated hemp seeds with thiram to control *G. debaryanum* and *Fusarium* spp. in Romania. Canada restricted the use of thiram in 2018, and the EU banned it in 2019. Clarke (1981) used **captan** ($C_9H_8Cl_3NO_2S$, a masked source of ammonia) to protect clone cuttings, but evidence of carcinogenicity led Frank (1988) to discourage its use in *Cannabis*. Web scuttlebutt suggests French hemp breeders treat seeds with captan to protect against damping off (captan also repels seed-eating birds). Captan residues still turn up on *Cannabis* in Europe (Fusari *et al.*, 2013).

Benomyl™ has been applied as seed treatment to control damping off in *Cannabis* (Mishra, 1987; Ashok, 1995). The sale of benomyl ended in 2002 in the USA. In fact, *all* synthetic fungicides are banned from use on medical and recreational *Cannabis* plants, in Holland, Canada, California, Colorado, Washington, and Oregon.

Robinson (1943a) included two non-synthetics in his study of ten fungicide seed treatments. Cuprous oxide (Cu_2O) at high rates—22.5 oz per bushel of seed (44 lb)—resulted in better germination rates than Ceresan™ or Thiram™. Today, cuprous oxide (e.g., Nordox®) is mostly used for treating foliar diseases, but there's one labeled use for damping off, in tobacco: dust the seed bed after planting, 3–4 lb/acre. Robinson also tested zinc oxide (ZnO), which outperformed Thiram™. Today, a zinc oxide seed treatment (Advanced Coating®) is labeled for maize, cotton, rice, and wheat (2–8 fluid ounces per 100 lb seed).

Copper sulfate pentahydrate has controlled *Pythium* spp. in other crops. A product containing this active ingredient is approved for use on *Cannabis* (EPA 2023), although no controlled trials have been performed to date.

In a case of botanical irony, Ji *et al.* (2014) showed that an extract of *C. sativa* could inhibit some of the pathogens that cause of *Cannabis* damping off. They tested the efficacy of 19 plant species as ethanol extracts mixed into PDA agar (4 mg/mL). *C. sativa* extract decreased *P. aphanideratum* colony size by 75%, second only to *Helianthus tuberosus* among 19 tested species. *C. sativa* extract also inhibited *Fusarium graminearum* (60%), *Phytophthora infestans* (51%), and *Rhizoctonia solani* (43%).

Taylor *et al.* (2019) applied biocontrol organisms or fungicides to 'Anka' seeds with a rotary pan seed coater. Seedling emergence rate improved after treatment with biocontrols (*Bacillus* and *Trichoderma* spp.), but marginally better after treatment with phosphorus and potassium salts (TKO-phosphite™, Prudent® 44, Nutrol®), or stacked synthetics (mefenoxam+ fludioxonil, or mefenoxam+ fludioxonil+ thiabendazole). Few statistical differences were seen—the researchers seeded too deeply, resulting in poor emergence overall.

(Prepared by J. McPartland; revised by S. Koike)

References

Adedokun O, Smith H, Anthony N, *et al.* (4 other authors). 2021. "Evaluation of the susceptibility of hemp (*Cannabis sativa*) to black shank disease of tobacco (*Phytophthora nicotianae*)," p. 14 in *Second annual Science of Hemp Conference*. Auburn University, Auburn, Alabama.

Andrén F. 1946. Betningsförsök med Lin- och Hampfrö. *Växtskyddsnotiser* 1946(1): 10–12.

Andrén F. 1947. Betningsförsök med Lin och Hampa. *Växtskyddsnotiser* 1947(6): 85–87.

Ashok K. 1995. Chemical control of root rot of ganja. *Current Research, University of Agricultural Sciences Bangalore* 24(9): 99–100.

Barloy J, Pelhate J. 1962. Premières observations phytopathologiques relatives aux cultures de chanvre en Anjou. *Annales des Épiphyties* 13:117–149.

Barna J, Lencsepeti JS, Rközy P, Zsombokos G. 1982. *Mezögazdasági Lexikon, vol 1 (A–K)*. Mezögazdasági Kiadó, Budapest.

Beckerman JL, Nisonson H, Albright N, Creswell T. 2017. First report of *Pythium aphanidermatum* crown and root rot of industrial hemp in the United States. *Plant Disease* 101: 1038.

Beckerman JL, Stone J, Ruhl GE, Creswell T. 2018. First report of *Pythium ultimum* crown and root rot of industrial hemp in the United States. *Plant Disease* 102: 2045.

Bócsa I, Karus M. 1997. *Der Hanfanbau: botanik, sorten, anbau und ernte*. Müller Verlag, Heidelberg.

Bredemann G. 1922. Versuche über Erhöhung der Keimkraft unserer Hanf—saat durch Beizung. *Faserforschung* 2: 58–63.

Burk H. 1923. Zur Steinbrandbekämpfung des Weizens. *Zeitschrift für Pflanzenkrankheiten und Gallenkunde* 23: 193–240.

California DPR (Department of Pesticide Regulation). 2017. *Legal pest management practices for Cannabis growers in California*. Available at: https://www.cdpr.ca.gov/docs/county/cacltrs/penfltrs/penf2015/2015atch/attach1502.pdf

California DPR (Department of Pesticide Regulation). 2018. *Cannabis pesticides that cannot be used*. Available at: https://www.cdpr.ca.gov/docs/cannabis/cannot_use_pesticide.pdf

Ceapoiu N. 1958. *Cinepa, studiu monografic*. Editura Academiei Republicii Populare Romine. Bucharest.

Chang Y, Rochon D, Sekimoto S, *et al.* (7 other authors). 2021. Genome-scale phylogenetic analyses confirm *Olpidium* as the closest living zoosporic fungus to the non-flagellated, terrestrial fungi. *Scientific Reports* 11: e3217.

Clarke RC. 1981. *Marijuana botany*. And/Or Press, Berkeley, California.

Colorado DOA (Department of Agriculture). 2023. *Pesticide use in Cannabis production information*. Available at: https://ag.colorado.gov/plants/pesticides/pesticide-use-in-cannabis-production-information

Coulibaly O, Mbila D, Sonwa DJ, Adesina A, Bakala J. 2002. Responding to economic crisis in Sub-Saharan Africa: new farmer-developed pest management strategies in cocoa-based plantations in southern Cameroon. *Integrated Pest Management Reviews* 7: 165–172.

Dippenaar MC, du Toit CLN, Botha-Greeff MS. 1996. Response of hemp (*Cannabis sativa* L.) varieties to conditions in Northwest Province, South Africa. *Journal of the International Hemp Association* 3(2): 63–66.

Drechsler C. 1952. Production of zoospores from germinating oospores of *Pythium ultimum* and *Pythium debaryanum*. *Bulletin of the Torrey Botanical Club* 79: 431–460.

Drechsler C. 1953. Development of *Pythium debaryanum* on wet substratum. *Journal of the Washington Academy of Sciences* 43: 213–225.

EPA. 2023. *Pesticide products registered for use on hemp*. Available at: https://www.epa.gov/pesticide-registration/pesticide-products-registered-use-hemp.

Ferri F. 1961b. Microflora dei semi di canapa. *Progresso Agricolo (Bologna)* 7(3): 349–356.

Ferri F, Noviello C. 1963. Prove di lotta contro *Sclerotium rolfsii* Sacc. sulla canapa. *Phytopathologia Mediterranea* 2: 72–75.

Flachs K. 1936. Krankheiten und schädlinge unserer gespinstpflanzen. *Nachrichten über Schädlingsbekämpfung* 11: 6–28.

Flores-Sanchez IJ, Pec J, Fei J, *et al.* (3 other authors). 2009. Elicitation studies in cell suspension cultures of *Cannabis sativa* L. *Journal of Biotechnology* 143: 157–168.

Frank M. 1988. *Marijuana grower's insider's guide*. Red Eye Press, Los Angeles, California.

Frezzi MJ. 1956. Especies de *Pythium* fitopatogenas identificadas en las República Argentina. *Revista de Investigaciones Agrícolas* 10 (2):113–241.

Fusari P, Rovellini P, Folegatti L, Baglio D, Cavalieri A. 2013. Olio e farina da *Cannabis sativa* L. analisi multiscreening di micotossine, ftalati, idrocarburi policiclici aromatici, metalli e fitofarmaci. *Rivista Italiana delle Sostanze Grasse* 90 (2013) 9–19.

Galloway LD. 1937. Report of the imperial mycologist. *Annual Science Report of the Agricultural Research Institute, New Delhi* 1935–1936: 105–111.

Gauthier N. 2020b. "Considering crop rotations and the potential for carry-over," p. 25 in Gauthier N, Leonberger K, Bowers C, *eds. Science of hemp: production and pest management*. University of Kentucky Agricultural Experiment Station, no. SR-112, Lexington, Kentucky.

Ghaderian YSM, Lyon AJE, Baker AJM. 2000. Seedling mortality of metal hyperaccumulator plants resulting from damping off by *Pythium* spp. *New Phytologist* 146: 219–224.

Ghani M, Basit A, Anwar M. 1977. *Investigations on the natural enemies of marijuana, Cannabis sativa L. and opium poppy, Papaver somniferum L.* Commonwealth Institute of Biological Control, Pakistan station, Rawalpindi.

Haney A, Kutscheid BB. 1975. An ecological study of naturalized hemp (*Cannabis sativa* L.) in east-central Illinois. *American Midland Naturalist* 93: 1–24.

Harvey JV. 1925. A study of the water molds and pythiums occurring in the soils of Chapel Hill. *Journal of the Elisha Mitchell Scientific Society* 41: 151–164.

Haskell RJ, Leukel RW, Otten CJ. 1943. Organized seed treatment to improve stands and conserve seed. A part of the government's wartime hemp program. *Plant Disease Reporter* 27: 252–253.

Health Canada. 2023. *Pest control products for use on cannabis*. Available at: https://www.canada.ca/en/health-canada/services/cannabis-regulations-licensed-producers/pest-control-products.html

Heuser O. 1927. "Die Hanfpflanze," pp. 1–102 in: Herzog, RO, *ed. Technologie der Textilfasern, Band 5, Teil 2: Hanf und Hartfasem*. Julius Springer, Berlin.

Hiltner L. 1892. *Einige durch Botrytis cinerea erzeugte Krankheiten gärtnerischer und landwirtschaftlicher Kulturpflanzen und deren Bekämpfung*. Doctoral thesis, University of Erlangen.

Hiltner L. 1904. Über neuere Erfahrungen und Problem auf dem Gebiet der Bodenbakteriologie und unter besonderer Berücksichtigung der Gründüngung und Brache. *Arbeiten der Deutschen Landwirtschaftlichen Gesellschaft* 98: 59–79.

Hiltner L. 1910. Über die Beizung des Wintergetreides mit Sublimatlösung. *Praktischen Blätter für Pflanzenbau und Pflanzenschutz* 8: 114.

Hiltner L, Gentner G. 1915. Ist das sog. Usvulun als Reizmittel für Getreide und andere Sämereien empfehlenswerter als die von der R. Agrikultur botanischen Anstalt abgegebenen Reizmittel? *Praktischen Blätter für Pflanzenbau und Pflanzenschutz* 13: 31–40.

Hiltner L, Gentner G. 1916. Ueber die Wirkung der Beizung der Samen von Hanf, Sonnenblumen, Buchweizen, Hierse, Mais und Mohar. *Praktischen Blätter für Pflanzenbau und Pflanzenschutz* 14: 85–90.

Ji YR, Yang QL, Wang YY, *et al.* (8 other authors). 2014. Studies on 19 plant extracts Including *Solanum nigrum* for antifungal activity in vitro. *Chinese Agricultural Science Bulletin* 30(25): 303–307.

Jian FJ, Al-Mamun MA, White NDG, *et al.* (3 other authors). 2019. Safe storage time of FINOLA® hemp (*Cannabis sativa*) seeds with dockage. *Journal of Stored Products Research* 83: 34–43.

Khan MRF, Bhuiyan MZR, Lakshman DK, *et al.* (3 other authors). 2023. First report of damping-off and seedling rot of hemp (*Cannabis sativa*) caused by *Fusarium solani* in North Dakota. *Plant Disease* 107: 232.

Kirchner HA. 1966. *Phytopathologie und Pflanzenschutz, vol. 2*. Akademie Verlag, Berlin.

Kirchner O. 1906. "Hanf, *Cannabis sativa* L." pp. 319–323 in *Die Krankheiten und Befehädigungen uhferer landwirtschaftlichen Kulturpflanzen*. E. Ulmer, Stuttgart.

Koehler B. 1946. Hemp seed treatments in relation to different dosages and conditions of storage. *Phytopathology* 36: 937–942.

Kotev S, Georgieva M. 1969. Химически средства за обеззаразяване семената на конопа срещу бактериолата. *Zashchita Rastenii* 17(1): 25–29.

Leukel RW. 1948. Recent developments in seed treatment. *Botanical Review* 14: 235–269.

Lévesque CA, de Cock AWAM. 2004. Molecular phylogeny and taxonomy of the genus *Pythium*. *Mycological Research* 108: 1363–1383

Lucke A von. 1925. *Die Ausbildung des Fasergehaltes beim Hanf (Cannabis sativa) in Abhangigkeit von verschiedenen Wachstumsbedingungen*. Friedrich-Wilhelms-Universitat, Breslau.

Marquart B. 1919. *Der Hanfbau*. Paul Parey, Berlin.

Maude RB. 1996. *Seedborne diseases and their control*. CAB International, Wallingford, UK.

McEno J. 1990. Hydroponic IPM. *The Growing Edge* 1(3): 35–39.

McGehee CS, Raudales RE. 2021. Pathogenicity and mefenoxam sensitivity of *Pythium*, *Globisporangium*, and *Fusarium* isolates from coconut coir and rockwool in marijuana (*Cannabis sativa* L.) production. *Frontiers in Agronomy* 3: e706138.

McPartland JM. 1996. A review of *Cannabis* diseases. *Journal of the International Hemp Association* 3: 19–23.

McPartland JM, Clarke RC, Watson DP. 2000. *Hemp diseases and pests – management and biological control.* CABI, Wallingford, UK.

Mishra D. 1987. Damping off of *Cannabis sativa* caused by *Fusarium solani* and its control by seed treatment. *Indian Journal of Mycology and Plant Pathology* 17(1): 100–102.

Niethammer A. 1930a. Über chemische Reizwirkungen an den Früchtschen von *Cannabis sativa* und den Samen von *Linum ustiatissimum. Faserforschung* 8: 312–315.

Niethammer A. 1930b. Histochemische Untersuchungen imd Permeabilitätsstudien an landwirtschattlichen Samereien im Hinblicke auf ihre Keimungsbiologie. *Wissenschaftliches Archiv für Pfanzenbau, Abteilung A: Archiv für Pfanzenbau* 3: 321–348.

Niethammer A. 1930c. Die dosis toxica und tolerate von Uspulun Universal für einzelne landwirtshaftliche Sämereien. *Zeitschrift für Pflanzenzüchtung (Pflanzenpathologie) und Pflanzenschutz* 40: 517–520.

Noble M, Richardson MJ. 1968. *An annotated list of seed-borne diseases.* Commonwealth Mycological Institute, Kew, UK.

Noble R, Coventry E. 2005. Suppression of soil-borne plant diseases with composts: a review. *Biocontrol Science and Technology* 15: 3–20.

Palmucci HE. 2015. *Caracterización de especies fitopatógenas de Pythium y Phytophthora (Peronospromycetes) en cultivos ornamentales del cinturón verde La Planta-Buenos Aires y otras áreas y cultivos de interés.* Doctoral thesis, University of La Plata, Argentina.

Patel JK, Joshi KR, Prajapati HN, Jage NP. 2014. Bait techniques for isolation of *Pythium aphanidermatum* causing damping-off of chilli from soil and efficacy of bio-agents in *vitro. Trends in Biosciences* 76: 474–476.

Patschke K, Gottwald R, Müller R. 1997. Erste Ergebnisse phytopathologischer Beobachtungen im Hanfanbau im Land Brandenburg. *Nachrichtenblatt des Deutschen Pflanxenschutzdienstes* 49: 286–290.

Paulitz TC, Adams K. 2003. Composition and distribution of *Pythium* communities in wheat fields in eastern Washington state. *Phytopathology* 93: 867–873.

Pellegrini M, Ercole C, Gianchino C, *et al.* (3 other authors). 2021. *Fusarium oxysporum* f. sp. *cannabis* isolated from *Cannabis sativa* L.: in vitro and in planta biocontrol by a plant growth promoting-bacteria consortium. *Plants* 10: e2436.

Pettitt TR, Wakeman AJ, Wainwright MF, White JG. 2002. Comparison of serological, culture, and bait methods for detection of *Pythium* and *Phytophthora* zoospores in water. *Plant Pathology* 51: 720–727.

Pietkiewicz TA. 1958. Mikroflora nasion konopi. Przeglad literaty. *Roczniki Nauk Rolniczych Seria A* 77(4): 577–590.

Plaats-Niterink AJ van der. 1981. Monograph of the genus *Pythium. Studies in Mycolology* 21: 1–242.

Plăcintă DD, Murariu D. 2016. Fungus evaluation from seeds germplasm before medium and long term storage. *Cercetări Agronomice în Moldova* 49: 71–82.

Pringsheim EG. 1928. Vergleichende Untersuchungen über Saatgutdesinfektion. *Angewandte Botanik* 10: 208–279.

Punja ZK, Scott C, Chen S. 2018. Root and crown rot pathogens causing wilt symptoms on field-grown marijuana (*Cannabis sativa* L.) plants. *Canadian Journal of Plant Pathology* 40: 528–551.

Punja ZK, Collyer D, Scott C, *et al.* (3 other authors). 2019. Pathogens and molds affecting production and quality of *Cannabis sativa* L. *Frontiers in Plant Science* 10: e1120.

Raper JR. 1939. Sexual hormones in *Achlya.* I. Indicative evidence for a hormonal coordinating mechanism. *American Journal of Botany* 26: 639–650.

Raper JR. 1959. Sexual versatility and evolutionary processes in fungi. *Mycologia* 51: 107–124.

Rataj K. 1957. Skodlivi cinitele pradnych rostlin. *Prameny literatury* 2: 1–123.

Robinson BB. 1943a. Seed treatment of hemp seeds. *Journal of the American Society of Agronomy* 35: 910–914.

Röder K, Krüger E. 1942. Zur Frage der Hanf- und Leinbeizung. *Nachrichtenblatt für den deutschen Pflanzenschutzdienst* 22: 9–11.

Rodríguez E, Sinobas J, Varés Megino L. 1999. Influencia del tipo de "cebo" y la temperatura en el aislamiento y desarrollo de *Pythium* spp. *Boletín de Sanidad Vegetal, Plagas* 25(2): 131–142

Rossouw SJ. 2015. *A novel organic substrate based on hemp (Cannabis sativa), or flax (Linum usitatissimum) bast fibre for NFT hydroponic systems.* MSc thesis, McGill University, Montreal, Canada.

Rothrock CS, Avanzato MV, Rupe JC. 2015. "Pythium seed rot, damping-off, and root rot," pp. 76-79 in Hartman GL, *et al.*, eds. *Compendium of soybean diseases and pests, 5th edn.* APS Press, St Paul, Minnesota.

Saccardo FA. 1886. *Syllogue Fungorum, vol. IV.* Typis Seminarii, Patavii.

Sánchez J, Sánchez-Cara J, Gallego E. 2000. Suitability of ten plant baits for the rapid detection of pathogenic Pythium species in hydroponic crops. *European Journal of Plant Pathology* 106: 209–214.

Sandoval-Denis M, Lombard L, Crous PW. 2019. Back to the roots: a reappraisal of *Neocosmospora. Persoonia* 43: 90–185.

Sands LB, Cheek T, Reyonlds J, Ma Y, Berkowitz GA. 2022. Effects of Harpin and Flg22 on growth enhancement and pathogen defense in *Cannabis sativa* seedlings. *Plants* 11(9): e1178.

Schoener J, Wilhelm R, Wang S. 2018a. Molecular identification of *Alternaria* species associated with imported industrial hemp seed. *Phytopathology* 108: S1.185

Schoener J, Wilhelm R, Yazbek M, Wang S. 2018b. First detection of *Pythium aphanidermatum* crown rot of industrial hemp in Nevada. *Phytopathology* 108: S1.187.

Schultz H. 1939. Untersuchungen über die rolle von *Pythium*-arten als erreger der fusskrankheit der lupine I. *Phytopathologische Zeitschrift* 12: 405-420.

Scott CA. 2021. *Disease management approaches for Cannabis sativa L.* Master's thesis, Simon Fraser University, British Columbia, Canada.

Scott CA, Punja ZK. 2023. Biological control of *Fusarium oxysporum* causing damping-off and *Pythium myriotylum* causing root and crown rot on cannabis (*Cannabis sativa* L.) plants. *Canadian Journal of Plant Pathology* 45: 238–252.

Seda A. 1935. Boj proti chorobam skuducum domacich olejnych ros t lin. *Casove Otazky Zemedelske* 45: 87–108

Serzane M. 1962. "Kanepju – *Cannabis sativa* L. Slimibas," pp. 366–369 *in Augu slimibas, praktiskie darbi.* Riga Latvijas Valsts Izdevnieciba, Lativa.

Sinobas J, Vares L, Rodriguez E. 1999. Influence of the type of bait and temperature in the isolation and development of Pythium spp. *Boletín de Sanidad Vegetal Plagas* 25: 131–142.

Sorrentino R, Lahoz E, Battaglia V, Sorrentiono MC, Cerrato D. 2023. First report of seedling damping-off of industrial hemp (*Cannabis sativa*) caused by seed-transmitted *Alternaria rosae* in Italy. *Plant Disease* 107: 1636.

Stephan J. 1929. Stimulationsversuche mit *Cannabis sativa.* I. (Wüttemberg. Landesanst. f. Samenprüf., Hohenheim b. Stuttgart.). *Faserforschung* 7: 292–298.

Taylor A, Mayton H, Bergstrom G, Fike J, Johnson B. 2019. *Evaluation of seed treatments on industrial hemp for management of damping off pathogens.* Cornell University IR-4 Project report. Available at: https://ir4.rutgers.edu/Fooduse/PerfData/4853.pdf

Trotuş E, Naie M. 2008. Cercetări privind reducerea atacului unor agenţi patogeni şi dăunători specifici culturilor de cânepă prin tratamentul chimic al seminţei. *Lucrari Stiintifice, Universitatea de Stiinte Agricole Si Medicina Veterinara "Ion Ionescu de la Brad" Iasi, Seria Agronomie* 51(2): 219–223.

Uzuhashi S, Tojo M, Kakishima M. 2010. Phylogeny of the genus *Pythium* and description of new genera. *Mycoscience* 51: 337–365.

Vakhrusheva TE, Davidian GG. 1979. Методические указания по инвентаризации болезней и микрофлоры льна и конопли. All-Union Vavilov Scientific Research Insititute of Crop Cultivation, Leningrad.

Wei GF, Ning K, Zhang GZ, *et al.* (5 other authors). 2021. Compartment niche shapes the assembly and network of *Cannabis sativa*-associated microbiome. *Frontiers in Microbiology* 12: e714993.

Wilsie CP, Reddy CS. 1946. Seed treatment experiments with hemp. *Journal of the American Society of Agronomy* 38: 693–701.

Wilsie CP, Black CA, Aandahl AR. 1944. *Hemp production experiments.* Iowa State College Agricultural Experiment Station, Bulletin P63. Ames, Iowa.

Zelenay A. 1960. Fungi of the genus *Fusarium* occurring on seeds and seedlings of hemp and their pathogenicity. *Prace Naukowe, Instytutu Ochrony Roślin, Poznan* 2: 248–249.

13 Phyllosphere/rhizosphere epiphytes and endophytes

Abstract

Microbes thrive in a zone around roots (the rhizophere) and leaves (phyllosphere). They live atop epidermal surfaces (epiphytes) or just below, in extracellular spaces between cells (endophytes). Their symbiotic relationships with plants are often mutualistic—where each species benefits. Other microbes are commensal—they benefit from plants (living off cellular leakage) while the plant is neither benefited nor harmed. Microbial symbiosis is the plant world's version of "gut health." Plant growth-promoting rhizobacteria (PGPRs) and fungal mycorrhizae are critical partners in an optimized crop system. PGPRs and mycorrhizae are commercially available—some are appropriate for *Cannabis*, others not. A subset of commensals have the ability to "go rogue." Über-nasty *Botrytis cinerea* can exist as a facultative endophyte, then flip a switch and become a necrotrophic pathogen—killing plant cells and feeding on the dead cells.

13.1 Introduction

This chapter concerns the *Cannabis* **microbiome**. A portmanteau of *microbe* and *biome*, the human microbiome is very trendy; dysbiosis has been associated with cancer, inflammatory bowel disease, multiple sclerosis, diabetes, asthma, allergies, and autism.

Perturbations in the vaginal microbiome may interfere with orgasm (Giovanetti *et al.*, 2023). The term was coined, however, in a plant disease context: Whipps *et al.* (1988) defined the microbiome "as a characteristic microbial community occupying a reasonably well-defined habitat which has distinct physiochemical properties. This term thus not only refers to the microorganisms involved but also encompasses their theatre of activity." Although the term is new, the underlying concept is much older.

Lorenz Hiltner (1862–1923) was a pioneer Bavarian agronomist active in hemp research. Hiltner (1892) made an early description of *Cannabis* gray mold disease, and he conducted the first controlled trial of a fungicide on hemp (see Fig. 12.7). Nobbe and Hiltner (1893) used pure cultures of *Rhizobium leguminosarum* to improve yields in legume crops, and commercialized the cultures (Nitragin®, still for sale).

Hiltner (1904) coined **rhizosphere** to describe the zone of soil surrounding a plant root inhabited by microorganisms. He postulated that these microbes were influenced by root growth and activity, and vice versa. Rhizosphere biodiversity surpasses that of any above-ground ecosystem on Earth, rainforests included. Rhizosphere microbes provide myriad benefits to our crop plants, and they benefit the planet by sequestering carbon. Pesticides and N-P-K fertilizers destroy the rhizosphere, thereby turning soil into dirt.

The **phyllosphere** is a comparable above-ground habitat (Last, 1955; Ruinen, 1956). It consists of a zone immediately above and below the **phylloplane** or leaf surface (Kerling, 1958, 1964). The phylloplane is not flat and two-dimensional—it's a stormy sea of bulging epidermal cells, cystolithic trichomes, glandular trichomes, epicuticular waxes, and stomata (Fig. 13.1).

The endophyllosphere, internal to the phylloplane, comprises mesophyll cells and the extracellular spaces between them.

An **epiphyte** is an organism that lives on plant surfaces. Epiphytes by definition are *not* parasitic, but either commensal or mutualistic. **Commensal** organisms benefit from the plant (e.g., living off cellular leakage from epidermal cells) but do not affect the plant. **Mutualistic** organisms benefit from the plant, and they provide benefits in return (e.g., supply plants with nutrients, or protect them from pathogenic organisms).

An **endophyte** is an organism that lives within a plant without causing apparent disease. Endophytes are either commensal or beneficial to the plant. The study of *Cannabis* epiphytes and endophytes received little attention until recently. Best known are mycorrhizal fungi, which are endophytes or epiphytes in the rhizosphere.

13.2 Mycorrhizae

A mycorrhiza (plural: mycorrhizae) is a mutualistic relationship between a plant and a fungus. The plant supplies the fungus with carbon and energy; the fungus supplies the plant with nutrients and other benefits. Fossil evidence of mycorrhizae dates to the Devonian Period, 407 million years ago. About 90% of plant species form mycorrhizal relationships. Most plants are facultative symbionts, i.e., they benefit from mycorrhizal fungi, but can also live without them. *Cannabis* is a facultative symbiont.

While trying to grow a truffle (*Elaphomyces granulatus*), Frank (1885) understood the fungus acted as a "wet nurse," feeding plant roots. He coined the word "mycorhiza" (one "r"). Frank (1887) recognized two types—ectomycorrhizae and endomycorrhizae. **Ectomycorrhizal** fungi do not penetrate plant cells. They form a hyphal mantle that covers the root tip, and a "Hartig net" of hyphae that surround plant cells within the plant cortex. Ectomycorrhizal fungi usually form associations with woody plants. Most forest mushrooms are ectomycorrhizal fungi.

DOI: 10.1079/9781836990352.0013

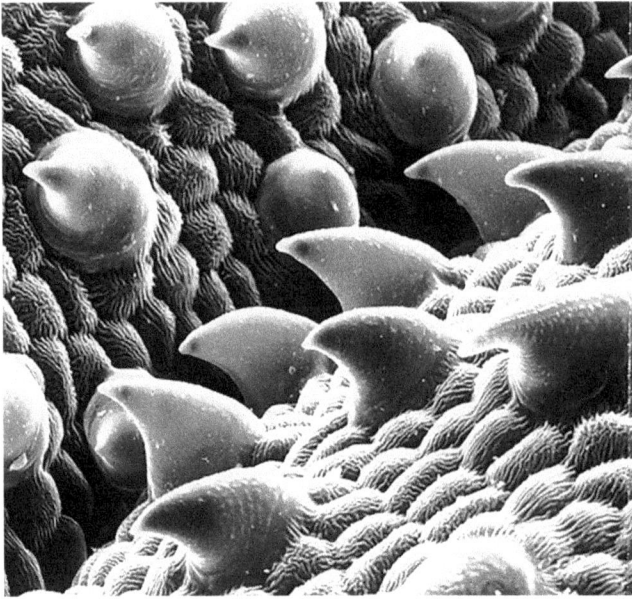

Fig. 13.1. *Cannabis* phylloplane (SEM ×2000)

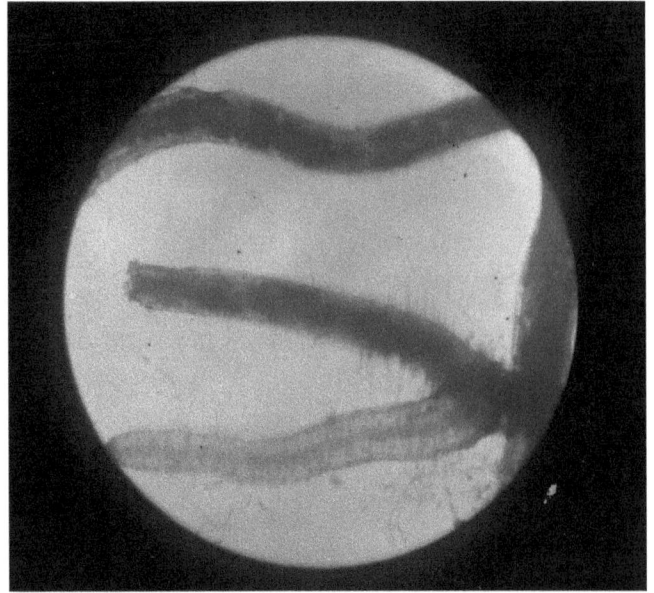

Fig. 13.2. Photomicrograph by E.G. Arzberger, labelled "Endotrophic mycorrhiza on *Cannabis sativa*" (lantern slide found in USDA archives).

No associations between *Cannabis* and ectomycorrhizal fungi have been reported. Yet many products marketed to *Cannabis* growers contain ectomycorrhizal fungi. They won't do any harm, and they might help in some mysterious way: Melin and Das (1954) showed that that metabolites produced by *Cannabis* roots enhanced the growth of ectomycorrhizal fungi.

Endomycorrhizal fungi penetrate plant cells. Within cells they form dichotomously branching invaginations (arbuscules) as well as balloon-like structures (vesicles), and are known as **arbuscular mycorrhizae** (AM). AM fungi constitute their own subphylum within the fungal kingdom, the Glomeromycotina, with about 250 species. The unsung mycologist **Emil Godfred Arzberger** (1877–1930) discovered AM fungi colonizing *Cannabis* roots. We found Arzberger's notes and a glass "lantern slide" in USDA archives (Fig. 13.2). Arzberger used the lantern slide in a presentation about mycorrhizae at the Cosmos Club in Washington, DC (Arzberger, 1920).

Barbara Mosse (1919–2010) and **Jim Gerdemann** (1921–2008) made great strides in AM fungal taxonomy. Mosse (1953) observed fungal hyphae in strawberry roots, and traced the microscopic hyphae to *Endogone* sp. sporocarps in the soil. She inoculated strawberry plants with the sporocarps and successfully formed an AM mycorrhizal relationship—a landmark, 3-paragraph *Nature* paper. Critics argued that the sporocarps were not sterile, and *Pseudomonas* bacteria clinging to sporocarps played a role.

Gerdemann (1964) shut down that argument: he inoculated maize seedlings with either *Endogone* sporocarps or sporocarp washings. Only the *Endogone*-inoculated seedlings became mycorrhizal, and they weighed 3.5-fold greater than control plants. Nicolson and Gerdemann (1968) named the fungus in Barbara's honor: *Endogone mosseae*. Later, Gerdemann and Trappe (1974) transferred *Endogone mosseae* to a new genus, *Glomus*, as *G. mosseae*.

Barrett (1947) used boiled hemp seed as "bait" to isolate an AM fungus from the root of a garden pea. He placed a section of colonized pea root in a petri plate with sterile water, and dropped the hemp seed beside it. After a few days, hyphae from the root penetrated the seed; he transferred the seed to another petri plate with a freshly boiled hemp seed, and the fungus sporulated. Over the course of a decade, he isolated AM fungi from the roots of 90 plant species (Barrett, 1961). Despite using hemp seeds as bait, he did not test *Cannabis*.

Gerdemann (1955) prepared "hemp seed agar" by adding two hemp seeds to 10 cc of water agar, and isolated "*Endogone* type b," now classified as ***Gigaspora gigantea*** (Nicolson & Gerd.) Gerd. & Trappe 1974. Gerdemann and his colleague Carol Shearer sourced hemp seeds from feral plants growing near the University of Illinois (pers. comm., C. Shearer, 1981).

Mosse (1961) created mycorrhizal *Cannabis* plants. She inoculated seedlings with sporocarps of an *Endogone* sp. harvested from soil. Compared with other plants in her experiment (clover, onion, strawberry, tomato), mycorrhizal anatomy differed within *Cannabis* roots—the fungus formed arbuscules, but also formed tightly coiled hyphae.

Gerdemann retired in 1981; McPartland was in his last class of students. McPartland and Cubeta (1997) documented naturally occurring *G. mosseae* in feral hemp. The fungus produced intracellular hyphal H-connections and tight hyphal coils, but few arbuscules—just like Mosse (1961) described.

With the advent of molecular systematics, *G. mosseae* is now classified as ***Funneliformis mosseae*** (Nicols. & Gerd.) Walker & Schüßler 2010. Several studies have artificially inoculated *Cannabis* with *F. mosseae* (Citterio *et al.*, 2005; Zarei *et al.*, 2014; Tadayon and Zarci, 2014). Naturally occurring *F. mosseae* has been confirmed in Chinese hemp (Gai *et al.*, 2006) and Indian *gañjā* (Chahar, 2022).

Fester *et al.* (2002) isolated *Glomus intraradices* Schenck & Small from hemp roots in Germany. The taxon is now *Rhizophagus intraradices* (Schenck & Small) Walker & Schüßler 2010, but most commercial products still call it *G. intraradices*. Over a dozen recent *Cannabis* studies on a range of topics have utilized Pro-Mix HP Mycorrhizae™, which contains *R. intraradices*.

R. intraradices was also identified in India by Chahar (2022), along with the aforementioned *F. mosseae* and *G. gigantea*, as well as three species new to *Cannabis*: *Rhizophagus fasciculatus* (Thaxter) Walker & Schüßler 2010, *Dentiscutata nigerita* Khade 2010, and *Racocetra coralloidea* (Trappe & Gerd.) Oehl *et al.* 2009.

Zielonka *et al.* (2019) used molecular methods (PCR with primers for ITS) to identify mycorrhizae colonizing roots of fiber-type plants in Poland. They identified *F. mosseae*, *Funneliformis geosporum* (Nicols. & Gerd.) Walker & Schüßler 2010, *Glomus caledonium* (Nicols. & Gerd.) Trappe & Gerd., *Paraglomus occultum* (Walker) Morton & Redecker, and *Diversispora* sp. Earlier, Hepper (1979) had shown that adding pieces of *Cannabis* seeds to soil stimulated *G. caledonium* spore germination and hyphal growth.

Zalaghi *et al.* (2021) artificially inoculated *Cannabis* with three fungi formerly classified as *Glomus* spp.: *F. mosseae*, *Claroideoglomus etunicatum* (Becker & Gerd.) Walker & Schüßler 2010, and *Septoglomus constrictum* (Trappe) Sieverding *et al.* 2011. Shoot and root dry weight of seedlings averaged 8.0 g with *C. etunicatum*, 7.0 g with *S. constrictum*, 6.7 g with *F. mosseae*, and 5.8 g in non-inoculated controls.

Several studies have compared different plant species regarding their colonization by AM fungi. *Cannabis* generally ranks low in the percentage of root tissues colonized by AM fungi, but ranks high in the concentration of AM fungal spores in surrounding soil. Betekhtin (2008) compared 36 steppe plants, ranked on a five-point scale. Ruderal *Cannabis sativa* ranked the lowest regarding the abundance of arbuscules in roots (0.3), but had the highest variation of fungal spores (0.0–1.4).

Zubek *et al.* (2012) surveyed 25 species of plants in Poland, and examined their roots for hyphae and arbuscules. *Cannabis* had the lowest intensity of colonization among all plants (1.2% of roots colonized). Yaseen *et al.* (2016) surveyed 20 species of plants in Pakistan; in terms of intensity of colonization, their results with *Cannabis* were inconsistent. However, *Cannabis* had the third highest concentration of AM fungal spores in soil surrounding roots (80 spores/100 g soil).

Tian *et al.* (2009) surveyed wild plants growing in the Mongolian steppe. Compared with other plants, *Cannabis ruderalis* was colonized lightly (1.2% of root length colonized) but supported a high concentration of AM fungal spores in soil (253 spores/50 g air-dried soil). Chahar (2022) surveyed six plants in Indian agricultural fields. In this case, *Cannabis sativa* ranked highest in percentage root colonization, but second lowest in spore density in surrounding soil (260 spores/100 g soil).

13.2.1 Application for crops

AM fungi serve as "biotic fertilizers," substituting for nutrient supplements. Gerdemann (1964) showed that mycorrhizal maize, compared with control plants, contained significantly greater phosphorus (P) content. P is immobile in soil, which causes a nutrient-depletion zone to form around roots. AM fungi grow beyond the nutrient-depletion zone and transport P back to the plant. AM fungi form an extensive hyphal network—Tisdall and Oades (1982) measured 150 ft (45.7 m) of mycelium per cm^2 of soil.

Menge (1983) estimated that mycorrhizal plants absorb 60 times more P than non-mycorrhizal plants in nutrient-depleted soil. Root absorption of K, Ca, Fe, Mg, Mn, and S improved as well. In the aforementioned study by Zalaghi *et al.* (2021), inoculation of *Cannabis* increased phosphorus concentration in shoot and root tissues, in the following order: *S. constrictum* > *C. etunicatum* > *F. mosseae* > non-inoculated controls.

The discovery of *Cannabis* mycorrhizae (McPartland and Cubeta, 1997) led to McPartland *et al.* (2000) predicting these species would be commercialized. Now there are dozens of mycorrhizal products being marketed to *Cannabis* growers (see section 20.6). Back then, McPartland *et al.* (2000) inoculated *Cannabis* potting soil with indigenous AMFs obtained from healthy field soil, a sort of "sourdough starter," based on research by Wilson *et al.* (1988).

The number of companies producing AMF products in Europe increased from fewer than 10 in the late 1990s to more than 75 in 2017 (Chen *et al.*, 2018). Chen estimated retail prices: for hobby and semi-professional users, the average price ranged from 10 to 50 cents per plant. The cost of mycorrhizal inoculation on an agricultural scale is considerably lower, about US$135 per hectare.

Seleiman *et al.* (2013) grew fiber-type 'USO-31' and measured the effects of soil amendments upon AM fungi. They counted the number of AM fungal spores in 10 g soil around roots: soil with synthetic N-P-K, 38 spores; sand with synthetic N-P-K, 42 spores; soil with sewage sludge, 42 spores; sand with sewage sludge, 52 spores; soil with digested (composted) sewage sludge, 160 spores. They also examined roots for hyphae and arbuscules: full colonization was seen in soil with digested sewage sludge, and no colonization was detected in soil or sand with synthetic N-P-K.

Zielonka *et al.* (2021) tested the effects of sewage sludge ("OŚ" treatment) upon AM fungi in 'Białobrzeskie', 'Tygra', and 'Beniko'. The frequency of AM colonization in roots of control plants averaged 18.60%, compared with 13.58% in OS-treated plants. Seemakram *et al.* (2022) tested the effects of two AM fungi, *Rhizophagus prolifer* and *R. aggregatus*, compared with N-P-K fertilizer. Plant biomass and cannabinoid content in plants inoculated with AM fungi was equal to plants receiving N-P-K.

Critics argue that inoculating soil with AM fungi is more expensive than fertilizing with N-P-K. However, AM fungi provide other benefits. They inhibit pathogenic fungi and nematodes. They enhance root branching, and plant ethylene production (reviewed in Koide and Moss, 2004). Particularly pertinent to *Cannabis*, AM fungi enhance CO_2 assimilation and photosynthetic capacity (Boldt *et al.*, 2011). *Cannabis* colonized by AM fungi tolerates heavy metals in the soil (Citterio *et al.*, 2005; Zalaghi *et al.*, 2021), drought and soil salinity (Zarei *et al.*, 2014; Tadayon and Zarei, 2014), and fertilization with sewage sludge (Zieolonka *et al.*, 2019).

Kakabouski *et al.* (2021) showed that AM fungi are aided by biocontrol fungi such as *Trichoderma harzianum*. They applied *T. harzianum* strain T-22 to two fiber-type cultivars, 'Fedora 17' and 'Felina', growing in 12-liter pots filled with soil and compost. They watered *T. harzianum* into pots at two rates: 2 g per pot (2×10^{12} CFU/kg) and 4 g per pot (4×10^{12} CFU/kg). Both treatments increased root colonization by AM fungi, and increased plant height, harvested dry weight, and CBD content.

13.3 Rhizosphere bacteria

Rhizosphere bacteria can be divided into two groups. **Rhizobia** are the bacterial counterparts of mycorrhizae. They invade root tissue and are truly mutualistic (both partners benefit). **Plant growth-promoting rhizobacteria** (PGPRs) colonize root surfaces but do not penetrate root tissues.

Rhizobia infect legumes (Fabaceae family) and form root nodules. Within nodules they convert atmospheric nitrogen into ammonia. Some nitrogen is used by the plant and some remains in the soil. Theophrastus (371–287 BC) noted that legumes "are not a burdensome crop to the soil, they even seem to manure it" (Theophrastus, 1916). Hellriegel and Wilfarth (1888) discovered rhizobia as the source of this "manure." Albert Frank, who discovered mycorrhizae, named the nitrogen-fixing bacterium *Rhizobium leguminosarum* (Frank, 1889).

PGPRs colonize root surfaces and do not form nodules. Kloepper and Schroth (1978) coined "PGPR" and elaborated the concept. PGPRs enhance plant growth in many ways: they improve access to soil nutrients, prevent plant diseases, and they produce plant growth hormones. PGPRs aggressively compete with pathogens for space on root surfaces. They ooze metabolites that suppress or kill pathogens. They also stimulate ISR (induced systemic resistance) in host plants. Pagnani *et al.* (2018) illustrated PGPRs colonizing root surfaces of the 'FINOLA' cultivar (Fig. 13.3).

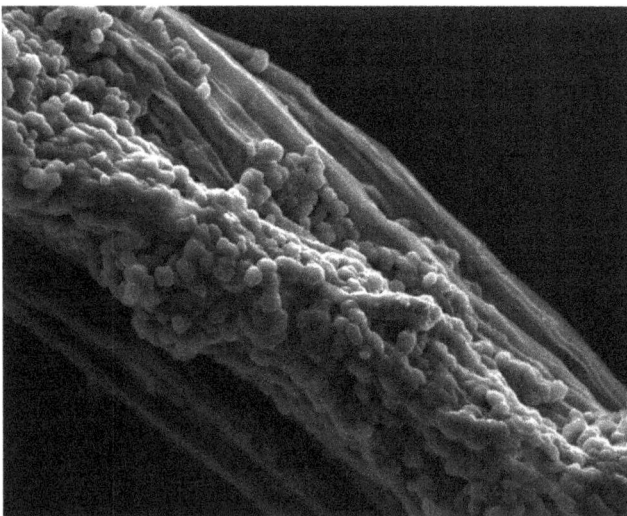

Fig. 13.3. PGPR bacteria colonizing a *Cannabis* root (courtesy Giancarlo Pagnani)

13.3.1 Rhizosphere microbiome

The *Cannabis* rhizosphere microbiome is being explored, and it's huge and complex. Analyzing the rhizosphere has become much easier using molecular methods, especially high-throughput NextGen sequencing.

Winston *et al.* (2014) analyzed five drug-type strains, grown in four different soils, and sampled three microbiomes from each: endorhiza (root samples), rhizosphere (soil adhering to roots), and bulk soil (10 cm away from roots). They counted the number of unique **operational taxonomic units** (OTUs), which are proxies for bacterial species—clusters of similar sequence variants (sharing >97% similarity of the 16S rDNA sequence). They counted 3400 OTUs—essentially, 3400 different bacterial species—in 27 *Cannabis* specimens. They found strain-specific differences in OTUs. OTUs varied between endorhiza, rhizosphere, and soil samples, and were influenced by different types of soils.

Afzal *et al.* (2015) identified endophytes in roots of wild-type *Cannabis* in Pakistan. They isolated endophytes and eliminated epiphytes by surface-sterilizing roots with ethanol and sodium hypochlorite. Roots were plated on agar, and bacterial colonies identified by cloning and sequencing their 16 rDNA genes. They identified 32 isolates, primarily *Acinetobacter*, *Chryseobacterium*, *Enterobacter*, *Microbacterium*, and *Pseudomonas* spp.

Comeau *et al.* (2020) extracted bacterial DNA (16S rDNA) from three chemotypes (CBD dominant, mixed CBD/THC, THC dominant), sampled over the course of their indoor life cycles—propagation stage, vegetative, early flowering, and late flowering stages. They sampled the rhizosphere (soil around roots), endorhizosphere (tissue within roots), and phyllosphere (aerial plant surfaces). Bacteria were identified at the family level (e.g., Burkholderiaceae, Streptomycetaceae, Rhizobiacea). Microbial diversity was greatest underground. When clones were grown in sterile media such as rockwool and coconut coir, their microbiome transferred from the mother plants.

Barnett *et al.* (2020) analyzed three microbiomes associated with field-cultivated 'Anka' in upstate New York: in root samples, the rhizosphere (soil adhering to roots), and bulk soil (a meter away from roots). OTUs were identified by sequencing 16S rDNA (bacteria) and ITS1 (fungi). Root tissue ITS1 amplicons were overwhelmed by reads from *Cannabis* ITS1, so they did not report fungal data ("N.D." in Table 13.1). Across all samples, they identified 8913 bacterial OTUs and 982 fungal OTUs. The most abundant OTUs were identified to species, genus, family, or class. Statistical analysis revealed little difference between bulk soil and rhizosphere communities. These surprising results suggest that *Cannabis* does not have a strong effect on overall microbial community composition in the rhizosphere. They sampled plants from six sites, which showed significant variations in microbiome composition, probably driven by variations in soil chemistry.

Wei *et al.* (2021) investigated microbiomes in four Chinese cultivars—a fiber-type ('Huoma #1'), fiber- and seed-type ('Yunma #1'), seed-type ('Gansuqingshui'), and fiber- and cannabinoid-type ('Yunmaza #1'). In each they investigated

Table 13.1. Bacterial and fungal organisms associated with soil and roots, identified by Barnett *et al.* (2020)

Sample	Most common bacterial OTUs	Most common fungal OTUs
Bulk soil	Pseudarthrobacter, Chloroflexi, Janibacter, Comamonadaceae, Lysobacter	Verticillium dahlia, Fusarium, Hypocreales
Rhizosphere	Pseudarthrobacter, Chloroflexi, Janibacter, Comamonadaceae, Acidobacteria	Verticillium dahlia, Fusarium, Tilletiopsis washingtonensis, Didymella dimorpha, Bullera alba
Root tissue	Streptomyces, Halieaceae, Comamonadaceae, Massilia, Aquabacterium, Rhizobium, Pseudomonas, Sphingomonas	N.D.

microbiomes in six "niches": bulk soil, rhizosphere, and endophytes of root, stem, leaf, and flower. OTUs were identified by sequencing 16S rDNA (bacteria) and ITS1 (fungi).

They showed that the microbiome community was determined by niche, not by cultivar type. Bacterial microbiome diversity was greatest in soil (bulk soil, OTUs n =1006; rhizosphere, n =1046), and decreased from root (n =872) to stem (n =731) to leaf (n =933) to flower (n =781). This trend was not as evident in fungi, with OTUs for bulk soil (n =220), rhizosphere (n =213), root (n =234), stems (n =262), leaf (n =287), and flower (n =146). Nevertheless they concluded, "According to source tracking analysis, hemp microbiota primarily originated from soil and were subsequently filtered in different plant compartments."

13.3.2 Application for crops

Pagnani *et al.* (2018) improved the growth of 'FINOLA' by inoculating seeds with *Azospirillum brasilense*, *Gluconacetobacter diazotrophicus*, *Burkholderia ambifaria*, and *Herbaspirillum seropedicae*. Bacteria colonized root surfaces, as well as vascular tissues of hemp seedlings (Fig. 13.3). Inoculated plants compared with controls produced more biomass (stem dry weight, leaf dry weight), equal to plants fertilized with 80 kg nitrogen/ha. THC and CBD content also significantly increased in PGPR-inoculated plants.

Lyu *et al.* (2019) reviewed the many benefits of PGRPs, and their application in *Cannabis*. PGRPs can fix nitrogen, solubilize phosphate, antagonize pathogens, and produce plant growth hormones. Commercially available PGPRs are detailed in section 20.6.

13.3.3 Nitrogen-fixing PGPRs

Azotobacter and *Azospirillum* spp. (*Azo-*, nitrogenous) can fix atmospheric nitrogen. Soviet researchers began testing *Azotobacter chroococcum* as a biofertilizer in 1928, and by 1937 released a commercial product, Азотобактерина, "Azotobakterin." Soviet researchers reported >50% increases in yields, although their claims were discounted by USA researchers (Allison, 1947). The Soviets proposed that *Azotobacter* protected plants from disease—also discounted by Allison (1947), but recently proven true (Wani *et al.*, 2016). Allison (1947) proposed that *Azotobacter* spp. stimulate plant growth

through the synthesis of growth-promoting substances— auxins, cytokinins, and gibberillins—also recently proven true (Wani *et al.*, 2016).

Hemp experiments with Azotobakterin began in 1937. Fiber yields increased 3.8% in a Central Russian variety, and 4.9% in an Italian variety (Gorodny, 1940). Stenin *et al.* (1942) published a booklet detailing the use of Azotobakterin on hemp. Studies on hemp by Tulaykova (1958) showed that Azotobakterin only worked in soils with a moderate amount of nitrogen. In severely deficient soils, the bacteria actually *sequestered* nitrogen, and hemp plants grew poorly. Azotobakterin is still available, under the trade name Азотовит® (Azotovit®).

Azospirillum spp. are distinct from *Azotobacter* spp. A genetic study by Young and Park (2007) demonstrated that they were unrelated; they proposed that *Azotobacter* was synonymous with *Pseudomonas*. In addition to fixing nitrogen, *Azospirillum* spp. also produce plant growth hormones (Wani *et al.*, 2016). Two species, *Azospirillum brasilense* and *A. lipoferum*, were isolated from *Cannabis* (Kosslak and Bohlool, 1983). This prompted McPartland *et al.* (2000) to predict these species could serve as *Cannabis* "biofertilizers." Our prediction has proven true (Pagnani *et al.*, 2018; O'Brien *et al.*, 2018).

13.3.4 Phosphate solubilizing bacteria (PSBs)

Soil phosphorus often exists as phosphate, frequently complexed with other elements to form mineral precipitates that are biologically unavailable. PSBs gained attention for their ability to "solubilize" phosphate, converting it from unavailable $[PO_4]^{3-}$ to available $[HPO_4]^{2-}$ and $[H_2PO_4]^-$.

Albert Caron in Ellenbach, Germany, isolated *Bacillus ellenbachensis* in pure culture, and patented it as Alinit® (Caron, 1897). He sold the rights for Alinit® to Friedrich Bayer & Co—now Bayer AG. The identity of what Caron called *B. ellenbachensis* has been debated. Lauck (1898) synonymized *B. ellenbachensis* under *Bacillus subtilis*, and Kellerman and Fawcett (1913) synonymized it under *Bacillus cereus*. Borriss (2011) assigned *B. ellenbachensis* to *B. subtilis*, but later assigned it to *B. cereus* (Borriss, 2017).

The Soviets considered Alinit® too costly, so they commercialized Фосфоробактерин, "Phosphorobakterin," *Bacillus megaterium* mixed in kaolin, 5 billion spores per gram. By 1942, about 10 million hectares in the Soviet Union were treated with Phosphorobakterin, Azotobakterin, or Nitragin (Smith *et al.*, 1962). Senchenko *et al.* (1963) stated that

Cannabis should only be treated with Phosphorobakterin in humus-rich peat-bog soils. Phosphorobakterin is still available, under the trade name Фосфатовит® (Phosphatovit®). Russian websites promote its use on *Cannabis*.

Other PSBs include *Rhizobium*, *Pseudomonas*, *Paenibacillus*, and *Pantoea* spp. Afzal *et al.* (2015) isolated *Pantoea vagans* and *Pseudomonas geniculata* from wild-type *Cannabis* in Pakistan and demonstrated their ability to solubilize phosphate. Corredor-Perilla *et al.* (2023) collected soil samples from *Cannabis* grown in Colombia and six *Bacillus* spp. significantly increased the bioavailability of unavailable tricalcium phosphate into a useable form of monophosphate.

Dumigan and Deyholos (2022) isolated phosphate-solubilizing *Paenibacillus mobilis* from all 15 *Cannabis* accessions collected across western Canada (fiber and drug strains, outdoor and indoor). They demonstrated "microbial inheritance"—seed-borne endophytes subsequently isolated from seedlings. Other common species included *Paenibacillus pabuli* and *Bacillus subtilis* (*n* =8), and *B. megaterium* (*n* =7). Two less common species, *Paenibacillus polyxma* and *Bacillus velezensis*, showed potent antagonism against seed-borne fungi.

Conant *et al.* (2017) inoculated *Cannabis* with Mammoth P™, a consortium of PGPRs (*Pseudomonas putida*, *Enterobacter cloacae*, *Citrobacter freundii*, *Comamonas testosteroni*) designed to increase phosphorus availability. Treatment with the PGRPs increased bud yields by 16.5%. Plant height also increased, but stem biomass did not.

Donald Smith and colleagues focused upon *Bacillus mobilis* and *Pseudomonas koreensis* isolated in Québec. A different strain of the latter species has been isolated from diseased *Cannabis* in North Carolina (see section 15.3). Lyu *et al.* (2022) inoculated "CBD Kush" with *B. mobilis* and *P. koreensis*, which increased flower weight by 5.13% and 11.45%, respectively, compared with controls. Tanney *et al.* (2023) applied *B. mobilis* and/or *P. koreensis* to "CBD Kush" that received either suboptimal or optimal nutrition. They measured the density of stalked glandular trichomes (SGTs) on perigonal bract (calyx) tissues. A combination of *B. mobilis* and *P. koreensis* increased SGT density 3.5% under optimal nutrition, and 10.3% under suboptimal nutrition. THCA% and CBDA% did not vary under optimal nutrition, but under suboptimal nutrition, THCA% increased by 30.7% (*B. mobilis*), 3.6% (*P. koreensis*), and 8.02% (combined) compared with controls. CBDA% increased by 20.5% (*B. mobilis*), 18.5% (*P. koreensis*), and 13.9% (combined) compared with controls.

13.3.5 Microbe-antagonizing bacteria

PGRPs elute metabolites that inhibit other bacteria and fungi: antibiotics, bacteriocins, siderophores, and cellulases, as well as hydrolytic enzymes such as β-1,3-glucanase and chitinase. Some PGPRs even produce hydrogen cyanide. Several bacteria isolated from *Cannabis* in the aforementioned study by Afzal *et al.* (2015) inhibited the growth of two fungal pathogens, *Aspergillus niger* and *Fusarium oxysporum*.

Pellegrini *et al.* (2021) tested several PGRPs for their ability to inhibit *F. oxysporum* f. sp. *cannabis*. First they tested

Fig. 13.4. Damage from inoculation with a PGPR product (Gianpaolo Grassi)

in vitro antagonistic activity. In their hands, growth inhibition of *F. o. cannabis* by *Azospirillum brasilense* (<20%) was less than three other PGPRs tested, *Gluconacetobacter diazotrophicus*, *Herbaspirillum seropedicae*, and *Burkholderia ambifaria* (inhibition 65–70%). Even in the absence of *F. o. cannabis*, PGPR-treated plants grew much better than control plants.

Dumigan and Deyholos (2022) isolated *Paenibacillus polyxma* and *Bacillus velezensis* from *Cannabis* plants in western Canada. They showed potent antagonism against seed-borne fungi (*Alternaria*, *Aspergillus*, *Penicillium*, and *Fusarium* spp.). Several of the phosphate-solubilizing *Bacillus* spp. collected by Corredor-Perilla *et al.* (2023) in Colombia inhibited the growth of *Fusarium oxysporum*. Gonsior *et al.* (2004) induced ISR (induced systemic resistance) in *Cannabis* by treating plants with non-pathogenic *Pseudomonas* sp., or with extracts from an alga, *Ascophyllum nodosum*. The treatment improved resistance against the parasitic plant *Phelipanche ramosa*.

Inoculating plants with PGRPs may inexplicably backfire. Gianpaolo Grassi (pers. comm., 2020) inoculated potting soil with a commercial PGPR product, after seedling emergence. The *Cannabis* seedlings were damaged, and the damage was dose-related (Fig. 13.4). Control plants (center column, red stakes) were healthy; plants receiving half-strength inoculation (right column, blue stakes) were damaged; plants receiving full-strength inoculation (left column, white stakes) were killed.

13.4 Phyllosphere Microbiome

Phyllosphere microbes gained attention because they occupy flowering tops, which are consumed by humans, and potentially cause human disease. Phyllosphere bacteria and fungi can exist as epiphytes or endophytes, like rhizosphere organisms. Some are obligate pathogens, such as powdery mildew fungi. Others appear to be obligate commensals, such as *Aureobasidium pullulans*, and live off cellular leakage from plant cells or detritus falling from the sky (e.g., pollen). Yet others are mutualistic, and provide benefits to plants (see below).

Then there are commensals that have the ability to "go rogue" and flip into a pathogenic state. *Botrytis cinerea* can exist as a facultative endophyte, then "flip a switch" and become a necrotrophic pathogen, by killing plant cells and feeding on the dead cells. Punja *et al.* (2019) isolated *B. cinerea* from symptomless *Cannabis* as well as diseased plants. They proposed that disease arose under conditions of high inoculum levels on plants approaching harvest.

Van Kan *et al.* (2014) investigated the "switch" that activates virulence in *B. cinerea*. They implicated mitogen-activated protein (MAP) kinase cascades, the calcineurin pathway, and trafficking of small RNAs (sRNAs). They also implicated highly inbred and hybridized crop plants, moved out of their native ecosystems, as tipping the balance in favor of *B. cinerea* virulence. That's certainly the case with *Cannabis*, where plants from arid Afghanistan—lacking resistance to *B. cinerea*—have been hybridized into nearly all commercially available drug-type strains (McPartland and Small, 2020).

13.4.1 Phyllosphere bacteria

Bacteria are the most numerous colonists of leaf surfaces; they may number in millions per cm^2 of leaf surface (Lindow and Barandl, 2003). Phyllosphere bacteria attract interest for Jekyll and Hyde reasons. On the Jekyll side, they help plants deal with abiotic stresses (e.g., drought, frost) and biotic stress (e.g., plant pathogens). On the Hyde side, some can invade human tissues (e.g., *Pseudomonas aeruginosa*, *Burkholderia cepacia*) or cause dysentery (*Salmonella*, *Shigella*, and *Escherichia* spp.). They have caused small epidemics in cannabis users (see section 18.3.5).

Phyllosphere populations differ widely between plant species, as well as within the same species, over the course of the growing season. Variations in population sizes are due, in part, to fluctuations in the physical and nutritional conditions of the phyllosphere. *Pseudomonas* and *Erwinia* (*Pantoea*) spp. are ubiquitous bacterial epiphytes. *Pseudomonas syringae* is another example of an epiphyte that can "go rogue" and cause plant disease (see section 15.2).

The community of *Cannabis* bacterial epiphytes is large. Thompson *et al.* (2017) used a molecular method to estimate the number of species in 20 samples of medical cannabis sold by dispensaries. They probed for bacteria using the 16S rDNA gene sequence, and counted the number of unique "operational taxonomic units" (OTUs), which are proxies for bacterial species. They recovered >5000 bacterial OTUs—essentially >5000 different bacterial species.

Two studies isolated **endophytes** by first killing epiphytes—they surface-sterilized leaves with ethanol and sodium hypochlorite, and then plated the leaves on nutrient agar (Fig. 13.5). Kusari *et al.* (2014) sampled plants grown by Bedrocan BV, a Dutch supplier of medicinal cannabis products. Endophytic bacteria were identified by cloning and sequencing their 16 rDNA genes. Nine isolates were identified: *Bacillus licheniformis*, *Bacillus megaterium*, *Bacillus pumilus*, *Bacillus subtilis*, *Brevibacillus borstelensis*, *Mycobacterium perigrinum*, and several unidentified species of *Bacillus* and *Mycobacterium*.

Fig. 13.5. Isolating endophytes (courtesy of Kusari *et al.*, 2014)

Scott *et al.* (2018) expanded sampling to leaves, petioles, and seed embryos, obtained from three industrial hemp cultivars growing in Canada. Ninety isolates were recovered from petioles, 25 from leaves, and 19 from seeds. More isolates were collected at early- and mid-growing periods (June and July) than late (August). Frequently isolated species belonged to the gram-negative genera *Pseudomonas* (35%), *Pantoea* (17%), and *Enterobacter* (6%), and the gram-positive genera *Staphylococcus* (16%) and *Bacillus* (9%). Other genera (1–2%) included *Acinobacter*, *Agrobacterium*, *Bevibacterium*, *Cedecea*, *Curtobactrum*, *Enterococcus*, *Erwinia*, *Microbacterium*, *Ochrobactrum*, *Paenibacillus*, *Rhizobium*, *Stenotrophomonas*, and *Xanthomonas*. The distribution of some endophytes was genotype-dependent. For example, *Acinobacter*, *Agrobacterium*, and *Enterococcus* were limited to 'Anka'.

Barnett *et al.* (2020) analyzed epiphytes. Cotton swabs dipped in sampling buffer were rubbed over leaf or flower surfaces. DNA was extracted and OTUs were identified by sequencing bacterial 16S rDNA and fungal ITS1. The most abundant OTUs were identified to species, genus, family, or class or other taxonomic levels (Table 13.2). Epiphyte populations differed considerably from rhizosphere populations (Table 13.1).

Application for crops

Phyllosphere bacteria benefit plants in ways similar to rhizosphere bacteria: they antagonize pathogens and produce plant growth hormones. They also fix nitrogen and solubilize phosphate—although these roles on leaf surfaces may not be important. The study by Barnett and colleagues (Table 13.2) identified several beneficial bacteria. One of them, *Lactococcus lactis*, is the Wisconsin state microbe. It is a fermentative organism associated with cheese, sauerkraut, and kimchi production.

Kusari *et al.* (2014) isolated several beneficial bacterial endophytes from *Cannabis* that are commercially available: *Bacillus licheniformis* is the active ingredient in EcoGuard™, a biofungicide. *B. licheniformis* and *B. subtilis* go into Quartzo®, a bionematicide, and Nemix C®, a plant "biostimulant." *B. pumilus* is a "mycorrhiza helper bacterium." *B. megaterium* was commercialized by the Soviets as "Phosphorobakterin."

Scott *et al.* (2018) screened *Cannabis* bacterial endophytes for beneficial attributes, and four isolates were standouts: BTC6-3 and BTC8-1 (*Pseudomonas fulva*) and BTG8-5 and BT14-4 (*P. orientalis*). They tested the isolates against ten fungi in dual culture assays. Four were significantly inhibited: *Sclerotinia sclerotiorum* (reduced 47.8% to 50.8%), *Botrytis cinerea* (reduced 19% to 22%), *Rhizoctonia solani* (reduced 28% by BT14-4), and binucleate *Rhizoctonia* (reduced 23.4%

to 31.3%). Other fungi were not significantly affected—*Colletotrichum gloeosporiodes*, *Fusarium solani*, *F. graminearum*, *Helminthosporium solani*, and the biocontrol organisms *Trichoderma virens* and *Stachybotrys elegans*. BTC6-3 (*Pseudomonas fulva*) and BT14-4 (*P. orientalis*) produced significant amounts of indole-acetic acid, a plant hormone. Other isolates (*Bacillus*, *Pseudomonas* spp.) had the ability to solubilize phosphate.

Balthazar *et al.* (2022a) reviewed the diversity of *Pseudomonas* spp. in the *Cannabis* microbiome, and then Balthazar *et al.* (2022b) screened beneficial *Pseudomonas* and *Bacillus* spp. for *in vitro* activity against several fungal pathogens. The beneficial bacteria were sourced from other crop species, the fungal pathogens were isolated from *Cannabis*: Bc (*Botrytis cinerea*), Ss (*Sclerotinia sclerotiorum*), Fo (*Fusarium oxysporum*), Fc (*F. culmorum*), Fs (*F. solani*), Aa (*Alternaria alternata*), Ps (*Phoma* sp.), and Cs (*Cercospora* sp.) (Fig. 13.6).

Balthazar and colleagues then conducted an *in vivo* test with the most promising bacteria against *Botrytis cinerea*. Plants in a growth chamber were sprayed with bacterial suspensions (10^5 CFU/ml) or water for controls. Two days later, plants were inoculated with *B. cinerea*, ~20 conidia/leaf, then kept under high humidity for 9 days (covered by clear plastic bags), and leaf symptoms assessed. Control leaves became chlorotic and necrotic with sporulating lesions. Best control was provided by *Bacillus velezensis* LBUM279 and *B. velezensis* FZB42 (33% of leaves showed no visible symptoms), less so by *B. velezensis* LBUM1082 (17% of leaves showed no symptoms), and *Pseudomonas synxantha* and *P. protegens* (8% showed no symptoms), and lastly *B. subtilis* (all leaves showed some symptoms and 42% were necrosed).

Some bacterial endophytes isolated from *Cannabis* have an ugly side. Intracellular bacterial endophytes may contribute to the **dudding** phenomenon usually attributed to a viroid, HLVd (see section 14.6). Bacterial contamination of plants grown in "sterile" plant tissue culture is particularly rife, due to the high nutrient availability in Murashige and Skoog culture medium (Cassells, 2012). This pertains to *in vitro Cannabis* explants (plantlets) grown in M&S agar. Ian Davidson (pers. comm. 2019) described a screening method: supplement M&S agar with 3% sucrose. In contaminated explants, bacteria "ooze out of the shoot like snot."

13.4.2 Phyllosphere fungi

The (leaf) phyllosphere concept has been extended to other above-ground parts, such as the caulosphere (stems), anthosphere

Table 13.2. Bacterial and fungal epiphytes associated with leaves and flowers, identified by Barnett *et al.* (2020)

Sample	Most common bacterial OTUs	Most common fungal OTUs
Leaves	*Sphingomonas, Methylobacterium, Hemenobacter, Microbacteriaceae, Pseudomonas*	*Tilletiopsis washingtonensis, Bullera alba, Epicoccum, Neoascochyta, Alternaria infectoria, Dioszegia hungarica, Basidiomycota, Sporobolomyces ruberrimus, Pleosporales, Filobasidium, Mycosphaerellaceae*
Flowers	*Pseudomonas, Lactococcus lactis, Pantoea, Enterobacter, Bacillis cereus, Ralstonia pickettii*	*T. washingtonensis, B. alba, Epicoccum, Sporobolomyces ruberrimus, Ascomycota*

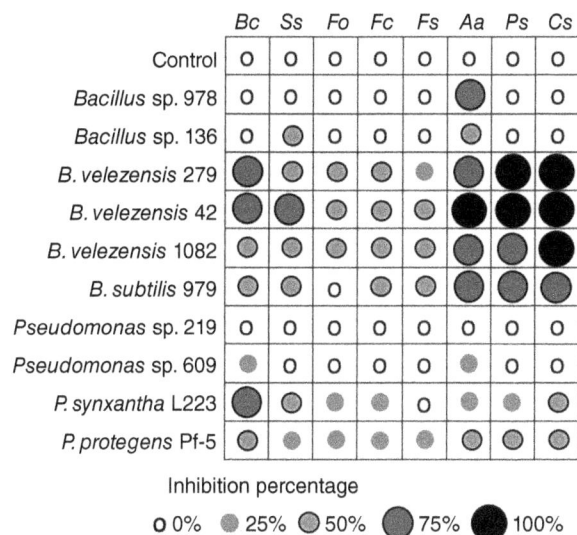

Fig. 13.6. *In vivo* confrontational assay between beneficial bacteria and fungal pathogens (adapted from Balthazar *et al.*, 2022b)

(inflorescences), and carposphere (fruits). Some studies have studied endophytes in pith tissue within the center of stalks.

Early work focused on caulosphere fungi. This is because hemp stalks are **retted** to release the fibers. Retting may lead to rotting, and hemp agronomists wanted to know what organisms were responsible. Hrebenyuk (1984) identified 79 fungal species, and Ondřej (1991) identified 41 fungal species. They are discussed under hemp retting (see section 18.1.1) and rotting (see section 18.1.2). Both studies identified a mixed bag of saprophytes and pathogens. Some of the organisms were undeniably epiphytes, such as the yeast *Aureobasidium pullulans*, and the biocontrol organism *Acremoniella atra*.

Analogous to harvested hemp stalks, researchers have studied the phyllosphere of harvested cannabis flowers. Thompson *et al.* (2017) estimated the number of fungal species in medical cannabis sold by dispensaries, using the same technique they used for estimating phyllosphere bacteria (see above). They recovered ~4000 fungal OTUs—essentially 4000 different fungal species. Some fungi that grow in harvested cannabis flowers cause human diseases (see section 18.3.5).

The phyllosphere of *living* plants has been surveyed by several groups. Kusari *et al.* (2013) sampled plants grown by Bedrocan BV, and surface-sterilized leaves, branches, and apical buds. Endophytic fungi growing out of samples were isolated and then identified by cloning and sequencing their ITS gene regions. The predominant endophyte was *Penicillium copticola*. Other species included *Penicillium meleagrinum*, *Penicillium sumatrense*, *Eupenicillium rubidurum*, *Chaetomium globosum*, *Paecilomyces lilacinus*, and *Aspergillus versicolor*. None of these fungi had been previously associated with *Cannabis* except for *C. globosum* (McPartland *et al.*, 2000).

Gautam *et al.* (2013) sampled plants in Himachal Pradesh. Leaves, petioles, and stems were surface-sterilized,

plated on potato dextrose agar, and endophytes identified by their morphological and cultural characteristics. Colony counts from leaves and stems were about equal, with fewer fungi cultured from petioles. They identified three *Aspergillus* spp. (*A. niger*, *A. flavus*, *A. nidulans*), two *Penicillium* spp. (*P. citrinum*, *P. chrysogenum*), and *Rhizopus stolonifer*. They also identified five other species known to be foliar pathogens of *Cannabis*: *Alternaria alternata* (the most frequently isolated fungus), *Curvularia lunata*, *Cladosporium* sp., *Colletotricum* sp., and *Phoma* sp.—facultative endophytes that can "go rogue."

Qadri *et al.* (2013) sampled plants in Lolab Valley, Kashmir. Plant material was surface-sterilized then cultured. Isolates were identified by their ITS gene sequences. This study screened ten species of plants, and from *Cannabis* only four endophytes were highlighted: *Alternaria alternata*, *Alternaria brassicae*, *Alternaria* sp., and *Schizophyllum commune*. Jin *et al.* (2014) sampled leaves of the fiber and seed cultivar 'Yún má No.1' in Yúnnán Province. They isolated *Plectosphaerella* sp., *Chaetomium* sp., *Nigrospora* sp., *Colletotrichum* sp., *Fusarium* sp., and *Graphium* sp., identified by morphology.

Scott *et al.* (2018) sampled three industrial hemp cultivars growing in Canada. Fifty-three fungal isolates were cultured and identified by ITS sequencing—but only to the level of genus or class. They consisted of *Aureobasidium* (24%), *Alternaria* (19%), *Cochliobolus* (19%), *Cladosporium* (15%), *Dreschlera* (5%), *Sordariomycetes* (4%), and 2% each of *Cryptococcus*, *Eutypella*, *Irpex*, *Stagonosporopsis*, *Dothideomycetes*, *Pezizomycetes*, and *Sordariomycetes*. The distribution of some endophytes was genotype-dependent, and specific to leaves, petioles, or seeds. Unlike the three aforementioned studies, no *Penicillium* spp. were identified. We re-examined some of these sequences (see Table 1.2, Fig. 1.12).

Lubna *et al.* (2019) isolated an endophyte colonizing healthy *Cannabis* plants in Pakistan. They considered the fungus a beneficial endophyte—it produced the plant growth hormones IAA and GA$_3$. They amplified ITS and sequenced it. BLAST analysis revealed the closest match in GenBank was a *Bipolaris* sp. We subjected the "*Bipolaris* sp.*" sequence to a BLAST "best-hit" analysis of GenBank, and it shared 100% identity with *Curvularia spicifera* (see Table 1.2, Fig. 1.12).

Punja *et al.* (2019) isolated epiphytes by gently wiping plant surfaces with cotton swabs, streaking the swabs on PDA plates, subculturing, and identifying them with ITS sequences. Inflorescences epiphytes included *Cladosporium westeerdijkieae*, *Botrytis cinerea*, *Alternaria alternata*, and four *Penicillium* spp.—*P. copticola*, *P. olsonii*, *P. simplicissimum*, and *P. spathulatum*. Endophytes were isolated from internal tissues by sterile dissection of pith, cortex, and vascular tissues. Culturing followed by ITS sequencing identified *Chaetomium globosum*, *Cladosporium westeerdijkieae*, *Fusarium oxysporum*, *Penicillium olsonii*, *Trametes versicolor*, *Trichoderma harzianum*, *Penicillium chrysogenum*, and *P. griseofulvum*. Frequency of isolation was greatest near the crown of the plant, and decreased as sampling progressed up the stem. They proposed that fungi found in growing media (coco fiber) colonized plants by growing in pith tissues.

Punja (2021a) extended his work on pith. A developmental study demonstrated that pith parenchyma cells disintegrated with age, resulting in the well-known hollow in the center of stems. Disintegration left behind layers of dead cells, suitable for fungal colonization. Some of the endophytes recovered from pith tissues, such as *T. globosum*, *T. versicolor*, *C. westeerdijkieae*, and *Fusarium oxysporum*, possess strong cellulolytic enzyme activities. They could "go rogue" given proper conditions (e.g., plant stress).

Comeau *et al.* (2020) extracted fungal DNA (ITS1) from three chemotypes (CBD dominant, mixed CBD/THC, THC dominant), sampled over the course of their indoor life cycles—propagation stage, vegetative, early-flowering, and late-flowering stages. Fungi were identified at the genus level (e.g., *Penicillium*, *Aspergillus*, *Fusarium*, *Zopfiella*). They expected to see differences between CBD- and THC-dominant plants, but did not.

Barnett *et al.* (2020) analyzed fungal epiphytes associated with 'Anka'. Across all samples, they identified 982 OTUs. The most abundant OTUs were identified to species, genus, family, or class or other taxonomic levels (Table 13.2). The yeast-like species *Tilletiopsis washingtonensis* and *Bullera alba* were new findings. Both these species produce antifungal compounds, and *T. washingtonensis* has been investigated as a biocontrol agent against powdery mildew (Urquhart and Punja, 2002).

Application for crops

Kusari *et al.* (2013) tested endophytic fungi against two *Cannabis* pathogens, *Botrytis cinerea* and *Trichothecium roseum*, in dual culture plate antagonistic assays. One or both pathogens were inhibited by many endophytes, which varied depending on the agar medium. *Paecilomyces lilacinus*, *Penicillium copticola*, and *Penicillium* sp. (isolate L3) were the most effective. Kusari provided color photographs of battles raging in petri plates.

Gautam *et al.* (2013) tested two endophytes, *Aspergillus flavus* and *A. niger*, in dual culture plate antagonistic assays against *Curvularia lunata* and *Colletotrichum gloeosporioides*. Both endophytes partially inhibited the growth of both pathogens.

Qadri *et al.* (2013) evaluated the antifungal potential of two *Cannabis* endophytes against several pathogens in dual culture plate antagonistic assays. *Alternaria alternata* inhibited the growth of *Fusarium solani* (by 52%), *Rhizoctonia solani* (50%), *Fusarium oxysporum* (25%), and *Aspergillus flavus* (23%). *Schizophyllum commune* inhibited the growth of *Fusarium solani* (60%), *Rhizoctonia solani* (90%), *Fusarium oxysporum* (75%), and *Aspergillus flavus* (31%). Neither endophyte inhibited *Verticillium dahlia*.

Punja (2021d) said these *in vitro* studies were encouraging. But he advised caution, until studies on living *Cannabis* plants have been conducted. However, *in vivo* results by Balthazar *et al.* (2022b) correlated well with their *in vitro* studies (Fig. 13.6).

Jin *et al.* (2014) inoculated plants with endophytes they isolated from 'Yún má No.1'. They sprayed plants with inoculum in evenings (without rain), four times at weekly intervals (control plants were sprayed with water). Two of the six endophytes, *Chaetomium* sp. and *Fusarium* sp., upregulated plant antioxidant enzymes (superoxide dismutase and peroxidase). Subtle improvements were seen in plant growth (no significant change in fresh weight or height, but improvements in dry weight of bast fiber).

Punja *et al.* (2019) inoculated plants with *Cannabis* endophytes to see if they might switch to a pathogenic state. When inoculated onto freshly excised buds and incubated in high humidity, *Fusarium oxysporum*, *Penicillium olsonii*, and *Botrytis cinerea* caused browning of tissues and decay. This circles back to our earlier discussion of endophytes "going rogue" and becoming pathogens. Maintaining plant health is imperative, with adequate irrigation and proper nutrition. If plants get stressed (an extreme example being Punja's excised buds), the endophytes can go rogue.

(Prepared by J. McPartland)

References

Afzal I, Shinwari ZK, Iqrar I. 2015. Selective isolation and characterization of agriculturally beneficial endophytic bacteria from wild hemp using canola. *Pakistan Journal of Botany* 47: 1999–2008.

Allison FE. 1947. *Azotobacter* inoculation of crops. 1. historical. *Soil Science* 64: 413–427.

Arzberger EG. 1920. Mycorrhiza, cytrids and related fungi in the roots of our common economic plants. *Journal of the Washington Academy of Sciences* 10: 211.

Balthazar C, Joly DL, Filion M. 2022a. Exploiting beneficial *Pseudomonas* spp. for *Cannabis* production. *Frontiers in Microbiology* 12: e833172.

Balthazar C, Novinscak A, Cantin G, Joly DL, Filion M. 2022b. Biocontrol activity of *Bacillus* spp. and *Pseudomonas* spp. against *Botrytis cinerea* and other *Cannabis* fungal pathogens. *Phytopathology* 112: 549–560.

Barnett SE, Cala AR, Lansen JL, et al. (5 other authors). 2020. Evaluating the microbiome of hemp. *Phytobiomes Journal* 4: 351–363.

Barrett JT. 1947. Observations on the root endophyte *Rhizophagus* in culture. *Phytopathology* 37: 359–360.

Barrett JT. 1961. Isolation, culture, and host relation of the phycomycetoid vesicular-arbuscular mycorrhizal endophyte *Rhizophagus*. *Recent Advances in Botany* 2: 1725–1727.

Betekhtin AA 2008. Изменчивость микотрофности травянистых растений. *Fundamental and Applied Botany at the Beginning of the XXI Century* 2: 108–109.

Boldt K, Pors Y, Haupt B, *et al.* (4 other authors). 2011. Photochemical processes, carbon assimilation and RNA accumulation of sucrose transporter genes in tomato arbuscular mycorrhiza. *Journal of Plant Physiology* 168: 1256–1263.

Borriss R. 2011. "Use of plant-associated *Bacillus strains* as biofertilizers and biocontrol agents in agriculture," pp. 41–76 in Maheshwari DK, *ed. Bacteria in agrobiology: plant growth responses.* Springer, Berlin.

Borriss R. 2017. Plant growth promoting bacteria—early investigations, present state and future prospects. *Vegetos* 30: 211–215.

Carbone D, Tombolato A. 1917. Sulfa macerazione rustica della canapa. *Stazioni Sperimentali Agrarie Italiane* 50: 563–575.

Caron A (Hartleb R, *ed.*). 1897. Ueber Alinit und den *Bacillus ellenbachensis alpha. Botanisches Centralblatt* 71: 229–231.

Cassells AC. 2012. Pathogen and biological contamination management in plant tissue culture: phytopathogens, vitro pathogens, and vitro pests. *Methods in Molecular Biology* 877: 57–80.

Chahar S. 2022. Diversity of arbuscular-mycorrhizal fungi in the agricultural fields of Kanhauri village, district Rewari, Haryana, India. *International Journal of Botany* Studies 7: 583–589.

Chen M, Arato M, Borghi L, Nouri E, Reinhardt D. 2018. Beneficial services of arbuscular mycorrhizal fungi—from ecology to application. *Frontiers in Plant Science* 9: e1270.

Citterio S, Prato N, Fumagalli P, *et al.* (5 other authors). 2005. The arbuscular mycorrhizal fungus *Glomus mosseae* induces growth and metal accumulation changes in *Cannabis sativa* L. *Chemosphere* 59: 21–29.

Comeau D, Novinscak A, Joly DL, Filion M. 2020. Spatio-temporal and cultivar-dependent variations in the *Cannabis* microbiome. *Frontiers in Microbiology* 11: e491.

Conant R, Walsh R, Walsh M, Bell C, Wallenstein M. 2017. Effects of a microbial biostimulant, Mammoth P, on *Cannabis sativa* bud yield. *Journal of Horticulture* 4: e191.

Corredor-Perilla IC, Cuervo Andrade JL, Olejar KJ, Park SH. 2023. Beneficial properties of soil bacteria from *Cannabis sativa* L.: Seed germination, phosphorus solubilization and mycelial growth inhibition of *Fusarium* sp. *Rhizosphere* 27.

Dumigan CR, Deyholos MK. 2022. *Cannabis* seedlings inherit seed-borne bioactive and anti-fungal endophytic bacilli. *Plants* 11: e2127.

Fester T, Hause B, Schmidt D, *et al.* (5 other authors). 2002. Occurrence and localization of apocarotenoids in arbuscular mycorrhizal plant roots. *Plant & Cell Physiology* 43: 256–265.

Frank AB.1885. Ueber die auf Wurzelsymbiose beruhende Ernährung gewisser Bäume durch unterirdische Pilze. *Berichte der Deutschen Botanischen Gesellschaft* 3: 128–145.

Frank AB. 1887. Ueber neue Mykorrhiza-Formen. *Berichte der Deutschen Botanischen Gesellschaft* 5: 395–409.

Frank AB. 1889. Über die Pilzsymbiose der Leguminosen. *Berichte Deutschen Botanischen Gesellschaft* 7: 332–346.

Gai JP, Christie P, Feng G, Li XL. 2006. Twenty years of research on community composition and species distribution of arbuscular mycorrhizal fungi in China: a review. *Mycorrhiza* 16: 229–239.

Gautam AK, Kant M, Thakur Y. 2013. Isolation of endophytic fungi from *Cannabis sativa* and study their antifungal potential. *Archives of Phytopathology and Plant Protection* 46: 627–635.

Gerdemann JW. 1955. Relation of a large soil-borne spore to phycomycetous mycorrhizal infections. *Mycologia* 47: 619–632.

Gerdemann JW. 1964. The effect of mycorrhiza on the growth of maize. *Mycologia* 56: 342–349

Gerdemann JW, Trappe JM. 1974. The Endogonaceae in the Pacific Northwest. *Mycologia Memoirs* 5: 1–76.

Giovanetti O, Tomalty D, Velikonja L, *et al.* (7 other authors). 2023. Pre- and post-LEEP: analysis of the female urogenital tract microenvironment and its association with sexual dysfunction. *Sexual Medicine* 11: 1–12.

Gonsior G, Buschmann H, Szinicz G, Spring O, Sauerborn J. 2004. Induced resistance: an innovative approach to managed branched broomrape (*Orobanche ramosa*) in hemp and tobacco. *Weed Science* 52: 1050–1053.

Gorodny NG. 1940. Применение бактериальных удобрений под коноплю. *Flax and Hemp* 1940(1): 45–47.

Hellriegel H, Wilfarth H. 1888. *Untersuchungen über die Stickstoffnahrung der Gramineon und Leguminosen.* Kayssler & Co., Berlin.

Hepper CM. 1979. Germination and growth of *Glomus caledonius* spores: the effects of inhibitors and nutrients. *Soil Biology and Biochemistry* 11: 269–277.

Hiltner L. 1892. *Einige durch Botrytis cinerea erzeugte Krankheiten gärtnerischer und landwirtschaftlicher Kulturpflanzen und deren Bekämpfung.* Doctoral thesis, University of Erlangen.

Hiltner L. 1904. Über neuere Erfahrungen und Problem auf dem Gebiet der Bodenbakteriologie und unter besonderer Berücksichtigung der Gründüngung und Brache. *Arbeiten der Deutschen Landwirtschaftlichen Gesellschaft* 98: 59–79.

Hiltner L, Gentner G. 1916. Ueber die Wirkung der Beizung der Samen von Hanf, Sonnenblumen, Buchweizen, Hierse, Mais und Mohar. *Praktischen Blätter für Pflanzenbau und Pflanzenschutz* 14: 85–90.

Hrebenyuk NV. 1984. Распространенне грибов на тресте конопли (The occurrence of fungi on hemp stems). *Микология и фитопатология* 18(4): 322–326.

Jin R, Yang MZ, Liu FH. 2014. Effects of endophytic fungi re-inoculation on physiological agronomic characters of hemp (*Cannabis sativa*). *Plant Diversity and Resources* 36: 65–69.

Kakabouski I, Tataridas A, Mavroeidis A, *et al.* (8 other authors). 2021. Effect of colonization of *Trichoderma harzianum* on growth development and CBD content of hemp (*Cannabis sativa* L.). *Microorganisms* 9: e518.

Kellerman KF, Fawcett KH. 1913. Studies of the *Bacillus subtilis* group. *Science* 38: 372–373.

Kerling LCP. 1958. De microflora op het blad van Beta vulgaris. *Tijdschrift over Plantezieken* 64: 402–410.

Kerling LCP. 1964. Fungi in the phyllosphere of leaves of rye and strawberry. *Mededelingen Landbhoogesch Opzoekstations Gent* 29: 885–895.

Kloepper JW, Schroth MN. 1978. Plant growth promoting rhizobacteria on radishes. *Proceedings of the 4th International Conference on Plant Pathogenic Bacteria* 2: 879–882.

Koide RT, Mosse B. 2004. A history of research on arbuscular mycorrhiza. *Mycorrhiza* 14: 145–163.

Kosslak RM, Bohlool BB. 1983. Prevalence of *Azospirillum* spp. in the rhizosphere of tropical plants. *Canadian Journal of Microbiology* 29: 649–652.

Kusari P, Kusari S, Spiteller M, Kayser O. 2013. Endophytic fungi harbored in *Cannabis sativa* L.: Diversity and potential as biocontrol agents against host plant-specific phytopathogens. *Fungal Diversity* 60: 137–151.

Kusari P, Kusari S, Lamshoeft M, *et al.* (3 other authors). 2014. Quorum quenching is an antivirulence strategy employed by endophytic bacteria. *Applied Microbiology and Biotechnology* 98: 7173–7183.

Last FT. 1955. Seasonal incidence of *Sporobolomyces* on cereal leaves. *Transactions of the British Mycological Society* 38: 221–239.

Lauck H. 1898. Welches sind die Bestandteil des als "Alinit" bezeichneten Inpfdüngers für Saatgetreide. *Centralblatt für Bakteriologie, Parasitenkunde, und Infektionskrankheiten, Zweite Abteilung* 4: 290–295.

Lindow SE, Brandl MT. 2003. Microbiology of the phyllosphere. *Applied and Environmental Microbiology* 69: 1875–1883.

Lubna SA, Khan AL, Waqas M, *et al.* (4 other authors). 2019. Growth-promoting bioactivities of *Bipolaris* sp. CSL-1 isolated from *Cannabis sativa* suggest a distinctive role in modifying host plant phenotypic plasticity and functions. *Acta Physiologiae Plantarum* 41: e65.

Lyu DM, Backer R, Robinson WG, Smith DL. 2019. Plant growth-promoting rhizobacteria for *Cannabis* production: yield, cannabinoid profile and disease resistance. *Frontiers in Microbiology* 10: e1761.

Lyu, DM, Backer R, Smith DL. 2022. Three plant growth-promoting rhizobacteria alter morphological development, physiology, and flower yield of *Cannabis sativa* L. *Industrial Crops Products* 178: e114583.

McPartland JM. 1984. Pathogenicity of *Phomopsis ganjae* on *Cannabis sativa* and the fungistatic effect of cannabinoids produced by the host. *Mycopathologia* 87: 149–153.

McPartland JM, Cubeta MA. 1997. New species, combinations, host associations and location records of fungi associated with hemp (*Cannabis sativa*). *Mycological Research* 101: 853–857.

McPartland JM, Small E. 2020. A classification of endangered high-THC cannabis (*Cannabis sativa* subsp. *indica*) domesticates and their wild relatives. *PhytoKeys* 144: 81–112.

McPartland JM, Clarke RC, Watson DP. 2000. *Hemp diseases and pests – management and biological control*. CABI Publishing, Wallingford, UK.

Melin E, Das VSR. 1954. Influence of root-metabolites on the growth of tree mycorrhizal fungi. *Physiologia Plantarum* 7: 851–858.

Menge JA. 1983. Utilization of vesicular-arbuscular mycorrhizal fungi in agriculture. *Canadian Journal of Botany* 61:1015–1024.

Mosse B. 1953. Fructifications associated with mycorrhizal strawberry roots. *Nature* 171: 974.

Mosse B. 1961. Experimental techniques for obtaining a pure inoculum of an *Endogone* sp. and some observations of the vesicular-arbuscular infections caused by it and other fungi. *Recent Advances in Botany* 2: 1728–1732.

Nicolson TH, Gerdemann JW. 1968. Mycorrhizal *Endogone* species. *Mycologia* 60: 313–325.

Nobbe F, Hiltner L.1893. Impfet den Boden! Eine aus der ungleichen Wirkungskraft der Knöllchenbakterien auf die verschiedenen Leguminosen sich ergebende praktische Schulussfolgerung. *Sächsische Landwirtschaftliche Zeitschrift* 51: 595–600.

O'Brien ER, Read E, Deyholos M, Nelson L. 2018. Effects of nitric oxide producing bacteria *Azospirillum brasilense* on microbial composition and secondary metabolite profile of *Cannabis*. *Planta Medica* 5(S 01): S4–S5.

Ondřej M. 1991. Výskyt hub na stoncích konopí (*Cannabis sativa* L.). *Len a Konopé* (Sumperk, Czech Republic) 21: 51–57.

Pagnani G, Pellegrini M, Galieni A, *et al.* (8 other authors). 2018. Plant growth-promoting rhizobacteria (PGPR) in *Cannabis sativa* 'Finola' cultivation: an alternative fertilization strategy to improve plant growth and quality characteristics. *Industrial Crops and Products* 123: 75–83.

Pellegrini M, Ercole C, Gianchino C, *et al.* (3 other authors). 2021. *Fusarium oxysporum* f. sp. *cannabis* isolated from *Cannabis sativa* L.: *in vitro* and *in planta* biocontrol by a plant growth promoting-bacteria consortium. *Plants* 10: e2436.

Punja ZK. 2021a. The diverse mycoflora present on dried cannabis (*Cannabis sativa* L., marijuana) inflorescences in commercial production. *Canadian Journal of Plant Pathology* 43: 88–100.

Punja ZK. 2021d. Emerging diseases of *Cannabis sativa* and sustainable management. *Pest Management Science* 77: 3857–3870.

Punja ZK, Collyer D, Scott C, *et al.* (3 other authors). 2019. Pathogens and molds affecting production and quality of *Cannabis sativa* L. *Frontiers in Plant Science* 10: e1120.

Qadri M, Johri S, Shah BA, *et al.* (6 other authors). 2013. Identification and bioactive potential of endophytic fungi isolated from selected plants of the Western Himalayas. *SpringerPlus* 2: e8

Ruinen J. 1956. Occurrence of *Beijerinckia* species in the "phyllosphere." *Nature* 177: 220–221.

Scott M, Rani M, Samsatly J, Charron JB, Jabaji S. 2018. Endophytes of industrial hemp (*Cannabis sativa* L.) cultivars: identification of culturable bacteria and fungi in leaves, petioles, and seeds. *Canadian Journal of Microbiology* 64: 664–680.

Seemakram W, Paluka J, Suebrasri T, Lapjit C. 2022. Enhancement of growth and cannabinoids content of hemp (*Cannabis sativa*) using arbuscular mycorrhizal fungi. *Frontiers in Plant Science* 13: e845794.

Seleiman MF, Santanen A, Kleemola J, Stoddard FL, Mäkelä P. 2013. Improved sustainability of feedstock production with sludge and interacting mycorrhiza. *Chemosphere* 91: 1236–1242.

Senchenko GI, Arinstein AI, Timonin MA. 1963. Конопля. Publishing House of Agricultural Literature, Moscow.

Smith JH, Allison FE, Soulides DA. 1962. *Phosphobacterin as a soil inoculant*. USDA Technical Bulletin no. 1263. Washington, DC.

Stenin VA, Dyukov R, Volkov T. 1942. За высокий урожай льна и конопли. Oblispolkoma, Novosibirsk.

Tadayon MR, Zarei M. 2014. Investigation of the symbiotic effect of the mycorrhiza fungus *Glomus mosseae* on salinity resistance in *shāh dānah* (*Cannabis*) ecotypes. *Journal of Plant Process and Function* 3(7): 105–114.

Tamburini E, Daly S, Steiner U, Vandini C, Mastromei G. 2001. *Clostridium felsineum* and *Clostridium acetobutylicum* are two distinct species that are phylogenetically closely related. *International Journal of Systematic and Evolutionary Microbiology* 51: 963–966.

Tanney CAS, Lyu DM, Schwinghamer T, *et al.* (3 other authors). 2023. Sub-optimal nutrient regime coupled with *Bacillus* and *Pseudomonas* sp. inoculation influences trichome density and cannabinoid profiles in drug-type *Cannabis sativa*. *Frontiers in Plant Science* 14: e1131346.

Theophrastus (Hort A, *trans*). 1916. *Enquiry into plants*. W. Heinemann, New York.

Thompson GR, Tuscano JM, Dennis M, *et al.* (8 other authors). 2017. A microbiome assessment of medical marijuana. *Clinical Microbiology and Infection* 23: 269–270.

Tian H, Gai JP, Zhang JL, Christie P, Li XL. 2009. Arbuscular mycorrhizal fungi associated with wild forage plants in typical steppe of eastern Inner Mongolia. *European Journal of Soil Biology* 45: 321–327.

Tisdall JM, Oades JM. 1982. Organic matter and water-stable aggregates in soils. *Journal of Soil Science* 33: 141–163.

Tulaykova KP. 1958. Роль бактерий типа радиобактера в корневом питании растений. *Агробиология* 1958(1): 93–101.

Urquhart EJ, Punja ZK. 2002. Hydrolytic enzymes and antifungal compounds produced by *Tilletiopsis* species, phyllosphere yeasts that are antagonists of powdery mildew fungi. *Canadian Journal of Microbiology* 48: 219–229.

Van Kan JAL, Saw MW, Grant-Downton RT. 2014. *Botrytis* species: relentless necrotrophic thugs or endophytes gone rogue? *Molecular Plant Pathology* 15: 957–961.

Wani SA, Chand S, Wani MA, Ramzan M, Hakeem KR. 2016. "*Azotobacter chroococcum*—a potential biofertilizer in agriculture: an overview," pp. 333–348 in Hakeem KR *et al.*, eds. *Soil science: agricultural and environmental prospectives*. Springer International, Switzerland.

Wei GF, Ning K, Zhang GZ, *et al.* (5 other authors). 2021. Compartment niche shapes the assembly and network of *Cannabis sativa*-associated microbiome. *Frontiers in Microbiology* 12: e714993.

Whipps JM, Lewis K, Cooke RC. 1988. "Mycoparasitism and plant disease control," pp. 161–187 in Burge MN, *ed. Fungi in biological control systems*. Manchester University Press, Manchester, UK.

Wilson GWT, Daniels-Hetrick BA, Gerschefske-Kitt B. 1988. Suppression of mycorrhizal growth response of big bluestem by non-sterile soil. *Mycologia* 80: 338–343.

Winston ME, Hampton-Marcell J, Zarraonaindia I, *et al.* (6 other authors). 2014. Understanding cultivar-specificity and soil determinants of the cannabis microbiome. *PLoS ONE* 9: e99641.

Yaseen T, Khan Y, Rahim F, *et al.* (4 other authors). 2016. Arbuscular mycorrhizal fungi spores diversity and AMF infection in medicinal plants of District Charsadda KPK. *Pure and Applied Biology* 4: 1176–1182.

Young JM, Park DC. 2007. Probable synonymy of the nitrogen-fixing genus *Azotobacter* and the genus *Pseudomonas*. *International Journal of Systematic and Evolutionary Microbiology* 57: 2894–2901.

Zalaghi R, Safari-Sinegani AA, Aliasgharzad N. 2021. The effects of inoculation with three *Glomus* species on growth and Pb uptake by hemp (*Cannabis sativa*) in a Pb-contaminated soil. *Systematic Bioscience* 1(1): e1.

Zarei M, Tadayon MR, Tadayyon A. 2014. Effect of biofertilizer, under salinity condition on the yield and oil content of three ecotype of hemp (*Cannabis sativa* L.). *Journal of Crops Improvement* 16: 517–529.

Zielonka D, Sas-Paszt L, Derkowska E, Lisek A, Russel S. 2019. Occurrence of arbuscular mycorrhizal fungi in hemp (*Cannabis sativa*) plants and soil fertilized with sewage sludge and phosphogypsum. *Journal of Natural Fibers* 18: 250–260.

Zubek S, Błaszkowski J, Buchwald W. 2012. Fungal root endophyte associations of medicinal plants. *Nova Hedwigia* 94: 525–540.

14 Virus, viroid, and phytoplasma diseases

Abstract

The causal agents in this chapter can only replicate within a living host, yet they cause serious disease symptoms and reduce yields. Their symptoms can be confused with genetic diseases or nutritional deficiencies. In fact the first viral diseases described on hemp—hemp streak and hemp mosaic virus—might not be caused by viruses. Ten viruses are described in detail. Viral diseases have exploded in the past 30 years, because of the increased use of vegetative clones to propagate *Cannabis*. Once acquired, they are nearly impossible to eradicate. Hop latent viroid (HLVd) was identified only five years ago, but HLVd now threatens the entire *Cannabis* industry. Putative control measures are described. Phytoplasmas are bacteria without cell walls, but they behave more like viruses. At least six *Candidatus* Phytoplasma spp. have been named on hemp, with several unknowns. We constructed a phylogenetic tree of *Cannabis* phytoplasmas to sort them out.

14.1 Introduction

In the 1880s, Louis Pasteur suspected some pathogens were too small to be seen through a microscope. He studied rabies, whose cause was invisible. Less than a decade later, a plant pathologist named **Dmitri I. Ivanovsky** (1864–1920) proved the existence of viruses. Ivanovsky (1892) filtered sap from diseased tobacco plants through a porcelain filter that removed all bacteria, and the sap remained infective. Beijerinck (1898) confirmed Ivanovsky's results. Beijerinck considered the cause a *contagium vivum fluidum*, "soluble living germ," and coined the word **virus**. He showed that viruses, unlike bacteria, could not be grown in petri dishes—they required the presence of living host cells in order to replicate.

Ivanovsky and Beijerinck studied the **tobacco mosaic virus** (TMV). TMV played a central role in establishing the field of virology. Bernal and Fankuchen (1941) discovered that crystals of TMV consisted of proteins and nucleic acids. Bernal's lab was joined by **Rosalind Franklin** (1920–1958), famous for her X-ray diffraction analysis of DNA. The DNA "double-helix" model proposed by Watson and Crick (1953) depended upon Franklin's data, which were yet unpublished, obtained without her consent, and uncited—a theft confessed later by Watson (1968).

Both Watson and Franklin shifted their focus from DNA to TMV. Watson (1954) proposed that TMV was arranged in a helical structure, but he got the structure wrong. A year later, Franklin (1955) corrected Watson's structure, by showing that TMV has 49 subunits per three turns of the helix (Fig. 14.1). She published a second landmark paper regarding the arrangement of RNA within TMV (Franklin, 1956). Then she died of ovarian cancer, denied credit for her key contributions to science (Creager and Morgan, 2008).

Phytoplasmas are also obligate parasites, like viruses, and difficult to culture *in vitro*. The comparison ends there. Phytoplasmas are members of Phylum Mycoplasmatota, formerly called Phylum Tenericutes (Latin *tener cutis*, soft skin). They are bacteria without cell walls. Phytoplasmas were discovered in a photo darkroom shared by Japanese botanists and veterinarians. The botanist, studying diseased plants, was surprised by a lack of viruses appearing in his electron microscope images. His veterinary colleague, who studied *Mycoplasma hyopneumonia* in pigs, replied, "But look at all those mycoplasmas" (Doi *et al.*, 1967). The pathogens were named mycoplasma-like organisms (MLOs). MLOs were renamed phytoplasmas in 1992.

Viroids are pathogens stripped down to the minimum required for reproduction: a naked, circular loop of ssRNA. They don't even have a protein coat, which makes them a hundred times smaller than the smallest viruses. Viroids were originally discovered by a botanist, Theodor Diener, in 1971. Since then a score of viroid diseases have been described, and *Cannabis* has joined the list of hosts—succumbing to hop latent viroid (HLVd). This viroid now threatens the entire *Cannabis* industry.

14.2 Viral diseases

Viral diseases have become a serious problem in the past 30 years, because of the increased use of vegetative clones to propagate *Cannabis*. The explosion of viral and viroid diseases in *Cannabis* has been likened to "the wild wild west" (Nachappa *et al.*, 2020). Viruses rarely kill *Cannabis*. They cause serious symptoms in foliage and flowers, and reduce yields. Once acquired, viruses are nearly impossible to eradicate.

Clonal propagation enables the mass production of phenotypically uniform plants. But clonal propagation allows the transmission of viruses and viroids from infected mother plants to infected clones. This is called **vertical transmission**—disease transmission from parent to offspring. Some vertical transmission occurs through viral-infected pollen or seed, but in *Cannabis* it primarily occurs through clonal propagation.

Vectors play a role in the transmission of plant viruses—the dissemination of viruses through insects moving from plant to plant. The hemp aphid (*Phorodon cannabis*) is often implicated (Röder, 1941b; Goidànch, 1955; Ceapoiu, 1958; Pitt

© John M. McPartland 2025. *Hemp Diseases and Pests* 2nd Edition (J.M. McPartland)
DOI: 10.1079/9781836990352.0014

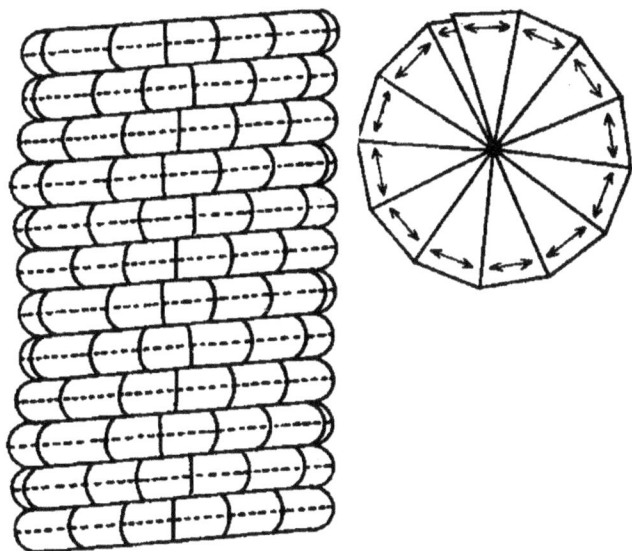

Fig. 14.1. Schematic drawing of the TMV helical structure (redrawn from Franklin, 1955)

et al., 2022). Mites, nematodes, fungi, and parasitic plants also vector viruses. Vectors, viruses, and plants have coevolved in complex ways.

Viruses are named by symptoms they cause and the organisms they infect. TMV causes a mosaic of yellow and green spots on tobacco leaves. Viruses are classified in a Linnaean hierarchy; TMV is in Phylum *Kitrinoviricota*, Class *Alsuviricetes*, Family *Virgaviridae*, Genus *Tobamovirus*. For a while, viruses were tagged with Latin binomials, and TMV was named *Marmor tabaci* (Holmes, 1939).

Most viruses that attack humans contain double-stranded (ds) DNA. In contrast, about 75% of plant viruses consist of single-stranded RNA (ssRNA). And a majority of these have positive-sense, (+) ssRNA, meaning that they are in the same sense orientation (i.e., "+") as messenger RNA. The genome of many plant viruses is split into multiple packets, encapsulated in separate particles. These "multipartite" genomes are unique to plant viruses.

Currently about 5000 viruses are recognized, of which 1200 are plant viruses. A dozen or so are known to infect *Cannabis*. Plant viruses rarely spread by direct contact between infected and uninfected individuals (e.g., natural root grafts, or plants rubbing in the wind). Spreading by direct contact is known as mechanical or sap transmission, and it is used in the laboratory, by rubbing infectious sap into the leaf of a healthy plant, or by grafting a diseased scion onto a healthy stock.

Plant viruses pose no threat to human health. A separate concern is the contamination of plant products by human viruses. An outbreak of hepatitis A virus (HAV) in Asotin County, Washington, was linked to Mexican marijuana fertilized with feces contaminated by HAV (Alexander, 1987). Outbreaks of HAV in Oklahoma (Harkess *et al.*, 1989) and Queensland, Australia (Shaw *et al.*, 1999) have also been linked with cannabis use—by handling fecally contaminated plants, or sharing contaminated paraphernalia. Sikora *et al.* (2017) investigated a

person with HAV, whose medical history showed no high-risk behaviors, but he smoked cannabis. They tested his weed, and RT-PCR amplified an HAV sequence identical to the one amplified from his blood.

Because viruses, viroids, and phytoplasma cannot be grown in pure culture, they cannot satisfy **Koch's postulates**, a protocol for establishing microbial causality (see section 1.5.4). *Adapted* Koch's postulates can be applied as follows. (1) Visualize or isolate the virus from disease plants, and demonstrate that it is absent in diseased plants. (2) Transmit the virus from disease plants to healthy plants, which subsequently develop disease symptoms.

Transmission can be performed by mechanical inoculation. (1) Grind lesioned tissue from a diseased plant in a mortar and pestle with sterile water or a buffer (e.g., sodium phosphate, pH 7.0). (2) Rub the slurry into healthy plant tissue with an abrasive such as carborundum. (3) Rinse off and wait for symptoms (up to 3 weeks). Many *Cannabis* studies have used this method (e.g., Röder, 1941a,b; Schmidt and Karl, 1969, 1970; Koziel, 2010; Righetti *et al.*, 2018; Warren *et al.*, 2019).

Some viruses are difficult to transmit mechanically, especially phloem-restricted viruses. In these cases, grafting between infected and virus-free plants is effective. This method has not yet been used to test *Cannabis* viruses. Dodder (*Cuscuta* spp.) has also been used to experimentally transmit viruses. Dodder is a parasitic vine that transmits viruses from plant to plant. Salehi *et al.* (2002) used *Cuscuta campestris* in a phytoplasma transmission experiment between *Citrus latifolia* and *Cannabis sativa*. Many phloem-restricted viruses are vectored by aphids; *Phorodon cannabis* has been used to experimentally vector viruses between diseased and healthy plants (Schmidt and Karl, 1970; Karl, 1971; Pitt *et al.*, 2022).

14.3 Hemp streak virus, hemp mosaic virus

These two viruses are named after their classic symptoms in leaves. Hemp streak virus (HSV) causes interveinal yellow streaks or chevron-stripes. Hemp mosaic virus (HMV) causes mottled light and dark green patches. A mystery surrounds HSV and HMV: the viruses have never been visualized in diseased tissues. There has been ample opportunity—viewing plant viruses with transmission electron microscopy (TEM) began with Markham *et al.* (1964). The International Committee on Taxonomy of Viruses does not recognize HSV and HMV (https://ictv.global/taxonomy).

14.3a. Hemp streak virus

Röder (1941a,b) described the first viral disease of *Cannabis*, in the Berlin region. He named its causal agent the *streifenvirus* (streak virus), later known as HSV. Röder (1941b) provided photographs of foliar symptoms. In addition to yellow streaks and chevron-stripes, brown necrotic flecks appeared, each fleck surrounded by a pale green halo. Flecks appeared along the margins and tips of older leaves, and often coalesced. Streak symptoms predominated in moist weather, whereas fleck symptoms appeared during dry weather. Eventually, leaf margins became wrinkled, leaf tips roll upward, and leaflets curled into spirals (Fig. 14.2).

Röder (1941a,b) expressed sap from diseased plants, and mechanically inoculated healthy plants. The inoculated plants developed symptoms within 6 days of inoculation, although, he said, "atypical" symptoms. Male plants suffered higher rates of infection than females. Röder (1941a) detected the hemp aphid, *Phorodon cannabis*, for the first time in the Berlin area, and Röder (1941b) implicated *P. cannabis* as a vector of the virus.

14.3b. Hemp mosaic virus

Mosaic symptoms were first described by Blattný *et al.* (1950), near Šumperk, Czech Republic. In addition to chlorotic mosaic spots, leaves developed a "warty" or puckered surface, followed by necrosis, and a twisting and torsions of leaflets. These symptoms were similar to those of HSV described by Röder (1941b), but Blattný and colleagues were unaware of Röder's publication, stating that "a viral disease has not been described." They implicated the aphid *Myzus persicae* as a viral vector. They stated that the disease's viral nature "has not been proven, but is very probable."

The same group reported that the mosaic virus could not be transmitted to tobacco (Svobodová *et al.*, 1954). The virus

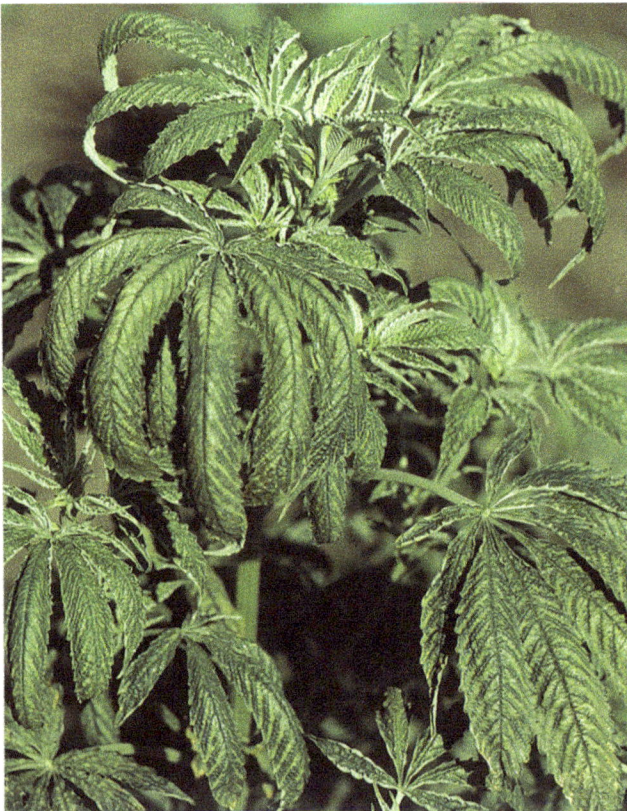

Fig. 14.2. Symptoms of putative hemp streak virus (R. Clarke)

was seed-borne, and decreased seed germination. They implicated *P. cannabis* as a vector of the virus. Rataj (1957), who worked with the group, said the viral disease was more common in the southern part of the Czech Republic, limited by the range of its aphid vector.

Ceapoiu (1958) described *mozaicul cînepea* (hemp mosaic) in Romania. Symptoms began as punctate, yellow-green leaf spots, which turned necrotic, coalesced, and finally the entire leaf wilted. He provided a photograph of symptoms. Ceapoiu attributed hemp mosaic to a virus, and differentiated it from Röder's "chlorosis" viral disease. He reported that nearly all Romanian and foreign varieties were susceptible, although some resistance was seen in Chinese hemp and Romanian wild hemp. Ceapoiu blamed *P. cannabis* as the vector, as well as green peach aphids (*M. persicae*), greenhouse whiteflies (*Trialeudodes vaporariorum*), and onion thrips (*Thrips tabaci*).

Goidànich (1955) reported "streak virus" in Italy; he noted a similar disease was described in Germany, Czech Republic, and the USA. The USA citation was based on Brierley (1945), which was actually a translation of Röder (1941b). Symptoms begin as a chlorosis at leaflet tips, then chlorotic stripes running between veins, and the entire leaf becomes striped and curls up. Later, brown spots surrounded by pale green halos develop, and leaf tissues become lacerated. Affected plants are stunted, and yield of fiber and seed decreases. *P. cannabis* vectors the virus.

Ferri (1963) described interveinal chlorotic stripes in the new 'Fibranova' cultivar in Italy, along with leaf curling, stunted growth, and low seed production. Seeds from affected plants gave rise to 40–50% seedlings with identical symptoms. Ferri noted the symptoms were similar to those described by Röder. He provided photographs of disease symptoms. Gitman (1968b) considered viral disease rare in Russia. Spaar *et al.* (1990) illustrated HSV symptoms in a color painting.

Righetti *et al.* (2018) addressed a central mystery of HSV and HMV: they had never been seen by electron microscopy or detected by serological tests. They studied two Italian cultivars, 'Chamaeleon' and 'Codimono', with symptoms of interveinal chlorosis and leaf margin wrinkling. Symptom intensity in plants ranged from highly symptomatic (severity index (**SI**) = 16) to symptomless (SI = 1). The plants were screened for 12 viruses, using virus-specific PCR primers. Next-generation (NextGen) sequencing of RNA extracted from symptomatic leaves revealed the presence of only one virus: Cannabis cryptic virus (CanCV), discovered earlier by Ziegler *et al.* (2012).

Righetti and colleagues quantified CanCV viral load in plants, and it did not correlate with disease severity—highly symptomatic plants carried the same viral load as symptomless plants. They concluded that CanCV was not responsible for interveinal chlorosis and leaf margin wrinkling. Finally, transmission electron microscopy revealed no virus-like structures in plants, except for CanCV.

Righetti therefore questioned whether hemp streak was caused by a virus. They proposed that hemp streak may be a genetic disease—they transmitted hemp streak vertically (through seeds of symptomatic plants), but not through mechanical inoculation or grafting. They also proposed that environmental stress might bring out symptoms—in young plantlets struggling to root in rockwool blocks.

Punja (2021a) cited Righetti's work, and stated that old reports of viral pathogens need to be reassessed using NextGen sequencing approaches. Nachappa *et al.* (2020) also cited Righetti's work, and used NextGen sequencing to identify viruses and viroids present in *Cannabis* at six sites in Colorado. CanCV was present at one site. Beet curly top virus was the most common. They identified a total of eight viruses and viroids, including three new to *Cannabis* (see below).

14.4 Arabis, alfalfa, and cucumber mosaic viruses

Identifying plant viruses by **serology** (antibodies and antigens) goes back to Chester (1935). Antibodies (antisera) to viruses are produced by injecting plant viruses into rabbits. The resulting antibodies are extracted and pipetted into wells in agar gel in a petri plate. Then the sap from a diseased plant (the antigen) is placed in an adjacent well. Antibodies and antigens diffuse into the agar, and if they match, a positive reaction (white lines of precipitate) arises between the wells (Van Slogteren and Van Slogteren, 1957). Schmidt and Karl (1969) first utilized serological tests for *Cannabis* research.

14.4a. Arabis mosaic virus (ArMV)

Schmidt and Karl (1969) observed hemp plants with chlorotic stripes and spots and deformed leaves near Potsdam, close to the original sighting of HSV by Röder. The viral nature of the disease was confirmed by mechanical transmission to a variety of test plants (*Chenopodium album, C. quinoa, Cucumis sativus, Petunia hybrida, Nicotiana megalosiphon*). Results with test plants, and the fact that experiments with *Phorodon cannabis* did not result in viral transmission, led them to suspect Arabis mosaic virus (ArMV). ArMV is primarily vectored by migratory dagger nematodes (*Xiphinema* spp.). Furthermore, they knew hop—closely related to hemp—was an ArMV host.

In an agar diffusion test, sap from infected plants formed a precipitate with ArMV antibodies. Schmidt and Karl proposed that Röder (1941b) may have worked with ArMV-infected material, because symptoms produced by ArMV were similar to those described by Röder. Koziel (2010) mechanically inoculated *Cannabis* plants with ArMV and illustrated symptoms.

ArMV is classified in the *Nepovirus* genus, which includes about 40 viruses. ArMV is icosahedral in shape, with ssRNA(+) nucleic material. ArMV has a wide host range; Schmelzer (1963) infected 93 plant species from 28 dicot families with ArMV via mechanical inoculation. Principal hosts are hop, strawberry, grape, and raspberry.

14.4b. Cucumber mosaic virus (CMV)

Schmidt and Karl (1970) observed hemp plants with symptoms that differed from those caused by ArMV. One symptom they called *Hanfscheckung* (hemp-spotting). Sap from *Hanfscheckung*-infected plants formed a precipitate with antibodies from cucumber mosaic virus (CMV) in agar diffusion tests. Inoculation tests by others confirmed that *Cannabis* was susceptible to CMV (Hartowicz *et al.*, 1971; Paulsen, 1971; Kegler and Spaar, 1997). Koziel (2010) mechanically inoculated *Cannabis* plants with CMV and illustrated symptoms. McKernan (oral comm., 2023) identified CMV when sequencing a *Cannabis* genome, "never before reported."

CMV may have the widest host range of any known plant virus (>1200 species in >100 families of dicots and monocots). It is transmitted by seed, through grafting, by over 60 aphid species, and by parasitic plants (*Cuscuta* spp.). CMV is a member of the *Cucumovirus* genus, which includes four viruses. CMV is icosahedral in shape, with ssRNA(+) nucleic material in a tripartite genome.

14.4c. Alfalfa mosaic virus (AMV)

Gelbstreifung des Hanfes (yellow streaking of hemp) was also observed by Schmidt and Karl (1970). The *Gelbstreifung* virus could be transmitted experimentally by *P. cannabis* and *M. persicae*. Their experiments with the panel of test plants listed above led them to suspect alfalfa mosaic virus (AMV). AMV was high on their suspect list because Schmelzer (1963) artificially infected *Cannabis* with AMV, which caused a systemic green mosaic. Sap from *Gelbstreifung*-infected hemp plants formed a precipitate with antibodies from AMV.

AMV is the sole member of the *Alfamovirus* genus, icosahedral in shape, with ssRNA(+) nucleic material in four particles. AMV infects >600 plant species in 70 families. Koziel (2010) mechanically inoculated *Cannabis* plants with AMV and illustrated symptoms with color photographs. Spaar *et al.* (1990) illustrated symptoms of ArMV, CMV, and AMV in color paintings. In their rendition, the symptoms look similar.

14.5 Assorted viruses

Traversi (1949) conducted inoculation studies with a sunflower virus in Argentina, and *Cannabis* was susceptible. Exactly what Traversi was studying has been debated—possibly sunflower mosaic virus, or sunflower chlorotic mottle virus.

Hartowicz *et al.* (1971) tested 22 viruses as potential biological control agents against illegal *Cannabis*. They stated that five viruses produced systemic symptoms and dwarfing, but none were considered effective biocontrol agents. The viruses were not named. However, they were named in an unpublished report by a team member (Paulsen, 1971). AMV, CMV, cucumber mosaic yellow strain (CMV-X), and euonymus ringspot virus (i.e., tomato ringspot virus, ToRSV) caused systemic mosaic and dwarfing. Necrotic local lesions were caused by elm mosaic (EMV).

Back-inoculations to indicator plants showed that symptomless infections were caused by tobacco mosaic virus (TMV) and foxtail mosaic virus (FMV). The remaining viruses did not infect *Cannabis*: barley stripe mosaic (BSMV), brome mosaic

(BMV), maize dwarf mosaic strain A (MDMV-A), panicum mosaic (PMV), wheat streak mosaic (WSMV), soil-borne wheat mosaic (SBWMV), bean pod mottle (BPMV), potato virus X (PVX), strawberry latent ringspot (SLRV), tobacco etch (TEV), tobacco rattle (TRV-C), soybean mosaic (SMV), and squash mosaic (SqMV) (Paulsen, 1971).

Kegler and Spaar (1997) tested three fiber cultivars ('USO-11', 'USO-14', 'USO-31') for susceptibility to eight viruses: AMV, CMV, and ArMV (previously tested by Schmidt and Karl, 1969, 1970), tomato ringspot (TomRSV), potato virus X (PVX), potato virus Y (PVY), and broad bean wilt (BBWV), and raspberry ringspot nepovirus (RRV). Their results are presented in Table 14.1.

AMV, CMV, and ArMV caused light green speckling or diffuse yellow-green mosaics and blotches ('USO-11' was not susceptible to AMV, 'USO-14' was not susceptible to ArMV). TomRSV caused light-green blotches in in one cultivar ('USO-31') and did not infect the other cultivars. PVX caused a mosaic between leaf veins in 'USO-31' and did not infect other cultivars. PYV caused a mosaic, as well as upturned leaf tips. BBWV caused a mosaic at the base of young leaves. RRV caused a mosaic and curling of leaf petioles.

14.5a. Potato virus Y (PVY)

Kegler and Spaar (1997) mechanically inoculated *Cannabis* with PVY and PVX (Table 14.1). Pitt *et al.* (2022) demonstrated that *P. cannabis* is an efficient vector of PVY in hemp (96% transmission rate) and potato (91%) using cohorts of 20 aphids. Even an individual aphid could transmit the virus, to hemp (63% transmission rate) and potato (19%). PVY is not phloem-limited. It is in epidermal cells and transmitted by aphids in a non-persistent manner (stylet-borne).

14.5b. Hop latent virus (HpLV)

Ziegler *et al.* (2012) found that *Cannabis* was a useful propagation host for HpLV. They artificially inoculated the fiber variety 'Fedora 17' with HpLV originating in hop (Fig. 14.3). Natural HpLV infection in *Cannabis* has not been reported in

peer-reviewed literature. It pops up in Wikipedia and other websites, probably confused with **hop latent viroid** (see below). HpLV occurs in commercial hop yards worldwide, and is transmitted by the hop aphid, *Phorodon humuli* (Adams and Barbara, 1982).

14.5c. Cannabis cryptic virus (CanCV)

While utilizing *Cannabis* as a propagation host for HpLV, Ziegler *et al.* (2012) accidently discovered CanCV. When they visualized HpLV using transmission electron microscopy, a second virus particle was seen (Fig. 14.3). It was spherical, with a diameter of 34 nm, and quite unlike HpLV—which is filamentous, measuring 675 × 14 nm. They named it Cannabis cryptic virus because it occurred in healthy plants. Cloning and sequencing of CanCV revealed a bipartite genome of dsRNA nucleic material, placing it in the *Betapartivirus* genus. Ziegler and colleagues searched for CanCV in six cultivars, 'Epsilon 68', 'Fedora 17', 'Felina 32', 'Futura 75', 'Ferimon', and 'USO-31'. All the cultivars except 'USO-31' contained the virus.

Righetti *et al.* (2018) isolated CanCV in Italy (see section 14.3). When they looked for CanCV in European fiber-type cultivars, they pronounced it "ubiquitous." Nachappa *et al.* (2020) isolated CanCV in Colorado.

14.5d. Beet curly top virus (BCTV)

Giladi *et al.* (2020) found plants in Colorado with stunted growth and a variety of foliar symptoms. Yellowing began at the leaf base, and expanded toward the tips, producing a yellow-green mosaic pattern. Within 10 days, the symptoms spread to the entire plant. In plants with advanced symptoms, newly developed leaves were pale green, narrow, and curled sideways, leading to a stunted, curled plant (Fig. 14.4).

The team used NextGen sequencing to identify beet curly top virus (BCTV). BCTV is a member of the *Curtovirus* genus, icosahedral in shape, with ssDNA(+) nucleic material in a

Table 14.1. Susceptibility of three cultivars to eight viruses (Kegler and Spaar, 1997).

Virus	Plants with symptoms (%)		
	USO-11	USO-14	USO-31
AMV	0	57.1	11.1
CMV	18.2	66.7	50.0
ArMV	10.0	0	7.6
BBWV	37.5	0	16.7
RRV	25.0	0	9.1
PVX	0	0	14.5
PVY	11.1	33.3	18.2
TomRSV	0	0	42.9

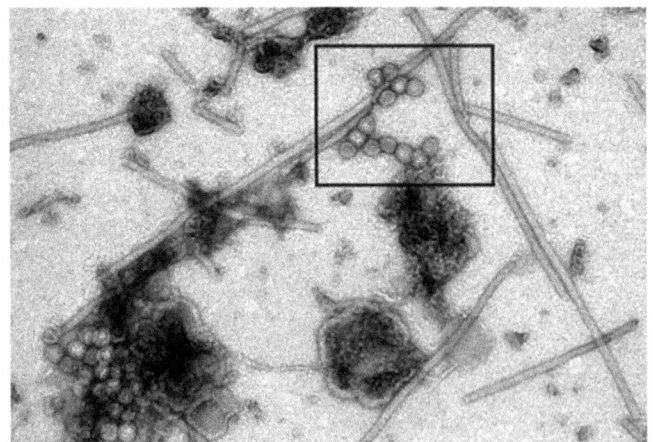

Fig. 14.3. TEM photomicrograph of CanCV (in box) (courtesy Ziegler *et al.*, 2012)

monopartite genome. The virus infects >300 plant species, with crop damage in sugar beets, beans, cucumber, peppers, spinach, and tomatoes. A dozen BCTV strains have been identified.

BCTV is limited to phloem tissue, and is vectored by the beet leafhopper, *Neoaliturus* (*Circulifer*) *tenellus* (Baker) 1896. *N. tenellus* feeds on a wide variety of hosts, including *Cannabis* (see Fig. 4.24). *N. tenellus* also vectors *Spiroplasma citri* (see section 14.8).

A severe BCTV disease outbreak occurred in Arizona, with up to 100% crop losses in some fields (Hu *et al.*, 2021). Symptoms varied by the stage of infection and plant growth stage. Early-stage symptoms included light green-to-yellowing of new growth, similar to sulfur or micronutrient deficiency, often combined with older leaves developing dark green "blotchy" mottling overlaying light green chlorosis. Curling and twisting of new leaves were also observed, with severe yellowing and witches' broom symptoms. Severely affected plants developed leaf necrosis, stunting, and died late in the season.

Two strains of BCTV that shared 97–99% sequence identity, named BCTV-WOR (Worland) and BCTV-CO (Colorado), were reported in Colorado (Nachappa *et al.*, 2020; Chiginsky *et al.*, 2021). In Oregon, Rivedal *et al.* (2022) identified BCTV-WOR and BCTV-CA/Logan. The authors provided oligonucleotide sequences for the primers and carefully described their protocol. They illustrated BCTV-WOR and BCTV-CA/Logan symptoms, as well as a plant coinfected with BCTV and HLVd (Fig. 14.4).

Melgarejo *et al.* (2022) also reported two strains in California, BCTV-CO and BCTV-WOR. The authors developed a rapid multiplex PCR test to detect the two strains. Mixed infections were present in 43% of samples. They coined "curly top disease" for the symptoms of distorted, upcurled, and yellowed leaves, along with stunting and busy growth. They fulfilled Koch's postulates by cloning BCTV-CO into a plasmid vector, which was transformed into *Agrobacterium*

tumefaciens. Hemp plants infected with transformed *A. tumefaciens* developed curly top disease. BCTV-CO and BCTV-WOR also affected crops in Washington State (Jarugula *et al.*, 2023).

14.5e. Cannabis sativa mitovirus 1 (CasaMV1)

CasaMV1 was an incidental discovery by plant geneticists who sequenced the *Cannabis* transcriptome (van Bakel *et al.*, 2011). CasaMV1 was sequenced from a drug-type strain, "Purple Kush". Another CasaMV1 sequence discovered in a fiber-type cultivar, 'FINOLA', had a 1-nucleotide insertion in its 5′ UTR region relative to the "Purple Kush" reference sequence.

Mitoviruses are small, ssRNA(+) viruses. They cannot be seen, because they do not code a capsid or any other structural proteins. Their genome is stripped down to one open reading frame, which encodes the RNA-dependent RNA polymerase (RdRp) enzyme—their method of replication. Fungi serve as their natural hosts, where they replicate in mitochondria. *Botrytis cinerea* is host for a mitovirus (Bruenn *et al.*, 2015).

At least 175 vascular plant genomes contain mitovirus-related sequences. These are called non-retroviral endogenized RNA virus elements (NERVEs). Bruenn *et al.* (2015) proposed that an ancestral fungal mitovirus integrated into the mitochondrial genome of the common ancestor of vascular plants—giving rise to today's NERVEs.

Nibert *et al.* (2018) presented evidence for the existence of genuine plant mitoviruses. They used a NERVE in *Arabidopsis thaliana* as a query sequence in a BLAST search of vascular plant transcriptomes. This search yielded hits in ten plant species, including *C. sativa* and *H. lupulus*. They named the sequences *Cannabis sativa* mitovirus 1 (CasaMV1) and *Humulus lupulus* mitovirus 1 (HuluMV1). Having found genuine plant mitoviruses, Nibert and colleagues wondered about their effects on plant health. Perhaps they are cryptic—causing no disease—but their association with mitochondria suggests otherwise, and Nibert called for more studies.

Bektaş *et al.* (2019) identified a transcript with high homology to CasaMV1 in their study of hop latent viroid in California. Chiginsky *et al.* (2021) found CasaMV1 in 11 of 12 counties sampled in Colorado.

14.5f. Lettuce chlorosis virus (LCV)

Hadad *et al.* (2019) observed plants in Israel showing symptoms of interveinal chlorosis, leaf brittleness, and occasional necrosis (Fig. 14.5). NextGen sequencing of RNA extracted from symptomatic leaves revealed the presence of **lettuce chlorosis virus (LCV)**.

LCV is a member of the *Crinivirus* genus, icosahedral in shape, with ssRNA(+) nucleic material in a bipartite genome. The whitefly *Bemisia tabaci* biotype MEAM1 (i.e., silverleaf whitefly, *Bemisia argentifolii*) transmitted the disease from symptomatic *Cannabis* to healthy *Cannabis*, *Lactuca sativa*, and *Catharanthus roseus*.

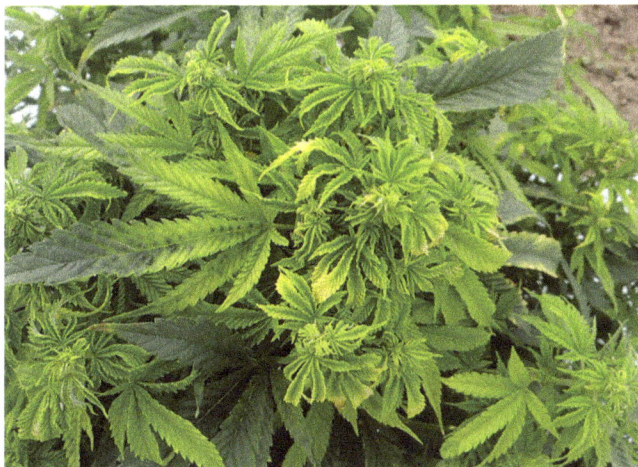

Figure 14.4. Symptoms of beet curly top virus: pale mottling of some leaves, chlorosis, and twisting of new growth (W. Cranshaw)

Fig. 14.5. Symptoms of lettuce chlorosis virus (kind courtesy of Aviv Dombrovsky)

14.5.1 Yet more viruses (YmVs)

Nachappa *et al.* (2020) and Chiginsky *et al.* (2021) used NextGen sequencing to survey the *Cannabis* "virome" at six sites in Colorado. In addition to CanCV, BCTV, HLVd, PVY, and CasaMV1, they found five new ones. **Plasmopara viticola-associated ourmia-like virus (PvOLv)** was isolated from two of six sites. PvOLv is a ssRNA(+) virus in the Botourmiaviridae family. Its vector is unknown.

Hop yellow virus (HYV) was isolated from one of six sites. HYV is a ssRNA(+) virus in family Bromoviridae, genus Anulavirus. It has been isolated in Italy on *Vitis vinifera*, and in China on *Humulus lupulus* and *Fritillaria thunbergia* (Elbeaino *et al.*, 2020). HYV is transmitted mechanically and via aphids.

Citrus yellow-vein-associated virus (CYVaV) was isolated from two of six sites. Two strains were isolated, sharing 98% identity, with 90% identity of CYVaV isolated from *Citrus floridana*. Citrus yellow-vein disease has been known in California since the 1950s, but CYVaV was only identified recently (Kwon *et al.*, 2021). In addition to Colorado (Chiginsky *et al.*, 2021), CYVaV has also been isolated from hemp in Washington State (Jarugula *et al.*, 2023). CYCaV is a ssRNA(+) virus, transmitted mechanically and via aphids.

Opuntia umbra-like virus (OULV) was isolated from two of six sites. Two strains were isolated, sharing 98% identity, and they shared 77% identity with CYVaV. OULV is another newly discovered ssRNA(+) virus. OULV was discovered in stunted prickly pear cactus, *Opuntia ficus-indica*, and spreads by mechanical transmission and by aphids (Felker *et al.*, 2019).

Tobacco streak virus (TSV) is a ssRNA(+) virus with a tripartite genome; Chiginsky *et al.* (2021) isolated it from Colorado *Cannabis*, with the three RNAs in its genome sharing 81% to 83% identity with TSV from soybean (RNA 1), *Dahlia* (RNA 1), and soybean (RNA 3) suggesting that it is a novel genotype/strain of the virus. TSV has a broad host range, infecting >200 plant species in 30 plant families. In other hosts TSV is transmitted by thrips (*Thrips tabaci, Frankliniella occidentalis*) that also infest *Cannabis*.

Tobacco mosaic virus (TMV) is a ssRNA(+) virus, and the original virus discovered by Iwanowsky and Beijerinck (see Fig. 14.1). It causes serious problems in tobacco, and it infects other solanaceous hosts and a few other plants. A grower in Colorado sent diseased samples for viral testing and TMV was identified (Nachappa, pers. comm., 2022).

14.5.2 VIGS viruses

Virus-induced gene silencing (VIGS) is a genetic tool that transiently reduces gene expression of a plant gene. VIGS exploits an antiviral mechanism in plants that interferes with mRNA after transcription, preventing gene translation. A VIGS vector is constructed by cloning a fragment of a targeted gene into a viral genome. The chosen virus should have a broad host range and cause only mild symptoms in an infected plant. VIGS was invented by Kumagai *et al.* (1995), who used the **tobacco rattle virus (TRV)** to silence *PDS* (phytoene desaturase) in tobacco plants. Silencing *PDS* reduces chlorophyll biosynthesis, resulting in an easily detected albino or "photobleached" phenotype.

Paulsen (1971) reported that TRV could not infect *Cannabis*, and initial VIGS experiments in *Cannabis* that used TRV did not successfully silence the olivetolic acid cyclase gene (Gagne *et al.*, 2012) or acyl-activating enzyme genes (Stout *et al.*, 2012). However, Alter *et al.* (2022) successfully used TRV to silence *PDS*. They noted that photobleaching was localized, because cell-to-cell movement by TRV in *Cannabis* was limited. Schachtsiek *et al.* (2019) used **cotton leaf crumple virus (CLCrV)** to create VIGS vectors that silenced *PDS* and *ChlI* (magnesium chelatase subunit I). *PDS* silencing was observed in newly emerging leaves that were not infiltrated, indicating a degree of systemic movement by CLCrV in *Cannabis*.

14.5.3 Differential diagnosis

Viral diseases may be confused with symptoms of bacterial blight, or early stages of some fungal disease (brown leaf spot, yellow leaf spot). Interveinal chlorosis might be confused with nitrogen deficiency. A genetic disease, chimeric leaf variegation arising from a somatic mutation, may resemble a viral infection (Fig. 14.6). But a leaf chimera usually arises on a single plant, and not across the field.

Differentiating between viruses is difficult. Using an electron microscope is helpful but expensive. Cross-inoculation studies with other plants are time-intensive. Serological tests for ArMV, CMV, and AMV have long been available. Serology has been replaced by **reverse transcriptase PCR** (RT-PCR) which can detect the presence of viruses in plant material. Recall that plant viruses and viroids have RNA genomes. RT-PCR is like real-time PCR (qPCR), but first an enzyme called reverse transcriptase converts the RNA to complementary DNA (cDNA). Virus- or viroid-specific oligonucleotide

Fig. 14.6. Leaf chimera

primers turn RT-PCR into a qPCR diagnostic assay. Laboratories offering RT-PCR diagnosis of *Cannabis* viruses and viroids are becoming a multimillion-dollar industry; see the HLVd section below.

14.5.4   Disease cycle

In nature, most plant viruses depend on sap-sucking insects for their survival, transmission, and spread. The period of time that insects retain the viruses—and therefore spread them—varies between viruses and between insects (Fereres and Raccah, 2015). In some cases, viruses circulate around the insect's haemocoel (the body cavity between organs through which blood circulates). This results in persistent transmission of viruses, often for the life of the insect. In non-persistent transmission, viruses stay in the stylet (mouthparts) and are retained in the vector for hours, whereas in semi-persistent transmission, virus is localized to the foregut of the insect, and the retention of viruses is limited to days.

AMV and CMV are primarily transmitted by aphids, where the viruses localize to stylets, and persistence is limited to a few hours. ArMV is vectored by nematodes (*Xiphinema* spp.), and can persist for years. *P. humuli* and *M. persicae* transmit hop viruses (e.g., hop mosaic virus, hop latent virus) in a non-persistent manner (Eppler, 1995). In an experimental

study, Karl (1971) demonstrated that *P. cannabis* transmitted the pea mosaic virus in a non-persistent manner.

Studies have shown that viruses can make infected plants more attractive to insects. Aphids are attracted to yellow sticky traps, and they are attracted to chlorosis arising in virus-infected leaves. Eppler (1995) demonstrated that *P. humuli* is attracted to chlorotic hop leaves. Jalapeño peppers infected by CMV emit more volatile compounds—odor cues for *M. persicae* (Safari *et al.*, 2019). Viruses are also known to modify vector feeding behavior. For instance, PVY-infected *P. cannabis* spent less time ingesting phloem compared with non-infected aphids on hemp. This suggests that virus-infected aphids are more likely to disperse, thereby increasing virus transmission (Pitt *et al.*, 2022).

Viruses may overwinter in their vectors. Some insects pass viruses to their offspring, through transovarial transmission, but this only occurs with viruses that circulate around the insect's hemocoel. Viruses also overwinter as seed-borne infections, and in diseased mother plants kept indoors.

14.5.5   Virus disease management

Breeding for virus-resistant plants is a possibility. Agronomists are gauging the resistance of different fiber cultivars to different viruses (Table 14.1). No antiviral chemicals have been effective. Controlling vectors, particularly *P. cannabis*, is a keystone for managing viruses (see section 4.6). Weeds may harbor symptomless infections, so eradicate them.

Preventive measures: Keep glasshouses and growrooms clean; limit visitors; quarantine new plants; destroy infected plants; do not clone virus-infected plants or harvest seeds from them; and minimize the mechanical spread of viruses by sanitizing hand tools. Virus-inactivating disinfectants include 10% solution of Chlorox Regular (5.25% sodium hypochlorite), 2% solution of Virkon® S, or a 20% solution (weight to volume) of non-fat dry milk.

Controlling viral diseases is a challenge in drug-type *Cannabis*, which is often propagated through vegetative cuttings. Vertical transmission of viruses from mother plants to clones is rampant, and difficult to control. Using sterile technique when taking cuttings is paramount, but not the solution.

Websites have proliferated that advertise "patent-pending cleaning processes" to eliminate viruses and viroids from plants. This involves *in vitro* meristem tip culture, a relatively new technology (Smýkalová *et al.*, 2019). The uninfected meristem tip can be excised from a plant, then transferred to a culture medium and regenerated into a plantlet (discussed below, Fig. 14.8).

Some researchers claim viruses can be eradicated from seeds using thermotherapy (see section 19.2.1). Heat therapy is a time-honored method of eliminating plant viruses (Cassells, 2012). It aims to denature the virus before killing the plant.

The certification of virus-free seeds and clones is discussed by Chiginsky *et al.* (2021). Certification requires the establishment, implementation, and monitoring of standards agreed upon by industry and government at the state level. Reliable sampling strategies and diagnostic protocols should be developed by certification agencies and diagnosticians. The authors suggest this might be easy, because hemp cultivators are familiar with the certification of THC content. However, the use of certified hemp *varieties* has not gained traction in the USA, never mind the certification of pathogen-free plants. In the EU, the use of certified hemp cultivars is required by the government, and not without difficulties (Montanari, 2017).

14.6   Hop latent viroid (HLVd)

Viroids such as HLVd consist of a naked, circular loop of ssRNA. The HLVd genome has only 256 nucleotides. It is one of the smallest replicating agents known, a "living fossil of the hypothetical RNA World" (Diener, 2016). HLVd was originally discovered in the hop-growing region around Hallertau in Germany (Puchta *et al.*, 1988). Puchta and colleagues called it "latent" because infected plants showed no symptoms. They tested samples from Oregon, England, Hungary, Yugoslavia, and Japan. HLVd was everywhere. Rather than symptomless, Patzak *et al.* (2001) demonstrated that HLVd caused a reduction in sesquiterpenoids in hop flowers.

Dudding is a *Cannabis* disease that causes stunted growth, small leaves and flowers, and reduced trichome production. The term was coined in 2015 by Rick Crum, a plant pathologist in Humboldt County. Crum called the causal agent a "putative *Cannabis* infectious agent" (Mann, 2016).

The cause of dudding, HLVd, was simultaneously reported by two groups in California (Bektaş *et al.*, 2019; Warren *et al.*, 2019). Their publications appeared on the same page in *Plant Disease*. Warren *et al.* (2019) began studying the disease in 2017, which appeared near Davis. Symptoms included stunting, malformation or chlorosis of leaves, brittle stems, and reduction in yields (Fig. 14.7A). Cuttings taken from symptomatic plants showed a reduced rooting success rate. NextGen sequencing of RNA extracted from symptomatic leaves revealed HLVd.

Bektaş *et al.* (2019) also began studying the disease in 2017. A separate outbreak across the state of California in Santa Barbara County led to renewed efforts. Symptoms included brittle stems, "an outwardly horizontal plant structure" (i.e., wider than tall—like plants were topped), and reduced flower mass and trichomes. They called the syndrome "cannabis stunting disease." Symptoms in some plants can be subtle: small, shriveled, yellow bracts, and a reduction of glandular trichomes on fan leaves compared to controls (Fig. 14.7B). NextGen sequencing of RNA extracted from symptomatic *and* asymptomatic plant revealed HLVd.

Nachappa *et al.* (2020) isolated HLVd from two sites in Colorado. Punja (2021d) identified HLVd as the cause of dudding in British Columbia. Joshi *et al.* (2022) detected HLVd in 70 of 389 diseased samples (18%) submitted to their plant clinic in British Columbia. Rivedal *et al.* (2022) isolated HLVd from 55% of all specimens submitted to their plant clinic in Oregon in 2019. Some plants were co-infected by HLVd and BCTV (Fig. 14.7C). Punja *et al.* (2023) surveyed HLVd incidence in nine provinces across Canada, which ranged from 5.3% to 92%, with a nationwide average of 25.6%.

Punja *et al.* (2023) reported new symptoms. Reductions of 12–42% in inflorescence stem lengths, fresh weights, and plant heights were observed, the reduction depending on the genotype. Flowering plants displayed yellowing or darkening of inflorescence leaves surrounding the pistillate flowers. The root systems of these plants were visibly reduced in volume by as much as 50%. Glandular trichomes in HpLV-infected plants were small and shriveled. THC content in infected plants decreased by 28.8% compared with controls. All these symptoms were profusely illustrated (including LM and SEM micrographs of trichomes). Insidiously, many plants in the vegetative stage were asymptomatic, yet harbored HpLV.

Fig. 14.7. Symptoms of HLVd and BCTV: **A.** HLVd (Steve Koike); **B.** HLVd vs control (Ali Bektaş); **C.** coinfection of HLVd and BCTV (Hannah Rivedal)

14.6.1   Disease cycle

Viroids, like viruses, are transmitted through clonal propagation—HLVd spreads from infected mother plants to infected clones (Bektaş *et al.*, 2019; Warren *et al.*, 2019). Punja *et al.* (2023) assumed seed and pollen transmission occurred in *Cannabis*, given the high incidence of HLVd detected in floral tissues. Atallah *et al.* (2024) reported seed-borne transmission at rates of 58–80%. They also revealed new hosts for HLVd: tomato, cucumber, chrysanthemum, *Nicotiana benthamiana*, and *Arabidopsis thaliana*. Earlier, Guček (2020) artificially inoculated *Cannabis* ('FINOLA' variety) with HpLV from hop using mechanical transmission.

Alternative hosts could pose a serious threat if a vector connected them with *Cannabis*. So far, no insects are known to vector HpLV. Adams *et al.* (1995) reported no transmission of HLVd in hop plants exposed to hop aphids (*Phorodon humuli*) that had fed on infected source plants. Knabel *et al.* (1999) found no evidence of HpLV vectoring by two hop pests, *P. humuli* and *Tetranychus urticae*, using PCR. HLVd has been detected in root aphids (another new plague, see Fig. 8.2), although this has not been demonstrated conclusively (Bektaş, 2024).

Punja *et al.* (2023) noted that the systemic spread of other viroids occurs through phloem, which explains the spread of HLVd from shoots to roots. Punja (publication forthcoming) artificially inoculated leaves with HLVd, and within 4 weeks the entire plant was infected, including roots. Flowering (switching to 12:12 lighting) increased viroid load in plants. Cuttings derived from HpLV-infected mother plants exhibited poor root development. HpLV-infected plants were more susceptible to diseases caused by *Fusarium oxysporum* and powdery mildew fungi.

In the process of sequencing HLVd, McKernan *et al.* (2022) identified a stretch of 19 nucleotides with 100% identity to the 5' coding region of a *Cannabis sativa* gene (*COG7*) involved in apical meristem growth. The authors proposed that HLVd may be involved in RNA interference of *COG7*; infected plants showed decreased *COG7* expression. Viroids are known to hijack the host's transcriptional machinery, specifically, the plant's DNA-dependent RNA polymerase enzymes. This contrasts with viruses, which hijack the host's translation system to produce proteins encoded in virus genomes.

14.6.2   HLVd management

Cannabis cultivators in British Columbia prevented the spread of HLVd "through good sanitation practices and strong adherence to biosecurity measures" (Drohan, 2021). Good sanitation included sterile pruning tools and clean growing medium. Disinfecting tools with a 10% bleach solution is effective, alcohol is not (Bektaş, 2024). Punja (publication forthcoming) described HLVd as very stable, and resistant to disinfectants that denature most viruses, such as Virkon® and ZerTol®. What Drohan meant by "biosecurity measures" was strict quarantine of new plants in a facility. Incoming plant material must be screened with molecular methods, primarily RT-PCR (see below).

Punja *et al.* (2023) evaluated 23 genotypes, five of which showed tolerance to infection, "but further research is needed to confirm this." Bektaş (2024) confirmed that

some infected genotypes are less prone to symptoms than others. The hop industry has developed HLVd-resistant cultivars.

For a mother plant with highly valued genetics, infection may spell disaster. Jones and Monthony (2022) recommended a high-temperature pre-treatment combined with meristem tip culture to eliminate HLVd from *Cannabis*. However, Podwyszńska *et al.* (2021) stated that this method was ineffective in eliminating HLVd in hop. They reported better results with cold-temperature pre-treatment and meristem tip culture, devised by Grudzinska *et al.* (2006). A patent for eliminating HLVd from *Cannabis* combined high-temperature pre-treatment with meristem tip culture (Grace and Roberts, 2019).

The apical meristem or growing tip of a plant is the last place a virus invades. It comprises the apical dome with young leaf primordia (red box in Fig. 14.8). Excised tissue must be free of xylem (Fig 14.8A), procambium (Fig. 14.8B), and phloem (Fig. 14.8C)—tissues that harbor HLVd. The uninfected meristem tip is transferred to MS agar and regenerated into clean plantlets. This method yielded between 77% and 100% HLVd-free plants, amongst 17 tested *Cannabis* strains (Grace and Roberts, 2019). For more about difficulties arising in MS culture and plantlet regeneration, as well as the specter of somaclonal variation, see text below Fig. 19.2.

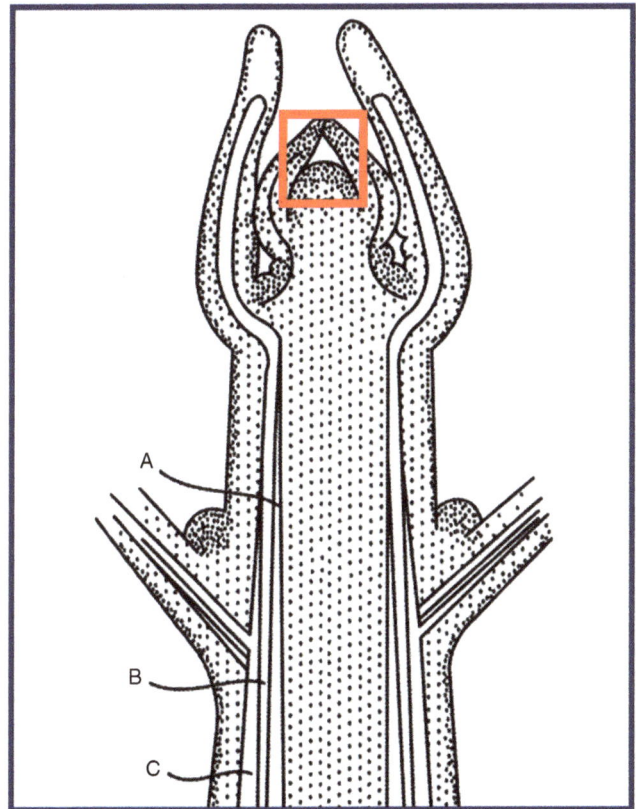

Fig. 14.8. *Cannabis* meristem tip and surrounding tissues (redrawn from Grace and Roberts, 2019)

As emphasized by Bektaş (2024), the best control of HLVd is exclusion. All plants coming into a grow facility should be tested for HLVd. Rivedal *et al.* (2022) described their RT-PCR protocol, and provided oligonucleotide sequences for HLVd-specific primers. Brunstein (2021) also described RT-PCR protocols. Fernandez i Marti *et al.* (2023) argued that RT-PCR was too technical and expensive; they developed a LAMP (loop-mediated isothermal amplification) assay to detect HLVd. However, sensitive and specific LAMP assays are hard to design, and they generate more false positives than RT-PCR.

RT-PCR sensitivity and specificity depend upon proper sampling and handling. Obtain plant samples with disinfected tools, ship material with cold packs, and minimize transport time between the sampling site and testing lab. Leaves are a poor sample source. Evidence suggests that roots act as a sink for HLVd—a good sample source, but this involves destructive sampling. Bektaş (2024) suggested sampling petioles and stems, which can be performed while excising material for clone production. He also highlighted two Quality Control steps include LODs and assaying for false-positive results.

Limit of detection (LOD) is the minimum amount of HLVd that can be detected with confidence in an assay. This is determined by testing the RT-PCR with quantified amounts of HLVd. The lower the LOD, the less likely is a false-negative test. Monitoring for false-positives is more difficult—the testing lab should run a parallel RT-PCR on *Cannabis* guaranteed to be HLVd-free. Bektaş recommended using pure *Cannabis* mRNA. Testing labs offering detection services for HLVd should practice both of these Quality Control steps.

Given the multimillion-dollar market for HLVd testing, shysters arise. The results of a "round-robin" audit of three labs showed that one lab reported negative results across the board. That lab's results could be due to "the use of subpar methods (at best) or cooking the books (at worst) to provide inflated results to their customers" (Bektaş, 2024). Similar scenarios arise in labs that provide THC testing (see section 2.8.5) and microbial testing for health requirements (see section 18.3.7). The era of honor-among-outlaws ended with legalization.

14.7   Phytoplasma diseases

Phytoplasmas are related to bacteria, but lack a cell wall. They are obligate parasites of plant phloem tissues. Phytoplasmas are spread by phloem-sucking insects, especially leafhoppers and planthoppers. In some cases, mites can also transmit them. Phytoplasmas are distributed worldwide, predominantly in warmer regions.

Being obligate parasites, phytoplasmas are difficult to culture in petri plates, making it difficult to measure traditional bacterial taxonomic characters. With the advent of molecular biology, Lee *et al.* (1998) devised a classification scheme based on restriction fragment length polymorphism (RFLP) analysis. They amplified the 16S rRNA gene, and then digested the amplicon with 17 restriction enzymes. The resulting RFLP

fragments, separated by gel electrophoresis, produced banding patterns unique for groups of phytoplasmas (Fig. 14.9).

The RFLP approach has identified 36 groups of phytoplasmas, designated 16Sr-I to 16Sr-XXXVI. Each group has subgroups designated by letters. For example, Spartium witches'-broom phytoplasma belongs to group 16Sr-Vb, and Buckthorn witches'-broom phytoplasma belongs to 16Sr-Vc.

The rise of inexpensive DNA sequencing has rendered RFLP largely obsolete. A new classification system is based on 16S rRNA nucleotide sequences. Phytoplasmas with <97.5% similarity in their 16S rRNA sequences are considered distinct species. "*Candidatus* Phytoplasma" was adopted as a genus name for prokaryotes lacking cell walls that colonize plant phloem ("*Candidatus*" is used for prokaryotes that cannot be cultured). The aforementioned species were renamed "*Candidatus* Phytoplasma spartii" and "*Candidatus* Phytoplasma rhamni," respectively (Marcone *et al.*, 2004).

Thanks to new molecular tools, research on phytoplasmas has grown exponentially. The first *Cannabis* phytoplasma was discovered in 1975. It took 32 years to discover the second one, and in the 15 years since then, six more have been discovered. Molecular identification has revealed that phytoplasmas have wide host ranges, spanning crop plants and neighboring weeds, and are vectored by polyphagous insects.

The 16S rRNA sequences of many newly discovered phytoplasmas are deposited at GenBank. We searched GenBank for these sequences, and constructed a phylogenetic tree of *Cannabis* phytoplasmas. The sequences of additional species, from other hosts, were added to the analysis, to frame the phylogenetic tree. Phylogenetic analysis was performed with the software platform at MABL (Méthodes et algorithmes pour la Bio-informatique LIRMM, www.phylogeny.fr). Results appear in Fig. 14.10, and are discussed over the next two pages.

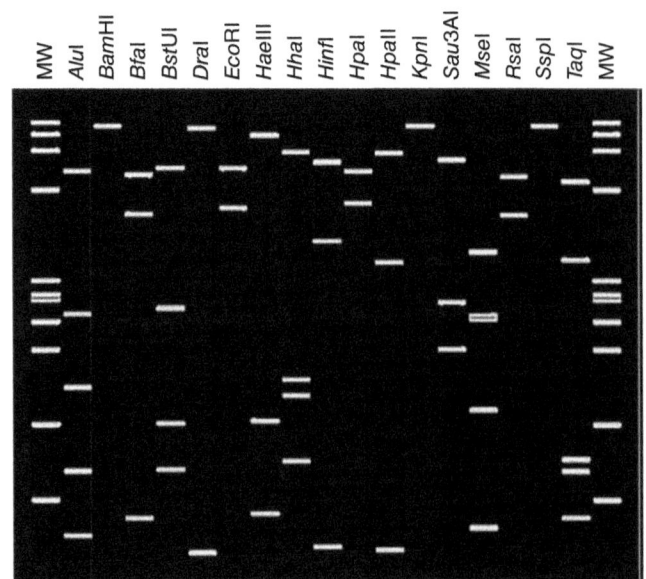

Fig. 14.9. RFLP restriction pattern of *Candidatus* Phytoplasma ziziphi

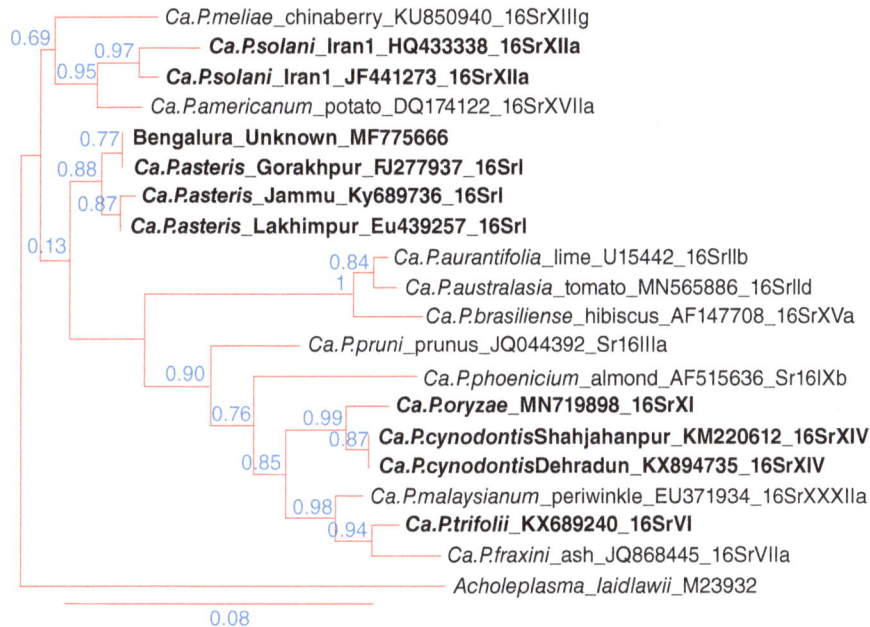

Fig. 14.10. Phylogenetic tree of some representative phytoplasma 16S rRNA sequences. *Cannabis* pathogens in bold print

14.7.1 Symptoms and signs

Plants afflicted with phytoplasmas show stunting (in plant height and leaf size) and leaf chlorosis. The classic symptom is **witches' broom**—a shortening of internodes, with a proliferation of axillary buds into branches, causing a bushy appearance. Flowers show **phyllody**, where floral organs (bracts, stamens) transform into small leafy structures, adding to the bushy appearance.

These symptoms are due to alterations in plant hormones—increases in auxin and cytokinin, and decreases in gibberellic acid. Now that we can classify phytoplasmas by their 16S rRNA sequences, it becomes apparent that phylogenetically distinct phytoplasmas induce the same symptoms in *Cannabis*.

14.7a. *Cannabis* phyllody MLO

Phatak *et al.* (1975) discovered an MLO in *Cannabis* growing near Delhi, India. They published the first-ever transmission electron microscope (TEM) photograph of a *Cannabis* pathogen (Fig. 14.11). Visualized by TEM, the MLO cells were pleomorphic, membrane-bound, roughly circular, and less than 0.5 μm in diameter. The cell membrane was electron dense (seen as dark, compared with cell contents). They were only observed in phloem elements of infected plants. Infected plants showed classic symptoms of witches' broom and phyllody, especially in male plants. A photo compared normal and diseased male inflorescences.

Sastra (1973) described witches' broom symptoms in *Cannabis* from northeastern India, but attributed the disease to viruses. Female plants were affected more than male plants.

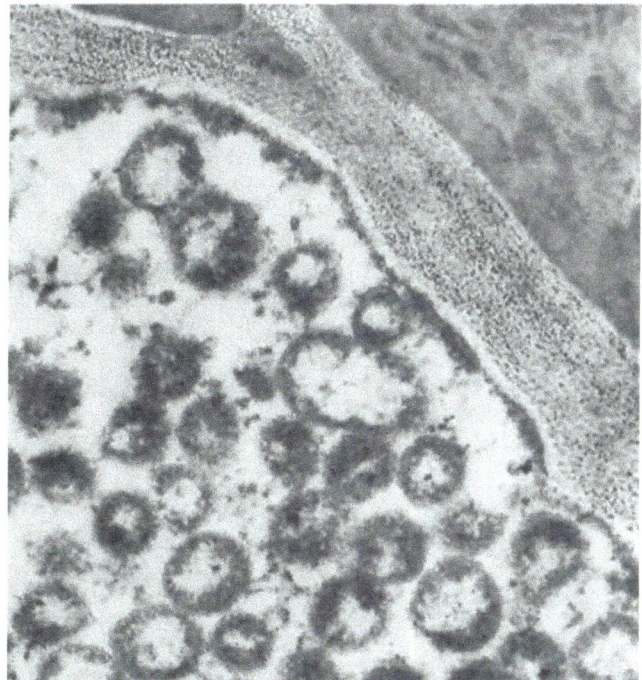

Fig. 14.11. TEM photomicrograph of phytoplasmas plugging a phloem vessel (Phatak *et al.*, 1975)

Ghani and Basit (1975) reported "bunchy top" affecting Pakistani plants, but also blamed viruses.

Scientists in India have led phytoplasma research in crop plants, reviewed by Rao *et al.* (2017). In India *Cannabis* grows like a weed, so most reports on *Cannabis* come from there.

Govind Pratap Rao (1960–) has led much of this work, beginning with his PhD research at Gorakhpur University.

14.7b. *Candidatus* Phytoplasma ziziphi

Zhao *et al.* (2007) observed witches' broom in feral hemp plants growing near Tài'ān in Shāndōng Province. Diseased plants developed clusters of proliferating branches with shortened internodes, and leaves on affected branches were significantly reduced in size. DNA was extracted and PCR carried out with 16S rDNA primers. RFLP restriction patterns indicated that the phytoplasma belonged to subgroup 16SrV-B of the elm yellows (EY) phytoplasma group.

The 16S rDNA sequence of the *Cannabis* pathogen was nearly identical (99.9%) to that of the jujube witches' broom (JWB) phytoplasma, which also occurs in Tài'ān. JWB is now called *Candidatus* Phytoplasma ziziphi. RFLP restriction patterns can be simulated with an online tool, creating an "*in silico* gel." We obtained the 16S rDNA sequence for *Candidatus* Phytoplasma ziziphi from PubMed (GenBank no. AB052876), and submitted it to virtual gel analysis at iPhyClassifier. The results appear in Fig. 14.9.

14.7c. *Candidatus* Phytoplasma asteris

Raj *et al.* (2008) observed symptoms of witches' broom in weedy roadside *Cannabis* growing in Lakhimpur Kheri, Uttar Pradesh, India, near the Nepal border. The plants exhibited proliferation of branches with shortened internodes and small leaves. DNA was extracted, and PCR carried out with 16S rDNA primers. RFLP analysis indicated that the phytoplasma belonged to 16SrI of the aster yellows group. The 16S rDNA sequence was nearly identical (99%) with *Candidatus* Phytoplasma asteris, associated with aster yellow disease.

Govind Pratap Rao's group in Gorakhpur, about 350 km east of Lakhimpur Kheri, observed the same symptoms, which they photographed (Mall *et al.*, 2011). Sequencing 16S rDNA sequence revealed 99% identity with *Candidatus* Phytoplasma asteris. They constructed a phylogenetic tree of 16S rDNA sequences, which separated the Gorakhpur and Lakhimpur Kheri isolates. They separated here as well (Fig. 14.10).

Govind Pratap Rao's group subsequently found weedy *Cannabis* growing in a sesame crop, and both had *Candidatus* Phytoplasma asteris (Un Nabi *et al.*, 2015). The authors implicated weedy *Cannabis* for perpetuating "sesame phyllody" disease. They traveled up to Kotbhalwal region of Jammu, over 1300 km northwest of Lakhimpur Gorakhpur, and found the same situation—weedy *Cannabis* and sesame crops sharing *Candidatus* Phytoplasma asteris (Singh *et al.*, 2018). Symptoms were observed in 5–12% of *Cannabis* plants and in 8% of sesame plants.

Gopala and Rao (2018) surveyed ornamental plants in India suspected of harboring phytoplasmas. Samples from a dozen plants, subjected to 16S rDNA amplification with RFLP analysis, revealed nearly a dozen different *Candidatus* species. Their sampling included weedy *Cannabis* in jasmine

fields at Bengaluru in Karnataka. The *Cannabis* phytoplasma was sequenced, but not characterized to species. We plugged their sequence into the phylogenetic tree (Fig. 14.10). The Bengaluru sequence fell into a clade with *Candidatus* Phytoplasma asteris sequences, with good bootstrap support. Thus it belongs to group 16Sr-I, which is the most common group of phytoplasmas worldwide (Rao *et al.*, 2017).

14.7d. *Candidatus* Phytoplasma solani

In Yazd Province, Iran, Vali Sichani *et al.* (2011) found plants with stunting and witches' broom in several hemp fields. They provided a color photograph of symptoms. PCR of 16S rDNA with RFLP analysis yielded two distinct sequences. One was 99% identical to *Candidatus* Phytoplasma solani, in group 16SrXIIa. The other sequenced shared only 96.79% identity—safely under the 97.5% boundary for naming a novel species. Vali Sichani declined to name a new species, saying other characters were necessary—either a unique host (*Cannabis* is a known host of phytoplasmas), a unique insect vector (unknown), or a unique chromosome size (unknown). In the phylogenetic tree (Fig. 14.10), the two Iranian sequences are sister, and form a clade with another member of group 16SrXIIa.

14.7e. *Candidatus* Phytoplasma cynodontis

Govind Pratap Rao's group observed symptoms of yellowing, little leaf, and witches' broom in plants they characterized as *C. sativa* subsp. *sativa* and *C. sativa* subsp. *indica* (Chaube *et al.*, 2015). Judging from their photos, they used the folk taxonomy of "Sativa" (narrow leaflet) and "Indica" (broad leaflet). Both populations were roadside weeds in Shahjahanpur, Uttar Pradesh, India. PCR of 16S rDNA with RFLP analysis revealed the phytoplasma belonged to 16SrXIV group. The sequence shared 99% similarity with *Candidatus* Phytoplasma cynodontis. The pathogen was also isolated from dūrvā grass (*Cynodon dactylon*), and the authors proposed that dūrvā grass could play a key role in the epidemiology of the disease.

The same research group observed yellowing, little leaf, and witches' broom in *Cannabis* growing at Dehradun, Uttarakhand, about 390 km from Shahjahanpur. PCR of 16S rDNA with RFLP analysis again identified the pathogen as *Candidatus* Phytoplasma cynodontis (Maury *et al.*, 2017).

14.7f. *Candidatus* Phytoplasma trifolii

Brinjal little leaf (BLL) is a widespread disease in India (brinjal is eggplant). Kumar *et al.* (2017) collected 18 eggplant samples with BLL symptoms in eight states (Uttar Pradesh, Haryana, Maharashtra, Odisha, Chhattisgarh, Assam, Bihar, and Delhi). They collected weeds in brinjal fields showing little-leaf, phyllody and witches' broom symptoms—including a *Cannabis* sample. PCR of 16S rDNA with RFLP analysis showed that the majority (16 eggplant samples and the *Cannabis* sample) belonged to group 16SrVI, with high sequence identity to *Candidatus* Phytoplasma trifolii.

Candidatus Phytoplasma trifolii was also described by two research groups in Nevada. Schoener and Wang (2019) reported 25–30% of plants affected; symptoms included excessive apical branching and crowded growth of underdeveloped leaves. Leaves in a cluster were yellowish, often distorted, and very small (Fig. 14.12). Feng *et al.* (2019) described symptoms of leaf curling, mottling, chlorosis, stunting, and node shortening. Symptomatic leaves showed "green islands," constrained by veins on the upper leaf surface, and interveinal discoloration on the lower surface, forming a pale "pustule-like" appearance. PCR of 16S rDNA with RFLP analysis revealed that the phytoplasma belonged to group 16SrVI. The sequence shared 99% similarity with *Candidatus* Phytoplasma trifolii, associated with alfalfa witches'-broom disease. Pair-wise BLAST comparison of their 16S rDNA sequence with the sequence from *Cannabis* in India (Kumar *et al.*, 2017) revealed 98.88% identity.

14.7g. *Candidatus* Phytoplasma phoenicium-related phytoplasma

Rasoulpour *et al.* (2017) observed severe little leaf, yellowing, shortened internodes, dwarfing, and witches' broom symptoms in hemp plants near Shiraz, Iran. PCR of 16S rDNA with RFLP analysis revealed that the phytoplasma belonged to

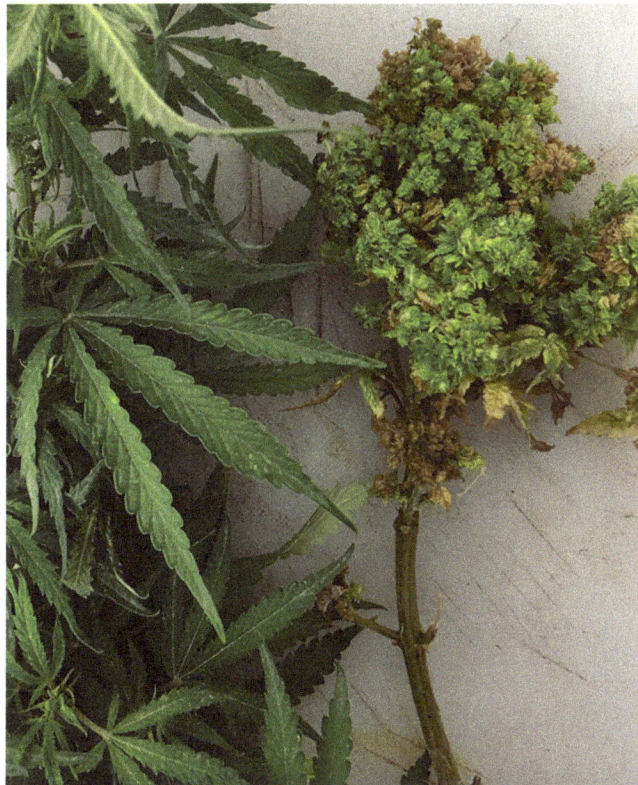

Fig. 14.12. Witches' broom symptoms caused by *Candidatus* Phytoplasma trifolii (courtesy of Shouhua Wang)

group 16SrIX-C. This phytoplasma is tentatively named Khafr almond witches' broom (KAlmWB), closely related to *Candidatus* Phytoplasma phoenicium, in group 16SrIX-B.

The *Cannabis* sequence showed 100% identity with other 16SrIX-C isolates from periwinkle (*Catharanthus roseus*), sesame, eggplant, and tomato; "hemp plants may serve as a phytoplasma reservoir for infection of other crops." Rasoulpour and colleagues established dodder (*Cuscuta campestris*) on periwinkle, and connected the dodder to diseased *Cannabis*. The phytoplasma was successfully transmitted from *Cannabis* to periwinkle.

14.7h. *Candidatus* Phytoplasma oryzae

Govind Pratap Rao's group searched for natural reservoirs of sugarcane grassy shoot phytoplasma (SGSP) among common weeds in Uttar Pradesh (Maurya *et al.*, 2020). *Cannabis* plants with symptoms of witches' broom were found in SGSP-afflicted sugarcane crops. DNA extracted from *Cannabis* revealed that the phytoplasma belonged to group 16SrXI, with 99.8% similarity with *Candidatus* Phytoplasma oryzae. Sequences from sugarcane were identical to those on *Cannabis*. The plants must share an insect vector.

14.7.2 Differential diagnosis

Witches' broom and phyllody are classic symptoms of phytoplasmas, rarely caused by other causal organisms. Some viruses may cause witches' broom, but not the viruses that affect *Cannabis*.

14.7.3 Disease cycle

The phytoplasma life cycle requires replication in two distinct and diverse hosts—plants (Kingdom Plantae) and insects (Kingdom Animalia). In plants, phytoplasmas survive and replicate in phloem sieve tubes and immature phloem cells. Phloem-sucking insects ingest phytoplasmas from the sap of infected plants. The pathogens pass through insect gut cells, and replicate in hemolymph and salivary glands. Reaching saliva, the phytoplasmas are injected into the next plant that a leafhopper feeds upon. Phytoplasmas negatively impact plants, but most leafhopper species do not show obvious negative effects from infection.

Phytoplasmas overwinter in their insect hosts. Infected female leafhoppers may pass the pathogen to their offspring, via transovarial transmission (Mittelberger *et al.*, 2017). Phytoplasmas may also overwinter in seeds; seed-borne infection has been documented in other plants (Satta *et al.*, 2019).

14.7.4 Phytoplasma management

Rogue infected plants and destroy them. Controlling insect vectors is a plausible way to limit the spread of phytoplasmas, but controlling leafhoppers is not easy (see section 4.9).

Eliminate weeds, because many are susceptible to phytoplasmas and provide a reservoir of infection. Developing cultivars resistant to either phytoplasmas or their vectors would be a long-lasting control measure.

Do not harvest seeds from infected plants. Assaying seeds for phytoplasmas is not required by plant protection quarantine protocols, but it should be. It would prevent the spread of disease into uncontaminated areas.

Phytoplasmas lack cell walls, which makes them resistant to antibiotics that target cell wall synthesis. Tetracycline binds to 16S rDNA, inhibiting protein synthesis, and has been used against phytoplasmas (Rao *et al.*, 2017). However, tetracycline and other antibiotics are not cost effective, because they must be reapplied every time a new flock of leafhoppers arrives. Chemical inducers of SAR (systemic acquired resistance) have been used, such as salicylic acid.

14.8   Spiroplasma diseases

Spiroplasmas lack cell walls and are members of the Mycoplasmatota, like *Candidatus* Phytoplasma spp., and they are vectored by insects. Spiroplasmas have a unique helical morphology and move in a corkscrew motion. Twenty-five *Spiroplasma* spp. are known, and one has been reported from *Cannabis*.

14.8a. *Spiroplasma citri* Saglio 1973

Saglio *et al.* (1973) originally described *S. citri* as a mycoplasma-like organism associated with "stubborn citrus disease" in California and Morocco. It was isolated from phloem sieve tubes. It could not be cultured *in vitro*. In addition to citrus, it infects a wide assortment of plants, including cabbage, horseradish, carrot, sesame, and many weed species. It is vectored by the beet leafhopper, *Neoalitarus* (*Circulifer*) *tenellus* (Liu *et al.*, 1983). *N. tenellus* feeds on a wide variety of hosts, including *Cannabis* (see Fig. 4.24).

Kenneth Frost observed a *S. citri*-infected cabbage field in Oregon growing adjacent to hemp fields; he predicted that *S. citri* would soon be reported in *Cannabis* (Frost, pers. comm., 2021). This happened quickly: Rivedal *et al.* (2024) isolated *S. citri* from hemp plants in Oregon with symptoms of leaf puckering, ringspots, interveinal chlorosis, and fasciation of flowers. Identification was confirmed using primers for the P89 adhesin gene and spiralin gene of *S. citri*. Leaf material from an infected plant successfully transferred the pathogen to healthy plants.

14.8.1   Management

Control the vector and you control *S. citri*. The incidence of beet leafhopper, and associated infections of *S. citri*, vary greatly from season to season, due largely to climatic conditions, such as temperature, rainfall, and wind patterns. Monitoring beet leafhopper populations can be accomplished by sweep netting and yellow sticky traps.

Beet leafhopper does not breed in *Cannabis*; eliminate weeds that serve as hosts, such Russian thistle (*Salsola tragus*) and kochia (*Bassia scoparia*). Biocontrol is difficult because of the high dispersal ability of beet leafhopper. Chemical control requires repeated foliar sprays of broad-spectrum insecticides throughout the growing season—which might be acceptable for potato crops, but not for *Cannabis*.

(Prepared by J. McPartland; revised by P. Nachappa)

References

Adams AN, Barbara DJ. 1982. Host range, purification and some properties of two carlaviruses from hop (*Humulus lupulus*): hop latent and American hop latent. *Annals of Applied Biology* 101: 483–494.

Adams AN, Barbara DJ, Morton A, Darby P, Green CP. 1995. Control of hop latent viroid in UK hops. *Acta Horticulturae* 385: 91–97.

Alexander T. 1987. Hepatitis outbreak linked to imported pot. *Sinsemilla Tips* 7(3): 22.

Alter H, Peer R, Dombrovsky A, Flaishman M, Spitzer-Rimon B. 2022. Tobacco rattle virus as a tool for rapid reverse-genetics screens and analysis of gene function in *Cannabis sativa* L. *Plants* 11(3): e327.

Atallah OO, Yassin SM, Verchot J. 2024. New insights into hop latent viroid detection, infectivity, host range, and transmission. *Viruses* 16(1): e30.

Beijerinck MW. 1898. Ueber ein Contagium vivum fluidum als Ursache der Fleckenkrankheit der Tabaksblätter. *Verhandelingen van de Koninklijke Akademie van Wetenschappen Te Amsterdam* 65: 1–22.

Bektaş A. 2024. Hop latent viroid: a guide to sampling, testing, and selection. *Cannabis Business Times*, available at: https://www.cannabisbusinesstimes.com/news/hop-latent-viroid-hlvd-guide-to-sampling-testing-lab-selection-to-prevent-virus-spread/

Bektaş A, Hardwick KM, Waterman K, Kristof J. 2019. The occurrence of hop latent viroid in *Cannabis sativa* with symptoms of cannabis stunting disease in California. *Plant Disease* 103(10): 2699.

Bernal JD, Fankuchen I. 1941. X-ray and crystallographic studies of plant virus preparations, I: Introduction and preparation of specimens; II: Modes of aggregation of the virus particles. *Journal of General Physiology* 25: 111–146.

Blattný C, Osvald CV, Novak J. 1950. Virosy a z viros podezřelé zjevy u konopí. *Ochrana Rostlin* 23: 5–9.

Brierley P. 1945. Viruses described primarily on ornamental or miscellaneous plants. II. *Plant Disease Reporter* 158(supplement): 167–200.

Bruenn JA, Warner BE, Yerramsetty P. 2015. Widespread mitovirus sequences in plant genomes. *PeerJ* 3: e876.

Brunstein J. 2021. *Issues and mitigations: hop latent viroid.* Segra International, Richmond, British Columbia.

Cassells AC. 2012. Pathogen and biological contamination management in plant tissue culture: phytopathogens, vitro pathogens, and vitro pests. *Methods in Molecular Biology* 877: 57–80.

Ceapoiu N. 1958. *Cinepa, studiu monografic.* Editura Academiei Republicii Populare Romine. Bucharest.

Chaube S, Kumar S, Dubey D, *et al.* (4 other authors). 2015. Identification of a novel phytoplasma (16Sr XIV-A subgroup) associated with little leaf and witches' broom of *Cannabis sativa* L. ssp. *sativa* and *C. sativa* L. ssp. *indica* in India. *Phytoparasitica* 43: 275–279.

Chester KS. 1935. Serological evidence in the study of the relationships of certain plant viruses. *Phytopathology* 25: 686–701.

Chiginsky J, Langemeier K, MacWilliams J, *et al.* (6 other authors). 2021. First insights into the virus and viroid communities in hemp (*Cannabis sativa*). *Frontiers in Agronomy* 3: e778433.

Creager AN, Morgan GJ. 2008. After the double helix: Rosalind Franklin's research on tobacco mosaic virus. *Isis* 99(2): 239–272.

Diener TO. 2016. Viroids: "living fossils" of primordial RNAs? *Biology Direct* 11: e15.

Doi YJ, Teranaka M, Yora K, Asuyama HF. 1967. クワ萎縮病, ジャガイモてんぐ巣病, Aster yellows 感染ペチュニアならびにキリてんぐ巣病の罹病茎葉篩部に見出された Mycoplasma 様 (あるいは PLT様)微生物について. *Annals of the Phytopathological Society of Japan* 33: 259–266.

Drohan J. 2021. Cannabis growing in Canada: an expanding market. *Crop & Soils Magazine* 54(6): 1–7.

Elbeaino T, Kontra L, Demian E, *et al.* (8 other authors). 2020. Complete sequence, genome organization and molecular detection of grapevine line pattern virus, a new putative Anulavirus infecting grapevine. *Viruses* 12: e602.

Eppler A. 1995. Zur Ökologie der Blattläuse an Hopfen und ihre Bedeutung bei der Ausbreitung von Hopfenviren. *Zeitschrift für Pflanzenkrankheiten und Pflanzenschutz* 102: 2–15.

Felker P, Bunch R, Russo G, *et al.* (6 other authors). 2019. Biology and chemistry of an umbravirus like 2989 bp single stranded RNA as a possible causal agent for Opuntia stunting disease (engrosamiento de cladodios). *Journal of the Professional Association for Cactus Development* 21: 1–13.

Feng X, Kyotani M, Dubrovsky S, Fabritius AL. 2019. First report of '*Candidatus* Phytoplasma trifolii' associated with a witches' broom disease in *Cannabis sativa* in Nevada, U.S.A. *Plant Disease* 103: 1763.

Fereres A, Raccah B. 2015. "Plant virus transmission by insects," pp. 1-12 in *eLS*, John Wiley and Sons, Chichester, UK.

Fernandez i Marti A, Parungao M, Hollin J, *et al.* (5 other authors). 2023. A novel, precise and high-throughput technology for viroid detection in *Cannabis* (MFDetect™). *BioRxiv Preprint*, https://doi.org/10.1101/2023.06.05.543818

Ferri F. 1963. Alterazioni della canapa trasmesse per seme. *Progresso Agricolo* 9: 346–351.

Franklin RE. 1955. Structure of tobacco mosaic virus. *Nature* 175: 379–381.

Franklin RE. 1956. Structure of tobacco mosaic virus: location of the ribonucleic acid in the tobacco mosaic virus particle. *Nature* 177: 928–930.

Gagne SJ, Stout JM, Liu E, *et al.* (3 other authors). 2012. Identification of olivetolic acid cyclase from *Cannabis sativa* reveals a unique catalytic route to plant polyketides. *Proceedings of the National Academy of Sciences USA* 109: 12811–12816.

Ghani M, Basit A. 1975. *Investigations on the natural enemies of marijuana, Cannabis sativa L. and opium poppy, Papaver somniferum L.* Annual Report, Commonwealth Institute of Biological Control, Pakistan station, Rawalpindi.

Giladi Y, Hadad L, Luria N, Cranshaw W, Lachman O, Dombrovsky A, 2020. First report of beet curly top virus infecting *Cannabis sativa* L., in Western Colorado. *Plant Disease* 104: 1656–1656.

Gitman LS. 1968. Малоизвестные болезни конопли (Little-known hemp diseases). *Защита растений (Plant Protection)* 13(3): 44–45.

Goidànich G. 1955. *Malattie crittogamiche della canapa.* Associazione Produttore Canapa, Bologna-Naples.

Gopala, Rao GP. 2018. Molecular characterization of phytoplasma associated with four important ornamental plant species in India and identification of natural potential spread sources. *3 Biotech* 8: e116.

Grace DJ, Roberts DJ. 2019. *Method of producing pathogen-free cannabis plants and pathogen-free plants and clones produced therefrom.* US Patent 20190387697A1.

Grudzinska M, Solarska E, Czubacka A, Przybys M, Fajbus A. 2006. Elimination of hop latent viroid from hop plants by cold treatment and meristem tip culture. *Phytopathologia Polonica* 40: 21–30.

Guček T. 2020. *Biologija viroida razpokanosti skorje agrumov (CBCVd) in razvoj metod za določanje viroidov v hmelju.* Doctoral thesis, Univerza v Ljubljani, Ljubljana, Slovenia.

Hadad L, Luria N, Smith E, *et al.* (3 other authors). 2019. Lettuce chlorosis virus disease: a new threat to *Cannabis* production. *Viruses* 11: e802.

Harkess J, Gildon B, Istre GR. 1989. Outbreaks of hepatitis A among illicit drug users, Oklahoma, 1984-87. *American Journal of Public Health* 79: 463–466.

Hartowicz LE, Knutson H, Paulsen A, Eaton BJ, Eshbaugh E. 1971. Possible biocontrol of wild hemp. *North Central Weed Control Conference, Proceedings* 26: 69.

Holmes FO. 1939. Proposal for extension of the binomial system of nomenclature to include viruses. *Phytopathology* 29: 431–436.

Hu J, Masson R, Dickey L. 2021. First report of beet curly top virus infecting industrial hemp (*Cannabis sativa*) in Arizona. *Plant Disease* 105: 1233.

Ivanovsky D. 1892. Über die Mosaikkrankheit der Tabakspflanze. *Bulletin Scientifique Publié Par l'Académie Impériale des Sciences de Saint-Pétersbourg. Nouvelle Série III* 35: 67–70.

Jarugula S, Wagstaff C, Mitra A, *et al.* (3 other authors). 2023. First reports of Bee curly top virus, Citrus yellow vein-associated virus, and Hop latent viroid in industrial hemp (*Cannabis sativa*) in Washington State. *Plant Disease*.

Jones M, Monthony AS. 2022. "Cannabis propagation," pp. 91–121 in Zheng YB, ed. *Handbook of Cannabis production in controlled environments.* CRC Press, Boca Raton, Florida.

Joshi V, Burlakoti P, Babu B. 2022. Diseases/symptoms on commercial crop samples submitted to the British Columbia Ministry of Agriculture, Food and Fisheries (BCMAFF) Plant Health Laboratory in 2021. *Canadian Plant Disease Survey* 102: 9–16.

Karl E. 1971. Neue Vektoren fur einige nichtpersistente Viren. *Archiv für Pflanzenschutz* 7: 337–342.

Kegler H, Spaar D. 1997. Zur virusanfälligkeit von hanfsorten (*Cannabis sativa* L.). *Archiv für Phytopathologie und Pflanzenschutz* 30: 457–464.

Knabel S, Seigner L, Wallnöfer PR. 1999. Nachweis des Hop Latent Viroids (HLVd) mit der Polymerase Kettenreaktion (PCR). *Gesunde Pflanzen* 51: 234–239.

Koziel SP. 2010. *Genetic analysis of lignification and secondary wall development in bast fibers of industrial hemp (Cannabis sativa).* Master's thesis, University of Alberta, Edmonton.

Kumagai MH, Donson J, Della-Cioppa G, Harvey D, Grill LK. 1995. Cytoplasmic inhibition of carotenoid biosynthesis with virus-derived RNA. *Proceedings of the National Academy of Sciences USA* 92: 1679–1683.

Kumar M, Madhupriya, Rao GP. 2017. Molecular characterization, vector identification and sources of phytoplasmas associated with brinjal little leaf disease in India. *3 Biotech* 7: e7.

Kwon SJ, Bodaghi S, Dang T, *et al.* (7 other authors). 2021. Complete nucleotide sequence, genome organization, and comparative genomic analyses of Citrus yellow-vein associated virus (CYVaV). *Frontiers in Microbiology* 12: e683130.

Lee IM, Gundersen-Rindal DE, Davis RE, Bartoszyk IM. 1998. Revised classification scheme of phytoplasmas based on RFLP analyses of 16S rRNA and ribosomal protein gene sequences. *International Journal of Systematic Bacteriology* 48: 1153–1169.

Liu HY, Gumpf DJ, Oldfield GN, Calavan EC. 1983. The relationship of *Spiroplasma citri* and *Circulifer tenellus*. *Phytopathology* 73(4): 585–590.

Mall S, Upadhyaya PP, Rao GP. 2011. Detection of a 16SrI phytoplasma associated with *Cannabis sativa* in India. *Indian Phytopathology* 43: 132–137.

Mandolino G, Ranalli P. 1999. "Advances in biotechnological approaches for hemp breeding and industry," pp. 185–212 in Ranalli P, *ed.*, *Advances in Hemp Research*. Haworth Press, New York.

Mann P. 2016. CEO: *Cannabis* needs mega research. *Mad River Union*, March 29, 2016 Available at: https://madriverunion.com/ceo-cannabis-needs-mega-research/

Marcone C, Gibb KS, Streten C, Schneider B. 2004. '*Candidatus* Phytoplasma spartii', '*Candidatus* Phytoplasma rhamni' and '*Candidatus* Phytoplasma allocasuarinae', respectively associated with spartium witches'-broom, buckthorn witches'-broom and allocasuarina yellows diseases". *International Journal of Systematic and Evolutionary Microbiology* 54: 1025–1029.

Markham R, Hitchborn JH, Hills GJ, Frey S. 1964. The anatomy of the tobacco mosaic virus. *Virology* 22: 342–359.

Maury R, Jadon V, Upadhaya PP. 2017. Witches'-broom disease of *Cannabis sativa* in India. *Agrica* 6(1): 72–75.

Maurya R, Mall S, Tiwari AK, Jadon V, Marcone C, Rao GP. 2020. *Cannabis sativa* L.: a potential natural reservoir of sugarcane grassy shoot phytoplasmas in India. *Sugar Tech* 22: 535–539.

McKernan K, Kane LT, McLaughlin S. 2022. Hop latent viroid chares a 19 nucleotide sequences with Cannabis sativa COG7. *OSF Preprints* https://osf.io/bwnmv.

Melgarejo TA, Chen LF, Rojas MR, Schilder A, Gilbertson RL. 2022. Curly top disease of hemp (*Cannabis sativa*) in California is caused by mild-type strains of beet curly top virus often in mixed infection. *Plant Disease*, Online ahead of print.

Mittelberger C, Obkircher L, Oettl S, *et al.* (6 other authors). 2017. The insect vector *Cacopsylla picta* vertically transmits the bacterium '*Candidatus* Phytoplasma mali' to its progeny. *Plant Pathology* 66: 1015–1021.

Montanari M. 2017. La certificazione del seme di canapa e le relative problematiche. *Dal Seme* 10: 38–41.

Nachappa P, Fulladolsa AC, Stenglein M. 2020. Wild wild west: emerging viruses and viroids of hemp. *Outlooks on Pest Management* 31(4): 175–179.

Nibert ML, Vong M, Fugate KK, Debat HJ. 2018. Evidence for contemporary plant mitoviruses. *Virology* 518: 14–24.

Patzak J, Matoušek J, Krofta K, Svoboda P. 2001. Hop latent viroid (HLVd)-caused pathogenesis: effects of HLVd Infection on lupulin composition of meristem culture-derived *Humulus lupulus*. *Biologia Plantarum* 44: 579–585.

Paulsen AQ. 1971. *Plant diseases affecting marijuana (Cannabis sativa)*. Unpublished manuscript, Kansas State University.

Phatak HC, Lundsgaard T, Verma VS, Singh S. 1975. Mycoplasma-like bodies associated with *Cannabis* phyllody. *Phytopathologische Zeitschrift* 83: 281–284.

Pitt WJ, Kairy L, Villa E, Nalam VJ, Nachappa P. 2022. Virus infection and host plant suitability affect feeding behaviors of cannabis aphid (Hemiptera: Aphididae), a newly described vector of Potato Virus Y. *Environmental Entomology* 20: 1010.

Podwyszńska M, Orlikowska T, Trojak-Goluch A, Wojtania A. 2021. Application and improvement of in vitro culture systems for commercial production of ornamental, fruit, and industrial plants in Poland. *Acta Societatis Botanicorum Poloniae* 91: e914.

Puchta H, Ramm K, Sänger HL. 1988. The molecular structure of hop latent viroid (HLV), a new viroid occurring worldwide in hops. *Nucleic Acids Research* 16: 4197–4216.

Punja ZK. 2021a. The diverse mycoflora present on dried cannabis (*Cannabis sativa* L., marijuana) inflorescences in commercial production. *Canadian Journal of Plant Pathology* 43: 88–100.

Punja ZK. 2021d. Emerging diseases of *Cannabis sativa* and sustainable management. *Pest Management Science* 77: 3857–3870.

Punja ZK, Wang K, Lung S, Buirs L. 2023. Symptomology, prevalence, and impact of Hop latent viroid on greenhouse-grown cannabis (*Cannabis sativa* L.) plants in Canada. *Canadian Journal of Plant Pathology*.

Raj SK, Snehi SK, Khan MS, Kumar S. 2008. '*Candidatus* Phytoplasma asteris' (group 16SrI) associated with a witches'-broom disease of *Cannabis sativa* in India. *Plant Pathology* 57: 1173.

Rao GP, Madhupriya, Thorat V, *et al.* (3 other authors). 2017. A century progress of research on phytoplasma diseases in India. *Phytopathogenic Mollicutes* 7: 1–38.

Rasoulpour R, Salehi M, Salehi E. 2017. Detection and partial characterization of a 16SrIX-C phytoplasma associated with hemp witches'-broom in Iran. *Iranian Journal of Plant Pathology* 99: 219–223.

Rataj K. 1957. *Skodlivi cinitele pradnych rostlin*. *Prameny Literatury* 2: 1–123.

Righetti L, Paris R, Ratti C, *et al.* (8 other authors). 2018. Not the one, but the only one: about *Cannabis* cryptic virus in plants showing 'hemp streak' disease symptoms. *European Journal of Plant Pathology* 150: 575–588.

Rivedal HM, Funke CN, Frost KE. 2022. An overview of pathogens associated with biotic stresses in hemp crops in Oregon, 2019 to 2020. *Plant Disease* 106: 1334–1340.

Rivedal HM, Temple T, Thomas WJ, *et al.* (8 other authors). 2024. First report of *Spiroplasma citri* associated with disease symptoms in field-grown hemp (*Cannabis sativa* L.) in the Pacific Northwest. *Plant Disease* 108: 202.

Röder K. 1941a. Untersuchungen über Hanfschädiger. *Mitteilungen aus der Biologischen Reichsanstalt für Land- und Forstwirtschaft in Berlin-Dalem* 63: 58–59.

Röder K. 1941b. Einige Untersuchungen über ein an Hanf (*Cannabis sativa* L.) auftretendes virus. *Faserforschung* 15: 77–81.

Safari M, Ferrari MJ, Roossinck MJ. 2019. Manipulation of aphid behavior by a persistent plant virus. *Journal of Virology* 93: e01781–18.

Saglio P, L'Hospital M, Lafleche D, *et al.* (4 other authors). 1973. *Spiroplasma citri* gen. and sp. n.: a mycoplama-like organism associated with "stubborn" disease of citrus. *International Journal of Systematic and Evolutionary Microbiology* 23:191–204.

Salehi M, Izadpanah K, Taghizadeh M. 2002. Witches' broom disease of lime in Iran: new distribution areas, experimental herbaceous hosts and transmission trials. *Proceedings of the Annual IOCV Conference* 15: 293–296.

Sastra KS. 1973. Studies on virus diseases of medical plants. *Indian Journal of Horticulture* 30: 562–566.

Satta E, Paltrinieri S, Bertaccini A. 2019. "Phytoplasma transmission by seed," pp. 131–147 in Bertaccini A *et al.*, *eds. Phytoplasmas: plant pathogenic bacteria, vol. II.* Springer Nature, Singapore.

Schachtsiek J, Hussain T, Azzouhri K, Kayser O, Stehle F. 2019. Virus-induced gene silencing (VIGS) in *Cannabis sativa* L. *Plant Methods* 15: e157.

Schmelzer K. 1963. Untersuchungen an Viren der Zier- und Wildgehölze. 1. Mitteilung Virosen an *Viburnum* und *Ribes. Phytopathologische Zeitschrift* 46: 17–52.

Schmidt HE, Karl E. 1969. Untersuchungen über eine flecken- und streifenbildung am hanf (*Cannabis sativa* L.). *Zentralblatt für Bakteriologie, Parasitenkunde, Infektionskrankheiten und Hygiene, Abteilung* 2 123: 310–314.

Schmidt HE, Karl E. 1970. Ein beitrag zur analyse der virosen des hanfes under berücksichtigung der hanfplattlaus als virusvektor. *Zentralblatt für Baktteriologie, Parasitenkunde, Infektionskrankheiten und Hygiene, Abteilung* 2 125: 16–22.

Schoener J, Wang S. 2019. First detection of a phytoplasma associated with witches' broom of industrial hemp in the United States. *Phytopathology* 109: S2.110.

Shaw DD, Whiteman DC, Merritt AD, *et al.* (4 other authors). 1999. Hepatitis A outbreaks among illicit drug users and their contacts in Queensland, 1997. *Medical Journal of Australia* 170: 584–587.

Sikora C, Tipples G, Pang X-L, Andonov A. 2017. Hepatitis A virus infection associated with cannabis use. *Canada Communicable Disease Report* 43: 245–256.

Singh AK, Gopala, Rao A, Goel S, Rao GP. 2018. Identification of '*Candidatus* Phytoplasma asteris' causing sesame phyllody disease and its natural weed host in Jammu, India. *Indian Phytopathology* 71: 143–146.

Ślusarkiewicz-Jarzina A, Ponitka A, Kaczmarek Z. 2005. Influence of cultivar, explant source and plant growth regulator on callus induction and plant regeneration of *Cannabis sativa* L. *Acta Biologica Cracoviensia Series Botanica* 47: 145–51.

Smýkalová I, Vrbová M, Cvečková M, *et al.* (7 other authors). 2019. The effects of novel synthetic cytokinin derivatives and endogenous cytokinins on the *in vitro* growth responses of hemp (*Cannabis sativa* L.) explants. *Plant Cell Tissue Organ Culture* 139: 381–394.

Spaar D, Kleinhempel H, Fritzsche R. 1990. *Öl- und Faserpflanzen.* Springer-Verlag, Berlin.

Stout JM, Boubakir Z, Ambrose SJ, Purves RW, Page JE. 2012. The hexanoyl-CoA precursor for cannabinoid biosynthesis is formed by an acyl-activating enzyme in *Cannabis sativa* trichomes. *Plant Journal* 71: 353–365.

Svobodová J, Blattný C, Pozdena J. 1954. "Virové choroby konopí," pp. 149–155 in Rataj K, *ed. Škůdci a choroby přadných rostlin.* Státní Zemědělské Nakladatelství, Praha.

Traversi BA. 1949. Estudio inicial sobre una enfermedad del girasol (*Helianthus annuus* L.) en Argentina. *Lilloa* 21: 271–278 and *Revista de Investigociones Agrícolas* 3: 345–351.

Un Nabi S, Madhupriya, Dubey D, Rao GP. 2015. Identification of *Cannabis sativa* L. ssp. *sativa* as putative alternate host of sesame phyllody phytoplasma belongs to 16SrI group in India. *Medicinal Plants* 7(1): 68–70.

Vali Sichani F, Bahar M, Zirak L. 2011. Characterization of stolbur (16SrXII) group phytoplasmas associated with *Cannabis sativa* witches'-broom disease in Iran. *Plant Pathology Journal (Faisalabad)* 10(4): 161–167.

Van Bakel H, Stout JM, Cote AG, *et al.* (4 other authors). 2011. The draft genome and transcriptome of *Cannabis sativa. Genome Biology* 12(10): R102.

Van Slogteren E, Van Slogteren DHM. 1957. Serological identification of plant viruses and serological diagnosis of virus diseases of plants. *Annual Review of Microbiology* 11: 149–164.

Warren J, Mercado J, Grace D. 2019. Occurrence of hop latent viroid causing disease in *Cannabis sativa* in California. *Plant Disease* 103(10): 2699.

Watson JD. 1954. The structure of tobacco mosaic virus, I: x-ray evidence of a helical arrangement of sub-units around the longitudinal axis. *Biochimica et Biophysica Acta* 13: 10–19.

Watson JD. 1968. *The double helix.* Atheneum, New York.

Watson JD, Crick FHC. 1953. A structure for deoxyribose nucleic acid. *Nature* 171: 737–738.

Zhao Y, Sun Q, Davis RE, Lee IM. 2007. First report of witches'-broom disease in a *Cannabis* spp. in China and its association with a phytoplasma of elm yellows group (16SrV). *Plant Disease* 91: 227.

Ziegler A, Matoušek J, Steger G, Schubert J. 2012. Complete sequence of a cryptic virus from hemp (*Cannabis sativa*). *Archives of Virology* 157: 383–385.

15 Bacterial and nematode diseases

Abstract

When diagnosing a plant disease, fungi are the usual suspects. We learn this in school, "When you hear hoof beats, look for horses, not zebras." The zebras are nematode and bacterial diseases. They're out there. Nematodes mostly limit their damage to roots, and the symptoms they cause are often misdiagnosed. Treatment can be difficult. Most problems are caused by root-knot nematodes (*Meloidogyne* spp.), cyst nematodes (*Heterodera* and *Globodera* spp.), and lesion nematodes (*Pratylenchus* spp.). Stem nematodes (*Ditylenchus dipsaci*) attack above-ground parts. Bacterial diseases are the opposite. They rarely attack roots (crown gall caused by *Agrobacterium* spp.), but cause all sorts of above-ground problems—stem diseases, blights, wilts, and leaf spots. The most common offenders are *Pseudomonas*, *Xanthomonas*, and *Serratia* spp. Treatment of bacterial diseases is relatively straightforward, if a proper diagnosis is made—the hard part.

15.1 Introduction

Nematodes attack roots, and the *Cannabis* root system is a large, juicy target (see Fig. 8.1). Because nematodes are mostly underground and unseen, their subterranean damage is underappreciated. Above-ground symptoms are non-specific and might be misdiagnosed as nutrient deficiencies or drought, mysteriously appearing despite adequate nutrients and moisture.

Koenning *et al.* (1999) estimated USA crop losses due to nematodes alone: maize up to 20%, soybean up to 15%, tobacco up to 5%, and miscellaneous vegetable crops up to 8%. Bernard *et al.* (2022) gave an excellent overview of *Cannabis*-parasitic nematodes—including several reports overlooked by us, for which we are indebted. Around 500 nematode species parasitize plants, but only 20 infest *Cannabis*: root-knot nematodes, and to a lesser degree, cyst nematodes, lesion nematodes, and others (needle, spiral, dagger, and spiral nematodes). One species attacks above-ground plant parts—the stem nematode.

Bacteria in the *Cannabis* literature that make the biggest splash are human pathogens such as *Salmonella* spp. They contaminate marijuana, do not cause any plant symptoms, but give rise to small epidemics in consumers (see section 18.3). Other bacteria invade hemp stalks after harvest, as retters and rotters (see section 18.1). Some mutualistic bacteria benefit *Cannabis*, such as PGPRs (plant growth-promoting rhizobacteria, see section 13.3). Subtracting these contaminants, saprophytes, and PGPRs leaves us with a handful of bacteria that cause disease in living plants. They produce leaf spots, blights, and stem ulcers. Most of them are *Pseudomonas* and *Xanthomonas* spp.

Traditionally, plant-pathogenic bacterial species were recognized by their *in vitro* biochemical–nutritional characters and pathogenicity. In 1980, pathogenicity became the most important character, and **pathovars** (abbreviation pv.) were named based on host range studies (Dye *et al.*, 1980). Pathovars can be difficult to differentiate, much like *formae speciales* among fungi. The advent of DNA–DNA hybridization (DDH) studies led to the concept of **genomospecies** and **genomovars** (Gardan *et al.*, 1999). Since then, DDH studies have been supplanted by 16S rRNA gene sequence analysis, or multilocus sequence typing analysis (with *gap1*, *gltA*, *gyrB*, and *rpoD*) (Bull *et al.*, 2010a).

15.2 Striatura ulcerosa

A bacterial disease called striatura ulcerosa (ulcerative streak) was first described in Italy, where monographs have been written about the disease (Ferri, 1957a,b; Goidànich and Ferri, 1960). Striatura ulcerosa was discovered in Rovigo by Peglion (1896, 1897), who identified the causal organism as *Bacillus cubonianus*, a pathogen of mulberry trees (*Morus alba*). It's still a problem in Rovigo; Gianpaolo Grassi provided us with photographs from there (Fig. 15.1A,B).

Julius Petri, of the eponymous petri dish, described the disease in Germany (Petri, 1908), as did others (Kirchner, 1906; Sorauer, 1908; Flachs, 1936). *Bacillus cubonianus* or its modern epithet, *Pseudomonas syringae* pv. *mori*, has also been described in Czechoslovakia (Rataj, 1957), Romania (Ceapoiu, 1958), Russia (Gitman and Boytchenko, 1934; Gitman, 1968a), and in the USA (McPartland and Hillig, 2004; Hillig, 2005a).

15.2.1 Symptoms and signs

Symptoms begin as light-green lesions, oval-shaped, that elongate along the length of the stem. Lesions rarely encircle more than half the stem circumference, but may run 10 cm along its length. They tend to first appear on mature plants, toward the base of the stem. Tiny whitish pustules (0.5–2 mm) arise within the lesions, filled with a mucilaginous white or yellow bacterial ooze. The pustules rupture and shred the stem epidermis, forming light-brown ulcerative streaking (Fig. 15.1).

Ferri (1957a) microscopically illustrated cellular disruption under the pustules, which extended to parenchymal cells between bast fibers. Parenchymal cells undergo hypertrophy

DOI: 10.1079/9781836990352.0015

Fig. 15.1. Striatura ulcerosa (**A**, **B**: Gianpaolo Grassi; **C**: Karl Hillig)

and distension, ruining the neighboring bast fibers. Secondary pustules may arise around the shredded remains of ruptured pustules. Placing stems with lesions in a humidity chamber (damp filter paper under a petri plate) gives rise to cloudy droplets emerging from the pustules, filled with motile bacteria seen under a microscope (Peglion, 1896, 1897).

15.2.2 Causal organism

15.2a. *Pseudomonas syringae* pv. *mori* (Boyer & Lambert 1893) Young, Dye & Wilkie 1978

≡*Bacterium mori* Boyer & Lambert 1893, ≡*Pseudomonas mori* (Boyer & Lambert 1893) Stevens 1913, ≡*Bacillus mori* (Boyer & Lambert) Holland 1920
=*Bacillus cubonianus* Macchiati 1892, ≡*Pseudomonas cuboniani* (Macchiati) Krasil'nikov 1949
Description: Colonies white, flat, circular, margins entire becoming undulate. Aerobic gram-negative rods, straight with rounded ends, exhibiting 1–7 polar flagella, 1.8–4.5 × 0.9–1.3 μm. Non-proteolytic, starch weakly hydrolysed, acid (but no gas) produced from xylose, dextrose, galactose, mannose, salicin, sucrose, raffinose, and glycerol. No sheaths or spores are produced. Bacteria often form in chains.
Comments: Smith (1920) synonymized Macchiati's taxon under Boyer and Lambert's taxon, on the rationale that Macchiati described the organism poorly, and did not conduct host inoculation studies. Elliott (1951) confirmed the synonymy. Young *et al.* (1978) reduced Boyer and Lambert's taxon to a pathovar of *Pseudomonas syringae*, coining *P. syringae* pv. *mori*.

In the ground-breaking DNA–DNA hybridization by Gardan *et al.* (1999), 48 pathovars of *P. syringae* were classified into nine genomospecies. *P. syringae* pv. *mori* belonged to genomospecies 2, along with 19 other pathovars. *Pseudomonas amygdali* is the earliest valid name for this genomospecies, thus *P. syringae* pv. *mori*

should be a synonym under *P. amygdali*. However, the taxon *Pseudomonas amygdali* **pv.** *mori* is not yet generally accepted (Krawczyk and Łochyńska, 2020).

Although the bacterium causing striatura ulcerosa in hemp has been identified as *P. syringae* pv. *mori*, it may be a case of mistaken identity. The bacterium has never been deposited in a culture collection amenable to DNA classification. Perhaps the cause of striatura ulcerosa on hemp stalks is actually *Pseudomonas cannabina*, a leaf pathogen (see below). Note that Peglion (1896, 1897) originally described *B. cubonianus* infecting hemp stalks, but he subsequently described *B. cubonianus* infecting leaves (Peglion, 1902).

15.2.3 Disease cycle and epidemiology

P. syringae pv. *mori* is found wherever mulberry trees grow. The pathovar also infects lima beans, *Phaseolus lunatus* (CABI, 2009). Plant-pathogenic bacteria are highly dependent on natural openings and wounds to invade plants. Peglion (1896, 1897) blamed epidemics of striatura ulcerosa on hailstone damage, which caused breaks in the epidermis—allowing bacteria to enter. Epidermal damage caused by frost, insects, and humans has also been implicated. After entering host tissue, the bacterium spreads through intercellular spaces, and forms bacteria-filled cavities. Pustules rupture and ooze bacteria, which spread by wind-driven rain to neighboring plants. People and contaminated equipment also spread the pathogen as they move through infested fields. Goidànich and Ferri (1960) reported that seed-borne infection was rare, but possible.

Hillig (2005a) conducted a taxonomic study of *Cannabis* germplasm from around the world, and striatura ulcerosa spread throughout his glasshouse in Indiana. Hillig visually rated symptoms on an ordinal scale of 1 (no or very few pustules) to 4 (many pustules). The most susceptible plants were fiber-type plants from East Asia. McPartland and Hillig (2004) proposed that *P. syringae* pv. *mori* originated in China. Italy would have been the bacterium's first stop in Europe—Italian agronomists have long imported Chinese germplasm (Amoretti, 1789; Re, 1806; de Montluisant, 1885).

15.2.4 Cultural and mechanical control

Pseudomonas plant pathogens share commonalities in their control. Exclusion is critical. Prevent the spread of seed-borne pathogens into new regions. Do not harvest seed from infected plants, and carefully clean seeds after threshing, to remove infected leaf fragments. Purchase certified clean seed where possible. Scout for disease after plant-damaging environmental events have occurred (e.g., hailstones). *Pseudomonas* spp. spread by splashing water, so avoid overhead irrigation. Indoors, keep foliage as dry as possible by increasing ventilation. Do not walk through crops or operate machinery when plants are wet.

Sanitation is key: collect and destroy crop debris at the end of the season, so the pathogen doesn't overwinter in the field. Shredding of stalks and roots exposes bacteria to the environment. Shredding hemp stalks requires special tillage equipment, because standard shredders cannot adequately cope with the fiber. Debris can be plowed under, although many *Pseudomonas* spp. can overwinter in soil. After crops are harvested, the pathogen can shift to weeds and multiply as an epiphyte—eliminate weeds from within and outside fields.

Grow healthy plants. Peglion (1902) said weakly growing crops in poor soil were predisposed to striatura ulcerosa, as well as crops sown late (in Italy, at the end of May). Ghillini (1951) added that striatura ulcerosa occurred primarily in crops grown in recently drained or previously unworked soil. As aforementioned, Hillig (2005a) measured the susceptibility of 135 worldwide *Cannabis* accessions to striatura ulcerosa. Fiber-type plants from East Asia were the most susceptible, followed by drug-type plants from South Asia and Central Asia. European plants—both cultivated and wild-type—were the least susceptible.

15.2.5   Biological and chemical control

Biological control of *Pseudomonas* spp. utilizes plant growth-promoting rhizobacteria (PGPRs), such as *Bacillus subtilis*, and *B. amyloliquefaciens*, other *Pseudomonas* spp. (*P. chlororaphis, P. fluorescens*), and *Streptomyces griseoviridis* and *S. lydicus* (see Chapter 20). *P. fluorescens* strain A506 has been used as a foliar spray against bacterial blight. Non-pathogenic strains of *P. syringae* are available (Bio-Save®). They exist as epiphytes on leaf surfaces and compete with fungal pathogens, but have not been tested against pathogenic *Pseudomonas* spp.

Bacteriophages are viruses that attack bacteria; they are the most common and diverse entities in the biosphere. Dion *et al.* (2020) estimated 10^{31} bacteriophages populate the planet (ten million trillion trillion), far greater than all other organisms combined, including bacteria—by tenfold. Bacteriophages destroy half the world's bacteria every 48 h. Buttimer *et al.* (2017) reviewed "bacteriophage biocontrol" of bacterial plant diseases, first tested in 1924, less than a decade after bacteriophages were discovered. Due to a number of issues, commercialization has been slow. Phage cocktails need to be updated constantly, in order to kill their relentlessly evolving bacterial hosts.

Antibiotics deployed against *Pseudomonas* plant pathogens include streptomycin, tetracycline, and kasugamycin. Bacterial resistance to antibiotics is rife, rendering them useless (Takahashi *et al.*, 2013). Public concern about pesticide residues has curtailed their use. Copper compounds (especially copper oxychloride) have been used since the 1950s. Chemicals that induce systemic acquired resistance (SAR) have found a place. Ishiga *et al.* (2019) applied a soil drench of acibenzolar-S-methyl (ASM), a well-known SAR activator, to control *P. cannabina* pv. *alisalensis* disease in cabbage.

15.3   Bacterial blight

Also known as bacterial leaf spot, this disease has been described in Serbia (Šutič and Dowson, 1959), Hungary (Klement and Kaszonyi, 1960), northern Italy (Peglion, 1902; Ferri, 1959a; Goidànich and Ferri, 1960), Russia (Gitman 1968a,b), and Bulgaria (Georgieva and Kotev, 1972).

Šutič and Dowson suspected that the Serbian epidemic began after infected seeds were imported from Turkey. They conducted host range testing: *Pseudomonas cannabina* caused leaf spots on beans (*Phaseolus vulgaris*), and a hypersensitivity reaction in hops. No other hosts were susceptible, in Moraceae (*Morus, Ficus* spp.), Urticaceae (*Urtica, Boehmeria* spp.), or other crop plants susceptible to *Pseudomonas* spp.—maize, tomato, pepper, lemon, walnut, and oleander.

15.3.1   Symptoms and signs

Symptoms begin as small water-soaked leaf spots. Spots enlarge along leaf veins (rarely crossing them) and turn brown or gray. Dead tissue breaks apart, causing leaf perforations. Floral leaves may also be affected, becoming yellow then brown. Growth of plants affected at an early age was stunted, with short internodes, and leaves compacted into rosettes. An Italian strain of the bacterium ("variety *italica*") produced "ulcerous striping"—small protuberances arising in rows between leaf veins, turning into dark, necrotic stripes (Goidànich and Ferri, 1960). The Italian strain also caused stem lesions.

15.3.2   Causal organisms

Bacterial blight is primarily caused by *Pseudomonas cannabina*, but a second species, *Pseudomonas koreensis*, has been isolated from plants with leaf spots.

15.3a.   *Pseudomonas cannabina* pv. *cannabis* Bull et al. 2010

≡*Pseudomonas cannabina* (ex Šutič & Dowson 1959) Gardan *et al.* 1999, ≡*Pseudomonas cannabina* Šutič & Dowson 1959; ≡*Pseudomonas cannabina* var. *italica* Dowson *apud* Goidànich and Ferri 1960, ≡*Pseudomonas syringae* pv. *cannabina* (Šutič & Dowson 1959) Young *et al.* 1978

Description: Colonies white, convex, circular, some strains produce green fluorescent pigment on King's medium B agar, other strains exude a non-fluorescent brown pigment (Smith *et al.*, 1988). Aerobic gram-negative rods, straight or slightly curved with rounded ends, exhibiting 1–4 polar flagella (see Fig. 3.2). Size was originally given as 1.5×0.3 µm (Šutič and Dowson, 1959), but recently as $3.0–4.0 \times 1.1–3.0$ µm (Gardan *et al.*, 1999; Bull *et al.*, 2010a). Non-proteolytic, starch weakly hydrolysed, acid (but no gas) produced from xylose, dextrose, galactose, mannose, sucrose, raffinose, and glycerol. No sheaths or spores are produced. Bacteria often form in chains.

Comments: Peglion (1902) thought the causal organism might be *Bacillus cubonianus*, the cause of striatura ulcerosa on stems. Šutič and Dowson (1959) named it *Pseudomonas cannabina*. Young *et al.* (1978) reduced *P. cannabina* to a pathovar of *Pseudomonas syringae*, coining *P. syringae* pv. *cannabina*. Takahashi *et al.* (1996) applied this taxon to the organism in Japan. In the ground-breaking DNA–DNA hybridization by Gardan *et al.* (1999), genomospecies 9 consisted of three isolates (from Hungary and Yugoslavia) of only one pathovar, *P. syringae* pv. *cannabina*. They considered it a separate species, and re-erected *P. cannabina*. An isolate of *P. syringae* pv. *mori* from *Morus alba* was not related, in genomospecies 2.

In the ground-breaking 16S rDNA and MLSA study by Bull *et al.* (2010a), *P. cannabina* did not genetically differ from *P. syringae* pv. *alisalensis*, a pathogen of broccoli raab, broccoli, and arugula. In host range testing, *P. syringae* pv. *alisalensis* was not pathogenic on *Cannabis*. Conversely, *P. cannabina* strains from *Cannabis* were not pathogenic on broccoli raab. Bull and colleagues recognized two pathovars: *P. cannabis* pv. *cannabis*, and *P. cannabis* pv. *alisalensis*. The latter pathovar has caused epidemics in broccoli (Cintas *et al.*, 2002), cauliflower (Koike *et al.*, 2006), rutabaga (Koike *et al.*, 2007), and brussels sprouts (Bull *et al.*, 2010b).

15.3b. *Pseudomonas koreensis* Kwon et al. 2003

P. koreensis was first isolated from farm soils in Korea. A Swedish strain isolated from the filter of a hydroponic system in which tomato was cultivated attracted attention. It produced biosurfactants that lysed zoospores of oomycete pathogens. Potato leaves pretreated with *P. koreensis* and then inoculated with *Phytophthora infestans* showed reduced disease compared with controls (Hultberg *et al.*, 2010).

Since then, other strains of *P. koreensis* have been applied as a PGPR to the rhizosphere to mitigate plant stress (drought, salinity, heavy metals) and provide antimicrobial activity against bacteria and fungi. Inoculation of soil with *P. koreensis* mitigated the effects of nutrient deficiency in *Cannabis* (Tanney *et al.*, 2023).

Thiessen *et al.* (2020) isolated a *Pseudomonas* sp. from *Cannabis* with leaf spots in North Carolina. Its 16S rDNA sequence was 99.85% identical with a *P. koreensis* strain isolated from disease-suppressive soil. Lesions were small (1–5 mm), medium brown to dark brown, angular, and limited by leaf veins. They fulfilled Koch's postulates.

15.3.3 Differential diagnosis

Peglion (1902) suggested that bacterial blight might be confused with diseases caused by foliar-infecting fungi, such as *Phyllosticta* and *Septoria* spp. More likely, Pseudomonas blight might be attributed to Xanthomonas blight—see next.

15.3.4 Disease cycle and epidemiology

P. cannabis pv. *cannabis* surged through Europe in the 1950s–1960s, and then disappeared from the literature. The epidemiology of *P. cannabis* is poorly understood, but it is closely related to *P. syringae*, which can exist as an epiphyte on leaf surfaces. It turns pathogenic after gaining access to mesophyll tissue, via natural openings or wounds, and colonizing the intercellular space within leaves.

P. syringae (as well as *Xanthomonas campestris*) often enters via **hydathodes**, which are water-secreting pores located along leaf margins. Hydathodes resemble stomata, but they are larger, have lost the power of movement, and connect to the plant's vascular system via xylem vessels. *Cannabis* does not have hydathodes, but each leaflet tip possesses a cluster of 6–15 stomata, probably modified to function as hydathodes (Dayanandan and Kaufman, 1976).

When introduced through a wound, *P. cannabis* causes symptoms 10 days after inoculation (Goidànich and Ferri, 1960). Šutič and Dowson (1959) stated that seed-borne infection "undoubtedly presents the main and most dangerous source of the disease." Conversely, Goidànich and Ferri (1960) blamed infected crop residues left in the field as the primary source of inoculum. Other authors suggested that seed was the primary source of inoculum (Noble and Richardson, 1968; Georgieva and Kotev, 1972). Cook (1981) presumed seed-borne infection, or seed contaminated with infected leaf fragments during threshing.

15.3.5 Disease management

See instructions under striatura ulcerosa, above. Given the limited distribution of *P. cannabina* pv. *cannabis* and *P. koreensis*, exclusion becomes paramount.

15.4   Xanthomonas blight

Xanthomonas blight of *Cannabis* has been reported in Japan and Korea (Watanabe, 1947; Okabe, 1949; Okabe and Goto, 1955; Kyokai, 1965; Netsu *et al.*, 2014) and in Romania (Sandru, 1977; Severin, 1978; Trotuş and Naie, 2008). A *Xanthomonas* sp. was isolated from retted hemp stalks in France (Riberio *et al.*, 2015), and a *Xanthomonas* sp. causing blight symptoms was isolated from industrial hemp in Kansas (Hollandbeck, 2019).

15.4.1 Symptoms and signs

Symptoms have been described as either leaf spots or a leaf blight. Leaf spots begin as pinhead-sized water-soaked lesions. They enlarge to 1–2 mm in diameter, and the central part becomes brown and necrotic, often accompanied by a yellow halo 2–3 mm wide (Fig. 15.2). Spots are generally circular, sometimes becoming polygonal, and delineated by leaf veins. Bacteria can also infect the shoot apex, causing bud blight (Severin, 1978; Netsu *et al.*, 2014).

15.4.2 Causal organism

15.4a. *Xanthomonas cannabis* pv. *cannabis* (Watanabe) Jacobs *et al.* 2015

≡*Pseudomonas cannabis* Watanabe 1947, ≡*Bacterium cannabis* (Watanabe) Okabe 1949, ≡*Xanthomonas cannabis* (Watanabe) Okabe & Goto 1965

Fig. 15.2. Xanthomonas blight. White arrows, necrotic lesions; black arrows, yellow halos (courtesy Netsu *et al.*, 2014)

= *Xanthomonas campestris* pv. *cannabis* Severin 1978
= *Xanthomonas cannabis* Mukoo *apud* Kishi 1988

Description: Colonies usually yellow, smooth and viscid. Aerobic gram-negative straight rods, with single polar flagella, usually $0.7–1.8 \times 0.4–0.7$ μm. No denitrification or nitrate reduction. Proteolytic, non-lipolytic; acid production from arabinose, glucose, mannose, galactose, trehalose, cellobiose.

Comments: Watanabe (1947) named the causal bacterium *Pseudomonas cannabis*. Okabe (1949) reclassified it as *Bacterium cannabis*, and Okabe and Goto (1955) reclassified it as *Xanthomonas cannabis*. In Romania, Severin (1978) named the bacterium *Xanthomonas campestris* pv. *cannabis*. He made no reference to the earlier Japanese publications. The Japanese publications were overlooked by sequential editions of *Bergey's Manual of Determinative Bacteriology* (6th: 1948; 7th: 1957; 8th: 1974). In 1980, international bacterial nomenclature was simplified with an "approved list of bacterial names," which cited *Xanthomonas campestris* pv. *cannabis* (Dye *et al.*, 1980).

Then the molecular era arrived. Based on DNA–DNA hybridization studies, Vauterin *et al.* (1995) limited *X. campestris* to only six pathovars that attack crucifers. They transferred 34 other *X. campestris* pathovars to new species, and listed 66 other *X. campestris* pathovars that awaited analysis, including *X. campestris* pv. *cannabis*. Parkinson *et al.* (2009) sequenced *gyrB* genes from over 200 *Xanthomonas* isolates, including *X. campestris* pv. *cannabis* from Romania. That isolate formed a species-level clade (SLC) with *X. campestris* pv. *zinniae* from *Zinnia elegans*, and *X. campestris* pv. *esculenti* from okra, *Hibiscus esculentus*.

Netsu *et al.* (2014) reisolated the hemp pathogen in Japan, and compared it with a culture of the Romanian isolate. Pathogenicity tests produced identical symptoms (Fig. 15.2). Phylogenetic analysis based on *gyrB* sequences placed both isolates in the same clade (SLC), separate from isolates of *X. campestris* pathovars. Netsu and colleagues made note of the long-overlooked taxon *Pseudomonas cannabis* by Watanabe (1947), which predated *Xanthomonas campestris* pv. *cannabis* by Severin (1978), and Netsu summarized that "they should be classified under a new species name in the future."

This was done by Jacobs *et al.* (2015). They sequenced the entire genomes of 24 *Xanthomonas* spp., including the Romanian and Japanese isolates. In their phylogenetic analysis, the Romanian and Japanese isolates formed a clade with a *Xanthomonas* that was pathogenic on beans in Rwanda identified by Aritua *et al.* (2015). Jacobs and colleagues renamed the Romanian and Japanese isolates *X. cannabis* pv. *cannabis*, and the Rwanda bean isolate *X. cannabis* pv. *phaseoli*.

The clade has gained members: Merda *et al.* (2017) published a whole-genome phylogenetic tree of 80 *Xanthomonas* isolates. The two hemp isolates and the Rwanda isolate formed a clade with a new bacterial pathogen of beans and Niger seeds (*Guizotia abyssinica*) in Zimbabwe. Meline *et al.* (2019) isolated a non-pathogenic *Xanthomonas* strain, CFBP 7698, that formed a clade with the two bean isolates from Rwanda and Zimbabwe, sister to the clade of two hemp isolates. Harrison *et al.* (2022) found that *X. cannabis* pv. *cannabis* formed a clade with the aforementioned okra pathogen, *X. campestris* pv. *esculenti*. Obviously more work remains to be done.

15.4.3   Disease cycle and epidemiology

Little has been published on the disease cycle and epidemiology of *X. cannabis* pv. *cannabis*. Symptoms first arise in spring. Water-splashed bacteria give rise to secondary infections, which continue until harvest time (Netsu *et al.*, 2014). Water-soaked lesions appear 7–10 days after inoculation. The authors report growth at 35°C, which is consistent with most *Xanthomonas* spp. (25–35°C).

Disease transmission in most *Xanthomonas* spp. occurs via infected crop residues. The *Xanthomonas* infection cycle includes an epiphytic stage and endophytic stage (An *et al.*, 2020). In the epiphytic stage, bacteria are rain-splashed onto the leaf surface. They excrete an extracellular polysaccharide known as xanthan gum (yes, the food additive) and form a biofilm. The endophytic stage begins when bacteria invade mesophyll tissue via natural openings or wounds. Bacteria colonize leaf tissue and vascular vessels. Xanthan gum plugs up xylem and disrupts the flow of fluids, causing wilt symptoms.

When bacterial populations reach high levels, the pathogens re-emerge onto the leaf surface. This is facilitated by warm and humid conditions, causing plants to open stomata and hydathodes. Bacteria are dispersed by wind and rain, as well as people and agricultural machinery. Seed-borne infection is reported for many *Xanthomonas* spp. Seed-borne transmission has not been analyzed in *X. cannabis*, but the species was formerly considered a pathovar of *X. campestris*, where seed contamination plays a dominant role in disease transmission.

15.4.4   Disease management

See instructions under striatura ulcerosa, above. Exclusion is paramount, given the limited distribution of *X. cannabis*. Seed-borne infection is likely; avoid purchasing seeds from places where *X. cannabis* is endemic.

15.5 Serratia blight

Serratia blight causes angular, dark brown-black leaf spots on leaves, stems, and flower parts (Schappe *et al.*, 2020). As the disease progresses, lesions coalesce to form larger regions of necrosis that engulf large portions of leaves, and whole plants are lost to disease (Fig. 15.3). Puncturing lesions result in red bacterial ooze emerging from plant tissues (Thiessen *et al.*, 2020).

15.5a. *Serratia marcescens* Bizio 1823

Description: Colonies on PDA are raised and dark-pink in color. Facultative anerobic gram-negative rods, with variable number of flagella, 0.9–2.0 × 0.5–0.8 μm. Proteolytic, uniquely casein hydrolytic; nitrate reduction to nitrite; acid production from arabinose, glucose, lactose, maltose, mannitol, sucrose, but not lactose.
Comments: Bizio (1823) identified *S. marcescens* as the cause of "red-colored polenta." The bacterium produces prodigiosin, a blood-red pigment, and its appearance on communion bread in Catholic Italy was considered miraculous. *S. marcescens* is an ecologically and genetically diverse species in the family Enterobacteriaceae. Various *S. marcescens* strains live as water or soil saprophytes, rhizobacteria, insect pathogens, plant endophytes and pathogens, and opportunistic human pathogens.

Schappe *et al.* (2020) isolated *S. marcescens* from four glasshouses in North Carolina. They identified the pathogen by culture characteristics and 16s rRNA and *rpoB* sequence analyses, and they fulfilled Koch's postulates. Because lesions were observed with inoculations in the absence of wounding, Thiessen *et al.* (2020) suggested that *S. marcescens* could invade plant tissues through hydathodes, trichomes, and/or stomata.

Fig. 15.3. Serratia blight (courtesy L. Thiessen)

Adding insult to injury, *S. marcescens* can metabolize and break down THC (Binder and Meisenberg, 1978). For disease management, see instructions under striatura ulcerosa, above. *S. marcescens* has been proposed as a biocontrol agent against soil-borne fungi (*Sclerotium rolfsii*, *Rhizoctonia solani*), oomycetes (*Pythium ultimum*), and insects (cutworms, termites, cockroaches). It has not been commercialized, because it may cause opportunistic infections in humans (Konecka *et al.*, 2019).

15.6 Other bacteria

Several other species of bacteria have been cited on *Cannabis*. They rarely pose problems, and some may have been misidentifications. We mentioned them briefly. See McPartland *et al.* (2000) for full descriptions of the first four organisms.

15.6a. Bacterial wilt *Erwinia tracheiphila* (Smith 1895) Bergey *et al.* 1923

Gitman (1968a) cited this bacterium infecting hemp. Ghani and Basit (1975) described a bacterial wilt of Pakistani *Cannabis* with symptoms similar to those described by Gitman, but they did not identify the causal agent. *E. tracheiphila* is transmitted via striped cucumber beetles (*Acalymma vittata*) and spotted cucumber beetles (*Diabrotica undecimpunctata*). Both of these insect species have been reported on *Cannabis* (see section 7.4h), but they show a strong preference for Cucurbitaceae hosts.

15.6b. Wildfire *Pseudomonas syringae* pv. *tabaci* (Wolf & Foster 1917) Young *et al.* 1978

≡*Pseudomonas tabaci* (Wolf & Foster 1917) Stevens 1925
This pathogen normally infects solanaceous crops (tobacco, tomato, potato). Johnson (1937) reported that it could infect *Cannabis*, but plants required stressful water-soaking for successful infection.

15.6c. Angular leaf spot *Pseudomonas syringae* pv. *mellea* (Johnson 1923) Young *et al.* 1978

≡*Pseudomonas mella* Johnson 1923
This pathogen normally infects tobacco; Gitman (1968a) reported it on hemp.

15.6d. Bacterial leaf blight *Sphingobium yanoikuyae* (Yabuuchi *et al.* 1990) Takeuchi *et al.* 2001

≡*Sphingomonas yanoikuyae* Yabuuchi et al. 1990
Thiessen *et al.* (2020) isolated this species from glasshouse-grown plants in North Carolina. Lesions were small

(1–5 mm), medium-brown to dark-brown, angular, and limited by leaf veins. In culture, *S. yanoikuyae* isolates produced yellow, convex colonies on nutrient agar, which could be easily confused with *Xanthomonas* spp. Identification was confirmed with 16S rDNA sequencing. *S. yanoikuyae* rarely causes plant disease, but Thiessen and colleagues completed Koch's postulates on *Cannabis* seedlings. *S. yanoikuyae* has a wide distribution across freshwater, seawater, and terrestrial habitats.

15.6e. Crown gall *Agrobacterium tumefaciens* (Smith & Townsend 1907) Conn 1942

≡*Bacterium tumefaciens* Smith & Townsend 1907
A. tumefaciens is a saprophytic soil organism, and can invade plants via wounds near the crown (root–shoot interface). *A. tumefaciens* inserts a segment of its DNA into plants, the Ti (tumor-inducing) plasmid. This induces gall formation, a nutrient-rich environment for the bacteria. Galls are round, white to light-green or light-tan in color, with an irregular surface (Fig. 15.4A).

Young *et al.* (2001) conducted a phylogenetic study of *Agrobacterium* and *Rhizobium* spp. using 16S rDNA sequences. They amalgamated *Agrobacterium* and *Rhizobium* into a single genus, and synonymized *A. tumefaciens* under *R. radiobacter*. Most workers still use the taxon *A. tumefaciens*, but phylogenetic studies recognize it as a species complex (Gan and Savka, 2018).

A. tumefaciens infects many dicotyledonous plants, and is found worldwide. Lopatin (1936) and Gitman (1968a) described *A. tumefaciens* infecting *Cannabis* in Russia. Lopatin placed *Cannabis* in the "extremely susceptible" category. Wahby *et al.* (2013) inoculated the hypocotyls of 5-day-old

Cannabis seedings with two wild-type strains of *A. tumefaciens* from Spain. They tested fiber-type cultivars ('Delta-405', 'Delta-Llosa', 'Futura77'), and two drug-type landraces (from the Rif in Morocco). After 3 weeks, they measured gall weight and diameter.

Moroccan plants were more susceptible. They provided raw data, from which we made statistical comparisons of this small dataset (three fiber-types × two bacterial strains, versus two drug-types × two bacterial strains). Gall weight (mean mg ±SEM): fiber-type 44.62 ±6.57, drug-type 73.08 ±6.55, two-tailed $P = 0.019$, statistically significant. Gall diameter (mean mm ±SEM): fiber-type 4.53 ±0.214, drug-type 5.90 ±0.187, two-tailed t test, $P = 0.002$, statistically significant.

David Watson (pers. comm. 2019) observed *Cannabis* crown gall in Amsterdam, but did not isolate the pathogen. His photos were consistent with *A. tumefaciens* infection. Holmes *et al.* (2023) observed symptoms of crown gall on indoor-grown plants in British Columbia. Culturing diseased tissues on MacConkey medium (selective for gram-negative bacteria) did not isolate the pathogen, but the presence of *A. tumefaciens* was confirmed following PCR amplification with primers for the *iaaH* gene. Artificial inoculation using an *A. tumefaciens* strain from ATCC (*Rhizobium radiobacter* strain TT134) resulted in gall formation in five *Cannabis* genotypes.

A. tumefaciens has been controlled by a non-pathogenic relative named *A. radiobacter* K84. It produces an antibiotic, agrocin 84, to which *A. radiobacter* K84 is resistant. The risk of transferring this resistance gene to *A. tumefaciens* led to the creation of a GMO strain, *A. radiobacter* K1026, released in 2001.

15.6f. Hairy root disease *Agrobacterium rhizogenes* (Riker *et al.* 1930) Conn 1942

≡*Bacterium rhizogenes* Riker *et al.* 1930
≡*Rhizobium rhizogenes* (Riker *et al.* 1930) Young *et al.* 2001.
This species resembles *A. tumefaciens*: a soil organism that invades dicotyledonous hosts through roots, with worldwide distribution. *A. rhizogenes* inserts a segment of its DNA into plants, the Ri (root inducing) plasmid. This induces an overabundant growth of adventitious roots at the site of infection. Young *et al.* (2001) conducted a phylogenetic study of *Agrobacterium* and *Rhizobium* spp., and transferred *A. rhizogenes* to *Rhizobium*. Most workers still use the taxon *A. rhizogenes*.

Wahby *et al.* (2013) reported *A. rhizogenes* infection in *Cannabis* (Fig. 15.4B). They inoculated 30 seedlings via wounds in the hypocotyl, cotyledonary node, cotyledon, and leaf—with infection rates of 88.3%, 58.7%, 0%, and 0%, respectively. Then they inoculated hypocotyls of three fiber-type cultivars and two drug-type landraces, with two strains of *A. rhizogenes* from Morocco, and measured root number (mean number of roots per infected site) and root size (mean length of roots (mm) emerging per infected site). They provided raw data, from which we made statistical comparisons of this small dataset. Root number: fiber-type mean 7.00 ±0.362, drug-type mean 8.28 ±0.704, two-tailed t test, $P = 0.114$, not statistically different. Root size: fiber-type mean 9.20 ±1.14, drug-type mean 8.15 ±0.566, two-tailed t test, $P = 0.502$, not statistically different.

Fig. 15.4. *Cannabis* seedlings inoculated with *Agrobacterium* spp.: **A**. crown gall caused by *A. tumefaciens*; **B**. hairy root caused by *A. rhizogenes* (courtesy Wahby *et al.*, 2013)

15.6.1   Assorted xylem pathogens

McKernan *et al.* (2016b) analyzed medical cannabis samples for bacterial contaminants, purchased at several medical dispensatories in New England. They used NextGen sequencing of PCR amplicons generated with 16S rRNA primers. This was performed directly off medical cannabis samples, without culturing. PCR amplifies all 16S rRNA sequences, whether the organisms are living or dead. They identified 29 bacterial species, including three xylem pathogens never before cited on *Cannabis*: *Ralstonia solanacearum*, *Xylella fastidiosa*, and *Leifsonia xyli*. All three species have been divided into subspecies. Unfortunately, McKernan and colleagues did not deposit voucher 16S rRNA sequences for subspecies analysis. Koch's postulates were not performed.

15.6g. *Leifsonia xyli* (Davis *et al.* 1984) Evtushenko *et al.* 2000

This species has been divided into several subspecies. *L. xyli* subsp. *xyli* causes a major disease of sugarcane.

15.6h. *Ralstonia solanacearum* (Smith) Yabuuchi *et al.* 1996

R. solanacearum is considered a species complex. Collectively the complex attacks many plants over a broad geographic range. *R. solanacearum* causes economic damage to solanaceous crops. An unspeciated *Ralstonia* sp. reportedly causes wilt symptoms in *Cannabis* (Munir *et al.*, 2023).

15.6i. *Xylella fastidiosa* Wells *et al.* 1987

X. fastidiosa has been divided into several subspecies. These subspecies cause alfalfa dwarf diseases, Pierce's disease of grapes, and citrus variegated chlorosis (CVC). Gibin *et al.* (2023) listed 535 plants host—including *Humulus scandens* and *Celtis occidentalis*, but not *Cannabis*. Many plant hosts remain asymptomatic, making them reservoirs for infection.

The rest of this chapter concerns **nematodes**. For an introduction to Phylum Nematoda, see text around Fig. 3.5. Methods of soil extraction and morphological identification are presented in section 1.5.1; molecular methods appear in section 1.5.2.

Nematodes are ubiquitous. Nathan Cobb (1859–1932), who laid the foundations of nematology, put it this way: "If all the matter in the universe except the nematodes were swept away, our world would still be dimly recognisable ... we should find its mountains, hills, vales, rivers, lakes, and oceans represented by a film of nematodes" (Cobb, 1915). A renaissance man, Cobb made improvements in microscope equipment, traveled around the world, published studies on bacteria and fungi, and was an accomplished artist.

15.7   Root-knot nematodes

Root-knot nematodes (*Meloidogyne* spp.) collectively attack thousands of plant species around the world. They cause greatest damage in warm regions, where summers are long and winters are short and mild. In the USA, root-knot nematodes get nasty south of the Mason-Dixon line (39° latitude), especially in sandy soils along coastal plains and river deltas. Larvae are known as **juveniles**; only the second stage (J2) is migratory in soil; later juvenile stages are sedentary in roots.

Root-knot nematodes cause additional havoc by predisposing plants to other pathogens, such as soil fungi. This is known as a **disease complex**, the result of "a synergistic interaction between two organisms" (Bernard *et al.*, 2022). Fungal root rot may arise around galls, especially late in the season.

15.7.1   Symptoms and signs

Above-ground symptoms are non-specific: stunting, chlorosis, and midday wilting with nightly recovery. Symptoms do not arise in plants uniformly across fields. Rather, they appear in patches of scattered infestations. Plants near the center of these patches have the worst symptoms, which blend to healthy plants at the periphery.

Below-ground symptoms and signs are more distinctive. After penetrating roots, *Meloidogyne* juveniles inject saliva into 5–7 cells located near xylem. Saliva proteins reorganize the cells into a placenta-like nutritional structure: 5–7 enlarged, multinucleated **giant cells**. Hypertrophy of tissues surrounding the giant cells and the nematode produce a **root gall**—a hard, white swelling in lateral roots, spherical to fusiform shaped (Fig. 15.5A). Egg sacs protrude from the gall, which can be stained pink (Fig. 15.5B), using a technique described in the epidemiology section below.

Root galls may coalesce to form larger **root knots**—conspicuous, hypertrophied roots with swollen surfaces. This rarely happens with *Cannabis*, even in susceptible varieties that are heavily galled (Bernard and Chaffin, 2020). Root knots may stimulate the formation of adventitious rootlets, creating roots with a bushy appearance. The excessive branching of rootlets may lead to false impressions, since diseased roots look more developed than healthy roots.

15.7.2   Causal organisms

Root-knot nematodes were unknown prior to Berkeley (1855), who classified them as *Vibrio* spp. This was followed by an era of collective names, such as *Anguillula radicicola* Greeff 1872, *Heterodera radicicola* Müller 1884, *Anguillula marioni* Cornu 1879, and *Heterodera marioni* (Cornu) Goodey 1932. The true identity of species tagged with these collective names cannot be known with certainty.

The first nematode identified on *Cannabis* was a collective name. Passon (1909) described *H. radicicola* on roots of *canhamo da India* (Indian hemp), along with an unidentified

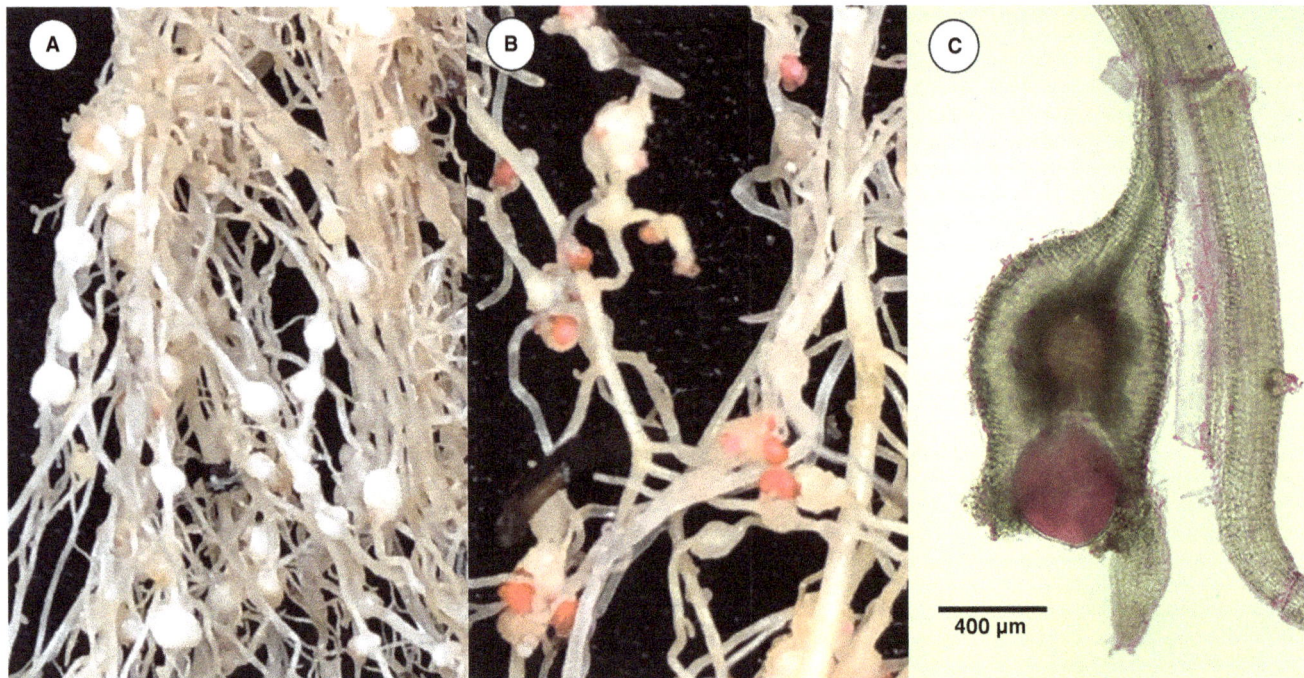

Fig. 15.5. *Meloidogyne incognita* infesting *Cannabis*: **A**. galled root system; **B**. protruding egg sacs stained pink; **C**. nematode stained pink, with giant cells in the center of gall (courtesy Ernest C. Bernard)

fungus, near São Paulo in Brazil. Passon made the first observation of a nematode–fungus disease complex, because he mentioned both. Passon's discovery of *H. radicicola* was repeated by Richter (1911). The collective name *H. marioni* was cited on *Cannabis,* but without a location record, by authors in the USA (Buhrer *et al.,* 1933) and co-authors from Holland and Russia (Filipjev and Schuurmans-Stekhoven, 1941). *H. marioni* was also cited in Japan (Watanabe, 1947). A root-knot nematode isolated from hemp in Tennessee was assigned two collective names, first *H. radicicola* (Whittle and Drain, 1935) and then *H. marioni* (Weiss, 1948).

Chitwood (1949) resurrected the genus *Meloidogyne* and stabilized the nomenclature in a major revision. He transferred the collective names to *Meloidogyne* and divided them into five *Meloidogyne* spp. Separating Chitwood's species by morphology can be difficult. Characters include patterns on female perineal shields (wrinkles or striae in the cuticle around the vulva and anus), the shape of male head caps, and stylet shape (Fig. 15.6). Chitwood (1949) established the use of perineal "fingerprint"patterns as a useful diagnostic aid.

15.7a. Southern root-knot nematode *Meloidogyne incognita* (Kofoid & White 1919) Chitwood 1949

Description: Morphological characters that differentiate *M. incognita* from other *Meloidogyne* spp. are presented in Fig. 15.6. Adult females are pear-shaped, 750 × 1300 µm; adult males are vermiform-shaped, 35 × 1300 µm. Females lay eggs in a jelly-like sac (gelatinous matrix) extruded from rectal glands at the posterior end and attached at the vulva; eggs are

laid into this matrix, which provides some protection from parasites. On roots the matrix is yellow-brown in color (Fig. 15.5). Eggs are ellipsoidal, colorless, 40 × 80 µm. Juveniles undergo their first molt while still in eggs, and emerge from eggs in the J2 stage. They are long and slender, 400 × 15 µm. After penetration of roots, J2s begin to swell (the "sausage" stage), undergo two more molts, and develop a sexual dimorphism (Fig. 15.7).

Comments: Most citations of "*Meloidogyne* sp." in the *Cannabis* literature likely refer to *M. incognita,* including Tennessee (Miller *et al.,* 1960), India (Johnston, 1964), and the former USSR (Goodey *et al.,* 1965). Correctly speciated *M. incognita* was first reported on *Cannabis* in Armenia (Pogosyan, 1960) and India (Nirula and Kumar, 1964). *Heterodera marioni* reported by Watanabe (1947) was later identified as *M. incognita* by Tsuge and Sato (2020).

M. incognita consists of physiological races, each with different host ranges. Six races are generally recognized (Subbotin *et al.,* 2021), but Sano and Iwahori (2005) recognized nine in Japan. When Whittle and Drain (1935) named hemp as a host of *Heterodera radicicola,* they assessed the nematode's ability to infest other plants (cotton was a good host, but not tobacco or peanut). This enabled Bernard *et al.* (2022) to identify it as *M. incognita* race 3.

Susceptibility to *M. incognita* races 2 and 4 was screened by Van Biljon (2017). All ten fiber-type cultivars were susceptible to race 4, whereas three cultivars were resistant to race 2 ('Diana', 'Kubanskaia Rannaja', 'Futura 75'). 'Futura 75', however, was susceptible to race 3 (Bernard *et al.,* 2022). Race 3 was identified by Lawaju *et al.* (2021) infesting CBD strains "Boax" and "Otto2".

Example of perineal pattern in *M. incognita*	Species	Dorsal arch of perineal pattern	Striae of perineal pattern	Stylet cone of females	Head cap of males
	M. incognita	squarish, high	coarse, smooth to wavy to zigzaggy	large, anterior half cylindrical, curved	flat to concave
	M. javanica	rounded, interrupted by lateral ridges	coarse, smooth to slightly wavy	anterior half tapered, less curved	convex, rounded, broad
	M. hapla	rounded, low	fine, smooth to slightly wavy	small, narrow and delicate, slightly curved	convex, rounded, narrow
DA, dorsal arch; A, anus; V, vulva	*M. enterolobii*	rounded, high	coarse and smooth	straight, narrow, sharply pointed	convex, rounded

Fig. 15.6. Some differential characteristics of *Meloidogyne* species (illustration of perineal pattern courtesy of Stirling *et al.*, 2002)

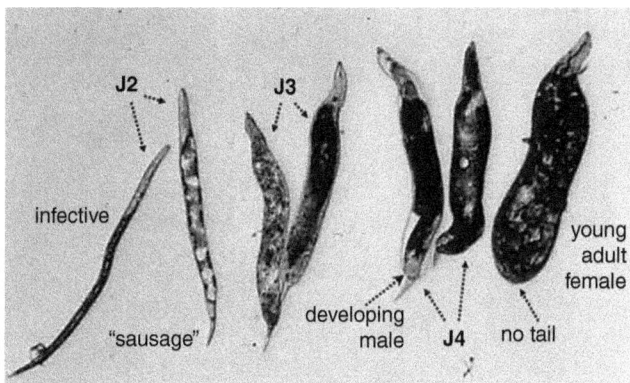

Fig. 15.7. Developmental stages of *Meloidogyne incognita* (courtesy EC Bernard and RS Hussey, University of Georgia)

Bernard and Chaffin (2020) and Hansen *et al.* (2020) provided excellent photographs of *M. incognita* symptoms on plants in Tennessee. Thiessen *et al.* (2020) isolated *M. incognita* on roots of stunted plants in North Carolina; identity was confirmed with primers for a SCAR (sequence characterized amplified region) specific for *M. incognita*. Lawaju *et al.* (2021) also used this approach to confirm the identity of *M. incognita*.

15.7b. Northern root-knot nematode *Meloidogyne hapla* Chitwood 1949

Description: This species resembles *M. incognita* in size and shape. *M. hapla* can be differentiated by perineal patterns, by the shape of male head caps, and by stylet shape (Fig. 15.6).
Comments: Inoculation tests proved *Cannabis* susceptibility to *M. hapla* (Norton, 1966; de Meijer, 1993; Hansen *et al.*, 2020). On a 0–4 scale, Norton reported a moderate gall rating of 1.9 on hemp. *M. hapla* generally causes less damage than *M. incognita*. Root galls formed by *M. hapla* remain small and

spherical; compound galls rarely develop. *M. hapla* causes serious damage in tomatoes and potatoes.

The eggs of *M. hapla* tolerate cold better than those of *M. incognita*, so the geographic range of *M. hapla* extends into temperate regions. In the USA, *M. hapla* is found in every state except Alaska (in Hawai'i it only occurs at higher elevations). It is also found in Canada, northern Europe, and southern Australia.

M. hapla can be confused with *Meloidogyne chitwoodi* Golden, a species distributed in the western USA. *Cannabis* is resistant to *M. chitwoodi*. Kok *et al.* (1994) quantified the effects of *Cannabis* grown in field soil heavily infested with *M. chitwoodi*. The initial soil population (1500 juveniles/100 ml soil) decreased to only 10 juveniles/100 ml soil after a crop of 'Kompolti'. In comparison, the soil population zoomed to 12,000 juveniles following a susceptible crop, ryegrass (*Lolium multiflorum*). A field left fallow (keeping the soil free of plants) resulted in 3 juveniles/100 ml soil. Korthals *et al.* (2000) also reported a reduction in *M. chitwoodi* soil populations after a season of 'Kompolti' cultivation—a reduction equivalent to a season of fallow soil.

15.7c. Javanese root-knot nematode *Meloidogyne javanica* (Treub 1885) Chitwood 1949

Description: *Meloidogyne* spp. are differentiated by patterns on female perineal shields, by the shape of male head caps, and by stylet shape (Fig. 15.6).
Comments: *M. javanica* has no cold tolerance; it lives in tropical and semi-tropical regions. In the USA the species has been reported in the Carolinas, the Gulf states, and southern California. *M. javanica* has infested *Cannabis* in Armenia (Pogosyan, 1960), India (Nirula and Kumar, 1964), the former USSR (Decker, 1972), and Iran (Esfahani and Ahmadi, 2010). South African researchers report various degrees of susceptibility to *M. javanica* in fiber-type cultivars (Van Biljon, 2007; Pofu *et al.*, 2010; Pofu and Mashela, 2014; Van Biljon, 2017).

In Yúnnán Province, Song *et al.* (2017) collected *M. javanica* from plants showing symptoms of slow growth, leaf chlorosis, leaf abscission, and root galls. Morphological identification was confirmed by sequencing the XSBN-1 SCAR marker (part of the ITS gene region). Koch's postulates were fulfilled using the 'Yún má 1' cultivar. Coburn and Desaeger (2020) inoculated six fiber-type cultivars with *M. javanica* ('Helena', 'Tygra', 'Fibranova', 'Eletta Campana', 'Carmagnola', 'Carmagnola Selezionata'). All were susceptible hosts, but damage was minimal—root galls were small and did not coalesce, and plant biomass was not affected. *M. javanica* consists of five races (Subbotin *et al.*, 2021). The race infesting *Cannabis* has not been identified.

15.7d. Guava root-knot nematode *Meloidogyne enterolobii* Yang & Eisenback 1983

Description: *Meloidogyne* spp. are differentiated by patterns on female perineal shields, by the shape of male head caps, and by stylet shape (Fig. 15.6).

Comments: *M. enterolobii* is a tropical or subtropical species found in southern China, central and southern Africa, Brazil, Venezuela, Puerto Rico, and southeastern USA. Its hosts include guava, eggplant, bell pepper, tomato, watermelon, and tobacco. Ren *et al.* (2021) described *M. enterolobii* infesting hemp in southern China. Morphological identification was confirmed by DNA—its ITS sequence was identical to other *M. enterolobii* sequences in GenBank. Pathogenicity testing determined a nematode reproduction factor (final population density/initial population density) of 18.2. Because of this high reproduction rate and its strong pathogenicity, the authors considered *M. enterolobii* a "major threat" to hemp production.

15.7.3   Differential diagnosis

The above-ground symptoms caused by *Meloidogyne* spp. are non-specific, such as wilting and stunting. They might be misinterpreted as symptoms of nutrition deficiencies or drought, or perhaps root damage from fungi or insects. Differentiating *Meloidogyne* spp. by morphology isn't easy. Photographs and line drawings of the characters can be found at http://nemaplex.ucdavis.edu.

Meloidogyne identification entered the molecular era when Guirao *et al.* (1995) used PCR amplification and RAPD markers to differentiate *M. incognita*, *M. hapla*, *M. javanica*, and *M. arenaria*. Nematode barcodes include rDNA (ITS gene region) and mDNA (COI and COII). Species-specific field kits are available that utilize real-type PCR (qPCR). Multiplex qPCR kits are available with primers for several *Meloidogyne* spp.

15.7.4   Disease cycle and epidemiology

M. incognita and *M. hapla* are slimmed-down land sharks. Their genomes are extremely compact, 86 Mb and 54 Mb,

respectively (Abad *et al.*, 2008; Opperman *et al.*, 2008). The latter is the smallest known genome for an invertebrate animal, 118-fold smaller than the human genome (6370 Mb, females). *M. incognita* and *M. hapla* shed metabolic pathways for which they can rely upon their hosts to provide. Conversely, they picked up genes involved in plant parasitism (pectate lyases and cell wall-degrading enzymes), which are not present in free-living nematodes, such as *Caenorhabditis elegans*. These genes were likely acquired via horizontal gene transfers from bacterial sources.

Meloidogyne spp. overwinter as eggs. The first molt occurs in eggs, so eggs hatch into second-stage juveniles (J2), which penetrate roots of susceptible plants. Once in roots, juveniles undergo three more molts and develop a sexual dimorphism. Males become migratory, whereas females stay sedentary and remain in roots. Most *Meloidogyne* spp. are obligately parthenogenetic—females produce eggs without mating. *M. hapla* is facultatively parthenogenetic. Females usually produce 300–500 eggs (rarely over 2000) in egg sacs. Females die and egg sacs decay, releasing eggs into the soil. Juveniles from the oldest eggs hatch and escape into the soil while new eggs are still being laid. Under optimal temperatures (25–30°C) this life cycle may turn in 20 days.

Root wounds by nematodes provide portholes for many root-pathogenic fungi, including *Rhizoctonia solani*, *Pythium ultimum*, *Fusarium oxysporum*, and *Verticillium* spp. Studies with tobacco show that weakly pathogenic fungi (*Curvularia*, *Botrytis*, *Penicillium*, *Aspergillus* spp.) also invade roots via nematode wounds (Lucas, 1975).

To estimate infestation levels, two methods can be used—root examination and soil assay. Both are done at the end of the season. **Root examination** estimates the percentage of the root system galled by nematodes. First, root tissue is cleared and stained: wash soil from a root sample, and place in a beaker containing 100 mL tap water with 40 mL household bleach (5.25% NaOCl). Soak for 4 min with occasional agitation, rinse in running water, and soak another 15 min in tap water to remove residual bleach. Drain and transfer to a beaker containing 400 mL water with 0.6 g phloxine B stain powder. Soak for 5–15 min until the egg matrix appears pink.

The "Gall Index 2" system by Hussey and Janssen (2002) is often used, where 0 = no galling; 1 = trace infection with a few small galls; 2 = ≤25% roots galled; 3 = 26–50%; 4 = 51–75%; and 5 = >75% roots galled.

In the **soil assay**, collect random samples across a field, 15–20 cm deep, mix the samples into a composite sample, and ship one pint of soil to an agricultural extension agent (if they still provide diagnostic services) or a diagnostic private lab. Samples must remain cool (<27°C) and moist.

No damage threshold for *Meloidogyne* spp. on *Cannabis* has been calculated (i.e., the density of eggs and juveniles in soil at which yield loss begins). Pofu *et al.* (2010) mentioned the Seinhorst model in their study of *M. javanica*, but did not determine a damage threshold. We propose the following index: 0–20 nematodes per pint of soil = very low infestation; 21–100/pint = low infestation; 101–300/pint = moderate infestation; >300/pint = high infestation.

15.7.5 Cultural and mechanical control

Growing vigorous plants will curb nematode damage. Healthy plants can regenerate new roots. Prevent water stress. Supplement soil with potassium. Add compost and manure to soil, to encourage the growth of organisms antagonistic to nematodes. Avoid planting in light, sandy soils favored by *Meloidogyne* spp.

Quarantine prevents the movement of nematodes into new areas. Juveniles left on their own only migrate 20 cm in a lifetime. But eggs can travel for miles on tractor tires, floating in streams, even blowin' in the wind. Quarantine must be rigorously enforced to be effective—exclude plants, soil, even dirty shoes and spades from clean areas.

For treating nematode-sick soil, extreme measures have been taken, such as deep plowing, flooding fields, soil solarization, or strict fallow (keeping the soil free of plants for a season). Glasshouse soils have been steam-sterilized.

Crop rotation is attractive, but options are limited for *M. incognita*, because of the existence of physiological races, each with different host ranges. Marigolds, *Tagetes* spp., are resistant to *Meloidogyne* spp., particularly African marigolds (*T. erecta*), French marigolds (*T. patula*), and South American marigolds (*T. minuta*). But they have limited value as alternative crops. Cereal crops—oat, wheat, and sorghum—are a good option, but their resistance to *M. incognita* varies among cultivars (Lima de Brida *et al.*, 2017).

Resistant *Cannabis* varieties should be used. Bernard *et al.* (2022) reviewed the literature, which revealed variability in *Cannabis* resistance to nematodes, due to the *Meloidogyne* species (or race) in question, as well as specific *Cannabis* cultivars or strains.

The first large-scale screening for resistance was undertaken by de Meijer (1993), who tested resistance to *M. hapla* in 148 worldwide *Cannabis* accessions. He conducted a glasshouse screening: seeds were planted in Conetainers (upright PVC tubes) with sandy soil, and 2-week-old seedlings were inoculated with J2s. Seven weeks later, plants were harvested, roots washed free of sand, and numbers of root galls and egg masses were counted.

Relatively resistant accessions came from Nepal, Spain, Hungary, and a hybrid drug strain from Holland. The most susceptible cultivars came from France (e.g., 'Futura 77') and Ukraine (e.g., 'USO-13'). De Meijer then conducted a field trial, using six accessions that spanned the range of resistance—susceptibility. Seeds were sown in naturally infested sandy soil (100 *M. hapla* J2/100 cm³ soil). At the end of the season, soil cores were taken, nematodes extracted with an elutriator and collected with a set of sieves, and juveniles counted per 100 cm³ soil. The relative resistance seen in Conetainers was not as evident in the field trial.

South Africans have conducted several resistance trials: Van Biljon (2007) evaluated 'Kompolti', 'Uniko-B', 'Novoadska', 'Fedora 17', 'Futura 75', and 'VIR-140'. Compared with controls, plants inoculated with *M. javanica* yielded less above-grown mass, except for relatively resistant 'VIR-140', a Russian landrace from Perm. Pofu *et al.* (2010) tested the susceptibility of four cultivars to *M. javanica*: 'Kompolti', 'Futura 75', 'Felina 34', and 'Ferimon' did not suffer significant yield losses, but they were susceptible to infection, and populations of *M. javanica* increased in glasshouse soil (particularly after 84 days of 'Kompolti' and 'Ferimon').

Pofu and Mashela (2014) repeated their study in a full-season time frame. They reported variable *rates* of infection: a peak at 56 days with 'Kompolti', 'Ferimon', and 'Futura 75', whereas 'Felina 34' required 151 days. Van Biljon (2017) screened ten fiber-type cultivars for resistance to *M. incognita* races 2 and 4 in glasshouse experiments. Most cultivars varied between tolerant to susceptible. Three cultivars ('Futura 75', 'Diana', 'Kubanskaia Rannaja') were resistant to *M. incognita* race 2. Kotcon *et al.* (2018) tested *M. hapla* and *M. incognita* against 'Canda', 'Delores', 'Futura', 'Fedora', and 'Felina 32'. *M. hapla* caused more root galls on 'Felina 32' than the others, but reproduction did not differ. *M. incognita* caused greater galling and reproduction on 'Canda' than the others.

Reproduction factor (Rf) is a more quantitative measure, calculated as Rf = Pf/Pi, where Pf is the final number of eggs and J2s, and Pi is the initial inoculum. Resistant or non-host plants show Rf <1, whereas susceptible plants show Rf >1. Bernard *et al.* (2022) noted that Rf has rarely been calculated in *Cannabis* literature, and methodological differences among different studies (time length of study, type of potting soil, size of pot) make comparisons between studies problematic.

Coburn and Desaeger (2020) calculated Rf in six fiber-type cultivars. They inoculated potted seedlings with 10,000 eggs of mixed *M. javanica* and *M. arenaria*, and measured Pf after 60 days. Rf values were high for all cultivars, ranging from 33 in 'Helena' to 52 in 'Tygra'. A positive control, cucumber, had Rf of 46. Plant growth (height and biomass) was not significantly reduced, but root dry weight was reduced by 44–52% in 'Helena', 'Tygra', and 'Eletta Campana', less so in 'Carmagnola', 'Carmagnola Selezionata', and 'Fibranova'.

Nematologists in Tennessee evaluated susceptibility to *M. incognita* and *M. hapla* in 15 fiber-type and CBD-type varieties (Bernard and Chaffin, 2020; Hansen *et al.*, 2020). Potted plants were inoculated with 5000 nematode eggs. After 8 weeks roots were harvested and rated for galling, and eggs extracted from roots and counted for Rf calculations. Susceptibility did not correlate with cannabinoid levels. Fiber-type cultivars ranged from 'Futura 75' (Rf = 20.4) to 'Fedora 17' (Rf = 10.4). CBD-type strains ranged from "Charlotte's Web" (Rf = 39.6) to "Wife" (Rf = 0.2). The authors noted that field studies are needed.

15.7.6 Biological and chemical control

Plant growth-promoting rhizobacteria (PGPRs) enhance nutritional health and induce systemic resistance, thereby protecting plants from root-knot nematodes (Xiang *et al.*, 2017). PGPRs also create a living barrier around roots to prevent infection, and exude nematicidal chemicals. Commercially available PGPRs include *Bacillus amyloliquefaciens*, *B. subtilis*, *B. firmus*, *Pseudomonas fluorescens*, and *P. chlororaphis* (see section 20.6).

Albifimbria verrucaria is a fungal equivalent of PGPRs; it enhances plant health, induces resistance, and acts synergistically with PGPRs to suppress nematodes (Fernández *et al.*, 2001). *A. verrucaria* produces a slew of compounds that inhibit the hatching of nematode eggs, inhibit feeding through muscle paralysis, and kill nematodes outright. Products are available consisting of heat-killed *A. verrucaria* broth brimming with nematicidal toxins (more in section 20.6).

Pochonia chlamydosporia, formerly *Verticillium chlamydosporium*, parasitizes the egg masses of sedentary nematodes (*Meloidogyne*, *Heterodera*, and *Globodera* spp.). The fungus is supplied as conidia (with mycelial fragments) or as chlamydospores (resting spores). Conidia have low survival rates in soil; chlamydospores cost more to produce, but survive longer in the absence of nematodes (Manzanilla-López *et al.*, 2013). Products can be stored for up to 3 months in cool temperatures, at low humidity (<50 RH), and away from sunlight. Hu *et al.* (2023) inoculated hemp seedlings with a field-collected isolate of *P. chlamydosporia*; the fungus readily colonized roots, and inoculated plants grew significantly larger compared with controls.

Two commercial varieties are available. *P. chlamydosporia* var. *catenulata* (KlamiC®) does best in soil temperatures of 15–32°C and pH 5.5–8.0. Drench soil at a rate of 30 g/m², or pour 1 g/plant into seedling transplantation holes. *P. chlamydosporia* var. *chlamydosporia* (PcMR-1 by Clamitec, IPP-21 by CNR in Italy) has been applied as a soil drench or in drip irrigation systems. Farming systems have a marked effect on *P. chlamydosporia*. The fungus performs better in humus-rich soils than in mineral soils. Moisture must be present. Prior to inoculation, partial soil sterilization (e.g., solarization) reduces competition from antagonistic soil microflora, enabling *P. chlamydosporia* to establish more readily. *P. chlamydosporia is compatible with* other beneficial fungi (mycorrhizae, *Trichoderma* spp.).

Purpureocillium lilacinum, formerly *Paecilomyces lilacinus*, parasitizes the egg masses of sedentary nematodes (*Meloidogyne*, *Heterodera*, and *Globodera* spp.). The fungus penetrates pregnant nematodes through their anus or vulva and destroys the eggs (Jatala *et al.*, 1979). Some isolates also infect insects, giving rise to "zombie bugs" whose behavioral changes improve the dispersal of *P. lilacinum* spores (Eberhard *et al.*, 2014).

Several products are available (MeloCon®, PAECILO®, BioACT®, Nemachek®, Pl Plus®). Supplied as spores and mycelial fragments, 10^7–10^{10} viable conidia/g, as a wettable powder or granule, or talc-based product. *P. lilacinum* can be applied as a soil drench, mixed into compost, or as a seed treatment. Read the label for dosages. Optimal soil conditions are rather warm and humid, 20–30°C and RH ≥65. Good farming practices improve its efficacy (see *Pochonia chlamydosporia* above). A combination of *P. lilacinus*, *Hirsutella rhossiliensis* (see below), and *Glomus mosseae* (a mycorrhizal fungus) reduced *Meloidogyne javanica* populations by 94.7%, compared with 93.5% for carbofuran—one of the most toxic carbamate pesticides (Abo-Korah, 2017).

Pasteuria penetrans, formerly *Bacillus penetrans*, is a bacterial parasite of nematodes (*Meloidogyne* spp.). Bacterial spores attach to the nematode's body, germinate, and form a germ tube that penetrates the body cuticle. Ovarian tissue in females is destroyed, and bacterial endospores replace the eggs, preventing the nematode from reproducing. *P. penetrans* is biotrophic (an obligate pathogen), thus cannot be grown on artificial media. Kokalis-Burelle (2015) successfully controlled

Meloidogyne incognita with *in vitro* produced *P. penetrans*, a project funded by Pasteuria Bioscience, Inc. But that company was acquired by Syngenta, who turned its attention to *Pasteuria nishizawae*.

Pasteuria nishizawae has the same life cycle as *P. penetrans*—it destroys eggs in females of *Heterodera* and *Globodera* spp. The species is relatively new to science (Sayre *et al.*, 1991), described 50 years after *P. penetrans*. It is also an obligate parasite, and difficult to manufacture. Syngenta devised an *in vitro* method, and launched Claiva® in 2013. It is formulated as a seed treatment.

Carnivorous fungi capture mobile nematodes and eat them; see the online review by Jiang *et al.* (2017). *Drechslerella* and *Arthrobotrys* spp. form constricting rings. When a nematode contacts the inner surface of ring cells, they rapidly triple their volume (within 0.1 second), tightening the noose and strangling the nematode (Fig. 15.8). The fungus penetrates the nematode through enzyme degradation and mechanical pressure, then kills the animal and absorbs its nutrients. *Drechslerella* spp. have not been commercialized; their numbers can be augmented by adding organic matter to soil (Gray, 1988).

Orbilia oligospora (*Arthrobotrys oligospora* until 2020) produces sticky reticulate traps (Fig. 15.9). *O. oligospora* was the first fungus documented to actively capture nematodes (Zopf, 1888). It secretes a substance that paralyzes the nematode, then secretes proteases and collagenases that enable it to consume the animal. *O. oligospora* is chemotropically attracted to plant roots, such as tomato and barley, where it exists as an endophyte (Bordallo *et al.*, 2002). This gives *O. oligospora* greater contact with plant-parasitic nematodes. Two commercial products are available, Royal 300® and Royal 350® (Jiang *et al.*, 2017).

Many *Orbilia* spp. produce nematode-trapping *Arthrobotrys* anamorphs as part of their lifecycle (Baral *et al.*, 2020). This is interesting, because *Orbilia luteola* was isolated from field-retted *Cannabis* stalks (McPartland and Cubeta, 1997). No anamorph was associated with *O. luteola* (see Fig. 18.3), and the species was not included in the review by Baral *et al.* (2020). Jiang *et al.* (2017) hypothesized that the *Arthrobotrys* stage

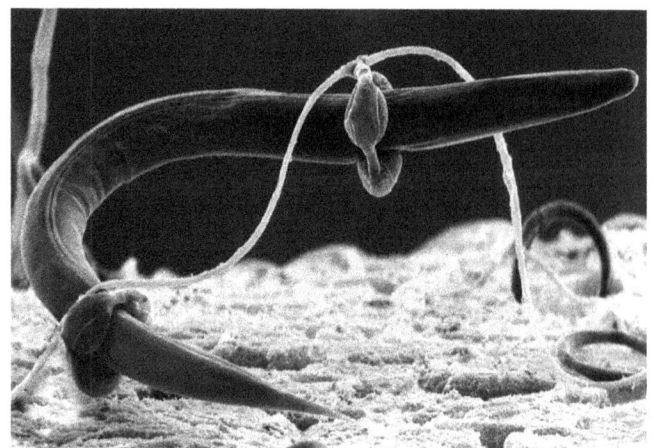

Fig. 15.8. *Drechslerella anchonia* trapping a nematode (SEM image courtesy of Nancy Allin and George L. Barron)

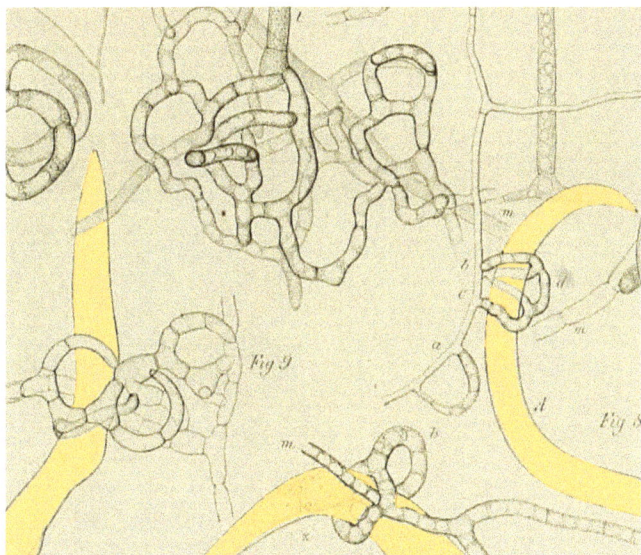

Fig. 15.9. Adhesive coils of *Orbilia oligospora* trapping three nematodes (Zopf, 1888)

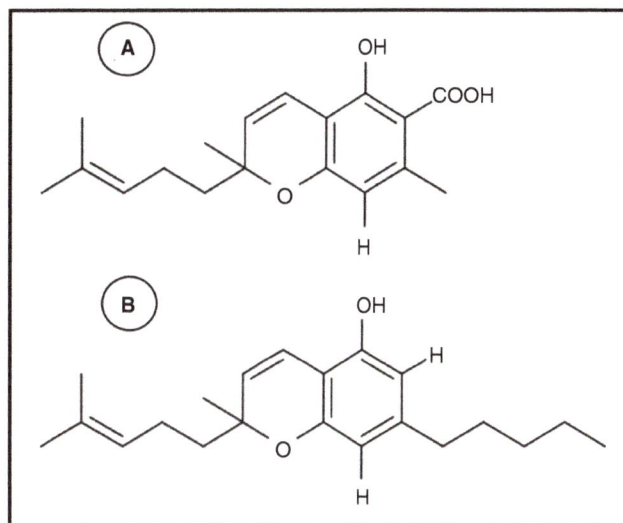

Fig. 15.10. Chemical structures: **A**. cannabiorcichromenic acid; **B**. cannabichromene

provides a supply of nitrogen (from nematode cadavers) for the *Orbilia* stage, which grows in nitrogen-deficient stalks.

Regarding nematicides, many researchers have ironically turned to *Cannabis* (see section 3.3.6). Dried leaves and flowers have been mixed into soil, as well as aqueous or ethanol extracts. Efficacy in petri dishes often disappeared in field studies. In general, the efficacy of *Cannabis* was middle-of-the-road in comparisons with other plants (e.g., marigold, *Tagetes erecta*; neem, *Azadirachta indica*; datura, *Datura metel*; tobacco, *Nicotiana tabacum*).

Cannabiorcichromenic acid (CCA) is a nematicidal compound produced by *Thelonectria olida* (formerly *Cylindrocarpon olidum*), a fungus that parasitizes eggs and juveniles of *M. incognita* and *H. schachtii* (Coosemans, 1991; Quaghebeur *et al.*, 1994). Some varieties of *T. olida* have been tested as biological controls (Yolageldi and Turhan, 2005), whereas other *T. olida* varieties are plant pathogens (Abreo *et al.*, 2010).

CCA was the first phytocannabinoid of non-plant origin to be discovered. It is an analog of cannabichromene (CBC) produced by *Cannabis* (Fig. 15.10). CBC is a minor constituent in the plant, usually <0.3% dry weight flowering tops, although immature plants have higher concentrations, and de Meijer *et al.* (2009) found an Afghan landrace that produced 2.2%. CBC is antibacterial and antifungal (Turner and ElSohly, 1981), its nematicidal potential has not been evaluated.

Nematodes are tough *hombres* and not easily killed. Shatilovich *et al.* (2023) revived 46,000-year-old nematodes from Siberian permafrost, crushing the longevity record for any plant or animal. Broad-spectrum chemical nematicides were the mainstay for reducing nematode populations in the second half of the 20th century. Most of them were fumigants and were applied under plastic sheets covering the soil. Environmental and human health concerns phased them out—dibromochloropropane in 1979, methyl bromide in 2017, and aldicarb anytime soon.

Some nitrogenous soil amendments are nematicidal, such as urea and calcium cyanamide. Urea is mixed into soil after harvest, and calcium cyanamide is mixed into soil 2 weeks before sowing. Calcium cyanamide killed soil nematodes nearly as well as über-nasty oxamyl and dazomet (Hussain *et al.*, 2017).

ClandoSan® combines urea with chitin from ground-up crab shells. Crab shell chitin encourages the growth of nematicidal soil organisms when mixed into soil at a rate of 2–4 kg/10 m², or about 1% by weight. Chitosan is a chemically deacetylated form of chitin that is soluble in water. The EPA added it to their list of Section 25(b) exempt pesticides in 2022, and EPA (2023) added a product with chitosan (Nemasan®) to its list of pesticides allowed for use on *Cannabis*.

A garlic-derived polysulfide product is approved for use in the EU (Anwar *et al.*, 2017). Britt *et al.* (2020) listed three biofriendly products for controlling nematodes in *Cannabis* crops: heat-killed *Burkholderia* spp. (Majestene®), *Chromobacterium subtsugae* strain PRAA4-1T (Grandevo® CG), and azadirachtin (Neemix®).

15.8 Cyst nematodes

Cyst nematodes, *Heterodera* and *Globodera* spp., are found worldwide but cause greatest damage in temperate regions. Their endoparasitic lifestyle is similar to that of root-knot nematodes (*Meloidogyne* spp.). *Heterodera* cysts are lemon-shaped with a prominent vulval cone; *Globodera* cysts are round without a vulval cone.

Heterodera cysts can be identified to species by fenestra (cuticle patterns) on their vulval cone—analogous to the identification of *Meloidogyne* spp. by their perineal patterns (Fig. 15.6). For example, *Heterodera schachtii* has an ambifenestrate vulval cone and *H. humuli* has a bifenestrate vulval cone. *H. schachtii* has a wide host range, and its pathogenicity on *Cannabis* seems certain, although *Cannabis* is not a preferred host. *H. humuli* has a narrow host range—primarily hops—and the host status of *Cannabis* is "contradictory" (Bernard *et al.*, 2022)

Above-ground symptoms caused by cyst nematodes are non-specific, like symptoms caused by root-knot nematodes. Below-ground symptoms consist of distorted roots—knobbed and bushy. Saliva injected into root cells by *Heterodera* spp. induce the formation of a **syncytium**—one large, polynucleated cell, formed through fusion with adjacent cells. The nematode enlarges into a white, lemon-shaped cyst. It breaks through the root surface, dies, and turns into a brown cyst (see Fig. 15.11). Roots damaged by cyst nematodes become predisposed to other soil pathogens, notably *Fusarium* and *Rhizoctonia* fungi.

15.8.1   Causal organisms

15.8a. Sugar beet cyst nematode *Heterodera schachtii* Schmidt 1871

Description: Cysts are lemon-shaped with a prominent vulval cone, colored white-yellow-brown (depending on their age), and less than a millimeter in length (average 500–800 μm). Fully developed eggs average 106 × 44 μm. Hatching J2 juveniles are 400–600 μm long. Once feeding begins, juveniles develop a sexual dimorphism. Males reach 1600 μm in length; females enlarge to the size and shape of cysts.
Comments: *H. schachtii* attacks hosts in the Chenopodiaceae (e.g., sugar beet) and Cruciferae (e.g., cabbage). It lives in temperate zones; in North America it is found from southern California and Florida north to Ontario and Alberta. The life cycle takes less than a month at an optimal temperature of 25°C.

Hollrung (1890) found cysts of *H. schachtii* on *Cannabis* roots in Germany. Kühn (1891) tested *Cannabis* as a trap crop to rid *H. schachtii* from soil where sugar beets were cultivated. After two seasons of hemp cultivation, yields of sugar beets increased. Hemp plants tolerated *H. schachtii* and grew well; Kühn considered hemp an economically viable rotational crop.

Vanha and Stoklasa (1896) listed plant species that are damaged by *H. schachtii,* and plant species that are resistant. *Cannabis* appeared in both lists. That paradox was noted by Marcinowski (1909) and Shaw (1915). Somma (1923) reported *H. schachtii* infesting hemp in Italy, and cited Neppi (1899), which we haven't seen. Filipjev and Schuurmans-Stekhoven (1941) suggested that reports of *H. schachtii* on hemp "must be considered as a separate species." Goodey *et al.* (1965) suggested that Hollrung's original record of *H. schachtii* on hemp might have been *H. humuli.*

15.8b. *Heterodera humuli* Filipjev 1934

Description: *H. humuli* cysts, like those of *H. schachtii*, are lemon-shaped and yellow to brown colored, but smaller (300–600 μm long). *H. humuli* J2 juveniles are smaller than those of *H. schachtii* (males reaching 1 mm long, females averaging 410 μm). Eggs are oval, 89–93 × 37–42 μm. Only one or two generations arises per year.
Comments: *H. humuli* normally infests hop and nettle. Filipjev and Schuurmans-Stekhoven (1941) stated that *H. humuli* "probably" occurred on hemp. In some host range studies, *H. humuli* reproduced on hemp (Franklin, 1951; Winslow, 1954; Sen and Jensen, 1969). In the latter study, hemp plants lost 63% weight compared with controls; hops

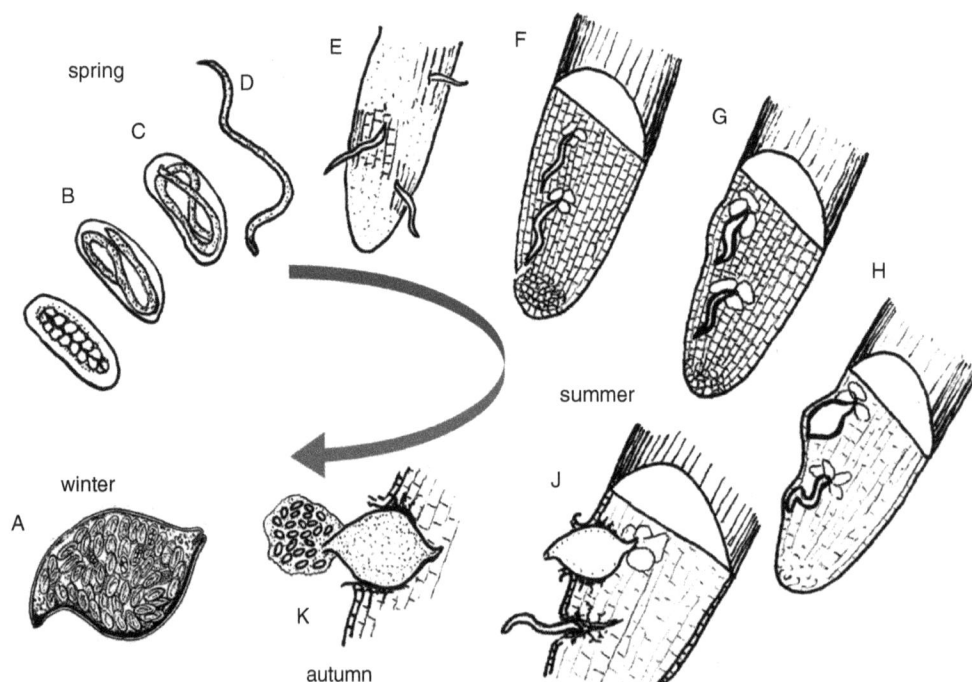

Fig. 15.11. Disease cycle of *Heterodera humuli* (redrawn from Agrios, 2005)

lost 68%. *H. humuli* has been reported on hemp in Russia (Decker, 1972) and Germany (Spaar *et al.*, 1990; Gutberlet and Karus, 1995).

Others found that *H. humuli* could not infect *Cannabis* (Kir'yanova and Krall, 1971; Subbotin, 1986). Hemp-root diffusates could not stimulate *H. humuli* egg hatch (De Grisse and Gillard, 1963). Bernard *et al.* (2022) summarized by saying that the ability of *H. humuli* to parasitize hemp seems variable; nematode populations may have evolved host selectivity, or hemp varieties may differ in susceptibility as they do with root-knot nematodes.

15.8c. Other Heterodera/Globodera spp.

Cannabis seems to be resistant to other cyst nematodes. Soil densities of soybean cyst nematode (*Heterodera glycines* Ichinohe 1952) decline after rotation with hemp (Scheifele *et al.*, 1997). Zhang *et al.* (2013) assessed soil populations of *H. glycines* following six different crops. Populations were predictably largest following soybeans. Following a crop of hemp the nematode population decreased by 29.8%, which was more than decreases following maize, sugar beet, wheat, or tobacco.

In a search for weed hosts of *H. glycines* in South Dakota, Basnet *et al.* (2020) examined several specimens of feral hemp, and no cysts were found. Kir'yanova and Krall (1971) rotated hemp with potatoes to suppress potato cyst nematode (*Globodera rostochiensis* Wollenweber 1923). Nettle cyst nematode (*Heterodera ripae* Subbotin 2003) did not produce cysts on *Cannabis* roots when artificially inoculated (López-Robles *et al.*, 2011).

Cannabis sativa was resistant to pigeon pea cyst nematode (*Heterodera cajani* Koshy 1967) in a host-range study (Koshy and Swarup, 1972). Mojumdar *et al.* (1989) killed *H. cajani* with *Cannabis* extracts. They ground up 100 g of fresh leaves in 25 ml water and filtered it through muslin cloth. This extract killed J2-stage juveniles of *H. cajani* in 6 h. A weaker 5% solution of the extract killed nematodes after 24 h.

15.8.2 Disease cycle and epidemiology

Heterodera spp. overwinter as eggs in cysts (Fig. 15.11A). Eggs develop into J1 juveniles (Fig. 15.11B), then molt into J2 juveniles (J2) while still within eggs (Fig. 15.11C). J2s hatch in the spring (Fig. 15.11D) and invade rootlets (Fig. 15.11E). Three molts later (Fig. 15.11F–H), females become more rotund while males revert to a vermiform shape. Growing females eventually rupture roots, their posteriors exposed to the soil, and males exit roots to inseminate the exposed females (Fig. 15.11J). Females may extrude a few eggs into a gelatinous matrix but the great majority (up to 500) are retained within the body (Fig. 15.11K), and the cycle may repeat. After females have produced their eggs, the body turns into a brown, leathery "cyst," protecting the eggs from the soil environment (Fig. 15.11A). Cyst nematode eggs typically have

a variable diapause or dormancy period that postpones hatching for one to several years.

15.8.3 Cyst nematode management

Follow instructions for root-knot nematodes. *Heterodera* spp. have narrower host ranges than *Meloidogyne* spp., so crop rotation is practical. If *H. schachtii* is a problem, do not rotate after Chenopodiaceae (e.g., sugar beet) or Cruciferae (e.g., cabbage). If *H. humuli* is a problem, stay away from hops, and eliminate nettles (*Urtica* spp.).

For biological and chemical control, the most vulnerable stages of *Heterodera* spp. (and all plant-parasitic nematodes) are the egg and J2 juveniles. These life stages exist outside of the plant, which brings them into contact with antagonistic biocontrol organisms and chemicals. Many biocontrol organisms listed in the root-knot section are effective against cyst nematodes.

Additionally, *Hirsutella rhossiliensis* parasitizes J2 juveniles of many species of *Heterodera*, *Globodera*, and *Rotylenchus*. It produces sticky conidia that attach to passing nematodes. The spores germinate, penetrate, and hyphae consume the host in about 3 days. New conidia sprout from the cadaver, launching another round of infections. Jaffee and Muldoon (1989) showed that natural populations of *H. rhossiliensis* reduced populations of *Heterodera schachtii* J2 juveniles by 50–77%. Constraints of *H. rhossiliensis* include its limited spread in soil, because it is a poor saprophytic competitor; *H. rhossiliensis* has not yet been commercialized.

The oomycete *Nematophthora gynophila* is an obligate parasite of *Heterodera* spp. Its flagellated zoospores attach to a J2, kill the animal with toxins, and then produce more zoospores to initiate new infections. Its life cycle takes 5 days at 13°C (Kerry, 1975). *N. gynophila* produces thick-walled oospores, which can live in the soil up to 5 years. It has not yet been commercialized.

15.9 Stem Nematodes

These unusual nematodes do *not* damage roots. They move out of soil and migrate up stems to branches and leaves, when plants are covered by a film of moisture. They enter through stomata or wounds, and feed on parenchymatous tissue.

Symptoms begin as inconspicuous thickenings of branches and leaf petioles. Sometimes even the middle veins of leaves become swollen. Stems subsequently become twisted and distorted, with shortened internodes, resulting in malformed, stunted plants (Fig. 15.12). Infected stems feel spongy. Leaves become curled and twisted, especially near the top. Plants often wilt due to severed xylem vessels. Lightly infected plants may send up new shoots from below infested areas.

15.9.1 Causal organism

15.9a. *Ditylenchus dipsaci* (Kühn) Filipjev 1936

≡*Anguillulina dipsaci* Kühn 1857
=*Tylenchus devastatrix* (Kühn) Oerley 1880

Fig. 15.12. *Incappucciamento* by *Ditylenchus dipsaci* (Aducco and Neppi, 1898)

Description: Females and males retain their filiform shape into maturity. Stylets are short and thin and difficult to see under a microscope; the stylet cone is about half of the total stylet length; stylet knobs are small and rounded. Females may reach 1.6 mm in length, with a long slender tail. Eggs are oval and relatively large, 60 × 20 μm; only a few eggs are found in the uterus. DNA-based methods have been developed to diagnose *D. dipsaci*. Marek *et al.* (2014) described a field kit using PCR and *D. dipsaci*-specific ITS primers.

Comments: *D. dipsaci* occurs in most temperate areas of the world, with a wide host range. In Italy, symptoms caused by *D. dipsaci* were known as *rachitismo* or *rachitide*, prior to the discovery of its cause. Berti Pichat (1866) attributed *rachitide* to torrential rains in early spring, which smeared seedlings with splashed soil, causing them to desiccate and die. Elsewhere in Italy the disease was called *incappucciamento* (Marconi, 1873). The cause of *incappucciamento* was identified as *Tylenchus devastatrix* by Aducco and Neppi (1898) in Ferrara. They provided microscopic illustrations of the nematode, and a classic illustration of symptoms (Fig. 15.12), copied by others (e.g., Mezzetti, 1951; Ciferri and Brizi, 1955).

Peglion (1901), also in Ferrara, elaborated on the epidemiology of *Tylenchus devastatrix*. Kirchner (1906) said hemp damage from *T. devastatrix* was limited to Italy. Navarro (1911) described *T. devastatrix* infecting hemp in Spain, where the disease was known as *porra* and *cinta*. In Napoli, Trotter (1938) attributed *rachitismo* to *Anguillulina dipsaci*. Steiner and Buhrer (1932) listed *Cannabis sativa* among crops attacked by *Tylenchus dipsaci* in the USA, but with a question mark.

Mezzetti (1951) first employed the taxon *D. dipsaci*, in a monograph on the nematode's damage on hemp, and its control. Sturhan and Brzeski (1991) gave the geographic range of *D. dipsaci* infesting hemp as Italy, Spain, Germany, Holland, Czechoslovakia, and USSR. Ceapoiu (1958) added Romania. Since then, *D. dipsaci* has been reported in Turkey (Tunçdemir, 1987) and Slovenia (Cizej and Poličnik, 2018). *D. dipsaci* has not yet been reported on hemp in North America, but the species is here.

Over 20 biological races of *D. dipsaci* have been identified. Some are polyphagous, but most have a limited host range. Kotthoff (1950) classified the hemp-infesting population in the "rye race." Kir'yanova and Krall (1971) recognized a flax–hemp race. Sturhan and Brzeski (1991) proposed that the rye and the flax–hemp races were identical. DNA-based taxonomy now recognizes *D. dipsaci* as a species complex. Vovlas *et al.* (2011) began using ITS sequences to cleave away new species. It's a matter of time before the rye–flax–hemp race gets a new name.

15.9.2 Disease cycle and epidemiology

The life cycle of *D. dipsaci* begins like root-knot nematodes and cyst nematodes. Eggs overwinter in the soil, and first molt occurs within eggs, which hatch into J2 juveniles. Thereafter, stem nematodes travel a different path —they migrate out of soil and infest above-ground plant parts. They enter stems through stomates, lenticels, or wounds. Invasion of plants is favored by cool, moist conditions. Mating with males is necessary for reproduction. The life cycle quickens to 19 days in wet weather and optimal temperatures (15–20°C). *D. dipsaci* is one of the few nematodes that spreads by seed, but this has only been reported in beans and onions.

15.9.3 Stem nematode management

Follow instructions for root-knot nematodes. Crop rotation is attractive, given the limited host range of the rye–flax–hemp race. Do not rotate hemp after flax (*Linum* species) or rye (*Secale cereale*). In cases of infestation, rotate out of hemp for 4–5 years (Sturhan and Brzeski, 1991). Early planting has been recommended for controlling *D. dipsaci* in various crops, but this was not helpful with hemp (Tunçdemir, 1987). Rogue infected plants from fields, remove leaves that have fallen to the ground, and destroy them. Ferraris (1935) burned hemp crop debris to eliminate *D. dipsaci*.

For biological control, PGRPs suppress newly hatched *D. dipsaci* juveniles while still in the soil. Mobile juveniles, while still in the soil, are killed by *Hirsutella rhossiliensis* and "carnivorous fungi" (see the root-knot section). For chemical control, Ferraris (1935) sprayed plants with 1% copper compounds. Tunçdemir (1987) reported success with Nemacur, Temac, and Vydate—now banned substances. A product with chitosan (Nemasan®) can be sprayed on plants or applied as a soil drench, approved for use on *Cannabis* (EPA, 2023).

15.10 Lesion nematodes

Genus *Pratylenchus* comprises ≥20 species, and five have been reported from *Cannabis*. Lesion nematodes are migratory endoparasites, meaning they penetrate roots, feed internally on cortex tissues, and then exit to migrate to new roots. This lifestyle can cause a lot of damage when populations are high. Their wounds predispose plants to secondary fungal infections, particularly by *Verticillium dahliae* and *Rhizoctonia solani*.

Symptoms arise as spots on small feeder roots, first appearing water-soaked or cloudy yellow in color. Soon the spots turn dark brown, and root tissues collapse. Collapsed lesions enlarge and coalesce as secondary fungi invade tissues. The entire root system may become stubby and discolored. Aboveground symptoms resemble those of drought stress or mineral deficiencies. Plants gradually appear less vigorous, wilt easily, turn yellow, and become stunted.

15.10.1 Causal organisms

15.10a. *Pratylenchus penetrans* (Cobb 1917) Filipjev & Schuurmans 1941

Description: *P. penetrans* adults are 400–800 µm long and 20–25 µm in diameter. They have blunt, broad heads and rounded tails. The stylet is relatively short (16–19 µm) with well-developed basal knobs. Eggs are elongate, 60–70 × 20–25 µm. A real-time PCR assay using *P. penetrans*-specific 28S rDNA primers has been developed (Baidoo *et al.*, 2017).
Comments: *P. penetrans* has a wide host range with >150 recorded hosts. It predominates in temperate regions, especially in areas with sandy soil. *P. penetrans* has been reported on hemp in the Netherlands (Kok *et al.*, 1994) and South Africa (Dippenaar *et al.*, 1996). Compared with other crops, *Cannabis* is a highly susceptible host; soil populations multiply 4- to 6-fold after a hemp crop (Kok and Coenen, 1996).

Brommer and Beers (2001) evaluated field reproduction of *P. penetrans* on 18 different crop plants, and hemp was the most susceptible. The initial field density was 424 nematodes per 100 ml soil; a season of hemp resulted in 3000 nematodes/100 ml soil. Núñez Rodríguez *et al.* (2023) first reported *P. penetrans* on hemp in the USA (Oregon and Washington); they reported a reproduction factor of 2.2 after 60 days post-inoculation.

15.10b. Other *Pratylenchus* spp.

Pratylenchus globulicola (Romaniko) 1960 infested hemp in Western Siberia (Romaniko, 1963). Loof (1978) synonymized *P. globulicola* under *P. penetrans*, but Ryss *et al.* (2015) recognized *P. globulicola* as a distinct species based on rRNA gene sequences.

Van Biljon (2017) recovered four other *Pratylenchus* spp. from hemp roots in South Africa: *Pratylenchus teres* Khan and Singh 1974, *Pratylenchus zeae* Graham 1951, *Pratylenchus scribneri* Steiner 1943, and *Pratylenchus brachyurus* (Godfrey 1929) Filipjev & Stekhoven 1941. Greco *et al.* (1988) conducted a host-range study with *Pratylenchus thornei* Sher & Allen 1953, a serious pathogen of chickpea, faba bean, lentil, and alfalfa. They included *Cannabis sativa*, and it was totally resistant.

15.10.2 Disease cycle and epidemiology

P. penetrans overwinters in infested plant debris or in soil at any life stage, although fourth-stage juveniles seem to be the optimal survival stage. In the spring they penetrate root epidermis either intra- or intercellularly, but once inside, they migrate intracellularly. The nematodes feed on cells until the cells lyse and cavities are formed, and then they migrate. Migration is rarely >1–2 m. The entire life cycle (egg to egg) can occur within roots, in 4–8 weeks.

15.10.3 Lesion nematode management

Follow instructions for root-knot nematodes. Population levels can be reduced by moldboard plowing (turning over the soil layer) to expose infected roots to the elements. Sanitize farm equipment after plowing infected fields. Weed control is important. Rotation to non-host plants is limited, because of the wide host range.

Dippenaar *et al.* (1996) tested the susceptibility of five cultivars to *Pratylenchus* sp. in naturally infested soil. In a 10 g root sample, 'Kompolti' harbored 158 nematodes, considerably more than 'Fedora 19' (n =13), 'Secuini' (4), 'Futura 75' (2), and 'Felina 34' (zero). Wijnholds and Hoek (2009) hypothesized that *P. penetrans* damage would be less if a short-season cultivar was cultivated ('USO-31'). They reported only 5% yield losses, but soil populations multiplied 5.7-fold, making hemp a poor choice as a rotation crop with potato—the primary host in Holland. Other susceptible crop plants included maize, alfalfa, and white clover (Brommer and Beers, 2001). Sugar beet, spring barley, and perennial ryegrass are resistant.

PGRPs and mycorrhizal fungi induce systemic resistance and inhibit *P. penetrans* activity (see the root-knot section). *P. penetrans* has been controlled with the soil fungus *Albifimbria verrucaria* and "carnivorous fungi." Multi-purpose fumigants (e.g., methyl bromide, chlorpicrin) and other nasties (Nemacur, Temac, Vydate, Furadan) used to be cornerstones of control, but they're history.

15.11 Other Nematodes

Several other groups of nematodes have been reported on *Cannabis*. Some are semi-endoparasitic, such as reniform nematodes. Others are ectoparasites, such as spiral-, stunt-, needle-, and dagger-nematodes—they live in soil and feed on plants without entering the root. Reniform nematodes include *Rotylenchulus parvus* (Williams) Sher 1961 and *Rotylenchulus unisexus* Sher 1965—both reported in South Africa (Dippenaar *et al.*, 1996; Van Biljon, 2017).

Spiral nematodes include Heliocotylenchus dihystera (Cobb) Sher 1961, Heliocotylenchus pseudorobustus (Steiner) Golden 1956, Heliocotylenchus paraplatyurus Siddiqi 1972, Heliocotylenchus serenus Siddiqi 1963, Scutellonema brachyurus (Steiner) Andrassy 1958, Scutellonema truncatum Sher 1963—all in South Africa (Dippenaar et al., 1996; Van Biljon, 2017). Scheifele et al. (1997) reported Heliocotylenchus sp. in Ontario, and Munir et al. (2023) reported Heliocotylenchus sp. in the USA.

Stunt nematodes have been cited on Cannabis: Tylenchorhynchus dubius (Buetschli) Filpjev 1936 was commonly collected from field soil samples in Russia (Skarbilovich, 1969). Unspeciated Tylenchorhynchus sp. was reported in the USA (Munir et al., 2023). Hemicriconemoides mangiferae Siddiqi 1961 was found in association with "Cannabis sativa, Indian hemp" at Jinju in South Korea (Choi and Geraert, 1975). Symptoms were not described in either report.

Several needle- and dagger-nematodes have been cited: Paralongidorus maximus (Bütschli) Siddiqi 1964 in Germany (Goodey et al. 1965), and Longidorus pisi Edwards 1965 in South Africa (Van Biljon, 2017). The latter also reported Xiphinema sp. in South Africa. Arabis mosaic virus has been recovered from Cannabis (Schmidt and Karl, 1969; Kegler and Spaar, 1997), which implies the plants were infected by the virus's vector, Xiphinema diversicaudatum (Micoletzky) Thorne 1939.

Lastly, stubby root nematodes, Trichodorus spp., ring nematodes, Mesocriconema spp., and lance nematodes, Hopolaimus spp., have been reported in the USA (Munir et al., 2023).

(Prepared by J. McPartland; revised by L. Thiessen (bacteria) and E. Bernard (nematodes))

References

Abad P, Gouzy J, Aury JM, *et al.* (51 other authors). 2008. Genome sequence of the metazoan plant-parasitic nematode *Meloidogyne incognita*. *Nature Biotechnology* 26: 909–915.

Abo-Korah MS. 2017. Biological control of root-knot nematode, *Meloidogyne javanica* infecting ground cherry, using two nematophagous and mychorrhizal fungi. *Egyptian Journal of Biological Pest Control* 27: 111–115.

Abreo E, Martinez S, Bettucci L, Lupo S. 2010. Morphological and molecular characterisation of *Campylocarpon* and *Cylindrocarpon* spp. associated with black foot disease of grapevines in Uruguay. *Australasian Plant Pathology* 39: 446–452.

Adegbite AA. 2011. Effects of some indigenous plant extracts as inhibitors of egg hatch in root-knot nematode (*Meloidogyne incognita* race 2). *American Journal of Experimental Agriculture* 1(3): 96–100.

Aducco A, Neppi C. 1898. L'incappucciamento della canapa. *L'Italia Agricola Giornale di Agricoltura* 35: 492–493.

Agrios GN. 2005. *Plant pathology, 5th edn.* Elseivier Academic Press, Amsterdam.

Amoretti C, *ed.* 1789. Canape della Cina. *Atti della Società Patriotica di Milano* 2:92.

An SQ, Potnis N, Dow M *et al.* (8 other authors). 2020. Mechanistic insights into host adaptation, virulence and epidemiology of the phytopathogen *Xanthomonas*. *FEMS Microbiology Reviews* 44: 1–32.

Anwar A, Gould E, Tinson R, Groom M, Hamilton CJ. 2017. Think yellow and keep green—role of sulfanes from garlic in agriculture. *Antioxidants* 6(3): e3.

Aritua V, Musoni A, Kabeja A, *et al.* (10 other authors). 2015. The draft genome sequence of *Xanthomonas* species strain Nyagatare, isolated from diseased bean in Rwanda. *FEMS Microbiolology Letters* 362: fnu055.

Baidoo R, Yan GP, Nagachandrabose S. 2017. Developing a real-time PCR assay for direct identification and quantification of *Pratylenchus penetrans* in soil. *Plant Disease* 101: 1432–1441.

Baral HO, Weber E, Marson G. 2020. *Monograph of Orbiliomycetes (Ascomycota) based on vital taxonomy*. National Museum of Natural History, Luxembourg.

Basnet P, Clay SA, Byamukama E. 2020. Determination of weed hosts of soybean cyst nematode in South Dakota. *Weed Technology* 34: 337–382.

Berkeley MJ. 1855. Vibrio forming cysts on the roots of cucumbers. *Gardener's Chronicle and Agricultural Gazette* 14: 220.

Bernard EC, Chaffin AG. 2020. Hemp cultivar susceptibility to the southern root-knot nematode, *Meloidogyne incognita*. *Journal of Nematology* 52: 1–2. Available at: https://figshare.com/articles/presentation/Bernard-Chaffin_SON2020_final_pdf/13366343/1

Bernard EC, Chaffin AG, Gwinn KD. 2022. Review of nematode interactions with hemp (*Cannabis sativa*). *Journal of Nematology* 54: e2022-2. Available at: https://www.ncbi.nlm.nih.gov/pmc/articles/PMC8975275/

Berti Pichat CB. 1866. "La pianta della canapa," pp. 396–499 in *Istituzioni scientifiche e tecniche, ossia Corso teorico e practico de Agricoltura, Libri XIII, Volume Quinto*. Presso l'Union Tipografico-Editrice, Torino.

Binder M, Meisenberg G. 1978. Microbial transformation of cannabinoids: Part 2; A screening of different microorganisms. *European Journal of Applied Microbiology and Biotechnology* 5: 37–50.

Bizio B. 1823. Lettera di Bartolomeo Bizio al chiarissimo canonico Angelo Bellani sopra il fenomeno della polenta porporina. *Biblioteca Italiana o sia Giornale di Letteratura Scienze e Arti* 30: 275–295.

Bordallo J, Lopez-Llorca L, Jansson HB, *et al.* (3 other authors). 2002. Colonization of plant roots by egg-parasitic and nematode-trapping fungi. *New Phytologist* 154: 491–499.

Britt K, Fike J, Flessner M, Johnson C, Kuhar T, McCoy T, Reed D. 2020. *Integrated pest management of hemp in Virginia*. Virginia Cooperative Extension publication no. ENTO-349NP, Blacksburg, Virginia.

Brommer E, Beers TG van. 2001. Wortellesieaaltje *Pratylenchus penetrans*. *PPO-Bulletin Akkerbouw* 2: 22–25.

Buhrer EA, Cooper C, Steiner G. 1933. A list of plants attacked by the root-knot nematode (*Heterodera marioni*). *Plant Disease Reporter* 17: 64–96.

Bull CT, Manceau C, Lydon J, *et al.* (3 other authors). 2010a. *Pseudomonas cannabina* pv. *cannabina* pv. nov., and *Pseudomonas cannabina* pv. *alisalensis* (Cintas Koike and Bull, 2000) comb. nov., are members of the emended species *Pseudomonas cannabina* (ex Sutic & Dowson 1959) Gardan, Shafik, Belouin, Brosch, Grimont & Grimont 1999. *Systematic and Applied Microbiology* 33: 105–111.

Bull CT, Mauzey SJ, Koike ST. 2010b. First report of bacterial blight of Brussels sprouts (*Brassica oleracea* var. *gemmifera*) caused by *Pseudomonas cannabina* pv. *alisalensis* in California. *Plant Disease* 94: 1375.

Buttimer C, McAuliffe O, Ross RP, *et al.* (3 other authors). 2017. Bacteriophages and bacterial plant diseases. *Frontiers in Microbiology* 8: e34.

CABI. 2009. *Pseudomonas syringae pv. mori. Distribution map no. 1069*. CABI, Wallingford, UK.

Ceapoiu N. 1958. *Cînepa, Studiu monografic*. Editura Academiei Republicii Populare Romine. Bucharest.

Chitwood BG. 1949. "Root-knot nematodes"—Part 1. A revision of the genus *Meloidogyne* Goeldi, 1887. *Proceedings of the Helminthological Society of Washington* 16: 90–104.

Choi YE, Geraert E. 1975. Criconematids from Korea with the description of eight new species (Nematoda: Tylenchida). *Nematologica* 21: 35–52.

Ciferri R, Brizi A. 1955. *Manuale di patologia vegetale, vol. III*. Societa Editrice Dante Alighieri, Rome.

Cintas NA, Koike ST, Bull CT. 2002. A new pathovar, *Pseudomonas syringae* pv. *alisalensis* pv. nov., proposed for the causal agent of bacterial blight of broccoli and broccoli raab. *Plant Disease* 86: 992–998.

Cizej MR, Poličnik F. 2018. Škodljivci industrijske konoplje (*Cannabis sativa* L.) v Sloveniji. *Hmeljarski Bilten* 25: 36–43.

Cobb NA. 1915. "Nematodes and their relationships," pp. 457–490 in *Yearbook of the Department of Agriculture for 1914*. Washington, DC.

Coburn JD, Desaeger J. 2020. "Root-knot nematode host status of different hemp cultivars," p. 19 in Gauthier N, *ed. Science of hemp: production and pest management*. University of Kentucky Agricultural Experiment Station no. SR-112. Lexington, Kentucky.

Cook A. 1981. *Diseases of tropical and subtropical field, fiber, and oil crops*. Macmillan, New York.

Coosemans J. 1991. Antagonistic activity of a *Cylindrocarpon olidum* (Wollenw.) isolate against nematodes. *Mededelingen van de Faculteit Landbouwwetenschappen, Rijksuniversiteit Gent* 56: 223–228.

Dayanandan P, Kaufman PB. 1976. Trichomes of *Cannabis sativa* L. (Cannabaceae). *American Journal of Botany* 63: 578–591.

De Grisse A, Gillard A. 1963. Morphology and biology of hop cyst eelworm (*Heterodera humuli* Filipjev 1934). *Nematologica* 9: 41–48.

de Meijer EPM. 1993. Evaluation and verification of resistance to *Meloidogyne hapla* Chitwood in a *Cannabis* germplasm collection. *Euphytica* 71: 49–56.

de Meijer EPM, Hammond KM, Micheler M. 2009. The inheritance of chemical phenotype in *Cannabis sativa* L. (III): variation in cannabichromene production. *Euphytica* 165: 293–311.

de Montluisant C. 1885. *Notice sur les produits des États Pontificaux à l'exposition universelle*. Bailly, Divry et Ce, Paris.

Decker H. 1972. *Plant nematodes and their control*. Kolos Publ. Co., Moscow.

Dion MB, Oechslin F, Moineau S. 2020. Phage diversity, genomics and phylogeny. *Nature Reviews Microbiology* 18: 125–138.

Dippenaar MC, du Toit CLN, Botha-Greeff MS. 1996. Response of hemp (*Cannabis sativa* L.) varieties to conditions in Northwest Province, South Africa. *Journal of the International Hemp Association* 3(2): 63–66.

Duraisamy S, Mishra AK, Kocábek T, Matoušek J. 2018. Activation of polyketide synthase gene promoter in *Cannabis sativa* by heterologous transcription factors derived from *Humulus lupulus*. *Biologia Plantarum* 62: 250–260.

Dye DW, Bradbury JF, Goto M, *et al.* (3 other authors). 1980. International standards for naming pathovars of phytopathogenic bacteria and a list of pathovar names and pathotype strains. *Review of Plant Pathology* 59:153–168.

Eberhard W, Pacheco-Esquivel J, Carrasco-Rueda F, *et al.* (5 other authors). 2014. Zombie bugs? The fungus *Purpureocillium* cf. *lilacinum* may manipulate the behavior of its host bug *Edessa rufomarginata*. *Mycologia* 106: 1065–1072.

Elliott C. 1951. *Manual of bacterial plant pathogens*. Chronica Botanica, Waltham, Massachusetts.

EPA. 2023. *Pesticide products registered for use on hemp*. Available at: https://www.epa.gov/pesticide-registration/pesticide-products-registered-use-hemp.

Esfahani MN, Ahmadi A. 2010. Field observations on the reaction of medicinal plants to root-knot nematodes in Isfahan, Iran. *International Journal of Nematology* 20: 107–112.

Fernández C, Rodríguez-Kábana R, Warrior P, Klopper JW. 2001. Induced soil suppressiveness to a root-know nematode species by a nematicide. *Biological Control* 22: 103–114.

Ferraris T. 1935. *Parassiti vegetali della canapa*. Rivista Agricola, Rome.

Ferri F. 1957a. La "striatura ulcerosa" della canapa. *Informatore Fitopatologica* 7(14): 235–238.

Ferri F. 1957b. La "striatura ulcerosa" flagello della canapa. *Progresso Agricolo* 3(10): 1194–1194.

Ferri F. 1959a. *Atlante delle avversità della canapa*. Edizioni Agricole, Bologna, Italia.

Filipjev IN, Schuurmans-Stekhoven JHS. 1941. *A manual of agricultural helminthology*. Brill, Leiden.

Flachs K. 1936. Krankheiten und schädlinge unserer gespinstpflanzen. *Nachrichten über Schädlingsbekämpfung* 11: 6–28.

Franklin MT. 1951. *The cyst-forming species of Heterodera*. Commonwealth Agricultural Bureaux, Farnham Royal, UK.

Gan HM, Savka MA. 2018. One more decade of *Agrobacterium* taxonomy. *Current Topics in Microbiology and Immunology* 418: 1–14.

Ganaie MA, Khan TA. 2017. Evaluation of some locally available plant leaves for the control of *Meloidogyne incognita* and *Rhizoctonia solani* disease complex. *Trends in Biosciences* 10: 8534–8539.

Gardan L, Shafik H, Belouin S, *et al.* (3 other authors). 1999. DNA relatedness among the pathovars of *Pseudomonas syringae* and description of *Pseudomonas tremae* sp. nov. and *Pseudomonas cannabina* sp. nov. (ex Sutic and Dowson 1959). *International Journal of Systematic Bacteriology* 49: 469–478.

Georgieva M, Kotev S. 1972. Причинителят на бактериозата по конопа в България. *Plant Protection* 20(9): 17–18.

Ghani M, Basit A. 1975. *Investigations on the natural enemies of marijuana, Cannabis sativa L. and opium poppy, Papaver somniferum L.* Annual Report, Commonwealth Institute of Biological Control, Pakistan station, Rawalpindi.

Ghillini CA. 1951. I parassiti nemici vegetali della canapa. *Notiziario sulle Malattie delle Piante* 15: 29–36.

Gibin D, Linares AG, Fasanelli E, Pasinato L, Delbianco A. 2023. *Xylella fastidiosa*. Update of the *Xylella* spp. host plant database. *EFSA Journal* 3;21: e8477.

Gitman LS. 1968a. "Бактериальные болезни конопли," pp. 264–267 in Beltykov KI, *ed. Бактериальные болезни растений и методы борьбы с ними.* Naukova Dumka, Kiev.

Gitman LS. 1968b. Малоизвестные болезни конопли. *Защита растений* 13(3): 44–45.

Gitman L, Boytchenko E. 1934. "*Cannabis*" pp. 45–53 in Справочник по болезням бовых лубианых культур. Institute New Bast Raw Materials, Moscow.

Goidànich G. 1959. *Manual di patologia vegetale.* Edizioni Agricole, Bologna.

Goidànich G, Ferri F. 1960. La batteriosi della canapa da *Pseudomonas cannabina* Sutic & Dowson var. *italica* Dowson. *Phytopathologische Zeitschrift* 37: 21–32.

Goodey JB, Franklin MT, Hooper DJ. 1965. *The nematode parasites of plants catalogued under their hosts.* Commonwealth Agricultural Bureaux, Farnham Royal, UK.

Goswami BK, Vijaylakshmi K. 1986. Effect of some indigenous plant materials and oilcake amended soil on the growth of tomato and root-knot nematode population. *Annals of Agricultural Research* 7: 363–366.

Gray N. 1988. Ecology of nematophagous fungi: effect of the soil nutrients N, P and K, and seven major metals on distribution. *Plant and Soil* 108: 286–290.

Greco N, Di Vito M, Sexena MC, Reddy MV. 1988. Investigation on the root lesion nematode *Pratylenchus thornei*, in Syria. *Nematologica Mediterranea* 16: 101–105.

Guirao P, Moya A, Cenis JL. 1995. Optimal use of random amplified polymorphic DNA in estimating the genetic relationship of four major *Meloidogyne* spp. *Pytopathology* 85: 547–551.

Gutberlet V, Karus M. 1995. *Parasitäre Krankheiten und Schädlinge an Hanf (Cannabis sativa).* Nova Institut, Köln, Germany.

Haidar MG, Askary TH. 2011. Management of plant parasitic nematodes through botanicals and growth of sugarcane (*Saccharum officinarum* L.). *Annals of Plant Protection Sciences* 19: 433–436.

Hansen Z, Bernard E, Grant J, *et al.* (4 other authors). 2020. *Hemp disease and pest management. University of Tennessee Extension publication* no. W916.

Harrison J, Hussain RMF, Greer S, *et al.* (5 other authors). 2022. Draft genome sequences for ten strains of *Xanthomonas* species that have phylogenomic importance. *Access Microbiology.*

Hillig KH. 2005a. *A systematic investigation of Cannabis.* Doctoral Dissertation, Department of Biology, Indiana University, Bloomington, Indiana.

Hollandbeck G. 2019. Plant diseases found in industrial hemp. *Plant Disease in Kansas* 45(2): 1–3.

Hollrung M. 1890. *Jahresbericht über die Tätigkeit der Versuchsstation für Nematoden vertilgung und Pflanzenshcutz.* Halle, Germany.

Holmes JE, Sanghera H, Punja ZK. 2023. Crown gall development on cannabis (*Cannabis sativa* L., marijuana) plants caused by *Agrobacterium tumefaciens* species-complex. *Canadian Journal of Plant Pathology* 45: 433–445.

Hu SS, Mojahid MS, Bidochka MJ. 2023. Root colonization of industrial hemp (*Cannabis sativa* L.) by the endophytic fungi *Metarhizium* and *Pochonia* improves growth. *Industrial Crops and Products* 198: e116716.

Hultberg M, Bengtsson T, Liljeroth E. 2010. Late blight on potato is suppressed by the biosurfactant-producing strain *Pseudomonas koreensis* 2.74 and its biosurfactant. *BioControl* 55: 543–550.

Hussain M, Zouhar M, Ryšánek P. 2017. Comparison between biological and chemical management of sugar beet nematode, *Heterodera schachtii. Pakistan Journal of Zoology* 49: 45–50.

Hussey RS, Janssen GJW. 2002. "Root-knot nematode: *Meloidogyne* species," pp. 43–70 in Starr JL, *et al.*, eds. *Plant resistance to parasitic nematodes.* CABI, Wallingford, UK.

Ishiga T, Iida Y, Sakata N, *et al.* (5 other authors). 2019. Acibenzolar-s-methyl activates stomatal-based defense against *Pseudomonas cannabina* pv. *alisalensis* in cabbage. *Journal of General Plant Pathology* 86: 48–54.

Jacobs JM, Pesce C, Lefeuvre P, Koebnik R. 2015. Comparative genomics of a *Cannabis* pathogen reveals insight into the evolution of pathogenicity in *Xanthomonas. Frontiers in Plant Science* 6: e431.

Jaffee BA, Muldoon AE. 1989. Suppression of cyst nematode by a natural infestation of a nematophagous fungus. *Journal of Nematology* 21: 505–510.

Jatala P, Kaltenbach R, Bocangel M. 1979. Biological control of *Meloidogyne incognita acrita* and *Globodera pallida* on potatoes. *Journal of Nematology* 11: 303.

Jiang XZ, Xiang MC, Liu XH. 2017. Nematode-trapping fungi. *Microbiology Spectrum* 5(1): 1–12.

Johnson J. 1937. Relation of water-soaked tissues to infection by *Bacterium angulatum* and *B. tabacum* and other organisms. *Journal of Agricultural Research* 55(8): 599–618.

Johnston A. 1964. *Additional records of plant nematodes in the south east Asia and Pacific region.* FAO Plant Protection Committee for the South East Asia & Pacific Region, Technical Document no. 30.

Kayani MZ, Mukhtar T, Hussain MA. 2012. Evaluation of nematicidal effects of *Cannabis sativa* L. and *Zanthoxylum alatum* Roxb. against root-knot nematodes, *Meloidogyne incognita. Crop Protection* 39: 52–56.

Kegler H, Spaar D. 1997. Zur virusanfälligkeit von hanfsorten (*Cannabis sativa* L.). *Archiv für Phytopathologie und Pflanzenschutz* 30: 457–464.

Kerry BR. 1975. Fungi and the decrease of cereal cyst-nematode populations in cereal monoculture. *EPPO Bulletin* 5: 353–361.

Kirchner O. 1906. "Hanf, *Cannabis sativa* L.," pp. 319–323 in *Die Krankheiten und Befehädigungen uhferer landwirtschaftlichen Kulturpflanzen.* E. Ulmer, Stuttgart.

Kir'yanova ES, Krall EL. 1971. *Plant-parasitic nematodes and their control, Vol. II.* Academy of Sciences of the USSR, Nauka Publishers, Leningrad.

Klement Z, Kaszonyi S. 1960. A kender új baktériumos betegsége. *Növénytermelés* 9(2): 153–158.

Koenning SR, Overstreet C, Noling JW, *et al.* (3 other authors). 1999. Survey of crop losses in response to phytoparasitic nematodes in the United States for 1994. *Journal of Nematology* 31(4S): 587–618.

Koike ST, Kammeijer K, Bull CT, O'Brien D. 2006. First report of bacterial blight of romanesco cauliflower (*Brassica oleracea* var. *botrytis*) caused by *Pseudomonas syringae* pv. *alisalensis* in California. *Plant Disease* 90: 1511.

Koike ST, Kammeijer K, Bull CT, O'Brien D. 2007. First report of bacterial blight of rutabaga (*Brassica napus* var. *napobrassica*) caused by *Pseudomonas syringae* pv. *alisalensis* in California. *Plant Disease* 91: 112.

Kok CJ, Coenen GCM. 1996. Host suitability of alternative oilseed and fiber crops to *Pratylenchus penetrans. Fundamental and Applied Nematology* 19: 205–206.

Kok CJ, Coenen GCM, de Heij A. 1994. The effect of fibre hemp (*Cannabis sativa* L.) on selected soil-borne pathogens. *Journal of the International Hemp Association* 1(1): 6–9.

Kokalis-Burelle, N. 2015. *Pasteuria penetrans* for control of *Meloidogyne incognita* on tomato and cucumber, and *M. arenaria* on snapdragon. *Journal of Nematology* 47: 207–2013.

Konecka E, Mokracka J, Krzyminnska S, Kazonwski A. 2019. Evaluation of the pathogenic potential of insecticidal *Serratia marcenscens* to humans. *Polish Journal of Microbiology* 68: 185–191.

Korthals GW, Nijboer H, Molendijk LPG. 2000. *Meloidogyne chitwoodi*: waardplantgeschiktheid van landbouw-en groenbemestinggewassen. PAV Bulletin Akkerbouw, April 2000: 1–3.

Koshy PK, Swarup G. 1972. Susceptibility of plants to pigeon-pea cyst nematode, *Heterodera cajani*. *Indian Journal of Nematology* 2: 1–6.

Kotcon J, Wheeler K, Cline RH, Carter S. 2018. Susceptibility and yield loss relationships of *Meloidogyne hapla* and *M. incognita* infecting *Cannabis sativa*. *Journal of Nematology* 50: 644.

Kotthoff P. 1950. Die Verbreitung von *Ditylenchus dipsaci* (Kühn) als Schadling an landwirtschaftlichen Kulturpflanzen in Westfalen. *Zeitschrift fur Pflanzenkrankheiten, Pflanzenpathologie und Pflanzenschutz* 57: 4–14.

Krawczyk K, Łochyńska M. 2000. Identification and characterization of *Pseudomonas syringae* pv. *mori* affecting white mulberry (*Morus alba*) in Poland. *European Journal of Plant Pathology* 158: 281–191.

Kühn J. 1891. Neuere Versuch zur Bekämpfung der Rübennematoden. *Neue Zeitschrift fur Rübenzucker-Industry* 26: 149–159.

Kyokai NSB. 1965. 農林病害虫名鑑. Japan Plant Protection Association, Tokyo.

Lawaju BR, Groover W, Kelton J, *et al*. (3 other authors). 2021. First report of *Meloidogyne incognita* infecting *Cannabis sativa* in Alabama. *Journal of Nematology* 53: e2021-52.

Lima de Brida A, Souza da Silva Correia EC, *et al*. (3 other authors). 2017. Oat, whet, and sorghum genotype reactions to *Meloidogyne incognita* and *Meloidogyne javanica*. *Journal of Nematology* 49: 386–389.

Loof PAA. 1978. The genus *Pratylenchus* Filipjev, 1936 (Nematoda: Pratylenchidae): a review of its anatomy, morphology, distribution, systematics and identification. *Växtskyddsrapporter* 5: 1–50.

Lopatin MI. 1936. Поражаемость растений возбудителем корневого рака растений *Bacterium tumefaciens* Sm. A. Town. Микробиология 5: 716–724.

López-Robles J, Sacristán-Pérez-Minayo G, Olalla-Gómez C. 2011. First report of *Heterodera ripae* on common nettle in Spain. *Plant Disease* 95: 883.

Lucas GB. 1975. *Diseases of tobacco, 3rd ed*. Biological Consulting Assoc., Raleigh, North Carolina.

MacKinnon L, McDougall G, Aziz N, Millam S. 2000. Progress towards transformation of fibre hemp. *Scottish Crop Research Institute Annual Report 2000/2001*. Scottish Crop Research Institute, Dundee, pp. 84–86.

Manzanilla-López RH, Esteves I, Finetti-Sialer MM, *et al*. (4 other authors). 2013. *Pochonia chlamydosporia*: advances and challenges to improve its performance as a biological control agent of sedentary endo-parasitic nematodes. *Journal of Nematology* 45: 1–7.

Marcinowski K. 1909. *Parasitisch und semiparasitisch an Pflanzen lebende Nematoden*. Paul Parey, Berlin.

Marconi F. 1873. La coltivazione della canapa. *Enciclopedia Agraria Italiana*. G. Cantoni, Torino.

Marek M, Zouhar M, Douda O, Maňasová M, Ryšánek P. 2014. Exploitation of FTA cartridges for the sampling, long-term storage, and DNA-based analysis of plant-parsitic nematodes. *Phytopathology* 104: 306–312.

McKernan K, Spangler J, Helbert Y, *et al*. (9 other authors). 2016b. Metagenomic analysis of medicinal cannabis samples; pathogenic bacteria, toxigenic fungi, and beneficial microbes grow in culture-based yeast and mold tests. *F1000Research* 5: e2471.

McPartland JM, Cubeta MA. 1997. New species, combinations, host associations and location records of fungi associated with hemp (*Cannabis sativa*). *Mycological Research* 101: 853–857.

McPartland JM, Hillig KW. 2004. Striatura ulcerosa. *Journal of Industrial Hemp* 9(1): 89-96.

McPartland JM, Clarke RC, Watson DP. 2000. *Hemp diseases and pests – management and biological control*. CABI Publishing, Wallingford, UK.

Meline V, Delage W, Brin C, *et al*. (13 other authors). 2019. Role of the acquisition of a type 3 secretion system in the emergence of novel pathogenic strains of *Xanthomonas*. *Molecular Plant Pathology* 20: 33–50.

Merda D, Briand M, Bosis, *et al*. (5 other authors). 2017. Ancestral acquisitions, gene flow and multiple evolutionary trajectories of the type three secretion system and effectors in *Xanthomonas* plant pathogens. *Molecular Ecology* 26: 5939–5952.

Mezzetti A. 1951. *Alcune alterazioni della canapa manifestatesi nella decorsa annata agraria*. Quaderni del Centro di Studi per le Ricerche sulla Lavorazione Coltivazione ed Economia della Canapa, Laboratorio Sperimentale di Patologia Vegetale di Bologna.

Miller PR, Weiss F, O'Brien MJ. 1960. *Index of plant diseases in the United States*. Agriculture Handbook No. 165. USDA, Washington, DC.

Mojumdar V, Mishra SD, Haque MM, Goswami BK. 1989. Nematicidal efficacy of some wild plants against pigeon pea cyst nematode, *Heterodera cajani*. *International Nematology Network Newsletter* 6(2): 21–24.

Mukhtar T, Kayani MZ, Hussain MA. 2013. Nematicidal activities of *Cannabis sativa* L. and *Zanthoxylum alatum* Roxb. against *Meloidogyne incognita*. *Industrial Crops and Products* 42: 447–453.

Munir M, Leonberger K, Kesheimer K, *et al*. (14 other authors). 2023. Occurrence and distribution of common diseases and pests of US *Cannabis*: a survey. *Plant Health Progress*.

Navarro L. 1911. *La enfermedad de los cáñamos en el término municipal de Ceheguín (Murcia), denominado vulgarmente porra y cinta*. Estación de Patología Vegetal, Ministerio de Fomento, Madrid.

Neppi C. 1899. *La canapa nel ferrarese*. Tipografia Nanicas, Modena.

Netsu O, Kijima O, Takikawa Y. 2014. Bacterial leaf spot of hemp caused by *Xanthomonas campestris* pv. *cannabis* in Japan. *Journal of General Plant Pathology* 80: 164–168.

Nirula KK, Kumar R. 1964. New host records of the root-knot nematodes. *Nematologica* 10(1): 184.

Noble M, Richardson MJ. 1968. *An annotated list of seed-borne diseases*. Commonwealth Mycological Institute, Kew, UK.

Norton DC. 1966. Additions to the known hosts of *Meloidogyne hapla*. *Plant Disease Reporter* 50: 523–524.

Núñez Rodríguez LA, Rivedal HM, Peetz A, Ocamb C, Zasada I. 2023. First report of the root-lesion nematode, *Pratylenchus penetrans*, parasitizing hemp (*Cannabis sativa*) in the United States. *Plant Health Progress*.

Okabe N. 1949. 植物細菌学. Asakura Shoten, Tokyo, p. 176.

Okabe T, Goto M. 1955. 日本に於ける植物細菌病.1.細菌病およびその病菌の種類について. Shizuoka University Agriculture Research Report 5: 63–71.

Oppermann CH, Bird DM, Williamson VM, *et al*. (16 other authors). 2008. Sequence and genetic map of *Meloidogyne hapla*: a compact nematode genome for plant parasitism. *Proceedings of the National Academy of Sciences USA* 105: 14802–14807.

Papp CS. 1976. *Manual of scientific illustration*. American Visual Aid Books, Sacramento, California.

Parkinson N, Cowie C, Heeney J, Stead D. 2009. Phylogenetic structure of *Xanthomonas* determined by comparison of *gyrB* sequences. *International Journal of Systematic and Evolutionary Microbiology* 59: 264–274.

Passon M. 1909. Relação—em resumo, does trabalhos e observações realizados durante o mez de Fevereiro, no Instituto Agronomico do Estado. *Boletim de Agricultura* 10: 135–140.

Peglion V. 1896. Una nuova malatti della canapa (bacteriosis dello stelo). *Malpighia* 10: 556–560.

Peglion V. 1897. Eine neue Krankheit des Hanfes (Bacteriosis des Stengels). *Zeitschrift für Pflanzenkrankheiten* 7: 81–84.

Peglion V. 1901. Intorno al cosiddetto "incappucciamento" della canapa. Annuario della R Stazione di Patologia Vegetale di Roma 1: 154–164.

Peglion V. 1902. La bacteriosi della canepa. *Rendiconti Lincei: Atti della Reale Accademia dei Lincei, serie quinta* 11: 32–34.

Petri J. 1908. Patologia vegetale. Ricerche sopra la batteriosi del fico. *Rendiconti Lincei: Atti della Reale Accademia dei Lincei, serie quinta* 15: 644–651.

Pofu KM, Mashela PW. 2014. Density-dependent growth pattern of on hemp cultivars: establishing nematode-sampling timeframes in host-status trials. *American Journal of Experimental Agriculture* 4: 639–640.

Pofu KM, Van Biljon ER, Mashela PW, Shimelis HA. 2010. Responses of selected hemp cultivars to *Meloidogyne javanica* under greenhouse conditions. *American-Eurasian Journal of Agriculture & Environmental Science* 8: 602–606.

Pogosyan EE. 1960. Галловые нематоды в Армянской ССР. *Izvestiya Akademii Nauk Armyanskoĭ SSR* 13(8): 27–34.

Quaghebeur K, Coosemans J, Toppet S, Compernolle F. 1994. Cannabiorci- and 8-chlorocannabiorcichromenic acid as fungal antagonists from *Cylindrocarpon olidum*. *Phytochemistry* 37: 159–161.

Rataj K. 1957. Skodlivi cinitele pradnych rostlin. *Prameny Literatury* 2: 1–123.

Re F. 1806. *Elementi d'agricoltura di Filippo Re. Edizione Terza. Volume Primo.* Stamperia Vitarelli, Venezia.

Ren ZH, Chen XJ, Luan MB, Guo B, Song ZQ. 2021. First report of *Meloidogyne enterolobii* on industrial hemp (*Cannabis sativa*) in China. *Plant Disease* 105: 230.

Riberio A, Pochart P, Day A, *et al.* (5 other authors). 2015. Microbial diversity observed during hemp retting. *Applied Microbiology and Biotechnology* 99: 4471–4484.

Richter L. 1911. In Brasilien beobachtete pflanzenkrankheiten. *Zeitschrift für Pflanzenkrankheiten* 21: 49–50.

Romaniko VI. 1963. Susceptibility of plants to the nematode *Pratylenchus globulicola*. *Scientific Conference of the All-Union Society of Helminthologists*, 1963: 59–60.

Ryss A, Pridannikov M, Subbotin SA. 2015. Characterisation of plant-parasitic nematodes described by V.I. Romaniko from Chelyabinsk Region, Russia. *Russian Journal of Nematology* 23: 168.

Sandru ID. 1977. Vestejirea bacteriana a frunzelor—o boala noua la cinepa. *Productia Vegetala Cereale si Plante Tehnice* 29(7): 34–36.

Sano Z, Iwahori H. 2005. Regional variation in pathogenicity of *Meloidogyne incognita* populations on sweetpotato in Kyushu Okinawa, Japan. *Japanese Journal of Nematology* 35: 1–12.

Saxena R, Gupta S. 2004. Efficacy of aqueous extract of some medicinal plants against second stage juveniles of *Meloidogyne incognita*. *Current Nematology* 15: 51–53.

Sayre RM, Wergin WP, Schmidt JM, Starr MP. 1991. *Pasteuria nishizawae* sp. nov., a mycelial and endospore-forming bacterium parasitic on cyst nematodes of genera *Heterodera* and *Globodera*. *Research in Microbiology* 142 (5): 551–564.

Schappe T, Richie DF, Thiessen LD. 2020. First report of *Serratia marcescens* causing a leaf spot disease on industrial hemp (*Cannabis sativa*). *Plant Disease* 104: 1248.

Scheifele G, Dragla P, Pinsonneault C, Laprise JM. 1997. *Hemp (Cannabis sativa) research report, Kent County, Ontario, Canada.* Ridgetown College of Agricultural Technology, Ontario, Canada.

Schmidt HE, Karl E. 1969. Untersuchungen über eine flecken- und streifenbildung am hanf (*Cannabis sativa* L.). *Zentralblatt für Bakteriologie, Parasitenkunde, Infektionskrankheiten und Hygiene, Abteilung* 2 123: 310–314.

Sen AK, Jensen HJ. 1969. Host-parasite relationships of various plants and the hop cyst hematode, *Heterodera humuli*. *Plant Disease Reporter* 53: 37–40.

Severin V. 1978. Ein neues pathogenes Bakterien an Hanf—*Xanthomonas campestris* pathovar. *cannabis*. *Archiv für Phytopathologie und Pflanzenschutz* 14: 7–15

Shah NH, Dar AR, Qureshi IA, Afroza A, Wani MR, Lubna A. 2018. Control of root-knot disease of brinjal (*Solanum melongena* L.) by the application of leaf extracts of certain medicinal plants. *Indian Journal of Agricultural Research* 52: 444–447.

Shatilovich A, Gade VR, Pippel M, *et al.* (11 other authors). 2023. A novel nematode species from the Siberian permafrost shares adaptive mechanisms for cryptobiotic survival with *C. elegans* dauer larva. *PLoS Genetics* 19: e1010798.

Shaw HB. 1915. The sugar-beet nematode and its control. *Sugar* 17: 58–61.

Singh SP, Singh V. 2002. Evaluation of some plant extracts against root-knot nematode, *Meloidogyne incognita*. *Indian Journal of Nematology* 32: 226–227.

Skarbilovich TS. 1969. Исследование фауны нематод конопли, *Cannabis sativa* L. Problems of Parasitology Pt. II, pp. 329–330. Naukova Dumka, Kyiv.

Smith EF. 1920. *An introduction to bacterial diseases of plants.* Saunders, Philadelphia.

Smith IM, Dunez J, Phillips DH, Lelliott RA, Archer SA. 1988. *European handbook of plant diseases.* Blackwell Scientific Publications, Oxford.

Somma U. 1923. *La canapa: coltura, lavorazione-commercio.* L. Cappelli, Bologna.

Song ZQ, Cheng FX, Zhang DY, Liu Y, Chen XW. 2017. First report of *Meloidogyne javanica* infecting hemp (*Cannabis sativa*) in China. *Plant Disease* 101: 842–843.

Sorauer P. 1908. *Handbuch der Pflanzenkrankheiten, zweiter band.* Paul Parey, Berlin.

Spaar D, Kleinhempel H, Fritzsche R. 1990. *Öl- und Faserpflanzen.* Springer-Verlag, Berlin.

Steiner G, Buhrer EM. 1932. A list of plants attacked by *Tylenchus dipsaci*, the bulb or stem nema. *Plant Disease Reporter* 16: 76–85.

Stirling G, Nicol J, Reay F. 2002. *Advisory services for nematode pests.* Rural Industries Research and Development Corporation, Moggill, Brisbane, Australia.

Sturhan D, Brzeski MW. 1991. "Stem and bulb nematodes, Ditylenchus spp.," pp. 423–464 in Nickel WR, ed. *Manual of agricultural nematology*. Marcel Dekker, New York.

Subbotin SA. 1986. Крапивная раса хмелевой нематоды *Heterodera humuli* Filipjev, 1934. *Bulletin of the All-Union Institute of Helminthology* 45: 98–99.

Subbotin SA, Palomares-Rius JA, Castillo P. 2021. *Systematics of root-knot nematodes (Nematoda: Meloidogynidae).* Brill, Leiden.

Šutič D, Dowson WJ. 1959. An investigation of a serious disease of hemp (*Cannabis sativa* L.) in Yugoslavia. *Phytopathologische Zeitschrift* 34: 307–314.

Takahashi K, Nishiyama K, Sato M. 1996. *Pseudomonas syringae* pv. *broussonetiae* pv. nov., the causal agent of bacterial blight of paper mulberry (*Broussonetia kazinoki* x *B. papyrifera*). *Annals of the Phytopathological Society of Japan* 62: 17–22.

Takahashi F, Ochiai M, Ikeda K, Takikawa Y. 2013. Streptomycin and copper resistance in *Pseudomonas cannabina* pv. *alisalensis*. *Japanese Journal of Phytopathology* 79: e35.

Tanney CAS, Lyu DM, Schwinghamer T, *et al.* (3 other authors). 2023. Sub-optimal nutrient regime coupled with *Bacillus* and *Pseudomonas* sp. inoculation influences trichome density and cannabinoid profiles in drug-type *Cannabis sativa*. *Frontiers in Plant Science* 14: e1131346.

Taylor AL, Sasser JN. 1977. *Biology, identification and control of root-knot nematodes (Meloidogyne spp.)*. North Carolina State University Graphics, Raleigh, North Carolina.

Thakur SK. 2014. Effect of green manuring and plant dry powder on soil properties and nematode infecting maize. *Agricultural Science Digest* 34: 56–59.

Thiessen LD, Schappe T, Cochran S, Hicks K, Post AR. 2020. Surveying for potential diseases and abiotic disorders of industrial hemp (*Cannabis sativa*) production. *Plant Health Progress* 21(4): 321.

Trotter A. 1938. L'incappucciamento e la *Peronospora* nei canapai della Campania. *Ricerche, Osservazioni e Divulgazioni Fitopatologiche, per la Campania ed il Mezzogiorno* 7: 57–63.

Trotuş E, Naie M. 2008. Cercetări privind reducerea atacului unor agenţi patogeni şi dăunători specifici culturilor de cânepă prin tratamentul chimic al seminţei. *Lucrari Stiintifice, Universitatea de Stiinte Agricole Si Medicina Veterinara "Ion Ionescu de la Brad" Iasi, Seria Agronomie* 51(2): 219–223.

Tsuge T, Sato T, eds. 2020. 日本植物病名目録. Japanese Society for Plant Pathology. Available at: https://www.ppsj.org/pdf/mokuroku/mokuroku202001.pdf

Tunçdemir U. 1987. Karadeniz bölgesínde kenevirlerde zararli Soğan ve sak nematodu (*Ditylenchus dipsaci* Kühn)'nun savas olanaklari üzerinde arastirmalar. *Bitki Koruma Bülteni* 27: 201–216.

Turner CE, ElSohly MA. 1981. Biological acitivity of cannabichromene, its homologs and isomers. *Journal of Clinical Pharmacology* 21: 283S–291S.

Van Biljon ER. 2007. "Important plant parasitic nematodes affecting fiber crops such as cotton and hemp in South Africa," pp. 63–69 in Anandjiwala R, *et al.*, eds. *Textiles for sustainable development*. Nova Science Publishers, New York.

Van Biljon ER. 2017. "Nematode pests of tobacco and fibre crops," pp. 285–310 in Fourie H, *et al.*, eds. *Nematology in South Africa: a view from the 21st century*. Springer International, Cham, Switzerland.

Vanha J, Stoklasa J. 1896. *Die Rübennematoden (Heterodera, Dorylaimus u. Tylenchus). Mit Anhang über die Enchytraiden*. Paul Barey, Berlin.

Vauterin L, Hoste B, Kersters K, Swings J. 1995. Reclassification of *Xanthomonas*. *International Journal of Systematic Bacteriology* 45: 472–489.

Vovlas N, Troccoli A, Palomares-Ruis JE, *et al.* (5 other authors). 2011. *Ditylenchus gigas* n. sp. parasitizing broad bean: a new stem nematode singled out from the *Ditylenchus dipsaci* species complex using a polyphasic approach with molecular phylogeny. *Plant Pathology* 60: 762–775.

Wahby I, Caba JM, Ligero F. 2013. *Agrobacterium* infection of hemp (*Cannabis sativa* L.): establishment of hairy root cultures. *Journal of Plant Interactions* 8: 312–320.

Watanabe T. 1947. 繊維作物病学. Asakura Shoten, Tokyo.

Weiss F. 1948. Check list revision. *Plant Disease Reporter* 32: 402–422.

Whittle WO, Drain BD. 1935. The root-knot nematode in Tennessee. *University of Tennessee Agricultural Experiment Station Circular No. 54*. Knoxville, Tennessee.

Wijnholds KH, Hoek H. 2009. *Effect van hennep op de populatiedichtheid van Pratylenchus penetran*. Praktijkonderzoek Plant & Omgeving B.V., Wageningen.

Winslow RD. 1954. Provisional lists of host plants of some root eelworms (*Heterodera* spp.). *Annals of Applied Biology* 41: 591–605.

Xiang N, Lawrence KS, Donald PA. 2017. Biological control potential of plant growth-promoting rhizobacteria suppression of *Meloidogyne incognita* on cotton and *Heterodera glycines* on soybean: A review. *Journal of Phytopathology* 166: 449–458.

Yolageldi L, Turhan G. 2005. Effect of biological seed treatment with *Cylindrocarpon olidum* var. *olidum* on control of common bunt (Tilletia laevis) of wheat. *Phytoparasitica* 33: 327–333.

Young JM, Dye DW, Bradbury JF, Panagopoulos CG, Robbs CF. 1978. A proposed nomenclature and classification for plant pathogenic bacteria. *New Zealand Journal of Agricultural Research* 21: 153–177.

Young JM, Kuykendall LD, Martínez-Romero E, Kerr A, Sawada H. 2001. A revision of *Rhizobium* Frank 1889, with an emended description of the genus, and the inclusion of all species of *Agrobacterium* Conn 1942 and *Allorhizobium undicola* de Lajudie *et al.* 1998 as new combinations: *Rhizobium radiobacter*, *R. rhizogenes*, *R. rubi*, *R. undicola* and *R. vitis*. *International Journal of Systematic and Evolutionary Microbiology* 51: 89–103.

Zhang QY, Li ZL, Han BJ, *et al.* (3 other authors). 2013. Immediate responses of cyst nematode, soil-borne pathogens and soybean yield to one-season crop disturbance after continuous soybean in Northeast China. *International Journal of Plant Production* 7: 341–353.

Zopf W. 1888. Ueber einen Nematoden fangenden Schimmelpilz. *Nova Acta der Kaiserlichen Leopoldinisch-Carolinischen Deutschen Akademie der Naturforscher* 52: 321–334.

16   Assorted parasitic plants, invertebrates, vertebrates, and weeds

Abstract

Other threats lie beyond insects, mites, fungi, nematodes, and bacteria. Parasitic plants can pose intractable problems, especially broomrape (*Orobanche* and *Phelipanche* spp.) and dodder (*Cuscuta* spp.). Vertebrate pests are not trivial (birds, rodents, rabbits, deer). Evidence suggests birds in particular coevolved with *Cannabis* and ate hemp seeds long before humans came along. The balance of protein and oil in hemp seeds seems optimized for bird nutrition, and they know it. Invertebrates other than insects and mites cause problems (slugs, snails, sowbugs, symphylans). Weed science is a separate specialty that has not received the attention it deserves from cannabinologists. Nearly all these assorted problems affect outdoor crops. Control measures are discussed and new ones proposed.

16.1   Parasitic plants: broomrape

Dodder (*Cuscuta* spp.), broomrape (*Orobanche* and *Phelipanche* spp.), and mistletoe (*Viscum* spp.) are obligate plant parasites. They have modified roots, called haustoria, that penetrate the host plant's vascular system, from which they withdraw nutrients. Albertus Magnus (AD 1200–1280) first applied a control measure against a parasitic plant. He pruned mistletoe out of an oak, and the sick tree got well (Agrios, 2005). It was a miracle; he was canonized and declared the patron saint of the natural sciences.

Broomrape was the first plant parasite recognized on *Cannabis* (Mattioli, 1544), prior to any descriptions of fungi or even insect pests. Broomrape's threat to hemp led to books and monographs on the subject. Pier Antonio Micheli (1679–1737), a founder of plant pathology, wrote the first monograph (Micheli, 1723). Many others followed (Guettard, 1751; Bonanome, 1766; Vaucher, 1827; Casali and Marconi, 1876; Beck-Mannagetta, 1890; Garman, 1903; Munerati, 1929; Nicot, 1959; Barloy, 1962).

16.1.1   Symptoms

Broomrapes parasitize roots and do most of their damage underground and unseen. Infested plants become stunted, sickly, and die prematurely. Surviving plants have weak stalks and yield imperfect fiber. Only briefly do broomrapes become visible above ground, when they send up flowers near the end of the season. Broomrapes starve their hosts and their wounds provide portholes for root-rot fungi. Barloy and Pelhate (1962) considered a combination of *Phelipanche ramosa* (broomrape) and *Fusarium solani* (root-rot fungus) as the greatest threat to hemp cultivation in southern France.

16.1.2   Causal organisms

Four broomrape species parasitize hemp. Phylogenetic studies show that *Orobanche sensu lato* is not monophyletic, but segregates into several clades, including *Orobanche sensu stricto* and *Phelipanche* spp. (Manen *et al.*, 2004). *Orobanche* spp. have unbranched inflorescences and their flowers lack bracteoles; *Phelipanche* spp. are usually branched and have bracteoles (Parker, 2013). Many broomrape species have narrow host ranges and perennial life histories, although the four species on *Cannabis* have wide host ranges and annual life histories.

16.1a. Branched broomrape *Phelipanche ramosa* (L.) Pomel 1874, Lamiales; Orobanchaceae

≡*Orobanche ramosa* L. 1753, ≡*Phelipaea ramosa* (L.) C.A. Meyer 1831, =*Orobanche du chanvre* Vaucher 1827, ≡*Orobanche cannabis* Vaucher ex Duby 1828, =*Orobanche cannabios* Schultz 1830

Description: Shoots arise near the base of hemp stems (Fig. 16.1). Prior to flowering the shoots resemble pale *Asparagus* but with a bulbous base. They usually branch, average 10–20 cm in height, and bear brownish-yellow scalelike leaves 3–8 mm long. They produce more than ten flowers, ranging in color from whitish to pale blue to deep blue, with a tubular corolla supported by a bract and two small, lateral and opposite bracteoles. Corolla usually <20 mm long, anther filaments not densely hairy, arising from base of corolla. Flowers mature into resinous capsules filled with 600–800 seeds. Seeds are tiny, ovoid, 300 µm long, dark brown, with reticulations dividing the surface into polygonal areas.

Comments: Dewey (1914) characterized *P. ramosa* as "the only really serious enemy to hemp." He wrote a 270-word summary of its disease cycle, control measures, and provided a

DOI: 10.1079/9781836990352.0016

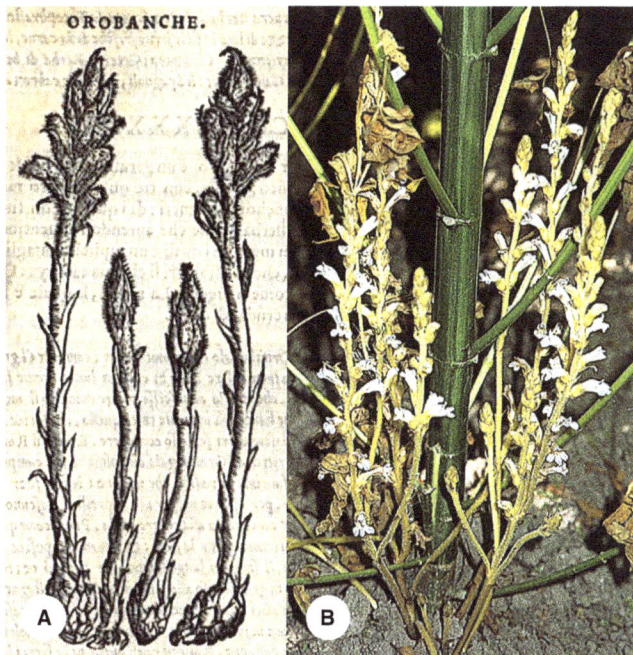

Fig. 16.1. *Phelipanche ramosa*: **A.** illustration in Mattioli (1554); **B.** growing on hemp (McPartland *et al.*, 2000).

photograph. Rataj (1957) described *P. ramosa* "raging" on hemp in Europe. Barloy (1962) reported 30% losses in fields that averaged 5–6 broomrapes per hemp plant, yet 15–20 broomrapes per hemp plant were not uncommon. In bad infestations, with 50–60 broomrapes per plant, the crop was abandoned, the land condemned, and cultivation banned.

P. ramosa is the most widespread broomrape in the world. China is often given as its native range (e.g., Garman, 1903; Wilhelm, 1958; Musselman, 1994). Mastromarino (1937) suggested that *Cannabis* was its original host in China. However, *P. ramosa* is not included in *Flora of China* (Zhang and Tzvelev, 1998). Sannong (2016) wrote that 大麻列当 (*dà má lièdāng*, great hemp broomrape) "has not yet been discovered in China." Sannong was referring to eastern China, because *P. ramosa* parasitizes tomatoes in far-western Xīnjiāng Region (Yao *et al.*, 2017).

Parker (2013) suggested a native range in the Middle East. Boissier (1879) documented early discoveries of *P. ramosa* by European botanists in Turkey and Syria, where it is still widespread (Gilli, 1982). Piwowarczyk (2012) proposed, "The species probably travelled to Central Europe from Asia with hemp crops *ca.* 500 BC." Gonsior *et al.* (2004) suggested the species spread to Europe in the 17th century.

Our literature review suggests *P. ramosa* spread from Turkey to Italy in the 1500s. Merchants from the Republic of Venice traded with the Ottoman Empire throughout the 15th–16th centuries (when they weren't fighting). *P. ramosa* spread to Europe more than once: Nicot (1959) described a new and more virulent race of *P. ramosa* in French hemp fields. He associated it with hemp seed imported from Turkey between 1920 and 1939.

Mattioli (1544) discovered *l'Orobanche* in the Anania (Non) Valley above Trento. "He kills legumes, hemp, and cereals who are born around him only with his presence without

his touching them, or clinging to their feet; the workers crown him *herba lupa* [wolf herb], who devours plants which are born next to him." Mattioli (1554) subsequently illustrated *l'Orobanche* (Fig. 16.1), and Sansovino (1560) of Venice echoed Mattioli's description of *l'Orobanche* killing hemp.

Herbaria, or collections of dried plants, were first assembled in Italy. An anonymous "Rome herbarium," which dates to 1550–1553 (Stefanaki *et al.*, 2019), has two specimens of *P. ramosa*, and one includes the lower stalk and roots of a *Cannabis* plant (Penzig, 1905). The anonymous En Tibi herbarium, made in Bologna around 1558, also includes a specimen of *P. ramosa* (Stefanaki *et al.*, 2018). In Pisa, Andrea Cesalpino collected *P. ramosa* in the 1560s (Caruel, 1858). He assigned a polynomial name to the species: *Orobanche altera brevior et ramosa* (Cesalpino, 1583).

Ulisse Aldrovandi collected several *P. ramosa* specimens for his herbarium in Bologna. They are undated, but labeled with Cesalpino's *ramosa* name. Aldrovandi (1600) wrote about *Orobanche*, "For rightly the farmers hate this herb since it draws away nourishment from the *Cannabi* in which they put all hope of profit, and because, taking birth stuck on its roots, keeps it from growing to its proper height." Soldano (2000) translated an unpublished manuscript by Aldrovandi: "*Orobanche herba taura* is called in Bologna *Erba scalogna* [bad luck herb] because of the strong damage it produces to hemp crops."

Leonhart Fuchs in Germany next mentioned broomrape. In a manuscript unpublished at his death (in 1566), Fuchs described *Orobanche aphyllos multicaule* on hemp (Fuchs, 2001). The illustration clearly depicts *P. ramosa*. The specimen may have come from the Italian botanist Luca Ghini, with whom Fuchs traded herbarium specimens and illustrations (Findlen, 2017).

Mattioli's text was edited and enlarged by Joachim Camerer, who provided new data: "This emaciating *Orobanche* is not universal, it is found in many places, and is called *ramosa* because it has a lot of stems. In Thüringia [central Germany] it is called *Hanffman* because it likes to grow under hemp" (Mattioli and Camerer, 1586). In France, d'Aléchamps and Bauhin (1587) used two names interchangeably, *Orobanche* and *Hamoderos* (Pliny's name for broomrape). "*Hamoderos* is found in fields of hemp. She grows close to the hemp root, glued against it, and takes away hemp's nourishment. Because *Hamoderos* is full of juice and succulent, she needs a lot of food, which she cheats from hemp." Bauhin (1622) coined the binomial *Orobanche ramosa* and said that it infested hemp "especially in Michelfeld" (Baden-Württemberg). Parkinson (1640) utilized Bauhin's binomial, *Orobanche ramosa*. He named the same three hosts as Mattioli: "corne, pulse, hempe."

The English botanist Ray (1693) said he saw *P. ramosa* in Italy, France, and Germany, implying it wasn't in England. But two decades later, Salmon (1710) described broomrape in England, "which is parted towards the top into several branches, whose flowers are either blew, purplish or white, and is sometimes found amongst hemp." Dillenius posthumously revised Ray (1724), and by then *P. ramosa* was "growing plentifully" in England. Hemp cultivation in England nosedived thereafter, and 70 years later, Sowerby (1794) described *P. ramosa* as a "rare plant."

Italian hemp, along with *P. ramosa*, spread to eastern Europe in the 1740s: Empress Maria Theresa sent Italian hemp growers to Bačka, in Hungary and Serbia (Berenji, 1996; Butter, 2001). In a Hungarian flora, Jósef (1775) said that *Orobanche ramosa*, which grew on hemp, was known as *Misegyertya* ("Mass-candle"). *P. ramosa* also reached Poland around then (Piwowarczyk, 2012). In Russia, an agricultural dictionary stated that коноплей заразиха, hemp broomrape, was first mentioned in 1769 (Dyigubsky, 1836). Beck-Mannagetta (1890) wrote a monograph on *P. ramosa*, and by then it grew throughout Europe, except Scandinavia and northern Russia.

P. ramosa reached Kentucky in the 1800s, and threatened "the end of hemp" (Garman, 1889). **Harrison Garman** (1856–1944) (Fig. 16.2) served as Kentucky's state entomologist from 1887 to 1920. He published many articles on hemp pests: cutworms (Garman, 1892, 1895), orthopterans (Garman, 1894a,b, 1904b), wireworms (Garman, 1892), root maggots (Garman, 1904a), flea beetles (Garmon, 1906), even grackles (Garman, 1907), but especially *P. ramosa*, "the bitterless foe" (Garman, 1889, 1890, 1892, 1903, 1905).

Because Chinese hemp seed was imported into Kentucky, Garman (1890) said it was "probably safe to assume" that *P. ramosa* came from China, but "I have been thus far unable to find any positive record of its occurrence there." A Kentucky landowner told Garman that "old colored men" observed *P. ramosa* in hemp fields 50 years prior to 1890, or about 1840. Around then, Clay (1837) acknowledged receipt of foreign hemp seed from David Aiken Hall—likely from Turkey, not China. Hall worked for the American Board of Commissioners for Foreign Missions, which was active in Turkey.

Fig. 16.2. Harrison Garman (courtesy of University of Kentucky)

A specimen of *P. ramosa* was collected near the border of Tennessee and North Carolina in 1884 (Musselman, 1984). Hemp growers in Illinois obtained seed from Kentucky, and *P. ramosa* grew in Illinois by 1890 (Garman, 1903) and became worse in 1895 (Garman, 1903). Subsequently it appeared on hemp in Wisconsin (Weiss, 1948). *P. ramosa* reached California by 1929, presumably via contaminated hemp seed (Anonymous, 1929).

Subspecies and pathovars: *P. ramosa* is morphologically variable. Some botanists split it into three subspecies: subsp. *nana*, subsp. *ramosa*, and subsp. *mutelii*. These differ in the length of their corollas, a gradient between subsp. *nana* (10–15 mm) and subsp. *mutelii* (22 mm) (Parker, 2013). The three subspecies form a clade in phylogenetic studies (Manen *et al.*, 2004). Together they attack over 60 plant species, the main hosts being hemp, tobacco, potato, tomato, crucifers, and beans.

P. ramosa may be split into pathovars—populations with distinctive pathogenicity towards one or more plant hosts. Vaucher (1827) first identified *P. ramosa* pathovars, by conducting cross-virulence tests. Broomrape collected from hemp could not infect beans. Vaucher thought this warranted differentiation at a species level, and erected *Orobanche cannabis* (now a synonym of *P. ramosa*).

Benharrat *et al.* (2005) recognized two pathovars on the basis of ISSR genetic markers and cross-virulence tests: pathovar A infested winter rape; pathovar B infested hemp and tobacco. They differed in life cycles: pathovar A attached to its host in autumn, pathovar B attached to its host in spring. Thus pathovar A had a 40-week life cycle, and pathovar B had a 16-week life cycle. Brault *et al.* (2007) recognized three pathovars, on hemp, winter rape, and tobacco, based on RAPD markers and cross-virulence tests.

The hemp pathovar may be divided into strains or races: Pasković (1941) said an Italian race could infest nearly all hemp cultivars, whereas a Russian race only caused problems in Russian cultivars, not southern Ukrainian or Italian cultivars. Khrennikov and Tollochko (1953) distinguished three physiological races or "biotypes" of *P. ramosa* infesting different Russian hemp cultivars.

16.1b. Egyptian broomrape *Phelipanche aegyptiaca* (Persoon) Pomel 1874

≡*Orobanche aegyptiaca* Persoon 1806
=*Orobanche indica* Buchanan-Hamilton ex Roxburgh 1832
Description: *P. aegyptiaca* closely resembles *P. ramosa*, but tends to be taller (20–45 cm) with larger leaves (5–12 mm). The corolla is longer, usually 20–35 mm; anther filaments densely hairy.

P. aegyptiaca ranges from the eastern Mediterranean region to India, and attacks the same plants as *P. ramosa*. Under the synonym *O. indica*, Prain (1903) reported *P. aegyptiaca* parasitizing *Cannabis* in Bengal, India. Parker (1986) reported *P. aegyptiaca* on *Cannabis*, but without a location record. Pohle (2001) listed both *C. sativa* and *P. aegyptiaca* as members of the "tropical" flora in Kāgbeni, Nepal.

16.1c. Tobacco broomrape *Orobanche cernua* Löfling 1758

=*Orobanche cumana* Wallroth

Description: Shoots usually not branched, 20–30 cm tall, leaves 5–10 mm long. Deep blue-purple flowers do not have bracteoles, corolla 12–20 mm long, anthers arise well up the corolla at least 4 mm distal from base.

O. cernua is a serious pest of tobacco, sunflower, and *Artemisia* spp. It is found around the Mediterranean region, east to the Arabian Sea, and Australia. *O. cernua* has been cited on *Cannabis* in Romania (Ceapoiu, 1958) and India (Marudarajan, 1950). Parker and Riches (1993) claimed that *Cannabis* served as a trap crop to induce *O. cernua* germination, and they cited Krishnamurthy *et al.* (1977), but that paper does not mention *Cannabis*.

Pujadas-Salva and Velasco (2000) addressed the relationship between *O. cernua* and *Orobanche cumana* Wallroth 1825. *O. cernua* and *O. cumana* are synonymous according to some but not others. *O. cumana* has not been reported on *Cannabis*, although extracts made of *Cannabis* roots can induce *O. cumana* seed germination (Yu and Ma, 2014). Sacchetti (1940) pulled into the *cernua–cumana* debate a third: *Orobanche crenata* Forsskål 1775. He claimed *O. crenata* infested *Cannabis*, and cited Chabrolin (1934), but Chabrolin did not mention *Cannabis*.

16.1d. Amethyst broomrape *Orobanche amethystea* Thuillier 1799

Description: This robust species produces large shoots, up to 50 cm tall, with a dense inflorescence. Flowers lack bracteoles, corolla 12–20 mm long, colored amethyst (light to dark violet), orange-yellow, or yellowish-white.

O. amethystea was collected on *Cannabis* in Morocco (Rumsey and Carine 2002). The species also infests *Eryngium campestre*, *Daucus carota*, *Vicia sativa*, and *Ballota* spp. (Parker and Riches, 1993). It grows around the Mediterranean, including southern Europe and northern Africa.

16.1.3 Disease cycle

Broomrape seeds germinate in the presence of host root exudates. *Cannabis* root tissues, aqueous extracts of roots, and even the soil surrounding roots are potent stimulators of seed germination (Yu and Ma, 2014). The seed germination tube (radical) is chemotactically guided towards host roots. This was illustrated by Garman (1903), who noted the hairpin turn in Fig. 16.3A. The radical drills a haustorium into the root, and penetrates phloem vessels.

As *P. ramosa* gains nutrients, it swells into a white, spherical nodule. The nodule becomes studded with prominences, which push out slender roots. An apical prominence grows into a "bulb" (Fig. 16.3B), from which the shoot emerges from soil. The shoot quickly flowers and sets seed. A single plant can produce 500,000 seeds (Barloy and Pelhate, 1962) or even a

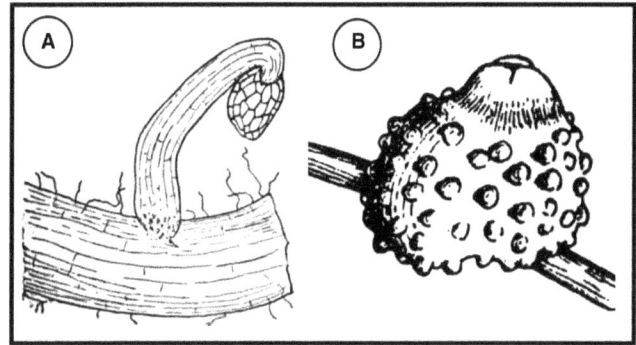

Fig. 16.3. *Phelipanche ramosa*: **A.** Garman (1903); **B.** Garman (1890) (not to same scale)

million (Piwowarczyk, 2012). One gram of *P. ramosa* seed consists of 200,000 individuals.

The flowers and seed capsules are sticky. Garman (1903) observed broomrape shoots stuck to workmen's shoes and harvest equipment, which contributed to its spread. The stickiness is sufficient that Fournier and Paris (1983) detected THC and CBD in flowers and seeds of *P. ramosa*—no doubt from tiny bits of *Cannabis* stuck to the broomrape sample. Garman proposed that *P. ramosa* spread around the world while stuck to hemp seeds transported by humans. *P. ramosa* seeds reach maturity the same time that hemp seed matures.

Seeds in soil can spread by flood water. In Kentucky, infestations spread downstream along the Kentucky River (Garman, 1903). Barloy (1962) said that 500 ha of floodplain in Maine-et-Loire was so heavily infested that hemp cultivation was banned. A century prior, Bertolini (1861) said that Bologna's most fertile land could no longer support hemp cultivation because of *P. ramosa*. The seeds can remain dormant in soil for up to 13 years (Garman, 1903). The power of *P. ramosa* lies in its ability to form a bank of seeds in the soil. Sustainable control requires a reduction of the seed bank. A combination of cultural, mechanical, biological, and chemical control may be needed.

16.1.4 Cultural and mechanical control

Micheli (1723) advised farmers to gather broomrape shoots before they set seed. Guettard (1751) noted the difficulty of plucking broomrape out of a densely sown fiber crop. Hand-pulling *P. ramosa* was not practical, because the parasite was firmly attached to roots. In loose soil the roots came up too. By the time *P. ramosa* sends up a shoot, its damage has already been done.

Garman (1889) stated emphatically that hemp shocks should not be threshed on ground tarps where seeds might get contaminated with broomrape seeds. Even then, hemp seed from fields infested with broomrape should not be planted for next year's crop. Vaucher (1827) suggested screening out broomrape seeds from hemp seeds. Garman (1890) emphasized the difficulty this presented. Screening requires meticulous attention, because broomrape seeds are tiny (a 20th the size of hemp seeds), and adhere to hemp seeds. Rather than screening, Garman (1903) killed broomrape seeds by immersing contaminated hemp seed in a 60°C water bath for 10 min; this treatment did not harm hemp seed.

Vaucher (1827) recommended crop rotation with non-host plants. Piwowarczyk (2012) recommended crop rotation for a period of at least 5 years. Garman (1890) bemoaned crop rotation, because *P. ramosa* seeds could live up to 13 years in soil. He reported a case where 34 acres of contaminated land was rotated between wheat and clover, and the fourth year back to hemp, and broomrape came back as well. In his opinion, farmers should burn badly infested fields and switch to wheat (Garman, 1889).

Broomrape does its worst in infertile soils; adding soil nutrients may reduce broomrape damage. Garman (1905) amended infested soil with muriate of potash (KCl), but at too high a rate—1.5 tons/acre. This decreased the broomrape population but was injurious to hemp. He also tested unslaked lime (CaO) at a high rate of 2.75 tons/acre, which suppressed broomrape and increased hemp yields. Abu-Irmaileh (1981) reduced *P. ramosa* infestation with two nitrogen supplements, ammonium nitrate (NH_4NO_3) and ammonium sulfate ((NH_4)$_2SO_4$). *P. ramosa* seed germination was reduced by ammonium nitrate (Abu-Irmaileh, 1994). Piwowarczyk (2012) claimed that goat manure suppressed broomrape.

Solar sterilization is effective in hot areas with clear skies. Jacobsohn *et al.* (1980) laid sheets of plastic (clear polyethylene 0.1 mm thick) on soil infested with broomrape. Drip irrigation under the plastic kept the soil moist to increase heat conductivity. After 36 days the plastic was removed, seeds were sown, and crops were free of broomrape (90% of control plants were infested).

The search for resistant varieties began with Casali and Marconi (1876*)*. 'Carmagnola' was more resistant than 'Bologna', which they attributed to faster seed germination. 'Carmagnola' also responded to fertilization (with nitrogen and phosphorus) more so than 'Bologna', which allowed it to outgrow *P. ramosa*. Samoggia (1898) claimed that roots of 'Carmagnola' contained more cellulose than 'Bologna', making it harder for *P. ramosa* to penetrate.

Zambrano (1910) noted that Turkish 'Pelosella' was more susceptible to *P. ramosa* than Italian landraces. Munerati (1929) measured susceptibility by the percentage of infected plants in three Italian landraces: 'Ortichina', 75–95% of plants; 'Bolognese', 51–88%; and 'Carmagnola', 22–33%. Savelli (1932) and Pasković (1941) also recommended 'Carmagnola'. Gikalov (1936) measured susceptibility by the number of plants infested with *P. ramosa* per square meter: Italian hemp, 0.2/m^2; Ukrainian hemp, 0.9–1.2/m^2; Russian hemp, 23.3/m^2. Ceapoiu (1958) reported that his Romanian cultivar 'ICAR 42-118' was resistant, as was 'G-3' from Bulgaria. 'ICAR 42-118' is a hybrid of Italian and Turkish hemp. 'G-3' is a selection of Italian hemp.

Bócsa and Beke (1956) reported that Chinese varieties grown in Hungary were highly susceptible to *P. ramosa*. Safra (1959) identified three resistant Russian cultivars: 'Yuzhnaya Krasnodarskaya', 'Yuzhnaya Pavlogradskaya', and 'Yuzhnaya Cherkasskaya'. All Yuzhnaya (Южная, "Southern") varieties were bred by Vasily A. Nevinnykh in the early 1950s, with various degrees of Italian ancestry (Nevinnykh and Schukina 1972).

Barloy (1962) conducted the largest resistance study ever done on hemp, for any pest or pathogen. His team assembled 120 landraces and cultivars, and tested them on naturally infested soil. French monoecious cultivars were susceptible (the 'Fibrimon' series), due to their narrow genetic base—ancestry from a susceptible monoecious Turkish variety ('Faza monoïque'). Conversely, a dioecious Turkish landrace, 'Tire Keneviri', was resistant. Four other varieties were very resistant: two Greek landraces, a Libyan landrace, and 'Fertodi' from Hungary. Barloy confirmed that Italian varieties were fairly resistant, as was 'ICAR 42-118'. Conversely, 'G-3' was moderately susceptible. The 'Kentucky' cultivar was very susceptible, as were Chinese varieties. All drug-type plants were very susceptible, from Morocco, Iran, India, and Belgian Congo.

Dempsey (1975) cited resistance in Hungarian 'Kompolti V-7L' and Russian 'JUS-1', 'JUS-87', and 'Juznoja Odnodomnoja'. Senchenko and Kolyadko (1973) tested 26 cultivars for resistance and ranked them accordingly. Most of those cultivars are no longer available (de Meijer, 1995).

16.1.5 Biological and chemical control

Crop rotations can be enhanced by sowing fields with trap-crops or catch-crops. **Trap-crops** are "false hosts." Root exudates from these plants stimulate germination of *P. ramosa* seeds, but the plants do not become infected ("suicide germination"). Trap-crops for *P. ramosa* include bean (*Phaseolus vulgaris*) and sorghum (*Sorghum bicolor*) (Parker and Riches, 1993), and white mustard (*Brassica alba*) (Lucas, 1975).

Catch-crops are plants that are susceptible to *P. ramosa*. Catch-crops are sown in a contaminated field, which induces broomrape seed generation. Then the crops are killed by tilling them into the soil. Guettard (1751) recommended *Xanthium* as a trap-crop to reduce *P. ramosa* in hemp, but *Xanthium* is somewhat susceptible to *P. ramosa*, so it is a catch-crop. Other *P. ramosa* catch-crops include flax (*Linum usitatissimum*), Asteraceae (*Sonchus* and *Senecio* spp.), and Brassicaceae (*Capsella*, *Arabidopsis*, and *Sinapis* spp.). *Cannabis* has been suggested as a catch-crop to control *Orobanche cernua* (Parker and Riches, 1993), and *Phelipanche aegyptica* and *Orobanche cumana* (Yu and Ma, 2014).

Inoculating tomatoes with *Glomus mosseae*, a beneficial mycorrhizal fungus, protected them from *P. ramosa* (López-Ráez *et al.*, 2012). *G. mosseae* also colonizes *Cannabis* roots (McPartland and Cubeta, 1997). Plant roots secrete strigolactones which attract mycorrhizal fungi. Seeds of *P. ramosa* eavesdrop on this plant-to-fungus communication to sense host roots. Once roots are colonized by mycorrhizal fungi, strigolactone production is downregulated. Therefore mycorrhizal colonization results in less *P. ramosa* germination (López-Ráez *et al.*, 2012).

Tobacco farmers in Bulgaria allowed geese to graze in *P. ramosa*-infected fields. Two geese/ha controlled broomrape (Lucas, 1975). Gonsior *et al.* (2004) controlled *P. ramosa* in hemp by drenching soil with Proradix®, which contains a rhizosphere bacterium, *Pseudomonas* sp. strain DSMZ. Treatment of potted plants reduced *P. ramosa* by 80% compared with controls, at a field rate of 1.6 kg/ha.

Insect pests of *Orobanche/Phelipanche* have attracted research. Forty contenders were reviewed by Klein and Kroschel (2002). They focused on the lead candidate, *Phytomyza orobanchia*, describing the fly and its life cycle. Kaltenbach (1864) first discovered *P. orobanchia* maggots boring into broomrape roots, shoots, and flowers. Garman (1903) suggested deploying *P. orobanchia* against broomrape in hemp. Russian scientists released *P. orobanchia* pupae at a rate of 500–600 pupae/ha, which reduced seed production by 80–90% (Kapralov, 1975). Unfortunately, the larvae are culled by their own hyperparasites.

Several fungi have been tested, reviewed by Boari and Vurro (2004). The lead candidates are *Fusarium orthoceras* (Appel and Wollenweber, 1910) and *Fusarium orobanches* (Jaczewski, 1912). Formin (1936) developed a product from *F. orobanches* in Russia called "Preparation F." It reportedly killed 60–80% of *P. ramosa*. Formin (1954) continued to promote Preparation F, but it was never commercialized. *Fusarium orthoceras* was synonymized under *Fusarium oxysporum* by Snyder and Hansen (1940). Bilaĭ (1955) recognized *F. oxysporum* var. *orthoceras*. Thomas *et al.* (1998) reported that *F. oxysporum* f. sp. *orthoceras* infected and killed shoots of *O. ceruna* before they produced seeds, and underground tubercles suffered 85% mortality. Boari and Vurro (2004) isolated *F. oxysporum* from *P. ramosa* (not designating a subspecies) that reduced the number of shoots by 70% compared with controls.

Turning to chemical control, Garman (1903) sprinkled 0.75 oz of salt (NaCl) into a box with 0.5 cubic ft (14.2 liters) of soil and planted hemp seed mixed with broomrape seed. It killed most of the *P. ramosa*, but "it was plain that the salt had a very injurious effect on the hemp." In a field trial, Garman (1905) mixed salt at a rate of 1.5 tons/acre into an infested field, which yielded 900 lb of hemp fiber per acre, compared with 640 lb/acre in the control.

Gonsior *et al.* (2004) induced resistance against *P. ramosa* in hemp by applying two biorational pesticides as soil drenches: a salicylic acid derivative (Bion®), and extracts of the algae *Ascophyllum nodosum* (Fruton®). Both reduced *P. ramosa* infestation in potted plants by 92% and 84%, respectively, compared with controls. Fruton® caused some chlorosis in hemp leaves, Bion® showed no phytotoxic effect.

Chemicals that induce suicide germination have been investigated. Johnson *et al.* (1976) synthesized analogues of strigol (a germination stimulant of *Striga* spp.) that stimulated *P. ramosa* germination. Blanco-Ania *et al.* (2019) demonstrated the potential of "suicide germination agents" against *P. ramosa* isolated from *Cannabis* plants. One commercialized strigol analogue, Nijmegen-1, induced suicide germination of *P. ramosa* when applied at a rate of only 6.25 g/ha. It was sprayed evenly on the soil surface, and then watered into the soil with irrigation. Within 2 weeks, many *P. ramosa* seeds germinated and died (Zwanenburg *et al.*, 2015).

16.2   Parasitic Plants: Dodder

Dodders are members of Convolvulaceae, the bindweed family. Bindweeds can spiral up fence posts, whereas dodders must twine around living plants. Unlike broomrapes, which live in soil and have roots, dodders escape from the soil and grow rootless. Dodders must therefore invade their host's xylem, in addition to phloem. Dodders drain plants of water and nutrients, and their tangles cause mechanical damage. Dodders can vector viruses as they grow from plant to plant. Rasoulpour *et al.* (2017) documented that a dodder (*Cuscuta campestris*) could vector a phytoplasma from *Cannabis* to periwinkle (*Catharanthus roseus*).

Hemp dodder went unmentioned by early botanists. They wrote about thyme dodder (*Cuscuta epithymum* L.) and flax dodder (*Cuscuta epilinum* Weihe). Both species epithets combine *epi-* (upon), with the host name. Thus Salmon (1710) coined the name *Epicannabis* for hemp dodder—a taxon ignored by Linnaeus, and Salmon's *Epicannabis* disappeared from the literature. Linnaeus (1749) referred to hemp dodder as *Cuscuta* (no species name).

16.2.1   Symptoms

Dodder arises as a conspicuous tangle of yellow filaments, bearing vernacular names such as "gold thread," "hair weed," and "love vine." They twine themselves around stems and branches (Fig. 16.4). Robust specimens girdle branches and pull down their hosts. *Cannabis* stalks sometimes form galls at the sites of hausorial penetrations, as illustrated by Dean (1937).

16.2.2   Causal organisms

Eight *Cuscuta* spp. have been pulled off hemp. *Cuscuta* spp. tend to have a wide host range, and they easily spread through shipments of host seeds, so most species are now worldwide.

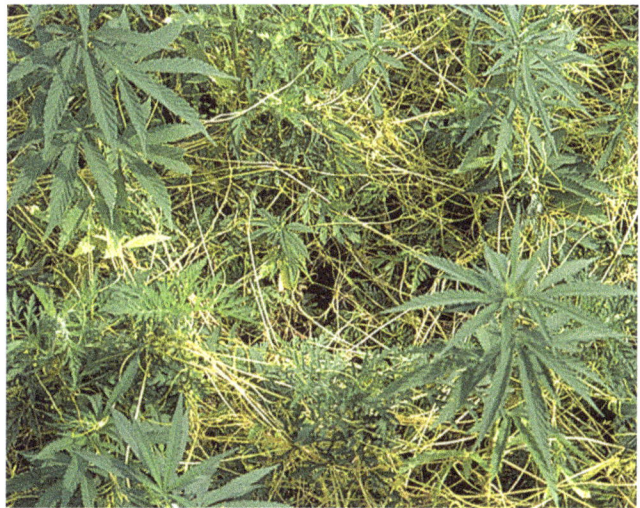

Fig. 16.4. Hemp infested by *Cuscuta europaea* (McPartland *et al.*, 2000)

16.2a. *Cuscuta europaea* L. 1753

= *C. major* Bauhin 1623, = C. *europaea* var. *indica* Engelmann 1859
Known as greater dodder, *C. europaea* has a much wider native distribution than its name implies: Europe, North Africa, West Asia, the Himalaya, and East Asia. Eastern populations go by *C. europaea* var. *indica*. As an invasive, *C. europaea* arrived recently in Maine, USA. It is the most commonly cited *Cuscuta* on *Cannabis* (Durande, 1782; Lamarck and de Candolle, 1805; Schmitz and Regel, 1841; Kirchner, 1906; Transhel *et al.*, 1933; Ferraris, 1935; Flachs, 1936; Ciferri, 1941; Dobrozrakova *et al.*, 1956; Barloy and Pelhate, 1962; Serzane, 1962; Vakhrusheva, 1979; Kojić and Vrbnicanin 2000).

    C. europaea entwines around many herbaceous plants with little host specificity, though nettles (*Urtica dioica*) and hops (*Humulus lupulus*) are particularly favored hosts. Schlechtendal *et al.* (1848) summarized, "A summer plant that feeds on *Urtica dioica*, *Humulus lupulus*, and *Cannabis sativa*, but then spreads from these locations and grows with other neighboring plants, *e.g.* with willows, chenopods, *etc.*" Due to its wide distribution and host range, *C. europaea* has many taxonomic synonyms, one of which was cited on *Cannabis* by Rataj (1957): *C. major* Bauhin. Salmon's "*Epicannabis*" was probably *C. europaea*. Peglion (1906) photographed a flowering top so tightly entwined by *C. europaea* that it looked like a Thai stick.

16.2b. *Cuscuta epithymum* (L.) L. 1774

Lesser dodder has been cited on hemp (Boisduval, 1828). It is native to Europe, West Asia, and northern Africa, but now occurs everywhere. Unlike other dodders, *C. epithymum* has red pigmentation. It is best known as a *Thymus* spp. parasite.

16.2c. *Cuscuta campestris* Yunker 1932

This is the common field dodder in North America. It has entwined drug-type *Cannabis* in Oregon and Pennsylvania, and feral hemp in Illinois (McPartland, pers. obs., 1987). *C. campestris* has spread east, to *Cannabis* in Romania (Ceapoiu, 1958), Syria (Malmuk, 1980), India (Bhellum and Rani, 1996), and Iran (Rasoulpour *et al.*, 2017). *C. campestris* infests a wide range of plants, especially herbaceous dicots such as clover (*Trifolium* spp.) and alfalfa (*Medicago* spp.).

16.2d. *Cuscuta pentagona* Engelman

This may be a subspecies of *C. campestris*. It is native to North America but now lives everywhere (Parker and Riches, 1993). *C. pentagona* has attacked hemp in Serbia (Stojanovic, 1959).

16.2e. *Cuscuta epilinum* Weihe 1824

Flax dodder is found wherever humans have transported *Linum usitatissimum* (Parker and Riches, 1993). It is a fast-growing plant; one individual can infest up to 120 flax plants (Larina, 2005). *C. epilinum* has been reported on *Cannabis* in Italy (Ferraris, 1935), Serbia (Kojić and Vrbnicanin, 2000), Russia (Larina, 2005), and Kenya (Nattrass, 1941).

16.2f. *Cuscuta racemosa* Martius 1823

Dewey, (1914) reported this "Chilean dodder" tangling hemp in California. The species normally infests *Trifolium* and *Medicago* spp. It is native to South America but now occurs in North America and Europe.

16.2g. *Cuscuta gronovii* Willdenow 1820

This species infested hemp in Iowa (Dean, 1937). It is native to North America east of the Rocky Mountains, and has spread to Europe.

16.2h. *Cuscuta reflexa* Roxburgh 1798

Sharma and Kapoor (2014) reported this species on Indian hemp in Uttar Pradesh. Its distribution extends from Afghanistan, through South Asia and Indonesia, to southern China.

16.2.3   Description

Dodder twines are threadlike, flexous, 1–2 mm wide, orange-yellow in color, sometimes tinged with red or purple, sometimes almost white. Tendrils arise at frequent intervals along the stems, twining around host stems. Leaves are reduced to minute scales, arising opposite tendrils. Differentiating *Cuscuta* spp. by flower structures is not easy. They all produce small (1–4 mm), short-lived flowers. The eight *Cuscuta* spp. cited on hemp produce two styles per flower; some stigmas have knobs, some do not. Seeds are reddish-brown, 1–2 mm long. For more information see Parker and Riches (1993).

16.2.4   Disease cycle

Each dodder plant produces many thousands of seeds, from late summer until frost. Seeds fall to soil, where they can survive for up to a decade. Or seeds fall into *Cannabis* flowering tops, and slip into the seed harvest. Dodder seeds germinate in the spring after soil is well-warmed. Seedlings grow a rudimentary root and send up a heliotropic (sun-seeking) tendril. They use plant volatiles as chemo-attractants to help find their hosts. Some chemo-attractants, such as β-myrcene, a-pinene, and β-phellandrene (Runyon *et al.*, 2006) are produced by *Cannabis*. After the tendril contacts a host, the root atrophies and the parasite escapes from the soil (Fig. 16.5).

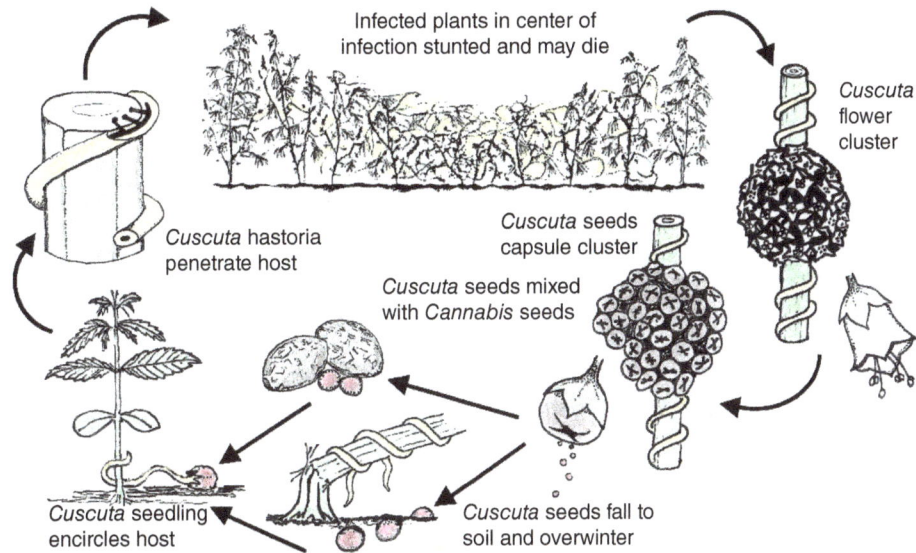

Fig. 16.5. Disease cycle of *Cuscuta europaea* (adapted from Agrios, 2005)

16.2.5   Control methods

Ceapoiu (1958) stated the obvious: hemp seed from infected crops should not be planted. If sowing such seed is necessary, carefully screen all alfalfa-sized dodder seed. Dodder seed also spreads via contaminated field equipment.

Once established in a field, getting rid of *Cuscuta* is difficult. Hand-pulling tendrils off plants always leaves behind fragments. Badly infested *Cannabis* crops, full of *Cuscuta* seed capsules, have been burned. Rotate with crops resistant to dodder. Depending on the *Cuscuta* sp., these may include beans (*Phaseolus vulgaris*), squash (*Cucurbita pepo*), tomato (*Lycopersicon lycopersicum*), or cotton (*Gossypium hirsutum*). Delay planting until later in the season, after the first spring flush of *Cuscuta* germination.

Chinese researchers have controlled *Cuscuta* by spraying a dodder-specific strain of *Colletotrichum*, a pathogenic fungus (Kendrick, 1985). Russians have controlled dodder with an *Alternaria* sp. (Parker and Riches, 1993). The use of herbicides, even in herbicide-resistant transgenic crops, has not been effective.

16.3   Parasitic plants: harland's balanophora

Balanophora harlandii Hooker 1859, Balanophoraceae
This species deserves a footnote as the most colorful parasite of *Cannabis*. Like broomrapes, *Balanophora* spp. are root parasites and live underground, emerging only to flower. The flower looks like a big mushroom, with a bright red stalk and cap. The hosts of *B. harlandii* are *Cannabis* and *Pueraria* (Huang and Jin, 2003). *B. harlandii* occurs in mountainous areas of southern China, India (Assam), and Thailand, where it is used medicinally. No control measures have been proposed.

16.4   Invertebrates other than insects and mites

16.4.1   Slugs and snails

Slugs and snails can be vexing in humid corners of the globe. Heuzé (1860) first complained about slugs, in France, destroying hemp seedlings shortly after they emerged. In Italy, Berti Pichat (1866) said "very fat snails" damaged seedlings, especially if the season was hot and humid. The future hemp expert Nikolai Vavilov published his undergraduate thesis on field slugs, *Deroceras agrestis* (Vavilov, 1910).

Slugs leave ragged holes in leaves with their rasping mouthparts. Traces of silvery slime appear on plants or along the ground. Seedlings may be completely eaten, with only a stem fragment remaining. Slugs usually feed at night. They cause two peaks of damage—in April–May (seedlings) and in September–November (flowering tops). Snails can be more aggressive than slugs, attack larger plants, and feed in flowering tops. One or two generations arise per year. Slugs and snails overwinter as eggs.

16.4a.   European field slug

Deroceras agreste (Linnaeus) 1758; Mollusca, Agriolimacidae
≡*Limax agrestis* Linnaeus 1758
Description: Body is slug-shaped, pale brown to orange-brown, with a darker head and tentacles; the sole of the foot is white; up to 5 cm in length.
Comments: *D. agreste* is native to Eurasia, and lives in moist grassy and marshy habitats. It lives about one year. It produces a clear mucus trail unless disturbed, and then forms a milky-colored mucus. Nördlinger (1869) named this species in hemp. It is likely the species photographed by Vogl (Fig. 16.6). Slugs

wiped out *Cannabis* grown by the famed botanist Miles Joseph Berkeley. He obtained 12 seeds of rare Egyptian *Cannabis sativa*, "from which is made the hashish." The slugs killed 11 of 12 seedlings (Berkeley, 1866).

16.4b. Other slugs and snails

The gray field slug, ***Deroceras reticulatum*** (Müller) 1774, destroyed seedlings in Pennsylvania (McPartland *et al.*, 2000). This European species now occurs in the Americas, Australia, and New Zealand. In humid New Zealand, McPartland and Rhode (2005) reported ***Deroceras panormitanum*** Lessona 1882 (now *Deroceras invadens* Reise 2011) and ***Arion distinctus*** Mabile 1868. There are many unidentified slugs and snails eating *Cannabis* in photos on the internet.

Snails are slugs with shells large enough to crawl into, which puts them in a different family, the Helicidae. Ottavi (1897) reported Italian hemp "ruined" by snails, and Del Guercio (1900a) said hemp may "suffer severely." In Russia, Zolotarev (1918) reported snails chewing up 7-week-old plants 40 cm tall.

Gainullin and Shikov (2012) identified ***Lucilla singleyana*** (Pilsbry) on hemp seedlings in the Republic of Georgia. This minute snail (2.0–2.5 mm diam.) is known as the smooth coil, and distributed worldwide. The fragile and translucent shell of *L. singleyana* has a glossy surface marked by growth wrinkles and radial grooves. *L. singleyana* damages the roots and small shoots of plants in many plant families. It caused ample damage on *Cannabis*, and Gainullin and Shikov recommended it for biological control against drug plants.

McPartland and Rhode (2005) encountered the brown garden snail, ***Cornu asperum*** Müller 1774 (≡*Helix aspersa* Müller). The snails crawled into flowering tops, coating everything with slime (Fig. 16.7). Adults wear a shell with four or five helical whorls, brownish-gold to beige in reticulated bands, 25–35 mm high. *C. asperum* is native to Europe and was introduced to the USA for use as food.

16.4.2 Control measures

Eliminate places under which slugs and snails hide: cardboard, stones, and woody debris. In small plots, hand-picking at night can be effective. Or place traps in the garden (as simple as carboard squares or inverted flower pots), where slugs and snails congregate as the sun comes up. Drop them in a bucket of soapy water. Switch from sprinkler irrigation to drip irrigation to reduce humidity. Protect small gardens from slugs with a circular strip of copper (the best), zinc, or steel. The metal interacts chemically with slug slime, and repels them.

Phasmarhabditis hermaphrodita is a soil nematode that kills slugs (*Deroceras* and *Arion* spp.) and snails (*Helix aspersa* and many others). The nematode hunts down its hosts by following trails of mucus, and penetrates an opening behind the slug's mantle. *P. hermaphrodita* vectors a bacterium, *Moraxella osloensis*, which kills the nematode in 7–21 days. Then *P. hermaphrodita* feeds on its bacteria-rich cadaver. Wilson *et al.*

(1993) demonstrated the efficacy of *P. hermaphrodita* against *Dercoceras reticulatum*. A year later, Nemaslug® was released in Europe. The product was not available in the USA, because *P. hermaphrodita* was not found there. Tandingan DeLey *et al.* (2017) found it in California, and their strain is being tested for its biological control potential.

Nemaslug® contains infective juveniles formulated as a wettable powder. They can be stored for a few days in a cool, dark place. *P. hermaphrodita* is applied as a soil drench, into moist soil, and not exposed to UV light (apply in evening). Soil must be kept moist for at least 2 weeks after application. Loamy

Fig. 16.6. Slug (*Deroceras* sp.) in Austria (courtesy C. Vogl)

Fig. 16.7. Snail (*Cornu asperum*) in New Zealand (McPartland and Rhode, 2005)

or sandy soil is best, not clay, with soil temperatures between 5°C and 20°C (15°C is optimal).

Rumina decollata is a decollate snail that preys on garden snails and their eggs. *R. decollata* resembles elongate versions of its prey, growing to 25–50 mm in length. It feeds at night and during the day hides under rocks. It does best in moderate humidity and warm temperatures. *R. decollata* is supplied as dormant adults. Release 1–10 snails/m^2 in moist, shaded areas with adequate organic matter. Occasionally *R. decollata* eats seedlings, so do not release in a freshly sown field. To protect native mollusks, the US Department of Fish & Game limits the release of *R. decollata* to Arizona, New Mexico, Texas, and parts of California. McDonnell *et al.* (2016) questioned the efficacy of *R. decollata* against the garden snail *Cornu asperseum;* they found it only attacked small snails, and preferred eating carrot roots.

Ocypus olens, the devil's coachman, is a large (20–32 mm long) black staphylinid beetle that eats garden snails. It has large jaws and breaks apart shells. Orth *et al.* (1975) reported that a single female consumed 20 small-sized snails (*Helix aspera*) in 22 days, eating almost her weight in snails daily. *O. olens* is native to Europe and North Africa, accidently introduced into California by 1920. Orth *et al.* (1975) considered *O. olens* an excellent biocontrol candidate for distribution. That never happened, but the species when seen should be protected (Albertini *et al.*, 2019).

Setting a pan of stale beer in the garden, edges flush with the ground, is a time-honored slug trap. Slugs wallow into the beer and drown. For slugs not falling for this old saw, set out toxic bait—bran mixed with metaldehyde, a molluscicide. Do not use metaldehyde baits where children or dogs might encounter them.

16.4.3 Pillbugs, sowbugs, symphylans

Pillbugs and sowbugs are similar-looking pests. They are arthropods, like insects and mites, but in the subphylum Crustacea, with crabs and shrimp. They have oval, segmented bodies with seven pairs of legs.

Pillbugs are also called woodlice, because they can be found hiding under logs or boards in damp areas. Another common name, armadillo bug, refers to the series of prominent plates that covers their bodies. The common pillbug *Armadillidium vulgare* (Latreille) 1804 usually feeds on decaying plant material—a beneficial activity—but sometimes injures crop seedlings.

Sowbugs look like pillbugs but with a flatter body. They have a pair of tail-like appendages projecting from the rear end of their bodies, absent in pillbugs. Sowbugs do not roll up into a tight ball when disturbed, like pillbugs. Sowbugs feed on decaying leaf litter, but may injure young plants. Frank (1988) described sowbugs killing *Cannabis* sprouts. He didn't name a species; the most common sowbugs are *Oniscus asellus* L. 1758, *Porcellio scaber* Latreille 1804, and *Philoscia muscorum* (Scopoli) 1763.

Garden symphylans are small (8 mm long), slender, segmented, pearly-white centipede-like arthropods. They have 12 pairs of legs (centipedes have 15 pairs). *Scutigerella immaculata*

(Newport) 1845 is the most common species worldwide. Symphylans feed on plant roots and can kill young seedlings, particularly in glasshouses. They have been reported on *Cannabis* in web blogs.

Control of sowbugs, pillbugs, and symphylans is rarely needed. Remove their hiding places on the soil surface—plant debris, boards, unused flower pots. Conversely, sowbugs and pillbugs can be trapped by placing cardboard on the soil, they congregate there during the heat of the day, and they're slow-moving.

16.5 Vertebrate Pests: Birds

Any ornithologist worth their weight in binoculars knows that birds devour *Cannabis* seeds. Birds scratch up seeds sown in field soil. If seed husks are still attached to newly emerged seedlings, birds will pluck the seedlings out of soil (Duhamel du Monceau, 1763). *Aesop's Fables*, passed down from ancient Greece, tells of a swallow advising other birds to eat all the hemp seed that a farmer sowed in his field, otherwise the hemp would grow up and be made into nets, to trap them (Aesop, 1894). Moral of the story: destroy the seed of evil, before it ripens into your ruin (Fig. 16.8). The swallow was right: Varro used hemp netting (*rete cannabina*) to trap birds around 36 BC (Cato and Varro, 1918).

Markham (1615) advised hemp farmers to keep a close eye on their freshly sown fields. "Especially an houre or two before Sunne rise, and as much before it set, from birds and other vermine, which will otherwise picke the seed out of the earth, and so deceive you of your profit."

At the other end of the season, when seeds are ripening on plants, birds will dive-bomb flowering tops. Doing so, birds shred flowering tops to their stalks (Fig. 16.9). They may shake flowering tops to dislodge seeds, dropping more than they eat. Birds will also rummage through harvested shocks drying in the field (Hicks, 1896).

Historically, many hemp breeding studies have been foiled by birds stripping seeds from plants (e.g., Auclerc, 1843; Fruwirth, 1922; Dewey ,1928; Surkov, 1932; Crescini, 1939; Callaway, 2004; McPartland *et al.*, 2004; Čeh and Čremožnik, 2016). Furthermore, birds shit where they eat. Pathogenic fungi found in bird excrement have been isolated from cannabis, and caused disease in consumers: *Histoplasma capsulatum* (Ramírez, 1990) and *Cryptococcus neoformans* (Shapiro *et al.*, 2018).

Sorauer (1958) identified several hemp pests in Europe: house sparrow, *Passer domesticus*; Eurasian tree sparrow, *Passer montanus*; European turtle dove, *Streptopelia turtur*; Eurasian nuthatch, *Sitta europaea*; linnet, *Linaria cannabina*; purple grackle, *Quiscalus quiscula*; starling, *Sturnus vulgaris*; and common magpie, *Pica pica*.

There are 23 Orders of birds, but members of only three Orders regularly dine on *Cannabis* seed: Passeriformes, Columbiformes, and Galliformes. (1) Passeriformes are perching birds, distinguished by the arrangement of their toes, which facilitates perching. (2) Columbiformes are stout-bodied seed-eaters—pigeons and doves. (3) Galliformes are heavy-bodied ground-feeding birds, such as chickens, turkeys, and quail.

Fig. 16.8. Woodcut illustration from *Aesop's Fables* (Steinhöwel and Brant, 1501)

Fig. 16.10. *Má què,* the hemp bird, *Passer montanus* (Wikimedia Commons)

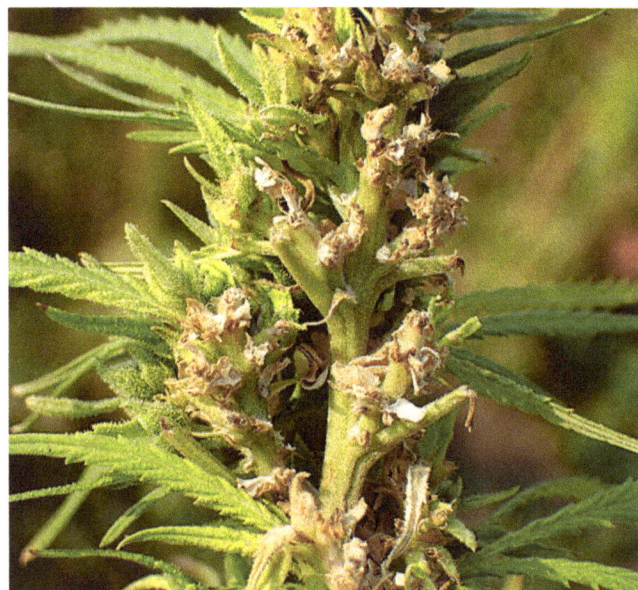

Fig. 16.9. Bird damage (W. Cranshaw)

16.5.1 Passeriformes

A perching bird was the earliest hemp pest of any kind described in the literature. Jiǎ Sīxié described the 麻雀 (*má què,* hemp bird) in AD 544. He reported *má què* eating seeds in Shāndōng Province (Jiǎ, 1962). *Má què* is the Eurasian tree sparrow, *Passer montanus* (Fig. 16.10). A related species, the russet sparrow, *Passer cinnamomeus,* is called 山麻雀 (*shān má què*). A Chinese name for a wild-type hemp is *shān má,* "mountain hemp".

Renaissance authors named several Passeriformes that ate hemp seed, but their pre-Linnaean names make identification tentative. William Turner wrote the first-ever printed book devoted wholly to ornithology. Turner (1544) named a bird

feeding on hemp seed and, handily for us, he gave the name in both Latin (*Rubicilla*) and English (bulfinche), which can be identified as the Eurasian bullfinch, *Pyrrhula pyrrhula.*

Gesner (1555b) named several lovers of hemp seed, identified best as *P. pyrrhula*; linnet, *Linaria cannabina*; common pigeon, *Columba livia*; and European goldfinch, *Carduelis carduelis.* Aldrovandi (1600) listed the same: *P. pyrrhula, L. cannabina, C. livia, C. carduelis,* plus two other species: the great tit, *Parus major* L. (which he illustrated next to a *Cannabis* plant), and the "voracious" hemp-eating common crossbill, *Loxia curvirostra.* He mentioned two birds that made nests in *Cannabe sylvestri* (feral hemp): the reed bunting, *Emberiza schoeniclus,* and the long-tailed tit, *Aegithalos caudatus.*

Rabelais (1546) said that hemp seeds were "delicious to all song birds, like linnets, finches, larks, canaries, serins, tarins, and others." Belon (1555) named hemp seed eaters: *P. pyrrhula, C. carduelis, C. chloris,* and the Eurasian siskin, *Carduelis spinus.* Coryate (1611) referred to finches in England as "hempseede birds." The English author Anonymous (1735) said that hemp fields were a good place to find wild birds. For capturing birds, Anonymous used hemp seed as bait, because it is "intoxicating of fowl, without tainting or hurting their flesh."

Linnaeus no doubt named *Linaria cannabina* L. for its food preference (Fig. 16.11). Germans and Swedes call the species "*Hänfling*" or "*Hämpling,*" reflecting its fondness for *hanf* or *hämp* seeds. Its common name in English, the linnet, refers to its other favorite food, flax seed. *L. cannabina* was rediscovered and assigned different names by ornithologists exploring the far reaches of Eurasia: *Passer cannabina* Pallas 1811, *Ligurinus cannabinus* Koch 1816, *Cannabina propria* Billberg 1828, *Cannabina arbustorum* Brehm 1828, *Linaria cannabina* Boie 1828, *Cannabina linota* Gray 1840, *Cannabina linaria* Rüpp 1845, *Acanthis cannabina* Blyth 1849, *Cannabis linaria* Blyth 1850, and *Cannabina major* Brehm 1855.

Bartram (1791) coined *Fringilla cannabina* for the "hemp bird" of Pennsylvania. Bartram no doubt considered the bird a pest—he operated a seed company, and published the first USA seed catalog that offered *Cannabis* for sale (Bartram & Son, 1807). Barton (1799) stated that Bartram's *Fringilla cannabina* was not

Fig. 16.11. Common linnet, *Linaria cannabina* (Wikimedia Commons)

the *Fringilla cannabina* of Linnaeus (now *Linaria cannabina*). Osbert Salvin, who edited Barton's book a century later, suggested they were the same species (Barton, 1883). Mitchill and Miller (1800) and Trotter (1909) voted for the redpoll, *Acanthis flammea*. Stone (1913) suggested that Bartram's species was the purple finch, *Haemorhous purpureus*.

The USA's first great ornithologist, Alexander Wilson (1766–1813), observed many species feeding on hemp seeds in the USA: American goldfinch, *Spinus tristis*; Baltimore oriole, *Icterus galbula*; tufted titmouse, *Baeolophus bicolor*; purple finch, *Haemorhous purpureus*; rose-breasted grosbeak, *Pheucticus ludovicianus*; and blue grosbeak, *Passerina caerulea* (Wilson, 1839). John James Audubon (1785–1851) wrote less about hemp. Audubon (1832) mentioned that its seeds were eaten by the American goldfinch, and its fiber went into nests built by indigo bunting (*Passerina cyanea*) and yellow warbler (*Setophaga petechia*).

Writing in 1856, Henry David Thoreau used the name "hemp-bird" for the American goldfinch, *S. tristis* (Thoreau, 1993), as did Flagg (1881). In a survey of birds in North Dakota hemp fields at the end of the season (Kotten *et al.*, 2022), *S. tristis* was the third most frequently encountered species (*n* = 40), bested by the house finch (*Haemorhous mexicanus*, *n* = 53), and mourning dove (*Zenaida macroura*, *n* = 116).

In Nepal, Matthiessen (1978) saw the Himalayan goldfinch, *Carduelis caniceps*, feeding on seeds of wild *Cannabis*. Whistler and Kinnear (1949) said the same about the Himalayan greenfinch, *Chloris spinoides*. "Their favourite food is the seed of the wild hemp which grows in large patches where buffaloes have been kept." Sokolov *et al.* (1990) named sparrows, finches, siskins, goldfinches, and crossbills as hemp pests in European Russia, Ukraine, Moldova, Kazakhstan, and Central Asia.

Passer domesticus, the house sparrow or English sparrow, is cited as a hemp pest (Barrows, 1889; Crescini, 1939; Sorauer, 1958; Samorini, 2000). It has been persecuted as an agricultural scourge. Farmers poisoned *P. domesticus* with hemp seeds soaked in strychnine (Fisher, 1889; Aston, 1903; Phillips, 1910). Given a choice of seeds, however, *P. domesticus* will choose seeds that are smaller than hemp and easier to crack open, "except when hungry" (Ziswiler, 1965). Some individuals

learned to crack hemp in 3 seconds. Others took 12 seconds, and tried to crack seeds along their lateral keel, "wherein the ejected shells have irregular fracture edges." Fryday *et al.* (1994) also experimented with *P. domesticus*. The birds found hemp seeds difficult to handle, and dropped about half while trying to dehusk them. They dropped even more when exposed to chlorfenvinphos, a pesticide that affects the nervous system. Pesticide dosage correlated with percentage of dropped seeds and bird weight loss.

P. domesticus evolved an obligate commensal relationship with humans about 10,000 years ago, at the dawn of agriculture. The species breeds in farmlands, villages, and cities, where it feeds on crops and grain spillings. One subspecies in Central Asia, *P. domesticus bactrianus*, lives away from humans, and breeds along rivers in grasslands. It likely represents a relict population of the ancestral house sparrow (Saetre *et al.*, 2012). Perhaps the ancestral sparrow ate wild hemp seeds prior to the dawn of agriculture.

Garman (1907) wrote, "I have no doubt about the evil character of the English sparrow," but he questioned the threat posed by the purple grackle, *Quiscalus quiscula*. He identified stomach contents in dozens of grackles, to ascertain whether they were graminivorous (destructive) or insectivorous (therefore useful). He determined the latter, and surmised they were ground-feeders. Therefore, seeds gleaned by them on the soil were of no crop value. Maize was the most common, followed by hemp seeds.

Bechstein (1868) commented upon the ease in which perching birds cracked hemp seeds. They manipulated hemp seeds with their beak and tongue, cracked the shell, removed the husk and spit it out, and swallowed the kernel. Kear (1962) elaborated on the process in finches: hemp seeds split easiest along definite lines of weakness. The birds rolled a seed around to its line of weakness, cracked its husk with their beaks, scooped out the kernel with their tongues, swallowed the kernel, and spit out the husk. If a seed was not held in the correct position before biting, it either shot from the bill, or the whole seed was crushed and the bird could not sort the kernel from the husk.

Ziswiler (1965) published a landmark paper that differentiated the Fringillidae (true finches) from the Estrildidae (weaver finches). He studied their bill structure and hemp seed-shelling techniques. Fringillidae used a "cutting" technique—a forward/backward (rostrocaudal) movement of the lower jaw during the cracking phase, and mediolateral movements of the lower jaw during the husking phase. Estrildidae used a "crushing" technique—opening and closing (dorsoventral) movements of the jaws, restricted to the vertical plane, to crack the seed coat.

Boaler (1919) suggested that birds spit out hemp seed husks because "this outer shell contains some quantity—small, of course—of an irritant poison." Van der Meij (2004) sensibly suggested that they removed the shell because of its poor digestibility and low nutritive value. Birds with delicate beaks, such as the marsh tit, *Poecile palustris*, cannot crack open hemp seeds. They clamp a seed in their claws, and laboriously hack it open to pick the kernel out of the shell (Naumann, 1824; Dorflein, 1914).

16.5.2 Galliformes

Brasavola (1536) wrote about *Gallus domesticus*, the domesticated chicken, that "hens fed with this seed [hemp] lay eggs for the whole winter." Brasavola's claim was repeated by others (Lusitanus, 1553; Mattioli, 1554; Gesner, 1555b; Piemontese, 1555; Cardano, 1550; Marcin of Urzędów, 1595; Aldrovandi, 1600; de Serres, 1600; Gerard, 1636; Parkinson, 1640; Bomare, 1764; Martyn, 1807a, etc.).

Liger (1700) questioned the idea from an economic standpoint—a regular diet of hemp seed would cost more than eggs are worth. Instead, Liger used hemp seed as veterinary medicine. He prescribed ground-up hemp seed to treat two poultry diseases: hens with "flux of the abdomen," and young chickens who molted their feathers in cold weather. Conversely, he withheld hemp seed from hens suffering from "festers," and hens who fell into "melancholy" from eating food that overheated them.

The "lay eggs aplenty" credo has been tested by three 21st century research groups (Silversides and Lefrançois, 2005; Gakhar *et al.*, 2012; Neijat *et al.*, 2014). These studies did not support the credo, although feeding hemp seed to egg-laying hens offered other benefits, such as larger eggs, richer in unsaturated fats, and healthier birds.

Non-domesticated Galliformes are known as gamebirds. Gamebirds in the USA Midwest rely upon feral hemp seed as a food source. In Iowa, bobwhite quail (*Colinus virginianus*) "relish" hemp seeds; Errington (1936) found a quail with its crop "full of hemp seeds." Feral hemp seed was a food source for California quail, *Callipepla californica* (Sumner, 1935). Hartowicz and Eaton (1971) documented the dependence of bobwhite quail and ringtail pheasant (*Phasianus colchicus*) on feral hemp seed. The dependence of these game birds on feral hemp led wildlife agencies to oppose police who eradicated feral hemp (Vance, 1971).

Gigstead (1939) inspected the crop contents of 157 ringtail pheasants in Wisconsin; feral hemp constituted 2.5% of their diet in October. Robel (1969) inspected the crop contents of 591 bobwhite quail in Kansas; feral hemp, despite being actively eradicated by police, still constituted 3.7% of their diet in January.

Wildlife conservationists grew hemp, even after *Cannabis* became illegal. Allen (1938) sowed a mixture of 17 crop plants in fallow farmland, which grew into "winter food patches" for birds. Hemp had key qualities: it produced agreeable food, with persistent fruits, and its stalks remained upright despite winter storms. In February the snow reached a depth of 26 inches (66 cm), and hemp projected above the snow: "When snow was deep and other foods most scarce, the only food patch plant that was available was hemp." Bobwhite quail and pheasants fed on the hemp.

Aldo Leopold, the revered environmentalist, had a day-job in wildlife management. Leopold *et al.* (1939) sowed hemp in Wisconsin food patches. They compared Kentucky hemp and Manchurian hemp. The seeds from both varieties were eaten by birds and depleted by October. Leopold bemoaned that fact that a federal permit was required. Wallace (1942) sowed hemp seed in "bird food patches" in Vermont. He reported success in 1938 and 1939, but in 1940 the hemp failed to germinate. "The reason becoming apparent when it was learned that hemp seed has to be sterilized in accordance with strict Federal regulations." Davison and Sullivan (1963) offered 200 food items to mourning doves, *Zenaida macroura*, classified as choice, fair, or uneaten. Hemp numbered among 64 choice foods.

Wild turkeys, *Meleagris gallopavo*, graze in freshly sown hemp fields (McPartland, pers. obs. 2019). They are quite brazen, seemingly cognizant that they are protected game birds. Crops should be sown during hunting season (May 1–31), when the turkeys vanish. In the late 19th century, wild turkeys were on the road to extinction, due to hunting and habitat destruction. The population dropped to 30,000 birds, a smaller number than exists today for orangutans, polar bears, or African elephants. To help turkeys survive, Gilbert (1891) noted that they loved hemp seeds: "Hemp is easily grown in the garden or any out of the way corner."

16.5.3 Columbiformes

The genera *Columba* (pigeons and doves) and *Streptopelia* (collared and turtle doves) evolved and diverged about 30 million years ago in Central Asia (Pereira *et al.*, 2007). This is the same time and place that *Cannabis* evolved, so it's likely that *Columba* and *Streptopelia* were early seed predators of *Cannabis*. Lapiedra *et al.* (2013) studied feeding behavior in *Columba* and *Streptopelia*. Most species searched for seeds on the ground, rather than foraging in canopies. Ground-foraging behavior would impart strong selection pressure on *Cannabis* to evolve camouflaged seeds covered by a persistent perianth (see Fig. 3.9).

The common pigeon, *Columba livia*, grazes on hemp seed, first noted by Gesner (1555b). Aldrovandi (1600) illustrated *C. livia* next to a hemp plant, with five seeds scattered on the ground (Fig. 16.12). De Serres (1600) extended the "hens lay eggs

Fig. 16.12. *Columba cum cannabe* (*Pigeon with hemp*) in Aldrovandi (1600)

aplenty" credo to domesticated pigeons—he fed hemp seeds to pigeons "for warmth," and to help them lay eggs in winter.

In *Les Misérables*, Victor Hugo (1862) wrote, "Every creature had its food or its fodder. The ring dove [*Streptopelia risoria*] found hemp seed …". Desportes *et al.* (1994) presented *S. risoria* with a choice of four seeds, and the doves chose hemp seed over canary seed (*Phalaris canariensis*), oats (*Avena sativa*), wheat (*Triticum vulgare*), and millet (*Panicum miliaceum*). Wilson and Korovin (2003) counted an astounding 793 *Cannabis* seeds in the crop of an Oriental turtle dove, *Streptopelia orientalis*. They caught the bird near the border of Russia and Kazakhstan. *S. orientalis* is migratory, and its range includes northern India, Afghanistan, Turkestan, and China.

Kabir (2014) said hemp seed was a "standard food" of the collared dove, *Streptopelia decaocto*, in Bangladesh. The turtle dove, *Streptopelia turtur*, is a pest of hemp crops in Europe (Sorauer, 1958). Over the past 40 years, *S. turtur* populations have declined by as much as 79%, making the species vulnerable to extinction. Young (2022) tracked this migratory species between Europe (France and Hungary) and Africa (Senegal), to determine its food sources. *Cannabis* seed was eaten in all three locations.

Abid (2019) assessed bird damage on crops in France. Common pigeons, *C. livia*, were the usual culprits, and hemp and soybean were the most affected—100% of hemp crops suffered some damage. Farmers sometimes had to reseed their fields. The wood pigeon, *Columba palumbus*, also causes problems in hemp crops (Vappula, 1965).

The American mourning dove, *Zenaida macroura*, thrives on feral hemp seed (King, 1883). McClure (1943) inspected the crop contents of 157 birds, and found an average of 61 hemp seeds per bird. Hemp was the most important single item in their diet: "Captive doves thrived for long periods of time on hemp alone." Mourning doves that fed on feral hemp in Iowa suffered a winter-kill after heavy snow covered the plants (Errington, 1935). Even on the grounds of Fort Leavenworth, Brumwell (1951) encountered mourning doves feeding "extensively" on feral hemp. In a survey of birds in North Dakota hemp fields at the end of the season, *Z. macroura* was the most frequently encountered species (Kotten *et al.*, 2022).

As a historical post-script, the now-extinct passenger pigeon, *Ectopistes migratorius*, added *Cannabis* to its diet after hemp cultivation began in North America. Wilson (1839) observed *E. migratorius* feeding on hemp seeds. Birds nested in branches of hemp plants, and wove nests of hemp fiber. Others reported passenger pigeons foraging in hemp fields (Wheaton, 1875; King, 1883; Butler, 1900; Allen, 1900; French, 1919; Pearson, 1925). The passenger pigeon population dropped from billions of birds to zero in 50 years. The crash was multifactorial. Its final demise matched the late 19th century collapse in hemp cultivation. Two of the last passenger pigeons shot in the wild were flushed from a hemp field in Kentucky, and from feral hemp in Indiana, both in 1898 (Greenberg, 2014).

16.5.4 Endozoochory

Birds obtain nutrition from seeds, and seeds are dispersed by birds—enabling plants to colonize new habitats. The dispersal mechanism is called **endozoochory**—dispersal of seeds via the animal digestive tract. We introduce endozoochory in text below Fig. 3.9. The vast majority of seeds consumed by birds, however, are cracked open in their beaks and do not survive.

Birds may swallow seeds whole, without cracking them open. This is a generalized feeding strategy in Columbiformes. Species observed swallowing hemp seeds whole include pigeon, *Columba livia* (Darwin, 1859; Zeigler *et al.*, 1980), American mourning dove, *Zenaida macroura* (McClure, 1943), ring dove, *Streptopelia risoria* (Desportes *et al.*, 1994), and Oriental turtle dove, *Streptopelia orientalis* (Wilson and Korovin, 2003). Galliformes also use this feeding strategy; intact hemp seeds found in the crop indicates they were swallow whole—bobwhite quail, *Colinus virginianus* (Errington, 1936; Robel, 1969), and ringtail pheasant, *Phasianus colchicus* (Gigstead, 1939).

Passeriformes have also been observed swallowing hemp seeds whole: blackbird, *Turdus merula* (Sonnini, 1799); leafbirds, *Chloropsis* spp. (Ruß, 1899); purple grackle, *Quiscalus quiscula* (Garman, 1907); dunnock, *Prunella modularis*; redbilled leiothrix, *Leiothrix lutea*; and several species of Illinois finches (Willson, 1971); several estrildid finch species (Van der Meij and Bout, 2006); and, despite its common name, the nutcracker *Nucifraga caryocatactes* (Fisher, 1845).

Birds also disperse seeds by "scatter-hoarding"—they hold seeds in their beaks and stash them in hiding places (Vander Wall, 2010). Scatter-hoarding of *Cannabis* was first documented by Pernau (1720) in Austria. He observed the *Hanfmeise*, "Hemp-tit," probably *Poecile palustris*, collecting three or four hemp seeds and carrying them away to a hoard. Brodin (2010) reported *P. palustris* hoarding hemp seed in Sweden. Bechstein (1868) described the Eurasian nuthatch (*Sitta europaea*) collecting as many hemp seeds as it could store in its beak and carrying them off to a food cache.

Researchers use the scatter-hoard instinct to study bird memory. Shettleworth and Krebs (1982) allowed birds to hide 12 hemp seeds at any of 97 storage sites in an aviary—small holes drilled into trees. After 2 hours without food, birds were returned to the aviary. They flew to the holes harboring seeds, ignoring most of the vacant storage sites. Shettleworth and Krebs (1986) used a similar method to compare retrieval rates in *P. palustris* and chickadees (*Poecile atricapillus*). The former species performed better. Urhan and Brodin (2015) used a similar method to compare retrieval rates in *P. palustris* and humans. The latter species performed better. Shiflett *et al.* (2004) conducted memory tests on *P. atricapillus*, using mealworms instead of hemp seeds. Injecting them with a synthetic antagonist of cannabinoid receptors (SR141716A) improved their memory.

Birds also transport pollen. Laursen *et al.* (1997) collected *Cannabis* pollen from the face of the lesser whitethroat, *Sylvia curruca*. This species winters in North Africa (e.g., Morocco) and Laursen caught the birds in Denmark. The chiffchaff, *Phylloscopus collybita*, has been caught with *Cannabis* pollen on its face, migrating north from Morocco (da Silva *et al.*, 2014).

Evidence of endozoochory began anecdotally: Spallanzani (1783) stated that certain seeds "do not lose their vegetative capacity by their stay in the stomach … bittersweet, mistletoe, and hemp, which sometimes grow upon trees, are produced by means of the excrement of birds." In 1856 Henry David

Thoreau observed feral hemp "sprung up" from dispersal of "bird's seed" (Thoreau, 1906). Feral *Cannabis* in Poland was attributed to seed transported by birds (Sanio, 1891). Harry Anslinger, architect of the 1937 Marihuana Tax Act, published a photograph of feral hemp rampaging through Philadelphia, "the result of bird-seed dissemination" (Fig. 16.13).

Hemp seed may directly improve bird reproduction—the core of coevolution. Brasavola (1536) observed that "hens fed with this seed [hemp] lay eggs for the whole winter." Sexual selection in birds is driven by females choosing males with sweet song and bright plumage (Darwin, 1871). Feeding hemp seed to males provokes song, for example the pigeon, *Columba livia* (Moore, 1735; Piper, 1871; Siegel, 1989), the Eurasian bullfinch, *Pyrrhula pyrrhula* (Günther, 1774), American robin, *Turdus migratorius* (Mann. 1848), European greenfinch, *Carduelis chloris* (Bechstein. 1868), European goldfinch, *Carduelis carduelis* (Dyson. 1889), chaffinch, *Fringilla coelebs* (Holden. 1903), and field sparrow, *Passer domesticus* (Samorini. 2002).

Shortly after the Spanish colonized the Canary Islands, they brought back **canaries** (*Serinus canaria*) for sale as caged birds. Canaries ate canary seed (*Phylaris canariensis*) from the Canary Islands, although de Herrera (1539) noted they loved *cañamones* (hemp seeds). Many experiments with hemp seed have been conducted with canaries. Feeding hemp seed to male canaries stimulates "sweeter song" (Samorini, 2000). Feeding hemp seed to female canaries makes them respond to sweeter song (Lerch *et al.*, 2011).

Robert Stroud (1890–1963) studied the effects of nutrition on canary reproduction. The "Birdman of Alcatraz" raised canaries during his 54 years in prison. Canary hens fed

MARIHUANA FOUND GROWING AS THE RESULT OF BIRD-SEED DISSEMINATION, PHILADELPHIA, PA.

Fig. 16.13. Dissemination via birdseed (US Treasury, 1937)

hemp seeds laid eggs with harder shells, which had a better hatch rate (Stroud, 1933). Chicks born to hemp-fed hens were larger and grew better. Young birds fed an unlimited supply gained too much weight, but this did not happen to older birds. Stroud noted that hemp seeds go rancid quickly. Seeds that smelled like stale butter could kill nestlings under 5 days old. "The evil narcotic properties often ascribed to hemp seed were not found in any of the tests that I have made. I am forced to conclude that they are largely imaginary."

Harry Anslinger pushed for a ban of hemp seed in bird food. The birdseed industry wrangled a concession—*heat sterilized* hemp seed was exempted from prohibition. The concession nearly got derailed over a nonsensical argument that "sterilization will not remove marihuana from the seed," raising the spectre of addicts smoking sterilized seeds (US Congress 1937). Stroud (1943) protested: "I have recently run some tests on this sterilized seed and have found it utterly unfit for canaries to eat. The stimulating effects of good hemp on song and reproduction were lacking." Stroud reiterated, "The oil of hemp seed becomes rancid very quickly and what was formerly a valuable food becomes a deadly poison."

On the evolutionarily negative side, hemp seed may *decrease* sexual selection, by darkening the plumage of male songbirds, as noted by Darwin (Stauffer, 1975). We traced this discovery to Frisch (1763). He said when *Feldlerche* (Eurasian skylark, *Alauda arvensis*) are fed a lot of hemp seed, "they often turn completely black." Frisch's observation was quoted by others (Günther, 1774; Buffon, 1793b).

Günther (1774) fed nothing but hemp seed to a Eurasian bullfinch (*Pyrrhula pyrrhula*), and his red plumage turned "gradually blackish" over four years. Günther also attributed a hemp seed-rich diet to melanism in the linnet (*Linaria cannabina*) and European goldfinch (*Carduelis carduelis*). He implicated the oily kernel rather than the pigmented husk of the seed, because birds shelled the seed and rejected the husk.

White (1789) did not cite Günther, but gave a similar account in England—*Pyrrhula pyrrhula* darkened over four years: "A few years ago I saw a cock bullfinch in a cage ... In about a year it began to look dingy; and, blackening every succeeding year, it became coal-black at the end of four. Its chief food was hempseed."

A Cambridge botanist stated, "A very singular effect is recorded, on very good authority, to have been sometimes produced by feeding Bullfinches and Goldfishes on Hemp-seed alone, or in too great quantity: *viz.* that of changing the red and yellow on those birds to a total blackness" (Martyn, 1807a). Stevenson (1866) owned a goldfinch that became black from a high hemp-seed diet. Newman (1856) described a similar experience with a hawfinch, *Coccothraustes coccothrauste*, fed hemp seed for 6 years. Staples (1948) attributed hemp seed's high oil content or high tyrosine content as the cause of melanism in bird feathers.

Other side effects from excess hemp seed were recorded: Estienne and Liébault (1583) reported *phthisis* (eye inflammation) and *flux de ventre* (liquid diarrhea). The aforementioned birder Johann Frisch said finches got too fat from excess hemp seed, and soon went blind (Frisch, 1763). Buc'hoz (1775) said excess hemp seed caused ophthalmia in chickens, blindness in chickadees, and "falling sickness" in goldfinches. Bechstein

(1868) warned that too free a supply of hemp seed caused songbirds to become "asthmatic, blind, and generally die of consumption."

16.5.5 Food preference studies

Birds have preferences. The *Boy Scout's Handbook* recommended hemp seed for feeding wild birds (Boy Scouts of America, 1911). Nearly a score of experimental studies regarding food preference have included *Cannabis* seeds. Rennie (1834) added to a canary's nest the fledgling of a cuckoo (*Cuculus canorus*), a species that feeds on insects. The canary fed caterpillars to the cuckoo, while the canary kept to her diet of hemp seeds.

Roessler (1936) offered the house finch, *Haemorhous mexicanus*, a dish of mixed seeds from 20 plant species. Roessler used many species that appeared in subsequent studies: niger seed, *Guizotia abyssinica*; canary seed, *Phalaris canariensis*; common millet, *Panicum miliaceum*; foxtail millet, *Setaria italica*; flax, *Linum usitatissimum*; oats, *Avena sativa*; poppy, *Papaver somniferum*; rape, *Brassica napus*; sunflower, *Helianthus annuus*; wild cherry, *Prunus avium*; and *Asteraceae* spp. Hemp seed ranked second in preference, behind niger seed.

Kear (1962) showed that seed choice was affected by seed size and husk thickness. Larger, harder seeds required more husking time, making them less desirable. She offered several finch species a choice of six of seeds, and hemp's husking time fell in the middle. For example, chaffinch, *F. coelebs*, took 5.0 seconds to husk a hemp seed; rape seed, canary seed, and millet took less time, and flax and sunflower took longer. In terms of efficiency (seed weight/min), hemp moved down to fifth place (5.35 g/min). Offered *pre-husked* kernels, five of seven finch species chose hemp over all other seeds: Eurasian bullfinch, *P. pyrrhula* (80.5% of all seeds eaten); greenfinch, *C. chloris* (48.3%); chaffinch, *F. coelebs* (42.3%); brambling, *Fringilla montifringilla* (37.6%); and canary, *S. canarius* (29.5%, higher than canary seed). Linnet, *L. cannabina*, preferred flax over hemp (agreeing with its common name rather than its Latin name).

Kear wondered why finches preferred hemp seed, since it took longer to husk. She proposed calorie count or nutrition, noting that hemp seed was highest in protein and oil content. During winter months, chaffinches ate a higher percentage of hemp seeds than in summer months. However, a forced diet of pure hemp seeds was detrimental. Chaffinches suffered from enteritis and opthalmitis (which caused vision loss and difficulty in feeding).

Ziswiler (1965) conducted seed choice experiments with ten finches in the Fringillidae family, and he gave them a choice of 18 seed species. Hemp was first choice with common linnet, *L. cannabina*; greenfinch, *C. chloris*; red crossbill, *Loxia curvirostra*; and Eurasian bullfinch, *P. pyrrhula*. Hemp was second choice with hawfinch, *C. coccothraustes*; and third choice with redpoll, *Acanthis flammea*.

Ziswiler also determined seed choice in estrildid finches. Hemp was first choice with red-headed parrotfinch, *Erythrura cyaneovirens* (45% of all consumed seeds), whereas it was only 10–30% in two other parrotfinch species, *E. trichroa* and *E. psittacea*. He also measured husking time in *E. cyaneovirens*: hemp took 4.0 seconds and foxtail millet took 1.8 seconds. Lastly Ziswiler tested four New World finches: painting bunting, *Passerina ciris*; white-throated seedeater, *Sporophila albogularis*; red crested finch, *Coryphospingus cucullatus*; and red-crested cardinal, *Paroaria coronata*. All rejected hemp seed except for the cardinal, who learned to crack it "after some exercise," and took 10 seconds.

Willson (1971) examined bird bill size in seed choice. She tested eight birds and eight seeds. Large-billed species chose large seeds (sunflower, hemp). Small-billed species chose small seeds (millet). The bird with the largest bill, cardinal, *Richmondena cardinalis*, husked hemp in 13.5 seconds (slower than millet, canary, or thistle seeds, but faster than oats or sunflower—the latter taking 36.5 seconds). Willson (1972) studied cardinals eating *feral* Illinois hemp seed (the previous experiment used commercial seed from Europe). Feral hemp seeds were husked faster than commercial seed, at 5.1 seconds by females and 6.6 seconds for males.

Willson (1971) also measured calories (cal) per seed. Hemp seed was in the middle range (59 cal/seed), between sunflower (440 cal) and millet (8 cal). Factoring calories per seed with husking time, cardinals could obtain more calories/min eating sunflower (744 cal/min) than hemp (262 cal/min). Thus seed preference was not based on maximizing the number of calories, because cardinals preferred hemp. Willson suggested the preference may be due to hemp seed containing 25% protein, compared with sunflower's 15%. Given husking time, fox sparrows (*Passerella iliaca*) and song sparrows (*Melospiza melodia*) could obtain maximal calories from hemp. The former preferred hemp, but the latter preferred millet.

Willson and Harmeson (1973) compared seed preference in cardinals and song sparrows at room temperature (26°C) and winter temperature (0°C). They measured calories per seed in four plants, in descending order: giant ragweed, *Ambrosia trifida* (141.8 cal/seed); feral hemp (37.4 cal./seed); smartweed, *Polygonum pensylvanicum* (10.7 cal/seed); and foxtail, *Setaria faberi* (7.6 cal/seed). Feral hemp seed contained one-third less calories than the commercial hemp seed used previously (Willson, 1971). They also measured husking time (Table 16.1).

For cardinals and sparrows at 26°C, preferences correlated more closely with seed-handling time than with calories. Cardinals at 0°C, however, preferred high-calorie hemp. Cardinals at 0°C must consume 3100 calories/hour to survive. Compared with the other three plants, hemp seed provided the best deal: active cardinals needed to eat 137 hemp seeds/h to meet their needs, yet they could husk 654 seeds/h. In comparison, cardinals needed 468 smartweed seeds/h to meet calorie needs, but could only husk 354 seeds/h.

Doherty and Cowie (1994) analyzed seed choice in young canaries, *S. canarius*, removed from the nest at the start of independent foraging. They were offered niger, common millet, flax, hemp, as well as a nutritional supplement, "egg-biscuit." In the first week they preferred egg-biscuit and niger, but thereafter switched to hemp seed. The authors attribute this to increased bill size during development. By the eighth week their diet consisted of hemp 39%, niger 32%, millet 12%, flax 6%, and egg-biscuit 10% (by weight).

Table 16.1. Feeding preference (number of seeds eaten) and husking time for four weed species, from Willson and Harmeson (1973)

Species, temperature	Feral hemp	Foxtail	Ragweed	Smartweed
Cardinal 26°C	30% (5.5 sec.)	46% (3.0 sec.)	13% (62.2 sec.)	11% (11.7 sec.)
Cardinal 0°C	52%	24%	11%	13%
Sparrow 26°C	9% (11.4 sec.)	45% (2.0 sec.)	44% n.m.	2% (10.1 sec.)
Sparrow 0°C	9%	53%	32%	6%

Van der Meij (2004) conducted a video analysis of serine, *S. serinus*, feeding on hemp seeds. Body size generally correlated with husking time and mandibulations (the number of seed positioning movements before cracking). The number of mandibulations was also affected by seed size. Also, the larger the seed, the greater the number of cracking attempts (*Digitaria* 1.1 attempts, hemp 1.8, sunflower 3.5). She provided scanning electron photomicrographs of an intact hemp seed and a cracked hemp seed, which split along the lateral keel. Van der Meij and Bout (2006) revisited the study by Ziswiler (1965). They tested seven species of estrildids and 11 species of fringillids. Bite force was measured by placing a pressure gauge at the tip of the beak. Fringillids bit harder (mean 11.9 Newtons ±3.89 SEM) and husked hemp seeds faster (mean 5.2 seconds ±1.72) than Estrildids (5.3 N ±0.90 SEM; 10.0 seconds ±1.72 SEM). The fastest Fringillid was the Chinese grosbeak, *Eophona migratoria*, 2.1 seconds. The slowest Estrildid was the zebra finch, *Taeniopygia guttata*, 16.0 seconds.

16.5.6   Food benefit studies

Hemp seed is rich in protein, oil, fiber, vitamins, and minerals. Anonymous (1733) wrote that "hemp-seed is *meat* to goldfinches, robins, nightingales, bullfinches, and canary birds." Newborn birds benefit from hemp seed nutrition: some species regurgitate "crop milk" to their young. The nestlings of pigeons gained more weigh when fed on a diet of hemp seed than fed a diet of corn seed (Carr and James, 1931). Juvenile canaries learned to pick up hemp seeds in 9.7 min when their parents were eating hemp seeds, compared with 13.3 min when parents were not eating hemp seeds (Cadieu *et al.*, 1995).

Almquist (1937) studied an erosive disease of the gizzard lining that occurred in young chicks on standard diets. He fed over 20 dietary supplements to chicks, and hemp seed meal (HSM) prevented the disease. Almquist (1938) proposed that the beneficial effects of HSM were due to the grit-like action of hemp seed husks, as well as a fat-soluble substance in HSM. Dam (1935) described a hemorrhagic disease in chickens. The disease was prevented by a fat-soluble substance in hemp seed. He named the substance *Koagulationsvitamin*—vitamin K.

Mary Willson, who conducted seed choice studies, also analyzed seed "digestibility." Willson and Harmeson (1973) used **apparent metabolizable energy** (AME) as their metric of digestibility. AME is the nutritional energy in a food minus the energy contained in excreta (feces and urine). The mean AME of four seeds in cardinals was: hemp 73%, foxtail 73%, ragweed 73%, smartweed 71%. In song sparrows it was: hemp 83%, foxtail 88.5%, pigweed 68.5%, smartweed 55%.

Hemp seed AME has also been measured in pigeons, chickens, and quails. AME in these species is lower than Passeriformes, possibly because they often swallow hemp seeds whole without husking them. Hemp seed husk contains a lot of undigestible fiber. In homing pigeons, hemp seed averaged 58.5% AME—the lowest of all seeds tested (Hullar *et al.*, 1999). However, correcting for zero nitrogen retention ("AME_n") improved hemp's ranking. Its AME_n value, 4308 kcal/kg, was less than sunflower, but more than corn, wheat, barley, millet, peas, lentils, sorghum, and canary seed.

Kalmendal (2008) measured AME in broiler chickens between 28 and 35 days old. She fed chickens with hemp seed cake (HSC), as 10%, 20%, and 30% of the diet. On average, AME equaled 43%, using a methodology similar to Willson and Harmeson (AME determined from excreta), but only 37% when determined from ileal digesta samples. She noted AME values for HSC were inferior to those for soybean meal and sunflower seed meal. She suggested dehulling seeds to improve HSC digestibility.

Hervé *et al.* (2014) compared hemp seed meal and dehulled hemp seed in broiler chicks. They reported AME_n values of 3135 kcal/kg and 6453 kcal/kg, respectively. Given the total gross energy in HSC (5299 kcal/kg) and dehulled HSC (7049 kcal/kg), these AME_n values represented 59% and an impressive 91.5%, respectively.

In bobwhite quail, *C. virginianus*, hemp seed AME was only 45.1% (Robel *et al.*, 1979a). Compared with other seeds, Robel considered hemp seed a poor source of metabolizable energy. Bobwhites require 74 kcal/day during the winter, and they consume about 17 g seed/day. At that rate, hemp seed could only provide 58% of their daily needs. When bobwhites were switched from a balanced poultry diet to pure hemp seed for 3 days, the birds lost 3.1 g—from 187 g to 184 g (Robel *et al.*, 1979b). Switched to corn, they gained 1.9 g. Switched to acorns they lost 20.1 g. Switched to switchgrass they lost 23.3 g. The birds consumed about the same number of calories in hemp and corn, but hemp's low AME caused them to lose weight.

Konca *et al.* (2014a) added hemp seed to the diet of 192 Japanese quail, *Coturnix coturnix japonica*, 7–42 days old. A maize–soybean diet was supplemented with 5%, 10%, or 20% hemp seed. Hemp seed supplementation did not change body weight, measured at three time points (days 7, 21, 42), except for a decrease at day 42 with 20% hemp seed. Relative weights of carcasses compared with internal organs steadily increased: 0% hemp: 67.1%; 5%: 70.9%; 10%: 70.6%; 20%: 72.0%.

Hemp seed supplementation improved measures of antioxidant potential.

Konca *et al*. (2014b) conducted a similar experiment with older female Japanese quail, ages 8–14 weeks. Hens on diets supplemented with 20% hemp seed gained 5.7% more weight than controls (non-significant difference). Hens supplemented with hemp seed laid 37.3 eggs while hens on control diets laid 35.8 eggs (non-significant difference). Egg weight showed the opposite trend—11.8 g on 20% hemp diets versus 12.0 g on control diets (non-significant difference). What was significant was that egg omega-3 fatty acid content increased linearly with dietary hemp seed content (P <0.01). Okay, time to move on.

16.5.7 Control of bird pests

Rataj (1957) lamented, "Birds cause the most serious damage to hemp seed. In spite of their harmfulness we lack protective measures against this danger. It is usually necessary to guard seed plots and drive the birds away."

Bird control methods include frightening (non-chemical fear inducers), exclusion (preventing access), repellents (irritating chemicals), and lethal force. Frightening devices may be visual (e.g., balloons, owl effigies, and "scarecrows" or human effigies), audio (distress calls, propane cannons), or audio-visual (firecrackers, 0.22 caliber rimfire CB cartridges). Frightening devices work best if employed early (before birds habituate to a site), used persistently, and different devices are rotated to preserve novelty.

Chairman Máo Zédōng initiated the Great Sparrow Campaign (打麻雀运动) against *má què* (麻雀) in 1958. A propaganda poster by Bì Chéng is reproduced in Fig. 16.14. It says 大家都来打麻雀, "Everyone comes to shoot the hemp bird." People shot at birds, banged pots and pans, ran around crop fields, and spread poison. Nests were torn down, eggs smashed, and nestlings killed. Bird populations dropped, resulting in an explosion of locusts and crop damage, which contributed to China's Great Famine of 1958–1961 (Shapiro, 2001).

Back in AD 544, Jiǎ Sīxié used scare tactics: "Disperse *má què* for several days in order to protect the seeds that have just germinated from being eaten by them" (Jiǎ, 1962). Nearly 1500 years later, Clarke (1995) observed Chinese hemp farmers keeping a constant daytime watch over their seed plots, and frightening birds by hanging brightly colored cloth or using firecrackers.

European farmers grew hemp in fields close to their homes, to keep an eye on the crops. They were worried about birds, not human thieves. Surveillance by French peasants in the 18th century was considered critical during the 7–12 days between sowing and the emergence of plants, to protect sown seeds from pigeons (Saint Jacob, 1960). In Irish hemp fields, Sigerson (1866) recommended "fluttering scare-crows or noisy little boys."

The earliest mention of scarecrows in hemp fields came from Martin Luther (1483–1546), but as a metaphor. Luther likened the Pope or the Catholic Church to a *Vogelscheuche* (scarecrow) in a hemp field. Sometimes Luther used the word *Hanfpoßens* (hemp-dummy). "If you are afraid of the

Fig. 16.14. Shooting hemp birds (Wikimedia Commons)

Hanfpoßens in the *hanf*, it will devour you; if you fear it not, it will not harm you" (Kruse, 1907).

Lisson and Mendham (1995) deployed hawk effigies and used "bioacoustics"—the broadcasting of bird distress calls over loudspeakers. Others scare tactics include hanging balloons with large owl eyes, ringing "bird bells," firing "bird cannons," or deploying scarecrows that spray water when set off by motion detectors. Those wacky inflatable men dancing outside of car dealerships have provided bird deterrence in *Cannabis* grain crops (M. Bolt, pers. comm., 2023). Also known as "tube guys" or "scare dancers," their limbs flail nonstop in random motion.

Auclerc (1843) said that *chanvre du Piémont* (Italian hemp), which was late-maturing, exposed seeds to "the voracity of birds," more so than early-maturing French landraces. Surkov (1932) also stated that late-maturing итальянской конопли (Italian hemp) got hit by sparrows. However, Callaway (2004) noted that 'FINOLA' matured early, but it still got hammered.

Romanian farmers lured away birds from hemp by scattering wheat and maize seeds. They did this religiously on Shrove Tuesday, and they spoke to the birds, "This will be your food for the year" (Colta, 2008). Grain-baited traps have also been used. They are inefficient, and sometimes trap non-target animals.

The Italian hemp breeder Crescini (1939) bemoaned seed losses from greenfinches (*C. chloris*) and sparrows (*P. domesticus*). To protect his breeding stock, Dewey (1928) covered individual plants with cheesecloth sacks. Dewey's notebooks (www.lysterdewey.com) indicate that he got this advice from Edwin R. Kalbach in 1920. Kalbach worked for the USDA, specializing in starlings, crows, and magpies. The cheesecloth didn't work. Notebook entry in 1921: "The work of an entire season is lost because of the birds which harvested all of the seed." Notebook entry in 1925: "The birds got nearly all of the hemp seed produced this last season at Arlington Farm except that which was bagged early."

The first hemp breeder to lament experimental variance due to bird predation was Fruwirth (1922). In 1907 he protected maturing plants with a gauze sack, but "it was chopped through by *Meisen* [titmice] trying to get to the fruits." The following year Fruwirth covered plants with glassine sacks and gauze. Finally in 1909 he covered the entire breeding garden in a large cage made of wire mesh.

The breeding of 'Aotearoa #1' in New Zealand required cultivation under steel wire mesh (McPartland *et al.*, 2004). Crops can be covered by framed canopies with "chicken wire" (polyethylene netting). Pochop *et al.* (1990) said that stretching wires, monofilament lines, or nylon strings above crops will repel birds. Spacing, height, and installation pattern of the lines depends on the species to be repelled.

The bird repellent Avitrol® (4-aminopyridine) is formulated as chemically treated grain bait. Poisoned birds emit distress calls, and scare off the flock. Some birds die. French hemp breeders treated seeds with anthraquinone, a distasteful bird repellent extracted from grapes (Gutberlet and Karus, 1995).

Monserrate *et al.* (2024) decreased bird predation in seedbeds at Cornell by running tractor-drawn field rollers (Fig. 16.15) after planting with a John Deere 1560 no-till seed drill. Seedling emergence was not hampered by the heavy machinery—they compared stand density in fields that were rolled pre- or post-planting and found no differences.

16.6 Vertebrate Pests: Mammals

Cannabis coevolved with mammals, adapting to endozoochory and becoming a nitrophile. Vavilov (1926) noted that wild-type *Cannabis* proliferated in soil manured by grazing cattle, and grew in gorges and ravines where the dung of wild animals accumulated. Soil disturbances provided by large mammals enabled wild-type *Cannabis* to penetrate established stands of perennial vegetation (Haney and Bazzaz, 1970; Yunusbaev *et al.*, 2003). Mammals eat *Cannabis* seeds, and some seeds survive passage through the digestive tract.

Giles (1939) identified hemp seeds in the scat of raccoons, *Procyon lotor*. Snyder (1974) wrote about hemp seed in horse manure. Okulova *et al.* (2015) found seeds in the feces of two Russian voles, *Microtus arvalis* and *M. levis*. Valentini *et al.* (2009) extracted *Cannabis* DNA from the feces of the golden marmot (*Marmota caudata*). De Barba *et al.* (2013) extracted *Cannabis* DNA from bear dung (*Ursus arctos*). Rodents, like birds, will cache seeds in hiding places, including hemp seed (Howard and Evans, 1961; Spengler, 2013).

Rodents are the worst mammalian seed predators. Buffon (1785) said the "Kazan souslik" gathered hemp seeds from sown fields and stored the seeds in their burrows. Buffon's drawing clearly illustrated the spotted ground squirrel, *Spermophilus suslicus* (Fig. 16.16). Albert Ivanov wrote a children's tale about the hemp-eating squirrel, "*How the souslik got*

Fig. 16.15. Field roller (McPartland at Cornell)

Fig. 16.16. The Kazan souslik (Buffon, 1785)

intoxicated." Ivanov was accused of drug propaganda (Aprelskaya, 2016).

Loiseleur Deslongchamps (1833) said wood mice, *Apodemus sylvaticus*, can "make a great failure" of hemp cultivation. Hemp seed sown in fields was avidly eaten by mice, voles, and *les rats* (Heuzé, 1860). Field voles (*Microtus* spp.) and hamsters (*Cricetus cricetus*) also feed on sown hemp seeds (Sorauer, 1958).

Eccard *et al.* (2021) conducted a food choice experiment with wild Norway rats (*Rattus norvegicus*), offering them seeds of eight species. Hemp seeds ranked third, behind sunflower and safflower seeds. The rats husked hemp seeds before eating them. The black rat (*Rattus rattus*) is a selective feeder—it refuses peas or beans, but readily eats hemp seeds (Mohr, 1928). Kangaroo rats (*Dipodomys* spp.) feed on hemp seed "with great avidity" (Shufeldt, 1902), and ground squirrels (*Ictidomys tridecemlineatus*) eat feral hemp seeds (Fitzpatrick, 1923). Groundhogs (woodchucks, *Marmota monax*) can quickly destroy a young *Cannabis* stand. They seem to feed on some plants and roll around in the rest.

Montana DoA (2019) cited several rodent pests: The house mouse (*Mus musculus*) and deer mouse (*Peromyscus* sp.) ate seeds and young sprouts. The Norway rat (*Rattus norvegicus*) and wood rat (*Neotoma* sp.) stripped bark from stalks to build nests, causing damage similar to Fig. 16.17. Voles (*Microtus* spp.) ate seeds, stripped bark, and tunneled through planted areas. Pocket gophers (*Thomomys* spp.) fed on plants, tunneled through planted areas, and gnawed on drip irrigation lines.

Rabbits are not rodents. Rabbit problems have been reported in Kansas (Hartowicz and Eaton, 1971), Texas (Brown, 1987), and New Zealand (McPartland and Rhode, 2005). Clarke (pers. comm., 1996) observed rabbits, *Sylvilagus* sp., stripping bark from stalks in search of moisture during droughts (Fig. 16.17). Note that he pruned away three lower limbs to better photograph the damage.

Moles (*Scalopus* and *Talpa* spp.) tunnel through fields, tearing up roots. They're after beetle grubs in the soil. They sometimes forage on the soil surface, and bring hemp seeds back to their burrows (Imaizumi, 1979). Chomel (1709) suggested repelling moles by burying a bundle of "green hemp" in a pit and watering it, until the "stench" drives them away.

Fig. 16.17. Stalk damage caused by rabbits (R. Clarke)

Latreille (1817) said the European hedgehog, *Erinaceus europaeus*, sometimes mixed its ordinary insect diet with fruits and seeds. *Touché*, he joked that a dead hedgehog's spiny pelt can be used to hackle (comb) hemp fiber. Wild boars (*Sus scrofa*) in Texas ate a diet dominated by feral hemp (Robeson *et al.*, 2018). Monkeys in South America ate hemp seed (Siegel, 1989), and dik-dik in Kenya did the same (Quinn, pers. comm., 1996).

Montana DoA (2019) reported black bears (*Ursus americanus*) knocking over plants. A Canadian grower fed black bears with dog food while they guarded his 1100-plant garden. He was fined $6000 for feeding dangerous wildlife, jail time for growing *Cannabis*, and lost his land (35 ha) to civil forfeiture (Keating, 2013).

Members of the deer family will browse on plants. This was particularly true back in the days of guerilla gardening on forest land. Many of these animals are protected by wildlife laws and cannot be killed except under rare circumstances. Montana DoA (2019) reported damage by mule deer (*Odocoileus hemionus*) and elk (*Cervus canadensis*).

Hamerstrom and Blake (1939) said white-tailed deer (*Odocoileus virginianus*) fed heavily on hemp "where available." The first report of deer damage on drug-type plants was by Frank and Rosenthal (1978). Deer tend to graze on younger plants, and shy away from mature plants, but they will browse on seeded flowering tops (Fig. 16.18). Trampling may cause more damage than their feeding.

Domesticated mammals rarely damage hemp crops. Markham (1616) protected grain crops from cattle by planting a border of *hempe*, "for upon it no cattell will bite." Jain and Arora (1988) mixed *gañjā* by-products (leaves, sticks, and seeds) into cattle feed. The cattle showed drowsiness and uncoordinated movements, and feed consumption dropped by 50% compared with the control diet. Cattle, sheep, and horses in Kazakhstan find wild-type *Cannabis* unpalatable. This leads to increased populations of wild-type *Cannabis* at the expense of native grasses, which are heavily grazed (Asanova, 2001). Palatability improves when cattle are fed hemp silage—the silage fermentation process presumably neutralizes repellent constituents (Campbell *et al.*, 2019).

Conversely, Siegel (1989) reported cattle and horses happily feeding on flowering tops in Hawai'i, "which gives them a case of mild staggers." Clarke and Merlin (2013) photographed a horse grazing on flowering tops. Patton (1998) reported cattle and horses eating hemp seed meal (a by-product of hemp seed oil production).

Goats devour *Cannabis* without ill effects (Clarke, pers. comm., 1996). A farmer in Guanajuato, Mexico, reported that his goats ate marihuana on land where former tenants cultivated it (Anonymous, 1909). "Some days after turning my goats into the new pasture I noticed that they had begun to act queerly … I have tried to drive the goats back into their old pasture, but they won't go."

Buffalo (*Bubalus bubalis*) graze freely on feral *Cannabis* in Rawalpindi, Pakistan (Ahmad and Ahmad, 1990). Urine tests revealed THC metabolites in six of ten buffalo, and in the milk

Fig. 16.18. Browsed plants with deer tracks

of five animals. Some children drinking milk from the animals also had THC metabolites in their urine.

Cats are not attracted to *Cannabis,* but may dig up transplanted seedlings, attracted to the fish emulsion in soil mixes (McPartland, pers. obs., 1993). Dogs only seem attracted to baked goods (Meriwether, 1969; Clarke *et al.*, 1971).

16.6.1   Control of mammal pests

Chomel (1709) advised storing hemp seed in a place "where the rats cannot damage it." Heuzé (1860) recommended close surveillance of fields after sowing seed. Discourage rodents and rabbits by removing refuge sites such as brush piles, stone and trash heaps, and weed patches.

Fencing works, but is exacting. Rodent fencing cannot be climbable. Use sheets of aluminum at least 18 inches (45 cm) tall. Tall tin cans, with tops and bottoms removed, keep rodents and rabbits from stripping bark off the bottoms of stalks. To exclude diggers (gophers, groundhogs, ground squirrels) requires a perimeter of underground fencing, such as chicken fencing or welded wire "hardware cloth," buried 2 ft (0.6 m) deep.

Deer fencing must be tall. Standard plastic deer fencing is 6 ft (1.8 m) tall—the minimum. When deer pressure is low, Dixon and Harper (2002) recommended single-strand deer fencing: ¼ inch (6.35 mm) nylon rope at a height of 24–30 inches (0.6–0.75 m). Fleming (1719) turned the tables on deer by using hemp twine to make *Hirsch Netze*, "deer nets."

Dixon and Harper (2002) recommended spraying fences with deer repellent. They used Deer Away® (containing putrescent egg solids) and Deer Stopper® (mint oil and rosemary oil). They're on lists of *Cannabis*-approved pesticides (California DPR, 2017; Montana DoA, 2019). Predator urine is commercially available, and repels rabbits and rodents. Dried blood in mesh bags may repel deer and rabbits. Repellents for deer and rabbits or rodents must be hung at different heights. Rotate scents every 2–3 weeks. Pests smaller than a breadbox can be captured in Havahart® cages, and then released at a distance—to become someone else's problem. Spring-traps kill rodents, but quickly, which is more humane than rodenticides.

Rodenticides were first used by Slator and Hall (1724) to rid stacks of hemp stalks of "vermin." The vermin were attracted to stacks by stray seeds, and then they made "beds to breed in." Slator and Hall mixed arsenic with flour and milk, and baited the paste with "canary wine and a few drops of oyl of annis-seed." They packed the paste into the hollow of water reeds, and laced the poisoned reeds in hemp stacks.

Anti-coagulant rodenticides such as warfarin and brodifacoum are on the list of pesticides that *cannot* be used in *Cannabis* crops (California DPR, 2018). The rodents bleed to death, as do predators who eat them. The use of brodifacoum in guerilla gardens killed northern spotted owls (*Strix occidentalis caurina*), an endangered species (Franklin *et al.*, 2018). Cholecalciferol (vitamin D$_3$) has replaced brodifacoum. But it can also kill dogs, and unlike brodifacoum has no antidote.

The Russians use biocontrol to kill rats and mice: the bacterium *Salmonella enteritidis* var. *issatschenko*. It causes a typhoid-like septicemia—no nicer than rodenticides, but it selectively kills rodents. Commercial products (Днисюк®, Бактороденцид®) are formulated as grain bait and scattered in crop fields or storage barns. Efficacy in crop fields averaged 83%, in stored grain averaged 89%. It remains active for six months at 4–22°C (Yakovlev, 2017).

16.7   Weeds

"HEMP DESTROYS WEEDS," wrote Dewey (1914) in uppercase letters. He wasn't first to say so. Teodosio (1553) wrote in Latin, *canabis alias herbas enecat*, "cannabis kills other plants." Tusser (1557) penned a poem about hemp with the couplet:
Where plots full of nettles be noisome to eie,
sowe thereupon hempseed, and nettle will die.

Markham (1615) said that hemp "is naturally of itself swift of growth, rough, and venomous to any thing that grows under it, and will sooner of its own accord destroy those unwholesome weeds." Markham (1616) recommended rotating wheat, barley or peas with "a full crop of hempe, which destroying the weeds and superfluous growths which spring from the fertilenesse of the soyles."

Henry (1771) stated, "Hemp is so great an enemy to weeds … every weed is choaked that cannot keep pace with hemp in its growth." Nevertheless, he added, "docks, sowthistles, thistles, and other rank growers should be hoed out or plucked up while the hemp is in its infancy." Norgord (1912) published a photograph, "Hemp kills Canada thistles" (Fig. 16.19). The figure legend stated, "On left, dead Canada thistle plants with rotten underground rootstocks killed by dense growth of hemp. On right, vigorous, healthy thistles from margin of same field."

Fig. 16.19. "Hemp kills Canada thistles" (Norgord, 1912)

How does hemp do it? Markham's remark offers an answer: it's the "full crop" or dense canopy formed by hemp. The dense canopy deprives other plants of sunlight, nutrients, water, even air: Bomare (1764) said that hemp "suffocates it [grass] by depriving it of air." Dewey (1914) surmised that the canopy blocked sunlight. This concept dates back to the celebrated French agronomist **François Rozier** (1734–1793), who said that hemp "destroys" weeds because "it smothers them with its shadow" (Rozier, 1787).

Depriving weeds of nutrients was proposed by Miller (1754). He wrote that hemp "is esteemed very good to destroy weeds … by robbing them of their nourishment."

Similarly, Wissett (1808) stated, "Hemp is esteemed very effectual for destroying weeds: but this it accomplishes by impoverishing the ground, and thus robbing them of their nourishment." Reisinger *et al.* (2005) determined that *Cannabis* competes with wheat (*Triticum aestivum* L.) by reducing the amount of nutrients *and water* available to wheat. They concluded this by comparing water and mineral content in the two plants. *Cannabis* hogged phosphorus and calcium the most.

Rather than resource competition, *Cannabis* may inhibit the growth of other plants through **allelopathy**: the chemical inhibition of one plant by another, through the release of substances that inhibit seed germination or growth. Black walnut (*Juglans nigra*) is famously allelopathic. Two millennia ago, Pliny (see *Naturalis Historia*, edited by Mayhoff in 1870) observed that "the shadow of a walnut tree … touching any plant whatever is undoubtedly poison."

Davis (1928) identified juglone, a water-soluble quinone, as the allelopathic poison produced by black walnut. Molisch (1937) coined the word *allelopathie*, although the toxicity of juglone was debated for decades, because of varying results obtained by different research groups. Some plants are tolerant of juglone, including *Cannabis* (Zubay *et al.*, 2021). The allelopathic toxicity of juglone is modulated by soil water, pH, texture, and organic matter.

Back in AD 544, Jiǎ Sīxié made an early allusion to allelopathy. He warned against planting hemp and soybean together, "they hate each other" (Zeng, 2008). Anonymous (1691) hinted at the "poison" of allelopathy, "hemp is so swift a grower, and such a *poison* unto all the Weeds, that it over-runneth, choaketh, and destoryeth them." Good (1953) reported that *Cannabis* can suppress the growth of neighboring plants, even when not grown in a dense canopy. He listed maize (*Zea mays*), lupine (*Lupinus luteus*), beets (*Beta vulgaris*), and brassicas (*Brassica oleracea*).

Vice versa, other plants exert allelopathy against *Cannabis*. Hemp grows poorly near *Spinacia oleracea, Secale cereale, Cicia sativa,* and *Lepidium sativum* (Good 1953). Muminović (1990) conducted a large study. Compared with control seeds (normalized to 100% germination), *Cannabis* germination decreased when exposed to *Rumex crispus* (44.3%), *Avena fatua* (45.1%), *Lamium purpureum* (63.5%), *Chenopodium album* (66.2%), *Elymus repens* (79.8%), *Ambrosia artemisiifolia* (80.3%), *Galinsoga parviflora* (81.4%), *Echinochloa crus-galli* (81.4%), *Juglans regia* (88.2%), *Sorghum halepense* (91.0%), and *Amaranthus retroflexus* (94.2%). However, some plants *improved* hemp seed germination: *Urtica dioica* (100.5%),

Taraxacum officinale (109.2%), and *Geranium dissectum* (112.6%).

The notoriously allopathic weed *Parthenium hysterophorus* can displace wild-type *Cannabis*, unless stands of *Cannabis* are well-established (Qureshi *et al.*, 2006). On the other hand, rainwater washing off *Cannabis* can inhibit the germination of *P. hysterophorus* seeds (Singh and Thapar, 2003).

Nearly 20 studies have tested the allelopathic effects of *Cannabis* against crop plants or weed species (see Table 3.4). Some degree of inhibition, of seed germination or growth, was seen in nearly all studies, although in one study, at minimal concentrations, *Cannabis* extracts *stimulated* weed seed germination (Stupnicka-Rodzynkiewicz, 1970).

Schlippe (1956) described an interesting application of allelopathy, invented by Zande people in the Congo. In September, the tops of *Cannabis* plants were covered with an inflorescence of *Desmodium* sp. (known to be an allelopathic species). This forced the plants to branch, and they produced numerous thick colas.

16.7.1 Weeds in hemp

Seed- and drug-type crops are sown less densely than fiber-type crops, and do not form a closed canopy. These crops have weed problems. This obvious reality was noted by Sandler and Gibson (2019). They made "a call for weed research." Their long trawl through the *Cannabis* literature found very few studies that addressed weed management. They found no publications that described relationships between *Cannabis* yields and weed pressure.

When Teodosio (1553) stated that "*Canabis alias herbas enecat*" and he named *gramen* (grass) as hemp's chief competitor. The first specific species named as a weed "infesting" hemp was wild mustard, *Sinapsis arvensis* (Villars and Chovin, 1789). Rowlandson (1849) stated that twitch (quackgrass, *Elymus repens*) and horse mint (*Mentha longifolia*) "are fatal to the growth of hemp." Although Dewey (1914) said hemp destroys weeds, he named two creeping perennials as troublesome weeds in poor stands of hemp: Canada thistle (*Cirsium arvense*) and quackgrass (*Elymus repens*). Conversely, these were species that Norgord (1912) eliminated with hemp (Fig. 16.19).

Zou *et al.* (2014) tried suppressing *E. repens* with four smother crops: a seed-type hemp cultivar ('FINOLA', sown at a dense rate of 160 seeds/m²), buckwheat (*Fagopyrum esculentum*), caraway (*Carum carvi*), and broad bean (*Vicia faba*). At 79 days after sowing (DAS), the weed dry mass (WDM) of *E. repens* was higher in 'FINOLA' fields than in the other crops. Correspondingly, the leaf area index (LAI) of 'FINOLA' was lower than the other crops. At 114 DAS, WDM was still highest in hemp. The authors concluded that dioecious hemp does not suppress the growth of *E. repens*. They proposed that a monoecious cultivar might do better.

Chiriță (2008) listed the worst weeds in Romanian hemp fields: *Echinochloa crus-galli*, *Setaria viridis*, *Chenopodium album*, *Amaranthus retroflexus*, *Raphanus raphanistrum*, and *Cirsium arvense*. Jankauskienė *et al.* (2014) identified *Chenopodium album*, *Polygonum aviculare*, and *Veronica arvensis* as the worst weeds in a 2-year study, with *Lamium purpureum*, *Thlaspi arvense*, and *Poa annua* causing problems one year.

The growth of wild-type hemp was inhibited by several perennials: Asanova (2001) named *Agropyron* (wheatgrass), *Eurotia* (white sage), and *Haloxylon* (saxaul). Conversely, wild-type hemp successfully competed with *Elymus* (wild rye), *Malva* (mallow), *Melilotus alba* (sweet clover), and *Artemisia* (sagebrush).

Johnson grass (*Sorghum halepense*) is a bad weed; its allelopathic powers are legendary. It was accidently introduced into South Carolina in the late 18th century, allegedly a contaminant in imported hemp seed (Tellman, 1996). However, *Cannabis* can turn the tables—it exerts allelopathic effects on Johnson grass (Konstantinović *et al.*, 2019).

Weedy *Cannabis* can even be a problem in hemp fields. "Volunteer hemp, which shall have come up from the seed of the previous year … will get so far ahead of the newly sowed hemp as to overshadow and injure it to some extent" (Beatty, 1844).

Bindweeds pose special problems. They are members of the Convolvulaceae family, along with dodder. Bindweeds are not parasitic, but they can spiral up hemp plants and topple them. Dewey (1914) named two Convolvulaceae causing problems in hemp, "wild morning-glory and bindweed." Today both of these common names refer to **Convolvulus arvensis** L., best known as field bindweed or small-flowered morning glory. The species is perennial, and creeps through the soil with an extensive rhizomatous root system.

To control *C. arvensis*, Franzke and Hume (1936) grew a smother crop of densely sown hemp. The fight wasn't close. "By Aug. 15, the bindweed had practically choked out the hemp. It was a very matted tangled mass." Franzke and Hume provided an ugly photo of hemp overtopped by bindweed.

Dempsey (1975) and Bócsa (2004) named three bindweed problems: *C. arvensis*, black bindweed (*Fallopia convolvulus* (L.) Löve), and hedge bindweed (*Calystegia sepium* (L.) Brown). *F. convolvulus* is an annual, and *C. sepium* is a perennial. Dewey noted that bindweed seeds are nearly the same size as hemp seeds, and difficult to separate by screening. Seed contamination commonly occurs in Chinese hemp seed (Clarke, pers. comm., 1999).

Cannabis has **weed mimics**. Weed mimics imitate crop plants, so farmers overlook them while weeding. Seedlings of hemp nettle (*Galeopsis* species), for instance, are difficult to distinguish from *Cannabis* seedlings. Male plants of false hemp, *Datisca cannabina* L., are also remarkable mimics of *Cannabis*, fooling even professional plant taxonomists, as pointed out by Small (1975a).

16.7.2 Cultural and mechanical control

Dewey (1914) may have acclaimed that "Hemp destroys weeds," but he still recommended weed control—a well-harrowed seedbed. Harrowing a field immediately prior to seeding will kill weeds that have already germinated. Lack of harrowing is why no- or low-till systems have problems with weeds; these systems often require a pre-seeding burn-off with herbicides.

In Dewey's time, *gañjā* growers in British India harrowed their seedbeds a minimum of four times, at intervals of 3–4 days. Chattopadhyaya (2018) obtained this information from a Bengali author who published in 1913. "Weeds sapped the fertility of the land and were considered by cultivators to substantially reduce the intoxicant power of the plant being grown."

Gañjā growers practiced what is now called the **stale seedbed technique**: harrow a seedbed, pause to allow weed seeds to germinate, then kill them by re-harrowing, and then plant hemp seeds. Unfortunately, re-harrowing triggers another flush of weed seed germination. **Flame-weeding** the seedbed is an alternative to re-harrowing.

Flame-weeding has been used with success in New York hemp fields to create a stale seedbed (L. Sosnoskie, pers. comm., 2024). Flaming is not burning—emerging weeds do not ignite. Temperatures between 55°C and 95°C kill seedlings by causing rapid protoplast expansion and rupture (Ascard *et al.*, 2007). Only 5% of hemp farmers in the Midwest use flame-weeding, compared with 30% using pre-plant tillage, and 60% using "hand removal," that is, hoeing (Zavala *et al.*, 2023).

Hemp needs hoeing early in the season. "It is therefore essential to get rid of the supernumerary plants, and to weed rigorously, if we do not want weeds to gain on hemp in the first days of its vegetation; past this time, hemp becomes their most assured destroyer" (Rozier, 1787). Aubin *et al.* (2015) reported good control with hand hoeing in a relatively large 2-year study. But it was economically feasible because they had grad students.

In row-cropped hemp, mechanical tillage can provide weed control early in the season. Row spacings can be adjusted to accommodate the tillage equipment being used. Post-emergence harrowing (in-crop tillage) is tolerated by some crops (e.g., canola) but not by densely sown fiber-type hemp. Commercially available tractor-towed machines combine computer vision with AI (machine learning) to identify crops versus weeds—and kill the weeds with lasers. This futuristic scenario dawned nearly 20 years ago; Slaughter *et al.* (2008) reported a crop vs weed classification accuracy of 90.3%.

Futuristic **weed electrocution** has even an older history. Lyster Dewey described several experiments, one at his alma mater, Michigan State: a boom carrying 24 kV (amps not stated) was towed through a field. Tops of weeds that brushed against the boom were killed, but roots of perennial weed were uninjured, such as Canadian thistle (Dewey, 1899a) and nutsedge (Dewey, 1899b). He preferred "the hoe."

Schreier *et al.* (2022) tested a modern implement, which delivered 15 kV and 7–20 amps to plants touching the boom. It was front-end mounted to a tractor, with boom height adjusted to weed height, and killed in-row and between-row weeds. This system worked in soybean, where a height differential existed between the weed and crop canopies. Tall crops like hemp present a different scenario.

Getting back to basics: early planting is essential for weed management. Hemp seed can germinate at cool temperatures, at least in studies of European fiber-type varieties (see section 17.6c). For early planting, Baxter and Scheifele (2000) recommended an optimal soil temperature of 8–10°C, at a depth of 2–3 cm, although hemp seed will germinate at a soil temperature of 4–6°C. Dewey (1914) noted that if hemp is planted late, or grows poorly due to drought or poor soil, it will not form a dense canopy. Neuer (1935) said weeds become a problem when hemp is planted in damp places and gets a slow start.

Sowing seeds at a high density results in earlier canopy closure, and therefore earlier weed suppression. Van der Werf (1997) provided dates of canopy closure in 'Kompoli Hibrid

TC', measured in growing degree days (GDDs, °Cd, base 2°C). In crops sown at 270 plants/m², the canopy closed at 265 °Cd, and crops sown at 10 plants/m² closed at 537 °Cd. Weeds only posed problems at the light seeding density.

Vera *et al.* (2006) measured weed density at three seeding rates, 20, 60, and 80 kg/ha, over three seasons. As seeding rates increased, weed density decreased, by 26%, 33%, and 41% in each year, respectively. They compared two dioecious seed-type cultivars 'FINOLA' (mean height 59 cm) and 'Fasamo' (112 cm). 'FINOLA' was shorter, yet it had a lower weed density than 'Fasamo', although it tended to be denser (plants/m²) at harvest time. Row spacing (18 vs 36 cm) had no effect on weed density.

Hall *et al.* (2014) evaluated 'BundyGem', a dioecious fiber-type cultivar, at seeding rates that resulted in 100, 200, 300, or 400 plants/m². Weed weight was inversely proportional to seeding rates: 23.2, 6.5, 2.6, and 1.5 g/m², respectively. Jankauskienė *et al.* (2015) grew monoecious 'Białobrzeskie' and 'USO 31' at two seeding rates, 45 and 70 kg/ha, which resulted in weed densities of 166 and 140 plants/m², respectively.

Jankauskienė *et al.* (2014) grew eight monoecious hemp cultivars at a sowing rate of 50 kg/ha and compared weed pressure. Weed density was highest in 'USO 31', followed by 'Beniko'. Since 'USO 31' was the shortest cultivar tested, the authors suggested that short plants had greater weed pressure (a result not seen in Vera *et al.*, 2006). 'USO 31' and 'Beniko' also matured early, so the authors proposed that early flowering limited canopy development.

There is a trade-off when fiber-type hemp is densely sown. Hemp sown at an excessive rate results in excessive self-thinning, plant mortality, and decreased yields. Rozier (1787) noted this: "if all the seeds sown germinated and prospered, the *chènevière* [hemp field] would be overfilled, and the stalks would be no larger than those of flax." Plant mortality resulting from self-thinning can be prevented by ensuring that the plant density at emergence does not exceed the maximum plant density possible at the expected yield. For a dry matter yield of 15 metric tons/ha, this plant density would be about 120 plants/m² (Van der Werf *et al.*, 1995a).

Controlling bindweeds can be difficult. Cutting or pulling bindweed has a negligible effect unless the rhizomes are dug out of the soil. The bindweed gall mite, *Aceria malherbae*, controls *C. arvensis* and *C. sepium*, but the availability of this biocontrol agent is limited. Kousta *et al.* (2020) controlled *C. arvensis* in hemp with heavy nitrogen fertilization, 140 mg/kg soil. This worked because *C. arvensis* is not a nitrophile like hemp.

Plasticulture, growing plants in strips of plastic, is frequently employed in the cultivation of drug-type *Cannabis* (see Fig. 19.4). White or black polyethylene, in strips 36" (0.9 m) wide, can be laid with tractor-towed equipment that forms raised beds, lays irrigation drip tape, and puts down plastic mulch, all in one pass. Plasticulture has downsides. It is expensive. Another machine is required to remove plastic at the end of the season. Plastic is a petrochemical, not biodegradable, and not consistent with claims of hemp as an ecofriendly crop. Plastic mulch can degrade into microplastic particles, which impact soil microorganisms, insects, and us (Qiang *et al.*, 2023). Furthermore, weeds between the strips of plastic mulch can still be an issue (Fig. 16.20).

Herbicides are an option—in the future. To date only one synthetic herbicide has been approved in the USA: ethalfluralin (EPA, 2023). The application of any other synthetic herbicide is off-label and therefore illegal.

Pre-emergence herbicides (PRE) are applied prior to seedling emergence. PRE herbicides are often applied on the day of sowing, or a few days before, or 1–3 days thereafter. **Post-emergence** herbicides (POST) are applied while hemp seedlings are young, often 3–4 weeks after they emerged, or 25–30 cm tall.

Corn gluten meal is approved by several states and Canada as a PRE herbicide, but see comments in section 21.6. The Organic Materials Review Institute approved several herbicides for use on other crops (OMRI, 2020). Their active ingredients include acetic acid, caprylic acid, ammonium nonanoate, and ammonium soap of fatty acids. They are non-selective and probably damage hemp. *Cannabis* is exquisitely sensitive to herbicide injury (see Fig. 17.8).

To control *Ambrosia artemisiifolia* weeds in hemp crops, Tarasov and Tarchokov (1974) applied dinitro-ortho-cresol and dinoseb. Tarchokov (1975) applied linuron and pebulate. Today they're all banned except linuron, and linuron is an endocrine disrupter. Linuron has been applied to hemp in agronomic studies (Mańkowski, 2003; Tang *et al.*, 2016), but these studies did not comment upon weed control or herbicide tolerance. Amaducci *et al.* (2015) reviewed Chinese literature regarding herbicides. They used PRE acetochlor, metolachlor, and pendimethalin. Kudryavtsev *et al.* (2017) reported success with

Fig. 16.20. CBD crop grown in rows of black plastic, without attention to weeds between the rows

highly potent sulfonylurea herbicides (chlorsulfuron, metsulfuron-methyl, thifensulfuron-methyl), which seems surprising.

In Canada two herbicides are labeled for use on fiber-type hemp, PRE ethalfluralin and POST quizalofop-p-ethyl (Beaton *et al.*, 2020). Quizalofop-p-ethyl has the unfortunate side-effect of stressing *Phorodon cannabis*, which actually increases aphid reproduction (Durak *et al.*, 2021). PRE ethalfluralin (Sonalan®) has been approved by EPA (2023) for use on *Cannabis*. Ethalfluralin primarily targets annual grasses, although some broadleaf weeds are susceptible (e.g., pigweeds, nightshades, lambsquarters).

Chiriță (2008) tested six herbicides on a monoecious hemp cultivar in a field trial. The best-tolerated herbicides were: PREs, acetachlor and metolachlor; and POST, fluozifop-p-butyl. The other three caused too much phytotoxicity: chlopyralid, trisulfuron, and tribenuron methyl.

Woosley *et al.* (2015) evaluated herbicide tolerance in 'FINOLA' in a field trial, measured as seed yield. Best tolerance was seen in POST bromoxynil, bispyribac-Na, and MSMA (monosodium methanearsonate), all of which produced greater yields than test plots with uncontrolled weeds. Less tolerated (in order of greatest yield reduction) were mesotrione (PRE), trifloxysulfuron (POST), fomesafen (PRE), rimsulfuron (POST), flazasulfuron (POST), pyroxasuflone (PRE), metalochlor (PRE), and pendimethalin (PRE).

Maxwell (2016) evaluated herbicide tolerance at two field sites, one with dioecious 'FINOLA', the other with a mix of unnamed Italian cultivars. Five PRE and five POST agents were evaluated. Tolerance was gauged by visual phytotoxicity, harvested biomass, and seed yield. With 'FINOLA', POSTs were better tolerated than PREs, and the opposite was seen with the Italian cultivars. Overall, MSMA (POST), bromoxynil (POST), and pendimethaline (PRE) were best tolerated.

Byrd (2019) screened monoecious 'Felina 32' for tolerance to 14 PRE and 13 POST agents, in glasshouse and field studies. Best-tolerated PREs were pendimethalin, *S*-metalochlor, and fomesafen; the best POSTs were sethoxydim, bromoxynil, clopyralid, and quizalofop. Flessner *et al.* (2020) tested monoecious 'Felina 32' in a glasshouse experiment and a field study. In the glasshouse study, less applicable to real life, the safest PREs were diuron, linuron, and pendimethalin, followed by sulfentrazone and flumioxazin. The safest POSTs were quizalofop, clopyralid, fomesafen, acifluorfen, sethoxydim, and halosulfuron. Their field study yielded different results: the safest PREs were diuron, linuron, and pendimethalin, followed by sulfentrazone and flumioxazin. The safest POSTs were quizalofop, clopyralid, fomesafen, acifluorfen, sethoxydim, and halosulfuron.

(Prepared by J. McPartland)

References

Aarti YD. 2018. The phytotoxic effect of aqueous extract of *Cannabis sativa* on the germination and growth of *Cicer arietinum*. *Research Journal of Pharmacy and Technology* 11: 5096–5100.

Abid S. 2019. *Evaluation et déterminants des dégâts provoqués aux grandes cultures par les oiseaux déprédateurs*. Master's thesis, University of Paris.

Abu-Irmaileh BE. 1981. Response of hemp broomrape (*Orobanche ramosa*) infestation to some nitrogenous compounds. *Weed Science* 29: 8–10.

Abu-Irmaileh BE. 1994. Nitrogen reduces branched broomrape (*Orobanche ramosa*) seed germination. *Weed Science* 42: 57–60.

Aesop (Jacobs J, *ed*). 1894. *The fables of Æsop*. MacMillan & Co., New York.

Agnieszka S, Magdalina R, Jan B, *et al*. (4 other authors). 2016. Phytotoxic effects of fiber hemp essential oil on germination of some weeds and crops. *Journal of Essential Oil Bearing Plants* 19: 262–276.

Agrios GN. 2005. *Plant pathology, 5th edn*. Elsevier Academic Press, Amsterdam.

Ahmad GR, Ahmad N. 1990. Passive consumption of marijuana through milk: a low level chronic exposure to delta-9-tetrahydro-cannabinol (THC). *Journal of Toxicology Clinical Toxicology* 28: 255–260.

Albertini A, Baptista P, Burgio G, *et al*. (7 other authors). 2019. The use of complementary approaches for conservation biological control: carabid and staphylinid beetles as natural enemies of the olive fruit fly. *IOBS-WPRS Bulletin* 141: 168–171.

Aldrovandi U. 1600. *Ornithologiae tomus alter*. W. Ritcheri, Frankfurt.

Allen DL. 1938. Some observations on fall and winter food patches for birds in southern Michigan. *Wilson Bulletin* 50: 42–46.

Allen JL. 1900. *The reign of law, a tale of the Kentucky hemp fields*. Macmillan Co., New York.

Almquist HJ. 1937. Sources and nature of the chick gizzard factor. *Journal of Nutrition* 14: 241–245.

Almquist HJ. 1938. The effect of hempseed preparations and of fineness of diet on the chick gizzard lining. *Poultry Science* 17: 155–158.

Amaducci S, Scordia D, Liu FH, *et al*. (4 other authors). 2015. Key cultivation techniques for hemp in Europe and China. *Industrial Crops and Products* 68: 2–16.

Anonymous. 1691. *England's improvement and seasonable advice to all gentlemen and farmers*. Booksellers in Westminster-Hall, London.

Anonymous. 1733. *A philosophical enquiry concerning the nature, use, and antiquity of hemp*. T. Thorton, Dublin.

Anonymous. 1735. *The sportsman's dictionary*, 2 vols, C. Hitch, C. Davis, and S. Austen, London.

Anonymous. 1909. Goats that feed on "dope." *New York Tribune*, April 11, 1909, part V, p. 2.

Anonymous. 1929. Broom rape in tomato. *California State Department of Agriculture Bulletin* 18(11): 626.

Appel O, Wollenweber HW. 1910. *Grundlagen einer Monographie der Gattung Fusarium (Link)*. Paul Parey and Julius Springer, Berlin.

Aprelskaya E. 2016. Автор книги про одурманившегося Суслика: "На моем бедолаге клеймо "наркоман." Available at: https://www.mk.ru/culture/2016/05/16/avtor-knigi-pro-odurmanivshegosya-suslika-na-moem-bedolage-kleymo-narkoman.html

Asanova DK. 2001. Fighting wild hemp using a method of biological action of perennial grasses and sub-shrubs in a system of radical pasture improvement. *Mezhdunarodnyĭ Sel'skokhozyaĭstvennyĭ Zhurnal* 2001(3): 63–64.

Ascard J, Hatcher PE, Melander B, Upandhyaya MK. 2007. "Thermal weed control," pp. 155–175 in Upandhyaya MK, *ed. Non-chemical weed management*. CABI Publishing, Wallingford, UK.

Aston BC. 1903. "The best method of poisoning small birds," pp. 27–29 in *Eleventh Report of the New Zealand Department of Agriculture*. Government Printer, Wellington.

Aubin, M, Seguin P, Vanasse A, *et al*. (3 other authors). 2015. Industrial hemp response to nitrogen, phosphorus, and potassium fertilization. *Crop, Forage & Turfgrass Management* 1(1): cftm2015.0159.

Auclerc L. 1843. Sur le chanvre du Piémont. *Bulletins de la Société d'Agriculture du Département du Cher* 21: 205–211.

Audubon JJ. 1832. *Ornithological biography, or an account of the habits of the birds of the United States of America*. Carey and Hart, Philadelphia.

Barloy J. 1962. Comportement en Anjou de diverses origines de chanvre vis-à-vis de l'orobanche (*Phelipea ramosa* Ca. Meyer). *Comptes rendus des séances de l'Académie d'Agriculture de France* 48: 194–201.

Barloy J, Pelhate J. 1962. Premières observations phytopathologiques relatives aux cultures de chanvre en Anjou. *Annales des Épiphyties* 13:117–149.

Barrows WB, *ed*. 1889. *The English sparrow (Passer domesticus) in North America*. USDA Division of Economic Ornithology, Bulletin no. 1. Government Printing Office, Washington, DC.

Barton BS. 1799. *Fragments of the natural history of Pennsylvania, part first*. Way & Groff, Philadelphia, Pennsylvania.

Barton BS (Salvin O, *ed*.). 1883. *Barton's fragments of the natural history of Pennsylvania*. Taylor and Francis, London.

Bartram W. 1791. *Travels through North & South Carolina, Georgia, East & West Florida, the Cherokee country, the extensive territories of the Muscogulges, or Creek confederacy, and the country of the Chactaws*. James & Johnson, Philadelphia, Pennsylvania.

Bartram & Son. 1807. *Catalogue of Trees, Shrubs & Herbaceous Plants*. Bartram & Reynolds, Philadelphia, Pennsylvania.

Bauhin C. 1622. *Catalogus plantarum circa Basileam sponte nascentium*. Typis J.J. Genathii, Basel.

Baxter WJ, Scheifele G. 2000. *Growing industrial hemp in Ontario*. Ontario Ministry of Agriculture, Food and Rural Affairs. Available at: http://www.omafra.gov.on.ca/english/crops/facts/00-067.htm

Beaton D, Chaput J, Cowbrough M, Obeid K. 2020. *Guide to weed control. Field crops*. Ministry of Agriculture, Food and Rural Affairs, Ottawa, Ontario.

Beatty A. 1844. *Essays on practical agriculture*. Collins & Brown, Maysville, Kentucky.

Bechstein JM. 1868. *The natural history of cage birds, new edition*. Groombridge & Sons, London.

Beck-Mannagetta GR. 1890. *Monographie der Gattung Orobanche*. Fischer, Cassel.

Belon P. 1555. *L'Histoire de la nature des oyseaux*. G. Corrozet, Paris.

Benharrat H, Boulet C, Theodet C, Thalouarn P. 2005. Virulence diversity among branched broomrape (*O. ramosa* L.) populations in France. *Agronomy for Sustainable Development* 25: 123–128.

Berenji J. 1996. Interview with Dr. Janoš Berenji. *Journal of the Industrial Hemp Association* 3(2): 69–70.

Berger J. 1969. *The world's major fibre crops: their cultivation and manuring*. Centre d'Etude de l'Azote. Zurich.

Berkeley MJ. 1866. Fortnightly meeting. *Journal of Horticulture, Cottage Gardener, and Country Gentleman* 11: 122–124.

Berti Pichat CB. 1866. "La pianta della canapa," pp. 396–499 in *Istituzioni scientifiche e tecniche, ossia Corso teorico e practico de Agricoltura, Libri XIII, Volume Quinto*. Presso l'Union Tipografico-Editrice, Torino.

Bertolini G. 1861. Descrizione di una nuova malattia della canepa nel Bolognese. *Memorie della Accademia delle Science dell'Istituto di Bologna* 12: 289–303.

Bhellum BL, Rani M. 1996. *Cuscuta campestris* Yuncker—a new record for the flora of Jammu and Kashmir State. *Indian Journal of Forestry* 19: 103–104.

Bilaĭ VI. 1955. Фузарии (биология и систематика). National Academy of Sciences of Ukraine SSR, Kiev.

Blanco-Ania D, Mateman JJ, Hýlová A, et al. (3 other authors). 2019. Hybrid-type strigolactone analogues derived from auxins. *Pest Management Science* 75: 3113–3121.

Boaler GH. 1919. "Bird seeds and all about them," pp. 7–42 in *Seeds, foods and wild plants for cage birds*. Cage Birds Press, London.

Boari A, Vurro M. 2004. Evaluation of *Fusarium* spp. and other fungi as biological control agents of broomrape (*Orobanche ramosa*). *Biological Control* 30: 212–219.

Bócsa I. 2004. *A kender és termesztése*. Agroinform Kiadó, Budapest.

Bócsa I, Karus M. 1997. *Der Hanfanbau: Botanik, Sorten, Anbau und Ernte*. C.F. Müller, Heildelberg. English translation: *The cultivation of hemp: botany, varieties, cultivation, and harvesting*. Hemptech, Sebastopol, California.

Bócsa L, Beke F. 1956. Néhány külföldi kenderfajta hazai termesztési és nemesítési értéke. *Növénytermelés* 5: 39–50.

Boisduval JA. 1828. *Flore française, ou description synoptique, tome premier*. Hautefeuille, Paris.

Boissier E. 1879. *Flora Orientalis sive enumeratio plantarum, volume quartum*. H. Georg, Genevae et Basileae.

Bomare JC de Valmont. 1764. "Chanvre," *Dictionnaire Raisonné Universel d'Histoire Naturelle* 1: 514–531.

Bonanome G. 1766. Sopra le piante parasite, che danneggiano gli alberi, le erbe de' prati, le canapaie, e le liniere. *Gionale d'Italia* 12: 89–93.

Boy Scouts of America. 1911. *The official handbook for boys*. Doubleday, Page & Co., NY.

Brasavola A. 1536. *Examen omnium simplicium medicamentorum: quorum in officinis usus est*. Bladus de Asula, Rome.

Brault M, Betsou F, Jeune B, Tuquet C, Sallé G. 2007. Variability of *Orobanche ramosa* populations in France as revealed by cross infestations and molecular markers. *Environmental and Experimental Botany* 61: 272–280.

Brodin A. 2010. The history of scatter hoarding studies. *Philosophical Transactions of the Royal Society B* 365: 869–881.

Brown D. 1987. Rodent problems. *Sinsemilla Tips* 7(1): 9–10.

Brumwell MJ. 1951. An ecological survey of the Fort Leavenworth military reservation. *American Midland Naturalist* 45: 187–231.

Buc'hoz PJ. 1775. *Dictionnaire vétérinaire et des animaux domestiques, 6 vols*. Brunet, Paris.

Buffon GLL compte de (Smellie W, *trans*). 1785. *Natural History, general and particular, 2nd edn, vol. VIII*. Strahan and Cadel, London.

Buffon GLL compte de (Smeillie W, *trans*). 1793b. *The natural history of birds, Vol. V*. Strahan, London.

Butler AW. 1900. Notes on Indiana birds. *Proceedings of the Indiana Academy of Science* 9: 149–151.

Butter M. 2001. *Hanf. Das weiße Gold der Batschka*. Books on Demand GmbH, Kirchbierlingen.

Byrd J. 2019. *Industrial hemp (Cannabis sativa L.) germination temperatures and herbicide tolerance screening*. Master's thesis, Virginia Polytechnic Institute and State University, Blacksburg, Virginia.

Cadieu JC, Cadieu N, Lauga J. 1995. Local enhancement and seed choice in the juvenile canary, *Serinus canaries*. *Animal Behaviour* 50: 793–800.

California DPR. 2018. *Cannabis pesticides that cannot be used*. California Department of Pesticide Regulation. Available at: https://www.cdpr.ca.gov/docs/cannabis/cannot_use_pesticide.pdf

Caligaris G. 2005. "Una coltura industriale: produzione, trasformazione, commercio della canapa piemontese tra il XVIII e il XX secolo," pp. 153–182 in: Poni C, Silvio Fronzoni S, eds. *Una fibra versatile: la canapa in Italia dal Medioevo al Novecento*. CLUEB, Bologna.

Callaway JC. 2004. Hemp seed production in Finland. *Journal of Industrial Hemp* 9(1): 97–103.

Campbell BJ, Berrada AF, Hudalla C, Amaducci S, McKay JK. 2019. Genotype x environment interactions of industrial hemp cultivars highlight diverse responses to environmental factors. *Agrosystems, Geosciences & Environment* 2: 180057.

Cardano G. 1550. *De Subtilitate, libri XXI*. Michaëlis Fezanddat & Robert Granion, Paris.

Carr RH, James CM. 1931. Synthesis of adequate proteins in the glands of the pigeon crop. *American Journal of Physiology* 97: 227–231.

Caruel T. 1858. *Illustratio in hortum siccum Andreae Caesalpini*. Typis Le Monnier, Florence.

Casali A, Marconi F. 1876. La canapa bolognese e quella Carmagnola di fronte all'*Orobanche*. *Giornale di Agricoltura Industria e Commercio del regno d'Italia, serie nuova* 1: 138.

Cato MP, Varro MP (Harrison F, *trans*.). 1918. *Roman farm management: the treatises of Cato and Varro done into English*. Macmillan Co., New York.

Ceapoiu N. 1958. *Cînepa, Studiu monografic*. Editura Academiei Republicii Populare Romine. Bucharest.

Čeh B, Čremožnik B. 2016. Vpliv sorte in količine semena za setev na pridelek vršičkov in stebel navadne konoplje (*Cannabis sativa* L.). *Hmeljarski Bilten* 23: 80–87

Cesalpino A. 1583. *De plantis libri XVI Andreae Caesalpini Aretini*. Georgium apud Marescott, Florence.

Chabrolin C. 1934. La germination des graines d'*Orobanche*. *Comptes Rendus Hebdomadaires des Séances de l'Académie des Sciences* 198: 2275–2277.

Chattopadhyaya U. 2018. *Naogaon and the world: intoxication, commoditization, and imperialism in South Asia and the Indian Ocean, 1840–1940*. Doctoral thesis, University of Illinois, Urbana-Champaign, Illinois.

Chiriță, N. 2008. Selectivity and efficiency of some herbicides in controlling weeds from monoecious hemp crops. *Cercetări Agronomice în Moldova* 41(3): 59–64.

Chomel N. 1709. *Dictionnaire œconomique, 2 vols*. Pierre Thened, Lyon.

Ciferri R. 1941. *Manuale di patologia vegetale*. Societa Editrice Dante Alighieri, Rome.

Clarke EGC, Greatorex JC, Potter R. 1971. *Cannabis* poisoning in the dog. *Veterinary Record* 88: 964.

Clarke RC. 1995. Hemp (*Cannabis sativa* L.) cultivation in the Tai'an district of Shandong Province, Peoples Republic of China. *Journal of the International Hemp Association* 2(2): 57, 60–65.

Clarke RC, Merlin MD. 2013. *Cannabis evolution and ethnobotany*. University of California Press, Berkeley, California.

Clay H. 1837. *Letter to D.A. Hall, Esq., 8 April 1837*. Available at: www.raabcollection.com/henry-clay-autograph/henry-clay-experiments-new-type-hemp-seed.

Colta ER. 2008 Cultura cânepii la românii din Ungaria. *Din Tradițiile Populare ale Românilor din Ungaria* 15: 79–96.

Coryate T. 1611 *Coryat's crudities*. Stansby, London.

Crescini F. 1939. Intorno alla produzione e alla conservazione del seme di canapa. *La canapa: Bollettino del Consorzio Industriali Canapieri dei Consorzi per la Canapa* 7(2): 14–18.

D'Aléchamps J, Bauhin J. 1587. *Historia generalis Plantarum, vol. 1*. Guillaume Rouille, Lyon.

da Silva LP, Ramos JA, Olesen JM, Traveset A, Heleno RH. 2014. Flower visitation by birds in Europe. *Oikos* 123: 1377–1383.

Dam H. 1935. The antihaemorrhagic vitamin of the chick. *Biochemical Journal* 29: 1273–1284.

Darwin CR. 1859. *On the origin of species*. John Murray, London.

Darwin CR. 1871. *The descent of man, and selection in relation to sex*. John Murray, London.

Darwin C (Stauffer RC, ed.). 1975. *Charles Darwin's natural selection*. Cambridge University Press, Cambridge, UK.

Davis EF. 1928. The toxic principle of *Juglans nigra* as identified with synthetic juglone, and its toxic effects on tomato and alfalfa plants. *American Journal of Botany* 15: 620–621.

Davison VE, Sullivan EG. 1963. Morning dove's selection of foods. *Journal of Wildlife Management* 27: 373–383.

De Barba M, Miquel C, Boyer F, *et al.* (4 other authors). 2013. DNA metabarcoding multiplexing and validation of data accuracy for diet assessment: application to omnivorous diet. *Molecular Ecology Resources* 14(2): 306–323.

de Herrera GA. 1539. *Obra de Agricultura*. Alcala, Spain. English edition: Arellano JE, trans. 2006. *Ancient agriculture*. Ancient City Press, Layton, Utah.

de Meijer EPM. 1995. Fiber hemp cultivars: a survey of origin, ancestry, availability and brief agronomic characteristics. *Journal of the International Hemp Association* 2(2): 66–73.

de Meijer EPM, Keizer LCP. 1996. Patterns of diversity in *Cannabis*. *Genetic Resources and Crop Evolution* 43: 41–52.

de Serres O. 1600. *Théâtre d'agriculture et mesnage des champs*. Jamet-Métayer, Paris.

Dean HL. 1937. Gall formation in host plants following haustorial invasion by *Cuscuta*. *American Journal of Botany* 24: 167–173.

Del Guercio G. 1900a. Osservazioni naturali sulle lumache dei campi e sulle varie esperienze fatte per allontanarle dale piante e per distruggerle. *Nuove relazioni intorno ai Lavori della R. Stazione di Entomologia Agraria di Firenze, Serie Prima* 2: 237–267

Dempsey JM. 1975. "Hemp," pp. 46–89 in *Fiber crops*. University of Florida Press, Gainesville, Florida.

Desportes JP, Gallo A, Cézilly F. 1994. Un indice de préférence normalisé pour la mesure des choix alimentaires individuels. *Canadian Journal of Zoology* 72: 552–555.

Dewey LH. 1899a. Weed-killing by electricity. *Electrical Engineer* 24: 131–132.

Dewey LH. 1899b. Electrical destruction of grass. *Scientific American* 81: 55.

Dewey LH. 1914. "Hemp," pp. 283–347 in: *USDA Yearbook 1913*. United States Department of Agriculture, Washington, DC.

Dewey LH. 1928. "Hemp varieties of improved type are result of selection," pp. 358–361 in: *USDA Yearbook 1927*. United States Department of Agriculture, Washington, DC.

Dixon CE, Harper CA. 2002. *Using single-strand fencing to manage deer*. Agricultural Extension Service, University of Tennessee, Knoxville, Tennessee.

Dobrozrakova TL, Letova MF, Stepanov KM, Khokhryakov MK. 1956. "*Cannabis sativa* L.," pp. 242-248 *in* Определитель болезней растений. State Publishing House for Pural-Economic Literature, Leningrad.

Doherty S, Cowie RJ. 1994. The effects of early feeding experience on long-term seed choice by canaries (*Serinus canaria*). *Ethology* 97: 177–189.

Dorflein F. 1914. *Tierbau und tierleben in ihrem zusammenhang betrachtet. II. Band: Das Tier als Glied des Naturganzen*. B.G. Teubner, Leipzig.

Duhamel du Monceau HL. 1763. *Éléments d'agriculture, tome second*. Guerin & Delatour, Paris.

Durak R, Jedryczka M, Czjka B, *et al.* (3 other authors). 2021. Mild abiotic stress affects development and stimulates hormesis of hemp aphid *Phorodon cannabis*. *Insects* 12: e420.

Durande JF. 1782. *Flore de Bourgogne, premiere partie*. L.N. Frantin, Dijon.

Dyigubsky I. 1836. *Лексикон городского и сельского хозяйства, том К. К. S.* Selivanovsky, Moscow.

Dyson CE. 1889. *Bird-keeping*. Frederick Warne & Co., London.

Eccard JA, Ferreira CM, Arce AP, Dammhahn M. 2021. Top-down effects of foraging decisions on local, landscape and regional biodiversity of resources (DivGUD). *Ecology Letters* 25: 3–16.

EPA. 2023. *Pesticide products registered for use on hemp*. Available at: https://www.epa.gov/pesticide-registration/pesticide-products-registered-use-hemp.

Errington PL. 1935. Winter killing of mourning doves in central Iowa. *Wilson Bulletin* 47: 159–160

Errington PL. 1936. The winter of 1934–35 and Lowa bob-whites. *American Midland Naturalist* 17: 554–568.

Estienne C, Liébault J. 1583. *L'Agriculture et maison rustique*. Iaques Du Pays, Paris.

Ferraris T. 1935. *Parassiti vegetali della canapa*. Rivista Agricola, Rome.

Findlen P. 2017. "The death of a naturalist: Knowledge and community in late Renaissance Italy," pp. 155–196 in Manning G, Klestinec C, eds. *Professors, physicians and practices in the history of medicine*. Springer Nature, Cham.

Fisher AK. 1889. "Destruction of the sparrow by poisons," pp. 174–183 in Barrows WB, ed. *The English sparrow (Passer domesticus) in North America*. USDA Division of Economic Ornithology, Bulletin no. 1. Government Printing Office, Washington, DC.

Fisher WR. 1845. On the two British species or varieties of the nutcracker. *The Zoologist* 3: 1073–1075.

Fitzpatrick FL. 1923. *The ecology and economic status of Citellus tridecemlineatus tridecemlineatus (Mitchill) in Iowa*. Doctoral thesis, University of Iowa, Iowa City.

Flachs K. 1936. Krankheiten und schädlinge unserer gespinstpflanzen. *Nachrichten über Schädlingsbekämpfung* 11: 6–28.

Flagg W. 1881. *A year with the birds*. Educational Publ. Co., Boston.

Fleming HF von. 1719. *Der Vollkommene Teutsche Jäger, Erster Haupf-Theil*. Martini, Leipzig.

Flessner ML, Bamber KW, Fike JK. 2020. Evaluating herbicide tolerance of industrial hemp (*Cannabis sativa* L.). *Crop Science* 60: 419–427.

Formin EE. 1936. *Новый биологический метод борьбы с заразихой*. Ukrainian Institute of Grain Farming, Dnepropetrovsk.

Formin EE. 1954. Микробиологический метод борьбы с заразихой на помидорах и капусте. *Scientific works of the Ukrainian Research Institute of Horticulture* 3: 217–219.

Fournier G, Paris M. 1983. Mise en evidence de cannabinoïdes chez *Phelipaea ramosa*, parasitant le chanvre, *Cannabis sativa*, Cannabinacees. *Planta Medica* 49: 250–251.

Frank M. 1988. *Marijuana grower's insider's guide*. Red Eye Press, Los Angeles, California.

Frank M, Rosenthal E. 1978. *Marijuana grower's guide*. And/Or Press, Berkeley, California.

Franklin AB, Carlson PC, Rex A, *et al.* (8 other authors). 2018. Grass is not always greener: rodenticide exposure of a threatened species near marijuana growing operations. *BMC Research Notes* 11: e94.

Franzke CJ, Hume AN. 1936. *Field bindweed*. South Dakota Agricultural Experiment Station Bulletin 305, Brookings, South Dakota.

French JC. 1919. *The passenger pigeon in Pennsylvania*. Altoona Tribune Company, Altoona Pennsylvania.

Frisch JL (Frisch JL, Frisch FH, *eds*). 1763. *Vorstellung der Vögel Deutschlandes.* Birnstiel, Berlin.

Fruwirth C. 1922. Zur Hanfzüchtung. *Zeitschrift für Pflanzenzüchtung* 8: 340–401.

Fryday SL, Hart ADM, Dennis NJ. 1994. Effect of exposure to an organophosphate on the seed-handling efficiency of the house sparrow. *Bulletin of Environmental Contamination and Toxicology* 53: 869–876.

Fuchs L (Baumann B, Baumann H, Baumann-Schleihauf S, *eds*). 2001. *Die Kräuterbuch-Handschrift des Leonhart Fuchs.* Ulmer, Stuttgart.

Gainullin RR, Shikov EV. 2012. "*Helicodiscus singleyanus* Pilsbry, 1890 (Mollusca, Gastropoda, Endodontidae) в Абхазии," p. 137 in: *Proceedings of the IV International Conference, Abkhazian State University, Russian Academy of Sciences Institute of Ecology of Mountain Territories.* Kotljarova Publishers, Nalchik, Russia.

Gakhar N, Goldberg E, Jing M, Gibson R, House JD. 2012. Effect of feeding hemp seed and hemp seed oil on laying hen performance and egg yolk fatty acid content: evidence of their safety and efficacy for laying hen diets. *Poultry Science* 91: 701–711.

Gallesio G. 1829. Sur le chanvre. *Annales Administratives et Scientifiques de l'Agriculture Française* 2e série, 3: 353–368 (continues in 4: 43–61).

Garman H. 1889. "The broomrape of tobacco and hemp," pp. 314–320 in *Eighth Annual Report from the Bureau of Agriculture, Horticulture and Statistics of the State of Kentucky.* Frankfort, Kentucky.

Garman H. 1890. *The broom-rape of hemp and tobacco.* Kentucky Agricultural Experiment Station, Lexington, Kentucky.

Garman H. 1892. "Observations on injurious insects and fungi," pp. 8–11 in *Second Annual Report of the Kentucky Agricultural Experiment Station.* Frankfort, Kentucky.

Garman H. 1894a. *Destructive locusts in Kentucky.* Kentucky Agricultural Experiment Station Bulletin no. 49. Lexington, Kentucky.

Garman H. 1894b. "The Orthoptera of Kentucky, pp. 23–30 in *Sixth Annual Report of the Kentucky Agricultural Experiment Station.* Frankfort, Kentucky.

Garman H. 1895. *Cutworms in Kentucky.* Kentucky Agricultural Experiment Station Bulletin no. 58. Lexington, Kentucky.

Garman H. 1903. *The broom-rapes.* Kentucky Agricultural Experiment Station Bulletin No. 105. Lexington, Kentucky.

Garman H. 1904a. "The fringed anthomyian *Pegomya fusciceps*," pp. 45–46 in *Insects injurious to cabbage.* Kentucky Agricultural Experiment Station Bulletin no. 114, Lexington Kentucky.

Garman H. 1904b. *On an injury to fruits by insects and birds.* Kentucky Agricultural Experiment Station Bulletin no. 116, Lexington, Kentuck.

Garman H. 1905. "Broom-rape family," pp. 316–322 in *Some Kentucky weeds and poisonous plants,* Kentucky Agricultural Experiment Station Bulletin no. 183. Lexington, Kentucky.

Garman H. 1906. *Observations and experiments on clover, alfalfa and soy beans.* Kentucky Agricultural Experiment Station Bulletin no. 125. Lexington, Kentucky.

Garman H. 1907. *The food of the crow blackbird.* Kentucky Agricultural Experiment Station Bulletin no. 130. Lexington, Kentucky.

Garrett ER, Hunt CA. 1974. Physiochemical properties, solubility, and protein binding of delta-9- tetrahydrocannabinol. *Journal of Pharmaceutical Sciences* 63: 1056–1064.

Gerard J (Johnson T, *ed.*). 1636. *The herball, or, Generall historie of plantes.* Adam Islip, Joice Norton and Richard Whitakers, London.

Gesner C. 1555b. *Historiae animalium liber III qui est de auium natura.* Froschauer, Zurich.

Gigstead G. 1939. Habits of Wisconsin pheasants. *Wilson Bulletin* 49: 28–34.

Gikalov SY. 1936. Борьба с заразихой на конопле. *Flax and Hemp* 1936(8): 20–23.

Gilbert ES. 1891. Bird seed. *Vick's Illustrated Monthly Magazine* 14: 130.

Giles LW. 1939. Fall food habits of the raccoon in central Iowa. *Journal of Mammalogy* 20: 68–70.

Gilli A. 1982. "Orobanchaceae," pp. 1–22 in Davis PH, *ed. Flora of Turkey and the East Aegean Islands, vol. 7.* Edinburgh University Press, Edinburgh.

Glivar T, Eržen J, Kreft S, *et al.* (4 other authors). 2020. Cannabinoid content in industrial hemp (*Cannabis sativa* L.) varieties grown in Slovenia. *Industrial Crops and Products* 145: e112082.

Gonsior G, Buschmann H, Szinicz G, Spring O, Sauerborn J. 2004. Induced resistance: an innovative approach to manage branched broomrape (*Orobanche ramosa*) in hemp and tobacco. *Weed Science* 52: 1050–1053.

Good R. 1953. *The geography of flowering plants, 2nd edn.* Longmans, Green & Co., London.

Green WT. 1899. *British birds for cages and aviaries.* Upcott Gill, London.

Greenberg J. 2014. *A feathered river across the sky: the passenger pigeon's flight to extinction.* Bloomsbury, New York.

Guettard M. 1751. Second memoire sur les piantes parasites. *Histoire de l'Academie Royale des Sciences* Année 1746: 189–206.

Günther FC. 1774. Gedanken über die Entstehungsart der anomalisch-schwarzen Farbe verschiedener sonst anders gefärbter Vögel. *Der Naturforscher* 2: 1–9.

Gutberlet V, Karus M. 1995. *Parasitäre Krankheiten und Schädlinge an Hanf (Cannabis sativa).* Nova Institut, Köln, Germany.

Hall J, Bhattarai SP, Midmore DJ. 2014. Effect of industrial hemp (*Cannabis sativa* L.) planting density on weed suppression, crop growth, physiological responses, and fibre yield in the subtropics. *Renewable Bioresources* 2(1): e1.

Hamerstrom FN, Blake J. 1939. Winter movements and winter foods of white-tailed deer in central Wisconsin. *Journal of Mammalogy* 20: 206–215.

Haney A, Bazzaz FA. 1970. "Some ecological implications of the distribution of hemp (*Cannabis sativa* L.) in the United States of America," pp. 39–48 in Joyce CRB, Curry SH, *eds. The botany and chemistry of Cannabis.* J & A Churchill, London.

Haney A, Kutscheid BB. 1973. Quantitative variation in the chemical constituents of marihuana from stands of naturalized *Cannabis sativa* L. in East-Central Illinois. *Economic Botany* 27: 193–203.

Hartowicz LE, Eaton BJ. 1971. Reducing the impact of wild hemp control on farm game. *North Central Weed Control Conference, Proceedings* 26: 70.

Henry D. 1771. *The complete English farmer.* F. Newbery, London.

Hervé J, Dalila F, Antoine R, Célia B. 2014. "Nutritional value of organic raw materials for poultry," pp. 291–294 in Rahmann G, Aksoy U, *eds. Proceedings of the 4th ISOFAR science conference.* Organic World Congress.

Heuzé G. 1860. *Les plantes industrielles, seconde partie.* Hachette et Cie, Paris.

Heuzé G. 1893. *Les plantes industrielles, troisième edition, Vol. 1.* Librairie Agricole de la Maison Rustique, Paris.

Hicks GH. 1896. "Oil-producing seeds," pp. 185–204 in *Yearbook of the United States Department of Agriculture 1895.* Washington, DC.

Holden GH. 1903. *Holden's new book on birds.* Holden Publishing, Boston, Massachusetts.

Howard WE, Evans FC. 1961. Seeds stored by prairie deer mice. *Journal of Mammalogy* 42: 260–263.

Huang SM, Jin MR. 2003. Balanophoraceae. *Flora of China* 5: 272–276.

Hugo V. 1862. *Les Misérables.* Lacroix, Verboeckhoven & Cie, Paris.

Hullar I, Meleg I, Fekete S, Romvari R. 1999. Studies on the energy content of pigeon feeds I. Determination of digestibility and metabolizable energy content. *Poultry Science* 78: 1757–1762.

Imaizumi Y. 1979. Observations of food obtaining behavior of the Japanese shew-mole, *Urotrichus talpoides*. *Mammalian Science* 39: 13–22.

Inam B, Hussain F, Bano F. 1989. *Cannabis sativa* L. is allelopathic. *Pakistan Journal of Scientific and Industrial Research* 32: 617–620.

Jacobsohn R, Greenberger A, Katan J, Levi M, Alon H. 1980. Control of Egyptian broomrape (*Orobanche aegyptiaca*) and other weeds by solar heating of the soil by polyethylene mulching. *Weed Science* 28: 312–316

Jaczewski AA. 1912. Fusarium orobanches. *Ежегодник сведений о болезнях и повреждениях культурных и дикорастущих полезных растений* 1910(6): 191.

Jain MC, Arora N. 1988. Ganja (*Cannabis sativa*) refuse as cattle feed. *Indian Journal of Animal Science* 58: 865–867.

Janischevsky DE. 1924. Форма конопли на сорных местах в Юго-Восточной России. Ученые записки Саратовского государственного университета имени Н. Г. Чернышевского 2(2): 3–17.

Jankauskiené Z, Gruzdevienė E, Lazauskas S. 2014. Potential of industrial hemp (*Cannabis sativa* L.) genotypes to suppress weeds. *Zemdirbyste Agriculture* 101: 265–270.

Jankauskiené Z, Gruzdevienė E, Burbulis N, Maumevičius E, Layko IM. 2015. Investigation of hemp (*Cannabis sativa* L.) crop weediness. *Rēzeknes Augstskola* 2: 120–123.

Jiǎ SX. 1962. *A preliminary survey of the book Ch'i min yao shu : an agricultural encyclopaedia of the 6th century*. Science Press, Beijing.

Johnson AW, Rosebery G, Parker C. 1976. A novel approach to *Striga* and *Orobanche* control using synthetic germination stimulants. *Weed Research* 16: 223–227.

Jósef C. 1775. *Uj füves es viragos magyar kert*. Landerer Mihaly, Budapest.

Kabir MA. 2014. Breeding biology of domesticated Eurasian Collared Dove (Columbidae) *Streptopelia decaocto* Frivaldszky 1838 in Saidpur, Bangladesh. *International Journal of Environment* 3: 48–54.

Kalmendal R. 2008. *Hemp seed cake fed to broilers*. Doctoral thesis, Swedish University of Agricultural Sciences, Uppsala.

Kaltenbach JH. 1864. Die deutschen Phytophagen aus der Klasse der Insekten (M-P). *Verhandlungen des naturhistorischen Vereins der preussischen Rheinlande und Westfalens* 21: 228–404.

Kapralov SI. 1975. Применение фитомизы в Ростовской области. Защита растений 20(6): 22–23.

Kear J. 1962. Food selection in finches with special reference to interspecific differences. *Proceedings of the Zoological Society of London* 138: 163–204.

Keating B. 2013. Could the tale of a grow op guarded by bears get any weirder? Yes. *Maclean's* January 21, 2013.

Kendrick B. 1985. *The fifth kingdom*. Mycologue Publications, Waterloo, Canada.

Khrennikov AS, Tollochko YM. 1953. Коноплеводство *(Hemp breeding)*. Research Institute of Bast Crops, Selkhozgiz, Moscow.

King FH. 1883. "Economic relations of Wisconsin birds," pp. 441-610 in Chamberlin TC, *ed. Geology of Wisconsin, Survey of 1873–1879, Vol* 1. David Atwood, Madison Wisconsin.

Kirchner O. 1906. "Hanf, *Cannabis sativa* L.," pp. 319–323 in *Die Krankheiten und Befehädigungen uhferer landwirtschaftlichen Kulturpflanzen*. E. Ulmer, Stuttgart.

Klein O, Kroschel J. 2002. Biological control of *Orobanche* spp. with *Phytomyza orobanchia*, a review. *BioControl* 47: 245–277.

Kojić M, Vrbnicanin S. 2000. Parazitski Korovi osnovne karakteristike, taksonomija, biodiversitet i rasprostanjenje. I. vilina kosico (*Cuscuta* L.). *Acta Herbologica* 9: 21–28.

Konca Y, Cimen B, Yalcin H, Kaliber M, Beyzi SB. 2014a. Effect of hempseed (*Cannabis sativa* sp.) inclusion to the diet on performance, carcass and antioxidative activity in the Japanese quail (*Coturnix coturnix japonica*). *Korean Journal for Food Science of Animal Resources* 34: 141–150.

Konca Y, Yalcin H, Karabacak M, Kaliber M, Durmescelebi. 2014b. Effect of hempseed *(Cannabis sativa* L.) on performance, egg traits and blood biochemical parameters and antioxidant activity in laying Japanese Quail (*Coturnix coturnix japonica*). *British Poultry Science* 55: 785–794.

Konstantinović B, Vidović S, Stojanović A, *et al.* (5 other authors). 2018. "Allelopathic effect of essential oil of *Cannabis sativa* L. on selected vegetable species," pp. 1212–1215 in *Proceedings, IX International Scientific Agriculturla Symposium "Agrosym 2018."* Jahorina, Bosnia and Herzogovina.

Konstantinović B, Šućur J, Kojić M, Samardzic N. 2019. Influence of extract *Cannabis sativa* L. on peroxidation activity *Sorghum halepense* (L.) Pers. *Knowledge International Journal* 34: 641–643.

Kotten EA, Hennessy I, Kluever BM, *et al.* (4 other authors). 2022. Industrial hemp as a resource for birds in agroecosystems: human–wildlife conflict or conservation opportunity? *Human–Wildlife Interactions* 16(3).

Kousta A, Papastylianou P, Cheimona N, *et al.* (3 other authors). 2020. Effect of fertilization and weed management on weed flora of hemp crop. *Bulletin UASVM Horticulture* 77(2): 44–51.

Krishnamurthy GVG, Lal R, Nagarajan K. 1977. Further studies on the effect of various crops on the germination of *Orobanche* seed. *Pest Article and News Summaries (PANS)* 23: 206–208.

Kruse WH. 1907. Dr. Martin Luther's treatise of confession. *Theological Quarterly* 11: 17–30.

Kudryavtsev N, Zaĭtseva L, Golubkov D, *et al.* (3 other authors). 2017. Efficacy of the high molecular weight preparation Artafit produced in Russia for cultivation of flax and hemp. *Mezhdunarodnyĭ Sel'skokhozyaĭstvennyĭ Zhurnal* 2017(3): 40–43.

Lamarck JB, de Candolle AP. 1805. *Flore française, troissieme edition, tome troissieme*. Agasse, Paris.

Lapiedra O, Daniel Sol D, Carranza S, Beaulieu JM. 2013. Behavioural changes and the adaptive diversification of pigeons and doves. *Proceedings of the Royal Society B* 280: 20122893.

Larina SY. 2005. *Cuscuta epilinum Weihe—flax dodder*. Agroatlas. Available at: http://www.agroatlas.ru/en/content/weeds/Cuscuta_epilinum/index.html

Latreille M. 1817. *Le règne animal, tome I*. Déterville, Paris.

Latta RP, Eaton BJ. 1975. Seasonal fluctuations in cannabinoid content of Kansas marijuana. *Economic Botany* 29: 153–163.

Laursen K, Holm E, Sørensen I. 1997. Pollen as a marker in migratory warblers, Sylviidae. *Ardea* 85: 223–231.

Leopold A, Moore EB, Sowls LK. 1939. Wildlife food patches in southern Wisconsin *Journal of Wildlife Management* 3: 60–69.

Lerch A, Rat-Fischer L, Gratier M, Nagle L. 2011. Diet quality affects mate choice in domestic female canary *Serinus canaria*. *Ethology* 117: 769–776.

Liger L. 1700. *Oeconomie générale de la campagne, ou Nouvelle maison rustique, 2 vols*. Charles de Sercy, Paris.

Linnaeus C. 1749. "Versuch von pflanzung der gewächse, wie solche auf der Natur gegründet," pp. 3–26 in *Der Königl. Schwedischen Akademie der Wissenschaften Abhandlungen, aus der Naturlehre, Erster band*. G.C. Grund, Hamburg.

Lisson SN and Mendham NJ. 1995. Tasmanian Hemp Research. *Journal of the International Hemp Association* 2(2): 82–85.

Loiseleur Deslongchamps L. 1833. "Canapa," pp. 9–15 in *Dizionario delle scienze naturali, volume quinto*. Batelli e Figli, Firenze.

López-Ráez JA, Jung SC, Fernandez I, *et al.* (3 other authors). 2012. Mycorrhizal symbiosis as a strategy for root parasitic weed control. *IOBC-WPRS Bulletin* 83: 59–63.

Lucas GB. 1975. *Diseases of tobacco, 3rd edn.* Biological Consulting Assoc., Raleigh, North Carolina.

Lusitanus A. 1553. *In Dioscoridis Anazarbei de medica materia libros quique enarrationes Amati Lusitani.* Scotus, Venetiis.

Mahmoodzadeh H, Ghasemi M, Zanganeh H. 2015. Allelopathic effect of medicinal plant *Cannabis sativa* L. on *Lactuca sativa* L. seed germination. *Acta Agriculturae Slovenica* 105: 233–239.

Makkizadeh TM, Farhoudi R, Rabii M, Rastifar M. 2011. Evaluation allelopathic effect of hemp (*Cannabis sativa* L.) on germination a growth of three kinds of weeds. *Crop Physiology (Iran)* 3(11): 77–88.

Malmuk OF. 1980. The spread of *Cuscuta* on alfalfa and on other plants in the "Ghouta" of Damascus. *Phytopathologia Mediterranea* 19: 64–66.

Manen JF, Habashi C, Jeanmonod D, Park JM, Schneeweiss GM. 2004. Phylogeny and intraspecific variability of holoparasitic *Orobanche* (Orobanchaceae) inferred from plastid rbcL sequences. *Molecular Phylogenetics and Evolution* 33: 482–500.

Mańkowski J. 2003. The effect of some agronomic factors on the amount and quality of homomorphic fibre. *Fibres & Textiles in Eastern Europe* 11: 20–25.

Mann J. 1848. *The American bird-keeper's manual.* Little and Brown, New York.

Marcin of Urzędów. 1595. *Herbarz Polski.* Drukarni Łazarzowey, Kraków.

Markham G. 1615. *Countrey contentments in two bookes, the second intituled the English Huswife.* Roger Jackson, London.

Markham G. 1616. *Maison Rustique, or, the countrey farme.* Adam Islip, London.

Martyn T. 1807a. *The gardener's and botanist's dictionary, Vol.* I, *Part I.* F.C. and J. Rivington, London.

Marudarajan D. 1950. Note on *Orobanche cernua* Loefl. *Current Science* 19: 64–65.

Mastromarino A. 1937. Osservazioni e ricerche sull'orobanche del tabacco (*Phelipaea ramosa* C. A. Mey.). *Bollettino Tecnico Istituto Sperimentale Coltivazioni Tabacchi* 15: 187–193.

Matthiessen P. 1978. *The snow leopard.* Viking Press, New York.

Mattioli PA, ed. (Dioscorides P). 1544. *Di Pedacio Dioscoride Anazarbeo libri cinque.* Niccolò Bascarini, Venice.

Mattioli PA, ed. (Dioscorides P). 1554. *Commentarii in sex libros Pedacii Dioscoridis de medica materia.* Vincent Valgrisum, Venice.

Mattioli PA, Camerer J. 1586. *Kreuterbuch.* Feyerabends, Fischers, & Dacken, Frankfurt.

Maxwell BA. 2016. *Effects of herbicides on industrial hemp (Cannabis sativa) phytotoxicity, biomass, and seed yield.* Master's Thesis, Western Kentucky University, Bowling Green, Kentucky.

McClure HE. 1943. Ecology and management of the morning dove in Iowa. *Iowa Agricultural Experiment Station Research Bulletin* 310: 353–415.

McDonnell RJ, Santangelo R, Paine T, Hoddle MS. 2016. The feeding behavior of *Rumina decollata* (Subulinidae: Gastropoda) raises questions about its efficacy as a biological control agent for *Cornu aspersum* (Helicidae: Gastropoda). *Biocontrol Science and Technology* 26: 331–336.

McPartland JM, Cubeta MA. 1997. New species, combinations, host associations and location records of fungi associated with hemp (*Cannabis sativa*). *Mycological Research* 101: 853–857.

McPartland JM, Rhode B. 2005. New hemp diseases and pests in New Zealand. *Journal of Industrial Hemp* 10(1): 99–108.

McPartland JM, Clarke RC, Watson DP. 2000. *Hemp diseases and pests – management and biological control.* CABI Publishing, Wallingford, UK.

McPartland JM, Cutler S, McIntosh DJ. 2004. Hemp production in Aotearoa. *Journal of Industrial Hemp* 9(1): 105–115.

Meriwether WF. 1969. Acute marijuana toxicity in a dog. *Veterinary Medicine* 64: 577–578.

Micheli PA. 1723. *Relazione dell'erba detta da' botanici orobanche e volgarmente succiamele.* Tartini e Franchi, Firenze.

Miller P. 1754. *The gardeners dictionary, 4th edn.* J & J Rivington, London.

Mitchill SL, Miller E. 1800. *The medical repository, vol. III.* Swords, New York.

Mohr EW. 1928. *Epimys rattus* in captivity. *Journal of Mammalogy* 9: 113–117.

Molisch H. 1937. *Der Einfluss einer Pflanze auf die Andere, Allelopathie.* Gustav Fischer, Jena.

Monserrate L, Wares A, Smart L. 2024. *Rolling before vs after planting.* Cornell Hemp. Available at: https://hemp.cals.cornell.edu.

Montana DoA (Department of Agriculture). 2019. Pest management practices for hemp growers in Montana. Available at: https://agr.mt.gov/Portals/168/Documents/IndustrialHemp/PestMgmtGuidelinesforHempinMT.pdf

Morimoto S, Tanaka Y, Sasaki K, *et al.* (5 other authors). 2007. Identification and characterization of cannabinoids that induce cell death through mitochondrial permeability transition in *Cannabis* leaf cells. *Journal of Biological Chemistry* 282: 20739–20751.

Moore J. 1735. *Columbarium: or, the pigeon-house.* Wilford, London.

Muminović Š. 1990. Alelopatski efekti ekstrakta nekih korova na klijavost sjemena usjeva. *Fragmenta Herbologica Jugoslavica* 19: 93–102.

Munerati O. 1929. Sul problema della canapa di fronte alla orobanche. *L'Italia Agricola* 66(4): 177–184.

Musselman LJ. 1984. An unusual specimen of *Orobanche* from North Carolina collected by John Ball in 1884. *Castanea* 49: 91–93.

Musselman LJ. 1994. "Taxonomy and spread of *Orobanche*," pp. 27–35 in Pieterse AH, *et al.*, eds. *Biology and management of Orobanche.* Royal Tropical Institute, Amsterdam.

Nattrass RM. 1941. Dodder. *East African Agricultural Journal* 6: 187–188.

Naumann JA. 1824. *Naturgeschichte der Vögel Deutschlands, Vierter Theil.* E. Fleischer, Leipzig.

Neijat M, Gakhar N, Neufeld J, House JD. 2014. Performance, egg quality, and blood plasma chemistry of laying hens fed hempseed and hempseed oil. *Poultry Science* 93: 2827–2840.

Neuer H. 1935. Der Anbau des Hanfes. *Berichte über Landwirtschaft: Sonderheft* 20: 16–30.

Nevinnykh VA, Schukina GN. 1972. Коноплеводство на Северном Кавказе. *Flax and Hemp* 1972(9): 5–8.

Newmann E. 1856. Black hawfinch. *The Zoologist* 14: 4994.

Nicot A. 1959. La culture du chanvre en France, problèmes techniques et économiques posés par l'*Orobanche*. *Association de Coordination Technique Agricole*, Paris.

Nördlinger H. 1869. *Die kleinen Feinde der Landwirthschaft, zweite auflage.* J.G. Cotta, Stuttgart.

Norgord CP. 1912. Hemp as a weed eradicator and money crop. *Annual Report of the Agricultural Experiment Station of the University of Wisconsin* 29: 18–19.

Okulova NM, Mironova TA, Sapel'nikov SF, *et al.* (3 other authors). 2015. Autumn diets of sibling species *Microtus arvalis* sensu lato and *M. agrestis* (Rodentia, Arvicolinae) in the forest-steppe of the central chernozem zone. *Russian Journal of Ecology* 46: 181–188.

Olsson M. 2015. *Allelopatisk effekt av industrihampa som förfrukt till åkerböna, ärt och lupin.* Doctoral thesis, Institutionen för Växtproduktionsekologi, Uppsala.

OMRI (Organic Materials Review Institute). 2020. OMRI products list. Available at: https://www.omri.org/sites/default/files/opl_pdf/CropByCategory-NOP.pdf

Orth RE, Moore I, Fisher TW, Legner EF. 1975. A rove beetle *Ocypus olens*, with potential for biological control of the brown garden snail, *Helix aspersa*, in California. *Canadian Entomolologist* 107: 1111–1116.

Ottavi E. 1897. *Il coltivatore*. Carlo Cassone, Casale.

Parker C. 1986. "Scope of the agronomic problems caused by *Orobanche* species," pp. 11–17 in Borg SJ, *ed. Biology and control of Orobanche*. Landbouwhogeschool Press, Wageningen, The Netherlands.

Parker C. 2013. "The parasitic weeds of the Orobanchaceae," pp. 313–344 in Joel DM, *et al.*, *eds. Parasitic Orobanchaceae: parasitic mechanisms and control strategies*. Springer, Heidelberg.

Parker C, Riches CR. 1993. *Parasitic weeds of the world*. CABI Publishing, Wallingford UK.

Parkinson J. 1640. *Theatrum botanicum*. Thomas Cotes, London.

Pasković F. 1941. *Konoplja*. Ministarstvo Seljačkog Gospodarstva, Zagreb, Yugoslavia.

Patton J. 1998. Hemp: from seed to feed. *Lexington Herald-Leader, Lexington, Kentucky, June* 2 1998.

Pearson TG. 1925. *Portraits and habits of our birds, vol. I*. National Association of Audubon Societies, New York.

Peglion V. 1906. La cuscuta parassita delle bietole e della canapa. *L'Italia Agricola* 43: 492–494.

Penzig OAJ. 1905. "Illustrazione degli erbarii di Gherardo Cibo," pp. 1–237 in Penzig O, *ed. Contribuzioni alla storia della botanica*. Ciminago, Genova.

Pereira SL, Kohnson KP, Clayton DH, Baker AJ. 2007. Mitochondrial and nuclear DNA sequences support a Cretaceous origin of Columbiformes and a dispersal-driven radiation in the Paleogene. *Systematic Biology* 56: 656–672.

Pernau FJ von. 1720. *Angenehme Land-Lust*. Peter Conrad Monath, Frankfurt.

Phillips JC. 1910. Notes on attracting birds. *Bird-Lore* 11(5): 175–178.

Piemontese A. 1555. *De' secreti del reuerendo donno Alessio Piemontese, prima parte, diuisa in sei libri*. Sigismondo Bordogna, Venetia.

Piper H. 1871. *Pigeons: their varieties, management, breeding and diseases*. Groombridge and Sons, London.

Pivano S. 1902. *Cartario della Abazia di Rifreddo fino all'anno 1300*. Chiantore-Mascarelli, Pinerolo.

Piwowarczyk R. 2012. A revision of distribution and historical analysis of preferred hosts of *Orobanche ramosa* (Orobanchaceae) in Poland. *Acta Agrobotanica* 65: 53–62.

Pliny (Mayhoff C., *ed.*). 1870. *Naturalis Historia [Latin edition]*. In aedibus B.G. Teubneri, Leipzig

Pochop PA, Johnson RJ, Aguero DA, Eskridge KM. 1990. "The status of lines in bird damage control—a review," pp. 317–324 in *Proceedings of the 14th Vertebrate Pest Conference*. University of California, Davis, California.

Pohle P. 2001. *Kāgbeni: contributions to the village's history and geography*. Institut für Geographie, Giessen.

Prain D. 1903. *Bengal plants, vol. II*. Botanical Survey of India, Calcutta.

Pudełko K, Majchrzak L, Narożna D. 2014. Allelopathic effect of fibre hemp (*Cannabis sativa* L.) on monocot and dicot plant species. *Industrial Crops and Products* 56: 191–199.

Pujadas-Salva AJ, Velasco L. 2000. Comparative studies on *Orobanche cernua* L. and *O. cumana* Wallr. (Orobanchaceae) in the Iberian peninsula. *Botanical Journal of the Linnean Society* 134(4): 513–527.

Qiang LY, Hu HB, Li GQ, *et al.* (4 other authors). 2023. Plastic mulching, and occurrence, incorporation, degradation, and impacts of polyethylene microplastics in agroecosystems. *Ecotoxicology and Environmental Safety* 263: e115274.

Qureshi RA, Gilani SA, Ghufran MA, Sultana KN. 2006. Effects of *Parthenium hysterophorus* L. on soil characteristics initially inhabited by *Cannabis sativa* L. *Pakistan Journal of Biological Sciences* 9: 2794–2797.

Rabelais F. 1546. *La Vie de Gargantua et de Pantagruel. Le Tiers Livre*. Edition Michel Fezandat, Paris.

Ramírez J. 1990. Acute pulmonary histoplasmosis: newly recognized hazard of marijuana hunters. *American Journal Medicine* 88 (Supplement 5): 60N–62N.

Ranalli P, *ed.* 1999. *Advances in Hemp Research*. Haworth Press, Binghamton, New York.

Rasoulpour R, Salehi M, Salehi E. 2017. Detection and partial characterization of a 16SrIX-C. *Journal of Plant Pathology* 99: 219–223.

Rataj K. 1957. Skodlivi cinitele pradnych rostlin. *Prameny literatury* 2:1–123.

Ray J. 1693. *Historia Plantarum, Tomus Secundus*. Smith and Walford, London.

Ray J (Dillenius JJ, *ed.*). 1724. *Synopsis methodica stirpium Britannicarum, editio tertia*. Innys, London.

Reisinger P, Lehoczky É, Komives T. 2005. Competitiveness and precision management of the noxious weed *Cannabis sativa* L. in winter wheat. *Communications in Soil Science and Plant Analysis* 36: 629–634

Rennie J, *ed.* 1834. Notes. *The Field Naturalist* 2: 11.

Robel RJ. 1969. Food habits, weight dynamics, and fat content of bob-whites in relations to food plantings in Kansas. *Journal of Wildlife Management* 33: 237–294.

Robel RJ, Bisset AR, Clement TM, Dayton AD. Morgan KL. 1979a. Metabolizable energy of important foods of bobwhites in Kansas. *Journal of Wildlife Management* 43: 982–987.

Robel RJ, Bisset AR, Dayton AD, Kemp KE. 1979b. Comparative energetics of bobwhites on six different foods. *Journal of Wildlife Management* 43: 987–992.

Robeson MS, Khanipov K, Glovko G, *et al.* (7 other authors). 2018. Assessing the utility of metabarcoding for diet analyses of the omnivorous wild pig (*Sus scrofa*). *Ecology and Evolution* 8: 185–196.

Roessler ES. 1936. Viability of weed seeds after ingestion by California linnets. *Condor* 38: 62–65.

Rowlandson T. 1849. On hemp. *Journal of the Royal Agricultural Society of England* 10: 172–182.

Rozier F. 1787. "Mémoire sur la culture et rouissage du chanvre," pp. 1–136 in *Recueil de mémoires sur la culture et le rouissage du chanvre*. Chez les frères Périsse, Lyon.

Rumsey F, Carine M. 2002. *Plant collecting in Morocco*. Plant Cuttings (Department of Botany On-Line Newsletter), e8. National History Museum, London.

Runyon JB, Mescher MC, De Moraes CM. 2006. Volatile chemical cues guide host location and host selection by parasitic plants. *Science* 313: 1964–1967.

Ruß K. 1899. *Die fremdländischen Stubenvögel*. Creutz, Magdeburg.

Sacchetti. M. 1940. "Nemici vegetali della canapa e del suo tiglio," pp. 93–109 in *Lezioni al corso di perfezionamento per la stima del tiglio di Canapa*. Fratelli Lega, Faenza.

Saetre GP, Riyahi S, Alibadian M, *et al.* (7 other authors). 2012. Single origin of human commensalism in the house sparrow. *Journal of Evolutionary Biology* 25: 788–796.

Safra R. 1959. The control of *Orobanche* and wild oats in the U.S.S.R. *Pest Articles & News Summaries, Section C. Weed Control* 5(2/3): 59–60.

Saint Jacob P de. 1960. *Les Paysans de la Bourgogne du Nord au dernier siècle de l'Ancien Régime*. Belles lettres, Paris.

Salmon W. 1710. *Botanologia. The English herbal: or, history of plants*. Dawks, London.

Samoggia M. 1898. Studi e ricerche sulla canapa. *Stazioni sperimentali Agrarie Italiane* 31(fasc. 4): 417–448.

Samorini G. 2000. *Animali che si drogano*, Telesterion, Vicenza.

Sanchis Serra A, Margalef CR, Morales Pérez JV, *et al.* (7 other authors). 2014. Towards the identification of a new taphonomic agent: an analysis of bone accumulations obtained from modern Egyptian vulture (*Neophron percnopterus*) nests. Quaternary International 330: 136–149.

Sandler L, Gibson K. 2019. A call for weed research in industrial hemp (*Cannabis sativa* L). *Weed Research* 59: 255–259.

Sanio C. 1891. Zahlenverhältnisse der Flora Preussens. II. *Verhandlungen des Botanischen Vereins für die Provinz Brandenburg* 32: 55–126.

Sannong ZH. 2016. 大麻病害. China Agricultural Press. Available at: http://www.pwsannong.com/c/2016-04-13/565118.shtml

Sansovino F. 1560. *Della agricoltura, libri cinque*. Sansovino et Co., Venetia.

Savelli R. 1932. Studien über der ferrarischen Hanf. *Der Züchter* 4: 286–290.

Schlechtendal DFL von, Langethal CG, Schenk E. 1848. *Flora von Deutschland, Band IX*. F. Mauke, Jena.

Schlippe P de. 1956. *Shifting cultivation in Africa: the Zande system of agriculture*. Routledge and Kegan Paul, London.

Schmitz JJ, Regel E. 1841. *Flora Bonnensis*. König, Bonnae, Germany.

Schreier H, Bish M, Bradley KM. 2022. The impact of electrocution treatments on weed control and weed seed viability in soybean. *Weed Technology* 36: 481–489.

Senchenko GI, Kolyadko IV. 1973. Устойчивость гибридов конопли к заразихе ветвистой. *Selection and Seed Production* 23: 27–33.

Serzane M. 1962. "Kanepju – *Cannabis sativa* L. Slimibas," pp. 366–369 *in Augu slimibas, praktiskie darbi*. Riga Latvijas Valsts Izdevnieciba, Latvia.

Shapiro BB, Hedrick R, Vanle BC, *et al.* (7 other authors). 2018. Cryptococcal meningitis in a daily cannabis smoker without evidence of immunodeficiency. *BMJ Case Reports* 2018: bcr2017221435.

Shapiro J. 2001. *Mao's war against nature*. Cambridge University Press, Cambridge UK.

Sharma S, Devkota A. 2014. Allelopathic potential and phytochemical screening of four medicinal plants of Nepal. *Scientific World* 12(12): 56–61.

Sharma YP, Kapoor V. 2014. Parasitic angiosperms and biology of *Cuscuta* species—an overview. *Review of Plant Pathology* 6: 577-608.

Shettleworth SJ, Krebs JR. 1982. How marsh tits find their hoards: the roles of site preference and spatial memory. *Journal of Experimental psychology: Animal Behavior Processes* 8: 354–375.

Shettleworth SJ, Krebs JR. 1986. Stored and encountered seeds: a comparison of two spatial memory tasks in marsh tits and chickadees. *Journal of Experimental Psychology: Animal Behavior Processes* 12: 248–257.

Shiflett MW, Rankin AZ, Tomaszycki ML, DeVoogd TJ. 2004. Cannabinoid inhibition improves memory in food-storing birds, but with a cost. *Proceedings of the Royal Society of London B* 271: 2043–2048.

Shoyama Y, Sugawa C, Tanaka H, Morimoto S. 2008. Cannabinoids act as necrosis-inducing factors in *Cannabis sativa*. *Plant Signaling and Behavior* 3: 1111–1112.

Shufeldt RW. 1902. On the habits of the kangaroo rats in captivity. *American Naturalist* 36: 47–51.

Siegel RK. 1989. *Intoxication: life in pursuit of artificial paradise*. E.P. Dutton, New York.

Sigerson G. 1866. *Cannabiculture in Ireland, its profit and possibility*. W. Kelly, Dublin.

Silversides FG. Lefrançois MR. 2005. The effect of feeding hemp seed meal to laying hens. *British Poultry Science* 46: 231–235.

Singh NB, Thapar R. 2003. Allelopathic influence of *Cannabis sativa* on growth and metabolism of *Parthenium hysterophorus*. *Allelopathy Journal* 12: 61–70.

Slator L, Hall R (Coote T, *ed*). 1724. *Instructions for the cultivating and raising of flax and hemp: in a better manner, than that generally practis'd in Ireland*. Grierson, Dublin.

Slaughter DC, Giles DK, Fennimore SA, Smith RF. 2008. Multispectral machine vision identification of lettuce and weed seedlings for automated weed control. *Weed Technology* 22: 378–384.

Small E. 1975a. The case of the curious "*Cannabis.*" *Economic Botany* 29: 254.

Small E. 2015b. Response to the erroneous critique of my *Cannabis* monograph by R.C. Clarke and M.D. *Merlin. Botanical Review* 81: 306–316.

Snyder G. 1974. *Turtle Island*. New Directions, New York.

Snyder WC, Hansen HN. 1940. The species concept in *Fusarium*. *American Journal of Botany* 27(2): 64–67

Sokolov VE, Ilyichev VD, Emelyanova IA. 1990. Млекопитающие и птицы, повреждающие технику и сооружения. Nauka, Moscow.

Soldano A. 2000. La provenienza delle Raccolte dell'erbario di Ulisse Aldrovandi volume I e II. *Atti del R. Istituto Veneto di Scienze, Lettere et Arti. Classe di Scienze fisiche, Matematiche e Naturali* 158(1): 1–245.

Sonnini CS. 1799. *Histoire naturelle des Oiseaux, par LeClerc de Buffon, Tome Dixième*. Dufart, Paris.

Sorauer P. 1958. *Handbuch der Pflanzenkrankheiten, Band 5*. Paul Parey, Berlin.

Sowerby J. 1794. *English botany, vol. 3*. Davis, London.

Spallanzani L (Senebier J, *trans*). 1783. *Expériences sur digestion de l'homme et de différentes espèces d'animaux*. Barthelemi Chirol, Geneva.

Spengler RN. 2013. *Botanical resource use in the Bronze and Iron Age of the Central Eurasian mountain/steppe interface: decision making in multiresource pastoral economies*. PhD thesis, Washington University in St Louis. St. Louis, Missouri.

Srivastava PP, Das LL. 1974. Effect of certain aqueous plant extracts on the germination of *Cyperus rotundus* L. *Science and Culture* 40: 318–319.

Staples CP. 1948. Further as to colour change without a moult. *Bulletin of the British Ornithology Club* 68: 80–88.

Stefanaki A, Thijsse G, van Uffelen GA, Eurlings MCM, van Andel T. 2018. The En Tibi herbarium, a 16th century Italian treasure. *Botanical Journal of the Linnean Society* 20: 1–31.

Stefanaki A, Porck H, Grimaldi IM, *et al.* (8 other authors). 2019. Breaking the silence of the 500-year-old smiling garden of everlasting flowers: the En Tibi book herbarium. *PLoS ONE* 14: e0217779.

Steinhöwel H, Brant S. 1501. *Esopi appologi sive mythologi*. Jacob von Pfortzheim, Basel.

Stevenson H. 1866. *The birds of Norfolk, Vol. I*. J. Van Voorst, London.

Stojanovic D. 1959. A contribution to the knowledge of species and varieties of *Cuscuta* in the territory of North Serbia. *Zashtita Bilja* 54: 21–27.

Stone W. 1913. Bird migration records of William Bartram. *Auk* 30(3): 325–358.

Stroud R (Sanborn HG, *ed.*). 1933. *Diseases of canaries*. T.F.H. Publications, Jersey City, New Jersey.

Stroud R. 1943. *Stroud's digest on the diseases of birds*. L.G. Marcus and Robert Stroud, Minneapolis, Minnesota.

Stupnicka-Rodzynkiewicz E. 1970. Zjawiska allelopatii między niektórymi roślinami uprawnymi i chwastami. *Acta Agraria et Silvestria (Series Agraria)* 10(2): 75–105.

Sumner EL. 1935. A life history study of the California quail, with recommendations for its conservation and management. *California Fish and Game* 21(2): 167–253, 275–342.

Surkov N. 1932. Сортоводство и семеноводство южных лубяных и районирование семенных рассадников. *За новое волокно* 1932(3): 12–19.

Tandingan DeLey I, McDonnell RJ, Paine T, DeLey P. 2017. "*Phasmarhabditis*: the slug and snail parasitic nematodes in North America," pp. 560–578 in Abd-Elgawad MM, *et al.*, eds. *Biocontrol agents: entomopathogenic and slug parasitic nematodes*. CABI, Wallingford, UK.

Tang K, Struik PC, Thouminot C, *et al.* (3 other authors). 2016. Comparing hemp (*Cannabis sativa* L.) cultivars for dual-purpose production under contrasting environments. *Industrial Crops and Products* 87: 33–44.

Tarasov AV, Tarchokov KS. 1974. Chemical control of common ragweed in hemp grown for seed production. *труды Всесоюзный Институт Лубяных Культур* 35: 91–96

Tarchokov KS. 1975. Herbicides in hemp. *Лен и Конопля* 1975(4): 22–23.

Tellman B. 1996. "Stowaways and invited guests: How some exotic plants reached the American southwest," pp. 144–149 in Tellman B, *et al.*, eds. *The future of arid grasslands*. USDA Rocky Mountain Research Station, Fort Collins, Colorado.

Teodosio G. 1553. *Medicinales epistolae LXVII*. Nicolaum Episcopium Juniorem, Basileae.

Thiébaut de Berneaud A. 1835. "Chanvre," pp. 87–89 in Guérin-Méneville FE, ed. *Dictionnaire pittoresque d'histoire naturelle et des phénomènes de la nature, Tome deuxième*. De Cosson, Paris.

Thomas H, Sauerborn J, Müller-Stöver D, *et al.* (3 other authors). 1998. The potential of *Fusarium oxysporum* f. sp. *orthoceras* as a biological control agent for *Orobanche cumana* in sunflower. *Biological Control* 13: 41–48.

Thoreau HD (Torrey B, *ed.*). 1906. *The writings of Henry David Thoreau, vol IV*. Houghton Mifflin & Co., Boston, Massachusetts.

Thoreau HD (Dean BP, *ed.*). 1993. *Faith in a seed: The dispersion of seeds and other late natural history writings*. Island Press, Washington, DC.

Transhel V, Gutner L, Khokhryakov M. 1933. Новые виды грибных паразитов на новых лубяных растениях. (A list of fungi found on new bast crops). *Болезни и вредители новых лубяных культур* 4: 127–140.

Trotter S. 1909. An inquiry into the history of the current English names of North American land birds. *Auk* 26: 346–363.

Turner W. 1544. *Avium praecipuarum, quarum apud Plinium et Aristotelem mentio est*. J. Cymnicus, Coloniae.

Tusser T. 1557. *A hundredth good pointes of husbandrie*. Richard Tottel, London.

Ulmer A, Yousaf Z, Khan F, *et al.* (4 other authors). 2010. Evaluation of allelopathic potential of some selected medicinal species. *African Journal of Biotechnology* 9: 6194–62006.

Urhan AU, Brodin A. 2015. No evidence for memory interference across sessions in food hoarding marsh tits *Poecile palustris* under laboratory conditions. *Animal Cognition* 18: 649–656.

US Congress. 1937. "Taxation on Marihuana," *Hearings before the Committee on Ways and Means, House of Representatives, on HR 6385, 75th Congress, 1st session*. Government Printing Office, Washington, DC.

US Treasury. 1937. *Traffic in opium and other dangerous drugs for the year ended December 31, 1936*. Government Printing Office, Washington, DC.

Vakhrusheva TE. 1979. *Methodological instructions for categorizing diseases and cultures of flax and hemp*. All-Union N. I. Vavilov Scientific Research Institute of Horticulture, Leningrad.

Valentini A, Miquel C, Ali Nawas M, *et al.* (9 other authors). 2009. New perspectives in diet analysis based on DNA barcoding and parallel pyrosequencing: the *trnL* approach. *Molecular Ecology Resources* 9: 51–60.

Van der Meij MAA. 2004. *A tough nut to crack. Adaptions to seed cracking in finches*. Doctoral thesis, Leiden University.

Van der Meij MAA, Bout RG. 2006. Seed husking time and maximal bite force in finches. *Journal of Experimental Biology* 209: 3329–3335.

Van der Werf HMG. 1997. The effect of plant density on light interception in hemp (*Cannabis sativa* L.). *Journal of the International Hemp Association* 4(11): 8–13.

Van der Werf HMG, Wijhuizen M, De Schutter JAA. 1995a. Plant density and self-thinning affect yield and quality of fibre hemp (*Cannabis sativa* L.). *Field Crops Research* 40: 153–164.

Vance JM. 1971. Marijuana is for the birds. *Outdoor Life* 147(6): 53–55, 96–100.

Vander Wall SB. 2010. How plants manipulate the scatter-hoarding behavior of seed-dispersing animals. *Philosophical Transactions of the Royal Society B* 365: 989–997.

Vappula NA. 1965. Pests of cultivated plants in Finland. *Annales Agriculturae Fenniana* 1 (Suppl.): 7–239.

Vaucher JPE. 1827. *Monographie des Orobanches*. Paschoud, Genève.

Vavilov NI. 1910. Голые слизни (улитки), повреждающие поля и огороды в Московской губернии [*Naked slugs (snails) damaging fields and vegetable gardens in the Moscow province*]. S.P. Yakovlev, Moscow.

Vavilov NI. 1926. происхождение культурной конопли и воэникновение культуры группы «первичных» растений. *Bulletin of Applied Botany and Plant-Breeding* 16(2): 221–233.

Vera CL, Woods SM, Raney JP. 2006. Seeding rate and row spacing effect on weed competition, yield and quality of hemp in the Parkland region of Saskatchewan. *Canadian Journal of Plant Science* 86: 911–915.

Villars D, Chovin JR. 1789. *Histoire des plantes de Dauphiné, tome troisieme*. Prevost, Paris.

Wallace GJ. 1942. A three-year trial feed patch for songbirds. *Journal of Wildlife Management* 6: 110–117.

Warmke HE, Davidson H. 1943. Polyploidy investigations. Annual Report of the Director of the Department of Genetics. *Carnegie Institute of Washington Yearbook* 42: 153–157.

Weiss F. 1948. Check list revision. *Plant Disease Reporter* 32: 404–424.

Wheaton JM. 1875. "The food of birds as related to agriculture," pp. 561–578 in *29th Annual Report of the Ohio State Board of Agriculture*. Columbus, Ohio.

Whistler H, Kinnear NB. 1949. *Popular handbook of Indian birds*. Gurney and Jackson, London.

White G. 1789. *The natural history of Selborne*. T. Bensley, London.

Wilhelm S. 1958. Parasitic seed plants with special reference to the broomrapes. *Proceedings Annual California Weed Conference* 10: 13–18.

Willson MF. 1971. Seed selection in some North American finches. *Condor* 73: 415–429.

Willson MF. 1972. Seed size preference in finches. *Wilson Bulletin* 84: 449–445.

Willson MF, Harmeson JC. 1973. Seed preferences and digestive efficiency of cardinals and song sparrows. *Condor* 75: 225–234.

Wilson A (Brewer TM, *ed.*). 1839. *Wilson's American ornithology*. Broaders & Co., Boston, Massachusetts.

Wilson MG, Korovin VA. 2003. Oriental turtle dove breeding in the Western Palearctic. *British Birds* 96: 234–241.

Wilson MJ, Glen DM, George SK. 1993. The rhabditid nematode *Phasmarhabditis hermaphrodita* as a potential biological control agent for slugs. *Biocontrol Science and Technology* 3: 503–511.

Wissett R. 1808. *A treatise on hemp*. J. Harding, London.

Woosley PB, Willian T, Williams DW, *et al.* (3 other authors). 2015. *Herbicide tolerance trail with industrial hemp*. Available at: https://hemp.ca.uky.edu/sites/hemp.ca.uky.edu/files/2015_herbicide_trial_0.pdf

Yakovlev AA. 2017. Значение бактороденцида (*Salmonella enteritidis* var. *issatschenko*, 29/1) в оптимальном ассортименте родентицидов для защиты растений. *IPRS-IDB Newsletter* 52: 339–341.

Yao ZQ, Cao XL, Fu C, Zhao SF. 2017. 新疆列当的种类、分布及其防治技术研究进展. *Journal of Biosafety* 26(1): 23–29.

Young RE. 2022. *Saving an iconic species from extinction in the UK: interactions between diet, parasites and environmental change*. Doctoral thesis, Cardiff University.

Yu R, Ma Y. 2014. Melon broomrape and sunflower broomrape seeds germination induced by hemp (*Cannabis sativa* L.) plants. *Journal of China Agricultural University* 19(4): 38–46.

Yu R, Ma YQ. 2014. 大麻对瓜列当和向日葵列当种子萌发诱导作用研究. *Journal of China Agricultural University* 19(4): 38–48.

Yunusbaev UB, Musina LB, Suyundukov YT. 2003. Dynamics of steppe vegetation under the effect of grazing by different farm animals. *Russian Journal of Ecology* 34: 43–47.

Zambrano G. 1910 (published 1911). Del seme canapa. Filogenesi, produzione e commercio [Of the hemp seed. Phylogeny, production and trade]. *Atti del Reale Istituto d'Incoraggiamento di Napoli, serie sesta* 62: 209–234.

Zavala J, Villier S, Philocles S, *et al.* (4 other authors). 2023. *Pest management of hemp*. Extension Bulletin H0-339-W, Purdue University, West Lafayette, Indiana.

Zeigler HP, Levitt PW, Levine RR. 1980. Eating in the pigeon (*Columba livia*): movement patterns, stereotypy, and stimulus control. *Journal of Comparative and Physiological Psychology* 94: 783–794.

Zeng RS. 2008. "Allelopathy in Chinese ancient and modern agriculture," pp. 39-59 in Zeng RS, Mallik AU, Luo SM, *eds. Allelopathy in Sustainable Agriculture and Forestry*. Springer, New York.

Zhang ZY, Tzvelev NN. 1998. Orobanchaceae. *Flora of China* 18: 231–239. Available at: http://www.efloras.org/florataxon.aspx?flora_id=2&taxon_id=123211

Ziswiler V. 1965. Zur kenntnis des Samenöffnens und der Struktur des hörnernen Gaumens bei körnerfressenden Oscines. *Journal für Ornithologie* 106(1): 1–47.

Zolotarev SA. 1918. Трудовая школа: статьи. All-Russian Teachers Union, Moscow.

Zou L, Santanen A, Tein B, Stoddard FL, Mäkela PA. 2014. Interference potential of buckwheat, fababean, oilseed hemp, vetch, white lupine and caraway to control couch grass weed. *Allelopathy Journal* 33: 227–236.

Zubay P, Kunzelmann J, Ittzés A, Zámboriné N, Szabo K. 2021. Allelopathic effects of leachates of Juglans regia L., Populus tremula L. and juglone on germination of temperate zone cultivated medicinal and aromatic plants. *Agroforestry Systems* 95: 431–442.

Zwanenburg B, Mwakaboko AS, Kannan C. 2015. Suicidal germination for parasitic weed control. *Pest Management Science* 72: 2016–2025.

17 Abiotic diseases and injuries

Abstract
Not all crop problems are caused by living organisms. Soils may have structures and textures suboptimal for *Cannabis*. Nutrient imbalances are rife: deficiencies of macronutrients (e.g., N-P-K) and toxicities from excess micronutrients (e.g., iron, manganese, molybdenum). Diagnosing symptoms can be confusing, nutrient testing is not easy, and much hinges on soil pH and salinity. Some soil amendments correct both soil texture and nutrient problems; crop rotation helps. Water problems are like nutrient problems: too little or too much. The same goes for temperature: too hot or too cold. Replicating sunlight indoors can be challenging—deficiencies and imbalances may arise. Pollutants, toxins, and pesticides cause abiotic injuries. And then there are genetic diseases, arising from inbreeding.

17.1   Introduction

Abiotic diseases and injuries have non-living causal factors. *Diseases* are caused by *persistent* factors—soil nutrient imbalances, light deficiencies, environmental toxins, and genetic factors. *Injuries* are caused by *transient* factors, such as extreme weather events (e.g., hailstones), pesticides, and four-wheelers. Spanning *transient—persistent* we have drought, salination, and air pollution.

Injuries can be large or small. When lightning strikes a hemp field, it leaves a dramatic patch of injured and dead plants. Conversely, a windstorm may cause microscopic injuries in the growing tips of bast fibers (Fig. 17.1). This injury is hidden within the stalk, but decreases the value of extracted fiber.

Abiotic injuries may predispose plants to diseases caused by living organisms. For example, drought-stressed plants become more susceptible to fungal invasions (McPartland and Schoeneweiss, 1984). Abiotic problems may be confused with diseases caused by living organisms. Early observers of Dutch elm disease in Europe, for instance, attributed elm decline to poison gas released during World War I.

17.2   Nutrients and soil classification

Soil texture and chemistry are introduced in section 2.5. The USDA created an elaborate taxonomy to classify soils, based on structure (inorganic and organic phases), soil horizon and depth, moisture content, temperature, and underlying bedrock. The classification is hierarchical, descending through the ranks of Order, Group, Family, and Series. There are twelve Orders of soils worldwide (USDA, 1999). They are often defined by a single dominant characteristic, such as the prevalent vegetation (Alfisols, Mollisols), the type of parent material (Andisols, Vertisols), or climate variables—such as lack of rainfall (Aridisols), or the presence of permafrost (Gelisols).

Of the twelve Orders of soils, two are optimal for commercial hemp production: Mollisols and Alfisols. **Mollisols** (Fig. 17.2) are dark, mineral-rich soils of the steppe—where

Cannabis likely evolved. They occupy the mid-latitude plains of Eurasia, North America, and South America. They generally lie between the Aridisols of arid climates and the Alfisols of humid climates. Worldwide, Mollisols cover 6.89% of ice-free land area (USDA, 1999). **Alfisols** (Fig. 17.2) generally arise under hardwood forest cover, ranging from semi-arid to humid regions. Alfisols cover 9.65% of ice-free land area worldwide. Alfisols occur in Europe and China, the USA (such as the Ohio River basin), and in the drier parts of peninsular India.

Hemp has also been cultivated on **Ultisols**—red clay soils found in the southeastern USA and southeastern China. They are typically deficient in potassium and calcium, and require lime and other fertilizers. **Inceptisols**, which are rare in the USA, are found in hemp-cultivating areas of China and peninsular India. Other countries use different soil classification systems. USDA (1999) translated its classification system into equivalent soil names in Canada, France, and the former USSR.

The lowest taxonomic rank in the USDA's system is the **Series**. Soil Series are named after a type location where they were first described. For example, the Gilpin Series was first described near Gilpin, Pennsylvania, but Gilpin Series soils occur from Indiana to Kentucky and New York. Gilpin Series was compared with 11 other soil Series by Coffman and Gentner (1975). They cultivated an Afghani drug-type landrace in these soils, and plants grown in Gilpin Series yielded the greatest THC content.

17.2.1   Improving soil texture

See section 2.5.1 for an introduction to soil texture. Soil texture is enhanced by its organic phase—humus or compost, technically **particulate organic material** (POM). POM improves the moisture-holding capacity of sandy soils. POM makes clay soils lighter, improves drainage, and makes root penetration and tillage easier. POM has negatively charged surfaces, which keeps K, Mg, Ca, and other cations from leaching out of soil.

Moorland is a damp habitat with acidic POM-rich soil. It supports grasses, bracken ferns, heather, and mosses.

© John M. McPartland 2025. *Hemp Diseases and Pests* 2ⁿᵈ Edition (J.M. McPartland)
DOI: 10.1079/9781836990352.0017

Farmers burned moorland to create new farmland, and they planted stress-tolerant crops such as *Cannabis* (Blith, 1652). In the run-up to WWII, the German agronomist Backe (1936) sought "national self-sufficiency," and recommended growing hemp in burned moorland. Baur and Lampe (1943) wrote a poem (translated from German):

> *Here springs the mighty hemp,*
> *the sole savior of the moorland.*
> *It grows quick and large and ...*
> *forces weeds to their knees.*

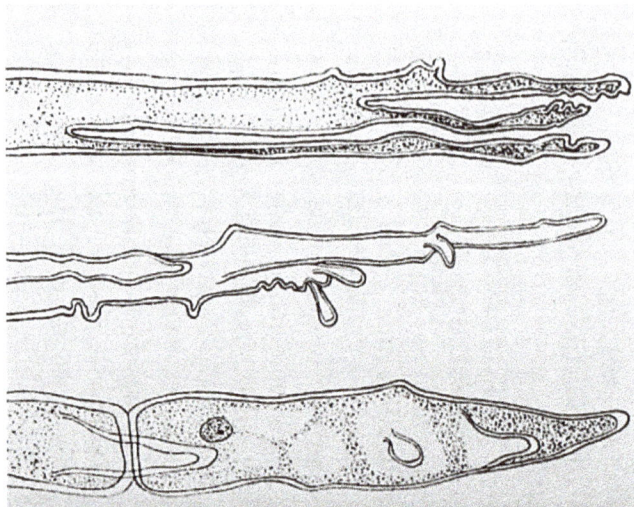

Fig. 17.1. Hemp bast fibers 10 days after stem buckling, ×300 (Schilling, 1923)

In fact, *Cannabis* growing on low-lying moorland in Germany developed nutrient deficiencies (Weigert and Fürst, 1939) and became predisposed to diseases (Röder, 1941b). Agricultural soils should consist of 4–10% POM. Soil with >10% POM is termed **muck**, and soil underlying moorland is **peat** with >25% POM. Muck and peat soils are not recommended for the production of high-quality fiber (Robinson, 1943a; Berger, 1969). Yield may be high, but the fiber is poor and lacks strength. Muck and peat contain too much nitrogen and not enough phosphorus and potassium.

Agricultural soils worldwide have lost POM, due to tillage practices and erosion. Dewey (1914) gave bad advice when he recommended tilling the soil *four times* before planting hemp, "twice rather deep, and twice with cultivators with fine teeth, merely stirring the surface." POM is best preserved by no-till farming (see section 19.2.2). Soil can be supplemented with POM by adding soil amendments. Several amendments have been tested by *Cannabis* agronomist: manure, sewage sludge, worm castings, humic acid, and seaweed.

Manure includes farmyard manure, farm slurry (liquid manure), or human excreta ("night soil"). Farmyard manure contains remnants of animal bedding (often straw), which has absorbed feces and urine—a great source of POM and nutrients.

The earliest recommendation for manuring hemp came from a Chinese agronomist named Fàn Shèngzhī in the 1st century BC. He fertilized hemp with silkworm excreta, three *shēng* (three liters) per plant, when plants were one *chǐ* high (36 cm). When silkworm excreta wasn't available, he recommended "well-ripened manure from a cesspit," one *shēng* per plant (Fàn Shèngzhī 1959).

Around AD 64 the Roman author Columella stated, "Hemp demands a rich, manured, well-watered soil" (Columella, 1941).

Fig. 17.2. Profiles of four soils, left to right: Mollisol, Alfisol, Ultisol, Inceptisol (Soil Survey Staff, 2015)

De Serres (1600) recommended fertilizing hemp with pigeon dung, but warned that the manure must be added to damp soil, otherwise it will burn the seedlings. Markham (1615) said that "a principal place to sow *hempe* is in old stockyards."

Whitworth (1758) described semi-nomadic herdsmen growing hemp in Ukraine. They grazed cattle near their villages. The manure "would ultimately yield up abundant crops of corn, flax and hemp." The hemp advocate Arthur Young (see Fig. 17.6) quantified the amount of manure in England: hemp fields should be manured with 16 horse-loads of dung per acre (Young, 1797). Farmers near Boston, USA, added ten tons of manure per acre, "good rotten stable dung" (Stevenson, 1812). Andrei T. Bolotov, writing in Russia in the 1820s, said that hemp required twice as much manure as rye. He suggested up to 800 wagon loads per *desyatina*, or 2.7 acres (Milov, 1998).

Thaer (1812) recommended fertilizing hemp with "hot" sheep and horse manure when grown on damp ground, and for dry ground, a heavy application of fermented cattle manure. Boussingault (1851) suggested fertilizing hemp with ten "slurry cars" of liquid manure per hectare. Gasparin (1842) cited two French agronomists: Crud in Boulogne-sur-Mer said 1500 kg manure had the power of producing 100 kg of *filasse de chanvre* (hemp fiber), whereas Leclerc in Anjou said 1500 kg manure produced 65 kg hemp fiber. In India, *ganja* required a "great quantity of manure" (Buchanan-Hamilton, 1833; Prain, 1893).

In the Flanders region, Rham (1842) said farmers applied 15 tons of "good rotten manure" per acre before sowing seed—or better yet, the autumn before. They also dressed fields with liquid manure, "15 hogsheads of the common tank liquor, which is chiefly cows' urine. This manure is allowed to sink into the soil for three or four days; the land is then harrowed, and about half a bushel of hemp-seed is sown per acre."

Sewage sludge has POM levels similar to natural humus, first noted by Albert Howard, one of the founders of organic farming (Howard, 1938). Polish agronomists have conducted many studies with hemp and sewage sludge (Czyżyk, 1985; Kutera and Czyżyk, 1992; Wiśniewski and Kołodziej, 1999; Piotrowska-Cyplik and Czarnecki, 2003; Badora and Filipek, 2004), followed by others (Alaru *et al.*, 2009, 2011; Seleiman *et al.*, 2012, 2013; Zielonka *et al.*, 2017, 2019). Sludge improves fiber yields more than its equivalent N-P-K delivered alone (Kutera and Czyżyk, 1992; Piotrowska-Cyplik and Czarnecki, 2003). Fiber content (percentage of bast fiber) also increases in comparison with N-P-K alone (Wiśniewski and Kołodziej, 1999). Sludge improves *Cannabis* Leaf Area Index and chlorophyll content, proxies for photosynthetic efficiency (Zielonka *et al.*, 2017).

Alaru *et al.* (2009) measured biomass in two fiber-type cultivars amended with four fertilizers, including nitrogen-fixing vetch (*Vinca caroliniana*) intercropped with hemp. Even though the four fertilizers were calibrated to deliver the same amount of nitrogen (100 kg N/ha), the yields differed (Table 17.1).

Alaru *et al.* (2011) repeated this on a larger scale, with more cultivars ('USO-31', 'Chameleon', 'FINOLA', 'Santhica-27'). They noted that the *quality* of sewage sludge was important: their source did not adequately condition the sludge, which was "gelatinous and gluey." This decreased the percentage of seedlings emerging from the soil, compared with other treatments. Nevertheless, biomass yield was greater with sludge than other

Table 17.1. Biomass yields, g/m² dry matter (Alaru *et al.*, 2009)

	'USO-31'	'Chameleon'
Sewage sludge	870	638
NH₄NO₃	582	664
Cattle slurry	519	504
Vetch	317	421
Control	410	418

treatments. Seleiman *et al.* (2012) reduced sludge stickiness by mixing it with peat, which increased yields.

Seleiman *et al.* (2013) grew 'USO-31' in sand or soil with various amendments: sand with synthetic N-P-K (Sa-N), sand with sewage sludge (Sa-SS), soil with synthetic N-P-K (So-N), soil with sewage sludge (So-SS), or soil with digested sludge (So-DS). Above-ground biomass in 14-day-old seedlings was greatest in So-SS (91.0 mg dry weight/seedling), followed by So-N (85.3 mg), So-DS (71.3 mg), Sa-SS (48.1 mg), and Sa-N (11.9 mg).

The downside of sewage sludge is heavy metal content. Urban sewage consists of domestic waste and storm-water runoff, as well as industrial effluents. *Cannabis* is a well-known bioaccumulator of heavy metals. Badora and Filipek (2004) measured the accumulation of nickel (Ni) in six plants grown in sludge-amended soil. Ni content varied: grass mixtures > rape > hemp > maize > poplar > willow.

Piotrowska-Cyplik and Czarnecki (2003) measured Ni, zinc (Zn), and copper (Cu) in 'Benico' amended with sludge (5% or 10% of soil content). Sludge-amended plants compared with controls produced more biomass, but accumulated more Ni, Zn, and Cu. Most of the Ni and Zn accumulated in roots, and most of the Cu accumulated in leaves. Stalks accumulated little Ni, Zn, and Cu. Seleiman *et al.* (2012) grew 'USO-31' in soil amended with sludge. The plants thrived, but they extracted Ni, Zn, and cadmium (Cd), more so than maize or oilseed rape. Plants grown in sludge mixed with peat accumulated less Ni, Zn, and Cd.

Worm castings are a popular high-POM soil additive, with no heavy metals. Ievinsh *et al.* (2017) grew seedlings of fiber-type 'Pūrini' in soil amended with worm castings (vermicompost). Control plants were amended with nutrient supplements equal to those in the vermicompost. At an optimal concentration of 5%, vermicompost stimulated hypocotyl and radical growth greater than control seedlings. They weighed more and had greater chlorophyll content. Seeds soaked for 4 hours in 5% vermicompost solution had a higher germination rate. Ievinsh also extracted humic acid from vermicompost, which did not provide the same beneficial results.

Humic acid (HA) is derived from humus, and known to improve the physical and chemical properties of soil. Berstein *et al.* (2019a) supplemented potting soil with daily irrigation of a liquid HA solution. Compared with control plants, HA-treated plants produced slightly more biomass, but not statistically different. Surprisingly, HA lowered cannabinoid concentration (THC% and CBD%).

Da Cunha Leme Filho *et al.* (2020) tested the effects of adding HA and/or a manure tea containing a consortium of mycorrhizal and beneficial bacteria (*Bacillus, Pseudomonas,*

Streptomycetes spp.). Compared with controls (which received only inorganic fertilizers), the application of HA, biofertilizer, or in combination, increased plant height, chlorophyll content, and photosynthetic efficiency by 55%, 8%, and 12%, respectively, after water stress. Biomass did not increase. The authors concluded that biostimulants may help mitigate *Cannabis* environmental stress.

Seaweed has long been used as a soil amendment. Pliny in AD 79 described the gathering of seaweed by "peoples of Britain and Gaul" to fertilize their fields (Monagail *et al.*, 2017). **Kelp meal** is dehydrated and pulverized seaweed, rich in POM, macronutrients, and micronutrients. Beyond nutrition, seaweed extracts serve as "biostimulants," improving root growth, nutrient uptake, and resistance to abiotic stress and disease organisms. Inter-tidal seaweeds such as *Ascophyllum nodosum* produce compounds (alginic acid and fucoidans) to cope with stresses of extreme variations in salinity, temperature, and light. *A. nodosum* extracts enhance root development in hydroponic *Cannabis* (Wise *et al.*, 2024). As with all good things, wild-harvesting of *A. nodosum* is not sustainable and resources are becoming overutilized and depleted (Monagail *et al.*, 2017).

Indoor-grown plants have their own special soil needs. According to Frank and Rosenthal (1978), "Most growers prefer to buy their soil, while some prefer to dig it." In fact, soil dug out of a field is too heavy for plants in containers. Soil conditioners must be added, to balance water retention, water drainage, and promote root growth. The same can be said for most bagged commercial soil. Popular choices for soil conditioners (in increasing cost) include sand, peat, vermiculite, perlite, and granular rockwool. Vermiculite can be problematic, because it decomposes into a clay sludge. Some examples of indoor soil mixtures are as follows:

- 5 parts soil/2 parts perlite/1 part composted manure
- 5 parts soil/1 part sand/1 part peat/half-part blood meal/half-part wood ash
- 5 parts soil/1 part vermiculite/1 part sand/1 part peat/quarter-part 12-12-12 chemical fertilizer.

17.3   Nutrient imbalances

We introduced soil nutrients in section 2.5.2. **Macronutrients** consist of nitrogen, phosphorus, and potassium (N-P-K), as well as calcium (Ca), sulfur (S), and magnesium (Mg). **Micronutrients** ("trace elements") include iron (Fe), zinc (Zn), boron (B), copper (Cu), manganese (Mn), chlorine (Cl), and molybdenum (Mo). See Table 2.6 and Table 2.10 for the physiological roles that nutrients play in plants.

Nutrient imbalances arise from too little or too much of these essential elements. For macronutrients, too little is more common than too much. For micronutrients, the reverse is true. As the poet William Blake said, "You never know what is enough unless you know what is more than enough" (Blake, 1790). Symptoms can be misleading—symptoms of macronutrient deficiencies are often mimicked by symptoms of micronutrient toxicities.

17.3.1   Soil pH and salinity

Soil acidity, measured as pH, ranges from 0 (very acidic) to 14 (very basic), and a pH of 7 is neutral. Duke (1982) summarized worldwide *Cannabis* data from 44 reports, and suggested a soil pH of 6.5 is best. Bócsa and Karus (1997) suggested a range of 5.8 to 7.0. Upton *et al.* (2013) proposed a range of 6.5 to 7.2 in soil, and 5.2 to 5.8 in hydroponic solutions.

Soil pH affects the availability of nutrients; a pH between 6.0 and 6.5 is optimal for most nutrients. In Fig. 17.3, the availability of each nutrient is indicated by the width of the bar. Above a pH of 7.5, the availability of several nutrients decreases. Hand-held meters for measuring pH are inexpensive and accurate.

Adjusting the pH of potted soil or small garden plots is relatively simple. To raise the pH, add limestone (calcium carbonate). The finer the limestone particles, the faster they raise the pH. Carefully blend pulverized, granular, or pelletized limestone into soil. The amount of limestone depends upon the soil's clay content. This needs to be done 2–3 months prior to planting, to allow time for lime to neutralize soil acidity. Hydrated lime (calcium hydroxide) works faster. To lower the pH, add elemental sulfur. To speed the effects of sulfur, use finely pulverized product, and keep soil moist. Aluminum sulfate has been recommended, but *Cannabis* bioaccumulates aluminum, which passes through smoke into lungs (Exley *et al.*, 2006). Frank and Rosenthal (1978) and Cervantes (2015) provided charts and tables for adjusting different soils to a proper pH.

Adjusting the pH of a large field may consume a lot of lime or sulfur. Soil pH depends upon the parent material from which the soil was formed. Shale, sandstone, and granite generally form acidic soils. Limestone forms basic soils. Application of fertilizers containing ammonium or urea increases acidity. The decomposition of organic matter also adds to soil acidity.

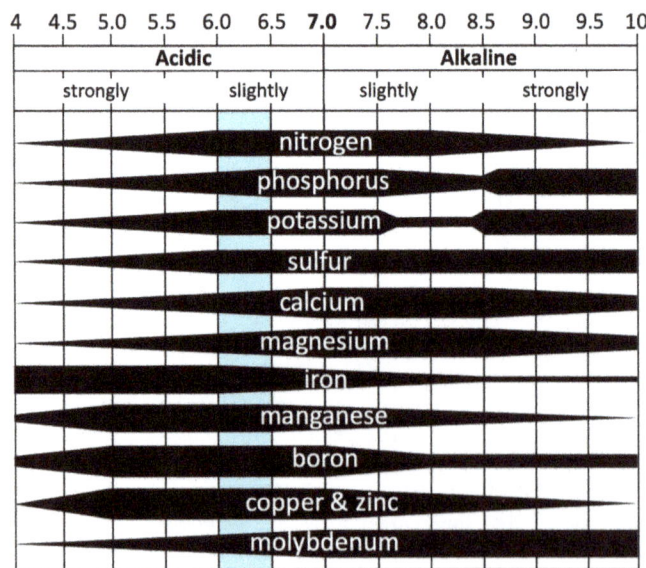

Fig. 17.3. Effect of soil pH on availability of plant nutrients (courtesy of Coolkoon, Wikimedia Commons)

Salinity reflects the amount of salt dissolved in water. Salt (sodium chloride, NaCl) is a natural component of soil. Rainfall leaches salt from soil, and groundwater drainage flushes it into the ocean. Seawater has a salinity of about 3.5% NaCl, or 35 g/L. This percentage can be expressed as molarity: the number of moles of NaCl per liter of solution. Seawater averages 559 mM NaCl.

Cannabis does not tolerate NaCl, even though chlorine (Cl) is an essential plant nutrient, and sodium (Na) may be beneficial in trace amounts. Garman (1903) mixed NaCl into soil, at a rate of 1.5 oz/ft^2, to control a parasitic plant that attacked hemp (*Phelipanche ramosa*). The treatment killed *P. ramosa*, but "it was plain that the salt had a very injurious effect on the hemp." Even salty air is detrimental: Hancer (1992) noted that "poisonous salty breezes" killed seedlings growing near the Hawaiian shoreline.

Elevated salt content in soil—**salinization**—is a worldwide problem. Salinization limits plant growth and crop productivity. Soil salinity increases soil water potential (ψ), which makes it harder for roots to absorb water. Excess Na interferes with many physiological processes (Acosta-Motos *et al.*, 2017). For more about ψ, see section 2.2.

Irrigation causes salinization. Irrigation water may have low levels of salt, but the water evaporates, leaving the salt behind. Salinization from irrigation dates back 6000 years, in ancient Mesopotamia. Acosta-Motos *et al.* (2017) estimated that a third of irrigated land worldwide is affected by salinity—800 million ha. Millions of acres of farmland have been permanently abandoned due to salinization in the lower Colorado River basin.

Growers who use **fertigation** (fertilizer added to water irrigation) may encounter salinization within a single growing season. Caplan *et al.* (2017a) blamed decreased *Cannabis* yields at high fertigation rates to salinization.

Salts other than NaCl contribute to salinization. Potassium chloride (KCl) and sodium sulfate (Na_2SO_4) are pH neutral, like NaCl. Alkaline soils contain alkaline salts, such as sodium carbonate (Na_2CO_3, pH 11) and sodium bicarbonate ($NaHCO_3$, pH 9). These are more damaging to plants than neutral salts.

To account for all salts—not just NaCl—marine scientists adopted a new standard for reporting salinity. The TEOS-10 standard measures electrical conductivity, quantified in units of milliSiemens per centimeter (mS/cm). Pure water equals 0.001 mS/cm and seawater averages 43 mS/cm. A soil is considered to be saline when it reaches 4.0 mS/cm (Acosta-Motos *et al.*, 2017).

Caplan *et al.* (2017a) estimated that *Cannabis* can tolerate salinity up to 3.0 mS/cm, whereas Alipour *et al.* (2014) report zero seed germination above 2.0 mS/cm. This classifies *Cannabis* as a **glycophyte**—salt-sensitive, along with most crop plants. In contrast, salt-tolerant plants (halophytes, such as mangroves) tolerate tenfold greater salinity. In a comparison of seed oil crops, Dadkhah (2010) showed that *Cannabis* was more sensitive to salinity than sesame, *Sesamum indicum*. Javadi *et al.* (2014) showed that *Cannabis* was more sensitive to salinity than fenugreek (*Trigonella foenum-graecum*), but more tolerant than black cumin (*Nigella sativa*).

A research group in Kūnmíng published a series of studies on salt stress. Na_2CO_3 impacted *Cannabis* more than $NaHCO_3$, NaCl, or Na_2SO_4 (Hu *et al.*, 2015). Rising salt concentration (0, 50, 100, 150, 200, 250 and 300 mmol/L) of all the salts linearly decreased seed germination rate, germination vigor index, and lengths of radicles and hypocotyls of seeds. A hormetic effect was seen with NaCl, where seeds benefited from mild stress—germination was greater at 50 mmol/L compared with zero (Hu *et al.*, 2015). Chinese cultivars varied in salt sensitivity. Seed-type 'Bama-huoma' showed more salt sensitivity than fiber-type 'Yun-ma 5' (Hu *et al.*, 2018).

Salt tolerance in hemp seedlings correlated with chlorophyll content, SOD activity (superoxide dismutase, an enzyme that converts O_2^- radicals into O_2), and water content in aboveground plant tissues (Hu *et al.*, 2016). A transcriptome analysis of 'Bama-huoma' and 'Yun-ma 5' subjected to salt stress showed cultivar-specific up- or down-regulation of genes, including transcription factors such as *MYB*, *NAC*, *GATA*, and *HSF* (Liu *et al.*, 2016).

Guerriero *et al.* (2017) conducted a similar transcriptome analysis of 'Santhica 25', a French monoecious cultivar. Salt stress upregulated several genes involved with cell wall biosynthesis (*CesA4*, *FLA10*, *FLA8*), lignification (*4CL*, *CAD*, *PAL*), and ethylene response production (*ERF1*). Microscopic analyses of hypocotyls in salt-stressed seedlings revealed changes in vascular tissues, where the lumen of xylem vessels was reduced.

Salinity in potted soil can be flushed out, using lots of fresh water. But this also removes mobile nutrients (e.g., N-P-K). Better to start fresh with new growing media. Salinity in field soils is a bigger problem. Deep tillage to break up claypans and hardpans will improve drainage and allow rainwater to leach the salinity. Reduce the surface evaporation of water by applying mulch to soil. Add soluble calcium to the soil, such as calcium sulfate (gypsum), to reduce soil sodium content. Limestone (calcium carbonate) does not dissolve in soils with high pH and cannot lower sodium levels. Adding sulfur, iron sulfate, or sulfuric acid will convert calcium carbonate to calcium sulfate.

Mycorrhizae (see Fig. 13.2) confer salt tolerance in plants. Under saline conditions, Zarei *et al.* (2014) inoculated *Cannabis* with mycorrhizae, and seed yields were greater than those of control plants. Tadayon and Zarei (2014) grew *Cannabis* inoculated with mycorrhizae in saline soil; the plants contained more potassium and less sodium, which conferred salt tolerance. The use of endophytic bacteria (see section 13.3) is another low-cost and eco-friendly method of enhancing crop productivity in saline soils. Afzal *et al.* (2015) isolated bacterial endophytes from feral *Cannabis*; the bacteria tolerated high salt concentrations, and promoted plant growth under salt stress. Breeding *Cannabis* for salt tolerance is beginning in Iran and China. Italian accessions are being tested for tolerance to saline water (G. Grassi, pers. comm., 2016).

17.3.2 Nutrient testing

After testing soil pH and salinity, the next step is nutrient testing. Nutrient testing may uncover "hidden hunger," a nutrient deficiency without symptoms that nevertheless affects the quantity and quality of crop yields. Soil tests are relatively inexpensive, and actually save money otherwise wasted on needless fertilizers. Cervantes (2015) estimated that for every US$10 spent on fertilizers by *Cannabis* gardeners, US$9 is wasted. Excess fertilizers wash into groundwater, where they cause environmental damage and health problems. Soil testing falls into two philosophies: SLAN and BCSR.

SLAN (Sufficiency Level of Available Nutrient) is the method used by university extension services. SLAN measures individual nutrients in a soil sample (N, P, K, Ca, Mg, S, micronutrients), and compares each with a predefined range of concentrations considered sufficient (or optimal) for plant nutrition (Black, 1993). Soils testing *high* or *very high* in a particular nutrient do not need supplemental fertilizer. Soils testing *medium*, *low*, or *very low* require fertilization, and a pro-rated amount of nutrients will be recommended.

BCSR (Base-Cation Saturation Ratio) measures the ratio of cationic nutrients (namely Ca^{++}, K$^+$, and Mg^{++}) required to balance the soil cation-exchange complex. Fertilizer recommendations are made to optimize the ratios. Soil balanced with BCSR leads to greater plant and human health, and increases soil biological activity and soil organic carbon content. The principles of BCSR were elucidated at Rothamsted Experiment Station in southern England and developed further by William Albrecht in Missouri (introduced in section 1.3). Given the emphasis on Ca^{++} (optimal ratios being 6.5Ca:1Mg and 13Ca:1K), BCSR is an efficacious approach for fiber-type hemp, which requires a fair amount of Ca. Many BCSR practitioners typically use both BCSR and SLAN in a hybrid approach (Culman *et al.*, 2021).

Leaf tissue analysis is based on the concept that nutrients present in foliar tissues correlate with nutrient concentrations in the soil. Foliar nutrient levels vary depending on the growth stage of the plant, so the sampling protocol must be standardized. Suchoff *et al.* (2021) recommended sampling "fully expanded, most recently mature or most recently expanded leaves—generally the third to fifth leaf down from the growing point." Suchoff sampled healthy *Cannabis* leaves in the vegetative stage prior to flowering.

There's a problem with leaf tissue analysis: the nutrient optima for *Cannabis* have not been definitively established. Fiber-, seed-, and drug-type plants have different nutrient optima. In the first edition of *HD&P* we proposed nutrient optima for *Cannabis* leaf tissue (McPartland *et al.*, 2000). These were based on general crop guidelines proposed by Bennett (1993), with alterations-based nutrient studies of fiber-type *Cannabis* (Table 17.2). Iványi and Izsáki (2009) proposed new guidelines, based on tissues samples obtained from healthy fiber-type hemp plants.

Bryson and Mills (2014) proposed nutrient optima based on leaf samples obtained from plants of an unreported *Cannabis* cultivar—likely a fiber-type plant, given legalities at that time. Landis *et al.* (2019) proposed guidelines for glasshouse-grown drug-type plants, based on leaf samples collected from five CBD strains (Table 17.2). The same group from North Carolina expanded their survey to 13 CBD strains, to refine their proposed nutrient optima for glasshouse-grown plants (Kalinowski *et al.*, 2020). They identified significant differences among the 13 strains in both macro- and micronutrients, and therefore widened the range of fertility requirements (Table 17.2).

Cannabis samples submitted to the North Carolina Department of Agriculture for disease diagnosis frequently fell outside these ranges. Thiessen *et al.* (2020) tested all samples (*n* = 572) received in 2017 and 2018, and applied the guidelines proposed by Bryson and Mills (Table 17.2). Nearly all samples were either deficient or toxic for one or more nutrients (Table 17.3). The authors suggested that the majority of plants fell out of range because the recommended guidelines were incorrect, not because of wholesale nutrient mismanagement—which led to the refined guidelines by Landis *et al.* (2019) and Kalinowski *et al.* (2020). Toxic levels of manganese (Mn) reported by Thiessen and colleagues may be due to cultivation in acidic Ultisols and Alfisols, which accumulate Mn in North Carolina (McDaniel and Buol, 1991).

Table 17.2. Proposed guidelines for nutrient levels in *Cannabis* leaf tissue.

Nutrient	McPartland *et al.* (2000) fiber crop	Iványi and Izsáki (2009) fiber crop	Bryson and Mills (2014) likely fiber	Landis *et al.* (2019) CBD crop	Kalinowski *et al.* (2020) CBD crop
N, %	2.0–5.0	5.0–6.0	3.3–4.76	2.56–4.47	3.29–4.98
P, %	0.2–0.5	0.5–0.6	0.24–0.49	0.31–0.44	0.26–0.43
K, %	1.0–5.0	2.7–3.0	1.83–2.35	1.84–2.98	2.08–3.41
Mg, %	0.1–1.0	0.6–0.8	0.40–0.81	0.25–0.46	0.34–0.69
Ca, %	0.1–1.0	2.4–3.0	1.47–4.42	0.53–1.40	1.50–5.34
S, %	0.1–0.3		0.17–0.26	0.19–0.29	0.25–0.37
Fe, ppm	50–250	65–105	100–150	59–132	83.5–169.0
Zn, ppm	20–100	25–40	24–52	23.2–46.2	26.4–54.9
Mn, ppm	20–300	85–130	41–93	24.3–71.9	66.5–264.0
Cu, ppm	5–20	2.0–5.0	5.0–7.1	1.83–11.4	1.6–7.0
B, ppm	10–100		56–105	22.6–57.3	25.8–90.5
Mo, ppm	0.1–0.5		0.50–1.5	0.53–2.36	
Cl, %	0.2–2.0				
Si, %	0.2–2.0				

There's another problem with leaf tissue analysis: results from an analysis might arrive 2–3 weeks after a sample is sent to a laboratory. For a *Cannabis* crop that reaches maturity within 3 months, the 2–3-week delay does not leave much time to effectively correct nutritional imbalances. Alternatively, leaf sap can be quickly tested in the field using a hand-held meter. But leaf sap tests are limited to N, P, and Ca.

17.3.3 Symptoms of nutrient imbalances

We gathered data from field reports as well as two experimental studies that manipulated nutrition. Tibeau (1936) grew *Cannabis* in pots with silica sand, and irrigated plants with Knop's solution, modified by a single nutrient—either absent, or eightfold greater concentration than normal. Cockson *et al.* (2019) replicated the study, but used Hoagland's solution.

Color photos by Cockson and colleagues improved upon Tibeau's black-and-white photos; they illustrated symptoms in early, intermediate, and advanced stages. Cervantes (2015) provided color drawings of deficiencies and toxicities, in early, intermediate, and advanced stages. Deficiency symptoms of mobile nutrients (N, P, K, Mg, B, Mo) usually begin in leaves near the bottom of plants. Deficiency symptoms from less mobile nutrients (Mn, Zn, Ca, S, Fe, Cu) usually begin in young leaves near tops.

17.3a. Nitrogen

N-deficiency: Lower leaves become pale and develop chlorosis (yellowing). Chlorosis intensifies with deepening deficiency, and moves up the plant from lower leaves to the middle part of the plant (Fig. 17.4A). In extreme cases, whole plants turn pale yellow-green; leaves may necrose and abscise. Petioles and stems may accumulate reddish anthocyanin pigment (Tibeau, 1936), a symptom more prevalent in P deficiency. The sex ratio may shift towards male plants (Tibeau, 1936; Frank and Rosenthal, 1978).

Lack of N is the most common deficiency in *Cannabis*. Because of N's high mobility in soil, N deficiency is more common in loose, sandy soils. Some popular drug strains, particularly "Skunk No. 1", produce lime green leaves, which may be mistaken for N or Fe deficiencies (Schoenmakers, 1986; Kees, 1988). Don't be fooled: "Skunk No. 1" is easily burned by N amendments (Alexander, 1988). For organic sources of N, see Table 17.4. Recovery time is fairly rapid—a matter of days (leaves too far gone and very yellow won't recover).

Anderson *et al.* (2021) diagnosed N deficiency using a SPAD (Soil Plant Analysis Development) meter by Minolta. SPAD meters make non-destructive measurements of chlorophyll content by measuring leaf absorbance in red and near-infrared regions. Anderson and colleagues demonstrated a high correlation

Table 17.3. Percent of plant tissue analyses (*n* = 572) that fell within or out of threshold recommendations for 11 nutrients (Thiessen *et al.*, 2020)

Level	N	P	K	Ca	Mg	S	Fe	Mn	Zn	Cu	B
Within threshold	41.78	32.52	18.88	40.38	37.76	60.66	29.90	14.69	21.85	39.51	33.39
Below minimum threshold	43.36	58.04	56.29	11.54	15.38	10.66	25.00	11.71	75.35	11.36	2.10
Above maximum threshold	14.86	9.44	24.83	48.08	46.85	28.67	45.10	73.60	2.80	49.13	64.51
Above 2-fold maxim. threshold	0.00	0.00	0.00	0.52	21.15	1.57	10.49	60.49	0.00	1.40	14.51

Fig. 17.4. Deficiency symptoms: **A.** nitrogen (Clarke); **B.** phosphorus (Cervantes); **C.** potassium (McPartland)

between SPAD measurements and actual chlorophyll content ($r = 0.82$) in five high-CBD strains. They determined that SPAD measurements <44 indicated N deficiency.

N-toxicity: Plants produce overly abundant foliage, leaves are dark green and may develop rolled edges with brown spots (Tibeau, 1936). *In extremis,* roots of N-overdosed plants discolor and decay, and above-ground parts turn a golden-copper color, wilt, and die. Ammonium tends to become toxic in soils with an acid pH; nitrate is a bigger problem in soils with a neutral to alkaline pH (Agrios, 2005). Potted plants with excess fertilizer can be rescued: leach out excess N by applying frequent, heavy irrigation. Cervantes (2015) recommended at least seven cycles of flushing and draining.

17.3b. Phosphorus

P-deficiency: Lower leaves initially develop olive-green spots in an irregular pattern (Cockson *et al.,* 2019). Olive-green spots enlarge with deepening deficiency, and leaves develop marginal and tip necrosis (Fig. 17.4B). Petioles may accumulate reddish anthocyanin pigment (Cervantes, 2015), but in plants with *afghanica* heritage this is normal (Frank, 1988). *In extremis,* much of the leaf becomes necrotic. Flowering is delayed and buds are small. P-deficiency symptoms mimic fungal diseases, and deficient plants become predisposed to fungal diseases. P-deficiency is relatively uncommon, but arises in heavy soil, in acidic soil, and in cold, wet soil. For organic sources of P, see Table 17.4. Recovery time is relatively fast.

P-toxicity: Excess P manifests as symptoms of Fe, Ca, and Zn deficiencies in new leaves (see below). It is fairly uncommon, *Cannabis* tolerates excessive P better than most plants (Westmoreland and Bugbee, 2022).

17.3c. Potassium

K-deficiency: Lower leaves develop a chlorotic or rusty-colored leaf margin, often limited to serrated edges (Fig. 17.4C). Chlorosis intensifies with deepening deficiency, and expands inward towards the midrib. *In extremis,* leaf margins and tips turn brown and curl, and brown spots appear between veins (Rataj, 1957). K deficiency is fairly common; it arises frequently in sandy soil because K leaches easily. Symptoms worsen in dry weather, cold weather, and under low light intensity. For organic sources of K, see Table 17.4. Recovery time is relatively slow, depending on the depth of the deficiency, and may take 2 weeks (Tibeau, 1936).

K-toxicity: Excess K manifests as symptoms of Ca, Fe, and Zn deficiencies in new leaves (see below). It is fairly uncommon, but may arise in coir-based growing media.

17.3d. Magnesium

Mg-deficiency: Lower leaves develop interveinal chlorosis— leaf tissue between veins turns yellow and the veins remain green. With deepening deficiency, the chlorosis intensifies and expands inward towards the midrib. *In extremis,* the entire leaf turns yellow except for the green midrib. Morad and Bernstein (2023) induced Mg deficiency by fertilizing plants with only 2 mg Mg/L. Symptoms (interveinal chlorosis) progressing from lower leaves to upper leaves, demonstrating the ability for Mg *in planta* remobilization. Plants were stunted with 28% of the biomass of plants fertilized with 35 mg/L. Mg deficiency is fairly common, especially in sandy, acidic soils in rainy seasons. Excess ammonia, K, or Ca lock up Mg in soil, making it unavailable. For organic sources of Mg, see Table 17.4. Recovery may take 2 weeks (Tibeau, 1936).

Mg-toxicity: Excess Mg causes tip scorch and brown spots on leaves, and internodes may lengthen (Tibeau, 1936). Mg toxicity is uncommon. Morad and Bernstein (2023) induced toxicity with a rate of 140 mg/L. Necrotic spots were seen on old leaves near the bottom of the plant.

17.3e. Calcium

Ca-deficiency: Upper leaves are affected first, because Ca is a less mobile nutrient. As newly expanding leaves develop, the basal portion of the leaflet is narrower and displays a lighter green coloration than the leaflet tip (Cockson *et al.,* 2019). As the deficiency progresses, uppermost leaves turn yellow and brown, their margins curl, and tips hook backwards. Apical meristems may wither and die, causing a proliferation of axillary shoot development. Plants have brittle stems and poor root development (Berger, 1969). Ca deficiency predisposes plants to stem and root pathogens (*Botrytis, Fusarium,* and *Rhizoctonia* fungi; *Ditylenchus* nematodes). Deficiencies are most common in acidic, sandy soils low in organic matter. For organic sources of Ca, see Table 17.2. Recovery time is slow, but plants can make a complete recovery (Tibeau, 1936).

Ca-toxicity: Excess Ca manifests as symptoms of K, Fe, and Zn deficiencies. Excess Ca in hydroponic solutions will combine with sulfur and form gypsum, which flocculates in the solution (Cervantes, 2015).

17.3f. Sulfur (Sul*ph*ur)

S-deficiency: Leaves in the upper and middle part of the plant turn lime-green to yellow. The symptoms resemble those of N deficiency, except symptoms from N deficiency arise in lower leaves. As the deficiency progresses, chlorosis intensifies on uppermost leaves. Stalks are thin, brittle, and etiolated. Drought worsens S deficiency. For organic sources of S, see Table 17.4. Spontaneous recovery may arise after rainfall washes S from the sky.

S-toxicity: Leaves are smaller and darker green than normal, the tips and margins may yellow and brown. This happens more often when using elemental S than magnesium sulfate. Symptoms can be partially corrected with Ca amendments.

17.3g. Zinc

Zn-deficiency: Compared with other micronutrients, Zn deficiency arises relatively frequently in *Cannabis*. Upper leaves develop chlorotic margins. As the deficiency progresses, chlorosis expands inward, causing an interveinal chlorosis, leaving only green veins and midrib. New leaves become twisted and deformed. Zn deficiency arises in alkaline soils low in organic matter. Excess P may precipitate Zn deficiency.

Zn-toxicity: Leaves are small with dark mottling. Zn overload is rare but very toxic; entire plants may die quickly (Cervantes, 2015).

17.3h. Manganese

Mn-deficiency: Leaves in the upper and middle part of the plant develop a bright yellow netted interveinal chlorosis (Cockson *et al.*, 2019). According to Weigert and Fürst (1939) and Wahlin (1944), necrotic spots may develop, and plants become extremely stunted (20–30 cm instead of 60 cm in height). They said Mn deficiencies worsen in hemp grown on reclaimed marshland.

Mn-toxicity: Coffman and Genter (1975) reported toxicity in *afghanica* plants cultivated in soil with 602 ppm Mn (soil derived from shale with pockets of Mn ore). They described symptoms as interveinal chlorosis of young leaves. Cockson *et al.* (2019) described marginal chlorosis in lower leaves. With deepening toxicity, the entire leaf becomes chlorotic, leaf margins become necrotic, and leaves sometimes abscised. Thiessen *et al.* (2020) illustrated interveinal chlorosis followed by interveinal necrosis.

Excess Mn arises in acidic soils and in soils that have been steam-sterilized.

17.3i. Iron

Fe-deficiency: Upper leaves develop yellow margins, which progresses to interveinal chlorosis, leaving only green veins. Cervantes (2015) noted the similarities between Fe, Mn, and Zn deficiencies. They likely arise concurrently, especially in alkaline soils with a pH >7.5 (see Fig. 17.3). Cervantes treated plants in alkaline soil with a chelated dose of all three micronutrients.

Fe-toxicity: Leaves become chlorotic to copper-colored. With progression, small brown spots appear between leaf veins. Fe toxicity is very rare.

17.3j. Copper

Cu-deficiency: Upper leaves are distorted and twisted, especially at the base of the leaflets. With deepening deficiency, new and expanding leaves display a pronounced basal narrowing, exhibit marginal interveinal chlorosis, and become limp—lacking turgidity (Cockson *et al.*, 2019). In extreme cases, roots become affected and whole plants may wilt. Kirchner (1966) referred to plants with Cu deficiency as "*gummihanf.*" The stalks lose their rigidity, become rubbery, and plants bend over. Conversely, Berger (1969) described Cu-deficient stalks as brittle and easily broken. Berger watched for Cu shortages in peat soils. Kirchner applied 50–100 kg copper sulfate/ha to soil in autumn to correct deficiencies.

Cu-toxicity: Leaves develop interveinal chlorosis, which can be confused with symptoms of Fe deficiency. Plants are stunted, with few branches (Cervantes, 2015). Cu toxicity may arise from repeated use of copper sulfate fungicides.

17.3k. Boron

B-deficiency: Upper leaves are distorted and twisted, especially at the base of leaflets.

With deepening deficiency, leaves curl at odd angles from the petiole, and leaf margins may become necrotic (Cockson *et al.*, 2019). The entire apical meristem looks burned, which may be confused with symptoms in tops growing too close to hot lights (Cervantes, 2015). Root tips die back, and larger roots look unhealthy. B deficiency arises in hemp grown on moorland soil (Weigert and Fürst, 1939).

B-toxicity: Lower leaves develop marginal chlorosis. As the deficiency progresses, the yellowing moves inward toward the midrib, leaf margins become necrotic, and leaves may abscise.

17.3l. Molybdenum (Mo, not Mb)

Mo-deficiency: Leaves in the lower and middle part of the plant develop interveinal chlorosis, which progresses to leaf distortion and marginal necrosis (Cervantes, 2015). Cervantes said that deficiency symptoms are rare; Cockson *et al.* (2019) reported a lack of visual symptoms in plants subjected to 9 weeks of deficiency. Mo toxicity is rare, and specific symptoms have not been described in the literature. Excess Mo causes Fe and Cu deficiency.

17.3m. Chlorine

Cl-deficiency: Upper leaves turn pale green and go limp; this deficiency rarely arises, and like all micronutrients, is easily over-corrected (Cervantes, 2015).

Cl-toxicity: Young leaves develop burned leaf tips and margins; as the toxicity progresses, leaves become copper-colored. Young seedlings and clones are the most susceptible to damage (Cervantes, 2015). Excess Cl has an adverse effect on cigarette combustibility in tobacco (Lucas, 1975).

17.3n. Silicon

Silicon is a non-essential nutrient, but it alleviates toxicities caused by heavy metals and salinity, as well as drought stress.

In rice, Si improves resistance to lodging, and enhances UV light tolerance. Plant species vary in their ability to absorb and accumulate silicon. Tomato (*Lycopersicon esculentum*) is a low accumulator, averaging 0.05% Si dry weight. Rice (*Oryza sativa*) is a high accumulator, averaging 10–15% Si. In addition to passive silica uptake (from groundwater into roots), rice *actively* absorbs silica via **aquaporins**, which are channel-forming proteins facilitating the passage of water, urea, as well as boron and silicon (Piperno, 2006).

Guerriero *et al.* (2019) identified aquaporin genes in the *Cannabis* genome, yet *Cannabis* appears to be a relatively low accumulator: Saijonkari-Pahkala (2001) reported Si content of 0.19% in dry stalks. Long ago, Jaffa (1891) showed that hemp stalks contained much less Si than leaves (see Table 2.7). Guerriero *et al.* (2019) found relatively little Si deposited in the cell walls of bast fibers. In contrast, many leaf structures were heavily silicified: cystolithic trichomes, stomata, and the waxy "wrinkles" on the surface of epidermal cells.

17.4   Nutrient supplementation

A primer on N-P-K and cannabinoids is provided in section 2.5.3. Here we discuss additional issues: organic vs chemical fertilizers, and the NH_4/NO_3 ratio.

17.4.1   Organic vs chemical

Natural (organic) fertilizers generally consist of animal effluvia and comprise an assortment of macro- and micronutrients.

Chemical (mineral) fertilizers are generally synthesized and comprise one or two macronutrients (see Table 17.4). Mineral fertilizers are rapidly assimilated by plants and easy to quantify. Organic fertilizers are not immediately available to plants; they require microbial degradation in the soil. The exact amount of organic nutrients provided to plants is difficult to quantify because the microbial rhizosphere is varied and complex. Massuela *et al.* (2023), who worked with both, recommended a combination of organic and mineral fertilizers for *Cannabis* cultivation, because "the higher availability of mineral fertilizer can be important for initial plant growth, while organic fertilizers can be employed as a more sustainable nutrient complementation during flowering."

Massuela's crucial word is "sustainable." Organic fertilizers balance soil nutrients, improve soil texture, and nurture soil microorganisms. Mineral fertilizers easily injure plants, harm soil texture, and kill soil organisms. Growers often use excessive amounts of mineral fertilizers, which only add economic costs and mineral leachates pollute water bodies (see section 2.5.3).

Surplus mineral salts decrease flower quality. The smoke is harsh and produces a black, greasy ash—instead of a smoother smoke with a light gray ash. Smoking organic cannabis causes less pharyngitis and laryngitis than smoking cannabis cultivated with mineral fertilizers (Clarke, unpublished research, 1996). Surplus mineral salts can be "flushed" by withholding nutrients before harvest. Irrigate with slightly acidic water (pH 6.0–6.5). The flushing period depends upon the growing medium, but 4–7 days is usually sufficient. Commercial flushing solutions that contain chelates are available.

Using an organic approach, macronutrient deficiencies are best corrected with a mix of rapidly absorbed fertilizers and slow-release fertilizers (Table 17.4). For N deficiencies, side-dress

Table 17.4. Some fertilizers, their N-P-K ratio, and nutrient availability

Fertilizer		Percentage (by weight)			Availability to plant
		N	P_2O_5	K_2O	
Natural:	Manure (dairy cow)	1	0.75	1	Medium
	Activated sludge	5	3	0.1	Medium
	Worm castings	3.5	1	1	Medium/rapid
	Blood meal	15	3	0.75	Medium/rapid
	Bird guano	12	8	3	Rapid
	Bat guano	6	9	1	Rapid
	Fish emulsion	8	3	3	Slow/medium
	Cottonseed meal	8	3	3	Slow/medium
	Bone meal (raw)	4	21	0.2	Medium
	Seaweed	4	21	0.2	Medium
	Wood ash	0	2	7	Rapid/medium
	Phosphate rock	0	30	0	Slow
	Greensand	0	1.5	5	Slow
Chemical:	Anhydrous ammonia	82	0	0	Rapid
	Urea	46	0	0	Rapid
	Ammonium nitrate	35	0	0	Rapid
	Ammonium sulfate	21	0	0	Rapid
	Ammo-phos (Grade A)	11	15	0	Rapid
	Potassium sulfate	0	0	39	Rapid
	triple Superphosphate	0	46	0	Rapid
	"Rose Food"	12	12	12	Rapid

plants with a mix of bird guano (rapid) and composted manure (slow). Supplement P with a mix of wood ash (fast) and bone meal (slow). Supplement K with wood ash (fast) and greensand (very slow).

Other macronutrients are not in Table 17.4. Supplement Mg with dolomitic (high-magnesiuim) limestone. If liming is contraindicated, use Epsom salt (hydrous magnesium sulfate, $MgSO_4 \cdot 7H_2O$) or sulfate of potash magnesia (sul-po-mag, 11% Mg, 22% K, 23% S). Ca can be supplemented by liming. If you need Ca without changing pH, apply gypsum (calcium sulfate—21% Ca, 16% S). Sulfur deficiency can be corrected with sul-po-mag or gypsum.

The relative merits of organic vs mineral fertilizers in *Cannabis* are gaining attention. Marshmann *et al.* (1976) studied this variable with regard to THC content in Jamaica. Their samples of organically grown plants ($n = 19$) contained more THC than their samples of plants grown conventionally ($n = 18$). They did not present statistical data, so we calculated means from their raw data: THC% in organic plants averaged 3.4 ±0.47 SE; conventionally grown plants averaged 2.4 ±0.27 SE, significantly different, $t = 0.031$. Their most potent sample came from an organic farm, 7.9% THC.

Hemp seeds from organically certified crops had slightly more omega-3 fatty acids than conventional crops (Sova *et al.*, 2018). Organic versus conventional sources of nitrogen fertilizer did not change fiber yield in 'Fédrina' (Forrest and Young, 2006). Laleh *et al.* (2021) measured seed yield in field crops amended with cow manure (0, 10, 20, 30 t/ha) and/or inorganic urea + triple superphosphate that delivered similar N-P-K. Applying a combination of manure (20 t/ha) and inorganic (100 kg N/ha) produced the highest seed yield.

Organic fertilizers need to be applied earlier and at higher rates than mineral fertilizers. For example, regarding N supplementation during flowering, Caplan *et al.* (2017b) fertigated plants with an organic fertilizer that required 389 mg N/L to optimize inflorescence yield and cannabinoid concentrations. Saloner and Bernstein (2021) fertigated plants with mineral fertilizers that required only 160 mg N/L for optimization.

Massuela *et al.* (2023) compared organic vs mineral fertilizers in high-CBD plants grown in a fiber and perlite substrate. The two liquid solutions delivered the same concentrations of macronutrients (N 240, P 96, K 240, Mg 18.5, Ca 74 mg/L) and micronutrients. At harvest, flower dry weight differed significantly: 19.6 g/plant for organic and 23.7 g/plant for mineral fertilizer treatments. Conversely, CBD% was greater for organic vs mineral fertilizer, 6.5% vs 5.8%, respectively. Total CBD yield (potency × flower biomass) followed the flower dry weight trend: 1197 mg/plant for organic, and 1319 mg/plant for mineral fertilizer.

Massuela also calculated **nutrient uptake efficiency** (NUpE) and **nutrient utilization efficiency** (NUtE) (see explanations near Fig. 2.11). NUpE was greater for mineral than organic fertilizers. NUtE was greater for mineral than organic fertilizers, but only for P, not N or K. **Nutrient harvest index** (N-HI) is the ratio between nutrient content in flowers to the nutrient content in whole plants. N-HI was greater for mineral than organic fertilizers, but only for P, not N or K. These results indicate the importance of P for building inflorescence biomass.

Agronomic use efficiency (AE) is the contribution of fertilizer toward yield, compared with a non-fertilized control. Massuela was interested in AE for inflorescences dry matter ($AE_{flowers}$) and CBD yield ($AE_{CBDyield}$). Plants under nutrient stress (fertilized at a rate of 160 mg/L) could increase their AE efficiency to produce CBD in comparison with well-fertilized plants (240 mg/L). $AE_{CBDyield}$ for N and K increased 34% for organic fertilizers, and 72% for mineral fertilizers. The same trends were seen for $AE_{flowers}$.

The availability of organic fertilizers can be enhanced by adding mycorrhizal fungi as well as nitrogen-fixing and phosphate-solubilizing bacteria. For example, Massuela *et al.* (2023) inoculated their growth substrate with a product that contained a slew of mycorrhizal species and beneficial bacteria. However, the science of mycorrhizae and PGPMs is in its infancy (see Chapter 13), and may even be *Cannabis* strain-specific (Winston *et al.*, 2014).

Many organic gardeners recommend spraying plants with fish emulsion and other leaf fertilizers (e.g., Frank and Rosenthal, 1978). But Farnsworth and Cordell (1976) warned against spraying marijuana plants with leaf fertilizers—virtually all liquid foliar fertilizers contain nitrates. *Cannabis* leaves convert nitrates into carcinogenic N-nitrosamines.

Bruce *et al.* (2022) cultivated a high-CBD strain and applied three regimens of organic fertilizers 1 week prior to transplanting clones to the field: (1) cow manure-based compost (CM); (2) standard fertilizer blend (ST: composted poultry manure, feathermeal, sulfate of potash); and (3) a boutique blend designed for cannabis production (BB: ST plus alfalfa–kelp–neem seed meal, Azomite®, Buddha grow and Buddha bloom). Compared with controls, biomass (dried flowering tops) was greater in all three regimens (Table 17.5), although statistically CM did not differ from controls (controls grew in a soil with relatively high background P and K levels), and ST did not differ from BB. CBD% and THC% were lowest in BB—probably a dilutional effect from increased biomass. CBD/THC ratio did not significantly vary.

17.4.2 NH_4/NO_3 ratios

Plants absorb two forms of nitrogen: nitrate (NO_3^-) and ammonium (NH_4^+). Following uptake by roots, nitrate (NO_3^-) is reduced to nitrite (NO_2^-) for further reduction to ammonium (NH_4^+). Ammonium is assimilated into glutamine and glutamate, which serve as N donors for the biosynthesis of all other amino acids. These combine into proteins, such as the enzyme THCA synthase.

Table 17.5. Results from Bruce *et al.*, 2022

	Biomass[a] (g/plant)	CBD%	THC%
None	517.3	8.03	0.31
CM	667.7	8.82	0.33
ST	689.8	8.06	0.31
BB	725.4	6.99	0.27

[a]Dried flowering tops

N also goes into nucleic acids, chlorophyll, some phytohormones (auxins, cytokinins), and other plant metabolites.

As stated in the title of an important paper by Saloner and Bernstein (2022a), *nitrogen source matters*. NO_3^- is an anion and NH_4^+ is a cation, so they have different impacts on the plant and its environment. When roots take up charged molecules, they release an identically charged molecule into the soil. Absorbing NH_4^+ will decrease (acidify) soil pH, and NO_3^- has a net effect of increasing soil pH.

Chemical fertilizers in Table 17.4 are NH_4^+-based, except for ammonium nitrate, which is 50% NH_4^+ and 50% NO_3^-. A century ago, farmers used a higher percentage of NO_3^--based fertilizers, such as sodium nitrate and calcium nitrate, but now they favor NH_4^+-based fertilizers (Cao *et al.*, 2018). Organic fertilizers require soil microbes to convert N into plant-available NH_4^+ (the "mineralization" process), and further converted into NO_3^- ("nitrification" process). NO_3^- can also be lost to microbe denitrification: NO_3^- converts to NO_2^- (nitrite) and N_2O (a greenhouse gas).

NO_3^- is water soluble and mobile in soil, so it is lost to leaching and runoff after rainfall. NH_4^+ can convert to ammonia (NH_3), and outgas from the soil. This loss is greater when urea, ammonium, or manures are applied to the soil surface and not incorporated into the soil. Kakabouki *et al.* (2021) applied urea with the urease inhibitor 2-NPT, which blocks the conversion of urea to NH_3. Plant height, seed yield, and CBD content was greatest in plants receiving urea + 2-NPT. Soil populations of mycorrhizal fungi were not affected by urea + 2-NPT compared with control plants.

Which form is best for plants? For most crop species, NO_3^- works best during most of the plant's growth period, although NH_4^+ is preferentially utilized by young plants (Bennett, 1993). NH_4^+ can interfere with root uptake of cationic macronutrients (K, Ca, Mg) as well as micronutrients (Fe, Mn, Cu, Zn, Co).

Saloner and Bernstein (2022a) grew plants under five NH_4/NO_3 ratios: 0:100, 10:90, 30:70, 50:50, and 100:0. Plant function (plant height, inflorescence yield, transpiration and photosynthesis rates, stomatal conductance, chlorophyll content) were highest with 100% NO_3 (0:100 ratio). At the 50:50 ratio, plants started to show toxicity symptoms—wilting, chlorosis, and necrosis. Pure NH_4 (100:0 ratio) induced substantial plant damage, resulting in plant death (Fig. 17.5). Extrapolation of data indicated that THCA% began to decline when $NH_4 \geq 20\%$. The authors did not recommend fertilizers containing >30% NH_4.

Massuela *et al.* (2023) stated that organic fertilizers have higher NH_4/NO_3 ratios than mineral fertilizers. It's an interesting claim, because they used a mineral fertilizer solution that contained ammonium (SSA, ammonium sulfate), nitrate (Plantaaktiv Fertilizer Type B™), and a mix of both (Wuxal Super). Regardless, they observed symptoms of NH_4^+ toxicity in plants at high rates of organic fertilizer treatment.

17.4.3 Crop rotation

Continuous monocropping depletes soil nutrients and injures soil structure. It causes a build-up of pests, pathogens, and weeds. *Cannabis* can exhaust soil nutrients, a fact first noted in

Fig. 17.5. NH_4/NO_3 ratios: **A**. 100% NO_3; **B**. 100% NH_4 (courtesy of Saloner and Bernstein, 2022a)

AD 544 by Jiǎ Sīxié, an agronomist from Shāndōng. Jiǎ (1962) recommended crop rotation: sow hemp after the soil has been enriched by a crop of *shōzu*—a legume (*Vigna angularis*). Jiǎ interplanted hemp with a perennial legume, *Styphnolobium japonicum*, to increase the growth of hemp and help the legume grow straighter (Willis, 2007). Lastly, Jiǎ recommended another legume, mung bean (*Vigna radiata*), as a green manure to enrich the soil (Bray, 1984).

Green manures are crop plants turned into the soil to provide organic matter, nutrients, and beneficial organisms. Green manures include legumes (clover, vetch, beans, peas) or grasses (annual ryegrass, oats, winter wheat, winter rye) as well as rapeseed or buckwheat. Fred (1916) recommended an interval of least 25 days between turning over clover and planting hemp seeds. This gave the clover time to decompose, which consumes soil oxygen.

Greeks and Romans practiced a simple crop rotation called the **two-field system**. Arable land was divided into two fields; one field was planted to wheat, barley, or rye, while the other was left fallow to recover its fertility. Some Roman authors—Cato, Columella, Varro, and Pliny—recommended two alternatives to this system: either continuous cropping with heavy manuring, or rotating grains crops with nitrogen-fixing legumes such as broad beans, vetch, or lupins (White, 1970). Columella, Varro, and Pliny promoted hemp cultivation, but did not describe its place in their crop rotation.

Charlemagne (AD 748–814) established the **three-field system**, which White (1962) called the greatest agricultural advance of the early Middle Ages. In the three-field system, only one-third of the land lay fallow. In autumn one-third of land was planted to rye, wheat, or barley, and in the spring another third of land was planted to oats, legumes, or other crops—including *lino vel canava* (flax or hemp), decreed by Charlemagne in AD 795 (Boretius, 1883).

Hemp was "a specialty of the fenlands" in England's Norfolk, Suffolk, and Lincolnshire counties. In late Tudor times hemp occupied as much as 15% of sown land in Lincolnshire, when it was continuously cropped. A century later, 1650–1700, the percentage of hemp dropped to 6%, when it became a part of crop rotations (Thirsk, 1957).

Farmers in Pays de Waes (between Antwerp and Ghent, Belgium) pioneered a **four-field rotation** in the early 16th century, which lacked a fallow year. This "Flemish-style" agriculture spread to neighboring France around 1630–1650, and included turnips, artificial meadows such as clover, and "the introduction on a larger scale of textile plants like hemp" (Delleaux, 2010).

Charles "Turnip" Townshend (1674–1738) introduced the system to England in the 1730s, where it became known as the Norfolk system. The sequence of four crops (wheat, turnips, barley, and clover) included a fodder crop (turnips) and a grazing crop (clover) for livestock. Better fed, the livestock yielded more manure, enabling better fertilization of the other crops. In the county of Norfolk, farmers grew "turnips, barley, wheat, and a little hemp" (Armstrong, 1781). Varro (1774) described the "new system of husbandry" involving turnips, and recommended planting hemp on "the most weedy ground you have," prior to starting the rotation with wheat, "as it will most effectually kill the weeds, and bring the ground into a good tilth."

Hemp crops in Ukraine were rotated with oats or turnips, then wheat or rye, then fields turned to meadow for 4–5 years (Marshall, 1772). Cattle grazed the meadow and fertilized the fields with manure. Marshall noted that fields turned to meadow grew grass and not weeds, and he attributed this to hemp, "a great cleaner of the land, and that no weeds can live among it."

In the French hemp heartland of Seine-et-Marne, Gaujac (1810) put an interesting spin on the turnip rotation: he sowed turnip seed after *male* hemp plants were harvested. "The turnip greens maintain a freshness in the earth which helps accelerate the maturity of the *chenevis* [female seed hemp]. After picking the female hemp, the land is covered with turnips." The hemp plants, in turn, protected turnips from aphids.

Arthur Young (1741–1820) advocated crop rotation (Fig. 17.6). Young (1774) stated that "improvements are wanting" in hemp

cultivation, and he undertook tours around Europe. He described hemp in France and Ukraine (Young, 1768), and hemp's importance for the economy in England (Young, 1772) and Ireland (Young, 1780). Young (1794) described experiments with Chinese hemp, and dedicated 17 pages to hemp in a book about agriculture (Young, 1797).

Young made the earliest reports of fiber yields/acre (Table 2.11), and he wrote up balance sheets—expenses and profits—for growing an acre of hemp in Suffolk (profit: 7 £. 8 *s.* 8 *d.*, Young 1797) and Lincolnshire (profit: 6 £. 12 *s.* 6 *d.*, Young 1799). He corresponded with President George Washington about hemp (Washington, 1801). Young condemned Great Britain's "breach of contract" with Ireland by taxing Irish-grown hemp (Young, 1780). He described France's rural poor living in grinding poverty and predicted the French Revolution (Young, 1794).

Unlike "Turnip" Townshend, Young added nitrogen-fixing beans (*Vicia faba*) into rotation. Young (1784) offered several "courses of crops" with beans in rotation. One course of crops included hemp: 1. beans; 2. hemp; 3. wheat; 4. clover; 5. beans; 6. wheat. Young (1792) observed farmers in Flanders practicing a complex 10-year rotation (Fig. 17.6). In Limagne some farmers sowed: 1. barley; 2. rye; 3. hemp followed by cabbage; 4. rye. Other Limagne farmers, interested in hemp as a cash crop, practiced a three-year rotation: 1. hemp; 2. rye; 3. add manure, and hemp again. Farmers in Alsace rotated wheat with a variety of legumes (beans, peas, clover, vetch) and other crops (hemp, potatoes, poppies for oil, maize, tobacco, cabbages).

Low (1834) advocated for a fallow rotation instead of turnips: "Hemp may be introduced into the rotation in such a manner as the following: 1. fallow or green crop; 2. wheat; 3. sown grasses; 4. hemp; 5. oats. Or: 1. fallow manured; 2. wheat or other corn-crop; 3. sown grasses; 4. oats; 5. hemp manured; 6. corn-crop." Dixon (1854) doubled-down on turnips: 1. turnips; 2. barley undersown with clover or grass; 3. hay or pasture; 4. flax; 5. turnips; 6. hemp. Dixon mentioned that farmers in Maine-et-Loire (France), who grew fiber plants as cash crops, used a triennial rotation of hemp, trefoil (i.e., clover), and flax. This system required rich land, and an abundance of manure, but was very remunerative.

Given all these expert opinions, the first person to actually conduct a quantitative experiment was Charles Daubeny (1795–1867), a botanist, chemist, geologist, and physician. He grew hemp in experiments at Oxford University's Botanical Garden from 1835 to 1844 (Daubeny, 1845). When hemp was continuously cropped, the yield continuously dropped, from 46.5 lb of dried stalks in 1836, down to 21.5 lb in 1844 (in a 100 ft^2 plot). In rotation, hemp yields were greater following clover, which fixes nitrogen (52.5 lb), than following turnip (37.6 lb), barley (42.5 lb), or buckwheat (39.5 lb).

In 19th century Kentucky, farmers monocropped hemp year after year until the crop no longer yielded a profitable return, then moved to new land. Clover was sown in the abandoned land and grown for two or three years to restore fertility (Hopkins, 1951). Russian peasants in the mid-19th century practiced the three-field system. Other crops sometimes replaced grains—nitrogen-fixing plants (legumes or vetches), lentils, turnips, millet, and lastly hemp or flax. The four-field

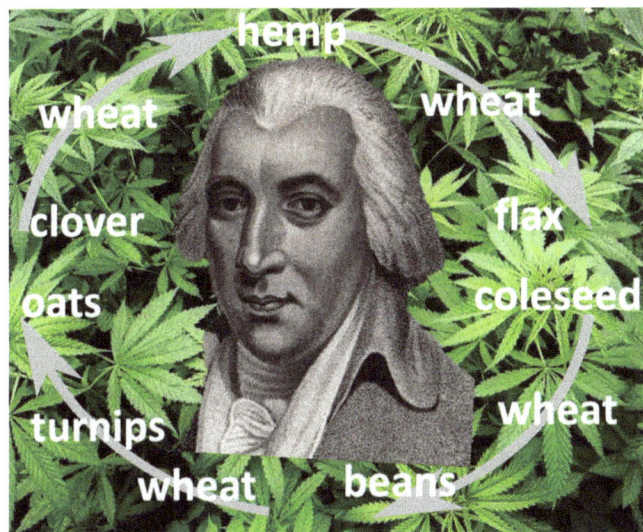

Fig. 17.6. Arthur Young (redrawn from Pell, 1893)

system did not become widespread in Russia until the early 20th century (Kahan, 1985).

Flint (1884) described a complex rotation in Italy: 1. wheat sown in November, with clover sown the following February; 2. wheat harvested in July, then clover cut or cattle pastured on it; 3. clover cut four times, and spread manure in winter; 4. clover again cut four times; 5. ground plowed and sown with hemp in March, harvested in June, land plowed again, and planted with maize, harvested in October, plowed again, and planted in wheat.

When industrial hemp returned to Canada and Europe in the 1990s, researchers sought its place in existing crop rotations. As a weed suppressant, Callaway (2004) recommended hemp followed by grain or root crops. Hemp's ability to suppress soil-borne pathogens made it attractive as a crop preceding suscep-tible crops (de Meijer, 1993; Kok *et al.*, 1994; Dippenaar *et al.*, 1996; Scheifele *et al.*, 1997). Gauthier (2020b) flipped the equa-tion and reviewed diseases in other crops that might carry over and infect hemp the following year. In 2017, the first year that Pennsylvania issued licenses for industrial hemp, the Rodale Institute jumped in. They slotted hemp into a 4-year rotation that included grains (wheat or maize), legumes (soybeans or vetch), and oats or hay (Rodale Institute, 2020).

17.5   Light deficiencies and imbalances

See section 2.1 for a heavy essay on light. *Cannabis* evolved in sunlight, but *Cannabis* cultivation often occurs indoors. Ever since Edison we have tried to replicate sunlight's intensity and spectrum with electric lights. Replicating sunlight is challen-ging, because plants respond to specific wavelength spectra. Photosynthesis is driven by visible light in the 400–700 nm spectrum (see Fig. 2.3). Other aspects of plant development are driven by specific wavelengths, and this is termed photo-morphogenesis. These aspects include phototropism, stalk elongation, stomata activity, and flowering.

Replicating sunlight is particularly challenging for *Cannabis* growers, because our plant of interest is a light hog— it has a very high light saturation point. *Cannabis* likely evolved in the steppe, "big sky" territory. Bazzaz *et al.* (1975) reported *Cannabis* approaching light saturation at 120,000 Lux, which is more intense than natural sunlight.

17.5.1   Electrical lighting systems

Cannabis growers have used a variety of electrical lighting systems. We briefly touch on a few systems here. For much more about lights, see Cervantes (2015). At the dawn of indoor growing, **fluorescent bulbs** were state-of-the-art (Stevens, 1975). Fluorescent bulbs are low-pressure mercury vapor dis-charge lamps that produce ultraviolet (UV) light, which causes a phosphor coating on the inside of the bulb to fluoresce in the visible spectrum. Now their use is limited to cloning rooms. Young cuttings have a relatively low light saturation point, and the wavelength spectrum of fluorescent bulbs is beneficial for young cuttings. Fluorescent bulbs are weak and inefficient, but they operate at a cool temperature and can be placed directly atop the plant canopy.

Next came high-intensity discharge (HID) bulbs. Frank and Rosenthal (1978) championed **metal halide** (MH) lamps. MH bulbs run an electric arc through a mixture of mercury and metal halides (e.g., sodium iodide). The color (wavelength) emitted from MH lamps can be optimized to deliver photosyn-thetically-active light in the 400–700 nm spectrum by adding different metals and inert gases. MH lamps are still popular, particularly for plants during vegetative stages.

High-pressure sodium (HPS) lamps soon gained popu-larity, because of their luminous efficiency (Frank, 1988). HPS bulbs run an electric arc through a mixture of sodium, mercury, and xenon gas. They deliver a rather monochromatic light in the yellow-orange wavelength band. HPS lamps are still popular, particularly for flowering-stage plants.

Light-emitting diodes (LEDs) are semiconductor light sources that emit light when current flows through a silicon-based substrate. Chandra *et al.* (2008) first used LEDs for *Cannabis* cultivation. LEDs are versatile; they can emit a monochromatic light, or a range of wavelengths by using mul-tiple semiconductors, or by placing a phosphor coating over the light source. "Quantum boards" array hundreds of small LED diodes in rows, attached to a flat board. They are more energy efficient than older LED arrays. LED technology continues to evolve rapidly; anything written here will be outdated soon.

17.5.2   Luminous efficiency

Indoor *Cannabis* cultivation is a very energy-intensive industry (Warren, 2015), and illumination consumes 35% of total elec-trical use in a growroom (Mills, 2012). Therefore, the luminous efficiency of artificial light is crucial. Luminous efficiency can be quantified as lumens/circuit watt.

A comparison of different electrical lighting systems is presented in Table 17.6. All the lights in Table 17.6 are more efficient than incandescent bulbs (18 lumens/circuit watt). The bulb–fixture price in Table 17.6 includes bulbs, fittings, reflective hoods, timers, and ballasts.

Nelson and Bugbee (2014) compared new HPS tech-nology (double-ended bulbs with electronic ballasts) with old HPS technology (single, mogul-base bulbs with magnetic ballasts), and ten different LED fixtures. They used a new metric for electrical efficiency: the number of photons per joule of energy input, measured as µmol/J. The new HPSs delivered 1.7 µmol/J, a dramatic improvement over old HPSs (mean 1.0 µmol/J). LED efficiency depended on color, ranging from 0.9 to 1.7 µmol/J. Factoring electrical costs with capital outlay (amortized over a 5-year period), they figured LED costs 2.3× more than new HPSs per µmol of photons.

Backer *et al.* (2019) conducted a meta-analysis of several studies, and reported that most strains grown under HPS lights produced higher flower yields/m² than plants grown under MH lamps. CBD/m² increased with light intensity at 400 W/m² compared with 600 W/m².

In addition to the data in Table 17.6, there are other con-siderations. The intensity of all bulbs declines with age, except HPSs, which just stop working. MH and HPS bulbs contain mercury, and their ballasts have electrical safety concerns. MH

Table 17.6. Aspects of indoor lighting (updated from Bush Doctor 1993b)

Type of bulb[a]	Initial lumens	Lumens per circuit watt	PPFD/PAR spectrum	Bulb and fixture price
FLO 32 W warm white	2200	50	Good (nearly full spectrum)	$
FLO 40 W Gro-Lux	2000	40	Excellent (full spectrum)	$
MH 400W standard	38,000	90–95	Average (blue)	$$
HPS 150 W	12,000	80	Below average (yellow-orange)	$$
LED array 200 W	20,000	100	Good (nearly full spectrum)	$$$

[a]FLO, fluorescent; HP, metal halide; HPS, high-pressure sodium; LED, light-emitting diodes; PPFD/PAR, photosynthetically active flux or radiation

and HPS bulbs generate a lot of heat. They cook seedlings, and must be suspended a distance above mature plants, which lowers their efficiencies.

LEDs generate less heat, but unlike the other bulbs, they are very directional—they emit light in a singular direction, necessitating an array of lights. Quantum boards provide a more even spread of light. Because LEDs are relatively cool, growers position them close to the canopy. But plant stress (chlorosis, leaf curling) may arise when Quantum boards are suspended 30 cm or less over the canopy. Aim for 40–45 cm when flowering (70–75 cm during vegetative stage). Hawley *et al.* (2018) deployed LEDs for sub-canopy lighting during the flowering stage. They positioned small, individual LEDs 15 cm on either side of the plant stalk and 2 cm off the soil surface, pointing up. Sub-canopy LEDs significantly increased flower yields, increased the flower-to-leaf ratio, and increased THC%.

17.5.3 Wavelength efficiency

In addition to light intensity, proper wavelength is equally important. Photosynthesis is driven by visible light in the 400–700 nm spectrum. The light source spectra (relative intensity at different wavelengths) of several light sources are presented in Fig. 17.7. LEDs are not included, because their wavelengths can be manipulated. Compare Fig. 17.7 with the chlorophyll absorption spectra (see Fig. 2.3).

In addition to harvesting light for photosynthesis, *Cannabis* utilizes specific wavelengths for other aspects of photomorphogenesis. See section 2.1.2 for a review of the specific photoreceptors that sense light at different wavelengths, and Bilodeau *et al.* (2019) for a more extensive review.

Red light (RL, wavelengths ~650–700 nm) and far-red light (FRL, ~700–740 nm) trigger phytochrome receptors. Plants growing in a closed canopy shade each other, and FRL predominates over RL. This FRL enrichment is how plants detect shade, and it accelerates flowering (in short-day conditions). In long-day conditions (during vegetative growth), a high FRL-to-RL ratio promotes stalk elongation. Seedlings are hard-wired for shade avoidance, and therefore do best when RL predominates over FLR. They do not elongate, produce a

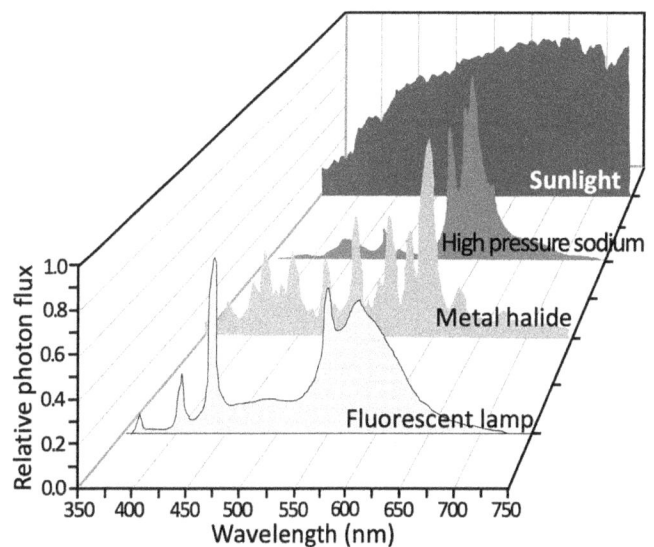

Fig. 17.7. Comparisons of brightness and wavelength spectrum (redrawn from Bilodeau *et al.*, 2019)

higher number of leaves, and an increase in root development (Bilodeau *et al.*, 2019).

Blue light (BL, ~420–520 nm) triggers cryptochome and phototropin receptors. Through these receptors, blue light regulates seed germination, chlorophyll synthesis, and stalk elongation and phototropism. Blue light also regulates stomatal activity, hence water transpiration and CO_2 exchange. Cryptochromes and phototropins also respond to ultraviolet UV-A light. The UVR8 photoreceptor responds to UV-B light. Cannabinoid production may be enhanced by UV light (see section 2.1.3).

17.6 Pollutants, toxins, and pesticides

Air pollution injures plants. Damage peaks during daylight hours, and worsens in warm, humid conditions. "Acid rain" doubles the damage by acidifying soil and creating various nutrient deficiencies. According to Agrios (2005), acid rain in Los Angeles (pH 1.7) was more acidic than vinegar (pH 3.0).

Sulfur dioxide (SO_2) is the smell of burnt matches. It is toxic at concentrations as low as 5 ppm. Burning sulfur-laden fossil fuel releases SO_2 into the atmosphere. According to the EPA, the amount of SO_2 released in the USA peaked in 1970, at 31.16 tons. SO_2 combines with water to form toxic droplets of sulfuric acid. SO_2 causes interveinal leaf chlorosis in *Cannabis* (Goidànich, 1959). **Nitrogen dioxide** (NO_2) is another consequence of our fossil fuel addiction, belching from internal combustion engines. NO_2 causes symptoms similar to those of SO_2 toxicity, at levels as low as 2–3 ppm. NO_2 combines with O_2 in bright sunlight to form ozone.

Ozone (O_3) arises from the action of light on O_2, and from electrical discharges. Indoor *Cannabis* cultivation generates ozone, as a result of high electrical use and oxidation reactions with outgassing terpenoids (Wang *et al.*, 2019). From a plant perspective, ozone is the most destructive air pollutant (Agrios, 2005). Concentrations >0.1 ppm cause tissue damage (humans can smell ozone at 0.1 ppm). Ozone initially causes chlorotic spots on leaves, primarily on upper surfaces. The spots enlarge and turn brown, leaves defoliate prematurely, and plants become severely stunted.

Peroxyacyl nitrates (PANs, also known as smog) arise when ozone combines with incompletely combusted gasoline. PANs may combine with atmospheric water to create nitric acid rain. Sharma and Mann (1984) studied *C. indica* growing next to a smoggy highway in India. Roadside plants suffered chlorosis and necrosis. Plants produced smaller leaves with shorter petioles than plants growing in the hills. Microscopically, polluted plants produced fewer stomates but more trichomes per leaf area.

Hydrogen fluoride (HF) is emitted by industry, especially petrochemical factories. It reacts with moisture to form hydrofluoric acid. HF causes interveinal chlorosis and necrosis of leaf tips and margins. Alternatively, leaves may develop a white speckling (Goidànich, 1959). Young plant growth is most susceptible, especially when wet.

Carbon dioxide (CO_2) is produced by automobiles, industries, and burning rainforests. Human activities currently emit about 29 billion tons of CO_2 per year. Rising CO_2 levels represent an "existential threat," as Bernie Sanders puts it. CO_2 promotes photosynthesis and increases carbon accumulation and plant growth (see section 2.4). But it dilutes the plant's N concentration, which forces insects and other N-seeking herbivores to eat more plant tissue. Roberts (1988) highlighted this fact in an early article on climate change.

17.6.1   Heavy metals

Some heavy metals are essential plant nutrients (iron, copper, zinc), whereas others are toxic: mercury, lead, cadmium, nickel, arsenic, and lithium. Source of heavy metal contamination include mining, tailings, industrial wastes, and agricultural runoff. *Cannabis* is a "bioaccumulator" of heavy metals. The plant easily absorbs heavy metals, and they are relatively well-tolerated.

These toxins deserve mention for their effects on us. Ingesting plants laced with lead (a contaminant in chemical fertilizers) causes anemia and brain damage. Chronic absorption of mercury causes neurologic deficits. Cadmium wrecks kidneys and causes breast cancer. Nickel and arsenic are carcinogenic. Lithium is a cumulative poison.

Siegel *et al.* (1988) measured 440 ng mercury per gram of *Cannabis* grown on Hawaii's mercury-rich volcanic soil. They noted that mercury is absorbed ten times more efficiently by the lungs than by the gut, and calculated that smoking 100 g of volcanic marijuana per week would lead to mercury poisoning. Volcanic soil also contains significant levels of cadmium. Grant *et al.* (2004) attributed volcanic soil to elevated levels of cadmium in Jamaican-grown tobacco and cannabis.

Tainted fertilizer is another source of heavy metal contamination. Safari Singani and Ahmadi (2012) showed that *Cannabis* readily takes up lead and cadmium from soils amended with contaminated cow and poultry manures. *Cannabis* accumulated cadmium from reputedly "clean" fertilizer (Ahmad *et al.*, 2015).

Phosphate ions are the main carriers of heavy metal contamination, and hydroponic fertilizers are particularly vulnerable to contamination (Karadjov, 2014). Phosphate fertilizers targeted at *Cannabis* growers ("bud blooms") have particular problems with arsenic, in some cases 10–50 ppm (N. Palmer, pers. comm., 2016). Rockwool, used as hydroponic growth medium, may also be contaminated.

In an analysis of hemp seeds, Mihoc *et al.* (2012) found cadmium contamination; they measured levels of 1.3–4.0 mg/kg. Eboh and Thomas (2005) showed that concentrations of arsenic, cadmium, chromium, iron, nickel, lead, and mercury were greater in leaf material than in seeds. Moir *et al.* (2008) measured heavy metals in marijuana smoke, including mercury, cadmium, lead, chromium, nickel, arsenic, and selenium. Deep inhalation, typical of marijuana smokers, doubled the exposure to heavy metals.

Health Canada (2008) mandated upper limits for arsenic (0.14 μm/kg body weight per day), cadmium (<0.09 μm/kg), lead (<0.29 μm/kg), and mercury (<0.29 μm/kg). Upton *et al.* (2013) proposed maximal limits for orally consumed cannabis products: mercury 2.0 μm/day, arsenic 10.0 μm/day, and cadmium 4.1 μm/day.

Individual states in the USA have also mandated testing for heavy metals (McPartland and McKernan, 2017). More recently, Jameson *et al.* (2022) examined regulatory documents by 36 states that legalized recreational and/or medical cannabis. Twenty-eight states mandated testing of arsenic, cadmium, lead, and mercury, with allowable tolerances ranging 0.00009 to 10 ppm. Other regulated elements included chromium (listed by six jurisdictions), nickel (two jurisdictions), and copper (two jurisdictions).

Cannabis is so efficient at absorbing heavy metals that it has gained attention as a "**bioremediation crop**." Bioremediation uses plants or microorganisms to remove pollutants from soil. Plants such as *Thlaspi caerulescens* (= *T. alpestre*) extract toxins from soil and store the toxins in their tissues. The plants are harvested and the toxins removed. *Cannabis* is a good candidate for bioremediation (Shi and Cai, 2009), although the amount of metal taken up by *Cannabis* pales in comparison with *T. caerulescens* (Giovanardi *et al.*, 2002; Löser *et al.*, 2002; Citterio *et al.*, 2003; Meers *et al.*, 2005).

Jurkowska *et al.* (1990) kicked off bioremediation research. They measured high levels of lithium in hemp (1.04 mg/kg), higher than the other crop plants tested, including barley, maize, mustard, oats, radish, rape, sorrel, spinach, sunflower, and wheat. *Cannabis* has been sown on toxic waste sites contaminated with cadmium and copper in Silesia. The metals are recovered by harvesting seed, and leaching seed with hydrochloric acid (Kozlowski, 1995). Hemp also accumulates heavy metals in its roots (Giovanardi *et al.*, 2002; Citterio *et al.*, 2003; Shi and Cai, 2009), and in leaf material (Giovanardi *et al.*, 2002; Arru *et al.*, 2004). Plants with mycorrhizal fungi growing in their roots show *greater* translocation of heavy metals from roots to shoots (Citterio *et al.*, 2005). Perhaps mycorrhizal-inoculated plants can better tolerate heavy metal stress.

Ciurli *et al.* (2002) exposed 'Fibranova' fiber-type plants to zinc chloride ($ZnCl_2$), zinc sulphate ($ZnSO_4$), and zinc nitrate ($Zn(NO_3)_2$). Seed germination rate decreased as zinc concentrations increased. Zinc chloride was particularly potent; 0.05 M reduced germinability. Seedlings grown in potting soil tolerated zinc chloride at 0.1 M, but that concentration in hydroponic culture killed plants in 8 days.

Cannabis can extract toxic polycyclic aromatic hydrocarbons from soil, such as benzo[a]pyrene and chrysene (Campbell *et al.*, 2002). *Cannabis* also extracts radioactive caesium-137 and strontium-90 from contaminated soil (Vandenhove and Van Hees, 2005; Hoseini *et al.*, 2012). Hemp crops were planted near the Chernobyl site for the purpose of removing radionucleotides (Anonymous, 2000).

Löser *et al.* (2002) were not impressed with the ability of *C. sativa* to uptake heavy metal-polluted river sediment. Although the plants took up zinc, cadmium, and nickel, about 95% of the plants died within a week. Apparently different cultivars vary in their ability and tolerance in taking up cadmium from contaminated soils (Shi *et al.*, 2012).

17.6.2 Pesticide injury

Some of the most dangerous toxins are pesticides. Insecticides, miticides, and fungicides can be phytotoxic. They are mixed with solvents, emulsifiers, diluents, and carriers, which can also cause plant injury.

Herbicides account for two-thirds of all pesticide usage. Nearly all herbicides (95%) are applied to just four crops: maize, soybeans, wheat, and cotton. Globally, usage of **glyphosate** has risen 15-fold since the 1996 introduction of "Roundup Ready" transgenic crops (genetically modified for glyphosate tolerance). In the USA alone, over 1.6 *billion* kg of glyphosate have been sprayed in the USA since 1974 (Benbrook, 2016).

That kind of tonnage generates a lot of **spray drift**—the unintended down-wind movement of herbicides from a target area. Off-target damage to neighboring crops occurs with as little as 1% drift, depending on plant sensitivity (Cederlund, 2017).

Another issue is **carryover** in soil—herbicide persistence into the next growing season. The potential for carryover injury is influenced by the amount of herbicide present in the soil and

soil conditions—high pH allows herbicides to persist, as does a high amount of organic matter or clay. Soil persistence varies amongst herbicides. Atrazine and simazine, for example, have long soil half-lives, and *Cannabis* is sensitive to them (Bócsa, 2004).

Cannabis is particularly sensitive to glyphosate—the most widely used herbicide in the world—with significant potential for drift (Córdova *et al.*, 2020) and carryover (Cornelius and Bradley, 2017). Horowitz (1977) tested nearly 50 herbicides and reported that "glyphosate at 820 g/ha was the only herbicide which killed both young and developed *Cannabis* plants." Symptoms of glyphosate injury included leaf burn, chlorosis, petiole malformation, and bud necrosis.

Toth *et al.* (2021) sprayed glyphosate on three CBD strains that had initiated flowering, and measured cannabinoid content for 3 weeks. Concentrations of CBD, THC, CBC, and CBG were significantly lower than those in control plants. CBD content, for example, dropped from 10% in controls to 6% in glyphosate-treated plants.

Herbicides are *purposely* sprayed on *Cannabis* for three reasons: (1) as defoliants and desiccants to simplify the harvest process; (2) to eliminate weedy *Cannabis* from crop fields; and (3) to eradicate illicit *Cannabis* crops.

Herbicides have been applied as defoliants and desiccants shortly before the harvest of either fiber- or seed-type crops (Goloborod'ko, 1986). Treflan is popular for this purpose (Nazirov and Tukhtaeva, 1981), as is glyphosate (Ranalli and Venturi, 2004; Serkov *et al.*, 2012; Rózańka *et al.*, 2023). Growth-retarding chemicals are also used (Keijzer *et al.*, 1990).

Eliminating feral/weedy *Cannabis* from crop fields is practiced in Eurasia. Feral hemp has been killed with herbicides in Bulgarian sunflower fields (Dochev *et al.*, 2016), Hungarian sunflowers (Kukorelli *et al.*, 2007), Slovakian sugar beets (Tóth *et al.*, 2015), Romanian wheat (Pascu, 2012), Hungarian wheat (Reisinger *et al.*, 2005), wheat in India (Kumar *et al.*, 2000), and onions in India (Prakash *et al.*, 2000) … and that's just a sample of recent 21st century publications.

Haney and Kutscheid (1975) estimated that feral hemp covered 5–10 million acres in the USA. This was mostly in the Midwest, where feral hemp encroached upon maize, soybean, and wheat fields. Soon after **2,4-D** became available, American agronomists used it to control feral hemp (Hanson *et al.*, 1946; Freed *et al.*, 1951; Dunham *et al.*, 1956; Klingman and McCarthy, 1958). Soviet agronomists also tested 2,4-D to control wild hemp growing in cereal crops (Dobretsov, 1957). Damage from 2,4-D includes small, narrow, deformed leaflets (Fig. 17.8). Chlorosis may arise, and leaf margins roll inward. Actively growing young leaves are the most sensitive, and damage decreases further down the shoot.

Bjerken (1958) shifted the dialogue from controlling feral hemp to eradicating illicit *Cannabis* crops. Bjerken called for an organized eradication program, to stop "the alarming spread of wild hemp or marijuana and the increasing use of the drug." Australia engaged in a systematic *Cannabis* eradication campaign prior to the USA. In 1964, police were alerted to a large area of feral *Cannabis* growing along 40 miles of the Hunter River. They sprayed it with herbicides, although "there has been no evidence of these plants being used for illicit purposes" (UN-CDN, 1964). They field-tested a variety of herbicides—2,4-D, atrazine, amitrole, simazine, bromacil, and

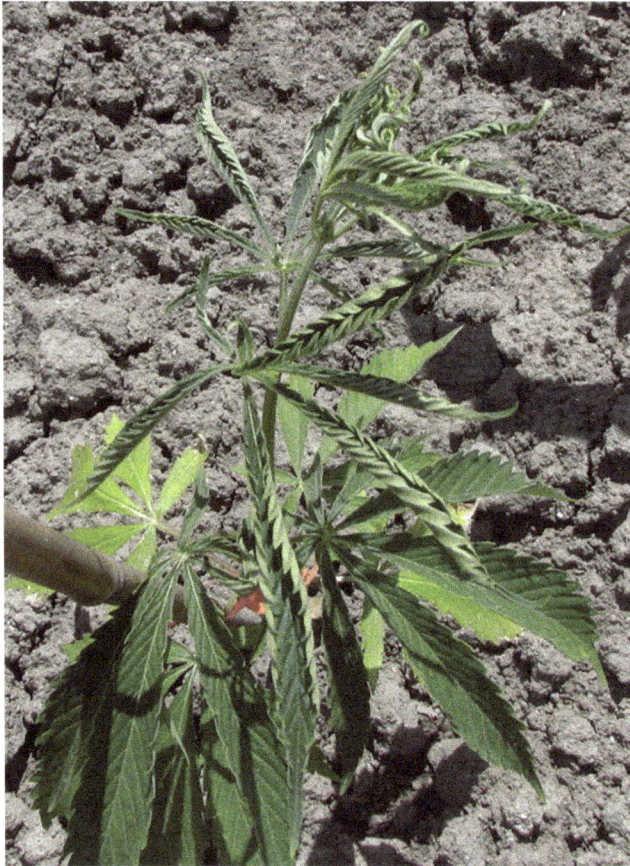

Fig. 17.8. 2,4-D herbicide injury (Gianpaolo Grassi).

ammonium thiocyanate. After 9 years, they contained the "outbreak" (Vane, 1973).

Entrepreneurs in the USA began wild-harvesting feral hemp in the Midwest after Nixon's "Operation Intercept" squeezed the supply of Mexican marijuana at border crossings in 1969. This drew the attention of law enforcement, who began eradicating feral hemp in earnest—the "Swat Pot" program (Drake, 1970b). McGlamery and Knake (1970) and US Department of Defense (1970) recommended a combination of 2,4-D and 2,4,5-T ("Agent Orange," used in Vietnam). Midwestern game birds feed on feral hemp (see section 16.5), which led wildlife agencies to oppose law enforcement eradication efforts (Vance, 1971).

Eaton *et al.* (1972) tested the effectiveness of eight herbicides against wild hemp. They also suggested mycoherbicides—spraying plants with killer fungi—and they identified several *Cannabis*-selective insects, as well as the hemp russet mite. Coffman and Genter (1978, 1980) demonstrated the effectiveness of paraquat against Jamaican *Cannabis* grown in a glasshouse. At low doses paraquat decreased THC and CBD content, and high doses killed the plants.

Horowitz (1977) tested nearly 50 herbicides against *Cannabis*, and listed 20 chemicals which killed or severely injured the crop. Most herbicides were absorbed by roots, limiting their effectiveness to young, actively growing plants. A few herbicides worked directly on foliage, in plants of all

ages—glyphosate (Roundup™), amitrole (Weedazol™), phenmedipham (Betanal™), 2,4-D, and paraquat.

Paraquat gained notoriety. In 1973 the Nixon administration covertly funded the Mexican government with US$50 million to spray cannabis crops with paraquat (Trux and Torry, 1978). Spraying in Mexico began in 1975, and in 1977 alone 9500 acres were destroyed by airborne spraying (Smith, 1978). The effort backfired. Plants harvested immediately after exposure looked normal, and paraquat is odorless and tasteless. The contaminated crop was harvested and sold by unscrupulous drug dealers. Mexican marijuana tainted with paraquat quickly appeared in the USA. By 1978, 21% of marijuana tested in the southwestern USA was contaminated with paraquat (Landrigan *et al.*, 1983). A public outcry ensued. Funding ceased in 1979 under the Carter Administration. The fear of poisoned Mexican marijuana gave birth to America's domestic cannabis industry.

The DEA hatched a new plan—spray paraquat on domestic *Cannabis* crops, beginning in Florida (Marshall, 1982). In 1983 the DEA began spraying crops in Georgia and Kentucky. Widespread concern over health hazards (and an EPA law prohibiting the use of paraquat in national forests) forced the DEA to stop spraying. The DEA quietly resurrected its herbicide agenda, and published environmental impact statements (DEA, 1986). Legal challenges by NORML and the Sierra Club curtailed spraying in the USA.

17.7   Climatic hazards

Hemp is susceptible to water stress—summarized in six words by Hackleman and Domingo (1943): either too little or too much. See section 2.2 for introductory concepts regarding moisture, such as water potential (ψ), field capacity (FC), volumetric water content (θ), evapotranspiration (ET), and water use efficiency (WUE).

17.7.1   Too little water

Symptoms caused by water deficits are obvious: wilting. Caplan *et al.* (2017b) considered a plant to be wilted when leaf angle (between the center of the middle leaflet and the branch from which it originated) increased by 50% from that in a well-watered plant. They measured this in leaves on a branch coming off the first internode.

They measured *temporary* wilting, from which plants can recover if watered. Beyond that lies *permanent* wilting, from which plants cannot recover, and die.

Mauz (1836) suggested that drought shifted the gender ratio in *Cannabis*, normally 50% female and 50% male, towards a greater percentage of female plants. Drought may exacerbate hermaphroditic flowering in genotypes predisposed to that condition (e.g. Thai landraces; Clarke, 1981).

Chronic water deficits cause chlorosis in leaves, stunting, and reduced yields. In the first month after sowing, seedlings exposed to more than ten consecutive days of drought showed more tip dieback (Gao *et al.*, 2018). Robinson (1943a) stated that plentiful moisture was critical during the first 6 weeks after sowing; thereafter *Cannabis* becomes well-rooted.

Abot *et al.* (2013) subjected monoecious French varieties to water stress (withholding water until leaves displayed the first sign of wilting). Drought-stressed plants compared with well-watered plants were shorter (stalk length 75 cm and 171 cm, respectively), and stem diameter at base was reduced by 32%. Maturation of phloem fibers was delayed. Drought stress had one positive effect—increased fiber tensile stiffness.

In Iranian plants subjected to drought stress, Bahador *et al.* (2017) reported an increase in canopy temperature (controls 41.5°C, drought-stressed 47.1°C), with greater leaf electrolyte leakage, as well as losses in leaf weight and area, and decreased seed yields (controls 1425 kg/ha, drought-stressed 260 kg/ha).

Estimates of *Cannabis* water consumption vary a great deal. The variation is due to a diversity of phenotypes, variable cropping densities, and the crop's wide geographic range. The first estimate was made in Vermont: Boyce (1900) estimated that an acre of fiber-type hemp required 450,000 gallons of water, or 20–25 inches of rainfall (508–635 mm) during its growing season. Dewey (1914) noted that Kentucky's hemp-growing region received 15.6 inches (396 mm) rainfall during its 4-month growing season.

Other estimates for optimal rainfall are 250–280 mm during the growing season (Lisson and Mendham, 1998), 650 mm rainfall per annum (Van Dam, 1995), seasonal irrigation of 535 mm (Lisson and Mendham, 1998), 305–381 mm during the growing season (BCMAF, 1999), 500–700 mm per growing season (Cherrett *et al.*, 2005), and 353 mm per growing season (Thevs and Aliev, 2022). Cosentino *et al.* (2013) estimated that a short, early-maturing French monoecious cultivar ('Futura 74') required 250 mm per season, and a tall, late-maturing Italian dioecious cultivar ('Fibranova') required 430 mm per season.

Estimates for drug-type crops are fewer in number. Cervantes (2015) stated that larger "Sativa" plants, with their extensive root systems, consumed more water than shorter "Indica" plants. At the dawn of the home-grown era, Drake (1970a) recommended watering each plant with 0.85 L/week, at 24°C (75°F).

In contrast, Carah *et al.* (2015) estimated that each plant consumes 22 L/day during the June–October growing season. Carah's widely quoted study is based on a tiny experiment: they cited a study from Humboldt Growers Association (2010), who calculated 22 L/day per plant from drip irrigation in a 5 × 5 ft (25 ft²) area. Carah multiplied 22 L/day × 130,000 plants/km² (their estimation of crop density), resulting in water consumption of 430 million L/km². This equals 430 mm rainfall during the June–October growing season (because 1 mm rainfall equals 1 L water over a 1 m² area).

Evapotranspiration (ET) is defined as "the combined loss of water from a given area, and during a specified period of time, by evaporation from the soil surface and by transpiration from plants" (Kirkham, 2014). Reference evapotranspiration (ET_0) is an estimate of the evapotranspiration from a reference crop, defined as a full-cover grass surface that is actively growing, well-watered, 0.12 m in height, and green (with an albedo of 0.23).

ET_0 is measured in units of mm H_2O/day, computed from meteorological data, including solar radiation, air temperature, air humidity, and wind speed. There are apps for that. ET_0 can also be estimated from pan evaporation—the amount of water evaporated from a pan filled with water (mm/day), where $ET_0 = E_{pan} \times K_p$. The pan coefficient ($K_p$) corrects for different types of pans, the "Class A evaporation pan" being standard.

Crop evapotranspiration, ET_C, is determined by a specific crop's characteristics, according to the formula $ET_C = ET_0 \times K_C$, where K_C is the crop coefficient. K_C values for fiber-type hemp were determined by García-Tejero *et al.* (2014), using the mostly monoecious cultivars 'Carma' and 'Ermes'. In the initial growth stage (from sowing to day 28), $K_C = 0.6$. In the development stage (days 29 to 59), $K_C = 0.8$. In mid-season (days 60 to 90), $K_C = 1.0$. In late season (days 61 to harvest), $K_C = 1.2$. Irrigation is needed when crop evapotranspiration ($ET_0 \times K_C$) exceeds rainfall.

Root-bound plants become predisposed to drought stress (Fig. 17.9A). Root-bound or pot-bound plants have outgrown their containers and don't have any more room to grow. Their roots get tangled and girdled, and self-prune. They become predisposed to root rot (Fig. 17.9B). Root-bound plants may never recover, even after removed from containers and transplanted into the field. In a survey of hemp growers, root binding was reported by 41.3% of respondents (Thiessen *et al.*, 2020).

Seedlings in rockwool cubes may become root-bound if transplanted too late (Fig. 17.9C). Peat pots (e.g., Jiffypot®) are reportedly biodegradable, allowing the entire pot with seedling to be transplanted into field soil. In fact the pots decompose too slowly, *Cannabis* roots never escape them, and the plants die of drought stress.

17.7.2 Irrigation

Areas with insufficient or unseasonable rainfall can import water from elsewhere. Water can be sourced from the surface (lakes and rivers) or underground (well water). Wells access two types of underground aquifers: an "unconfined" aquifer (a shallow water table at atmospheric pressure) or a "confined" aquifer (a deeper layer of water under impermeable material and under pressure). Unconfined aquifers are impacted by drought; confined aquifers ("fossil water") are often a nonrenewable resource.

The oldest reliably dated well, in Cyprus, was driven through limestone to reach an aquifer at a depth of 8 m, and dated to 8400 BC (Peltenburg, 2012). Hemp irrigation is described in the world's oldest surviving agricultural treatise, *Fàn Shèngzhī Shū*, written in the 1st century BC. Fàn Shèngzhī (1959) recommended irrigation with water, five *shēng* (five liters) per plant. Archaeological evidence is older: Shaw *et al.* (2007) unearthed 2nd century BC irrigation dams in Madhya Pradesh, India. They recovered pollen from rice, cereals, spinach, amaranth, mustard, and *Cannabis sativa*.

Kārēz irrigation systems in Turpan (Xīnjiāng Region, China) are even older. These ingenious systems utilized subterranean channels, with vertical shafts periodically dug for ventilation. The oldest *kārēz* systems in Turpan may date to 600 BC (Hu *et al.*, 2012), and Turpan's residents were the oldest-known cultivators of drug-type *Cannabis*, 800–500 BC, based on archaeological studies (Jiang *et al.*, 2006; Russo *et al.*, 2008; Jiang *et al.*, 2016).

Hemp was cultivated under irrigation in France and Italy (Young, 1794). Nöschel (1856) said Kazakhs used "tedious irrigation" to convert steppe to arable land, so they could grow hemp, millet, barley, and melons. English botanists found *Cannabis* growing under irrigation in Xīnjiāng Region, China (Price and Simpson, 1913).

In a book about irrigation in the USA, Steward (1893) stated that hemp "is peculiarly adapted to irrigation, its yield and quality being both improved." He recommended growing hemp in strips 3 ft wide, separated by irrigation channels.

Dewey (1914) claimed that hemp was first irrigated on a "commercial scale" in 1912, near Bakersfield. George Schlichten, inventor of the Schlichten hemp decorticator, wrote a 15-page primer on hemp irrigation. "The land should be level within the checks and with good solid borders, so there will be no trouble to irrigate properly and thoroughly" (Schlichten, 1921).

Methods of distributing water in fields can be categorized as surface irrigation, overhead irrigation, or drip irrigation. Surface or "furrow irrigation" is an ancient technology dating back to Mesopotamia (Fig. 17.10A). It is labor intensive: fields must be leveled and sloped, furrows dug and firmed, and pumping systems adjusted. Surface irrigation requires expertise to avoid water loss from field runoff and deep percolation.

Fig. 17.9. Root-bound plants: **A.** wilting and leaf chlorosis; **B.** root rot; **C.** rockwool cube transplanted into soil (Gianpaolo Grassi)

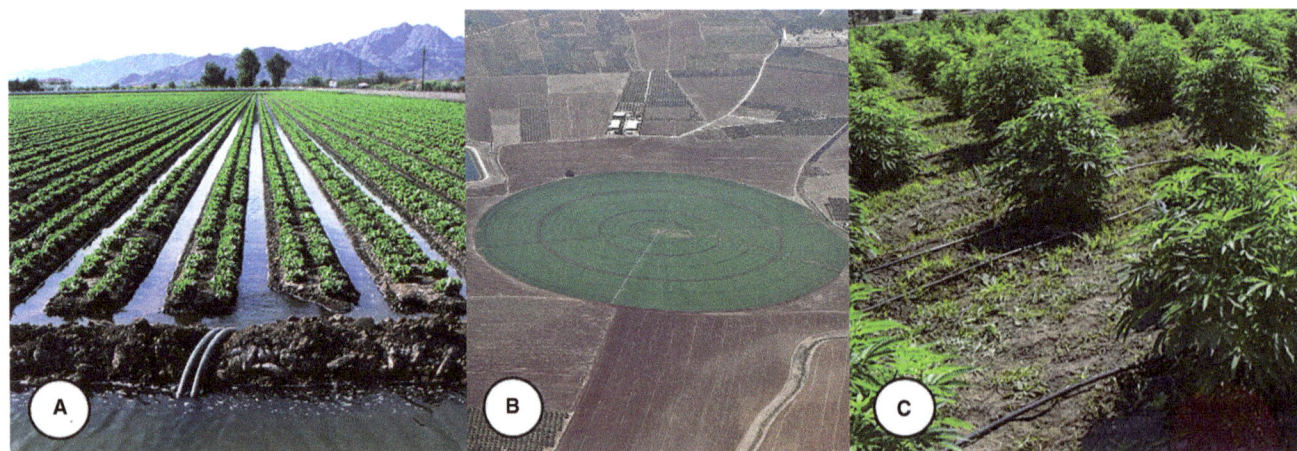

Fig. 17.10. Irrigation systems: **A.** Furrow irrigation with siphon tubes; **B.** Center pivot system; **C.** Drip irrigation. (Photo credits: A. USDA, Wikimedia Commons; B. Asaf Sagi, Wikimedia Commons; C. Southern Irrigation, https://southernirrigation.com)

"Surge" irrigation intermittently applies water to an irrigation furrow. This reduces water loss (less percolation loss), but requires greater engineering and closer monitoring.

Overhead irrigation utilizes high-pressure sprinklers. They can be mounted on permanently installed risers, or on mobile (wheeled) towers, which move on center pivots or linear machines (Fig. 17.10B). They reduce labor costs and improve irrigation performance, at the cost of greater capital outlay. They work better than furrows in sandy soil, but consume a lot of water.

Drip irrigation utilizes a network of perforated tubes (Fig. 17.10C). They can be placed on the surface or dug subsurface. Drip irrigation reduces water consumption and energy usage, and works well on hilly land, or small and odd-shaped fields. Drip irrigation involves capital outlay (40–50% greater than center pivot systems), labor for installation (and fixing leaks and clogs), and has a limited lifespan—especially if water contains particles that clog the emitters. Drip irrigation compared with overhead sprinkler irrigation lessens foliar diseases in *Cannabis* (Thiessen *et al.*, 2020).

In Fort Collins, Colorado, Campbell *et al.* (2019) compared crop yields under "dry" versus "wet" conditions. Their experiment was performed during a particularly dry year—only 53 mm rainfall fell during the growing season. In "dry" plots, seedlings were irrigated if wilting was observed, and they received 147 mm of supplemental irrigation. In "wet" plots, plants received 398 mm of supplemental irrigation throughout the season. Seed yield in "dry" plots averaged 404 kg/ha, and in "wet" plots averaged 1123 kg/ha. In dry and wet conditions, total plant dry biomass averaged 2482 and 6239 kg/ha, respectively. In dry and wet conditions, CBD% averaged 1.431% and 2.244%, respectively. THC% averaged 0.136% and 0.158%, respectively.

There are many approaches to determining when irrigation is needed, and how much. For small gardens, Cervantes (2015) offered three simple approaches. One is the "50% watering rule" for potted plants, based on volumetric water content (θ). Weigh plants in 3-gallon (1.1 L) pots after watering to field capacity. When the weight drops to half, it's time to irrigate. Alternatively, in small outdoor plots, irrigate when soil is dry 1.3 cm below the surface. The third approach is to use a soil moisture sensor.

Calculating needs for large fields is more complicated. Cosentino *et al.* (2012) determined irrigation needs for fiber-type hemp on the basis of maximum available soil water content in the top 0.6 m of soil, calculated with the formula: V = (FC - WP) × θ × D × 10^3, where V = water amount, FC = soil field capacity, WP = soil wilting point, θ = bulk density (1.1 g/cm^3), and D = rooting depth (0.6 m). See section 2.2 for these metrics.

Automated water management systems estimate irrigation needs by monitoring meteorological conditions (wind run, relative humidity, rainfall, temperature, solar radiation). Automated systems adjust irrigation pump rates on an hourly basis (e.g., Rain Bird ET Manager™, Hunter Pro-C Hydrawise™, Irritrol Climate Logic®).

Drought resistance varies by genotype. Barbieri (1943) noted that Turkish landraces were more drought resistant than Italian landraces. Amaducci *et al.* (2008) reported that monoecious 'Futura 75' yielded more fiber than dioecious 'Tiborszállási' under drought stress. Mihoc *et al.* (2012) found that dioecious 'Denise' outperformed monoecious 'Diana' under drought stress.

Campbell *et al.* (2019) measured yields in 13 European fiber-type cultivars, and monoecious cultivars were the most responsive to irrigation (e.g., 'Félina 32', 'Fédora 17', 'Bialobrzeskie'). Babaei and Ajdanian (2020) screened a number of Persian landraces for drought resistance. Among Chinese cultivars, 'Yúnmá 1' was drought-sensitive, and 'BM001' was drought-resistant (Gao *et al.*, 2011). Gao *et al.* (2018) identified 1292 genes in drought-stressed 'Yúnmá 1' plants that were down- or up-regulated. Many genes were in the abscisic acid (ABA) signaling pathway.

Traditional drug-type landraces in Morocco, *kif* or *beldiya*, were cultivated as rain-fed crops. Increased adoption of drought-intolerant hybrid strains ("Critical" and "Amnesia") has necessitated the widespread use of drip irrigation (Chouvy and Macfarlane, 2018). Affluent *kif* farmers rip up arable land to build water reservoirs out of packed earth, lined with polyethylene (R. Lee, pers. comm., 2016). The reservoirs supposedly capture rainwater, but in reality they collect groundwater. This depletes underground aquifers, leaving other farmers without water during the summer.

17.7.3 Too much water

Cannabis does not tolerate "wet feet." Neuer (1935) stated, "Although hemp with its enormous leaf forest evaporates huge amounts of water, it is extremely sensitive to waterlogging." Seedlings are particularly sensitive to waterlogging, especially in clay soils (Asquer *et al.*, 2019). In a survey of hemp growers in North Carolina, problems with excess water were reported by 32.5% of respondents (Thiessen *et al.*, 2020), but it was a bad hurricane season.

Soggy soil damages roots by reducing oxygen availability. A sudden excess—flooding—causes the same symptoms as drought: wilting. Wilting may be temporary, pending quick correction of the problem. Asphyxia of plant roots from overwatering may also cause chlorosis in emerging growth (Fig. 17.11). Excess water may exacerbate nutritional imbalances in *Cannabis*, such as excess manganese (Mihoc *et al.*, 2012). Excess soil moisture predisposes roots to fungal pathogens, such as *Pythium aphanidermatum*—which may occur in drip-irrigated field plants (Hu and Masson, 2021).

Field soil must have adequate drainage. Drainage rate can be estimated with a percolation test. Dig a hole in soil at least 30 cm (1 ft) deep and 30 cm wide, with straight sides. Fill the hole with water, and as soon as it has drained, refill it. Lay a straight edge across the top of the hole, use a yardstick to measure the water level, and set a timer. A drainage rate of <5 cm (<2 inches) per hour indicates poor drainage.

Poorly draining soil can be improved by adding organic matter (e.g., compost). Raised beds (berms or mounds) promote drainage. The height of the berm depends upon the density of the soil; 15 cm is often adequate. Deep tillage to break up claypans and hardpans improves drainage.

Laying tile drainage is the ultimate solution—a recommendation dating back to Cato the Elder around 170 BC (Cato, 1934). Wilsie *et al.* (1944) illustrated the benefits of tile drainage: a strip of hemp plants growing over a tile line was considerably taller than surrounding plants (Fig. 17.12). Overwatered plants can be revived with Oxygen Plus™, a hydrogen peroxide product. Jiang *et al.* (2021) rescued waterlogged Chinese hemp with uniconazole, a plant growth retardant.

In addition to soil moisture, consider atmospheric moisture: relative humidity. Extremely dry air causes leaf tips to turn brown (Frank, 1988). This problem mostly arises indoors, under hot lights.

17.7.4   Hot and cold

Temperature maxima and minima vary by genotype. In general, drug-type plants have higher temperature optima than fiber-type plants; see section 2.3, which also introduces growing degree days (GDD)—the amount of heat accumulation or "thermal time" that *Cannabis* needs to reach maturity.

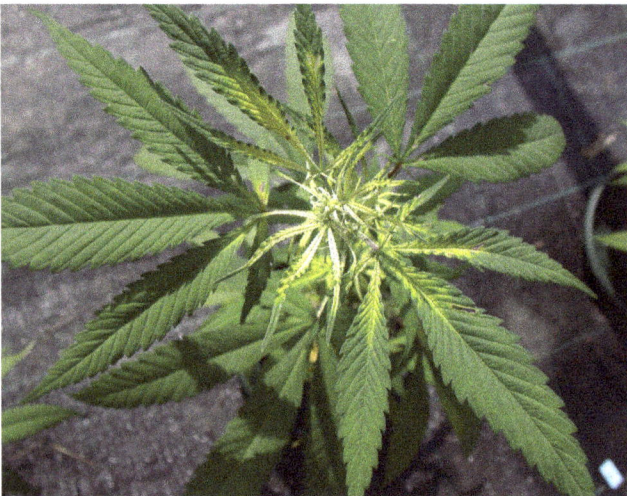

Fig. 17.11. Symptoms of root asphyxia from overwatering (Gianpaolo Grassi)

Fig. 17.12. Benefits of tile drainage, from Wilsie *et al.* (1944)

The minimum temperature at which seeds germinate became a focus of attention, with the goal of starting a crop as early in the springtime as possible. Haberlandt (1874) measured seed germination in 48 crops, including hemp. For a lower limit he tested seed germination at 3.8°C, and all hemp seeds germinated. At the upper end, he measured percent germination and mean germination time at 25, 30, 35, and 40°C. For hemp, germination was 100% at 25, 30, and 35°C. The shortest germination time was at 30°, which he took as optimal. The length of germinating radicals (roots) was greatest at 25°C, however.

Uloth (1875) tested seed germination in 25 species of plants at a colder temperature. He mixed seeds in a little soil and sandwiched them between blocks of ice. After 4 months, seeds of six species germinated: hemp, wheat, barley, rye, oats and pea. Uloth did not provide percentages.

Haberlandt (1879) investigated Uloth's claim by testing germination in insulated cases that kept temperatures constant at 0–1°C. After 45 days, eight of 25 species germinated, led by rye with nearly 100% germination; 14% of hemp seeds germinated. After 120 days, he transplanted the eight species to soil at 16°C. Four of the species grew, but no hemp seeds emerged from the soil. Haberlandt (1879) summarized measurements on minimum, maximum, and optimal temperatures for seed germination. For hemp he gave 1–2°C, 45°C, and 35°C, respectively. Hemp had one of the widest ranges of any plants measured. Haberlandt proposed that wide ranges were only seen in crop plants cultivated over a wide range of latitudes, like hemp.

Kirchner (1883) tested ten crop species that he assumed were cold tolerant. He used a different way to measure minima: seed were germinated at room temperature, then placed in cold chambers to see if their radicles (roots) continued to grow. He placed five germinated hemp seeds in a chamber maintained between 0.75°C and 1.0°C. After 7 days, two seedlings died, but the others grew an average of 0.05 mm per day. After 17 days the seedlings grew an average of 2.39 mm. Out of the ten species, only three (white mustard, rye, wheat) grew better than hemp at cold temperatures.

The thermal time (growing degree days, °Cd) required by seeds to germinate and emerge from the soil has been calculated. Tamm (1933) measured 96 °Cd, with a base temperature of 0°C. Van der Werf *et al.* (1995c) measured 88.3 °Cd, with a base temperature of 0°C. This was the mean of three cultivars: 'Kompolti Hibrid TC', 84.4 °Cd; 'Fédrina 74', 89.1 °Cd; and 'Kompolti Sárgaszárú', 91.3 °Cd. Lisson *et al.* (2000) summarized the results of several studies with 'Kompolti'. In two glasshouse studies, seedling emergence required 81.2 °Cd and 91.3 °Cd above a base temperature of 1°C.

Honek *et al.* (2014) measured the mean time it took for 50% of seeds to germinate at 5, 9, 13, 17, 21, 25, 29, 33, and 37°C. Using a complex formula, they calculated three temperatures for germination: minimum (–4.16°C), optimal (29.87°C), and maximum (37.03°C). Honek explained that "negative" minimums do not mean a seed can germinate below zero, but the seed is metabolically active at temperatures close to 0°C. Their study included 125 plant species, and only seven had minimum temperatures less than hemp. They were wild herbaceous species, with high-altitude *Rumex alpinus* L. setting the record of –8.30°C.

Seed germination is an exothermic (heat-releasing) event, as seeds metabolize their stored oil. Hemp seeds produce more heat upon germination than most other crops plants. Darsie *et al.* (1914) measured the average heat yield in 10 g of seeds; hemp generated 1.84°C; barley, wheat, oats, corn, and clover yielded 0.49–0.88°C.

Seedlings can survive down to –5 to –7°C for short periods of time (Urban, 1979; Hanson, 1990). Surviving seedlings had chlorotic leaves. Dzhaparidze (1967) claimed that female seedlings tolerate spring frosts better than male seedlings.

Cannabis has less tolerance for late-season frosts than for early-season frosts. Even wild-type plants cannot tolerate hard frosts (Bócsa and Karus, 1997). For mature cultivated plants in autumn, temperatures below –3°C damage flowering tops (Hanson, 1990). Grigory'ev (2000) linked cold resistance with early flowering; he reasoned that early harvesting reduced losses from frost and cold. He screened Russian accessions for early maturation, and for cold tolerance during germination.

Theimer and Mölleken (1995) proposed that varieties from high latitudes, hence cold tolerant, produced seeds with an elevated ratio of unsaturated/saturated fatty acids. Deferne and Pate (1996) reasoned that unsaturated fatty acids remain more mobile at cooler temperatures. A Finnish variety, 'FINOLA', is frost tolerant to –5°C in all stages of growth; Callaway (2002) attributed this to early maturity, and a high percentage (80%) of unsaturated fatty acids in the seeds.

Shelenga *et al.* (2012) measured percentages of nine fatty acids in the seeds of 20 landraces collected across Russia. From their data we plotted latitude against the sum of unsaturated fatty acids (SDA+GLA+ALA), and found no correlation ($r^2 = 0.07$, $p = 0.27$). Perhaps the percentage of unsaturated fatty acids is *not* a predictor of cold tolerance.

In drug-type plants exposed to late-season cold, frosted flowers blacken, and the smoked flowers yield a harsh taste (Frank and Rosenthal, 1978). Landraces from Afghanistan (*a.k.a.* "Indica") have more late-season cold tolerance than South Asian landraces (*a.k.a.* "Sativa"). Clarke (1981) first noted early maturation in Afghani landraces, and Clarke (1987) first noted cold tolerance. Drug-type strains advertised as cold tolerant come from Central Asia (e.g., "Afghani No. 1"; Schoenmakers, 1986), and they mature early (e.g., "Early Bird", Kees, 1988).

Late-season cold triggers **anthocyanin** accumulation in Afghani flowers and leaves—the sought-after "purple weed." However, this also arises in some populations of European fiber-type hemp, first noted by Dewey (1913). Figure 17.13 illustrates fiber-type 'Białobrzeskie' with anthocyanin accumulation after a frost—and quite a variable response in two individuals growing side-by-side.

17.7.5 Other environmental stresses

Insufficient light causes problems; see section 2.1. Plants become spindly and chlorotic. Too much light too quickly causes **sun scald**. This may happen when indoor seedlings are transplanted outside. Undersides of fast-growing leaves are vulnerable if wind turns them over to exposure in brilliant sunlight. Scalded leaf tissues become dry and brown and resemble "hopperburn" symptoms caused by potato leafhoppers.

Hailstone damage tends to be less of a problem with *Cannabis* than with large-leaved plants, but it does happen. Both leaves and stems are damaged (Fig. 17.14). For some reason, in world *Cannabis* literature, *grandine* (hailstone) damage is primarily reported in Italy (Casaletti, 1787; Berti Pichat, 1866; Levi, 1882; Peglion, 1897; Fabbri, 1949; Goidànich, 1959;

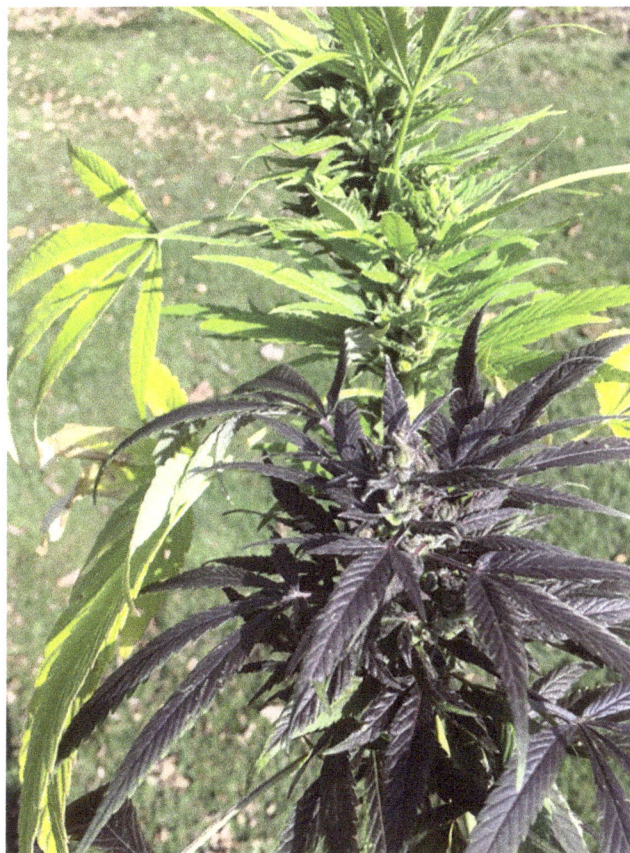

Fig. 17.13. Anthocyanin accumulation in cold-stressed plants

Fig. 17.14. Hailstone damage in Italy (Gianpaolo Grassi)

Capasso, 2001). Young plants, 40–60 cm tall, are the most sensitive. Their stems snap, and the plants either die or develop side shoots (Bócsa, 2004). Surviving plants are susceptible to fungal infections.

Heavy rainfall can damage plants. Heavily flowered branches snap under the load. This happens more often in Afghani landraces, because of their brittle (woody), menorah-shaped branching habitus (Fig. 17.15). This does not occur in South Asian plants, which evolved tolerance to monsoonal rainfall. Their branches are more flexible, and come off the stalk at more acute angles (McPartland and Small, 2020). Repair split branches and broken stems with splints, tape, and aloe gel. Prevent rainstorm damage by hanging a horizontal trellis of netting above plants, so they can grow up and through the grid for support.

17.8   Genetic diseases

Cannabis is normally **dioecious**, with separate female (pistillate) and male (staminate) individuals. Separate genders are normal in animals, but not in plants. Only about 5% of angiosperm species are dioecious. Individual plants that express both female and male flowers are either monoecious or hermaphroditic.

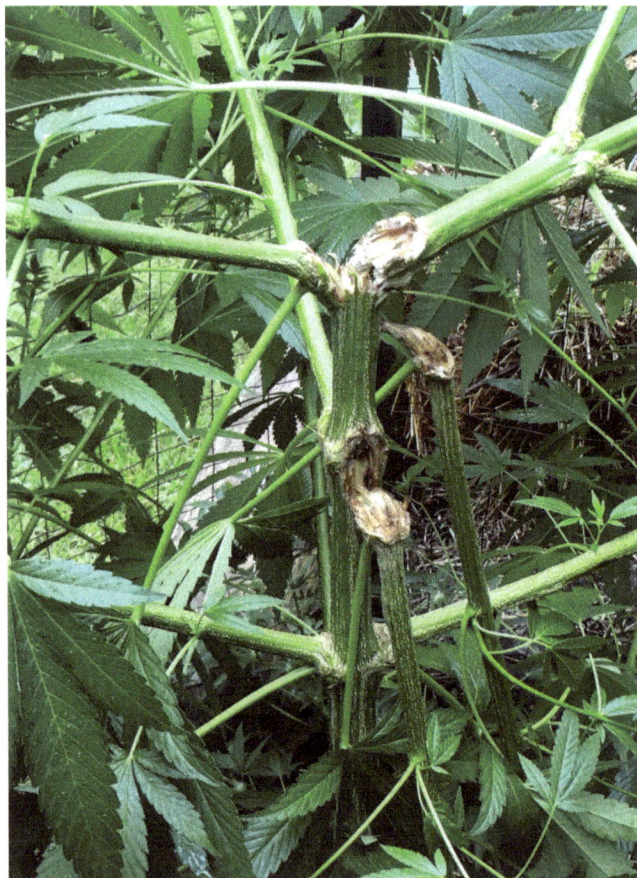

Fig. 17.15. Rainstorm damage

Monoecious (i.e., intersexual) individuals produce separate male and female flowers on the same plant. Characteristically, male flowers are located in whorls at the base of primary and secondary branches, while female flowers are found in branch terminals. Monoecious *Cannabis* arises spontaneously in fiber-type plants (Mauz, 1836; Holuby, 1878; Borthwick and Scully, 1954).

The 20th century saw a concerted effort to breed monoecious hemp. This is because staminate plants flower *weeks* before staminate plants in dioecious hemp. The desire to mechanize agriculture, and harvest all plants at uniform maturity in one pass through a field, led to breeding for "simultaneously maturing hemp" (Grishko, 1937; Hoffmann, 1941). Monoecious plants are good seed crops, because every plant bears seeds.

Monoecious plants are unserviceable as *gañjā* (sinsemilla) crops, yet drug-type landraces from India express a significant incidence of monoecy (Prain, 1904). Tendency towards monoecy is a trait that separates landraces of South Asian provenance from Central Asian landraces (*C. afghanica*), where monoecious plants are rarely seen (McPartland and Small, 2020). Landraces from Thailand are notorious for turning monoecious or hermaphroditic—expose a female plant to drought stress, and male flowers will appear (Clarke, 1981).

Hermaphroditic (i.e., androgynous) plants are relatively rare compared with monoecious plants. Hermaphroditic individuals express both stamens and pistils in an individual flower. Autenrieth (1821) induced hermaphroditic flowers by pruning a male plant, and some of the male flowers produced pistils (Fig. 17.16A). This was repeated by Gasparrini (1862), who provided a better image (Fig. 17.16B).

Prain (1904) illustrated hermaphroditic *mória* Indian hemp (Fig. 17.16C)—bane to the existence of *gañjā* cultivators, who were trying to grow sinsemilla. They painstakingly removed all male plants, but female *mória* plants produced male flowers, whose pollen ruined the sinsemilla. Drake (1970a) illustrated

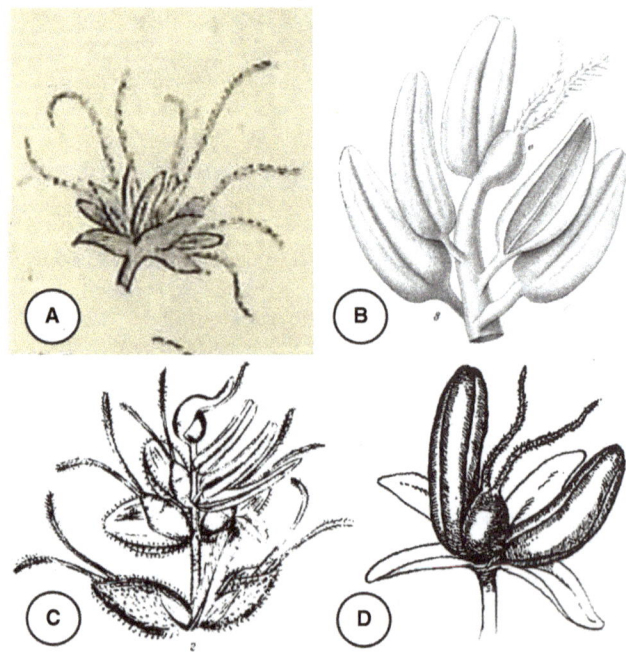

Fig. 17.16. Hermies through the years. See text for letters.

four types of hermaphroditic abnormalities, including a "female showing maleness by developing stamens" (Fig. 17.16D). Punja and Holmes (2020) provided color close-up photographs of hermaphroditic inflorescences.

17.8.1   Inbreeding and mutations

Humans inbreed *Cannabis* to fix desirable traits. Inbreeding may also lead to the expression of recessive or deleterious traits. Lyster Dewey described several kinds of mutations. Dewey (1913) found a purple-leaved mutation arising in 'Kymington', a Chinese landrace he inbred for 9 years. Dewey crossed 'Kymington' with plants from Ferrara (northern Italy) to create 'Ferramington.' Among the progeny he found mutant plants with their leaflets webbed together, forming a palmately lobed leaf that looks a bit like hops. Others referred to the 'Kymington' × Ferrara mutants as a "simple-leaved mutation" (Malloch, 1922) or the "pinnatifid-leaf variety" (Schaffner, 1923a).

Italian breeders described *pinnatifidofilla* plants from Ferrara (Savelli, 1929b; Allavena, 1961). In addition to *pinnatifidofilla*, Crescini (1956) described plants with fasciation, stalk ramification, and half-albino leaves. Jannaccone (1947) assigned the name *C. sativa* var. *albicata* to albino plants he discovered amongst 'Paesana' in Napoli. Lower leaves were completely white, becoming pale green towards the plant apex.

Palmately lobed leaf mutations also arise in drug plants, named "Ducksfoot" and "Hawaiian webbed indica" varieties (de Meijer *et al.*, 2003). Perhaps the most aberrant leaf mutant goes by the name "Australian Bastard Cannabis" (Larsen, 1999). Leaflets are non-serrated and small (<5 cm length), often in a whorled arrangement.

Fasciation is an inbreeding mutation, where normally round stems developed a flat or ribbon shape, with thickening and twisting of stems and leaves (Fig. 17.17). It was first

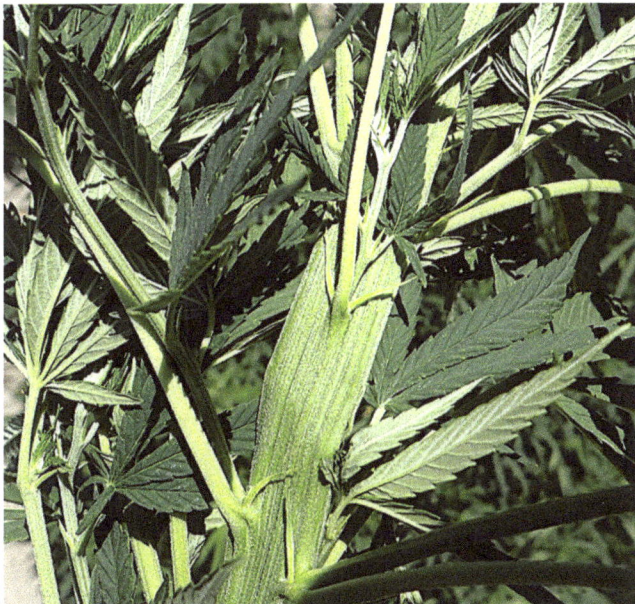

Fig. 17.17. Stem fasciation (W. Cranshaw)

described and illustrated in German hemp by Muth (1906). Crescini (1934c) determined that the mutation was inheritable, a recessive trait, and affected 4–10% of the 'Carmagnola' population. Lyster Dewey's herbarium contains 'Kymington' specimens with fasciation (McPartland, pers. obs., herb BPI). Flower fasciation and other strange teratologies were illustrated by Borodina and Migal (1987).

Bócsa (1958) identified other symptoms of inbreeding: small and sterile seed, reduced plant height, and increased susceptibility to fungal diseases. Lai (1985) described the deleterious effects of inbreeding on yield of fiber and seed. Unwanted characteristics from inbreeding included short stature (only 68% the height of normal hemp), and shortened lifespan (vegetative growth 9 weeks shorter than normal plants).

Some genetic mutations turn out to be beneficial. Hoffmann (194crossed an early-ripening Finnish landrace with a late-ripening Italian landrace, and selected a chlorophyll-deficient mutant with yellow stalks. Its lighter-colored stalks led to a more even coloration when fibers were processed. Bócsa (1972) crossed Hoffmann's mutant with 'Kompolti' and released the yellow-stalked cultivar 'Kompolti Sárgaszárú'. Its fiber had excellent spinning properties, was suitable for mechanized harvesting, and could provide fiber for paper pulp—where chlorophyll is undesirable (Kiss, 1973).

'Kompolti Sárgaszárú' was acclimated to the Glukhov (Hlukhiv) region by Ukrainian breeders, who renamed it 'Глуховская-10' (Glukhovskaya-10) (Virovets and Sitnik, 1975). The monogenic recessive mutation had pleiotropic effects—and not all beneficial, such as reduced yield. But crossing 'Глуховская-10' with Ukrainian cultivars (e.g., 'ЮС-9' and 'ЮСО-1') produced high-yielding yellow-stemmed hybrids, including monoecious forms (Sitnik, 1981). Hoffmann's yellow mutant lives on—the new Italian variety 'Carmaleonte' is a descendant of 'Kompolti Sárgaszárú'. Its fiber is easily separated by dew retting, thereby avoiding water retting, which can be environmental harmful (Grassi and McPartland, 2017).

17.8.2   Chimeras and somatic mutations

A **chimera** is a single organism composed of cells with more than one genotype. One well-known type of plant chimera gives rise to **variegation**—leaves exhibiting a mosaic of green and albino cells. The albino (actually yellow) cells do not produce chlorophyll, due to a somatic mutation.

A **somatic mutation** is a genetic change that occurs in a somatic cell during mitosis. Somatic mutations are caused by internal factors (e.g., DNA replication error, DNA repair error, transposable elements), and environmental factors (e.g., ultraviolet radiation, high temperature, drought). Studies with *Arabidopsis* estimate that one somatic mutation occurs approximately every six cell divisions (Frank and Chitwood, 2016). These mutations may be deleterious and result in cell death. Or they occur in non-coding regions and have no obvious effect on plant phenotype (74.75% of the *Cannabis* genome consists of non-coding elements; Gao *et al.*, 2020). However, rare mutation events create striking shifts in the phenotype, such as chimeric variegation.

Variegation arises in various patterns due to the location of where the albino cell first arose in the **shoot apical meristem** (SAM). The *Cannabis* SAM is a small, domed structure, 0.5 mm at its base, and the site of virus eradication techniques (see Fig. 14.8). SAM cells are stratified, consisting of different cell layers—the **tunica–corpus** arrangement (Schmidt, 1924). The tunica is the outer layer of SAM; it grows in discrete layers by anticlinal cell division (at right angles to the surface). The corpus lies beneath the tunica. Corpus cells divide in various directions and so become irregular in arrangement. The thickness of the tunica varies amongst plant species, between one and four layers. Majd *et al.* (2019) determined that *Cannabis* has a two-layered tunica, labeled L1 and L2 (layer L3 is the corpus) (Fig. 17.18).

Chimeric plants can be classified as periclinal, mericlinal, or sectorial, based on the cell layer in which the initial albino cell arose. In a **periclinal chimera**, a single cell in the tunica undergoes a mutation, and by anticlinal cell divisions the entire layer becomes differentiated from the other layers. A periclinal chimera arising in L2 is illustrated in Fig. 17.18. In a **mericlinal chimera**, a single cell in the tunica undergoes a mutation, but in this case only part of the layer becomes differentiated (L1 in Fig. 17.18). In a **sectorial chimera**, the altered genotype involves all layers of the meristem (L1, L2, and L3).

Periclinal chimeras are stable, and the most common cause of variegation. They tend to be uniform in color and often arise in symmetrical patterns around the periphery of a leaflet. Mericlinal chimeras are unstable and often a transition stage, where the meristem will eventually form a stable periclinal chimera or lose the mutation. Sectorial chimeras are unstable, with atypical patterns consisting of different ratios of green and albino areas that vary from leaflet to leaflet, and entire albino leaves are common (Frank and Chitwood, 2016).

Malloch (1922) first described and illustrated a *Cannabis* sectorial chimera (Fig. 17.19A). It arose on a plant in which several leaves were half green and half albino/yellow—all on the same side of the stem, and the stem rib from which they branched was also half green and half albino/yellow. A better example of a sectorial chimera appears in Fig. 17.19B. For symptoms of a specific type of sectorial chimera, a marbled or non-patterned sectorial chimera, see Fig. 14.6. Marbled sectorial chimeras can be confused with viral infections. Plants with the latter are more prevalent in a field, however, and are not restricted to a single plant or even a single leaf.

Half-albino plants with lost chlorophyll grow poorly. These undesirable plants can be rogued, "especially when a crop is produced for seed" (Wang, 2021). Nevertheless, Casano (2019) received a patent for plant breeders' rights for the new cultivar 'Davina', a chimeric plant that arose from 'Pilar', with "a distinctive mottled yellow and green leaf phenotype." Casano mentioned that 'Davina' cuttings were more prone to fungal disease during *in vivo* rooting stage than 'Pilar' cuttings.

Chimeras go beyond variegated leaves. If a somatic mutation occurs in the L2 level of SAM cells destined to become floral parts of the plant, a flower chimera could arise. Savelli (1929b) proposed a case of *chimera bisessuale* for separate male and female flowers on the same plant. Grassi (pers. comm., 2019) attributed monoecy in one branch of a plant to a flower chimera. Explaining this phenomenon's mechanism is beyond our pay grade. Suffice to say *Cannabis* is a monopodially branching plant (Spitzer-Rimon *et al.*, 2019), which enhances the transmission of a somatic mutation to subsequently produced plant tissues (Salomonson, 1996).

Whitham and Slobodchikoff (1981) proposed that genetic diversity arising through somatic mutations may play a role in disease or herbivore resistance. Cuttings from individual branch bearing these adaptive traits could be clonally reproduced.

The root apical meristem (RAM) is another site predisposed to somatic mutations and chimeras. Polyploid chimeras— where some cells were diploid, and others were tetraploid—have been observed in the RAM of *Cannabis* (Litardière, 1925; Langlet, 1927). Galán-Ávila *et al.* (2020) found a mix of diploid and tetraploid cells in hypocotyls and cotyledons of 7-day-old seedlings, but not in leaves.

The accumulation of somatic mutations in vegetatively maintained plants, in the absence of sexual recombination, is known as Muller's rachet (McKey *et al.*, 2010). This phenomenon arises in crop species maintained for extended periods of time through vegetative clonal propagation (e.g., apple, citrus, potato, cassava, banana, grape, coffee, hops). Theoretically,

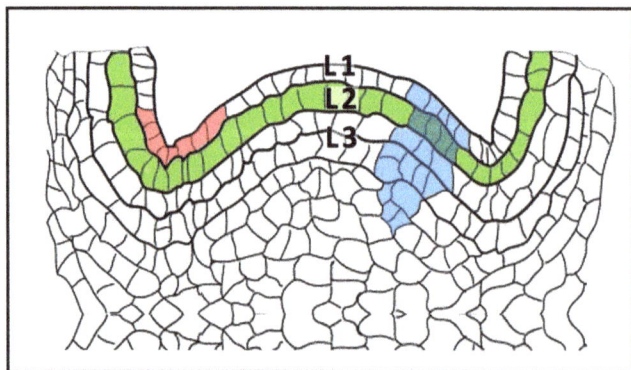

Fig. 17.18. Chimeras: periclinal (green), mericlinal (red), sectorial (blue) (adapted from Geneve *et al.*, 2023)

Fig. 17.19. Cannabis chimeras: **A.** from Malloch (1922); **B.** courtesy of Neta Rimmerman

clonal cuttings produce plants that are genetically identical to mother plants. However, the accumulation of somatic mutations in a purely clonal line ultimately leads to a decline in the quality of clones, associated with reduced cannabinoid production and plant vigor.

Adamek et al. (2022) sequenced the genomes of cuttings taken from "Honey banana", a high-THC hybrid strain. The mother plant was relatively young (1.5 years), a clonally propagated plant. Three cuttings were taken from the bottom, middle, or apical part of the plant, at 59 cm, 151 cm, and 226 cm, respectively. Compared with a reference genome (Grassa et al., 2021), a total of 3.8 million nucleotide variants were detected amongst the three clones. This equated to a variant arising every ~425 bases in reputably "clonal" lines, which was only sixfold less than the average found in a study of 40 diverse Cannabis genomes, with a variant arising every ~73 bases (McKernan et al., 2020).

A subset of "high-impact mutations" (HIMs, those predicted to disrupt protein coding) were smaller in number, and heterogeneously distributed: Shared HIMs, inherited from the mother plant, totaled 845. Additionally the apical cutting harbored 1234 unique HIMs, the bottom cutting had 247 unique HIMs, and the middle cutting had 63 unique HIMs. The authors attributed this disparity to increased cell division arising from frequent pruning of mother plants: bottom limbs were pruned to encourage more apical growth, and apical growth was pruned to obtain cuttings. Adamek and colleagues identified nucleotide variants in genes encoding enzymes for terpenoid and cannabinoid synthesis. Although none were HIMs, their impact was difficult to predict.

17.8.3 Somaclonal variation

Somaclonal variation is the genetic and epigenetic variability arising in plants regenerated from in vitro micropropagation. Lata et al. (2010) first discussed somatoclonal variation, and they used 15 inter-simple sequence repeat (ISSR) markers to evaluate the genetic stability of micropropagated Cannabis plants over 30 passages in tissue culture.

Cannabis plants regenerated from callus tissue possess more somaclonal variation than plants regenerated from nodal explants (Chandra et al., 2017b). Putting positive spin on this, ElSohly et al. (2017) suggested that somaclonal variation arising in callus-regenerated plants could be used for new product development and crop improvement. In reality, the majority of somaclonal variations are deleterious (Roux et al., 2021).

Hesami et al. (2023) evaluated the genetic stability of explants (stem segments) from the bottom, middle, or apical part of a plant. They were taken through four passages in tissue culture. Plantlets regenerated from explants obtained near the basal part of plants were taller, developed more nodes (branching points), leaves with more leaflets, and produced a larger canopy than plantlets from middle- or apical-sourced explants. The authors proposed that a reduction in the number of nodes of Cannabis plantlets after four subcultures served as a morphological sign of somaclonal variation.

The same group described Cannabis **hyperhydricity** (a.k.a. vitrification), a physiological disorder associated with in vitro micropropagation, with symptoms of vitreous (glassy, fragile) shoots and leaves, short internodes, and plantlet death. They found higher rates of hyperhydricity when using Murashige and Skoog medium (the industry standard) than with Driver and Kuniyuki Walnut medium (Page et al., 2021).

(Prepared by J. McPartland)

References

Abot A, Bonnafous C, Touchard F, et al. (4 other authors). 2013. Effects of cultural conditions on the hemp (Cannabis sativa) phloem fibres: Biological development and mechanical properties. Journal of Composite Materials 47: 1067–1077.

Acosta-Motos JR, Ortuño MF, Bernal-Vicente A, et al. (3 other authors). 2017. Plant responses to salt stress: adaptive mechanisms. Agronomy 7: 18.

Adamek K, Jones AMP, Torkamaneh D. 2022. Accumulation of somatic mutations leads to genetic mosaicism in cannabis. Plant Genome 15: e20169.

Afzal I, Shinwari ZK, Iqrar I. 2015. Selective isolation and characterization of agriculturally beneficial endophytic bacteria from wild hemp using canola. Pakistan Journal of Botany 47: 1999–2008.

Agrios GN. 2005. Plant pathology, 5th edn. Elseivier Academic Press, Amsterdam.

Ahmad A, Hadi F, Ali N. 2015. Effective phytoextraction of cadmium (Cd) with increasing concentration of total phenolics and free proline in Cannabis sativa (L.) plant under various treatments of fertilizers, plant growth regulators and sodium salt. International Journal of Phytoremediation 17(1-6): 56–65.

Alaru M, Noormets M, Raave H, et al. (3 other authors). 2009. Farming systems and environmental impacts. Agronomy Research 7: 3–10.

Alaru M, Lukk L, Olt J, et al. (4 other authors). 2011. Lignin content and briquette quality of different fibre hemp plant types and energy sunflower. Field Crops Research 124: 332–338.

Alexander T. 1988. Rating the strains. Sinsemilla Tips 8(2): 31–33.

Alipour SH, Darbandi MC, Ejirloo SJ, Bellehgh AB. 2014. Effect of sodium chloride induced salinity on germination and vigor indices of shāh dānah seed (Cannabis sativa L.). Third National Congress on Organic and Conventional Agriculture, 20–21 August 2014, Ardabil, Iran.

Allavena D. 1961. Fibranova, nuova varieta di canapa ad alto contenuto di fibra. Sementi Elette 5: 34–44.

Amaducci S., Zatta A., Pelatti F., Venturi G. 2008. Influence of agronomic factors on yield and quality of hemp (Cannabis sativa L.) fibre and implication for an innovative production system. Field Crops Research 107(2): 161–169.

Anderson SL, Pearson B, Kjelgren R, Brym Z. 2021. Response of essential oil hemp (*Cannabis sativa* L.) growth, biomass, and cannabinoid profiles to varying fertigation rates. *PLoS ONE* 16: e0252985.

Anonymous. 2000. *Phytoremediation: using plants to clean soil.* McGraw Hill Botany. Available at: www.mhhe.com/biosci/pae/botany/botany_map/articles/article_10. html

Armstrong MJ. 1781. *History and antiquities of the County of Norfolk. Volume IX: containing the hundreds of Smithdon.* J. Crouse, Norwich.

Arru L, Rognoni S, Baroncini M, Bonatti PM, Perata P. 2004. Copper localization in *Cannabis sativa* L. grown in a copper-rich solution. *Euphytica* 140: 33–38.

Asquer C, Melis E, Scano EA, Carboni G. 2019. Opportunities for green energy through emerging crops: biogas valorization of *Cannabis sativa* L. residues. *Climate* 7: e142.

Autenrieth HF. 1821. *Disquisitio quaestionis academicae de discrimine sexuali jam in seminibus plantarum dioïcarum apparente.* H. Laupp, Tübingen.

Babaei M, Ajdanian L. 2020. Screening of different Iranian ecotypes of *Cannabis* under water deficit stress. *Scientia Horticulturae* 260: e108904.

Backe H, *ed.* 1936. *Der hanfbau, seine wirtschaftliche bedeutung.* Reichsministerium für Ernährung und Landwirtschaft, Berlin.

Backer R, Schwinghamer T, Rosenbaum P, *et al.* (8 other authors). 2019. Closing the yield gap for *Cannabis*: a meta-analysis of factors determining *Cannabis* yield. *Frontiers in Plant Science* 10: 495.

Badora A, Filipek T. 2004. The influence of applying purified municipal sewage on nickel content in plants. *Polish Journal of Environmental Studies* 13: 435–437.

Bahador M, Tadayon MR, Rafie-Alhoseini M, Salehi MH. 2017. Changes of canopy temperature and some physiological traits of hemp (*Cannabis sativa*) under deficit water stress and zeolite rates. *Environmental Stresses in Crop Sciences* 10: 269–279.

Barbieri R. 1943. Aspetti e soluzione del problema del semi di canapa in Campania. *Annali della Facoltà di agraria di Portici della R. Università di Napoli* 15: 85–127.

Baur V, Lampe F von. 1943. *Die Lustige Hanffibel.* Reichsnährstand Verlag, Berlin.

Bazzaz FA, Dusek D, Seigler DS, Haney AW. 1975. Photosynthesis and cannabinoid content of temperate and tropical populations of *Cannabis sativa. Biochemical Systematics and Ecology* 3: 15–18.

BCMAF (British Columbia Ministry of Agriculture and Food). 1999. *Industrial Hemp (Cannabis sativa L.). Specialty Crops Factsheet.* Kamloops, BC, Canada.

Benbrook CM. 2016. Trends in glyphosate herbicide use in the United States and globally. *Environmental Sciences Europe* 38: e3.

Bennett WF. 1993. *Nutrient deficiencies and toxicities in crop plants.* American Phytopathological Society, St Paul, Minnesota.

Berger J. 1969. *The world's major fibre crops: their cultivation and manuring.* Centre d'Etude de l'Azote, Zurich.

Berstein N, Gorelick J, Zerahia R, Koch S. 2019a. Impact of N, P, K, and humic acid supplementation on the chemical profile of medical cannabis (*Cannabis sativa* L). *Frontiers in Plant Science* 10: 736.

Berti Pichat CB. 1866. "La pianta della canapa," pp. 396–499 in *Istituzioni scientifiche e tecniche, ossia Corso teorico e practico de Agricultura, Libri XIII, Volume Quinto.* Presso l'Union Tipografico-Editrice, Torino.

Bilodeau SE, Wu BS, Rufyikiri AS, MacPherson S, Lefsrud M. 2019. An update on plant photobiology and implications for Cannabis production. *Frontiers in Plant Science* 10: e296.

Bjerken S. 1958. Correlation and cooperation between agencies in weed control in Minnesota. *Weeds* 6: 465–467.

Black CA. 1993. *Soil fertility evaluation and control.* CRC Press, Boca Raton, Florida.

Blake W. 1790. *The marriage of heaven and hell.* W. Blake, London (copy at Houghton Library, Harvard University).

Blith W. 1652. *The English improver improved.* John Wright, London.

Bócsa I. 1958. A kender beltenyésztésémek újabb jelenségei. *Növénytermelés* 7: 1–10.

Bócsa I. 1972. Különleges célok elérése a kendernemesítésben. *Rostnövények* 1972: 29–34

Bócsa I. 2004. *A kender és termesztése.* Agroinform Kiadó, Budapest.

Bócsa I, Karus M. 1997. *Der Hanfanbau: Botanik, Sorten, Anbau und Ernte.* C.F. Müller, Heidelberg. English translation: *The cultivation of hemp: botany, varieties, cultivation, and harvesting.* Hemptech, Sebastopol, California.

Boretius A, *ed.* 1883. *Monumenta Germaniae historica, Legum sectio II, Capitularia regum Francorum I.* Hahnsche Buchhandlung, Hannover.

Borodina EI, Migal ND. 1987. Flower teratology in intersexual hemp plants. *Soviet Journal of Developmental Biology* 17(4): 262–269.

Borthwick HA, Scully NJ. 1954. Photoperiodic responses of hemp. *Botanical Gazette* 116: 14–29.

Boussingault JB. 1851. *Economie rurale considérée dans ses rapports avec la chimie, la physique et la meteorology, deuxième edition, Tome Premier.* Béchet Jeune, Paris.

Boyce SS. 1900. *Hemp (Cannabis sativa). A practical treatise on the culture of hemp for seed and fibre.* Orange Judd, New York.

Bray F. 1984. "Agriculture," Volume 6, Part 2 of Needham J, *ed. Science and Civilisation in China.* Cambridge University Press, Cambridge, UK.

Bredemann G. 1952. Weitere Beobachtungen bei Züchtung des Hanfes auf Fasergehalt. *Der Züchter* 22: 257–269.

Bruce D, Connelly G, Ellison S. 2022. Different fertility approaches in organic hemp (*Cannabis sativa* L.) production alter floral biomass yield but not CBD:THC ratio. *Sustainability* 14: e6222.

Bryson GM, Mills HA, *eds.* 2014. *Plant analysis handbook IV.* Micro-Macro Publishing, Inc., Athens, Georgia.

Buchanan-Hamilton F. 1833. *A geographical, statistical and historical description of the district, or zila, of Dinajpur.* Baptist Mission Press, Calcutta.

Bush Doctor. 1993b. A heavy essay on lights. *High Times* no. 220, pp. 36–38.

Callaway JC. 2002. Hemp as food at high latitudes. *Journal of Industrial Hemp* 7: 105–117.

Callaway JC. 2004. Hemp seed production in Finland. *Journal of Industrial Hemp* 9: 97–103.

Campbell S, Paquin D, Awaya JD, Li QX. 2002. Remediation of benzo[a]pyrene and chrysene-contaminated soil with industrial hemp (*Cannabis sativa*). *International Journal of Phytoremediation* 4: 157–158.

Campbell BJ, Berrada AF, Hudalla C, Amaducci S, McKay JK. 2019. Genotype x environment interactions of industrial hemp cultivars highlight diverse responses to environmental factors. *Agrosystems, Geosciences & Environment* 2: e180057.

Cao PY, Lu CQ, Yu Z. 2018. Historical nitrogen fertilizer use in agricultural ecosystems of the contiguous United States during 1850–2015: application rate, timing, and fertilizer types. *Earth System Science Data* 10: 969–984.

Capasso S. 2001. *Canapicoltura: passato, presente e futuro.* Istituto di Studi Atellani, Napoli.

Caplan D, Dixon M, Zheng YB. 2017a. Optimal rate of organic fertilizer during the vegetation-stage for *Cannabis* grown in two coir-based substrates. *HortScience* 52: 1307–1312.

Caplan D, Dixon M, Zheng YB. 2017b. Optimal rate of organic fertilizer during the flowering stage for *Cannabis* grown in two coir-based substrates. *HortScience* 52: 1796–1803.

Carah JK, Howard JK, Thompson SE, *et al.* (12 other authors). 2015. High time for conservation: adding the environment to the debate on marijuana liberalization. *BioScience* 65: 822–829.

Casaletti A. 1787. *Istruzione per ben esercitare l'impiego di Provisioniere dei lini e canape filabili.* Massimi, Roma.

Casano S. 2019. Development of ornamental *Cannabis sativa* L. cultivars: phytochemical, morphological, genetic characterization and propagation aspects. *Acta Horticulturae* 1263: 283–290.

Cato the Elder (Ash HB, Hooper WD, *trans*). 1934. *On Agriculture.* Harvard University Press, Cambridge, Massachusetts.

Cederlund H. 2017. Effects of spray drift of glyphosate on nontarget terrestrial plants—a critical review. *Environmental Toxicology and Chemistry* 36: 2879–2886.

Cervantes J. 2015. *The Cannabis encyclopedia.* Van Patten Publishing, Vancouver, Washington.

Chandra S, Lata H, Khan IA, ElSohly MA. 2008. Photosynthetic response of *Cannabis sativa* L. to variations in photosynthetic photon flux densities, temperature and CO2 conditions. *Physiology and Molecular Biology of Plants* 14: 299–306.

Chandra S, Lata H, ElSohly MA, Waler LA, Potter D. 2017b. *Cannabis* cultivation: methodological issues for obtaining medical-grade product. *Epilepsy & Behavior* 70: 302–312.

Cherrett N, Barrett J, Clemett A, Chadwick M, Chadwick MJ. 2005. *Ecological footprint and water analysis of cotton, hemp, and polyester.* Stockholm Environmental Institute, Stockholm.

Chiriță, N. 2008. Selectivity and efficiency of some herbicides in controlling weeds from monoecious hemp crops. *Cercetări Agronomice în Moldova* 41(3): 59–64.

Chouvy PA, Macfarlane J. 2018. Agricultural innovations in Morocco's cannabis industry. *International Journal of Drug Policy* 58: 85–91.

Citterio S, Santagostino A, Fumagalli P, *et al.* (3 other authors). 2003. Heavy metal tolerance and accumulation of Cd, Cr and Ni by *Cannabis sativa* L. *Plant and Soil* 256: 243–252.

Citterio S, Prato N, Fumagalli P, *et al.* (5 other authors). 2005. The arbuscular mycorrhizal fungus *Glomus mosseae* induces growth and metal accumulation changes in *Cannabis sativa* L. *Chemosphere* 59: 21–29.

Ciurli A, Alpi A, Perata P. 2002. Impiego della canapa nella fitodepurazione da metalli pesanti. *Agroindustria* 1(1): 64–60.

Clarke RC. 1981. *Marijuana botany.* And/Or Press, Berkeley, California.

Clarke RC. 1987. *Cannabis evolution.* MS thesis, Indiana University, Bloomington, Indiana.

Cockson P, Landis H, Smith T, Hicks K, Whipker BE. 2019. Characterization of nutrient disorders of *Cannabis sativa.* *Applied Sciences* 9: 4432.

Coffman CB, Genter WA. 1975. Cannabinoid profile and elemental uptake of *Cannabis sativa* L. as influenced by soil characteristics. *Agronomy Journal* 67: 491–497.

Coffman CB, Genter WA. 1978. *Cannabis sativa* L. response to paraquat. *Proceedings of the Northeastern Weed Science Society* 32: 9.

Coffman CB, Genter WA. 1980. Biochemical and morphological responses of *Cannabis sativa* L. to postemergent applicatiions of paraquat. *Agronomy Journal* 72: 535–537.

Columella LIM (Ash HB, *trans*). 1941. *On agriculture: with a recension of the text and an English translation, Vol 1.* Harvard University Press, Cambridge, Massachusetts.

Córdova RA, Tomazetti M, Refatti JP, *et al.* (3 other authors). 2020. Drift distance in aircraft glyphosate application using rice plants as indicators. *Planta Daninha* 38: e020223422.

Cornelius CD, Bradley KW. 2017. Carryover of common corn and soybean herbicides to various cover crop species. *Weed Technology* 31: 21–31.

Cosentino SL, Testa G, Scordia D, Copani V. 2012. Sowing time and prediction of flowering of different hemp (*Cannabis sativa* L.) genotypes in southern Europe. *Industrial Crops and Products* 37: 20–33.

Cosentino SL, Riggi E, Testa G, Scordia D, Copani V. 2013. Evaluation of European developed fibre hemp genotypes (*Cannabis sativa* L.) in semi-arid Mediterranean environment. *Industrial Crops and Products* 50: 312–324.

Crescini F. 1934c. Sulla fasciazione della canapa (*Cannabis sativa* L). *Archivio Botanico per la Sistematica, Fitogeografia e Genetica* 10: 387–389.

Crescini F. 1956. La fecondazione incestuosa processo mutageno in *Cannabis sativa* L. *Caryologia* 9(1): 82–92.

Culman SW, Brock C, Doohan D, *et al.* (6 other authors). 2021. Base cation saturation ratios vs. sufficiency level of nutrients: a false dichotomy in practice. *Agronomy Journal* 113: 5623–5634.

Czyżyk W. 1985. Nawadnianie ściekami komunalnymi gruntów ornych w okresie pozawegetacyjnym. *Wiadomości Instytutu Melioracji i Użytków Zielonych* 15: 55–77.

Da Cunha Leme Filho JF, Thomason WE, Evanylo GK, *et al.* (4 other authors). 2020. Biochemical and physiological responses of *Cannabis sativa* to an integrated plant nutrition system. *Agronomy Journal* 112: 5237–5248.

Dadkhah A. 2010. Salinity effect on germination and seedling growth of four medicinal plants. *Iranian Journal of Medicinal and Aromatic Plants* 26: 358–369.

Darsie ML, Elliott C, Peirce GJ. 1914. A study of the germinating power of seeds. *Botanical Gazette* 58: 101–136.

Daubeny C. 1845. Memoir on the rotation of crops, part I. *Philosophical Transactions of the Royal Society of London* 1: 179–211.

de Meijer EPM. 1993. Evaluation and verification of resistance to *Meloidogyne hapla* Chitwood in a *Cannabis* germplasm collection. *Euphytica* 71: 49–56.

de Meijer EPM, Bagatta M, Carboni A, Crucitti P, Cristiana Moliterni VM, Ranalli P, Mandolino G. 2003. The inheritance of chemical phenotype in *Cannabis sativa* L. *Genetics* 163: 335–346.

de Serres O. 1600. *Théâtre d'agriculture et mesnage des champs.* Jamet-Métayer, Paris.

DEA. 1986. *Final environmental impact statement: Cannabis eradication on non federal and Indian lands in the contiguous United States and Hawaii.* DEA-EIS-2. Drug Enforcement Administration, U.S. Department of Justice. Washington, DC.

Deferne JL, Pate DW. 1996. Hemp seed oil: a source of valuable essential fatty acid. *Journal of the International Hemp Association* 3(1): 1–7.

Delleaux F. 2010. Diffusion et application des méthodes culturales flamandes dans les anciens Pays-Bas méridionaux au XVIIIᵉ siècle. *Revue Historique* 653: 27–58.

Dewey LH. 1913. A purple-leaved mutation in hemp. *USDA Plant Industry Circular* 113: 23–24.

Dewey LH. 1914. "Hemp," pp. 283–347 in: *USDA Yearbook 1913.* United States Department of Agriculture, Washington, DC.

Dippenaar MC, du Toit CLN, Botha-Greeff MS. 1996. Response of hemp (*Cannabis sativa* L.) varieties to conditions in Northwest Province, South Africa. *Journal International Hemp Association* 3(2): 63–66.

Dixon ("Delamer") ES. 1854. *Flax and hemp: their culture and manipulation.* Routledge & Co., London.

Dobretsov AN. 1957. Гербицид 2,4-Д уничтожает дикую коноплю. *Защита растений от вредителей и болезней* 1957(2): 36–37.

Dochev C, Mitkov A, Yanev M, Neshev N, Toney T. 2016. "Herbicide control of wild hemp (*Cannabis sativa* L.) at sunflower grown by 'Express Sun' technology," pp. 1339–1344 in *Proceedings of the VII International Scientific Agricultural Symposium*, Sarajevo.

Drake B. 1970a. *The cultivator's handbook of marijuana, Revised edn.* Agrarian Reform Co., Eugene, Oregon.

Drake G. 1970b. The battle against marihuana. *Congressional Record—Senate.* March 13, 1970, p. 7308.

Duke JA. 1982. "Ecosystematic data on medicinal plants," pgs 13-23 *in* Atal CK, Kapur BM, *eds. Utilization of medicinal plants.* United Printing Press, New Delhi.

Dunham RS, Nylund RE, Hansen HL, Jensen EH. 1956. *Cultural and chemical weed control in Minnesota.* University of Minnesota Agricultural Extension Service, Extension Folder no. 191. St Paul, Minnesota.

Dzhaparidze LI. 1967. *Sex in plants.* Israel Program for Scientific Translations, Jerusalem.

Eaton BJ, Hartowicz LE, Latta RP, *et al.* (3 other authors). 1972. Controlling wild hemp. *Kansas Agricultural Research Station, Report of Progress* 188: 1–10.

Eboh LO Thomas BE. 2005. Analysis of heavy metal content in *Cannabis* leaf and seed cultivated in southern part of Nigeria. *Pakistan Journal of Nutrition* 4: 349–351.

ElSohly MA, Radwan MM, Gul W, Chandra S, Galal A. 2017. Phytochemistry of *Cannabis sativa* L. *Progress in the Chemistry of Organic Natural Products* 103: 1–34.

Exley C, Begum A, Woolley MP, Bloor RN. 2006. Aluminum in tobacco and cannabis and smoking-related disease. *American Journal of Medicine* 119: P276.E9–P276.E11.

Fabbri A. 1949. *La stima dei danni della grandine alla canapa e al granoturco.* Edizioni Agricole, Bologna.

Fàn Shèngzhī (Shih SH, *trans, ed*). 1959. *On "Fan sheng-chih shu": an agriculturistic book of China written by Fan Shêng-chih in the first century B.C.* Science Press, Beijing.

Farnsworth NR, Cordell GA. 1976. New potential hazard regarding use of marijuana—treatment of plants with liquid fertilizers. *Journal of Psychedelic Drugs* 8: 151–155.

Flint CL. 1884. *The American farmer, vol. I.* Park & Co., Hartford, Connecticut.

Forrest C, Young JP. 2006. The effects of organic and inorganic nitrogen fertilization on the morphology and anatomy of Cannabis sativa 'Fédrina' (industrial fibre hemp) grown in Northern British Columbia. *Journal of Industrial Hemp* 11(2) :3–24.

Frank M. 1988. *Marijuana grower's insider's guide.* Red Eye Press, Los Angeles, California.

Frank M, Rosenthal E. 1978. *Marijuana grower's guide.* And/Or Press, Berkeley, California.

Frank MH, Chitwood DH. 2016. Plant chimeras: the good, the bad, and the 'Bizzaria'. *Developmental Biology* 419: 41–53.

Fred EB. 1916. Relation of green manures to the failure of certain seedlings. *Journal of Agricultural Research* 5: 1161–1176.

Freed VH, Warren R, Leach CM. 1951. *Selective weed control in grain and grass crops.* Oregon State College, Extension Bulletin no. 719, Corvallis, Oregon.

Galán-Ávila A, García-Fortea E, Prohens J, Herraiz FJ. 2020. Development of a direct *in vitro* plant regeneration protocol from *Cannabis sativa* L. seedling explants: developmental morphology of shoot regeneration and ploidy level of regenerated plants. *Frontiers in Plant Science* 11: e645.

Gao CS, Cheng CH, Zhao LN, *et al.* (7 other authors). 2018. Genome-wide expression profiles of hemp (*Cannabis sativa* L.) in response to drought stress. *International Journal of Genomics* 2018: 3057272.

Gao S, Wang BS, Xie SS, *et al.* (6 other authors). 2020. A high-quality reference genome of wild *Cannabis sativa. Horticulture Research* 7: e73.

Gao Y, Wang YF, Qiu CS, *et al.* (3 other authors). 2011. Preliminary study on effects of drought stress on physiological characteristics and growth of different hemp cultivars (*Cannabis sativa* L.). *Plant Fiber Sciences in China* 33(5): 235–239.

García-Tejero IF, Durán-Zuazo VH, Pérez-Álvarez R, *et al.* (4 other authors). 2014. Impact of plant density and irrigation on yield of hemp (*Cannabis sativa* L.) in a Mediterranean semi-arid environment. *Journal of Agricultural Science and Technology* 16: 887–895.

Garman H. 1903. *The broom-rapes.* Kentucky Agricultural Experiment Station Bulletin No. 105. Lexington, Kentucky.

Gasparin A de. 1842. Mémoire sur la valeur des engrais. *Mémoires publiés la Société Royale et Central d'Agriculture* 1842: 122–164.

Gasparrini G. 1862. *Ricerche sulla embriogenia della canape.* Stamperia e Carterie del Firreno, Napoli.

Gaujac C. 1810. Suite de l'extrait d'un Mémoire de M. Gaujac, Propriétaire-Cultivateut à Dagny, près Coulommiers (Seine-et-Marne), sur la culture compare des plantes oléagineuses. *Bulletin Société d'Encouragement pour l'Industrie Nationale* 9: 247–252.

Gauthier N. 2020b. "Considering crop rotations and the potential for carry-over," p. 25 in Gauthier N, Leonberger K, Bowers C, *eds. Science of hemp: production and pest management.* University of Kentucky Agricultural Experiment Station, no. SR-112, Lexington, Kentucky.

Geneve RL, Wilson SB, Davies FT. 2023. *Chimeras.* Available at: https://propg.ifas.ufl.edu/03-genetic-selection/04-genetic-chimera.html

Giovanardi R, Marchiol L, Mazzocco GT, Zuliani F. 2002. Possibilità di impiego della canapa nella fitoestrazione di metalli pesanti in terreni contaminati: primi risultati. *Agroindustria* 1(1): 69–73.

Goidànich G. 1959. *Manual di patologia vegetale.* Edizioni Agricole, Bologna.

Goloborod'ko PA. 1986. Дефолиация и десикация конопли (Defoliation and desiccation of hemp). *Zashchita Rastenii "Agropromizdat"* 1986(8): 53.

Grant CN, Lalor GC, Vutchkov MK. 2004. Trace elements in Jamaican tobacco. *West Indian Medical Journal* 53(2): 66–70.

Grassa CJ, Weiblen GD, Wegner JP, *et al.* (5 other authors). 2021. A new *Cannabis* genome assembly associates elevated cannabidiol (CBD) with hemp introgressed into marijuana. *New Phytologist* 230: 1665–1679.

Grassi G, McPartland JM. 2017. "Chemical and morphological phenotypes in breeding of *Cannabis sativa* L.," pp. 137–160 in Chandra S, Lata H, ElSohly MA, *eds. Cannabis sativa: Botany and Biotechnology.* Springer International Publishing, Cham, Switzerland.

Grigory'ev SV. 2000. "Изучение холодостойкости конопли," pp. 38-49 in *Collection of scientific papers dedicated to the 100th anniversary of V.A. Nevinnykh.* Adygea, Krasnodar. Available at: http://www.demeter.org.es/pdf/investi_a/Cold_resitance_of_hemp.pdf

Grishko NN. 1937. Одновременно созревающая конопля (*Simultaneously maturing hemp*). State Publishing House of Agricultural and Collective Farm Literature, Moscow.

Guerriero G, Behr M, Hausman JF, Legay S. 2017. Textile hemp vs. salinity: insights from a targeted gene expression analysis. *Genes* 8(10): e242.

Guerriero G, Deshmukh R, Sonah H, *et al.* (6 other authors). 2019. Identification of the aquaporin gene family in *Cannabis sativa* and evidence for the accumulation of silicon in its tissues. *Plant Science* 287: e110167.

Haberlandt F. 1874. Die oberen und unteren Temperaturgrenzen für die Keimung der wichtigeren landwirtschaftlichen Sämereien. *Landwirtschaftlichen Versuchs-Stationen* 17: 104–116.

Haberlandt F. 1875. Ueber die untere Grenze der Keimungstemperatur der Samen unserer Getreidepflanzen. *Wissenschaftlich-Praktische Untersuchungen auf dem Gebiete des Pflanzenbaues* 1: 109–117.

Haberlandt F (Hecke W, *ed.*). 1879. *Der allgemeine Landwirtschaftliche Pflanzenbau.* Faesy & Frick, Wien.

Hackleman JC, Domingo WE. 1943. *Hemp: an Illinois war crop.* University of Illinois Agricultural Experiment Station Circular No. 547. Urbana, Illinois.

Hancer JH. 1992. Buds of paradise. *High Times* No. 199: 48–50.

Haney A, Kutscheid BB. 1975. An ecological study of naturalized hemp (*Cannabis sativa* L.) in east-central Illinois. *American Midland Naturalist* 93: 1–24.

Hanson AA. 1990. *CRC practical handbook of agricultural science.* CRC Press, Boca Raton, Florida.

Hanson NS, Smith CW, Keim FD. 1946. "2,4-D—a new chemical for weed control," pp. 35-37 in *59th annual report of the Agricultural Experiment Station.* University of Nebraska Agricultural Experiment Station, Lincoln, Nebraska.

Hawley D, Graham T, Stasiak M, Dixon M. 2018. Improving *Cannabis* bud quality and yield with subcanopy lighting. *HortScience* 53: 1593–1599.

Health Canada. 2008. *Product information sheet on dried marihuana (cannabis).* Available at: www.hc-sc.gc.ca/dhp-mps/marihuana/about-apropos/dried-information-sechee-eng.php

Hesami M, Adamek K, Pepe M, Phineaus Jones AM. 2023. Effect of explant source on phenotypic changes of in vitro growth *Cannabis* plantlets over multiple subcultures. *Biology* 12: e443.

Hoffmann W. 1941. Gleichzeitig reifender Hanf. *Der Züchter* 13: 277–283.

Hoffmann W. 1947a. Helle Stengel—eine wertvolle Mutation des Hanfes (*Cannabis sativa* L.). *Der Züchter* 17/18: 56–59.

Hoffmann W. 1947b. Die Vererbung der Geschlechtsformen des Hanfes (*Cannabis sativa* L.). *Der Züchter* 17/18: 257–277.

Holuby JL. 1878. Cannabis sativa monoica "sverepá konopa" der Slovaken. *Öesterreichische Botanische Zeitschrift* 28: 367–369.

Honek A, Martinkova Z, Lukas J, Dixon AF. 2014. Plasticity of the thermal requirements of exotherms and adaptation to environmental conditions. *Ecology and Evolution* 4: 3103–3114.

Hopkins JF. 1951. *A history of the hemp industry in Kentucky.* University of Kentucky Press, Lexington, Kentucky.

Horowitz M. 1977. Herbicidal treatments for control of *Cannabis sativa* L. *Bulletin on Narcotics* 29(1): 75–84.

Hoseini PS, Poursafa P, Moattar F, Amin MM, Rezaei AH. 2012. Ability of phytoremediation for absorption of strontium and cesium from soils using *Cannabis sativa. International Journal of Environmental Health Engineering* 1: 17.

Howard A. 1938. The manufacture of humus from the wastes of the town and the village. *Journal of the Royal Sanitary Institute* 58: 279–286.

Hu HR, Liu H, Deng G, Du GH, Xu Y, Liu FH. 2015. Effects of different salt-alkaline stresses on seed germination and seedling growth of *Cannabis sativa. Journal of Plant Resources and Environment* 24: 61–68.

Hu HR, Du GH, Xu Y, Chen ZY, Liu H, Liu FH. 2016. Effects of salt-alkaline stress on seedling growth and physiological indexes of two hemp varieties. *Journal of Yunnan University* 38: 974–981.

Hu HR, Liu H, Liu FH. 2018. Seed germination of hemp (*Cannabis sativa* L.) cultivars responds differently to the stress of salt type and concentration. *Industrial Crops and Products* 123: 254–261.

Hu JH, Masson R. 2021. First report of crown and root rot caused by *Pythium aphanidermatum* on industrial hemp (*Cannabis sativa*) in Arizona. *Plant Disease* 105: 2257.

Hu WJ, Zhang JB, Liu YQ. 2012. The qanats of Xinjiang: historical development, characteristics and modern implications for environmental protection. *Journal of Arid Land* 4: 211–200.

Humboldt Growers Association. 2010. *Humboldt County 314-55.1 medical marijuana land uses draft proposal.* Available at: http://library.humboldt.edu/humco/holdings/HGA2.pdf

Ievinsh G, Vikmane M, Iirse A, Karlsons A. 2017. Effect of vermicompost extract and vermicompost-derived humic acids on seed germination and seedling growth of hemp. *Proceedings of the Latvian Academy of Sciences, Section B* 71: 286–292.

Iványi I, Izsáki Z. 2009. Effect of nitrogen, phosphorus, and potassium fertilization on nutritional status of fiber hemp. *Communications in Soil Science and Plant Analysis* 40: 1–6.

Jaffa ME. 1891. *Composition of the ramie plant.* Bulletin of the University of California Agricultural Experiment Station, Bulletin No. 94. Berkeley, California.

Jameson LE, Conrow KD, Pinkhasova DV, *et al.* (9 other authors). 2022. Comparison of state-level regulations for cannabis contaminants and implications for public health. *Environmental Health Perspectives* 130: e97001.

Jannaccone A. 1947. Una nuova razza di canapa (*C. sativa* L. var. *albicata* Jannaccone). *L'Agricoltura Italiana* 48: 8–9.

Javadi H, Seghatoleslami MJ, Mosavi SG. 2014. The effect of salinity on seed germination and seedling growth of four medicinal plant species. *Iranian Journal of Field Crops Research* 12: 53–64.

Jiǎ SX. 1962. *A preliminary survey of the book Ch'i min yao shu: an agricultural encyclopaedia of the 6th century.* Science Press, Beijing.

Jiang HE, Li X, Zhao YX, *et al.* (7 other authors). 2006. A new insight into *Cannabis sativa* (Cannabaceae) utilization from 2500-year-old Yanghai Tombs, Xinjiang, China. *Journal of Ethnopharmacology* 108: 414–422.

Jiang HE, Wang L, Merlin MD, Clarke RC, *et al.* (4 other authors). 2016. Ancient *Cannabis* burial shroud in a Central Eurasian cemetery. *Economic Botany* 70: 213–221.

Jiang Y, Sun YF, Zheng DF, *et al.* (6 other authors). 2021. Physiological and transcriptome analyses for assessing the effects of exogenous uniconazole on drought tolerance in hemp (*Cannabis sativa* L.). *Scientific Reports* 11: e14476.

Jurkowska H, Rogóz A, Wojciechowicz T. 1990. The content of lithium in some species of plants following differentiated doses of nitrogen. *Polish Journal of Soil Science* 23: 195–199.

Kahan A. 1985. *The plow, the hammer, and the knout: an economic history of eighteenth-century Russia.* University of Chicago Press, Chicago.

Kakabouki I, Kousta A, Folina A, *et al.* (4 other authors). 2021. Effect of fertilization with urea and inhibitors on growth, yield, and CBD concentration of hemp (*Cannabis sativa* L.). *Sustainability* 13: e2157.

Kalinowski J, Edmisten K, Davis J, *et al.* (5 other authors). 2020. Augmenting nutrient acquisition ranges of greenhouse grown CBD (cannabidiol) hemp (*Cannabis sativa*) cultivars." *Horticulturae* 6(4): e98.

Karadjov J. 2014. *Eliminating heavy-metal toxicity in medical marijuana.* Available at: www.advancednutrients.com/heavymetal//Eliminating-Heavy-Metal-Toxicity-in-Medical-Marijuana.pdf

Kees. 1988. *The marijuana seed catalog, '88–'89.* Super Sativa Seed Club, Amsterdam.

Keijzer P, Lubberts JH, Stripper H. 1990. Evaluation of the growth retardant triapenthol on hemp. *Annals of applied Biology* 116 (Supplement): 72–73.

Kirchner HA. 1966. *Phytopathologie und Pflanzenschutz, band 2.* Akademie Verlag, Berlin.

Kirchner O. 1883. Ueber das Längenwachstum von Pflanzenorganen bei niederen Temperaturen. *Beiträge zur Biologie der Pflanzen* 3: 335–364.

Kirkham MB. 2014. *Principles of soil and plant water relations.* Academic Press, Amsterdam.

Kiss T. 1973. Az 1968–1970. évi kender-fajtakísérletek eredményei, adatok az elsö sárgaszárú fajtáról (Results of the 1968–1970 hemp variety trials, and data on the first yellow-stemmed variety). *Fajtakísérletezés–Fajtaminősítés (Országos Mezőgazdasági Fajtakísérleti Intézet) évi 1971:* 283–299.

Klingman DL, McCarty MK. 1958. *Interrelations of methods of weed control and pasture management at Lincoln, Nebr., 1949–55.* USDA Technical Bulletin no. 1180. Washington, DC.

Kok CJ, Coenen GCM, de Heij A. 1994. The effect of fibre hemp (*Cannabis sativa* L.) on selected soil-borne pathogens. *J International Hemp Association* 1(1): 6–9.

Kozlowski R. 1995. Interview with Professor R. Kozlowski, director of the Institute of Natural Fibres. *Journal of the International Hemp Association* 2(2): 86–87.

Kudryavtsev N, Zaĭtseva L, Golubkov D, *et al.* (3 other authors). 2017. Efficacy of the high molecular weight preparation Artafit produced in Russia for cultivation of flax and hemp. *Mezhdunarodnyĭ Sel'skokhozyaĭstvennyĭ Zhurnal* 2017(3): 40–43.

Kukorelli G, Nagy S, Reisinger P. 2007. Comparative experiments with imidazolinone and tribenuron-methyl tolerant sunflower hybrids. *Magyar Gyomkutatás és Technológia* 8(1): 67–73.

Kumar A, Singh S, Singh KK. 2000. Weed management through cropping systems and herbicides in wheat under north Bihar ecosystem. *Annals of Agricultural Research* 21: 580–583.

Kumar V, Chopra AK. 2013. Reduction of weeds and improvement of soil quality and yield of wheat by tillage in Northern Great Plains of Ganga River in India. *International Journal of Agricultural Science Research* 2: 249–257.

Kutera J, Czyżyk W. 1992. Wpływ nawodnień ściekami z zakładów utylizacyjnych na plonowanie łąk i roślin uprawnych. *Wiadomości Instytutu Melioracji i Użytków Zielonych* 17: 447–463.

Lai T van. 1985. Effects of inbreeding on some major characteristics of hemp. *Acta Agronomica Academiae Scientiarum Hungaricae* 34: 77–84.

Laleh S, Jami Al-Ahmadi M, Parsa S. 2021. Response of hemp (*Cannabis sativa* L.) to integrated application of chemical and manure fertilizers. *Acta Agriculturae Slovenica* 117(2): 1–15.

Landis H, Hicks K, Cockson P, *et al.* (3 other authors). 2019. Expanding leaf tissue nutrient survey ranges for greenhouse cannabidiol-hemp. *Crop, Forage & Turfgrass Management* 5: 1–3.

Landrigan PJ, Rowell KE, James LM, Taylor PR. 1983. Paraquat and marijuana: epidemiologic risk assessment. *American Journal of Public Health* 73: 784–788.

Langlet O. 1927. Zur Kenntnis der polysomatischen. *Zellkerne im Wurzelmeristem. Svensk Botanisk Tidskrift* 21: 397–422.

Larsen D. 1999. Growing down under. *Cannabis Culture* no. 19. Available at: www.cannabisculture.com/articles/1485.html.

Lata H, Chandra S, Techin N, Khan IA, ElSohly MA. 2010. Assessment of the genetic stability of micropropagated plants of *Cannabis sativa* by ISSR markers. *Planta Medica* 76: 97–100.

Levi IA. 1882. Gli effetti della grandine sulla canapa. *Giornale d'Agricoltura, Industria e Commercio* 19: 368.

Lisson SN, Mendham NJ. 1995. Tasmanian hemp research. *Journal of the International Hemp Association* 2(2): 82–85.

Lisson S, Mendham N. 1998. Response of fiber hemp (*Cannabis sativa* L.) to varying irrigation regimes. *Journal of the International Hemp Association* 5(1): 9–15.

Lisson SN, Mendham NJ, Carberry PS. 2000. Development of a hemp (*Cannabis sativa* L.) simulation model 1. General introduction and the effect of temperature on the preemergent development of hemp. *Australian Journal of Experimental Agriculture* 40: 405– 411.

Litardière R de. 1925. Sur l'existence de figures didiploides dans le meristème radiculaire du *Cannabis sativa* L. *La Cellule* 35: 21–25.

Liu JJ, Qiao Q, Cheng X, *et al.* (4 other authors). 2016. Transcriptome differences between fiber-type and seed-type *Cannabis sativa* variety exposed to salinity. *Physiology and Molecular Biology of Plants* 22: 429–443.

Löser C, Zehnsdorf A, Fussy M, Stärk HJ. 2002. Conditioning of heavy metal-polluted river sediment by *Cannabis sativa* L. *International Journal of Phytoremediation* 4: 27–45.

Low D. 1834. *Elements of practical agriculture.* Bell and Bradfute, Edinburgh.

Lucas GB. 1975. *Diseases of tobacco, 3rd edn.* Biological Consulting Assoc., Raleigh, North Carolina.

Majd A, Roustaee Z, Sattari TN, Arbabian S. 2019. The developmental study vegetative and generative organs of *Cannabis sativa* L. *Quarterly Journal of Developmental Biology* 9(3): 63–73.

Magagnini G, Grassi G, Kotiranta S. 2018. The effect of light spectrum on the morphology and cannabinoid content of *Cannabis sativa* L. *Medical Cannabis and Cannabinoids* 2018(1): 19–27.

Malloch WS. 1922. Value of the hemp plant for investigating sex inheritance. *Journal of Heredity* 13: 277–283.

Markham G. 1615. *Countrey contentments in two bookes, the second intituled the English Huswife.* Roger Jackson, London.

Marquart B. 1919. *Der Hanfbau.* Paul Parey, Berlin.

Marshall E. 1982. Pot-spraying plan raises some smoke. *Science* 217: 429.

Marshall J. 1772. *Travels through Holland, Flanders, Germany, Denmark, Sweden, Lapland, Russia, the Ukraine, and Poland.* J. Almon, London.

Marshmann JA, Popham RE, Yawney CD. 1976. A note on the cannabinoid content of Jamaican ganja. *Bulletin on Narcotics* 28(4): 63–68.

Massuela DC, Munz S, Hartung J, Nkebiwe PM, Graeff-Hönninger S. 2023. Cannabis Hunger Games: nutrient stress induction in flowering stage—impact of organic and mineral fertilizer levels on biomass, cannabidiol (CBD) yield and nutrient use efficiency. *Frontiers in Plant Science* 14: e1233232.

Mauz EF. 1836. "Versuch und Beobachtungen über das Geschlecht der Pflanzen und die Veränderung derselben durch äussere Einflüsse," pp. 1–15 in Babo F von, ed. *Kurze Anleitung zur Anlage und Behandlung der Wiesen.* Osswald, Heidelberg.

McDaniel PA, Buol SW. 1991. Manganese distributions in acid soils of the North Carolina Piedmont. *Soil Science Society of America Journal* 5: 152–158.

McGlamery MD, Knake EL. 1970. Marihuana. *Insect, Weed, and Plant Disease Survey Bulletin* 17: 4.

McKernan KJ, Helbert Y, Kane LT, *et al.* (12 other authors). 2020. Sequence and annotation of 42 cannabis genomes reveals extensive copy number variation in cannabinoid synthesis and pathogen resistance genes. *BioRxiv* 2020.01.03.894428.

McKey D, Elias M, Pujol B, Duputié A. 2010. The evolutionary ecology of clonally propagated domesticated plants: Tansley review. *New Phytologist* 186: 318–332.

McPartland JM, McKernan KJ. 2017. "Contaminants of concern in cannabis: microbes, heavy metals and pesticides," in Chandra S, Lata H, ElSohly MA, eds. *Cannabis sativa: Botany and Biotechnology.* Springer International Publishing, Cham, Switzerland.

McPartland JM, Schoeneweiss DF. 1984. Hyphal morphology of *Botryosphaeria dothidea* in vessels of unstressed and drought-stressed *Betula alba. Phytopathology* 74: 358–362.

McPartland JM, Small E. 2020. A classification of endangered high-THC cannabis (*Cannabis sativa* subsp. *indica*) domesticates and their wild relatives. *PhytoKeys* 144: 81-112.

McPartland JM, Clarke RC, Watson DP. 2000. *Hemp diseases and pests – management and biological control.* CABI Publishing, Wallingford, UK.

Medwedewa GB. 1935. The climatic influences upon the pollen development of the Italian hemp. *Zeitschrift für Induktive Abstammungs- und Vererbungslehre* 70: 170–176.

Meers E, Ruttens A, Hopgood M, Lesage E, Tack FM. 2005. Potential of *Brassica rapa, Cannabis sativa, Helianthus annuus* and *Zea mays* for phytoextraction of heavy metals from calcareous dredged sediment derived soils. *Chemosphere* 61: 561–572.

Mihoc M, Pop G, Alexa E, Radulov I. 2012. Nutritive quality of Romanian hemp varieties *(Cannabis sativa* L.) with special focus on oil and metal content of seeds. *Chemistry Central Journal* 6: 122.

Mills E. 2012. The carbon footprint of indoor *Cannabis* production. *Energy Policy* 46: 58-67.

Milov LV. 1998. Великорусский пахарь и особенности российского исторического процесса. Rosspen, Moscow.

Ming R, Bendahmane A, Renner SS. 2011. Sex chromosomes in land plants. *Annual Review of Plant Biology* 62: 485–514.

Mitkov A, Yanev M, Neshev N, Tonev T. 2018. Evaluation of low herbicide rates of gardoprom® Plus Gold 550 SC and Spectrum® 720 EC at conventional sunflower (*Helianthus annuus* L.). *Scientific Papers Series A, Agonomy* 61(2): 94–97.

Moir D, Rickert WS, Levasseur G, *et al.* (4 other authors). 2008. A comparison of mainstream and sidestream marijuana and tobacco cigarette smoke produced under two machine smoking conditions. *Chemical Research in Toxicology* 21 (2):494–502.

Monagail M, Cornish L, Morrison L, Araújo R, Critchley AT. 2017. Sustainable harvesting of wild seaweed resources. *European Journal of Phycology* 52: 371–390.

Morad D, Bernstein N. 2023. Response of medical cannabis to magnesium (Mg) supply at the vegetative growth phase. *Plants* 12: e2676.

Muth F. 1906. Untersuchungen über die Früchte des Hanfes (*Cannabis sativa* L.). *Jahresbericht der Vereinigung für Angewandte Botanik* 3: 76–121.

Nazirov Kh, Tukhtaeva S. 1981. Трефлан на кенафа (Application of treflan to hemp). *Len i Konoplya* 1981(2): 34.

Nelson JA, Bugbee B. 2014. Economic analysis of greenhouse lighting: light emitting diodes vs. high intensity discharge fixtures. *PLoS ONE* 96(6): e99010.

Neuer HV. 1935. Der Anbau des Hanfes. *Berichte über Landwirtschaft: Sonderheft* 20: 16–30.

Nobbe F, Hiltner L.1893. Impfet den Boden! Eine aus der ungleichen Wirkungskraft der Knöllchenbakterien auf die verschiedenen Leguminosen sich ergebende praktische Schulussfolgerung. *Sächsische Landwirtschaftliche Zeitschrift* 51: 595–600.

Nöschel A (Helmerson G, *ed*). 1856. Bemerkungen über die naturhistorischen, insbesondere die geognostische-hydrographischen Verhältnisse der Steppe zwischen den Flüssen Or und Turgai, Kumak und Syr-Darja. *Beiträge zur Kenntnis des Russischen Reiches und der angränzenden Länder Asiens* 18: 117–196.

Page SRG, Monthony AS, Jones AMP. 2021. DKW basal salts improve micropropagation and callogenesis compared with MS basal salts in multiple commercial cultivars of *Cannabis sativa*. *Botany* 99(5): 269–279.

Pagnani G, Pellegrini M, Galieni A, *et al.* (8 other authors). 2018. Plant growth-promoting rhizobacteria (PGPR) in *Cannabis sativa* 'Finola' cultivation: an alternative fertilization strategy to improve plant growth and quality characteristics. *Industrial Crops and Products* 123: 75–83.

Pascu G. 2012. Research regarding weed infestation in wheat crop at ecologic prod farm, Mihail Kogalniceanu-Constanta county. *Scientific Papers, Series A, Agronomy* 55: 212–215.

Peglion V. 1897. Eine neue Krankheit des Hanfes (Bacteriosis des Stengels). *Zeitschrift für Pflanzenkrankheiten* 7: 81–84.

Pell A. 1893. Arthur Young. *Journal of the Royal Agricultural Society of England (third series)* 4: 1–23.

Peltenburg E. 2012. "East Mediterranean water wells of the 9th–7th millennium BC," pp. 69-82 in Klimsha F, *et al.*, eds. *Wasserwirtschaftliche Innovationen im archäologischen Kontext.* Leidorf GmbH, Rahden/Westfalia.

Piotrowska-Cyplik A, Czarnecki Z., 2003. Phytoextraction of heavy metals by hemp during anaerobic sewage sludge management in the non-industrial sites. *Polish Journal of Environmental Studies* 12: 779–784.

Piperno DR. 2006. *Phytoliths. A comprehensive guide for archaeologists and paleoecologists.* Altamira Press, Lanham, Maryland.

Prain D. 1893. *Report on the cultivation and use of gánjá.* Bengal Secretariat Press, Calcutta.

Prain D. 1904. On the morphology, teratology and diclinism of the flowers of *Cannabis*. *Scientific Memoirs by Officers of the Medical and Sanitary Departments of the Government of India* 12: 51–82.

Prakash V, Pandey AK, Singh RD, Mani VP. 2000. Integrated weed management in winter onion (*Allium cepa*) under mid-hill conditions of north-western Himalayas. *Indian Journal of Agronomy* 45: 816–821.

Price MP, Simpson ND. 1913. An account of the plants collected by Mr. M. P. Price on the Carruthers-Miller-Price Expedition through North-West Mongolia and Chinese Dzungaria in 1910. *Journal of the Linnean Society of London, Botany* 41: 385–456.

Punja ZK, Holmes JE. 2020. Hermaphroditism in marijuana (*Cannabis sativa* L.) inflorescences – impact on floral morphology, seed formation, progeny sex ratios, and genetic variation. *Frontiers in Plant Science* 11: e718.

Ranalli P, Venturi G. 2004. Hemp as a raw material for industrial applications. *Euphytica* 140: 1–6.

Rataj K. 1957. Skodlivi cinitele pradnych rostlin. *Prameny literatury* 2: 1–123.

Reisinger P, Lehoczky É, Komives T. 2005. Competitiveness and precision management of the noxious weed *Cannabis sativa* L. in winter wheat. *Communications in Soil Science and Plant Analysis* 36: 629–634.

Rham WL. 1842. Agriculture of the Netherlands, part II. *Journal of the Royal Agricultural Society of England* 3: 240–263.

Roberts L. 1988. Is there life after climate change? *Science* 242: 1010–1012.

Robinson BB. 1943a *Hemp* USDA Farmer's Bulletin No. 1935. Washington, DC.

Rodale Institute. 2020. *Industrial hemp trials.* Available at: https://rodaleinstitute.org/science/industrial-hemp-trial/

Röder K. 1941b. Einige untersuchungen über ein an hanf (*Cannabis sativa* L.) auftretendes virus. *Faserforschung* 15: 77–81.

Roux N, Chase R, Ven den Houwe I, *et al.* (6 other authors). 2021. "Somatoclonal variation in clonal crops: containing the bad, exploring the good," pp. 355-365 in Sivasankar S, *et al.*, eds. *Mutation breeding, genetic diversity and crop adaptation to climate change.* CABI, Wallingford UK.

Rózańka W, Romanowska B, Rojewski S. 2023. The quantity and quality of flax and hemp fibers obtained using the osmotic, water- and dew-retting processes. *Materials* 16: e7436.

Russo EB, Jiang HE, Li X, *et al.* (15 other authors). 2008. Phytochemical and genetic analyses of ancient cannabis from Central Asia. *Journal of Experimental Botany* 59: 4171–4182.

Safari Singani AA, Ahmadi P. 2012. Manure application and *Cannabis* cultivation influence on speciation of lead and cadmium by selective sequential extraction. *Soil and Sediment Contamination* 21: 305–321.

Saijonkari-Pahkala K. 2001. *Non-wood plants as raw material for pulp and paper.* Doctoral thesis, Helsinki University.

Salomonson A. 1996. Interactions between somatic mutation and plant development. *Vegetatio* 127: 71–75.

Saloner A, Bernstein N. 2021. Nitrogen supply affects cannabinoid and terpenoid profile in medical cannabis (*Cannabis sativa* L.). *Industrial Crops and Products* 167: e113516.

Saloner A, Bernstein N. 2022a. Nitrogen source matters: high NH_4/NO_3 ratio reduces cannabinoids, terpenoids, and yield in medical cannabis. *Frontiers in Plant Science* 13: e830224.

Savelli R. 1929b. Canapa monofilla e canapa pinnatifidofilla. *L'Italia Agricola* 66: 629–646.

Schaffner JH. 1923a. The influence of relative length of daylight on the reversal of sex in hemp. *Ecology* 4: 323–334.

Scheifele G, Dragla P, Pinsonneault C, Laprise JM. 1997. Hemp (*Cannabis sativa*) research report, Kent County, Ontario, Canada.

Schilling E. 1923. Zur Morphologie, Physiologie und diagnostischen Bewertung der Bastfasern von *Cannabis sativa*. *Berichte der Deutschen Botanischen Gesellschaft* 41: 121–127.

Schlichten GW. 1921. *Hemp growing by irrigation.* H.S. Crocker Co., Sacramento, California.

Schmidt A. 1924. Histologische Studien an phanerogamen Vegetationspunkten. *Botanisches Archiv* 8: 345–404.

Schoenmakers N. 1986. *The Seed Bank 1986/1987 Catalogue.* Drukkerij Dukenburg Printers, Nijmegen, The Netherlands.

Seleiman MF, Santanen A, Stoddard FL, Mäkelä P. 2012. Feedstock quality and growth of bioenergy crops fertilized with sewage sludge. *Chemosphere* 89: 1211–1217.

Seleiman MF, Santanen A, Kleemola J, Stoddard FL, Mäkelä P. 2013. Improved sustainability of feedstock production with sludge and interacting mycorrhiza. *Chemosphere* 91: 1236–1242.

Serkov VA, Pluzhnikov AI, Smirnov AA. 2012. Эффективность приёма десикации на однодомной конопле. *Advances in Science and Technology APK* 2012(4): 29–30.

Sharma GK, Mann SK. 1984. Variation in vegetative growth and trichomes in *Cannabis sativa* L. (marihuana) in response to environmental pollution. *Journal of the Tennessee Academy of Science* 59: 38–40.

Shaw J, Sutcliffe J, Lloyd-Smith L, *et al.* (4 other authors). 2007. Ancient irrigation and Buddhist history in Central India: Optically stimulated luminescence dates and pollen sequences from Sanchi dams. *Asian Perspectives* 46: 166–201.

Shelenga TV, Grigory'ev SV, Illarionova KV. 2012. Биохимическая характеристика семян конопли (*Cannabis sativa* L.) из различных регионов России. *Proceedings in Botany, Genetics, and Selection* 170: 212–219.

Shi G, Cai Q. 2009. Cadmium tolerance and accumulation in eight potential energy crops. *Biotechnology Advances* 27: 555–561.

Shi G, Liu C, Cui M, Ma Y, Cai Q. 2012. Cadmium tolerance and bioaccumulation of 18 hemp accessions. *Applied Biochemistry and Biotechnology* 168: 163–173.

Siegel BZ, Garnier L, Siegel SM. 1988. Mercury in marijuana. *BioScience* 38: 619–623.

Sitnik VP. 1981. Наследование признаков обусловленных плейотропным действем гена желтостебельности конопли. *Breeding and Seed Production, Ukraine SSR* 47: 46–49.

Smith RJ. 1978. Spraying of herbicides on Mexican marijuana backfires on US. *Science* 199: 861–864.

Soil Survey Staff. 2015. *Illustrated guide to soil taxonomy, version 2.* USDA Natural Resources Conservation Service, National Soil Survey Center, Lincoln, Nebrasla.

Sova N, Lutsenko M, Korchmaryova A, Andrusevych K. 2018. Research of physical and chemical parameters of the oil obtained from organic and conversion hemp seeds varieties "Hliana". *Ukrainian Food Journal* 7: 244–552.

Spitzer-Rimon B, Duchin S, Bernstein N, Kamenetsky R. 2019. Architecture and florogeneis in female *Cannabis sativa* plants. *Frontiers in Plant Science* 10: 350.

Stevens M. 1975. *How to grow marijuana indoors under lights, 3rd edn.* Sun Magic, Seattle.

Stevenson W. 1812. *General view of the agriculture of the County of Dorset.* McMillan, London.

Steward H. 1893. *Irrigation for the farm, garden, and orchard.* Orange Judd Co., New York.

Suchoff D, McGinnis M, Davis J, Whipker B, Hicks K. 2021. *Hemp leaf tissue nutrient ranges: refinement of reference standards for floral hemp.* North Carolina Cooperative Extension AG-904.

Tadayon MR, Zarei M. 2014. Investigation of the symbiotic effect of the mycorrhiza fungus *Glomus mosseae* on salinity resistance in *shāh dānah* (*Cannabis*) ecotypes. *Journal of Plant Process and Function* 3(7): 105–114.

Tamm E. 1933. Weitere Untersuchungen über die Keimung und das Auflaufen landwirtschaftlicher Kulturpflanzen. *Pflanzenbau* 10: 297–313.

Thaer A. 1812. *Grundsätze der rationellen Landwirthschaft, Vierter Band.* G. Reimer, Berlin.

Theimer RR, Mölleken H. 1995. "Analysis of the oil from different hemp (*Cannabis sativa* L.) cultivars—perspectives for economical utilization," pp. 536–543 in *Bioresource Hemp, Proceedings of the Symposium, Frankfurt, Germany.* nova-Institute, Köln, Germany.

Thevs N, Aliev K. 2022. Water consumption of industrial hemp (*Cannabis sativa* L.) from a site in northern Kazakhstan. *Central Asian Journal of Water Research* 8(2): 19–30.

Thiessen LD, Schappe T, Cochran S, Hicks K, Post AR. 2020. Surveying for potential diseases and abiotic disorders of industrial hemp (*Cannabis sativa*) production. *Plant Health Progress* 21(4): 321–332.

Thirsk J. 1957. *English peasant farming; the agrarian history of Lincolnshire from Tudor to recent times.* Routledge & Kegan Paul, London.

Tibeau ME. 1936. Time factor in utilization of mineral nutrients by hemp. *Plant Physiology* 11: 731–747.

Toth JA, Smart LB, Smart CD, *et al.* (4 other authors). 2021. Limited effect of environmental stress on cannabinoid profiles in high-cannabidiol hemp (*Cannabis sativa* L.). *GCB Bioenergy* 13: 1666–1674.

Tóth Š, Sikora V, Kovalová L, Harcár M, Porvaz P. 2015. Konopa rumovisková *Cannabis ruderalis* Janisch a cukrová repa. *Listy Cukrovarnické a Řepařské* 131: 292–294

Trux J, Torry L. 1978. Poison pot probe. *New Scientist* 78: 242–243.

Tulaykova KP. 1958. Роль бактерий типа радиобактера в корневом питании растений. *Агробиология* 1958(1): 93–101.

Uloth W. 1875. Ueber die Keimung von Pflanzensamen in Eis. *Flora* 33: 266–268.

UN-CDN (United Nations Commission on Narcotic Drugs). 1964. *Summary of Annual Reports of Governments.* CDN, Economic and Social Council. New York, New York.

Upton R, Craker L, ElSohly M, *et al.* (5 other authors). 2013. *Cannabis inflorescence, Cannabis spp., standards of identity, analysis, and quality control.* American Herbal Pharmacopoeia, Scotts Valley, California.

Urban L. 1979. Fagyhatás és fagyvédelem. *Növéntermelés* 28(5): 473–478.

USDA. 1999. *Soil taxonomy, a basic system of soil classification for making and interpreting soil surveys, 2nd edn.* Agriculture Handbook 436. Government Printing Office, Washington, DC.

US Department of Defense. 1970. *Herbicide Manual for noncropland weeds.* Superintendent of Documents, Washington, DC.

Van Dam JEG. 1995. "Potentials of hemp as an industrial fibre crop," pp. 405–422 in *Proceedings of First Bioresource Hemp Symposium*, nova-Institut, Frankfurt, Germany.

Van der Werf HMG, Brouwer K, Wijhuizen M, Withagen JMC 1995c. The effect of temperature on leaf appearance and canopy establishment in fibre hemp (*Cannabis sativa* L.). *Annals of Applied Biology* 126: 551–561.

Vance JM. 1971. Marijuana is for the birds. *Outdoor Life* 147(6): 53–55, 96–100.

Vandenhove H, Van Hees M. 2005. Fibre crops as alternative land use for radioactively contaminated arable land. *Journal of Environmental Radioactivity* 81: 131–141.

Vane NW. 1973. The eradication of *Cannabis sativa* L. with herbicides in the Hunter River Valley, N.S.W., Australia. *Bulletin on Narcotics* 3: 49–50.

Varro C. 1774. *A new system of husbandry, vol. II.* Printed by the author, London.

Virovets VG, Sitnik VP. 1975. Изучение некоторых физиологических особенностей желтостебельъной конопли (Study of some physiological features of yellow-stemmed hemp). *Proceedings of the All-Russian Research Institute for Bast Processing* 38: 29–33.

Wahlin B. 1944. Några fall av manganbrist sommaren 1943. *Växtskyddsanst Stockholm* 8(1): 11–15.

Wang CT, Wiedinmyer C, Ashworth K, *et al.* (5 other authors). 2019. Potential regional air quality impacts of cannabis cultivation facilities in Denver, Colorado. *Atmospheric Chemistry and Physics* 19: 13973–13987.

Wang SH. 2021. *Diagnosing hemp and cannabis crop diseases.* CABI, Wallingford, UK.

Warren GS. 2015. Regulating pot to save the polar bear: energy and climate impacts of the marijuana Industry. *Columbia Journal of Environmental Law* 40: 385–432.

Washington G. 1801. *Letters from His Excellency General Washington, to Arthur Young, Esq. F.R.S.* McMillian, London.

Weigert J, Fürst F. 1939. Die wirkung von spurenelementen in den randgeibieten südbayerischer moore. *Praktische Blätter für Pflanzenbau und Pflanzenschutz* 17: 117–140.

Westmoreland FM, Bugbee B. 2022. Sustainable *Cannabis* nutrition: elevated root-zone phosphorus significantly increases leachate P and does not improve yield or quality. *Frontiers in Plant Science* 13: e1015652.

White KD. 1970. Fallowing, crop rotation, and crop yields in Roman times. *Agricultural History* 44: 281–290.

White L. 1962. *Medieval technology and social change.* Oxford University Press, Oxford, UK.

Whitham TG, Slobodchikoff CN. 1981. Evolution by individuals, plant–herbivore interactions, and mosaics of genetic variability: the adaptive significance of somatic mutations in plants. *Oecologia* 49: 287–292.

Whitworth C. 1758. *An account of Russia as it was in the year 1710.* Strawberry-Hill, London.

Willis RJ. 2007. *The history of allelopathy.* Springer, Dordrecht, The Netherlands.

Wilsie CP, Black CA, Aandahl AR. 1944. *Hemp production experiments. Cultural practices and soil requirements.* Agricultural Experiment Station, Iowa State College, Bulletin P63, Ames, Iowa.

Winston ME, Hampton-Marcell J, Zarraonaindia I, *et al.* (6 other authors). 2014. Understanding cultivar-specificity and soil determinants of the cannabis microbiome. *PLoS ONE* 9: e99641.

Wise K, Selby-Pham J, Chai XY, Simovich T, Gupta S, Gill H. 2024. Fertiliser supplementation with a biostimulant complex of fish hydrolysate, *Aloe vera* extract, and kelp alters cannabis root architecture to enhance nutrient uptake. *Scientia Horticulturae* Epub before print.

Wiśniewski J, Kołodziej B. 1999. Wpływ nawadniania ściekami komunalnymi na plonowanie i skład chemiczny konopi siewnych. *Folia Universitatis Agriculturae Stetinensis, Agricultura* 77: 379–385.

Yamamoto A, Akiyama H, Naokawa T, Yagi K. 2013. Lime-nitrogen application reduces N_2O emission from a vegetable field with imperfectly-drained sandy clay-loam soil. *Soil Science and Plant Nutrition* 59: 442–449.

Yep B, Zheng YB. 2020. Potassium and micronutrients fertilizer addition in aquaponic solution for drug-type Cannabis sativa L. cultivation. *Canadian Journal of Plant Science* 101: 341–352.

Young A. 1768. *The farmer's letters to the people of England.* W. Nicoll, London.

Young A. 1772. *Political essays concerning the present state of the British Empire.* Strahan and Cadell, London.

Young A. 1774. *Political arithmetic.* W. Nicoll, London.

Young A. 1780. *A tour of Ireland, 2 vols.* H. Goldney, London.

Young A. 1784. "The cultivation of beans, recommended to the farmers of Scotland," pp. 678–681 in *Present State of Husbandry in Scotland, Vol. IV, Part II.* William Creech, Edinburgh.

Young A. 1792. *Travels during the years 1787, 1788, and 1789.* 1st edn. J. Rackham, London

Young A. 1794. *Travels during the years 1787, 1788, and 1789, 2nd edn.* J. Rackham, London.

Young A. 1797. *General view of the agriculture of the County of Suffolk.* Macmillan, London.

Young A. 1799. *General view of the agriculture of the County of Lincoln.* W. Bulmer & Co., London.

Zarei M, Tadayon MR, Tadayyon A. 2014. Effect of biofertilizer, under salinity condition on the yield and oil content of three ecotype of hemp (*Cannabis sativa* L.). *Journal of Crops Improvement* 16: 517–529.

Zielonka D, Nierebiński M, Kalaji HM, *et al.* (3 other authors). 2017. Efficiency of the photosynthetic apparatus in *Cannabis sativa* L. fertilized with sludge from a wastewater treatment plant and with phosphogypsum. *Ecological Questions* 28(4): 55–61.

Zielonka D, Sas-Paszt L, Derkowska E, Lisek A, Russel S. 2019. Occurrence of arbuscular mycorrhizal fungi in hemp (*Cannabis sativa*) plants and soil fertilized with sewage sludge and phosphogypsum. *Journal of Natural Fibers* 18: 250–260.

Abstract

Much of *Cannabis* can be utilized: stalks for bast fibers and hurds, achenes for seeds and seed oil, female flowers for marijuana (with seeds) or without seeds (sinsemilla, *gañjā*). Resin glands go into *hashīsh* or *charas*. These products are vulnerable to decay and destruction by fungi, insects, and other organisms. Two kinds of organisms cause problems after plants are harvested. The first group consists of field organisms that invade living plants and then carry over as storage problems. The second group consists of storage organisms that only infest products after harvest. Post-harvest problems pose dangers to humans: textile workers are susceptible to byssinosis, a disease caused by inhaling bacteria and fungi. Rancid hemp seed can be poisonous. Flowering tops harboring *Aspergillus* or *Penicillium* fungi pose all kinds of dangers—opportunistic infections in humans, aflatoxin poisoning, and loss in cannabinoid shelf life. Control methods for post-harvest problems are described.

18.1   Hemp fiber

Bast fibers are the plant's phloem or sap-conducting cells. An individual fiber is a single plant cell, 10–20 μm wide but up to 7500 μm long (3 inches). The fiber has a thick cell wall, making it long and strong. Bast fibers constitute about 13% of dried stalk in wild-type plants (de Meijer and Keizer, 1996). They arise in bundles, typically 10–40 fiber cells per bundle, and the bundles reside in a ring of parenchyma cells. Collectively this layer is known as the cortex or inner bark, and lies beneath the epidermis. A hemp stalk in cross-section consists of concentric rings—epidermis, cortex, cambium, xylem, and pith (Fig. 18.1).

Cambium consists of unspecialized meristem cells, that give rise to both phloem (outwards) and xylem (inwards). Internal to the cambium is the ring of **xylem**. Xylem fiber cells transport water. They are woody (with heavily lignified walls) and short, reaching only 0.2–0.6 mm in length. This woody component is known as the **hurd** (*a.k.a.* core or shive). It thins out toward the center of the stalk, becoming pith. The pith may thin to a hollow in the center of the stalk. This simplified description elides some complexities, such as the presence of primary and secondary fibers, procambium and collenchyma (Snegireva *et al.*, 2015).

The cell walls of adjacent bast fiber cells are glued together, and this interface is known as the **middle lamella**. The middle lamella is composed of pectins and hemicelluloses, which prevent fiber cells from separating or sliding against each other. Removing these glues, and loosening fibers from the rest of the stalk, is called **retting**. Retting is accomplished by microbial, mechanical, or chemical means.

Microbial retting is a technical term for *controlled rotting*. Microbes dissolve pectins and hemicelluloses in the middle lamella—ideally without attacking cellulose in the fiber. A distinction is made between **water retting** and **land retting**, depending on the moisture source. Water retting uses a pond or the eddy of a quiet river. Water retting is mentioned in a Chinese book of poetry, *Shī Jīng*, that dates to 750–600 BC. "The Pond by the Eastern Gate / Is good for steeping hemp" (Waley, 1996). Archaeological evidence (massive quantities of

Cannabis pollen in pond sediment) dates to 2000 BC in Garhwal Himalaya near Māna Pass, along an ancient trade route between India and China (Demske *et al.*, 2016).

Land retting utilizes atmospheric moisture; different microbes are involved. Land-retted stalks are spread in thin layers on the stubble of fields or on a mowed pasture. A seasonal distinction is made between **dew retting** (in the autumn) or less commonly **snow retting** (in the winter). Water-retted fiber is superior to dew-retted fiber. The retting is more consistent, and the fiber is not discolored (dew-retted fiber is often stained gray). Modern pond retting utilizes concrete or steel tanks. The water is heated and aerated with air compressors, and tanks are inoculated with pure cultures of retting organisms. Approximately 50 tons of water are used per ton of fiber (Kozlowski, 1995).

Mechanical retting utilizes machines to strip and separate fiber from stalks. Mechanical "decorticators" were first invented in Spain (Salvá Campillo and Santponç Roca, 1784). Mechanical decorticators yield low-quality fibers. **Steam retting** is traditionally practiced in Japan, Korea, and China (Clarke, 2010a). The new "steam explosion" method is much quicker than traditional steam retting, but the fiber is shredded and unsuitable for spinning.

Chemical retting employs acids or alkalis to accelerate the separation of fiber from the stalk. A Jesuit missionary observed the Chinese mixing quicklime (calcium oxide) into retting water in the 1740s. "This process provides them with a white thread, but a little deranged" (Pigeonneau and Foville, 1882). **Enzymatic retting** employs various proteins that catalyze the breakdown of pectins and hemicelluloses. Smythe (1951) proposed using hemicellulose enzymes for the retting of hemp, but Akkawi (1987) obtained the first patent for enzymatic retting of hemp. Akkawi used SPS-ase, a wide-spectrum polysaccharidase patented in Denmark 3 years prior. Akkawi also retted hemp with pectinase and *β*-glucanase.

Amarasinghe *et al.* (2022) compared three retting methods. Chemical retting pre-treated stalks with 0.3% HCl, followed by 6% NaOH at 70°C for 1 h, then post-treated with 1% acetic acid. Enzymatic retting immersed stalks in 0.1% pectinase at

DOI: 10.1079/9781836990352.0018

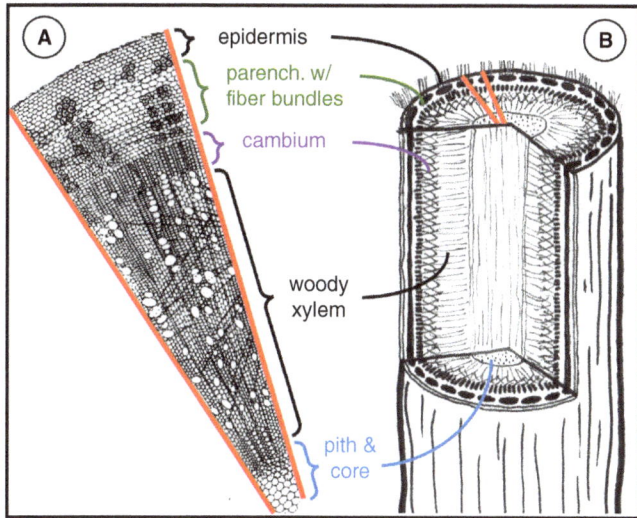

Fig. 18.1. Cross-section of hemp stalk: **A**. microscopic view (Mirbel, 1809); **B**. macroscopic view (McPartland *et al.*, 2000)

30°C for 30 min at pH 4, transferred to 10% cellulase at 50°C for 30 min at pH 6, transferred to 0.05% xylanase and 1% laccase at 70°C for 30 min at pH 7, and then heated at 80°C for 15 min to deactivate enzymes. Microbial retting utilized an uncharacterized microbial solution obtained from rotted okra (*Abelmoschus esculentus*), by immersing stems in the microbial solution at room temperature until fibers separated from the woody core. Chemical retting yielded fibers with the highest tensile strength, and enzymatic retting yielded fibers with the lowest tensile strength.

18.1.1   Fiber retting organisms

Kolb (1868) showed that pectin was the glue that held together fibers of flax and hemp. Fermentation of pectin yielded pectic acid, and he found pectic acid in retting ponds. Pasteur (1857) determined that fermentation was caused by microorganisms.

Van Tieghem (1877), a student of Pasteur, discovered *Bacillus amylobacter* (=*Clostridium butyricum*) fermenting vegetable matter under water. Two years later he proposed that bast fibers were retted by the action of *B. amylobacter* on cellulose (Van Tieghem, 1879). Praźmowski (1879) refuted the ability of *B. amylobacter* to dissolve cellulose; rather, it acted on a "more digestible substance." De Bary (1886) showed that fungi could dissolve the middle lamella in hemp fibers. He studied a pathogen (*Sclerotinia sclerotiorum*), and not retting organisms, but his results were applicable.

Mangin (1888) cited the work by Kolb and Van Tieghem, and proposed that *B. amylobacter* retted stalks by dissolving pectin in the middle lamella. This was demonstrated experimentally by Winogradski and Fribes (1895): an anaerobic bacterium fermented pectin in flax stalks, and did not break down cellulose. They described the bacterium as relatively large (10–15 μm long), and formed a large, terminal spore ("tadpole form")—no doubt a *Clostridium* sp.

Julius Wiesner (1838–1916), an Austrian plant anatomist, investigated retting and its microscopic effects on bast fiber. Wiesner (1873) dedicated five pages to hemp in his pioneering research. He recognized *Thauröste* (dew retting) as the dissolving of an intercellular substance in bast tissue. Based on Kolb's work, Wiesner surmised the substance was pectin. Microscopic examination of retted stalks revealed "fermentation organisms of a yeast-like and bacterial-like nature or else in the form of fungal mycelia, which are definitely formed from atmospheric germs."

Wiesner's work was rediscovered by Behrens (1902, 1903), who studied bacteria and fungi involved in hemp retting. He identified *B. amylobacter* (=*C. butyricum*) as the primary agent in water-retted hemp. Sterilized hemp stalks inoculated with pure cultures of *B. amylobacter* required the addition of nitrogen (ammonium salt) for robust fermentation. He also identified *Bacillus asterosporus* (=*Paenibacillus polymyxa*), which we now know is capable of fixing nitrogen.

In dew-retted hemp, Behrens identified *Mucor stolonifer* as the primary agent. He cultured *M. stolonifera* on calcium pectinate (extracted from beetroot). He also identified fungi that attacked cellulose instead of pectin, which rotted and ruined hemp fiber (discussed below). *M. stolonifera* could not ret hemp at 2–5°C, the temperature of winter retting. Under these conditions, he identified another *Mucor* species, which a colleague named *Mucor hiemalis* (Wehmer, 1903). Weisner's student Gerhard Ruschmann published a book about retting and that summarized work by all the aforementioned researchers. Ruschmann (1923) argued that Behrens misidentified *Mucor stolonifer*, it was actually *Rhizopus nigricans*.

Italian researchers began isolating bacteria from hemp retting ponds. Giacomo Rossi of Napoli identified *Bacterium cannabinum* (Rossi, 1902), a taxon that has disappeared from the literature. Subsequently, Rossi and Guarnieri (1907) identified *Bacillus comesii*. Rossi obtained a patent for hemp retting: cultures of *B. comesii* were added to an aerated vat of water kept at 28–35°C (US Patent 977,133). The retting could be completed in 3–20 days, depending upon the material under treatment and the degree of retting desired.

Rossi's former collaborator Dominico Carbone isolated *Bacillus felsineus* (= *Clostridium felsineum*) from the mud of a retting pit (Carbone and Tombolato, 1917). Soon the "Carbone method" supplanted the "Rossi method." Because *C. felsineum* was an anaerobe, it didn't require Rossi's aeration machinery. *C. felsineum* retted hemp in 60 h at 37°C. The microbe also retted flax and ramie, and it was adopted all over Europe.

Wallace H. Fuller (1915–2006) researched dew retting in Iowa during the WWII "war hemp" era. He wrote poetry and was a good artist (Fig. 18.1B is based on his drawing in Fuller *et al.*, 1946). He quantified microbial populations by plating sections of stalks on agar and counting colonies. In 1943, a dry year, colony counts obtained from unretted hemp showed that bacteria outnumbered fungi by 250 colonies to 32 colonies. After retting, bacterial counts increased up to 30-fold, and fungal counts increased up to 23-fold. In 1944, a wet year, bacterial counts increased up to 1050-fold, and fungal counts increased up to 25-fold (Fuller and Norman, 1944; Fuller *et al.*, 1946).

Fuller proposed that large bacterial populations in the wet year were responsible for more rapid retting that year. He made no attempt to identify the bacteria (he was a chemist), and he identified fungi to genus: *Alternaria*, *Hormodendrum* (=*Cladosporium*), *Fusarium*, *Aspergillus*, *Phoma*, and *Cephalosporium* spp. When brought into the laboratory (which is what Behrens had done), *Mucor* sp. and *Rhizopus arrhizus* appeared.

Fuller *et al.* (1946) evaluated dew-retting parameters—the effects of stalk moisture content and temperature, the effects of piling depth, turning over piles, and sprinkling with water. He also experimented with water retting, both aerobic and anaerobic (Fuller and Norman, 1946a). Instead of inoculating with pure cultures of *B. comesii* or *B. felsineus*, he made a "starter inoculum" by covering chopped hemp stalks in a crock with water, and incubated the crock in a warm place. He tested different pH levels and temperatures. Lastly, he measured biochemical changes arising during retting (Fuller and Norman, 1945, 1946b).

Research on retting resumed with relegalization—using 21st century molecular methods. Tamburini *et al.* (2004) revisited *Clostridium felsineum* and *C. acetobutylicum* in a 16S rRNA sequence analysis. They were phylogenetically related, but DNA–DNA hybridization indicated that they were distinct species.

Tamburini *et al.* (2003) used pectin-enriched agar to culture bacteria sampled from green hemp stalks, retted hemp stalks, and retting pits. They isolated 104 anaerobic and 23 aerobic strains, identified by their 16S rRNA sequences. A subset had ≥99.8% sequence identity with known species, so they could be identified. From anaerobic substrates: *C. felsineum*, *C. acetobutylicum*, *C. saccharobutylicum*, *C. aurantibutyricum*. From aerobic substrates: *Paenibacillus amylolyticus* and *Bacillus subtilis* ssp. *spizizenii*.

Tamburini *et al.* (2004) assayed these bacteria for polygalacturonase activity and for retting efficiency (which correlated). They selected two standouts, *C. felsineum* L1/6 and *B. subtilis* ROO2A, and illustrated hemp stalks retted in tanks inoculated with the two species. Di Candilo *et al.* (2010) conducted more tests with L1/6 and ROO4B. Inoculation of retting tanks with these bacteria, besides speeding up the process, significantly improved fiber quality. Quality was judged by color, homogeneity, softness, ease of hackling, and analysis by scanning electron microscopy.

Yadav *et al.* (2008) extracted enzymes from fungi as a source of chemical retting. They chose a strain of *Aspergillus terricola* from 20 candidates isolated in Gorakhpur, India, selected for its ability to secrete pectin lyase enzymes. Green hemp stalks were completely retted after 24 h in 0.24 IU of the purified enzyme.

Riberio *et al.* (2015) analyzed microbe populations of dew retted hemp in France. They compared samples from different soil types, harvested at different dates. Culturable organisms were identified by 16S rRNA (bacteria) or 18S rRNA (fungi).

The most frequent bacteria were *Escherichia coli*, *Pantoea agglomerans*, *Pseudomonas rhizosphaerae*, *P. fulva*, *Rhizobium huautlense*, *Massilia timonae*, and *Rhodobacter* sp. Fungal species were principally *Cladosporium* spp. (*C. herbarum*, *C. cladosporiodes*, *C. bruhnei*), *Cryptococcus carnescens*, *Aspergillus versicolor*, *A. glaucus*, and *A. fumigatus*. Stalks retted on "red ground" (clay soil) had higher bacterial and fungal counts than "white ground" (chalky soil); retting was more efficient on red ground.

Law *et al.* (2020) simulated dew retting in a greenhouse environment. They determined the effects of different moisture levels on the retting process. Microbes were directly sequenced from hemp and soil samples, using NextGen sequencing of 16S rRNA amplicons. Species diversity was lower at low moisture levels compared with high moisture levels; when sampled over six time points, bacterial populations shifted at high moisture levels, but not at low moisture levels. Dominant bacteria, identified to genus, were *Pseudomonas*, *Sphingomonas*, *Pantoea*, *Aureomonas*, *Methylobacterium*, *Massilia*, and *Enterobacteriaceae*.

18.1.2   Fiber *rotting* organisms

As pectin becomes depleted, many microbes begin to digest cellulose. At this point, retting of the middle lamella switches to rotting of bast fibers. When Tamburini *et al.* (2004) assayed bacteria for polygalacturonase activity, they also searched for strains with low cellulolytic activity. They found a few. Law *et al.* (2020) observed a shift towards *Chryseobacterium* spp. late in the retting season. This was associated with a depletion of pectin, increased cellulolytic activity, and rotted fibers.

It might be mentioned that insects can damage retted stalks, as well as finished hemp fiber. Olivier (1789) mentioned that cockroaches (*Periplaneta americana*) gnaw on hemp fabric. In Italy, larvae of the Mediterranean flour moth (*Ephestia kühniella*) posed a threat to stored fiber (Ferri, 1959a). The silverfish (*Lepisma saccharinum*) may chew holes in hemp fabric (Raven, 1987). In his famous book *Micrographia*, Robert Hooke (1665) described silverfish eating holes in books, "finding, perhaps, a convenient nourishment in those hulks of hemp and flax."

The greatest threat, however, comes from fungi. Behrens (1902) implicated fungi as a "damaging influence" in dew-retted hemp. He identified **Cladosporium herbarum**, which was "already present everywhere" on dried stalks *prior* to retting—seen as black spots and stripes (Fig. 18.2). Behrens demonstrated

Fig. 18.2. *Cladosporium* sp. causing black spots at an early stage of dew retting

that *C. herbarum* could grow on pure cellulose (Swedish filter paper), given adequate moisture and mineral supplements. Under a microscope, Behrens observed *C. herbarum* growing between and within bast cells. This explained the gray discoloration of dew-retted hemp. Behrens also identified *Botrytis cinerea* and *Aspergillus glaucus* attacking cellulose in hemp stalks.

Fuller in Iowa also identified *Cladosporium* ("*Hormodendrum*") in over-retted hemp. However, *Trichothecium roseum*, *Cephalosporium* sp., *Alternaria* sp., and *Fusarium* sp. occurred more frequently in over-retted hemp (Fuller and Norman, 1944; Fuller *et al.*, 1946). Earlier, Agostini (1927) identified *Alternaria alternata* in rotted hemp fabrics, along with *Stachybotrys lobulata*.

Hrebenyuk (1984) identified 79 species of fungi growing on dew-retted stalks in Ukraine. The three most common were *Cladosporium cladosporioides*, *Alternaria cheiranthi* (possibly misidentified *A. alternata*), and *Gonatobotrys simplex*. Further down the list appeared *C. herbarum*, *B. cinerea*, and *T. roseum*. It's quite a rogue's gallery, including bad pathogens (e.g., *Verticillium albo-atrum*, *Sclerotium rolfsii*), and even a slime mold, *Didymium clavus*. In Czechoslovakia, Ondřej (1991) compiled a shorter list, with 41 species. *A. alternata*, *C. herbarum*, and *Stemphylium botryosum* were among the abundant species, with *B. cinerea*, *T. roseum*, and *C. cladosporiodes* less abundant. Clearly, some of these bad actors are double trouble: causing disease in living plants, then rotting fiber after plants die.

18.1.3 Special fungal species

Some fungi have only been reported on dead and rotten stalks—and not on living plants. A subset of these fungi has only been reported on *Cannabis*. They are listed alphabetically.

18.1a. *Calycina herbarum* (Persoon) Gray 1821

≡*Hymenoscyphus herbarum* (Pers.) Dennis 1964, ≡*Helotium herbarum* (Pers.) Fries 1849
Description: *Apothecia* gregarious, arising on a very short stalk, pale yellow to ochraceous, minutely downy, flat to slightly convex, disc up to 3 mm in diameter. *Asci* cylindrical to clavate, stalked, 8-spored, up to 90 × 8 µm. *Paraphyses* hyaline, filiform, 90 × 2 µm. *Ascospores* 1-septate, hyaline, fusiform to cylindrical, 13–17 × 2.5–3.0 µm.
Comments: Saccardo found *C. herbarum* fruiting on *Cannabis* stalks near Padova, Italy. He distributed exsiccati as *Mycotheca italica* No. 119 (BPI!). *C. herbarum* infests many herbaceous hosts, especially *Urtica* spp. The fungus occurs worldwide in temperate climates. It sporulates on dead stalks in October, spreading ascospores by wind and water.

18.1b. *Chaetomium* spp.

Hrebenyuk (1984) isolated five *Chaetomium* spp. from hemp stalks in Ukraine: **Chaetomium globosum** Kunze 1817, **Chaetomium elatum** Kunze 1818, **Chaetomium murorum** Corda 1837 (now **Botryotrichum murorum** (Corda) Wei *et al.* 2016), **Chaetomium trilaterale** Chivers 1912, and **Chaetomium piluliferum** Daniels 1961. Morphological characters of the five species are provided in Table 18.1. Gwinn *et al.* (2023) isolated three *Chaetomium* spp. from products sold by licensed cannabis facilities in British Columbia: *C. globosum*, *C. elatum*, and *C. brasiliensis*.

18.1c. *Cyathicula coronata* (Bulliard) P. Karst 1865

≡*Peziza coronata* Bulliard 1789
Description: *Apothecia* gregarious, arising on a short stalk, 2–4 mm, yellow to ochraceous, concave with the disc margin exhibiting a series of teeth like a crown, disc up to 3 mm in diameter. *Asci* cylindrical, 8-spored, 90–110 × 7–9 µm. *Paraphyses* hyaline, cylindrical, 90–120 × 2 µm. *Ascospores* 1-celled rarely 2-celled, hyaline, biseriate, often biguttulate, cylindrical with rounded ends, 15–20 × 3–4 µm.
Comments: Bulliard (1789) named *Cannabis* as a host in his original description, but no one else has. This well-known Eurasian species is found on fallen stems of herbaceous plants in the Urticaceae, Asteraceae, and especially the Asteraceae.

Table 18.1. Characteristics of five *Chaetomium* species described on harvested hemp stalks

Species	Perithecium size (µm)	Ornamental hairs	Ascospore size (µm)	Ascospore shape
C. globosum	200–300 tall × 200–280 diam.	Undulating, tapering to hyaline tips	9–13 × 6–10	Lemon-shaped with apiculate ends
C. elatum	400–500 tall × 335–450 diam.	Dichotomous branches tapering to points	11–14 × 8–10	Lemon-shaped with apiculate ends
C. murorum	240–340 tall × 200–345 diam.	Circinate tips with graceful arches	13–17 × 7–9	Ellipsoid with a longitudinal furrow
C. piluliferum	280–560 tall × 222–480 diam.	Circinate tips sinuous and smooth	13–17 × 8–10	Ellipsoid with apiculate ends
C. aureum	100–140 tall × 100–130 diam.	Yellowish brown with spiral ends	9–12 × 5–7	Ovate and flattened on one side

18.1d. *Diplodiella ramentacea* Tassi 1900

Description: *Pycnidia* scattered or gregarious, black, more or less spherical, 200–250 μm in diameter. *Conidiogenous cells* not described. *Conidia* elliptical, initially unicellular, almost hyaline, finally with a transverse wall, slightly constricted, with two oil drops, pale yellow-brown color, 6–8 × 2.3 μm.

Comments: Tassi (1900) isolated this fungus from rotting hemp stalks near Siena in Italy, where it had the common name *Fiammifero* because the rotting stalks smelled sulfurous like match sticks. Allescher (1903) reported *D. ramentacea* in Germany. Zambettakis (1955) synonymized it under *Diplodiella fibricola* (Berk.) Saccardo 1884, although Mycobank still considers *D. ramentacea* a legitimate taxon.

18.1e. *Leptospora rubella* (Persoon) Rabenhorst 1857

Description: *Pseudothecia* appear on blackened or purple-red stalks, scattered or gregarious, immersed then more-or-less erumpent, globose to conical, often laterally compressed, black, glabrous, papillate; contains periphyses, paraphyses, and asci; 200–300 μm in diameter. *Asci* bitunicate, borne upon short stalks, cylindrical, 8-spored, 140–160 × 4.5–6.0 μm (Holm says asci grow to 225 μm long). *Ascospores* slightly spiraled within the ascus, filiform, yellow, with many obscure septa and guttules, nearly as long as asci and 1 μm wide.

Comments: Holm (1957) listed ten synonyms for this species. Hrebenyuk (1984) reported two of those synonyms on Ukrainian hemp stalks, *Ophiobolus porphyrogonus* (Tode) Saccardo 1883, and *Ophiobolus vulgaris* (Saccardo) Saccardo 1881.

18.1f. *Micropeltopsis cannabis* McPartland 1997

Description: *Ascomata* thyriothecioid, 45–130 μm in diameter, flattened ampulliform, 25–46 μm high, dark brown to black, ostiolate, margin entire; upper layer composed of radially arranged quadrangular cells; basal layer of similar construction to upper layer but paler. *Ostiole* central, raised, composed of small cells and bearing a crown of divergent setae (Fig. 18.3B). *Setae* straight, subulate, thick-walled, non-septate, smooth, dark brown, 12–22 μm long. *Asci* bitunicate, ovoid to obclavate, 4–8-spored, 21–40 × 4–9 μm. *Ascospores* hyaline, guttulate, ellipsoid, with a single septum, 11–12 × 2.5–3.0 μm.

18.1g. *Ophiobolus cannabinus* Passerini 1888

Description: *Pseudothecia* scattered, globose or conical, black, immersed with conical apex and ostiole barely erumpent. *Asci* cylindrical, short-stalked, 8-spored, 85 × 5 μm. Paraphyses cylindrical. *Ascospores* filiform, aseptate, hyaline, 65–85 × 1.0–1.25 μm.

Comments: Passerini (1888) found *O. cannabinus* on hemp stalks near Parma, Italy. His description of aseptate ascospores

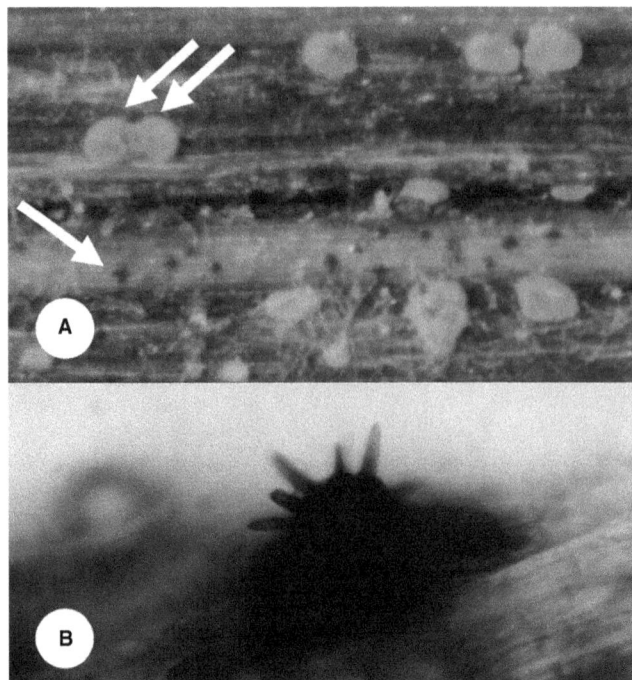

Fig. 18.3. *Orbila* and *Micropeltopsis* on a hemp stalk: **A.** macroscopic view; **B.** microscopic view of *Micropeltopsis*

would preclude this fungus from *Ophiobolus*. Three other *Ophiobolus* spp. have been described on hemp stalks. Preston and Dosdall (1955) collected *O. anguillides* (Cooke) Saccardo from hemp near Judson, Minnesota (see Fig. 3.3). Hrebenyuk (1984) identified two species on hemp stalks in Ukraine, *O. porphyrogonus* and *O. vulgaris*, both of which have been synonymized under *Leptospora rubella* (see above).

18.1h. *Orbilia luteola* (Roumeguere) McPartland 1997

≡*Calloria luteola* Roumeguere 1881

Description: *Apothecia* superficial, sessile, translucent yellow-orange, margin spherical to ellipsoidal, up to 0.6 mm in diameter and 100 μm in thickness. *Excipulum* consists of hyaline thin-walled textura globulosa. *Asci* cylindrical, 8-spored, 26.0 × 4.5 μm. Paraphyses hyaline, filiform, slightly enlarged at the apex. *Ascospores* hyaline, single-celled, fusiform, some indistinctly guttulate, 6.5 × 1.5 μm.

Comments: Roumeguere (1881) found *C. luteola* on rotten hemp stalks near Villernur, France. He distributed specimens in his *Fungi Gallici exsiccati*. Roumeguere's holotype specimen (herb. BR), as well as two isotypes (BPI, CUP) were examined by McPartland and Cubeta (1997), who transferred *C. luteola* to *Orbilia*. In the isotype at BPI, *O. luteola* apothecia (Fig. 18.3A, double arrow, white cups) cohabitated with tiny thyriothecioid fungi (Fig. 18.3A, single arrow, black dots). The spiny little fungi proved to be a new species, *Micropeltopsis cannabis* (McPartland and Cubeta 1997).

18.1i. *Periconia byssoides* Persoon 1801

≡*Cephalotrichum byssoides* (Persoon) Kuntze 1898
=*Periconia pycnospora* Fresenius 1850
Description: *Conidiophores* erect, straight or slightly flexous, 2–3-septate, dark brown at base and paling to subhyaline at the apex, 12–23 μm in diameter at base tapering to 9–18 μm below the head, then forming a septum with an apical head swelling to 11–28 μm in diameter, entire length 200–1400 μm (up to 2 mm). *Conidiogenous cells* hyaline, ellipsoidal to spherical, monoblastic to polyblastic. *Conidia* cantenate, forming chains, spherical, brown, verrucose, 10–15 μm in diameter.
Comments: Gitman and Malikova (1933) and Gitman and Boytchenko (1934) described blackened hemp stalks with *P. byssoides* near Moscow. Saikia and Sarbhoy (1982) collected the fungus from dead branches and stalks in Assam, India. Fuller and Norman (1944) found a *Periconia* sp. on retted hemp in Iowa. Hrebenyuk (1984) reported a related species, *Periconia cookei* Mason & Ellis 1953, on hemp stalks near Kiev. *P. cookei* differs little from *P. byssoides*—it lacks a septum at the apical cell, and produces slightly larger conidia (13–16 μm in diameter).

18.1j. *Torula herbarum* (Persoon) Link 1809

Description: *Colonies* olivaceous when young, dark brown to black when old, velvety. *Conidiophores* septate, pale olive, 2–6 μm wide, enlarging at conidiogenous cells to 7–9 μm in diameter. *Conidia* cylindrical, resembling a chain of attached spheres, mostly 4–5-septate (range 2–10), constricted at septa, straight or curved, pale olive to brown, verruculose to finely echinulate, 20–70 × 5–9 μm.
Comments: Behrens (1902) identified *T. herbarum* retting hemp in Germany. Hrebenyuk (1984) reported it colonizing harvested hemp stalks in Ukraine. Clarke collected *T. herbarum* from hemp stalks previously parasitized by *Colletotrichum dematium* in China (McPartland, unpublished data 1995). This species also colonizes hemp leaves as a secondary parasite, and causes a post-harvest storage mold of marijuana (McPartland, unpublished data).

18.1.4 Occupational hazards

After retting, microbes persist in hemp stalks, and workers who process the stalks can get sick. **Byssinosis**, also called "Monday syndrome," is a disease common to all textile workers exposed to fiber dust. After a weekend away from work, renewed exposure causes chest tightness, shortness of breath, wheezing, and coughing. These symptoms usually dissipate as the work week progresses, only to reappear the following Monday (McPartland, 2003). Ramazzini (1713) first described hemp workers suffering from chronic coughing and wheezing in northern Italy. French mill owners blamed these symptoms on their workers' weekend "moral excesses" (Mareska and Heyman, 1845). Cazin and Cazin (1868) blamed dust arising from the beating and carding of hemp. They noted, without irony, that inhaling the fumes of nitrated hemp leaves could treat the symptoms.

Salomon (1893), a physician in France, wondered why asthmatics gained relief by smoking *chanvre indien*, yet hemp factory workers became asthmatic. Other acute symptoms included headache, dyspnea, chills, fever, sweating, conjunctivitis, stomatitis, eczema, general malaise, various neuralgias, and "sleep peopled with fantastic dreams." Chronic exposure led to bronchitis and pneumonia, postural abnormalities, muscle atrophy, and weight loss. Salomon compared the physique of a hemp factory worker to another manual laborer in Sarthe (Fig. 18.4A). Twelve hemp factory workers weighed an average of 101.6 pounds, compared with the average manual laborer's weight of 140 pounds. Salomon blamed hemp dust, breathed constantly in confined air. He invented a filter mask for workers to wear (Fig. 18.4B).

In Spain, respiratory disease in hemp workers was called "cannabosis" (Jimenez and Lahoz, 1944). The British physician Schilling (1956) equated cannabosis to byssinosis, a syndrome shared by flax, cotton, jute, and sisal workers. Schilling's breakthrough studies in the 1950s and 1960s were followed by brilliant work published by Arend Bouhuys (Bouhuys *et al.*, 1967; Bouhuys and Zuskin, 1976), and Eugenija Zuskin (Valic *et al.*, 1968; Zuskin *et al.*, 1976, 1990, 1992). Among all textile workers, hemp processors suffer the worst (Zuskin *et al.*, 1976). Thomas *et al.* (1988) reported that 15% of hemp workers suffered from byssinosis ("pousse" in Ireland), whereas only 2.8% of sisal workers were afflicted.

Bacterial endotoxins have been implicated as the cause of byssinosis, rather than fungal toxins or constituents in hemp itself (Castellan *et al.*, 1984). Endotoxins are lipopolysaccharides produced by gram-negative bacteria. Er *et al.* (2016) reported highest endotoxin levels in fine hemp dust (605 EU/mg), compared with coarse hemp dust (336 EU/mg), and hemp fibers (114 EU/mg).

Caminita *et al.* (1943) blamed "hemp fever" on gram-negative *Enterobacter cloacae* isolated from hemp dust. They isolated *E. cloacae* from retted hemp and not from unretted hemp. Nicholls *et al.* (1991) found that hemp fibers contain 10 times more bacteria than cotton, flax, or jute fibers. Nicholls isolated *E. cloacae* and four other species of Enterobacteriaceae, three species of *Pseudomonas*, and assorted *Bacillus*,

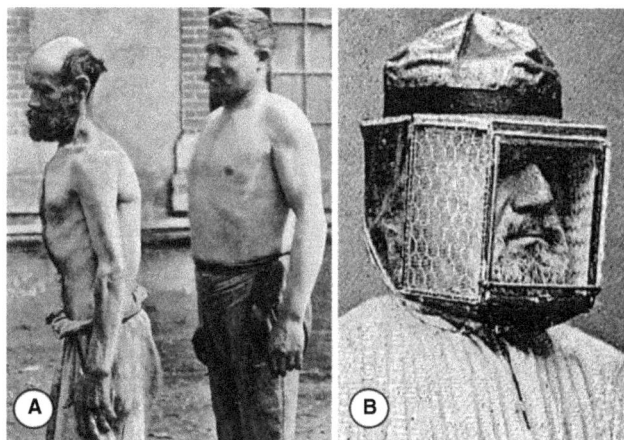

Fig. 18.4. A. Emaciated hemp factory worker compared with another worker; **B.** mask for filtering hemp dust (Salomon, 1893)

Corynebacterium, *Staphylococcus*, and *Acinetobacter* spp.—a nasty bunch, left over from hemp retting. Fungi are not exonerated. *Pantoea agglomerans*, a hemp-retting organism, produces endotoxins (Riberio *et al.*, 2015), and causes byssinosis in cotton workers (Dutkiewicz *et al.*, 2015). In Bulgaria, Dimitrov *et al.* (1990) found a high concentration of fungal spores in hemp mill air (mostly *Cladosporium*, *Alternaria*, *Aspergillus*, *Penicillium*, and *Fusarium* spp.).

Within hemp mills, the prevalence of byssinosis varies by mill occupation. Valic *et al.* (1968) measured disease rates in workers involved in hackling, carding, and spinning of hemp (Table 18.2). Workers tested with hemp dust extracts (skin-prick tests) also varied in their response to dust collected from different parts of the mill (Zuskin *et al.*, 1992).

After 10–20 years of exposure, workers with byssinosis may develop chronic obstructive lung disease (COPD), and begin to lose lung tissue (Bouhuys and Zuskin, 1976). COPD includes emphysema and chronic bronchitis. Bouhuys and Zuskin stated emphatically, "We believe that deterioration of lung function among hemp workers begins before the age of 45 and it continues even if further exposure to dust ceases." In Turkey, Er *et al.* (2016) reported a byssinosis rate of 28.2% in hemp workers, with highest prevalence in tobacco smokers. The disease becomes lethal if not treated. In one surveyed Spanish town, the average lifespan of hemp workers was 39.6 years, whereas non-hemp farm workers in the same town lived for 67.6 years (Schilling, 1956).

Keeping down dust levels is difficult. Improving machinery and ventilation can decrease dust exposure. But even in a modern hemp facility in Australia, dust levels exceeded those recommended by law (Gardner *et al.*, 2020). Masks are an option, but never a popular one.

18.2   Hemp seeds

Hemp seeds have a long history of use as human food. Some scholars believe that humans were attracted to seeds first, prior to the plant's fiber or cannabinoids (Liger, 1700; Vavilov, 1926; Reininger, 1946; Schultes, 1973). The oldest archeological evidence of *Cannabis* usage consists of seeds found in kitchen middens, rather than hemp fibers. They date to cal. ^{14}C 8000 BC (Kobayashi *et al.*, 2008; Kudo *et al.*, 2009).

Harvested seeds are lost to many kinds of organisms. As neatly summed in an olde English proverb:

> *One for the rat, one for the crow,*
> *One to rot, and one to grow.*

Rats and mice are notorious marauders of all stored foodstuffs. They're the reason we domesticated cats 9500 years ago (see Fig. 20.1). Chomel (1709) stated that *chènevis* (hempseed) should be stored in a place "where the rats cannot damage it." For more about rodents, see section 16.6.

18.2.1   Insect pests of stored grain

Christensen (1957) first made the distinction between *field organisms* (Group 1) and *storage organisms* (Group 2) in a study of deteriorated grain. Group 1 organisms invade living plants and then carry over as storage problems. Group 2 organisms only infest products after harvest.

Group 1 organisms do poorly in properly stored products, which lack the high levels of moisture and oxygen they require. Group 1 insects include budworms (*Heliothis* and *Helicoverpa* spp.), which specialize in destroying flowers, fruits, and seeds (see section 5.2). **Group 2** organisms have evolved tolerance for low moisture and oxygen levels. Other attributes of Group 2 organisms include short life cycles, high reproductive rates, good dispersal ability, and rapid mutation and selection. Prior to humans storing food products, Group 2 insects lived in bird nests, rodent burrows, or anthills.

The granary weevil *Sitophilus granarius* has completely adapted to life in piles of seeds. Larvae complete their entire development hidden within seeds, and the adults have lost the ability to fly—their wing covers have fused. Loss of flight prevents water loss in a dry environment, and fused wing covers protect them from the rough-and-tumble among seeds as they are transported by humans to new places. *S. granarius* no longer exists in nature outside of stored seed (Plarre, 2010).

S. granarius infests hemp seed (Ormerod, 1898). Conversely, its cousin, *Sitophilus zeamais*, did not feed on hemp seeds in a host-range study (Obata *et al.*, 2011). Yet another cousin, the worldwide curse *Sitophilus oryzae* (rice weevil), can feed on hemp seeds as an adult, but the larvae cannot develop and reproduce in hemp seeds (Rusynov *et al.*, 2019). Another

Table 18.2. Prevalence of byssinosis within hemp mills

Working section	Disease prevalence (%)[a]	Positive test prevalence (%)[b]
Hacklers (combing machines)	62.5	41%
Carders (carding machines)	57.1	38%
Strippers and grinders	50.0	Not tested
Spinners (spinning machines)	40.0	33%
Drawers	30.2	Not tested
Weaving machines	Not tested	33%

[a]Prevalence of byssinoisis symptoms as reported by Valic *et al.* (1968)
[b]Positive skin-prick tests using extracts of hemp dust collected from different mill machines by Zuskin *et al.* (1992)

well-known Group 2 insect, the fruit fly *Drosophila melano-gaster*, normally feeds on rotting vegetative matter. However, adding hemp seed meal to their diet increases their body weight by 20%, and females lay 16% more eggs than controls (Lee *et al.*, 2010). Hemp seed is good food.

Other notorious Group 2 insects infest stored hemp seeds. Glover (1869) reported larvae of a lined flat beetle, *Laemophloeus modestus*, in stored hemp seed. Now named *Placonotus modestus*, it is closely related to *Cryptolestes* spp., which are common pests of stored grain products. The confused flour beetle, *Tribolium confusum,* has been found in stored seed (Marlatt, 1921; Smith and Olson, 1982).

Arnaud (1974) identified the foreign grain beetle (*Ahasverus advena*), rusty grain beetle (*Cryptolestes ferrugineus*), flat grain beetle (*Cryptolestes pusillus*), saw-toothed grain beetle (*Oryzaephilus surinamensis*), the Mediterranean flour moth (*Ephestia kühniella*), and the Indian meal moth (*Plodia interpunctella*) from hemp seeds in storage in California. The latter two species, *E. kühniella* and *P. interpunctella*, were also reported by Meyer (1974) in Indiana.

Stadnyk *et al.* (2018) tested hemp seeds as a food source for ten species of storage insects. Four species reproduced on hemp: red flour beetle, *Tribolium castaneum*; drugstore beetle, *Lasioderma serricorne*; saw-toothed grain beetle, *Oryzaephilus surinamensis*; and warehouse beetle, *Trogoderma variabile*. Six species did not reproduce: confused flour beetle, *Tribolium confusum*; rusty grain beetle, *Cryptolestes ferrugineus*; rice weevil, *Sitophilus oryzae*; lesser grain borer, *Rhyzopertha dominica*; flour mill beetle, *Cryptolestes turcicus*; and cigarette beetle, *Stegobium paniceum*.

Hamilton *et al.* (2021) studied reproduction of 11 species stored with hemp seed for 9 weeks, at two moisture contents (dry 9% or damp 15%), and two dockage levels (chaff-free or 15% chaff). *Ephestia kühniella* did best, especially at 9% moisture and 15% chaff. Four beetle populations increased over the 9 weeks: *Tribolium castaneum* (red flour beetle), *Lasioderma serricorne* (cigarette beetle), *Oryzaephilus surinamensis* (sawtoothed grain beetle) and *Trogoderma variabile* (warehouse beetle). The following species did not increase their populations: *Cryptolestes ferrugineus* (rusty grain beetle), *Rhyzopertha dominica* (lesser grain borer), *Sitophilus oryzae* (rice weevil), *Cryptolestes turcicus* (flour mill beetle), *Tribolium confusum* (confused flour beetle), and *Stegobium paniceum* (drugstore beetle).

Rusynov *et al.* (2019) named six other storage beetles infesting hemp seed in Ukraine: *Ophonus griseus*, *O. rufipes*, *Amara apricaria*, *A. ingenua*, *Phyllopertha horticola*, and *Amphimallon solstitiale*. See their publication for descriptions. All six species are native to Eurasia.

Mites (not insects) can infest stored grains and cause problems in stored hemp seeds. Dombrovskaya (1940) identified the cereal mite (*Tyrophagus noxius*, =*Tyrophagus putrescentiae*) in stored hemp seed in Ukraine. Dombrovskaya found larvae of a predacious fly feeding on the mite, which she named as a new species, *Silvestrina tyrophagi*. Chmielewski and Filipek (1968) also found *T. putrescentiae* infesting hemp seeds. Boczek *et al.* (1960) reported two notorious mite pests feeding on hemp seeds, the flour mite (*Acarus siro*, =*Tyroglyphus farinae*) and the hairy grain mite (*Glycyphagus destructor*). *G. destructor* was also reported by Piao (1990), along with

Tydeus kochi and a *Tarsonemus* sp. Chmielewski (1984) identified the grain mite *Tyroglyphus longior* in stored seed, which reduced seed germination rates.

18.2.2 Hemp seed microbes

Hemp seed microbes helped debunk the theory of spontaneous generation. Wrisberg (1765) placed hemp seeds in warm water, from which microscopic "animalcules" materialized 4 days later—transmutated into life by a mysterious vegetative force. Spallanzani (1765) tried destroying the "vegetative force" by mincing hemp seeds to bits, yet animalcules still emerged from minced seeds when placed in water. Needham (1769) argued that mincing seeds did not eliminate their vegetative force. He boiled seeds in a flask, sealed the flask with a cork, and animalcules still emerged (no doubt from the contaminated cork). Spallanzani (1776) boiled hemp seeds in a glass flask, and then hermetically sealed the flask by melting the glass—no animalcules emerged.

Instead of spontaneous generation, Ellis (1769) proposed that animalcules arose through "spontaneous division," later called binary fission. Ellis observed microscopic animalcules emerging from hemp seeds soaked in water. They multiplied by dividing in two. Different animalcules also appeared, larger and more elongate than the others. They opened their trunks to swallow the smaller ones (Fig. 18.5). Ellis obtained various samples of hemp seed from all over London, soaked them in distilled water, and the same animalcules emerged.

Some fungi invade seeds while still attached to the plant. Paulsen (1971) isolated *Penicillium* sp. and *Aspergillus* sp. localized at the hilum end of feral hemp seeds in Kansas, and *Alternaria alternata* colonized the entire seed coat, as well as bracts. Haney and Kutscheid (1975) estimated that 20.5% of seeds were destroyed by *A. alternata* in feral hemp populations in Illinois.

After seeds are harvested and placed in storage, fungi may ruin entire lots of poorly stored hempseed. Pietkiewicz (1958) and Ferri (1961b) investigated the microflora of deteriorating

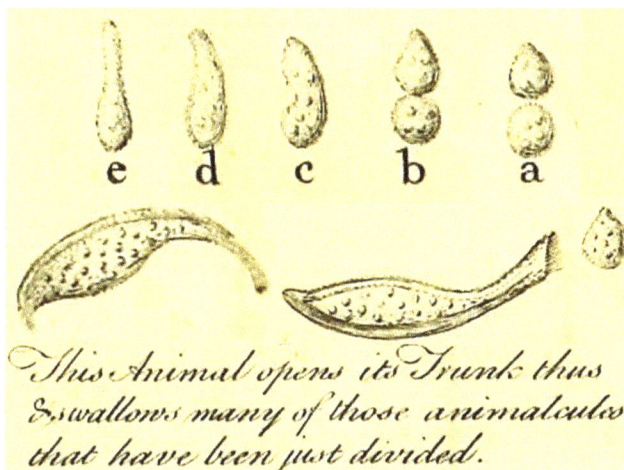

Fig. 18.5. Animalcules in a hemp seed infusion (Ellis, 1769)

seeds. Both researchers isolated *Alternaria alternata*, *Rhizopus stolonifera*, *Cladosporium herbarum*, and several *Fusarium* spp.

Babu *et al.* (1977) identified 16 fungi and a *Streptomyces* sp. growing on the surfaces of seeds. More importantly, they uncovered 10 organisms from internal parts of seeds: *Penicillium chrysogenum*, *P. frequentans*, *P. chermesinium*, *P. lævitum*, *P. fellutanum*, *P. chrlichii*, *Aspergillus niger*, *A. sulphureus*, *Cladosporium herbarum*, and *Cephalosporium curtipes*. Abdulla and Khalaf (2016) isolated *Aspergillus* sp. and *Rhizopus stolonifer* from hemp seeds sold as bird seed, sealed in polyethylene bags.

18.2.3   Control of seed pests and pathogens

Hemp seeds should be stored in cool, dry, and secure conditions. Chomel (1709) simply said, "keep it [hemp seed] in a place where it does not mold." Seeds harboring bacteria and fungi can be disinfested with thermotherapy (see section 19.2.1). Stored seeds have been protected with pesticides (Robinson, 1943a; Koehler, 1946; Ferri, 1961b).

Crescini (1943) proposed that seed viability improves if seeds are allowed to "after-ripen" on plants, by placing harvested plants in a shady place for 7–10 days prior to threshing. In the USA researchers began studying ways to store hemp seed during WWII, when a shortage of germplasm thwarted "war hemp" cultivation. Koehler (1946) reported that hemp seeds with high moisture content deteriorated more rapidly than seeds with low moisture content when stored at 21°C. Seed moisture content was less of a factor when seeds were stored at 0°C.

Toole *et al.* (1960) described studies they began in 1947. Fresh, undried seeds of Kentucky hemp contained 8.3% water; when dried at 47°C, water content decreased to 5.7%. Dried and undried seeds were placed in cloth bags or sealed in glass jars, and subjected to three storage conditions under variable relative humidity (RH) and temperatures (Table 18.3). For storage over 2 years (24 months), seeds dried to 5.7% and sealed in glass jars germinated the best, at any temperature. Dried seeds in cloth bags reverted to ambient RH. They repeated the glass jar study in 1950, with dried and undried seeds at four temperatures (–10°C, 0°C, 10°C, 21°C). Dried

seeds germinated well at all four temperatures, and undried seeds did best at –10°C.

Sengbusch (1955) reported studies initiated in 1946. Low temperatures preserved germination rates. Hemp seed stored at 20°C lost viability after 3 years, but seed stored at –20°C retained viability for 5 years. He exposed seeds to –40°C without damage. Sengbusch said that extending preservation for several years would help germplasm banks, like Leningrad and Gatersleben, which grew out their hemp collections every year—at a high cost, and a risk of cross-fertilization of accessions.

Crocioni (1950) analyzed the effects of moisture and temperature on germination rates. He preserved viability for 16 months by lowering seed moisture to 6.0%, stored slightly above 0°C. In a meta-analysis of seed longevity studies concerning 92 plant species, Priestley *et al.* (1985) summarized three studies on hemp (from France, Germany, and the USSR, without citing references). The three studies measured loss of germinability over several years and calculated P_{50} values (half-viability period): 5.48, 6.25, and 7.25 years. Priestley omitted a study in Japan by Kondo *et al.* (1950), who reported that hemp seeds could germinate after 19 years under desiccated conditions.

Small and Brookes (2012) tested fiber-type hemp ('ESTA-1', 'Carmen', 'USO-31'), a drug-type cultivar ('Medicinal'), and Canadian feral hemp, stored in glass jars. Storage for 18 months under ambient conditions (room temperature, seed moisture 11%) reduced seed germination to zero. Then they tested four temperatures and three moisture levels, and measured percentage germination after 1, 6, 12, 18, and 66 months. Results at 66 months are presented in Table 18.4. They concluded that seed desiccated to 6% and refrigerated at 5°C is sufficient for several years. Colder temperatures (–20°C) improved germinability, which began to decline at –80°C.

Parihar *et al.* (2014) also tested the interaction of seed moisture × storage temperature, in seeds of Himalayan provenance. In seeds stored at 15°C for 36 months, seed moisture made a big difference: at 5% seed moisture the germination rate was 84%; it was 88% at 7%, 63% at 8%, 43% at 10%, and 0% at 12%. In seeds stored at –20°C for 36 months, seed moisture made little difference: at 5% seed moisture the germination rate was 90%, it was 90% at 7%, 85% at 8%, 86% at 10%, and 85% at 12%.

Table 18.3. Germination rate of seeds stored under different conditions (Toole *et al.*, 1960)

Months storage	Seed dried to 5.7% and stored in cloth bags			Undried seed at 8.3% and stored in cloth bags		
	10°C and 50% RH	–10°C and "high RH"	Ambient conditions, unheated building	10°C and 50% RH	–10°C and "high RH"	Ambient conditions, unheated building
0	98%	98%	98%	100%	100%	100%
24	99%	98%	7%	98%	99%	4%
75	99%	91%	0%	99%	96%	0%
	Seed dried to 5.7% and stored in sealed glass jars			**Undried seed at 8.3% and stored in sealed glass jars**		
0	98%	98%	98%	100%	100%	100%
24	98%	99%	99%	98%	98%	80%
75	98%	98%	96%	99%	99%	0%

Table 18.4. Percentage seed germination after 66 months under different conditions (Small and Brookes, 2012)

Moisture content	+20°C	+5°C	−20°C	−80°C
11%	0	34.6	66.0	65.4
6%	56.4	71.0	78.0	69.8
4%	52.8	73.6	75.8	70.4

Ironically, stored wheat has been protected from insect pests with *Cannabis* leaves and flowers. Equally ironically, *Cannabis* has served as both a repellent and an attractant (to trap pests). Société d'Agriculture du Marne (1813) used hemp as an attractant to draw *charançons* (weevils) from piles of wheat. Piles were stirred with a shovel to awaken the weevils, then "newly picked hemp" was placed on a sheet nearby. After attracting weevils, the sheet with hemp was shaken out in a poultry yard, where chickens were "very eager for this insect." Perris (1851) proposed the opposite strategy: repel *calandre du blé* (wheat weevils) by placing bundles of fresh *Cannabis* near heaps of wheat. Carpentier (1862) said the same, and identified the pest: *Curculio granarius* (now *S. granarius*). Flament (1874) was more exacting: sweep out the storage attic, and place handfuls of fresh *female* hemp. "A little before harvest, it exhales enough odor to be put in the attic." Valyn (1885) recommended doing this every year, as a preventive.

This French thing spread to the USA. Riley and Howard (1892) scattered hemp leaves in stored grain to repel grain weevils. MacIndoo and Stevers (1924) scattered hemp leaves among bags and heaps of grain to protect against grain weevils. Hemp extracts were also used to protect stored seeds from bacteria and fungi. Zelepukha (1960) described a preparation made from wild hemp that protected potato and tomato seeds from bacterial diseases. Pandey (1982) protected millet seed from over 25 species of fungi with extracts of Indian *Cannabis*.

18.3 Flowering tops

Until the 1980s, cannabis in the USA was usually seeded. Thereafter, unseeded products gained market share—*gañjā* and sinsemilla. These products may not reach consumers for months after harvest, which is unusual for a fungible commodity. This interlude provides a wide berth for other "consumers" to intrude. To prevent decay and destruction requires attention to drying, curing, packaging, and storage.

18.3.1 Drying and manicuring

Proper drying begins with proper harvesting. A few days before harvest, prune away the last of large fan leaves. Inspect plants for mold-infected buds, and remove them too, with sanitized clippers or scissors. Harvested plants are hung upside-down, mostly for ease in hanging, but also because sugar leaves curl around the bud. Cervantes (2015) discussed alternatives (horizontal rack drying) and several pages of finesse points regarding proper harvesting, manicuring, and drying. He recommended slow drying, for 5–14 days, at 60–70°F (16–21°C), 45–55% relative humidity, with adequate air circulation.

Slow drying enhances aroma and burning characteristics, via the degradation of chlorophyll, and the conversion of starches → reducing sugars → CO_2 + H_2O. This is mostly an enzymatic process, but partly fermentative. Adequate air circulation restrains fermentation, which you can monitor by smell (avoid musty hay odors), but monitoring with a hygrometer is less subjective.

Slow drying is called "air-cured" by tobacco growers. "Flue-cured" tobacco is faster and more economical—hang plants in a shed and heat to 130°F (54.4°C), at 30% RF, for 3–4 days. Cannabis dried this way is harsh, with chlorophyll intact, and terpenoids gone. "Sun-cured" tobacco is prepared by spreading leaves on racks in direct sunlight. Sun-curing oxidizes chlorophyll; eliminating the green reveals other colors. Sun-cured cannabis used to be common: Acapulco gold, Panama red, Columbian gold, and red Thai stick.

Living *Cannabis* plants contain about 80% water or moisture content (MC). Optimally dried cannabis inflorescences contain about 10–12% MC. Fungi cannot grow below 15% MC. Material below 10% MC becomes brittle and disintegrates. Dishonest commercial agents market their crop above 15% MC. This is because cannabis, like corn flakes, is sold by weight, not by volume.

Manicuring (trimming) removes leaf material from inflorescences. Trimmed buds lend an appearance and consistency preferred by connoisseurs. They burn better, and have a higher cannabinoid content than buds with sugar leaves still attached. **Wet trimming** is done immediately after harvest. Wet trimming speeds drying time, and when done correctly, reduces the chance of buds molding on a drying rack. **Dry trimming** is performed after the harvested plant has been dried. Dry trimming requires greater manual dexterity, but yields a higher-quality end product.

Large operations use **trimming machines**, which yield lower-quality buds than hand trimming. Furthermore, machine-trimmed buds, compared with hand-trimmed buds, have a higher frequency of *Penicillium olsonii* colonization (Punja *et al.*, 2019). Some machines are designed for use on wet material, other for dry material, and some machines handle either. Wounding of flowers while wet trimming might cause even greater *Penicillium* colonization. The fungi grow and spread while the wet flowers are drying (Punja *et al.*, 2021a). The machines trap bits of plant debris in nooks and crannies, glued in place by resin, where *Penicillium* spp. can hide (Scott and Punja, 2022).

18.3.2 More on curing

Curing or "aging" reduces chlorophyll, sugars, and moisture. Curing is not essential—some people like the green, minty taste of uncured cannabis. Done incorrectly, curing allows fungi to flourish. "Air curing" or "burping" is simple. Place dried buds in airtight containers (canning jars, turkey-baster bags) packed loosely, with lots of air between them. Store sealed containers in a cool, dry, dark place. Moisture deep inside buds migrates out and into the air. Open the containers two or three times a day to release moisture, for 1–3 weeks. Monitor for the smell of ammonia—bacterial fermentation shouldn't verge into anaerobiosis.

"Fermentation curing" accelerates the degradation of chlorophyll and sugars. It is known as sweat-curing in the tobacco world. Tobacco is packed into closed, heated rooms with humidifiers. Anaerobic bacteria do their job, then die, and the process is completed. Fermentation curing is fraught with microbial dangers—it's a step away from composting.

Africans invented fermentation curing of cannabis (they also invented pipe smoking). In Angola, *diamba* was compressed in a conical mass, 2–4 inches (5–10 cm) in diameter and 2 ft (60 cm) long, wrapped in "some dried vegetable" and secured with strings (Daniell, 1850). Flückiger (1867) clarified that Angolan herb was wrapped in palm leaves. In Tanzania, dried tops were compressed and wrapped in bark or banana leaves, "65 cm long and thick as your arm" (Engler, 1895). Dahl (1915) gave the Tanzanian name for these bundles as *kihango*. In Malawi, bundles of *chamba*, called *mpaka*, were stuffed into sections of bamboo (Scott, 1892).

To assure fermentation, bundles were sometimes buried. In South Africa, DuToit (1980) reported that "ancestors" buried bundles in grain pits, which "assured sweating and slow aging like a good wine." Growers moistened *dagga*, rolled it tightly in paper, and allowed mold to grow on it. Malawians sealed *chamba* in corn husks, which were left in the sun, or buried in warm earth, or inserted into a thatched roof.

African slaves brought *Cannabis* to Brazil, along with sweat-curing and water pipes. "The parts of the plant used by the habitués are cut at their maturity, dried in the shade, and exposed to the evening dew for several nights in order to cure them" (Dória, 1917). Native Brazilians, the Tenetehára, adopted sweat-curing. They packed buds tightly in gourds or bamboo and buried them, "where they are said to become better cured" (Henman, 1980). Larger quantities were cured by laying dried buds outside at night to dampen with dew, then piled indoors during the day to ferment. "After a few days they turned from their original 'home-grown' lime colour to a much darker brown or black."

Sweat-curing was widely practiced in Columbia. Harvested colas were layered up to 0.5 meter deep, in piles 3 × 4.5 m². Piles were unpacked daily, to remove colas that had changed color (Frank and Rosenthal, 1978). Sweat-cured Colombian had a "tradition" of *Aspergillus* contamination (Bush Doctor, 1993a).

18.3.3   Packaging and storing

Package well-dried cannabis in an airtight container. Store it in a dry, dark, cold place. Containers with rigid walls prevent buds from being crushed and rupturing resin gland heads. Our favorite storage container is a wide-mouthed glass jar with an airtight seal, like a mason jar or canning jar. Containers made of plastic may outgas into buds, giving them an off-taste, although some people swear by polycarbonate plastic (e.g., Nalgene). Plastic surfaces generate a static charge, and hold resin gland heads that have separated from buds.

Frank and Rosenthal (1978) recommended storing cannabis with an orange peel to impart a "pleasant bouquet." Bush Doctor (1993a) isolated *Penicillium italicum* from cannabis stored with an orange peel. *P. italicum*, the "blue citrus mold," is notorious for its ability to spread by contact (i.e., "one bad apple spoils the whole bunch").

All rules were broken when cannabis was smuggled, such as Mexican "brick weed" of the 1970s and 1980s. Speed being essential, plants were incompletely dried. Small volumes were essential, so plants were compressed into bricks using hydraulic presses. Manicuring was minimal—whole branches were crushed into 1 kg bricks. (Recall the 1971 Cheech and Chong song, "No stems no seeds that you don't need.") Bricks were shipped in odor-proof (but unbreathable) wrapping, such as cement bag paper or polyethylene plastic—perfect conditions for fungi.

DEA agents said that Mexican marijuana was frequently moldy, because it was stored in the open air on the Mexico side while awaiting transportation across the border (Smith and Olson, 1982). There are photos on the web showing border agents digging up Mexican bricks buried close to the border. Columbian sweat-cured cannabis was also compressed with hydraulic presses. The product was wrapped in burlap, and shipped across the Caribbean as ballast in leaky boats.

18.3.4   Storage pests

As we said earlier, insects in stored products can be categorized into two groups. Group 1 consists of field organisms that invade living plants and then carry over as storage problems. Group 2 consists of storage organisms that only infest products after harvest. Group 1 insects that infest living *Cannabis* plants, such as aphids and spider mites, do not damage cannabis in storage. Like other fastidious organisms (e.g., viruses, nematodes, phytoplasmas, etc.), they stop feeding after their host plants die at harvest.

Group 2 insects—storage pests—are relatively rare in herbal cannabis. There is no cannabis equivalent of the tobacco beetle, *Lasioderma serricorne*, which infests stored tobacco bales, hogsheads, even cigarettes. The paucity of storage pests may be due to cannabinoid and terpenoids, which repel insects. See "*Cannabis* as insect killer" in section 3.3.6.

Back in the heyday of seeded Mexican marijuana, storage pests were attracted to seeds, not the herb. Arnaud (1974) named several beetles feeding on seeds in confiscated cannabis (see previous subsection). Smith and Olson (1982) analyzed seized cannabis at the request of border agents whose offices were overrun by insects. They found 20% of the seeds had been hollowed out by the confused flour beetle, *Tribolium confusum*. They also found two carpet beetles, *Trogoderma variabile* and *Attagenus megatoma*, which feed on almost anything—possibly the herb, but more likely they too were eating seeds. The seized cannabis was moldy, which attracted fungus-feeding beetles, *Adistermia watsoni* and *Microgramme arga* (now *Dienerella argus*). Confiscated cannabis analyzed by Arnaud (1974) was probably moldy too. He reported fungus gnats (*Bradysia* sp.), scavenger flies (*Scatopse fuscipes* and *Desmometopa* species), filth flies (*Desmometopa* sp.), and a fruit fly (*Drosophila busckii*).

Criminal investigators have used insects to identify a drug seizure's location of origin. New Zealand investigators identified 20 different species in a seizure, whose geographic distributions overlapped in the Tenasserim Hills of Burma and Thailand (Crosby *et al.*, 1986). Macedo *et al.* (2013) conducted a similar study in Brazil. Three unusual insects identified the seizure's place of origin as southern Brazil (Mato Grosso) or adjoining Argentina and Paraguay.

18.3.5 Storage microbes

Group 1 fungi are species that infest living plants. Some of these fungi carry over in storage. Punja (2021a) analyzed the mycoflora present in dried cannabis inflorescences, by gently swabbing sample surfaces, and streaking the swabs on PDA agar. Isolated colonies were sequenced with ITS primers. Among 34 different fungal species identified, four were Group 1 fungi: *Cladosporium westeerdijkieae*, *Botrytis cinerea*, *Fusarium oxysporum*, and *Alternaria alternata*. Punja's swab method identified viable spores. For the spores of Group 1 fungi to actually *grow* in harvested products would require a lot of moisture. A typical consumer would reject that kind of damp, degraded product.

Punja (2021a) also isolated several *Aspergillus* and *Penicillium* spp. These fungi rarely cause disease in living plants, but reports exist (Westendorp, 1854b; Babu *et al.*, 1977; Jerushalmi *et al.*, 2020a). A few Group 1 outliers have been reported: a case of sinusitis caused by *Curvularia lunata* may have come from a patient's cannabis (Schwartz, 1987; Brummond *et al.*, 1987). Ramírez (1990) reported four policemen contracting histoplasmosis, caused by *Histoplasma capsulatum*, after destroying a *Cannabis* field in Puerto Rico. *Cryptococcus neoformans* was isolated from cannabis consumed by a patient with cryptococcal meningitis (Shapiro *et al.*, 2018).

Group 2 fungi are storage organisms that infest products after harvest. *Aspergillus* and *Penicillium* spp. are quintessential Group 2 fungi in grain (Christensen, 1957) and tobacco (Welty, 1972). They can grow in conditions of limited moisture, low temperatures, and low oxygen levels. *Aspergillus* and *Penicillium* spp. appear to be resistant to the anti-fungal effects of cannabinoids that inhibit other fungi (Dahiya and Jain, 1977; Grewal, 1989).

Storage molds reported from cannabis are illustrated in Fig. 18.6. From left to right: *Rhizopus nigrans*, *Mucor hiemalis*, *Penicillium chrysogenum*, *P. italicum*, *Aspergillus flavus*, *A. fumigatus*, *A. niger*. Top row: sporangia or conidial heads cross-sectioned to reveal internal structures, with *M. hiemalis* demonstrating spore release after sporangium dehiscence (×400). Bottom row: natural habit of these fungi (×25).

18.3a. *Aspergillus* spp.

Aspergillus was common back in the 1970s and 1980s, when cannabis was smuggled in a sullied state. The first case of *Aspergillus* contamination was published by Dr Anthony Fauci,

later famous for AIDS and Covid-19 research. He described pulmonary aspergillosis in a 17-year-old, and *Aspergillus fumigatus* was cultured from the patient's marijuana (Chusid *et al.*, 1975).

Inhaling *Aspergillus*-contaminated cannabis has caused a spectrum of diseases, from acute wheezing (Kagen *et al.*, 1983) to systemic infections (Kagen *et al.*, 1983), bronchopulmonary aspergillosis (Llamas *et al.*, 1978; Schwartz and Hollick, 1981; Kouevidjin *et al.*, 2003), sinus infection requiring sphenoidotomy (Schwartz, 1985), invasive pulmonary fungus balls (Marks *et al.*, 1996; Cescon *et al.*, 2008; Gargani *et al.*, 2011), necrotizing pulmonary granulomata (Cunnington *et al.*, 2000; Bal *et al.*, 2010), disseminated infection from lung to brain (Salam and Pozniak, 2017), and invasive pulmonary aspergillosis, sometimes leading to systemic dissemination and death (Sutton *et al.*, 1986; Hamadeh *et al.*, 1988; Denning *et al.*, 1991; Szyper-Kravitz *et al.*, 2001).

These published studies may be the tip of an unreported iceberg. In a review of 225 organ transplant specialists, 43% reported cases of "cannabis-associated fungal infections" in transplant recipients (Levi *et al.*, 2019). In a USA-wide review of medical records associated with health insurance claims, Benedict *et al.* (2020) determined that people who used cannabis were 3.5 times more likely to have a fungal infection than people who did not use cannabis. Some cases (aspergillosis, mucormycosis) likely involved contaminated cannabis, other cases did not involve known cannabis contaminants (blastomycosis, coccidioidomycosis) or rare contaminants (cryptococcosis, histoplasmosis).

In these reports the most frequently encountered species were ***Aspergillus fumigatus***, ***Aspergillus niger***, and ***Aspergillus flavus*** (Fig. 18.6). Several studies measured the incidence of *Aspergillus* spores in street samples: 2 out of 10 (Chusid *et al.*, 1975), 11 of 12 (Kagen, 1981), 13 of 14 (Kagen *et al.*, 1983), and 3 of 7 (Verweij *et al.*, 2000). Kurup *et al.* (1983) analyzed 24 samples: *A. flavus* in 12, *A. fumigatus* in seven, and *A. niger* in six.

McKernan *et al.* (2015, 2016a,b) analyzed samples purchased at medical dispensaries in Massachusetts, Maine, and Rhode Island. Their metagenomic study, utilizing sensitive NextGen sequencing, identified *A. fumigatus*, *A. niger*, and *A. flavus*.

Additionally, they identified *A. candidus*, *A. leporis*, *A. puniceus*, *A. oryzae*, *A. ostianus*, *A. sepultus*, *A. terreus*, *A. unguis*,

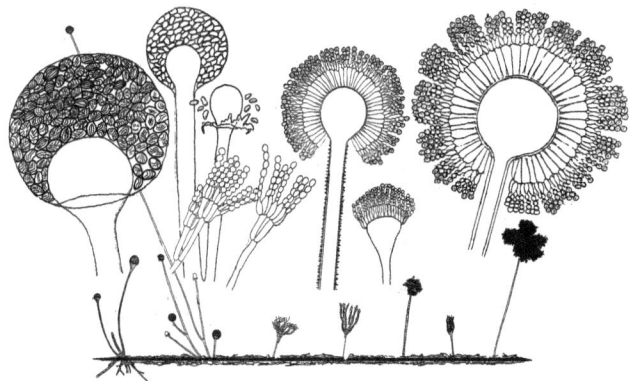

Fig. 18.6. Storage molds associated with cannabis (McPartland, 2000)

A. ustus, A. versicolor, A. filifer (=*Emericella filifer*), *A. nidulans* (=*Emericella nidulans*), *A. rugulosus* (=*Emericella rugulosa*), *A. fruticulosa* (=*Emericella fruticulosa*), and *A. nivioglacus* (=*Eurotium niveoglaucum*).

Punja (2021a) analyzed the mycoflora in dried inflorescences by gently swabbing sample surfaces, isolating colonies, and sequencing them with ITS primers. *Aspergillus niger* was moderately prevalent, *A. flavus* and *A. ochraceus* less so.

Rumors circulated in the 1970s that moldy marijuana buried underground gained potency. Folb and Fromkin (1972) recorded three slang terms in Los Angeles for "a particularly potent form of marijuana": *black mota, black moat,* and *black mold.* Folb (1980) clarified that *black mo* was "buried an' moldy an' it turn black." This erroneous rumor gained wide credence (Margolis and Clorfene, 1975; *Green and Miller, 1975;* Starks, 1977; DuToit, 1980). Medical case reports described patients burying marijuana in soil for "aging," or knowingly smoking moldy marijuana (Chusid *et al.,* 1975; Llewellyn and O'Rear, 1977; Llamas *et al.,* 1978). No doubt they hoped for a potency-enhancing mold, but *Aspergillus* found them instead.

Forensic scientists cited hazards faced by police who seized and dismantled indoor grow operations. Air samples revealed heavy contamination by *Aspergillus* spp. and *Penicillium* spp. in Canada (Johnson and Miller, 2012; Gwinn *et al.,* 2023), Colorado (Martyny *et al.,* 2013; Root *et al.,* 2020), and Belgium (Vanhove *et al.,* 2018). In the Belgian report, police complained of headache, nose and/or eye irritation, and skin irritation.

Aflatoxins are toxic metabolites exuded by *Aspergillus* spp. They are acutely poisonous and mutagenic; chronic exposure increases the odds of liver cancer. Llewellyn and O'Rear (1977) provided proof-of-concept: they inoculated "street marijuana" with *A. flavus* and *A. parasiticus,* which yielded aflatoxin B$_1$ after 14 days (average 0.323 and 0.207 µm/g cannabis, respectively). But first they added 25 mL water to 5 g samples and sterilized them prior to inoculation. Incidentally, the samples averaged 1.5% THC after removal of stems and seeds (sign of the times).

Concern over aflatoxins has prompted jurisdictions to require aflatoxin testing of cannabis products. Health Canada (2008) set a limit of <20 µm/kg cannabis for aflatoxins B$_1$, B$_2$, G$_1$, and G$_2$. Jameson *et al.* (2022) examined regulatory documents by 36 states that legalized recreational and/or medical cannabis, and 27 jurisdictions mentioned aflatoxin testing. We found only two reports of aflatoxin contamination in world literature: Di Nardo *et al.* (2020) tested inflorescences sold legally in Italy, and detected aflatoxin B$_1$ in an astounding 50% of 14 samples. In comparison, Jameson *et al.* (2022) reported only one case of "mycotoxin" detection out of 9414 samples submitted to California. In a small study, Narváez *et al.* (2020) did not recover aflatoxins from ten samples of CBD gelatin capsules.

This paucity may be due to the anti-*Aspergillus* activity of *Cannabis* compounds seen in several studies (see text below Table 3.3). Al Khoury *et al.* (2021) went the extra step and showed that extracts from inflorescences inhibited both *A. flavus* growth and aflatoxin B$_1$ production.

The uncharacterized "mycotoxin" reported by Jameson *et al.* (2022) may have been **ochratoxin A** (OTA), a toxic metabolite produced by *Aspergillus* and *Penicillium* spp. OTA is probably a contaminant of greater concern than aflatoxins in cannabis (McPartland and McKernan, 2017). Gwinn *et al.* (2023) listed seven *Aspergillus* and six *Penicillium* spp. isolated from cannabis with the capacity to synthesize OTA. Twenty-six jurisdictions require OTA testing (Jameson *et al.* (2022), and Health Canada (2008) set a limit of <20 µm cannabis/kg for OTA.

Narváez *et al.* (2020) tested ten CBD gelatin capsules obtained from online shops across Europe. They found toxic metabolites produced by *Fusarium* spp. in 60% of samples: trichothecene (T-2) toxin, zearalenone, and enniatins (enniatin A, A1, and B1). No jurisdictions mandate testing for these metabolites (Jameson *et al.* 2022).

18.3b. *Penicillium* spp.

Some *Penicillium* spp. can colonize living plants (see section 9.4). They carry over as storage molds. Others *Penicillium* spp. do not contaminate cannabis until after it is harvested (Fig. 18.6). Three studies measured the incidence of *Penicillium* spores in seized samples. Kagen *et al.* (1983) reported 2 out of 28 samples. Kurup *et al.* (1983) reported 2 out of 24. Verweij *et al.* (2000) reported 6 out of 7, producing very high plate counts of 10^4–10^7 CFU/g material. In contrast, only 3 of 98 tobacco cigarettes were contaminated, with plate counts of 200–300 CFU/g material.

McKernan *et al.* (2015, 2016a,b) analyzed samples purchased at medical dispensaries in New England. Their metagenomic approach, utilizing NextGen sequencing with ITS primers, identified the four aforementioned species. Additionally, they identified *P. camerberti, P. canescens, P. citrinum, P. citreonigrum, P. commune, P. concentricum, P. corylophilum, P. decaturense, P. digitatum, P. glabrum, P. miczynskii, P. raistrickii, P. spinulosum, P. steckii, P. soppii, P. sumatrense, P. toxicarium, P. egyptiacum* (=*Eupenicillum crustaceum*), *P. javanicum* (=*Eupenicillum javanicum*), *P. euglacum* (=*Eupenicillium anatolicum*).

Punja (2021a) analyzed the mycoflora in dried inflorescences by gently swabbing sample surfaces, isolating colonies, and sequencing them with ITS primers. The most frequently recovered fungi were *Penicillium* spp.—*P. citrinum, P. olsonii, P. simplicissimum,* and *P. spathulatum.* Punja *et al.* (2023) analyzed total yeast and mold counts (TYMC) in >2000 fresh and dried specimens in British Columbia. Most dried commercial cannabis samples contained <1000–5000 CFU/g. Sequencing with ITS1-2 identified 21 species, dominated by *Penicillium* (*P. olsonii, P. citrinum, P. brevicompactum*), *Aspergillus* (*A. flavus, A. niger, A. ochraceus*), *Fusarium* (*F. oxysporum, F. graminearum, F. sporotrichoides*), and *Cladosporium cladosporioides.*

Other molds

Mucor spp. and *Rhizopus* spp. are occasionally reported (Fig. 18.6). They are usually indicative of cannabis in a deteriorated condition. Kurup *et al.* (1983) reported that 10 out of 24 samples contained *Mucor* spp. Kagen *et al.* (1983) isolated *Mucor* spp. in 12 out of 28 samples. Gwinn *et al.* (2023) isolated *Mucor circinelloides* and *M. racemosus* from products sold by licensed cannabis facilities in British Columbia. Punja (2021a) isolated

Rhizopus solani from 2–10% of samples of dried inflorescences. Punja also isolated two fungi that were used for biological control when the plants were alive: *Beauveria bassiana* and *Trichoderma harzianum*.

Thermophilic actinomycetes are indicative of highly deteriorated herb. Kurup *et al.* (1983) reported three species: *Thermoactinomyces candidus* (7 out of 24 samples), *T. vulgaris* (3 out of 24), and *Micropolyspora faeni* (1 out of 24). Kagen *et al.* (1983) isolated *T. candidus* (4 out of 28 samples) and *T. vulgaris* (3 out of 28). Inhaling thermophilic actinomycetes causes hypersensitivity pneumonitis, "farmer's lung."

Bacteria

Field bacteria cannot grow or replicate in harvested and dried plants, as noted by Christensen (1957), because they require free water to grow. Bacteria may play a part during curing, but when curing stops, they die. Zhao *et al.* (2007) identified two bacteria in flue-cured tobacco: *Bacillus megaterium* and *Bacteriovorax* sp.

Accidental contaminants may cause human disease. Ungerleider *et al.* (1982) isolated a rogue's gallery of human pathogens from NIDA-sourced medical cannabis: *Klebsiella pneumoniae*, *Enterobacter cloacae*, and group D *Streptococcus*. It should be noted that NIDA's cannabis at that time was sweat-cured by placing harvested material on concrete floors (B. Thomas, pers. comm., 1999)—an unacceptable method today. An epidemic caused by *Salmonella muenchen* that involved 85 people was traced to contaminated marijuana (Taylor *et al.*, 1982). The investigators concluded that the plant material, sourced from Mexico, was contaminated by untreated manure— another unacceptable practice today.

McKernan *et al.* (2016b) analyzed samples purchased at medical dispensaries in New England. They generated PCR amplicons of 16S rRNA and used NextGen sequencing. PCR amplifies all sequences, whether the organisms are living or dead. They identified an astonishing 29 species. They highlighted *Clostridium botulinum*, but its significance in cannabis is questionable. *C. botulinum* grows in canned foods and produces a toxin under anaerobic conditions. Other bacteria they identified, but didn't comment upon, are serious human pathogens: *Salmonella enterica*, *Pseudomonas aeruginosa*, *Escherichia coli*, *Corynebacterium diphtheriae*, and *Streptococcus pneumoniae*.

Other 16S rRNA sequences likely came off the hands of people trimming buds—*Propionibacterium acnes*, *Staphylococcus aureus*, and *Acinetobacter pitii*. Maybe they were eating yogurt, the possible source of *Bacillus coagulans* and *Lactococcus lactis*. They licked their lips: *Anaerococcus prevotii* and *Finegoldia magna*. Other identified bacteria—*Pseudomonas putida*, *P. stutzeri*, *Kocuria rhizophila*, and *Maricaulis maris*, are soil saprophytes—which doesn't say much for harvesting methods at the medical dispensaries sampled.

18.3.6 Microbial degradation of THC and CBD

THC and CBD are antifungal (see section 3.3.4) and antibacterial (see Table 3.1). Fungi and bacteria have evolved the ability to metabolize and break down these compounds.

The literature is large, and reviewed by Akhtar *et al.* (2016). Here we provide some highlights: Robertson and Lyle (1975) provided proof-of-concept: they incubated *Syncephalastrum racemosum* with THC or CBD, which yielded hydroxylated metabolites: 4'-OH-THC and 4'-OH-CBD. Binder (1976) did the same with *Cunninghamella blakesleeana*. Robertson *et al.* (1978) showed that a bacterium, *Rhodococcus* (*Mycobacterium*) *rhodochrous*, could metabolize THC and CBD. None of these organisms naturally occur on *Cannabis*. Christie *et al.* (1978) isolated *Chaetomium globosum* from *Cannabis*; it metabolized THC into 3'-hydroxy-, 11-hydroxy-, and 3',11-dihydroxy- derivatives.

Binder and Meisenberg (1978) screened 163 strains of fungi and bacteria for their ability to metabolize THC. Fifty-one strains were capable, on a scale of +++ (metabolites >5%), ++ (metabolites >2% to 5%), and + (metabolites 0.05% to <2.0%). These included the *Cannabis* pathogens *Botrytis cinerea* (+++), *Sclerotinia sclerotiorum* (+++), *Chaetomium globosum* (++), *Fusarium moniliforme* (++), *F. sulphurem* (+), and *Serratia marcenscens* (+). Saprophytic storage fungi included *Aspergillus niger* (+) and *Rhizopus stolonifera* (+). Beneficial biocontrol organisms also made the list: *Lecanicillium* (*Verticillium*) *lecanii* (++), *Gliocladium roseum* (+), *Paraphaerosphaeria* (*Coniothyrium*) *minitans* (+), *Bacillus subtilis* (+), and *B. cereus* (+).

18.3.7 Microbial testing requirements

Because some fungi and bacteria are human pathogens, federal or state governments have mandated microbial testing. This has led to testing recommendations by government-based or industry-based organizations, who developed and published *Cannabis*-specific standards.

Microbial testing was initiated by the Office of Medicinal Cannabis in Holland (Hazekamp, 2006, 2016). Bedrocan BV, the primary supplier of medical cannabis in Holland, tests harvested plants as well as final packaged products. They use two petri plate-based screening tests—one for **total aerobic microbial count** (TAMC), the other for **total yeast and mold count** (TYMC). Degree of contamination is quantified by counting the number of colony-forming units (CFU) arising per gram of plated cannabis (CFU/g). The Dutch placed limits of <100 CFU/g for TAMC, and <10 CFU/g for TYMC—which is close to sterility. Certain pathogens must be completely absent—*Staphylococcus aureus*, *Pseudomonas aeruginosa*, and bile-tolerant gram-negative bacteria such as *Escherichia coli*.

Health Canada (2008) mandated similar tests, with different limits: <100 CFU/g for TAMC, and <100 CFU/g for TYMC, as well as specific tests for Coliform bacteria (<3 MPN/g), and *E. coli* (absent). In the USA, *American Herbal Pharmacopoeia* (Upton *et al.*, 2013) developed protocols based on cannabis assays used in Holland: TAMC, TYMC, bile-tolerant gram-negative bacteria, plus a specific petri plate screen for *E. coli* and *Salmonella* spp. Foundation of Cannabis Unified Standards (FOCUS) proposed testing cannabis products with TAMC and TYMC, with upper limits of <1 CFU/g of *E. coli* or *Salmonella* spp. (FOCUS, 2016).

McPartland and McKernan (2017) reviewed microbial testing in the USA, first mandated by New Jersey (NJMMP,

2011). California didn't mandate testing until 2018, which was 22 years after its legalization of medical cannabis. Protocols varied from state to state, often petri plate- or film-based assays proposed by the *American Herbal Pharmacopoeia*, with variable action levels—from zero tolerance to <100 CFU/g.

It bears mentioning that tests based on CFU/g detection are prone to sampling bias due to subsampling (Holmes *et al.*, 2015). Because of this, and to accelerate testing turn-around time, McKernan *et al.* (2016b) proposed using molecular PCR-based methods. There are two drawbacks. First, PCR-based methods using NextGen sequencing may identify dozens of species, amplified from miniscule samples, such as *Trichophyton rubrum* (ringworm fungus) coming off the hands of people trimming buds (McKernan *et al.*, 2016a). Second, PCR will amplify DNA from both living and dead organisms. This was demonstrated by Frink *et al.* (2022), who spiked 1000 *Aspergillus* spores into 1 g dried cannabis, and used X-rays to sterilize the samples. Subsequent culturing revealed no growth, but qPCR tests were positive.

The first drawback can be avoided by using a multiplex qPCR assay, which detects a predefined panel of the most frequent offenders (e.g., *Aspergillus flavus*, *A. fumigatus*, and *A. niger*). To mitigate the second drawback, an enrichment step can be performed, where samples are incubated overnight in TSB broth prior to qPCR detection. This ensures only live organisms are measured, but raises questions over preferential culture conditions for broader total yeast and mold tests. To address this conundrum, some labs perform a qPCR on total yeast and molds, and positive results are confirmed with an additional test extracted 24 h later to ensure the signal from the pre-incubation test was from live organisms (McPartland and McKernan, 2017).

McKernan *et al.* (2015) compared results between qPCR and three petri plate- or film-based detection systems: 3M Petrifilm™ (3M-P), Simplate-Biocontrol Systems™ (S-BS), and BioLumix™ (BLX). They tested 17 dispensary-obtained cannabis samples. Six samples tested positive with qPCR, five samples tested positive with S-BS (>10,000 CFU/g), four samples test positive with 3M-P (>10,000 CFU/g), and only one sample tested positive with BLX, which is a simple pass–fail test.

The USA Farm Bill of 2018 legalized CBD hemp, opening the big-money floodgates. Standards organizations that previously ignored the public health concerns of cannabis consumers quickly began feeding at the money trough—ASTM International, American National Standards Institute, and Association of Official Analytical Chemists. Their lobbyists are positioning them as chieftains to certify and regulate *Cannabis* cultivators, processors, and dispensaries. For them, *Cannabis* is not a matter of lived experience but a lucrative gravy train.

First on the bandwagon, *United States Pharmacopeial Convention* (USP) convened an expert panel regarding microbial testing of cannabis (Sarma *et al.*, 2020). Limits for inflorescences included TAMC (<10^5 CFU/g), TYMC (<10^4 CFU/g), and the absence of *E. coli* and *Salmonella* spp. They proposed more stringent limits for inhalable products sold to immunocompromised people: TAMC (<100 CFU/g), TYMC (<10 CFU/g), and the absence of *E. coli*, *Salmonella* spp., *Staphylococcus aureus*, and *Pseudomonas aeruginosa*.

Turning to fungi, Sarma *et al.* (2020) recommended culture-based tests for *Aspergillus niger*, *A. flavus*, *A. fumigatus*, and *A. terreus*, but the accurate identification of *Aspergillus* spp. could be difficult. They noted that qPCR for specific *Aspergillus* spp. was more sensitive than culture-based methods, but may lead to false positives due to cross-reactivity with non-specified *Aspergillus* spp. They recommended an enrichment step in culture media prior to qPCR testing.

Interlab differences in TAMC and TYMC counts have been found, possibly due to the economic benefits of "undercounting"—labs that consistently reported lower counts attracted producers who required testing. Punja *et al.* (2023) determined TYMC counts in six specimens that were also analyzed by two (unnamed) commercial labs. Both labs underreported counts compared to Punja—undercounting by 22% (lab A) or 76% (lab B). This illegitimate activity was also seen among labs in Michigan, where blinded "round-robin" audits of identical products revealed consistent discrepancies (J. Crookston, pers. comm., 2022).

In the latest review of microbial testing in the USA, Jameson *et al.* (2022) examined regulatory documents for 36 states that legalized recreational and/or medical cannabis, and all jurisdictions mentioned microbial contamination, but only 32 jurisdictions listed specific species, such as *E. coli*, *Salmonella*, *Candida*, *Listeria*, and *Klebsiella* spp.

18.3.8 Controlling storage microbes

Prevention is the best way to avoid microbial contamination. See the previous instructions regarding proper harvesting, drying, air-curing, packaging, and storage. Punja (2021a) demonstrated the ubiquity of airborne fungal spores in trimming and drying rooms. Scott and Punja (2022) recommended using air filters and purifiers to reduce the inoculum load floating through these spaces. Technologies include HEPA filter mats, photocatalytic oxidation, UV light, or ozone. Trimming machines must be scrupulously cleaned on a regular basis, using food-safe degreasers.

Punja *et al.* (2023) assayed TYMC for 3 years to identify factors that contribute to yeast and mold counts. Factors that *decreased* TYMC included the absence of leaf litter in the glasshouse, running fans 24 h/day for a week prior to harvest, adequate drying of buds (moisture content ≤14%), drying method (hanging whole plants upside-down was better than wet-trim/rack-drying), and harvest season (November–April was better than May–October). Plant with dense inflorescences (more stigmatic tissues and unifoliate "sugar leaves") held humidity and generated higher TYMC counts than plants with thready inflorescences (e.g., "Kush" vs "Jack Herer").

It should be mentioned that water pipes do not prevent the transmission of fungal spores from contaminated cannabis into smoke (Moody *et al.* 1982), not even water pipes with filters (Sullivan *et al.*, 2013). Fungi and bacteria are capable of passing through vaporizers (Ruchlemer *et al.*, 2015; Sopovski *et al.*, 2023).

18.3.9 Biological control

Fungi associated with post-harvest contamination can be controlled with other fungi. Kusari *et al.* (2013) isolated *Penicillium copticola* from flowering tops, and *in vitro* assays showed that it antagonized *Trichothecium roseum*. Punja and Ni (2021) tested *Trichoderma asperellum* and it antagonized *B. cinerea*, both *in vitro* and *in vivo*. Punja et al. (2023) tested *Trichoderma harzianum* and *Gliocladium catenulatum* and they antagonized *Penicillium* spp. McPartland et al. (2000) suggested spraying harvested flowers with *Candida oleophila* (Nexy®) before hanging to dry. *Candida* spp. and other yeast biocontrol agents were also discussed by Mahmoud et al. (2023). However, in products subjected to microbial testing for regulatory purposes, surviving *Candida* may trigger a positive test. This is true for all the aforementioned biocontrols (Punja and Ni, 2021; Punja et al., 2023).

18.3.10 Sterilization

To kill microbial contaminants in harvested cannabis, growers have used heat, gas sterilization, and ionizing radiation.

Heat sterilization

Levitz and Diamond (1991) killed conidia of *A. fumigatus*, *A. flavus*, and *A. niger* by baking cannabis in a dry oven at 350°F (150°C) for 15 min. This temperature no doubt decarboxylated THCA into THC, and accelerated THC degradation into CBN; many terpenoids probably outgassed. Herb blended into a brownie mix and baked at nearly the same temperature (325°F, 163°C) loses less THC. The brownies' interior will not exceed 100°C as long as water remains in the mix (Iffland et al., 2016). Ruchlemer et al. (2015) autoclaved herb for 50 min at 135°C and a pressure of 316 kPA (3.1× atmospheric pressure). This treatment decreased THC content by 22.6% (from 15.0% to 11.6%).

Gas sterilization

Heat-sensitive materials (spices and medical devices) have been gas-sterilized with ethylene oxide since 1938. Ungerleider et al. (1982) gassed cannabis for 4.5–5 h, which killed pathogenic bacteria (*Bacillus*, *Klebsiella*, *Enterobacter*, *Streptococcus* spp.). They performed this at 140°F (60°C) under reduced pressure, 8.5–10 psi (586–689 mbar). Ruchlemer et al. (2015) exposed cannabis to ethylene oxide for 165 min, at 51°C, and 892 mbar. This treatment decreased THC content by 26.6%.

Chlorine dioxide sterilization allows for shorter exposure times and is generally safer than ethylene oxide. Chlorine dioxide gas has successfully reduced microbe levels in cannabis, according to a USA patent application (US 2020/0221701 A1). Its effects on THC were not analyzed.

Ionizing radiation

Gamma rays are extremely lethal because of their high energy and penetrative power. Radioactive sources such as cobalt-60 (^{60}Co) emit gamma rays as they decay, and this source has been used to irradiate food. Gamma radiation is controversial. Only 60 countries permit food irradiation. Many EU countries limit irradiation to dried herbs and spices. The amount of radiation energy absorbed by food is measured in gray (Gy) or kiloGray (kGy) units. Packaged meat and poultry may be irradiated with up to 70 kGy. Ungerleider et al. (1982) irradiated cannabis with a dose of 15–20 kGy, which killed pathogenic bacteria (*Bacillus*, *Klebsiella*, *Enterobacter*, *Streptococcus* spp.).

Dutch and Canadian medical cannabis is treated with 10 kGy (Hazekamp, 2006; Health Canada, 2008). Microbial counts in Dutch cannabis are tested before and after irradiation, because "bad quality" cannabis should not be rescued by irradiation (Hazekamp, 2016). Jerushalmi et al. (2020b) used gamma rays to kill *Botrytis cinerea* in cannabis. A dose of 7.5 KGy reduced *B. cinerea* inoculum levels 6-fold and 4.5-fold, in naturally infected and artificially inoculated flowers, respectively.

Hazekamp (2016) evaluated the effects of 10 KGy on cannabinoids and terpenoids. Quantification with GC-FID and UPLC showed that THC and/or CBD levels were not altered by irradiation treatment, compared with controls. Irradiation decreased four monoterpenoids: α-guaiene (10%), *cis*-ocimene (7–23%), β-myrcene (8–18%), terpinolene (16–38%). Irradiation decreased seven sesquiterpenoids—guaiol (6%), nerolidol (7%), trans-β-farnesene (7–10%), β-caryophyllene (6–10%), γ-selinene (13–17%), eudesma-3,7(11)-diene (14%), and γ-elemene (8–19%).

X-rays are similar to gamma rays but have less energy and can be produced by machines using electricity. Frink et al. (2022) used X-ray irradiation at a dose of 2.5 kGy to kill *Aspergillus niger*, *A. flavus*, *A. fumigatus*, and *A. terreus* in cannabis flower and cannabis inhalable products. The treatment did not alter THCA, THC, CBDA, or CBD content. Some terpenoids showed a downward trend (β-caryophyllene, humulene) but "these changes were not tested for statistical significance."

Beta radiation, also known as electron beam (e-beam) radiation, consists of electrons or positrons. Beta radiation lacks the lethality of gamma rays; it can be used for treating cancer. Rhenium-188 (^{188}Re) is a common source, but beta radiation can also be created using an electron accelerator, rather than a radioactive source. Electron accelerators can be turned on and off, unlike radioactivity.

Jerushalmi et al. (2020b) e-beamed cannabis with 15 kW and an energy capacity of 5.25 MeV. They treated cannabis with natural *B. cinerea* infection to a range of doses, and determined an ED$_{50}$ dose of 3.55 KGy. For cannabis artificially inoculated with *B. cinerea*, the ED$_{50}$ was 5.18 KGy. Punja (2021a) e-beamed dried buds contaminated with *Penicillium* spp. The treatment markedly reduced *Penicillium* populations, in one experiment dropping from 100 CFUs to 2 CFUs per swabbed specimen (e-beam dosage not stated).

Cold atmospheric plasma

Adding energy to a gas will form a plasma. It is analogous to adding energy to a liquid, boiling it into a gas. Cold atmospheric plasma (CAP) is generated from atmospheric air by exposing it to a dielectric discharge between two plate electrodes. This creates reactive oxygen and nitrogen species, such as ozone, hydroxyl radicals ('OH) and amino radicals ('NH$_2$). Adding hydrogen peroxide (H$_2$O$_2$) increases the formation of 'OH. Ruchlemer et al. (2015) exposed cannabis to CAP with

H_2O_2 for 49 min, at a temperature of 51°C and a pressure of 40 mbar. This decreased THC content by 12.6% (compared with 22.6% with autoclaving, and 26.6% by ethylene oxide). Jerushalmi *et al.* (2020b) exposed cannabis to CAP with H_2O_2 for 10 min at an operating voltage of 6 kV. This reduced naturally infected *B. cinerea* levels to nearly zero, and decreased artificially inoculated *B. cinerea* levels 4-fold.

18.4   Cannabinoid shelf life

Cannabinoid shelf life and its effects on pharmacology have focused on THCA, THC, CBN, and CBD (see Fig. 1.1). *Cannabis* biosynthesizes THCA, called "Δ⁹-THC acid" in older studies. There are two isomers, THCA-A (Fig. 1.1) and rarely encountered THCA-B.

THCA is not psychoactive—chewing freshly harvested plant material imparts no "high." THCA has other therapeutic benefits (Russo and Marcu, 2017). Rarely in medicinal chemistry does the pharmacology of a substance become significantly altered by simple decarboxylation; Filer (2022) gave only one other example. When exposed to heat or light (or time), THCA decarboxylates to THC. With further exposure to heat or light (and oxygen), THC dehydrogenates to CBN.

The relative psychoactivity of cannabinoids can be gauged by their K_i values (Fig. 1.1). K_i measures "binding affinity" at cannabinoid receptor 1 (CB_1). K_i is a metric similar to potency (e.g., EC_{50}), measured as molar concentration (nM). A small K_i indicates high binding affinity. The K_i of THCA is very large, about 100-fold greater than the K_i of THC (McPartland *et al.*, 2017). CBN is about 20-fold less potent than THC as gauged by its K_i (McPartland *et al.*, 2007d).

CBD (see Fig. 1.1) has little binding affinity at CB_1. It has gained popularity for myriad other therapeutic benefits (McPartland and Pruitt, 1999). CBD has a longer shelf life than THCA or THC. Turner *et al.* (1973a) reported no loss in CBD content in dried flowering tops, stored for two years at 4°C. *Hashīsh* stored for two years at 4°C lost 25% of its CBD content; THC losses were greater (Trofin *et al.*, 2012b). CBD decay in hash oil was approximately a quarter of THC's rate of decay (Trofin *et al.*, 2012a).

Terpenoids also have limited shelf life and are lost in storage. This is a process of outgassing. Lighter-weight monoterpenoids outgas faster than sesquiterpenoids (see section 2.8.6).

18.4.1   THCA → THC

The instability of THCA hampers its clinical application as well as its pharmacological exploration. "How can anybody do an experiment if the compound likes to convert into something else just by sitting around, and the 'something else' has all kinds of activities?" (R. Mechoulam, pers. comm., 2017). Measuring decarboxylation *in planta* is difficult, but it occurs in living plants. Happyana *et al.* (2013) micro-dissected gland heads from trichomes and analyzed them with liquid chromatography and cryogenic mass spectrometry. Between weeks 4 and 8 of flowering, the THCA/THC ratio decreased, nearly linearly: 157:1 to 110:1 to 91:1 to 94:1 to 56:1.

Taschwer and Schmid (2015) dried fresh plants at 25°C for 24 h on a drying rack, and the THCA/THC ratio dropped to 5.5. Drying at 50°C under the same conditions was essentially the same (THCA/THC ratio ~5.4), but drying at 100°C decarboxylated all the THC (THCA/THC ratio ~0). Several studies, reviewed by Iffland *et al.* (2016), suggest that 100°C is an inflection point for THCA decarboxylation. Iffland *et al.* (2016) claimed that only 33% of THCA is decarboxylated in a cake baked at 180°C for 45 min, because the cake's interior will not exceed 100°C as long as water remains in the cake.

Storage conditions affect decarboxylation rates. Yamauchi *et al.* (1967) stored dried herb for 2 months at 35°C and reported 70% decarboxylation of THCA. In an Oxford museum at room temperature (*ca.* 20°C), Harvey (1990) found trace amounts of THCA in 90-year-old *gañjā* specimens stored on shelving in the dark. Turner *et al.* (1973a) stored dried flowering tops using an industry standard: amber glass bottles with aluminum caps, with limited exposure to light. They monitored cannabinoid levels at 10-week intervals for 25 months. At room temperature (22°C), in a refrigerator (4°C), or a freezer (–18°C), THCA dropped to 80%, 95.8%, or 94.7% (respectively) of initial levels. Taschwer and Schmid (2015) also tested freezer storage (–25°C), for 4 months, and THCA content remained stable.

Storing cannabis under nitrogen reduces THCA decarboxylation: Doorenbos *et al.* (1971) heated herbal material under nitrogen, at seven temperatures between 25° and 100°C, for 75 min. There was no significant THCA decarboxylation below 80°C.

Cannabis cigarettes burn at 500–600°C (Fehr and Kalant, 1972; Pomahacova *et al.*, 2010), sufficient to decarboxylate all THCA. However, the temperature proximal to a cigarette's combustion point is much lower, allowing THCA to vaporize rather than pyrolyze. Consistent with this, Pomahacova and colleagues showed that cigarette smoke yielded some THCA, although its THCA/THC ratio (1:95) was much less than the ratio generated by a vaporizer at 170°C (1:14).

Solvents vary in their ability to extract THCA from herbal material. Romano and Hazekamp (2013) reported that ethanol (22°C, 20 min) extracted a THCA/THC ratio of 42:2, and olive oil (98°C, 60 min) extracted a slightly lower THCA/THC ratio (42:6). Citti *et al.* (2016) also compared solvents. Supercritical CO_2 (90 min) extracted the best THCA/THC ratio, followed by olive oil (110°C, 2 h), then ethanol (25°C, 3 h). The latter two were tested for stability; after 10 days in amber-colored bottles at 25°C, olive oil retained 78.0% of initial THCA levels, and ethanol retained only 33.4%.

Peschel (2016) extracted flowering tops in hydroethanolic solvents, and stored them in amber-colored bottles. After 3 months at 20°C (ambient light, no direct sunlight), THCA levels dropped to 50% of initial levels—identical to 15 months in a refrigerator at 2–6°C (he obtained similar results with 40% or 90% ethanol). Other studies have tested the stability of THCA in laboratory reagents—methanol, chloroform, petroleum ether, and n-hexane (Smith and Vaughan, 1977; Veress *et al.*, 1990).

Hazekamp *et al.* (2007) demonstrated short-term THCA stability in "cannabis tea." They added quantified amounts of THCA and THC to heated water. After 15 min of simmering above 55°C, they recovered 63% of THCA and only 17% of THC. However, THCA loss was substantial in cannabis tea stored at 4°C—decreasing to 71% of initial levels after one day. THCA rapidly decarboxylates if water is boiled *after* cannabis is added to it (De Zeeuw *et al.*, 1972).

18.4.2 THC → CBN

THC can oxidize to CBN *in planta*, as inflorescences reach maturity. Most growers are aware of this and monitor the color of gland heads as flowers mature. Young gland head contents are generally clear or translucent to cloudy. With maturity, yellow → amber → brown colors indicate the conversion of THC to CBN. These color changes parallel the color of stigmas as they change from whitish-yellow to reddish-brown.

Potter (2009) rated gland head color on a 1 to 8 scale, and measured their cannabinoid content (Fig. 18.7). 1 = completely clear, 2 = misty white, 3 = translucent, 4 = opaque white, 5 = yellow to slightly brown, 6 = light brown, 7 = mid-brown, 8 = dark brown. Note that gland head maturity and color can vary in a single plant, even in a single leaf, because of the asynchronous onset of trichome formation.

The correlation between gland head color and cannabinoid content by Potter (2009) confirmed earlier studies by others (Turner *et al.*, 1977; Mahlberg and Kim, 2004), and has been confirmed since then (Gjorgievska *et al.*, 2023). Sutton *et al.* (2023) reported the time-course for this color change in four drug-type hybrids. Clear gland heads were most abundant 25 days after flowering began (DAF). Milky gland heads peaked around 50–55 DAF. Brown gland heads began appearing 50 DAF and peaked 65–75 DAF.

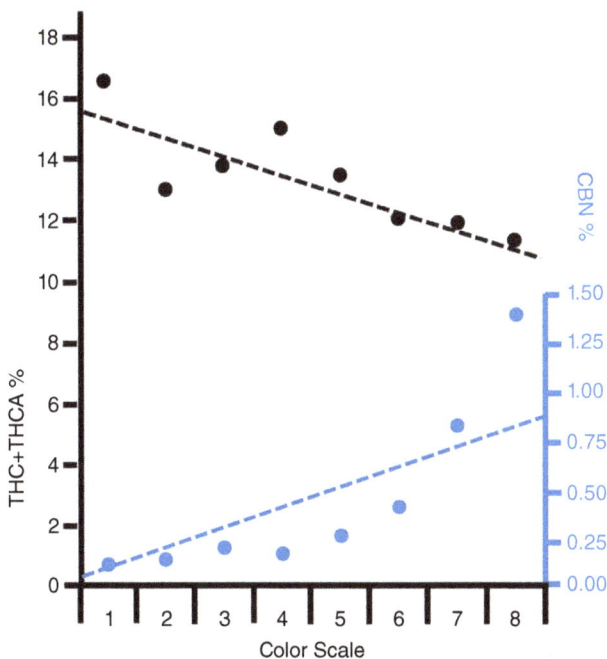

Fig. 18.7. Cannabinoid content in gland heads rated on a color scale (adapted from Potter, 2009)

For herb in storage, the rate at which THC decays to CBN is a function of temperature, exposure to oxygen and light, and time. This was recognized as a loss in potency before chemists identified THC and CBN. Prain (1893) estimated *gañjā* psychoactivity using monkeys and cats. At room temperature in India, *gañjā* lost two-thirds of its potency after 1 year in storage; after 2 years, *gañjā* produced weak effects; after 3 years, *gañjā* lost all effectiveness.

Marshall (1898) assayed psychoactivity in dogs, cats, rabbits, and himself. He bubbled oxygen through a cannabis extract, and it lost all psychoactivity. Passing a stream of CO_2 through the extract scarcely changed its activity. Marshall surmised that changes were due to oxidation, "the obvious remedy is to keep cannabis preparations in air-tight vessels until they can be used."

Eckler and Miller (1917) used the dog ataxia assay to monitor deterioration of bulk herb, stored for 50 months in a hot attic or in a cool basement. Attic-stored herb lost all activity, estimated as 2% loss per month. Basement-stored herb lost 60% activity, estimated as 1% loss per month. Eckler and Miller also tested herb soaked in ethanol, and a *USP* cannabis extract (also in ethanol), which retained their activity over time.

Levine (1944) monitored changes in a 100 lb "lump of *charas*" from India, stored in goatskin for 3 years. The outer surface of the *charas* developed a hard crust and lost its characteristic odor. The center of the lump of *charas* retained a fresh appearance and smell. Levine isolated CBN from the crust, but none from the center of the lump (the center contained THC—not discovered yet). Drug potency of the crust fell to one-twentieth of its original value, measured in the dog ataxia assay. One-twentieth the potency is commensurate with K_i differences between THC and CBN (see Fig. 1.1).

Lerner (1969) stored "manicured marihuana" at different temperatures, and measured the loss of THC using gas chromatography. At room temperature (24°C), THC was lost at a rate of 3–5%/month over 61 months. At 100°C, all THC was gone after a month. Coffman and Gentner (1974) showed how THC content (in parts per million) declined in oven-dried leaf tissue at three different temperatures, after 1, 4, 16, and 64 hours, measured in ppm (parts per million):

65°C: *1 h:* 763; *4 h:* 775; *16 h:* 695; *64 h:* 676
85°C: *1 h:* 720; *4 h:* 690; *16 h:* 580; *64 h:* 378
105°C: *1 h:* 680; *4 h:* 610; *16 h:* 445; *64 h:* 147

Turner *et al.* (1973a) stored dried flowering tops in amber glass bottles with aluminum caps, with limited exposure to light, for 25 months. At room temperature (22°C), THC decomposed at a rate of 6.9% loss/year. In a refrigerator at 4°C, the loss rate was 5.4%/year. In a freezer at −18°C, the loss rate was 3.8%/year.

Fairbairn *et al.* (1976) ground herbal material through a 1 mm mesh prior to storage. This ruptured resin gland heads, which accelerated THC oxidation. At 20°C in ambient light, THC decreased to 37% of starting levels (loss of 33.4%/year). At 20°C in darkness, THC decreased to 75% of starting levels (loss of 13.3%/year). At 5°C in darkness, THC decreased to 91% of starting levels (loss of 4.8%/year). Ross and ElSohly (1997) measured THC in "coarsely manicured" marijuana stored in closed barrels in an air-conditioned vault (20–22°C). THC decreased by 16.6% ±7.4 of its original value after 1 year,

26.8% ±7.3 after 2 years, 34.5% ±7.6 after 3 years, and 41.4% ±6.5 after 4 years.

Trofin *et al.* (2011) monitored THC content in ten samples of dried flowering tops. When stored at 22°C in ambient light, 34.1% of THC was lost the first year; the calculated THC half-life was 20.7 months over a 4-year span. Storage at 4°C in darkness resulted in 27.8% loss the first year, with a half-life of 28.6 months (summary statistics calculated by us from their raw data).

18.5   Resin glands and concentrates

Cannabinoids and terpenoids are sequestered in the resin heads of glandular trichomes (see Fig. 3.1). Products containing mechanically separated resin glands are known as *hashīsh*, *charas*, and *nasha*. These products will mold if improperly prepared. When a block of moldy *hashīsh* is broken in half, the mold can be seen as gray veins (Clarke, 1998). Cherniak (1979) illustrated moldy hash—see his figure 8.17. Pérez-Moreno *et al.* (2019) reported that 10% of *hashīsh* in Madrid was contaminated by *Aspergillus* spp. They quantified the amount of contamination by grinding up *hashīsh* and culturing it, and then counting the number of CFUs/g of plated material. They reported an average of 3833 CFUs/g of material—quite high.

Clarke (1998) reviewed mechanical methods for separating resin glands from the rest of the plant. Briefly: **hand-rubbed resin** was traditionally made in India, Pakistan, and Nepal. Hand-rubbing of flowering tops causes resin to accumulate upon the palms and fingers, which is scraped off and collected (Fig. 18.8).

Sieved resin utilizes finely woven fabric (traditionally silk) as a screen or sieve, to separate gland heads from trichome stalks and leaf detritus. Sieving requires cool and dry conditions, which prevail during winters in Turkestan, Afghanistan, and Persia.

Sieving is a more efficient extraction method than handrubbing, in terms of resin yield per acre (see Table 2.14). Hooper (1908) measured the percentage of ethanol-extracted resin in *charas* samples collected from South and Central Asia. Clarke (1998) made an educated guess which samples were sieved vs hand-rubbed. Sieved samples averaged 36.1% extractable resin, and rubbed samples averaged 38.1% extractable resin.

Hooper's ethanol-extracted resin included THC, CBN, and CBD. Rubbing likely causes resin glands to rupture, accelerating the deterioration of THC → CBN, but this has not been explicitly tested experimentally. Hanuš *et al.* (2016b) tested *hashīsh* from India (possibly rubbed), Morocco, and Lebanon (probably sieved), but did not describe the products, other than "disk, chocolate, sole," respectively. Their THC/CBN ratios were 1.9, 4.8, and 1.2, respectively. The concentration of THC in sieved *hashīsh* tops out around 60% (Clarke, 1998; Potter, 2009). Chouvy and Macfarlane (2018) reported Moroccan *hashīsh* containing 68% THC, which seems unusually high.

Ice water separation is like dry sieving, except plant material is submerged in a bucket of ice water and agitated. Ice-cold resin heads readily detach from trichomes and sink to the bottom. The bucket is lined with a series of sieves that separate resin glands from other plant material. Russo *et al.* (2021) described an alternative—using "dry ice" (solid CO_2) to freeze and detach resin heads. These are sifted through 150 μm screens in a freezer, maintaining the "cold chain."

Solvent extraction and the manufacturing of concentrates is beyond the remit of this text. Suffice to say a variety of solvents can be employed—non-polar solvents (e.g., hexane, chloroform, petroleum ether, butane, naphtha, supercritical CO_2) or polar solvents (methanol, ethanol, butanol, acetonitrile). Concentrates are known as butane hash oil (BHO), dabs, wax, and shatter. They often contain >90% THC. Concentrates have birthed a 21st century industry: vape pens, which emulsify the concentrates in propylene glycol or polyethylene glycol.

Jameson *et al.* (2022) examined regulatory documents by 36 states that legalized recreational and/or medical cannabis, and 25 jurisdictions mandated testing for solvent residues—butane, benzene, toluene, hexane, and heptane. Allowable residue levels varied widely, from zero tolerance to 5000 ppm. Several jurisdictions encouraged the use of supercritical CO_2, which leaves behind no solvent residues.

Rosin extraction utilizes heat and pressure to squeeze resin out of plant material. The product is similar to solvent extracts, but without the presence of residual solvents, and terpenoids are better preserved than with solvent extraction.

18.5.1   *Hashīsh* THCA → THC

THCA may be more stable in sieved *hashīsh* than in herbal cannabis. The gland heads are detached and compacted to minimize exposure to light and oxygen. De Zeeuw *et al.* (1972) noted that "large amounts" of THCA could be detected in *hashīsh*. Smith and Vaughan (1977) analyzed 2-year-old *hashīsh*, and it contained more THCA (3.5%) than THC (2.6%). Sectioning the block revealed a higher percentage of THCA in inner layers than outer surfaces.

Baker *et al.* (1981) measured THCA and THC levels in seized samples of herbal cannabis and resin. The THCA/THC ratio was higher in resin (mean 3.08) than herbal material (mean 1.96). Sieved *hashīsh* samples (Morocco, Lebanon, Turkey) had a higher THCA/THC ratio than hand-rubbed *hashīsh* (India, Pakistan), means 4.0 and 0.65, respectively.

Fig. 18.8. Hand-rubbed resin (courtesy of Clarke, 1998)

Lindholst (2010) determined that THCA had a half-life of 462 days in *hashīsh* stored at room temperature in darkness for 4 years. When *hashīsh* was homogenized, extracted in methanol:chloroform, and stored under the same conditions, THCA half-life plummeted to 32 days—probably because homogenization ruptured the gland heads.

18.5.2   *Hashīsh* THC → CBN

Der Marderosian and Murthy (1974) proposed that THC retains longer shelf life within resin gland heads, whose contents are "hermetically sealed and thus less prone to oxidation." Terpenoids in resin glands may serve as antioxidants to protect THC from oxidizing into CBN (McPartland *et al.*, 2017).

Fairbairn *et al.* (1976) described resin gland heads as "well filled, well closed containers." They made *hashīsh* by freezing dried tops, "to make the glands hard and brittle," then sieved the resin glands through a 125 μm mesh, and compressed them into blocks. When stored at 20°C in darkness, THC content dropped from 11.6% to 11.0% in a year—a loss of 5% per annum. Decomposition was much greater when stored at 20°C in ambient light—THC content in the surface layer of the *hashīsh* dropped to 5.2%—a loss of 55% per annum.

Trofin *et al.* (2012b) stored *hashīsh* at 22°C in ambient light. THC levels dropped 25.87% the first year; they calculated a THC half-life of 24.9 months over a 4-year span. Storage at 4°C in darkness was nearly the same—25.87% loss the first year, with a half-life of 25.4 months. These losses were less than Trofin *et al.* (2011) reported from herbal cannabis. Trofin *et al.* (2012a) monitored THC content in two samples of "hash oil." These were police seizures and the solvent was not identified. When stored at 22°C in ambient light, 32.2% of THC was lost the first year; the calculated half-life was 26.2 months over a 4-year span. Storage at 4°C in darkness was similar—32.7% loss the first year, with a half-life of 27.9 months. These losses are greater than Trofin and colleagues reported in *hashīsh*, but less than losses in herbal cannabis.

The stability of THC in various solvents has been tested. Prior to 1937, nearly all medicinal cannabis was dispensed as an ethanol extract. Fetterman *et al.* (1971) analyzed a 40-year-old *USP* extract made by Mulford Co. It contained 0.43% THC, 5.2% CBN, and 2.7% CBD. Kubena *et al.* (1972) analyzed a 43-year-old *USP* cannabis extract made by Parke-Davis & Co. It contained 0.4% THC, 0.04% CBN, and 0.1% CBD. Harvey (1985) detected "traces" of THC in a 140-year-old bottle of Peter Squire's ethanol tincture.

Fairbairn *et al.* (1976) emphasized the deleterious effects of light when THC is extracted in ethanol. Ethanolic extracts stored at room temperature (20°C) in the dark did not lose THC, but ethanolic extracts stored in clear glass bottles and ambient light lost nearly half their potency in just 20 days. They obtained similar results with THC extracted in petroleum ether or chloroform—it quickly decomposed when stored in light. Lindholst (2010) extracted THC in 9:1 methanol:chloroform. It quickly decomposed when stored at room temperature in ambient light. He reported a half-life of 35 days, with THC becoming undetectable after 140 days. When stored in darkness, however, THC concentration remained stable (after an initial increase as THCA decarboxylated into THC).

CBD loses stability when extracted and stored in non-polar solvents. Parker *et al.* (1974) showed that CBD was more stable in ethanol than chloroform, stored in tightly stoppered bottles at room temperature in the dark. After 8 days in ethanol, CBD reduced to 84% of the starting concentration, and in chloroform only 5% remained. Fairbairn *et al.* (1976) tested CBD in chloroform at room temperature. After just 6 days, 32% of CBD remained in dark storage, and only 5% remained when stored in ambient light.

(Prepared by J. McPartland; revised by E. Russo)

References

Abdulla SA, Khalaf AS. 2016. A comprehensive study on bird seed mixtures entered to Iraqi Kurdistan region markets. *Journal of the University of Duhoh* 19: 218–222.

Agostini A. 1927. Observazioni informa a due ifomiceti saprofiti dannosi di tessuti di canapa. *Atti della Reale Accademia dei Fisiocritici* 1(3): 25–33.

Akhtar MT, Shaari K, Verpoorte R. 2016. Biotransformation of tetrahydrocannabinol. *Phytochemistry Reviews* 15: 921–934.

Akkawi JS. 1987. *Procédé de rouissage biochimique des plantes fibreuses libériennes. SPS-ase enzyme polysaccharidase.* European patent no. WO/1987/002390.

Al Khoury A, Sleiman R, Atoui A, *et al.* (4 other authors). 2021. Antifungal and anti-aflatoxigenic properties of organs of *Cannabis sativa* L.: relation to phenolic content and antioxidant capacities. *Archives of Microbiology* 203: 4485–4492.

Allescher A. 1903. "*Diplodiella ramentacea,*" p. 930 in *Dr. L. Rabenhorst's Kryptogamen-Flora von Deutschland, Oesterreich und der Schweiz. Die Pilze, VII Abtheilung: Fungi imperfecti.* E. Kummer, Leipzig.

Amarasinghe P, Pierre C, Moussavi M, *et al.* (3 other authors). 2022. The morphological and anatomical variability of the stems of an industrial hemp collection and the properties of its fibres. *Heliyon* 14: e09276.

Arnaud PH. 1974. Insects and mites associated with stored *Cannabis sativa* Linnaeus. *Pan-Pacific Entomologist* 50: 91–92.

Babu R, Roy AN, Gupta YK, Gupta MN. 1977. Fungi associated with deteriorating seeds of *Cannabis sativa* L. *Current Science* 46(20): 719–720.

Baker PB, Taylor BJ, Gough TA. 1981. The tetrahydrocannabinol and tetrahydrocannabinolic acid content of cannabis products. *Journal of Pharmacy and Pharmacology* 33: 369–372.

Bal A, Agarwal AN, Das A, Vikas Suri, Varma SC. 2010. Chronic necrotising pulmonary *Aspergillosis* in a marijuana addict: a new cause of amyloidosis. *Pathology* 42: 197–200.

Behrens J. 1902. Untersuchungen über die Gewinnung der Hanffaser durch natürliche Röstmethoden. *Centralblatt für Bakteriologie, Parasitenkunde, und Infektionskrankheiten, Zweite Abteilung* 8: 114–120, 131–137, 264–268, 295–299.

Behrens J. 1903. Ueber die taurotte von flachs und hanf. *Centralblatt für Bakteriologie, Parasitenkunde, und Infektionskrankheiten, Zweite Abteilung* 10: 524–530.

Benedict K, Thompson GR, Jackson BR. 2020. Cannabis use and fungal infections in a commercially insured population. *Emerging Infectious Diseases* 26: 1308–1310.

Binder M. 1976. Microbial transformation of tetrahydrocannabinol by *Cunninghamella blakesleeana* Lender. *Acta Helvetica Chimica* 63: 1674–1684.

Binder M, Meisenberg G. 1978. Microbial transformation of cannabinoids: Part 2; A screening of different microorganisms. *European Journal of Applied Microbiology and Biotechnology* 5: 37–50.

Boczek J, Golebiowska Z, Krzeczkowski K. 1960. Roztocze szkodliwe w przechowalniach siemienia lnianego i konopi w Polsce. *Prace Naukowe Instytutu Ochrony Roślin* 2(1): 57–86.

Bouhuys A, Zuskin E. 1976. Chronic respiratory disease in hemp workers. *Annals Internal Medicine* 84: 398–405.

Bouhuys A, Barbero A, Lindell SE, Roach SA, Schilling RS. 1967. Byssinosis in hemp workers. *Archives Environmental Health* 14: 533–544.

Brummond W, Kurup VP, Zablocki CJ, *et al.* (3 other authors). 1987. Fungal sinusitis and marijuana, in reply. *JAMA* 257: 2915.

Brygadyreko VV, Reshetniak DY. 2014. Trophic preferences of *Harpalus rufipes* (Coleoptera, Carabidae) with regard to seeds of agricultural crops in conditions of laboratory experiment. *Baltic Journal of Coleopterology* 14: 179–190.

Bulliard P. 1789. La Pezize couronnée, Peziza coronata. *Herbier de la France* 9: plate 416. Didot le jeune, Paris.

Bush Doctor. 1993a. Stash alert: how to preserve pot potency. *High Times* 213, 75–79.

Caminita BH, Schneiter R, Kolb RW, Neal PA. 1943. Studies on strains of *Aerobacter cloacae* responsible for acute illness among workers using low-grade stained cotton. *Public Health Reports* 58: 1165–1183.

Carbone D, Tombolato A. 1917. Sulfa macerazione rustica della canapa. *Stazioni Sperimentali Agrarie Italiane* 50: 563–575.

Carpentier JBA. 1862. Du charençon. *Bulletin de la Société d'Agriculture de l'arrondissement de Boulogne-Sur-Mer* 3: 256.

Carter DJ. 1984. *Pest Lepidoptera of Europe, with special reference to the British Isles.* W. Junk, Dordrecht.

Castellan RM, Olenchock SA, Hankinson JL, *et al.* (5 other authors). 1984. Acute broncho-constriction induced by cotton dust: dose related responses to endotoxin and other dust factors. *Annals Internal Medicine* 101: 157–163.

Cazin FJ, Cazin H. 1868. *Traité des plantes médicinales indigènes, troisième edition.* Asselin, Paris.

Cervantes J. 2015. *The Cannabis encyclopedia.* Van Patten Publishing, Vancouver, Washington.

Cescon DW, Page AV, Richardson S, *et al.* (3 other authors). 2008. Invasive pulmonary aspergillosis associated with marijuana use in a man with colorectal cancer. *Journal of Clinical Oncology* 26: 2214–2215.

Cherniak L. 1979. *The great books of hashish, vol. 1, book 1.* And/Or Press, Berkeley, California.

Chmielewski W. 1984. *Tyrophagus longior*—bioekologia, wystepwanie i szkodliwosc. *Prace Naukowe Instytutu Ochrony Roślin* 26: 69–85.

Chmielewski W, Filipek P. 1968. Wpływ zaprawiania nasion lnu i konopi panogenem na roztocze. *Roczniki Nauk Rolniczych* 93(A) 4: 701–710.

Chomel N. 1709. *Dictionnaire œconomique, 2 vols.* Pierre Thened, Lyon.

Chouvy PA, Macfarlane J. 2018. Agricultural innovations in Morocco's cannabis industry. *International Journal of Drug Policy* 58: 85–91.

Christensen CM. 1957. Deterioration of stored grains by fungi. *Botanical Review* 23: 108-134.

Christie RM, Rickards RW, Watson WP. 1978. Microbial transformation of cannabinoids. I. Metabolism of (–)-Δ^9-6a,10a-trans-tetrahydrocannabinol by *Chaetomium globosum. Australian Journal of Chemistry* 31: 1799–1807.

Chusid MJ, Gelfand JA, Nutter C, Fauci AS. 1975. Pulmonary aspergillosis, inhalation of contaminated marijuana smoke, chronic granulomatous disease. *Annals of Internal Medicine* 82: 682–683.

Citti C, Ciccarella G, Braghiroli D, *et al.* (3 other authors). 2016. Medicinal cannabis: principal cannabinoids concentration and their stability evaluated by a high performance liquid chromatography coupled to diode array and quadrupole time of flight mass spectrometry method. *Journal of Pharmaceutical and Biomedical Analysis* 128: 201–209.

Clarke RC. 1998. *Hashish!* Red Eye Press, Los Angeles, California.

Clarke RC. 2010a. Traditional fiber hemp (*Cannabis*) production, processing, yarn making, and weaving strategies—functional constraints and regional responses, part 1. *Journal of Natural Fibers* 7: 118–153.

Coffman CB, Gentner WA. 1974. *Cannabis sativa* L.: Effect of drying time and temperature on cannabinoid profile of stored leaf tissue. *Bulletin on Narcotics* 26(1): 67–70.

Crescini F. 1943. Forme di canapa (*Cannabis sativa* L.), III. la germinabilità. *Nuovo Giornale Botanico Italiano* 50(3/4): 209–231.

Crocioni A. 1950. *Durata del potere germinativo del seme di canapa in rapporto alle condizioni di conservazione.* Istituto di Agronomia Generale e Coltivazioni Erbacee, Università di Bologna.

Crosby TK, Watt JC, Kistemaker AC, Nelson PE. 1986. Entomological identification of the origin of imported *Cannabis. Journal of the Forensic Science Society* 26(1): 35–44.

Cunnington D, Teichtahl H, Hunt JM, Dow C, Valentine R. 2000. Necrotizing pulmonary granulomata in a marijuana smoker. *Chest* 117: 1511–1514.

Dahiya MS, Jain GC. 1977. Inhibitory effects of cannabidiol and tetrahydrocannabinol against some soil inhabiting fungi. *Indian Drugs* 14(4): 76–79.

Dahl E. 1915. *Nyamwesi-Wörterbuch.* L. Friederichsen, Hamburg.

Daniell WF. 1850. On the d'amba, or dakka, of southern Africa. *Pharmaceutical Journal and Transactions* 9: 363–365.

De Bary A 1886. Ueber einige Sclerotinien und Sclerotinienkrankheiten (fortsetzung). *Botanische Zeitung* 44: 449–461.

de Meijer EPM, Keizer LCP. 1996. Patterns of diversity in *Cannabis. Genetic Resources and Crop Evolution* 43: 41–52.

De Zeeuw RA, Malingre TM, Merkus FW. 1972. Delta-1-tetrahydrocannabinolic acid, an important component in the evaluation of cannabis products. *Journal of Pharmacy and Pharmacology* 24: 1–6.

Demske D, Tarasov PE, Leipe C, *et al.* (3 other authors). 2016. Record of vegetation, climate change, human impact and retting of hemp in Garhwal Himalaya (India) during the past 4600 years. *The Holocene* 26: 1661–1675.

Denning DW, Follansbee SE, Scolaro M, *et al.* (3 other authors). 1991. Pulmonary aspergillosis in the acquired immunodeficiency syndrome. New England Journal of Medicine 324: 654–662.

Der Marderosian A, Murthy S. 1974. Analysis of old samples of *Cannabis sativa* L. *Journal of Forensic Sciences* 19: 670–675.

Di Candilo M, Bonatti PM, Guidetti C, *et al.* (4 other authors). 2010. Effects of selected pectinolytic bacterial strains on water-retting of hemp and fibre properties. *Journal of Applied Microbiology* 108: 194–203.

Di Nardo FD, Cavalera S, Baggiani C, *et al.* (3 other authors). 2020. Enzyme immunoassay for measuring aflatoxin B1 in legal cannabis. *Toxins* 12: e265.

Dimitrov M, Ivanova-Dzhubrilova S, Nikolcheva M, Drenska E. 1990. Study of mycotoxical and dust air pollution in enterprises for preliminary processing of cotton and hemp. *Problemi na Khigienta* 15: 121–127.

Dombrovskaya EV. 1940. Описание нового вида галлицы *Silvestrina tyrophagi* sp. m. (сем. Cecidomyidae), уничтожающей клеща *Tyrophagus noxius* Zachv. Leningrad Bull. Plant Protection 1940(3): 87–88.

Doorenbos NJ, Fetterman PS, Quimby MW, Turner CE. 1971. Cultivation, extraction and analysis of *Cannabis sativa* L. *Annals of the New York Academy of Sciences* 191: 3–14.

Dória R. 1917. Os fumadores de maconha: effeitos e males do vicio. *Proceedings of the Second Pan American Scientific Congress* Section 8 (Part 1): 151–162.

Dutkiewicz J, Mackiewicz B, Lemieszek MK, Golec M, Milanowski J. 2015. *Pantoea agglomerans*: a mysterious bacterium of evil and good. Part I. Deleterious effects. *Annals of Agricultural and Environmental Medicine* 22: 576–588.

DuToit BM. 1980. *Cannabis in Africa.* A.A. Balkema Press, Rotterdam.

Eckler CR, Miller FA. 1917. On the deterioration of crude Indian cannabis. *Journal American Pharmaceutical Association* 6: 872–875.

Ellis J. 1769. Observations on a particular manner of increase in the Animalcula of vegetable infusions, with the discovery of an indissoluble salt arising from hemp-seed put into water till it becomes putrid. *Philosophical Transactions of the Royal Society of London* 79: 138–152.

Engler A. 1895. *Die pflanzenwelt Ost-Afrikas und der nachbargebiete. Theil A.* Dietrick Reimer, Berlin.

Er M, Emri SA, Demir AU, Thorne PS, *et al.* (3 other authors). 2016. Byssinosis and COPD rates among factory workers manufacturing hemp and jute. *International Journal of Occupational Medicine and Environmental Health* 29: 55–68.

Fairbairn JW, Liebmann JA, Rowan MG. 1976. The stability of cannabis and its preparations on storage. *Journal of Pharmacy and Pharmacology* 28: 1–7.

Fehr KO, Kalant H. 1972. Analysis of cannabis smoke obtained under different combustion conditions. *Canadian Journal of Physiology and Pharmacology* 50: 761–767.

Fetterman PS, Seith ES, Waller CW, *et al.* (3 other authors). 1971. Mississippi-grown *Cannabis sativa* L.: preliminary observation on chemical definition of phenotype and variations in tetrahydrocannabinol content versus age, sex, and plant part. *Journal of Pharmaceutical Sciences* 60: 1246–1249.

Ferri F. 1959a. *Atlante delle avversità della canapa.* Edizioni Agricole, Bologna, Italia.

Ferri F. 1961b. Microflora dei semi di canapa. *Progresso Agricolo (Bologna)* 7(3): 349–356.

Filer CN. 2022. Acidic cannabinoid decarboxylation. *Cannabis and Cannabinoid Research* 7(3): 262–273.

Flament A. 1874. Moyen de chasser les charancons. *Le Cosmos* 35: 53–54.

Flückiger FA. 1867. Pharmaceutische Reissindrucke. *Pharmaceutische Centralhalle für Deutschland* 8: 436–442.

FOCUS. 2016. *Cannabis retail. FS-3001-1.* Foundation of Cannabis Unified Standards, Scottsdale, Arizona.

Folb EA. 1980. *Runnin' down some lines.* Harvard University Press, Cambridge, Massachusetts.

Folb EA, Fromkin V. 1972. *A comparative study of urban black argot.* Project report no. 0-1-055, University of California, Los Angeles.

Frank M, Rosenthal E. 1978. *Marijuana grower's guide.* And/Or Press, Berkeley, California.

Frink S, Marjanovic O, Tran P, *et al.* (6 other authors). 2022. Use of X-ray irradiation for inactivation of *Aspergillus* in cannabis flower. *PLoS ONE* 17(11): e0277649.

Fuller WH, Norman AG. 1944. The nature of the flora on field-retting of hemp. *Proceedings of the Soil Science Society of America* 9: 101–105.

Fuller WH, Norman AG. 1945. Biochemical changes involved in the decomposition of hemp bark by pure cultures of fungi. *Journal of Bacteriology* 50: 667–671.

Fuller WH, Norman AG. 1946a. *The retting of hemp. II. controlled retting of hemp.* Iowa Agricultural Experiment Station, Bulletin no. 343. Ames, Iowa.

Fuller WH, Norman AG. 1946b. *The retting of hemp. III. Biochemical changes accompanying retting of hemp.* Agricultural Experiment Station, Bulletin no. 344. Ames, Iowa.

Fuller WH, Norman AG, Wilsie CP. 1946. *The retting of hemp I. Field retting of hemp in Iowa.* Agricultural Experiment Station, Bulletin no. 342. Ames, Iowa.

Gardner M, Reed S, Davidson M. 2020. Assessment of worker exposure to occupational organic dust in a hemp processing facility. *Annals of Work Exposures and Health* 64: 745–753.

Gargani Y, Bishop P, Denning DW. 2011. Too many mouldy joints—marijuana and chronic pulmonary aspergillosis. *Mediterranean Journal of Hematology and Infectious Diseases* 3: e2011005.

Gitman L, Boytchenko E. 1934. "*Cannabis*" pp. 45–53 *in* Справочник по болезням бовых лубианых культур. Institute New Bast Raw Materials, Moscow.

Gitman LS, Malikova TP. 1933. "Пораженность стеблей конопли по сортам," pp. 51–57 *in* Болезни и вредители новых лубяных культур, вып. 3. Institute of New Bast Raw Materials, Moscow.

Gjorgievska VS, Karanfilova IC, Trajkovska A, *et al.* (4 other authors). 2023. Monitoring of *Cannabis* cultivar technological maturity by trichome morphology analysis and HPLC phytocannabinoid content. *Pharmacognosy Research* 15: 94–100.

Glover T. 1869. "The food and habit of beetles," pp. 78–117 *in* *Report of the Commissioner of Agriculture, 1868.* Government Printing Office, Washington, DC.

Green M, Miller RD. 1975. "*Cannabis* use in Canada," pp. 497–520 in Rubin V, ed. *Cannabis and culture.* Mouton, The Hague.

Grewal PS. 1989. Effects of leaf-matter incorporation on *Aphelenchoides composticola* (Nematoda), mycofloral composition, mushroom compost quality and yield of *Agaricus bisporus*. *Annals of Applied Biology* 115: 299–312.

Gwinn KD, Leung MCK, Stephens AB, Punja ZK. 2023. Fungal and mycotoxin contaminants in cannabis and hemp flowers: implications for consumer health and directions for further research. *Frontiers in Microbiology* 14: e1278189.

Hamadeh R, Ardehali A, Locksley RM, York MK. 1988. Fatal aspergillosis associated with smoking contaminated marijuana, in a marrow transplant recipient. *Chest* 94: 432–433.

Hamilton K, White NDG, Jian FJ, Fields PG. 2021. Hemp (*Cannabis sativa*) seed for reproduction of stored-product insects. *Journal of Stored Products Research* 92: e101787.

Haney A, Kutscheid BB. 1975. An ecological study of naturalized hemp (*Cannabis sativa* L.) in east-central Illinois. *American Midland Naturalist* 93: 1–24.

Hanuš LO, Levy R, La Vega D, *et al.* (3 other authors). 2016b. The main cannabinoids content in hashish samples seized in Israel and Check Republic. *Israel Journal of Plant Sciences* 63: 182–190.

Happyana N, Agnolet S, Muntendam R, *et al.* (3 other authors). 2013. Analysis of cannabinoids in laser-microdissected trichomes of medicinal *Cannabis sativa* using LCMS and cryogenic NMR. *Phytochemistry* 87: 51–59.

Harvey DJ. 1985. "Examination of a 140 year old ethanolic extract of cannabis: identification of new cannabitriol homologues and the ethyl homologue of cannabinol," pp. 23–30 in Harvey DJ, Paton W, Nahas GG, *eds. Marihuana 84—Proceedings of the Oxford Symposium on Cannabis.* IRL Press, Oxford, UK.

Harvey DJ. 1990. Stability of cannabinoids in dried samples of cannabis dating from around 1896–1905. *Journal of Ethnopharmacology* 28: 117–128.

Hazekamp A. 2006. An evaluation of the quality of medicinal grade cannabis in the Netherlands. *Cannabinoids* 1(1): 1–9.

Hazekamp A. 2016. Evaluating the effects of gamma-irradiation for decontamination of medicinal cannabis. *Frontiers in Pharmacology* 7: e108.

Hazekamp A, Bastola K, Rashidi H, Bender J, Verpoorte R. 2007. Cannabis tea revisited: a systematic evaluation of the cannabinoid composition of cannabis tea. *Journal of Ethnopharmacology* 113: 85–90.

Health Canada. 2008. *Product information sheet on dried marihuana (cannabis).* Available at: www.hc-sc.gc.ca/ dhp-mps/marihuana/about-apropos/dried-information-sechee-eng.php

Henman A. 1980. War on drugs is war on people. *The Ecologist* 10: 282–289.

Holm L. 1957. Etudes taxonomiques sur les Pléosporacées. *Symbolae Botanicae Upsalienses* 14: 1–188.

Holmes M, Vyas JM, Steinbach W, McPartland J. 2015. *Microbiological safety testing of cannabis.* Available at: http://cannabissafetyinstitute.org/wp-content/uploads/2015/06/Microbiological- Safety-Testing-of-Cannabis.pdf

Hooke R. 1665. *Micrographia.* Martyn and Allestry, London.

Hooper D. 1908. Charas of Indian hemp. *Year-Book of Pharmacy* 1908: 435–444.

Hrebenyuk NV. 1984. Распространенне грибов на тресте конопли (The occurrence of fungi on hemp stems). *Микология и фитопатология* 18(4): 322–326.

Huang HT. 2000. "Fermentation and Food Science," Volume 6, Part 5 of Needham J, *ed. Science and Civilisation in China.* Cambridge University Press, Cambridge, UK.

Iffland K, Carus M, Grotenhermen F. 2016. *Decarboxylation of tetrahydrocannabinolic acid (THCA) to active THC.* European Industrial Hemp Association web site. Available at: http://eiha.org/media/2014/08/16-10-25-Decarboxylation-of-THCA-to-active-THC.pdf

Jameson LE, Conrow KD, Pinkhasova DV, *et al.* (9 other authors). 2022. Comparison of state-level regulations for cannabis contaminants and implications for public health. *Environmental Health Perspectives* 130: e97001.

Janischevsky DE. 1924. Форма конопли на сорных местах в Юго-Восточной России. Ученые записки Саратовского государственного университета имени Н. Г. *Чернышевского* 2(2): 3–17.

Jerushalmi S, Maymon M, Dombrovsky A, Freeman S. 2020a. Fungal pathogens affecting the production and quality of medical cannabis in Israel. *Plants* 9(7): e882.

Jerushalmi S, Maymon M, Dombrovsky A, Freeman S. 2020b. Effects of cold plasma, gamma and e-beam irradiations on reduction of fungal colony forming unit levels in medical cannabis inflorescences. *Journal of Cannabis Research* 2: e12.

Jimenez DC, Lahoz G. 1944. Cannabosis (enfermedad de los trabajadores del canamo). *Rev Clin Esp* 14: 366–376.

Johnson LI, Miller JD. 2012. Consequences of large-scale production of marijuana in residential buildings. *Indoor and Built Environment* 21: 595–600.

Kagen SL. 1981. *Aspergillus:* an inhalable contaminant of marihuana. *New England Journal of Medicine* 304: 483–484.

Kagen SL, Kurup VP, Sohnle PG, Fink JN. 1983. Marijuana smoking and fungal sensitization. *Journal of Allergy and Clinical Immunology* 71: 389–393.

Kobayashi M, Momohara A, Okitsu S, Yanagisawa S, Okamoto T. 2008. Fossil hemp fruits in the earliest Jomon period from the Okinoshima site, Chiba Prefecture. *Shokuseishi kenkyū (Japanese Journal of Historical Botany)* 16(1): 11–18.

Koehler B. 1946. Hemp seed treatments in relation to different dosages and conditions of storage. *Phytopathology* 36: 937–942.

Kolb J. 1868. Recherches sur le blanchiment des tissus. *Bulletin de la Société industrielle de Mulhouse* 38(suppl. Juillet): 37–57.

Kondo M, Kasahara Y, Akita S. 1950. Germination of hemp seeds stored for 19 years and their growth. *Nōgaku Kenkyū* 39: 37–39.

Kouevidjin G, Mazieres J, Sayas S, Didier A. 2003. Aspergillose bronchopulmonaire allergique aggravée par la consommation de cannabis. *Revue Française d'Allergologie et d'Immunologie Clinique* 43: 192–194.

Kozlowski R. 1995. Interview with Professor R. Kozlowski, director of the Institute of Natural Fibres. *Journal of the International Hemp Association* 2(2): 86–87.

Kubena RK, Barry H, Sagelmann AB, Theiner M, Farnsworth NR. 1972. Biological and chemical evaluation of a 43-year-old sample of cannabis fluid extract. *Journal of Pharmaceutical Sciences* 61: 144–145.

Kudo Y, Kobayashi M, Momohara A, *et al.* (5 other authors). 2009. Radiocarbon dating of fossil hemp fruits in the earliest Jomon period. *Shokuseishi kenkyū (Japanese Journal of Historical Botany)* 17(1): 27–32.

Kurup VP, Resnick A, Kagen SL, Cohen SH, Fink JN. 1983. Allergenic fungi and actinomycetes in smoking materials and their health implications. *Mycopathologia* 82: 61–64.

Kusari P, Kusari S, Spiteller M, Kayser O. 2013. Endophytic fungi harbored in *Cannabis sativa* L.: diversity and potential as biocontrol agents against host plant-specific phytopathogens. *Fungal Diversity* 60: 137–151.

Law AD, McNees CR, Moe LA. 2020. The microbiology of hemp retting in a controlled environment: steering the hemp microbiome towards more consistent fiber production. *Agronomy* 10: e492.

Lee MJ, Park MS, Hwang SJ, *et al.* (13 other authors). 2010. Dietary hempseed meal intake increases body growth and shortens the larval stage via the upregulation of cell growth and sterol levels in *Drosophila melanogaster. Molecules and Cells* 30: 29–36.

Lerner P. 1969. The precise determination of tetrahydrocannabinol in marihuana and hashish. *Bulletin on Narcotics* 21: 39–42.

Levi ME, Montague BT, Thurstone C, *et al.* (3 other authors). 2019. Marijuana use in transplantation: a call for clarity. *Clinical Transplantation* 33: e13456.

Levine J. 1944. Origin of cannabinol. *Journal of the American Chemical Society* 66: 1868–1870.

Levitz SM, Diamond RD. 1991. Aspergillosis and marijuana. *Annals of Internal Medicine* 115: 578–579.

Liger L. 1700. *Oeconomie générale de la campagne, ou Nouvelle maison rustique, 2 vols.* Charles de Sercy, Paris.

Lindholst C. 2010. Long term stability of cannabis resin and cannabis extracts. *Australian Journal of Forensic Sciences* 42: 181–190.

Llamas R, Hart DR, Schneider NS. 1978. Allergic bronchopulmonary aspergillosis associated with smoking moldy marihuana. *Chest* 73: 871–872.

Llewellyn GC, O'Rear CE. 1977. Examination of fungal growth and aflatoxin production on marihuana. *Mycopathologia* 62: 109–112.

Macedo MP, Kosmann C, Pujol-Luz JR. 2013. Origin of samples of *Cannabis sativa* through insect fragments associated with compacted hemp drug in South America. *Revista Brasileira de Entomologia* 57: 197–201.

MacIndoo NL, Stevers AF. 1924. *Plants tested for or reported to possess insecticidal properties.* USDA Department Bulletin no. 1201, Washington, DC.

Mahlberg PG, Kim ES. 2004. Accumulation of cannabinoids in glandular trichomes of *Cannabis* (Cannabaceae). *Journal of Industrial Hemp* 9(1): 15–36.

Mahmoud M, BenRejeb I, Punja ZK, Buirs L, Jabaji S. 2023. Understanding bud rot development, caused by *Botrytis cinerea*, on cannabis (*Cannabis sativa* L.) plants grown under greenhouse conditions. *Botany* 101: 200–231.

Mangin L. 1888. Sur la constitution de la membrane des végétaux. *Comptes Rendus Hebdomadaires des séances de l'Académie des Sciences* 107: 144–146.

Mareska J, Heyman J. 1845. Enquete sur le travail et la condition physique et morale des ouvriers employes dan les manufactures de cotton, a Gand. *Ann Soc Med Gand* 16 (pt 2): 5–245.

Margolis JS, Clorfene R. 1975. *A child's garden of grass.* Ballantine Books, New York.

Marks WH, Florence L, Leiberman J, *et al.* (4 other authors). 1996. Successfully treated invasive aspergillosis associated with smoking marijuana in a renal transplant recipient. *Transplantation* 61: 1771–1774.

Marlatt CL, ed. 1921. *Pests collected from imported plants and plant products from January 1, 1921, to December 31, 1921.* Annual letter of information, no. 35. Federal Horticulture Board, USDA, Washington, DC.

Marshall CR. 1898. A contribution to the pharmacology of cannabis indica. *Journal of the American Medical Association* 31: 882–891.

Martyny JW, Serrano KA, Schaeffer JW, Van Dyke MV. 2013. Potential exposures associated with indoor marijuana growing operations. *Journal of Occupational and Environmental Hygiene* 10: 622–639.

McKernan K, Spangler J, Zhang L, *et al.* (4 other authors). 2015. Cannabis microbiome sequencing reveals several mycotoxic fungi native to dispensary grade cannabis flowers. *F1000Research* 4: 1422.

McKernan K, Spangler J, Zhang L, *et al.* (6 other authors). 2016a. Cannabis microbiome sequencing reveals *Penicillium paxilli* and the potential for paxilline drug interactions with cannabidiol. *Proceedings of the 26ᵗʰ Annual Symposium on the Cannabinoids.* International Cannabinoid Research Society, Research Triangle Park, North Carolina, pp. P1–28.

McKernan K, Spangler J, Helbert Y, *et al.* (9 other authors). 2016b. Metagenomic analysis of medicinal cannabis samples; pathogenic bacteria, toxigenic fungi, and beneficial microbes grow in culture-based yeast and mold tests. *F1000Research* 5: e2471.

McPartland JM. 2000. The *Cannabis* pathogen project: report of the fourth 5-year plan. *Inoculum (Mycological Society of America Newsletter)* 51(3): 46.

McPartland JM. 2003. Byssinosis in hemp mill workers. *Journal of Industrial Hemp* 8(1): 33–44.

McPartland JM, Cubeta MA. 1997. New species, combinations, host associations and location records of fungi associated with hemp (*Cannabis sativa*). *Mycological Research* 101: 853–857.

McPartland JM, McKernan KJ. 2017. "Contaminants of concern in cannabis: microbes, heavy metals and pesticides," in Chandra S, Lata H, ElSohly MA, *eds. Cannabis sativa: Botany and Biotechnology.* Springer International Publishing, Cham, Switzerland.

McPartland JM, Pruitt PL. 1999. Side effects of pharmaceuticals not elicited by comparable herbal medicines: the case of tetrahydrocannabinol and marijuana. *Alternative Therapies Health and Medicine* 5: 57–62.

McPartland JM, Clarke RC, Watson DP. 2000. *Hemp diseases and pests – management and biological control.* CABI Publishing, Wallingford, UK.

McPartland JM, Glass M, Pertwee RG. 2007d. Meta-analysis of cannabinoid ligand binding affinity and cannabinoid receptor distribution: interspecies differences. *British Journal of Pharmacology* 152: 583–589.

McPartland JM, MacDonald C, Young M, *et al.* (3 other authors). 2017. Affinity and efficacy studies of tetrahydrocannabinolic acid A at cannabinoid receptor types one and two. *Cannabis and Cannabinoid Medicine* 2(1): 87–95.

Meyer RW. 1974. Insects and other arthropods of economic importance in Indiana during 1974. *Proceedings of the Indiana Academy of Science* 84: 313–321.

Mirbel CF Brisseau de. 1809. *Exposition de la théorie de l'organisation végétale, Seconde Édition.* Dufart, Paris.

Moody M, Wharton RC, Schnaper N, Schimpff SC. 1982. Do water pipes prevent transmission of fungi from contaminated marijuana? *New England Journal of Medicine* 306: 1492–1493.

Narváez A, Rodríguez-Carrasco Y, Castaldo L, Izzo L, Ritieni A. 2020. Ultra-high-performance liquid chromatography coupled with quadrupole orbitrap high-resolution mass spectrometry for multi-residue analysis of mycotoxins and pesticides in botanical nutraceuticals. *Toxins* 12: e114.

Needham JT. 1769. *Nouvelles recherches sur les découvertes microscopiques, et la geenéraation des corps organisés, premiere partie.* Lacombe, Dauphine.

Nicholls PJ, Tuxford AF, Hould B. 1991. Bacterial content of cotton, flax, hemp and jute. *Proceedings, 15th Cotton Dust Research Conference,* pp. 289–292.

NJMMP. 2011. *Medicinal marijuana program rules.* New Jersey Medicinal Marijuana Program. Available at: www.nj.gov/health/medicalmarijuana/documents/final_rules.pdf

Obata H, Manabe A, Nakamura N, Onishi T, Senba Y. 2011. A new light on the evolution and propagation of prehistoric grain pests: the world's oldest maize weevils found in Jomon potteries, Japan. *PLoS ONE* 6(3): e14785.

Olivier GA. 1789. *Histoire naturelle. Tome quatreime. Insectes.* Panckoucke, Paris.

Ondřej M. 1991. Výskyt hub na stoncích konopí (*Cannabis sativa* L.). *Len a Konopí* (Sumperk, Czech Rep) 21: 51–57.

Ormerod EA. 1898. *Report of observations of injurious insects and common farm pests during the year 1897.* Simpkin & Co., London.

Pandey KN. 1982. Antifungal activity of some medicinal plants on stored seeds of *Eleusine coracana. Journal of Indian Phytopathology* 35: 499–501.

Parihar SS, Dadlani M, Lal SK, *et al.* (3 other authors). 2014. Effect of seed moisture content and storage temperature on seed longevity of hemp (*Cannabis sativa*). *Indian Journal of Agricultural Sciences* 84: 1303–1309.

Parker JM, Borke ML, Black LA, Cochran TG. 1974. Decomposition of cannabidiol in chloroform solution. *Journal of Pharmaceutical Sciences* 63: 970–971.

Passerini G. 1888. Diagnosi di funghi nuovi. Nota III. *Atti della Reale Accademia dei Lincei. Rendiconti di classe di Scienze Fisiche, Matematiche e Naturale* 4(2): 55–66.

Pasteur L. 1857. Mémoire sur la fermentation appelée lactique. *Comptes Rendus Chimie* 45: 913–916.

Paulsen AQ. 1971. *Plant diseases affecting marijuana (Cannabis sativa).* Unpublished manuscript, Kansas State University.

Pérez-Moreno M, Pérez-Lloret P, González-Soriano J, Santos-Álvarez I. 2019. Cannabis resin in the region of Madrid: Adulteration and contamination. *Forensic Science International* 298: 34–38.

Perris E. 1851. Quelques considérations sur les insectes nuesibles a l'agriculture. *Bulletin de la Société agricole, scientifique & littéraire des Pyrénées-Orientales* 8: 310–335.

Peschel W. 2016. Quality control of traditional cannabis tinctures: pattern, markers, and stability. *Scientia Pharmaceutica* 84: 567–584.

Piao XG. 1990. Studies on the mites infesting herb medicine in China and Japan and their derivation. *Japanese Journal of Sanitary Zoology* 41: 1–7.

Pietkiewicz TA. 1958. Mikroflora nasion konopi. Przeglad literatury. *Roczniki Nauk Rolniczych Seria A* 77(4): 577–590.

Pigeonneau H, Foville A de, *eds.* 1882. *L'Administration de l'Agriculture au Contrôle général des Finances (1785-1787) Procès-verbaux et rapports.* Librairie Guillaumin, Paris.

Plarre R. 2010. An attempt to reconstruct the natural and cultural history of the granary weevil, *Sitophilus granarius* (Coleoptera: Curculionidae). *European Journal of Entomology* 107: 1–11.

Pomahacova B, Van der Kooy F, Verpoorte R. 2010. Cannabis smoke condensate III: the cannabinoid content of vaporised *Cannabis sativa*. *Inhalation Toxicology* 21: 1108–1112.

Potter DJ. 2009. *The propagation, characterisation and optimisation of Cannabis sativa L. as a phytopharmaceutical*. Doctoral thesis, King's College, London.

Prain D. 1893. *Report on the cultivation and use of gánjá*. Bengal Secretariat Press, Calcutta.

Praźmowski A. 1879. Zur Entwickelungsgeschichte und Ferment wirkung einiger Bakterienarten. *Botanische Zeitung* 37: 409–424.

Preston DA, Dosdall L. 1955. *Minnesota plant diseases*. Horticultural Crops Research Branch, Special Publication no. 8. USDA, Beltsville, Maryland.

Priestley DA, Cullinan VI, Wolfe J. 1985. Differences in seed longevity at the species level. *Plant, Cell and Environment* 8: 557–562.

Punja ZK. 2021a. The diverse mycoflora present on dried cannabis (*Cannabis sativa* L., marijuana) inflorescences in commercial production. *Canadian Journal of Plant Pathology* 43: 88–100.

Punja ZK. 2021d. Emerging diseases of *Cannabis sativa* and sustainable management. *Pest Management Science* 77: 3857–3870.

Punja ZK, Ni L. 2021. The bud rot pathogens infecting cannabis (*Cannabis sativa* L., marijuana) inflorescences: symptomology, species identification, pathogenicity and biological control. *Canadian Journal of Plant Pathology* 43: 827–854.

Punja ZK, Collyer D, Scott C, *et al*. (3 other authors). 2019. Pathogens and molds affecting production and quality of *Cannabis sativa* L. *Frontiers in Plant Science* 10: e1120.

Punja ZK, Li N, Lung S, Buirs L. 2023. Total yeast and mold levels in high THC-containing cannabis (*Cannabis sativa* L.) inflorescences are influenced by genotype, environment, and pre-and post-harvest handling practices. *Frontiers in Microbiology* 14: e1192035.

Ramazzini B. 1713. *De Morbis Artificum Diatriba*. Patavii, Padua.

Ramírez J. 1990. Acute pulmonary histoplasmosis: newly recognized hazard of marijuana hunters. *American Journal Medicine* 88 (Supplement 5): 60N–62N.

Raven L. 1987. *Hands on spinning*. Interweave Press, Loveland, Colorado.

Reininger W. 1946. "Remnants from prehistoric times," pp. 374–404 in: Reininger W, *ed*. Hashish. *Ciba Symposia* 8 (5/6), Ciba Pharmaceutical Products, Summit, New Jersey.

Riberio A, Pochart P, Day A, *et al*. (5 other authors). 2015. Microbial diversity observed during hemp retting. *Applied Microbiology and Biotechnology* 99: 4471–4484.

Riley CV, Howard LO. 1892. Hemp as a protection against weevils. *Insect Life* 4: 223.

Robertson LW, Lyle MA. 1975. Biotransformation of cannabinoids by *Syncephalastrum racemosum*. *Biomedical Mass Spectrometry* 2: 266–271.

Robertson LW, Koh SW, Huff SR, Malhotra RK, Ghosh A. 1978. Microbiological oxidation of pentyl side-chain of cannabinoids. *Experientia* 34: 1020–1022.

Robinson BB. 1943a. Seed treatment of hemp seeds. *Journal of the American Society of Agronomy* 35: 910–914.

Romano LL, Hazekamp A. 2013. Cannabis oil: chemical evaluation of an up- coming cannabis-based medicine. *Cannabinoids* 1: 1–11.

Root KS, Magzamen S, Sharp JL, *et al*. (3 other authors). 2020. Application of the environmental relative moldiness index in indoor marijuana grow operations. *Annals of Work Exposures and Health* 64: 728–744.

Ross SA, ElSohly MA. 1996. The volatile oil composition of fresh and air-dried buds of *Cannabis sativa*. *Journal of Natural Products* 59: 49–51.

Ross SA, ElSohly MA. 1997. CBN and Δ^9-THC concentration ratio as an indicator of the age of stored marijuana samples. *Bulletin on Narcotics* 49: 139–147.

Rossi G. 1902. Primo contributo allo studio della macerazione della canapa. *Stazioni Sperimentali Agarie Italiane* 35: 241–278.

Rossi G, Guarnieri G. 1907. *Bacillus comesii* e le su proprietà. *Atti del VI Congresso di Chimica Applicata*, Roma.

Roumeguere C. 1881. Fungi Gallici exsiccati, Centuria XVII, Index. *Revue Mycologique* 3(12): 6–7.

Ruchlemer R, Amit-Kohn M, Raveh D, Hanuš L. 2015. Inhaled medicinal cannabis and the immunocompromised patient. *Supportive Care in Cancer* 23: 819–822.

Ruschmann G. 1923. *Grundlagen der Röste: ein wissenschaftlich-technische Einführung*. S. Hirzel, Leipzig.

Russo EB, Marcu J. 2017. *Cannabis* pharmacology: the usual suspects and a few promising leads. *Advances in Pharmacology* 80: 67–134.

Russo EB, Plumb J, Witely VL. 2021. Novel solventless extraction technique to preserve cannabinoid and terpenoid profiles of fresh *Cannabis* inflorescence. *Molecules* 26: e5496.

Rusynov VI, Martynov VO, Kolombar TM. 2019. "Coleoptera pests of stored food supplies and field crops," pp. 34–133 in Gristan YI, *et al*., *eds*. Current problems of agrarian industry in Ukraine. Accent Graphics Publishing, Vancouver, Canada.

Saikia UN, Sarbhoy AK. 1982. Hyphomycetes of India. V. the genus *Periconia*. *Indian Phytopathology* 35: 277–281.

Salam AP, Pozniak AL. 2017. Disseminated aspergillosis in an HIV-positive cannabis user taking steroid treatment. *Lancet Infectious Diseases* 17: 882.

Salomon L. 1893. *Essai sur une intoxication aiguë et chronique observée chez les peigneurs de chanvre*. Steinheil, Paris.

Salvá Campillo F, Santponç Roca F. 1784. *Disertación sobre la explicación y uso de una nueva máquina para agramar cáñamos y linos inventada por los doctores en medicina*. Imprenta Real, Madrid.

Sarma ND, Waye A, ElSohly MA, *et al*. (13 other authors). 2020. Cannabis inflorescence for medical purposes: USP considerations for quality attributes. *Journal of Natural Products* 83: 1334–1351.

Schilling RSF. 1956. Byssinosis in cotton and other textile workers. *Lancet* 271: 261–265, 319–324.

Schmidt M. 1929. Blattlausfliegen (Syrphidae) als vorratsschädlinge. *Mitteilungen Gesellschaft für Vorratsschutz* 5(6): 80–81.

Schultes RE. 1973. Man and marijuana. *Natural History* 82(7): 58–63, 80, 82.

Schwartz I. 1985. Marijuana and fungal infection. *American Journal of Clinical Pathology* 84: 256.

Schwartz IS. 1987. Fungal sinusitis and marijuana. *JAMA* 257: 2914–2915.

Schwartz RH, Hollick GE. 1981. Allergic bronchopulmonary aspergillosis with low serum immunoglobulin E. *Journal of Allergy and Clinical Immunology* 68: 290–294.

Scott C, Punja Z. 2022. "Management of diseases on *Cannabis* in controlled environment production," pp. 231–251 in Zheng YB, *ed*. Handbook of Cannabis production in controlled environments. CRC Press, Boca Raton, Florida.

Scott DC. 1892. *A cyclopaedic dictionary of the Mang'anja language spoken in British Central Africa*. Foreign Mission Committee of the Church of Scotland, Edinburgh.

Sengbusch RV. 1955. Die erhaltung der keimfähigheit von samen bei tiefen temperature (Preserving the germination of seeds by low temperature). *Züchter* 25: 168–169.

Shapiro BB, Hedrick R, Vanle BC, *et al.* (7 other authors). 2018. Cryptococcal meningitis in a daily cannabis smoker without evidence of immunodeficiency. *BMJ Case Reports* 2018: bcr2017221435.

Small E. 1975. Morphological variation of achenes of *Cannabis*. *Canadian Journal of Botany* 53: 978–987.

Small E, Brookes B. 2012. Temperature and moisture content for storage maintenance of germination capacity of seeds of industrial hemp, marijuana, and ditchweed forms of *Cannabis sativa*. *Journal of Natural Fibers* 9: 240–255.

Smith RL, Olson CA. 1982. Confused flour beetle and other coleoptera in stored marijuana. *Pan-Pacific Entomologist* 58: 79–80.

Smith RN, Vaughan CG. 1977. The decomposition of acidic and neutral cannabinoids in organic solvents. *Journal of Pharmacy and Pharmacology* 29: 286–290.

Smythe CV. 1951. Microbiological production of enzymes and their industrial applications. *Economic Botany* 5: 126–144.

Snegireva A, Chernova T, Ageeva M, Lev-Yadun S, Gorshkova T. 2015. Intrusive growth of primary and secondary phloem fibres in hemp stem determines fibre-bundle formation and structure. *AoB Plants* 7: plv061.

Société d'Agriculture du Marne. 1813. *Séance publique de la Société d'agriculture, Commerce, Sciences et Arts du Département de la Marne*. Boniez, Chalons.

Sopovski DS, Han J, Stevens-Riley M, *et al.* (6 other authors). 2023. Investigation of microorganisms in cannabis after heating in a commercial vaporizer. *Frontiers in Cellular and Infection Microbiology* 12: e1051272.

Spallanzani L. 1765. *Saggio di osservazioni microscopiche concernenti il sistema della generazione dei signori di Needham e Buffon*. B. Soliani, Modena.

Spallanzani L. 1776. *Opuscoli di fisica animale e vegetabile*. Società tipographica, Modena.

Stadnyk K, White NDG, Jian F, Gields PG. 2018. Suitability of hemp seed for reproduction of stored-product insects. *Julius-Kühn-Archiv* 463: 172.

Starks M. 1977. *Marijuana potency*. And/Or Press, Berkeley, California.

Sullivan N, Elzinga S, Raber JC. 2013. Determination of pesticide residues in cannabis smoke. *Journal of Toxicology* 2013: e378168.

Sutton DB, Punja ZK, Hamarneh G. 2023. Characterization of trichome phenotypes to assess maturation and flower development in *Cannabis sativa* L. (cannabis) by automatic trichome gland analysis. *Smart Agricultural Technology* 3: e10011.

Sutton S, Lum BL, Torti FM. 1986. Possible risk of invasive aspergillosis with marijuana use during chemotherapy for small cell lung cancer. Drug Intelligence & Clinical Pharmacy 20: 289–291.

Szyper-Kravitz M, Lang R, Manor Y, Lahav M. 2001. Early invasive pulmonary aspergillosis in a leukemia patient linked to *Aspergillus* contaminated marijuana smoking. Leukemia and Lymphoma 42: 1433–1437.

Tamburini E, Daly S, Steiner U, Vandini C, Mastromei G. 2001. *Clostridium felsineum* and *Clostridium acetobutylicum* are two distinct species phylogenetically closely related. *International Journal of Systematic and Evolutionary Microbiology* 51: 963–966.

Tamburini E, Gordillo León AG, Perito B, Mastromei G. 2003. Characterization of bacterial pectinolytic strains involved in the water retting process. *Environmental Microbiology* 5: 730–736.

Tamburini E, León AG, Perito B, Di Candilo M, Mastromei G. 2004. Exploitation of bacterial pectinolytic strains for improvement of hemp water retting. *Euphytica* 140: 47-54.

Taschwer M, Schmid MG. 2015. Determination of the relative percentage distribution of THCA and Δ(9)-THC in herbal cannabis seized in Austria—impact of different storage temperatures on stability. *Forensic Science International* 254: 167–171.

Tassi F. 1900. Novae micromycetum species descriptae et iconibus illustratae. *Bollettino del Laboratorio de Orto Botanico Reale Universita Siena* 3: 117–132.

Taylor DN, Wachsmuth IK, Sangkuan YH, *et al.* (7 other authors). 1982. Salmonellosis associated with marijuana: a multistate outbreak traced by plasmid fingerprinting. *New England Journal of Medicine* 306: 1249–1253.

Thomas HF, Elwood JH, Elwood PC. 1988. Byssinosis in Belfast ropeworks: an historical note. *Annals Occupational Hygiene* 32: 249–251.

Toole EH, Toole VK, Nelson EG. 1960. *Preservation of hemp and kenaf seed*. USDA Agricultural Research Service, Technical Bulletin no. 1215. Government Printing Office, Washington, DC.

Traill GW. 1828. Statistical sketch of Kamaon. *Asiatick Researches* 16: 137–236.

Trofin IG, Vlad CC, Dabija G, Filipescu L. 2011. Influence of storage conditions on the chemical potency of herbal cannabis. *Revista de Chimie Bucharest* 63: 293–297.

Trofin IG, Dabija G, Vaireanu DI, Filipescu L. 2012a. Long-term storage and cannabis oil stability. *Revista de Chimie Bucharest* 62: 639–645.

Trofin IG, Dabija G, Vaireanu DI, Filipescu L. 2012b. The influence of long-term storage conditions on the stability of cannabinoids derived from cannabis resin. *Revista de Chimie Bucharest* 63: 422–427.

Turner CE, Hadley KW, Fetterman PS, *et al.* (3 other authors). 1973a. Constituents of *Cannabis sativa* L. IV: Stability of cannabinoids in stored plant material. *Journal of Pharmaceutical Sciences* 62: 1601–1605.

Turner JC, Hemphill JK, Mahlberg PG. 1977. Gland distribution and cannabinoid content in clones of *Cannabis sativa*. *American Journal of Botany* 64: 687–693.

Ungerleider JT, Andrysiak T, Tashkin DP, Gale RP. 1982. Contamination of marihuana cigarettes with pathogenic bacteria—possible source of infection in cancer patients. *Cancer Treatment Reports* 66: 589–590.

Upton R, ed. 2013. *Cannabis inflorescence, Cannabis spp. Standards of identity, analysis, and quality control*. American Herbal Pharmacopoeia, Scotts Valley, CA.

Upton R, Craker L, ElSohly M, *et al.* (5 other authors). 2013. *Cannabis inflorescence, Cannabis spp., standards of identity, analysis, and quality control*. American Herbal Pharmacopoeia, Scotts Valley, California.

Valic F, Zuskin E, Walford J, Kersic W, Paukovic R. 1968. Byssinosis, chronic bronchitis, and ventilatory capacities in workers exposed to soft hemp dust. *British Journal Industrial Medicine* 25(3): 176–186.

Valyn M. 1885. Tablettes du travail. Le chanvre destructeur des charançon. *Le Petit Journal*, 16 mars 1885.

Van Tieghem P. 1877. Sur le *Bacillus amylobacter* et son rôle dans la putrefaction des tissues végétaux. *Bulletin de la Société Botanique de France* 24: 128–134.

Van Tieghem P. 1879. Sur la fermentation de la cellulose. *Comptes Rendus Hebdomadaires des séances de l'Académie des Sciences* 88: 205–210.

Vanhove W, Cuypers E, Bonneure AJ, *et al.* (4 other authors). 2018. The health risks of Belgian illicit indoor *Cannabis* plantations. *Journal of Forensic Sciences* 63(6): 1783–1789.

Vavilov NI. 1926. The origin of the cultivation of "primary" crops, in particular cultivated hemp. *Bulletin of Applied Botany and Plant-Breeding* 16(2): 221–233.

Vavilov NI, Bukinich DD. 1929. Konopli. Труды по прикладной ботанике, генетике и селекции 33 (Suppl.): 380–382.

Veress T, Szanto JI, Leisztner L. 1990. Determination of cannabinoid acids by high-performance liquid chromatography of their neutral derivatives formed by thermal decarboxylation: I. Study of the decarboxylation process in open reactors. *Journal of Chromatography A* 520: 339–347.

Verweij PE, Kerremans JJ, Voss A, Meis JF. 2000. Fungal contamination of tobacco and marijuana. *JAMA* 284: 2875.

Waley A. 1996. *The Book of Songs.* Grove Press, New York.

Wehmer CW. 1903. Der *Mucor* der hanfrötte, *M. hiemalis nov. spec. Annales Mycologici* 1: 39–41.

Welty RE. 1972. Fungi isolated from flue-cured tobacco sold in southeast United states, 1968-1970. *Applied Microbiology* 24: 518–520.

Westendorp GD. 1854b. *Les Cryptogames.* I.S. Van Doosselaere. Gand, Belgium.

Wiesner J. 1873. *Rohstoffe des Pflanzen reiches.* W. Engelmann, Leipzig.

Winogradski S, Fribes V. 1895. Sur le rouissage du lin et son agent microbien. *Comptes Rendus Hebdomadaires des séances de l'Académie des Sciences* 121: 742–745.

Wrisberg HA. 1765. *Observationum de animalculis infusoriis satura.* Vandenhoek, Göttingen.

Yadav S, Yadav PK, Yadav D, Yadav KD. 2008. Purification and characterization of pectin lyase produced by *Aspergillus terricola* and its application in retting of natural fibers. *Applied Biochemistry and Biotechnology* 159: 270–283.

Yamauchi T, Shoyama Y, Aramaki H, Azuma T, Nishioka I. 1967. Tetrahydrocannabinolic acid a genuine substance of tetrahydrocannabinol. *Chemical & Pharmaceutical Bulletin* 15: 1075–1076.

Zambettakis CE. 1955. Recherches anatomiques et biologiques sur les Sphaeropsidales-Phaeodidymae des fungi imperfecti. *Archives du Muséum National d'Histoire Naturelle, 7ème Série* 3(1): 43–145.

Zelepukha SI. 1960. Третя нарада з проблеми фітонцидів. *Mikrobiolohichnyi Khurnal (Kiev)* 22(1): 68–71.

Zhao MQ, Wang BX, Li FX, *et al.* (4 other authors). 2007. Analysis of bacterial communities on aging flue-cured tobacco leaves by 16S rDNA PCR–DGGE technology. *Applied Microbial and Cell Physiology* 73: 1435–1440.

Zuskin E, Valic F, Bouhuys A. 1976. Byssinosis and airway responses due to exposure to textile dust. *Lung* 154: 17–24.

Zuskin E, Kanceljak B, Pokrajac D, Schachter EN, Witek TJ. 1990. Respiratory symptoms and lung function in hemp workers. *British Journal Industrial Medicine* 47(9): 627–632.

Zuskin E, Kanceljak B, Schachter EN, *et al.* (5 other authors). 1992. Immunological findings in hemp workers. *Environmental Research* 59: 350–361.

Abstract
Cultural and mechanical methods of pest and disease control can be preventive or curative. Proper sanitation is key, particularly indoors—the use of disease-free seeds and clones, establishing a quarantine area, and at the high end: meristem tip culture. Hydroponic solutions need to be sanitized. Field sanitation includes the destruction of infested crop residues, attention to weeds, and crop rotation. Mechanical controls include insect traps, mechanical barriers, judicious pruning, and optimizing plant growth.

19.1 Introduction

Cultural controls are practices that alter the landscape and make it less favorable for pest establishment, reproduction, dispersal, and survival. Examples include crop rotation, resistant or tolerant cultivars, regulating moisture levels, and tilling the soil (Fig. 19.1). Cultural controls are usually preventive, and consist of ordinary farm practices that encourage healthy plant growth.

Mechanical controls kill pests and pathogens directly, or make the environment unsuitable for them. Mechanical controls can be preventive (e.g., steam sterilization of glasshouse soil, screens over greenhouse vents) or curative (e.g., pruning away fungus-infected branches, hoeing weeds). Mechanical methods can be as simple as hand-picking pests off plants, or as complicated as *in vitro* meristem tip culture.

Useful cultural and mechanical methods are listed below. Many of these methods are ancient. Many were abandoned with the advent of chemical pesticides, but are enjoying a resurgence in popularity.

19.2 Sanitation

Sanitation is a cornerstone of cultural control, and key in a crop like *Cannabis* with limited pesticide options. Sanitation involves the removal of infested materials, as well as *potential* sources of infestation, and the disinfection of equipment. Field sanitation and greenhouse/growroom sanitation will be treated separately. They share some strategies, such as the use of clean starting material.

19.2.1 Clean start

Purchase seeds that are pest- and pathogen-free. Seeds can harbor fungi, bacteria, and viruses, as well as weed seeds, insects, and mites. Seeds may be marketed as "certified" disease-free, but currently there are no government programs or independent certifying agencies that test these claims for *Cannabis*. Many

growers save their own seeds to grow next year's crop. Do not save seeds from plants infested by diseases or pests.

To prevent the importation of seed-borne diseases and pests, some governments have implemented "phytosanitary requirements." New Zealand stipulated that all imported *Cannabis* seeds undergo a "hot water treatment" (Thomson, 2019), based on a hot water dip described in the first edition of *HD&P*. Subsequently, Thomson (2020) specified requirements for the hot water dip: 50°C for 30 min, or at 60°C for 10 min. Unfortunately, hot water causes thermal damage—it may kill seeds, delay seed germination, or stunt seedlings. Thermal damage can be mitigated by plunging seeds into cold water after treatment, then carefully drying them.

Seed surfaces can be sterilized by rinsing seeds in a solution of 1 part Clorox® bleach (sodium hypochlorite) in 9 parts water (resulting in 0.5% NaOCl) for 15 min, with gentle agitation, followed by a water rinse, then drying. The addition of a second rinse in 70% ethanol can be detrimental—it may reduce seed generation by half (M. Bolt, unpublished results, 2023).

Clonal propagation has become the norm in drug-type *Cannabis* crops—taking vegetative cuttings from mother plants and rooting them. Clones infected by viruses have become pandemic (see section 14.2). In the absence of certifying agencies, growers must trust their clone sources, and quarantine new clones.

Establish a **quarantine area** in your facility. Cultivate new arrivals in their own grow space, separate from other plants. Limit movement of people and tools between the quarantine area and the rest of the facility. Keep quarantined clones separated until you are confident that they are not infected—which entails flowering a few clones to watch for disease symptoms. Researchers have developed assays to test for common *Cannabis* pathogens; these include assays performed by the grower and assays performed in laboratories.

Micropropagation

The clonal propagation of plants, when performed in sterile conditions on synthetic media, is termed **micropropagation** or plant tissue culture. Small bits of plant tissue, called **explants,**

DOI: 10.1079/9781836990352.0019

Fig. 19.1. Row cultivation of hemp (adapted from Lesik, 1958)

are cultured into whole plants. Morel and Martin (1952) first used this *in vitro* approach to propagate disease-free *Dahlia* plants. Unfortunately, *Cannabis* is notoriously recalcitrant to micropropagation techniques. Two approaches have been used: indirect and direct organogenesis.

Indirect organogenesis regenerates explants into plant organs (roots, shoots) using an intermediate "callus" step: Explants are plated on Murashige and Skoog (MS) medium, and exposed to a cocktail of plant growth regulators, which turns them into **callus tissue**—a mass of undifferentiated parenchyma cells. This is also known as somatic embryogenesis. The next step is notoriously difficult with *Cannabis*: regenerating callus into whole plants (roots, shoots). Callus tissue is treated with another cocktail of plant growth regulators. Ślusarkiewicz-Jarzina *et al.* (2005) made the breakthrough. They regenerated shoots from callus (Fig. 19.2), then rooted plants, and acclimatized the plantlets for life outside petri plates.

Direct organogenesis refers to the formation of plant organs (roots, shoots) directly from explants, without a callus stage. Similar cocktails of plant growth regulators are used to regenerate explants and callus, but explants are easier. Sources of explants for regenerating *Cannabis* plantlets include seedling hypocotyls, cotyledons, and primary leaves, as well as apical and axillary meristems (reviewed by Chandra *et al.*, 2017; Monthony *et al.*, 2021). Explants growing *in vitro* can be cleansed of bacteria and fungi by culturing them with antibiotics and fungicides.

Champions of direct organogenesis criticize indirect organogenesis for somaclonal variation that sometimes arises in callus tissue (Chandra *et al.*, 2017). Lata *et al.* (2010) took micropropagated plants through 30 cycles of direct organogenesis, and showed no genetic variance using 15 ISSR markers. On the negative side, direct organogenesis has a higher risk of yielding chimeric plants (Feeney and Punja, 2017).

Meristem tip culture utilizes explants excised from the growing tip of a plant. Morel and Martin (1952) used meristem tip culture to propagate virus-free *Dahlia* plants. This is possible because vascular tissues, within which viruses migrate, do not extend into the apical meristem or growing tip of a plant.

Fig. 19.2. *Cannabis* plantlet regenerated from callus (courtesy of Aurelia Ślusarkiewicz-Jarzina)

Their method was further refined, by using heat treatment (thermotherapy) of meristem cultures to denature any viruses present (Quak, 1961).

The shoot apical meristem (SAM) of *Cannabis* is small—0.5 mm at its base—and difficult to dissect away from nearby leaf tissue, which contains vascular tissues (see Fig. 14.8). Spitzer-Rimon *et al.* (2019) provided photomicrographs of its structure. The SAM is frequently the site of somatic mutations that give rise to chimeras (see Fig. 17.18).

Small size, coupled with fidgety organogenesis, makes meristem tip culture difficult. Nodal segments with apical or axillary buds have been regenerated into plantlets (Ślusarkiewicz-Jarzina *et al.*, 2005; Casano and Grassi, 2009; Lata *et al.*, 2009). But these studies utilized SAM + leaf tissue. Smýkalová *et al.* (2019) excised "isolated apical meristem" from seedlings, and regenerated plantlets using a novel cytokinin derivative. So the technology is there. Websites have proliferated that advertise "patent-pending cleaning processes" to eliminate *Cannabis* viruses. The entrepreneurs are ahead of the scientists.

19.2.2 Field sanitation

Avoid moving from infested fields to clean fields. This includes equipment (heavy machinery, hand tools) and yourself. Work in clean areas first, then move to diseased areas, and then disinfect equipment. Sodium hypochlorite (bleach) is a good disinfectant. Especially avoid moving through fields when plants are wet, to minimize the spread of bacteria and fungi.

Rogue (pull out) plants that are heavily infected or infested. Properly dispose of them by burial, burning, or composting. Composting effectively destroys pests and pathogens, and it provides an end-product: compost. But it must be done properly, otherwise the compost pile turns into a pest nursery.

Keep fields free of weeds. Weeds suck nutrients and water from crops. Weeds attract many pests and pathogens that subsequently move into *Cannabis*. Weeds create humid microenvironments that favor mold development. Keeping the soil surface clean eliminates insects who cannot lay eggs on bare ground. Hand-hoe small plots, and machine-till row-cropped hemp. Row spacings can be adjusted to accommodate the tillage equipment being used (Fig. 19.1).

Immediately after harvest, destroy infested crop residues. If harvesting is done at different dates, because of different cultivars, treat those crop blocks separately—do not wait until all the blocks have been harvested to destroy crop residues. Shredding of stalks and roots exposes overwintering pests and pathogens. Burying or burning residues is also effective. Remove, bury, or burn anything that may shelter overwintering pests, such as weeds and surface trash (dead leaves, brush heaps, boxes, equipment, etc.). Clean all harvesting equipment.

Tillage or "working the soil" exposes overwintering pests, pathogens, and weed seeds to their natural enemies and to the weather. Tillage can be done after harvest or prior to planting. **Ards** or scratch-plows were the earliest tillage equipment. Other than use in loose alluvial soil, ards were arduous. They were not suited for heavy soil or clearing new land. For these situations, the ancient Chinese invented **turnplows**, which turned over the soil (Bray, 1984). The turnplow consists of a **plowshare**, which cuts through the soil, connected to a **moldboard**, which inverts the soil. Europeans used turnplows by early Medieval times, but their moldboards were straight and required four to eight oxen to pull them. Chinese turnplows with curved moldboards required only one or two animals. The Dutch acquired Chinese turnplows in the early 17th century that had iron plowshares and curved moldboards, whose depth could be adjusted. In 1837 John Deere of Middlebury, Vermont, used smooth, self-cleaning steel rather than cast iron—the modern moldboard.

The English author Gervase Markham (1568–1637) said that hemp fields should be tilled twice, at the end of February and the latter half of April immediately before sowing (Markham, 1615). Clods of soil must be broken up with a "clotting-beetle" or wooden mallet. He emphasized that plowed soil with its furrows must be smoothed with a harrow before sowing hemp seeds.

A **harrow** is an implement dragged across plowed land to level and smooth the soil. In Markham's time, a harrow was a wooden lattice with spikes, dragged by a horse. Later these were armed with curved and flexible iron teeth mounted in rows—**spring-tooth harrows**. These were replaced by **disc harrows**, whose cutting edges are a row of metal discs. Disc harrows with added horsepower—petrol-powered tractors and hydraulic controls—became disc plows.

The **seed drill** was invented in China in the 2nd century BC and introduced to Italy in the mid-16th century (Bray, 1984). In England it was further refined by Jethro Tull (1674–1741). He invented a horse-drawn seed drill that sowed seeds at the correct depth and spacing, and then covered the seed with soil. Crops planted in rows could be more easily weeded. Tull (1731) advocated horses over oxen because they were lighter and caused less soil compaction. Tull's "horse-hoeing husbandry" was recommended for hemp by Henry (1771), to facilitate the separate harvest of males (for best fiber) and females (for fiber and seeds). Wissett (1808) seconded that opinion. In 1937 Walter Huhnke invented the first hemp-specific seed drill (Fig. 19.3) for breeding studies near Berlin (Neuer *et al.*, 1946).

Field history related to the previous years' crops will inform your selection of light tilling or deep disking. Tillage comes at a price: fractured soil structure, soil moisture depletion, and accelerating soil erosion. Intensive tillage releases large amounts of CO_2 into the atmosphere.

No-till or **low-till** are alternatives to routine plowing. The benefits are many: reduced soil erosion (especially the loss of particulate organic material, POM), improved soil quality (less soil compaction), and increased soil microbiome diversity (thereby reducing pathogens and soil-dwelling insects). Fuel usage and labor costs are reduced. The dust bowl era of over-plowing gave rise to the first book about no/low-till, *Plowman's Folly*. Faulkner (1943) stated in his opening paragraph, "The truth is that no one has ever advanced a scientific reason for plowing." Faulkner was a proponent of low-till methods, such as disc harrowing, rather than moldboard plowing.

Disc harrowing slices-and-dices crop stubble on the soil surface. This preserves POM, but may hamper the sowing of seed. Hemp seed is sensitive to planting depth, and achieving a precise depth can be challenging in a no-till field. A lot of force is required to cut through hemp stubble: 243 N for each hemp stalk (Chen *et al.*, 2004). In comparison, cutting through maize stubble requires 23–80 N. Grabowska (2005) proposed no-till as a viable option for growing fiber-type hemp. Zengin *et al.* (2018) reported that a disc-harrowed crop of fiber-type 'Futura 75' produced high biomass with a high yield of essential oil. Medicinal *Cannabis* has also been grown in no-till systems (Bennett, 2018).

No/low-till has a downside: increased weed pressure, necessitating greater reliance on herbicides. Researchers are seeking ways to decrease herbicide usage in no-till crops. Rodale Institute (2020) grew fiber-type hemp in a no-till system that utilized cover crops and roller-crimpers instead of herbicides and tillers. Ironically, no-till systems have trouble controlling *Cannabis ruderalis* where it poses as a troublesome weed, such as in India (Kumar and Chopra, 2013). A no-till sunflower crop in Bulgaria utilized low levels of herbicides to control *C. ruderalis* (Mitkov *et al.*, 2018).

Plasticulture—growing plants in strips of plastic—is frequently employed in the field cultivation of drug-type *Cannabis*. This is because crops of drug-type *Cannabis* are planted in widely spaced rows, providing ample space for weeds. Plasticulture reduces weed pressure (Fig. 19.4), and limits diseases that spread from soil by rain splash (e.g., *Septoria* spp.). *Cannabis* easily becomes a weed, and volunteer plants growing from seeded crops are potent sources of

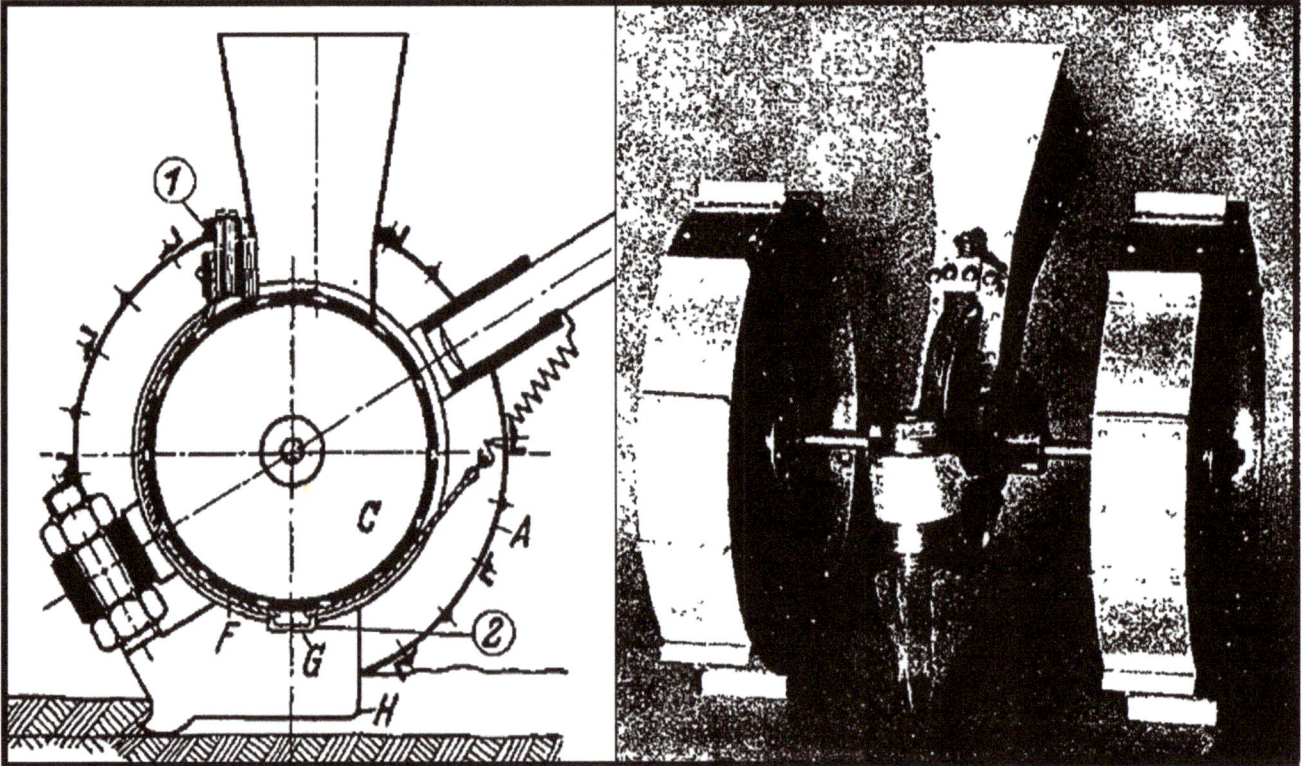

Fig. 19.3. Huhnke hemp seed drill (with courtesy, Neuer *et al.*, 1946).

Fig. 19.4. Plasticulture (courtesy M. Bolt)

inoculum and must be eradicated. Plasticulture has downsides; see Fig. 16.20.

Field soil can be partially sterilized with **soil solarization**, by covering the soil with strips of plastic and heating under the sun for 4–6 weeks. Solarization works best in sunny climates, obviously. The field must be carefully disked and its surface smoothed. Remove any rocks or clods that might raise the plastic or puncture it. Treated soil must be kept moist so that the heat will penetrate evenly. Consider laying drip irrigation lines before laying plastic.

Plastic sheets should be polyethylene, transparent, and at least 76 cm wide. Proper thickness ranges from 1 mil (one-thousandth of an inch, or 0.025 mm) to 2 mils, depending on the wind. Plastic sheets must be laid tightly against the soil; row edges are anchored by burying the edges in a shallow trench. Solarization controls many soilborne fungal and bacterial pathogens, as well as some nematodes, weed seeds, and insects. Earthworms survive by burrowing to lower soil depths. Done properly, solarization kills pathogenic fungi while maintaining a rich microflora of beneficial organisms. Beneficial fungi, such as *Trichoderma* and mycorrhizae, are very heat resistant. Ristaino *et al.* (1991) mixed beneficial *Gliocladium virens* into soil, then covered it with clear polyethylene 1 mil thick (0.025 mm) for 6 weeks during the warmest months of summer. Soil temperatures reached 41–49°C at 10 cm soil depth.

19.2.3 Glasshouse/growroom sanitation

Indoors, sanitation and exclusion are key. Cover air vents and windows with fine screens or mesh, and make sure doorframes shut tight. Keep areas around glasshouses as weed-free as possible. Limit the entry of visitors into facilities. People and equipment can move from clean areas to infested areas, but never the reverse (Wainwright-Evans, 2017). Sanitize footwear by placing foot baths at entryways. Have personnel change out of street clothes into work attire (lab jackets, hairnets).

All work surfaces, such as propagation benches, potting tables, etc., must be disinfected regularly. Sodium hypochlorite, hydrogen peroxide, peroxyacetic acid, and quaternary ammonium chloride compounds are described below, in the section on hydroponic disinfectants. Dodecyl dimethyl ammonium chloride (DDAC) is a hospital-grade disinfectant used for sanitizing work surfaces and tools, and employed in footbath floor mats. Scott and Punja (2022) compared Chemprocide® (with DDAC) to ZeroTol® (with hydrogen peroxide) against three *Cannabis* pathogens—*Botrytis cinerea*, *Fusarium oxysporum*, *Penicillium olsonii*—and DDAC was slightly more effective.

Prune, rogue, and dispose of infested plants. Prune with sharp scissors, and dip in isopropyl alcohol or DDAC between cuts. Dab prune sites with alcohol, or with biocontrol organisms. Remove all clippings, fallen leaves, and other debris throughout the season. Eliminate all plant residues after harvest.

Facilities with catastrophic problems, such as spider mites, broad mites, russet mites, root aphids, and powdery mildew, should be completely emptied and sanitized. Wash down walls, and clean nutrient tanks with an enzyme product (e.g., Hygrozyme®) to facilitate the breakdown of dead plant matter

in circulating systems. Trays, flats, pots, and other containers should be discarded, or if re-used, thoroughly washed and sanitized. Heat-sterilize hand tools. Pests and pathogens are killed at different temperatures (Table 19.1).

Potting soil can be sterilized or pasteurized. Re-using spent rockwool is risky because it cannot be completely sanitized, and may harbor inoculum. Heating soil to 80°C for 30 min kills most soil organisms (Table 19.1). Steam heat is economical. Batches of soils and compost can be steamed in autoclaves, which come in various sizes, including walk-in models. Steam can be injected into soil through hollow-tined soil injection devices, or through perforated pipes buried permanently underground.

Soil sterilization has drawbacks—it eliminates beneficial organisms and creates a biological vacuum. Another drawback is nutrient toxicity. Manganese in soil heated above 80°C becomes available to plants in toxic amounts, particularly in acidic, clay soils (Jarvis, 1992). Unless this free manganese is leached from the soil, it may remain toxic for 60 days or more and contribute to iron deficiency. Nitrogen in steamed soil also undergoes changes. Nitrifying bacteria are killed, so ammonia may build up to phytotoxic levels (Jarvis, 1992).

Soil pasteurization is better than sterilization—it kills pathogenic organisms while maintaining a rich microflora of beneficial organisms. Pasteurization is analogous to soil solarization, described above. Pasteurization is accomplished by mixing steam with air at 60°C. Instead of pasteurizing soil, many small-time gardeners choose to purchase pre-pasteurized, ready-mixed potting soil.

Hydroponic facilities can be completely shut down by water molds (McEno, 1990). Facilities with catastrophic problems should be dismantled and sanitized before another crop is introduced. Equipment amenable to heat treatment can be sterilized. Remove reservoir tanks and pipes from the room and do a deep clean. Bleach down walls and floors. Reassemble, and "shock treat" the nutrient solution with disinfectants prior to adding plants.

Hydroponic nutrient solution that recirculates can act as both a primary inoculum source and a dispersal mechanism for oomycetes and fungi. Methods of nutrient solution disinfection were reviewed by Stewart-Wade (2011); we'll provide

Table 19.1. Thermal inactivation of selected pathogens and pests (from Jarvis, 1992)

Organism	Temperature (°C)	ET[a]
Botrytis cinerea	55	15
Sclerotinia sclerotiorum	50	5
Pythium ultimum	46	20–40
Fusarium oxysporum	60	30
Rhizoctonia solani	53	30
Most other fungi	60	30
Most bacteria	60–70	10
Most viruses	100	15
Most nematodes	48–56	10–15
Most insects and mites	60–70	30
Most weed seeds	70–80	15

a.ET = exposure time, minutes

summaries below. A disadvantage shared by nearly all disinfection methods is the elimination of beneficial microbes. Biocontrol organisms and rhizobacteria added to growing media normally recirculate in nutrient solution.

Heat treatment

Flash-pasteurization is common in the Netherlands. Recirculating nutrient solution is heated to 95°C, for 30 seconds, and rapidly cooled. Two sets of heat exchangers do the job. Disadvantages include an expensive infrastructure and high energy costs (and carbon footprint—most systems use natural gas). Ancillary steps are needed to keep heat exchangers from corroding, such as prefiltration and adjusting pH levels.

Ozone

Ozone injected into nutrient solution is a potent and rapid-acting disinfectant. The ozone is generated on-site from oxygen in the air, and reverts to oxygen in minutes. Compared with heat treatment, ozone systems are economically feasible for smaller-scale operations. Ozone systems require a qualified professional for installation and maintenance. Several companies market their systems to *Cannabis* cultivators. Ozone efficacy declines in solutions that are high in pH and organic matter. Ozone causes iron to precipitate, knocking that nutrient out of solution. Ozone levels are hard to measure, and excessive levels damage plant roots.

Ultraviolet (UV) light

UV-C light at an intensity of 100 mJ/cm^2 kills pathogenic fungi, but higher doses are need to kill viruses and nematodes. UV light takes longer to kill pathogens than ozone. It doesn't work if water is turbulent and contains suspended organic material or sediment, so it requires a pre-filtration step. The simplest apparatus consists of a UV-C lamp mounted centrally within a steel cylinder. Nutrient solution flows past the lamp and the cylinder wall, its flow rate controlled by a solenoid valve. Getting the flow rate correct is the tricky part. Monitoring light intensity is important, because UV lamps degrade with time.

Filtration systems

Slow sand filtration (SSF) has been used for a century in sewage treatment plants (Fig. 19.5). Greenhouse operations began using SSF 30 years ago in Europe. Efficacy depends on sand particle size, water flow rate, and the concentration of the pathogen. In addition to filtration, the top layer of sand, termed *schmutzdecke*, is colonized by beneficial microbes that kill pathogens. Granulated rockwool (e.g., Grodan®) or pumice have also been used. SSF is slow and cumbersome, and sequesters lots of nutrient solution. Small, in-line membrane filters can remove *Pythium* zoospores from recirculating solutions. They need to be micropore-sized, and require prefiltration with a 100 µm strainer along with a 20 µm filter, otherwise they rapidly clog. Schuerger and Hammer (2009) tested four membrane filter brands (Honeycomb, Polypure, Polymate, and Absolife). They predicted that a porosity of 5 µm would be sufficient. But they found that a double-filter configuration with pore sizes of 1 µm and 0.5 µm was required to suppress *Pythium* root rot.

Ionization

Ionizers pass an electrical charge through nutrient solution to release copper ions (Cu^{2+}) from copper anodes in the water. Cu^{2+} in nutrient solution at 0.5–1.0 ppm kills most pathogens. Copper ionization has proven efficacy against *Pythium*, algae, and bacteria, but has not worked for *Fusarium* and viruses. Ionizers are simple to install, and anodes last 2–6 years. Cu^{2+} at excessive levels is a cumulative phytotoxin.

Chemicals

Chemical disinfectants used here are oxidizing agents. **Chlorine** can be used in liquid, solid, or gaseous forms. Inline injection of chlorine gas into water is the mainstay disinfectant of municipal water supplies. The EPA allows chlorine levels up to 4 ppm (i.e., mg/L) in tap water. Cayanan *et al.* (2009) tested chlorine concentration and contact time required to kill pathogens in recycled irrigation water: *Phytophthora infestans* (1 ppm for 3 min), *Pythium aphanidermatum* (2 ppm for 3 min), *Fusarium oxysporum* (14 ppm for 6 min), *Rhizoctonia solani* (12 ppm for 10 min).

Chlorine is an essential plant nutrient, but excess levels cause toxicity, especially in young seedlings and clones. *Cannabis* is fairly chlorine-tolerant; Cervantes (2015) estimated that ≥20 ppm causes toxicity (leaf tip burn) in some varieties, others tolerate up to 140 ppm. Many companies sell filters to remove chlorine from water intended for *Cannabis* hydroponics. They emphasize chlorine's damage to beneficial microbes in the growth medium. Chlorine-based disinfectants may corrode

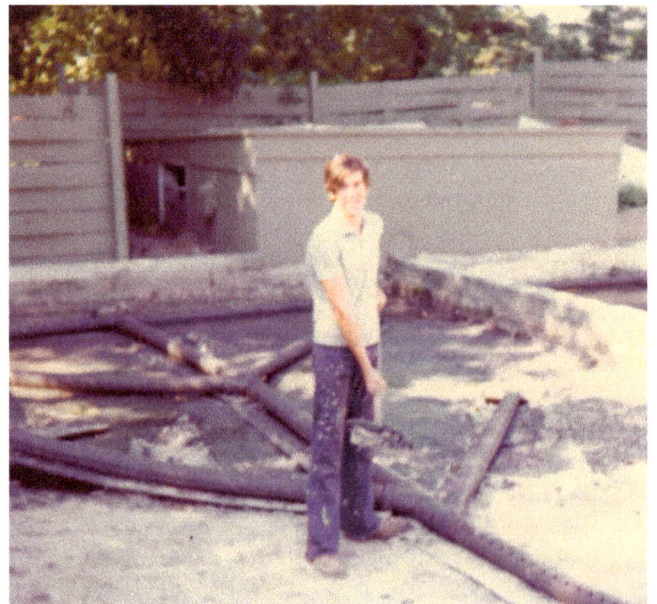

Fig. 19.5. Slow sand filtration: shoveling *schmutzdecke* (McPartland, 1972 photo)

some components of hydroponic systems. They are tricky to dose, because their efficacy depends on the pH (6.5–7.0 is optimal), and on the concentration of nitrogenous compounds dissolved in solution. Chlorine will combine with ammonia and other nitrogenous compounds to form chloramines.

Monochloramine is marketed as Pythoff™, for controlling *Pythium* in nutrient solutions at a dose of 1 mL per 10 L. Chloramine compounds are more phytotoxic than chlorine. They persist in solution (whereas chlorine outgasses), and concentrate over time. Pythoff™ has fans in the hobby grower community.

Quaternary ammonium chlorides (Quats) are potent disinfectants commonly found in hand-wipes. Physan 20™ is marketed for disinfecting hydroponic systems between crops. The company's website says it can be added to nutrient solutions, but gives no guidance regarding dosage. A dilute solution (1 tablespoon per gallon of water) can be used as a foliar spray for pathogen control. The Physan 20™ website states that "Physan 20 leaves no harmful residue if foliage is dried and smoked."

Liquid chlorine, as **sodium hypochlorite** (Clorox® bleach), is used for sterilizing equipment. When diluted, bleach can also disinfect nutrient solution without phytotoxicity. In water, sodium hypochlorite forms hypochlorous acid (HOCl), which is responsible for biocidal activity. Hydroxyl ions (OH⁻) are also formed, which raises the pH. When pH goes over 7.5, HOCl breaks down and loses efficacy. Adding Clorox® bleach (6% sodium hypochlorite) at a dose of 0.1 mL per liter of nutrient solution is effective and safe for plants (but kills beneficial microbes).

Solid chlorine, as **calcium hypochlorite** ("powdered bleach" or "pool shock") can be added to nutrient solution. In water it forms HOCl and OH⁻, as well as calcium, available for uptake by plants. Calcium hypochlorite is less phytotoxic than bleach, and less corrosive to pipes and equipment. Slow-release systems are available that use compressed blocks or pellets in disposable cartridges. Commercial hydroponic products contain HOCl (e.g., Clear Rez™, UC Roots™). Other products contain mixtures of chlorine dioxides (e.g., Ultra-Shield™, Selectrocide®). Read the labels. They are more potent than sodium hypochlorite or calcium hypochlorite, and last longer.

Hydrogen peroxide is a good disinfectant but expensive. It has a short half-life when added to nutrient solution, about 4 days. Food-grade hydrogen peroxide (33–35% H_2O_2) contains no additives, but should be diluted before being added to a hydroponic reservoir. Make a 3% stock solution by diluting one part food-grade H_2O_2 to eleven parts water. The recommended dose of 3% solution is 3 mL per liter of nutrient solution, or about 2–3 teaspoons per gallon. Higher concentrations damage roots. Quadruple that dose can be used to flush the system, 4 days before adding plants. However, H_2O_2 dissociates into ozone, which may cause minerals, especially iron, to fall out of solution.

Products that combine H_2O_2 with **peroxyacetic acid** (PAA) are widely available (e.g., Sanidate, Terraclean, Zertol). In water, PAA dissociates into H_2O_2 and acetic acid. PAA is widely used for used for sanitizing surfaces and equipment. As a disinfectant added to nutrient solutions, the proper dose to avoid *Cannabis* phytotoxicity has not been clearly elucidated.

19.3   Rotation, rotation, rotation

Crop rotation keeps pests moving, a lethal inconvenience. It eliminates many pests and pathogens that attack *Cannabis*. Crop rotation is less effective against generalist feeders, migratory pests, and pathogens that survive for long periods in the soil. Some soil-borne pathogens require a very long rotation. Crop rotation schemes for *Cannabis* are ancient. An agronomist from Shāndōng named Jiǎ Sīxié, writing around AD 544, recommended planting *Cannabis* after legumes, such as azuki beans and mung beans (Jiǎ, 1962). Dewey (1914) recommended rotating *Cannabis* after beans. These rotations utilized the nitrogen-fixing capacities of leguminous plants.

Crop rotation to control pests and pathogens requires a knowledge of previous cropping history. For example, Koike *et al.* (2019) reported that lettuce fields with *Sclerotinia minor* infection resulted in white mold disease in a subsequent hemp crop. Gauthier (2020b) provided other examples of fungal pathogen "carry over" in hemp crops in Kentucky, including:

- *Fusarium graminearum* following cereals, maize, pasture grasses
- *Fusarium solani* following beans, cucurbits, potatoes
- *Colletotrichum coccodes* following maize, cucurbits, and solanaceous vegetables
- *Colletotrichum dematium* following sorghum, soybean, pepper
- *Drechslera gigantea* following cereals, maize, grasses, other monocots
- *Golovinomyces spadiceus* following hosts in the Asteraceae family
- *Pythium ultimum* following brassicas, cereals, maize, soybean, tobacco
- *Cercospora flagellaris* following aster, amaranth, johnsongrass, soybean
- *Rhizoctonia solani* following beans, brassicas, cereals, cucurbits, rice, potato, soybean, tobacco, turf
- *Sclerotium rolfsii* following vegetable crops, peanut, cotton, grasses, soybean, sunflower, tobacco.

Similar advice applies to insect pests: European corn borers, corn earworms, and fall armyworms follow maize crops; cotton bollworms and cotton mirids follow cotton crops; cabbage moths, cabbage curculios, and flea beetles follow vegetable crops; cutworms and white root grubs follow pasture and turf crops.

Practice "escape cropping"—vary the time of planting and/or harvesting. Early planting permits susceptible seedlings to harden before pests arrive. Late planting avoids cool temperatures that predispose seedlings to damping off and root rots. Late planting also eludes the egg-laying period of some pests. Early harvest avoids the endless autumn rains that predispose plants to gray mold in the Pacific Northwest and the Netherlands, or Fusarium flower blight in the Midwestern USA.

19.4   Optimize plant growth

Healthy plants are more resistant to pathogens and pests. Maintain proper moisture levels that are optimal for plant health.

- Drought kills plants outright and also predisposes them to pathogens and pests.
- Overwatering also kills plants and predisposes them to pathogens and pests. It is a common problem with novice growers. Allow soil

to dry between irrigations, and do not plant in poorly draining soils. If poorly draining soils are unavoidable, plant in raised rows.

- Avoid excess atmospheric humidity. Excess humidity is a common problem and permits many fungi and some pests to flourish.

Outdoors, increase air circulation between plants by not overcrowding them. Humidity always increases after canopy closure, which is when leaves of adjacent plants touch each other and shade the soil. This happens in seedling beds, and when fiber-type crops approach maturity. Properly spaced plants keep the canopy open, increase air circulation, and reduce humidity and leaf wetness. This is difficult to achieve in fiber production systems, where dense seedings are the ideal production practice.

Plant rows in the direction of prevailing winds, or in an east–west or northeast–southwest orientation to promote solar drying. Air circulates better on sloped fields or high points than on low or flat ground. Avoid overhead irrigation during flowering.

Indoors, proper ventilation is critical so humidity can escape. In glasshouses, turn on heat (electric, propane, butane or natural gas, *not* kerosene or gasoline) before sundown. This slows the drop in temperature and prevents moisture condensation on plants.

Heat all night and keep all vents open—although a "waste of heat," this is the only way to drive moisture out of the night air. During the day use air conditioning—cool air holds less water and lowers the atmospheric humidity. The lower the temperature differential between night and day, the lower is the chance of dew formation. Avoid dew at all costs during flowering. Do not irrigate plants late in the day or at night. Irrigate at the base of the plant, if possible, and not with misters or overhead irrigation.

Humidity increases in low-ceiling structures such as Dutch Venlo™ glasshouses. Never allow plants to fill more than one-third of the volume inside a glasshouse or growroom. Have a high roof and short plants. Humidity also increases in glasshouses using polythene blackouts (which restrict daylength and induce flowering), and in glasshouses using thermal screens to reduce heat losses. Using blackout shades made of porous woven cloth is better than plastic sheeting.

Optimize soil structure and nutrition. Balance nutrients carefully (see section 17.3). Excess nitrogen predisposes plants to bacteria, fungi, and leaf-eating insects. Indiscriminate use of phosphorus to promote flowering is ill-advised for the same reason.

Improve soil structure (see section 17.2). Adding organic amendments to soil will augment naturally occurring biological control organisms. This tactic works particularly well against soil-borne fungi and nematodes. Adding *well-composted* material is safe and effective. Conversely, poorly composted materials may contain pests and pathogens. Adding municipal sludge to soil has *increased* the incidence of root-rot diseases (Windels, 1997). Patented formulations of organic materials and minerals are available for the suppression of soil pathogens. Some formulations also contain urea and other pesticidal compounds (e.g., Clandosan™) and are described under chemical control.

19.5   Resistant varieties

Although most *Cannabis* cultivars are bred for high yields, a few have been selected for pathogen and pest resistance. In most cases, the mechanism of resistance is poorly understood. But sometimes it is simple—for instance, slimmer buds hold less moisture, which protects them from gray mold. Landraces from Central Asia (i.e., "Indica" or *C. sativa* ssp. *indica* var. *afghanica*) evolved in an arid environment; they produce dense, water-retaining inflorescences, and have little resistance to gray mold and powdery mildew (Clarke, 1987; McPartland and Small, 2020).

Evaluating fiber-type landraces and cultivars for resistance dates back to Dewey (1914). His co-workers Charles and Jenkins (1914) noted that a Chinese landrace with purple-colored foliage was particularly susceptible to the fungus *Botryosphaeria marconii*.

Fiber-type plants have been evaluated for resistance to insects (Dempsey, 1975; Virovets and Lepskaya, 1983; Grigoryev, 1998; Bòcsa, 1999), nematodes (de Meijer, 1993; Pofu *et al.*, 2010), fungi (Ferri, 1959b; McCain and Noviello, 1985; de Meijer *et al.*, 1992; Mediavilla *et al.* 1997), bacteria (Sandru, 1977; McPartland and Hillig, 2004a), viruses (Kegler and Sparr, 1997), and parasitic plants (Senchenko and Kolyadko, 1973; Bòcsa and Karus, 1997).

Drug-type plants have received less attention. They have been evaluated for resistance to insects (McPartland *et al.*, 2000), fungi (McCain and Noviello, 1985; de Meijer *et al.*, 1992; Krishna, 1996; Mediavilla *et al.*, 1997), nematodes (de Meijer, 1993), and bacteria (Hillig, 2005a).

Chief Seven Turtles (1988) described traditional breeding techniques well-suited for *Cannabis*. He was particularly fond of reciprocal recurrent selection. This strategy maintains two select populations while crossing them to produce hybrid offspring.

Genetic engineers have created genetically modified *Cannabis* with resistance to the gray mold fugus, *Botrytis cinerea* (see Fig. 1.5). McKernan *et al.* (2020) identified a gene that imparted resistance to powdery mildew in a CBD strain, "Jamaican Lion". They cloned the gene into *Escherichia coli*, and it exerted antifungal effects.

19.6   Mechanical control

Mechanical control can be as simple as removing insects by hand—eggs, larvae, pupae, adults. Crush them or drop them into a bucket of soapy water. Chilly mornings are a good time to shake beetles and bugs off plants and onto a ground cloth. Stem borers can also be removed by hand—carefully split galls lengthwise, remove borers, then bind the stem back together.

Branches with isolated fungal infections can be pruned from otherwise healthy plants. Many molds establish a foothold in senescent plant tissues, so pruning yellow leaves and injured branches is preventive. Dab wounds and pruning cuts with alcohol, or a fungicidal tree sealer, or a biocontrol agent (e.g., *Trichoderma harzianum* or *Gliocladium roseum*). Whole plants can be rogued from otherwise healthy fields or glasshouses. Place infected/infested material in a plastic bag, and cinch the bag tight before removing the material. This prevents the spread of air-borne fungi or insects.

19.6.1 Insect traps

Mechanically trap or repel insects with light and color. **Sticky traps** can be used to attract and kill winged insects. Commercial sticky traps are square or rectangular cards coated with Tanglefoot™, mineral oil, vaseline, or other sticky and water-proof materials. Sticky ribbons (tapes) are also commercially available. Stir up insects by shaking plants periodically. Sticky traps lose effectiveness in windy locations when they get covered by dust. Homemade sticky traps can be made of plastic, glass, masonite, thin plywood, heavy cardboard, or even upside-down soda/water bottles. Hang traps vertically on plants, near the top of the crop. Or attach them to wooden stakes, and raise them as the crop grows taller.

The Moericke pan trap is a dish, bowl, or bucket, often made of yellow plastic, and filled with water to the rim (with a drop or two of soap to lessen the surface tension of the water). It attracts flying insects, who alight on the water and get trapped in it.

Yellow works best against aphids, whiteflies, leafhoppers, and fungus gnats. Yellow Solo® cups reflect a wavelength of *550–560 nm, which is perfect. For painting homemade traps, use school-bus yellow or sunset yellow (*Rustoleum No. 659, for taxi cabs). For thrips, blue (460 nm) is better than yellow; blue also attracts leafminers. The color of commercial thrips traps is hard to match in a paint store (baby boy blue, dixie belle blue). Tiny LED bulbs (10 lumen) of the proper wavelength can be clipped to traps to enhance attraction. White traps attract flea beetles and true bugs.

Sticky traps are primarily used for monitoring pest populations, but they can be used for mass-trapping. For monitoring purposes, distribute one trap/200 m² crop area. For mass-trapping, distribute a minimum of one trap/20 m², up to one trap/2 m², and replace the traps frequently. Massive sticky traps are commercially available (Fig. 19.6).

Unfortunately, colored traps attract beneficial insects (Broughton and Harrison, 2012). Yellow attracts parasitic wasps

(e.g., *Encarsia formosa*, *Aphidius* spp., *Dacnusa siberica*), winged stages of aphid predators (*Aphidoletes aphidimyza*, *Chrysoperla carnea*), lady beetles (*Coccinella* spp., *Hippodamia convergens*, *Rodolia cardinalis*, *Stethorus picipes*), even honey bees. Yellow sticky traps should be removed before releasing parasitic wasps, and thereafter only hung for 2 days per week for monitoring purposes.

Trap selectivity can be enhanced by adding artificial sex pheromones (see Fig. 21.5). Sex pheromones only attract specific pests. For example, Lurem-TR® attracts two thrips species, *Frankliniella occidentalis* and *Thrips tabaci*. Food attractants have also been added to sticky traps. For example, Lösel *et al.* (1996) lured *Phorodon humuli* with a mix of essential (volatile) oils from *Prunus* and *Humulus* plants. But essential oils also attract beneficial parasitoids and predators. Food attractants and sex pheromones wash off the sticky traps.

Light traps attract nocturnal insects. Night-flying moths are particularly attracted to light, "the fate of the moth in the flame" (an often-quoted line from Aeschylus in *Fragments*, written in 480 BC). In the past, farmers lit bonfires in cotton fields to attract and kill corn earworm moths (Comstock, 1879). Light traps are useful for monitoring purposes and perhaps pest control in enclosed spaces.

An electric light (mercury vapor bulb or fluorescent tube) can be suspended above a bucket with soapy water, or surrounded by an electric grid or adhesive boards. Nocturnal entomologists hang a white sheet next to a light and collect specimens that come to rest on the sheet. Ultraviolet (blacklight) traps became commercially available in the 1950s, which are more attractive to insect photoreceptors. Low-wattage LED lights can be solar powered (Fig. 19.7).

Mechanical barriers protect plants from insects and other pests. Yepsen (1976) aptly described mechanical barriers as "traps without the power of attraction." Super-sticky Tanglefoot® is a popular barrier against crawling insects. Do not apply Tanglefoot directly on stems—the product is a

Fig. 19.6. Catchy™ sticky glue trap (courtesy Davik Industries)

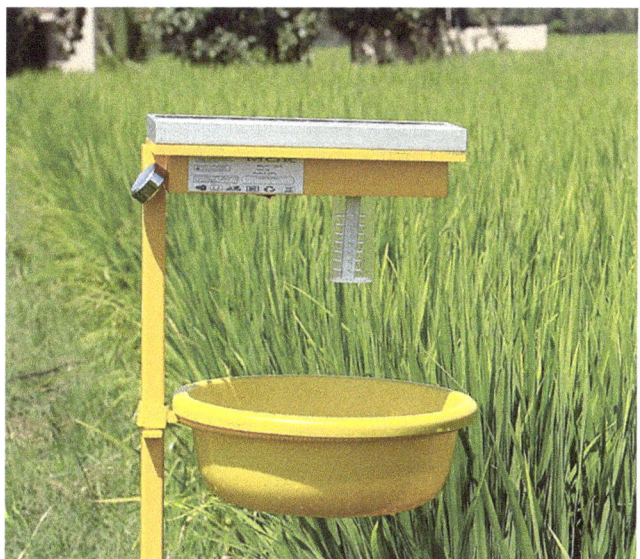

Fig. 19.7. MGK™ light trap (courtesy MGK, Wikimedia Commons)

mixture of castor oil, gum resins, and wax—it stresses plants and may translocate to flowers. Instead, wrap stems with a sheath of aluminum foil and coat the foil with Tanglefoot.

Protect seedlings from flying pests and their egg-laying mothers. Outdoors, use tent netting, tightly woven cloth, polyethylene screens (Green-Tek®), or floating row covers (Harvest-Guard®, Reemay®). They can be fashioned into row cover tunnels with steel hoops. Indoors, be sure to screen vents, open windows, and exhaust fans. A wire fence keeps out most large mammals. Australians have taken mechanical barriers to the extreme by stretching "vermin fences" across their continent to slow the migration of rabbits.

(Prepared by J. McPartland; revised by M. Bolt)

References

Bennett EA. 2018. Extending ethical consumerism theory to semi-legal sectors: insights from recreational cannabis. *Agriculture and Human Values* 35: 295–317.

Bòcsa I. 1999. "Genetic improvement: conventional approaches," pp. 153–184 in: Ranalli P, *ed., Advances in Hemp Research*. Haworth Press, Binghamton, New York.

Bòcsa I, Karus M. 1997. *Der Hanfanbau: Botanik, Sorten, Anbau und Ernte*. C.F. Müller, Heildelberg. English translation: *The cultivation of hemp: botany, varieties, cultivation, and harvesting*. Hemptech, Sebastopol, California.

Bray F. 1984. "Agriculture," Volume 6, Part 2 of Needham J, *ed. Science and Civilisation in China*. Cambridge University Press, Cambridge, UK.

Broughton S, Harrison J. 2012. Evaluation of monitoring methods for thrips and the effect of trap colour and semiochemicals on sticky trap capture of thrips (Thysanoptera) and beneficial insects (Syrphidae, Hemerobiidae) in deciduous fruit trees in Western Australia. *Crop Protection* 42: 156–163.

Casano S, Grassi G. 2009. Valutazione di terreni di coltura per la propagazione *in vitro* della canapa (*Cannabis sativa*). Italus Hortus 16(2):109–112.

Cayanan DF, Zhang P, Liu W, Dixon M, Zheng Y. 2009. Efficacy of chlorine in controlling five common plant pathogens. *HortScience* 44: 157–163.

Cervantes J. 2015. *The Cannabis encyclopedia*. Van Patten Publishing, Vancouver Washington.

Chandra S, Lata H, Khan IA, ElSohly MA. 2017. "*Cannabis sativa* L.: botany and horticulture," pp. 79–100 in Chandra *et al.*, eds. *Cannabis sativa L.—botany and biotechnology*. Springer International, Cham, Switzerland.

Charles VK, Jenkins AE. 1914. A fungous disease of hemp. *Journal of Agricultural Research* 3: 81–84.

Chen Y, Gratton JL, Liu J. 2004. Power requirements of hemp cutting and conditioning. *Biosystems Engineering* 87: 417–424.

Chief Seven Turtles. 1988. "Reciprocal recurrent selection in *Cannabis*," pp. 34–37 *in* Alexander T, *ed. The best of Sinsemilla Tips*. Full Moon Publishing, Corvallis, Oregon.

Clarke RC. 1987. *Cannabis evolution*. Master's thesis, Indiana University, Bloomington, Indiana.

Comstock JH. 1879. *Report upon cotton insects*. Government Printing Office, Washington, DC.

de Meijer EPM. 1993. Evaluation and verification of resistance to *Meloidogyne hapla* Chitwood in a *Cannabis* germplasm collection. *Euphytica* 71: 49–56.

de Meijer EPM, van der Kamp, HJ, van Eeuwijk FA. 1992. Characterization of *Cannabis* accessions with regard to cannabinoid content in relation to other plant characters. *Euphytica* 62: 187–200.

Dempsey JM. 1975. "Hemp" pp. 46–89 in *Fiber crops*. University of Florida Press, Gainesville.

Dewey LH. 1914. "Hemp," pp. 283–347 in *USDA Yearbook 1913*. United States Department of Agriculture, Washington, DC.

Faulkner EH. 1943. *Plowman's folly*. University of Oklahoma Press, Norman, Oklahoma.

Feeney M, Punja ZK. 2017. "The role of *Agrobacterium*-mediated and other gene-transfer technologies in *Cannabis* research and product development," pp. 343–363 in Chandra S, *et al.*, eds. *Cannabis sativa: Botany and Biotechnology*. Springer International Publishing, Cham, Switzerland.

Ferri F. 1959b. La Septoriosi della canapa. *Annali della sperimentazione agraria* N.S. 13:6, Supplement pg. CLXXXIX–CXCVII.

Gauthier N. 2020b. "Considering crop rotations and the potential for carry-over," p. 25 in Gauthier N, Leonberger K, Bowers C, *eds. Science of hemp: production and pest management*. University of Kentucky Agricultural Experiment Station, no. SR-112, Lexington, Kentucky.

Grabowska L. 2005. *Perspektywy uprawy konopi przemysłowych w Polsce*. IBMiER, Warsaw.

Grigoryev SV. 1998. Survey of the VIR Cannabis collection: the resistance of accessions to corn stem borer. *Journal of the International Hemp Association* 5(2): 72–74.

Henry D. 1771. *The complete English farmer*. F. Newbery, London.

Hillig KH. 2005a. *A systematic investigation of Cannabis*. Doctoral Dissertation, Department of Biology, Indiana University, Bloomington, Indiana.

Jarvis WR. 1992. *Managing diseases in greenhouse crops*. APS Press, St Paul, Minnesota.

Jiǎ SX. 1962. *A preliminary survey of the book Ch'i min yao shu: an agricultural encyclopaedia of the 6th century*. Science Press, Beijing.

Kegler H, Spaar D. 1997. Kurzmitteilung zur Virusanfälligkeit von Hanfsorten (*Cannabis sativa* L.). *Archives Phytopathologie und Pflanzenschutz* 30: 457–464.

Koike ST, Stanghellini H, Mauzey SJ, Burkhardt A. 2019. First report of sclerotinia crown rot caused by *Sclerotinia minor* on hemp. *Plant Disease* 103: 1771.

Krishna A. 1996. Resistance in ganja cultivars to *Sclerotium rolfsii*. *Indian Journal Mycology and Plant Pathology* 26: 307.

Kumar V, Chopra AK. 2013. Reduction of weeds and improvement of soil quality and yield of wheat by tillage in Northern Great Plains of Ganga River in India. *International Journal of Agricultural Science Research* 2: 249–257.

Lata H, Chandra S, Khan I, ElSohly MA. 2009. Propagation through alginate encapsulation of axillary buds of *Cannabis sativa* L. – an important medicinal plant. *Physiology and Molecular Biology of Plants* 15: 79–86.

Lata H, Chandra S, Techen N, Khan IA, Elsohly MA. 2010. Assessment of the genetic stability of micropropagated plants of *Cannabis sativa* by ISSR markers. *Planta Medica* 76: 97–100.

Lesik BV. 1958. *Приемы повышения качества лубяного волокна. Конопля, кенаф и джут*. Selkhozgiz, Moscow.

Lösel PM, Lindemann M, Scherkenback J, *et al.* (4 other authors). 1996. Effect of primary-host kairomones on the attractiveness of the hop-aphid sex pheromone to *Phorodon humuli* males and gynoparae. *Entomologia Experimentalis et Applicata* 80: 79–82.

Markham G. 1615. *Countrey contentments in two bookes, the second intituled the English Huswife*. Roger Jackson, London.

McCain AH, Noviello C. 1985. Biological control of *Cannabis sativa*. *Proceedings, 6th International Symposium on Biological Control of Weeds*, pp. 635–642.

McEno J. 1990. Hydroponic IPM. *The Growing Edge* 1(3): 35–39.

McKernan KJ, Helbert Y, Kane LT, *et al.* (12 other authors). 2020. Sequence and annotation of 42 *Cannabis* genomes reveals extensive copy number variation in cannabinoid synthesis and pathogen resistance genes. *BioRxiv Preprint*, Available at: https://doi.org/10.1101/2020.01.03.894428

McPartland JM, Hillig KW. 2004a. Striatura ulcerosa. *Journal of the International Hemp Association* 9(1): 89–96.

McPartland JM, Small E. 2020. A classification of endangered high-THC cannabis (*Cannabis sativa* subsp. *indica*) domesticates and their wild relatives. *PhytoKeys* 144: 81–112.

McPartland JM, Clarke RC, Watson DP. 2000. *Hemp diseases and pests – management and biological control*. CABI Publishing, Wallingford, UK.

Mediavilla V, Spiess E, Zürcher B, *et al.* (6 other authors). 1997. Erfahrungen aus dem Hanfanbau 1996," pp. 253–262 in *Bioresource Hemp 97, Proceedings of the Symposium, Frankfurt, Germany*. Nova-Institute, Köln, Germany.

Metcalf CL, Flint WP, Metcalf RL. 1951. *Destructive and useful insects: their habits and control*. McGraw-Hill, New York.

Mitkov A, Yanev M, Neshev N, Tonev T. 2018. Evaluation of low herbicide rates of gardoprom® Plus Gold 550 SC and Spectrum® 720 EC at conventional sunflower (*Helianthus annuus* L.). *Scientific Papers Series A, Agronomy* 61(2): 94–97.

Monthony AS, Page SR, Hesami M, Jones AMP. 2021. The past, present, and future of *Cannabis sativa* tissue culture. *Plants* 10: e185.

Morel G, Martin C. 1952. Guérison de dahlias atteints d'une maladie á virus. *Comptes Rendus Hebdomadaires des Séances de l'Académie des Sciences* 235: 1324–1325.

Neuer HV. Prieger E, Sengbusch RV. 1946. Hanfzüchtung. I. Die Steigerung des Faserertrages von Hanf. *Der Züchter (Zeitschrift für theoretische und angewandte Genetik)* 17/18(2): 33–39.

Pofu KM, Van Biljon ER, Mashela PW, Shimelis HA. 2010. Responses of selected hemp cultivars to *Meloidogyne javanica* under greenhouse conditions. *American-Eurasian Journal of Agricultural and Environmental Science* 8: 602–606.

Quak F. 1961. Heat treatment and substances inhibiting virus multiplication, in meristem culture to obtain virus-free plants. *Advances in Horticultural Science and Their Applications* 1: 144–148.

Ristaino JB, Perry KB, Lumsden RD. 1991. Effect of solarization and *Gliocladium virens* on sclerotia of *Sclerotium rolfsii*, soil microbiota, and the incidence of southern blight of tomato. *Phytopathology* 81: 1117–1124.

Rodale Institute. 2020. *Industrial hemp trials*. Available at: https://rodaleinstitute.org/science/industrial-hemp-trial/

Sandru ID. 1977. Vestejirea bacteriana a frunzelor—o boala noua la cinepa. *Productia Vegetala, Cereale si Plante Tehnice* 29: 34–36.

Schuerger AC, Hammer W. 2009. Use of cross-flow membrane filtration in a recirculating hydroponic system to suppress root disease in pepper caused by *Pythium myriotylum*. *Phytopathology* 99: 597–607.

Scott C, Punja Z. 2022. "Management of diseases on *Cannabis* in controlled environment production," pp. 231–251 in Zheng YB, *ed. Handbook of Cannabis production in controlled environments*. CRC Press, Boca Raton Florida.

Senchenko GI, Kolyadko IV. 1973. A method for producing broomrape-resistant varieties of hemp. Селекция и семеноводство 23: 27–33.

Ślusarkiewicz-Jarzina A, Ponitka A, Kaczmarek Z. 2005. Influence of cultivar, explant source and plant growth regulator on callus induction and plant regeneration of *Cannabis sativa* L. *Acta Biologica Cracoviensia Series Botanica* 47: 145–151.

Smýkalová I, Vrbová M, Cvečková M, *et al.* (7 other authors). 2019. The effects of novel synthetic cytokinin derivatives and endogenous cytokinins on the *in vitro* growth responses of hemp (*Cannabis sativa* L.) explants. *Plant Cell Tissue Organ Culture* 139: 381–394.

Spitzer-Rimon B, Duchin S, Bernstein N, Kamenetsky R. 2019. Architecture and florogeneis in female *Cannabis sativa* plants. *Frontiers in Plant Science* 10: 350.

Stewart-Wade SM. 2011. Plant pathogens in recycled irrigation water in commercial plant nurseries and greenhouses: their detection and management. *Irrigation Science* 29: 267–297.

Thomson P. 2019. *Import health standard: seeds for sowing*. Ministry for Primary Industries, Wellington, New Zealand.

Thomson P. 2020. *Approved biosecurity treatments*. Ministry for Primary Industries, Wellington, New Zealand.

Tull J. 1731. *The horse-hoeing husbandry*. A. Rhames, Dublin.

Virovets VG, Lepskaya LA. 1983. "Сортовая устойчивость конопли к стеблевому мотыльку (*Ostrinia nubilalis* Hb.)," pp. 53–58 in Биологические особенности, технология возделывания и первичная обработка лубяных культур. Glukhov, Ukraine.

Wainwright-Evans S. 2017. Take control of mites on cannabis crops. *Greenhouse Grower*, Available at: https:// www.greenhousegrower.com/production/insect-control/ take-control-of-mites-in-cannabis-crops/.

Windels CE. 1997. "Altering community balance: organic amendments, selection pressures, and biocontrols," pp. 282–300 in Andow DA, *et al.*, eds. *Ecological interactions and biological control*. Westview Press, Boulder, Colorado.

Wissett R. 1808. *A treatise on hemp*. J. Harding, London.

Yepsen RB. 1976. *Organic plant protection*. Rodale Press, Emmaus, Pennsylvania.

Zengin G, Menghini L, Di Sotto A, *et al.* (13 other authors). 2018. Chromatographic analyses, *in vitro* biological activities, and cytotoxicity of *Cannabis sativa* L. essential oil: a multidisciplinary study. *Molecules* 23: 3266.

Abstract

It's a bug-eat-bug world. This chapter provides a long list of commercially available biocontrols—their descriptions, shelf lives, release rates, and compatibilities. The list includes lacewings, predatory mirid and pirate bugs, lady beetles, *Aphidoletes* and *Feltiella* flies, parasitic wasps (*Trichogramma* and *Encarsia* spp.), and last but not least, a bevy of predatory mites. This chapter includes what the EPA classifies as "microbial pesticides," such as *Bacillus thuringiensis* (Bt), fungi that attack other fungi or insects, predatory nematodes, and insecticidal viruses. We also discuss mycorrhizal fungi and rhizobacteria. The most common strategy is "inundative" release of biocontrol organisms. "Conservation biocontrol" encourages naturally occurring predators and parasites, through the use of indoor "banker plants" and outdoor companion planting (intercropping). Interestingly, *Cannabis* has a long history of use for repelling pests from other crops.

20.1   Introduction

Untold numbers of pests and pathogens are culled by their natural enemies. Encouraging this carnage is biological control. Our planet prefers biocontrol to pesticides. Concern over pesticides has prompted government funding for biocontrol research, and has funded a flourishing sector of entrepreneurial start-ups.

The EPA (Environmental Protection Agency, USA) classifies *Bacillus thuringiensis* (Bt) as a "microbial pesticide." We cover it in this chapter, rather than the chemical pesticides chapter. The EPA has streamlined the development of microbial pesticides. To navigate EPA registration, a microbial pesticide (for use on a food crop) takes about 19 months and costs about US$48,621. In comparison, a conventional pesticide averages 24 months and US$590,000 (Leahy *et al.*, 2014).

Agrochemical corporations employed 419 lobbyists and spent US$43 million to lobby the US Congress in 2022 alone (Center for Responsive Politics, 2024). Among those lobbyists, 58% were former government employees. There is a "revolving door" of former agrochemical executives and attorneys who now serve in government agencies tasked with industry oversight. Consumers are also targeted. The industrial food and agriculture industry spent US$126 million in "tobacco-style" public relations tactics between 2009 and 2013; separately, Monsanto alone spent US$95 million on marketing in 2013 (Hamerschlag *et al.*, 2015).

Biocontrols have been used longer than pesticides—if one considers the taming of cats for rodent control, which began at least 9500 years ago (Vigne *et al.*, 2004). Cats killed rats and mice in stored grain, and we elevated them to goddess status (Fig. 20.1). The dispersal of cats followed human trade routes. A DNA analysis of 352 ancient cats traced their dispersal from Egypt to ancient Greece and Rome, all the way to Viking ports by the 7th century AD (Ottoni *et al.*, 2017). When Roman Legions marched north of the Alps, they brought cats (Peters,

1998). In the "Age of Sails" it was compulsory to keep cats onboard ships (Johansson and Hüster, 1987).

Around 1762 a team of French colonials shipped mynah birds (*Acridotheres tristis*) from India to Mauritius to control red locusts (*Nomadacris septemfasciata*). Within 8 years the birds controlled the locusts—a feat characterized as the first international transfer of a biocontrol organism (Cheke and Hume, 2009).

Carl Linnaeus loved plants, but his second love was insects. When the Swedish Academy of Science requested proposals for controlling insects, Linnaeus submitted a proposal under the pseudonym Nelin (1763): "Nobody has thought of getting rid of insects with insects. Every insect has its predator which follows and destroys it. Such predatory insects should be caught and used for disinfesting crop plants" (translation by Hörstadius, 1974).

The entomologist Charles De Greer, Linnaeus's contemporary, was quoted by Stillingfleet (1775): "We shall never be able to guard ourselves against them [insects] but by their means [other insects]." De Greer was a great admirer of René Antoine de Réaumur, an early advocate of biocontrol. To control aphids in glasshouses, Réaumur (1737) recommended collecting the eggs of predators. He illustrated three aphid predators: (1) syrphid larvae and an adult (Fig. 20.2A); (2) lady beetles (Fig. 20.2B); and (3) lacewing eggs, larva, and adult (Fig. 20.2C).

Charles Valentine Riley (1843–1895) (Fig. 20.3) is the founder of modern biological control (Wheeler *et al.*, 2010). Born in London, Riley was the illegitimate child of an Anglican clergyman who died in a debtors' prison. Riley hand-wrote his first "book" as a schoolboy, *Natural History of Insects*. He emigrated to the USA at age 16 and landed a job at the new journal *Prairie Farmer* in Chicago. Riley's expertise led to his appointment as the first State Entomologist of Missouri in 1868 (Sorensen *et al.*, 2018). His reports from Missouri included two hemp pests, the corn earworm *Helicoverpa zea* (Riley, 1871) and the Rocky Mountain locust *Melanoplus spretus* (Riley, 1877b).

© John M. McPartland 2025. *Hemp Diseases and Pests* 2nd Edition (J.M. McPartland)

DOI: 10.1079/9781836990352.0020

Fig. 20.1. Egyptian statue of Bastet, the cat goddess, 600 BC. Museum of Fine Arts, Boston (Photo by John McPartland)

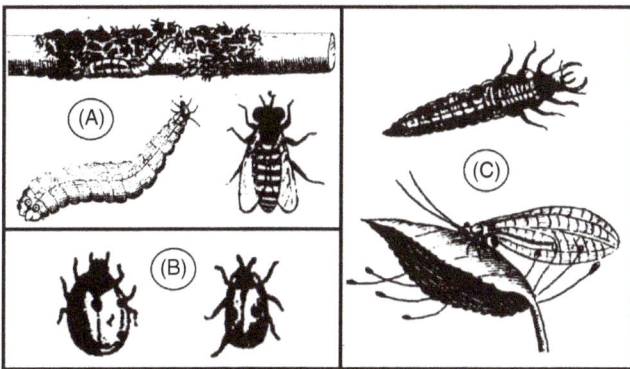

Fig. 20.2. Aphid predators (Réaumur, 1737)

An avid field worker and taxonomist, Riley eventually named 203 new species. Riley (1892) estimated there were 10 million insect species in the world—quite a leap from the earlier estimate of 600,000 species by Kirby and Spence (1826b). In 1873 Riley organized the first transfer of an arthropod biocontrol agent from one country to another. He exported *Tyroglyphus phylloxerae* to France to help fight the grapevine phylloxera. Next he imported a parasitoid, *Cotesia glomerata*, from England to the USA for control of the imported cabbageworm *Pieris rapae* (see Fig. 20.5).

The USDA hired Riley in 1878. He continued to write about hemp pests, such as *Helicoverpa zea* (Riley, 1885). Riley and Howard (1892) tested *Cannabis* as a botanical pesticide, by scattering hemp leaves in stored grain to repel grain weevils. Riley gained international acclaim for importing the vedalia lady beetle (*Novius cardinalis*) from Australia to California in 1888. Riley died six years later, due to skull injuries from a bicycle accident.

Biocontrol lost market share after World War II, thanks to the availability of inexpensive insecticides. Fifteen years later,

Fig. 20.3. Charles Riley (courtesy of Special Collections, USDA National Agriculture Library)

Rachel Carson's *Silent Spring* put a damper on DDT. Interest in biocontrols was renewed. By the late 1960s, biocontrol became a core component of integrated pest management (IPM). Now it is here to stay.

20.2 Biocontrol agents

Natural enemies of pests and pathogens are classified as predators, parasitoids, pathogens, and competitors/antagonists. *Cannabis* can present unique challenges that hamper the efficacy of biocontrol agents. Glandular trichomes may restrict hunting by predators and parasitoids, and short daylengths during flowering induce diapause in many predators (Lemay *et al.*, 2022). Nevertheless, insect surveys in *Cannabis* crops have revealed large and diverse populations of naturally occurring predators and parasitoids (Batra, 1976; Lago and Stanford, 1989; Trotuş and Naie, 2008; Schreiner, 2019; Britt *et al.*, 2022).

20.2.1 Predators

Predators feed upon other animals (prey), that are either smaller or weaker than themselves. Many predators are generalists—they do not target specific prey species. Some predators are specialists. Lady beetles often limit their diet to aphids, and *Stethorus* beetles specialize in *Tetranychus* spider mites. Both adults and larvae may be predatory (e.g., lady beetles), or just the larvae are predatory (e.g., lacewings).

Commercially available predators include **beetles** (*Adalia, Cryptolaemus, Delphastus, Dalotia, Stethorus* spp.), **lacewings**

(*Chrysopa*, *Chrysoperla* spp.), **flies** (*Aphidoletes*, *Feltiella* spp.), **pirate bugs** (*Orius* spp.), **mirid bugs** (*Macrolophus*, *Dicyphus* spp.), mantids, and others. **Predatory mites** are also available (*Phytoseiulus*, *Neoseiulus*, *Amblyseius*, *Galendromus*, *Stratiolaelaps* spp.).

The written history of farmers using predators dates to AD 304 in present-day Vietnam. A Chinese botanist named Jī Hán was posted there, and wrote *Nánfāng Căomù Zhuàng*, *"Plants of the Southern Regions."* He described citrus growers purchasing bags of predatory ants in the market, which were hung in orange trees, "if the trees do not have this kind of ant, the fruits will be damaged by many harmful insects" (Li, 1979). Thus biocontrols were already articles of commerce, purchased in the market. USDA visitors in China met citrus growers who still reared ants (*Oecophylla smaragdina*) for release in their orchards (Groff and Howard, 1925).

In England, Jewell (1612) placed bags of ants in fruit trees, and coated the base of trees with a ring of tar. Trapped in the tree and hungry, the ants "eate and destroye all the caterpyllers there, without hurting any of the fruite." Forsskål (1775), during his fatal journey to Yemen, described the use of ants (*Formica animosa*) to "persecute" pests of date palms (*Phoenix dactylifera*). Ant nests were moved from surrounding hills and placed in trees. El-Haidari (1981) dated this Yemeni practice to AD 1200.

The first predator collected on *Cannabis* was a syrphid fly. Scopoli (1763) named it *Musca cannabina*. The syrphid was cited by Vallot (1836), "Hemp provides the observer with many means of study; the large quantity of aphids, which are sometimes found on the lower side of its leaves, attracts several insects whose larvae are aphidivores." *M. cannabina* is a widespread species, and the subject of hundreds of research publications. Due to a series of poor taxonomic decisions, *M. cannabina* is now called *Episyrphus balteatus* (see text above Fig. 20.12).

Erasmus Darwin, the grandfather of Charles, recommended protecting and encouraging syrphid flies, as well as ichneumonid wasps (Darwin, 1800). He was cited by Kirby and Spence (1826a), who wrote the founding text of entomology in English. They penned a lengthy chapter on predatory insects for pest control, including *M. cannabina*.

Wilson *et al.* (2019) surveyed *Cannabis* growers in California regarding their methods of pest and disease management. Growers used a variety of predators (Fig. 20.4). Other biocontrols were more popular (e.g., microbial pathogens), as were organic pesticides (e.g., azadirachtin).

20.2.2   Parasitoids

Parasitoids are parasitic insects that develop on or within another organism (the host), eventually killing it. The larval stages are parasitic; the adults are nectar feeders or predators. Unlike predators, parasitoids are smaller than their host. Parasitoids complete their life cycle on a single host, whereas predators consume prey in large numbers. Parasitoids are selective—they often target a single species, or several closely related species. Selectivity extends to the host's life stage—parasitoids target only eggs or larvae, sometimes pupae.

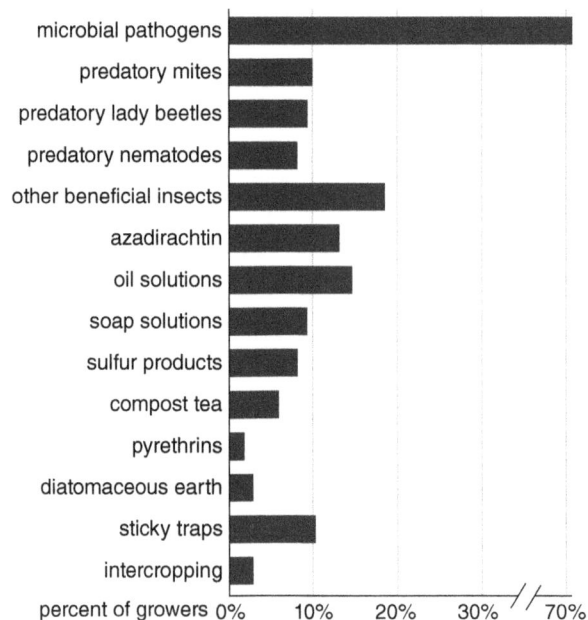

Fig. 20.4. Survey of methods used by California *Cannabis* growers for pest and disease management (adapted from Wilson *et al.*, 2019)

A single female parasitoid usually inserts individual eggs into multiple hosts. Larvae hatch and feed on bodily fluids, leaving vital organs for last. The host dies when its fluids have been sucked dry or its internal organs cease to function. Death is timed perfectly, occurring when larvae are fully developed. Larvae pupate in the cadavers, emerge as adults, and go off to lay more eggs.

Parasitoids are more efficient at finding prey than predators. Parasitoids aggressively hunt prey until they are nearly eradicated. Predators, on the other hand, prefer being surrounded by prey. When the prey population abates a bit, predators migrate away in search of happier hunting grounds, and leave many pests behind. Parasitoids stick around. They are better suited for preventive control, but tend to work too slowly for large infestations.

There are four major groups of parasitoids. **Chalcid wasps** (*Trichogramma*, *Encarsia*, *Eretmocerus*, *Aphelinus*, *Anagyrus*, *Leptomastix* spp.) parasitize eggs and larvae of many insect species. **Ichneumonid wasps** (*Ichneumon*, *Eriborus*, *Hyposoter* spp.) mainly target caterpillars and aphids as hosts. **Braconid wasps** (*Aphidius*, *Cotesia*, *Microplitis* spp.) differ from ichneumonids only by their wing veins; they parasitize eggs and larvae of many insect species. **Tachinid flies** (*Tachina*, *Trichopoda*, *Erynnia*, *Voria* spp.) are often generalist parasitoids. In an insect survey of *Cannabis* crops in southern California, the number of chalcids (*n* = 437 specimens) were greater than braconids (*n* = 107), ichneumonids (*n* = 5), and tachinids (*n* = 3) (Britt *et al.*, 2022).

Parasitoids came to human attention later than predators, due to their small size and complex life cycles. In AD 1096, Lù Diàn described a tachinid fly that attacked a domesticated insect—the silkworm, *Bombyx mori*. "A fly lays eggs on the body of silkworms. The maggots come out when the silkworm is cocooning. Then the maggots hide in the soil and become flies" (Cai *et al.*, 2005). A contemporary book, *Wù lèi xiāng gǎn zhì*, written by either Zàn Níng or Sū Shì, stated, "When being bitten by flies, silkworm will have maggots inside their body."

Van Lenteren and Godfray (2005) reviewed the European discovery of parasitoids, with three major players. **Maria Sibylla Merian** (1647–1717) (Fig. 20.5) was a German-born Swiss who moved to Amsterdam and conducted field work in Surinam (Dutch Guiana). She made close observations of insects and etched careful copper engravings. Merian (1679) observed larvae of the wasp *Cotesia glomerata* parasitizing a larva of the cabbage moth *Pieris brassicae*. After killing the caterpillar, wasp larvae changed into cocoons (highlighted in pink in Fig. 20.5), and "after twelve days a fly emerged out of each cocoon."

Jan Swammerdam (1637–1680) was a Dutch contemporary of Merian. Swammerdam (1669) described the development of ichneumon wasps ("*Vespa ichneumon*") inside their hosts. He observed pupation and emergence, and wondered how they came to be inside the host insect. Swammerdam figured it out around 1675, when his friend Otto Marsilius mentioned that he'd seen caterpillars being stung by "flies" (Van Lenteren and Godfray, 2005). Swammerdam deduced that eggs were laid in the caterpillars, which was how they came to be inside the host insect. Swammerdam solved the riddle of the parasitoid life cycle, which he described in *Bybel der Natuur*. He wrote the book around 1679, but it wasn't published until long after his death (Swammerdam, 1737).

In the meantime he was scooped by a Dutchman with new technology: **Antoni van Leeuwenhoek** (1632–1723) and his microscope. Van Lenteren and Godfray (2005) examined Leeuwenhoek's correspondence: In 1687 Leeuwenhoek described the emergence of "flies" from pupae of moths. He reasoned that fly eggs must have been laid on the pupae by adult flies. In 1695 he described mummified aphids on roses; he dissected the mummies and found larvae (probably *Aphidius*

rosae larvae in *Macrosiphum rosae*). By 1700 he had unraveled the whole parasitoid life cycle (Leeuwenhoek, 1702), complete with microscopic drawings (Fig. 20.6).

Darwin (1800) described an ichneumon wasp attaching its eggs to a cabbage caterpillar, "and thus perishes instead of changing into a butterfly. This ichneumon fly should therefore be encouraged, if his winter habitat could be discovered." Darwin's grandson Charles was horrified: "I cannot persuade myself that a beneficent and omnipotent God would have designedly created the Ichneumonidae with the express intention of their feeding within the living bodies of caterpillars" (Darwin, 1860).

The first parasitoid associated with hemp was discovered by Scopoli (1763), who named the species *Ichneumon cannabis*. From our perspective, *I. cannabis* is detrimental, because it parasitizes a beneficial insect, *Musca cannabina*. Vallot (1836) also observed *I. cannabis* parasitizing *M. cannabina*, which in turn was preying upon aphids, which were feeding on hemp.

20.2.3   Pathogens

If handling predators and parasitoids makes you squeamish, try *biocontrol in a can*. A packet of **microbial pathogens** (MPs) contains millions-to-trillions of beneficial microbes. MPs can be deployed like pesticides—mix with water and spray onto foliage. MPs can be applied to soil in five different ways (see Table 21.4). MPs have been used by a high percentage of California *Cannabis* growers, according to a recent survey (Fig. 20.4).

MPs are relatively specific to particular groups of hosts, during specific life stages. They rarely infect non-target hosts, such as beneficial insects. They may precipitate allergies in us, or the carriers in some commercial formulations cause allergies. Some MPs, such as *Bacillus thuringiensis* (Bt), only work if they are swallowed by pests. They do poorly against sucking insects (e.g., aphids, whiteflies, leafhoppers).

Not all MPs kill their hosts. Instead they slow growth, shorten lifespans, or reduce reproduction. Some fungi take

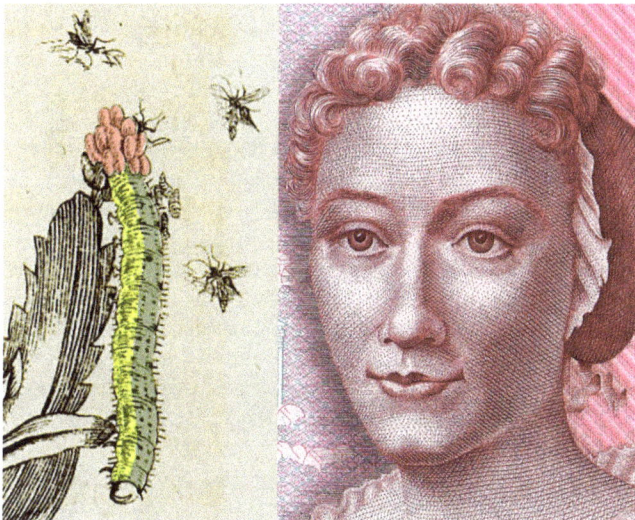

Fig. 20.5. Illustration of parasitism from Merian (1679); her portrait on a 500 DM banknote, Deutsche Bundesbank, public domain

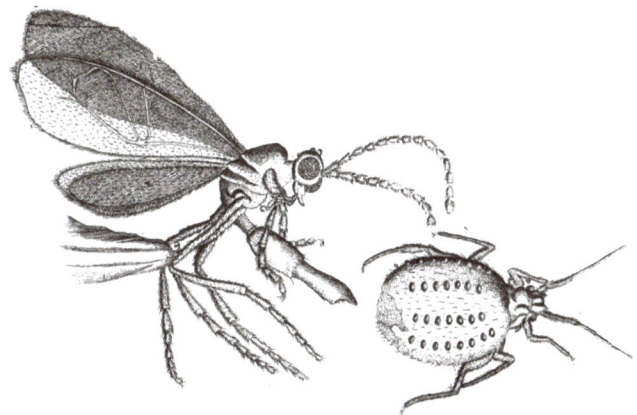

Fig. 20.6. A braconid parasitoid, likely *Aphidius ribus*, in oviposition over an aphid, likely *Cryptomyzus ribis* (Leeuwenhoek, 1702)

control of insect brains and manipulate their behavior. Grasshoppers infected by *Entomophaga grylli*, for example, climb to high places, zombie-like, and die at locations optimal for fungal spore dispersal (Roy *et al.*, 2006). The abdomen of the *Euchorthippus* cadaver in Fig. 20.7 is bulging with spores of *E. grylii*.

Steinhaus (1975) wrote a historical review of MPs. The earliest MPs were discovered in domesticated insects. Aristotle (384–322 BC) described two honey bee diseases. "Clerus" consisted of "a growth of little worms." Another disease caused "a lassitude on the part of the bees and malodorousness of the hive" (Aristotle, 1910). A fungus found on silkworm larvae, *Cordyceps militaris*, was called the "plant worm" by ancient Chinese. The first illustrated description of a diseased insect was published by Réaumur (1726). A Jesuit in China sent Réaumur a dead caterpillar from which emerged a long, club-shaped fungus. Réaumur's drawing is good enough to identify the fungus as *Ophiocordyceps* (*Cordyceps*) *sinensis*.

The Italian entomologist **Agostino Bassi** (1773–1856) studied a white growth on silkworm larvae called *mal de segno* (in Italian) or *muscardine* (in French). He identified it as a fungus (Bassi, 1835), now called *Beauveria bassiana* in his honor. *B. bassiana* is approved for use on *Cannabis* in many jurisdictions. Bassi proposed that microorganisms cause many diseases in plants, animals, and humans—prior to Louis Pasteur and Robert Koch, who got all the credit.

Fig. 20.7. Dead *Euchorthippus* sp. infected with *Entomophaga grylli* (courtesy Hélène Rival, Wikimedia Commons)

The first deployment of an MP was by **Ilya I. Metchnikov** (1845–1916) in Russia. Metchnikov (1879) discovered the fungus *Metarhizium anisopliae* and cultured it on artificial medium (sterilized beer mash). He poured this into soil, and the fungus killed grubs of its namesake—*Anisoplia austriaca* (wheat grain beetle). Metchnikov went on to win a Nobel Prize, but not for that.

Available MPs include **bacteria** (*Bacillus*, *Acinetobacter*, *Pantoea*, *Pseudomonas* spp.), **fungi** (*Beauveria*, *Cordyceps*, *Lecanicillium*, *Metarhizum*, *Trichoderma*, *Clonostachys*, *Gliocladium*, *Verticillium*, *Arthrobotrys*, *Ampelomyces* spp.), **nematodes** (*Heterorhabditis*, *Steinernema*, *Phasmarhabditis* spp.), **protozoans** (*Nosema*, *Brachycentrus*, *Vairimorpha* spp.), and **viruses** (nuclear polyhedrosis virus).

Bacteria are rock stars in the MP world. They too were discovered in silkworms. Ishiwata (1901) initiated a series of publications on *sottō-byō*, "collapse disease." He referred to the causal agent as "*sottō* bacillus," without assigning a scientific name. His student erected the taxon *Bacterium sotto* Ishiwata (Iwabuchi, 1908). Aoki and Chigasaki (1915) found that the filtrate of a *B. sotto* culture was lethal to silkworms—the first demonstration of a Bt toxin.

Meanwhile in Germany, Berliner (1915) named an organism *Bacillus thuringiensis* (Bt). He suggested Bt could be used to control insect pests. Husz (1928) used Bt to control the European corn borer *Ostrinia nubilalis*. This led to the commercialization of Bt in 1938, a product called Sporéine in France. Further development was disrupted by World War II and the introduction of broad-spectrum insecticides.

Research on *B. sotto* continued; a landmark *Nature* paper by Angus (1954) demonstrated that crystalline inclusions produced by *B. sotto* were responsible for killing caterpillars. Heimpel and Angus (1958) recognized that *B. thuringiensis* and *B. sotto* were the same species. They chose the name *B. thuringiensis* over *B. sotto*, arguing that Berliner (1915) published in April, and Aoki and Chigasaki (1915) published in July. They were unaware that Iwabuchi (1908) had coined *Bacterium sotto*, documented later by Steinhaus (1961). As the older name, *B. sotto* should have priority (Dooling, 1978). We should be writing about Bs, not Bt.

Sanchis (2011) and Sansinenea (2012) reviewed the rest of the story: Steinhaus (1956) rekindled interest in Bt with a popular article, "Living pesticides," published in *Scientific American*. He convinced Pacific Yeast Products to produce a Bt-based product. By 1957 Thuricide® was available for testing, and 4 years later it was registered by the EPA. In 1962 Edouard Kurstak isolated a new variety of Bt, now named *B. thuringiensis* serovar *kurstaki*. The new serovar entered the market as Dipel® in 1970. It replaced the old strain, *B. thuringiensis* serovar *thuringiensis*. In 1963 Keio Aizawa isolated *B. thuringiensis* serovar *aizawai*. It was soon marketed as Certan®, and now sells as Xentari®. All these products are limited to lepidopteran pests.

In 1977 a Bt strain toxic to mosquitos was discovered, *B. thuringiensis* serovar *israelensis*. By 1992 *israelensis*-based products were sold as Vectobac®, Bactimos®, and Teknar®. A Bt strain toxic to beetle larvae was discovered in 1983, *B. thuringiensis* serovar *tenebrionis*. The first *tenebrionis*-based product was released in 1988, M-One®. The first transgenic Bt crop plant was a *tenebrionis*-based product, Monsanto's NewLeaf® potato, registered by the EPA in 1995.

Toxins produced by *B. thuringiensis*, collectively known as delta-endotoxins, are classified into a crystal (Cry) group and a cytolytic (Cyt) group. Insects have alkaline digestive tracts, which proteolytically activate Cry and Cyt toxins. Activated toxins bind to the insect cell membrane, leading to gut disruption. Hundreds of *cry* genes have been cloned and sequenced from insects; they have their own nomenclature (Crickmore *et al.*, 1998). *B. thuringiensis* genes encoding Cry1A toxins are classic caterpillar killers. Their genes have been genetically engineered into maize and cotton. Cry3A toxins kills beetle grubs, and their gene went into transgenic potatoes. Cyt toxicity is mostly limited to dipterans, although Cyt1Aa kills beetle grubs as well as fly larvae.

Fungi have one major advantage over bacteria: they do not have to be ingested. Fungi work on contact, by infecting insects right through their skin. In order to do this, however, fungal MPs require high humidity. This is a disadvantage, because high humidity encourages gray mold (*Botrytis cinerea*). In glasshouses, cycling two nights with elevated humidity and two nights with normal humidity is adequate for fungal MPs, without increasing mold problems (Van Lenteren, 1995). Fungal MPs work well in cloning chambers, which are always humid.

The taxonomic names of most fungal MPs have changed in the past 20 years, often repeatedly, roiled by molecular systematics. These name changes annoy end-users, who want stable nomenclature. However, DNA research has revealed the existence of "cryptic" species, which will improve the use of these MPs, tailored to specific hosts and specific meteorological conditions.

Nematode parasites of insects were described by Réaumur (1742). While dissecting the abdomen of a bumble bee, he observed a mass of threads. Placed in water under a microscope, the threads began moving, each a "small white worm in the shape of an eel." They were juveniles of *Sphaerularia bombi*, named a century later by Dufour (1837). Fifteen years later, Diesing (1851) listed no fewer than 175 insect species as nematode hosts. Many of Diesing's "nematodes" (e.g., *Gordius* spp.) are now classified as Nematomorpha, or horsehair worms. They are big, measured in centimeters rather than millimeters. The true nematodes in Diesing's list (*Sphaerularia*, *Mermis* spp.) were also big. None of them are commercially-available biocontrol agents.

Nematodes we use today are much smaller. They were not discovered until Steiner (1923) described *Aplectana kraussei* (now *Steinernema kraussei*). Six years later he described *Neoaplectana glaseri*, the first nematode to be marketed for biocontrol. Glaser (1932) pioneered the artificial culture of *N. glaseri*, which he poured into soil for control of Japanese beetle grubs. The grubs were actually killed by bacterial symbionts that lived within nematodes (Poinar and Thomas, 1967). The nematodes act as vectors, transporting bacteria into the insect hosts. In a survey of *Cannabis* growers in California, nematodes were nearly as popular as lady beetles (see Fig. 20.4).

Viruses arrived last in the biocontrol arsenal. Again they were discovered in silkworms. Cornalia (1856) observed refractive crystals in blood cells of diseased silkworms: "The crystals decrease the blood cells to the point of disappearance." The crystals, now known as occlusion bodies (OBs), were aggregates of the **nuclear polyhedrosis virus** (NPV). Glaser and Chapman (1913) deduced that "wilt disease" of spongy (gypsy) moth larvae was caused by a virus, because it could pass through a filter that removed larger microorganisms. After the invention of electron microscopes, Bergold (1947) published the first images of NPVs. He demonstrated that OBs were large protein shells, containing up to 200 rod-shaped viruses.

Not until 1973 did the EPA register the first NPV, Viron-H® (later Elcar®), for control of *Heliocoverpa zea*. After new pyrethroids came on the market a decade later, Elcar® sales plummeted, and manufacturing ceased. When off-target costs of pyrethroids became known, Elcar® returned, marketed as Gemstar®, along with Spod-X® (*Spodoptera exigua* NPV).

20.2.4   Competitors and antagonists

Rather than acting as predators, parasitoids, or pathogens, organisms in this group interact with pests through competition or antagonism. Competition arises for nutrients, space, and even sex.

Autocidal control utilizes competition for sex, by sterilizing male insects and releasing them. If the ratio of sterile males to normal males is 2:1, then 67% of females will mate with sterile insects, and produce infertile eggs (Knipling, 1992). Knipling and colleagues sterilized massive populations of screwworms (*Cochliomyia hominivorax*) with gamma radiation. This "peaceful use of atomic energy" successfully eradicated the screwworm from the USA by 1966.

Researchers at Oxitec (Oxford University) created sterile males using genetic modification (GM). These included GM mosquitos (*Aedes* and *Anopheles* spp.), GM fruit flies (*Ceratitis capitata*), and GM diamondback moths (*Plutella xylostella*). Releasing them into the environment has faced steep opposition. The Bill and Melinda Gates Foundation funded field trials of GM *Aedes aegypti*. The foundation threw its weight behind GM technology after it purchased 500,000 shares of Monsanto stocks in 2010. Between 2014 and 2019 the foundation donated US$15 million to fund research on Monsanto's GM crop plants. Its shares in Bayer-Monsanto are valued at US$23 million (Bassey-Orowuje *et al.*, 2019).

Oxitec convinced Brazil to release their mosquitoes. Genetic sampling by independent researchers (from Yale) revealed that GM *Aedes aegypti* hybridized with indigenous Brazilian mosquitos, transferring their transgenes to the native population. "These results highlight the importance of having in place a genetic monitoring program during such releases to detect un-anticipated outcomes" (Evans *et al.*, 2019).

Microbial competitors and antagonists (C&As) are non-pathogenic bacteria and fungi that suppress plant pathogens and pests. They do this by simply overgrowing the other organisms, or producing chemicals that antagonize growth and reproduction. C&As colonize above-ground portions of plants (the phyllosphere), and below-ground portions of plants (the rhizosphere). We dedicate Chapter 13 to them.

Some C&As are commercially available—see sections later in this chapter on *Bacillus cereus*, *Bacillus megaterium*, *Bacillus subtilis*, *Burkholderia* spp., *Fusarium oxysporum* Fo47, *Pseudomonas chlororaphis*, *Pseudomonas fluorescens*, *Pseudomonas syringae*, mycorrhizal fungi, and plant growth-promoting rhizobacteria.

20.2.5   Companion plants

Intercropping describes the sowing of two or more plant species in close proximity. Other names for this practice are **polyculture** and **companion planting**. Intercropping mimics the biodiversity of natural ecosystems. In a survey of *Cannabis* growers in California, 3% used intercropping (Fig. 20.4).

Intercropping may involve two or more cash crops, or a cash crop intercropped with a cover crop. A **cover crop** is a plant species grown for its environmental benefits rather than its own crop yield. A cover crop improves soil fertility and quality, lessens soil erosion, suppresses weeds, controls diseases and pests, and promotes biodiversity. Cover crops are typically grasses or legumes.

A classic example of intercropping is the Native American "Three Sisters" system: beans (*Phaseolus vulgaris*) climbed up cornstalks (*Zea mays*), while squash (*Cucurbita pepo*) covered the ground. Beans fixed soil nitrogen, cornstalks provided a trellis, and squash smothered weeds. By AD 544 the Chinese intercropped *Cannabis* with a legume (Willis, 2007). The legume fixed nitrogen for hemp, and the hemp helped the legume grow straighter.

Intercropping can decrease pests and pathogens. Intercropping performs this function via four mechanisms: (1) reducing the number of susceptible hosts (a dilution effect); (2) resistant plants provide an obstacle to susceptible plants (a barrier effect); (3) resistant plants produce chemicals that repel pests (a repellent effect); and (4) provide a haven for natural enemies (a reservoir effect).

Trap-crops attract pests away from main crops. Trap-crops and the pests they attract can be sprayed with pesticides, or removed from the field, taking the pests with them. An effective trap-crop will attract 70–85% of a pest population while covering only 1–10% of the total crop area (Hokkanen, 1991). Trap-crops require careful monitoring, otherwise they turn into *pest nurseries*.

Intercropping works. In a meta-analysis of 21 studies that compared intercropping with monocropping, Tonhasca and Byrne (1994) reported a 60% reduction in herbivorous insect populations. Langellotto and Denno (2004) conducted a meta-analysis of 43 studies, which showed that "increased habitat complexity" (i.e., intercropping) significantly increased populations of natural enemies. Boudreau (2013) reviewed 111 studies regarding fungi and oomycetes, and intercropping reduced disease by 71% compared with monocropping. Mechanisms included: alterations of wind, rain, and vector dispersal; modification of microclimate (especially temperature and moisture); changes in host morphology and physiology; and direct pathogen inhibition.

20.2.6   *Cannabis* as insect repellent

None of the aforementioned reviews mentioned the benefits of intercropping *Cannabis* upon neighboring crop plants (Fig. 20.8). *Cannabis* harbors robust populations of predators and parasitoids (Batra, 1976; Lago and Stanford, 1989; Trotuş and Naie, 2008; Schreiner, 2019; Britt *et al.*, 2022). The dense canopy formed by crops of fiber-type hemp provides an ideal microclimate for natural predators (McPartland *et al.*, 2000). Gupta *et al.* (2012) described an "entomophage park" of natural vegetation growing near crop fields in Kashmir. The park supported large populations of natural enemies, and feral *Cannabis sativa* grew there. Lots of old research claimed that hemp plants repelled many types of insects from neighboring crop plants: caterpillars, aphids, phylloxera, and beetles.

20.2.7   Caterpillars

Chomel and La Marre (1767) recommended planting hemp around cabbage fields to drive away the cabbage moth, *Pieris brassicae*. "Cover the entire edge of the hemp in which you want to plant cabbage with hemp. Even if the whole neighborhood is infested with caterpillars, there will not be a leaf (it is said) in the space enclosed by the hemp." A flurry of publications followed.

Pratje (1768) claimed, "If one wants to drive the cabbage to safety from the caterpillars, one should sow hemp around the land on which it has been sowed. It will be astonishing to realize that, although all the land lying around is covered with caterpillars, [cabbage] on which hemp is surrounded, not a single one [caterpillar] will be seen." Pratje's claim was repeated, often without attribution, by authors across western Europe. They included Willich, Hamm, Jentink, D'Arenberg, Blanchard, and Linsbauer (full citations in McPartland and Sheikh, 2018).

Alletz (1773) posed an interesting hypothesis regarding this remedy: "The cause comes, either from the aversion that the caterpillars have for this plant, or from birds who are very fond [of hemp seed], and by probing the hemp, destroy at the same time the caterpillars which also makes one of their foods."

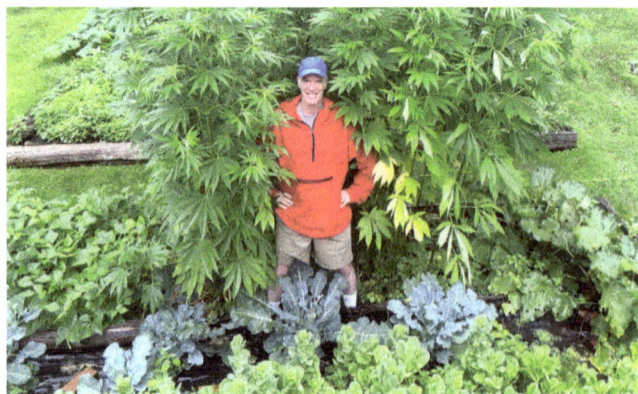

Fig. 20.8. Intercropping with *Cannabis* (courtesy P. Pruitt)

Grosier (1781) proposed that hemp's odor caused caterpillars to wander, hence perish.

Köllar (1837) coined the term "natural control." Regarding the hemp-around-cabbage story, he wrote, "Certainly it is not the smell that is acting against the insects, since several [insects] even feed on its leaves. It should be the case that various birds, especially sparrows, attracted to its seeds and protected by its bushy habit, like to stay on the fields bordered by hemp, and help to reduce the harmful insects."

Boisduval (1867) wrote about the oak leafroller, *Tortrix viridana*, whose numbers were increasing in Paris. He attributed their rise to the decline of birds who ate them, such as *mésanges* (*Parus major*), *friquets* (*Passer montanus*), and *moineaux* (*Passer domesticus*). Boisduval sowed "bird seed" plants, hemp and millet, in forest clearings to bolster those bird populations. In Spain, Ramirez (1865) wrote, "To exterminate aphids and others of legumes, it is indicated to surround them with a hemp border." We're guessing that didn't work: one of the worst aphids in legumes, *Aphis fabae*, also infests hemp.

Experts repeated the hemp-around-cabbage story for nearly 200 years, until someone conducted an actual experiment. Herold (1949) grew different types of cabbage (white, red, savoy, rose, and cauliflower) on their own, or surrounded by hemp. He hypothesized that volatile substances that outgassed from hemp temporarily blocked olfaction in egg-laying cabbage moths, "maybe even numb them." Alternatively, hemp masked the smell of cabbage plants. Disappointingly, hemp had no deterrent effect on egg-laying moths. On the contrary, cabbage plots enclosed by hemp were preferred for oviposition, because they were protected from wind.

Fahlpahl (1949) argued that Herold did not grow plants in moorland soil. In this soil, hemp grew more vigorously and emitted a greater smell, making egg-laying moths *biologische Vergrämung* (biologically deterred). Fahlpahl added that hemp's volatiles inhibited the growth of pathogens, so plants growing in the vicinity of hemp suffered fewer diseases. Ziarkiewicz and Anasiewicz (1961), unaware of Herold's study, repeated it. They found no protective effect against *P. brassicae* by planting hemp around cabbage plants.

French grape vineyards needed protection. Anonymous (1783) grew *Cannabis* around grape vines to protect them from "caterpillars," probably the vine moth, *Eupoecilia ambiguella* (Hübner) 1796. *Cannabis* worked as a trap crop: "the caterpillars left the vine to attach themselves to it [*Cannabis*]." The author added that the technique did not work in all locations. Roberjot (1787) exposed the vine moth to *feuille de chenevi* (leaves of hemp), to no effect. Fahnenberg (1815) planted *Cannabis* near grape vineyards to repel the vine moth. However, an editorial attached to Fahnenberg commented that "enlightened men in France" believed that the odor of hemp did *not* repel insects.

20.2.8 Aphids

Gaujac (1810) intercropped hemp with turnips in France. He claimed the odor of hemp protected turnip plants from *pucerons* (aphids). The best known "turnip aphid" in France is *Lipaphis erysimi*, which, possibly erroneously, has been reported as a *Cannabis* pest (Prasad, 1997).

Hemp planted near fava beans (*Vicia faba*) protected the beans from two species of aphids: the "black aphid" *Aphis fabae* (Scheidweiler, 1847) and the dock aphid, *Aphis rumicis* (Feldt, 1919). Scheidweiler said protection did not begin until hemp started flowering, when it "began to exhale its penetrating odor, and the aphids disappeared as if by enchantment." Conversely, Mackiewicz (1962) reported that hemp planted near beets (*Beta vulgaris*) had no effect on *Aphis fabae*. In fact, *A. fabae* is a well-known *Cannabis* pest.

Harand (2000) observed that potatoes bordering a hemp field were not infested by aphids (species not stated), whereas the rest of the potato field was heavily infested. He used gas chromatography to analyze hemp plants, and discovered that they produce (E)-β-farnesene, an aphid alarm pheromone.

20.2.9 Grape phylloxera

Viticulturists were hit by the "Great French Wine Blight" in the 1860s. A North American aphid, the grape phylloxera (*Daktulosphaira vitifoliae*), began devastating French vineyards. Chemicals and pesticides proved useless. Grape growers turned to *Cannabis*.

Intercropped hemp reportedly repelled phylloxera from vineyards (Du Mont, 1873; De Montlaur, 1873; Naudin, 1874; Türckheim, 1874; Roger-Dobos, 1875; Jamet, 1875; Garreau, 1876). Intercropped tobacco reportedly worked the same way, although De Céris (1874) noted that "hemp is within the reach of everyone, while the planting of tobacco is subject to restrictions." Naudin (1874) argued that the "insecticide or repellency properties" of *Cannabis* "are perhaps more pronounced than those of tobacco." Reverchon-Chamussy (1874) described *Cannabis* as an "ancient, formidable enemy opposed to the phylloxera." He cut down hemp plants and mixed them into soil around vine roots. "The smell of buried hemp will kill or chase away the phylloxera." Dalmagne (1882) suggested sowing *Cannabis*, as well as watering vineyard soil with water from hemp retting ponds.

Others were skeptical. Trimoulet (1876), the leading entomologist in Bordeaux, reviewed four studies on the effects of *Cannabis* on phylloxera: a study by LeComte-Tridan produced no results; two studies by Peyraut and Tourette "remain to be verified," and the results of watering vine roots with an infusion of hemp by Docoux and Pouilb "are unknown."

Jules-Émile Planchon, who originally identified phylloxera, doubted the ability of hemp to drive away the pest, "a property which is lent to it [hemp] free of charge as a result of a prejudice as false as it is widespread" (Planchon, 1875). Dumas (1876) reported that decoctions of hemp did not kill phylloxera. Sauzey (1879) tried using hemp and tobacco, "and that of all these attempts nothing serious has resulted." Chadeuil (1882) bemoaned claims regarding the "supposed" effects of hemp against phylloxera. Goulet (1886) referred to the effects as an "illusion." Clément (1887) reported no success when intercropping vineyards with *Cannabis*. Laboulbéne (1887) summarized, "Botanical procedures are unfortunately ineffective."

Intercropping with hemp, tobacco, pyrethrum, belladonna, datura, castor beans, lupine, spurges, etc. had little or no effects on the recalcitrant phylloxera.

Hopes persisted. Béguy (1885) advocated for the use of hemp to destroy phylloxera. He credited the toxic odor given off by green plants; the odor could even kill a human trapped in an enclosure with green hemp. The last gasp came from Maret (1904), who said the "bad smell" from hemp drove off winged phylloxera aphids, preventing egg-laying. "Would the development of phylloxera not be due to the fact that we gave up planting hemp in the vineyards?"

20.2.10   Other pests

The cinch bug, *Blissus leucopterus*, attacks grain crops in the grass family—wheat, rye, oats, and maize. To control cinch bugs, Charles Valentine Riley (see Fig. 20.3) surrounded grain crops with a ring of hemp plants. Unfortunately, "as soon as they [cinch bugs] become winged, they fly over it [hemp]" (Riley, 1875).

Riley (1885) noted that *Cannabis* growing near cotton exerted a "protective influence" against the cotton leafworm (*Alabama argillacea*, he called it *Aletia xylina*). Balfour (1887) suggested intercropping *Cannabis* to repel spider mites from tea, *Camellia sinensis*. Given the susceptibility of *Cannabis* to spider mites, this seemed like a bad idea, unless the *Cannabis* was used as a trap crop.

Stratii (1976) claimed that when potato plants were surrounded by hemp plants, plants nearest to hemp were free from infestation by the Colorado potato beetle (*Leptinotarsa decemlineata*). However, Mackiewicz (1962) grew hemp around the edges of a potato field and within the field, and found no effect on *L. decemlineata*. Hemp also had no effect on black bean aphid (*Aphis fabae*) in beet fields.

Nematodes are difficult to control through intercropping (Boudreau, 2013). Intercropping with marigolds (*Tagetes* spp.) has shown some success. Marigolds repel root-knot nematodes, *Meloidogyne* spp. (Hooks *et al.*, 2010). A flurry of 20th century publications investigated the effects of *Cannabis* on nematodes. That literature is reviewed in section 3.3.6. Collectively it suggests that *Cannabis* can suppress *Meloidogyne* and *Heterodera* spp.

20.3   Biocontrol strategies

There are three basic strategies for biological pest control: classical, inundative, and conservation. Each has advantages and disadvantages.

20.3.1   Classical biocontrol

Classical biocontrol targets exotic insect pests by importing their natural enemies from their native lands. Optimally this results in permanent establishment of the natural enemy and long-term control in an area where the pest has invaded.

We previously mentioned some classical success stories, such as Charles Riley's importation of the vedalia lady beetle from Australia to California to control *Icerya purchasi*, also from Australia. Only 129 vedalias were released in 1888, but within 18 months their offspring nearly eradicated *I. purchasi* from Californian orange groves.

However, serious off-target impacts may arise. Non-native biocontrols may themselves become "invasive" in the targeted ecosystem. They may switch hosts and prey on native species, or compete with native species, or disperse far beyond their intended range. The French shipped mynah birds from India to Mauritius to control locusts. Success led to the mynah bird being shipped around the world. Later it was perceived as a "great nuisance," because it destroyed the nests and eggs of domestic birds (Palmer, 1894). Now the mynah numbers among the world's 100 worst invasive species. It's a member of the starling family (Sturnidae).

The British shipped the Indian mongoose (*Herpestus auropunctatus*) to Jamaica in 1872 to control rats. Within 10 years, Jamaican naturalists noted a sharp decrease in the number of ground-nesting birds, fowl, and reptiles. By then, however, the mongoose had been released in Hawai'i, resulting in the extermination of native birds (Palmer, 1894).

Vertebrate biocontrol organisms are notorious for going off-target. There are over 100 cases of biocontrols driving non-target hosts to extinction (McPartland and Nicholson, 2003). Ironically, far more species have been driven to extinction by biocontrols than by pesticides. Insects imported for biocontrol may become invasive. Jennings *et al.* (2017) highlighted the Asian lady beetle, *Harmonia axyridis*. It feeds on many non-target insects, as well as some plants, and it has displaced native lady beetles. Since its introduction to the USA in 1916, *H. axyridis* has spread everywhere, and its bodies are piling up in my attic at this moment.

In most cases classical biocontrol has not been ecologically disruptive, and natural biodiversity has been maintained. Risk assessments and cost–benefit analyses are now implemented before biocontrols are introduced. The advantages of classical biocontrol include low cost, an inability of pests to develop resistance to biocontrol agents, and an almost total absence of harmful effects on humans. Disadvantages of classical biocontrol include off-target effects, pest populations continue to exist at low levels, risk assessments can be demanding in terms of research expertise, and results are not predictable.

20.3.2   Inundative biocontrol

Inundative release (*a.k.a.* augmentative or inoculative biocontrol) liberates a large population of natural enemies for quick pest control. This tactic is used in situations where natural enemies are lacking—growrooms and glasshouses, as well as monocropped field systems. This expensive approach makes economic sense for high-value crops.

Inundative biocontrol under glass began in 1926 with *Encarsia formosa*, the whitefly parasitoid. But growers abandoned *E. formosa* for cheap DDT in the 1940s. The appearance of DDT-resistant spider mites around 1968 revived interest in biocontrol, using predatory mites (*Phytoseiulus persimilis*).

Van Lenteren (1995) estimated that glasshouses covered 150,000 ha of the globe, and the area treated with *P. persimilis* was >7500 ha The area treated with *E. formosa* was >2500 ha (it had been reintroduced in 1970).

Cannabis cultivators in California began using inundative biocontrol in the 1970s, releasing lady beetles and mantids (D. Watson, pers. comm., 1995). After moving to Amsterdam, Watson collaborated with Koppert Biological Systems BV to devise biocontrol methods. With Watson's urging, Koppert helped underwrite the first edition of *HD&P*.

To work effectively, inundative biocontrols should be introduced at an early stage of pest infestation. Biocontrols often reproduce faster than pests, but if pests get a head start, crop damage arises before biocontrols can catch up. Optimally, biocontrols should be introduced preventively (i.e., prophylactically) before pests even appear. Several preventive releases are scheduled per season ("dribble release"), because biocontrols will die out if no pests are available.

Delivery systems for biocontrols vary. Many are sold loose, in shaker bottles or bags, mixed with carrier material (vermiculite or wood chips). They can be broadcast by hand or with machines that dispense them via fans or compressed air (e.g., Koppert's Airbug, BioBest's Entomatic). Biocontrols can be dispensed into cardboard distribution boxes hung from plants (e.g., Koppert's Dibox, BioBest's Bio-Box, BioBee's D-Box). These are particularly useful for treating pest hotspots, and they prevent carrier material from coming into contact with plants (Fig. 20.9A).

Sachets are small containers designed to be hung from plants for slow-release of biocontrols. They contain a breeding colony of the biocontrol, along with a food source such as *Tyrophagus putrescentiae* ("mold mites") that do not harm plants (Fig. 20.9B). Advantages over broadcasting include less frequent introductions (every 4–6 weeks) and more consistent levels of biocontrols.

Application rates for inundative biocontrol depend on many factors. First in importance is the number of pests, measured

as the Infestation Severity Index (ISI) for each pest. Other factors include crop biomass, crop location (field, glasshouse, or growroom), local environmental conditions (do temperature and humidity favor the biocontrol or the pest?), and the longevity of biocontrol organisms versus the longevity of pests.

If biocontrols are released in *field* crops, more care must be taken in timing, and providing an immediate food source for them. Otherwise the biocontrols will exit crops in search of shelter and food. This is especially true with winged biocontrols. Many non-winged biocontrols, such as predatory mites, have the opposite problem. They cannot disperse enough. If adjacent plants are not touching each other, it may be necessary to disperse biocontrols onto each individual plant.

Some predators go dormant (they enter diapause) when crops begin to flower in autumn. This is triggered by a short photoperiod, and is affected by temperature. Researchers have overcome this trait by selective breeding. Non-diapausing breeds of biocontrols are becoming available. Winged biocontrols may ultimately fly away from crops. Hopefully they leave behind eggs which hatch into more biocontrols. Entice them to stay and lay eggs by providing water and artificial honeydew or nectar.

Advantages of inundative biocontrol include commercial availability, minimal disruption of indoor ecosystems, and an almost total absence of harmful effects on humans. Disadvantages include cost, and the possibility of pest resistance to microbial biocontrols if confronted with sufficient selection pressure (Gould, 1991).

20.3.3 Conservation biocontrol

Conservation biocontrol (CBC) differs from classical and inundative biocontrol in one big way: no natural enemies are released. Instead, CBC takes measures to support and enhance populations of naturally occurring beneficial organisms. CBC was introduced by Réaumur (1737), who protected the eggs of insect predators. Darwin (1800) called for the "encouragement" of ichneumon parasitoids. CBC took a back seat when Charles Riley initiated the era of classical biocontrol.

CBC regained prominence after *Silent Spring*. Scientists recognized that natural enemies had been decimated by synthetic pesticides. IPM practitioners established the strategy of scouting populations of natural enemies. Scouting is the regular, systematic inspection of crops and surrounding areas, and keeping careful records. Naturally occurring biocontrols are best enhanced by increasing plant diversity that would otherwise be absent in a monocropped field. Plant diversity provides biocontrols with shelter, favorable microclimates, and a supply of alternative prey.

Meadow vegetation provides a concentration of biodiverse resources for natural enemies. Meadows provide perennial habitats, whereas annual crop habitats are transient. Strips of native vegetation within crop fields, known as "beetle banks," harbor predators that disperse into the crop (Hajek and Eilenberg, 2018). Native flowering plants provide nectar and pollen. These floral resources are used by syrphid and tachinid flies, some parasitoid wasps, lacewing adults, lady beetles, and

Fig. 20.9. Biocontrol dispensers in *Cannabis*: **A.** boxes; **B.** sachets (courtesy Koppert Biocontrols)

predatory bugs (mirids and anthocorids). Healthy soil provides a reservoir of beneficial organisms, such as microbial pathogens and predatory insects. Many parasitoids pupate in the soil, and their populations are higher in no-till or low-till fields. Decreasing the use of broad-spectrum pesticides is key for CBC.

Indoors growers use **banker plants** as a miniature version of CBC. Banker plants provide alternative food sources for natural enemies. The goal is to sustain a reproducing population of natural enemies that will provide long-term suppression. For example, the cereal aphid (*Rhopalosiphum padi*) is maintained on cereal plants (wheat, barley, oats) as a host for the aphid-killing parasitoid *Aphidius colemani*. Because *R. padi* only infests monocotyledonous plants, it poses no risk to *Cannabis* in the glasshouse. *A. colemani* spreads from banker plants and kills hemp aphids (*Phorodon cannabis*) in *Cannabis* growing nearby (Fig. 20.10).

Banker plants can be inoculated with actual pests of a crop, but this strategy is more controversial. For example, banker plants can be inoculated with greenhouse whiteflies, to support populations of the parasitoid *Encarsia formosa* (Frank, 2010). Many people reject this "pest-in-first" concept, but it works (Van Lenteren, 1995).

Advantages of CBC include low-cost outlay, and a healthier, more biodiverse environment. Disadvantages include the fact that cost–benefit analyses are difficult with CBC. The loss of land to "beetle banks" and meadows, even banker plants, can be a significant consideration in high-value crops. CBC is not generally considered a "stand-alone" type of control strategy; it needs to be integrated into a larger IPM system (Hajek and Eilenberg, 2018).

20.3.4   Biocontrols and pesticides

Large pest populations may require chemical control before biocontrols can work effectively. Theiling and Croft (1988) summarized 1000 publications that studied the effects of

Fig. 20.10. Banker plant in *Cannabis* (courtesy Sami Simmons, Biobee USA)

pesticides on biocontrol arthropods. The effects of pesticides on a few biocontrol agents are presented in Table 20.1. Pesticides permitted for use on *Cannabis* are contrasted with pesticides that are banned. Table 20.1 shows that banned pesticides are quite toxic to biocontrols. Some pesticides stay toxic for a long time (e.g., deltamethrin, a synthetic pyrethroid). However, a few permitted pesticides also hammer biocontrols, such as insecticidal soap. After insecticidal soap dries, its toxicity is gone. These products have a very short re-entry period (days after spraying insecticidal soap, biocontrols can be introduced).

20.4   Pollinators 1: Bees

Pollinators are a related topic in this chapter on biocontrols. Flowering plants and their animal pollinators are one of nature's most striking examples of mutualism. Animals transport male pollen to female stigmas—bees, wasps, hoverflies, flower beetles, some ants and moths, even bats and birds. The pollinator family Apidae (honey bees, bumble bees, digger bees, etc.) (Fig. 20.11) radiated during the mid-Cretaceous, 100–70 million years ago, simultaneously with the Eudicot clade of flowering plants. Many Eudicots evolved an "entomophily syndrome," where flowers grow large and colorful, and offer nectar as pollinator rewards.

Honey bees obtain nutrition from plant nectar (carbohydrates, which they convert into honey) and from pollen (for carbohydrates, lipids and protein). Carbohydrates in pollen are in the form of starch and sugars. Lipids in pollen are converted into beeswax.

Cannabis is an **anemophilous** plant: it is wind pollinated. Its flowers do not produce showy floral parts or produce nectar. It has not evolved a mutualistic relationship with pollinators. *Cannabis* pollen contains relatively little starch—1.14%, compared with 16.56% in *Zea mays* (T'ai and Buchmann, 2000). Briosi and Tognini (1894) noted, "hemp flowers are visited by insects, but what they look for in the flowers we do not know; their visits do not seem to have any connection with fertilization."

Cannabis does produce copious pollen, and its pollen is sought by bees. In a literature review of anemophilous plants visited by pollinators, Saunders (2018) included *Cannabis*. He reported *Cannabis* being visited by honey bees (*Apis mellifera, Apis cerana*) and *Lipotriches* spp. (ground-nesting solitary bees).

The earliest mention we could find of honey bees in hemp was by Butler (1609), in Oxford, England. Alefeld (1856) included *Cannabis* in his definitive German text on *Bienen-flora*, "bee-flora." Alefeld noted that in July and August, honey bees "wear the dust [*Cannabis* pollen] very strongly." In the USA, Wagner (1867) wrote, "Hemp is a favorite with bees, though seldom met with since cotton has supplanted sail duck and hemp has gone out of fashion." Wilder (1879) monitored plants visited by bees in Illinois: "the common hemp stands far ahead of anything I ever saw. Bees literally swarm on it, from early daylight till dark. It is a wonder that others have not noticed it and made it known." Wilder planted 0.75 acre of hemp for his bees.

Table 20.1. Effects of some pesticides on biocontrol organisms, on a scale of 4 = very harmful (75% affected), 3 = moderately harmful (51–75% affected), 2 = slightly harmful (25–50% affected), 1 harmless (<25% affected), Also the amount of time (*n* = weeks) a pesticide remains harmful after application. ND = no data.[a]

Pesticide,type	Phytoseiulus persimilis Larva/adult	Chrysoperla carnea Larva/adult	Orius laevigatus Larva/adult	Encarsia formosa Larva/adult	Aphidius colemani Larva/adult	Lecanicillium lecanii fungal spores
Banned						
Abamectin	4/4 (>1)	4/3 (>1)	3/4 (1-6)	4/1 (3)	4/ND (>1)	1 (ND)
Deltamethrin	4/4 (8–12)	4/4 (8–12)	4/4 (8–12)	4/4 (8–12)	4/4 (8–12)	1 (ND)
Diazinon	2/2 (ND)	2/2 (4)	3/2 (ND)	4/4 (4–6)	4/4 (ND)	4 (ND)
Imidacloprid	2/2 (2)	4/4 (0)	4/4 (4-6)	4/4 (>12)	4/4 (ND)	ND
Rotenone	4/4 (ND)	4/2 (ND)	4/4 (2)	4/4 (2)		ND
Permitted						
Bacillus thuringiensis	1/1 (0)	1/1 (0)	1/1 (0)	1/1 (0)	1/1 (0)	1 (ND)
Hort oil	3/3 (ND)	1/1 (0)	2/3 (0)	1/1 (0)	1/1 (0)	1 (ND)
Insecticidal soap	4/4 (0)	4/4 (0)	4/4 (0)	4/ND (2)	4/ND (ND)	1 (ND)
Neem oil	2/2 (0)	1/1 (0)	1/2 (0)	2/1 (ND)	1/1 (0)	ND
Pyrethrum	4/4 (1)	2/2 (1)	4/ND (ND)	4/2 (ND)	4/1 (ND)	ND
Spinosad	2/2 (1)	2/1 (ND)	2/2 (1)	2/2 (2)	1/1 (ND)	ND
Sulfur dust	2/2 (1)	1/1 (0)	3/2 (ND)	3/ND (3-4)	2/2 (ND)	3 (ND)

[a]Data from Koppert, https://sideeffects.koppert.com/side-effects/
The data presented here are only estimations and do not guarantee safety if followed.

Fig. 20.11. Pollinators gathering hemp pollen: **A.** honey bee; **B.** bumble bee (courtesy of Heather Grab)

Hamet (1861) blamed hemp pollen on "vertigo disease" in honey bees, where they lost the ability to fly, and ran around until exhausted. Other authors have said *Cannabis* yields "toxins" that affect honey bees (e.g., Bista and Shivokoti, 2000). *Cannabis* pollen can be contaminated with THC and CBD (Paris *et al.*, 1975). But THC wouldn't cause vertigo or other neurological changes in honey bees, because the brains of *Apis mellifera*—like all insects—lack the cannabinoid receptors found in vertebrates (McPartland *et al.*, 2001).

In South Asia (India, Pakistan, Nepal), *Cannabis* is visited by *Apis mellifera* and *Apis cerana* (Suryanarayana *et al.*, 1992; Bista and Shivokoti, 2000). Less common bee species, *Apis dorsata* and *Apis florea*, also collect pollen from Indian hemp (Schmalzel, 1980; Sharma *et al.*, 2017). In India, *Cannabis* becomes an important source of pollen during periods of floral scarcity—in the late summer, when major pollen flora are not flowering (Dalio, 2012). Dalio reported *Cannabis* pollen loads of 4 mg/bee. During summer months, *Cannabis* pollen was the most frequently encountered pollen in honey samples (Bibi *et al.*, 2008).

In Mississippi, pollinators visiting *Cannabis* included *A. mellifera*, the eastern bumble bee *Bombus impatiens*, and the sweat bee *Lasioglossum imitatum* (Lago and Stanford, 1989). In upstate New York, Flicker *et al.* (2020) identified 16 species, led by *A. mellifera* (59% of total abundance), *Bombus impatiens* (30%) and the sweat bee *Lasioglossum hitchensi* (3%). The remainder were ten other *Lasioglossum* spp., *Augochlora pura*, and *Xylocopa virginica*. The height of hemp plants influenced the abundance of bees, with tall cultivars attracting more bees and a richer species diversity. Hemp provided vital subsistence to bees in late summer and early autumn, coinciding with a dearth of other pollinator-friendly crop plants in New York. Similar results were reported in two surveys conducted in Colorado. O'Brien and Arathi (2019) collected 1937 individuals from 23 different genera, led by *A. mellifera* (38% of total abundance). Next came longhorn bees *Melissodes bimaculata* and *Peponapis pruinosa* (25% and 16%, respectively). Three bumble bees, *Bombus griseocollis*, *B. huntii*, and *B. pensylvanicus*, collectively contributed 6%, and 19 genera constituted the remaining 15% (*Svastra, Anthophora, Lasioglossum, Megachile, Agopostemon, Halictus, Triepeolis* spp.).

Schreiner (2019) surveyed all pollen-feeding insects, and 70% were bees from two families. Members of family Apidae dominated, thanks to *A. mellifera*, followed by *M. bimaculata* and *M. agilis*, then *B. griseocollis, B. huntii, B. pensylvanicus*, and *B. fervidus*, then *Peponapis pruinose, Svastra obliqua, Anthophora montana*, and *Andrena helianthi*. Members of family Halicidae (sweat bees) included *Agopostemon, Halictus, Lasioglossum*, and *Augochlorella* spp.

Honey bees collect plant resins and convert them into **propolis**, which they use as a sealant within hives. Schreiner (2019) proposed that honey bees she found in sinsemilla CBD crops were collecting resins for propolis—there were no male plants or pollen in the sinsemilla crops. Alternatively, the honey bees were attracted to sinsemilla crops by aphid "honeydew," which they feed upon (Kunkel and Kloft, 1977). Based on a pollen analysis, Özkök *et al.* (2023) found that honey bees prefer *Cannabis* to other plants when collecting resin for propolis.

Resins collected from *Cannabis* and converted to propolis contain cannabinoids. French (2016) analyzed propolis in bee hives placed in 70 acres of dioecious hemp. Cannabinoid content in propolis averaged 9.30 mg/g, or 0.93% by weight. The hives contained very little honey, because *Cannabis* does not produce nectar. French pulled hives from hemp fields before the bees starved for lack of nectar. THC or THCA has been extracted from propolis collected in Greece and Cyprus (Kalogeropoulos *et al.*, 2009), Iran (Abdulkhani *et al.*, 2015), and Turkey (Türk *et al.*, 2022; Özkök *et al.*, 2023).

The importance of pollen from wind-pollinated plants was emphasized by Saunders (2018). Bee populations are declining worldwide, due to pesticides, diseases, air pollution, climate change, and the effects of monoculture. Avoiding the use of pesticides in hemp crops would protect these pollinators (O'Brien and Arathi, 2019; Schreiner, 2019). Imidacloprid and other neonicotinoid pesticides have been linked with honey bee colony collapse disorder (Lu *et al.*, 2014; Tsvetkov *et al.*, 2017).

Cannabis might provide other benefits to honey bees: Surina and Stolbov (1981) partially controlled the parasitic honeybee mite, *Varroa jacobsoni*, with vapors from fresh leaves and stems, reduced to a powder. They rubbed the inner walls of the hive with 10–12 g powder per bee family. Skowronek *et al.* (2022) fed honey bees sugar syrup mixed with *Cannabis* extract, and they lived longer than controls (52 days and 35 days, respectively). Honey bees fed the *Cannabis* extract had higher levels of antioxidant enzymes (catalase, peroxidase, glutathione, superoxide dismutase). The *Cannabis* extract was not well-characterized: "hemp paste extract + 3 mL water-glycerine solution."

20.5 Pollinators too: Syrphids

Syrphids or hoverflies provide dual services for crops. The flies are often seen hovering around flowers, feeding on nectar and pollen. Syrphid flies are pollinators in natural landscapes and commercial crops, second in importance behind bees. Bees carry more pollen than syrphid flies, because of their larger size and hairiness. Fivefold more syrphids (*Episyrphus balteatus*) than wild bees (*Osmia rufa*) were required to reach a similar fruit set (and seed yield) in a trial with rapeseed plants (Jauker *et al.*, 2011). *Cannabis* does not require syrphid pollination services, because it is wind pollinated. Nevertheless, syrphids, like honey bees, collect *Cannabis* pollen (Mutin, 2017).

Syrphids provide a second service for commercial crops: the larvae feed voraciously on aphids and other pests. Réaumur (1737) recommended collecting the eggs of syrphid flies for controlling aphids (see Fig. 20.2A). A single larva of the aforementioned species *Episyrphus balteatus* can consume up to 1322 medium-sized aphids (Tenhumberg, 1995). We'll return to this species shortly.

Syrphids are flies but their appearance mimics wasps and bees (Fig. 20.12). They may look scary, but they cannot sting or bite. Schmidt (1929) found alarmingly high numbers of syrphid larvae in stored hemp seed in Germany. He noted that the crop was heavily infested with aphids, which led to a great syrphid boom—until autumn, when the aphid population crashed. Remaining syrphid larvae, unable to pupate, hibernated in *Cannabis* flowers, and ended up in seed during threshing.

Two surveys of syrphids in *Cannabis* have been published. Danuta Malinowska surveyed populations in Poland. She listed 14 syrphid species collected in colonies of *Phorodon cannabis* (Malinowska, 1973). This list appeared in her subsequent quantitative survey of syrphids in eight crop plants

Fig. 20.12. *Episyrphus balteatus* larvae and adults. (Drawings by De Geer, 1776; larva photo courtesy of Syrio; adult photo courtesy of Hans Hillewaert)

(Malinowska, 1979). In terms of the number of larvae collected, *Cannabis* ranked highest among eight crops she surveyed (*n* = 9186, 14 species). In terms of the number of adults collected, *Cannabis* ranked third (*n* = 386), but richest in the number of species (*n* = 17).

The predominant species in summer was *Episyrphus balteatus* (De Geer) 1776. Two species predominated in autumn, *Syrphus vitripennis* Meigen 1822 and *Platycheirus peltatus* (Meigen) 1822. The other "accessory" species include: *Sphaerophoria scripta* (L.) 1758, *Eupeodes* (*Metasyrphus*) *corollae* Fabricius 1794, *Syrphus ribesii* L. 1758, *Sphaerophoria rueppelli* Wiedemann 1830, *Melanostoma mellinum* L. 1758, *Platycheirus clypeatus* Meigen 1822, *Scaeva pyrastri* L. 1758, *Eupeodes luniger* Meigen 1822, *Platycheirus scutatus* Meigen 1822, *Pipiza noctiluca* L. 1758, *Platycheirus angustatus* Zetterstedt 1843, *Eupeodes nitens* Zetterstedt 1843, *Syrphus tarvus* Osten Sacken 1875, *Sphaerophoria menthastri* (L.) 1758, *Sphaerophoria picta* Meigen 1822, *Sphaerophoria dubia* Zetterstedt 1849. Among syrphids in the eight crops, the community on *Cannabis* was least similar to the others.

The other list of *Cannabis* syrphids was collated in India by Mitra and Banerjee (2007). Most of their data came from a study in Jammu and Kashmir by Datta and Chakraborti (1983). They listed *Episyrphus balteatus* (De Geer) 1776, *Metasyrphus latifasciatus* (Macquart) 1829, *Ischiodon scutellaris* (Fabricius) 1805, *Syritta pipiens* (L.) 1758, *Sphaerophoria scripta* (L.) 1758, and unidentified *Ischyrosyphus* sp. and *Melanostoma* sp.

Despite vastly different terrain—the Lublin lowlands versus the Himalaya highlands—the studies in Poland and India had several species in common.

Foremost was *Episyrphus balteatus* (Fig. 20.12), a species distributed across Eurasia. The species has a common name: the marmalade hoverfly. De Geer (1776) named the species *Musca balteata*. Its larvae targeted *Cannabis* aphids in France (Vallot, 1836) and Germany (Gistel, 1856). Prior to De Geer, Scopoli (1763) discovered *Musca cannabina*. Scopoli's species was encountered by others, such as Gmelin in 1788, Petagna in 1808, Olivier in 1811, and Wilkes in 1819. Meigen (1822) recognized that *M. cannabina* was identical to De Geer's species. Meigen erroneously believed that *M. cannabina* was coined by Gmelin in 1788, so he synonymized *M. cannabina* under De Geer's "older" name.

Schiner (1862) noted that that Scopoli coined *Musca cannabina*, not Gmelin, and Scopoli's taxon was older than De Geer's taxon, but "the interpretation of Scopoli's *Musca canabina* [*sic*] is not entirely certain; the De Geer name of the species must therefore remain." Schmid (1996) argued for De Geer's taxon in the interests of nomenclatural stability. The *International Commission on Zoological Nomenclature* officially recognized De Geer's taxon and rejected *Musca cannabina* as a "disused subjective synonym" (Tubbs, 2000). By all rights, this well-known insect should be called *Episyrphus cannabina* (Scopoli).

The adult flies feed on pollen. Jia *et al.* (2022) investigated pollen sources for the flies in China, and *Cannabis* was a "preferred host plant," along with *Humulus scandens* and *Helianthus annuus*. The larvae hold the world record in aphid consumption

(1322 aphids; Tenhumberg, 1995), and prey on over 230 aphid species (Leroy *et al.*, 2010).

Introducing *E. balteatus* as augmentative biocontrol has been stymied by raising the species in artificial conditions. Dong *et al.* (2004) successfully reared adults on hemp pollen. The adults are strong flyers, which also stymies biocontrol; keeping released adults in glasshouses is difficult. Pineda and Marcos-García (2008) tried retaining *E. balteatus* by placing "banker plants" in glasshouses (barley with the aphid *Rhopalosiphum maidis*), but with marginal success. They suggested inundative release of eggs, rather than the commercially available pupae. Leroy *et al.* (2010) induced adults to lay eggs on artificial surfaces by stimulating oviposition with limonene and (E)-β-farnesene (common terpenoids produced by *Cannabis*).

The *Cannabis* surveys in Poland and India shared other species in common. **Sphaerophoria scripta** has also been recovered in Italian hemp (Goidànich, 1928). The species has a common name, the long hoverfly, has a worldwide distribution, and has attracted attention as a biocontrol against aphids. Another species in common, *Melanostoma mellinum*, has also been recovered from hemp in Germany (Berdrow, 1907; Dziock, 2002), Italy (Goidànich, 1928), and Colorado (Schreiner, 2019). *M. mellinum* is distributed across Eurasia and North America, with grasslands its preferred habitat. It too has attracted attention as a biocontrol agent.

Gardeners and farmers attract and retain egg-laying syrphid females by creating florally rich habitats along field margins and hedgerows. Yarrow, parsley, chamomile, and alyssum spp. are excellent for this purpose.

20.6   Biocontrol list

We provide an alphabetical list of biocontrol organisms mentioned in the text, with brief descriptions. Their commercially availability can fluctuate. Some require permits for shipment to certain areas (like Hawai'i).

For updates on commercially availability go to producers' websites. Three helpful to us are: Koppert (www.koppert. com); Biobee USA (https://biobee.us); and Biobest (www. biobest.be). Or go to distributors websites, such as Applied Bionomics (www.appliedbio-nomics.com/), Arabico (www. arbico-organics.com/), Beneficial Insectary (https://green-methods.com/), Gardener's Supply (www.gardeners.com/), IPM Labs (www.ipmlabs.com/), Nature's Control (www. naturescontrol.com/), Natural Enemies (www.naturalenemies. com), and Rincon-Vitova Insectaries (www.rinconvitova.com).

Besides producers and distributors, there are many retail suppliers. A directory is provided by the Bio-Integral Resource Center (www.birc.org/). BIRC also publishes a journal, *IPM Practitioner.* They've been in the business since 1979. Biocontrol specialists have suggested that growers buy directly from an insectory and not through a distributor, to obtain the highest quality product.

For each commercially availability species we follow a format with the following abbreviations:

Supplied as	Suppl	A. All live stages in shaker bottles or sachets
		B. Mummies or pupae in shaker bottles, tubes, or cards
		C. Conidia or spores in water-dispersible granules or wettable powder
		D. Conidia or spores in liquids (emulsifiable or flowable concentrates)
Shelf life	SL	A. Release immediately
		B. Store maximum 1–2 days in a cool (12–14°C), dark place
		C. Store for weeks or months in a cool (8–10°C), dark place
Release rate	RR	Prevention release (PreR); light infestation release (LiR); heavier infestation release (HiR)
Compatibility	C	Compatible with (Cw); not compatible with (NCw)

Albifimbria verrucaria, formerly *Myrothecium verrucaria*, is a soil fungus that controls many nematodes. Its mycotoxins inhibit the hatching and development of nematode eggs, and paralyze muscles that control feeding and locomotion. *A. verrucaria* also produces enzymes—chitinase and β-1,3-glucanase—that kill insects (e.g., *Helicoverpa armigera*) and pathogenic fungi (e.g., *Sclerotium rolfsii*) (Chavan *et al.*, 2017). Natural populations of *A. verrucaria* were isolated from Ukrainian hemp plants (Hrebenyuk, 1984). **Suppl**: C (DiTera® G™) or D (DiTera® ES™) of a heat-killed *A. verrucaria* broth. **SL**: C. Incorporated into soil as a pre-plant mix, or injected into the root zone of established plants. Approved for use on *Cannabis* by Colorado DOA (2023).

Amblyseius swirskii is a mite that preys on thrips (*T. tabaci*, *F. occidentalis*, *Scirtothrips dorsalis*), whiteflies (*T. vaporariorum*, *B. tabaci*), and spider mites (Fig. 20.13). It is native to the eastern Mediterranean region, and does best at 25–29°C and 65% RH. Its life cycle spins weekly in optimal conditions; adults consume about ten thrips/day or 20 whitefly eggs/day. **Suppl**: A; **SL**: B; **PreR**: 25/m² weekly, **LiR**: 50/m² weekly, **HiR**: 100–300/m² weekly. **NCw**: *Neoseiulus cucumeris* (they eat each other). *A. swirskii* does not enter diapause under short daylengths. Its need for high humidity makes it suboptimal for cultivars susceptible to gray mold.

Ampelomyces quisqualis is a fungus that hyperparasitizes powdery mildew fungi. It infects hyphae, conidiophores, and cleistothecia. It reduces hyphal growth, decreases the production of spores, and may eventually kill the mildew colony. Parasitized mildew colonies become flattened and dull gray within a week of infection. In optimal conditions (15–25°C, moderate humidity), *A. quisqualis* produces pycnidia which ooze tiny cylindrical conidia (see Fig. 9.3C). **Suppl**: C (AQ-10®) or D (AQ-SF®). **SL**: C. Conidia sprayed on plants require free water to germinate. **Cw**: horticultural oil; *A. quisqualis* actually works better when mixed with oil and sprayed.

Anagyrus pseudococci is an encyrtid wasp that parasitizes mealybug larvae (*Planococcus* and *Pseudococcus* spp.). In optimal conditions (25°C, moderate humidity) the life cycle may take only 3 weeks, females lay up to 50 eggs. Parasitized mealybugs turn into swollen mummies. Wasps emerge from mummies via small, ragged holes. Females are brown, with black and white banded antennae, 1.5–2 mm long. Male wasps are smaller, without the distinctive antenna bands. **Suppl**: B; **SL**: B; **PreR**: 0.10/m² every 14 days; **LiR**: 0.25/m² weekly, **HiR**: 0.50/m² weekly; **Cw**: the predator *Cryptolaemus montrouzieri*.

Fig. 20.13. *Amblyseius swirskii* preying on thrips (courtesy Steven Arthurs)

Antonospora locustae, formerly *Nosema locustae*, is a microsporidian parasite that infects locusts and grasshoppers, especially members of the Acrididae family (*Melanoplus* and *Opeia* spp.). After eating *A. locustae*, the host stops feeding, its activity decreases, and it slowly dies (1st or 2nd instars die in 1–3 weeks, older insects take longer). *A. locustae* is effective for long-term suppression of populations, rather than stopping sudden outbreaks. Henry (1971) applied *A. locustae* spores to bait (wheat bran) which he spread across rangeland, and reduced grasshopper populations by 50% within 4 weeks. **Suppl**: bran bait; **SL**: C. Apply in the morning after the temperature reaches 16°C, and when no rain is forecast (rain or heavy dew makes the bran unpalatable).

Aphelinus abdominalis is a chalcid wasp that controls aphids. It is native to temperate Europe, and does best in moderate humidity and temperatures. Adults are 2.5–3 mm long, with a black thorax, yellow abdomen, and short antennae. Female wasps deposit eggs in adult aphids. Parasitized aphids die and turn into black mummies about 2 weeks after adults are released. Larvae pupate within mummies and emerge as wasps, leaving behind a ragged exit hole at the rear of the mummy. **Suppl**: A; **SL**: B; **PreR** or **LiR**: 50 wasps/infected plant; **Cw**: other parasitoids. *A. abdominalis* migrates poorly and should be released close to aphids. It can handle higher temperatures than *Aphidius* spp.

Aphidius colemani*, *A. matricariae*, and *A. ervi are braconid parasitoids with aphids as their hosts. Adults are tiny black wasps, 2 mm long (Fig. 20.14). Females lay around 150 eggs, one egg per aphid, in either adults or nymphs. Eggs hatch into larvae, feed within their hosts, and pupate there. Aphids swell and stiffen into papery, light-brown mummies. Emerging wasps leave small, round exit holes at the back of the abdomen (see Fig. 4.14). The life cycle takes 2–3 weeks in optimal temperatures, and adults live another 2 weeks.

Aphidius females track at least three scents: (E)-*β*-caryophyllene, (E)-*β*-farnesene, and the smell of aphid honeydew. The first two compounds are produced by *Cannabis*; Heuskin *et al.* (2012) encapsulated the scents into slow-release beads as attractants for *Aphidius* females. *Aphidius* spp. work best as preventives. They reproduce relatively slowly and cannot respond quickly to heavy infestations. Heavy honeydew hinders them. On the positive side, they do not diapause during flowering.

Suppl: B; **SL**: B; **PreR**: 0.25 pupa/m2 weekly, **LiR**: 1/m2 weekly for 3 weeks, **HiR**: 2/m2 weekly for 6 weeks. Release in shaded areas, from open boxes hanging in plants or on rockwool slabs. *A. colemani* and *A. matricariae* are marketed for control against "smaller" aphids, such as *Aphis gossypii* and *Myzus persicae*, whereas *A. ervi* controls larger *Macrosiphum euphorbiae* and *Aulacorthum solani*. All three reportedly attack *Phorodon cannabis*, and mixes of all three are commercially available. According to the Biobest website, *A. matricariae* is better adapted to lower temperatures. But *A. matricariae* naturally parasitizes *Phorodon cannabis* in warm places, such as India (Khan and Shah, 2017) and Pakistan (Starý *et al.*, 1998).

Aphidoletes aphidimyza is a cecidomyiid fly whose larvae prey on aphids. Adult flies search for aphid colonies and deposit eggs there; they look like tiny mosquitos, with a wingspan of 2.5 mm. They are night fliers, feed on aphid honeydew, and lay 100–250 eggs. Larvae are white upon hatching but turn orange as they mature, reaching 3 mm in length (Fig. 20.15). Each larva eats up to 65 aphids per day, and it kills more aphids than it can eat when aphid populations are high. It injects a paralyzing poison into aphids, allowing it to feed on hosts much larger than itself. The life cycle takes about 4 weeks. Native to northern Europe and North America, the species does best in moderate humidity and temperatures (70% RH, 18–27°C). *A. aphidimyza* does poorly in hot, dry, windy locations. Short photoperiods send maggots into diapause, ending their effectiveness when *Cannabis* flowers.

Suppl: B; **SL**: B; **LiR**: 1-2 pupae/m², **HiR**: 10 pupae/m² in heavy infestations, repeating every week. **Cw**: *Aphidius matricariae,* **NCw**: parasitic nematodes. The pupae do not need to be sprinkled from the bottle, just opening the bottle near aphid colonies is sufficient. *A. aphidimyza* larvae eat less than lacewing larvae, but they do not migrate away, and they work better in light infestations. Adult females require an established aphid population before they lay eggs. *A. aphidimyza* larvae need soil to pupate. In soilless hydroponic glasshouses with concrete or plastic-covered floors, sprinkle peat moss between rows of plants for pupation sites. *A. aphidimyza* adults are not attracted to bright light (unlike lacewings and lady beetles), so they won't fly into bulbs and windows.

Aphytis melinus is a chalcid wasp that feeds on hard scales (*Aonidiella, Aspidiotus, Pseudaulacaspis* spp.). It is native to

Fig. 20.14. *Aphidius ervi* probing *Phorodon cannabis* (courtesy Melissa Schreiner)

Fig. 20.15. *Aphidoletes aphidimyza* larva feeding on the *Phorodon cannabis* (courtesy Melissa Schreiner)

India and Pakistan and tolerates a wide range of temperatures (13–33°C) and moderate humidity (40–60% RH). Females are golden-brown wasps, 1.2 mm long, and lay eggs in about 30 scales. The life cycle takes 2–3 weeks in optimal conditions, then adults live another 3 weeks. **Suppl**: A; **SL**: A; **LiR**: 24-48 wasps/m² every week.

Assassin bugs are predatory Hemipterans; at least 170 species are known to prey on insect pests. They are big—adults may reach 40 mm in length. Their size enables them to feed upon a wide variety of prey. Ambrose (2003) detailed their excellent search efficiency and fecundity. Assassin bugs are not commercially available. Protect them. Assassin bugs are recognizable by their elongated head with a distinct narrowed "neck," long legs, and a long, tubular proboscis. Nymphs are covered with short spines which aid in holding onto their prey. The nymph in Fig. 20.16 is on a *Cannabis* branch with the head of a hemp borer impaled on its proboscis. It is likely *Sinea diadema* (Fabricius) 1776, which naturally breeds in hemp crops (Schreiner, 2019).

Bacillus cereus is a gram-positive bacterium, classified as a plant growth-promoting rhizobacterium (PGPR), discussed

Fig. 20.16. Assassin bug nymph (M. Bolt)

in section 13.3. *B. cereus* was the first PGPR to be commercialized, back in 1897. Comparisons show that *B. cereus* is inferior to *Bacillus subtilis* (García-Gutiérrez *et al.*, 2012; Almaghrabi *et al.*, 2013), and few *B. cereus* products are currently available. Some products combine *B. subtilis* and *B. cereus*, such as BioStart™ Defensor, a soil inoculant and foliar spray. **Suppl**: C and D; **SL**: C. The long shelf life of spore-forming *Bacillus* spp. is an advantage over *Pseudomonas*-based biocontrol agents.

Bacillus mycoides (isolate J) colonizes the phyllosphere and induces host resistance against bacterial and fungal pathogens. Isolate J was originally isolated from healthy sugar beet foliage in a field with a severe *Cercospora* outbreak (Jacobsen *et al.*, 2004). It is marketed as LifeGard® and shows efficacy against several *Cannabis* pathogens (see Table 9.1).

Bacillus megaterium is a PGPR that controls soil-borne fungal pathogens. Dumigan and Deyholos (2022) isolated it from *Cannabis* seeds and seedlings. Serkov and Pluzhnikov (2012) applied *B. megaterium* as a seed treatment to protect 'Surskaya', a fiber-type cultivar. The test soil contained *Fusarium oxysporum*, *Botrytis cinerea*, *Verticillium albo-atrum*, and *Pythium* sp. Seeds were soaked in *B. megaterium* solution (Albit TPS®) for 3 h. Seed treatment increased the germination of hemp seeds by 18%, and decreased root rot by 7.1%. *B. megaterium* treatment achieved a better seed germination rate than treatment with fungicides (benomyl and thiram).

Bacillus subtilis species complex refers to several PGPR spp. that impact pathogens by competing for space on plant surfaces, and by oozing antifungal metabolites. They indirectly impact pathogens by promoting root growth ("disease escape"), and by inducing systemic resistance in plants. Although the term "PGPR" was coined for *Pseudomonas* spp. (Kloepper *et al.*, 1980), the PGPR concept was discovered by Kloepper's older colleagues at Berkeley working with *B. subtilis* (Weinhold and Bowman, 1968).

The number of species allocated to this species complex has waxed and waned. In 1974, the species complex consisted of *Bacillus subtilis*, *Bacillus licheniformis*, and *Bacillus pumilus* (Fritze, 2004). Since then, genetic studies using 16S rRNA sequences have uncovered many "cryptic" species. These include

Bacillus amyloliquefaciens (named in 1987), *Bacillus atrophaeus* in 1989, *Bacillus mojavensis* in 1994, *Bacillus vallismortis* in 1999, *Bacillus spizizenii* in 2001, *Bacillus sonorensis* in 2001, and *Bacillus velezensis* in 2005. Two members of the species complex naturally occur in *Cannabis* seeds and seedlings, *B. subtilis* and *B. velezensis* (Dumigan and Deyholos, 2022).

Cawoy *et al.* (2011) listed 20 products, their species names, and target pathogens. Perhaps best-known is Serenade® (*B. subtilis*). It can be applied to soil for control of *Pythium*, *Rhizoctonia*, or *Fusarium* damping off. It can be sprayed on foliage to control powdery mildew, downy mildew, gray mold, and leaf spots caused by *Cercospora*, *Colletotrichum*, and *Xanthomonas* spp. Similar indications are given for other *B. subtilis* products (Ballad®, Cease™, Companion®, Rhapsody®, Subtilex®), as well as *B. pumilus* products (Sonata®), *B. amyloliquefaciens* products (Double Nickel 55®) and a mix of *B. amyloliquefaciens* and *B. subtilis* (Microflora™).

Some products specifically target root rot by *Pythium*, *Fusarium*, and *Rhizoctonia* spp., such as Kodiak® (*B. subtilis*), Yield Sheild® (*B. pumilus*), and EcoGuard™ (*B. licheniformis*). Botrybel® (*B. velezensis*) specifically targets gray mold caused by *Botrytis cinerea*. According to Fan *et al.* (2017), all of these products should be labeled *B. velezensis*. The authors state that the best products contain both living *Bacillus* and their antimicrobial metabolites (e.g., Serenade®, Double Nickel 55®). But even in these products, concentrations of the metabolites are not indicated. Balthazar *et al.* (2022b) screened several *B. velezensis* isolates against *Botrytis cinerea* on *Cannabis* leaves, and they performed well (see Fig. 13.6).

Bacillus spp. also control nematodes. Abd-Elgawad and Askary (2018) listed 14 products, but many—made in Egypt, India, or China—are not readily available. Well-known products include BioStart™ Defensor (*B. subtilis* and *B. cereus*), NemaControl® (*B. amyloliquefaciens*), Quartzo® and Nemix C® (*B. subtilis* and *B. licheniformis*), and BioStart™ RhizoBoost (*B. licheniformis*, *B. laterosporus*, *B. chitinosporus*).

Bacillus thuringiensis, "Bt," has a long history (section 20.2). Bt produces a spore toxin, which becomes toxic when it is cleaved by a stomach enzyme. After eating Bt, insects stop feeding within the hour. A foul-smelling fluid—liquified digestive organs—trickles from their mouths and anuses. They shrivel, blacken and die in several days (Fig. 20.17).

Bt products do not contain living bacteria. Most products contain the toxin and spore, but some contain only the toxin. **Suppl**: C or D; **SL**: C (up to a year); **Cw**: most beneficial insects. Bt toxin is harmless to people, although some individuals may develop allergic reactions. Immediately before spraying, mix with water that is slightly acidic (pH 4–6.5). For better results, add a spreader-sticker and UV inhibitor. For best results add a caterpillar feeding stimulant, such as Entice™ or Pheast™. Spray all foliage uniformly and completely. Bt on foliage is degraded by UV light within 1–3 days, so spray outdoors in late afternoon or on cloudy (not rainy) days.

B.t. var. *kurstaki* kills Lepidopterans, such as ECBs, hemp borers, and budworms. It works best against young, small larvae—spray at the first sign of caterpillar damage. Spraying Bt every 10–14 days prevents infestations. Surface sprays will not kill borers already inside stalks. Bacterial genes encoding *B.t. kurstaki* Cry1A toxins have been genetically engineered

Fig. 20.17. European corn borer larvae killed and discolored by *Bacillus thuringiensis* (courtesy Koppert Biocontrols)

into maize and cotton. The transgenic crops were released in 1996, and McPartland *et al.* (2000) predicted that organic farmers, who have relied on Bt for decades, may lose their best weapon against caterpillars. This has come to pass: Britt *et al.* (2021a) sprayed corn earworms (*Helicoverpa zea*) infesting hemp with several *B.t. kurstaki* products. They were only moderately effective.

B.t. **var.** *aizawai* is effective against Lepidopterans, such as armyworms, budworms, and some borers. It contains different endotoxins than *B.t. kurstaki*, which have not been bioengineered into crop plants. Britt *et al.* (2021a) showed that *B.t. aizawai* worked better than *B.t. kurstaki* against corn earworms in controlled field trials. Several products combine toxins from *B.t. aizawai* and *B.t. kurstaki*.

B.t. **var.** *israelensis* selectively kills Dipterans in the suborder Nematocera. *B.t. israelensis* products have been marketed for control of mosquitoes, fungus gnats, blackflies, and crane flies (Ben-Dov, 2014).

B.t. **var.** *tenebrionis* was isolated from the beetle *Tenebrio molitor* (Krieg *et al.*, 1983). Krieg had a research contract with Boehringer Mannheim GmbH, whose lawyers convinced the publishers of *Zeitschrift für Angewandte Entomologie* to delay Krieg's publication until a patent was filed (Charles, 2002). In the late 1990s there were seven *B.t. tenebrionis* products commercially available, but only Trident® and Novodor® are available today. The *B.t. tenebrionis* delta endotoxin is only effective against the youngest 1st and 2nd larval instars. Susceptibility to *B.t. tenebrionis* varies greatly in beetles, presumably because of variation in their gut wall, where the *B.t. tenebrionis* toxin must attach.

B.t. **var.** *san diego* has an interesting history. A new biotech corporation, Mycogen of San Diego, was flush with venture capital and looking for new genes to patent. Mycogen contacted Boehringer to obtain a license to probe *B.t. tenebrionis*, but couldn't agree on terms. A Mycogen scientist, Corinna Herrnstadt, visited Aloysius Krieg, the discoverer of *B.t. tenebrionis*. He recalled, "I showed her everything, I wanted to help her so she could get ahead" (Charles, 2002). Two years later, Herrnstadt *et al.* (1986) announced discovery of *B.t. san diego*, from an undisclosed source of isolation. Krieg cried foul. His team compared *B.t. tenebrionis* and *B.t. san diego* biochemically,

serologically, and genetically, and rejected *B.t. san diego* as a "novel organism" (Krieg *et al.*, 1987). Boehringer sued Mycogen. Testimony revealed that Herrnstadt slid her hand along the surface of tables in Krieg's laboratory, then wiped her fingers on a petri dish hidden in her purse. Mycogen agreed to pay Novo Nordisk (the new owner of *B.t. tenebrionis*) US$4 million, and disclaim its patents on *B.t. san diego* (Charles, 2002).

Beauveria bassiana is a fungus that kills sap-sucking pests (whiteflies, aphids, thrips, planthoppers, bugs, mites) as well as chewing insects (grasshoppers, beetles, termites, ants, European corn borers). Under optimal conditions, insect cadavers sprout more spores, which cause secondary infections. *B. bassiana* is approved for use on *Cannabis* by several US states. California does not specify a *B. bassiana* strain. Washington and Colorado specify strain GHA (e.g., Botanigard® MAXX, with pyrethrin), Oregon specifies two strains: GHA (e.g., Botanigard® ES, Mycotrol® WPO) and PPRI 5339 (*e.g.*, Velifer®). Health Canada specifies two strains: GHA (see above) and ANT-03 (e.g., BioCeres®).

Suppl: D; **SL**: C; **Cw**: parasitic wasps (e.g., *Aphidius matricariae*) and Bt—which works best against young larvae, whereas *B. bassiana* kills older larvae. **NCw**: lady beetles and green lacewings. Mix spores with cool water and a wetting agent, then spray on all surfaces of pest-infested plants. The spores can also be poured into soil. Spray around sunset—dew and darkness facilitate spore germination, and UV light kills them. Being a fungus, *B. bassiana* does best at moderate temperatures (18–29°C) and high humidity (≥60% RH). Thus *B. bassiana* cannot be used when plants are flowering and humidity renders them susceptible to gray mold. *B. bassiana* is not as deadly as *Lecanicillium* spp., nor is it as selective.

Beauveria brongniartii (=*B. tenella*) is marketed for control of the European cockchafer, *Melolontha melolontha*. Products include Betel (France), Engerlingspilz (Switzerland), Schweizer Beauveria (Switzerland), and Melocont (Austria).

Burkholderia vietnamiensis is another PGPR soil bacterium, sold as Botrycid® and marketed for control of fungi (*Botrytis cinerea*, *Rhizoctonia*, *Thielaviopsis*, *Verticillium*, *Fusarium*, *Pythium*), as well as *Xanthamonas* bacteria and nematodes. *B. vietnamiensis* belongs to the *Burkholderia cepacia* species complex. *B. cepacia* was registered as a biopesticide between 1992 and 1998 (Deny®, Blue Circle®, and Intercept®). Infections of *B. cepacia* in people with cystic fibrosis led Holmes *et al.* (1998) to call for a moratorium on *B. cepacia* agricultural use, and those products were pulled. Cawoy *et al.* (2011) referred to Botrycid® as *B. cepacia*, not *B. vietnamiensis*.

***Burkholderia* spp. strain A396**, also known as *Burkholderia rinojensis*, is a bacterium isolated from Japanese soil, with strong insecticidal and miticidal activities. **Suppl**: D; **SL**: C; **NCw**: honey bees, possibly predatory beetles and lacewing larvae. Commercial products (e.g., Venerate®) contain heat-killed bacteria (60°C for 2 h) and their thermostable toxins—so it's more akin to chemical control than biological control. The toxins cause exoskeleton degradation, and interfere with molting and pupation. Cordova-Kreylos *et al.* (2013) demonstrated efficacy against an insect (*Spodoptera exigua*) and a mite (*Tetranychus urticae*). California DPR (2017) approved its use against mites, leafhoppers, aphids, whiteflies, thrips, and moth larvae infesting *Cannabis*.

Chromobacterium subtsugae is a bacterium found to be orally toxic to foliar-feeding insects (Martin *et al.*, 2007). In other insects and mites it reduced feeding or reduced fecundity. **Suppl**: D; **SL**: C; **NCw**: honey bees and predatory beetles. Commercial products (e.g., Grandevo®) contain heat-killed bacteria with spent fermentation medium solids—so it's more akin to chemical control than biological control. It is marketed against armyworms, leaf beetles, aphids, whiteflies, thrips, and mites. Britt and Kuhar (2020) showed good efficacy against hemp russet mites (see Fig. 4.7). Pace (2019) found poor efficacy against root aphids. Grandevo® is approved for use on *Cannabis* in many jurisdictions.

Chrysoperla carnea is the common green lacewing. Adults feed on nectar, pollen, and aphid honeydew. Larvae are voracious predators of aphids, whitefly nymphs, mealybug larvae, thrips, spider mites, and the eggs of leafhoppers, leaf miners, and budworms. *C. carnea* does best at moderate temperatures (24–27°C) and moderate humidity (55% RH), but adapts to a wide range of humidity (35–75% RH). Adults have slender pale green bodies 12–20 mm long, with delicate, lacy, light-green wings and bright eyes. They can turn grayish-brown during the fall when *Cannabis* is flowering. Larvae ("aphid lions") have yellowish-gray bodies with brown marks, tufts of hair, and large jaws. They look like tiny alligators, 2–10 mm long (Fig. 20.18). Eggs are oval, white or light green. They are suspended from undersides of leaves by slender threads 1 cm long (see Fig. 20.2C).

C. carnea overwinters as pupae or adults. Adults have strong migratory instincts and may fly for 3–4 h before settling down on plants. Females lay up to 400 eggs. Larvae consume 300–400 aphids during their 2-week larval period. In optimal conditions the life cycle takes 25 days, then adults live another 40 days. Short days induce diapause in some populations of *C. carnea*, rendering them ineffective in flowering crops.

Suppl: A (larvae); **SL**: B; **Cw**: Bt, other *Chrysoperla* spp., lady beetles, *Trichogramma* wasps, and predatory mites, although some of these biocontrols may be eaten by lacewing larvae in the absence of aphids. Recommendations for release rates are presented in Table 20.2. Place larvae at the base of plants, or in distribution boxes hanging from plants. Spread them out to avoid cannibalism. Some suppliers use buckwheat as a carrier to help prevent cannibalism. Do not irrigate plants for a couple days, to avoid washing lacewings away (misting is okay and may help young larvae survive). Entice adults to stay and lay eggs by providing artificial honeydew—mix honey and brewer's yeast on sticks or wax paper and distribute among plants (replace when moldy); commercial products include Wheast™ or Biodiet™.

Chrysoperla rufilabris is difficult to tell apart from *Chrysoperla carnea* (see photos in Tauber *et al.*, 2000). Their life cycles are similar, and they eat the same prey. *C. rufilabris* performs better than *C. carnae* in humid environments (≥75% RH), may be a more aggressive predator than *C. carnae*, produces more eggs per female, may be found in higher numbers towards the end of the season, and it lacks the strong migratory wanderlust of *C. carnea*, so it stays in release areas longer (Hoffmann and Frodsham, 1993; Tauber *et al.*, 2000). *C. rufilabris* is supplied like *C. carnea*, and released at the same rates.

Clonostachys rosea is a soil fungus that kills other soil fungi as well as foliar pathogens. It consists of two subspecies, *C. rosea* f. *rosea* and *C. rosea* f. *catenulatum* (Schroers, 2001), formerly called called *Gliocladium roseum* and *Gliocladium catenulatum*, each with demonstrated efficacy against fungal pathogens, by Moody and Gindrat (1977) and Huang (1978), respectively. *G. catenulatum* strain J1446 is commercially available as PreStop® WG or Lalstop® G46. **Suppl**: C; **SL**: C; applied as a foliar spray, soil drench, and hydroponic solution, or incorporated into potting media and drip irrigation.

If there's such a thing as one-stop shopping in fungal biocontrol, this is it. PreStop/Lalstop is EPA-labeled for seed-borne and soil-borne diseases (damping-off, root and stem rot, and wilt), caused by *Botrytis*, *Sclerotinia*, *Fusarium*, *Verticillium*, *Cladosporium*, *Colletotrichum*, *Mycosphaerella*, *Phytophthora*,

Fig. 20.18. *Chrysoperla carnea* larvae feeding on hemp aphids (R. Clarke)

Table 20.2. *C. carnea* release rates for control of aphids

ISI[a]	Number of predators released per plant or m² of glasshouse crop[b]
Preventive	5 larvae/m² every 3 weeks
Light	5 larvae/plant OR 10/m² every 2 weeks
Moderate	25 larvae/m² on trouble spots, 10/m² elsewhere every week
Heavy	50 larvae/m² on trouble spots, 25/m² elsewhere every week
Critical	100 larvae/m² on trouble spots, 50/m² elsewhere every week

[a]ISI = Infestation Severity Index for aphids, see Table 4.2.
[b]Control of aphids with *C. carnea* is difficult during flowering (light cycles below 12 h/day)

and *Pythium*. Foliar applications control *Botrytis*, *Alternaria*, *Colletotrichum*, *Bipolaris*, *Cladosporium*, *Didymella*, *Mycosphaerella*, and "also suppresses powdery mildew symptoms." The manufacturer's website claims that "university tests showed superior to Rootshield, PlantShield, Actinovate, Cease, and Serenade." PreStop® proved superior to other biocontrols at controlling *Fusarium oxysporum* and *Pythium myriotylum* in *Cannabis* seedlings (Scott and Punja, 2023).

Cordyceps fumosorosea, formerly *Isaria fumosoroseus* (and before that *Paecilomyces fumosoroseus*), is an entomopathogenic fungus. It parasitizes whiteflies, with reported success against spider mites, aphids, mealybugs, thrips, caterpillars, rootworms, and beetle grubs—both foliar and soil-dwelling pests. *C. fumosorosea* is a species complex: Fargues *et al.* (2002) sequenced 48 samples of *C. fumosorosea* collected from insects worldwide, which split into three monophyletic groups. Samples from two groups were isolated from whiteflies. Samples from the third group were isolated from the aforementioned wide host range. A commercially available strain, Apopka 97 (PFR 97®, Preferal®, Ancora™), fell into the third group. The study did not include another commercially available strain, FE 9901 (Isarid™, NoFly®).

Suppl: C; **SL**: C; **Cw**: other fungi and beneficial parasitoid wasps. Generalist predators such as lady beetles are likely infected by the fungus. The NoFly® manufacturer claims no adverse effects on honey bees, mirid bugs (*Macrolophus pygmaeus*), pirate bugs (*Orius laevigatus*), and the mites *Phytoseiulus persimilis* and *Amblyseius swiskii* (Lara and Fernández, 2014).

Commercial products are supplied as conidia (produced in solid culture) or yeast-like blastospores (produced in liquid fermentation tanks). Blastospores germinate faster than conidia, so they are preferable. Many product labels don't specify conidia or blastospores—they list "CFUs" (colony-forming units). The product needs to be well-mixed in water for 30 min and periodically agitated, to ensure a well-dispersed suspension. Spores can be sprayed directly on plants, applied as a soil drench, or added to irrigation systems. Foliar pests require good underside leaf coverage. UV light exposure will decrease spore survival, so spray in the evening.

Spores germinate best at moderate temperatures (18–28°C) and high humidity (≥80% RH). These conditions preclude usage in *Cannabis* varieties predisposed to gray mold in high humidity. The fungus directly penetrates the insect cuticle, and kills the host in 4–5 days. In humid conditions, dead insects sprout more spores, which spread secondary infections.

Cryptolaemus montrouzieri is a lady beetle

known as the mealybug destroyer. The beetles also eat aphids, but do not reproduce on them (they lay eggs in *egg masses*, and aphids give birth to *live larvae*). *C. montrouzieri* does best at 27°C (range 17–32°C) and 70–80% RH. Adults are 3–4 mm long, with red-orange heads and thoraxes, and shiny, black-brown wing covers (Fig. 20.19A). Larvae are long (up to 13 mm) and resemble disheveled mealybugs (Fig. 20.19B).

Adults feed on mealybug eggs and young larvae; females lay up to 500 eggs, one per host egg mass. Young larvae eat mealybug eggs and young larvae; older larvae eat all mealybug stages. The larvae eat up to 250 mealybugs and eggs, then pupate in sheltered places on plant stems or greenhouse structures. In optimal conditions the life cycle takes 29–38 days, then adults live another 50–60 days.

C. montrouzieri works best in heavy infestations and tends to fly away before its job is done, like most lady beetles. **Suppl**: A (adults); **SL**: B; **HiR**: 20–50/m² or 5/plant; **Cw**: the parasitoids *Leptomastix dactylopii* and *Anagyrus pseudococci*. Release in the early morning or late at night to decrease dispersal. Exercise caution when using white sticky traps—the beetles are attracted to white. The adults have a bad habit of flying into lights.

Dacnusa sibirica is a braconid wasp that parasitizes

Liriomyza larvae. This Siberian native does best in cool to moderate temperatures (optimally 14–24°C) and moderate humidity (70% RH). Adult wasps are dark brown-black, 2–3 mm in length, and have long antennae (Fig. 20.20). Wasps lay eggs in all ages of leafminer larvae, but prefer 1st and 2nd instars. Larvae develop and pupate within leafminer larvae. The life cycle takes 16 days under optimal conditions; adults live

Fig. 20.19. *Cryptolaemus montrouzieri*: **A**. adult (Koppert Biocontrols); **B**. larva (W. Cranshaw)

Fig. 20.20. *Dacnusa sibirica* ovipositing in a leafminer larva (courtesy Koppert Biocontrols)

another 7 days, females lay 60–90 eggs. **Suppl**: A (adults); **SL**: B; **PreR**: 2/m² weekly, **LiR**: 5/m² weekly; **Cw**: *Diglyphus isaea*, which may be more effective than *D. sibirica*. Exercise caution with yellow sticky traps, the wasps may be attracted to yellow.

Delphastus pusillus is a lady beetle that preys on whiteflies—*Bemisia tabaci*, *B. argentifolii*, and *Trialeurodes vaporarium*. Lacking whiteflies, it will eat spider mites and baby aphids. This Florida native does best in moderate humidity and moderate to warm temperatures (19–32°C). Adults are shiny black, small, 1.3–2.0 mm long, with brown heads (Fig. 20.21). Larvae are pale yellowish-white, 3 mm long. Females lay up to 75 eggs on leaves alongside whitefly eggs. The life cycle takes 21 days in optimal conditions, then adults live another 5–9 weeks. Both adults and larvae consume whitefly eggs and larvae. Adults are voracious—during their lifetime, they devour as many as 10,000 whitefly eggs or 700 larvae.

Suppl: A (adults); **SL**: B; **LiR**: 1-2 beetles/m² weekly, **HiR**: 4/m² weekly; **Cw**: green lacewings; the beetles avoid eating *Encarsia-* or *Eretmocerus*-parasitized whiteflies, making these biocontrols compatible. *D. pusillus* works best in moderate outbreaks. At low whitefly densities the beetles disperse. In heavy infestations *D. pusillus* bogs down in heavy honeydew. It moves poorly across hairy tomato leaves—and may find hairy *Cannabis* leaves difficult.

Dicyphus hesperus is a North American mirid that preys on whiteflies—*Bemisia tabaci* and *Trialeurodes vaporarium*. In the absence of whiteflies, it eats aphids, spider mites, moth eggs, and thrips. Adults are black and green with red eyes, an elongate shape, 6 mm long, with long antennae and legs. Both adult bugs and nymphs are predators. **Suppl**: A; **SL**: B; **LiR**: 0.25–0.5 bugs/m² weekly, **HiR**: 2/m² weekly. In the absence of prey *D. hesperus* may feed on crop plants. *D. hesperus* should be combined with other biocontrols, such as *Eretmocerus eremicus*. Development time from egg to adult takes 5 weeks in optimal conditions of 25°C and RH ≤50%.

Diglyphus isaea is a chalcid wasp that controls leafminers. It is native to northern Eurasia and does best in moderate humidity and temperatures (80% RH, 24–32°C). Adults are black with a metallic greenish tinge, 2–4 mm long, with short antennae (Fig. 20.22). The female drums her antennae on the leaf mine to locate the miner, and kills it in the process of inserting an egg. Larvae feed off the carcass and pupate there. The life cycle takes 17 days under optimal conditions; adults live another 3 weeks, females kill 200–300 leafminers. **Suppl**: A (adults); **SL**: A; **LiR**: 1 wasp/m² weekly, **HiR**: 5/m² weekly; **Cw**: *Dacnusa sibirica* and beneficial nematodes. Release in the early morning or evening. *D. isaea* prefers warmer temperatures than *D. sibirica*, and works faster and more effectively than *D. sibirica*. Exercise caution with yellow sticky traps.

Encarsia formosa, a chalcid wasp, slayer of whiteflies, caught the attention of Maxwell-Lefroy (1915), "Up to the present, no parasite destructive of this pest [*T. vaporariorum*] has been known in Great Britain, and we are anxious to establish this one in as many places as possible." Maxwell-Lefroy was killed in a pesticide mishap, and the distribution of *E. formosa* did not restart until after his tragic death (Speyer, 1927). *E. formosa* parasitizes *T. vaporariorum* best, less so *B. tabaci* and *B. argentifolii*. The species occurs throughout the northern hemisphere in temperate regions, but was almost wiped out by DDT.

Fig. 20.21. *Delphastus pusillus* feeding on whitefly larvae (courtesy USDA)

Fig. 20.22. *Diglyphus isaea* searching for leafminers (courtesy Koppert Biocontrols)

E. formosa reproduces asexually via parthenogenesis and males are rarely seen. Females deposit eggs individually inside the bodies of whitefly larvae (2nd–4th instars) and pupae. The larvae develop in about 2 weeks at optimum temperatures. *T. vaporariorum* larvae and pupae—normally white—turn black when parasitized, making them easy to spot (Fig. 20.23B). Parasitized *B. tabaci* turns amber-brown. Adult wasps emerge from carcasses, leaving behind an exit hole. They live 30 days and lay 50–100 eggs. *E. Formosa* does best in bright light, 50–80% RH, and temperatures of 18–28°C (optimal 24°C).

Suppl: A (pupae in whitefly larvae attached to cards); **SL**: B; **Cw**: green lacewings (*Chrysoperla carnea*), *Delphastus pusillus*, and *Eretmocerus eremicus*. Some products sell *E. formosa* and *E. eremicus* together. Release rates are presented in Table 20.3. Hang cards in shade near the bottoms of plants. *E. formosa* is not effective in heavy infestations and works best as a preventive. The percentage of parasitized larvae can be monitored easily (black vs white) and should be ≥80% for effective control. Dense leaf hairs covering *Cannabis* may interfere with the wasp's ability to locate prey. This problem also arises on cucumbers. *E. formosa* has a bad habit of flying into lights, which

Fig. 20.23. *Encarsia formosa*: **A**. adult female (courtesy Koppert Biocontrols); **B**. parasitized larvae and pupae of *T. vaporariorum* (courtesy Goldlocki, Wikimedia Commons)

Table 20.3. *E. formosa* release rate for control of glasshouse whiteflies

ISI[a]	Number of predators released per plant or m^2 of glasshouse crop[b]
Preventive	1–5 pupae per plant or 10/m^2 every 2 weeks
Light	20/m^2 every 2 weeks until 80% of whitefly larvae have turned black
Moderate	30/m^2 on trouble spots, 20/m^2 elsewhere, then 10/m^2 every 2 weeks
Heavy	50/m^2 on trouble spots, 30/m^2 elsewhere, then 20/m^2 every 2 weeks

[a]ISI = Infestation Severity Index of greenhouse whitefly, see Table 4.4.

limits its use in growrooms. Yellow sticky traps may attract the wasps. Most *E. formosa* strains tolerate some pesticides (see Table 20.1).

Entomophthoraceae is a family of fungi (*Entomophthora, Entomophaga, Zoophthora, Erynia, Pandora, Batkoa* spp.) comprising *ca.* 180 species. They are exclusively entomopathogenic, having coevolved with insects for 225 to 387 million years (Gryganskyi *et al.*, 2012). Along the way they picked up striking reproductive strategies. Some species exert mind-control over their victims (see Fig. 20.7). Others forcibly discharge ballistospores from the cadaver, surrounding it with a halo of sticky spores (Fig. 20.24). *Entomophthoraceae* specialize in killing specific lineages of insect—aphids, whiteflies, thrips, flies, grasshoppers, or true bugs.

At least two species infest the aphid *Phorodon cannabis*: *Entomophthora planchoniana* (Mattirolo, 1898; Barta and Cagáň, 2006) and *Pandora neoaphidis* (Keller, 2006). Considering *P. cannabis* is *Cannabis*-specific, this association of *Cannabis*, *P. cannabis*, and the two fungi is an ancient relationship.

Fig. 20.24. Housefly killed by *Entomophthora muscae*

Unfortunately, *Entomophthoraceae* are not easily cultured or stored in a viable state, making mass production difficult. Instead, Messelink *et al.* (2014) used a banker plant system (see Fig. 20.10). Grain aphids (*Sitobion avenae*—not a *Cannabis* pest) naturally infected with *P. neoaphidis* were maintained on banker plants (oats), thereby inducing epizootics in other aphids in the glasshouse.

Eretmocerus eremicus is a chalcid wasp that parasitizes whiteflies—*Bemisia tabaci, B. argentifolii*, and *Trialeurodes vaporarium*. The wasps are tiny (1 mm long), with yellow bodies, green eyes, and clubbed antennae. Females lay a solitary egg under the whitefly nymph. The hatching larva chews into its host, then lays dormant, until the whitefly pupates. At that point the larva releases digestive enzymes and ingests liquified body parts. Pupae turn from white to beige-yellow, and adults emerge in about 12 days. Adult females live for 6–12 days at 27°C, laying 3–5 eggs per day.

Suppl: B (pupae attached to cardboard strips, or in shaker bottles); **SL**: B; release rates same as *Encarsia formosa*; **Cw**: same as *E. formosa*. Optimal conditions are 23–35°C with ≤60% RH. *E. eremicus* handles temperatures as high as 45°C, much higher than *Encarsia formosa*. Yellow sticky traps should be removed prior to releasing the wasps. Female wasps also kill whitefly nymphs by stabbing them with ovipositors, and feeding on hemolymph that exudes from the wound.

Feltiella acarisuga is a cecidomyiid fly whose larvae prey on spider mites. The adults search for spider mite colonies and deposit eggs there. Each larva consumes 15 adult mites or 80 eggs per day. Adults look like mosquitos, with a wingspan of 1.6–3.2 mm. Larvae are maggot-like, yellow to brick-red (color depends on body contents), up to 2.0 mm long (Fig. 20.25). They spin a white cocoon and pupate on undersides of leaves. *F. acarisuga* occurs worldwide, under optimal conditions of 80% RH and 20–27°C. The life cycle takes about 2–4 weeks. **Suppl**: A (pupae); **SL**: B; **LiR**: 0.25/m^2 weekly; **HiR**: 10/m^2 every weekly; **Cw**: *Phytoseiulus persimilis*. The adults fly well and can find spider mites overlooked by growers. *F. acarisuga* works well in combination with *P. persimilis*. It reproduces more slowly than *P. persimilis*, but eats more, and does not diapause—a welcome advantage in flowering crops.

Galendromus occidentalis, a phytoseiid mite, is sometimes classified in either *Metaseiulus* or *Typhlodromus*. It preys on spider mites. Native to western North America, it tolerates a range of humidity (40–80% RH) and temperatures

Fig. 20.25. *Feltiella acarisuga* larva and *Tetranychus urticae* (courtesy Koppert Biocontrols)

(26–35°C). *G. occidentalis* resembles *Neoseiulus fallacis*. It is a small predator, so some spider mite adults may be too big for it. The life cycle takes 7–14 days, depending on temperature. Adults eat 1–3 spider mites per day or 6 mite eggs per day. *G. occidentalis* goes into diapause with short days and cold temperatures, so its effectiveness ends when *Cannabis* begins flowering. **Suppl**: A; **SL**: B; **PreR** and **LiR**: 50/m² every 2 weeks. *G. occidentalis* eats less than *Phytoseiulus persimilis* and reproduces more slowly, but lives longer without food, disperses more rapidly (including aerial dispersal), and tolerates semiarid conditions (it does not tolerate high humidity).

Geocoris punctipes, the "big-eyed bug," eats any host smaller than itself, including whiteflies, aphids, spider mites, eggs and nymphs of plant bugs (e.g., *Lygus* spp.), and eggs and young larvae of lepidopterans. Adults have large, bulging eyes and somewhat flattened brown bodies, covered with black specks, 3–4 mm long (Fig. 20.26). Nymphs look similar but without wings. Under optimal conditions (moderate humidity and temperatures), the life cycle takes about 30 days. In shaker bottles of 500 nymphs when commercially available. In areas where it occurs naturally (southern USA and Mexico), *G. punctipes* benefits from conservation efforts (see section 20.3.3).

Heterorhabditis bacteriophora (=*H. heliothidis*) is a tiny nematode, 1–1.5 mm long, that parasitizes soil insects, including beetle grubs, caterpillars, root maggots, ants, and thrips. *Heterorhabditis* spp. and other nematodes (*Steinernema* spp.) kill pests more quickly than most other biological control agents. *H. bacteriophora* uses a "cruiser" foraging strategy: it is highly mobile, and find hosts by tracking CO_2 and excretory products. It penetrates hosts through natural openings (mouth, spiracles, anus). *Heterorhabditis* spp. can also penetrate the insect cuticle (body wall). Cuticular penetration is important regarding white root grubs, because they have sieve plates protecting their spiracles, and they defecate so often that most nematodes entering the anus are pushed back out.

Nematodes carry gram-negative anaerobic bacteria (*Photorhabdus* spp.), which do the actual killing. Once nematodes penetrate their hosts, they release toxins which inhibit the insect immune system, then they release bacteria. Bacteria liquefy host organs within 48 h. Bacteria also produce antibiotics, which prevent

Fig. 20.26. *Geocoris punctipes* feeding on whitefly larvae, *Bemisia tabaci* (USDA)

Fig. 20.27. *Heterorhabditis bacteriophora* emerging from cadaver of *Galleria mellonella* (courtesy Peggy Greb, Wikipedia Commons

the putrefaction of dead insects, allowing nematodes to use the insect's body to reproduce and multiply. A new generation of nematodes emerges from the carcass in 2 weeks (Fig. 20.27).

Suppl: 3rd-stage larvae in polyethylene sponge packs or bottles; **SL**: B (sealed containers remain viable for up to 3 months at 2–6°C). Gently stir contents of package into 5 liters of water (15–20°C), let stand for five 5 min, then transfer to a watering-can or spray tank. Some gel preparations must be mixed with warm water (<32°C) and allowed to dissolve for 30 min before they are released. Release within 3 h, otherwise the nematodes drown. Apply in early evening, to avoid direct sunlight. To apply with a sprayer, remove filters smaller than 500 µm, use a spray nozzle at least 0.5 mm wide, spray with a pressure under 8 Bars, i.e. under 116 lb per square inch (PSI), and agitate the spray tank frequently so nematodes remain suspended. The usual recommended dose is 50,000 nematodes per plant or 500,000–1,000,000/m². Water the soil before and after application (at least 0.64 cm of irrigation), or apply during a rainstorm, and keep the soil moist for at least 2 weeks after application. *H. bacteriophora* does best in soil temperatures of 15–32°C.

Cw: Bt, *Paenibacillus* (*Bacillus*) *popilliae*, and predatory mites (including *Stratiolaelaps scimitus*). Lacey and Georgis (2012) reported synergy between nematodes (*Heterorhabditis* and *Steinernema* spp.) and Bt, and between nematodes and the biocontrol fungus *Metarhizium anisopliae*. **NCw:** incompatible with any biocontrol that spends part of its life in the soil (e.g., pupating *Aphidoletes aphidimyza*). *Heterorhabditis* and *Steinernema* spp. have similar host ranges, and similar release rates. *Heterorhabditis* spp. live deeper in the soil (down to 15 cm) than *Steinernema* spp., and they have superior host-seeking abilities. *Heterorhabditis* spp. also have better host-penetrating abilities—*Steinernema* spp. cannot penetrate insect cuticles. On the other hand, *Steinernema* products store better than *Heterorhabditis* products, and they survive in soil for longer periods of time in the absence of hosts. *Heterorhabditis* spp. infect hosts from 7°C to 35°C, *Steinernema feltiae* can be infective from 2°C to 30°C, and *Steinernema carpocapsae* has a tighter range, 10–30°C (Lacey and Georgis, 2012).

Heterorhabditis megidis was discovered on an Ohio golf course parasitizing Japanese beetles. It is larger than *H. bacteriophora*, 2.0–3.6 mm in length. *H. megidis* is relatively cold-tolerant and active below 15°C. It has the same host range as *H. bacteriophora* (except for small insects like aphids and thrips), and similar release rates. In Europe *H. megidis* has been marketed for control of weevil grubs.

Hirsutella thompsonii is a fungus that controls spider mites. Conidia and mycelial fragments quickly germinate, infect mites, and kill them in 1–2 weeks. Mite cadavers sprout conidiophores and conidia in moderate to high humidity and warm temperatures. **Suppl:** C and D; **SL:** B; **LiR:** spray 2–4 lb/acre (2.3–4.6 kg/ha); **Cw:** insect biocontrols (reportedly). Commercial sources of *H. thompsonii* come and go. Mycar™ was discontinued in 1985 due to product iinstability (McCoy, 1996). Currently No-Mite™ is sold in Europe, and ATCC® 24874™ is available, permit pending.

Iphiseius degenerans is a predatory phytoseiiid mite. *I. degenerans* preys on the young larval stages of thrips (*T. tabaci* and *F. occidentalis*). In the absence of thrips, it eats spider mites and can survive on pollen. The species does well in low to moderate humidity (55–80% RH), and warm temperatures (21–32°C). Adults are dark reddish-brown, oval, 0.7 mm long, and mobile. Eggs are transparent and turn brown before hatching, laid along veins on undersides of leaves. The life cycle takes 10 days, then adults live another month. Adults consume 4–5 thrips/day and lay 1–2 eggs/day. **Suppl:** A; **SL:** B; **PreR:** 2–4 mites/m²/month, **LiR:** 10 mites/m² every 2 weeks, **HiR:** 20 mites/m² weekly; **Cw:** *Orius* spp., **NCw:** may feed on *Neoseiulus cucumeris*. *I. degenerans* eats less than *N. cucumeris*, but can outperform *N. cucumeris* during flowering—short daylength does not send it into diapause, and it tolerates low humidity. *I. degenerans* tolerates neem, but is more sensitive to insecticidal soap and sulfur than *N. cucumeris*.

Lady beetles, a.k.a. "ladybugs," are coccinellid beetles (Fig. 20.28). The adults have red to yellowish-orange wing covers with black spots, dome-shaped bodies, 4–10 mm long. Larvae look like tiny alligators, usually gray-black with orange markings. Both adults and larvae are voracious predators of aphids. Lady beetles like to *wallow* in aphids—they may curb heavy infestations, but fly away before their job is done. Lady beetles overwinter as adults, often in large aggregations. Commercial sources are wild-harvested from aggregation sites, and we no longer recommend wild-harvested lady beetles. Native populations of lady beetles have declined in the past 20 years, due to several factors, such as the spread of non-native lady beetles (Harmon *et al.*, 2007). Wild-harvested lady beetles may harbor parasites and pathogens, which infect native lady beetles in release areas.

Lady beetles are worthy beneficiaries of conservation biocontrol (see section 20.3.3). Nearby meadows provide shelter and high humidity, needed by adults. Lady beetles eat *Cannabis* pollen as a food source (Smith, 1960, 1961), but *Cannabis* does not provide nectar. A source of flower nectar (or an artificial substitute) attracts adult beetles, induces them to lay eggs, and reduces dispersal. Mix honey and brewer's yeast on sticks or wax paper, and distribute among plants (replace when moldy); commercial products include Wheast™ or Biodiet™.

Adalia bipunctata, the "two-spotted lady beetle" (Fig. 20.28A), is native to North America. *A. bipunctata* is the most common lady beetle that is reared in commercial insectaries, rather than wild-harvested. Native populations are disappearing and its native range is shrinking (Harmon *et al.*, 2007).

Coccinella septempunctata and *C. undecimpunctata* are the "seven-spotted lady beetle" and "eleven-spotted lady beetle," respectively. Both are native to Eurasia, and repeatedly introduced to North America as biocontrol agents. *C. septempunctata* preys on *Phorodon cannabis* in Eurasia (Khan and Shah, 2017) and in Colorado (Schreiner, 2019). In an olfactory

Fig. 20.28. Lady beetles: **A.** *Adalia bipunctata* (W. Cranshaw); **B.** *Harmonia axyridis* (McPartland); **C.** *Hippodamia convergens* larva (R. Clarke); **D.** *Hippodamia convergens* adult (R. Clarke)

behavior study, *C. septempunctata* adults were attracted to (E)-β-farnesene, an aphid alarm pheromone (which is also produced by *Cannabis*). This attraction was inhibited by another terpenoid, (E)-β-caryophyllene—also produced by *Cannabis* (Al Abassi *et al.*, 2000).

Harmonia axyridis, the "multicolored Asian lady beetle" (Fig. 20.28B), is a voracious aphid predator. It also preys on thrips, scale insects, mites, and moth eggs. Under optimal conditions (21–30°C, 70% RH), the life cycle takes only 25 days. Adults live up to 3 years, and females lay up to 700 eggs. The USDA introduced *H. axyridis* from Japan, and the species has become invasive throughout the USA. Its competitive advantages over native lady beetles include feeding breadth and reproduction rate. Unlike native lady beetles, *H. axyridis* feeds on aphids killed by the fungus *Pandora neoaphidis* (Roy *et al.*, 2008). *H. axyridis* thereby spreads *P. neoaphidis*, which is a parasite of *Phorodon cannabis* (Keller, 2006). When winter approaches, *H. axyridis* seeks shelter indoors, making it a stinky household pest.

Hippodamia convergens, the "convergent lady beetle" (Fig. 20.28C,D), is native to North America. It is the most common lady beetle in commerce, but all supplies are wild-harvested from California's Sierra Nevada mountains. The Association of Natural Biocontrol Producers does not endorse wild-harvesting of biocontrol organisms.

Lecanicillium **spp.** are fungi formerly classified under the single species *Verticillium lecanii*. Molecular DNA analyses now recognize *L. lecanii*, *L. muscarium*, *L. longisporum*, and *L. nodulosum* (Zare and Gams, 2001). This has confused yet clarified commercial applications. The host range of *Verticillium lecanii* was considered quite wide, but that's because it consisted of several species. Today, Mycotal® is based upon *L. muscarium*, for use against whiteflies and thrips. Vertalec® is based upon *L. longisporum*, for use against aphids. It works best against *Myzuz persicae* and worst against *Aphis fabae*. *L. longisporum* has potential use against powdery mildew fungi, and other *Lecanicillium* spp. kill nematodes (Goettel *et al.*, 2008).

Suppl: C; **SL**: C; **Cw**: *Encarsia formosa* is slightly susceptible to *Lecanicillium* spp. Most other biocontrol organisms are unharmed. Soak in water (15–20°C) for 2–4 h prior to spraying. Spray around sunset—dew and darkness facilitate spore germination, and UV light kills them. Keep the spray tank agitated. Spores germinate and directly penetrate the insect cuticle. After infection the fungus takes 4–14 days to kill the pest. Under ideal conditions, dead insects sprout a white fluff of conidiophores bearing slimy, 1-celled conidia, which cause secondary infections. Being fungi, *Lecanicillium* spp. do best at moderate temperatures (18–28°C) and high humidity (≥80% RH) for at least 24 h after spraying. Thus *Lecanicillium* spp. cannot be used when plants are flowering and susceptible to gray mold.

Leptomastix dactylopii is a chalcid wasp that lays eggs in 3rd-stage larvae of *Planococcus citri* and other mealybugs. It is native to Brazil and does best at 25°C and 60–65% RH. Females are yellow-brown, 3 mm long; male wasps are smaller and have hairy antennae. Females lay up to 200 eggs, one per mealybug. Parasitized mealybugs turn into dark, swollen mummies. Wasps emerge from small, round holes. The life cycle takes about 3 weeks in optimal conditions. **Suppl**: A; **SL**: A (dispersed the day of receipt, either in the early morning

or evening); **PreR**: 2 wasps/m² weekly, **LiR**: 5/plant weekly; **Cw**: predators such as *Cryptolaemus montrouzieri*. The wasps work well at low pest densities.

Lindorus lophanthae, a coccinellid lady beetle, preys on nymphs of soft scales (*Saissetia* and *Parasaissetia* spp.). In the absence of soft scales, *L. lophanthae* eats mealybugs. Native to Australia, *L. lophanthae* was imported into California by the USDA, and has spread through the southern half of the USA. *L. lophanthae* does best in warmer temperatures (23–31°C) and moderate humidity (65% RH). Adults are small, pubescent, black beetles with red-orange highlights, 2.5 mm long. Females lay over 100 eggs, one at a time, on plants among scales. Larvae are gray, reach 3 mm long, and prey on young scales and eggs. The life cycle takes 3 weeks, then adults live another 6–8 weeks, eating constantly. **Suppl**: A; **SL**: B; **LiR**: 3–6 beetles/m² every 3 weeks, **HiR**: 6–12 beetles/m² weekly. Release in the early morning or late at night.

Macrolophus pygmaeus is a mirid bug that feeds on whiteflies—*Bemisia tabaci*, *B. argentifolii*, and *Trialeurodes vaporarium*. Nymphs and adults prefer whitefly eggs and larvae, which they pierce to suck out body fluids, feeding on 30–40 eggs per day. In the absence of whiteflies the bugs eat aphids, spider mites, moth eggs, and thrips. Adults are light green, 6 mm long, with long antennae and legs. The entire life cycle takes about a month. Each female lays about 250 eggs, inserted into plant tissue. **Suppl**: A; **SL**: B; **LiR**: 5 adults or nymphs/m² weekly, **HiR**: 5 adults or nymphs/m² weekly; **Cw**: *Encarsia formosa* and *Delphastus pusillus*. Gently sprinkle onto infested leaves.

Mesoseiulus longipes, a phytoseiid mite, preys on spider mites and possibly other mites. A native of South Africa, it tolerates conditions that are warm (up to 38°C) and dry (down to 40% RH)—conditions that wilt *Phytoseiulus persimilis* and even *Neoseiulus californicus*. Adults resemble *P. persimilis*. The life cycle turns slightly slower than that of *P. persimilis*. Female mites lay 3–4 eggs/day, and live up to 34 days. **Suppl**: A; **SL**: A; release rate same as *P. persimilis*. This species migrates better than most predatory mites, making it suitable for taller plants.

Metaphycus helvolus is an encyrtid wasp that controls soft scales (*Saissetia*, *Parasaissetia*, *Coccus* spp.). The species is native to semi-tropical regions and does best in warm glasshouses (24–32°C). Adults are yellow and black, 1.3 mm long. Female wasps lay up to 100 eggs, one at a time, under immature scales. Females also feed on scales. Larvae burrow into scales, kill them, pupate therein, and emerge as adults. The life cycle takes 24 days, then adults live another month. **Suppl**: A; **SL**: B; **PreR**: 1–2 wasps/m²/month, **LiR**: 5–10 wasps/plant or 10 wasps/m² every 2 weeks. *M. helvolus* is not useful in heavy infestations. It should be released in the evening, or when powering-down lamps (the wasps fly towards bright lights). The wasps are attracted to yellow sticky traps.

Metarhizium anisopliae was the first fungal agent deployed to control insect pests (Metchnikov, 1879). *M. anisopliae* has a wide host range (*ca.* 200 insect species) because it is a species complex. This explains why a generic "*M. anisopliae*" product doesn't work equally well on all insects in its host range. Bischoff *et al.* (2009) split the species complex into eight species.

Four species are generalists: *M. robertsii* has the widest host range, including Coleoptera, Lepidoptera, Diptera, Orthoptera, Hemiptera, and Isoptera. Three or four of those insect orders are infected by *M. anisopliae* (strict sense), *M. brunneum*, and *M. pingshaense*. Two species, *M. majus* and *M. guizhouense*, have more limited host ranges, against Coleoptera and some Lepidoptera. *M. flavoviride* is virulent against Hemiptera and some Coleoptera (*M. flavoviride* itself is a species complex; see Montalva *et al.*, 2016). *M. album* is specific for Hemiptera, and *M. acridum* only infects Acrididae (locusts and most grasshoppers).

In addition to killing insects, *Metarhizium* spp. colonize the roots of plants and boost plant growth by enhancing the uptake of soil nutrients, as well as increasing root resistance to plant pathogens (Hu *et al.*, 2014). Hu *et al.* (2023) inoculated hemp seedlings with *M. robertsii*, *M. anisopliae*, *M. brunneum*, and *M. acridum*. All four species readily colonized roots, and positively influenced shoot length, stem weight, and root weight compared with control plants. *Metarhizium* spp. differ in their habitat specificities, regarding soil temperature and aridity. The deployment of specific species should be tailored to field conditions (Kryukov *et al.*, 2017). **Suppl**: C or D; **SL**: C; **Cw**: lady beetles, green lacewings, parasitic wasps, honey bees, or earthworms (according to studies submitted by manufacturers). Most commercial products are registered as soil drenches, but some are approved as foliar sprays or even in drip irrigation systems. See instructions for *Beauveria bassiana*.

Metarhizium rileyi appears as *Nomuraea rileyi* in most biocontrol literature and products (Keppler *et al.*, 2014). *M. rileyi* is restricted to noctuid caterpillars, and has been deployed against serious problems in *Cannabis*, such as budworms (*Helicoverpa zea*, *H. armigera*, *H. viriplaca*), cutworms/armyworms (*Spodoptera litura*, *S. exigua*, *Agrotis ipsilon*), and defoliators (*Mamestra brassicae*). **Suppl**: C or D; **SL**: B; dosage on label instructions usually 300–600 g/ha. See instructions for *Beauveria bassiana*. Spores germinate and directly infect caterpillars through their skins. Death occurs in about a week. Under humid conditions the cadavers become covered with greenish conidia. The cycle is repeated when conidia contact new hosts.

Mycorrhizal fungi colonize plant roots and enter symbiotic relationships with their host plants. Over a dozen species of mycorrhizal fungi have been isolated from *Cannabis* under natural conditions (see section 13.2). All belong to one group: arbuscular mycorrhizal fungi (**AMFs**). Most AMFs have wide host ranges. A second group, ectomycorrhizal fungi (**EMFs**), primarily colonizes tree species, not herbaceous plants.

AMFs prevent infection by soilborne pathogens (*Pythium*, *Rhizoctonia*, *Fusarium*, *Botrytis cinerea*). AMFs interact with beneficial bacteria that colonize plant roots. Mosse (1962) first showed that mycorrhizal roots enhance the survival of beneficial *Pseudomonas fluorescens*. Reciprocally, "mycorrhiza helper bacteria" stimulate mycorrhizal symbioses. These helpers include species treated in this chapter—*Azotobacter*, *Azospirillum*, *Bacillus*, *Burkholderia*, *Pseudomonas*, and *Streptomyces* spp. (Frey-Klett *et al.*, 2007).

Mycorrhizal products usually contain a mix of AMF species. Products may also contain EMFs (useless for *Cannabis*), as well as *Bacillus*, *Pseudomonas*, and other plant growth-promoting rhizobacteria (PGPRs). In one *Cannabis* study, Massuela *et al.*

(2023) used a product that contained nine AMF species, 11 EMF species, 18 PGPR species, three *Trichoderma* species, and *Bacillus thuringiensis*.

The plethora of commercial products leave us at a loss for specific recommendations. Individual plant species respond best to specific AMF species (Klironomos, 2003; Rouphael *et al.*, 2015). Optimal AMF species for *Cannabis* have not been determined. Utilizing AMFs isolated from *Cannabis* under natural conditions (see section 13.2) would be a good start. Some caveats are in order: Soils containing adequate phosphorus will inhibit the growth of AM fungi (Koide and Moss, 2004). The benefits of supplementing soil with AM fungi are context-dependent, because AM fungi interact synergistically or antagonistically with other organisms in the soil.

Several stages in *Cannabis* cultivation benefit from AMF inoculation. Clones and micropropagated plantlets face a "weaning" stage as they adopt to growth media or soil. Applying AMFs at transplantation improves root development and nutrient uptake (Chen *et al.*, 2018). The benefits of AMFs in hydroponic cultivation have been debated. The constant flow of soluble nutrients in aeroponics systems eliminates the nutrient depletion zone around roots. When organic substrates are used (e.g., coir), AMFs can successfully colonize roots. This becomes more difficult with mineral substrates, such as rockwool (Rouphael *et al.*, 2015). Commercial potting soils—often steam-sterilized—usually lack AMFs, and benefit from inoculation. Potting soils with pre-mixed AMFs are available. Pro-Mix® contains *Glomus intraradices* as well as *Bacillus pumilus*, a "mycorrhiza helper bacterium."

Agricultural soils are AMF-impoverished, particularly in terms of numbers of species. The success of AMFs in field applications depends on many factors, reviewed by Rouphael *et al.* (2015). Crop rotation enhances AMFs. Rotating with legumes is particularly beneficial, rotating with non-mycorrhizal crops is not (e.g., brassicas, spinach, sugar beets, lupins). Judicious use of organic fertilizers (e.g. manure, compost), and slow-release minerals (e.g. rock phosphate) do not seem to suppress AMFs, and may even stimulate them. Organic farming methods as a whole benefit AMFs, as reported in many papers.

Hayman (1982) found that adding organic matter to soil improved AMF development, whereas ammonium nitrate (i.e., petrochemical fertilizer) decreased AMF populations and root colonization. High rates of phosphorus application also imparted negative effects, more so in potted plants than field soil. Other detrimental practices included monoculture and deep soil tillage, which disrupts the mycorrhizal network.

Menge (1983) tested the effects of eight synthetic pesticides—none approved for use on *Cannabis*—and all were lethal to AMFs. Copper sulfate caused little damage, and may have improved mycorrhizae by eradicating mycorrhizal parasites and predators. Ipsilantis *et al.* (2012) evaluated the impact of biorational pesticides on AMFs. Azadirachtin inhibited some AMFs, whereas spinosad, pyrethrum, and terpenes did not affect AMFs or root colonization.

Neoseiulus barkeri, formerly *Amblyseius barkeri* (=*A. mackenziei*), is a phytoseiid mite that preys on young thrips but also eats broad mites. The species is distributed worldwide, and does best at 65–72% RH and moderate temperatures (19–27°C). Adults resemble *Neoseiulus cucumeris*, perhaps a deeper tan color. *N. barkeri* can reproduce every 6.2 days at 25°C. Adults consume 2–3 thrips/day and lay 1–2 eggs/day. Each mite eats about 85 thrips larvae. **Suppl**: A; **SL**: B; release rates the same as *N. cucumeris*; **Cw**: some products combine *N. barkeri* and *N. cucumeris*. Commercial strains diapause under short daylengths.

Neoseiulus californicus

*is a phytoseiid mite that preys on spider mites, broad mites, and aphid eggs. A native of southern California, it does best in moderate humidity (≥60% RH) and moderate to high temperatures (18–35°C). Adult mites are pear-shaped and translucent to beige in color, depending on what they eat. The life cycle may take only 6 days in optimal conditions. Females lay 3 eggs/day for up to 14 days. Adults eat 5–15 adult spider mites/day, along with some eggs and larvae. **Suppl**: A; **SL**: B; release rates the same as *Phytoseiulus persimilis*; **Cw**: *P. persimilis* and everything compatible with *P. persimilis*. This species lives longer without prey than *P. persimilis*, so it works better as a preventive control. But *N. californicus* eats more slowly than *P. persimilis* and its populations build more slowly. *N. californicus* migrates fairly well and has been used outdoors.*

Neoseiulus cucumeris

*is a generalist predatory mite, feeding on spider mites, whiteflies, aphids, and psyllids. MacGill (1939) observed it preying on thrips larvae—the first predator mite known to do so—and *N. cucumeris* is still marketed for thrips control. The species does best at moderate temps (19–27°C) and high humidity (≥75% RH). *N. cucumeris* adults are pear-shaped, 0.5 mm long, beige to pink, with legs shorter than those of *Phytoseiulus persimilis*. The life cycle takes 6–9 days at 25°C, then adults live another month. Adults consume 6 thrips/day and lay 2–3 eggs/day.*

Suppl: A; **SL**: B (mites in sachets come with a food source and can be stored longer); **PreR**: 100 mites/m²/month, **LiR**: 100–500 mites/plant every 2–3 weeks, or 200–300 mites/m²/week until the infestation subsides. **Cw**: the absence of hosts, *N. cucumeris*, *P. persimilis* and *I. degenerans* prey upon each other, but they can coexist (Hussey and Scopes, 1985). **NCw**: some *N. cucumeris* strains are susceptible to biocontrol fungi (*Lecanicillium lecanii* and *Beauveria bassiana*), so they are not compatible. *N. cucumeris* can be sprinkled directly on plants, sprinkled into distribution boxes, or heaped on rockwool blocks. The sachets are hung from plants. *N. cucumeris* survives well without prey, much better than *P. persimilis*, which makes it a good preventive. Many strains of *N. cucumeris* enter diapause under short photoperiods, while plants are flowering. Some new strains do not, but their need for high humidity makes them unsuitable for *Cannabis* cultivars susceptible to gray mold. *N. cucumeris* tolerates insecticides better than *I. degenerans* and most other predatory mites, but is very sensitive to pyrethroids.

Neoseiulus fallacis

*preys on spider mites and possibly hemp russet mites. Native to the northwestern USA, it does best in moderate to high humidity (60–90% RH) and moderate temperatures (optimally 10–27°C, but up to 38°C in high humidity). Adults are pear-shaped and white until they feed, and then take on the color of their prey, usually pale red or brown. They are slightly smaller than *P. persimilis*. The life cycle takes 10 days in optimal conditions, then adults live another 20–60 days. Adult females eat about 15 mites/day, and lay a total of 40–60 eggs. **Suppl**: A; **SL**: B; **PreR**: 10 mites/m² every 2 weeks,*

LiR: 20–40 mites/m² weekly. Hops growers have used *N. fallacis* to control spider mites outdoors, released at 50,000/acre. *N. fallacis* migrates vigorously on plants of all sizes, and uses aerial dispersal to blow from plant to plant. Short days and cold temperatures send *N. fallacis* into diapause. Diapause is delayed by warm temperatures, and completely averted above 27°C.

Neozygites floridana

*is a pathogenic fungus that targets twospotted spider mite, *Tetranychus urticae*. Time between infection and death is 3.4 days at 25°C, and 11.2 days at 15°C (Carner, 1976). Males have a "fatal attraction" for *N. floridana* (Trandem et al., 2015). Males normally guard quiescent females, to increase their chances of fathering offspring. Males preferred *N. floridana*-killed females to live healthy females, which promoted transmission of the fungus. *N. floridana* is biotrophic (an obligate pathogen) and difficult to produce on artificial media. Brazilian and Norwegian isolates are under development for commercial production. *N. floridana* does not infect *Phytoseiulus persimilis*; this predator facilitates spread of the fungus, and a combination of fungus and predator was superior to either alone (Trandem et al., 2015).*

Nosema spp.

*are microsporidian parasites, formerly considered protozoans, but more closely related to unicellular fungi. One species, *Nosema locustae*, has been renamed *Antonospora locustae* (see there). *Nosema apis* and *Nosema ceranae* cause devastating diseases in honey bee hives. *Nosema pyrausta* was isolated from European corn borers in the 1920s in Europe. It appeared in the USA by 1951, possibly arriving via infected parasitoids introduced from Europe. Lewis and Lynch (1978) reduced ECB populations by as much as 44% by spraying maize with *N. pyrausta*. The pathogen has not been commercialized because it is widely distributed and epidemics occur naturally.*

Nuclear polyhedrosis viruses

*are a family of ~25 viruses. NPVs infect and kill noctuid larvae (e.g., budworms, cutworms, armyworms). Caterpillars that ingest NPV soon stop feeding, become flaccid, darken, then die. Sometimes the cadavers hang from plants, and sometimes they totally liquify (Fig. 20.29). Either way, they spread more viruses. **Suppl**: D (viral occlusion bodies); **SL**: C; **Rate**: 4–10 fl. oz of product/acre. Since NPV must be eaten to kill caterpillars, spray all*

Fig. 20.29. Beet armyworm (*Spodoptera exigua*) killed by NPV, compared with normal larva (courtesy USDA)

foliage uniformly and completely. NPV on foliage is degraded by UV light, so spray in late afternoon or on cloudy (not rainy) days. To slow degradation, combine NPV with Pheast™ or Coax™, which are lepidopteran feeding stimulants.

***Helicoverpa zea* NPV** kills *H. zea* and *H. armigera*. HzNPV was the first commercialized NPV strain, registered by the EPA in 1973. It was discontinued a decade later, but now it's back (Elcar®, Gemstar®, Heligen®), for use in cotton and vegetable crops. Britt *et al.* (2021a) showed that Heligen® worked better than Gemstar® for controlling corn earworms in hemp. Both are permitted for use on *Cannabis* by Colorado DOA (2023).

***Spodoptera exigua* NPV** kills the beet armyworm, *S. exigua* (Fig. 20.29). SeNPV is marketed as Spod-X® for use in vegetable crops.

***Autographa californica* NPV** was isolated from the alfalfa looper, *A. californica*. The host range of AcNPV is limited to that species, although Kondo and Maeda (1991) expanded its range to *Spodoptera litura* and *Heliothis virescens* after a little genetic engineering. The commercial product (Gusano®) is not transgenic.

***Mamestra brassicae* NPV** was originally isolated from the cabbage moth, *M. brassicae*. MbNPV has a wide host range of over 30 species, mostly in the Noctuidae (Doyle *et al.*, 1990). Many susceptible species are *Cannabis* pests: *M. brassicae*, *Helicoverpa armigera*, *Spodoptera exigua*, *Agrotis segetum*, and *Melanchra persicariae*. Sold in France (Mamestrin®) and Russia (Virin-EKS®), but not the USA.

***Orius* spp.**, "minute pirate bugs," prey upon any host smaller than themselves, such as thrips, whiteflies, aphids, psyllids, and spider mites. *Orius* spp. are perhaps the only biocontrol that kill *adult* thrips, so they are marketed for thrips control. Adults eat 5–20 thrips/day. They are good for preventive control, but in the absence of pests, they eat other biocontrols, then each other. They bite humans. In an insect survey conducted in southern California *Cannabis* crops (Britt *et al.*, 2022), minute pirate bugs were more common (*n* = 240 specimens) than lacewings (*n* = 46), lady beetles (*n* = 23), and assassin bugs (*n* = 10).

Adults are yellow-brown to black with white wing patches, 1.5–3 mm long (Fig. 20.30). *Nymphs* are teardrop-shaped, bright yellow to brown, with red eyes. Adults overwinter; females lay 30–130 eggs in plant tissue. The life cycle takes as little as 3 weeks, then adults live another month. Nymphs and adults move quickly and catch prey with their forelegs. *Orius* spp. diapause under short daylengths. Young *Orius* nymphs migrate with difficulty across leaves and flowers with prominent trichomes; adults do better. In mixed pest infestations, *Orius* spp. eat the largest pests first. They may be swimming in baby aphids, but they would rather eat a large thrips.

Orius insidiosus is native to North America and the most common pirate bug commercially available. *O. insidiosus* eats thrips, whitefly larvae and pupae, spider mites, aphids, and the eggs of budworms (*Helicoverpa zea*, *H. virescens*). **Suppl**: A (adults and nymphs); **SL**: B; **PreR**: 0.5/m², **LiR**: 1/m², **HiR**: 10/m² weekly; **Cw**: reportedly compatible with *Neoseiulus cucumeris*; **NCw**: eats the aphid predator *Aphidoletes aphidimyza*.

O. laevigatus hails from the Mediterranean region. It feeds primarily on *F. occidentalis* and *T. tabaci*, but also eats spider mites, aphids, and even small larvae of ECBs (*Ostrinia*

Fig. 20.30. *Orius insidiosus* feeding on whiteflies (courtesy Jack Dykinga, Wikimedia Commons)

nubilalis) and cutworms (e.g., *Spodoptera litura*). **Suppl**: A (adults and nymphs); **SL**: B; release rates same as *O. insidiosus*; **Cw**: reportedly coexists with lady beetles (*Coccinella unidecimpunctata*) and lacewings (*Chrysopa carnea*). *O. laevigatus* may survive without thrips longer than *O. insidiosus*.

O. majusculus comes from central Europe, but arrived in North America by 2008. This species prefers eating thrips and whiteflies. *O. majusculus* produces more eggs than any other *Orius* sp. However, it is more photoperiod sensitive (entering diapause in daylengths under 16 h) than *O. insidiosus* or *O. laevigatus* (under 11 h).

O. tristicolor is native to western North America. It feeds on thrips (*F. occidentalis*, *T. tabaci*, *Caliothrips fasciatus*), spider mites, aphids, and eggs of budworms (*Helicoverpa zea*). Twenty years ago several western USA suppliers carried it (e.g., ARBICO, Rincon-Vitova, Applied Bio-nomics, EcoSolutions, etc.), but no longer.

Paenibacillus popilliae, formerly *Bacillus popilliae*, is a soil bacterium that causes "milky spore disease" in grubs of Japanese beetles (*Popilia japonica*). After grubs eat spores of *P. popilliae*, the bacteria multiply, the host turns white to nearly translucent, and dies. Under optimal conditions (28°C, pH 7), this happens in 2 or 3 weeks. As the cadaver decomposes, it releases 5×10^9 bacteria into the soil. *P. popilliae* was the first bacterial insecticide available in the USA, registered in 1948, and marketed as Doom™ and Japidemic™. **Suppl**: C (100 million spores/g in talc powder); **SL**: B. Broadcast soil with spores, about 5 g/m², and follow with 20 min of watering. In stable soils the bacteria become established, providing long-term control. Temperature is a critical factor—*P. popilliae* needs a minimum soil temperature of 16°C, so it doesn't work well north of New Jersey (where it was discovered). *P. popilliae* is a highly selective pathogen and can combined with other biocontrols (e.g., *Heterorhabditis bacteriophora* and *Steinernema glaseri*).

Paraphaerosphaeria minitans, formerly *Coniothyrium minitans*, is a soil fungus that parasitizes the sclerotia of *Sclerotinia sclerotiorum*, *S. minor*, and *Botrytis* *cinerea*. The fungus naturally occurs in temperate zones worldwide, at optimal temperatures of 20–30°C. It penetrates sclerotia through tiny cracks, or by secreting chitinase and β-1, 3-glucanase enzymes. Once inside, it kills and then reproduces. **Suppl**: C (Contans®, Intercept®); **SL**: B. Applied as a spray to the soil (1–4 lb/acre), followed by mechanical mixing into upper 5 cm of topsoil, with adequate irrigation or timely rainfall. The product should be applied to soil after harvest, or 3–4 months prior the typical onset of disease, so that *P. minitans* can kill sclerotia before they germinate and attack the crop. Use higher rates (3–4 lb/acre) to control *S. minor* or if disease pressure is high.

Phytoseiulus persimilis, a phytoseiid mite, is the

most popular mail-order predator of spider mites. Adults are orange-red in color, pear-shaped, 0.5–0.7 mm long, with long legs and no spots (Fig. 20.31). Compared with spider mites, *P. persimilis* is more elongate and moves more quickly. Eggs are oblong and twice the size of spider mite eggs. Eggs survive best in high humidity (70–95% RH).

Adults work best at moderate temperatures (20–30°C) and >70% RH (feeding nearly stops in humidity <30% RH). *P. persimilis* avoids high temperatures, such as flowering tops basking in bright light. The life cycle takes 7–8 days in optimal conditions, much faster than spider mites (see Fig. 4.3). Adults live another 30–40 days, and consume up to 24 immature spider mites or 30 eggs/day. A week after hatching, females start laying 4–5 eggs/day, for a total of 60 eggs/lifetime.

Suppl: A (adults and nymphs mixed with vermiculite, bran, corn cob grit); **SL**: A or B (*P. persimilis* is cannibalistic); **Release rates**: see Table 20.4; increase rates when plants are taller than 1 m, planted more densely than 6/m², or when humidity is low (<45% RH). Encourage *P. persimilis* by misting plants with water during periods of low humidity (this also discourages spider mites). **Cw**: Bt, *Feltiella acarisuga*, and most parasitic wasps. *P. persimilis* can coexist with *Neoseiulus cucumeris* but they prey on each other in the absence of hosts. *P. persimilis* has been mixed with lacewings (*Chrysoperla* spp.), but lacewings

may eat predatory mites. *P. persimilis* is often combined with *Neoseiulus californicus* and *Mesoseiulus longipes* because these mites live longer without food. *P. persimilis* can be sprinkled into distribution boxes hanging on plants, or sprinkled directly on plants with spider mites. Plants should be touching each other in a closed canopy for effective dispersion.

Some strains of *P. persimilis* enter diapause when plants begin flowering. Diapause can be delayed by elevating the temperature. Importing new predators *weekly* might work—*P. persimilis* feeds for about a week before entering diapause. Regardless of daylength sensitivity, the effectiveness of *P. persimilis* decreases when *Cannabis* begins flowering, because predators find it difficult to move across sticky flower resins. Avoid most insecticides, miticides, and even fungicides while utilizing *P. persimilis* (Table 20.1). Allow previously applied pesticides to break down for 2–3 weeks before introducing predators.

Plant growth-promoting rhizobacteria

(PGPRs) are soil bacteria that flourish in the rhizosphere surrounding roots of plants. The bacteria flourish, and the plants flourish—it's a mutualistic symbiosis. Some PGPRs are included in this chapter because of their powerful biocontrol activities (e.g., *Bacillus cereus*, *B. subtilis*, *Burkholderia cepacia*, *Paenibacillus lentimorbus*, *Pseudomonas chlororaphis*, *Pseudomonas fluorescens*, *Streptomyces griseoviridis*). Notable nitrogen-fixing PGPRs are described in section 13.3 for their use as biofertilizers (*Azotobacter* and *Azospirillum* spp.).

Pseudomonas chlororaphis is a PGPR. *P. chlororaphis* and other fluorescent pseudomonads (*P. fluorescens*, *P. syringae*), together with *Bacillius* and *Streptomyces* spp., constitute a substantial fraction of the bacterial community in the rhizosphere. Their presence correlates with the suppression of plant diseases caused by fungi and other pathogens. Johnsson *et al.* (1998) demonstrated the efficacy of *P. chlororaphis* against a variety of seed-borne fungi in barley, oats, wheat, and rye. Bacterial liquid culture was directly applied to pathogen-infested seeds, and treated seeds could be stored for 2 years without losing the disease-suppressing effects of the bacterial treatment.

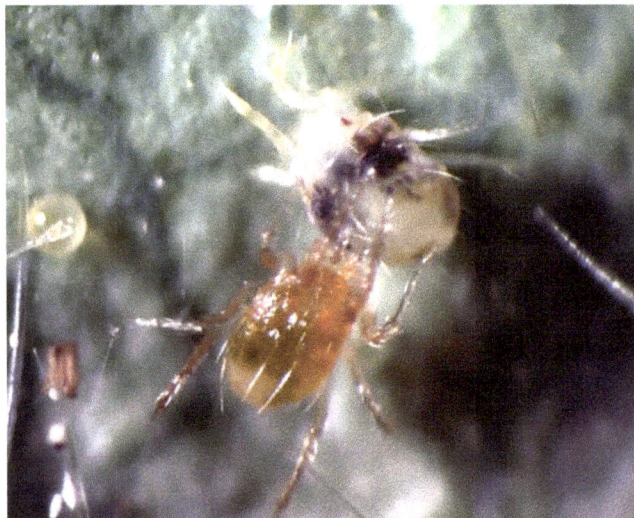

Fig. 20.31. *Phytoseiulus persimilis* preying on *Tetranychus urticae* (courtesy Koppert Biocontrols)

Table 20.4. *P. persimilis* release rate for control of twospotted spider mites

ISI[a]	Number of predators released per m² of glasshouse crop[b]
Preventive	5/m² every 3 weeks
Light	10/m² initial release, then 5/m² every 3 weeks
Moderate	25/m² on trouble spots, 10/m² elsewhere, then 5/m² every 3 weeks
Heavy	mechanical & chemical control of trouble spots OR 200/m² on trouble spots, 25/m² elsewhere, then 10/m² every 3 weeks
Critical	mechanical & chemical control of trouble spots, 25/m² elsewhere, then 10/m² every 3 weeks

[a]ISI = Infestation Severity Index of two-spotted spider mite, see Table 4.1.
[b]Effectiveness of *P. persimilis* decreases during flowering

Howler® is approved for use on seeds, seedlings, and flowering plants (including tobacco) to control *Botrytis*, *Sclerotinia*, *Fusarium*, *Rhizoctonia*, *Colletotrichum*, *Pythium*, and *Phytophthora*. NematoKill™ demonstrates efficacy against the root-knot nematode *Meloidogyne hapla* (Nam *et al.*, 2018). *P. chlororaphis* is also lethal to caterpillars, such as *Spodoptera littoralis* and *Heliothis virescens* (Ruffner *et al.*, 2013). **Suppl**: C or D; **SL**: B; applied as a seed dressing, soil drench, hydroponic irrigant, or foliar spray.

Pseudomonas fluorescens is another PGPR.

P. fluorescens directly impacts pathogens by aggressively competing for space on plant surfaces. It suppresses the growth of pathogens by producing antibiotics, bacteriocins, siderophores, and hydrolytic enzymes such as β-1,3-glucanase and chitinase. These benefits of *P. fluorescens* led to Kloepper *et al.* (1980) coining the term "PGPR" in a *Nature* paper. The commercialization of *P. fluorescens* has been hampered by difficulties in preparing stable and long-lived formulations. Products with *P. fluorescens* come and go (e.g., Pseudon-F®, Ecomonas®). They have been labeled for controlling damping-off pathogens such as *Pythium*, *Fusarium*, and *Rhizoctonia* spp. Some mixtures of PGPRs and mycorrhizal fungi contain *P. fluorescens*.

Pseudomonas syringae has a Jekyll-and-Hyde

reputation. Some isolates cause *Cannabis* diseases (see section 15.2), yet non-pathogenic isolates live as epiphytes, and they protect plants from disease agents. The epiphytes produce antibacterial and antifungal compounds (Sinden *et al.*, 1971). Janisiewicz and Marchi (1992) used *P. syringae* to control fruit storage rots caused by gray mold (*Botrytis cinerea*) and blue mold (*Penicillium expansum*). *P. syringae* has been genetically modified, and it was the first GMO released into the environment. Rogue professors made the releases without authorization. Steven E. Lindow (1951–) at UC-Berkeley injected GMO *Pseudomonas syringae* into trees, 2 years after a court injection blocked his experiment. His biotech firm was fined $20,000 by the EPA (Sun, 1986). A year later, Gary A. Strobel (1938–) at Montana State injected GMO *Pseudomonas syringae* into trees. Strobel was a repeat-offender, having illicitly released GMO *Rhizobium meliloti* in 1984, in Montana, South Dakota, Nebraska, and California (Schneider, 1987; Anderson, 1987).

Pythium oligandrum parasitizes fungal pathogens,

particularly other *Pythium* spp. Dreschler (1943) inoculated petri plates with *P. oligandrum* alongside *P. ultimum*, *P. debaryanum*, or *P. irregulare*, and those species were attacked "in a violently parasitic manner." Thirty-five years later, Veselý (1977) demonstrated the ability of *P. oligandrum* to control damping off in sugar beets caused by *P. ultimum* and *P. debaryanum*.

Thirty years later, the EPA registered *P. oligandrum* as Polyversum®. **Suppl**: C; **SL**: C. Apply as a seed dressing (0.1–0.3 oz/60–100 lb seed), or as an overhead spray or soil drench (1.5–3.0 oz/acre every 2 weeks). Target species on the label include soil-borne pathogens (*Pythium* spp., *Phytophthora* spp., *Fusarium* spp., *Verticillium* spp. *Rhizoctonia solani*, *Sclerotinia sclerotiorum*, *Botrytis cinerea*), and foliar pathogens (*Botrytis cinerea*, *Alternaria* spp., *Phoma* spp., *Ascochyta* spp., *Peronosplasmopara* spp.).

Steinernema carpocapsae is a small nematode (males

1–1.7 mm long, females 2.8–5.1 mm) that parasitizes insects in soil, including beetle grubs, many lepidopterans (cutworms,

armyworms, earworms, borers), root maggots, ants, and thrips. Like *Heterorhabditis* nematodes, *Steinernema* spp. contain mutualistic gut bacteria (*Xenorhabdus* spp.) that actually do the killing. *S. carpocapsae* is cosmopolitan, and does best in soil temperatures at 15–30°C. The life cycle of *S. carpocapsae* is like *Heterorhabditis bacteriophora*, but quicker—hosts begin dying within 24 h of infection, and the next generations emerges within 10 days. *Steinernema*-infected larvae become limp and turn a cream or brown color (Fig. 20.32).

Suppl: 3rd-stage larvae in polyethylene sponge packs or bottles (NemAttack™, Millenium®, Ecomask, Exhibitline Sc, many others); **SL**: C (sealed containers remain viable for 6–12 months at 2–6°C); apply to soil at the same rate as *Heterorhabditis bacteriophora*. Unlike that species, *S. carpocapsae* has been used to control insects above ground (Lacey and Georgis, 2012). Spray onto foliage or inject into borer holes. In these applications, nematode survival is short. It can be lengthened by coating nematodes with antidesiccants (e.g., Folicote, Biosys 627). Repeat applications every 3–6 weeks for the duration of infestations.

S. carpocapsae is the hardiest *Steinernema* sp. in storage, and it survives the longest in soil without hosts. *S. carpocapsae* acts as an "ambusher"—it tends to lie and wait for hosts, while nictating (the nematode stands on its tail and waves its head). Thus it works well with mobile hosts above the ground, or at the ground level (cutworms and fungus gnat maggots). *Steinernema* spp. are easier to mass produce than *Heterorhabditis* spp., and last longer in storage. For more comparisons, see the section on *H. bacteriophora*.

Steinernema feltiae and *Steinernema glaseri*

are related to *S. carpocapsae*, but not as commercially popular. Dutky's experiments with *S. feltiae* in the 1950s rekindled interest in nematodes as biocontrol agents. *S. feltiae* was originally marketed for control of dipteran maggots—immature fungus gnats and craneflies. But *S. feltiae* also kills cutworms, beetle grubs, and thrips. *S. feltiae* has an intermediate foraging strategy between the ambush and cruiser type. The *S. feltiae* Scanmask™ strain, native to northern Sweden, is active down to 10°C.

S. glaseri is especially effective against beetle larvae, particularly scarabs (Fig. 20.32). This species uses a "cruiser" foraging

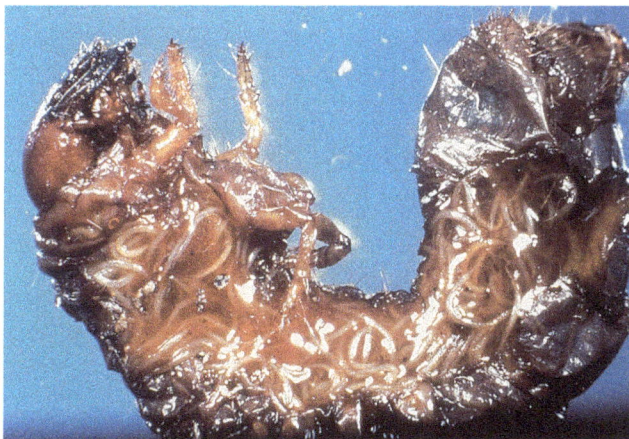

Fig. 20.32. *Steinernema glaseri* nematodes visible through the cuticle of a *Popillia japonica* cadaver (courtesy USDA)

strategy: it is highly mobile and tracks host volatiles. *S. glaseri* does not nictate, and does not attach well to passing hosts. Cruising makes it effective against hosts with low mobility, further down in the soil profile.

Steinernema riobrave is new (NemAttack™).

Isolated from the Rio Grande Valley in Texas, *S. riobrave* is tolerant of hot (up to 35°C) and semi-arid conditions. Target species on the label include budworms (*Helicoverpa zea*, the original host), fall armyworm (*Spodoptera frugiperda*), beet armyworm (*Spodoptera exigua*), fungus gnats (*Bradysia* spp.), and various beetle grubs. *S. riobrave* uses both ambush and cruiser foraging strategies.

Stethorus punctillum is a lady beetle known as the

"spider mite destroyer." Two related species, *S. punctum* and *S. picipes*, are sometimes confused, but not commercially available. All three are native to North America, and do best in moderate humidity (60% RH) and temperatures (20–32°C). Adults are oval, black, shiny, pubescent, 1.5–3 mm long, with brown or yellow legs. Larvae are cylindrical, dark gray to black (turning reddish before pupation), covered with fine pale-yellow spines, 1–2 mm long. The life cycle takes 20–40 days, then adults live another 30–40 days. Adults consume all stages of mites and their eggs, as many as 75–100/day. **Suppl**: A; **SL**: B; **PreR**: 1–2 mites/m^2 monthy, **LiR**: 3–4 mites/m^2 every 2 weeks, **HiR**: 4–8/m^2 weekly; **NCw**: *S. punctillum* may eat predatory mites. *S. punctillum* beetles are good fliers and find infestations easily. They feed greedily but migrate before their job is finished. To reduce migration, use strategies described in the "Lady beetles" section. Dense mite webbing on leaves with large trichomes may impede newly hatched larvae.

Stratiolaelaps scimitus, formerly *Hypoaspis*

miles, is a soil-dwelling predatory mite that preys on small insects: fungus gnat pupae, thrips pupae, root aphids, symphylans, even overwintering spider mites. *S. scimitus* is native to the northern USA, and does best in moist (not soaked) greenhouse soils at 15–30°C. Adults are brown, oblong, and slightly larger than *Phytoseiulus persimilis*. Adults and nymphs feed on pests, consuming about 5 pests/day. The life cycle takes 10–13 days. *S. scimitus* does not diapause. When pests are not present, the mites survive on algae and plant debris. *S. scimitus* inhabits the top layer of soil (1–3 cm), and can colonize the surface of rockwool. **Suppl**: A; **SL**: A (or up to 2 weeks when supplied with a food source such as *Tyrophagus putrescentiae*); **PreR** against fungus gnats: 100 adults/m^2 soil area, **LiR**: 200/m^2, **HiR**: 300/m^2 weekly. **LiR** against thrips: 200–500/m^2. **Cw**: Bt, *Beauveria bassiana*, *Aphidoletes aphidimyza*, and beneficial nematodes (*Steinernema* spp.). Thanks to selective breeding and a subterranean lifestyle, *S. scimitus* tolerates neem and many fungicides.

Streptomyces griseoviridis is an actinomycete

(filamentous bacterium) found in soil. Together with *Bacillius* and *Pseudomonas* spp., *Streptomyces* spp. constitute a substantial fraction of the bacterial community in the rhizosphere. Tahvonen (1982) isolated *S. griseoviridis* from disease-suppressive Finnish peat moss. It inhibited damping-off caused by *Rhizoctonia*, *Alternaria*, and *Fusarium* spp. Ten years later, Tahvonen's strain was registered by the EPA as Mycostop®. **Suppl**: C; **SL**: C; apply as a seed dressing (0.3 oz/lb seed), soil drench (0.035–0.07 oz/10 m^2 soil with sufficient water), or leaf spray (0.18 oz/13

gal) (1 oz = 28.35 g; 1 lb = 454 g; 1 gal = 3.8 liters). Target species on the label include aforementioned the fungi as well as *Pythium* spp. and *Botrytis cinerea*.

Streptomyces lydicus came along about a decade

after *S. griseoviridis*. Yuan and Crawford (1995) isolated *S. lydicus* from a flax field in England. It prevented damping off caused by *Pythium ultimum* and *Rhizoctonia solani*. Ten years later, their strain was registered by the EPA as Actinovate®. **Suppl**: C; **SL**: C; apply as a seed dressing (6–18 oz/100 lb seed), soil drench (4–6 oz/100 gal, apply at a rate of 1 gal/ft^3), or leaf spray (6–12 oz/50–100 gal weekly) (1 oz = 28.35 g; 1 lb = 454 g; 1 gal = 3.8 liters; 1 ft = 0.3 m). It is labeled for the aforementioned fungi, as well as *Phytophthora*, *Fusarium*, *Verticillium*, *Botrytis cinerea*, powdery mildew (*Erysiphe*, *Podosphaera*), and downy mildew (*Pseudoperonospora*, *Peronospora*).

Talaromyces flavus is a rhizosphere-living fungus

that kills fungal pathogens, even species with thick-walled sclerotia, via cell wall-degrading enzymes (chitinase, β-1,3-glucanase, cellulase), with hydrogen peroxide (via glucose oxidase), and by direct invasion (mycoparasitism). *T. flavus* provided the best control against *Verticillium dahliae* in a screen of 34 biocontrol fungi (Marois *et al.*, 1982). Since then, *T. flavus* has proven efficacy against *V. albo-atrum*, *Rhizoctonia solani*, *Sclerotium rolfsii*, and *Sclerotinia sclerotiorum*. A commercial product is available in Europe (Protus®) as wettable granules, applied as a seed treatment or soil drench. In a comparison test for treating *Rhizoctonia solani* disease in peas, Protus® tied in first place with *Trichoderma virens* (SoilGuard®), better than *Trichoderma harzianum* (Supresivit®), and *T. viride* (Ecofit®) (Koch 1999).

Townesilitus spp. are braconids wasps that parasitize

flea beetles (*Phyllotreta*, *Psylloides*, *Chaetocnema* spp.). Haeselbarth and Loan (1983) split ten *Townesilitus* spp. away from the genus *Microctonus* on the basis of egg-laying behavior. Female adults of *Townesilitus* and *Microctonus* are morphologically similar: a dark brown thorax and lighter brown abdomen separated by a thin, curved petiole (waist), 2.0–3.0 mm long, with yellow or brown head, prominent black eyes, beige wings, and antennae nearly the length of the body.

Females lay a solitary egg in an adult flea beetle, as the host emerges from its pupa in late summer. First-instar larvae overwinter within the body of their host. Larval development resumes in spring, and it kills its host by summer. Larvae pupate in the soil, and emerge 2–3 weeks later as adult wasps. This coincides with a new generation of flea beetles, which the wasp parasitizes (Stigenberg, 2017). *Townesilitus* and *Microctonus* spp. parasitize various *Cannabis* pests. The list in Table 20.5 was collated from Jolivet (1950), Cox (1994), Lesage and Majka (2010), and Stigenberg (2017). *T. bicolor* has the widest host range. It is native to Europe, and was introduced into Canada for biological control (Wylie, 1988).

Thripoctenus javae, formerly *Thripobius javae*

(=*T. semileuteus*), is a eulophid wasp that parasitizes the greenhouse thrips (*Heliothrips haemorrhoidalis*). Adults have a black head and thorax, a yellow abdomen with dark bands, and yellow legs and antennae, 0.6 mm long. They crabwalk across leaf surfaces in search of 1st- and 2nd-stage thrips larvae. Females lay individual eggs in 40 or more larvae. Parasitized thrips larvae turn black. The entire life cycle takes 22 days in optimal conditions. Between 1986 and 1990 more than 500,000 *T. javae* parasitoids

Table 20.5 Townesilitus and Microctonus spp. on Cannabis pests

Host species	Parasitoid species
Psylloides	*Townesilitus bicolor*
attenuata	*Townesilitus psylliodis*
	Microctonus cerealium
	Microctonus labilis
Psylliodes	*Townesilitus psylliodis*
punctulata	Microctonus punctulatae
Phyllotreta	*Townesilitus bicolor*
atra	Microctonus areolatus
Phyllotreta	*Townesilitus bicolor*
nemorum	Microctonus aethiops
	Microctonus areolatus
	Microctonus vittatae
Chaetocnema	*Townesilitus bicolor*
hortensis	Microctonus cerealium

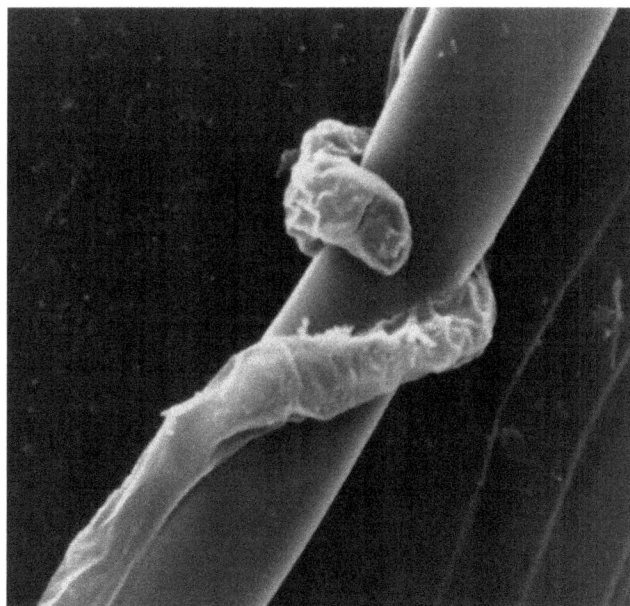

Fig. 20.33. *Trichoderma virens* coiling around hypha of *Rhizoctonia solani* (SEM, courtesy USDA)

were released in southern California; they overwintered and dispersed from release sites, with a decline in *H. haemorrhoidalis* populations (McMurty, 1995). A German company sells *T. javae* for inundative control, supplied as pupae attached to cards.

Trichoderma **spp.** are soil fungi formerly classified under the single species *Trichoderma viride*. Rifai (1969) split it into nine morphological "species aggregates," including five cited in biocontrol research: *T. viride*, *T. harzianum*, *T. koningii*, *T. pseudokoningii*, and *T. polysporum*. Phylogenetic research has further split those aggregates into 89 species. Kullnig *et al.* (2001) detected many misidentified species reported in the literature as *T. harzianum*. The same goes for species reported as *T. viride* (Jaklitsch *et al.*, 2006), and *T. koningii* (Samuels *et al.*, 2006). Thus, species identified in research publications have to be questioned.

Trichoderma virens differs significantly from other *Trichoderma* species, insofar as Rifai (1969) considered it **Gliocladium virens**. That's the name that appears in many studies, but phylogenetic research has moved it to *T. virens* (Rehner and Samuels, 1994). Howell (1982) showed that *T. virens* controlled damping off caused by *Pythium ultimum* and *Rhizoctonia solani*, by different mechanisms—an antifungal metabolite and mycoparasitism, respectively (Fig. 20.33).

T. virens was registered as GlioGard® in 1990, formulated as an alginate-wheat bran prill. Five years later it changed to a granular formulation, SoilGard®, with improved water dispersibility and efficacy (Lumsden *et al.*, 1996). Koch (1999) compared SoilGuard® with *Trichoderma harzianum* (Supresivit®), and *T. viride* (Ecofit®), and *Talaromyces flavus* (Protus®). SoilGuard® performed the best against *Rhizoctonia solani* disease in peas. The label now includes *Rhizoctonia*, *Pythium*, *Phytophthora*, *Fusarium*, *Sclerotinia*, and *Sclerotium* spp., always applied as a soil drench.

Trichoderma harzianum showed promise for biocontrol of *Sclerotium rolfsii* and *Botrytis cinerea* (Wells *et al.*, 1972). *T. harzianum* does best in warm regions, and can be combined with partial soil sterilization (e.g., solarization). *T. harzianum* strain T-39 (Trichodex®) gained clearance in 1996 as a

foliar spray to control *Botrytis cinerea* in grapes. *T. harzianum* strain T-22 is RootShield®, PlantShield®, and Trianum® (soil or foliar applications, labeled for *Botrytis*, *Pythium*, *Rhizoctionia*, *Fusarium*, *Thielaviopsis*, and *Cylindrocladium* spp.). Treating fiber-type hemp with *T. harzianum* strain T-22 increased plant height, harvested dry weight, and CBD content (Kakabouski *et al.*, 2021).

Trichoderma atroviride was overlooked by Rifai (1969), and misidentified as *T. harzianum* in many studies (Kullnig *et al.*, 2001). Weindling (1934) may have actually worked with *T. atroviride* when he demonstrated the lethal effects of "*Trichoderma lignorum*" against *Rhizoctonia solani*. Binab®, a blockbuster registered in 1989 as *T. viride*, was later identified as *T. harzianum*, and now as *T. atroviride*. Various *T. atroviride* products are labeled for use against soil pathogens (*Botrytis*, *Rhizoctonia*, *Fusarium*, *Verticillium*, *Pythium*, *Phytophthora* spp.), and foliar pathogens (*Didymella* spp.).

Trichoderma asperellum was also misidentified as *T. harzianum* (Kullnig *et al.*, 2001) or *T. viride* (Sriram *et al.*, 2013). Today, various *T. asperellum* products are labeled for use against soil pathogens (*Botrytis*, *Sclerotinia*, *Fusarium*, *Rhizoctonia*, *Pythium*, *Phytophthora* spp.) and anthracnose (*Colletotrichum* spp.).

With the correct identification of *T. atroviride* and *T. asperellum*, other species have dropped from view, *Trichoderma koningii* and *Trichoderma pseudokoningii* in particular. *T. koningii* still appears in products that combine *Trichoderma* species. Combinations also add cold-tolerant *Trichoderma polysporum*, never seen as a stand-alone product.

Woo *et al.* (2014) named 273 *Trichoderma*-based products sold worldwide (published on the internet). Each trade name was listed with its active ingredient(s), targeted pathogens, and geographic availability. They calculated that most *Trichoderma*-combined products contain *T. harzianum* (83%), of which 55%

are combined with *T. viride*, and 28% with *T. koningii*. No taxonomic names appeared on 33 combination products. Other microorganisms are added to mixtures, such as mycorrhizal fungi and plant growth-promoting rhizobacteria.

Which *Trichoderma* is the best biocontrol for *Cannabis*? For targeting foliar pathogens, such as *Botrytis* and *Alternaria*, the product must be formulated for that application (e.g., PlantShield®). For damping off and root rots, any of the other *Tricoderma*-based products are appropriate. Even in the absence of pathogens, *Trichoderma* spp. in the rhizosphere promote root growth and enhance nitrogen uptake by plants, resulting in higher crop yields (Woo *et al.*, 2014).

Trichogramma **spp.** are tiny wasps that parasitize eggs of lepidopterans. At least 20 *Trichogramma* spp. have been mass-reared for field use. Adult wasps are 0.3–1.0 mm long, with brown or orange-colored bodies, red eyes, and short antennae (Fig. 20.34). *Trichogramma* wasps are most effective at parasitizing young eggs, so repeated releases are necessary. *Trichogramma* spp. can be combined with predator biocontrols. *Trichogramma* spp. are rarely affected by entomopathogenic fungi or Bt. Watch out for ants, which eat *Trichogramma* pupae.

Suppl: pupae within parasitized eggs, attached to cards made of cardboard, paper, or bamboo; **SL**: up to a week in a cool (6–12°C), dark place. Cold-hardy species may tolerate longer storage. *Trichogramma* pupae are manually distributed by hanging cards from plants in a warm, humid place out of direct sunlight. This approach takes about 30 min per hectare (Smith, 1996). *Trichogramma* pupae have been broadcast from tractors or airplanes. They are attached to carriers, such as bran. Pupae and carriers can be coated with acrylamide sticky gel so they adhere to plant surfaces. Avoid releasing in cold, rainy, or windy conditions. The wasps cannot fly against winds >7 km/h, which blow them out of release fields (Bigler *et al.*, 1997).

The life cycles of *Trichogramma* spp. are similar. Females oviposit individual eggs into up to 200 pest eggs, which turn black when parasitized (Fig. 20.35A,B). Larvae feed within the eggs (Fig. 20.35C), and pupate there (Fig. 20.35D). This development takes about 10 days, and adults emerge from exit holes in the eggs (Fig. 20.35E). Adult males (Fig. 20.35F) have longer antennae than adult females (Fig. 20.25G). Adults live 7–14 days. *Trichogramma* wasps search for target eggs by

walking across leaf surfaces. Leaf trichomes slow them down. On tomato leaves the wasps get entangled in trichome exudates and die, especially if trichome exudates contain methyl ketones (Kashyap *et al.*, 1991). *Cannabis* trichomes exude methyl ketones (Turner *et al.*,1980), and probably hinder *Trichogramma* wasps.

T. minutum was the first *Trichogramma* sp. described in North America. "Though scarcely more than 0.02 inch long [0.5 mm], it can jump to the distance of several inches" (Riley, 1871). Riley (1879) reared the second species, *T. pretiosum*, from eggs of the cotton budworm, *Alabama argillacea*. Flanders (1930) mass-reared *T. minutum* on eggs of *Sitotroga cerealella*. He attached 150,000 *S. cerealella* eggs onto sticky carboard discs. Each *T. minutum* female laid a solitary egg in up to 300 *S. cerealella* eggs. Soon the 150,000 eggs were parasitized, and Flanders hung the cardboard discs in crop fields. Within 3–6 days, 150,000 adult *Trichogramma* wasps emerged, and flew off to parasitize field pests. Not much has changed with this methodology.

T. minutum and *T. pretiosum* are polyphagous, and parasitize the eggs of over 200 species of Lepidoptera. Mixtures of both species are available. Clarke (unpublished research, 1996) used a combination of *T. minutum*, *T. pretiosum*, and *T. evanescens* for best control in glasshouses. *T. pretiosum* works best on plants under 5 ft (1.5 m) tall, while *T. minutum* protects taller plants. Neither species is very cold-hardy.

T. evanescens is European and parasitizes the eggs of over 60 Lepidopteran species. These include major *Cannabis* pests: *Grapholita delineana*, *Ostrinia* spp., *Heliocoverpa* spp., *Agrotis* spp., *Spodoptera* spp., *Mamestra* spp., *Chilo* spp., *Autographa gamma*, etc. Adults are weak fliers and usually move less than 3 m from release sites. At the first sight of ECB or budworm moths, release 10 pupae/m² in the glasshouse. *T. evanescens*, which is cold-hardy, can be released at wider intervals than other *Trichogramma* spp.—every 1–3 weeks, using 200,000–300,000 wasps/hectare (Smith, 1996).

T. dendrolimi is an East Asian species, commercialized in China. It parasitizes the eggs of over 72 Lepidopteran

Fig. 20.34. *Trichogramma dendrolimi* parasitizing eggs (courtesy Victor Fursov, Wikimedia Commons)

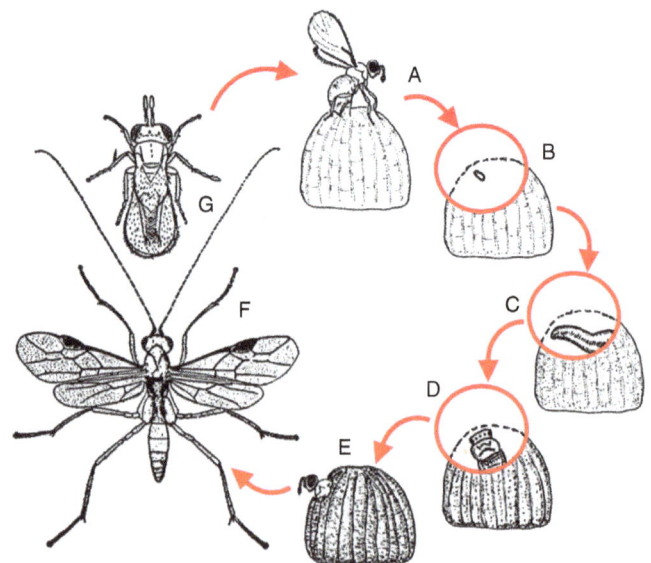

Fig. 20.35. Life cycle of *Trichogramma minutum*

pests—basically the same as *T. evanescens*. The species is cold-hardy, but is a poor flier, and does not do well in windy conditions.

T. brassicae is a Eurasian species, marketed for pest control of brassica crops, but works well on any low-growing plant (i.e., young *Cannabis*). *T. brassicae* is polyphagous on over 150 lepidopteran species. It is cold-hardy, but susceptible to mechanical injury during preparation and shipping. Depending on the number of *Ostrinia nubilalis* generations per season, a total of 4–9 releases is required (Smith, 1996). Bigler *et al.* (1997) found that *T. brassicae* heavily parasitized ECB eggs within an 8 m radius of each release site.

T. ostriniae is native to China, where it parasitizes *Ostrinia furnacalis*, and has been introduced to the USA to control *Ostrinia nubilalis*. It is not cold-hardy and does not overwinter. Releases of 30,000 *T. ostriniae* per acre have resulted in season-long parasitism of *O. nubilalis* egg masses. In maize, *T. ostriniae* prefers *Ostrinia* eggs found in the lower and middle parts of plants—not flowering tops.

T. nubilale is a North American parasitoid of *O. nubialis*, and has been introduced into China for control of *Ostrinia furnacalis*. However, it may not reduce *O. nubialis* populations consistently (Andow *et al.*, 1995). One study showed *T. nubilale* was 15% less effective than *T. ostrinia*. Simultaneous release of both species resulted in worse control—so *T. nubilale* and *T. ostrinia* should not be used together.

Typhlodromus cannabis is a phytoseiid mite that preys on its fellow spider mites. This new species was described in Yúnnán Province, China. Ke and Xin (1983) collected it on *Cannabis sativa*; it was presumably feeding on two-spotted spider mites, *Tetranychus urticae*.

Tree frogs are not commercially available, but they turn up in *Cannabis* fields cultivated along forest edges or near breeding ponds. They're another benefit of practicing conservation biocontrol (see section 20.3.3). The species *Hyla versicolor* (Fig. 20.36), an avid insectivore, was found in Vermont hemp (courtesy John Bruce).

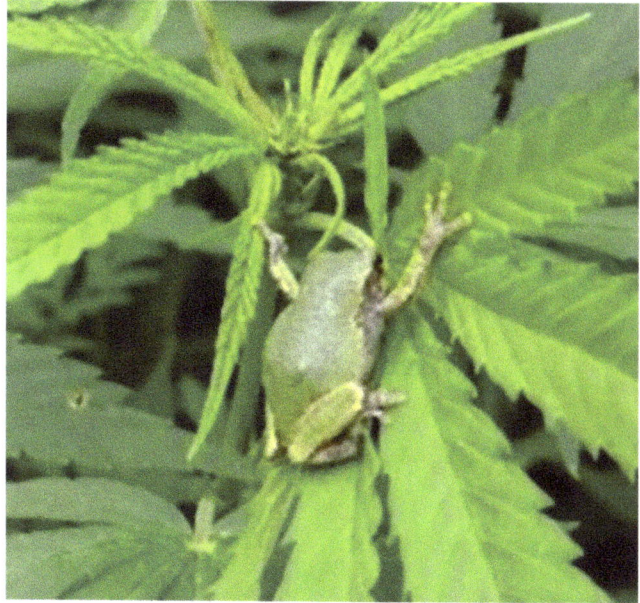

Fig. 20.36. Unexpected biocontrol (courtesy John Bruce)

(Prepared by J. McPartland; revised by M. Bolt, Purdue; F. van Elswijk, Koppert)

References

Abd-Elgawad MMM, Askary TH. 2018. Fungal and bacterial nematicides in integrated nematode management strategies. *Egyptian Journal of Biological Pest Control* 28: e74.

Abdulkhani A, Hosseiinzadeh J, Ashori A, Esmaeeli H. 2015. Evaluation of the antibacterial activity of cellulose nanofibers/polylactic acid composites coated with ethanolic extract of propolis. *Polymer Composites* 38: 13–17.

Al Abassi S, Birkett MA, Pettersson J, *et al.* (3 other authors). 2000. Response of the seven-spot ladybird to an aphid alarm pheromone and an alarm pheromone inhibitor is mediated by paired olfactory cells. *Journal of Chemical Ecology* 26: 1765–1771.

Alefeld F. 1856. *Die Bienenflora Deutschlands und der Schweiz*. Rüchler, Darmstadt.

Alletz PA. 1773. *L'Albert moderne, ou nouveaux secrets éprouvés et licites, recueillis d'après les découvertes les plus récentes*. Duchesne, Paris

Almaghrabi OA, Massoud SI, Abdelmoneim. 2013. Influence of inoculation with plant growth promoting rhizobacteria (PGPR) on tomato plant growth and nematode reproduction under greenhouse conditions. *Saudi Journal of Biological Sciences* 20: 57–61.

Ambrose DP. 2003. Biocontrol potential of assassin bugs (Hemiptera: Reduviidae). *Journal of Experimental Zoology, India* 6: 1–44.

Anderson I. 1987. Rumpus over rogue release of microbes. *New Scientist* 115 (no. 1575): 15.

Andow DA, Klacan GC, Bach D, Leahy TC. 1995. Limitations of *Trichogramma nubilale* (Hymenoptera: Trichogrammatidae) as an inundative biological control of *Ostrinia nubilalis* (Lepidoptera: Crambidae). *Environmental Entomology* 24: 1352–1357.

Angus TA. 1954. A bacterial toxin paralyzing silkworm larvae. *Nature* 173: 545–546.

Anonymous. 1783. Suite des Mémoires sur les vers des Vignes. *Journal de l'Agriculture, du Commerce et des Finances et Arts* 1783(1): 1–33.

Aoki K, Chigasaki Y. 1915. Über die Pathogenität der sog. Sottô-Bacillen (Ishiwata) bei Seidenraupen. *Mitteilungen aus der Medicinischen Facultät der Kaiserlich-Japanischen Universität zu Tokyo* 13: 419–440.

Aristotle (Thompson DW, *trans*). 1910. *A history of animals.* Clarendon Press, Oxford, UK.

Balfour E. 1887. *The agricultural pests of India.* Bernard Quaritch, London.

Balthazar C, Novinscak A, Cantin G, Joly DL, Filion M. 2022b. Biocontrol activity of *Bacillus* spp. and *Pseudomonas* spp. against *Botrytis cinerea* and other *Cannabis* fungal pathogens. *Phytopathology* 112: 549–560.

Barta M, Cagáň L. 2006. Aphid-pathogenic Entomophthorales (their taxonomy, biology and ecology). *Biologia, Bratislava* 62(suppl. 21): S543–S616.

Bassey-Orowuje M, Thomas J, Wakeford T. 2019. Exterminator genes: the right to say no to ethics dumping. *Development* 62: 121–127.

Bassi A. 1835. *Del mal del segno, calcinaccio o moscardino.* Tipografia Orcesi, Lodi.

Batra SW. 1976. Some insects associated with hemp or marijuana (*Cannabis sativa* L.) in northern India. *Journal of the Kansas Entomological Society* 49: 385–388.

Béguy M. 1885. Une lettre de M. Béguy préconisant l'emploi du chanvre pour détruira lo phylloxéra. *Le Courrier de la Rochelle*, 13 juin 1885 [no page number].

Ben-Dov E. 2014. *Bacillus thuringiensis* subsp. *israelensis* and its dipteran-specific toxins. *Toxins* 6: 1222–1243.

Benhamou N, Garand C, Goulet A. 2002. Ability of nonpathogenic *Fusarium oxysporum* strain Fo47 To induce resistance against *Pythium ultimum* infection in cucumber. *Applied and Environmental Microbiology* 68: 4044–4060.

Berdrow H, ed. 1907. Lebensrätsel im Pflanzenreich. *Illustriertes Jahrbuch der Naturkunde* 5: 131–158.

Bergold G. 1947. Die isolierung des Polyeder-Virus und die Natur der Polyeder. *Zeitschrift für Naturforschung B* 2: 122–143.

Berliner E. 1915. Über die Schlaffsucht der Mehlmottenraupe (*Ephestia kuhniella* Zell) und ihren Erreger *Bacillus thuringiensis* n. sp. *Zeitschrift fur angewandte Entomologie Berlin* 2: 29–56.

Bibi S, Husain SZ, Malik RN. 2008. Pollen analysis and heavy metals detection in honey samples from seven selected countries. *Pakistan Journal of Botany* 40: 507–516.

Bigler F, Suverkropp BP, Cerutti F. 1997. "Host searching by Trichogramma and its implications for quality control and release techniques," pp. 240–253 *in* Andow DA, *et al.*, eds. *Ecological interactions and biological control.* Westview Press, Boulder, Colorado.

Bischoff JF, Rehner SA, Humber RA. 2009. A multilocus phylogeny of the *Metarhizium anisopliae* lineage. *Mycologia* 101: 512–530.

Bista S, Shivokoti GP. 2000. Honeybee flora at Kabre, Dolakha District. *Nepal Agricultural Research Journal* 4/5: 18–25.

Boisduval JB. 1867. *Essai sur l'entomologie horticole.* Donnaud, Paris.

Boudreau MA. 2013. Diseases in intercropping systems. *Annual Review of Phytopathology* 51: 499–519.

Briosi G, Tognini F. 1894. Intorno alla anatomia della canapa (*Cannabis sativa* L.). Parte prima, organi sessuali. *Atti della Instituto Botanico di Pavia* (Serie II) 3: 91–209.

Britt KE, Kuhar TP. 2020. Evaluation of miticides to control hemp russet mite on indoor hemp in Virginia, 2019. *Arthropod Management Tests* 45(1): 1–2.

Britt KE, Reed TD, Kuhar TP. 2021a. Evaluation of biological insecticides to managed corn earworm in CBD hemp, 2020. *Arthropod Management Tests* 46(1): 1–2.

Britt KE, Meierotto S, Morelos V, Wilson H. 2022. First year survey of arthropods in California hemp. *Frontiers in Agronomy* 4: e901416.

Butler C. 1609. *The feminine monarchie, or a treatise concerning bees and the due ordering of them.* Joseph Barnes, Oxford.

Cai WZ, Yan YH, Li LY. 2005. The earliest records of insect parasitoids in China. *Biological Control* 32: 8–11.

California DPR (Department of Pesticide Regulation). 2017. *Legal pest management practices for Cannabis growers in California.* Available at: https://www.cdpr.ca.gov/docs/county/cacltrs/penfltrs/penf2015/2015atch/attach1502.pdf

Carner GR. 1976. A description of the life cycle of *Entomophthora* sp. in the two-spotted spider mite. *Journal of Invertebrate Pathology* 28: 245–254.

Cawoy H, Bettiol W, Fickers P, Ongena M. 2011. "Bacillus-based biological control of plant diseases," pp. 273–302 in Stoytcheva M, ed. *Pesticides in the modern world.* InTech, Rijeka, Croatia.

Center for Responsive Politics. 2024. *Industry profile: agricultural services and products.* Available at: https://www.opensecrets.org/federal-lobbying/industries/summary?cycle=2022&id=A07

Chadeuil G. 1882. Le phylloxera. *Le XIXe Siècle*, 7 avril 1882 [no page number].

Charles D. 2002. *Lords of the harvest: biotech, big money, and the future of food.* Perseus Publishing, Cambridge, Massachusetts.

Chavan SB, Vidhate RP, Kallure GS, *et al.* (3 other authors). 2017. Stability studies of cuticle degrading and mycolytic enzymes of *Myrothecium verricaria* for control of insect pests and fungal phytopathogens. *Indian Journal of Biotechnology* 16: 404–412.

Cheke A, Hume JP. 2009. *Lost land of the dodo. The ecological history of Mauritius, Réunion and Rodrigues.* Bloomsbury, London.

Chen M, Arato M, Borghi L, Nouri E, Reinhardt D. 2018. Beneficial services of arbuscular mycorrhizal fungi—from ecology to application. *Frontiers in Plant Science* 9: e1270.

Childers CC, Rodrigues JC, Welbourn C. 2002. Host plants of *Brevipalpus californicus*, *B. obovatus*, and *B. phoenicis* (Acari: Tenuipalpidae) and their potential involvement in the spread of viral diseases vectored by these mites. *Experimental and Applied Acarology* 30: 29–105.

Chomel N, La Marre LH de. 1767. *Dictionnaire oeconomique, nouvelle édition, tome premier.* Taneau, Paris.

Clément JHM. 1887. *Lettres sur le phylloxera, adressées aux vignerons de la paroisse d'Huriel.* Librairie Prot, Montlucon.

Colorado DOA (Department of Agriculture). 2023. *Pesticide use in Cannabis production information.* Available at: https://ag.colorado.gov/plants/pesticides/pesticide-use-in-cannabis-production-information.

Coppel HC, Mertins JW. 1977. *Biological insect pest suppression.* Springer-Verlag, Berlin.

Cordova-Kreylos AL, Fernandez LE, Koivunen M, *et al.* (3 other authors). 2013. Isolation and characterization of *Burkholderia rinojensis* sp. nov., a non-*Burkholderia cepacia* complex soil bacterium with insecticidal and miticidal activities. *Applied and Environmental Microbiology* 79: 7669–7678.

Cornalia E. 1856. Monografia del bombice del gelso. *Memorie dell I.R. Instituo Lombardo di Scienze, Lettere ed Arti* 6: 3–387.

Cox ML. 1994. "The hymenoptera and Diptera parasitoids of Chrysomelidae," pp. 419–468 in Jolivet PH, *et al.*, eds. *Novel aspects of the biology of Chrysomelidae.* Springer Science, Dordrecht.

Crickmore N, Zeigler DR, Feitelson J, *et al.* (5 other authors). 1998. Revision of the nomenclature for the *Bacillus thuringiensis* pesticidal crystal proteins. *Microbiology and Molecular Biology Reviews* 62: 807–813.

Dalio JS. 2012. *Cannabis sativa*—an important subsistence pollen source for *Apis mellifera*. *IOSR Journal of Pharmacy and Biological Sciences* 1(4): 1–3.

Dalmagne H. 1882. Le Phyloxera. *Le Pays*, 28 Mai 1882 [no page number].

Darwin, C. (1860). Letter no. 2814, to Asa Gray. *Darwin Correspondence Project.* Available at: https://www.darwinproject.ac.uk/letter/DCP-LETT-2814.xml.

Darwin E. 1800. *Phytologia.* J. Johnson, London.

Datta M, Chakraborti M. 1983. On a collection of flower flies (Diptera: Syrphidae) with new records from Jammu and Kashmir. *Records of the Zoological Survey of India* 81: 237–253.

De Céris A, *ed.* 1874. Chronique Agricole. *Journal d'Agriculture Pratique* 2: 227–232.

De Geer C. 1776. *Mémoires pour servir à l'histoire des insects, tome sixieme.* Hesselberg, Stockholm.

De Montlaur E. 1873. Réveil du phylloxera après son engourdissement hivernal. *Journal d'Agriculture Pratique* 2: 118–121.

Diesing KM. 1851. *Systema helminthum, vol.* II. Braumüller, Vindobonae.

Dong K, Dong Y, Luo YZ. 2004. 黑带食蚜蝇成虫室内的饲养方法. *Entomological Knowledge* 41(2): 157–160.

Dooling OJ. 1978. "A note on bacterial nomenclature," p. 110 in *Proceedings of the twenty-sixth Annual Western International Forest Disease Work Conference.* USDA Forest Service, Davis, California.

Doyle CJ, Hirst ML, Cory JS, Entwistle PF. 1990. Risk assessment studies: detailed host range testing of wild-type cabbage moth, *Mamestra brassicae* (Lepidoptera: Noctuidae), Nuclear Polyhedrosis Virus. *Applied and Environmental Microbiology* 56(9): 2704–2710.

Dreschler C. 1943. Two species of *Pythium* occurring in southern States. *Phytopathology* 33: 261–299.

Du Mont G. 1873. Note relative à l'influence que pourrait avoir la culture du chanvre pour éloigner des vignobles le Phylloxéra. *Comptes rendus hebdomadaires des séances de l'Académie des Sciences* 77: 715.

Dufour L. 1837. Recherches sur quelques Entozoaires et larves parasites des insectes orthoptères et hyménoptères. *Annales des Sciences Naturelles Zoologie, seconde série* 7: 5–20

Dumas JB. 1876. Études sur le phylloxera et sur les sulfocarbonates. *Annales de Chimie et de Physique, cinquième série* 7: 5–112.

Dumigan CR, Deyholos MK. 2022. *Cannabis* seedlings inherit seed-borne bioactive and anti-fungal endophytic bacilli. *Plants* 11: e2127.

Dziock F. 2002. *Überlebensstrategien und Nahrungsspezialisierung bei räuberischen Schwebfliegen (Diptera, Syrphidae).* UFZ-Bericht, Leipzig.

Eilenberg J. 2002. *Biology of fungi from the order Entomophthorales with emphasis on the genera Entomophthora, Strongwelsea and Eryniopsis.* Doctoral thesis, Royal Veterinary and Agricultural University, Copenhagen.

El-Haidari HS. 1981. The use of predator ants for the control of date palm insect pests in the Yemen Arab Republic. *Date Palm Journal* 1(1): 129–130.

Evans BR, Kotsakiozi P, Costa-da-Silva AL, *et al.* (7 other authors). 2019. Transgenic *Aedes aegypti* mosquitos transfer genes into a natural population. *Scientific Reports* 9: e13047.

Fahlpahl H. 1949. Schütz Hanf gegen Kohlweißlingsschaden? *Nachrichtenblatt für den deutschen Pflanzenschutzdienst* 3: 221.

Fahnenberg C. 1815. Correspondance. *Bulletin de la Société d'Encouragement pour l'Industrie Nationale* 12: 213–220.

Fan B, Blom J, Klenk HP, Borriss R. 2017. *Bacillus amyloliquefaciens, Bacillus velezensis,* and *Bacillus siamensis* form an "Operational Group *B. amyloliquefaciens*" within the *B. subtilis* Species Complex. *Frontiers in Microbiology* 8: e22.

Fargues J, Bon M, Maguin S, Couteaudier Y. 2002. Genetic variability among *Paecilomyces fumosoroseus* isolates from various geographical and host insect origins based on rDNA-ITS regions. *Mycological Research* 106: 1066–1074.

Feldt W. 1919. Vorbeugungsmittel gegen Bohnen-Blattläuse und einige andere Erfahrungen mit Acker- und Puffbohnen in Ostpreussen. *Mitteilungen des Verein zur Förderung der Moorkultur im Deutschen Reiche* 37: 37–40.

Flanders SE. 1930. Mass production of egg parasites of the Genus *Trichogramma. Hilgardia* 4: 465–501.

Flicker NR, Poveda K, Grab H. 2020. The bee community of *Cannabis sativa* and corresponding effects of landscape composition. *Environmental Entomology* 49: 197–202.

Forsskål P (Niebuhr C, *ed.*). 1775. *Descriptiones animalium.* Mölleri, Copenhagen.

Frank SD. 2010. Biological control of arthropod pests using banker plant systems: past progress and future directions. *Biological Control* 52: 8–16.

French NJ. 2016. *Industrial hemp as forage for honey bees.* Available at: https://coloradohemphoney.com/blogs/news/industrial-hemp-as-forage-for-honey-bees

Frey-Klett P, Garbaye J, Tarkka M. 2007. The mycorrhiza helper bacteria revisited. *New Phytologist* 176: 22–36.

Fritze D. 2004. Taxonomy of the genus *Bacillus* and related genera: the aerobic endospore-forming bacteria. *Phytopathology* 94: 1245–1248.

García-Gutiérrez L, Romero D, Zeriouh H, *et al.* (4 other authors). 2012. Isolation and selection of plant growth-promoting rhizobacteria as inducers of systemic resistance in melon. *Plant and Soil* 358: 201–212.

Garreau M. 1876. La destruction du Phylloxera par la culture de plantes parasiticides. *Comptes rendus hebdomadaires des séances de l'Académie des Sciences* 83: 388.

Gaujac C. 1810. Suite de l'extrait d'un Mémoire de M. Gaujac, Propriétaire-Cultivateut à Dagny, près Coulommiers (Seine-et-Marne), sur la culture compare des plantes oléagineuses. *Bulletin Société d'Encouragement pour l'Industrie Nationale* 9: 247–252.

Gistel J. 1856. *Die mysterien der europäischen insectenwelt.* Dannheimer, Kempten.

Glaser RW. 1932. *Studies on Neoaplectana glaseri, a nematode parasite of Japanese beetles (Popillia japonica).* New Jersey Department of Agriculture, Circular no. 211, New Brunswick

Glaser RW, Chapman JW. 1913. The wilt disease of gypsy caterpillars. *Journal of Economic Entomology* 8: 140–150.

Goettel MS, Koike M, Kim JJ, *et al.* (3 other authors). 2008. Potential of *Lecanicillium* spp. for management of insects, nematodes and plant diseases. *Journal of Invertebrate Pathology* 98: 256–261.

Goidànich A. 1928. Contributi alla conoscenza dell'entomofauna della canapa. I. Prospetto generale. *Bollettino del Laboratorio di Entomolgia del R. Istituto Superiore Agrario di Bologna* 1: 37–64.

Gould F. 1991. The evolutionary potential of crop pests. *American Scientist* 79: 496–507.

Goulet C, *ed.* 1886. Chronique de la Quinzaine. *La Vigne Française* 7 (14): 209–211.

Groff GW, Howard CW. 1925. The cultured citrus ant of South China. *Lingnan Agricultural Review* 2(1): 108–114.

Grosier JBG. 1781. Remèdes pour détruire les insectes qui dévorent les bleds à conserve, & garantir des chenilles les choux & les arbres-fruitiers. *Journal de Littérature, des Sciences et des Arts* 4: 408–409.

Gryganskyi AP, Humber RA, Smith ME, *et al.* (6 other authors). 2012. Molecular phylogeny of the Entomophthoromycota. *Molecular Phylogenetics and Evolution* 65: 682–694.

Gupta RK, Srivastava K, Bali K. 2012. An entomophage park to promote natural enemy diversity. *Biocontrol Science and Technology* 22: 1442–1464.

Haeselbarth E, Loan CC. 1983. *Townesilitus,* a new genus for a species group in *Microctonus* (Hymenoptera: Braconidae, Euphorinae). *Contributions of the American Entomological Institute* 20: 384–387.

Hajek AE, Eilenberg J. 2018. Natural enemies: an introduction to biological control, 2nd edition. Cambridge University Press, Cambridge, UK.

Hamerschlag K, Lappé A, Malkan S. 2015. *Spinning food.* Available at: https://foe.org/resources/spinning-food-how-food-industry-front-groups-and-covert-com

Hamet HL. 1861. *Cours Pratique d'apiculture (culture des Abeilles), deuxième edition*. Bureaux de l'apiculteur, Paris.

Harand W. 2000. Hanf (*Cannabis sativa*) gegen Blattläuse. *Pflanzenschutz* 16(4): 8–9.

Harmon JP, Stephens E, Losey J. 2007. The decline of native coccinellids (Coleoptera: Coccinellidae) in the United States and Canada. *Journal of Insect Conservation* 11: 85–94.

Hayman DS. 1982. Influence of soils and fertility on activity and survival of VA mycorrhizal fungi. *Phytopathology* 72: 1119–1125.

Heimpel MA, Angus TA. 1958. The taxonomy of insect pathogens related to *Bacillus cereus* Frankland and Frankland. *Canadian Journal of Microbiology* 4, 531–541.

Henry JE. 1971. Experimental application of *Nosema locustae* for control of grasshoppers. *Journal of Invertebrate Pathology* 18: 389–394.

Herold W. 1949. Schützt Hanf gegen Kohlweißlingsschäden? *Nachrichtenblatt für den deutschen Pflanzenschutzdienst* 3: 155–156.

Herrnstadt C, Soares GG, Wilcox ER, Edwards DL. 1986. A new strain of *Bacillus thuringiensis* with activity against coleopteran insects. *Bio/Technology* 4: 305–308.

Heuskin S, Lorge S, Godlin B, *et al.* (8 other authors). 2012. Optimisation of a semiochemical slow-release alginate formulation attractive towards *Aphidius ervi* Haliday parasitoids. *Pest Management Science* 68: 127–136.

Hoffmann MP, Frodsham AC. 1993. *Natural enemies of vegetable insect pests*. Cooperative Extension, Cornell University, Ithaca, New York.

Hokkanen HMT. 1991. Trap cropping in pest management. *Annual Review of Entomology* 36: 119–138.

Holmes A, Govan J, Goldstein R. 1998. Agricultural use of *Burkholderia* (*Pseudomonas*) *cepacia*: A threat to human health? *Emerging Infectious Diseases* 4: 221–227.

Hooks CRR, Wang K, Ploeg A, McSorley R. 2010. Using marigold (*Tagetes* spp.) as a cover crop to protect crops from plant-parasitic nematodes. *Applied Soil Ecology* 46: 307–320.

Hörstadius S. 1974. Linnaeus, animals and man. *Biological Journal of the Linnean Society* 6: 269–275.

Howell CR. 1982. Effect of *Gliocladium virens* on *Pythium ultimum*, *Rhizoctonia solani*, and damping-off of cotton seedlings. *Phytopathology* 72: 496–498.

Hrebenyuk NV. 1984. Распространенне грибов на тресте конопли (The occurrence of fungi on hemp stems). *Микология и фитопатология* 18(4): 322–326.

Hu SS, Mojahid MS, Bidochka MJ. 2023. Root colonization of industrial hemp (*Cannabis sativa* L.) by the endophytic fungi *Metarhizium* and *Pochonia* improves growth. *Industrial Crops and Products* 198: e116716.

Hu X, Xiao GH, Zheng P, *et al.* (7 other authors). 2014. Trajectory and genomic determinants of fungal-pathogen speciation and host adaptation. *Proceedings of the National Academies of Science USA* 111: 16796–16801.

Huang HC. 1978. *Gliocladium catenulatum*: hyperparasite of *Sclerotinia sclerotiorum* and *Fusarium* species. *Canadian Journal of Plant Pathology* 56: 2243–2246.

Hussey NW, Scopes N. 1985. *Biological pest control: the glasshouse experience*. Blandford Press, Poole, UK.

Husz B. 1928. *Bacillus thuringiensis* Berl., a bacterium pathogenic to corn borer larvae. *International Corn Borer Investigations. Scientific Reports* 1: 191–193.

Ipsilantis I, Samourelis, C, Karpouzas DG. 2012. The impact of biological pesticides on arbuscular mycorrhizal fungi. *Soil Biology and Biochemistry* 45: 147–155.

Ishiwata S. 1901. 劇烈なる一種の軟化病(卒倒病)に就て. 大日本蚕糸会報 (*Great Japanese Silk thread Newsletter Bulletin*) 114: 1–5.

Iwabuchi H. 1908. 通俗蚕体病理学. Meibundo, Tokyo.

Jacobsen BJ, Zidack NK, Larson BJ. 2004. The role of *Bacillus*-based biological control agents in integrated pest management systems: plant diseases. *Phytopathology* 94: 1272–1275.

Jaklitsch WM, Samuels GJ, Dodd SL, Lu BS, Druzhinina IS. 2006. *Hypocrea rufa/Trichoderma viride:* a reassessment, and description of five closely related species with and without warted conidia. *Studies in Mycology* 56: 135–177.

Jamet L, *ed.* 1875. Le Phylloxera. *La Charente*, 25 Mai 1875 [no page number].

Janisiewicz WJ, Marchi A. 1992. Control of storage rots on various pear cultivars with a saprophytic strain of *Pseudomonas syringae*. *Plant Disease* 76: 555–560.

Jauker F, Bondarenko B, Becker HC, Steffan-Dewenter I. 2011. Pollination efficiency of wild bees and hoverflies provided to oilseed rape. *Agricultural and Forest Entomology* 14: 81–87.

Jennings DE, Duan JJ, Follett PA. 2017. "Environmental impacts of arthropod biological control: an ecological perspective," pp. 105–129 in Coll M, Wajnberg E, eds, *Environmental pest management*. John Wiley & Sons, New York.

Jewell W. 1612. *The golden cabinet of true treasure*. John Crosley, London.

Jia HR, Liu YQ, Li XK, *et al.* (6 other authors). 2022. Windborne migration amplifies insect-mediate pollination services. *eLife* 0: e76230.

Johansson F, Hüster H. 1987. *Untersuchungen an Skelettresten von Katzen aus Haithabu (Ausgrabung 1966–1969)*. Wachholtz, Neumünster.

Johnsson L, Hökeberg M, Gerhardson B. 1998. Performance of the *Pseudomonas chlororaphis* biocontrol agent MA 342 against cereal seed-borne diseases in fields experiments. *European Journal of Plant Pholology* 104: 701–711.

Jolivet P. 1950. Les parasites, prédateurs et phorétiques des Chrysomeloidea

(Coleoptera) de la faune franco–belge. *Bulletin de l'Institut royal des Sciences naturelles de Belgique* 26(34): 1–39.

Kakabouski I, Tataridas A, Mavroeidis A, *et al.* (8 other authors). 2021. Effect of colonization of *Trichoderma harzianum* on growth development and CBD content of hemp (*Cannabis sativa* L.). *Microorganisms* 9: e518.

Kalogeropoulos N, Konteles SJ, Troullidou E, Mourtzinos I, Karathanos VT. 2009. Chemical composition, antioxidant activity and antimicrobial properties of propolis extracts from Greece and Cyprus. *Food Chemistry* 116: 452–461.

Kashyap RK, Kennedy GG, Farrar RR. 1991. Behavioral response of *Trichogramma pretiosum* and *Telenomus sphingis* to trichome/methyl ketone mediated resistance in tomato. *Journal of Chemical Ecology* 17: 543–556.

Ke LS, Xin JL. 1983. Notes on three new species of the genus *Typhlodromus* (Acari: Phytoseiidae). *Entomotaxonomia* 5: 185–188.

Keller S. 2006. Entomophthorales attacking aphids with a description of two new species. *Sydowia* 58: 38–74.

Kepler RM, Humber RA, Bischoff JF, Rehner SA. 2014. Clarification of generic and species boundaries for *Metarhizium* and related fungi through multigene phylogenetics. *Mycologia* 106(4): 811–829.

Kerry BR. 1975. Fungi and the decrease of cereal cyst-nematode populations in cereal monoculture. *EPPO Bulletin* 5: 353–361.

Khan AA, Shah MA. 2017. Records of aphids and their natural enemies in agro-ecosystem with special reference to horticultural ecosystem of Kashmir. *Journal of Entomology and Zoology Studies* 5(4): 189–203.

Kirby W, Spence W. 1826a. "Diseases of insects," pp. 197–232 in *An introduction to entomology, vol. IV*. Longman, Rees, Orme, and Green, London.

Kirby W, Spence W. 1826b. "Geographical distribution of insects," pp. 474–508 in *An introduction to entomology, vol.* IV. Longman, Rees, Orme, and Green, London.

Klironomos JN. 2003. Variation in plant response to native and exotic arbuscular mycorrhizal fungi. *Ecology* 84(9): 2292–2301.

Kloepper JW, Schroth MN. 1978. Plant growth-promoting rhizobacteria on radishes. *Proceedings of the 4th International Conference on Plant Pathogenic Bacteria* 2: 879–882. INRA, Angers, France.

Kloepper JW, Leong J, Teintze M, Schroth MN. 1980. Enhanced plant growth by siderophores produced by plant growth-promoting rhizobacteria. *Nature* 286: 885–886

Knipling EF. 1992. *Principles of insect parasitism analyzed from new perspectives.* USDA Agriculture Handbook no. 693, Washington, DC.

Koch E. 1999. Evaluation of commercial products for microbial control of soil-borne plant diseases. *Crop Protection* 18: 119–125.

Koide RT, Mosse B. 2004. A history of research on arbuscular mycorrhiza. *Mycorrhiza* 14: 145–163.

Kokalis-Burelle, N. 2015. *Pasteuria penetrans* for control of *Meloidogyne incognita* on tomato and cucumber, and *M. arenaria* on snapdragon. *Journal of Nematology* 47: 207–2013.

Köllar V. 1837. *Naturgeschichte der schadlichen insecten in beziehung auf landwirthschaft und forstcultur.* Landwirtshafts Gesellschaft, Vienna.

Kondo A, Maeda S. 1991. Host range expansion by recombination of the baculoviruses *Bombyx mori* nuclear polyhedrosis virus and *Autographa californica* nuclear polyhedrosis virus. *Journal of Virology* 65: 3625–3632.

Krieg A, Huger AM, Langenbruch GA, Schnetter W. 1983. *Bacillus thuringiensis* var. tenebrionis: ein neuer, gegenüber Larven von Coleopteren wirksamer Pathotyp. *Zeitschrift für Angewandte Entomologie* 96: 500–508.

Krieg A, Huger AM, Schnetter W. 1987. "Bacillus thuringiensis var. san diego" Stamm M-7 ist identisch mit dem zuvor in Deutschland isolierten käferwirksamen *B. thuringiensis* subsp. *tenebrionis* Stamm BI 256-82. *Zeitschrift für Angewandte Entomologie* 104: 417–424.

Kryukov V, Yaroslavtseva O, Tyurin M, *et al.* (6 other authors). 2017. Ecological preferences of *Metarhizium* spp. from Russia and neighboring territories and their activity against Colorado potato beetle larvae. *Journal of Invertebrate Pathology* 149: 1–7.

Kullnig CM, Krupica T, Woo SL, *et al.* (4 other authors). 2001. Confusion abounds over identities of *Trichoderma* biocontrol isolates. *Mycological Research* 105: 770–771.

Kunkel H, Kloft WJ. 1977. Progress in honeydew research. *Apidologie* 8: 369–391.

Laboulbéne A. 1887. "Phylloxera," pp. 823–836 in *Dictionnaire encyclopédique des sciences médicales, tome vingt-quatrième.* G. Masson, Paris.

Lacey LA, Georgis R. 2012. Entomopathogenic nematodes for control of insect pests above and below ground with comments on commercial production. *Journal of Nematology* 44: 218–225.

Lago PK, Stanford DF. 1989. Phytophagous insects associated with cultivated marijuana, *Cannabis sativa*, in Northern Mississippi. *Journal of Entomological Science* 24: 437–445.

Langellotto GA, Denno RF. 2004. Responses of invertebrate natural enemies to complex-structured habitats: a meta-analytical synthesis. *Oecologia* 139: 1–10.

Lara JM, Fernández C. 2014. Compatibility of the entomopathogenic fungus *Isaria fumosorosea* with pollinators. Ecoletter #10. Available at: https://www.futurecobioscience.com/storage/pdf/compatibility-nofly-pollinators-282.pdf

Leahy J, Mendelsohn M, Kough J, Jones R, Berckes N. 2014. "Biopesticide oversight and registration at the U.S. Environmental Protection Agency," pp. 3–18 in Gross A, *et al.*, eds. *Biopesticides: state of the art and future opportunities.* American Chemical Society, Washington, DC.

Leeuwenhoek A van. 1702. A letter from Mr. Anthony van Leeuwenhoek, F.R.S concerning some insects observed by him on fruit trees. *Philosophical Transactions of the Royal Society of London* 22: 659–672.

Lemay J, Zheng YB, Scott-Dupree C. 2022. Factors influencing the efficacy of biological control agents used to manage insect pests in indoor cannabis (*Cannabis sativa*) cultivation. *Frontiers in Agronomy* 4: e795989.

Leroy PD, Verheggen FJ, Capella Q, Francis F, Haubruge E. 2010. An introduction device for the aphidophagous hoverfly *Episyrphus balteatus* (De Geer) (Diptera: Syrphidae). *Biological Control* 54, 181–188.

Lesage L, Majka CG. 2010. Introduced leaf beetles of the Maritime Provinces, 9: *Chaetocnema concinna* (Marsham, 1802) (Coleoptera: Chrysomelidae). *Zootaxa* 2610: 27–49.

Lewis LC, Lynch RE. 1978. Foliar application of *Nosema pyrausta* for suppression of populations of European corn borer. *Entomophaga* 23: 83–88.

Li HL. 1979. 南方超木状 : *a fourth century flora of Southeast Asia.* Chinese University Press, Hong Kong.

Loomans AJM, Lenteren JC van. 1995. Biological control of thrips pests: a review on thrips parasitoids. *Wagenigen Agricultural University Papers* 95(1): 92–193.

Lu CS, Warchol KM, Callahan RA. 2014. Sub-lethal exposure to neonicotinoids impaired honey bees winterization before proceeding to colony collapse disorder. *Bulletin of Insectology* 67: 125–130.

Lumsden RD, Walter JF, Baker CP. 1996. Development of Gliocladium virens for damping-off disease control. Canadian Journal of Plant Pathology 18: 463–468.

MacGill EI. 1939. A gamasid mite (*Typhlodromus thripsi* n. sp.), a predator of *Thrips tabaci* Lind. *Annals of Applied Biology* 26: 309–317.

Mackiewicz S. 1962. Wpływ wsiewek konopi w ziemniakach i burakach na gęstość porażenia przez stonkę ziemniaczaną i mszyce. *Biuletyn Instytutu Ochrony Roślin* 16: 101–131.

Malinowska D. 1973. Larwy bzygowatych (Diptera) w koloniach mszyc na niektórych roslinach uprawnych. *Polskie Pismo Entomologiczne* 43: 607–613.

Malinowska D. 1979. Communities of aphidophagous syrphids (Diptera, Syrphidae) in the Lublin region. *Memorabilia Zoologica* 30: 37–62.

Maret H, *ed.* 1904. Le phylloxera et le chanvre. *Le Rappel,* 23 août 1904, p. 1.

Marois JJ, Johnston SA, Dunn MT, Papavizas GC. 1982. Biological control of *Verticillium* wilt of eggplant in the field. *Plant Disease* 66: 1166–1168.

Martin PAW, Gundersen-Rindal D, Blackburn M, Buyer J. 2007. *Chromobacterium subtsugae* sp. nov., a betaproteobacterium toxic to Colorado potato beetle and other insect pests. *International Journal of Systematic and Evolutionary Microbiology* 57: 993–999.

Massuela DC, Munz S, Hartung J, Nkebiwe PM, Graeff-Hönninger S. 2023. Cannabis Hunger Games: nutrient stress induction in flowering stage—impact of organic and mineral fertilizer levels on biomass, cannabidiol (CBD) yield and nutrient use efficiency. *Frontiers in Plant Science* 14: e1233232.

Mattirolo O. 1898. Sulla comparsa in Italia della *Entomophora planchoniana* Cornu. *Le Stazioni Sperimentali Agrarie Italiane* 31: 315–326.

Maxwell-Lefroy HM. 1915. The control of white fly and soft scale. *The Gardeners' Chronicle* 58: 154.

McCoy CW. 1996. "Pathogens of eriophyid mites," pp 481–492 in Lindquist EE, *et al.*, eds. *Eriophyoid mites: their biology natural enemies and control.* Elsevier, Amsterdam.

McMurty JA. 1995. "Greenhouse thrips," pp. 77–78 in Nechols JR, ed. *Biological control in the western United States.* University of California, Oakland.

McPartland JM. 2002. Epidemiology of the hemp borer, *Grapholita delineana* Walker, a pest of *Cannabis sativa* L. *Journal of Industrial Hemp* 7(1): 25–42.

McPartland JM, Cubeta MA. 1997. New species, combinations, host associations and location records of fungi associated with hemp (*Cannabis sativa*). *Mycological Research* 101: 853–857.

McPartland JM, Hegman W. 2018. *Cannabis* utilization and diffusion patterns in prehistoric Europe: a critical analysis of archaeological evidence. *Vegetation History and Archaeobotany* 27: 627–634.

McPartland JM, Nicholson J. 2003. Using parasite databases to identify potential nontarget hosts of biological control organisms. *New Zealand Journal of Botany* 41(4): 699–706.

McPartland M, Sheikh Ź. 2018. A review of *Cannabis sativa*-based insecticides, miticides, and repellents. *Journal of Entomology and Zoology Studies* 6: 1288–1299.

McPartland JM, Clarke RC, Watson DP. 2000. *Hemp diseases and pests – management and biological control.* CABI Publishing, Wallingford, UK.

McPartland JM, Di Marzo V, De Petrocellis L, Mercer A, Glass M. 2001. Cannabinoid receptors are absent in insects. *Journal of Comparative Neurology* 436: 423–429.

Meigen JW. 1822. *Systematische Beschreibung der bekannten Europäischen zweiflügeligen Insekten, dritter theil.* Schultz-Wundermann'schen Buchhandlung, Hamm.

Menge JA. 1983. Utilization of vesicular-arbuscular mycorrhizal fungi in agriculture. *Canadian Journal of Botany* 61: 1015–1024.

Merian MS. 1679. *Der Raupen wunderbare Verwandelung.* J.A. Graffen, Nürnberg.

Messelink GJ, Bennison J, Alomar O, *et al.* (5 other authors). 2014. Approaches to conserving natural enemy populations in greenhouse crops: current methods and future prospects. *BioControl* 59: 377–393.

Metchnikov E. 1879. Болезни личинок хлебного жука. P. Frantsova, Odessa.

Mitra B, Banerjee D. 2007. Fly pollinators: assessing their value in biodiversity conservation and food security in India. *Records of the Zoological Survey of India* 107: 33–48.

Montalva C, Collier K, Rocha LFN, *et al.* (4 other authors). 2016. A natural fungal infection of a sylvatic cockroach with *Metarhizium blattodeae* sp. nov., a member of the *M. flavoviride* species complex. *Fungal Biology* 120: 655–665.

Moody AR, Gindrat D. 1977. Biological control of cucumber black root rot by *Gliocladium roseum. Phytopathology* 67: 1159–1162.

Mosse B. 1962. The establishment of vesicular-arbuscular mycorrhiza under aseptic conditions. *Journal of General Microbiology* 27: 509–520.

Mutin VA. 2017. Питание мух-журчалок (Diptera: Syrphidae) пыльцой ветроопыляемых растений. *Чтения памяти Алексея Ивановича Куренцова* 28: 87–94.

Nam HS, Anderson AJ, Kim YC, 2018. Biocontrol efficacy of formulated *Pseudomonas chlororaphis* O6 against plant diseases and root-knot nematodes. *Plant Pathology Journal* 34: 241–249.

Naudin C. 1874. Chanvre contre le phylloxéra. *Le Cosmos* 35: 40.

Nelin CN. 1763. *Svar på Frågan, huru kunna Maskar, som göra skada på Frukt-träd, medelst bloomornas och lösvens affrätande, bäst förekommas och fördisvas?* Lars Salvius, Stockholm.

O'Brien C, Arathi HS. 2019. Bee diversity and abundance on flowers of industrial hemp (*Cannabis sativa* L.). *Biomass and Bioenergy* 122: 331–335.

Ottoni C, Van Neer W, De Cupere B, *et al.* (26 other authors). 2017. The palaeogenetics of cat dispersal in the ancient world. *Nature Ecology & Evolution* 1: e0139

Özkök A, Karlıdağ S, Keskin, *et al.* (5 other authors). 2023. Palynological, chemical, antimicrobial, and enzyme inhibition properties of *Cannabis sativa* L. propolis. *European Food Research and Technology* 149: 2175–2187.

Pace R. 2019. *Studying biocontrols for root aphids.* Cannabis Horticultural Association of Humbolt County. Available at: https://www.chascience.com/horticultural-blog/tag/Beauveria+bassiana

Palmer TS. 1894. "The danger of introducing noxious animals and birds," pp. 87–110 in *Yearbook of United States Department of Agriculture for 1893.* Washington, DC.

Paris M, Boucher F, Cosson L. 1975. The constituents of *Cannabis sativa* pollen. *Economic Botany* 29: 245–253.

Peters J. 1998. *Römische Tierhaltung und Tierzucht. Eine Synthese aus archäozoologischer Untersuchung und schriftlich-bildlicher Überlieferung.* Leidorf, Rahden.

Pineda A, Marcos-García MA. 2008. Introducing barley as aphid reservoir in sweet-pepper greenhouses: Effects on native and released hoverflies (Diptara: Syrphidae). *European Journal of Entomology* 105: 531–535.

Planchon JE. 1875. La défense contre le phylloxera. *Annales Agronomiques* 1: 74–97.

Poinar GO, Thomas GM. 1967. The nature of *Achromobacter nematophilus* as an insect pathogen. *Journal of Invertebrate Pathology* 9: 510–514.

Prasad SK. 1997. Hemp *Cannabis sativa*, a new place to harbour mustard aphid, *Lipaphis erysimi. Indian Journal of Entomology* 59: 232.

Pratje JH. 1768. Mittel wider die Kohl-Raupen. *Landwirthschaftliche Erfahrungen zum Besten des Landmannes: eine Wochenschrift.* Erstes Quartal (5): 552.

Ramirez BA. 1865. *Diccionario de Bibliografía agronómica.* Rivadeneyra, Madrid.

Réaumur RA de. 1726. Remarques sur la plante appellée à la Chine Hia Tsao Tom Tschom, ou plante ver. *Mémoires de l'Académie Royale des Sciences* 1726: 302–306.

Réaumur RA de. 1736. *Mémoires pour servir a l'histoire des insectes, tome second.* Imprimerie Royale, Paris.

Réaumur RA de. 1737. "Histoire des vers mangeurs de pucerones," pp. 363-412 in *Mémoires pour servir a l'histoire des insectes, tome troisieme.* Imprimerie Royale, Paris.

Réaumur RA de. 1742. *Mémoires pour servir a l'histoire des insectes, tome sixième.* Imprimerie Royale, Paris.

Rehner SA, Samuels GJ. 1994. Taxonomy and phylogeny of *Gliocladium* analyzed from nuclear large subunit ribosomal DNA sequences. *Mycological Research* 98: 625–634.

Reverchon-Chamussy P. 1874. La mort du phylloxera. *Annales Agricoles et Littéraires de la Dordogne* 35: 804–807.

Rifai MA. 1969. A revision of the genus *Trichoderma. Mycological Papers* 116: 1–56.

Riley CV. 1871. Third annual report on the noxious, beneficial and other insects of the State of Missouri. *Annual Report of the Missouri Board of Agriculture* 3: 157-158.

Riley CV. 1875. *Seventh annual report on the noxious, beneficial, and other insects of the State of Missouri.* Jefferson City, Missouri.

Riley CV. 1877a. *Ninth annual report on the noxious, beneficial, and other insects of the State of Missouri.* Jefferson City, Missouri.

Riley CV. 1879. Parasites of the cotton worm. *Canadian Entomologist* 11: 161–162.

Riley CV. 1885. *On the cotton worm together with a chapter on the boll worm.* Fourth Report of the US Entomological Commission, USDA. Government Printing Office, Washington, DC.

Riley CV. 1892. *Directions for collecting and preserving insects.* Smithsonian Institution, United States National Museum Bulletin no. 39 part F. Government Printing Office, Washington, DC

Riley CV, Howard LO. 1892. Hemp as a protection against weevils. *Insect Life* 4: 223.

Roberjot C. 1787. Mémoire sur un moyen propre à détruire les chenilles qui ravagent la vigne. *Mémoires de la Société d'Agriculture,* trimestre de printemps, p. 193–206.

Roger-Dobos M. 1875. Extraits des procès-verbaux. *Bulletin de la Société d'Acclimation,* 3e série 2: 240–241.

Rouphael Y, Franken P, Schneider C, *et al.* (6 other authors). 2015. Arbuscular mycorrhizal fungi act as biostimulants in horticultural crops. *Scientia Horticulturae* 196: 91–108.

Roy HE, Steinkraus DC, Eilenberg J, Hajek AE, Pell JK. 2006. Bizarre interactions and endgames: entomopathogenic fungi and their arthropod hosts. *Annual Review of Entomology* 51: 331–357.

Roy HE, Baverstock J, Ware RL, *et al.* (4 other authors). 2008. Intraguild predation of the aphid pathogenic fungus Pandora neoaphidis by the invasive coccinellid Harmonia axyridis. *Ecological Entomology* 33: 175–182.

Ruffner B, Péchy-Tarr M, Ryffel F, *et al.* (5 other authors). 2013. Oral insecticidal activity of plant-associated pseudomonads. *Environmental Microbiology* 15: 751–763.

Sakimura K. 1937. Introduction of *Thripoctenus brui* Vuillet, a parasite of *Thrips tabaci* Lind. from Japan to Hawai. *Journal of Economic Entomology* 30: 799–802.

Samuels GJ, Dodd SL, Lu BS, *et al.* (3 other authors). 2006. The *Trichoderma koningii* aggregate species. *Studies in Mycology* 56: 67–133.

Sanchis V. 2011. From microbial sprays to insect-resistant transgenic plants: history of the biospesticide *Bacillus thuringiensis*. A review. *Agronomy for Sustainable Development* 31: 217–231.

Sansinenea E. 2012. "Discovery and description of *Bacillus thuringiensis*," pp. 3–18 in Sansinenea E, *ed. Bacillus thuringiensis biotechnology*. Springer Science+Business Media, Dordrecht.

Saunders MA. 2018. Insect pollinators collect pollen from wind-pollinated plants: implications for pollination ecology and sustainable agriculture. *Insect Conservation and Diversity* 11: 13–31.

Sauzey A. 1879. Procès-Verbaux. *Annales de la Société d'Agriculture, Histoire naturelle et Arts utiles de Lyon, cinquième série* 2: xxvii.

Scheidweiler M, *ed.* 1847. Moyen de garantir les choux des chenilles et d'autres insects. *Journal d'Horticulture Pratique* 4: 147.

Schiner JR. 1862. *Fauna Austriaca. Die Fliegen (Diptera), Teil 1*. Carl Gerold's Sohn, Wien.

Schmalzel RJ. 1980. *The diet breadth of Apis (Hymenoptera: Apidae)*. Master's thesis, University of Arizona, Tuscon AZ.

Schmid U. 1996. Rettet *Episyrphus balteatus* (De Geer, 1776). *Volucella* 2: 101–103.

Schmidt M. 1929. Blattlausfliegen (Syrphidae) als vorratsschädlinge. *Mitteilungen Gesellschaft für Vorratsschutz* 5(6): 80–81.

Schneider K. 1987. Gene scientist freed germs in 1984 tests. *New York Times* Sept. 2, A1, D19.

Schontzler G. 1999. Decade of classified MSU research sought. *Billings Gazette*, 8 November 1999.

Schreiner M. 2019. *A survey of the arthropod fauna associated with hemp (Cannabis sativa L.) grown in eastern Colorado*. Master's thesis, Colorado State University, Fort Collins.

Schroers HJ. 2001. A monograph of *Bionectria* (Ascomycota, Hypocreales, Bionectriaceae) and its *Clonostachys* anamorphs. *Studies in Mycology* 46: 1–214.

Scopoli GA. 1763. *Entomologia carniolica*. Trattner, Vindobonae.

Scott CA, Punja ZK. 2023. Biological control of *Fusarium oxysporum* causing damping-off and *Pythium myriotylum* causing root and crown rot on cannabis (*Cannabis sativa* L.) plants. *Canadian Journal of Plant Pathology* 45: 238–252.

Serkov VA, Pluzhnikov II. 2012. Эффективность предпосевной обработки семян однодомной конопли посевной (Efficiency of presowing seed treatment in monoecious hemp). Достижения науки и техники АПК 2012(2): 46–46.

Sharma HK, Chauhan A, Katna S, *et al.* (3 other authors). 2017. Nesting attributes of dwarf bee, *Apis florea* Fabricius in Himachal Pradesh. *International Journal of Farm Sciences* 7(2): 22–24.

Sinden SL, DeVay JE, Backman PA. 1971. Properties of syringomycin, a wide spectrum antibiotic and phytotoxin produced by *Pseudomonas syringae*, and its role in the bacterial canker disease of peach trees. *Physiological Plant Pathology* 1: 199–213.

Skowronek P, Wójcik Ł, Strachecka A. 2022. Impressive impact of hemp extract on antioxidant system in honey bee (*Apis mellifera*) organism. *Antioxidants* 11: e707.

Smith BC. 1960. A technique for rearing Coccinellid beetles on dry foods, and influence of various pollens on the development of *Coleomegilla maculata lengi* Timb. (Coleoptera: Coccinellidae). *Canadian Journal of Zoology* 38: 1047–1049.

Smith BC. 1961. Results of raring some coccinellid (Coleoptera: Coccinellidae) larvae on various pollens. *Proceedings of the Entomological Society of Ontario* 91: 270–271.

Smith SM. 1996. Biological control with *Trichogramma*. *Annual Reviews of Entomology* 41: 375–406.

Sorensen WC, Smith EH, Smith JR, Weber DC. 2018. *Charles Valentine Riley: founder of modern entomology*. University of Alabama Press, Tuscaloosa.

Speyer ER. 1927. An important parasite of the greenhouse white-fly (*Trialeurodes vaporariorum*, Westwood). *Bulletin of Entomological Research* 17: 301–308.

Sriram S, Savitha MJ, Rohini HS, Jalali SK. 2013. The most widely used fungal antagonist for plant disease management in India, *Trichoderma viride* is *Trichoderma asperellum* as confirmed by oligonucleotide barcode and morphological characters. *Current Science* 104: 1332–1340.

Starý P, Naumann-Etienne K, Remaudière G. 1998. A review and tritrophic associations of aphid parasitoids (Hym., Braconidae, Aphiliinae) of Pakistan. *Parasitica* 54(1): 3–21.

Steiner G. 1923. *Aplectana kraussei* n. sp., eine in der Blattwespe *Lyda* sp. parasitierende Nematodenform, nebst Bemerkungen über das Seitenorgan der parasitischen Nematoden. *Zentralblatt für Bakteriologie, Parasitenkunde, Infektionskrankheiten und Hygiene, Abteilung II* 59: 14–18.

Steinhaus E. 1956. Living pesticides. *Scientific American* 195(2): 96–104.

Steinhaus EA. 1961. On the correct author of *Bacillus sotto*. *Journal of Insect Pathology* 3: 97–100.

Steinhaus EA. 1975. *Disease in a minor chord*. Ohio State University Press, Columbus, Ohio.

Stigenberg J. 2017. Review of the genus *Townesilitus* (Hymenoptera, Braconidae) in Sweden, with description of a new species and a molecular characterization. *Entomologisk Tidskrift* 138 (2): 137–150.

Stillingfleet B. 1775. *Miscellaneous tracts relating to natural history, husbandry, and physick*. J. Dodsley, London.

Stratii YI. 1976. Hemp and the Colorado beetle. *Zashchita Rasteniĭ* 5: 61.

Sun M. 1986. EPA suspends biotech permit. *Science* 232: 15.

Surina LN, Stolbov NM. 1981. конопля при варроатоз (Hemp and varroatosis). *Пчеловодство (Beekeeping)* 1981(8): 21–22.

Suryanarayana MC, Rao GM, Singh T. 1992. Studies on pollen sources for *Apis cerana* Fabr and *Apis mellifera* L bees at Muzaffarpur, Bihar, India. *Apidologie* 23: 33–46.

Swammerdam JJ. 1669. *Historia Insectorum Generalis, ofte Algemeene Verhandeling van de Bloedeloosen Dierkens*. Meinardus van Dreunen, Utrecht

Swammerdam JJ (Boerhaave H, *ed*). 1737–1738. *Bybel der Natuure, 2 vols*. Severinius, Boudewyn, Van der Aa, Leyden.

Tähvonen R. 1982. The suppressiveness of Finnish light coloured Sphagnum peat. *Journal of the Scientific Agricultural Society of Finland* 54: 345–356.

T'ai HR, Buchmann SL. 2000. A phylogenetic reconsideration of the pollen starch–pollination correlation. *Evolutionary Ecology Research* 2: 627–643.

Tauber MJ, Tauber CA, Daane KM, Hagen KS. 2000. Commercialization of predators: recent lessions from green lacewings (Neuroptera: Chrysopidae: *Chrysoperla*). *American Entomologist* 46(1): 26–38.

Tenhumberg B. 1995. Estimating predator efficiency of *Episyrphus balteatus* (Diptera: Syrphidae) in cereal fields. *Environmental Entomology* 24: 687–691.

Theiling KM, Croft BA. 1988. Pesticide side-effects on arthropod natural enemies: a database summary. *Agriculture Ecosystems and Environment* 21: 191–218.

Tonhasca A, Byrne DN. 1994. The effects of crop diversification on herbivorous insects: A meta-analysis approach. *Ecological Entomology* 19: 239–244.

Trandem N, Bhattarai UR, Westrum K, Knudsen GK, Klingen I. 2015. Fatal attraction: male spider mites prefer females killed by the mite-pathogenic fungus *Neozygites floridana*. *Journal of Invertebrate Pathology* 128: 6–13.

Trandem N, Berdinesen R, Pell JK, Klingen I. 2016. Interactions between natural enemies: effect of a predatory mite on transmission of the fungus *Neozygites floridana* in two-spotted spider mite populations. *Journal of Invertebrate Pathology* 134: 35–37.

Trimoulet AH. 1876. *La maladie de la vigne.* Degréteau, Bordeaux.

Trotuş E, Naie M. 2008. Cercetări privind reducerea atacului unor agenţi patogeni şi dăunători specifici culturilor de cânepă prin tratamentul chimic al seminţei. *Lucrari Stiintifice, Universitatea de Stiinte Agricole Si Medicina Veterinara "Ion Ionescu de la Brad" Iasi, Seria Agronomie* 51(2): 219–223.

Tsvetkov N, Sampson-Robert O, Sood K, *et al.* (5 other authors). 2017. Chronic exposure to neonicotinoids reduces honey bee health near corn crops. *Science* 356: 1395–1397.

Tubbs PK, *ed.* 2000. Precedence of names in wide use over disused synonyms or homonyms in accordance with Article 23.9 of the Code. *Bulletin of Zoological Nomenclature* 57: 6–10.

Türckheim R de. 1874. Du "Phylloxera vastatrix" ou puceron de la vigne. *Bulletin de la Société Académique du Bas-Rhin* 8: 20–32.

Türk MU, Şahinler N, Dinler H. 2022. Chemical structure and antifungal activity of Agean region of propolis in Türkiye. *Turkish Journal of Agriculture* 10: 2571–2582.

Turner CE, Elsohly MA, Boeren EG. 1980. Constituents of *Cannabis sativa* L. XVII. A review of the natural constituents. *Journal of Natural Products* 43: 169–234.

Vallot JN. 1836. Sur une Phrygane. *Mémoires de l'Académie des Sciences, Arts et Belles-lettres de Dijon* 1836(2): 237–246.

Van Lenteren JC. 1995. "Integrated pest management in protected crops," pp. 311-343 *in* Dent D, *ed. Integrated pest management.* Chapman & Hall, London.

Van Lenteren JC, Godfray HCL. 2005. European science in the Enlightenment and the discovery of the insect parasitoid life cycle in the Netherlands and Great Britain. *Biological Control* 32: 12–24.

Veselý D. 1977. Potential biological control of damping-off pathogens in emerging sugar beet by *Pythium oligandrum* Drechsler. *Phytopathologische Zeitschrift* 90: 113–115.

Vigne JD, Guiliane J, Debue K, Haye L, Gérard P. 2004. Early taming of the cat in Cyprus. *Science* 304: 259.

Wagner S, *ed.* 1867. Bee pasturage. *American Bee Journal* 3: 29–30.

Weindling R. 1934. Studies on a lethal principle effective in the parasitic action of *Trichoderma lignorum* on *Rhizoctonia solani* and other soil fungi. *Phytopathology* 24: 1153–1179.

Weinhold AR, Bowman T. 1968. Selective inhibition of the potato scab pathogen by antagonistic bacteria and substrate influence on antibiotic production. *Plant and Soil* 28:12–24.

Wells HD, Bell DK, Jaworski CA. 1972. Efficacy of *Trichoderma harzianum* as a biocontrol for *Sclerotium rolfsii*. *Phytopathology* 62: 442–447.

Wheeler AG, Hoebeke ER, Smith EH. 2010. Charles Valentine Riley: taxonomic contributions of an eminent agricultural entomologist. *American Entomologist* 56: 14–30.

Wilder A. 1879. Letter to the editor. *American Bee Journal* 15: 84.

Willis RJ. 2007. *The history of allelopathy.* Springer, Dordrecht, The Netherlands.

Wilson H, Bodwitch H, Carah J, *et al.* (4 other authors). 2019. First known survey of cannabis production practices in California. *California Agriculture* 73: 119–127.

Woo SL, Ruocco M, Vinale F, *et al.* (7 other authors). 2014. *Trichoderma*-based products and their widespread use in agriculture. *Open Mycology Journal* 8: 71–126.

Wylie HG. 1988. Release in Manitoba, Canada of *Townesilitus bicolor* (Hym.: Braconidae), an European parasite of *Phyllotreta* spp. (Col.: Chrysomelidae). *Entomophaga* 33: 25–32.

Yuan WM, Crawford DL. 1995. Characterization of *Streptomyces lydicus* WYEC108 as a potential biocontrol agent against fungal root and seed rots. *Applied and Environmental Microbiology* 61: 3119–3128.

Zare R, Gams W. 2001. A revision of *Verticillium* section *Prostrata*. IV. The genera *Lecanicillium* and *Simplicillium* gen. nov. *Nova Hedwigia* 73: 1–50.

Zarei M, Tadayon MR, Tadayyon A. 2014. Effect of biofertilizer, under salinity condition on the yield and oil content of three ecotype of hemp (*Cannabis sativa* L.). *Journal of Crops Improvement* 16: 517–529.

Ziarkiewicz T, Anasiewicz T. 1961. Badania nad wplywem konopi na wystepowanie bielinka kapustnika (*Pieris brassicae* L.) na kapuscie (*Brassica alba* v. *capitata* L.). *Roczniki nauk rolniczych Seria A* 83 (A): 641–649.

21 Biorational chemical control

Abstract

Pesticides have been used since the dawn of monoculture in Mesopotamia. A biorational pesticide is a chemical that is relatively non-toxic to people, with few environmental side effects, a modicum of selectivity, and will not persist in the environment. Pesticide regulations, formulations, application strategies, and safety requirements are summarized. Pesticides pose costs to human health and to the environment. Costs vary, ranging from naturally occurring substances (e.g., neem seed oil, hort oil, insecticidal soap, sulfur), to biorational synthetic pesticides (e.g., heat-killed bacterial cultures, insect pheromones), to a rogues' gallery of chemicals banned from use on *Cannabis*. Naturally occurring products draw from the EPA's list of "Section 25(b) exempt pesticide products." What can be used on *Cannabis* depends upon the intended product: fiber, seed, or cannabinoids. Many pesticide residues in herbal cannabis pass into smoke, which poses a unique risk. This has necessitated the passage of residue testing regulations.

21.1 Introduction

A **pesticide** is a chemical that kills, incapacitates, or deters a pest or pathogen. This definition offers options beyond lethal force. Baiting a field with an artificial sex pheromone, for instance, confuses male insects and makes reproduction impossible. Pesticides are named by their targets: **miticides (acaricides)** kill mites, **insecticides** kill insects, **fungicides** kill molds, **antibiotics** kill bacteria, **herbicides** kill plants, **nematicides** kill nematodes, **avicides** kill birds, **rodenticides** kill rodents.

Use pesticides only when all else fails (Fig. 21.1). They cause off-target damage—*acute* toxicity or *chronic* debility, the latter of which may transcend generations. Pesticides trigger unpredictable side effects and have hidden costs. Costs to the environment and costs to human health often go unrecognized at first. We get blindsided later—wading knee-deep in the chemical, facing a crisis.

Stern *et al.* (1959) emphasized that pesticides should be *selective*—targeting pests, while sparing other species. In fact, few chemicals are truly pest-selective; "pesticide" is a misnomer. They are really **biocides**, and harm many living things. The casualties of collateral damage are called "non-target organisms." Honey bees and bald eagles are famous non-target victims. Mycorrhizae are less famous—they live underground and are microscopic—but they are killed by fungicides, insecticides, and nematicides. Pesticides may even damage the plants they are supposed to protect.

Ideally a pesticide does not persist in the environment. After serving its purpose, it should degrade. Rachel Carson brought our attention to the dangers of persistent pesticides, which "biomagnify" in the food chain (Carson, 1962). The poster-child of persistence is DDT. One ppm DDT in pond water might accumulate to 100 ppm in plankton, and accumulate to 1000 ppm in plankton-feeding fish. Eagles eat an accumulation of fish, and become an endangered species.

A **biorational pesticide** is a chemical that is relatively non-toxic to people, with few environmental side effects, a modicum of selectivity, and will not persist in the environment. The term was coined by Djerassi *et al.* (1974), who listed pheromones, insect hormones, and insect growth regulators as examples. These examples show that "biorational" is open to interpretation; after all, some insect growth regulators are not very selective (e.g., azadiractin). Similarly, insecticidal soaps and horticultural oils are considered biorational because they lack persistence, but they are broadly biocidal.

Experts on the internet define biorational pesticides as "natural." This fuzzy concept encompasses nasty toxins produced by living things (e.g., nicotine), or—in a total misuse of the word—synthetic analogues (e.g., neonicotinoids). *Natural* does not means these chemicals are safe or belong in baby food. Horowitz *et al.* (2009) had difficulty defining "biorational pesticides." The term is not interchangeable with "organic pesticides" as defined by USDA standards, or with "biopesticides" as regulated by the EPA (see section 21.4). Horowitz concluded that "the term biorational should be used descriptively as an adjective denoting compatibility with living systems within specific contexts."

Our specific context is *Cannabis*, with three sub-contexts: crops grown for fiber, or for seeds and seed oil, or for pharmaceuticals (THC, CBD, etc.). Regarding the latter, we have another sub-context: cannabis is often smoked and inhaled. Few pesticides have been tested in this route of delivery. Tobacco researchers have been interested in pesticide residues in smoke, and the effects of pyrolysis upon pesticides. Studies on pesticide residues in cannabis smoke have only begun (e.g., Sullivan *et al.*, 2013). We discuss the issue at length in section 21.4.

21.1.1 Banned: a brief history of pesticides

Pesticides began with the advent of civilization. Plant domestication in the Fertile Crescent, *ca.* 10,000 BC, gave rise to Sumer, founded around 4000 BC. The civilization's urban center, Uruk, had a population of perhaps 50,000 people by

DOI: 10.1079/9781836990352.0021

Fig. 21.1. Signage required for commercial applicators, purchased at local Walmart store

2900 BC. Sumerians learned to use sulfur as an acaricide and insecticide by 2500 BC (Glass, 1979). In China, wood ashes and lime were used as early as 1200 BC. The Chinese also made early use of botanical pesticides, and by 200 BC their pesticide arsenal included arsenic. In ancient Rome, Pliny (AD 23–79) recorded the use of arsenic, sulfur, and olive oil.

Cannabis has been used as an insecticide, first appearing nearly simultaneously in an Arabic text *ca.* AD 904–931 (Rodgers, 1980), in a Byzantine text *ca.* AD 945–959 (Beckh, 1895), and in a Chinese text *ca.* AD 980 (Needham *et al.*, 1986). Perhaps 200 research publications have shown that *Cannabis* can repel or kill insects, mites, nematodes, fungi, and bacteria (see Chapter 3). These pesticidal capabilities are marginal, otherwise we wouldn't need to write this book.

During Europe's agricultural revolution, which began 300 years ago, new pesticides came from exotic plants, inorganic compounds (e.g., copper sulfate), and heavy metals (e.g., arsenic, mercury, lead). The first synthetic insecticide—a substance that does not occur naturally—was Antinonin™ (4,6-dinitro-*o*-cresol), marketed by Germany's Friedrich Bayer & Co. in 1899. It was not banned by the EU until 1999.

The Bayer corporation (founded in 1863) and BASF (founded in 1865, also in Germany) are now the 2nd and 3rd largest agrochemical corporations worldwide. Syngenta in Basel became the largest in 2000, after the merger of Novartis and AstraZeneca. In the USA, Dow Chemical (founded 1897), Monsanto (founded 1901, now part of Bayer), and DuPont (founded 1802) ranked 4th, 5th, and 6th worldwide.

Other corporations poisoned themselves to death. Chemische Fabrik Marktredwitz (CFM) produced a fungicide made from mercuric chloride, Fusariol®, beginning in 1912. Fusariol® was the first fungicide tested on hemp (see Fig. 12.7). Mercury contamination overtook CFM, and the factory closed in 1985. The site became the world's first large-scale mercury remediation project; between 1992 and 1996, a government contractor processed 50,000 metric tons of mercury-contaminated wastes.

The Golden Age of pesticides began after Paul Müller discovered the insecticidal powers of DDT in 1939. He was awarded a Nobel Prize. The postwar period heralded an unprecedented surge in pesticide usage, accompanied by the adoption of monocultural crop regimes. *Cannabis* crops, being illegal in most Western countries, did not board that runaway train.

Public attitudes pivoted after Rachel Carson's *Silent Spring*, which dampened the celebratory and triumphalist rhetoric. Paradoxically, since the publication of Carson's book in 1962, the use of pesticides has tripled in the USA. Mart (2015) addressed the paradox: Increased usage is primarily due to the rise of pesticide-resistant pests. Pimentel (2005) claimed that insecticide usage in the USA has grown tenfold since 1945, yet at the same time, crop losses from insects have nearly doubled, from 7% to 13%. The more we spray, the more we accelerate evolution towards resistant pests.

The first case of pesticide resistance was documented by Melander (1914): populations of San Jose scale (*Quadraspidiotus perniciosus*) were untouched by sulfur-lime sprays. By 1946 resistance was identified in 11 insect pests (Metcalf *et al.*, 1962). The raw power of DDT gave hope that resistance wouldn't arise. But by 1947 housefly resistance to DDT had evolved—just 2 years after DDT was approved for public sale. With the introduction of every new insecticide class (organochlorines, carbamates, imidazolines, organophosphates, pyrethroids, neonicotinoids), resistance has surfaced within 2–20 years. Today, according to various estimates, there are 500–1000 resistant pest species.

21.2   Pesticide formulations

Pesticide formulations are summarized in Table 21.1. **Fumigants** are gases. They are very effective but they must be used in enclosed areas, they require expensive equipment, and most are carcinogenic. **Sprays** are liquids and much easier to apply. Spray formulations include emulsifiable concentrates (ECs), flowable concentrates (FCs), soluble powders (SPs), dry flowables (DFs), and wettable powders (WPs). Sprays may be ineffective if they bead up and roll off plant surfaces. This occurs on plants whose foliar surfaces have waxy layers and a pubescence of trichomes (i.e., *Cannabis*).

Dusts adhere firmly to plant surfaces but are susceptible to wind drift during application. **Granules** are similar to dusts but consist of larger particles, with less wind-drift. Granules pose a serious hazard to birds, who mistake them for food or grit and eat them. **Baits** are often formulated as granules but have the power of attraction (e.g., food attraction, color attraction, sex attractant).

The amount of active ingredient (a.i.) present in a pesticide is listed on the label. Solid formulations list the amount as a percentage of weight. For instance, a 50 WP contains 50% a.i. by weight. Liquid formulations indicate the amount of a.i. in pounds per gallon. For instance, a 4 EC contains 4 pounds of the a.i. per gallon of emulsifiable concentrate.

Pesticides work when pests ingest them (**stomach poisons**) or touch them (**contact poisons**). Stomach poisons work against pests who chew and swallow their food. They are not very effective against insects with piercing-sucking mouth parts (e.g., aphids), who do not ingest poisons adhering to plant surfaces. Sap-sucking insects can be controlled with contact poisons.

Table 21.1. Chemical formulations

Formulation	Abbrev.	Description
Aerosol	A	Pesticide dissolved in volatile solvent and pressurized in a can by a propellant gas like CO_2, a convenient but expensive formulation.
Bait	B	Pesticide impregnated into a substrate attractive to pests, such as food or sex pheromones.
Dry flowable	DF	Dry pesticide formulated as granules, to be mixed with water and sprayed; much like a WP, but the heavier DF granules reduce a person's exposure to airborne particles during handling and mixing.
Dust	D	Nearly microscopic particles of powder-dry pesticides diluted in a dry carrier; dusts adhere well to plant surfaces, useful for treating small indoor areas but tend to drift outdoors.
Emulsifiable concentrate	EC	Petroleum-based liquid plus emulsifier, mix with water and spray; like all sprays, may ineffectively roll off plant surfaces. Phytotoxicity hazard usually greater than other liquid formulations, may leave visible residue on plants.
Flowable concentrate	FC	Solid or semi-solid pesticide wet-milled into a pudding-like consistency to be mixed in water and sprayed; requires frequent agitation to remain suspended, may clog spray equipment, may leave visible residue on plants.
Fumigant	F	Vapor usually stored under pressure in metal bottles as liquified gas; vaporizes when released, quickly dissipates, dangerous, expensive.
Granule	G	Prepared by applying liquid pesticides to coarse particles of porous material; granules are like dusts, but larger so they wind-drift less.
Plant-incorporated protectant	PIP	A category proposed by the EPA for bioengineered crop plants yielding toxins normally produced by other organisms, such as plants that produce Bt—a toxin ordinarily produced by *Bacillus thuringiensis*.
Slow-release	SR	Pesticide embedded in polychlorovinyl resin (e.g., No-Pest Strip®) to slow the rate of volatilization; other SRs include paint-on pesticides and microencapsulation in semi-permeable membranes.
Soluble powder	SP	Soluble powder that dissolves in water to form a true solution; packaged as powders or concentrated solutions; unlike FCs, no further agitation is needed after SPs are dissolved.
Wettable powder	WP	Insoluble pesticide mixed with mineral clay (WPs look like Ds but are more concentrated); suspended in water with a surfactant; act like FCs (i.e., sprays may clog), safer to plants than ECs, may leave visible residue on plants.

A **systemic pesticide** moves inside a plant following absorption. This movement is usually upward (through xylem) and outward, into the plant's tissues. Systemics often remain in plant tissues without breaking down. For example, aldicarb, a systemic pesticide used to treat spider mites on hops, produces a peak concentration in hops flowers *60 days* after application (Duke, 1985). You can't wash off systemic pesticides. Systemics that move into pollen and nectar may kill honey bees and other pollinators.

The Environmental Protection Agency (EPA) proposed a new pesticide category in 1994. Originally called "plant-pesticides," the category now has the less alarming name "**plant-incorporated protectants**" (PIPs). PIPs are GMOs (genetically modified organisms)—plants genetically manipulated to biosynthesize toxins normally produced by other organisms. We introduce GMOs in section 1.2.3, with a photo of the world's first GMO *Cannabis* (see Fig. 1.5). In 1995 the EPA registered the first PIP plant: Monsanto's transgenic NewLeaf® potato. The plant expressed Cry 3A proteins, normally produced by *Bacillus thuringiensis* (Bt). Six years later, Monsanto quietly mothballed NewLeaf®. Farmers didn't want to pay Monsanto's

premium prices, and McDonald's, the fast-food giant, refused to use GMO potatoes for its french fries.

Calgene's Flavr Savr™ tomato was registered around the same time. Genetic engineers added an antisense polygalacturonase gene which prolonged shelf life. Campbell Soup Company did not to use the GMO tomato because of public opposition, plus it didn't taste any better (as promised), it was more expensive, and the product was pulled from the market (Bruening and Lyons, 2000).

Since 1995, the EPA has registered 12 PIPs, primarily GMO cotton and maize. Many Bt crops now express multiple transgenes, including two or more Cry proteins. A new class of PIPs express double-stranded RNA (dsRNA), which suppresses the expression of specific proteins in pests. The first dsRNA-expressing GMO was approved in 2017, Monsanto's SmartStax® corn. Unanticipated problems have cropped up with transgenic plants. Insect resistance has arisen. Transgenes from GMOs have introgressed into their wild relatives. Corporations that create GMO plants seek to monopolize seed sales. This leads to losses in crop biodiversity and food sovereignty. For more see Chapter 22, available on the web.

21.3   Pesticide costs

Pesticides pose costs to human health and costs to the environment. *Most pesticides harm humans.* Farmers face acute poisoning as a job hazard. Organophosphates cause the majority of acute poisonings, because of their potency and widespread use. Most are EPA Class I toxins. The EPA's toxicity rating system, based on rodent studies, appears in Table 21.2. A lethal dose of organophosphates can be absorbed through the skin, eyes, lungs, or stomach (never use your mouth to prime a siphon or clear a spray line). In one remarkable report, parathion spilled on a pair of coveralls hospitalized three workers over a 2-week period, despite repeated launderings (Clifford and Nies, 1989).

Pimentel (2005) cited an EPA estimate of 300,000 pesticide poisonings per year in the USA. Most poisonings involved children. He estimated that poisonings cost $228 million per year in the USA, including hospitalizations, outpatient treatments, lost work, and cost of fatalities. Additionally, Pimentel counted over 8 million cases of domestic animal pesticide poisonings in the USA, costing $30.7 million per year. To quote Cynthia Westcott (1964), "The use of chemicals by amateurs is hazardous in any event." Always read the label and never act in haste.

Besides acute toxicity, many pesticides cause chronic toxicity as a result of long exposure. Pesticides may be carcinogenic (cause cancer), mutagenic (cause gene damage), teratogenic (cause birth defects), or oncogenic (cause tumors). Agricultural workers who handle pesticides suffer an increased incidence of cancer. Mills and Kwong (2001) linked data from the California Cancer Registry with a membership list of the United Farm Workers. They found an elevated incidence of leukemia, stomach cancer, uterine cancer, and brain tumors.

Pimentel (2005) implicated pesticides in Mills and Kwong's findings, and extrapolated to the general population: if everyone suffered the same pesticide exposure as farm workers, there would be 83,000 new cancer cases per year. Instead, Pimentel estimated the incidence of new cancer cases in the USA population due to pesticides as "only" 10,000 to 15,000 per year. He reckoned a cost of $1 billion per year (10,000 cases × US$100,000 per case—the latter a considerable underestimate).

Estrogenic pollutants, *a.k.a.* endocrine-disrupting chemicals (EDCs), are an increasing problem. Estrogenic EDCs may trigger estrogen-dependent breast cancer, or harm developing fetuses. *In utero* EDC exposure may cause a boy to be born with undescended testicles, have a low sperm count at puberty,

testicular cancer in middle age, or prostate cancer as an old man—all from prenatal exposure (Colborn *et al.*, 1996).

Levels of EDCs are rising in the water supply (Lecompte *et al.*, 2017). EDCs include insecticides (especially neonicotinoids) and plasticizers (phthalates, bisphenol A, 4-nonylphenol). Plasticizers are added to tubing used in hydroponic systems, and they leach into the water (Colborn *et al.*, 1996). In 2013 the FDA finally banned bisphenol A from infant formula packaging.

The environmental costs of pesticides are large. Some costs are quantifiable, and some are not—what price do we place on the loss of bird song, of bats eating mosquitos, of soil microbes decomposing soil litter, or a world without frogs? Can we place a cost on pesticides disrupting entire ecosystems?

Pimentel (2005) quantified costs due to the destruction of beneficial natural enemies in 16 USA crops, which summed to US$257 million in added *control* costs (i.e., more pesticides). He could not calculate added control costs due to pesticide resistance.

Honey bees and wild bees are vital for pollination of fruit and vegetable crops. Pesticides have a major impact on pollinators. Pimentel (2005) parsed out damage from pesticides, and estimated annual costs of $13.3 million for colony losses, $25.3 million for honey and wax losses, $8.0 million for increased bee rentals, and $210 million for agricultural losses due to reduced pollination.

Pesticides may damage the very plants they are intended to protect. Factors include dosage, pesticide formulation, droplet size, and temperature. Some plants suffer collateral damage from pesticides not aimed at them; herbicide drift is a well-known problem. Pimentel (2005) quantified these costs: $136 million for direct crop losses, $1000 million for crop products destroyed because of excess pesticide contamination, $245 million for crop applicator insurance, and $10 million for government investigations and testing.

21.4   Pesticide regulations

The EPA was established by President Nixon in 1970. The EPA was charged with pesticide regulation under the Federal Insecticide, Fungicide and Rodenticide Act (FIFRA). For a pesticide to be registered, studies must show that it poses no "unreasonable risk" to human health or the environment. **Tolerances** (maximum allowable residue limits) for pesticides on agricultural commodities were established under the Food Additives Amendment

Table 21.2. EPA pesticide acute toxicity classification

Class	Oral LD_{50} (mg/kg)	Dermal LD_{50} (mg/kg)	Inhalation LD_{50} (mg/L)	Eye effects	Skin effects
I	50 or less	200 or less	0.2 or less	Corrosive, corneal opacity not reversible	Corrosive
II	50–500	200–2000	0.2–2.0	Corneal opacity reversible within 7 days, irritation persisting for 7 days	Severe irritation at 72 hours
III	500–5000	2000–20,000	2.0–20	No corneal opacity, irritation reversible within 7 days	Moderate irritation at 72 hours
IV	>5000	>20,000	>20	No irritation	Mild irritation at 72 hours

of 1958 (FAA). The FAA's Delaney Clause ruled a zero-tolerance for pesticides shown to cause cancer, but the Delaney Clause was overturned by provisions signed into law by President Clinton.

Few pesticides have been labeled in the USA for use on *Cannabis*. "The label is the law." Using pesticides in a manner inconsistent with their labeling brings penalties, up to US$1000 per offense. Biorational pesticides are regulated by two additional sets of federal regulations in the USA: the USDA regulates "organic pesticides," and the EPA regulates "biopesticides."

21.4.1   USDA organic pesticides

The Organic Foods Production Act of 1990 authorized a National Organic Program, administered by the US Department of Agriculture (USDA). Because federal laws have prohibited *Cannabis* cultivation, the USDA refused to certify cannabis products as "organic." Thanks to the Farm Bill Act of 2018, which lifted restrictions on *Cannabis sativa* with THC <0.3%, that is now changing. In the meantime, third parties offer unofficial certification, such as Clean Green Certified (since 2004), Dragonfly Earth Medicine (since 2007), Certified Kind (since 2014), and the Cannabis Certification Council (since 2016).

Organic farmers certified by the USDA must restrict pesticide usage to items appearing on *The National List*. Individual states may adopt additional requirements after review and approval by the USDA. *The National List* approves some synthetic materials, such as micronutrient fertilizers, humic acid, fish emulsions (which contain synthetic pH stabilizers), plastic mulch and row covers, disinfectants (bleach, hydrogen peroxide, isopropanol), and synthesized sources of CO_2.

Biorational pesticides on *The National List* include insecticidal soap (potassium salts of fatty acids), horticultural oil (paraffinic oils from petroleum), minerals (sulfur, lime sulfur and sulfurous acid gas, copper sulfate, fixed copper, potassium bicarbonate, boric acid), sticky traps, and insect pheromones. No botanical extracts or essential oils are approved. *The National List* specifically *prohibits* some natural substances, such as rotenone, tobacco dust (nicotine sulfate), sodium nitrate, arsenic, and strychnine.

21.4.2   EPA biopesticides

The EPA has a biopesticide website (www.epa.gov/pesticides/biopesticides). The EPA defines biopesticides as naturally occurring substances that control pests, as well as microorganisms that control pests (e.g., *Bacillus thuringiensis*, *Pseudomonas fluorescens*, *Streptomyces lydicus*, *Trichoderma* spp.). We cover these separately as **biocontrol microorganisms** in section 20.2. Biopesticides, according to the EPA, include PIPs—pesticidal substances produced by GMO plants. The EPA's list includes insecticidal soaps, horticultural oils, plant-based botanicals (clove, thyme, and garlic oils), and minerals (sulfur, kaolin, cryolite, potassium bicarbonate). A few plant extracts make the list (azadirachtin, linalool, piperidine), as do insect pheromones. The EPA regularly updates its list of registered "biopesticide active ingredients," currently 390 entries and counting.

Many EPA biopesticides are "minimum risk" pesticides, otherwise known as **Section 25(b) exempt pesticide products**. The EPA created the Section 25(b) category in 1994 to encourage the adoption of minimum risk pesticides. Products meeting Section 25(b) criteria are Generally Recognized As Safe (GRAS) by the FDA: present in food, widely available to the public, with no evidence of damage to humans or the environment, and not likely to persist in the environment.

Biopesticides meeting Section 25(b) criteria are exempt from EPA registration. The active ingredients in this list include essential oils (rosemary, thyme, cinnamon, clove, and garlic oils, plus geraniol), seed oils (corn, soybean, cottonseed, sesame, and linseed oils), and others (citric acid, potassium sorbate, lauryl sulfate, and sodium lauryl sulfate).

Until recently, the EPA refused to register pesticides for *Cannabis*, or set tolerance levels on products (Stone, 2014). This failure to act in the interests of the American public, the EPA claimed, was simply because it "has yet to receive any applications for pesticide use on marijuana and, therefore, we have not evaluated the safety of any pesticide on marijuana" (EPA, 2016).

Thanks to the Farm Bill Act of 2018, the times they are a-changing. On December 19, 2019, the EPA approved ten biopesticide products for low-THC hemp production. The agency received applications from ten companies seeking label approval, opened a 30-day public comment period, and approved products made by three companies. Since then, the EPA has approved one conventional pesticide (ethalfluralin, a herbicide) and 98 biopesticides (EPA, 2023, available online).

Now that the EPA had begun regulating *Cannabis* pesticides, Cranshaw *et al.* (2019) foresaw several regulatory paths for different products. Pesticides for *Cannabis* grown strictly for fiber or seed might be regulated under the EPA's Oilseed Group (crop group 20). This group includes cottonseed, flax seed, sunflower seed, and rapeseed (canola). There is no separate fiber crop group. Inclusion in the Oilseed Group would streamline pesticide registration through label changes, or through state-by-state Special Local Needs (SLN) registrations. President Trump's EPA boss, Scott Pruitt, rejected efforts by California, Washington, Nevada, and Vermont to approve SLN registrations for *Cannabis*.

Cranshaw noted that hemp grown for CBD or THC will have a more difficult path to pesticide registration. Among existing EPA groups, this crop is closest to the Herbs and Spices Group (crop group 19), which includes 68 commodities, such as basil, chamomile, chives, lavender, lemon balm, rosemary, and wormwood. There is no "herbal medicines" group. Former mainstays of the pharmaceutical industry, such as foxglove, goldenseal, belladonna, monkshood, and ipecac, do not appear in any EPA groups.

Complicating registration, CBD is consumed by several routes of administration: orally, topically, vaporized, and smoked. Furthermore, CBD is often extracted using solvent-based methods, further concentrating pesticide residues that may have been present at low or even undetectable concentrations on the plant. All concentrations and routes of administration must be tested separately—probably a deal-breaker for the industry. Punja (2021a) made a helpful proposal: register pesticides (particularly fungicides) that are limited in application to the propagative stage of cultivation. Pesticides applied early in the growth cycle could meet "zero residue limits" at harvest time.

21.4.3   State regulations

In the absence of federal regulations, individual states have issued regulations. The Maine legislature permitted the application of Section 25(b) pesticides on *Cannabis* (State of Maine, 2013). Maine was followed by the Washington Department of Agriculture (WSDA, 2016), Colorado Department of Agriculture (Colorado DOA, 2016), and the California Department of Pesticide Regulation (California DPR, 2017). California DPR (2018) followed up with a list of pesticides specifically *prohibited* from use on *Cannabis* under any circumstances, with civil or criminal penalties.

Washington, Colorado, and California released lists of allowable pesticides beyond Section 25(b) pesticides. Most of them appear in the USDA's *The National List* or the EPA's list of biopesticides. According to Washington's regulations, pesticides had to be: (1) labeled for use on unspecified food crops; (2) exempt from residue tolerance on all food crops; and (3) not denied by the EPA for use on tobacco.

Tobacco pesticides have an interesting history, reviewed by Stephenson *et al.* (2003). Since 1950, over 200 pesticides and growth regulators have been used on tobacco crops. Most of these chemicals have fallen by the wayside, such as DDT. In the decade leading up to Stephenson's report, more than 30 pesticides had been cancelled for use on tobacco. The EPA has not set pesticide residue tolerances on tobacco, because tobacco is not classified as a food crop. The EPA has not assessed long-term risks of pesticides to tobacco smokers, reasoning that tobacco itself is a health risk.

The EPA has not required the evaluation of combustion products in smoke. Manufacturers nevertheless provided pyrolysis residue studies of 13 pesticides, which the EPA evaluated, and Stephenson *et al.* (2003) reviewed the EPA's documentation. In most cases, residues were ≤0.1 ppm, "below the agency's level of concern." The USDA stepped in, and placed residue limits on 15 pesticides. Rodgman and Perfetti (2013) provided a more recent review of tobacco pesticides, the transfer rates of pesticides to smoke, and thermal degradation products in smoke.

The list of active ingredients allowed in *Cannabis* by Washington, Colorado, and California included Section 25(b) pesticides, botanicals (e.g., neem oil, azadirachtin, garlic oil, capsaicin, pyrethrins, extract of *Reynoutria sachalinensis*), biological control organisms (e.g., *Bacillus thuringiensis*, *Streptomyces griseoviridis*, *Beauveria bassiana*, *Isaria fumosorosea*, *Gliocladium virens*, *Burkholderia* spp.), and others (e.g., insecticidal soaps, horticultural oils, copper, sulfur, kaolin, iron phosphate, potassium silicate). Washington and Colorado allow piperonyl butoxide, a synthetic compound linked with human disease.

Regarding state-by-state regulations, Cranshaw (2020) wrote, "It is hard to imagine any crop ever grown in the United States that has had such a confused approach to the use of pesticides than with hemp." In addition to the pioneering states,

Cranshaw detailed pesticides allowed elsewhere. Several states permit the same pesticides as Washington and Colorado (Arizona, Hawaii, Oregon, South Carolina, Virginia). Other states restrict their lists to Section 25(b) pesticides (Kentucky, Massachusetts, Montana, Nebraska, and North Carolina)

Other states add items. Indiana adds castor oil, mint oil, and hypochlorous acid; their list includes products by FoxFarm, with crazy labels (Fig. 21.2, compare with Fig 21.1). FoxFarm's miticide contains cottonseed oil, clove oil, garlic oil, sodium bicarbonate, oleic acid, and lauric acid. New York adds *Chromobacterium subtsugae*, *Helicoverpa armigera* NPV virus, diatomaceous earth, and boric acid. Vermont adds diatomaceous earth and ammonium salts of fatty acids.

Wisconsin, Illinois, and Michigan have the largest lists, adding *Gliocladium virens*, *Bacillus pumilus*, *B. subtilis*, *Chromobacterium* sp., *Metarhizium anisopliae*, *Streptomyces lydicus*, *Trichoderma harzianum*, *T. virens*, *T. gamsii*, *T. asperellum*, *Myrothecium verrucaria*, citronella oil, diatomaceous earth, and farnesol. Wisconsin uniquely adds *Muscodor albus*. Nevada has a "List of pesticides that are not legally prohibited for use on

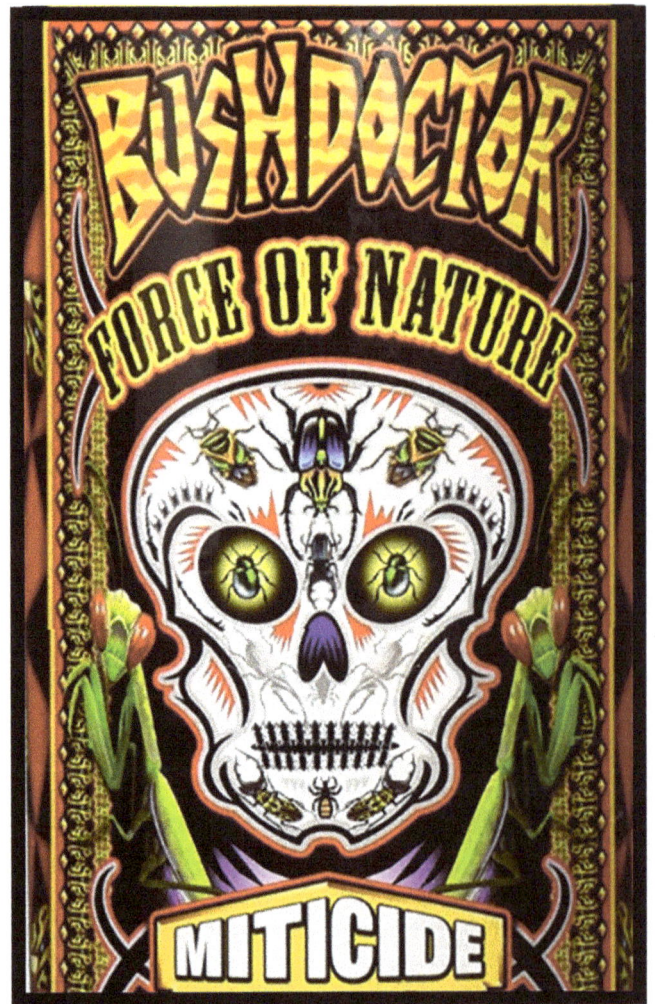

Fig. 21.2. Label of a miticide made by Foxfarm, Humboldt County, California. (Author's collection)

industrial hemp," which lists ingredients in three categories: Section 25(b) pesticides, "EPA registered pesticides—exempt from tolerance," and "EPA registered pesticide—tolerance is monitored." The latter group includes items specifically banned by other states.

21.4.4 Canadian regulations

Pesticides in Canada are regulated by the *Pest Control Products Act*, under the auspices of the Pest Management Regulatory Agency (PMRA), a branch of Health Canada. Under the *Cannabis Act of 2018*, growers can only use pesticides approved for *Cannabis* under the PMRA. A list of approved pesticides is updated on the PMRA's online "Pesticide Label Search" site (Health Canada, 2023).

Currently 25 pesticide products are approved for *Cannabis* intended for medical or recreational use. Active ingredients include insecticidal soap, horticultural oil, botanical poisons (e.g., garlic powder, pyrethrins, canola oil, extract of *Reynoutria sachalinensis*, fermentation products of *Lactobacillus casei*), minerals (e.g., potassium bicarbonate, copper, wettable sulfur), and biological control organisms (e.g., *Bacillus thuringiensis*, *Beauveria bassiana*, *Streptomyces lydicus*, *Gliocladium catenulatum*, *Trichoderma harzianum*, *Trichodema virens*), as well as a rooting hormone, indole-3-butyric acid. Bifenazate and abamectin were formerly on the list, but removed. For industrial fiber-type hemp, no insecticides or fungicides are registered.

21.4.5 EU regulations

The European Union has a two-tiered system of pesticide regulation. Firstly, before an actual pesticide can be sold on the European market, the active ingredient (Plant Protection Product, PPP) needs to be approved by the EU. After the active ingredient is approved, a procedure of approval of the PPP can begin in individual Member States. The EU's pesticide policies are very stringent, and no pesticides are approved for use on *Cannabis*.

21.5 Pesticide residues

Pesticide residues pose a unique risk to smokers, because up to 69.5% of residues in herbal cannabis pass into the smoke (Sullivan *et al.*, 2013). Furthermore, Sullivan and colleagues showed that when pesticides were pyrolyzed, they created toxic mixtures of newly formed compounds. Our boom in legalization has its unregulated dark side. The use of illegal pesticides is a rising crisis, and a breakdown in ethics.

Thirty years ago, when indoor grows were small, and dominated by *au naturel* hippie ethics, a grower might rogue infested plants, rather than blast his space with poisons. The stakes are higher now: Large corporations have stockholders to please. *A crop cannot go down.* For example, glasshouses filled with 114 acres (46.1 ha) of *Cannabis* are operated by Glass House Farms at a single location near Los Angeles—the equivalent of 86 football fields (Wilson, 2023).

Already 25 years ago, a witches' brew of pesticides was catalogued by McPartland *et al.* (2000), obtained from anecdotal reports and the literature. The list included acephate, benomyl, carbaryl, carboxin, chlorpyrifos, chlorothalonil, diazinon, dichlorvos, dicofol, dimethoate, fenbutatin oxide, iprodione, malathion, maneb, parathion, vinclozolin, and a slew of synthetic pyrethroids.

Upton *et al.* (2013) collated a similar list of pesticides illegally used on *Cannabis*, including 12 insecticides or miticides (abamectin, acequinocyl, bifenazate, etoxazole, fenoxycarb, imidacloprid, spinosad, spiromesifen, and several synthetic pyrethroids), four fungicides (imazalil, myclobutanil, trifloxystrobin, paclobutrazol), and three plant-growth regulators (daminozide, paclobutrazol, chlormequat chloride).

Tests by forensic chemists show more of the same. Samples from Oregon dispensaries contained residues of abamectin, azadirachtin, bifenazate, bifenthrin, carbaryl, chlorfenapyr, chlordane, chlorpyrifos, coumaphos, cypermethrin, diazinon, dichlorvos, ethoprophos, imidacloprid, malathion, metalaxyl, mevinphos, myclobutanil, paclobutrazol, permethrin, piperonyl butoxide, propoxur, and 4-4'-DDE (Voelker and Holmes, 2015). Nearly half of the 389 samples had pesticide residues, and 2% of samples contained >100,000 ppm pesticides.

Testing of legally grown medical cannabis in California identified pesticides in 49.3% samples, and 12 compounds were identified (Wurzer, 2016). Myclobutanil led the list (40% of samples), followed by bifenazate (20%), spiromesifen (15%), imidacloprid (4.6%), and spinodad (1.3%), as well as abamectin, acequinocyl, bifenazate, daminozide, fenoxycarb, pyrethrum, and spirotetramat.

Russo (2016) tested 26 samples from legal dispensaries in Washington. Pesticide residues were detected in 84.6% of samples, including 24 agents of every class: insecticides (organophosphates, organochlorides, carbamates, neonicotinoids), miticides, fungicides, an insecticidal synergist, and growth regulators. One sample was contaminated with nine agents, include boscalid (112,033 ppb) and extremely toxic carbaryl (25,483 ppb). Samples obtained from indoor grows had a higher risk of contamination than samples obtained from outdoor grows.

Wylie *et al.* (2020) tested 21 seized samples, and 11 had pesticides, with one sample smeared with seven: dieldrin, atrazine, mevinphos, fenarimol, hexazinone, DEET, and tentatively identified DDE (a degradation product of DDT). The others were nicotine, myclobutanil, chlorpyrifos, chlordane, and malathion. The latter was present at 82 ppm, which is 164 times greater than California maximum residue limits for inhalable products, and 4100 times greater than Canadian maximum residue limits for dried cannabis flower.

Many of these pesticides are "over the counter" products, available in garden supply stores. Some are only approved for landscape plants, not food plants. For example, Rosenthal (1998) recommended spraying plants with Wilt-Pruf® to suffocate spider mites. This is a tree nursery product, whose active ingredient is polyvinyl chloride, the same as Saran Wrap. McPartland and McKernan (2017) documented reports of hydroponic shops repackaging pesticides for ornamental plants, such as bifenazate and abamectin, and selling them to *Cannabis* cultivators. A dubious corporation marketed Guardian, a "100% natural" miticide, which contained abamectin.

Disturbingly, some of the chemicals listed above are "restricted use pesticides," a designation the EPA assigns to chemicals considered too hazardous for sale to the general public. This means *Cannabis* growers obtained them from government-certified applicators. Banned insecticides, no longer for sale in the USA, have been encountered, such as carbofuran (Thompson *et al.*, 2017), and dieldrin and chlordane (Wylie *et al.*, 2020).

In Canada, pesticides confiscated from illicit indoor grow operations included chlorpyrifos, diazinon, and 11 synthetic pyrethroids (NCCEH, 2009). Moulins *et al.* (2018) made unannounced inspections of licensed producers, and found unauthorized pesticides in 26 of 144 samples: myclobutanil, bifenazate, boscalid, fludioxonil, fluopyram, tebuconazole, and piperonyl butoxide.

In Europe, Fusari *et al.* (2013) detected pesticides (amitraz, chlorpyrifos, disulfoton, captan, fenvalerate, methiocarb, phosalone, trifluralin) as well as estrogenic phthalates, probably absorbed from plastic packaging. Schneider *et al.* (2014) detected fungicides (propamocarb, tebuconazole, propiconazole, tolylfluanid), and less frequently insecticides or acaricides (imidacloprid, bifenthrin, hexythiazox) in 50 seized samples. Cuypers *et al.* (2017) analyzed plants and carbon filter swipe samples in illicit grow sites. They detected pesticides in 64.3% of 72 cannabis plant samples, and in 65.2% of 46 carbon filter cloths. Nineteen pesticides were identified, primarily o-phenylphenol, bifenazate, imidacloprid, abamectin, propamocarb, propiconazole, chlorpyrifos, and tebuconazole. Furthermore, labeled pesticide containers found at crime scenes implicated deltamethrin, cyermethrin, permethrin, and parathion.

In Uruguay, Pérez-Parada *et al.* (2016) tested seized samples and detected residues of diazinon, tebuconazole, and teflubenzuron. Analysis of seizures in Rio de Janeiro, with possible provenance from Uruguay, identified imidacloprid, metazachlor, buprofezin, and metalaxyl (Daniel *et al.*, 2019).

21.5.1   Residue testing regulations

Detecting pesticide residues requires expensive analytical methods, using GS-MS and HPLC (Upton *et al.*, 2013). Assaying *Cannabis* for pesticides is more difficult than microbial testing (see section 18.3.7). Valdes-Donoso *et al.* (2019) estimated laboratory costs for pesticide testing in California. Itemizing capital investment (interest plus depreciation), equipment maintenance, office rental, labor, and consumables, the cost ranges from US$334 to $750 per sample, depending on the size of the operation. This estimate bundled testing for pesticides, microbial impurities, heavy metals, and cannabinoid content.

Each pesticide compound must be individually targeted for testing, and the black-market usage of pesticides keeps regulators in the dark regarding which compounds should be targeted (Stone, 2014). For example, banned pesticides no longer sold in most of the world, such as carbofuran, dieldrin, and chlordane, are omitted by typical testing procedures, and would therefore escape detection.

A suspected pesticide found at a grow site was analyzed by Wilson and Graff (2013). Identifying the mysterious liquid with infrared spectroscopy and GC-MS proved difficult, because common spectral reference libraries (e.g., NIST) had no matching spectra. After an extensive literature search, the unknown compound was identified as spiromesifen, a recently introduced miticide. Regulators are now testing for spiromesifen (Valdes-Donoso *et al.*, 2019; Atapattu and Johnson, 2020).

Unethical growers are turning to new subterfuge, instigating an "arms race" with regulatory chemists. Maguire *et al.* (2019) used a solid phase extraction method (QuEChERS), coupled with LC-MS/MS (triple quadrupole mass spectrometry) and GS-MS/MS. This system was capable of detecting 367 targeted compounds. They sampled legal products in Oregon, previously tested for pesticides and declared clean. They detected three pesticides on Oregon's list of targeted compounds—myclobutanil, spinosad, and piperonyl butoxide (indicating lapses by testing laboratories), as well as three pesticides new to *Cannabis*: fluopyram, pendimethalin, and diethyltoluamide (DEET).

Wylie *et al.* (2020) used QuEChERS extraction with GC coupled to a high-resolution accurate mass quadrupole time-of-flight mass spectrometer (GC/Q-TOF). To identify mass ion spectra, they developed a find-by-fragments software tool, and a separate peak-finding algorithm, which enabled them to screen for over 1000 pesticides.

Perhaps due to costs and other complications, states have been slow to mandate testing. Medical cannabis was first legalized by California in 1996. It took 22 years for the state to mandate pesticide residue testing (Valdes-Donoso *et al.*, 2019). California's law was preceded by mandatory testing in Oregon (Farrer, 2015) and Washington (Washington Administrative Code, 2016).

Atapattu and Johnson (2020) compared these regulations with mandatory testing by Health Canada. Oregon mandated testing of 59 pesticides, with limits of quantification (LOQ) set between 0.2 and 2.0 μg/g, depending on the target pesticide. California required testing for 66 pesticides, with an LOQ of 0.1–5.0 μg/g for inhalable products and 0.1–4.0 μg/g for other cannabis goods. Valdes-Donoso *et al.* (2019) clarified that California now has zero tolerance levels for a subset of 21 pesticides.

The latest review by Jameson *et al.* (2022) examined regulatory documents by 36 states that legalized recreational and/or medical cannabis. Testing has ramped up. All 36 jurisdictions mentioned pesticide testing, although five did not mention any specific contaminants in their regulatory documents. Among the remaining, 25 states listed ≤100 specific pesticides requiring testing, and four states listed >400 pesticides, which is the full gamut of pesticides listed by the EPA. Allowable levels varied among states; for instance, chlorpyrifos, dichlorvos, and dimethoate ranged from zero tolerance in California to 1 ppm in several states. The authors noted a mismatch between testing regulations and prevalent contaminants detected in cannabis. For example, herbicides accounted for 24% of all regulated pesticides, but herbicides are rarely used in *Cannabis* cultivation.

Jameson *et al.* (2022) also reported California testing records in 5654 flower and 3760 samples: a failure rate of 2.3% for flowers and 9.2% for extracts. Insecticides accounted for more than half of the contaminated samples (77), followed by fungicides (40 samples), and miticides (18 samples). In flowers, the insecticide chlorpyrifos was detected in the highest number of samples (19), followed by the fungicide myclobutanil (16 samples), boscalid (11 samples), and the miticide spiromesifen (10 samples).

Health Canada (2008) mandated testing for 96 pesticides, with an LOQ of 0.02–3.0 µg/g for dried flowers, 0.01–2.5 µg/g for oils, and 0.01–1.5 µg/g for fresh plants. Atapattu and Johnson (2020) provided a list of Canada's 96 pesticide targets, some of which must be assayed by GC-MS, or LC-MS/MS, or both. One pesticide is actually on Canada's list of approved pesticides (pyrethrin), the rest are banned compounds.

There have been several high-profile cases of cannabis removed from sale due to contamination. In 2011 California issued a cease-and-desist order against the sale of cannabis contaminated with daminoside and paclobutrazol (Upton *et al.*, 2013). In 2012, a whistleblower at Maine's largest medical cannabis dispensary revealed that nine types of insecticides and fungicides were being applied to *Cannabis* (Shepard, 2013). Colorado regulators quarantined thousands of plants grown by a dispensary chain that used myclobutanil, a turfgrass fungicide (Wyatt, 2015). This was one of nine marijuana recalls in Denver that year (Baca and Migoya, 2015). A quick internet search reveals recalls in Arizona, California, Colorado, Michigan, Missouri, Nevada, Oregon, Vermont, Washington, and Canada.

More disturbing are cases of cannabis testing labs caught falsifying results for pesticides. Sequoia Analytical Lab in Sacramento was shut down for this, and millions of dollars' worth of product were recalled and destroyed (Fimrite, 2018). Cannabis spiked with known amounts of pesticides were submitted to five testing laboratories in California, who returned a false negative rate of 78%, "indicating that these laboratories were falsely passing pesticide-laden material" (Smith *et al.*, 2019). This fraudulent activity also occurred in Michigan, revealed by blinded "round-robin" audits (J. Crookston, pers. comm., 2022). Fraud by laboratory testing facilities is lucrative, thanks to collusion between testing labs and producers requiring certificates of analysis. Industrial Bio-Test Laboratories in Illinois operated the largest toxicology lab in the USA, overseen by a former Monsanto scientist, until it was shut down in 1983 for fabricating safety data on pesticides.

21.6   List of epa section 25(b) exempt pesticides

Several US states and Canada permit *Cannabis* cultivators to use Section 25(b) exempt pesticide products, which are exempt from EPA registration. Some products have proven useful for small-acre farmers. For industrial-scale farming, however, Cranshaw (2020) opined, "as a class of products, Minimum Risk Pesticides appear to provide close to zero potential value for help in managing the hemp insect/mite problems."

Chitosan is a polysaccharide derived from the shells of crustaceans, and the EPA added it to their list of Section 25(b) exempt pesticides in 2022. Recently two products containing chitosan were approved for use on *Cannabis* (EPA, 2023). Chitosan acts as an extrinsic elicitor in plants and triggers pathogen resistance. Suwanchaikasem *et al.* (2023a) supplemented hydroponically grown *Cannabis* with 0.1–0.5% chitosan. After 8 days plants were harvested; treated plants showed reduced root growth; and shoot growth was unchanged. They called this a "growth-defense tradeoff," because treated plants upregulated myriad defense mechanisms in roots: chitinase and peroxidase levels increased, as did defense proteins (Thaumatin-like protein 1, Mulatexin-like protein, PR protein 1). Subsequently, Suwanchaikasem *et al.* (2023b) primed plant roots with chitosan and exposed them to *Athelia rolfsii*, and no improvement was seen. *A. rolfsii* attacks plants at the base of stalks, and the authors proposed that chitosan might work better with root pathogens.

Citric acid and **malic acid** are simple carboxylic acids, long known to be fungicidal. When sprayed on plants, they induce disease resistance against gray mold, powdery mildew, downy mildew, and the cause of hemp canker (e.g., Elad, 1992; Hassan *et al.*, 2006; Jafari and Hadavi, 2012). Blogs on the web report success with *Cannabis*.

Corn gluten meal inhibits the germination of both grassy and broad-leaf weeds, and may be used as a pre-emergent herbicide. However, extremely high rates must be applied (e.g., 2 t/ha), which may be cost-prohibitive.

Dried blood repels vertebrate pests, such as deer. Mesh bags filled with bloodmeal can be placed around the perimeter of small fields. To keep deer out of *Cannabis*, bloodmeal must be replaced every week or two (McPartland *et al.*, 2000).

21.6.1   Essential oils

Essential oils (EOs) are extracted from aromatic plants via steam distillation or hydrodistillation. EOs contain terpenoids and phenols that impart biological activity. Citronella oil, for example, contains citronellal, citronellol, geraniol, camphene, pinene, dipenthene, limonene, linalool, and borneol. Section 25(b) includes EOs extracted from cedarwood (*Cedrus* spp.), cinnamon (*Cinnamonum* spp.), citronella (*Cymbopogon* spp.), clove (*Syzygium aromaticum*), cornmint (*Mentha arvensis*), garlic (*Allium sativum*), geranium (*Geranium* spp.), lemongrass (*Cymbopogon* spp.), peppermint (*Mentha piperita*), rosemary (*Salvia rosmarinus*), spearmint (*Mentha spicate*), and thyme (*Thymus* spp.). Terpenoids purified from EOs are also permitted, such as citronella, eugenol, and geraniol.

EOs act as insect repellents, and have some larvicidal and adulticidal activity. Citronella oil has been used as a repellent since 1882, making it one of the first botanical pesticides (Gerberg and Novak, 2007). EcoSMART®, the first mixture of EOs marketed as an insect repellent, was released in 1998 (Isman, 2015). Since then, the EPA has registered two EO-based repellents, Requiem® (from *Chenopodium ambrosioides*) and Captiva® (from *Capsicum* oleoresin and garlic oil).

Some terpenoids, such as eugenol and geraniol, are actually insect *attractants*. Fleming *et al.* (1940) impregnated a device with eugenol and geraniol to attract and trap Japanese beetles. Their familiar device is still used today (Fig. 21.3).

The downside of EOs is lack of persistence (they are also called volatile oils); EOs must be reapplied frequently. Mitra *et al.* (2020) tested Section 25(b) EOs against the mosquito *Aedes aegypti*. All EOs showed repellency when first released, but after 90 minutes, only cinnamon oil showed any efficacy. The need for frequent reapplication makes EOs an expensive proposition. Cranshaw and Baxendale (2013) noted that EOs may cause plant injury, notably cinnamon oil, which is the primary active ingredient in Ed Rosenthal's Zero Tolerance™ herbal pesticide.

No experimental studies have demonstrated the efficacy of EOs against hemp pests (Cranshaw, 2020). The evidence is entirely anecdotal. Exactly which EO might work best against a given pest is anybody's guess. Proprietary products often combine several EOs (e.g., Foxfarm's Bush Doctor miticide, Fig. 21.2). Combining EOs may provide **additive** (1+1 = 2), or **synergist** (1+1 = 3) benefits. Synergistic interactions amongst EOs have been demonstrated (e.g., Zibaee and Khorram, 2015; Pumnuan and Insung, 2016; Esmaeily *et al.*, 2017), but not always (*e.g.*, Behi *et al.*, 2017).

The compositional complexity of EOs makes interactions difficult to predict. In a study with *Aedes aegypti*, Muturi *et al.* (2017) demonstrated synergy between oregano and manuka EOs, but an antagonistic interaction between clove and manuka EOs. Muturi proposed that carvacrol in oregano EO contributed to its synergism with manuka EO, whereas eugenol in clove EO contributed to its antagonism with manuka EO. Sarma *et al.* (2019) took this work further. They tested the efficacy of five EOs against *Aedes aegypti*: garlic, peppermint, basil, eucalyptus, and myrtle. Then they extracted major constituents from each, and tested the adulticidal effects of 28 binary combinations. Seven combinations showed synergism, six showed no interaction, and fifteen showed antagonistic effects. The best combination was carvone and limonene.

Miresmailli *et al.* (2006) tested the toxicity of rosemary EO against spider mites, *Tetranychus urticae*, as well as the comparative toxicities of rosemary's individual constituents (limonene, α-pinene, β-pinene, 1,8-cineole, p-cymene, α-terpineol, borneol, camphor, camphene). Binary comparisons showed synergy among the constituents, with the presence of all constituents necessary to equal the toxicity of the natural oil. Wu *et al.* (2017) obtained the same results with thyme oil and its individual constituents against *Tetranychus cinnabarinus*.

Synergy between EOs and biocontrol fungi has been reported. Sabbour and Abd-El-Aziz (2010) tested three EOs combined with *Lecanicillium muscarium*, *Metarhizium riley*, and *Isaria fumosorosea*. The combinations increased pathogenicity against the beetle *Bruchidius incarnatus*. EOs damaged the epidermis of insects, permitting better penetration of fungal germ tubes. Additive effects were seen by Razmjou *et al.* (2016), who combined EOs with *Lecanicillium muscarium* against *Aphis gossypii*. Synergy between EOs and synthetic pesticides has also been shown, against the aphid *Myzus persicae* (Faraone *et al.*, 2015), the locust *Schistocera gregaria* (Mansour *et al.*, 2015), the budworm *Spodoptera litura* (Radhika and Sahayaraj, 2014), and the armyworm *Spodoptera frugiperda* (Fazolin *et al.*, 2016). This opens opportunities to use EOs in pest management, while reducing the amount of synthetic insecticide needed to suppress pest populations.

Lauryl sulfate and **sodium lauryl sulfate (SLS)** are non-ionic surfactants in soaps, derived from fatty acids and fatty alcohols. They have weak insecticidal activity against thin-skinned aphids and whitefly nymphs (Cory and Langford, 1935). SLS was shown to be less effective against whitefly nymphs than comparable doses of insecticidal soap, horticultural oil, cottonseed oil, and synthetic surfactants (Liu and Stansly, 2000). *Cannabis* web blogs report success with products that combine SLS with EOs. However, the combination of SLS and EOs may be phytotoxic (Cloyd *et al.*, 2009). Lauryl sulfate combined with soybean oil and other surfactants controls powdery mildew in dogwood (Deyton *et al.*, 2001).

PeP (2-phenylethyl propionate) is a natural insect attractant that is applied in traps. It is primarily used as a Japanese beetle lure.

Potassium sorbate is a fungicide, primarily used to prevent post-harvest rot of fruits and vegetables, as well as tobacco spoilage fungi. Soliman and El-Mohamedy (2017) showed that

Fig. 21.3. Japanese beetle trap (Fleming *et al.*, 1940)

potassium sorbate could induce resistance against powdery mildew, but its effect was more powerful when combined with chitosan. Numerous *Cannabis* web blogs report success against mites and aphids with Nuke 'Em® (potassium sorbate and citric acid). Potassium sorbate is also insecticidal when combined with SLS (e.g., RestAsure® against bed bugs)

Putrescent whole egg solids (PWESs) are vertebrate repellents, useful against deer. Palmer *et al.* (1983) found PWESs were more efficacious than Meat Meal (bloodmeal plus bone meal), Hinder (ammonium soaps of fatty acids), Hot Sauce (capsicum), Spotrete-F (thiram), and Feather Meal (poultry feathers). PWESs have less efficacy against rabbits and squirrels.

21.6.2 Seed oils

Section 25(b) includes seed oils (SOs) extracted from corn (*Zea mays*), castor bean (*Ricinus communis*), cottonseed (*Gossypium* spp.), linseed (*Linum usitatissimum*), sesame (*Sesamum indicum*), and soybean (*Glycine max*). Proprietary products combine several SOs, such as PureCrop1™, which combines corn and soybean oils, with hemp listed on its label (for control of piercing-sucking insects and mites, as well as leaf fungi). SOs work by blocking respiratory openings (spiracles), thereby suffocating insects and mites. In some cases, SOs also may act as poisons, interacting with fatty acids and interfering with normal metabolism (Cranshaw and Baxendale, 2013).

Unlike EOs, Cranshaw (2020) noted that SOs have demonstrated value in controlling spider mites and piercing-sucking insects. After SOs are applied to plants and dried, they are compatible with beneficial insects. SOs are sometimes lumped with horticultural oils (HOs), but HOs are paraffinic (petroleum)-based, and covered in the next section. Researchers began experimenting with SOs a century ago. Staniland (1926) tested several SOs (cottonseed, linseed, rapeseed, sesame seed, castor, olive) emulsified with 1% soap against "various insects" on apple foliage. Aphids and capsids were controlled better than caterpillars, and rapeseed SO 0.5% worked the best. It compared favorably with the cost-benefits of nicotine sprays.

Balachowsky (1933) showed that peanut oil and rapeseed oil, used as 2% emulsions, were equal to or superior to petroleum-based HOs against scale insects. Martin and Salmon (1933) controlled powdery mildew (*Sphaerotheca humuli*) in glasshouse-grown hops with SOs (cottonseed, olive, and sesame oil at 0.5%, rapeseed oil at 2%, emulsified with 0.5% soft soap), without causing foliage injury. The SOs worked better than HOs, which caused foliar damage.

SOs, unlike EOs, are not very repellent. Mitra *et al.* (2020) tested the repellency of Section 25(b) active ingredients against *Aedes aegypti*, on a scale ranging from 100% DEET (score 34) to mineral oil (negative control, score 74). The mean score for EOs was 58.8, and for SOs was 74.2. Castor oil, with its strong smoky odor, was the most repellent SO (score 62). Castor oil applied as a soil drench in small *Cannabis* gardens repelled burrowing pests (moles, voles), as well as gophers (California DPR, 2017). Soil drenches also suppress nematode populations.

SOs may inhibit plant transpiration—do not apply to drought-stressed plants, or above 100°F (37.8°C). Paradoxically, applying SOs in conditions of low temperatures and high humidity may also damage plants, because these conditions inhibit oil evaporation. Products that combine SOs and EOs are efficacious against spider mites, aphids, whiteflies, and mealybugs, but they cause phytotoxicity in some plants (Cloyd *et al.*, 2009). SOs are less effective against robust-bodied insects, such as caterpillars. However, Westgate *et al.* (2017) successfully reduced feeding damage by budworms (*Heliocoverpa zea*) and European corn borers (*Ostriania nubilalis*) in maize. Corn, canola, safflower, and soybean oil were equipotent; adding emulsifiers provided a modest additional benefit.

21.7 List of organic-friendly biopesticides

Several pesticides that didn't make the cut as Section 25(b) agents nevertheless appear on the USDA's *National List*, or the EPA's list of biopesticides, or lists of approved pesticides by individual states or Health Canada. Wilson *et al.* (2019) surveyed California *Cannabis* growers regarding their pest management approaches, and organic-friendly biopesticides were prominent (see Fig. 20.4). "Oil solutions" led the way, a category that included EOs, SOs, and HOs. Next came azadirachtin, followed by insecticidal soap.

Azadirachtin is the primary active ingredient in neem seed oil (see below), extracted from the neem tree (*Azadirachta indica*) or the chinaberry tree (*Melia azedarach*). Azadirachtin acts as a broad-spectrum insect growth regulator, feeding deterrent, and repellent. It is most effective against caterpillars, but also controls immature whiteflies, leafminers, fungus gnats, mealybugs, leafhoppers, and some thrips and beetles. It is less effective against aphids, although commercial formulations vary in efficacy (Shannag *et al.*, 2014). The EPA approved the use of azadirachtin products in 1990, the first major market outside India (Isman, 2015). In 2019 the EPA approved azadirachtin products for low-THC hemp production. Azadirachtin is available as an EC (Table 21.1) from several manufactures. Azadirachtin and neem oil decompose rapidly in water, so don't mix more EC than needed. Diluting water should be acidic (pH 3–7). Use a spreader/sticker. Apply when humidity is high (morning or evening). Azadirachtin persists for 1–2 weeks on plant surfaces. In some plants it is absorbed *systemically*.

Boric acid (sodium tetraborate) is a stomach poison useful against ants, crickets, roaches, and earwigs. It is available as a dust, granule, or bait. These formulations are persistent in dry conditions, but wash out rapidly after rainfall. Boric acid is phytotoxic. It is not dangerous to honey bees or birds. Boric acid works slowly; ant populations aren't reduced for a week or more.

Capsicum oleoresin is an ethanol or methanol extract derived from hot peppers, *Capsicum annum* or *C. frutescens*. Hot Shot™ Insect Killer was originally approved for treating pests in poultry (USDA, 1960). Walter (1995) first proposed using capsicum oleoresin against plant pests, particularly spider mites, as well as Hemiptera (aphids, whiteflies, thrips).

The EPA has registered many products with capsicum oleoresin, usually combined with garlic oleoresin or EOs. Its efficacy alone has been debated. Edelson *et al.* (2002) reported no significant mortality with capsicum oleoresin at any dose. Koleva Gudeva *et al.* (2013) tested purified capsicum against *Myzus persicae*, and reported an LC_{50} dose of 0.2934 mg/mL—pretty good—but only at high concentrations (diluted 1:50 in water).

Antonious *et al.* (2006) screened 24 *Capsicum* accessions for their toxicity and repellency to the spider mite *Tetranychus urticae*. Only one accession (*C. annuum* Grif-9169) showed significant toxicity (60% mortality after 24 h); other accessions (*C. annuum, C. frutescens, C. chinensis, C. baccatum*) were down around 20% mortality. Mixing 1.5% capsicum oleoresin into soil reportedly repels gophers, but Sterner *et al.* (2005) questioned its efficacy. Although capsicum oleoresin is water-soluble, a spreader/sticker should be added for better dispersal of spray droplets on leaf surfaces.

Copper is a fungicide, useful against gray mold, powdery mildew, downy mildew, leaf spots, blights, anthracnose, and also some foliar bacterial diseases. There are many forms of copper; some appear on *The National List,* and some are prohibited (copper nitrate, copper chloride). Fixed copper (all the formulations listed below, except Bordeaux mixture) is less soluble than the copper used in Bordeaux mixture, so fixed copper is less phytotoxic, but also less effective. Copper is not toxic to birds and bees, but is very toxic to fish, and acts as an eye and skin irritant in mammals. Copper can bioaccumulate in food chains. Copper works best as a preventive because it inhibits spore germination. Copper persists on foliage unless washed off. Every time fixed copper gets wet (a heavy dew), it releases a little metallic copper onto plants.

Copper sulfate ($CuSO_4$) is the most popular form of fixed copper, with the least phytotoxicity. It forms hydrates ($CuSO_4 \cdot nH_2O$), and the pentahydrate ($n = 5$) is most common. Prévost (1807) first discovered that copper sulfate prevented fungal spore germination. But it took 70 years for copper sulfate to gain wide usage (Millardet, 1883). Farahmand *et al.* (2023) used copper sulfate to inhibit foliar growth of *Penicillium olsonii* in drug-type plants. A product with copper sulfate pentahydrate (Sysstem®-Cal Blue) is approved for use on *Cannabis* (EPA, 2023). Copper sulfate is moderately toxic to mammals and fish, and persistent in soil.

Copper oxychloride ($H_3ClCu_2O_3$) is sold as a dust or wettable powder. A sulfated version (copper oxychloride sulfate) is also available. Copper oxychloride is more popular in Europe, copper oxychloride sulfate is more popular in the USA.

Cupric hydroxide (CuH_2O_2) is sold as a dust or wettable powder. This formulation dissolves in water better than most fixed coppers, causing less clogging of spray nozzles.

Cuprous oxide (Cu_2O) is sold as a dust, granule, or wettable powder. It is less phytotoxic than other copper compounds, but more toxic to us. Residual period and reapplication interval is 7–10 days. Cuprous oxide is also used as a seed treatment.

Copper octanoate is fixed copper combined with octanoate (a fatty acid) to form a copper soap. On plant surfaces it dissociates into soluble copper ion (Cu^{+2}) and octanoate. Marketed as Cueva®, it shows efficacy against several *Cannabis* pathogens (see Table 9.1), and is permitted for use on *Cannabis* by Colorado DOA (2023).

Bordeaux mixture is an "unfixed" mix of copper sulfate and calcium hydroxide (quick lime). After 2 years of research, Millardet and Gayon (1885) announced that Bordeaux mixture was superior to copper sulfate alone. It became the first fungicide to be used worldwide. Only the copper is fungitoxic; the lime serves as a "safener" to protect plants. Preparations of Bordeaux are labeled as a ratio of copper sulfate (in pounds) and quick lime (in pounds) to water (in gallons). The most popular mixture is 8:8:100. It is not very soluble; agitate the mix frequently to keep it from settling in the spray tank. Phytotoxicity arises when Bordeaux mixture is applied in cool, wet weather. To reduce phytotoxicity, decrease copper or increase lime (mix 2:6:100 or 8:24:100), and spray on a warm, dry day. Bordeaux is slightly toxic to honey bees, practically non-toxic to mammals, but highly toxic to fish.

Dead microbial pesticides are new products that span the realm of biocontrol and chemical control. They consist of insecticidal bacteria along with remnants of fermentation media, with no viable cells in the end product. They are approved by the USDA's National Organic Program and the Organic material Review Institute, and are listed by the EPA as biopesticides. Several states (Colorado, Washington, Oregon, Nevada, New York) approve their use on *Cannabis*, and they are listed in university extension guides (e.g., Britt *et al.*, 2020). The two products listed below function as stomach poisons, killing caterpillars with chewing mouthparts such as *Spodoptera exigua*. Their effects resemble those of Bt-based insecticides, with rapid cessation of feeding, and death in 3–5 days. They also work on contact, killing piercing-sucking insects (aphids, whiteflies, thrips, lygus bugs, stink bugs, even mites).

Heat-killed *Burkholderia* spp., *Burkholderia rinojensis* strain A396 (Venerate® XC or CG, Nemorax®, Majestene®), is labeled for use on many fruit and vegetable crops, to control armyworms, cutworms, and piercing-sucking insects. "Nematicide" was added to its label in 2015. Active ingredients include templazole (a phenylurea) and templamide (a mysin antibiotic).

Chromobacterium subtsugae strain PRAA4-1T cells, and spent fermentation media (Grandevo® CG), deters feeding, and reduces fecundity and oviposition. Active ingredients include violacein, an indol-derived pigment, and chromamide, a peptide. EPA (2023) approved its use on *Cannabis*.

Diatomaceous earth (DE) consists of the fossilized remains of diatoms (microalgae) that are millions of years old (Fig. 21.4). DE works as a contact insecticide. It tears microscopic holes in the surfaces of soft-bodied insects, and slices between the exoskeleton plates of hard-bodied insects. Injured insects rapidly dehydrate and die in hot, dry weather. DE is formulated as a talc-like dust. It is persistent in dry weather. The dust can irritate eyes. It should not be applied to drug crops destined for inhalation, because DE is a known respiratory hazard.

Horticultural oils (HOs) are petroleum-based mixtures of paraffinic hydrocarbons. Seed oils are sometimes included under the HO umbrella, but we treat them separately. Petroleum-based HOs are approved for use on *Cannabis* by Canada and several states. HOs were traditionally sprayed on trees in winter as a "dormant oil" to kill aphid eggs and mites. HOs used to be very heavy and phytotoxic to herbaceous

Fig. 21.4. SEM photomicrograph of diatomaceous earth (courtesy Dawid Siodłak, Wikimedia Commons)

plants, with their use limited to wood plants in winter ("dormant oil"). Improved refining techniques have produced lighter HOs ("summer oils") with flashpoints below 345°F (171°C), and these can be sprayed onto foliage.

HOs rarely cause phytotoxicity if sprayed in a 1% solution. A 3% solution may cause damage—always test-spray. HOs are also ranked by their *UR rating* (unsulfonated residue)—look for 92% or greater. SuffOil-X® has been used on *Cannabis*; its pre-emulsified formulation enables the spraying of small droplets—an essential consideration with HOs (UR rating 92%).

HOs plug up spiracles (breathing tubes), and suffocate mites, whiteflies, aphids, mealybugs, scales, and other small insects. To reduce phytotoxicity, plants should be well-watered, and then sprayed on a warm day (21–38°C) with low humidity (<50% RH), so oil does its work, then evaporates quickly. HOs can be used against fungi if mixed with 0.5% baking soda (3 tsp/gal). HOs have low mammalian toxicity, with little impact against beneficial insects (lady beetles, parasitic wasps, honey bees) *after* they have dried on the plant.

Iron phosphate and **sodium ferric EDTA** are slug and snail killers. They are formulated as pellets or in baits. California DPR (2017) approved their use on *Cannabis*, and EPA (2023) approved several products for use on *Cannabis*. Iron damages the digestive system of slugs and snails; they stop eating and die. Iron is non-toxic to mammals and birds, earthworms are not affected, and because it is applied to soil, beneficial insects and honey bees don't encounter it. Iron is less effective than metaldehyde—a common synthetic used against slugs—but a far safer choice (Speiser and Kistler, 2002).

Insecticidal soaps (ISs) consist of potassium salts of fatty acids. ISs suffocate many insects (but not their eggs). On *Cannabis* crops, Safer's™ Insecticidal Soap caused 98.6% mortality in aphids and 91% mortality in two-spotted spider mites (Puritch, 1982). Dr Bronner's soap also works well (mix 0.5 tsp/gal). M-Pede® is marketed as an insecticide, miticide (including root aphids), and fungicide (against powdery mildew).

ISs work best if mixed in soft water; soften the water if hardness exceeds 300 ppm (or 17.5 grains/gallon). Plants should be sprayed at 5-day intervals. IS loses activity after drying, so spray early in the morning when temperatures are cool. Follow each soap spray after several hours by a water rinse, especially in warm weather. High concentrations of ISs strip the cuticle off plant leaves. ISs are not selective, and they kill many beneficial insects. They kill predatory mites but their eggs survive, so ISs and predatory mites can work together. ISs can be used on *Cannabis* up to a week before harvest without any distasteful residues discerned on finished dry flowers.

Insect pheromones are volatile hydrocarbons emitted by a species to communicate with other members of the same species. Pheromones can be dispersed into the open air, to control pests by creating confusion, disrupt mating, and to prevent further egg laying. Dispensers are either passive (pheromone-impregnated polymers, Fig. 21.5A), or active (pheromone aerosol "puffers"). More commonly, pheromone-impregnated lures are used in traps, such as tanglefoot (Fig. 21.5B) or funnel traps (Fig. 21.5C).

Soon after pheromones were chemically elucidated and synthesized, they were used in pest management, first against the codling moth (Riedl and Croft, 1974). European hemp growers quickly adopted pheromones for use against Eurasian hemp borers, European corn borers, cutworms, armyworms, and budworms (Nagy, 1979).

Reproductive (sex) pheromones are emitted by female insects to attract males. They were intuited by the great naturalist (and *Cannabis* taxonomist) John Ray—or rather by his wife, Margaret Oakley. Ray caged a female moth, and Margaret caught two males that had flown into an open window and were attempting to reach the female. "They were attracted, as it seems to me, by the scent of the female" (Ray, 1710). In 1874, Jean-Henri Fabre caged a female peacock moth, *Paturnia pyri*, and over the course of 8 days he captured, marked, and released 150 males. He removed antennae from some males, and they were no longer attracted to the female (Fabre, 1949).

Traps with reproductive pheromones are used for pest monitoring purposes, or for control—either a mass-capturing strategy (many traps, coupled with sticky glue or funnels), or an "attract and kill" strategy (lures coupled with insecticides or biocontrol agents). Alternatively, scatter dispensers across a field—the males fly in a pheromone fog, confused and overwhelmed, and mating ceases.

Some pheromones appear on the USDA's *National List,* and the EPA's website of allowed biopesticides. Japanese beetle traps, for instance, contain (R)-(Z)-5-(dec-1-enyl)-oxacyclopentan-2-one (Japonilure™). (Z,E)-9,12-tetradecadienyl acetate attracts three *Cannabis* pests: the beet armyworm (*Spodoptera exigua*), the fall armyworm (*Spodoptera frugiperda*), and the tobacco cutworm (*Spodoptera litura*). (Z,E)-9,12-tetradecadienyl acetate also acts as a **kairomone**—a chemical signal detected by a parasite seeking a host, such as the braconid wasp *Bracon brevicornis*, the ichneumonid wasp *Venturia canescens*, and commercially available *Trichogramma evanescens* (www.pherobase.com).

Aggregation pheromones are often produced by male insects, attract both sexes, and are often released after feeding or pooping. Aggregation pheromones attract insects from a distance (attractants) or induce them to remain at the pheromone source (arrestants). Aggregated pests can be trapped and removed from the environment. The gregariousness of cockroaches is due to aggregation pheromones. Hardee *et al.* (1972) produced the first synthetic aggregation pheromone,

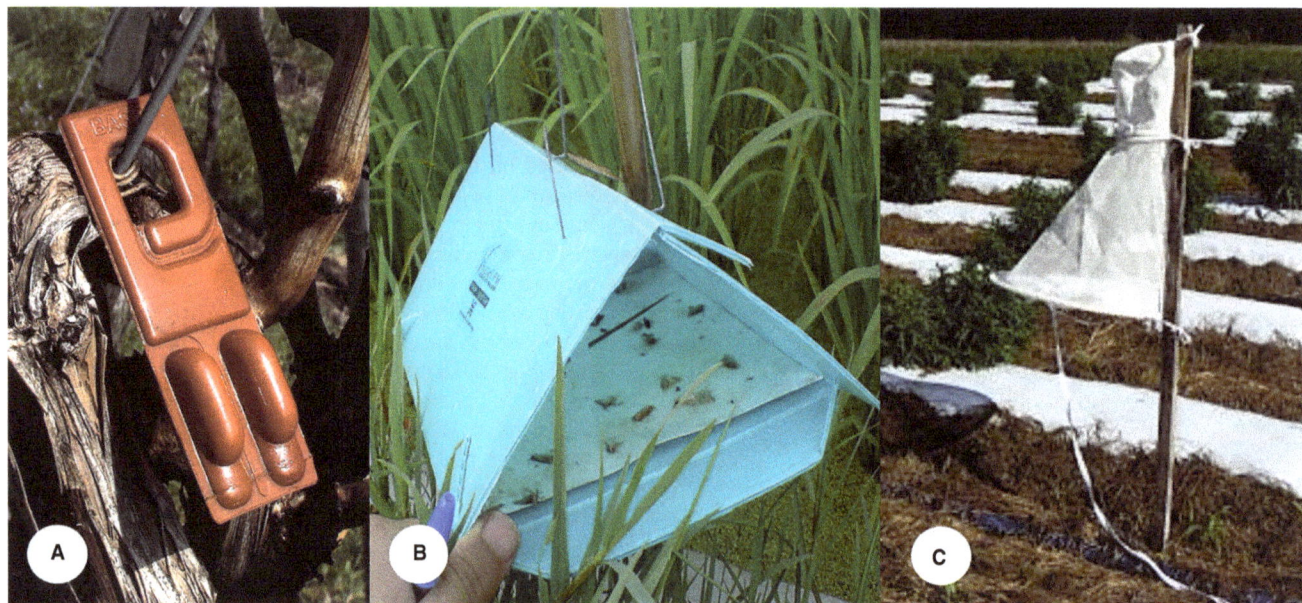

Fig. 21.5. Pheromone products: **A**. pheromone-impregnated polymer by BASF; **B**. pheromone-baited sticky trap; **C**. Sentry® pheromone-baited net trap. Credits: (A) Roger Kreja, Wikimedia Commons; (B) Mehdi, Wikimedia Commons; (C) Kadie Britt (Britt *et al.*, 2019).

Grandlure™, to trap boll weevils (*Anthonomus grandis*). Some terpenoids, such as eugenol and geraniol, act as aggregation attractants (Fig. 21.3).

Alarm pheromones or *dispersal agents* such as β-farnesene cause some aphids to drop to the ground. Other aphids disperse across the surfaces of plants. Dispersal increases their exposure to pesticides or biocontrol organisms. β-farnesene is not on the EPA's list of biopesticides, but its alcohol conjoiner is: farnesol. Both farnesol and β-farnesene are produced by *Cannabis* (Turner *et al.*, 1980).

Neem oil is a clarified hydrophobic extract obtain from seeds of the neem tree, *Azadirachta indica*, native to the Indian subcontinent. Indian farmers observed that neem trees were left uneaten after locust swarms. Indeed, the only healthy vegetation at Sarnath, where the Buddha first taught the Dharma, was a neem tree (Fig. 21.6). But even *A. indica* is damaged by dozens of insects, mites, and fungi, such as soft and hard scales, aphids, whiteflies, thrips, eriophyid mites, caterpillars, grubs, fire ants, termites, *Fusarium oxysporum*, *F. solani*, *Rhizoctonia solani*, *Sclerotium rolfsii*, and powdery mildew (Boa, 1995).

Neem oil is similar to other horticultural oils, but contains 0.03% azadirachtin (an insecticidal antifeedant), and repellent terpenoids (1.4% salannin, 0.5% nimbin, 0.2% azadiradione). As such, it has greater potency than other horticultural oils against insects, as well as miticidal activity. Neem oil is not a Section 25(b) exempt pesticide, but is registered as a biopesticide by the EPA. It is non-toxic to birds, bees, and earthworms, causes minimal adverse effects in predatory insects, and readily biodegrades in soil and water (EPA, 2012). It is formulated as an emulsifiable concentrate; spray in the evening, and repeat every 10–15 days.

Potassium bicarbonate is approved for use on *Cannabis* in Canada and several states to treat powdery mildew (Kaligreen®, Armicarb®, Milstop®). Non-toxic, low persistence, little phytotoxicity. Runoff enters the soil and provides plants with potassium. Potassium bicarbonate is used in baking, is added to wine and soda, and provides a buffering agent in pharmaceutical formulations. Sutton *et al.* (2023) warned that spraying Milstop® twice on maturing *Cannabis* flowers (at 5 and 6 weeks after flowering began) caused trichome gland heads to brown, and reduced trichome density, due to degradation of the gland head cuticle.

Pyrethrin is a mixture of six closely related compounds (pyrethrin I and II, cinerin I and II, jasmolin I and II) extracted from flowers of *Chrysanthemum* spp., predominantly *Chrysanthemum* (*Tanacetum*) *cinerariifolium*. The crude extract made from *Chrysanthemum* spp. is known as **pyrethrum**, and containing pyrethrin. Pyrethrum originated in the Middle East and was introduced into the USA around 1870. It is now our most popular botanical pesticide. Insect resistance has appeared.

Pyrethrin is a contact insecticide. It works best against lepidopterans (caterpillars) and coleopterans (beetles), with less effectiveness against aphids, whiteflies, bugs, beetles, and mites. Pyrethrin exerts a rapid paralytic action and can "knock down" flying insects, but they may recover. Pyrethrin is on Health Canada's approved pesticides list. Several US states limit pyrethrin residues in *Cannabis* to ≤1 μg/g. Disadvantages of pyrethrin include high cost and its rapid deterioration in air (about 1 day). Sunlight accelerates its deterioration, so apply in early evening or during cloudy weather. To slow down degradation, combine pyrethrin with Pheast™, a lepidopteran feeding stimulant. Do not mix pyrethrum in alkaline water; it is not compatible with Bordeaux, lime, and soaps.

Formulations include pyrethrum powder (powdered flowers, 1% pyrethrins), pyrethrin dust (20–60 D, mixed with clay or talc carriers), pyrethrin sprays (1–20 EC, WP), and

Fig. 21.6. Neem tree at Sarnath

aerosol products. Insect resistance has appeared. Many formulations contain synergists, which lack pesticidal effects, but enhance the activity of pyrethrin. Common synergists are semi-synthetic **piperonyl butoxide** (PBO), or the synthetic compound N-octyl bicycloheptene dicarboximide (MGK 264). Colorado and Washington allow very low PBO residues, Canada has zero-tolerance for PBO residues.

Edelson *et al.* (2002) included pyrethrin+PBO in a test of four biorational pesticides against *Myzus persicae*. They determined LC_{50} levels, which were unfortunately large: pyrethrin+PBO, 0.18 g/L; insecticidal soap, 6.8 g/L; azadirachtin, 193 g/L; capsicum oleoresin, no significant mortality at any dose. In field studies, there was no evidence that the density of aphids was reduced by any insecticide treatment when applied at manufacturers' labeled rates. However, combinations of biorationals produced synergistic effects, particularly insecticidal soap + capsicum oleoresin.

Pyrethrin is not phytotoxic. It is lethal to most biocontrol organisms. It kills honey bees if sprayed directly on them, but foraging bees who encounter sprayed plants are repelled and not killed. Pyrethrin is highly toxic to fish. Most mammals safely ingest pyrethrin. Chevalier (1930) prescribed pyrethrin pills to combat intestinal worms. This practice continued into the 1970s, and topical pyrethrin is still used for treating lice. Some people develop skin allergies, and exposure at high levels may induce asthmatic attacks, nausea, and facial flushing.

Reynoutria sachalinensis **extract** (RSE) is a 5% ethanol extract made from *Reynoutria* (*Fallopia*) *sachalinensis* (giant knotweed). Herger *et al.* (1988) discovered that RSE is a phytoalexin elicitor, and induces plant resistance to powdery mildew and gray mold. It was approved by the EPA a decade later. RSE prevents powdery mildew, and is approved for use on *Cannabis* by California DPR (2017) and other states. The product is made from cultivated plants, which is a pity, because *R. sachalinensis* is an invasive weed. Using RSE isn't easy: foliar application should start at the 4–6-leaf stage, repeated at 7–10-day intervals.

Ryania is extracted from roots and stems of *Ryania speciosa*. It is on the USDA's *National List* of organic substances. Ryania is a slow-acting stomach poison used against European corn borers and other caterpillars. Rutgers University researchers "discovered" ryania by screening extracts of plants used by Amazonian people (Pepper and Carruth, 1945). Ryanodine, an alkaloid in ryania, targets eponymous ryanodine receptors (RyRs), a class of calcium channels. Caffeine also works on RyRs.

Sodium bicarbonate, baking soda, is a fungicide akin to potassium bicarbonate, but with much less efficacy. It kills mycelium but not the conidia. Furthermore, sodium may burn foliage, whereas potassium is a plant nutrient. Spray a 0.5% solution of baking soda (3 tsp/gal). Some growers double the concentration. Sodium bicarbonate works better when combined with light horticultural oil (also mixed 3 tsp/gal).

Sodium hypochlorite is sold as household bleach in a 5% solution. Bleach disinfects viruses, bacteria, fungi, and small arthropods from equipment, glasshouse walls, and the air. A solution of 1 part bleach in 9 parts water (0.5% NaOCl) is used as a seed soak or soil drench against fungi. Diluted further, it can be sprayed on plants. Undiluted bleach is toxic to plants and caustic to our eyes and skin.

Spinosad is produced by *Saccharopolyspora spinosa*, a soil-dwelling actinomycete. It is relatively new, first registered by the EPA in 1997 (McPartland *et al.*, 2000), and now on the *National List*. Spinosad works as both a contact and stomach poison against thrips, caterpillars, beetles, and even leafminer maggots (Entrust®, Conserve®, SpinTor®). Entrust® showed good efficacy against *Cannabis* budworms (Britt and Kuhar, 2020; Britt *et al.*, 2021a). On plant surfaces spinosad is broken down relatively rapidly, with no phytotoxicity. Its half-life in soil, in the absence of sunlight, is much longer (with moderate toxicity to earthworms). Spinosad exhibits low toxicity to mammals, fish and birds. It is relatively safe to beneficial organisms and honey bees.

Sulfur in several formulations appears on *The National List*. Specific formulations allowed on *Cannabis* differ on a state-by-state level, with different regulations in Canada. Wilson *et al.* (2019) surveyed *Cannabis* growers regarding their pest management approaches, and "sulfur products" ranked just behind insecticidal soap (see Fig. 20.4). Sulfur should not be combined with seed oil (SO) or horticulture oil (HO); they interact to create a herbicidal stew. Once sulfur is applied, you cannot follow up with SO or HO applications.

Elemental sulfur, applied as a dust or spray, is a fungicide (especially powdery mildew) and pesticide (mites, aphids, and other small insects). Sulfur is not water-soluble, so it must be formulated as a wettable powder or granule (e.g., Kumulus®, Cosavet®, Thiolux®, Microthiol®). Sulfur persists on plants until wash-out. Elemental sulfur converts to SO_4^{2-} in soil, which can be absorbed by plants as a nutrient. Elemental sulfur is totally unavailable to plants, and can injure foliage, especially in hot, dry weather. Do not apply if the temperature is expected to exceed 86°F (30°C). Sulfur has low mammalian toxicity, but can irritate the eyes and skin, and should not be inhaled. It is non-toxic to birds and fish, but lethal to most biocontrol organisms.

Lime sulfur or calcium polysulfide, CaS_5 (e.g., Sulforix®) was first used at Versailles to control powdery mildew on grapes (Heuzé, 1852). By the 1880s it was used as an insecticide in California. Lime sulfur is more effective than elemental sulfur. But it causes greater phytotoxicity, especially in hot, dry weather. Cherian (1932) sprayed *gañjā* with lime sulfur, "which was tested by veteran smokers who gave their verdict against it."

Sulfur dioxide is created when sulfur pellets are vaporized in a sulfur burner. Although the pellets and burners can be purchased online, they are not registered with the EPA, and therefore illegal. Health Canada (2023) *does* permit the use of vaporized sulfur, specifically Agrotek Ascend™ Vaporized Sulphur. In high humidity, burning sulfur produces sulfuric acid, which causes chlorosis in *Cannabis* (Goidànich, 1959). Sulfur dioxide also damages lights, eradicates biocontrol organisms, and is toxic to humans. See more downsides in the chemical control section on spider mites (section 4.2).

Swinglea glutinosa **extract** (SGE) is an antifungal essential oil extract made from *Swinglea glutinosa*, a small tree from SE Asia. Labeled uses include treatment of gray mold, powdery mildew, downy mildew, and *Alternaria alternata*. New York includes SGE on its list of approved *Cannabis* pesticides (EcoSwing™).

Urine produced by predators repels prey animals from a garden perimeter. Fox urine repels small mammals like rabbits and rodents. Coyote urine reportedly works better against larger pests, like ground hogs and deer. Reapply predator urine products after rainfall and every week or so, depending on the product. Human urine is said to work the best. Mark your territory.

Versitude peptide, also known as GS-omega/kappa-Hxtx-Hv1a (Spear T®), is derived from the venom of an Australian spider, *Hadronyche versuta*. It is labeled against thrips, whiteflies, spider mites, and some lepidopteran caterpillars. It is approved for use on *Cannabis* by Colorado DOA (2023). Britt *et al.* (2021a) tested Spear T® against the budworm *Helicoverpa zea*, and it performed about the same as Bt *aizawai* (XenTari®).

The next four agents—horsetail, polygodial, quassia, and sabadilla—do not appear on lists of approved biopesticides by the USDA or the EPA. Perhaps they were overlooked; these agents are relatively safe and effective.

Horsetail (*Equisetum hyemale* or *E. arvense*) has been used medicinally since the time of Dioscorides. Steiner (1924a,b) prevented foliar plant diseases with a "quite concentrated tea" of *E. arvense*, which he diluted and applied as a "liquid manure" to soil around plants. Yepsen (1976) provided specifics: boil 1.5 ounces of dried leaves and stems in one gallon of water for 20 minutes. Diver (2002) prepared a fermented tea of horsetail, nettle, and comfrey, "stuff a barrel about three-quarters full of fresh green plant material, then top off the barrel with tepid water." After fermenting for 3–10 days, strain the fluid, dilute to 1:10 or 1:5, and apply as a foliar spray or soil drench.

The primary active ingredient is likely silicone. Horsetail extract also contains phenolic acids, flavonoids, and terpenoids, which likely contribute to its antifungal effects (Garcia *et al.*, 2012). Those authors prepared a hydro-alcoholic extract (50 g dried material in 1000 ml of 70% ethanol), evaporated to 50% concentration, which inhibited *Aspergillus flavus* and *Fusarium verticillioides* in maize seeds. Rodino *et al.* (2014) extracted 10 g dried material in 100 ml of 70% ethanol, diluted to 4%, to inhibit the *in vitro* growth of *Rhizoctonia solani*. De Queiroz *et al.* (2015) determined a minimum fungicidal concentration of 0.62 mg/mL against *Trichophyton rubrum* and *Microsporum canis* (human ringworm fungi).

Polygodial is a sesquiterpene extracted from *Polygonum* and *Warburgia* spp., first shown to have antifeedant and deterrent effects against *Spodoptera littoralis* and *S. exempta* (Kubo and

Ganjian, 1981). Rather than killing pests, polygodial deters them from feeding, so they migrate elsewhere. Analysis of video-recordings indicated that aphids (*Myzus persicae*) were repelled following contact of their antennal tips with polygodial-treated leaves (Powell *et al.*, 1995). Aphids with surgically removed antennae showed no apparent response to polygodial, suggesting that polygodial may induce noxious signals in aphids. In mammals it activates the TRPA1 ion channel—the same mechanism as allyl isothiocyanate (in wasabi and horseradish).

Prota *et al.* (2014) compared polygodial to pyrethrin in antifeedant assays with two insects. Against green peach aphid (*Myzus persicae*), polygodial had an ED$_{50}$ of 54 µg/g, about half the effectiveness of pyrethrin (28 µg/g). Against silverleaf whitefly, *Bemisia tabaci*, polygodial (25 µg/g) was far less effective than pyrethrin (1.4 µg/g). Prota proposed combining polygodial and pyrethrin; their different mechanisms of action may give rise to synergy. Polygodial likely has additive effects with CBD, which is also a potent TRPA1 agonist (De Petrocellis *et al.*, 2011).

Quassia or bitterwood (*Quassia amara*) is native to South America. Linnaeus coined *Quassia* in honor of Kwasi, an African slave brought to Suriname by the Dutch. He learned about quassia's medicinal uses, which he shared with the Dutch. *Cloëz (1869) placed* 100 g of quassia wood shavings and 20 g of lice-bane (*Delphinium staphisagria*) seeds in 3 liters of water, and boiled it down to 2 liters. He sprayed dozens of plant species, and rid them of aphids and flea beetles (compared with control plants). He is also known for publishing an early chemical analysis of hemp seeds (Cloëz, 1866).

Parker (1913a) combined quassia with soap to kill 96% of *Phorodon humuli* and other aphids on hops. Sallam *et al.* (2009) compared quassia with Neemazal T/S (azadirachtin plus limonoids) and Trifolio S-forte (vegetable oil with emulsifiers) against *Rhopalosiphum padi* and *Metopolophium dirhodum*. Quassia killed more aphids than Trifolio S-forte, but less than Neemazal T/S. Aggarwal *et al.* (2006) made a similar comparison between quassin (a triterpene lactone extracted from quassia), Neemazal T/S, and *Bacillus thuringiensis* ssp. *aizawai* against the budworm *Helicoverpa armigera* and the armyworm *Spodoptera exigua*. Quassin took third place. It works better as a feeding deterrent than a biocide. Quassia reportedly spares lady beetles and bees, and is relatively nontoxic to mammals.

Sabadilla is extracted from seeds of *Schoenocaulon officinale*. Its active ingredients are alkaloids, cevadine and veratridine. Explorers in Mexico discovered the ancient use of "cebadilla" for treating body vermin (Mondardes, 1574). By 1726 sabadilla became part of "Capuchin powder," a French remedy for lice (Zörnig, 1909). Its use as a crop pesticide revived after Allen *et al.* (1944) discovered that heating sabadilla enhanced its insecticidal activity. Sabadilla acts as both a contact and stomach poison against thrips, bugs, beetle grubs, caterpillars, and grasshoppers (Veratran D®). Once dusted on plants, sabadilla breaks down quickly in sunlight (about 2 days). The dust irritates mucous membranes and may cause sneezing. Mammalian toxicity has dampened enthusiasm for sabadilla, and it is also toxic to biocontrol organisms. Although it remains on the *National List*, sabadilla products are hard to find.

Calcium cyanamide (CaCN$_2$) is abhorred because of its method of synthesis (burnt limestone fed into a furnace with coal charcoal and atmospheric N$_2$), but we have used it to kill soil-borne organisms that cause damping off, such as *Pythium*, *Phytophthora*, *Rhizoctonia*, and *Fusarium*, when mixed into soil at a rate of 10–12 g/ft² (108–130 g/m²). CaCN$_2$ also kills nematodes (Hussain *et al.*, 2017) and root-damaging insects such as wireworms (Ritter *et al.*, 2014). CaCN$_2$ is a slow-release fertilizer that decomposes into urea, then ammonium, then nitrate (not nitrite). Some CaCN$_2$ is converted to hydrogen cyanamide, which does the killing. Hydrogen cyanamide is also phytotoxic, so CaCN$_2$ must be mixed into soil in early spring, at least 2 weeks before sowing seed. CaCN$_2$ has low mammalian toxicity, but hydrogen cyanamide may cause contact dermatitis.

21.8 Failed "non-toxic" pesticides

As the *Cannabis* industry has grown, we've learned that some "non-toxic" pesticides are really toxic. A few "non-toxics" that appeared in our first edition (McPartland *et al.*, 2000) are no longer recommended, and we explain why below.

Abamectin (avermectin) exhibits translaminar activity in some plants, meaning it is systemic. Abamectin was blamed for allergic reactions and neurological symptoms in Jane Weirick, who ran a trimming and processing service (Mikuriya *et al.*, 2005). She eventually died—the first documented casualty of sloppy and unscrupulous pesticide usage by *Cannabis* growers (Gardner, 2005). California DPR (2018) included abamectin on its list of pesticides that *cannot* be used on *Cannabis* under *any* circumstances.

Imidacloprid is a chlorinated analog of nicotine: a neonicotinoid. The EPA first approved its use in 1992, and within 20 years imidacloprid became the most widely used insecticide in the world. It is systemic and penetrates every cell in a plant. Imidacloprid has neurotoxic and mutagenic effects. The European Union banned it in 2018. Imidacloprid and other neonicotinoids have been linked with honey bee colony collapse disorder (Lu *et al.*, 2014; Tsvetkov *et al.*, 2017). Imidacloprid is often used as a seed dressing, which is seen as environmentally friendly, compared with spraying crops. However, imidacloprid leaches from coated seeds into the soil. It poisons groundwater, killing earthworms and other invertebrates crucial to soil health (Van Dijk *et al.*, 2013).

Juvenoids (juvenile hormones) arrest insect growth and prevent pests from maturing and reproducing. Two juvenoids are on the EPA's list of biopesticides (methoprene and pyriproxyfen, described below). Applying juvenoids results in a *gradual* reduction of pest populations. Some juvenoids kill immature insects and eggs, thus they work more quickly. When first introduced in 1967, juvenoids were considered "resistance proof." Unfortunately, insects have developed resistance to their own hormone mimics.

Methoprene is a wide-spectrum juvenoid, for controlling aphids, leafhoppers, caterpillars, beetles, and leafminers in a variety of cereal and vegetable crops. It is approved for use on *Cannabis* by Colorado DOA (2023). Formulations include emulsifiable concentrates, aerosols, baits, and flea collars for pets. Methoprene has a half-life of 2 days in water, 2 weeks on tobacco leaves in sunlight, and persists longer on tobacco in storage. It is moderately toxic to fish, non-toxic to honey bees, birds, and mammals, but may be irritating to mucous membranes.

Pyriproxyfen targets aphids, whiteflies, scales, and thrips, on fruit and vegetable crops. Soil drenches control fungus gnats, baits are used for fire ants, and pet collars to control fleas. Pyriproxyfen residues are persistent and can harm lady beetles and other predatory beetles for 2–3 months. It is moderately toxic to fish, and non-toxic to honey bees, birds, and mammals, but may be irritating to mucous membranes.

Other juvenoids are not recommended. *Kinoprene* is on the EPA's list of biopesticides, but its registration is limited to ornamentals, not vegetables or tobacco. *Fenoxycarb* acts as a juvenoid, but it is a carbamate insecticide. It kills beneficial insects; pet use (flea control) has been curtailed due to possible carcinogenic effects on dogs. *Buprofezin* is banned in some countries due to its negative environmental impacts. Some of its metabolites are carcinogenic.

Nicotine sulfate (NS) extracted from tobacco (*Nicotiana tabacum*) is no longer recommended. Tobacco has long been used as an insecticide, beginning with the King's gardener at Versailles (La Quintinye, 1690). Pichl (1902) found a 1% aqueous extract "very effective" at controlling aphids and spider mites infesting *Cannabis*. Instead of an aqueous extract, the King's gardener at Kew fumigated tobacco in a glasshouse to kill aphids and spider mites on cucumbers (Green, 1771). We've found only one scientific report of *Cannabis* fumigated with tobacco smoke: Schaffner (1928) fumigated hemp plants in a glasshouse infested by scale insects, mealybugs, and aphids. Tobacco fumigation had a "very severe effect" on the plants, and most died. In 1910 the Kentucky Tobacco Company began selling concentrated NS, containing 40% nicotine by weight (Black Leaf 40®). That's when the real problems started. Within a decade, reports of accidental poisonings from NS appeared, including five deaths (McNally, 1920).

Piperonyl butoxide (PBO) does not kill insects; rather, it enhances the killing power of pesticides, by inhibiting cytochrome P-450 (CYP) enzymes in insects that metabolize pesticides. Canyon *et al.* (2010) estimated that 1700 products, or 8% of all USA-registered pesticides, contained PBO. It is frequently added to pyrethrins and synthetic pyrethroids, for agriculture as well as medical uses (e.g., lice-killing shampoo).

PBO is moderately persistent in the environment. PBO exposure may cause skin irritation, asthma attacks, nausea, vomiting, and diarrhea (Canyon *et al.*, 2010). The EPA classifies PBO as a group C substance, "possibly carcinogenic to humans." PBO residues are commonly detected in contaminated *Cannabis*. Voelker and Holmes (2015) tested 389 samples, and PBO was the most common contaminant, with an astounding 407,000 ppm in one sample.

Pyrethroids are synthetic pyrethrins (permethrin, cypermethrin, deltamethrin, bifenthrin). Several states and Canada allow the use of natural pyrethrins on *Cannabis*, but nobody permits the use of synthetic pyrethroids. They are counter-productive: pests rapidly gain resistance to pyrethroids, while natural predators are killed off, allowing pest populations to explode. Pyrethroids are more toxic to mammals and more persistent in the environment than their natural cousins.

Synthetic pyrethroids are known to disrupt the endocrine system (Colborn *et al.*, 1996). Pyrethroid residues are considered an occupational health hazard for workers harvesting and processing cannabis (Couch *et al.*, 2017).

Rotenone is no longer recommended. Rotenone is derived from roots of *Lonchocarpus* and *Derris* spp. Millions of pounds of derris and cubé root were imported annually into the USA, sourced from Malaya, Indonesia, and South America. Rotenone was the most popular pesticide in the USA until DDT in the 1950s. Rotenone use has been linked with Parkinson's disease in at least five epidemiology studies. The Netherlands banned rotenone back in 1980, the UK banned it in 2009, and its use in organic farming is now banned in the USA.

Sodium nitrate ($NaNO_3$) became popular as a fertilizer, "Peruvian saltpetre," in the mid-1800s. The Saltpeter War (1879–1884) between Chile and Peru was fought over sodium nitrate deposits. Sodium nitrate kills soil-borne damping-off fungi and nematodes, and suppresses fungal production of mycotoxins (Covarelli *et al.*, 2004). However, it can no longer be recommended. *The National List* specifically prohibits sodium nitrate. It is water soluble, highly mobile, and quickly leaches into groundwater. This results in water pollution and the contamination of water supplies. Following ingestion, the body reduces nitrate to nitrite, or to nitrosamines, known to be carcinogens.

Streptomycin is an antibiotic oozed by *Streptomyces griseus*. Streptomycin has been applied as a soil drench or foliar spray on orchard trees, tobacco, and many other crops since the 1950s (Stockwell and Duffy, 2012). Streptomycin resistance in plant pathogens is now fairly widespread. It may cause phytotoxicity, is toxic to fish, and applicators may develop allergenic reactions. *Streptomyces griseus* is approved as a biocontrol agent for use on *Cannabis*, which is a viable alternative to spraying streptomycin. A related antibiotic, **oxytetracycline** (Terramycin®), is produced by *Streptomyces rimosus*. It is sprayed on orchard trees and vegetable crops. Its use is limited to only a few countries (Stockwell and Duffy, 2012).

21.9   Rogues' gallery

Extremely nasty pesticides continue to be used on crops. Many turn up in studies of *Cannabis* pesticide residues. In the spirit of *harm reduction*, we describe some here. European hemp researchers have tested toxic pesticides on fiber crops. These chemicals should never be sprayed on crops destined for human consumption, such as seed oil or herbal cannabis. Placing synthetics in baited traps is a relatively safe application. Baited traps (such as Japanese beetle traps) keep chemicals off plants, and are containerized for effective disposal. Using synthetics for seed treatment (coating seeds before they are planted) is no longer considered safe—the pesticides wash off seeds and into groundwater.

Synthetic pesticides are classified by their chemical structures. The major groups include organochlorines, organophosphates, carbamates, dipyridilium herbicides, and the most popular group, "miscellaneous."

21.9.1 Organochlorines

Organochlorines are nerve poisons. They are relatively safe to mammals, and rarely cause acute poisoning in humans. But their long residual (i.e., resistance to degradation), initially a desired feature, makes them environmentally dangerous. Most organochlorines have been banned in the USA, such as DDT, chlordane, dieldrin, and lindane. In 1993 the American Public Health Association urged a phase-out of all chlorinated organic compounds.

Chloropicrin was developed as a poison gas in World War I, but found use as an agricultural fumigant by 1919. Soil fumigation with chloropicrin is expensive, technically difficult, and generally nasty. It is acutely toxic and carcinogenic. Chloropicrin reduces *Cannabis* seed germination by at least 30% (Miège, 1921). Mysteriously, Health Canada (2023) includes chloropicrin on its list of approved *Cannabis* pesticides.

Pentachloronitrobenzene (PCNB) was introduced by Bayer in the 1930s as a substitute for mercurial fungicides. Ferri and Noviello (1963) drenched soil with PCNB, and decreased the incidence of *Sclerotium rolfsii* in hemp plants. Ashok (1995) found it helpful for controlling *gañjā* root diseases, but Mishra (1987) did not. PCNB is phytotoxic, slightly toxic to honey bees and birds, and highly toxic to fish. PCNB may disrupt the endocrine system (Colborn *et al.*, 1996). In 2013 many of its pesticide registrations were cancelled.

Dicofol is structurally similar to DDT, and dicofol products contain traces of DDT. It causes eggshell thinning in various bird species, which is why DDT was banned. Dicofol is carcinogenic to male (not female) mice, and may disrupt the endocrine system (Colborn *et al.*, 1996).

21.9.2 Organophosphates

Organophosphates are nerve poisons, like organochlorines, but degrade relatively quickly in the environment. Organophosphates are far more acutely poisonous to mammals than organochlorines. Organophosphates have not been banned for agricultural uses, but the EPA banned most of them for residential uses.

Parathion kills insects and careless farmers. It is half-strength Sarin (a chemical warfare agent). Cuypers *et al.* (2017) found a container of parathion at an illicit grow site. Parathion is moderately toxic to fish, and extremely toxic to birds, bees, and mammals. California DPR (2018) included parathion on its list of pesticides that *cannot* be used on *Cannabis* under *any* circumstances.

Malathion remains the most widely used organophosphate. It turns up frequently in studies of *Cannabis* pesticide residues. Back in the day, grower's guides recommended it (Frank and Rosenthal, 1978; Rosenthal, 1999). It is highly toxic to honey bees, and moderately toxic to mammals and birds. Malathion is a probable carcinogen. In an epidemiology study of 119 children, Bouchard *et al.* (2010) found that elevated levels of dimethylthiophosphate (a metabolite of malathion) doubled their odds of having ADHD (attention-deficit hyperactivity disorder).

Diazinon also turns up in studies of *Cannabis* pesticide residues, and formerly appeared in grower's guides. It is highly toxic to honey bees and birds (who mistake granular formulations for food). Symptoms of acute poisoning include dizziness, headache, salivation, lacrimation, rhinorrhea, nausea, vomiting, diarrhea, muscle twitching, weakness, and tremor. Diazinon, like malathion, has been linked with non-Hodgkin's lymphoma (Guyton *et al.*, 2015).

Chlorothalonil formerly appeared in grower's guides. Products with chlorothalonil contain traces of dioxin, a notorious toxin. The EU has banned it since 2019. It is moderately toxic to birds and mammals (especially eye and skin irritation). Off-target victims of chlorothalonil include fish, amphibians, and pollinator insects.

Chlorpyrifos residues have been blamed on "toxic marijuana syndrome" (Gillet, 2016). In 2015 the Obama administration proposed banning chlorpyrifos, but the ruling was appealed by the Trump administration. California unilaterally cancelled the registration of chlorpyrifos in 2019. Chlorpyrifos is among commonly used pesticides for attempted suicide.

Dimethoate is also used for suicide attempts, with a higher rate of success than chlorpyrifos, 23% compared to 8% (Eddleson *et al.*, 2005). It turns up in studies of *Cannabis* pesticide residues, and formerly appeared in grower's guides. Dimethoate is banned in the EU. **Dichlorvos** or DDVP is the insecticide in No-Pest™ strips, formerly recommended in grower's guides and still appears all over the internet. Dichlorvos is carcinogenic, and the EU has banned it since 1998. The EPA currently has dichlorvos under review.

21.9.3 Carbamates

Carbamates are similar to organophosphates in their chemical structures, their targets in insects (acetylcholinesterase), and their high toxicity. Symptoms of acute poisoning go by the "DUMBBELS" mnemonic: defecation, urination, miosis, bronchospasm, emesis, lacrimation, salivation. Sulfur-containing carbamates (e.g., maneb, mancozeb, thiram) are biodegraded by plants into ethylene thiourea, a potent carcinogen.

Maneb is a sulfur- and manganese-containing fungicide, with a wide spectrum of activity. Bósca and Karus (1997) recommended maneb against gray mold, but it was banned by the EU in 2009, ruled a probable carcinogen. **Zineb** is a sulfur- and zinc-containing fungicide, recommended by Bósca and Karus (1997), but Mishra (1987) did not find zineb helpful for controlling *gañjā* diseases. All zineb registrations were voluntarily cancelled following an EPA review. **Mancozeb** is a mix of maneb and zineb; its registrations were also voluntarily cancelled by its manufacturers.

Benomyl was formerly used to control fungal diseases of *Cannabis* (Mishra, 1987; Ashok, 1995), but many fungi are resistant to it. Benomyl works systemically, persists in the environment, and is classified as a probable carcinogen. **Carbaryl** is the most commonly-used carbamate. It formerly appeared in grower's guides, and turns up in studies of *Cannabis* pesticide residues. Russo (2016) found one sample, obtained from a legal dispensary, with 25,483 ppb. The residue limit for carbaryl in tobacco is 100 ppb (Stephenson *et al.*, 2003). The Bhopal disaster occurred at a carbaryl factory—8000 people died within 2 weeks, and another 8000 died thereafter (Eckerman, 2005). The EPA classifies carbaryl as a probable carcinogen. It is toxic to birds and fish, very toxic to honey bees, and kills earthworms.

21.9.4   Dicarboximides

Dicarboximides are ammonia derivatives that target triglyceride biosynthesis in sclerotia-forming fungi. They are systemic in plants, toxic to birds, very toxic to fish, and act as endocrine disruptors (Colborn *et al.*, 1996). Two closely related dicarboximides, **vinclozolin** and **iprodione**, were tested on fiber-type hemp (de Meijer *et al.*, 1995). The treatment reduced gray mold, but did not significantly increase crop yield—11.2 t/ha for control plots versus 12.3 t/ha for treated plots (Van der Werf, 1994). Since then, vinclozolin has been banned in many European countries. It has been phased out in the USA, limited to canola and turf. Iprodione's approval was not renewed in Europe in 2017.

21.9.5   Miscellaneous synthetics

Metalaxyl is an acylalanine fungicide with systemic action. It is used to control oomycetes such as *Pythium*, *Phytophthora*, and *Pseudoperonospora* spp., although resistance is commonplace. Metalaxyl frequently appears in studies of pesticide residues. California DPR (2018) included metalaxyl on its list of pesticides that *cannot* be used on *Cannabis* under *any* circumstances. **Triforine** is a piperazine fungicide used for controlling powdery mildew. Frank (1988) recommended it for *Cannabis*, and web scuttlebutt suggests wide use. Triforine is not permitted on tobacco (Stephenson *et al.*, 2003). The EU banned it in 2007. It is moderately toxic to fish, but highly toxic to mammals.

Myclobutanil is a triazole fungicide with systemic action. It turns up frequently in studies of *Cannabis* pesticide residues. When heated, myclobutanil releases cyanide gas, so it is banned from use on tobacco. It may contribute to vaping-associated lung injury (Xantus, 2020). Myclobutanil is banned in Canada, Colorado, Washington, and Oregon for the production of medical and recreational cannabis. **Bifenazate** is used to control mites on ornamental plants. It has not yet received a human carcinogen classification, because its use is limited to non-food plants. Yet it turns up frequently in studies of *Cannabis* pesticide residues. In one study of legally grown medical cannabis in California, bifenazate was the second most common pesticide (Wurzer, 2016). It is moderately toxic to mammals and birds, and highly toxic to fish and aquatic invertebrates.

Boscalid is a carboxamide fungicide with systemic activity. The EPA classified boscalid as "suggestive evidence of carcinogenicity, but not sufficient to assess human carcinogenic potential." Allowable residue tolerances are as low as 0.05 ppm. Russo (2016) found 112 ppm in a cannabis sample obtained from a legal dispensary in Washington. California DPR (2018) included boscalid on its list of pesticides that *cannot* be used on *Cannabis* under *any* circumstances.

Paclobutrazol is a triazole fungicide, used on turfgrass and ornamentals, and not labeled for use on food crops. Because of this, paclobutrazol has not been subject to cancer studies. Paclobutrazol is also a plant growth regulator (PGR), and it replaced another PGR, **daminozide** (Alar™), removed from the market in 1989. Both paclobutrazol and daminozide have turned up in studies of *Cannabis* pesticide residues. In 2011 California issued a cease-and-desist order against the sale of cannabis contaminated with paclobutrazol and daminoside (Upton *et al.*, 2013). California DPR (2018) included paclobutrazol on its list of pesticides that *cannot* be used on *Cannabis* under *any* circumstances.

Tebuconazole is a triazole fungicide, used on turfgrass and vegetable crops. It is moderately toxic to mammals, fish, and aquatic invertebrates, and rated as a possible carcinogen by the EPA. Tebuconazole leaches into groundwater and contaminates water supplies. Due to its endocrine-disrupting effects, it is being reviewed by the EU for removal from the market. Tebuconazole is a frequent offender in studies of *Cannabis* pesticide residues. California DPR (2018) included tebuconazole on its list of pesticides that *cannot* be used on *Cannabis* under *any* circumstances.

Metaldehyde causes slugs to secrete a heavy trail of mucus, so they dehydrate and die. Rainfall leaches metaldehyde into groundwater, which harms non-target species (fish, amphibians), and contaminates drinking water. The pellets attract children and pets, they have a sweet flavor. The LD_{50} in dogs is only 100 mg/kg; symptoms of toxicosis include drooling, seizures, coma, and death from respiratory failure. Heating metaldehyde yields acetaldehyde, an irritant of the skin, eyes, throat, and respiratory tract; strong exposure causes nausea and vomiting, and it's a carcinogen. A ban on metaldehyde in the UK was overturned after a legal challenge in 2019.

21.9.6   Metalloid-based pesticides

Inorganic pesticides made of arsenic, lead, and mercury were major breakthroughs in the 19th century. Now they are specifically prohibited in the USDA's *National List* of organic pesticides.

Arsenic is acutely toxic, "the poison of kings," and carcinogenic. The EPA ranked arsenic as #1 in its priority list of hazardous substances at Superfund sites. Yet arsenical compounds are still prescribed today for medical uses (Firth, 2013). A variety of arsenical compounds have been used for pest control, beginning with Slator and Hall (1724), who laced arsenic into stacks of hemp stalks to "prevent the Vermin from doing Mischief." Copper acetoarsenite gained the name Paris green, for its use as a rat poison in the sewers of that city. Its insecticidal properties were discovered by farmers in Illinois and Wisconsin (Walsh and Riley, 1868). Calcium arsenite became known as "London purple" after a London chemist, Henry Hemingway, sent a sample to Charles Bessey at Iowa State University. Paris green and London purple burned foliage, and farmers worried about toxicity.

Lead arsenate, a white powder with no place name, came to the attention of entomologists in Boston (Kirkland, 1896). Cherian (1932) dusted *gañja* plants with lead arsenate. He measured 8.3 mg of lead arsenate residues in 100 g of harvested *gañjā*. Lead arsenate, Paris green, and London purple became obsolete when DDT arrived. Lead arsenate was formerly the most common cause of accidental honey bee poisonings, implicated in 81% of surveyed cases (Shaw and Mendal, 1940).

Mercurial pesticides seem medieval, but they were still around in the 1980s. Naaman Goodsell, a garden marketing pioneer in Rochester, first used mercuric chloride ($HgCl_2$) as an insecticide (Goodsell, 1831). Hiltner and Gentner (1916) treated hemp seeds with mercuric chloride in a controlled trial, complete with photographs (Fig. 12.7). Organomercurials, less toxic, supplanted mercuric chloride. The hazards of cumulative contamination with mercury led to the phasing out of organomercurials during the 1980s. All mercurial pesticides were finally banned in 1992, due to adverse toxicology (Maude, 1996).

21.10   Pesticide safety

A pesticide container must be labeled with a signal word that matches the product's acute toxicity classification (Table 21.2). Those blazed with "Danger" and "Poison" are extremely toxic Class I poisons. Simply tasting these chemicals can be lethal. A "Warning" label marks pesticides of moderate toxicity (EPA Class II); a teaspoon to a tablespoon will kill. Chemicals carrying the word "Caution" (EPA Class III & IV) are also lethal, but at larger doses.

When working with pesticides, follow these common-sense precautions:

1. Read the pesticide label and follow instructions.
2. Handle all used spraying/dusting equipment with as much caution as the pesticide itself.
3. Store pesticides and equipment in lockable storage away from children and animals, and away from water, food, and feed.
4. Keep pesticides in original containers. Discard unlabeled pesticide containers; do not guess at contents. Dispose of pesticide wastes properly. Empty pesticide containers should be triple-rinsed before throwing them away or recycling them. Do not re-use containers in the home.
5. Prepare pesticides and spray equipment in a protected area away from children and animals, where spills can be cleaned up easily. Clean up spilled chemicals with absorbent materials such as sawdust. Do not hose down the area. Decontaminate pesticides with household bleach and hydrated lime.
6. Do not combine two pesticides unless you are certain they are compatible. To avoid cross-contamination of pesticides, use separate labeled sprayers for each class of pesticide (fungicide, insecticide, herbicide, etc.).
7. Do not work alone when handling dangerous pesticides, especially in enclosed areas like storage sheds, preparation areas, or glasshouses and growrooms.
8. Always wear protective clothing, including long-sleeved shirts and long-legged pants made from tightly woven fabric. Protect your scalp with a waterproof, brimmed hood or hat. Rubber or neoprene footwear should fit over (or replace) leather or fabric shoes. Wash or discard protective clothing immediately following use.
9. Some pesticides require additional **Personal Protective Equipment** (PPE): an apron, coveralls, or chemical-resistant suit (especially when pouring and mixing chemicals or cleaning pesticide equipment). Gloves should be made of polyvinyl chloride or rubber (nitrile, neoprene, etc.), unlined (fabric lining absorbs pesticides), and extend to mid-forearm. Protect eyes with safety glasses, goggles, or a full-face shield. Wear a well-fitting dust mask or respirator. Take steps to avoid heat illness when wearing respirators and protective clothing. Even botanical pesticides and biocontrol products should be applied while wearing PPE (Table 21.3).
10. Never eat, drink, chew gum, or smoke while working with pesticides or in treated areas. Wash your hands before using the toilet. Avoid inhaling pesticide fumes; if pesticides drift nearby, get away. If eyes or skin become contaminated, quickly flush with large volumes of water. Remove contaminated clothing; handle clothing cautiously; wash in hot water with maximum detergent. Wash work clothes separate from the family laundry. Be sure that cleaners know clothes are contaminated and dangerous.
11. Know the symptoms of pesticide poisoning—irritated skin/throat/eyes, nausea, vomiting, dizziness, headache, trouble breathing, muscle pains, pinpoint pupils. Post telephone numbers of the local hospital and the National Center for Poison Control, (800) 222-1222. On line: https://www.poison.org, there's an app: webPOISONCONTROL®.

Table 21.3. Personal Protective Equipment recommended for some botanicals and biocontrols

Active Ingredient	Recommended PPE
Bacillus species	Gloves (latex is sufficient), dust mask, goggles, LSS, LLP
Copper	Gloves (latex or rubber), dust mask, goggles for mixing powders or safety glasses for mixing liquids, coveralls or LSS, LLP
Diatomaceous Earth	Goggles, dust mask
Neem	Safety glasses, dust mask, LSS, LLP
Oil, vegetable or horticultural	Safety glasses, mask
Pyrethrins	Gloves (latex or rubber), safety glasses, dust mask or respirator, coveralls or LSS, LLP
Ryania	Safety glasses, dust mask, LSS, LLP
Sabadilla	Gloves (latex or rubber), safety glasses, dust mask, LSS, LLP
Soaps	Gloves (latex or rubber), safety glasses, respirator, LSS, LLP
Sulfur	Gloves (latex or rubber), goggles for powders or safety glasses for liquids, dust mask (if using micro-sul formulation, use respirator with HEPA filter), coveralls or LSS, LLP

LSS=long-sleeved shirt; LLP=long-legged pants with shoes and socks

21.10.1   Worker Protection Standard

The common-sense precautions listed above are now law. Employers who hire agricultural workers must meet federal guidelines for pesticide safety. (Some states have additional requirements.) In 2015 the EPA revised its **Worker Protection Standard (WPS '15)** for the first time in 23 years. The provisions in WPS '15 are complicated. Employers should contact their regional EPA office or a state agricultural extension agent for full information regarding compliance. In summary:

1. WPS '15 is label-driven. Requirements are pesticide-specific and listed on labels. Both general-use and restricted-use pesticides are covered by WPS '15. If you are using a pesticide with labeling that refers to the Worker Protection Standard, you must comply. Many home-use pesticides and biocontrols are exempt, but we recommend using precautions even in the absence of specific label requirements (Table 21.3).

2. All workers *potentially* exposed to pesticides—not just pesticide handlers—must be trained in pesticide safety on an annual basis. The EPA provides a checklist of 36 required topics, including the precautions listed above.

3. Before spraying, a fact sheet describing the pesticide must be posted. Required information included its EPA number and active ingredients, the area treated, date of application, and **restricted-entry interval** (REI). The REI is the period of time after a pesticide has been applied when entry into the treated area is restricted. Generally, Class I poisons (Table 21.2) have an REI of 48 h, Class IIs have an REI of 24 h, and Class IIIs have an REI of 12 h. Some pesticides also require "Keep Out" signs be posted during the REI (in English and Spanish).

4. Information posters on pesticide safety and emergency assistance must also be displayed at a central location.

5. Establish decontamination stations within 1/4 mile of all agricultural workers. These should be mobile, so field workers can clean up before meals and at the end of the workday. Supply water (1 gallon for each worker and 3 gallons for each handler) for routine and emergency washing, plenty of soap, and paper towels.

6. Workers who apply pesticides require additional training, decontamination supplies, and personal protective equipment (PPE). PPE must be clean and in operating condition, worn and used correctly, and washed or replaced as needed.

REI posters and warning signs, pesticide fact sheets, worker training manuals, compliance record books, PPE equipment, and decontamination equipment can be purchased at farm supply stores. For one-stop shopping, contact Gempler's based in Janesville, Wisconsin (1-800-382-8473), on line: https://gemplers.com/pages/wpsstandard

21.11   Pesticide application

Pesticides can be applied directly on plants, or applied near plants. Pesticides are applied on plants as foliar treatments, seed treatments, or onto roots as soil drenches. Off-plant applications include baits, barriers, and pre-season soil treatments. Off-plant applications are safest for *Cannabis* consumers, but still contaminate the environment.

21.11.1   Baits, barriers, and soil treatments

Most baits are sold ready-to-use. Examples include pheromone baits for moths, food baits for cutworms, Japanese beetle traps, and rodent baits. Tanglefoot is a popular barrier, useful against crawling insects. Soil treatments can be applied by any method in Table 21.4. Furrow, root zone, and seed treatment methods are most commonly used.

21.11.2   Seed treatments

Patschke *et al.* (1997) recommended seed dressings as the best way to control some hemp diseases. Sowing pesticide-coated seeds is safer than applying soil drenches. But the pesticides nevertheless wash off seeds and into groundwater. Some IPM practitioners cannot justify seed treatment, because it is a form of *preventive* chemical control. IPMers normally withhold pesticide application until an economically damaged threshold has been reached.

Seed dressings work at different levels within seeds. Some only work on the surface. Other pesticides penetrate seeds and eliminate internally pathogens. Some systemic pesticides persist in young seedlings. Seeds may be coated with a variety of pesticide formulations, such as D, WP, FC, and SP (Table 21.1). The machinery used for coating seeds is reviewed and illustrated by Maude (1996). Briefly:

Dusts are mixed with seeds in drum mixers or auger mixers. Mixers can be modified to apply liquid formulations (WPs and FCs). Improved methods of applying liquids to seeds include perforated drums, fluidized beds, spinning disc techniques, and film-coating techniques. The trouble with dust is poor adhesion to seeds, resulting in the environmental and health hazards of loose pesticide dust in bags of seed. Spraying seeds with a fast-drying liquid results in better adherence. But the liquid may adhere irregularly, with too little on some seeds and heavy phytotoxic loads on other seeds.

A second approach is seed *immersion*—steeping seeds in liquids or fumigant gases for various periods of time, at ambient or elevated temperatures. Many hemp researchers have utilized this method. Soaking seeds in hot water (50°C) eradicates some pathogens. Soaking seeds in 10% household bleach for 1 min kills many fungi. Rinse seeds in water at least three times before drying or planting. To eradicate bacteria, Kotev and Georgieva

Table 21.4. Methods of applying pesticides to soil

Method	Description
Broadcast	Pesticide broadcast on the surface and mixed into soil before sowing seed
Furrow	Pesticide selectively poured into soil behind plow or drill, before sowing seed
Root zone	Pesticide mixed into soil of intended root zone, before transplanting seedlings
Irrigation	Pesticide added to trickle irrigation system
Seed treatment	Seeds coated with pesticide using an adhesive

(1969) soaked hemp seed in 0.3% formalin. Ferri (1961b) soaked hemp seeds in 0.2% sulfuric acid for up to 30 min, or in 90% isopropyl alcohol for 10 min, apparently without seed damage. Mishra (1987) soaked *gañjā* seeds with seven fungicides and tested their efficacies, but most of the fungicides he tested are now banned.

21.11.3 Foliar treatments

Dusts adhere firmly to plant surfaces, particularly if electrostatically charged or applied to dew-covered plants. Dusts are more susceptible to wind drift than are sprays. Do not dust when wind speed exceeds 5 mph. (Sprays tolerate breezes up to 12 mph.)

Sprays are easier to apply than dusts, but spray droplets often roll off plant surfaces, especially off leaves covered with trichomes such as *Cannabis*. **Spray adjuvants** must be added to improve the effectiveness of spray applications. There are many types of spray adjuvants (over 150 adjuvants are registered in 17 categories in the USA). Here are a few.

Spreaders (i.e., *wetting agents*) keep sprays from beading up and rolling off leaves. They reduce the surface tension of liquids, so the sprayed droplets spread out and hold onto sprayed surfaces. High rounded drops on leaf surfaces indicate the need for a spreader. Flat drops that slide off leaves indicate too much spreader. There are *non-ionic, anionic,* and *cationic* spreaders.

Non-ionic spreaders are the most common. They do not ionize in water, so they do not react with most pesticides (but they tend to remain as residues on plant surfaces). Two common non-ionic spreaders are alkyl-aryl-poly-oxy-ethylenate (AAPOE) and alcohol-poly-oxy-ethylenate (APOE). Triton AG-44M® is a popular non-ionic spreader. Therm X-70®, derived from *Yucca schidigera*, is a USDA-approved organic spreader. About 10% of spreaders fall into the *anionic* category; they are ionized with a strong negative charge. Examples include fatty acids and linear alkyl sulfonate (LAS). *Cationic* spreaders (ionized with a strong positive charge) are rarely used.

Stickers provide an adhesive effect after the spray has dried on the plant. Otherwise, the pesticide would quickly wash off from heavy dew, irrigation, or rain, as well as from wind erosion and leaf abrasion. Stickers work three ways: by increasing adhesion, slowing evaporation, or providing a waterproof coating. Some stickers are also spreaders, such as AAPOE, fatty acids, and LAS. Waterproofing stickers include latex, polyethylene, resins, and menthene polymers—not meant for human consumption.

Extenders (also called *stabilizing agents*) protect spray residues against UV radiation and heat, which degrade many pesticides. **Emulsifiable oil** adjuvants enhance the penetration of sprays into waxy plant surfaces. They are also called *plant penetrants* or *translocators*. **Activator** is a general term indicating any adjuvant that increases the effectiveness of a pesticide; all of the aforementioned adjuvants are activators.

Buffers include *acidifying* adjuvants and *softening* agents. Many pesticides require mixing in water that is slightly acidic (optimal pH = 6.0). The most common acidifying adjuvant is phosphoric acid. Muriatic acid and vinegar may not be suitable. Hard water contains metal ions (Ca, Fe, Mg, etc.) which precipitate many pesticides (especially fatty acids and soaps) into a scum, curd, or "ring around the tub." Water hardness is measured in ppm or grains; if water hardness exceeds 300 ppm or 18 grains, softening may be required. Softening agents include Spray-Aide, Blendex®, and Triton AG-44M®.

Defoamers keep foam from forming in spray tanks. The most common defoamer is a silicone/carbon polymer called dimethylpolysiloxane. **Drift retardants** keep sprays from breaking into ultrafine droplets that float away. **Compatibility agents** facilitate the mixing of pesticides to keep them from separating or curdling in spray tanks.

21.11.4 Foliar application equipment

Spray equipment

The popularity of pesticides in the late 19th century would not have been possible without the introduction of spray equipment. Prior to the invention of pressured sprayers and fine spray nozzles, pesticides were brushed onto plants with whisk brooms. The Vermorel corporation near Lyon invented a sprayer for use in vineyards in 1880. They made backpack models and horse-drawn machines (Fig. 21.7). Their brass piston-plunger was fitted with hemp packing to keep the pneumatic seal.

Pesticide application devices are sold at hardware stores and garden centers. Selecting equipment adapted to your individual needs makes the choice less overwhelming. Here is a partial list of spraying and dusting equipment, with comments.

1. *Aerosol can:* prepackaged, very convenient, high cost.
2. *Trigger-pump sprayer:* inexpensive, reusable, Windex-type hand sprayers work best with water-soluble pesticides for small problems. They are too small and tiring for large gardens.

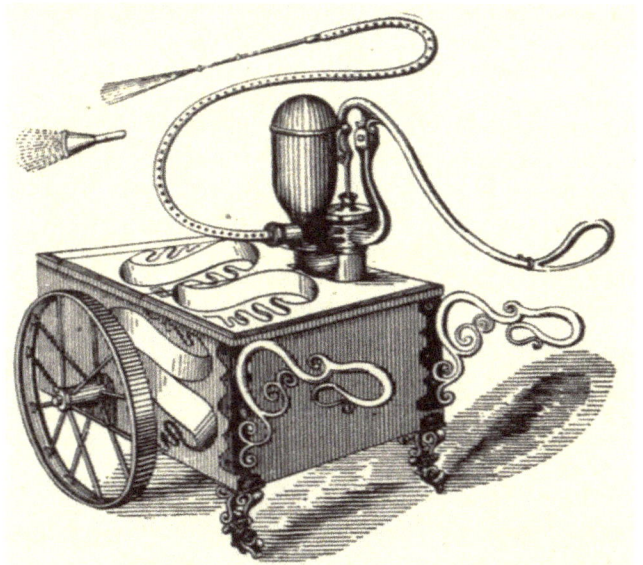

Fig. 21.7. A horse-drawn pesticide sprayer (Lodeman, 1896)

3. *Slide-pump sprayer:* similar to trigger-pump sprayers, but generate pressure via a telescoping plunger. More applicable to larger gardens. Usually limited to one nozzle, and most slide-pump sprayers only discharge during half of the pump cycle. They are leaky and often contaminate the applicator.

4. *Compression sprayer:* a manual pump compresses air above liquid in a tank. Air pressure forces the liquid out through an opening at the bottom of the tank to a hand-held nozzle. The tank can be carried in the other hand or in a backpack. Air pressure slowly discharges as pesticides are applied, requiring frequent repumping. Compression sprayers clog easily, and require frequent cleaning and maintenance to prevent corrosion of working parts.

5. *Power compressed-air sprayer:* a gas or battery-powered engine pumps air pressure into a tank, and the rest works like a compression sprayer. Usually mounted on a tractor. Higher maintenance costs.

6. *Hydraulic sprayer:* a hydraulic pump that draws liquid up from a tank and forces it under pressure through multiple discharge tubes (spray booms) and out of any number of nozzles. Usually mounted on a tractor—expensive maintenance.

7. *Bulb or bag duster:* squeeze by hand to propel dust onto plants. Simple, easy, and many are reusable.

8. *Hand-crank dust or granule applicator:* dust or granules are stored in a hopper, beneath which spins a fan, powered by a hand-crank. Air currents or the fan itself discharge the granules. Hole size at the bottom of the hopper regulates application rate, and fan speed controls the range of discharge. The granules often contaminate the applicator.

9. *Power duster:* similar to the above, but an engine generates compressed air to blow dust. Usually tractor-mounted, generating considerable drift and collateral contamination.

10. *Power granule applicator:* like the power duster, but spreads granules, much less drift.

Spray nozzles come as cones, flat fans, whirl chambers, and many other configurations. Liquid pesticides can be atomized into different droplet sizes. The smaller the droplet the better the coverage, and smaller droplets are more toxic to insects. But smaller droplets are more susceptible to drift.

Fogs and smokes are the smallest droplets, with a size of <5µm diameter. Fogs and smokes almost act as fumigants—they cover all plant surfaces (including the undersides of leaves). Fogs are produced by ultrasonic or thermal fogging machines, and smokes are generated by combustion. Aerosol droplets (5–50 µm in diameter) are produced by blasting pressurized air over a liquid to atomize the spray. Mists (50–100 µm) also require high pressure. Fine sprays (100–400 µm) and coarse sprays (>400 µm) can be generated at low pressures.

For calibration and use of **tractor-drawn power equipment**, refer to Matthews (2008) for information concerning droplet size, nozzle types, pump capacity, and other aspects of application technology. Here is an example of calibration. Using tractor-drawn power sprayers, the amount of insecticide applied per acre depends on the height of the crop. For plants under 18 inches tall (45 cm), only one hollow-cone spray nozzle is required per row. Spraying rows 36 inches (90 cm) apart at a pressure of 60 pounds/inch^2 (4.2 kg/cm^2), while moving 4 miles per hour (6 km/h) will use 7 gallons/acre (65 L/ha).

For taller plants, use 3 hollow cone nozzles per row, same pressure, same tractor speed, and apply 21 gallons/acre (196 L/ha). Always position the center nozzle 6–12 inches (15–30 cm) above plant height. The two side nozzles should be positioned on either side of the row, low to the plant, with the nozzles angled upward—so sprays hit pests that congregate on undersides of leaves.

Just 40 years after the invention of the Vermorel sprayer (Fig. 21.7), Neillie and Houser (1922) reported that entomologists in Ohio used an airplane to spread lead arsenate (Fig. 21.8). This was the first aerial application of a pesticide.

Fig. 21.8. First aerial application of a pesticide (Neillie and Houser, 1922)

(Prepared by J. McPartland; revised by E. Russo)

References

Aggarwal N, Holaschke M, Basedow T. 2006. Evaluation of bio-rational insecticides to control *Helicoverpa armigera* (Hübner) and *Spodoptera exigua* (Hübner) (Lepidoptera: Noctuidae) fed on *Vicia faba* L. *Mitteilungen der Deutschen Gesellschaft für Allgemeine und Angewandte Entomologie* 15: 245–250.

Allen TC, Dicke RJ, Harris HH. 1944. Sabadilla, *Schoenocaulon* spp., with reference to its toxicity to houseflies. *Journal of Economic Entomology* 37: 400–407.

Antonious GF, Meyer JE, Syner JC. 2006. Toxicity and Repellency of Hot Pepper Extracts to Spider Mite, *Tetranychus urticae* Koch. *Journal of Environmental Science and Health Part B* 41: 1383–1391.

Ashok K. 1995. Chemical control of root rot of ganja. *Current Research, University of Agricultural Sciences Bangalore* 24(9): 99–100.

Atapattu SA, Johnson KRD. 2020. Pesticide analysis in cannabis products. *Journal of Chromatography A* 1612: e460656.

Baca R, Migoya D (2015) Denver issues ninth pot product recall in 10 weeks. *The Denver Post*, November 18, 2015.

Backer R, Schwinghamer T, Rosenbaum P, *et al.* (8 other authors). 2019. Closing the yield gap for *Cannabis*: a meta-analysis of factors determining *Cannabis* yield. *Frontiers in Plant Science* 10: 495.

Balachowsky A. 1933. Propriété insecticide des huiles végétales. *Le Progrès Agricole et Viticole* 100: 382–388.

Beckerman J, Stone J, Ruhl G, Creswell T. 2018. First report of *Pythium ultimum* crown and root rot of industrial hemp in the United States. *Plant Disease* 102: 2045.

Beckh H. 1895. *Geoponica Sive Cassiani Bassi scholastici de re rustica ecologae*. Teubner, Lipsiae.

Behi F, Bachrouch O, Ben Fekih I, Boukhris-Bouhachem S. 2017. Insecticidal and synergistic activities of two essential oils from *Pistacia lentiscus* and *Mentha pulegium* against the green peach aphid *Myzus persicae*. *Tunisian Journal of Plant Protection* 12: 53–65.

Boa ER. 1995. *A guide to the identification of diseases and pests of neem (Azadirachta indica)*. Food and Agriculture Organization of the United Nations, Bangkok.

Bócsa I, Karus M. 1997. *Der Hanfanbau: Botanik, Sorten, Anbau und Ernte*. Müller Verlag, Heildelberg.

Bouchard MF, Bellinger DC, Wright RO, Weisskopf MG. 2010. Attention-deficit/hyperactivity disorder and urinary metabolites of organophosphate pesticides. *Pediatrics* 125(6): e1270–1277.

Britt KE, Kuhar TP. 2020. Laboratory bioassays of biological/organic insecticides to control corn earworm on hemp in Virginia, 2019. *Arthropod Management Tests* 45(1): 1–2.

Britt KE, Pagani MK, Kuhar TP. 2019. First report of brown marmorated stink bug (Hemiptera: Pentatomidae) associated with *Cannabis sativa* (Rosales: Cannabaceae) in the United States. *Journal of Integrated Pest Management* 10(1): 1–3.

Britt K, Fike J, Flessner M, Johnson C, *et al.* (3 other authors). 2020. *Integrated pest management of hemp in Virginia*. Virginia Cooperative Extension publication no. ENTO-349NP, Blacksburg, *Virginia*.

Britt KE, Reed TD, Kuhar TP. 2021a. Evaluation of biological insecticides to manage corn earworm in CBD hemp, 2020. *Arthropod Management Tests* 46(1): 1–2.

Bruening G, Lyons JM. 2000. The case of the Flavr Savr tomato. *California Agriculture* 54: 6–7.

California DPR (Department of Pesticide Regulation). 2017. *Legal pest management practices for Cannabis growers in California*. Available at: https://www.cdpr. ca.gov/docs/county/cacltrs/penfltrs/penf2015/2015atch/attach1502.pdf

California DPR. 2018. *Cannabis pesticides that cannot be used*. California Department of Pesticide Regulation. Available at: https://www.cdpr.ca.gov/docs/ cannabis/cannot_use_pesticide.pdf

Canyon DV, Heukelbach J, Shaalan ES, Speare R. 2010. "Piperonyl butoxide," pp. 2306–2309 in Grayson ML, Mills J, *eds. Kucer's use of antibiotics, 6th edn*. Hodder and Stoughton, London.

Carson R. 1962. *Silent spring*. Mifflin, Boston, Massachusetts.

Cherian MC. 1932. Pests of ganja. *Madras Agricultural Journal* 20: 259–265.

Chevalier J. 1930. Lepyrèthre (chrysantheme insecticide). Activité pharmacodynamique et thérapeutique. *Bulletin des Sciences Pharmacologiques* 37: 154–164.

Clifford NJ, Nies AS. 1989. Organophosphate poisoning from wearing a laundered uniform previously contaminated with parathion. *JAMA* 262: 3035–3036.

Cloëz FS. 1866. *Observations et expériences sur l'oxydation des matières grasses d'origine végétale*. Thunot, Paris.

Cloëz M (Carrière EA, *ed.*). 1869. Destruction des pucerons, altises, etc. *Revue Horticole* 41: 288–289.

Cloyd RA, Galle CL, Keith SR, Kalschuer NA, Kemp KE. 2009. Effect of commercially available plant-derived essential oil products on arthropod pests. *Journal of Economic Entomology* 104: 1567–1579.

Colborn T, Dumaroski D, Peterson-Myers J. 1996. *Our stolen future*. Dutton Books, NY.

Colorado DOA (Department of Agriculture). 2016. *Pesticide use in Cannabis production information*. *Colorado Department of Agriculture*. Available at: https://www. colorado.gov/pacific/agplants/pesticide-use-cannabis-production-information

Colorado DOA (Department of Agriculture). 2023. *Pesticide use in Cannabis production information*. Available at: https://ag.colorado.gov/plants/pesticides/ pesticide-use-in-cannabis-production-information.

Cory EN, Langford GS. 1935. Sulfated alcohols in insecticides. *Journal of Economic Entomology* 28(2): 257–260.

Couch J, Victory K, Lowe B, *et al.* (5 other authors). 2017. *Evaluation of potential hazards during harvesting and processing cannabis at an outdoor organic farm*. US Department of Health and Human Services, Report no. 2015-0111-3271. Washington, *DC*.

Covarelli L, Turner AS, Nicholson P. 2004. Repression of deoxynivalenol accumulation and expression of Tri genes in *Fusarium culmorum* by fungicides *in vitro*. *Plant Pathology* 53: 22–28.

Cranshaw W. 2020. *State guidelines for the use of pesticides to control insects and mites: a review of the present situation*. Colorado state hemp insect website. Available at: https://hempinsects.agsci.colostate.edu/

Cranshaw WS, Baxendale B. 2013. *Insect control: horticultural oils*. Colorado State University Extension. Available at: https://extension.colostate.edu/docs/ pubs/insect/05569.pdf

Cranshaw W, Schreiner M, Britt K, Kuhar TP, McPartland JM, Grant J. 2019. Developing insect pest management systems for hemp in the United States: a work in progress. *Journal of Integrated Pest Management* 10(1): 26.

Cuypers E, Vanhove W, Gotink J, *et al.* (3 other authors). 2017. The use of pesticides in Belgian illicit indoor cannabis plantations. *Forensic Science International* 277: 59–65.

Daniel D, Lopes FS, do Lago CL. 2019. A sensitive multiresidue method for the determination of pesticides in marijuana by liquid chromatography-tandem mass spectrometry. *Journal of Chromatography A* 1603: 231–239.

De Meijer WJM, van der Werf HMG, Mathijssen EWJM, van den Brink PWM. 1995. Constraints to dry matter production in fibre hemp (*Cannabis sativa* L.). *European Journal of Agronomy* 4: 109–117.

De Petrocellis L, Ligresti A, Moriello AS, *et al.* (5 other authors). 2011. Effects of cannabinoids and cannabinoid-enriched cannabis extracts on TRP channels and endocannabinoid metabolic enzymes. *British Journal of Pharmacology* 163: 1479–1494.

De Queiroz GM, Politi FAS, *et al.* (5 other authors). 2015. Phytochemical characterization, antimicrobial activity, and antioxidant potential of *Equisetum hyemale* L. (Equisetaceae) extracts. *Journal of Medicinal Food* 18: 830–834.

Deyton DE, Sams CE, Cannon AL, Cummins JC, Windham MT. 2011. Management of powdery mildew on flowering dogwood with soybean oil. *Journal of Environmental Horticulture* 29: 185–192.

Diels J, Cunha M, Manaia C, Sabugosa-Madeira B, Silva M. 2011. Association of financial or professional conflict of interest to research outcomes on health risks or nutritional assessment studies of GM products. *Food Policy* 36: 197–203.

Diver S. 2002. *Notes on compost teas.* ATTRA Bulletin. Available at: file:///Users/johnmcpartland/Downloads/compost-tea-notes%20(1).pdf

Djerassi C, Shih-Coleman C, Diekman J. 1974. Insect control of the future: operational and policy aspects. *Science* 186: 596–607.

Duke JA. 1985. *CRC handbook of medicinal herbs.* CRC Press, Boca Raton, Florida.

Eckerman I. 2005. *The Bhopal saga—causes and consequences of the world's largest industrial disaster.* Universities Press, Hyderabad, India.

Eddleston M, Eyer P, Worek F, *et al.* (11 other authors). 2005. Differences between organophosphorus insecticides in human self-poisoning: a prospective cohort study. *Lancet* 366: 1452–1459.

Edelson JV, Duthie J, Roberts W. 2002. Toxicity of biorational insecticides: activity against the green peach aphid, *Myzus persicae* (Sulzer). *Pest Management Science* 58: 255–260.

Elad Y. 1992. The use of antioxidants (free radical scavengers) to control grey mould (*Botrytis cinerea*) and white mould (*Sclerotinia sclerotiorum*) in various crops. *Plant Pathology* 41: 417–426.

Ellstrand NC. 2003. *Dangerous liaisons? When cultivated plants mate with their wild relatives.* Johns Hopkins University Press, Baltimore, Maryland.

EPA. 2012. *Cold pressed neem oil.* US Environmental Protection Agency. Available at: https://www3.epa.gov/pesticides/chem_search/reg_actions/registration/decision_PC-025006_07-May-12.pdf

EPA. 2016. *Pesticide use on marijuana.* US Environmental Protection Agency. Available at: https://www.epa.gov/pesticide-registration/pesticide-use-marijuana

EPA. 2023. *Pesticide products registered for use on hemp.* US Environmental Protection Agency. Available at: https://www.epa.gov/pesticide-registration/pesticide-products-registered-use-hemp.

Esmaeily M, Bandani A, Zibaee I, Sharifian I, Zare S. 2017. Sublethal effects of *Artemisia annua* L. and *Rosmarinus officinalis* L. essential oils on life table parameters of *Tetranychus urticae* (Acari: Tetranychidae). *Persian Journal of Acarology* 6(1): 39–52.

Fabre JH (Teixeira de Mattos A, *trans*, Teal EA, *ed*). 1949. *The insect world of J. Henri Fabre.* Beacon Press, Boston, Massachusetts.

Farahmand H, Robinson GI, Gerasymchuk M, Kovalchuk I. 2023. Copper sulphate inhibits *Penicillium olsonii* growth and conidiogenesis on *Cannabis sativa.* *Journal of Plant Pathology.*

Faraone N, Hillier NK, Gutler GC. 2015. Plant essential oils synergize and antagonize toxicity of different conventional insecticides against *Myzus persicae* (Hemiptera: Aphididae). *PLoS ONE* 10(5): e0127774.

Farrer DG. 2015. *Technical report: Oregon Health Authority's process to determine which types of contaminants to test for in Cannabis products and levels for action.* Available at: https://www.oregon.gov/oha/ph/preventionwellness/marijuana/documents/oha-8964-technical-report-marijuana-contaminant-testing.pdf

Fazolin M, Estrela JLW, Monteiro Medeiros AF, *et al.* (3 other authors). 2016. Potencial sinérgico do óleo essencial rico em dilapiol para inseticidas piretroides sintéticos frente à lagarta-do-cartucho. *Ciência Rural* 46: 383–388.

Ferri F. 1961b. Microflora dei semi di canapa. *Progresso Agricolo (Bologna)* 7(3): 349–356.

Ferri F, Noviello C. 1963. Prove di lotta contro *Sclerotium rolfsii* Sacc. sulla canapa. *Phytopathologia Mediterranea* 2: 72–75.

Fimrite P. 2018. Big recall ordered for pot products after lab is caught faking tests. *San Francisco Chronicle*, Dec 21, 2018.

Firth J. 2013. Arsenic—the "poison of kings" and the "saviour of syphilis." *Journal of Military and Veterans' Health* 21: 11–17.

Fleming WE. Burgess ED, Maines WW. 1940. *The use of traps against the Japanese beetle.* USDA Circular no. 594. Washington, DC.

Frank M. 1988. *Marijuana grower's insider's guide.* Red Eye Press, Los Angeles.

Frank M, Rosenthal E. 1978. *Marijuana grower's guide.* And/Or Press, Berkeley, California.

Fusari P, Rovellini P, Folegatti L, Baglio D, Cavalieri A. 2013. Olio e farina da *Cannabis sativa* L. analisi multiscreening di micotossine, ftalati, idrocarburi policiclici aromatici, metalli e fitofarmaci. *Rivista Italiana delle Sostanze Grasse* 90 (2013): 9–19.

Garcia D, Ramos AJ, Sanchis V, Marín S. 2012. Effect of *Equisetum arvense* hydro-alcoholic extract: phenolic composition and antifungal and antimycotoxigenic effect against *Aspergillus flavus* and *Fusarium verticillioides* in stored maize. *International Journal of Food Microbiology* 153: 21–27.

Gardner F. 2005. Jane Weirick: death of an organizer. *CounterPunch* October 29, 2005.

Gerberg EJ, Novak RJ. 2007. "Considerations on the use of botanically-derived repellent products," pp. 305–310 in Debboun M, Frances S, Strickman D, *eds. Insect repellents: principles, methods and uses.* CRC Press, Boca Raton, Florida.

Gillet S. 2016. *True bud: understanding toxic marijuana syndrome.* Bird, Briarcliff Manor, New York.

Glass EH. 1979. *Pest management strategies in crop protection, vol.* I. Office of Technology Assessment, Washington, DC.

Goidànich G. 1959. *Manual di patologia vegetale.* Edizioni Agricole, Bologna.

Goodsell N. 1831. Gardens. *Genesee Farmer (Rochester)* 1(22): 169.

Green T. 1771. "Account of a machine or bellows for fumigating and destroying insects on melons, cucumbers, and fruit-trees," reproduced in "Extracts from the Society's records," *Journal of the Royal Society of Arts* 86 (1936): 97–98.

Guyton KZ, Loomis D, Gross Y, *et al.* (6 other authors). 2015. Carcinogenicity of tetrachlorvinphos, parathion, malathion, diazinon, and glyphosate. *Lancet Oncology* 16: 490–491.

Hardee DD, McKibben GH, Gueldner RC, *et al.* (3 other authors). 1972. Boll weevils in nature respond to grandlure, a synthetic pheromone. *Economic Entomology* 65: 97–100.

Hassan MEM, Abd El-Rahman SS, El-Abbasi IH, Mikhail MK. 2006. Induced resistance against faba bean chocolate spot disease. *Egyptian Journal of Phytopathology* 34: 69–79.

Health Canada. 2008. *Product information sheet on dried marihuana (cannabis).* Available at: www.hc-sc.gc.ca/ dhp-mps/marihuana/about-apropos/dried-information-sechee-eng.php

Health Canada. 2023. *Pest control products for use on cannabis.* Available at: https://www.canada.ca/en/health-canada/services/cannabis-regulations-licensed-producers/pest-control-products.html

Herger G, Klingauf F, Mangold D, Pommer EH, Scherer M. 1988. Die Wirkung von Auszügen aus dem Sachalin-Staudenknöterich, *Reynoutria sachalinensis* (F. Schmidt) Nakai, gegen Pilzkrankheiten, insbesondere Echte Mehltau-Pilze. *Nachrichtenblatt des Deutschen Pflanzenschutzdienstes* 40: 56–60.

Heuzé G. 1852. Maladie de la vigne, 1. Procédé Grison. *Revue Horticole, quatrième série* 1: 168–170.

Hiltner L, Gentner G. 1916. Ueber die Wirkung der Beizung der Samen von Hanf, Sonnenblumen, Buchweizen, Hierse, Mais und Mohar. *Praktischen Blätter für Pflanzenbau und Pflanzenschutz* 14: 85–90.

Horowitz AR, Ellsworth PC, Ishaaya I. 2009. "Biorational pest control—an overview," pp. 1-20 in Ishaaya I, Horowitz AR, eds. *Biorational control of arthropod pests.* Springer Netherlands, Dordrecht.

Hussain M, Zouhar M, Ryšánek P. 2017. Comparison between biological and chemical management of sugar beet nematode, *Heterodera schachtii. Pakistan Journal of Zoology* 49: 45–50.

Isman MB. 2015. A renaissance for botanical insecticides? *Pest Management Science* 71: 1587–1590.

Jafari N. Hadavi E. 2012. Growth and essential oil yield of dill (*Anethum graveolens*) as affected by foliar sprays of citric acid and malic acid. *Acta Horticulturae* 995: 287–290.

Jameson LE, Conrow KD, Pinkhasova DV, *et al.* (9 other authors). 2022. Comparison of state-level regulations for cannabis contaminants and implications for public health. *Environmental Health Perspectives* 130: e97001.

Kirkland AH. 1896. "Insecticides," pp. 363–371 in *Forty-third Annual Report of the Secretary of the Massachusetts State Board of Agriculture.* Boston, Massachusetts.

Koleva Gudeva L, Mitrev S Maksimova V, Spasov D. 2013. Content of capsaicin extracted from hot pepper (*Capsicum annuum* ssp. *microcarpum* L.) and its use as an ecopesticide. *Hemijska Industrija* 67: 671–675.

Kotev S, Georgieva M. 1969. Химически средства за обеззаразяване семената на конопа срещу бактериолата. *Zashchita Rastenii* 17(1): 25–29.

Kubo I, Ganjian I. 1981. Insect antifeedant terpenes, hot-tasting to humans. *Experientia* 37: 1063–1064.

La Quintinye J de. 1690. *Instruction pour les jardins fruitiers et potages, Tome II.* Claude Barbin, Paris.

Lecompte S, Habauzit D, Charlier TD, Pakdel F. 2017. Emerging estrogenic pollutants in the aquatic environment and breast cancer. *Genes* 8(9): e229.

Lintner JA. 1882. "Hellebore as an insecticide," pp. *41–43 in First annual report on the injurious and other insects of the State of New York.* Weed, Parsons & Co., Albany.

Liu TX, Stansly PA. 2000. Insecticidal activity of surfactants and oils against silverleaf whitefly (*Bemisia argentifolii*) nymphs (Homoptera: Aleyrodidae) on collards and tomato. *Pest Management Science* 56: 861–866.

Lodeman EG. 1896. *The spraying of plants.* Macmillan & Co., New York.

Lu CS, Warchol KM, Callahan RA. 2014. Sub-lethal exposure to neonicotinoids impaired honey bees winterization before proceeding to colony collapse disorder. *Bulletin of Insectology* 67: 125–130.

Maguire WJ, Call CW, Cerbu C, *et al.* (2 other authors). 2019. Comprehensive determination of unregulated pesticide residues in Oregon cannabis flower by liquid chromatography paired with triple quadrupole mass spectrometry and gas chromatography paired with triple quadrupole mass spectrometry. *Journal of Agricultural and Food Chemistry* 67: 12670–12674.

Mansour SA, El-Sharkawy AZ, Abdel-Hamid NA. 2015. Toxicity of essential plant oils, in comparison with conventional insecticides, against the desert locust, *Schistocerca gregaria* (Forskål). *Industrial Crops and Products* 63: 92–99.

Mart M. 2015. *Pesticides, a love story.* University Press of Kansas, Lawrence, Kansas.

Martin H. 1948. *The scientific principles of plant protection, 3rd edn.* Arnold & Co., London.

Martin H, Salmon ES. 1933. The fungicidal properties of certain spray-fluids. X., glyceride oils. *Journal of Agricultural Science (England)* 23: 228–251.

Matthews GA. 2008. *Pesticide application methods, 3rd edn.* Blackwell Science, London.

Maude RB. 1996. *Seedborne diseases and their control.* CAB International, University Press, Cambridge, UK.

McNally WD. 1920. A report of five cases of poisoning by nicotine. *Journal of Laboratory and Clinical Medicine* 5: 213–217.

McPartland JM, McKernan KJ. 2017. "Contaminants of concern in cannabis: microbes, heavy metals and pesticides," in Chandra S, Lata H, ElSohly MA, *eds. Cannabis sativa: Botany and Biotechnology.* Springer International Publishing, Cham, Switzerland.

McPartland JM, Clarke RC, Watson DP. 2000. *Hemp diseases and pests – management and biological control.* CABI Publishing, Wallingford, UK.

Melander AL. 1914. Can insects become resistant to sprays? *Journal of Economic Engineering* 7: 167–173.

Metcalf C, Flint W, Metcalf R. 1962. *Destructive and useful insects, 4th edn.* McGraw-Hill Book Co., New York.

Miège E. 1921. Action de la chloropicrine sur la faculté germinative des graines. *Comptes Rendus de l'Académie des Sciences* 172(3):170–173.

Mikuriya TH, North W, Lucido F, Jaffee R, Weirick J. 2005. Cerebrospinal delayed allergic reaction to pesticide residue. *Conference Proceedings, International Association for Cannabis as Medicine,* p. 14.

Millardet A. 1883. Le mildiou et le rot. *Zeitschrift für Wein-, Obst-, und Gartenbau für Elsasz-Lothringen,* numéros des 1 et 15 mars, 1883.

Millardet A, Gayon LU. 1885. Traitement du mildiou par le mélange de sulfate de cruivre et de chaux. *Journal d'Agriculture Pratique* 49: 707–710.

Mills PK, Kwong S. 2001. Cancer incidence in the United Farmworkers of America (UFW), 1987-1997. *American Journal of Industrial Medicine* 40: 596–603.

Milne GWA, ed. 2002. *Ashgate handbook of pesticides and agricultural chemicals.* Ashgate Publishing, Aldershot, UK.

Miresmailli S, Bradbury R, Isman MB. 2006. Comparative toxicity of *Rosmarinus officinalis* L. essential oil and blends of its major constituents against *Tetranychus urticae* Koch (Acari: Tetranychidae) on two different host plants. *Pest Management Science* 62: 366–371.

Mishra D. 1987. Damping off of *Cannabis sativa* caused by *Fusarium solani* and its control by seed treatment. *Indian Journal of Mycology and Plant Pathology* 17(1): 100–102.

Mitra S, Rodriguez SD, Vulcan J, *et al.* (5 other authors). 2020. Efficacy of active ingredients from the EPA 25(b) list in reducing attraction of *Aedes aegypti* (Diptera: Culicidae) to humans. *Journal of Medical Entomology* 57: 477–484.

Monardes N. 1574. *Primera y segunda y tercera partes de la Historia medicinal, de las cosas que se traen de nuestras Indias Occidentales, que sirven en medicina.* Alonso Escriuano, Seville.

Moulins JR, Blais M, Montsion K, *et al.* (7 other authors). 2018. Multiresidue method of analysis of pesticides in medical cannabis. *Journal of AOAC International* 101: 1948–1960.

Muturi EJ, Ramirez JL, Doll KM, Bowman MJ. 2017. Combined toxicity of three essential oils against *Aedes aegypti* (Diptera: Culicidae) larvae. *Journal of Medical Entomology* 54:1684–1691, 2017.

Nagy B. 1979. Different aspects of flight activity of the hemp moth, *Grapholitha delineana* Walk., related to integrated control. *Acta Phytopathologica Academiae Scientiarum Hungaricae* 14: 481–488.

NCCEH. 2009. *Recommendations for safe re-occupancy of marijuana grow operations.* National Collaborating Centre for Environmental Health. Available at: www.ncceh.ca/documents/guide/recommendations-safe-re-occupancy-marijuana-grow-operations

Needham J, Lu GD, Huang HT. 1986. "Botany," Volume 6, Part 1 of Needham J, *ed. Science and Civilisation in China.* Cambridge University Press, Cambridge, UK.

Neillie CR, Houser JL. 1922. Fighting insects with airplanes. *National Geographic Magazine* 41: 232–238.

Palmer WL, Wingard, RG, George JL. 1983. Evaluation of white-tailed deer repellents. *Wildlife Society Bulletin* 11(2): 164–166.

Parker WB. 1913a. The hop aphis in the Pacific region. *USDA Entomology Bulletin* 111: 9–39.

Patschke K, Gottwald R, Müller R. 1997. Erste Ergebnisse phytopathologischer Beobachtungen im Hanfanbau im Land Brandenburg. *Nachrichtenblatt des Deutschen Pflanzenschutzdienstes* 49: 286–290.

Pepper BP, Carruth LA. 1945. A new plant insecticide for control of the European corn borer. *Journal of Economic Entomology* 38: 59–66.

Pérez-Parada A, Alonso B, Rodríguez C, *et al.* (8 other authors). 2016. Evaluation of three multiresidue methods for the determination of pesticides in marijuana (*Cannabis sativa* L.) with liquid chromatography-tandem mass spectrometry. *Chromatographia* 79: 1069–1083.

Pichl J. 1902. Über die Geschlechts- und Blütenbildung beim Hanf. *Sitzungsberichte des Deutschen Naturwissenschaftlich-Medicinischen Vereines für Böhmen "Lotos" in Prag* 22: 143–146.

Pimentel D. 2005. Environmental and economic costs of the application of pesticides primarily in the United States. *Environment, Development and Sustainability* 7: 229–252.

Powell G, Hardie J, Pickett JA. 1995. Behavioral evidence for detection of the repellent polygodial by aphid antennal tip sensilla. *Physiological Entomology* 20: 141–146.

Prévost B. 1807. *Mémoire sur la cause immediate de la carie ou charbon des blés.* Bernard, Paris.

Prota N, Bouwmeester HJ, Jongsma MA. 2014. Comparative antifeedant activities of polygodial and pyrethrins against whiteflies (*Bemisia tabaci*) and aphids (*Myzus persicae*). *Pest Management Science* 70: 682–688.

Pumnuan J, Insung A. 2016. Fumigant toxicity of plant essential oils in controlling thrips, *Frankliniella schultzei* (Thysanoptera: Thripidae) and mealybug, *Pseudococcus jackbeardsleyi* (Hemiptera: Pseudococcidae). *Journal of Entomological Research* 40: 1–10.

Punja ZK. 2021a. The diverse mycoflora present on dried cannabis (*Cannabis sativa* L., marijuana) inflorescences in commercial production. *Canadian Journal of Plant Pathology* 43: 88–100.

Puritch G. 1982. An inside look at insecticidal soap. *Sinsemilla Tips* 3(3): 34.

Radhika SA, Sahayaraj K. 2014. Synergistic effects of monocrotophos with botanical oils and commercial neem formulation on *Spodoptera litura* (Fab.) (Lepidoptera: Noctuidae). *Journal of Biopesticides* 7(Suppl): 152–159.

Ray J. 1710. *Historia Insectorum.* Churchill, London.

Razmjou J, Davari M, Ebadollahi A. 2016. Effect of two plant essential oils and the entomopathogenic fungus, *Lecanicillium muscarium* (Zare & Gams) on the cotton aphid, *Aphis gossypii* Glover. *Egyptian Journal of Biological Pest Control* 26: 775–779.

Riedl H, Croft BA. 1974. A study of pheromone trap catches in relation to codling moth (Lepidoptera: Olethreuidae) damage. *Canadian Entomologist* 106: 525–537.

Ritter C, Richter E, Knölck I, Katroschan KU. 2014. Laboratory studies on the effect of calcium cyanamide on wireworms (*Agriotes ustulatus*, Coleoptera: Elateridae). *Journal of Plant Diseases and Protection* 121: 133–137.

Robinson BB. 1943a. Seed treatment of hemp seeds. *Journal of the American Society of Agronomy* 35: 910–914.

Rodgers RH. 1980. Hail, frost, and pests in the vineyard: Anatolius of Berytus as a source for the Nabataean Agriculture. *Journal of the American Oriental Society* 100: 1–11.

Rodgman A, Perfetti TA. 2013. *The chemical components of tobacco and tobacco smoke, 2nd edn.* CRC Press, Boca Raton, Florida.

Rodino S, Butu A, Butu M, Petruta Cornea C. 2014. *In vitro* efficacy of some plant extracts against damping off disease of tomatoes. *Journal of International Scientific Publications* 2: 240–244.

Rosenthal E. 1998. *Marijuana grower's handbook.* Quick American Archives, Oakland, California.

Rosenthal E. 1999. *Closet cultivator.* Quick American Archives, Oakland, California.

Russo EB. 2016. Current therapeutic cannabis controversies and clinical trial design issues. *Frontiers in Pharmacology* 7: e309.

Sabbour MM, Abd-El-Aziz SE. 2010. Efficacy of some bioinsecticides against *Bruchidius incarnatus* (Boh.)(Coleoptera: Bruchidae) infection during storage. *Journal of Plant Protection Research* 50: 28–34.

Sallam AA, Volkmar C, El-Wakeil NE. 2009. Effectiveness of different bio–rational insecticides applied on wheat plants to control cereal aphids. *Journal of Plant Diseases and Protection* 116: 283–287.

Sarma R, Adhikari K, Mahanta S, Khanikor B. 2019. Combinations of plant essential oil based terpene compounds as larvicidal and adulticidal agent against *Aedes aegypti* (Diptera: Culicidae). *Scientific Reports* 9: e9471.

Schaffner JH. 1928. Further experiments in repeated rejuvenation in hemp and their bearing on the general problem of sex. *American Journal of Botany* 15: 77–85.

Schneider S, Bebing R, Dauberschmidt C. 2014. Detection of pesticides in seized illegal cannabis plants. *Analytical Methods* 6: 515–520.

Shannag HS, Capinera JL, Freihat NM. 2014. Efficacy of different neem-based biopesticides against green peach aphid, *Myzus persicae* (Hemiptera: Aphididae). *International Journal of Agricultural Policy and Research* 2: 61068.

Shaw FR, Mendal SC. 1940. A survey of bee poisonings. *Gleanings in Bee Culture* 68: 221, 265.

Shepard M. 2013. Maine marijuana company fined $18K for pesticides. *Portland Press Herald*, September 6, 2013.

Slator L, Hall R (Coote T, *ed*). 1724. *Instructions for the cultivating and raising of flax and hemp: in a better manner, than that generally practis'd in Ireland.* Grierson, Dublin.

Smith BC, Lessard P, Pearson R. 2019. Inter-laboratory variation in cannabis analysis: pesticides and potency in distillates. *Cannabis Science and Technology* 2(1): 48–53.

Soliman MH, El-Mohamedy RSR. 2017. Induction of defense related physiological and antioxidant enzyme response against powdery mildew disease in okra (*Abelmoschus esculentus* L.) plant by using chitosan and potassium salts. *Mycobiology* 45: 409–420.

Speiser B, Kistler C. 2002. Field tests with a molluscicide containing iron phosphate. *Crop Protection* 21: 389–394.

Staniland LN. 1926. Oil sprays for spring and summer use. *Annual Report of the Agricultural and Horticultural Research Station, Long Ashton, University of Bristol* 1926: 78–81.

State of Maine. 2013. *An Act to maintain access to safe medical marijuana.* Available at: www.mainelegislature.org/legis/bills/bills_126th/billtexts/SP057801.asp

Steiner R. 1924a. *Geisteswissenschaftliche Grundlagen zum Gedeihen der Landwirtschaft. [trans: Spiritual-Scientific Foundations for the Prosperity of Agriculture (Agriculture Course)]* R. Steiner Verlag, Dornach.

Steiner R. 1924b. "Report to members of the Anthroposophical Society after the Agriculture Course, Dornach, Switzerland, June 20, 1924," pp. 1–12 in Creeger CE, Gardner M, *trans, ed.* 1984. *Spiritual foundations for the renewal of agriculture.* Biodynamic Farming and Gardening Association, Junction City, Oregon.

Stephenson JB, Fishkin C, Crothers N, *et al.* (6 other authors). 2003. *Pesticides on tobacco.* United States General Accounting Office, Washington, DC.

Stern VM, Smith RF, van den Bosch R, Hagen KS. 1959. The integration of chemical and biological control of the spotted alfalfa aphid. *The integrated control concept. Hilgardia* 29: 81–101.

Sterner RT, Shumake SA, Gaddis SE, Bourass JB. 2005. Capsicum oleoresin: development of an in-soil repellent for pocket gophers. *Pest Management Science* 61: 1202–1208.

Stockwell VO, Duffy B. 2012. Use of antibiotics in plant agriculture. *Revue Scientifique et Technique, Office International des Épizooties* 31: 199–210.

Stone D. 2014. Cannabis, pesticides and conflicting laws: the dilemma for legalized States and implications for public health. *Regulatory Toxicology and Pharmacology* 69: 284–288.

Sullivan N, Elzinga S, Raber JC. 2013. Determination of pesticide residues in cannabis smoke. *Journal of Toxicology* Epub 378168.

Sutton DB, Punja ZK, Hamarneh G. 2023. Characterization of trichome phenotypes to assess maturation and flower development in *Cannabis sativa* L. (cannabis) by automatic trichome gland analysis. *Smart Agricultural Technology.*

Suwanchaikasem P, Nie SA, Idnurm A, *et al.* (3 other authors). 2023a. Effects of chitin and chitosan on root growth, biochemical defense response and exudate proteome of *Cannabis sativa. Plant–Environment Interactions* 4: 115–133.

Suwanchaikasem P, Nie SA, Selby-Pham J, *et al.* (3 other authors). 2023b. Hormonal and proteomic analyses of southern blight disease caused by *Athelia rolfsii* and root chitosan priming on *Cannabis sativa* in an *in vitro* hydroponic system *Plant Direct* 7: e528.

Thompson CM, Gabriel MW, Purcell KL. 2017. An ever-changing ecological battlefield: marijuana cultivation and toxicant use in western forests. *Wildlife Professional* 11(3): 42–46.

Tsvetkov N, Sampson-Robert O, Sood K, *et al.* (5 other authors). 2017. Chronic exposure to neonicotinoids reduces honey bee health near corn crops. *Science* 356: 1395–1397.

Turner CE, Elsohly MA, Boeren EG. 1980. Constituents of *Cannabis sativa* L. XVII. A review of the natural constituents. *Journal of Natural Products* 43: 169–234.

Upton R, Craker L, ElSohly M, *et al.,* eds. 2013. *Cannabis inflorescence, Cannabis spp. Standards of identity, analysis, and quality control.* American Herbal Pharmacopoeia, Scotts Valley, California.

USDA. 1960. *List of chemical compounds and additives approved for use under USDA poultry and poultry product inspection and grading programs.* USDA Agricultural Marketing Service Bulletin 419, Washington, DC.

Valdes-Donoso P, Sumner DA, Goldstein R. 2019. Costs of mandatory cannabis testing in California. *California Agriculture* 73(3-4): 154–160.

Van der Werf HMG. 1994. *Crop physiology of fibre hemp (Cannabis sativa L.).* Doctoral thesis, Wageningen Agricultural University, the Netherlands.

Van Dijk TC, Van Staalduinen MA, Van der Sluijs JP. 2013. Macro-invertebrate decline in surface water polluted with imidacloprid. *PLoS ONE* 8(5): e62374.

Voelker R, Holmes M. 2015. *Pesticide use on Cannabis.* Cannabis Safety Institute. Available at: http://cannabissafetyinstitute.org/wp-content/uploads/2015/06/CSI-Pesticides-White-Paper.pdf

Walsh BD, Riley CV, *eds.* 1868. To destroy Colorado potato bugs. *American Entomologist* 1: 219.

Walter WR. 1995. *Wax and capsaicin based pesticide.* Wilder Agricultural Product Co Inc., New Wilmington, Pennsylvania. US Patent 5466459.

Washington Administrative Code. 2016. *WAC 246-70-050. Quality assurance testing.* Available at: https://apps.leg.wa.gov/wac/default.aspx?cite=246-70-050

Westcott C (Horst RK). 1964 (1990 revised). *Westcott's plant disease handbook, 5th edn.* Van Nostrand Reinhold, New York.

Westgate PJ, Schultz BB, Hazzard RV. 2017. Effects of carriers, emulsifiers, and biopesticides for direct silk treatments on caterpillar feeding damage and ear development in sweet corn. *Journal of Economic Entomology* 110: 507–516.

Wilson DK, Graff CL. 2013. The identification of spiromesifen, a recently introduced pesticide, using approaches to chemical unknown analysis. *Journal of Forensic Sciences* 58: 220–223

Wilson H, Bodwitch H, Carah J, *et al.* (4 other authors). 2019. First known survey of cannabis production practices in California. *California Agriculture* 73: 119–127.

Wilson S. 2023. Giant California greenhouse signals a bet on cannabis legalization. *Washington Post,* Feb 28, 2023.

WSDA. 2016. *Pesticide and fertilizer use on marijuana.* Washington State Department of Agriculture. Available at: http://agr.wa.gov/pestfert/pesticides/pesticideuseonmarijuana.aspx

Wu LP, Huo Z, Zhou XL, *et al.* (6 other authors). 2017. Acaricidal activity and synergistic effect of thyme oil constituents against carmine spider mite (*Tetranychus cinnabarinus* (Boisduval)). *Molecules* 22: e1873.

Wurzer J. 2016. Safety testing of medicinal cannabis. *Proceedings of the 2006 Cannmed Conference.* Medical Genomics, Woburn, Massachusetts.

Wyatt K. 2015. Marijuana subject to product liability claim in Colorado. *US News & World Report,* October 5, 2015.

Wylie PL, Westland J, Wang M, *et al.* (3 other authors). 2020. Screening for more than 1,000 pesticides and environmental contaminants in cannabis by GC/Q-TOF. *Medical Cannabis and Cannabinoids* 3: 14–24.

Xantus GX. 2020. Vaping-associated lung injury—VALI facts, assumptions and opportunities: review of the present situation. *Postgraduate Medical Journal* 96: 61–63.

Yepsen RB. 1976. *Organic plant protection.* Rodale Press, Emmaus, Pennsylvania.

Zibaee I, Khorram P. 2015. Synergistic effect of some essential oils on toxicity and knockdown effects, against mosquitos, cockroaches and housefly. *Arthopods* 4(4): 107–123.

Zörnig H. 1909. *Arzneidrogen, Teil I.* Werner Klinkhardt, Leipzig.

Notes: Page numbers in *italics* refer to figures and page numbers in **bold** refer to tables.